Geotechnical and Geophysical Site Characterization

Volume 1

PROCEEDINGS OF THE SECOND INTERNATIONAL CONFERENCE ON SITE CHARACTERIZATION
ISC-2, Porto, Portugal, 19-22 September 2004

Geotechnical and Geophysical Site Characterization

EDITED BY

ANTÓNIO VIANA DA FONSECA

Faculdade de Engenharia da Universidade do Porto (FEUP), Portugal

PAUL W. MAYNE

School of Civil & Environmental Engineering, Georgia Institute of Technology, Atlanta, GA, USA

VOLUME 1

MILLPRESS ROTTERDAM NETHERLANDS 2004

This Conference was organised in the University of Porto under the auspices of:
International Society of Soil Mechanics and Geotechnical Engineering - ISSMGE, TC16 on In-Situ Testing in collaboration with TC10 on Geophysics, and the Portuguese Geotechnical Society.

Cover design: Mariana Serra and Millpress

Disclaimer: The Organizing Committee of the Second International Conference on Site Characterization and Millpress Science Publishers accept no responsibility for errors or omissions in the papers. The Organizing Committee of the Second International Conference on Site Characterization and Millpress Science Publishers shall not be liable for any damage caused by errors or omissions in the papers.

All rights reserved.
This publication may not be reproduced in whole or in part, stored in a retrieval system or transmitted in any form or by any means without permission from the publisher, Millpress Science Publishers.
info@millpress.com

Published and distributed by Millpress Science Publishers, P.O. Box 84118, 3009 CC Rotterdam, Netherlands
Tel.: +31 (0) 10 421 26 97; Fax: +31 (0) 10 209 45 27; www.millpress.com

ISBN 90 5966 009 9 set of 2 volumes
ISBN 90 5966 010 2 volume 1
ISBN 90 5966 011 0 volume 2

© 2004 Millpress Rotterdam
Printed in the Netherlands

Volume 1

Keynote lectures
1. Mechanical *in-situ* testing methods
2. Geophysical methods applied to geotechnical engineering
3. Innovative technologies and equipment
4. New developments in interpretation of *in-situ* data

Volume 2

5. Case studies involving practical projects
6. Characterization of non-textbook geomaterials
7. Applications to geotechnical structures
8. Enhanced characterization by combined *in-situ* testing
9. Laboratory and field comparisons

Table of contents

Volume 1

Introduction	XXI
Reviewing Committee	XXIII
Reference documents on *in-situ* testing	XXV
Acknowledgements	XXVIII

KEYNOTE LECTURES

The James K. Mitchell Lecture: "*In situ* soil testing: from mechanics to interpretation" H-S. Yu	3
Characterization of granite and the underground construction in Metro do Porto, Portugal S. Babendererde, E. Hoek, P. Marinos & A. Silva Cardoso	39
In-situ test characterization of unusual geomaterials F. Schnaid, B.M. Lehane & M. Fahey	49
Routine and advanced analysis of mechanical in-situ tests. Results on saprolitic soils from granites more or less mixed in Portugal A. Gomes Correia, A. Viana da Fonseca & M. Gambin	75
The contributions of *in situ* geophysical measurements to solving geotechnical engineering problems K.H. Stokoe II, S-H. Joh & R.D. Woods	97
Deformation properties of fine-grained soils from seismic tests K.R. Massarsch	133
Advanced calibration chambers for cone penetration testing in cohesionless soils A-B. Huang & H-H. Hsu	147
In-situ seismic survey in characterizing engineering properties of natural ground S. Shibuya, S. Yamashita, Y. Watabe & D.C.F. Lo Presti	167
Characterization of the Venice lagoon silts from *in-situ* tests and the performance of a test embankment P. Simonini	187
Characterization of soft sediments for offshore applications M.F. Randolph	209
Evaluating soil liquefaction and post-earthquake deformations using the CPT P.K. Robertson	233

1. MECHANICAL IN-SITU TESTING METHODS

General report: Mechanical *in-situ* testing methods 253
M. Devincenzi, J.J.M. Powell & N. Cruz

Comparison between plate load test results and dilatometer test results - Case study, 265
Seymareh dam and Sazbon dam projects
A. Agharazi & M. Moradi

Influence of vane size and equipment on the results of field vane tests 271
H. Åhnberg, R. Larsson & C. Berglund

Determination of *in-situ* deformation moduli with a penetrometer and static loading test 279
H. Arbaoui, R. Gourvès, Ph. Bressolette & L. Bodé

Characterization of soft marine clay using the flat dilatometer 287
A. Arulrajah, M.W. Bo & H. Nikraz

Estimating the SPT penetration resistance from rod penetration based on instrumentation 293
E.H. Cavalcante, F.A.B. Danziger & B.R. Danziger

Characterization of SPT grain size effects in gravels 299
C.R. Daniel, J.A. Howie, R.G. Campanella & A. Sy

International and European standards on geotechnical investigation and testing for site characterization 307
V. Eitner, R. Katzenbach & F. Stölben

CPTU site characterization: offshore peninsular and East Malaysia experience 315
S.M. Fairuz, J. Rohani & T. Lunne

The repeatability in results of Mackintosh Probe test 325
A. Fakher & M. Khodaparast

Application of the CPT to the characterization of a residual clay site 331
S.G. Fityus & L. Bates

Energy ratio measurement of SPT equipment 339
D-S. Kim, W-S. Seo, E-S. Bang

CPT-DMT interrelationships in Piedmont residuum 345
P.W. Mayne & T. Liao

Energy measurements for standard penetration tests and the effects of the length of rods. 351
E. Odebrecht, F. Schnaid, M.M. Rocha & G.P. Bernardes

SPT-T: test procedure and applications 359
A.S.P. Peixoto, D. de Carvalho & H.L. Giacheti

Dynamic cone penetrometer to estimate subgrade resilient modulus for low volume roads design 367
A.M. Rahim & K.P. George

Soil profile interpretation based on similarity concept for CPTU data 373
F. Saboya Jr & D.J.G. Balbi

Centrifuge penetration tests in saturated layered sands 377
M.F. Silva & M.D. Bolton

Sampling by drilling and excavations and groundwater measurements according to 385
EN ISO 22475 for geotechnical investigations and site characterization
F. Stölben, V. Eitner & H. Hoffmann

Undrained shear strength and OCR of marine clays from piezocone test results 391
F.A. Trevor & P.W. Mayne

Results of a comparative study on cone resistance measurements 399
V. Whenham, N. Huybrechts, M. De Vos, J. Maertens, G. Simon & G. Van Alboom

2. GEOPHYSICAL METHODS APPLIED TO GEOTECHNICAL ENGINEERING

General report: Geophysical methods applied to geotechnical engineering   409
S. Foti & A.P. Butcher

ISC'2 experimental site investigation and characterization - Part 2:   419
from SH waves high resolution shallow reflection to shallower GPR tests
F. Almeida, H. Hermosilha, J.M. Carvalho, A. Viana da Fonseca & R. Moura

Quantitative resistivity methods for marine and fluvial site investigations   427
P. Brabers

ISC'2 experimental site investigation - Part 1: conventional and tomographic P and   433
S waves refraction seismics vs. electrical resistivity
J.M. Carvalho, A. Viana da Fonseca, F. Almeida & H. Hermosilha

The role of crosshole seismic tomography for site characterization and grout injection evaluation   443
on Carmo convent foundations
M.J. Coelho, F.M. Salgado & L. Fialho Rodrigues

Geophysical characterization for seepage potential assessment along the embankments of the Po River   451
C. Comina, S. Foti, L.V. Socco & C. Strobbia

The use of multichannel analysis of surface waves in determining G_{max} for soft clay   459
S. Donohue, M. Long, K. Gavin & P. O'Connor

The use of electrical resistivity for detection of leacheate plumes in waste disposal sites   467
V.R. Elis, G. Mondelli, H.L. Giacheti, A.S.P. Peixoto & J. Hamada

Wavelets in surface wave cavity detection - theoretical background and implementation   475
N. Gucunski, P. Shokouhi & A. Maher

Investigation of bridge foundation sites in karst terrane via multi-electrode electrical resistivity   483
D.R. Hiltunen & M.J.S. Roth

Monitoring water saturation gradient in a silty sand site using geophysical techniques   491
E. Kalantarian & D. Doser

Characterizing ground hazards induced by historic salt mining using ground-penetrating radar   497
B. Kulessa, A. Ruffell & D. Glynn

Hydrogeological characterization by electrical resistivity surveys in granitic terrains   505
A.S. Lima & A.C.V. Oliveira

Development of a TDR dielectric penetrometer   513
C-P. Lin, C-C. Chung & S-H. Tang

Joint acquisition of SWM and other seismic techniques in the ISC'2 experimental site   521
I. Lopes, I. Moitinho, C. Strobbia, P. Teves-Costa, G.P. Deidda, M. Mendes & J.A. Santos

Decomposed rock mass characterization with crosshole seismic tomography at the Heroísmo   531
station site (Porto)
M. Oliveira & M.J. Coelho

Shear velocity model appraisal in shallow surface wave inversion   539
A. O'Neill

Full waveform reflectivity for inversion of surface wave dispersion in shallow site investigations   547
A. O'Neill

Offshore geosciences for coastal infrastructure   555
J. Peuchen & W.M. NeSmith

A framework for inversion of wavefield spectra in seismic non-destructive testing of pavements   563
N. Ryden, P. Ulriksen & C.B. Park

Geoelectrical resistivity imaging of domestic waste disposal sites: Malaysian case study *A.R. Samsudin, U. Hamzah & W.Z.W. Yaacob*	571
The small strain stiffness of gold tailings *M. Theron, G. Heymann & C.R.I. Clayton*	575
An assessment of heavy metal contamination using electromagnetic survey in the solid waste disposal site of Campos dos Goytacazes, Brazil *S. Tibana, L.A.C. Monteiro, E.L. dos Santos Júnior & F.T. de Almeida*	581
Geotechnical site characterization using surface waves, case studies from Belgium and the Netherlands *V. van Hoegaerden, R.S. Westerhoff, J.H. Brouwer & M.C. van der Rijst*	585
Geophysical methods and identification of embankment dam parameters *O.K. Voronkov, A.A. Kagan, N.F. Krivonogova, V.B. Glagovsky & V.S. Prokopovich*	593
Application of advanced geophysical technologies to landslides and unstable slopes *R.J. Whiteley*	601
Resistivity imaging as a tool in shallow site investigation – a case study *R. Wisén, T. Dahlin & E. Auken*	607
Evaluation of grouting effect with seismic tomography considering grouting mechanism *Y. Yamaguchi & H. Satoh*	615

3. INNOVATIVE TECHNOLOGIES AND EQUIPMENT

General report: Innovative technologies and equipment *M. Long & K.K. Phoon*	625
The use of microtremors for soil and site categorization in Thessaloniki city, Greece *P. Apostolidis, D. Raptakis, M. Manakou & K. Pitilakis*	635
New de-coupled shear wave source for the SCPT test *L. Areias, W. Haegeman & W.F. Van Impe*	643
Comparing 360° televising of drill hole walls with core logging *R. T. Baillot, N. Barton, R. Abrahão & A. Ribeiro Jr.*	647
Developments of a full displacement pressuremeter for municipal solid waste site investigations in Brazil *L.A.L. Bello, T.M.P. de Campos, J.T. Araruna & B.G. Clarke*	657
Rock characterization using drilling parameters *J. Benoît, S.S. Sadkowski & W.A. Bothner*	665
Penetration resistance in soft clay for different shaped penetrometers *S.F. Chung & M.F. Randolph*	671
Evaluation of the undrained shear strength profile in soft layered clay using full-flow probes *J.T. DeJong, N.J. Yafrate, D.J. DeGroot & J. Jakubowski*	679
Rapid site assessment with the electrical conductivity probe flexible approach for rapid and deep site investigation *M.P. Harkes, E.E. van der Hoek & J.J. van Meerten*	687
A framework for using textured friction sleeves at sites traditionally problematic for CPT *G.L. Hebeler, J.D. Frost & J.D. Shinn II*	693
Thin layer and interface characterization by VisCPT *R.D. Hryciw & S. Shin*	701
Evaluation of shear wave velocity profile using SPT based uphole test *D-S. Kim, E-S. Bang, W-S. Seo*	707

Suitability of using Iwasaki's P_L in characterization of liquefaction damages during the 1999 Chi-Chi Earthquake, Taiwan — 713
C.P. Kuo, M. Chang, S.H. Shau & R.E. Hsu

T-bar testing in Irish soils — 719
M. Long & G.T. Gudjonsson

Site characterization and QA/QC of deep dynamic compaction using an instrumented dilatometer — 727
H.J. Miller, K.P. Stetson & J. Benoît

Soil-rock sounding with MWD – a modern technique to investigate hard soils and rocks — 733
B. Möller, U. Bergdahl & K. Elmgren

A new *in-situ* apparatus: the geomechameter — 741
J. Monnet & S.M. Senouci

Permeability of sand sediment soil determined by the pressure infiltrometer test — 749
T. Morii, Y. Takeshita & M. Inoue

Adapted T-bar penetrometer versus CPT to determine undrained shear strengths of Dutch soft soils — 757
O. Oung, J.W.G. Van der Vegt, L. Tiggelman & H.E. Brassinga

Development of seismic site characterization method using harmonic wavelet analysis of wave (HWAW) method — 767
H-C. Park & D-S. Kim

Use of dilatometer and dual dilatometer test for soft soils and peats — 775
P.P. Rahardjo, Y. Halim & L. Sentosa

Field evaluation of the LCPC *in-situ* triaxial test — 783
Ph. Reiffsteck

CPTWD (Cone Penetration Test While Drilling) a new method for deep geotechnical surveys — 787
M. Sacchetto, A. Trevisan, K. Elmgren & K. Melander

Enhanced access penetration system: a direct push system for difficult site conditions — 795
J.D. Shinn II & J.W. Haas III

Use of a radioisotope cone to characterize a lumpy fill — 801
T.S. Tan, M. Karthikeyan, K.K. Phoon, G.R. Dasari & M. Mimura

Development of a carrier tool for a multi-method investigation of brownfield sites — 809
P. Wotschke & S. Friedel

4. NEW DEVELOPMENTS IN INTERPRETATION OF IN-SITU DATA

General report: New developments in the interpretation of *in-situ* test data — 819
B. Lehane & M. Fahey

Discussion report: Benefit of new developments in interpretation of *in-situ* data — 825
J. Peuchen

Calibration chamber size and boundary effects for CPT q_c measurements — 829
M.M. Ahmadi & P.K. Robertson

State-space seismic cone minimum variance deconvolution — 835
E.J. Baziw

The effects of vibration on the penetration resistance and pore water pressure in sands — 843
J. Bonita, J.K. Mitchell & T.L. Brandon

In-situ testing and foundation engineering: recent contributions — 853
J-L. Briaud

Disturbance effects of field vane tests in a varved clay *A.B. Cerato & A.J. Lutenegger*	861
Assessment of cyclic stability of cohesive deposits using cone penetration *R. Debasis*	869
High resolution stratigraphic and sedimentological analysis of Llobregat delta nearby Barcelona from CPT & CPTU tests *M.J. Devincenzi, S. Colàs, J.L. Casamor, M. Canals, O. Falivene & P. Busquets*	877
Fractal analysis of CPT data *J.A. Díaz-Rodríguez & P. Moreno-Carrizales*	885
Neural networks in soil characterization *W. Ding & J.Q. Shang*	889
Cylindrical cavity expansion modeling for interpretation of cone penetration tests *F. Elmi & J.L. Favre*	897
Stratigraphic profiling by cluster analysis and fuzzy soil classification from mechanical cone penetration tests *J. Facciorusso & M. Uzielli*	905
Site variability, risk, and beta *R.A. Failmezger, P.J. Bullock & R.L. Handy*	913
A case study using in-situ testing to develop soil parameters for finite-element analyses *E. Farouz, J-Y. Chen & R.A. Failmezger*	921
Using a non linear constitutive law to compare Menard PMT and PLT E-moduli *A. Gomes Correia, A. Antão & M. Gambin*	927
Computational and physical basis for dynamic characterization using Love waves *B.B. Guzina, A.I. Madyarov & R.H. Osburn*	935
Effect of ageing on shear wave velocity by seismic cone *J.A. Howie & A. Amini*	943
Dissipation test evaluation with a point-symmetrical consolidation model *E. Imre & P. Rózsa*	951
Feasibility of neural network application for determination of undrained shear strength of clay from piezocone measurements *Y-S. Kim*	957
Measurement of *in-situ* deformability in hard rock *D. Labrie, B. Conlon, T. Anderson & R.F. Boyle*	963
Estimation of shear strength increase beneath embankments by seismic cross-hole tomography *R. Larsson & H. Mattson*	971
Statistical evaluation of the dependence of the liquidity index and undrained shear strength of CPTU parameters in cohesive soils *J. Liszkowski, M. Tschuschke, Z. Młynarek & W. Tschuschke*	979
CPTU – replication test in post flotation sediments *Z. Młynarek, W. Tschuschke & T. Lunne*	987
Evaluation of the coefficient of subgrade reaction for design of multipropped diaphragm walls from DMT moduli *P. Monaco & S. Marchetti*	993
Evaluation of SCPTU intra-correlations at sand sites in the Lower Mississipi River Valley, USA *J.A. Schneider, A.V. McGillivray & P.W. Mayne*	1003

Risk assessment for contaminated tropical soil *A.C. Strava Corrêa & N. Moreira de Souza*	1011
An examination of the engineering properties and the cone factor of soils from East Asia *M. Tanaka & H. Tanaka*	1019
The regional information system 'Databank Ondergrond Vlaanderen - DOV' *I. Vergauwen, P. De Schrijver & G. Van Alboom*	1025

INDICES

Keyword index

Author index

Volume 2

5. CASE STUDIES INVOLVING PRACTICAL PROJECTS

General report: Case studies involving practical projects   1033
P.W. Mayne & D. Hight

Site characterization and site planning for the reclamation of brown fields of former open-pit mining sites with help of cone penetration tests   1035
W. Al Hamdan & R. Azzam

Pre-reclamation characterization of marine clay using *in-situ* testing methods   1041
A. Arulrajah, M.W. Bo & H. Nikraz

Estimating the *in-situ* cohesion of Tehran cemented alluvium using plate load test along the edge of trench   1047
E. Asghari & S.S. Yasrebi

In-situ tests for the geotechnical characterization of airship hangar soil in the city of Augusta   1053
A. Cavallaro, M. Maugeri & A. Ragusa

Characterization of material strata and slide mechanism of slopes based on *in-situ* ground monitoring   1061
M. Chang, G.C. Doo, Y.F. Chiou, S.Y. Lin, S.C. Chien, & J.W. Chen

Geology and geotechnics of Alqueva and Pedrogão dam sites and their importance on the foundation zoning   1067
J.M. Cotelo Neiva, J. Neves & C. Lima

Site characterization of five hazardous waste landfills   1075
A. De, N. Matasovic & R.J. Dunn

Slope degradation and analysis of Mokattam plateau, Egypt   1081
M. EL-Sohby, M. Aboushook & O. Mazen

Casa da Música do Porto: site characterization   1089
A.R. Gaba, A.C. Pickles & R. Oliveira

Analysis of the geotechnical behaviour of tailings disposal systems using CPTU tests   1097
R.C. Gomes, L.H. Albuquerque Filho, L.F.M. Ribeiro & F.M. Pereira

Integral geotechnical investigation – A case study   1105
N.Sh. Guirguis

Ouiqui dyke characterization and rehabilitation   1111
Y. Hammamji, P. Vannobel, J.P. Tournier, R. Piché, G. Lefebvre & M. Karray

The soil investigation and interpretation for a complex metro project   1119
J. Herbschleb

The application of shallow seismic refraction and penetration tests for site investigation for an airport runway extension in Calabar, South Eastern Nigeria   1125
A.O. Ilori, C.S. Okereke & A.E. Edet

Site characterization of paleoliquefaction features in Missouri   1131
H. Jadi, R. Luna, D. Hoffman & P.W. Mayne

Characterization of granular test sites for grillage load testing   1139
F.H. Kulhawy & H.E. Stewart

Application of electrical resistivity tomography to leak detection in a geomembrane at A55 Conwy Tunnel, North Wales   1147
D. Nichol, J.K. Ferris & J.M. Reynolds

A risk assessment method of rain-induced slope failure and a portable dynamic cone penetration test 1155
O. Nunokawa, T. Sugiyama, T. Fujii & K. Okada

A historical case in the Bolivia-Brazil natural gas pipeline: slope on the Curriola River 1161
H.R. Oliveira

Settlements of a building with shallow foundations in Recife, Brazil 1165
J.T.R. Oliveira, R.Q. Coutinho & A.D. Gusmão

Control of subsidence with borehole extensometer and surveying measurements in Murcia, Spain 1171
F. Peral, A. Rodríguez & J. Mulas

The use of geophysical surveying for the investigation of slope failures in southern Brazil 1177
R.J.B. Pinheiro, A.V.D. Bica, L.A. Bressani & A. Strieder

In situ methods and their quality assessments in ground improvement projects 1185
A.J. Puppala, A. Porbaha & V. Bhadriraju

Site characterization for settlement analyses of a large diameter cryogenic tank 1191
P.M. Rao, K.M. Vennalaganti, S.N. Endley & K. Sreerama

Liquefaction potential mapping in Memphis, Tennessee using stochastically simulated CPT data 1199
S. Romero-Hudock & G.J. Rix

Wali Al Ahed flyover – long distance assistance in site characterization 1207
J.S. Steenfelt

Application of *in-situ* geotechnical tests in characterization of slope failures 1213
J. Vakili

Drilling process monitoring for ground characterization during soil nailing in weathered soil slopes 1219
Z.Q. Yue, J.Y. Guo, L.G. Tham & C.F. Lee

Site investigation for unstable slope instrumentation and assessment 1225
J. Záleský, J. Schröfel, J. Pruška & J. Salák

6. CHARACTERIZATION OF NON-TEXTBOOK GEOMATERIALS

General report: Characterization of non-textbook geomaterials 1233
R.Q. Coutinho, J.B. de Souza Neto & K.C. de Arruda Dourado

Discussion report: Characterization of non-textbook geomaterials 1259
J.A. Howie

Determination of optimum sampling locations in a solid waste landfill using PCSV technique 1263
S. Avsar, A. Bouazza & E. Kavazanjian

Evaluation of the collapsibility of a sand by Ménard pressuremeter 1267
R.Q. Coutinho, K.C.A. Dourado & J.B. Souza Neto

Evaluation of effective cohesive intercept on residual soils by DMT and CPT 1275
N. Cruz, A. Viana da Fonseca & E. Neves

Penetrometer testing in residual soils from granitic rocks in the South of Portugal 1279
I.M.R. Duarte, A.B. Pinho & F.L. Ladeira

Physical, mechanical, and hydraulic properties of coal refuse for slurry impoundment design 1285
Y.A. Hegazy, A.G. Cushing, & C.J. Lewis

Example of erratic distribution of weathering patterns of Porto granite masses and its implication 1293
on site investigation and ground modelling
E. Marques, A. Viana da Fonseca, P. Carvalho & A. Gaspar

In-situ geotechnical characterization of the Brasilia porous clay 1301
F.E.R. Marques, J. Almeida e Sousa, C.B. Santos, A.P. Assis & R.P. Cunha

Bearing capacity evaluation of drilled piles into a non-saturated soil profile *M.M.A. Mascarenhas, R.C. Guimarães, J. Camapum de Carvalho & P.R.F. Falcão*	1311
Effects of partial saturation on CPTU readings in Burleson clay *J.B. Nevels, Jr. & N. Khoury*	1319
Geotechnical site investigation of municipal solid waste landfills *D.A.F. Oliveira & P. Murrieta*	1325
Density measurements in a tailings dam using penetrologger tests *L.F.M. Ribeiro, L.H. Albuquerque Filho, E.L. Pereira, R.C. Gomes & A.L. Leite*	1331
The contamination level through an organic soil of Gramacho MSW landfill *E. Ritter, J.C. Campos & R.L. Gatto (deceased)*	1339
SPT, CPT and CH tests results on saprolitic granite soils from Guarda, Portugal *C.M.G. Rodrigues & L.J.L. Lemos*	1345
Interpretation of pressuremeter tests in a gneiss residual soil from São Paulo, Brazil *F. Schnaid & F.M. Mántaras*	1353
Geotechnical characterization of a residual soil profile: the ISC'2 experimental site, FEUP *A. Viana da Fonseca, J. Carvalho, C. Ferreira, E. Costa, C. Tuna & J.A. Santos*	1361
The quality control of post flotation reservoir dam by determination of relative compaction index in various methods *J. Wierzbicki, A. Niedzielski, M. Waliński & W. Wołyński*	1371

7. APPLICATIONS TO GEOTECHNICAL STRUCTURES

General report: Applications to geotechnical structures *C.W.W. Ng*	1379
Discussion report: Applications to geotechnical structures *R.P. Cunha*	1391
The engineering geology assessment of Sabzkuh water conveyance tunnel route, Central Iran *R. Ajalloeian, M. Hashemi & Sh. Moghaddas*	1397
Simulation of transient water flow through Robert Bourassa Main Dam, Quebec, Canada *V. Alicescu, J.-P. LeBihan (deceased) & S. Leroueil*	1403
Geotechnical investigations on the Zagrad location in Rijeka, Croatia *Ž. Arbanas, M.S. Kovačević & B. Jardas*	1415
Bearing capacity assessment for an Omega pile in a Brazilian tropical soil through cone penetration test *E. Beira Fontaine & D. de Carvalho*	1421
Study of drilled piles behaviour, modelling by finite-element method *A. Bekkouche & A. Ras*	1425
Pressuremeter test for evaluating load transfer characteristics along bored piles in residual soil *M-F. Chang & H. Zhu*	1431
Individual foundation design for column loads *R.A. Failmezger & P.J. Bullock*	1439
Prediction of the ground conditions ahead the TBM face in the tunnels of Guadarrama (Spain), using Electrical Resistivity Tomography (ERT) *J.M. Galera, F. Peral & A. Rodríguez*	1443
Ground characterization for railroad embankments by seismic methods *W. Gardien, H.G. Stuit, F.H. Drossaert & G.G. Drijkoningen*	1449

A case history: dock enlargement at Barcelona harbour R. Gómez Escoubès, M. Arroyo, J.M. González Herrero, M. Devincenzi & R. Sáenz de Navarrete	1457
Drilled shafts impedance evaluation on Doremus Avenue Bridge N. Gucunski, M. Balic, A. Maher & H.H. Nassif	1465
Site characterization for the investigation of tunnel 8 collapse A-B. Huang, C.P. Lin, J.J. Liao, Y.W. Pan & J. Tinkler	1473
Results of surface waves testing in the investigation of the Mont-Blanc tunnel after the 1999 fire M. Karray, G. Lefebvre & R.M. Faure	1481
Determination of unit skin friction in a sand deposit from *in situ* test results S.P. Kelley & A.J. Lutenegger	1489
CPT-based load capacity of closed- and open-ended pipe piles J.-H. Lee, R. Salgado & K-H. Paik	1499
Evaluation of pile behaviour using piezocone tests A. Mahler, E. Imre & A. Kocsis	1507
DMT-predicted vs measured settlements under a full-scale instrumented embankment at Treporti (Venice, Italy) S. Marchetti, P. Monaco, M. Calabrese & G. Totani	1511
Advantages and equations for pile design in Brazil via DPL tests T. Nilsson & R. Cunha	1519
Pile settlement predictions using theoretical load transfer curves and seismic CPT data 1525 M.A. Pando, A.L. Fernández & G.M. Filz	
Site characterization and project-specific rock mass classification, Rio Tinto Tunnel, Metro do Porto I. Pöschl & R. Pais	1533
Clay seams below a pile tip – how much do they affect pile performance? H.G. Poulos	1543
Behavior of Omega piles, subjected to compression instrumented load tests P.J. Rocha de Albuquerque, D. de Carvalho & F. Massad	1551
Geotechnical site investigation for tunneling design. A step-by-step approach: Case study J.M.C. Roxo	1557
The monitoring of pile construction to assess integrity and load carrying capacity of CFA piles M. Rust, C.R.I. Clayton, N. Mure, J. Scott, D. Seward & S. Quale	1565
Site investigation by trial embankments J. Škopek & J. Boháč	1571
Reliability of shallow foundation design using the standard penetration test D.P. Zekkos, J.D. Bray & A. Der Kiureghian	1575

8. ENHANCED CHARACTERIZATION BY COMBINED IN-SITU TESTING

General report: Enhanced characterization by combined *in-situ* testing V. Fioravante	1585
A brief study of the repeatability of *in-situ* tests at the Florida Department of Transportation deep foundations research site in Orlando, Florida, USA J.B. Anderson, F.C. Townsend & E. Horta	1597
CPTu-DMT performance-based correlation for settlement design M. Arroyo, M. Devincenzi, M.T. Mateos, R. Gómez-Escoubès & J.M. Martínez	1605

Two and three dimensional imaging utilizing the seismic cone penetrometer *E. Baziw*	1611
The importance of geological investigation in geotechnical characterization of Pusan clays *S.G. Chung, C.K. Ryu, S.H. Baek & S.W. Kim*	1619
Integration of testing data in a site investigation program: examples from tropical soils sites *G. De Mio & H.L. Giacheti*	1627
Site characterization for tunnel-building interaction problem in Brazil *S.B. Foá, P.G.O. Passos, A.P. Assis & M.M. Farias*	1635
Use of piezocone tests to characterize the silty soils of the Venetian lagoon (Treporti test site) *G. Gottardi & L. Tonni*	1643
Critical train speed and the benefit of seismic profiling *V. Hopman & P. Hölscher*	1651
In-situ measurement of material damping with seismic cone penetration test *L. Karl, W. Haegeman & G. Degrande*	1655
Comparison of *in-situ* tests to determine engineering properties of a deltaic sand *S.P. Kelley & A.J. Lutenegger*	1663
Using SCPT and DMT data for settlement prediction in sand *B. Lehane & M. Fahey*	1673
Geotechnical correlations in Madrid soil from SPT and Pressuremeter tests *A. López Carrasco, A.Rodríguez Soto & M. Torres*	1681
Shear wave velocity-penetration resistance correlation for Holocene and Pleistocene soils of an area in central Italy *C. Madiai & G. Simoni*	1687
Seismic piezocone and seismic flat dilatometer tests at Treporti *A. McGillivray & P.W. Mayne*	1695
Stiffness of cemented gravel of Tehran from pressuremeter and other *in-situ* tests *B. Pahlavan, A. Fakher & M. Khamehchian*	1701
Use of DMT and DPL tests to evaluate ground improvement in sand deposits *P.G.O. Passos, M.M. Farias, & R.P. Cunha*	1709
Small strain stiffness assessments from *in-situ* tests *J.J.M. Powell & A.P. Butcher*	1717
Comparison of cross-hole, seismic cone penetrometer, spectral wave (SASW) to characterize bridge sites in the New Madrid Seismic Zone *R.W. Stephenson, R. Luna, N. Anderson & W. Liu*	1723
Experimental *in-situ* test sites *G. Togliani & G. Beatrizotti*	1731
Characterization of marine clay soil for a mass rail transit system in Singapore *S.Y. Tong, S.K. Tang & S. Sugawara*	1739
The subsoil – Crisis or risk management? *M.Th. van Staveren*	1747
Tentative evaluation of K_0 from shear waves velocities determined in Down-Hole (V_s^{vh}) and Cross-Hole (V_s^{hv}) tests on a residual soil *A. Viana da Fonseca, C. Ferreira & J.M. Carvalho*	1755
Evaluation of *in-situ* K_0 for Ariake, Bangkok and Hai-Phong clays *Y. Watabe, M. Tanaka & J. Takemura*	1765

9. LABORATORY AND FIELD COMPARISONS

General report: Laboratory and field comparisons D.J. DeGroot & R. Sandven	1775
Comparison of CPTU and laboratory tests interpretation for Polish and Norwegian clays Z. Bednarczyk & R. Sandven	1791
Deriving geotechnical parameters of residual soils from granite by interpreting DMT+CPTU tests N. Cruz, S. Figueiredo & A. Viana da Fonseca	1799
Dynamic characterization of alluvial deposit in urban area V. Fioravante	1805
Evaluation of Bryozoan limestone properties based on *in-situ* and laboratory element tests P.G. Jackson, J.S. Steenfelt, N. Foged & J. Hartlén	1813
Geotechnical characterization for slope failure analysis A. Menkveld & U.F.A. Karim	1821
Geotechnical characterization of an overconsolidated Pliocene clay by field and laboratory tests F. Pelli, D. Minuto, E. Isetta & G. Lombardi	1827
Variability of OCR-Q_t correlation in Singapore upper marine clay K.K. Phoon, H.E. Low & T.S. Tan	1835
Comparison between *in-situ* and laboratory tests results on undisturbed frozen samples for a natural coarse sand D. Porcino & V.N. Ghionna	1843
Undrained shear strength obtained from *in situ* and laboratory tests R. Sandven & J. Black	1851
Sample disturbance in highly sensitive clay R. Sandven, T. Ørbech & T. Lunne	1861
Characterizing artificial and natural bonding of clays S. Shibuya, D.J. Li & H. Tanaka	1869
A study on disturbance effect of clay by block sampling Y-S. Shin, Y-J. Kim & H-H. Chang	1875
Laboratory and *in-situ* measurement of attenuation in soils Y.H. Wang, W.M. Yan & K.F. Lo	1883
Numerical modeling considering nonlinear deformability of soft rock foundations Y. Yamaguchi & M. Nakamura	1891

INDICES

Keyword index	1901
Author index	1907

Introduction

Site characterization is the initial required step towards the proper analysis and design of all types of geotechnical projects, ranging from foundations, tunnels, excavations, earth dams, embankments, and offshore structures. It is paramount that a proper and thorough site investigation be completed in order to fully realize the likelihood of construction difficulties, long term performance, and optimal economy for the underground design. Towards the betterment of subsurface exploration practices, the Second International Conference on Site Characterization (ISC-2) was held in Porto, Portugal from 19-22 September 2004 to bring together researchers, users, and practicing engineers & geologists involved in geotechnical & geophysical site characterization. The concept of ISC-2 was initiated by ISSMGE Technical Committee TC16 on In-Situ Testing in collaboration with TC10 on Geophysics.

These two volume proceedings contain 11 invited keynote lectures that were prepared by outstanding experts in the field, as well as some 219 contributed technical papers from over 40 different countries on various aspects of site characterization. Papers were refereed by at least two outside reviews to obtain a high quality publication and technical standards. Specifically, the conference addressed the use and benefits of geotechnical site characterization, including *in-situ* tests, field methods, interpretation, and geophysics for the exploration of the subsurface environment. The main objectives were used to create a forum for the exchange of ideas and experiences between researchers, equipment manufacturers, and practicing civil engineers & geologists regarding the application of geotechnical and geophysical methods for the solution of ground-related problems in soil and rock. The scope included conventional geotechnical issues, as well as geo-environmental, seismic risk, and soil dynamics. Emphasis was placed on the development of new equipment & field data acquisition, improved interpretation of data, cross-comparisons of methods, and the application of *in-situ* tests in engineering practice.

Accordingly, the papers have been sorted into nine general themes that were established for presentation, including: 1. Mechanical *in-situ* testing methods; 2. Geophysical methods applied to geotechnical engineering; 3. Innovative technologies & equipment; 4. New developments in the interpretation of *in-situ* data; 5. Case studies involving practical projects; 6. Characterization of "nontextbook" geomaterials; 7. Applications to geotechnical structures; 8. Enhanced characterization by combined *in-situ* testing; 9. Laboratory and field comparisons.

The conference would not have come together without the dedicated work and commitment of many fine individuals of the organizing committee and international societies, as well as the varied international authors of papers who contributed their innovation, creativity, experience, and findings in technical writing & presentation. We are most grateful for their help and assistance in preparing these two-volume proceedings for your reference.

António Viana da Fonseca	Paul W. Mayne
University of Porto	Georgia Institute of Technology

Reviewing Committee

CORE MEMBERS OF TC16 - ISSMGE

Prof. P. W. Mayne *(Chair - GIT, USA)*
Prof. An-Bin Huang *(NCTU, Taiwan)*
Prof. Fernando Schnaid *(UFRGS, Brazil)*
Dr. John M. Powell *(BRE, UK)*
Prof. Martin Fahey *(UWA, Australia)*
Dr. Tom Lunne *(NGI, Norway)*
Prof. Zbigniew Mlynarek *(Poznan Univ., Poland)*

Chair of TC10
Dr. K. Rainer Massarsch *(SGS, Sweden)*

CORE MEMBERS OF TC16 AND TC10

Prof. A. Viana da Fonseca *(president ISC'2, FEUP, Portugal)*

Prof. J. Almeida e Sousa *(FCTUC, Portugal)*
Ing. Ricardo Andrade Cruz *(Mota-Engil, Portugal)*
Dr. Tony Butcher *(BRE, UK)*
Prof. Jorge Carvalho *(FEUP, Portugal)*
Prof. Abílio Cavalheiro *(FEUP, Portugal)*
Prof. Roberto Coutinho *(UFP, Brazil)*
Ing. Nuno Cruz *(Mota-Engil, Portugal)*
Prof. Renato Cunha *(UFB, Brazil)*
Prof. Luís de Sousa *(LNEC, FEUP, Portugal)*
Prof. Don DeGroot *(UMass, USA)*
Dr. Marcelo Devincenzi *(Igeotest, Spain)*
Prof. Vicenzo Fioravante *(Unife, Italy)*
Prof. Sebastiano Foti *(Politecnico Torino, Italy)*
Dr. David Hight *(GCG, UK)*
Prof. John Howie *(UBC, Canada)*

Prof. Mike Jamiolkowski *(Politecnico Torino, Italy)*
Prof. Barry Lehane *(UWA, Australia)*
Prof. Silvano Marchetti *(Univ. L'Aquila, Italy)*
Prof. M. Matos Fernandes *(FEUP, Portugal)*
Prof. Charles Ng *(UST, Hong Kong, PRC)*
Dr. Marília Oliveira *(LNEC, Portugal)*
Ing. Joek Peuchen *(Fugro, Netherlands)*
Prof. K.K. Phoon *(NUS, Singapore)*
Prof. Peter Robertson *(Univ. Alberta, Canada)*
Prof. Rolf Sandven *(NTNU, Norway)*
Prof. Jaime Santos *(IST-UTL, Portugal)*
Prof. Jørgen Steenfelt *(COWI A/S, Denmark)*
Dr. Hiroyuki Tanaka *(IAI, Japan)*
Prof. Hai-Sui Yu *(Univ. Nottingham, UK)*

Reference documents on *in-situ* testing
(Geotechnical Site Characterization)

Field Testing of Soils (1962). ASTM Special Technical Publication 322, ASTM 65th Annual Meeting, American Society for Testing & Materials, West Conshohocken, Pennsylvania, 318 pages: www.astm.org

Symposium on Soil Exploration (1964). ASTM Special Technical Publication 351, ASTM 66th Annual Meeting, American Society for Testing & Materials, West Conshohocken, Pennsylvania, 156 pages: www.astm.org

In-Situ Investigations in Soils & Rocks (1970). Proceedings of Conference by British Geotechnical Society, William Clowers & Sons Ltd, London, 324 pages.

Special Procedures for Testing Soil & Rock for Engineering Purposes (1970). ASTM Special Technical Publication 479, American Society for Testing & Materials, West Conshohocken, Pennsylvania: www.astm.org

Determination of the In-Situ Modulus of Deformation of Rock (1970). ASTM Special Technical Publication 477, American Society for Testing & Materials, West Conshohocken, Pennsylvania: www.astm.org

Field Testing and Instrumentation of Rock (1973). ASTM Special Technical Publication 554, American Society for Testing & Materials, West Conshohocken, Pennsylvania: www.astm.org

Capacite Portante et Tassements des Fondations à Partir d' Essais In Situ (Nuyens, J. (1973). Presses Universitaires de Bruxelles (Belgium) or Eyrolles, Paris (France)

European Symposium on Penetration Testing (ESOPT-l, 1974, Amsterdam), two volumes, published by the Swedish Geotechnical Society, Stockholm.

In-Situ Measurement of Soil Properties (1975), Proceedings of the Conference on In-Situ Measurement of Soil Properties held at North Carolina State University, Raleigh, Vol. I and II, ASCE, Reston/Virginia.

Dynamic Geotechnical Testing (1978), Special Technical Publication (STP) No.654, American Society for Testing & Materials, West Conshohocken, Pennsylvania.

The Pressuremeter and Foundation Engineering (Baguelin, F., Jezequel, J., and Shields, D., 1978), Trans Tech Publications, Clausthal, Germany.

Les Essais In Situ en Mecanique des Sols (Cassan, M., 1978), Volumes 1 and 2, Eyrolles, Paris, France (Volume 1 re-published in 1980)

First International Symposium on Pressuremeter Testing (ISP-1, 1982), Editions Technip, 27 rue Ginoux, 75737, Paris, France.

Cone Penetration Testing and Experience (1981), Proceedings from ASCE National Convention, St. Louis, Missouri; published by American Society of Civil Engineers, Reston, Virginia.

European Symposium on Penetration Testing (ESOPT-2, 1982), Amsterdam, The Netherlands; published as Penetration Testing, Volumes 1 and 2, Balkema, Rotterdam.

Reconnaissance Géologique et Géotechnique des Traces de Routes et Autoroutes (Rat, M., editor, 1982). Laboratoire Central des Ponts et Chaussées, 75732 Paris, France, purchase from www.lcpc.fr

Proceedings, Soil and Rock Investigations by in Situ Testing (1983), three volumes (bilingual), published by Revue Française de Geotechnique, 75343 Paris, France

Second International Symposium on Pressuremeter Testing (ISP-2, 1986) College Station, Texas, USA; published as The Pressuremeter and Its Marine Applications, Special Technical Publication No.950, American Society for Testing & Materials, West Conshohocken, Pennsylvania.

Use of In-Situ Tests in Geotechnical Engineering (In-Situ '86), Blacksburg, Virginia, USA; published as Geotechnical Specialty Publication (GSP) No.6, American Society of Civil Engineers, Reston, Virginia, 1986.

Vane Shear Strength Testing in Soils: Field & Laboratory Studies (1987), Special Technical Publication No.1014, American Society for Testing & Materials, West Conshohocken, Pennsylvania.

Essais de Penetration (Amar, S., editor, 1987), Laboratoire Central des Ponts et Chaussées, 75732 Paris, France, purchase from www.lcpc.fr

Penetration Testing in the UK, Proceedings of the Geotechnology Conference on Penetration Testing in the United Kingdom (PTUK), Birmingham (1988); published by Thomas Telford, London.

First International Symposium on Penetration Testing (ISOPT -1, 1988) Orlando, Florida, USA; published as Penetration Testing 198, Volumes 1 and 2, Balkema, Rotterdam.

Third International Symposium on Pressuremeter Testing (ISP-3, 1990) Oxford, UK; published as Pressuremeters by Thomas Telford and the British Geotechnical Society, London.

Proceedings First International Conference on Calibration Chamber Testing (ISOCCT-1, 1991), Potsdam, New York; ed. by A-B. Huang, published as Calibration Chamber Testing, Elsevier, New York.

The Pressuremeter (Briaud, J.L., 1992), A.A. Balkema, Rotterdam and Brookfield, Vermont.

Dynamic Geotechnical Testing II (1994), Special Technical Publication No.1213, American Society for Testing & Materials, West Conshohocken, Pennsylvania.

Pre-Failure Deformation of Geomaterials (1994, Sapporo). Proceedings, International Symposium on Pre-Failure Deformation Characteristics of Geomaterials, Vol. 1 and 2, edited by S. Shibuya, T. Mitachi, and S. Miura. Balkema, Rotterdam.

Fourth International Symposium on Pressuremeters (ISP-4, 1995) Sherbrooke, Quebec, Canada; published as The Pressuremeter and Its New Avenues by A.A. Balkema, Rotterdam.

International Symposium on Cone Penetration Testing (CPT'95), Linköping, Sweden; published by the Swedish Geotechnical Society, Report No. 3-95.

Pressuremeters in Geotechnical Design (Clarke, B. G., 1995), Blackie Academic and Professional (Chapman & Hall), Glasgow, Scotland

Advances in Site Investigation Practice (1996), Proceedings of the Institution of Civil Engineers, London, UK, March 1995; published by Thomas Telford, London.

Cone Penetration Testing in Geotechnical Practice (1997) by T. Lunne, P.K. Robertson, and J.J.M. Powell, SPON/Blackie/London, Rutledge Publishing, New York.

Proceedings International Site Characterization (1998, ISC-1, Atlanta), published as Geotechnical Site Characterization, Volumes 1 and 2, editors P.K. Robertson and P.W. Mayne, Balkema, Rotterdam.

Proceedings of the International Symposium on Characterization of Soft Marine Clays (1997, Yokosuka). edited by T. Tsuchida and A. Nakase, published as Characterization of Soft Marine Clays (1999), Balkema, Rotterdam.

Pre-Failure Deformation Characteristics of Geomaterials (1999), Volumes 1 and 2, Proceedings of the Second Intl. Conference, Torino, edited by M. Jamiolkowski, R. Lancellotta, and D. LoPresti, Balkema, Rotterdam, 1419 pages.

Innovations & Applications in Geotechnical Site Characterization (2000) editors P.W. Mayne and R.D. Hryciw, GSP 97, ASCE, Reston/VA, 247 pages.

Proceedings, International Conference on Soil Properties and Case Histories (In-Situ 2001), ed. P. Rajardjo, Bali, Indonesia.

Soils and Waves (2001) by J.C. Santamarina, K.A. Klein, and M.A. Fam, John Wiley & Sons, New York, 488 pages.

Characterization and Engineering Properties of Natural Soils (2002, Singapore), edited by T.S. Tan, K.K. Phoon, D. Hight and S. Leroueil, two volumes, Swets & Zeitlinger, Lisse.

A Short Course in Geotechnical Site Investigation (2002). by N. Simons, B. Menzies, and M. Matthews, Thomas Telford, London, 353 pages.

Symposium International on Identification & Determination of Soil & Rock Parameters for Geotechnical Design (PARAM 2002), Paris, 2-3 September 2002; Laboratoire Central des Ponts et Chaussées (LCPC), 620 pages.

Deformation Characteristics of Geomaterials / Comportement des Sols et des Roches Tendres Proceedings of the 3rd International Symposiumon Deformation Characteristics of Geomaterials, two volumes, edited by Di Benedetto, H., Doanh, T., Geoffroy, H., Sauzéat, C. (IS Lyon 03), Lyon, France, 22-24 September, Swets & Zeitlinger, Lisse.

Acknowledgements

The Organizing Committee of the 2nd International Conference on Site Characterization is grateful to:

Faculdade de Engenharia
Universidade do Porto
International Society of Soil Mechanics and Geotechnical Engineering
ISRM National Group from Portugal
International Association for Engineering Geology and the Environment
Sociedade Portuguesa de Geotecnia
Câmara Municipal do Porto
ASCE Geo-Institute
Ordem dos Engenheiros
British Council
Fundação Luso-Americana
Fundação Oriente
Instituto da Construção

Grupo Mota-Engil
Administração dos Portos do Douro e de Leixões, Igeotest, Instituto de Estradas de Portugal, Rodio, Casa da Música, Somague-Mesquita, ACE, Sopecate, Tecnasol FGE, Teixeira Duarte
APvdBerg, Atlas Copco, ETECLda, Fundasol, Metro do Porto

The publication of the Proceedings has been partly funded by the "Portuguese Science and Technology Foundation"

Keynote lectures

James K. Mitchell Lecture
In situ soil testing: from mechanics to interpretation

Hai-Sui Yu
Nottingham Centre for Geomechanics, The University of Nottingham, UK

Keywords: in situ soil testing, mechanics, interpretation, analysis, strength, stiffness, state, clay, sand

ABSTRACT: This paper reviews and evaluates the current use of fundamental mechanics in developing rational interpretation methods for deriving soil properties from in situ test results. The focus is on some of the most widely used in situ test devices including cone penetrometers with and without pore pressure measurements (CPTU and CPT), self-boring and cone pressuremeters (SBPMT and CPMT), and flat dilatometers (DMT). In situ tests in both cohesive and frictional soils for measuring strength and stiffness properties, in situ state parameters, consolidation coefficients, stress history and in situ stresses are considered in detail.

1 INTRODUCTION

In his foreword to the Author's book 'Cavity Expansion Methods in Geomechanics' (Yu, 2000), Professor James K. Mitchell stated:

'The ability to treat the results of cone penetration and pressuremeter tests in sand and clay on a realistic theoretical basis enhances their value for site characterisation and determination of relevant soil mechanical properties'.

The preparation of this inaugural Mitchell Lecture therefore provides a good opportunity to conduct a brief review of the current use of both continuum and particle mechanics in the interpretation of in situ soil tests for measuring design parameters. Because of time and space constraints, the review will be selective, and is organised in terms of different in situ tests and their related interpretation methods. The focus will be on the interpretation of cone penetration tests (CPT/CPTU), self-boring and cone pressuremeter tests (SBPMT/CPMT) and flat dilatometer tests (DMT) that is based on a sound understanding of the mechanics of these tests. The selected topics cover more recent developments and, to some extent, also reflect the Author's own research interests in the area.

In situ testing serves a number of purposes in geotechnical engineering, which include (Ladd et al., 1977; Wroth, 1984; Jamiokowski et al., 1985):

- Site classification and soil profiling.
- Measurement of a specific property of the ground.
- Development of empirical rules for foundation design.
- Control of construction.
- Monitoring of performance and back analysis.

Whilst all these operations will benefit from a good understanding of the mechanics of in situ tests, it is essential if an accurate measurement of a specific property of the ground is to be made. This is because, unlike laboratory testing, in situ testing is generally an indirect technique as soil properties cannot be obtained directly from measured response without solving it as a boundary value problem.

In a most comprehensive review on the measurement of soil properties in situ, Mitchell et al. (1978) identified the following main reasons for the increased use of field testing:

- To determine properties of soils, such as continental shelf and sea floor sediments and sands, that cannot be easily sampled in the undisturbed state.
- To avoid some of the difficulties of laboratory testing, such as sample disturbance and the proper simulation of in situ stresses, temperature and chemical and biological environments.
- To test a volume of soil larger than can conveniently be tested in the laboratory.

- To increase the cost effectiveness of an exploration and testing programme.

However it has also long been realised (Wroth, 1984) that the interpretation of in situ tests is beset with difficulties especially if their results are needed to assess the stress-strain and strength characteristics of the tested soils. Jamiokowski (1988) highlighted the following difficulties that could form the major sources of uncertainty:

1) With the exception of the self-boring pressuremeter tests (SBPMT) and some geophysical tests, all other in situ tests represent complex boundary value problems rendering their theoretical interpretation very difficult.
2) The drainage conditions during in situ tests are usually poorly controlled and present the problem of determining whether the tests have been performed in undrained, drained or partially drained conditions.
3) Frequently, during the execution of in situ tests the soil involved is subjected to effective stress paths (ESP) which are very different from those representative of the relevant engineering problem. Hence, the measured soil stress-strain response is ESP dependent, and reflects its anisotropic and elastoplastic behaviour. This represents one of the most difficult problems when interpreting in situ test results.
4) Due to the highly pronounced nonlinear behaviour of all soils, even at small strains, it is difficult to link the stress-strain characteristics obtained from in situ tests to the stress or strain level relevant to the solution of the specific design problem.

Despite a large amount of empiricism and oversimplified assumptions involved with interpretation, in situ testing has and will continue to play a key role in the characterisation of natural soil deposits. Table 1 lists some of the fundamental soil properties that in situ testing can measure with a sound theoretical basis.

Table 1: Some current in situ testing capabilities for measuring soil properties

Test	Measured properties	Selected references
Cone penetration tests (CPT/CPTU)	Soil profiling Stress history (OCR) Consolidation coefficient In situ state parameter for sand Undrained shear strength Hydrostatic pore pressure	Robertson (1986) Wroth (1984), Mayne (1993) Baligh and Levadoux (1986) Teh (1987) Been et al. (1987) Yu and Mitchell (1998) Lunne et al. (1997)
Self-boring pressuremeter tests (SBPMT)	Horizontal in situ stress Shear modulus Shear strength Stress-strain curve In situ state parameter for sand Consolidation coefficient Small strain stiffness	Jamiolkowski et al. (1985) Wroth (1982) Gibson and Anderson (1961) Hughes et al. (1977) Palmer (1972) Manassero (1989) Yu (1994, 1996, 2000) Clarke et al. (1979) Byrne et al. (1990) Jardine (1992) Fahey and Carter (1993) Bolton and Whittle (1999)
Cone pressuremeter tests (CPMT)	Horizontal in situ stress Shear modulus Shear strength In situ state parameter for sand	Houlsby and Withers (1988) Schnaid (1990) Yu (1990) Yu et al. (1996)
Flat dilatometer tests (DMT)	Soil profiling Horizontal in situ stress Stress history (OCR) Shear strength In situ state parameter for sand	Marchetti (1980) Mayne and Martin (1998) Finno (1993) Huang (1989) The present paper - Yu (2004)

Given the interpretation of an in situ test requires the analysis of a corresponding boundary value problem, some simplifying assumptions will have to be made as in the case of solving any other boundary value problems. In particular, assumptions will have to be made with respect to the following three aspects:

1) Geometry and boundary conditions.
2) Soil behaviour.
3) Water drainage conditions.

2 SELF-BORING PRESSUREMETER TESTS IN CLAY

2.1 Overview

The self-boring pressuremeter (shown in Figure 1) has been established as one of the best in situ testing devices for measuring soil properties since its initial development over three decades ago in France (Baguelin et al., 1972) and the UK (Wroth and Hughes, 1972). Almost all the theoretical interpretation methods developed for it were based on the fundamental assumption that the pressuremeter geometry is such that the test can be simulated as the expansion and/or contraction of an infinitely long, cylindrical cavity in the soil. The advantage of this fundamental assumption is that the pressuremeter problem becomes one-dimensional for which many analytical solutions exist even for complex soil models (Yu, 2000). For tests in clay, it is often assumed that the test is carried out fast enough so that the undrained condition may be valid. With respect to soil behaviour, many models of various complexities (e.g., linear or nonlinear elastic together with perfectly plastic or strain hardening plastic models) have been used in the interpretation.

In recent years, the validity of these earlier assumptions with respect to pressuremeter geometry, water drainage and soil behaviour has been assessed in detail by numerical methods. It is now known that some of these simplifying assumptions could lead to significant errors in the derived soil properties.

2.2 Undrained shear strength

Self-boring pressuremeters are frequently used to determine undrained shear strengths of clays. Most interpretation methods take the following steps: First, a complete stress-strain relation is assumed for the soil, based on which the theoretical pressuremeter curves can be obtained by analysing the test as a cylindrical cavity expansion problem, either analytically or numerically. Then by matching some key or all parts of the theoretical pressuremeter curves with those of a real pressuremeter test curve, the undrained shear strength may be estimated. Examples of the interpretation methods include Gibson and Anderson (1961), Jefferies (1988), and Yu and Collins (1998).

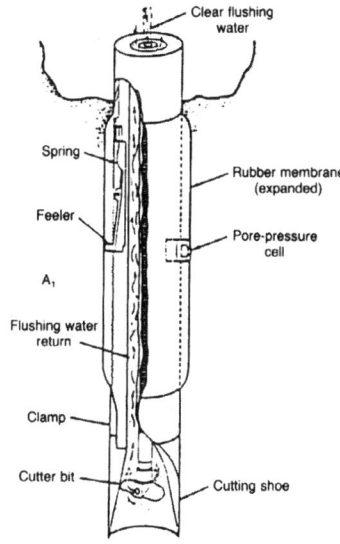

Figure 1: The Cambridge self-boring pressuremeter

2.2.1 Total stress loading analysis

Gibson and Anderson (1961) were among the first to use cavity expansion theory to develop interpretation methods for deriving soil properties from pressuremeter test results. In their analysis, the clay was assumed to behave as a linear elastic-perfectly plastic Tresca material obeying the following failure criterion:

$$\sigma_1 - \sigma_3 = 2S_u \qquad (1)$$

where σ_1 and σ_3 are the major and minor principal stresses, and S_u is the undrained shear strength, whose value is not unique for a clay but depends on stress conditions imposed by a particular test (Wroth, 1984).

The pressuremeter test was idealised as the expansion of an infinitely long, cylindrical cavity in soil under undrained conditions. For simplicity, a total stress formulation was used in the analysis of Gibson and Anderson (1961). With the above assumptions, cavity expansion theory can be used to give the following theoretical pressuremeter expansion curve at the stage of plastic loading:

$$P = \sigma_{h0} + S_u\left\{1 + \ln\left(\frac{G}{S_u}\right)\right\} + S_u \ln\frac{\Delta V}{V} \quad (2)$$

where $\Delta V/V = (a^2 - a_0^2)/a^2$ is the volumetric strain; a and a_0 are the current and initial cavity radii respectively; P and σ_{h0} are the total pressuremeter pressure and total in situ horizontal stress; and G is the shear modulus of the soil.

The theoretical pressuremeter curve, as defined by equation (2), suggests that if pressuremeter results are plotted in terms of cavity pressure against the logarithm of the volumetric strain, the slope of the plastic portion (which is a straight line) is equal to the undrained shear strength of the soil S_u.

2.2.2 Total stress unloading analysis

Jefferies (1988) and Houlsby and Withers (1988) independently extended Gibson and Anderson's solution to include unloading. Jefferies (1988) derived the unloading solution for application to self-boring pressuremeter tests, where some small strain assumptions were used to simplify the mathematics. On the other hand, Houlsby and Withers (1988) were concerned with cone pressuremeter tests for which a large strain analysis is necessary.

The small strain cavity unloading solution, as derived by Jefferies (1988), can be expressed in the following form:

$$P = P_{max} - 2S_u\left\{1 + \ln\left(\frac{G}{S_u}\right)\right\}$$
$$- 2S_u \ln\left\{\frac{a_{max}}{a} - \frac{a}{a_{max}}\right\} \quad (3)$$

where a_{max} is the cavity radius at the end of the loading stage, P_{max} is the cavity pressure at the end of the loading stage and a denotes cavity radius at any stage of pressuremeter unloading.

The theoretical pressuremeter unloading solution, as defined by equation (3), suggests that if the pressuremeter unloading results are presented as the pressuremeter pressure versus $-\ln(a_{max}/a - a/a_{max})$, the slope of the plastic unloading portion (which is a straight line) is equal to twice the soil undrained shear strength.

2.2.3 Total stress analysis with a hyperbolic soil model

If the stress-strain behaviour of clay can be described by a hyperbolic equation, then closed form solutions can be obtained for cavity expansion curves if elastic strains are ignored (Prevost and Hoeg, 1975; Denby and Clough, 1980). Both strain hardening and strain softening may be considered. For example, in the case of strain hardening, the stress-strain relation may be described as follows:

$$q = \frac{\gamma}{D + \gamma} q_u \quad (4)$$

where q is defined as $\sqrt{3}/2$ times the difference of the major and minor principal stresses and γ is shear strain (i.e. the difference between the major and minor principal strains). D is a soil constant and the second soil constant q_u is the ultimate shear stress (i.e. $\sqrt{3}$ times the undrained shear strength). It then follows that the pressuremeter loading curve can be described as a function of two soil constants D and q_u in the following closed form:

$$P = \sigma_{h0} + \frac{q_u}{\sqrt{3}} \ln\left(1 + \frac{2}{\sqrt{3}D}\varepsilon_c\right) \quad (5)$$

where $\varepsilon_c = (a - a_0)/a_0$ is the cavity strain. In practice, the constant D may be easily chosen for a given soil. If this is the case, pressuremeter loading curves may be used to estimate the ultimate shear stress (strength). This can be achieved by plotting the pressuremeter loading results in terms of cavity pressure P versus $\ln\left(1 + \frac{2}{\sqrt{3}D}\varepsilon_c\right)$. The slope of the pressuremeter curve in this plot should be equal to the undrained shear strength S_u.

2.2.4 Effective stress analysis with critical state models

The self-boring pressuremeter test in clay is usually interpreted using undrained cavity expansion theory based on total stresses. This is reasonably accurate for normally and lightly overconsolidated clays where the shear resistance of the soil does not change significantly during the pressuremeter test. For heavily overconsolidated clay, however, the shear resistance may vary considerably with deformation history and this cannot be easily accounted for by the total stress approach with a perfectly plastic soil model.

Collins and Yu (1996) were the first to derive analytical solutions for large strain cavity expansion in critical state soils. Using these analytical solutions, Yu and Collins (1998) showed that the direct application of the total stress-based interpretation method of Gibson and Anderson (1961) is accurate for soils with low overconsolidation ratio (*OCR*) values. However the total stress approach tends to underestimate undrained shear strength of the soil for heavily overconsolidated soils. As shown in Figure 2, the underestimate could be as high as 50% for soils with a very high *OCR* value. A more detailed discussion can be found in Yu and Collins (1998) and Yu (2000).

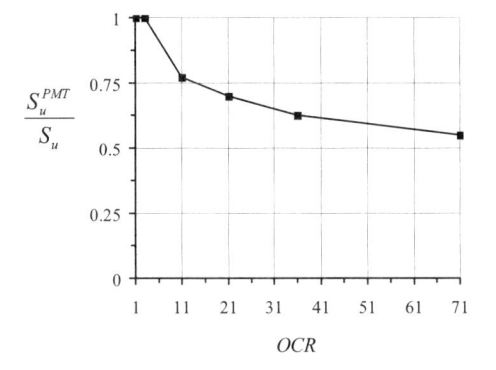

Figure 2: Ratio of pressuremeter strength S_u^{PMT} to triaxial strength S_u versus *OCR* (after Yu and Collins, 1998)

2.3 Complete shear stress-strain curve

The development of a method for deducing a complete shear stress-strain curve from measured pressuremeter expansion results was generally attributed to Palmer (1972), Baguelin et al. (1972) and Ladanyi (1972). As noted by Hill (1950), however, the same procedure for deriving shear stress-strain relation from a known cavity expansion curve had been outlined many years earlier by W.M. Shepherd, as reported in Morrison (1948). The key feature of this alternative approach lies on the fact that there is no need to assume a specific form of stress-strain relations apart from a plastic flow rule (i.e., incompressibility for undrained loading).

It can be shown (Yu, 2000) that the shear stress definition and the incompressibility condition lead to the following cavity expansion curve:

$$P = \sigma_{h0} + \int_0^{\varepsilon_c} \frac{\tau}{\varepsilon} d\varepsilon \qquad (6)$$

where ε_c is the cavity strain and τ is shear stress. The above equation can be used to derive the following shear stress-strain relation:

$$\tau = \varepsilon_c \frac{dP}{d\varepsilon_c} \qquad (7)$$

in which the derivative $dP/d\varepsilon_c$ is readily obtained from the measured pressuremeter results in terms of P versus ε_c.

It has now been recognised that a serious limitation of this method is that the derived stress-strain curve appears to be very sensitive to initial disturbance and the datum selected for the strain (Wroth, 1984; Mair and Wood, 1987).

2.4 Consolidation coefficients

Another soil property that can be measured with a self-boring pressuremeter is the horizontal consolidation coefficient c_h. Such a measurement can be made by conducting either a 'strain holding test' (Clarke et al., 1979) or a 'pressure holding test' (Fahey and Carter, 1986).

When a pressuremeter is expanded in clay under undrained conditions, excess pore pressures are generated in the surrounding soil which is deformed plastically. If at this stage the diameter of the pressuremeter is held constant, relaxation of soil is observed by the decrease of the measured excess pore pressure and the total cavity pressure. This is called a strain-holding test. On the other hand, if the total pressure is held constant, relaxation occurs as the decrease of the measured excess pore pressure and the continuing increase in cavity diameter. This is called a pressure-holding test.

If the pressuremeter expansion is modelled as a cylindrical cavity expansion process in a Tresca soil, then it can be shown (Gibson and Anderson, 1961) that the excess pore pressure takes a maximum at the cavity wall, which is linked to the cavity volumetric strain by

$$\Delta U_{max} = S_u \ln\left(\frac{G}{S_u}\right) + S_u \ln\left(\frac{\Delta V}{V}\right) \qquad (8)$$

If the cavity radius is held constant (i.e., strain holding test), the excess pore pressure dissipates. The consolidation coefficient may be estimated using a dimensional time factor $T_{50} = c_h t_{50}/a^2$, where t_{50} is the time taken for the excess pore pressure to fall to half its maximum value.

By assuming that soil behaves as an entirely elastic material during consolidation, a closed form solution for the time dependence of the excess pore pressures around a cavity was derived by Randolph and Wroth (1979). A subsequent elastoplastic consolidation analysis carried out by Carter et al. (1979) using the finite element method confirmed that the elastic consolidation analysis of Randolph and Wroth (1979) is sufficiently accurate. In particular, a relationship between the normalised maximum excess pore pressure $\Delta U_{max}/S_u$ and the time factor T_{50} was obtained and is plotted in Figure 3.

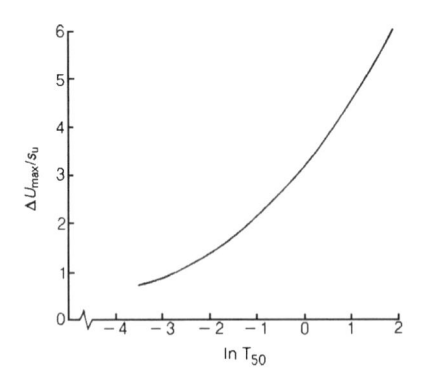

Figure 3: Time for 50% pore pressure decay at the cavity wall (after Randolph and Wroth, 1979)

It then follows that with the actual time t_{50} and the normalised maximum excess pore pressure at the cavity wall measured in pressuremeter strain-holding tests, the correlation, as shown in Figure 3, can be used to determine the horizontal consolidation coefficient c_h.

2.5 Shear modulus and non-linear stiffness

Wroth (1982) noted that a major use of the self-boring pressuremeter is to measure soil stiffness. As stressed by Jamiolkowski et al. (1985) and Mair and Wood (1987), the measurements obtained at the initial stage of a pressuremeter expansion test are not usually reliable. Therefore emphasis is more often placed on using small unloading-reloading loops at later stages of the tests for estimating soil stiffness. For a linear-elastic/plastic soil, cylindrical cavity expansion theory suggests that the shear modulus of the soil is equal to half the slope of an unloading-reloading loop of a pressuremeter curve. However it is well known that soil behaviour is often highly nonlinear even at small strains (e.g., Burland, 1989). In other words, the secant shear modulus of a soil is the highest at very small strains and tends to decrease considerably with increasing shear strain.

If the soil being tested is linear-elastic/plastic then in theory the unloading curve should coincide with reloading curve for a pressuremeter unloading-reloading loop. The slope of such an elastic pressuremeter loop is twice the shear modulus of the soil. Otherwise the small strain behaviour of the soil would be nonlinear and in this case, the interpretation of the pressuremeter results would be more complex because of the strain dependence of the stiffness (Jardine, 1992). For a nonlinear-elastic stress-strain behaviour, a number of theories can be used to describe it. Simple and well-known examples include those based on a power law (Bolton and Whittle, 1999) and a hyperbolic equation (Denby and Clough, 1980).

If an elastic soil stress-strain relationship can be described by the following power law:

$$\tau = G_I \gamma^\beta \tag{9}$$

where τ is the shear stress (i.e. half of the difference between the major and minor principal stresses) and γ is the shear strain. G_I and β are two nonlinear elastic constants and the value of β is between 0 and 1. Obviously G_I is the shear modulus for a linear elastic material when $\beta = 1$. It would be useful if we could derive the soil constants G_I and β from pressuremeter test results.

Cavity expansion theory suggests that for an elastic material governed by equation (9), the following initial elastic cavity stress-strain relation from rest may be derived (Ladanyi and Johnston, 1974; Bolton and Whittle, 1999):

$$P = P_0 + \frac{G_I}{\beta}\gamma_c^\beta \tag{10a}$$

or

$$\ln(P - P_0) = \ln\left(\frac{G_I}{\beta}\right) + \beta \ln(\gamma_c) \tag{10b}$$

where the shear strain at the cavity wall is defined as $\gamma_c = 2\ln(a/a_0)$ with a and a_0 being the current and initial cavity radii respectively. P and P_0 are the current and initial cavity pressures respectively.

Now consider the case when the cavity pressure has increased to a value say, P_{max}, then it is gradually reduced. In this case (i.e., pressuremeter unloading), we can derive the following theoretical relationship between cavity pressure and contraction (note the initial condition is at the end of cavity loading test):

$$P = P_{max} - \frac{G_I}{\beta}\gamma_c^\beta \tag{11a}$$

or

$$\ln(P_{max} - P) = \ln\left(\frac{G_I}{\beta}\right) + \beta \ln(\gamma_c) \tag{11b}$$

where the shear strain at the cavity wall is defined as $\gamma_c = 2\ln(a_{max}/a)$ with a and a_{max} being the current and maximum cavity radius at the end of the loading stage respectively.

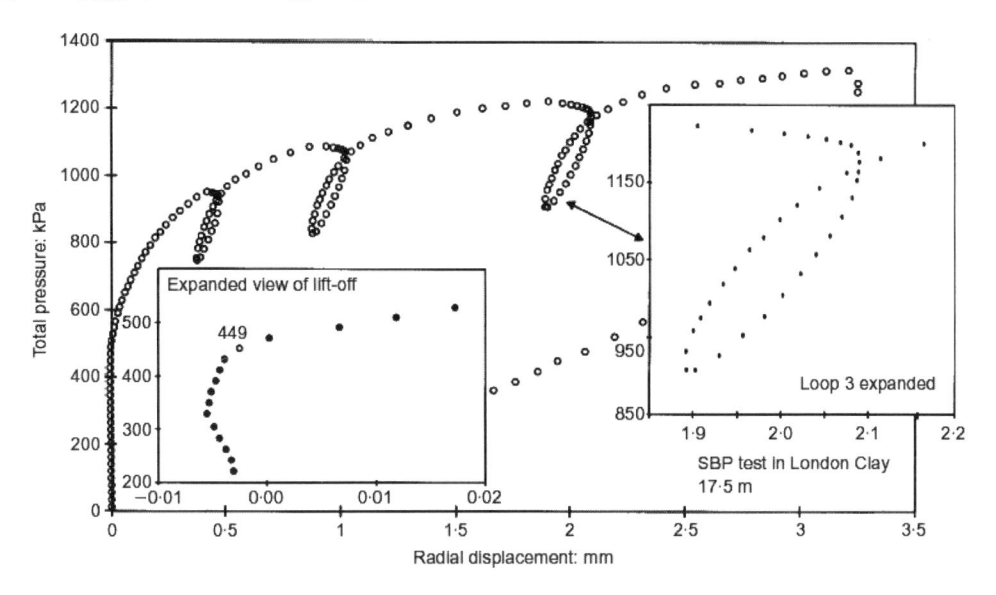

Figure 4: A self-boring pressuremeter test in London clay with three unloading-reloading loops (after Bolton and Whittle, 1999)

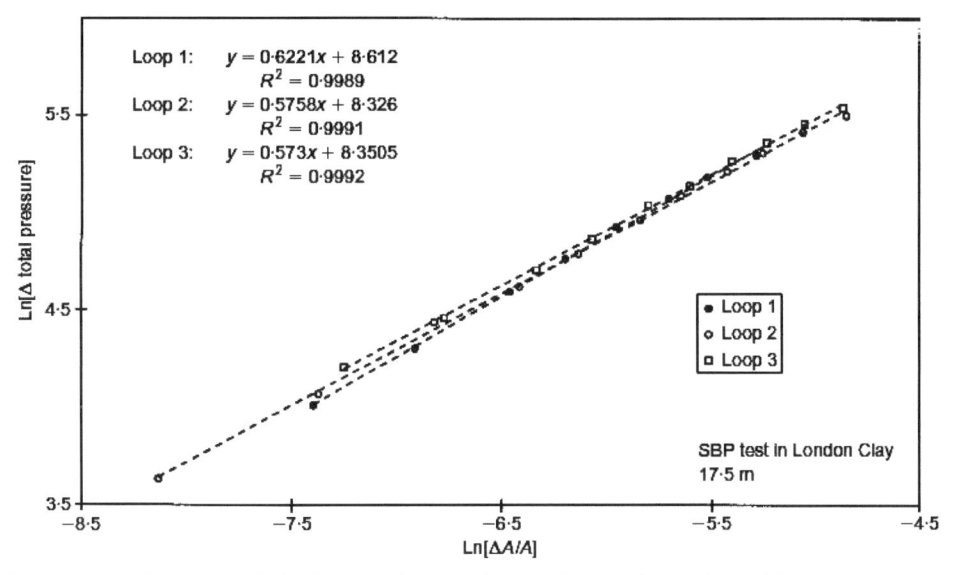

Figure 5: Deriving the nonlinear elastic relationship from unloading-reloading loops (after Bolton and Whittle, 1999)

Equation (11b) suggests that if the pressuremeter unloading results are plotted in terms of $\ln(P_{max} - P)$ versus $\ln(\gamma_c) = \ln(2(\varepsilon_c)_{max} - 2\varepsilon_c)$, then the slope is equal to the nonlinear constant β. The other nonlinear elastic constant G_l can be derived from the fact that the intercept of the plot is equal to $\ln(G_l/\beta)$. Figure 4 shows the results of a self-boring pressuremeter test in London clay with three unloading/reloading loops. By using equation (10b), Bolton and Whittle (1999) suggested a similar method that can be used to derive the elastic constants G_i and β (as shown in Figure 5).

Alternatively, a nonlinear-elastic stress-strain relation may be described by a hyperbolic relation of the following type:

$$\tau = \frac{\gamma}{1/G_i + \gamma/\tau_{max}} \quad (12)$$

where again two material constants are required. They are the initial shear modulus G_i and the maximum shear stress τ_{max} at infinite shear strain.

Using the stress-strain relation (12), cavity expansion theory can be used to give the following theoretical cavity contraction curve for elastic pressuremeter unloading:

$$P = P_{max} - \tau_{max} \ln\left(1 + \frac{G_i}{\tau_{max}}\gamma_c\right) \quad (13)$$

Although not as easy to use as equation (11b), equation (13) may also be used to match a measured pressuremeter unloading curve in order to estimate the nonlinear elastic constants G_i and τ_{max} (e.g., Ferreira and Robertson, 1992).

2.6 Sources of inaccuracy

The interpretation methods described in the preceding sections were developed based on simplified assumptions about pressuremeter geometry, water drainage and initial disturbance. Possible inaccuracies of soil properties caused by these assumptions can be assessed numerically.

2.6.1 Effect of finite pressuremeter length

All the interpretation methods described so far were based on the fundamental assumption that the pressuremeter is sufficiently long so that its expansion can be simulated as the expansion of an infinitely long, cylindrical cavity. In reality, however, the pressuremeter length varies depending on the type of pressuremeter used. A typical example of a pressuremter is the Cambridge self-boring pressuremeter which had a length to diameter ratio of 6. It is therefore necessary to use numerical methods (such as finite elements) to assess the validity of using one-dimensional cavity expansion analysis to solve the two-dimensional pressuremeter problem. Research in this area was first undertaken by Yu (1990, 1993a) and Yeung and Carter (1990), who were later followed by many other researchers.

The most important conclusion of these numerical studies was that ignoring the two-dimensional pressuremeter geometry would significantly overestimate the undrained shear strength. For example, by using a linear elastic perfectly plastic model obeying the von Mises failure criterion, Yu (1990, 1993a) obtained the following correction factor:

$$F_c = \frac{S_u}{S_u^6} = 1 - 0.02\ln\left(\frac{G}{S_u^6}\right) \quad (14)$$

where S_u^6 is the undrained shear strength derived from pressuremeters with a length to diameter ratio of 6 (as for the Cambridge self-boring pressuremeter). The actual undrained shear strength can then be estimated by multiplying the undrained shear strength S_u^6 by a reduction factor F_c given by equation (14). A more recent study of the presssuremeter geometry effect, reported by Yu et al. (2003) using a critical state model, suggests that effective stress analysis gives a smaller geometry effect. In addition, the effect is found to decrease with the OCR value of the soil.

It was also found that the two-dimensional pressuremeter geometry has a quite small effect on the measurement of stiffness (Houlsby and Carter, 1993) and consolidation coefficients (Jang et al., 2003).

2.6.2 Effect of partial drainage and strain rate

The validity of the undrained assumption for pressuremeter analysis in clay has been assessed by Fioravante et al. (1994) and Jang et al. (2003) amongst others. These studies indicate that the pressuremeter expansion can be assumed to occur under the undrained condition at a 1%/min rate, only if the coefficient of permeability of the clay is less than $10^{-9} m/s$. Otherwise the effect of partial drainage would become significant and the undrained condition is no longer a valid assumption.

The effect of strain rate on pressuremeter test results was investigated in detail by Pyrah and Anderson (1990) and Prapaharan et al. (1989). From a parametric study in the latter paper, it was concluded that if laboratory results at a strain rate of 0.01%/min are the reference, then the usual pressuremeter test gives an overestimate of the undrained shear strength. The strain rate effects are most significant for soils with a strain softening behaviour. For a strain hardening soil, the pressuremeter test can yield a derived stress-strain curve similar to that of a material curve corresponding to the reference strain rate.

2.6.3 Effect of disturbance during pressuremeter installation

Although it was commonly assumed that the installation of a self-boring pressuremeter causes no disturbance to the surrounding soil, in reality some disturbance would inevitably occur. As mentioned earlier, the method for deriving stress-strain relations from pressuremeter curves is particularly sensitive to initial disturbance.

A theoretical study of the possible effects of initial disturbance has been reported by Aubeny et al. (2000) using strain path analysis. This study indicates that disturbance induced during ideal self-boring penetration (i.e., where the volume of soil extracted exactly balances the volume of soil displaced by the device) causes a reduction in lift-off pressures compared to the in situ horizontal stress and a higher peak undrained shear strength. The analysis also shows that more reliable undrained shear strengths can be derived from pressuremeter unloading tests.

3 SELF-BORING PRESSUREMETER TESTS IN SAND

3.1 Overview

As in clay, cavity expansion theory forms the main theoretical basis for the interpretation of self-boring pressuremeter tests in sand. For simplicity, the tests are assumed to be carried out under a fully drained condition so that excess pore pressures will be zero throughout the test. The main difference in behaviour between clay and sand lies in the significant volume change occurred in sand during shear, and this must be captured by any realistic sand model. Over the last two decades, significant advances have been made in the analysis of pressuremeter tests in sand using realistic stress-strain equations of various complexities (Yu, 2000).

3.2 Drained shear strength

Hughes et al. (1977) modified the analysis of Gibson and Anderson (1961) to account for the effect of dilation during drained pressuremeter tests in sand. To derive a closed from solution, they assumed that the angles of friction and dilation were constant during the pressuremeter test. From the analysis, a simple procedure was suggested for deriving the value of friction and dilation angles from the pressuremeter loading results. Subsequently an interesting drained analysis, similar to that of Palmer's undrained analysis in clay, has been proposed by Manassero (1989). With this analysis, a stress-strain relationship can be derived from the pressuremeter loading results provided that a plastic flow rule is assumed.

3.2.1 Angles of friction and dilation

Hughes et al. (1977) developed a small strain cavity expansion solution that can be used to deduce the angles of friction and dilation from the pressuremeter loading test results. In their analysis, the sand was assumed to behave as an elastic-perfectly plastic Mohr-Coulomb material obeying the following failure criterion in terms of effective stresses:

$$\frac{\sigma_1'}{\sigma_3'} = \frac{1+\sin\phi}{1-\sin\phi} \tag{15}$$

where ϕ is the angle of internal soil friction, which, like the undrained shear strength of clay, also depends on stress conditions imposed by a particular test (Wroth, 1984). By ignoring elastic deformation in the plastically deforming zone, the analytical solution for the cavity expansion curve in the plastic range can be approximated as follows:

$$\ln(P') = s \ln \varepsilon_c + A \tag{16}$$

where P' is the effective cavity pressure, $s = (1+\sin\psi)\sin\phi/(1+\sin\phi)$, A is a constant and ψ is the angle of soil dilation.

The theoretical pressuremeter loading curve, as defined by equation (16), indicates that if the pressuremeter results are plotted as the effective cavity pressure P' versus the cavity strain on a logarithmic scale, the slope of the plastic portion (which is a straight line) is equal to s, which is a function of the friction angle ϕ and dilation angle ψ. If Rowe's stress-dilatancy equation is used to link the angles of friction and dilation, we can obtain the following formula for deducing them from the pressuremeter loading slope and the angle of soil friction at the critical state ϕ_{cs}:

$$\sin\phi = \frac{s}{1+(s-1)\sin\phi_{cs}} \tag{17}$$

$$\sin\psi = s + (s-1)\sin\phi_{cs} \tag{18}$$

3.2.2 Complete stress-strain curve

As shown by Manassero (1989) and Sousa Coutinho (1989), a pressuremeter loading curve can also be used to deduce a complete soil stress-strain curve, provided that a plastic flow rule can be assumed.

For dilatant sand, the relationship between the radial and hoop strains may be assumed to be related by an unknown function f such as $\varepsilon_r = f(\varepsilon_\theta)$ with a condition that $\varepsilon_r = 0$ when $\varepsilon_\theta = 0$. The function f must be determined numerically from the pressuremeter loading test results. Yu (2000) shows that the equations of equilibrium, strain compatibility condition and Rowe's stress-dilatancy relation can be combined to give the following equation:

$$\frac{1+\frac{f'}{K}}{f-\varepsilon_\theta} = -\frac{1}{\sigma'_r} \times \frac{d\sigma'_r}{d\varepsilon_\theta} \quad (19)$$

in which $K = (1+\sin\phi_{cs})/(1-\sin\phi_{cs})$ and $f' = d\varepsilon_r/d\varepsilon_\theta$. The above equation cannot be integrated analytically. However when applying it at the cavity wall, the finite difference method can be used to solve for a numerical function f and therefore the relationship between the radial and hoop strains. This is possible because at the cavity wall both the effective radial stress σ'_r and $d\sigma'_r/d\varepsilon_\theta$ are given from the pressuremeter curve. The stress ratio is linked to the function f as follows:

$$\frac{\sigma'_r}{\sigma'_\theta} = -\frac{K}{f'} \quad (20)$$

Further application and extension of this approach were given by Ajalloeian and Yu (1998) and Silvestri (2001). Presented in Figure 6 are derived stress ratio-shear strain curves using this approach from the results of model pressuremeter tests in a large chamber obtained by Ajalloeian and Yu (1998) with three different pressuremeter length to diameter (L/D) ratios.

It is stressed that the above analysis is valid only when elastic deformation can be ignored in the plastically deforming zone. As will be discussed later, this assumption could have a significant effect on the derived soil strength properties (Yu, 1990).

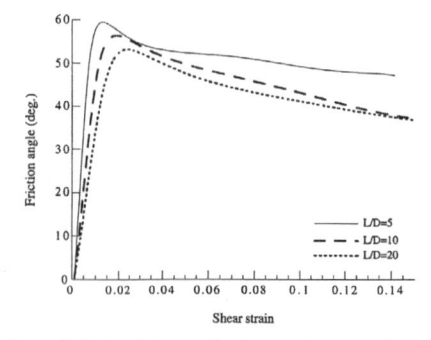

Figure 6: Derived stress ratio-shear strain curves from laboratory pressuremeter tests (after Ajalloeian and Yu, 1998)

3.3 In situ state parameter

A state parameter (defined as the vertical distance between the current state and the critical state line in the usual $v - \ln p'$ plot) was introduced by Wroth and Bassett (1965) and Been and Jefferies (1985) to combine the effects of both relative density and stress level on soil behaviour in a rational way. The state parameter concept represents an important step forward from the conventional relative density concept in characterising sand behaviour. It has been demonstrated that many commonly used sand properties, such as the angles of friction and dilation, normalise well to the state parameter. The practical application of the state parameter concept is dependent upon the ability to measure it in situ. To meet this demand, Yu (1994,1996) developed a procedure to deduce the in situ (or pre-shear) state parameter from either loading or unloading curves of a self-boring pressuremeter test in sand.

3.3.1 State parameter from loading results

Using a state parameter-based critical state soil model, Yu (1994) developed an interpretation method by which the results of a self-boring pressuremeter test can be correlated with the in situ sand state parameter. It was found that for a particular sand, a linear correlation exists between the pressuremeter loading slope s and the pre-shear (or in situ) state parameter of the soil. In addition, this correlation was found to be largely independent of initial stress state and soil stiffness, and may therefore be considered to be unique for a given soil.

The numerical results obtained for six different sands suggest that the following linear correlation may be used for practical purposes:

$$\xi_0 = 0.59 - 1.85s \quad (21)$$

where ξ_0 is the in situ sand state parameter and s is the measured pressuremeter loading slope. Once the state parameter is known, the angles of friction in situ can then be estimated using an average correlation between the angle of friction and state parameter (Been et al., 1987).

The validity of Yu's analysis was further confirmed by Hsieh et al. (2002) using the more advanced sand model MIT-S1.

3.3.2 State parameter from unloading results

As pointed out by Jamiolkowski et al. (1985), soil disturbance during the installation of a self-boring pressuremeter may have a significant effect on the shape of the initial loading portion of the pressuremeter curve. It is therefore necessary, whenever possible, to place less reliance on interpretation methods that are purely based on the initial portion of the test results.

Thus Yu (1996) developed an interpretation method for the unloading stage of a pressuremeter test in terms of the state parameter. The method uses the unloading pressuremeter curve to deduce the in situ state parameter, and thus represents an attractive alternative to the loading analysis. Using this unloading analysis, the pressuremeter results are plotted as $\ln(P')$ versus $-\ln((\varepsilon_c)_{max} - \varepsilon_c)$, and the slope of the pressuremeter unloading curve s_d in this plot is then estimated. The numerical study with six different sands again confirms that there is a largely unique correlation between in situ state parameter and pressuremeter unloading slope, that is given by

$$\xi_0 = 0.53 - 0.33 s_d \tag{22}$$

3.4 Shear modulus and small strain stiffness

Because of sampling difficulties, one of the most common uses of self-boring pressuremeter tests in sand is for the measurement of shear modulus (Wroth, 1982). However the interpretation and application of the soil stiffness derived from the pressuremeter unloading-reloading loops requires special care, and this is largely due to the strong dependence of soil stiffness on both stresses and strains (Bellotti et al., 1989).

3.4.1 Interpretation of unloading-reloading shear modulus

If the soil is linear elastic and plastic, then cavity expansion theory would suggest that the unloading-reloading loop of a pressuremeter test should be a straight line. The slope of the loop is twice the shear modulus of the tested soil. In reality, however, most soils exhibit a nonlinear elastic stress-strain feature even at very small strains. Therefore actual pressuremeter unloading and reloading sections do not coincide. Nevertheless some average slope of the loop is still widely measured to give the so-called pressuremeter unloading-reloading shear modulus G_{ur}, which may be regarded as a secant shear modulus for a nonlinear soil.

For a rational interpretation of soil moduli, it is crucial to note the fact that they are dependent on both stress and strain levels. Given that the stress level at which the unloading-reloading modulus is measured is different from that of an in situ state (i.e. a pre-shear state), it is useful to estimate the equivalent in situ shear modulus G_{ur}^i, at a particular shear strain level (as represented by the size of the unloading-reloading cycle performed). A simple equation that can be used for this estimation is:

$$G_{ur}^i = G_{ur}\left(\frac{p'_0}{p'_m}\right)^n \tag{23}$$

where p'_0 and p'_m are the in situ mean effective stress and the mean effective stress at the cavity wall when the unloading-reloading cycle is performed. For sand, the value of n is generally in the range of 0.4-0.5, with a tendency to increase with increasing level of strains (Wroth et al., 1979).

3.4.2 Estimate of small strain (or maximum) shear modulus

At very small strains (say less than $10^{-4}\%$), the soil modulus is at peak and tends to decrease with increasing strain levels. This peak modulus is often termed as the maximum or small strain shear modulus G_0. Unfortunately the small strain modulus is not a constant for a given soil and rather it is a function of the void ratio, mean stress level as well as stress ratios (Hardin, 1978; Yu and Richart, 1984). The following equation has been frequently used to describe this dependence (Hardin, 1978):

$$\frac{G_0}{p_a} = BF(e)\left(\frac{p'_m}{p_a}\right)^{0.5}\left(1 - 0.3k^{1.5}\right) \tag{24}$$

where p_a is the atmospheric pressure used as a reference pressure. p'_m is the effective mean stress and the stress ratio effect is expressed in terms of $k = (\sigma'_1/\sigma'_3 - 1)/[(\sigma'_1/\sigma'_3)_{max} - 1]$. The parameters B and $F(e)$ depend on particle shape and void ratio e. Equation (24) has been shown to be in agreement with quality laboratory measurement of the small strain modulus such as those obtained using resonant column tests reported by Byrne et al. (1990).

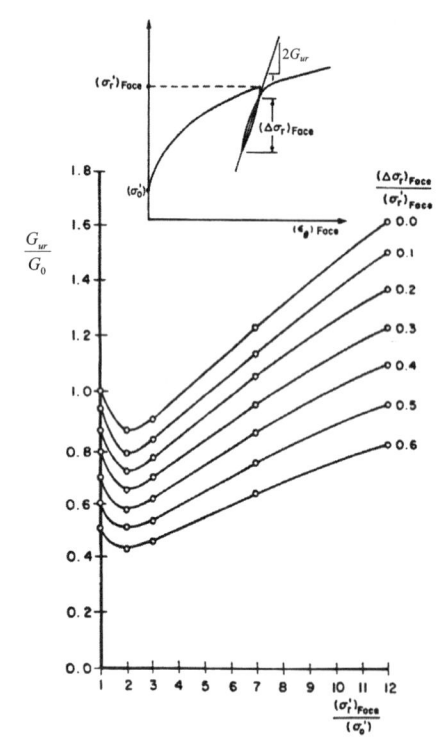

Figure 7: A chart for determination of G_0 from measured G_{ur} (after Byrne et al., 1990)

1) An elastic-plastic cavity expansion analysis to determine the stress field and volume change caused by pressuremeter expansion. These stresses allow the in situ small strain modulus values to be computed prior to pressuremeter unloading tests using equation (24).
2) A nonlinear elastic analysis to determine the displacement at the pressuremeter face upon unloading. These displacements are used to compute the equivalent elastic pressuremeter unloading-reloading shear modulus G_{ur}.
3) By comparing the unloading-reloading shear modulus with the in situ small strain shear modulus for various levels of applied cavity stress prior to unloading, and for various amounts of unloading, a chart is generated from which the ratio of G_{ur}/G_0 can be obtained depending on the applied pressuremeter loading and unloading conditions.

Figure 7 presents such a chart developed by Byrne et al. (1990) for determining the in situ small strain shear modulus from a pressuremeter unloading-reloading modulus. A further study has been presented by Fahey and Carter (1993).

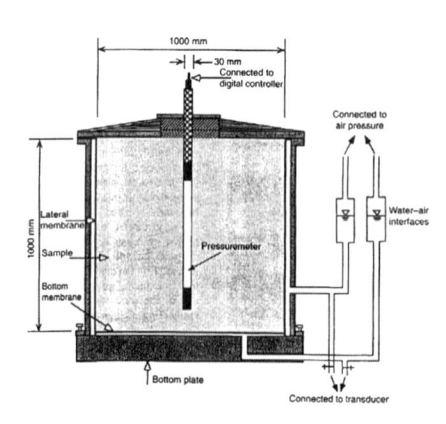

Figure 8: The chamber used by Ajalloeian and Yu (1998)

To derive the small strain shear modulus G_0, Byrne et al. (1990) proposed a numerical procedure to correlate it with the pressuremeter unloading-reloading modulus. The procedure takes the following steps:

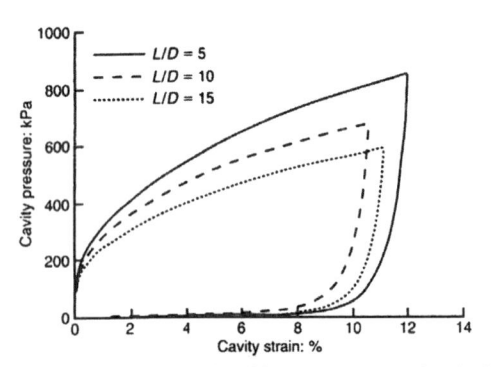

Figure 9: Laboratory results of finite pressuremeter length effects (after Ajalloeian and Yu, 1998)

3.5 Sources of inaccuracy

As in the case for tests in clay, the possible effects of the simplified assumptions used in developing the above mentioned interpretation methods can be assessed either using numerical methods and/or more realistic soil models.

3.5.1 Effect of finite pressuremeter length

The effect of ignoring the finite pressuremeter length on drained pressuremeter analysis was assessed in detail by Yu (1990) using finite element methods. The result of this numerical study was confirmed by a comprehensive chamber study of finite pressuremeter length effects reported by Ajalloeian and Yu (1998) – see Figures 8 and 9.

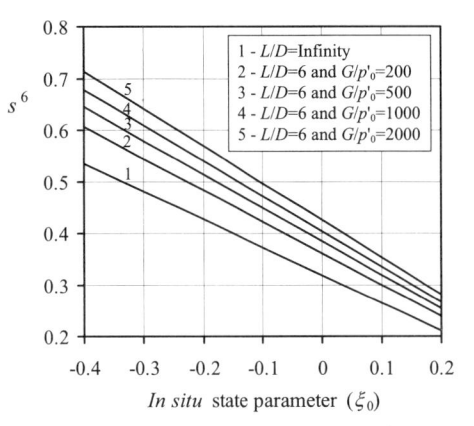

Figure 10: A chart to derive in situ state parameter by accounting for the effect of finite pressuremeter length

As expected, both numerical and laboratory chamber studies suggest that a finite pressuremeter length results in a stiffer pressuremeter loading response. In particular, the pressuremeter loading slope s^6 for a length to diameter ratio of 6 was found to be 10-20% higher than those from the cylindrical cavity expansion theory. The overestimate is slightly dependent upon the soil stiffness index (defined as the shear modulus G over the initial mean effective stress, p'_0), as given by the following equation:

$$F_C = \frac{s}{s^6} = 1.19 - 0.058 \ln\left(\frac{G}{p'_0}\right) \leq 1 \qquad (25)$$

In practice, the effect of finite pressuremeter length can be simply taken into account by determining the correction factor F_C from equation (25). This can then be applied to the measured pressuremeter loading slope before correlating with soil properties such as the angles of friction and dilation (equations (17) and (18)) and the in situ state parameter (equation (21)). Figure 10 presents a chart that can be used to derive in situ state parameter from the pressuremeter loading results by accounting for the effect of finite pressuremeter length.

The experimental data obtained by Ajalloeian and Yu (1998) suggests that finite pressuremeter length has a smaller effect on unloading results than on the loading section of the test. This is to be expected since the unloading involves a very small cavity contraction.

3.5.2 Effect of elastic deformation in the plastically deforming zone

It was noted earlier that elastic deformation in a plastically deforming region was ignored by both Hughes et al. (1977) and Manassero (1989). The effect of this simplifying assumption was assessed by Yu (1990). It was shown that neglecting elastic strain in plastic zones tends to give a softer pressuremeter response, and therefore underestimates the measured angle of friction. The study also suggests that the effect of elastic deformation in the plastic zone is particularly marked for dense soil with a high stiffness index. Using a numerical study, Yu (1990, 1993a) suggested the following single equation for the corrected angle of friction ϕ^c to account for the effect of both finite pressuremeter length and elastic deformation in the plastically deforming zone:

$$\frac{\phi^c}{\phi^6} = 1.36 - 0.078 \ln\left(\frac{1 - \sin\phi^c}{\sin\phi^c} \times \frac{G}{\sigma'_{h0}}\right) \qquad (26)$$

where ϕ^6 is the friction angle derived from the method of Hughes et al. (1977) for pressuremeters with a length to diameter ratio of 6.

3.5.3 Effect of sand particle crushing

It is now established that sand particles crush at high stresses (McDowell and Bolton, 1998). One important feature common to all the findings of recent studies in this area is the distinct steepening of the compression line at elevated stresses (Konrad, 1998).

The possible effect of particle crushing on cavity expansion solutions in sands was studied recently by Russell and Khalili (2002). In their work, a single function for a nonlinear critical state line was introduced which is able to capture the main features of sand behaviour for stresses that are lower and higher than those needed for particle crushing. Limited pressuremeter expansion calculations given by Russell and Khalili show that ignoring particle crushing may lead to a stiffer pressuremeter loading response. This would be particularly true for tests performed in sands with high initial density and/or mean effective stresses.

4 CONE PENETRATION TESTS IN CLAY

4.1 Overview

Over the last few decades, cone penetration testing (with or without pore pressure measurement, CPTU/CPT) has been established as the most widely used in situ testing device for obtaining soil profiles worldwide. This has been achieved mainly by developing empirical correlations and soil classification charts (Robertson, 1986; Lunne et al., 1997; Mitchell and Brandon, 1998). In addition, good progress has also been made, though slowly, in the understanding of the fundamental mechanics of the cone penetration tests in undrained clay. This progress provides confidence in derived soil properties from CPTU test results. Yu and Mitchell (1996, 1998) noted the great difficulties of carrying out a rigorous analysis of cone penetration problems and gave a brief review and evaluation of the theoretical methods that may be used for such an analysis. The most widely used theories are:

1) Bearing capacity methods (BCM)
2) Cavity expansion methods (CEM)
3) Strain path methods (SPM)
4) Finite element methods (FEM)

While each of these four theories may be used alone for cone penetration analysis (Yu and Mitchell, 1996, 1998), better predictions of the cone penetration mechanism may be achieved if some of them are used in combination. Successful examples are SPM-FEM (Teh and Houlsby, 1991), CEM-SPM (Yu and Whittle, 1999), CEM-FEM (Abu-Farsakh et al., 2003), and CEM-BCM (Salgado et al., 1997).

Apart from the above theories that have been the main approaches currently used for cone penetration analysis, other methods such as the discrete element method (DEM) may also be useful for cone penetration analysis in granular materials (e.g., Huang and Ma, 1994; Yu et al., 2004).

4.2 Undrained shear strength

If cone penetration tests in clay are assumed to occur under undrained conditions, cone tip resistance q_c (with the correction for porewater effects on the back of the cone tip) may be related to the undrained shear strength S_u as follows:

$$q_c = N_c S_u + \sigma_0 \Rightarrow S_u = \frac{q_c - \sigma_0}{N_c} \quad (27)$$

where σ_0 denotes the in situ total stress (either vertical or mean total stress depending on the type of theory used for cone penetration analysis). The theory of cone penetration can be used to give the so-called cone factor N_c.

4.2.1 Cavity expansion combined with steady penetration of infinite cone

Based on the rigorous plasticity solutions of steady penetration of a rigid cone in a von Mises soil (Durban and Fleck, 1992; Sagaseta and Houlsby, 1992), Yu (1993b) derived the following expression for the cone factor:

$$N_c = \frac{2}{\sqrt{3}}\left[\pi + \alpha + \arcsin(\lambda_c) + \lambda_c \cot\frac{\alpha}{2} - \sqrt{1-\lambda_c^2} + \frac{H}{2}\right]$$
$$+ \frac{2}{\sqrt{3}}\ln\left(\frac{\sqrt{3}}{2}I_r\right) \quad (28)$$

where $I_r = G/S_u$ is known as the rigidity or stiffness index and the parameter H is defined as

$$H = \frac{\sin\frac{\beta}{2} + \lambda_c \sin\beta}{\cos\frac{\beta}{2} - \cos\beta}; \quad \beta = 180° - \frac{\alpha}{2}$$

in which α is the cone apex angle and λ_c is used to indicate a smooth cone ($\lambda_c = 0$) or a rough cone ($\lambda_c = 1$).

Yu's analytical solution (28) has been extended recently by Su and Liao (2002) to include the effect of shear strength anisotropy (see Figure 11). The cone factor for soil obeying an anisotropic failure criterion is:

$$N_c = \frac{1+A_r}{\sqrt{1+2A_r}}\ln I_r + \frac{1-A_r}{3} + R\left\{1 + \frac{1+A_r}{\sqrt{1+2A_r}}\right.$$
$$\left. + 0.52 A_r^{1/8}(1+A_r)\right\} \quad (29)$$

where $R = 3.13$ for a rough cone and $R = 1.39$ for a smooth cone. The shear strength anisotropy is defined by the parameter A_r, which is the ratio between the undrained shear strength from extension triaxial tests and that from compression tests.

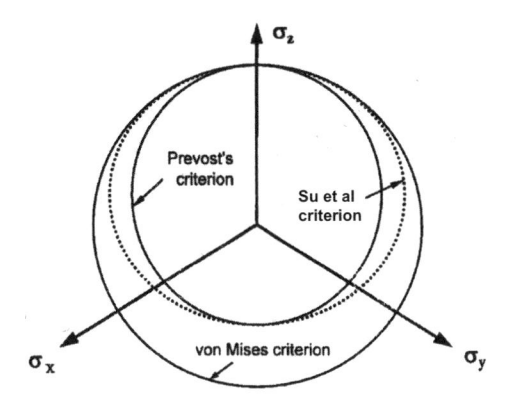

Figure 11: The anisotropic strength criteria (after Su and Liao, 2002)

A simple comparison between equations (28) and (29) indicates that the effect of strength anisotropy of clay will become significant only when the strength anisotropy ratio A_r is less than 0.6.

4.2.2 Strain path analysis combined with finite element methods

The analysis of cone penetration in a von Mises soil by combining strain path analysis and finite element calculations was used by Teh and Houlsby (1991) to overcome the inequilibrium problem of a pure strain path analysis. This combined analysis gives a slightly higher cone factor than that from a pure strain path analysis. The resulting cone factor is:

$$N_c = \left(1.67 + \frac{I_r}{1500}\right)(1 + \ln I_r) + 2.4\lambda_c$$

$$- 0.2\lambda_s - 1.8\Delta \qquad (30)$$

where λ_s and λ_c are used to indicate either rough (with a value of 1) or smooth interfaces (with a value of 0) for the shaft and cone respectively. The parameter $\Delta = (\sigma_{v0} - \sigma_{h0})/(2S_u)$ is used to include the effect of anisotropic in situ stress states.

4.2.3 Steady state finite element analysis

Yu et al. (2000) developed a novel finite element formulation for the analysis of steady state cone penetration in undrained clay modelled by both the von Mises and modified Cam clay models. The proposed finite element analysis focuses on the total displacements experienced by soil particles at a particular instant in time during the cone penetration test. This is possible because, with the steady state assumption, the time dependence of stresses and strains can be expressed as a space-dependence in the direction of penetration (see Figure 12). As a result, the finite element solution of steady cone penetration can be obtained in one step. This new analysis offers the following advantages over the strain path method:

1) All equations of soil equilibrium are fully accounted for.
2) Cone and shaft roughness can be taken into account in a more rigorous manner.
3) It can be more easily adapted to analyse cone penetration in dilatant soils.

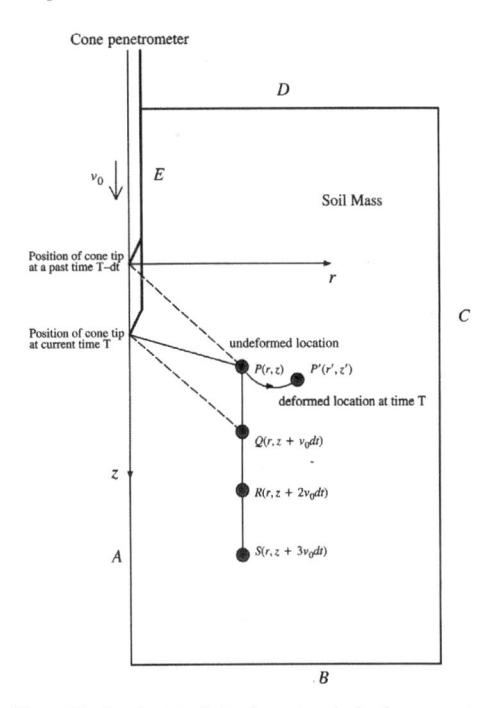

Figure 12: Steady state finite element analysis of cone penetration (after Yu et al., 2000)

The cone factor obtained by Yu et al. (2000) for a von Mises soil is given by the following equation:

$$N_c = 0.33 + 2\ln I_r + 2.37\lambda - 1.83\Delta \qquad (31)$$

where λ (ranging between 0 and 1) is used to indicate roughness of the cone/shaft and soil interface.

4.2.4 Cavity expansion combined with finite element analysis

Most recently, a numerical model has been presented by Abu-Farsakh et al. (2003) for the analysis of cone penetration in clay. As shown in Figure 13, the penetration problem is numerically simulated in two stages. First, the cone penetrometer is expanded radially from a small initial radius to its radius and this is similar to a cylindrical cavity expansion process. Second, the continuous penetration of the penetrometer is simulated by imposing incremental vertical displacements on the nodes along the cone and soil interface. The cone factor from this combined cavity expansion and finite element analysis using the modified Cam clay model is given as follows:

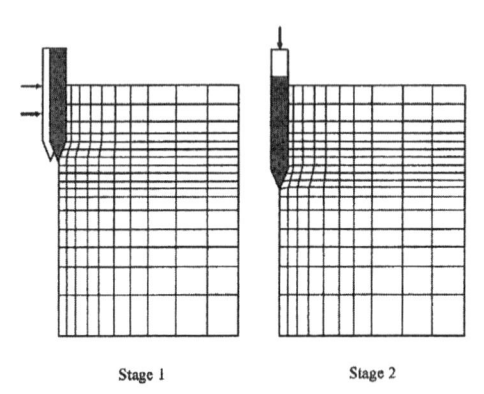

$$N_c = 2.45 + 1.8\ln I_r - 2.1\Delta \quad (32)$$

Figure 13: Combined cavity expansion and finite element analysis (after Abu-Farsakh et al., 2003)

4.2.5 Strain path analysis combined with cavity expansion methods

Yu and Whittle (1999) presented a novel approach to estimate the cone factor by making use of both strain path analysis and cavity expansion methods. With this new method, the strain path solution of a simple pile developed by Baligh (1986) for a von Mises soil was used to estimate the size of the plastic zone in the soil caused by cone penetration. Once the plastic region is established, spherical cavity expansion theory was then used to determine the stress distribution and therefore cone resistance. The cone factor for smooth cone and shaft derived from this hybrid method is:

$$N_c = 1.93 + 2\ln I_r \quad (33)$$

which gives slightly higher values than those from a pure strain path analysis. For example, Baligh's strain path solution for a simple pile geometry is (van den Berg, 1994):

$$N_c = 1.51 + 2\ln I_r \quad (34)$$

and the strain path solution of Teh and Houlsby (1991) for an actual cone geometry is:

$$N_c = 1.25 + 1.84\ln I_r \quad (35)$$

4.2.6 Adaptive finite element analysis

Most recently, Lu (2004) presented a finite element analysis of cone penetration in clay using the adaptive remeshing technique proposed by Hu and Randolph (1998). The adaptive remeshing technique was first used for modelling metal forming processes (Cheng, 1988) and localisation problems (Lee and Bathe, 1994) to overcome the severe distortion in large deformation finite element analysis.

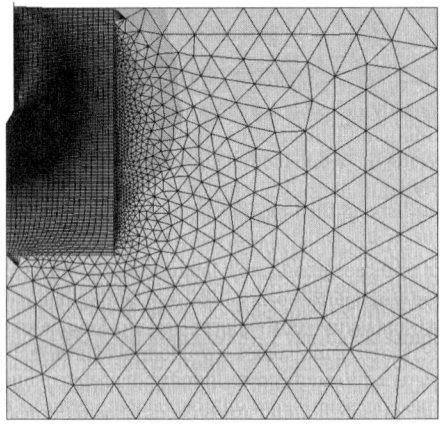

Figure 14: Deformed finite element mesh and plastic region due to cone penetration in clay

A similar finite element study was also carried out by the Author and his student Mr. J. Walker using the commercial finite element program, ABAQUS, with the option of adaptive meshing techniques. The adaptive meshing in ABAQUS is often referred to as Arbitrary Lagrangian-Eulerian (ALE) analysis. A deformed finite element mesh and the plastic region (represented by the dark area) generated by cone penetration in clay are shown in Figure 14. Soils were modelled by the von Mises criterion. In this approach, remeshing and remapping of the field variables from an old mesh to a new one are carried out at a prescribed frequency. A preliminary solution obtained for the cone factor for a smooth cone and shaft/soil interface can be written as follows:

$$N_c = 0.27 + 1.915\ln I_r \quad (36)$$

The influence of in-situ stress states and the roughness of soil-shaft/cone interface can be readily accounted for using adaptive finite element analysis and is currently being studied at the University of Nottingham.

4.3 Consolidation coefficients

The coefficient of consolidation is one of the most difficult soil properties to measure in geotechnical engineering. As mentioned earlier, it can be measured in situ using self-boring pressuremeter holding tests to observe the excess pore pressure decay with time. The interpretation of the pressuremeter holding tests was based on the initial excess pore pressure derived from cavity expansion theory and one-dimensional consolidation solution.

A similar procedure has been used to measure the coefficient of consolidation using cone penetrometer with pore pressure measurement (i.e. CPTU or piezocone) by interrupting the cone penetration and observing the excess pore pressure decay with time. The interpretation of piezocone consolidation can be carried out by using either of the following two methods:

- One-dimensional cavity expansion methods (Torstensson, 1977; Randolph and Wroth, 1979).
- Two-dimensional strain path methods (Levadoux and Baligh, 1986; Baligh and Levadoux, 1986; Teh and Houlsby, 1991).

4.3.1 Cavity expansion approach

Torstensson (1977) developed an interpretation method based on cavity expansion theories. With this method, the initial excess pore pressures prior to consolidation were estimated using cavity expansion theories with an elastic-plastic soil model. It is noted in passing that more accurate solutions are now available with critical state models (Collins and Yu, 1996; Yu, 2000). The consolidation stage of the test was predicted using a one-dimensional, linear, uncoupled consolidation theory (i.e., neglecting the coupling between total stresses and pore pressures during consolidation).

As is the case in the pressuremeter holding tests, Torstensson (1977) suggested that the coefficient of consolidation should be interpreted at 50% dissipation from the following equation:

$$c = \frac{T_{50}}{t_{50}} r^2 \qquad (37)$$

where T_{50} is a time factor which can be obtained from cavity expansion theory (Figure 3), r is the penetrometer radius, and t_{50} is the actual time taken for 50% consolidation (i.e. the excess pore pressure reduces to half of its initial value).

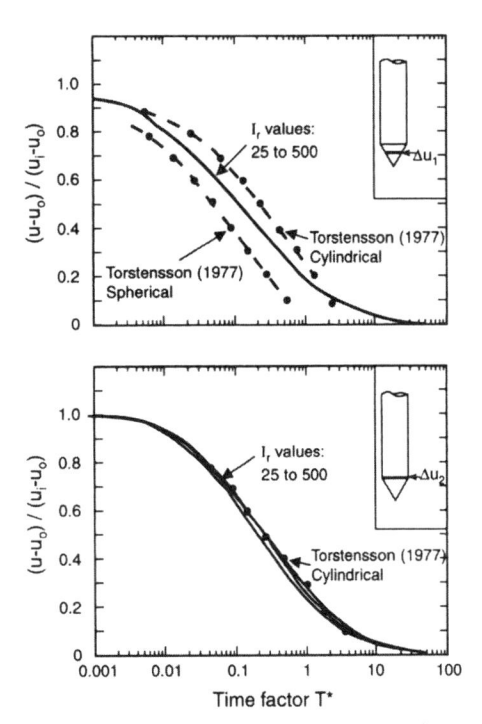

Figure 15: Theoretical solutions for consolidation around cones (after Teh and Houlsby, 1991 and Lunne et al., 1997)

It seems obvious that if the filter element for measuring pore pressures is located on the cone face the spherical cavity expansion solution would be more applicable. On the other hand, the cylindrical cavity expansion solution would be more suitable if the filter element is located on the shaft (Lunne et al., 1997).

4.3.2 Strain path approach

To account for the effect of the two-dimensional nature of cone penetration, Levadoux and Baligh (1986) and Baligh and Levadoux (1986) have used strain path methods (Baligh, 1985) to predict the excess pore pressures generated by the cone installation. Then a finite element method was used to carry out the subsequent coupled and uncoupled linear consolidation analysis. Their study led to some important conclusions including:

1) The effect of the coupling between total stresses and pore pressures is not very significant.
2) The initial distribution of the excess pore pressures has a significant influence on the dissipation process.
3) Dissipation is predominantly in the horizontal direction.

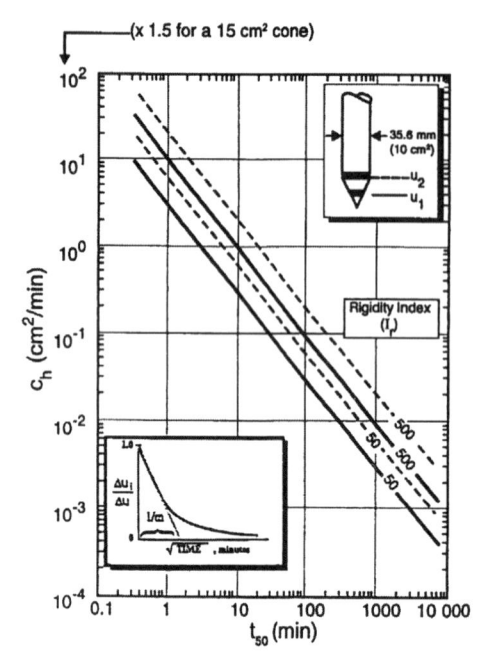

Figure 16: A chart for finding c_h from t_{50} (after Robertson et al., 1992)

By using a method similar to that of Baligh and Lavadoux (1986), Teh and Houlsby (1991) reported the results of a parametric study on cone penetration and consolidation. In the study of Teh and Houlsby, strain path analysis was used to determine the initial excess pore pressures and the subsequent uncoupled, linear consolidation was modelled by the finite difference method. To account for the effect of the stiffness index, $I_r = G/S_u$, Teh and Houlsby (1991) suggested the use of a modified dimensionless time factor, T^*, defined as

$$T^* = \frac{c_h t}{r^2 \sqrt{I_r}} \qquad (38)$$

Figure 15 shows the strain path solutions of a normalised excess pore pressure versus the modified dimensionless time factor obtained by Teh and Houlsby (1991). For comparison, the cavity expansion solutions of Torstensson (1977) are also shown in the figure for two filter element locations with one immediately behind the cone and another on the cone face. It is most interesting to note that for the case with the filter element located immediately behind the cone, the one-dimensional cavity expansion solutions are practically the same as the two-dimensional strain path solutions.

Based on the above theoretical solutions, Robertson et al. (1992) produced a chart (shown in Figure 16) that may be readily used to obtain the coefficient of consolidation from the actual time taken for 50% consolidation t_{50}.

4.4 Stress history - overconsolidation ratio

For clay, the overconsolidation ratio (OCR) is a key property that is needed to define its mechanical behaviour. Several approaches have been proposed to estimate the OCR from CPTU data (Lunne et al., 1997). In particular, Mayne (1993) proposed an analytical method based on cavity expansion theory and critical state soil mechanics. Mayne's method includes the following elements:

1) Use of Vesic's cavity expansion solution to estimate the cone factor (Vesic, 1977).
2) Use of critical state soil mechanics to link the undrained shear strength to the OCR (Wroth, 1984).
3) Use of spherical cavity expansion solutions and critical state soil mechanics to estimate excess pore pressures.

Based on the above assumptions, Mayne (1993) showed that the value of the OCR can be derived from CPTU data using

$$OCR = 2\left\{\frac{1}{1.95M+1}\left(\frac{q_c - u_2}{\sigma'_{vo}}\right)\right\}^{1/\Lambda} \qquad (39)$$

for the case with the filter element located behind the cone, and

$$OCR = 2\left\{\frac{1}{1.95M}\left(\frac{q_c - u_1}{\sigma'_{vo}} + 1\right)\right\}^{1/\Lambda} \qquad (40)$$

for the case with the filter element located on the cone face. In equations (39) and (40), u_1 and u_2 are pore pressures at the cone face and behind the cone respectively; M is the slope of the critical state line in the usual $q - p'$ plot; and Λ is a soil property typically in the range of 0.75-0.85 (Wroth, 1984).

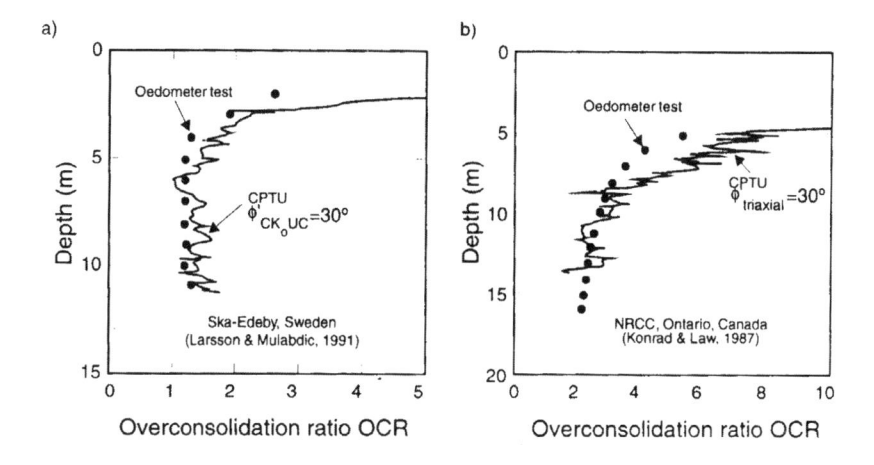

Figure 17: Measured and predicted *OCR* for sites in (a) Sweden and (b) Ontario (after Mayne, 1993)

Figure 17 demonstrated that the estimated values of the *OCR* using equation (39) from CPTU data are consistent with those measured using laboratory odometer tests.

5 CONE PENETRATION TESTS IN SAND

5.1 Overview

Because of the dilatant characteristics of sand during shear, cone penetration in sand is much more difficult to analyse than that in undrained clay. Over the last two decades, good progress has been made in understanding the mechanics of cone penetration in undrained clay. By contrast, progress has been slow in developing rigorous methods to analyse cone penetration in cohesionless soil. This is why large laboratory chamber testing was widely used to develop empirical correlations between cone results and sand properties (e.g., Parkin and Lunne, 1982; Been et al., 1987; Houlsby and Hitchman, 1988; Ghionna and Jamiolkowski, 1991).

Most existing methods for the analysis of deep cone penetration in sand are based on either bearing capacity theory (Durgunoglu and Mitchell, 1975) or cavity expansion theory (Vesic, 1977; Yu and Mitchell, 1998; Salgado et al., 1997). In addition, attempts have also been made in using finite element and discrete element methods to simulate deep penetration problems in sand (van den Berg, 1994; Huang et al., 2004; Huang and Ma, 1994; Yu et al., 2004).

Cone penetration testing in sand is generally drained and therefore the analysis methods presented here are based on the assumption that there would be no excess pore pressures generated as a result of cone penetration.

5.2 Drained shear strength

Cone tip resistance in sand is often used to derive soil friction angle. Various correlations have been proposed in this aspect and most of them were based on either bearing capacity analysis or cavity expansion theory.

5.2.1 Bearing capacity approach

Durgunoglu and Mitchell (1975) presented a well-known bearing capacity solution for deep cone penetration problems. A major advantage of this approach is its relative simplicity. This approach can be easily accepted by the engineer who is already familiar with bearing capacity calculations. As pointed out by Yu and Mitchell (1998), however, the major limitation of bearing capacity theory for cone penetration modelling in sand is its inability of accounting for soil stiffness and volume change.

In the study of Durgunoglu and Mitchell (1975), a failure mechanism was used to give a plane strain solution first (i.e. for wedge penetration). Then an empirical shape factor was used to account for the axisymmetric geometry of cone penetration problems. For the case when the soil-cone interface friction angle is half of the soil friction angle, the solution of Durgunoglu and Mitchell (1975) may be expressed by a simple expression:

$$N_q = \frac{q'_c}{\sigma'_{v0}} = 0.194 \exp(7.63 \tan \phi) \quad (41)$$

where N_q is the cone factor in sand and ϕ is drained soil friction angle.

(a). Ladanyi & Johnston (1974) (b). Vesic (1977)

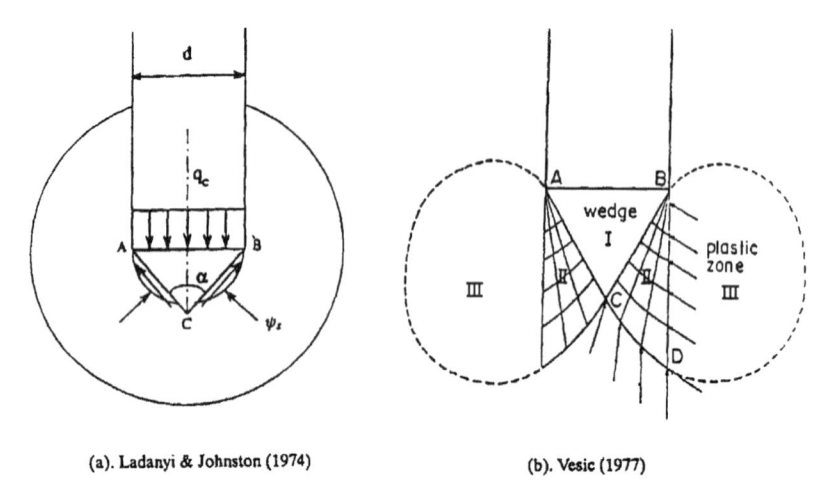

Figure 18: Mechanisms linking cone resistance with cavity limit pressures (after Yu and Mitchell, 1998)

5.2.2 Cavity expansion approach

The analogy between cavity expansion and cone penetration was first pointed out by Bishop et al. (1945) after observing that the pressure required to produce a deep hole in an elastic-plastic medium is proportional to that necessary to expand a cavity of the same volume under the same conditions. As discussed by Yu and Mitchell (1996, 1998), proposals were made by many researchers to relate cone tip resistance with cavity (mainly spherical cavities) limit pressures, which include those by Ladanyi and Johnston (1974) and Vesic (1977) - see Figure 18.

For example, Vesic (1977) assumed that cone tip resistance is related to the spherical cavity limit pressure by a failure mechanism shown in Figure 18(b). This assumption leads to the following simple expression of the cone factor:

$$N_q = \left(\frac{1+2K_0}{3-\sin\phi}\right) \exp\left[\left(\frac{\pi}{2}-\phi\right)\tan\phi\right]$$
$$\times \tan^2\left(\frac{\pi}{4}+\frac{\phi}{2}\right)(I_{rr})^n \quad (42)$$

in which $K_0 = \sigma'_{h0}/\sigma'_{v0}$, and $I_{rr} = I_s/(1+I_s\varepsilon_v)$ is the reduced rigidity index where ε_v is the average volumetric strain estimated in the plastically deforming region, and the rigidity index I_s and parameter n are given by $I_s = G/(p'_0\tan\phi)$ and $n = 4\sin\phi/[3(1+\sin\phi)]$.

After applying the Vesic correlation to the results of a number of chamber tests, Mitchell and Keaveny (1986) concluded that measured cone resistances may be closely modelled for sands with a low value of the reduced index (i.e. more compressible soils). Since dilation was not accounted for in Vesic's solution, this approach cannot be used to model cone penetration in medium dense to very dense sands where dilation is significant.

To extend Vesic's approach, Salgado (1993) and Salgado et al (1997) used a stress rotation analysis to relate cone resistance to a cylindrical cavity limit pressure (see Figure 19). Based on a number of simplifying assumptions, cone resistance is linked to the cylindrical cavity limit pressure as follows:

$$q'_c = 2\exp(\pi\tan\phi)\frac{\left[(1+C)^{1+l}-(1+l)C-1\right]}{C^2 l(1+l)} P'_{lc} \quad (43)$$

where P'_{lc} denotes the effective cylindrical cavity limit pressure, l is determined numerically and C is linked to soil dilation angle ψ by

$$C = \sqrt{3}\exp\left(\frac{\pi}{2}\tan\psi\right)$$

Salgado et al. (1997) applied the theoretical correlation (43) to predict measured cone resistances for a large number of cone tests in large calibration chambers and concluded that the correlation worked well. Typically the measured cone resistances can be predicted to within 30%.

2) Use of spherical cavity expansion theory to determine the cone tip resistance from the estimated plastic region.

This approach was motivated by a recent finite element study of cone penetration in sand (Huang et al., 2004), which suggests that the plastic zone behind the cone and around the shaft is similar to that predicted by the cylindrical cavity expansion theory. Around the cone tip and face, the elastic-plastic boundary may be assumed to be circular or elliptical in shape (see Figure 20).

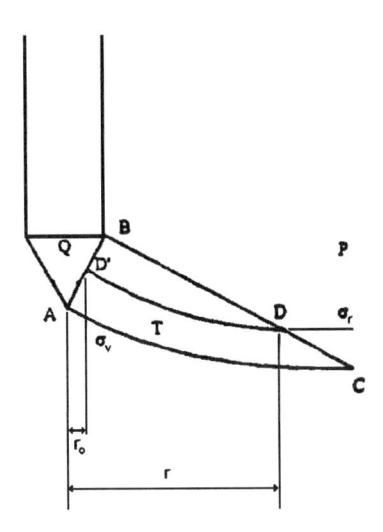

Figure 19: Linking cone resistance with cylindrical cavity limit pressure (after Salgado, 1993)

By using both numerical cavity expansion solutions and chamber data for cone tip resistance, Cudmani and Osinov (2001) recently proposed the following average equation to link cone tip resistance q'_c with the spherical cavity limit pressure P'_{ls}:

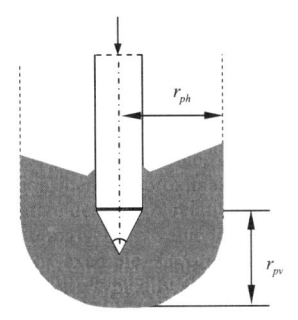

Figure 20: Plastic zone around a cone in sand (after Huang et al., 2004)

$$q'_c = \left[1.5 + \frac{5.8(D_r)^2}{(D_r)^2 + 0.11}\right] P'_{ls} \qquad (44)$$

where D_r is the relative density ranging between 0 and 1. Note that in the study of Cudmani and Osinov, a slightly different, pressure-dependent relative density was considered. Cudmani and Osinov (2001) showed that equation (44) is able to predict 85% of their chamber test data of cone tip resistances to within 25%.

5.2.3 Combined cylindrical-spherical cavity expansion method

All the cavity expansion methods described in the previous section assumed that cone tip resistance is related, through theoretical or semi-analytical considerations, to either spherical cavity limit pressure or cylindrical cavity limit pressure. Here a new method is proposed to estimate cone tip resistance by using both cylindrical and spherical cavity expansion solutions. The basic idea of the new method consists of two steps:

1) Estimate of the size of the plastically deforming zone around the cone using the cylindrical cavity solution for the size of the plastic region.

By following the above procedure and using the cavity expansion solutions in Mohr-Coulomb materials, as derived by Yu (2000) and Yu and Carter (2002), cone tip resistance for a purely frictional soil is given by:

$$\frac{q'_c}{p'_0} = \frac{3\alpha'}{2+\alpha'}\left(F\frac{c}{a}\right)^{\frac{2(\alpha'-1)}{\alpha'}} \qquad (45)$$

where F is a plastic zone shape factor that takes a value of unity if the plastic zone around the cone is a circle (i.e. $r_{ph} = r_{pv}$) and otherwise would be less than 1. Pending more numerical studies, F may be assumed to be between 0.7-0.8 and (c/a) denotes the relative size of the plastic zone generated by the expansion of a cylindrical cavity from zero radius. Yu (2000) derived an analytical solution for this quantity, which can be readily obtained by solving the following simple non-linear equation for a purely frictional soil:

$$1 = \gamma\left(\frac{c}{a}\right)^{\frac{\alpha'-1}{\alpha'}} + [2\delta - \gamma]\left(\frac{c}{a}\right)^{\frac{\beta'+1}{\beta'}}$$

in which

$$\gamma = \frac{\alpha'\beta's'}{\alpha'+\beta'}, \quad \delta = \frac{(\alpha'-1)p_0'}{2(1+\alpha')G}, \quad s' = \frac{\chi(1-\alpha')}{\alpha'\beta'}$$

$$\alpha' = \frac{1+\sin\phi}{1-\sin\phi}, \quad \beta' = \frac{1+\sin\psi}{1-\sin\psi},$$

$$\chi = \frac{(1-\nu)\alpha'p_0'}{((\alpha')^2-1)G}\left\{\left[\beta'-\frac{\nu}{1-\nu}\right]+\frac{1}{\alpha'}\left[1-\frac{\nu\beta'}{1-\nu}\right]\right\}$$

and ν is Poisson's ratio. It is noted that following the same procedure, the solution for the cone tip resistance in a cohesive-frictional soil can also be obtained in closed form using the cavity expansion solutions derived by Yu (2000).

5.3 *In situ state parameter*

Based on the results of a large number of cone tests in calibration chambers, Been et al. (1987) were the first to observe that cone tip resistance may correlate with the initial (in situ or pre-shear) state parameter. Their empirical correlation between the cone resistance and the initial state parameter is given in the following form:

$$\frac{q_c'}{p_0'} = k\exp(-m\xi_0)+1 \qquad (46)$$

After a more detailed analysis of chamber test data on Ticino sand, Sladen (1989) later showed that this correlation is not unique, rather the relationship varies significantly with mean stress level. In other words, the coefficients k and m in equation (46) are not constants even for the same sand.

Yu and Mitchell (1996, 1998) provided a theoretical explanation for such a pressure-dependent, cone resistance-state parameter relationship. Using a state parameter soil model, Collins et al. (1992) found that the spherical cavity limit pressure is linked to the initial mean stress and void ratio as follows:

$$\frac{P_{ls}'}{p_0'} = m_1(p_0')^{(m_2+m_3v_0)}\exp(-m_4v_0) \qquad (47)$$

where $v_0 = 1+e_0$ is the initial specific volume of the soil and e_0 is the initial void ratio. For Ticino sand, the constants are: $m_1 = 2.012\times10^7$, $m_2 = -0.875$, $m_3 = 0.326$, $m_4 = 6.481$.

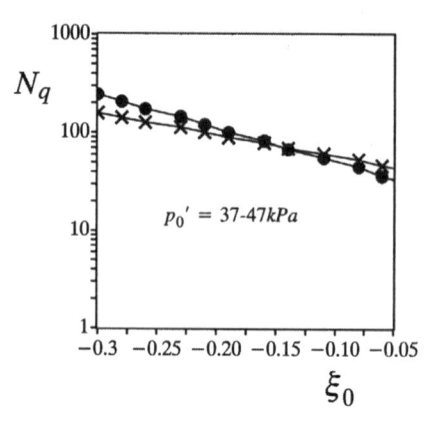

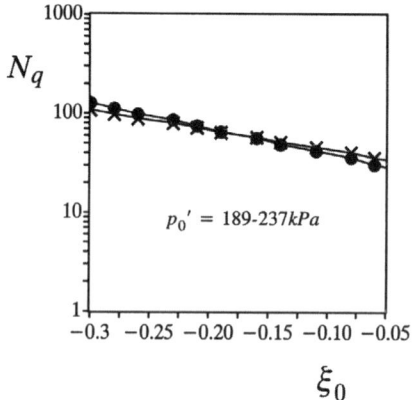

Figure 21: Measured (cross) and predicted (solid circle) cone factor-state parameter relations (after Yu and Mitchell, 1998)

Yu and Mitchell (1998) then used the correlation of Ladanyi and Johnston (1974) to estimate cone tip resistance from the spherical cavity limit pressure determined from equation (47). Presented in Figure 21 are comparisons between cavity expansion predictions and experimental data for chamber cone test results in Ticino sand at two different stress levels. The experimental curves were obtained by applying a chamber size correction factor (given by Been et al., 1987) to the best-fit lines presented by Sladen (1989). It is clear from the figure that a good agreement was obtained.

Russell and Khalili (2002) recently extended the cavity expansion solution of Collins et al. (1992) and showed that the theoretical correlation between spherical cavity limit pressure and initial state parameter is also strongly affected by particle crushing which was reflected by a steeper critical state line at high stress level. Particle crushing was also shown experimentally by Konrad (1998) as an important factor in the interpretation of state parameters from cone penetration tests in a calibration chamber.

6 CONE PRESSUREMETER TESTS IN CLAY AND SAND

6.1 Overview

The cone pressuremeter (also known as the full-displacement pressuremeter) is an in situ testing device that combines a standard cone penetrometer with a pressuremeter module incorporated behind the cone tip. The idea of mounting a pressuremeter module behind the cone tip was first introduced in the early 1980s. The development aims to combine the merits of both the standard cone and the pressuremeter into a single instrument. The cone pressuremeter can be installed by standard CPT jacking equipment and this enables pressuremeter tests to be carried out as part of routine CPT operations (see Figure 22).

Cone pressuremeter tests are difficult to analyse because the tests are carried out in a soil that has already been disturbed by the penetration of the cone (Withers et al., 1989). As a result, a rigorous interpretation of cone pressuremeter tests must account for the effect of installation process. This is why the development of equipment for the cone pressuremeter was, for a long time, more advanced than its interpretation methods. So far, the analytical interpretation methods for cone pressuremeter tests have been mainly based on cavity expansion/contraction theory (Houlsby and Withers, 1988; Yu et al., 1996).

6.2 Cone pressuremeter tests in undrained clay

Using cavity expansion theory, Houlsby and Withers (1988) developed the first theoretical interpretation method for deriving soil properties from cone pressuremeter tests in undrained clay. In the analysis, the initial installation of the instrument was modelled as the expansion of a cylindrical cavity in the soil. The expansion phase of cone pressuremeter tests was modelled as a continuous expansion of the same cavity, and the unloading phase of the tests as the cavity contraction.

This one-dimensional simulation of cone penetration is somewhat in error because it ignores the two-dimensional nature of the problem. However more rigorous, two-dimensional analyses of the cone penetration problem (e.g., Baligh, 1986; Teh and Houlsby, 1991; Yu et al., 2000) show that the stress distribution far behind the cone is similar to that obtained from the expansion of a cylindrical cavity from zero radius. Given the pressuremeter module is located some distance from the cone, it seems sensible to use a simple cavity expansion theory as the basis for the interpretation of cone pressuremeter test results.

Since the installation and subsequent expansion of the cone pressuremeter is simulated as the expansion of a cylindrical cavity from zero radius, it can be easily shown that the cavity pressure remains constant during any stage of installation and loading tests. The constant pressure is the same as the limiting pressure obtained from the expansion of a cavity from a finite radius (Yu and Houlsby, 1991). For a Tresca soil, the limiting pressure is well known (Gibson and Anderson, 1961):

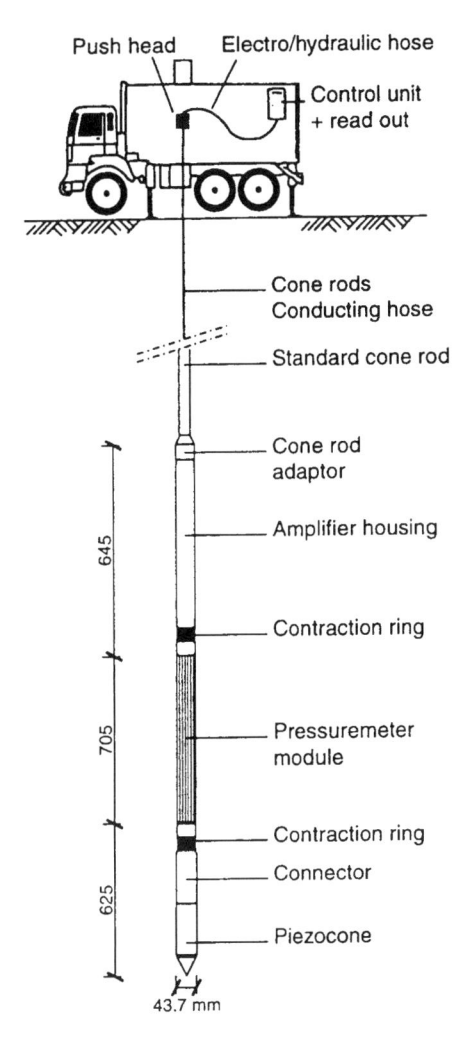

Figure 22: A cone pressuremeter (after Withers et al., 1989)

$$P_{max} = \sigma_{h0} + S_u(1 + \ln I_r) \qquad (48)$$

The complete analytical solution for pressuremeter unloading curves is defined by the following equation:

$$P = P_{max} - 2S_u \{1 + \ln[(\varepsilon_c)_{max} - \varepsilon_c] + \ln I_r\} \qquad (49)$$

where $(\varepsilon_c)_{max}$ is the maximum cavity strain at the start of cone pressuremeter unloading tests. The above solution is plotted in Figure 23, which shows that the slope of the unloading plastic curve in a plot of P versus $-\ln[(\varepsilon_c)_{max} - \varepsilon_c]$ is equal to $2S_u$. From the figure, both shear modulus and initial horizontal total stress may also be estimated.

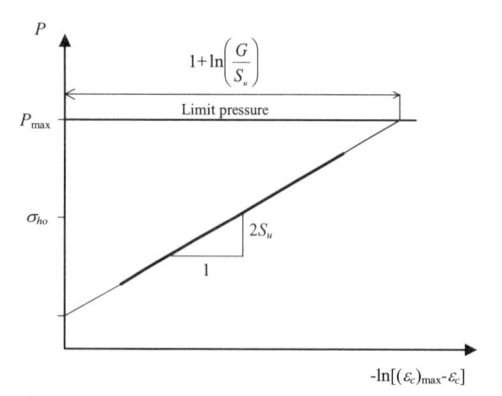

Figure 23: The interpretation method of Houlsby and Withers (1988)

6.3 Cone pressuremeter tests in sand

For obvious reasons, rigorous analysis of cone pressuremeter tests in sand is extremely difficult. Using a non-associated Mohr-Coulomb model, Yu (1990) derived a large strain cavity expansion/contraction solution for sand, equivalent to that of Houlsby and Withers (1988) for clay (see also Yu and Houlsby,1991 and 1995). However, limited applications of this solution in the interpretation of cone pressuremeter tests in sand suggested that it could give unrealistic soil properties. This indicates that soil behaviour during cone pressuremeter tests may be too complex to be modelled accurately by a perfectly plastic Mohr-Coulomb model.

Using both cone tip resistance and pressuremeter limit pressure measured with a cone pressuremeter, Yu et al. (1996) proposed a semi-analytical method for deriving the soil friction angle and the in situ state parameter. In this approach, it was assumed that pressuremeter limit pressure can be estimated by the limit pressure from cylindrical cavity expansion.

The cone tip resistance was estimated from the limit pressure of spherical cavity expansion using the correlation of Ladanyi and Johnston (1974). Therefore the theoretical ratio of cone resistance and pressuremeter limit pressure (ψ'_l) can be expressed in terms of the ratio of spherical cavity limit pressure to cylindrical cavity limit pressure as follows:

$$\frac{q'_c}{\psi'_l} = (1 + \sqrt{3} \tan \phi) \frac{P'_{ls}}{P'_{lc}} \qquad (50)$$

For the determination of the friction angle, cavity expansion solutions in a perfectly plastic Mohr-Coulomb soil were used. In the evaluation of the in situ state parameter, cavity expansion solutions using a state parameter-based, critical state model were used (Collins et al., 1992; Yu, 2000).

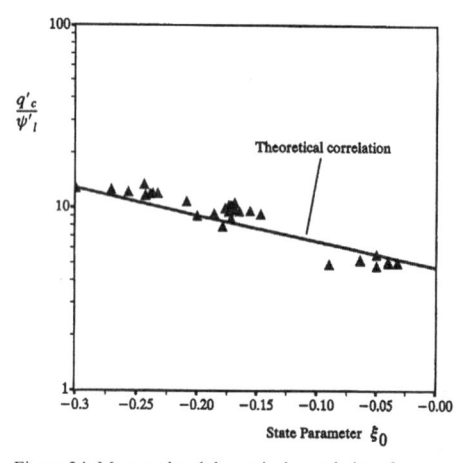

Figure 24: Measured and theoretical correlations for cone pressuremeter tests in Leighton Buzzard sand (after Yu et al., 1996)

6.3.1 Drained shear strength

Yu et al. (1996) used the analytical cavity limit pressures of Yu and Houlsby (1991) to correlate the ratio of cone tip resistance to pressuremeter limit pressure with the angle of soil friction. After a parametric study, Yu et al. (1996) proposed the following correlation:

$$\phi = 22.7 + \frac{14.7}{\ln(G/p'_0)} \times \frac{q'_c}{\psi'_l} \qquad (51)$$

which may be used to derive friction angles from measured ratio of q'_c/ψ'_l, provided a reasonable estimate can be made for stiffness index G/p'_0.

6.3.2 In situ state parameter

Using a state parameter-based, critical state soil model, Collins et al. (1992) presented the limit pressure solutions for the expansion of both spherical and cylindrical cavities in six different sands that have been widely used for calibration chamber testing. The results of these numerical solutions suggest that the ratio of spherical and cylindrical cavity limit pressures may be estimated by the following equation:

$$\frac{P'_{ls}}{P'_{lc}} = C_1(p'_0)^{C_2+C_3(1+e_0)} \exp[C_4(1+e_0)] \quad (52)$$

where e_0 is the initial void ratio and the constants C_1, C_2, C_3, C_4 for the six reference sands are given in Table 2.

Table 2. Material constants (Collins et al., 1992)

Sand	C_1	C_2	C_3	C_4
Monterrey No 0	1087	-0.47	0.225	-3.214
Hokksund	560	-0.424	0.195	-2.84
Kogyuk	237	-0.359	0.167	-2.485
Ottawa	1163	-0.469	0.24	-3.483
Reid Bedford	342	-0.385	0.172	-2.521
Ticino	376	-0.387	0.175	-2.604

Yu et al. (1996) showed that the theoretical correlation between the ratio q'_c/ψ'_l and the in situ state parameter ξ_0 is largely independent of initial stress level. In addition its dependence on sand type was also found to be small. The following average correlation was therefore proposed by Yu et al. (1996) for practical applications:

$$\xi_0 = 0.4575 - 0.2966 \ln \frac{q'_c}{\psi'_l} \quad (53)$$

This can be used to derive the in situ state parameter from the measured ratio of q'_c/ψ'_l.

The theoretical correlation (53) was supported by experimental results presented in Yu et al. (1996) for cone pressuremeter tests in sand (see Figure 24). Its relevance has also been demonstrated by Robertson et al. (2000) using both cone and self-boring pressuremeter test data from the Canadian Liquefaction Experiment (CANLEX) project. Powell and Shields (1997) applied this theoretical correlation to obtain in situ state parameters from field cone pressuremeter tests in sand.

6.4 Effect of finite pressuremeter length

Like self-boring pressuremeters, cone pressuremeters have a finite length to diameter ratio (typically around 10) and therefore the one-dimensional analysis of Houlsby and Withers (1988) may lead to errors in the derived soil properties. To quantify these errors, Yu (1990) carried out a large strain finite element analysis of cone pressuremeter tests. In this analysis, the installation of the cone pressuremeter was modelled as the expansion of a cylindrical cavity. The stress field at the end of the installation can be obtained from analytical cavity expansion solutions. Then starting from this initial stress state, a large strain finite element formulation was used to analyse the expansion and contraction of the pressuremeter membrane. The parametric study reported by Yu (1990, 1993a) for a pressuremeter length to diameter ratio of 10 leads to the following conclusions:

1) The one-dimensional analysis of Houlsby and Withers (1988) overestimates the undrained shear strength and this overestimate could be as high as 10% for a high stiffness index.
2) The neglect of finite pressuremeter length underestimates the shear modulus and this underestimate may increase to 20% for a high stiffness index.
3) The one-dimensional analysis leads to very significant errors in the derived in situ total horizontal stress. The corrected in situ total horizontal stress after accounting for finite pressuremeter length is:

$$(\sigma_{h0})^c = \sigma_{h0} - 0.63 S_u - 0.073 S_u \ln I_r \quad (54)$$

where σ_{h0} is the in situ total horizontal stress derived directly from the one-dimensional analysis of Houlsby and Withers (1988).

Yu (1990) applied equation (54) to the cone pressuremeter test data reported by Houlsby and Withers (1988) and found that the measured total horizontal stresses with finite length corrections are consistent with those measured from self-boring pressuremeter tests (see Figure 25). This has been further confirmed recently by Powell (2004) after applying equation (54) to a large number of cone pressuremeter tests in other clays.

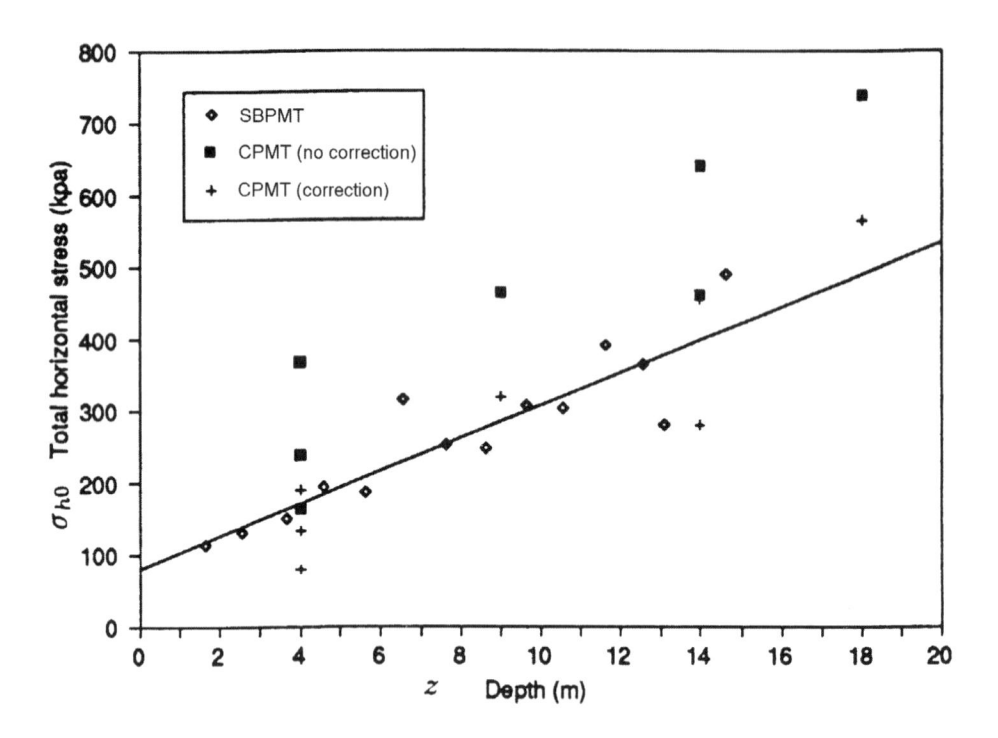

Figure 25: Measured in situ horizontal stresses with various methods for tests at Madingley, Cambridge (after Yu, 1990)

7 FLAT DILATOMETER TESTS IN CLAY

7.1 Overview

The flat dilatometer (shown in Figure 26) is being used increasingly in geotechnical practice to obtain design parameters for a variety of soils (Marchetti, 1980; Marchetti et al., 2001). This is because

1) It is simple to operate and maintain.
2) It does not rely on minimizing disturbance during insertion.
3) It provides a repeatable and continuous profile of the measured parameters.

To date, however, the interpretation of the test has been performed almost exclusively using empirical methods (Marchetti, 1980; Lutenegger, 1988; Campanella and Robertson, 1991; Mayne and Martin, 1998). Research aiming at a better understanding of the fundamental mechanics of the dilatometer test is very limited and seems to be only related to tests in undrained clay. These existing studies were based on either strain path analysis (Huang, 1989; Finno, 1993; Whittle and Aubeny, 1993) or flat cavity expansion methods (Yu et al., 1993; Smith and Houlsby, 1995).

7.2 Total stress flat cavity expansion analysis

As a simple model, Yu et al. (1993) proposed that the installation of a flat dilatometer can be simulated as a flat cavity expansion process. This is consistent with the usual practice of modelling cone pressuremeter installation as a cylindrical cavity expansion process (Houlsby and Withers, 1988). The difference is that no analytical solutions are available for the expansion of a flat cavity in soils. Therefore numerical methods must be used for modelling dilatometer tests. Whilst it is expected that the simple two-dimensional flat cavity expansion modelling approach will introduce errors in the calculated stresses close to the tip of the dilatometer blade, the stresses predicted at some distance behind the dilatometer tip would be reasonably accurate (Finno, 1993).

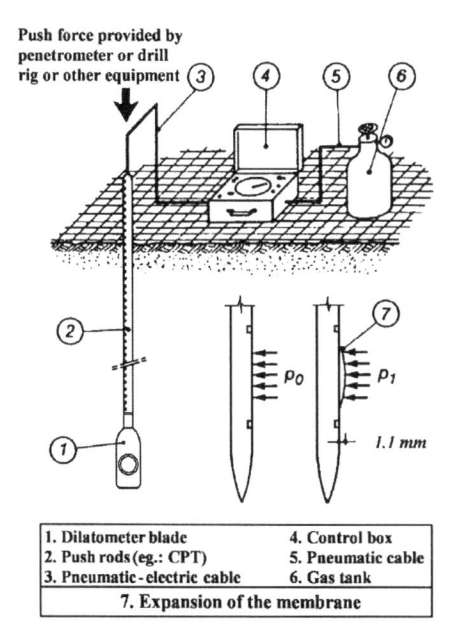

Figure 26: Setup and procedure of the flat dilatometer testing (after Marchetti et al., 2001)

Legend:
1. Dilatometer blade
2. Push rods (eg.: CPT)
3. Pneumatic-electric cable
4. Control box
5. Pneumatic cable
6. Gas tank
7. Expansion of the membrane

By using a linear elastic-perfectly plastic Tresca soil model, Yu et al. (1993) conducted a finite element analysis of the dilatometer installation. The numerical results showed that the first pressure reading (i.e. lift-off pressure) of the dilatometer P_0 can be linked to the in situ total horizontal stress σ_{h0} and stiffness index $I_r = G/S_u$ in terms of a dilatometer factor N_{P_0} as follows:

$$N_{P_0} = \frac{P_0 - \sigma_{h0}}{S_u} = 1.57 \ln I_r - 1.75 \quad (55)$$

The numerical study indicates that the lift-off pressure of the dilatometer is similar to that of a cone pressuremeter. This theoretical finding is in agreement with experimental observation (Lutenegger and Blanchard, 1990).

In addition, Yu et al. (1993) noted that the dilatometer factor N_{P_0} may be usefully linked to the dilatometer horizontal stress index K_D and the coefficient of earth pressure at rest $K_0 = \sigma'_{h0}/\sigma'_{v0}$ as follows:

$$K_D - K_0 = N_{P_0} \times \frac{S_u}{\sigma'_{v0}} \quad (56)$$

7.3 Effective stress flat cavity expansion analysis

To account for the effect of soil stress history, the installation of the dilatometer into undrained clay can be analysed using an effective stress formulation in conjunction with a critical state model. Together with his student Mr. C. Khong, the Author has carried out a parametric study using the critical state model CASM which was developed by Yu (1998). The model CASM has been implemented into the finite element programme CRISP, which was then used in dilatometer analyses. The material constants used are relevant to three different clays as given in Table 3.

Table 3: Clay constants used in CASM

Clay	London clay	Weald clay	Speswhite kaolin clay
M	0.89	0.9	0.86
λ	0.161	0.093	0.19
κ	0.062	0.025	0.03
μ	0.3	0.3	0.3
Γ	2.759	2.06	3.056
n	2.0	4.5	2.0
r	2.718	2.718	2.718

For a given clay, it is possible numerically to relate the dilatometer factor, N_{P_0}, with the overconsolidation ratio of the soil. It is well known that OCR is used to denote the overconsolidation ratio defined in terms of vertical effective stress. Overconsolidation ratios can also be defined in terms of mean effective stress, which is usually denoted by R (Wroth, 1984). The exact relationship between these two overconsolidation ratios is complex and depends on the actual consolidation history of the soil. For example, they become identical for an isotropically consolidated soil sample (Wroth, 1984) and for a one-dimensionally consolidated sample it may be shown that R tends to be somewhat smaller than OCR (Wood, 1990).

The preliminary numerical studies reported here refer to the plane strain analysis of a dilatometer installation into an isotropically consolidated clay, and in this case the two definitions of overconsolidation ratio become identical. As shown in Figure 27, the finite element results indicate that the dilatometer factor may be linked to the initial state (overconsolidation ratio) of the clay by the following form:

$$N_{P_0} = \frac{P_0 - \sigma_{h0}}{S_u} = c_1(OCR)^{c_2} \quad (57)$$

where c_1 and c_2 are constants depending on material type. For the three clays used, their values are: $c_1 = 6.17$ and $c_2 = -0.086$ for London clay; $c_1 = 7.24$ and $c_2 = -0.121$ for Weald clay; and $c_1 = 6.65$ and $c_2 = -0.046$ for Kaolin clay.

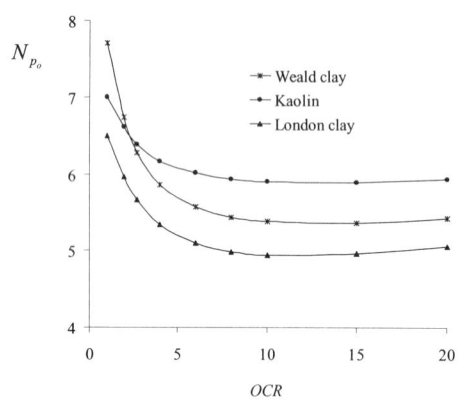

Figure 27: Theoretical correlation between dilatometer index and OCR

By combining equations (57) and (56), we can obtain the following relationship:

$$K_D - K_0 = c_1 (OCR)^{c_2} \times \frac{S_u}{\sigma'_{v0}} \tag{58}$$

Wroth (1984) showed that for isotropically consolidated soils the critical state theory links the undrained strength ratio to the OCR in an elegant form:

$$\frac{S_u}{\sigma'_{v0}} = \frac{M}{2} \left(\frac{OCR}{r} \right)^\Lambda \tag{59}$$

where r is the spacing ratio (Yu, 1998) and $\Lambda = (\lambda - \kappa)/\lambda$. The validity of this theoretical prediction has been confirmed by experimental data (Ladd et al., 1977). In addition, Mayne and Kulhawy (1982) showed that K_0 may be empirically related to the OCR as follows:

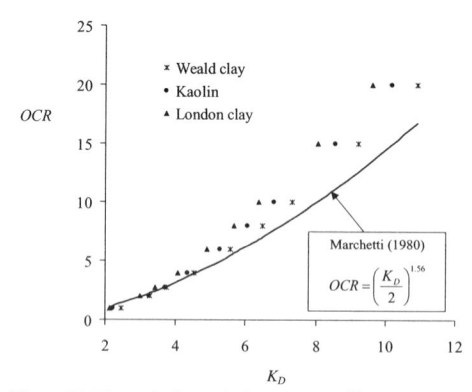

$$K_0 = (1 - \sin\phi)(OCR)^{\sin\phi} \tag{60}$$

Figure 28. Theoretical correlation between K_D and OCR

By using equations (59) and (60) and noting $\sin\phi = 3M/(6+M)$, equation (56) gives a theoretical correlation between K_D and the OCR:

$$K_D = \frac{6-2M}{6+M}(OCR)^{\frac{3M}{6+M}} + \frac{c_1 M}{2r^\Lambda}(OCR)^{c_2+\Lambda} \tag{61}$$

which is shown in Figure 28 for the three clays considered.

Furthermore, equations (60) and (61) can be combined to give the following correlation between K_D and K_0:

$$K_D = K_0 + \frac{c_1 M}{2r^\Lambda} \left[\frac{(6+M)}{(6-2M)} K_0 \right]^{\frac{6+M}{3M}(c_2+\Lambda)} \tag{62}$$

which is shown in Figure 29 for the three clays considered.

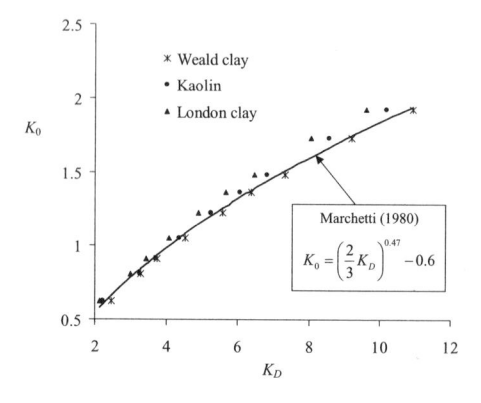

Figure 29: Theoretical correlation between K_D and K_0

It is clear from these comparisons that whilst the empirical correlations of Marchetti (1980) may be reasonable for some clays, they could be very inaccurate for others depending on their mechanical properties. In particular, the theoretical $OCR - K_D$ correlations for the three clays considered show considerable differences from the Marchetti correlation. This difference was also observed by Powell and Uglow (1988) when comparing the Marchetti's correlations with field dilatometer test data obtained in several UK clays.

7.4 Strain path analysis

In an important contribution, Huang (1989) implemented a numerical technique to conduct strain path analysis for arbitrary three-dimensional penetrometers. Further strain path analyses of the installation of flat dilatometers in clays were reported by Whittle and Aubeny (1993) and Finno (1993).

Whilst these studies have provided useful insights, their scopes were rather limited and no theoretical correlations were produced for direct use in practice. The parametric study reported by Finno (1993), using a bounding surface soil model for relatively low OCR values, seems to support the empirical correlation between K_D and the OCR proposed by Marchetti (1980).

8 FLAT DILATOMETER TESTS IN SAND

8.1 Overview

Very little work has been published on the analysis of dilatometer tests in sand. The existing correlations are almost entirely empirical in nature. Due to volume changes, it is not straightforward to extend strain path analysis to sand. However, the approach of simulating dilatometer installation as a flat cavity expansion process can be equally used for both clay and sand.

Presented below are the results of finite element simulations of the installation of a dilatometer in sand performed by the Author and his students Mr. C.D. Khong and Mr. X. Yuan.

8.2 Drained shear strength

Following the study of Yu et al. (1993) in clay, the insertion of a dilatometer in sand has been modelled as a flat cavity expansion process. First we model the sand using a linear elastic, perfectly/plastic Mohr-Coulomb theory. The aim of the study is to theoretically link the dilatometer horizontal index K_D with the fundamental soil properties. The commercial finite element package, ABAQUS, was used in the numerical simulations with the Mohr-Coulomb model.

In the parametric study reported here, the friction angle ϕ varies from 30 to 50 degrees. In addition, soil stiffness index (G/p_0') is varied from 200 to 1500. The dilation angle ψ is derived from the angle of friction using Rowe's stress dilatancy relation (Bolton, 1986) by assuming a value of 30° for the critical state friction angle.

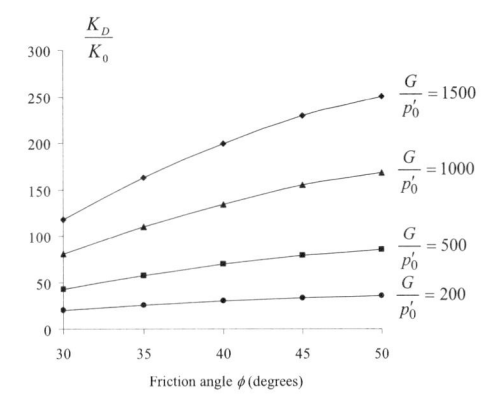

Figure 30: Theoretical correlation for deriving friction angle

Figure 30 shows that whilst the normalised dilatometer horizontal index K_D/K_0 increases with soil friction angle, the influence of the soil stiffness index G/p_0' is also very significant. This is because a large initial part of the dilatometer insertion process occurs when soil is in an elastic state. As a result, the first reading of the dilatometer is a strong function of soil stiffness.

$$\frac{K_D/K_0}{G/p'_0} = -0.0001\phi^2 + 0.0125\phi - 0.1967$$

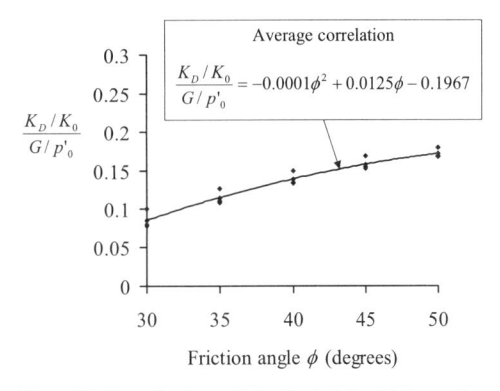

Figure 31: Normalised correlation for deriving friction angle

The numerical results presented in Figure 30 are re-presented in Figure 31 so that a single equation may be used to relate the normalised horizontal index with friction angle and stiffness index, namely

$$\phi = 1013\left[\frac{K_D/K_0}{G/p_0'}\right]^2 - 42.4\left[\frac{K_D/K_0}{G/p_0'}\right] + 26.5 \quad (63)$$

which clearly shows that estimates for both the stiffness index and the in-situ horizontal stress coefficient must be made before the angle of soil friction can be deduced from K_D values measured from the dilatometer tests.

8.3 *In situ state parameter*

As a better alternative to the perfectly plastic Mohr-Coulomb theory, the unified state parameter model CASM (Yu, 1998) can be used to model sand behaviour. A previous section reported that CASM has been used successfully to model dilatometer tests in undrained clay. Here we report the results of a finite element analysis of the dilatometer installation in sand modelled by CASM. Like most other critical state models, CASM uses a pressure-dependent shear modulus. The parametric study reported here uses material model constants relevant to four well-known reference sands, listed in Table 4.

Table 4. Sand constants used in CASM

Sand	Hokksund sand	Kogyuk sand	Ticino sand	Reid Bedford sand
M	1.29	1.24	1.24	1.29
λ	0.024	0.029	0.04	0.028
κ	0.01	0.01	0.01	0.01
μ	0.3	0.3	0.3	0.3
Γ	1.934	1.849	1.986	2.014
n	2	2	2	2
r	10	10	4	10

The correlation does depend on the soil type. For practical application, however, an average correlation may be useful (Figure 33). This is given below:

$$\frac{K_D}{K_0} = -185.4\xi_0^2 - 68.2\xi_0 + 7.3 \quad (64)$$

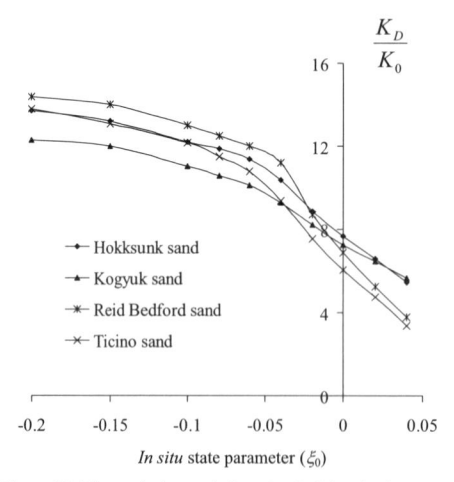

Figure 32: Theoretical correlations for deriving in situ state parameter

The numerical results are plotted in Figure 32 in terms of the normalised dilatometer horizontal index K_D/K_0 against the in situ state parameter ξ_0 prior to the dilatometer insertion. As expected, the normalised dilatometer horizontal index increases when the in situ state parameter decreases from a positive value (i.e., looser than critical state) to a negative value (i.e., denser than critical state).

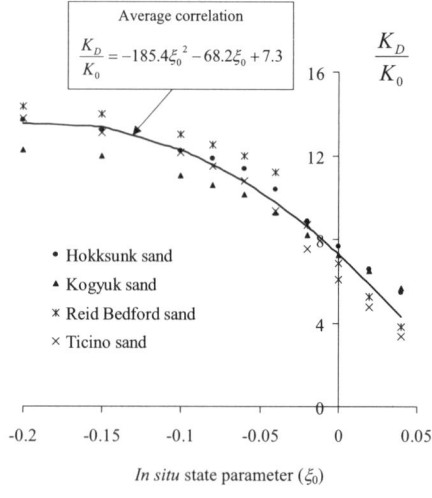

Figure 33: Average correlation for deriving in situ state parameter

Alternatively, the in situ state parameter may be estimated from the normalised dilatometer horizontal index using the following equation:

$$\xi_0 = -0.002\left(\frac{K_D}{K_0}\right)^2 + 0.015\left(\frac{K_D}{K_0}\right) + 0.0026 \quad (65)$$

9 PARTICLE MECHANICS APPROACH

9.1 *Overview*

So far almost all the analyses of in situ tests have been based on continuum mechanics by treating soils as a continuous medium. A useful alternative, particularly for granular material, would be to treat it as a system of discrete particles. The theory of this approach is known as particle or discontinuous mechanics (Harr, 1977; Cundall and Strack, 1979). Application of this approach to the analysis of real soil mechanics problems is still limited because it requires a large number of particles to be used and therefore demands extensive computer resources.

9.2 DEM modelling of deep penetration in sand

Huang and Ma (1994) were among the first to apply the discrete element method (DEM) to simulate deep penetration in sand. In their study, a plane strain penetrometer was pushed into a ground made of a large number of particles. However, the number of simulations reported by them was very limited.

To gain further insights, a study using DEM to simulate deep wedge penetration in sand has been carried out most recently by Yu et al. (2004), who used a two-dimensional, plane strain DEM code that was an extended version of Jiang et al. (2003). The cohesionless soil chosen has a particle size distribution as shown in Figure 34. Due to the geometric

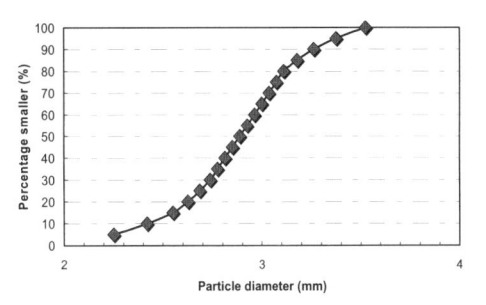

Figure 34: Particle size distribution of cohesionless soil

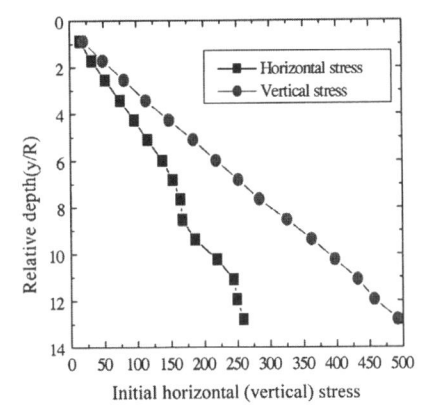

Figure 35: Initial stress state of cohesionless soil

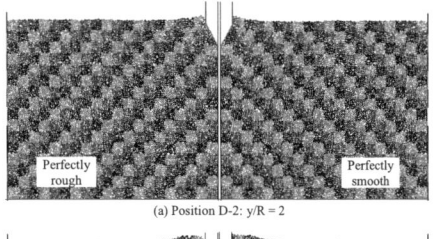

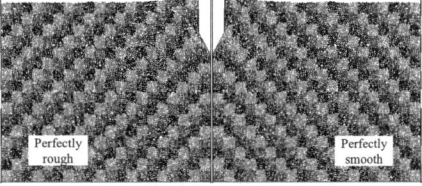

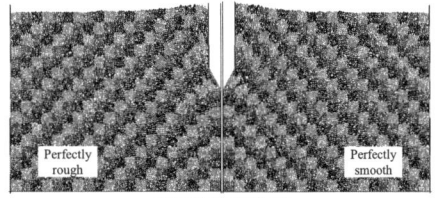

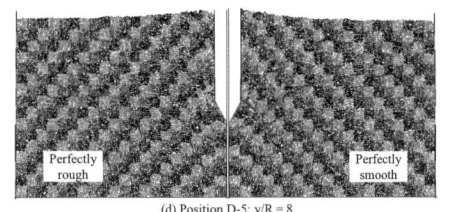

Figure 36: Process of deep penetration modelled by DEM

symmetry of the problem, only half of the medium-dense granular ground with a void ratio of 0.24 was considered. The penetrometer used in the simulations has a half-width of $R=18$ mm with an apex angle of 60° and is composed of 3 rigid walls, i.e. frictional tip wall, frictional and frictionless sleeve walls. A DEM-based simulation of deep penetration takes the following main steps:

1) A soil layer of 10,000 particles was first generated using the undercompaction method (Jiang et al. 2003) with depth and width as $16R$ and $17.5R$ respectively.
2) The soil layer was then allowed to settle under an amplified gravity field of 1000g.
3) The top wall was removed to simulate a free boundary, and the remaining walls are kept as frictionless.
4) The outside boundary was divided into 10 small sections of the same height and the pressure on

each section was measured, and kept as a constant during the penetration to simulate a K_0 stress boundary condition, see Figure 35.
5) By choosing different values of the tip (sleeve)-particle friction, between 0 and 1.0, the penetrometer was pushed downward at 2 mm/s and several aspects of the test results were analysed.

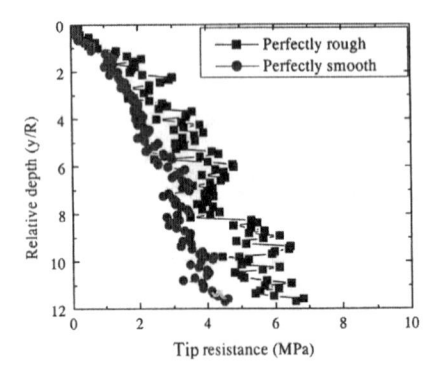

Figure 37: Penetration resistance versus penetration depth

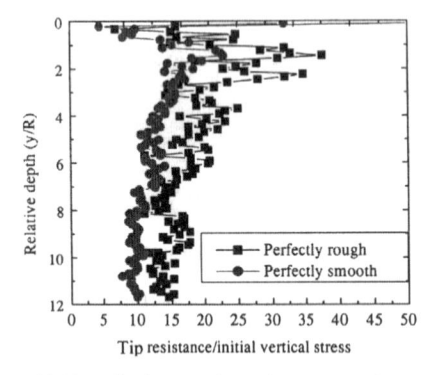

Figure 38: Normalised penetration resistance (cone factor) versus penetration depth

The continuous penetration process of a wedge penetrometer from the ground surface is shown in Figure 36 for both smooth and rough soil-penetrometer interfaces. As shown in Figure 37, the penetration resistance increases steadily with penetration depth and as expected a rough penetrometer generates a higher resistance. Plotted in Figure 38 are normalised penetration resistances (equivalent to the cone factor for cone penetrometers) against penetration depth. The pattern predicted with the discrete element method (DEM) is consistent with what was observed in both centrifuge testing of a cone penetrometer (Bolton et al., 1999) and calibration chamber testing of a plane strain pile (White, 2002) and a cone penetrometer (Houlsby and Hitchman, 1988).

10 CONCLUSIONS

10.1 Overview

The rational interpretation of in situ tests depends on the successful analysis of corresponding boundary value problems. As in the solution of most other soil mechanics and geotechnical engineering boundary value problems, continuum mechanics forms the main theoretical basis, although particle mechanics-based discrete element methods have the potential to further advance our understanding of in situ testing processes in granular soil.

Incomplete as this review had to be, the Author hopes that it has conveyed an idea of the tremendous development that has occurred in this field during the last two decades. In particular, significant progress has been made in developing the rational theoretical basis for the interpretation of pressuremeter tests in soils. Good progress has also been achieved in understanding the mechanics of cone penetration and dilatometer tests in undrained clay. These achievements justify the expectation that the next decade will see a more rapid development of mechanics-based, rigorous interpretation methods for in situ tests in some of the geomaterials that have so far proved intractable.

Mitchell et al. (1978) correctly pointed out that the refinement of existing procedures and further development of new methods of interpretation is an on-going process. Indeed, much research is still needed in further enhancing our understanding in the following key areas:

1) The mechanics of cone penetration/cone pressuremeter tests in granular soil.
2) The mechanics of flat dilatometer tests in soil.
3) The effect of layered soils on in situ test results.
4) The effect of partially saturated soils on in situ test results.
5) The interpretation of in situ tests in soils other than clay and sand.
6) The interpretation of in situ tests in granular soil by accounting for the effects of particle crushing and non-coaxial behaviour.

10.2 Self-boring pressuremeter tests

1) The one-dimensional cavity expansion theory proves to be a useful theoretical framework for the interpretation of self-boring pressuremeter tests.

2) The two-dimensional pressuremeter geometry effects appear to be significant but can be easily accounted for by applying the correction factors derived from finite element analysis.
3) The undrained condition assumed for tests in clay is valid only when the coefficient of permeability is less than 10^{-9} m/s.

10.3 Cone penetration/cone pressuremeter tests

1) The one-dimensional cavity expansion theory (applicable to both clay and sand) and the two-dimensional strain path method (applicable to undrained clay only at the present time) prove to be useful theoretical frameworks for the interpretation of cone penetration/cone pressuremeter tests.
2) The newly developed steady-state finite element technique and large strain finite element methods with adaptive remeshing are more general methods and potentially should provide a more accurate theoretical basis for the understanding of cone penetration/cone pressuremeter in soils.
3) The discrete element method (DEM) has the potential to be a useful theoretical tool for advancing our understanding of cone penetration/cone pressuremeter tests in granular soil.

10.4 Flat dilatometer tests

1) The two-dimensional flat cavity expansion method (applicable to both clay and sand) and the three-dimensional strain path method (applicable to undrained clay only at the present time) prove to be useful theoretical frameworks for modelling the installation of the flat dilatometer in soils.
2) The discrete element method (DEM) should provide a useful numerical tool for modelling the installation of a dilatometer into granular soils.
3) Other numerical techniques, such as three-dimensional finite element methods, will be required to model the expansion of the dilatometer following its insertion into the ground. No work of this type has been reported.

ACKNOWLEDGEMENTS

The Author is greatly honoured to have been selected by the Scientific Committee of ISC'2 (led by TC16 and TC10 of the International Society of Soil Mechanics and Geotechnical Engineering) to present the Inaugural James K. Mitchell Lecture. He is immensely privileged to have known and worked under Professor Mitchell, who is not only a giant of his generation but also truly a teacher in the highest sense of the word.

The Author wishes to acknowledge support and contributions from many people during the preparation of this lecture: first and foremost his wife, Xiu-Li, daughter, Christina and son, Thomas for their encouragements and sacrifices; mentors at key stages of his career, Ted Brown, Peter Wroth, Jim Mitchell, Kerry Rowe and Steve Brown for their inspiration and guidance; colleagues and staff at the Nottingham Centre for Geomechanics (NCG), particularly Caroline Dolby, Glenn McDowell, Ed Ellis and Guoping Zhang, for their valuable assistance; collaborators, Guy Houlsby, Fernando Schnaid, John Carter, Ian Collins, Mark Randolph, Andrew Whittle, Len Herrmann, Ross Boulanger, Rodrigo Salgado, Ken Been, An-Bin Huang, and John Powell for many useful discussions; students and assistants of past and present including Rassoul Ajalloeian, Mark Charles, Cuong Khong, Xun Yuan, James Walker, Wenxiong Huang, and Mingjing Jiang for choosing to undertake research with him on in situ soil testing.

REFERENCES

Abu-Farsakh, M., Tumay, M. and Voyiadjis, G. 2003. Numerical parametric study of piezocone penetration test in clays. *International Journal of Geomechanics*, 3(2):170-181.

Ajalloeian, R. and Yu, H.S. 1998. Chamber studies of the effects of pressuremeter geometry on test results in sand. *Geotechnique*, 48(5): 621-636.

Aubeny, C.P., Whittle, A.J. and Ladd, C.C. 2000. Effects of disturbance on undrained strengths interpreted from pressuremeter tests. *Journal of Geotechnical and Geoenvironmental Engineering*, ASCE, 126(12):1133-1144.

Baguelin, F, Jezequel, J.F., Lemee, E. and Mehause, A. 1972. Expansion of cylindrical probe in cohesive soils. *Journal of the Soil Mechanics and Foundations Division*, ASCE, 98(11):1129-1142.

Baligh, M.M. 1985. Strain path method. *Journal of Geotechnical Engineering*, ASCE, 111(3), 1108-1136.

Baligh, M.M. 1986. Undrained deep penetration. I: shear stresses. *Geotechnique*, 36(4):471-485.

Baligh, M.M and Levadoux, J.N. 1986. Consolidation after undrained piezocone penetration. II: Interpretation. *Journal of Geotechnical Engineering*, ASCE, 112(7):727-745.

Been, K., Crooks, J.H.A., Becker, D.E. and Jefferies, M.G. 1987. The cone penetration test in sand: II general inference of state. *Geotechnique*, 37(3):285-299.

Been, K. and Jefferies, M.G. 1985. A state parameter for sands. *Geotechnique*, 35(2):99-112.

Bellotti, R., Ghionna, V., Jamiolkowski, M., Robertson, P.K. and Peterson, R.W. 1989. Interpretation of moduli from self-boring pressuremeter tests in sand. *Geotechnique*, 39(2):269-292.

Bishop, R.F., Hill, R. and Mott, N.F. 1945. The theory of indentation and hardness tests. *Proceedings of Physics Society*, 57:147-159.

Bolton, M.D. 1986. The strength and dilatancy of sands. *Geotechnique*, 36(1):65-78.

Bolton, M.D., Gui, M.W., Garnier, J., Corte, J.F., Bagge, G., Laue, J. and Renzi, R. (1999). Centrifuge cone penetration tests in sand. *Geotechnique*, 49(4):543-552.

Bolton, M.D. and Whittle, R.W. 1999. A non-linear elastic/perfectly plastic analysis for plane strain undrained expansion tests. *Geotechnique*, 49(1):133-141.

Burland, J. 1989. Small is beautiful - the stiffness of soils at small strains. *Canadian Geotechnical Journal*, 26:499-516.

Byrne, P.M., Salgado, F.M. and Howie, J.A. 1990. Relationship between the unload shear modulus from pressuremeter tests and the maximum shear modulus for sand. *Proceedings of ISP3*, Oxford, 231-241.

Campanella, R.G. and Robertson, P.K. 1991. Use and interpretation of a research dilatometer. *Canadian Geotechnical Journal*, 28(1):113-126.

Carter, J.P., Randolph, M.F. and Wroth, C.P. 1979. Stress and pore pressure changes in clay during and after the expansion of a cylindrical cavity. *International Journal for Numerical and Analytical Methods in Geomechanics*. 3:305-323.

Cheng, J.H. 1988. Automatic adaptive remeshing for finite element simulation of forming processes. *International Journal for Numerical Methods in Engineering*, 26:1-18.

Clarke, B.G., Carter, J.P. and Wroth, C.P. 1979. In situ determination of the consolidation characteristics of saturated clays. *Proceedings of 7th European Conference on Soil Mechanics*, Vol. 2:207-213.

Collins, I.F. and Yu, H.S. 1996. Undrained cavity expansion in critical state soils. *International Journal for Numerical and Analytical Methods in Geomechanics*. 20(7):489-516.

Collins, I.F., Pender, M.J. and Wang, Y. 1992. Cavity expansion in sands under drained loading conditions. *International Journal for Numerical and Analytical Methods in Geomechanics*, 16(1):3-23.

Cudmani, R. and Osinov, V.A. (2001). The cavity expansion problem for the interpretation of cone penetration and pressuremeter tests. *Canadian Geotechnical Journal*, 38:622-638.

Cundall, P.A. and Strack, O.D.L. 1979. A discrete numerical model for granular assemblies. *Geotechnique*, 29(1):47-65.

Denby, G.M. and Clough, G.W. 1980. Self-boring pressuremeter tests in clay. *Journal of Geotechnical Engineering*, ASCE, 106(12):1369-1387.

Durban, D. and Fleck, N.A. 1992. Singular plastic fields in steady penetration of a rigid cone. *Journal of Applied Mechanics*, ASME, 59:1725-1730.

Durgunoglu, H.T. and Mitchell, J.K. 1975. Static penetration resistance of soils. *Proceedings of the ASCE Specialty Conference on In-Situ Measurements of Soil Properties*, Vol 1: 151-189.

Fahey, M. and Carter, J.P. 1986. Some effects of rate of loading and drainage on pressuremeter tests in clays. *Proceedings of Speciality Geomechanics Symposium*, Adelaide, 50-55.

Fahey, M. and Carter, J.P. 1993. A finite element study of the pressuremeter test in sand using a nonlinear elastic plastic model, *Canadian Geotechnical Journal*, 30:348-362.

Ferreira, R.S. and Robertson, P.K. 1992. Interpretation of undrained self-boring pressuremeter test results incorporating unloading, *Canadian Geotechnical Journal*, 29:918-928.

Finno, R.J. 1993. Analytical interpretation of dilatometer penetration through saturated cohesive soils. *Geotechnique*, 43(2):241-254.

Fioravante, V., Jamiolkowski, M. and Lancellotta, R. 1994. An analysis of pressuremeter holding tests. *Geotechnique*, 44(2):227-238.

Ghionna, V.N. and Jamiolkowski, M. 1991. A critical appraisal of calibration testing of sands. *Proceedings of 1st International Symposium on Calibration Chamber Testing*, Potsdam, 13-39.

Gibson, R.E. and Anderson, W.F. 1961. In situ measurement of soil properties with the pressuremeter. *Civil Engineering Public Works Review*, Vol. 56:615-618.

Hardin, B.O. 1978. The nature of stress-strain behaviour of soils. *Proceedings of ASCE Geotechnical Engineering Specialty Conference*, California, 3-90.

Harr, M.E. 1977. *Mechanics of Particulate Media*. McGraw-Hill, New York.

Hill, R. 1950. *The Mathematical Theory of Plasticity*. Oxford University Press.

Hsieh, Y.M., Whittle, A.J. and Yu, H.S. 2002. Interpretation of pressuremeter tests in sand using advanced soil model. *Journal of Geotechnical and Geoenvironmental Engineering*, ASCE, 128(3):274-278.

Houlsby, G.T. and Carter, J.P. 1993. The effect of pressuremeter geometry on the results of tests in clays. *Geotechnique*, 43:567-576.

Houlsby, G.T. and Hitchman, R. 1988. Calibration chamber tests of a cone penetrometer in sand. *Geotechnique*, 38:575-587.

Houlsby, G.T. and Withers, N.J. 1988. Analysis of the cone pressuremeter test in clay. *Geotechnique*, 38:575-587.

Hu, Y. and Randolph, M.F. 1998. A practical numerical approach for large deformation problem in soil. *International Journal for Numerical and Analytical Methods in Geomechanics*, 22(5):327-350.

Huang, A.B. 1989. Strain path analysis for arbitrary three dimensional penetrometers. *International Journal for Numerical and Analytical Methods in Geomechanics*.13:551-564.

Huang, A.B. and Ma, M.Y. 1994. An analytical study of cone penetration tests in granular material. *Canadian Geotechnical Journal*, 31:91-103.

Huang, W., Sheng, D., Sloan, S.W. and Yu, H.S. 2004. Finite element analysis of cone penetration in cohesionless soil. *Computers and Geotehnics* (accepted).

Hughes, J.M.O., Wroth, C.P. and Windle, D. 1977. Pressuremeter tests in sands. *Geotechnique*, 27(4):455-477.

Jamiolkowski, M. 1988. Research applied to geotechnical engineering. James Forrest Lecture. *Proceedings of Institution of Civil Engineers,*, London, Part 1, 84:571-604.

Jamiolkowski, M., Ladd, C.C., Germaine, J.T. and Lancellotta, R. 1985. New developments in field and laboratory testing of soils. Theme Lecture. *Proceedings of the 11th International Conference on Soil Mechanics and Foundation Engineering*, Vol 1:57-154.

Jang, I.S., Chung, C.K., Kim, M.M. and Cho, S.M. 2003. Numerical assessment on the consolidation characteristics of clays from strain holding, self-boring pressuremeter test. *Computers and Geotechnics*, 30:121-140.

Jardine, R.J. 1992. Nonlinear stiffness parameters from undrained pressuremeter tests. *Canadian Geotechnical Journal*, 29(3):436-447.

Jefferies, M.G. 1988. Determination of horizontal geostatic stress in clay with self-bored pressuremeter. *Canadian Geotechnical Journal*, 25:559-573.

Jiang, M.J., Konrad, J.M. and Leroueil, S. 2003. An efficient technique for generating homogeneous specimens for DEM studies. *Computers and Geotechnics*, 30(7):579-597.

Konrad, J.M. 1998. Sand state from cone penetrometer tests: a framework considering grain crushing stress. *Geotechnique*, 48(2):201-215.

Ladanyi, B. 1972. In situ determination of undrained stress-strain behaviour of sensitive clays with the pressuremeter. *Canadian Geotechnical Journal*, 9(3):313-319.

Ladanyi, B. and Johnston, G.H. 1974. Behaviour of circular footings and plate anchors embedded in permafrost. *Canadian Geotechnical Journal*, 11:531-553.

Ladd, C.C., Foott, R., Ishihara, K., Schlosser, F. and Poulos, H.G. 1977. Stress-dormation and strength characteristics. Theme Lecture. *Proceedings of 9th International Conference on Soil Mechanics and Foundation Engineering*. Vol.2:421-497.

Lee, N.S. and Bathe, K.J. 1994. Error indicators and adaptive remeshing in large deformation finite element analysis. *Finite Element Analysis in Design*, 16:99-139.

Levadoux, J.N. and Baligh, M.M. 1986. Consolidation after undrained piezocone penetration. I: prediction. *Journal of Geotechnical Engineering*, ASCE, 112(7):707-726.

Lu, Q. 2004. *A Numerical Study of Penetration Resistance in Clay*. PhD Thesis, The University of Western Australia.

Lunne, T., Robertson, P.K. and Powell, J.J.M. 1997. *Cone Penetration Testing*. E&FN Spon, London.

Luttenegger, A.J. 1988. Current status of the Marchetti dilatometer test. Special Lecture. *Proceedings of 1st International Symposium on Penetration Testing* (ISOPT-1), Orlando, 1:137-155.

Lutenegger, A.J., Blanchard, J.D. (1990). A comparison between full displacement pressuremeter tests and dilatometer tests in clay., *Proceedings of ISP3*, Oxford, 309-320.

Mair, R.J. and Wood, D.M. 1987. *Pressuremeter Testing, Methods and Interpretation*. CIRIA Report, Butterworths, London.

Manassero, M. 1989. Stress-strain relationships from drained self-boring pressuremeter tests in sand. *Geotechnique*, 39(2):293-308.

Marchetti, S. 1980. In situ tests by flat dilatometer. Journal of Geotechnical Engineering, ASCE, 106(GT3):299-321.

Marchetti, S, Monaco, P., Totani, G. and Calabrese, M. 2001. The flat dilatometer test (DMT) in soil investigations. *A Report by the ISSMGE Committee TC 16. Proceedings of International Conference on In situ Measurement of Soil Properties*, Bali, 41pp.

Mayne, P.W. 1993. In situ determination of clay stress history by piezocone. In: *Predictive Soil Mechanics*, Thomas Telford, London, 483-495.

Mayne, P.W. and Kulhawy, F.H. 1982. K_0-OCR relationships in soils. *Journal of Geotechnical Engineering*, ASCE, 108(6):851-872.

Mayne, P.W. and Martin, G.K.1998. Commentary on Marchetti flat dilatometer correlations in soils. ASTM *Geotechnical Testing Journal*, 21(3):222-239.

McDowell, G.R. and Bolton, M.D. 1998. On the micromechanics of crushable aggregates. *Geotechnique*, 48(5):667-679.

Mitchell, J.K. and Brandon, T.L. 1998. Analysis and use of CPT in earthquake and environmental engineering. Keynote Lecture, *Proceedings of ISC'98*, Vol.1:69-97.

Mitchell, J.K., Guzikowski, F. and Villet, W.C.B. 1978. The measurement of soil properties in-situ: present methods – their applicability and potential. *Lawrence Berkeley Laboratory Report 6363*, University of California at Berkeley.

Mitchell, J.K. and Keaveny, J.M. 1986. Determining sand strength by cone penetrometer. *Proceedings of the ASCE Specialty Conference*, In Situ'86, Blacksburg, 823-839.

Morrison, J.L.M. 1948. The criterion of yield of Gun steel. *Proceedings of the Institution of Mechanical Engineers*, 159:81-94.

Palmer, A.C. 1972. Undrained plane strain expansion of a cylindrical cavity in clay: a simple interpretation of the pressuremeter test. *Geotechnique*, 22(3):451-457.

Parkin, A.K. and Lunne, T. 1982. Boundary effects in the laboratory calibration of a cone penetrometer in sand. *Proceedings of 2nd European Symposium on Penetration Testing*, Amsterdam, 2:761-768.

Powell, J.J.M. 2004. Personal communication.

Powell, J.J.M. and Shields, C.H. 1997. The cone pressuremeter - a study of its interpretation in Holmen sand. *Proceedings of 14th International Conference on Soil Mechanics and Foundation Engineering*, 573-576.

Powell, J.J.M. and Uglow, I.M. 1988. The interpretation of the Marchetti dilatometer test in UK clays. *Proceedings of Penetration Testing in the UK*, Thomas Telford, Paper 34:269-273.

Prapaharan, S., Chameau, J.L. and Holtz, R.D. 1989. Effect of strain rate on undrained strength derived from pressuremeter tests. *Geotechnique*, 39(4):615-624

Prevost, J.H. and Hoeg, K. 1975. Analysis of pressuremeter in strain softening soil. *Journal of Geotechnical Engineering*, ASCE, Vol. 101(GT8):717-732.

Pyrah, I.C. and Anderson, W.F. 1990. Numerical assessment of self-boring pressuremeter tests in a clay calibration chamber. *Proceedings of ISP3*, 179-188.

Randolph, M.F. and Wroth, C.P. 1979. An analytical solution for the consolidation around a driven pile. *International Journal for Numerical and Analytical Methods in Geomechanics*, 3:217-229.

Robertson, P.K. 1986. In situ testing and its application to foundation engineering. *Canadian Geotechnical Journal*, 23(4):573-594.

Robertson, P.K., Sully, J.P., Woeller, D.J., Luune, T., Powell, J.J.M. and Gillespie, D.G. 1992. Estimating coefficient of consolidation from piezocone tests. *Canadian Geotechnical Journal*, 29(4):551-557.

Robertson, P.K. et al. 2000. The Canadian Liquefaction Experiment: an overview. *Canadian Geotechnical Journal*, 37:499-504.

Russell, A.R. and Khalili, N. 2002. Drained cavity expansion in sands exhibiting particle crushing. *International Journal for Numerical and Analytical Methods in Geomechanics*, 26:323-340.

Sagaseta, C. and Houlsby, G.T. 1992. Stresses near the shoulder of a cone penetrometer in clay. *Proceedings of 3rd International Conference on Computational Plasticity*, Vol.2:895-906.

Salgado, R. 1993. *Analysis of Penetration Resistance in Sands*. PhD Thesis, University of California at Berkeley.

Salgado, R., Mitchell, J.K. and Jamiolkowski, M. 1997. Cavity expansion and penetration resistance in sand. *Journal of Geotechnical and Geoenvironmental Engineering*, ASCE, 123(4):344-354.

Schnaid, F. 1990. *A Study of the Cone Pressuremeter Test in Sand*. DPhil Thesis, Oxford University.

Silvestri, V. 2001. Interpretation of pressuremeter tests in sand. *Canadian Geotechnical Journal*, 38:1155-1165.

Sladen, J.A. 1989. Problems with interpretation of sand state from cone penetration test. *Geotechnique*, 39(2):323-332.

Smith, M.G. and Houlsby, G.T. 1995. Interpretation of the Marchetti dilatometer in clay. *Proceedings of 11th ECSMFE*, Vol 1:247-252.

Sousa Coutinho, A.G.F. 1990. Radial expansion of cylindrical cavities in sandy soils: application to pressuremeter tests. *Canadian Geotechnical Journal*, 27:737-748.

Su, S.F. and Liao, H.J. 2002. Influence of strength anisotropy on piezocone resistance in clay. *Journal of Geotechnical and Geoenvironmental Engineering*, ASCE, 128(2):166-173.

Teh, C.I. and Houlsby, G.T. 1991. An analytical study of the cone penetration test in clay. *Geotechnique*, 41(1):17-34.

Torstensson, B.A. 1977. The pore pressure probe. Norskjord- og fjellteknisk forbund. Fjellsprengningsteknikk – bergmekanikk – geoteknikk, Oslo, Foredrag, 34.1-34.15, Trondheim, Norway, Tapir.

Van den Berg, P. 1994. *Analysis of Soil Penetration*, PhD Thesis, Delft University.

Vesic, A.S. 1977. Design of pile foundations. *National Cooperation Highway Research Program, Synthesis of Highway Practice* 42. TRB, National Research Council, Washington DC.

White, D.J. 2002. *An Investigation Into Behaviour of Pressed-in Piles*. PhD Thesis, Cambridge University.

Whittle, A.J. and Aubeny, C.P. 1993. The effects of installation disturbance on interpretation of in situ tests in clays. In: *Predictive Soil Mechanics*, Thomas Telford, London, 742-767.

Withers, N.J., Howie, J, Hughes, J.M.O. and Robertson, P.K. 1989. Performance and analysis of cone pressuremeter tests in sands. *Geotechnique*, 39(3):433-454.

Wood, D.M. 1990. *Soil Behaviour and Critical State Soil Mechanics*, Cambridge University Press.

Wroth, C.P. 1982. British experience with the self-boring pressuremeter. *Proceedings of the Symposium o Pressuremeter and its Marine Applications*, Paris, Editions Technip, 143-164.

Wroth, C.P. 1984. The interpretation of in situ soil tests. *Geotechnique*, 34:449-489.

Wroth, C.P. and Bassett, N. 1965. A stress-strain relationship for the shearing behaviour of sand. *Geotechnique*, 15(1):32-56.

Wroth, C.P. and Houlsby, G.T. 1985. Soil mechanics – property characterization and analysis procedures. Theme Lecture No. 1. *Proceedings of 11th International Conference on Soil Mechanics and Foundation Engineering.* Vol 1:1-56.

Wroth, C.P. and Hughes, J.M.O. 1972. An instrument for the in situ measurement of the properties of soft clays. *Report CUED/D, Soils TR13*. University of Cambridge.

Wroth, C.P., Randolph, M.F., Houlsby, G.T. and Fahey, M. 1979. A review of the engineering properties of soils with particular reference to the shear modulus. *Cambridge University Report CUED/D Soils TR75.*

Yeung, S.K. and Carter, J.P. 1990. Interpretation of the pressuremeter test in clay allowing for membrane end effects and material non-homogeneity. *Proceedings of 3rd International Symposium on Pressuremeters*, Oxford, 199-208.

Yu, H.S. 1990. *Cavity Expansion Theory and its Application to the Analysis of Pressuremeters*. DPhil Thesis, Oxford University.

Yu, H.S. 1993a. A new procedure for obtaining design parameters from pressuremeter *tests. Australian Civil Engineering Transactions*, Vol. CE35 (4):353-359.

Yu, H.S. 1993b. Discussion on: singular plastic fields in steady penetration of a rigid cone. *Journal of Applied Mechanics*, ASME, 60:1061-1062.

Yu, H.S. 1994. State parameter from self-boring pressuremeter tests in sand. *Journal of Geotechnical Engineering*, ASCE, 120(12):2118-2135.

Yu, H.S. 1996. Interpretation of pressuremeter unloading tests in sands. *Geotechnique*, 46(1):17-31.

Yu, H.S. 1998. CASM: A unified state parameter model for clay and sand. *International Journal for Numerical and Analytical Methods in Geomechanics*, 22:621-653.

Yu, H.S. 2000. *Cavity Expansion Methods in Geomechanics*. Kluwer Academic Publishers, The Netherland.

Yu, H.S. and Carter, J.P. 2002. Rigorous similarity solutions for cavity expansion in cohesive-frictional soils. *International Journal of Geomechanics*, 2(2):233-258.

Yu, H.S., Carter, J.P. and Booker, J.R. 1993. Analysis of the dilatometer test in undrained clay. In: *Predictive Soil Mechanics*, Thomas Telford, London, 783-795.

Yu, H.S., Charles, M. and Khong, C.D. 2003. Analysis of pressuremeter geometry effects using critical state models. *International Journal for Numerical and Analytical Methods in Geomechanics* (under review).

Yu, H.S. and Collins, I.F. 1998. Analysis of self-boring pressuremeter tests in overconsolidated clays. *Geotechnique*, 48(5):689-693.

Yu, H.S., Herrmann, L.R. and Boulanger, R.W. 2000. Analysis of steady cone penetration in clay. *Journal of Geotechnical and Geoenvironmental Engineering*, ASCE, 126(7):594-605.

Yu, H.S. and Houlsby, G.T. 1991. Finite cavity expansion in dilatant soil: loading analysis. *Geotechnique*, 41(2):173-183.

Yu, H.S. and Houlsby, G.T. 1995. A large strain analytical solution for cavity contraction in dilatant soils. *International Journal for Numerical and Analytical Methods in Geomechanics*, 19(11):793-811.

Yu, H.S., Jiang, M. and Harris, D. 2004. DEM simulation of deep penetration in granular soil. *Journal of Geotechnical and Geoenvironmental Engineering*, ASCE (under review).

Yu, H.S. and Mitchell, J.K. 1996. Analysis of cone resistance: a review of methods. *Research Report No. 142.09.1996*, The University of Newcastle, NSW, 50pp.

Yu, H.S. and Mitchell, J.K. 1998. Analysis of cone resistance: review of methods. *Journal of Geotechnical and Geoenvironmental Engineering*, ASCE, 124(2):140-149.

Yu, H.S., Schnaid, F. and Collins, I.F. 1996. Analysis of cone pressuremeter tests in sand. *Journal of Geotechnical Engineering*, ASCE, 122(8):623-632.

Yu, H.S. and Whittle, A.J. 1999. Combining strain path analysis and cavity expansion theory to estimate cone resistance in clay. *Unpublished Notes*.

Yu, P. and Richart, F.E. 1984. Stress ratio effects on shear moduli of dry sands. *Journal of Geotechnical Engineering*, ASCE, 110(3):331-345.

Characterization of granite and the underground construction in Metro do Porto, Portugal

S. Babendererde
Am Lotsenberg 8, D-23570 Lubeck-Travemunde, Germany
contact@bab-ing.com

E. Hoek
3034 Edgemont Boulevard, P.O. Box 75516, North Vancouver, B.C, V7R 4X1, Canada
ehoek@attglobal.ne

P. Marinos
National Technical University of Athens, 9 Iroon Polytechniou str. 15780, Greece
marinos@central.ntua.gr

A. Silva Cardoso
Faculdade de Engenharia, Rua Roberto Frias, 4200-465 Porto, Portugal
scardoso@fe.up.pt

Keywords: rockmass classification, granite, tunnelling

ABSTRACT: The characterization of the granitic mass of Porto for the design and construction of the Metro works of the city was based on weathering grades and structural features which were used for the derivation of the design parameters. The highly variable nature of the deeply weathered Oporto granite posed significant challenges in the driving of the 2.3 km long C line and the 4 km long S line of the project. Two 8.7 m diameter Herrenknecht EPB TBMs were used to excavate these tunnels but the nature of the rock mass made it extremely difficult to differentiate between the qualities of the mass and apply an open or a closed mode operation of the TBM accordingly. Thus early problems were encountered due to over excavation and face collapse. The matter was finally resolved by the introduction of an Active Support System, which involves the injection of pressurized bentonite slurry to compensate for deficiencies in the face support pressure when driving in mixed face conditions. Both the C and S lines have now been completed with minimal surface subsidence and no face instability.

1. INTRODUCTION

In late 1998 the Municipality of Porto took a decision to upgrade its existing railway network to an integrated metropolitan transport system with 70 km of track and 66 stations. Seven kilometres of this track and 10 stations are located under the picturesque and densely populated city of Porto, an UNESCO world heritage site. A map of the surface and underground routes is presented in Figure 1. Metro do Porto SA, a public company, is implementing the project. The design, construction and operation of this concession were awarded to Normetro, a joint venture. The civil works design and construction was awarded to Transmetro, a joint venture of Soares da Costa, Somague and Impregilio.

The underground tunnel, driven by two Earth Pressure Balance (EPB) TBMs, has an internal diameter of 7.8 m and accommodates two tracks with trains. Line C stretches 2,350 m from Campanhã to Trindade and has five underground stations, a maximum cover of 32 m and a minimum of 3m before reaching Trinidad station. Line S is 3,950m long and runs from Salgueiros to São Bento with 7 stations and a maximum overburden of 21 m.

Figure 1: Map of Metro do Porto routes. Underground tunnels are Line C from Campanhã to Trindade and Line S from Salgueiros to São Bento.

Tunnel driving was started in August 2000 with the drive from Campanhã to Trindade. It was originally planned that the EPB TBM would be run with a partially full, unpressurized working chamber in the better quality granite in order to take advantage of the higher rates of advance in this mode as compared with operating with a fully pressurized working chamber. It was soon found that the highly variable nature of the rock mass made it extremely difficult to differentiate between the better quality rock masses in which the working chamber could be operated safely with no pressure and the weathered material in which a positive support pressure was required on the face. There were indications of over-excavation and two collapses reached the surface. The second occurred on 12 January 2001, almost a month after the passage of the TBM on 16 to 18 December 2000. This collapse resulted in the death of a citizen in a house overlying the tunnel.

At the invitation of Professor Manuel de Oliveira Marques, Chief Executive Officer of Metro do Porto S.A., one of the authors (E.H) visited Porto from in early February 2001 to review the geotechnical and tunnelling issues of the C Line tunnel. As a result of this visit a Panel of Experts, consisting of the authors of this paper, was established in order to provide advice to Metro do Porto.

2. GEOLOGICAL CONDITIONS

The underground portion of the line passes through the granite batholith which was intruded into the Porto-Tomar regional fault in the late Hercinian period (Figure 2). The Porto Granite, a medium grained two mica granite, is characterized by deep weathering and the tunnel passes unevenly through six grades of weathering and alteration ranging from fresh granite to residual soil. The granite is crossed randomly by aplitic/pegmatitic dykes which display much less weathering, following tectonically determined tension joints.

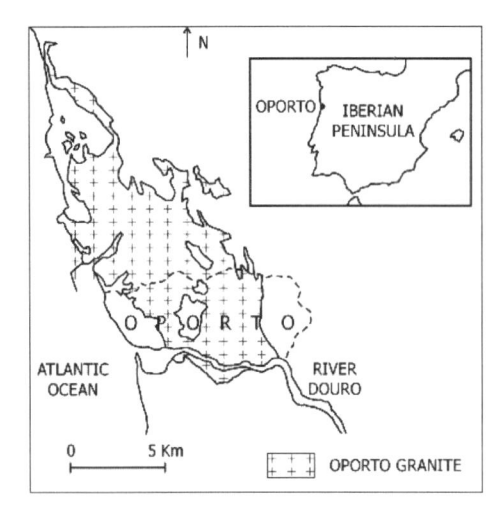

Figure 2: Distribution of granite in the City of Oporto (from A. Begonha and M. A. Sequeira Braga, 2002)

3. CHARACTERISATION OF WEATHERING

The particular feature of most engineering significance of the rock mass is its weathering. All weathering grades (W1 to W6, as established in the engineering geological classification according to the scheme proposed by the Geological Society of London, 1995, and the recommendations of ISRM) can be encountered. Through analyzing the associated geomechanical properties from laboratory tests, the designers developed a re-classification of the degree of weathering aiming to better define the characteristic values of each class and to reduce the overlap between classes (Table 1, Russo et al., 2001).

Table 1: Weathering classes over the uniaxial compressive strength range (clear bars indicate classification based only on qualitative evaluation, shaded bars indicate re-classification after statistical analysis, from Russo et al., 2001)

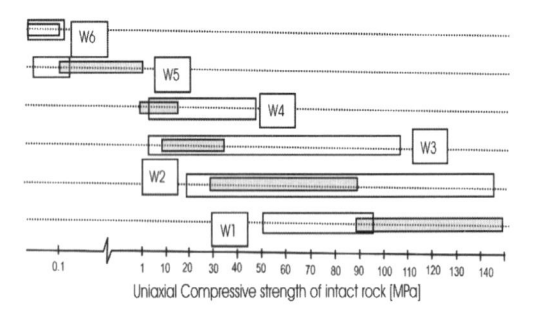

The depth of weathering is of the order of few tens of meters as weathering was assisted by the stress relief regime due to the deepening of Duro valley. Depths of weathering of 30m are reported by Begonha and Sequeira Braga, 2002. Hence, the ground behaviour varies from a strong rock mass to a low cohesion or even cohesionless granular soil. The granularity and frictional behaviour is retained, as the kaolinitisation of feldspaths is not complete and the clay part not important. Furthermore, the spatial development of the weathered rock is completely irregular and erratic.

The change from one weathered zone to another is neither progressive nor transitional. It is thus possible to move abruptly from a good granitic mass to a very weathered soil like mass. The thickness of the weathered parts varies very quickly from several meters to zero. Blocks of sound rock, "bolas", of various dimensions can "float" inside a completely decomposed granite. Weathered material, either transported or in situ, also occurs in discontinuities.

A particularly striking feature is that, due to the erratic weathering of the granite, weathered zones of considerable size well beyond the size of typical "bolas" can be found under zones of sound granite (see Figure 3). While this phenomenon is an exception rather than the rule and it was expected to disappear with depth, it could not be ignored in the zone intersected by the construction of the metro works. A typical case of such setting is in Heroismo station where weathered granite with floating cores of granite occurs under a surficial part of a sound granitic rock mass (Figure 5).

Figure 3: Appearance of different degrees of weathering in granite in a core recovered from a site investigation borehole on the tunnel alignment. Note that the weathered granite in the left box is at a depth of about 24m under the sound granite of the right box. This must therefore correspond to a huge boulder (core).

4. CHARACTERIZATION OF GRANITIC ROCK MASSES

The definition of rock mass properties for use in the face stability analyses and the machine selection, in the design of stations and the settlement- risk analysis, was based on a geotechnical characterization of the granitic mass in various groups. The approach applied in the design is illustrated in Table 2 and the values of the geotechnical parameters after statistical analysis are shown in Table 3 (Russo et al., 2001, Quelhas et al., 2004). Groups g5, g6 and g7 refer to material with soil-like behaviour. Thus it was generally possible to apply principles of soil mechanics to define the geotechnical parameters and the design values of the soil mass were based on sample properties, taking into account the results of the available in situ tests (SPT, etc).

Deformations modulus for groups g2 and g3 was derived from empirical correlations and the results of the 136 Menard tests conducted in the boreholes. It is worth noting that the values of the pressiometric modulus showed significant variability when only associated with the weathering class. On the other hand, when the structure of the mass was considered variability and discrepancies were significantly reduced (Russo et al., 2001).

Figure 4: Appearance of Oporto granite in the face of an excavation for the new (2002) football stadium. Fracturing of the rock mass and heterogeneity in weathering is obvious

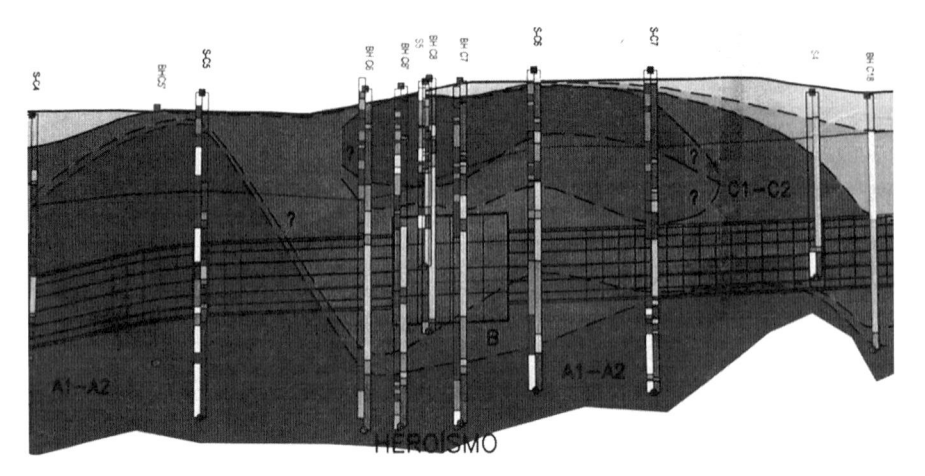

Figure 5: Predicted geology for the Heroismo mined station (Assessment by Transmetro, documents of Metro do Porto). Heterogeneity in weathering and its erratic geometry is evident.

Table 2. Conceptual procedure for the geotechnical characterization of the granitic rock mass and for design (from Russo et al., 2001)

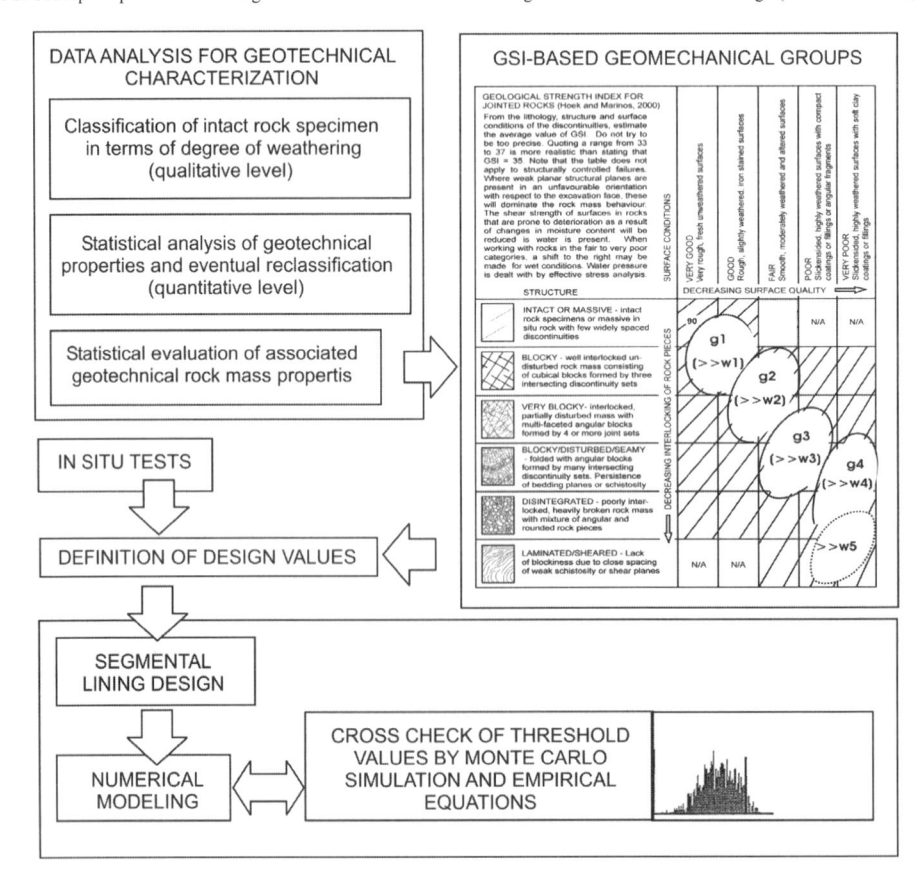

Table 3 Geotechnical parameters (average values, with brackets are given the standard deviations, from Russo et al., 2001 and from Quelhas et al., 2004)

Geotechnical groups	σ_{ci}	γ (KN/m³)	Hoek-Brown criterion parameters mb	s	Ed (GPa)
g1	90-150	25-27	7.45 (1.15)	6.9E-2 (3.2E-2)	35 (10)
g2	30-90	25-27	3.2 (0.5)	7.5E-3 (3.4E-3)	10.7 (3.0)
g3	10-35	23-25	0.98 (0.07)	7.5E-4 (1.7E-4)	1.0 (0.5)
g4	1-15	22-24	0.67 (0.12)	0	0.4 (0.2)

Geotechnical groups	N_{SPT}	γ (KN/m³)	c′ (MPa)	φ′ (°)	Ed (GPa)
g5	>50	19-21	0.01-0.05	32-36	0.05-0.20
g6	<50	18-20	0-0.02	30-34	0.02-0.07
g7	Var.	18-20	0	27-29	<0.05

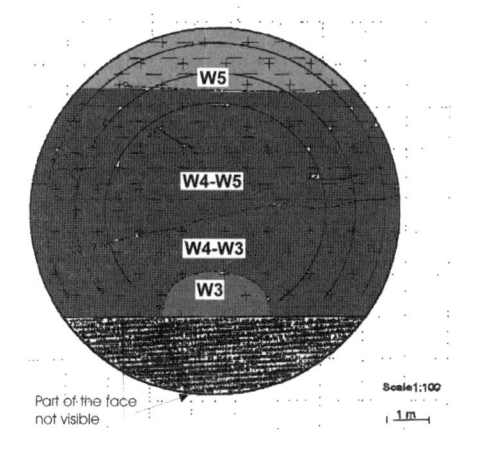

Figure 6: As typical distribution of weathered granite in the face of the EPB driven Tunnel.

It is clear that this characterization cannot be integrated in the design for the selection of parameters, without taking into account the spatial development and variation of geotechnical groups along the alignment or in the area around the stations.

The significance of this comment was shown dramatically soon after boring with the EPB TBM has started. Thus, for the needs of this specific mechanized excavation such a characterization was meaningless and the mode of operation of the TBM had to be selected in such a way that the worst anticipated conditions could be dealt with at any time.

5. PERMEABILITY

The permeability of the rock mass is dependent upon the weathering grade and the associated fractures. In the less weathered rock the flow is related primarily to the fracture system while, in the more heavily weathered material, the ground behaves more like a porous medium. Porosity in the latter case may have been increased from leaching and this together with the highly variable permeability of the rock mass, has resulted in a very complex groundwater regime. The overall permeability is rather low; of the order of 10^{-6} m/s or lower. However higher permeabilities were measured in pumping tests. We consider that preferential drainage paths exist within the granite mass. The very weathered material, having little or no cohesion may be erodible under high hydraulic gradients.

The frequent occurrence of old wells connected by drainage galleries was a hazard for tunnelling. Opinion was expressed that long term exploitation of these wells had led to the washing out of fines increasing permeability and formation of an unstable soil structure (Grasso et al., 2003)

6. EPB TBM CHARACTERISTICS

The complex geological and hydrogeological conditions described above resulted in a decision by Transmetro to utilize an 8.7 m diameter Herrenknecht EPB TBM (see Fruguglietti et al. 1999, and 2001). Initially, only one machine was to be used to drive both lines but following start-up problems, a second machine was added in order to make it possible to complete the tunnel drives on schedule.

The TBMs are equipped with a soil conditioning system capable of injecting foam, polymer or bentonite slurry into the working chamber. Muck removal is by continuous belt conveyor from the TBM back-up to the portal and then by truck to the muck disposal areas. Tunnel lining is formed from 30 cm thick, 1.4 m wide pre-cast concrete segments. The lining comprises six segments and a key and dowel connectors are used in the radial joints while guidance rods are used in the longitudinal joints. The features of the EPB TBM are illustrated in Figure 7. In a review paper by N. Della Valle (Tunnels and

Tunnelling, 2002) details are presented. Gugliementi et al. (2004), in a recent paper, offer a full presentation of the control of ground response and face stability during excavation. In those papers issues proposed by the authors of the present paper and discussed here are described.

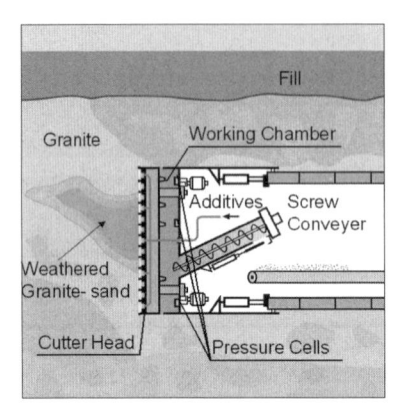

Figure 7: Characteristics of the Herrenknecht EPB TBM used in Oporto.

7. CHARACTERIZATION OF GEOLOGICAL CONDITIONS IN TERMS OF THE TBM OPERATION

The geological conditions discussed above can be translated to the following geological models in front, at the face and immediately above the TBM:
1. Granitic mass of sound or slightly weathered rock, no weathered material in the discontinuities;
2. Granitic mass of sound or slightly weathered rock but with very weathered material (filled or in situ) in substantial fractures; these fractures may communicate with overlaying parts of completely weathered granite;
3. Very weathered or completely weathered granite, W5 (almost granular soil with little or no cohesion);
4. Very weathered or completely weathered granite with blocks of the rock core;
5. Mixed conditions with both sound mass and completely weathered granite appearing in the face.

In all cases the water table is above the tunnel crown

Only the first of these geological models can be excavated using an EPB TBM operating in an open mode. However, because of the unpredictable changes in the geological conditions described above, we considered that the risk of operating in an open mode was unacceptable unless there was unambiguous evidence that this condition persisted for a considerable length of tunnel drive. This was not the case in this tunnel and we recommended that the entire drive should be carried out with the TBM operating in a closed mode.

Indeed in all other models, uncontrolled over-excavation could occur unless the chamber of the machine was full of appropriately conditioned excavated material with the necessary support pressure and control of the evacuation of the muck through the screw conveyor. Lack of adequate face support could result in piping of the weathered material in the fractures that could, in turn, induce collapse of the overlying weathered granite. The mixed face conditions described in item 5 above were considered to be particularly difficult because of the uneven pressure distribution on the face induced by the different stiffness of the rock and soil masses. The successfully handling of this problem is discussed in a following section.

A significant number of wells and old galleries exist in the area and, while most were located on old city maps and by inspection of existing properties, there remained the possibility that some unpredicted wells and galleries could be encountered. The wells usually end above the tunnel but some were deep enough to interfere with the construction. The crossing of such features clearly involved some risk but this was substantially lower when operating the TBM in a fully closed and pressurised mode than in an open or partially open mode.

8. FACE SUPPORT PRESSURE

The face support pressure of EPB - TBMs was controlled by measuring the pressure at the bulkhead with pressure cells, approximately 1.5 m from the face, as shown in Figure 8. In closed mode operation, the working chamber is completely filled with conditioned excavated material, the earth paste. The earth paste is pressurized by the advancing forces induced by the advance jacks via the bulkhead. The pressure level is controlled by the effectiveness of the excavating cutter head in relation to the discharging screw conveyor.

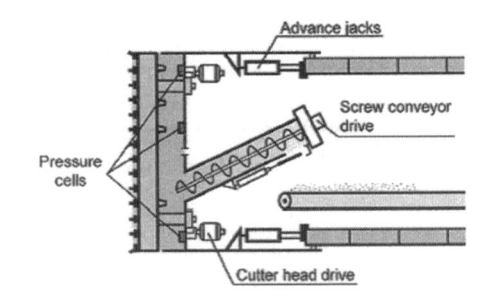

Figure 8: Measurement devices for face support pressure

To verify complete filling of the working chamber, the density of the earth paste in the working chamber was controlled by pressure cells on the bulkhead at different levels. This method satisfies the demand of preventing a sudden instability of the face caused by a partially empty working chamber but it does not guarantee a reliable face support pressure.

Pressure measurement at the bulkhead, 1.5 m behind the face, provides only partial information about the support pressure at the face. The support medium, the earth paste created from excavated ground, conditioned by a suspension with different additives, must have the physical properties of a viscous liquid. However, the shear resistance in that viscous liquid reduces the support forces which can be transferred onto the face. The shear resistance of the earth paste depends on the excavated ground and the conditioning, which is a complex and sensitive procedure. Consequently, the shear resistance of the support medium often varied considerably.

Therefore, the fluctuation of the face support pressure could exceed 0.5 bars. This fluctuation may be acceptable in homogeneous geology but in mixed ground, as found in the Oporto granite, the variable support pressure entailed the danger of significant over excavation.

One of the processes which can cause a drop in the face support pressure is illustrated in Figure 9 which shows a situation in which the lower part of the face is in unweathered granite while the upper part of the face is in residual soil. A major part of the thrust of the machine is consumed by the cutter forces required to excavate the unweathered granite and there is a deficiency in the forces available to generate the pressure in the earth paste in the upper part of the working chamber. This results is an imbalance between the soil and water pressure in the unweathered granite and the support pressure in the upper part of the working chamber. If this deficiency is too large, the face will collapse inwards into the working chamber and this will result in progressive over excavation ahead and above the face.

The deficiency of face support pressure can be compensated for by the addition of an Active Support System, proposed by Dr Siegmund Babendererde (one of the authors of this paper) and shown in Figure 9. This system is positioned on the back-up train and consists of a container filled with pressurized bentonite slurry linked to a regulated compressed air reservoir. The Bentonite slurry container is connected with the crown area of the working chamber of the EPB TBM. If the support pressure in the working chamber drops below a predetermined level, the Active Support System automatically injects pressurized slurry until the pressure level loss in the working chamber is compensated. The addition of this Active Support System to the EPB TMB results in an operation similar to that of a Slurry TBM. This automatic pressure control system reduces the range of fluctuations of the face support pressure to about 0.2bar.

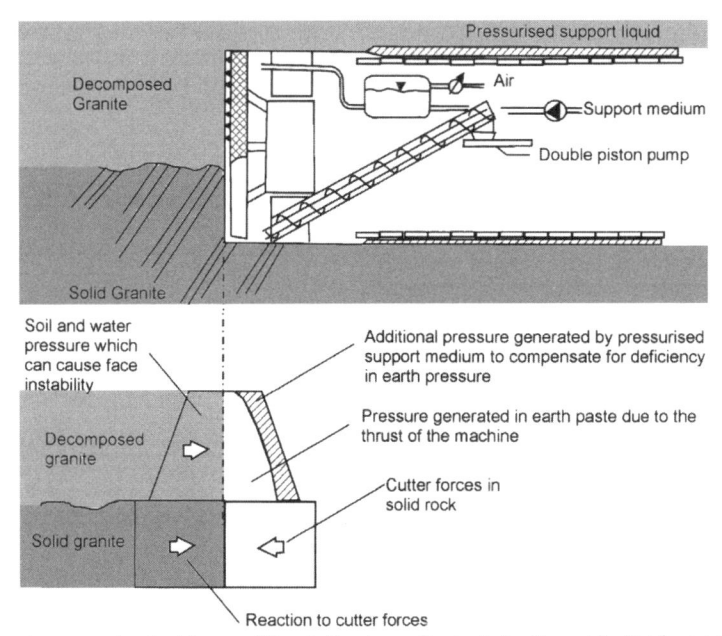

Figure 9: Face support pressures in mixed face conditions in Oporto granite. An Active Support System for overcoming the support pressure deficiency is also illustrated.

In the case of an open and potentially collapsible structure in the weathered granite surrounding the wells, resulting from leaching of the fines, we considered that stable face conditions can be maintained by the correct operation of the TBM in fully closed EPB mode with supplementary fluid pressure application. However, care was required in the formulation and preparation of the pressurizing fluid in order to ensure that an impermeable filter cake was formed at the face. This was necessary in order to prevent fluid loss into the open structure of the leached granite mass.

The application of the Active Support System in the Metro do Porto project was the first time that this system had been used. There was initial concern that the addition of the bentonite slurry would alter the characteristics of the muck to the point where it could no longer be contained on the conveyor system and that an additional slurry muck handling facility may be required. This concern proved to be unfounded since the volume of bentonite slurry injected proved to be very small and there was no discernable change on the characteristics of the muck.

The predetermined support pressure was determined from calculations using the method published by Anagnostou and Kovari (1996) which proved to be reliable for these conditions. The Active Support System was extremely effective in maintaining the predetermined support pressure and no serious face instability or over excavation problems were encountered after it was introduced. In fact, the system permitted the 8.7 m diameter tunnel to pass under old houses with a cover of 3 m to the foundations, without any pre-treatment of the ground. Surface settlements of less than 5 mm were measured in this case. The boring of the section under this shallow cover is described in a paper of Diez and Williams, 2003.

The Active Support System was also connected to the steering gap abound the shield and the filling of this gap with bentonite slurry provided a reliable means of maintaining a predetermined pressure in this gap.

CONCLUSIONS

The highly variable characteristics of the weathered granite in Oporto and their sudden changes imposed substantial risks on the driving of the C and S lines by means of EPB TBMs. The impossibility of accurately predicting and maintaining the correct face support pressure resulted in significant over excavation and two collapses to surface during the first 400 m of the C line drive. Characterization in different geotechnical groups for the selection of the mode of operation of the EPB was almost meaningless and the mode of operation of the TBM had to be selected in such a way that the worst anticipated conditions could be dealt with at any time.

The introduction of the Active Support System, which involves the injection of pressurized bentonite slurry to compensate for deficiencies in the face support pressure when driving in mixed face conditions, proved to be a very effective solution. The remaining C and S line drives have now been completed without further difficulty although the rate of progress was less than that originally projected when the project was planned.

The final breakthrough of the C line drive is illustrated in Figure 13.

Figure 13: Final breakthrough of the TBM S-203 on the completion of the drive from Salgueiros to Trindade on Thursday 16 October 2003.

ACKNOWLEDGEMENTS

The authors wish to acknowledge the permission of Metro do Porto to publish the details contained in this paper. The cooperation of Transmetro and particularly of Ing. Giovanni Giacomin in working with the Panel of Experts is also acknowledged. Part of this paper was presented in a workshop in Aveiro, Portugal, April 2004.

REFERENCES

Anagnostou, G and Kovari, K. 1996. "Face stability conditions with earth-pressure balanced shields". *Tunnelling and Underground Space Technology*. Vol 11, No 2, pp 163-173.

Begonha, A. and Sequeira Braga, M. A. 2002. "Weathering of the Oporto granite: geotechnical and physical properties". *Caten*a. Vol. 49, pp. 57-76

Della Valle, N. 2002. "Challenging soil conditions at Oporto". *Tunnels and Tunnelling International*. December 2002, pp. 16-19.

Diez R. and Williams R. "TBM tunneling under very low cover- Approach to Trindade station, Porto Metro (Portugal)"

Fruguglietti, A., Ferrara, G, Gasparini, M and Centis, S. 2001 "Influence of geotechnical conditions on the excavation methods of Metro do Porto project".*Proceedings, Congress ITA*. Milan, pp. 135-141

Fruguglietti, A., Guglielmetti, V., Grasso, P., Carrieri, G and Xu, S. 1999. "Selection of the right TBM to excavate weathered rocks and soils". *Proceedings Conference: Challenges for the 21 st Century*, Allen et al (eds), Balkema Publ.. pp. 839-947.

Geological Society of London 1995. "The description and classification of weathered rocks for engineering purposes". *QJEG*, pp 28

Grasso P., Xu Sh., Fedele M., Russo G. and Chiriotti E. 2003. "Particular failure mechanisms of weathered granite observed during construction of metro tunnels by TBM". *Proc. ITA World Tunnelling Congress,* 2, Amsterdam

Guglielmetti V., Grasso P., Gaj F. and Giacomin G. 2003. "Mechanized tunneling in urban environment: control of ground response and face stability, when excavating with an EPB Machine". *Proc. ITA World Tunnelling Congress,* 2, Amsterdam

Quelhas, J., Gomez, J., Ferreira, P., Pires, A., Andrade, c., Baiao, C., Maia, C., Pistone, R. 2004. "Estacoes mineiras do metro do Porto – Metodos construtivos". *9th Portugeuse Congress of Geotechnics (9th CNG),* Aveiro.

Russo, G., Kalamaras, G.S., Origlia, P and Grasso, P. 2001. "A probabilistic approach for characterizing the complex geological environment for the design of the new Metro do Porto". *Proceedings, Congress ITA*. Milan, pp 463-470.

In situ test characterisation of unusual geomaterials

F. Schnaid
Federal University of Rio Grande do Sul, Porto Alegre, Brazil

B.M. Lehane & M. Fahey
School of Civil and Resource Engineering, The University of Western Australia, Australia

Keywords: *in situ* tests, soil structure, small strain stiffness, stiffness non-linearity, shear strength, partial drainage, unsaturated soil conditions

ABSTRACT: This paper presents a review and discussion of recent and current developments in the interpretation of *in situ* tests conducted in 'unusual' geomaterials. Insights are provided into the assessment of stress-strain-time and strength characteristics of a number of soils that are still relatively poorly understood. Soil characterization and classification, the role of hydraulic conductivity and partial drainage effects, the strength and stiffness of both natural and man-made deposits, the influence of structural effects, and the significance of unsaturated soil mechanics are topics selected for discussion. Advantages and limitations of currently adopted interpretative techniques and analyses are highlighted. Since the *in situ* behaviour of unusual soils is complex, emphasis has been placed on correlations with mechanical properties that are based on the combination of measurements from independent tests such as the ratio of the elastic stiffness to ultimate strength (G_o/q_c, G_o/N_{60}), the ratio of cone resistance and pressuremeter limit pressure (q_c/Ψ) and the association of strength and energy measurements (N_{60} and energy).

1 INTRODUCTION.

In situ testing is a vital component of a site investigation programme, and one in which the complexity of nature starts to be revealed. Such importance can be measured by the variety of available *in situ* test equipment, techniques and procedures, as well as by the diversity of interpretative methods designed to characterize soils and assess their mechanical properties. International conferences, some of which have been devoted specifically to *in situ* testing, have provided the forum for publication of major pioneering developments. These started with the development of the SPT, which arose from the need to supplement geological information with quantitative data. Since then, notable contributions include: the standardisation of SPT results by a reference energy value (Schmertmann & Palacious, 1979); growth in understanding of the basis and applicability of cone penetration tests (e.g. Lunne *et al.*, 1997), framed by theoretical solutions such as the strain path method (Baligh, 1985); recognition of the importance of cavity expansion theory and its application to penetration tests (e.g. Ladanyi, 1963; Vesic, 1972); development of correlations between test results and soil properties from large laboratory calibration chamber tests in sand (Schmertann, 1975; Bellotti *et al.*, 1989; Baldi *et al.*, 1996); appreciation of the importance of geophysical methods (e.g. Stoke *et al.*, 1995) and the recent achievements on the understanding of the role of small strain stiffness (Burland, 1989).

These new developments set the standards for interpretation, using *in situ* tests, of the stress-strain-strength behaviour of both sand and clay to assist in the solution of a variety of geotechnical problems. They also provided the background necessary for the assessment of other aspects of soil behaviour in more complex environments, including the effects of microstructure (fabric and bonding), small strain stiffness, anisotropy, weathering and destructuration, partial saturation, large scale volume changes and viscosity. These important features of natural soil (and man-made) ground behaviour are now recognized and are addressed on the basis of a framework that has been established from a comprehensive characterization of laboratory tests on reconstituted soils and a number of well known natural clays and sands.

The considerable body of knowledge accumulated from laboratory tests now has to be extended to the field of *in situ* testing, and a necessary step in this direction is to develop a new generation of interpretation methods and constitutive models that capitalizes on our existing experience. Models can either attempt to capture the full nature of the observed mechanical behaviour of geomaterials or simply aim at only reproducing the essential features related to a given soil or soil condition. The challenge is therefore threefold: to evaluate the applicability of existing theoretical and empirical approaches in order to extend the experience of 'standard' clays and sands to other geomaterials, to develop interpretative methods that incorporate new constitutive models whenever required, and to gather experimental data that justifies the applicability of proposed interpretation methods to engineering applications.

This paper begins by reviewing key aspects of the behaviour of unusual geomaterials, and then examines if given *in situ* test interpretation methods can lead to a rational selection of soil parameters and hence economical designs in such materials. Only a selection of aspects are discussed here as the topic is clearly too extensive to be covered in a single paper.

2 UNUSUAL GEOMATERIALS

For the purpose of this paper, an 'unusual geomaterial' will be defined as one that satisfies any one or more of the following criteria:
- classical constitutive models do not offer a close approximation of its true nature;
- it is difficult to sample or to be reproduced in the laboratory (interpretation is therefore solely based on *in situ* test data);
- very little systematic experience has been gathered and reported;
- values of parameters are outside the range that would be expected for more commonly encountered soils such as sand and clay;
- the soil state is variable due to complex geological conditions.

The term 'unusual' employed in this paper can therefore only be interpreted within the context of the currently accepted state-of-the-art; when more experience and data are gathered, materials that are now seen as unusual may be in the future regarded as ordinary.

The need to focus research efforts on unusual soils and unusual soil conditions is justified by the fact that soil mechanics has largely evolved from research on the drained and undrained mechanical behaviour of sedimentary clays and the drained response of reconstituted sands. This research has led to the development of a family of complete mathematical models based on the critical state concept (Schofield & Wroth, 1968, Bolton, 1986). The undrained response of sands has also been the focus of much research in recent years, after the recognition that the high hydraulic conductivity of these soils does not necessarily prevent pore pressure accumulation under dynamic loading. Although departing from this well established background is not a simple task, the efforts of many researchers, over the past twenty years, has resulted in remarkable progress in the integration of many geomaterials into a consistent and unified framework. This link has been realized by the recognition that soil structure is a common feature of all geomaterials (e.g. Vaughan, 1985; Burland, 1989; Tatsuoka *et al.,* 1997; Leroueil and Hight, 2003). A widespread approach has been to compare the response of the natural soil with that of the corresponding reconstituted material (e.g. Burland 1990) to identify features of behaviour emerging from structure from those related to changes in state.

A fundamental understanding of soil behaviour is developed in the laboratory in tests carried out under strictly controlled boundary conditions. A preliminary requirement is to retrieve good quality samples with minimal disturbance. Unfortunately, our ability to achieve such samples is limited due to a variety of factors including the sensitivity of soil structure to stress relief (e.g. Hight, 1998). In hard bonded soils, Shelby tubes can only be used if the soil is sufficiently soft to permit driving, and even in this case rock fragments can obstruct driving or result in a sample that experiences considerable disturbance (de Mello, 1972; Sandroni, 1988). Even block samples, which do not suffer from the relatively severe disturbances effects inherent in tube sampling, can experience considerable destructuration due to stress relief and exhibit stiffness characteristics that contrast with observed field behaviour. In granular materials, tube sampling takes place under drained conditions and the volumetric and shear strains produced during sampling are sufficient to completely destructure the sand. The only viable option for these soils is to freeze the ground prior to sampling. Natural silt and silt sandy soils are also subjected to unavoidable disturbance. These and other geomaterials listed in Table 1 offer great challenge to engineering characterization as design is based primarily on interpretation of *in situ* test results.

Within this very broad subject area, some specific themes have been selected for discussion:
- evaluation of the *mechanical properties* of natural soils and some man-made geomaterials from *in situ* tests;
- delineation of *soil profiles* from *in situ* measurements, with emphasis on geophysics and penetration tools; the role of *hydraulic conductivity* in unusual geomaterials;
- evaluation of *partial drainage* and uncertainties introduced to *in situ* test interpretation;

Table 1: Unusual soils and their general characteristics.

Geomaterials	Features of behaviour
Natural soils — Bonded soils (hard soils, soft rocks, residual soils)	Bonding and structure are important components of shear strength Cohesive-frictional nature (characterized by ϕ' and c') Anisotropy derived from relic structures of the parent rock Structure and fabric may be developed *in situ* by weathering process Very variable fabric and mineralogy Destructuration in shear Stress-history relatively unimportant
Calcareous sand	Primarily a product of biological activities and therefore more susceptible to post-depositional physical and chemical alterations Poorly graded, wide range of particle shapes and sizes Carbonate content, crushability, interparticle cementation, index properties, stress history, geological processes affect engineering behaviour Fairly compressible, their compressibility results from grain-crushing and the collapse of grain-structure
Intermediate soils	Conditions of drainage are difficult to determine for given loading conditions Lenses of both finer and coarser materials often encountered *In situ* test interpretation methods are related to either sand (drained) or clay (undrained)
Coarse-grained cemented aged materials	Soils with sufficient high permeability to ensure drained conditions Strength and stiffness higher than fine grained soils Less likely to be strongly time dependent Important influence of aging and cementation Susceptible to erosion and liquefaction
Volcanic soils	Characterized by low specific gravity and angular crushable grains Often high compressibility
Difficult soft soil conditions	Peat layers impart considerable spatial variations in water content and index properties Organic content has a strong impact on soil fabric and mechanical properties Extremely soft and compressible
Man-made — Earth-fills and improved ground	Void ratio and structure controlled by the mode of placement and/or by compaction Anisotropy and stress history determined from compaction Evaluation of consolidation and settlement in cohesive soils Assessment of changes in density in cohesionless soils
Tailings	Very stratified and layered, the particle size ranging from coarse rock to clay size Mechanical properties vary with ore type, method of placement, location, exposure to evaporation, ageing, among other effects
Waste repositories	Rheological effects due to physical and chemical alterations
State — Partial saturation	Effects of partial consolidation in the soil ahead of an advancing probe
Unsaturated soils	Frame of reference is described by four variables – net mean stress $(p - u_a)$, deviator stress q, suction s $(u_a - u_w)$ Suction measurements and their practical significance on controlling soil collapsibility and hydraulic conductivity

- interpretation of *in situ* tests in *coarse grain geomaterials* incorporating acknowledged effects of age and cementation;
- analysis of the response of *cemented materials* where the threshold yield stress controls soil response;
- the practical significance of suction measurements in *partially saturated soils* and their influence on testing interpretation and prediction of soil collapsibility.

To cover these topics, the paper examines data from coarse-grained materials, intermediate soils (including natural silty deposits and tailings) and bonded geomaterials (including residual soils and carbonated sands), under both saturated and unsaturated conditions. Space constraints impose a number of restrictions on this report: the continuum mechanics approach is assumed as valid in all calculations, rheological effects produced by both physical and chemical alterations will not be considered, and problematic soils (e.g. peats) that are now covered in specific publications (under TC36) will not be addressed. Soil improvement techniques are also outside the scope of the present review.

3 TESTING TECHNIQUES

Some of the most common *in situ* tests available for routine investigation are listed in Table 2. Changes or developments to *in situ* testing equipment and testing procedures have not been significant in recent years and, for geotechnical applications, engineers can rely on a variety of commercial tools, International Reference Test Procedures and well established national codes of practice. Although development has slowed, some recent exciting trends include (a) the combination of different sensors in a single test device (e.g. Mayne, 2001), (b) adaptation of additional sensors to penetration tools to enhance and expand their capabilities for geo-environmental purposes (e.g. Robertson *et al.*, 1998), (d) use of T-bar and ball penetrometers because of the more predictable nature of the strain paths that they induce to the soil (e.g. Randolph*, 2004) and (d) changes in execution procedures and testing for advanced off-shore engineering (e.g. Van Impe & Van der Broeck, 2001; Randolph*, 2004).

This paper attempts to identify the applicability of existing techniques for characterisation of unusual soils, a task that necessarily has to cover a critical revision of available empirical and theoretical methods. In particular, correlations with mechanical properties that are based on the combination of measurements from independent tests are explored. These combinations are:

- the ratio of the 'small strain' elastic stiffness to ultimate strength (G_o/q_c, G_o/N_{60});
- the ratio of cone resistance and pressuremeter limit pressure (q_c/ψ_L);
- the association of strength and energy measurements (N_{60} and energy).

Thereafter, the field investigation methods are briefly addressed, together with a critical appraisal on how results can be compiled to obtain a ground model and appropriate geotechnical parameters. The discussion focuses on data from the four most popular *in situ* testing techniques: geophysics (for general ground characterization and measurement of the small strain stiffness), penetration tools (SPT and CPT for soil profiling and property assessment) and pressuremeter tests (for rational prediction of soil properties).

4 GROUND CHARACTERIZATION

A ground investigation programme aims to determine the ground and groundwater conditions relevant to a given construction site, profiling and classification of soils being a primary step. Profile description and geotechnical classification can be rather complex, difficult to understand and generalize and demand a considerable number of trial pits and boreholes to facilitate detailed geological logging (e.g. Clayton *et al.*, 1995). The Unified Soil Classification System can only be used provided that shortcomings in relation to unusual geomaterials are recognised. For example, drainage conditions during penetrometer installation are difficult to establish in intermediate and partially saturated soils.

A geophysical survey is regarded as a powerful technique for subsurface exploration. Despite due recognition of its risks and limitations, there has been a steady increase in the perceived value of geophysics in representing complicated subsurface conditions involving large spatial variability, stratified soils, weathered profiles, among others. In papers presented to this conference, these complexities are reported many times, as for example, in the paper by Marques *et al** (2004) which discusses the strong inhomogeneities and erratic weathering grades of the Porto Granite weathered profiles.

To enhance its consistency, a site investigation campaign should encompass a combination of geophysical surveys with a mesh of boreholes and/or penetration tests. Among the many forms of *in situ* penetration tests used worldwide, the two most common penetration tools designed for soil profiling are the SPT and CPT. Whereas the CPT has proved to be a very efficient technique in a wide variety of materials, the SPT remains one of the few viable options where very stiff soils and the presence of very coarse-grained and rocky materials may limit the penetration of hydraulic tools.

Table 2: Commercial *in situ* testing techniques

Test	Designation	Measurements	Common Applications
Geophysical Tests:			
Seismic Refraction	SR	P-waves from surface	Ground characterization
Surface Waves	SASW	R-waves from surface	Small strain stiffness, G_o
Crosshole Test	CHT	P & S waves in boreholes	
Downhole Test	DHT	P & S waves with depth	
Standard Penetration Test	SPT	Penetration (N value)	Soil profiling
			Internal friction angle, ϕ'
Cone penetration Test			Soil profiling
Electric	CPT	q_c, f_s	Undrained shear strength, s_u
Piezocone	CPTU	q_c, f_s, u	Relative density/internal friction angle, ϕ'
Seismic	SCPT	$q_c, f_s, V_p, V_s, (+u)$	Consolidation properties
Resistivity	RCPT	q_c, f_s, ρ	Stiffness (seismic cone)
Pressuremeter Test			
Pre-bored	PMT	G, ($\psi \times \varepsilon$) curve	Shear modulus, G
Self-boring	SBPMT	G, ($\psi \times \varepsilon$) curve	Undrained shear strength, s_u
Push-in	PIPPMT	G, ($\psi \times \varepsilon$) curve	Internal friction angle, ϕ'
Full-displacement	FDPMT	G, ($\psi \times \varepsilon$) curve	In situ horizontal stress
			Consolidation properties
Flat Dilatometer Test			
Pneumatic	DMT	p_o, p_1	Stiffness
Seismic	SDMT	p_o, p_1, V_p, V_s	Shear strength
Vane Shear Test	VST	Torque	Undrained shear strength, s_u
Plate loading test	PLT	(L × δ) curve	Stiffness and strength
Combined Test			Shear modulus, G
Cone pressuremeter	CPMT	$q_c, f_s, (+u), G, (\psi \times \varepsilon)$	Shear strength

4.1 Cone Penetration Test (CPT)

The CPT, with the possible inclusion of pore water pressure, shear wave velocity and resistivity measurements is now recognized worldwide as an established, routine and cost-effective tool for site characterization and stratigraphic profiling, and a means by which the mechanical properties of the subsurface strata may by assessed. CPTs were particularly popular in sands and in marine and lacustrine sediments in costal regions, but are now also commonly used in peats, silt, residual soils, a variety of hard materials (chalk, cemented sands) and reclaimed land formed by hydraulic fills, dredgings and mine tailings. For a general review on the subject the reader is encouraged to refer to Lunne *et al.* (1997) – *CPT in Geotechnical Practice,* and the Proceedings of the International Symposia on Penetration Testing (1988; 1998).

Routine penetrometers have employed either one midface element for pore water pressure measurement (designated as u_1) or an element positioned just behind the cone tip (shoulder, u_2). The ability to measure pore pressure during penetration greatly enhances the profiling capability of the CPTU, allowing thin lenses of material to be detected. Additionally, geotechnical site characterization is enhanced by independent seismic measurements, adding the downhole shear wave velocity (V_s) to the measured tip cone resistance (q_t), sleeve friction (f_s) and pore water pressure (u). The combination of different measurements into a single sounding provides a particular powerful means of assessing the characteristics of unusual materials.

Profiling and soil classification are fairly well-established practices, but there is still a need for careful planning to ensure that the required information is obtained. Two examples are presented here as a reminder of the different aspects of site characterization that are achieved by pore pressure measurement.

The first case study, extracted from a paper submitted to the present conference by Costa Filho* *et al.* (2004), in collaboration with the first author, relates to a gold tailing deposit. Figure 1 shows a typical CPT profile representative of the deposit obtained for underflow conditions. The differences in the waste characteristics (ore type, mine processing,

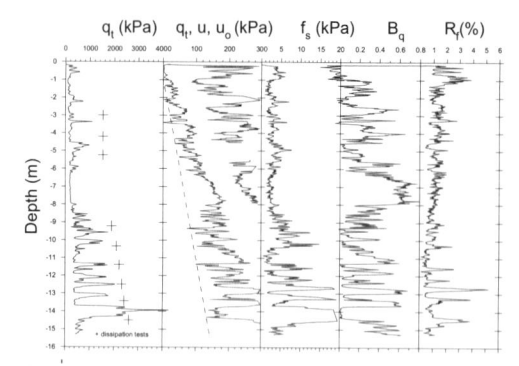

Figure 1. CPT probes obtained for underflow gold tailings

mineralogy) as well as the placement process during disposal tend to affect the geomechanical behaviour of tailings and produce a highly stratified and layered profile. Classification charts presented in Figure 2 illustrate the considerable dispersion in material type indicated by pore pressure measurements. As later discussed in section 6, pore pressure measurements are paramount for the evaluation of partial drainage during penetration in intermediate soils and therefore for the assessment of representative soil parameters. The result is that for silts, pore water pressure measurements do not reflect undrained conditions at a standard penetration rate and soil classification from q_t versus u (or B_q) charts may become unreliable.

The second case is reported by Schneider *et al.* (1999) in a residual deposit. This Piedmont test site in the USA is composed of silty to sandy residual soils grading eventually to partially-weathered schist and gneiss. The water table lies 3 m below ground level under an unsaturated crust that is thought to result from groundwater fluctuations. A representative piezocone profile is represented in Figure 3, which indicates a profile with increasing tip resistance and relatively constant sleeve friction and shear wave velocity with depth. Of particular interest are the negative pore water pressure measurements at the shoulder filter exhibiting a value of $u_2 \approx -100$ kPa recorded throughout the 15m depth. Measurements

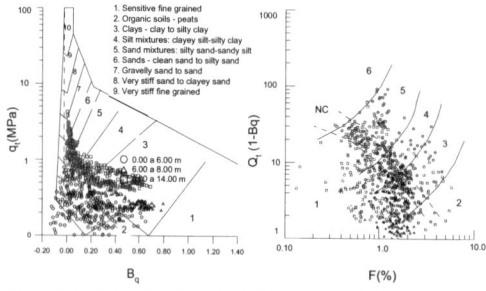

Figure 2. Soil classification chart (Robertson, 1990)

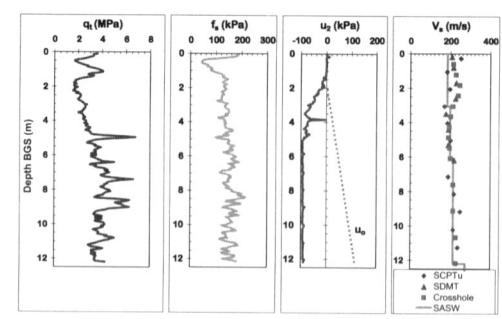

Figure 3. Piezocone penetration in a residual soil at the Opelika Test Site, USA (Schneider *et al*, 1999)

of pore pressure, in this case, reveal a particular 'unusual' characaristic of the *in situ* material. According to Sowers (1994) the Piedmont residuum has relic features and qualities of the parent rock, including remnant bonding of the intact rock itself, as well as discontinuities and fissures of the rock mass. It is likely that the shoulder u_2 reading is negative (and limited to -100 kPa because of cavitation of the sensor fluid) as it primarily reflects shear induced pore pressures related to the remnant discontinuities within the deposit.

Bearing in mind that soil classification using CPTU data is indirect and relies entirely on empirical charts developed for interpretation of strata, u_2 measurements cannot always be considered useful to ensure a proper soil classification in unusual geomaterials. Since classification charts should rely on at least two independent measurements, in the absence of pore pressure measurements, it is suggested that q_c should be compared with the small strain stiffness G_o. The G_o/q_c ratio provides a measure of the ratio of the elastic stiffness to ultimate strength and may therefore be expected to increase with sand age and cementation, primarily because the effect of these on G_o is stronger than on q_c. For sands, work reported by Bellotti *et al.* (1989), Rix & Stokes (1992), Lunne *et al.* (1997) and Fahey *et al.* (2003) provides some new insights by correlating G_o/q_c versus q_{c1}, where q_{c1} is defined as:

$$q_{c1} = \left(\frac{q_c}{p_a}\right)\sqrt{\frac{p_a}{\sigma'_v}} \qquad (1)$$

and where p_a is the atmospheric pressure (note that the normalized parameter q_{c1} is dimensionless). Once profiles of q_c and G_o are determined, these values can be used directly to evaluate the possible effects of stress history, degree of cementation and ageing for a given profile, as already recognised by Eslaamizaad & Robertson (1997). Data points are shown in Figure 4 for CPT tests carried out in residual soils (artificially cemented Monterey soils also included). Since residual soils always exhibit some bond structure, the data fall outside and above the

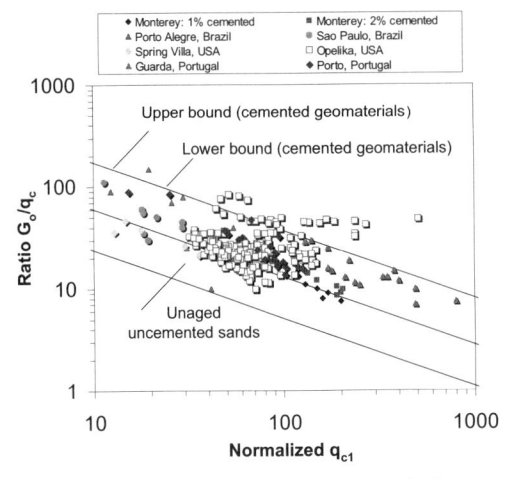

Figure 4 Relationship between G_o and q_c for residual soils

band proposed by Eslaamizaad & Robertson as indicated in the figure.

The variation of G_o with q_c observed in the range of sand deposits by Robertson (1997) was expressed by upper and lower bounds. The upper bound for uncemented material can be assumed as a lower bound for cemented soils and a tentative new upper bound for cemented materials can be expressed as:

$$\left.\begin{array}{l} G_0 = 800\sqrt[3]{q_c \sigma'_v p_a} \text{ upper bound : cemented} \\ G_0 = 280\sqrt[3]{q_c \sigma'_v p_a} \text{ lower bound : cemented} \\ \qquad\qquad\qquad\qquad\text{upper bound : uncemented} \\ G_0 = 110\sqrt[3]{q_c \sigma'_v p_a} \text{ lower bound uncemented} \end{array}\right\} \quad (2)$$

An examination of the potential of employing a G_o/q_c ratio to assess such structural effects is made in the following using *in-situ* data obtained in a variety of sand types in Perth, Australia. The stratigraphy in the Perth area includes the following sand deposits, which are listed in order of increasing age:
- Siliceous sand fill placed hydraulically for reclamation works in the 1950s and 1960s;
- 'Safety Bay Sand', which was deposited under littoral and aeolian conditions in the mid-Holocene; this sand contains many shells and has a calcium carbonate content in excess of 50%;
- 'Spearwood dune sand' which was laid down in the late Pleistocene as a limestone but was subsequently leached of virtually all its calcium carbonate content;
- Alluvial 'Upper Guildford (siliceous) sand', which was laid down by streams flowing from the pre-Cambrian Darling Ranges to the east of Perth during the early Pleistocene;
- 'Lower Guildford (siliceous) sand', which was formed in the same way as the Upper Guildford sand but prior to the deposition of Guildford clay which underlies the upper sand.

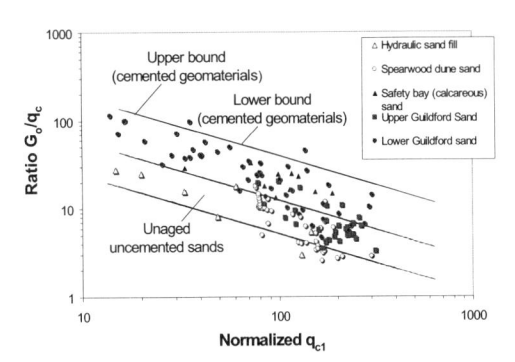

Figure 5. Relationship between G_o and q_c for Perth sands, Australia

The G_o/q_c ratios for these Perth sands are plotted against corresponding q_{c1} values on Figure 5, where q_{c1} may be considered approximately proportional to the sand relative density. A relatively clear trend for G_o/q_c ratios at a given q_{c1} value to increase with age is evident. For example, G_o/q_c ratios in the lower Guildford sand are typically about five times higher than those recorded in the hydraulic fill. It is also apparent that the relatively young calcareous sand in this database indicates G_o/q_c ratios which are as high as those of the much older Guildford siliceous sands. This is presumably because of the likelihood of stronger cementation effects in calcareous sands.

The scatter in Figures 4 and 5 is thought to be principally a result of the influence of the horizontal stress on both initial stiffness and tip cone resistance. Calibration chamber data in sand have clearly shown that, for a given density, cone resistance depends primarily on the *in situ* horizontal stress and therefore σ'_{ho} must be accounted for in a rational interpretation of field tests (Schnaid & Houlsby, 1992). Equation 2 should ideally be referred to horizontal stress or mean *in situ* stress rather than to vertical stress. The preference for σ'_{vo} is justified by the impossibility of determining with reasonable accuracy the value of the horizontal stress in most natural deposits, because they have undergone complex stress history, cementation and aging effects that are difficult to reconstruct.

4.2 *Standard Penetration Test (SPT)*

The SPT is the most widely used *in situ* testing technique, primarily because of its simplicity, robustness and its ability to cope with difficult ground conditions in addition to providing disturbed soil samples. A comprehensive review of procedures and applications of the SPT is given by Decourt *et al.* (1988) and Clayton (1993). There is a range of types of SPT apparatus in use around the world (e.g. those employing manual and automatic trip hammers) and, consequently, variable energy losses cannot be avoided. Variability due to unknown values of energy delivered to the SPT rod system can now be properly accounted for by standardizing the measured N value to a reference value of 60% of the potential energy of the SPT hammer (N_{60}), as suggested by Skempton (1986). In many countries, however, this recommendation has not been incorporated into engineering practice. Moreover, even an SPT N value normalized to a given reference energy is not 'standard' because of the presently contentious issue of the influence of the length of the rod string; this effect is discussed later in this paper.

As for the CPT, SPT N values can also be combined with seismic measurements of G_o to assist in the assessment of the presence of a deposit's bonding structure and its variation with depth. Such a combination is provided on Figure 6, which plots G_o/N_{60} vs $(N_1)_{60}$ in residual soils (Barros, 1997; Schnaid, 1997), where $(N_1)_{60} = N_{60} (p_a/\sigma'_{vo})^{0.5}$ and is analogous to q_{c1} on Figures 4 and 5. The bond structure is seen to have a marked effect on the behaviour of residual soils, producing values of normalised stiffness (G_o/N_{60}) that are considerably higher than those observed in fresh cohesionless materials. A guideline formulation to compute G_o from SPT tests is given by the following equations:

$$\frac{(G_0/p_a)}{N_{60}} = \alpha N_{60} \sqrt{\frac{p_a}{\sigma'_{vo}}} \quad \text{or} \quad \frac{(G_0/p_a)}{N_{60}} = \alpha (N_1)_{60} \quad (3)$$

where α is a dimensionless number that depends on the level of cementation and age as well as the soil compressibility and suction. The variation of G_o with N can also be expressed by upper and lower boundaries, similarly to the cone penetration data:

$$\left.\begin{array}{l} G_0 = 1200 \sqrt[3]{N_{60} \sigma'_v p_a^2} \quad \text{upper bound : cemented} \\ G_0 = 450 \sqrt[3]{N_{60} \sigma'_v p_a^2} \quad \text{lower bound : cemented} \\ \qquad\qquad\qquad\qquad\qquad\; \text{upper bound : uncemented} \\ G_0 = 200 \sqrt[3]{N_{60} \sigma'_v p_a^2} \quad \text{lower bound uncemented} \end{array}\right\} \quad (4)$$

The process of *in situ* weathering of parent rocks (which creates residual soils) gives rise to a profile containing material ranging from intact rocks to completely weathered soils. Rock degradation generally progresses from the surface and therefore there is normally a gradation of properties with no sharp boundaries within the profile. Lateritic and saprolitic residual soils are distinguished on Figure 6 because of their different geological history. Lateritic soils are formed under hot and humid conditions involving a high permeability profile which often results in a bond structure with high contents of oxides and hydroxides of iron and aluminium. In saprolitic profiles the original disposition of the decomposed crystals of the parent rock is retained.

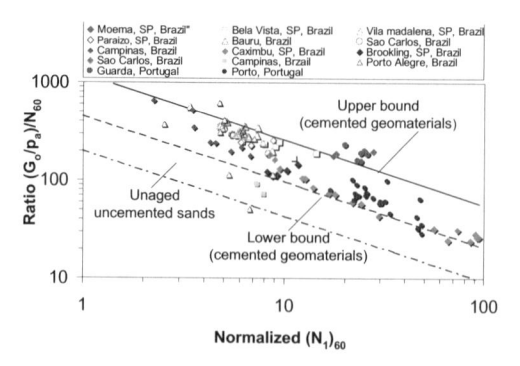

Figure 6. Correlation between G_o and N_{60} for residual soils (Schnaid, 1997).

This gives to the soil a peculiar relic structure where the soil grains are well arranged and orientated (e.g. Novais & Ferreira, 1985; Vaughan, 1985). The small strain stiffness to strength ratio embodied within the G_o/N_{60} term is seen on Figure 6, at a given $(N_1)_{60}$ (or relative density), to be generally appreciably higher for lateritic soils than that of the saprolites, primarily because the latter generally exhibit higher N_{60} (or strength) values.

It follows from the foregoing that a bonded/cemented structure produces G_o/q_c and G_o/N_{60} ratios that are systematically higher than those measured in cohesionless soils. These ratios therefore provide a useful means of assisting site characterization.

5 SOIL STIFFNESS

Soil stiffness depends upon complex interactions of state (bonding, fabric, degree of cementation, stress level), strain level (and effects of destructuration), stress history and stress path, time dependent effects (aging and creep) and type of loading (monotonic or dynamic). Whereas the initial shear modulus G_o is considered to be a fundamental soil property, a knowledge of the non-linear and inelastic stress-strain response of geomaterials is now fully recognized as being critical to the prediction of ground movements. Comprehensive reviews of this topic are reported at the International Conferences of Pre-failure Deformation Behaviour of Geomaterials" in 1995, 1997, 1999 and 2003 and only a short description of the main features relevant to *in situ* testing interpretation is provided in the following:

- Soils display non-linear stress strain behaviour that can be broadly characterized by the linear threshold strain, ε_{Y1}, the strain marking the limit to recoverable behaviour ε_{Y2} and the strain denoting the onset of large scale yielding, ε_{Y3} (Jardine, 1985; Tatsuoka et al., 1997).
- At very small strains, within the limit state curve defined by ε_{Y1}, soils are believed to be-

have as elastic materials represented by the initial elastic stiffness G_0. The magnitude of G_0 is preferably measured using seismic field tests, or alternatively using laboratory bender elements or resonant column tests.
- For clean sands, the effects on initial stiffness of strain rate and stress history have been found to be insignificant. For carbonate and crushable sands, the effects of over-consolidation are more pronounced (Tatsuoka et al., 1997). In clays, G_0 depends on mean effective stress, void ratio and OCR (e.g. Hardin & Drnevich, 1972; Jamiolkowski et al., 1995a).
- Whilst in reconstituted soils the shear behaviour is controlled solely by a combination of deviator stress, mean effective stress and specific volume (and as a consequence the shear stiffness at any strain is expressed as a function of their current state), natural soils exhibit a structural behaviour that does not conform with the framework developed for reconstituted materials (Burland, 1990; Leroueil & Vaughan, 1990).
- In strongly bonded materials, the zone of elastic behaviour is also enlarged (e.g. Tatsuoka et al., 1997; Matthews et al., 2000; Cuccovillo & Coop, 1999) and the value of G_0 becomes particularly important as a bench-mark for engineering applications.
- Stiffness anisotropy (inherent and stress induced) is a common feature of geomaterials, over a wide range of strains (e.g. Tatsuoka & Shibuya, 1991; Belloti et al., 1996; Hight et al., 1997; Jardine et al., 1995). However, stiffness anisotropy of residual soils and other unusual geomaterials is not presently well understood.

Small-strain shear stiffness is determined in situ from the shear wave velocity measured in conventional cased boreholes using crosshole (CHT) and downhole (DHT) techniques, or with surface techniques such as spectral analysis of surface waves (SASW), seismic refraction (SR) and reflection surveys. Particularly attractive are the downhole methods of SCPT and SDMT obtained from the piezocone and dilatometer, respectively, as well as the combination of velocity measurements with the SPT. Propagation velocity should preferably be computed from the measured difference in travel times between two geophones mounted in the penetration tools, to avoid problems with trigger delays and uncertainties on travelling paths. The CHT and DHT techniques enable the velocity of horizontally propagating, vertically polarized (S_{hv}), vertically propagating, horizontally polarized (S_{vh}) and horizontally propagating, horizontally polarized (S_{hh}) shear waves to be measured. Information of the anisotropy of small strain stiffness from these seismic measurements is becoming more readily available. An example of S_{vh} records in an SCPT carried out in a residual soil has been previously given in Figure 4.

In the preceding section, the variation of G_0 with q_c and G_0 with N_{60} observed for residual soils and natural sands was summarised in Figures 4 and 5. Equations 2 and 4 can match the range of recorded G_0 values and despite the fact that these equations have originally been proposed to distinguish cemented and uncemented soils, it is likely that practitioners may be tempted to employ them to estimate G_0. In the absence of direct measurements of shear wave velocities, the proposed lower bounds are recommended for a preliminary evaluation of the small strain stiffness from q_c or N_{60}:

$$\left.\begin{array}{l} G_0 = 280\sqrt[3]{q_c \sigma'_v p_a} \\ G_0 = 450\sqrt[3]{N_{60} \sigma'_v p_a^2} \end{array}\right\} \text{lower bound, cemented} \\ \left.\begin{array}{l} G_0 = 110\sqrt[3]{q_c \sigma'_v p_a} \\ G_0 = 200\sqrt[3]{N_{60} \sigma'_v p_a^2} \end{array}\right\} \text{lower bound, uncemented} \quad \begin{array}{l} 2,4 \\ (bis) \end{array}$$

These equations predict values of G_0 that are not far from previously published relationships developed for sands (Baldi et al., 1996; Rix & Stoke, 1992; Jamiolkowski et al., 1995b). However the effect of natural cementation and ageing is quantified here and is shown to produce a marked increase in both G_0/q_c and G_0/N_{60} ratios. Given the considerable scatter observed for different soils, correlations such as given in equations (2) and (4) are only approximate indicators of G_0 and do not replace the need for *in situ* shear wave velocity measurements.

Whereas G_0 represents the small-strain stiffness, the non-linear stiffness characteristics of both 'usual' and 'non-usual' soils have a major influence on the performance of geotechnical structures. However, as it is rarely possible to obtain good quality intact samples of 'unusual' soils to study these characteristics in the laboratory, *in situ* testing should offer a means for their evaluation. Most of our common *in situ* test profiling devices, such as the CPT and SPT, cannot fulfil this requirement and consequently practitioners continue to employ single site/soil specific operational stiffness values that are derived from linear elastic backanalyses of full scale foundation performance in local soil conditions.

The ever-increasing popularity of the shear wave velocity and hence G_0 measurement is encouraging, but, in its own right, is of limited value because governing strain levels in the vicinity of geotechnical structures are 'small-intermediate' and far greater than the very small strain pertaining to seismic measurements. The pressuremeter and plate load tests are currently the only *in situ* tests that can provide a measure of the *in situ* non-linear stiffness of soils. These tests do not, however, measure an 'element stiffness' and reliable interpretation of (non-

(a)

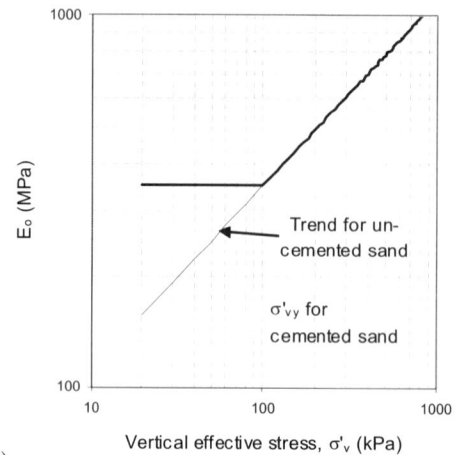

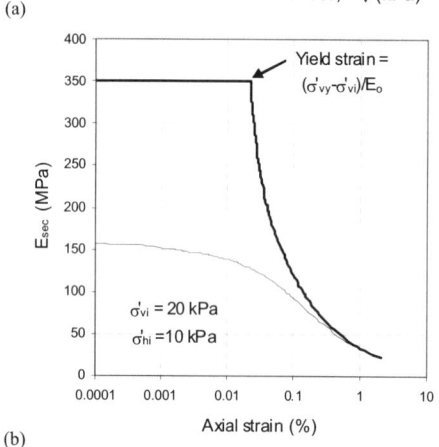

(b)

Figure 7. Typical variations on the small strain Young's modulus

linear) elemental stiffness characteristics relies on an appropriate numerical backanalysis method coupled with a realistic soil constitutive model incorporating representative parameters.

The Plate Load Test (PLT) can provide a good estimate of an average operational stiffness for use in shallow foundation design. However, as shown by the following example, interpretation of non-linear stiffness soil properties from a drained PLT is more problematic. Typical (simplified) variations of the very small strain Young's modulus, E_o (inferred from shear wave velocities with a Poisson's ratio of 0.1) of uncemented and cemented Perth sand are shown on Figure 7a, while Figure 7b shows the expected variation in triaxial compression of the secant Youngs modulus (E_{sec}) with initial vertical and horizontal effective stresses (σ'_{vi} and σ'_{hi}) of 20 kPa and 10 kPa respectively. The E_{sec} value of the cemented sand is seen to reduce from a high initial value (E_o) of 350 MPa, which prevails until a presumed yield stress (σ'_{vy}) of 100 kPa is exceeded, to a stiffness comparable to that of the uncemented sand at axial strains in excess of 0.4%.

The data on Figure 7 and the simplified non-linear settlement prediction method proposed by Lehane & Fahey (2002) were employed to predict the applied stress (q_{app})-settlement (s) response of a 300mm diameter (D) plate on the Perth sands. This method incorporates the strain and stress level dependence of stiffness in a computer program, and although it does not model plastic flow and assumes a Boussinesq stress distribution, it has been shown to be a reliable predictive tool at typical working settlements (i.e. s/D <2%). Parameters to match the stiffness characteristics shown on Figure 7 were derived using the procedures described in Lehane & Fahey (2002). The predictions assumed that the stiffness of the cemented and uncemented sands were identical at initial effective stress levels in excess of σ'_{vy} (=100 kPa).

The q_{app}-s/D predictions are shown on Figure 8. It is apparent that, despite the significant differences in stiffness seen on Figure 7b, the curves are almost linear and not dissimilar. It appears that the increase in E_o with stress level in the un-cemented sand almost compensates for its much lower E_o value at the beginning of loading. Significant softening that may be expected on inspection of Figure 7b when the applied stresses exceeded σ'_{vy} is also not apparent for the same reason. This latter observation is consistent with that observed by Viana da Fonseca et al. (1997) in a footing test on a cemented saprolitic soil, and it would appear that backanalysis of PLTs under drained conditions can lead to the inference of a range of markedly different non-linear stiffness characteristics. Clearly, PLT interpretation would benefit from unload-reload loops.

The non-linear behaviour of soils can be also estimated from the measured pressure-expansion curve

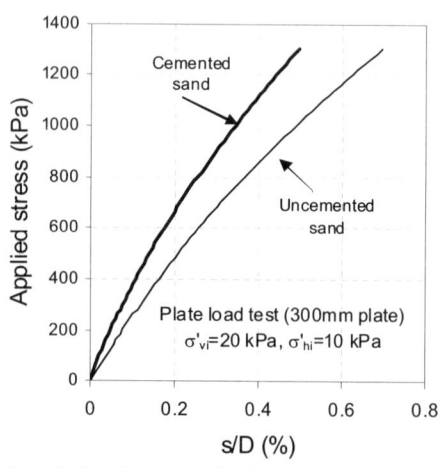

Figure 8. Calculated stress (q_{app})-settlement (s) response of a 300mm diameter (D) plate on the Perth sands

58

© 2004 Millpress, Rotterdam, ISBN 90 5966 009 9

and particularly from the information extracted from small unload-reload cycles (Fahey, 1998; Whittle, 1999). Regardless of the method adopted for modelling the non-linear response of the pressuremeter, there is always an uncertainty that cannot be solved. The strains undergone by soil elements at different distances from the pressuremeter probe vary strongly with radius and therefore an arbitrary choice of a representative value of shear strain (γ) has to be chosen from a range of representative values.

A finite element analysis of the pressuremeter tests in sand using the non-linear elastic model proposed by Fahey and Carter (1993) was adopted to produce a set of results similar to those of the PLT. The analysis employed a modified hyperbolic-type expression for shear stiffness:

$$\frac{G}{G_0} = 1 - f\left(\frac{\tau}{\tau_{max}}\right)^g \quad (5)$$

where f controls the strain to peak strength (τ_{max}) and g determines the shape of the degradation curve as a function of mobilised stress level (τ). Two soils were examined: an uncemented sand represented by f=0.85 and g=0.2, and a cemented sand with the same f value but with g =2.0; a G_o value of 150 MPa at an initial mean effective stress of 100 kPa was assumed for both materials. The resulting G/G_o versus shear strain (γ) curves shown in Figure 9 demonstrates that g=2 gives high G_o values until just after 0.01% strain. These two values of g yielded different SBP curves and unload–reload cycles, as illustrated in Figure 10. The expansion curve up to a cavity strain of 0.6% gives a direct comparison with the plate test. A 35% difference in stiffness was measured by the pressuremeter, which is a little higher than the difference predicted for the PLT. Soil elements in the vicinity of the pressuremeter are subject to shear and experience some minor increase in mean stress due to non-linearity, which may explain its slightly higher sensitivity to variations in stiffness. As for the PLT, there is no clear sign of yield.

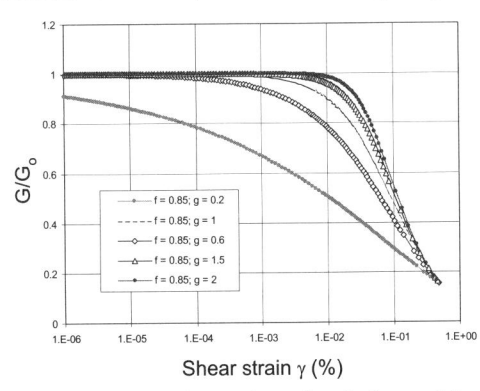

Figure 9. Typical variations on the small strain shear modulus

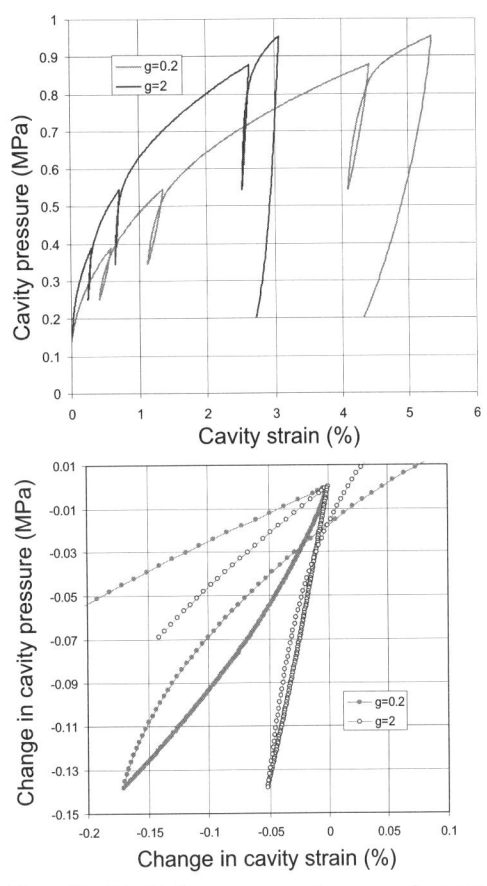

Figure 10. Calculated pressure-expansion curves from non-linear model and unload-reload cycles.

It appears that unload-reload loops are good indicators of *in situ* small strain stiffness and are therefore an essential component of SBPTs to improve the accuracy of inverse modelling to derive soil stiffness parameters. Indeed, Fahey (1998) has suggested that combining seismic G_o measurements and SBP tests incorporating multiple unload-reload loops is currently the only accurate method of obtaining non-linear stiffness parameters from *in situ* tests.

Comparisons between field and laboratory non-linear characteristics of some unusual soils have been extensively reported in recent publications (e.g. Schneider *et al.*, 1999; Schnaid, 1997; Ng and Wang, 2001). Despite its limitations, the pressuremeter has been able to describe trends of behaviour of G_o/G versus γ curves. Discrepancies between field and laboratory results are attributed to differences in loading and boundary conditions, anisotropy and sampling disturbance. However, it is clear from the foregoing analyses that great care should be exercised in back-figuring *in situ* stiffness non-linearity.

5.1 Combining data from various in-situ tests

Fahey et al. (2003) and Lehane & Fahey* (2004) describe an approximate means of assessing stiffness non-linearity by combining G_o measurements with trends indicated in SBPTs and DMTs. The approach, which was developed for Perth sands using a relatively large database of *in situ* test results in Perth, had the following features:

- Correlations such as those given in equation (2) and (4) were derived for sands of various ages and converted to equivalent E_o values assuming a Poisson's ratio of 0.1.
- Dilatometer E_D data were found to vary in a similar way with q_c and σ'_v to the G_o data but indicated a relatively low sensitivity to sand age and stress history. Best-fit correlations indicated E_o/E_D ratios of 11 ±3 and 7 ±2 for overconsolidated and normally consolidated sands respectively. Lehane & Fahey (2004) indicate that, as a consequence of a number of compensatory factors, E_D values approximate to about 70% of the in situ vertical operational stiffness at s/D =1.8%.
- G_{ur} values measured using the procedure suggested by Fahey & Carter (1993) were found to be typically 0.40 ±0.05 of G_o at cavity strains of ≈0.1%.
- Correlations proposed by Baldi et al. (1989) for the Young's modulus of a sand measured in triaxial compression at an axial strain of 0.1% were converted to a format similar to that of equation (2); these correlations were shown to predict average ratios of E_o to operational stiffness of 5 for both normally and overconsolidated sand.

These features are then combined in Figure 11, with due acknowledgement to the strain levels induced in each test, to provide the practitioner with an approximate means of deriving a strain dependent Young's modulus (E_{eq}) operational at a vertical stress level of σ'_v. Fahey et al. (2003) show that this approach provides a good estimate of stiffnesses backfigured from footings in Perth sand and Lehane & Fahey* (2004) show that a refinement of the method describes the non-linear response to load of the Texas experimental footings on sand reported by Briaud & Gibbens (1994), even if only V_s and DMT data are available.

6 PARTIAL DRAINAGE AND UNDRAINED SHEAR STREGTH

The assessment of the flow and consolidation characteristics of unusual materials is a crucial first step in design. Many of the unusual soils referred to in Table 1 exhibit a complex macro and micro structure and may have very scattered grain/aggregate size distribution, and variations in mineralogy and clay content. These features have a dominant effect on soil permeability and hence on *in situ* behaviour at given loading rates.

First consider the hydraulic conductivity of residual soils, where excellent reviews have been published by Deere & Patton (1971), Costa Filho & Vargas (1985), Garga & Blight (1997). The permeability of residual soils is controlled to a large extent by the relic structure of the parent rock. Typical values of permeability measured both in the laboratory and in the field are given in Figure 12, but generalization for various types of residual soils can be misleading and must therefore be avoided. Clearly, many residual soils are of medium to high permeability, particularly young saprolite strata due to the presence of sandy and silty grains aggregated in large pore sizes and the highly fractured structure inherited from the matrix rock.

In residual soils, relationships between hydraulic conductivity (k) and void ratio (or porosity) should be treated with great caution because of the well known effects of scale effects on permeability measurement e.g. see Garga & Blight (1997).

Assessed values of k in other unusual soils also often show enormous variability. Calcareous materials encompass a wide range of soil and rock types and may vary from well-cemented limestone to moderately cemented and lightly cemented sands and silts to completely uncemented calcareous muds (Fahey, 1997). Natural silty deposits or tailing materials are often of intermediate permeability with k values within the range in which partial drainage is likely to occur during CPTs and SPTs. For soils with k in the range of 10^{-5} to 10^{-8} m/s, the simplest accepted approach of a broad distinction between drained (gravel and sand) and undrained (clay) conditions cannot be applied to the interpretation of *in situ* tests without a great deal of uncertainty.

Assessment of the possible effects of partial drainage is essential and, at present, it seems far from being satisfactorily solved. There are a number

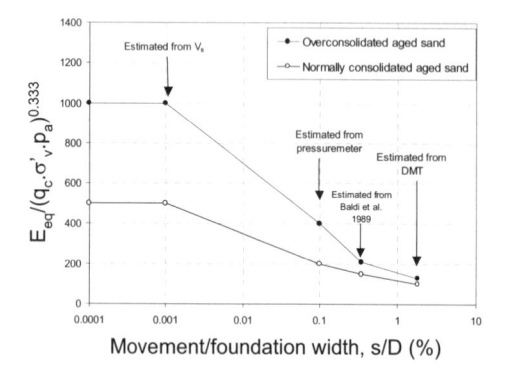

Figure 11. Proposed degradation curve for E_{eq}

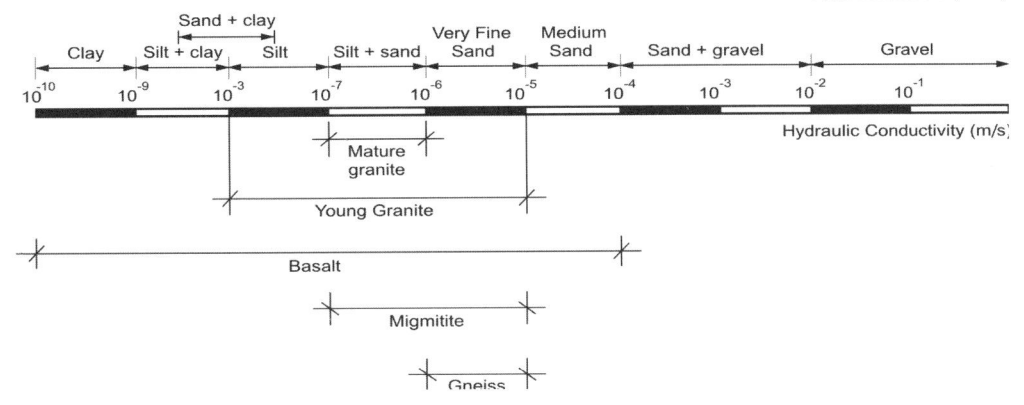

Figure 12. Tentative values of saturated permeability for residual soils (database from Costa Filho & Vargas Jr, 1985; Garga & Blight, 1997; Leong et al, 2003; Viana da Fonseca, 2003; Schnaid et al, 2004)

of important questions to be addressed before deriving the undrained strength for soils, such as:
- how is partial drainage avoided during penetration? If it happens, how is it recognized and what are the possible consequences on the derived s_u values?
- how are *in situ* test results normalised in a sensible way for engineering applications?

Field vane tests provide a direct measure of the s_u in clay for the particular mode of shearing imposed. However, in practice, the following criteria need to be satisfied: (a) the soil must be weak enough to facilitate insertion and rotation of the vane blades without damage, (b) the standard rate of rotation should be fast enough to avoid the possibility of partial drainage and (c) the effects of disturbance, testing rate, progressive failure and stress history can be accounted for. These criteria cannot be satisfied, even approximately, for soils containing a significant quantity of silt, sands, shell, intact organic matter and sulphides (Bergdahl et al., 2003). Vane tests are, nonetheless, frequently performed in materials other than clay.

Difficulties in the interpretation of *in situ* test measurements when the penetration mode changes from drained to partially drained to undrained requires continuous monitoring of pore pressures throughout the test. The CPTU thus becomes a natural choice to assess the likely drainage conditions. It is largely accepted that under a standard rate of penetration (20mm/s), undrained response will occur if the permeability of the soil is less than about 10^{-7} m/s. Two approaches offer guidelines for evaluating drained conditions during penetration in intermediate soils having permeabilities between 10^{-6} to 10^{-3} m/s:
- Hight et al. (1994) found that a relationship between B_q, $(q_t-\sigma_{vo})/\sigma'_{vo}$ and clay content could be a useful approach for interpreting results under fully undrained conditions. Their analyses appear to suggest that penetration is fully undrained for values of B_q greater than 0.5.
- House et al. (2001) showed that the degree of partial consolidation can be expressed as a function of the rate of penetration v, probe diameter, d, and coefficient of consolidation c_v, expressed in non-dimensional form as V= vd/c_v. Penetration tests conducted using kaolin in a centrifuge have suggested a rather sharper transition between drained and undrained conditions. Approximate limits of V < 0.2 for drained conditions and V > 20 for undrained conditions are suggested as qualitative boundaries.

Despite the fact that Hight's approach is expressed in terms of the normalized parameter $(q_t-\sigma_{vo})/\sigma'_{vo}$, a small adaptation to the original method is recommended here by examining trends of B_q with the ratio of undrained strength (or directly to OCR) and to couple this analysis to the non-dimensionalised velocity V. This adjustment facilitates engineering judgment when analyzing CPTU penetration data and is in line with the view that the identification of complex patterns of behaviour may be assisted by combining sets of independent measurements.

Ladd et al. (1977) proposed the following empirical equations based on the concepts of critical state soil mechanics:

$$\frac{(s_u/\sigma'_{vo})_{oc}}{(s_u/\sigma'_{vo})_{nc}} = (OCR)^{0.8} \quad (6)$$

and

$$\frac{s_u}{\sigma'_{vy}} = 0.2 \cdot \text{to} \cdot 0.3 \quad (7)$$

A normally consolidated Cam Clay type of soil should yield a ratio $s_u/\sigma'_{vy} = s_u/\sigma'_{vo}$ of between 0.25 to 0.30 depending on shearing mode. Deviation from this pattern is related to (a) overconsolidation, (b) partial drainage or (c) a characteristic behaviour of silty soils that does not fully comply with Cam Clay models. Separating out these effects is not a straightforward task, though.

Stiffness and strength characteristics of materials containing fine and coarse particles have been investigated by Hight et al. (1994), Lehane & Faulkner (1998), Zdravkovic & Jardine (2000), among others. The response in undrained triaxial compression of low OCR silts shows a reduction in mean effective stress (p'), typical of a lightly overconsolidated clay, followed by a significant increase in p' as failure approaches. This tendency for dilation at large strains leads to the mobilisation of large undrained shear strength ratios. For example, Lehane & Faulkner (1998) measured s_u/σ'_{vy} ratios of 0.45 ±0.05 for reconstituted glacial till.

CPTU data for a natural silty deposit and a tailings deposit from a gold mine are presented on Figures 13 and 14 in terms of the variation of the undrained strength ratio and the normalized cone resistance with the pore pressure parameter, B_q. Both deposits are essentially normally consolidated[1] with a c_v value in the range 0.01 to 0.2 cm^2/s, a cone factor, N_{kt}, of 15 was assumed to convert the CPT end resistance to undrained strength (e.g. Lunne et al., 1997).

It is evident that, for both deposits considered, the s_u/σ'_{vo} ratio reduces significantly with increasing B_q value, and at a B_q value of approximately 0.5 the *undrained strength ratio* reaches a plateau at a constant value of about 0.25; this value is coincidentally of the same order as that given by equation (7). Since variations in B_q do not follow any pattern with respect to depth, the observed changes are not related to overconsolidation. For values of B_q ranging between 0.3 to 0.5, s_u/σ'_{vo} ratios vary within the range of 0.3 to 0.4 with a tendency to increase slightly with reducing B_q. The calculated values of s_u/σ'_{vo} are consistent with the measured range of ϕ' values, suggesting that these investigated layers are predominantly silty soils and that partial drainage is not dominant. These data are in close agreement with observations made in the centrifuge (House et al. 2001): values of B_q greater than 0.5 are associated with non-dimensionalised velocities (V) greater than 200, whereas for B_q within the range of 0.3 to 0.5, V lies between 20 and 50. A B_q value of 0.3 seems to give a lower boundary below which the undrained strength ratio exhibits considerable scatter and a marked increase to unrealistic values in a normally consolidated deposit. It is then reasonable to

[1] It is worth mentioning that despite the fact that chemical cementation can occur and variations in water table can produce some overconsolidation, tailing disposal produce relatively young deposits that are in general consolidated under their own overburden pressure, *except* when successive layers are exposed to evaporative drying.

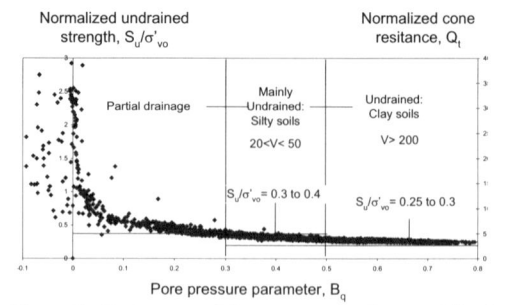

Figure 13: Drainage conditions on a normally consolidated silty deposit.

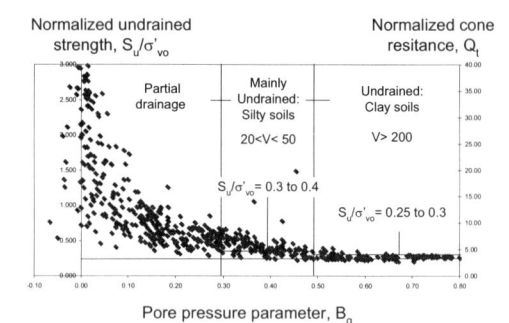

Figure 14: Drainage conditions on a silt gold tailing (data from Costa Filho et al*, 2004).

consider that partial drainage prevails and that the derived values of undrained shear strength are overestimated.

The observed patterns enable the following conclusions to be made:

- Normalization of undrained strength is useful in interpreting CPT results in silty soils.
- The combination of B_q, V, $(q_t-\sigma_{vo})/\sigma'_{vo}$ and s_u/σ'_{vo} provides general guidance for evaluating soil stratification and associated drainage conditions during cone penetration. For a standard rate of penetration, $B_q > 0.5$ and V > 200 are representative of a more clayey type of material tested under undrained conditions, whereas B_q from 0.3 to 0.5 and V from 20 to 50 are characteristic of silty soils also tested under predominantly undrained conditions.
- Partial drainage is likely to prevail for B_q values lower than 0.3 in normally consolidated silty soils tested at a standard rate of penetration of 20mm/s. Partial drainage should be detected if an accurate assessment of the undrained shear strength is to be made from CPTU test data. Non-recognition of this effect can result in a gross overestimation of s_u values, which can lead to unsafe design.

Assessment of strain rate effects and the influence on partial drainage in intermediate soils is an area in which further research is necessary before attempting to generalize the above experimental observations.

7 DRAINED SHEAR STRENGTH

7.1 Cohesionless soils

Bishop (1971) defined cohesionless soils as geomaterials in which intrinsic interparticle forces or bonds make a negligible contribution to their mechanical behaviour. In cohesionless soils, the strength parameter of major interest is the internal friction angle ϕ'. Whereas the strength-dilation theory and critical state concepts apply, uncertainty arises from the shear strength envelope non-linearity (which increases with increasing relative density and grain crushability) and the complicated strain field produced during CPT/SPT penetration in frictional dilative soils. *In situ* penetration tests remain the only viable option to characterize granular soils and such characterization is based primarily on empirical evidence. In major engineering projects, additional field geophysics, pressuremeter and dilatometer tests, as well as laboratory tests on reconstituted samples, may be available to assist development of site-specific correlations.

There are two possible approaches for interpretation of the CPT in sand: (a) analysis based on bearing capacity theories which, given the complexities of modelling penetration in sand, can only be regarded as approximate (Vesic, 1972; Durgunoglu & Mitchell, 1975; Salgado et al., 1997) and (b) methods based on results from large laboratory calibration chamber tests (e.g. Bellotti et al., 1996, Jamiolkowski et al., 1985). A recent trend has been to interpret cone penetration results in terms of the state parameter, although this approach detracts from the simplicity of estimating properties from field data only.

A widespread approach in engineering practice is to estimate relative density D_r from cone tip resistance (e.g. Jamiolkowski et al., 1985; Houlsby, 1998). Values of D_r can later be combined with operational stress levels to produce an estimate of peak friction angles (e.g. Bolton, 1986). Limitations to this approach are that (a) the database is predominantly based on tests carried out on un-aged, clean fine to medium, uniform silica sands and (b) most available correlations are referred to effective overburden stress instead of mean stress and are therefore applicable only in normally consolidated deposits.

7.1.1 Application of the SPT to ϕ' measurement

Alternatively, numerous methods have been proposed to estimate friction angle from the SPT adopting inverse application of bearing capacity theories. Assuming an elastic-perfectly plastic medium, solutions are known to be dependent on shape factor, rigidity index and plastic strain in the failure zone and are sensitive to soil crushability and cementation. Although heavily criticized, the SPT remains as a prominent tool in geotechnical engineering practice. There are, however, many unanswered aspects in the basic interpretation that have not been studied critically. Wave equation studies (Schmertmann & Palacios, 1979) suggest that the theoretical energy reaching the sampler decreases with decreasing rod length. This is a controversial statement supported by some researchers (e.g. Skempton, 1986) and contested by others (e.g. Aoki & Cintra, 2000).

The energy transferred to the composition of SPT rods and sampler was recently investigated in a calibration chamber testing programme (Odebrecht, 2003; Odebrecht et al., 2004). A building under construction was used to house a calibration chamber at ground level over which SPT rods and the SPT hammer were located. The hammer was positioned at elevations corresponding to several different floors to allow rod lengths varying from 5.8m up to 35.8m to be employed; a casing prevented the rods from buckling. Instrumentation was located immediately below the anvil, at the mid-height of the total rod length and immediately above the SPT sampler.

A typical record of the force measured by the load cell and the force calculated from the accelerometer readings, measured in the calibration chamber, is presented in Figure 15. The maximum energy transmitted to the rod stem is also shown, calculated by the F-V method and known as the Enthru energy:

$$E = \int_0^\infty F(t)V(t)\,dt \qquad (8)$$

Take the example of signals recorded at the top of a 35.8m long rod. This long length is selected here as, in theory, all energy is transmitted by the first compression wave. The concepts postulated by Schmertmann and Palacios (1979) can therefore be fully applied to the interpretation of the test data. The figure illustrates the penetration in very loose sand (N ≈ 3). Although the energy was expected to be fully transmitted to the rods during a time interval

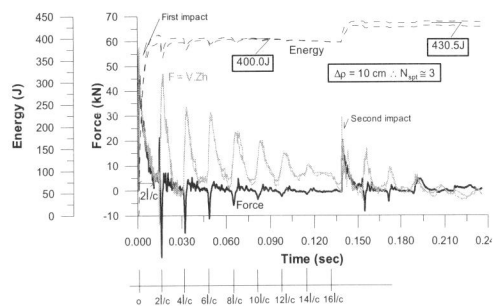

Figure 15 - Typical measured force-time relationships for a 35.80m long rod steam measured below anvil (Odebrecht el al*, 2004)

of 2ℓ/c, it is evident that for the large penetration induced by the hammer a second and late impact (Δt > 100 ms >> 2ℓ/c) produces a further increase in energy that eventually contributes to the penetration. The energy effectively transmitted to the rod stem therefore appears to be affected by the permanent penetration of the sampler and is not only a function of the so called nominal potential energy E* (474 J).

In a short string of rods, the second and third impacts also produce a significant increase in the transmitted energy that cannot be disregarded when interpreting the measured data. In conclusion, the energy transferred to the soil is a function of the nominal potential energy E*, the permanent penetration of the sampler, the rod length and rod weight. A rational method of interpreting the SPT should take into account the combined effect of these four variables. The *maximum potential energy*, PE_{h+r}, delivered to the soil should therefore be expressed as a function of the *nominal potential energy* E^* (= 474 J - ASTM, 1986), permanent sampler penetration and weight of both hammer and rods (Odebrecht et al., 2004);

$$PE_{h+r} = \eta_3[\eta_1(0.76+\Delta\rho)M_h g + \eta_2 \Delta\rho M_r g] \quad (9)$$

where:

M_h = hammer weight;
M_r = rod weight;
g = gravity acceleration;
Δρ = Sample penetration under one blow;
E^* = nominal potential energy
= 0.76m x 63.5kg x 9.8m²/s = 474 J

$$\eta_1 = \text{hammer efficiency} = \frac{\int_0^\infty F(t)\,V(t)\,dt}{(0.76+\Delta\rho)M_h g}$$

$\eta_2 = \beta_2 + \alpha_2 \ell$
$\eta_3 = \beta_3 + \alpha_3 \ell$

The estimation of η_2 and η_3, as well as of the corresponding α and β coefficients, is not a trivial task and requires some additional simplifying hypotheses. After several attempts, the experimental data were adjusted by keeping $\eta_2 = 1$ and allowing η_3 to be expressed as a function of the length of the rods:

η_3 = energy efficiency = $1 - 0.0042\ell$

The effect of the rod length on measured values of penetration may be examined using Equation (9) and is summarised on Figure 16, which plots the ratio of the energy effectively delivered to the soil, PE_{h+r}, and the ENTHRU energy, E^*, against the rod length. As indicated, a greater amount of energy is delivered to soil with a lower shear resistance (i.e. SPT N values). The influence of the length of the rod is twofold: the longer the rod length the greater

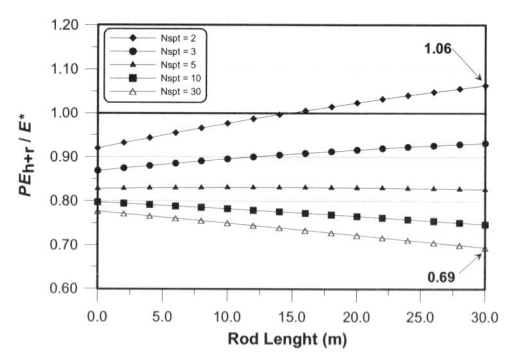

Figure 16. The ratio between the energy delivered to the soil and the ENTHRU energy (Odebrecht el al*, 2004).

the energy losses observed during propagation along the drilling rods; however for low resistance soils (low N-values), the gain in energy from the weight of the rods can sometimes be greater that the energy losses resulting from wave propagation. The combination of a very long rod and a significant sampler penetration can result in an energy ratio (PE_{h+r}/E^*) greater than unity.

It follows from the foregoing that normalization to a reference value of N_{60} is no longer sufficient to fully explain the mechanism of energy transfer to the soil and it is proposed that the system energy should be calculated using equation (9). Furthermore, it interesting to recall that the maximum potential energy can be transformed into work by the non-conservative forces (W_{nc}) acting on the sampler during penetration, and since the work is proportional to the measured permanent penetration of the sampler, it is possible to calculate the dynamic force transmitted to the soil during driving:

$$PE_{h+r} = W_{nc} = F_d\,\Delta\rho \quad \text{or} \quad F_d = PE_{h+r}/\Delta\rho \quad (10)$$

The dynamic force F_d can be considered as a fundamental measurement for the prediction of soil parameters from SPT results. One example to illustrate possible applications is given in Figure 17, in which the F_d is directly related to the friction angle by combining equations (9) and (10) to bearing capacity theory assuming a rigid-plastic stress-strain relationship and Vesic's bearing capacity factors (Vesic, 1972). The correlation results in a simple set of relationships where the combined values of N_{60} and σ'_{vo} are directly related to φ' values. The proposed approach depicts the trends obtained from the database of the *United States Bureau of Reclamation* (Gibbs & Holtz, 1957), reproduces the correlation proposed by de Mello (1971) and incorporates into the analysis the effect of the rigidity index.

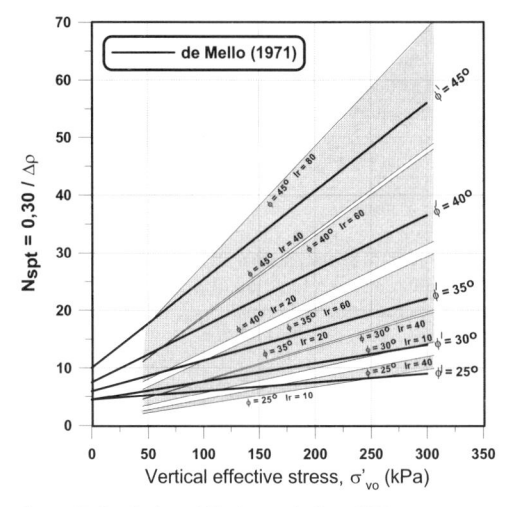

Figure 17. Prediction of friction angle from SPT energy measurements.

7.1.2 Application of the cone pressuremeter to φ' determination

The CPMT is an *in situ* testing device that combines the 15 cm² cone with pressuremeter module mounted behind the cone tip. Since the pressuremeter test is not carried out in undisturbed ground, the effects of installation have to be accounted for and large strain analysis is required. This technique is perceived as having a great potential that has not yet been fully recognized in practice. Analysis of the test in clay is achieved by a simple geometric construction of the curve to determine the undrained shear strength, the shear modulus and the *in situ* horizontal stress (Houlsby and Withers, 1988). Analysis in sand is, however, significantly more complex and interpretation is largely empirical based (Schnaid & Houlsby, 1992; Nutt & Houlsby, 1992).

Experimental results of calibration chamber testing of the Fugro cone pressuremeter revealed that the ratio of cone tip resistance (q_t) and to the pressuremeter limit pressure (ψ_L) correlates well with many soil properties such as relative density and friction angle. It is worth noting that both cone resistance and pressuremeter limit pressure are dependent on the size of the calibration chamber used (Schnaid & Houlsby, 1991), but the ratio of these two quantities is relatively unaffected by chamber size, and therefore correlations established in the laboratory may be applied directly to field conditions. Approximate empirical expressions for relative density, D_r (Schnaid and Houlsby, 1992; Nutt & Houlsby, 1992), expressed as a percentage, are:

$$D_r = 9.6 \frac{q_c - \sigma_{h0}}{\psi_L - \sigma_{h0}} - 30.5 \qquad (11)$$

$$D_r = \frac{1}{3} \frac{q_c - \sigma_{h0}}{\sigma_{h0}} + 10 \qquad (12)$$

The basis of these correlations is that both the cone resistance and the limit pressure depend on the combined effects of horizontal stress and relative density. The combination of equation (11) and (12) give σ'_{h0} as a function of q_c and Ψ_L, expressed as the root of a quadratic equation.

A theoretically sound correlation based on CPTM data has been proposed by Yu et al. (1996). In the theoretical development, the authors have assumed that both the cone resistance q_t and the pressuremeter limit pressure ψ_L are strongly related to the limit pressure of spherical and cylindrical cavities respectively. Solutions for cavity expansion in an elastic perfectly plastic Mohr-Coulomb soil have been used to correlate the ratio of q_c/ψ_L to the peak friction angle of the soil. In addition, the limit pressure solutions for cavity expansion in a strain hardening/softening soil using a state-parameter-based soil model are used to correlate q_c/ψ_L to the *in situ* sand state parameter. This recognizes the idea that prior to the achievement of the critical state, the behaviour of granular materials is largely controlled by the state parameter ψ and that ψ can be directly correlated with triaxial friction angles (Been & Jefferies, 1985). From this background, Yu et al. (1996) demonstrated that the ratio of q_c/ψ_L is found to be mainly dependent on the initial state parameter of the soil and that it can be conveniently expressed as:

$$\psi = 0.4575 - 0.2966 \ln \frac{q_c}{\psi_L} \qquad (13)$$

and the (plane strain) friction angle (in degrees) as (Yu & Houlsby, 1991):

$$\phi'_{ps} = \frac{14.7}{\ln I_s} \frac{q_c}{\psi_L} + 22.7 \qquad (14)$$

where the stiffness index $I_s = G/p'_o$; and G and p'_o = operational shear modulus and initial effective mean pressure, respectively. Further verification of the proposed interpretation method from field tests is still needed to enhance the confidence of these correlations in engineering practice.

Both the pressuremeter and the dilatometer can also be useful in deriving strength in saturated drained cohesionless soils. These techniques will be addressed in more detail in the following section, devoted to the prediction of properties in cohesive-frictional soil, where penetration tests offer less reliable information.

7.2 Cohesive-frictional materials

Although most geomaterials are recognized as being 'structured', the natural structure of bonded soils has a dominant effect on their mechanical response

(Vaughan, 1985; 1999). Soil improvement, especially with the addition of other materials such as fibres, cement or lime, can also produce marked structural changes and a cohesion/cementation component that dominates the resultant material's shear strength. Geotechnical problems involving slope stability, excavations, road pavements and other applications involving low stress levels in over-consolidated and weathered soils cannot be addressed without accounting for a cohesion/cementation component in the maintenance of long-term shear strength.

The debate on an appropriate level of site investigation for any particular project continues. A limited ground investigation based on penetration tests (CPT, CPTU or SPT) will not produce the necessary database for any rational assessment of soil properties, for the simple reason that two strength parameters cannot be derived (ϕ', c') from a single measurement (q_c or N_{60}). Limited investigations are, however, often the preferred option. In such cases involving cohesive frictional soil, engineers tend to (conservatively) ignore the c' component of strength and correlate the *in situ* test parameters with the internal friction angle ϕ'. Average c' values may be later assessed from previous experience and backanalyses of field performance.

The pressuremeter offers the possibility of characterizing the mechanical properties of a cohesive-frictional material, although the analysis is complicated by a number of factors such as the influence of bonding on the stress-dilatancy response of soils and the effects of destructuration. Ideally, the c' and ϕ' should be coupled to stiffness, dilatancy and mean stress level.

7.2.1 *Stress-dilatancy*

Various stress-dilatancy relationships have been proposed in the literature. In the most commonly used approach for sands, Rowe (1962) postulated that dilatancy is related only to the effective stress ratio and internal friction angle. Rowe's theory has been inappropriately considered as general and extended to other geomaterials. Experimental research has shown that dilation is also a function of voids ratio and degree of cementation, as well as the magnitude of suction (Cubrinovski & Ishihara, 1999; Li & Dafalias, 2000; Mantaras & Schnaid, 2002).

For cohesive-frictional materials, Rowe (1962, 1963) showed that equilibrium of the shear and normal forces applied to a contact of a sliding plane is obtained from:

$$D = \frac{\sigma'_1/\sigma'_3}{\left[\tan\left(\frac{\pi}{4}+\frac{\phi'_{cv}}{2}\right)\right]^2 + \frac{2.c'}{\sigma'_\theta}\tan\left(\frac{\pi}{4}+\frac{\phi'_{cv}}{2}\right)} \quad (15)$$

where $D=1-(d\varepsilon_v/d\varepsilon_1)$, $d\varepsilon_v$ and $d\varepsilon_1$ are plastic volumetric and principal strain increments and σ'_θ the effective circumferential stress. For a purely frictional material, equation (15) is simply expressed as:

$$\frac{\sigma'_1}{\sigma'_3} = \tan^2\left(\frac{\pi}{4}+\frac{\phi'_{cv}}{2}\right)\cdot\left(1-\frac{d\varepsilon_v}{d\varepsilon_1}\right) \quad (16)$$

Equation (16) is often expressed as R=K.D, where R is the ratio of the principal stresses, K is a constant and D is a function of the ratio of the plastic strain increments. It is usual to represent the dilation response of the soil through the parameter D, so that when D is unity, a soil is at the critical state and shearing occurs at constant volume.

Recent studies have shown that dilation of the intact soil is inhibited by the presence of the cement component (e.g. Cuccovillo and Coop, 1999) To illustrate this aspect, Schnaid *et al.* (2001) carried out a programme of triaxial compression tests on artificially cemented soil samples derived from weathered sandstone. Typical results are plotted in Figure 18, in which the stress ratio R is plotted against the dilation component D, and the ratio R/D is plotted against axial strain ε_a. It is evident that, prior to peak, the dilatancy experienced by the cemented samples at a given stress ratio is smaller than that typical of a reconstituted sample (for which K≈3.39). The rate of dilation increases with increasing shear strain amplitude in a continuous pattern that goes up to peak stresses. Peak states are accompanied by dilation and plastic strains that developed after the soil had yielded and the bonds started to degrade. This process starts at very small strains. Strain measurements after peak are very unreliable due to strain localisation, but it appears that an ultimate state has been reached when the experimental data curves down towards the reconstituted line and reach a value of dilatancy D approximately equal to 1 in all tests.

In Figure 18, Rowe's stress-dilatancy relationship has been drawn for comparison considering a soil characterised by an angle of friction and a true cohesion (equation 15). Note that Rowe assumes that the cohesion between the particles is constant while the value of K is directly proportional to friction and cohesion and inversely proportional to the stress level. Despite the fact that Rowe's equation does not depict trends from structure degradation, the average lines drawn from equation (15) represent the soil flow rule with a greater accuracy than lines obtained from equation (16) for describing the pre-peak behaviour of structured soils. It is therefore concluded that, from both the theoretical and experimental points of view, there is little justification in adopting a flow rule in a cohesive frictional material that does not incorporate the effect of cohesion on the dilatancy response of the soil.

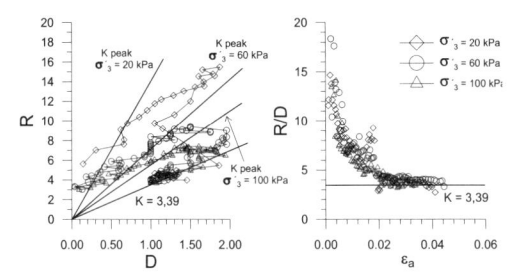

Figure 18. Stress dilatancy relationship for silty sand mixed with Portland cement (Mantaras & Schnaid, 2002).

7.2.2 Destructuration

Degradation of structured materials has been a much researched topic in recent years (e.g Aversa et al., 1993; Ishihara et al., 1989). The variation of shear strength employing a Mohr-Coulomb criterion can only be modelled by a reduction in the cohesion intercept during the shear process. A simple strength reduction idealization during softening has already been proposed (Carter et al., 1986) in which the post-peak softening is described as linear and eventually reaches residual strength behaviour. It is here recommended that variation in cohesion should be expressed as a function of plastic shear strains (γ_p). Experimental results suggest that, as a first approximation, the reduction in interparticle cohesion could be expressed as the following simple hyperbolic function of γ_p, with c´ tending to zero at large strains:

$$c'_f = f[(1+\gamma_p)^{-n}] \qquad (17)$$

A new cavity expansion model that incorporates the effects of structure degradation into cylindrical cavity expansion theory, as well as the influence of cohesion on the stress-dilatancy relationships of soils, was introduced by Mantaras & Schnaid (2002) and Schnaid & Mantaras (2003). The Euler Method is applied to solve simultaneously two differential equations that lead to the continuous variations of strains, stresses and volume changes produced by cavity expansion. Despite the mathematical complexity, an explicit expression for the pressure-expansion relation is derived without any restriction imposed on the magnitude of deformations. For materials with no cohesion, the given equations simplify to the equations adopted for describing a frictional material and the proposed solution approaches Yu and Housby's (1991) pressure-expansion relation. For undrained expansion the solution converges to that proposed by Jefferies (1988).

It is now fully recognized that strength, stiffness and *in situ* stresses interact to produce a particular pressuremeter expansion curve. Parameters that produce an analytical curve fitting the experimental results satisfactorily are, in theory, representative of the soil behaviour. This approach is mainly used for cohesive-frictional soils where the strength parameters cannot be extracted directly from the experimental curve. Input parameters are always kept within the limits defined by independent test data, but engineering judgment is required to avoid the selection of a set of *doubtful* parameters values that may produce a good fit to the data.

A case study summarised in a paper submitted to the present conference by Schnaid & Mantaras* (2004) can illustrate the applicability of the proposed approach. An extensive site investigation programme comprising laboratory triaxial tests, SPT and a large number of high quality pressuremeter tests in a residual gneiss soil profile has been reported by Pinto & Abramento (1997). A typical example of the fit provided by the analytical solution is shown on Figure 19 and a summary of the shear strength data obtained from the interpretation of 15 such tests is presented in Figure 20, which also plots SPT N_{60} and the pressuremeter limit pressures(ψ_L). The SBPM yielded ϕ'_{ps} from 27° to 31° with considerable data scatter but within the range measured from laboratory testing data. The curve fitting applied to the loading portion of the SBPM tests gave results which are rather consistent, being slightly above the assumed critical state values and compatible with laboratory data. The presence of mica at given locations has yielded a lower boundary for predicted ϕ'_{ps} values, compatible with evidence provided by N values.

Figure 19. Typical example of a pressuremeter test carried out in the saprolite gneiss residual soil of Sao Paulo.

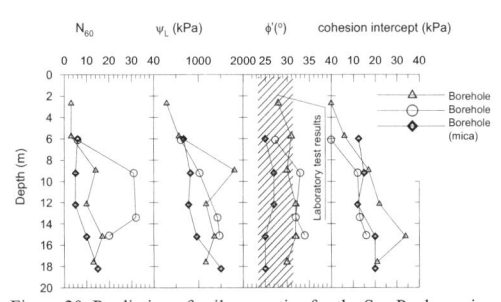

Figure 20. Prediction of soil properties for the Sao Paulo gneiss residual soil (Schnaid & Mantaras*, 2004).

8 UNSATURATED SOIL CONDITIONS

In the interpretation of *in situ* tests, it is also necessary to recognise that various geomaterials, such as hard soils and soft rocks, may not be saturated. In this case, the role of matrix suction and its effect on soil permeability has to be acknowledged and accounted for. The constitutive relationship to describe flow in unsaturated soils is Darcy's Law, as it is for saturated soil conditions. However, whereas the velocity and hydraulic gradient are directly proportional in saturated soils (with the constant of proportionality equal to the saturated permeability, k_s), a non-linear permeability needs to be employed for flow predictions in unsaturated soil. Numerous analyses have been proposed to express this permeability function (e.g. Huang et al., 1998; Fredlund, 2000), which for practical purposes is taken as a function of k_s and the matric suction, $u_a - u_w$ (u_a being the pore air pressure and u_w the pore water pressure). Common to all methods is the existence of a mathematical relationship between the coefficient of permeability at a given suction value $k(\Psi)$ and the soil water characteristic curve, which suggests that $k(\Psi)$ remains relatively constant until the air-entry value of the soil is reached and decreases rapidly with increasing matric suction beyond this point.

The influence of partial saturation imparts a very distinct behaviour to a soil and, for this reason, unsaturated soils are treated here as unusual. The discussion presented here describes how the recent body of research conducted in this area may be used to assist interpretation of *in situ* tests in unsaturated soil conditions. Given space restrictions, two aspects of significance are briefly addressed: (a) suction measurement and its practical significance and (b) suction control in field tests and soil collapsibility.

An important contribution in the analysis of unsaturated soils has been the extension of the elastic-plastic critical state concepts to unsaturated soil conditions by Alonso et al. (1990). In this method, the frame of reference is described by four variables – net mean stress ($p - u_a$), deviator stress q, suction s ($u_a - u_w$) and specific volume v, where u_a is the air pressure and u_w the pore water pressure. Several constitutive models have subsequently been proposed following these same concepts (Josa et al. 1992; Wheeler and Sivakumar 1995). These constitutive models allow derivation of the yield locus in the (p, q, s) space, an analysis that requires nine soil parameters. Model parameters are assessed from laboratory suction controlled testing such as isotropic compression tests and drained shear strength tests. For isotropic conditions, the model is characterized by the loading-collapse (LC) yield curve whose hardening laws are controlled by the total plastic volumetric deformation. A third state parameter has to be incorporated to include the effect of the shear stress q. The yield curve for a sample at

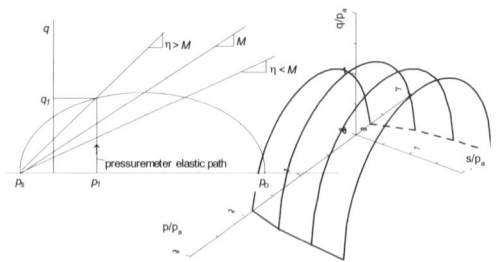

Figure 21: Three dimensional yield surface in unsaturated soils (after Alonso et al, 1990).

a constant suction s is described by an ellipse, in which the isotropic preconsolidation stress is given by the previously defined p_0 value that lies on the loading-collapse yield curve. The critical state line (CSL) for non-zero suction is assumed to result from an increase in (apparent) cohesion, maintaining the slope M of the CSL for saturated conditions, as illustrated in Figure 21.

These concepts have to be incorporated into the analysis of *in situ* tests in unsaturated soil conditions and, for that purpose, the first necessary step is to measure the *in situ* matric suction. The negative pore water pressure, u_w, or matric suction ($u_a - u_w$), when referenced to the pore air pressure, u_a, has been found to play a significant role in the behaviour of residual soils. Geotechnical engineering problems in tropical and subtropical regions are therefore often associated with variations of suction due to the unsaturated nature of these soils. Several techniques have been developed recently to measure the matric suction, such as the non-flushable vacuum tensiometer, the flushable piezometer and the miniature non-flushable tensiometer (e.g. Ridley and Burland, 1995). However, these techniques are normally used in the laboratory in compacted soils or in the field in sedimentary clay deposits.

Suction measurements in a granular granite residual soil site in southern Brazil, in which experience of such measurements is scarce, have been presented by Kratz de Oliveira et al. (1999). At this site, the measured suction ranged from 25 to 70 kPa with a trend of increasing suction during any dry period, which can be attributed to evaporation processes. A typical result is illustrated in Figure 22 for measurements of up to 50 kPa. There is no marked difference between the readings recorded from the different instruments, which implies that any of the three techniques can be used with some confidence in engineering practice for suctions less than 100 kPa in coarse grained soils. The relationship between matric suction and gravimetric water content for *in situ* and laboratory specimens is shown in Figure 23, over the range of suctions being considered. In general, the data agree well with the general equation proposed by Fredlund and Xing (1994). This relationship is the so-called soil water characteristic

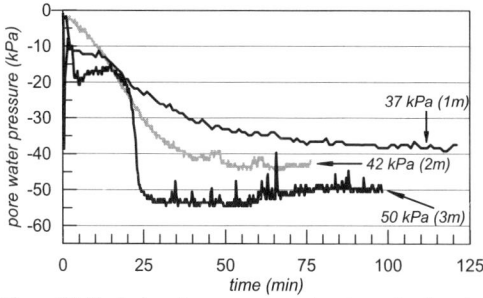

Figure 22. Typical suction measurements using a tensiometer for the granite residual soil (Kratz de Oliveira et al, 1999).

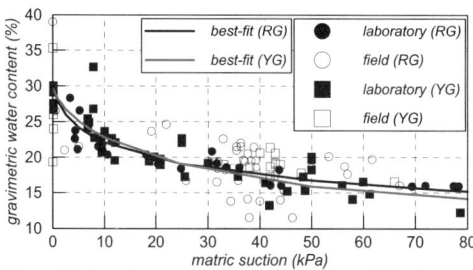

Figure 23. Suction-water characteristic curve (Schnaid et al, 2004).

curve and provides vital information concerning the hydraulic and mechanical behaviour of partially saturated soils.

The recognition that matric suction produces an additional component of effective stress suggests the need to link the magnitude of *in situ* suction to the observed response of field tests. This led to the development of monitored suction pressuremeter tests, *SMPMT*, in which the *in situ* suction is monitored throughout the test by tensiometers positioned close to the pressuremeter probe. First it is necessary to recognize that the standard self-boring technique cannot be applied to unsaturated soil conditions. The drilling technique using either a flushing fluid or compressed air would produce changes in the pore water pressure u_w or in the pore air pressure u_a, affecting the *in situ* soil suction $u_a - u_w$ in the vicinity of the pressuremeter probe. The pre-bored technique appears to be a viable option despite its well known limitations.

Typical pressuremeter curves in a granite residual soil are illustrated in Figure 24 (Schnaid et al., 2004). The first test was performed at an *in situ* suction of 40 kPa. After soaking the area, another test was carried out, producing a marked reduction in both pressuremeter initial stiffness response and cavity limit pressure. A straightforward conjecture is that stiffness degradation with shear strain is likely to be shaped by changes in matric suction. The response of tensiometers installed at the same depth of the pressuremeter tests at 30 and 60 cm from the centre of the *SMPMT* borehole are also shown in the

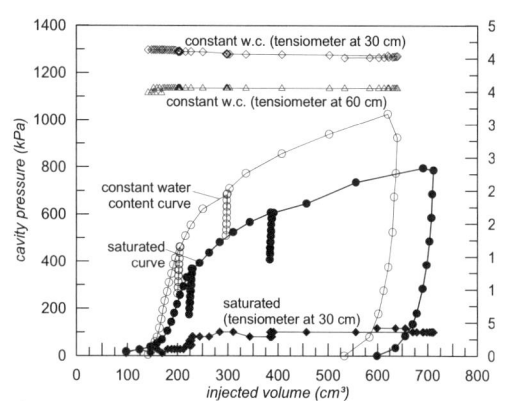

Figure 24. Typical suction monitored pressuremeter tests..

figure. Suction measurements remained approximately constant throughout the expansion phase in the tests carried out both in soaked and unsaturated soil conditions.

Similar patters are observed during wetting-induced collapse investigated using both conventional suction controlled oedometer tests and plate loading tests. Data suggest that shear strains induced by loading do not produce significant changes in matric suction and this enables cavity expansion theory to be extended to accommodate the framework of unsaturated soil behaviour in the interpretation of pressuremeter tests. As a consequence, it is possible to demonstrate that the pressuremeter system is not only suitable for estimating the potential collapse of soils but also for assessing the constitutive parameters that are necessary to describe the 3D-yield surfaces in a (p, q, s) space in unsaturated soils (Schnaid et al., 2004). The same cavity expansion theoretical background discussed in the previous sessions for saturated drained materials remains valid (Schnaid et al., 2004; Gallipoli et al., 2000).

9 CONCLUSIONS

Although we believe we are making steady progress in our ability to measure properties in a wide range of geomaterials, interpretation of *in situ* tests is still very problematic. However, in the light of recent recognition of aspects such as structural effects, soil non-linearity and unsaturated soil conditions, there is a new opportunity to make significant improvement in methods for predicting behaviour from *in situ* tests. This is of fundamental importance in soils where constitutive parameters cannot be measured from laboratory tests in routine site investigation practice.

Despite the multiplicity of factors controlling the interpretation of *in situ* tests in complex soil conditions, some concluding remarks can be drawn from the topics discussed in this paper:

- Soil classification from *in situ* test results should rely on at least two independent measurements. Combinations of G_o/q_c, G_o/N_{60}, q_c/ψ_L can provide additional insights to current interpretation methods. In particular, a measure of the ratio of the elastic stiffness to ultimate strength, expressed as G_o/q_c and G_o/N_{60}, has shown to be fairly sensitive to cementation and ageing and is therefore useful for identifying unusual geomaterials.
- At a qualitative level, the variation of the small-strain stiffness G_o with both q_c and N_{60} was shown to be particularly sensitive to soil structure. A set of equations expressed as lower and upper bounds have been proposed to derive G_o values in both residual soils and sands. These equations cover the range of experimental data of CPT and SPT tests summarized in Figures 4, 5 and 6.
- The reduction in the ratio of G/G_o with shear stress and shear strain is known to be sensitive to degradation of cementation and structure, among several other factors (Tatsuoka *et al.*, 1997). Research is still necessary to envisage forms of expressing the degradation curve from *in situ* tests.
- A reliable evaluation of the mass permeability is a preliminary requirement for interpretation of *in situ* tests, assumed to be undertaken under fully drained or fully undrained conditions. Very frequently, unusual materials exhibit coefficients of hydraulic conductivity in the range of transitional soils, a range over which partial drainage is often observed for currently adopted rates of loading in various tests. Implications are that the undrained shear strength derived from penetration tests can be grossly overestimated. Specific recommendations have been made to evaluate partial drainage from CPTU data. Whereas engineering judgment is guided by a combined evaluation of the undrained strength ratio, normalized cone resistance and non-dimensional velocity, efforts should be made to test silty soils both in the laboratory (centrifuge or calibration chamber) and *in situ* to supplement data in unusual soil conditions.
- Sands and gravels cannot be regarded as unusual soils. However the recognition of the influence of light cementation and ageing in coarse-grained materials is relatively recent and the assessment of these effects in both laboratory and *in situ* tests is still controversial. *In situ* penetration testing remains as the most viable option for assessing soil parameters in sand. CPT and CPMT correlations from calibration chamber tests are recognized as useful in normally consolidated deposits. A new method is proposed to estimate the angle of internal friction that combines the SPT blow count number to the energy generated by the non-conservative forces produced during penetration.
- Interpretation of *in situ* tests in cohesive-frictional geomaterials is a function of the angle of internal friction, cohesion intercept, soil stiffness, dilation and mean effective stress. There are also specific features to be accounted for, such as the influence of bonding on the stress-dilatancy response and the effects of destructuration. The pressuremeter is the only *in situ* testing technique that may take account of all these effects in everyday engineering projects.
- Unsaturated soil conditions impose a further degree of complexity in testing interpretation. The influence of suction measurements and its practical significance to assess soil parameters and soil collapsibility from pressuremeter tests has been demonstrated.

ACKNOWLEDGMENTS

The Brazilian Research Council CNPq has funded a part of the research projects described in this paper. This paper was written while the first author was a visitor to the University of Western Australia, funded by a Gledden Senior Visiting Fellowship from the University. This support is gratefully acknowledged.

REFERENCES

Alonso, E.E., Gens, A. & Josa, A. 1990. A constitutive model for partially saturated soils. Géotechnique, 36(3), 405-430.
Aoki, N. & Cintra, J.C.A. 2000. The application of energy conservation Hamilton's principle to the determination of energy efficiency in SPT test. Proceedings of the VI International Conference on the Application of Stress-Wave Theory to Piles, São Paulo, pp 457 – 460.
Aversa, S.; Evangelista, A.; Leroueil, S. & Picarelli, L. 1993. Some aspects of the mechanical behaviour of structured soils and soft rocks. Geotechnical engineering of hard soils – soft rocks. Anagnostopulos *et al.* (eds). 359-366
Baldi, G.; Ghionna, V.N.; Jamiolkowski, M., & Lo Presti 1989. Modulus of sand from CPTs and DMTs. 12th Int. Conf. Soil Mech. Found. Engng, Rio de Janerio, 1: 165-170.
Baligh, M.M 1985. Strain path method. J. Soil Mech. Fdn. Engng. Div., ASCE. 11(7), 1108-1136.
Baligh, M.M & Levadoux, J.N 1986. Consolidation after undrained piezocone penetration. II: Interepretation. J. Soil Mech. Fdn. Engng. Div., ASCE. 11(7), 112(7), 727-745.
Barros, J.M.C. 1997. Dinamic shear modulus in tropical soils, PhD Thesis, São Paulo University, In portuguese.
Been, K. & Jefferies, M.G. 1985. A state parameter for sands. Géotechnique, 35(2), 99-112.
Bellotti, R.; Ghionna, V.N.; Jamiolkowski, M., Robertson, P.K. & Peterson, R.W. 1989. Interpretation of moduli from self-boring pressuremeter tests in sand, Geotechnique, 39(2): 269-292.
Bellotti, R.; Jamiolkowski, M., Lo Presti, D.C.F. & O'Neill, D.A. 1996. Anisotropy of small strain stiffness in Ticino sand. Geotechnique, 46 (1): 155-131.

Bergdahl U., Larsson R. and Viberg L. (2003). Ground Investigations and parameter assessment for different geological deposits in Sweden. In situ characterisation of soils., 119-170, Ed. Saxena K.R. and Sharma V.M., Balkema.

Bishop, A.W. 1971 Shear strength parameters for undisturbed and remolded soil specimens. Proc. of the Roscoe Memorial Symp., Cambridge, UK, 3-58.

Bolton, M.D. 1986. The strength and dilatancy of sands. Géotechnique 36 (1), 65-78.

Briaud, J.-L. and Gibbens, R.M. (Eds.) (1994). Predicted and Measured Behaviour of Five Spread Footings on Sand. Proc. ASCE Prediction Symposium, Texas A&M University, June, ASCE Geotechnical Special Publication No. 41.

Burland, J.B. 1989. Small is beautiful: the stiffness of soils at small strains, Can. Geotch. J., 26: 499-516.

Burland, J.B. 1990. On the compressibility and shear strength of natural clays, Géotechnique 40 (3), 329-378.

Carter, J.P.; Booker, J.R. & Yeung, S.K. 1986. Cavity expansion in cohesive frictional soils, Géotechnique 36 (4), 349-358.

Clayton, C.R.I 1993. SPT Energy Transmission: Theory Measurement and Significance, Ground Engineering, 23 (10), pp. 35-43

Clayton, C.R.I.; Matthews, M.C. & Simons, N.E. 1995. Site Investigation. Blackwell Science, 584pp.

Costa Filho, L.M. & Vargas, Jr. E. 1985. Hydraulic properties. Peculiarities of geotechnical behaviour of tropical lateritic and saprolitic soils. Progress Report (1982-1985). Brazilian Society of Soil Mechnics. 67-84.

Cubrinovski, M. & Ishihara, K. 1999. Modelling of sand behaviour based on the state concept. Soils and Foundations. 38 (3), 115-127.

Cuccovillo, T. & Coop, M. 1999. On the mechanics of structured sands. Géotechnique 49, No. 4, 349-358.

de Mello, V.F.B. 1971. The standard penetration test. 4th Panamerican Conf. Soil Mech. Found. Engng, Porto Rico, 1: 1-87.

de Mello, V.F.B. 1972. Thoughts of soil engineering applicable to residual soils. Proc. 3rd Symposium Southeast Asian Conf., 1 ,15-34, Hong Kong.

Decourt, L; Muromachi, T; Nixon, I.K.; Schmertmann, J.H.; Thorbum, C.S. & Zolkov, E. 1998. Standard Penetration test (SPT): International reference test procedure Proc. Int. Symp. on Penetration Testing, ISOPT-1, USA, 1: 37-40.

Deere, D.V. & Patton, F.D. 1971. Slope stability in residual soils. Proc. 4th Pan American Conf. Soil Mech. and Found. Engng., Puerto Rico, 1, 87-170.

Durgunoglu, H.T. & Mitchell, J.K. 1975. Static penetration resistance of soils. I Proc. of the ASCE Specialty Conf. on In Situ Measurement of Soil Properties. Raleigh, 1, 151-189.

Eslaamizaad, S. and Robertson, P.K. (1996) A Framework for In-situ determination of Sand Compressibility" 49th Canadian Geotechnical Conference; St John's Newfoundland.

Fahey, M. (1997) Some unusual calcareous soils. 14th Int. Conf. Soil Mech. Found. Engng, Hamburg, 4: 2197-2198.

Fahey, M. 1998. Deformation and in situ stress measurement, Geotech. Site Charact., ISC'98, Balkema, 1: 49-68.

Fahey, M. & Carter, J.P. 1993. A finite element study of the pressuremeter test in sand using a non-linear elastic plastic model, Can. Geotch. J., 30: 348-362.

Fahey, M., Lehane, B. and Stewart, D.P. (2003). Soil stiffness for shallow foundation design in the Perth CBD. Australian Geomechanics, Vol. 38, No. 3, 61–89.

Fredlund, 2000. The implementation of unsaturated soil mechanics into geotechnical enginenering. The 1999 R.M. Hardy Lecture. Canadian Geotechnical J., 37: 963-986.

Fredlund, D.G. and Xing, A. (1994). Equations for the soil-water characteristic curve. Canadian Geotech. J., 31(4), 521-532.

Gallipoli, D.; Wheller, S.J. & Karstunen, M. 2000. Importance of modelling degree of saturation variation: the pressuremeter test. Proc. 8th Int. Symp. Numer. Models Geomech., Rome, 627-633.

Garga, V.K. & Blight, G.E. 1997. Permeability. Mechanics of Residual Soils. Balkema. Chapter 6, 79-94.

Gibbs, H. J. & Holtz, W.G. 1957. Research on determining the density of sands by spoon penetration testing, 4th Int. Conf. Soil Mech. Found. Engng, London, 1: 35-39.

Hardin, B.O. & Drnevich, J.H. 1972. Shear modulus and damping in soils: design equations and curves. J. Soil Mech. Fdn. Engng. Div. Am. Soc. Civ. Engrs. 98, No. 7, 667-692.

Hight, D.W. 1998. Soil characterization: the importance of structure and anisotropy. 38th Rankine Lecture.

Hight, D.W.; Georgiannou, V.N. & Ford, C.J. 1994. Characterization of clayey sand. Proc. 7th Int. Conf. of Offshore Structures, USA, 1, 321-340.

Hight, D.W.; Bennell, J.B.; Chana, B.; Davis, P.D. Jardine, R.J. and Porovic, E. (1997) Wave velocity and stiffness measurements on the Crag and Lower London Terceraries and Sizewell. Geotechnique, 47 (3): 451-474.

Houlsby, G.T. 1998. Advanced interpretation of field tests, , Geotech. Site Charact., ISC'98, Balkema, 1, 99-112 Geotech. Site Charact., ISC'98, Balkema, 1: 99-109.

Houlsby, G.T. & Withers, N.J. 1988. Analysis of the cone pressuremeter test in clays. Geotechnique, 38 (4): 575-587

House, A.R.; Oliveira, J.R.M.S. & Randolph, M.F. (2001) Evaluating the coefficient of consolidation using penetration tests. Int. J. of Physical Modelling in Geotechnics. 1(3): 17- 26.

Huang, S.Y.; Barbour, S.L. & Fredlund, D.G. 1998. Development and verification of a coefficient of permeability function for a deformable, unsaturated structure. Canadian Geotech. J., 35(4), 411-425.

Ishihara, K.; Kokusho, T. & Silver, M.L. 1989. Earthquakes: Influence of local conditions on seismic response, State-of-the-Art Report. 11th Int. Conf. Soil Mech. Found. Engng, San Francisco, 4: 2719-2734.

Jamiolkowski, M.; Ladd, C.C.; Germaine, J.T. & Lancellotta, R. 1985. New developments in field and laboratory testing of soils. State-of-the-Art Report. 11th Int. Conf. Soil Mech. Found. Engng, San Francisco, 4: 57-153.

Jamiolkowski, M.; Lancellotta, R.; Lo Presti, D.C.F. & Pallara, O. 1995a. Stiffness of Toyoura sand at small and intermediate strains, 13th Int. Conf. Soil Mech. Found. Engng, Hamburg, 1: 169-172.

Jamiolkowski, M.; Lancellotta, R. & Lo Presti. 1995b. Remarks on the stiffness at small strains os six Italian clays, Keynote Lecture- IS Hokkaido, 2: 817-836.

Jardine, R.J.; Fourie, A.B.; Maswoswse, J. & Burland, J.B. 1985. Field and laboratory measurements of soil stiffness. 11th Int. Conf. Soil Mech. Found. Engng, San Francisco, 2: 511-514.

Jardine, R.J. Kuwano, R.; Zdravkovic, L. & Thornton, C. 2001. Some fundamental aspects of pre-failure behaviour of granular soils. 2nd Int. Conf. on Pre-failure Behaviour of Geomaterials. Torino, 2: 1077-1111.

Jefferies, M. G. 1988. Determination of horizontal geostatic stress in clay with self-bored pressuremeter. Canadian Geotechnical Journal, 25, pp. 559-573.

Josa, A., Balmaceda, A., Gens, A. & Alonso, E. E. 1992. An elastoplastic model for partially saturated soils exhibiting a maximum of collapse. Proc. 3rd Int. Conf. Comput. Plasticity, Barcelona 1, 815-826.

Kratz de Oliveira, L.A.; Ridley, A.M.; Schnaid, F.; Gehling, W.Y.Y. & Bressani, L.A. 1999. Rainfall and evaporation effects on matric suctions in a granite residual soil, XI Panamerican Conf. Soil Mech. Geotech. Engng., 2: 985-990.

Ladanyi, B. 1963. Evaluation of pressuremeter tests in granular soils. Proc. 2nd Pan American Conference on Soil Mechanics and Foundation Engineering, São Paulo, 1, 3-20.

Ladd, C.C.; Foott. R.; Ishihara, K.; Schlosser, F. & Poulos, H.G. 1997. Stress-deformation and strength characteristics. State-of-the-Art Report. 9th Int. Conf. Soil Mech. Found. Engng, *Tokyo, 2: 421-494.*

Lehane B.M. and Faulkner A. Stiffness and strength characteristics of a hard lodgement till. *Proc. 2nd Int. Symp. on the Geotechnics of hard soils and soft rocks*, Naples, 1998, 2, 637-646.

Leong, E.C.; Rahardjo, H. & Tang, S.K. 2003. Characterisation and engineering properties of Singapore residual soils. Characterization and Engineering Properties of Natural Soils, Balkema, 2, 1279-1304.

Leroueil, S. & Vaughan, P.R. 1990. The general and congruent effects of structure in natural soils and weak rocks. Géotechnique 40, No. 3, 467-488.

Leroueil, S. & Hight, D.W. 1990. Behaviour and properties of natural and soft rocks. Characterization and Engineering Properties of Natural Soils, Balkema, 1, 29-254.

Li, S.S. & Dafalias, Y.F. 2000. Dilantancy for cohesionless soils. Geotechnique, 52 (3): 449-460.

Lunne, T.; Robertson, P.K. & Powell, J.J.M. 1997. Cone penetration testing in geotechnical practice, Blackie Academic & Professional, 312p.

Mantaras, F.M. & Schnaid, F. 2002. Cavity expansion in dilatant cohesive-frictional soils. Geotechnique, 52 (5), 337-348.

Manassero, M. 1989. Stress-strain relationships from drained self-boring pressuremeter test in sand. Géotechnique 39 (2), 293-308.

Mayne, P.W. 2001. Stress-strain strength flow characteristics of enhanced in situ testing. Int. Conference of In Situ Measurement of Soil Properties and Case Histories, Indonesia, 29-48.

Matthews, M.C.; Clayton, C.R.I. & Own, Y. 2000. The use of field geophysical techniques to determine geotechnical stiffness Parameters, Geotechn. Engng, 143(1) 31-42.

Ng, C.W.W. & Wang, Y. 2001. Filed and laboratory measurements of small strain stiffness of decomposed granites. Soils and Foundations, 41 (3), 57-71.

Novais Ferreira, H. 1985. Characterisation, identification and classification, of tropical lateritic and saprolitic soils for geotechnical purposes. General Report, Int. Conf. Geomechanics in Tropical Lateritic and Saprolitic Soils, Brasilia, 3: 139-170.

Nutt, N.R.F. & Houlsby, G.T. 1992. Calibration tests on the cone pressuremeter in carbonate sand. Int. Symp. Calib. Chamber Testing. Potsdam, New York, 265-276.

Odebrecht, E. 2003. Energy Measurements in SPT Test, PhD Thesis, Porto Alegre Federal University of Rio Grande do Sul, In portuguese. pp203.

Odebrecht, E.; Schnaid, F.; Rocha, M.M. & Bernardes, G.P. 2004. Energy efficiency for Standard Penetration Tests. J. Geotech. and Geoenvironmental Engng, ASCE, Submitted for publication.

Pinto, C.S. & Abramento, M. 1997. Pressuremeter tests on gneissic residual soil in São Paulo, Brazil. Proc. of the 14th. Int. Conf. on Soil Mech. and Found. Engng, Hamburg, 1, 175-176.

Ridley, A. M. & Burland, J. B. 1995. A pore water pressure probe for the in situ measurement of a wide range of soil suctions. Proc. 1st Int. Conf. on Advances In Site Investigation Practice, London. 510-520.

Rix, G.J. & Stokes, K.H. 1992. Correlation of initial tangent modulus and cone resistance. Int. Symp. Calibration Chamber Testing, Potsdam, USA: 351-362.

Robertson, P.K. 1990. Soil classification using the cone penetration test. Can. Geotch. J., 27(1): 151-158.

Robertson, P.K.; Lunne, T. & Powell, J.J.M. 1998. Geoenvironmental applications of penetration testing. Proc. Int. Conf. Site Characterization, ISC'2, Atlanta, USA, 1, 35-48.

Rowe, P.W. 1962 The stress-dilatancy relation for static equilibrium of an assembly of particles in contact. Proc. Royal Soc. London, A269, 500-527.

Rowe, P.W. 1963. Stress-dilatancy, earth pressure and slopes. J. Soil Mech. Fdns. Div. ASCE 89, SM3, 37-61.

Rowe, P.W. 1971. Theoretical meaning and observed values of deformation parameters for soil. Stress-strain behaviour of soils, Proc. Roscoe Memorial Symposium, Cambridge University, 143-195.

Salgado, R.; Mitchell, J.K. & Jamiolkowski, M. 1997. Cavity expansion and penetration resistance in sand. J. Geotch. and Geoenvironmental Engng., 123 (4), 344-354.

Sandroni, S.S.. 1988. Sampling and testing of residual soils in Brazil. In sampling and Testing of Residual Soils. A Review of Int. Practice.31-48, E.W. Brand & H.B. Phillipson, Scorpion Press, Hong Kong.

Schmertmann, J.H 1975. Measurement of in situ shear strength. Proc. Specialty Conf. In Situ Measur. of Soil Properties, ASCE, 2, 57-138.

Schmertmann, J.H. and Palacios, A. 1979. Energy dynamics of SPT. Journal of the Soil Mechanics and Foundation Division, ASCE, 105 (8), 909-926.

Schnaid, F. 1997. Panel Discussion: Evaluation of in situ tests in cohesive frictional materials, 14th Int. Conf. Soil Mech. Found. Engng, Hamburg, 4: 2189-2190.

Schnaid, F. & Houlsby, G.T. 1991. An assessment of chamber size effects in the calibration of in situ tests in sand; Geotechnique, 41 (4), 437-445.

Schnaid, F.; Houlsby, G. T. 1992. Measurement of the proprieties of sand in a calibration chamber by cone pressuremeter test, Geotechnique, 42 (4), 578-601.

Schnaid, F. & Houlsby, G.T. 1994. Interpretation of shear moduli from cone-pressuremeter tests in sand; Geotechnique, 44 (1), 147-164.

Schnaid, F.; Prietto, P.D.M.; & Consoli, N.C. 2001. Characterization of cemented sand in triaxial compression. J. Geotech. and Geoenvironmental Engng, ASCE, 127 (10), 857-868.

Schnaid, F.; Kratz de Oliveira, L.A. and Gehling, W.Y.Y. 2004. Unsaturated constitutive surfaces from pressuremeter tests. J. Geotech. and Geoenvironmental Engng, ASCE, 130 (2), 174-185.

Schnaid, F. & Mantaras, F.M. 2003 Cavity expansion in cemented materials: structure degradation effects. Geotechnique, 53 (9): 797-807.

Schneider, J.A.; Hoyos, L.Jr.; Mayne, P.W.; Macari, E.J. and Rix, G.J. 1999. Field and laboratory measurements of dynamic shear modulus of Piedmont soils. Behavioral Characteristics of Residual Soils, ASCE, Reston, USA, 12-25.

Schofield, A. and Wroth, P. 1968. Critical State Soil Mechanics. Mc Graw-Hill, London, 310pp.

Skempton, A.W. 1986. Standard penetration test procedures and effects in sands of overburden pressure, relative density, particle size, aging and over consolidation. Géotechnique, 36 (3), 425-447.

Stoke, K.H.II, Hwang, S.K.; Lee, J.N.K. & Andrus, R.D. 1995. Effects of various parameters on the stiffness and damping of soils at small to medium strains. Proc. Int. Conf. on Prefailure Deformation Characteristics of Geomaterials. Vol. 2, 785-816.

Tatsuoka, F. and Shibuya, S. 1991. Deformation characteristics of soils and rocks from field and laboratory tests. 9th Asian Reg. Conf. on Soil Mech. and Found. Engng., Bangkok, 2: 101-170.

Tatsuoka, F.; Jardine, R.J.; Lo Presti, D.; Di Benedetto, H. & Kodaka, T. 1997. Theme Lecture: Characterising the pre-failure deformation properties of geomaterials. 14th Int. Conf. Soil Mech. Found. Engng, Hamburg, 4: 2129-2164.

Vaughan, P.R. 1985. Mechanical and hydraulic properties of tropical lateritic and saprolitic soil, General Report, Int. Conf. Geomechanics in Tropical Lateritic and Saprolitic Soils, Brasilia, 3: 231-263.

Vaughan, P.R. 1999. Problematic soils or problematic soil mechanics. Proc. Int. Symp. on Problematic Soils, Sendai, Japan, 2: 803-814.

Van Impe, W.F. & Van den Broeck, M. (2001) Geotechnical characterisation in offshore conditions. Conferenze di Geotecnica di Torino. Italy, 1-15.

Vesic, A.S. 1972. Expansion of cavities in infinite soil mass. J. of Geotch. Eng. Div., ASCE, 98 (3): 265-290.

Viana da Fonseca, A.; Matos Fernandes, M. & Cardoso, A.S. 1997. Interpertation of a footing load test on a saprolitic soil from granite. Géotechnique, 47(3), 633-651.

Viana da Fonseca, A. 2003. Characterization and deriving engineering properties of a saprolitic soil from granite in Porto. Characterization and Engineering Properties of Natural Soils, Balkema, 2, 1341-1378.

Wheeler, S. J. and Sivakumar, V. 1995. An elasto-plastic critical state framework for unsaturated soil. Géotechnique, 45(1), 35-53.

Whittle, R.W. 1999. Using non-linear elasticity to obtain the engineering properties of clay – a new solution for the self-boring pressuremeter test. Ground Engineering: 30-34.

Wroth, C.P. 1984. The interpretation of in situ soil test. 24th Rankine Lecture. Géotechnique 34 (4): 449-489.

Yu, H.S. & Houlsby, G. 1991. Finite cavity expansion in dilatant soils: loading analysis. Géotechnique 41 (2): 173-183.

Yu, H.S. & Houlsby, G.T. 1995. A large strain analytical solution for cavity contraction in dilatant soils. Int. J. Numer. Anal. Methods Geomech. 19, 793-811

Yu, H.S.; Schnaid, F. & Collins (1996) Analysis of cone pressuremeter tests in sands. J. Geotech. Engng. 122 (8): 623-632.

Zdravkovic, L. & Jardine R.J. 2000. Some anisotropy stiffness characteristics of a silt under general stress conditions. Geotechnique, 47-3, 407-437.

Proceedings ISC-2 on Geotechnical and Geophysical Site Characterization, Viana da Fonseca & Mayne (eds.)
© 2004 Millpress, Rotterdam, ISBN 90 5966 009 9

Routine and advanced analysis of mechanical in situ tests. Results on saprolitic soils from granites more or less mixed in Portugal

A. Gomes Correia
Dept. of Civil Engineering, School of Engineering, University of Minho, Portugal

A.Viana da Fonseca
Dept. of Civil Engineering, Faculdade de Engenharia da Universidade do Porto, Portugal

M. Gambin
Scientific adviser of Apageo, France, member of the ISSMGE Board

Keywords: stress-strain of soils, small shear modulus, moduli, angle of shearing resistance, transported soils, saprolitic granite soils, seismic tests, penetration tests, pressuremeter tests, plate load test, elastic-plastic model

ABSTRACT: This paper covers the more recent findings in the interpretation of different in-situ tests, such as SPT, CPT, PMT, SBPT, DMT and PLT to obtain geotechnical parameters of significant use in engineering practice. It concerns mainly shearing resistance properties and stiffness properties with special emphasis on the importance of stress and strain dependency. In this context some practical rules are presented for using parameters at two levels of design: routine and advanced levels. These practical rules concern transported soils (unaged and uncemented) are compared with those established in this paper for residual saprolitic soils from granite from different regions of Portugal.

It was noticed that the bonded structure and fabric of residual saprolitic soils from granite have a significant influence on their geomechanical behaviour. Consequently, the structural peculiarities of these residual soils influence the pattern of their non-linear constitutive behaviour. Deformability modulus derived from robust but relatively crude tests, such as SPT, CPT, DPT or even PMT, are compared with reference values taken from seismic survey (CH) and load tests, such as PLT. They can be situated on stress-strain levels defined from laboratory triaxial tests over high quality undisturbed samples. Several parametric correlations were established, which agree well with other correlations proposed for residual soils of the same nature. Significant differences are apparent between those correlations and the ones established for transported soils with identical grading curves, which may be explained by the weak bonded structure, inherited from the parent rock.

1 INTRODUCTION

The geotechnical site investigation is a function of the specific project and the associated risk. In general it takes into account the construction conditions and covers the following aspects (Hight and Higgins, 1995; Roberstson, 2001):
a. Geological regime: nature and sequence of the subsurface strata, stress history;
b. Groundwater regime: hydrogeological regime;
c. Soil and rock properties and behaviour: stress-strain-strength-creep-hydro properties and behaviour of the subsurface strata;
d. Geo-environmental regime: composition, distribution and flow of contaminants.

There are a variety of in-situ tests available to meet these objectives. A list of these major tests can be found in Lunne et al. (1997), where their applicability and usefulness for different type of ground conditions are described.

The best approach to use and choose the appropriate field tests is an integration approach involving structural engineer and geotechnical engineer. The approach should consider firstly the nature of the construction and the proposed methods of analysis and secondly the nature of the ground. Consequently the best in-situ tests for each given project and ground conditions will then be the ones which give the required information regarding the understanding of geological and geotechnical ground conditions and the relevant geotechnical properties and behaviour of the ground such as their constitutive laws to be used by the proposed methods of analysis used in the structural design. They should also give this information with an acceptable degree of accuracy at the lowest cost.

In this context it is interesting to classify the different types of ground behaviour used at different levels of structural analysis matching proper codes:
- Level No.1: routine calculation, assuming pseudo-elastic parameters for the ground (servi-

ceability stiffness) obtained by routine analysis of test results;
- Level No.2: advanced calculation, assuming non-linear soil stiffness obtained by advanced analysis of tests results and;
- Level No.3: research calculation, using complex soil models obtained by analysing test results with complex constitutive laws of soil behaviour (hydro-stress-strain-strength-creep).

However, the use of these codes addresses a major practical difficulty which is related to the choice of the characteristic values to give to the relevant parameters of the constitutive laws. These values are usually obtained from laboratory tests where distribution of stresses and strains are homogeneous and boundary conditions are well defined. However, it is always difficult to obtain undisturbed samples; moreover, the selection of samples and their size can lead to uncertainty regarding the way they represent the soil. Consequently, in-situ tests could then become an alternative means to obtain these values. In any case it is always interesting to be able to compare the field test results with laboratory tests, and, if possible, combine both results. Unfortunately it must be pointed out that the drawback of most routine in-situ tests lies in the fact that the stress and strain distribution necessary for the identification of constitutive laws is unknown.

In this respect, it is also interesting to classify in-situ tests in three categories, regarding the method by which the stress-strain parameters are calculated:
- Category A: includes field measurements by seismic tests in which the small strain shear modulus G_0 is determined by using a sound theoretical basis;
- Category B: includes field tests such as pressuremeter tests, and plate load tests (PLT) which can yield deformation parameters using also a sound theoretical approach, but with a few or more assumptions or approximations. In this last context cone penetration test (CPT) and Marchetti dilatometer test (DMT) can also be included;
- Category C: includes field tests such as standard penetration test (SPT), CPT and DMT, for which the soil reaction cannot be easily modelled by a theory and deformation parameters are obtained by empirical correlations.

The importance of this classification is that only categories A and B can be expected to be used universally, while category C will apply only for the cases for which they were established and different correlations may be required for different soils (Atkinson and Sallfors, 1991).

This paper describes the most recent findings of tests of category B and C the results of which are suitable to characterise the stress-strain behaviour of soils, particularly stiffness, and that can be used for levels Nos.1 and 2 of design, mostly used by practising engineers. The tests of category A and the analysis involving level No.3 of design are covered in other papers (Stokoe et al., 2004, Yu, 2004). However, test results of category A will be used in this paper, since they are a fundamental parameter of the ground, considered as a benchmark value (Tatsuoka et al., 1997).

Emphasis is given to results obtained at experimental sites of residual saprolitic soils of the Centre (Guarda) and North of Portugal (Porto sites Nos.1and 2, and FEUP site). In fact, these soils originating from granite constitute the main geotechnical ambient for foundation design in most urban areas. Their cemented structure and fabric influence the engineering behaviour, particularly the stiffness, often estimated from in situ tests.

These experimental sites have been chosen in different regions of the Portuguese territory in order to establish fundamental correlations between simple parameters, such as cone-penetration resistances or pressuremeter data with deformability moduli (Viana da Fonseca, 1996, 1998, 2003, Viana da Fonseca et al. 1994, 1997, 1998, 2001, 2003, 2004, Duarte, 2002, Rodrigues, 2003). The use of seismic in situ tests for the evaluation of shear and compression wave velocities has enabled more precise reference values for stiffness. Some of these surveys included load tests on prototype footing or plates with different sizes, as well as laboratory tests on high quality samples, for the purpose of predicting foundation settlements using more or less complex models. The comparison between derived moduli is most relevant, in order to rely on the premises of design, mainly for service conditions.

Geotechnical characterization of typical granitic residual soils profiles from the metropolitan area of Porto based on extensive *in situ* testing allowed the discussion of some particularities for the terms used to derive geotechnical design parameters.

A first synthesis of two experimental sites (Porto sites Nos.1 and 2) was reported on Viana da Fonseca et al. (2001), a third survey (FEUP site) was included in Viana da Fonseca et al. (2004). Another carefully tested site (Guarda) will be considered herein, this one based on the work from Rodrigues (2003).

These soils were classified (by laboratory identification on undisturbed samples) as silty sands and sandy silts, more rarely as clayey siltes, being in agreement with DMT and CPT based classifications.

2 BACKGROUND

2.1 Basic soil behaviour, parameters and properties

Damage of civil engineering structures interests the pre-failure behaviour of the ground (Burland, 1989, Burland and Wroth, 1974). Presently there is a general consensus that the range of strains interesting the serviceability of structures is between 0,001% and 0,5% (Burland, 1989, Gomes Correia and Biarez, 1999, Biarez et al., 1999, Jardine, 1995, Simpson, 2001). Consequently ground behaviour in this deformation range (from small to medium strains) must be accurately characterised. Figure 1 summarises the main soil features with respect to strain level, including the positioning of in-situ tests in this context.

It is well recognised that soil exhibits an approximate elastic behaviour at very small and small strains and a non-linear behaviour at medium strains. This non-linear pre-failure behaviour complicates in-situ test interpretation and may conflict with simplifying assumptions made in the past. It is then crucial to define and identify the type of modulus that will be adopted. Figure 2 defines different modulus that can be associated with the pre-failure deformation of the soil.

It is obvious, that any correlation with a soil modulus must specify the type of equipment used and also the level of strain. The direct use of this modulus in practical applications will only be suitable if it is defined for the magnitude of strain that the soil shall exhibit at the site under working conditions (serviceability modulus or stiffness).

In routine design (level N° 1) the serviceability state can also be reached indirectly by using a global safety factor to the stresses obtained by ultimate limit state analysis. Furthermore, in advanced design (level N° 2) the commercial FEM geotechnical codes use simple elastoplastic models, needing information about post-peak shearing resistance. This, involves large and very large strains, for which range it is also recognised the post peak differences in strength due to the influence of dilatancy (Fig. 3).

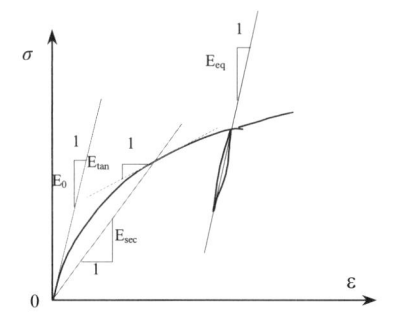

Figure 2. Schematic definition of the different moduli on a stress-strain curve

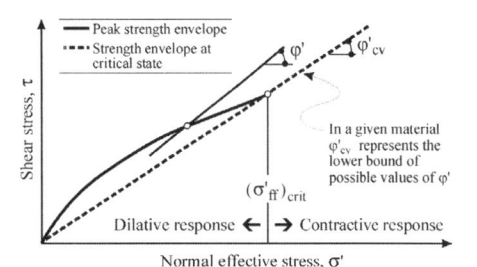

Figure 3. Schematic definition of different angle of shearing resistance (Randolph et al., 2004)

As a consequence it is necessary to identify different frictions angles: (1) peak angle of shearing resistance, ϕ'_p, (2) angle of shearing resistance at critical state, ϕ'_{cv}, and (3) angle of shearing at residual state, ϕ'_r.

Following the same attitude as for moduli, any correlation to be established with an angle of shearing resistance must specify which one is used. Present updated and detailed information about the pressure dependency of the angle of shearing resistance useful for foundation design can be found in Randolph et al. (2004).

Shear strain (%)	0,001	0,01	0,1	1	10
Strain level	Small		Medium		Large - failure
Soil behaviour	Quasi-elastic		Elastoplastic	-	failure
					Critical state, residual state
		No dilatancy			Dilatancy
Analysis for design		Deformation analysis		Ultimate state: bearing capacity and stability analisys	
Structures (serviceability state)	Roads, airports, rail track		Foundations Tunnels Retaining walls		
In-situ test operation	Seismic tests: CH, DH, SASW, SCPT, SDMT		LPT, SBPT (unload-reload)		SPT, CPT, PMT, SBPT, DMT

Figure 1. Important aspects related with soil strain level

Table 1. Values of Q and ϕ'_{cv} for different uncemented sands (Randolph et al., 2004)

SAND		MINERALOGY	Q	ϕ'_{cv} (°)	REFERENCE
TICINO		Siliceous (**)	10,8	33,5	
TOYOURA		Quartz	9,8	32	Jamiolkowski et al.
HOKKSUND		Siliceous (**)	9,2	34	(2003)
MOL		Quartz	10	31,6	Yoon (1991)
OTTAWA	Quartz	Fines 0%	9,8	30	
		Fines 5%	10,9	32,3	
		Fines 10%	10,8	32,9	Salgado et al. (2000)
		Fines 15%	10	33,1	
		Fines 20%	9,9	33,5	
ANTWERPIAN		Quartz and Glauconite	7,8 to 8,5	31,5	Yoon (1991)
KENYA		Calcareous	8,5	40,2	Jamiolkowski et al.
QUIOU		Calcareous	7,5	41,7	(2003)

In practical terms ϕ'_{cv} and ϕ'_p can be related by the empirical strength-dilatancy relationship proposed originally by Bolton (1986):

$$\phi'_p - \phi'_{cv} = m\{D_R[Q - \ln(\sigma'_{mf})]\} - R \quad ; \phi'_p \geq \phi'_{cv} \quad (1)$$

where: m is a coefficient respectively equal to 3 and 5 for axisymmetric and plane strain conditions ; R is a term, in first approximation function of (ϕ'_{cv} - ϕ'_μ), for sands ≅ 1; Q is a logarithmic function of grains compressive strength (Table 1) and σ'_{mf} the mean effective stress at failure. The importance of the relative density index becomes evident. Its value can only be obtained by field tests and this will be addressed in this paper.

2.2 G_0 a benchmark value obtained by seismic tests

The small strain shear modulus G_0 is the initial stiffness of the stress-strain curve for a given soil. In isotropic conditions, is related with the Young's modulus E_0 (Fig. 2) by $G_0=E_0/[2(1+\nu)]$.

This modulus, if properly normalised with respect to void ratio and effective stress, is in practical terms independent of the type of loading, number of loading cycles, strain rate and stress/strain history. It is then a fundamental parameter of the ground, considered as a benchmark value, which reveals the true elastic behaviour of the ground.

The first expression relating small strain shear modulus with void ratio and effective stress was derived in the early sixties by Hardin and Richard (1963) from field and laboratory tests on granular materials. This expression has been modified by different authors to accommodate clays and non isotropic conditions too. In a simplified form the expression can be written as:

$$G_0 = S \cdot p_a^{1-n} \cdot F(e) \cdot p'^n \quad (2)$$

where p_a is a reference stress, generally assumed equal to 100 kPa;

p' is the effective mean stress;
S and n are experimental constants and
$F(e)$ the void ratio function generally adopted as:

$$F(e) = \frac{(C-e)^2}{1+e} \quad (3)$$

where C is a constant, function of the shape and nature of grains.

More recently, based on results of six soft clays, Jamiolkowski et al. (1995) proposed the following equation:

$$F(e) = e^{-x} \quad (4)$$

In the following when the reference pressure (p_a) will be not used in formulae, then the value of S will expressed in pressure unities.

In the field the small strain shear modulus can be obtained by seismic tests (category A) which have in the last years opened new perspectives for the interpretation and use of their results in geotechnical engineering. This has been a consequence of the improvement of the testing equipment, signal processing and interpretation. On top of the evaluation of the small strain shear modulus, it is now well established that some more relevant information can be obtained by this category of tests:

1. evaluation of anisotropy by using polarised shear waves;
2. estimation of k_0;
3. evaluation of material damping;
4. evaluation of modulus degradation curve with strain;
5. evaluation of undrained behaviour and of the susceptibility of in situ materials to static or cyclic liquefaction.

When interpreting the results of in situ seismic tests it should be kept in mind that they are influenced by aging. This can explain why velocity of body waves of natural deposits of some age differ from that of same soil reconstituted in laboratory with the same state of effective stress and void ratio. Jamiolkowski et al. (1995) quantified the influence of aging on G_0 (see Table 2) by means of the following empirical formula (Anderson and Stokoe, 1978; Mesri, 1987):

$$G_0(t) = G_0(t_p)\left[1 + N_G \cdot \log\left(\frac{t}{t_p}\right)\right] \quad (5)$$

where $G_0(t)$ is the small shear modulus;
$G_0(t_p)$ as above at $t = t_p$, t is any generic time larger that t_p, t_p is the time to the end of primary consolidation and
N_G is a dimensionless parameter indicating the rate of increment of G_0 per log cycle of time (see Table 2).

Table 2. NG values to quantify aging in small shear modulus (Jamiolkowski et al. 1995)

Soil	d_{50} (mm)	PI (%)	N_G (%)	Notes
Ticino sand	0,54	-	1,2	Predominantly silica
Hokksund	0,45	-	1,1	Predominantly silica
Messina sand and gravel	2,10	-	2,2 to 3,5	Predominantly silica
Messina sandy gravel	4,00	-	2,2 to 3,5	Predominantly silica
Glauconite sand	0,22	-	3,9	50% quartzo, 50% glauconite
Quiou sand	0,71	-	5,3	Carbonatic
Kenya sand	0,13	-	12	Carbonatic
Pisa clay		23 to 46	13 to 19	
Avezzano silty clay		10 to 30	7 to 11	
Taranto clay		35 to 40	16	

Consequently comparison between in situ and laboratory measured values of velocity of seismic waves offers insight into the quality of the undisturbed samples. As already referred this type of tests will be considered in another lecture and will be not developed in this paper.

For a natural alluvial sands, aged and cemented, Ishihara (1986) proposed the following empirical equation to estimate G_0:

$$\frac{G_0(\text{MPa})}{F(e)} = [3.16 \text{ to } 5.72] \cdot \left[p'_0 (\text{MPa}) \cdot 10^3\right]^{0.4} \quad (6)$$

where:

$$F(e) = \frac{(2.17-e)^2}{1+e} \quad (7)$$

Viana da Fonseca (1996) from results of cross-hole tests in a saprolitic soil of granite in Portugal (Porto) showed a very small stress dependency of the small strain shear modulus:

$$\frac{G_0(\text{MPa})}{F(e)} = 110 \cdot \left[p'_0 (\text{kPa})\right]^{0.02} \quad (8)$$

where the void ratio function $F(e)$ was calculated using equation (7) based on results obtained on undisturbed samples recovered at the experimental site.

These results show that the constant value of the small strain shear modulus expression is much higher for these residual soils where S=110 kPa (eq. 2 without p_a) than for sandy transported soils where S=3.2 to 5.7, while the exponent n, reflecting the influence of the mean effective stress, is substantially lower. These different values of power n could be a consequence of different types of binding between grains (or glue) affecting the Hertz type of behaviour existing in particulate materials (Biarez et al., 1999).

More recent data for a Porto silty sand, Viana da Fonseca at el. (2004) found different constants, as illustrated in equation (9). This may result from the fact that the weathering conditions of the investigated soils are different. The comparison of these trends is presented in Figure 4.

$$\frac{G_0(\text{MPa})}{F(e)} = 65 \cdot \left[p'_0 (\text{kPa})\right]^{0.07} \quad (9)$$

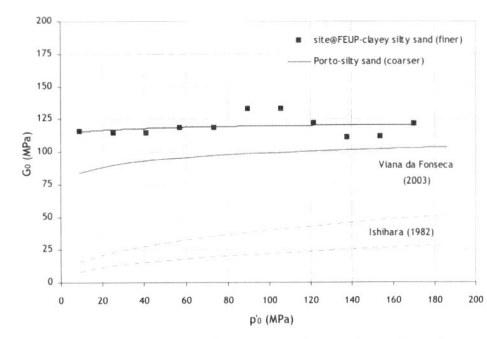

Figure 4. Comparison between observed and reference proposals of G_0 variation with effective stresses

This same analysis was made for Guarda's soils, another site in Portugal at Guarda, considering the void ratio that corresponds to the mean effective stress at that depth (Fig. 5). The two extreme values of the parameter S = 7,9 MPa and 14,3 MPa proposed by Ishihara (1982) were used in order to frame the results.

The results show that:
1. The magnitude of S and n parameters (eq. 2) reflects the reference value of the shear modulus to a certain degree. For the saprolitic granite under study, these values almost coincide with those found for the Porto granite by Viana da Fonseca (1996). These are both a great deal higher, however, than those indicated by Ishihara for the case of sedimentary materials. This express that the interparticular bonds present in the structured materials of residual origin have a predominant role in defining stiffness.
2. The value of parameter n exhibits significant differences in the case of the Guarda, as opposed to the Porto, saprolitic granite. These soils clearly show that a distinct dependence exists between the shear modulus and the in situ stress reflected in p'_0. In the case of the Guarda saprolite, the value of n is much closer to that proposed by Ishihara (1982) for aged cemented sands of sedimentary origin.

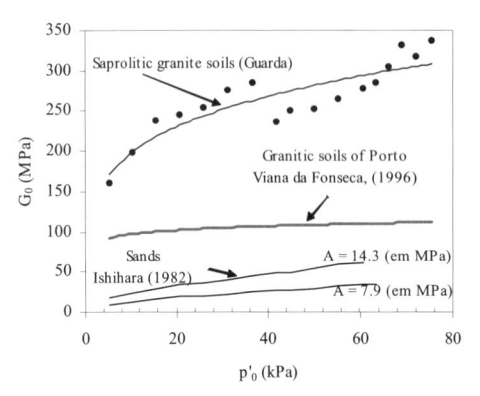

Figure 5. Relation between G_0 and p'_0 for the Guarda and Porto saprolitic granite and its conformity with the relation defined for granular sedimentary soils.

3 ROUTINE ANALYSIS OF MECHANICAL IN-SITU TESTS

3.1 Elastic stiffness

3.1.1 SPT

In many countries, as in Portugal, the Standard Penetration Test (SPT) is still a common in-situ test for geotechnical investigation.

A standardisation effort of SPT values has been done in relation with energy and depth influence, mainly for site liquefaction evaluation; this application of SPT values is not discussed here. Nowadays standardised SPT values for a energy ratio of 60% (N_{60}) is common.

Correlations between of SPT results and stiffness are very sensitive to different factors, while those relations between penetration parameters and small strain shear modulus (G_0) are somewhat independent of misleading factors, such as scale effects, non-linearity, etc (Jamiolkowski et al., 1988).

From the many empirical correlations it is presented the one relating standardised SPT values with shear wave velocity, from which the small shear modulus is obtained (Seed et al., 1986):

$$Vs = 69 \cdot N_{60}^{0,17} \cdot Z^{0,2} \cdot F_A \cdot F_G \qquad (10)$$

$$G_0 = \rho_t \cdot V_s^2 \qquad (11)$$

where: Vs is the wave velocity (m/s), N_{60} is the number of blow/feet for a energy ratio of 60%, Z is the depth (m), F_G is a geological factor (clays=1; sands=1,086), F_A is the age factor (Holocene=1; Pleistocene=1,303), ρ is the total mass density.

Following Stroud's (1988) suggestion, a simple and very useful power law between G_0 and N_{60} is:

$$G_0(\text{MPa}) = C \cdot N_{60}^n \qquad (12)$$

For the case of Porto granites the constant values were the following: $C = 63$; $n = 0.30$, for the first surveys (Viana da Fonseca, 2003), and, $C = 57$; $n = 0.20$, for the very last survey (Viana da Fonseca et al., 2004) – the former in a clear silty sand and the latter a clayey-sand with silt.

The variation of G_0 versus effective mean stress (p'_0) is very small when its variation versus other parameters, such as N_{60}, is analyzed. Correlations between G_0 and N_{60} for relevant values of p'_0 on shallow foundations are shown strongly underestimating elastic stiffness of these soils (Stroud, 1988).

Another methodology proposed by Schnaid (1999), is to establish the relation between G_0 and N_{60} by using normalised values as the following law:

$$\frac{\left(G_0/p_a\right)}{N_{60}} = \alpha \cdot N_{60} \sqrt{\frac{p_a}{\sigma'_{v0}}} \qquad (13)$$

in which, p_a – atmospheric pressure; α – adimensional value that reflects the dependence on interparticular bonds of the soil.

This will be displayed in a way such as Figure 6, where results from Brazilian saprolitic and lateritic soils, as well as those from Porto are included (Schnaid, 1999 & 2004).

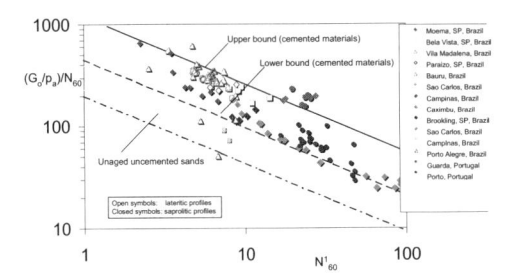

Figure 6. Initial stiffness normalized values predicted by the SPT test (Schnaid, 2004)

The results yielded by Porto (sites Nºs. 1 and 2) and Guarda saprolitic granite soils are above those corresponding to the Brazilian saprolitic soils. They are the same order of magnitude as those found for the lateritic soils discussed by Schnaid (1999). This would explain the distinct regional peculiarities already suggested by other indicators, such as weathering characteristics, as revealed by various chemical and petrographic weathering indices (Viana da Fonseca, 1996, and Rodrigues, 2003), or even by the relationship established between tip resistance of CPT (q_c) and N. This value is also influenced by varying weathering conditions, tectonic history and parent rocks. Values obtained are clearly above those for the uncemented granular soils. The effect that interparticular bonds have on the behaviour of residual soils is thus confirmed. These bonds generate normalised stiffness values that are considerably higher than those for destructured soils with similar grading characteristics, void ratio and stresses.

3.1.2 CPT

For a number of years engineers have attempted to correlate the cone resistance with different deformation moduli in order to predict settlements of structures. Most methods involve the estimation of some parameter linked to a certain calculation method. The bases of these equations are empirical and they all attempt to link observed settlements to cone resistance measured before construction. In some cases researchers have tried to correlate the cone resistance to deformation modules obtained in the laboratory. The deformation parameters of a soil arc strongly dependent on stress history. Since the cone resistance, as well as the friction ratio are rather insensitive to stress history. Consequently, the deformation modulus cannot be expected to correlate well with the cone resistance, except for normally consolidated soils.

The only modulus which seems to be reasonably insensitive to stress history, as already mentioned, is the small strain shear modulus G_0. Consequently G_0 is the more appropriate to obtain a reliable correlation with cone resistance q_c. According to Jamiolkowski et al. (1988), this correlation can be established as a function of mean effective stress. Figure 7, based on field and calibration chamber data can be used to predict small shear modulus from CPT for uncemented sands.

Other correlations between q_c and G_0 for uncemented and unaged cohesionless soils, such as those given by Robertson (1991) and also Rix and Stokoe (1992) are represented in Figure 8. The last relationship relative to uncemented siliceous sands is expressed by the following equation:

$$\frac{G_0}{q_c} = 290.57 \left[\frac{q_c}{(\sigma'_{vo} p_a)^{0.5}} \right]^{-0.75} \tag{14}$$

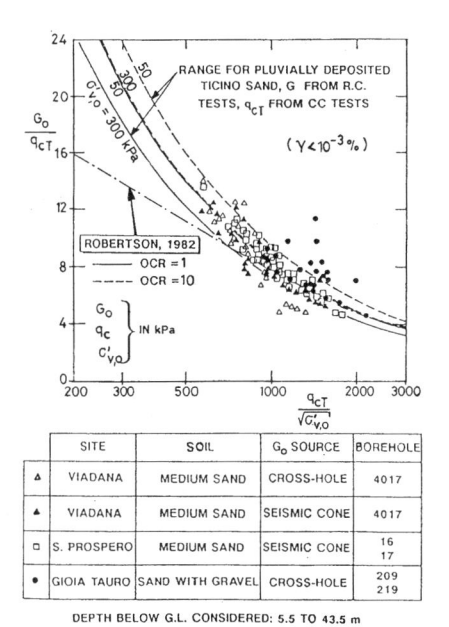

Figure 7. Normalized q_c versus G_0 correlation for uncemented predominantly quartz sand (Jamiolkowski et al, 1988)

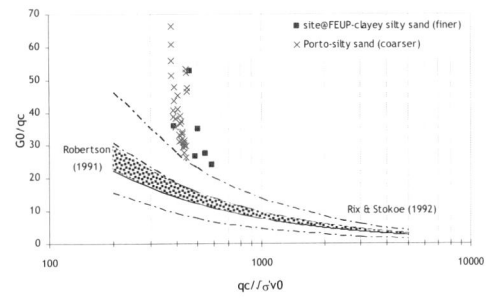

Figure 8. G_0/q_c versus $q_c/\sqrt{\sigma'_{vo}}$ from in-situ tests at FEUP experimental site, compared with other regional data and with reference curves

However, the small shear strain modulus G_0 is better determined by geophysical methods. Information can now be obtained by the seismic cone SCPT, which can also include pore water pressure filter to become a SCPTu. Using this device, Mayne and Rix (1993) proposed the following empirical correlation between G_0 and cone penetration resistance based on a database gathering 31 clay sites results:

$$G_0 = \frac{99.5(p_a)^{0.305}(q_c)^{0.695}}{e_0^{1.130}} \quad (15)$$

More recently Jamiolkowski (2004), for sand and gravel of Pleistocene age at Messina straits obtained the following empirical equation:

$$\frac{G_0}{q_c} = 144.04\left[\frac{q_c}{(\sigma'_{vo}p_a)^{0.5}}\right]^{-0.631} \quad (16)$$

It must be pointed out that these purely empirical correlations, should only be applied to sites similar from those considered in the original database.

In fact, for cemented materials these correlations follow a completely different trend as is shown in Figure 8. In these materials results of CPT denote an approximately linear increase of q_c with σ'_{v0} (and depth), as shown in Figure 9. Robertson's (1990) classification chart identifies this material as cemented, aged or very stiff natural soil, with a grain size distribution typical of sands or silt/sand mixtures, although its density index values are low.

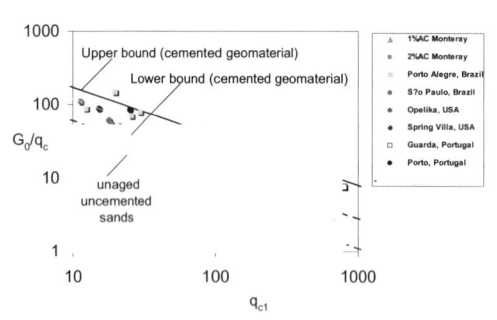

Figure 9. Results of CPT and CH over depth on Porto silty sand

The results of Figure 8 for the Porto saprolitic soil (silty sand), were obtained by means of CPT (q_c values) and cross hole (G_0 values) (Viana da Fonseca et al., 1998 and 2004).

Another way of doing this representation is by means of the proposals on Schnaid (1999), as shown in Figure 10. Again, the results yielded by Porto sites Nºs 1 and 2 and Guarda saprolitic granite soils are above those corresponding to the Brazilian saprolitic soils.

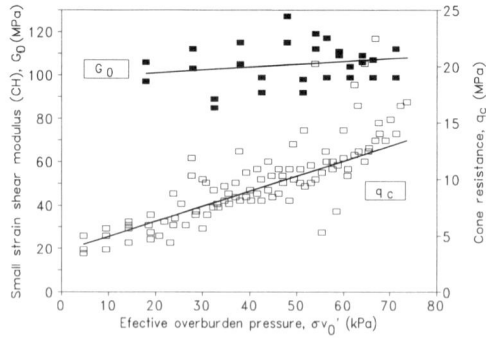

Figure 10. Relationship between G_0 and q_c for residual soils (Schnaid, 1999).

3.1.3 PMT/SBPT

Many kinds of pressuremeters probes are currently in use (Briaud, 1992, Clarke, 1995). Their differences are mostly related to the way they are inserted into the ground: predrilled hole (PMT), self-bored (SBPT), pushed-in (CPMT). It is obvious that the SBPT is the one that probe insertion causes a limited soil disturbance, contrary to the other types that cause an unavoidable stress relief. Consequently, the SBPT is the only one that can allow the measurement of the geostatic total horizontal stress σ_{h0}. It also offers a better interpretation of test results from small to large strains levels. Jamiolkowski and Manassero (1995) summarized the different geotechnical parameters that can be obtained by the three types of pressuremeters.

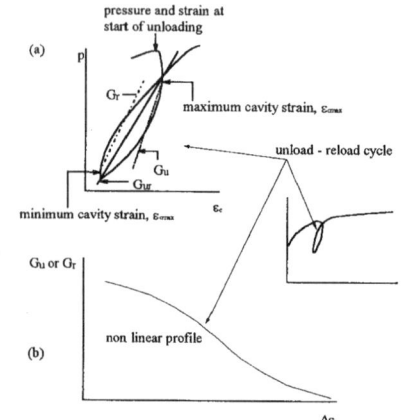

Figure 11. Selection of shear moduli (Clarke, 1995)

In Figure 11 are represented the different modulus that can be obtained by the SBPT.

Theoretically, the initial slope of a SBPT yields the G_0 value. However, in practice there is still some disturbance (Wroth, 1982) and, therefore, the modulus must be taken from an unload-reload cycle (G_{ur}). For very overconsolidated soils and cemented geomaterials it could be assumed that $G_{ur}=G_0$ if the strain of one cycle is less than 0,01%.

The use of G_{ur} in practice can be done by two approaches:
- To link G_{ur} to G_0 using a determined stress-strain relationship (Bellotti et al., 1989; Ghionna et al., 1994);
- To compare G_{ur} values to the degradation modulus curve G/G_0 versus shear strain - γ from laboratory, taking into account the average values of shear strain and mean plane effective stress associated with the soil around the expanded cavity (Bellotti et al., 1989).

The PMT is not appropriated to obtain directly G_0 because of the unavoidable disturbance during predrilling.

The SCPMT obtained by incorporating velocity geophones to a CPMT can directly measure G_0. As emphasized by Mayne (2001) new directions for enhanced geotechnical site characterization might optimize the amounts and types of data recorded. In this context the seismic piezocone pressuremeter (SPCPMT) seems to be an interesting tool (Fig. 12).

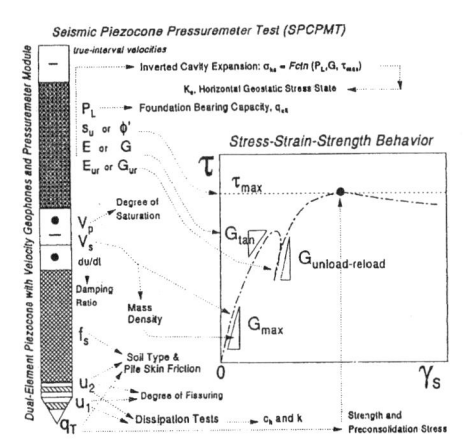

Figure 12. SPCPMT and recorded data availability (Mayne, 2001)

3.1.4 DMT

It is not the conventional dilatometer modulus E_D modulus which yields the best correlation between small strain shear modulus and dilatometer test results but the horizontal stress index (K_D), as pointed out by Marchetti (1997). Recently Tanaka and Tanaka (1998) for three sand sites found that G_0/E_D decreases as K_D increases. They observed the following trend: G_0/E_D decreases from ≈ 7,5 at small K_D (1,5 – 2) to ≈ 2 for $K_D > 5$.

It is also possible to incorporate velocity geophones to a DMT equipment and directly measure G_0 (Mayne, 1999).

3.2 Serviceability Stiffness

3.2.1 Factoring G_0

For practical purposes is necessary to extrapolate the results of small strains to the range of strain of engineering significance, generally 0,001% to 0,5%. This need arises from the recognition that the displacements of well designed civil engineering structures are generally quite small and overpredicted when using soil parameters that are inferred from conventional soil tests in theoretical settlement solutions (Burland, 1989; Tatsuoka et al., 1997; Simpsom 2001; Jardine et al., 2001).

Naturally the case of using a settlement solution based on the measured parameter is not considered here (Menard, 1962, Schmertmann, 1970).

A modified hyperbola can be used as a simple means to reduce the small strain shear modulus to secant values of G at working strain levels, in terms of shear strain γ, or at working load levels, in terms of the mobilized strength (q/q_u).

The generalized form may be given, in terms of γ, as:

Jardine et al. (1986):

$$\frac{G_s}{p'} = A + B \cdot \cos\left\{\alpha \cdot \left[\log_{10}\left(\frac{\varepsilon_D}{\sqrt{3} \cdot C}\right)\right]^\gamma\right\} \quad (17)$$

where: A, B, C, α, γ are constants; p' is the mean effective normal stress $p' = (\sigma_1' + \sigma_2' + \sigma_3')/3$ and:

$$\varepsilon_D = \left\{(2/3) \cdot \left[(\varepsilon_1 - \varepsilon_2)^2 + (\varepsilon_2 - \varepsilon_3)^2 + (\varepsilon_3 - \varepsilon_1)^2\right]\right\}^{1/2} \quad (18)$$

or Gomes Correia et al. (2001):

$$\frac{G_s}{G_0} = \frac{1}{\left[1 + a \cdot \left(\frac{\gamma}{\gamma_{0,7}}\right)\right]} \quad (19)$$

where γ is shear strain;
$\gamma_{0,7}$ is the shear strain for a stiffness degradation factor of $G/G_0=0.7$ and
a is a constant (a ≈ 0,385, for the database used).

This relationship between G/G_0 and $\gamma^*=\gamma/\gamma_{0,7}$ seems to be very promising as a reference stiffness degradation curve, since, for the range of shear strain tested, it seems scarcely affected by the kind of soils (temperate or tropical soils), plasticity index, confining pressure, degree of saturation and overconsolidation ratio (Fig. 13).

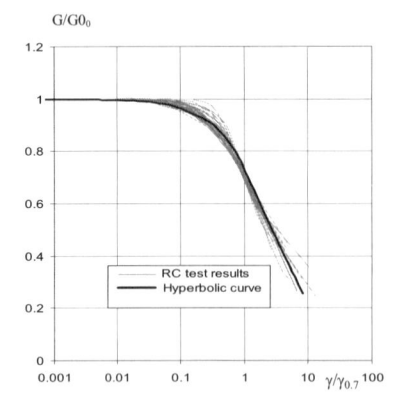

Figure 13. Relationship between normalized secant shear modulus and normalized shear strain ($\gamma/\gamma_{0.7}$) and hyperbolic fitting

Based on laboratory experimental results from 37 tests, by resonant column, of lateritic and saprolitic soils it was possible to establish the relationship presented in Figure 14 for lateritic and saprolitic Brazilian soils, allowing a practical use of these results.

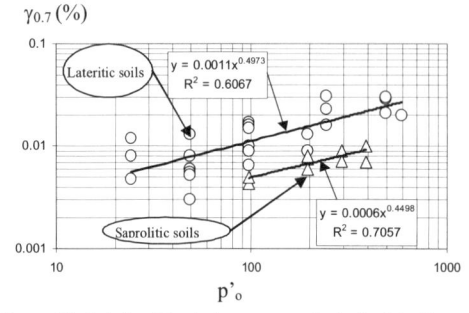

Figure 14. Relationships between $\gamma_{0.7}$ and p'$_0$ for lateritic and saprolitic soils

Puzrin and Burland (1998) proposed a more fundamental approach covering the full range of strains from small to large strains, through medium strains.

In terms of mobilized strength or stress level (q/q_u), the generalized form is (Fahey and Carter, 1993):

$$\frac{E}{E_0} = 1 - f\left(\frac{q}{q_{ult}}\right)^g \qquad (20)$$

where f and g are fitting parameters. Values of $f=1$ and $g=0,3$ appear reasonable first order approximation for unstructured and uncemented geomaterials (Mayne, 2001).

The mobilized strength or stress level (q/q_u) can also be considered as an inverse factor of safety (FS), i.e. a stress level half of the ultimate corresponds to a FS = 2.

Using these functions factoring G_0 is possible to obtain a serviceability shear modulus, or serviceability stiffness, to be used in routine calculations to obtain settlements. As a very rough approach a factoring value of 0,5 can be used.

3.2.2 SPT/CPT

Viana da Fonseca and Almeida e Sousa (2001), for the Porto silty sand, using a crossed interpretation of footing and plate loading tests with the SPT values in the settlement influence zone, for a service level of $q_s/q_{ult} \cong 10\text{-}20\%$, obtained an average ratio between the serviceability secant Young's modulus and SPT values of:

$$E(\text{MPa})/N_{60} \cong 1 \qquad (21)$$

This relationship is similar to the proposal of Stroud (1988) for normally consolidated soils, in identical stress levels.

The analysis of a large scale loading test (circular concrete footing 1.20m in diameter) and of two other plates of smaller diameter (0.30m and 0.60m), performed in the Porto silty sand, lead to the Young's moduli values presented in Table 3, for different loading stages. These results were obtained by back-analysis of the footing loading test (rigid footing), considering a linear elastic layer with constant modulus underlain by a rigid base at 6.0m depth.

Table 3. Secant Young's modulus, E_s, from loading test for different service criteria

Loading Tests	q (s/B= 0.75%)	q/q_{ult}[*] (F_S=10)	q/q_{ult} (F_S = 4)	q/q_{ult} (F_S = 2)
Footing	17.3	20.7	16.0	11.0
Plate (0,6)	11.9	11.2	12.5	12.7
Plate (0,3)	6.7	6.9	5.9	5.7

[*] Corresponding to the allowable pressure for serviceability limit state design.

The intermediate stress level (FS = 4) corresponds approximately to the allowable pressure for residual soils, from Décourt's (1992) criterion.

Correlations between q_c and Young's modulus, established for different stress-strain levels by triaxial tests (CID and CAD) with local strain measurements, confirmed the very strong influence of non-linearity on E/q_c ratios as well as a singular pattern of that variation when compared to proposals for transported soils (Viana da Fonseca et al., 1997).

The possibility of inferring design values for Young's modulus to predict the behavior of load tests, on plates and on a prototype footing, conducted in the vicinity of was these penetration tests was developed and thoroughly discussed elsewhere (summarized in Viana da Fonseca and Ferreira, 2002). The main conclusions drawn for the most common methods are as follows: the Burland and Burbidge (1985) equation based on SPT results led to an overestimation of the observed settlement by a factor of 2 to 3, while the application of the Schmertmann et al.'s (1978) method reproduced accurately the footing settlement for $\alpha = E/q_c$ values in the range of 4.0 to 4.5. Both methods identify this saprolitic soil in the global typology of cemented or overconsolidated granular soils.

3.2.3 PMT/SBPT

The routine analysis of PMT tests follows the method originally developed by Menard (1955). It gives design parameters directly obtained from the pressuremeter test curve (ASTM, 2004). Figure 15 shows the interpretation of the curve and Figure 16 exemplifies the procedure to obtain the pressuremeter modulus (E_m), based on the present ASTM (2004) standard. It must be pointed out that this modulus is related with the average stiffness exhibited by the ground associated with a determined strain level. Consequently the use of this value must be only applied in settlement formulae developed by Ménard (Ménard, 1963, 1965), as this is done in the French Code for foundation design (MELT, 1993, Gambin and Frank, 1995). Consequently Menard modulus must be considered as a test-specific design parameter.

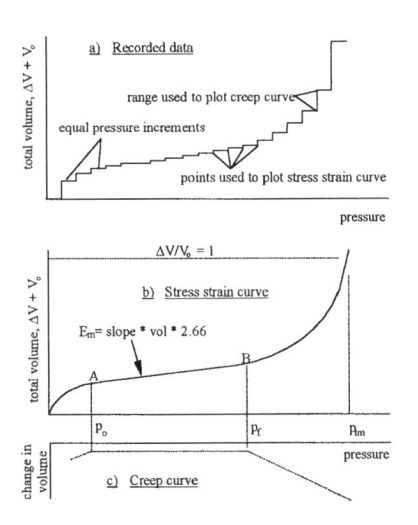

Figure 15. Interpretation of Menard test (PMT) according ASTM standard (Clarke and Gambin, 1998)

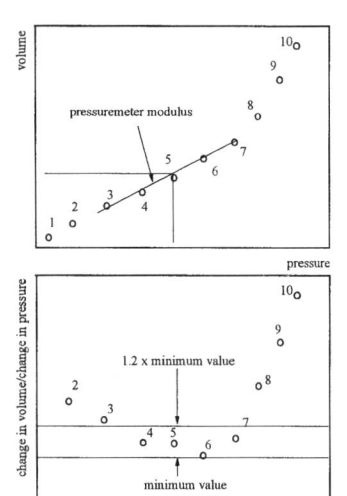

Figure 16. Selection of the pressure range to calculate E_m according ASTM standard (Clarke and Gambin, 1998)

Concerning SBPT, it is possible to analyse several moduli from SBPT results (Fig. 10): the initial shear modulus G_i (G_0 if enough precision is obtained) and the secant modulus from a unload-reload loop G_{ur}. This last modulus is judged to be more reliable and suitable to engineering practice, since G_i is strongly influenced by disturbance, even when small, and to the compliance of the measuring system (Fahey and Carter, 1993; Ghionna et al., 1994). However, even if this loops are suitable to measure shear stiffness, it is recognised the difficulty to associate this value to a strain level. According to Jamiolkowski and Manassero (1995) the value of G_{ur} measured in coarse grained soils represents the drained stiffness at intermediate strain level, between 1.10^{-4} to 1.10^{-3}, relatively insensitive to soil disturbance caused by probe insertion.

In practice, only the strain at the pressuremeter rubber cover surface is known, which means that the stiffness will be a little higher because stiffness is increasing further away from the pressuremeter. This assumption will allow obtaining degradation stiffness curve of the tested soil by varying the amplitude of the loop, which could be useful to compare with other test results.

Experimental in-situ work described by Viana da Fonseca (2003) revealed stiffness from reload-unload cycles of PMT (E_{pmur}) and SBPT tests in saprolitic granite soils apparently very different. In fact, for PMT it were found the following relations: $E_{pmur}/E_{pm} \cong 2$ and $E_0/E_{pm} \cong 18\text{-}20$, with E_0 determined on seismic survey (G_0 - CH), while for SBPT $G_0/G_{ru} \cong 2,6$ to $3,0$. It must be noticed that these last values are substantially lower than the ratio ($\cong 10$), reported by Tatsuoka & Shibuya (1992) on Japanese residual soils from granite. The non-linearity model of Akino - cited by the previous authors - developed for a high range of soil types, including residual soils, is expressed simply by:

$$E_{sec} = E_0, \quad \varepsilon \leq 10^{-4} \quad (22)$$

$$E_{sec} = E_0 \cdot (\varepsilon/10^{-4})^{-0.55}, \quad \varepsilon \geq 10^{-4} \quad (23)$$

SBPT unload-reload modulus correspond to secant values for shear strain of about 6×10^{-4}, which agrees very well with the above indicated trends for this test (Viana da Fonseca, 2003).

It must be pointed that the comparison of results of the two types of tests can only be properly discussed if the mean effective stress during the cycle (p') is well estimated and the strain level of the cycle of each test reported. These aspects will be analyzed in item 4.

3.2.4 DMT

The modulus determined by the Marchetti's Flat Dilatometer (DMT), designated M_{DMT}, is the vertical confined (one dimensional) tangent modulus at σ'_{v0} and is said to be the same as E_{oed} ($=1/m_v$) obtained from an oedometer test in the same range of strains. This modulus can be converted to the Young's modulus (E) via the theory of elasticity. For $\nu=0,25-0,30$ it is possible to write: $E \approx 0,8 \, M_{DMT}$.

This empirical Marchetti's modulus is applied to predict settlements in sand and clays (Marchetti et al., 2001) and it was validated by different researchers (Schmertmann, 1986 in Marchetti et al. 2001 and Hayes, 1990, in Mayne, 2001).

Viana da Fonseca and Ferreira (2002) for characterization of the soil stiffness for shallow foundations settlement assessment, used correlations between the moduli E_{DMT} and $E_{s10\%}$ (secant modulus corresponding to 10% of peak shear strength). The following correlations were obtained:

$$G_0/E_{DMT} \cong 16.7 - 16.3 \cdot \log_{10}(p_{0N}) \quad (24)$$

$$E_{s10\%}/E_{DMT} = 2.35 - 2.21 \cdot \log_{10}(p_{0N}) \quad (25)$$

These formulae are situated between those that are used for NC and OC transported soils.

3.3 Angle of shear resistance

In saturated geomaterials, drained and undrained conditions can prevail during in-situ testing. For penetration tests it is common to assume fully drained penetration in clean sands (drained conditions - ϕ') and for clays with very low permeability fully undrained conditions (s_u).

The undrained shear resistance (s_u) is greatly affected by several factors such as initial stress state, anisotropy, stress history, boundary conditions, strain rate, ...) and consequently it is generally normalized to the preconsolidation stress (σ'_p).

3.3.1 SPT/CPT

SPT can be used to predict the peak angle of shear resistance in granular soils when normalised to a reference energy (60%) and a stress-level of $p_a = 100$ kPa $(N_1)_{60}$, by:

$$(N_1)_{60} = \frac{N_{60}}{(\sigma'_v/p_a)^{0,5}} \quad (26)$$

Hatanaka and Uchida (1996) obtained the following equation, also corroborated by Mayne (2001) for residual silty sand in Atlanta and Georgia:

$$\phi'_p = [15,4(N_1)_{60}]^{0,5} + 20° \quad (27)$$

CPT is recognised primarily as a strength-measuring device (Houlsby, 2001).

Robertson and Campanela (1983) recommended for unaged, uncemented quartz sands the following correlation:

$$\phi'_p = \arctan[0,1 + 0,38 \cdot \log(q_c/\sigma'_{v0})] \quad (28)$$

An alternative equation considering the non linear normalization of qc with the stress level has been proposed by Kulhawy and Mayne (1990) in Marchetti et al. (2001):

$$\phi'_p = 17,6° + 11,0 \cdot \log(q_{c1}) \quad (29)$$

where q_{c1} is calculated by the following expression:

$$q_{c1} = \frac{q_c}{(\sigma'_v/p_a)^{0,5}} \quad (30)$$

A more general approach consist of estimating a secant friction angle. In fact, considering the non-linear shear resistance envelop defined in Figure 2 and equation 1, the secant friction angle can be estimated knowing the relative density D_r. This parameter can be estimated by means of equation 31 represented in Figure 17 (Jamiolkowski et al., 2003).

$$D_R = \frac{1}{C_2} \ell n \left(\frac{q_c}{C_o p'^{C_1}_o} \right) \quad (31)$$

where q_c and p'_o are both in kPa, and the various parameters are: $C_o = 300$ (dimensional); $C_1 = 0.46$; $C_2 = 2.96$.

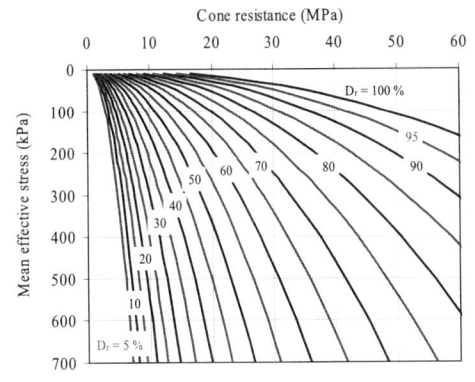

Figure 17. Relationship between cone resistance, relative density and mean effective stress for coarse-grained soils (Jamiolkowski et al. 2003)

In sands the pressure influences the peak and the dilatancy angles. However, as these angles are related (see Bolton, 1986 or Schanz and Vermeer, 1996), then as long as the angle of critical state is known, the friction angle can be deducted.

3.3.2 *PMT/SBPT*

Theoretically a peak and a post peak resistance can be obtained by pressuremeter tests. However, because the influence of disturbance during installation the peak resistance is usually ignored for PMT.

An usual prediction of undrained shearing resistance is obtained by the Ménard limit pressure - p_{lm} (Amar et al., 1975):

$$s_u = \frac{(p_{lm} - \sigma_h)}{5,5} \quad for \quad (p_{lm} - \sigma_h) < 300\,kPa \quad (32)$$

$$s_u = 25 + \frac{(p_{lm} - \sigma_h)}{10} \quad for \quad (p_{lm} - \sigma_h) > 300\,kPa \quad (32a)$$

where p_{lm} is the applied pressure required to double the cavity diameter and
p_0 is the estimated in-situ horizontal stress.

In the SBPT the following relationship can be used:

$$s_u = \frac{(p - \sigma_h)}{[1 + \ln(G/s_u) + \ln(\Delta V/V)]} \quad (33)$$

For drained conditions the angle of shearing resistance can also be estimated as follow:

$$sen\,\phi' = \frac{s}{[1 + (s-1)sen\,\phi'_{cv}]} \quad (34)$$

It is also possible to estimate both, drained and undrained shearing resistance by CPMT, but the method will be not presented here. For more details, see (Clarke and Gambin, 1998)

3.3.3 *DMT*

For undrained conditions the original correlation established by Marchetti (1980) is:

$$s_u = 0,22 \cdot \sigma'_{v0} \cdot (0,5 \cdot K_D)^{1,25} \quad (35)$$

This estimation of undrained shearing resistance seems according Marchetti et al (2001), to be quite accurate for design, at least for everyday practice.

Regarding the estimation of the drained angle of shearing resistance in sands two methods were proposed by Marchetti (1997). They both use the horizontal stress index k_D calculated by:

$$k_D = \frac{(p_0 - u_0)}{\sigma'_{v0}} \quad (36)$$

The use of a wedge plasticity solution relate I_D as a function of ϕ' and lateral stress state, including active pressure, at-rest k_0 (NC) value and passive pressure. Mayne (2001) found the following expression for the k_p case:

$$\phi' = 20° + \frac{1}{0,04 + 0,06/k_D} \quad (37)$$

This solution was later cross-correlated for CPT-DMT relationships by Campanella and Robertson (1991). Durgunoglu and Mitchel (1975), cited by Marchetti et al. (2001) presented a chart (Fig. 18) that allows the estimation of ϕ', in function of q_c, σ'_{vo} and k_0 (NC).

3.4 *Correlation between in-situ tests in residual soils*

The difficulties of sampling residual soils, which cause a number of problems for the characterization of stress-strain behaviour of soils through laboratory tests, make in situ tests very important tools in geotechnical practice. The most common tests are by far, the dynamic penetration tests, the classical SPT and in specific conditions dynamic probing (DPSH), but other more limited in penetration capacities (such as CPT and DMT) or more time consuming, such as PMT or PLT (plate load tests) are becoming more frequent as they give a more fundamental parametrical information. More recently a special attention is being made to seismic tests for the evaluation of initial shear modulus (G_0), regarded as a highly important benchmark parameter. Although in situ tests suffer serious limitations in terms of interpretation of their results, they nevertheless make a valuable contribution to geomechanical characterization.

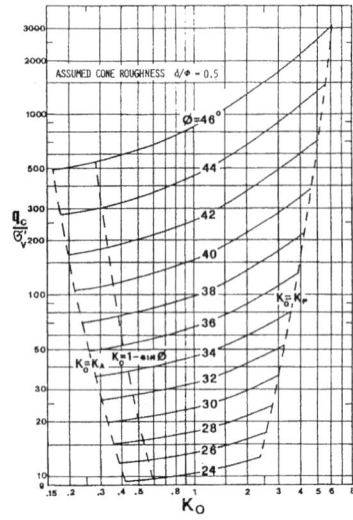

Figure 18. Estimation of ϕ', in function of q_c, σ'_{vo} and k_0 (NC), proposed by Durgunoglu and Mitchel (1975), in Marchetti et al. (2001)

It is interesting to examine if among the many correlations established between the results given by the various in situ tests some of them are applicable to residual soils, as a preference for evaluating strength and stiffness parameters, giving emphasis to the importance of the specific strain level associated with the deformability modulus derived from each of them.

3.4.1 CPT – SPT correlations

From their collection of data of different parent rocks of residual soils, Danziger et al. (1998), concluded that correlations between CPT ad SPT present a large scatter due to intrinsic heterogeneity.

These authors have concluded that different parent rocks generally produce different correlations for the same particle size distribution (a pattern of soil type). It is assumed that from Brazilian data, there is a general trend of decreasing values of q_c/N_{SPT} with D_{50} and generally lower values than those expressed by Robertson and Campanella (1983) average line. Results from Porto granites, corroborate the very high sensitivity to the type of matrix, as it is expressed in Figure 18. In this figure, very recent data from the experimental FEUP site of the University of Porto is also included. Parts of these results were reported at this ISC'2 in the paper by Viana ad Fonseca et al. (2004). Results obtained are presented in Figure 19 including correlations proposed by other authors.

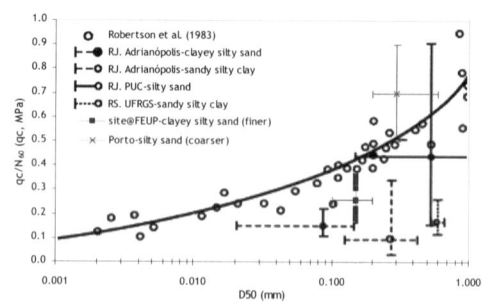

Figure 19. Ranges of q_c/N versus D_{50} on Brazilian residual soils, compared with the experimental site results (based on Danziger et al, 1998)

It is remarkable that data from this experimental site vary significantly with the dominant matrix of each soil. The results of the more silty sand Porto matrix shown in the previous data, exhibit a large contrast with the results of the more clayey soil in the last experimental site.

Corroborating this, Rodrigues & Lemos (2004) presented additional data obtained for the saprolitic granite soils from Guarda, a much coarser matrix, and plotted on the same graph (Figure 20).

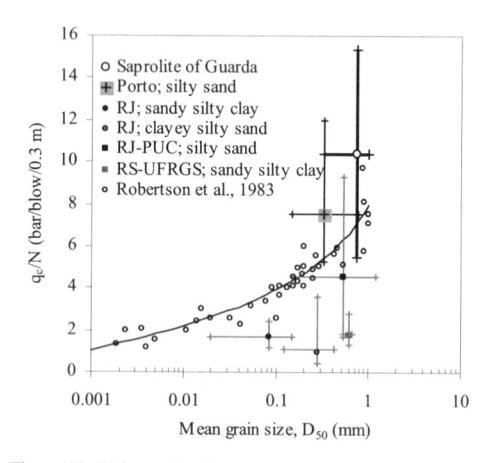

Figure 20. Values of q_c/N_{60} versus D_{50} for Guarda saprolitic granite and other residual soils.

The results obtained and presented on Figure 20 clearly show that, in the case of saprolitic soils from Guarda and the former Porto sites, q_c/N relations are conspicuously higher than those proposed for the granular sedimentary soils. This fact should be related to the greater sensitivity of the q_c parameter of the CPT test, than the value of N of the SPT test, concerning the cohesive part of the resistance, due to the existence of weak inter-particulate bonding and significant quartz coarse grains. It is indeed reasonable to accept that grain size distribution plays an important part in controlling stress-strain behaviour, since the coarser grain size of Guarda's saprolitic granite soils exhibits a higher q_c/N ratio than the saprolitic granite soils from Porto, whose grain size is finer. The Brazilian residual soils and the those from FEUP site have a q_c/N ratio that is lower than that predicted for sedimentary soils. Both theses soils have a similar mineralogical nature due to the original rock: the Brazilian rock being made of gneiss and sandstone, and the FEUP rock made of granite at the interface of gneiss and schist.

These findings lend further support to the idea that grain size properties do not in themselves explain the behaviour of the residual soils. They mean that other parameters must be incorporated into the analysis of the behaviour of these soils, namely, weathering indices, chemical and mineralogical ones.

An important aspect is the link between the drainage conditions during cone penetration and that expected in the design problem (Lunne et al., 1995). Takesue et al. (1995) showed this aspect for a volcanic soil, pointing out that the change in drainage, function of the penetration rate, has a larger effect on the sleeve friction than on the cone resistance. This is consistent with the fact that cone resistance is a total stress measurement contrarily to sleeve friction that is controlled by the effective stresses. The drainage also affects the relationship between CPT and SPT. In fact SPT is the summation of cutting shoe resistance and friction along the outside wall (and to a less extend along the inside wall) of the SPT sampler.

3.4.2 *PMT versus CPT/SPT correlations*

For the first two experimental sites on Porto granite saprolitic soils, there are some derived ratios between PMT and SPT or CPT parameters, which were reported in a paper by Viana da Fonseca et al. (2003). Theses correlations are included in the table 4.

Table 4. Ratios between SPT, CPT and PMT parameters

q_c/p_l^*	f_s/p_l^*	N_{60}/p_l^* (MPa)	N_{60}/E_m (MPa)	E_m/p_l^*	E_{mur}/E_m
14.3	0.390	14.6	1.4	10.6	1.4 - 1.9

f_s is the friction sleeve of CPT; p_l^* net limit pressure and the other symbols already defined.

3.4.3 *A synthesis of correlations obtained between in situ tests parameters and ratios between moduli*

In Table 5 comparative parameters between in situ tests are presented.

Table 5. Ratios obtained from in situ tests

q_d/q_c	0,75 – 1,25	E_m/p_l	12
N_{20}/q_c (MPa^{-1})	0,6 – 0,8	p_{0DMT}/p_{0m}	2 – 3
q_c/p_l^*	4 – 6	P_{1DMT}/P_{fPMT}	≅ 1
f_s/p_l^*	0,10 – 0,25	E_D/E_{PMT}	≅ 1,5

q_d is the dynamic tip resistance; N_{20} is the number of blows in 0,20m penetration of DPSHT; p_{0DMT} – Lift-off pressure of DMT; p_0 - Lift-off pressure of PMT (see Fig. 15); p_{1DMT} – Limit pressure of DMT; p_f – Creep pressure of PMT (see Fig. 15); p_l^* net limit pressure

Ratios between distinct values of Young's moduli inferred from the investigations conducted have the obvious interest of fulfilling the needs of geotechnical designers to obtain data from different origins for each specific purpose.

Viana da Fonseca et al. (2003) reported some interesting correlations from the data available at the experimental sites:
– values of Young's moduli determined directly, with no empirical treatment, or even, no deriving assumptions;
– common constant ratios that are assumed to correlate SPT (DP) or CPT parameters with Young's modulus, comparing them with transported soils;
– relative values of moduli can be summarized in the way that is expressed in Table 6a, while some relations could be pointed out between in situ tests, as expressed in Table 6b.
– In what respects the relative position of the values deduced from the tri-axial tests on undisturbed samples, the data can be also summarized by some ratios presented in Table 6c.

Table 6a. Ratios between Young's modulus

$\dfrac{E_0(CH)}{E_{s1\%}(PLT)}$	$\dfrac{E_0(CH)}{E_{ur}(PLT)}$	$\dfrac{E_0(CH)}{E_m}$
≅ 8 – 15	≅ 2 - 3	≅ 20 - 30

Table 6b. Average ratios between Young's modulus and in situ "gross" tests

$\dfrac{E_0(CH)}{N_{60}(SPT)}$	$\dfrac{E_0(CH)}{q_c}$	$\dfrac{E_0(CH)}{q_d(DPL)}$	$\dfrac{E_0(CH)}{p_l}$
≅ 10 (MPa)	≅ 30	≅ 50	≅ 8

Table 6c. Ratios between Young's moduli obtained in tri-axial tests and in situ CH tests

$\dfrac{E_0(CH)}{E_0(BE)_{tx}}$	$\dfrac{E_0(CH)}{E_{el}(LI)_{tx}}$	$\dfrac{E_0(CH)}{E_{ur}(LI)_{tx}}$	$\dfrac{E_0(CH)}{E_{ti}(LI)_{tx}}$
≅ 2,0	≅ 2,4	≅ 3,1	≅ 4,5

Triaxial tests (tx): seismic waves velocities determined by bender elements (BE) and modulus in elastic loops (el) or between vertices on unload-reload cycles (ur), and secant to 10% of failure ($s_{10\%}$), using local instrumentation (LI).

3.4.4 A synthesis of shearing resistance obtained between in situ tests

The application of the proposal of Robertson and Campanella's (1983) to the first two surveys in the residual soils from Porto conducted to higher values of ϕ' than those derived both by the application of Décourt (1989) proposal based on SPT, or that one from DMT, following Marchetti (1997) correlation established for sandy soils. It should be noted that this correlation is assumed to underpredict ϕ', since the accepted value results from the lower limit of 3 curves based on Marchetti's assumptions, who considers K equal to K_{0nc} or to the square root of passive earth pressure coefficient (K_p). The reason of the discrepancy between SPT and CPT derived values of ϕ' (single resistance parameter that can be derived from tests that generated only "one" parameter) was discussed in Viana da Fonseca et al. (2003). This is a consequence of the high effective intercept on the vertical axis in the compression-shear domain, a peculiarity of residual soils. This cannot be identified by a dynamic test, which reflects large strain strength, mostly ruled by the friction component, while the less destructive testing procedure of CPT is more sensitive to this low strain strength component.

This natural important deviations towards to the behaviour detected in transported soils modelled by the classical theories of Soils Mechanics are, to a great extent, due to a structural cementation inherited from the original rock mass and are, in terms of strength, essentially characterized by the existence of this effective cohesive intercept (c') and the development of a yielding behavior induced by the break of the cementation structure, independently from the failure corresponding to the plastific yield of the soil matrix component. The quantification of the cohesive resistance component (c') has been achieved mainly by triaxial tests and, less often, by back-analysis of load tests with plate or footing of different sizes. Getting undisturbed samples on these soils is extremely difficult, usually implying the partial or even complete loss of the cemented natural structure. Cruz et al. (2004) present an experimental conceptual approach, aiming at quantifying the effective cohesive component (c') of resistance by means of Marchetti's DMT. Since this test allows the determination of two basic parameters (p_0 and p_1), it is stated generating the possibility of evaluating both the angle of shear resistance and cohesive intercept. Assuming that K_D reflects the overall resistance of soil, it can be expected that either c' and ϕ' may affect this parameter. Then, if ϕ' from tri-axial testing is assumed, the corresponding K_D may be back-calculated. The difference between the two values of K_D (measured and back-calculated) will reveal the effective cohesive intercept. More detailed information can be found in Cruz et al. (1997).

Another issue that is also very pertinent for these residual soils is their particular sensitivity to sampling, since their behaviour is strongly controlled by the structure inherited from the parent rock. This issue was discussed in detail in other papers (Viana da Fonseca and Ferreira, 2000; 2002). It is also relevant to emphasizes the influence that the stress-path, mainly when the test is carried out in compression or in extension, has on the values of resistance parameters. As a clear illustration of this issue the derived values from in situ and laboratory tests results from the FEUP site are presented in Table 7 (extracted from Viana da Fonseca et al. 2004)

Table 7 Resistance parameters from in situ and laboratory tests

TESTS		ϕ' [°]	c' [kPa]
In situ	SPT	38	n/a
	CPT	37	n/a
	DMT	39	n/a
Laboratory	TX compression	45.8	4.5
	TX extension	28	12

4 ADVANCED ANALYSIS

The result of in situ test is either a penetration resistance or a relationship between some load (or stress) and the induced deflection (or strain); this result reflects the integrated and complex response of many soil elements around the instrument. It is then necessary to convert the test result by back analysis into soil parameters that at this level of analysis should be related with some soil model, which may be used in engineering design. However, this will require full understanding of the theories and models for which parameters are required (Atkinson and Sallfors, 1991). In this paper only simple soil models and structural models common in engineering practice are addressed.

4.1 Simplified soil modelling – modulus degradation curve

Elhakim and Mayne (2003) showed an approach to represent nonlinear stiffness soil behaviour based on CPT with seismic transducers, i.e. SCPT (Fig. 12). In this approach a modulus degradation graph is needed, like the one proposed before (see 3.2.1). In their work they choose Fahey and Carter (1993) degradation modulus (equation 20).

This same concept can be applied to PMT and SBPT with the incorporation of direct measurement of G_0.

SBPT offers also the possibility to assess the entire shear stress τ versus shear strain γ relationship in sands (drained) and in fine grained soils, undrained (Palmer, 1972; Manassero, 1989).

For undrained pressuremeter tests Wood (1990) obtained theoretical relationship between a non-linear pressuremeter test curve and a non-linear elastic stiffness-strain curve. This opened the possibility of using curve fitting methods to obtain non-linear undrained stiffness moduli from pressuremeter tests.

4.2 *Simple elastic-plastic soil model to derive test curves*

The common theories for the derivation of the test curves:
− cavity expansion curve, i.e. pressure versus strain curve in PMT, SBPT and DMT,
− load (or stress) versus displacement in plate load test (PLT))
mainly use simple elastic-plastic modelling of soil, considering stiffness as stress dependent.

Of course if a more realistic soil model is used, then it is obvious that a different test curve will be obtained.

It must be noticed that cavity expansion tests and PLT do not provide enough information to derive a unique solution of parameters of a simple elastic-plastic soil model. It is then useful to assess some of the parameters by other tests, mainly laboratory tests in order to narrow the range of possibilities. However, these parameters must be appropriated for the boundary conditions of the problem.

Among all the above-mentioned in-situ tests, the pressuremeter tests are the more appropriate for back analysis since they provide a complete pressure-strain curve for which elements of soil exhibit the same strain history at different radii distances from the surface. This last aspect is not the case for the PLT where each element of soil under the plate undergoes different strain history (Houlsby, 2001).

Gomes Correia et al. (2004) back-analysed PMT and PLT results on a silty sand (residual soil of Granite, close to Porto) using a simple model developed by PLAXIS and called HSM (hardening soil model) which assumes a non linear elastic response of the soil during loading and a isotropic hardening during unloading. The main features of this model are:
- Stress dependent stiffness.
- Plastic straining due to both primary deviatoric and compression loading.
- Elastic unloading and reloading.
- Failure according to the Mohr Coulomb model.

The stress-strain geotechnical parameters of the model are well known by professionals: stress dependent Young's modulus, Poisson's ratio, angle of shearing resistance at critical state, dilatancy angle and effective cohesion.

Some of the relevant results of this study are also pointed out here (Gomes Correia et al. 2004).

Using the previous modeling technique where the friction angle was derived from tri-axial tests (Fleureau et al., 2002), the following conclusions were drawn:
- The HSM in PLAXIS could be a good compromise to back-analyze PMT results, while for PLT the identification of the parameters of this non linear law from a load-settlement curve is more difficult without having other information. A possible solution to this is the measurement of deformations in depth under the plate with strain gauges, as it was proposed by Burland (1989) and Tatsuoka et al. (1989).
- Menard modulus (E_m) obtained in the routine analysis according to the ASTM (2004) standard (see Fig. 16) is associated to a strain level near 1 % (Biarez et al 1998; Gomes Correia et al. 2004). The secant modulus of the unload-reload cycle is around 2,2 times Ménard's modulus.
- The secant modulus of an unload-reload cycle of the plate load test is rather close to a strain level of 0,1 %. Furthermore, the unload-reload modulus of plate load test is about three times the unload-reload modulus of pressuremeter, as a consequence of the associated different strain levels, assuming that the representative stresses in the two tests are identical.
- These strain levels are in good agreement with the E-moduli values obtained for both types of tests.
- In the non linear elastic behaviour domain of the soil, the curve which expresses the variation of the applied pressure during a PLT versus the ratio settlement over diameter is close to the curve which shows the function of the vertical stress versus the vertical strain in a tri-axial test. In the numerical modeling with a power law equal to 0.5 (Hillier and Woods, 2001), the ratio between the relative plate deformations or "relative strains" δ/D and the tri-axial strains was about 0.5 (see Gomes Correia et al. 2004).

The routine interpretations of PMT and PLT led to very different values of modulus. Besides, the Ménard modulus is a tangent modulus (Fig. 2), in the sense that it is obtained by the slope of the pseudo-elastic zone of the pressuremeter curve, while the modulus obtained by the plate load test is generally a secant modulus. In addition it is obvious that these moduli will be modified if test procedure or interpretation is modified. This is a consequence of the non-linear material behaviour, where the modulus depends on the level of stress and strain, among others. The main point is to know in practice how to use correctly these values. In fact, modelling geotechnical structures is being more and more popular, and consequently the results of category B tests must be more and more used for the identification of the model parameters. It is evident that the correctness of this identification is a function of the model adopted. So, the appropriateness of the model must be carefully analysed and confirmed.

5 CONCLUSIONS

It is nowadays well established in the geotechnical community that soils exhibits non linear behaviour giving place to the definitions of different modules and angles of shearing resistance. This non-linear behaviour complicates in-situ test interpretation and may conflict with simplified assumptions made in the past. To clarify this, the following directions were presented in this paper:
– A direct use of moduli in practical application will be only suitable if it is defined for the magnitude of strain and stress that the soil shall exhibit at the site under working conditions. Otherwise it must be associated either with correction factors or by using design rules well calibrated by the real behaviour of structures, as is the case of Ménard's Modulus.
– Any correlation of in-situ test results should specify, furthermore the type of equipment and test procedure, the type of modulus or angle of shearing resistance;
– The small shear strain modulus (G_0), if normalised with respect to void ratio and effective stress, is in practical terms independent of the type of loading, number of loading cycles, strain rate and stress/strain history. Consequently it is the most appropriate parameter to establish correlations with in-situ tests (SPT, CPT, DMT). A great improvement will be the incorporation of seismic transducers with these equipments for routine site investigation work. It must be stressed that these correlations are only valid for the materials tested, but they are very useful for regional and country applications in order to create databanks. This will be of great help during design phase, since it will allow analysis to be developed, while specific tests results are not yet available.
– At routine design level, G_0 could be adapted to strain level of engineering significance (0,001 % to 0,5 %). Some rules are presented acting in terms of shearing strain or of the mobilized strength;
– The peak angle of shearing resistance seems to be well correlated with SPT and CPT results. It is also possible to correlate that value with the non linear secant angle knowing mainly the relative density;
– In advanced design level soil parameters of engineering significance to be used in soil modelling (non linear behaviour) can be obtained by two approaches:
  • By using results of category C tests (SPT, CPT, DMT, PMT, SBPT) simultaneously with G_0, or G_{ur} (SBPT) with different strain amplitudes, to obtain a modulus degradation curve – simplified soil model;
  • By back-analysing results of category C tests, stress-strain results of PMT, SBPT, or load-displacements results from PLT.
– The comparison of different modulus obtained by category C tests in the same soil can be done using some kind of analysis in order to situate the stress and strain associated to each modulus. In an experimental study on residual soils of granite it was obtained that E_m (PMT) according ASTM (2004) is associated to an strain level around 1%, while the unload-reload modulus of PLT according ASTM (1993) is close to 0,1%.

In this paper it also pointed out the peculiar behaviour of saprolitic granitic soils (aged and cemented) of some regions of Portugal putting in evidence the differences related with transported soils (unaged and uncemented). Several correlations between different tests of categories A, B and C, following the directions mentioned previously, are presented. They are very useful for day to day design practice and are being collected to update a knowledge based system already implemented covering a variety of geomaterials from rock to soils.

REFERENCES

Amar, S., Baguelin, F., Jézéquel, J.F., Le Mèhauté, A. 1975. In situ shear resistance of clays. *Proc. ASCE Spec. Conf. on In situ Measurements of Soil Properties*, Raleigh, Vol. 1, pp. 22-45.

Anderson, D.G., Stokoe, K.H. II. 1978. Shear modulus: A time dependent soil property. Dynamic Geotechnical Testing, *ASTM STP* 654, pp. 66-90.

Atkinson, J.H. and Sallfors, G. 1991. Experimental determination of stress-strain-time characteristics in laboratory and in situ tests. *Proc. X ECSMFE, Deformation of Soils and Displacements of Structures*, Vol. III, pp. 915-956, Associazione Geotecnica Italiana.

ASTM, 1993, Standard test method for repetitive static plate tests of soils and flexible pavements components, for use in evaluation and design of airport and highway pavements. *Annual Book of ASTM Standards, D 1195-93, American Society for Testing and Materials*, West Conshohocken, Pensylvania

ASTM, 2004, Standard test method for prebored pressuremeter testing in soils. *American Society for Testing and Materials, vol. 04.08, Soil and Rocks (I): D420-D5779*, West Conshohocken, Pensylvania.

Bellotti, R., Ghionna, V.N., Jamiolkowski, M.B., Robertson, P.K., Peterson, R.W. 1989. Interpretation of moduli from self-boring tests in sand. *Géotechnique* 39 (2), pp. 269-292.

Biarez, J.; Gomes Correia, A.; Liu, H.; Taibi, S. 1999. Stress-strain characteristics of soils interesting the serviceability of geotechnical structures. *2nd International Symposium on Pre-Failure Deformation Characteristics of Geomaterials.* Balkema, Rotterdam, Vol. 1, pp. 617-624.

Biarez, J., Gambin, M., Gomes Correia, A., Flavigny, E.,Branque, D. 1998. Using pressuremeter to obtain parameters to elastic plastic models for sand. *Geotechnical Site Characterisation (ISC'98)*, Robertson and Mayne editors, Balkema, Rotterdam, Vol 2, pp.747-752

Bolton, M.D. 1986. The strength and dilatancy of sands. *Géotechnique*, 36 (1), 65-78.

Briaud, J.-L. 1992. *The pressuremeter*. Balkema, Rotterdam.

Burland, J. 1989, The 9th Bjerrum memorial lecture: Small is Beautiful, the Stifness of Soils at Small Strain, *Can. Geotech. Journal,* Vol 26, pp.499-516.

Burland, J.B., Burbidge M.C. 1985. *Settlement of foundations on sand and gravel.* Proc. Inst. of Civil Eng. 78: 1325-1381. London: Thomas Tellford.

Burland, J.B. and Wroth, C.P. 1974. Settlement of buildings and associated damage: State of the art review. *Proc. Conf. on Settlement of Structures*, Cambridge, 611-654.

Campanella R.G., Robertson, P.K. 1991. Use and interpretation of a research dilatometer. *Canadian Geotechnical Journal*, Vol. 28, pp. 113-126.

Clarke, B. 1995. The pressuremeter in geotechnical Design,,*Blackie Academic and Professionnal, Glasgow.*

Clarke, B.G. and Gambin, M.P. 1998. *Pressuremeter testing in onshore ground investigations: A report by the ISSMGE Committee TC 16.* Geotechnical Site Characterisation (ISC'98),Robertson and Mayne editors, pp. 1429-1468, Balkema, Rotterdam.

Cruz, N., Viana da Fonseca, A., Neves, E. 2004. Evaluation of effective cohesive intercept on residual soils by DMT data. *In situ Conference, ISC2*. Porto, 2004.

Cruz, N., Viana, A., Coelho, P., Lemos, J. 1997. Evaluation of geotechnical parameters by DMT in Portuguese soils. *XIV Int. Conf. on Soil Mechanics and Foundation Engineering*, pp 77-80.

Danziger, F.A.B.; Politano, C.F.; Danziger, B.R. 1998. CPT-SPT correlations for some Brazilian residual soils. *Proc. 1st Int. Conf. on Site Characterization ISC'98* Atlanta, USA. E^{ds} Robertson & Mayne, Vol. 2, pp. 907-912.

Décourt, L. 1992. SPT in non classical material. Applicability of classical soil mechanics principles in structured soils. *Proc. US/Brazil Workshop*, Belo Horizonte, pp. 67-100. Univ. Fed. Viçosa,MG, Brazil.

Décourt, L. 1989. The standard penetration test. State of the art report. *Proc. XII ICSMFE*, Rio de Janeiro. Balkema, Rotterdam. Vol. 4: 2405-2416.

Duarte, I. M. R. 2002. *Residual soils of granitoid rocks from the South of Portugal. Geological and geotechnical characteristics*. Ph. D. Thesis, University of Évora, 373 p. (in Portuguese).

Elhakim, A.F. and Mayne, P.W. (2003). "Derived stress-strain-strength of clays from seismic cone tests". *Proceedings, Deformation Characteristics of Geomaterials*, Vol. 1., Lyon, France.

Fahey, M. and Carter, J.P. 1993. A finite element study of the pressuremeter test in sand using nonlinear elastic plastic model. *Canadian Geotechnical Journal* 30 (2): 348-362.

Fleureau, J-M.; Gomes Correia, A.; Hadiwardoyo, S.; Dufour-Laridan, E. & Langlois, V. 2002. Influence of suction on the dynamic properties of a silty sand. *Third International Conference on Unsaturated Soils*, Recife, Brasil.

Gambin, M.P., Frank, R.A. 1995. The present design rules for foundations based on Ménard PMT results. *Proc. 4th Int. Symp. On Pressuremeters*, Sherbrooke, Canada.

Ghionna, V.N., Jamiolkowski. M.B., Pedroni. S. and Sangado, R. 1994. Tip displacement of drilled shafts in sands. in vertical and horizontal deformations of foundations and embankments. Ed. A.T. Yeung and G.Y. Felio, *ASCE, GSP40*, New York, 2, 1039-1057.

Gomes Correia, A., Barros, J.M.C., Santos, J.A., Sussumu, N. 2001. An approach to predict shear modulus of soils in the range of 10^{-6} to 10^{-2} strain levels. *4th International Conference on Recent Advances in Geotehcnical Earthquake Engineering and Soil Dynamics*, 26-31 March 2001, San Diego, California.

Gomes Correia, A. and Biarez, J. 1999. *Stiffness properties of materials to use in pavement and rail track design.* Proc. XIIth ECSMGE, vol.1, pp. 1245-1250.

Gomes Correia, A., Antão, A., Gambin, M. 2004. Using a non linear constitutive law to compare Menard PMT and PLT E-moduli. *Second Int. Conf. on Site Characterization – ISC'2*, Porto. Ed. Viana da Fonseca & Mayne Millpress, Rotterdam.

Hardin, B.O., Richart, F.E. 1963. Elastic wave velocities in granular soils. *J. Soil Mech. Found. Div., ASCE*, 89, Nº SM1: 33-65.

Hatanaka, M. and Uchida, A. 1996. *Empirical correlation between penetration resistance and φ of sandy soils. Soil and Foundations* 36 (4), pp. 1-9.

Hight, D.W. and Higgins, K.G. 1995. *An approach to the prediction of ground movements in engineering practice: Background and application.* Keynote Lecture 7, IS Hokkaido: 2, pp. 909-946, Balkema, Rotterdam.

Hillier, R.P. and Woods, R.I. 2001. Characterisation of non-linear elastic behaviour from field plate load tests. *Proc. XV th ICSMGE*, Istanbul, Balkema, Rotterdam, vol. 1, pp. 421-424.

Houlsby, G. T. 2001. In situ tests and pre-failure deformation behaviour of soils. *Proc. 2nd Int. Symp. on Pre-failure Deformation Characteristics of Geomaterials* – IS Torino 99. Torino, Italy. E^{ds} Jamiolkowski, Lancellota & Lo Presti, Vol. 2, pp. 1319-1324.

Ishihara, K. 1982. Evaluation of soil properties for use in earthquake response analysis. *Proc. Inter. Symp. Num. Models in Geomechanics*, Zurich, pp. 237-259.

Ishihara, K. 1986. *Evaluation of soil properties for use in earthquake response analysis.* Geot. Mod. Earthq. Eng. Pr.: pp. 241–275. Balkema, Rotterdam.

Jamiolkowski, M. and Manassero, N. 1995. *The role of in situ testing in geotechnical engineering – thoughts about the future.* Advances in Site Investigation Practice, pp. 929-951, Thomas Telford, U.K.

Jamiolkowski, M., Ghionna, V.N., Lancellotta, R., Pasqualini, E. 1988. New correlations of penetration tests for design practice. *Proc. of Int. Symp. On Penetration Testing*, Orlando

Jamiolkowski, M., Lancellotta, R., Lo Presti, D.C.F. 1995. *Remarks on the stiffness at small strains of six Italian clays.* Keynote Lecture 3, IS Hokkaido: 2, pp. 817-836, Balkema, Rotterdam.

Jamiolkowski, M., Ghionna, V.N, Lancellotta, R. & Pasqualini, E. 1988. New correlations of penetration tests for design practice. *Proc. ISOPT-1*, Orlando. Ed. De Ruiter. Balkema, Rotterdam. Vol. I: 263-296.

Jamiolkowski, M., Lo Presti, D.C.F., Pallara, O. 1995. *Role of in situ testing in Geotechnical Earthquaque Engineering and Soil Dynamics*, St. Louis.

Jamiolkowski, M.B. 2004. *Soil properties evaluation from static cone penetration test.* Private communication.

Jamiolkowski, M.B., Lo Presti, D.C.F., Manassero, M. 2003. *Evaluation of relative density and sheer strength of sands from cone penetration test (CPT) and flat dilatometer (DMT)*. Soil Behaviour and Soft Ground Construction, Eds. J.T. Germain, T.C. Sheahan and R.V. Whitman, ASCE, GSP 119, 201-238.

Jardine, R.J. 1995. One perspective of the pre-failure deformation characteristics of some geomaterials. *IS Hokkaido '94*, Vol. 2, pp. 855-885.

Jardine, R.J., Potts, D.M., Fourie, A.B. and Burland, J.B. 1986. Studies of the influence of non-linear stress strain characteristics in soil-structure interaction. *Géotechnique*, 36(3), 377-396.

Jardine R.J., Kuwano, R., Zdravkovic, L., Thornton, C. 2001. Some fundamental aspects of the pre-failure behaviour of granular soils. *Proc. 2nd Int. Symp. on Pre-failure Deforma-*

tion Characteristics of Geomaterials – IS Torino 99. Torino, Italy. E^{ds} Jamiolkowski, Lancellota & Lo Presti, Vol. 2, pp. 1077-1111.

Lunne, T., Robertson, P.K., Powell, J.J.M. 1997. *Cone penetration testing in geotechnical practice*. London, U.K. Chapman and Hall.

Manassero, M. 1989. Stress-strain relationship from drained self-boring pressuremeter tests in sand. *Géotechnique*, 39(2), 293-307.

Marchetti, S. 1997. The flat dilatometer design applications. *III Geotechnical Engineering Conference*, Cairo University.

Marchetti, S. 1980. In situ tests by flat dilatometer. *ASCE JGED*, Vol. 106, GT3, pp. 299-321.

Marchetti, S. Monaco, P., Totani, G., Calabrese, M. 2001. The flat dilatometer test (DMT) in soil investigations. Report by the ISSMGE Committee TC 16. *Conf. On In Situ Measurement of Soil Properties and Case Histories*, Rahardjo and Lunne eds., Bali, pp. 95-131.

Mayne, P.W. 2001. Stress-strain-strength parameters from enhanced in-situ tests. *International Conference on In situ Measurement of Soil Properties and Case Histories*, Bali: 27-47.

Mayne, P.W., Scneider, J.A., Martin, G.K. 1999. Small-and large-strain soil properties from seismic flat plate dilatometers tests. Proc. *2nd Int. Symp. on Pre-failure Deformation Characteristics of Geomaterials – IS Torino 99*. Torino, Italy. E^{ds} Jamiolkowski, Lancellota & Lo Presti, Vol. 1, pp. 419-426.

Mayne, P.W. and Rix, G.J. 1993. $G_{max} - q_c$ relationships for clays. *ASTM Geotechnical Testing Journal* 16 (1), pp. 54-60.

MELT, 1993, Fascicule 62, Titre V du CCTG, *Imprimerie des Journaux Officiels, Paris*

Ménard, L., 1955. *Pressiomètre, brevet français d'invention*, n° 1.117.983, 19.01.1955

Ménard, L. 1963. *Calcul de la force portante des fondations sur la base des essais pressiométriques*, Sols Soils, 5, No. 5, pp. 9-24.

Ménard, L. 1965. Règles pour le calcul de la portante et du tassement des fondations en fonction des résultats pressiométriques. *Proc. 6th ICSMFE*, Montreal, pp. 295-299.

Mesri, G. 1987. The fourth law of soil mechanics; the law of compressibility. *Proc. Int. Symp. On Geotechnical Engineering of Soft Soils*, Mexico City.

Palmer, A.C. 1972. Undrained plane-strain expansion of a cylindrical cavity in clay: a simple interpretation of the pressuremeter test. *Géotechnique*, 22 (3), pp. 451-457.

Palmer, A.C. 1990. Undrained plane-strain expansion of a cyclindrical cavity in clay. *Géotechnique*, **22**(3), 451-457.

Puzrin, A.M. and Burland, J.B. 1998. Nonlinear model of small-strain behaviour of soils. *Geotechnique* 48 (2): 217-233.

Randolph, M.P., Jamiolkowski, M.B and Zdravković, L. 2004. *Load carrying capacity of foundations*. Keynote Lecture for Skempton Memorial Conference, Vol. 1: 207-240, Imperial College London, UK.

Rix, G.J. & Stokoe, K.H. 1992. Correlations of initial tangent modulus and cone resistance. *Proc. Int. Symp. Calibration Chamber Testing*. Potsdam, New York: 351-362. Elsevier.

Robertson, P.K. 1990. *Soil classification using the cone penetration test*. Canadian Geot. J. Vol. 27: 151-158.

Robertson, P.K. 1991. Estimation of foundation settlements in sand from CPT. 'Vertical and horizontal deformation of foundation and embankments', *Geot. Special Pub., ASCE*, Vol. II, N° 27, pp. 764-778.

Robertson, P.K. 2001. *Sixty years of CPT – What advances have we made?* Special Lecture on the Int. Conf. On In Situ Measurement of Soil Properties and Case Histories, Rahardjo and Lunne eds., Bali, pp. 1-16.

Robertson, P.K.; Campanella, R.G. 1983. Interpretation of cone penetrometer test, Part I: Sand. *Canadian Geotech. J.*, Vol. 20, N° 4, pp. 718-733.

Rodrigues, C.M.G. 2003. *Geotechnical characterization and geomechanic behaviour study an saprolitic granite soils from Guarda*. PhD thesis, Coimbra University, Coimbra, Portugal (in Portuguese).

Rodrigues, C.M.G.; Lemos, L.J.L. 2003. Strength and stress-strain behaviour of saprolitic granite soils from Guarda – sampling effects. *Proc. Int. Symp. Deformation Characteristics of Geomaterials – IS Lyon 2003*. Lyon, France. E^{ds} Benedetto et al, pp. 663-668.

Rodrigues, C.M.G., Lemos, L.J.L. 2004. SPT, CPT and CH tests results on saprolitic granite soils from Guarda, Portugal. *Second Int.Conf.on Site Characterization – ISC'2*, Porto. Ed. Viana da Fonseca & Mayne, Millpress, Rotterdam.

Salgado, R., Bandini, P. and Karim, A. 2000. Shear strength and stiffness of silty sand. *J. of Geotech and GeoEnviron Eng.*, ASCE, 126(5), 451-462.

Schanz, T. and Vermeer, P.A. 1996. Angles of friction and dilatancy of sand. *Geotechnique* 46 (1): 145-151.

Schnaid, F. 1999. On the interpretation of in situ tests in unusual soil conditions. *Proc. 2nd Int. Symp. on Pre-failure Deformation Characteristics of Geomaterials – IS Torino 99*. Torino, Italy. E^{ds} Jamiolkowski, Lancellota & Lo Presti, Vol. 2, pp. 1339-1345.

Schnaid, F., Lehane, B.M. & Fahey M. 2004. *In situ test characterisation of unusual geomaterials*. Keynote Lecture on Second Int. Conf. on Site Characterization – ISC'2, Porto. Ed. Viana da Fonseca & Mayne Millpress, Rotterdam.

Schmertmann, J. 1970. Static cone to compute static settlement over sand. *ASCE*, JSMFD, SM 3.

Schmertmann, J.H.; Hartman, J.P. & Brown, P.R. 1978. Improved strain influence factor diagram, *J. Geot. Eng. Div.*, Vol. 104, GT8, pp. 1131-1135. ASCE, New York

Seed, H.B., Wong, R.T., Idriss, L.M., Tokimatsu, K. 1986. Moduli and damping factors for dynamic analysis of cohesionless soils. *JGED, ASCE*, 112 (11), pp. 1016-1032.

Simpson, B. 2001. Engineering needs. *Proc. 2nd Int. Symp. on Pre-failure Deformation Characteristics of Geomaterials – IS Torino 99*. Torino, Italy. E^{ds} Jamiolkowski, Lancellota & Lo Presti, Vol. 2, pp. 1011-1026.

Stokoe, K. H. II, Chung-Ang, S-H, Woods, R.D. 2004. *Some contributions of in situ geophysical measurements to solving geotechnical engineering problems*, Second Int. Conf. on Site Characterization – ISC'2, Porto. Ed. Viana da Fonseca & Mayne Millpress, Rotterdam.

Stroud, M.A. 1988. The standard penetration test - its application and interpretation. *Proc. Geot. Conf. Penetration Testing in U.K.*, Birmingham. Thomas Telford, London. : 24-49.

Tanaka, H. and Tanaka, M. 1998. Characterization of sandy soils using CPT and DMT. *Soils and Foundations*, Japanese Geotechnical Society 38 (3), pp. 55-65.

Tatsuoka, F. & Shibuya, S. 1992. Deformation characteristics of soils and rocks from field and laboratory tests. *Proc. 9th Asian Reg.CSMFE*, Bangkok, 2: 101-170. Rotterdam: Balkema.

Tatsuoka, F., Jardine, R.J., Lo Presti, D.C.F., Di Benedetto, H., Kohata, T. 1997. Characterizing the pre-failure deformation properties of geomaterials". *Proc 14th ICSMFE*, Hamburg, 4, 2129-2164.

Viana da Fonseca, A. 1996. *Geomechanics of residual soils from Porto granite. Design criteria for shallow foundations*. PhD thesis, Porto University, Porto (in Portuguese).

Viana da Fonseca, A. 1998. Identifying the reserve of strength and stiffness characteristics due to cemented structure of a saprolitic soil from granite. *Proc. 2nd International Sympo-*

sium on Hard Soils – Soft Rocks. Naples. Vol.1: pp.361-372. Balkema, Rotterdam.

Viana da Fonseca, A. 2003. *Characterizing and deriving engineering properties of a saprolitic soil from granite, in Porto. Characterization and Engineering Properties of Natural Soils*. Eds. Tan et al., pp.1341-1378. Swets & Zeitlinger, Lisse.

Viana da Fonseca, A., M. Matos Fernandes, A.S. Cardoso & J.B. Martins 1994. Portuguese experience on geotechnical characterization of residual soils from granite. *Proc. 13th I.C.S.M.F.E.*, New Delhi 1: 377-380.

Viana da Fonseca, A., M. Matos Fernandes, A.S.Cardoso 1997. Interpretation of a footing load test on a saprolitic soil from granite. *Géotechnique*, Vol. 47, 3: 633-651.

Viana da Fonseca, A., Matos Fernandes, M. & Cardoso, A. S. 1998. Characterization of a saprolitic soil from Porto granite by in situ testing, *First Int. Conf. on Site Characterization –ISC'98*. Atlanta: 2, 1381-1388. Balkema, Rotterdam.

Viana da Fonseca, A. & Ferreira, C. 2000. Management of sampling quality on residual soils and soft clayey soils. Comparative analysis of in situ and laboratory seismic waves velocities. (in Portuguese) *Proc. Workshop Sampling Techniques for Soils and Soft Rocks & Quality Control*. FEUP, Porto.

Viana da Fonseca, A. & Almeida e Sousa, J. 2001. At rest coefficient of earth pressure in saprolitic soils from granite. *Proc. XIV ICSMFE*, Istambul, Vol. 1: 397-400.

Viana da Fonseca, A. & Ferreira, C. 2002. The application of the Bender Elements technique on the evaluation of sampling quality of residual soils. (in Portuguese). *Proc. XII COBRAMSEG*, Vol.1: 187-199. ABMS, Sao Paulo.

Viana da Fonseca, A., Vieira de Sousa, F. & Ferreira, C. 2003. Deriving stiffness parameters from "gross" in situ tests and relating them with "noble" reference values on saprolitic soils from granite. *Proc. XII Pan-American CSMGE, MIT*, Vol.1, pp 321-328, Verlag Guckuf GmbH Essen, Germany.

Viana da Fonseca, A., Carvalho, J., Ferreira C., Tuna, C., Costa, E., Andrade, R., Cruz, N., Santos, J. 2004. Geotechnical characterization of a residual soil profile: the experimental site of the University of Porto. *Second Int. Conf. on Site Characterization – ISC'2*, Porto. Ed. Viana da Fonseca & Mayne Millpress, Rotterdam.

Wroth, C.P. 1982. British experience with the self-boring pressuremeter. *Proc. Int. Symp. Pressuremeter and its Marine Appl.*, Paris, pp. 143-164.

Wood, M. 1990. Strain-dependent moduli and pressuremeter tests. *Géotechnique*, 40 (3), pp. 509-512.

Yoon, Y. 1991. *Static and Dynamic Behaviour of Crushable and Non-Crushable Sands*. Ph.D.Thesis, Ghent University.

Yu, H-S 2004. In situ testing: from mechanics to interpretation. First James K. Mitchell Lecture, *Second Int. Conf. on Site Characterization – ISC'2*, Porto. Ed. Viana da Fonseca & Mayne Millpress, Rotterdam.

Proceedings ISC-2 on Geotechnical and Geophysical Site Characterization, Viana da Fonseca & Mayne (eds.)
© 2004 Millpress, Rotterdam, ISBN 90 5966 009 9

Some contributions of in situ geophysical measurements to solving geotechnical engineering problems

Kenneth H. Stokoe, II
University of Texas, Austin, Texas, USA.

Sung-Ho Joh
Chung-Ang University, Seoul, South Korea

Richard D. Woods
University of Michigan, Ann Arbor, Michigan, USA

Keywords: geophysics, geotechnics, seismic testing, in situ tests, body waves, surface waves, case histories

ABSTRACT: This paper focuses on one in situ geophysical method, seismic measurements. Seismic (stress wave) measurements have been used for more than 50 years in geotechnical engineering, primarily in the areas of soil dynamics and geotechnical earthquake engineering. In the past 30 years, their role has steadily increased to the point where they also play an important part in characterizing sites, materials and processes for non-dynamic problems. Case histories and applications are presented to highlight some examples.

1 INTRODUCTION

The geotechnical engineer has always been faced with the problem of characterizing near-surface materials. The near-surface region is often within 10 to 100 m of the ground surface. Traditionally, field exploration programs have involved boring, sampling, and penetration testing. In the 1960s, in situ geophysical measurements began to be employed in geotechnical engineering. This work primarily involved seismic (stress wave) measurements which were adapted from exploration geophysics. Seismic measurements were used to characterize geotechnical sites (e.g. layering, top of bedrock, depth to water table) and geotechnical materials (e.g. stiffnesses in shear and compression). The real demand for seismic measurements grew out of the need to evaluate the dynamic properties of near-surface soils, specifically the shear-wave velocity, V_s. V_s is a key parameter in soil dynamics and geotechnical earthquake engineering. Today, however, in situ seismic measurements are used in many more applications as discussed herein.

The discipline of geophysics and in situ geophysical measurements encompass much more than seismic methods. Other geophysical methods include electrical, magnetic, electromagnetic, ground-penetrating radar, and gravity. All of these methods offer the geotechnical engineer new and improved in situ techniques to characterize sites, materials and processes. These opportunities arise from the strong theoretical bases upon which geophysical methods are founded, the complementary physical principles that support various field tests, and the ability to perform the same basic measurement in the laboratory and in the field. Furthermore, many geophysical methods are noninvasive which make them well suited and cost effective in profiling spatially and temporally.

The geotechnical engineering profession has not adopted other geophysical methods as rapidly as seismic methods. One reason is that seismic methods directly measure a mechanical property, the initial slope of the stress-strain relationship, which is used in the solution of many geotechnical engineering problems. However, new demands in geoenvironmental, geotechnical, and military applications are constantly increasing the need for improved and higher-resolution site characterization methods. Geophysical methods have an important role to fill in these areas. Technical papers in the Proceedings of the First International Conference on Site Characterization (Robertson and Mayne, 1998) emphasize this point.

1.1 *Organization*

The information and examples presented in this paper focus on the use of in situ seismic measurements. However, this material demonstrates the relevance of geophysical methods in geotechnics. A brief background on stress waves is presented in Section 2 to facilitate subsequent discussions. An overview is presented in Section 3 of four noninvasive surface-wave methods. The reason is that the use of this generalized method is rapidly increasing in geotechnical engineering. Case histories and applications are presented in Section 4 that involve geosystems loaded statically as well as dynamically. The use of in situ V_s measurements in evaluating sample disturbance and in predicting nonlinear soil response is discussed in Section 5 followed by conclusions in Section 6.

2 BACKGROUND ON STRESS WAVES AND TRADITIONAL SEISMIC METHODS

2.1 Types of Stress Waves

Traditionally, in situ seismic testing has been conducted by initiating a mechanical disturbance at some point in the earth and monitoring the resulting motions (stress waves) at other points in the earth. The modes of propagation most often used are body waves, compression and shear waves, and one type of surface wave, the Rayleigh wave. These waves, in terms of their far-field particle motions, are illustrated in Figure 1. Compression waves (P waves) have particle motion parallel to the direction of wave propagation, while shear waves (S waves) have particle motion perpendicular to the direction of wave propagation. Rayleigh waves (R waves) exist because of the exposed ground surface. R waves have particle motions that are a combination of vertical (shear) and horizontal (compression) motions. Near the surface of a uniform material, R waves create particle motion that follows a retrograde elliptical pattern as illustrated in Figure 1c. The decay with depth of the vertical and horizontal components of R-wave particle motion is illustrated in Figure 2. The depth axis is normalized by the Rayleigh wavelength, λ_R. It is interesting to see in Figure 2 that the horizontal component changes sign at a normalized depth around 0.15. The meaning of this change in sign is that R-wave particle motion changes from a retrograde ellipse to a prograde ellipse in a uniform half-space.

2.2 Stress Wave Velocities

Stress waves are non-dispersive in a uniform elastic medium. The term non-dispersive indicates that the propagation velocity is independent of frequency. These waves are also considered non-dispersive in low-loss homogeneous soil and rock at small strains and low frequencies. However, stratigraphy and other forms of heterogeneity cause frequency-dependent velocity. (This dependency is the fundamental premise on which surface-wave testing is based as noted in Section 3.) The far-field velocities of stress waves depend on the stiffness and mass density of the material as,

P-wave velocity:

$$V_P = \sqrt{\frac{M}{\rho}} = \sqrt{\frac{B + \frac{4}{3}G}{\rho}} = \sqrt{\frac{E}{\rho} \frac{(1-\nu)}{(1+\nu)(1-2\nu)}} \quad (1)$$

S-wave velocity:

$$V_S = \sqrt{\frac{G}{\rho}} \quad (2)$$

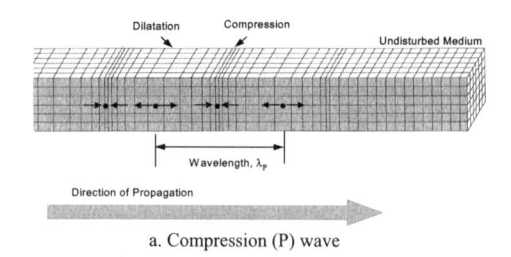

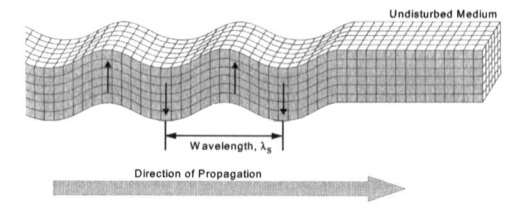

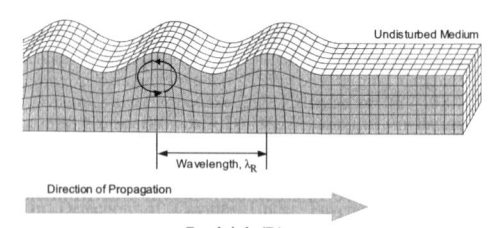

Figure 1. Wave propagation modes: body waves within a uniform, infinite medium and Rayleigh waves along the surface of a uniform half space (after Bolt, 1976)

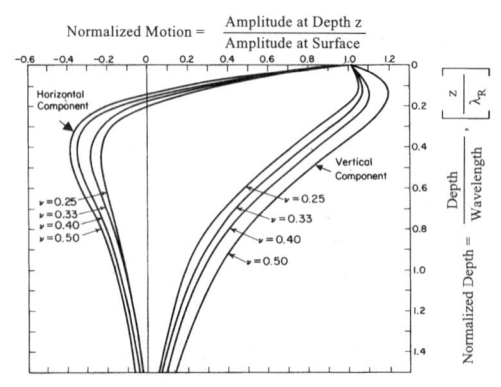

Figure 2. Variation in normalized particle motions with normalized depth for Rayleigh waves propagating along a uniform half space (from Richart et al., 1970)

where ρ is the mass density and M, B, G and E are the constrained, bulk, shear, and Young's moduli, respectively, and ν is Poisson's ratio. For a homogeneous, isotropic material, compression and shear wave velocities are related through Poisson's ratio, ν, as,

$$V_P = V_S\sqrt{\frac{1-\nu}{0.5-\nu}} \qquad (3)$$

The far-field velocity of the Rayleigh wave, V_R, is related to the velocities of P and S waves as (Achenbach, 1975),

$$\left[2-\left(\frac{V_R}{V_S}\right)^2\right]^2 - 4\cdot\left[1-\left(\frac{V_R}{V_P}\right)^2\right]^{1/2}\left[1-\left(\frac{V_R}{V_S}\right)^2\right]^{1/2} = 0 \qquad (4)$$

A good approximation for V_R in terms of V_s and Poisson's ratio is (modified from Achenbach, 1975),

$$V_R \cong \frac{0.874 + 1.117\nu}{1+\nu} V_S \qquad (5)$$

These equations permit computing the relative values of V_P, V_s and V_R as a function of Poisson's ratio, as shown in Figure 3. At $\nu=0$, $V_P=\sqrt{2}\,V_s$ and $V_R = 0.874\,V_s$. At $\nu= 0.5$ (which theoretically represents an incompressible material; hence, an infinitely stiff material), $V_P=\infty$ so that $V_P/V_s=\infty$. At $\nu=0.5$, $V_R=0.955\,V_s$. The ratios of body wave velocities (V_P/V_s) typically determined with small-strain seismic tests on unsaturated soil and rock are around ~1.5 to 2.0, which corresponds to Poisson's ratio ~0.10 to 0.33; therefore, the small-strain Poisson's ratio is relatively low.

It is important to note that the S-wave velocity is the same in an infinite medium as in a rod (torsional motion). However, the longitudinal P-wave velocity is different, being $V_P=\sqrt{(M/\rho)}$ in an infinite medium and $V_L=\sqrt{(E/\rho)}$ in a rod. The "L" denotes a longitudinal wave. The relationship between V_L and E is for tests in which wavelengths are much greater than the radius of the rod. For shorter wavelengths, V_L decreases as frequency increases. Also, wave velocity, V, wavelength, λ, and frequency of excitation, f, are related for any type of stress wave as,

$$V = f\lambda \qquad (6)$$

It is also worth mentioning that the terms "elastic" and "small strain" are often used to describe stress waves and associated propagation velocities and moduli when dealing with in situ seismic measurements. These terms are used because transient mechanical disturbances created in situ during testing generate stress waves in geotechnical materials that have maximum strain amplitudes less than 0.0001%. As a result, the stress waves exhibit propagation behavior that is independent of strain amplitude and possess only a minor amount of energy dissipation due to material damping.

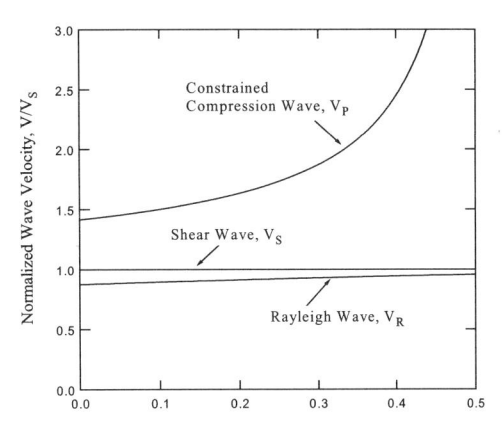

Figure 3. Relationship between stress wave velocities in a uniform half space and Poisson's ratio (from Richart et al., 1970)

2.3 Wave Velocities and Degree of Saturation

The shear wave velocity is related to the shear stiffness of the soil skeleton. In clean coarse sands, where capillary effects are negligible, the effective stress controls the shear stiffness, and the effect of saturation on shear wave velocity is only related to changes in mass density ρ, through $V_s=\sqrt{(G/\rho)}$. The relevance of capillary forces at interparticle contacts on shear stiffness increases with fines content. And, the lower the degree of saturation, the higher G and V_s become (Cho and Santamarina, 2001).

On the other hand, P-wave velocity is controlled by the constrained modulus, M=B+4G/3. Therefore, the fluid and the granular skeleton contribute to V_P. For degrees of saturation, S_r, less than about 99 percent, P-wave velocity is controlled by the stiffness of the soil skeleton in constrained compression in the same fashion as shear waves; that is, the main influence of water on V_P over this range in S_r comes from unsaturated conditions which impact the soil skeleton stiffness. However, if the degree of saturation equals 100 percent, the constrained modulus of this two-phase medium is dominated by the relative incompressibility of the water in comparison to the soil skeleton. The resulting value of V_P varies with the void ratio or porosity, n, the bulk stiffness of the material that makes the grains, B_g, and the bulk stiffness of the fluid, B_f. The bulk stiffness of the fluid phase is very sensitive to the presence of air. Therefore, when the degree of saturation S_r is about 99.5 to 100 percent, the value of V_P is very sensitive to S_r. Figure 4 shows the typical influence of degree of saturation on V_P over this very small change in degree of saturation (the shear wave velocity remains unaffected by such a small change in saturation). For completeness, it is also noted that the impact of S_r on V_s and V_P of rock is very small (only a few percent change) for S_r going from zero to 100%.

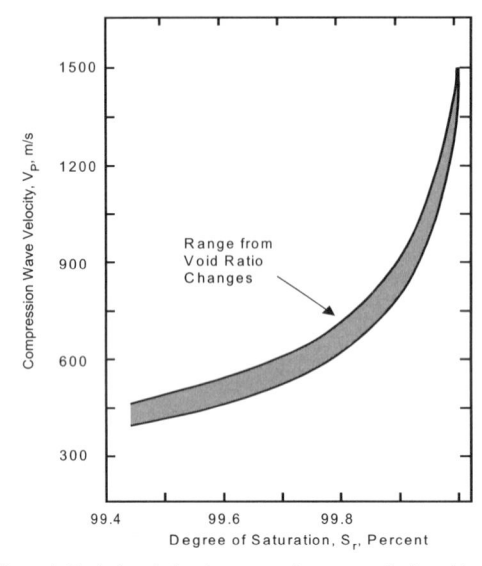

Figure 4. Typical variation in compression wave velocity with degree of saturation changing from 99.4 to 100 % for sand (after Allen et al., 1980)

2.4 Traditional Field Methods

Field testing methods can be classified as active or passive. Active-type methods are generally employed in geotechnical engineering. In this case, a wave is radiated into the medium from a source that is energized as part of the test. Passive-type methods are used less frequently. However, a passive system can be selected when background noise can be used as the excitation source. Field testing methods can also be classified as nonintrusive if all instrumentation is placed on the ground surface, or intrusive when boreholes or penetrometers are used. The most common stress-wave based methods in field use today are briefly reviewed below.

2.4.1 Nonintrusive, active methods

Nonintrusive methods have many advantages including: (1) elimination of the time and cost of drilling, (2) avoidance of potential environmental consequences of drilling, and (3) effective coverage of large areas. These methods include surface refraction, surface reflection, and surface waves as illustrated in Figure 5.

Surface Reflection Method - The surface reflection method is one of the oldest and most common seismic methods. This method is well documented in numerous textbooks in geophysics (e.g., Dobrin and Savit, 1988, and Burger, 1992). The main principle of the seismic reflection method is illustrated in Figure 5a which shows one arrangement of the source and receivers. Both the source and receivers are placed on the ground surface. Typically, compression wave measurements are performed using either mechanical sources that are vertically oriented or explosive sources. Waves reflected from interfaces at depth are monitored with vertically-sensitive geophones. The main purpose of testing is typically to identify and approximately locate key interfaces at depth.

Surface Refraction Method - The surface refraction method is an established geophysical method for nonintrusively identifying sediment stiffnesses and layer interfaces at depth. The method is based on the ability to detect the arrival of wave energy that is critically refracted from a higher velocity layer which underlies lower velocity sediment. Seismic signals are generated with an active source, and wave arrivals are detected on the surface with an array of receivers as shown in Figure 5b. As with the surface reflection method, compression wave measurements are typically performed using vertical mechanical sources or explosives. The arrivals of refracted waves on the ground surface are monitored with vertically-sensitive geophones.

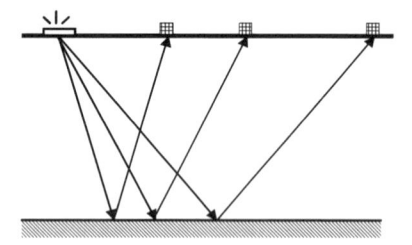

a. One surface reflection arrangement: normal moveout (NMO)

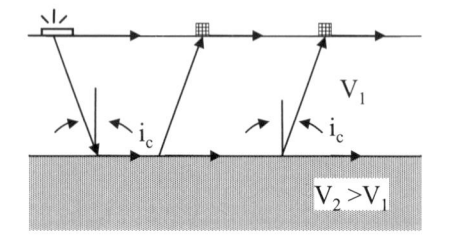

b. Surface refraction testing

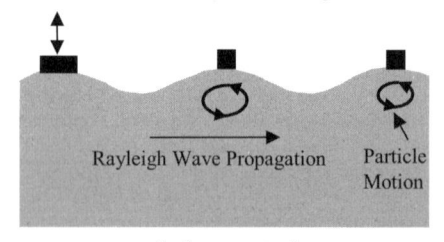

c. Surface-wave testing

Figure 5. Generalized field arrangements used in noninstrusive, active seismic methods

Surface-Wave Method – The surface-wave method can involve Rayleigh and Love waves, and testing has been conducted on land and offshore (Stokoe et al., 1994, Stoll et al., 1994, Tokimatsu, 1995, and Luke and Stokoe, 1998). Most testing in geotechnical engineering involves R waves. Several variations of the generalized method are currently being used or are under development as discussed in Section 3. The most common approach used on land is called the spectral-analysis-of-surface-waves (SASW) method. This test method involves actively exciting Rayleigh wave energy at one point and measuring the resulting vertical surface motions at various distances (receiver points) away from the source as illustrated in Figure 5c. Measurements are performed at multiple source-receiver spacings along a linear array. The generalized method and variations under development have tremendous potential in geotechnical engineering and are therefore discussed in more detail in Section 3.

2.4.2 *Intrusive, active methods*

Intrusive active methods have been widely used in geotechnical engineering, particularly in soil dynamics and geotechnical earthquake engineering beginning in the 1960s. The crosshole and downhole methods were initially developed/adapted for use followed by development of the seismic cone penetrometer (SCPT) and suspension logger. Each method is briefly discussed below.

Crosshole Method - Shear and compression wave velocities are determined from time-of-travel measurements between a source and one or more receivers in the crosshole method. Testing is generally conducted by placing the active source and receivers at the same depth in adjacent boreholes, as illustrated in Figure 6a. The times of travel from the source to the receivers, called direct travel times, and the times of travel between receivers, called interval travel times, are measured. Vertically oriented impacts with mechanical sources are usually applied to the borehole wall using a wedged source. Vertically oriented receivers are used to monitor horizontally propagating shear waves with vertical particle motion; hence SV waves. Radially oriented receivers are used to monitor horizontally propagating P waves. Compression and shear wave velocities are determined by dividing the borehole spacings at the testing depth by the respective travel times. The test is repeated at multiple depths to compile a complete profile of shear and compression wave velocities versus depth.

There are several strengths associated with crosshole testing. First, the source and receivers are placed closed to the material/target to be evaluated, thus enhancing resolution. Second, measurements can also be gathered along multiple inclined ray paths which can be processed together to render a tomographic image of the cross section (Menke, 1989, and Santamarina and Fratta, 1998). Third, P, SV and SH waves can be generated and measured. (SH waves are shear waves with particle motions in the horizontal direction.) The main disadvantage in crosshole testing is the time and cost associated with drilling boreholes; however, ongoing developments in penetrometer-deployed sources combined with effectively deployed receivers promise efficient crosshole implementations under the appropriate soil conditions (e.g., Fernandez, 2000).

Downhole Method - In the downhole method, the times for compression and shear waves to travel between a source on the surface and points within the soil mass are measured. Wave velocities are then calculated from the corresponding travel times after travel distances have been determined. Travel distances are typically based on assuming straight ray paths between the source and receivers, although the analysis may sometimes account for refracted travel paths. Figure 6b shows a conventional setup which requires the drilling of only one borehole. One of the main advantages of the downhole method in comparison to the crosshole method is the need for only one borehole, so the cost is less. However, the disadvantage is that wave energy has to travel increasingly larger distances as the depth of testing increases. In the writers' experience, the optimum testing depths range from about 10 m to 50 m unless specialized personnel are involved. This depth is, of course, dependent on the energy developed by the source (various high-energy, mechanical sources have been constructed, e.g., Liu et al., 1988).

Seismic Cone Penetrometer - The cone penetrometer (CPT) is a well established tool for characterizing soil properties by measuring tip and side resistances on a probe pushed into the soil (Lunne et al., 1997). The SCPT test is a modification of the cone penetrometer test that allows measurement of shear wave velocities in a downhole testing arrangement (Campanella et al., 1986). Seismic energy is generated at the surface near the insertion point of the cone. Usually a horizontal impact on an embedded anvil is used to generate the SH waves. Travel times of the shear wave energy, either direct or interval, are measured at one or more locations above the cone tip as shown in Figure 6c. After testing at one depth, the cone is penetrated further into the soil, and the test is repeated. One of the important benefits of this method is that the seismic data can be combined with the cone resistance values to build a clearer picture of both soil type, strength, stiffness, and layering. This is an excellent example of using multiple techniques to investigate sites.

Suspension Logger - Logging tools can be lowered into a borehole to determine material properties with stress waves, electromagnetic waves, gamma radiation, and other physical principles. The main limitations in borehole logging are the effect of casing and drilling fluids on the measured response and

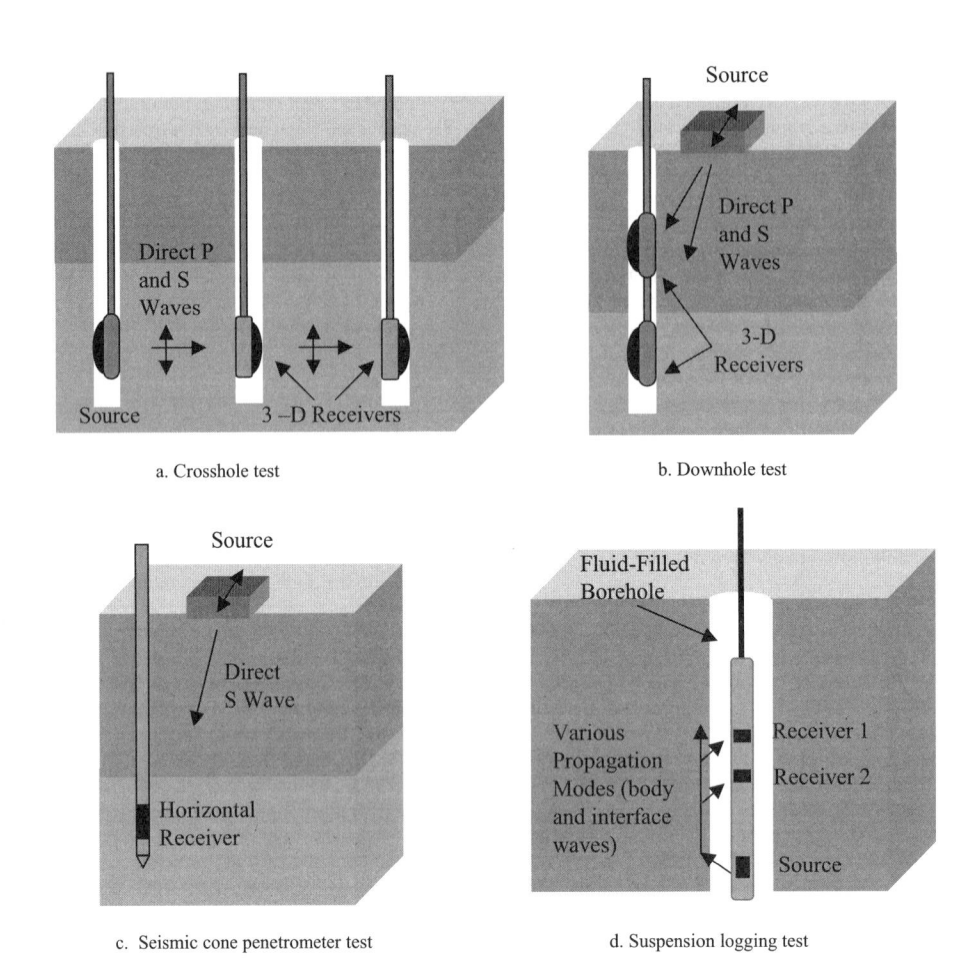

Figure 6: Field arrangements used to perform intrusive seismic tests (from Stokoe and Santamarina, 2000)

the depth scanned by the technique relative to the zone affected by drilling the borehole. One of the more recent advances in borehole shear wave methods is the suspension logger (Kitsunezaki, 1980, Toksoz and Cheng, 1991, and Nigbor and Inai, 1994). This test is performed in a single, mud-filled borehole. The device is lowered on a wire line into the borehole, and seismic energy is generated and received by a receiver array in the borehole as shown in Figure 6d. The shear and compression wave velocities of the surrounding material are determined from the arrival times of these waves following standard travel-time procedures. One of the advantages of this method is that the wire-line nature of the test allows for measurements at significant depths (hundreds of meters). Two drawbacks of the method are that it generally can not be performed in a steel or thick plastic casing if soft soils are to be tested and it does not work well within about 7 m of the ground surface.

2.5 *Additional Information*

Most of the information presented above was extracted from the article by Stokoe and Santamarina, 2000. The information is briefly presented to facilitate the following discussion. However, much more information is available in the article and in the literature because of the strong theoretical bases upon which seismic and other geophysical measurements are founded. Textbooks such as Richart et al. (1970), Aki and Richards (1980), Ward (1990), Sharma (1997), and Santamarina et al. (2001) are excellent references. Manuals such as ASCE Press (1998) and NRC (2000) are also good references. Many important topics such as material damping, geotechnical spreading, near-field effects, mode conversions, effects of stress state on wave velocities, inherent and stress-induced anisotropies could not be covered herein but can be important in properly applying these seismic methods.

3 OVERVIEW OF EVOLVING SURFACE-WAVE METHODS

There is one field seismic method that is under active development and deployment today. That method is the surface-wave method. The great interest in this method arises, in large part, from the noninstrusive nature of the method combined with the capabilities of imaging softer layers beneath stiffer materials and testing large areas rapidly and cost effectively. One test configuration is described in Section 2.4.1 and illustrated in Figure 5c. The method, in terms of a generalized method, can be divided into two basic parts: (1) monitoring Rayleigh-wave propagation along the ground surface ("field testing"), and (2) empirical or numerical modeling of the field measurements to yield the subsurface V_s profile. The objective of field testing is to determine the phase-velocity dispersion curve for the test site. This objective is discussed in detail below as are the different approaches to meeting this objective. Modeling of the field measurements, either empirical or numerical, can vary significantly from one analysis procedure to another. The strengths and limitations of the modeling procedures are discussed in Section 3.2. However, since the forward modeling theory of surface-wave propagation was introduced by Thomson (1950) and Haskell (1953), empirical procedures should no longer be used to analyze the field measurements.

3.1 *Surface-Wave Techniques: Determination of Phase-Velocity Dispersion Curves and Associated Modeling Approaches*

The steady-state, Rayleigh-wave method (Richart et al., 1970) is one of the initial surface-wave methods that was used in geotechnical engineering, which directly measure the wavelengths of Rayleigh waves for the determination of phase velocities (velocities associated with wavelengths or frequencies). The two-station method (Landisman et al., 1968 and Sato, 1971) is another surface-wave method based on the inter-station phase difference. Sato used a transfer function to determine phase velocities of surface waves for a range of frequencies, and Landisman et al. used an inter-station cross-correlogram to eliminate the adverse effects of low-energy noises. In the 1980s, the University of Texas at Austin (Heisey, et al., 1982, Nazarian and Stokoe, 1984, and Stokoe and Nazarian, 1985) established the SASW method to determine phase-velocities of Rayleigh waves. The SASW method was an innovative method to make faster and more efficient measurements than any previous methods. In addition, the dynamic stiffness matrix method (Kausel and Roësset, 1981) became the theoretical basis for modeling the phase-velocity dispersion curves to yield the V_s profile of the site. Advances continue to occur in this aspect of the test (Roësset, et al., 1991, Gucunski and Woods, 1991, Al-Hunaidi, 1994, Al-Hunaidi and Rainer 1995, and Joh, 1996).

Recently many other methods have been developed for determination of phase-velocity dispersion curves in the generalized surface-wave method. The four most widely used methods are the spectral-analysis-of-surface-waves (SASW) method, the frequency-wave number (f-k) spectrum method, the multi-channel analysis of surface wave (MASW) method, and the continuous surface wave (CSW) method. Currently, the SASW method is used around the world including Asia and Europe, the MASW method is mostly used in the America and some Asian countries, and the CSW method is actively used in the United Kingdom, Australia and some Asian countries. In Table 1, general features of each method are compared. Advantages and disadvantages of each method are summarized in Table 2. In the following sections, the fundamental principles of these methods are described and some important issues are discussed.

3.1.1 *SASW method*

In the SASW method, the dispersive characteristics of Rayleigh waves propagating through a layered material are measured and then used to evaluate the S-wave profile of the material (Stokoe et al., 1994). SASW measurements involve generating waves at one point on the ground surface and recording them as they pass by two or more locations, as illustrated in Figure 7a. All measurement points are arranged along a single radial path from the source. Measurements are performed with several (typically six or more) sets of source-receiver spacings. In each set, the distance between the source and first receiver is kept equal to the distance between receivers. The phase shift versus frequency relationship is measured for surface waves propagating between the receivers for each receiver spacing. A typical phase plot is shown in Figure 7b. From each phase plot, the phase velocity of the surface wave is calculated at each frequency knowing the frequency, phase angle and distance between the receivers. The result is a plot of phase velocity versus frequency for a given receiver spacing, called an individual dispersion curve (Figure 7c). This procedure is repeated for all source-receiver spacings used at the site and typically involves significant overlapping in the dispersion data between adjacent receiver sets. The individual dispersion curves from all receiver spacings are combined into a single composite dispersion curve called the experimental or "field" dispersion curve (Figure 8). Once the composite dispersion curve is generated for the site, an iterative forward modeling procedure or an inversion analysis algorithm is used to determine a shear-wave velocity profile by matching the field dispersion curve with the theoretically-determined dispersion curve (Figure 8).

Table 1. Key features of four, widely used surface-wave methods

Surface-Wave Method	Key Features
SASW method	• phase velocities from phase differences • two to four receivers typically used • superposed-mode phase velocity (apparent phase velocity) • global property over receiver-spread area • shear-wave velocity profile from the apparent phase velocities • comprehensive forward modeling or inversion analysis • impulsive source, swept-sine source, or random vibration source
f-k spectrum method	• phase velocities from frequency-wave number spectrum • multiple receivers (e.g. 128, 256, etc. receivers) • fundamental and higher-mode phase velocities • global property over receiver-spread area • shear-wave velocity profile from fundamental and higher modes • impulsive source
MASW method	• limited number of receivers (usually 24 receivers) • fundamental and higher-mode phase velocities • walk-away measurement • same measurement configuration as common-midpoint reflection survey • global property over receiver-spread area • shear-wave velocity profile from the fundamental mode • impulsive source or swept-sine source
CSW method	• phase velocity from the average phase-angle slope over receiver-spread area • four to six receivers used • superposed-mode phase velocity (apparent velocity) • global property over receiver-spread area • shear-wave velocity profile from the apparent velocities • steady-state harmonic source

Table 2. Advantages and disadvantages of four, widely used surface-wave methods

Surface-Wave Method	Advantages	Disadvantages
SASW method	• good sampling of shallow material • more sensitive measurements for layer stiffness contrast, using apparent velocity inversion analysis	• multiple measurements using different source-receiver configurations are required • expertise required for phase unwrapping and forward modeling
f-k method	• dispersion curves separated for fundamental and higher modes • body-wave effect extracted • dispersion curve global to the receiver-spread area	• aliasing problem in wave number domain • inaccurate mode separation in case of poor resolution in f-k spectrum • large number of traces required for good resolution in wave-number domain • limitation due to topographic constraint and instrumentation capability • long measurement time
MASW method	• mode separation of surface waves	• aliasing problem in wave-number domain • use of the fundamental mode only in inversion analysis
CSW method	• the effects of local anomalies minimized with the use of average phase-angle slope • no expertise required to calculate phase velocity • reliable measurements with controlled source	• dedicated inversion analysis required but not used • near-field effects included • exploration depth limited • frequency-content of vibrator is limited

The SASW method is a simple technique that is easily implemented in terms of field testing. The requirement of several measurements using different source-receiver configurations is time and labor intensive. However, the multiple source-receiver configurations employ multiple sources which are selected appropriately for the measured wavelength range at each source-receiver configuration, Therefore, time- and labor-intensive measurements preferably lead to high-quality results.

The SASW method measures apparent phase velocities, which correspond to the superposed mode of higher-mode surface waves and body waves. Determination of apparent phase velocities incorporates phase unwrapping. In a complicated multi-layered system, phase unwrapping may be non-systematic and sometimes requires expertise, which is cumbersome to inexperienced personnel. However, the non-systematic nature of phase unwrapping can be improved by a signal processing technique such as the impulse-response filtration technique (Joh et al., 1997) and Gabor spectrum.

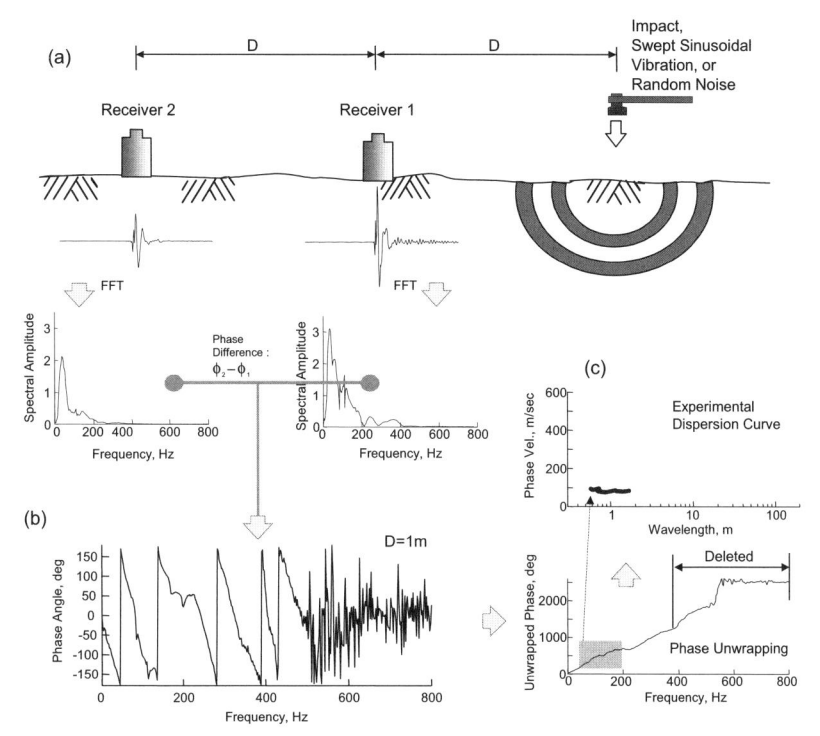

Figure 7. Spectral-analysis-of-surface-waves (SASW) method: Calculation of phase velocities

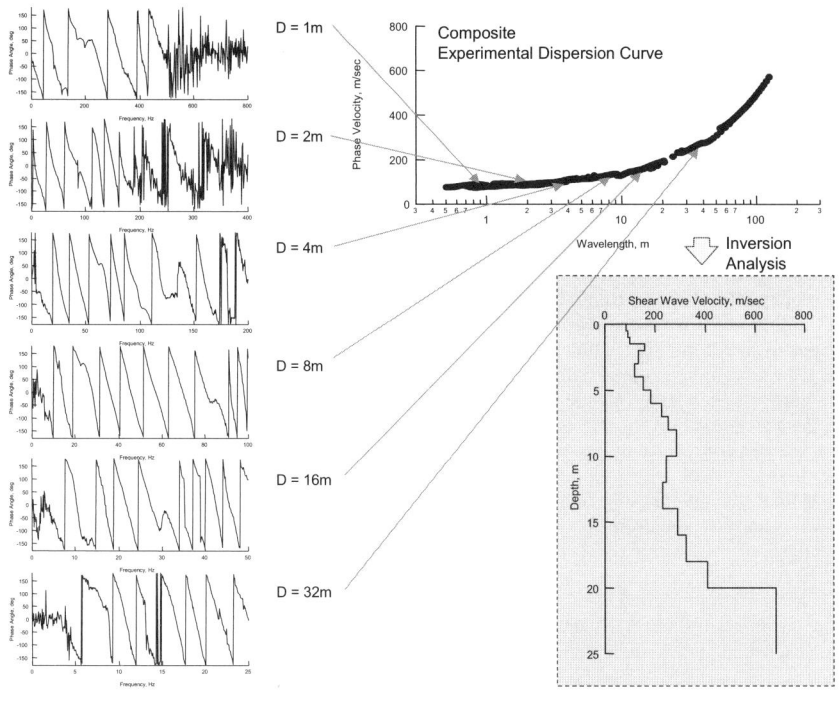

Figure 8. Spectral-analysis-of-surface-waves (SASW) method: Determination of a dispersion curve and shear-wave velocity profile

Importantly, the SASW method uses the apparent phase velocity dispersion curve along with source and receiver locations in the forward modeling or inversion analysis. The dynamic stiffness matrix method (Kausel and Roësset, 1981), which is the forward modeling algorithm used in the matching or inversion process, can simulate the apparent phase velocity specific to the source-receiver configuration. The inversion analysis based on apparent phase velocities and the dynamic stiffness matrix method are key features of the SASW method, which improves the reliability and accuracy of the shear-wave velocity profile.

3.1.2 *Frequency-wave number (f-k) spectrum method*

The frequency-wave number (f-k) method is another method which has been widely used in the geophysical area and recently adopted for geotechnical engineering applications. In the f-k method, the propagating surface waves are measured at a significant number of locations in a line with the source. The measurement of propagating surface waves at many sequential locations in a line can reveal the wavelengths of the surface waves, which are basically the reciprocals of the wave numbers. Along with the frequency information obtained from the time-domain waveform, the wave number information is used to determine phase velocities.

To describe the f-k spectrum method, a total of 256 synthetic seismograms were generated using the dynamic stiffness matrix method. The layered system shown in Figure 9a was used as the model profile. The stacked traces in the time-space domain are shown in Figure 9b. These results can be transformed to the frequency-wave number domain by means of a 2-D FFT or slant-stack analysis (McMechan and Yedlin, 1981). Figure 9c displays the frequency-wave number (f-k) contour plot transformed from the time-space domain data in Figure 9b. The fundamental and higher modes of the surface-wave propagation are identified by the ridge analysis of the frequency-wave number contour plot. In the frequency-wave number contour plot, the modes of the surface-wave propagation refer to different wave numbers for a given frequency, and correspond to the loci identified by the ridge analysis. Figure 9d is the phase-velocity dispersion curve determined from the modes identified in Figure 9c. This approach to determine phase velocities from the frequency-wave number spectrum is called the f-k spectrum analysis (Gabriels et al., 1987).

The f-k spectrum method is superior to any other method in characterizing the fundamental and higher modes from the measured surface wave. However, the required use of numerous receivers is the main disadvantage both for repetitiveness problems and for required testing time. Also, the data acquisition required for a large number of traces may be expensive, and the topographic constraints may limit the reliability of the measurements.

3.1.3 *Multi-channel analysis of surface waves (MASW) method*

In the MASW method (Park et al., 1999, and Miller et al., 1999), a large array of time traces is measured using a swept-sine vibratory source or an impulsive hammer, using the walk-away method (Figure 10). The basic field configuration and acquisition procedure for the MASW measurements is generally the same as the one used in conventional common midpoint (CMP) body-wave reflection surveys. In the MASW method, the dispersion curve can be determined in two approaches: the swept-frequency record approach and the frequency-wave number spectrum approach. In the swept-frequency record approach shown in Figure 11a, the linear slope of each component of a swept-frequency record is determined and used to calculate the phase velocity. The frequency-wave number spectrum shown in Figure 11b is almost the same approach as the frequency-wave number spectrum method described in the Section 3.2. In the frequency-wave number spectrum method, the ridge of the frequency-wave number contour plot is identified and used to determine the phase velocity from the relationship among frequency, wave number and phase velocity. On the other hand in the MASW method, the phase velocity-frequency contour plot is first determined from the frequency-wave number contour plot and then the ridge of the phase velocity-frequency contour plot is identified for the calculation of the phase velocity corresponding to a frequency.

The MASW method uses only the fundamental mode for the inversion analysis. For the site with a normally dispersive dispersion curve, in which phase velocities increase with increasing wavelength, the fundamental mode alone may be enough to resolve the layer stiffness reliably. However, for a typical geotechnical site with a more complex stiffness profile, where the measured dispersion curve may be inversely dispersive or heavily fluctuating with a up-and-down pattern, the inversion analysis using the fundamental-mode only can not work well (Tokimatsu et al., 1992). To make the MASW method a reliable exploration method, it is crucial to incorporate higher modes as well as the fundamental mode in the inversion analysis. Recently, an effort to use higher modes in the inversion analysis was made (Kansas Geological Survey, 2003).

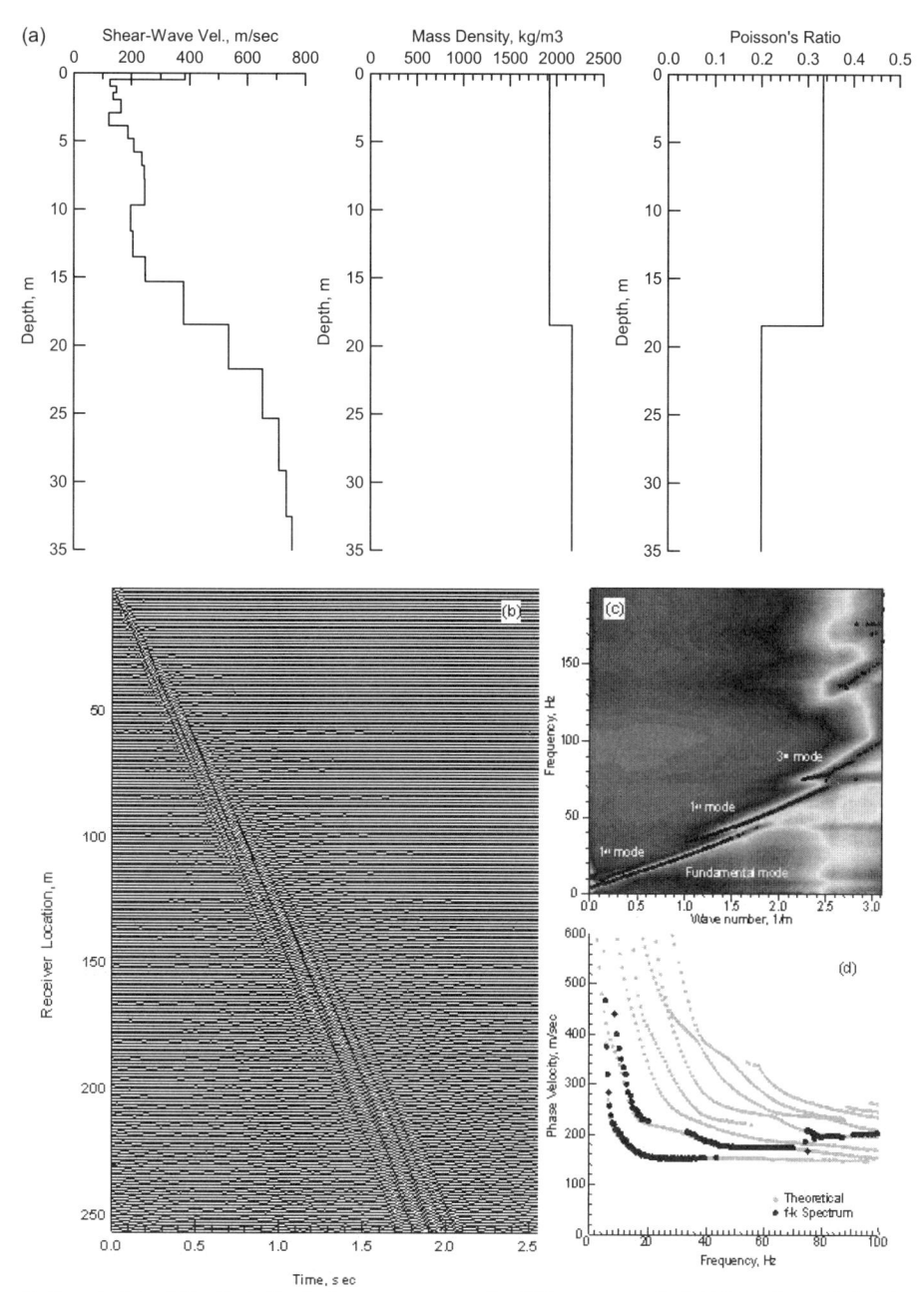

Figure 9. Numerical simulation illustrating the frequency-wave number (f-k) spectrum method: (a) layered geotechnical site, (b) synthetic seisograms, (c) f-k contour plots, and (d) phase-velocity dispersion curve.

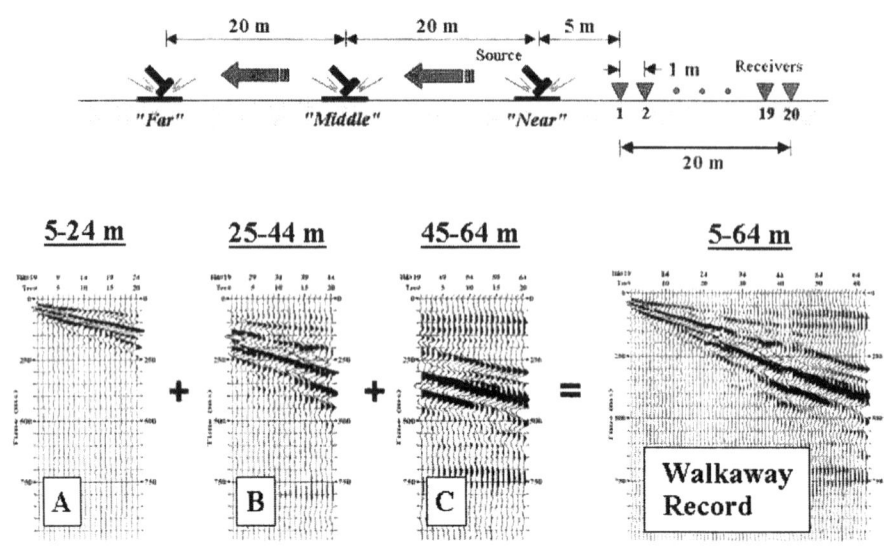

Figure 10. Multichannel analysis of surface waves (MASW) method: walk-away method for measuring a large array of traces (Kansas Geological Survey, 2003)

3.1.4 *Continuous surface wave (CSW) method*
The continuous surface-wave (CSW) method is a geophysical exploration technique to evaluate the subsurface stiffness structure using a vibrator and more than four receivers, as depicted in Figure 12. Since the CSW method was initiated by British researchers (Matthews et al, 1996, and Menzies and Matthews, 1996), it has been used in Europe, Australia and some Asian countries. Unlike other surface-wave methods, the CSW testing only uses a vibrator to generate surface waves. The application of the CSW method is limited to shallow stiffness profiling like compaction-quality control, because the vibrator source does not generate enough energy for sampling deep material.

The CSW testing shown in Figure 12 uses four geophones to measure the particle-velocity history of the ground for sinusoidal vibration induced by the vibrator. The geophones are placed in a linear array with an equal spacing. Sometimes five or six geophones are used to improve the accuracy of the measurement. The time history of particle velocity that is measured at each geophone is transformed into the frequency domain by Fourier transformation. And the phase angle is determined for an excitation frequency at each geophone. Then, the phase angles are plotted against the location of the geophone, as shown in the lower portion of Figure 12. If the soil is homogeneous, the phase angle should have the tendency to linearly increase with the distance from the source. In some cases, the phase angle goes over 180 degree or below -180 degree because, the phase angle is wrapped to fall between -180 degree and 180 degree due to the nature of Fourier transformation. Usually the phase wrapping can be easily identified in the plot of the source-to-receiver distance versus the phase angle, which has more than two parallel lines. In the case with wrapped phase angles, the phase unwrapping operation can be applied to recover the original phase angles.

After the phase velocities are determined for all the excitation frequencies, the shear-wave velocity profile can be determined from an empirical relationship or an inversion analysis like the one for the SASW method. Presently an empirical analysis is used. One advantage of the CSW method is to use an average phase-angle slope. The average of the phase-angle slop eliminates the local anomalies which may mislead the evaluation of the global S-wave velocity profile. The other advantage of using the average phase-angle slope is that expertise is not needed in determining the phase velocities, which enables the automation of the phase-velocity calculation. The controlled source like a electromechanical vibrator allows reliable measurements only in the frequency range compliant to the vibrator specification, and measurements of frequencies out of the vibrator specification lose reliability. This indicates that very shallow and very deep materials can not be sampled, which turns out to be a disadvantage of the CSW method. Also, the measurement time is usually long compared with other surface-wave methods. Finally, the CSW method needs to be more refined in that an inversion analysis specific to this method needs to be developed for reliable use in the future.

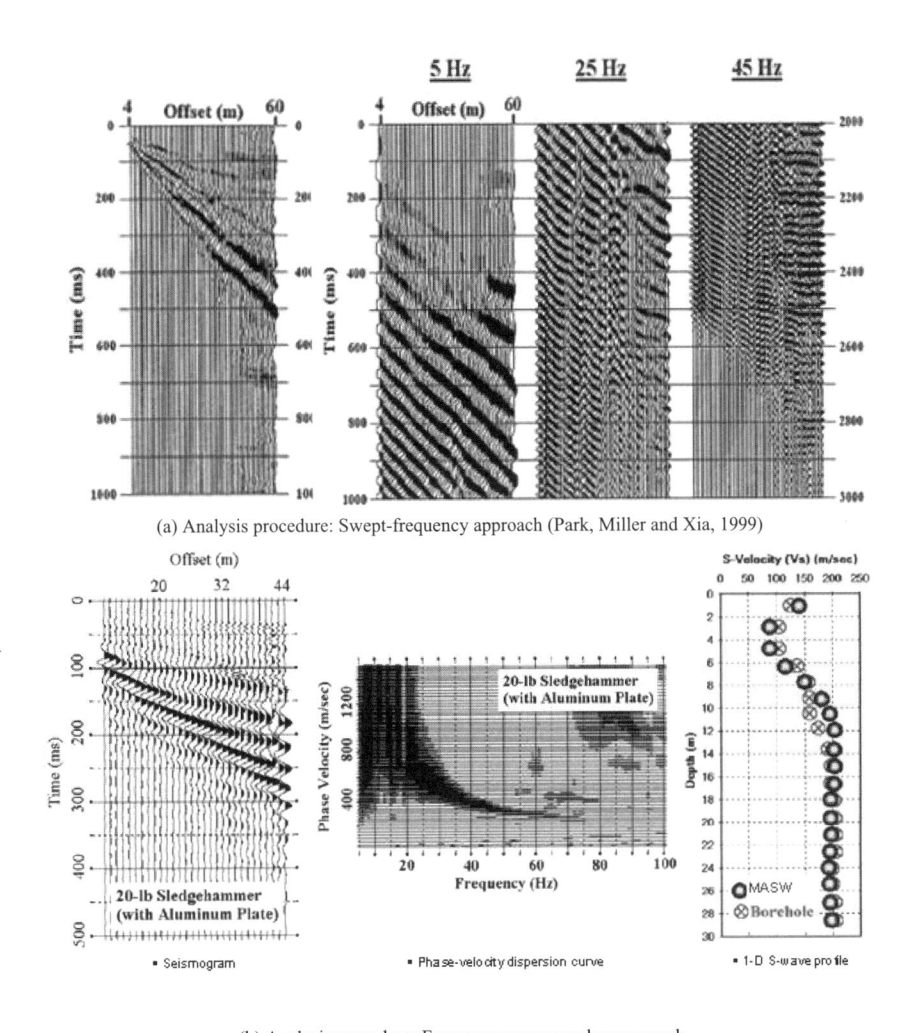

(a) Analysis procedure: Swept-frequency approach (Park, Miller and Xia, 1999)

- Seismogram
- Phase-velocity dispersion curve
- 1-D S-wave profile

(b) Analysis procedure: Frequency-wave number approach

Figure 11 Multichannel analysis of surface waves (MASW) method: Analysis Procedure (Kansas Geological Survey, 2003)

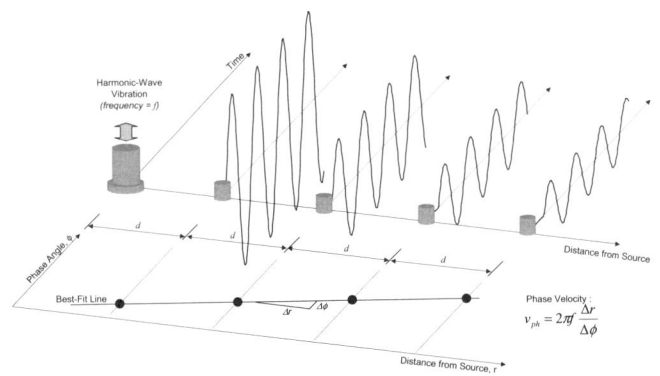

$$v_{ph} = 2\pi f \frac{\Delta r}{\Delta \phi}$$

Figure 12 Continuous surface wave (CSW) method

3.2 Theoretical Aspects Associated with the Surface-Wave Methods

Most surface-wave methods as applied today are sophisticated in measurement and analysis, and therefore give rise to important issues in terms of theoretical background aspects. In this section, two major issues related to the surface-wave methods are discussed. These issues are the forward modeling procedure and the inversion analysis; that is, how to theoretically calculate phase velocities for a given layered system and how to evaluate a shear-wave velocity profile from a measured dispersion curve. These issues are important topics for better and more reliable profiling of subsurface stiffness.

3.2.1 Higher-mode velocities and apparent velocity

Fundamental and higher modes in surface-wave propagation are the distinctive features in a multi-layered system. When surface waves propagate through a multi-layered system, the stiffness of each layer affects the propagation of the surface waves (Gucunski and Woods, 1992, and Al-Hunaidi and Rainer, 1995). The different stiffnesses in the layers may confine the stress waves in some layers or cause multiple refractions and reflections, leading to different ray paths, which result in different propagation velocities even for the same frequency. Both the transfer matrix method (Thomson, 1950, and Haskell, 1953) and the dynamic stiffness matrix method (Kausel and Roësset, 1981) can determine fundamental- and higher-mode velocities. These modes correspond to plane waves in 2-D space, and can be calculated from the eigenvector analysis of the transfer matrix or the dynamic stiffness matrix.

The superposed mode in surface-wave propagation is also an important feature, because this mode is actually generated during testing. The superposed mode corresponds to the 3-D wave propagating in a cylindrical pattern, not like the planar pattern of a 2-D wave. This propagation is often observed when the source is close to the receivers and the wavefront still has a significant cylindrical pattern. The superposed mode does not fall into specific normal modes, but is somewhere between the normal modes. The superposed mode is often called an apparent velocity or an effective velocity. Calculated or measured apparent phase velocities are dependent on the actual locations of the source and receivers.

In surface-wave measurements, there are two different approaches in terms of using surface-wave modes. The SASW and CSW methods use the superposed mode, because the source is close to the receivers and mode separation of the measured surface waves is not practical. The f-k spectrum and MASW methods use the fundamental and higher modes, because the source is far enough from the receivers and the modes are well separated. Specifically, the f-k spectrum and MASW methods use a large array of receivers which helps the separation of surface-wave modes.

Figure 13 shows the differences among the normal-mode solution, 2-D solution and 3-D solution of propagating surface waves. The 2-D solution is different from the 3-D solution in that the 2-D velocity is a superposed-mode velocity of plane Rayleigh-wave modes without body-wave interference (Roësset et al, 1991). The dynamic stiffness matrix method was used to calculate theoretical phase velocities for the layered systems in Figure 13. Case 1 is a soil system with increasing stiffness with depth, and Case 2 is a soil system with a soft layer trapped between a harder surface layer and a half-space. Phase velocities were calculated for: (1) different modes of plane Rayleigh waves, (2) the 2-D solution of a plane Rayleigh wave, and (3) the 3-D solution, which is an apparent dispersion curve for the cylindrical Rayleigh wave. The contribution of different modes to the simulated dispersion curve is presented in Figure 13b. In the case of the soil system with increasing stiffness with depth, the apparent dispersion curve essentially coincides with the fundamental mode of the Rayleigh wave over the complete frequency (hence wavelength) range. However, in Case 2, the 2-D and 3-D solutions in the higher-frequency region (smaller-wavelength region) are not just from one mode but a superposition of several modes. Also, it is important to realize that: (1) the 2-D solution resides between the modes of a plane Rayleigh wave, and (2) the 3-D solution may become lower than the fundamental mode at low frequencies. This phenomenon is probably due to the multiple reflections and refractions of body waves. The comparison of the normal-mode solution, 2-D solution and 3-D solution in Figure 13b implies that the 3-D solution may be the closest to the actual measurements contaminated with body waves and higher-mode Rayleigh waves.

3.2.2 Inversion analysis to evaluate a shear-wave velocity profile

In the surface-wave methods, two different categories of inversion analysis are available, dependent on the type of experimental dispersion curve. The first one is to use the normal-mode solution. The f-k spectrum method and the MASW method belong to this category. Most of the available inversion techniques are based on this approach (Hossian and Drnevich, 1989, Addo and Robertson, 1992, Yuan and Nazarian, 1993, and Xia, et al., 1999). In this category, the most crucial step is to well separate fundamental and higher modes. Specially in the high-frequency range, the phase velocities for

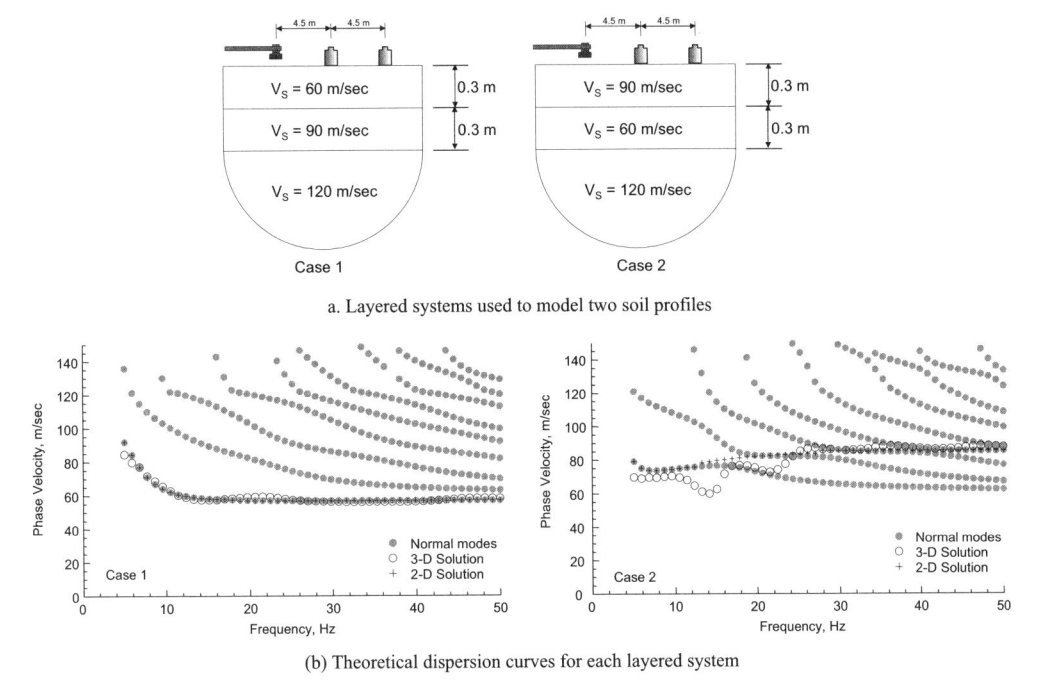

Figure 13 Contribution of different modes of the Rayleigh wave to the 3-D solution of wave propagation

normal modes are very close to each other. Therefore, if the field measurement configuration is not good enough to differentiate modes, the inversion analysis may end up with misleading results. Figure 9c is a good example of difficulty in resolving lower modes in the high-frequency region. In the high-frequency region, the evaluated mode is not necessarily the fundamental mode, but is one of the higher modes. In this case, it is almost impossible to identify which higher-mode the measured mode belongs to. Therefore, the inversion analysis using the normal-mode solution needs to focus on only the low-frequency region to avoid the problem in miscounting the normal-mode number.

In the second category of inversion analysis, the apparent phase-velocity dispersion curve is (or should be) used. The SASW and CSW methods belong to this category (Gucunski and Woods, 1991, Rix and Leipski, 1991, Tokimatsu et al, 1992, Joh, 1996, and Ganji et al., 1998). In this case, it is very important to calculate the apparent theoretical phase velocity. The apparent theoretical phase velocity should be calculated using the exact locations of source and receivers, which can have a significant influence on the resulting phase-velocity dispersion curve. Several sets of the experimental dispersion curves from different sets of receiver combinations should be included to evaluate the layer stiffness contrast more reliably.

In inverting the dispersion curves determined in the SASW or CSW method, it is more beneficial to incorporate information on the source and receiver locations rather than to neglect them by assuming the measured phase velocities are far-field velocities. Figure 14 compares the resulting shear-wave velocities of two approaches: (1) the global inversion analysis, and (2) the array inversion analysis (Joh, 1996).

In the global inversion analysis, information on the source and receiver locations is ignored, and it is assumed that the receivers are located in the far field. In this inversion analysis, the theoretical phase velocities are calculated for receivers deployed at virtual locations of 2λ and 4λ (λ is wavelength for a specific frequency) and optimized to match the general trend of the dispersion curve. On the other hand, the array inversion analysis uses the phase velocities specific to the source and receiver locations, and finds the optimum shear-wave velocity profile to match all the individual experimental dispersion curves with theoretical dispersion curves corresponding to each source-receiver configuration. As shown in Figure 14b, the array inversion analysis made a fit between five experimental dispersion curves with the corresponding theoretical dispersion curves, while the global inversion analysis made a fit to follow the general trend of the experimental dis-

a. Global inversion

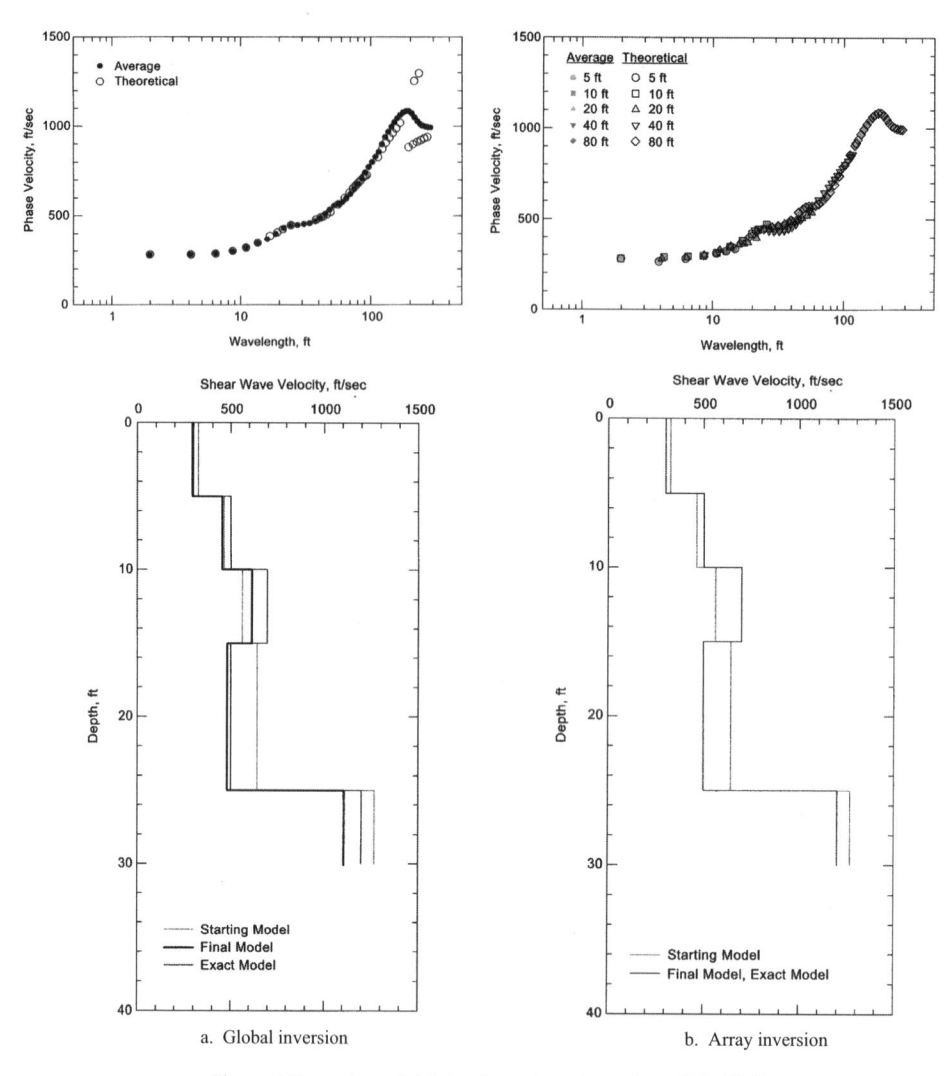

b. Array inversion

Figure 14 Comparison of global and array inversion analyses (Joh, 1996)

persion curve. The resulting shear-wave velocities also show the superiority of the array inversion analysis. The array inversion analysis was able to produce the shear-wave velocity profile almost the same as the exact model assumed to generate the synthetic dispersion curves. However, in some cases, environmental noise and undesirable effects due to lateral geologic variability may intervene into real measurements so that this approach may not work perfectly and needs to be applied with care.

4 CASE HISTORIES AND APPLICATIONS

The purpose of this section is to present some case histories and applications that demonstrate the importance of in situ geophysical methods to the solution of geotechnical engineering problems. The examples focus on the use of in situ seismic measurements, but demonstrate the relevance of geophysical measurements in geotechnical engineering. The examples include problems that involve geosystems loaded statically as well as dynamically and loaded in the linear (small strain) and nonlinear ranges.

4.1 Soil Modulus for Settlement Analysis and Soil Structure Interaction

Settlement predictions/calculations based on the ultimate strength of soils have been made for a century or more. However, beginning in the late 1960s in the construction of nuclear power plants and other very large and heavy structures, it was evident that geotechnical engineers had to find better ways of analyzing the deformation behavior of soils including analyses of settlement and soil structure interaction. However, the geotechnical community has made very slow progress in pursuing a rational approach to these analyses. The following four case histories describe some attempts to use elastic modulus derived from seismic wave velocity measurements to estimate the settlement of foundations on sands, gravels, heavily overconsolidated clays, and soft rock.

4.1.1 Case history 1- settlement analysis of large and heavy structures

In an early recognition of the prevailing irrational approach to settlement prediction based on ultimate strength, William Swiger of Stone and Webster in 1974 (Swiger, 1974) suggested an improvement to the geotechnical practice of settle-ment prediction using elastic moduli derived from seismic waves. His description of the problem and potential solution is presented here as the basis for a modest advancement over the past quarter century in use of elastic modulus for settlement prediction.

In a rational approach to settlement (deformation) prediction, stress and strain should be related through modulus. The complication for soils is that soil is a nonlinear material, starting from very low strain levels, so application of any approach using modulus has to recognize and accommodate the nonlinear behavior. Swiger determined, based on calculations for five power plant structures, that the average strain causing settlement under large structures was on the order of 10^{-3} throughout a depth about equal to the minimum dimension of the loaded area. He then outlined an approach for determining a soil modulus at an appropriate strain level.

Swiger pointed out correctly that methods of modulus determination requiring sampling of soils and testing specimens in the laboratory suffered from a major inescapable drawback, sample disturbance. (The subject of sample disturbance is presented in more detail in Section 5.0). He also noted that techniques exist by which modulus could be measured in situ and without disturbance, namely seismic wave velocity measurements. Crosshole and downhole seismic tests were well established by the mid-1970s and Swiger used them to determine small-strain (10^{-6}) soil moduli at multiple depths in the ground.

Swiger acknowledged the benefits of using shear wave velocity (V_s) over compression wave velocity (V_p) because V_s is unaffected by the water table and calculation of shear modulus from shear wave velocity requires only an estimate or measurement of soil density. Shear modulus is calculated using Equation 2. Furthermore, for isotropic materials, shear modulus can be converted to other moduli including Young's modulus or constrained modulus using Equation 1. The simplified equation relating shear and Young's moduli is,

$$E = 2G(1+\nu) \qquad (7)$$

Swiger further pointed out that shear modulus measured at low strain can be adjusted to larger strains through relationships provided by Hardin and Drnevich (1972a and 1972b) or as shown by Seed (1969) in Figure 15. Based on the first cycle of a large load test at the Brookhaven National Laboratories and on modulus derived from crosshole shear wave velocity at the site, he found reasonable agreement in back calculated Young's modulus and E determined from seismic wave velocity at a strain level of 4×10^{-5} as shown in Table 3 for two values of ν. At the time, Swiger decided to use a range in ν but noted that $\nu = 0.3$ seemed more reasonable. Today, we realize that ν in the range of 0.15 to 0.35 is appropriate for the soil skeleton.

Table 3 Moduli calculated from load test and crosshole shear wave velocity (from Swiger, 1974)

Poisson's Ratio	$\nu = 0.3$	$\nu = 0.45$
Seismic Modulus (E)	3.7×10^6 psf	4.2×10^6 psf
Load Test Modulus (E)	3.9×10^6 psf	3.4×10^6 psf

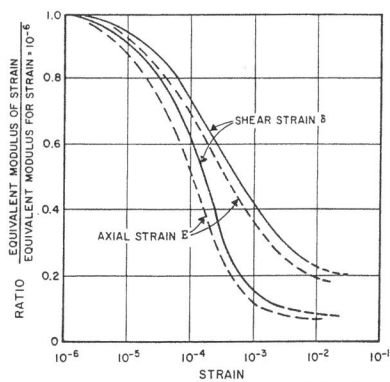

Figure 15. Modulus-strain relations used by Swiger, 1974 (from Seed, 1969)

At the site of the turbine room of the Shippingport nuclear power station, a site underlain by about 60 ft (18m) of medium dense to dense sand and

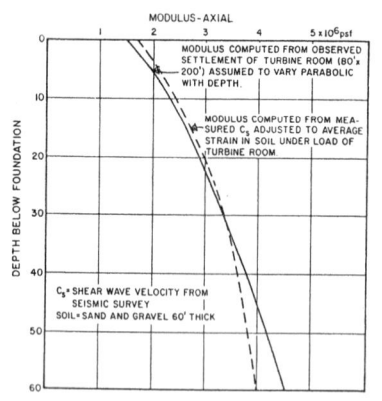

Figure 16. Modulus profiles from observed settlements and seismic measurements (from Swiger, 1974)

gravel, strain-adjusted modulus determined from seismic wave velocity was compared to modulus computed from observed settlement by Swiger as shown in Figure 16. These moduli show very good agreement.

Since the time of Swiger's paper, another important seismic method has become available permitting modulus profiles to be determined without the boreholes or any ground disturbance, namely surface-wave testing as discussed in Section 3. With this nondestructive and nonintrusive method, elastic moduli profiles for homogeneous and layered soil sites can be readily obtained for the purpose of settlement analysis and soil structure interaction. Surface-wave testing mitigates one of the high-cost elements of crosshole and downhole seismic testing, namely boreholes.

4.1.2 *Case history 2 – settlement analysis of a power plant*

Konstantinidis et al. (1986) reported a case study in which moduli developed for prediction of settlement of a power plant were based on the approach suggested by Swiger, 1974. The site consisted of both sand and clay layers of about equal thicknesses to a depth of 200 ft (61 m). Seismic wave velocities measured at the site are presented in Figure 17. Other methods of estimating moduli for settlement prediction considered by Konstantinidis et al. (1986) included the CPT, pressuremeter testing (PMT) and laboratory tests including consolidation and triaxial compression. Short-term settlement measurements (initial elastic settlement) showed about 1 inch (0.68 to 1.19 inch) [17 to 30 mm] of settlement for Unit 1 of a two-unit plant. Unit 2 did not have the same sequence of settlement measurements so it could not be compared.

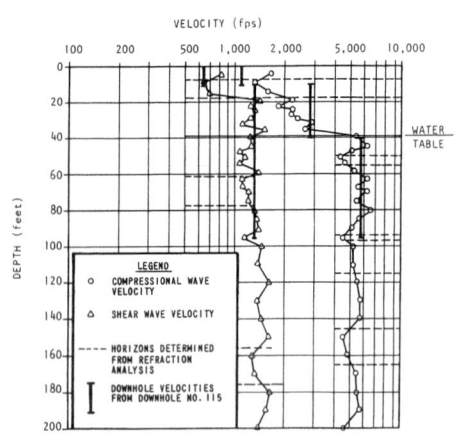

Figure 17. Typical seismic survey results at power plant site (from Konstantinidis et al., 1986)

Moduli from all methods used except moduli from seismic crosshole tests overestimated the measured settlement by a factor of at least two. Using the moduli from seismic crosshole tests, the estimated settlements were within +/- 15 % of the measured settlements.

It is noteworthy that the laboratory tests, although performed with special refinements designed to eliminate sample disturbance and conducted on carefully sampled specimens, consistently produced unrealistically low estimates of soil stiffness. It was postulated that the highly overconsolidated soils at this site were more susceptible to disturbance during sampling than "average" soils. The overconsolidated state of this site, compared to a soft soil site, may have also added to the applicability of seismically determined moduli in this case.

The authors conclude that field-determined moduli produce better estimates of soil compressibility than laboratory tests, and that the modified version of Swiger's suggested method based on seismic wave velocity measurements produced the most comprehensive and realistic assessment of settlement.

4.1.3 *Case history 3 – settlement of a water tank*

John Burland, in his Bjerrum Lecture, focused on the need for small-strain soil properties for many geotechnical problems, including soil structure interaction (Burland, 1989). He emphasized the nonlinear behavior of soils and promoted the use of strain-appropriate strength or stiffness for analysis. Since soil strains in most geotechnical problems fall in the range of 0.1% or smaller, Burland urged that geotechnical engineers recognize the importance of strain-appropriate soil properties. The development of laboratory techniques capable of precise measurement of small strains convinced him that the gap between dynamic and static measurements of soil stiffness was being closed. Previously, dynamic measurements of soil stiffness gave values so much higher than static measurements that many engineers discounted the dynamic measurements. However, Burland cited cases where accurately determined static small-strain values of stiffness were compatible with seismically measured stiffness, giving greater credibility to dynamically measured values. As an example, he presented a case where Young's modulus deduced from a static water tank loading test on Mundford Chalk produced stiffness very nearly the same as those determined from a seismic refraction survey. Figure 18 shows Young's modulus versus depth determined by three methods; 0.86-m diameter plate loading tests at seven depths, finite-element back calculation, and seismic refraction.

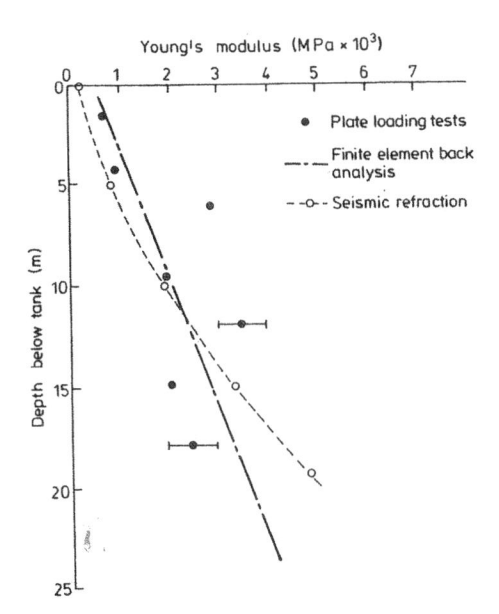

Figure 18. Young's modulus determined from three methods for settlement analysis of a water trunk on Mundford Chalk (from Burland, 1989)

Burland concluded that these results along with others open up the way for a whole new area of study linking dynamic and static deformation properties of soils. He speculated that studies of this kind would lead to wider application of geophysical measurements for determining in situ stiffness properties of geotechnical materials. The writers agree wholeheartedly.

4.1.4 *Case history 4 – settlement analysis of footings*

In 1994, the Geotechnical Division of ASCE held a settlement prediction symposium in conjunction with an ASCE Specialty Conference at Texas A&M University (Briaud and Gibbens, 1994). Full-scale footings of five sizes / configurations were tested to failure with detailed measurements of load-settlement. Thirty-one predictors were bold enough to make first class predictions of settlement based on detailed characterization of the site. Three of the 31 predictors used seismic wave velocities from cross-hole tests to determine modulus for settlement prediction.

The predictors used 22 different settlement prediction methods based on soil properties determined by five field tests and two laboratory tests. Several measures of accuracy of prediction were compared with predictions including settlement at several stages of loading and the factor of safety at ultimate load. Table 4 is a compilation of factors of safety for the five footings computed from predictions by the 31 predictors. Those who used seismically determined moduli were numbers 22, 23 and 28. Two of those (numbers 22 and 28) were consistently better than the mean of all predictors. The other predictor, number 23, was better than the mean for the three larger footings. Although this prediction event did not specifically showcase settlement predictions based on seismically determined moduli, it provides further evidence that this method of prediction has potential for future application.

Table 4. Factors of Safety $F = Q_f / Q_d$ (measured design load*/predicted design load) [*ultimate with FS =3] (from Briaud and Gibbens, 1994)

No.	Authors	Q(f)/Q(d) 1m	Q(f)/Q(d) 1.5m	Q(f)/Q(d) 2.5m	Q(f)/Q(d) 3.0m(s)	Q(f)/Q(d) 3.0m(n)
1	Wiseman	4.66	3.92	2.70	2.39	2.72
2	Poulos	4.66	0.90	2.77	3.23	2.78
3	Siddiquee	29.49	29.31	24.07	22.11	24.70
4	Silvestri	4.81	4.76	4.77	4.67	5.31
5	Horvath	3.16	2.43	2.27	2.00	2.28
6	Thomas	11.42	15.69	7.28	16.53	6.34
7	Surendra	6.53	3.86	4.54	1.93	2.20
8	Chang	11.60	20.40	8.19	7.50	6.15
9	Brahma	4.75	5.13	3.84	5.81	3.39
10	Floess	5.22	4.25	2.77	2.35	2.65
11	Boone	6.87	4.74	3.04	5.14	3.42
12	Cooksey	5.80	5.10	3.38	3.00	3.42
13	Scott	6.14	4.25	2.98	3.00	3.11
14	Townsend	4.31	1.62	2.03	1.67	1.69
15	Foshee	4.11	6.03	5.97	13.74	5.58
16	Mesri	1.88	2.31	2.56	2.71	3.08
17	Ariemma	1.58	2.92	1.89	7.20	1.87
18	Tand	5.44	4.70	3.54	3.06	3.52
19	Funegard	5.44	4.70	3.54	3.06	3.52
20	Deschamps	3.48	3.00	3.94	2.50	2.85
21	Altaee	2.90	3.09	3.74	3.91	4.46
22	Decourt	2.23	2.63	2.59	1.93	2.39
23	Mayne	13.22	8.10	4.14	3.20	3.64
24	Kuo	12.43	10.52	6.09	5.00	5.59
25	Shahrour	6.33	7.56	9.59	5.88	6.70
26	Abid	3.26	3.40	2.73	2.50	3.11
27	Utah State	5.80	4.81	2.89	1.87	2.21
28	Gottardi	4.78	3.25	1.95	1.63	1.88
29	Chua	10.42	10.89	8.83	8.13	9.19
30	Bhowmik	4.75	4.25	3.55	3.21	3.42
31	Dyaljee	8.52	6.47	4.03	3.70	4.26
Mean		6.64	6.29	4.72	4.99	4.43
Standard Deviation		5.23	5.90	4.12	4.63	4.13
Measured Value		3	3	3	3	3

4.1.5 *Summary*

A few successful demonstrations of the use of seismically determined soil modulus for settlement predictions have been presented. Other case histories can be found in conferences dealing with the pre-failure deformation characteristics of geomaterials such as Shibuya et al. (1994), Jardine et al. (1998), Jamiolkowski et al. (2001), and Di Benedetto et al. (2003). The potential for use of this more rational approach to determining soil stiffness has been confirmed, but not yet widely adopted. In some cases, engineers report that obtaining the seismic wave velocities is too expensive (Konstantinidis et al, 1986), but with the development of surface-wave methods, the cost of boreholes has been eliminated. The writers hope that elimination of this cost impediment will allow broader application of seismically determined moduli for geotechnical engineering purposes.

4.2 *Crosshole Seismic Velocity for Grouting Control*

When soil improvement in the form of grouting is selected in geotechnical applications, some means of confirming the expected improvement is necessary. Seismic wave velocity can be used to quantify soil improvement by measuring wave velocities before and after grouting. Following are three case histories describing the successful use of seismic wave velocity to confirm the extent of ground improvement by grouting.

4.2.1 *Case history 1 - forge foundations*

Seismic shear wave velocities determined from crosshole tests were used to confirm the degree and extent of soil improvement from combined compaction and chemical grouting, (Woods and Partos, 1981). New forging machines were to be installed in a forge shop located on deep beach deposits near the Atlantic coast. Figure 19 shows blow count versus depth for two locations at this site, B1 and B3. The blow count ranged from 2 to about 20 in the upper 5 meters. The new forge machines were considerably larger than the old forges and vibrations within the plant from the new, larger installations were of concern as were differential settlements of the forges. Based on preliminary calculations, it was clear that the loose sand deposits were susceptible to shakedown settlement if large-amplitude vibrations occurred. There was also a potential for transmission of large-amplitude vibrations through the plant, so soil improvement in the form of grouting was selected to mitigate these concerns. Both compaction and chemical grouting techniques were chosen to be performed in series, with compaction grouting being performed as a first stage and then chemical grouting in a second state if sufficient stiffness had not been achieved in the first stage.

The plan view of the site, Figure 20, shows the footprint of four new foundation blocks and locations for compaction grouting, chemical grouting and boreholes for crosshole tests. Figure 21 shows the shear wave velocity versus depth profiles determined from crosshole seismic tests after each of the two stages of grouting. Compaction grouting achieved the minimum shear wave velocity at this location, but chemical grouting was performed as an added factor of safety. The crosshole tests confirmed a significant increase in shear wave velocity leading to successful operation of forging machines at this site. No excessive settlements were observed and vibration levels throughout the plant were not noticeable.

Fig. 19. Composite soil profile (from Woods and Partos, 1981)

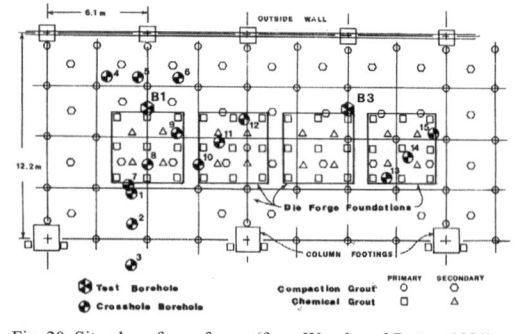

Fig. 20. Site plan of new forges (from Woods and Partos, 1981)

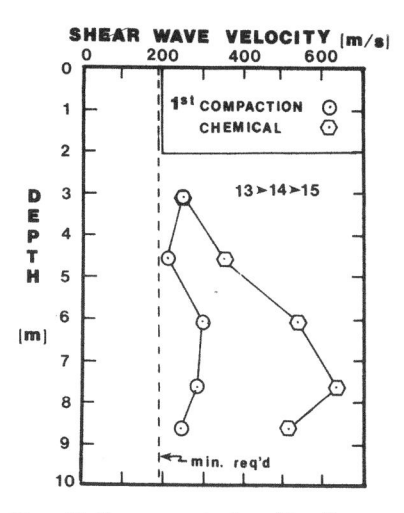

Figure 21. Shear wave velocity profiles after compaction grouting and then after chemical grouting (from Woods and Partos, 1981)

4.2.2 Case history 2 - subway construction

One route of the subway in Pittsburgh, Pennsylvania was constructed under a part of Sixth Avenue, a narrow street with old, heavy masonry structures on both sides (ENR, 1982). The invert of the subway was well below the lower elevation of the spread footings supporting the heavy buildings. The foundation material consisted of coarse sand, gravel and cobbles. Construction of the subway called for excavation in a cut-and-cover process, but stability of the adjacent building foundations was in question. Chemical grouting was chosen to improve (stabilize) the soil, and crosshole seismic tests were used to confirm achievement of sufficient improvement over the un-grouted condition.

Crosshole equipment was fabricated that made use of the grout pipes for placement of the source and receivers. Shear wave velocities were used to characterize the soil before and after chemical grouting. A target shear wave velocity was determined in the laboratory using resonant column tests, and a test section of crosshole tests was performed at the site to confirm expectations from the laboratory study. Before-grouting shear wave velocities ranged from about 500 ft/sec to 1000 ft/sec (150 m/s to 305 m/s) and after-grouting velocities ranged from about 1400 ft/sec to 3000 ft/sec (425 m/s to 915 m/s). The criteria for satisfactory soil improvement by grouting was either: (1) a doubling of the shear modulus (1.41 times increase in shear wave velocity) over the before-grouting condition, or (2) a minimum of 1400 ft/sec (425 m/s). The Sixth-Avenue section of the subway was successfully completed without disturbance of the adjacent buildings.

4.2.3 Case history 3 - old bridge support

New design loads on a railway line in Italy required structural and ground improvements along the entire line and particularly on two XIX century masonry arch bridges (Volante et al., 2004). A need for careful ground movement control during upgrading of the old bridges led to design of a multistage-multiport, low-pressure grouting technique. Careful grout control was exercised during the injection process, but in the long run it was important to determine the strength and deformation parameters of the newly grouted soil. Figure 22 shows P-wave and S-wave velocities before and after grouting at one of the bridge sites. In this case, the modulus of the ground under the piers was increased by a factor of about 2.5.

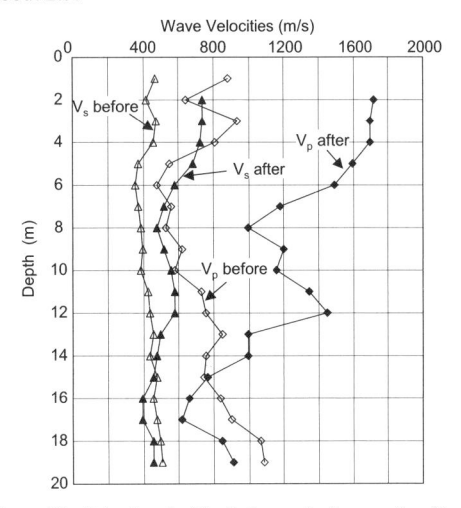

Figure 22. Seismic velocities before and after grouting (from Volanta et al., 2004)

4.2.4 Summary

The three case histories presented here clearly show that seismic wave velocities can be used to advantage in confirming quality and extent of grouting operations. While the examples cited all used the crosshole seismic method, applications of surface wave methods may provide economies where there is sufficient lateral extent to apply them. For a broad-area dynamic compaction project, the writers have also successfully used the SASW method to confirm ground improvement by showing increased shear wave velocities. Stokoe and Santamarina, 2000 have also shown evaluation of blast densification by the SASW method.

4.3 Underground Cavity Detection

Many engineering situations require the determination of the existence or absence of underground obstacles, solid or void, as well as their locations and depths. Probing for these obstacles with penetrometers is a time consuming and expensive process. Several currently available geophysical techniques have been proposed and used to identify underground anomalies. The following example and two case histories describe the use of some of these techniques.

4.3.1 *Example 1 – use of GPR, SASW and crosshole testing for cavity detection*

Three geophysical methods were studied for the detection of buried cavities (Al-Shayea, 1994, and Al-Shayea et al., 1994). A soil bin, 7 m in diameter and 2 m in depth, was used in this study. The soil bin is shown in Figures 23 and 24. A three-cell cavity was buried in the bin at a depth to center of the cavity of 614 mm. SASW data were collected along five lines identified by the source and receiver symbols and the skew lines marked a-c on Figure 23. The general SASW test arrangement is shown in Figure 24. Ground-penetration radar (GPR) data were also collected along the grid lines identified on Figure 23. Crosshole shear-wave tests were performed across both long and short axes of the buried cavity and in the free-field. GPR gave the most obvious identification of the cavity as indicted on Figure 25. In this figure, it is very clear where the electromagnetic wave field produced by GPR was distorted, (Figure 25b) compared with the wave field of the free-field (Figure 25a). Approximate depth and size of the cavity were calculated from Figure 25b knowing the frequency of the GPR source and the dielectric constant of the sand.

SASW tests were performed directly over the centerline of the cavity with various states of filled and empty cells. (The cells were filled using the same sand as in the bin.) Differences in cavity filling

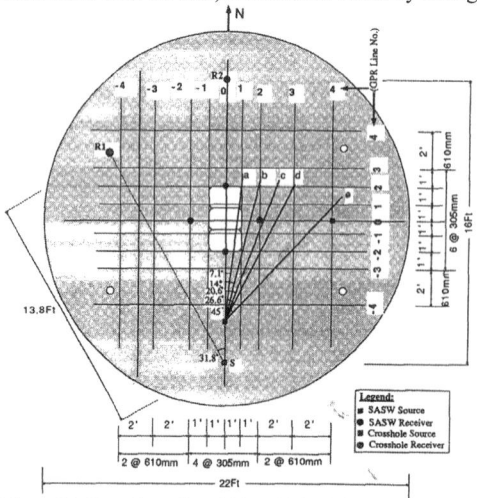

Figure 23. Plan view of sand bin with buried three-cell cavity (from Al-Shayea et al, 1994)

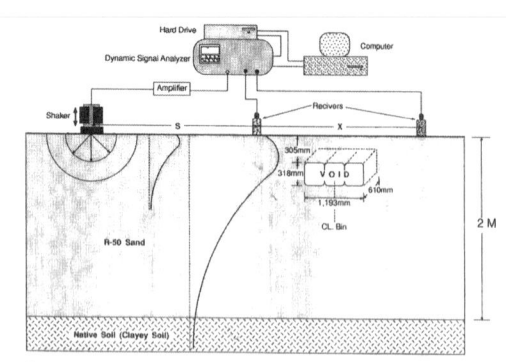

Figure 24. Cross section of sand bin with SASW setup for cavity detection (from Al-Shayea et al, 1994)

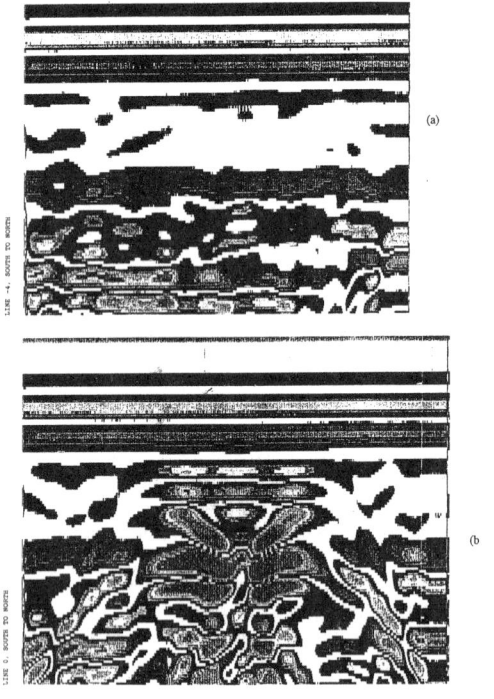

Figure 25. GPR Scans: (a) free-field and (b) parallel to and over the long axis of the cavity with all cells empty (from Al-Shayea et al, 1994)

showed substantially different dispersion curves, Figure 26. A smoothed free-field dispersion curve is shown as a solid line while the empty and partially empty void dispersion curves are shown with other symbols. For this discussion, the symbols representing cavity conditions need not be identified because the key result is that the existence of a void and the size of the void both influenced the shape of the dispersion curve.

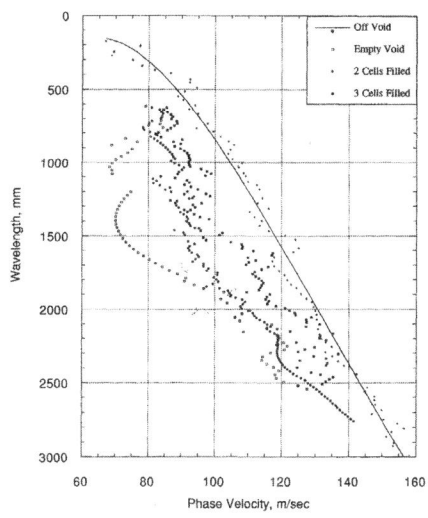

Figure 26. Dispersion curves for free-field and for lines over the long axis of the cavity with various cells filled (from Al-Shayea et al., 1994)

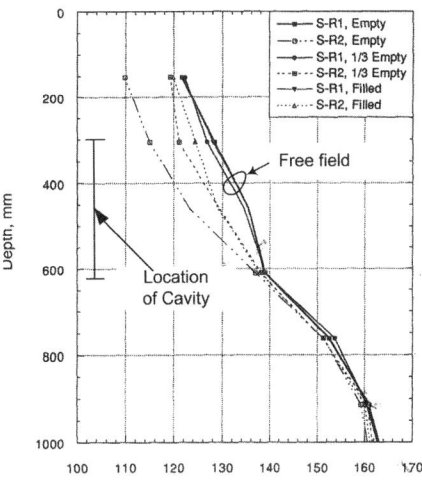

Figure 27. Crosshole velocity profiles along two paths, free-field and along centerline of the cavity (from Al-Shayea, 1994)

Other SASW tests were performed directly over and on lines skewed to the centerline of the cavity. These varying lines also showed substantially different dispersion curves.

Results of SV-wave seismic crosshole tests performed on lines S-R1 (free-field) and S-R2 (over cavity) in Figure 23 are presented in Figure 27. Here it can be seen that the shear wave profile for the free-field direction is quite consistent for three conditions of the void, all cells empty, 1 cell filled with sand, and all cells filled with sand. The cases of all cells filled did not exactly match the free-field condition because the cell filling process could not duplicate the free field density of the sand. In the case of line S-R2, the average shear wave velocity in the depth region of the cavity was clearly reduced by the existence of the cavity. Had the crosshole boreholes been closer together, shear wave velocity differences would have been more dramatic at depths representative of the cavity.

All three of the geophysical techniques used in this study have characteristics that allow identification of cavities or anomalies in the underground, but each have their limitations. To mention just the most salient drawbacks, GPR suffers from ability to penetrate clay, SASW is limited by the need to run multiple lines of data, and crosshole boreholes need to strategically located to straddle the cavity. Some of these limitations may be minimized through future study while others are inherent to the basic principles of the geophysical methods. Some of these drawbacks simply point to the necessity of performing suites of complementary geophysical tests for cavity detection.

It is worth noting that some of the drawbacks of the SASW method that were cited above for cavity detection may be mitigated by using recently developed wavelet theory (Shokouhi and Gucunski, 2003, and Gucunksi and Shokuohi, 2004). Continuous wavelet transforms (CWT) are a new class of transformations that can produce a time-frequency map of the ground surface from which indications of near-surface cavities can be derived. The data collection required for CWT can be achieved simultaneously with SASW data collection.

4.3.2 *Case history 1- mine collapse under a highway*

High resolution SH-wave reflection tests were preformed along the right-of-way of Interstate Highway I-70 in southeastern Ohio, (Guy et al., 2003). A portion of the east bound lane of I-70 at this location collapsed into old underground coal-mine workings. A plan view of the east bound lanes of I-70, the location of the collapsed highway, and a projection of the old underground mine workings are shown in Figure 28. High-resolution SH-reflection surveys were performed along two lines straddling the east bound lanes of I-70, lines GUE-I70-1 and EBPassYY. The interface between soil and rock was clear for most of the lengths of both of these survey lines, but in the interval between stations 48320 and 48360 on line EBPassYY the SH-wave stacking velocity plot showed a discontinuous segment of the soil/rock interface in the region indicted by the angled bars in Figure 29 and on the geologic cross section shown in Figure 30. The continuous soil/rock interface as interpreted is indicated on Figure 29 by the dash-dot line across the plot. Soil borings and rock coring

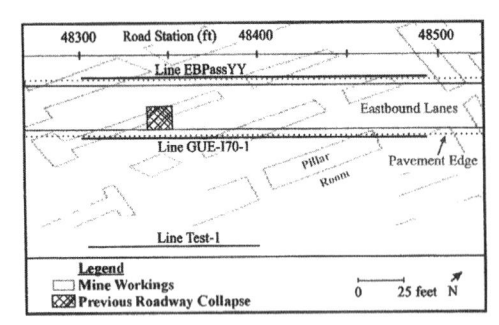

Figure 28. Plan view of site showing eastbound lanes of I-70, seismic reflection lines and projection of Murray Hill No. 2 Mine Workings (from Guy et al., 2003)

were performed at six locations selected by interpreting the soil/rock interface reflector in Figure 29. These voids indicate stopping from the former coal mine upwards, but that stopping had not progressed to the pavement level.

This seismic method penetrated relatively deep, about 20 meters in this case. The results of the tests show a potential for future subsidence and sinkhole development along this highway. Based on this kind of information, remedial efforts could be applied to stop progression of the stopping or stabilize the ground above the current voids.

Because high resolution SH reflection utilizes steady-state ground excitation, wavelengths can be controlled for good wavelength/cavity size ratios and may permit better cavity size and depth determination than other seismic wave based cavity detection methods.

4.3.3 Case history 2 – other geophysical testing at the mine collapse under highway I-70

All of the geophysical techniques cited in Section 4.3.1 (except CWT) were applied by Hiltunen et al., (2004) at the mine-collapse site on I-70 described above. The results confirmed the drawbacks cited previously. However, the work by Hiltunen et al. confirmed that quality geophysicalmeasurements could be made in close proximity to an active interstate highway with heavy truck traffic. Both

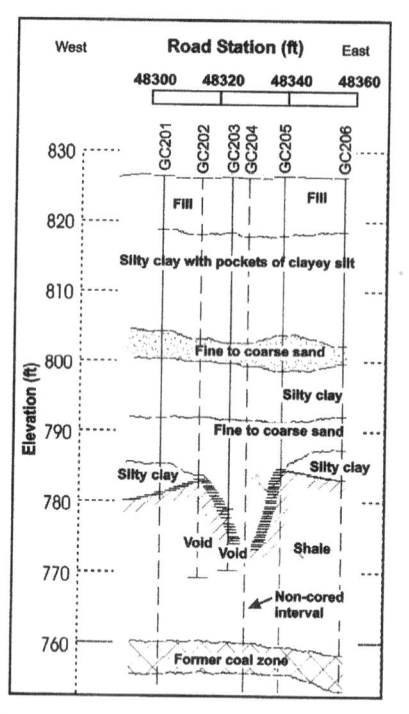

Figure 30. Geologic cross section for line EBPassYY for stations 48300 thru 48360 including exploratory borings and identification of voids (from Guy et al., 2003)

GPR and SASW testing at this site could not probe deep enough to explore the soil/rock interface, GPR because of clay in the soil and SASW because of the lack of a sufficiently energetic excitation source. The crosshole test did not happen to encounter a void or loose soil. A major conclusion from these tests was that no single technique could unambiguously detect voids or other anomalies throughout a wide range of depths.

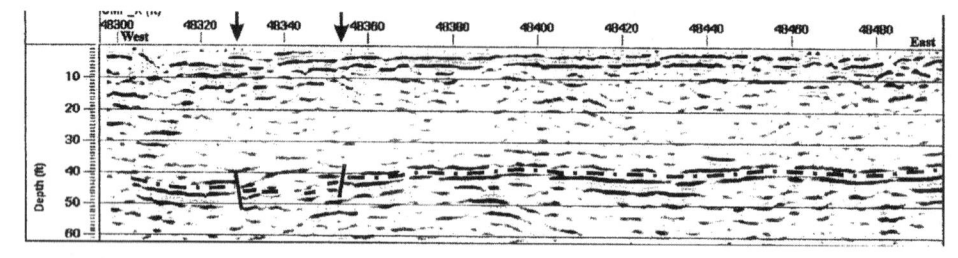

Figure 29. Interpreted stacked time record for line EBPassYY for Stations 48300 thru 48480 (from Guy et al., 2003)

4.3.4 Summary

While geophysical methods, including several seismic wave propagation methods, have considerable potential for locating and sizing buried objects, no single method is appropriate for all sites. In the past, engineers and geophysicists have often chosen one method in an attempt to locate cavities or buried objects and have been disappointed with the results. Indications from the cases cited herein are that, while any one method may provide some parts of the identification puzzle, a suite of tests can provide more confidence in finding cavities or buried objects

4.4 Tunnel Investigation

Sometimes, V_s measurements are used to profile constructed systems and their geotechnical foundation materials to assist in forensic studies. A forensic study of a concrete-lined tunnel in rock is described below (Stokoe and Santamarina, 2000). A generalized cross section of the tunnel is shown in Figure 31a. The tunnel is approximately 3 m in diameter, with a concrete liner that has a nominal thickness of 30 cm.

An extensive investigation was conducted in which SASW testing was performed at more than 100 locations along the longitudinal axis of the tunnel. SASW testing was performed with hand-held hammers as sources and accelerometers as receivers. The accelerometers were held magnetically to metal disks attached to the liner. This general configuration is shown in Figure 31b. Testing was conducted to profile along two planes into the liner-rock system. One profile was along the springline, and the other profile was near the crown as illustrated in Figure 31b.

The SASW testing program was designed to investigate the following: 1. thickness and quality of the concrete liner in the springline and crown areas, 2. thickness and quality of any grout in the area of the crown, 3. identification of any voids in the crown area, and 4. stiffness and variability of the rock behind the liner. (Grouting in the crown area was done some time after construction of the liner.) The program successfully answered these questions. Examples showing how some of the questions were answered follow.

An interpreted V_s profile at one springline location is shown in Figure 32a. The profile shows a high-quality concrete liner ($V_s > 2500$ m/s) that is about 35 cm thick. At this location, the liner is in direct contact with the rock, and the rock is stiffer (and presumably stronger) than the concrete.

Results from one crown location are shown in Figure 32b. In this case, the liner is thicker than 40 cm, and there is grout between the liner and the rock. Based on the V_s values, both the concrete and grout are high quality. The concrete-grout-rock interfaces have intimate contact; hence, no voids. Also, the rock is less stiff than the concrete at this location.

Clearly, SASW testing was successfully and cost-effectively applied in the tunnel investigation. The writers have had other successful projects in many other underground applications (for instance, Madianos et al., 1990, Olson et al., 1993, and Luke et al., 1998).

4.5 Offshore Shear-Wave Velocity Profiling Using Seismic Interface Waves

As offshore construction moves into deeper water (depths greater than 1.6 km), traditional drill-and-sample geotechnical site investigations become expensive and less reliable. The expense of drilling in deep water often dictates the extraction and testing of only a few samples. Furthermore, the quality of these samples can be severely compromised when extracted through great water depths. Other geotechnical site investigation methods, such as the seismic cone penetrometer (e.g. Robertson et al., 1986), are effective on land and in shallow water, but become more difficult and costly to apply in the deep-water environment. One seismic method that has potential for deep-water seafloor investigation is the surface-

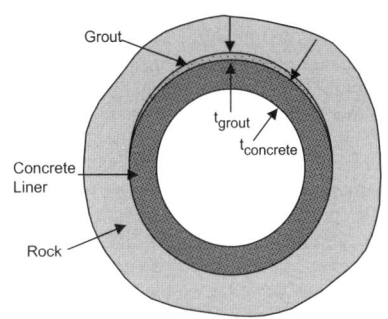

a. Generalized tunnel cross section

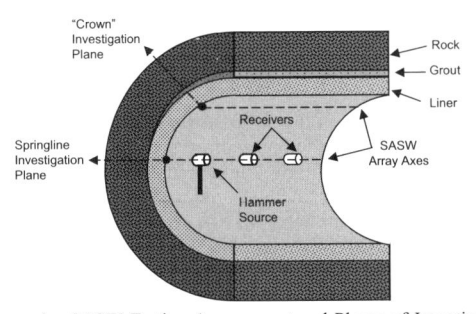

b. SASW Testing Arrangement and Planes of Investigation (from Stokoe and Santamarina, 2000)

Figure 31. SASW testing performed inside a concrete-lined tunnel (from Stokoe and Santamarina, 2000)

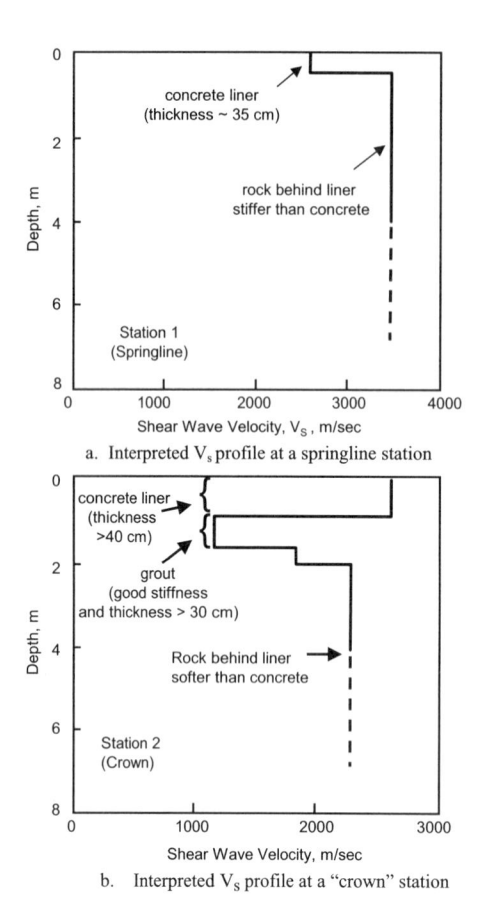

Figure 32. Examples of V_s profiles measured inside a concrete-lined tunnel (from Stoke and Santamarina, 2000).

wave method for V_s measurements of the sediment. Shear-wave velocity is used because it is essentially unaffected by the presence of water (and air if the sediment is unsaturated) and because, in a saturated soil, V_s is far better correlated with shear strength than V_p wave, which has been used to predict shear strength in previous offshore investigations (Blake and Gilbert, 1997).

The SASW method is being applied to the offshore environment to determine shear-wave velocity profiles of the seafloor. In this application, a soil-water interface wave, called a Sholte wave (Wright et al., 1994), is measured. The measured Sholte-wave dispersion curve is used to determine a shear-wave velocity profile. Figure 33 shows the results from SASW testing performed by ConeTec, Inc. off the coast of Vancouver, B.C. (Rosenblad and Stokoe, 2001). The soils in this region were composed of loose silty-sands and sands that were unsaturated (gaseous). Testing was conducted in shallow water depths, ranging from 12.2 to 76.2 m, to demonstrate the potential of the method. Personnel from ConeTec also conducted SCPT measurements at the site. The SASW results agree reasonably well with the SCPT values at depths between 5.5 and 10 m as seen in Figure 33. The differences that are observed are likely due to the localized versus global nature of the SCPT and SASW tests, respectively. Based on the writers' experience on land, development of a lower-frequency source and longer arrays would allow SASW profiling to significantly greater depths at this site, certainly to depths on the order of 30 to 50 m. However, the results do demonstrate the feasibility of the surface-wave method offshore. Stoll et al. (1994), Luke and Stokoe (1998) Rosenblad (2000), and Rosenblad et al. (2003) are among others who have also shown this surface-wave application.

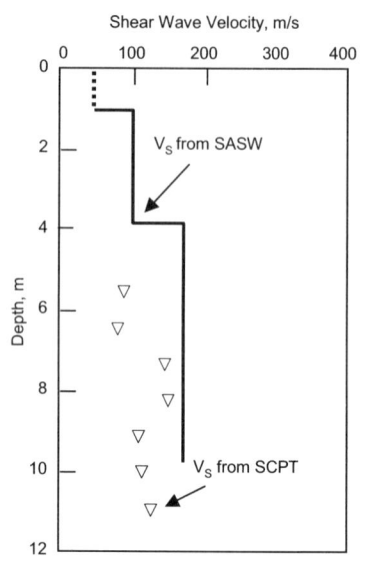

Figure 33. Results from the SASW testing performed at the offshore near Vancouver, B.C. (Rosenblad and Stokoe, 2001)

4.6 Geotechnical Earthquake Engineering

The importance of the shear stiffness of geotechnical materials in calculating their response during dynamic loading initially stimulated the development of in situ seismic methods tailored to measure V_s. This development began in earnest in the 1960s with modifications/refinements to the crosshole and downhole methods. It is continuing today with improvements to surface-wave methods. The following two case histories and one application describe some recent work in geotechnical earthquake engineering.

4.6.1 Case history 1 - profiling hard-to-sample alluvium

Yucca Mountain, Nevada, was approved as the site for development of the geologic repository for high-level radioactive waste and spent nuclear fuel in the United States. The U.S. Department of Energy has been conducting studies to characterize the site and assess its future performance as a geologic repository. As part of these studies, a comprehensive program of in situ seismic investigations was performed at the proposed site of the Waste Handling Building (WHB). The purpose of these investigations was to characterize the velocity structure of the subsurface for seismic design of the WHB facilities.

In situ seismic velocity measurements were performed by three different methods at this site. The seismic methods include two borehole methods, downhole and suspension logging, and one surface-wave method, spectral-analysis-of-surface waves (SASW). The borehole surveys were conducted in 16 cased boreholes to a maximum depth of 198 m. SASW surveys were performed at 34 locations around much of the proposed area which was about 300 m by 450 m in plan dimensions. The SASW surveys were aimed at evaluating the top 50 m of the site and investigating lateral variability. The SASW surveys provided greater spatial coverage of the site while the borehole surveys added critical deeper information.

Stokoe et al., 2003 presented a comparison of the V_s profiles determined by the three seismic methods in the material where the most overlap in measurements existed. This material is a hard-to-sample Quaternary alluvium/colluvium (Qal) which ranges from a poorly graded gravel (GP) to a silty gravel (GW). The alluvium contains varying amounts of sand, cobbles and boulders, and it varies in thickness, depth, and amount of cementation over the WHB site. The alluvium was measured in 15 of the 16 boreholes. The results form the most comprehensive set of V_s measurements, in terms of multiple seismic methods in a localized area and in one material type, that has ever been compared. In addition, this comparison represents a "blind comparison" overseen by URS personnel, in that each measurement team was not aware of the other's results until after they were all submitted to URS. Therefore, not only did the seismic tests at the WHB site provide the information needed for the seismic design of the facility, but they also provided an interesting comparison of V_s profiles measured by different methods as described below.

The comparison of the V_s profiles is presented as an average profile for each method. The average profiles were determined by first dividing the 1.5-to-30-m depth range into four intervals, with the smallest near the surface where the V_s gradient was the greatest and largest at depth where the gradient was the smallest. This division allowed the overall characteristic V_s profile of the alluvium to be preserved. The four depth intervals were: 1.5 to 4.6 m (layer no. 1), 4.6 to 9.2 m (layer no. 2), 9.2 to 18.3 m (layer no. 3), and 18.3 to 30.5 m (layer no. 4). An example of the results from this procedure at one borehole is presented in Figure 34. Figure 34a shows the profiles determined from downhole, suspension logging and SASW measurements in and near borehole RF-19. Figure 34b shows the averaged V_s profiles.

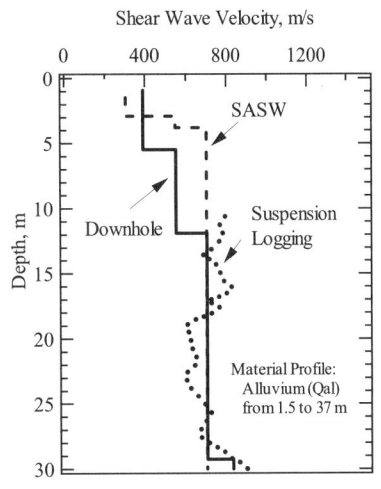

a. V_s profile measured with each seismic method

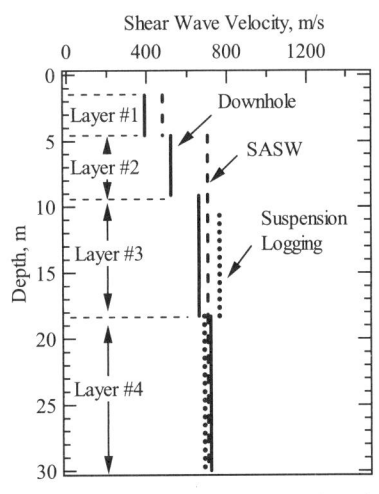

b. Averaged V_s profile computed for each seismic method from Figure 34a above

Figure 34 Example of measured and averaged V_s profiles used in comparing the seismic methods; Measurements in or near Borehole RF-19 (from Stokoe et al., 2003)

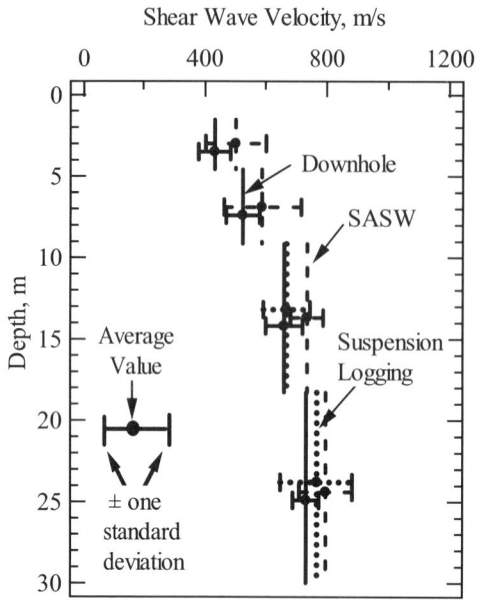

Figure 35. Comparison of average V_s profiles from 15 locations in the WHB area (Stokoe et al., 2003)

Comparison of the average V_s profiles is presented in Figure 35. Also shown in the figure is the variability in the V_s values measured with each method, expressed by ± one standard deviation (±σ). These data are presented numerically in Table 5, including the coefficient of variation (COV = σ/Avg V_s). The comparison is shown in Figure 35 and demonstrates the strength and robustness of S-wave velocity measurements today. First, identical trends of increasing V_S with depth were measured with all three methods. Second, differences in average V_s values in each layer are small. The two largest differences are 16% and 12% and are found between the downhole and SASW measurements in layers no. 1 and no. 2, respectively. These layers are the shallowest layers (less than 9.2 m deep) and are the ones which should be expected to show the largest differences due to: (1) the SASW method having the most resolution of these methods near the surface, (2) the averaging effect of placing straight-line segments through the measured travel times in downhole data reduction, and (3) wave refraction and lateral variability in the Qal affecting each method differently. The standard deviations determined from the measurements and the COV values support points (1) and (2) above. The COV values are about 0.21 for SASW measurements in layers no. 1 and no. 2 and about 0.11 for the downhole measurements in the same layers. It should be noted that the variability shown by ±σ includes both measurement uncertainty and variability in material properties.

4.6.2 *Case history 2 - deep Vs profiling*

Another part of the seismic investigations at the Yucca Mountain site discussed above involved deep V_s profiling along the top of Yucca Mountain (Stokoe et al., 2004). Deep profiling is defined as evaluating the shear-wave velocity structure to depths of about 200 m. This work involved the SASW method and required the use of a Vibroseis (see Figure 36) to generate the low-frequency (hence long wavelength) waves necessary to profile to 200 m. Yucca Mountain consists of stacked layers of tuffs with V_s generally above 900 m/s. Therefore, the lowest excitation frequency was in the range of 3 Hz and the farthest measurement point from the source was around 500 m.

SASW measurements were performed at 22 array sites along the top of Yucca Mountain. These sites were spread over a distance of about 5 km. The SASW surveys were aimed at evaluating: (1) the top 150 to 200 m of the mountain, (2) an apparent V_s gradient in the near surface (within about 5 to 15 m), and (3) any lateral variability over the 5-km distance. The mean V_s profile that was determined from the 22 profiles is presented in Figure 37 along with the 16th and 84th percentile V_s values. The coefficient of variation (COV) about the mean profile was calculated by assuming the V_s values follow a log-normal distribution.

Table 5 Numerical Analyses of Average V_S Profiles (from Stokoe et al., 2003)

Depth Interval (m)	Downhole Surveys					SASW Surveys					Suspension Logging Surveys				
	No. of Meas.	Avg. V_S (m/s)	St. Dev., σ (m/s)	$\frac{\sigma}{\text{Avg. } V_S}$		No. of Meas.	Avg. V_S (m/s)	St. Dev., σ (m/s)	$\frac{\sigma}{\text{Avg. } V_S}$		No. of Meas.	Avg. V_S (m/s)	St. Dev., σ (m/s)	$\frac{\sigma}{\text{Avg. } V_S}$
1.5 – 4.6	10	432	53	0.12		10	502	101	0.20		0[1]	—	—	—
4.6-9.2	12	526	56	0.11		12	590	126	0.21		2[1]	549	—[2]	—
9.2-18.3	13	662	59	0.09		12[3]	736	54	0.07		9	669	75	0.11
18.3-30.5	8	732	43	0.06		7[3]	798	88	0.11		8	768	118	0.15

1 Suspension logging was not successfully performed at shallow depths due to high attenuation, backscattering and tube-wave interference.
2 Insufficient data to perform meaningful calculations.
3 SASW surveys were performed at one less borehole than the downhole surveys due to surface obstructions.

Figure 36. Photograph of Vibroseis truck in operation on the top of Yucca Mountain (from Stokoe et al., 2001)

Several interesting trends are evident in Figure 37. First, it is observed that the near-surface V_s gradient is quite abrupt. The values of V_s change from approximately 300 m/s in the top meter to over 800 m/s at a depth of only 5 m. Below 5 m, the mean V_s value increases gradually from about 900 to 1000 m/s at a depth of 150 m. A gradient near the surface is expected due to the effects of weathering on the near-surface rock. The abruptness of the gradient, which has important implications in terms of the ground motion hazard, can not be predicted without field measurements of this kind. Another important result shown in Figure 37 is the nearly constant value of the mean V_s below the 5-m-thick, near-surface zone. The measurements show a mean value of approximately 1000 m/s in the depth range of 10 to 150 m.

4.6.3 *Application 1 - liquefaction resistance*

Evaluation of the liquefaction resistance of soils can be a critical factor in many geotechnical engineering investigations. Such an evaluation is typically performed with field test such as the standard penetration tests (SPT) or cone penetration test (CPT). The general procedure, called the "simplified procedure,"

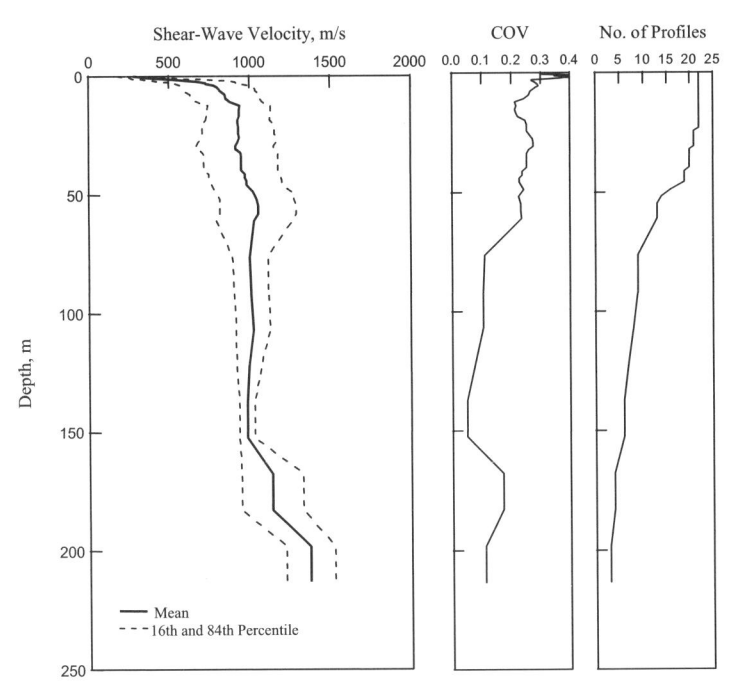

Figure 37 Statistical analysis of V_S profiles from SASW measurements on top of Yucca Mountain (Stokoe et al., 2004)

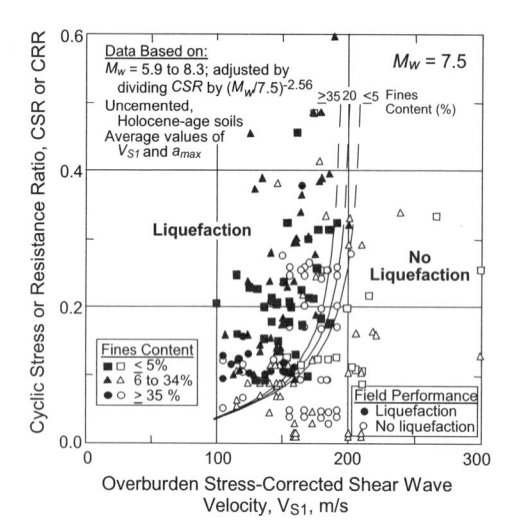

Figure 38. Curves recommended by Andrus and Stokoe (2000) for delineating liquefiable and nonliquefiable granular soils based on field V_s measurements

cedure," was initiated by Seed and Idriss (1971) using SPT blow counts correlated with a parameter called the cyclic stress ratio that represents the earthquake loading. This procedure has been updated over the years. CPT measurements have been added, initially by Robertson and Campanella (1985). A national workshop was convened in 1996 which further updated the SPT and CPT procedures and added an in situ geophysical method to the suite of field tests (Youd et al., 2001). The geophysical method is in situ seismic measurements of V_s. Values of V_s are correlated with earthquake loading in the same manner as done in the SPT and CPT procedures.

The procedure involving V_s was presented by Andrus and Stokoe, (2000). The procedure is based on field performance data from 26 earthquakes and in situ V_s measurements at over 70 sites. The case history data from this procedure, adjusted to an earthquake moment magnitude (M_W) of 7.5, is shown in Figure 38. Of the 90 liquefaction case histories shown in the figure, only two incorrectly lie in the no-liquefaction region. These two points are, however, very near the boundary. Clearly, the procedure based on field V_s measurements can be used as a supplement or in lieu of the SPT and CPT procedures. The procedure is especially important for use with hard-to-sample soils such as soils containing gravel and/or cobbles. The nonintrusive nature of the SASW and other surface-wave methods make them especially well suited for this application in hard-to-sample soils.

5 IMPACT OF DISTURBANCE FROM SAMPLING ON PREDICTED NONLINEAR SOIL RESPONSE IN THE FIELD

One of the strengths of seismic measurements, as well as other geophysical measurements, is that the same basic measurement can be performed in the field and in the laboratory. Field measurements of V_s are performed in the small-strain or elastic range as discussed earlier. Therefore, laboratory measurements of V_s also have to be evaluated in this small-strain range if they are to be compared directly with the field values. This comparison presently forms the way sample disturbance is evaluated in geotechnical earthquake engineering when dealing with nonlinear deformational characteristics. Field and laboratory values of V_s at small strains are used to adjust the nonlinear response of soil measured in the laboratory to field conditions. The nonlinear response is typically shown in terms of the nonlinear variation in shear modulus with shearing strain (G – log γ). This comparison, the adjustment procedure, and the impact on the G – log γ and stress-stain (τ – γ) curves are discussed below.

5.1 Comparison of Small-Strain Field and Laboratory Values of V_s

Invariably, when field and laboratory values of V_s are compared, values of $V_{s,\ lab}$ range from slightly less to considerably less than the in situ values, $V_{s,\ field}$ (Anderson and Woods, 1975, Long, 1980, Yasuda and Yamagushi, 1985, Yokoa and Konno, 1985, and Chiara, 2001). A just-completed project dealing with the resolution of site response issues in the 1994 Northridge, CA earthquake, called the ROSRINE project, involved numerous field and laboratory investigations. Sixty-three intact samples were recovered and tested in the laboratory at the University of Texas using combined resonant column and torsional shear equipment (Darendeli, 2001, and Choi, 2003). Additionally, in situ seismic measurements were performed during the field investigation phase, mainly be GeoVision Geophysical Services, Corona, CA using a suspension logger. Therefore, the ROSRINE project afforded an excellent opportunity to investigate further the relationship between field and laboratory values of V_s.

An example field V_s profile measured in this study is presented in Figure 39. At this site, called La Cienega, in situ seismic tests (shallow crosshole testing and deep suspension logging) were performed. A depth of nearly 300 m was logged. Intact samples were recovered from depths ranging from 4 to about 240 m. The laboratory values of V_s, shown by the solid circular symbols, are plotted at the corresponding sample depths. There is considerable

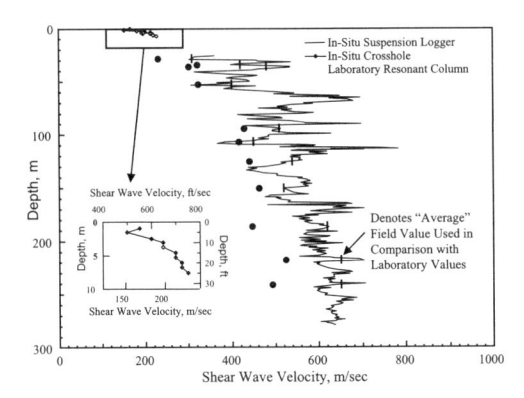

Figure 39. Example profile of small-strain field and laboratory shear wave velocities evaluated at a strong-motion earthquake site on the ROSRINE project (from Stokoe and Santamarina, 2000)

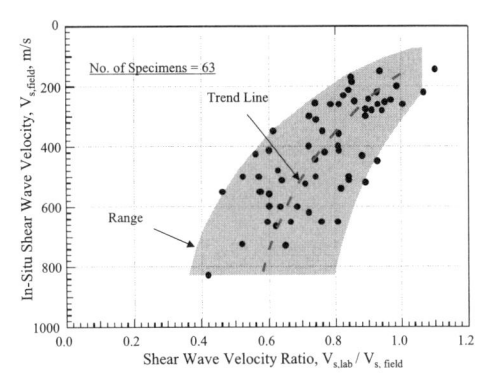

Figure 40. Variation in the ratio of laboratory-to-field shear wave velocities ($V_{s,\,lab}/V_{s,\,field}$) with the in-situ value of $V_{s,\,field}$ determined in the ROSRINE project

variability in the field V_s profile. The "average" field values associated with the laboratory values are shown by the short vertical lines through the field V_s profile in the vicinity of the sample depth.

A summary of all field-lab V_s comparisons from the ROSRINE project is presented in Figure 40. A total of 63 samples were tested in the laboratory. There is a clear trend in the data, with the velocity ratio ($V_{s\,lab}/V_{s\,field}$) decreasing as the in situ value of V_s increases. (There was essentially no correlation with sample depth.) In general terms, the velocity ratio is around one at $V_s \cong 160$ m/s. However, at $V_s \cong 725$ m/s, the velocity ratio is about 0.6, which means that the small-strain shear modulus from laboratory testing is on the order of 1/3 of the value in the field. This comparison strongly supports the need to perform field seismic tests, certainly in studies dealing with siting and retrofitting of important facilities.

5.2 Estimated Field G - log γ Curves from Field and Laboratory Measurements

Once the V_s profile has been determined at important or high-risk sites, the next step in the geotechnical earthquake engineering investigation is determination of the nonlinear characteristics of the soil. This step typically involves cyclic and/or dynamic laboratory testing of intact specimens. In terms of nonlinear shear modulus, these results are presented in the form of the variation in normalized modulus, G/G_{max}, with shearing strain amplitude, γ, or simply G – log γ. Typical examples of G – log γ curves from the ROSRINE project for soft, moderately stiff and very stiff soils are shown in Figures 41a, 41b, and 41c, respectively. The laboratory G - log γ curves are shown by the solid lines in the figures.

With the in situ V_s value and the laboratory G – log γ curve, the final step is to estimate the field G – log γ curve. This step is accomplished by scaling the laboratory G – log γ curve using G_{max} determined from the field seismic tests as,

$$G_{\gamma,\,field} = \left(\frac{G_{\gamma,\,lab}}{G_{max,\,lab}}\right) G_{max,\,field} \qquad (8)$$

where,

$G_{\gamma,\,field}$ = in situ shear modulus at a shearing strain of γ,
$G_{\gamma,\,lab}$ = shear modulus determine in the -laboratory with an intact specimen at a shearing strain of γ,
$G_{max,\,lab}$ = small-strain shear modulus determined in the laboratory, and
$G_{max,\,field}$ = in situ shear modulus measured by seismic testing.

It is assumed, of course, that evaluation of the G – log γ curve in the laboratory was performed at a confinement state, excitation frequency, number of loading cycles, drainage condition, etc. that represent the field conditions. Also, $G_{max,\,field}$ was calculated from $V_{s,\,field}$ using Equation 2.

The estimated field G – log γ curves are shown by the dashed lines in Figures 41a, 41b and 41c for the soft, medium stiff and very stiff soil examples taken from the ROSRINE project. Clearly, adjustment of the laboratory curve is critical to correctly predicting the earthquake ground motions at the site.

5.3 Laboratory and Field Stress-Strain (τ – γ) Curves

The laboratory shear stress-shear strain (τ – γ) curve can be calculated from the laboratory G – log γ curve as,

$$\tau = G * \gamma \qquad (9)$$

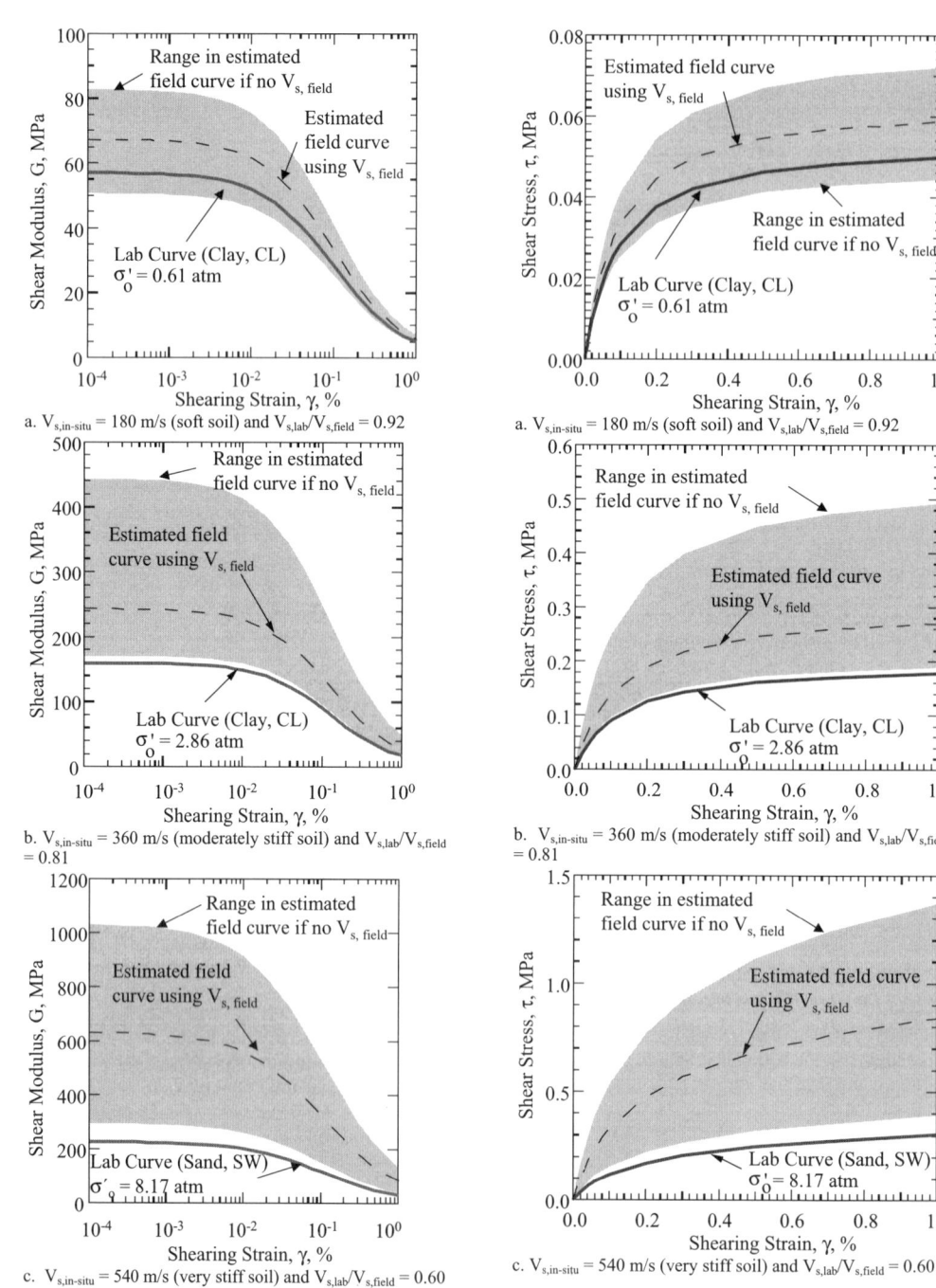

Figure 41. Measured laboratory G – log γ curves, estimated field G – log γ curves using $V_{s,\,field}$, and possible range in field G – log γ curves using range in Figure 40 but no measurement of $V_{s\,field}$; Examples for soils from the ROSRINE project with a range in shear stiffness

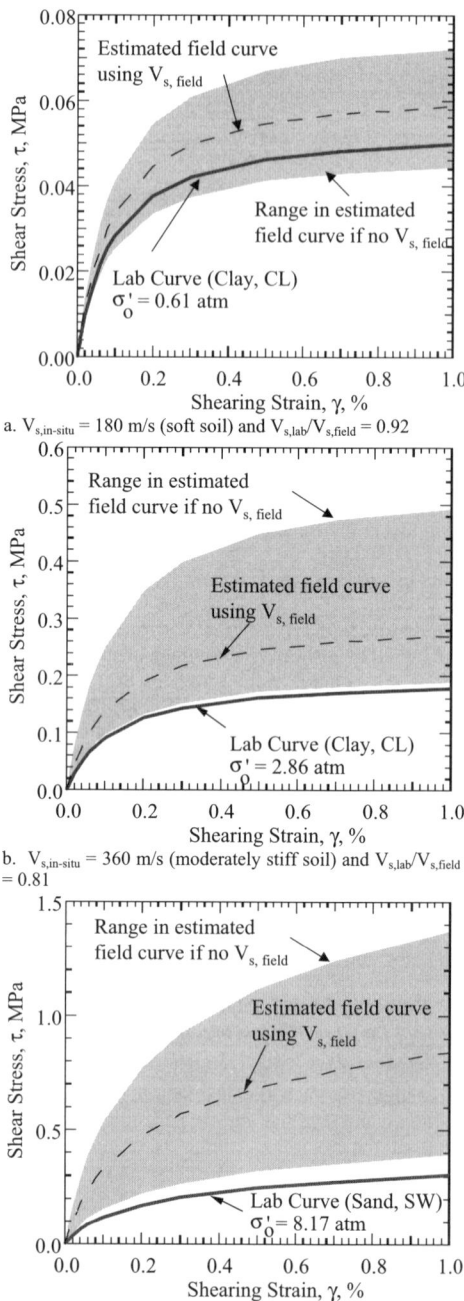

Figure 42. Calculated laboratory τ – γ curves, estimated field τ – γ curves using $V_{s\,field}$, and possible range in field τ – γ curves using range in Figure 40 but no measurement of $V_{s\,field}$; Examples for soils from the ROSRINE project with a range in shear stiffness

with companion sets of G – γ values taken from the dynamic laboratory curve. The laboratory τ – γ curves that were derived from the laboratory G – log γ curves for the soft, medium stiff and very stiff soils in Figures 41a, 41b and 41c are shown by the solid lines in Figures 42a, 42b and 42c, respectively. The estimated field τ – γ curves for these soils were determined following the procedure expressed by Equation 9, except that the estimated field G – log γ curves were used. The estimated field curves are shown by the dashed lines in Figure 42.

This example is presented to show the importance of $V_{s,\,field}$ in predicting the field τ – γ curves which are used for deformational analyses like the ones presented in Section 4.1. In such deformational analyses, shear strains rarely exceed 1 % which is the reason why the τ – γ curves are shown with a scale of 0 to 1 %. Unfortunately, many geotechnical engineers are not aware of this adjustment procedure for sample disturbance or the importance of V_s.

5.4 What If No In Situ V_s Values Are Measured?

At times, the owner or client may elect to test only intact samples and not perform in situ V_s measurements. This decision may be based on cutting cost, incomplete understanding of the importance of $V_{s,\,field}$ or other reasons. In any case, the field G – log γ and τ – γ curves can not be estimated using the adjustment procedure discussed above. Therefore, a wide range in the estimated nonlinear field curve will result. Figure 40 can be used in reverse to find the range in the expected field curves if one only had laboratory G – log γ curves. These ranges are shown by the shaded zones in Figure 41 for the three different soil stiffnesses. The ranges are quite large, exceeding factors of two and three for the moderately stiff and very stiff soils, respectively. The same relative comparison is shown by the shaded zones in Figure 42 for the ranges in expected τ – γ curves.

5.5 What If No Laboratory G – log γ Curves Are Measured?

Obviously, the other situation that the geotechnical engineer might face is having only the in situ V_s measurements. In the writers' opinion, this situation may not be as troublesome as the case above with no $V_{s,\,field}$ values. Hopefully, the engineer has a boring log and the soil types identified. In this case, empirical soil models can be substituted for the laboratory G – log γ curves (assuming no unusual or difficult soils). The empirical models should include variables such as soil type, confinement state, some measure of uncertainty, etc. such as the model by Darendeli (2001). However, this approach would only be used if no laboratory G – log γ curves were measured and will result in wide ranges for the estimated field G – log γ and τ – γ curves.

6 CONCLUSIONS

Geophysical methods have an important and ever-increasing role to play in the solution of geotechnical engineering problems. Seismic methods have been embraced by geotechnical engineers over the past 50 years. They have been heavily used in the solution of soil dynamics and geotechnical engineering problems, especially in the evaluation of small-strain shear and compression stiffnesses. Today, V_s, measurements in the field and laboratory form a critical link in evaluating sample disturbance and in predicting nonlinear G – log γ and τ – γ curves.

The adoption of geophysical methods in the solution of non-dynamic problems has occurred more slowly in geotechnics, excluding geoenvironmental and military applications. Seismic testing is still the most widely used method, particularly for evaluating site characteristics (layering, top of bedrock, voids, etc.) and monitoring processes (grouting, damaged or changed zones from construction activities, etc.). The use of strain-adjusted moduli in settlement and other deformational analyses offers a rational approach to the solution of many of these problems. This approach will continue to grow.

The seismic method that continues to evolve in geotechnical engineering is the surface-wave method. The nonintrusive nature of the method makes its application very cost effective, and its usefulness will continue to increase. It would be very beneficial to this method, as well as other geophysical methods, to incorporate increased automation. Adoption by the profession would also benefit from increased coverage of geophysical methods in the civil engineering curriculum. Finally, the engineer also needs to consider that the robustness of the solution is significantly enhanced in many applications by the use of a suite of geophysical measurements.

ACKNOWLEDGMENTS

The writers sincerely appreciate the opportunity given by the organizers of this conference to present these results. The patience and understanding of Prof. António Viana da Fonseca is especially appreciated.

Support from the California Department of Transportation, the National Science Foundation, the United States Geological Survey, the ROSRINE project, and the U.S. Department of Energy through a subcontract to Bechtel SAIC is gratefully acknowledged. Interaction, encouragement and guidance from many colleagues is appreciated as is the work of many excellent graduate students. Several of the case histories involved work with geotechnical consulting firms and their clients. Permission to publish the results is appreciated. Finally, a special thanks is given to Ms. Alicia Zapata for assisting in preparation of this paper.

REFERENCES

Achenbach, J. D. 1975. Wave Propagation in Elastic Solids. North Holland, 425p.
Addo, K.O., and Robertson, P.K. 1992. Shear-wave velocity measurements of soils using Rayleigh waves. Canadian Geotechnical Journal, 29, pp. 558-568.
Aki, K., and Richards, P.G. 1980. Quantitative Seismology: Theory and Methods, W.H. Freeman and Co., Vol. I.
Al-Hunaidi M.O. 1994. Analysis of dispersed multi-mode signals of the SASW method using the multiple filter/crosscorrelation technique. Soil Dynamics and Earthquake Eng., Vol. 13, Elsevier, pp. 13-24.
Al-Hunaidi M.O., and Rainer J.H. 1995. Analysis of multi-mode signals of the SASW method. Proc. 7th Int. Conf. Soil Dynamics and Earthquake Eng., pp.259-266.
Allen, N. F., Richart, F.E., Jr., and Woods, R.D. 1980. Fluid wave propagation in saturated and nearly saturated sands. Journal of Geotechnical Engineering, ASCE, Vol. 106, No. GT3, pp. 235-254.
Al-Shayea, N. 1994. Detection of subsurface cavities using the spectral-analysis-of- surface waves method. Ph.D. Dissertation, University of Michigan, April, 269 pp.
Al-Shayea, N., Woods, R.D., and Gilmore, P. 1994. SASW and GPR to detect buried objects. Proceedings of the Symposium on the Application of Geophysics to Engineering and Environmental Problems (SAGEEP), Vol. 1, pp. 543-560.
Andrus, R.D. and Stokoe, K.H., II. 2000. Liquefaction resistance of soils from shear-wave velocity. Journal of Geotechnical and Geoenvironmental Engineering, American Society of Civil Engineers, Vol. 126, No. 11, pp. 1019-1025.
ASCE Press. 1998. Geophysical exploration for engineering and environmental investigations. Adapted from the US Army Corps of Engineers, No. 23, Technical engineering and design guides, 204 p.
Blake, W. D., and Gilbert, R. B. 1997. Investigation of possible relationship between undrained shear strength and shear wave velocity for normally consolidated clays. Offshore Technology Conf., Houston, 411-420.
Bolt, B.A. 1976. Nuclear Explosions and Earthquakes, W.H. Freeman and Company.
Briaud, J-L., and Gibbens, R.M., Editors 1994. Predicted and Measured Behavior of Five Spread Footings on Sand. Geotechnical Special Publication No. 41, ASCE. Results of Spread Footing Prediction Symposium, Sponsored by the Federal Highway Administration, June, 255 p.
Burger, H.R. 1992. Exploration Geophysics of the shallow subsurface, Prentice Hall, Englewood Cliffs, New Jersey.
Burland, J.P. 1989. Small is beautiful: the stiffness of soils at small strains. Canadian Geotechnical Journal, Vol. 26, No. 4, pp. 499-516.
Campanella, R.G., Robertson, P.K., and Gillespie, D. 1986. Seismic cone penetration test, Use of in situ tests in geotechnical engineering. Proceedings, In Situ '86, ASCE Geotechnical Specialty Publication No. 6, Samuel P. Clemence (ed.), Blacksburg, VA, June, pp. 116-130.
Chiara, N. 2001. Investigation of small-strain shear stiffness measured in field and laboratory geotechnical studies. M.S. Degree, University of Texas.
Cho, G. C., and Santamarina, J. C. 2001. Unsaturated particulate materials: particle-level Studies. Journal of Geotechnical and Geoenvironmental Engineering, January Vol. 127, Issue 1, pp. 84-96.

Choi, W.J. 2003. Linear and nonlinear dynamic properties from combined resonant column and torsional shear tests of ROSRINE phase-II specimens. Masters Thesis, University of Texas.
Darendeli, M.B. 2001. Development of a new family of normalized modulus reduction and material damping curves. Ph.D. dissertation, University of Texas.
Di Benedetto, H., Doanh, T., Geoffroy, H., and Sauzeat, C. (Eds.) 2003. Deformation Characteristics of Geomaterials. A.A. Balkema Publishers, Tokyo, 1425 p.
Dobrin, M.B., and Savit, C.H. 1988. Introduction to Geophysical Prospecting. Fourth Edition, McGraw-Hill Book Company.
ENR. 1982. Shoehorning Pittsburgh's subway. Cover Story, Engineering News Record, McGraw-Hill, N.Y., July 22, pp. 30-34.
Fernandez, A. 2000. Tomographic imaging stress fields in discrete media. Ph.D. Dissertation, Georgia Institute of Technology, Atlanta.
Gabriels, P., Snieder, R., and Nolet, G. 1987. In situ measurements of shear-wave velocity in sediments with higher-mode Rayleigh waves. Geophys. Prospect., Vol. 35, pp. 187-196.
Ganji V., Gukunski N., and Nazarian S. 1998. Automated inversion procedure for spectral analysis of surface waves. J. Geotech. and Geoenv. Eng., Vol. 124, ASCE, pp. 757-770.
Gucunski N., and Woods R.D. 1991. Inversion of Rayleigh wave dispersion curve for SASW test. 5th Int. Conf. on Soil Dynamics and Earthquake Eng., Kalsruhe, pp. 127-138.
Gucunski N., and Woods R.D. 1992. Numerical simulation of SASW test. Soil Dynamics and Earthquake Eng., Vol. 11 (4), Elsevier, pp. 213-227.
Gucunski, N., and Shokouhi, P. 2004. Application of Wavelets in Detection of Cavities Under Pavement by Surface Waves. Proceedings of the 5th International Conference on Case Histories in Geotechnical Engineering, New York, April 13-17, paper 10.09.
Guy, E.D., Nolen-Hoeksema, R.C., Daniels, J.J., and Lefchick, T. 2003. High-resolution SH-wave seismic reflection investigations near a coal mine-related roadway collapse feature. Journal of Applied Geophysics, Vol. 54, pp. 51-70.
Hardin, B.O. and Drnevich, V.P. 1972a. Shear modulus and damping in soils – I. measurement and parameter effect. Journal of Soil Mechanics and Foundation Engineering, ASCE, Vol. 98, June, pp. 603-624.
Hardin, B.O. and Drnevich, V.P. 1972b. Shear modulus and damping in soils – II. design equations and curves. Journal of Soil Mechanics and Foundation Engineering, ASCE, Vol.98, June, pp. 667-692.
Haskell N.A. 1953. The dispersion of surface waves on multi-layered media. Bulletin of the Seismological Society of America, Vol. 43 (1), pp. 17-34.
Heisey, S., Stokoe, K.H., II and Meyer, A.H. 1982. Moduli of pavement systems from spectral analysis of surface waves. Transportation Research Record 852, pp. 22-31.
Hiltunen, D.R., Nolen-Hoeksema, R.C., and Woods, R.D. 2004. Characterization of abandoned mine sites beneath I-70 via crosshole and SASW seismic wave methods. Proceedings of 5th International Conference on Case Histories in Geotechnical Engineering, New York, N.Y, April 13-17, paper 7.08.
Hossian, M.M., and Drnevich, V.P. 1989. Numerical and optimization techniques applied to surface waves for backcalculation of layer moduli. Nondestructive testing of pavements and backcalculation of moduli. ASTM STP 1026, A.J. Bush III, and G.Y. Baladi, eds., Am. Soc. for Testing and Mat., Philadelphia, PA., pp. 649-669.
Jamiolkowski, M., Lancellotta, R., and Lo Presti, D. (Eds.). 2001. Pre-Failure Deformation Characteristics of Geomaterials. Vols. 1 and 2, A.A. Balkema Publishers, Tokyo

Jardine, R.J., Davies, M.C.R., Hight, D.W., Smith, A.K.C., and Stallebrass, S.E. (Eds.). 1998. Pre-Failure Deformation Behavior of Geomaterials. Thomas Telford Publishing, London, 417 pp.

Joh, S.-H. 1996. Advances in interpretation and analysis techniques for spectral-analysis-of-surface-waves (SASW) measurements. Ph.D. dissertation, University of Texas at Austin.

Joh, S.-H., Rosenblad, B. L., and Stokoe, K. H., II. 1997. Improved data interpretation method for SASW tests at complex geotechnical sites. International Society of Offshore and Polar Engineering.

Kansas Geological Survey 2003. Handout of the MASW workshop: Symposium on the application of geophysics to engineering and environmental problems, San Antonio, TX.

Kausel E., and Roësset J.M. 1981. Stiffness matrices for layered soils. Bulletin of the Seismological Society of America, Vol. 71 (6), pp. 1743-1761

Kitsunezaki, C. 1980. A new method for shear wave logging. Geophysics, Vol. 45, pp. 1489-1506.

Konstantinidis, B., Van Riessen, G., and Schneider, J.P. 1986. Structural settlements at a major power plant. Settlement of Shallow Foundations on Cohesionless Soils: Design and Performance, Geotechnical Special Publication No. 5, ASCE, Seattle Washington, pp.54-73.

Landisman, M., Dziewonski, A, Sato, Y., and Masse, R. 1968. Cited in Geophys. J. Roy. Astron. Soc., Vol. 17, p.369.

Liu, H.-P., Warrick, R.E., Westerlund, R.E., Fletcher, J.B., and Maxwell, G.L. 1988. An air-powered impulsive shear-wave source with repeatable signals. Bulletin Seismological Society of America, Vol. 78, No. 1, pp 355-369.

Long, L.G. 1980. Comparison of field and laboratory dynamic soil properties," M.S. Thesis, University of Texas at Austin.

Luke, B.A., and Stokoe, K.H., II. 1998. Application of SASW method underwater. Journal of Geotechnical and Geoenvironmental Engineering, Vol. 124, No. 6, June, pp. 523-531

Luke, B.A., Stokoe, K.H., II, Bay, J.A., and Nelson, P.P. 1999. Seismic Measurements to investigate disturbed rock zones. 3rd National Conference of Geo-Institute of ASCE, Urbana-Champaign, IL.

Lunne, T., Robertson, P.K., and Powell, J.M. 1997. Cone Penetrometer Testing in Geotechnical Practice. 1st Edition, London, NY, Blackie Academic & Professional Press, 312 p.

Madianos, M., Nelson, P.P., and Stokoe, K.H., II. 1990. Evaluation of discrete discontinuities and repeated monitoring of rock mass characteristics with SASW testing,. 8th Annual Workshop, Generic Mineral Technology Center, U.S. Bureau of Mines, Reno, Nevada, pp. 15-23.

Matthews M.C., Hope V.S., and Clayton C.R.I. 1996. The use of surface waves in the determination of ground stiffness profiles. Geotechnical Eng., Vol. 119, Proc. Inst. Civil Eng, pp. 84-95

McMechan, G.A., and Yedlin, M.J. 1981. Analysis of dispersive waves by wave field transformation. Geophysics, Vol. 46, pp. 869-874.

Menke, W. 1989. Geophysical Data Analysis: Discrete Inverse Theory. Academic Press, 289 p.

Menzies B., and Matthews M. 1996. The continuous surface-wave system: a modern technique for site investigation. Special Lecture: Indian Geot. Conf., Madras

Miller, R.D., Xia, J., Park, C.B., and Ivanov, J. 1999. Multichannel analysis of surface waves to map bedrock. *The Leading Edge*, Vol. 18, no. 12, pp. 1392-1396.

Nazarian, S., and Stokoe, K.H., II. 1984. Nondestructive testing of pavements using surface waves. Transportation Research Record 993, pp. 67-79.

Nazarian, S., Baker, M., and Crain, K. 1995. Use of seismic pavement analyzer in pavement evaluation. Transportation Research Record 1505, pp.1-8.

Nigbor R.L., and Imai T. 1994. The suspension P-S velocity logging method. Geophysical Characteristics of Sites, ISSMFE, Technical Committee 10 for XIII ICSMFE, International Science Publishers, New York, pp. 57-63.

NRC. 2000. Seeing into the earth. Committee for Noninvasive Characterization of the Shallow Subsurface for Environmental and Engineering Applications, P.R. Romig, Chair, 129 p.

Olson, L.D., Sack, D.A., and Stokoe, K.H., II. 1993. Stress-wave nondestructive testing of tunnels and shafts,. Transportation Research Record, No. 1415, pp. 95-99.

Park, C.B., Miller, R.D., and Xia, J. 1999. Multichannel analysis of surface waves (MASW): Geophysics, Vol. 64, pp. 800-808.

Richart, F.E., Jr., Hall, J.R. and Woods, R.D. 1970. Vibrations of Soils and Foundations. Englewood Cliffs, New Jersey, Prentice Hall, 414 p.

Rix G.J., and Leipski, A.E. 1991. Accuracy and resolution of surface wave inversion. Recent advances in instrumentation, data acquisition and testing in soil dynamics. Geotechnical special publication, No. 29, N.Y., ASCE 1991, pp. 17-23.

Robertson, P.K., and Campanella, R.G. 1985. Liquefaction potential of sands using the CPT. Journal of Geotechnical Engineering, ASCE, Vol. 111, No. 3, pp. 384-403.

Robertson, P.K., Campanella, R.G., Gillespie, D., and Rice, A., 1986. Seismic CPT to measure in situ shear wave velocity. Journal of Geotechnical Engineering, American Society of Civil Engineers, Vol. 112, No. 8, pp 791-803.

Robertson, P.K., and Mayne, P.W. (Eds.). 1998. Geotechnical Site Characterization. Vols. 1 and 2, A.A. Balkema Publishers, Rotterdam.

Roësset J.M., Chang D.W., Stokoe K.H. 1991. Comparison of 2-D and 3-D models for analysis of surface wave tests. Proc. 5th Int. Conf. on Soil Dyn. and Earthq. Eng., Kalsruhe, Vol. 1, pp. 111-126.

Rosenblad, B.L. 2000. Experimental and Theoretical studies in support of implementing the spectral-analysis-of-surface-waves (SASW). Ph.D., University of Texas.

Rosenblad, B. L., Stokoe, K.H., II, Kalinski M.E. and Kavazanjian E., 2003 Shear wave velocity profiles of sediments determined from surface wave measurements," 13th International Offshore and Polar Engineering Conference, Honolulu, Hawaii.

Rosenblad, B.L. and Stokoe, K.H., II. 2001. Offshore shear wave velocity profiling using interface waves. OTRC 2001 Conference honoring Prof. Wayne Dunlap, Houston, TX.

Santamarina, J. C., and Fratta, D. 1998. Introduction to Discrete Signals and Inverse Problems in Civil Engineering. ASCE Press, Reston, 327p.

Santamarina, J. C., Klein, K.A., and Fam, M.A. 2001. Soils and Waves. John Wiley & Sons, Ltd., Rexdale, Ontario, Canada, 488 p.

Sato, Y. 1971. Bull. Earthq. Res. Inst. Tokyo, Vol. 33, p.33., Cited in Dziewonski A.M. and Hales, A.L., Numerical Analysis of Dispersed Seismic Waves. Methods in computational physics, Vol. 11, pp.39-85.

Seed, H.B. 1969. Influence of Local Soil Conditions on Earthquake Response. Proceedings of Specialty Session 2 – Soil Dynamics, Seventh International Conference on Soil Mechanics and Foundation Engineering, Mexico City, Mexico.

Seed, H.B., and Idriss, I.M. 1971. Simplified procedure for evaluating soil liquefaction potential. Journal of Soil Mechanics and Found. Div., ASCE, Vol. 97, No. 9, pp. 1249-1273.

Sharma, P.V. 1997. Environmental and Engineering Geophysics, Cambridge University Press, Cambridge, U.K., 475 pp.

Shokouhi, P. and Gucunski, N. 2003. Application of Wavelet Transform in Detection of Shallow Cavities by Surface Waves. Proceedings of the Symposium on the Application of Geophysics to Engineering and Environmental Problems (SAGEEP), EEGS, San Antonio, TX.

Sibuya, S., Mitachi, T., and Miura, S. (Eds.) 1994. Symposium on Pre-Failure Deformation Characteristics of Geomaterials, Two Volumes, Sapporo, Japan.

Stokoe, K.H., II and Santamarina, J.C. 2000. Seismic-wave-based testing in geotechnical engineering. Plenary Paper, International Conference on Geotechnical and Geological Engineering, GeoEng 2000, Melbourne, Australia, pp. 1490-1536.

Stokoe, K.H., II, Rosenblad, B.L., Bay, J.A., Redpath, B., Diehl, J.G., Steller, R.A., Wong, I.G., Thomas, P.A. and Luebbers, M., 2003. Comparison of V_S profiles from three seismic methods at Yucca Mountain. Volume I, Soil and Rock America 2003,Cambridge, MA, pp. 299-306.

Stokoe, K.H.II, Rosenblad, B.L., Wong, I.G., Bay, J.A. Thomas, P.A., and Silva, W.J. 2004. Deep V_s profiling along the top of Yucca Mountain using a vibroseis source and surface waves. 13th World Conference on Earthquake Engineering, Vancouver, B.C., Canada (accepted for publication).

Stokoe, K.H., II, and Nazarian, S. 1985. Use of Rayleigh waves in liquefaction studies. Proceedings, Measurement and Use of Shear Wave Velocity for Evaluating Dynamic Soil Properties. Geotechnical Engineering Division, ASCE, pp. 1-17.

Stokoe, K.H., II, Wright, S.G., Bay, J.A. and J.M. Roesset 1994. Characterization of geotechnical sites by SASW method. Geophysical Characteristics of Sites, ISSMFE, Technical Committee 10 for XIII ICSMFE, International Science Publishers, New York, pp. 15-25.

Stoll, R.D., Bryan, G.M., and Bautista, E.O. 1994. Measuring lateral variability of sediment geoacoustic properties. Journal Acoustical Society of America, Vol. 96, No. 1, pp. 427-438.

Swiger, W.F. 1974. Evaluation of Soil Moduli. Proceedings of Conference on Analysis and Design in Geotechnical Engineering, Vol. II, Austin, TX, June, pp. 79-92.

Thomson W.T. 1950. "Transmission of elastic waves through a stratified solid medium. J. Applied Physics, Vol. 21 (1), pp. 89-93

Tokimatsu, K. 1995. Geotechnical site characterization using surface waves. First International Conference on Earthquake Geotechnical Engineering, Vol. 3, Kenji Ishihara, editor, Tokyo, pp. 1333-1368.

Tokimatsu, K., Tamura, S., and Kojima, H. 1992. Effects of multiple modes on Rayleigh wave dispersion. J. Geotech. Engrg., ASCE, 118(10), 1529-1543.

Toksoz, M.N., and Cheng, C.H. 1991. <u>Wave Propagation in a Borehole.</u> in J.M. Hovem, M.D. Richardson, and R.D. Stoll (eds.), Shear Waves in Marine Sediments, Kluwer Academic Publishers, Dordrecht, The Netherlands.

Volante, M., Scattolini, E., and Pizzarotti, E.M. 2004. Injection Consolidation Under the Piers of the Railway Bridges for the Rehabilitation of the Line Merano-Malles. Proceedings of the 5th International Conference on Case Histories in Geotechnical Engineering, New York, N.Y., April 13-17, Paper No. 10.07.

Ward, S. H. 1990. <u>Geotechnical and Environmental Geophysics.</u> Investigations in Geophysics No. 5, Society of Exploration Geophysics, 3 volumes.

Woods, R. D., and Partos, A. 1981. Control of Soil Improvement by Crosshole Testing. Proc. of the 10th International Conference of the Inter. Soc. for Soil Mech. and Found. Engrg. Stockholm, Sweden, Vol. 3, June, pp. 793-796.

Wright, S.G., Stokoe, K.H., II and Roesset, J.M. 1994. SASW Measurements at geotechnical site overlaid by water. Dynamic Geotechnical Testing: Second Volume, ASTM STP 1213, R.J. Ebelhar, V.P. Drnevich and B.L. Kutter, Eds., American Society for Testing and Materials, Philadelphia, pp. 39-57.

Xia, J., Miller, R.D., and Park, C.B. 1999. Estimation of near-surface velocity by inversion of Rayleigh wave. Geophysics, 64, 691-700.

Yokota, K. and Konno, M. (1985). Comparison of soil constants obtained from laboratory tests and in situ tests. Symposium on Evaluation of Deformation and Strength of Sandy Grounds. Japanese Society of Soil Mechanics and Foundation Engineering, pp. 111-114 (in Japanese).

Yasuda, S. and Yamaguchi, I. 1985. Dynamic shear modulus obtained in the laboratory and in situ. Symposium on Evaluation of Deformation and Strength of Sandy Grounds. Japanese Society of Soil Mechanics and Foundation Engineering, pp. 115-118 (in Japanese).

Youd, T.L., Idriss, I.M., Andrus, R.D., Arango, I., Castro, G., Christian, J.T., Dobry, R., Finn, W.D.L., Harder, L.F., Jr., Hynes, M.E., Ishihara, K., Koester, J.P., Liao, S.S.D., Marcuson W.F., III, Martin, G.R., Mitchell, J.K., Moriwaki, Y., Power, M.S., Robertson, P.K., Seed, R.B., and Stokoe, K.H., II. 2001. Liquefaction resistance of soils: summary of report from the 1996 NCEER and 1998 NCEER/NSF workshops on evaluation of liquefaction resistance of soils. Journal of Geotechnical and Geoenvironmental Engineering, Vol. 127, No. 10, pp. 817-833.

Yuan D., and Nazarian S. 1993. Automated surface wave method: inversion technique. J. Geotechnical Eng., Vol. 119, No.7, ASCE, pp. 1112-1126

Deformation properties of fine-grained soils from seismic tests

K.R. Massarsch
Geo Engineering AB, Stockholm, Sweden

Keywords: clay, dynamic testing, fine-grained soils, elastic modulus, rate of loading, seismic testing, shear modulus, shear strain, shear strength, silt, spring constant, strain rate, subgrade reaction

ABSTRACT: Geotechnical design requires the prediction of soil structure interaction, for which the deformation properties of the soil are needed. Little guidance can be found in the literature for estimating the soil modulus during undrained loading. Therefore, over-simplified methods are frequently used even for the analysis of complex problems. The concepts used to describe the deformation behavior of fine-grained, normally consolidated soils are presented and critically reviewed. The deformation properties (shear modulus) at small and large strain are discussed. Based on a comprehensive survey of seismic field and laboratory data, it is possible to predict the shear modulus at small strain and the variation of the shear modulus with increasing shear strain. A relationship is proposed which can be used to predict the variation of the normalized shear modulus as a function of shear strain. It can be shown that the strain rate at seismic small-strain testing is slow and comparable to that of conventional geotechnical laboratory tests. The starting point of the stress-strain curve (at low shear strain level) can be accurately established from seismic tests, and its end point (at high strain) by conventional shear tests. The variation of the shear modulus with strain can be determined from resonant column tests. A numerical model is presented which makes it possible to predict the variation of shear modulus as a function of shear strain. The practical application of the concept is illustrated by a case history, where good agreement was obtained between predicted and measured deformation properties.

1 INTRODUCTION

1.1 *International Conference on Geotechnical Site Characterization, ISC in 1998*

The prediction of the deformation behavior of soils has been an important task in geotechnical research and has become increasingly important as more sophisticated analytical methods have become available. At the first International Conference on Geotechnical Site Characterization, ISC in 1998, several papers were presented, which addressed this topic. In this context, geophysical testing – and in particular seismic testing – can play an important role. Many valuable concepts were presented in one of the Theme Lectures "Deformation and in situ stress measurements", Fahey (1998). The paper outlined a generally applicable framework for establishing deformation parameters, which are required for deformation analysis of geotechnical structures. In his conclusions, Fahey stated that: *"predictions of soil deformations under the influence of foundation loads have been generally found to be of very limited accuracy. A major reason for this has been that the non-linearity of the stress-strain response in this strain range has not been taken into account until recently"*. He concluded that *"a number of questions needed to be answered with the seismic methods, particularly how the effect of soil "fabric" on shear wave velocity can be differentiated from the effects of various principle stresses"*.

The present paper addresses the same issues, but with emphasis on the undrained deformation behavior of normally consolidated fine-grained soils. The objective is to present a practically applicable concept for establishing the stress-strain behavior from very low to large strains. Although the paper focuses on normally consolidated soils at undrained conditions, the basic concept presented herein could be expanded to other soil types and loading conditions.

1.2 *Static versus Dynamic Soil Behavior*

Research on the stress-strain behavior of soils has been an important issue in earthquake and off-shore engineering. Major progress has been made in developing laboratory and field testing methods which have become routine tools for practicing engineers.

Today it is possible to solve even complex dynamic soil-structure interaction problems. However, these advances have not been recognized by geotechnical engineers, and surprisingly crude soil models are still being used for analyzing static soil-structure interaction problems. One reason for this gap of knowledge between soil dynamics and traditional geotechnical engineering was – and in many cases still is – the notion that dynamic (and cyclic) soil properties can not be used for the analysis of static geotechnical problems.

This paper aims to demonstrate that the rate of loading during seismic small-strain testing is comparable to – or even slower than – most conventional geotechnical field testing. Thus, geotechnical engineers can use the information obtained from seismic tests for the analysis of conventional geotechnical soil-structure interaction problems.

1.3 Serviceability limit state (SLS)

Geotechnical design is based on the fundamental requirement that a structure and its components are safe under maximum loads and forces. However, the structure must also be capable to serve the designed functions without excessive deformations. The onset of excessive deformations is called serviceability limit state (SLS). SLS is defined as the state beyond which specified service requirements are no longer met. The evolving standard on which geotechnical design in Europe will be based, Eurocode 7 establishes the principles and requirements for safety and serviceability of structures. Traditionally, geotechnical engineers were trained to, and capable of analyzing and designing stability and bearing capacity problems, which require information about the strength of foundation materials. In contrast, designing structures for normal operating conditions (SLS) requires often new, more complex analytical concepts, and more sophisticated soil models. These must account for the variation of soil stiffness (soil modulus) over a wide strain range.

During the recent past, important progress has been made in the development of analytical methods, which can treat even complex loading situation and accommodate sophisticated soil models. The main limitation in the past has been – and in many cases still is – the difficulty to select realistic deformation parameters for soils. Little effort is often spent on verifying that the chosen soil parameters realistically represent the actual foundation conditions, even in the case of important and complex projects. One of the most difficult soil parameters to assess is soil stiffness (modulus), and its variation with stress (strain).

Under undrained conditions, deformations in fine-grained soils occur quickly. However, the assumed soil stiffness has an important effect on the calculated response, i.e. influences the interaction between the construction element (e.g. a pile or sheet pile) and the surrounding soil.

1.4 Simplified Soil Models

In the early days of soil mechanics, deformation properties of soils were chosen based on practical experience, i.e. from the observation and back-analysis of actual projects. It was known early on that almost all soils behave "non-linearly" even at low stress levels. However, suitable investigation methods (in the field and laboratory) did not exist. Therefore, empirical correlations were developed between the elastic modulus, E (Young's modulus) of the soil and soil parameters obtained from various testing methods, such as the Standard Penetration Test (SPT) or the cone penetration test (CPT).

In order to calculate the contact pressure and stress distribution below footings, it was necessary to develop simplified soil models, e.g. by replacing the supporting soil by a bed of equally spaced and equally compressible springs, Terzaghi & Peck (1948). In spite of this crude assumption, the concept has found wide-spread acceptance and is still used by many geotechnical engineers. The ratio between the applied stress and the corresponding settlement is known as the "coefficient of subgrade reaction", k_s which is defined as

$$k_s = \frac{p}{s} \tag{1}$$

where p, kg/cm^2 = load and s, cm = deformation of the subsoil (subgrade). In an elastic material, the settlement below the center of a rigid plate can be calculated from

$$s_0 = \frac{r\pi p}{2E}(1-v^2) \tag{2}$$

where r = plate radius, E = modulus of elasticity and v = Poisson's ratio. With the definition of the coefficient of subgrade reaction according to equation 1, the following relationship between k_s and E can be obtained

$$k_s = \frac{E}{2r(1-v^2)} \tag{3a}$$

Note that this relationship depends on the plate size, which normally is 50 to 70 cm. For fine-grained soils it can be assumed that $v = 0.5$, which gives the following expression

$$k_s = \frac{E}{1.5r} \tag{3b}$$

The modulus of subgrade reaction is equivalent to the spring constant, which is commonly used to analyze the dynamic response of foundations on elastic material. The spring constant represents the load required to move the foundation block in the direction of the force, exerted by the load through a distance 1.

In Sweden, it is frequently assumed that $k_s = 80\tau_{fu}$ where τ_{fu} = the undrained shear strength determined by the field vane test and corrected for plasticity. Broms (1963) proposed the following, still widely used, relationship for the calculation of the lateral resistance of piles in clay. The modulus of subgrade reaction, k_0 for a rigid plate with a side length of 1.0 m, and assuming $v = 0.5$, can be estimated from

$$k_0 = 1.67 E_s \quad (4)$$

where E_s is the equivalent modulus of elasticity. The value of E_s depends on the stress level. At 50 % of the failure load (factor of safety = 2) at short-term loading (undrained conditions), E_s is equal to 50 – 200 times the undrained shear strength, τ_{fu}.

2 ESTIMATION OF SOIL MODULUS

2.1 Elastic Modulus

At undrained loading, the elastic modulus, E_u reflects the immediate settlements which occur before consolidation starts. Due to difficulties of determining the deformation characteristics by laboratory tests, empirical relationships are frequently used. It is often assumed that E_u is related to the undrained shear strength. Bjerrum (1972) has proposed the ratio E_u / τ_{fu} ranges from 500 to 1500, where τ_{fu} is determined by the vane shear test. The lowest value is for highly plastic clays, where the applied load is large. The highest value is for clays of low plasticity, where the added load is relatively small. A wide range of values has been proposed in the literature, cf. Holtz & Kovac (1981).

In Fig. 1, the ratio of the elastic modulus E_f normalized by the undrained shear strength, τ_f is plotted against plasticity index, *PI*. There is much scatter for *PI* below 50 and not much data available for higher *PI* values. The scatter is not surprising, considering the different methods used to measure the undrained shear strength and the stress level, at which the modulus values were determined. The above given range of values and the data shown in Fig. 1 are of little benefit for design.

Figure 2 shows for the case of normally consolidated clays the variation of the normalized modulus E_u/s_u as a function of the applied shear stress, τ_n / s_u, after Ladd et al. (1977). The elastic modulus decreases with increasing shear stress and this effect can explain to some extent the large scatter of values in Fig. 1. The normalized modulus decreases with increasing plasticity index. At low shear stress level (0.2), the E_u / s_u ratio varies between 100 – 1500, and decreases at higher shear stress level (0.8) to 25 – 700.

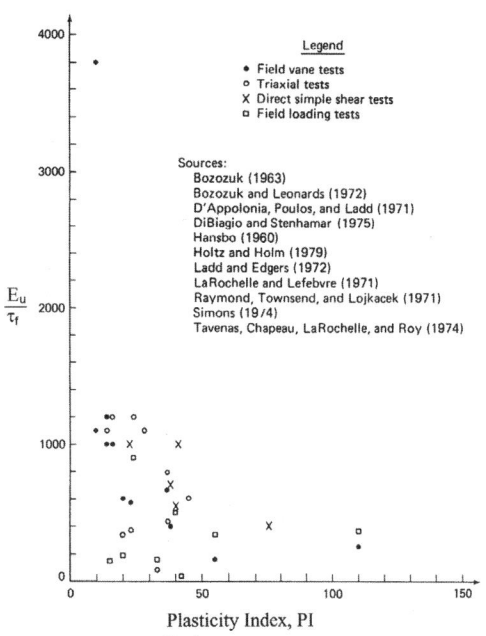

Figure 1. The ratio E_u / τ_{fu} versus plasticity index, PI as reported by several authors, Holtz & Kovacs (1981).

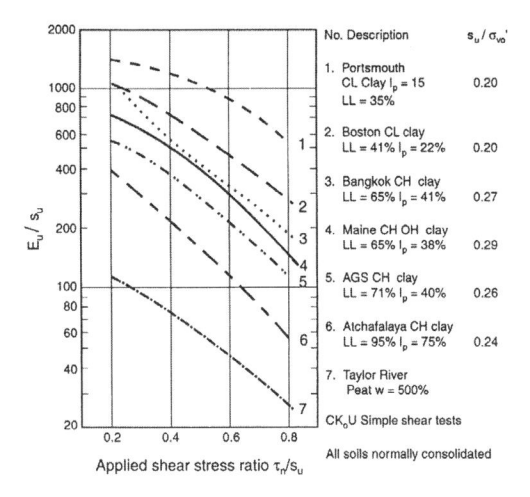

Figure 2. Modulus ratio as a function of the shear stress ratio, from Lunne et al. (1997), after Ladd (1977). Note the semi-logarithmic scale.

2.2 Definitions

For the case of an elastic material, Hooks law applies, which defines the relationship between the vertical compression ε_z and the axial stress σ_z

$$\varepsilon_z = \frac{\sigma_z}{E} \qquad (5)$$

where E is Young's modulus of elasticity. The ratio between strains in the three directions is given by

$$\varepsilon_x = \varepsilon_y = -\nu \varepsilon_z \qquad (6)$$

where ε_x and ε_x are the strains in the directions x and y, respectively and ν is Poisson's ratio. If the shear stress τ_{zx} is applied to an elastic cube, shear distortion γ_{zx} is related to the shear stress according to

$$\gamma_{zx} = \frac{\tau_{zx}}{G} \qquad (7)$$

where G is the shear modulus. From equation 5 and 7, the relationship between the modulus of elasticity E and the shear modulus G is obtained

$$G = \frac{E}{2(1+\nu)} \qquad (8)$$

The relationship depends thus on Poisson's ratio, ν which needs to be assessed. It is commonly assumed that for undrained conditions in fine-grained soils, $\nu = 0.5$. However, this assumption is not necessarily valid at small strain levels, where ν can be significantly lower (0.15 – 0.3). This aspect can have important consequences when interpreting the results of small-strain tests, but it is usually not recognized when applying equation 8.

2.3 Definitions of Shear Modulus, G

The value of the shear modulus depends on the strain level (or the applied shear stress level, i.e. the factor of safety), cf. Fig. 2. In Fig. 3 a typical shear stress-shear strain relationship is shown for fine-grained soils at undrained loading. Three commonly used definitions of the shear modulus are indicated. At very low stress levels (very low strains), the shear modulus is called the maximum shear modulus, G_{max}. With increasing stress level, the shear modulus decreases, cf. Fig. 2. At a stress level corresponding to 50 % of the failure stress the term G_{50} is frequently used, which corresponds to a factor of safety typical for normal operating conditions. At failure, the shear modulus is defined as G_f.

The stress-strain relationship for the case of repeated loading is shown in Fig. 4. The initial loading curve (G_{max}) and the unloading-reloading curves are shown. It is common practice to define the stress-strain relationship of soils by the secant modulus, G_s. Note that at unloading and re-loading, the modulus is often assumed to correspond to the modulus at initial loading, G_{max}.

Frequently, the soil modulus is normalized by the undrained shear strength, cf. Fig. 2. It is implicitly assumed that a linear relationship exists between soil stiffness and soil strength. However, this assumption is incorrect. For normally consolidated, fine-grained soils, a close correlation exists between the ratio τ_f / σ_v' and PI (Bjerrum, 1972)

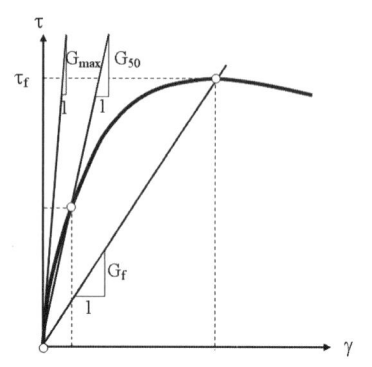

Figure 3. Shear stress – shear strain relationship for fine-grained soil at undrained loading.

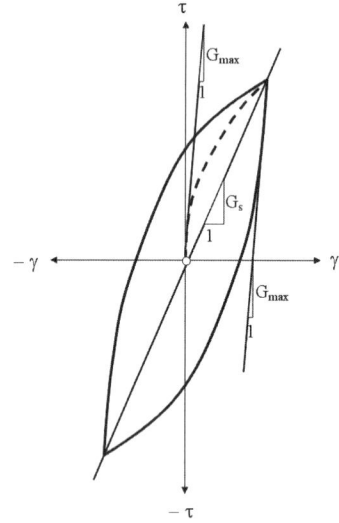

Figure 4. Stress-strain relationship during shear for soils at repeated loading.

$$\frac{\tau_{fu}}{\sigma_v'} = 0.0029 PI + 0.13 \qquad (9)$$

where σ_v' is the vertical effective stress. On the other hand, the shear modulus at small strains is not related linearly to the effective (overburden) stress, as will be shown in the next section.

3 SHEAR MODULUS AT SMALL STRAINS

3.1 Empirical Correlations

Hardin (1978) has proposed the following semi-empirical relationship for the estimation of the shear modulus at small strains, G_{max}.

$$G_{max} = \frac{625}{0.3 + 0.7e^2} OCR^k \left(\sigma_0' p_a\right)^{0.5} \tag{10}$$

where e = void ratio, OCR = overconsolidation ratio, k = empirical constant, which depends on PI, σ_0' is the mean effective stress and p_a is a reference stress (98.1 kPa). The shear modulus at small strains is thus a function of the square root of the mean effective stress and thus also of the vertical effective stress. Therefore, the assumption of a linear relationship, G/τ_f is not justified.

The mean effective stress σ_0' can be determined from

$$\sigma_0' = \frac{(1+2K_0)}{3} \sigma_v' \tag{11}$$

where K_0 is the coefficient of lateral earth pressure at rest (effective stress). An empirical relationship has been proposed for normally consolidated clays, Massarsch (1979)

$$K_0 = 0.0042 PI + 0.44 \tag{12}$$

Hardin (1978) has suggested the following relationship for estimating the parameter k from PI

$$k = 0.006 PI + 0.045 \tag{13}$$

Equation 10 for estimating G_{max} is in reasonable agreement with measured values for soft clay and silt (Andreasson, 1979, Bodare, 1983, Langö, 1991, Länsivaara, 1999, Larsson & Mulabdic, 1991, Massarsch, 1985, Vucetic & Dobri, 1991).

3.2 Determination of Shear Modulus at Small Strains

The shear modulus at small strain can be determined accurately in the field and in the laboratory. In the field, the seismic down-hole and/or cross-hole test have become routine methods, while the SASW-method is becoming increasingly popular. A description of the different seismic methods and their practical application is given by Stokoe & Santamarina (2000). In Scandinavia, the dynamic plate load test has been used by several investigators, (Andreasson, 1979, Bodare, 1983).

Seismic and dynamic laboratory tests have been described in the literature, e.g. Woods and Henke (1981) and Woods (1994). An interesting development is the bender element measuring technique, which can be combined with conventional laboratory testing methods, such as the triaxial and oedometer test, (Dyvik & Olsen, 1989).

The measuring accuracy of conventional laboratory tests has also improved and stress-strain measurements can now be performed at very low strain levels, during triaxial, simple or direct shear tests.

A high-precision torsional shear test was developed at the University of Kentucky, Drnevich & Massarsch (1979). The unique feature of this device at that time was that the shear modulus could be measured with high accuracy at shear strains as low as 0.001%. The strain rate of the torsional shear test was 0.1 Hz, thus more than one order of magnitude lower than that of a resonant column test. Comparative tests on clays, silty sands and sands have shown that at small strains ($\leq 0.001\%$) the modulus is almost independent of frequency and thus of strain rate. This aspect will be discussed below in more detail.

3.3 Correlation between G_{max} and τ_{fu}

Döringer (1997) analyzed data from seismic field and laboratory measurements on fine-grained soils. The tests were evaluated, using the concept presented in the previous section. Substituting equation 9 into equation 10, and inserting appropriate values for k and K_0, as well as by replacing void ratio, e by water content w_n (assuming saturated conditions), the relationship given in equation 14 is obtained. Note in this relationship G_{max} is normalized by the square root of the undrained shear strength, τ_{fu}.

$$\frac{G_{max}}{\sqrt{\tau_{fu} p_a}} =$$

$$= \frac{625}{0.3 + 0.7 e^2} OCR^k \sqrt{\frac{1+2K_0}{3(0.0029 PI + 0.13)}} \tag{14a}$$

$$\frac{G_{max}}{\sqrt{\tau_{fu} p_a}} =$$

$$= \frac{625}{0.3 + 0.7 \left(w_n \dfrac{\rho_s}{\rho_w}\right)^2} OCR^k \sqrt{\frac{1+2K_0}{3(0.0029 PI + 0.13)}}$$

$$\tag{14b}$$

where w_n = natural water content, ρ_s = density of solid particles and ρ_w = density of water. The normalized shear modulus is shown in Fig. 5 as a function of the water content, for different values of the plasticity index, PI and assuming normally consolidated soil (OCR = 1). It is apparent that the normalized shear modulus decreases markedly when the water content of the soil increases. The reduction of shear modulus is less pronounced at higher water content.

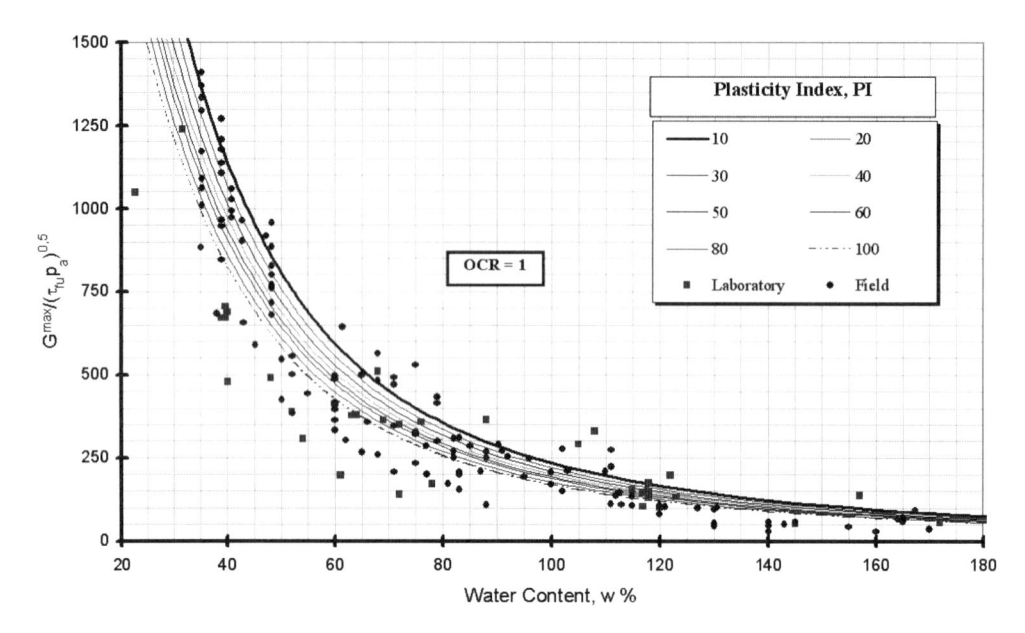

Figure 5. Relationship between the normalized shear modulus at small strains, G_{max} and the water content, cf. equation 14b; from Döringer (1997).

In Fig. 5 are also shown the results of seismic measurements from field and dynamic laboratory tests in a wide variety of soils, reported in the literature, Döringer (1997). In spite of the fact that different methods were used to determine the undrained shear strength, the data follow the semi-empirical relationship. Modulus values from field measurements are generally about 10 to 20% higher than those from laboratory measurements.

It is apparent that water content (and thus void ratio) has a strong influence on the small-strain modulus. The normalized shear modulus (at small shear strain) is much higher in silty clays and silts than in clays, and can range from 1000 – 2000. In the case of low-plastic clays ($w_n = 20$ %), the ratio is in excess of 1500 but decreases to 200 when w_n approaches 100 %. The value can be even lower in organic soils. The large scatter of values shown in Fig. 1 and Fig. 2 is thus not surprising.

Equation 14 can be used to estimate the shear modulus and thus also the shear wave velocity. In the case of normally consolidated soft clay with $\tau_{fu} = 15$ kPa, $w_n = 80\%$, and $PI = 60$, the normalized shear modulus ratio $G_{max}/(\tau_{fu}p_a)^{0.5} = 280$. In this case, the rigidity index $G_{max}/\tau_{fu} = 230$. Assuming that at small strains $\nu = 0.3$ then $E_{max}/\tau_{fu} = 600$. This value is in reasonable agreement with the range of modulus values at low shear stress level, shown in Fig. 2.

4 EFFECT OF STRAIN ON SHEAR MODULUS

4.1 General trends

The shear modulus is affected by stress level and thus by strain level. The measuring accuracy of conventional laboratory testing devices is limited and these can usually not measure G_{max}. On the other hand, the Resonant Column (RC) test can measure shear strain levels down to 10^{-4} % or lower with high precision. Figure 6 shows the results of a RC test on a reconstituted sample of silty clay, Drnevich & Massarsch (1979). The test was performed at a vibration frequency of approximately 30 Hz. At shear strains lower than 10^{-3} %, the shear modulus is almost constant ($G_{max} = 77$ MPa). However, with increasing shear strains, the modulus decreases markedly and is at 0.1% shear strain 24 MPa, i.e. only 30 % of the maximum value. In conventional laboratory tests, the first data readings would usually be taken at this strain level!

It is thus not surprising that conventional laboratory tests grossly underestimate soil stiffness.

Massarsch (1985) reported results from resonant column tests on a variety of fine-grained soils. Figure 7 shows these results with the normalized shear modulus G_s/G_{max} as a function of shear strain in linear scale. It can be seen that PI has a strong influence on the degradation of the shear modulus. The shear modulus decreases more rapidly in low-plastic soils.

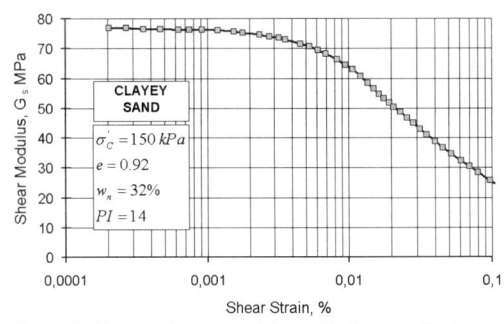

Figure 6. Change of shear modulus with shear strain determined from resonant column test, after Drnevich & Massarsch (1979).

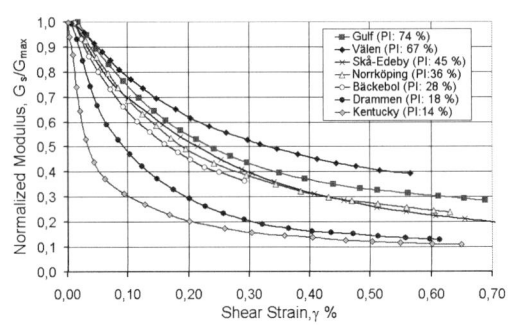

Figure 7. Normalized stress-strain relationship of silts and clays, determined from RC test, (Massarsch, 1985).

4.2 Stress-Strain Behavior

The stress-strain behavior of fine-grained soils has been investigated extensively in the areas of soil dynamics and earthquake engineering. Recommendations have been given for estimating the shear modulus as a function of shear strain (Kovacs et al., 1971, Seed & Idriss, 1970). The most widely used correlation between was proposed by Vucetic and Dobry (1991).

Based on a review of laboratory test data published in the literature, they published stress-strain curves, which are shown in Fig. 8. The effect of soil plasticity and number of loading cycles on the stress-strain relationship are also indicated. It can be concluded that in fine-grained soils the plasticity index, PI is the most important parameter for the stress-strain behavior and thus the modulus reduction curve. Soils with higher plasticity generally exhibit a more linear stress-strain behavior. The number of strain cycles also affects the soil modulus, which decreases as the number of cycles increases. Since the stress-strain curves are given in chart form this complicates their application in numerical analysis.

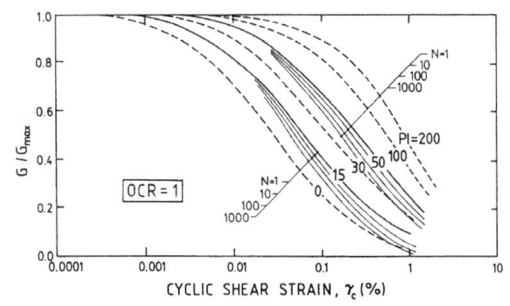

Figure 8. Variation of the normalized shear modulus for normally consolidated soils as a function of the cyclic shear strain, (Vucetic & Dobri, 1991). Indicated is the effect of the plasticity index, PI and the number of loading cycles..

Döringer (1997) analyzed stress-strain data published in the literature (mainly RC tests) and performed a regression analysis. A modulus reduction factor, $R_m = G_s/G_{max}$ was used to define the reduction of the shear modulus G_s at three shear strain levels, 0.1, 0.25 and 0.5 %, cf. Fig. 9.

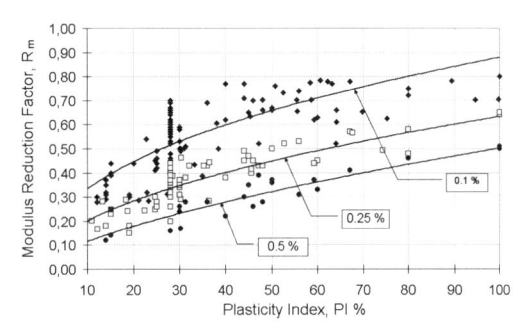

Figure 9. Modulus reduction factor, $R_m = G_s/G_{max}$ as function of the plasticity index, PI at three strain levels, (Döringer, 1997).

In spite of the variation in quality of the background material and the wide range of tested soils, a reasonable correlation was obtained. The test results are also in good agreement with those reported by Vucetic & Dobri (1991). The modulus reduction factor R_m decreases rapidly in the case of silty soils. For a soil with PI = 20 % at γ = 0.1 %, the shear modulus is 0.45 G_{max}, and at γ = 0.5 % the value is 0.15 G_{max}, respectively.

4.3 Proposed Stress-Strain Model

Rollins et al. (1998) have compiled stress-strain data for sandy and gravely soils and proposed the following relationship for the variation of the normalized shear modulus G_s/G_{max} with shear strain γ(%)

$$\frac{G_s}{G_{max}} = \frac{1}{\left[1.2 + 16\gamma\left(1 + 10^{-20\gamma}\right)\right]} \quad (15)$$

The data shown in Fig. 9 for fine-grained soils were analyzed using a modified relationship

$$\frac{G_s}{G_{max}} = \frac{1}{\left[1 + a\gamma\left(1 + 10^{-\beta\gamma}\right)\right]} \quad (16)$$

where the coefficients α and β were determined empirically from Fig. 9. The correlation between these coefficients and *PI* is shown in Fig. 10.

For clays with $PI = 20 - 40$, a typical range of values for $\alpha = 6.5 - 4.0$, and $\beta = 0.75 - 0.9$, respectively. Equation 16 defines the stress-strain behavior of fine-grained soils numerically, which facilitates its use in analytical models.

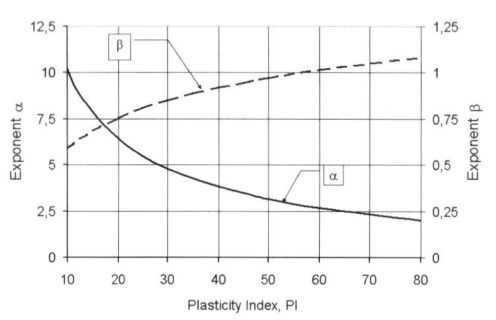

Figure 10. Correlation between PI and parameters α and β, cf. equation 15.

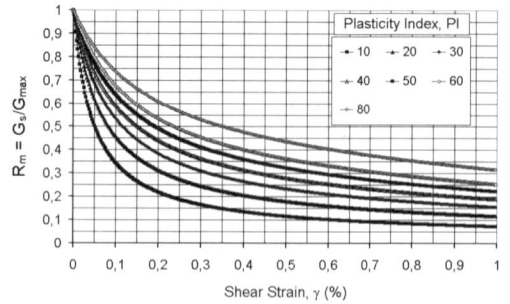

a) Linear scale

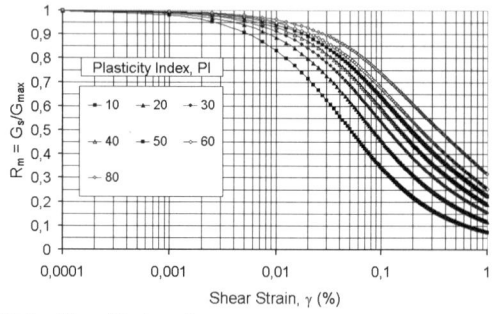

b) Semi-logarithmic scale

Figure 11. Variation of the normalized shear modulus as a function of shear strain for different values of *PI*, cf. equation 16.

The variation of the normalized shear modulus is shown as a function of shear strain, for different values of *PI*, both in linear as well as in semi-logarithmic scale, cf. Fig. 11 a and b. The data presented in Fig. 11 are in good agreement with previously proposed correlations by Vucetic and Dobry (1991). This is not surprising as the present investigation used part of the same database. However, the present investigation includes additional data, mainly from Scandinavia.

5 EFFECT OF STRAIN RATE ON SOIL STIFFNESS

5.1 *Some Observations Concerning Strain Rates*

One reason why geotechnical engineers have been reluctant to use stress-strain data from seismic investigations and RC tests was their notion of a "seismic modulus", which is only applicable for dynamic problems (i.e. at high strain rates). In the following, it will be shown that this is a misconception, and that in the case of seismic tests at small strains, the loading rate is slow, and in many cases slower than during conventional geotechnical testing.

In order to illustrate the problem, typical values of loading rate during different types of construction activities and for commonly used geotechnical investigation methods were estimated, Fig 12 a & b. It should be pointed out that these estimates are not very accurate and merely intended to illustrate the point. In order to assess the loading rate, it has been assumed that deformations (shear strain, %) occurs with time according to a sinusoidal relationship (1/4 of a sine wave). The secant from origin to peak strain was then calculated and this value was used as the average loading rate (either %/min or %/s).

It is noteworthy that the average loading rates stretch across more than 10 orders of magnitude (%/min). Even more interesting is that the loading rate at conventional field and laboratory tests, which are used on a daily basis for "static" design purposes, is significantly higher than that of seismic and cyclic soil tests. For instance, the loading rate of the three most common geotechnical field tests (SPT, CPT and vane test) is significantly higher than that of a seismic cross-hole test.

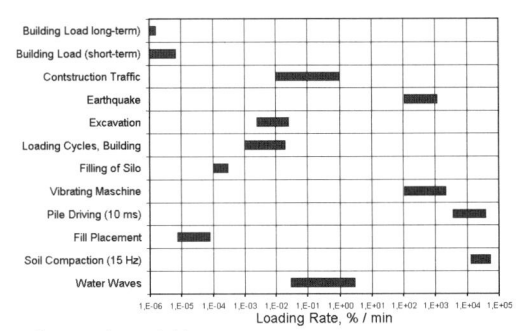

0.0002 % shear strain, the measured shear modulus was 76 MPa. The threshold value, where the shear modulus started to decrease, was 0.002 %. Since the test was performed at a constant vibration frequency, the rate of loading increased between these two values by one order of magnitude, without any discernable effect on soil stiffness. This fact has been demonstrated by numerous investigations using the RC test. Thus strain rate has little or negligible influence on the shear modulus at low strain level (~ <0.001%).

It is interesting to compare the loading rate of a "dynamic" resonant column (RC) test on soft clay with a "static" direct shear test. In Fig. 13 a & b the rate of loading of the two tests is compared.

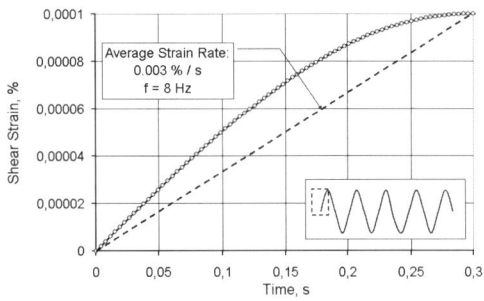

Figure 12. Estimated range of average loading rate during construction activities (a) and geotechnical investigations (b). The loading rate was estimated using a ¼-sine curve.

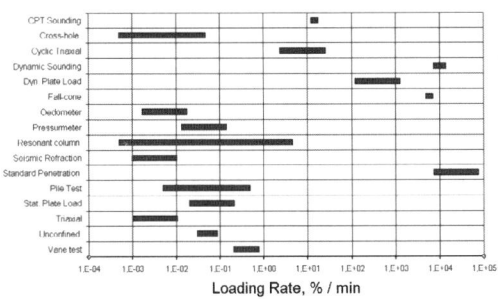

The loading rate at small strain levels of the laboratory RC test is comparable to that of an undrained shear or triaxial tests, and significantly lower than that of the fall-cone or the laboratory vane test.

5.2 Loading Rates during Seismic Testing

The effect of the rate of loading on the undrained shear strength has been discussed in the literature, e.g. Bjerrum (1972). However, the effect of loading rate on soil stiffness (shear modulus) has not been the focus of much attention. More than 20 years ago, at the ICSMFE in Stockholm in 1981, the following topic was discussed - *The shear modulus determined from "seismic tests" is generally referred to as a "dynamic modulus". Its significance for static geotechnical engineering is not yet generally appreciated. It can be shown, however, that at small shear strain, the "dynamic" shear modulus actually is determined at a strain rate which corresponds to static loading conditions.*- (Massarsch, 1982). After more than 20 years, this fact is still not appreciated, although it has potentially very important consequences. To illustrate this important point, reference is made to Fig. 6, which shows the results of a RC test on silty clay. The vibration frequency was approxmately 30 Hz. At a shear strain level of 0.0002 %, the rate of loading is 0.024%/s. This is comparable to a conventional undrained compression test. At

a. "Dynamic" Resonant Column test

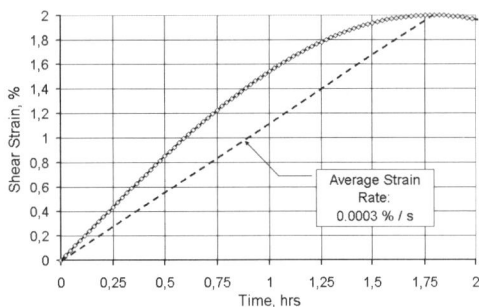

b. "Static" Direct Shear test

Figure 13. Comparison of average shear strain rate from dynamic RC and static shear test.

Shear strain is plotted against time (seconds and hours, respectively). In the case of the "dynamic" RC test, the average rate of loading at a vibration frequency of 8 Hz and a shear strain level of 0.0001 % is 0.0003 %/s. In the case of a "static" direct shear test, which is performed typically during 1 - 2 hours to failure (2 – 5% shear strain), the average shear strain rate is almost the same (2% during 1.75 hrs: 0.003 %/s). The shear modulus determined at a strain level one order of magnitude lower would remain essentially unchanged if performed at the same strain rate as the laboratory test. Thus the start (at low strain level) and the end of the stress-strain curve (at high strain level) can be established reliably by two tests, which are carried out at the same strain rate.

It is then possible to combine the results of a dynamic and a static test to establish the stress-strain relationship over a large strain range. This concept is illustrated in Fig. 14.

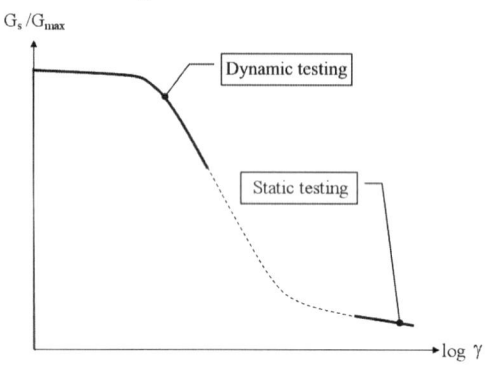

Figure 14. The results of a dynamic test and a static test can be used to establish the stress-strain curve, as they are determined at the same strain rate.

6 CASE HISTORY BÄCKEBOL

6.1 Geotechnical Conditions

The Bäckebol site, located north of Gothenburg on the Swedish west coast, has been the subject of numerous detailed geotechnical investigations, (Andreasson, 1979, Fellenius, 1972, Larsson & Mulabdic, 1991, 1991, Massarsch, 1976, Sällfors, 1975, Torstensson, 1973). The depth of the soft, plastic clay exceeds 40 m. Below an approximately 1 m thick dry crust a relatively homogeneous deposit of marine, post-glacial clay is found down to 10 m. The ground water level is located about 1 m below the ground surface and the pore water pressure is hydrostatic below that level. The water content in this layer varies between 70 – 90 % and is slightly above the liquid limit. The plastic limit is around 35 % and the plasticity index around 50 %. At 4 to 5 m depth, the clay is slightly overconsolidated. The coefficient of lateral earth pressure at rest, K_0 has been investigated in the field as well as in the laboratory, (Massarsch & Broms, 1976), cf. equation 12 and at 5 m depth K_0 = 0.6. The density of the clay is 15.5 kN/m^3.

The undrained shear strength has been determined in a comprehensive testing program involving both field vane tests and model pile tests, Torstensson (1973). The sensitivity of the clay at 4 to 5 m depth is around 20. The undrained shear strength at 4 to 5 m depth is 15 kPa, determined from field vane tests and corrected for plasticity. Figure 15 shows the results of field vane tests, which were performed at a depth of 3.75 m at different loading rates. Note that the standard loading rate during a field vane test is normally 1 min to failure.

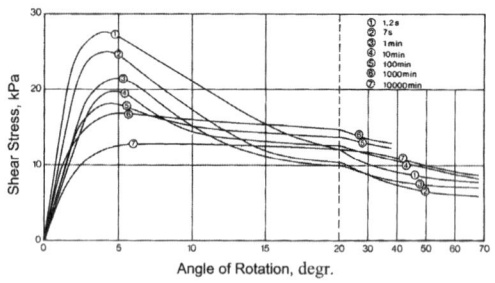

Figure 15. Shear stress from field vane tests as a function of angle of rotation for different loading rates, Torstensson (1973).

If it is assumed that failure occurs at about 1.0 % shear strain during rapid loading, and at about 5 % during slow loading, it is possible to estimate the influence of shear strain rate on the undrained shear strength from Fig. 16. The strain rate during a standard vane test is 0.03 %/s (1.6 %/min). The undrained shear strength in Bäckebol clay increases by about 15 % per log cycle. Thus also the secant shear modulus determined at failure, G_f, will be affected by the rate of loading.

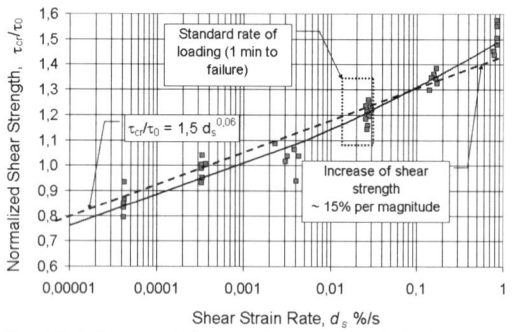

Fig. 16. Influence of shear strain rate on undrained shear strength from field vane tests. The approximate standard loading rate is indicated. Evaluation of results shown in Fig. 16.

Consolidated, undrained triaxial tests were also performed by Torstensson (1973). The undrained shear strength, determined from the deviator stress, is in good agreement with the field vane tests. The stress-strain curve of the tests on a sample from 4.5 m depth was digitized and is shown in Fig. 17.

Failure occurred at 1.25 % axial strain. Shear strain γ is related to axial strain ε_a according to the following relationship, cf. also equation 8

$$\Delta \gamma = (1+\nu)\Delta \varepsilon_a \qquad (17)$$

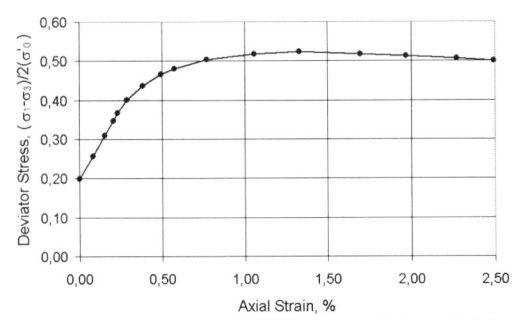

Fig. 17. Stress-strain curve of consolidated, undrained triaxial test from sample at 4.5 m depth, evaluated from test results reported by Torstensson (1973).

where ν is Poisson's ratio. Assuming undrained conditions ($\nu = 0.5$) $\gamma = 1.5 \, \varepsilon_a$. It is then possible to calculate the equivalent shear modulus, which at failure (1.9 % shear strain) is $G_f = 790$ kPa.

6.2 Seismic Tests

Different types of seismic and dynamic tests were performed at Bäckebol, (Andreasson, 1979, Bodare, 1983. Larsson & Mulabdic, 1991). These comprised dynamic screw plate tests and different types of seismic DH- and CH-tests. In addition, Andreasson (1979) performed RC tests on solid and hollow soil specimens. Figure 18 shows the results of seismic CH tests with two different types of energy sources. The shear wave velocity at 4 and 5 m depth ranged from 70 to 80 m/s.

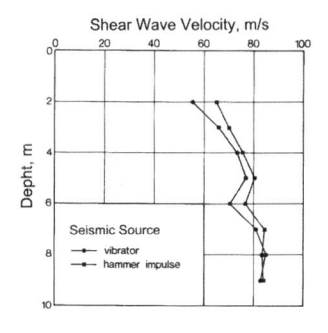

Fig. 18. Results of seismic CH-tests, Andreasson (1979).

Resonant column tests were performed on undisturbed samples from different depths, using a fixed-free device and following standard test procedures, Andreasson (1979). The measuring range of shear strains was 0.0003 – 0.1 %. The undisturbed samples were reconsolidated to the in-situ stress state and thereafter the test was started. Figure 19 a & b show the results of RC tests at 4 and 5 m depth, respectively. The maximum shear modulus was 5.5 and 6.7

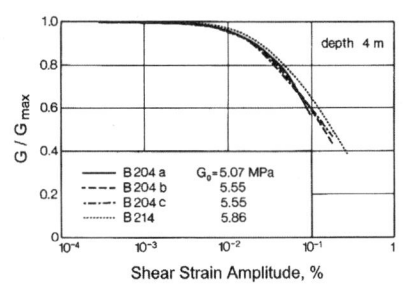

a. RC-test at 4 m depth

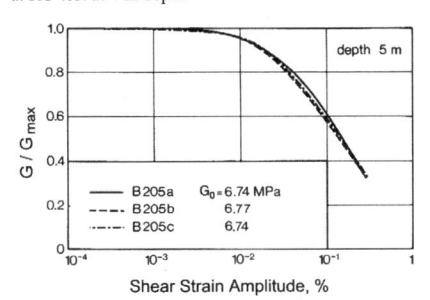

b. RC-test at 5 m depth

Figure 19. Results of resonant column tests, Andreasson (1979).

MPa at 4 and 5 m depth, respectively. From the RC tests, the modulus reduction curve could be established, Fig. 19.

6.3 Application of Stress-Strain Concept

The shear wave velocity determined at small strains from RC tests (65 m/s) was slightly lower than the field values from CH tests (70 - 80 m/s). The normalized stress-strain curves determined at both depths were in good agreement. It is now possible to combine the results from the seismic tests (CH and RC tests), Fig. 18 and 19 with the triaxial tests shown in Fig. 17. Axial strain was converted to shear strain according to equation 17. The data at 4 m depth were used to establish the stress-strain curve according to the concept presented in Fig. 14. The results are shown in Figure 20, which demonstrates that it is possible to determine reliably a stress-strain relationship over a large strain range, from very small strains (0.0001 %) to large (5 %) shear strains. In Fig. 20 is also shown the stress-strain relationship as determined from equation 16, and Fig. 10, assuming $PI = 50$ ($a = 3.17$, $\beta = 0.97$).

6.4 Pore Pressure during Pile Driving

In the vicinity of the Bäckebol test area, comprehensive investigations were performed, aiming to predict the excess pore water pressure during driving of

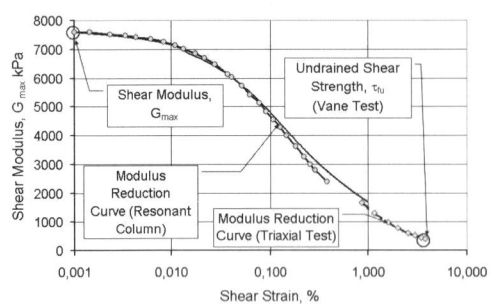

Figure 20. Combination of stress-strain measurements from resonant column and triaxial tests. Fig. 18, 19 and 20.

prefabricated concrete piles, Massarsch (1976) and Massarsch & Broms (1981). The prediction model was based on cavity expansion theory and Fig. 21 shows the relationship used in the investigation.

As is the case in many theoretical prediction models, the results are strongly influenced by the assumed soil parameters. An important in-put parameter is the stiffness ratio (G/τ_f). Considering the uncertainty of soil modulus values as described in the earlier part of this paper (Fig. 1 & 2), it is often difficult to make realistic assumptions on which to base predictions. The stress-strain relationship for Bäckebol, as shown in Fig. 20 suggests that at failure (i.e. close to the pile), the stiffness ratio G/τ_f can be as low as 50 (~790/15), based on triaxial tests. However, at small shear strain level (0.0001%), further away from the pile the ratio increases $G/\tau_f = 500$ (~7600/15). This variation is significant as it corresponds to a factor of 10, which is not negligible even in the case of preliminary studies.

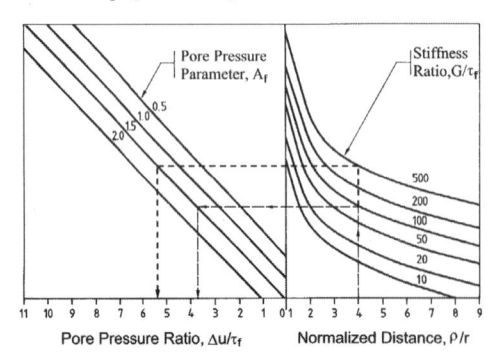

Figure 21. Relationship between the excess pore water pressure in the vicinity of an expanding cavity for different stiffness ratios, Massarsch (1976).

The soil stiffness (G/τ_f) can be established using the stress-strain relationship shown in Fig. 20. The concept presented in this paper made it possible to predict excess pore water pressure more reliably, Massarsch & Broms (1981). The shear strain level at different distances from the pile can be calculated theoretically. In Fig. 21, the range of stiffness ratios (100 – 500) used in the analysis is indicated.

7 CONCLUSIONS

The assessment of deformation properties is an important part of geotechnical design. Simplified concepts, which were developed many decades ago, such as the coefficient of subgrade reaction or spring constants, are still used in practice. These empirically determined values do not take into account fundamental geotechnical concepts, such as the effect of confining pressure or strain level.

Information about the soil modulus published in the literature shows large scatter and is not acceptable for reliable design. In spite of this, many complex projects are analyzed by sophisticated analytical methods, where geotechnical input parameters are chosen based on crude assumptions.

Major progress has been made in earthquake engineering and soil dynamics, and reliable methods exist for the determination of deformation properties of soils even at very low strain levels (down to 0.0001 % shear strain). Based on a comprehensive analysis of published seismic data a surprisingly good correlation was obtained between the shear modulus at small strain and water content (void ratio).

It is well-known that shear strain affects soil stiffness. The reduction of the shear modulus for a wide range of fine-grained soils was determined at three strain levels (0.1, 0.25 and 0.5 % shear strain). The modulus reduction factor is strongly affected by the plasticity index. The shear modulus decreases more rapidly in silts and silty clays, while the effect of shear strain is less pronounced in soils with high plasticity. A numerical relationship is proposed, which makes it possible to establish the stress-strain relationship over a large strain range (from 0.0001 – 0.5 % shear strain).

The soil modulus, determined by seismic or dynamic methods, is often termed the "dynamic" modulus. It is shown that the loading rate during dynamic testing (for instance the resonant column test) at small shear strain levels is slow (0.001 – 0.01 %/sec) and comparable to that of conventional laboratory tests (triaxial and shear tests). Thus, the starting point and the end-point of a stress-strain curve can be established by a seismic or dynamic test (preferably performed in the field) and a static shear test, respectively. The modulus reduction curve can be determined accurately by the resonant column test. In the absence of such test data, the relationship proposed in this paper can be used.

The application of the proposed concept was exemplified using data from a well-documented test site in Bäckebol, Sweden. Good agreement was obtained between measured and predicted soil deformation data.

ACKNOWLEDGEMENTS

The investigations presented in this paper were performed during the past three decades at the University of Kentucky (UK) and the Royal Institute of Technology (KTH) in Stockholm, Sweden. Prof. Vincent Drnevich, then at UK, has initiated the concept of strain rate effect, and has had a profound influence on the author. His astute and constructive comments have always been highly appreciated. The collaboration with the members of the geotechnical group at UK was stimulating and characterized by enthusiasm and innovative thinking.

The work started at UK was continued in the following years at KTH. Several masters students, and in particular Heike Döringer, have spent many hours performing tests and analyzing data. Without their commitment it would have been difficult to carry out this research successfully.

Research funds were made available by several organizations, such as the Swedish Council for Building Research and the Åke och Greta Lisshed's Foundation. Their support is gratefully acknowledged.

Some of the theoretical studies, on which this paper is based, were performed as part of the PrognosVib research project, sponsored by the Swedish National Rail Administration – Banverket. The contributions by the members of the PrognosVib research team, and in particular the support and encouragement by its project manager, Mr. Alexander Smekal is acknowledged with gratitude.

The valuable comments made by Dr. Bengt H. Fellenius and Dr. John Powell are highly appreciated.

REFERENCES

Andreasson, B. 1979. Deformation characteristics of soft, high-plastic clay under dynamic loading conditions. Chalmers Tekniska Högskola (CTH). Geoteknik med grundläggning. 242 p.
Bjerrum, L. 1972. Embankments on soft ground. Proc. ASCE Specialty Conference on Performance of Earth and Earth-Supported Structures, Purdue University, Vol. II, pp. 1-54.
Bodare, A. 1983. Dynamic screw plate for determination of soil modulus in situ. Doctoral thesis, Uppsala University, UPTEC 83 79 R, 273 p.
Broms. B. B. 1963. Allowable bearing capacity of initially bent piles. Proc. ASCE, Vol. 89, No. SM5, p. 73 – 90.
Döringer, H. 1997. Veformungseigenschaften von bindigen Böden bei kleinen Deformationen (Deformation properties of cohesive soils at small deformations). Masters thesis, Dept. Soil and Rock Mechanics, Royal Institute of Technology (KTH), 54 p.
Drnevich, V.P. and Massarsch, K.R., 1979. Sample Disturbance and Stress - Strain Behaviour. ASCE Journal of the Geotechnical Engineering Division, Vol. 105 No GT 9, pp. 1001-1016.
Dyvik, R. & Olsen, T.S. 1989. G_{max} measured in oedometer and DSS tests using bender elements. International conference on soil mechanics and foundation engineering, 12, Rio de Janeiro, Aug. 1989. Proceedings, Vol. 1, pp. 39-42.
Fahey, M. 1998. Deformation and in situ stress measurement. International conference on site characterization, ICS '98, 1, Atlanta, GA, 19-22 April 1998. Proceedings, Vol. 1, pp. 49-68.
Fellenius,B.H., 1972. Downdrag on piles in clay due to negative skin friction. Canadian Geotechnical Journal, Vol.9, No.4, pp.323 – 337.
Holtz R.D. & Kovac, W.D. 1981. An Introduction to Geotechnical Engineering. Prentice-Hall, Inc. 733 pp.
Kovacs, W.D., Seed, H.B. & Chan, C.K. 1971. Dynamic moduli and damping ratios for a soft clay. ASCE. Soil Mechanics and Foundations Division. Journal 1971, Vol 97, No SM1, pp 59-75
Ladd, C.C., Foott, R., Ishihara, K., Schlosser, F & Poulos, H. G. 1977. Stress-deformation and strength characteristics. ICSMFE 9, Tokyo, July 1977. Proceedings, Vol. 2, pp. 421-494.
Langö, H.V. 1991. Cyclic shear modulus of natural intact clay. Doctoral thesis, University of Trondheim (NTH). Report 1991:51.
Länsivaara, T.T. 1999. A study of the mechanical behavior of soft clay. Norwegian University of Science and Technology. NTNU. Doktor ingeniöravhandling 1999:85, 191 p.
Lanzo, G, Vucetic, M, Doroudian, M. 1997. Reduction of shear modulus at small strains in simple shear. ASCE. Journal of Geotechnical and Geoenvironmental Engineering. Vol 123, No 11, pp 1035-1042.
Larsson, R. & Mulabdic, M. 1991. Shear moduli in Scandinavian clays. Measurement of initial shear modulus with seismic cones. Empirical correlations for the initial shear modulus in clay. Swedish Geotechnical Institute, SGI Rapport 40, 127 p.
Larsson, R. & Mulabdic, M., 1991. Shear Modulus in Scandinavian Clays. Swedish Geotechnical Institute, Report No. 40, 127 p.
Lunne, T., Robertson, P.K. & Powell, J.J.M. 1997. Cone Penetration Testing in Geotechnical Practice. Blackie Academic & Professional, 352 p.
Massarsch, K.R. 1979. Lateral earth pressure in normally consolidated clay. European conference on soil mechanics and foundation engineering, 7, Brighton, Sept. 1979. Proceedings, Vol. 2. pp. 245-249.
Massarsch, K.R. and Drnevich, V. P., 1979. Deformation Properties of Normally Consolidated Clay, 7th European Conference on Soil Mechanics and Foundation Engineering, Brighton, Proceedings, Vol. 2, pp. 251-255.
Massarsch, K.R., 1976. Soil Movements Caused by Pile Driving in Clay, Inst. för jord-och bergmekanik, KTH, Thesis in partial fulfillment of the requirements for the Degree Doctor of Engineering, Job-Rapport No 6, 261 p.
Massarsch, K.R., 1982. Dynamic and Static Shear Modulus, Discussion Session 10. Soil Dynamics, Proc. 10th Int. Conference on Soil Mechanics and Foundation Engineering, Stockholm, 15-19 June, 1981, Proceedings, Vol. 4, pp. 880-881.
Massarsch, K.R., 1985. Stress-Strain Behaviour of Clays, 11th International Conference on Soil Mechanics and Foundation Engineering, San Francisco, Proceedings, Volume 2, pp. 571 - 574.
Massarsch, K. R. & Broms, B.B. 1976. Lateral Earth Pressure at Rest in Soft Clay, Proc. ASCE Journal of the Geotechnical Engineering Division, 0ctober 1976, pp.1041-1047.
Massarsch, K. R. & Broms, B. B., 1981. Pile Driving in Clay Slopes, Proceedings of the Tenth International Conference

on Soil Mechanics and Foundation Engineering, Stockholm, (also VBB Special Report 21:80.7), Vol. 3. p 469-474.

Rollins, K.M., Evans, M.D., Diehl, N.B. and Daily, WD, III., 1998. Shear modulus and damping relationships for gravels. ASCE. Journal of Geotechnical and Geoenvironmental Engineering, Vol. 124, No 5, pp. 396-405.

Sällfors, G. 1975. Preconsolidation pressure of soft, high-plastic clays. Chalmers University of Technology (CTH) Doctoral Thesis, 231p.

Seed, H. B. & Idriss, I. M. 1970. Soil moduli and damping factors for dynamic response analysis. Rep. No. EERC 70-10. Earthquake Engineering Res.Ctr. Coll Engrg. University of California, Berkeley,

Stokoe, K.H, & Santamarina, J.C. 2000. Seismic-wave-based testing in geotechnical engineering. GeoEng 2000. Proceedings, Vol. 1. pp 1490-1536.

Terzaghi, K. 1946. Theoretical Soil Mechanics. 3rd Edition, John Wiley and Sons, Inc. 510 p.

Terzaghi, K. & Peck, R.B. 1948. Soil mechanics in engineering practice. Wiley & Sons. 1 Ed, 566 p.

Torstensson, B.A., 1973. Kohesionspålar i lös lera. En fältstudie i modellskala (Cohesion piles in soft clay. A field study in model scale). Doctoral Thesis. Chalmers tekniska högskola. Geoteknik med grundläggning, 185 p.

Vucetic, M. & Dobry, R. 1991. Effect of Soil Plasticity on Cyclic Response. Journal of the Geotechnical Engineering Division, ASCE Vol. 117, No. 1. Jan, pp. 89 - 107.

Woods, R.D. & Henke, R. 1981. Seismic techniques in the laboratory. ASCE Journal, Geotechnical Engineering Division. Vol 107, nr GT10, pp. 1309-1325.

Woods, R.D. 1994. Laboratory measurement of dynamic soil properties. American Society for Testing and Materials, ASTM. Special technical publication STP 1213, pp. 165-190.

Advanced calibration chambers for cone penetration testing in cohesionless soils

A.B. Huang
Department of Civil Engineering, National Chiao Tung University, Hsin Chu, Taiwan

H.H. Hsu
Department of Civil Engineering, Chien Kuo Institute of Technology, Chang Hua, Taiwan

Keywords: sand, CPT, physical modeling, calibration chamber, centrifuge, numerical simulation

ABSTRACT: The calibration chamber as it was first introduced had its main purpose to calibrate cone penetration test (CPT) in dry, uniformly graded clean sand. The use of an ingenious cavity-wall design enabled the soil specimen be subjected to either a constant stress or zero strain boundary condition. The calibration chamber allowed full scale CPT to be performed in a uniform specimen with known stress and density state. This unique capability was ideal for establishing correlations between CPT and soil design parameters. Empirical rules based on these correlations have routinely been used for the interpretation of CPT in cohesionless soils. The cavity-wall design, with its large but limited specimen, could not properly duplicate the field conditions where soil extends laterally to infinity. CPT performed in the cavity-wall chamber, regardless of the applied boundary conditions, could yield cone tip resistance values that were significantly lower than what would be expected in the field. Much of the calibration chamber related research in the last three decades has evolved around new techniques to either understand or overcome the boundary effects. Various forms of numerical simulations have been implemented and/or coupled into the physical model for CPT calibration. The test material has extended to silty sand under saturated conditions. Centrifuge has also been experimented as an alternative and may be more efficient way to calibrate CPT in cohesionless soils. The paper reviews these various techniques and discusses their implications in the use of calibration test data and future developments in CPT calibration techniques.

1 INTRODUCTION

The first calibration chamber is believed to have been established at Materials Research Division, Country Road Board (CRB), Melbourne, Victoria, Australia in late 1960's (Holden, 1991). Its main purpose was to simulate the field conditions in the laboratory and calibrate the CRB electrical friction cone penetrometer. The original concept of a calibration chamber test was to prepare a large sand specimen in the laboratory, consolidated to a desired stress level, and then perform the experiment under given boundary conditions. Since the entire experiment was conducted in the laboratory, the test quality could be readily controlled. The large sand specimen, with uniform deposition and known engineering properties, provided reference values for the interpretation and thus calibration of the cone penetration test (CPT). Different from test pits, the calibration chamber had a flexible lateral boundary and partial or full flexible boundary in the vertical direction. In addition, the stress and displacement on the boundary of a chamber specimen could be controlled and/or measured to provide information for a more rigorous analysis of the calibration test data.

According to statistics by Ghionna and Jamiolkowski (1991), there were at least 19 calibration chambers in the world in 1991. More calibration chambers have been built (e.g., Peterson and Arulmoli, 1991; Hsu and Huang, 1998; Ajalloeian and Yu, 1998; and Tan et al., 2003) since then. The applications of calibration chamber have been extended to cohesive soils and in situ testing methods other than CPT. These applications have included Marchetti dilatometer (Borden, 1991; Huang et al., 1991; and Wang, 1998), pressuremeter (Huang et al., 1991; Manassero, 1991; Anderson et al., 1991: and Ajalloeian and Yu, 1998), hydraulic fracture (Been and Kosar, 1991) and the calibration of pile foundations (Kulhawy, 1991; O'Neill, 1991; and Foray, 1991). Jamiolkowski and his co-workers (Jamiolkowski et al., 1988) have proposed a series of empirical equations to interpret various in situ test results based on calibration chamber tests. Based on load tests conducted in a calibration chamber, Kulhawy (1991) proposed design methods for drilled shaft foundations. Lee and Stokoe (1988) used the calibration chamber to measure shear wave velocities in sand and correlated them with stress states. Tan et al. (2003) reported their preliminary chamber CPT results in unsaturated silt.

Among all types of in situ test methods that can be calibrated in a laboratory set up, CPT is probably the most challenging one. The main difficulty in CPT calibration stems from the large strains induced by CPT and its sensitivity to conditions imposed on a finite chamber specimen. The CRB chamber was designed to house a 0.76 m diameter and 0.91 m high soil specimen. The relatively large size was used with an intention to minimize the boundary effects. Unfortunately, even with its large dimensions, significant boundary effects remained with the original CRB chamber design. The development of calibration chambers in the last three decades has evolved around new techniques to either understand or overcome these boundary effects. The CPT calibration has extended beyond the realm of physical modeling. Various forms of numerical simulations have been implemented and/or coupled into the physical model for CPT calibration. The use of centrifuge has been explored as an alternative and may be more efficient way to calibrate CPT in cohesionless soils.

Parkin (1988) has provided a comprehensive state-of-the-art report on the basic concept of calibration chamber testing and correlations established from CPT calibration tests. This paper will concentrate on the experimental techniques of CPT calibration in cohesionless soils, their values and challenges in future developments.

2 THE CAVITY-WALL CALIBRATION CHAMBERS

In addition to its large specimen, the most significant accomplishment in the CRB chamber was probably the utilization of an ingenious cavity-wall design to simulate the zero lateral strain (K_o) conditions. By maintaining a cavity pressure equal to the chamber cell pressure, an average rigidity of the inner-wall was effectively established. Figure 1 provides a schematic view of the cavity-wall design.

The vertical and lateral chamber boundaries were controlled independently to be constant stress or zero average deformation. A typical cavity-wall calibration chamber was capable of creating four types of boundary conditions as shown in Table 1. The terminology used in Table1 is essentially universal amongst the chamber users. Been et al. (1988) have indicated that boundary conditions on top and bottom of the chamber specimen had little effect on CPT results. Parkin (1988) stated that of the four boundary conditions, the most significant were B1 and B3.

Holden (1991) summarized the advantages of the calibration chamber as follows:

1. The lateral boundary was flexible. It was able to produce the normally consolidated (NC) or overconsolidated (OC) specimens under K_o conditions.
2. The boundary stresses were known and controlled.
3. The chamber specimen was large enough to perform a full scale CPT.
4. By means of pluvial methods, the chamber specimens were uniform and reproducible.

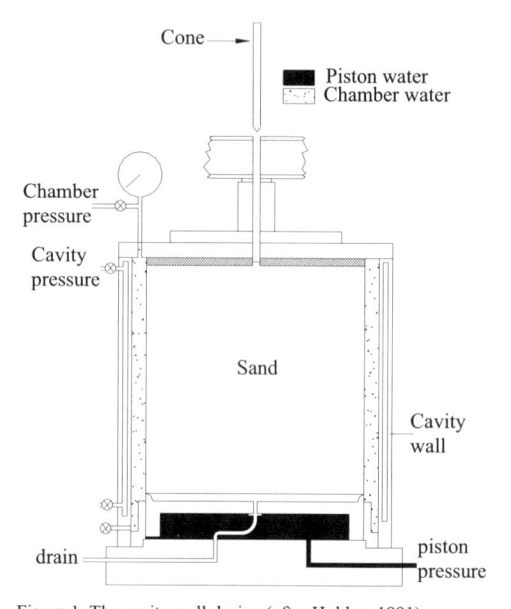

Figure 1. The cavity wall design (after Holden, 1991).

Table 1. Boundary conditions in calibration chamber tests.

Boundary conditions	Top and bottom boundary		Lateral boundary	
	stress	strain	stress	strain
B1	constant	-	constant	-
B2	-	0	-	0
B3	constant	-	-	0
B4	-	0	constant	-

2.1 Fabrication of a Clean Sand Specimen

Creating a uniform sand specimen with the desired density and fabric of a field deposit in a repeatable fashion has been the ultimate goal in chamber specimen preparation. Most of the CPT calibration tests in clean sand have been performed under dry conditions. Two types of pluviator systems have been reported for the preparation of sand specimens:

the stationary and traveling sand pluviators (Salgado, 1993; Fretti et al., 1995; and Salgado et al., 1998).

The stationary sand pluviator consists of a hopper, a rainer plate, a shutter plate, and diffuser meshes. The holes in the rainer plate are initially closed by the shutter plate. Upon opening the rainer plate, the sand falls from the hopper through the holes of the rainer plate and down to the diffuser meshes. The diffuser meshes are positioned at a constant distance over the surface of sand deposition. The relative density (D_r) is mainly affected by the deposition intensity (DI) and falling height (FH) which is the distance between the deposition surface and the bottom of diffuser meshes (Rad and Tumay, 1987). The weight of sand falling per unit area and per unit time is referred to as deposition intensity (DI) (Lo Presti et al., 1993). The quantity, diameter, and arrangement of holes on the rainer plate are the controlling factors of DI. A high DI tends to create a looser specimen (i.e., lower D_r). Upon exiting the rainer plate, sand particles pass through diffuser meshes before reaching the deposition surface. When the sand particles finally impact on the deposition surface, the velocity is determined by FH. The velocity of sand particles, that controls the impact force of the deposition particles, also affects D_r. The particle velocity is a function of gravity force and air resistance. A terminal velocity is reached when FH is far enough to balance these two forces acting on the particle. Thus, D_r is independent from FH only if FH is higher than the "terminal" FH. Lo Presti et al. (1992) compiled a series of experimental data and recommended that the "terminal" FH is approximately 500mm for sand and 750mm for gravel.

In a traveling pluviator, a motor driven hopper traversing over the deposition is used to spread the sand. DI is determined mainly by the width of the opening and moving speed of the hopper (Lo Presti et al., 1993). The test sand is loaded in the hopper and drops from the opening.

Both the stationary and traveling pluviators are known to create repeatable, uniform, dry and uniformly graded clean sand specimens. These methods are not suitable for graded sand however, as they could result in particle segregation. Experience gained from shearing tests on soil element has indicated that the sand specimen preparation method can significantly alter the stress-strain behavior (Ishihara, 1993). The dry pluviation method is not likely to duplicate a fabric of natural alluvial soil deposit (Ishihara, 1993; and Vaid et al., 1999) where soil particles are sedimented under water. Ghionna and Jamiolkowski (1991) have provided convincing evidence that shows significant differences in soil fabric between a fresh reconstituted sand specimen and an aged, natural sand deposit. If and how the differences in soil fabric can be correlated to variations in CPT results is not clear. These potential drawbacks should be kept in mind when using the interpretation methods developed from CPT calibration tests.

2.2 The Boundary Effects

A standard cone penetrometer has a diameter of 35.7mm (ASTM D3441). The diameter ratio (D/B) of the chamber specimen diameter (D) over that of a standard cone (B) is approximately 42 even for a relatively large 1.5m-diameter chamber specimen. Ideally, D/B is infinite in the field.

Previous experiences in performing CPT in a cavity-wall calibration chamber as they relate to boundary effects are summarized as follows:

1. The field condition for CPT where soil extends laterally to infinity is expected to be between B1 and B3 (Veismanis, 1974; Parkin, 1988). The finite specimen dimensions tend to cause the cone tip resistance (corrected for unequal end area where applicable) q_T to be lower than those expected in the field under similar density and stress conditions.
2. The q_T under B3 conditions continues to increase with depth and does not reach a "plateau" in dense sand (Parkin and Lunne, 1982; Parkin, 1988).
3. For tests under B1, q_T is mostly a function of initial (prior to cone penetration) horizontal effective stress σ'_{ho}, at least up to over consolidation ratio (OCR) of 8 (Veismanis, 1974; Chapman and Donald, 1981; Parkin, 1988; Houlsby and Hitchman, 1988).
4. Figure 2 shows the CPT data compiled by Parkin and Lunne (1982) under different boundary conditions, D_r and D/B. Experimental data show that the boundary effects are more apparent in dense sand than in loose sand. The influence decreases with the compressibility of sand. For loose sand, chamber results are relatively independent of boundary conditions, even when D/B is as low as 21 (Parkin, 1988). For dense sand, all calibration chamber results are affected by boundary conditions, even for D/B of 60 or greater (Parkin, 1988).

Recognizing the existence of boundary effects, Baldi et al. (1982) reported a series of chamber size correction factors, r for Ticino sand at D/B = 34.2 and

$$r = \frac{q_T(\text{expected field value})}{q_T(\text{measured in chamber})} \quad (1)$$

The value of r increases with D_r and OCR. Based on the analysis of a database, Mayne and Kulhawy (1991) suggested an empirical equation that shows r was a function of D/B and D_r only and not related to stress history. Salgado (1993) proposed a more elaborate analytical scheme to correct the boundary effects. The expected field q_T was computed assuming cone penetration as a plane strain cylindrical cavity expansion. The method considered the effects of sand dilatancy, boundary conditions, D/B, and regarded the horizontal stress as a predominant influence factor on q_T.

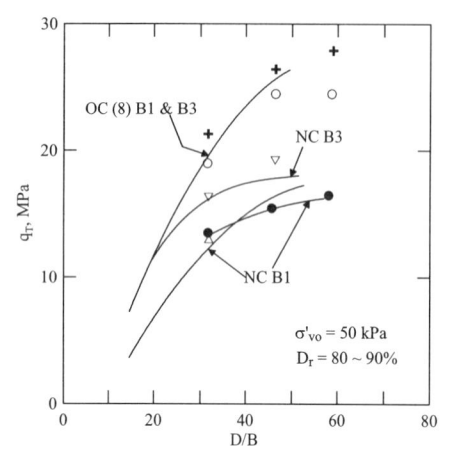

Figure 2. Effects of boundary conditions, Dr and D/B on q_T (after Parkin and Lunne, 1982).

There have been reports (e.g., Schmertmann, 1976; Villet and Mitchell, 1981; Baldi et al., 1982; Jamiolkowski et al., 1988) that q_T should increase with effective overburden stress (σ'_{vo}) and/or effective mean normal stress (σ'_{oo}). Parkin (1988) has suggested that q_T responds to the minor principal stress (σ_3), it could be σ'_{vo} or σ'_{ho} depends on the K (ratio of σ'_{ho} over σ'_{vo}) values. Houlsby and Hitchman (1988) conducted a series of CPT in a calibration chamber under various K values and concluded that q_T is mostly a function of σ'_{ho}. Figures 3 to 5 show different correlations between q_T and the initial stress component as proposed by various researchers.

The above-mentioned approaches on the interpretation of CPT contain significant differences in relating q_T to soil properties. The conflicting conclusions could be a result of differences in laboratory setups, in analytical procedures, or both.

3 A NUMERICALLY SIMULATED CALIBRATION CHAMBER

How the boundary conditions affect the cone tip resistance is an intriguing problem. In order to tackle this problem, it is apparent that some understanding of what happens within the soil mass as a result of cone penetration would be of great help. Calibration chamber or other types of physical model tests usually offer measurements on the boundary only. A numerical analysis in this regard would be much more effective as it can compute stress and strain distributions within the soil mass.

CPT is a large strain problem and is difficult to simulate using continuum mechanics techniques such as the Finite Element Method. Huang and Ma (1994) reported a numerical scheme using a two-dimensional distinct element method (DEM) to simulate cone penetrations in granular material. A granular medium was treated as an assembly of circular disks. The disks had the ability to break and reform contacts. Movements of the individual disks followed Newton's second law and physical properties at the contacts. Thus, the deformation of the particle assembly under a given boundary condition was treated as a transient problem with states of equilibrium developing whenever the internal forces balanced. Readers are referred to Cundall (1974) and Ting et al. (1989) on the details of DEM and its applications in geotechnical engineering.

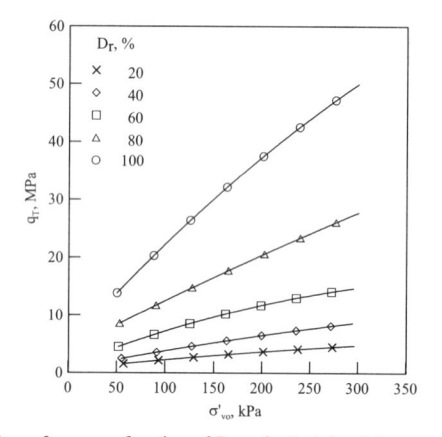

Figure 3. q_T as a function of Dr and σ'_{vo} (after Schmertmann, 1976).

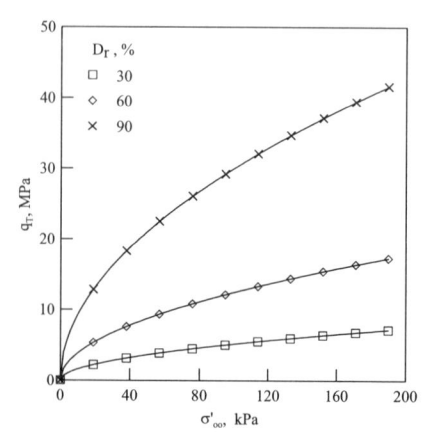

Figure 4. q_T as a function of Dr and σ'_{oo} (after Jamiolkowski et al., 1988).

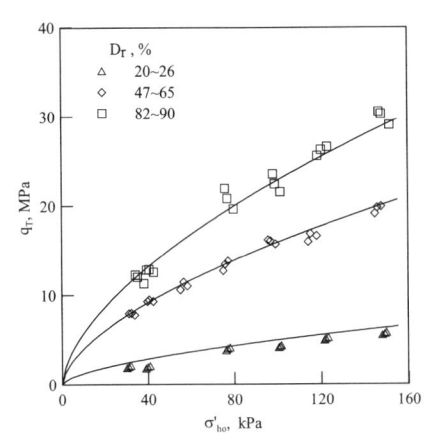

Figure 5. q_T as a function of Dr and σ'_{ho} (after Houlsby and Hitchman, 1988).

In the numerical simulations by Huang and Ma (1994), physical properties of the DEM disks included: mass density = 2670 kg/m^3, normal contact stiffness = 300 MN/m^2, shear contact stiffness = 210 MN/m^2, contact coefficient of restitution = 0.25, damping ratio = 0.2 and contact friction angle = 25°. These properties in many ways emulated the behavior of natural sand particles. The DEM specimen for cone penetration tests was created by a numerically simulated pluviation process. DEM disks with the size distribution shown in Figure 6 were aligned in a regular pattern and contained in a bin under zero gravity (g = 0) initially. The bin had an opening at the bottom. The disk pluviation process was started by changing gravity from 0 to 1 g. As there was no air friction in the numerical scheme, particle segregation was not possible.

Taking advantage of symmetry, only half of the penetrometer and soil mass were simulated. All linear dimensions were normalized with respect to the cone radius, R (R = 5 mm). A total of 12000 disks could simulate a specimen that was 16R wide and approximately 19.5R deep. The center vertical boundary was rigid and frictionless. The penetrometer had a 60° apex angle which was the same as a standard cone penetrometer. A contact friction angle of 25° between the cone surface and neighboring DEM disks was used in the computations. The sum of contact forces in vertical direction between the inclined cone tip surface and DEM disks were taken as the tip resistance. To simulate K$_o$ consolidation, the top boundary was set to stress controlled, all other boundaries were set to rigid and frictionless. An equivalent of B1 condition was materialized by converting the lateral boundary to stress controlled upon consolidation. The B3 was simulated by applying the same boundary conditions as in K$_o$ consolidation. CPT was simulated by forcing the cone into the disk assembly at a constant penetration rate of 20 cm/sec. The relatively high penetration rate was used to keep the computation time manageable. The penetration rate was not expected to have a significant influence on CPT results in dry sand (Dayal and Allen, 1975).

In the cases to be presented, the σ'_{vo} was 1200 kPa for all the DEM assemblies. To create an overconsolidated (OC) assembly, a σ'_{vo} of 12000 kPa was applied first and then unloaded to 1200 kPa and thus resulted in an overconsolidation ratio (OCR) of 10. The K$_O$ values were 0.34 and 1.24 for the NC and OC assembly, respectively. The 12000 disk assembly had a void ratio of approximately 0.22 for both the NC and OC assemblies. The void ratio was not sensitive to the static loading and unloading as is the case of low compressible natural granular soil.

3.1 *Numerical Simulation of Field Conditions*

Large strains during a cone penetration were expected only in close vicinity of the penetrometer (Huang, 1989). Thus, in the far field, where the strain was expected to be small, the soil mass could be reasonably assumed as a continuous material. The Boundary Element Method (BEM) was capable of simulating semi-infinite soil mass. The material was treated as linear elastic in the BEM region that extended outwards to infinity as shown in Figure 7. The combined DEM/BEM regions simulated a semi-infinite, dry granular soil mass without a lateral boundary. The contact forces between the DEM disks and the DEM/BEM boundary were linearly distributed among the BEM nodal points under an explicit numerical scheme. The forces distributed on the nodal points were used as input to the BEM. The computed BEM boundary movement was then entered in the DEM computation in the next time step. Monitoring zones as numbered 1 and 2 in Figure 7 could be set up within the DEM region where the disk movement and contact force variations could be traced during simulated CPT.

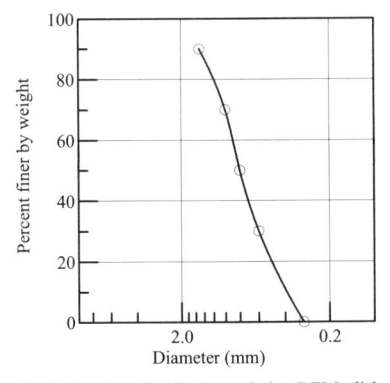

Figure 6. Grain size distribution of the DEM disks (from Huang and Ma, 1994).

The BEM region during cone penetration was expected to experience a predominant shearing mode that was similar to lateral compression. The Young's modulus, E and Poisson's ratio, υ needed to characterize the BEM region were determined by simulated lateral compression biaxial tests performed on the same DEM disk assemblies used in CPT simulations.

Figure 8 shows the q_T profiles from three CPT (i.e., CPT-1, CPT-2 and CPT-3) under simulated field conditions, to be referred to as B5, in a normally consolidated (NC) assembly with DEM region containing 8000, 12000 and 18000 disks. The corresponding lateral dimensions of the DEM region were 8R, 16R and 24R, respectively. The q_T values were normalized with respect to σ'_{vo}.

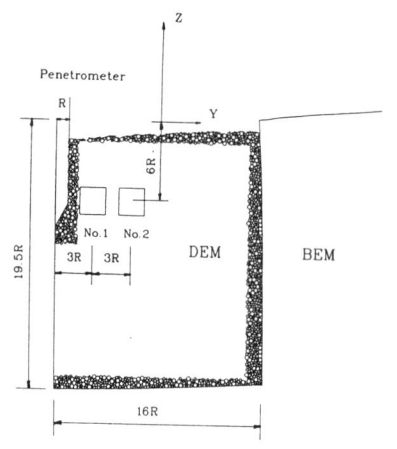

Figure 7. Numerical scheme of CPT simulation (from Huang and Ma, 1994).

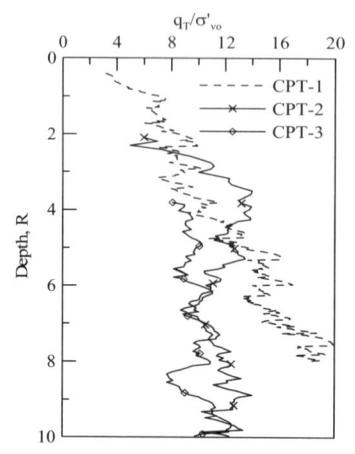

Figure 8. Simulated cone penetration tests (from Huang and Ma, 1994).

As in field tests, the simulated CPT was rather "noisy". In the 8000 disk assembly (CPT-1), there was a trend of q_T increasing with depth. This was likely due to the proximity of the BEM region. The mean q_T/σ'_{vo} values among tests CPT-1, CPT-2 and CPT-3 varied by a maximum of 12% as the lateral dimension of the DEM region increased from 8R to 24R. The 12% difference was roughly the same as the noise level. If a sufficiently large lateral dimension for the DEM region was selected, any further increase of the DEM lateral dimension should have little effect on q_T values. Physical chamber tests in dense sand under either a stress controlled or rigid lateral boundary (Parkin, 1988) generally showed an increase of approximately 100% of q_T as the chamber specimen radius changed from 16R to 24R. Physical chamber tests have indicated (e.g., Parkin, 1988) that q_T in OC sands were less sensitive to the rigidity of the lateral boundary than those in NC sands. Hence, if the DEM/BEM numerical scheme worked for NC assemblies, it was expected to work for OC assemblies as well.

3.2 *Boundary and Stress History Effects according to Numerical Simulations*

For each disk, the DEM computed its current position, contacts with the neighboring disks and their contact forces in every time step. The data allowed the results to be statistically or graphically analyzed from micromechanical point of view. Or, the displacement and contact forces could be statistically related to stress and strain and then analyzed from continuum mechanics point of view. Figure 9 demonstrates the distribution of normal contact forces and disk movements within the DEM disk region as the cone tip reached 8R when simulated under B1 and B3 conditions. The DEM region consisted of 12000 disks and was 16R wide and 19.5R deep. The line thickness was proportional to the magnitude of contact force. The vectors associated with disk movements in Figure 9(b) were established by comparing the current disk positions with those when the cone tip was at 7R. The corresponding distribution of normal contact forces and disk movements within the DEM disk region under B5 are shown in Figure 10.

Results in Figures 9 and 10 are based on CPT simulations in the NC assembly. According to CPT under B5, the contact forces were concentrated in the vertical direction below and around the face of the penetrometer tip. The disk movements followed a pattern that was similar to the deep failure mechanism described by Vesic (1963) where large soil movements concentrated in areas below the base of penetrometer. The force chains were much less defined and the disk movements more wide spread in the case of CPT in B1. The rigid boundary applied in B3 forced the contact forces to be more concentrated

B1   B3
(a) Normal contact force

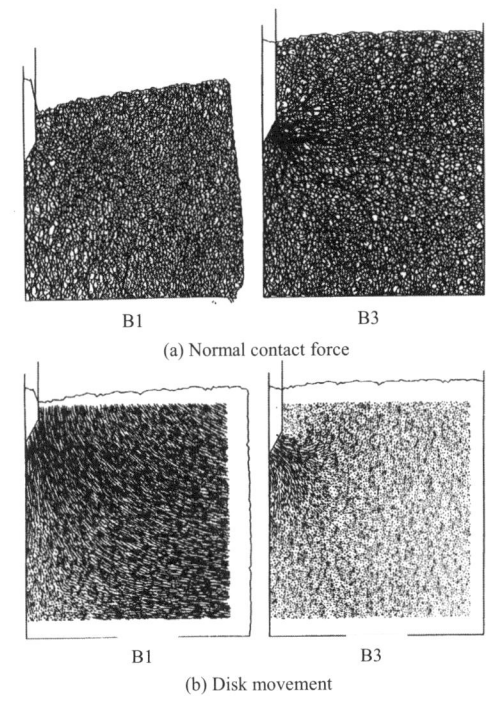

B1   B3
(b) Disk movement

Figure 9. Normal contact force and disk movement distribution in NC assembly under B1 and B3 (from Ma, 1994).

Normal contact force   Disk movement

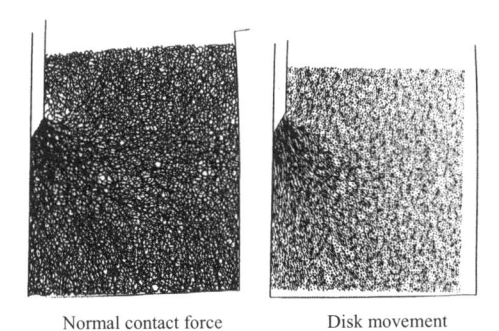

Figure 10. Normal contact force and disk movement distribution in NC assembly under B5 (from Ma, 1994).

in horizontal direction and disk movements in vertical direction.

The distribution of normal contact forces and disk movements of the OC assembly (OCR = 10) under B5 are shown in Figure 11. Due to higher stiffness in lateral compression and initial lateral stress, the contact forces in the OC assembly were more concentrated in the horizontal direction and disk movements in vertical direction. The magnitude of disk movements in the OC assembly was less than that in the NC assembly. In OC assembly, the disk movements were confined in a smaller area but the disk

Normal contact force   Movement

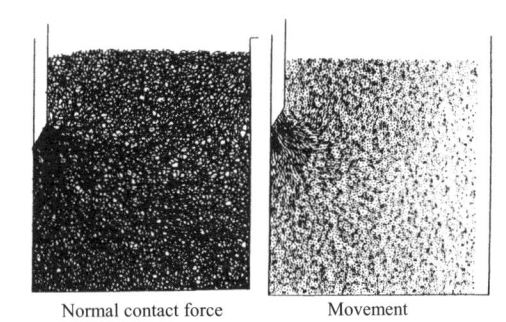

Figure 11. Normal contact force and movement distributions in OC assembly under B5 (from Ma, 1994).

movements extended above the cone base. There was no evidence indicating a slip surface that reached a vertical tangency in either of the NC or OC assemblies.

The CPT induced disk movements can be statistically converted into a distribution of strains. If the movements of a group of disks are described by a set of linear functions as

$$d_y = ay + bz \tag{2}$$

$$d_z = cy + dz \tag{3}$$

The strain tensor is derived by taking derivatives of Eqs. 2 and 3.

$$\varepsilon_{yy} = \frac{\partial (dy)}{\partial y} = a \tag{4}$$

$$\varepsilon_{zz} = \frac{\partial (dz)}{\partial z} = d \tag{5}$$

$$\varepsilon_{yz} = 0.5 \left(\frac{\partial (dy)}{\partial z} + \frac{\partial (dz)}{\partial y} \right) = 0.5(b + c) \tag{6}$$

And the volumetric strain, ε_v is

$$\varepsilon_v = \varepsilon_{yy} + \varepsilon_{zz} \tag{7}$$

Compressive strain is negative. In computing the strain tensor using Eqs. 4 to 6, the DEM assembly was divided into small "computation zones." Each zone had a dimension of 1.5R by 1.5R and contained approximately 100 disks. The displacement for each disk was computed by comparing its current position with that in the previous time step. For each zone, the coefficients a, b, c and d were determined by curve fitting Eqs. 2 and 3 using the least square method. The corresponding strain increments were computed using Eqs. 4 and 5. Accumulating the strain increments from the beginning of cone penetration (i.e., strain components are zero) gived the current state of strain.

Figure 12 shows the ε_v contours within the 12000 dsik assembly under B1 as the cone tip reached 8R following the above procedure. For the NC assembly the results showed signifcant dilation thoroughout the DEM assembly with the maximum dilative ε_v occuring at 2 to 4R ahead of the cone tip. In the OC assembly, the dilation was less significant and some contraction occurred above the cone base. The dilation was apparently associated with the wide spread lateral disk movements under B1 as shown in Figure 9.

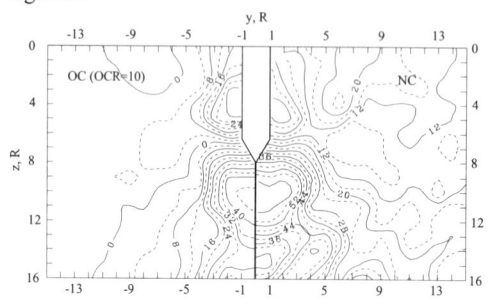

Figure 12. Volumetric strain contours under B1 (from Ma, 1994).

Figure 13 shows the ε_v contours within the DEM region under B5. In the NC assembly, dilation remiained below the cone tip but was less significant then those in B1 conditions. In the OC assembly, where the σ'_{ho} was 3.6 times that in the NC assembly, cone penetration induced contraction below the cone tip. The maximum contraction located at approximately 3R below the cone tip. In both the NC and OC assemblies under B5, contraction occurred after the disks passed the cone base.

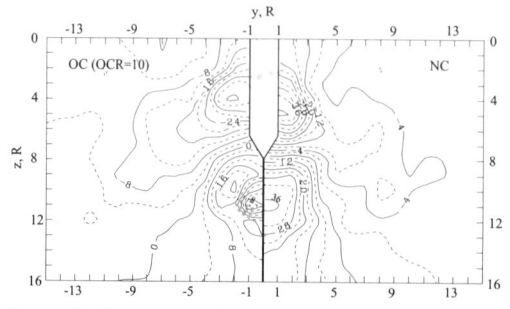

Figure 13. Volumetric strain contours under B5 (from Ma, 1994).

Based on the computed contact forces, the stress tensor, σ_{ij} could be computed for a given group of disks (Christoffersen et al. 1981) as

$$\sigma_{ij} = \frac{1}{V}\sum_{\alpha=1}^{N}\frac{1}{2}\left(f_i^\alpha d_j^\alpha + f_j^\alpha d_i^\alpha\right) \qquad (8)$$

where

f_i = contact force between disks
d_i = vector connecting the centers of disks in contact
V = total volume
N = number of disks

By dividing the DEM region into computation zones and using Eq. 8, the distribution of stresses around the cone can be computed. Figure 14 shows the contours of the normalized vertical stress, σ_{zz}/σ'_{vo} when the cone tip reached 8R, under B1. In NC assembly, apparently the lack of confinement caused the σ_{zz}/σ'_{vo} to be significantly less than 1.0 with their maximum value at 0.7. In the OC assembly, the peak σ_{zz}/σ'_{vo} occurred at approximately 1R below the cone tip. The value of σ_{zz}/σ'_{vo} attenuated rapidly along the face of the cone tip. Above the cone base, σ_{zz}/σ'_{vo} dropped below 1.0. Similar reduction in vertical stress was also suspected by Wesley (1998), who has inidcated that this reduction of vertical stress above the cone base was the main cause for the low q_T from CPT performed in cavity-wall calibration chambers with limited lateral dimensions.

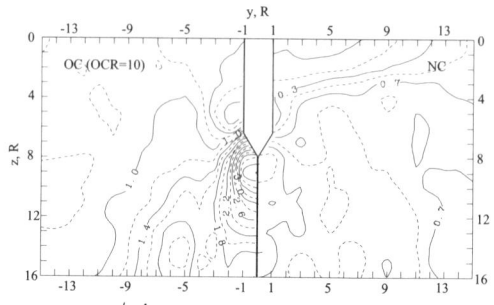

Figure 14. σ_{zz}/σ'_{vo} contours under B1 (from Ma, 1994).

Figure 15 shows the contours of σ_{zz}/σ'_{vo} under B5, when the cone tip reached 8R. In both NC and OC assemblies, the peak σ_{zz}/σ'_{vo} occurred at approximately 1R below the cone tip. Above the cone base, σ_{zz}/σ'_{vo} in both the NC and OC assemblies stabilized at values below 1.0. Apparently, the B1 boundary effects were more significant on the NC assembly in reducing σ_{zz} above the cone base.

The numerical simulations have indicated that by coupling DEM/BEM schemes, it was possible to simulate boundary conditions typically applied in cavity-wall chambers as well as field conditions where soil extended outward to infinity. The DEM/BEM system can be viewed as a different form of calibration chamber. The response due to CPT within the soil mass can be readily evaluated from micromechanical as well as continuum mechanics point of view. The computations showed that boundary conditions could alter the dilatancy, failure mechanism and stress field within the soil mass around the penetrating cone tip. Thus, for a correction factor as described in Eq. 1 to be valid, it should properly address all these issues.

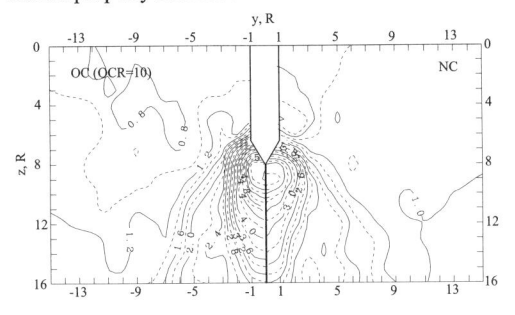

Figure 15. σ_{zz}/σ'_{vo} contours under B5 (from Ma, 1994).

The numerically simulated CPT calibration tests can be conducted without the cost of an elaborate laboratory system. This numerical tool can be a valuable compliment to the physical calibration tests. Issues such as the effects of cementation between particles and gravity as in the case of centrifuge can be studied and evaluated using the numerical simulation.

4 SERVO CONTROLLED CALIBRATION CHAMBERS

For a chamber specimen with limited lateral dimension, it is possible that the lateral boundary deforms in response to the calibration test conducted within the soil specimen. The deformation should then induce variation of stress on the lateral boundary due to the reaction of soil from beyond the lateral boundary, if under field conditions. In that case, the lateral boundary is neither constant stress nor zero strain. By introducing some intelligence to the lateral stress control scheme, it is possible to significantly reduce, if not eliminate the boundary effects. Some of the available "smart" calibration chambers are introduced in the following sections.

4.1 The Modified ISMES Calibration Chamber

By allowing the lateral boundary stress to vary in accordance to the amount of strain induced at the boundary (ε_b), it was possible to simulate the infinite soil mass beyond the lateral boundary. Ghionna and Jamiolkowski (1991) modified the ISMES calibration chamber (specimen diameter = 1.2m, height = 1.5 m) by implementing a computer controlled feedback at the lateral boundary as shown in Figure 16 to perform self-boring pressuremeter (SBPT) tests in dense Ticino sand where a cavity expansion pressure (p') versus strain (ε_c) curve was measured at the center of the specimen. The lateral boundary stress (σ'_h) was adjusted during SBPT according to a constitutive model ($\Delta\sigma'_h = p'(\varepsilon_b)$) that emulated soil from the physical boundary to infinity and SBPT was considered an axisymmetric and plane strain problem. Consequently, the lateral stress on the boundary varied with time as SBPT expanded. The boundary strain was considered uniform and a uniform stress was applied on the lateral boundary throughout the specimen.

4.2 The IMG Calibration Chamber

The IMG (Institut de Mécanique de Grenoble) calibration chamber (Foray, 1991) shown in Figure 17 consisted of three cylindrical segments of 50 cm height and 1.2 m internal diameter jointed together by flanges with O ring seals. Each cylindrical segment was equipped with an independent, 50 cm high rubber membrane. Depending on the number of cylindrical segments used, the total height could be in increments of 50 cm. The system allowed the stress on the lateral boundary to be servo controlled. The IMG chamber was used to perform axial load tests on model piles. The stress paths measured by Foray (1991) in the sand during the loading stage of a model pile, at a distance of 3.5 diameters from the pile are shown in Figure 18. These measured stress paths were substantially similar to a pressuremeter stress path in its elastic stage.

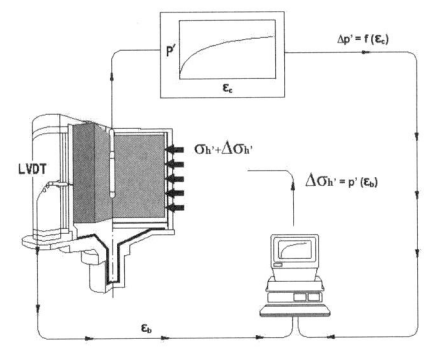

Figure 16. The SBPT calibration chamber (Ghionna and Jamiolkowski, 1991).

Based on his observations, Foray (1991) concluded that the lateral boundary of the chamber specimen was submitted to a loading-unloading pressuremeter stress path during the cone penetration or model pile installation. Thus the ideal lateral boundary conditions should reproduce such a loading path. Using the servo controlled boundary and stiffness from pressuremeter unload-reload tests in the calibration chamber with the same stress and density states, Foray (1991) has demonstrated that it was possible to significantly reduce the boundary effects in the case of a model pile pull-out test.

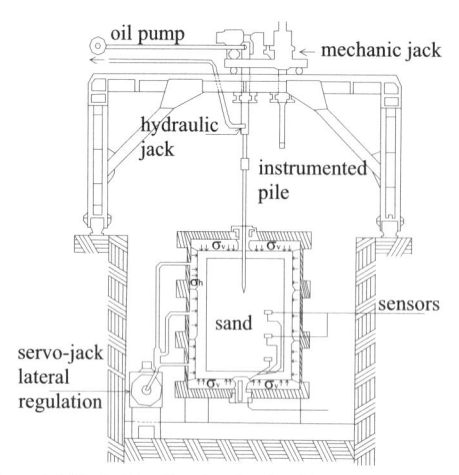

Figure 17. The IMG calibration chamber (Foray, 1991).

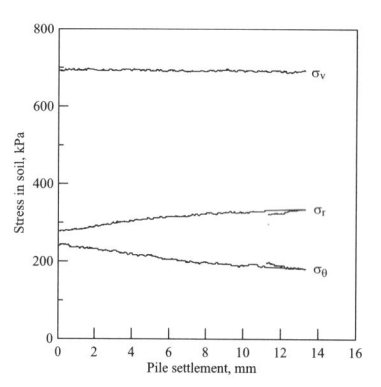

Figure 18. Stresses measured at a distance of 3.5 diameters from the pile (Foray, 1991).

4.3 The NCTU Field Simulator

The servo control systems reported by Ghionna and Jamiolkowski (1991) and Foray (1991) allowed the stress to vary but remained uniform throughout the lateral boundary. The technique was a step forward in providing a potential capability to simulate the field conditions. In CPT calibration however, the lateral boundary response to cone penetration was expected to vary with depth. Hsu and Huang (1998) at National Chiao Tung University developed a servo controlled calibration chamber that allowed the boundary response to be individually controlled at different depths. The system, referred to as the NCTU field simulator had a physical cylindrical specimen, as in the conventional cavity-wall calibration chamber, and a numerically-simulated soil mass that extended laterally from the physical boundary to infinity. The NCTU field simulator design took advantage of the earlier studies on numerical simulations which indicated that it was possible to explicitly couple discrete and continuum schemes to effectively minimize, if not eliminate the boundary effects for CPT simulations. The physical sand specimen, equivalent to the DEM disks in numerical simulations, was housed in a stack of twenty rings. These rings were lined with an inflatable silicone rubber membrane on the inside to facilitate boundary displacement measurement and stress control. The membrane expansion measuring system consisted of a wax lubricated, heavy duty fishing line wrapped around the membrane. The ends of the fishing line were attached to an extensometer that sensed the circumferential displacement of the rubber membrane. The specimen, 0.79 m in diameter and 1.6 m high, could be handled efficiently because of the relatively small dimensions and light weight.

During a cone penetration, the boundary displacement at each ring level was monitored and entered into a computer program. The stress for each ring level was adjusted pneumatically, according to the desired boundary conditions. The lateral boundary could be set as constant stress (B1), rigid (B3), or simulated field conditions, B5 (constant stress applied at top and bottom of specimen). When performing calibration tests under B5, the numerical scheme that simulated soil mass from beyond the physical boundary was executed repeatedly to determine the pressure at each ring level with the updated extensometer readings, as the cone penetration continued. Figure 19 provides a schematic view of the design and coupling concept of the NCTU field simulator. The experience gained in numerical simulations indicated that it should be possible to use an axisymmetric BEM scheme to simulate the lateral boundary under B5. However, BEM was too slow and complicated to be repeatedly executed during cone penetration. If the soil at certain distance away from the cone tip should experience a stress path that was similar to a pressuremeter expansion as shown in Figure 18, then it would be possible to simply simulate the field soil response using the cylindrical cavity expansion theory. Instead of performing a pressuremeter test in the specimen to determine the soil stiffness as suggested by Foray (1991), the radial strain (ε_r) and stress ($\sigma_r = \phi(\varepsilon_r)$) relationship could be determined upon boundary stress application, by a lateral compression test on the sand specimen. In the lateral compression test, σ'_{vo} remained constant while σ_r increased beyond σ'_{ho}. The lateral boundary was set to rigid (B3) during sand pluviation. The vertical and horizontal stresses were than adjusted following a stress path set by the desired K value to reach the target vertical (σ'_{vo}) and horizontal (σ'_{ho}) stress level, a procedure similar to that reported by Houlsby (1998). The main reason to use stress as the controlling factor was based on an argument that the influence of horizontal stress was much more important than the influence of overconsolidation ratio (Houlsby, 1998).

4.3.1 *CPT Calibration Tests in Da Nang Sand*

A series of cone penetration tests have been performed in dry Da Nang sand from Vietnam using the simulator. Da Nang sand was a clean quartz sand with average grain size, D_{50} = 1.13 mm, specific gravity G_s = 2.61, minimum void ratio, e_{min} = 0.515 and maximum void ratio, e_{max} = 0.808. Figure 21 shows the steady state line of Da Nang sand according to a series of isotropically consolidated drained (CID) triaxial tests and the initial states under which the CPT calibration tests were performed.

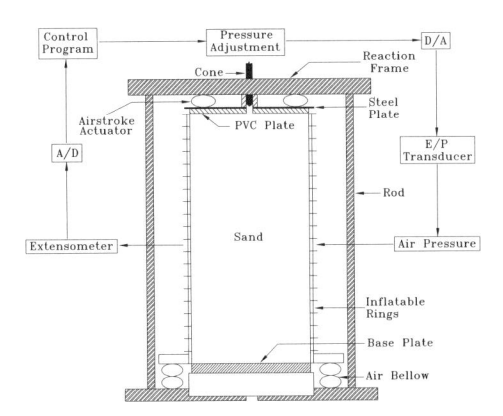

Figure 19. Physical /Numerical coupling of the simulator.

The relationship between stress (P_{ro}) and radial strain (ε_{ro}) at the physical-simulated soil mass interface, considering a cylindrical cavity expansion in an infinite soil mass, was derived by the following integration

$$P_{ro} = \int_0^{\varepsilon_{ro}} \frac{\phi(\varepsilon_r)}{2\varepsilon_r} d\varepsilon_r + \sigma_{ho} \qquad (9)$$

Figure 20 shows the $\phi(\varepsilon_r)$ obtained from the lateral compression test and the corresponding P_{ro}-ε_{ro} curve. The relationship of P_{ro}-ε_{ro} was stored in the computer. During cone penetration, the circumferential displacement at the boundary of every ring level, ΔC was measured and converted to ε_{ro} as

$$\varepsilon_{ro} = \frac{\Delta C}{\pi D} \qquad (10)$$

where
D = diameter of the physical specimen.

The corresponding P_{ro} in response to ε_{ro} for every ring, under simulated field conditions was then determined in accordance with the P_{ro}-ε_{ro} relationship stored in the computer. During cone penetration, the P_{ro} for each ring level was adjusted pneumatically and continuously updated with the change of ΔC.

Eq. (9) is essentially an inverse of interpreting a pressuremeter test with no soil dilatancy. P_{ro} and ε_{ro} may be viewed as the pressure and radial strain of the cavity expansion in a pressuremeter test. In the interpretation of a pressuremeter test, the soil $\phi(\varepsilon_r)$-ε_r relationship was obtained by taking derivative of the P_{ro}-ε_{ro} curve. In order for this scheme to be valid, the simulated sand mass outside of the cavity wall should experience insignificant volume change during cavity expansion to satisfy the no dilatancy assumption.

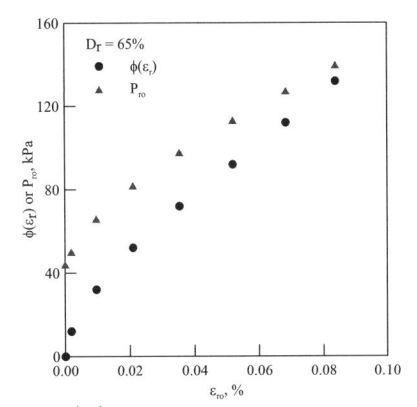

Figure 20. $\phi(\varepsilon_r)$ and the P_{ro}-ε_{ro} curves.

Figure 22 shows the q_T profiles under B5, where σ'_{vo} = 56 kPa and σ'_{ho} = 22 kPa, and D_r of 65 and 84%. Two types of cone diameters D_{cone} = 17.8 and 35.7 were used for tests in Figure 22. For a physical specimen diameter of 790 mm, these D_{cone} correspond to D/B of 44 and 22, respectively. The D_r of 65 and 84% under the given stress states corresponded to state parameters (ψ) of -0.19 and -0.24, respectively. Results show that upon reaching a stable value, q_T of the two types of D/B agree within 11.4% for D_r = 65% and 5.2% for D_r = 84%. As described earlier, dilatancy is a predominant factor in chamber boundary effects. The similarity of q_T from CPT with such a wide range of D/B reflects the success of reducing boundary effects with the simulator. The simulation scheme should remain valid as long as the CPT is performed in Da Nang sand with less dilatant state, or ψ larger than -0.24. Essentially all CPT calibration tests to be presented herein had their initial ψ close or larger than -0.24 as shown in Figure 21.

A comparison of q_T profiles under B1, B3 and B5 conditions (σ'_{vo} = 56 kPa and σ'_{ho} = 22.0 kPa), ΔC and P_{ro} measured/applied to ring No. 10 (depth at 800 mm) are shown in Figure 23. The result confirms our general belief that under field conditions, the physical lateral boundary is neither constant stress nor zero lateral strain. P_{ro} started at σ'_{ho} and gradually increased as the cone tip approached, and then stabilized after the cone tip passed. The amount of P_{ro} in B5 was higher than that under B3 but the maximum P_{ro} value was reached at a later stage in cone penetration. The q_T under B5 was higher than that of B3, confirming the fact that q_T under B3 was not an upper bound (Parkin, 1988). For the case shown in Figure 23, ΔC under B5 was just slightly larger than 0. Or, the ε_{ro} under B5 was very small and the assumption of no dilatancy in soil beyond the lateral boundary could be justified. In B3 conditions, significant increase in P_{ro} occurred at a very early stage of cone penetration to maintain the fixed boundary or $\Delta C = 0$.

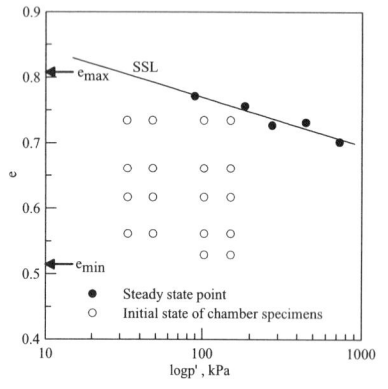

Figure 21. Steady state line of Da Nang sand and initial states of chamber specimens.

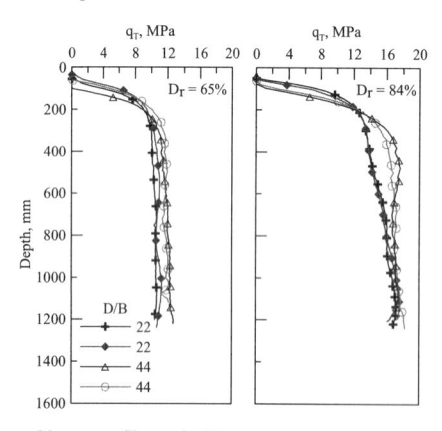

Figure 22. q_T profiles under B5.

Figure 24 depicts the ΔC measurement and P_{ro} applied to ring No. 10 (midheight of the specimen) at the physical boundary during cone penetration under the same initial stress conditions ($\sigma'_{vo} = 160$ kPa and $\sigma'_{ho} = 74$ kPa) but five different relative densities (25, 50, 65, 84 and 95%). The depth in Figure 24 is in reference with the cone tip level and normalized with respect to the cone diameter D_{cone}. The corresponding q_T profiles are shown in Figure 25. The q_T values continued to increase with depth in dense specimens, as observed in conventional chamber (Parkin and Lunne, 1982; and Parkin, 1988) under B3. The variation of ΔC followed a similar pattern as P_{ro} except that ΔC started at 0. In general, ΔC reached a maximum at approximately 3 to 5 D_{cone} ahead of the cone tip, and then remained more or less constant. The relative magnitudes of ΔC and P_{ro} reflected the combined effects of sand stiffness and dilatancy. The increases of P_{ro} and ΔC were positively related to D_r as a result of higher stiffness of the sand. In some cases, the correlations among P_{ro}, ΔC and D_r tend to offset one another and became less apparent.

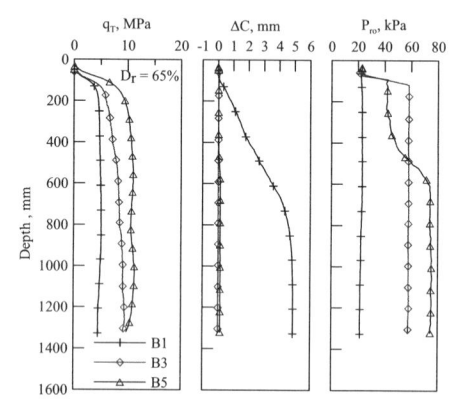

Figure 23. Comparison of q_T profiles under B1, B3 and B5 conditions.

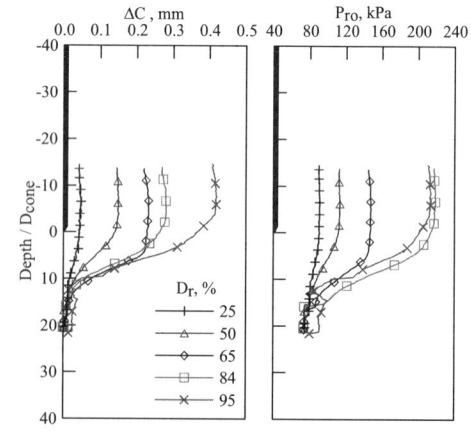

Figure 24. ΔC and P_{ro} measurements at ring No. 10.

Figure 26 plots cone tip resistance normalized with respect to the horizontal stress, Q_h (= $(q_T - \sigma_{ho})/\sigma'_{ho}$ versus Dr from a series of CPT in Da Nang sand under B5. These CPT were performed with $\sigma'_{vo} = 38$ to 161 kPa and K = 0.4 to 1.8. While the data follow the general trend of the empirical

158

© 2004 Millpress, Rotterdam, ISBN 90 5966 009 9

equation proposed by Houlsby (1998), the data do not support a consistent correlation between q_T and σ'_{ho}. As shown in Figure 27, for a given mean normal stress (σ'_{oo}), q_T can have a negative relationship with σ'_{ho}. This negative correlation becomes more significant as Dr increased.

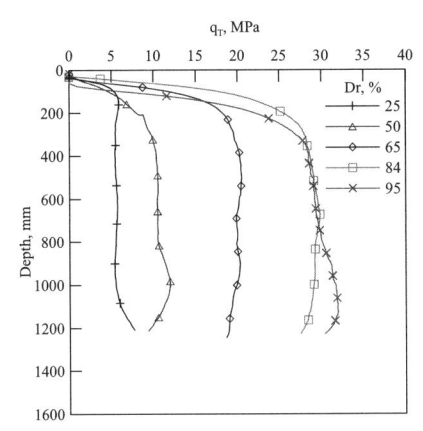

Figure 25. The q_T Profiles of different Dr under B5.

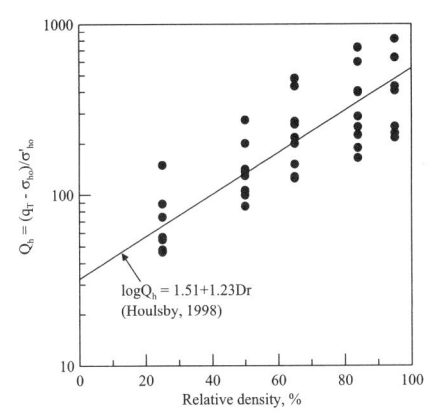

Figure 26. Normalized cone tip resistance versus Dr.

According to the CPT data in Da Nang sand, a relatively consistent relationship was possible only between q_T and σ'_{oo} as shown in Figure 28. Using the format of Jamiolkowski et al. (1988) to curve fit the test data in Da Nang sand, the following equation was obtained:

$$q_T = 369 p_a \left(\frac{\sigma'_{oo}}{p_a} \right)^{0.5} \exp(2.34 D_r) \quad (11)$$

where q_T and stress values are in kPa and p_a is the atmospheric pressure. Eq. (11) depicted in Figure 28 as solid curves had a coefficient of correlation of 0.98 with the available test data. For comparison purpose, the equation by Jamiolkowski et al. (1988) is also plotted in Figure 28 using dashed lines.

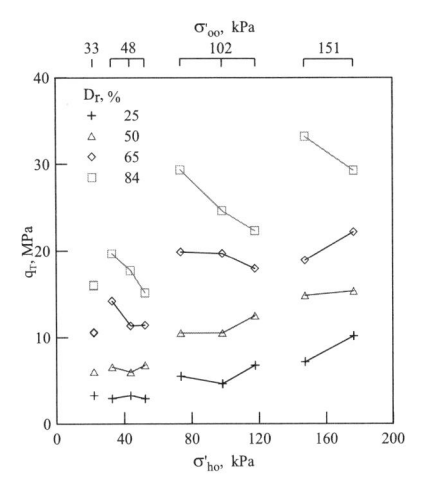

Figure 27. q_T versus horizontal stress.

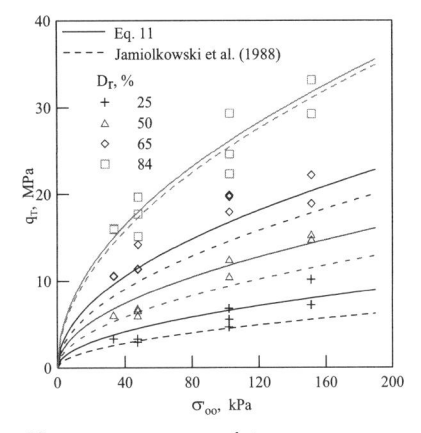

Figure 28. q_T versus mean normal stress.

The servo-controlled calibration chambers are practical and effective in reducing the boundary effects without the use of a large specimen. There is no need to involve correction factors for CPT results under B5.

5 CALIBRATION TESTS IN SILTY SAND

Natural sand often contains various amounts of fines (particles passing #200 sieve). The effects of plastic or non-plastic fines on the behavior of sand have been a subject of research for many decades. In spite of these research efforts, there is still a lack of consensus as to how the fines should affect the monotonic or cyclic shear strength of sand (Thevanayagam, 1998; Polito and Martin, 2001). The adjustment or consideration for fines content is an important aspect when q_T is used for assessing liquefaction potential. According to empirical charts compiled by Ishihara (1993), for a given q_T the corresponding cyclic strength increased with fines content. An opposite trend was reported by Carraro et al. (2003) according to their estimate of q_T in silty sand using the cavity expansion theory and cyclic triaxial tests on reconstituted specimens. A systematic study on the effects of fines on q_T would be very helpful in clarifying these controversies.

As the fines content increases, CPT becomes partially drained. Thus, CPT calibration in saturated specimen is highly desirable if not necessary for silty sand. Except for limited data reported by Rahardjo (1989) and Rahardjo et al. (1995), information on CPT calibration in silty sand is scarce. It is believed that difficulties associated with specimen preparation and conducting CPT calibration in full saturation are the main reasons for the lack of earlier attempts.

5.1 Calibration Chamber for Saturated Specimens

Figure 29 shows a schematic view of a medium sized calibration chamber developed at NCTU. It was designed for a 525 mm diameter and 760 - 815 mm high specimen. The relatively small chamber size was utilized for ease of specimen saturation, back pressuring and handling. The chamber was designed to provide constant stress lateral boundary conditions only. For compressible sand, which is often the case for silty sand (Ishihara, 1993), these chamber dimensions should be sufficient in offsetting the boundary effects (Bellotti and Pedroni, 1991). The chamber provided top and bottom drainage and six open-ended piezometers to monitor the pore pressure development within the specimen. These piezometers were made of 1.65mm OD and 0.23mm wall stainless steel tubing filtered with a piece of nonwoven geotextile. The void within the piezometers and pressure transducer ports were filled with deaired glycerin. During saturation of the specimen, the cone penetrometer was inserted in an O-ring sealed cone container at the top of the specimen. The container, made of Delrin, also served as a low friction bushing.

5.2 Fabrication of a Silty Sand Chamber Specimen

Dry pluviation could not be used to prepare graded or silty sand specimens as it would inevitably create segregation. Research has indicated that specimens prepared by water sedimentation (Høeg et al. 2000; Vaid et al. 1999) could most likely duplicate the soil fabric of an alluvial silty sand. Water sedimentation however, tends to cause particle segregation or layering in the soil specimen. CPT is known for its effectiveness in detecting thin soil non-uniformities. A layered silty sand specimen would inevitably result in erratic penetration test profiles and thus defeat the purpose of calibration test. Kuerbis and Vaid (1988) developed a slurry deposition procedure to prepare silty sand specimens for laboratory shearing tests. They have demonstrated that slurry deposition can produce silty sand specimens with similar stress-strain behavior as those from water sedimentation, but with much improved uniformity. The slurry deposition procedure according to Kuerbis and Vaid (1988) is rather tedious, its practicality as a means to create chamber specimens which involves large amounts of soil, is highly questionable. Rahardjo (1989) applied a much simplified slurry consolidation procedure where the silty sand slurry was either mixed by hand or a concrete mixer, and then consolidated into a chamber specimen. The CPT results as presented by Rahardjo (1989) however, were not consistent with those performed in specimens with acceptable uniformity.

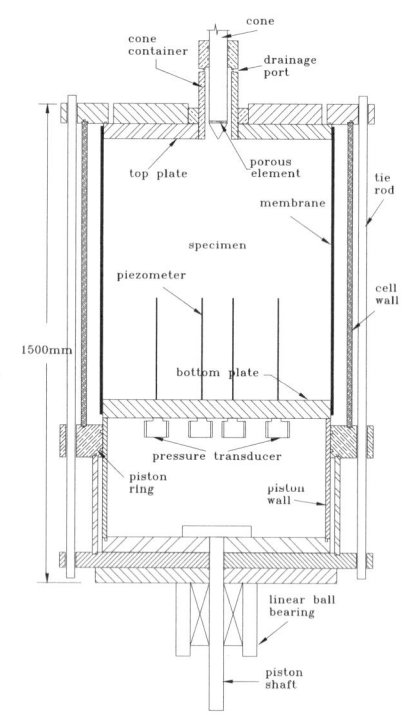

Figure 29. The NCTU medium sized calibration chamber.

A dry deposition method was used at NCTU to prepare the silty sand specimens for CPT calibrations. This procedure had the advantage of allowing CPT calibration to be performed in dry and saturated specimens. Dry silty sand mixture was placed in a 500-kg bag and lifted by a crane. The sand was slowly released from the bottom of the bag immediately above the surface of deposition. This process minimized the height of deposition, and thus the chance of segregation. The initial relative density upon deposition was approximately 50%. A steel rod was inserted in the sand and repeatedly rocked sideways to densify the sand. The densification stopped when the desired specimen height was reached. This sand placement and densification process was repeated to create a specimen in four equal layers. Upon completion of deposition, a 20 kPa vacuum was applied to the sealed specimen and the membrane jacket removed. The chamber cell was then assembled and the desired stress conditions applied to the specimen, if the CPT was performed under dry conditions.

If the CPT was to be performed in a saturated specimen, the saturation process started by infiltrating CO_2 from the bottom of the specimen. The piezocone tip was placed in a sealed container filled with glycerin and deaired under vacuum prior to installation in the chamber. The chamber specimen was permeated with deaired water from the bottom until the piezocone container was flooded with water. The specimen was then saturated under a backpressure of 300 kPa. The status of saturation was monitored by the six piezometers installed in the specimen.

The confining stress was applied in steps following a given K value in a zigzag fashion. For each stress increment, the vertical (piston) and lateral (cell) stresses were independently controlled so that $(\sigma'_v + \sigma'_h)/2$ increased to a desired level while $(\sigma'_v - \sigma'_h)/2$ remained constant. In this stage the effective stress path moved horizontally. Drainage was allowed at top and bottom of the specimen. Upon pore pressure dissipation, the effective stress path moved vertically by adjusting $(\sigma'_v - \sigma'_h)/2$ and $(\sigma'_v + \sigma'_h)/2$ remained constant until the desired K value was reached. The next stress increment was applied upon dissipation of the excess pore pressure according to the piezometer readings. This consolidation procedure was necessary to avoid specimen failure during stress application for saturated specimens.

5.3 CPT Calibration Tests in Mai Liao Sand

The authors performed a series of CPT calibration tests in Mai Liao Sand (MLS) from Central Western Taiwan. The specific gravity of MLS had an average value of 2.69. X-ray diffraction analysis of MLS showed significant amounts of muscovite and chlorite, in addition to quartz. The fines had a liquid limit of 32 and a plasticity index less than 8. Figure 30 shows the grain size distribution of MLS. The natural MLS was sieved first to separate the fines and then remixed to create specimens with desired gradation.

Figure 31 shows the post CPT fines content (FC) distributions from three specimens prepared initially with 15% fines content. Upon calibration test and removal of the cone, the chamber specimen was sliced vertically into two halves from the center to expose the cylindrical cavity created by the cone penetration. Soil samples were taken within 20mm from the wall of the central cavity and on the edge of the specimen away from areas affected by CPT. These samples were then wash sieved to determine their fines contents. The fines contents near the center cavity wall increased significantly as a result of cone penetration and this fines content increase had a positive relationship with the confining stress. The fines content from the edge of the specimens remained close to 15%. The uniformity of fines contents away from the center of the specimen indicated that the chamber specimens were reasonably uniform in terms of gradation.

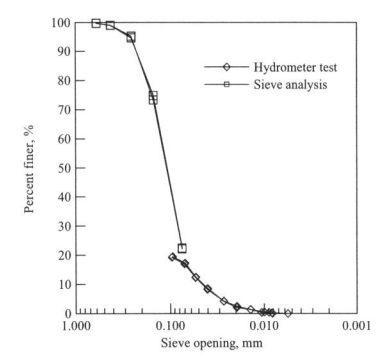

Figure 30. Grain size distribution of Mai Liao Sand.

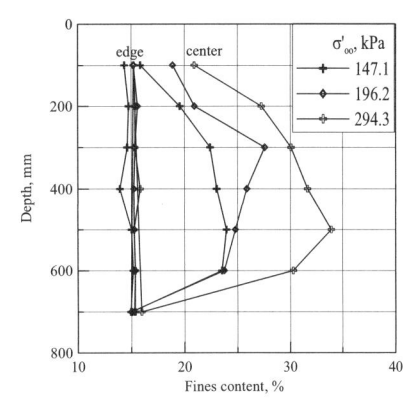

Figure 31. Fines contents at various locations in the specimen.

CPT calibrations were performed in dry and saturated specimens. For CPT under saturated conditions, the cone penetration test was conducted with specimen's top and bottom drainage valves open. A standard piezocone (diameter = 35.7 mm) with the porous element behind the cone tip (U_2) was used for CPT in saturated specimens. A half-size cone (17.8 mm diameter) was used in most of the dry specimens as a means to verify or minimize the boundary effects. All CPT's were performed with a penetration rate of 20 mm/sec, under stress controlled vertical and horizontal boundary conditions.

Figure 32 compares the profiles of q_T from tests using the standard piezocone (D/B = 15) and the half-size cone (D/B = 30). For a given FC, good agreement in q_T was obtained from tests with two different D/B values. The insensitivity of q_T to such a large variation of D/B was a reflection of the compressible nature of MLS. An important additional implication is that the chamber specimen was large enough even for a standard cone penetrometer and the boundary effects should be insignificant.

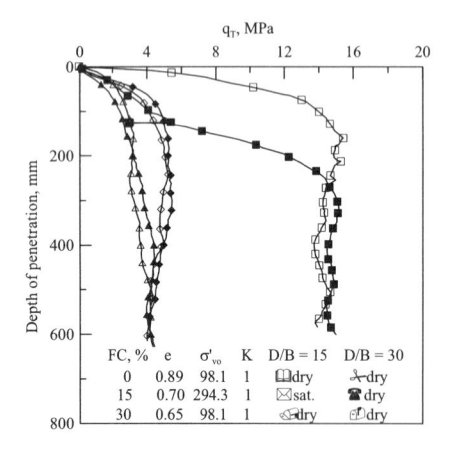

Figure 32. Comparisons of q_T with different diameter ratios.

For some of the CPT in saturated specimens, the penetration was stopped at a depth of approximately 350mm to perform a pore pressure dissipation test. Immediately upon resumption of cone penetration, there was a sudden increase in q_T before a stable value was reached again. A comparison of cone tip resistance normalized with respect to the mean normal stress, Q_o ($Q_o = (q_T - \sigma_{oo})/\sigma'_{oo}$) from CPT performed in dry and saturated specimens with FC = 15, 30 and 50% are presented in Figure 33. Other than saturation, the test pairs had the same void ratio and stress conditions. When FC = 15%, the Q_o in dry and saturated specimens agreed within 6%, after reaching a penetration depth of 300mm. Apparently CPT performed in MLS with fines contents less than 15% may be considered a drained test. When FC = 30%, Q_o in a dry specimen could be more than twice the value as in a saturated specimen. When FC = 50%, the Q_o in a dry specimen was over three times that in a saturated specimen. It is apparent that as fines content exceeded 30%, the CPT in MLS could no longer be considered a drained test.

Figure 34 plots Q_o versus e from 94 CPT calibration tests in MLS. The tests in MLS covered a σ'_o ranging from 50 to 500 kPa and K varied from 0.5 to 2.0. Also included are results in Da Nang sand by the authors and those in Quiou and Ticino sand reported by Almeida et al. (1991). Quiou sand was a clean crushable calcareous sand and Ticino sand was a clean uniformly graded quartz sand. The data points for CPT in MLS with FC=0% fell between those of Quiou and Ticino sand. As fines content increased to 15%, the Q_o in MLS were mostly smaller than those of Ticino or Da Nang sand under the same e. For MLS, occupying the voids among larger particles by the fines apparently did not significantly increase its resistance to cone penetration. As the fines contents exceeded 30%, Q_o became extremely small as CPT became a partially drained test.

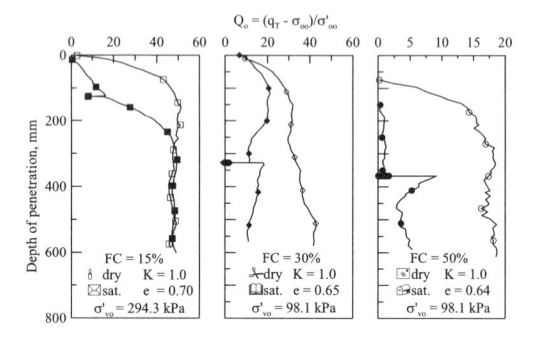

Figure 33. Comparisons of Q_o in dry and saturated specimens.

By curve fitting the available q_T from calibration tests in MLS with fines contents of 0 and 15% using a slightly modified format proposed by Fioravante et al. (1991), the following equations were obtained:

For FC = 0%:

$$\frac{q_T}{P_{a2}} = 383 \left[\frac{\sigma'_{vo}}{P_a}\right]^{0.03} \left[\frac{\sigma'_{ho}}{P_a}\right]^{0.42} \exp[-2.02e] \quad (12)$$

and for FC=15%:

$$\frac{q_T}{P_{a2}} = 236 \left[\frac{\sigma'_{vo}}{P_a}\right]^{0.23} \left[\frac{\sigma'_{ho}}{P_a}\right]^{0.44} \exp[-1.63e] \quad (13)$$

where p_{a2} is the atmospheric pressure in the same unit as q_T. Since q_T was much larger than σ_{oo} under which the CPT was conducted, the potential error associated with back pressure was insignificant. The equations have a minimum coefficient of correlation of 0.94 with the data points.

Rearranging Eqs. (12) and (13) so that,

$$383 \left[\frac{\sigma'_{ho}}{\sigma'_{vo}}\right]^{0.42} \exp[-2.02e] = \left[\frac{q_T}{P_{a2}}\right] \left[\frac{P_a}{\sigma'_{vo}}\right]^{0.03+0.42} \quad (14)$$

$$236 \left[\frac{\sigma'_{ho}}{\sigma'_{vo}}\right]^{0.44} \exp[-1.63e] = \left[\frac{q_T}{P_{a2}}\right] \left[\frac{P_a}{\sigma'_{vo}}\right]^{0.23+0.44} \quad (15)$$

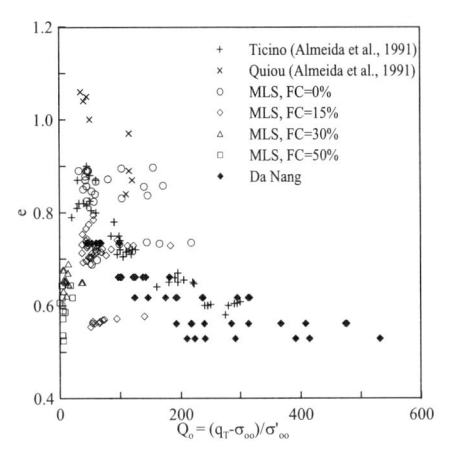

Figure 34. Void ratios versus Q_o.

When applying CPT for soil classification or liquefaction potential assessment, q_T is often corrected for overburden stress (Youd et al., 2001), or converted to q_{T1N} as

$$q_{T1N} = \left(\frac{q_T}{P_{a2}}\right)\left(\frac{P_a}{\sigma'_{vo}}\right)^n \quad (16)$$

The stress exponent n was believed to be related to the type or behavior of soil. A comparison among Eqs. (14) to (16) would show that $n = 0.45$ and 0.67 for FC=0 and 15% in MLS, respectively. This trend was consistent with those reported by Olsen and Malone (1988). The stress exponent of σ'_{ho} was much larger than that of σ'_{vo}, suggesting that σ'_{ho} was a much more significant contributing factor to n than σ'_{vo} for MLS.

The available simplified methods (Youd et al., 2001) in assessing soil liquefaction potential hypothesized that for the same cyclic resistance ratio, the standard penetration test (SPT) or CPT penetration resistance in silty sand was smaller than clean sand. The calibration data in MLS offered an opportunity to verify this hypothesis from q_T point of view. It would be very useful if q_T values from tests with different fines contents under a given state of stress and density could be compared. For silty sand, there was another issue as to what is the appropriate index to represent the state of density, which may include global void ratio, e or the intergranular void ratio, e_s defined as:

$$e_s = \frac{e + FC/100}{1 - FC/100} \quad (17)$$

The e_s is a void ratio that considers fines as voids (Thevanayagam (1998); and Polito and Martin, 2001). The ratio of q_T at FC=15% ($q_{T(FC=15\%)}$) over that at FC=0% ($q_{T(FC=0\%)}$) under different stress conditions according to Eqs. 14 and 15 are shown in Figure 35. The stress and density parameters used in the computations were chosen to reflect CPT within a depth of 20 m. The comparison of computed q_T was kept within or near the boundaries of stress and density conditions applied in the chamber tests. The minimum void ratios were 0.646 and 0.589 and maximum void ratios were 1.125, 1.058 for MLS with 0 and 15% of fines content, respectively. The $e = 0.75$ corresponds to D_r of 78% and 66% for FC = 0 and 15%, respectively.

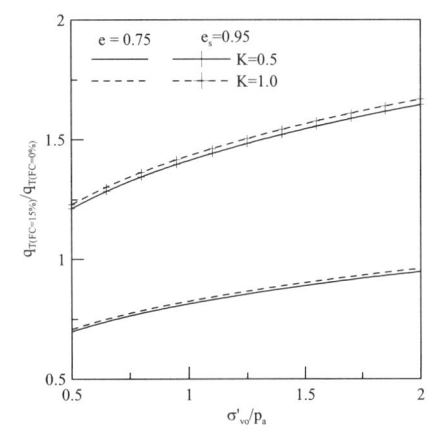

Figure 35. Effects of fines contents on q_T.

According to Figure 35, there was a general decrease in q_T as fines content increased from 0 to 15% under the given conditions. These differences diminish however, as the overburden stress became higher. The relationship was not sensitive to K. For $e_s = 0.95$, the corresponding D_r under FC=0, and 15% were 36 and 85%, respectively. Under these circumstances, $q_{T(FC=15\%)}/q_{T(FC=0\%)}$ was considerably larger than 1, indicating that ignoring the existence of fines would over-compensate the effects of fines.

The above series of CPT calibrations tests were performed in silty sand specimens prepared by dry deposition. If and how the specimen preparation method could affect the penetration test results have not been resolved.

6 CPT IN CENTRIFUGE

Centrifuges have been widely used in modeling geotechnical problems. In-flight CPT has been used to determine the uniformity and soil parameters of centrifuge specimens. Due to its limited vertical dimension, the vertical stress within a calibration chamber specimen is more or less uniform. In field CPT however, the cone penetration is under the effects of a significant stress increase with depth. This stress gradient effect can be readily simulated in centrifuge using a scaled physical model. A collaborative research effort has been carried out among five European centrifuge centers (Bolton et al., 1999) where in-flight miniature CPT's were performed in Fontainebleau sand. The Fontainebleau sand was a clean, uniform silica sand with $d_{50} = 0.22$mm. Cone diameters (B) ranging from 10 to 12 mm were used. Dry sand specimens prepared by pluviation were held in cylindrical or rectangular containers imposing a rigid lateral boundary. The lateral dimension (D) of these containers varied from 100 to 1200 mm, resulting in diameter ratios (D/B) ranging from 10 to well over 80. Profiles of normalized cone tip resistance ($Q_v = (q_T - \sigma_{vo})/\sigma'_{vo}$) versus normalized depth (Z = (depth of penetration)/B) from centrifuge CPT performed at these laboratories as shown in Figure 36, agreed within ±10% according to Bolton et al. (1999). For comparison, profiles from chamber CPT calibrations in dense (Dr = 84%) Da Nang sand under B5 are also included in Figure 36. The chamber CPT's were performed under σ'_{oo} ranging from 34 to 151 kPa and K from 0.4 to 1.8. For Da Nang sand, higher Q_v generally corresponded to lower σ'_{oo}. A similar trend was also reported by Bolton et al. (1999) in their centrifuge CPT. The Q_v values from chamber CPT in Da Nang sand were consistently higher than those of centrifuge tests. It is not clear whether these inconsistencies are due to differences in test conditions (i.e., 1g versus high g and rigid boundary versus B5 condition) and/or particle characteristics.

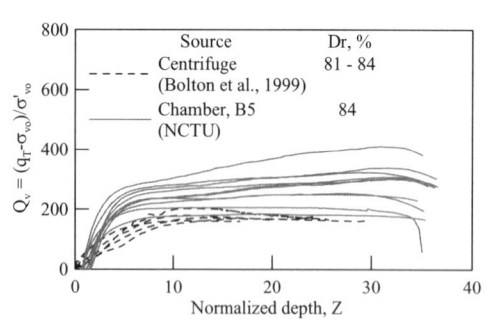

Figure 36. Comparison of cone tip resistance between centrifuge and chamber tests.

Centrifuge CPT results performed in Fontainebleau sand (Bolton and Gui, 1993; and Bolton et al., 1999) with Dr ranging from 54 to 89% showed that Q_v generally decreased as D/B became larger. Figure 37 shows the relationship between Q_v and D/B according to centrifuge CPT in dense Fontainebleau sand reported by Bolton et al. (1999). Bolton and Gui (1993) have suggested that a minimum D/B of 40 would be required to reach a stable cone tip resistance for centrifuge CPT in dense sand. The size effect for centrifuge CPT had a trend that was opposite to that normally observed in chamber CPT calibrations. As described in Figure 2, except for loose sand, cone tip resistance generally increased with the lateral dimension of the chamber specimen. A diameter ratio well in excess of 60 would be required before the boundary effect diminished in dense sand for chamber CPT. The reasons for such drastically different boundary effects between centrifuge and calibration chambers are not clear. It is strongly suggested that further studies be carried out to ascertain the mechanisms that are responsible for such differences, before centrifuge is regularly accepted as a means to calibrate CPT in sand.

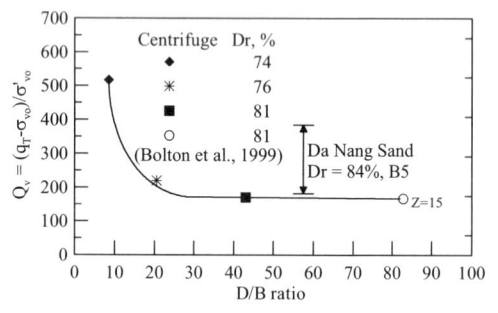

Figure 37. Effect of D/B on Q_v.

7 CONCLUSIONS

The experimental and numerical techniques developed so far have enabled us to perform CPT calibration tests that can substantially duplicate the field conditions. With the help of servo control and numerical simulations, the physical dimension is no longer a dominant factor in chamber design. If necessary, the calibration test can be performed in a saturated specimen under a back pressure. Experience in CPT calibration however, has been concentrated in dry, uniform clean sand. The need to calibrate CPT in silty sand is obvious. There has been substantial evidence indicating that the specimen preparation method can significantly affect the stress-strain relationships of reconstituted sand. How the specimen preparation method affects the CPT results in clean and silty sand should be an important issue to be addressed in the future. Practical methods to prepare a uniform silty sand specimen that can duplicate naturally deposited silty sand would be another challenge to be undertaken.

Centrifuge can be a useful part of the solution to the above challenges. Experience has indicated that it may be feasible to perform scaled CPT calibration tests in centrifuge. For a 10 mm diameter miniature cone, a centrifuge specimen with lateral dimension less than 500 mm may be sufficient in minimizing the boundary effects. The lateral dimension can be further reduced if a container with servo controlled lateral boundary is used. With this set up, the specimen preparation would be much more efficient because of the reduced dimensions. Problems of soil fabric and specimen preparation effects can then be practically studied. Before this potential is materialized, it may be necessary to first evaluate the soil failure mechanisms induced by CPT under elevated gravity and how these failure mechanisms compare with CPT under 1 g. The DEM simulation or similar numerical techniques may help in providing at least a qualitative description as to how the gravity field affects CPT in a granular material in centrifuge environment.

REFERENCES

Ajalloeian, R. and Yu, H.S. (1998). "Chamber Studies of The Effects of Pressuremeter Geometry on Test Results in Sand," Geotechnique, Vol.48, No.5, pp.636.

Almeida, M.S.S., Jamiolkowski, M., and Peterson, R.W. (1991) "Preliminary Results of CPT Tests in Calcareous Quoiu Sand." Proceedings, *First International Symposium on Calibration Chamber Testing*, Potsdam, New York, pp.41-54.

Anderson, W.F., and Pyrah, I.C. (1991). "Pressuremeter Testing in a Clay Calibration Chamber." Proceedings, *First International Symposium on Calibration Chamber Testing*, Potsdam, New York, pp.55-66.

Baldi, G., Bellotti, R., Ghionna, V., Jamiokowski, M., and Pasqualini, E. (1982). "Design Parameters for Sands from CPT." Proceedings, *Second European Symposium on Penetration Testing*, Amsterdam, Vol. 2, pp.425-432.

Been, K., Crooks, J.H.A., and Rothenburg, L. (1988). "A Critical Appraisal of CPT Calibration Chamber Tests." Proceedings, *First International Symposium on Penetration Testing, ISOPT-1*, Vol.2, pp.651-660.

Been, K. and Kosar, K.M. (1991). "Hydraulic Fracture Simulations in a Calibration Chamber." Proceedings, *First International Symposium on Calibration Chamber Testing*, Potsdam, New York, pp.67-78.

Bellotti, R., and Pedroni, S. (1991). "Design and Development of a Small Calibration Chamber for Compressible Sands." Proceedings, *First International Symposium on Calibration Chamber Testing*, Potsdam, New York, pp.91-100.

Bolton, M.D. and Gui, M.W. (1993). "The Study of Relative Density and Boundary Effects for Cone Penetration Tests in Centrifuge." Report No. CUED/D-SOILS/TR256, Cambridge University, 30p.

Bolton, M.D., Gui, M.W., Garnier, J., Corte, J.F., Bagge, G., Laue, J., and Renzi, R. (1999). "Centrifuge Cone Penetration Tests in Sand." Geotechnique, Vol.49, No.4, pp.543-552.

Borden, R.H. (1991). "Boundary Displacement Induced by DMT Penetration." Proceedings, *First International Symposium on Calibration Chamber Testing*, Potsdam, New York, pp.101-118.

Carraro, J. A. H., Bandini, P., and Salgado, R. (2003). "Liquefaction Resistance of Clean and Nonplastic Silty Sands Based on Cone Penetration Resistance." *Journal of Geotechnical and Geoenvironmental Engineering*, ASCE, Vol.129, No.11, pp.965-976.

Chapman, G.A., and Donald, I.B. (1981). "Interpretation of Static Penetration Tests in Sand." Proceedings, *10th International Conference on Soil Mechanics and Foundation Engineering*, Stockholm, Vol.2, pp.455-458.

Christoffersen, J., Mehrabadi, M.M., and Nemat-Nasser, S. (1981). "A Micromechanical Description of Granular Material Behavior." *Journal of Applied Mechanics*, Vol.48, pp.339-344.

Cundall, P.A., (1974). "A Computer Model for Rock-Mass Behavior Using Interactive Graphics for the Input and Output of Geometric Data." United States National Technical Information Service Report AD/A-001 602.

Dayal, U., and Allen, J.H. (1975). "The Effect of Penetration Rate on the Strength of Remolded Clay and Sand Samples." *Canadian Geotechnical Journal*, Vol.12, No.3, pp.336-348.

Fioravante, V., Jamiolkowski, M., Tanizawa, F., and Tatsuoka, F. (1991). "Results of CPT's in Toyoura Quartz Sand." Proceedings, *First International Symposium on Calibration Chamber Testing*, Potsdam, New York, pp.135-146.

Foray, P. (1991). "Scale and Boundary Effects on Calibration Chamber Pile Tests" Proceedings, *First International Symposium on Calibration Chamber Testing*, Potsdam, New York, pp.147-160.

Fretti, C., Lo Presti, D.C.F., and Pedroni, S. (1995). "A Pluvial Deposition Method to Reconstitute Well-Graded Sand Specimens." *Geotechnical Testing Journal*, Vol.18, No.2, pp.292-298.

Ghionna, V.N., and Jamiolkowski, M. (1991). "A Critical Appraisal of Calibration Chamber Testing of Sands," Proceedings, *First International Symposium on Calibration Chamber Testing*, Potsdam, New York, pp.13-39.

Høeg, K., Dyvik, R., and Sandbaekken, G. (2000). "Strength of Undisturbed versus Reconstituted Silt and Silty Sand Specimens." *Journal of Geotechnical and Geoenvironmental Engineering*, ASCE, Vol.126, No.7, pp.606-617.

Holden, J.C. (1991). "History of the First Six CRB Calibration Chambers." Proceedings, *First International Symposium on Calibration Chamber Testing*, Potsdam, New York, pp.1-12.

Houlsby, G.T., and Hitchman, R. (1988). "Calibration Chamber Tests of a Cone Penetrometer in Sand." Geotechnique, Vol.38, No.1, pp.39-44.

Houlsby, G.T. (1998). "Advanced Interpretation of Field Tests." Proceedings, *First International Conference on Site Characterization*, ISC'98, Atlanta, Georgia, Vol.1, pp.99-112.

Hsu, H.H., and Huang, A.B. (1998). "Development of an Axisymmetric Field Simulator for Cone Penetration Tests in Sand." *Geotechnical Testing Journal*, Vol.21, No.4, pp.348-355.

Huang, A.B. (1989). "Strain Path Analyses for Arbitrary 3-D Penetrometers." *International Journal for Numerical and Analytical Methods in Geomechanics*, Vol.13, pp.551-564.

Huang, A.B., Holtz, R.D., and Chameau, J. L. (1991). "A Laboratory Study of Pressuremeter Tests in Clays." *Journal of Geotechnical Engineering*, ASCE, Vol.117, pp.1549-1567.

Huang, A.B., and Ma, M.Y. (1994). "An Analytical Study of Cone Penetration Tests in Granular Material." *Canadian Geotechnical Journal*, Vol.31, No.1, pp.91-103.

Ishihara, K. (1993). "Liquefaction and Flow Failure during Earthquakes." Geotechnique, Vol.43, No.3, pp.351-415.

Jamiolkowski, M., Ghionna, V.N., Lancellotta, R., and Pasqualini, E. (1988). "New Correlations of Penetration Tests for Design Practice." Proceedings, *First International Symposium on Penetration Testing, ISOPT-1*, Orlando, Florida, Vol.1, pp.263-296.

Kuerbis, R., and Vaid, Y.P. (1988). "Sand sample preparation - the slurry deposition method." *Soils and Foundations*, Vol.28, No.4, pp.107-118.

Kulhawy, F.H. (1991). "Fifteen Years of Model Foundation Testing in Large Chambers." Proceedings, *First International Symposium on Calibration Chamber Testing*, Potsdam, New York, pp.185-196.

Lee, S.H.H., and Stokoe, K.H. (1988). "Investigation of Low-Amplitude Shear Wave Velocity in Anisotropic Material."

Geotechnical Engineering Report, No.CR86-6, Civil Engineering Department, University of Texas at Austin, 343p.

Lo Presti D., Pedroni S., and Crippa V. (1992). "Maximum Dry Density of Cohesionless Soils by Pluviation and by ASTM D 4253-83: a Comparative Study." *Geotechnical Testing Journal*, Vol.15, No.2, pp.180-189.

Lo Presti, D.C.F., Berardi, R., Pedroni, S., and Crippa, V. (1993). "A New Traveling Pluviator to Reconstitute Specimens of Well-Graded Silty Sands." *Geotechnical Testing Journal*, Vol.16, No.1, pp.18-26.

Ma, Y. (1994). *A Numerical Study of Cone Penetration Tests in Granular Assemblies*, Ph.D. thesis, Department of Civil and Environmental Engineering, Clarkson University, Potsdam, New Yor, 166p.

Manassero, M. (1991). "Calibration Chamber Correlations for Horizontal in Situ Stress Assessment Using Self-Boring Pressuremeter and Cone Penetration Tests." Proceedings, *First International Symposium on Calibration Chamber Testing*, Potsdam, New York, pp.237-248.

Mayne, P.W., and Kulhawy, F.H. (1991). "Calibration Chamber Database and Boundary Effects Correction for CPT Data." Proceedings, *First International Symposium on Calibration Chamber Testing*, Potsdam, New York, pp.257-264.

Olsen, R.S., and Malone, P.G. (1988) "Soil Classification and Site Characterization Using the Cone Penetration Test," Proceedings, *First International Symposium on Penetration Testing, ISOPT-1*, Orlando, Florida, Vol.2, pp.887-893.

O'Neill, M.W. (1991). "Houston's Calibration Chamber: Case History." Proceedings, *First International Symposium on Calibration Chamber Testing*, Potsdam, New York, pp.277-288.

Parkin, A.K. (1988). "The Calibration of Cone Penetrometers." Proceedings, *First International Symposium on Penetration Testing, ISOPT-1*, Orlando, Florida, Vol.1, pp.221-243.

Parkin, A.K., and Lunne, T. (1982). "Boundary Effects in the Laboratory Calibration of a Cone Penetrometer for Sand." Proceedings, *Second European Symposium on Penetration Testing*, Amsterdam, Vol.2. pp.761-768.

Peterson, R.W., and Arulmoli, K. (1991). "Overview of a Large Stress Chamber System." Proceedings, *First International Symposium on Calibration Chamber Testing*, Potsdam, New York, pp.329-338.

Polito, C.P., and Martin, J.R. (2001). "Effects of Nonplastic Fines on the Liquefaction Resistance of Sands." *Journal of Geotechnical and Geoenvironmental Engineering*, ASCE, Vol.127, No.5, pp.408-415.

Rad, N.S. and Tumay, M.T. (1987) "Factors Affecting Sand Specimen Preparation by Raining." *Geotechnical Testing Journal*, Vol.10, No.1, pp.31-37.

Rahardjo, P.P. (1989). *Evaluation of Liquefaction Potential of Silty Sand Based on Cone Penetration Test*, Ph.D. Thesis, Department of Civil Engineering, Virginia Polytechnic Institute, Blacksburg, Virginia.

Rahardjo, P.P., Brandon, T.L., and Clough, G.W. (1995). "Study of Cone Penetration Resistance of Silty Sand in the Calibration Chamber." Proceedings, *the International Symposium on Cone Penetration Testing*, CPT '95, Swedish Geotechnical Society, Vol.2, pp.577-582.

Salgado, R. (1993). *Analysis of Penetration Resistance in Sands*, Ph.D. Thesis, Department of Civil Engineering, University of California at Berkeley, 357p.

Salgado, R., Mitchell, J.K., and Jamiolkowski, M. (1998). "Calibration Chamber Size Effects on Penetration Resistance in Sand." *Journal of Geotechnical and Geoenvironmental Engineering*, ASCE, Vol.124, No.9, pp.878-888.

Schmertmann, J. (1976). "An Updated Correlation between Relative Density D_r, and Fugro-Type Electric Cone Bearing q_c." *Contract Report No. DACW 39-76 M6646*, Waterways Experiment Station.

Tan, N.K., Miller, G.A., and Muraleetharan, K.K. (2003) "Preliminary Laboratory Calibration of Cone Penetration in Unsaturated Silt." Proceedings, *12th Pan American Conference on Soil Mechanics and Geotechnical Engineering*, Cambridge, Massachusetts, Vol.1, pp.391-396.

Thevanayagam, S. (1998)."Effects of Fines and Confining Stress on Undrained Shear Strength of Silty Sands." *Journal of Geotechnical and Geoenvironmental Engineering*, ASCE, Vol.124, No.6, pp.479-491.

Ting, J.M., Corkum, B.T., Kauffman, C.R., and Creco, C. (1989)."Discrete Numerical Model for Soil Mechanics." *Journal of Geotechnical Engineering*, ASCE, Vol.115, pp.379-398.

Vaid, Y.P., Sivathayalan, S., and Stedman, D., (1999). "Influence of Specimen-reconstituting Method on the Undrained Response of Sand." *Geotechnical Testing Journal*, Vol.22, No.3, pp.187-196.

Veismanis, A. (1974). "Laboratory Investigation of Electrical Friction Cone Penetrometers in Sands." Proceedings, *European Symposium on Penetration Testing*, Stockholm, Vol.2.2, pp.407-419.

Vesic, A.S. (1963). "Bearing Capacity of Deep Foundations in Sand." *Highway Research Board Record No. 39*, Stresses in Soils and Layered System, pp.112-153.

Villet, W.C.B., and Mitchell, J.K. (1981). "Cone Resistance, Relative Density and Friction angle" Proceedings, *ASCE Symposium on Cone Penetration Testing and Experience*, St. Louis, pp.178-208.

Wang, P.Y. (1998). *Calibration of Dilatometer Test in Silty Sand*, Master Thesis, Department of Civil Engineering, National Chiao Tung University, Hsin Chu, Taiwan (in Chinese).

Wesley, L.D. (1998). "Interpretation of Calibration Tests Involving Cone Penetrometers in Sands." *Geotechnique*, Vol.52, No.4, pp.289-293.

Youd, T.L., Idriss, I.M., Andrus, R.D., Arango, I., Castro, G., Christian, J.T., Dobry, R., Liam Finn, W.D., Harder Jr., L.F., Hynes, M.E., Ishihara, K., Koester, J.P., Liao, S.S.C., Marcuson III, W.F., Martin, G.R., Mitchell, J.K., Moriwaki, Y., Power, M.S., Robertson, P.K., Seed, R.B., and Stokoe II, K.H. (2001). "Liquefaction Resistance of Soils: Summary Report from the 1996 NCEER and 1998 NCEER/NSF Workshops on Evaluation of Liquefaction Resistance of Soils," *Journal of Geotechnical and Geoenvironmental Engineering*, ASCE, Vol.127, No.10, pp.817-833.

In-situ seismic survey in characterising engineering properties of natural ground

S. Shibuya
Dept. of Architecture and Civil Engineering, Kobe University (formerly Hokkaido University), Japan

S. Yamashita
Dept. of Civil Engineering, Kitami Institute of Technology, Japan

Y. Watabe
Port and Airport Research Institute, Yokosuka, Japan

D.C.F. Lo Presti
Dept. of Structural and Geotechnical Engineering, Technical University of Turin, Italy

Keywords: in-situ test, laboratory test, elasticity, shear modulus, case history

ABSTRACT: Quasi-elastic stiffness (QES) in shear behaviour of geomaterials is discussed. Recent developments in measurement of QES of natural ground by a means of in-situ seismic survey, together with comparable measurement of elastic wave velocities in the laboratory are outlined. Application of combined in-situ/lab measurements is discussed with reference to the interlink between QES and strengths, the quality of laboratory sample and the in-situ 'structure' of natural sedimentary clay. Two case records, each demonstrating the importance of in-situ seismic survey are described: these are i) prediction of soft clay ground deformation associated with deep excavation in Bangkok Metro project, and ii) geotechnical site investigation performed at leaning tower of Pisa in September 2003.

1 INTRODUCTION

In-situ seismic survey enables us to manifest the depth-profile of elastic wave velocities (i.e., P-wave & S-wave) in natural ground. In addition, when these data are combined with the relevant profile of soil density from borehole data, elastic properties such as shear modulus (G), Young's modulus (E) and Poisson's ratio can directly be explored by applying the theory of elasticity. Ground strains induced by in-situ seismic survey are usually small of the order from 10^{-7} to 10^{-5}. Therefore, this non-destructive survey provides a unique opportunity to acquire geo-information regarding the current state of natural ground.

In-situ seismic tests, however, give us very limited information about in-situ elastic properties, which are far insufficient for everyday design in geotechnical engineering. To clarify the link with conventional design parameters at intermediate/large strains such as strengths is therefore of importance with which the value of seismic survey will be greatly enhanced. Laboratory element tests play an important role of making up for the missing link between the in-situ small-strain properties and other properties at intermediate/large strains. To achieve this purpose, the laboratory tests should be capable of measuring engineering properties of geomaterials over a wide strain range, including comparable elastic properties at very small strains. Moreover, the sample disturbance in the laboratory ought to be minimized, and it must be properly evaluated.

Regarding interpretation of the result of in-situ survey, the Authors believe that geotechnical engineers must be 'trinitarians'. 'Trinitarianism' for seismic survey refers to three kinds of geo-information. Surface wave measurement using Layleigh wave manifests spatial (3-D) variation of wave velocity in natural ground that is cited (e.g., Stokoe et al. 1988, Tokimatsu 1997 among others). On the other hand, seismic cone (or down/cross hole) test (e.g., Robertson and Campanella 1986, Tanaka et al. 1994 among others) reveals 1-D in-formation of this kind, whereas the wave velocity measurement in the laboratory deals with more comprehensive examination at a 'point'. Ideally, these 3-D, 1-D and 'point' information of QES should be synthesized by which the result of in-situ seismic survey with natural variability of elastic stiffness can be understood with confidence.

In this paper, some details regarding in-situ/lab measurements of wave velocity, together with their comparisons are outlined, bearing the above-mentioned 'trinitarianism' in mind. Application of in-situ seismic tests in understanding engineering properties of various geomaterials is discussed. Lastly, two case histories, each demonstrating the importance of in-situ seismic survey in geotechnical site investigation scheme are described in detail.

2 QUASI-ELASTIC STIFFNESS OF GEOMATERIALS

Elasticity refers to materials' property involved with no energy dissipation for any closed stress cycle. It is irrelevant whether the stress-strain relationship is linear or non-linear. In addition, the elastic properties ought to be time-independent, for example, the stiffness must not be influenced by the rate of stressing or straining.

In theory, when shear (or tangential) loading is induced at the interparticle contacts, the elastic response will not be observed for an assembly of elastic particles like soil grains, which in turn brings about energy dissipation in a closed stress-strain cycle (e.g., Johnson 1985). In reality, however, soil elements both in-situ and in the laboratory have inevitably been subjected to over-consolidation (OC), cyclic pre-straining and creep strain.

Quasi-elastic geomaterial properties may be defined for the behaviour at small strains because of the above-mentioned strain (or stress) history with time. For example, the stress-strain response at extremely small strains is quasi-elastic in respect that it shows a hysteretic loop involved with a small amount of energy loss, but the stiffness is unaffected by the rate of shearing a over normal range encountered in geotechnical engineering (Tatsuoka and Shibuya 1992, Shibuya et al. 1992).

For a hypo-elastic medium like soils, the compliance matrix, $[D_{ijkl}^{e}]$ in use for expressing the stress-strain relationship is stress-state dependent (Tatsuoka et al. 1997). Therefore, only a strain increment tensor, $[d\varepsilon_{ij}^{e}]$, can be related to a stress increment tensor, $[d\sigma_{kl}]$; that is;

$$[d\varepsilon_{ij}^{e}] = [D_{ijkl}^{e}] \cdot [d\sigma_{kl}] \qquad (1)$$

In a classical framework of elasto-plastic theory in soil mechanics, the yielding characteristics are described by defining a single yield locus. The stress relevant to the threshold in e-log p' diagram or in stress-strain curve as subjected to consolidation or shearing is normally regarded as "yield stress". The yield stress in conventional terms is apparent, and relates to the geomaterial behaviour at relatively large strains. The stress-strain behaviour inside the supposed elastic region is thus neither fully recoverable nor completely time-independent. Accumulation of excess pore pressure when subjected to low-amplitude cyclic loading is a typical example to verify it.

Figure 1 shows a simplified scheme of multiple kinematic yield surfaces applicable to clay behaviour (Jardine 1985, 1992). Both of sub-yield surfaces, Y1 and Y2 are mobile along the current stress point, whereas the larger scale yield surface Y3 remains relatively immobile. The quasi-elastic behaviour can exclusively be observed inside the inner-

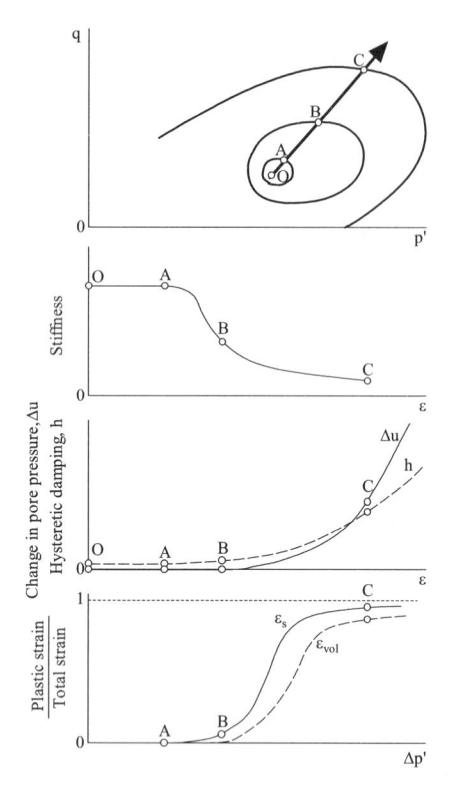

Figure 1. Scheme of multiple yield surfaces (Jardine 1985, 1992, Hight and Higgins, 1994).

most sub-yield surface, Y1. The stiffness when subjected to shearing remains constant so long as the current stress point stays within Y1 surface. Some reduction in stiffness takes place when the current stress point traverses Y1 surface.

The shear behaviour when the current stress point remains inside the Y2 surface is characterised by no accumulation of pore pressure subjected to undrained cyclic shear. The Y2 boundary is thus related to threshold cyclic shear strain proposed by Dorbry et al. (1982) (Shibuya et al. 1995a). Monotonous stress changes that remain within Y3 develop only small to moderate strains. The order of strains at the boundaries of Y1, Y2 and Y3 are approximately 0.001% (10^{-5}), 0.01% (10^{-4}) and 0.1-1% (10^{-3} - 10^{-2}), respectively. It has also been demonstrated that the Y1 and Y2 surfaces grow in size due to ageing and to the increase in rate of shearing (Tatsuoka et al. 1997).

Figure 2 illustrates in-situ and laboratory measurement of ground stiffness in a common practice of geotechnical site investigation (Yamashita et al. 1996). In-situ seismic survey involved with the measurement of shear body wave velocity, V_S,

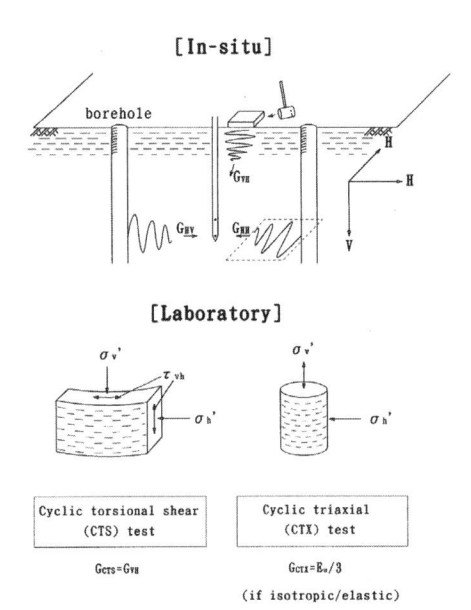

Figure 2. Measurement of quasi-elastic shear modulus in geotechnical site investigation (Yamashita et al. 1996).

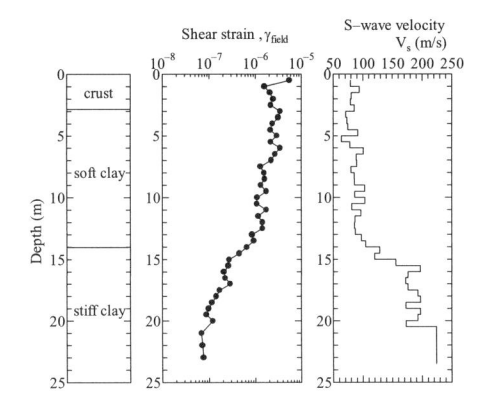

Figure 3. Ground strain in seismic cone test; n.b., a case record of Bangkok clay at Suttisan Station site along North Section of (Bangkok Metro) (Nishida et al. 1999).

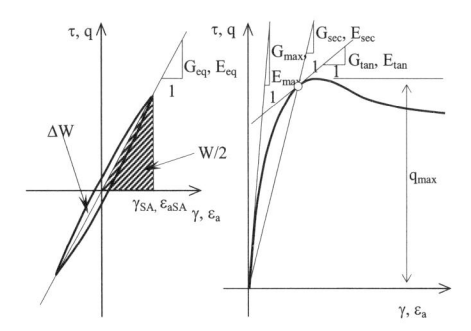

Figure 4. Definitions for Young's and shear moduli, together with hysteretic damping ratio.

manifests the profile of shear modulus, G_f, with depth, the estimate of which is based on;

$$G_f = \rho_t \cdot V_s^2 \quad (2)$$

where ρ_t denotes bulk density of subsoil at a cited depth.

Ground deformation in the geophysical survey is usually very small on the order of shear strain between 0.00001% and 0.0001% (10^{-7} and 10^{-6}). An example of geophysical survey performed in Bangkok is shown in Fig. 3 (Nishida et al. 1999). The results are obtained by performing a down-hole seismic cone test at Suttisan Station site on Northern Line of Bangkok Metro (Shibuya and Tamrakar 1997, 2003). Shear wave is generated at ground surface by a means of plank hammering. On the basis of the travel time record of shear wave, the magnitude of shear strain, γ_{field}, is estimated by using the following equation (e.g., Robertson and Campanella 1986);

$$\gamma_{field} = v_{max} / V_s \quad (3)$$

in which v_{max} denotes a maximum vibration velocity in the horizontal direction, $(\partial u / \partial t)_{max}$ (n.b., u: horizontal displacement), whereas V_S stands for the shear wave velocity in the vertical direction (i.e., z-direction), $\partial z / \partial t$, at pinpoint depth. The γ_{field} value is on the order of 10^{-6} in the upper soft clay extending down to 15 m depth, and of 10^{-7} in the lower stiff clay. In the light of the level of induced strain, the G_f value is practically free of any sources of disturbance, implying that geophysical survey has an important role to play as part of an integrated approach to stiffness investigations in geotechnical site characterization and in contributing towards checking the validity of laboratory stiffness data.

In the laboratory measurement, three types of tests are commonly practiced for measuring the QES in Y1 zone. Wave propagation techniques, for example in bender element test are employed in the laboratory by which G_{max} at strain level equivalent to G_f can be measured on the basis of Eq.2.2 (Dyvik and Madhus 1985). Alternation in stiffness over a wide strain range can be examined in other two tests; i.e., triaxial and torsional shear tests by using specimens of cylindrical and hollow cylindrical in shape, respectively (see Fig.2).

Figure 4 illustrates typical of stress-strain relation-ship of geomaterials in monotonic loading (*ML*) and cyclic loading (*CL*) tests. Young's modulus (*E*) can be defined by using the relationship between deviatoric stress (*q*) and axial strain (ε_a) in triaxial (or uniaxial) test. Similarly, shear modulus (*G*) can be

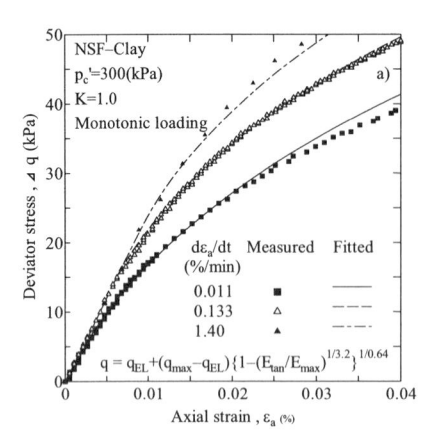

Figure 5. Results of undrained triaxial tests on *NSF* clay; a) stress-strain relationship at small strains, b) Young's moduli with strain, and c) hysteretic damping with strain (Shibuya et al. 1997a).

directly measured of the relationship between shear stress (τ) and shear strain (γ) in torsional (or simple) shear test.

Descriptions of the non-linear and irrecoverable stress-strain response may be made by using the following stiffness parameters;

Secant shear and Young's moduli;

$$G_{sec} = \tau/\gamma \text{ and } E_{sec} = q/\varepsilon_a \quad (4)$$

Tangent shear and Young's moduli;

$$G_{tan} = d\tau/d\gamma \text{ and } E_{tan} = dq/d\varepsilon_a \quad (5)$$

Equivalent (peak-to-peak) shear and Young's moduli;

$$G_{eq} = \tau_{SA}/\gamma_{SA} \text{ and } E_{eq} = q_{SA}/\varepsilon_{aSA} \quad (6)$$

Hysteretic damping ratio;

$$h = (1/2\pi) \cdot (\Delta W/W) \quad (7)$$

The subscript "*SA*" in Eq.6 is abbreviation of 'single amplitude' that means a half of double amplitude of cyclic strain or stress. The stress parameters, together with the corresponding strains are each defined against the initially equilibrium state prior to shearing.

Results of a series of *ML* and *CL* triaxial tests are shown in Fig.5. Specimens of a reconstituted clay are each subjected to undrained shear under a strictly controlled rate of axial straining, $d\varepsilon_a/dt$ in each *ML* test and in each cycle of a *CL* test using a fixed frequency *f* (Shibuya et al 1997a). Despite the marked difference in the rate of straining, E_{sec} in the *ML* tests coincides with E_{eq} in the *CL* tests at ε_a of 0.001% (10^{-5}). The damping ratio is not exactly zero by showing the values of 2-3 %. The trend is similar

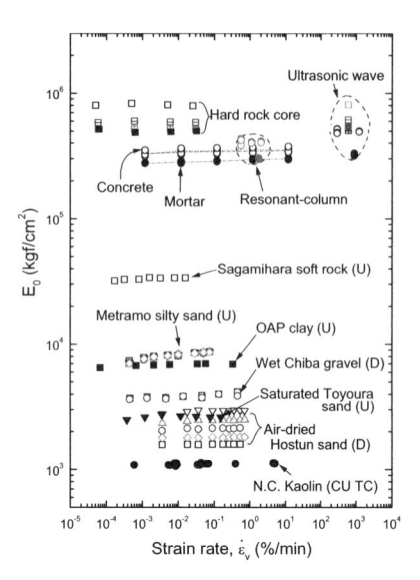

Figure 6. Dependency of E_{max} on strain rate (N.B., reproduction of Fig.4.6 of Tatsuoka et al. 1997).

for G_{sec} and G_{eq} when defined at shear strain of about 0.001% in a series of undrained *ML* and *CL* torsional shear tests on the reconstituted clay (Shibuya et al 1995a).

The small-strain stiffness is a maximum value at a given stress condition. The quasi-elastic shear and Young's moduli, G_{max} and E_{max}, of laboratory samples are therefore described as;

$$G_{max} = G_{sec} = G_{tan} \text{ and } E_{max} = E_{sec} = E_{tan} \quad (8)$$

When the stress-strain isotropy is postulated, E_{max} in undrained triaxial test on the fully saturated sample is related to;

$$G_{max} = E_{max}/2(1+\nu_u) = E_{max}/3 \quad (9)$$

where ν_u stands for undrained Poisson's ratio of 0.5.

A comprehensive summary of rate effect on E_0 (=E_{max}) has been reported by Tatsuoka et al. (1997) (Fig.6). It should be mentioned that experimental data are now abundant to show the existence of quasi-elastic shear (or Young's) modulus of a wide spectrum of geomaterials from soft soils to hard rocks, for which the G_{max} (or E_{max}) value is scarcely affected by rate of shearing and the type of loading (e.g., Tatsuoka and Shibuya 1992, Shibuya et al. 1992, 1995a, Porovic and Jardine 1994, Tatsuoka et al. 1997, Jamiolkowski et al. 1995, Lo Presti et al. 1993, Tanaka, Y. et al. 1994, 1995 and others). The coincidence between G_f and G_{max} has been observed by many researchers with which the existence of quasi-elastic stiffness is corroborated (e.g., Tatsuoka and Kohata 1995 among others).

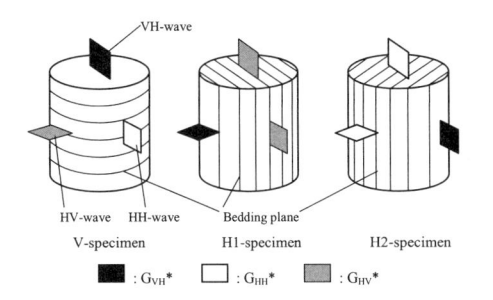

Figure 7. Relations between propagating direction and bedding plane.

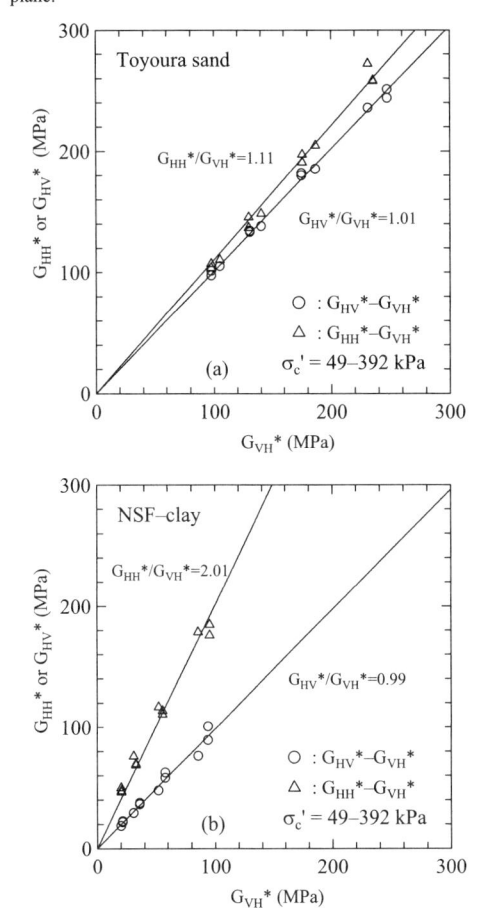

Figure 8. Relation of G_{HH}^* or G_{HV}^* to G_{VH}^*; (a) Toyoura Sand, (b) NSF-clay (Yamaguchi et al. 2004).

Provided that the subsoil exhibits a cross-anisotropy in the stiffness, the G_{vh} value from a down-hole survey where the shear wave propagates in the vertical direction with the particle motion in a horizontal direction differs from the comparable G_{hh} value from a cross-hole survey. The value of G_{hh}/G_{vh} in the literature shows a wide spread. The ratio was close to unity in Holocene sand deposits (Nishio and Katsura 1994, Lunne et al. 1992), in reconstituted clays (Lo Presti et al. 1999) and also in sedimentary soft rocks (Tatsuoka et al. 1997, Hayano 2001). It ranged between 1.2 and 1.5 in natural sedimentary clays in Italy (Jamiolkowski et al. 1995), and it was far greater than unity in over-consolidated fissured clays (Butcher and Powell 1995).

Figure 7 shows a schematic diagram of a triaxial specimen in which three pairs of bender elements (*BEs*) were instrumented in order to measure cross-anisotropy of quasi-elastic shear moduli; i.e., G_{hh}, G_{vh} and G_{hv}. In tests on air-pluviated Toyoura sand, a vertically-cut specimen (V-sample) and a horizontally-cut specimen (H-sample) defined relative to the bedding plane were prepared by using a freezing technique. Similarly, both of V- and H- samples relative to the 1-D preconsolidation (vertical) stress were prepared for a reconstituted clay. As can be seen in Fig.8, the G_{hh}/G_{vh} was 1.1 and 2.0 for Toyoura sand and the reconstituted clay, respectively (Yamaguchi et al. 2004). Authors are still puzzled with the fact that G_{hh} is twice G_{vh} even for the isotropically consolidated clay. Yet, a better understanding is needed into anisotropy in quasi-elastic shear modulus.

3 RECENT DEVELPMENTS IN SEISMIC TESTS

3.1 *In-situ test*

Geophysical survey has been developed for detecting underground resources, such as coal and petroleum. Afterwards, this technique has been gradually applied in geotechnical engineering for describing stratigraphy deep in the ground. In these applications, the point is not to measure the properties of each layer but to identify a specific layer having properties different from the others using reflection of the elastic wave(s). The dynamic response analysis of soil-structure interaction is common in regions where the seismic activity is intense. In this situation, the shear wave velocity, V_S, in the foundation is needed for the calculation. Compared to these applications, a geophysical survey for getting QEP requires more strict resolution of V_S because the profile of G_f is calculated as square of V_S as indicated in Eq.2.

As described earlier, 'trinitarianism' in understanding the outcome of seismic survey refers to three kinds of geo-information. Surface wave measurement using Layleigh wave manifests spatial (or 3-D) variation of wave velocity in natural ground that is cited (e.g., Stokoe et al. 1988, Tokimatsu 1997 among others). In this paper, a case history involved with the surface wave survey is afterwards

Figure 9. Twin-receiver type seismic cone developed at Port and Horbour Research Institute (Tanaka H. et al. 1994).

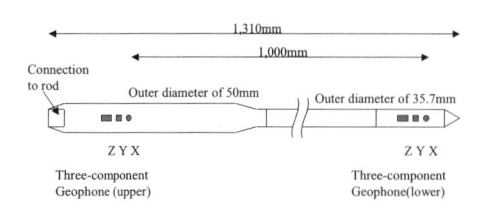

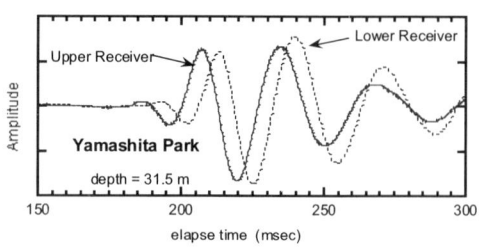

Figure 10. A typical of shear wave velocity measurement using dual-receiver seismic cone penetrometer (Tanaka H. et al, 1994).

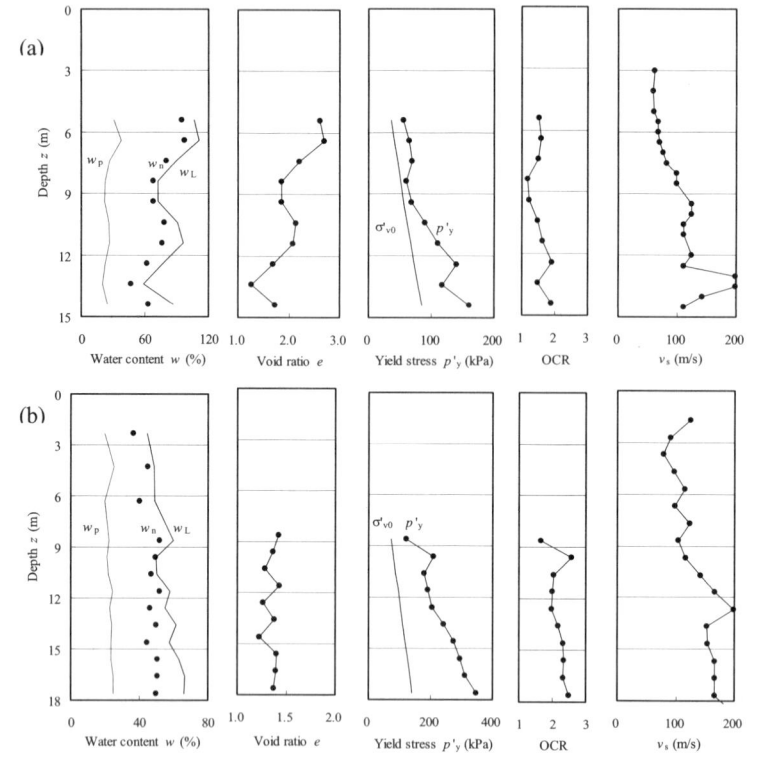

Figure 11. Two examples of seismic cone test; a) Bangkok (Ekachai) and b) Hanoi (Hai-Phong) (Watabe et al. 2004).

described in relation to geotechnical site investigation recently performed at leaning tower of Pisa.

Seismic cone penetration (SCP) test (e.g., Robertson and Campanella 1986) is a useful tool for obtaining 1-D information of quasi-elastic shear modulus. In advent of small but sufficiently sensitive receivers, it has been possible to develop a seismic cone comprising accelerometer(s) at the cone tip. The supreme advantages of using the seismic cone compared to other conventional geophysical methods, refer to low cost as well as promptness because boreholes are not necessary in performing the SCP test. In addition, when making a borehole in soft clayey or sandy ground, a casing is normally installed to keep the borehole open properly. This casing sometimes interrupts measurement of the arrival time of the shear wave. The cross-hole measurement is possible in the SCP test, for which the shear wave is generated at a cone penetrated at some depth and it is received by another cone comprising a receiver (Baldi et al. 1988). However, the down-hole method is a prevailing practice: that is, the shear wave is generated on the ground surface and the propagated wave is received by the in-soil cone (see Fig.2).

Figure 9 shows a prototype cone developed at Port and Harbour Research Institute (PHRI)

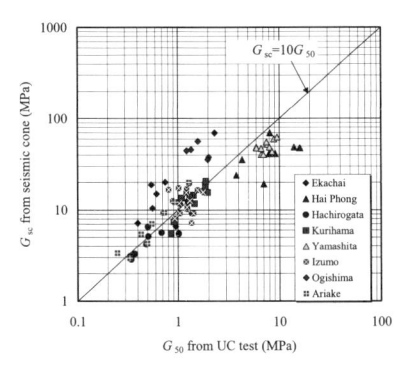

Figure 12. Comparison of shear moduli between G_{50} from UC test and G_{SC} from SCPT of clays (data from Tanaka et al. 1996, Watabe et al. 2004).

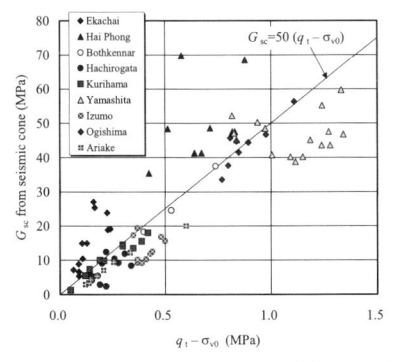

Figure 13. Relationship between G_{SC} and the cone tip resistance for clays (Tanaka et al. 1994, Tanaka, H. and Tanaka, M. 1996, Watabe et al. 2004).

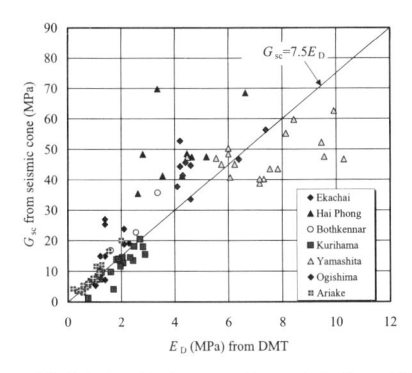

Figure 14. Relationship between G_{SC} and E_D from DMT for clays (Tanaka et al. 1994, Tanaka, H. and Tanaka, M. 1996, Watabe et al. 2004).

(currently Port and Air port Research Institute, PARI) (Tanaka H. et al. 1994). This seismic cone is instrumented with two receivers mounted at the cone tip to get high accuracy for measuring V_S (Tanaka, H. et al. 1996). By having the dual receiver system, the arrival time is determined by the time lag for the shear wave to propagate between the two receivers 1 m apart from one to the other (Fig.10). Two case records of the down-hole SCP test recently performed are presented in Fig.11 (Watabe et al. 2004).

Correlations between the shear modulus from SCP test, G_{SC} ($=G_f$) and other soil parameters such as from G_{50} from unconfined compression (UC) test, the corrected tip resistance (q_t) from cone penetration test (CPT) and E_D from dilatometer test (DMT) have been manifested in normally or slightly overconsolidated clays worldwide (Tanaka, H. et al. 1994, Tanaka, H. and Tanaka, M. 1996, Watabe et al. 2004). Figure12 shows the relationship between G_{50} measured by the UC test and G_{SC}. Bearing in mind that G_{50} from the UC test accounts for merely

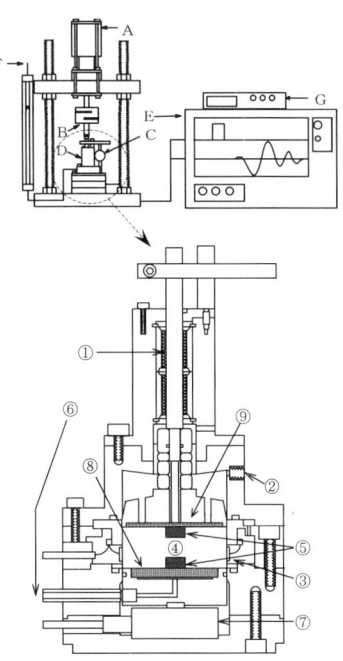

A Bellofram cylinder
B Axial load cell (upper)
C Displacement transducer for vertical displacement
D Consolidometer
E Oscilloscope
F Back pressure
G Function generator

1 Thrust bearing
2 Back pressure
3 Strain gauges for lateral stress mesurement
4 Specimen
5 Bender elements
6 Pore pressure
7 Axial load cell (lower)
8 Ceramin disk
9 porous stone

Figure 15. Consolidometer equipped with bender elements developed at Geotechnical Engineering Laboratory, Hokkaido University (Shibuya et al. 1997b).

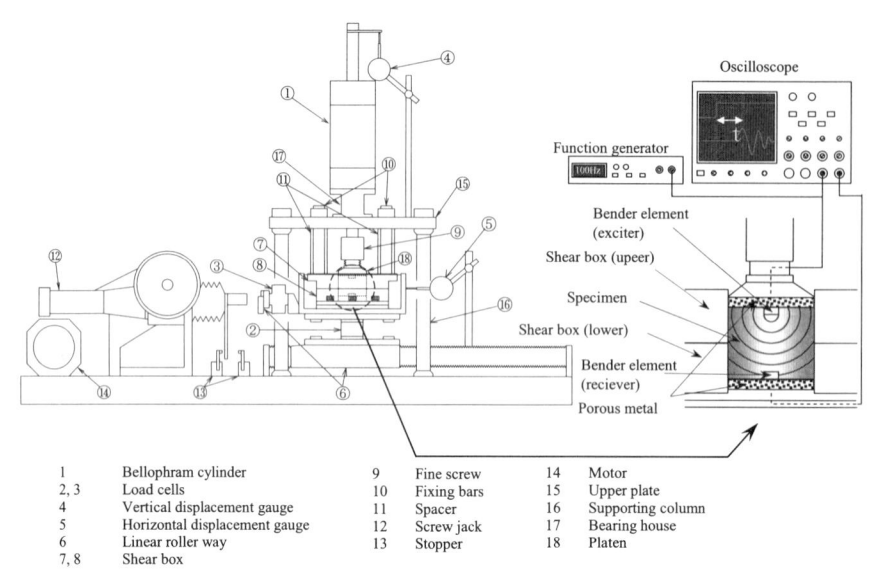

Figure 16. Direct shear box apparatus equipped with bender elements developed at Geotechnical Engineering Laboratory, Hokkaido University (Ogino et al. 1999).

1	Bellophram cylinder	9	Fine screw	14	Motor
2, 3	Load cells	10	Fixing bars	15	Upper plate
4	Vertical displacement gauge	11	Spacer	16	Supporting column
5	Horizontal displacement gauge	12	Screw jack	17	Bearing house
6	Linear roller way	13	Stopper	18	Platen
7, 8	Shear box				

1/10 of G_f, G_{50} from UC test roughly corresponds to the strain level of 1 %. As seen in Fig.13, the G_f value may be correlated to the tip resistance, q_t, deduced from CPT, that is;

$$G_{SC}(=G_f) = 50(q_t - p_{vo}) \quad (10)$$

where p_{vo} denotes the total overburden pressure.

Alternatively, G_{SC} in Fig.14 may be correlated to dilatometer modulus, E_D (Marchetti 1980) as;

$$G_{SC} = 7.5 E_D \quad (11)$$

3.2 Laboratory test

The bender element (BE) is handy, inexpensive and durable. The instrument suits well for measuring V_S in the laboratory by which G_{max} of soft soils is measured under controlled stress or strain rate (Dyvik and Madshus 1985). In fact, the BEs have been employed into a few testing devices such as consolidometer (Jamiolkowski et al. 1994, Shibuya et al. 1995b among others), triaxial apparatus (Tanizawa et al. 1994, Viggiani and Atkinson 1995, Jovicic et al. 1996, Kuwano and Jardine 1997 among others), and also applied to in-situ measurement of G_f (Nishio and Katsura 1994). Strength as well as G_{max} can be measured simultaneously in both the DSB and triaxial devices. Moreover, altering stiffness with strain over the entire pre-failure strain range may be measured in the triaxial apparatus.

Figure 15 shows a simple consolidometer equipped with a pair of BEs (Shibuya et al. 1997b). Figures 16 and 17 show similar cases of direct shear box (DSB) apparatus and triaxial apparatus, respectively (Ogino et al. 1999, Hwang et al. 1998).

In the consolidometer developed at Hokkaido University, the circular soil specimen of nominal dimensions 60 mm in diameter and 20 mm high undergoes 1D consolidation. It features by extra capabilities of

i) measurement of σ_h as well as σ_v,
ii) back pressure application,
iii) pore pressure monitoring at the base, and
iv) dual measurement of the vertical stress σ_v at the base and over the apparatus,

The measurement of two principal stresses enables us to determine K_0-value of the sample. Application of back pressure enhances the degree of saturation in the sample, which in turn improves accuracy of void ratio determination, hence e-$\log\sigma'_v$ curve on reconsolidation. Monitoring pore pressure at the base assists in minimizing development of creep strain when incremental loading is imposed. In the BE test performed at Hokkaido University, the drainage is allowed at the specimen's top surface alone (see Fig.15). In each loading step, the principal stresses, together with shear wave velocity, are measured at an instant when the excess pore water pressure at the bottom of the specimen is fully dissipated.

When soil moves relative to the rigid wall, friction force develops at the interface. Error in the σ_v measurement due to the wall friction has been found quite significant in conventional testing devices in which soil specimen is confined in a rigid box or ring (Shibuya et al. 1997b). Like the consolidometer

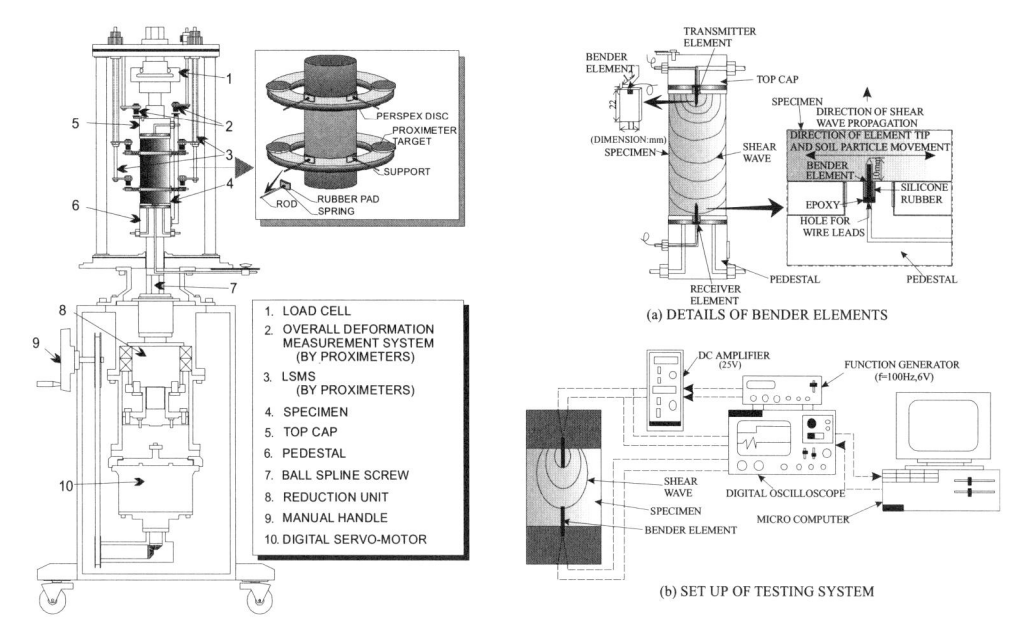

Figure 17. A multi-function triaxial apparatus equipped with bender elements developed at Geotechnical Engineering Laboratory, Hokkaido University (Hwang et al. 1998).

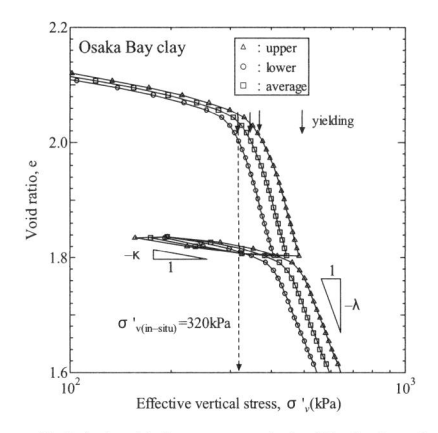

Figure 18. Relationship between e and σ'_v of Osaka Bay clay in CRS test (Li et al. 2004).

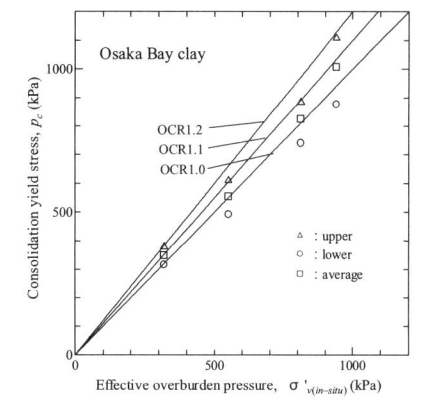

Figure 19. Determination of OCR of Osaka Bay clay in CRS test. (Li et al. 2004).

already shown in Fig.15, the vertical load at the base, W_{lower} as well as the measurement at the top, W_{upper} should therefore be measured by which the effect of wall friction is properly accounted for. Averaged σ_v in consolidometer test may be given by

$$\sigma'_{(oedometer)} = (W_{upper} + W_{lower})/2A \quad (12)$$

where A stands for cross-section area of specimen.

Figure 18 shows the relationship between void ratio e and σ'_v of Osaka Bay clay (Li et al. 2004) The λ value refers to the slope of NC line in terms of e-σ'_v relationship. The effect of interface friction acting on the lateral surface of the specimen was significant for the σ'_v measurement. The conventional measurement at the top grossly overestimated the σ'_v as compared to the average of the upper and lower measurements in compression, and vice versa in swelling. As seen in Fig.19, the conventional σ'_v measurement yielded the well-known design value of the overconsolidation ratio (OCR) of 1.2, whereas the OCR value from the averaged σ'_v yielded approximately unity. Accordingly, we have chosen the averaged vertical stress in order to determine the

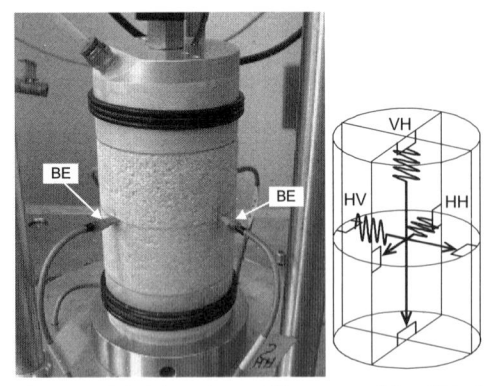

Figure 20. Triaxial specimen of Toyoura sand in which the shear wave velocity is measured using three sets of BEs (Yamashita et al. 2003).

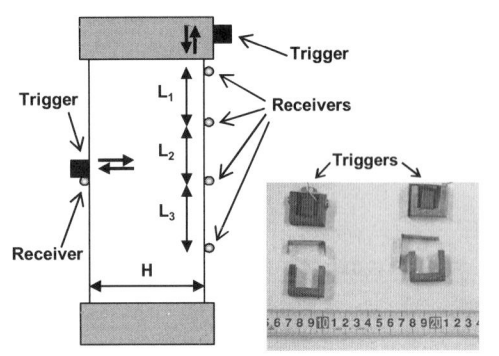

Figure 21. Measurement of P-S wave velocities in vertical and horizontal directions (AnhDan et al. 2002).

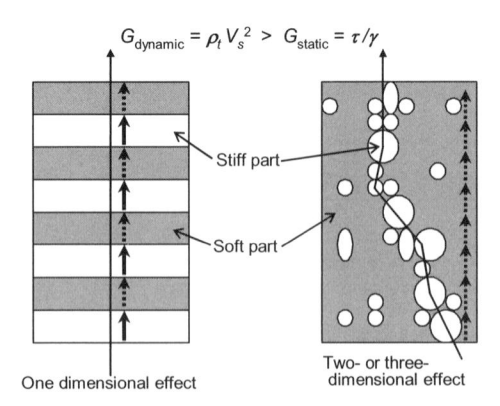

Figure 22. Effects of sample heterogeneity (modified from Tanaka et al. 1994).

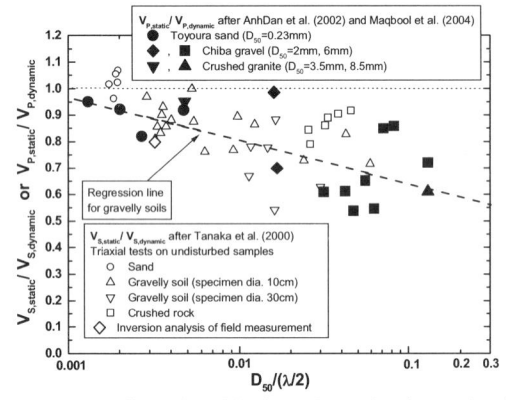

Figure 23. Effects of particle size and wave length on ratio of equivalent elastic wave velocities by static and dynamic measurements (modified from Tanaka et al. 2000, AnhDan et al. 2002, Maqbool et al. 2004).

yield stress, hence the *OCR* value. We should be more careful about the effects of interface friction even in conventional oedometer test using a specimen of merely 2 cm high.

An unsolved issue regarding the BE test is how to determine the correct travel time of the shear wave. Near field (NF) effects are a real nuisance which makes it difficult (Salinero et al. 1986, Kawaguchi et al. 2001 among others). Yet, no international consensus has been made on this issue. Therefore, TC29 has organized international parallel test on BE test and the latest information regarding BE test is available on the official website of TC29 (http://www.jiban.or.jp/e/tc29/index.htm, 2003).

Anisotropy of elastic stiffness is an interest for practical engineers when they interpret the result of in-situ seismic survey performed using different methods such as cross-hole and down-hole methods (for example, Butcher and Powell 1995). Anisotropy in shear modulus can also be measured using *BEs* (refer to Jamiolkowski et al. 1994, Fioravante 2000 among others). Figure 20 shows a triaxial specimen of Toyoura sand in which the shear wave velocity is measured using three sets of *BEs* (Yamashita et al. 2003). A set of *BEs* mounted on top cap and the pedestal enables us to measure shear wave velocity with the wave propagating in the vertical direction involved with the soil grain movement in the horizontal direction, VH. Similarly, the velocities associated with HV and HH shear can be measured with other two sets of *BEs* (refer also to Figs.7 and 8).

Figure 21 shows the measurement of *P-S* wave velocities (AnhDan et al. 2002). In this system, a multi-piezo-ceramic unit, which is more powerful than a single bender, is employed as the trigger to generate not only *S*-wave but also *P*-wave. The *P* and *S* wave velocities are measured using accelerometers mounted on the lateral surface of the specimen. This system seems most suitable for the elastic wave velocity measurement of large size soil specimen.

It is well known that in tests on coarse-grained soil or soft rocks, the elastic wave velocity meas-

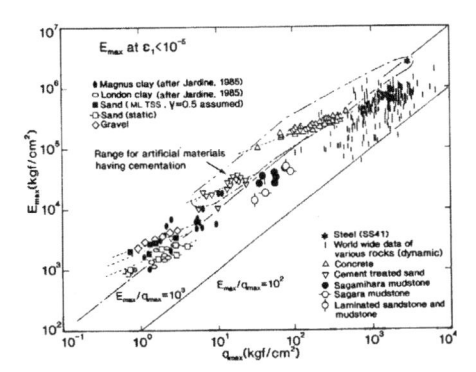

Figure 24. Relationship between E_{max} and q_{max} of various geomaterials (Tatsuoka and Shibuya 1992).

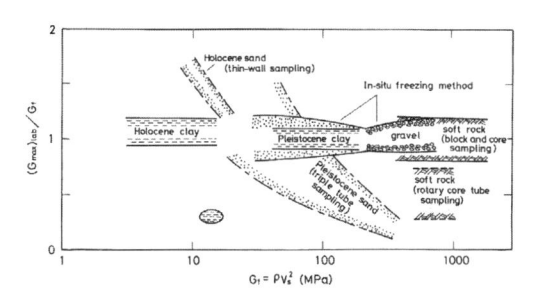

Figure 25. Ratio of G_{max} from laboratory tests to G_f from geophysical surveys (Toki et al. 1994, Shibuya et al. 1996).

urement undergoes the effects of sample heterogeneity (for example, refer to Tatsuoka and Shibuya 1992). As illustrated in Fig.22, the stiffness from the stress-strain curve in the laboratory refers to the overall deformation behaviour, whereas the elastic wave measurement reflects predominantly on those of the stiff part (Tanaka. Y. et al. 1994).

Figure 23 shows the ratio of the equivalent elastic wave velocity from static test to the measured velocity from dynamic test plotted against the ratio of the mean grain diameter to the half of wave length (Tanaka et al. 2000, AnhDan et al. 2002, Maqbool et al. 2004). As it can be seen, the ratio using both S and P waves tends to decrease as the wave length gradually approaches to the grain size. The results strongly suggest that we should use the wave length, say roughly 1,000 times or more, the D_{50} of the soil specimen in order to obtain the overall elastic stiffness on average.

4 APPLICATION OF ELASTIC SHEAR MODULUS MEASUREMENT

4.1 Interlink with strength (rigidity index)

The stiffness/strength ratio is usually termed rigidity index. The index is frequently employed for numerical analysis in geomechanics, and for interpretation of in-situ tests such as pressuremeter test.

Geomaterial that is stiffer in deformability may be stronger in strength. Figure 24 shows a summary of the relationship between E_{max} and q_{max} of various geomaterials ranging from soft soils to hard rocks (Tatsuoka and Shibuya 1992). In a logarithmic scale of consideration, the ratio can be the order of

$$E_{max}/q_{max} \approx 1,000 \pm 500 \quad (13)$$

As the E_{max}/q_{max} value is roughly constant for sedimentary soft rocks in particular (Barla et al. 1999), it is possible to evaluate the spatial variability of E_{max} from q_{max}, noting that the strength is readily obtained in conventional laboratory testing.

Accordingly, the ratio of shear modulus to undrained shear strength, S_u (n.b., $S_u = \tau_{max}$ is assumed) is roughly equal to

$$G_{max}/S_u \approx 660 \pm 330 \quad (14)$$

The rigidity index with G_{max} is far greater than the value of

$$G_{50}/S_u \approx 75 \pm 25 \quad (15)$$

that is often employed for numerical analysis in geomechanics.

4.2 Assessing disturbance of laboratory samples

The coincidence between G_f in natural ground and G_{max} of laboratory sample is normally considered to promise the quality of laboratory sample (Yasuda and Yamaguchi 1985, Tokimatsu and Oh-hara 1990, Hight 1996 among others).

Figure 25 shows a summary from well-described case records in Japan (Toki et al. 1995, Shibuya et al. 1996a). A few characteristic features have been stated as follows:

i) The G_{max}/G_f values fall within a narrow band between 0.8 and 1.2 for most of Holocene and Pleistocene clays recovered using thin-wall samplers, sands and gravels by means of in-situ freezing technique (Yoshimi et al. 1989) and soft rocks by block and core sampling, and

ii) the G_{max}/G_f value is greater than unity for loose sands retrieved by thin wall samplers, and conversely far smaller than unity for dense sandy soils sampled by a triple-tube sampler.

In the samplings of sands and gravels, the overestimate of G_f is attributed to some densification during sampling, whereas destructuration is responsible for the underestimate (Tokimatsu and Oh-hara 1990). On the other hand, hair-cracks developed during sampling are responsible for the underestimate in soft rock samples retrieved by rotary core tube sam-

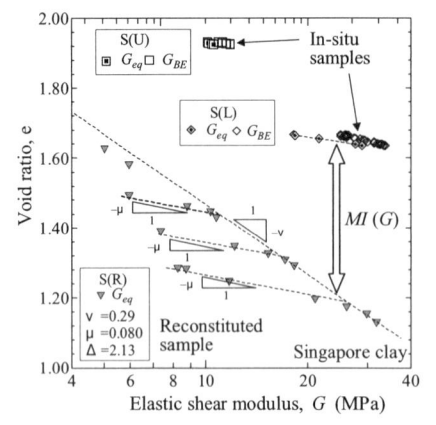

Figure 26. Relationship between void ratio and elastic shear modulus of Singapore clay when subjected to 1D compression (Li et al. 2003).

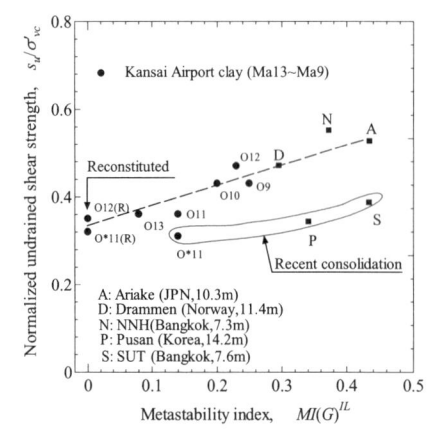

Figure 27. Relationship between s_u/σ_v and $MI(G)^{I_L}$ for Kansai Airport clay (Li 2003).

pling (Tatsuoka and Kohata 1994, Tatsuoka et al. 1996).

Interpretation of the data will be enhanced with the following descriptions;

a) density state between laboratory sample and the source ground at the cited depth,
b) details of reconsolidation method, including the consolidation path and period, and the stress system when G_{max} is measured, and
c) heterogeneity (or variability) of source ground reflecting to G_f.

As stated earlier, G_{max} of laboratory samples is strongly influenced by factors listed in a) and b). In addition, the comparison between G_{max} and G_f should preferably be made by using the averaged

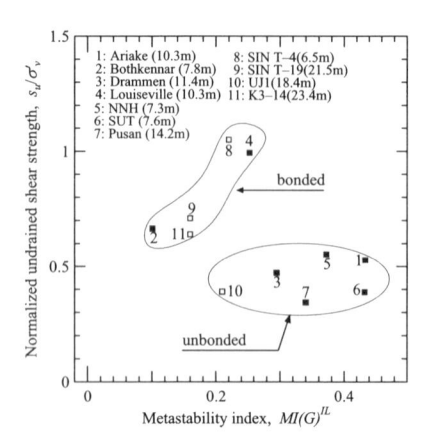

Figure 28. Relationship between s_u/σ_v and $MI(G)^{I_L}$ for Holocene clays worldwide (Temma et al. 2000, 2001).

values since the properties of natural ground vary with depth even in an allegedly uniform layer.

Unlike what it has been previously said, "the coincidence between G_f and G_{max} is a compulsory, not satisfactory, condition to guarantee the quality of laboratory samples" reflects the Authors' up-to-date understanding.

4.3 Evaluating in-situ structure of natural sedimentary clay

Soil 'structure' may be estimated by knowing the quasi-elastic stiffness. Figure 26 shows the relationship between void ratio, e, and the elastic shear modulus, G when subjected to 1-D compression. The triangular symbol refers to the behaviour of reconstituted clay, whereas the comparable results of two natural samples, each recompressed to in-situ geostatic stresses, are shown using squares (Li et al. 2003). As it can be seen in this figure, the 'structure' of the aged natural samples may be evaluated quantitatively by metastability index, $MI(G)$, which refers to the difference of void ratio between the non-structured reconstituted sample and the natural sample at a common G (Shibuya 2000).

Figure 27 shows the relationship between the ratio of undrained shear strength/consolidation pressure, s_u/σ'_{vc}, and metastability index, $MI(G)^{IL}$. In a series of consolidated-undrained (CU) compression tests, each sample of Osaka Bay clay was recompressed to in-situ effective overburden pressure (Li 2003). A trend is clear for s_u/σ'_{vc} to increase with $MI(G)^{IL}$; which may be expressed as;

$$\frac{S_u}{\sigma'_{vc}} = \left(\frac{S_u}{\sigma'_{vc}}\right)_{RC} + m \cdot MI(G_{max})^{I_L} \quad (16)$$

where 'RC' stands for 'reconstituted' (=non-structured). The results suggest the capability of

Table 1. Elastic deformation modulus employed for FE analysis in the past.

$E = \alpha S_u$	Author	
Embankment loading analysis		
$E_u = 200–500 S_u$	Bjerrum (1964)	
$E' = 70–250 S_{u\,FVS}$	Balasubramaniam et al. (1981)	
$E_u = 15–40 S_{u\,FVS}$	Bergado et al. (1990)	(Back-calculation)
Excavation works analysis		
$E_u = 200–500 S_u$	Bowels (1988)	
$E_u = 280–500 S_u$	Hock (1997)	(soft clay)
$E_u = 1200–1600 S_u$	Hock (1997)	(stiff clay)
$E' = E'_{max}/2$	Simpson (2000)	

$MI(G)^{IL}$ in expressing the effect of in-situ structure due to ageing that brought about noticeable increase in undrained strength. Figure 28 shows similar relationship of Holocene clays worldwide (Temma et al. 2000, 2001). Interparticle bonding of natural clays may be identified by using this kind of diagram.

5 CASE HISTORY

5.1 Predicting ground deformation in deep excavation of soft clay in Bangkok

A case history in Thailand on deep excavation in soft clay with concrete diaphragm wall is described with a particular attention paid to the ground deformation behind the diaphragm wall. As summarized in Table 1, there have been several empirical proposals how we should select the soil stiffness employed for *FE* analysis. For embankment loading, the use of several hundreds or tens times of undrained shear strength s_u has been proposed for the equivalent Young's modulus. For excavation works, on the other hand, much higher factors have been proposed by Bowels (1988) and Hock (1997) for the conversion from s_u, or the use of half of the small strain stiffness E_{max} obtained by in-situ seismic survey has been proposed by Simpson (2000).

Figure 29 shows the variations of secant Young's modulus normalized using s_u and E_{max} with axial strain, that were obtained in undrained triaxial compression test on normally consolidated clays collected from worldwide (Temma et al. 2000, 2001). In the present case history, the ground strains behind the diaphragm wall was on the order of 0.1 %, which is consistent with the proposals by Simpson and Hock. Similar result has been also reported by Ou et al. (2000) for deep excavation with diaphragm wall in Taipei, Taiwan. It should be pointed out that the axial strain associated with $E=E_{max}/2$ corresponds to a narrow range from 0.03 % to 0.3 %. The use of $E=E_{max}/2$, therefore, matches well the ground strains of soft clay behind diaphragm wall induced by deep excavation. Note also that the E_{sec}/s_u value

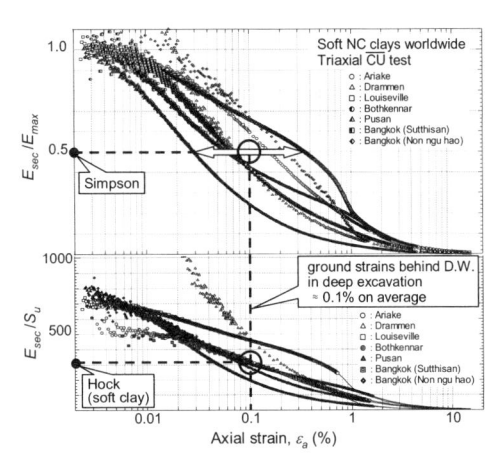

Figure 29. Normalized secant Young's modulus of normally consolidated clays (Temma et al. 2000, 2001).

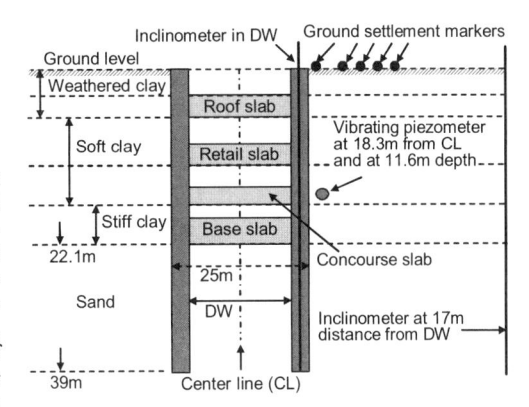

Figure 30. Cross-section of deep excavation with diaphragm wall in Bangkok (Tamrakar et al. 2001).

corresponding to the axial strain of 0.1% ranges from 200 to 400.

Figure 30 shows cross-section of deep excavation with diaphragm wall for the construction of subway tunnels in Bangkok by open-cut method. The concrete diaphragm walls were pre-installed to a depth of 39 m, and deep excavation was afterwards carried out in soft and stiff clays down to a depth of 22 m (Tamrakar et al. 2001). Instrumentations were a piezometer, two series of inclinometers set along the wall axis and the borehole located at a horizontal distance of 17 m from the diaphragm wall, and a series of markers for monitoring ground settlement.

Figure 31 shows the stratigraphy with representative profiles of *OCR*, compression and swell indices, λ and κ, and geostatic stresses and pore pressures.

Figure 32 shows the results of a series of undrained triaxial compression tests on undisturbed clay specimens retrieved from several depths.

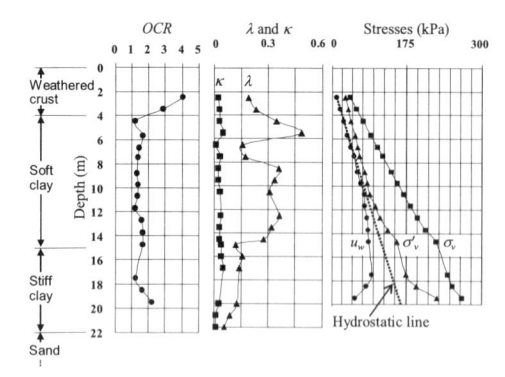

Figure 31. Soil profiles for deep excavation work in Bangkok (Tamrakar et al. 2001).

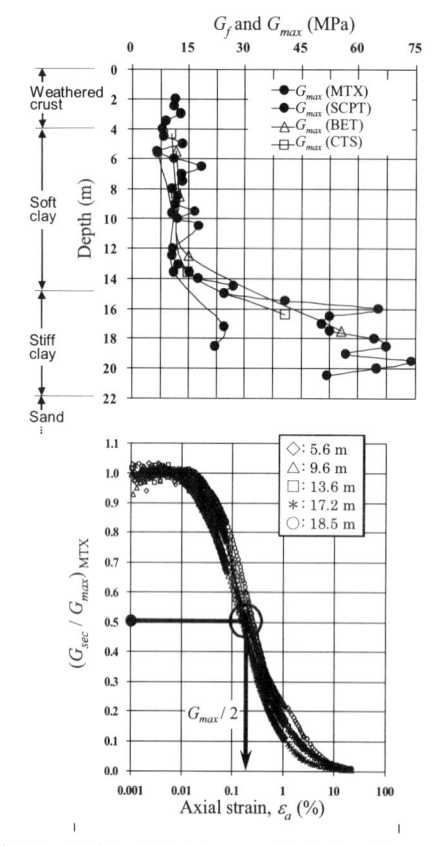

Figure 32 Undrained triaxial compression test results on undisturbed clay specimens (Shibuya et al. 2001).

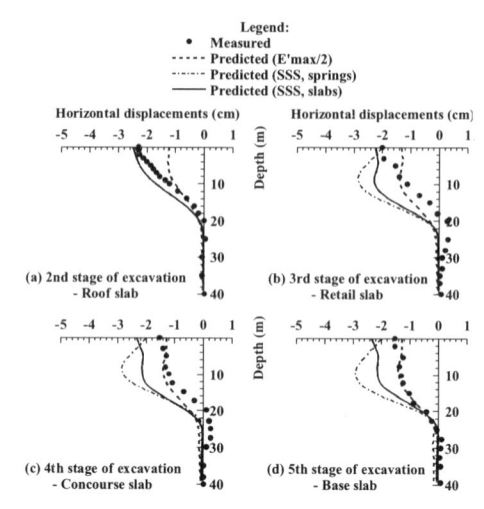

Figure 33. Horizontal deformation of diaphragm wall at four excavation stages (Kovacevic et al. 2003).

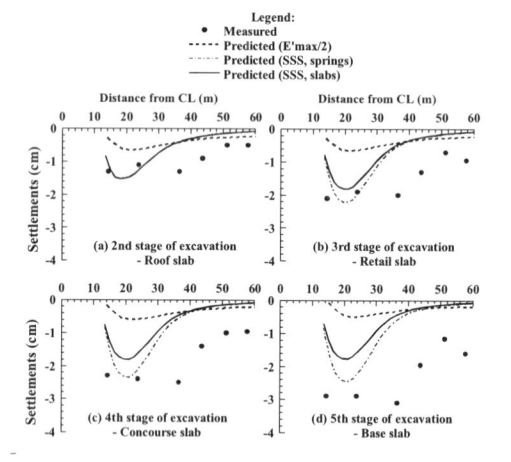

Figure 34. Ground surface settlement at four excavation stages (Kovacevic et al. 2003).

Although the small strain stiffness values vary with depth, the degradation curves of the normalized stiffness were almost similar to each other.

By considering the proposal by Simpson (2000), half of the small strain stiffness corresponding to the strain level of about 0.1 % was employed in the numerical simulation of the full scale behaviour using an equivalent-linear elastic approach. In addition, Kovacevic et al. (2003) has performed a non-linear elastic analysis using a model called the small strain stiffness or SSS model. The results of boundary values predicted using these two kinds of analysis are compared in Figs. 33 and 34. The horizontal deformation of the diaphragm wall was compared at several excavation stages (see Fig.33). From a practical point of view, the result from equivalent-linear analysis presented using dash lines was not bad, while the deformation profile could be better captured by the non-linear analysis shown in solid lines assuming full moment connection between the

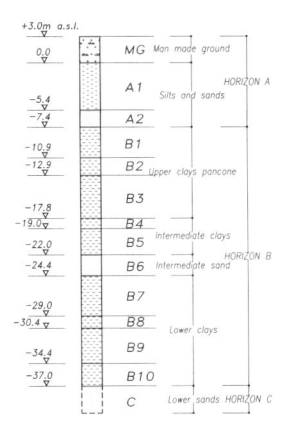

Figure 35. Thickness of soil strata (Lo Presti et al. 2003).

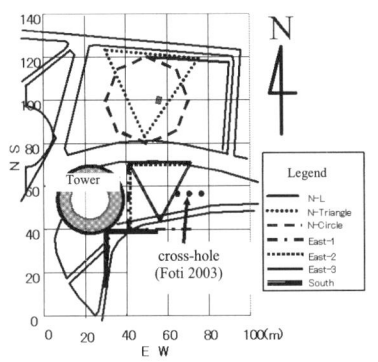

Figure 36. Location of surface wave measurement

roof slab and the diaphragm wall. Similarly, the nonlinear analysis using a small-strain stiffness (*SSS*) model could better capture the ground settlement profile than the linear analysis (see Fig.34), suggesting an importance of using proper soil stiffness considering its strain level and stress state dependencies.

5.2 Site investigation at leaning tower of Pisa

Distinguishing characteristics of geotechnical site investigation for conservation of antiquities may be outlined in the following;

i) sustainable monitoring is of great importance since extremely long time-span is usually considered,
ii) investigation should be non-destructive not to disturb the cited antiquity and surrounding environments, and
iii) spatial (3-D) geo-information over a wide area is needed, since an area of engineering interest is normally unknown.

Surface wave measurement (or SASW) would satisfy all of these three requirements.

Figure 37. A scene of the measurement.

The site investigation at leaning tower of Pisa described in this paper was carried out in September 2003 through bi-country research co-operation between Italy and Japan. Research organizations participated in the research project were Technical University of Turin (Italy), and Hokkaido University/ Kitami Institute of Technology/ Port and Airport Research Institute (Japan). Preceding site investigation at Pisa tower has been described in detail by Lo Presti et al. (2003). The conservation work of underexcavation completed in 2000 has been reported in detail by Jamiolkowski (2001).

The objective of the 2003 investigation was to examine properties change of subsoils, if any, underneath the foundation due to underexcavation work in 1998-2000. In the investigation, the practice of 'trinitarianism' in seismic survey as described earlier was attempted. Since the project is still on-going, the result of surface wave survey (or SASW) to examine spatial (3-D) variation of wave velocity only is described in this paper.

Figure 35 shows typical of soil stratigraphy with alternate layers of sand and clay (Lo Presti et al. 2003). Locations of the surface wave measurement are shown in Fig.36. Figure 37 shows a scene of the measurement. The variation of shear wave velocity with depth is shown in Fig.38. It should be mentioned that the V_s profile from the current investigation matched well with that of SASW measurement previously performed by Foti (2003) and also with the result of down-hole PS-logging data (Lo Presti et al. 2003). It is also interesting to note that the V_S in the North side at 8-16 m depth where the underexcavation was carried out exhibits relatively low values as compared to the rest (see Fig.38). The observation suggests a notion that soft clay in the zone may have been disturbed in the event of underexcavation. In near future, the Authors will report more com-

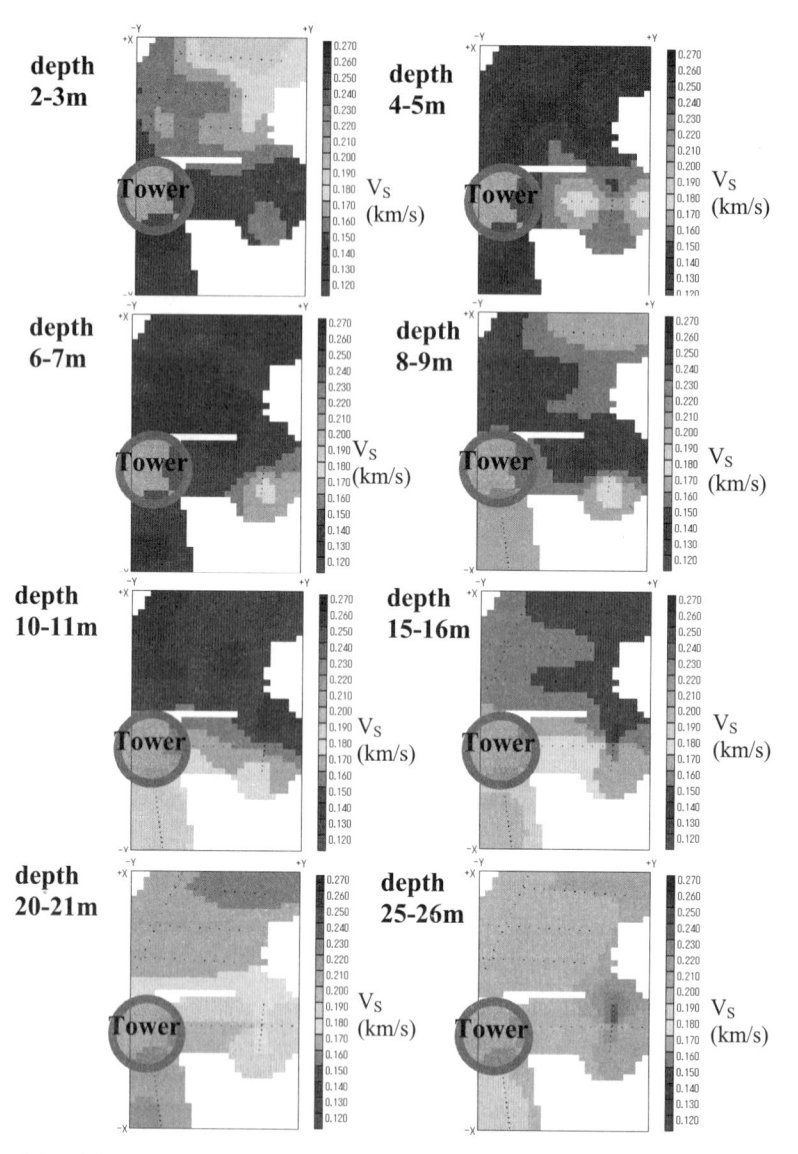

Figure 38. The variation of shear wave velocity with depth.

prehensive and conclusive interpretation with the results of laboratory tests.

6 CONCLUDING REMARKS

In geotechnical site investigation, in-situ and laboratory tests should be complementary to each other. A single property; i.e., quasi-elastic stiffness of natural ground seems a key to find out a missing link between in-situ and laboratory.

Quasi-elastic stiffness of geomaterials involved with strains from 10^{-7} to 10^{-5} is unaffected by rate of shearing as well as the type of loading (i.e., so-to-speak dynamic and static loadings). However, the stiffness sometimes shows strong anisotropy. The G_{hh}/G_{vh} value was 1.1 and 2.0 for Toyoura sand and an isotropically consolidated clay, respectively.

In-situ seismic test is non-destructive to source ground so that it certainly provides geo-information regarding the current state of natural ground as well demonstrated in geotechnical site investigation performed in soft clays worldwide.

In this paper, importance of 'trinitarianism' in properly understanding the outcome of seismic survey is demonstrated. Spatial (3-D) survey of wave velocities from surface wave measurement, 1-D survey from down/cross-hole measurement and more comprehensive 'point' examination in laboratory element tests should be synthesized. A case history as such recently performed at leaning tower of Pisa suggests that a portion of subsoil subjected to under-excavation may have been disturbed by showing relatively low profile of V_s.

A case history of deep excavation in Bangkok demonstrates the significance of G_f from in-situ seismic survey when coupled with G-γ relationship from laboratory test. Stiffness relevant to averaged ground strains may be employed for predicting ground deformation by FE analysis using equivalent linear or non-linear stress-strain model.

Laboratory tests should be capable of measuring deformation of geomaterials over a wide strain range, including comparable elastic properties at very small strains. Stiffness anisotropy should also be clarified.

Comparison of quasi-elastic stiffness between in-situ and laboratory is useful for understanding disturbance of laboratory sample. But, the coincidence between G_f in-situ and G_{max} in the laboratory is a compulsory, not satisfactory, condition to guarantee the quality of laboratory samples.

The G_f of soft soils from seismic test is larger by ten hold compared to G_{50} from UC test. The G_f value may be correlated to the tip resistance, q_t of CPT and to dilatometer modulus, E_D as shown in Eqs. 10 and 11, respectively. In a logarithmic scale of consideration, the E_{max}/q_{max} value is about one thousand, hence the G_{max}/S_u of about 660 (see Eqs. 14 and 15, respectively).

In-situ 'structure' of natural sedimentary clay may be evaluated quantitatively by metastability index, $MI(G)$ defined using G_f.

ACKNOWLEDGEMENTS

The Authors appreciate their colleagues and research partners Professor T. Mitachi (Hokkaido University, HU), Dr D.J. Li (HU), Dr T. Kawaguchi (Hakodate National College of Technology), Dr M. Tanaka (PARI), Mr. K. Hayashi (OYO Co.) and Y. Shoji (OYO Co.) and Dr S. Foti (Technical University of Turin) for their continuous supports without which most of the work described in this paper was not possible. The Authors are also indebt to Professor M. Jamiolkowski, Technical University of Turin, who made the site investigation at leaning tower of Pisa possible. The collaborative research performed at leaning tower of Pisa was supported by Grant-in-aid No.15404012, Ministry of Education & Science, Japan.

REFERENCES

AnhDan, L.Q., Koseki, J. and Sato, T. 2002. Comparison of Young's moduli of dense sand and gravel measured by dynamic and static methods, *Geotechnical Testing Journal* 25(4), ASTM: 349-368.

Baldi, G., Bruzzi, D., Superbo, S., Battaglio, M. and Jamiolkowski, M. 1988. Seismic cone in Po River sand, *Proc. of the First International Symposium on Penetration Testing*, Vol.2, 643-650.

Barla, M. Barla, G., Lo Presti, D.C.F. and Pallara, O. 1999. Stiffness of soft rocks from laboratory tests, *Proc. of Second International Symposium on Pre-failure Deformation of Geomaterials (IS-Torino'99)*, Balkema, Vol.1, 43-50.

Bowels, J.E. 1988. Foundation analysis and Design, *4th Edition, McGraw-Hill*.

Butcher,A.P. and Powell,J.M. 1995. Practical considerations for field geophysical techniques used to assess ground stiffness, *Advances in Site Investigation Practice (C. Craig edn.)*, Thomas Telford, 701-714.

Dyvik, R. and Madshus, C. 1985. Laboratory measurements of G_{max} using bender elements, *Proc. of ASCE Convention*, Detroit, 186-196.

Dobry,R., Ladd,R.S., Yokel,F.Y., Chung,R.M. and Powell,D. 1982. Prediction of pore pressure buildup and liquefaction of sand during earthquakes by cyclic strain method, *National Bureau of Standards Building Science 138*, Washington DC.

Fioravante V. 2000. Anisotropy of small strain stiffness of Ticino and Kenya sands from seismic wave propagation measured in triaxial testing, *Soils and Foundations* 40(4): 129-142.

Foti, S. 2003. Small-strain stiffness and damping ratio of Pisa clay from surface wave tests, *Géotechnique*, 53(5): 455-461.

Hight, D.W. 1996. Moderator's report on session 3:drilling, boring, sampling and description, *Advances in Site Investigation Practice (C. Craig edn.)*, Thomas Telford, 337-360.

Hock, G.C. 1997. Review and analysis of ground movements of braced excavation in Bangkok subsoil using diaphragm walls, *M.Eng.Thesis*, Bangkok Thailand.

Hwang, S.C., Mitachi, T., Shibuya, S. and Tateichi, K. 1998. Stress-strain characteristics in the wide strain rane from small strain to failure state and undrained shear strength of natural clays, *Journal of JSCE*, No.589/III-42, 305-319 (in Japanese).

Hayano, K. 2001. Pre-failure deformation characteristics of sedimentary soft rocks, *PhD thesis*, The University of Tokyo (in Japanese).

Jamiolkowski, M., Lancellotta, R. and Lo Presti, D.C.F. 1995. Remarks on the stiffness at small strains of six Italian clays, *Pre-failure Deformation of Geomaterials (Shibuya, S. et al edns)*, Balkema, Vol.2, 817-836.

Jamiolkowski, M. 2001. The leaning tower of Pisa: end of an Odyssey, Terzhaghi Oration, *Proc. 15th ICSMGE*, Vol.5, 2979-2996.

Jardine,R.J. 1985. Investigations of pile-soil behaviour with special reference to the foundations of offshore structures, *PhD thesis, Univ. of London*.

Jardine, R.J. 1992. Some observations on the kinematic nature of soil stiffness, *Soils and Foundations*, 32(2): 111-124.

Johnson, K.L. 1985. Contact mechanics, *Cambridge University Press*.

Jovicic, V., Coop, M.R. and Simic, M. 1996. Objective criteria for determining G_{max} from bender elements, *Géotechnique*, 46(2): 357-362.

Kawaguchi, T., Mitachi, T. and Shibuya, S. 2001. Evaluation of shear wave travel time in laboratory bender element test

Proc. of the 15th ICSMGE, Istanbul, Balkema, Vol.1, 155-158.

Kovacevic, N., Hight, D.W. and Potts, D.M. 2003. A comparison between observed and predicted behaviour of a deep excavation in soft Bangkok clay, *Deformation Characteristics of Geomaterials*, IS-Lyon, Balkema, Vol.1, 983-989.

Kuwano, R. and Jardine, R.J. 1997. Stiffness measurements in a stress-path cell, Pre-failure Deformation Behaviour of Geomaterials, *Geotechnique Symposium In Print*, 391-394.

Li, D.J., Shibuya, S., Mitachi, T. and Kawaguchi, T. 2003. Judging fabric bonding of natural sedimentary clay, *Deformation Characteristics of Geomaterials*, IS-Lyon, Balkema, Vol.1, 203-209.

Li, D.J. 2003. Study on Structure and Cementation Effects of Natural Sedimentary Clay, *Doctor Eng. Thesis*, Hokkaido University.

Li, D.J., Shibuya, S. and Mitachi, T. 2004. Engineering properties of Osaka bay clay, *Proc. of the International Symposium on Engineering Practice of Soft Deposits*, IS-Osaka. (in print)

Lo Presti,D.C.F., Pallara,O., Lancellotta,R., Armandi,M. and Maniscalco,R. 1993. Monotonic and cyclic loading behaviour of two sands at small strains, *Geotechnical Testing Journal*, 14(4), 409-424.

Lo Presti, D.C.F., Pallara, O. and Jamiolkowski, M. 1999. Anisotropy of small strain stiffness of undisturbed and reconstituted clays, *Proc. of Second International Symposium on Pre-failure Deformation of Geomaterials (IS-Torino'99)*, Balkema, Vol.1, 3-10.

Lo Presti, D., Jamiolkowski, M. and Pepe, M. 2003. Geotechnical characterisation of the subsoil in Pisa, *Characterisation and Engineering Properties of Natural Soils*, Balkema, Vol.1, 909-949.

Lunne, T., Robertson, P.K. and Powell, J.J.M. 1992. Cone penetration testing in geotechnical practice, *Blackie Academics and Professional*, London.

Nishida, K., Takamura, T., Nakajima, M., Tanaka, H. and Tanaka, M. 1999. Case studies of seismic cone test abd its amplitude characteristics on in-situ data, *Journal of JSCE*, No.631/III-48, 329-338 (in Japanese).

Nishio,S. and Katsura,Y. 1994. Shear wave anisotropy in Edogawa Pleistocene deposit, *Pre-failure Deformation of Geomaterials*, Vol.1, 169-174.

Maqbool, S., Koseki, J. and Sato, T. 2004. Effects of compaction on small strain Young's moduli by dynamic and static measurements, *Bulletin of ERS* 37, Institute of Industrial Science, University of Tokyo. (in print)

Marchetti, S. 1980): In-situ tests by flat dilatometer, *Journal of the GE Division*, ASCE, 106(GT3): 299-321.

Ogino, T., Mitachi, T., Shibuya, S. and Ikegame, Y. 1999. Evaluation of deformation and strength characteristics of clay by using direct shear apparatus with bender elements. *Technical Report No.39, Hokkaido Branch of JGS*, 1-10.

Ou, C.Y., Liao, J.T. and Cheng, W.L. 2000. Building response and ground movements induced by a deep excavation, *Géotechnique* 30(3): 209-220.

Porovic, E. and Jardine, R.J. 1994. Some observations on the static and dynamic shear stiffness of Ham River sand, *Pre-failure Deformation of Geomaterials (Shibuya, S. et al edns)*, Balkema, Vol. 1, 25-30.

Robertson, P.K., Campanella, R.G. and Rice, A. 1986. Seismic CPT to measure in situ shear wave velocity, *Proc. of GE Div., ASCE*, 112(8): 791-803.

Salinero, I. S., Roesset, J. M. and Stokoe, K. H. 1986. Analytical studies of body wave propagation and attenuation, *Report GR86-15*, University of Texas at Austin.

Shibuya, S., Tatsuoka, F., Teachavorasinskun, S., Kong, X.J., Abe, F., Kim, Y.S. and Park, C.S. 1992. Elastic deformation properties of geomaterials, *Soils and Foundations*, 32(3): 26-46.

Shibuya, S., Mitachi, T., Fukuda, F. and Degoshi, T. 1995a. Strain rate effects on shear modulus and damping of normally consolidated clays, *Geotechnical Testing Journal*, 18(3): 365-375.

Shibuya, S., Mitachi, T., Yamashita, S., and Tanaka, H. 1995b. Effects of sample disturbance on G_{max} of soils - a case study, *Proc. First Int. Conf. on Earthquake Geotechnical Engineering (K. Ishihara edn)*, Vol.1, 77-82.

Shibuya, S. and Tanaka, H. 1996a. Estimate of elastic shear modulus in Holocene soil deposits, *Soils and Foundations*, 36(4): 45-55.

Shibuya, S., Mitachi, T., Yamashita, S. and Tanaka, H. 1996b. Recent Japanese practice for investigating elastic stiffness of ground, *Advances in Site Investigation Practice (C. Craig edn.)*, Thomas Telford, 875-886.

Shibuya, S., Hwang, S.C. and Mitachi, T. 1997. Elastic shear modulus of soft clays from shear wave velocity measurement, *Géotechnique*, 47(3): 593-601.

Shibuya, S. and Mitachi, T. 1997a. Panel discussion: Effects of interparticle bonding on stiffness of geomaterials, *Proc. of 14th ICSMGE*, Vol.4, 2181-2182.

Shibuya, S., Mitachi, T., Fukuda, F. and Hosomi, A. 1997b. Modelling of strain-rate dependent deformation of clay at small strains, *Proc. of 12th. ICSMGE*, Hamburg, Vol.1, 409-412.

Shibuya, S. and Tamrakar, S.B. 1997. In-situ and laboratory investigations into engineering properties of Bangkok clay, *Proc. of International Symposium on Characterization of Soft Marine Clays*, Yokosuka, Balkema, 107-132.

Shibuya, S. 2000. Assessing structure of aged natural sedimentary clays, *Soils and Foundations* 40(3): 1-16.

Shibuya, S. and Tamrakar, S.B. (2003). Engineering properties of Bangkok clay, *Characterisation and Engineering Properties of Natural Soils (Tan T.S. et al. edns)*, Singapore, Balkema, Vol.1, 645-692.

Simpson, B. 2000. Engineering needs. *Prefailure Deformatioms Characteristics of Geomaterials*, Theme and Keynote Lectures, IS-Torino, Balkema, Vol.2, 1011-1026.

Stokoe, K.H.II, Nazarin, S., Rix, G.J., Sanchez-Salinero, I., Sheu, J.C. and Mok, Y.J. 1988. In situ seismic testing of hard-to-sample soils by surface wave method. *Earthquake Engineering and Soil Dynamics II; Recent Advances in Ground Mortion Evaluation*, ASCE, 264-278.

Tamrakar, S.B., Shibuya, S. and Mitachi, T. 2001. A practical FE analysis for predicting deformation of soft clay subjected to deep excavation, *Proc. of 3rd International Conference on Soft Soil Engineering*, Hong Kong, Vol.1, 377-382.

Tanaka, H., Tanaka, M., Iguchi, H. and Nishida, K. 1994. Shear modulus of soft clay measured by various kinds of test. *Pre-failure Deformation of Geomaterials (Shibuya, S. et al edns)*, Balkema, Vol. 1, 235-240.

Tanaka, H., Sharma, P., Tsuchida, T. and Tanaka, M. 1996. Comparative study on sample quality using several types of samplers, *Soils and Foundations*, 36(2): 57-68.

Tanaka, H. and Tanaka, M. 1996. A site investigation method using cone penetration and dilatometer tests, *Technical Note of the Port and Harbour Research Institute, Ministry of Transport, Japan*, No.837 (in Japanese).

Tanaka, H. and Tanaka, M. 1997. Applicability of the UC test to two European clays, *Proc. of 14th ICCSMGE*, Vol. 1, 209-212.

Tanaka, H., Hamouche, K., Tanaka, M., Watabe, Y., Leroueil, S. and Founier, I. 1998. Use of Japanese sampler in Champlain sea clay, *Geotechnical Site Characterization (Robertson, P.K. and Mayne, P.W. edns)*, Balkema, Vol.1, 439-444.

Tanaka, Y., Kudo, K., Nishi, K. and Okamoto, T. 1994. Shear modulus and damping ratio of gravelly soils measured by several methods, *Pre-failure Deformation of Geomaterials*, IS-Hokkaido, Balkema, Vol.1, 47-53.

Tanaka,Y., Kokusho,T., Okamoto,T. and Kudo,K. 1995. Evaluation of initial shear modulus of gravelly soil by laboratory test and PS-logging, *Proc. First Int. Conf. on Earthquake Geotechnical Engineering (K. Ishihara edn)*, Vol.1, 101-106.

Tanaka, Y., Kudo, K., Nishi, K., Okamoto, T., Kataoka, T. and Ueshima, T. 2000. Small strain characteristics of soils in Hualien, Taiwan. *Soils and Foundations* 40(3): 111-125.

Tanizawa, F., Teachavorasinskun, S. Yamaguchi, J., Sueoka, T. and Goto, S. 1994. Measurement of shear wave velocity of sand before liquefaction and during cyclic mobility, *Pre-failure Deformation of Geomaterials*, IS-Hokkaido, Balkema, Vol.1, 63-68.

Tatsuoka, F. and Shibuya, S. 1992. Deformation characteristics of soils and rocks from field and laboratory tests, Keynote Paper, *Proc. of 9th ARC on SMFE*, Vol.2, 101-170.

Tatsuoka,F. and Kohata,Y. 1995, Stiffness of hard soils and soft rocks in engineering applications, Keynote Lecture, *Pre-failure Deformation of Geomaterials (Shibuya, S. et al., eds.)*, Balkema, Vol.2, 947-1063.

Tatsuoka, F., Kohata, H., Tsubouchi, K., Murata, K., Ochi, K. and Wang, L. 1996. Sample disturbance in rotary core tube sampling of soft rock, *Advances in Site Investigation Practice (C. Craig edn.)*, Thomas Telford, 281-292.

Tatsuoka, F., Jardine, R.J., Lo Presti, D.C.F., Di Benedetto, H. and Kodaka, T. 1997. Characterizing the pre-failure properties of geomaterials, Theme Lecture for Plenary Session 1, *Proc. of 13th ICSMGE*, Hamburg, Vol.4, 2129-2164.

Temma, M., Shibuya, S. and Mitachi, T. 2000. Evaluating ageing effects of undrained shear strength of soft clays, *Proc. of International Symposium on Coastal Geotechnical Engineering in Practice*, Yokohama, Vol.1, 173-179.

Temma, M., Shibuya, S., Mitachi, T. and Yamamoto, N. 2001. Interlink between metastability index, MI(G) and undrained shear strength in aged holocene clay deposit, *Soils and Foundations*, 41(2): 133-142 (in Japanese).

Toki, S., Shibuya, S. and Yamashita, S. 1995. Standardization of laboratory test methods to determine the cyclic deformation properties of geomaterials in Japan, *Pre-failure Deformation of Geomaterials (Shibuya, S. et al edns)*, Balkema, Vol.2, 741-784.

Tokimatsu, K. 1997. Geotechnical site characterization using surface waves, *Earthquake Geotechnical Engineering (K. Ishihara edn), Balkema*, Vol.2, 1333-1368.

Tokimatsu, K. and Oh-hara, J. 1990. 8.2. In-situ freezing sampling, *Tsuchi-to-Kiso, JGS*, 38(11): 61-68 (in Japanese).

Viggiani, G. and Atkinson, J.H. 1995. Stiffness of fine grained soil at very small strains, *Géotechnique*, 45(2): 249-265.

Yamaguchi, T., Yamashita, S, Hori, T. and Suzuki, T. 2004. Shear wave velocity of three directions by bender element test on clay, *Technical Report No.44, Hokkaido Branch of JGS*, 281-286 (in Japanese).

Yamashita, S., Shibuya, S. and Tanaka, H. 1996. A case study for characterizing undrained cyclic deformation properties in young sand deposit from in-situ and laboratory tests, *Soils and Foundations*, 37(2), 117-126.

Yamashita, S., Hori, T. and Suzuki, T. 2003. Effects of fabric anisotropy and stress condition on small strain stiffness of sands, *Deformation Characteristics of Geomaterials*, IS-Lyon, Balkema, Vol.1, 187-194.

Yasuda, S. and Yamaguchi, I. 1985. Dynamic shear modulus obtained in the laboratory and in situ, *Proc. Sympo. on Evaluation of Deformation and Strength of Sandy Grounds*, JSSMFE, 115-118 (in Japanese).

Yoshimi,Y., Tokimatsu,K. and Hosaka,Y. 1989. Evaluation of liquefaction resistance of clean sands based on high quality undisturbed samples, *Soils and Foundations*, 29(1), 94-104.

Watabe, Y., Tanaka, M. and Takemura, J. 2004. Evaluation of in-situ K_0 for Ariake, Bangkok and Hai-Phong clays, *Proceedings of the 2nd International Conference on Geotechnical Site Characterization* (in print).

Characterization of the Venice lagoon silts from in-situ tests and the performance of a test embankment

P. Simonini
Department IMAGE, University of Padova, Italy

Keywords: silts, mechanical behavior, Venice lagoon, piezocone test, dilatometer test, heterogeneous soil

ABSTRACT: To protect the historic city of Venice against recurrent flooding, a huge project has been undertaken, involving the design and construction of movable gates located at the three lagoon inlets for controlling tidal flow. The selection of the movable gate foundations demands very extensive and low cost geotechnical characterization of the Venice lagoon soils. These are composed of a predominantly silty fraction combined with clay and/or sand forming an erratic interbedding of various types of sediments, whose basic mineralogical characteristics vary narrowly. To this end two typical test sites in the lagoon area - the Malamocco Test Site and the Treporti Test Site – were selected for the mechanical characterization of Venice soils and a site-specific calibration of the most widely used geotechnical investigation tools. The main outcome of the research carried out at the Malamocco Test Site and some preliminary results from the Treporti Test Site are here presented and discussed.

1 FOREWORD

To protect the city of Venice and the surrounding lagoon against recurrent flooding, a huge project has been undertaken, involving the design and construction of movable gates located at the three lagoon inlets (Figure 1). These gates, controlling the tidal flow, temporarily separate the lagoon from the sea at the occurrence of particularly high tides, which have increased notably and in frequency (Harleman et al., 2000).

Therefore, comprehensive geotechnical studies were carried out to characterize the Venetian soil and achieve a suitable design of the movable gates foundations.

The main characteristic of the lagoon soils is the presence of a predominant silt fraction, combined with clay and/or sand (Cola and Simonini, 2000). These form a chaotic interbedding of different sediments, whose basic mineralogical characteristics vary narrowly, as a result of unique geological origins and a common depositional environment.

This latter feature, together with the relevant heterogeneity of soil layering, seemed to suggest concentrating the main research efforts on selected test sites, considered as representative of typical soil profiles, where relevant in-situ and laboratory investigations could be carried out in the careful characterization of the Venetian lagoon soils.

The first test site, namely the *Malamocco Test Site* (MTS) was located at the Malamocco inlet. Within a limited area, a series of investigations that included boreholes, piezocone (CPTU), dilatometer (DMT), self-boring pressuremeter (SBPM) and cross hole tests (CHT) were performed on contiguous verticals.

On the MTS samples a comprehensive laboratory test program was completed (Cola and Simonini,

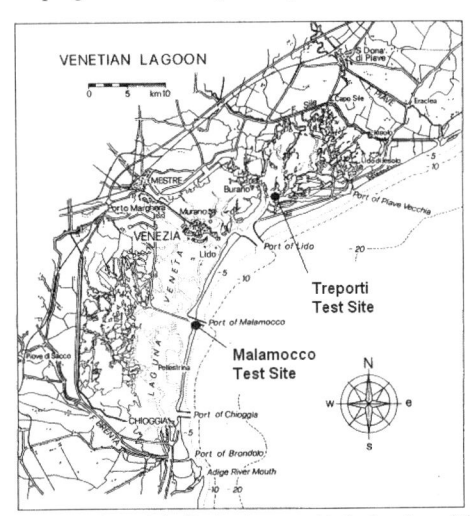

Figure 1. View of the Venice lagoon with the location of the test sites.

1999, 2002; Biscontin et al., 2001). It should be emphasized that the very heterogeneous nature of the Venice soils required a relatively large number of tests, in order to define even the simplest properties with a certain degree of accuracy.

One of the main aims of the MTS was to evaluate the reliability of the most widely used charts or correlative equations for the CPTU/DMT interpretation on the basis of the comparison with the results of the laboratory tests, with the aim of characterizing soil profile and the main geotechnical properties in an economical way (Ricceri et al., 2002).

Due to the influence of high soil heterogeneity, relevant difficulties were however encountered in comparing the relevant soil properties with those directly estimated from site tests. The absence of any clear horizontal layering provided no absolute certainty in carrying out a comparison of material properties exactly within the same type of soil.

A new test site was therefore selected, i.e. the *Treporti Test Site* (TTS), which is situated on the mainland, close to the Lido inlet. The goal of TTS is to measure directly in-situ the stress-strain-time properties of the heterogeneous Venetian soils.

At the TTS a vertically-walled circular embankment, loading up the ground to around 100 kPa, was very recently constructed, measuring, along with and after the construction, the relevant ground displacements together with the pore pressure evolution. To this end, the ground beneath the embankment was therefore heavily instrumented using plate extensometer, differential micrometers, GPS, inclinometers, piezometers and load cells.

Boreholes with undisturbed sampling, traditional CPTU (Gottardi and Tonni, 2004) and DMT (Marchetti et al., 2004), seismic SCPTU and SDMT (Mayne and McGillivray, 2002) were employed to characterize soil profile and estimate relevant soil properties for comparison with those directly measured in situ.

The main outcome of the research carried out at the MTS and some preliminary results from the TTS, where the site monitoring is still running, are here presented and discussed.

2 GEOLOGY OF THE VENICE LAGOON

During the Quaternary period the lagoon area underwent alternating periods of marine transgression and regression, as a result of which both marine and continental sediments coexist. More particularly, the deposits forming the upper 50-60 m are characterized by a complex system of interbedded sands, silts and silty clays with inclusions of peat (Bonatti, 1968; Rowe, 1973), deposited during the last glacial period of the Pleistocene, when the rivers transported fluvial material down from the Alpine ice fields. The Holocene is only responsible for the shallowest lagoon deposits, about 10-15 m thick.

The top layer of Würmian deposits is composed of a crust of highly over consolidated clay, commonly referred to as *caranto*, where many historical Venetian buildings are founded. It was subject to a process of overconsolidation as a result of oxidation during the 10000 year emergence of the last Pleistocenic glaciation (Gatto and Previatello, 1974).

The depositional patterns of the shallower Venetian sediments are rather complex due to the combined effects of geological history and human action, which have significantly modified the morphology of the lagoon, inlets and channels over the centuries.

Sands show two types of mineralogical composition, namely carbonatic and siliceous, the former in the forms of detrital calcite and dolomite crystals, especially at higher depths (Favero et al., 1973).

When carbonate and quartz-feldspar fractions decrease, the clay minerals increase alongside variation of the grain-size distribution from sands to clays. Clayey minerals, not exceeding beyond 20%, are mainly composed of illite or muscovite with chlorite, kaolinite and smectites as secondary minerals (Curzi, 1995).

3 THE MALAMOCCO TEST SITE

3.1 *Soil profile, basic properties*

Figure 2 shows the basic soil properties as a function of depth. For classification purposes the soil classes have been reduced to three: medium to fine sand (SP-SM), silt (ML) and very silty clay (CL).

From Figure 2, features to note are:
- The predominance of silty and sandy fraction can be clearly appreciated: the three soil classes occur approximately in proportions of 35% SM-SP, 20% ML, 40% CL and 5% medium plasticity clays and organic soils (CH, OH and Pt). Percentages of silt exceeding 50% are present in 65% of samples analyzed;
- Sands are relatively uniform but moving towards finer materials the soils become more graded: the coarser the materials, the lower the coefficient of non-uniformity U, whose range increases with the decreasing mean particle diameter D_{50}.
- Atterberg limits are characterized by average values of liquid limit $LL = 36\pm9\%$ and of plasticity index $PI = 14\pm7\%$. Activity $A=PI/CF$ (CF=clay fraction) is low, the great majority of samples falling in the range $0.25<A<0.50$.
- Significant variations have been observed in saturated unit weight γ_{sat}, even in cores a few

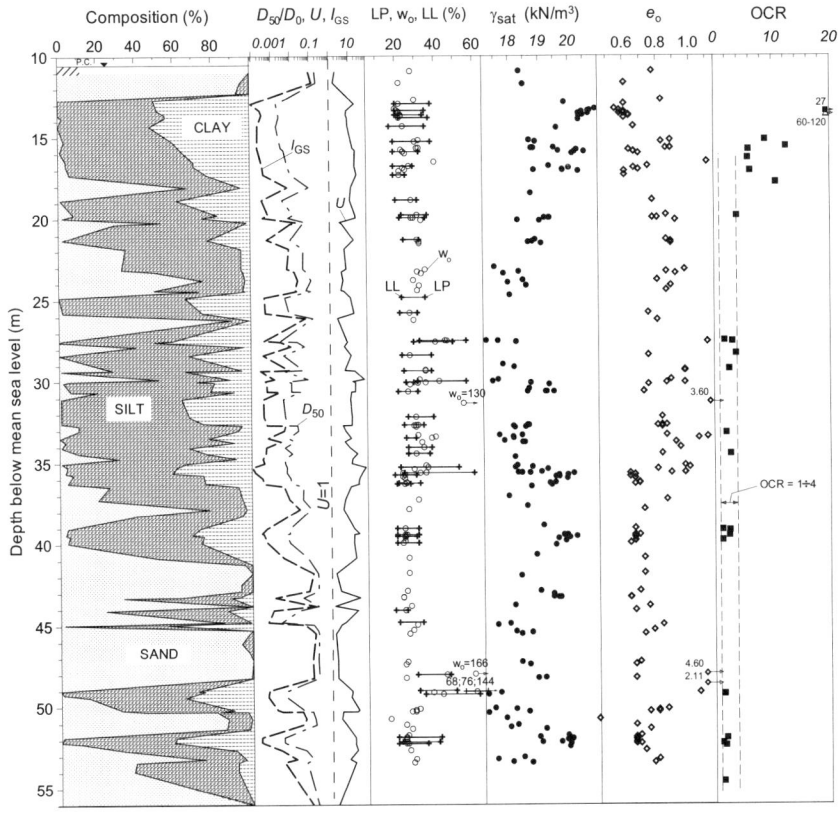

Figure 2. Soil profile, basic properties and stress history at the MTS.

centimeters long (specific gravity $G_s = 2.77 \pm 0.03$);
- Void ratio e_0 lies in the range between 0.7 and 1.0 from 19 m to 36 m below m.s.l.; at greater depths, it is lower and falls in the range 0.6-0.75 (higher values are due to laminations of organic soils).

From column 2 of Figure 2 it can be clearly noted that the profiles of D_{50} and of U are generally characterized by the opposite trends with depth, namely, when D_{50} decreases the U increases and vice-versa.

In order to take into account the coupled and opposite variation of D_{50} and U into a single parameter, a new grain-size index I_{GS} was proposed by Cola and Simonini (2002). I_{GS} is defined as:

$$I_{GS} = \frac{D_{50}/D_0}{U} \qquad (1)$$

where D_0 is a reference diameter (1 mm).

In the laboratory characterization of the Venice lagoon soils, it is assumed that significant time-independent material parameters can be related to the grain size index I_{GS}, accounting for their grading characteristics. The significance of I_{GS} is limited here to the range $8 \cdot 10^{-5} \leq I_{GS} \leq 0.12$ and $CF < 25\%$; when electrochemical effects assume a relevant role, the grain size distribution must be associated with another index such as clay fraction, Atterberg limits or mineral activity.

3.2 *Stress history*

Preconsolidation stress σ'_c was determined from the oedometric test using the Casagrande method, whose application turned out cumbersome due to low-pronounced yielding curvature. No particular improvement in estimating σ'_c was unfortunately obtained by other methods (e.g. Grozic et al., 2003).

A general decrease of OCR with depth (Figure 2, last column) can be noted, with the deeper formations remaining slightly OC. The higher values are due to *caranto* and to some deeper layers. Note that the main part of the cohesive soil layers or laminations are slightly OC (OCR ≈ 1.2-3.7).

3.3 *Mechanical behavior from laboratory tests*

The laboratory tests program consisted of one-dimensional tests, drained and undrained triaxial compression tests (TX-D/U), resonant column (RC)

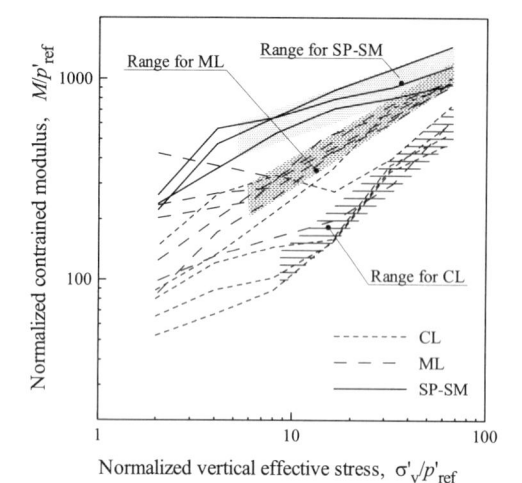

Figure 3. Normalized constrained modulus versus vertical effective stress.

tests and bender element (BE) measurements in the triaxial cell.

The main features of the mechanical behavior of the Venice lagoon soils are here briefly summarized. A more detailed description can be found in Cola and Simonini (2002).

3.3.1 *One dimensional compression*
The response in 1-D compression has been interpreted in terms of constrained modulus M as a function of vertical effective stress σ'_v (Janbu, 1963),

$$\frac{M}{p'_{ref}} = C \cdot \left(\frac{\sigma'_v}{p'_{ref}}\right)^m \qquad (2)$$

where C and m are two experimental constants and p'_{ref} a reference stress ($p'_{ref}=100$ kPa).

Figure 3 shows the trend of M as a function of σ'_v/p'_{ref} for the three classes of soil. The lowest values correspond to the CL class whereas the highest correspond to the SM-SP classes.

The experimental constants C and m were related to the grain size index I_{GS} through the following equations:

$$C = (270 \pm 30) + 56 \cdot \log I_{GS} \qquad (3)$$

$$m = (0.30 \pm 0.1) - 0.07 \cdot \log I_{GS} \qquad (4)$$

Figures 4a and 4b show the trend of C and m as a function of I_{GS}.

3.3.2 *Critical state*
Typical results of TX-CK$_0$U tests for the extreme classes SM-SP and CL are shown on Figure 5a in terms of deviatoric stress q vs. axial strain ε_a and pore pressure u vs. ε_a. Figure 5b sketches the

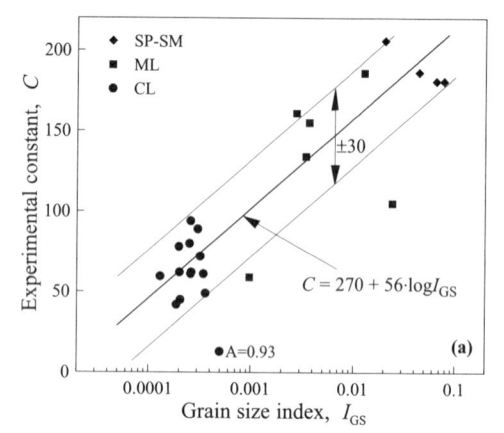

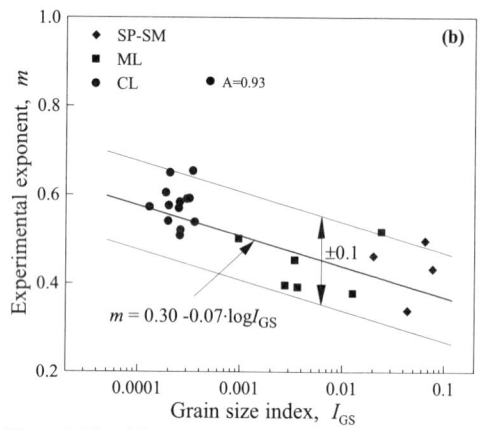

Figure 4. Material constants C (*a*) and m (*b*) as a function of grain-size index.

corresponding stress-paths in the p'-q plane (p'=mean effective stress).

The evolution of u indicates the presence of both contractant and dilatant phases: the Phase Transformation Point (Tatsuoka and Ishihara, 1974; Ishihara, et al., 1975) marks the end of the contractant phase, after which the sample exhibited dilatant behavior.

Non-dilative response was observed only in a few of the tests performed on samples with higher clay content, but this behavior cannot be considered as representative of typically Venetian soils.

The critical state angle ϕ'_c is reported in Figure 6 as a function of I_{GS}, distinguishing among the three soil classes and between drained (D) and undrained (U) tests. As expected, higher values are due to sands (34°- 39°) whereas silty clays show lower values (30°-34°). Critical angle of silts covers a larger range, reaching also the upper values of sands.

The parameters e_{ref} = void ratio at p'_{ref} and λ_c = slope of the critical state line in the plane e-logp' were also estimated and two regression lines, one for SM-SP-ML the other for CL, determined: the pair

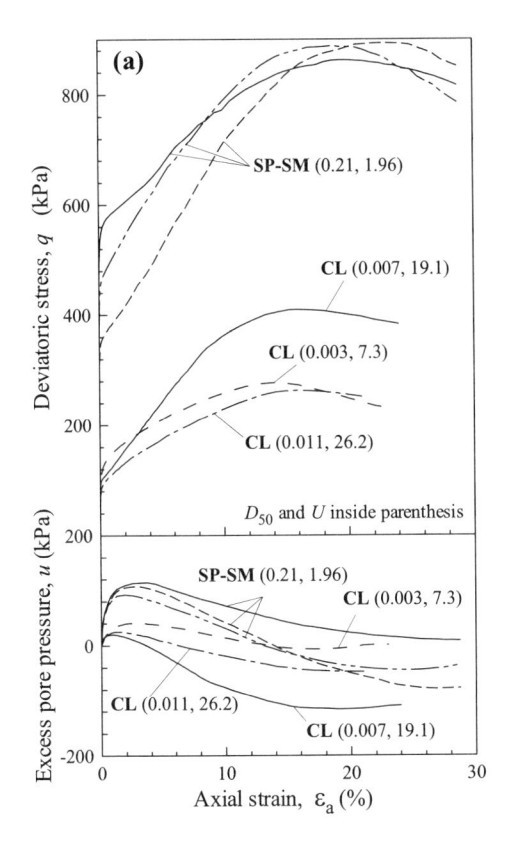

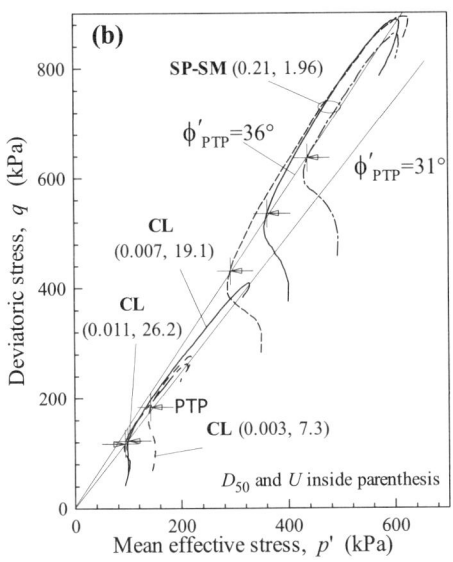

Figure 5. Typical undrained triaxial behaviour for SP-SM and CL soils: (a) q and u versus axial strain, (b) p'-q stress paths.

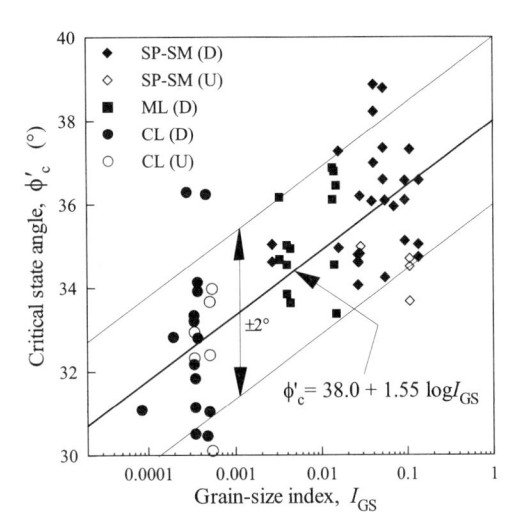

Figure 6. Critical state angle as a function of grain-size index.

(e_{ref}, λ_c) turned out to be equal to 0.918 and 0.066 for SM-SP-ML and 0.818 and 0.107 for CL, respectively.

The trends of critical state parameters ϕ'_c, λ_c and e_{ref} as a function of I_{GS} may characterized by the following expressions:

$$\phi'_c = (38.0 \pm 2) + 1.55 \cdot \log I_{GS} \quad (°) \tag{5a}$$

$$\lambda_c = (0.152 \pm 0.04) - 0.016 \cdot \log I_{GS} \tag{5b}$$

$$e_{ref} = (1.13 \pm 0.01) + 0.10 \cdot \log I_{GS} \tag{5c}$$

3.3.3 Peak strength and state parameter

The difference between peak and critical friction angle ($\phi'_p - \phi'_c$) measured in TX tests is reported against the mean effective stress at failure p'_f on Figure 7.

The equation suggested by Bolton (1986, 1987) for deriving ϕ'_p in triaxial compression was considered to interpret the triaxial test results:

$$\phi'_p - \phi'_c = 3D_R [(Q - \ln p'_f) - 1] \tag{6a}$$

where $D_R = (e_{max} - e)/(e_{max} - e_{min})$ is the relative density and Q an experimental constant (Bolton suggested Q = 10 for quartz and feldspar sands and 8 for limestone). The equation was adapted to the Venice soils as follows:

$$\phi'_p - \phi'_c = 3.9 D_R [(9 - \ln p'_f) - 1] \tag{6b}$$

Eq. (6b) is plotted on Figure 7 for $D_R = 0.5$ and $D_R = 0.7$, a typical range of density for Venice sands. The value 3.9, used here instead of 3 suggested by Bolton, may be partially justified by considering that the Venetian sands and silts are composed of angular

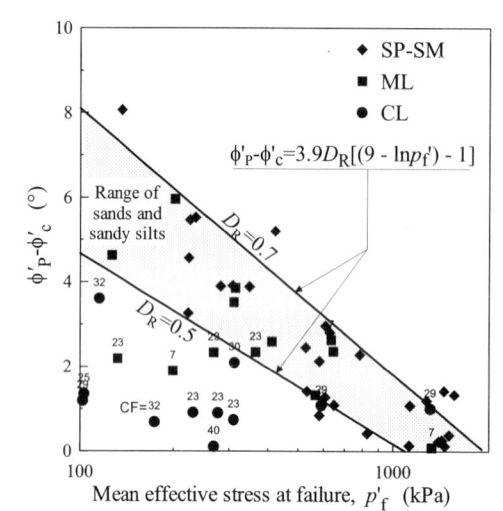

Figure 7. Difference between peak and critical angles versus mean effective stress at failure in triaxial tests.

grains, thus providing a higher degree of interlocking, whereas the data reported by Bolton mostly referred to sands characterized by a lower degree of angularity (e.g. Lee and Seed, 1967).

The state parameter ψ (Wroth and Basset, 1965; Been and Jefferies, 1985; Been et al. 1991) defined here as:

$$\psi = e - e_{ref} + \lambda_c \log \frac{p'}{p'_{ref}} \quad (7)$$

was also calculated and related to ϕ'_p. The data are represented rather well by the expression proposed by Wood et al. (1994):

$$\phi'_p = \phi'_c - \beta \cdot \psi \quad (8)$$

with $\beta = 40$ for sands and silts.

3.3.4 Very-small strain behavior

It is assumed that the soil behaves elastically at very small strains. The response can be interpreted in terms of maximum stiffness G_{max}, the latter being influenced by p', OCR and e (Hardin and Black, 1969). For sands, low-plasticity and slightly OC cohesive soils, the effect of OCR can be considered negligible (Hardin and Drnevich, 1972). Therefore, G_{max} can be related to e and p' by the expression:

$$\frac{G_{max}}{p'_{ref}} = D \cdot f(e) \cdot \left(\frac{p'}{p'_{ref}}\right)^n \quad (9)$$

where D is a material constant, $f(e)$ is a void ratio function, n is an exponent.

Among the relationships proposed to express $f(e)$ (Hardin and Drnevich, 1972; Jamiolkowski et al., 1994; Biarez and Hicher, 1994; Shibuya and Tanaka,

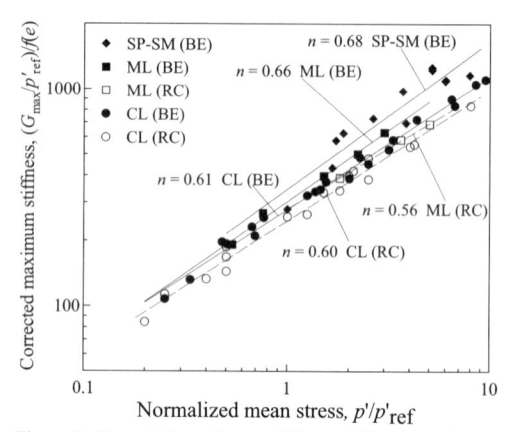

Figure 8. Corrected maximum stiffness as function of mean effective stress.

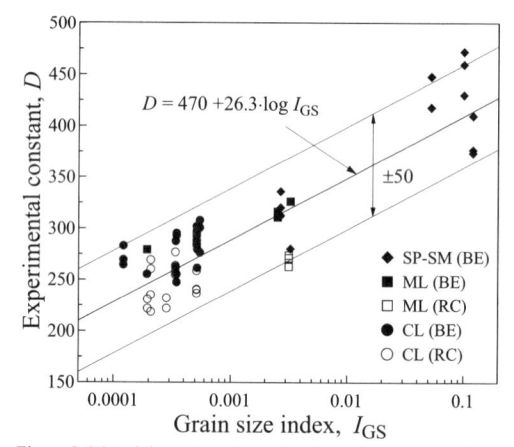

Figure 9. Material constant D as a function of grain-size index.

1996), the original expression proposed by Hardin and Drnevich (1972):

$$f(e) = (2.97 - e)^2 / (1 + e) \quad (10)$$

was used here to analyze experimental results.

Figure 8 depicts $G_{max}/(f(e)p'_{ref})$ as a function of p'/p'_{ref}. The exponent $n=0.60$ could be reasonably assumed as representative of the trend for all data.

The dependence of the material constant D on I_{GS} was assessed and fitted by the equation (Figure 9):

$$D = (470 \pm 50) + 26.3 \cdot \log I_{GS} \quad (11)$$

Note that, I_{GS} is once more capable of expressing the dependence of G_{max} on the three classes of soils.

3.3.5 Consolidation and secondary compression

Figure 10 shows the values of the primary consolidation coefficient c_v and secondary compression c_α estimated from the oedometric tests as a function of overburden stress. Note that c_v is

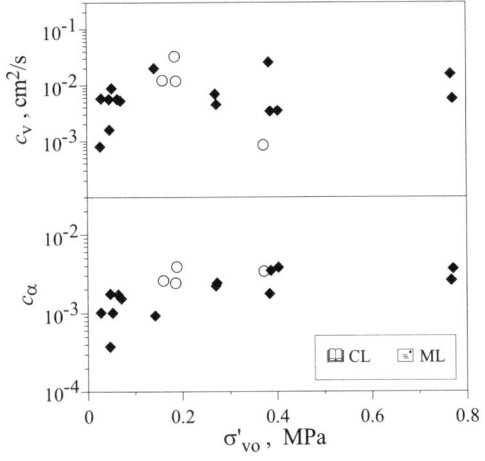

Figure 10. Consolidation coefficients at MTS.

rather high and seems to be relatively insensitive to variation of vertical effective stress whereas c_α tends to increase slightly at small stress levels and to remain constant at higher ones.

3.4 Calibration of in-situ tests

One of the aims of the MTS was to provide a site-specific calibration of CPTU and DMT, for extensive soil characterization at the three inlets, thus avoiding the use of more expensive borehole and laboratory testing as well as the scale effects inherent in these highly heterogeneous soils.

In the tests carried out at the MTS, the piezocone was the standard Dutch cone provided by the pore pressure transducer located along the shaft just behind the cone (Torstensson, 1975, 1977) and with the ratio a between net and total area equal to 0.83.

The following symbols have been used in the interpretation of the CPTU:

q_c = cone resistance;
u_2 = pore pressure measured at the cylindrical extension of the cone;
f_s = measured sleeve friction;
q_t = $q_c + (1-a) u_2$ = corrected cone resistance;
q_n = $q_t - \sigma_{vo}$ = net corrected cone resistance (σ_{vo} = total overburden stress);
Δu_2 = $u_2 - u_0$ = excess pore pressure, where u_0 is water pore pressure in a hydrostatic condition;
R_f = $100\, f_s/(q_t - \sigma_{vo})$ = normalized sleeve friction ratio;
B_q = $\Delta u_2 /(q_t - \sigma_{vo})$ = pore pressure ratio.

The standard dilatometer was utilized at the MTS (Marchetti, 1975, 1980). In the standard test procedure, two values of gas pressure are recorded:

p_0 = pressure to lift the membrane off the sensing device located just beneath the membrane
p_1 = pressure to cause 1 mm deflection.

From the two corrected pressure readings

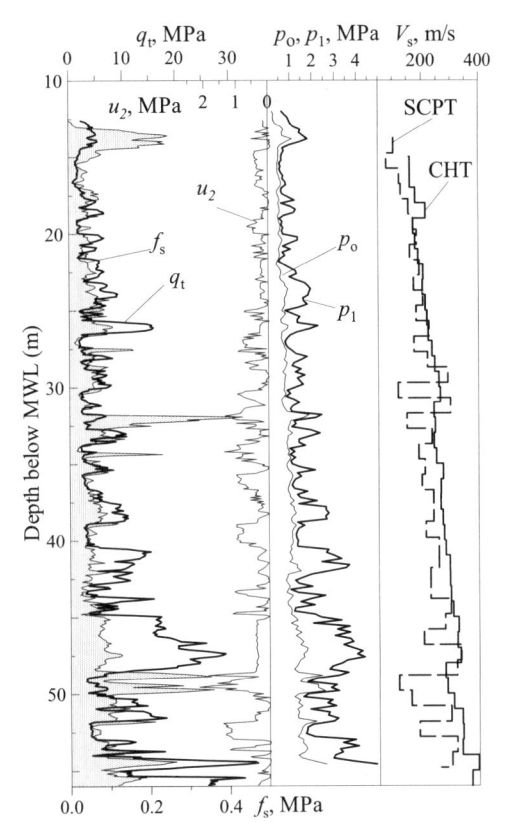

Figure 11. Profiles of q_t, f_s, u_2, p_0, p_1 and V_s at MTS.

Marchetti derived the following index parameters, namely:

The material index: $I_D = (p_1 - p_0)/(p_0 - u_0)$
The horizontal-stress index: $K_D = (p_0 - u_0)/\sigma'_{vo}$
The dilatometric modulus: $E_D = 34.7(p_1 - p_0)$

where σ'_{vo} is the in-situ vertical effective stress.

The standard DMT output is proposed in terms of corrected pressure readings of the three index parameters I_D, K_D and E_D and of a series of empirically derived properties through correlations based on the above index parameters (e.g. Totani et al., 1999).

The results of two typical tests SCPTU19 and MDMT2, carried out at closer distance, are shown in Figure 11 in terms of the profiles of significant quantities q_t, f_s, u_2, p_0, p_1 and V_s, where the latter is the shear wave velocity measured at various depths by the SCPTU and by the CHT.

Sharp variations of q_t, f_s and u_2 are observed; more particularly u_2 remains below the hydrostatic level not just in the *caranto* but also in some cases in the sandy layers. This effect has been observed in the special circumstances of dense cohesionless soil that exhibits dilation when sheared (Wroth, 1984).

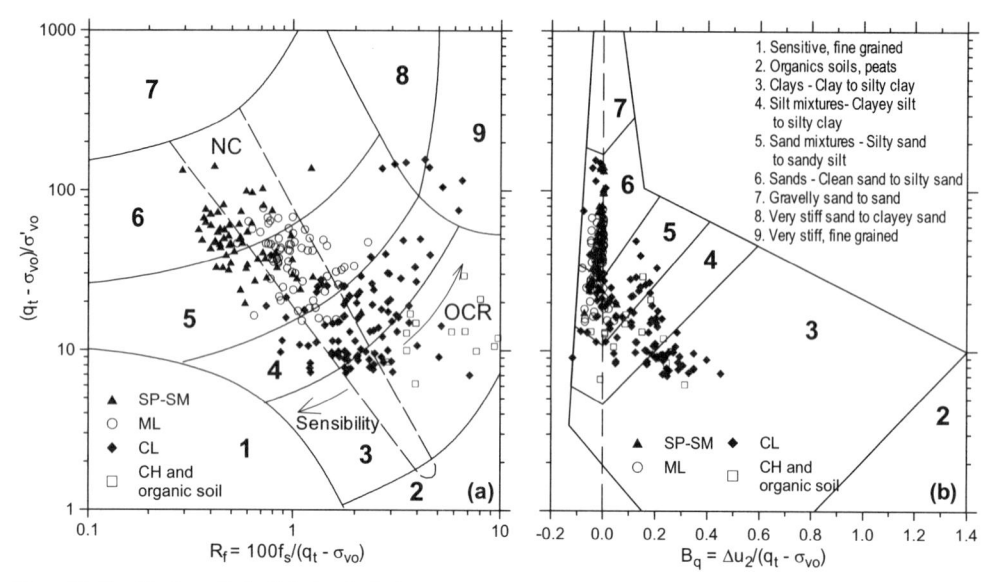

Figure 12. Classification with the Robertson's charts.

B_q rarely exceeds 0.4 at MTS, whereas in other sites higher values have been observed.

The p_0 and p_1 profiles are characterized by the continuous and coupled oscillations of the two pressures, thus confirming the high degree of heterogeneity already shown by SCPTU19.

An increasing trend of V_s with depth, from 150 m/s up to 300 m/s at 50 m below m.s.l., is noted with very small variations recorded in CHT among the different soil formations. Some flapping of V_s, measured with SCPTU at depths from 29 to 34 m and 45 to 52 m, may be due to the presence of some thin peaty layers, which could influence the propagation of V_s across horizontal soil layering.

On the basis of laboratory test results, a calibration of CPTU and DMT to estimate the main geotechnical properties is considered and discussed in the following sections.

3.5 Piezocone calibration

3.5.1 Soil classification

Soil classification with CPTU can be obtained by using q_t and f_s (Olsen and Farr, 1986; Robertson, 1990) and/or with u_2 measured during penetration (Senneset et al., 1982; Robertson, 1990).

All the classification methods are based on empirical charts, with the most widely accepted being Robertson's where the soils are grouped into 9 classes, accounting also for OCR. Figures 12a and 12b report the CPTU data superimposed onto Robertson's charts, dividing the soils into the four classes, SM-SP, ML and CL and organic soils.

From Figures 12a and 12b it can be noted that the proposed subdivision is relatively suitable for the classification of the Venetian soils, which mostly belong to the groups 4, 5 and 6.

3.5.2 Grain-size index

An attempt to relate the grain-size index I_{GS}, the parameter accounting for grain size distribution, to CPTU quantities was performed. The best result was obtained by linking I_{GS} to R_f, whose correlation, shown on Figure 13, is interpreted by the equation:

$$I_{GS} = 0.0059 R_f^{-3.33} \qquad (12)$$

Materials with $R_f > 4$ have been excluded from the regression analysis.

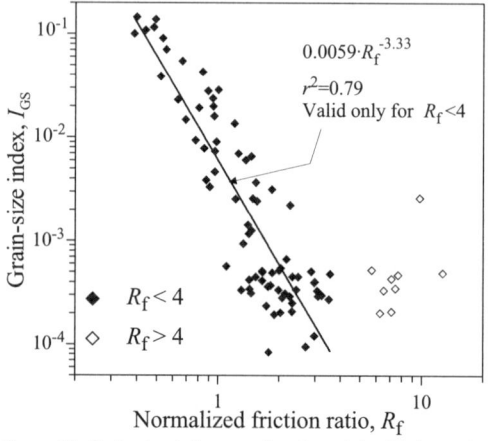

Figure 13. Grain-size index as a function of the friction ratio from CPTU.

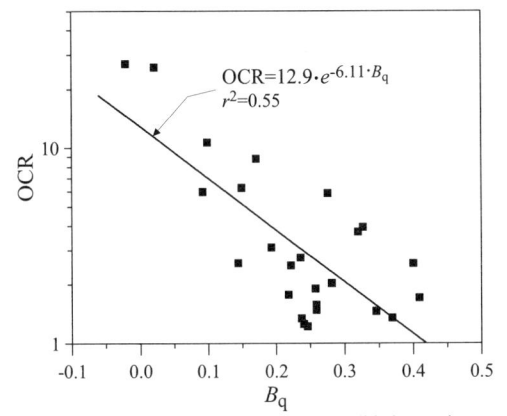

Figure 14. Relationship between overconsolidation and pore pressure ratios.

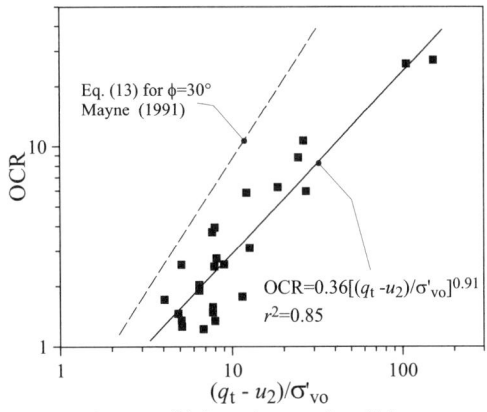

Figure 15. Overconsolidation ratio versus $(q_t-u_2)/\sigma'_{vo}$.

3.5.3 Overconsolidation ratio

The level of u_2 generated during cone penetration, and consequently the value of B_q, are functions of the type of soil and of OCR.

In order to verify if a possible relationship between OCR and B_q exists, the OCR's values determined from 1-D compression tests were plotted versus B_q in Figure 14. A general decrease of OCR along with an increase of pore pressure ratio can be observed, but no reliable correlation between these two quantities can be established, the level of u_2 being much more dependent on soil type than on OCR.

Another attempt to link OCR with CPTU results could be performed as suggested by Mayne (1991), who proposed a relationship between OCR and $(q_t-u)/\sigma'_{vo}$ on the basis of the coupled use of cavity expansion theory and of the MCC model:

$$OCR = 2\left\{\left(\frac{q_t - u_2}{\sigma'_{vo}}\right) \bigg/ \left[1.95\left(\frac{3\sin\phi'_c}{6-\sin\phi'_c}\right)+1\right]\right\}^{1.33} \quad (13)$$

where $3\sin\phi'_c/(6-\sin\phi'_c)$ is the slope of critical state line in the (p',q) plane.

The data reported in Figure 15 confirm the increase of OCR with $(q_t-u_2)/\sigma'_{vo}$: however the curve given by Mayne's theoretical approach for a friction angle of 30°, characteristic of the Venetian clayey silts, represents approximately an upper limit of OCRs.

Figure 15 also reports the best regression fit obtained on experimental data, with the expression

$$OCR = 0.36\left(\frac{q_t - u_2}{\sigma'_{vo}}\right)^{0.91} \quad (14)$$

which could be used to predict tentative profiles of OCR in the slightly over consolidated range.

3.5.4 Undrained strength

From q_t the undrained shear strength c_u can be determined by applying the bearing capacity equation to the cone resistance. Hence, we have:

$$c_u = \frac{q_c - \sigma_{vo}}{N_k} \quad (15)$$

where N_k is the cone bearing capacity factor. Several theoretical approaches are available for N_k (Yu and Mitchell, 1998), the latter being theoretically characterized by a large range of variation, between 11 and 19 for NC clays, approaching 25 for OC ones.

On the basis of c_u measured in the laboratory, N_k was calculated for CL formations. The result is shown in Figure 16. The scatter around the average value of 8.5 is relatively large; eq. (15) can be therefore used only for a rough estimate of c_u.

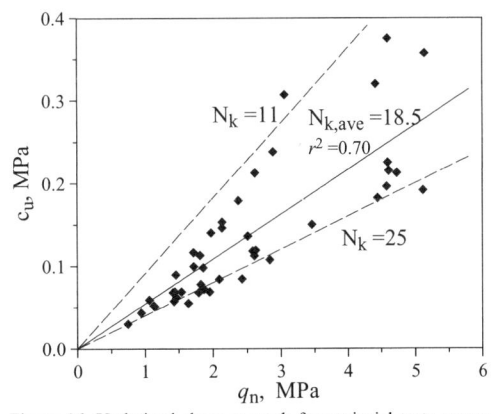

Figure 16. Undrained shear strength from triaxial tests versus net cone resistance.

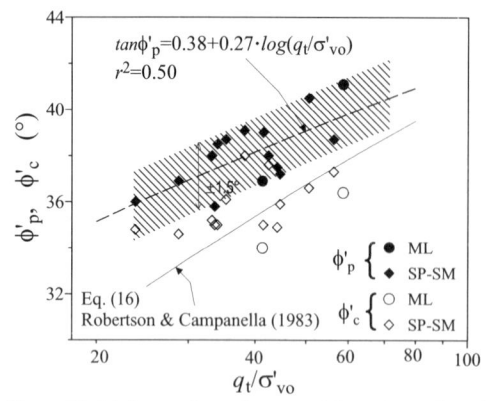

Figure 17. Friction angles and cone resistance in sands and silty sands.

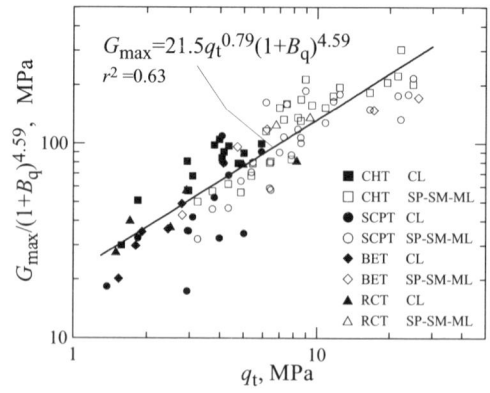

Figure 18. Relationship between maximum stiffness and CPTU parameters.

3.5.5 *Friction angle*

The simplest and most widely known relationship to estimate ϕ' is due to Durgunoglu and Mitchell (1975), modified by Robertson and Campanella (1983). This is based on the correlation between ϕ', q_t and σ'_{vo} with the following analytical expression:

$$\phi' = \arctan\left[0.10 + 0.38\log\left(\frac{q_t}{\sigma'_{vo}}\right)\right] \quad (16)$$

A similar relationship was used in our case and the regression analysis

$$\phi' = \arctan\left[0.38 + 0.27\log\left(\frac{q_t}{\sigma'_{vo}}\right)\right] \pm 1.5 \quad (°) \quad (17)$$

is proposed in Figure 17, where the shaded area defines the range of possible excursion between the upper and lower values of ϕ'.

3.5.6 *Maximum shear stiffness*

Correlations were proposed between CPT resistance and G_{max} for a large variety of soils (Baldi et al., 1989, Mayne and Rix, 1993, Hegazy and Mayne, 1995). The major criticism to all these correlations is that G_{max} is a parameter determined at very small shear strain levels whereas q_t is a quantity measured at large deformations involving yielding and failure of the soil surrounding the cone. However, as pointed out by Mayne and Rix (1993), both quantities, G_{max} and q_t, show, for a given soil, similar dependence on the same parameters, namely the mean effective stress level and void ratio.

The reconstruction of the e profile is a difficult task. The advantage of using the CPTU is given by the continuous profile of B_q which, in these soils, is mainly a function of pore size distribution rather than OCR. Therefore, the B_q profile may act as a simple substitute of e profile, in order to take into account soil type and its structural condition.

To analyze the dependence of G_{max} on q_t and B_q we used the data from the SCPT, CHT and from laboratory measurements performed with BE system and RC equipment (Simonini and Cola, 2000).

The following multiple regression function applied on data from all types of soils was selected:

$$G_{max} = 21.5 q_t^{0.79}(1+B_q)^{4.59} \quad (MPa) \quad (18)$$

The data on Figure 18 show that an association between Gmax, qt and Bq is reasonably possible for Venetian soils.

3.5.7 *Coefficient of consolidation*

Among the various methods for predicting the initial excess pore pressure distribution, generated by driving the piezocone, and its subsequent dissipation (Torstensson, 1977; Baligh and Levadoux, 1980; Gupta and Davidson, 1986; Teh and Houlsby, 1988), the method due to Teh and Houlsby was used to calculate the coefficient of consolidation.

Figure 19 shows the trend of c_h with overburned stress determined by the dissipation tests. Note the high values of c_h, always above those provided by laboratory calculations and characterized by large oscillations with depth.

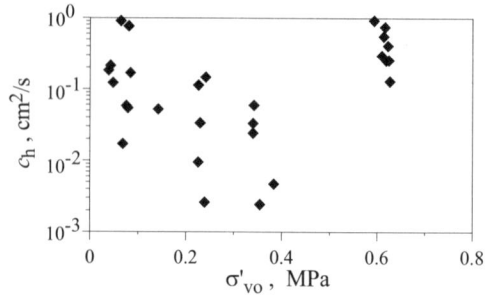

Figure 19. Coefficient of horizontal consolidation versus overburned stress.

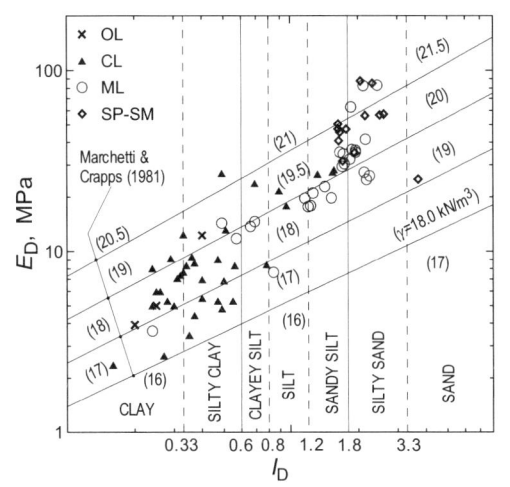

Figure 20. Classification by means of Marchetti and Crapps chart.

3.6 Dilatometer calibration

3.6.1 Soil classification

The material index I_D was originally used as approximate parameter for identifying the soil type, i.e. $I_D \leq 0.6$ for clay, $0.6 < I_D < 1.8$ for silt and $I_D \geq 1.8$ for sand. However, it was later recommended to combine the material index I_D with E_D for better soil classification.

Figure 20 summarizes the data obtained from the MTS and plotted on the improved chart proposed by Marchetti and Crapps (1981).

The two extreme classes of soil, namely the silty clay (CL) and silty sand (SM-SP) appear to be better defined than those of silts (ML).

Marchetti and Crapps soil classification chart provides indications about the unit weight of soil. Comparing the γ predicted with that measured in the laboratory, the range of maximum scatter around the average values does not exceed 15%. Nevertheless, the chart tends to overestimate γ of CL while underestimating that SM-SP.

3.6.2 Overconsolidation ratio

The estimation of overconsolidation ratio for clays was proposed by Marchetti relating K_D to OCR from oedometer tests with the following correlation:

$$\text{OCR} = (0.5 K_D)^{1.56} \qquad (19)$$

The use of correlation (19) is restricted to materials with $I_D < 1.2$, free of cementation which have experienced simple one-dimensional stress histories.

Among the improved relationships available in literature, that proposed by Lacasse and Lunne (1988):

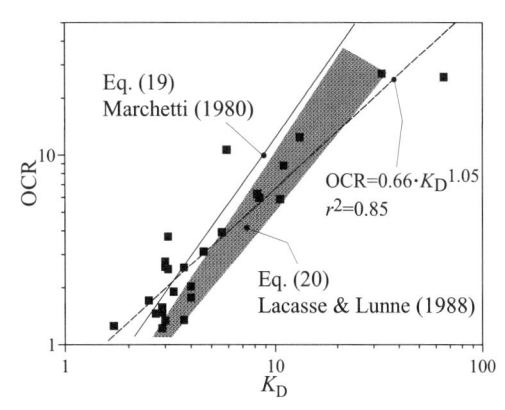

Figure 21. Overconsolidation ratio versus horizontal stress index.

$$\text{OCR} = 0.225 K_D^{1.35 \div 1.67} \qquad (20)$$

takes into account a large range of soil plasticity in the exponent, that varies from 1.35 for plastic clays and up to 1.67 for low plasticity materials.

The applicability of correlations (19) and (20) to MTS data was verified in Figure 21. The Marchetti correlation seems to provide a rather good estimation of OCR at the lowest values of K_D. Large differences compared to laboratory results occur at higher horizontal stress indexes ($K_D > 10$), that is, for the highly OC *caranto* at shallow depths.

No improvement of data relating to any of the K_D range was obtained by adopting the Lacasse and Lunne relationship.

If we apply the power regression to all our data independently of I_D and stress history we have the following relationship:

$$\text{OCR} = 0.66 K_D^{1.05} \qquad (21)$$

which could provide, as similar to the piezocone, an OCR prediction in the range $1.7 < K_D < 70$.

3.6.3 Coefficient of earth pressure at rest

DMT appears to be particularly suitable for measuring K_0. The original correlation between at-rest coefficient K_0 and K_D is:

$$K_o = (K_D / 1.5)^{0.47} - 0.6 \qquad (22)$$

Lacasse and Lunne (1988), on the basis of direct K_0 measurements, argued that the Marchetti relationship tends to overestimate K_0.

To determine K_0 in clays, they suggested the following correlation:

$$K_o = 0.34 K_D^{0.44 \div 0.64} \qquad (23)$$

where the lower exponent value is associated with highly plastic clays.

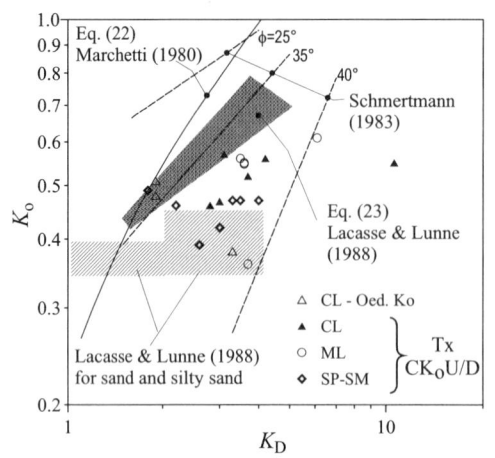

Figure 22. At-rest coefficient versus horizontal stress index.

To estimate K_0 from DMT in granular materials Schmertmann (1983) suggested the use of a chart based on CPT experience, that relates ϕ', σ'_{vo} and q_t.

In Figure 22, K_0 values measured in TX-CK$_0$U/D, and K_0-oedometer are plotted, distinguishing both the different types of soils. Correlations (22) and (23) together with the Schmertmann curves for granular materials are also sketched in Figure 22.

No significant differences in K_0 values are noted among the data from different tests or from the three classes of soils, nor according to the material index I_D. The points are located below both the Marchetti, and Lacasse and Lunne curves and seem to be more in accordance with Schmertmann's suggestions. This is probably due to the prevalently silty nature of Venetian cohesive soil, characterized by low chemical bonds between soil particles, together with a high sensitivity to the stress relief inherent in the sampling disturbance. A correlation between K_0 and K_D does not seem possible in this case using the data collected at MTS.

3.6.4 Undrained strength

Considering the dependence of c_u/σ'_{vo} on overconsolidation ratio, Marchetti proposed a correlation between c_u and K_D:

$$\frac{c_u}{\sigma'_{vo}} = 0.22(0.5K_D)^{1.25}. \qquad (24)$$

Correlation (24) was recommended, as similarly done for correlation (19), only for material with $I_D<1.2$ and simple loading history.

Undrained strength from TX tests is plotted against K_D in Figure 23: a general increase of c_u as a function of K_D is appreciated. It may therefore be suggested that the original correlation could be utilized for a preliminary estimate of c_u for all the range of K_D.

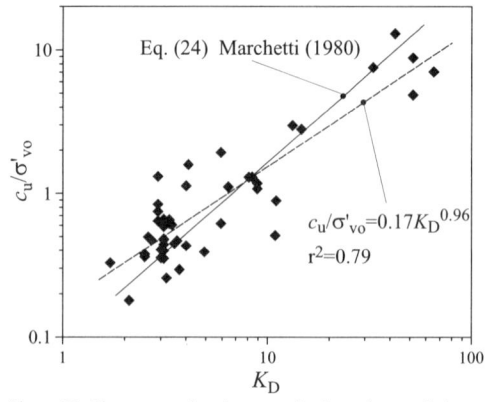

Figure 23. Shear strength ratio versus horizontal stress index.

3.6.5 Friction angle

Observing a linear correlation between K_D and q_t, Marchetti adapted the relationship (16), thus presenting a new chart that provides ϕ'_p of sand from K_D (if K_0 is known or determined with correlation (22)). Recently Marchetti (from Totani et al., 1999) suggested two direct empirical correlations, which could be used to give lower and upper bounds of the range of possible friction angles:

$$\phi'_{max} = 31 + K_D/(0.236 + 0.066K_D) \qquad (25a)$$

$$\phi'_{min} = 28 + 14.6 \cdot \log K_D - 2.1 \cdot (\log K_D)^2 \qquad (25b)$$

A comparison between laboratory values of peak and critical friction angles from triaxial tests on SP-SM and ML soils and the value predicted by eqs. (25a,b) is plotted in Figure 24: the fitting is rather limited showing a poor dependence of friction angle on K_D.

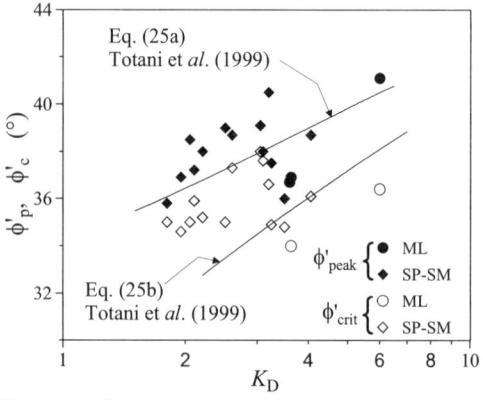

Figure 24. Friction angles versus horizontal stress index in sands and silty sands.

3.6.6 Constrained modulus

To estimate one-dimensional constrained modulus M, the correlation with E_D is used in the form:

$$M = R_M E_D \quad (26)$$

where R_M is a function of K_D and I_D as a result of some relationships not reported here for the sake of brevity. Comparisons between constrained modulus from DMT (M_{DMT}) and oedometer (M_{oed}), proposed by Lacasse and Lunne (1988) for various types of soil, showed that the ratio M_{DMT}/M_{oed} could vary from 0.5 for high-plasticity clays to 2.0 for sandy or silty soils.

In Figure 25 the M_{DMT} estimated with relationship (26) is plotted against M_{oed} determined in laboratory for all the types of Venetian soils, including sands. The ratio M_{DMT}/M_{oed} is always greater then the unity and is not influenced by the type of soil or by its I_D value. Even if R_M appears to decrease with both K_D and I_D, it does not seem possible to find a correlation between R_M, K_D and I_D.

3.6.7 Maximum stiffness

A very small-strain modulus can also be estimated from E_D. The ratio:

$$R_G = G_{max}/E_D \quad (27)$$

is, in some cases, expressed as a function of K_D or relative density (Jamiolkowski et al., 1988), but these relationships cannot be applied straightforward to all soils. Hryciw (1990) expressed G_{max} as a more general function of K_0, γ_{DMT} and σ'_{vo}, in which K_0 is given by correlation (22) and γ_{DMT} is the total unit weight determined from DMT data by means of Marchetti and Crapps' chart.

As presented above the errors in estimating γ_{DMT} and K_0 for Venetian soil are not negligible. Consequently, we preferred to propose a direct relationship between R_G and I_D: the best fitting was found in the form of the logarithmic relationship between R_G and I_D, as shown in Figure 26. Using G_{max} data measured both in-situ and in the laboratory, the relationship can be written as:

$$R_G = 6.71 - 14.2 \log I_D \quad (28)$$

3.7 Remarks on the calibration of CPTU and DMT at MTS

The relevant difficulties encountered at the MTS in comparing the relevant properties from laboratory tests with estimations from in-situ tests must be pointed out: the absence of any clear horizontal layering in fact provided no absolute certainty to carry out a comparison of material properties within the exact same type of soil. Nevertheless, some interesting remarks can be tentatively drawn:
- The CPTU represents an indispensable investigation tool to characterize the highly interbedding of Venice soil, thus allowing for a very accurate description of soil profile. Reliable classification can be also performed with the dilatometer, but with some uncertainties in defining the intermediate class of Venetian soils, namely the silts. No reliable value of unit soil weight seems to be determinable with the DMT;
- Both CPTU and DMT can be used to reasonably reconstruct stress history, especially for the slightly OC layers. For these, approximate values of the OCR can be determined to sketch tentative profiles of OCR with depth;
- For similar reasons, the method of determination of K_0 with DMT cannot be definitely validated on the basis of the comparison with SBPM and laboratory results, the latter particularly influenced by the typical stress-relief due to the non-structured nature of Venetian sediments.
- The applicability of both CPTU and DMT to calculate the c_u is rather limited, with a preference for CPTU.
- With reference to ϕ' of sands and silty sands, the

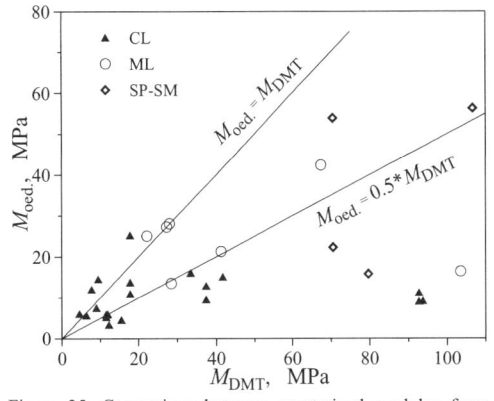

Figure 25. Comparison between constrained modulus from oedometric tests and dilatometric tests.

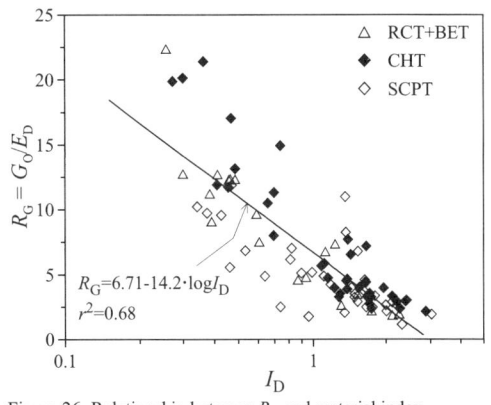

Figure 26. Relationship between R_G and material index.

well accepted Durgunoglu and Mitchell relationship was calibrated for CPTU, to suit these soils: the friction angle could be therefore reasonably calculated with a scatter of ±1.5°. Improved empirical relationships for DMT recently proposed by Marchetti to determine the maximum and minimum friction angles seem unfortunately not particularly applicable to these materials.
- CPTU dissipation tests are useful for calculating the coefficient of consolidation. Comparison with laboratory data obtained from vertical consolidation tests showed higher in-situ values of c_h typical of silty clay, but no particular comment is possible with respect to the influence of the inherent anisotropy.
- Constrained modulus from DMT was rather different from that measured in the laboratory on small oedometric samples. However, as discussed in the previous sections, the noticeable dispersion of data may probably be attributed to the sensitivity of this parameter even to small variations of grain size composition.
- In the absence of any direct measurement of V_s G_{max} can be tentatively estimated from both CPTU and DMT. In the first case the better result was obtained using an association between G_{max}, q_t and pore pressure factor B_q. Compared to CPTU, the DMT leads to a more reliable prediction of G_{max} profile.

4 THE TREPORTI TEST SITE

In order to measure in-situ the stress-strain-time properties directly and verify reliability of the correlations for SCPTU and DMT discussed in the previous sections, a comprehensive research program, coordinated by the University of Padova with the collaboration of the University of Bologna and L'Aquila was very recently launched.

The research program was concerned with the construction of a full-scale earth-reinforced embankment over a typical soil profile in the Venice lagoon and with the measurement of relevant displacements of ground together with pore pressure evolution.

The project has been developed in the following steps:
a) selection of the typical test site in the lagoon area;
b) execution of site investigations;
c) installation of the monitoring system;
d) construction of the embankment;
e) monitoring the ground displacements and pore pressure evolution during the bank realization;
f) execution of laboratory investigations;
g) repetition of some in-situ tests after the completion of the bank;
h) prosecution of monitoring for almost five additional years after bank completion.

The test site is located just outside Treporti, an old fishing village near the Cavallino shoreline, facing the North-Eastern lagoon. The soil profile was selected bearing in mind that embankment had to be placed on the most compressible profile, in order to develop vertical displacements as large as possible.

4.1 Site investigations

Figure 27 sketches the location of site investigations consisting of:
- 10 CPTU with several dissipation tests and 2 mechanical CPT (performed by the research group of the University of Bologna);
- 10 DMT with dissipation tests (by the research group of the University of l'Aquila);
- 6 seismic SCPTU / SDMT (performed in cooperation with Paul Mayne and Alec McGillivray of GeorgiaTech, Atlanta);
- 4 boreholes with undisturbed sampling (by the research group at the University of Padova).

Two additional CPTU and four DMT have been also performed after completion of the bank.

4.2 Monitoring instrumentation

The instrumentation installed at TTS was designed to keep the following main quantities under control:
- Surface vertical displacements using 7 settlement plate and 12 bench marks, the latter located outside the embankment area. The displacements

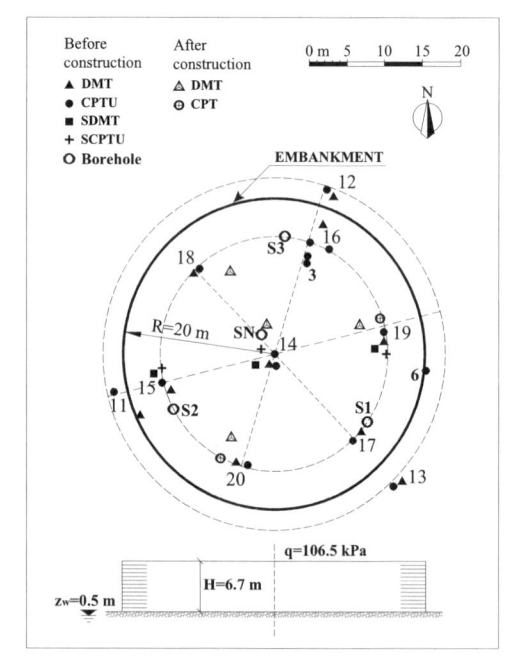

Figure 27. Embankment plan with location of site investigation tests.

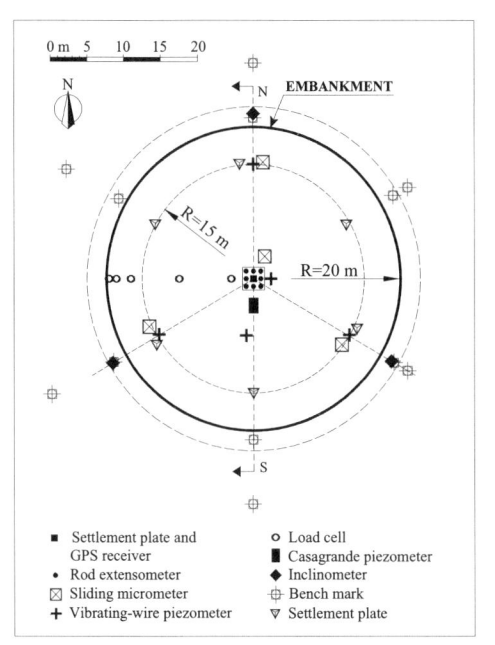

Figure 28. Embankment plan with location of monitoring devices.

of these instruments were monitored by high-precision surveying methods;
- Surface vertical displacement by one GPS antenna, located in the centre of the embankment area and fixed to the central settlement plate;
- Vertical deep displacements by means of 8 borehole rod extensometers;
- Vertical strains, along with four verticals, using 4 special multiple micrometers;
- Horizontal displacements by means of 3 inclinometers.
- Pore water pressure in fine-grained soils by means of 5 Casagrande as well as 10 vibrating wire piezometers.
- Total vertical stress beneath the loading embankment by means of 5 load cells.

Figures 28 and 29 show, respectively, the positioning of the instrumentation and a schematic soil profile, which is described in the next sections.

To measure the vertical displacements throughout the foundation ground very precisely, multiple micrometers, capable of measuring vertical displacements at 1 m intervals, were selected so that they could measure large movements with an adequate degree of accuracy (0.03 mm/m) (Kovari and Amstad, 1982).

4.3 Embankment construction

Short prefabricated vertical drains were first installed to speed up the drainage of a shallow clay layer (about 1.5 thick) and to prevent possible lateral soil spreading during construction.

The cylindrical sand bank construction started in September, 12th, 2002 and ended in March, 10th, 2003. The bank is formed by 13 geogrid-reinforced sand layers with 0.5 m thickness. As reinforcement, a stiff polypropylene geogrid was used; to avoid any sand flow through the holes of the grid, additional sheets of geotextile were placed onto the grids before placing the sand. Each sand layer was dynamically compacted to give an average dry unit weight of 15.6 kN/m^3.

After completion, the embankment was covered by an impermeable membrane on which 0.2 m medium-fine gravel was posed, thus reaching a final height of 6.7 m.

4.4 Preliminary results

The research program is still in progress and only a few of the preliminary results are shown here. Other information can be found in the papers written by Marchetti et al., (2004) and Gottardi and Tonni (2004), presented at this conference.

4.4.1 Soil profile, basic properties and stress history

The approximate ground sequence, whose first 40 meters are shown on Figure 29, consists of:
A) 0-2 m: soft silty clay, whose drainage has been improved by prefabricated drains. This thin layer contributes significantly to the total settlement of

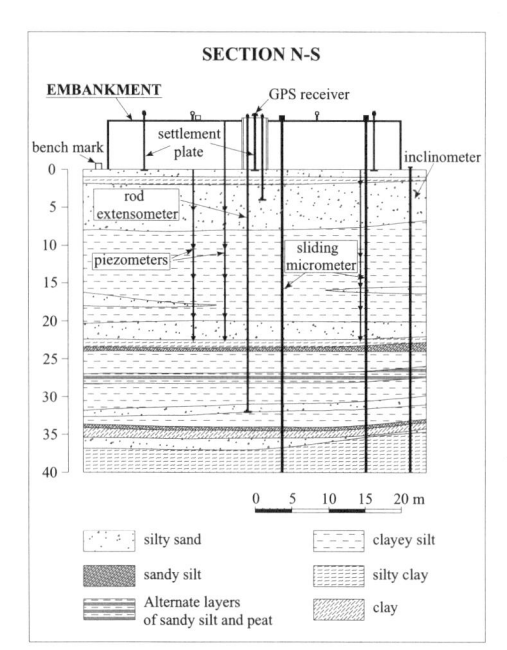

Figure 29. Cross-section of embankment with soil profile and monitoring devices.

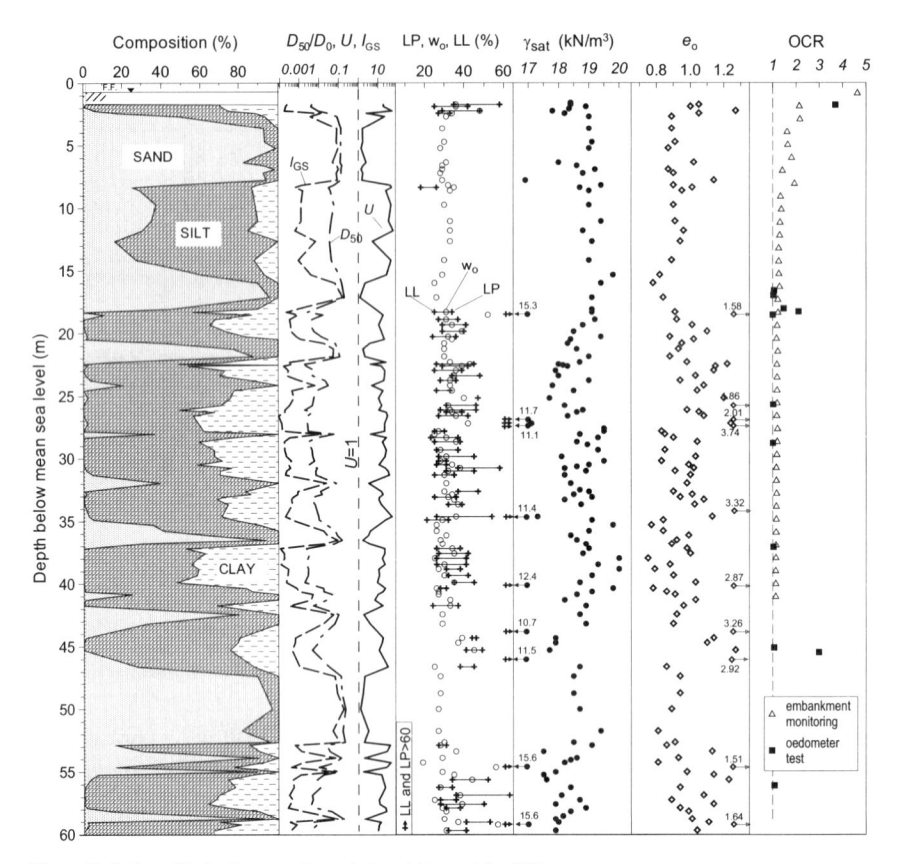

Figure 30. Soil profile, basic properties and stress history at the TTS.

the embankment, even though it cannot be considered as representative of the Venice lagoon cohesive formations;
B) 2-8 m: medium-fine silty sand;
C) 8-20 m: clayey silt with a sand lamination between 15 and 18 m, covering 270° in plan between North and East. This layer constitutes the fine-grained deposit that gave rise to most part of the vertical displacements;
D) 20-23 m: medium-fine silty sand;
E) 23-45 m: alternate layers of clayey and sandy silt;
F) 45-55 m: medium-fine silty sand.

Frequent laminations of peat are present below 25 m, in layers E and F.

Figure 30 reports the basic soil properties as a function of depth. The profile at TTS is very similar to that at MTS, thus confirming the validity of the selected site to reproduce a typical Venice lagoon ground profile.

Additional features to note are:
- From the soil grading reconstruction provided by borehole n. 2, the various types of soil occur up to 60 m, approximately, in the proportion; SM-SP: 22%; ML: 32%; CL: 37%, CH-Pt: 9%.
- Upper and deeper sands are relatively uniform; finer materials are more graded, the coarser the materials, the lower the coefficient of non-uniformity U. Grain size index I_{GS} lies in the same range observed at MTS.
- Except for the organic soils, Atterberg limits of cohesive fraction are similar to those determined at MTS;
- The unit weight γ_{sat}, showing large oscillations with depth, is somewhat lower than that measured at MTS, void ratio e_o is slightly higher and lying approximately in the range between 0.8 – 1.1, with higher values due to laminations of organic material.

The last column of Figure 30 sketches the profile of overconsolidation ratio, estimated from oedometric tests as well as from the interpretation of in-situ stress strain behaviour, as described in section 4.3.3. Note that the soils appear to be slightly overconsolidated.

Figure 31 reports also the compression and swelling coefficients C_c and C_s, the coefficient of

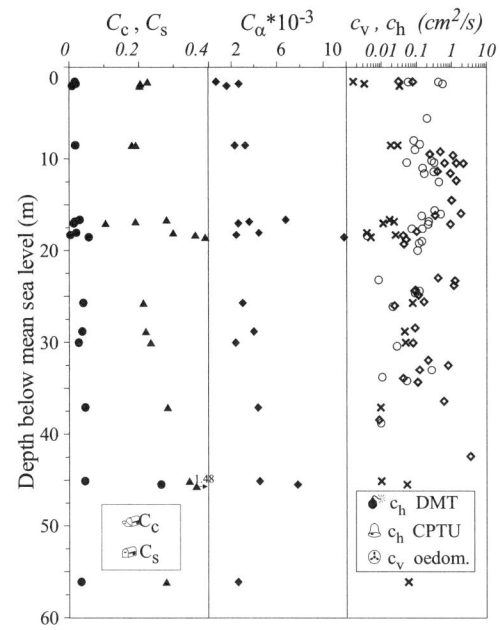

Figure 31. Profiles of compression indexes and consolidation coefficients at TTS.

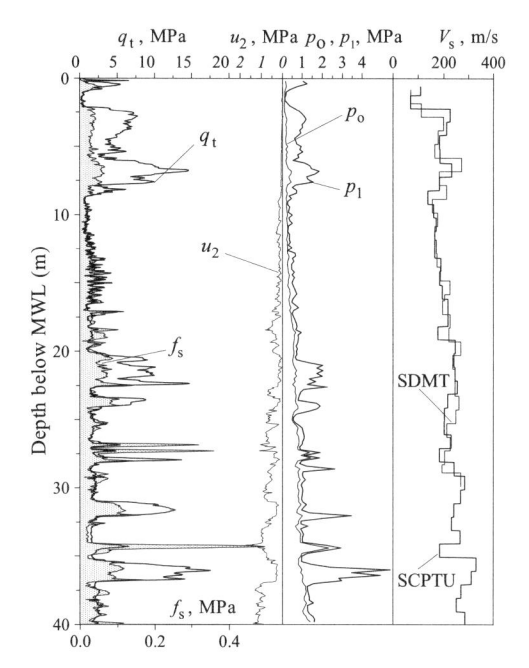

Figure 32. Comparison among the profiles of q_t, f_s, u_2, p_o, p_1 and V_s at TTS.

secondary compression c_α as well as the coefficient of consolidation estimated from laboratory and in-situ tests. A profile of the coefficient of consolidation proves once more the soil heterogeneity characterized by such a relevant variation of c_h, whose values are however in the same range measured at MTS.

The results of two typical tests CPTU14 and DMT14 (carried out near the center of the embankment at closer distance) are shown in Figure 32 in terms of profiles of significant quantities q_t, f_s, u_2, p_0, p_1 and V_s, where the latter is the shear wave velocity measured by SCPTU and SDMT. Relevant oscillations of u_2 are clearly evident, thus suggesting large differences in drainage conditions provided by a continuous alternation of different grain size composition in the whole deposit. Pore pressure parameter B_q lies in the same range of that measured at MTS with some values beyond 0.5. Similarly to MTS, V_s increases with depth from about 150 m/s to 300 m/s at higher depths.

Other relevant information on the mechanical behavior of soils at TTS is planned for publication in the very near future.

4.4.2 *Pore pressure evolution during construction*
The high values of the consolidation coefficient and the presence of prefabricated drains in the upper silty clay layer, let us suppose that primary consolidation should have been quite rapid and contemporary with the bank construction. In other words, no delayed deformation due to consolidation should have been observed, considering that the rate of load increase, required by the earth-reinforcement construction technique, was low compared to the drainage conditions of the deposit. This hypothesis has been confirmed by the application of the theory of consolidation under linearly increasing load (Viggiani, 1967; Olson, 1977). Therefore no significant and measurable pore pressure rise should have been monitored throughout the loading process.

This hypothesis was confirmed by the piezometer readings which gave no detectable pore pressure increase in any layer: the variation of pore pressure in the piezometers seemed, in fact, to be controlled only by the daily oscillations of the sea tide in the channel facing the embankment area.

4.4.3 *Vertical displacements*
Comparing the vertical and horizontal displacements throughout the loading process, it appeared that the total vertical displacement is one order of magnitude greater than the maximum horizontal displacement, that is, the deformation process developed prevalently in the vertical direction.

Figure 33 shows the evolution at the time of the maximum ground settlement measured under the center of the embankment. The total settlement at the completion of the embankment (around 180 days) was about 36 cm. This value includes, besides

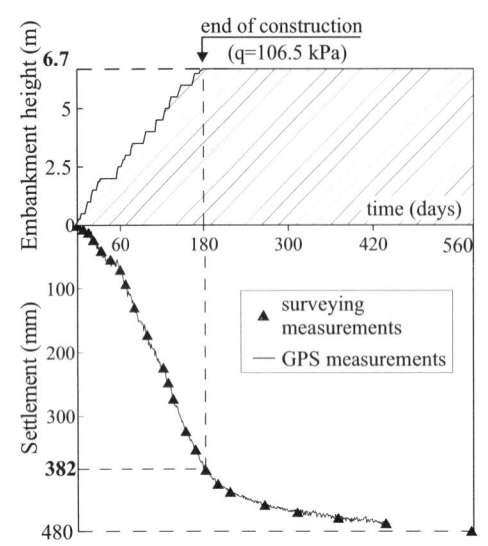

Figure 33. Evolution with time of maximum ground settlement under the embankment.

immediate and consolidation settlement, the secondary settlement that also occurred during the time of construction. Until the last measurement (25.03.04), an additional secondary settlement at constant load of 12 cm was measured, thus giving a total settlement of 48 cm.

The performance of GPS is also superimposed on the same plot: excluding some small oscillations, note the excellent agreement between the two methods of settlement measurement.

It is interesting to analyze the distribution of vertical displacement with depth provided by the multiple extensometers. These measurements are given in Figures 34 and 35, both in differential and accumulated form (for the extensometer installed close to the centre).

The relevant contribution of the upper silty clay A and particularly of the silt layer C is clearly evident. The influence of embankment reduces with depth and beyond 35 m the displacements are very small and not appreciable with the type of instruments installed at TTS.

The differential displacement representation allows for an appreciation of the strains, that are mostly concentrated, excluding the upper layer A, in the silty formation C. Note that the maximum vertical deformation ε_z does not exceed 4%.

The differential displacements plotted as a function of stress increments applied to the ground (the latter approximately estimated using an elastic finite element analysis using a realistic distribution of stiffness throughout the soil profile) allowed for a tentative evaluation of preconsolidation stress.

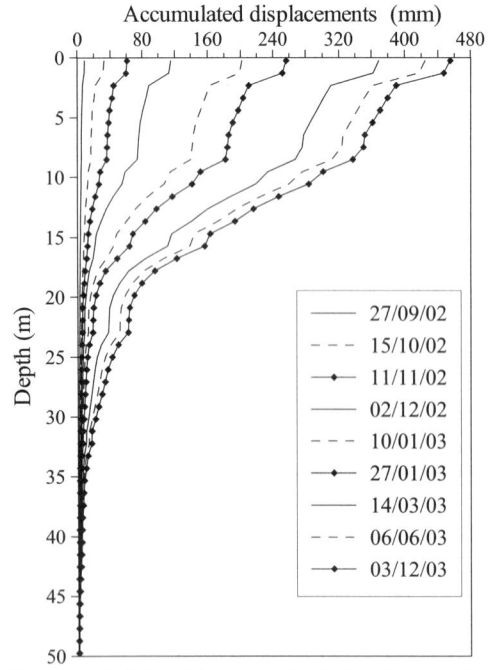

Figure 34. Evolution with time and depth of settlement under the embankment.

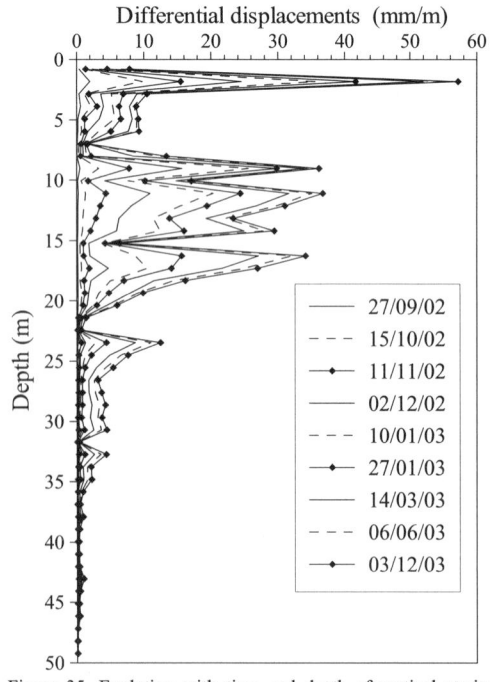

Figure 35. Evolution with time and depth of vertical strain under the embankment.

Figure 36 depicts a few of the typical stress-strain responses for some 1-m thick layers, recorded using the multiple micrometer located near the embankment centre. Note the variation of curvature throughout the loading process, characterized by a much stiffer response at the beginning of the loading phase. Hypothesizing that no delayed deformation due to the consolidation has occurred along with the loading process, the sharp variation of curvature has been interpreted in terms of yielding stress σ'_y. Since, strains in the ground developed prevalently in the vertical direction, it was tentatively assumed that the classical preconsolidation stress was $\sigma'_p \approx \sigma'_y$.

Therefore, OCR calculated by comparing σ'_y with σ'_{vo} have been plotted in Figure 30, last column, together with the OCR evaluated from oedometric tests. This slight OC state may essentially be due to ageing effects (the deposit is relatively young from a geological point of view) coupled with the influence cyclic pore pressure oscillations induced by the tide.

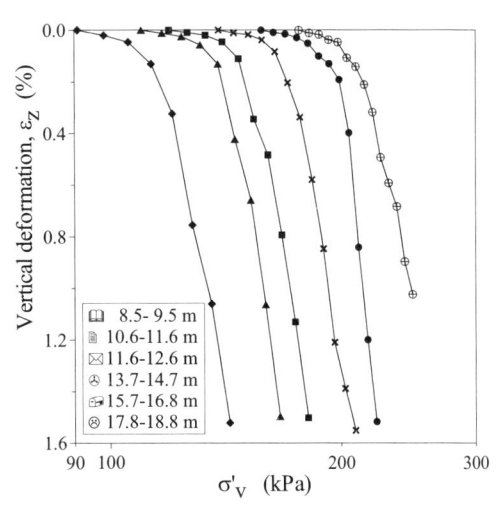

Figure 36. Vertical strain of the most compressive strata, developed during construction.

4.5 Preliminary remarks from TTS experience

As explained above, the analysis, interpretation and discussion on the results of the study carried out so far at TTS, is still in progress. Nevertheless, some preliminary remarks can here be pointed out:
- The selected site is highly representative of typical soil layering in the Venice lagoon, thus allowing the application of the findings based on the TTS study to other sites in the lagoon area;
- The embankment construction technique, namely the earth reinforcement system, has proved to be a suitable method to load the ground up to relevant loads and depths;
- The instrumentation installed at TTS provided high quality measurements of most relevant quantities, and particularly, of the vertical strains at 1 m intervals, throughout the ground profile. GPS applicability to measure vertical settlements has been also verified;
- In relation to the particular soil layering, drainage characteristics and geometric ratios between relevant dimensions, soil compression occurred mostly in a vertical direction.
- Bearing in mind the above observations, the micrometers readings plotted as a function of increasing stress, induced into the ground by the bank, allowed for an approximate estimation of a preconsolidation state in the ground;
- The high drainage properties of the soils at TTS together with the relatively low rate of loading ensured approximately a drained soil behavior, that is, consolidation took place along with the embankment construction. Confirmation of this hypothesis has been corroborated by no detectable increase in pore pressure due to the consolidation process: in the analysis and interpretation of displacement trends in time, a key role will be, therefore, assumed by the secondary compression whose evolution should be carefully investigated.

5 CONCLUSIONS

The paper outlined some relevant results of the research carried out so far to characterize the mechanical behavior of Venice lagoon soils.

The information obtained from the investigation carried out at the Malamocco Test Site is now being improved by a new experimental study at the Treporti Test Site, whose results, based on the measurement directly in situ the stress-strain–time soil properties, appear to represent a unique and valuable tool for a suitable calibration of the piezocone and dilatometer and for the formulation of a reliable geotechnical model aimed at an appropriate design of the mobile gate foundations.

ACKNOWLEDGEMENTS

The author wishes to extend special thanks to: the Consorzio Venezia Nuova, Venezia; the Magistrato alle Acque, Venezia; the research group of the University of Bologna; the research group of the University of L'Aquila; Prof. P.W. Mayne, Georgia Institute of Technology, USA; everybody at the geotechnical group of the University of Padova.
This research was supported partially by the Italian University Government Department as Scientific Research Program of National Relevance.

REFERENCES

Baldi, G., Bellotti, R., Ghionna, V.N., Jamiolkowski, M. and Lo Presti, D.C.F. (1989) Modulus of Sands from CPT and DMT. *Proc. XII Int. Conf. on Soil Mech. Found. Eng.*, Rio de Janeiro, Vol. 1, pp. 165-170, Balkema, Rotterdam.

Baligh, M. M. and Levadoux, J. N. (1980) Pore pressure dissipation after cone penetration. *Research Report R.80-11*, MIT, Cambridge, Mass.

Been, K., Jefferies, M.G., and Hachey, J. 1991. The critical state of sands. *Géotechnique*, Vol. 41, No. 3, pp. 365-381.

Been, K., and Jefferies, M.G. 1985. A state parameter for sands. *Géotechnique*, Vol. 35, No .2, pp. 91-112.

Biarez, J., and Hicher, P.I. 1994. *Elementary soils mechanics: saturated remoulded soils*. Balkema, Rotterdam.

Bolton, M.D. 1986. The strength and dilatancy of sands. *Géotechnique*, Vol. 36, No. 1, pp. 65-78.

Bolton, M.D. 1987. Discussion. *Géotechnique*, Vol. 37, No. 2, pp. 225-226.

Biscontin, G., Pestana, J., Cola, S. & Simonini, P. (2001). Influence of grain size on the compressibility of Venice Lagoon soils. *XV ICSMGE*, Istanbul, Vol.1, pp. 35-38, Balkema, Rotterdam.

Bonatti E. (1968). Late-Pleistocene and Postglacial Stratigraphy of a Sediment Core from the Lagoon of Venice (Italy). *Mem. Biog. Adr.*, Vol. VII (Suppl.), pp. 9-26, Venezia.

Cola, S., and Simonini, P. (1999). Some remarks on the behavior of Venetian silts, *Second Int. Symposyum on Prefailure behaviour of geomaterials*, IS Torino 99, pp. 167-174, Balkema, Rotterdam.

Cola S. and Simonini P. (2000). Geotechnical characterization of Venetian soils: basic properties and stress history. *Research Report*, Dept. of Hydraulic, Maritime and Geotechnical Engineering, University of Padova, July 2000.

Cola, S, and Simonini, P. (2002). Mechanical behaviour of silty soils of the Venice lagoon as a function of their grading properties. *Canadian Geotechnical Journal*, 39, pp. 879-893.

Curzi, P.V. (1995). *Sedimentological-environmental study of the Malamocco inlet: Final report*. Consorzio Venezia Nuova, Venezia.

Durgunoglu, H.T. and Mitchell, J.K. (1975) Static penetration resistance of soils: I. Analysis. *Proc. ASCE, Specialty Conference on In Situ Measurements of Soil Properties*, Raleigh, NC, Vol.1, pp. 151-172.

Favero V., Alberotanza L., Serandrei Barbero R. 1973. *Aspetti paleoecologici, sedimentologici e geochimici dei sedimenti attraversati dal pozzo VE1bis*. CNR, Laboratorio per lo studio della dinamica delle grandi masse, Rapporto tecnico n.63, Venezia.

Gatto P., Previatello P. (1974). Significato stratigrafico, comportamento meccanico e distribuzione nella Laguna di Venezia di un'argilla sovraconsolidata nota come "Caranto". CNR, Laboratorio per lo studio della dinamica delle grandi masse, *Rapporto tecnico n.70*, Venezia.

Gottardi, G, and Tonni, L. (2004). A comparative study of piezocone tests on the silty soils of the Venice lagoon (Treporti Test Site). *ISC-2*, Porto.

Grozic, J.L.H., Lunne, T. and Pande, S. (2003). An oedometer test study on the preconsolidation stress of glaciomarine clays. *Can. Geot. J.*, 40, pp. 857-872.

Gupta, R.C. and Davidson, J.L. (1986) Piezoprobe determined coefficient of consolidation. *Soils and Foundations*, 36 n.3, pp. 12-22.

Hardin, B.O., and Black, W.L. (1969). Vibration modulus of normally consolidated clay. *J. of SMFE Div., Proc. ASCE*, Vol. 95, No. SM1, pp. 33-65.

Hardin, B.O., and Drnevich, V.P. (1972). Shear modulus and damping in soils: design equations and curves. *J. of SMFE Div., Proc. ASCE*, Vol. 98, No. SM7, pp. 667-692.

Harleman, D.R.F., Bras, R.L., Rinaldo, A. & Malanotte, P. (2000). Blocking the Tide, *Civil Enginering, ASCE*, 52-57.

Hegazy, Y. A. and Mayne, P.W. (1995) Statistical correlation between V_s and cone penetration data for different soil types. *Int. Symposium on Penetration Testing*, CPT '95, Linkoeping, Vol. 2, pp. 173-178, Swedish Geotechnical Society.

Hryciw, R.D. (1990) Small-strain-shear modulus of soil by dilatometer. *Journal of Geotechnical Engineering*, ASCE, Vol.116, No.11, pp. 1700-1716.

Ishihara, K., Tatsuoka, F., and Yashuda, S. (1975). Undrained strength and deformation of sand under cyclic stresses. *Soils and Foundations*, Vol. 15, No. 1, pp. 29-44.

Jamiolkowski, M., Ghionna, V.N., Lancellotta, R. and Pasqualini, E. (1988) New correlations of penetration tests for design practice, *Penetration Testing 88, ISOPT-1*, pp. 263-296, Balkema, Rotterdam.

Jamiolkowski, M., Lancellotta, R., and Lo Presti, D.C.F. (1994). Remarks on the stiffness at small strains of six Italian clays, *Proc. of the Symposium on Prefailure Deformation Characteristics of Geomaterials*, Vol. 2, pp.817-836, Balkema, Rotterdam.

Janbu, N. (1963). Soil compressibility as determined by oedometer and triaxial tests. *Proc. 3rd ECSMFE*, Wiesbaden, Vol. 1, pp. 245-251.

Kovari, K and Amstad, C. (1982). A new method of measuring deformations in diaphragm walls and piles. *Géotechnique*, 22, No. 4, pp. 402-406.

Lacasse, S. and Lunne, T. (1988) Calibration of dilatometer correlations, *Penetration Testing 88, ISOPT-1*, pp. 539-548, Balkema, Rotterdam.

Lee, K.L., and Seed, H.B. (1967). Drained strength characteristics of sands. *J. Soil Mech. Found. Div.*, ASCE, Vol. 99, SM6, pp. 117-141.

Mayne, P.W. (1991) Determination of OCR in clays by piezocone tests using cavity expansion and critical state concepts, *Soils and Foundations*, Vol. 31, No. 2, pp. 65-76.

Mayne, P.W. and Rix, G.J. (1993) G_{max}-q_c relationships for clays. *Geotechnical Testing Journal*, Vol. 16, No.1, pp. 54-60.

Mayne, P. and McGillivray, A. (2002). Report of seismic flat dilatometer and seismic piezocone tests. Treporti Embankment Study, *GTRC Project E20-851*, Georgia Institute of Technology.

Marchetti, S. (1975) A new tn-situ test for the measurement of horizontal soil deformability. *Proc. Am. Soc. Civ. Engrs. Spec. Conf. In-situ measurements of soil properties*, Vol. 2, pp. 255-259.

Marchetti, S. (1980) In situ tests by flat dilatometer, *J. of Geot. Eng.*, ASCE, Vol. 106, GT3, pp. 299-321.

Marchetti, S. and Crapps, D.K. (1981) Flat Dilatomer Manual, *Schmertmann and Crapps*, Inc. Gainsville.

Marchetti, S., Monaco, P., Calabrese, M. and Totani, G. (2004). DMT- predicted vs. measured settlements under a full-scale instrumented embankment at Treporti (Venice, Italy). *ISC-2*, Porto.

Olsen, R.S. and Farr, J.V. (1986) Site characterization using the Cone Penetrometer Test, *Proc. Int. Symp. IN-SITU '86*, Blacksburg (USA), pp. 854-868.

Olson, R.E. (1977). Consolidation under time dependent loading, *J. of Geot. Eng. Div.*, ASCE, GT1, pp. 55-60.

Ricceri, G., Simonini, P., and Cola, S. (2002). Applicability of piezocone and dilatometer to characterize the soils of the Venice lagoon. *Geotechnical and Geological Engineering*, 20, pp. 89-121, Kluwer Academic Publishers.

Robertson, P.K. and Campanella, R.G. (1983). Interpretation of cone penetration test, Part I: Sands. *Can. Geotech. Journal*, Vol. 20, No. 4, pp. 718-734.

Robertson, P.K. (1990) Soil classification using the cone penetration test, *Can. Geotechnical Journal*, Vol. 27, pp. 151-158.

Rowe, P.W. (1973). Soil mechanic aspect of the cores of the deep borehole VE 1 in Venice. A critical analysis and recommended future investigations. *Technical Report N. 57. National Research Council*, Venice.

Schmertmann, J.H. (1983) Revised procedure for calculating K_o and OCR from DMTs. *DMT Workshop*. Gainsville.

Senneset, K., Janbu, N. and Svano, G. (1982) Strength and deformation parameters from cone penetration tests. *Proc. 2nd Eur. Symp. on Penetration Testing*, Amsterdam, Vol. 2, pp. 863-870, Balkema, Rotterdam.

Shibuya, S., and Tanaka, H. (1996). Estimate of elastic shear modulus in Holocene soil deposits. *Soils and Foundations*, Vol. 36, No. 4, pp. 45-55.

Simonini P. and Cola S. (2000). Use of piezocone to predict maximum stiffness of Venetian soils. *Journal of Geotechnical and Geoenvironmental Engineering*, ASCE, Vol. 126, No. 4, pp. 378-381.

Tatsuoka, F., and Ishihara, K. (1974). Drained deformation of sand under cyclic stress reversing direction. *Soils and Foundations*, Vol. 14, No. 3, pp. 51-65.

Teh, C.I. and Houlsby, G.T. (1991) An analytical study of the cone penetration test in clay. *Geotéchnique* 41, No. 1, pp. 17-34.

Torstensson, B.A. (1975) Pore pressure sounding instrument. *Proc. Am. Soc. Civ. Engrs. Spec. Conf. In-situ measurements of soil properties*, Vol. 2, pp. 48-54.

Torstensson, B.A. (1977) The pore pressure probe. Nordiske Geoteknisk Mote, Oslo, Paper N. 34.

Totani, G, Marchetti, S, Monaco, P. and Calabrese, M. (1999) Impiego della prova dilatometrica (DMT) nella progettazione geotecnica. *XX National Geotechnical Congress*, Parma, Italy, Vol. 1, pp. 301-308.

Viggiani, C. (1967). Su alcuni problemi di teoria della consolidazione. *Fondaz. Polit. Mezzogiorno*, Quad. 29.

Wood, D., Belkeir, K., and Liu, D.F. (1994). Strain softening and state parameter for sand modelling. *Géotechnique*, Vol. 44, No. 2, pp. 335-339.

Wroth, C.P., and Bassett, R.H. (1965). A stress-strain relationship for the shearing behaviour of sand, *Géotechnique*, Vol. 15, No. 1, pp. 32-56.

Wroth, C.P. (1984) The interpretation of in situ soil tests. *Geotechnique*, Vol. 34, No. 4, pp. 449-489.

Yu, H.S. and Mitchell, J.K. (1998) Analysis of cone resistance: review of methods. *J. of Geotechnical and Geoenviromental Engineering*, ASCE, Vol. 124, No. 2, pp. 140-149.

Characterisation of soft sediments for offshore applications

M.F. Randolph
Centre for Offshore Foundation Systems, The University of Western Australia

Keywords: anisotropy, clay, consolidation coefficient, penetration testing, sensitivity, shear strength, strain rate, vane test

ABSTRACT: Offshore facilities are being developed in increasingly deep water, with consequent challenges in obtaining high quality samples for laboratory testing and a growing reliance on in situ testing. Offshore design is dominated by assessment of capacity or ultimate limit state, rather than deformations and serviceability, and hence the shear strength profile of the seabed sediments is the most critical design parameter. This paper reviews current methods for in situ determination of shear strength in soft seabed sediments, including interpretation of different types of test and the influence of rate effects, soil sensitivity, strength anisotropy and consolidation characteristics. Experience with different types of penetrometer drawn from a recent joint project between the Norwegian Geotechnical Institute (NGI) and the Centre for Offshore Foundation Systems (COFS) is illustrated and ideas presented for new developments in penetration testing techniques.

1 INTRODUCTION

Characterisation of offshore sediments for the design of foundation and anchoring systems represents an ever increasing challenge as hydrocarbon fields are developed in water depths now approaching 3000 m. In most cases, sediments encountered in water depths greater than a few hundred metres are relatively fine-grained and essentially normally consolidated from a geological standpoint. Shear strengths may be no more than 2 to 10 kPa at seabed level and increase with depth at gradients typically in the range 1 to 2 kPa/m, although higher gradients may occur in silty deposits. More complex stratigraphies will arise near shelf-breaks, or where major submarine slides have occurred, and other features, including rocky outcrops and gas hydrates, may also be encountered (Quirós & Little, 2003).

The difficulties in obtaining undisturbed soil samples for laboratory testing are exacerbated offshore by reduced control of the robotic sampling devices and much greater relief of total stress. Where gas is dissolved in the pore fluid, the stress relief will cause the gas to come out of solution, leading to fracturing of the sample on horizontal planes and the possible formation of voids within the sample. Saturation levels of gas may be measured using devices such as the BAT probe (Rad & Lunne, 1994), which may give early warning of potentially disturbed samples. Even without obvious disturbance, however, the presence of gas in the pore fluid can distort the response in undrained tests (Lunne et al., 2001).

The increased potential for sample disturbance necessitates greater reliance on in situ testing in order to characterise offshore sediments. The stratigraphic advantages of penetrometer testing are similar to onshore situations, but there is increased requirement for accurate determination of an appropriate (absolute) shear strength profile, which is fundamental to the geotechnical design of offshore facilities. A good illustration of the problem is given by Quirós & Little (2003) for a lightly overconsolidated clay profile in the Gulf of Mexico, where a design shear strength profile based on laboratory tests varied essentially linearly from 75 kPa at 20 m depth to 250 kPa at 120 m. In order to fit the profile of net cone resistance (which varied with depth in a distinctly non-linear manner), the cone factor was varied from around 17 at shallow depths to 11 at 120 m.

This example raises a number of questions. For example, how reliable were the different strength measurements (UU, Shansep triaxial, miniature vane and in situ vane)? Are there theoretical reasons to expect the cone factor to decrease with depth, as the overconsolidation ratio decreases? Can other characteristics of the clay, such as rigidity index or in situ stress ratio, be gleaned from the cone factor, and associated pore pressure parameter and friction ratio?

Ideally, in situ testing by means of field vane and different forms of penetrometer should provide the most reliable guide to *absolute* shear strengths (albeit some form of average among the different modes of shearing involved in the given test), while laboratory testing would give *relative* information on the strength anisotropy and other aspects such as stiffness and consolidation characteristics. In practice, however, in almost all cases field data are calibrated to the laboratory data by adopting a given cone factor or vane correction factor. The question then is what relative reliability should be assigned to each form of testing, and what range of factors is plausible for each type of in situ test.

An excellent review of in situ testing for offshore soils was provided by Lunne (2001) and there is little point in duplicating that review here. Instead, this paper will focus mainly on some of the new penetrometer probes that have started to be used quite widely around the world (Randolph et al., 1998; Hefer & Neubecker, 1999). Penetrometers such as the T-bar, ball or plate have a projected area that is 5 to 10 times that of the shaft containing the load cell. Plasticity solutions give the ratio of penetration resistance to the soil shear strength and thus, in principle, avoid the need for calibration against laboratory strength data at each site.

The 'holy grail' of in situ testing in the offshore field would be for an economic test that gave an accurate absolute measurement of the average shear strength relevant for the design of foundation and anchoring systems. However, the potential for the new types of penetrometer to provide such a gold standard is muddied (literally) by factors such as the strain rate dependency of shear strength and the gradual reduction in strength exhibited by lightly overconsolidated soils as they are sheared.

The paper will explore these issues, illustrating the potential for such probes to evaluate factors such as strain rate effects, sensitivity and also to assess consolidation characteristics of the soil.

2 RECENT OFFSHORE SI EQUIPMENT

Before taking a more detailed look at in situ methods of characterizing seabed sediments, a brief summary is presented here of recent developments in offshore equipment. Perhaps the most exciting of these is the so-called PROD (Portable Remotely Operated Drill) developed and operated by Benthic Geotech Pty Ltd.

The PROD is a fully independent sea-bed unit, connected to the support vessel via an electrical umbilical, that is able to take samples and carry out penetration tests (Figure 1). Carter et al. (1999) gave an overview of the original device, but its current capabilities include:

- operating in up to 2000 m water depth;
- rotary drilling and sampling;
- piston sampling, 44 mm diameter and maximum core length of 2.75 m, down to maximum depth of 125 m;
- cone penetrometer device, 36 mm diameter with 2 m pushes to maximum depth of 100 m, using acoustic transmission from the CPT to the drilling module;
- ball penetrometer (under development) with 60 mm to 70 mm ball diameter and similar capability to cone penetrometer.

(a) Launching PROD off the stern of vessel

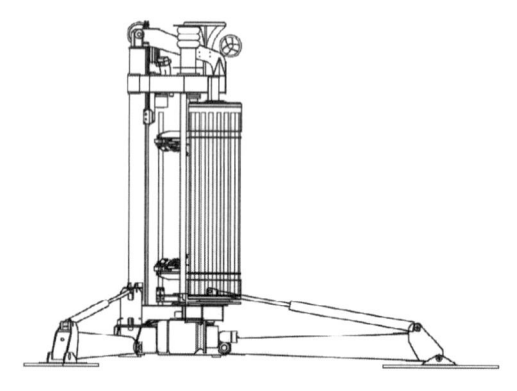

(b) Schematic of PROD after deployment

Figure 1 Portable Remotely Operated Drill (PROD)

The complete PROD weighs 10 tonnes in air, or 6 tonnes in water, and when collapsed can fit within a standard 6 m shipping container. Although it is still at a relatively early stage of development, it has already shown far-reaching potential for robotic deepwater investigations. Limitations of drilling from a surface vessel, with associated high costs and vulnerability to the weather, are reduced and the depth range for sampling and penetrometer testing exceeds that of other seabed frames or remote coring devices.

Other forms of remote seabed sampler have been described by Young et al. (2000) (Jumbo Piston Corer or JPC) and Borel et al. (2002) (STACOR sampler). These devices have a steel barrel, with typically a 100 mm diameter PVC liner. They penetrate the seabed under their own weight, and can retrieve samples of up to 20 to 30 m long. Recovery rates in excess of 90 % are achieved through the use of a piston that, in the case of the STACOR, is maintained at seabed level by a pulley system as the corer penetrates (Borel et al., 2002). Disturbance assessed from radiographs appears to be limited to the edges of the core, and strength measurements show similar normalised parameters to those obtained from conventional high quality sampling. For soft sediments, corers penetrating by self-weight have clear advantages in terms of the speed with which the samples can be obtained without the need for sophisticated drilling vessels, and in the continuous large diameter samples obtained.

Where pile foundations are anticipated, conventional drilling will generally be necessary since the soil characteristics must be investigated to great depths, typically up to 150 m below the seabed or even deeper if information is needed for installation of well conductors. Here, as described by Lunne (2001), developments have focused on sampling and testing tools that can be released on a wire line down the drill pipe, latching in at the base. Generally, such as in Fugro's XP system, data from a penetration or vane test will be recorded locally, and then retrieved with the instrument.

There will inevitably be some disturbance from the drilling operation and slight motion of the drill pipe due to only partial heave compensation. This disturbance may extend to 200 or 300 mm below the base of the hole. As such, penetrometer strokes of at least 1.5 to 3 m are generally adopted, thus minimising the proportion of the test through disturbed material. Equally, vane testing should not be attempted above 0.5 m below the base of the hole, and more typically 2 levels of testing, at 0.75 m and 1.5 m below the base, are achieved.

Although the geometry of the T-bar does not lend itself to operation below the base of a drill-pipe, Fugro are in the process of developing a ball penetrometer for wire-line deployment in soft sediments. This is likely to be around 80 mm diameter, attached to a standard 36 mm shaft, giving an area ratio (ratio of projected penetrometer area to shaft) of 5.

In most deepwater sites, where shallow anchoring systems are anticipated, seabed frames such as Fugro's Wheeldrive Seacalf (Peuchen, 2000; Lunne, 2001) provide the best approach (see Figure 2). The frame sits on the seabed and is thus independent of any motion of the vessel, and can also provide a continuous push for cone or other penetrometer, down

(a) Photograph of unit being prepared for launching

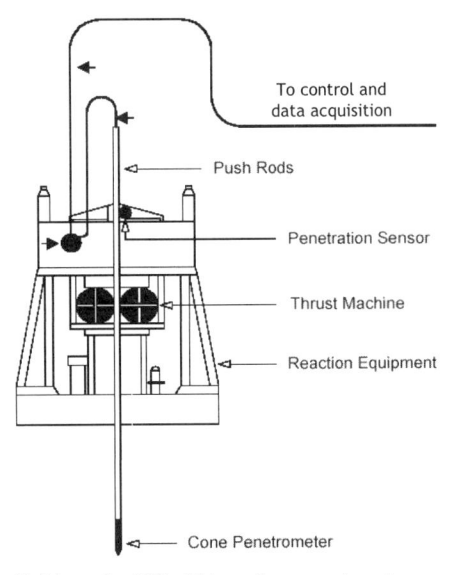

(b) Schematic of Wheeldrive and cone penetrometer

Figure 2 Schematic of Fugro's Wheeldrive Seacalf system

to depths of 30 to 60 m. The cone rods are pre-assembled and kept under tension from the vessel prior to penetrating the soil. Compressible studs may be mounted on the wheel-drives to provide a distributed grip on the cone rods, minimizing any slippage.

Figure 3 shows a comparison of profiles of net cone resistance obtained (a) through a wire-line deployment down a drill pipe, and (b) using a seabed frame. The latter profile shows less variability, and generally higher values of cone resistance; in particular it gives much better definition of the significant strength intercept at the seabed.

2.1 Full-Flow Penetrometers

Another advantage of a seabed frame is that there is less restriction on the use of alternative penetrometers, since these no longer have to pass through a drill pipe. Figure 4 shows alternative 'full-flow' penetrometers that have been used in research. These have been designed to give a projected area that is 10 times that of the cone shaft, and can be substituted for the cone merely by unscrewing the cone head and replacing it with the chosen device.

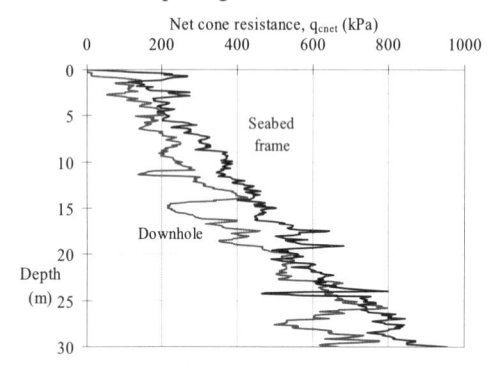

Figure 3 Comparison of downhole and seabed frame CPTs

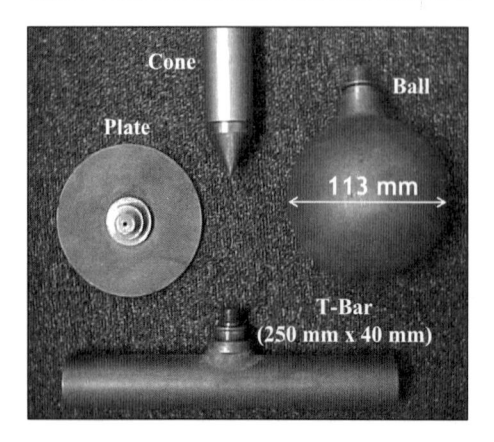

Figure 4 Alternative full-flow penetrometers with cone

The T-bar penetrometer was first developed as a laboratory device to give improved definition of the shear strength profile, and only later scaled up for field use (Stewart & Randolph, 1991, 1994). The first offshore T-bar penetrometer tests were conducted in Australian waters in late 1997 (Randolph et al., 1998) with a purpose built device incorporating its own load cell and also two pore pressure transducers within the cylindrical bar. This device was used 9 months later, again in Australian waters (Hefer & Neubecker, 1999). Subsequently, T-bar tests have been incorporated in several offshore site investigations from West Africa to the North Sea.

The axisymmetric ball penetrometer was introduced in order to reduce the potential for the load cell to be subjected to bending moments arising from non-symmetric resistance along the T-bar (Watson et al., 1998). The circular plate was an obvious extension of this idea, although the tendency for material to be carried down with the plate renders this a less attractive geometry than the ball. During extraction, the ball is also unaffected by any rotation of the rods (which would force a T-bar out of the remoulded zone created during penetration).

Recently a 78 mm diameter ball penetrometer attached to a 25 mm diameter shaft has been used offshore for the first time in conjunction with a seabed frame. Although the data are still proprietary, comparable resistance profiles from cone, T-bar and ball penetrometers were successfully obtained and showed excellent consistency. As mentioned previously, detailed plans for a wire-line ball penetrometer and also a similar device for the PROD are at an advanced stage.

A detailed discussion of the analysis and interpretation of full-flow penetrometers is presented later. However, it is useful to summarise the rationale behind their introduction, which rested on three main arguments:

(1) The measured penetration resistance requires minimal correction to provide net resistance, by contrast with potentially significant adjustments to the measured cone resistance to obtain a net value.

(2) Improved accuracy is obtained in soft soils due to the larger projected area; this gives a higher penetration force (particularly relative to the force on the load cell arising from hydrostatic water pressures of up to 20 MPa), so improved resolution and also reduced sensitivity to any load cell drift.

(3) Accurate plasticity solutions exist relating the net penetration resistance to the shear strength of the soil, within certain idealizations that are explored later in the paper.

Peuchen (2000) has provided a detailed discussion of the accuracy of cone resistance measurements in deepwater. In practice, minor corrections to the load cell zero may need to be considered, due to zero drift or bending effects for a T-bar. A possible way of assessing this is to consider the ratio of extraction resistance to penetration resistance. This is illustrated for T-bar resistance profiles in Figure 5, where a 20 kPa offset has been applied in the lower plot. This offset gives (arguably) a better fit with the cone profile, and a more rational pattern for the ratio of extraction resistance to penetration resistance.

It has not been customary to log data during cone extraction, and indeed this is currently not possible in most downhole offshore cone systems. However, it can be a useful check in soft sediments that the adjustments from measured to net penetration resistance have been made correctly. Figure 6 shows an example of this, from a site in Western Australia (see Chung & Randolph, 2004). The cone data have been corrected in the normal way (Lunne et al., 1997), first adjusting the measured resistance, q_c, to allow for pore pressure acting on the back of the cone shoulder, and then subtracting the overburden stress, σ_{vo}, from the resulting total cone resistance, q_t, to give

$$q_{cnet} = q_t - \sigma_{vo} \qquad (1)$$

For the T-bar or ball penetrometers, it has been suggested that the equivalent correction should be

$$q_{net} = q_m - [\sigma_{vo} - u_o(1-\alpha)] A_s / A_p \qquad (2)$$

where α is the area ratio of the load cell (or pore pressure filter) as for the cone, u_o is the hydrostatic pore pressure prior to conducting the test and A_s and A_p are respectively the shaft and (projected) areas of the penetrometer. As shown by Chung & Randolph (2004), the correction is essentially negligible (at least in comparison with that for the cone), largely because the ratio A_s/A_p is typically 0.1 to 0.2.

However, it should also be noted that the correction given in Eq. 2 ignores complications due to the shaft penetrating behind the penetrometer, which will give rise to a differential total pressure across the upper and lower surfaces of the penetrometer. This is difficult to quantify (although represents a small positive correction), and in practice data from full-flow penetrometers have tended to be interpreted taking the measured resistance as the net value, which without any correction.

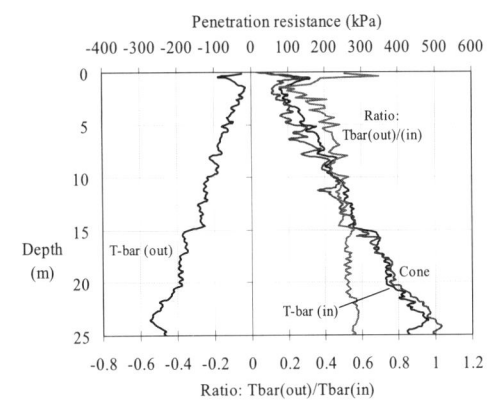

(a) Without any adjustment

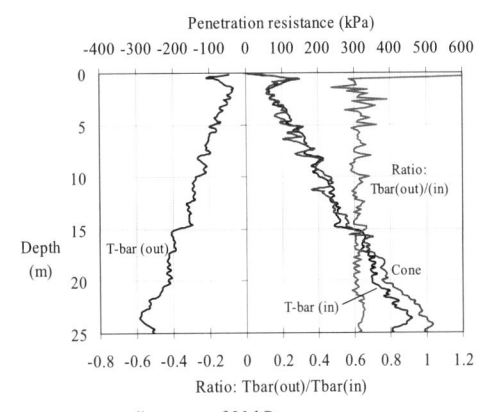

(b) After zero adjustment of 20 kPa

Figure 5 Effect of zero adjustment on T-bar resistance profiles

The CPT and T-bar data shown in Figure 6 are particularly interesting in respect of the ratios of extraction to penetration resistance. This ratio is relatively uniform for the T-bar, ranging between 0.6 and 0.67, but shows a marked trend to decrease with depth for the cone (ignoring results above 3 m and below 17 m, where the stratigraphy changes). This perhaps suggests either that the adjustments to the cone data are deficient in some way, or that other aspects of soil response, such as the rigidity index or certain aspects of the undisturbed structure of the clay, are varying strongly with depth. By contrast, Figure 7 shows data from centrifuge model tests on reconstituted kaolin, where the net cone resistance follows the T-bar resistance closely during both penetration and extraction.

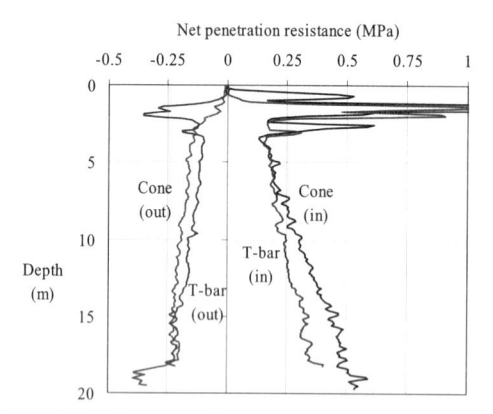

(a) Profiles of net penetration and extraction resistance

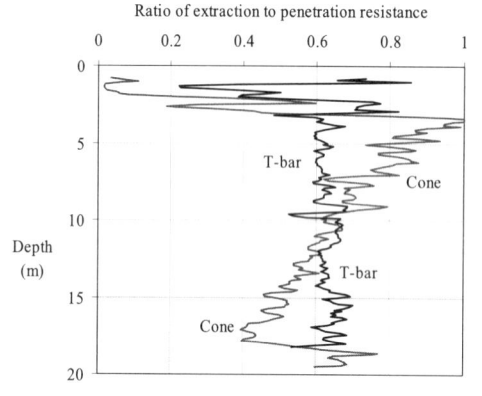

(b) Ratio of extraction to penetration resistance

Figure 6 Cone and T-bar resistance during penetration and extraction in a soft clay site in Western Australia

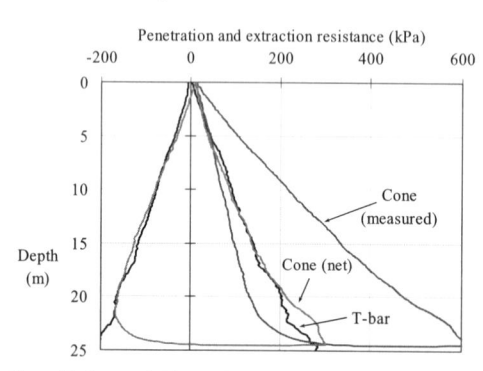

Figure 7 Cone and T-bar resistances from centrifuge model tests (after Watson et al., 2000)

2.2 *In Situ Vane Test*

In situ vane tests have been conducted routinely offshore since the mid 1980s (Geise et al., 1988; Johnson et al., 1988; Kolk et al., 1988; Quiros & Young, 1988; Young et al., 1988). During that time the equipment has been continuously improved, both in respect of wire-line operations and deployment from seabed frames. In soft clays, the (mostly) standard procedures involve a vane of 65 mm diameter and 130 mm high, with net area ratio of around 9 to 11 % and perimeter ratio of 3 to 4 % (Cerato & Lutenegger, 2004). The vane is pushed to the required depth at a rate of 20 mm/s and then left for 5 minutes before being rotated at 0.1 or 0.2 °/s (the onshore standard being 0.1 °/s or 6 °/min).

The strength measured in a vane test is sensitive to the precise testing procedure, in particular the delay between insertion and testing, and the rotation rate (Chandler, 1988). Standardising the test helps in consistency of results in a given soil type but, as is discussed in more detail later, the various effects of disturbance, partial consolidation (leading to strength recovery) and enhanced strength due to the high strain rates compensate each other to some extent but may not prove consistent among soils of different plasticity and consolidation characteristics. Current practice, however, is not to apply any correction factor to offshore vane strengths.

3 THEORETICAL SOLUTIONS

It is useful to summarise here the theoretical solutions of the different shapes of penetrometer and for the vane test. These solutions are based on simple elastic perfectly-plastic idealizations of soil response, and results are given here for a Tresca failure criterion. For all penetrometers, the relevant resistance factor, N, is defined as the ratio of net penetration resistance to shear strength. Traditionally, the factor for the cone has been labelled N_{kt}.

3.1 *Cone Penetrometer*

A number of different theoretical solutions for the cone penetrometer have been presented, based either on a strain path approach (Teh & Houlsby, 1991) or on some form of large displacement finite element analysis (Yu et al., 2000; Lu et al., 2004). All solutions have shown that, even for a simple Tresca soil model, the theoretical cone factor is affected by the rigidity index, $I_r = G/s_u$ where G is the shear modulus and s_u the shear strength of the soil, by the in situ stress ratio, $\Delta = (\sigma_{vo} - \sigma_{ho})/2s_u$, where σ_{vo} and σ_{ho} are the in situ vertical and horizontal stresses, and by the roughness coefficient, α_c, where α_c is the ratio of limiting shear stress on the cone-soil interface to the shear strength.

The expression obtained by Lu et al. (2004), which is similar to that of Teh & Houlsby (1991), is

$$N_{kt} = \frac{q_{cnet}}{s_u} \approx 3.4 + 1.6\ln(I_r) - 1.9\Delta + 1.3\alpha_c \quad (3)$$

For a typical roughness coefficient of around 0.3, the theoretical cone factor varies from 10.2 to 13.9 for $100 \le I_r \le 300$ and $-0.5 \le \Delta \le 0.5$.

3.2 T-bar and Ball Penetrometers

A theoretical solution for the T-bar (or any cylindrical object) moving through the soil was first presented by Randolph & Houlsby (1984). At the time, this was considered an exact solution, but subsequently an error was found in the upper bound calculation, due to a region where the shear strain rate derived from the kinematic mechanism was of opposite sign to the shear stress from the lower bound solution. Defining $\Delta = \sin^{-1}(\alpha)$ where α is the friction ratio at the T-bar-soil interface, the lower bound solutions for the T-bar factor is

$$N_{Tbar} = \pi + 2\Delta + 4\cos\left(\frac{\pi}{4} - \frac{\Delta}{2}\right)\left[\sqrt{2} + \sin\left(\frac{\pi}{4} - \frac{\Delta}{2}\right)\right] \quad (4)$$

Although there is a belief that the lower bound solution may indeed be exact, at present this can only be demonstrated for the case of a fully rough interface.

The lower bound, and corresponding upper bound, solutions are based on a characteristic mesh that comprises involute curves 'unwrapping' from an evolute cylinder inside the T-bar, which then change to concentric circular arcs. Martin (private communication) showed that a very simple mechanism, comprising counter-rotating rigid crescents centred at a radius, λR (where R is the T-bar radius) on the horizontal axis of symmetry (Figure 8), gave much improved upper bounds for low values of α. The upper bound from this mechanism is given by

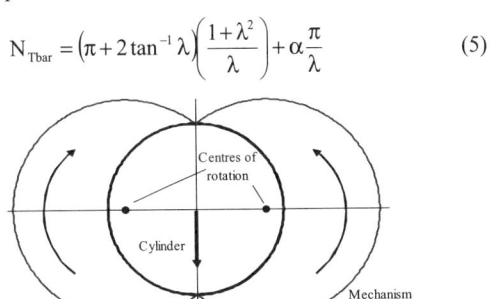

$$N_{Tbar} = \left(\pi + 2\tan^{-1}\lambda\right)\left(\frac{1+\lambda^2}{\lambda}\right) + \alpha\frac{\pi}{\lambda} \quad (5)$$

Figure 8 Rotational mechanism (after Dr Chris Martin)

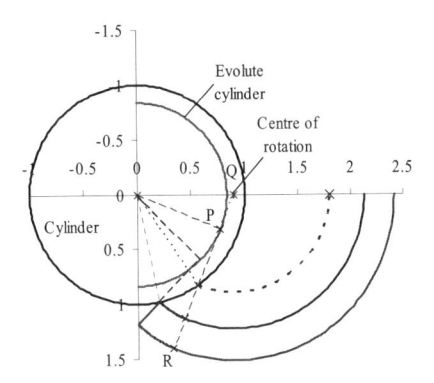

Figure 9 Combined involute and rotational mechanisms for plane strain T-bar and axisymmetric ball

The optimum upper bound, which is essentially indistinguishable from the lower bound for most values of α, involves a combination of a small wedge of non-deforming soil at the front and back of the T-bar, a region of the involute mechanism, and then a transition to the Martin-style rotational mechanism, as illustrated in Figure 9. This gives an upper bound of 9.20 for a perfectly smooth T-bar, compared with the lower bound of 9.14.

Lower and upper bound solutions for a spherical (ball) penetrometer were presented by Randolph et al. (2000). These gave bounds of 10.97 and 11.60 for a perfectly smooth ball, increasing to 15.10 and 15.31 for a fully rough ball. Again, the rotational mechanism of Figure 8 and the combined mechanism of Figure 9 provide improved upper bound for low values of α. Important differences between the plane strain cylinder and the axisymmetric ball are that in the latter case (a) the rotational region now contains internal shearing (due to hoop strains) and (b) the mechanism boundary is no longer a velocity discontinuity since the soil velocity on the bounding streamline is zero.

The theoretical solutions for the cone, T-bar and ball penetrometers are summarised in Figure 10. It may be seen that the range due to variations in interface friction ratio is around ±7 % for both T-bar and ball, although in practice friction ratios will rarely exceed about 0.3 during undrained penetration. The theoretical ball penetration resistance is 23 to 27 % greater than for the T-bar. Note also that the range shown for the cone factor is consistent with the lower limit of around 10 quoted by Aas et al. (1986) (relative to field vane tests) but falls well short of their upper limit of around 18. A possible implication is that the higher values reported for cone factors arise from comparative tests where disturbance has led to underestimation of the true shear strengths. In deepwater soft sediments, correlations for the cone resistance against laboratory strengths rarely yield cone factors in excess of 15.

3.3 Vane Test

For convenience, the classical relationship between torque and undrained shear strength is given here for the vane test. For a vane of height, h and diameter, d, (see Figure 13 later) the torque is given by

$$T = \frac{\pi d^3}{6}\left(1 + 3\frac{h}{d}\right)s_u \qquad (6)$$

For the usual aspect ratio, h/d, of 2 the contribution from the sides of the cylindrical shearing region is around 86 % of the total (and higher if allowance is made for non-uniform shear stress mobilization across the ends of the cylinder – Chandler, 1988).

4 ALLOWING FOR REAL SOIL BEHAVIOUR

The theoretical solutions presented in the previous section are based on a simple, rate-independent, perfectly plastic soil response. In practice, natural soils show significant rate-dependence, both for stiffness and strength, and the lightly overconsolidated sediments encountered in deep water offshore tend to be susceptible to disturbance, with sensitivities of around 3 and higher. In addition, the shear strengths measured by different apparatus in the laboratory reveal anisotropy, typically with $(s_u)_{tc} > (s_u)_{ss} > (s_u)_{te}$, where the subscripts tc, ss and te denote triaxial compression, simple shear and triaxial extension respectively.

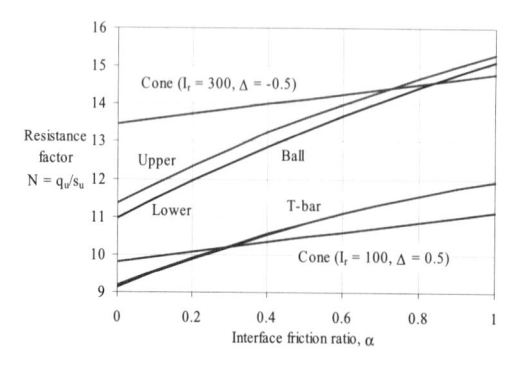

Figure 10 Theoretical factors for cone, T-bar and ball

4.1 Strength Anisotropy

Shear strengths deduced from field tests reflect an average shear strength and increasingly the shear strength measured in simple shear is being adopted as this average for correlation purposes and in design. This is partly because the simple shear strength is generally close to the numerical average from the three different types of test. A further practical reason, however, is that offshore samples are very expensive, and much less sample is required for a simple shear test than for a triaxial test.

The importance of strength anisotropy in offshore design has been emphasised (Lauritzsen & Schjetne, 1976), particularly in the detailed assessment of the response of gravity-base platforms under cyclic loading (Andersen & Lauritzsen, 1988). However, in many applications different shearing modes compensate and the net effect is that capacity based on the average shear strength provides a sufficient (if marginally conservative) estimate of that computed from a more sophisticated treatment of anisotropy.

The effect of anisotropic strengths was explored by Zdravkovic et al. (2001) for a skirted foundation, 17 m diameter with 12 m deep skirts, subject to tensile loading at an angle, θ, to the vertical. The soil was modelled using the MIT-E3 model for clay, where the triaxial extension and simple shear strengths were 52 and 66 % of that in triaxial compression (average strength 73 % of triaxial compression strength). As shown in Figure 11, the failure envelope using anisotropic strengths was some 22 % smaller than that using isotropic strengths equivalent to the triaxial compression profile. In this case, therefore, estimating the capacity using the average strength would have been conservative by 6 %.

Separate studies by Randolph (2000) and Aubeny et al. (2003) provide an instructive divergence in the calculated effects of strength anisotropy. The former paper suggested a reduction in capacity (relative to the average shear strength) by 11 % for a strength ratio, $(s_u)_{te}/(s_u)_{tc}$, of 0.5, and 17 % for a strength ratio of 0.33. By contrast, the study by Aubeny et al. showed no reduction in capacity with decreasing strength ratio (again, relative to the average shear strength of the soil).

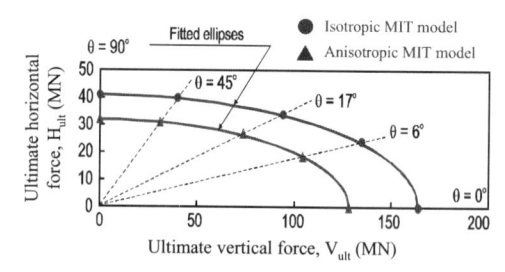

Figure 11 Failure envelopes for skirted foundation (after Zdravkovic et al., 2001)

An important shearing mode for 3-dimensional problems is that for shearing within a horizontal plane (such as occurs in a pressuremeter test). This strength is never measured in routine laboratory studies, and indeed would require careful hollow cylinder tests to quantify it. Randolph (2000) adopted a conservative assumption that this strength was equal to that in triaxial extension, which led to a marked reduction in caisson capacity with increasing strength anisotropy. If the alternative assumption was made (as by Aubeny et al., 2003) that this strength was equal to the average value, then the effect of strength anisotropy disappeared.

The above discussion is relevant to field results obtained with the T-bar and ball penetrometers. As noted from centrifuge model test data (Watson et al., 1998) and from recent field results (DeJong et al., 2004; Chung & Randolph, 2004), penetration resistances measured with the ball appear similar to, or even lower than, the T-bar resistance. Figure 12 shows this from the average of two ball tests and two T-bar tests conducted at a soft clay site in Western Australia. Experimental data are therefore inconsistent with the theoretical factors presented earlier, where the ball factor was more than 20 % greater than the T-bar factor.

Randolph (2000) commented that strength anisotropy may account, at least in part, for the discrepancy between theory and experimental data, with the theoretical ratio reducing to 1.07 for a strength ratio of 0.5. A numerical study conducted by Lu (2003) using the MIT-S1 model, with parameters chosen to simulate a calcareous silt, confirmed that the ratio of ball to T-bar resistance could drop to unity for a smooth interface ($\alpha = 0$). This was for a strength ratio, $(s_u)_{te}/(s_u)_{tc}$, of 0.69, but an unexpected feature of the soil model for the adopted parameters was that the predicted horizontal (or pressuremeter) strength was only 80 % of the triaxial extension strength.

Further study is needed of the effects of strength anisotropy, and in particular the value and influence of the pressuremeter-style strength. From the field data alone, however, it is clear that the conventional shape factor of 1.2 assumed for circular or square foundations, relative to strip foundations, may be significantly optimistic.

4.2 Strain Rate

There is extensive literature on the effect of strain rate on the measured or deduced shear strength of clays. Typically the peak strength increases by between 5 and 20 % per decade increase in shearing rate, which may be expressed as:

$$s_u = s_{u,ref}\left[1+\mu\log\left(\frac{\dot{\gamma}}{\dot{\gamma}_{ref}}\right)\right] \quad (7)$$

where μ is the rate of increase per decade and $s_{u,ref}$ is a reference strength measured at a strain rate of $\dot{\gamma}_{ref}$. Note that in other publications, the coefficient μ has been written variously as α or λ, but μ is adopted here to avoid confusion with alternative use of the other two symbols.

There are strong arguments (both from physical principles, Mitchell (1976), and to avoid problems with low strain rates) for preferring an alternative version of the above relationship, based on an inverse hyperbolic sine function, expressed as

$$s_u = s_{u0}\left[1+\mu'\sinh^{-1}\left(\frac{\dot{\gamma}}{\dot{\gamma}_o}\right)\right] \quad (8)$$

Taking $\mu' = \mu/\ln(10)$, this function reverts closely to Eq. 7 for strain rates in excess of the reference rate, but leads to rapidly decaying strain rate effects for strain rates below a threshold of $\dot{\gamma}_o$ so that, for strain rates below $0.1\dot{\gamma}_o$, a minimum strength is reached that is about 4 % lower than s_{u0}. The concept of a threshold strain rate below which the rate effect disappears has been noted by Sheahan et al. (1996).

In geotechnical design, the strain rates applicable to in situ tests, laboratory tests and operational conditions cover an extremely wide range, typically up to 6 to 8 orders of magnitude. As an example, suction caissons designed for deep water facilities in the Gulf of Mexico must contend with what are known as 'loop' currents that may last for several days or even weeks (Clukey et al., 2004). The (nominal) strain rate associated with base failure of a 5 m diameter caisson would thus be $\sim 10^{-5}$ %/s. This com-

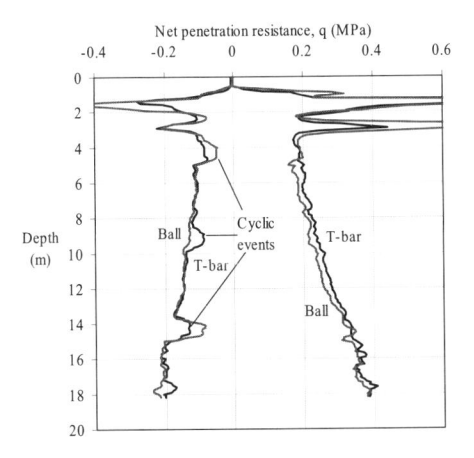

Figure 12 Comparison of T-bar and ball resistance profiles (after Chung & Randolph, 2004)

pares with typical strain rates in laboratory testing of 1 %/hour (or 0.3×10^{-3} %/s) and strain rates for in situ tests (vane, cone or full-flow penetrometer) that may reach 10^3 %/s.

With such differences in strain rate, we may wonder why we might expect laboratory strengths to match those deduced from in situ tests using factors derived from simple rate-independent perfectly plastic soil models. Indeed, strain-rate effects lie at the root of Bjerrum's vane correction factors, which range from 0.6 (high plasticity) to 1 (low plasticity). Aas et al. (1986) presented a detailed review of vane correction factors and concluded that the best correlation was with the (vane) strength ratio, $(s_u)_{fv}/\sigma'_{vo}$ with the correction factor decreasing from unity for a strength ratio of 0.2, down to 0.6 at a ratio of 0.6.

In offshore practice, however, vane shear strengths are rarely adjusted, even for strength ratios in excess of 0.4 (Kolk et al., 1988). A partial justification for this is the inevitable disturbance due to insertion of the vane. Cherato & Lutenegger (2004) have demonstrated very nicely how the degree of disturbance various almost linearly with the thickness of the vane blades. For a net area ratio of ~10 % (perimeter ratio of ~3 %), shear strains in the cylindrical shear zone around the vane caused by its insertion may be estimated by considering cavity expansion caused by a solid cylinder of diameter d_{eq} = 0.32d (where d is the vane diameter, see Figure 13).

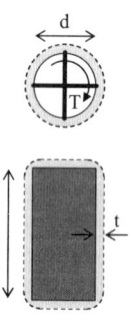

Figure 13 Schematic of vane and surrounding shear band

The maximum shear strain will be around 10 % (based on a cavity expansion strain of 5 %, and tensile circumferential strain of equal magnitude). Such strains will cause some reduction in strength for the lightly overconsolidated clays of interest. Regain in strength may be expected during subsequent consolidation. Radial consolidation solutions show that pore pressures start to change significantly at that radius for values of $T = c_v t/d_{eq}^2$ greater than about 0.5 ($c_v t/d^2 \sim 0.05$). For typical (soft clay) radial consolidation coefficients c_v in the range 10 to 30 m^2/yr (0.3 to 1 mm^2/s), this represents 200 to 700 s for d =
65 mm. Thus excess pore pressure changes in the shear zone will be relatively small for low permeability clays during the 5 minute wait period, but may be more significant in silty clays, perhaps leading to vane strengths that are too high.

Biscontin & Pestana (2001) presented results from a detailed study of the vane rotation rate (or peripheral velocity) on the deduced shear strength. They presented their results in terms of the peripheral velocity, relative to a standard velocity of 0.057 mm/s (equivalent to a 65 mm diameter vane rotating at 0.1 °/s). Their data from tests conducted in an artificial reconstitute soil, comprising 72 % kaolinite, 24 % bentonite and 4 % fly ash, showed a tendency for the rate of increase in vane strength with the logarithm of the peripheral velocity to increase as the velocity increased. In fact a power law relationship was found to give the best fit. If expressions equivalent to Eq. 7 were used to fit the data (with peripheral velocity replacing strain rate), then a µ value of 0.1 gave a lower bound fit, although the gradient rose to 0.2 or greater at peripheral velocities in excess of 1 mm/s.

For a rate-independent material it is not possible to define a strain rate associated with the (assumed) velocity discontinuity at the edges of the vane, without invoking some arbitrary shear band thickness. However, if the shear strength of the material increases with strain rate, there will be an optimum shear band thickness that minimises the internal plastic work as the material attempts to reduce the average shear strain rate.

Focusing just on the cylindrical section (which dominates the torque), the average shear strain rate in a shear band of width, t, (Figure 13) is

$$\dot{\gamma} = \frac{\omega d}{2t} \tag{9}$$

Assuming a rate-dependent shear strength as given by Eq. 7, the contribution to the torque, per unit height of vane, may then be written

$$\Delta T = \Delta T_{ref}\left[1+\mu \log\left(\frac{\omega d}{2t\dot{\gamma}_{ref}}\right)\right]\left(1+\frac{t}{d}\right) \tag{10}$$

where $\Delta T_{ref} = \pi d^2 s_{u,ref}/2$.
Differentiating this with respect to t, the minimum torque is found for a shear band thickness of

$$\frac{t}{d} = \frac{1}{\ln(10)\left[\frac{1}{\mu}+\log\left(\frac{\omega d}{2t\dot{\gamma}_{ref}}\right)\right]-1} \tag{11}$$

which converges very rapidly. As an example, for the standard vane rotation rate of 0.1 °/s and diameter of 65 mm, a rate coefficient of $\mu = 0.1$ and reference strain rate of 1 %/hour (typical laboratory rate), lead to $t/d = 0.032$ (t ~ 2 mm) and $\Delta T = 1.44 \Delta T_o$. The shear strain rate across the shear band is then 2.7 %/s, which is 10^4 times the reference rate.

The above analysis is over-simplified, as equilibrium is only satisfied approximately (in terms of a balance of internal and external work). A more rigorous analysis, based on the hyperbolic sine relationship of Eq. 8, allows deduction of a shear strain distribution around the vane that maintains torsional equilibrium. This may be expressed as

$$\dot{\gamma} = \frac{dv}{dr} = \dot{\gamma}_o \sinh\left(\frac{Ar_{vane}/r - 1}{\mu'}\right) \quad (12)$$

with v the circumferential velocity and A given by

$$A = 1 + \mu' \sinh^{-1}\left(\frac{(dv/dr)_{r=d/2}}{\dot{\gamma}_o}\right) \quad (13)$$

The shear strain rate at the vane edge may then be adjusted in order to satisfy a boundary condition at the outer edge of the shear band, where the velocity v and strain rate dv/dr must both decay to zero. The resulting pattern of strain rates is compared with that from the simple analysis in Figure 14 for the given set of parameters. As might be anticipated, the approximate analysis gives a sensible average shear strain rate and nominal shear band thickness.

Perhaps more surprising is the quite close agreement in magnitudes of $\Delta T/T_{ref}$ for a given value of μ from the two analyses. Figure 15 compares the variation of $\Delta T/\Delta T_{ref}$ with vane rotation rate and rate coefficient, μ, taking both $\dot{\gamma}_{ref}$ and $\dot{\gamma}_o$ as 1 %/hour. The slight increase in shear band thickness as the rotation increases has little effect on the rate of increase in torque for a given rate parameter. As noted earlier, for the standard vane rotation rate of 0.1 °/s and typical rate parameters, the torque is enhanced by 40 to 60 % compared with the reference torque.

This type of analysis is applicable to any shear band (or velocity jump) where broadening the band would lead to increased internal plastic work, either because of the greater volume of soil involved, or a less than optimum mechanism. An obvious example is the axial loading of a pile, where a similar shear strain distribution to that shown in Figure 14 would arise. Interestingly (and contrary to what was assumed by Biscontin & Pestana (2001)), it is the rotation rate of the vane, rather than the peripheral velocity, that controls the rate dependency of the torque. For the pile case, the rotation rate (expressed in radians/s) should be replaced directly by the axial displacement rate normalised by the pile radius, so that

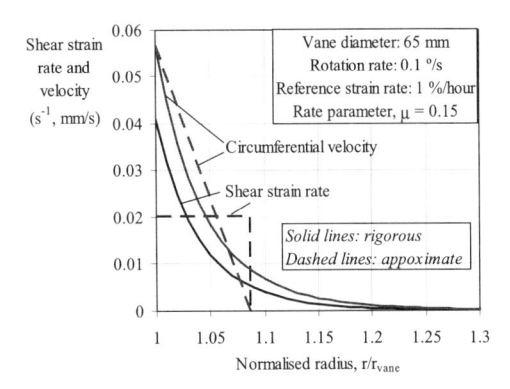

Figure 14 Profiles of circumferential velocity and strain rate for simple and rigorous analyses

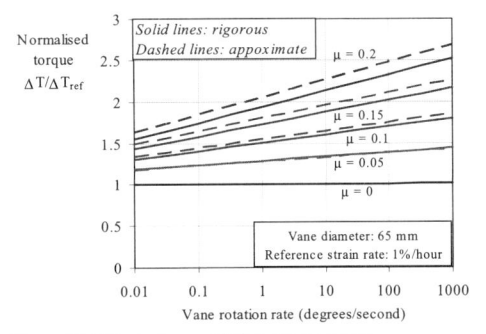

Figure 15 Effect of vane rotation rate on torque

for a 2 m diameter offshore pile, a displacement rate of 1.7 mm/s would lead to the same enhancement of shaft capacity as for a vane rotated at 0.1 °/s.

The above analysis for the cylindrical shear surface around the vane cannot be extended easily for shearing on the top and bottom surfaces of the vane, since the shear band width can be increased without penalty (apart from at the edges). This implies that correction factors for low aspect ratio vanes might be lower than for a standard vane. This agrees with results presented by Watson et al. (2000), who showed that for a vane of $h/d = 1.5$, tests at different rates led to a rate coefficient (μ) of 0.15 to 0.2, but that for a given rate the closest agreement with strengths from T-bar tests (which in turn were calibrated to laboratory simple shear tests) was for a very low aspect ratio vane with $h/d = 0.33$.

A further finding by Biscontin & Pestana (2001) was that the residual vane strength was essentially independent of rotation rate. By contrast, Watson et al. (2000) found that the residual vane strength decreased with increasing rotation rate. These conclusions may be complicated by partial consolidation occurring at standard rotation rates, with increasingly undrained conditions prevailing at higher rates of rotation.

The effect of shearing rate on the stress ratio mobilised under residual conditions has been studied extensively at Imperial College, with key results summarised by Fearon et al. (2004) and Lemos & Vaughan (2004). They found that both 'positive' and 'negative' rate effects could occur under residual conditions, with the former confined to soils with clay content in excess of 30 to 40 %, but with negative rate effects occurring for silty soils when sheared under fully saturated conditions.

The effect of strain rate on the T-bar resistance has been explored in the NGI-COFS project on characterisation of deep water sediments, the results from which will be presented in a forthcoming paper (Randolph et al., 2005). The mechanism associated with T-bar penetration includes both velocity jumps (or concentrated shear bands) and regions of diffuse shear. However, since this is a moving boundary problem, with the T-bar advancing at a penetration rate, v_p, the magnitude of the strain jump across the velocity discontinuity is finite. This may be visualised by imagining the soil approaching and moving through the T-bar mechanism (with the T-bar stationary). For a given velocity jump, Δv, and angle, θ, between the discontinuity and the approaching streamline, the strain jump is equal to $\Delta v/v_p\cos\theta$.

Thus, for the purely rotational mechanism shown in Figure 8 with centres of rotation at λR, the strain jump may be expressed as

$$\gamma = \frac{\sqrt{1+\lambda^2}}{\lambda \cos\theta} \tag{14}$$

This remains finite except at the very edge of the mechanism. However, while the strain increment is finite, the strain rate is infinite unless a shear band of finite thickness, t, is invoked.

In order to quantify the strain rates in the T-bar mechanism, the simplest approach is to compare the magnitudes of velocity jumps in the T-bar and vane tests, and to invoke a relative strength on the T-bar discontinuities given by

$$s_u = s_{uvane}\left[1+\mu\log\left(\frac{\Delta v_{Tbar}}{\Delta v_{vane}}\right)\right] \tag{15}$$

where Δv_{vane} is 0.057 mm/s (based on the standard rotation rate of 0.1 °/s and diameter of 65 mm).

For the T-bar, the average velocity jump for the rotational mechanism of Figure 8 is

$$\Delta v_{Tbar} = \frac{\sqrt{1+\lambda^2}}{\lambda}v_p \tag{16}$$

which is ~32 mm/s for the standard penetration rate of 20 mm/s (taking $\lambda = 0.8$). Thus the velocity jump in the T-bar mechanism is around 500 times that for a standard vane test. For a typical rate parameter, μ, in the range 0.1 to 0.2, the shear strength mobilised along the discontinuity in the T-bar mechanism should be 27 to 54 % greater than the vane strength.

A similar analysis as undertaken for the vane may be applied to the rotational discontinuity in the T-bar mechanism. As noted previously, rather than compare velocities directly (as in Eq. 15), it is more appropriate to compare velocities normalised by the radius of the discontinuity. However, these are similar in vane (32.5 mm) and T-bar (from the mechanicsm in Figure 8, the discontinuity radius is ~26 mm for a 40 mm diameter bar) and the ratio of normalised velocities is then ~700; this is not very different from the ratio of 500 for the absolute velocities.

Essentially, rate effects for the T-bar may be gauged from Figure 15, but taking an equivalent rotation rate of 700 times 0.1 °/s, or 70 °/s. For rate parameters in the range 0.1 to 0.2, the resulting 'nominal' shear band thickness is 5 to 8 % of the discontinuity radius (i.e. 1.3 to 2 mm) and the mobilised shear strength would be 1.7 to 2.5 times the reference shear strength.

There are a number of simplifications in the above discussion, but the key point is that the strain rates in regions of most concentrated distortion in the T-bar mechanism are some 3 orders of magnitude greater than in a vane test, and potentially 7 orders of magnitude greater than in a standard laboratory test. By all accounts, therefore it might be expected that the theoretical T-bar factors of around 10 to 11 should be too small, by a factor of up to 2, in relating penetration resistances measured in the field to shear strengths measured in the laboratory.

Whereas the major portion of plastic work in the T-bar mechanism occurs (at least theoretically) on discontinuities in the velocity field, the reverse is true for the ball mechanism. For low interface friction ratios, the simple rotational mechanism of Figure 8 is within 1 % of the optimum solution obtained from the general mechanism of Figure 9. In the rotational mechanism there is no velocity discontinuity for the ball (apart from on the ball surface itself), as the velocity on the bounding surface is zero.

The mechanism therefore lends itself to assessing, in principle, the effects of strain rates. Adopting a shear strength that follows Eq. 7, with a reference shear strain rate of 1 %/hour, it is found that for a typical interface friction ratio of $\alpha \sim 0.2$ the ball resistance varies according to

$$q_{ball} \approx q_{ball0}(1+4.7\mu) \quad (17)$$

where q_{ball0} is the resistance for rate-independent material; this relationship is based on a ball diameter of 100 mm, penetrating at 20 mm/s.

This result indicates that the average strain rate within the ball mechanism is just under 5 orders of magnitude higher than the reference laboratory strain rate (compared with about 7 orders of magnitude for the T-bar). This difference in average strain rate in the two types of full-flow penetrometer may be a contributing factor in the relative magnitudes of T-bar and ball resistances measured in the field.

5 STRAIN SOFTENING

The effect of high strain rates in the soil surrounding a penetrometer must be considered in conjunction with the absolute magnitude of strain, and the gradual reduction in strength as soil passes through the mechanism. This may be achieved by adopting a version of the strain path method, but incorporating this within the postulated upper bound mechanism (Einav & Randoph, 2004).

Figure 16 shows an example of the streamlines of soil flowing through the T-bar mechanism. The streamlines are derived by superimposing the upward flow of soil (or downward movement of the T-bar) on the velocities from the mechanism shown previously. The overall dissipation is calculated by integrating up each streamline, allowing for the volume of material (or current separation of the streamlines) and the current shear strength of the material (adjusting for strain rate and cumulative softening).

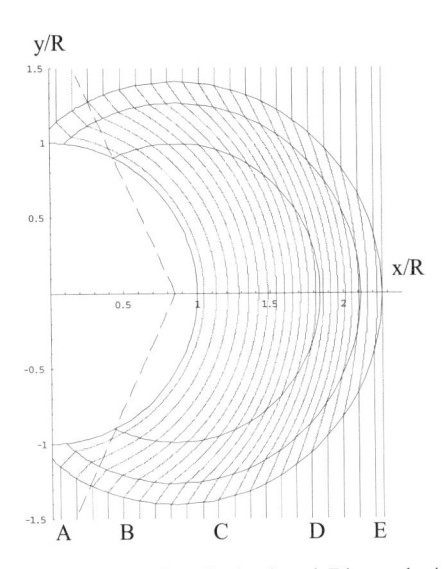

Figure 16 Streamlines flowing through T-bar mechanism optimised for α = 0.2 (Einav & Randolph, 2004)

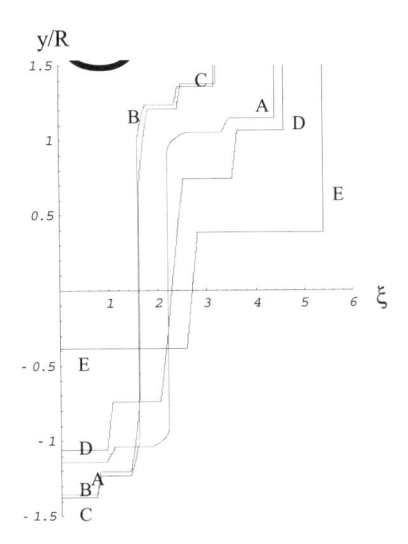

Figure 17 Cumulative strain for soil flowing through T-bar mechanism of Figure 16 (Einav & Randolph, 2004)

A simple isotropic softening law was adopted by Einav & Randolph (2004), of the form

$$\delta(\xi) = s_{us}/s_{ui} = \delta_{rem} + (1-\delta_{rem})e^{-3\xi/\xi_{95}} \quad (18)$$

where s_{ui} is the initial shear strength and δ is a degradation parameter, increasing from an initial value of 0, to a final value of δ_{rem} representing the fully remoulded condition. The accumulation of degradation is controlled by the parameter, ξ, which is the cumulative (plastic) shear strain defined as

$$\xi = \int_t |\dot{\gamma}_{max}|dt \quad (19)$$

where $\dot{\gamma}_{max}$ is defined according to Tresca or von Mises yield criterion as

$$\dot{\gamma}_{max} = 2\dot{\varepsilon}_{max} \quad \text{(Tresca)}$$
$$\dot{\gamma}_{max} = \sqrt{2\dot{\varepsilon}_{ij}\dot{\varepsilon}_{ij}} \quad \text{(von Mises)} \quad (20)$$

The final parameter in Eq. 18 is ξ_{95}, which represents the cumulative shear strain for 95 % of the shear strength degradation (since $e^{-3} \sim 0.05$). Figure 17 shows cumulate strain, ξ, for five different streamlines passing through the mechanism. The final cumulative strain is similar for all streamlines within the main body of the mechanism, although streamlines close to the T-bar (A) or close to edge of the mechanism (E) undergo greater strain. The strain jump tends to infinity at the edge of the mechanism, but the length of discontinuity there tends to zero, so the plastic work remains finite.

For low interface friction ratios, the average cumulative plastic strain throughout the mechanism is around 400 %. This is useful in assessing the average strength during first passage of the T-bar, and also further degradation caused by subsequent cycling of the T-bar up and down. Indeed, cyclic T-bar tests offer a useful way of calibrating the model adopted by Einav & Randolph (2004).

Figure 18 shows results from a cyclic T-bar test carried out at a soft clay site in Western Australia, while Figure 19 shows a calibration of the proposed degradation model against the measured degradation. The best fit corresponds to a sensitivity $(1/\delta_{rem})$ of 5, and a cumulative strain for 95 % degradation of 20 (2000 %). Note that the final resistance ratio is greater than δ_{rem} since the resistance has been normalised by that for initial penetration of the T-bar, during which degradation occurs. For the model parameters here, the average strength during first passage of the T-bar is 79 to 92 % of the peak strength as ξ_{95} varies between 20 and 50.

It is possible to combine the two (partly compensating) effects of high strain rates and gradual softening in order to arrive at an estimate of what T-bar factor might be expected in the field. At this stage, the strength increase due to high strain rates has been modelled using Eq. 15, since the T-bar mechanism is dominated by localised shear bands. As such, the resulting T-bar factor will be that appropriate for comparison with results from vane tests, and not necessarily laboratory strengths.

Figure 20 shows the net effect on T-bar resistance for different combinations of rate parameter, μ, and degradation parameter, ξ_{95}. The curves are based on a soil with sensitivity, 5 (hence $\delta_{rem} = 0.2$). It may be seen that for rate parameters in the range 0.1 to 0.2, the T-bar factor varies from 10.2 to 12.5 for rapidly softening soil ($\xi_{95} = 20$), and 11.7 to 14.3 for very gradually softening soil ($\xi_{95} = 50$).

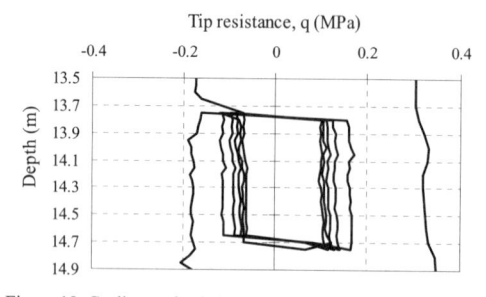

Figure 18. Cyclic test for field T-bar Test 3 at depth 14.3 m (Chung & Randolph, 2004)

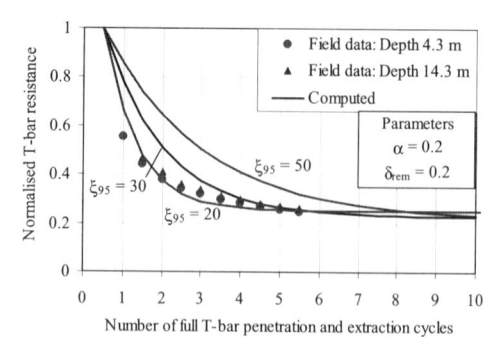

Figure 19. Calibration of degradation model from cyclic T-bar tests

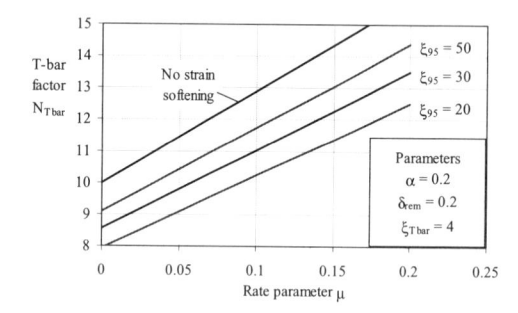

Figure 20 Combined effect of strain rate and strain-softening for T-bar (Einav & Randolph, 2004)

It is of interest to compare the shape of streamlines and the pattern of strain accumulation for the ball mechanism. Although the optimum mechanisms for a given interface friction ratio will be different for the plane strain and axisymmetric problems, for ease of comparison Figures 21 and 22 show streamlines and cumulative shear strain for identical mechanism parameters as for the T-bar in Figures 16 and 17.

A key difference in the shape of the streamlines is a greater decrease in curvature towards the edge of the mechanism, since the velocity field decays inversely with distance from the ball axis. An even stronger difference is evident from the shear strains accumulated along the streamlines A to E. For the ball, there is a monotonic decrease in strain magnitude for successive streamlines moving away from the ball. The cumulative shear strain is very large immediately adjacent to the ball, but the average shear strain throughout the whole mechanism is only between 200 and 250 % for low values of interface friction ratio, compared with ~400 % for the T-bar.

It might be expected that the lower average shear strain for the ball mechanism would lead to more gradual softening during cyclic penetration compared to a T-bar. However, results reported by Chung

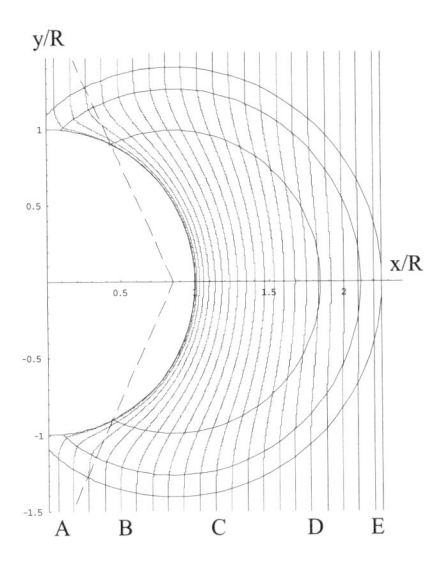

Figure 21 Streamlines flowing through ball mechanism with identical parameters to Figure 15 (Einav & Randolph, 2004)

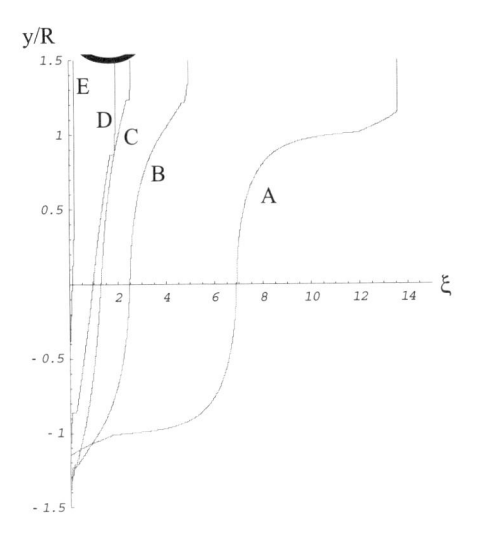

Figure 22 Cumulative strain for soil flowing through ball mechanism of Figure 21 (Einav & Randolph, 2004)

& Randolph (2004) show that the rate of softening is very similar, if anything with a slight tendency for the resistance to drop more rapidly for ball and plate compared to the T-bar. A possible explanation for this lies in the 3-dimensional nature of the strain fields for the axisymmetric penetrometers, and thus a greater degree of softening for a given magnitude of accumulated shear strain.

6 OFFSHORE DATA IN SOFT SEDIMENTS

Cone and T-bar data, together with vane and laboratory strength measurements, have been collected recently for a number of soft clay sites in a joint NGI-COFS project, the results of which will be presented by Lunne et al. (2005). Some of those data are presented here. Figures 23 to 28 show examples of cone, T-bar and shear strength data in four separate lightly overconsolidated deposits, two from offshore (Australia and West Africa) and two from onshore (Australia and Norway). These four sites cover material ranging from a calcareous clay (offshore Australia) with a plasticity index mainly in the range 27 to 38 %, to a highly plastic clay (offshore West Africa) with plasticity index typically 80 to 120 %.

The profiles of net cone resistance (Figure 23) and T-bar resistance (Figure 24) are remarkably similar with gradients of 19 to 23 kPa/m (cone) and 17 to 18 kPa/m (T-bar). Although the average ratio of T-bar to net cone resistance is mostly between 0.8 to 0.9, the ratio profiles (Figure 25) show some very interesting patterns.

With the exception of the onshore Norwegian site, the trend is for the ratio of T-bar to net cone resistance to decrease with depth, from a value of unity (or just greater) near seabed level, down to 0.8 to 0.95 at 30 m depth. By contrast, the ratio from onshore Norway lies mostly between 0.7 and 0.8 with an increasing trend below a depth of 10 m.

The theoretical cone and T-bar factors shown in Figure 10 would support an average ratio of ~0.8, with lower values for clays with high rigidity index, or high values of K_0 (negative value of the in situ stress parameter, Δ). T-bar resistances in excess of the cone resistance at very shallow depths suggest very low values of the rigidity index, while the trend in three of the sites for the ratio of T-bar to net cone resistance to decrease with depth implies increasing rigidity index or reducing in situ stress parameter, Δ. This would be consistent with a trend of rigidity index reducing with overconsolidation ratio, (Keavey & Mitchell, 1986).

Data from in situ vane strengths and laboratory simple shear tests for the four sites are shown in Figures 26 and 27. The vane strengths show significant scatter, with trendline gradients in the range 1.1 to 2.2 kPa/m. The simple shear strengths are more consistent, with gradients of 1.5 to 2.0 kPa/m. However, the absolute values of shear strength are extremely low, and laboratory strengths measured from offshore samples in the upper 10 to 15 m may well have been affected by disturbance.

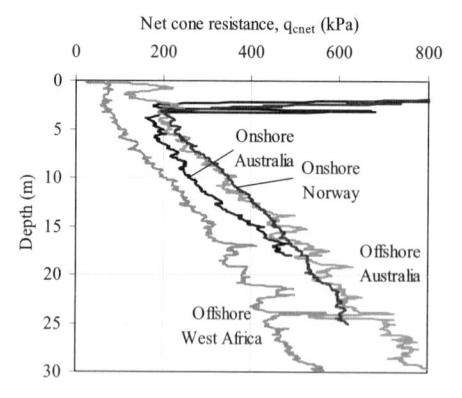

Figure 23 Comparison of net cone resistance profiles in lightly overconsolidated clays

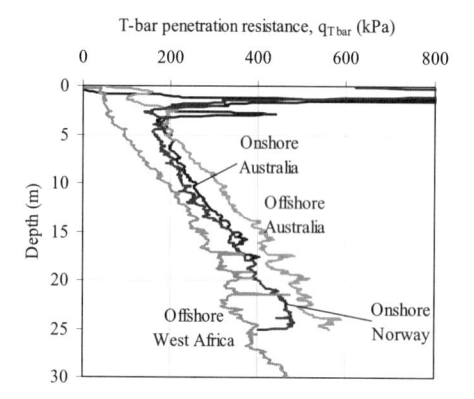

Figure 24 Comparison of T-bar resistance profiles in lightly overconsolidated clays

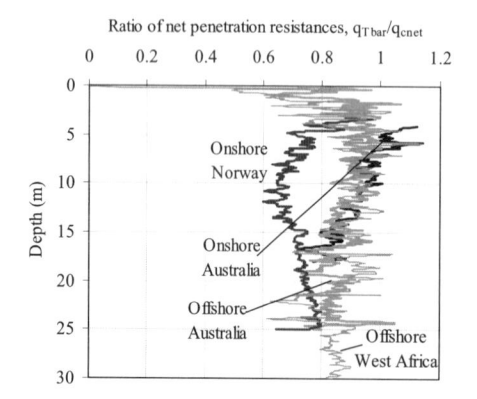

Figure 25 Ratio of T-bar and net cone resistances in lightly overconsolidated clays

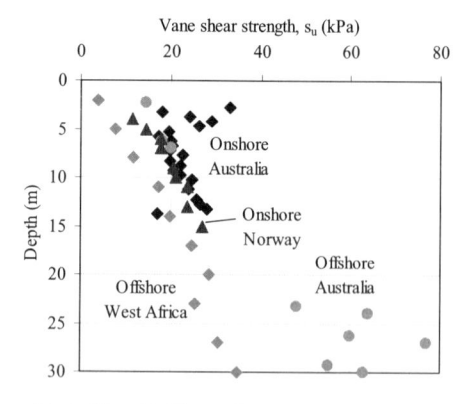

Figure 26 Profiles of field vane shear strengths in lightly overconsolidated clays

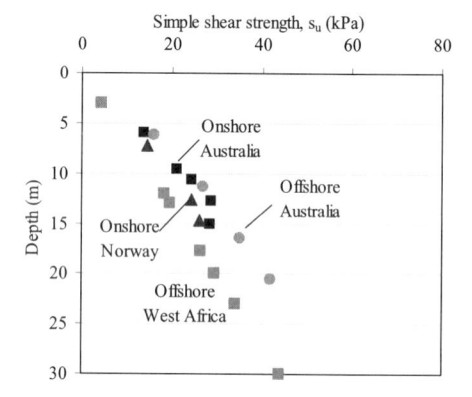

Figure 27 Profiles of shear strengths measured in simple shear in lightly overconsolidated clays

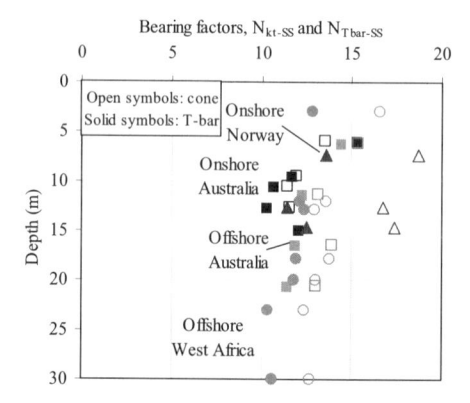

Figure 28 Cone (N_{kt}) and T-bar (N_{Tbar}) factors related to simple shear strengths in lightly overconsolidated clays

Profiles of cone factor, N_{kt}, and T-bar factor, $N_{T\text{-}bar}$, are shown in Figure 28. These do indeed show an overall trend to reduce slightly with depth, perhaps confirming underestimation of shear strengths at shallow depths. Of particular note in Figure 28 is the tendency for the cone factor to show a wider range among the four sites than the T-bar factor.

It is difficult to distinguish the effects of variability, due to minor stratigraphic changes or different degrees of sample disturbance, from any underlying trend for the cone and T-bar factors. One approach is to consider average factors for each site, within the depth range explored. These average values are summarised in Table 1.

Table 1 Average cone and T-bar factors for the four data sets

Site	$N_{kt\text{-}SS}$	$N_{kt\text{-}vane}$	$N_{Tbar\text{-}SS}$	$N_{Tbar\text{-}vane}$
Onshore Australia	12.6	11.2	11.9	10.9
Onshore Norway	17.6	16.3	12.5	11.6
Offshore Australia	13.7	12.3	12.4	11.3
Offshore W.Africa	13.6	15.0	12.2	12.7

Although the database is limited, the ratio of standard deviation in the average factor for each site to the average value for all four data sets is 2 % for the T-bar and 14 % for the cone (for simple shear strengths), and 7 % and 17 % respectively for vane strengths. This suggests a greater range for the cone factor compared with the T-bar factor.

Confirmation of this trend is needed urgently from additional offshore sites. Ultimately the aim is to understand what additional soil characteristics may be inferred from sites showing particularly high or low cone factors. By conducting parallel T-bar and cone penetration tests, there is the potential to interpret differences in penetration resistances quantitatively, perhaps to indicate differences in overconsolidaton (or yield stress) ratio or in rigidity index. Of particular importance is to understand the significance of any variation with depth of the ratio of T-bar and cone resistance.

7 PARTIAL CONSOLIDATION

So far the paper has concentrated on fine-grained soils where field tests may be considered undrained. However, many offshore deposits include silt zones, either in relative thick layers, or as thinner lenses interbedded amongst finer grained material. These zones are revealed by an increase in penetration resistance and decrease in excess pore pressure ratio, B_q. In order to interpret the penetration resistance in such intermediate soils, a first step is to assess the degree of drainage during the test.

The effect of soil consolidation characteristics has been studied numerically in a recent publication by Abu-Farsakh et al. (2003). Unfortunately, the transition points from undrained to partially drained, and then to drained behaviour, were expressed in terms of a normalised permeability, $K_N = k/v$ (where k is the hydraulic conductivity and v the penetration rate). This neglects both probe size and soil compressibility, both of which will affect the transition penetration rates.

A more appropriate non-dimensional parameter is

$$V = \frac{vd}{c_v} \quad (21)$$

where c_v is the coefficient of consolidation. Finnie & Randolph (1994) presented results from constant rate of penetration tests on shallow circular foundations and suggested transition points of V < 0.01 for drained response, and V > 30 to ensure undrained response.

In order to verify the transition point from undrained to partially drained response, the simplest approach is to conduct penetration tests at different rates in a single material, where the consolidation coefficient can be evaluated carefully from independent tests. Randolph & Hope (2004) presented data from model piezocone and T-bar tests conducted in normally consolidated kaolin on a centrifuge. The piezocone was 10 mm diameter (78.5 mm² area) while the T-bar was 5 mm diameter, and tests were conducted at rates ranging from 3 mm/s down to 0.005 mm/s.

Figure 29 shows the measured excess pore pressure ratio as a function of normalised penetration rate. The pore pressure was measured just behind the cone tip, at the u_2 position, and for comparison a curve attributed to Baligh (presented by Hight & Leroueil, 2003) for excess pore pressures at the cone tip (u_1 position) is also shown. Note that the Baligh curve was expressed as a variation of $\Delta u/\Delta u_{max}$ by Hight & Leroueil (ranging between 1 for undrained conditions and 0 for drained conditions) and has been adjusted on the vertical axis here in order to match the measured undrained B_q value of ~0.6.

Baligh et al. (1981) proposed the piezocone with pore pressure measurement at the cone tip, in order to maximise its sensitivity to stratigraphic changes. However, the alternative (u_2) position at the cone shoulder is more commonly adopted these days, in order to allow appropriate correction for pore pressures acting behind the cone. This is partly a matter of expediency, but also reduces problems associated with filter compression and wear. However, the results in Figure 29 suggest it results in some loss of

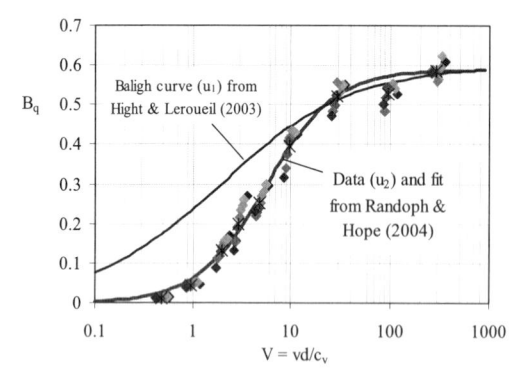

Figure 29 Effect of normalised penetration rate on excess pore pressure ratio

sensitivity to changes in consolidation characteristics (or penetration rate).

For the u_2 position, the transition from undrained to partially drained response starts at a normalised velocity of V ~ 100. However, a corresponding increase in penetration resistance appears to occur at a slightly lower rate, partly due to balancing the effects of reduced viscous enhancement of the cone resistance. This is shown in Figure 30 for cone and T-bar tests. The fitted curves through the data are of the form

$$\frac{q}{q_{ref}} = \left(a + \frac{b}{1+cV^d}\right)\left\{1 + \frac{\mu}{\ell n(10)}\left[\sinh^{-1}(V/V_o) - \sinh^{-1}(V_{ref}/V_o)\right]\right\}$$
(22)

where the first part represents the transition from undrained (or reference value, q_{ref}) to drained penetration resistance and the second part represents the effects of strain rate. Here, the reference rate has been taken as $V_{ref} = 100$, the rate at which viscous effects start to decay as $V_o = 10$, and the rate parameter as $\mu = 0.1$.

Two additional 'backbone' curves are shown in Figure 30(b), from unpublished T-bar tests by Watson & Suemasa (2000) undertaken at the University of Western Australia, and drum centrifuge tests reported by House et al. (2001). The latter tests may have been affected by small vibrations transmitted from the central turntable of the drum centrifuge, leading to additional excess pore pressures generated at the T-bar, and hence delayed effects of consolidation. Certainly the more recent test data of Randolph and Hope (2004) appear to lie more towards the Watson & Suemasa curve. The parameters in Eq. 22 are summarised in Table 2 for the different curves shown in Figures 29 and 30.

The net cone resistance shows a gradual transition for non-dimensionalised velocities of V < 30. The T-bar shows a minimum resistance for V ~10 but the resistance then doubles within an order of magnitude change (to V ~ 1). At the drained end of the spectrum, excess pore pressures decay essentially to zero by V ~ 0.1. The slowest tests reported by Randolph & Hope (2004) still showed increasing resistances, but the unpublished penetration data from Watson & Suemasa (2000) suggested that limiting (drained) resistances were reached also for V ~ 0.1.

Table 2 Parameters in Eq. 22 for curve fits in Figures 29, 30

Type of test and source	a	b	c	d
Cone Δu (u_1) (Baligh)	0	0.60	1.54	-0.65
Cone test, B_q	0	0.59	8.64	-1.26
Cone resistance, q_{cnet}	1	2.65	0.90	0.83
T-bar resistance, q_{Tbar}	1	2.77[a]	0.84	1.23
T-bar: House et al. (2001)	1	2.77[a]	2.47	1.30
T-bar: Watson & Suemasa	1	2.77	0.57	1.45

[a] Drained resistance of 3.77 times undrained resistance assumed, based on Watson & Suemasa (2000) (unpublished)

A number of observations may be made in relation to Figures 29 and 30. It appears that the excess pore pressure ratio gives the 'earliest' warning of partial consolidation (i.e. the highest transition penetration rate or lowest transition consolidation coefficient). This is helped by the fact that viscous effects enhance both q_c and Δu at high penetration rates, while the B_q value appears largely unaffected within the undrained zone.

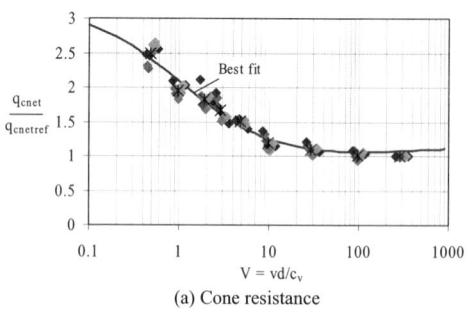

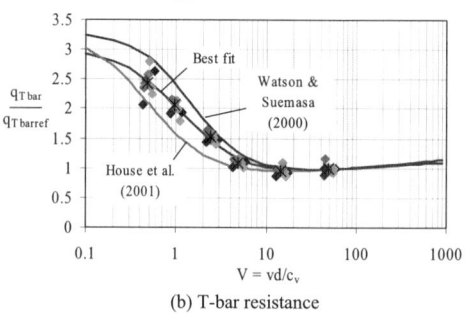

(a) Cone resistance

(b) T-bar resistance

Figure 30 Effect of normalised penetration rate on cone and T-bar resistance (Randolph & Hope, 2004)

The data in Figures 29 and 30 were plotted using c_v values obtained from Rowe cell tests in order to normalise the penetration rates. The resulting curve fits may then be used either to assess whether a given penetration test in, say, a silt is partially drained or not, or to deduce a value for the consolidation coefficient of a particular soil.

Where a single penetration test is conducted, some independent assessment of consolidation coefficient is needed in order to calculate a normalised velocity, $V = vd/c_v$, and hence assess the degree of partial consolidation that may be occurring during the test. Conversely, multiple tests conducted at different penetration rates are needed to assess a transition in the measured penetration resistance or excess pore pressure ratio, and hence deduce a value for c_v, by matching the transition point to an appropriate backbone curve (such as those in Figures 29 and 30).

8 VARIABLE PENETRATION RATE TESTS

Rather than carry out multiple penetration tests, the transition point for partially drained conditions may be assessed by varying the penetration rate over a limited depth range. A procedure to achieve this, referred to as a 'twitch test', was suggested by House et al. (2001). At a given depth, the penetration rate is successively halved, and the penetrometer advanced by 1 to 2 diameters of the probe at each new rate. A number of tests of this nature have been undertaken in the field recently (Lunne et al., 2005).

An example twitch cone test over the depth range 6.4 to 7.1 m at a soft clay site in Western Australia is shown in Figure 31. The penetration rate is (nominally) halved after each 75 mm of penetration (~2 cone diameters), from an initial rate of 20 mm/s down to a slowest rate of 0.02 mm/s before reverting to the standard rate. For comparison, results from a standard rate test are also shown. Note that spikes in the data correspond to pauses for rod changes.

As the penetration rate is reduced, initially the cone resistance decreases, owing to reduced viscous effects, but the excess pore pressure ratio remains similar to that from the standard test (average value around 0.5). At a penetration rate of 0.2 mm/s and below (depth of 6.9 m) the B_q value starts to reduce significantly, and the cone resistance starts to increase again towards the profile for the standard test.

A corresponding variable rate T-bar test at the same site is shown in Figure 32. Again, there is a gradual reduction in resistance as the penetration rate is reduced, before a marked increase for rates of 0.04 mm/s and below. Note also the significant local strengthening of the soil, and thus transient high resistance, once the penetration rate reverts to 20 mm/s just below 7 m depth.

Interpretation of variable rate tests in the field is complicated by any slight variability in the measured penetration resistance. It helps, therefore, to compare results against those from a test conducted at a standard rate. Analysis is then achieved by normalizing the penetration resistance by the corresponding value from the standard-rate test. A first stage is to take the data from both tests, and evaluate local average data at defined depth intervals (for example, every 5 mm for typical field probes).

Penetration resistance and excess pore pressure data normalised in this way are shown in Figure 33 for tests conducted at Burswood in Western Australia. The data are mainly within the undrained regime (penetration rates in excess of 0.2 mm/s, from B_q da-

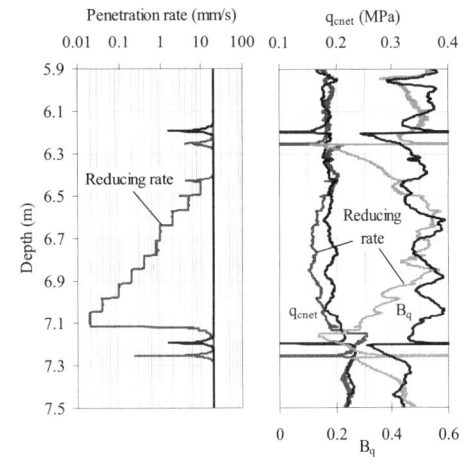

Figure 31 Standard and 'twitch' cone tests at Burswood

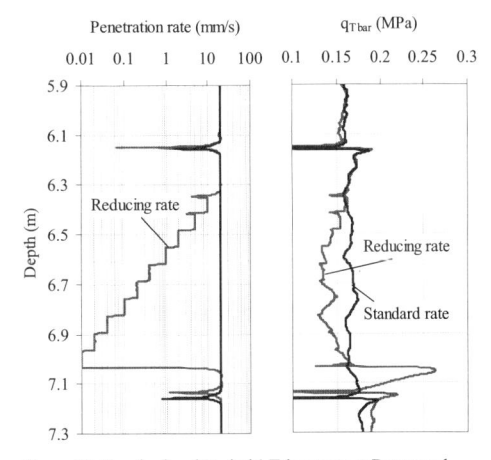

Figure 32 Standard and 'twitch' T-bar tests at Burswood

ta in Figure 31) and follow a reasonably linear relationship in semi-logarithm space. The average rate parameter, µ, is 0.13, although separate best fits for cone, T-bar and ball resistances give µ values of 0.10, 0.16 and 0.13 respectively.

Even before the transition point to partial drainage, there is a tendency for the rate effect to reduce at penetration rates below 1 mm/s. An improved fit to the data is obtained using a hyperbolic sine relationship (the curly bracket term in Eq. 22, but with actual velocities, v), with µ = 0.13, v_{ref} = 20 mm/s and v_o = 1 mm/s. This function will lead to decreasing rate effects for v between $0.1v_o$ and v_o, and negligible effects for rates lower than $0.1v_o$.

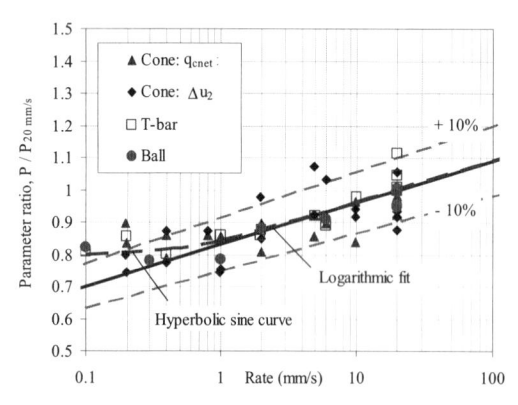

Figure 33 Effect of penetration rate for undrained conditions

The full relationship in Eq. 22 has been used to assess a coefficient of consolidation for the partially drained regime in each test. Figure 34 shows the data from variable rate cone tests conducted over the depth ranges 6 to 7 m and 8 to 9 m. Corresponding results from a T-bar test over the depth range 6 to 7 m are shown in Figure 35.

All results have been fitted using Eq. 22, with the parameters a, b, c and d based on those proposed by Randolph & Hope (2004) and given in Table 2. To fit the B_q data, the value of parameter, b, has been reduced from 0.59 to 0.50 (the measured B_q value for undrained conditions). For each test, the horizontal position of the data (or non-dimensional velocity, V) has been adjusted by varying c_v in order to obtain a fit with the curves shown.

The deduced c_v values compare with values of 0.03 to 0.13 mm²/s from constant rate of strain consolidation tests, and 0.3 to 0.6 mm²/s from piezocone dissipation tests (interpreted using the method of Teh & Houlsby, 1991). The higher values from the piezocone tests are consistent with the part-consolidating, part-swelling regimes in dissipation tests, which typically leads to c_v values that are 3 to 5 times those applicable for consolidation after yield (Fahey & Lee Goh, 1995).

The data from the T-bar test in Figure 35 fit more closely the shape from the cone backbone curve from Randolph & Hope, rather than the curve derived from their T-bar tests. The rather flatter variati-

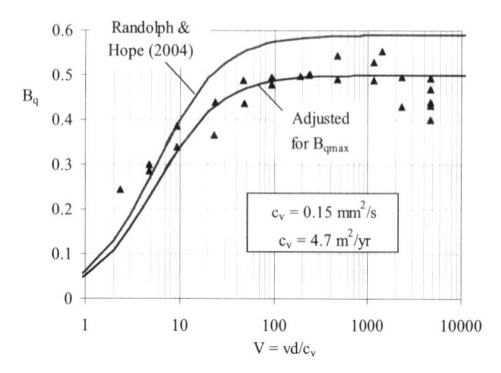

(a) Excess pore pressure ratio, B_q

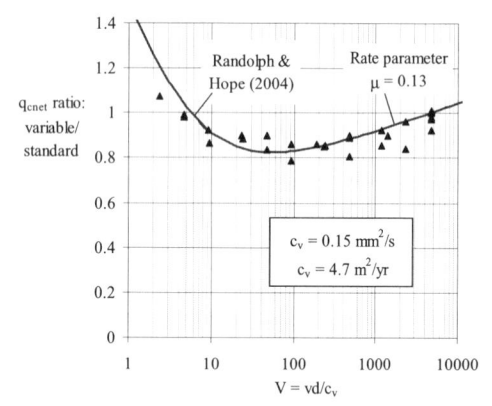

(b) Normalised net cone resistance

Figure 34 Fitting results of cone twitch test

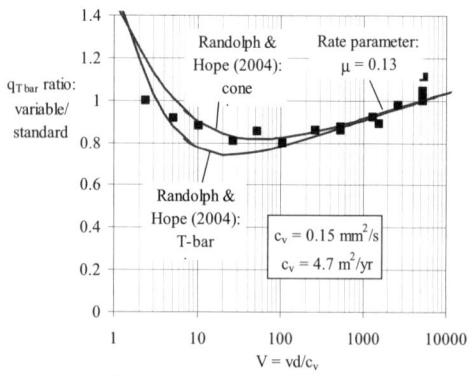

Figure 35 Fitting results of T-bar twitch test

on with penetration rate from the field data may be due to the slightly higher rate effects for the natural clay, compared with the reconstituted kaolin that was used to develop the backbone curves. Arguably, the minimum T-bar resistance for the field data should be matched with the minimum of the T-bar backbone curve (despite this giving a less good overall fit than using the cone backbone curve). That would have led to a deduced c_v up to 3 times higher (around 0.4 mm^2/s).

9 DISCUSSION

In some respects, offshore design is more straightforward than onshore design, since the focus is primarily on capacity (and hence strength of the soil) rather than deformations. On the other hand, greater challenges exist in evaluating the strength of offshore sediments, particularly as the industry moves towards ever greater water depths. There are significant difficulties in recovering high quality samples from the upper 20 or 30 m of the sea floor, where shear strengths will mostly range from 0 to 50 kPa, and this leads to a much greater reliance on in situ testing methods. It is therefore somewhat illogical that the absolute value of shear strengths derived from in situ penetrometer tests must rely on correlations with results from laboratory tests, with consequential uncertainty in respect of sample quality.

The practice of correlating shear strength data against the net cone resistance in order to arrive at an appropriate N_{kt} factor (or profile, varying with depth) for a given site arose from a number of contributing difficulties:
- scatter in shear strength data leads to uncertainty in the cone factor on any given site;
- variations in testing standards and procedures contributes to the range of average cone factors for different sites, even for soils of apparently similar characteristics;
- theoretical cone factors show a relatively broad range, and depend on additional soil properties such as rigidity index and in situ stress ratio.

The net result is the practice of site-by-site correlation we currently follow.

The advent of full-flow penetrometers such as the T-bar or ball has the potential to change this state of practice. The penetrometers are amenable to detailed analysis, where variations of strength due to strain rate, strain softening and anisotropy may all be incorporated. Initial correlations with shear strength data show a much reduced range of the average T-bar factor for each site. The potential is there to adopt a standard factor as a default, adjusting the factor only where there is strong evidence, supported by analysis, that particular soil characteristics require such adjustment.

From the data presented earlier, the average T-bar factor appears to lie in the range 12 to 12.5. While this range is higher than the proposed factor of 10.5 arising from earlier studies on reconstituted soils (Stewart & Randolph, 1991; Watson et al., 1998), it may be justified on the basis of greater strain rate dependency of the strength of natural soils.

With experience in different soil types (for example, DeJong et al., 2004; Long & Gudjonsson, 2004), the range of appropriate resistance factors for full-flow penetrometers may broaden. However, the onus should be on first ensuring that the shear strength data against which the resistance is correlated are of high quality, and secondly identifying the particular soil characteristic that might explain resistance factors outside the norm.

It would of course be heresy to suggest that a field penetrometer test might be regarded as the gold standard to which laboratory tests are correlated. On the other hand the design objective should be borne in mind. If this is estimation of the vertical and horizontal resistance provided by the soil to pipeline or riser motions, then a 'model' T-bar resistance might well provide better guidance than a set of laboratory strength tests. A similar argument could be made for estimation of anchor penetration resistance through seabed soils by comparison with a ball penetrometer.

9.1 *Improvements in Field Testing*

There are a number of measures that may be taken to improve the confidence in field penetrometer data, both in their acquisition and their interpretation. Parallel testing with two different types of penetrometer, such as cone and T-bar, or cone and ball, has obvious advantages, since it enables comparison of the two resistance profiles and, of particular importance, the trend with depth of the ratio of resistances. Odd features or trends in this resistance ratio may reflect an incorrect load cell zero, or errors introduced in the sequence of converting the measured cone resistance to a net resistance. Transition through different soil strata may lead to changes in the ratio of T-bar and cone (net) resistance, and the challenge is then to interpret what additional information that yields about the soil characteristics.

Mention was made earlier of the importance of comparing extraction and penetration resistance. It is essential to improve field equipment and procedures, and the state of practice, so that monitoring cone and other penetrometers during extraction becomes standard. Comparison of extraction and penetration resistance provides a further check on quality of the data (and interpretation), and also a first indication of the sensitivity of the soil.

The remoulded strength of the soil, and the amount of strain required to achieve a fully remoulded state, may be assessed by cyclic penetrometer tests, as shown previously. Lack of symmetry in the resulting cyclic extraction and penetration response may also indicate load cell drift, although minor asymmetry in favour of the penetration resistance appears common and could have an alternative explanation.

The ability to vary penetration rate has a number of potential benefits, ranging from assessment of the strain rate dependency of the peak and remoulded strengths, to measuring the consolidation coefficient of the soil. This would require development in field equipment, but controlled penetration rates spanning two or three orders of magnitude are not difficult to achieve.

This discussion has focused so far on penetrometer testing, but improvements in vane testing might also be considered. Varying the rotation rate for a vane would also allow assessment of strain rate effects, and would also have operational benefits. For example, increasing the vane rotation to 1 or 10 °/s would shorten the time for the additional rotations needed to reach residual conditions. Measuring the strain rate dependence of the vane torque would aid in interpreting the test, separating out enhancement of strength due to the high strain rates from the effects of disturbance due to vane insertion.

10 CONCLUSIONS

This paper has reviewed current practice and potential advances in characterising the soft sediments typically encountered in deepwater offshore developments. Particular focus has been on the new generation of full-flow penetrometers, and the extent to which these provide an absolute measurement of the soil strength, or merely a relative one.

All field penetrometer and vane tests involve strain rates in the soil that are many orders of magnitude greater than standard laboratory shearing rates. They also create disturbance of the soil during insertion, and this compensates to some degree the effects of the high strain rates.

Analytical assessment of strain rates and the resulting increase in resistance has been presented, both for the vane and for full-flow cylindrical (T-bar) and spherical (ball) penetrometers. The latter offer an exciting potential, in parallel with cone penetrometer tests, to provide a check on the quality of field testing and interpretation, and to evaluate soil characteristics with much greater insight.

While the focus of the paper has been offshore, the content is equally applicable to soft onshore soils, particularly where the ratio of shear strength to effective overburden pressure is less than unity. The challenge now is to capitalise on the greater extent to which we are able to analyse these tests, incorporating effects of strain rate, softening and strength anisotropy, in order to decrease current reliance on empirical correlations.

ACKNOWLEDGEMENTS

This paper has built on some of the current activities of the Centre for Offshore Foundation Systems (COFS), established and supported under the Australian Research Council's Research Centres Program. Particular thanks are due to BP Exploration Operating Company, Statoil, Norsk Hydro, Woodside Engineering, Petroleo Brasileiro and ConocoPhillips, who were sponsors of the joint industry project: Characterization of Soft Soils in Deep Water by In Situ Tests, undertaken jointly by NGI (project leader Mr Tom Lunne) and COFS, for permission to publish certain results from that project. The author would also like to acknowledge helpful review comments and advice from James Schneider and Marc Senders, PhD students at COFS, but both with extensive field experience from former employment with Fugro.

REFERENCES

Aas, G., Lacasse, S., Lunne, T. & Høeg, K. (1986). Use of in situ tests for foundation design on clay. *Proc. In Situ '86: Use of In situ Tests in Geotechnical Engineering*, ASCE, New York, 1-30.

Abu-Farsakh, M., Tumay, M. & Voyiadjis, G. (2003). Numerical parametric study of piezocone penetration test in clays. *Int. J. of Geomechanics*, ASCE, 3(2), 170-181.

Andersen, K.H. & Lauritzsen, R. (1988). Bearing capacity for foundations with cyclic loads. *J. Geotech. Eng. Div.*, ASCE, 114(5), 540-555.

Aubeny, C.P., Han, S-W. & Murff, J.D. (2003). Suction caisson capacity in anisotropic, purely cohesive soils. *Int.. J. of Geomechanics*, 3(3/4), 225-235.

Baligh, M.M., Azzouz, A.S., Wissa, A.Z.E., Martin, R.T. & Morrison, M.H. (1981). The piezocone penetrometer: cone penetration testing and experience. *Proc. ASCE Conference on Cone Penetration Testing*, St Louis, 247-263.

Biscontin, G. & Pestana, J.M. (2001). Influence of peripheral velocity on vane shear strength of an artificial clay. *Geotechnical Testing Journal*, 24 (4), 423-429.

Borel, D., Puech, A. * de Ruijter, M. (2002). High quality sampling for deep water geotechnical engineering: the STACOR experience. *Proc. Conf. on Ultra Deep Engineering and Technology*, Brest.

Carter, J.P., Davies, P.J. & Krasnostein, P. (1999). The future of offshore site investigation – robotic drilling on the seabed. *Australian Geomechanics*, 34(3), 77-84.

Cherato, A.B. & Lutenegger, A.J. (2004). Disturbance effects of field vane tests in a varved clay. *Proc. 2nd Int. Conf. on Site Characterisation*, Porto.

Chandler, R.J. (1988). The in-situ measurement of the undrained shear strength of clays using the field vane. *Vane Shear Strength Testing of Soils: Field and Laboratory Studies*, ASTM STP 1014: 13-45.

Chung, S.F. & Randolph, M.F. (2004). Penetration resistance in soft clay for different shaped penetrometers. *Proc. 2nd Int. Conf. on Site Characterisation*, Porto.

Clukey, E.C., Templeton, J.S., Randolph, M.F. & Phillips, R.A. (2004). Suction caisson response under sustained loop-current loads. *Proc. Offshore Tech. Conf.*, Houston, Paper OTC 16843.

DeJong, J.T., Yafrate, N.J., DeGroot, D.J. & Jakubowski, J. (2004). Evaluation of the undrained shear strength profile in soft layered clay using full-flow probes. *Proc. 2nd Int. Conf. on Site Characterisation*, Porto.

Einav, I. & Randolph, M.F. (2004). Combining upper bound and strain path methods for evaluating penetration resistance. *Int. J. Num. Methods in Eng.*, forthcoming paper.

Fahey, M. & Lee Goh, A. (1995). A comparison of pressuremeter and piezocone methods of determining the coefficient of consolidation. *Proc. 4th Int. Symp. on The Pressuremeter and Its New Avenues*, Quebec, 153-160.

Fearon, R.E., Chandler, R.J. & Bommer, J.J. (2004). An investigation of the mechanisms which control soil behaviour at fast rates of displacement. *Proc. Skempton Conf., Advances in Geotechnical Engineering*, ICE, London, 441-452.

Finnie, I.M.S. & Randolph, M.F. (1994). Punch-through and liquefaction induced failure of shallow foundations on calcareous sediments. *Proc. Int. Conf. on Behaviour of Offshore Structures, BOSS '94*, Boston, 217-230.

Geise, J.M., Hoope, J. & May, R. (1988). Design and offshore experience with an in situ vane. *Vane Shear Strength Testing of Soils: Field and Laboratory Studies*, ASTM STP 1014: 318-338.

Hefer, P.A & Neubecker, S.R. (1999). A recent development in offshore site investigation tools: the T-bar. *Proc. Australasian Oil and Gas Conf.*, Perth.

Hight, D.W. & Leroueil S. (2002). Characterisation of soils for engineering purposes. In *Characterisation and Engineering Properties of Natural Soils*, Eds Tan, T.S., Phoon, K.K., Hight, D.W. and Leroueil, S., Balkema, 1, 255-360.

House, A.R., Oliveira, J.R.M.S. & Randolph, M.F. (2001). Evaluating the coefficient of consolidation using penetration tests, *Int. J. of Physical Modelling in Geotechnics*, 1(3), 17-25.

Johnson, G.W., Hamilton, T.W., Ebelhar, R.J., Mueller, J.L. & Pelletier, J.H. (1988). Comparison of in situ vane, cone penetrometer and laboratory test results for Gulf of Mexico deepwater clays. *Vane Shear Strength Testing of Soils: Field and Laboratory Studies*, ASTM STP 1014: 293-305.

Keaveny, J. & Mitchell, J.K. (1986). Strength of fine-grained soils using the piezocone. *Use of In-Situ Tests in Geotechnical Engineering*, GSP 6, ASCE, Reston/VA, 668-685.

Kolk, H.J., Hoope, J. & Imms, B.W. (1988). Evaluation of offshore in situ vane test results. *Vane Shear Strength Testing of Soils: Field and Laboratory Studies*, ASTM STP 1014: 339-353.

Ladd, C.C. & DeGroot, D.J. (2003). Recommended practice for soft ground site characterization: Arthur Casagrande Lecture. *Proc. 12th Panamerican Conf. on Soil Mech. and Geotech. Engng*, Cambridge, USA.

Lauritzsen, R. & Schjetne, K. (1976). Stability calculations for offshore gravity structures. *Proc. 8th Offshore Tech. Conf.*, Houston, 1, 75-82.

Lemos, L.J.L. & Vaughan, P.R. (2004). Shear behaviour of pre-existing shear zones under fast loading. *Proc. Skempton Conf., Advances in Geotechnical Engineering*, ICE, London, 510-521.

Long, M. & Gudjonsson, G.T. (2004). T-bar testing in Irish soils. *Proc. 2nd Int. Conf. on Site Characterisation*, Porto.

Lu Q. (2003). *A Numerical Study of Penetration Resistance in Clay*. PhD Thesis, The University of Western Australia.

Lu, Q., Randolph, M.F., Hu, Y. & Bugarski, I.C. (2004). A numerical study of cone penetration in clay. *Géotechnique*, 54 (in press).

Lunne, T. (2001). In situ testing in offshore geotechnical investigations. *Proc. Int. Conf. on In Situ Measurement of Soil Properties and Case Histories*, Bali, 61-81.

Lunne, T., Randolph, M.F., Sjursen, M.A. & Chung, S.F. (2005). Comparison of cone and T-bar resistance factors in a range of offshore and onshore soft sediments. Proc. Int. Symp. on Frontiers in Offshore Geotechnics, Perth.

Lunne, T., Robertson, P.K. & Powell, J.J.M. (1997). *Cone Penetration Testing in Geotechnical Engineering*, Blackie Academic and Professional, London.

Lunne, T. Berre, T., Strandvik, S., Andersen, K.H. & Tjelta, T.I. (2001). Deepwater sample disturbance due to stress relief, *Proc. OTRC Int. Conf. on, Geotechnical, Geological and Geophysical Properties of Deepwater Sediments*, OTRC, 64-85.

Mitchell, J. K. (1976). *Fundamentals of Soil Behavior*, Wiley, New York.

Peuchen, J. (2000). Deepwater cone penetration tests. *Proc. Offshore Technology Conf.*, Houston, Paper OTC 12094.

Quirós, G.W. & Young, A.G. (1988). Comparison of field vane, CPT and laboratory strength data at Santa Barbara Channel site. *Vane Shear Strength Testing of Soils: Field and Laboratory Studies*, ASTM STP 1014: 306-317.

Quirós, G.W. & Little, R.L. (2003). Deepwater soil properties and their impact on the geotechnical program, *Proc. Offshore Technology Conf.*, Houston, Paper OTC 15262.

Rad, N.S. & Lunne, T. (1994). Gas in soil. I: Detection and η-profiling, *J. of Geotech. Eng.*, ASCE, 120(4), 697-715.

Randolph, M.F. (2000). Effect of strength anisotropy on capacity of foundations. *Proc. John Booker Memorial Symposium*, Sydney, 313-328.

Randolph, M.F., Andersen, K.H, Einav, I & Lunne, T. (2005). Numerical analysis of T-bar penetration in soft clay. Paper in preparation.

Randolph, M.F., Hefer, P.A., Geise, J.M. & Watson, P.G. (1998). Improved seabed strength profiling using T-bar penetrometer. *Proc Int. Conf. Offshore Site Investigation and Foundation Behaviour - "New Frontiers"*, Society for Underwater Technology, London, 221-235.

Randolph, M.F. & Hope, S. (2004). Effect of cone velocity on cone resistance and excess pore pressures. *Proc. Int. Symp. On Engineering Practice and Performance of Soft Deposits*, Osaka.

Randolph, M.F. & Houlsby, G.T. (1984). The limiting pressure on a circular pile loaded laterally in cohesive soil. *Géotechnique*, 34(4), 613-623.

Randolph, M.F., Martin, C.M. & Hu, Y. (2000). Limiting resistance of a spherical penetrometer in cohesive material. *Géotechnique*, 50(5), 573-582.

Sheahan, T.C., Ladd, C.C. & Germaine, J.T. (1996). Rate-dependent undrained shear behavior of saturated clay. *J. Geotech. Eng.*, ASCE, 122(2), 99-108.

Stewart, D.P. & Randolph, M.F. (1991). A new site investigation tool for the centrifuge. *Proc. Int. Conf. On Centrifuge Modelling – Centrifuge 91*, Boulder, Colorado, 531-538.

Stewart, D.P. & Randolph, M.F. (1994). T-bar penetration testing in soft clay. *J. Geot. Eng. Div.*, ASCE, 120(12), 2230-2235.

Teh, C. I. & Houlsby, G. T. (1991). An analytical study of the cone penetration test in clay. *Géotechnique*, **41**(1), 17-34.

Watson, P.G., Newson, T.A. & Randolph, M.F. (1998). Strength profiling in soft offshore soils. *Proc. 1st Int. Conf. On Site Characterisation - ISC '98*, Atlanta, 2, 1389-1394.

Watson, P.G., Suemasa, N. & Randolph, M.F. (2000). Evaluating undrained shear strength using the vane shear apparatus. *Proc. 10th Int. Conf. On Offshore and Polar Engng, ISOPE 00*, Seattle, 2, 485-493.

Young, A.G., Honganen, C.D., Silva, A.J. & Bryant, W.R. (2000). Comparison of geotechnical properties from large diameter long cores and borings in deep water Gulf of Mexico. *Proc. Offshore Technology Conf.*, Houston, Paper OTC 12089.

Young, A.G., McClelland, B. & Quirós, G.W. (1988). In situ vane testing at sea. *Vane Shear Strength Testing of Soils: Field and Laboratory Studies*, ASTM STP 1014: 46-67.

Yu, H.S., Herrmann, L.R. & Boulanger, R.W. (2000). Analysis of steady cone penetration in clay. *J. Geotechnical and Geoenvironmental Eng.*, ASCE, 126(7), 594-605.

Zdravkovic, L., Potts, D.M. and Jardine, R.J. (2001). A parametric study of the pull-out capacity of bucket foundations in soft clay. *Géotechnique*, 51(1), 55-67.

Evaluating soil liquefaction and post-earthquake deformations using the CPT

P.K. Robertson
Dept. of Civil and Environmental Engineering, University of Alberta, Edmonton, Canada

Keywords: soil liquefaction, ground deformations, site investigation, cone penetration test

ABSTRACT: Soil liquefaction is a major concern for many structures constructed with or on sand or sandy soils. This paper provides an overview of soil liquefaction, describes a method to evaluate the potential for cyclic liquefaction using the Cone Penetration Test (CPT) and methods to estimate post-earthquake ground deformations. A discussion is also provided on how to estimate the liquefied undrained shear strength following strain softening (flow liquefaction).

1 INTRODUCTION

Soil liquefaction is a major concern for structures constructed with or on sand or sandy soils. The major earthquakes of Niigata in 1964 and Kobe in 1995 have illustrated the significance and extent of damage caused by soil liquefaction. Soil liquefaction is also a major design problem for large sand structures such as mine tailings impoundment and earth dams.

To evaluate the potential for soil liquefaction it is important to determine the soil stratigraphy and in-situ state of the deposits. The CPT is an ideal in-situ test to evaluate the potential for soil liquefaction because of its repeatability, reliability, continuous data and cost effectiveness. This paper presents a summary of the application of the CPT to evaluate soil liquefaction. Further details are contained in a series of papers (Robertson and Wride, 1998; Youd et al., 2001; Zhang et al., 2002; Zhang et al., 2004).

2 DEFINITION OF SOIL LIQUEFACTION

Several phenomena are described as soil liquefaction, hence, a series of definitions are provided to aid in the understanding of the phenomena.

2.1 Cyclic (softening) Liquefaction

- Requires undrained cyclic loading during which shear stress reversal occurs or zero shear stress can develop.
- Requires sufficient undrained cyclic loading to allow effective stresses to reach essentially zero.
- Deformations during cyclic loading can accumulate to large values, but generally stabilize shortly after cyclic loading stops. The resulting movements are due to external causes and occur mainly during the cyclic loading.
- Can occur in almost all saturated sandy soils provided that the cyclic loading is sufficiently large in magnitude and duration.
- Clayey soils generally do not experience cyclic liquefaction and deformations are generally small due to the cohesive nature of the soils. Rate effects (creep) often control deformations in cohesive soils.

2.2 Flow Liquefaction

- Applies to strain softening soils only.
- Requires a strain softening response in undrained loading resulting in approximately constant shear stress and effective stress.
- Requires in-situ shear stresses to be greater than the residual or minimum undrained shear strength
- Either monotonic or cyclic loading can trigger flow liquefaction.
- For failure of a soil structure to occur, such as a slope, a sufficient volume of material must strain soften. The resulting failure can be a slide or a flow depending on the material characteristics and ground geometry. The resulting movements are due to internal causes and can occur after the trigger mechanism occurs.
- Can occur in any metastable saturated soil, such as very loose fine cohesionless deposits, very sensitive clays, and loess (silt) deposits.

Note that strain softening soils can also experience cyclic liquefaction depending on ground geometry.

Figure 1 presents a flow chart (Robertson and Wride, 1998) to clarify the phenomena of soil liquefaction.

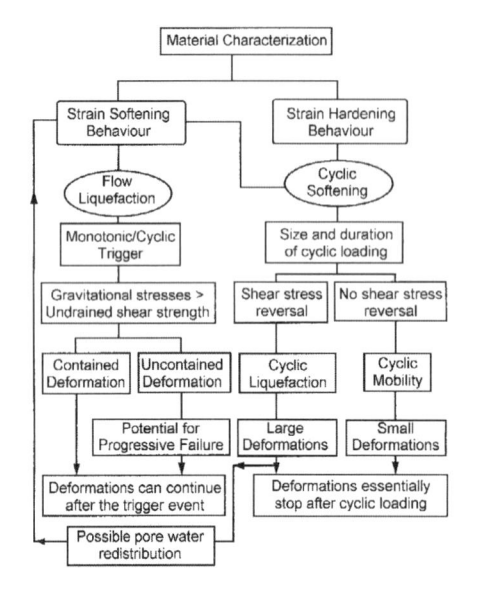

Figure 1. Flow chart to evaluate liquefaction of soils (After Robertson and Wride, 1998)

If a soil is strain softening, flow liquefaction is possible if the soil can be triggered to collapse and if the gravitational shear stresses are larger than the minimum undrained shear strength. The trigger can be either monotonic or cyclic. Whether a slope or soil structure will fail and slide will depend on the amount of strain softening soil relative to strain hardening soil within the structure, the brittleness of the strain softening soil and the geometry of the ground. The resulting deformations of a soil structure with both strain softening and strain hardening soils will depend on many factors, such as distribution of soils, ground geometry, amount and type of trigger mechanism, brittleness of the strain softening soil and drainage conditions. Examples of flow liquefaction failures are the Aberfan flow slide (Bishop, 1973), Zealand submarine flow slides (Koppejan et al., 1948) and the Stava tailings dam failure. In general, flow liquefaction failures are not common, however, when they occur, they typically take place rapidly with little warning and are usually catastrophic. Hence, the design against flow liquefaction should be carried out cautiously.

If a soil is strain hardening, flow liquefaction will not occur. However, cyclic liquefaction can occur due to cyclic undrained loading. The amount and extent of deformations during cyclic loading will depend on the density of the soils, the magnitude and duration of the cyclic loading and the extent to which shear stress reversal occurs. If extensive shear stress reversal occurs and the magnitude and duration of the cyclic loading are sufficiently large, it is possible for the effective stresses to essentially reach zero with resulting possible large deformations. Examples of cyclic liquefaction were common in the major earthquakes in Niigata in 1964 and Kobe in 1995 and manifest in the form of sand boils, damaged lifelines (pipelines, etc.) lateral spreads, slumping of embankments, ground settlements, and ground surface cracks. If cyclic liquefaction occurs and drainage paths are restricted due to overlying less permeable layers, the sand immediately beneath the less permeable soil can loosen due to pore water redistribution, resulting in possible subsequent flow liquefaction, given the right geometry.

The three main concerns related to soil liquefaction are generally:

- Will an event (e.g. a design earthquake) trigger significant zones of liquefaction?
- What displacements will result, if liquefaction is triggered?
- What will be the resulting residual (minimum) undrained shear strength, if a soil is potentially strain softening (and is triggered to liquefy)?

This paper describes how the CPT can be used to evaluate the potential for an earthquake to trigger cyclic liquefaction, and, then how to estimate post-earthquake displacements (vertical and lateral). Finally, a method is described on how to estimate the minimum undrained shear strength, using the CPT, which could result if the soil is potentially strain softening (flow liquefaction).

3 CYCLIC LIQUEFACTION

Most of the existing work on cyclic liquefaction has been primarily for earthquakes. The late Prof. H.B. Seed and his co-workers developed a comprehensive methodology to estimate the potential for cyclic liquefaction due to earthquake loading. The methodology requires an estimate of the cyclic stress ratio (CSR) profile caused by the design earthquake and the cyclic resistance ratio (CRR) of the ground. If the CSR is greater than the CRR cyclic liquefaction can occur. The CSR is usually estimated based on a probability of occurrence for a given earthquake. A site-specific seismicity analysis can be carried out to determine the design CSR profile with depth. A simplified method to estimate CSR was also developed by Seed and Idriss (1971) based on the maximum ground surface acceleration (a_{max}) at the site. The simplified approach can be summarized as follows:

$$\text{CSR} = \frac{\tau_{av}}{\sigma'_{vo}} = 0.65 \left[\frac{a_{max}}{g}\right]\left(\frac{\sigma_{vo}}{\sigma'_{vo}}\right) r_d \quad (1)$$

Where τ_{av} is the average cyclic shear stress; a_{max} is the maximum horizontal acceleration at the ground surface; $g = 9.81 \text{m/s}^2$ is the acceleration due to gravity; σ_{vo} and σ'_{vo} are the total and effective vertical overburden stresses, respectively; and r_d is a stress reduction factor which is dependent on depth. The factor r_d can be estimating using the following bi-linear function, which provides a good fit to the average of the suggested range in r_d originally proposed by Seed and Idriss (1971):

$$r_d = 1.0 - 0.00765z \quad (2)$$

if $z < 9.15$ m

$= 1.174 - 0.0267z$

if $z = 9.15$ to 23 m

Where z is the depth in metres. These formulae are approximate at best and represent only average values since r_d shows considerable variation with depth.

Seed et al., (1985) also developed a method to estimate the cyclic resistance ratio (CRR) for clean sand with level ground conditions based on the Standard Penetration Test (SPT). Recently the CPT has become more popular to estimate CRR, due to the continuous, reliable and repeatable nature of the data (Youd et al., 2001).

In recent years, there has been an increase in available field performance data, especially for the CPT (Robertson and Wride, 1998). The recent field performance data have shown that the existing CPT-based correlation by Robertson and Campanella (1985) for clean sands is generally good. Based on discussions at the 1996 NCEER workshop (NCEER, 1997), the curve by Robertson and Campanella (1985) has been adjusted slightly at the lower end. The resulting recommended CPT correlation for clean sand is shown in Figure 2 and can be estimated using the following simplified equations:

$$\text{CRR}_{7.5} = 93\left[\frac{(q_{c1N})_{cs}}{1000}\right]^3 + 0.08 \quad (3)$$

if $50 \leq (q_{c1N})_{cs} \leq 160$

$$\text{CRR}_{7.5} = 0.833\left[\frac{(q_{c1N})_{cs}}{1000}\right] + 0.05$$

if $(q_{c1N})_{cs} < 50$

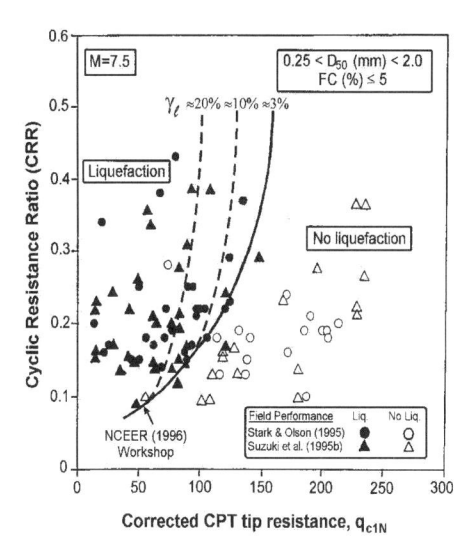

Figure 2. Cyclic resistance ratio (CRR) from the CPT for clean sands. (After Robertson and Wride, 1998).

Where $(q_{c1N})_{cs}$ is the equivalent clean sand normalized cone penetration resistance (defined in detail later).

The field observations used to compile the curve in Figure 2 are based primarily on the following conditions:
- Holocene age, clean sand deposits
- Level or gently sloping ground
- Magnitude M = 7.5 earthquakes
- Depth range from 1 to 15 m (85% is for depths < 10 m)
- Representative average CPT values for the layer considered to have experienced cyclic liquefaction.

Caution should be exercised when extrapolating the CPT correlation to conditions outside the above range. An important feature to recognize is that the correlation is based primarily on average values for the inferred liquefied layers. However, the correlation is often applied to all measured CPT values, which include low values below the average. Therefore, the correlation can be conservative in variable deposits where a small part of the CPT data can indicate possible liquefaction.

It has been recognized for some time that the correlation to estimate $\text{CRR}_{7.5}$ for silty sands is different than that for clean sands. Typically a correction is made to determine an equivalent clean sand penetration resistance based on grain characteristics, such as fines content, although the corrections are due to more than just fines content, since the plasticity of the fines also has an influence on the CRR.

One reason for the continued use of the SPT has been the need to obtain a soil sample to determine the fines content of the soil. However, this has been offset by the generally poor repeatability and reliability of the SPT data. It is now possible to estimate grain characteristics directly from the CPT. Robertson and Wride (1998) suggest estimating an equivalent clean sand normalized cone penetration resistance, $(q_{c1N})_{cs}$ using the following:

$$(q_{c1N})_{cs} = K_c (q_{c1N}) \quad (4)$$

where K_c is a correction factor that is a function of grain characteristics of the soil.

Robertson and Wride (1998) suggest estimating the grain characteristics using the soil behavior chart by Robertson (1990) (see Figure 3) and the soil behavior type index, I_c, where;

$$I_c = \left[(3.47 - \log Q)^2 + (\log F + 1.22)^2\right]^{0.5} \quad (5)$$

where $Q = q_{c1N} = \left(\dfrac{q_c - \sigma_{vo}}{P_{a2}}\right)\left(\dfrac{P_a}{\sigma'_{vo}}\right)^n$

is the normalized CPT penetration resistance, dimensionless; n = stress exponent; $F = f_s/[(q_c - \sigma_{vo})] \times 100\%$ is the normalized friction ratio, in percent; f_s is the CPT sleeve friction stress; σ_{vo} and σ'_{vo} are the total effective overburden stresses, respectively; P_a is a reference pressure in the same units as σ'_{vo} (i.e. $P_a = 100$ kPa if σ'_{vo} is in kPa); and P_{a2} is a reference pressure in the same units as q_c and σ_{vo} (i.e. $P_{a2} = 0.1$ MPa if q_c and σ_{vo} are in MPa). Robertson and Wride (1998) used a form of q_{c1N} that did not subtract the total vertical stress (σ_{vo}) from q_c and used n = 0.5. The more correct approach is the full form shown in equation 5. In general there is little difference between q_{c1N} and Q for most sandy soils at shallow depth ($\sigma'_{vo} < 300$ kPa).

The soil behaviour type chart by Robertson (1990) shown in Figure 3, uses a normalized cone penetration resistance (Q) based on a simple linear stress exponent of n = 1.0, whereas the chart recommended here for estimating CRR in sand is based on a normalized cone penetration resistance (q_{c1N}) based on a stress exponent n = 0.5.

Olsen and Malone (1988) correctly suggested a normalization where the stress exponent (n) varies from around 0.5 in sands to 1.0 in clays. The procedure using n = 1.0 was recommended by Robertson and Wride (1998) for soil classification in clay type soils when $I_c > 2.6$. However, in sandy soils when $I_c \leq 2.6$, Robertson and Wride (1998) recommended

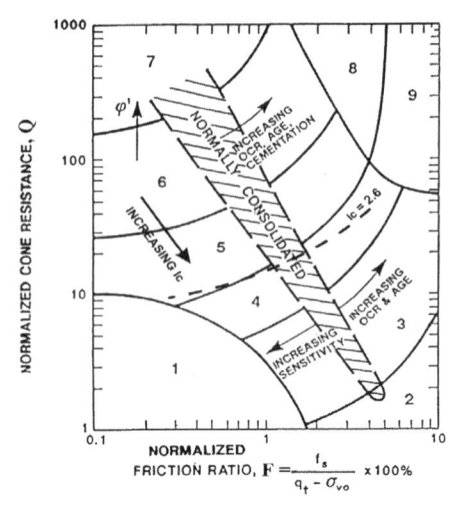

Figure 3. Normalized CPT soil behaviour type chart. (After Robertson, 1990).

Zone	Soil Behaviour Type	I_c
1	Sensitive, fine grained	N/A
2	Organic soils – peats	> 3.6
3	Clays – silty clay to clay	2.95 - 3.6
4	Silt mixtures – clayey silt to silty clay	2.60 – 2.95
5	Sand mixtures – silty sand to sandy silt	2.05 – 2.6
6	Sands – clean sand to silty sand	1.31 – 2.05
7	Gravelly sand to dense sand	< 1.31
8	Very stiff sand to clayey sand*	N/A
9	Very stiff, fine grained*	N/A

* Heavily overconsolidated or cemented
Note: Soil behaviour type index (I_c) is given by
$I_c = [(3.47 - \log Q)^2 + (\log F + 1.22)^2]^{0.5}$

that data being plotted on the chart be modified by using n = 0.5.

The simplified normalization suggested by Robertson and Wride (1998) is easy to apply but produces a somewhat discontinuous variation of the stress exponent, n. To produce a smoother variation of the stress exponent the following modified method is recommended.

Assume an initial stress exponent n = 1.0 and calculate Q and F and then I_c. Then:

If $I_c < 1.64$ n = 0.5   (6)

If $I_c > 3.30$ n = 1.0

If $1.64 < I_c < 3.30$ n = (I_c –1.64) 0.3 + 0.5

Iterate until the change in the stress exponent, Δn < 0.01.

When the in-situ vertical effective stress (σ'$_{vo}$) exceeds 300 kPa assume n = 1.0 for all soils. This avoids the need to make further corrections for high overburden stresses (i.e. K$_σ$).

The recommended relationship between I$_c$ and the correction factor K$_c$ is given by the following:

$$K_c = 1.0 \quad \text{if} \quad I_c \leq 1.64 \tag{7}$$

$$K_c = -0.403\, I_c^4 + 5.581\, I_c^3 - 21.63\, I_c^2 + 33.75\, I_c - 17.88 \quad \text{if } I_c > 1.64$$

The correction factor, K$_c$, is approximate since the CPT responds to many factors, such as, soil plasticity, fines content, mineralogy, soil sensitivity, age and stress history. However, in general, these same factors influence the CRR$_{7.5}$ in a similar manner. Caution should be used when applying the relationship to sands that plot in the region defined by 1.64 < I$_c$ < 2.36 and F < 0.5% so as not to confuse very loose clean sands with sands containing fines. In this zone, it is suggested to set K$_c$ = 1.0, to provide some added conservatism.

Soils with I$_c$ > 2.6 fall into the clayey silt, silty clay and clay regions of the CPT soil behavior chart and, in general, are essentially non-liquefiable in terms of cyclic liquefaction. Samples should be obtained and liquefaction evaluated using other criteria based on index parameters (Youd et al., 2001). Soils that fall in the lower left region of the CPT soil behavior chart defined by Ic > 2.6 but F < 1.0% can be sensitive fine-grained soils and hence, possibly susceptible to both cyclic and flow liquefaction. Fine-grained, cohesive (clay) soils can develop significant strains and deformations during earthquake loading if the level and duration of shaking are sufficient to overcome the peak undrained resistance of the soil. If that were to happen and sufficient movement were to accrue, the strength of the soil could reduce to it's residual or remolded strength depending on the amount of strain required to soften the soil.

The full methodology to estimate CRR$_{7.5}$ from the CPT is summarized in Figure 4.

Recent publications have attempted to update the relationship between normalized cone resistance (q$_{c1N}$) and CRR (Figure 2). However, these efforts have limited additional value since the relationship is based on average values in layers that were thought to have liquefied. This requires considerable judgment and is subject to much uncertainty. It is now time to move away from the boundary curve diagrams represented in Figure 2. The preferred approach is to continue to refine the total integrated CPT-based approach based on case histories using the full CPT and CRR profiles. Ideally case histories should include extensive CPT data, samples and observations of ground deformations (both surface and subsurface) so that a full evaluation can be made of the impact of the earthquake loading and where the deformations occurred. This can be important, since deformations can occur where full liquefaction may not have taken place (i.e. factor of safety greater than 1). Currently most publications take case history data and estimate the layer that experienced liquefaction and than assign an average penetration resistance value to that layer. In this way each case history can be represented by one data point. However, each case history should be represented by the many hundred data points contained in the full CPT profile. Often the ground profiles are complex and liquefaction may not have occurred in one well-defined single layer.

Liquefaction often occurs in multiple layers, which can only be observed in the full CPT profile. It can be overly simplistic to represent a complex ground profile and case history by one data point on a boundary curve like Figure 2. These curves have served as an excellent starting point in the development of the current simplified CPT and SPT liquefaction methods now available, however, it is now

q$_c$: tip resistance, f$_s$: sleeve friction
σ$_{vo}$, σ$_{vo}$': in-situ vertical total and effective stress
units: all in kPa

initial stress exponent : n = 1.0 and calcualte Q, F, and I$_c$
if I$_c$ <= 1.64, n = 0.5
if 1.64 < I$_c$ < 3.30, n = (I$_c$ −1.64)*0.3 + 0.5
if I$_c$ >= 3.30, n = 1.0
iterate until the change in n, Δn < 0.01
if σ$_{vo}$' > 300 kPa, let n = 1.0 for all soils

$$C_n = \left(\frac{100}{\sigma_{vo}'}\right)^n$$

$$Q = \frac{(q_c - \sigma_{vo})}{100} \cdot C_n, \quad F = \frac{f_s}{(q_c - \sigma_{vo})} \cdot 100$$

$$I_c = \sqrt{(3.47 - \log Q)^2 + (1.22 + \log F)^2}$$

if I$_c$ <= 1.64, K$_c$ = 1.0
if 1.64 < I$_c$ < 2.60, K$_c$ = -0.403I$_c^4$ + 5.581 I$_c^3$ − 21.63 I$_c^2$ + 33.75 I$_c$ − 17.88
if I$_c$ >= 2.60, evaluate using other criteria; likely non- liquefiable if F > 1%
BUT, if 1.64 < I$_c$ < 2.36 and F < 0.5%, set K$_c$ = 1.0

$$(q_{c1N})_{cs} = K_c \cdot Q$$

$$CRR_{7.5} = 93 \cdot \left(\frac{(q_{c1N})_{cs}}{1000}\right)^3 + 0.08, \text{ if } 50 <= (q_{c1N})_{cs} < 160$$

$$CRR_{7.5} = 0.833 \cdot \left(\frac{(q_{c1N})_{cs}}{1000}\right) + 0.05, \text{ if } (q_{c1N})_{cs} < 50$$

if I$_c$ >= 2.60, evaluate using other criteria; likely non- liquefiable if F > 1%

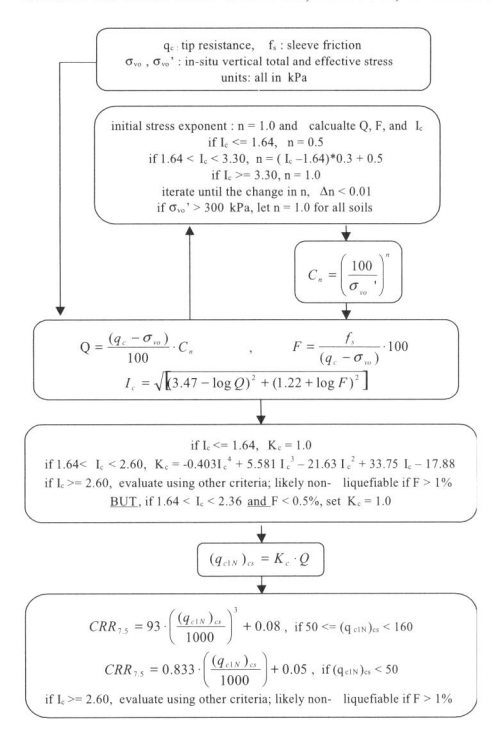

Figure 4. Flow chart to evaluate cyclic resistance ratio (CRR) from the CPT. (Modified from Robertson and Wride, 1998).

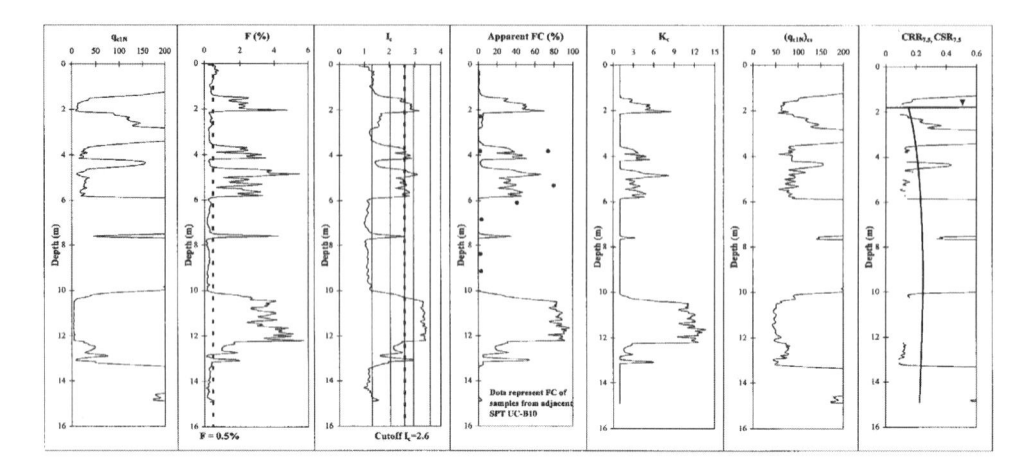

Figure 5. Example of CPT to evaluate cyclic liquefaction at Moss Landing Site. (After Robertson and Wride, 1998).

time to leave these highly simplistic curves and progress using data captured in the full soil profile.

The factor of safely against liquefaction is defined as:

$$\text{Factor of Safety, FS} = \frac{\text{CRR}_{7.5}}{\text{CSR}} \text{MSF} \quad (8)$$

here MSF is the Magnitude Scaling Factor to convert the $\text{CRR}_{7.5}$ for $M = 7.5$ to the equivalent CRR for the design earthquake. The recommended MSF is given by:

$$\text{MSF} = \frac{174}{M^{2.56}} \quad (9)$$

The above recommendations are based on the NCEER Workshop in 1996 (Youd et al., 2001).

An example of the CPT method to evaluate cyclic liquefaction is shown on Figure 5 for the Moss Landing site that suffered cyclic liquefaction during the 1989 Loma Prieta earthquake in California (Boulanger et al., 1997).

A major advantage of the CPT approach is the continuous and reliable nature of the data. CPT data are typically collected every 5 cm (2 inches). This means that data points are collected at the interface between layers, such as between clay and sand. During this transition, the CPT data points do not accurately capture the correct soil response since the penetration resistance is moving from either low to high values or vis-a-versa. The CPT penetration resistance represents an average response of the ground within a sphere of influence that can vary from a few cone diameters in soft clay to 20 cone diameters in dense sand. In these thin transition zones the CPT-based liquefaction method can predict low values of CRR. This is illustrated in Figure 5 at depths of around 6m and 10m, where there are clear interface boundaries between sand and clay. At these locations there a few data points that indicate low values of CRR and hence liquefaction. These thin interface zones are easy to identify and account for in the interpretation. In the following sections, methods to estimate post-earthquake ground deformations will be presented and discussed. Using these CPT-based methods, it is simple to identify the thin interface transition zones and remove, where appropriate.

A key advantage of the CPT based liquefaction method is that continuous profiles can be calculated quickly, which allows the engineer time to study the profile in detail and apply engineering judgment where appropriate. The CPT-based liquefaction method is a simplified approach and is hence conservative. The method was developed from the limit boundary curve in Figure 2 that was developed using average values but the resulting method is applied using all data points. Also, the continuous CPT data predicts low values of CRR in the thin interface transitions zones, as described above.

Juang et al. (1999) has shown that the Robertson and Wride CPT-based liquefaction method has the same level of conservatism as the Seed et al SPT-based liquefaction method, both represent a probability of liquefaction of between 20% to 30% (i.e. more conservative than the expected 50% probability). This conclusion was supported by the NCEER workshop (Youd et al., 2001).

4 POST-EARTHQUAKE DEFORMATIONS

The CPT-based method described above, can provide continuous profiles of CRR and Factor of Safety for given design earthquake loading. However, Factor of Safety is not always the most meaningful means to evaluate liquefaction potential. For most projects a more meaningful evaluation of the effect of a design earthquake on a given project is to estimate the ground deformations that may result from the earthquake. Ground deformations that follow earthquake loading are either vertical settlements or lateral deformations. Although the Factor of Safety due to a design earthquake may be less than 1.0, the resulting deformations may be either acceptable for the project or can be accommodated with appropriate design of the structures.

5 LIQUEFACTION INDUCED VERTICAL GROUND SETTLEMENTS

Liquefaction-induced ground settlements are essentially vertical deformations of surficial soil layers caused by the densification and compaction of loose granular soils following earthquake loading. Several methods have been proposed to calculate liquefaction-induced ground deformations, including numerical and analytical methods, laboratory modeling and testing, and field-testing-based methods. The expense and difficulty associated with obtaining and testing high quality samples of loose sandy soils may only be feasible for high-risk projects where the consequences of liquefaction may result in severe damage and large costs. Semi-empirical approaches using data from field tests are likely best suited to provide simple, reliable and direct methods to estimate liquefaction-induced ground deformations for low to medium risk projects, and also to provide preliminary estimates for higher risk projects. Zhang et al. (2002) proposed a simple semi-empirical method using the CPT to estimate liquefaction induced ground settlements for level ground.

For sites with level ground, far from any free face (e.g., river banks, seawalls), it is reasonable to assume that little or no lateral displacement occurs after the earthquake, such that the volumetric strain will be equal or close to the vertical strain. If the vertical strain in each soil layer is integrated with depth using Equation (10), the result should be an appropriate index of potential liquefaction-induced ground settlement at the CPT location due to the design earthquake,

$$S = \sum_{i=1}^{j} \varepsilon_{vi} \Delta z_i \qquad (10)$$

Where: S is the calculated liquefaction-induced ground settlement at the CPT location; ε_{vi} is the post-liquefaction volumetric strain for the soil sub-layer i; Δz_i is the thickness of the sub-layer i; and j is the number of soil sub-layers.

The method suggested by Zhang et al (2002) is based laboratory results (Ishihara and Yoshimine, 1992) to estimate liquefaction induced volumetric strains for sandy and silty soils, as shown on Figure 6. The procedure can be illustrated using a CPT profile from the Marina District site in California. Figure 7 illustrates the major steps in the CPT-based liquefaction potential analysis and shows the profiles of measured CPT tip resistance q_c, sleeve friction f_s, soil behavior type index I_c, cyclic resistance ratio CRR & cyclic stress ratio CSR, and factor of safety against liquefaction FS, respectively. The data in Figures 7a and 7b can be directly obtained from the CPT sounding.

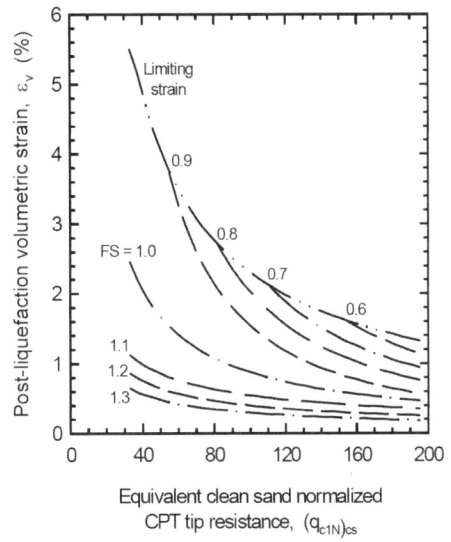

Figure 6. Relationship between post-liquefaction volumetric strain and equivalent clean sand normalized CPT tip resistance for different factors of safety (FS).
(After Zhang et el., 2002)

Figures 7c, 7d and 7e show the results calculated based on the Robertson and Wride procedure shown in Figure 4. Note that, according to Robertson and Wride's approach, CRR is not calculated when the soil behavior type index is greater than 2.6. These soils are assumed to be non-liquefiable in Robertson and Wride's approach.

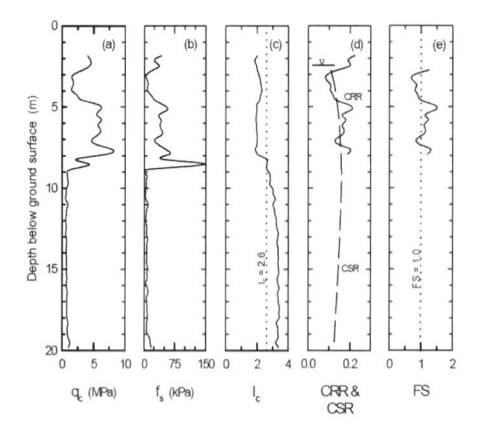

Figure 7. Example plots illustrating the major procedures in performing liquefaction potential analysis using the CPT based Robertson and Wride's (1998) method.
(After Zhang et al., 2002)

The four key plots for estimating liquefaction induced ground settlements by the CPT-based method are presented in Figure 8. Figures 8a to 8d show the profiles of equivalent clean sand normalized tip resistance $(q_{c1N})_{cs}$, factor of safety FS, post-liquefaction volumetric strain ε_v, and liquefaction induced ground settlement S, respectively. Data in Figures 8a and 8b are from the liquefaction potential analysis. Data in Figure 8c are calculated from the results presented by Ishihara and Yoshimine, (1992) and shown in Figure 6. The settlement shown in Figure 8d is obtained using Equation (10) and the volumetric strains from Figure 8c.

The method by Zhang et al (2002) was evaluated using liquefaction-induced ground settlements from the Marina District and Treasure Island case histories damaged by liquefaction during the 1989 Loma Prieta earthquake. Good agreement between the calculated and measured liquefaction-induced ground settlements was found. Although further evaluations of the method are required with future case history data from different earthquakes and ground conditions, as they become available, it is suggested that the CPT-based method may be used to estimate liquefaction-induced settlements for low to medium risk projects and also provide preliminary estimates for higher risk projects.

6 EFFECTS OF OTHER MAJOR FACTORS ON CALCULATED SETTLEMENTS

6.1 Maximum surface acceleration

The amplification of earthquake motions is a complex process and is dependent on soil properties, thickness, frequency content of motions and local geological settings. For a given earthquake and geological setting, the amplification increases with the increase of soil compressibility and with soil thickness. Maximum surface acceleration at a site is one important parameter used in evaluating liquefaction potential of sandy soils. However, its determination is difficult without recorded accelerograghs for a given earthquake because it likely varies with soil stratigraphy, soil properties, earthquake properties, the relative location of the site to the epicenter and even ground geometry. Ground response analysis may help to solve the problem but still leave some uncertainty in the results. Zhang et al. (2002) illustrated the importance of maximum surface acceleration and showed that calculated settlements are not linearly proportional to maximum surface acceleration beyond certain values since the calculated volumetric strains reach limiting values (Fig. 6). Using the Zhang et al (2002) approach it is possible to calculate settlements as a function of surface acceleration as a means to evaluate the sensitivity of the approach to the design earthquake.

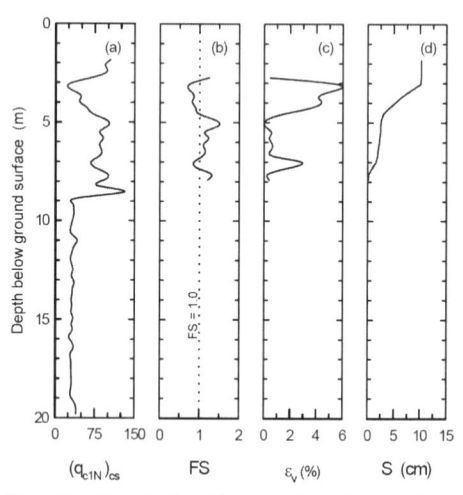

Figure 8. Example plots illustrating the major procedures in estimating liquefaction-induced ground settlements using the Zhang et al (2002) CPT based method.

6.2 Fines content or mean grain size

Although there are many practical situations where liquefaction settlements need to be estimated for sands with little silt to silty-sands, the volumetric strains shown in Figure 6 are applicable to saturated clean sands only. It is therefore necessary to consider the effects of fines content on post-liquefaction volumetric strains and liquefaction potential.

Volumetric strains tend to increase with increasing mean grain size at a given relative density (Lee and

Albaisa 1974). Since increasing the fines content of a sand will result in a decrease of the mean grain size, it is postulated that post-liquefaction volumetric strains would decrease with increasing fines content in sands at a given relative density.

Silty sands have been found to be considerably less vulnerable to liquefaction than clean sands with similar SPT blow-counts (Iwasaki et al. 1978; Tatsuoka et al. 1980; Tokimatsu and Yoshimi 1981; Zhou 1981; et al.). Consequently, modification factors to SPT blow-counts or cyclic resistance ratio for sands with different fines contents or mean grain sizes had been widely used in liquefaction potential analyses (Seed and Idriss 1982; Robertson and Campanella 1985; Seed et al. 1985; Robertson and Wride 1998).

Zhang et al (2002) used the equivalent clean sand normalized CPT tip resistance $(q_{c1N})_{cs}$ to account for the effects of sand grain characteristics and apparent fines content on CRR. $(q_{c1N})_{cs}$ is also used to estimate post-liquefaction volumetric strains for sands with fines. This approach assumes that both the liquefaction resistance and post-liquefaction deformations of silty sandy soils can be quantified using the same method and formulas as for clean sands provided that the equivalent clean sand normalized cone penetration resistance, $(q_{c1N})_{cs}$, is used. This implies that no further modification is required for the effects of fines content or mean grain size if $(q_{c1N})_{cs}$ is used to estimate the liquefaction induced settlements of sandy soils including silty sands. Because $(q_{c1N})_{cs}$ will increase with increase of fines content with sands for a given cone tip resistance, the calculated post-liquefaction volumetric strains will decrease with increase of fines content for a given factor of safety. This approach appears to indirectly account partially or wholly for the effect of grain characteristics on post-liquefaction volumetric strains and provides the same trend as observed by Lee and Albaisa (1974).

6.3 Transitional zone or thin sandy soil layers

It is recognized that transitional zones between soft clay layers and stiff sandy soil layers influence the calculated liquefaction-induced settlements. However, the influence of the transitional zones on calculated $(q_{c1N})_{cs}$, and FS has been partially counteracted implicitly in the Robertson and Wride method. Generally, the measured tip resistance in a sandy soil layer close to a soft soil layer (usually a clayey soil layer) is smaller than the "actual" tip resistance (if no layer interface existed) and the resultant friction ratio is greater than the "actual" friction ratio due to the influence of the soft soil layer. As a result, the calculated value of I_c will increase, and therefore the correction factor K_c, $(q_{c1N})_{cs}$, and FS will increase as well. $(q_{c1N})_{cs}$ and FS may be close to the "true" values in the same sandy soil layer that is not influenced by the adjacent soft soil layer. Therefore, the calculated ground settlements would be close the "actual" values because of this implicit correction incorporated within the Robertson and Wride method.

Zhang et al. (2001) took no correction in an attempt to quantify the influences of both the transitional zones and thin sandy layers on the tip resistance, yet achieved good agreement with the limited case history results. Making no correction for transitional zones and thin layers is conservative when estimating liquefaction potential and liquefaction related deformations. Further investigation is required to quantify the influence of transitional zones or thin sandy soil layers on calculated FS and liquefaction-induced ground settlements.

6.4 Three dimensional distribution of liquefied soil layers

The thickness, depth and lateral distribution of liquefied layers will play an important role on ground surface settlements. Liquefaction of a relatively thick but deep sandy soil (see Figure 9a) may have minimal effect on the performance of an overlying structure founded on shallow foundations. However, liquefaction of a near surface thin layer of soil (Figure 9b) may have major implications on the performance of the same structure.

Ishihara (1993) provided some guidance on the effect of thickness and depth to the liquefied layer on potential settlements that may be reasonable provided that the site is not susceptible to ground oscillation or lateral spread. Gilstrap (1998) concluded that Ishihara's relationship for predicting surface effects may be oversimplified. As well, the application of Ishirara's criteria in practice for cases with multiple liquefied layers (Figure 9c) is not clear.

The lateral extent of liquefied layers may also have an effect on ground surface settlements. A small locally liquefied soil zone with limited lateral extent (Figure 9d) would have limited extent of surface manifestation than that for a horizontally extensive liquefied soil zone with the same soil properties and vertical distribution of the liquefied layer. On the other hand, the locally liquefied soil zone may be more damaging to the engineered structures and facilities due to the potential large differential settlements. However, no quantitative study has been reported for the effect of lateral extent of liquefied layers on ground surface settlements.

Neglecting the effect of three-dimensional distribution of liquefied layers on ground surface settlements may result in over-estimating liquefaction-induced ground settlements for some sites. Engineering judgement is needed to avoid an overly conservative design. Case histories from previous earthquakes have indicated that little or no surface manifestation was observed for cases where the depth from ground surface to the top of the liquefied layer was greater than 20 m. Care is required to detect local zones of soil that may liquefy and to estimate the potential differential settlements that may occur.

6.5 Factor K_c

Robertson and Wride (1998) recommended the factor K_c be set equal to 1.0 rather than using K_c of 1.0 to 2.14 when the CPT data plot in the zone defined by $1.64 < I_c < 2.36$ and $F < 0.5\%$ to avoid confusion of very loose clean sands with denser sands containing fines. However, if the CPT data of a dense sand with fines plots in the zone ($1.64 < I_c < 2.36$ and $F < 0.5\%$), the calculated $(q_{c1N})_{cs}$ value for the dense sand could be reduced by one-half. Although this recommendation is conservative for evaluating liquefaction potential of sandy soils, it may result in over-estimating liquefaction-induced ground settlements for sites with denser sands containing fines that fit in that zone. This seems to be true for some of the CPT soundings in the two case histories studied by Zhang et al (2002). For example, based on soil profiles, CPT profiles, and engineering judgment, a portion of the soil that should have been assessed as dense sand containing fines, was classified as very loose clean sand with K_c equal to 1.0.

Zhang et al (2002) showed that when the settlements for the two case histories were recalculated without following the recommendation of K_c equal to 1.0 for $1.64 < I_c < 2.36$ and $F < 0.5\%$, there was almost no effect for the western and central parts of Marina District and only small (up to 14%) effects for Treasure Island. However the calculated settle-

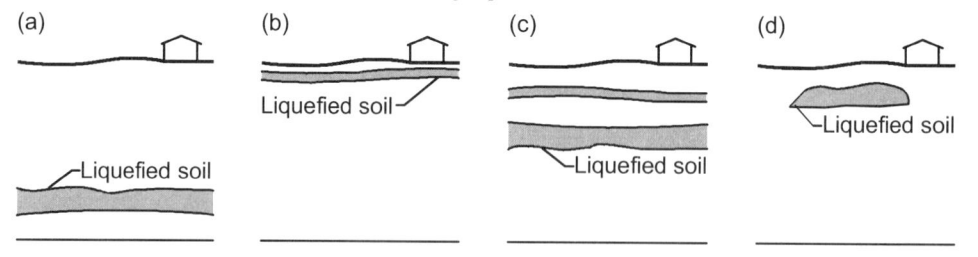

Figure 9. Four hypothetical cases showing importance of three-dimensional distribution of liquefied layers.

ments for the eastern part of Marina District were reduced by a factor of about 2 without the recommendation for K_c. The effect of this recommendation on calculated ground settlements depends on the amount of the soils that fit in the zone defined by $1.64 < I_c < 2.36$ and $F < 0.5\%$ within a soil profile for a site studied. If a large amount of the soils fit in this zone, the effect could be more significant than that for the two case history sites studied above. Soil sampling is therefore recommended to clarify soil properties for sites where a large amount of soil plots in the zone $1.64 < I_c < 2.36$ and $F < 0.5\%$.

6.6 Cutoff of I_c equal to 2.6

A cutoff of I_c equal to 2.6 is used to distinguish sandy and silty soils from clayey soils, which are believed to be non-liquefiable (Robertson and Wride, 1998). Gilstrap (1998) concluded that the I_c cutoff of 2.6 recommended by Robertson and Wride (1998) is generally reliable for identifying clayey soils, but noticed that 20% to 50% of the samples with I_c between 2.4 to 2.6 were classified as clayey soils based on index tests. This implies that the cutoff of I_c equal to 2.6 appears slightly conservative.

Zhang et al (2002) investigated the sensitivity of the calculated settlements to this cutoff for the two case histories using a cutoff of I_c equal to 2.5. The calculated settlements with the cutoff of I_c equal to 2.5 were slightly smaller than with the cutoff of 2.6. For the two cases, only a small portion of the soil in the profiles had I_c ranging from 2.5 to 2.6, thus the use of a cutoff of I_c equal to 2.6 does not greatly overestimate the settlements.

Neglecting the influence of the recommendation for K_c and the cutoff line of I_c equal to 2.6 on the calculated ground settlements is conservative. However, soil sampling is recommended to avoid unnecessary overestimation of liquefaction-induced ground settlements for some sites where a large amount of the soils have a calculated I_c close to 2.6 or/and fit in the zone defined by $1.64 < I_c < 2.36$ and $F < 0.5\%$.

7 LIQUEFACTION INDUCED LATERAL DISPLACEMENTS

Generally, liquefaction-induced ground failures include flow slides, lateral spreads, ground settlements, ground oscillation, and sand boils. Lateral spreads are the pervasive types of liquefaction-induced ground failures for gentle slopes or for nearly level (or gently inclined) ground with a free face (e.g., river banks, road cuts).

Several methods have been proposed to estimate liquefaction-induced lateral ground displacements including numerical models, laboratory tests, and field-test-based methods. Challenges associated with sampling loose sandy soils limit the applications of numerical and laboratory testing approaches in routine practice. Field-test-based methods are likely best suited to provide simple direct methods to estimate liquefaction-induced ground deformations for low- to medium-risk projects and to provide preliminary estimates for high-risk projects.

One-g shake table tests have been conducted to investigate the mechanisms of liquefaction-induced ground lateral spreads. These tests support the hypothesis that lateral spreads result from distributed residual shear strains throughout the liquefied layers. The residual shear strains in liquefied layers are primarily a function of: (a) maximum cyclic shear strains γ_{max}, and (b) biased insitu static shear stresses. In this paper, γ_{max} refers to the maximum amplitude of cyclic shear strains that are induced during undrained cyclic loading for a saturated sandy soil without biased static shear stresses in the direction of cyclic loading. Biased in situ static shear stresses are mainly controlled by ground geometry at the site (e.g., ground slope, free face height, and the distance to a free face). The thickness of liquefied layers will also influence the magnitude of lateral displacements, with greater lateral displacements for thicker liquefied layers. Both γ_{max} and the thickness of liquefied layers are affected by soil properties and earthquake characteristics.

Zhang et al (2004) suggested a semi-empirical method based on CPT results to estimate the lateral displacement using the combined results from laboratory tests with case history data from previous earthquakes. The method captures the mechanisms of liquefaction-induced lateral spreads and characterizes the major factors controlling lateral displacements. Application of the method is simple and can be applied with only a few calculations following the CPT-based liquefaction potential analysis. The approach may be suitable to estimate the magnitude of lateral displacements associated with liquefaction-induced lateral spread for gently sloping (or level) ground with or without a free face for low to medium-risk projects, or to provide preliminary estimates for higher risk projects.

The Zhang et al (2004) method uses the laboratory results presented by Ishihara and Yoshimine (1992) to estimate the maximum shear strains, γ_{max}, as shown on Figure 10. Zhang et al (2004 used the correlation between D_r and normalized cone tip resistance (q_{c1N}) suggested by Tatsuoka et al. (1990):

$$D_r = -85 + 76 \log(q_{c1N}) \qquad (q_{c1N} \leq 200) \qquad (11)$$

where: q_{c1N} is the normalized CPT tip resistance corrected for effective overburden stresses corresponding to 100 kPa (Robertson and Wride, 1998).

This correlation provides slightly smaller and more conservative estimates of relative density than the correlation by Jamiolkowski et al. (1985) when q_{c1N} is less than about 100.

Integrating the calculated γ_{max} values with depth will produce a value that is defined as the lateral displacement index, (LDI):

$$LDI = \int_0^{Z_{max}} \gamma_{max} dz \qquad (12)$$

where Z_{max} is the maximum depth below all the potential liquefiable layers with a calculated FS < 2.0.

Using case history results and ground geometric parameters to characterize the ground geometry, the lateral displacement (LD) can be estimated.

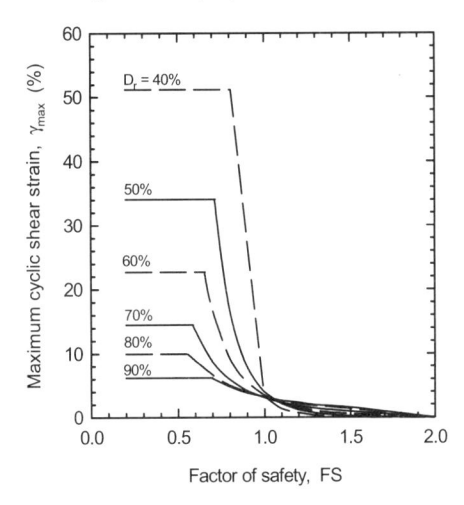

Figure 10. Relationship between maximum cyclic shear strain and factor of safety for different relative densities D_r for clean sands (after Zhang et al., 2004).

For gently sloping ground without a free-face; Zhang et al (2004) suggested the following relationship based on the available case histories;

$$\frac{LD}{LDI} = S + 0.2 \qquad \text{(for } 0.2\% < S < 3.5\%\text{)} \qquad (13)$$

Where: S is the ground slope as a percentage.

For level ground conditions with a free-face; Zhang et al (2004) suggested the following relationship based on the available case histories:

$$\frac{LD}{LDI} = 6 \cdot \left(\frac{L}{H}\right)^{-0.8} \quad \text{(for } 4 < L/H < 40\text{)} \quad (14)$$

Where: L is the horizontal distance from the toe of a free-face,
H is the height of the free-face.

Level ground is taken as ground with a slope less than 0.15%.

The Zhang et al (2004) approach is recommended for use within the ranges of earthquake properties, moment magnitude (M_w) between 6.4 and 9.2, and, peak surface acceleration (a_{max}) between 0.19g and 0.60 g, and free face heights less than 18 m. The case history data used for developing the approach, especially for gently sloping ground without a free face, were dominantly from two Japanese case histories associated with the 1964 Niigata and 1983 Nihonkai-Chubu earthquakes, where the liquefied soils were mainly clean sand only. The values for the geometric parameters used in developing the approach were within limited ranges, as specified in Equations (13) and (14). It is recommended that the approach not be used when the values of the geometric parameters go beyond the specified ranges.

The approach by Zhang et al (2004) is suitable to estimate the magnitude of lateral displacements associated with liquefaction-induced lateral spread for gently sloping (or level) ground with or without a free face for low to medium-risk projects, or to provide preliminary estimates for higher risk projects. Given the complexity of liquefaction-induced lateral spreads, considerable variations in magnitude and distribution of lateral displacements are expected. Generally, the calculated lateral displacements using the proposed approach for the available case histories showed variations between 50% and 200% of measured values. The accuracy of "measured" lateral displacements for most case histories is about ± 0.1 to ± 1.92 m. Therefore, it is unrealistic to expect the accuracy of estimated lateral displacements be within ± 0.1 m. The reliability of the proposed approach can be fully evaluated only over time with more available case histories.

The approach by Zhang et al (2004) was developed using case history data with limited ranges of earthquake parameters, soil properties, and geometric parameters. Therefore, it is not recommended that the approach be applied for values of input parameters beyond the specified ranges. Engineering judgement and caution should be always exercised in applying the proposed approach and in interpreting the results. Additional new data are required to further evaluate and update the proposed approach.

8 FLOW LIQUEFACTION

When a soil is strain softening there is the potential for instability. The key response parameter required to estimate if a flow slide would occur is the minimum (residual or liquefied) undrained shear strength, s_{min} following strain softening. Methods have been suggested to estimate the minimum (liquefied) undrained shear strength of clean sands from penetration resistance based on case histories (Seed and Harder, 1990; Stark and Mesri, 1992). A recent re-evaluation of these case histories (Wride et al., 1998) has questioned the validity of the proposed correlation. Recent research has also shown the importance of direction of loading on the minimum undrained shear strength of sands. The minimum undrained shear strength in triaxial compression (TC) loading is higher than that in simple shear (SS), which in turn, is larger than that in triaxial extension (TE). The difference is less as the sand becomes looser. Hence, the appropriate undrained shear strength to be used in a stability analyses will be a function of direction of loading. This has been recognized for some time in clay soils (Bjerrum, 1972).

The possibility of instability in undrained shear is also linked to the brittleness or sensitivity of the soil. Brittleness index (Bishop, 1967) is an index of the collapsibility of a strain softening soil when sheared undrained, which is defined as follows:

$$I_B = \frac{S_{peak} - S_{min}}{S_{peak}} \quad (15)$$

Where:
S_{peak} = the peak shear resistance prior to strain softening
S_{min} = the minimum undrained shear strength

A value of $I_B = 1$ indicates zero minimum undrained shear strength. If there is no strain softening then $I_B = 0$.

Figure 11 shows a link between the response characteristics of brittleness and minimum undrained strength ratio with only a small influence of direction of loading. When the minimum undrained strength ratio decreases, the brittleness increases. The sands are essentially non-brittle ($I_B = 0$) when the minimum undrained strength ratio (S_{min}/p') is greater than about 0.2 for TE, 0.25 for SS and 0.40 for TC. These values are similar to those observed for fine grained (clay) soils of low plasticity (Jamiolkowski et al., 1985). When the undrained strength ratio is less than about 0.1 the brittleness is usually high.

Yoshimine et al. (1998) proposed a relationship between normalized cone penetration resistance and the minimum undrained shear strength in simple shear loading for clean sands based on a combination of laboratory testing and case histories, as shown in Figure 12. The method suggested by Robertson and Wride (1998), can be used to calculate the equivalent clean sand normalized penetration resistance, $(q_{c1N})_{cs}$, which can then be used to estimate the minimum undrained shear strength ratio using Figure 12. The resulting values of minimum undrained shear strength ratio are approximate and apply primarily to young, normally consolidated, uncemented soils.

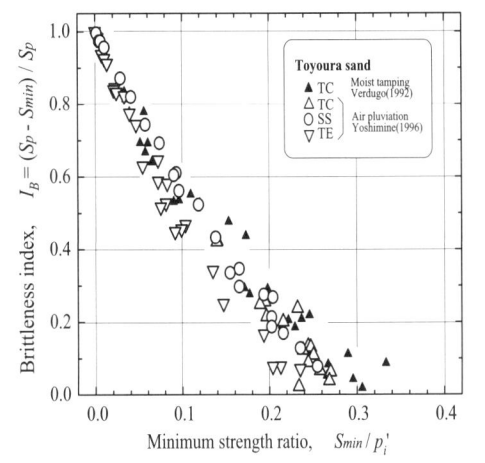

Figure 11. Relationship between minimum undrained strength ratio and brittleness index for clean sand (Yoshimine et al., 1998)

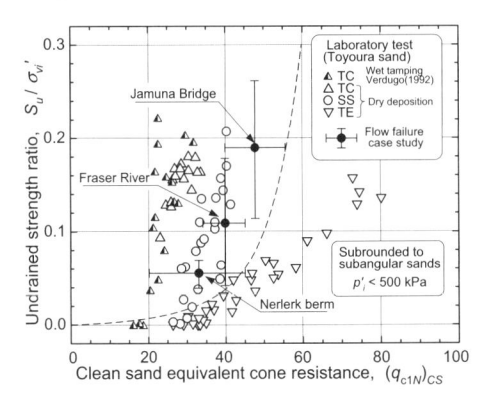

Figure 12. Undrained strength ratio from CPT for clean sand. (After Yoshimine et al., 1998).

Sandy soils with angular grains and aged soils would likely have higher strengths. The actual value of s_{min} to be applied to any given problem will depend on the ground geometry, but in general, the simple shear direction of loading, $(s_{min})_{SS}$, represents a reasonable average value for most problems.

Soils that have a minimum undrained shear strength ratio in simple shear of around 0.30 or higher are generally not brittle. Soils that have a minimum undrained shear strength ratio of around 0.10 or less are often highly brittle. Hence, the value of $s_{u(min)}/\sigma'_{vo} = 0.10$ represents the approximate boundary between soils that can show significant strain softening in undrained simple shear and soils that are general not strain softening. For simple shear direction of loading, this boundary can be represented by a normalized clean sand equivalent CPT value of $(q_{c1N})_{cs} = 50$ (see Figure 12). Hence, a value of $(q_{c1N})_{cs} < 50$ can be used as an approximate criteria to estimate if a soil may be brittle and strain softening in undrained simple shear. If the soil is cohesionless (i.e. non-plastic, $I_c < 2.6$) the loss of strength could occur rapidly over small strains (brittle). If the soil is cohesive the loss of strength could occur more slowly, depending on the degree of sensitivity and plasticity of the soil. Highly sensitive clays (quick clays) can loose their strength rapidly resulting in flow slides. Less sensitive clays tend to deform gradually without a flow slide.

Olsen and Stark (2002) reviewed case histories of flow liquefaction and showed that there was no evidence of flow liquefaction where the mean normalized cone resistance (q_{c1N}) was greater than 65, although only two data points (out of 32) exceeded a value of 50. They also showed that when the normalized cone resistance was less than 50 the minimum (liquefied) shear strength ratio was less than 0.1. No case history had a back-calculated liquefied shear strength greater than 0.12. Only 7 case histories had measured CPT values, the remaining had values either converted from measured SPT, or had estimated CPT values. Olsen and Stark (2002) suggested an average relationship described by;

$$s_{u(LIQ)} / \sigma'_{vo} = 0.03 + 0.00143 (q_{c1N}) \pm 0.03 \quad (16)$$

$$\text{for } q_{c1N} < 65$$

The data presented by Olsen and Stark (2002) made no adjustment for fines content, although the case histories indicated a range in fines contents from 0 to 100%.

The approach suggested by Yoshimine et al. (1998) tends to produce conservative low values compared to the Olsen and Stark (2002) method. Both methods are appropriate for young uncemented, normally consolidated soils where the vertical effective stress is less than about 300 kPa and $I_c \leq 2.6$. In sandy soils where the sand grains are angular the methods may under predict the undrained shear strength. Where the in-situ effective stress is greater than 300 kPa, the minimum undrained shear strength could be smaller depending on the compressibility of the soil. Soils that fall in the lower left region of the CPT soil behavior chart (F<1%, I_c>2.6) may be sensitive and may be susceptible to both cyclic and flow liquefaction depending on soil plasticity, sensitivity (brittleness). Clay soils, where I_c>2.6 and F>1%, are generally non-liquefiable. For high-risk projects, it is advisable to obtain samples to evaluate their liquefaction (strain softening) potential using other criteria.

9 REPRESENTATIVE VALUES

When evaluating the potential for flow liquefaction, there is little guidance given on what value of penetration resistance can be taken as representative of the deposit. In the SPT and CPT based methods for cyclic liquefaction suggested by the NCEER Workshop (Youd et al., 2001), the average values were taken from the case histories to develop the method. However, the CPT-based approach is generally applied using all the continuous CPT data. In general, if all values of the measured penetration resistance are used with a relationship that was based on average values, the resulting design will generally be somewhat conservative.

Seed and Harder (1990), Stark and Mesri (1992) and Olson and Stark (2002) used average values from case histories to develop the relationship between either $(N_1)_{60}$ or q_{c1N} and minimum (liquefied) undrained shear strength to estimate the potential for flow liquefaction. Wride et al. (1998) argued that the minimum value of penetration resistance should be more appropriate. A disadvantage of defining a criteria based on minimum values (especially for the SPT) is the uncertainty that the measured values represent the minimum. In practice, a lower bound relationship is often applied to all measured penetration resistance values. Popescue et al. (1997) suggested that the 20-percentile value would be appropriate as the representative value for cyclic liquefaction. The 20-percentile value is defined as the value at which 20 percent of the measured values are smaller (i.e. 80 percent are larger).

Olson and Stark (2002) used mean values of penetration resistance because most flow failure case histories had insufficient test results to reasonably estimate 20-percentile values. They suggest that if minimum values of penetration resistance are used for design in conjunction with their empirical correlation (equation 16), engineers should consider selecting a liquefied strength ratio greater than the average relationship (i.e. the upper bound values). The boundary suggested by Yoshimine et al. (1998) is essentially a lower bound relationship that applies to mean values and appears to be more conservative than the relationship suggested by Olsen and Start (2002).

In the authors' opinion, the mean value is likely the more representative value for any given deposit for the evaluation of flow liquefaction potential based on the criteria of $(q_{c1N})_{cs}$ = 50, although factors such as, ground geometry, soil profile and the probability of a triggering event can influence the representative value. Olsen and Stark (2002) suggest using mean values and the average relationship (equation 16). If the mean value of the CPT $(q_{c1N})_{cs}$ > 50 than flow liquefaction is unlikely. Olsen and Stark (2002) suggest that to incorporate a strength ratio in a post-triggering stability analysis, a liquefied soil layer can be separated into a number of sublayers of equal σ'_{vo} (stress contours) and (or) equal penetration resistance (penetration contours). For example, each vertical effective stress contour would have an equal value of $s_{u\ (LIQ)}$, and $s_{u\ (LIQ)}$ would increase as σ'_{vo} contours increase.

10 SUMMARY

For low-risk, small-scale projects, the potential for cyclic liquefaction can be estimated using penetration tests such as the CPT. The CPT is generally more repeatable than the SPT and is the preferred test, where possible. The CPT provides continuous profiles of penetration resistance, which are useful for identifying soil stratigraphy and for providing continuous profiles of estimated cyclic resistance ratio (CRR). Corrections are required to both the SPT and CPT results for grain characteristics, such as fines content and plasticity. For the CPT, these corrections are best expressed as a function of soil behavior type, I_c, which is affected by the full range of grain characteristics.

For medium- to high-risk projects, the CPT can be useful for providing a preliminary estimate of liquefaction potential in sandy soils. For higher risk projects, and in regions where there is little previous CPT experience, it is also preferred practice to drill sufficient boreholes with selective sampling adjacent to the CPT soundings to verify various soil types encountered and to perform index testing on disturbed samples. A procedure has been described to correct the measured cone resistance for grain characteristics based on the CPT soil behavior type, Ic. The corrections are approximate, since the CPT responds to many factors affecting soil behavior. Expressing the corrections in terms of soil behavior index is the preferred method of incorporating the effects of various grain characteristics, more than just fines content. When possible, it is recommended that the corrections be evaluated and modified to suit a specific site and project. However, for small-scale, low-risk projects and in the initial screening process for higher risk projects, the suggested general corrections provide a useful conservative guide and provide continuous profiles that capture the full detail of the soil profile.

It is also useful to evaluate CRR using more than one method. For example, the seismic CPT (SCPT) can provide a useful technique to independently evaluate liquefaction potential, since it measures both the usual CPT parameters, as well as shear wave velocity, within the same soil profile. The CPT provides detailed profiles of cone resistance, but the penetration resistance is sensitive to grain characteristics, such as fines content, soil plasticity and mineralogy, and hence corrections are required. The seismic part of the SCPT provides a shear wave velocity profile typically averaged over 1m intervals and, therefore contains less detail than the cone tip resistance profile. However, shear wave velocity is less influenced by grain characteristics and few or no corrections are required (Andrus and Stokoe, 1998). Shear wave velocity should be measured with care to provide the most accurate results, since the estimated CRR is sensitive to small changes in shear wave velocity. There should be consistency in the liquefaction evaluation using either method. If the two methods provide different predictions of CRR profiles, samples should be obtained to evaluate the grain characteristics of the soil. Differences between the shear wave velocity and penetration resistance methods can be caused by aging, cementation and grain compressibility (Lunne et al., 1997).

A key advantage of the integrated CPT method is that the algorithms can easily be incorporated into a spreadsheet or software, as illustrated by the Moss Landing example. The result is a straightforward method for analysing the entire CPT profile in a continuous manner. This provides a useful tool for the engineer to review the potential for cyclic liquefaction across a site using engineering judgement.

An extension to the CPT-based method to evaluate liquefaction potential was suggested by Zhang et al (2002 and 2004) and allows estimates of vertical settlements and lateral spread deformations to be made using the continuous CPT profile.

The method by Zhang et al (2002) to estimate post-earthquake vertical settlements was evaluated using liquefaction-induced ground settlements from the Marina District and Treasure Island case histories damaged by liquefaction during the 1989 Loma Prieta earthquake. Good agreement between the calculated and measured liquefaction-induced ground settlements was found. Although further evaluations of the method are required with future case history data from different earthquakes and ground conditions, as they become available, it is suggested that the CPT-based method may be used to estimate liquefaction-induced settlements for low to medium risk projects and also provide preliminary estimates for higher risk projects.

The method by Zhang et al (2004) to estimate lateral displacements was developed using case history data with limited ranges of earthquake parameters, soil properties, and geometric parameters. Therefore, it is not recommended that the approach be applied for values of input parameters beyond the specified ranges. Engineering judgement and caution should always be exercised in applying the proposed approach and in interpreting the results. Additional new data are required to further evaluate and update the proposed approach.

Soils that have a minimum undrained shear strength ratio of around 0.10 or less are often highly brittle. Hence, the value of $s_{u(min)}/\sigma'_{vo} = 0.10$ represents the approximate boundary between soils that can show significant strain softening in undrained simple shear and soils that are general not strain softening. Research has clearly shown that the undrained shear strength of soils is usually a function of direction of loading, with undrained shear strengths in compression loading often being higher than those in simple shear and triaxial extension. The resulting average minimum undrained shear strength is therefore a function of the slope geometry. Although all projects should be evaluated based on their actual geometry, often the average undrained shear strength is close to that in simple shear loading, which is consistent with the observations made by Bjerrum (1972) for slopes and embankments in clays. For simple shear direction of loading, this boundary can be represented by a normalized clean sand equivalent CPT value of $(q_{c1N})_{cs} = 50$. Hence, a limit of $(q_{c1N})_{cs} < 50$ can be used as an approximate criteria to estimate if a soil may be brittle and strain softening in undrained simple shear. The recent review of flow failure case histories by Olsen and Stark (2002) confirm this value. If the soil is cohesionless (i.e. non-plastic, $I_c < 2.6$) the loss of strength could occur rapidly over small strains (brittle). If the soil is cohesive the loss of strength could occur more slowly, depending on the degree of sensitivity and plasticity of the soil. Highly sensitive clays (quick clays) can loose their strength rapidly resulting in flow slides. Less sensitive clays tend to deform gradually without a flow slide.

Sands that have angular grains may have a minimum (liquefied) undrained strength ratio higher than predicted using the suggested CPT methods by Yoshimine et al (1998) and Olsen and Stark (2002). Aged soils (age > 1,000 years) may also be somewhat stronger. For high-risk projects, the proposed CPT criteria (based on $(q_{c1N})_{cs} < 50$) provides a useful screening technique to identify potentially critical zones where flow liquefaction may be possible. For low risk projects, the proposed CPT methods will generally provide a conservative estimate of the minimum undrained shear strength ratio in simple shear loading. The proposed relationship conservatively estimates the minimum (liquefied) undrained shear strength ratio for a soil structure that contains extensive amounts of loose sandy soils with impeded drainage, such as thick deposits of loose interbedded sands and silts. In soil structures where drainage and consolidation of the liquefied layer can occur during and immediately after the earthquake, higher values of undrained shear strength will likely exist. Such conditions may exist in a thin deposit with free drainage to the ground surface or a deposit interbedded with extensive pervious gravel layers.

ACKNOWLEDGEMENTS

This paper is a summary of many years of research that have involved many colleagues and graduate students as well as support by the Natural Science and Engineering Research Council of Canada (NSERC).

REFERENCES

Andrus, R.D., and Stokoe, K.H. 1998. Guidelines for evaluation of liquefaction resistance using shear wave velocity. Proceedings of the NCEER (National Center for Earthquake Engineering Research) workshop on evaluation of liquefaction resistance of soils, Salt Lake City, Utah, January 1996, T.L. Youd and I.M. Idriss (eds.), NCEER-97-0022, 89-128.

Bishop, A.W. 1967. Progressive failure – with special reference to the mechanism causing it. Panel discussion. Proceedings of the Geotechnical Conference, Oslo, Norway, 2: 142-150.

Bjerrum, L., 1972, Embankments on soft ground. Proceedings of Specialty Conference on Performance of Earth and Earth-supported structures, Lafayette, Indiana, 2: 1-54.

Bartlett, S.F., and Youd, T.L. 1995. Empirical prediction of liquefaction-induced lateral spread. Journal of Geotechnical Engineering, ASCE, 121(4): 316-328.

Boulanger, R.W., Mejia, L.H., and Irdiss, I.M. 1997. Liquefaction at Moss Landing during Loma Prieta earthquake. Journal of Geotechnical and Geoenvironmental Engineering, ASCE, 123(5): 453-468.

Ishihara, K., 1993. Liquefaction and flow failure during earthquakes. 33rd Rankine Lecture, Geotechnique, 43(3): 349-415.

Ishihara, K., and Yoshimine, M. 1992. Evaluation of settlements in sand deposits following liquefaction during earthquakes. Soils and Foundations, 32(1): 173-188.

Iwasaki, T., Tatsuoka, F., Tokida, K., and Yasuda, S. 1978. A practical method for assessing soil liquefaction potential based on case studies at various sites in Japan. Proceeding of the Second International Conference of Microzonation, San Francisco, CA, Vol. 2, pp.885-896.

Jamiolkowski, M., Ladd, C.C., Germaine, J.T., and Lancellotta, R. 1985. New developments in field and laboratory testing of soils. In Proceedings of the Eleventh International Conference on Soil Mechanics and Foundation Engineering, San Francisco, 12-16 August, Vol. 1, pp. 57-153.

Juang, C.H., Chen, C.J., and Tien, Y.M. 1999a. Appraising CPT-based liquefaction resistance evaluation methods -- artificial neural network approach. Canadian Geotechnical Journal, 36(3): 443-454.

Juang, C.H., Rosowsky, D.V., and Tang, W.H. 1999b. Reliability-based method for assessing liquefaction potential of soils. Journal of Geotechnical and Geoenvironmental Engineering, ASCE, 125(8): 684-689.

Lee, K.L., and Albaisa, A. 1974. Earthquake induced settlements in saturated sands. Journal of Geotechnical Engineering, ASCE, 100(GT4): 387-406.

NCEER 1997. Proceeding of the NCEER Workshop on Evaluation of Liquefaction Resistance of Soils, Edited by Youd, T.L., and Idriss, I. M., Technical Report NCEER-97-0022, Salt Lake City, Utah, Decemeber 31, 1997.

Olsen, R.S., and Malone, P.G. 1988. Soil classification and site characterization using the cone penetrometer test. Penetration Testing 1988, ISOPT-1, Edited by De Ruiter, Balkema, Rotterdam, 2: 887-893.

Olsen, S.M. and Stark, T.D., 2002. Liquefied Strength ratio from liquefaction flow failure case histories. Canadian Geotechnical Journal, 39: 629-647

Popescu, R., Prevost, J.H., and Deodatis, G., 1997. Effects od spacial variability on soil liquefaction: some design recommendations. Geotechnique, 47(5): 1019-1036

Robertson, P.K., 1990. Soil Classification using the CPT. Canadian Geotechnical Journal. 27(1),151-158.

Robertson, P.K., and Wride C.E. 1998. Evaluating cyclic liquefaction potential using the CPT. Canadian Geotechnical Journal, 35(3): 442-459.

Robertson, P.K., and Campanella, R.G. 1985. Liquefaction potential of sands using the cone penetration test. Journal of Geotechnical Engineering, ASCE, 22(3): 298-307.

Seed, H.B. 1979. Soil liquefaction and cyclic mobility evaluation for level ground during earthquakes. Journal of Geotechnical Engineering Division, ASCE, 105(GT2): 201-255.

Seed, H.B., and Idriss, I.M. 1971. Simplified procedure for evaluation soil liquefaction potential. Journal of the Soil Mechanics and Foundations Division, ASCE, 97(SM9): 1249-1273.

Seed, H.B., and Idriss, I.M. 1982. Ground motions and soil liquefaction during earthquakes. Earthquake Engineering Research Institute, p.134.

Seed, H.B., Tokimatsu, K., Harder, L.F., and Chung, R.M. 1985. Influence of SPT procedures in soil liquefaction resistance evaluations. Journal of Geotechnical Engineering, ASCE, 111(12): 1425-1440.

Seed, R. B., and Harder, L. F. 1990. SPT-based analysis of cyclic pore pressure generation and undrained residual strength. Proceedings H. B. Seed Memorial Symp., Vol. 2, BiTech Publishers, Vancouver, BC, May, pp.351-376.

Stark, T.D., and Mesri, G.M., 1992. Undrained shear strength of liquefied sands for stability analysis. Journal of Geotechnical Engineering, ASCE, 118 (11): 1727-1747.

Tatsuoka, F., Iwasaki, T., Tokida, K., Yasuda, S., Hirose, M., Imai, T., and Kon-no, M. 1980. Standard penetration tests and soil liquefaction potential evaluation. Soils and Foundations, JSSMFE, 20(4): 95-111.

Tokimatsu, K. and Yoshimi, Y. 1981. Field correlation of soil liquefaction with SPT and grain size. Proceedings of Eight World Conference on Earthquake Engineering, San Francisco, CA, 95-102.

Yoshimine, M., Robertson, P.K., and Wride, C.E., 1998, Undrained shear strength of clean sands, Accepted for publication in the Canadian Geotechnical Journal.

Wride, C.E., McRoberts, E.C. and Robertson, P.K., 1998. Reconsideration of Case Histories for Estimating Undrained Shear strength in Sandy soils. Accepted for publication in the Canadian Geotechnical Journal.

Youd, T.L., Idriss, I.M, Andrus, R.D., Arango, I., Castro, G., Christian, J.T., Dobry, R., Finn, W.D.L., Harder, L.F., Hynes, M.E., Ishihara, K., Koester, J.P., Liao, S.S.C., Marcuson, W.F., Martin, G.R., Mitchell, J.K., Moriwaki, Y., Power, M.S., Robertson, P.K., Seed, R.B., Stokoe, K.H. 2001. Liquefaction resistance of soils: summary report from the 1996 NCEER and 1998 NCEER/NSF Workshops on evaluation of liquefaction resistance of soils. Journal of Geotechnical and Geoenvironmental Engineering, **127**(4): 297 – 313.

Zhang, G., Robertson, P.K. and Brachman, R.W.I, 2002. Estimating liquefaction-induced ground settlements from CPT for level ground. Canadian Geotechnical Journal, 39: 1168-1180

Zhang, G., Robertson, P.K. and Brachman, R.W.I, 2004. Estimating liquefaction-induced lateral displacements using the SPT or CPT. Journal of Geotechnical and Geoenvironmental Engineering.

Zhou, S.G. 1981. Influence of fines on evaluating liquefaction of sand by CPT. Proceedings of International Conference on Recent Advances in Geotechnical Earthquake Engineering and Soil Dynamics, St. Louis, MO, 1, 167-172.

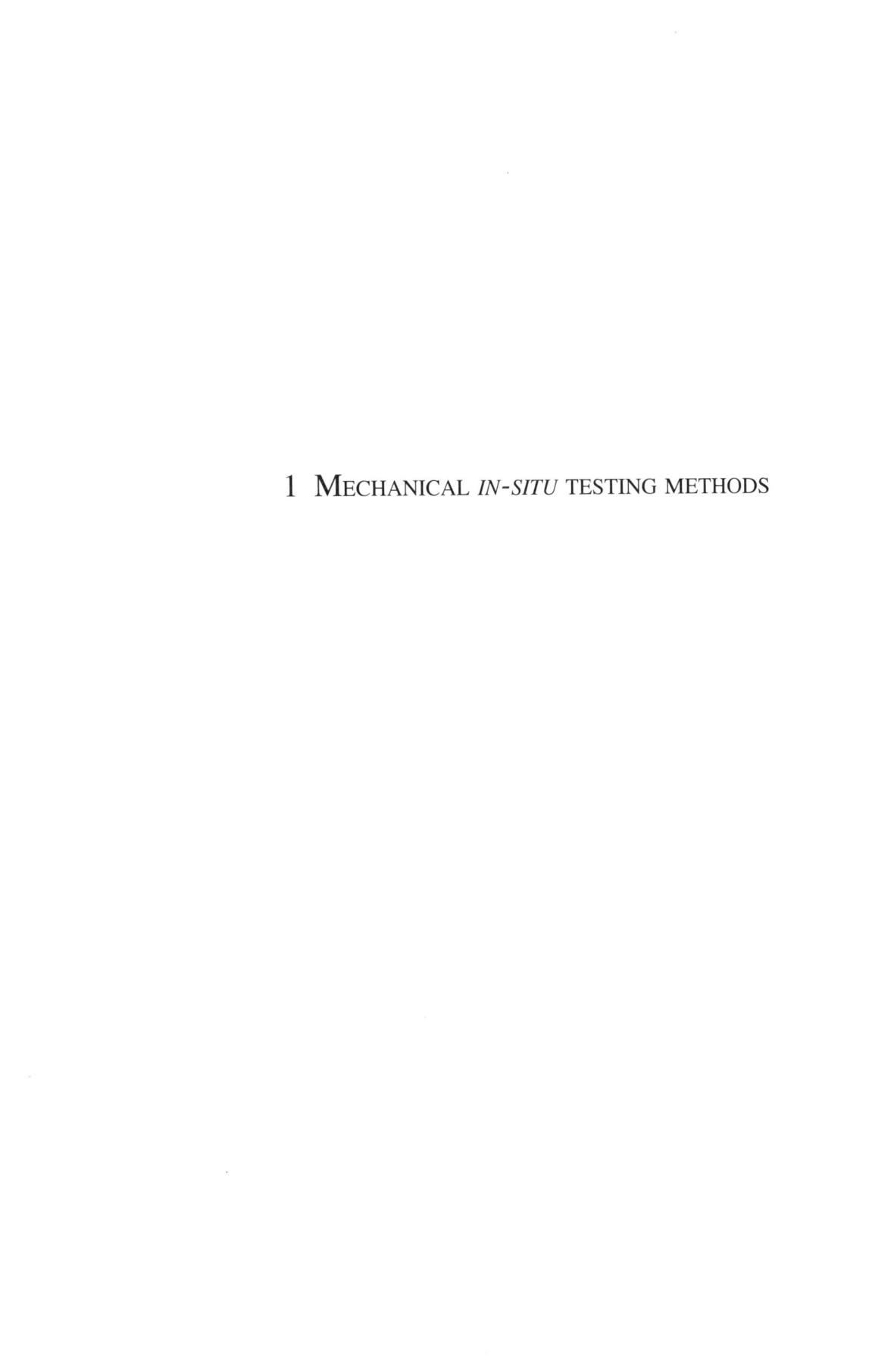

1 MECHANICAL *IN-SITU* TESTING METHODS

General report: Mechanical in-situ testing methods

Marcelo Devincenzi
Igeotest, SL, Spain

John J. M. Powell
Centre for Structural and Geotechnical Engineering, Building Research Establishment, Watford, UK

Nuno Cruz
Mota-Engil, Engenharia e Construções, SA; Faculdade de Ciências e Tecnologia da Universidade de Coimbra, Portugal

Keywords: in situ test, characterisation, investigation programme, standards

ABSTRACT: The objective of this report is to introduce ISC'2 Session 1 "Mechanical in-situ testing methods". A general prologue is made stressing the potential of geotechnical/geoenvironmental in situ testing and the importance of global standards that will ensure that tests are carried out with consistency, a subject of prime importance to transfer experience from one country to another or develop generic correlations. A wide range of parameters can be reliably obtained from in situ testing and it is noticeable that the ratings of some tests have improved as the database of experience has built up, but others have declined as the initial predictions of their capability have been found to be wanting.

Effort is required in order to improve the situation and get the best from existing tests. The papers presented to Session 1 are an example of this fact. The 20 papers, from 13 different countries around the world, cover a wide range of topics and uses of in situ tests and this variability is a good example of their versatility and usefulness for a ground investigation programme provided the right test is selected for the situation and the equipment and procedures allow repeatable and quality data to be obtained.

1 INTRODUCTION

Successful geotechnical design and construction require a good knowledge of the mechanical behaviour of the ground including its spatial variability. The requisite information is gathered as part of a ground investigation programme.

The objective of any subsurface exploration programme is to determine the following:

1. Nature and sequence of the subsurface strata (geological regime)
2. Groundwater conditions (hydrogeological regime)
3. Physical and mechanical properties of the subsurface strata (engineering regime)

For geo-environmental site investigations of ground contamination there is the additional requirement to determine the

4. Distribution and composition of contaminants (geoenvironmental regime)

These requirements vary in volumetric extent depending on the nature of the proposed project and the perceived ground related risks. There are many techniques available to meet the objectives of a ground investigation and these include both field and laboratory testing of the ground. Laboratory tests include those that test elements of the ground, such as triaxial tests and those that test prototype models, such as centrifuge tests. Field tests include drilling, sampling, in-situ testing, full-scale testing and geophysical tests.

An ideal ground investigation (GI) should include a combination of laboratory tests, primarily to classify the ground, and field tests, primarily to determine engineering parameters.

Field and laboratory techniques should be viewed as complementary rather than competitive. In situ tests can, however, often offer significant advantages over laboratory tests, for example:

- they can be quicker, easier and cheaper than sampling and laboratory testing,
- the soil can be assessed in its natural environment without the potential problems of sample disturbance,
- and the spatial variability of the deposit can be more fully investigated.

Table 1 (from Lunne et al 1997) is an updated version of a table originally devised in the early

Table 1: The applicability and usefulness of in situ tests. Lunne et al., 1997

Group	Device	Soil Type	Profile	u	*φ'	s_u	I_D	m_v	c_v	k	G_0	σ_h	OCR	σ–ε	Hard rock	Soft rock	Gravel	Sand	Silt	Clay	Peat
Penetrometers	Dynamic	C	B	-	C	C	C	-	-	-	C	-	C	-	-	C	B	A	B	B	B
	Mechanical	A	A/B	-	C	C	B	C-	-	-	C	C	C	-	-	C	C	A	A	A	A
	Electric (CPT)	B	A	-	C	B	A/B	C	-	-	B	B/C	B	-	-	C	C	A	A	A	A
	Piezocone (CPTU)	A	A	A	B	B	A/B	B	AB	B	B	B/C	B	C	-	C	-	A	A	A	A
	Seismic (SCPT/SCPTU)	A	A	A	B	A/B	A/B	B	AB	B	A	B	B	B	-	C	-	A	A	A	A
	Flat Dilatometer (DMT)	B	A	C	B	B	C	B	-	-	B	B	B	C	C	C	-	A	A	A	A
	Standard (SPT)	A	B	-	C	C	B	-	-	-	C	-	C	-	-	C	B	A	A	A	A
	Resistivity probe	B	B	-	B	C	A	C	-	-	-	-	-	-	-	C	-	A	A	A	A
Pressure-meters	Pre-Bored (PBP)	B	B	-	C	B	C	B	C	-	B	C	C	C	A	A	B	B	B	A	B
	Self boring (SBP)	B	B	A[1]	B	B	B	B	A[1]	B	A[2]	A/B	B	AB[2]	-	B	-	B	B	A	B
	Full displacement (FDP)	B	B	-	C	B	C	C	-	-	A2	C	C	C	-	C	-	B	B	A	A
Others	Vane (FVT)	B	C	-	-	A	-	-	-	-	-	-	B/C	B	-	-	-	-	-	A	B
	Plate load	C	-	-	C	B	B	B	C	C	A	C	B	B	B	A	B	B	A	A	A
	Screw plate	C	C	-	C	B	B	B	C	C	A	C	B	-	-	-	-	A	A	A	A
	Borehole permeability	C	-	A	-	-	-	-	B	A	-	-	-	-	A	A	A	A	A	A	B
	Hydraulic fracture	-	-	B	-	-	-	-	C	C	-	B	-	-	B	B	-	-	C	A	C
	Crosshole /Downhole / Surface seismic	C	C	-	-	-	-	-	-	-	A	-	B	-	A	A	A	A	A	A	A

Applicability: A = high, B = moderate, C = low, -= none

*φ' = will depend on soil type; [1] = only when pore pressure sensor fitted; [2] = only when displacement sensor fitted.

u = in situ static pore pressure; φ' = effective internal friction angle; s_u = undrained shear strength; m_v = constrained modulus; c_v = coefficient of consolidation; k = coefficient of permeability; G_0 = shear modulus at small strains; OCR = overconsolidation ratio; σ-ε = stress-strain relationship; I_D = density index

1980s and presents a list of some of the major in-situ tests and their applicability for use in different ground conditions. If this table were compared with the earlier versions it would be noticeable that the ratings of some tests have improved as the database of experience has built up, but others have declined as the initial predictions of their capability have been found to be wanting. *There is little doubt that the levels of applicability given in Table 1 can now be attained or exceeded provided the tests are selected and used correctly.* This applies to all levels of in situ testing from the simplest basic tests such as dynamic probing and the SPT, through tests such as the CPT and piezocone (CPTU), to the more complex devices such as the self-boring pressuremeter (SBP). Table 1 shows that a wide range of parameters can be reliably obtained from in situ testing. Ideally we would want all assessments in Table 1 to be 'A' and researchers and others continue to strive to raise the levels of applicability however this may not always be achievable if there is a fundamental weakness in the appropriateness of the test.

It would appear that many practitioners have felt that too often the capabilities of in situ tests have been over-sold or inappropriately applied. This has resulted in dissatisfaction when the tests failed to deliver what was promised. The fact that we continue to see papers with more and more correlations for in situ devices often does little to improve this feeling; however it is often not the test that is as fault but its operation and application. Before anything else the results from and in situ test must be repeatable within the bounds of ground variability. There must be consistency in both equipment and operation wherever it is specified and used not just in one country but around the world and this is particularly important if we are to transfer experience from one country to another or develop generic correlations.

We have seen with the CPT how the realisation of the effects of porewater pressure on measured cone resistance has enabled both consistency of results between devices and reduced scatter in developed correlations.

In the SPT we have over the years seen how the results can show significant scatter resulting not simply from the simplicity of the test and operator influences but more importantly from how equipment variations, and particularly the energy input from different equipment, influences the results. Several papers covering this specific topic are presented in Session 1 and commented upon below. The latest CEN/ISO specification for the test now requires knowledge of the actual energy being delivered by the equipment not just a theoretical value.

2 REALISING THE FULL POTENTIAL OF IN SITU TESTING

So how can we improve the situation and get the best from existing tests without developing new ones?
- Firstly, whoever is specifying a ground investigation programme should always consider in situ testing.
- they should at least have a basic understanding of the various tests and their strengths (and weaknesses or limitations).
- they should be able to select the right test for the situation and once a decision has been made then be able to specify the correct equipment and procedures for achieving the desired results. They need standards available to them that will ensure that the tests are carried out with consistency (NB. For confidence in the results there must always be a way of checking the quality of the data determined).
- they should also consider whether additional information might be required later, for example as a result of design changes. (It may well cost very little extra to gather that data at the same time thus avoiding having to make the best of the original data or incurring remobilisation costs later).

We are seeing increasingly in some countries and in CEN and ISO standards currently in preparation, that specifications for test procedures are now trying to help guide the specifier, for example having various specified classes of accuracy for CPT based on soil type and data use (profiling or soil parameters). Furthermore, we should be encouraging accreditation procedures for in situ testing. If current practice has resulted in cost cutting and bad practice, then this should be firmly discouraged even if some small additional costs are incurred. Two detailed papers dealing with ISO/CEN standards are presented in Session 1 and commented upon below.

With in situ testing we have at our disposal very powerful tools that can yield a great deal of valuable information as part of a well planned GI provided they are specified and used correctly. We should not be specifying them without due thought to the end result and the reliability we can put on the data gathered. The lessons learnt from the past must be used to ensure that as other in situ tests are developed they are validated with reliable databases and that specifications and procedures allow their full potential to be developed.

We do not all have to be experts in in situ testing but a sound understanding of the test methods and equipment, coupled with improved specifications and better guidance will ensure that the engineer is able to realise the full potential of these tests.

By selecting the right configuration of tests, in situ testing will give 4 main advantages over the traditional combination of borings, sampling and other testing, namely:
1. continuous or near continuous data
2. repeatable and reliable data
3. speed of operation (potential for shorter GI time-scales)
4. cost savings.

It should also be remembered that the power of in situ tests is not restricted to soil parameter determination; there are also many examples of their use in indirect design applications where parameters unique to a particular test type can be used directly in design procedures.

3 REVIEW OF PAPERS

Twenty papers from 13 different countries are presented in Session 1 covering a wide range of topics and uses of in situ tests as can be seen in Table 2. This breadth is a good example of their versatility and usefulness for a ground investigation programme. Some of the more relevant items are presented below.

3.1 *International and European Standards*

International and European standards currently in preparation are the main subject of two papers presented to Session 1. As stated above, international standardisation is of prime importance in order to harmonise the quality requirements for equipment as well as methodologies in order to enable comparable results to be obtained. Table 3 briefly summarises the Technical Committees involved in CEN and ISO and the ongoing standards and technical specifications (TS) they are producing.

The paper by Eitner et al. describes in detail the general framework of committees belonging to the International Organisation for Standardisation (ISO) and the European Committee for Standardisation (CEN) and their labours to prepare common standards on equipment and methods used for soil and rock identification, drilling, sampling, field and laboratory testing as well as groundwater measurements for geo-engineering practice. The paper by Stölben et al. deals exclusively with standards on sampling by drilling and groundwater measurements.

Table 2: Session 1 by tests & subject– summary of papers

Test	N° of papers	Authors	Main subject & Notes
Standards	2	V. Eitner, R. Katzenbach & F. Stölben	ISO & EN standards on Geotechnical investigation and testing for site characterisation: current framework in European countries. See Table 3.
		F. Stölben, V. Eitner & H. Hoffmann	ISO & EN Standards on Geotechnical investigation and testing for site characterisation: sampling by drilling and groundwater measurements. See Table 3.
SPT	5	E. Odebrecht, F. Schnaid, M.M. Rocha & G.P. Bernardes	Energy ratio with instrumentation of system. Different location of sensors. Effect of rod length analysed. New approach to define Energy efficiency. See Table 4.
		Dong-Soo Kim, Won-Seok Seo & Eun Scok Bang	Energy ratio with instrumentation of system. Different hammers. Force waveforms and factors affecting efficiency analysed. See Table 4..
		E.H. Cavalcante, F.A.B. Danziger & B.R. Danziger	Energy ratio with instrumentation of system. Manual donut hammer. Different ranges of Nspt blow count with different behaviours. See Table 4.
		A.S.P. Peixoto, D. de Carvalho & H.L. Giacheti	CPT-T: CPT with torque measurement. Electrical torquemeter. State of the art in Brazilian practice. Suggestion of test procedure. Comparison of 4 methods to predict pile capacity based on CPT-T.
		C.R. Daniel, J.A. Howie, R.G. Campanella & A. Sy	Characterisation of SPT grain size effects in gravels.
CPT	6	V. Whenham, N. Huybrechts, M. De Vos, J. Maertens, G. Simon & G. Van Alboom	CPT case study with comparative analysis of results using 3 types of mechanical cones (mantle cone, Begemann cone and simple cone with closing nut) and electrical cone. Ratios of cone resistance $q_{c\,mech}$ / $q_{c\,elec}$ presented.
		S.G. Fityus & L. Bates	CPT case study assessing parameters on expansive residual clay at Maryland, where others data available. Thickness and identification of expansive soil as well as degree of weathering of basal rock.
		D.J. Balbi & F. Saboya Jr.	CPTU case study with statistical treatment of data to identify soil groups in Quaternary soils north Rio de Janeiro. Cluster analysis using B_q and Q_t (normalised net q_T) as variables as Hegazy & Mayne (2003).
		M.F. Silva & M.D. Bolton	Centrifuge model to investigate sensitivity of CPTU to layering effects and grain size at different penetration rates under different loading conditions & permeabilities.
		S.M. Fairuz, J. Rohani & T. Lunne	CPTU off shore case study. Characterisation of sands, NC clays and OC clays at four locations in South China Sea. N_{kt} of fine grained soils 15 to 20 recommended.
		F.A. Trevor & P.W. Mayne	CPTU & lab tests case study in marine clays in Southeast Asia to estimate OCR and Su using Mayne (1991) method based on cavity expansion theory and critical state concepts. Determination of correction factors ξ for this method.
CPT&DMT	1	P.W. Mayne & T. Liao	CPT-DMT interrelationships in Piedemont Residuum: I_D f(FR%), E_D f(q_T). Equivalent CPT method developed for obtaining constrained modulus.
DMT	1	A. Arulrajah, M.W. Bo & H. Nikraz	DMT test case study in soft marine clays under pre-load improvement treatment. Dissipation testing with DMT to determine c_h and k_h as a tool to evaluate degree of consolidation of pre-loaded area using Bo et al. (1977) method.
DP	2	A. Fakher & M. Khodaparast	Dynamic penetration in cohesive soft soils. Correlation N – E.
		A.M. Rahim & K.P. George	Dynamic penetration to estimate subgrade resilient modulus.
DP & PLT	1	A. Arbaoui, R. Gourvès, Ph. Bressolette, L. Bodé	Determination of in situ deformation modulus with a penetrometer. Method. Deformation, at depth load test described.
PBP & PLT	1	A. Agharazi & M. Moradi	Comparison between plate load test results and dilatometer test results.
FVT	1	H. Åhnberg, R. Larsson & C. Berglund	Influence of vane size and equipment on the results of field vane tests. Method. Different equipment used and commented. Disturbance effects.

3.2 Standard Penetration Test

As can be seen in Table 2, special interest is paid by several papers to the instrumentation of the SPT test in order to determine the energy transmitted from the falling hammer, which is in accordance with the current CEN ISO procedures (EN ISO 22476, Part 4 in Table 3).

3.2.1 Instrumentation of SPT: Energy Efficiency

Transmitted energy to the rods by SPT hammer blows is highly dependent on different aspects such as the type of hammer, rod length, blow rate and energy calculation methods. Three papers to ISC'2 – two from Brazil and one from Korea- report investigations in this field using nearly the same

Table 3: ISO and CEN framework standards in geotechnical site investigation. Compiled after Eitner et al. and Stölben et al.

Technical Committees (TC) &Sub-committees (SC)	Goals	Standards
ISO/TC 182/SC 1 "Geotechnical Investigation and Testing"	Prepares standards on identification, description and classification of soils and rocks. International "mirror" of CEN/TC 341.	ISO 14688-1 to 3 "Geotechnical investigation and testing – Identification, description and classification of soil" Part 1: Identification and description. 2002. Part 2: Classification principles. 2003. Part 3: Electronic exchange of data of identification and description of soil. Technical Specification: TS. Under development ISO 14689-1 to 2 "Geotechnical investigation and testing – Identification and classification of rock" Part 1: Identification and description". 2003. Part 2: Electronic exchange of data of identification and description of rock. Technical Specification (TS): Under development
CEN/TC 341 "Geotechnical Investigation and Testing"	Standardization in equipment and methods used for drilling, sampling, field and laboratory testing of rock and soil as well as groundwater measurements. In co-operation with ISO/TC 182/SC 1. Five working groups: • Drilling and sampling methods and groundwater measurements. Germany (DIN). See Stölben et al. • Cone and piezocone penetration Tests. Netherlands (NEN) • Dynamic probing and SPT. Germany (DIN) • Testing of geotechnical structures. France (AFNOR) • Borehole expansion tests. France (AFNOR)	Several standards on field investigation in preparation will be published: EN ISO 22475-1 to 3 "Geotechnical investigation and testing – Sampling by drilling and excavation and groundwater measurements" Part 1: Technical principles for execution Part 2: Technical qualification criteria for enterprises and personnel Part 3: Conformity assessment of enterprise and personnel EN ISO 22476 1 to 16 "Geotechnical investigation and testing – Field testing" Part 1: Electrical cone and piezocone penetration tests Part 2: Dynamic probing Part 3: Standard penetration test Part 4: Ménard pressuremeter test Part 5: Flexible dilatometer test Part 6: Self-boring pressuremeter test Part 7: Borehole jack test Part 8: Full displacement pressuremeter test Part 9: Field vane test Part 10: Weight sounding test (TS) Part 11: Flat dilatometer test (TS) Part 12: Lefranc permeability tests Part 13: Water pressure tests Part 14: Pumping tests Part 15: Mechanical cone penetration test Part 16: Plate loading tests
CEN/TC 250/SC 7 "Geotechnical Design"	Eurocode 7, the base code for geotechnical design. Test results are used for evaluation and interpretation leading to derived values.	ENV 1997-1 (European pre-standard), soon available as EN. ENV 1997-2, design assisted by laboratory tests. ENV 1997-3, design assisted by field-testing.

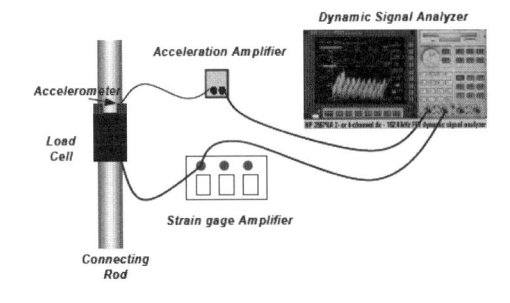

Figure 1: Configuration of instrumented SPT test. Dong-Soo Kim et al.

instrumentation consisting of strain gauges & accelerometers with a data acquisition system to measure force and velocity in order to evaluate the maximum energy delivered to the rods. Typical configuration is shown in Figure 1. Table 4 summarises the main topics of these investigations.

Odebrecht et al. observe that the *theoretical maximum potential energy* PE_{h+r}^* delivered to the soil should be expressed as a function of the *nominal potential energy* (E^* = $0.76.M_h.g$ = $0.76m*63.5kg*9.801m/s$ = 474J), the permanent sampler penetration and the weight of both hammer and rods, that is:

$$PE_{h+r}^* = (0.76 + \Delta\rho)M_h g + \Delta\rho M_r g \qquad (1)$$

Table 4: Instrumented SPT for energy ratio assessment

Author	Odebrecht, Schnaid, Rocha & Bernardes	Cavalcante, Danziger & Danziger	Dong-Soo Kim, Won-Seok Seo & Eun Scok Bang
Type of instrumentation	Load cell with 4 strain gauges sensitive to axial strains only. 2 accelerometers of different range	Strain gauges and piezoelectric accelerometers	8 strain gauges in full bridge and 2 accelerometers
Analysed parameters	Force and acceleration (integrated to velocities)	Force and velocity	Force and velocity
Location of sensors	Just below the anvil Above the sampler Attached to rod, middle position between anvil & sampler	0,5 m below the anvil some near the sampler	Just below the anvil
Computational method	FV	FV	FV & V^2 compared
Type of hammer	Manual pin guided	Donut hand operated with rope	4 types compared: donut hydraulically lifted (WD) chain automatic (CA) manual safety w/rope (RS) manual donut w/rope (RD)
Theoretical energy	478 J	478 J	474 J
Type of soil	Granular with different DR%	Different types	10 to 15 m of weathered soil
Nspt blow range		10 to 60	1 to 60
Main results / findings / contributions	Instrumentation at depth, above the sampler. Energy delivered to the spoon is influenced by both test depth and actual penetration. New concept and definition of efficiency accounted by 3 empirical coefficients, η_1, η_2, η_3 being η_1 function of rod length and inversely proportional to the length of the rods	Transmitted energy to the rod fount to be 83% of 474 J nominal energy (ISSMFE, 1989). Previous works in Brazil showed ≈70% efficiency (energy ratio). Ranges of Nspt with different behaviour found, 1-6, 7-16, 18-60. Correlations Ncorr/N for these values.	Wave shapes of each hammer. Effects of rod length & blow speed. Energy ratios for each hammer type. Energy ratios from FV method larger than F^2 one: ≈70% and ≈60% respectively for CA hammer. Efficiency depends on type of hammer, rod length, blow rate and energy calculation method.

where M_h is the hammer weight, M_r the rod weight, g the gravity acceleration and $\Delta\rho$ sampler penetration under one blow.

The first part of equation (1) represents the hammer potential energy (nominal + additional) and second the rod potential energy. In other words, both length of the rods and permanent spoon penetration contribute to the theoretical energy delivered to the sampler. This becomes significant in loose soils where $\Delta\rho$ is high. This factor is independent of the length of the rods.

Energy losses take place during the penetration process. The energy ratio ER_r or efficiency is traditionally defined as the ratio between the actual maximum delivered energy E_r (i.e. instrumentation) and the nominal potential energy, E^*:

$$ER_r = \frac{E_r}{E^*} \quad (2)$$

It is generally accepted to correct the Nspt result to a reference energy value of 60% of the potential nominal energy of the SPT hammer using standard correction factors (Seed et al., 1985; Skempton, 1986; ISSMFE 1989). However, these corrections assume a theoretical energy input and by using instrumented tests the correct adjustment of the measured Nspt value to the value with a reference energy ratio such as N_{60} can be made.

A new concept of efficiency is postulated by Odebrecht et al. These authors present experimental results of energy measurement in SPT tests carried out under controlled boundary conditions in a calibration chamber and the effective energy measured at the top and the bottom of the rods during impact of hammer. They conclude that efficiency is accounted for by three coefficients, η_1, η_2, η_3 using:

$$PE_{h+r} = \eta_3 [\eta_1 (0.76 + \Delta\rho) M_h g + \eta_2 \Delta\rho M_r g] \quad (3)$$

where η_1, the hammer efficiency, is obtained from measurement at the top of the rod stem, η_2 can be assumed as unity and η_3, the energy efficiency, is expressed as a function of the length of the rods l, being inversely proportional to them:

$$\eta_3 = 1 - 0.0042l \quad (4)$$

3.2.2 SPT-T procedure & practice

The state of the art in Brazilian practice and procedure suggestion for SPT-T test is presented by Peixoto et al. The use of the SPT with torque measurement, that required to turn the sampler, was proposed by Ranzini (1988), being the torque as a kind of *static* component of a *dynamic* test. The torque is empirically used to estimate pile skin friction – mainly for driven piles, being the T/N$_{spt}$ ratio an interesting index for practical purposes, though specific local correlations are needed.

General suggestions for test procedure are reviewed in Table 5; Figure 2 presents an example of a test with an electric device and data acquisition system. Oscillations in the curves are due to lateral movements of the rods.

Four methods for predicting bearing capacity of piles based on SPT-T are summarised and discussed by Peixoto et al. (Decourt, 1996; Alonso, 1996a, 1996b; Carvalho et al., 1998 and Peixoto, 2001).

Table 5: SPT-T test general procedure according Peixoto et al.

Topic	Recommendations	Notes
Procedure	After penetration, an adapter is placed on the anvil for the torquemeter to twist. A centralising device needed to avoid rod lateral movements.	Torquemeter of proper capacity must be used. Torquemeter horizontal during rotation. Alignment of rods during application of torque important key.
Measurements	Maximum and residual torque must be logged.	Mechanical creeping pointer torquemeter provides enough information. Readings immediately after driving sampler.
Rotation speed	5 turns/minute	

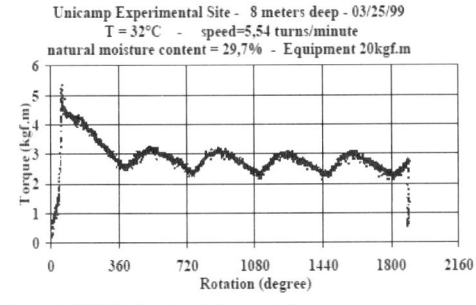

Figure 2: SPT-T plot after Peixoto et al.

3.3 Dynamic penetration

Fakher & Khodaparast present a study about the repeatability of Mackintosh probe at three sites in Iran. This equipment is a lightweight portable handy penetrometer which is still widely used. The authors also present a local correlation between N$_{10cm}$ and s$_u$. (errata: they do however incorrectly quote the correlation of Butcher et al, quoting the relationship between s$_u$ dynamic point resistance for stiff clays and not soft). Their correlation is lower than Butcher's i.e. more blows for the same shear strength and this may be the result of torque and friction not being eliminated with this simple device. It should be noted that the Mackintosh probe is not covered by the new dynamic probing standard in Table 3.

Another paper by Rahim & George presents a correlation to obtain subgrade resilient modulus from N.

3.4 CPT and DMT tests

A total of 8 papers dealing with CPT(u), flat dilatometer (DMT) and CPT-DMT combined investigation are presented to Session 1 (Table 1).

3.4.1 Mechanical CPT

Although electrical CPT(u) is nowadays more commonly used -and recommended by current standards- mechanical devices are still in use in some countries. Whenham et al. present a comparative analysis in Belgium soils establishing ratios q$_{c\ mech}$/q$_{c\ elec}$ for different mechanical devices (mantle cone, friction sleeve mantle cone and simple cone) and different types of soils. Care is needed in extrapolation to other electrical cones and soil types as q$_c$ is used and not q$_t$ and so the results from the electrical cone in clays will be specific to that cone type as they are not corrected for pore pressure effects (Lunne et al 1997).

3.4.2 Soil profile interpretation

The excellent profiling capability of the CPT(u) test is a well recognised feature in the Geotechnical industry. Recent research has also shown that the CPT(u) is a powerful tool for the sedimentological study of delta sediments, even allowing an understanding of how climatic changes and associated sea level changes during Late Quaternary took place (Amorosi & Marchi., 1999; Devincenzi et al., 2003, 2004).

Three papers related to this subject are presented to Session 1 describing investigations on statistical analysis as a helpful tool to use in layered strata, location and depth of expansive soils and evaluation of the degree of weathering in residual massifs.

Balby & Saboya present a case study using the statistical method based on similarity criteria on normalised Q and B$_q$ (Hegazy & Mayne, 2003) to identify soil groups in Quaternary soils. The interpretation is based on cluster analysis, which has a better capacity in detecting changes that could not be detected by direct observation or classification charts, having three important advantages:

1. Organise soils by similarity.
2. Locate boundaries between layers.
3. Identify lenses and mixed soils.
4. Identify systematic errors as rod changes or casual pauses during the test.

Fytius and Bates present another interesting application of CPT in residual expansive clays. In fact, residual soils have a characteristic behaviour, which is not yet well understood. However, characterisation of these soils usually is no more than the basic one. The data usually required for foundation design in expansive soils are:
1. Thickness and position of the layers.
2. Swell potential.
3. Depths of suction and cracks (active layer).
4. Magnitude of suction.

In the proposed approach the CPT results are used to give either direct or indirect estimates of the thickness of the top soil, the position and thickness of expansive clay layers and the depth and degree of weathering of underlying rock.

3.4.3 Centrifuge penetration
Silva & Bolton present research on the use of CPTu tests carried out in centrifuge models to observe the effect of layering of sand layers at different penetration rates and under different loading conditions and permeabilities. The data are used to discuss the methodology of characteristic grain size interpretation and the use of viscous fluid in centrifuge tests.

3.4.4 Assessing undrained shear strength and OCR of marine clays from CPTu
Trevor & Mayne present a modification of Mayne's 1991 method based on cavity expansion theory and critical state concepts to predict both the overconsolidation ratio (OCR) and undrained shear strength from CPTu tests.

The predicted values of s_u and OCR for marine clays soils using Mayne (1991) method are higher than reference triaxial compression tests values, though the interpreted CPTu profiles and the laboratory test profiles show similar trends.

The correction factors to improve the piezocone data interpretation method incorporating the critical state parameter (M) were derived by Trevor (2001) on the basis of statistical evaluation and are designated as:

$\xi_{OCR} = (0.029+0.409M)$ and :

$\xi_{SU} = (0.56+0.095M)$

Modified equations that can be used to predict OCR and s_u in these soils from CPTu tests can be written as:

$$OCR = 2\xi_{OCR}\left[\frac{1}{1.95M+1}\left(\frac{q_t - u_2}{\sigma'_{v0}}\right)\right]^{1.33} \quad (5)$$

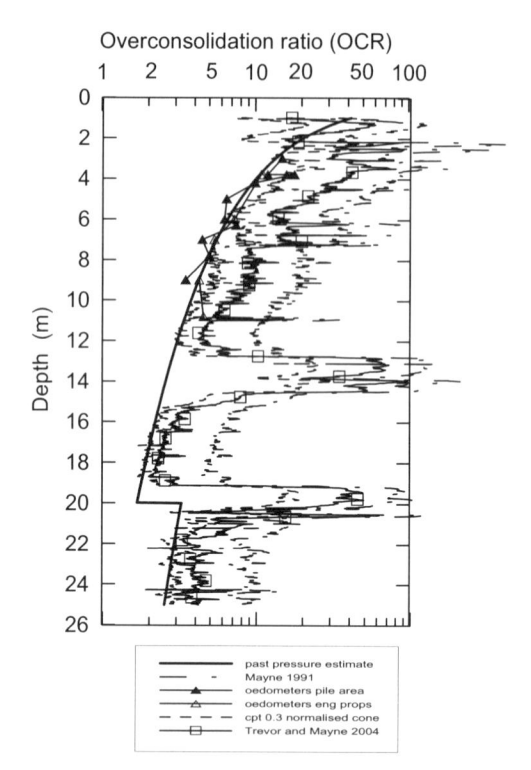

Figure 3. OCR profile for glacial till site using Trevor and Mayne method.

$$s_u = \frac{M}{2}\xi_{su}\left(\frac{q_t - u_2}{1.9M+1}\right) \quad (6)$$

These corrections imply limited sensitivity to M for the correction for s_u, the range is 0.48 – 0.43 as φ' changes from 20 to 35°, whilst for OCR the change is more significant ranging from 0.35 – 0.61. Powell (2001) showed very good agreement for OCR with the original equation i.e. a correction factor of 1, but Figure 3 shows results for a glacial till site (Powell and Butcher, 2002) with both the old and new correlations. It can be seen that the correction improves the predicted OCR profile but is still too high; the use of a simplified normalised cone resistance approach is seen to give a very good fit. Trevor and Mayne recommended to re-evaluate the proposed correction factors for other types of soil and this will be important in understanding when to apply it.

3.4.5 CPTu – DMT interrelationships
The main advantages of tests like CPT/CPTU and DMT are related to cost effectiveness and fast collection of data which permits statistical and numerical analysis. Thus, it is possible to quickly get in-

formation to be used in the powerful numerical models with good enough quality to give a compatible input to the models that does not lower their efficiency.

It should be stressed that combined (DMT+CPT/U) characterisation campaigns offer an approach to new possibilities to explore situations unresolved by each test in isolation, both in sedimentary and residual soils. Both tests, when used together in regular campaigns, can offer an improvement both in geological and geotechnical determinations. The combination of both tests presents the following advantages:

1. Cross checking of the same parameters.
2. Complementary data resulting from both tests.
3. Combination of basic parameters from both tests, offering the possibility of assessing data that otherwise could not be deduced.

With regards to geotechnical information, the combined tests can allow the determination of parameters related to stress history, strength and deformability, permeability and consolidation properties, liquefaction potential, etc. in a continuous manner.

The quantity and quality of information obtained from both tests leads to a better knowledge of soil characteristics, as well as numerical information for design practice.

Interrelationships between CPTu and flat dilatometer (DMT) in Piedemont residual soils that are comprised of silty fine sands to fine sandy silts are investigated by Mayne & Liao who propose:

$$E_D = 5q_t \quad (7)$$

and

$$I_D = 2.0 - 0.14 FR\% \quad (8)$$

E_D being the dilatometer modulus and I_D the material index obtained with DMT test.

The same subject was studied in a research program based on the results from 7 investigations performed in the region between Porto and Braga (Portugal), with a total of 30 borings, 22 CPTU profiles and 23 DMT profiles (Cruz et al, 2004). The region under study is characterized by granitic residual saprolitic soils with a flocculated structure, loose to medium compacted and with a grain size distribution that varies from sandy silts to sands (very similar to Mayne experimental site). The results obtained led to the following conclusions, related to this subject:

- M/q_t (where M is the constrained modulus calculated from the DMT as $M = R_M E_D$) seems to have more potential than E_D/q_t, since the R_M (correlation factor between E_D and M could be seen as a selection parameter for the type of soil; in fact, M parameter is calculated based on 3 basic DMT test parameters, i.e. the calculation is dependent on the type of soil (I_D) and on K_D (which seems to reflect the cementation structure), besides the dilatometer modulus E_D; this seems to be very helpful to work with a wider range of soil types;

- M/q_t is not a constant value, but tends to increase with the cementation structure in a similar manner to the one proposed by Marchetti (1997) to distinguish NC (M/q_c = 5 -12) from OC sands (M/q_c = 12 – 24); Cruz et al (2004) found two different M/q_t correlations in residual soils when working on "NC" and "OC" sides equal to 6,4 and 16,8, respectively; of course these can be sub-divided, but this is a basic well known frontier of mechanical behaviours, which can be used as a reference (Figure 4);

- The overall E_D data in the study ranges from 0,1 to 6,0 MPa, while E_D/q_t falls within the interval 1 – 20; in general E_D/q_t around 5 to 6 represents the division between NC – OC; Analysing data individually (from each of the 7 experimental sites) it becomes clear that the relation E_D/qt changes with the level of cementation, as well as M/q_t, and the results range from 2 to 7;

In conclusion it could be said that both relations are clearly not a constant value but increases with the strength due to cementation effects; the relation M/q_t shows a high correlation factor when used to deduce cohesive intercept due to cementation structure (Cruz et al, 2004). In that sense Mayne's data would be a local correlation that reflects a particular level of cementation structure. The use of M/q_t seems to have a higher potential than E_D/q_t since integrates the numerical definition of type of soil through identification index from DMT (I_D).

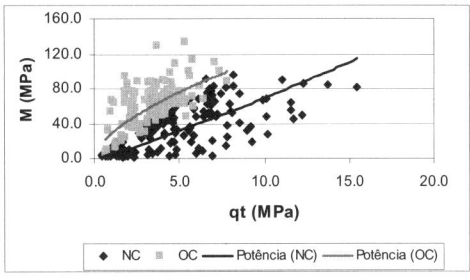

Figure 4: M/q_t for NC and OC (Cruz et al, 2004).

Considering the I_D - Fr relation, the general data given by Cruz et al. (2004) are not as well fitting as Mayne & Liao's. Nevertheless, results seem to converge, since the global analysis gives 1,92 – 0,05 Fr and if intersection is forced to 2 then it is 2 – 0,06 Fr. Analysing data separately and forcing the intersection at 2, then 3 of the experimental sites show a variation of Fr dependent term varying between 0,06 and 0,1. At other sites it is not possible to find any particular trend line.

Data from OC clays slightly cemented in Girona (Spain) also show a good agreement with Mayne & Liao's I_D-Fr proposed relationship, OCR ranging from to 3 to 6 and $M/q_t \approx 25$.

On the contrary, no relation could be found between I_D and Fr in Quaternary pro-delta silty clays (OCR $\approx$ 1.1) with silt intercalations as can be seen in Fig. 5. It must be stressed that these sediments are not of the same nature as those studied by Mayne & Liao.

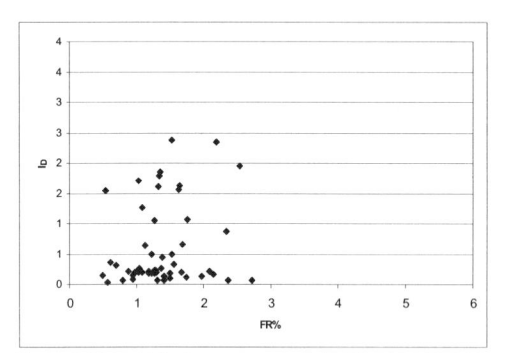

Figure 5: ID vs. Fr in Quaternary pro-delta silty clays in Barcelona, Spain.

3.5 *Influence of Vane size on results of vane tests*

Vane test results are affected by several factors such as the effect of disturbance at insertion, the effect of waiting before starting the test, the shape of the rupture surface, influences of the rate of rotation, the length of the vane shaft protruding below the casing for the vane, the number of blades and the relation between height and diameter of the vane, among other aspects

Research on the influence of the size of the vane on the results of the vane test -an important fact that has no effect on the standards yet- and a comparative analysis of different equipment available on the market today are presented by Åhnberg et al. A detailed listing of different factors affecting the results is also outlined.

The results of the investigation showed that for tests in clay the larger the vane size, the higher the shear strength, provided that the thickness of the shafts and wings are the same. The small vane (55x110mm^2) showed shear strengths that were only about 85% of those obtained with the normal vane (65x130mm^2). Figure 6 shows the comparison.

This can be related to the fact that the relative influence of the disturbance at installation of the vane increases with decreasing vane size.

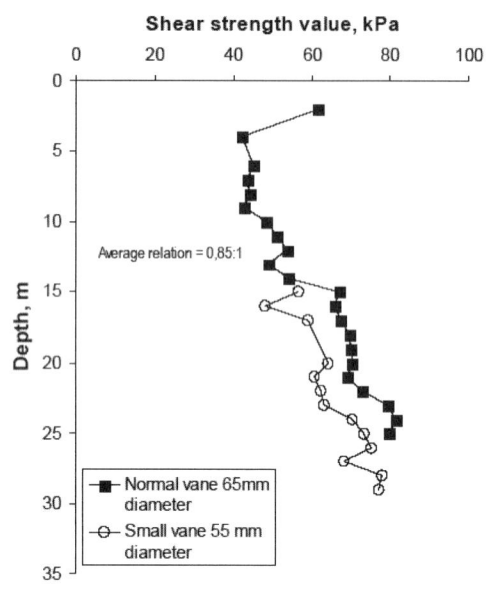

Figure 6: example of results from parallel vane tests in clay with different vane sizes. Åhnberg et al.

This effect is counteracted in organic soil owing to the fibre content, this may result in the opposite trend.

3.6 *Deformability tests*

Arbaoui et al. present an experimental device which permits performing static loading tests when the penetration is paused. The test allows the establishment of monotonic or cyclic stress-displacement curves to obtain deformation modulus.

A case study making a comparative analysis of plate load tests (PLT) and dilatometer (i.e. rock pressuremeter) tests results in rock masses are presented by Agharazi & Moradi. These two methods, the two prevailing for rock deformability assessment, do not result in an equal value of deformation modulus. The authors present empirical ratios between them.

4 CONCLUSIONS

Successful geotechnical design requires a good knowledge of the mechanical behaviour of the ground and in situ tests are no doubt powerful tools that can yield a great deal of valuable information. A wide range of parameters can be reliably obtained from in situ testing and it is noticeable that the ratings of some tests have improved as the database of experience has built up, but others have declined as the initial predictions of their capability have been found to be wanting.

Effort is required in order to improve the situation and get the best from existing tests. International (i.e. global) standards are a subject of prime importance that will ensure that tests are carried out with consistency in order to enable comparable results to be obtained, a key matter to transfer experience from one country to another or develop generic correlations. There must be consistency in both equipment and operation wherever it is specified and used not just in one country but around the world. Session 1 detailed the present efforts being made by the International Organization for Standardisation (ISO) and the European Committee for Standardisation (CEN).

Furthermore, we should be encouraging accreditation procedures for in situ testing. If current practice has resulted in cost cutting and bad practice, then this should be firmly discouraged even if some small additional costs are incurred.

On the other hand, it would appear that many practitioners have felt that too often the capabilities of in situ tests have been over-sold or inappropriately applied. This has resulted in dissatisfaction when the tests failed to deliver what was promised. The fact that we continue to see papers with more and more correlations for in situ devices often does little to improve this feeling. However it is often not the test that is as fault but its operation and application.

The papers presented to Session 1, from 13 different countries around the world, cover a wide range of topics and uses of in situ tests. This wide range is a good example of their versatility and usefulness for a ground investigation programme provided the right test is selected for the situation and the equipment and procedures allow repeatable and quality data to be obtained.

The papers have shown that there is still value in improving our understanding and operation of existing tests before indulging in evermore sophisticated techniques.

REFERENCES

Alonso, U.R. 1996a. *Estimativa da adesão em estacas a partir do atrito lateral medido com o torque no ensaio SPT-T.* Solos e Rochas, V18 (No 1): 191-194.

Alonso, U.R. 1996b. *Estacas hélice continua com monitoração electrônica, Previsão da capacidade de carga através do SPT-T.* Proc. SEFE III V2: 141-151.

Amrosi, A. & Marchi, N. 1999. *High resolution sequence stratigraphy from piezocone tests: an example from the Late Quaternary deposits of the Southeastern Po plain.* Sedimentary Geology, 128: 67-81.

Carvalho, J.C., Guimarães, R.C., Pereira, J.H.F., Paulocci, H.V.N. & Araki, M.S. 1998. *Utilizasão do ensaio SPT-T no dimensionamiento de estacas.* Proc. COBRAMSEG XI, V2: 973-982.

Cruz, N., Viana da Fonseca, A., Neves, E. 2004. *Evaluation of effective cohesive intercept on residual soils by DMT data.* Proc. 2nd International Conference on Site Characterization ISC'2, Porto

Decourt, L. 1996. *Investigações Geotécnicas.* In Hachich et al., Fundações, teoria e práctica: 119-162.

Devincenzi, M., Colas, S., Falivene, O., Canals, M & Busquets. P. 2003. *Aplicación del piezocono para el estudio sedimentológico de detalle de los sedimentos cuaternarios del delta del Llobregat, Barcelona.* Actas III Congreso Andaluz de Carreteras, Sevilla. V2: 937-954.

Devincenzi, M. Colas, S., Casamor, J.L., Canals, M., Falivene, O. & Busquets, P. 2004. *High resolution stratigraphic and sedimentological analysis of Llobregat delta nearby Barcelona from CPT & CPTU tests.* Proc. 2nd International Conference on Site Characterization ISC'2, Porto.

Hegazy, Y.A. & Mayne, P.W. 2003. *Objective site characterization using clustering of piezocone data.* Journal of Geotechnical and Geoenvironmental Engineering, 128 (No 12): 986-996.

ISSMFE 1989. Report of the ISSMFE Technical Committee on Penetration Testing of soils – TC 16, with reference to Test Procedures. Swedish Geotechnical Institute, Linkoping. Information, 7.

Marchetti, S. 1997. *The Flat Dilatometer Design Applications.* III Geotechnical Engineering Conference, Cairo University.

Mayne, P.W. 1991. *Determination of OCR in clays by piezocone tests using cavity expansion and critical state concepts.* Soils and Foundations, Vol. 31 (No 2): 65-76.

Lunne, T, Robertson, P.K. & Powell, J.J.M. 1997. *Cone Penetration Testing in Geotechnical Practice.* Blackie Academic & Professional and SponPress. London.

Peixoto A.S.P. 2001. *Estudo do ensaio SPT-T e sua aplicação na práctica de Engenharia de Fundações.* Thesis PhD. Unicamp, Brazil.

Powell, J.J.M. 2001. *In situ testing and its value in characterising the UK National soft clay testbed site, Bothkennar.* Proc. Int. Conf In situ 2001, Bali. 365-372.

Powell, J.J.M. and Butcher, A.P.2002. *Characterisation of a glacial till at Cowden, Humberside.* Proc.. Int. Syposium on Characterisation and engineering properties of natural soils, Singapore, (eds Tan et al), Singapore, December 2002. Balkema, Vol 2: 983-1020.

Seed, H.B., Takimatsu, K., Harder, L.F. & Chung, R. 1985. *Influence of SPT procedures in soil liquifaction resistance evaluations.* Journal of Geotechnical Engineering, ASCE. Vol. 111 (No 12): 1425-1445.

Skempton, A.W. 1986. *Standard penetration test procedures and the effects in sands of overburden pressure, relative density, particle size, aging and overconsolidation.* Geotechnique, Vol. 36(No 3): 425-427.

Trevor, F. 2001. *Evaluation of soil parameters from results of piezocone tests in clay.* MSc Thesis, Nanyang Technological University, Singapore.

Comparison between plate load test results and dilatometer test results Case study, Seymareh dam and Sazbon dam projects

A. Agharazi
KHAK & SANG Co., Tehran, Iran

M. Moradi
University of Tehran and KHAK & SANG Co., Tehran, Iran

Keywords: in-situ test, deformation modulus of rock mass, Plate Load Test, dilatometer test

ABSTRACT: Under effects of in situ conditions, such as in situ stress field and existing discontinuities, the plate load tests and dilatometer tests, usually, do not result in the same values as modulus of deformation for an individual rock mass. This subject has been focused on, with respect to results of executed plate load tests and dilatometer tests, in two large dam sites in Iran. Considering the blasting and stress release damages to rock mass located around the test gallery, a value between 1-2 has been evaluated for the ratio of plate load test modulus to dilatometer modulus.

1 INTRODUCTION

The deformation modulus of a rock mass is one of the most important parameters in analyzing the mechanical behavior of rock masses under stress. There are different methods for the determination of deformability characteristics of rock masses, but execution of in situ tests are the most reliable methods.

Among in situ tests, Plate Load Test (PLT) and the dilatometer test are more conventional methods and both of them, from a theoretical point of view, are based on the theory of elasticity, considering the rock mass as an elastic, isotropic and homogeneous medium. Hence, theoretically in the same rock mass, the calculated modulus should be similar for both methods. But in real conditions, this is not the case and the modulus determined from PLT is usually greater than values calculated based on dilatometer tests.

PLT is an expensive and time consuming in situ test for the determination of deformability characteristics of a rock mass but with regard to its stress distribution, method of loading and large volume of rock mass involved in the test, this method is one of the most reliable methods for the determination of the rock mass modulus of deformation. Hence in geotechnical site investigations, especially in large dam sites, execution of this test is one of the inalienable parts of rock mechanics tests program. Having an estimation of the ratio between PLT modulus and dilatometer modulus for an individual rock mass, execution of dilatometer test is a reasonable option for determining the deformability characteristics of rock masses in a wider range and with sufficient accuracy.

The following factors are the most predominant parameters causing the difference between these two tests results:
1. in situ stress field
2. existence of discontinuities
3. blasting and stress release damages on PLT
4. propagation of tensile cracks during dilatometer test

The existence of a compressive in situ stress field in a rock mass (such as gravitational stresses) limits the deformation of the rock, compared to that being predicted by theory of elasticity. Hence an error arises when the actual deformations in a rock mass are measured directly in a PLT or dilatometer test, where a stress distribution in accordance with the theory of elasticity is assumed, which is not the representative of the actual stress field existing in the rock mass. This results in higher modulus values and the higher the order of magnitude of in situ compressive stress field, the higher the modulus values. Although this parameter has a similar effect on both PLT and dilatometer test but severity of its effect is not the same, so a ratio greater than unity between PLT modulus and dilatometer modulus will occur. Moreover, the presence of discontinuities is another parameter affecting in situ tests results. The effect of this parameter is a function of orientation and mechanical characteristics of discontinuities and the severity of its effect depends on the number of discontinuities existing in the rock mass volume affected by the testing stresses. However this parameter has a decreasing effect on both methods and considering the larger volume of rock mass being affected during PLT, this parameter has a greater effect on PLT modulus than dilatometer modulus.

Blasting and its consequent weakening of rock mass can cause a serious reduction in modulus values being obtained from a PLT. This parameter has a greater effect on surface measurements especially in the first cycle of loading. Hence the modulus values obtained in the first cycle of loading are not suitable representatives of the rock mass deformability characteristics. This effect can be recompensed by a correction factor whose value is determined on the basis of comparison between slope of load-deformation curves in various loading cycles.

Development of tensile cracks around boreholes, during a dilatometer test, is the most predominant reason for the occurrence of a more than unity ratio between PLT modulus and dilatometer modulus. Considering this parameter as an inherent feature of dilatometer test, there is no correction factor for compensating its effect on the resulted deformation modulus. Figure 1 shows the development of a tensile crack during the third loading cycle of a dilatometer test in "Azadi" dam foundation in Iran. The curve shows the generation and development of a tensile crack, at a stress level of 4MPa and nearly perpendicular to the sensor 1 direction, which has caused a high and sudden increase in measured deformation. As a result, the secant modulus calculated in this loading cycle obviously decreases.

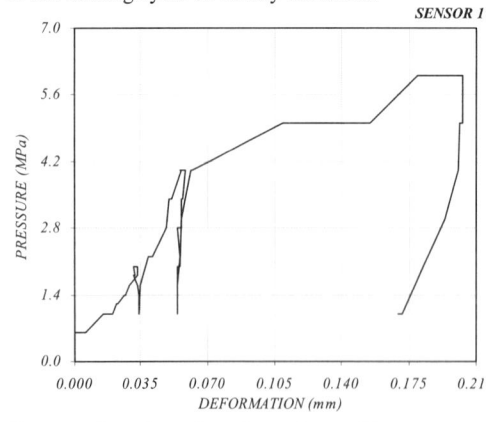

Figure 1. The effect of tensile cracks on dilatometer test. (Azadi Dam, KHAK & SANG Co.)

But paying attention to an important point is essential in relation to this subject, that is, one should not always consider this parameter as a source of error, because it may be the case in some circumstances, such as loading a rock mass by a pressure tunnel. In other words, the deformation that occurs around a pressure tunnel in the surrounding rock mass is more compatible with the deformability characteristics determined from a dilatometer test than those resulted from PLT. This is due to the similarity between dilatometer loading and pressure tunnel loading that cause a similar reflection in the rock mass. In both cases propagation of tensile cracks play an important role in the increase of deformation around the borehole or tunnel, whereas by execution of a PLT there is no possibility for estimating the effects of this parameter on the deformation modulus of the rock mass. So the validity of a test is not only a function of the prementioned factors, but the application of the results is an effective parameter in selecting a suitable test for a specified study plan.

In the following sections the results of test series executed in two large dam sites in Iran will be analyzed and the ratio between PLT modulus and dilatometer modulus will be evaluated.

2 SEYMAREH DAM AND POWERHOUSE

Seymareh dam is a double-curved concrete dam with a height of 180 m (from foundation level) and has been aimed to control flowing water at surface and to produce hydropower energy. This dam is located on "Asmary-Shahbazan" formation in the folded "Zagros" zone in Ilam province to the south-west of Iran. The rock mass of dam foundation consists mainly of thick-layered karstic or massive limestone.

In order to determine the deformability characteristics of Seymareh dam foundation, a total number of 13 PLTs including 7 vertical tests and 6 horizontal tests, and 49 dilatometer tests in 10 vertical boreholes have been carried out by KHAK & SANG Co.

During the plate load tests, the rock surface has been subjected to an incremental cyclic load (5 cycles) up to a stress level of 10MPa by a rigid plate of 1m in diameter and deformations in rock mass have been measured by a multiple positioned borehole extensometer (five points) installed in the central borehole. All the test procedures have been in accordance with ISRM suggested method.

The dilatometer tests were performed by an IF096 dilatometer made by "INTERFELS" (Germany) with 1m length and 98mm diameter, capable of exerting a maximum pressure of 10 MPa. The expansion of the borehole diameter is measured by three Linear Variable Differential Transformers (LVDTs) built inside the sleeve. The measurement devices were arranged at an angle of 120° relative to each, other and so the anisotropy of the rock mass can also be measured. Table1 shows the results of dilatometer tests and their statistical parameters in Seymareh dam.

Considering the direction of loading in the dilatometer tests, the six horizontal plate load tests results are comparable with deformation modulus determined on the basis of dilatometer tests. Table 2 shows the results of plate load tests, only for surface

Table 1. results of dilatometer tests in Seymareh dam site

	MAX. LOAD (MPa)	MODULUS OF DEFORMATION (GPa)			
		MAX.	MIN.	MEAN.	ST.DEV.	VAR.
CYC.1	3	11.98	0.89	5.24	2.96	8.76
CYC.2	5	22.04	2.41	10.33	5.05	25.48
CYC.3	7	30.15	2.15	13.17	6.60	43.60
TDM	7	14.48	0.39	6.26	3.02	9.14

readings, in Seymareh dam site. The modulus values calculated on the basis of deformations of the rock mass in depths are unreasonably high, such that in some cases its value is higher than those determined by uniaxial compressive strength tests of the intact rock samples taken from drilling cores of extensometer boreholes. Hence these results were withdrawn in comparing the dilatometers modulus and plate load tests modulus.

Table 2: Results of plate load tests of Seymareh dam site

Dir.	Cycle	Load (MPa)	Modulus of Deformation (GPa)				Ratio Between Modulus			
			Max.	Min.	Mean	St. Dev.	Var.	R_{21}	R_{32}	R_{43}	R_{54}
Horizontal	C1	3	7.75	1.74	4.36	2.00	4.01	3.24	1.30	1.11	1.31
	C2	5	35.43	4.13	14.13	8.08	65.21				
	C3	7	37.28	5.13	17.60	8.35	69.74				
	C4	10	33.58	5.41	19.04	8.42	70.91				
	C5	10	49.27	6.29	24.51	10.34	106.9				
Vertical	C1	3	5.44	0.9	2.4	1.38	1.92	2.7	1.44	1.16	1.26
	C2	5	12.16	2.49	5.4	2.98	8.85				
	C3	7	19.34	2.72	8.72	5.38	28.96				
	C4	10	19.74	2.81	9.91	5.65	31.89				
	C5	10	30.52	3.39	13.34	8.26	68.20				

According to table 2, the ratio of deformation modulus of cycle two to cycle one, R_{21}, is higher than the same ratios between other cycles of loading. This is mainly because of the blasting damages to the near surface rocks and consequently the closure of cracks during the first cycle of loading. Fig.1 shows such effect on the load – deformation curve of the PLTLH3 test in Seymareh dam site. As can be seen, the slope of the curve in cycle 1 is smaller than the slope of the curve in cycle 2. Similarly this applies to cycle 3 and cycle 2. But for cycles 3-5 the slope of the curve is nearly the same. It shows that the damages of blasting are limited to the near surface rock and the rock mass located at depth is not disturbed.

One reasonable method for the compensation of this effect, is to determine a correction factor on the basis of the ratio of modulus values in consecutive loading cycles in which no blasting damages to their resultant modulus are assumed. In Figure 2, cycles 3,4 and 5 show such a condition. According to this method, a correction factor of Bc=2.5 has been assumed to compensate for blasting damages to cycle one modulus in horizontal tests.

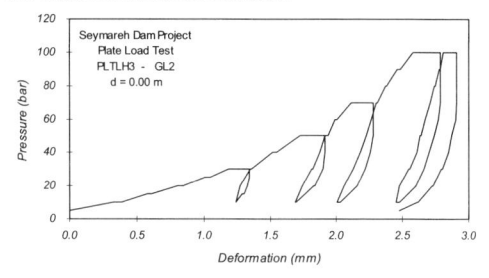

Figure2. A Load – Deformation curve showing blasting damage effects on PLT results in Seymareh Dam Project

Multiplying the cycle one modulus by this correction factor, the ratio between PLT modulus and dilatometer modulus in Seymareh Dam Site would be evaluated as shown in Table3.

As can be seen in this table, the ratio for cycle one is greater than the ratios obtained in the two other cycles. With regard to the low stress level of cycle one, this high ratio is not due to the propagation of tensile cracks during dilatometer test, but due to deflection of the weakened rock, located in the borehole wall periphery. This weakening is mainly because of the borehole wall convergence under the effect of in situ stress field or swelling of the rock, so in the first cycle of loading, a major portion of deflection belongs to the recovery of this induced deformation and accordingly leads to a lower modulus of deformation in this cycle. In addition, a bad drilling procedure can cause crushing and weakening of the rock and so a similar effect can be expected.

Table 3. The ratio between PLT Modulus and Dilatometer Modulus in Seymareh Dam Site

	CYC.1	CYC.2	CYC.3
E_{PLT} (GPa)	10.87 (*Bc)	14.13	17.60
$E_{DILATOMETER}$ (GPa)	5.24	10.33	13.17
R	2.07	1.38	1.37

3 SAZBON DAM PROJECT:

This dam is located on "Asmari" formation in Ilam Province of Iran. The foundation rock is limestone and almost similar to the Seymareh dam foundation rocks. For determining the rock mass deformation characteristics in this site, a total number of 10 PLTs, including five vertical tests and five horizontal tests, and 49 dilatometer tests in 10 vertical boreholes were performed by KHAK & SANG Co. All tests have been executed using similar equipments and procedures as in Seymareh Dam. Figure 3, shows the typical load – deformation curves of a horizontal PLT executed in this site.

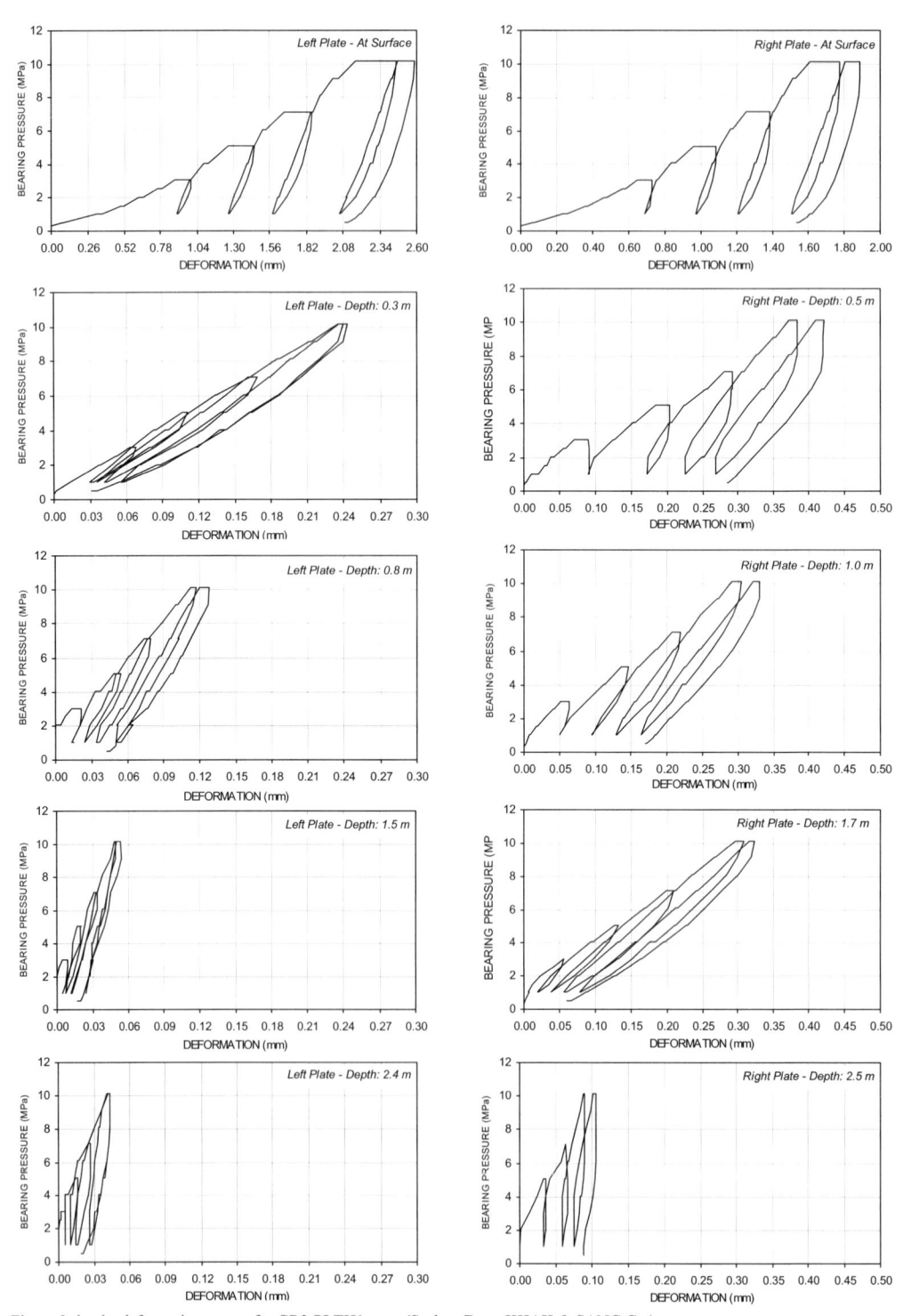

Figure 3: load – deformation curves for GR2-PLTH1 test. (Sazbon Dam, KHAK & SANG Co.)

Table 4: dilatometer tests results and their relevant statistical parameters in Sazbon Dam Site

	MAX. LOAD (MPa)	MODULUS OF DEFORMATION (GPa)			
		MAX.	MIN.	MEAN	ST. DEV.	VAR.
CYC.1	3	20.70	0.70	4.95	3.41	11.61
CYC.2	5	18.97	1.02	8.90	4.15	17.18
CYC.3	8	21.61	1.08	10.10	4.54	20.57
TDM	8	14.87	0.66	6.25	3.04	9.26

Table 4 and Table 5 show the dilatometer test results and PLT results (only surface readings) and their relevant statistical parameters in Sazbon Dam site.

Again a correction factor is needed to compensate for the effect of blasting damages on cycle one modulus of deformation. Comparing the modulus in this cycle with other cycles, a minimum correction factor of Bc=2.37 can be determined for horizontal tests. Table 6 shows the resultant ratios between PLT modulus and dilatometer modulus in Sazbon Dam site.

Table 5. plate load tests results in Sazbon Dam Site

Dir.	Cycle	Load (MPa)	Modulus of Deformation (GPa)				Ratio Between Modulus			
			Max.	Min.	Mean	St. Dev.	Var.	R_{21}	R_{32}	R_{43}	R_{54}
Horizontal	C1	3	6.50	1.84	3.59	1.49	2.22	2.93	1.15	1.04	1.46
	C2	5	20.06	6.54	9.78	4.04	16.36				
	C3	7	21.91	6.03	12.54	4.76	22.69				
	C4	10	20.93	6.22	12.62	4.47	19.97				
	C5	10	33.45	9.95	18.56	6.80	46.21				
Vertical	C1	3	5.66	0.91	2.41	1.56	2.44	3.05	1.40	1.02	1.40
	C2	5	16.17	2.41	6.97	4.80	23.02				
	C3	7	20.99	3.57	9.80	6.39	40.79				
	C4	10	19.93	3.82	9.79	5.88	34.59				
	C5	10	29.97	4.82	14.12	9.28	86.18				

Table 6. The ratio between PLT modulus and dilatometer modulus in Seymareh Dam Site

	CYC.1	CYC.2	CYC.3
E_{PLT} (GPa)	8.51 (*Bc)	9.78	12.54
$E_{DILATOMETER}$ (GPa)	4.95	8.90	10.10
R	1.72	1.10	1.24

Similarly, the cycle one ratio is higher than the other two ratios which is due to prementioned reasons.

4 CONCLUSIONS

In situ tests are the most reliable methods for the determination of rock mass deformability characteristics. The plate load test and the dilatometer test are two prevailing methods among in situ tests but under effect of in situ conditions, such as in situ stress field and discontinuities, in addition to other affecting parameters such as propagation of tensile cracks during dilatometer test and blasting damages in PLT, these two methods do not result in an equal value as deformation modulus of an individual rock mass. Considering the loading method and volume of rock being affected by loading, PLT is superior to dilatometer test, but it is an expensive and time consuming method. Therefore having a correlation between results of this test with the dilatometer modulus, it is possible to execute more dilatometer tests and reduce the number of PLTs in a site investigation program.

For test series executed in two large dam sites in Iran, the ratio between plate load tests modulus and dilatometer modulus have been evaluated at a range of 1.37<R<2.07 and 1.10<R<1.72 respectively for Seymareh Dam and Sazbon Dam. But this range can vary proportional to the existing conditions such as in situ stress field, discontinuities, applied stress level and mechanical characteristics of the rock mass. Hence this value should be determined individually for each specific rock mass.

REFERENCES

Agharazi A., 2003, "The determination of the different conditions effects on the results of in situ tests for determining the deformation modulus of rock masses", M.Sc. thesis, Department of mining engineering, Tehran University, Iran

ASTM, 1999, "Standard test method for determining the in situ modulus of deformation of rock mass using the flexible plate loading method"

ASTM, 1973, "Field Testing and Instrumentation of Rock", ASTM STP 554

International Society for Rock Mechanics, 1979, Commission on Standardization of Laboratory and Field Tests, "Suggested Methods for Determination In Situ Deformability of Rock," International J. Rock Mechanics Min. Sci. and Geomechanics Abstract, Vol 16, No.2, pp. 143-146

International Society for Rock Mechanics, 1986, Commission on Standardization of Laboratory and Field Tests, "Suggested Method for Deformability Determination Using a Flexible Dilatometer with Radial Displacement Measurements," International J. Rock Mechanics Min. Sci. and Geomechanics Abstract, Vol 3, No.3

KHAK & SANG Co., 2002, "Rock mechanics tests report: plate load tests of Sazbon Dam", Tehran, Iran

KHAK & SANG Co., 2001, "Rock mechanics tests report: dilatometer tests of Sazbon Dam", Tehran, Iran

KHAK & SANG Co., "Rock mechanics tests report: dilatometer tests of Seymareh Dam", Tehran, Iran

KHAK & SANG Co., "Rock mechanics tests report: plate load tests of Seymareh Dam", Tehran, Iran

Lama R.D. & Vutukuri V.S., 1978, "Handbook on Mechanical Properties of rocks", Vol.III, Trans Tech Publications

Palmstrom A. & Singh R., 2001, "The Deformation Modulus of Rock Masses – Comparisons Between In Situ Tests and Indirect Estimates", Tunnelling and Underground Space Technology, Pergamon Press, pp. 115-131

Influence of vane size and equipment on the results of field vane tests

H. Åhnberg, R. Larsson & C. Berglund
Swedish Geotechnical Institute, Linköping, Sweden

Keywords: field vane test, clay, equipment, vane size, disturbance, shear strength, fibre effect

ABSTRACT: The field vane test is commonly used for determination of undrained shear strength in fine-grained soils throughout the world and the test has been standardised in many countries. The standards prescribe demands for the equipment and execution of the test. They are based on comprehensive research regarding the effect of different factors on the results. In most standards, a certain number of vanes of different sizes are specified for use depending on the shear strength of the soil. The tests are interpreted in the same way and are expected to yield the same result regardless of the size of the vane used. However, in spite of the numerous investigations of the effect of different factors on the test results, the effect of the actual size of the vane has been studied only to a limited extent. A new investigation in Sweden has shown that the size of the vane can affect the results in different ways depending on the type of soil. It is therefore recommended to use as far as possible the standard size of the vane on which the bulk of the experience is based and, in cases where larger or smaller vanes are to be used, to check whether there is any effect of size on the results.

1 DEVELOPMENT OF THE FIELD VANE TEST

The field vane test was originally developed by the Geotechnical Commission of the Swedish State Railways between 1914 and 1922. Similar devices were used in Germany in the 1920s and by the British Army in the 1940s. The development of the modern field vane test started at the Swedish Geotechnical Institute (SGI) in the late 40s (Carlsson 1948), and Skempton (1948) used similar equipment in the U.K. The investigations comprised studies of the shape of the rupture surface and influences of the rate of rotation, the length of the vane shaft protruding below the casing for the vane, the number of wings (vane blades) and the relation between height and diameter of the vane, among other aspects.

A recommended equipment and test procedure were presented by Cadling and Odenstad (1950). This equipment consists of a set of three vanes with four wings having a height-diameter ratio of 2:1. During installation, the vane is protected in a robust casing and the turning rods are also encased in hollow drilling pipes. The whole system is pushed down until the tip of the casing is about half a metre above the test level. The vane and inner rod system are then pushed out of the casing and advanced until the free length between the casing and the upper end of the vane is at least 5 times the diameter of the vane. The rotation of the vane should be slow, resulting in failure between 2 and 4 minutes after start of the test, with an ideal time of 3 minutes (Flodin 1958). After the test, the vane is retracted into the casing, whereby any soil sticking to

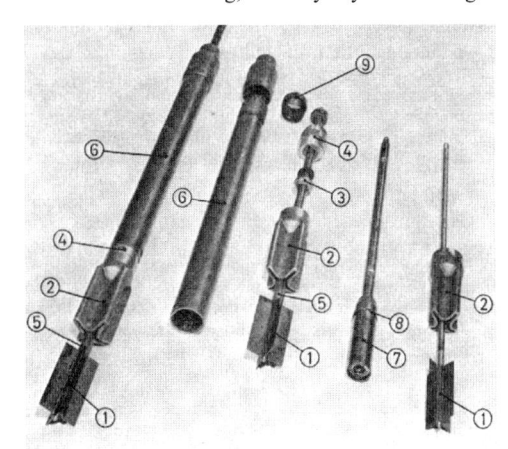

Fig. 1. The field vane equipment, (from Andrésen and Bjerrum 1957)

the wings is scraped off and the vane is again protected during the advance of the equipment to the next test level.

The use of the field vane test spread rapidly and similar equipment was also used mainly in Canada, Norway, the UK and the USA. Experience from practical applications and research was gathered and the American Society for Testing and Materials, ASTM, arranged special conferences on the field vane test in 1956, 1966 and 1987. Practical experience showed that the undrained shear strength evaluated from the field vane tests has to be corrected, particularly in high-plastic soils. The Swedish Geotechnical Institute (1969), Andréasson (1974), Helenelund (1977) and Larsson et al. (1984) successively proposed corrections with respect to the liquid limit of the soil, and Bjerrum (1973) presented similar correction factors based on the plasticity index. Experience has also shown that the evaluated undrained shear strength from the field vane tests should also be corrected with respect to the overconsolidation ratio (Aas et al. 1986, Larsson and Åhnberg 2003).

Other research on the field vane test has been aimed at estimating the anisotropy of the undrained shear strength and the mobilisation of the shear strength in various parts of the shear surface (e.g. Silvestri et al. 1993). In this context, vanes with different height-diameter ratios and vanes with a diamond shape have been used. La Rochelle et al. (1973) and Roy and Leblanc (1988) studied the effect of the design of the wings and particularly the thickness of their tips. The effect of disturbance at insertion of the vane and the effect of a waiting time before starting the test, as well as that of rate of rotation, were studied by Aas (1965), Wiesel (1973), Torstensson (1977), and Roy and Leblanc (1988). The effects of the latter factors may be assumed to be involved in the correction factors (Larsson et al. 1984). The latter research has not significantly affected the design of the field vane equipment, the performance of the test or the existing standards, but has emphasised the need to conform with the standards.

Apart from an investigation by Arman et al. (1975), research concerning the influence of the actual size of the vane has mainly been performed in organic soils (Golebiewska 1976, Landva 1980). The results showed that there is a significant size effect in such soils. However, this has similarly had no effect on the standards.

2 NEW EQUIPMENT

Relatively simple lightweight equipment was developed in Sweden during the 60s. In this equipment, no casing is used, but there is a slip coupling between the vane and the rods. The vane is pushed directly to the test level. The rod friction is first measured and then separated from the total torque required to rotate the equipment. This equipment has proved to yield results compatible with the original SGI equipment in the soft homogeneous clays for which it was designed, but great care has to be taken to pre-drill any dry crust and stiffer layers above the soft clay. The lightweight equipment may also cause greater disturbance in layered soils and the results have often been found to differ from those obtained with the heavier equipment in deeper profiles, particularly in varved and layered soils. The slip coupling is often incorporated in the old type of equipment in order to avoid any influence of internal friction in the rod and casing system.

A new generation of field vane equipment has recently been introduced. This is adapted to modern drill rigs and incorporates casings, electrical rotation of the vane, electronic measurement of the time and torque, and automatic storage of the measured data. In general, the geometric design and dimensions of the equipment are the same as for the traditional type. However, for manufacturing reasons the previous cast bronze casing for the vane, which was shaped to exactly accommodate the vane, has been replaced by a solid cylindrical casing with two slots at right angles to each other. This has caused some concern for a larger disturbed zone in front of the casing because of a larger displaced soil volume.

3 GEOMETRY OF THE FAILURE SURFACE

The geometry of the shear surface in vane tests with four or more wings has been found to be cylindrical and the evaluation of the test is based on that assumption. However, is has also been found that the diameter of the cylinder is not necessarily the same as that of the vane. Skempton (1948) found that the diameter of the shear surface was somewhat larger than the vane and proposed the relation $D_{shear\ surface} = 1.05\ D_{vane}$. However, this fixed relation does not affect the ratio between the shear strengths evaluated from different vanes with the same H/D value. Golebiewska (1976) found that the shear surface in organic soils is located at a distance outside the perimeter of the vane, which is constant for a particular type of soil. This was also found by Landva (1980), who showed that the shear zone in peat extends far outside the perimeter of the vane. This has a large effect on the evaluated shear strength and the smaller the vane is the higher the evaluated shear strength becomes. This effect is assumed to be related to the fibrous nature of the organic soils and the longer the fibres are, the larger the effect can be expected to be. In principle, this introduces major uncertainties regarding the relevance of the field vane test in fibrous peat. The possible influence should be considered also in more decomposed organic soils and other soils with a significant fibre content.

4 DISTURBANCE EFFECTS

Several types of disturbance effect occur when the vane is inserted in the ground. When using a casing, there is a heavily disturbed zone just below the tip of the casing. This effect is intended to be avoided by pushing the vane a sufficiently longer distance to be outside this disturbed zone. The current specification in the procedure recommended by the Swedish Geotechnical Society (1993) is that the free distance between the vane and the casing must be between 0.35 and 0.5 m depending on the size of the vane and its casing. This figure is based on investigations performed with the original design of the casing.

At insertion of the vane itself, there are two types of disturbance effect. Depending on the volume of soil displaced by the vane, the soil in its vicinity becomes more or less disturbed and the soil adjacent to the blades becomes almost completely remoulded, Fig. 2.

If the excess pore pressure created at insertion is allowed to dissipate and the soil reconsolidates, most of these effects are eliminated, at least in normally consolidated soils. Investigations by Aas (1965), Wiesel (1973) and Torstensson (1977) have shown that these disturbance effects are significant, particularly in low-plastic clays. They can also vary greatly between different soils of the same plasticity, Fig. 3.

The average disturbance at insertion is incorporated in the correction factors for the test (Larsson et al. 1984). The disturbance is related to the cross-sectional area of the vane and particularly to the thickness of the wings towards their edges. La Rochelle et al. (1973) and Roy and Leblanc (1988) found a linear relation between the disturbance effects and the thickness of the wing edge in tests performed with the normal procedure. The vanes are therefore often manufactured with wings that decrease in thickness towards the edge. However, the vanes have to be designed also with consideration to maximum torque and wear. The thicknesses of both the vane shafts and the wings are

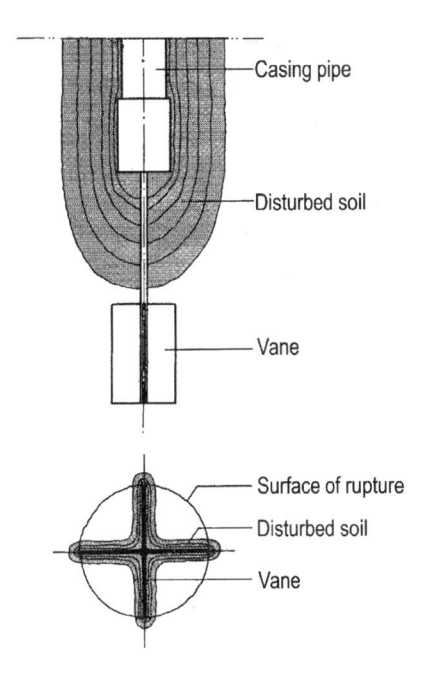

Fig.2. Schematic picture of heavily disturbed zones in vane tests (Cadling and Odenstad 1950).

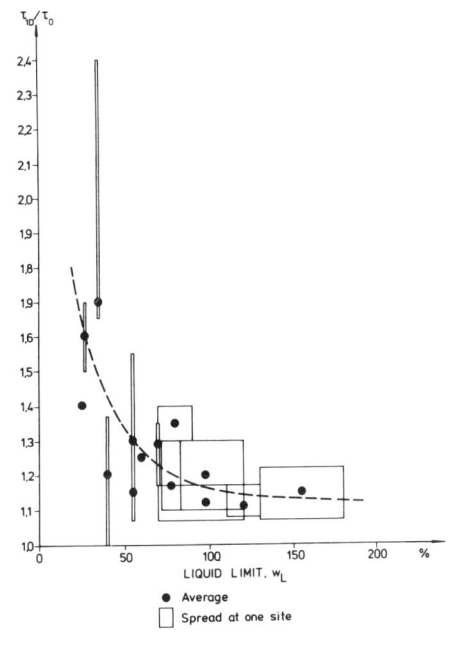

Fig. 3. Degree of disturbance evaluated as the relation between shear strengths measured directly after installation of the vane and strengths measured after 1 day of reconsolidation before the test (Larsson et al. 1984).

therefore often the same for vanes of different sizes. This entails that vanes with smaller diameters can be expected to disturb the involved soil volume more than vanes with larger diameters.

5 BACKGROUND TO THE NEW INVESTIGATION

There were mainly three reasons for the new investigation (Åhnberg et al. 2001):
- In tests in deep soil profiles in Sweden, it had been observed that a change to a smaller vane, necessitated by the increase in shear strength with depth, may be associated with a sudden decrease in measured shear strength at the depth for the change and a considerably lower increase in measured shear strength with depth thereafter.
- Most of the Swedish experience refers to vanes of the normal size of 65 x 130 mm. However, larger vanes are sometimes used in order to increase the accuracy and resolution of tests in soft soils. A discussion has thereby emerged as to whether the results from the two vane sizes are fully compatible.
- New equipment with somewhat altered design and new measuring devices had been introduced on the market and questions had arisen as to whether the results from these are entirely compatible with those from the traditional equipment.

6 SCOPE OF THE INVESTIGATION

The largest part of the investigation was performed in a deep clay profile at Munkedal where apparently different results had been obtained with vanes of normal and small size. Here, four types of equipment were tested: original SGI equipment, lightweight Nilcon equipment, a new Geotech equipment and a prototype of new ENVI equipment. For the SGI and the Nilcon equipment, both normal and small-size vanes of 55 x 110 mm were used. Parallel tests were also performed with mechanical and electronic measurement of the applied torque and rotation.

Supplementary tests were then performed using the same equipment with normal and small vanes at a location with very homogeneous soil conditions and repeatable test results.

Tests with normal and large-size vanes were performed in three areas, two with high-plastic clays and one with organic clay. The large vane had a size of 80 x 160 mm. The three vane sizes used are those normally supplied with the equipment and used in Sweden. The diameters of the shafts and the thicknesses of the wings are the same for all of them.

7 RESULTS

The first part of the investigation showed that there was a large difference in results depending on the size of the vane. For the tests with the SGI equipment, the small vane showed shear strengths that were only about 85% of those obtained with the normal vane, Fig. 4. The corresponding relation for the Nilcon equipment was about 0.89:1. This difference is larger than could be expected from reported Canadian investigations, which indicated differences of 2–3% for a corresponding change in vane size. However, this figure relates only to a change in thickness of the wing tip in relation to the diameter of the vane.

On the other hand, there was no significant difference in the results obtained with the different types of equipment. The new design of the casing therefore did not appear to have any influence on the results even in this obviously easily disturbed clay. The apprehension about the new design causing significantly greater disturbance could thereby be largely dismissed.

As expected, there was no difference between the mechanical and electronic measurements except for a better resolution in the latter. However, the electronic

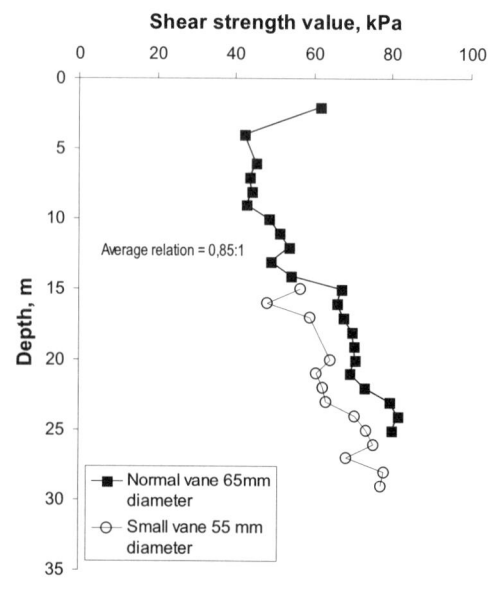

Fig. 4. Example of results from parallel tests with normal and small vanes in Munkedal clay.

measurement of the torque and rotation of the rods showed another aspect of the lightweight equipment that had not been previously taken into consideration. In tests with this equipment, measurements are made first of the torque required to overcome the rod friction and then the total torque. It is then assumed that the rod friction remains constant throughout the test. This is a fair assumption as long as the rod friction constitutes only a small fraction of the total torque. In tests in deep clay profiles, particularly when using small vanes, the rod friction may amount to more than half of the total torque, in which case a significant uncertainty may arise. This is illustrated by a test curve from a deep level in Munkedal in Fig. 5. The curve shows that there is a small peak in the rod friction and that this decreases with further rotation. It is impossible to judge whether the curve has flattened out when the slip coupling is engaged or the rod friction continues to decrease, but this question has a significant effect on the evaluated shear strength.

The supplementary tests with normal and small vanes were performed in a profile of high-plastic clay with organic soil in the upper layer at Sundholmen. Repeated test series with both SGI and Nilcon equipment had here shown that the results are very even and fully compatible when the normal vane is used (Larsson and Åhnberg 2003). The results obtained with the small vane are here about 4% lower in the clay but somewhat higher in the organic soil on top. The average relation is fairly consistent in the clay, although there is a certain scatter in the results obtained with the small vane, Fig. 6.

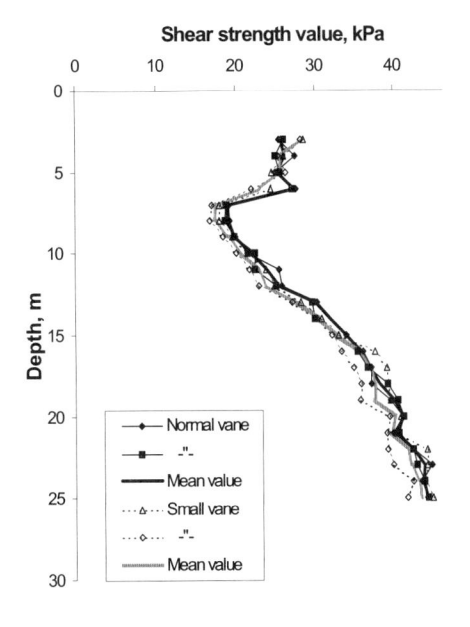

Fig. 6. Results from tests with normal and small vanes at Sundholmen

One investigation using normal and large vanes has been performed by Flygfältsbyrån AB in high-plastic clay at Kvibergslänken in Gothenburg. A large number of comparative tests were performed and all results fell within a narrow band without any significant difference between the two vane sizes.

Another such investigation was performed in high-plastic clay at Nödinge north of Gothenburg. The results here showed a consistent difference with about 5% higher values for the large vane, Fig. 7.

A further investigation was performed in clayey organic soil at Söderleden in southern Gothenburg. The results here had a larger scatter than in the previously investigated clays, partly because of a significant shell content. The average relations differed by up to ± 5% at individual test levels, but the curves were intertwined and the average difference in the total results was less than 0.5%. In this soil, it may be assumed that a possible greater disturbance by the normal vane is more or less offset by a larger fibre effect from the organic matter than for the large vane.

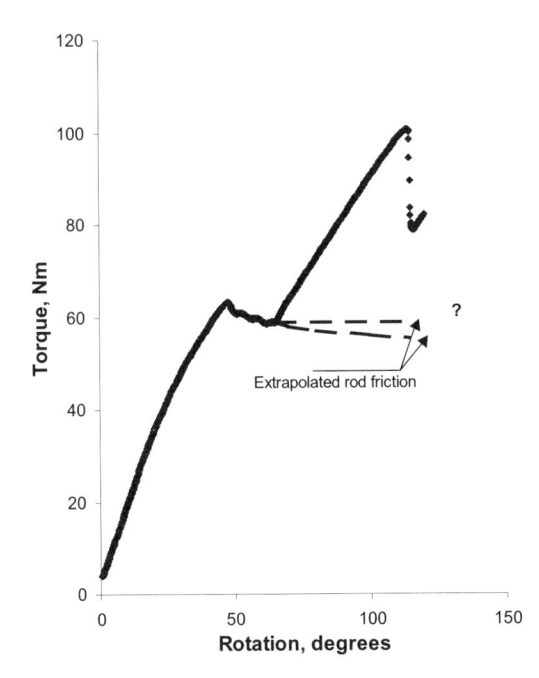

Fig. 5. Uncertainty in the evaluation of the results from lightweight field vane equipment at high rod friction.

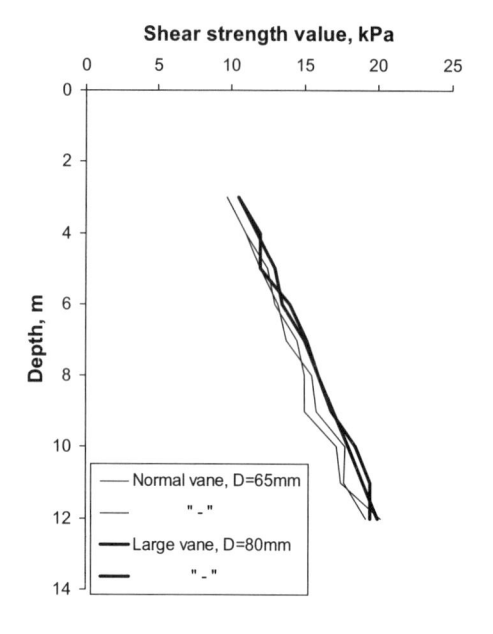

Fig. 7. Results from tests with normal and large vanes at Nödinge.

8 DISCUSSION

The results in this investigation in principle show that for tests in clay the larger the vane the higher the shear strength values are, provided that the thicknesses of the shafts and wings are the same. This can be related to the fact that the relative influence of the disturbance at installation of the vane increases with decreasing vane size. The difference in evaluated shear strength is often insignificant but can be considerable in easily disturbed soils, particularly for small vanes.

This effect is counteracted in organic soil by the fact that the fibre content will cause the shear surface to be located a certain distance outside the perimeter of the vane. This may result in the opposite trend that small vanes yield higher shear strength values than larger ones.

The measured shear strength values are converted to undrained shear strength by empirical correction factors. These factors include the effect of normal disturbance and possible discrepancies from the assumed shear surface. However, they are based on empirical experience from mainly the normal vane and do not compensate for possible unusual effects, such as effects of vane size.

9 CONCLUSIONS AND RECOMMENDATIONS

The investigation has shown that the new types of field vane equipment in principle yield results that are compatible with the traditional types. However, for one make only a prototype was tested.

A change to a smaller vane may result in considerably lower measured shear strength. This should be considered and, if possible, avoided.

High rod friction calls for an early change of vane size and is also a considerable source of uncertainty in evaluating the results. Use of lightweight equipment without casings should therefore be avoided in deep profiles and stiff clays.

A change to a larger vane normally entails that the spread in the test results decreases and that the accuracy of the measured values increases. However, it may also cause somewhat different results to be obtained. Depending on the type of soil, they may be both higher and lower than those obtained with the normal vane. It is therefore recommended that some kind of calibration be performed when using a larger vane, particularly in clays. This can be done by parallel tests with the normal vane and/or direct simple shear tests in the laboratory, depending on the character of the investigation and the demand for highly accurate determinations of the shear strength.

ACKNOWLEDGEMENT

This investigation was performed by the Swedish Geotechnical Institute in cooperation with the manufacturers Geotech AB and ENVI AB, the consulting companies Flygfältsbyrån AB, Forsgrens Konsultbyrå, Geogruppen AB and Jacobsson & Widmark AB, and Chalmers University of Technology.

REFERENCES

Aas, G. (1965). Study of the Effect of Vane Shape and Rate of Strain on Measured Values of In Situ Shear Strength of Clays. Proc. Conf. on Shear Strength of Soils, Oslo, Vol. 1, pp. 141-145.

Aas, G., Lacasse, S., Lunne, T. and Hoeg, K. (1986). Use of in situ tests for foundation design on clay. Proc. Conf. on Use of In Situ Tests in Geotechnical Engineering, ASCE STP 6, 1986, pp. 1-30.

Åhnberg, H., Larsson, R. and Berglund, C. (2001). Nygamla vingar, stora som små. Swedish Geotechnical Institute, Varia No. 509, Linköping. (In Swedish)

Andréasson, L. (1974). Förslag till ändrade reduktionsfaktorer vid reduktion av vingborrbestämd skjuvhållfasthet med ledning av flytgränsvärdet. Int. report. Chalmers University of Technology, Gothenburg. (In Swedish)

Arman, A., Poplin, J.K. and Ahmad, N. (1975) Study of the Vane Shear. Proc. Conf. on In Situ Measurement of Soil Properties, Raleigh, Vol. 1, pp. 93.120.

Bjerrum, L. (1972). Problems of Soil Mechanics and Construction on Soft Clays. Proc. 8th ICSMFE, Moscow, Vol. 3, pp. 111-159.

Cadling, L. and Odenstad, S. (1950). The vane borer. Royal Swedish Geotechnical Institute, Proceedings No. 2, Stockholm.

Carlsson, L. (1948) Determination in situ of the shear strength of undisturbed clay by means of a rotating auger. Proc. 2nd ICSMFE, Vol. 1, 1948, pp. 265-270.

Flodin, N. (1958). Anvisningar för geotekniska institutets fältundersökningar, Del 1. Royal Swedish Geotechnical Institute, Meddelanden No. 4, Stockholm. (In Swedish)

Golebiewska, A. (1976). Analisa stosowalnosci sony obtrowej do badania wytrzymalosci gruntow organicznych. Praca doktorska. SGGW-AR, Warzaw Agricultural University. (In Polish)

Helenelund, K.V. (1977). Methods of reducing undrained shear strength of soft clays. Swedish Geotechnical Institute, Report No. 3. Linköping.

Landva, A.O. (1980). Vane testing in peat. Canadian Geotechnical Journal, Vol. 17, No. 1, pp. 1-19.

La Rochelle, P., Roy, M. and Tavenas, F. (1973). Field Measurements of Cohesion in Champlain Clays. Proc. 8th ICSMFE, Moscow, Vol. 1, 1973, pp. 229-236.

Larsson, R., Bergdahl, U. and Eriksson, L. (1984). Evaluation of undrained shear strength in cohesive soil. Swedish Geotechnical Institute, Information No. 3E, Linköping.

Larsson, R. and Åhnberg, H. (2003). Long-term effects of excavations of slope crests. Swedish Geotechnical Institute, Report No. 61, Linköping.

Roy, M. and Leblanc, A. (1988). Factors Affecting the Measurement and Interpretation of the Vane Strength in Soft Sensitive Clays. ASTM Special Technical Publication STP 1014, pp. 117-130.

Silvestri, V., Aubertin, M. and Chapuis, R.P. (1993). A Study of Undrained Shear Strength Using Various Vanes. ASTM Geotechnical Testing Journal, Vol. 16, No. 2, pp. 228-237.

Skempton, A.W. (1948). Vane tests in the alluvial plane of the River Forth, near Grangemouth. Geotechnique, Vol. 1, No. 2, pp. 111-124.

Swedish Geotechnical Institute (1969). Reducering av skjuvhållfasthet med avseende på finlekstal och sulfidhalt. Swedish Geotechnical Institute. Int. report, Stockholm. (In Swedish).

Swedish Geotechnical Society (1993). Recommended Standard for Field Vane Shear Test. Swedish Geotechnical Society, SGF Report 2:93E, Linköping.

Torstensson, B-A. (1977) Time-Dependent Effects in the Field Vane Test. International Symposium on Soft Clay, Bangkok, 1977, pp. 387-397.

Wiesel, C-E. (1973). Some Factors Influencing In Situ Vane Test Results. Proc. 8th ICSMFE, Moscow, Vol. 1.2, pp. 475-479.

Determination of in-situ deformation moduli with a penetrometer and static loading test

H. Arbaoui, R. Gourvès, Ph. Bressolette & L. Bodé
Laboratoire de Génie Civil – C/U/S/T, Université Blaise Pascal, Clermont-Ferrand, France

Keywords: deformation moduli, settlements, static loading, penetrometer, Mohr-Coulomb model

ABSTRACT: Deformability characterized by a modulus is recognized as one of the most important parameter governing soils settlements. This deformation modulus can be estimated considering an in situ static loading test on a cone penetrometer. The description of the test and the interpretation of the test results are given in this paper. In addition, on further examination, the three mechanical parameters of the Mohr-Coulomb model can be obtained by a numerical simulation of the deformability test.

1 INTRODUCTION

Coupling a pile static loading test with a Gouda 25 kN static penetrometer, Faugeras (1979) and Faugeras et al. (1983) have developed a test based on the principle of a pile static loading test. A static loading is applied on a cone penetrometer driven to a chosen depth. After waiting few minutes for the soil to become stable (this is quite immediate in sand formations), the loading is carried out on the cone step by step waiting 60 s between the different steps. A 10 cm² truncated cone has been used in these tests. The graph of pressure versus cone's displacement was called by these authors the deformability curve. On this curve, the first part corresponds to the pseudo-elastic response of the soil while the second part depicts the elasto-plastic behaviour of the soil.

Here, another test with a light dynamic penetrometer with variable energy, called Panda, in place of the Gouda is proposed and described. This test may be implemented on experimental sites, using a light device, or in the laboratory; in that case, the loading is then applied using an hydraulic jack. The static loading is applied on a small and extended penetrometric point (section of 4 cm²). A few number of cheap and fast in situ tests can be realized with this light device. The deformation modulus can then be obtained from the deformability curve resulting from the test. A numerical simulation of the test is made too. Finally, a relation can be defined between the observed results and the most important parameters of a soil mechanical model. In this paper, we present first the experimental device, then the different tests performed and their results, and finally some conclusions are given about the finite element model of the deformability test.

2 SOIL DEFORMABILITY TEST: TESTING EQUIPEMENT, PROCEDURES, TESTED MATERIAL AND IDENTIFICATION OF DEFORMATION MODULI

The experimental apparatus comprises a dynamic penetrometer device and a static loading device. The dynamic point resistance q_d at different depths z is measured with the penetrometer. The details of the in situ static loading device are shown on figure 1. The cone displacements are recorded with three digital comparators and stored on a computer. Two typical pressure-displacement curves obtained in Allier's sand are shown on figure 2: a monotonous and a cyclic loading curve.

Even if this measuring system has given satisfactory results on different sites, the capacity of the static loading equipment was sometimes insufficient, in particular in firm or hard soils. This is due to the limited load applied with steel masses on the cone penetrometer. Moreover, the application of the load by steel masses – with no shock and no vibration – is difficult. Technological ameliorations are then needed with regard to the reliability and the precision of the test, and also to the automatic application of the load.

In laboratory, the experimental studies can be more reliable. The material used has a higher capacity and the laboratory environment allows a better measurement and a precise control of the different phases of the test. Thus, we present in this paper the tests performed in our laboratory. They are made in a sandy medium built into a big tank.

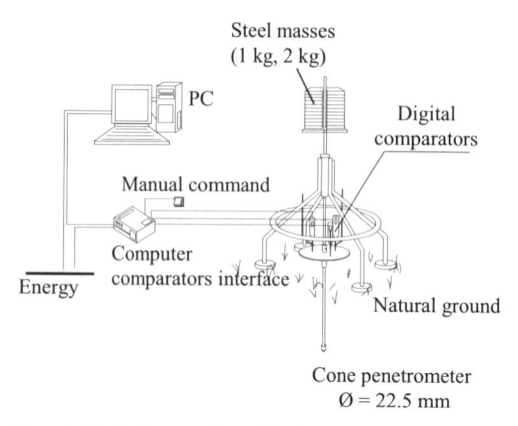

Figure 1. The in situ experimental device

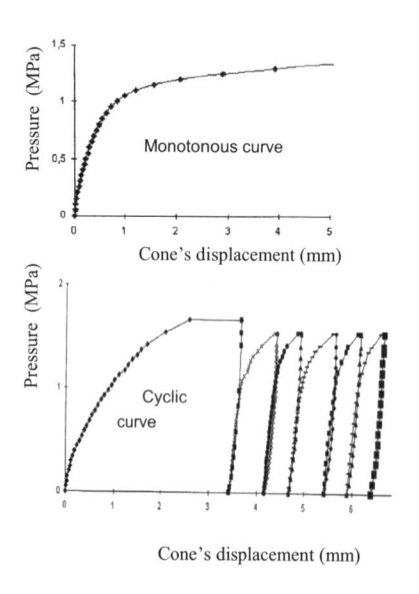

Figure 2. Typical deformability curves (Allier's sand)

2.1 The tank and the sand characteristics

The test is carried out on sand placed into a tank with a constant density (Fig. 3). The sand is an Allier's one with a continuous granulometry 0/5 (Fig. 4). The sand is put in the tank in two layers. The top layer is dense (a layer compaction has been performed). The bottom zone is loose: in that case the sand has been placed with a dipper from a low height.

The value of the friction angle φ' measured with triaxial tests is between 38° and 42°.

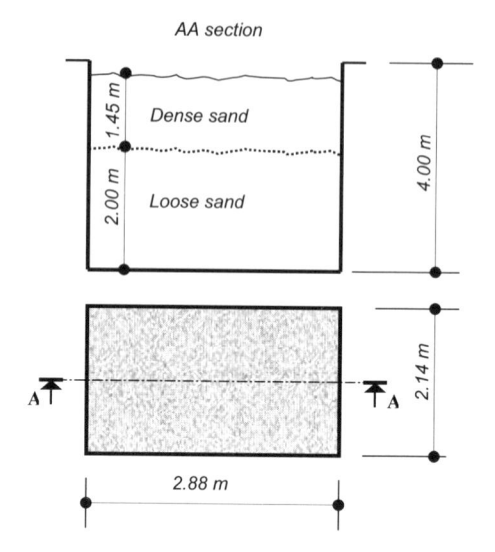

Figure 3. Tank for sand tests

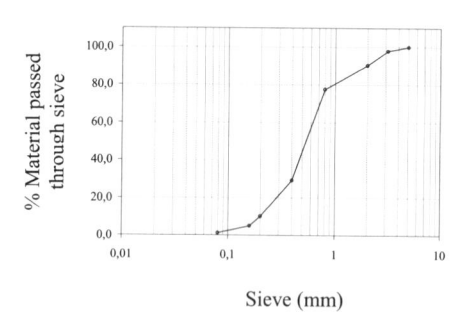

Figure 4. Granulometric curve of Allier's sand

2.2 Laboratory test procedure and tests positions

2.2.1 Monotonous static loading test

Considering one vertical profile in the proposed tank, several static loading tests are carried out at different depths. All these depths are superior to the critical depth denoted z_c, and equal here to 0.25 m (in a purely frictional dense or medium sand soil, the dynamic point resistance become constant from a depth called « critical depth »).

A test consists therefore in driving a 4 cm² cone penetrometer section (22.5 mm diameter) to a desired depth h by dynamic penetration (the rod train equipped with the cone is driven into the soil by a hammer about 2.350 kg and the depth e due to one blow of hammer ranges from 0.5 to 2 cm); then the static loading at very low speed can begin. The dynamic penetrometer device is then removed and replaced by the static loading test device shown on figure 5. A reaction frame is used (Fig. 5). The static loading of the cone is performed with a speed of 0.01 mm/s using a 10 tons hydraulic jack (Fig. 5).

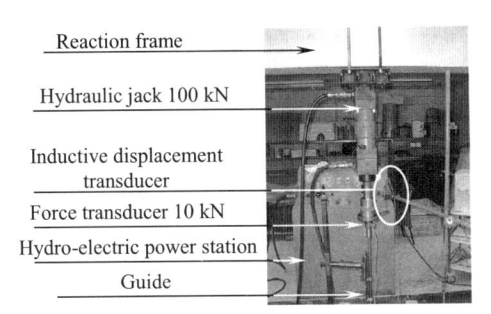

Figure 5. Test's instrumentation

A displacement type control is performed. The reaction force is measured by a 10 ± 0.01 kN force transducer, while the cone displacement is measured by a LVDT inductive displacement transducer of 20 mm nominal displacement span. All the data are recorded at a rate of one measure per second by a measuring unit. The force transducer is fixed under a loading flange which receives the hydraulic cylinder. The loading flange and the force transducer are connected to a guide who is related to the penetrometer rod train. Each static loading test is conducted up to a displacement Δh about 20 mm.

2.2.2 Cyclic loading test
The following unloading-reloading procedure is adopted: the cone penetrometer is first driven to a depth denoted h (the dynamic point resistance at this depth q'_{dm} corresponds to the mean of q_d calculated between h and $h - 5$ cm using the obtained penetrogram). Then, the static loading is conducted from a starting pressure $p_c = q'_{dm}/2$ assumed to be close to the creep pressure. We start a cycle at p_c and unload until a p_d pressure is reached such that $p_d < p_c/2$ so in practice $p_d < q'_{dm}/4$. The range of the cycle is $\Delta p = p_c - p_d$. This unloading-reloading loop is repeated in order to obtain the maximum number of cycles. The same device as above is employed. The cyclic pressuremeter tests conducted by Combarieu and Canépa (2001) show that there is no discernable difference between slow or rapid unloading mode when measurement uncertainty is taken into account. So we recommend to decompress the soil rapidly, in a single step.

2.2.3 Positions of different tests
The static loading tests are made at different depths in the tank, along four profiles S_1, S_2, S_3 and S_4 shown in figure 6. Two cyclic tests (C_1 and C_2) and four classical dynamic penetrometer tests Pd_1, Pd_2, Pd_3 and Pd_4 were also conducted.

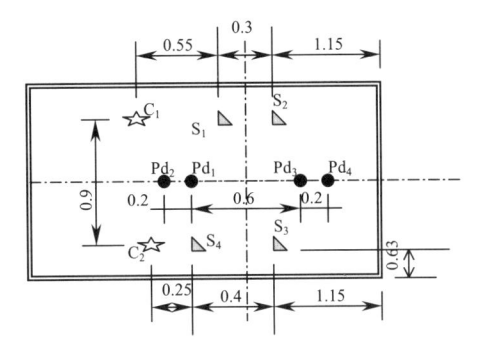

Figure 6. Tests positions in the tank

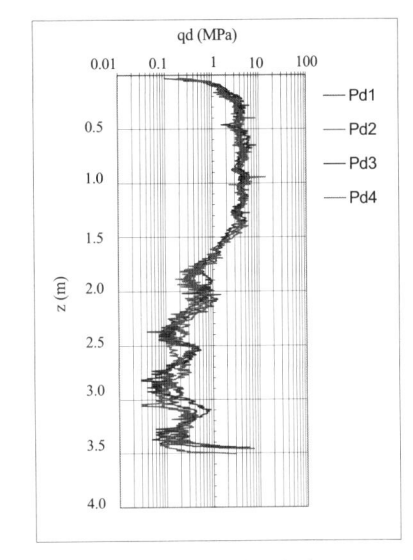

Figure 7. Penetrograms Pd_1, Pd_2, Pd_3 and Pd_4

The penetrograms (Fig. 7) show that the laboratory massif is correctly built and guarantees a good homogeneity of the material: the in situ value of γ_d is between 16.5 kN/m³ and 17.5 kN/m³ for the dense zone and between 13 kN/m³ and 14.5 kN/m³ for the loose one. These values are calculated from γ_d and q_{dl} with equation (1), given by Chaigneau (2001) for this sand:

$$\gamma_d = 1.27 \ln(q_{dl}) + 14.63 \tag{1}$$

2.3 Deformability curve interpretation (moduli)

We consider the penetrometer cone as a small rigid circular plate. If E is the Young's modulus and ν the Poisson's ratio of a semi-infinite elastic medium, it is known that the settlement w under a rigid circular and superficial foundation of radius R is given by Boussinesq's formula:

$$w = \frac{1-\nu^2}{E} \times \frac{N}{\pi R^2} \times \frac{\pi R}{2} \qquad (2)$$

where N is the vertical load applied to the centre of the plate.
If σ is the contact pressure, it is obvious that:

$$E = (1-\nu^2) \frac{\pi R}{2} \times \frac{\sigma}{w} \qquad (3)$$

In our case, the plate is embedded in the medium at a depth h; then it is necessary to take into account the Mindlin's factor of reduction in this formula (Mindlin, 1935). If k_M is this factor and considering an infinite depth h (in this case $k_M = 2$), it is possible to write:

$$E = (1-\nu^2) \frac{\pi R}{2} \times \frac{\sigma}{w} \times \frac{1}{k_M} = (1-\nu^2) \frac{\pi R}{4} \times \frac{\sigma}{w} \qquad (4)$$

Then, the previous formula is applied to the tests of the present studies, with $\nu = 0.33$, that is a value generally allowed in soil mechanics:

$$E = 0.7 R \frac{\Delta P_p}{\Delta h} \quad (MPa) \qquad (5)$$

Different deformation moduli could be obtained from the deformability curve: tangent and secant modulus from monotonous curves and secant moduli from a cyclic curve. They are calculated with equation (5), where ($\Delta P_p/\Delta h$), in MPa/mm, is an approximation of the gradient value of the part of a curve considered to define a deformation modulus.

■ The zero-point deformation modulus (elasticity modulus), defined from a monotonous deformability curve, is the quantity ($R = 11.25$ mm) defined by:

$$E^0_{a_1 a_2 a_3} = \frac{\Delta P_p}{\Delta h} \times 7.87 \qquad (6)$$

where ($\Delta P_p/\Delta h$) is the zero-point gradient value of the curve obtained mathematically. In this context, it has already been shown by Zhou (1997) that the stress-displacement curve deduced from the deformability test can be satisfactorily fitted by a three coefficients mathematical function $f(x)$:

$$P_p = f(x) = \frac{x}{a_1 x + a_2} + \frac{1}{a_3} x \qquad (7)$$

The observed curves are then adjusted using a non-linear regression method. The three coefficients noted a_1, a_2 and a_3 are illustrated on figure 8.

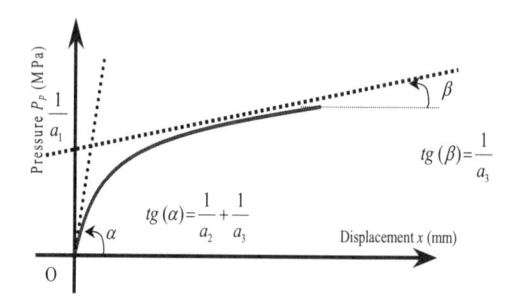

Figure 8. Deformability curve: interpretation of a_1, a_2, a_3 mathematical parameters

So, the elasticity modulus deduced from the proposed test denoted by $E^0_{a_1 a_2 a_3}$ is given by the expression (8) below:

$$E^0_{a_1 a_2 a_3} = \frac{dP_p}{dx}(0) \times 7.87 = tg(\alpha) \times 7.87$$

$$E^0_{a_1 a_2 a_3} = \left(\frac{1}{a_2} + \frac{1}{a_3} \right) \times 7.87 \qquad (8)$$

■ The secant deformation modulus, derived from a monotonous deformability curve at a deformation rate $\varepsilon = \Delta h /2R$, is the quantity:

$$E_{secant} = \left(\frac{\Delta P_p}{\Delta h} \right)_{\Delta h = 2R\varepsilon} \times 7.87 \qquad (9)$$

We consider particularly in this studies the three following deformation rates: 1 %, 2 % and 5 %.

■ Two other deformation moduli, obtained from cyclic curves, could be also calculated: an unloading secant modulus ($p_c \rightarrow p_d$) denoted by E^u_{secant} and a reloading secant modulus ($p_d \rightarrow p_c$) denoted by E^r_{secant}.

3 DETAILED RESULTS AND DISCUSSION

3.1 *Monotonous static loading tests*

Figures 9a and 9b illustrate some examples of deformability curves obtained with the proposed device.

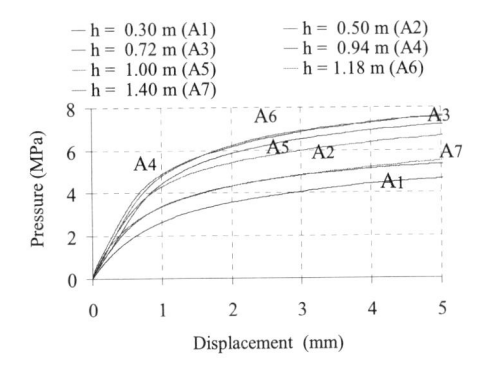

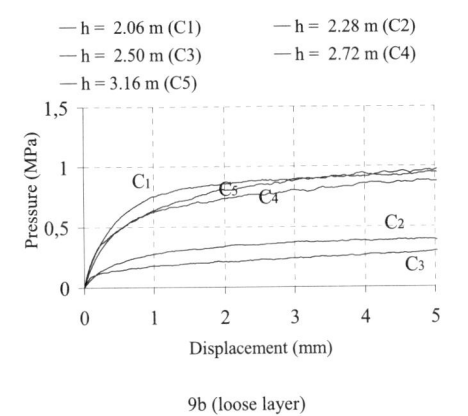

Figure 9. Examples of deformability curves in the tank (profile 2)

Figure 9 shows that all the obtained curves have the same shape. The failure pressure value denoted by P_{pl}, defined as the ordinate of the intersection point between the infinite asymptote and the axis of pressures (equal to $1/a_1$ see fig 8), is close to the dynamic cone resistance q_d. This observation was also made by Chaigneau (2001) during other researches performed on this type of penetrometer. This author point out that the measured values q_d and q_c using a Panda penetrometer are similar (q_c is the static cone resistance under 20 mm/s penetration's speed).

For each test, the three parameters a_1, a_2 and a_3 have been estimated. For all soundings, the initial tangent modulus (or initial instantaneous elasticity modulus) E^0_{a1a2a3} and the secant deformation moduli E_{secant} (about 1 %, 2 % and 5 % of deformation rate) in MPa have been calculated using equations (8) and (9). The values of these deformation moduli, with q_{dm} in MPa, at a depth h are given in tables 1 and 2. As an example, the secant deformation moduli values are only given for profile #2. The mean point resistance q_{dm} is the mean value of the dynamic point resistances q_d that is measured between $h - 5$ cm and $h + 5$ cm.

Table 1. Deformation modulus (dense layer)

		h (m)	0.3	0.5	0.72	0.94	1.00	1.18	1.40
S_1		q_{dm}	3.7	3.8	4.0	6.0	4.9	4.7	4.8
		E^0_{a1a2a3}	68	108	178	159	108	–	95
Profile S_2		q_{dm}	3.6	4.5	5.6	6.1	5.4	6.0	5.1
		E^0_{a1a2a3}	40	66.5	68	83.5	72.5	74.5	56.5
	E_{secant} 1%		33	50.5	42.5	58	50	44.5	44.5
	2%		28	39	36	52	46	38	37
	5%		20	25	33	21	32	31	25
S_3		q_{dm}	3.5	2.8	5.4	5.2	5.8	5.5	5.3
		E^0_{a1a2a3}	–	78	82	84	86	67.5	66
S_4		q_{dm}	5.2	6.6	5.3	5.8	4.5	5.3	4.3
		E^0_{a1a2a3}	–	71	55	73.5	73	54	45

Table 2. Deformation modulus (loose layer)

		h (m)	1.84	2.06	2.28	2.50	2.72	2.94	3.16
S_1		q_{dm}	0.7	0.5	0.2	0.3	0.2	0.1	0.52
		E^0_{a1a2a3}	14	9.5	–	5.5	–	–	–
Profil S_2		q_{dm}	0.7	0.5	0.3	0.4	0.6	0.2	0.5
		E^0_{a1a2a3}	–	21	6.5	12.5	21	–	16.5
	E_{secant} 1%		–	12.5	4.5	4.0	12.5	–	10
	2%		–	9.5	3.5	2.5	8.0	–	8.0
	5%		–	5.5	2.0	1.5	5.0	–	4.5
S_3		q_{dm}	0.9	0.6	0.3	0.4	0.3	0.2	0.8
		E^0_{a1a2a3}	25.5	22	18.5	30.5	27	17.5	6.0
S_4		q_{dm}	0.7	0.8	0.3	0.3	0.4	0.3	0.5
		E^0_{a1a2a3}	24.5	–	–	–	34.5	–	7.0

The cells without value indicate that the experimental curve has not been interpreted (questionable result). From the tables, first of all it can be seen that the secants moduli at 1 %, 2 % and 5 % of deformation rate are less than the initial modulus. In other words, the initial modulus measured with the proposed testing procedure could be a characteristic of in situ soil's behaviour undergoing small deformations (deformations less than 0.1 %). In the second place, the initial modulus is proportional to the point resistance. When we compare that modulus to q_{dm}, we observe a ratio denoted by $\lambda = E^0_{a1a2a3}/q_{dm}$ bounded between 10 and 25 for the dense sand and between 20 and 100 for the loose sand. We noticed finally that the testing procedure gives similar values of deformation moduli about each considered sounding (one by one). However, an important dispersion is observed between each sounding at a same depth for a considered zone (dense layer in the one hand and loose layer in the other hand). Tables show that in order to reach a repeatability of measurements (vertically and horizontally) we need a more homogenous layer (a medium with a more constant density) which is built at this moment.

3.2 Cyclic loading tests

Figure 10 shows a standard curve obtained for cyclic test C$_2$ performed at a depth of 1.20 m. The q'_{dm} value is about 6 MPa. We can observe a similar behaviour with other material under cyclic tests: thus, the displacement (Fig. 10) stabilizes after a few number of cycles which denotes the cyclic densification phenomena.

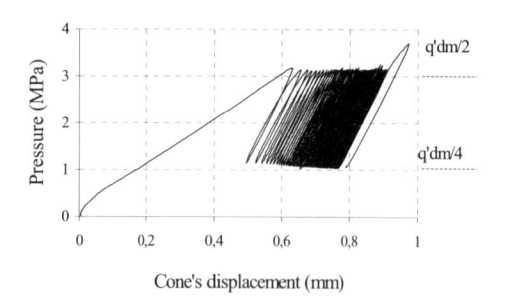

Figure 10. Example of cyclic deformability curve (50 cycles)

Table 3 gives some values of the reloading and unloading secant moduli (E^r_{secant}, E^u_{secant}) for this test.

Table 3. Reloading and unloading secant modulus

Number of cycle N	E^r_{secant} (MPa)	E^u_{secant} (MPa)
1	65.0	116.5
2	97.5	116.5
3	101.5	115.5
4	105.5	117.0
5	107.5	117.0
6	107.0	116.0
7	109.5	117.0
8	109.5	116.0
9	110.0	115.5
10	110.0	117.0
15	110.0	115.0
20	110.5	115.5
30	111.0	115.5
40	111.5	117.0
50	111.0	115.0

As can be seen, the measured unloading secant modulus denoted by E^u_{secant} is relatively constant. The reloading secant modulus E^r_{secant} gradually increases as early as the first cycles and becomes practically constant after a few number of cycles (9 cycles here). This typical cyclic test is very useful to reach a soil modulus which describes the purely elastic soil behaviour. Moreover, this usefulness was pointed out by Combarieu & Canépa (2001). Indeed, we should mention the hope that a cyclic test will be able to reduce the effects of soil disturbance related to the deformability test procedure. Some other geotechnical test procedures (plate bearing test, deep foundation loading test, oedometer test, etc.) include an unloading-reloading phase and the deformation behaviour measured during this loop is exploited. Ménard (1962) introduced this procedure (unloading and reloading are performed during the expansion of the pressuremeter probe). He mentioned that the deformation modulus measured over the cycle is practically identical to the purely elastic modulus or micro strain modulus denoted by E_ε. Theoretical studies of the expansion of spherical and cylindrical cavities in an elastic perfectly plastic soil have provided additional support for Ménard's ideas, notably the analyses conducted by Mestat (1994). Lastly, when this test procedure is followed, the factors mentioned here allow us to assume that it is possible to measure a strain modulus over an unloading-reloading cycle during cyclic deformability test which is characteristic of the behaviour of soils undergoing small deformations. It would be possible, for a whole range of problems, to use a modulus of this kind (secant or hysteretic modulus) directly for the calculations that assume a linear elastic behaviour of the soil.

4 NUMERICAL SIMULATION

In order to conduct a more accurate analysis of the in situ soil deformability test (study of soil-cone interaction, of the soil volume mobilized by the static loading of the cone penetrometer, etc.) a finite element model of the proposed test was built using the code CÉSAR-LCPC (Arbaoui, Bodé, Bressolette & Gourvès, 2002). At term, the three in situ parameters E, c and φ usually used in geotechnical studies, could be related to the three precedent mathematical coefficients a_1, a_2 et a_3 introduced in Eq (7). The soil is assumed to be isotropic, homogenous and infinite. Its behaviour is modelized by an elastic perfectly plastic Mohr-Coulomb law with 5 parameters (Young's modulus E, Poisson's ratio ν, cohesion c, friction angle φ and dilatancy angle ψ). ν was taken equal to 0.33 and ψ has been assumed to be equal to 0°. The 22.5 mm diameter cone may be considered to be rigid in comparison with the soil. For simplicity, the cone was modelized here as an elastic body with a large Young's modulus (equal to 21.10^4 MPa) and a null Poisson's ratio. Figure 11 shows the axisymmetric finite element discretization (5458 nodes and 2402 quadratic finite elements).

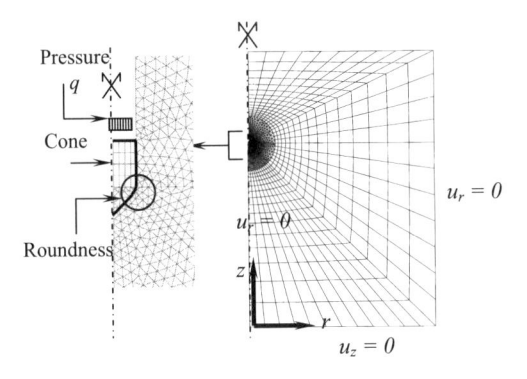

Figure 11. Finite element discretization

The kinematic boundary conditions are $u_z = 0$ on the bottom nodes ($z = 0$) and $u_r = 0$ on the lateral sides nodes ($r = 0$, $r = 2$ m). The boundaries of the soil domain are chosen far from the cone so as to avoid interactions with the predicted failure area of the soil (Fig. 11). The pressure q is equal to 2 MPa, and is applied on the cone by step of $0.1 \times q$. Neither the effects of dynamic penetration nor the water conditions have been taken into account, but the initial geostatic stresses has been considered. A study of the influence of the soil-cone interface conditions (perfectly sliding, Coulomb friction) on the soil's answer, (described by the pressure-displacement curve) has been realized. It has been shown that this influence was low on the global answer (deformability curve) of the considered model. The result of the confrontation between observed and calculated curves is satisfying and positive: a similar order of magnitude is especially observed (Fig. 12).

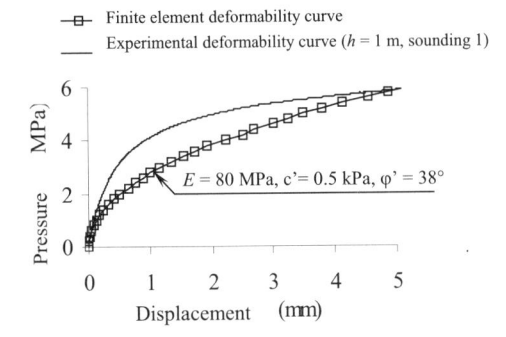

Figure 12. Example of a comparison between experimental and finite element pressure-displacement curves

However, the model is still simplified and didn't represent yet all the phases of the in situ test protocol: for example, the preliminary cone's installation into the soil is not modelized at the moment. Consequently, the model will have to be improved and its reliability examinated.

5 CONCLUSION AND PERSPECTIVES

The penetrometer test (static or dynamic) allows to obtain an in situ failure characteristic. The idea is to obtain at the same time with this traditional apparatus an estimation of one of the most important deformability parameter, the deformation modulus E. Thus, a new experimental device (light or with large capacity) is proposed to carry out a simple compression test at a desired depth with a penetrometer. This is a static loading test on the penetrometer's cone. This test, with its light device, allows fast execution of a large number of tests at low cost.

The test allows to draw a monotonous or cyclic stress-displacement curve. The test procedure has been performed on a sand placed into a tank with two layers (loose and dense) using a light dynamic penetrometer with variable energy type Panda. The monotonous curve shows a linear part, related to the pseudo-elastic response of the soil where a deformation modulus similar to the Young's modulus is defined. Moreover, traditional secant moduli are also obtained. We obtain a reliable and relevant estimation of the initial modulus for displacement levels corresponding to small deformations of the soil (typically under 0.1 %).

However, in order to obtain a modulus which represents better the purely elastic behaviour of soil (micro deformations modulus), it is noticed that the cyclic deformability test is the adequate testing. Cyclic tests have been realized and other moduli are proposed. In sum, the recommended test provides then to consider different categories of moduli (monotonous tangent and secants, cyclic unloading-reloading).

We can say that the Young's modulus determination problem with this approach may be summarised to a choice between a modulus which is estimated from a loading phase (in a domain assumed elastic) and a modulus from an unloading-reloading cycle. We must also keep in mind that, for engineering computations, only linear elasticity is used, which requires the Young's modulus estimation.

The cohesion and the friction angle of soil are others deformability parameters which can be deduced from the stress-cone's displacement curve. That is why a numerical simulation of the deformability test has been done. The results are encouraging and interesting. As perspective, others soil types would be studied (clay, silt): it should give further validation of the experimental test. Moreover, the reloading modulus law evolution according cycle numbers could be an interesting approach to reach a soil's micro deformation modulus.

Finally, the technique developed here could be adapted to another penetrometer with high capacity, for example a normalized CPT or CPTu.

ACKNOWLEDGEMENTS

This work is conducted in collaboration with the LCPC of Paris. We would like to acknowledge it for its financial and technical supports.

REFERENCES

Arbaoui, H., Bodé L., Bressolette Ph. & Gourvès, R. 2002. Finite element simulation of an in situ soil deformability test. *5th European Conference in Numerical Methods in Geotechnical Engineering*, Presses de l'école nationale des Ponts et Chaussées, Paris, France, 113-118.

Balaam, N.P., Poulos, H.G. & Booker J.R. 1975. Finite element analysis of the effects of installation on pile load-settlement behaviour. *Geotechnical Engineering* volume 6 :33-48.

Chaigneau, L. 2001. Caractérisation des milieux granulaires de surface à l'aide d'un pénétromètre. *PhD thesis : Blaise Pascal University, Clermont-Ferrand II*, France, 204 p.

Combarieu, O. & Canépa, Y. 2001. The unload-reload pressuremeter test. *Bulletin de liaison des laboratoires des Ponts et Chaussées* 233: 37-67.

Faugeras, J.C. 1979. L'essai de compressibilité des sols au pénétromètre statique et son interprétation sur modèle analogique. *PhD thesis: Paul Sabatier University, Toulouse*, France, 125 p.

Faugeras, J.C., Fortuna, G. & Gourvès, R. 1983. Soils compressibility measurement by static penetrometer. *In situ testing, International Symposium* volume 2: 269-274, Paris.

Ménard, L. & Rousseau J. 1962. L'évaluation des tassements. Tendances nouvelles, *Soils*, 1, 13-30.

Mestat, Ph. 1994. Validation du progiciel CESAR-LCPC en comportement mécanique non linéaire. *Fondations superficielles et tunnels*. Etudes et recherches des laboratoires des Ponts et Chaussées volume 1, *série géotechnique - GT 58*: 193 p.

Mindlin, R.D. 1935. A preliminary statement of the Galerkin vectors; force at a point in the interior of a semi-indefinite solid. *Physics*, t. 7, 195-202.

Zhou, S. 1997. Caractérisation des sols de surface à l'aide du pénétromètre dynamique léger à énergie variable type Panda. *PhD thesis: Blaise Pascal Univesity, Clermont-Ferrand II*, France, 179 p.

Characterization of soft marine clay using the flat dilatometer

A. Arulrajah & H. Nikraz
Department of Civil Engineering, Curtin University of Technology, Perth, Australia

M. W. Bo
Bullen Consultants Ltd., United Kingdom

Keywords: consolidation, dissipation testing, hydraulic conductivity, in-situ testing, shear strength

ABSTRACT: In-situ testing on Singapore marine clay was carried out with the flat dilatometer at an In-Situ Test Site in the Republic of Singapore. The flat dilatometer (DMT) penetration and dissipation tests were first conducted prior to reclamation at various elevations in the marine clay. Following the subsequent completion of land reclamation and ground improvement works with vertical drains and preloading, another series of flat dilatometer penetration and dissipation tests were conducted in the Vertical Drain Area as well as an adjacent untreated Control Area. The results obtained in the treated Vertical Drain Area and untreated Control Area were then compared with each other and the prior to reclamation results. In-situ penetration testing with the flat dilatometer enable the shear strength, overconsolidation ratio and degree of consolidation of the marine clay to be determined. In-situ dissipation testing with the flat dilatometer provide a means of evaluating the in-situ coefficient of horizontal consolidation and horizontal hydraulic conductivity of marine clays.

1 INTRODUCTION

The Changi East Reclamation Project in the Republic of Singapore comprises of the ground improvement of marine clay with the installation of prefabricated vertical drains and subsequent surcharge placement. The In-Situ Testing Site was located in the Northern area of the project where the thickest compressible layers existed.

The original seabed level in the site was 3.29 meters below Admiralty Chart Datum (−3.29 mCD). A series of pre-reclamation in-situ flat dilatometer penetration and dissipation tests were first conducted. Land reclamation was next carried out by the hydraulic placement of sand until the vertical drain platform level of 4 meters above Admiralty Chart Datum (+4 mCD). Vertical drains were installed at this elevation at 1.5 meter square spacing to depths of up to 35 meters in the Vertical Drain Area. Surcharge was next placed until the design elevation of 10 meters above Admiralty Chart Datum (+10 mCD) and was left in place for a period of 23 months.

The In-Situ Testing Site consists of two adjacent sub-areas namely the Vertical Drain Area where vertical drains were installed and the Control Area where no vertical drains were installed. The two sub-areas are located adjacent to each other and are subject to the same construction sequence and surcharge height and as such this enabled a comparison to be made between the two areas. Another series of post-improvement flat dilatometer tests were carried out in both the Vertical Drain Area and Control Area after a surcharge period of about 23 months. A comparison was then made between the vertical drain treated and untreated areas. In-situ testing works in this research study comprises the use of the flat dilatometer (DMT) penetration and dissipation tests.

2 SITE DESCRIPTION

Fig. 1 shows the location of the In-Situ Testing Site. Field tests carried out prior to reclamation were denoted as FT-2. Field tests carried out in the Vertical Drain Area after improvement were denoted as FT-8 while tests carried out in the untreated Control Area after improvement were denoted as FT-9.

The In-Situ Testing Site comprises two distinct layers of marine clay, which are the "Upper Marine Clay layer" and the "Lower Marine Clay layer". The "Intermediate Stiff Clay layer" separates these two distinct marine clay layers. The upper marine clay is soft with undrained shear strength values ranging from 10 to 30 kPa. The lower marine clay is lightly overconsolidated with an undrained shear strength varying from 30 to 50 kPa. Bo et al. (1998a, 1998b) has described the characteristics of the marine clay found at Changi.

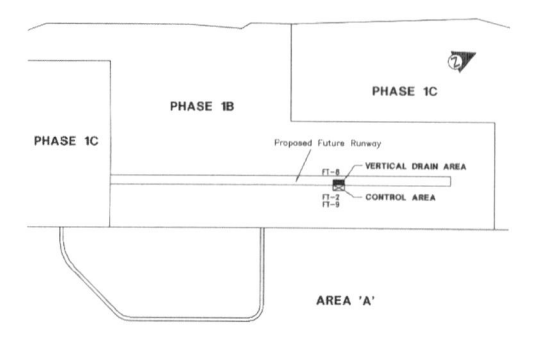

Figure 1. Location of In-Situ Test Site comprising Vertical Drain Area and Control Area (Bo et al. 2000).

3 FLAT DILATOMETER PENETRATION TESTING

A Marchetti flat dilatometer was used for the tests, which consist of a steel membrane on one side of the blade. The dilatometer blade is 96 mm in width and 230 mm in length. The diameter of the membrane is 60 mm. The dilatometer requires certain specialised skills and technical knowledge to operate. Fig. 2 shows the geometry and dimensions of the Marchetti dilatometer blade used in the study.

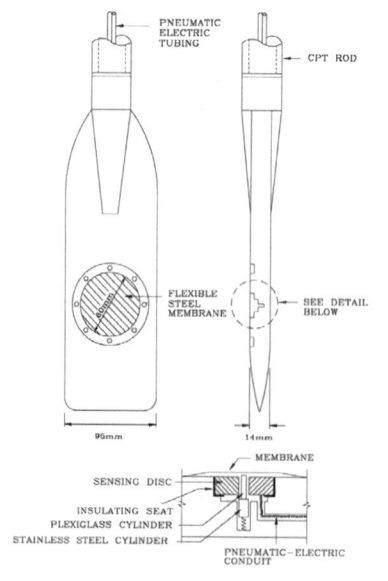

Figure 2. Geometry and dimension of Marchetti dilatometer blade (Chang, 1986).

3.1 Flat dilatometer penetration test procedure

The testing procedure followed that described by Marchetti and Crapps (1981). The testing consisted of pushing the flat dilatometer blade gradually into the soil at a prescribed rate of 20 mm per second with the use of a 20 ton Gouda dutch cone rig. The pushing was temporarily stopped at each of the proposed testing levels at which the two pressure readings A and B corresponding to two prefixed states of expansion of the membrane was recorded. The first pressure reading that is A-reading (p_o) corresponds to the membrane lift-off pressure while the second pressure reading that is B-reading (p_1) corresponds to the pressure required for the center of the membrane to deflect by a preset distance of 1 mm into the soil.

From the two pressure readings, three dilatometer indices are obtained being the material index (I_D), horizontal stress index (K_D) and dilatometer modulus (E_D):

$$I_D = (p_o - p_1) / (p_o - u_0) \quad (1)$$

$$K_D = (p_o - u_0) / (\sigma_{vo} - u_0) \quad (2)$$

$$E_D = 34.7 \, (p_1 - p_o) \quad (3)$$

Marchetti (1980) proposed the following correlation between the undrained shear strength, C_u, with the horizontal stress index, K_D:

$$C_u = 0.22 \, \sigma_{vo}' \, (0.5 \, K_D)^\eta \quad (4)$$

$$K_D = (P_0 - U_0) / \sigma_{vo}' \quad (5)$$

where σ_{vo}' is vertical effective stress; K_D is the horizontal stress index; η is a variable which may vary from one clay to another; P_0 is the A reading from dilatometer; U_0 is pre-inserting water pressure.

For Singapore Marine Clay at Changi, η can be taken as 1 for upper marine clay and intermediate clay while η can be taken as 0.7 for lower marine clays (Bo et al. 2000).

Marchetti (1980) proposed the following correlation for the estimation of OCR for clay:

$$OCR = (0.5 \, K_D)^n \quad (6)$$

where for Singapore marine clay at Changi, n can be taken as 1 for upper and lower marine clays and 0.8 for intermediate clays (Bo et al. 1997, 1998a).

3.2 Comparison of penetration testing results

Care should be taken in determining the horizontal stress index, K_D, when soil is still undergoing consolidation. In that case current pore pressure values from the piezometer should be used for equilibrium pore pressure. The method of determining degree of consolidation from DMT has been discussed by Bo et al. (1997). The comparison of the DMT results after soil improvement as compared to that before reclamation is shown in Fig. 3 to 6.

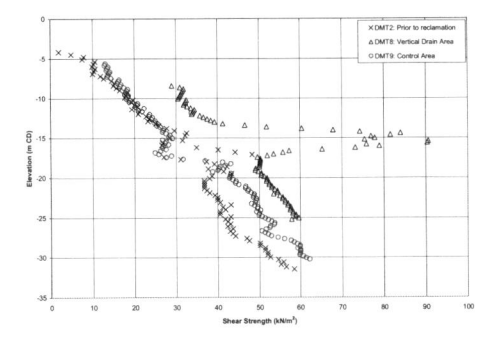

Figure 3. Variation of DMT shear strength with elevation.

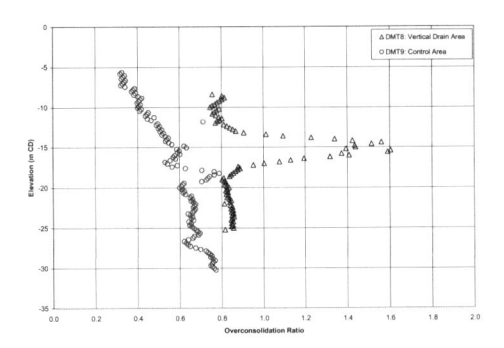

Figure 4. Variation of DMT OCR with elevation.

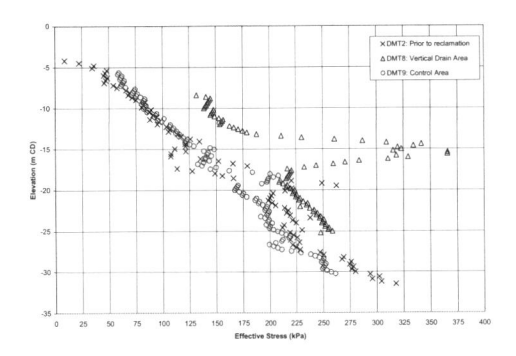

Figure 5. Variation of DMT effective stress with elevation.

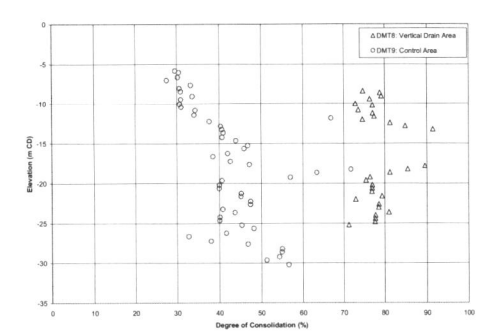

Figure 6. Variation of DMT degree of consolidation with elevation.

The DMT tests indicate that the degree of consolidation of the Vertical Drain Area had attained a degree of consolidation of about 70-80% while the Control Area had attained a degree of consolidation of about 40%. The degree of consolidation was estimated by the method of Bo et al. (1997).

4 FLAT DILATOMETER DISSIPATION TEST

The dilatometer test has the potential of providing estimates of the in-situ coefficient of consolidation due to horizontal flow from dissipation tests. The common dilatometer dissipation test involves two different procedure, one by recording the change of A-reading with time and the other the change of C-reading with time. The C-reading is the pressure reading, which corresponds to the resumption of the lift-off position of the membrane during deflation subsequent to taking the B-reading.

4.1 Flat dilatometer dissipation test procedure

The dissipation test which makes use of the A reading is called the DMTA dissipation test and can be performed at any depth. In this method, the A-reading is taken at different time intervals and plotted against log time. The time corresponding to the point of reverse curvature on the A-decay curve, T_{flex} is used as a basis for the interpretation of the C_h. The following expression was used as proposed by Marchetti and Tottani (1989):

$$C_h(DMTA) \times T_{flex} = 5 \text{ cm}^2 \qquad (7)$$

where C_h is in units of cm²/min.

In the second method, the C-reading is plotted against square root time and the time corresponding to 50% consolidation, t_{50} is determined and used in the interpretation of C_h (Schmertmann, 1988). Gupta et al. (1983, 1986) procedure, developed for piezocone dissipation analysis was modified and used in the interpretation of C_h. The dissipation test which makes use of the C-reading is called the DMTC dissipation test and can be performed at any depth. The procedure involves estimating rigidity index, E_u/C_u, and pore pressure at failure, A_f, for the clay and determining the time factor corresponding to 50% pore pressure dissipation, T_{50}, from the dissipation curves for $A_f = 0.9$ (Schmertmann, 1988). An adjustment of the time factor may be required if A_f is different from 0.9. The T_{50} can then be used in the following equation as proposed by Marchetti and Tottani (1989), which assumes $R^2 = 600 mm^2$:

$$C_h (DMTC) = 600 (T_{50} / t_{50}) \quad (8)$$

where C_h is in units of mm^2/min.

Similar to CPTU tests, the C_h values determined from either DMTA or DMTC corresponds to the unloading/reloading range. Corrections have to be made to obtain the in-situ C_h value (Chu et al. 2002). For foundation clays consolidated in the normally consolidated range, estimates of the coefficients of consolidation can be obtained from $C_h(probe)$ by means of the following expression published by Baligh and Levadoux (1980):

$$C_h (NC) = (Cr / Cc) [C_h(probe)] \quad (9)$$

where Cr is the recompression index and Cc is the compression index.

In order to obtain the hydraulic conductivity in the normally consolidated condition, a correction taking the recompression ratio into account needs to be applied. The horizontal hydraulic conductivity is given by the following equation:

$$k_h = (\gamma_w / 2.3\sigma'_v) (RR) C_h \quad (10)$$

where k_h is horizontal hydraulic conductivity; γ_w is unit weight of water in kN/m^3; RR is recompression ratio and σ'_v is effective vertical stress in kPa.

Fig. 7 and 8 show the prior to reclamation testing results. Fig. 9 and 10 show testing results after ground improvement at the Vertical Drain Area and Control Area respectively. Fig. 11 and 12 show typical dissipation test plots used for determination of C_h at the Vertical Drain Area and Control Area respectively.

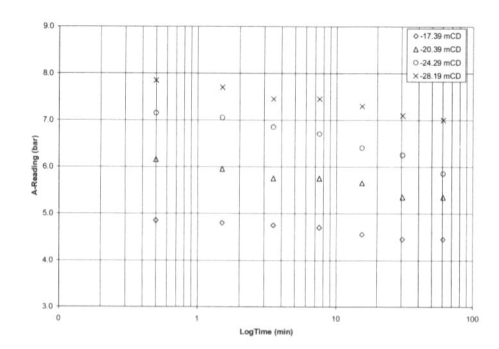

Figure 7. DMTA dissipation tests prior to reclamation.

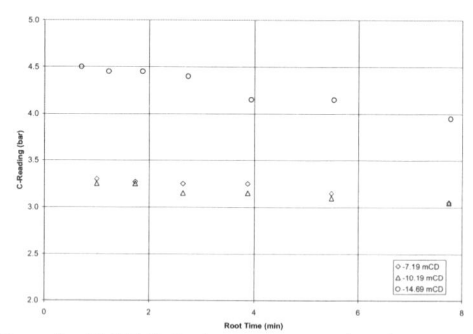

Figure 8. DMTC dissipation test prior to reclamation.

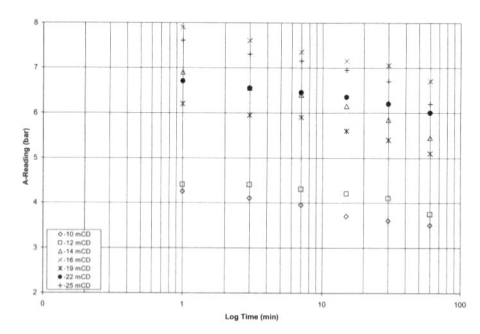

Figure 9. DMTA dissipation test at Vertical Drain Area.

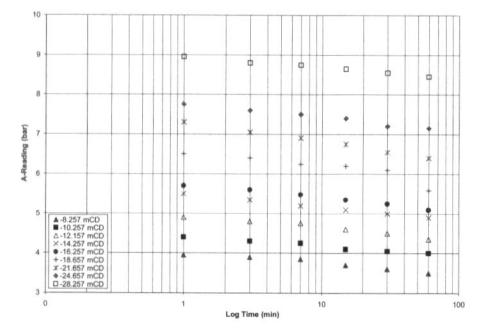

Figure 10. DMTA dissipation test at Control Area.

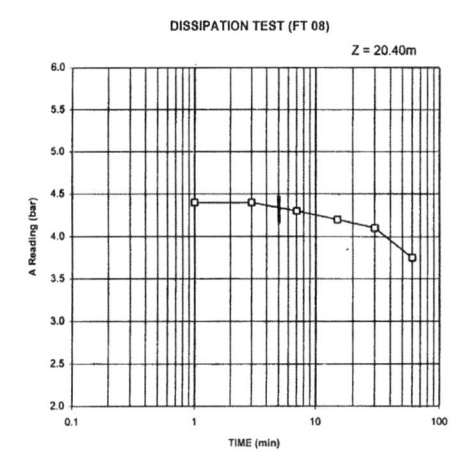

Figure 11. Typical DMTA dissipation test plot at Vertical Drain Area (elevation -12 mCD).

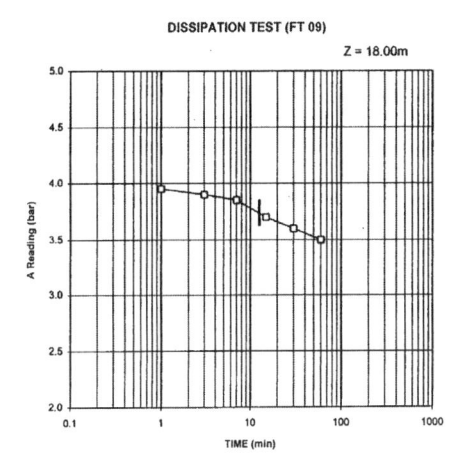

Figure 12. Typical DMTA dissipation test plot at Control Area (elevation -8.257 mCD).

4.2 Comparison of dissipation testing results

The comparison of the C_h results for the Vertical Drain Area and the Control Area is presented in Fig. 13 whereas Fig. 14 shows the k_h results.

C_h value is seen to be higher in the Vertical Drain Area as compared to the Control Area. The C_h could be higher due to greater ratio of reduction in the coefficient of volume change, m_v. C_h value may not be reduced to the degree expected due to the use of correction factors and change in compressibility. The in-situ results in the upper and lower marine clay layers indicate C_h values of 4-6 m²/yr in the Vertical Drain Area as well as in the Control Area. It is observed that all the method indicates large C_h values in the intermediate stiff clay layer.

The C_h values for the laboratory tests were obtained from radial flow Rowe cell of 75 mm diameter and 30 mm thickness and also from horizontally cut 63.5 mm oedometer test samples. The laboratory results for C_h is noted to be generally lower than that obtained from the DMT.

It also seems that k_h values in the Vertical Drain Area is lower than that in the Control Area, which should be the case. It is noted however that k_h values are indirectly obtained from C_h values. k_h values ranging from 10^{-9} to 10^{-10} m/s were obtained in the Vertical Drain Area while values of 10^{-9} m/s were obtained in the Control Area.

It is evident that the prior to reclamation DMT dissipation test has encountered the intermediate stiff layer strata at the lower elevations and hence the high initial C_h values. The DMT dissipation tests in the Vertical Drain Area and the Control Area also indicate higher C_h values in the intermediate stiff layer.

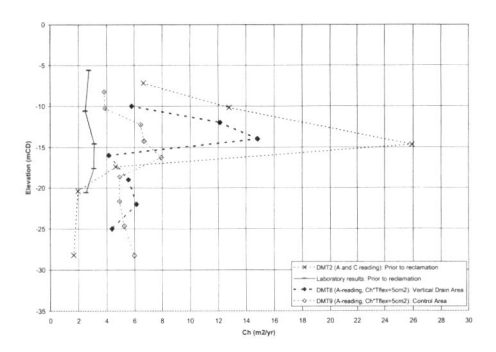

Figure 13. Comparison of coefficient of consolidation due to horizontal flow from DMTA dissipation test.

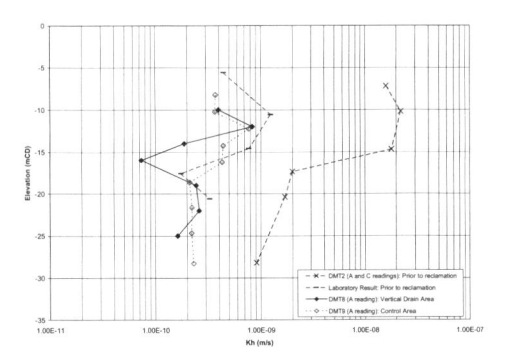

Figure 14. Comparison of horizontal hydraulic conductivity from DMTA dissipation test.

5 DISCUSSIONS

The DMT result seems high till depths of 20 meters and this could be due to the variation in the profile of the intermediate layer in the lower depths. However, the DMT results are reasonable in the lower marine clay layer. The method based on A readings needs to be calibrated before its use as it involves the selection of a constant between 5 and 10 cm^2. Unless the selection of this constant is verified, the DMT method is not recommended (Bo et al. 1998b). The selection of a constant of 5 cm^2 has been found to be suitable for the Singapore marine clay at Changi. (Bo et al. 2003, Chu et al. 2002).

The reason for the variations in the in-situ dissipation tests with depths could be due to variations in the soil stratigraphy at each of the in-situ testing locations. For instance, the actual depths of the intermediate stiff clay layers could differ slightly at each in-situ dissipation locations. Other factors could be slight differences in the seabed elevations at the test locations. Another obvious difference would be the differing assumptions and methods of analysis of the different tests.

6 CONCLUSIONS

The DMT test indicates that the degree of consolidation of the Vertical Drain Area had attained a degree of consolidation of about 70-80% while the Control Area had attained a degree of consolidation of about 40%.

C_h value is seen to be higher in the Vertical Drain Area as compared to the Control Area. Despite the k_h being lower in the Vertical Drain Area, the C_h could be higher due to greater ratio of reduction in the coefficient of volume change, m_v.

The in-situ results in the upper and lower marine clay layers indicate C_h values of 4-6 m^2/yr in the Vertical Drain Area as well as in the Control Area.

It also seems that k_h values in the Vertical Drain Area is lower than that in the Control Area, which should be the case. It is noted however that k_h values are indirectly obtained from C_h values. k_h values ranging from 10^{-9} to 10^{-10} m/s were obtained in the Vertical Drain Area while values of 10^{-9} m/s were obtained in the Control Area.

ACKNOWLEDGEMENTS

The authors are most grateful to Dr. A. Vijiaratnam the Former Chairman of SPECS Consultants (Singapore) Pte. Ltd. for his encouragement and full support in the submission of these findings.

REFERENCES

Baligh, M.M. and Levadoux, J. N. 1980. Pore Pressure Dissipation After Cone Penetration, *Massachusetts Institute of Technology Research Report*, Cambridge, Massachusetts.

Bo Myint Win, Arulrajah, A., Choa, V. 1997. Assessment of Degree of Consolidation in Soil Improvement Project, *Proceedings of the International Conference on Ground Improvement Techniques*, May, Macau: 71-80.

Bo Myint Win, Arulrajah, A., Choa, V. and Chang, M. F. 1998a. Site characterization for land reclamation project at Changi in Singapore, *Proceedings of the 1st International Conference on Site Characterization*, April 1998, Roberson & Mayne (eds), Balkema, Rotterdam, Atlanta, USA: 333-338.

Bo Myint Win, Arulrajah A. and Choa. V. 1998b. The hydraulic conductivity of Singapore Marine Clay at Changi, *Quarterly Journal of Engineering Geology*, 31: 291-299.

Bo Myint Win, Chang, M.F., Arulrajah, A., Choa, V. 2000. Undrained Shear Strength of the Singapore Marine Clay at Changi from In-Situ Tests, *Geotechnical Engineering, Journal of the Southeast Asian Geotechnical Society*, Vol. 31, Number 2, August 2000.

Bo Myint Win, Chu, J, Low, B. K. and Choa, V. 2003. Soil Improvement – Prefabricated Vertical Drain Techniques, Thomson Learning, Singapore.

Chang, M.F. 1986. The Flat Dilatometer and Its Application to Singapore Clays, *Proceedings of the 4th International Seminar Field Instrumentation and In-situ Measurements*, Nanyang Technological Institute, Singapore.

Chu J., Bo Myint Win, Chang M. F., and Choa V. 2002. Consolidation and permeability properties of Singapore marine clay, *Journal of Geotechnical and Geoenvironmental Engineering*, September 2002.

Gupta, R.C. 1983. Determination of the In-Situ Coefficient of Consolidation and Permeability of Submerged Soils Using Electrical Piezoprobe Sounding, *PhD Thesis*, University of Florida.

Gupta, R.C. and Davidson, J.L. 1986. Piezoprobe Determined Coefficient of Consolidation, *Soils and Foundations*, Vol. 26, No. 3, pp. 12-22.

Marchetti, S. 1980. Insitu Tests by Flat Dilatometer, *Journal of the Geotechnical Engineering Division*, ASCE, Vol. 106, No. GT3, Proc Paper 15290: 299-321.

Marchetti, S., Crapps, D.K. 1981. Flat Dilatometer Manual. Schmertmann and Crapps Inc., Gainsville, Florida, USA.

Marchetti, S., Totani, G. 1989. Ch Evaluation form DMTA Dissipation Curves, *Proceedings of the 12th International Conference on Soil Mechanics and Testing Engineering*, Vol. 1: 281-286.

Schmertmann J. H. 1988. The coefficient of consolidation obtained from p2 dissipation in the DMT, *Proceedings of the Geotechnical Conference*, Pennsylvania. Department of Transportation, Pennsylvania.

Estimating the SPT penetration resistance from rod penetration based on instrumentation

E.H. Cavalcante
Federal University of Sergipe, Aracaju, SE, Brazil

F.A.B. Danziger
Federal University of Rio de Janeiro, Rio de Janeiro, RJ, Brazil

B.R. Danziger
Fluminense Federal University, Niterói, RJ, Brazil

Keywords: Standard Penetration Test, SPT instrumentation

ABSTRACT: The SPT penetration resistance N is obtained as the number of hammer blows necessary to penetrate a standardized sampler 300 mm, after an initial setting of 150 mm. The common practice consists in considering the N value as an integer number, irrespective of the differences in penetration to the reference 300 mm. This paper presents the results of instrumented SPTs where the N value provided by the test in the usual way is compared with a corrected Ncorr value based on the penetrations obtained from accelerometer data. The N values analysed ranged from 1 to 60. Three ranges with different behaviours were found. In the first one, for N varying from 1 to 6, a significant scatter was found, with a quite clear trend of Ncorr higher than N. In the second range, for N varying from 7 to 16, a small scatter was found, and the correlation lines were very close to the Ncorr = N line. In the third range, for N varying from 18 to 60, the same scatter as in the second range was found. A trend of Ncorr 7% greater than N was found in this case. As far as the nominal 450 mm penetration is concerned, a trend of penetration smaller than 450 mm was found.

1 INTRODUCTION

The Standard Penetration Test (SPT) is one of the most widely used in situ test all over the world (Décourt, 1989). It provides information regarding soil classification, index parameters and a number of geotechnical applications through semi-empirical procedures, like e.g. evaluation of settlements of shallow foundations, bearing capacity of piles and evaluation of liquefaction potential.

The SPT penetration resistance N is obtained as the number of hammer blows necessary to penetrate the sampler 300 mm, after an initial setting of 150 mm. The common practice consists in considering the N value as an integer number, irrespective of the differences in penetration to the reference 300 mm.

This paper presents the results of instrumented SPTs where the N value provided by the test in the usual way is compared with a corrected N_{corr} value based on the penetrations obtained from accelerometer data.

2 ENERGY TRANSMISSION ON THE SPT SYSTEM

Energy transmission from the hammer to the rod stem has deserved the interest of many authors in several publications in the past (e.g. De Mello, 1971, Kovacs et al., 1977). The research carried out by Palacios (1977) and Schmertmann and Palacios (1979) has shown the energy transmission mechanism from the wave equation theory. Those authors established that the number of blows N depends on the amount of energy delivered to the rod stem according to

$$N_I E_I = N_{II} E_{II} \qquad (1)$$

where N_I and N_{II} are the number of blows corresponding to the energies E_I and E_{II}, respectively.

Other contributions to this subject have followed, emphasizing the need to standardize an energy level to be used internationally (e.g. Robertson et al., 1983, Seed et al., 1985, Skempton, 1986).

The International Geotechnical Society (ISSMFE, 1989) has established a reference value of 60 % of the theoretical free fall of 474 J, E_{60}, for comparisons among tests performed by different equipments. The corresponding N value is designated as N_{60} and can be obtained from expression (2) as:

$$N_{60} = N \frac{E}{E_{60}} \quad (2)$$

where N corresponds to the energy E, which has to be known. The energy is to be obtained below the driving head (ISSMFE, 1989).

3 THE BRAZILIAN SPT

The Brazilian SPT is described e.g. by Décourt et al. (1989) and is performed according to the Brazilian Standard NBR 6484 (2001).

The sampler used is very similar to the one recommended by ISSMFE (1989). A 65 kg pinweight cylindrical hammer falling from a free fall height of 0.75 m provides a nominal energy of 478 J, almost the same as the one in ISSMFE (1989), 474 J.

In early times different samplers have been used in the SPT system in Brazil (e.g. Belincanta, 1998, Belincanta and Cintra, 1998). From the beginning of the 70's only the standard sampler has been used. Correlations among the results obtained from the different samplers have been developed (e.g. Teixeira, 1977, Belincanta and Cintra, 1998).

Although there are nowadays few automatic hammers in operation in Brazil, in almost all cases the hammer is still hand-operated. Figure 1 illustrates an old picture from Vargas (1945), where Figure 2 shows one of the systems used in the research herein presented.

On the one hand, the use of such an old practice does represent a non evolution of the SPT technology in Brazil. On the other hand, it can be assumed that the energy level in the SPT system has been basically the same throughout the years. As a consequence, it seems that Brazilian foundations have not been overdesigned due to changes in SPT energy, an issue raised by Kovacs (1994) in respect to EUA practice.

NBR 6484 (2001) recommends the use of a hard wood cushion in the lower part of the hammer but in many cases it is not used at all.

According to the NBR 6484 (2001), the anvil mass is 3.5 kg to 4.5 kg. Instead a much lighter anvil is very often used (from 0.5 kg to 1.0 kg).

Brazilian rods are lighter than the ISSMFE (1989) ones. They have an internal diameter of 25.4 mm, an external diameter of 33.4 mm and a mass of 3.3 kg/m.

Figure 1. Rig used with hand-operated hammer (Vargas, 1945).

Figure 2. Hand-operated hammer still commonly used in Brazil (Cavalcante, 2002).

4 TEST RESULTS

4.1 Instrumentation used

The instrumentation used consisted in strain-gauges and piezoelectric accelerometers. The instruments were fixed in a rod section nearly 0.5 m below the anvil. The instrumented section is 0.23 m long, as shown in Figure 3.

The SPT Analyzer (Figure 4) was used for data acquisition. The SPT Analyzer transforms the instrumented data in force (F1 and F2) and velocity (v1 and v2) records with time. The equipment evaluates, at real time, the maximum energy delivered to the rod at each blow. The vertical displacement of the rod stem and the energy delivered versus time are also obtained (PDI, 1999, Maultsby, 1999).

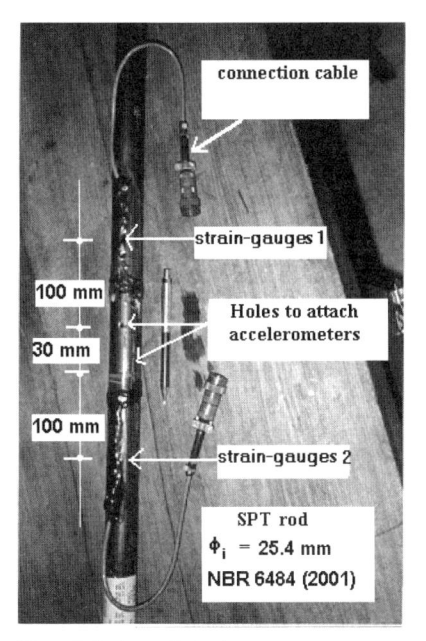

Figure 3. Instrumented rod.

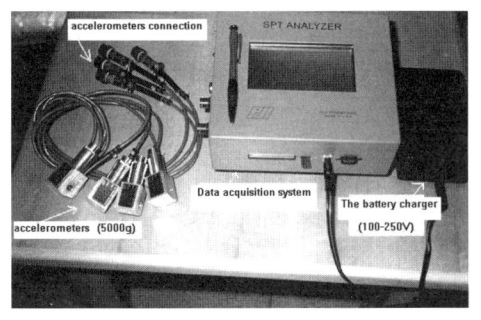

Figure 4. The SPT Analyzer.

Details about the instrumentation and data quality can be found in Cavalcante (2002) and Cavalcante et al. (2003).

4.2 Tests performed

Three series of tests have been performed in the period May 2000 – December 2001 in ten sites in two Brazilian cities, Rio de Janeiro and João Pessoa. Three companies involving five rigs performed the SPTs. Different kinds of soils in 13 borings have been tested. Boring depth ranged from 7 to 31 m, but most of the tests were limited to 14 m.

A total of 1393 blows were recorded, including both force and velocity records. In one of the sites the sensors were located not only close to the anvil, but also near (0.29 m above) the sampler.

The third series of tests aimed at the evaluation of the free fall height and rate of the hammer impact. Further details can be found in Cavalcante (2002).

4.3 The energy delivered to the rod stem

The tests performed by Cavalcante (2002) have shown that the average transmitted energy to the rod stem is about 82.3% of the nominal energy of 478 J (Brazilian nominal energy) or 83.0 % in respect to the ISSMFE (1989) nominal energy of 474 J, independently of the rod length. Most of the data were obtained from 2 m to 14 m depth. The standard deviation of the average 82.3 % efficiency was 5.6% for this rod interval. Figure 5 illustrates the obtained results.

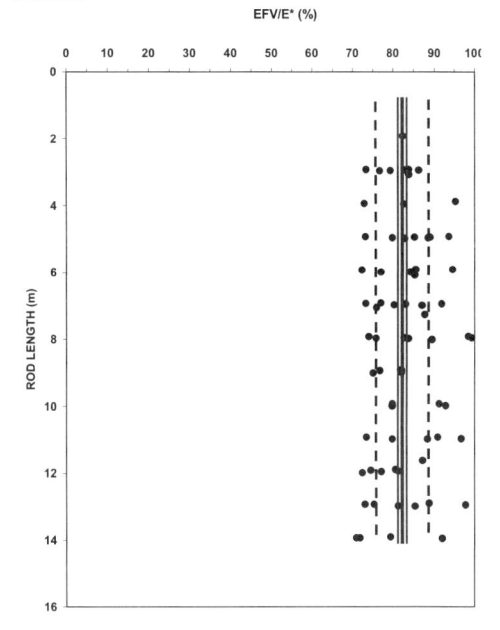

Figure 5. Energy efficiency transferred to the rod stem, EFV from equation (4), E*=478 J. Data in the figure represents the average of all blows in each SPT (Cavalcante, 2002, Danziger et al., 2004).

It was also found that multiple impacts in a single blow play an important role on the total energy delivered to the rod, especially in the case of short depths and low penetration values. Details on this subject can be found in Cavalcante (2002) and Danziger et al. (2004).

Previous results obtained for Brazilian data indicated efficiencies of 72% (Décourt et al., 1989), 66.7 to 72.3 % (Belincanta and Cintra, 1998) in respect to 478 J, or 72.6%, and a range of 67.3 to 72.9% in respect to 474 J, respectively.

4.4 The correction of N

The penetration resistance, N, is obtained as the number of hammer blows necessary to penetrate the sampler 300 mm, after an initial setting of 150 mm. The common practice consists in considering the N value as an integer number, irrespective of the differences in penetration to the reference 300 mm.

The instrumented data has allowed a comparison between the N value obtained in the usual way with the one inferred from accelerometer data. From the 1393 blows recorded, 1158 have been considered here.

The following procedure has been used in the analysis. The penetration for each blow has been evaluated from double integration of the accelerometer data, and included in a table. Then the total penetration was accumulated from the last blow until the value closest to 300 mm has been obtained. In most cases the corresponding number of blows was the same as the one recorded by the operator. However, in some cases this has not happened, and there have been few cases where a significant difference was found, showing a rough error. Once the accumulated penetration was obtained, a linear relation was assumed in order to evaluate N_{corr}, the corrected number of blows. Thus N_{corr} was obtained as

$$N_{corr} = \frac{300\,N}{d} \qquad (3)$$

where

N = number of blows used for obtaining d

d = accumulated penetration (in mm) closest to 300 mm

Table 1 illustrates the above procedure. The 26 blows necessary to drive the sampler the nominal 450 mm in one case is shown in Table 1. Accumulating the penetration from the last blow, the penetration closest to 300 mm is 306.1 mm, which is reached for 20 blows. Therefore, from expression (3) with N=20 and d=306.1mm, N_{corr} is obtained as 19.6. The N value recorded by the operator was 19. The energy transmitted to the rod stem, EFV, obtained from force (F) and velocity (v) records (equation 4) is also included in the table.

$$EFV = \int F\,v\,dt \qquad (4)$$

The N_{corr} versus N values are plotted in Figure 6, N values ranging from 1 to 60. N_{corr} was calculated by expression (3) while N was the value recorded by the operator. It must be pointed out that in all tests analysed in the paper the nominal 450 mm penetration was reached.

From Figure 6 one can depict three distinct N ranges with different behaviour. In order to get a better insight into the behaviour of the whole data, Figure 6 was split into three figures (7 to 9), for N ranging from 1 to 6, 7 to 16 and 18 to 60 respectively. There have been no records with N=17. Although the N values correspond to the above limits in each figure, the axes limits of all figures correspond to different values, in order to fit with the N_{corr} values, which are generally beyond those limits.

Table 1. Blows for 450 mm nominal penetration. Illustrative case.

Blow	Penetration (mm)	Accumulated penetration from the last blow (mm)	EFV (J)
1	37.7	442.6	385
2	26.9	404.8	411
3	21.3	377.9	389
4	20.7	356.6	397
5	15.0	335.8	383
6	14.8	320.9	356
7	14.8	306.1	342
8	15.1	291.3	350
9	13.4	276.2	327
10	15.9	262.8	351
11	14.7	246.9	353
12	16.3	232.2	363
13	16.3	215.9	379
14	15.1	199.6	401
15	16.6	184.5	416
16	17.8	168.0	438
17	17.7	150.2	450
18	15.8	132.4	438
19	14.8	116.6	428
20	14.7	101.8	415
21	15.2	87.1	406
22	14.6	71.9	411
23	15.2	57.3	402
24	15.3	42.1	398
25	13.0	26.8	363
26	13.9	13.9	403

Linear correlations have been included in the figures for two situations: the first one allowing $N_{corr} = f(N)$ to have an intercept at the y axis (dashed line); in the second one the line passes through the origin (continuous line). Table 2 summarizes the obtained correlations. The equation $N_{corr}=N$ is also represented in all figures.

It can be seen from Figure 7 that there is a significant scatter of the data, and Table 2 indicates that the squared correlation coefficients for both lines are very low. Moreover, both lines show very different angular coefficients, see Table 2. There is a quite clear trend of N_{corr} higher than N.

On the other hand, for N ranging from 7 to 16, it can be seen from Figure 8 and Table 2 a quite different trend. In fact, the squared correlation coefficients are high in both cases, and the correlation lines are very close to the $N_{corr}=N$ line.

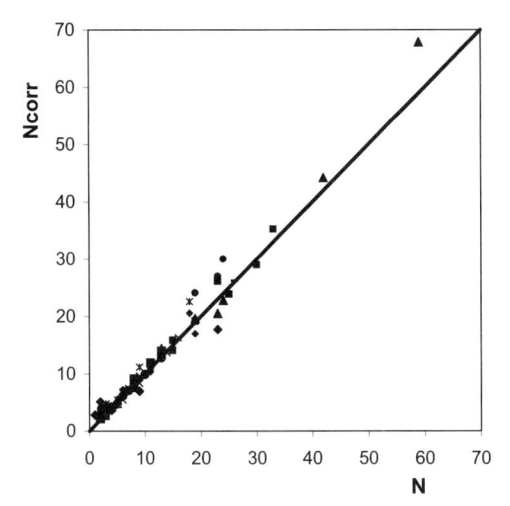

Figure 6. N_{corr} versus N for N ranging from 1 to 60.

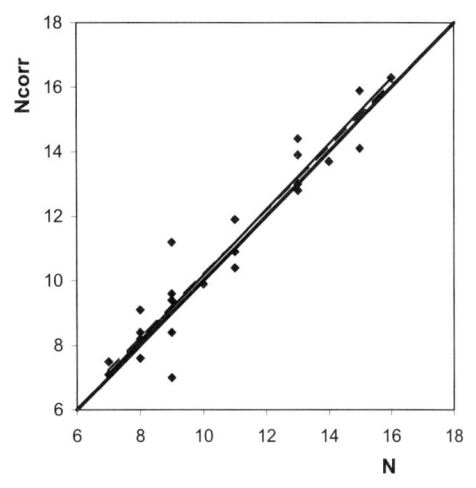

Figure 8. N_{corr} versus N for N ranging from 7 to 16.

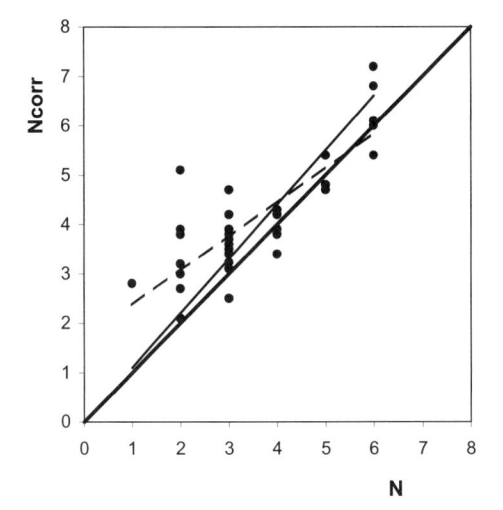

Figure 7. N_{corr} versus N for N ranging from 1 to 6.

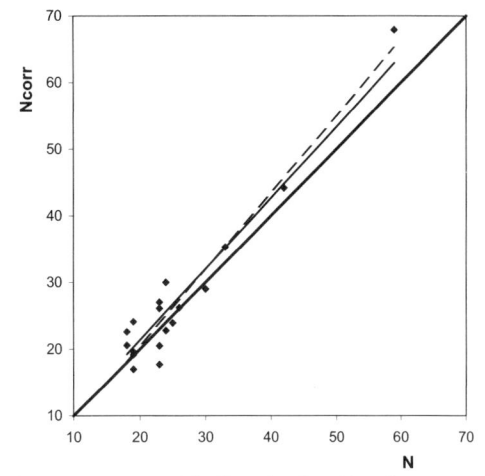

Figure 9. N_{corr} versus N for N ranging from 18 to 60.

The last range analysed, for N in the interval 18-60 (with few data for the higher values), about the same correlation coefficients were found for both lines as in the 7-16 N range. However, there is a trend for N_{corr} being about 7% greater than N, considering the correlation line passing through the origin.

It must be pointed out that N_{corr} only takes into consideration the errors related to the penetration length, i.e. N_{corr} should replace N in expression (2) to obtain N_{60}.

If the total nominal 450 mm penetration is now analysed, the picture shown in Table 3 is found. Data in Table 3 does not include those cases where it was not possible to reach the nominal 450 mm due to high soil resistance. This has eliminated the data relative to one of the instrumented sites.

It can be seen from Table 3 that in only 37% of the tests performed the penetration was in the range 420 mm – 480 mm. In 50% of the tests the penetrations did not reach 420 mm whereas in 13% of the tests they exceeded 480 mm. Such results clearly re-

Table 2. Correlations for the N ranges analysed.

N range	$N_{corr} = a + bN$			$N_{corr} = cN$	
	a	b	r^2	c	r^2
1 – 6	1.68	0.70	0.67	1.10	0.41
7 – 16	0.25	1.00	0.92	1.02	0.92
18 – 60	-2.17	1.15	0.93	1.07	0.93

r^2 = squared correlation coefficient

vealed a trend of smaller penetrations than the nominal 450 mm. When a shorter interval is considered, it can be seen that only 10% of the tests reached the penetration range 440 mm – 460 mm.

Table 3. Amount of tests performed in respect to the 450 mm nominal penetration.

Site	Instrumented tests	<420 mm	>480 mm	Penetration values in the range 420mm-480mm	in the range 440mm-460mm
1	10	2	5	3	3
2	10	7	1	2	2
3	12	3	0	9	1
4	9	5	0	4	2
5	14	12	0	2	0
6	10	1	4	5	0
7	7	3	0	4	0
8	1	1	0	0	0
9	5	5	0	0	0
total	78	39	10	29	8
%	100	50	13	37	10

5 CONCLUSIONS

This paper presents the results of instrumented SPTs where the N value provided by the test was compared with a corrected N_{corr} value based on the penetrations obtained from accelerometer data. The N values analysed ranged from 1 to 60. Three ranges with different behaviours were found. In the first one, for N varying from 1 to 6, a significant scatter was found, with a quite clear trend of N_{corr} higher than N. In the second range, for N from 7 to 16, a small scatter was found, and the correlation lines were very close to the $N_{corr} = N$ line. In the third range, for N varying from 18 to 60, the same scatter as in the second range was found. A trend of N_{corr} 7% greater than N was found in this case. As far as the nominal 450 mm penetration is concerned, a trend of penetrations smaller than 450 mm was found. Moreover, only 10% of the tests reached the penetration range 440 mm – 460 mm.

ACKNOWLEDGEMENTS

To Fundação José Bonifácio, for the financial support to purchase the PDI Analyser. To CNPq, for the financial support to the first author.

REFERENCES

Belincanta, A. 1998. *Evaluation of Factors Affecting the Penetration Resistance*. D.Sc. Thesis, EPUSP, São Paulo, Brasil, 362 p. (In Portuguese).
Belincanta, A. & Cintra, J.C.A. 1998. Factors Intervening Variants of ABNT Method for Performing SPT. *Solos e Rochas – Brazilian Geotechnical Journal* Vol. 21 (No. 3): 36-40 (In Portuguese).
Cavalcante, E.H. 2002. *Theoretical-Experimental Investigation on SPT*. D.Sc. Thesis, COPPE/UFRJ, 410 p. (in Portuguese).
Cavalcante, E.H., Danziger, F.A.B., Danziger, B.R. & Bezerra, R.L. 2003. Recent experience on SPT instrumentation in Brazil. *Proceedings XII Panamerican Conference on Soil Mechanics and Geotechnical Engineering. Cambridge, Mass.*: Vol. 1, 423-428.
Danziger, F.A.B., Cavalcante, E.H. & Danziger, B.R. 2004. Length effect on SPT. In preparation.
De Mello, V.F.B. 1971. Standard Penetration Test. *Proceedings IV Panamerican Conference on Soil Mechanics and Foundation Engineering. Porto Rico*: Vol. 1, 1-86.
Décourt, L. 1989. The Standard Penetration Test – State-of-the-Art Report. *Proceedings 12th International Conference on Soil Mechanics and Foundation Engineering, Rio de Janeiro*: Vol. 4, 2405-2416.
Décourt, L., Belincanta, A. & Quaresma Filho, A.R. 1989. Brazilian experience on SPT. *In Supplementary Contributions by the Brazilian Society of Soil Mechanics, published on the occasion of the 12th ICSMFE, Rio de Janeiro*: 49-54.
ISSMFE 1989. *Report of the ISSMFE –Technical Committee on Penetration Testing of Soils – TC 16. With Reference Test Procedures – CPT – SPT – DP – WST. International Reference Test Procedure for the Standard Penetration Test (SPT)*: 17-19.
Kovacs, W.D., Evans, J.C. & Griffith, A.H. 1977. Towards a More Standardized SPT. *Proceedings 9th International Conference on Soil Mechanics and Foundation Engineering, Tokyo*: Vol. 2, 269-276.
Kovacs, W.D. 1994. Effects of SPT equipment and procedures on the design of shallow foundations on sand. *Proceedings Settlement '94, ASCE, New York*: Vol. 1, 121-131.
Maultsby, J. P. 1999. *Evaluation of the Standard Penetration Test (SPT) Energy Transfer Performance for 58 Selected Florida-Based SPT Hammer Systems*. Report, University of Florida.
NBR 6484. 2001. *Soil – Simple Reconnaissance Borings with SPT – Testing Procedure*. ABNT, 17 p. (In Portuguese).
Palacios, A. 1977. *Theory and Measurements of Energy Transfer During Standard Penetration Test Sampling*. Ph.D. Thesis, University of Florida, Gainesville: 390 p.
PDI. 1999. *SPT ANALYZER - SPT Users Manual*. Pile Dynamics Inc. Ohio, 30 p.
Robertson, P.K., Campanella, R.G. & Wightman, A. 1983. SPT-CPT Correlations. *ASCE Journal of Geotechnical Engineering* Vol. 109 (No. 11): 1449-1459.
Schmertmann, J. H. & Palacios, A. 1979. Energy Dynamics of SPT. *ASCE Journal of the Geotechnical Engineering Division* Vol. 105 (GT8): 909-926.
Seed, H.B., Tokimatsu, K., Harder, L.F. & Chung, R.M. 1985. Influence of SPT Procedures in Soil Liquefaction Resistance Evaluations. *ASCE Journal of Geotechnical Engineering* Vol. 111 (No. 12): 1425-1445.
Skempton, A.W. 1986. Standard Penetration Test Procedures and the Effects in Sands of Overburden Pressure, Relative Density, Particle Size, Ageing and Overconsolidation. *Géotechnique* Vol. 36 (No. 3): 425-447.
Teixeira, A. H. 1977. Improvement of the SPT. Solos do Interior de São Paulo. ABMS/EESC/USP São Carlos, SP, Brazil: Chapter 4, 75-93 (In Portuguese).
Vargas, M. 1945. Soil exploration for foundation design. *Revista Politécnica*, São Paulo, Brazil, Vol. 41 (No. 149): 167-180 (In Portuguese).

Characterization of SPT grain size effects in gravels

C.R. Daniel, J.A. Howie & R.G. Campanella
Dept. of Civil Engineering, University of British Columbia, Vancouver, BC, Canada

A. Sy
Klohn-Crippen Consultants Ltd., Vancouver, BC, Canada

Keywords: SPT, LPT, gravel, grain size effects, Distinct Element Method (DEM), CASE method

ABSTRACT: The effects of gravel particles on the results of penetration test results, including those of the widely used Standard Penetration Test (SPT), are poorly understood. Existing empirical grain size effect databases are reviewed herein and shown to be ambiguous in the gravel range of grain sizes. Preliminary results of an ongoing investigation of the mechanisms underlying grain size effects are presented. Two-dimensional Distinct Element Method (DEM) computer modeling results demonstrate the relationship between the energy required to penetrate two parallel platens and the ratio of the platen spacing to the mean grain size (D_{50}). Decreasing the platen spacing ratio leads to the formation of stress arches in the soil ahead of the sampler and reduced penetration energies. Very low platen spacing results in plugging of the two-dimensional sampler. The CASE method of stress wave analysis is applied to stress wave data recorded during SPT in sands and gravels. Resulting back-calculated energy versus displacement plots show promise for identifying differences between sand and gravel responses recorded during tests of comparable $(N_1)_{60}$ values.

1 INTRODUCTION

Measurement of the engineering properties of gravels is an unique challenge for geotechnical engineers. The effects of gravel-sized particles on penetration test results, including those of the widely used Standard Penetration Test (SPT), are poorly understood. Concern about this shortcoming is increasing as evidence supporting the occurrence of gravel liquefaction during seismic events continues to accumulate (e.g. Valera et al., 1994). Various approaches have been proposed to minimize these grain size effects, including the use of tools of larger diameter than the SPT sampler ("Large Penetration Tests", LPT). Ideally, the LPT sampler would maintain a sampler to soil particle diameter ratio similar to that of the SPT in sand, but this can generally only be achieved in fine gravels due to practical limitations on sampler size. Thus, it would be advantageous to understand the mechanisms behind and be able to account for unavoidable grain size effects.

This paper presents preliminary results of studies being carried out to investigate apparent ambiguities in existing SPT grain size effect databases. Two such databases that extend into the gravel range of grain sizes ($D_{50} > 2$ mm) are presented first, followed by preliminary results from two approaches to analysis of sampler-soil interaction. The first consists of computer modeling of sampler penetration using Distinct Element Method (DEM) software in which the interaction between individual soil particles and the sampler are simulated. The second approach uses the CASE method of stress wave analysis to back-calculate the energy versus displacement relationship at the sampler-soil interface during field tests. It is demonstrated that both approaches have potential to improve our understanding of SPT sampler-soil interaction.

2 BACKGROUND INFORMATION

In addition to grain size effects, SPT blow counts (N) are known to vary with dynamic energy, effective overburden pressure, soil density, overconsolidation ratio and age (Skempton, 1986). It is widely accepted that (N) values should be corrected to a standard dynamic energy of 60% of the hammer potential energy (475 J) using the equation:

$$N_{60} = N \cdot \frac{\text{ENTHRU}}{0.6 \cdot 475 \text{ J}} \qquad (1)$$

where ENTHRU is the measured dynamic energy transmitted from the hammer to the drill rods. Similarly, (N_{60}) values can be corrected to a standard vertical effective stress of 98 kPa using the equation:

$$(N_1)_{60} = N_{60} \cdot C_N \qquad (2)$$

where (C_N) is the overburden stress correction factor, values for which have been proposed by several researchers (e.g. Liao and Whitman, 1986). Meyerhof (1957) demonstrated that, for a given soil and overburden stress, the blow count is proportional to the square of the relative density (D_r), defined by:

$$D_r = \frac{e_{max} - e}{e_{max} - e_{min}} \qquad (3)$$

where e_{max}, e and e_{min} are the maximum, measured and minimum void ratios, respectively. Skempton (1986) compiled a database of both calibration chamber and field SPT results for which estimates of (D_r) and the overburden pressure were available. Energy corrections were applied using an empirical database relating (ENTHRU) to test details such as the method of hammer release and the country in which the testing was performed. He noted that $(N_1)_{60}$ values normalized to the square of the relative density correlated reasonably well with the logarithm of the mean grain size (D_{50}). The database was later reviewed by Kulhawy and Mayne (1990) who proposed simple equations to describe the observed relationships, taking into account overconsolidation ratio and age.

Daniel et al. (2003b) summarize recent additions of calibration chamber data (Kokusho and Yoshida, 1998) and field data (Cubrinovski and Ishihara, 1999; Kokusho and Tanaka, 1994; Tanaka et al., 1991) to Skempton's original database. Dynamic energy was unfortunately not measured during these investigations but the new data are valuable because they extend well into the gravel range of grain sizes (Fig. 1a). Note that the KJ-gravel data are actually Japanese LPT (JLPT) blow counts converted to SPT blow counts using the correlation factor (N_{SPT} / N_{JLPT}) of 1.58 recommended by Daniel et al. (2003a). The best fit line in Fig. 1a was proposed by Cubrinovski and Ishihara (1999) and is described by the equation:

$$\frac{N_1}{D_r^2} = 9 \cdot \left(0.23 + \frac{0.06}{D_{50}}\right)^{-1.7} \qquad (4)$$

Equation (4) represents the sand data reasonably well, considering the lack of energy corrections. The scatter in the gravel data, however, is considerable. Some of this scatter is apparently related to differences between the KJ and T-gravel. The normalized blow counts appear to increase in the KJ-gravel but

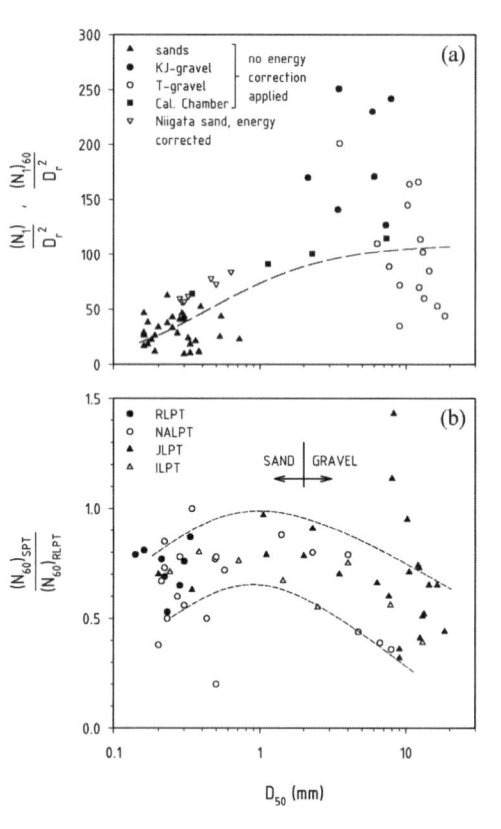

Figure 1. Existing grain size effect databases based on (a) normalized SPT blow counts and (b) SPT-RLPT blow count ratios (*after:* Daniel et al. 2003b).

Table 1. Details of KJ and T-gravels.

Parameter	KJ-gravel	T-gravel
Percent Fines [1]	1–14	1–8
Percent Gravel [1]	51–70	35-86
D_{50} (mm) [1]	2.1–7.8	1.1–18.5
D_r (%) [1]	32–70	21–81
G_o, frozen samples (MPa) [2]	260–560	~150–290
G_o, field (MPa) [2]	~850	~250–450

[1]. Cubrinovski and Ishihara (1999).
[2]. Kokusho and Tanaka (1994).

reach a plateau or even decrease in the T- gravel. Table 1 demonstrates that the two gravels are very similar except that the small strain shear modulii (G_o) of the KJ-gravel are significantly higher than those of the T-gravel (Kokusho and Tanaka, 1994). These higher (G_o) values can likely be attributed to cementation of the KJ-gravel, the effects of which would not be accounted for by normalizing blow counts to the square of the relative density. For this reason, the effect of increasing mean grain size demonstrated by the T-gravel data is likely the more representative response for non-cemented gravels.

Table 2. Test details for the SPT and four types of LPT (*after:* Daniel et al., 2003a)

Parameter	SPT	Japanese LPT (JLPT)	Italian LPT (ILPT)	North American LPT (NALPT)	Reference LPT (RLPT)
Sampler Outer Diameter, cm (in.)	5.08 (2)	7.3 (2.9)	14.0 (5.5)	7.62 (3)	11.43 (4.5)
Sampler Inner Diameter:					
Open shoe, cm (in.)	3.49 (1.375)	5 (2)	10 (3.9)	6.1 (2.4)	9.84 (3.875)
Barrel, no liner, cm (in.)	3.81 (1.5)	5.4 (2.13)	11 (4.3)	6.4 (2.52)	10.16 (4)
Hammer energy, kJ (ft-kip)	0.475 (0.35)	1.472 (1.084)	2.796 (2.062)	1.02 (0.75)	0.81 (0.6)
$(N_{60})_{LPT} / (N_{60})_{RLPT}$	-	0.44	0.73	0.50	1.00

The scatter demonstrated by the T-gravel data alone is considerable. Gravel deposits exhibit greater vertical and lateral variability than sands due to the higher energy of the depositional environment in which they are formed. This type of variability would contribute to the scatter in Fig. 1a, for example, if the frozen samples used to measure the relative density were not always representative of the soil tested during SPT in adjacent test holes. There is a second source of scatter that is related to the fixed size of the SPT sampler. The forces acting between soil particles vary considerably over short distances, and the concept of stress is only valid when a sufficiently large sample is considered. As the mean grain size increases, the influence of individual soil particles on the total penetration resistance also increases. Because the interparticle forces are variable relative to the larger scale stresses within the soil, the variability of the penetration resistance should also increase with increasing grain size. For these reasons, it is to be expected that a greater quantity of data will be required to reliably characterize grain size effects in gravels than in sands.

Daniel et al. (2003b) compiled a complementary database of SPT-LPT blow count ratios collected from the literature and during several field investigations. As the LPT is a non-standardized test, it was necessary to account for equipment variations (Table 2). Blow counts recorded using the various LPT's were converted to "Reference LPT" (RLPT) blow counts using the conversion factors listed in Table 2, which were estimated using a variation of the SPT-LPT correlation procedure described by Daniel et al. (2003a). The resulting database is shown in Fig. 1b. The original LPT's used to record each of the data points are indicated on the figure.

Normalizing SPT to RLPT (N_{60}) values is advantageous in that it eliminates the need for field measurements of relative density, reliable values of which are expensive to obtain. As a result, it is possible to collect a greater amount of this type of data at a lower cost. The obvious limitation of this database is that it does not directly relate measured SPT or RLPT blow counts to a useful soil property, such as density. The database can only be used to assess the validity of assumptions made about SPT and RLPT grain size effects.

The data in Fig. 1b follow a concave down trend that primarily reflects grain size variations but will also include effects of overconsolidation ratio, age and soil gradation, which have not been explicitly accounted for. Preliminary trendlines outlining the concave down trend have been sketched in Fig. 1b. For comparison, Fig. 2 presents three hypothetical SPT and RLPT grain size effect responses (variations on Fig. 1a) and the corresponding SPT-RLPT blow count ratios that would result. It has been assumed that the RLPT will exhibit the same pattern of grain size effects as the SPT but changes of slope will occur at coarser grain sizes, due to the larger dimensions of the RLPT sampler.

Fig. 2a demonstrates that the trend of decreasing SPT-RLPT blow count ratios observed in the field could not occur if the SPT and RLPT normalized blow counts increased in gravels at the same rate. Jamiolkowski and Lo Presti (2003) postulate that SPT and LPT samplers begin to penetrate in a plugged fashion in coarser soils. They note that the increase in end-bearing area would be much larger for an LPT sampler than an SPT sampler, and suggest that this would cause the LPT blow counts to increase at a greater rate than the SPT blow counts, a situation similar to that shown in Fig. 2b. This hypothesis recreates the observed SPT-RLPT blow count ratio trend and is in agreement with the widely held opinion that normalized SPT or LPT blow counts continue to increase in gravels, and is supported by the KJ-gravel data in Fig. 1a. As discussed above, however, the grain size effects demonstrated by the T-gravel data are likely more representative of non-cemented gravels.

The T-gravel data in Fig. 1a support the hypotheses of normalized SPT and RLPT blow counts becoming constant (not shown in Fig. 2) or beginning to decrease (Fig. 2c) with increasing grain size. Either of these scenarios would generate the trend of decreasing SPT-RLPT blow count ratios with increasing mean grain size observed in the field. Daniel et al. (2003b) suggested that this response could be related to the inability of coarser soils to generate frictional resistance on the internal surfaces of the sampler. This would lead to a reduction of the energy required to achieve penetration and to lower blow counts. It is apparent that additional work is

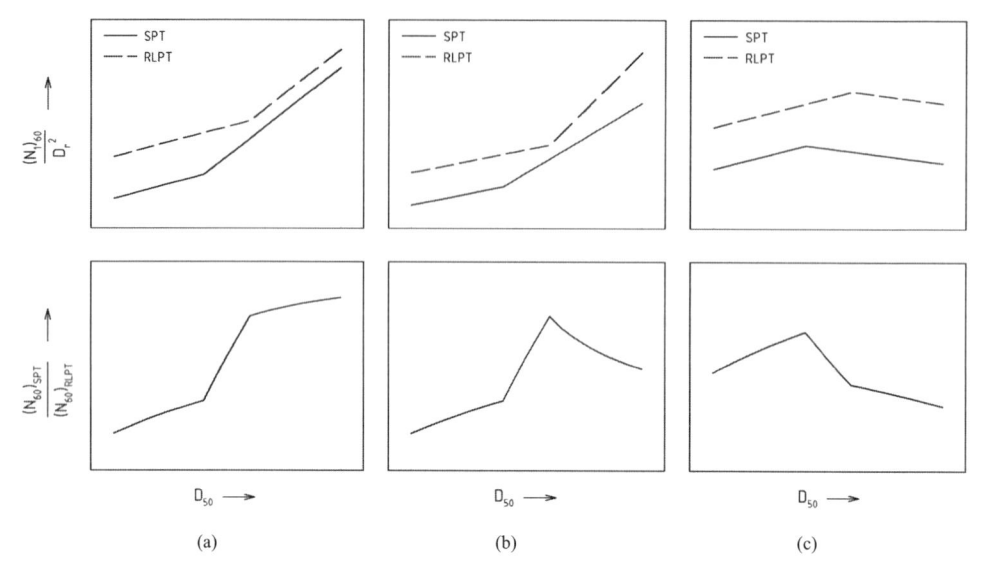

Figure 2. Three hypothetical grain size effect relationships for SPT and RLPT, and resulting SPT-RLPT blow count ratios.

required to accurately characterize the effect of gravel sized particles on SPT and LPT blow counts. Two approaches to this are described below, both of which are attempts to understand the sampler-soil interaction during the test.

3 DEM COMPUTER MODELING

The finite element and finite difference approaches to computer modeling treat soil as a continuum. An investigation of grain size effects must consider the effect of soil grains that are comparable in diameter to the sampler wall thickness or inner diameter, however, which clearly violates the continuum assumption. The Distinct Element Method (DEM) described by Cundall and Strack (1979) can be used to model soils as an assemblage of independent particles, thereby avoiding the continuum assumption. The authors have initiated an SPT grain size effect study using the DEM software "Particle Flow Code in 2 Dimensions" (PFC2D).

PFC2D models assemblages of circular particles and walls by tracking the location, spin and velocity of each particle and the location and orientation of each wall over thousands of very small time steps. At each time step, particle-particle and particle-wall contacts are identified and both normal and shear contact forces are calculated using a simple linear elastic contact model. Slip occurs between particles when the shear force exceeds a maximum allowable shear contact force equal to the normal force multiplied by a friction coefficient. Local non-viscous damping is employed to enhance model stability. The net force acting on each particle is used to calculate acceleration using Newton's second law. These particle specific accelerations are then double integrated over the length of the time step to determine new locations, at which point the process is repeated. Wall velocities are defined by the user and are not affected by contact forces. A comprehensive description of PFC2D is provided by Itasca (1999).

In addition to the ability to model soil as a discontinuous material, advantages of the DEM approach include the ability to model large strain problems such as sampler penetration and recreate complex soil behaviour such as shear induced volumetric strain using simple particle contact models. The primary limitations of the approach are its computationally intensive nature, which limits the number of particles that can be modeled within a reasonable period of time, and the inability to consider the effect of interstitial pore fluids. In addition, it is unlikely that results from a 2 dimensional analysis will be quantitatively comparable to 3 dimensional field results. It is anticipated, however, that they will provide useful insight into the mechanisms underlying grain size effects in gravels.

A particle assemblage consisting of 28,712 particles randomly distributed within a 1.0 m wide by 0.6 m high area was generated in PFC2D (Fig. 3). If the particles were assumed to be spherical, 50% of the particles by mass would be retained on each of the #4 and #5 U.S. standard sieves, corresponding to

(a)   (b)

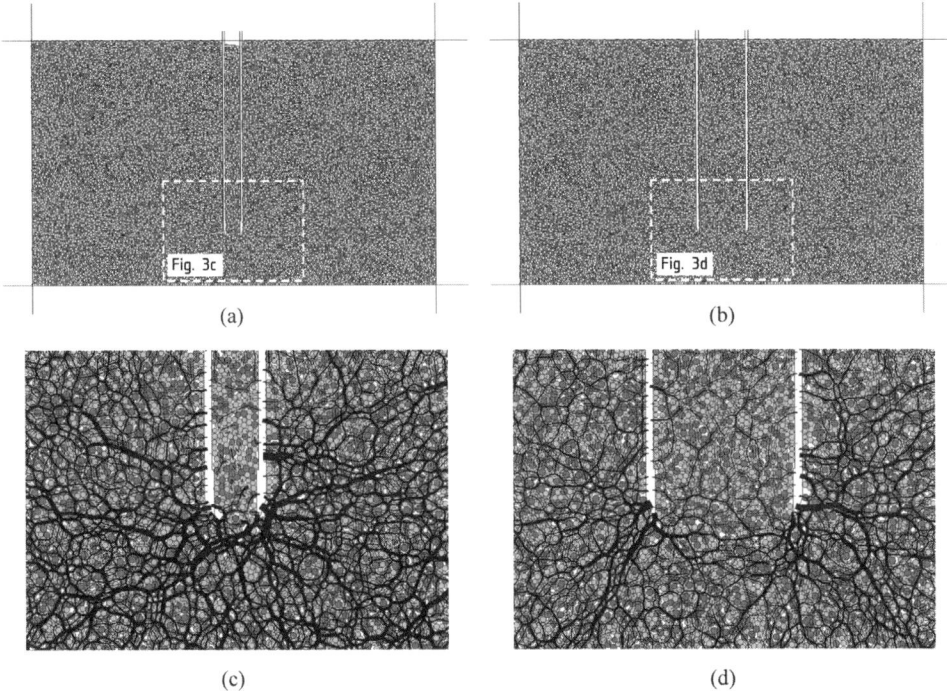

(c)   (d)

Figure 3. DEM sampler penetration simulations for (a, c) 50.8 mm (2") and (b, d) 127 mm (5") samplers. The superimposed black lines in (c) and (d) are proportional in thickness to the magnitude of the interparticle forces.

particle diameter ranges of 4.00 - 4.75 mm and 4.75 – 5.60 mm, respectively. Hence the 2 dimensional (D_{50}) of the assemblage is 4.75 mm. For all other aspects of program execution, the particles are treated as cylinders of unit length. The boundary walls were programmed to maintain constant vertical and horizontal stresses of 100 kPa along the edges of the sample. Sampler penetration was simulated by pushing two platens into the sample at a rate of 0.2 m/s. The dimensions of the platens are based on the cross section of a standard SPT sampler. Simulations were performed with sampler "outer diameters" of 25.4, 50.8, 76.2, 102 and 127 mm (1, 2, 3, 4 and 5 inches). Fig.'s 3a and 3b show the 50.8 and 127 mm (2 and 5 inch) trials, respectively. Fig.'s 3c and 3d provide greater detail of the area around the sampler opening. The superimposed black lines in the latter are proportional in thickness to the magnitude of the interparticle forces.

The net vertical forces acting on the platens were recorded at regular intervals. Integrating the total force as a function of sampler penetration yields a profile of penetration energy. Of particular interest is the energy required to penetrate the platens from 152.4 to 457.2 mm (6 to18 inches), the interval over which the blow count is measured during an SPT or LPT in the field. This penetration energy (E) is plotted as a function of the platen spacing normalized to the mean grain size in Fig. 4. The energy that would be required to penetrate a very large sampler was estimated by doubling the energy required to penetrate a single platen. Six of these single platen simulations were executed with the horizontal position of the platen shifted slightly for each simulation. The range of energies recorded in this manner is shown on Fig. 4. For comparison, the platen spacing to (D_{50}) ratio corresponding to an SPT sampler in soil with a mean grain size of 1 mm is also indicated. This ratio was selected because the trend of SPT-RLPT blow count ratios reverses at roughly that grain size in Fig. 1b.

The data in Fig. 4 demonstrate that the required penetration energy decreases with platen spacing, despite the fact that the platen dimensions were kept constant. This reduction in energy is likely related to the formation of stress arches in the soil ahead of the sampler opening, the effectiveness of which increases as the platen spacing decreases. Comparison of the patterns of interparticle force in Fig.'s 3c and 3d suggests that arching is prevalent ahead of the 50.8 mm (2") sampler but largely absent ahead of the 127 mm (5") sampler. The presence of stress arches reduces the quantity of internal friction required for plug formation, leading to some plugging of the 50.8 mm (2") sampler and very early plugging of the 25.4 mm (1") sampler, as indicated by the sample recovery values listed on Fig. 4.

The results of this preliminary investigation demonstrate the potential of the DEM approach to modeling sampler penetration. The complex phenomenon of sampler plugging is predicted automatically when conditions are appropriate and the effect of plugging on the penetration energy (E) is readily demonstrated. Additional studies are underway to better characterize the effect of platen dimensions, mean grain size and soil gradation on the penetration energy. The (E) values determined during such studies can be loosely related to field (N_{60}) values, which are also a measure of the energy required to penetrate the sampler. In order to have greater confidence in the application of the results of the DEM study to field evidence of grain size effects, it is important to improve our understanding of sampler-soil interaction during SPT performed in the field. One approach to this objective is described below.

4 CASE ANALYSIS OF STRESS WAVE DATA

Sample classifications and measured blow counts are the standard output of the SPT. The CASE method of stress wave analysis, originally developed for pile driving applications (Rausche et al., 1985), provides an unique opportunity to retrieve additional information about the soil response during an SPT. For the CASE method, measurements of axial force (F) and particle velocity (V) at a point on the drill rods, which are primarily recorded to calculate

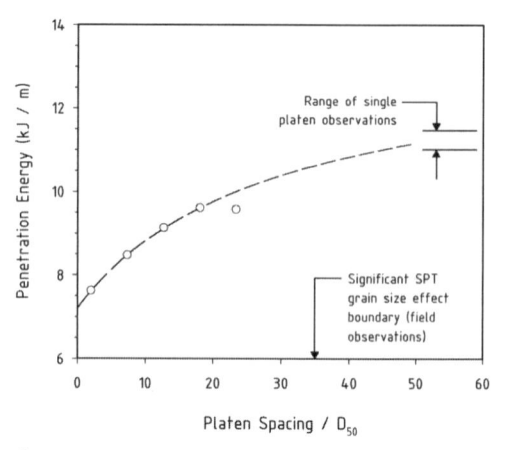

Figure 4. DEM modeling results demonstrating effect of platen spacing on required penetration energy. Recovery was 100% unless otherwise indicated.

ENTHRU, are used to back-calculate time histories of soil resistance and sampler velocity. Goble and Abou-matar (1992), Abou-matar et al. (1996) and Daniel et al. (2003c) have applied the CASE method to SPT stress wave data. There remains, however, considerable scope for a broad-based investigation of soil response. The effect of varying grain size on the observed soil response would be one obvious objective for such an investigation.

Soil resistance and sampler velocity time histories can be used to calculate the energy absorbed by the soil and the sampler displacement. Fig. 5 shows selected energy versus displacement plots from two sites: Patterson Park, located on the Fraser River Delta, near Vancouver, BC, Canada and Resurrection River, located near Seward, Alaska, USA.

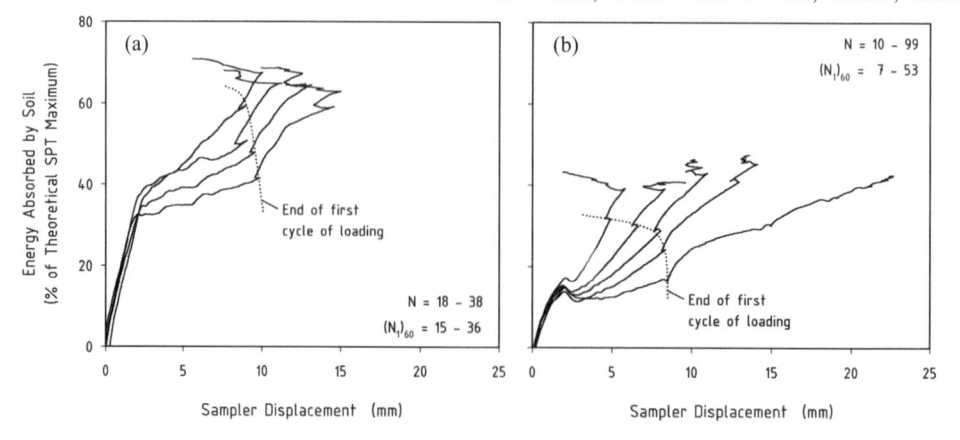

Figure 5. Average soil responses during selected SPT in (a) fine to medium grained sands (Patterson Park) and (b) fine sand to medium gravel (Resurrection River).

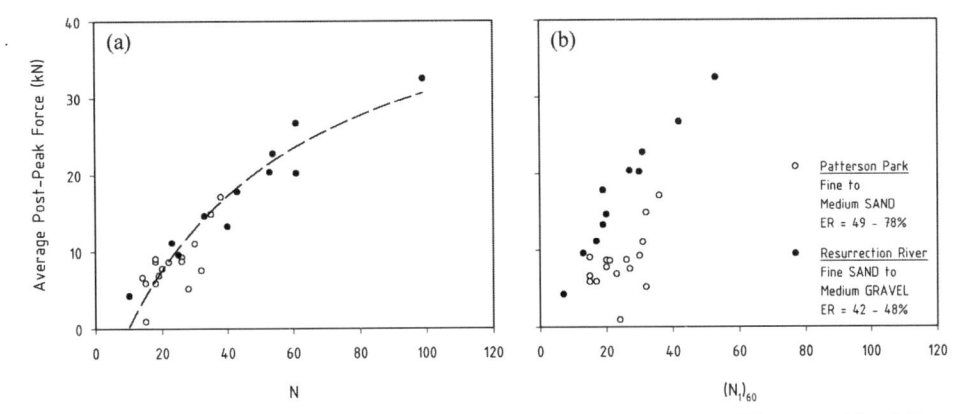

Figure 6. Average post-peak forces measured during SPT plotted as a function of (a) uncorrected blow counts (N) and (b) energy and overburden corrected blow counts $(N_1)_{60}$.

This plot format simplifies comparison of soil responses and helps to reduce the rod coupling effects noted by Daniel et al. (2003c). The slope of the curve at any point is equal to the instantaneous soil resistance. The soils at Patterson Park are poorly graded fine to medium sands while those at Resurrection River range from fine sands to medium gravels. Tests at both sites were conducted using safety hammers and "A" series rods but significantly different dynamic energies were provided (42 – 48% at Resurrection River, 49 – 78% at Patterson Park). More information is required to interpret the data in Fig. 5, however, because the shapes of the curves reflect both the nature of the applied load and the nature of the soil response to that load. Thus, it must first be determined which features of the curve contain the desired soil response information.

During an SPT, energy is transferred from the hammer to the sampler-soil interface by a stress wave that propagates down the drill rods following impact. Upon arrival at the sampler-soil interface, a portion of the wave will be absorbed by the soil and the remainder will be reflected back up the drill rods. The reflected portion will eventually return to reload the soil. These loading cycles continue until all of the energy of the original stress wave has been either absorbed by the soil or otherwise dissipated, a process that typically concludes 50 to 100 ms after impact. The nature of the loading that occurs during each cycle is controlled by the shape of the original stress wave, which typically consists of a sharp rise of axial force and particle velocity at the leading edge of the wave followed by roughly exponential decay to zero (e.g. Goble and Abou-matar, 1992).

The end of the first cycle of loading, corresponding to the second arrival of the stress wave at the sampler-soil interface, has been identified on each of the curves in Fig. 5. As a first approximation, each of these first cycles of loading can be represented by a simple bi-linear relationship, and hence can be further divided into a peak loading stage (first linear section) and a post-peak stage (second linear section). The data in Fig. 5 demonstrate that the soil response during the peak loading stage is essentially independent of soil conditions, generating almost identical responses for blow counts ranging from 18 to 38 in Fig. 5a and 10 to 99 Fig. 5b. In contrast, the slope and hence the average force of the post-peak stage of loading varies considerably between tests.

Figures 6a and 6b compare the average post-peak force to the uncorrected blow count (N) and the corrected blow count $(N_1)_{60}$, respectively for a total of 27 tests performed at the two sites. There is very little scatter in Fig. 6a, considering the wide range of soil type, dynamic energy, rod length and test depth represented, suggesting that there is a reasonably unique relationship between the post-peak force generated by the soil and the uncorrected blow count that will be recorded. Fig. 6b demonstrates that the data from the two sites become distinct when the blow counts are corrected for dynamic energy and overburden pressure using Equations (1) and (2). For a given value of $(N_1)_{60}$, a higher post peak force was measured at the Resurrection River site. This may be related to the generally coarser grain sizes encountered at Resurrection River. Alternatively, the greater intensity of the peak loading stage at Patterson Park may have resulted in greater softening of the soils and lower post-peak forces.

The brief analysis described above demonstrates the potential of the CASE method for SPT applications. At the least, the method can be used to generate sampler displacement histories for use in PFC2D analyses. It is hoped that additional data will also clarify the effect of grain size variations on elements of the soil response such as the post-peak force.

5 SUMMARY AND CONCLUSIONS

Two SPT grain size effect databases have been reviewed and shown to be ambiguous in gravels. Greater variability of blow counts is expected in gravels but the available data do not even consistently indicate whether blow counts will be higher, lower or roughly the same in gravels as in sands. The authors have initiated an investigation to further the characterization of grain size effects in gravels using the Distinct Element Method (DEM) of computer modeling and the CASE method of stress wave analysis. Two dimensional DEM software has been used to model the penetration of two rigid platens into an assemblage of circular particles. Decreasing the platen spacing led to the development of stress arches ahead of the sampler and eventually to plugging, with an associated decrease in penetration energy. Additional studies are planned to isolate the relationship between grain size and sampler plugging. The CASE method is a potentially useful tool for the investigation of grain size effects, providing time histories of soil resistance and sampler velocity that can be compared for soils of varying grain size. A relationship between post-peak soil resistance and the measured blow count has been demonstrated. Additional data are required to clarify the factors affecting this relationship, which may include dynamic energy and grain size.

ACKNOWLEDGEMENTS

The authors thank the British Columbia Ministry of Transportation and Highways and Foundex Explorations Ltd., of Surrey, BC for financial support of the Patterson Park field work. The Resurrection River field work was conducted in association with Dr. Joseph Koester under the Earthquake Engineering Research Program of the United States Army Corp of Engineers. Mr. Scott Jackson of UBC Civil Engineering built the stress wave monitoring equipment.

REFERENCES

Abou-matar, H., Rausche, F., Thendean, G., Likins, G.E., and Goble, G.G. 1996. Wave equation soil constants from dynamic measurements on SPT. *In* Proceedings, Fifth International Conference on the Application of Stress-Wave Theory to Piles. *Edited by* Townsend, Hussein and McVay. Thomson-shore, Dexter, Michigan, pp. 163-175.

Cubrinovski, M., and Ishihara, K. 1999. Empirical correlation between SPT N-value and relative density for sandy soils. Soils and Foundations, **39**(5): 61-71.

Cundall, P.A., and Strack, O.D.L. 1979. A discrete numerical model of granular assemblies. Geotechnique, **29**: 47-65.

Daniel, C.R., Howie, J.A., and Sy, A. 2003a. A method for correlating large penetration test (LPT) to standard penetration test (SPT) blow counts. Canadian Geotechnical Journal, **40**(1): 66-77.

Daniel, C.R., Howie, J.A., and Sy, A. 2003b. Compilation of an SPT-LPT grain size effects database for gravels. *In* Proceedings of the International Conference on Problematic Soils. *Edited by* Jefferson and Frost. Nottingham, United Kingdom. July 29-30, Vol.1, pp. 227-234.

Daniel, C.R., Howie, J.A., and Taylor, B. 2003c. Analysis of SPT stress wave data using the CASE method. *In* Proceedings of the 56th Canadian Geotechnical Conference. Winnipeg, Manitoba, Canada. Sept. 29 - Oct. 1.

Goble, G.G., and Aboumatar, H. 1992. Determination of wave equation soil constants from the standard penetration test. *In* Proceedings, Fourth International Conference on the Application of Stress-Wave Theory to Piles. *Edited by* F.B.J. Barends. The Hague, The Netherlands. A.A. Balkema, pp. 99-103.

Itasca 1999. Particle Flow Code in 2 Dimensions (PFC2D) User's Guide. Itasca Consulting Group, Inc., Minneapolis, Minnesota.

Jamiolkowski, M., and Lo Presti, D.C.F. 2003. Geotechnical characterization of Holocene and Pleistocene Messina sand and gravel deposits. *In* Proceedings, Characterisation and Engineering Properties of Natural Soils. *Edited by* Swets and Zeitlinger, pp. 1087-1119.

Kokusho, T., and Tanaka, Y. 1994. Dynamic properties of gravel layers investigated by in-situ freezing sampling. *In* Proceedings, Ground Failures under Seismic Conditions. Atlanta, Georgia. ASCE Geotechnical Special Publication No. 44, pp. 121-140.

Kokusho, T., and Yoshida, Y. 1998. SPT N-value and S-wave velocity of gravelly soils with different particle gradings. *In* Proceedings, Geotechnical Site Characterization. *Edited by* Robertson and Mayne. Balkema, Rotterdam, pp. 933-938.

Kulhawy, F.H., and Mayne, P.W. 1990. Manual on estimating soil properties for foundation design, Report EL-6800 Research Project: 1493-6, Electrical Power Research Institute.

Liao, S.C., and Whitman, R.V. 1986. Overburden correction factors for SPT in sand. Journal of Geotechnical Engineering Division, ASCE, **112**(3): 373-377.

Meyerhof, G.G. 1957. Discussion on Research on determining the density of sands by spoon penetration testing. *In* Proceedings, 4th International Conference on Soil Mechanics and Foundation Engineering. London, Vol.3, p. 110.

Rausche, F., Likins, G.E., and Goble, G.G. 1985. Dynamic determination of pile capacity. Journal of the Geotechnical Engineering Division, ASCE, **111**(3): 367-383.

Skempton, A.W. 1986. Standard penetration test procedures and the effects in sands of overburden pressure, relative density. particle size, ageing and overconsolidation. Geotechnique, **36**(3): 425-447.

Tanaka, Y., Kudo, K., Yoshida, Y., and Kokusho, T. 1991. Evaluation method of dynamic strength and residual settlement for gravelly soils, Report U90063, CRIEPI.

Valera, J.E., Traubenik, M.L., Egan, J.A., and Kaneshiro, J.Y. 1994. A practical perspective on liquefaction of gravels. *In* Geotechnical Special Publication No. 44 - Ground Failures under Seismic Conditions. ASCE, pp. 241-258.

International and European standards on geotechnical investigation and testing for site characterisation

V. Eitner
DIN Deutsches Institut für Normung e.V., Berlin, Germany

R. Katzenbach
Institut und Versuchsanstalt für Geotechnik, Technische Universität Darmstadt, Germany

F. Stölben
Stölben Bohrunternehmen GmbH, Zell/Mosel, Germany

Keywords: standards, investigation, sampling, groundwater measurements, field testing, laboratory testing

ABSTRACT: Three committees of the International Organisation for Standardisation (ISO) and the European Committee for Standardisation (CEN) are preparing common standards on equipment and methods used for soil and rock identification, drilling, sampling, field and laboratory testing of rock and soil as well as groundwater measurements for geo-engineering practice. The aim of this international standardisation work is to harmonise the quality requirements for equipment and to achieve comparable results whenever standardised equipment is used and standardised methods are applied. Manufacturers and users of equipment and methods of ground investigation and testing should find international standards valuable for common use as they enable comparable results to be obtained and reduce the variability of the results of geotechnical data which are the basis for geotechnical design.

These international standards for ground investigation and testing are an important basis for the safety of structures and risk analysis in geotechnical design and construction. They also make a substantial contribution to sustainable resource development around the world. Moreover, they facilitate the development of a common global market for free trade in services and equipment for ground investigation and testing. International standardisation in this field also leads to an noticeable cost reduction in the building market.

1 INTRODUCTION

The world is getting smaller since the continuous progress of technical and economical globalization, which is also present in the building and construction business. Therefore standards are necessary to solve problems due to communication difficulties or technical differences among clients, consultants, contractors etc. that often lead to a reduction of safety and serviceability. Standards on identification and classification, sampling, field and laboratory testing for geotechnical purposes are currently prepared by the European Committee for Standardization (CEN) and the International Organization for Standardization (ISO) according to the Vienna Agreement.

The World Trade Organization (WTO) is the international organization dealing with the global rules of trade between nations. Its main function is to ensure that trade flows as smoothly, predictably and freely as possible. ISO (International Organization for Standardization) has built a strategic partnership with WTO. The political agreements reached within the framework of WTO require underpinning by technical agreements. ISO has the complementary scopes, the framework, the expertise and the experience to provide this technical support for the growth of the global market.

The Agreement on Technical Barriers to Trade (TBT) – sometimes referred to as the Standards Code – aims to reduce impediments to trade resulting from differences between national regulations and standards. As far as international consensus-based standards are concerned, the Agreement invites the signatory governments to ensure that the standardizing bodies in their countries accept and comply with a "Code of good practice for the preparation, adoption and application of standards.

The TBT Agreement recognizes the important contribution that international standards and conformity assessment systems can make to improving efficiency of production and facilitating international trade. Where international standards exist or their completion is imminent, therefore, the Code of Good Practice says that standardizing bodies should use them, or the relevant parts of them, as a basis for standards they develop. It also aims at the harmonization of standards on as wide a basis as possible, encouraging all standardizing bodies to play as full a part as resources allow in the preparation of international standards by the relevant international body, including the ISO.

2 TECHNICAL COMMITTEES FOR STANDARDISATION IN GEOTECHNICAL INVESTIGATION AND TESTING

2.1 General

The standardization in the field of geotechnical investigation and testing is done in three committees of ISO International Organization for Standardization and CEN European Committee for Standardization, partly according to the Vienna Agreement (1991) that stipulates the general exchange of information at Central Secretariat level, the co-operation on standards drafting between ISO and CEN, the adoption of existing International standards as European standards and the parallel approval of standards (see Figure 1):
- ISO/TC 182/SC 1"Geotechnical Investigation and Testing";
- CEN/TC 250/SC 7 "Geotechnical Design";
- CEN/TC 341 "Geotechnical Investigation and Testing".

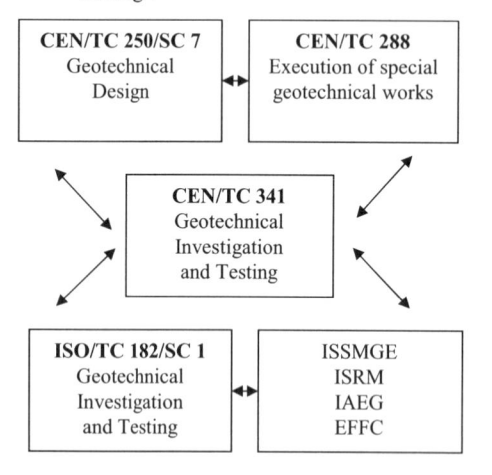

Figure 1. Technical committees on geotechnical engineering (from Eitner et al. 2002).

2.2 ISO/TC 182/SC 1 Geotechnical investigation and testing

ISO/TC 182/SC 1 was established together with its parent technical committee ISO/TC 182 "Geotechnics" in 1982. It was named "Classification and Presentation" according to its scope including all matters facilitating the communication with the aid of documents in the geotechnical field, such as
- classification of soil and rock;
- terminology, both standardization of terms and definitions not treated in connection with specific standards of other subcommittees, and in connection with specific standards of other subcommittees, and the co-ordination of the terminology the other subcommittees;
- symbols to be used in calculations;
- symbols to be used on drawings and graphs, expressing, data on soil and rock.
- The secretariat and the chairmanship were held by Sweden (SIS) but changed to Germany (DIN) in 1988.

ISO/TC 182/SC 1 prepares standards on identification, description and classification of soil and rock. It is also the international mirror committee of CEN/TC 341. Currently, ISO/TC 182/SC 1 has: 14 participating members (P): Austria, China, Czech Republic, Finland, France, Germany, Italy, Japan, Luxembourg, Netherlands, Norway, Republic of Korea, Sweden and United Kingdom and 10 observers (O-members): Argentina, Australia, Belgium, Iceland, India, Ireland, Israel, Portugal, South Africa, Turkey. There are several liaisons with other committees and organizations such as ISO/TC 82 "Mining", CEN/TC 341, CEN/TC 288 "Execution of special geotechnical works", ISRM, ISSMGE and IAEG.

In 2001 the committee resolved to add some new work items to its working program and to change its name and scope due to the formation of the new CEN/TC 341 "Geotechnical Investigation and Testing". Name, scope and working program are identical now. There is a close co-operation between both committees according to the Vienna Agreement.

2.3 CEN/TC 341 Geotechnical investigation and testing

The scope of CEN/TC 341 comprises standardization in the field of geotechnical investigation and testing pertains to equipment and methods used for drilling, sampling, field and laboratory testing of rock and soil as well as groundwater measurements as part of the ground and site investigation services.

The goal of this standardization work is to harmonize the quality requirements for equipment and to achieve comparable results whenever standardized equipment is used and standardized methods are applied. Manufacturers and users of equipment and methods of geotechnical investigation and testing should find European standards valuable for common use as they enable comparable results to be obtained and reduce the variability of the results of geotechnical data which are the basis for the geotechnical design. European standards for investigation and testing are important for the safety of infrastructures, e. g. such as roads, bridges, canals, railways, airfields, harbors, tunnels, sewer and communication lines and other structures, e. g. buildings dams. These European standards also facilitate the development of a common European market for the trade of services and equipment for geotechnical investigation and testing. The standardization in the field of geotechnical investigation and testing will also lead to an apparent cost reduction in the building market. Currently CEN/TC 341 has five working groups (see Table 2).

Other working groups on field and laboratory testing are planned for the future.

Table 2. Working groups of CEN/TC 341 Geotechnical investigation and testing.

WG	Title	Convenor/ Secretariat
1	Drilling and sampling methods and groundwater measurements	Germany (DIN)
2	Cone and piezocone penetration Tests	Netherlands (NEN)
3	Dynamic probing and standard Penetration test;	Germany (DIN)
4	Testing of geotechnical structures	France (AFNOR)
5	Borehole expansion tests	France (AFNOR)

2.4 CEN/TC 250/SC 7 Geotechnical Design

Since the mid-seventies, working groups have prepared European harmonized technical specifications for the design of buildings on behalf of the European commission. These documents also called "Eurocodes" should be available for different construction types and building type-independent problems, foundations and permissible loading of building. After the Council of the European Community agreed upon a new concept concerning technical harmonization and standardization, the Eurocodes were given to CEN for further preparation and publication as European standards or pre-standards.

The Eurocodes are a series of standards for the design of buildings, geotechnical structures and earth quake resistance. They are prepared by CEN/TC 250 "Structural Eurocodes" which has 9 subcommittees (see Table 2).

Table 2. Subcommittees of CEN/TC 250 Structural Eurocodes.

SC	Title
1	Basis of design and actions on structures
2	Design of concrete structures
3	Design of steel structures
4	Design of composite steel and concrete structures
5	Design of timber structures
6	Design of masonry structures
7	Geotechnical design
8	Design provisions for earth quake resistance
9	Design of aluminium structures

Every Eurocode consists of more than one part. Altogether there are more than 50 parts.

Eurocode 7, prepared by CEN/TC 250/SC 7, is the basic code for the geotechnical design having two parts finally Part 1 on general rules and Part 2 on design assisted by testing

Eurocode 7 Part 1 was finalized and available as ENV 1997-1 (European pre-standard) in November 1994. After two years all CEN members were asked to comment and to decide about its future development. Currently, Eurocode 7 Part 1 is under revision and will be soon available as EN. Then all CEN members are required to give this European standard the status of a national standard without any alteration beside a National Annex with the safety values.

CEN/TC 250/SC 7 created two further project teams PT 2 and PT 3 in autumn of 1994 to prepare two standards on design assisted by laboratory and field-testing. After finalization in summer 1997 they were available as ENV 1997-2 and ENV 1997-3. In March 2001 CEN started the 2-year enquiry asking for comments. Both parts will be merged to one part; hereby all specifications on test equipment and test execution will be deleted and transferred to the work program of CEN/TC 341. Only design related matters such as interpretation and evaluation of test results and the planning of geotechnical investigation and testing will be left.

3 STANDARDISATION PROJECTS

3.1 Identification and classification of soil and rock

ISO/TC 182/SC 1 has finalized one standard which was published by CEN and ISO in 2002: ISO 14688-1 Geotechnical investigation and testing – Identification and classification of soil – Part 1: Identification and description. Two other standards were published for parallel formal voting in July 2003:
- ISO 14688-2 Geotechnical investigation and testing – Identification and classification of soil – Part 2: Classification principles;
- ISO 14689-1 Geotechnical investigation and testing – Identification and classification of rock – Part 1: Identification and description;

Two new work items resulting in Technical Specifications (TS) were decided in 2002:
- ISO/TS 14688-3 Geotechnical investigation and testing – Identification and classification of soil – Part 3: Electronic exchange of data of identification and description of soil;
- ISO/TS 14689-2 Geotechnical investigation and testing – Identification and classification of rock – Part 2: Electronic exchange of data of identification and description of rock.

ISO 14688-1 and ISO 14688-2 establish the basic principles for the identification and classification of soils on the basis of those material and mass characteristics most commonly used for soils for engineering purposes.

The general identification and description specified by ISO 14688-1 of soils is based on a flexible system for immediate (field) use by suitably experienced persons covering both material and mass characteristics by visual and manual techniques. Details are given of the individual characteristics for identifying soils and the descriptive terms in regular use, including those related to the results of tests from the field. The field of application of this standard is natural soils in-situ and similar man-made materials in-situ and soils redeposited by man. It generally permits soil to be identified with adequate accuracy for general or preliminary characterization more accurate identification and classification based on grading, plasticity or organic content often require laboratory tests that are not covered. In addition to identifying soils, the condition in which a soil is encountered, any particular secondary constituents, other features of a soil, such as carbonate content, particle shape, surface roughness of particles, odor, any common names and the geological classification should all be indicated. ISO 14688-1 specifies several methods for the determination of these soil characteristics. It uses for example particle size as the fundamental basis for designating mineral soils.

The classification principles of ISO 14688-2 support the soil grouping into classes of similar composition and geotechnical properties. Soils shall be therefore classified into soil groups on the basis of their nature that is the composition only, irrespective of their water content or compactness, taking into account the following characteristics:

- particle size distribution (grading),
- plasticity,
- organic content.

The most common approach in classification is to divide the soils on the basis of particle size grading and plasticity. The division is made on the relative size fractions present for the coarser soil fractions, determined on the whole sample, and on the plasticity of the finer fractions.

ISO 14688-1 is based on international practice (IAEG 1981, ISRM 1977a, 1977b, 1978, 1980, 1984, ISSMGE 1994) and relates to the identification and the description of rock material and mass on the basis of mineralogical composition, predominant grain size, genetic groups, discontinuities, structure and other components.

The standard also provides rules for the description of other characteristics as well as for their designation and applies to the description of rock for geotechnics in civil engineering. The description is carried out on cores and other samples of natural rock and on rock masses. The lithological identification of rocks is based on the determination of: genetic group, structure, grain size, mineralogical composition and void content.

ISO 14688-1 and ISO 14689-1 both recommend to use the symbols of ISO 710 series to represent soils on borehole legends or on engineering geological maps. In a report it has to be clearly stated that the descriptions are based on visual and manual identification. Further the description of any soil shall at least contain:

- Author's name;
- date of description;
- details of origin of samples;
- condition of described soil and rock respectively;
- soil and rock type respectively;
- secondary soil fractions;
- color;
- legend of symbols and additional terms used.

Following the previous items any appropriate descriptions shall be added according to this standard.

ISO/TS 14688-3 and ISO/TS 14689-2 will cover requirements for the electronic exchange of data on identification and description of soil and rock. They will provide a data exchange format (XML) that facilitates the data exchange independently from a certain hardware of software system.

CEN will also vote on these three standards according to the Vienna Agreement and adopt them as European standards, if the majority of CEN members agrees.

3.2 *Drilling and Sampling Methods and Groundwater Measurements*

CEN/TC 341 is preparing standards dealing with the investigation of soil, rock and groundwater measurements for use as subsoil and construction materials as part of the geotechnical investigation services. It defines concepts and specifies requirements relating to exploration by excavation, drilling, sampling and groundwater measurements. The aims of such exploratory work are:

- to obtain information on the sequence, thickness and orientation of strata;
- to establish the type, composition and condition of strata;
- to obtain an indication of local groundwater conditions and recover water samples;
- to recover soil and rock samples of a quality sufficient to assess the general suitability of a site for civil engineering purposes and to determine the required soil and rock mechanical parameters for design purposes.

More information in detail is given in Stölben et al. (2004). ISO will also vote on this standard according to the Vienna Agreement and adopt them as an International standard, if the majority of p-members of ISO/TC 182/SC 1 agree.

3.3 *Field Testing*

CEN/TC 341 prepares in co-operation with ISO/TC 182/SC 1 several standards specifying the requirements for indirect ground investigations of soil and rock by field tests. These standards will be published under the common title "ISO 22476 Geotechnical investigation and testing – Field testing":
- Part 1: Electrical cone and piezocone penetration tests;
- Part 2: Dynamic probing;
- Part 3: Standard penetration test;
- Part 4: Menard pressuremeter test;
- Part 5: Flexible dilatometer test;
- Part 6: Self-boring pressuremeter test;
- Part 7: Borehole jack test;
- Part 8: Full displacement pressuremeter test;
- Part 9: Field vane test;
- Part 10: Weight sounding test (TS);
- Part 11: Flat dilatometer test (TS);
- Part 12: Permeability tests;
- Part 13: Water pressure tests
- Part 14: Pumping tests;
- Part 15: Mechanical cone penetration test;
- Part 16: Plate loading tests.

The common purpose of these standards is to eliminate as far as possible erroneous assessments of subsoil conditions as well as to limit scatter when repeating tests and improve reproducibility when undertaking field testing.

All standards on field-testing will have an identical list of content:
Foreword
1 Scope
2 Normative References
3 Terms and definitions
4 Equipment
5 Test procedure
6 Test results
7 Test report

ISO 22476-1 and ISO 22476-15 deal with the execution and reporting on cone penetration tests. These tests are performed with a cylindrical penetrometer with conical tip, or cone and if applied, the friction sleeve, are measured. The results from a cone penetration test can in principle be used to evaluate soil stratification, soil type, soil density and in situ stress conditions and mechanical soil properties.

ISO 22476-2 on dynamic probing covers the determination of the resistance of soils and soft rocks in-situ to the dynamic penetration of a cone. A hammer of a given mass and given falling height is used to drive the cone. The penetration resistance is defined as the number of blows required driving the penetrometer over a defined distance. A continuous record is provided with respect to depth but no samples are recovered.

Four procedures are included, covering a wide range of specific work per blow:
- Dynamic probing light (DPL): test representing the lower end of the mass range of dynamic equipment;
- Dynamic probing medium (DPM): test representing the medium mass range of dynamic equipment;
- Dynamic probing heavy (DPH): test representing the medium to very heavy mass range of dynamic equipment;
- Dynamic probing super-heavy (DPSH): test representing the upper end of the mass range of dynamic equipment.

The test results of this standard are specially suited for the qualitative determination of a soil profile together with direct explorations (e.g. drilling) or as a relative comparison of other in-situ tests. They may also be used for the determination of the strength and deformation properties of soils, generally of the cohesionless type but also possibly in fine-grained soils, through appropriate correlations. The results can also be used to determine the depth to very dense ground layers indicating the length of end bearing piles.

ISO 22476-3 on the "Standard Penetration Test" covers the determination of the resistance of soils at the base of a borehole to the dynamic penetration of a split barrel sampler and the obtaining of disturbed samples for identification purposes. The standard penetration test is used mainly for the determination of the strength and deformation properties of cohesionless soils, but some valuable data may also be obtained in other types of soils. The basis of the test consists in driving a sampler by dropping a hammer of 63,5 kg mass on to an anvil or drive head from a height of 760 mm. The number of blows (N) necessary to achieve a penetration of the sampler of 300 mm (after its penetration under gravity and below a seating drive) is the penetration resistance.

3.4 *Testing of geotechnical structures*

CEN/TC 341 started to prepare a set of standards for testing geotechnical structures (ISO 22477) such as
- Pile load tests;
- testing of anchorages;
- testing of soil nailing;
- testing of reinforced fills

to eliminate as far as possible erroneous assessments of these geotechnical structures and the uncertainties of design methods as well as to limit scatter when repeating tests and improve reproducibility when testing geotechnical structures.

3.5 Laboratory Testing

Based on international recommendations for laboratory testing (DIN, ISSMGE ed. 1998) a set of Technical Specifications were prepared under the common title "ISO/TS 17892 Geotechnical investigation and testing – Laboratory testing"
- Part 1: Determination of water content;
- Part 2: Determination of density of fine grained soils;
- Part 3: Determination of density of solid particles;
- Part 4: Particle size distribution;
- Part 5: Oedometer test;
- Part 6: Fall cone test;
- Part 7: Compression test;
- Part 8: Unconsolidated triaxial test;
- Part 9: Consolidated triaxial test;
- Part 10: Direct shear test;
- Part 11: Permeability test;
- Part 12: Determination of Atterberg limits.

3.6 Design assisted by field and laboratory testing (geotechnical investigation)

As already mentioned CEN/TC 250/SC 7 revises ENV 1997-2 and ENV 1997-3 in order to convert them into European standards (EN). Both standards will merge to a single part defining the concepts and specifying requirements relating to planning of geotechnical investigations and testing and evaluation of test results in order to provide the geotechnical data necessary for the design of buildings and civil engineering works, the aim being to prevent damage to the new structures and to maximize cost-effectiveness during the planning and construction stages. The final draft will be probably available at the end of 2004.

4 CONCLUSIONS

Figure 2 illustrates the links of the standards prepared by the different Technical committee on geotechnical design, investigation and testing.

The standards of CEN/TC 341 specify the test equipment and the correct test procedure leading to comparable test results. These test results are used by Eurocode 7 Part 2 (EN 1997-1) for evaluation and interpretation leading to derived values. Eurocode 7 Part 1 (EN 1997-1) uses these values for the determination of characteristic values and design values.

Together with standards for the execution of special geotechnical works, an almost complete set of geotechnical standards are currently in preparation.

This set of standards will facilitate trade, exchange and technology transfer through:

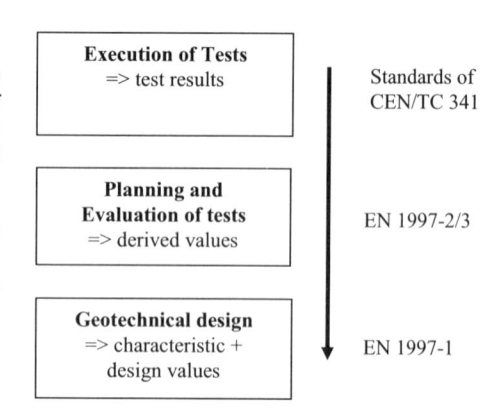

Figure 2. Standardization in geotechnical engineering (from Eitner et al. 2002).

a) enhanced product and service quality and reliability at a reasonable price;
b) improved health, safety and environmental protection;
c) greater compatibility and interoperability of goods and services;
d) simplification for improved usability;
e) reduction in the number of models and procedures, and thus reduction in costs;
f) increased distribution efficiency, and ease of maintenance.

Users will have more confidence in geotechnical products and services that conform to International Standards.

REFERENCES

DIN (ed.) (2000). Economic benefits of standardization - Summary of results. Final report and practical examples (Executive Summary). Beuth Verlag, Berlin (in German, English and Spanish, www.din.de)

DIN, ISSMGE (eds.) (1998). Recommendations of the ISSMGE for Geotechnical Laboratory Testing. Beuth Verlag, Berlin (in English, French and German)

Eitner, V., Katzenbach, R. & Stölben, F. (2002): Geotechnical investigation and testing - An outlook on European and international standardization. - In: Honjo, Y. et al. (eds.) (2002): Foundation design codes and soil investigation in view of international harmonization and performance based design. - Proceed. International Workshop IWS Kamakura, Japan, 10-12 April 2002, pp. 211-215, Rotterdam (Balkema)

IAEG (1981). Rock and Soil Description and Classification for Engineering Geological Mapping, Bulletin of the International Association of Engineering Geology, No.24, pp.235-274.

ISRM (1977 a). Suggested Method for Petrographic Description of Rocks, Int. J. Rock Mech. Min. Sci. & Geomech. Abstr. Vol.15, pp.41-45.

ISRM (1977 b). Suggested Methods for the Quantitative Description of Discontinuities in Rock Masses, Int. J. Rock Mech. Min. Sci. & Geomech. Abstr. Vol.15, pp.319-368.

ISRM (1978). Suggested Methods for Determining the Uniaxial Compressive Strength and Deformability of Rock Materials, Int. J. Rock Mech. Min. Sci. & Geomech. Abstr. Vol.16, pp.135-140.

ISRM (1980). Basic Geotechnical Description of Rock Masses, Int. J. Rock Mech. Min. Sci. & Geomech. Abstr. Vol.18, pp.85-110.

ISRM (1984). Suggested Method for Determining Point Load Strength, Int. J. Rock Mech. Min. Sci. & Geomech. Abstr. Vol.22, No.2, pp.51-60.

ISSMGE (1994). Testing Method of Indurated Soils and Soft Rocks - Suggestions and Recommendations, ISSMFE Technical committee on Indurated Soils and Soft Rocks, pp.65-69.

Stölben, F., Eitner, V. & Hoffmann, H. (2004). Sampling by drilling and excavations and groundwater measurements according to EN ISO 22475 for geotechnical investigations and site characterization. Proc. ISC'2, 20-22 September 2004; Porto.

CPTU site characterisation: offshore peninsular and East Malaysia experience

Syed M. Fairuz
TL Geotechnics Sdn Bhd, Kuala Lumpur, Malaysia

Joehan Rohani
TL Geotechnics Sdn Bhd, Kuala Lumpur, Malaysia

Tom Lunne
Norwegian Geotechnical Institute, Oslo, Norway

Keywords: CPT, CPTU, characterisation, cone resistance, sleeve friction, pore pressure

ABSTRACT: Modern cone penetrometer allows for the soil type to be determined from the measured values of cone resistance, sleeve friction and pore pressure. The paper discuss soil characterisation methods proposed by Robertson (1990) and compare them against four cases containing sand, normally consolidated clay, and overconsolidated clay for offshore Peninsular and East Malaysia marine soils. Net cone resistance, q_t versus undrained shear strength correlation are also plotted for the sites specified in this paper, providing N_{kt} factors required for determining undrained shear strength of fine grained soils. N_{kt},15 to 20 gives an undrained shear strength that corresponds to what is measured in a UU test on a reasonably good quality sample.

1 INTRODUCTION

With extensive offshore site investigations being carried out for offshore Peninsular Malaysia and East Malaysia soils, data compiled through comparing CPTU results and laboratory testing can enable more accurate characterisation of these soils.

The authors are all involved directly with offshore soil investigations. The experience in gathering CPTU data offshore Peninsular Malaysia and East Malaysia as shown in Figure 1.0 over more than ten years have provided a degree of confidence in the interpretation of CPTU data to build a database to characterize the offshore soils.

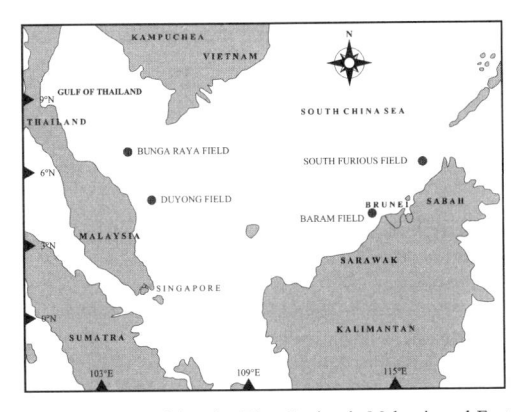

Figure 1.0 General location Plan, Peninsula Malaysia and East Malaysia

2 SOIL CHARACTERISATION METHOD

Robertson (1990) proposed plotting a "normalized cone resistance", Q_t, against a "normalized friction ratio", F_r in a cone resistance chart. The accompanying pore pressure ratio chart plots the "normalized cone resistance" against the pore pressure ratio, B_q.

The normalized cone resistance is defined by:

$$Q_t = \frac{q_t - \sigma_{vo}}{\sigma'_{vo}} \quad (1)$$

Where
q_t = net cone resistance corrected for pore water pressure on shoulder (Lunne et al., 1997)
σ_{vo} = total overburden stress
σ'_{vo} = effective overburden stress
$q_t - \sigma_{vo}$ = net cone resistance

The normalized friction factor is defined as the sleeve friction over the net cone resistance, as follows:

$$F_r = \frac{f_s}{q_t - \sigma'_{vo}} \times 100\% \quad (2)$$

Where
f_s = sleeve friction

The pore pressure ratio, B_q, is defined as follows:

$$B_q = \frac{u_2 - u_0}{q_t - \sigma_v} \quad (3)$$

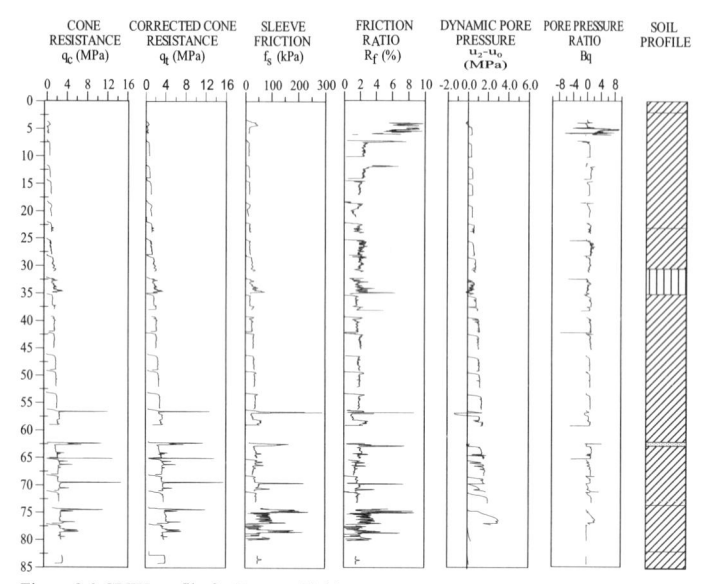

Figure 2.0 CPTU profile for Duyong Field

Table 1.0 Site Information for Duyong Field

No	Depth	Description	Su (kPa)/ φ	Water Content (%)	Soil Fractions Clay (%)	Silt (%)	Sand (%)	CPTU Data q$_t$ (MPa)	f$_s$ (kPa)	u$_2$ (MPa)	Atterberg Limits LL (%)	PL (%)	PI (%)
	(m)												
BH 1A, Duyong B, Water Depth: 76.6m MSL
1	0.0	Very soft CLAY	5 -8	69.5	28	53	19	-	-	-	65	26	39
2	2.0	Firm siltyCLAY	30 -45	40.8	34	63	3	0.6 -1.0	15.0 -20.0	0.4 -0.6	60	25	34
3	23.0	Firm to stiff silty CLAY	45 -62	-	-	-	-	1.0 -1.2	17.5 -30.0	0.6 -1.0	-	-	-
4	30.5	Stiff clayey SILT	62 -70	21.0	21	53	26	1.2 -1.3	30.0 -33.0	1.0 -0.6	30	16	14
5	35.0	Stiff to very stiff silty CLAY	70 -122	40.5	53	45	2	1.3 -1.8	33.0 -33.0	0.6 -1.5	59	27	32
6	62.0	Medium dense fine SAND	25°	-	-	-	-	6.0 -5.5	150 -125	0.0	-	-	-
7	62.7	Very stiff silty CLAY	125 -145	36	41	53	6	1.8 -2.5	33.0 -42.5	1.2 -2.0	58	27	31
8	73.5	Very stiff silty CLAY	145 -165	33	32	60	8	2.5 -3.0	42.5 -50.0	2.0 -1.8	50	22	28

Where

u_2 = pore pressure measured at cone shoulder
u_o = in-situ pore pressure
σ_v = total overburden stress

The normalisation was proposed to compensate for the cone resistance dependency on the overburden stress and, herefore, when analyzing deep CPTU soundings (i.e., deeper than about 30 m) apply quite well as seen in Figures 2.1, 2.2, 3.1, 3.2, 4.1, 4.2, 5.1 and 5.2.

Robertson, 1990 refers to the numbered areas by:
1. Sensitive, fine grained
2. Organic soils and peat
3. Clays to silty clays
4. Silty clay to clayey silt
5. Sandy silt to silty sand
6. Silty to clean sand

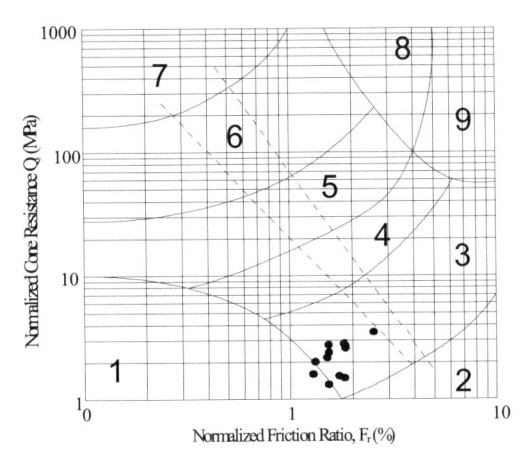

Figure 2.1 F_r chart by Robertson (1990) - BH 1A, Duyong B Field

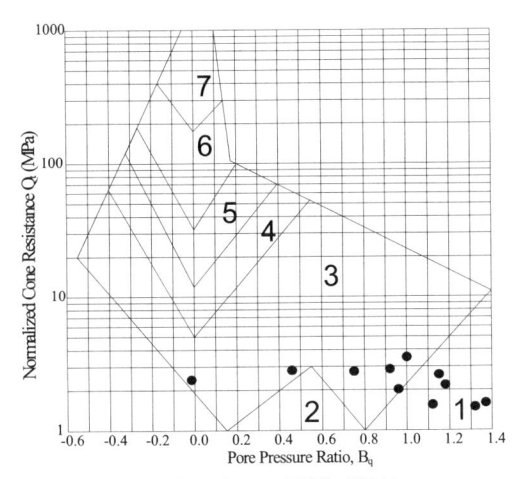

Figure 2.2 B_q chart by Robertson (1990) - BH 1A, Duyong B Field

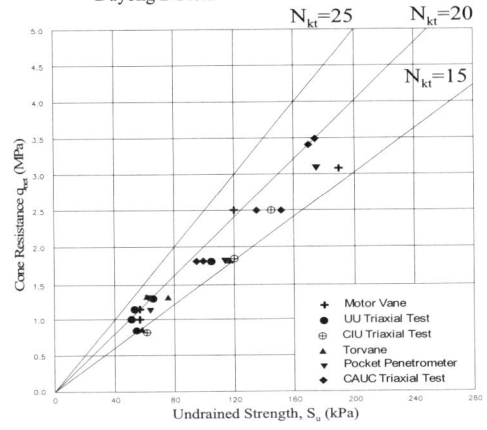

Figure 2.3 N_kt chart - BH 1A, Duyong B Field

7. Sand to gravelly sand
8. Clayey to 'very stiff' sand
9. Very stiff, overconsolidated or cemented soils

Estimation of Su from the CPTU data are derived from the following equation:

$$S_u = \frac{q_t - \sigma_{vo}}{N_{kt}} \qquad (4)$$

where N_{kt} is the empirical cone factor

3 SITES CONSIDERED

Table 1 to table 4 represents a range of data presented in this paper performed at four (4) fields, offshore Peninsular and East Malaysia. The soil conditions for Duyong, Bunga Raya and Baram fields ranged from soft, normally consolidated to stiff overconsolidated clays with sand and cemented sand conditions found at South Furious Field. Cone penetration results are available at all sites indicated in this paper.

4 SOIL CHARACTERISATION AND UNDRAINED SHEAR STRENGTH

4.1 Duyong Soil Description

The subsoil encountered at BH 1A, Duyong B Field constitutes generally of Clays with Silt/Sand inclusions at various borehole penetrations. From mudline to 2.0m, very soft Clays are encountered underlain by a 21.0m thickness firm Clay layer. Firm to stiff Clays are encountered from 23.0m to 30.5m underlain by a 4.5m thickness stiff clayey Silt layer. Stiff to very stiff Clays are encountered to the final depth explored of 85.0m interspersed by a medium dense Sand layer of 0.7m thickness at 62.0m penetrations. Figure 2.0 shows an example of a CPTU profile for the Duyong Field. Figure 2.3 indicates that the N_{kt} factor of 15 to 20 is generally acceptable for this site. In Table 1.0, values for Su, q_t, f_s and u_2 are usually expressed as a range, e.g. for very soft clay near the seabed, the Su range is between 5 to 8 kPa.

It can be seen that there is more scatter in the Bq data when compared between the F_r and B_q chart. The data tends to indicate finer materials in the B_q chart. The F_r chart predicts more reliably the soil types when compared to the B_q chart.

4.2 Bunga Raya Soil Description

The subsoil encountered at this BH BRC, Bunga Raya Field consists of surficial loose to medium sand to about 1.4m underlained by stiff Clay to a 1.0m thick 6.0m. This is followed by medium dense

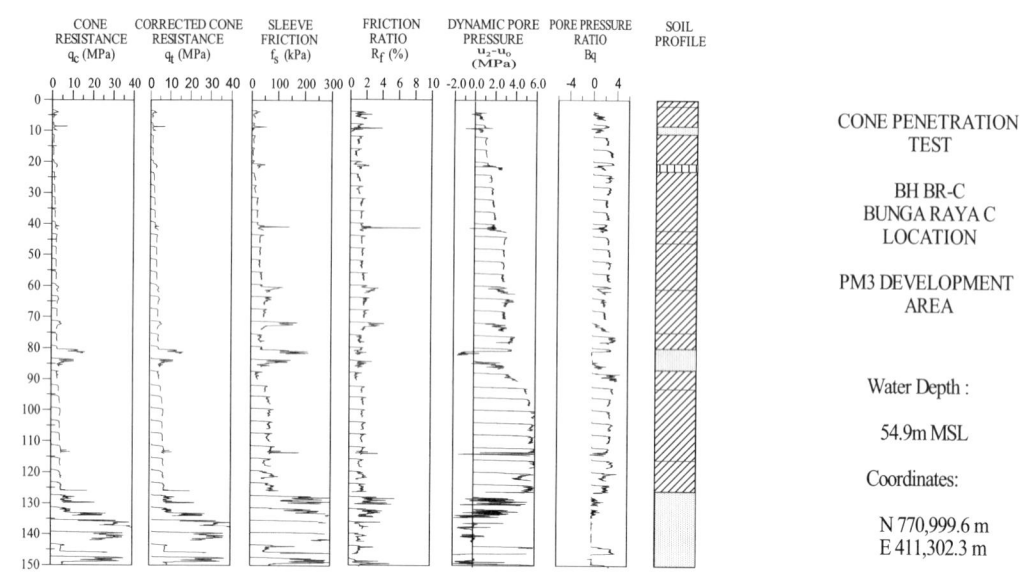

Figure 3.0 CPTU Profile of Bunga Raya Field

Table 2.0 Site Information for Bunga Raya Field

No	Depth (m)	Description	Su (kPa)/ φ	Water Content (%)	Clay (%)	Silt (%)	Sand (%)	q_t (MPa)	f_s (kPa)	u_2 (MPa)	LL (%)	PL (%)	PI (%)
\multicolumn{14}{	l	}{BH BR-C, Bunga Raya C Field, Water Depth: 54.9m MSL}
1	0.0	Loose to medium dense SAND	25°	54	-	-	-	-	-	-	-	-	-
2	1.4	Stiff silty CLAY	75	31	54	39	7	-	-	-	-	-	-
3	6.0	Medium dense silty SAND.	25°	39	15	24	61	1.0	13	0.0 / -0.4	-	-	-
4	7.0	Firm silty CLAY	30 -45	42	-	-	-	-	-	-	-	-	-
5	16.0	Very stiff CLAY	115 -125	33	35	60	5	3.0	117	0.6 / -1.7	58	26	32
6	61.0	Very stiff CLAY	180	40	-	-	-	3.0	104	1.6 / -1.8	-	-	-
7	73.0	Hard silty CLAY	240	29	-	-	-	3.0	52	1.8 / -2.4	-	-	-
8	86.0	Very stiff to hard silty CLAY	160 -220	31	-	-	-	3.5	65	2.4 / 1.2	63	31	32
9	122.0	Dense silty SAND	30°	27	0	10	90	30	234 -250	0.0	-	-	-
10	137	Hard silty CLAY	220 -240	30	-	-	-	5.0	78	(-0.6) / -4.2	-	-	-

Sand layer underlained by a firm Clay layer to 16.0m followed by very stiff Clays to 61.0m penetrations. Very stiff clay is encountered from 61.0m to 73.0m penetrations becoming hard from 73.0m to 86.0m penetrations. Very stiff to hard Clays are encountered from 86.0m penetration to the final depth explored of 152.0m. In between the final hard clays, a 15.0m thick dense Sand layer exists from 122.0m to 137.0m penetrations.

Figure 3.0 shows an example of a CPTU profile for the Bunga Raya Field. Figure 3.3 indicates that the N_{kt} factor of 15 to 20 is generally acceptable for this site. Negative readings are shown in brackets in Table 2.0. Scatter is seen again in the B_q chart. Again it is seen that the F_r chart is seen to predict the soil characteristics more reliably when compared to the B_q chart.

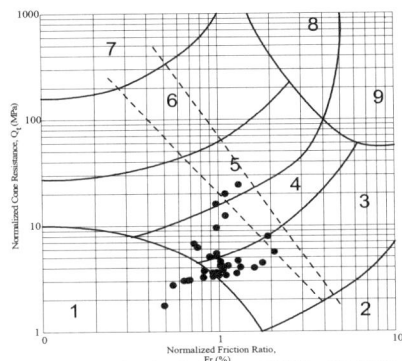

Figure 3.1 Fr chart by Robertson (1990) - BH BRC, Bunga Raya Field

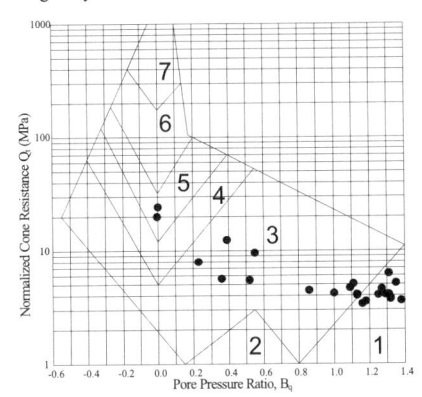

Figure 3.2 Bq chart by Robertson (1990) - BH BRC, Bunga Raya Field

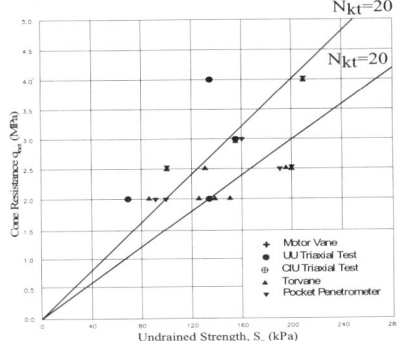

Figure 3.3 Nkt chart - BH BRC, Bunga Raya Field

4.3 Baram Soil Description

The subsoil encountered at BH BAKPB location, Baram Field generally consists of silty clays interlayered with sands along the depth explored. From mudline to 5.0m, very soft Clays are encountered. A 10.5m, 5.4m and 10.2m thickness layer of Sands were found at 5.0m, 31.0m and 45.9m penetrations respectively. From 62.0m to the final depth explored of 150.0m penetrations, dense to very dense sands were encountered.

Local variance to shear strength may exist due to inclusions of silt pockets, silt partings, shell fragments and organic matter. Figure 4.0 shows an example of a CPTU profile for the Baram Field. Figure 4.3 indicates that the N_{kt} factor of 15 to 20 is generally acceptable for this site. Here the F_r chart does indicate existence of clean sands but the B_q do not.

4.4 South Furious Soil Description

The subsoil encountered by drilling, sampling and CPTU testing indicates that the BH SFJT-C generally comprises of Sands. The top surficial layer of 6.0m thickness was found to be of medium dense gravelly sand/coral that is underlain by loose and medium dense gravelly silty sands to 50.0m penetrations. A stiff clay layer was encountered at 50.0m to about 69.3m penetrations followed by cemented sands/silts identified by cone resistance to be medium dense to very dense in condition. Thin-cemented sand/silt layers or seams are widely dispersed in the soil profile. Figure 5.0 shows an example of a CPTU profile for the South Furious Field. Figure 5.3 indicates that the N_{kt} factor of 15 to 20 is generally acceptable for this site.

Again in the camparison for the F_r and B_q chart clean sand is not distinguished in the B_q chart.

4.5 Discussion on Soil Data

The results of the soil characterisation using CPTU data for the four sites mentioned above using the Robertson (1990) method are shown in Figures, 2.1, 2.2, 3.1 3.2, 4.1, 4.2, 5.1 and 5.2. Tables 1 to 4 are only site specific. It does not necessarily represent the overall site characterization. However the sites have been chosen to represent the span of the South China Sea and provide a broad understanding of the scope of the soils offshore Peninsular Malaysia and East Malaysia.

Laboratory test conducted on the soil samples are carried out using British Standard (BS) as outline in BS 1377. In general soils with clay content above 20% are considered to have Clay as a major constituent, unless CPTu classification signifies a profound Silt profile. Description of soils is done according to American Society of Testing and Materials (ASTM) standard.

Due to an important need to obtain surface mudline sample for shallow foundation design, there is limited CPTU data presented in this paper for mudline material. Visual classifications are usually taken for mudline samples, where geotechnical engineers can have a better understanding of the soils.

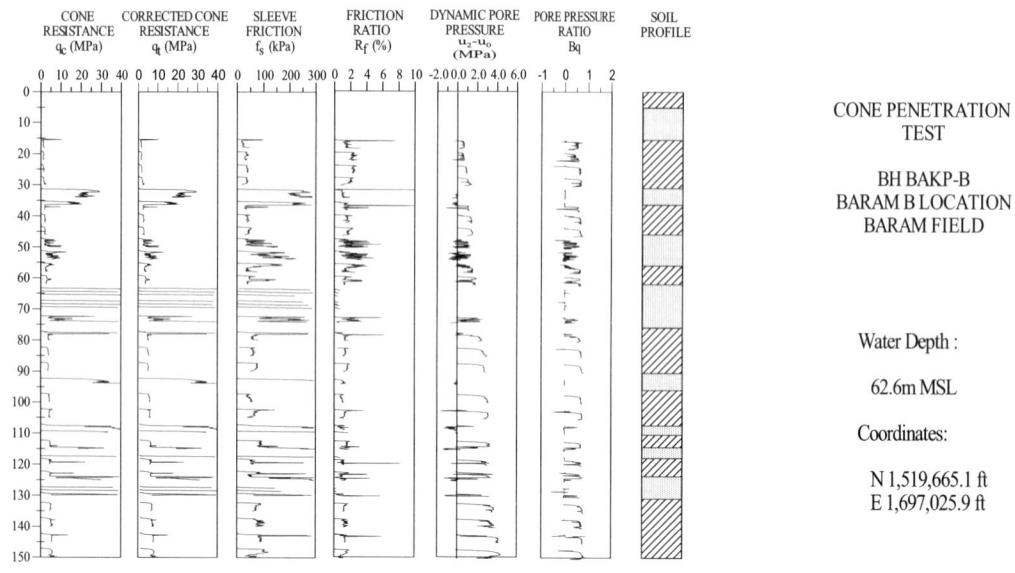

Figure 4.0 CPTU profile for Baram Field.

CONE PENETRATION TEST

BH BAKP-B
BARAM B LOCATION
BARAM FIELD

Water Depth:

62.6m MSL

Coordinates:

N 1,519,665.1 ft
E 1,697,025.9 ft

Table 3.0 Site Information for Baram Field

No	Depth (m)	Description	S_u (kPa)/ ϕ	Water Content (%)	Clay (%)	Silt (%)	Sand (%)	q_t (MPa)	f_s (kPa)	u_2 (MPa)	LL (%)	PL (%)	PI (%)
					Soil Fractions			CPTU Data			Atterberg Limits		
BH BAKP-B, BARAM B Field, Water Depth: 62.6m MSL													
1	0.0	Very soft silty CLAY	4	90	40	57	3	-	-	-	82	27	55
2	5.0	Medium dense fine SAND	20°	27	2	27	71	-	-	-	-	-	-
3	15.5	Stiff to very stiff silty CLAY	58 -95	37	55	45	0	1.4 -2.4	17.0 -18.0	0.7 -1.3	60	24	36
4	31.0	Medium dense fine SAND	25°	18	0	5	95	22.0 -16.0	240 -15.0	0.0 -(-0.3)	-	-	-
5	36.4	Very stiff silty CLAY	125	27	-	-	-	2.0 -2.1	36 -42	1.1 -1.5	50	23	28
6	45.9	Medium dense silty very fine SAND	20°	28	-	-	-	2.1 -5.2	42 -84	(-0.5) -1.2	-	-	-
7	56.1	Very stiff silty CLAY	138 -162	34	-	-	-	3.6 -2.4	30 -36	0.5 -1.75	-	-	-
8	62.0	Very dense very fine SAND	35°	26	0	0	100	28	180	0.0	-	-	-
9	76.0	Hard dense silty CLAY	212	28	-	-	-	4.0	60	1.5	44	23	21
10	90.5	Dense silty fine SAND	30°	28	-	-	-	30	360	0.0	-	-	-
11	96.0	Hard silty CLAY	225	24	23	64	12	4.1	48 -54	3.0	32	19	13
12	107.5	Very dense silty fine SAND	35°	-				36	360	(-0.5) -0.0	-	-	-
13	110.5	Hard silty CLAY	250	28	13	74	13	4.4	90	3.0	-	-	-
14	114.4	Dense silty fine SAND	30°	24	4	15	81	24	288	0.0	-	-	-
15	118.0	Hard silty CLAY	250	25				4.4	60	2.75	47	21	26
16	123.8	Very dense silty fine SAND	35°	-	0	3	97	28	180	2.8	-	-	-
17	131.0	Hard silty CLAY	250 -300	27	13	55	32	4.4	72	3.0	40	21	19

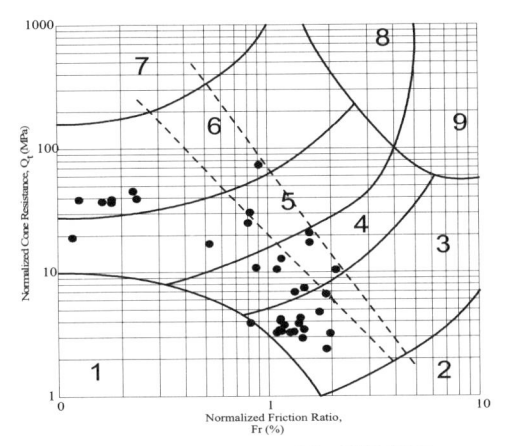

Figure 4.1. F_r chart by Robertson (1990) - BH BAKPB location, Baram Field

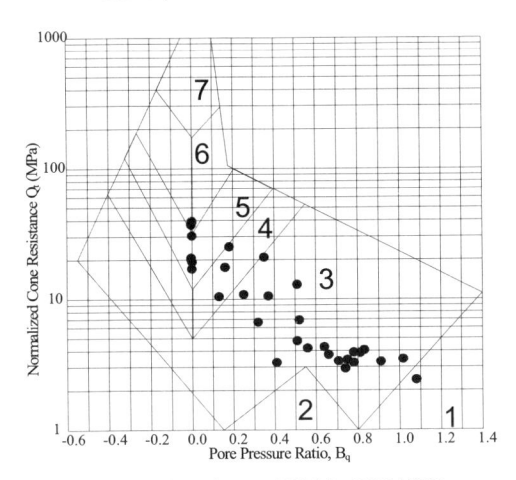

Figure 4.2 B_q chart by Robertson (1990)) - BH BAKPB location, Baram Field

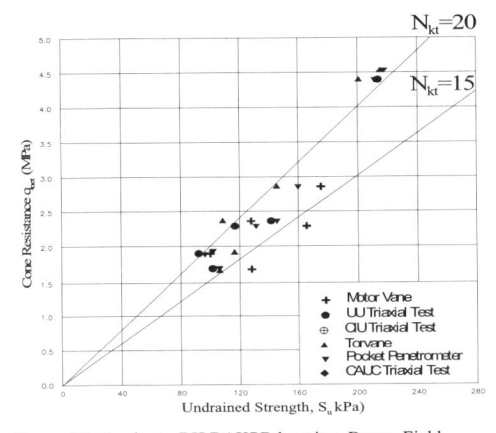

Figure 4.3 N_{kt} chart - BH BAKPB location, Baram Field

Even offshore laboratory tests undertaken on mudline samples can be difficult due to sampling techniques that impose disturbances to the samples collected. It needs to be noted that the authors do have in its unpublished database CPTU mudline results for other locations.

CPTU results presented in the tables indicate values that have been averaged over a depth of 3.0m. Typically in silty and sandy layers, local variance occurs to values of q_c, f_s and u_2. Thus engineering judgement is placed in the selection of the CPTU data in silty and sandy soils. In silt, penetration of CPTU can be partially drained due to permeability properties of the soils. Thus for the interpretation of CPTU data in silty soils it is important to identify the drainage conditions expected in the design problem and during cone penetration. Negative pore pressure data are indicated in bracket in the Table 1 to 4.

Generally, cone resistances, q_t and skin friction, f_s increases as the shear strength of saturated clay increases. The soils investigated in this paper are normally consolidated to moderately consolidated as can be observed from the overconsolidation ratio (OCR) in Table 5 and when contractive in nature are seen to generate positive pore pressures. In silt and sandy soils, which are dilative in nature as shown in Tables 1 to 4, negative pore pressures can develop.

Table 5 shows the summarised values of Fr and Bq used in the normalized Robertson (1990) plots. OCR and Ko in the table is estimated from the correlation provided by Andresen et. al. (1979) by first estimating Su from CPTU results and effective vertical stress from both soil profile and plasticity index relationships. N_{kt} values are representative at selected depths for all the locations selected.

The data tabulated in this paper only indicate the general classification of the soils offshore Peninsular and East Malaysia soils. There are other locations that portray special characteristics such as being over consolidated at shallow depths and having shallow gas influencing soil characteristics.

Corrected cone resistance results are presented in Tables 1 to 4 but both corrected and uncorrected cone resistance data are graphically shown for the subject fields. Although corrected and uncorrected cone resistance values represented in Figures 2.0, 3.0, 4.0 and 5.0 for Duyong, Bunga Raya, Baram, and South Furious Fields respectively are seen to indicate minor differences, it is necessary for cone resistances to be corrected for pore pressure effects for the purpose of characterisation of offshore soils. The pore pressure correction is especially important in soft fine-grained saturated soils, where pore pressures can be large relative to the cone resistance.

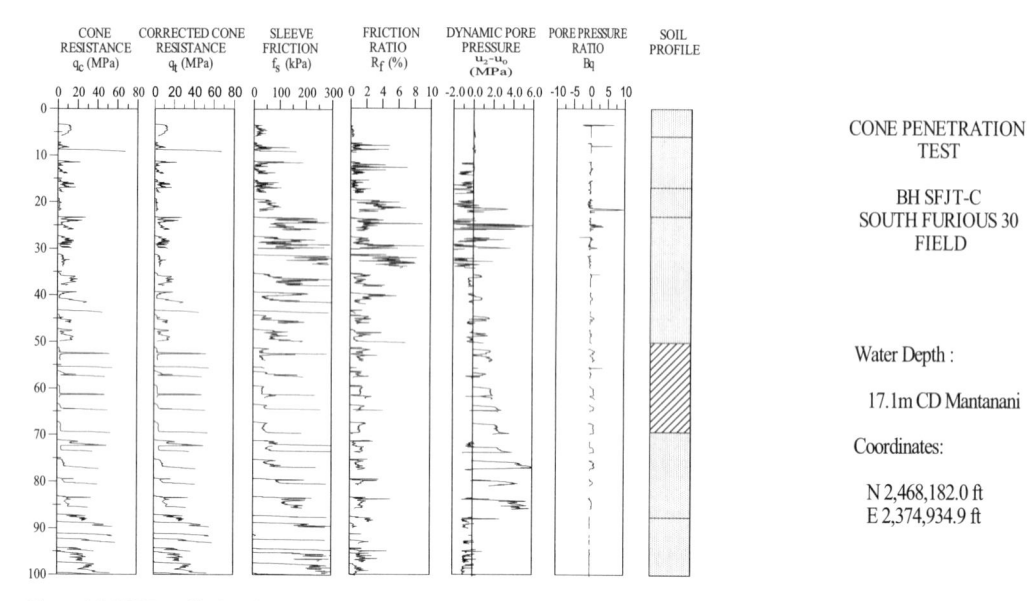

Figure 5.0 CPTU profile South Furious Field.

Table 4.0 Site Information for South Furious Field

No	Depth (m)	Description	Su (kPa)/ϕ	Water Content (%)	Soil Fractions Clay (%)	Silt (%)	Sand (%)	CPTU Data q_t (MPa)	f_s (kPa)	u_2 (MPa)	Atterberg Limits LL (%)	PL (%)	PI (%)
BH SFJT-C South Furious Field, Water Depth: 17.1m CD Mantanani														
1	0.0	Medium dense SAND	35°	31	0	2	98	11.0	26	0	-	-	-
2	6.0	Loose SAND.	25°	20	3	23	74	4.0	52	0	-	-	-
3	17.0	Loose SAND.	25°	-	13	15	72	3.0	39	(-1.0) -0.0	-	-	-
4	23.0	Medium dense SAND.	30°	18	12	26	62	5.0 -6.0	91 -130	(-1.2) -0.0	-	-	-
5	50.0	Stiff CLAY	175	26	24	66	10	3.0	26	1.2	43	20	23
6	69.3	Loose to medium dense silty SAND	26°	19	3	45	52	5.0 -20.0	13 -380	-0.6 -3.0	-	-	-
7	87.5	Medium dense SAND	32°	24	3	13	84	21.0 29.0	630 250	(-1.2) -0.0	-	-	-

Table 5.0 Summary of Characterisation Data

Locations	Depth (m)	γ'(kN/m3)	PI (%)	Su, uu (kPa)/ϕ	Qt (MPa)	Fr (%)	Nkt	OCR	Ko
Duyong									
	20	7.5	34	40	0.7	1.2	15-20	1.0	0.6
	41	7.5	59	105	2.2	1.4	15-20	1.2	0.7
	63	7.5	31	128	1.8	1.7	15-20	1.0	0.65
Baram									
	27	8.5	36	85	2.8	2.0	15-20	1.3	0.63
	125	9.0	-	35°	21.5	1.3	-	-	-
	145	8.5	19	275	4.1	1.3	15-20	1.0	0.59
Bunga Raya									
	6.0	8.0	-	25°	6.1	1.3	-	-	-
	18	7.5	32	115	2.9	0.6	15-20	3.1	0.82
	86	7.5	32	240	4.7	1.0	15-20	1.2	0.63
	130	9.0	-	30°	9.5	2.6	-	-	-
South Furious									
	3.0	8.4	-	35°	391	0.1	-	-	-
	10.0	8.4	-	25°	182.6	1.1	-	-	-
	30.0	8.4	-	30°	26.0	1.9	-	-	-
	60	8.4	23	175	5.0	1.0	15-20	1.0	0.61
	90.0	9.0	-	32°	60.6	0.5	-	-	-

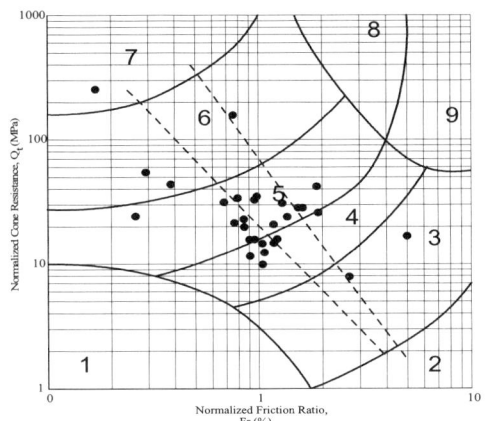

Figure 5.1 F_r chart by Robertson (1990) - BH SFJT-C South Furious Field

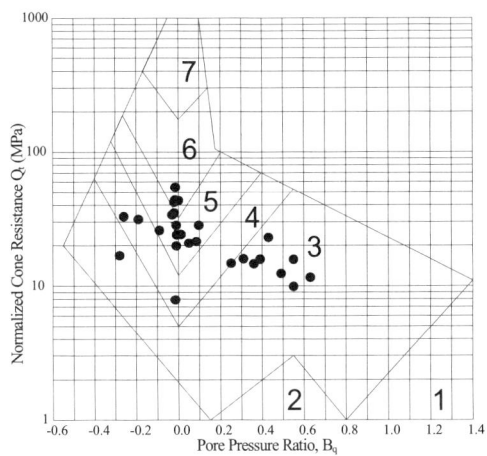

Figure 5.2 B_q chart by Robertson (1990) - BH SFJT-C South Furious Field

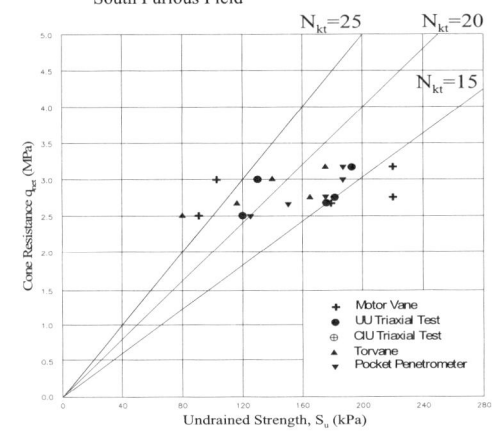

Figure 5.3 N_{kt} chart - BH SFJT-C, South Furious Field

Proper maintenance, calibration and preparation of cones are required for the CPT and CPTU data to be trustworthy. With a defined operating procedure the CPT and CPTU becomes an excellent equipment to enable the accurate classification of multi-layered soils.

5 CONCLUSIONS

The authors believe that work on characterisation of soils offshore Peninsular Malaysia and East Malaysia needs to be meticulously collected and processed to achieve a more elaborate database able to provide first insights into locations offshore, prior to planning and performing offshore site investigations.

The Su obtained in this paper is through UU triaxial test. This is mainly suitable for selecting parameters for design of fixed structure deep piled foundation or on shallow foundation such as for the purpose of spudcan penetration analyses.

N_{kt}, 15- 20 fits certain depths in all the locations presented and is recommended as first estimate for Su correlation at offshore Peninsular and East Malaysia soils.

It is observed by the authors that for the CPTU data used to characterize the offshore Peninsular and East Malaysian soils against published works the following points are salient:

1. Robertson (1990), include normalisation computations of the CPT data. Comparison between CPTU characterisation using the cone resistance and pore pressure methods indicate that the pore pressure chart needs careful understanding of drainage conditions in locations with pure sand such as Baram and South Furious. Experience offshore indicate that pore pressure charts implies finer soil type than is necessary the case.
2. The tables presented in this paper would provide a platform for geotechnical engineers to compare CPT/CPTU data gathered from other sites both onshore and offshore Peninsular Malaysia and East Malaysia and would act as references where more complex judgement of soil characterization is required.

REFERENCES

Andresen, T., T. Berre, A. Kleven and T. Lunne (1979) "Procedures used to obtain soil parameters for foundation engineering in the North Sea". Marine Geotechnology, Vol.3, No.3, pp.201-266. Also published in: Norwegian Geotechnical Institute, Publication 129.

Lunne, T., P.K. Robertson and J.J.M. Powell, (1997) "Cone Penetration Testing In Geotechnical Practice". Spon Press.

Robertson, P.K. (1990) "Soil classification using the cone penetration test". Canadian Geotechnical Journal, Vol. 27, No. 1, pp. 151 – 158.

The repeatability in results of Mackintosh Probe test

A. Fakher
Associate Professor, Civil Eng. Dept., Tehran University, Tehran, Iran

M. Khodaparast
PhD Student, Civil Engineering Department, University of Tehran, Iran

Keywords: Dynamic Probing, Mackintosh Probe, repeatability, and site investigation

ABSTRACT: Dynamic Probing is a continuous soil investigation technique and could have an important role in geotechnical site investigation. Mackintosh Probe is a lightweight, portable and handy penetrometer and considerably a faster and cheaper tool than boring equipment especially when the depth of exploration is moderate and the soils are soft or loose. It is important to study the repeatability of Mackintosh Probe results. The paper aims to present the capability of Mackintosh Probe for the investigation of cohesive soft soils. In addition, the repeatability of its results is studied. The data has been obtained from high quality site investigation at three sites, Emamie port, Khamir port and Emam-Khomeini port (all in South of Iran). The paper encourages the application of Mackintosh Probe for site investigation in soft soils.

1 INTRODUCTION

Dynamic Probing is a continuous soil investigation technique, which is probably the simplest soil penetration test. The apparatus for dynamic probing comprises a sectional rod with a cone fitted at the end whose base is of slightly greater diameter than rod. It is driven into the ground by a constant mass that is allowed to fall on to the rod through a constant distance, and the arrangement should be such that the mass falls through the constant distance without the operator having to use his judgment in any way. Testing is carried out continuously from ground level to the final penetration depth.

The continuous sounding profiles may enable easy recognition of dissimilar layers and even thin strata by observed variations in the penetration resistance (Shokrani, 2000). Mackintosh Probe is a lightweight dynamic penetrometer and a considerably fast and cheap tool especially when the depth of exploration is moderate and the soils penetrated are soft or loose (Sabtan and Shehata, 1994). The purpose of the presented paper is to describe the capability and limitations of Mackintosh Probe.

2 MACKINTOSH PROBE

Mackintosh Probe is a lightweight and highly portable tool consisting of a 1.10in (27.94-mm) diameter cone with its 30°-apex angle; 0.5in (12.7-mm) diameter solid rods and a 4.5kg (10-Lb) dead weight with standard drop height of 300mm. The cone is advanced into the soil by standard blows from the drop weight and the number of blows for 100mm penetration is counted (M). Friction losses on the rods are minimized through the use of enlarged conical couplings. Mackintosh Probe is illustrated in Fig. 1. British Standard Code of practice for site investigations, BS5930 (1981), recognized Mackintosh Probe.

The main advantages of Mackintosh Probe include:

1) The speed of operation is high.
2) Minimal equipment and personal are required.
3) It is a very low cost tool.
4) It is very easy to use.
5) It is can be used for the interpolation of soils between trail pits and boreholes and reduces the number of boreholes.

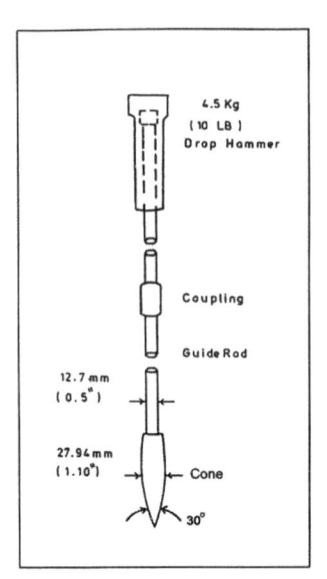

Fig. 1: The Set-up and dimensions of Mackintosh Probe (Sabtan and Shehata, 1994)

3 EXPERIMENTAL STUDY

The data of the paper has been obtained from high quality site investigation under supervision of the authors at three sites, Emamie port (South of Iran and 140km East of Mahshahr), Khamir port (South of Iran and 60km East of Bandar-Abbas port) and Emam-Khomeini port (South of Iran). In addition to probing, conventional boring and testing were undertaken in all three sites. The principal soil properties of these sites are given in Table (1). The soils investigated ranged from soft to medium clays. Plasticity Indices vary from 7 to 22.
In addition to above-mentioned data, some published data has been also considered for discussion.

4 REPEATABILITY OF MACKINTOSH PROBE RESULTS

It is very important to study the repeatability of Mackintosh Probe results. For this purpose, series of tests have been carried out at different site. In each group of test, two, three, or four Mackintosh testing have been repeated at very close proximity (less than 0.5m).

For illustration, the M (Mackintosh value) variations versus depth for some tests, undertaken in each three sites, Emamie port, Khamir port and Emam-Khomeini port, are shown in Fig. 2.

To study the repeatability of results, it is important to choose a suitable parameter that represents the repeatability. To analysis the results, mean, variance, standard deviation, and coefficient of variation have been considered.

The mean is the numerical average of the observed values, while the others are closely related measures of the dispersion of the data about the mean. The standard deviation value (s) is not appropriate for this purpose because (s) is large for the large values of M and it cannot be independently used for judgment about repeatability.

The coefficient of variation is dimensionless, and is expounded as a percentage. It is particularly useful for comparing groups with widely different means because it measures the degree of dispersion in terms of mean.

The coefficient of variation (C_v) can be considered as an indicative parameter because it somehow represents a normalized variance. It is calculated using the following formula:

$$C_v = s/\bar{x} \qquad (1)$$

Where $\bar{x}$ = average of M at each depth and s = standard deviation of M at each depth.

In Table (2) some of soil properties, as measured by the various standard tests, are listed together with their coefficient of variation reported by various researchers. Because of sources of variability in soils, it is not expected that these coefficients of variation to be the same for all properties tested.

Table 1: Soil properties of the investigated sites

Site	Depth (m)	Soil Description	Density (t/m³)	W %	PL %	PI %	C_u (kPa)	Date
Emamie port	2.5	Soft to very soft brown silty Clay	1.88	32	20	7	18	2000
	5		1.96	41	18	7	23	
	7.5		1.89	37	23	16	32	
	10		1.88	36	19	19	46	
Khamir port	2	Lean Clay with silt	1.85	31	20	9	10	1999
	5		1.91	35	22	22	23	
	7		1.95	39	25	17	30	
Emam-Khomeini port	3	Soft Clay with silt	1.95	30	20	19	28	2000
	5		2.00	33	20	12	34	
	8		1.90	31	18	16	45	
	12		1.99	27	16	9	48	

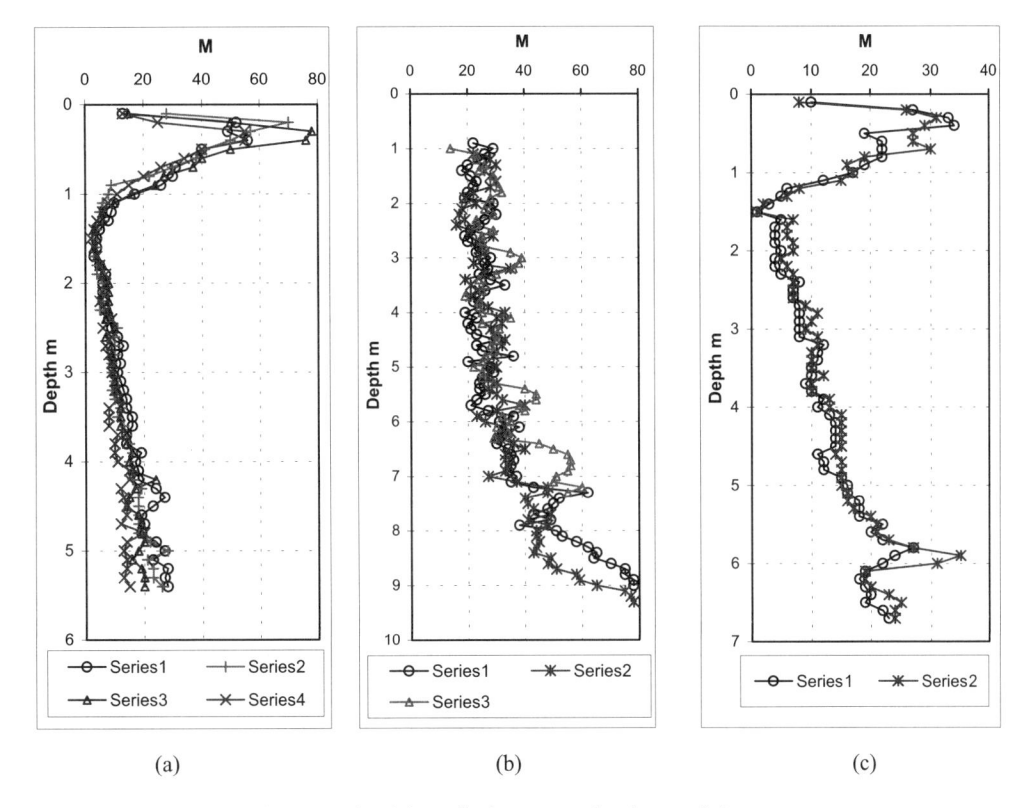

Fig. 2: Examples of the result of tests repeated at close proximity
(a) At Emamie port site
(b) At Emam-Khomeini port site
(c) At Khamir port site

Table 2: Coefficient of variation for soil engineering tests (Lee et al., 1983)

Test	Reported Cv (%)	Recommended Standard
Angle of friction (sands)	5-15	10
CBR	17-58	25
Undrained cohesion (clays)	20-50	30
Standard penetration test (SPT)	27-85	30
Unconfined compressive strength (clays)	6-100	40

It could be seen the variation of C_v for the results of Standard Penetration Test (N), that is a super heavy dynamic probing, reported between 27 to 85 % and recommended 30% by researchers (Lee et al., 1983).

As the repeatability of SPT test results is conclusive, so it could be used for judgment about repeatability of Mackintosh Probe by comparison between Cv of Mackintosh Probe results and Cv of Standard Penetration Test. In presented research, the value of C_V has been calculated for each depth in each group of tests performed at very close proximity.

The Cv variations versus depth in each site for repeated tests are shown in Fig. 3.

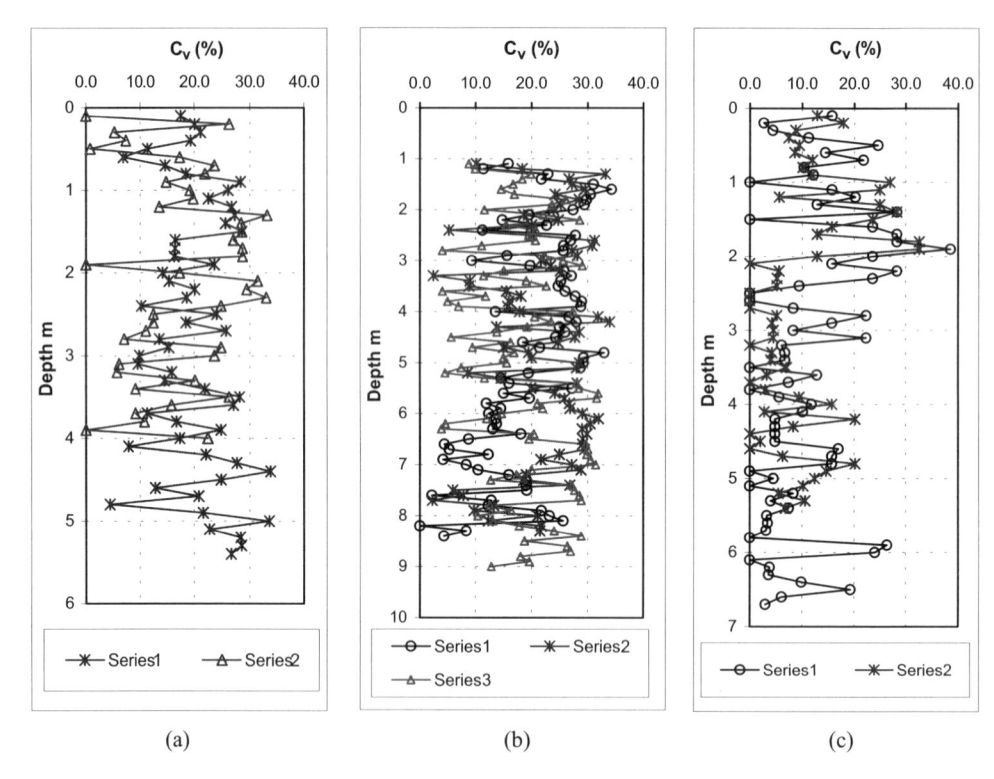

Fig. 3: The coefficient of variation (C_V) versus depth
(a) At Emamie port site
(b) At Emam-Khomeini port site
(c) At Khamir port site

Seeing that for the lighter weight hammer of dynamic probing, the values of M (the number of blows per 10 cm) become more in the same soil profile and properties. So it is expected, the value of C_V is large adv for results of Mackintosh Probe in comparison with SPT results.

In the tests, the C_V varies between 0 to 38% and in most case is less than 30%. Therefore the results of Mackintosh Probe tests could be considered as repeatable results in comparison with SPT, which has a proposed C_V of 30%.

5 RELATIONSHIP BETWEEN M AND C_U

Butcher et al. (1996) reported a correlation between M and c_u (undrained shear strength) for soft clays (c_u below 50 kPa):

$$c_u = \frac{q_d}{22} \qquad (2)$$

Where q_d = dynamic point resistance; which is resulted using the following:

$$q_d = \frac{W}{(W+W')} r_d \qquad (3)$$

$$r_d = \frac{W \cdot g \cdot h}{A \cdot e} \qquad (4)$$

Where r_d = the unit point resistance value in Pa; W = the mass of the hammer (4.5 kg); g = is the acceleration due to gravity (9.81 m/sec^2); h= the height of fall of the hammer (0.3 m); A= the area at the base of the cone (6.1E-4 m^2); e = the average penetration in meter per blow (0.1/M); M= the number of blows per 10 cm and W' = the total mass of the extension rods, the anvil and the guiding rods in kg.

Sabtan and Shehata (1994) presented the following Equation between M and c_u (kPa) using the results of tests in Saudi Arabia and also the data presented by Chan and Chin (1972) for Kuala Lumpur:

$$c_u = 0.96 M^{0.91} \qquad (5)$$

Using the data obtained in the presented research and the data presented by Sabtan and Shehata (1994), a new Equation is derived as follows:

$$c_u = 1.38 M^{0.85} \quad (6)$$

The Equations 2, 5 and 6 have been plotted in Fig. 4. The results of other tests such as vane shear test, UU triaxial test and unconfined compression test have been also presented in Fig. 4 for the three sites investigated.

of a correlation between M and c_u. The value of M for soft clays is higher than the value of N (SPT results), so M value is more sensitive than N to the variation of soil properties. It means that the variations of Mackintosh Probe testing seem to be more indicative than SPT in soft grounds.

If good correlations have been already established in an area, Mackintosh testing could be independently used for ground investigation. However, it could be always be used for interpolation of soils between boreholes to reduce the cost of investigation.

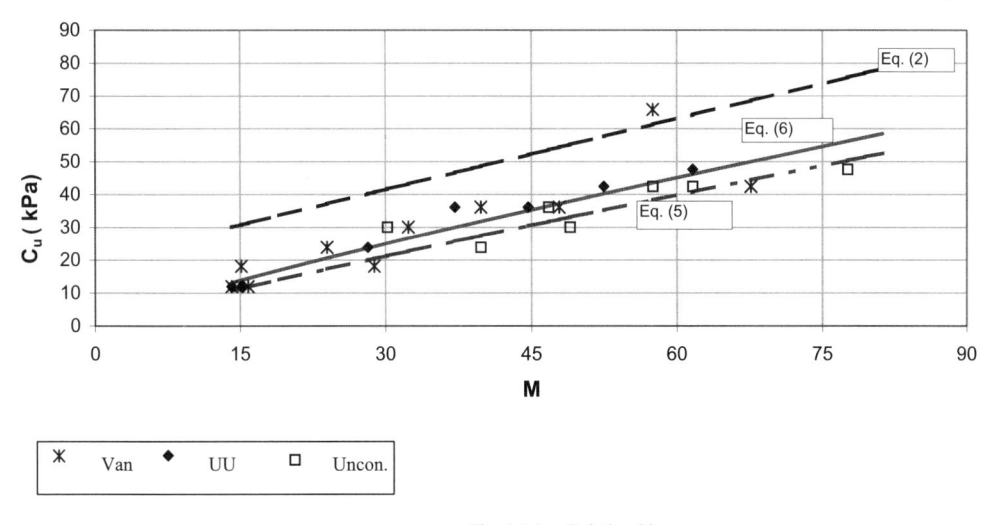

Fig. 4: M-c_u Relationship

As it can be seen in Fig. 4, Equations 5 and 6 better represent the data obtained by the authors for Emamie, Khamir and Emam-Khomeini sites. However, there is not much difference between Equations 5, 6 and 2.

Therefore it could be concluded that a good correlation between M and c_u could be established for each site to achieve a good estimation of c_u using the results of Mackintosh Probe.

6 DISCUSSION

The use of the dynamic probing in conjunction with trial pits and boreholes can produce information at a low cost. Mackintosh Probe can be widely used in site investigation in soft grounds because it is a promising device due to low cost and reasonably repeatable results. Mackintosh Probe testing in soft soil has been reported for depth of 0 to 10m (Kong, 1983). The authors experienced up to 9.5m depth in their experimental studies (Fakher et al., 2001). Therefore, it could be used for mentioned depths.

In addition, M is a good indication of soil parameters. The presented equations confirm the existence

7 CONCLUSIONS

Mackintosh Probe is a lightweight and handy device, which could be easily used for the investigation of soft soils up to a depth of about 10m. The results of Mackintosh Probing are:
    1-repeatable
    2-indicative

Good correlation could be established between M and c_u for soft clays. Therefore Mackintosh Probe can be used to quickly assess the variability of soil condition, allowing different conditions to be identified. This probe in soft soils could be used for depth to 10m.

This allows effective targeting of any subsequent boreholes or tests that may be required and also interpolation of soils between boreholes.

However, due to the relatively low energy hammer, the Mackintosh Probe is not a feasible device for investigation of dense to very dense soils.

ACKNOWLEDGEMENT

The authors wish to thank Sahel Consulting Engineers for cooperation in experimental study undertaken in Emamie, Khamir and Emam-Khomeini sites.

REFERENCES

BS 5930 (1981), *Code of Practice for Site Investigation*, British Standard Institution, pp.147.

Butcher, A.P., McElmeel, K. and Powel, J.J.M. (1996), "*Dynamic probing and its use in clay soils*", Advance in Site Investigation Practice, ed. Thomas Telford, London, pp. 383-395.

CHAN, S. F. and CHIN, F. K. (1972), "*Engineering characteristics of the soil along the federal highway in Kuala Lumpur*", Proceeding of the Third Southeast Asian Conference on Soil Engineering, Hong Kong, pp. 41-45.

FAKHER, A. and KHODAPARAST, M. and PAHLAVAN, B. (2001), "*Coastal soft clay improvements using preloading - A case study*", Proceeding of 3rd International Conference on Soft Soil Engineering.

Kong, T. B. (1983), "*In-situ soil testing at the Bekok dam site, Johor, Peninsular Malaysia*", Symposium international of In Situ Testing, Paris, Vol. 2, pp. 403-408.

LEE I. K. and WHITE W. and INGLES O. G. (1983), "*Geotechnical Engineering*", Copp Clark Pitman, Inc., pp.62.

SABTAN, A. A. and SHEHATA, W.M. (1995), "*Mackintosh Probe as an exploration tool*", Bulletin of the International Association of Engineering Geology, Paris, No. 50, pp. 89-94.

SHOKRANI, H. (2000), "*Calibration and implication of Mackintosh Probe for site investigation*", MSc Thesis, Civil Engineering Department, University of Tehran (in Persian Language).

Application of the CPT to the characterization of a residual clay site

S.G. Fityus & L. Bates
School of Engineering, The University of Newcastle, Australia.

Keywords: CPT, residual, expansive, clay, Maryland, mudstone

ABSTRACT: This paper presents a case study in which standard Cone Penetrometer equipment was applied in the assessment of important foundation parameters on a site underlain by shallow, expansive, residual clay soils. It is based on a testing program in which a grid of 7 by 7 CPT tests, at 5m spacings, was undertaken at the Maryland expansive soils test site, where a large quantity of site data has already been collected by a variety of other means. All of the CPT tests were pushed to refusal in weathered rock. An approach is proposed, in which data from the standard CPT test can be used to give either direct or indirect estimates of important parameters in expansive soils engineering such as the thickness of the topsoil, the position and thickness of expansive clay layers, and the depth and degree of weathering. Data extracted from the CPT tests includes depth of refusal, and tip resistances and friction ratios of major soil layers. The CPT data is compared with site parameter data measured using other site assessment techniques, and the ability of particular derived CPT quantities to either qualitatively or quantitatively describe important site parameters is discussed. Emphasis is placed on the value of the CPT as an economical and efficient tool for the characterization of relatively shallow soil deposits that display high variability. In particular, this paper demonstrates that, in having the ability to provide detailed information at a large number of discrete locations in a relatively short time, the CPT is an ideal tool for the assessment of site variability, where large datasets of spatial data are needed to give statistical confidence in a site model.

1 INTRODUCTION

Despite its widespread use in the geotechnical industry, the Cone Penetration Test or CPT has found relatively little use in the in situ testing of residual clay soils. The reasons for this are many. Finke et al, (1999) point out that the general difficulties in penetrating stiff and hard residual materials, and the possibility of encountering rock fragments, may be major reasons why CPT testing is not commonly employed in residual materials. They further suggest there are difficulties in interpreting the results obtained, although their results show that where residual materials are mostly saturated, and predominantly granular or silty granular, the measurement of pore pressures can be used as an aid in the interpretation of results.

Schneider et al. (2001) also studied the results of CPTU tests in residual soils, and they suggest that u_2 pore pressure response can be used to differentiate between stratigraphic breaks in the weathering profile.

However, ambiguity in CPT results is a particular problem in unsaturated residual soils of a clayey nature, where the stiffness of the soil is a function of the prevailing moisture content. In stiff and very stiff residual clays, the cone resistance can vary significantly depending upon the soil moisture condition and u_2 pore pressures are typically negative.

Another reason why CPT methods have not found frequent use in shallow, stiff, residual clays is because such soils are usually not problem soils, and they are easily explored by simpler methods. The properties of interest for stiff clay soils are readily explored by more direct techniques such as auger drilling and direct sampling. Indeed, in Australia, it would usually be considered excessive to deploy a 20 tonne CPT rig to investigate a site where weathered rock structure became evident within about 2m of the surface.

Reports in the geotechnical literature of applications of the CPT as an investigative tool for residual soils are scarce. Finke et al (1999) lists 12 reports, mostly dealing with the piedmont residual soils of the US, the porous residual clays of Brazil or so-called Tropical Residual Soils. The papers by Puechen et al. (1996) and Schneider (2001) are notable in that they consider CPT testing in a range of different residual soil materials including granite saprolites, weathered schists, igneous breccias, sandstones and siltstones.

In the present paper, consideration is given to the application of CPT testing to the expansive residual clay soils of eastern Australia, which are derived from a range different parent rocks including sandstones, mudstones and basalt. It seeks to demonstrate that several properties of interest in the modeling of expansive soil behaviour can be inferred from the results of the standard CPT test.

2 APPROACH

2.1 Expansive Clay Soils and the CPT

Expansive clay soils realize significant changes in volume in response to changes in suction (Nelson and Miller, 1992). Where substantial thicknesses of expansive clay exist in soil profiles which experience suction changes, this volume change can lead to ground movements as great as 160mm (Dhowian, 1990). Such movements often occur differentially beneath lightly loaded structures, leading to unacceptable levels of structural distress.

The engineering of foundations to accommodate such movements can be achieved by estimating the differential foundation movement on the basis of predicted characteristic ground surface movement for the site, and the application of empirically based mound shape models (Walsh and Cameron, 1998).

Estimation of the characteristic ground surface movement may be achieved in several ways, but is commonly determined by calculation, employing soil profile and laboratory test data (Fityus, 1999).

The aim of this paper is to consider the potential for using the CPT to estimate some of the important parameters used in ground movements predictions for expansive clay sites. In the tests undertaken for this study, the measured u_2 pore pressures were consistently negative and showed little consistency. Consequently, u_2 pore pressures have not been used in this work. Despite this, it will be shown that the CPT is particularly useful in assessing the variability in certain expansive soils parameters on a site wide basis.

2.2 Expansive Soil Movement Prediction

The characteristic value of the ground surface movement at a site can be estimated on the basis of the data obtained from a soil profile, and the prevailing environmental conditions. Data required for such an estimate comprises:
- the thickness of the expansive soil layer(s)
- the position of the expansive soil layer(s) present
- the expansiveness of the clay soils present
- the depth to which suction changes occur
- the depth to which cracking occurs
- the magnitude of the suction change at the surface
- the maximum magnitude of suction change as a function of depth

The first three of these are variables are mostly controlled by the localized susceptibility of the parent rock mass to weathering, and each may exhibit great variability. Even in areas underlain by a single geological unit, the composition and structure of residual soils can vary greatly due to localized differences in the degree and depth of weathering. Such differences are often due to the influence of primary geological structures in the parent rock mass such as joints and faults.

The next four parameters are largely controlled by the local environment, and so, they should be relatively spatially consistent across a single site, provided there are no significant changes in climatic exposure or anomalous influences such as structures, impervious ground covers or trees.

Variation in the characteristic ground surface movement is thus most likely to result from localized differences in the effects of weathering, causing differences in the amount, position and expansiveness of the clay soils formed.

Schneider et al. (2001) have plotted the CPT responses at a range of residual soil types and for different degrees of weathering, on the soil behaviour type charts of Robertson et al. (1986). These include results for mudstones from a range of depths in deeply weathered tropical soil profiles in Singapore. The interpretation of Robertson et al. (1986) will also be employed here.

In this study, aspects of the standard CPT response will be use to infer differences in the weathering of different soil profiles, and these will in turn be used to infer values of expansive soil profile parameters of interest in making ground movement predictions.

A number of CPT derived quantities have been considered. These are:
- the overall extent of weathering (in m) taken to be indicated by the depth of refusal (Q_t>70MPa).
- the depth of topsoil present, which typically exhibits reduced expansive potential.
- the depth of significantly expansive residual clay soil.

Consideration will also be given to the identification of residual clays and weathered rock on the basis of CPT test results, and the findings compared with the findings of Schneider et al. (2001).

3 FIELD SITE AND CPT TESTING.

Data for this study was collected from the Maryland expansive soils research site near Newcastle, in Australia (Fityus et al. 2004). From the results of a 10 year monitoring program, it was possible to assess the actual variability in ground movements for this site.

The residual clay soil profile at Maryland is described in detail in Fityus and Smith (2004). A summary of the soil layer details is shown in Figure 1. It is apparent from Figure 1 that the boundaries between layers are indistinct, and soil changes are generally gradational. What is less apparent is that the thickness of the identified soil layers exhibits significant lateral variation. For example, borehole data from the site indicates that the topsoil can vary in thickness from 250 to 400mm. Further, the depth at which rock structure is clearly discernable varies between 800 and 1300mm. Figure 1 does show that structures such as relict joints cause localized variation in the style and degree of weathering.

Figure 1. The Maryland expansive soil profile.

The instrumented field site at Maryland and the research activities undertaken there, are described in detail in Fityus et al., (2004). In summary, the activities included the construction of two ground slab covers, the establishment and monitoring of 154 surface movement indicators, positioned in open ground and adjacent to, and upon the covers, 28 subsurface movement indicators installed at various depths, 25 aluminium access tubes to 1.5 or 3m, installed to accommodate a neutron probe for in situ measurement of water content and a range of other devices for the in situ measurement of moisture content and suction.

Results of the Maryland study have provided high quality data on the soil profile characteristics, and on the ground movement behaviour at the site.

The site and the distribution of movement monitoring points used in this study are shown in Figure 2. Although there are around 150 surface movement measuring stations on the site, this study will only consider the results of the 29 of these which are situated in open ground areas, at sufficient distance from trees and ground covers to avoid being significantly affected by them. These are shown in black in Figure 2.

As the site is still being monitored, the CPT assessment was conducted within a 30m x 30m area beside the monitoring area. This involved 49 CPT tests on a 5m grid spacing, pushed to refusal. These are shown in Figure 2.

The CPT tests were performed using the University of Newcastle's 20 tonne, truck mounted, NEWSYD in situ testing facility. The facility, employing Hogentogler CPT equipment, performed the tests reported in this study using a 4 channel, 10 ton cone.

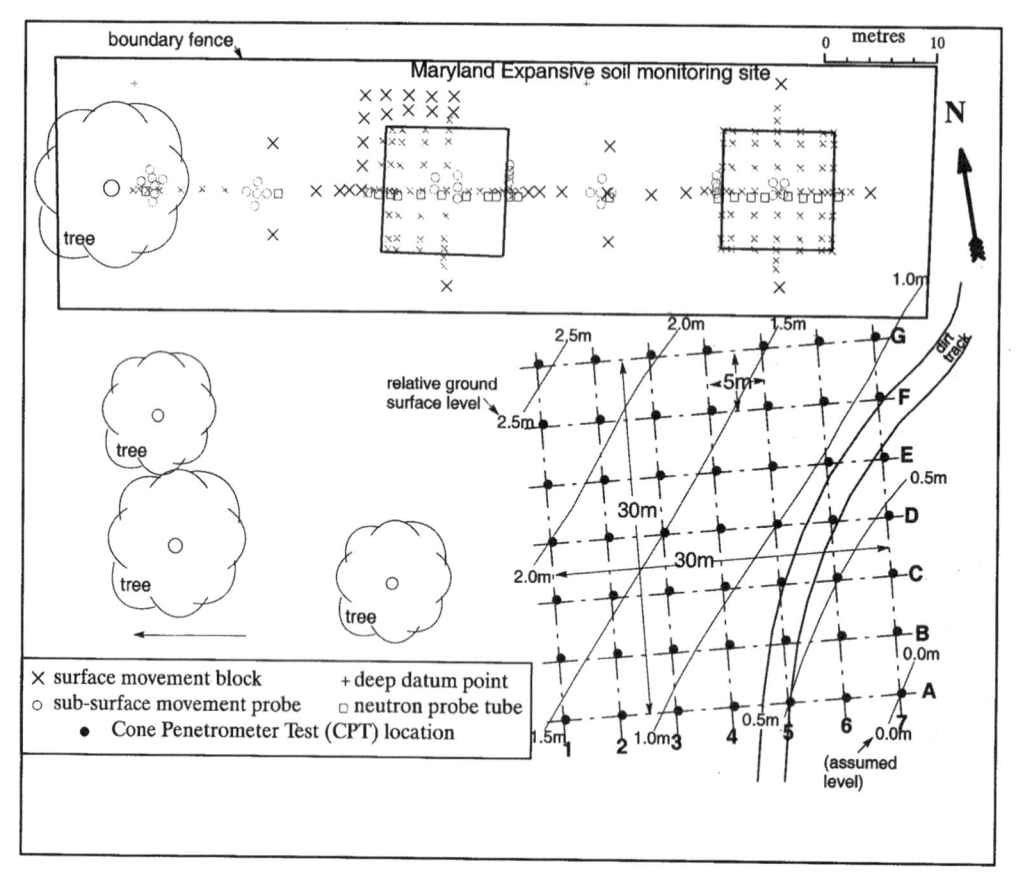

Figure 2. Site map of CPT test locations and the Maryland field site.

4 TYPICAL RESULTS

Two typical CPT test results are shown in Figure 3, corresponding to the deepest and shallowest profiles, encountered at positions A2 and A7 (respectively). Despite a significant difference in the depth to refusal, each of the results shows a number similar features, including

- A generally increasing tip resistance with depth
- A zone of low friction ratio just below the surface
- A lower zone with an elevated friction ratio
- Two zones (less distinct) with successively reduced friction ratio below the zone of elevated friction ratio.
- A zone of significantly elevated tip resistance, encountered just prior to refusal.

Whilst all of these features were evident in most of the measured CPT profiles, there were a number of anomalous profiles where one or more of the features could not be as readily discerned.

5 INFERRENCE OF TOPSOIL DEPTH

Figure 1 indicates that the topsoil is considerably less plastic than the underlying clay soils due to a significantly increased silt and sand content. It is also found to be significantly less expansive. As a means of estimating its thickness from the CPT results, the Robertson et al. (1986) interpretation was used, assuming that the topsoil would correspond to soil types 4 or 5.

Table 1 shows the statistics of the inferred topsoil depths, as well as observations from the Maryland study. The agreement is good.

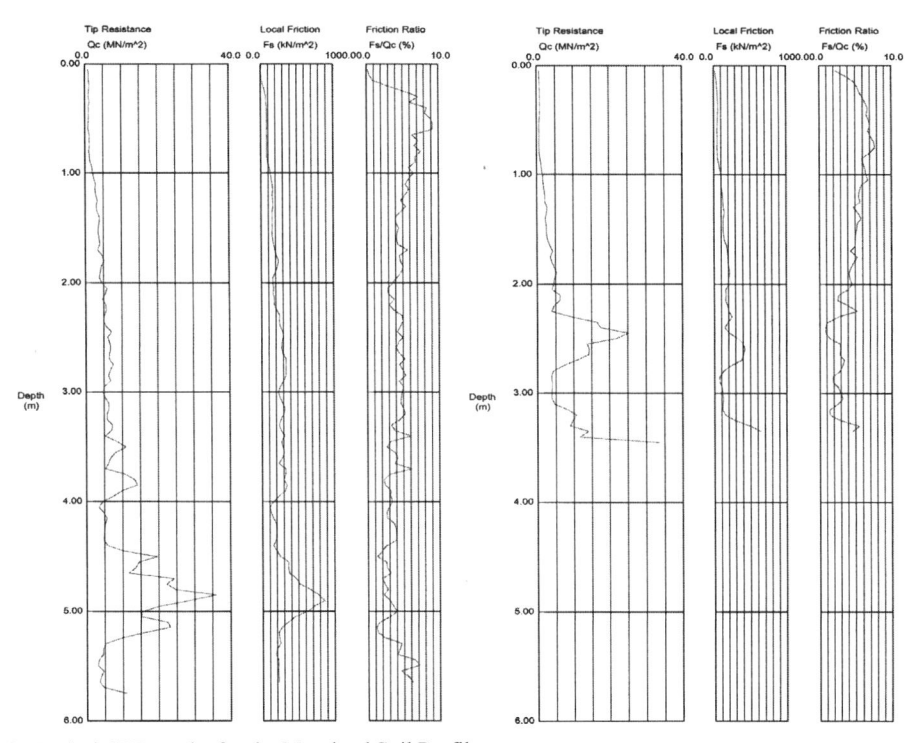

Figure 3 Typical CPT results for the Maryland Soil Profile

6 INFERRENCE OF EXPANSIVE CLAY THICKNESS

Figure 1 indicates that there is a layer beneath the topsoil (shown from 0.25-1.25m) that is considerably more plastic and expansive than the remainder of the soil in the profile. Although shown as clay in the figure on the basis of its measured liquid limit and plasticity index, it is could reasonably be described as a silty clay on the basis of particle size determinations (Fityus and Smith, 2004). Unlike the underlying weathered rock, this layer of the residuum has lost all of its original sedimentary structure through repeated swell and desiccation, in response to a large number of naturally occurring wetting and drying cycles. As a means of estimating its thickness from the CPT results, the Robertson et al. (1986) interpretation was used, assuming that the topsoil would correspond to soil type 3.

Table 1 shows the statistics of the inferred expansive clay thickness, as well as observations from the Maryland study. The agreement between inferred and directly measured value is again good. The average tip resistance in this layer is 1.1Mpa. The average and maximum friction ratios of this layer are 7.2 and 9.2%.

7 INFERRENCE OF ROCK

The material underlying the expansive clay layer is mudstone rock, weathered to greater or lesser extents. Whilst rock structure is clearly evident in this material, it is also expansive, but to a reduced extent. Two layers are evident within the first 4m of the profile, and they can be described as extremely and highly weathered rock. They are taken to correspond to two different layers on the CPT results, discerned by a slight reduction in the average friction ratio in the lower layer. The tip resistance in the lower layer (highly weathered rock) is also increased, with both layers being discerned from the expansive clay layer by tip resistances exceeding 3Mpa.

The results of Table 1 show that the depth at which moderately weathered to fresh rock (taken as refusal) is encountered varies from 3.5 to 6.75 with an average depth of 4.2m. The average properties of the inferred Extremely Weathered rock are a tip resistance of 4.7 MPa, and a friction ratio of 4.9%. The average properties of the inferred Highly Weathered rock are a tip resistance of 6.9MPa, and a friction ratio of 4.0%.

Table 1 Comparison of CPT derived statistics and directly measured values for the Maryland site.

criteria	n	CPT inferred values				Directly obtained values		
		mean	std. dev.	max	min	mean	max	min
Thickness of soil; category >3 (m)	49	0.3	0.13	0.6	0.0	0.32	0.45	0.25
Thickness of expansive soil; category 3 (m))	49	1.0	0.25	1.6	0.5	0.95	1.45	0.75
Average Tip Resistance in clay layer (MPa)	49	1.1	0.54	3.0	0.3			
Average Friction Ratio in clay layer (%)	49	7.2	1.61	11	4			
Maximum Friction Ratio in clay layer (%)	49	9.2	2.37	14	4			
Tip Resistance below clay (EW rock; MPa)	49	4.7	1.16	7	3			
Friction Ratio below clay (EW rock; %)	49	4.9	0.76	9	4			
Tip Resistance in HW rock above refusal (MPa)	49	6.9	1.06	12	3			
Friction Ratio in HW rock above refusal (MPa)	49	4.0	1.04	6	2			
Depth to Refusal (m)	49	4.2	0.48	6.75	3.5			

It is interesting to plot the data for the average tip resistance and friction ratio in the expansive clay and weathered rock materials on the soil behaviour chart of Robertson et al. (1986), including the data of Schneider et al. (2001). This is presented in Figure 4. The average CPT values for the expansive clay layers plot well within the soil category 3 or clay region on the diagram. The values from Schneider et al. (2001) for mudstone derived residual soils all plot as silty clays and clayey silts in regions 4, 5 and 6. The Schneider et al. (2001) data, however, was obtained from considerable depth (9-19m) below overlying sediments in a tropical environment. It is considered that the residuum under such conditions is more likely to be a weathered rock material and not a mature, desiccated residual clay, as occurs under near surface conditions under temperate climatic conditions.

Indeed, if the results of this study are compared with the data of Schneider et al. (2001), relatively good agreement is observed. The results from the Maryland CPT study show a well defined trend in the evolution of expansive clay from mudstone rock.

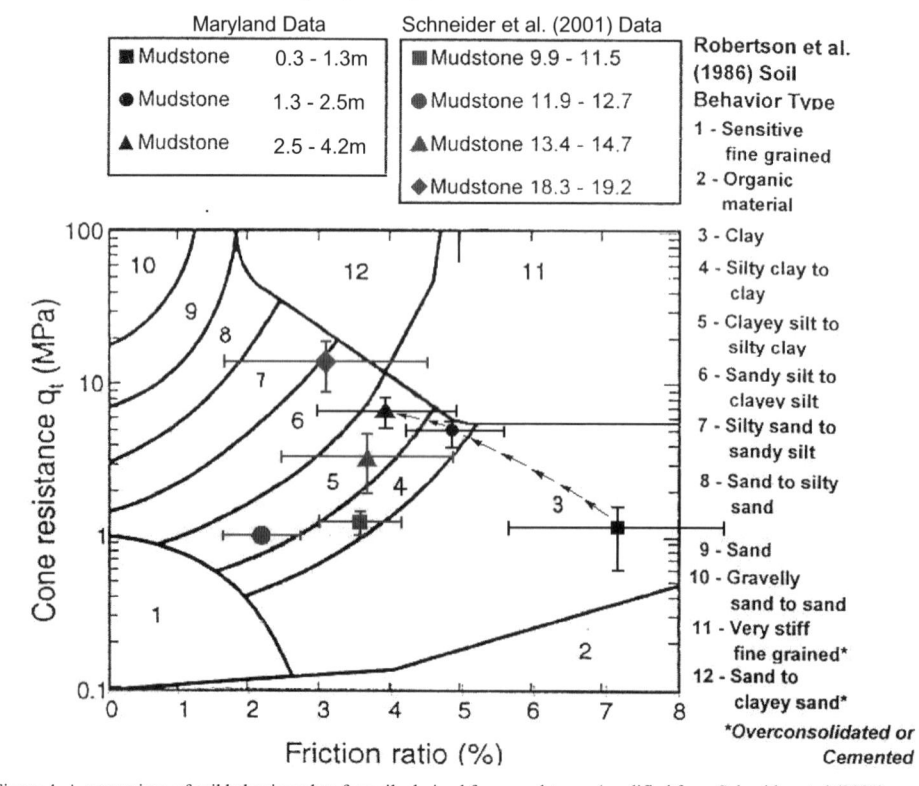

Figure 4 A comparison of soil behaviour data for soils derived from mudstones (modified from Schneider et al (2001)

8 THE ASSESSMENT OF SITE VARIABILITY

The above results show that clay layer depth and thickness details can be reliably inferred from CPT test results. Whilst such data is also readily obtained by shallow drilling, it cannot be obtained as quickly or as efficiently using such methods. If the statistics of Table 1 are considered, it is evident that soil layer characteristics are spatially variable, and it follows that a considerable amount of drilled site investigation would be required to confidently evaluate the magnitude of the variation. Alternatively, such an assessment of site variability can be undertaken using the CPT with considerably improved efficiency.

The Australian standard to guide the design of Residential Slabs and Footings (AS2870, 1996) guides the assessment of foundations for all soil conditions, including those of expansive soils. It requires a minimum of one borehole per housing site, but allows for a reduction in the number of boreholes if soil conditions are deemed to be consistent across an entire housing subdivision. This sometimes means that as few as 10 boreholes may be used as the basis for estimating foundation movements for up to 50 housing sites.

The use of a single estimate of characteristic surface movement on a house site, based on a single borehole log, can be misleading if the soil conditions are laterally variable over a small scale. In such cases, the potential differential foundation movement may be much greater than that caused by differential suction changes. Further, extrapolations of estimated soil profile characteristics over wide areas may lead to the adoption of inappropriate values, even where the localized conditions are consistent.

Fityus et al. (2002) showed that the CPT can be used as an efficient tool in the assessment of soil profile variability across expansive soil sites. They further demonstrated that the use of CPT-derived parameters is not limited to the inference of variability in the soil layer thicknesses, and that an appropriately chosen parameter might be used directly, to infer the variability in the expected ground movement. After considering 10 different, arbitrarily defined, CPT-derived parameters, it was found that only the depth of refusal displayed a similar degree of variation to the ground surface movements measured at the Maryland site. A comparison of distributions of CPT refusal depths and measured ground surface movements is presented in Figure 5.

9 CONCLUSIONS

The CPT test is a convenient and efficient tool for use in the assessment of soil stratigraphy in residual soil sites. In the study presented here, values of top

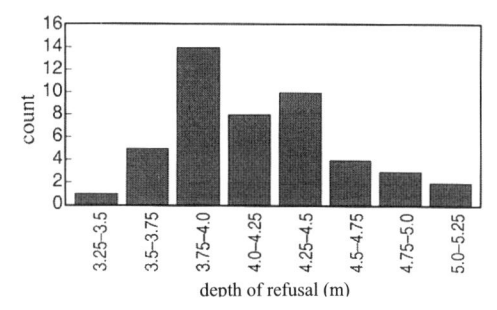

a) variation in depth of refusal for 49 CPT tests at Maryland

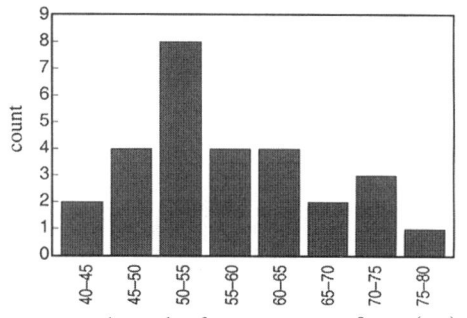

b) variation in 29 ground surface movement measurements at Maryland

Figure 5. A comparison of the variability of ground movements and CPT refusal.

soil thickness and expansive clay thickness were reliably inferred from CPT profiles, to give average, maximum and minimum values that are consistent with directly measured values.

The studied soil profile contains 4 main soil layers that can be identified both in drilled boreholes and CPTs. The characteristic values of tip resistance and friction ratio of these layers are consistent with those that have previously been reported for residual soils derived from mudstones.

The CPT is an ideal too for assessing site variability as it enables stratigraphic characteristics to be determined quickly and repeatedly at a large number of locations across a particular site, for a relatively low cost and small effort.

ACKNOWLEDGEMENTS

Thanks and acknowledgement are given to Mr. Nigel Dobson, who undertook much of the testing to obtain the CPT data used in this work.

Thanks also, to the Mine Subsidence Board of New South Wales, for their financial support of many of the projects that provided data for this study.

REFERENCES

AS2870-1996 *Residential Slabs and Footings.* Standards Australia.

Dhowian, A.W. 1990. Simplified Heave Prediction Model for Expansive Shale. *Geotechnical Testing Journal*, 13: 323–333.

Finke, K.A., Mayne, P.W. and Klopp, R.A. 2001 Piezocone testing in the Atlantic Piedmont residuum. *Journal of Geotechnical and Geoenvironmental Engineering*, 126: 307-316.

Fityus, S.G. 1999 *Transport processes in partially saturated soils.*, PhD. Thesis, The Department of Civil, Surveying and Environmental Engineering, The University of Newcastle, unpublished.

Fityus, S.G, Dobson, N. and Smith, D.W, 2002 'A study of the variability in ground movements for an expansive clay site.' *Proceedings of the 3rd int. conference on unsaturated soils*, Recife, Brazil, 611-617.

Fityus, S.G. and Smith, D.W. 2004 Characteristics of a residual soil profile developed from a mudstone in a temperate climate, Accepted for publication in *Engineering Geology*.

Fityus, S.G., Smith, D.W. and Allman M.A. 2004. An expansive soils research site near Newcastle Australia, accepted for publication in the June edition of the *Journal of Geotechnical and Geoenvironmental Engineering*, ASCE.

Nelson, J.D., and Miller D.J., 1992 *Expansive soils: problems and practice in foundation and pavement engineering.* John Wiley & Sons Inc., New York.

Peuchen, J, Plasman, S.J. and Steveninck, R. 1996, Cone penetration testing in tropical residual soils. *Ground Engineering*, 29: 37-40.

Robertson, P.K., Campanella, R.G. Gillespie, D. and Greig, J. 1986 Use of Piezometer cone data. *Use of In situ tests in Geotechnical Engineering.* GSP6: 1263-1280. Reston, VA. ASCE.

Schneider, J.A., Peuchen, J, Mayne, P.W. and McGillivray, A. V. 2001 *Proceedings of the International Conference on In Situ measurement of Soil Properties and Case Histories*, Bali, 593-598.

Walsh, P.F., and Cameron, D.A., 1997 '*The design of residential slabs and footings.*' Standards Australia, SAA HB28–1997.

Energy ratio measurement of SPT equipment

Dong-Soo Kim, Won-Seok Seo, Eun-Seok Bang
Department of Civil & Environmental Engineering, Korea Advanced Institute of Science and Technology, Daejeon, Korea

Keywords: SPT, N-Value, energy efficiency, types of hammer, F^2 method, FV method

ABSTRACT: The standard penetration test (SPT) used all over the world is a representative in situ method in ground investigation and N-value is employed as almost unique parameters in geotechnical design and analysis. However, the N-value in the SPT is affected by the magnitude of the rod penetration energy transmitted from the falling hammer as well as the geotechnical characteristics of the ground. In this study, the instrumented rod was devised with load cell consisted of 8 strain gauges in full bridge and accelerometer, and the force and velocity of elastic wave in rod were measured at between anvil and the top of drill rods system. Four types of hammer system including donut hammer with the rope pulley system, donut hammer with the hydraulic lift system, the automatic trip hammer, and the safety hammer with the rope pulley system, were used and test results obtained by both computational procedures F^2 and FV methods were compared. Energy ratio measurements were performed continuously with depth and the effects of tensile wave cutoff and the rod length were studied. At a given depth, the wave signals were continuously monitored and the repeatability of the applied energy for various types of equipment was investigated. The effects of various parameters such as blow rate of hammer, operating depth, type of hammer, on the energy ratio were also studied. Finally, the representative energy ratios of various types of SPT equipments frequently used in Korea are suggested.

1 INTRODUCTION

The standard penetration test (SPT) has been applied widely in practice for site investigations and for design and analysis of geotechnical structures in most countries since it has been developed. SPT has many advantages such as; 1) the experimental correlation with many soil parameters was suggested using N-value, 2) test equipment is relatively simple, 3) it is possible to acquire soil samples simultaneously, etc.

The *N*-value in the SPT is affected by the magnitude of the transmitted energy to the rod from the falling hammer as well as the geotechnical characteristics of the ground. As a matter of fact, it may not be guaranteed to reliably apply N-value in the design because the shapes and dimensions of the hammer and other parts of the SPT system are not standardized. This may affect the magnitude of the penetration of the rod/sampler even if the testing site is identical. Several investigators have measured the hammer energy in various SPT systems and found considerable variabilities(Schmertmann and Palacios, 1979; Kovacs and Salomone, 1982; Robertson and Woller, 1991). Therefore, each SPT equipment has its own impact energy delivered to the drill rod and is required to measure, in order to adjust the measured N-value to the value with reference energy ratio as like N_{60}.

In this paper, two methods of determining the energy ratio of SPT equipment, force-velocity(FV) and force-integration(F^2) methods, were discussed. Using the instrumented rod equipped with load cell and accelerometer, four different types of SPT hammer system -donut hammer with the hydraulic lift system, the chain type automatic hammer(CME), and the safety/donut hammer with the rope pulley system- were tested and the measured energy ratios obtained by both computational procedures, F^2 and FV methods, were compared. Parameters as like types of hammer, rod length, hammer falling speed, repeatability of hammer impact that affect to the transmitted energy are investigated. The importance of measuring the SPT energy ratio and adjusting the N value based on energy ratio was also addressed.

2 METHODS OF ENERGY CALCULATION

The standard procedure of SPT is that a 63.5kg weight hammer falling at a height of 0.76m impacts drilled rods connected sampler at the end of rods. On this process, the original potential energy of hammer, $E_{n100\%}$, is changed to E_r, the delivered energy to drill rods in compression wave. The delivered energy, E_r, is significantly influenced by operators and working environments as well as characteristics of equipment, and N-value is linearly decreased as increasing E_r. Therefore, the energy ratio, ER_r, defined as ratio of delivered energy (E_r) over theoretical potential energy ($E_{n100\%}$) is a crucial parameter for obtaining reliable N-value. Eq. 1 expresses the energy ratio of hammer:

$$ER_r = \frac{E_r}{E_{n100\%}} = \frac{E_r}{W \cdot h} \quad (\times 100\%) \quad (1)$$

where, W = weight of hammer and h = hammer fall height.

2.1 Force Squared (F^2) method

The delivered energy is defined as the total work done in the force-displacement space. Since the hammer impact force in a rod is variable with time, it is more convenient to express the work done, or energy, as a function of time, i.e.

$$E_r(t) = W = \int F(t)dx = \int F(t)V(t)dt \quad (2)$$

where, F(t) is the force and V(t)=dx/dt is the particle velocity, all with reference to a point in the rod. Eq. 2 is the fundamental equation describing energy delivered in a rod as a function of time, and can be calculated if the force and velocity time histories are known at a point in the rod. In the past, however, efforts to use accelerometers were not successful, the square of the force integrating was used. For one-dimensional impact wave propagation in a uniform unsupported elastic rod, it can be assumed that :

$$F(t) = V(t) \cdot \frac{EA}{c} \quad (3)$$

in which the quantity EA/c is refered to as the impedance of the rod.

The proportionality relationship in Eq. 3 will hold only from the time of impact to the time of arrival of the wave reflection from the sampler, i.e. for the time duration of the first compression pulse. The F^2 method is defined as follows;(ASTM, 1992)

$$E_n(t) = \frac{cK_1K_2K_c}{EA} \int_0^{\Delta t} [F(t)]^2 dt \quad (4)$$

where A = cross-sectional area of the drill rods; c = velocity of the compression wave in the drill rod, approximately 5120m/s for steel; E = modulus of elasticity of the drill rods; $E_n(t)$ = maximum energy transmitted to the drill rod; F(t) = dynamic force in the drill rod as a function of time; K_1 = a theoretical correction factor to account for closeness of the measuring device to the impact surface; K_2 = a theoretical correction factor to correct for the fact that some first wave compression energy is cut off prematurely; K_c =factor to correct the theoretical velocity, c, to the actual velocity; Δt = time duration of the first compression pulse starting at t=0. The factors, K_1 and K_2, are similar in principle to Schmertmann and Palacios' (1979) correction factors.

2.2 Force Velocity (FV) Method

The force velocity method which measures the force and velocity simultaneously is defined as Eq. 2. The maximum value calculated from Eq. 2 is the maximum transferred energy in the rod. Proponents of the FV method state the advantages of this method such as, 1) The integration is through the complete waveform capturing complete energy in the system, and 2) If proportionality is not present in the first compression interval, the F^2 method assumptions are not correct and integration of FV gives the correct energy of the system(Sy and Campanella, 1991).

3 FIELD WORKS

The hammer impact energies were measured with instrumented rod and calculated using both F^2 and FV methods. The instrumented rod was devised with a load cell consisted of 8 strain gauges in full bridge attached to the regular rod and two accelerometers, and the force and velocity due to elastic impact wave were measured just below the anvil. The measuring system is shown on Fig. 1.

In this study, the measurements of energy ratio were performed with four types of SPT hammer including donut hammer in hydraulic lift system (WD), the chain type automatic trip hammer (CA), the safety

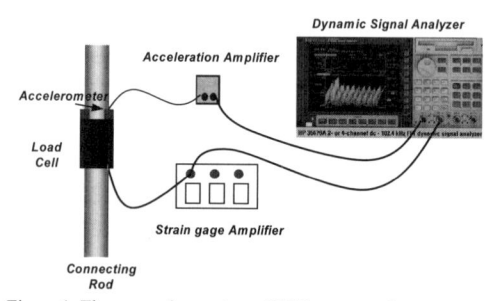

Figure 1. The measuring system of SPT energy ratio

hammer with the rope pulley system(RS), and donut hammer with the rope pulley system(RD). The field works were conducted at Kimje site in Korea and the soil conditions at the site consist of 10~15m of weathered soil. A series of measuring energy ratio of each hammer were performed at borehole as close as possible in order to compare N-values of each equipment as shown in Fig. 2. SPT equipment and operation methods used in this study are summarized in Table 1.

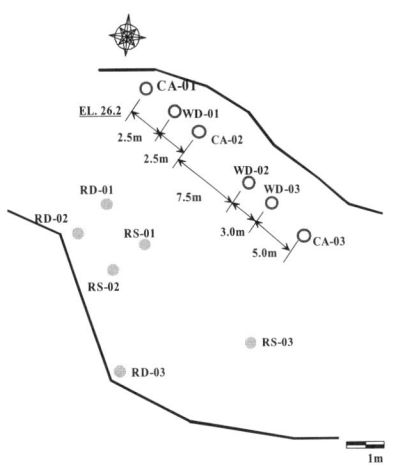

Figure 2. Test place of each type of equipment

Table 1 SPT Equipments and Types of Hammer

Hammer Type	Boring Method	Hammer Release	Blow Rate
WD (Donut)	Wash boring	Hydraulic lift Steel wire	20~30/min
CA (CME,Auto)	Continuous - flight auguring	Automatic Chain	25/min, 60/min
RS (Safety)	Wash boring	Rope and pulley	30~40/min
RD (Donut)	Wash boring	Rope and pulley	30~40/min

4 TEST RESULTS

4.1 Wave Shapes of Each Hammer

The initial compression wave in the rods by hammer impact reflects at the sampler end of SPT rod system of finite length and returns as a tension wave. When this tension wave reaches the hammer-rod contact point its tension stress magnitude exceeds the existing contact compressive stress between the hammer and rods. This process, so called tension cutoff, is important to analysis wave energy induced to the rod system.

Stress waveforms generated by three types of hammer (WD, CA, and RS) with same rod length are shown in Figure 3. It can be noticed that the tension cutoff times are almost identical irrespective of hammer type if the rod lengths are the same. The maximum compression stress, however, depends on hammer type and it could be expected that energy transmitted to rod system is different. In this study, maximum compression stress of CA hammer is the largest while that of WD hammer is smallest. The maximum compression stress induced in hammer-rod system could be estimated using Eq. 5 as

$$\sigma_{max} = \rho c (\frac{V_{hi}}{1+r}) \quad (5)$$

in which ρ = mass density of material; c = compression wave velocity in rods; V_{hi} = velocity of hammer at impact; and r = ratio of the cross-sectional area of the rods to the area of the hammer.

The maximum stress magnitude depends on hammer type and hammer releasing method or their combination because it is affected by falling velocity of hammer and anvil size. SPT equipments used in this study have various energy ratios because of the differences in hammers, releasing method, and friction during impact.

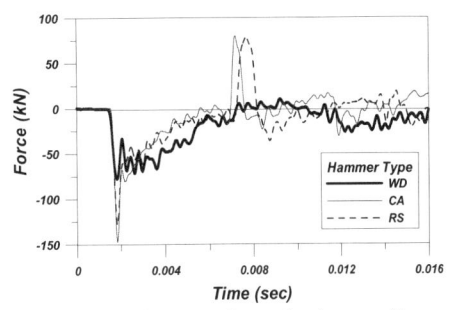

Figure 3. Comparing force waveforms of each types of hammer and equipment

Many countries have standardized only weight and falling height of hammer except hammer types, falling releasing method and any other factors affecting energy transmitted to rod. These problems produce variable N values used in design, so it is important to determine energy ratio for each types of hammer and characteristic equipments.

4.2 Effects of Rod Length and Blow Speed

The facts that energy induced in rod by hammer impacting is affected by rod length, are well known (Skempton, 1986). The effect of rod length was studied by measuring waveforms as varying rod length. Stress waveforms with various rod lengths for chain type automatic (CA) hammer are shown in Fig. 4 and 5. CA hammer has advantage in that it can reduce the effect of operator relative to other types of hammer.

Fig. 4 shows the force waveforms of short rod lengths of 3.3m, 4.7m, 6.2m. Although the maximum compression forces are almost identical, tension cutoff time is delayed as length of rod is increased. The first compression area (A) is decreased as rod length is getting short. On the other hand, it can be noticed that the second compression area (B1, B2, B3) increases as rod length decreases. Therefore, if the energy transmitted to rod is calculated using first compression area, it could be underestimated

For the cases of relatively longer rod lengths; 10.7m, 12.2m, 13.7m, the second compression areas (B1, B2, B3) are reduced considerably relative to first compression as shown in Fig. 5. In particular, the second compression area for the red length of 13.7m (B3) is almost vanished because tension cut-off reaches after the first compression wave almost fully generated in the rod length of 13.7m. Therefore, the suggestion of energy calibration is reasonable in rod length under 10m (Skempton, 1986). It is more reliable to determine energy ratio using 10m of rod length or more if possible and calibration is needed for the rod length shorter than 10m.

Hammer blow rate is also affecting parameter in calculating energy induced to rod. CA hammer can control the with blow speed, and it was operated with slow blow speed (25blows per minute) and fast blow speed (60 blows per minute) in order to compare the energy ratios. The variations in energy ratios with blow speed and rod length are shown in Fig. 6. Energy ratios of fast blow cases were considerably larger than those of slow blow cases and the difference was up to about 15%. Also, as expected, energy ratio is relatively small at short rod length regardless of blow speed.

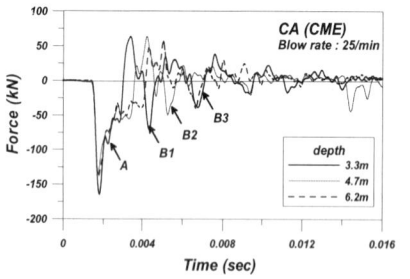

Figure 4. Comparing force waveforms of short rod lengths: L=3.3m, L=4.7m, L=6.2m

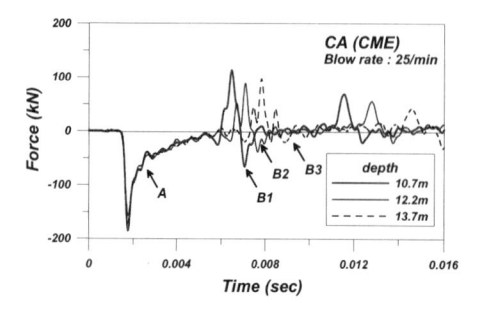

Figure 5. Comparing force waveforms of long rod lengths: L=10.7m, L=12.2m, L=13.7m

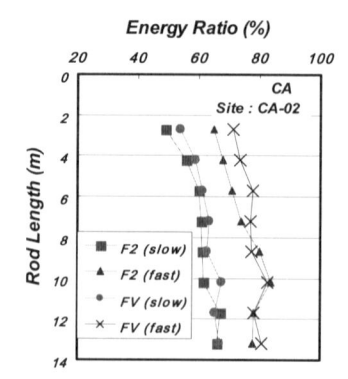

Figure 6. Energy ratios of CA with rod length and blow speed

4.3 Repeatability of Hammers

In order to assess the repeatability of SPT equipment, the force signals of 10 consecutive blows at the same depth were monitored for three different SPT equipments, and the results are shown in Fig. 7. The randomness of the force measurement for donut hammer(WD and RD) was greater than those for chain automatic (CA) or safety (RS) hammers, showing the poor repeatability of impact in donut hammer. However, the consecutive force waveforms of CA or RS hammer match almost exactly.

4.4 Energy Ratios of Each Equipment

The magnitudes of forces delivered to the rods were obtained by four types of hammers from the signals of both load cell and accelerometer. The typical signals obtained by CA are presented in Fig. 8. The sign conventions used are positive force for tension wave and negative force for compression wave. The two force traces, from load cell and accelerometer, are approximately proportional within the first compression pulse except for some local separations of the two wave traces caused by wave reflections in the system.

The proportionality is considerably broken from the tension cutoff point in which the polarity of velocity is reverse to that of force. It was shown that the area of force diagram obtained from the signal of accelerometer was larger than that from load cell, and therefore, the energy from FV method was larger than that from F^2 method. The energy ratio of the equipment (ER_r) was calculated by diving the E_r by theoretical falling energy ($E_{n100\%}$), and the energy ratios obtained by F^2 and FV method were also shown in Fig. 8.

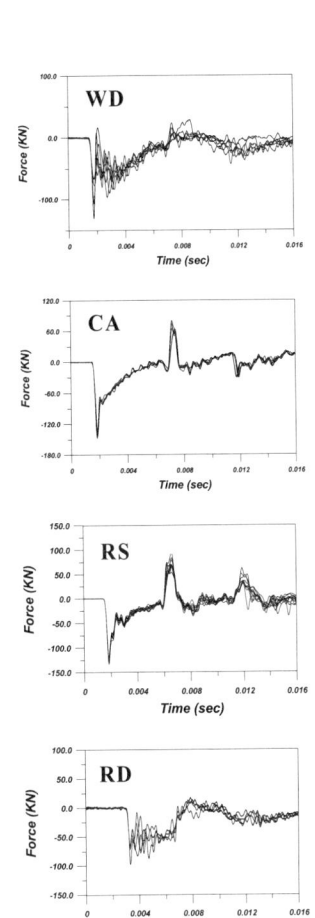

Figure 7. The continuous records from each hammer

Table 2 shows the energy ratios obtained by two different methods (F^2 and FV) of four types of SPT equipments. It was revealed that the energy ratios measured by different equipment are varied depending on types of SPT hammers and therefore the measured N values would be varied, even though test site has almost identical soil condition. The standard deviation measured by donut hammer was the largest.

The variations in N values with depth are shown in Fig. 9. At the almost same test site with depth, 2.5m distance between two boreholes, both WD and CA hammer provide the N value. Test results without correction and with correction to the reference energy ratio(E_r=60%) are shown in the figure. As shown in Fig. 9, the N values of donut hammer and auto trip hammer in each depth were relatively different before correction, however the N values between two types of hammer were well accordant after correction by N_{60}.

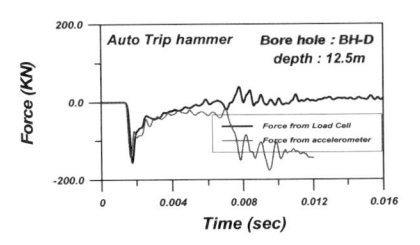

(a) Force signals from load cell and accelerometer (CA)

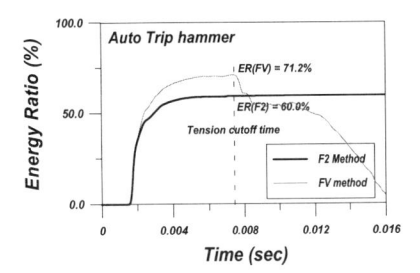

(b) Energy ratios from F^2 and FV methods (CA)

Figure 8. Stress waves measured in instrumented rod and energy ratios

Table 2. Energy ratios of each equipment

Hammer Type		F2 Method		FV Method	
		Mean (%)	Standard Deviation	Mean (%)	Standard Deviation
WD	<10m	44.9	9.1	57.9	7.4
	>10m	58.9	7.3	60.2	7.8
CA	<10m	57.4	3.1	60.0	4.5
	>10m	64.9	2.6	66.4	3.4
RS	<10m	54.7	3.4	58.2	3.9
	>10m	58.0	3.1	59.0	4.1
RD	<10m	36.9	9.5	38.0	10.8
	>10m	38.6	6.7	41.4	11.7

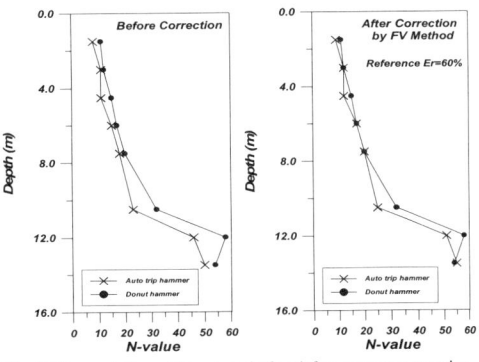

Fig. 9 The variation of N values before/after energy correction

5 CONCLUSIONS

Measurements of SPT hammer energy delivered to the rod were performed using 4 types of hammers : donut hammer in hydraulic lift system, chain automatic hammer, safety hammer with the rope pulley system and safety hammer with the rope pulley system. The energy efficiencies are quiet different depending on the types of hammer, rod length, blow rate, and energy calculating methods. Therefore, it is somewhat difficult to apply fixed energy ratio on each type of hammer. The energy ratios of automatic and safety hammer have small standard deviations relative to those of donut hammer, showing good repeatability. Energy measurements of other types of SPT equipment is underway for the reliable engineering application of N-value.

ACKNOWLEDGEMENTS

This study was sponsored by the SISTEC, which are gratefully acknowledged

REFERENCES

American Society for Testing and Materials 1992. Standard test method for stress wave energy measurement for dynamic penetrometer testing systems. *ASTM Designation: D 4633-86.* Annual Book of ASTM Standards, Section 4, 04.08: 943-946.

Kovacs, W.D. and Salomone, L.A.(1982), "SPT hammer energy measurement", *Journal of the Geotechnical Engineering Division, ASCE, Vol.108, No.GT4,* pp.599-620.

Robertson, P.K. and Woller, D.J.(1991), "SPT energy measurements using a PC based system", *44th Canadian Geo. Conference, Vol.1, Paper No.8,* pp.1-10.

Schmertmann, J.H. and Palacios, A.(1979), "Energy Dynamics of SPT", *Journal of the Geotchnical Engineering Division, Vol. 105, No. 8,* pp. 909-926.

Skempton, A.W.(1986), "Standard penetration test procedures and the effects in sands overburden pressure, relative density, particle size, ageing and overconsolidation", *Geotechnique, Vol.36, No.3,* pp.425-447.

Sy, A. and Campanella R.G.(1991), "An Alternative Method of Measurement SPT Energy", *Proceeding of the 2nd International Conference on Recent Advances in Geotechnical Engineering and Soil Dynamics,* pp. 499-505.

CPT-DMT interrelationships in Piedmont residuum

Paul W. Mayne & Tianfei Liao
School of Civil & Environmental Engineering, Georgia Institute of Technology, Atlanta, GA USA

Keywords: cone, constrained modulus, dilatometer, elastic modulus, in-situ tests, piezocone, residual soils

ABSTRACT: Interrelationships are investigated between the piezocone penetration test (CPTu) and flat plate dilatometer test (DMT) in Piedmont residual soils that are comprised of silty fine sands to fine sandy silts. Data from test sites in Alabama, Georgia, and North Carolina are used for these purposes. The flat plate dilatometer test is well-recognized for its ability to calculate settlements of shallow foundations. Within the Piedmont geology, an equivalent CPT method has been developed for obtaining constrained moduli. An interrelationship between CPTu midface porewater pressures and initial DMT contact pressures is also explored.

1 INTRODUCTION

Over two decades of calibration between the DMT and measured foundation performance records have shown its value & reliability in settlement computations (e.g., Schmertmann, 1986; Mayne & Frost, 1988; Marchetti, et al. 2001). The measured dilatometer modulus (E_D) is converted to a constrained modulus (M') per the procedures established by Marchetti (1980). For each sublayer, the uniaxial strain can be calculated as $\varepsilon = \Delta\sigma_v/M'$ and the resulting settlement in that sublayer is simply $\rho = \varepsilon \Delta z$. The change in stress at each sublayer can be obtained from classical elastic theory solutions (e.g., Poulos & Davis, 1974; Mayne & Poulos, 1999). Settlements from all sublayers are summed to find the total foundation settlement:

$$\rho_{Total} = \Sigma \frac{\Delta\sigma_v}{M'} \cdot \Delta z \qquad (1)$$

An earlier & well-received method for foundation settlement calculations (Schmertmann, 1970) related the settlement modulus directly to the measured cone tip resistance (q_c), particularly in fine sandy soils from northern Florida (e.g., M' = 2 q_c). This also utilized elastic theory solutions, but combined the influence of the modulus and stress distribution to form a simplified strain influence diagram, known as the 0.6-2B triangle. At that time (circa 1970), it was necessary to approximate the distributions because engineers relied on slide rules for their calculations. With electronic calculators and computers, there is no longer need for dependence upon approximate distributions (Mayne & Poulos, 1999).

Another clarification that needs addressing is the use of a one-dimensional modulus (M'), that corresponds to consolidation or oedometer testing, versus the three-dimensional problem associated with footings & mats that require an elastic modulus (E'). In fact, the two moduli are related via elastic theory (e.g., Poulos & Davis, 1974):

$$E' = \frac{(1+v')(1-2v')}{(1-v')} \cdot M' \qquad (2)$$

where v' is the drained Poisson's ratio. For v' = 0, in fact, the moduli are equal: E' = M', and for the normal drained case corresponding to primary consolidation where v' = 0.2 (Jamiolkowski, et al. 1994; LoPresti, et al. 1995), the elastic modulus E' is 90 percent of M', so in practical circles they are used somewhat interchangeably.

It is of interest to revisit both approaches and develop a methodology by which the advantages of the CPT can be appreciated for foundation settlement evaluations. In this case, the Piedmont geology comprised of silty residuum will be addressed.

2 PIEDMONT GEOLOGY

The Piedmont geology is comprised of "nontextbook geomaterials" including residual silty fine sands, clayey silts, and fine sandy silts derived from the weathering of old gneiss and mica schist of Precambrian Z-age and granites of Paleozoic age. Details on the formation and characteristics have been discussed elsewhere (e.g., Sowers & Richardson, 1983; Sowers, 1994; Martin, 1977; Mayne, 1999).

At several well-documented sites, data from both DMT and CPT have been collected to interrelate the measurements. For example, extensive testing has been reported for the National Geotechnical Experimentation Site (NGES) in Piedmont soils near Opelika, Alabama (Vinson & Brown, 1997; Brown & Vinson, 1998; Schneider, et al. 1999; Mayne, et al. 2000; Finke, et al., 2001). Similar data sets in the Piedmont have been reported for a test site on the Georgia Tech campus (Mayne & Harris, 1993; Harris & Mayne, 1994). Published and unpublished data also exist in Piedmont soils of North Carolina (e.g., Wang & Borden, 1996; Mayne et al. 2002).

A summary of CPT results derived from 22 piezocone soundings advanced at the Opelika NGES is presented in Figure 1. Here the mean profiles of cone tip stress (q_t) and sleeve friction (f_s) are plotted with depth and upper and lower limits given by plus and minus one standard deviation. The tip stress increases from 2.5 to 4.0 MPa in the upper 10 meters, while sleeve resistance increases less dramatically from 120 to 170 kPa. Porewater pressures are unusual in that they are positive on the cone face and negative at the shoulder (Finke, et al. 2001). Figure 2 shows the mean profiles of face penetration porewater pressures (u_1) from four soundings and shoulder readings (u_2) from the 22 standard type-2 piezocones. Illustrating a complete set, the measured lift-off pressures (p_0) and expansion pressures (p_1) from DMTs at the Opelika site are shown in Figure 3.

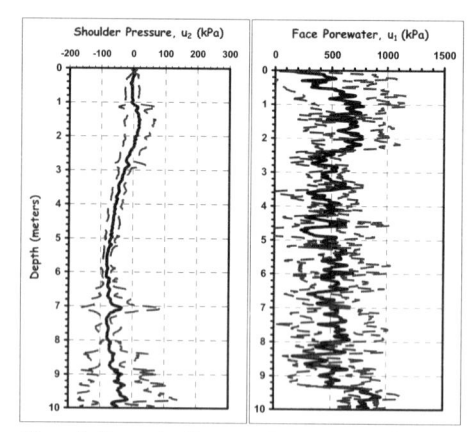

Figure 2. Penetration Porewater Pressures at Opelika.

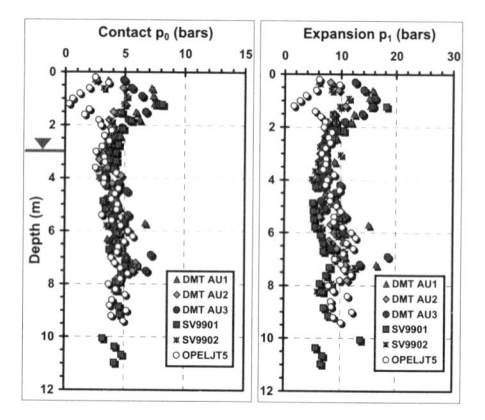

Figure 3. Flat Dilatometer Results at Opelika NGES.

3 CPT-DMT CORRELATIONS

Three sets of CPT-DMT data have been compiled from varied locations in the Piedmont residuum (Georgia, North Carolina, Alabama), for development of intracorrelative trends between the two in-situ test methods. The field tests were conducted in general accordance with ASTM D 5778 and D 6635 guidelines for the CPT and DMT, respectively.

From the cross-comparative analyses, Figure 4 shows that elastic modulus correlates well with the cone tip stress, as suggested by Schmertmann (1970). Later studies by Kulhawy & Mayne (1990) showed that the factor $\alpha_c = E'/q_t$ could range from 2 to 8 for NC sands and 7 to 30+ for OC sands. In this case, $\alpha_c = E'/q_t$ is about 5 for the Piedmont silts.

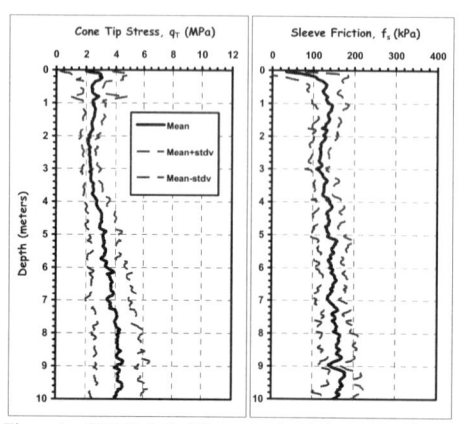

Figure 1. CPT Statistical Summary in Piedmont Residuum at Opelika Test Site, Alabama.

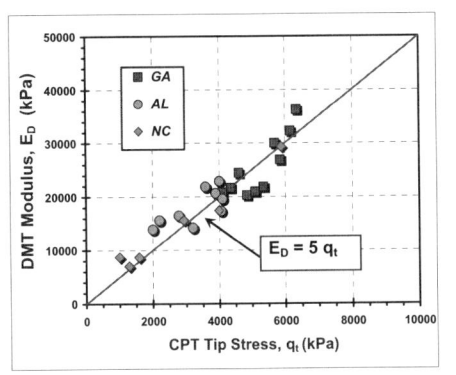

Figure 4. Relationship between the DMT Elastic Modulus and CPT Tip Stress in Piedmont Soils.

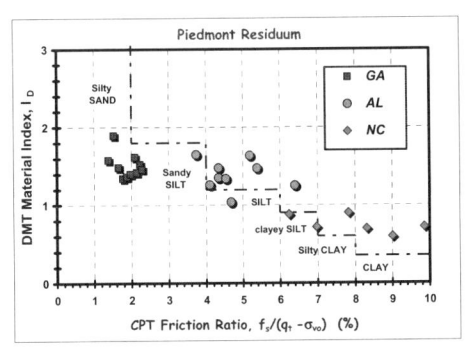

Figure 5. Trend between DMT material index and CPT friction ratio in Piedmont residuum. Soil behavior type from DMT.

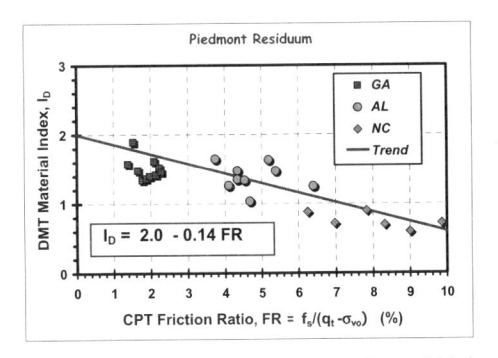

Figure 6. Relationship adopted between DMT material index and CPT friction ratio in Piedmont residual soils.

In the Marchetti procedure (1980) for obtaining a constrained modulus, the dilatometer material index (I_D) and horizontal stress index (K_D) are also required in order to attain the modulus ratio ($R_M = M'/E_D$). The DMT material index relates to the grain size of the soil, as does the CPT friction ratio (e.g., Robertson & Campanella, 1983). Thus, we can expect a trend between the DMT I_D and cone parameter, FR = $f_s/(q_t-\sigma_{vo})$. Figure 5 shows that an approximate trend is evident between these two parameters using the data sets from the Piedmont. The adopted relationship for obtaining an equivalent I_D from FR is presented in Figure 6.

As the DMT obtains two independent measurements (p_0 and p_1), the third index is readily obtained from the first two indices. The horizontal stress index is found from:

$$K_D = \frac{p_o - u_o}{\sigma_{vo}'} = \frac{E_D}{34.7 \cdot I_D \cdot \sigma_{vo}'} \quad (3)$$

Thus, in summary, the CPT data can be converted to equivalent DMT indices via the following expressions for Piedmont residual sandy silts:

Dilatometer Modulus: $E_D = 5\,q_t$ \quad (4)

Material Index: $I_D = 2.0 - 0.14\,(FR)$ \quad (5)

Standard DMT data reduction procedures are then employed to obtain the constrained modulus (D' = M' = $1/m_v$):

$$M' = R_M\,E_D \quad (6)$$

where R_M = fctn (I_D, K_D), as detailed in the following table (Marchetti, 1980; Schmertmann, 1986; Mayne & Martin, 1998; Marchetti, et al. 2001).

Table D-1. Constrained Modulus Parameter (R_M) for Settlement Calculations (Marchetti, et al. 2001)

Conditions	Relationship for R_M = M'/E_D	Notes
If $I_D < 0.6$	$R_M = 0.14 + 2.36 \log K_D$	Clay soils
If $I_D > 3$	$R_M = 0.50 + 2.0 \log K_D$	Clean (quartz) Sands
If $0.6 < I_D < 3$	$R_M = R_{M0} + (2.5-R_{M0}) \log K_D$ and $R_{M0} = 0.14 + 0.15(I_D-0.6)$	Silts to silty Sands
If $K_D > 10$	$R_M = 0.32 + 2.18 \log K_D$	High Values
If $R_M < 0.85$	Set $R_M = 0.85$	Minimum

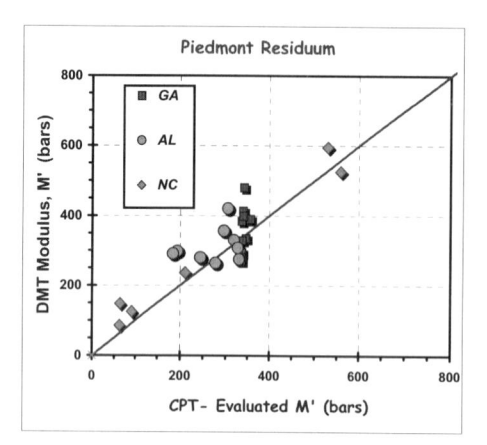

Figure 7. Validation Check on CPT-DMT M' Conversion.

As a cross-check on the procedure, the forward evaluation of constrained modulus M' determined from CPT data in the Piedmont using the new approach is seen to compare well with the DMT-interpreted values of M' in Figure 7, thus validating the proposed approach. Prior calibration efforts in the Piedmont using DMT results to evaluate foundation displacements have been reported elsewhere for footings, mats, and piles (Mayne & Frost, 1988).

4 OPELIKA TEST SITE

The CPTu data at Opelika represent a statistical compilation of soundings using a variety of penetrometers, cone rigs, and personnel. Readings have been acquired using equipment manufactured by Hogentogler (type 1, type 2, dual), Fugro BV (type 1 and 2), Vertek (type 2), and A.P. van den Berg (type 2). These data were used to develop the mean profiles of tip resistance, sleeve friction, and porewater pressures shown in Figures 1 and 2. More recent studies here have utilized Geotech AB and Envi memocone systems, as well as resistivity, dielectric, and seismic penetrometers. Most soundings have been advanced to depths of 15 to 20 m, although a few have gone as deep as 32 meters.

The DMT data at Opelika were collected using a standard flat blade system obtained from GPE of Gainesville, FL, yet the results were obtained in many separate trips to the site by varied personnel over a six-year period (Martin & Mayne, 1998; Schneider, et al. 1999; Mayne & Brown, 2003).

Site conditions are not uniform across the Opelika NGES, however, as rock outcrops are located at the other end of the 150-hectare test region on the opposite end of the pavement racetrack. As a result, foundation load testing has been conducted on a variety of piles and shafts in both soil and in weathered rock (e.g., Brown, 2002). Nevertheless, the Opelika NGES is an established testing ground to explore the interesting aspects of the residual soils of the Piedmont physiographic province. These silty soils have characteristics that give the appearance of "loose sand" in some instances, or behavior as "stiff clay" in other situations. Thus, site characterization on their behavior in construction and geotechnical concerns has been challenging. In certain cases, these geomaterials apparently act in both undrained and drained response (Mayne, et al. 2000).

5 POREWATER PRESSURES

Penetration porewater pressures at the midface and shoulder positions are quite different in the Piedmont soils. At the face position, positive values are recorded around 500 to 600 kPa with major spiked values of highs and lows (see Fig. 2). In contrast, at the shoulder position, variable readings are encountered in the vadose zone above the groundwater table, yet once the phreatic surface is encountered, the readings are primarily negative around -95 kPa. Returning to Fig. 2, the values go negative at depths of 2.5 to 3.0 m which is the depth of groundwater at Opelika at the times of testing. Similar types of paired positive-negative readings are encountered in stiff fissured clays (e.g., Lunne, et al. 1997). Though the marked differences, both u_1 and u_2 readings in the Piedmont dissipate quickly in about 1 to 2 minutes and reach the same hydrostatic value u_0 (Finke, et al. 2001).

For soft to stiff clays, the CPTu penetration porewater pressures interrelate with initial DMT contact pressure (Figure 8). An expression derived from cavity expansion theory (Mayne & Bachus, 1989)

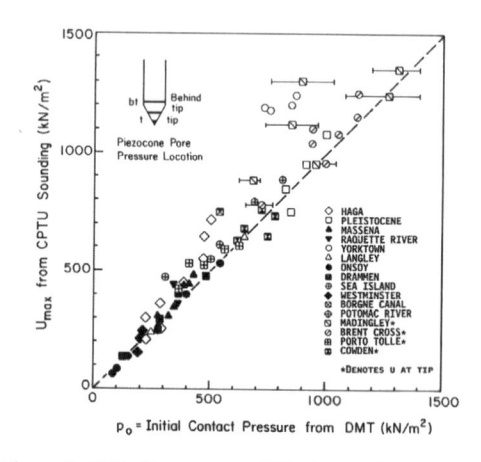

Figure 8. CPTu Porewater vs. DMT Contact Pressures in Clays. (after Mayne & Bachus, 1989)

was used to interrelate the readings as a first approximation:

$$u_{max} \approx p_0 \qquad (7)$$

For fissured overconsolidated clays, the above holds true provided that porewater pressures are measured at the cone tip or midface.

As the Piedmont sometimes acts as a "stiff fissured clay" during relatively fast rates of loading, it is of interest to investigate the above correlation. Mean values of midface porewater pressures from Opelika (Fig. 2) are compared with contact or lift-off pressures (Fig. 3) in Figure 9. Both readings can be seen to be comparable magnitude, thus confirming the interrelationship between u_1 and p_0.

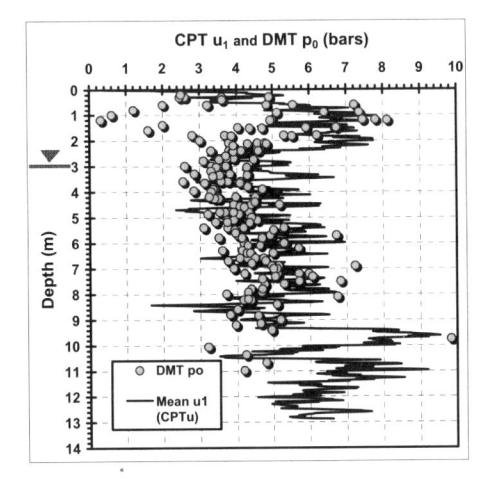

Figure 9. Comparison of CPTu Midface Readings and DMT Contact Pressures in Piedmont Residuum at Opelika NGES.

6 CONCLUSIONS

Site-specific correlations can be developed in the Piedmont residual silts that interrelate the cone penetration test parameters and flat plate dilatometer readings. In particular, the cone tip stress relates to the elastic modulus and friction ratio correlates with material index. This permits the use of CPT data for estimating constrained modulus (M') for calculating foundation settlements in these soils and offers a parallel approach to the DMT procedure well validated in practice. An additional interrelationship based on cavity expansion theory couples the midface porewater pressures from CPTu with the measured DMT A-contact readings.

ACKNOWLEDGMENTS

The authors appreciate the financial support of the National Science Foundation (CMS-0338445) and the Mid-America Earthquake Center (EEC-9701785). Any opinions, findings and conclusions or recommendations expressed herein are provided by the authors and do not necessarily reflect those of NSF or MAE.

REFERENCES

ASTM (2002). Standard Test Method for Electronic Friction Cone and Piezocone Penetration Testing of Soil, ASTM D 5778, American Society for Testing & Materials, West Conshohocken, PA.

ASTM (2002). Standard for Flat Dilatometer Test, ASTM D 6635, American Society for Testing & Materials, West Conshohocken, PA.

Brown, D.A. & Vinson, J.(1998). Comparison of strength and stiffness parameters for a Piedmont residual soil, Geotechnical Site Characterization (2), Balkema, 1229-1234.

Brown, D.A. (2002). Effect of construction on axial capacity of drilled foundations in Piedmont soils. Journal of Geotechnical & Geoenvironmental Engrg. 128 (12), 967-973.

Finke, K.A., Mayne, P.W. & Klopp, R.A. (2001). Piezocone penetration testing in Atlantic Piedmont residuum. Journal of Geotechnical & Geoenvironmental Engrg 127 (1), 48-54.

Harris, D.E., & Mayne, P.W. (1994). Axial compression behavior of two drilled shafts in Piedmont residual soils, Proceedings, International Conference on Design and Construction of Deep Foundations, Vol. 2, Federal Highway Administration, Washington, D.C., 352-367.

Jamiolkowski, M., Lancellotta, R., LoPresti, D.C.F., and Pallara, O. (1994). Stiffness of Toyoura sand at small and intermediate strains. Proc. 13th Intl. Conf. on Soil Mechanics & Foundation Engrg., Vol. 3, New Delhi, 169-173.

Kulhawy, F.H. and Mayne, P.W. (1990). Manual on Estimating Soil Properties for Foundation Design. Report EL-6800, Electric Power Research Institute, Palo Alto, 306 pages.

LoPresti, D.C.F., Pallara, O., and Puci, I. (1995). A modified commercial triaxial testing system for small strain measurements. ASTM Geotechnical Testing J. 18 (1), 15-31.

Lunne, T., Robertson, P.K., and Powell, J.J.M. (1997). Cone Pentration Testing in Geotechnical Practice, Blackie Academic/EF Spon/Rutledge Publishing Company.

Marchetti, S. (1980). In-situ tests by flat dilatometer. Journal of Geotechnical Engineering 106 (GT3), 299-321.

Marchetti, S., Monaco, P., Totani, G. & Calabrese, M. (2001). The flat dilatometer test (DMT) in soil investigations. Proceedings, International Conference on In-Situ Measurement of Soil Properties & Case Histories, Bali, 95-131.

Martin, R.E. (1977). Estimating foundation settlements in residual soils, Journal of the Geotechnical Engineering Division (A4 SCE) 103 (GT3), 197-212.

Martin, G.K. & Mayne, P.W. (1998). Seismic flat dilatometer tests in Piedmont residual soils. Geotechnical Site Characterization (2), Balkema, Rotterdam, 837-843.

Mayne, P.W. & Frost, D.D. (1988). Dilatometer experience in Washington, D.C. Transportation Research Record 1169, National Academy Press, Washington, D.C., 16-23.

Mayne, P.W. & Bachus, R.C. (1989). Penetration pore pressures in clays by CPTu, DMT, and SBP. Proceedings, ICSMFE, Vol. 1, Rio de Janeiro, 291-294.

Mayne, P.W. & Harris, D.E. (1993). Axial load-displacement behavior of drilled shaft foundations in Piedmont residuum.

Report E-20-X19 to Federal Highway Administration by Georgia Tech Research Corp, 162 p

Mayne, P.W. and Martin, G.K. (1998). Commentary on Marchetti flat dilatometer correlations in soils. ASTM Geotechnical Testing Journal 21 (3), 222-239.

Mayne, P.W. (1999). Site characterization aspects of Piedmont residual soils in eastern US. Proceedings, 14th International Conference on Soil Mechanics & Foundation Engineering, Vol. 4, Balkema, Rotterdam, 2191-2195.

Mayne, P.W. and Poulos, H.G. (1999). Approximate displacement influence factors for elastic shallow foundations. Journal of Geotechnical & GeoEnvironmental Engineering 125 (6), 453-460.

Mayne, P.W., Brown, D.A., Vinson, J., Schneider, J.A. and Finke, K.A. (2000). Site characterization of Piedmont residual soils at Opelika, Alabama. National Geotechnical Experimentation Sites, GSP No. 93, ASCE, Reston, Virginia, 160-185.

Mayne, P.W., Zavala, G., Liao, T. and McGillivray, T. (2002). Report of piezocone, seismic cone, and flat dilatometer tests, Winston-Salem, North Carolina, GTRC Project E20-850 to Trigon Engineering Consultants, Greensboro, NC.

Mayne, P.W. and Brown, D.A. (2003). Site characterization of Piedmont residuum of North America. Characterisation & Engineering Properties of Natural Soils, Vol. 2, Swets & Zeitlinger, Lisse, 1323-1340.

Poulos, H.G. and Davis, E.H. (1974). Elastic Solutions for Soil & Rock Mechanics, Wiley & Sons, New York; now available from University of Sydney Press.

Robertson, P.K. and Campanella, R.G. (1983). Interpretation of cone penetration tests: sand. Canadian Geotechnical Journal 20 (4), 718-733.

Schmertmann, J.H. (1970). Static cone to compute static settlement over sand. ASCE Journal of Soil Mechanics & Foundations Division 96 (3), 1011-1043.

Schmertmann, J.H. (1986). Dilatometer to compute foundation settlement. Use of In-Situ Tests in Geotechnical Engineering, GSP 6, ASCE, Reston/VA, 303-321.

Schneider, J.A., Hoyos, L., Mayne, P.W., Macari, E.J., & Rix, G.J. (1999). Field and lab measurements of dynamic shear modulus of Piedmont residual soils. Behavioral Characteristics of Residual Soils (GSP 92), ASCE, Reston, 12-25.

Sowers, G.F. (1994). Residual soil settlement related to the weathering profile. Vertical and Horizontal Deformations of Foundations & Embankments, GSP No. 40, Vol. 2, ASCE, Reston, Virginia, 1689-1702.

Sowers, G.F. & Richardson, T.L. (1983). Residual soils of the Piedmont and Blue Ridge, Transportation Research Record No. 919, National Academy Press, Wash, D.C., 10-16.

Vinson, J.L & Brown, D.A. (1997). Site characterization of the Spring Villa geotechnical test site and a comparison of strength and stiffness parameters for a Piedmont residual soil, Report No. IR-97-04, Highway Research Center, Harbert Engineering Center, Auburn University, AL, 385 p.

Wang, C.E. & Borden, R.H. (1996). Deformation characteristics of Piedmont residual soils. Journal of Geotechnical Engineering 122 (10), 822-830.

Energy measurements for Standard Penetration Tests and the effects of the length of rods

E. Odebrecht
Dept. of Civil Engineering, State University of Santa Catarina, Brazil
Formerly: Dept. of Civil Engineering, Federal University of Rio Grande do Sul, Brazil

F. Schnaid & M.M. Rocha
Dept. of Civil Engineering, Federal University of Rio Grande do Sul, Brazil

G.P. Bernardes
Dept. of Civil Engineering, UNESP, Brazil

Keywords: SPT, *in-situ* tests; energy measurements

ABSTRACT: The need to measure the energy transmitted to the rod composition to correct and uniform the number of SPT blows is now fully recognized by the engineering community. The present paper extends the existing theoretical approach based on the energy content of the first compression wave and introduces an analysis for large displacements that aims at determining the influence of the length of rod in the measured N_{spt} values. The study is based on calibration chamber tests with energy measurements in three different positions: immediately below the anvil, at the mid-height of the composition and immediately above the sampler. Both experimental and analytical results demonstrate that the energy efficiency is inversely proportional to the length of rod. However, it also demonstrates that for long rod composition penetrating a soil with low resistance, the contribution of both rod weight and the permanent sampler penetration should be considered when computing the total energy delivered to the soil-sampler system.

1. INTRODUCTION

The SPT has been used worldwide for preliminary exploration and complementary for prediction of soil properties, foundation performance and liquefaction potential by means of empirical correlations. These correlations relay on the blow count number N which in recent years has been subjected to various "corrections" to account for the lack of standardization in test procedures, effects of overburden pressure and influence of rod length. Scatter produced by the highly variable and unknown values of energy delivered to the SPT rod system can now be properly accounted for by standardizing the measured blow count (N-value) to a reference value of 60% of the potential energy of the SPT hammer, as suggested by Seed et al. (1985) and Skempton (1986).

However, even a N value normalized to a given reference energy is not a guarantee of an efficient engineering design. There are still some important aspects of interpretation that are open to debate:
a) The energy delivered to the rod stem has been consistently monitored by measurements of force and acceleration and is used to quantify by the so called "rod energy ratio". However, very few attempts have been made to produce measurements at depth, just above the sampler, to determine the actual loss in energy during wave propagation in the rod stem.

b) Wave equation studies (Schmertmann & Palacios, 1979) suggest that the theoretical energy reaching the sampler decreases with decreasing rod length. This is a controversial statement supported by some researchers (e.g. Skempton, 1986) and contested by others (e.g. Aoki & Cintra, 2000).

c) The energy delivered to the SPT can be very low for an SPT performed above a certain depth since the energy reaching the sampler is small due to energy losses and to the large mass of the drill rods. These factors are not taken into account in current interpretation methods and lead to conservative prediction of constitutive parameters derived from empirical formulae.

The present work provides new insights for the analysis of SPT testing data from which all the above aspects can be rationally interpreted by a combination of calibration chamber test results and a numerical finite difference code. Some experimental results are presented and analysed in the paper in an attempt to clarify aspects of the mechanism of wave propagation along the rods to the sampler and soil. In particular, it has been shown that the effects of the weight of the rod stem cannot be disregarded if discrepancies between field and laboratory are to be resolved.

The work described herein is part of a comprehensive research programme reported by Odebrecht (2003).

2. INSTRUMENTATION

The equipment developed for measuring force and acceleration signals comprises a notebook provided with an A/D conversion card (PCM-DAS-16D/16 (Computerboard)), a signal conditioner/amplifier, two 12V lead-acid batteries, an oscilloscope for real time signal monitoring, cables, connectors, and adaptors. The load cell is provided with four 350Ω strain gages, assembling a complete Whetstone bridge. The strain gages are disposed to be sensitive to axial strains only, by means of an eccentricity compensation scheme. To measure accelerations (that will be integrated to velocities), a couple of Brüel&Kjaer 4375S piezoelectric accelerometers are symmetrically mounted in the load cell. These accelerometers present a dynamic range from 100μG to 10kG, suitable for the 5kG accelerations produced by a hammer impact. A signal conditioner/amplifier was specifically designed and built for the load cell. This device provides three channels of information: one channel for the Whetstone bridge signal (axial force), and two channels for the piezoelectric accelerometers (averaged and integrated to velocity).

3. EXPERIMENTAL SETUP

In order to allow for a perfect understanding of the mechanism involved on driving a sampler into the ground and evaluate the energy effectively transferred to the composition of rods and sampler, a calibration chamber testing programme has been carried out. In this paper, experimental data are presented in order to evaluate the influence of rod length, keeping as constants all other factors that might affect penetration such as test equipment (hammer mechanism, anvil mass, rod size, sampler geometry) and operator, as well as the soil and boundary conditions in the calibration chamber. Four distinct lengths have been tested corresponding to 5.80m, 11.8m, 18.8m and 35.8m. Every rod length has been instrumented at three positions: immediately below the anvil, at the mid-height of the composition and immediately above the sampler.

A building under construction has been used to support the testing configuration in which the driving system was mounted at several different floors, whereas the calibration chamber was placed at underground level, as illustrated in Figure. 1. A casing prevented the rods from buckling.

The chamber allows a known granular material prepared at a given density to be tested under strictly controlled boundary conditions. The calibration chamber designed and constructed to calibrate the SPT can test a cylindrical sample of sand 0.60 m in diameter and 1.0 m high. A complete description of the equipment has been given by Odebrecht (2003). The sample is confined vertically at the top by a flexible rubber membrane, which is used to apply the vertical stress by means of water pressure. Both the bottom and lateral of the sample are confined by rigid boundaries. The vertical pressure is controlled by a pressurization system regulated by a self-relieving air pressure valve driving an air-water interface system. The sample is prepared in the tank by compacting thin layers at a given density. A major difference from previously reported data from calibration tests (e.g. Schnaid & Houlsby, 1992) is that penetration in the chamber is not constant during driving since soil resistance is alternately controlled by friction on the outside and inside of the spoon and by end-bearing. Penetration that results from a single blow reduces gradually as the sampler is driven into the soil from about 10cm at the first blow to about 1.5cm at final stages.

All tests have been carried out strictly according to the Brazilian Standard (NBR6484/2001) that is in accordance to the IRTP (1988). A Terzaghi-Peck split barrel connected to Schedule 80 (3.23kg/m) rods was used. A manual release of a 65 kg pin-guided hammer produced an impact on a 3.72 kg steel anvil through which the hammer energy passes into the drill rods.

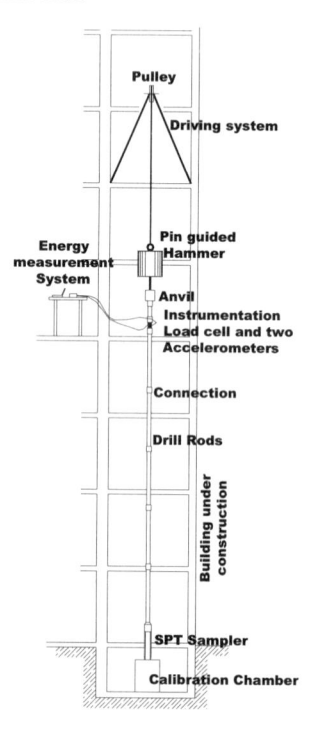

Figure. 1 - Illustrative representation of test setup in a construction building.

Interpretation of test data should be undertaken with great care. The diameter of the calibration chamber is only 12 times the diameter of the sampler and results are affected significantly by chamber size effects (e.g. Schnaid & Houlsby, 1992). These effects are not taken into account since calibration tests have been carried out here to produce a direct comparison between tests in which most of the variables that affect results are properly controlled.

4. EXPERIMENTAL RESULTS

Typical records of the force measured by the load cell and the force calculated from the readings of the accelerometers, measured in the calibration chamber, are presented in Figures 2 and 3. The maximum energy transmitted to the rod stem is also shown in these figures, calculated by the F-V method and known as the Enthru energy:

$$E = \int_0^\infty F(t)V(t)\,dt \qquad (1)$$

Both signals are recorded at the top of a 35.80m long rod. A long composition has been selected because in theory all energy is transmitted by the first compression wave and, as a consequence, the concepts postulated by Schmertmann and Palacios (1979) can be fully applied to the interpretation of test data. The difference between these two figures is that the first case shows a condition that represents a loose sample with an average N approximately 8, whereas the second case represents a very loose specimen where the blow produces a large permanent penetration ($N \cong 3$).

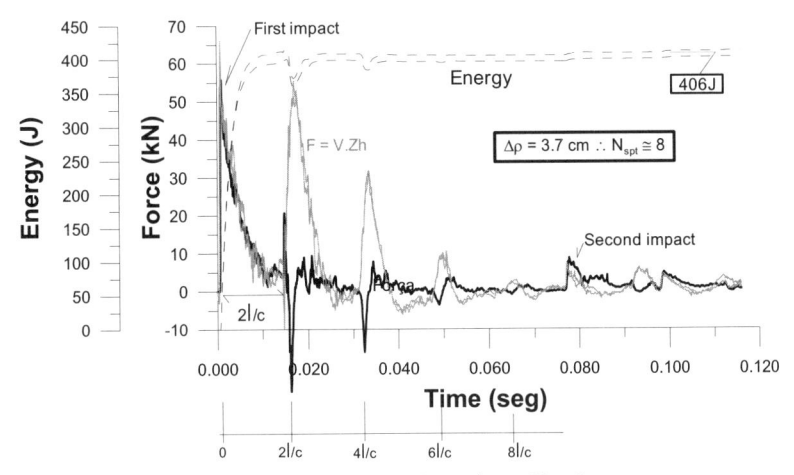

Figure. 2 – Typical measured forcer-time relationships for a 35.80m long rod stem ($N_{spt} \approx 8$).

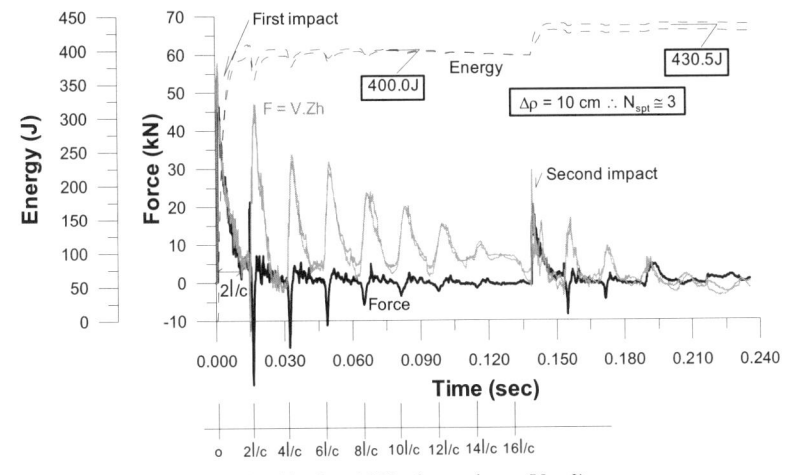

Figure. 3 - Typical measured forcer-time relationships for a 35.80m long rod stem ($N_{spt} \approx 3$).

Figure 2 illustrates the characteristics of a standard result that displays features already discussed by previously reported data (Schmertamann & Palacios, 1979; Skempton, 1986; Kovacs et al, 1982). The energy is fully transmitted to the rods during a time interval of $2\ell/c$ (where ℓ = rod length) and in this case the energy transmitted to the rod can be calculated by equation 2:

$$E = \int_{0}^{2\ell/c} F(t)V(t)\,dt \quad (2)$$

However, an interesting and perhaps surprising pattern is observed in Figure 3. For the large penetration induced by the hammer, a second and late impact ($\Delta t > 100$ ms $\gg 2\ell/c$) produces a further increase in energy that eventually contributes to the penetration of the sampler. The most unexpected feature is that energy delivered to this 35.80m long composition of rods is different to these two reported cases, suggesting that energy effectively transmitted is not only a function of the so called nominal potential energy E* (474J) but is also affected by the actual displacement that results from the permanent penetration of the sampler.

In a short composition of rods, the second and third impacts also produce a significant increase in the transmitted energy that cannot be disregarded when interpreting the measured data (see Figure 4). The impact of the falling hammer produces a first compression wave that propagates downwards to the sampler and reflects upwards as a tension wave arriving at the hammer at a time corresponding to $2\ell/c$. At this time the anvil is pulled away from the hammer causing a temporarily interruption until a second impact is produced. The same mechanism is then reproduced in the rods giving rise to a third impact. Every impact produces an increase in the energy delivered to the rod-sampler system. The obvious conclusion from all these evidences is that the integration time that is necessary to calculate the actual energy transmitted to the rods should be sufficient to allow the various impacts to be considered.

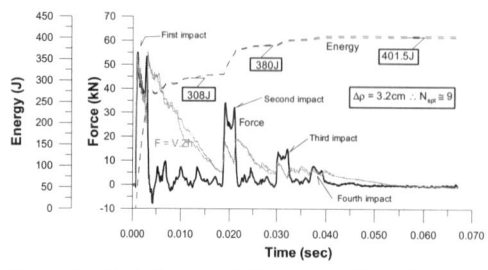

Figure. 4 – Typical measured forcer-time relationships for a 5.80m long rod stem.

The integration of the various experimental signals measured in the calibration chamber, adopting equation 1, produced values of energy which are greater that the nominal potential energy E*. This is not the case when using equation 2 (integration interval from 0 to $2\ell/c$) which is a particular solution of the stress wave equation and does not take into account additional ram impact due to large downward displacements of the rod stem.

Several authors have already mentioned the need to take into account the various impacts produced by the SPT hammer (Farrar, 1988; Butler et al, 1998; Aoki & Cintra, 2000). Implications to this recommendation have not been fully understood however. In calculating the energy delivered to sampler from equation 1 one should necessarily consider the overall displacement of the hammer, which is the sum of the initial height of fall ρ_o plus the penetration of the sampler $\Delta\rho$. For example, in Figure 3 $\Delta\rho$ is equal to 10 cm and therefore the overall height of fall is equal to 86 cm.

Based on the recognition that the energy transmitted to the rod is a function of ρ_o and $\Delta\rho$ and that it could also be affected by the rod length, an attempt was made to express the measured energy as a function of permanent penetration to all tested lengths. Results are plotted in Figure 5, which summarizes the data for rods lengths of 5.80m, 11.8m, 18.8m and 35.8m. The greater the permanent penetration the greater the energy transmitted to the rod with no specific pattern emerging as a function of the length of the rods. Furthermore, great penetrations can produce energy values greater than the theoretical E* value. None of these evidences can be explained from the theoretical framework proposed by Schmertmann & Palacios (1979).

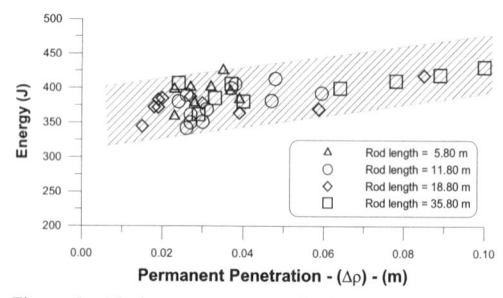

Figure. 5 – Maximum energy transmitted to the rod stem versus permanent penetration for different lengths of rods.

Whereas the transmitted maximum energy measured at the top of the rod stem depends solely on the hammer-anvil system, both the length of the rods and the permanent penetration of the sampler also influence the energy delivered to the sampler. Comparison between experimental values measured immediately below the anvil and immediately above the sampler is given later in this paper.

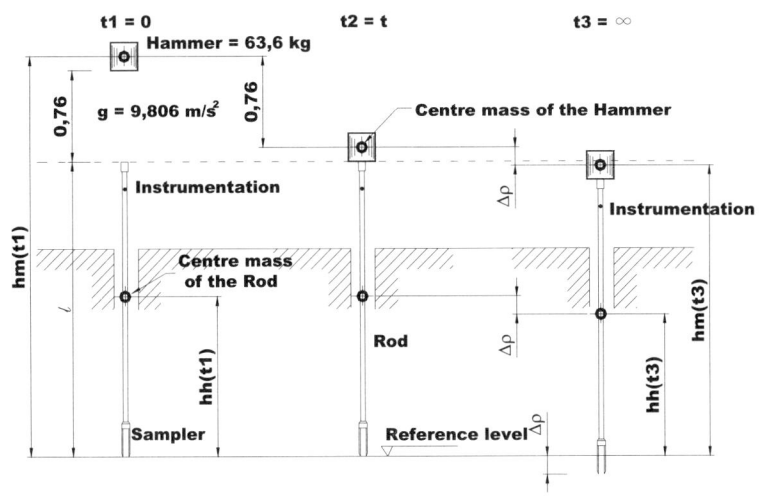

Figure. 6 – Illustrated representation of the potential energy at different stages of the SPT test.

5. ANALYTICAL FRAMEWORK

Figure 6 summarizes all the main features of behaviour that emerged from the analysis of the experimental data. Three distinct stages are represented in the figure. First t1=0 corresponds to the time immediately after the liberation of the hammer in free fall. A second stage is represented by t2=t in which the hammer is about to reach the top of the rod, and finally t3= ∞ that represents the end of the penetration process.

The *maximum theoretical potential energy*, PE^*_{h+r}, delivered to the soil should therefore be expressed as a function of the *nominal potential energy* E*, permanent sampler penetration and weight of both hammer and rods.

$$PE^*_{h+r} = (E^* + M_h \, g \, \Delta\rho) + \Delta\rho \, M_r \, g \quad (3)$$

where: M_h = hammer weight;
M_r = rod weight;
g = gravity acceleration;
$\Delta\rho$ = Sample penetration under one blow;
E* = nominal potential energy = 0.76m 63.5kg 9.801m/s = 474 J

The *nominal potential energy* E*= 474 J (ASTM, 1986) represents a part of the hammer potential energy that is transmitted to the soil. An *additional hammer potential energy* is given by $M_h \, g \, \Delta\rho$. The other part is transmitted by the *rod potential energy* $M_r \, g \, \Delta\rho$ which cannot be disregarded for tests carried out at great depths in soft soils, i.e. conditions in which $\Delta\rho$ and M_r are significant. It is here interesting to note that the dynamic formulae of driven piles also take in account the combined effect of weight of the hammer and the weight of the pile (e.g. Smith, 1960).

For convenience, equation 3 can be re-written so that the first term represents the *hammer potential energy* (*nominal + additional*) and the second the *rod potential energy*:

$$PE^*_{h+r} = (0.76 + \Delta\rho) \, M_h \, g + \Delta\rho \, M_r \, g \quad (4)$$

Finally, it should be clear from Figure 6 that an external fixed reference is adopted to compute energy. Previous published works consider an internal reference that is only acceptable when computing the effect of the first compression wave from equation 2.

6. EFFICIENCIES

Equation 4 deals with an ideal condition, where energy losses during the energy transference process are not taken into account. However, it is well known in engineering practice that these losses occur and may not be disregarded.

6.1 Instrumentation above the anvil

Figure 7 summarizes all results obtained with the instrumentation positioned immediately below the anvil. A comparison between energy calculated from equations 1 and 4 is shown in this figure, expressed as a function of measured displacements and rod lengths. As clearly observed, linear regression lines for measured energy and hammer potential energy are parallel. This is a strong indication that there is a fixed relation between both energies; consequently equation 4 may be rewritten to account for an efficiency factor η_1:

$$PE_{h+r} = \eta_1 (0.76 + \Delta\rho) M_h g + \Delta\rho M_r g \qquad (5)$$

where: PE_{h+r} = maximum potential energy

$$\eta_1 = \text{hammer efficiency} = \frac{\int_0^\infty F(t)V(t)dt}{(0.76 + \Delta\rho)M_h g}$$

It is important to emphasize that this definition of η_1 is not the SPT efficiency proposed by Schmertmann and Palacios (1979). In equation 1 the integration limits (from zero to infinity) are set to account for all compression waves and consequently to include the large displacements of the rod stem.

Regression lines estimated from SPT results, presented in Figure 7, exhibit a large coefficient of variation due to scatter. The efficiency factor, η_1, comprises all energy losses experienced during a SPT blow which includes the type of hammer, method of releasing, type of anvil, among others.

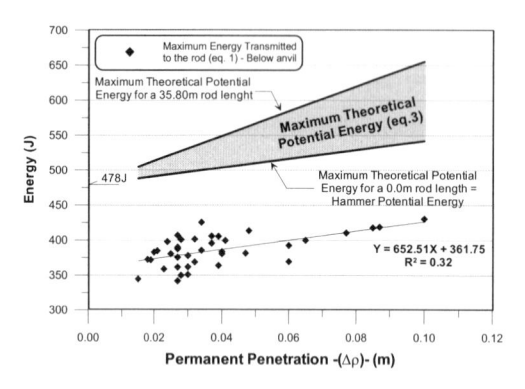

Figure 7 – Variation of potential energy versus permanent penetration.

Another important point to be noticed from Figure 7 is that energy increases with the permanent penetration of the sampler. It is possible to assume a linear relation between the energy measured close to the anvil (equation 1) and the hammer potential energy. Considering that both regression lines are parallel, it could be claimed that the efficiency increases with penetration. However, changes in η_1 for different penetration values are so small that may be disregarded. In this work, the efficiency factor mean value, η_1, has been estimated as 0.764 (constant for practical purposes), with a standard deviation of 0.036. Finally, regression lines of the hammer potential energy are not parallel to the regression line of the maximum potential energy (equation 4) which indicates that the contribution of rod mass to total energy is relevant.

6.2 Instrumentation above the sampler

Experimental results for the energy transfer obtained from the instrumentation positioned immediately above to the sampler are presented in Figure 8. This figure also shows energy measurements at the top of the rod stem, which are referred to as *maximum potential energy* and *hammer potential energy*. The nominal potential energy (474J) is also presented for the sake of comparison.

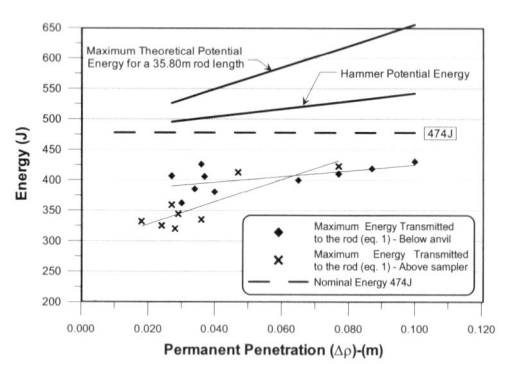

Figure. 8 – Energy as a function of penetration – 35.80m rod.

In this case, an additional energy variation occurs between the anvil and the sampler. This indicates the need to multiply the maximum potential energy by another factor. The following general form is proposed:

$$PE_{h+r} = \eta_3 [\eta_1(0.76 + \Delta\rho) M_h g + \eta_2 \Delta\rho M_r g] \qquad (6)$$

where: $\eta_2 = \beta_2 - \alpha_2 \ell$

$\eta_3 = \beta_3 - \alpha_3 \ell$

Estimation of η_2 and η_3, as well as of the corresponding α and β coefficients, is not a trivial task and requires some additional simplifying hypotheses. After several attempts, the experimental data was adjusted by keeping $\eta_2 = 1$ and allowing η_3 to be expressed as a function of the length of the rods:

η_3 = energy efficiency = $1 - 0.0042\ell$

A least square algorithm was used for η_3 model fitting. The equation above was found to minimize the error, defined here as the square of the difference between the maximum potential energy PE_{h+r} and the maximum energy transmitted to the rod stem calculated by the F-V method (instrumentation close to the sampler). A value of $\alpha = 0.0042$ obtained for the database indicates that efficiency is inversely proportional to the length of rods.

Finally, Figure 9 shows the differences between the energy calculated from equations 2 and 5. Differences are plotted against both permanent penetration and length of rods to allow the verification that equation 5 indeed cleans up the influence of these two parameters. Scatter is in the order of 47.4 J corresponding to an error of less than 10% of the nominal potential energy (474J).

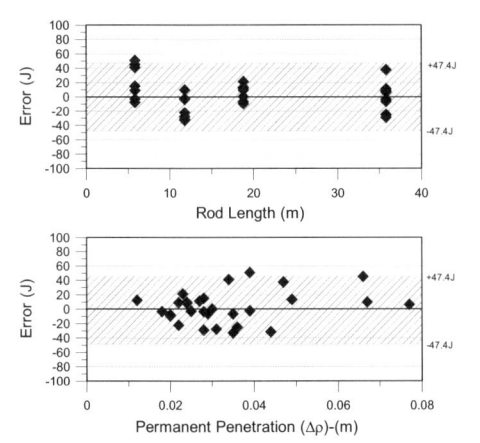

Figure. 9 – Model error versus (a) rod length and (b) permanent penetration.

Table 1 gives an example of the impact produced by the proposed approach, where the ratio of the *maximum potential energy* PE_{h+r} and the *nominal potential energy* E* is shown to vary with both the permanent penetration and rod length. For the given combinations, the ratio is consistently greater that unit and increase with increasing rod length. This suggests that the energy effectively reaching the sampler is significantly greater than the reference value of 60% of the potential energy.

Table 1 – Energy ratio PE_{h+r}/E^*.

N_{spt}	Rod Length (m)					
	2	8	14	20	26	32
2	1.22	1.28	1.34	1.40	1.46	1.52
4	1.11	1.14	1.17	1.20	1.23	1.26
8	1.06	1.07	1.09	1.10	1.12	1.13
12	1.04	1.05	1.06	1.07	1.08	1.09
16	1.03	1.04	1.04	1.05	1.06	1.07
20	1.02	1.03	1.03	1.04	1.05	1.05
24	1.02	1.02	1.03	1.03	1.04	1.04

The authors foresee many applications in engineering practice to the principles discussed in the present paper. Normalization to a reference value of N_{60} is no longer sufficient to fully explain the mechanism of energy transfer to the soil; the system energy calculated from equation 5 appears to produce a more sounding representation. Furthermore, is interesting to recall to that the maximum potential energy is transformed into work by the non conservative forces (W_{nc}) acting on the sampler during penetration which can be conveniently expressed as:

$$PE_{h+r} = W_{nc}$$

Since the work is proportional to the measured permanent penetration of the sampler, it is now possible to calculate the dynamic force transmitted to the soil during driving:

$$PE_{h+r} = W_{nc} = F_d \Delta\rho \text{ or } F_d = PE_{h+r}/\Delta\rho$$

The dynamic force can be considered as a fundamental measurement for the prediction of soil parameters from SPT results as already demonstrated by Odebrecht (2003). For example, F_d can be directly related to soil parameters from bearing capacity theories.

7. CONCLUSIONS

This paper presents experimental results of energy measurements in SPT tests carried out under controlled boundary conditions in a calibration chamber. From the basis of these experiments a new interpretation method is proposed, supported by stress wave theory and the effective energy measured at the top and bottom of the rod composition during the impact of a SPT hammer. Based on experimental results it is possible to draw some conclusions, which are summarized as fallows:

a) The energy transferred to the rod and to the sampler due a hammer impact should be obtained through the integration of equation 1, with an upper limit of integration equal to infinity. This recommendation is valid even for a long rod stem (greater than 15m) as clearly demonstrated throughout this paper.

b) The maximum potential energy can be conveniently expressed as a function of three components, nominal potential energy E*, sampler permanent penetration and weight of both hammer and rods. Therefore the greater the length of the rods the greater the energy transmitted to the sampler-soil system for large displacements.

c) Efficiency is accounted for by three coefficients η_1, η_2 and η_3 that should be obtained from calibration. The hammer efficiency η_1 is obtained from measurements at the top of the rod stem. Efficiency factor η_2 can be assumed as unit. The energy efficiency η_3 is inversely proportional to the length of rods.

d) Effects and corrections for rod lengths are embedded into the equation proposed to compute energy. Previous empirical propositions are no longer required.

ACKNOWLEDGEMENTS

The first author received a scholarship from the Brazilian Financial Agency CAPES. The writers wish to express their gratitude to the Federal University of Rio Grande do Sul and State University of Santa Catarina, and for all people who have contributed to this project.

REFERENCES

About-Matar. H & Coble,G.G.(1997), SPT Dynamic Analysis and Measurements, Journal of Geotechnical and Geoenviromental Engineering – ASCE – vol 123. n. 10, pp 921-928.

Aoki, N & Cintra J.C.A. (2000) The application of energy conservation Hamilton's principle to the determination of energy efficiency in SPT test. Proceedings of the sixth international conference on the application of stress-Wave Theory to Piles – São Paulo – September 2000, pp 457 – 460.

Butler, J.J.; Caliendo, J.A.; goble, G.G. (1998) Comparison of SPT energy measurement methods. In: International Conference on Site Characterization, ISC'98, Atlanta. Proceeding, v.2. p.901-906.

Farrar, J.A. (1998) Summary of Standard Penetration Test (SPT) energy measurement experience. In: International Conference on Site Characterization, ISC'98, Atlanta. Proceeding, v.2, p.919-926.

ISSMFE, (1989) Technical Committee on Penetration Testing of Soils, TC 16. Report on Reference Test Procedure CPT-SPT-DP-WST. Rio de Janeiro.

Kovacs, W.D.; salomone, L.A. (1982) SPT hammer energy measurements. Journal of the Geotechnical Engineering Division, ASCE, v.108, n.GT4, p.559-620, apr.

NBR 6484 – Solo – Sondagem de simples reconhecimento com SPT – Método de Ensaio. – Fev. 2001.

Odebrecht (2003), Energy measurements on SPT tests, Porto Alegre, Ph.D thesis, Federal University of Rio Grande do Sul, Porto Alegre, Brazil, (in Portuguese).

Schmertmann, J.H. & Palacios, A. (1979) Energy dynamics of SPT. Journal of the Soil Mechanics and Foundation Division, ASCE, v.105, n.GT8, p.909-926, aug.

Schnaid, F.; Houlsby, G.T. (1992) Measurement of the proprieties of sand in a calibration chamber by cone pressuremeter test – Géotechnique, Vol. 42, No. 04; pag 578-601.

Seed, R.B.; Tokimatsu, K.; Harder, L.F.; Chung, R.M. (1885) Influence of SPT procedures in soil liquefaction resistance evaluations, Journal of Geotechnical Engineering, Vol. 111, No. 12, pag 1425 , 1445.

Smith, E.A.L., (1960) Pile driving analysis by the wave equation. Journal Soil Mech. Found, Div., ASCE, No. SM4, august, pp 2574.

Skempton, A.W. (1986) Standard penetration test procedures and effects in sands of overburden pressure, relative density, particle size, aging and over consolidation. Géotechnique, v.36, n.3, p.425-447.

SPT-T: test procedure and applications

A.S.P. Peixoto
São Paulo State University, Unesp-Bauru, SP and University of Campinas, Unicamp, Brazil

D. de Carvalho
University of Campinas, Unicamp, Brazil

H.L. Giacheti
São Paulo State University, Unesp-Bauru, SP, Brazil

Keywords: Standard Penetration Test, torque measurements, soil classification, lateral skin friction of piles

ABSTRACT: Ranzini (1988) proposed the Standard Penetration Test with torque measurement (SPT-T) and some geotechnical engineers in Brazil have been using it since 1991. This paper presents the state of the art on SPT-T testing, emphasizing what is already established as common knowledge in Brazilian engineering practice, besides a suggestion for test procedure, including equipment and practical aspects. In addition, the study of the shape of the "torque versus rotation-degree angle" curve obtained by an electric torquemeter used in several SPT-T tests carried out on six experimental research sites in the southeast region of Brazil is discussed here. Four different methods to predict pile capacity based on SPT-T test results are briefly presented and a comparison with load tests carried out on different types of piles, on those six experimental research sites, is presented.

1 INTRODUCTION

SPT is still the most common test for site characterization of soils. The test is simple and provides a disturbed sample of the soil while the penetration resistance (*N-value*) is measured. The *N-value* can be empirically related to geotechnical design parameters but it has been found to be very dependent on the test equipment and procedure. Mayne (2001) inquires how just a numerical value is sufficient to estimate a great number of different soil parameters. He advocated the use of *in situ* testing with hybrid devices.

In Brazil, the Standard Penetration Test is commonly used in foundation practice since 1940 and this experience cannot be dismissed.

So, Ranzini (1988) suggested supplementing the conventional SPT with measurement of the torque required to turn the split spoon after driving. This is called the standard penetration test with torque measurement (SPT-T). The torque (*T*) provides a 'static' component for a 'dynamic' test. According to Ranzini (1994), the adhesion between the soil and the sampler, obtained by torque measurement, could be used to calculate the lateral skin friction of piles.

First, this paper will present the state of the art on SPT-T testing, emphasizing what is already established as common knowledge in Brazilian engineering practice.

After this, a suggestion for test procedure, including equipment and practical aspects, will be presented. Mechanical torquemeters have been currently used, but Peixoto (2001) developed an electric torquemeter and a data acquisition system that permits plotting "torque *versus* rotation-degree angle". This equipment was used in several SPT-T tests carried out on six experimental research sites in the southeast region of Brazil, where many geotechnical test data and pile load tests were available (Fig. 1). The shape of that curve and the definition of maximum and residual torque values will be presented and discussed in this paper.

Figure 1 Cities of Brazil where the research sites are located

Four different methods to predict bearing capacity of piles based on SPT-T test results were proposed by Decourt (1996), Alonso (1996a and 1996b), Carvalho et al. (1998) and Peixoto (2001). Those methods will be briefly presented and discussed, based on a comparison with a load test carried out on different types of piles. All methods led to a good prediction of bearing capacity for the six different sites.

The torque ratio (*T/N*) was strongly discussed by Decourt (1998), Peixoto and Carvalho (1999) and Peixoto (2001). This paper will present the major aspects related to *T/N* ratio. This ratio appears to be an interesting index to estimate the behavior of driven piles because it is affected by soil response to dynamic loads. It is important to emphasize that there are many factors affecting this index and it is difficult to identify the most prominent. So, any correlations using *T/N* ratio cannot be generalized and it has to be previously tested for a specific site.

As a conclusion, SPT has the advantage of producing a resistance index, while sampling the soil. Introducing torque measurement (SPT-T) is a simple, interesting and inexpensive option for site characterization. The torque has been empirically used to estimate pile skin friction, mainly for driven piles. The use of *T/N* to identify tropical and collapsible soils neglects the fact that the test procedure can destroy the soil structure of lateritic soil before applying the torque.

2 STATE OF THE ART

When Ranzine (1988) proposed the use of torque in SPT, he suggested the possibility of using that value to obtain lateral resistance of piles. Ranzine (1994) presented the following equation:

$$fT = \frac{T}{(41.336 \times h - 0.032)} \quad (1)$$

where fT = sampler-soil adhesion, kPa; T = torquemeter measurement, m.kN; and h = length of penetration of the sampler, m.

According to Ranzini (1994), the sampler-soil adhesion, fT, coud be used as lateral skin friction of piles.

Three years after Ranzini's (1988) proposal, some foundation consultants began to use that new technique. Some engineers, as Alonso (1996a and b), Carvalho et al (1998) and Peixoto (2001) use the value of the adhesion, fT, in the prediction of lateral resistance piles.

Decourt (1991), studying the soils of Tertiary Sedimentary Basin in the city of São Paulo (TSBSP), found a *T/N* ratio of 1,2. So, for that author, for soils out of that basin, it could be used a *Neq* (=*T*/1,2), besides *N-blow*, to turn the utilization of correlations established for TSBSP soils possible. After that, Decourt (1996) suggested an adaptation of the Decourt and Quaresma (1978) method to predict bearing of piles where, again for soils out of TSBSP, one would substitute N-blow by Neq.

Peixoto (2001) developed an electric torquemeter that allowed the plotting of the "torque versus rotation degree" curves. Using this, analyzing the curve, it could infer: the behavior of the curve; the definition of maximum and residual torques; the variability of *T/N* ratio; a procedure suggestion for the execution of the SPT-T test.

2.1 *T/N ratio*

Decourt (1998) presented a study of soils classification according to the degree of structuring, through the torque *T/N* ratio. Although during the execution of SPT, the structure of the soil is broken inside the sampler, the torque measures the skin friction in an area that, despite being partially disturbed, still preserves its original structure. Thus, structured soils tend to have larger *T/N* (Fig. 2).

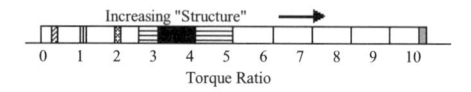

Soil	T/N
Sedimentary Sands, Lower Bound	~ 0.3
Soil from the TSESP	~ 1.2
Saprolitic Soils - SP	~ 2.0
Collapsible Soils - SP	2.5/5.0
Soft Marine Clays of Santos City	3.0/4.0
Sedimentary Sands, Upper Bound	~ 10.0

Note: TSESP = Tertiary Sedimentary Basin of São Paulo

Figure 2. Soil Classification based on torque ratio (*T/N*) *apud* Decourt (1998)

Peixoto and Carvalho (1999) presented a detailed study of seven thousand *T/N* ratio data addressing how grain size, geological aspects and presence of organic matter, gravel, mica and limonite concretions affect that value. In the majority of cases, the analyses showed great dispersions. As a conclusion, the many factors affecting the relationship and the difficulty to know the most prominent one in each problem make it difficult to propose a soil classification.

However, Peixoto (2001) also demonstrated the great variability of the *T/N* ratio. Nevertheless, in spite of that, it can be verified that the medium values show some tendencies. For instance: for *T/N* ratio smaller than value one, the soil is always collapsible. But, there are collapsible soils with *T/N* ratio greater than value one.

On the other hand, Peixoto (2001) concluded that the *T/N* ratio can be used as a parameter to predict the behavior of the soil, for example, when implanting a foundation element. Here, the *T/N* ratio is not used as a correlation, but to infer the soil behavior of the location where the data were obtained.

2.2 *Bearing capacity prediction*

Mayne (2001) inquires how just a numeric value (*N-blow*) is enough to estimate a great number of different soil parameters and advocated the use of *in situ* testing with hybrid devices.

In Brazil, SPT is commonly used in foundation practice since 1944 and this experience cannot be dismissed. Ranzini (1988) suggested supplementing conventional SPT with measurement of the torque required to turn the split spoon after driving. This is called the standard penetration test with torque measurement (SPT-T). The torque (T) provides a 'static' component for a "dynamic" test.

Nowadays, there are four methods to predict the bearing capacity of piles based on SPT-T methods: Decourt (1996), Alonso (1996a and 1996b), Carvalho et al. (1998) and Peixoto (2001). Below, a brief review of each of them is presented and then a comparison of the methods with the load test results.

The ultimate bearing capacity of a single pile based on empirical methods using SPT-T test results (Q_u), can be expressed as the ultimate tip resistance (Q_{ut}), and ultimate shaft resistance (Q_{us}), as follows:

$$Q_u = Q_{ut} + Q_{us} = x_1.(q_t.A_t) + x_2.U.\Sigma(q_{si}.L_{si}) \quad (2)$$

where x_1 and x_2 = empirical factors; q_t = tip resistance; A_t = area of pile tip; U = pile perimeter; q_{si} = shaft or skin resistance within a single soil type, labeled i, penetrated by the pile; L_{si} = shaft length interfacing with layer i.

For all Brazilian methods, the N value refers to a measurement obtained with 72% efficiency.

Table 1. x_1 coefficient *apud* Decourt (1996)

Pile Soil	Precast	Bored	Bentonitic bored	Continuos Flight Auger	Root	Injected
Clay	1.00	0.85	0.85	0.30	0.85	1.00
Intermediate Soil	1.00	0.60	0.60	0.30	0.60	1.00
Sand	1.00	0.50	0.50	0.30	0.50	1.00

* Just oriented values because few data

Table 2. x_2 coefficient *apud* Decourt (1996)

Pile Soil	Precast	Bored	Bentonitic bored	Continuos Flight Auger	Root	Injected
Clay	1.00	0.80	0.90	1.00	1.50	3.00
Intermediate Soil	1.00	0.65	0.75	1.00	1.50	3.00
Sand	1.00	0.50	0.60	1.00	1.50	3.00

* Just oriented values because few data

Table 3. C coefficient apud Decourt and Quaresma (1978)

Soil Type	C (kN/m²)
Clay	120
Clayey Silt (Residual Soil)	200
Sandy Silt (Residual Soil)	250
Sand	400

2.2.1 Decourt (1996) method

Decourt (1996) proposed a modification of the Decourt and Quaresma (1978) method adding the x_1 and x_2 coefficients as functions of soil and pile type (Tab. 1 and Tab. 2).

The author also suggested the use of "N_{eq}" (=T/1.2), besides N-blow, for soils located out of Tertiary Sedimentary Basin in the city of São Paulo, eq. (3) and eq. (4):

$$q_s = 10 \times \left(\frac{Neq}{3} + 1\right) \quad (kN/m^2) \quad (3)$$

$$q_t = C \times \overline{Neq} \quad (4)$$

where: $\overline{Neq}$ = Mean of $T(kgf.m)/1,2$ value at pile tip, one meter up, one meter down; C = coefficient that depends of ttype of soil (Tab. 3).

For Terciary Sedimentary Basin of São Paulo City, use N-blow.

2.2.2 Alonso (1996a and 1996b) method

Alonso (1996a) suggested the point resistance could be obtained by Aoki and Velloso (1975) or Decourt and Quaresma (1978). Alonso (1996b) suggested the following equation for Continuous-flight-auger:

$$qt = \beta \times \frac{T_{res}^1 + T_{res}^2}{2} \quad (6)$$

where: T_{res}^1 = average of values of the residual torque (kgf.m) along a length 8D over the pile tip; T_{res}^2 = same, for length 3D under the pile tip; D = pile diameter; β = coefficient that depends of soil type.

Here, the T_{res}^1 and T_{res}^2 values is considered as maximum value of 40 kgf.m.

The x_1 and x_2 values of eq. (2) are considered equal the value one.

This author suggested the q_s equations based on fT value for different pile types, Tab. 4.

Table 4. q_s based on fT In Alonso (1996a and 1996b)

Pile Type	Equation
Root	$q_s = 1.15*fT$
Precast Concrete	$q_s = fT/1.5$
Bentonitic Auger	$q_s = fT/1.7$
Continuous-flight-auger	$q_s = 0.65*fT (\leq 200kPa)$

2.2.3 Carvalho et al (1998) method

Carvalho et al.. (1998) presented the following equation to foresee the lateral load capacity of piles in porous clays of the Federal District, Brazil.

The authors proposed that Qu could be calculated basically by the Aoki and Velloso (1975) or Decourt and Quaresma (1978) methods with modified values of x_1 and x_2.

On the other hand, the q_s value is calculated as eq. (7):

$$q_s = fT.\alpha_T \qquad (7)$$

where: α_T = skin friction portion coefficients (Tab. 5)

Table 5. Coefficients to Lateral Portion α_T apud Carvalho et al. (1998)

Pile Type	Aoki and Velloso (1975)	Decourt and Quaresma (1978)
Strauss	0.85	1.10
Precast	0.85	0.90
Auger	1.30	1.35
Bored	0.90	0.95

2.2.4 Peixoto (2001) method

Q_{ut} is here calculated by Decourt and Quaresma (1978) using the N-blows value instead of Neq.

For Q_{us}, the x_2 factor is a function of pile type (Tab. 6). The shaft resistance (q_s) is as follows:

$$qs = F_1.fT \qquad (8)$$

where: F_1 = Correction value function of pile type and $T(kgf.m)/N$ ratio, (Tab. 6).

The T/N ratio is an interesting index to estimate the behavior of driven piles because it is affected by soil response to dynamic loads. In this manner, the use of a correction factor based on T/N ratio appears to be an interesting approach, since it is not a general correlation, because it is obtained directly on the site where the foundation will be installed.

Table 6. x_2 and F_1 coefficients

Pile Type	x_2	F_1 $T/N<1$	F_1 $T/N>1$
Precast concrete	0.8	1.0	1.0
Omega	3.0	1.0	1.0
Steel	0.3	1.0	1.0
Small Diameter Injected	2.0	1.0	1.0
Root	1.5	1.0	1.0
Strauss	0.8	1.3	0.7
Franki	0.8	0.7	0.5
Socketed	3.5	0.7	0.5
Continuous-Flight-Auger	2.0	1.0*	0.3*
Auger and Bored	1.4	1.3	0.7
Diaphragm Panel Bored	0.7	1.0	1.0

Note: when $\overline{fT} > 80$ kPa, use F_1=0.3 to any T/N

The x_2 and F_1 values are just a first suggestion and need improvement to be used in foundation practice.

3 SUGGESTION FOR TEST PROCEDURE

The test is very simple: the sounding is executed according to NBR 6484 (1980). After each penetration, the sampler is pulled out after measuring the torque value. According to this, the test procedure is as follows:

After penetration of the sampler, with the counting of blows, an adapter is placed on the anvil that allows joining the torquemeter (Fig. 3). A centralizing device should be placed either in the mouth of the hole or in the pipe to prevent the rod from leaving the center of the hole during application of the torque (Fig. 4 and 5).

Afterwards, rotation on the rod-sampler assembly is applied, using the torquemeter. The maximum torque is measured and the rotation continues without interruption until the torque remains constant, when the residual torque is obtained. The two measurements are logged in the sounding bulletin. The centralizing device and the adapter-torquemeter assembly are removed and the soil boring continues until the next meter, when N and torque values are measured again. This procedure continues till reaching impenetrable sounding or the maximum torquemeter capacity.

Many SPT-T tests were observed, which made the following recommendations possible:
- an adaptation in the anvil should be made to fit the adapter because it was observed that, no matter how careful the removal is, a small torque is applied to the rod-sampler group, affecting the reading of maximum torque.
- the centralizing device is an important part of the equipment, to maintain the rod alignment during application of torque. The test should not be conducted without that equipment.
- the minimum and maximum capacity should also be observed at each torquemeter reading (Table 7).
- reading the maximum torque is very fast. Therefore, it is difficult to be done visually. It is recommended to use torquemeters with a creeping pointer or, if it is missing, to advise the operator to perform the reading only while the equipment is rotating.
- during rotation, the torquemeters should stay in the horizontal position to prevent the rods from arching.

Peixoto (2001) performed many tests in six experimental research sites in the southeast region of Brazil and got the rotation speed for each torque measurement. It was concluded that, for the level of speed rotations obtained (between 3 and 5 turns/

minute), it does not influence the results. So, it is recommended to measure with 5 turns/minute.

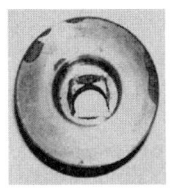

Figure 3. Adaptation made at the anvil

Figure 4. Centralizing device to be placed in the mouth of the hole

Figure 5. Centralizing device to be placed in the pipe

Table 7. Torquemeter capacity control apud Peixoto (2001)

N values	Maximum torquemeter capacity
0 – 10	265 N.m
11 – 30	471 N.m
31 – 45	785 N.m

* considering N value equal N_{72} that means the N with eficiency of 72%.

4 TORQUE VERSUS ROTATION CURVES

An electric torquemeter and a data acquisition system were developed to furnish aid to the proper use of torque measurement in the SPT. This equipment allowed the plotting of the "torque versus rotation degree" curve.

The SPT-T tests using that equipment were carried out on seven locations where much geotechnical data was available. The test results of the "torque versus rotation degree" allowed the following analyses:
 - behavior of the torque versus rotation degree curve;
 - definition of the maximum and residual torques;
 - suggestion of procedure to execute the SPT-T test.

4.1 Curve Shape

In each curve, the following factors were observed: shape; rotation speed; N-blow; moisture; equipment capacity; comparison between the mechanical and electrical torques obtained at the same time.

The most representative curves are presented below. The general curve is represented in Fig. 6.

The oscillation shows the difficulty to maintain the rod straight inside the hole, even using the centralizing device. Small oscillations indicate the relief caused by the operator's footsteps.

469 curves were analyzed, performed in seven locations, with different geological and geotechnical characteristics. The analyzes are presented by soil type.

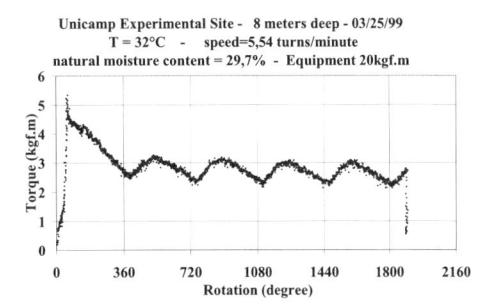

Figure 6. Typical curve

4.1.1 Clayey and Silty Soils

Figure 7 and 8 show the typical curves for clayey and silty soils respectively.

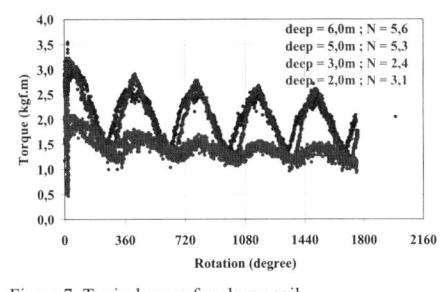

Figure 7. Typical curve for clayey soil

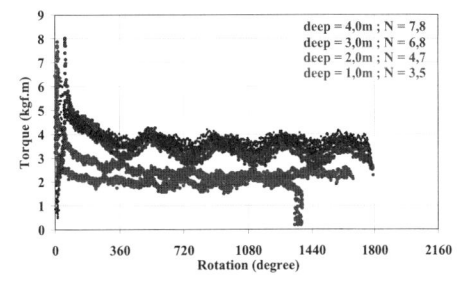

Figure 8. Typical curve for silty soil

Larger oscillations were observed for clayey soils than silty soils, indicating a major stiffness of those soils.

Torque reached a maximum value before rotation of 60° and the residual torque was well defined before the second complete turn.

4.1.2 *Sandy Soils*

The curve behavior for the sandy soils is greatly variable and the residual torque is difficult to obtain. But this problem is solved by greater N-blows (Fig. 9).

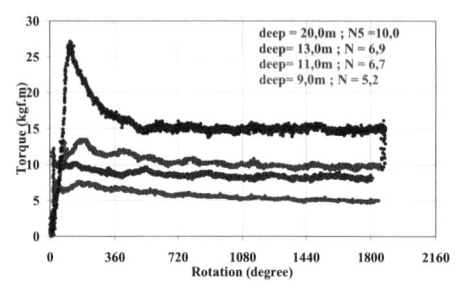

Figure 9. Typical curve for sandy soil

Figure 10 shows the continuous decrease of torque value for low N-blows. It could be explained by the opening of the hole, probably caused by the smaller cohesion in these soils.

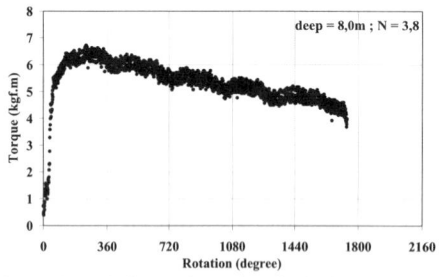

Figure 10. Typical curve for sandy soil with low N-blow

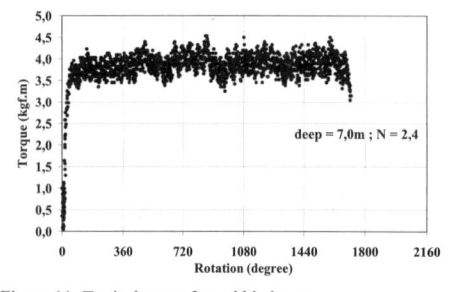

Figure 11. Typical curve for pebble layer

4.1.3 *Pebble Layers*

Figure 11 represents a curve obtained in a pebble layer. The not defined maximum and residual torque can be observed, probably because of the great friction in the soil-sampler wall. The same occurs in gravel layers

4.1.4 *Residual Soils*

The residual soils analyzed here are originated from Diabase and Migmatite rock, and they present a great amount of micaceous mineral. The maximum torque is instantaneously reached, but the residual value seems to be defined after the fourth turn (Fig. 12).

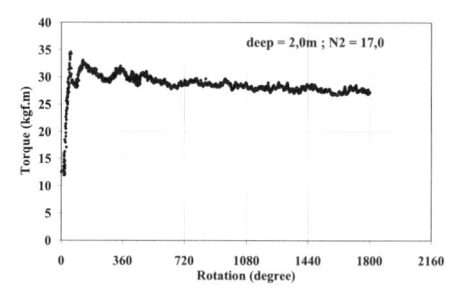

Figure 12. Residual soil typical curve

4.1.5 *Organic Clay*

The N-blow is too insignificant and there is no resistance to soil-sampler rotation. The curve shape does not have the typical oscillation of other types of soil. Both maximum and residual torque are well defined (Fig.13).

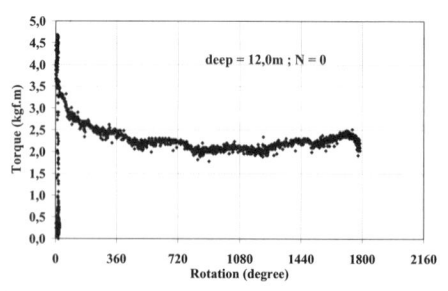

Figure 13. Typical curve for residual soil

4.2 Rotation Speed

This study consisted in allowing two samplers driven in two different depths (5m and 15m) and to perform rotation once a day at different rotation speeds. Figures 14 and 15 represent these measurements. It could be concluded that, for the level of rotation speeds (between 3 and 5 turn/minute), there is no need for correction.

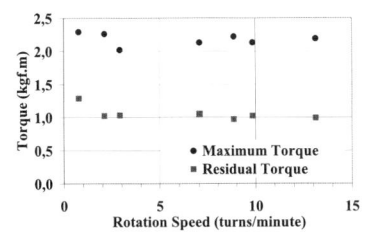

Figure 14. Torque variation at different rotation speeds (data obtained at five meters depth)

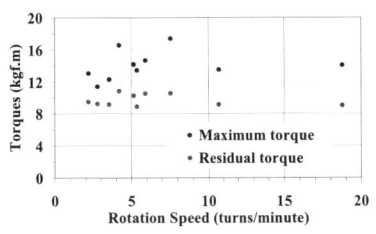

Figure 15. Torque variation at different rotation speeds (data obtained at fifteen meters depth)

4.3 Driven time of the sampler

The influence of driven time of the sampler in recuperating torque measurement was analyzed for the five experimental sites. These sites have soils with different geotechnical characteristics: residual, organic and collapsible soils.

For the residual soil of Migmatite rock, both residual and maximum torques decreased in a second measure immediately after the first test. However, after some hours, the torque increases to a higher value. Probably, with the first rotation, the test hole was open to a higher diameter, decreasing the adhesion and, after some hours; the soil structure was recuperated, but it is remoulded.

The same happened with organic soil and also probably because of the porous-pressure dissipation in time.

For collapsible soils, there was no recuperation of maximum torque with time and the residual torque remained constant. It indicates that soil structure does not recover after deformation for this type of soil.

So, for a standard procedure, it recommends that torque measurement needs to be executed immediately after driving the sampler in the SPT test.

4.4 Comparison between electric and mechanical torque measurements

In all tests performed by Peixoto (2001) the electrical and mechanical torque measurements were obtained simultaneously. The comparison proved that it is possible to use mechanical torquemeters, which are not expensive.

The reading of maximum torque is very fast. Therefore, it is difficult to be done visually. The use of torquemeters with creeping pointer is recommended or, if this is missing, advise the operator to perform the reading only while the equipment is rotating.

Although residual torque is difficult to be obtained in sandy soils, the difference between the reading executed on the second turn and the reading on the fifth turn is not significant, in practice. It is recommended to do the measurement at the end of the second turn, while the equipment is rotating.

5 COMPARISON BETWEEN LOAD TESTS RESULTS AND ULTIMATE LOADS PREDICTED BY EMPIRICAL METHODS BASED ON SPT-T TESTS

Comparative study of the empirical methods was based on the SPT-T results and load tests executed in six experimental research sites in the southeast region of Brazil, as shown in Figure 1. These sites are located in public universities as follows: Unicamp in the city of Campinas (SP); Unesp in the city of Bauru (SP), Unesp in the city of Ilha Solteira (SP), USP in the city of São Carlos (SP), USP in the city of São Paulo (SP) and Ufla in the city of Lavras (MG). Soil stratigraphy of these places and the pile characteristics were described in Peixoto et al (2003).

Table 8 summarizes the comparison between ultimate load determined, based on load tests (Q_{ul}) and estimated failure load (Q_u) based on SPT-T test methods: Decourt (1996), Alonso (1996a and 1996b), Carvalho et al.. (1998) and Peixoto (2001).

The Carvalho et al. (1998) method was not recommended by the author for socketed and continuous flight auger pile, so it was not applied.

The Peixoto (2001) method was proposed, based on the six experimental research data, so, the statistical evaluation presented on Table 8, just for this method, was calculated considering also other data of Brazilian references, which is indicated in that thesis.

For precast concrete pile, the methods using SPT-T data appear to be interesting, since this *in situ* test simulates pile execution.

For auger and continuous flight auger piles, the "Q_u/Q_{ul} ratio" based on SPT-T methods led to satisfactory results. But, the high coefficient of variation values indicate this method already needs to be improved.

On the other hand, the results of bored piles were acceptable, which means the behavior of the auger and continuous flight auger piles can be influenced by a scale factor (compared with SPT sampler diameter) or execution type.

Again, the results for the socketed pile are conservative.

The satisfactory results by Peixoto (2001) methods probably can be explained by the utilization of the empirical factor, "x_2", that depends on T/N ratio and represents the type and structure of the soil where the foundation was executed.

Table 8 Statistical evaluation of methods dispersion using SPT-T data

Method	Q_u/Q_{ul} mean	sd	cv (%)
Precast Concrete Pile			
Decourt (1996)	1.08	0.16	14.8
Alonso (1996 a and b)	0.89	0.25	28.1
Carvalho et al.. (1998)	0.78	0.10	12.9
Peixoto (2001)	1.02	0.16	15.3
Auger Pile			
Decourt (1996)	0.71	0.32	45.1
Alonso (1996 a and b)	0.89	0.38	42.7
Carvalho et al.. (1998)	1.26	0.53	42.1
Peixoto (2001)	1.05	0.31	28.9
Continuous Flight Auger Pile			
Decourt (1996)	1.29	0.64	49.6
Alonso (1996 a and b)	1.41	0.91	64.5
Peixoto (2001)	0.86	0.33	38,5
Bored Pile			
Decourt (1996)	0.63	0.09	14.9
Alonso (1996 a and b)	0.92	0.18	19.5
Carvalho et al.. (1998)	0.89	0.12	13.5
Peixoto (2001)	0.91	0.14	15.9
Socketed Pile			
Decourt (1996)	0.58	0.16	27.6
Alonso (1996a and b)	0.54	0.09	16.7
Peixoto (2001)	1.01	0.18	17.4

6 CONCLUSION

The amount of papers about SPT-T demonstrates the Brazilian geotechnical interest on the subject, especially by engineers since it is very simple and not inexpensive procedure.

The development of an electric torquemeter and the data acquisition system allowed plotting "torque versus rotation angle". The interpretation of the curves defined: maximum and residual torque, identification of susceptible soils to vibration; studying the soil behavior after deformation as function time.

The T/N ratio can be used as a parameter to predict the soil behavior after driven foundation construction.

As a final conclusion, torque measurement is additional information that improves a traditional in situ test, SPT, which is well-known in foundation practice. New methods based on SPT-T to predict the bearing capacity of piles lead to good correlation with the load test results, for the soils studied. It is also important to emphasize that the results are preliminary and more adjustment will be necessary with its application in practice.

ACKNOWLEDGEMENT

The authors gratefully acknowledge FAPESP (São Paulo State Research Support Foundation) and CNPq (National Council of Research) for the financial support.

REFERENCES

Alonso, U.R., (1996a). "Estimativa da Adesão em Estacas a Partir do Atrito Lateral Medido com o Torque no Ensaio SPT-T". Solos e Rochas. Vol. 18, no 1, p.191-194. 1996a

Alonso, U.R., (1996b). Estacas Hélice Contínua com Monitoração Eletrônica. Previsão da Capacidade de Carga através do SPT-T. In: SEFE, III. Proceedings .São Paulo. Vol 2, p. 141-151.

Aoki, N. and Velloso, D. A., (1975). "An Approximate method to EStimate the Bearing Capacity of Piles" . In: Cong. Panamericano Mec. Solos e Eng. Fund., V. Anais. Buenos Aires.

Bandini, P and Salgado, R., (1998). "Methods of Pile Design Based on CPT and SPT results". Geotechnical Site Characterization.Vol 2: 967-976.

Carvalho, J.C.; Guimarães, R.C.; Pereira, J.H.F; Paulocci, H.V.N. & Araki, M.S., (1998). Utilização do Ensaio SPT-T no Dimensionamento de Estacas . In: COBRAMSEG, XI. Proceedings. Vol. II, p.973-982.

Decourt, L., (1998). "A More Rational Utilization of Some Old In Situ Tests" .In: Geotechnical Site Cahracterization. Proceedings. Balkema. Atlanta,USA. p. 913 a 918.

Decourt, L; Quaresma, A R., (1978). "Capacidade de Carga de Estacas a partir de Valores de SPT", In: CBMSEF, V. Proceedings. Rio de Janeiro.

Decourt, L., (1996). "Investigações Geotécnicas". In: Hachich et al, Fundações, teoria e Prática.. pp 119-162.

Decourt,L., (1991). Previsão dos deslocamentos horizontais de estacas carregadas transversalmente com base em ensaios penetrométricos In: Seminário de Engenharia de Fundações Especiais, II. Anais. ABMS/ABEF. São Paulo. Vol 2, p. 340-362.

Mayne, P.W., (2001). "Stree-Strain-Strenght-Flow Parameters from Enhanced In-Situ Tests". In-Situ 2001. Bali. Pp. 21-48.

Meyerhof, G.G., (1976). "Bearing Capacity and Settlement of Pile Foundations". Journal of the Geotechnical Engineering. ASCE, 103(3): 197-228.

NBR 6484, (1980). "Execução de Sondagens de Simples Reconhecimento dos Solos" . Associação Brasileira de Normas Técnicas. 12p.

Peixoto, A.S.P.; Carvalho, D. & Giacheti, H.L., (2003). "SPT-T and CPT Tests to Predict Bearing Capacity of Piles in Brazilian Practice. XII PCSMGE. Cambridge/MA. Soil and Rock 2003. Soil and Rock 2003. V. I, p 282-286.

Peixoto, A.S.P., (2001)"Estudo do Ensaio SPT-T e sua Aplicação na Prática de Engenharia de Fundações". Thesis (doctor degree). Unicamp. 510p.

Peixoto, A.S.P. and Carvalho, D. (1999). "Standard Penetration Test with Torque Measurement (SPT-T) and Some Factors that affect the T/N Ratio" ". XI PCSMGE. Foz do Iguaçu. Brazil. Vol 3, p 1605- 1612.

Ranzini, S.M.T. (1988). "SPT-F". Solos e Rochas, vol.11, p. 29-30.

Ranzini, S.M.T., (1994). "SPTF : 2ª parte". Solos e Rochas, v. 17, p. 189-190.

Dynamic Cone Penetrometer to estimate subgrade resilient modulus for low volume roads design

A.M. Rahim
California State Polytechnic University, San Luis Obispo, USA

K.P. George
University of Mississippi, MS, USA

Keywords: DCP, Resilient Modulus, Shelby tube, pavement design, subgrade characterization

ABSTRACT: The subgrade soil characterization, in terms of Resilient Modulus (M_R), has become important in order to provide a structurally sound and economical pavement structure. The 1986 and 1993 *AASHTO Guide for Design of Pavement Structures* recommended laboratory testing procedure AASHTO T274-83 for determining laboratory M_R values. Because of the complexities encountered with the laboratory test, in-situ tests are desirable. This becomes even more important in the design of Low Volume Roads (LVR), managed by local or county agencies, responsible for managing extensive pavement network with limited resources. This paper presents an experimental study to investigate the viability of using an Automated Dynamic Cone Penetrometer (referred to as DCP throughout the paper) for subgrade characterization through correlation between DCP index (penetration per blow) and M_R. Twelve as-built subgrade test sections reflecting a range of typical subgrade materials of Mississippi are selected and tested with DCP. Undisturbed samples are extracted using Shelby tube and tested in a repeated triaxial machine employing TP 46-94 test protocol. Resilient modulus is now correlated to DCP index. The results suggest two relations—one for fine- and another for coarse-grain soil—in correlating DCP index to laboratory M_R. The predictability of the models are substantiated by using an independent set of data at different locations throughout the State of Mississippi comparing predicted resilient modulus with the actual laboratory values.

1 INTRODUCTION

In the original AASHTO Guide for design of pavement structures, published in 1961 and revised in 1972, the subgrade stiffness was accounted for by assigning a Soil Support Value (SSV) that has a scale ranging from 1 to 10. In 1986, the AASHTO Guide was substantially revised to include replacement of SSV with subgrade Resilient Modulus, (M_R) (AASHTO 1986). M_R values may be estimated directly from laboratory testing, indirectly through correlation with other laboratory/field tests, or back-calculated from deflection measurements.

For a new design, M_R values are generally obtained by conducting repeated triaxial test on reconstituted/undisturbed cylindrical specimens. The laboratory test is tedious, costly and time consuming procedure. Large number of samples needs to be collected and tested for reasonably accurate results. Even then, it is difficult to reproduce the in-situ sample conditions (Houston *et al.* 1992). Considering the complexity of the repeated load triaxial test, field testing procedures have been proposed to estimate subgrade moduli. Of particular interest in this paper is the Dynamic Cone Penetrometer (DCP), an automated version, as a potential device for estimating M_R through correlation. The importance of these correlations lies in the simplicity of conducting DCP test on prepared subgrade to predict the resilient modulus for low volume road design at the county/local road level where resources are limited and demands are competing.

The DCP consists of a steel rod with a cone at one end that is driven into the subgrade by means of a sliding hammer. The material's resistance to penetration is measured in terms of DCP index (DCPI) in mm/blow (Hammons *et al.* 1996). Figure 1 presents the fully portable trailer-mounted device designated Automated DCP (ADCP) during operation in the field. The DCP was originally designed and used to determine the strength profile of the subgrade due to its ability to provide a continuos record of relative soil strength with depth. Numerous studies have been conducted investigating the use of DCP and factors affecting its results (Hassan 1996, Livneh *et al.* 1995, Livneh 2000, Ayers 1989, Chen *et al.* 2001).

The direct use of DCP in pavement design is yet to be established; however, it has been correlated to commonly used soil parameters, for example, CBR (Livneh et al. 1995). Burnham et al. (1993) stated that DCP could be used to provide a reasonable estimate of unconfined compressive strength of soil-lime mixtures. The shear strength of soils was correlated with DCPI, for different confining pressures, in a laboratory study conducted by Ayers et al. (1989). Employing a DCPI-CBR relation followed by another one between CBR and resilient modulus, DCP results were converted to resilient modulus which showed good agreement with both laboratory-measured and Falling Weight Deflectometer (FWD) backcalculated values (Chen et al. 2001).

Only a few studies had attempted to correlate resilient modulus to DCPI. Hassan (1996) developed a simple regression model correlating M_R with DCPI for fine-grain soils at optimum moisture content. Chai et al. (1998) used the results of DCP tests and CBR-DCP relationships developed in Malaysia during the 1987 National Axle Load study to determine in-situ subgrade elastic modulus. Jianzhou et al. (1999) analyzed the FWD deflection data and DCP results of six pavement projects in Kansas to develop a relationship between the DCPI and backcalcualted subgrade moduli. Rahim and George (2002) developed two models correlating M_R to DCPI and other soil properties for both fine- and coarse-grain soils.

The objective of this paper is to explore the feasibility of using DCP characterizing subgrade soil resilient modulus, via correlating DCP index to resilient modulus. For two different soil groups, namely fine-grain and coarse-grain soils, two different regression models are attempted correlating M_R (as dependent variable) with DCPI.

2 TESTING PLAN

2.1 Test Sections

Twelve test sections of 244 m in completed grading projects, reflecting a range of soil types in Mississippi, were selected. Table 1 lists the locations, soil classifications, and other properties of the soils in each section.

2.2 Field Testing and Sampling

2.2.1 Dynamic Cone Penetrometer Test

Prior to the experimental program in the field, side-by-side tests were conducted with a manual DCP and an Automated DCP (ADCP), establishing that they provide statistically similar results. For purpose of discussion, therefore, the penetration test results

Table 1. Locations and other physical properties of tested sections.

Roadway	County	AASHTO Class.[a]	Proctor Test Max. Dry Density[b], kN/m^3 / Optimum Moisture, %	Shelby Tube Samples Dry Density[c], kN/m^3 / Moisture[b], %
SR25	Rankin	A-6	17.4 / 14.0	18.0 / 16.8
SR25	Rankin	A-6	18.2 / 12.0	19.0 / 12.5
SR25	Rankin	A-6	17.1 / 14.3	17.0 / 16.9
SR25	Rankin	A-6	18.00 / 13.0	18.4 / 12.8
SR25	Leake	A-6	18.4 /14.0	19.2 / 11.7
US45	Monroe	A-2-4	16.7 / 15.0	17.5 / 14.2
US45	Monroe	A-2-4	16.3 / 16.0	17.4 / 16.5
US45	Monroe	A-2-4	17.1 / 14.5	18.4 / 12.4
US45	Monroe	A-2-4	15.7 / 17.5	16.8 / 16.3
US45	Monroe	A-6	17.4 / 14.5	18.1 / 16.6
US45	Monroe	A-6	17.1 / 15.5	18.2 / 13.7
US45	Monroe	A-3	15.7 / 15.5	16.5 / 17.2

a Based on the majority of samples from each section (samples from each section might have different classification)

b Standard Proctor

c Average dry density/moisture determined from the undisturbed samples

will be referred to as DCP results, though most of the results were gathered employing an ADCP as in Figure 1. DCP consists of a steel rod with a steel-cone at one end that is driven into the subgrade be means of sliding hammer. The angle of cone tip is 60°, and its base diameter is 20 mm. The hammer weighs 8 kg, and its sliding height is 575 mm. The scheme for ADCP investigation consisted of testing at 30 m intervals to a depth of 1 m in the subgrade.

Graphing the results of penetration vs. number of blows, as shown in Figure 2, possible subgrade layering was sought by investigating slope changes in the trend line. By visually inspecting the plots, points of slope change were identified thus determining different layers, each with a finite slope. The slope of different segments is designated Dynamic Cone Penetration Index (DCPI) in units of mm/blow. As shown in Figure 2, most of the 60 sample locations exhibited three layers, the top layer between 200 mm and 300 mm thick, and the second layer between 300 mm and 400 mm. The calculated DCPIs were used for correlation with laboratory measured M_R, as will be discussed in detail in a later section.

Figure 1. Automated Dynamic Cone Penetrometer in operation in the field

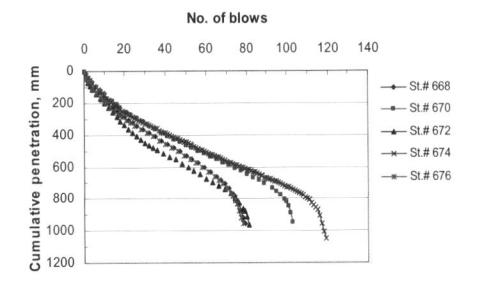

Figure 2. ADCP test results in section #12, US45-Monroe county.

2.2.2 Soil sampling and testing

For M_R testing, Shelby tube samples were obtained from five locations at 61 m intervals to a depth of 1.5 m. Retrieved from each foot was one test cylinder 71 mm diameter by 142 mm height, with the top three samples tested for M_R in the laboratory, accumulating 15 M_R values from each test section.

2.3 Laboratory tests

2.3.1 Resilient modulus test

Laboratory M_R test, following the AASHTO TP46 Protocol, was conducted employing the MDOT repeated triaxial machine furnished by Industrial Process Control (IPC). The deformation in the samples was monitored employing two Linear Variable Differential Transformers (LVDTs) mounted outside of the testing chamber. The average M_R values for the last five loading cycles of a 100-cycle sequence yielded the resilient modulus. Figures 3 and 4 are plots of M_R versus deviator stress for a fine- and coarse-grain soil samples, respectively. In total, 180 plots were prepared from which M_R of each sample was interpolated for the stress state representative of field conditions.

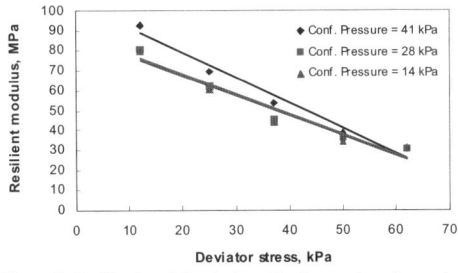

Figure 3. Resilient modulus test results, fine-grain soil sample

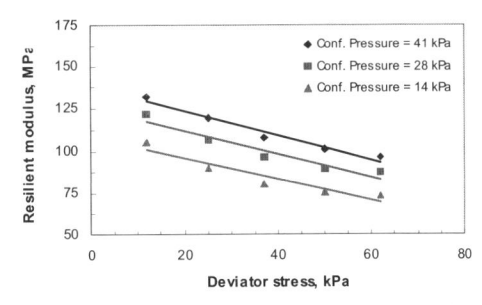

Figure 4. Resilient modulus test results, coarse-grain soil sample

3 RESILIENT MODULUS RELATED TO DCPI

3.1 Resilient modulus determination

The plan called for correlating M_R to DCPI. At each Shelby tube location, each sample from first, second, and third foot yielded M_R-stress state curves, such as in Figures 3 and 4. From penetration vs. blows plot (as in Figure 2), three DCPI values were extracted corresponding to the same sample location. Since M_R is a function of stress state, the question arises as to selecting an M_R from M_R-stress state relation. Relying on the results of Thompson and Ronett (1976), Elliot (1992) suggested using a zero confining pressure and a 41.6-kPa deviator stress when selecting an M_R value from laboratory test data. Recognizing that in place subgrade has to sustain the overburden of pavement layers, in addition to the standard 18-kips axle load, in situ stresse of a typical pavement is combined with stress due to a 20 kN wheel load at a tire pressure of 690 kPa. Stresses calculated by KENLAYER (Huang 1993), yielded a stress state of 37 kPa deviator stress and 14 kPa lateral stress. Making use of the above stress combination, one M_R value for each sample was interpolated from plots such as in Figures 3 and 4.

3.2 M_R – DCPI correlation

By necessity, the 180 test cylinders from 12 test sections were classified into two groups: fine-grain and coarse-grain soil in accordance with AASHTO M145-87 (AASHTO 1990). For each group, one regression model relating M_R to corresponding DCPI was attempted.

The Statistical Package for the Social Science (SPSS) (Norusis 1993) was used to perform the regression analysis in this study. Table 2 presents the ranges of modulus and DCPI values for both fine- and coarse-grain soils. In developing regression models, the curve estimation option in SPSS was employed investigating the best forms of relation between M_R and DCPI, employing coefficient of determination (R^2) as the best-fit criterion.

Table 2. Ranges of both dependent and independent variables for fine- and coarse-grain soil.

Soil Type	Variable Notation	Description	Range
Fine-grain	M_R	Resilient modulus, MPa	31– 269
	DCPI	Penetration index, mm/blow	3.7– 66.7
Coarse-grain	M_R	Resilient modulus, MPa	28– 158
	DCPI	Penetration index, mm/blow	5.6 – 40.0

The nonlinear regression option in SPSS was employed to determine the regression coefficients. After an exhaustive search, examining many different forms the following model forms were selected.

Fine-grain soil:

$$M_R = 532.1(DCPI)^{-0.492} \quad \ldots\ldots\ldots\ldots\ldots\ldots(1)$$

$R^2 = 0.4 \quad RMSE = 35.3$

Coarse-grain soil:

$$M_R = 235.3(DCPI)^{-0.475} \quad \ldots\ldots\ldots\ldots\ldots\ldots(2)$$

$R^2 = 0.4 \quad RMSE = 18.5$

where: R^2 = Coefficient of determination, and

$RMSE$ = Root Mean Square Error

3.3 Model Verification

To verify the predictability of the developed models, an independent set of data was made available from a recently conducted study sponsored by the Mississippi Department of Transportation (MDOT) to investigate the viability of using FWD in characterizing subgrade soils (Bajrachayara 2003). Nine test sections were tested using DCP and undisturbed Shebly tube samples were extracted and tested employing TP 46-94 protocol.

The laboratory MR values were determined, for the three samples from each location, at stress combination of 37 kPa deviator stress, and 14 kPa confining pressure. Other soil properties were determined and samples were grouped into two groups; fine- and coarse-grain soils according to AASHTO soil classification.

The predicted moduli were plotted against actual modulus values as presented in Figures 5.a and 5.b for fine- and coarse-grain soils, respectively. The points clustering along the line of equality is another indication the powerful predictability of developed models.

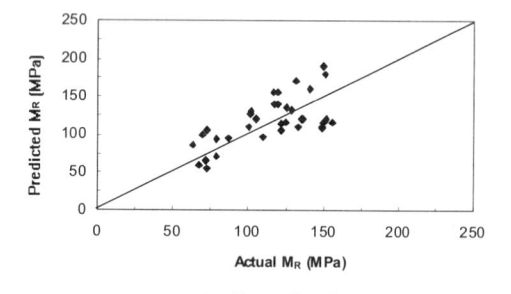

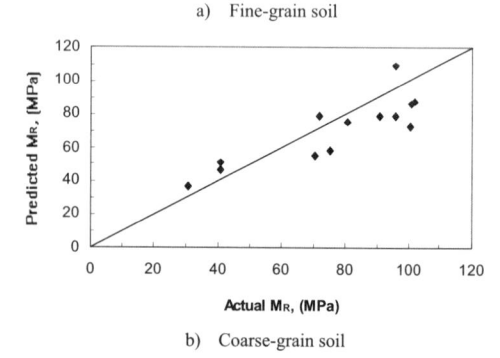

a) Fine-grain soil

b) Coarse-grain soil

Figure 5. Actual vs. predicted resilient modulus for two soil groups

4 SUMMARY AND CONCLUSION

The focus of this study was to investigate the use of Dynamic Cone Penetrometer (automated version) for subgrade soil characterization. In 12 sections of prepared subgrade, DCP test was conducted and undisturbed samples were collected for MR testing. MR values were regressed against DCP index obtaining two models, one each for fine- and coarse-grain soil. The models are verified by repeating the tests in another site and comparing measured and predicted moduli. Summarized herein are the significant conclusions:

1- Automated DCP is a simple and expedient device for field testing of soils and particulate material.
2- As mandated by data, two models were developed—one for fine- and another for coarse-grain soil.
3- The model predictability is substantiated by testing 9 independent test sections throughout the State of Mississippi with excellent comparison between measured and predicted M_R values.

By way of general conclusion/recommendation, it is advanced that for the range of Soils tested, the developed M_R-DCPI models provide useful predictions of resilient modulus especially for low volume roads.

ACKNOWLEDGEMNT

This paper includes some results from the two studies, " Subgrade Characterization for Highway Pavements", and "Falling Weight Deflectometer for Estimating Subgrade Resilient Moduli", both sponsored by the Mississippi department of Transportation (MDOT) and the U.S. department of Transportation, Federal Highway Administration. Assistance received from MDOT staff in the filed work is acknowledged. The opinions, findings and conclusions expressed in this report are those of the authors and not necessarily those of the MDOT or the Federal Highway Administration. This does not constitute standard specification or regulation.

REFERENCES

AASHTO Guide for Design of Pavement Structures, American Association of State Highway and Transportation Officials 1986, Washington, DC.

AASHTO Standard Specifications for Transportation Materials and Methods of Sampling and Testing, 1990, 15th Edition.

Ayers, M., Tomspson, M., and Uzaraki, D. 1989 "Rapid Shear Strength Evaluation of In-Situ Granular Materials," Presented at the 68th Transportation Research Board Annual Meeting, Washington, DC.

Bajracharya, M. 2003 "In-situ Subgrade Soil Characterization using Falling Weight Deflectometer", Master Thesis, The University of Mississippi, Oxford, Mississippi.

Burnham, T. and Johnson, D. 1993 "In-Situ Foundation Characterization Using the Dynamic Cone Penetrometer," Report MN-93/05, Minnesota Department of Transportation, Maplewood.

Chai, G. and Roslie, N. 1998 "The Structural Response and Behavior Prediction of Subgrade Soils using Falling Weight Defelectometer in Pavement Construction", 3rd International Conference on Road & Airfield Pavement Technology.

Chen, D-H., Wang, J-N. and Bilyeu, J. 2001 "Application of the DCP in Evaluation of Base and Subgrade", Paper Presented at the 80th Annual Meeting of Transportation Research Board, Washington, D.C.

Elliot, R.P. 1992 "Selection of Subgrade Modulus for AASHTO Flexible Pavement Design", In Transportation Research Record 1354, TRB, National Research Council, Washington, D.C.

Hammons, M.I., Parker, F., Malpartida, A.M. and Armaghani, J.M. 1996 "Development and Testing of an Autoimated Dynamic Cone Penetrometer for Florida Department of Transportation", Draft Report, Contract FLDOT-ADCP-WPI #0510751.

Hassan, A. 1996 "The Effect of Material Parameters on Dynamic Cone Penetrometer Results for Fine-grained Soils and Granular Materials," Ph.D Dissertation, Oklahoma State University, Stillwater, Oklahoma.

Houston, W.N., Mamlouk, M.S., and Perera, R.W.S. 1992 "Laboratory versus Nondestructive Testing for Pavement Design," ASCE Journal of Transportation Engineering, Vol. 118, No. 2, pp. 207-222.

Huang, Y.H. 1993 *Pavement Analysis and Design*, by Prentice-Hll, Inc., New Jersey.

Jianzhou, C., Mustaque, H. and LaTorella, T.M. 1999 "Use of Falling Weight Deflectometer and Dynamic Cone Penetrometer in Pavement Evaluation", Paper Presented in the Transportation Research Board, Washington, D.C.

Livneh, M. 2000 "Friction Correction Equation for the Dynamic Cone Penetrometer in Subsoil Strength Testing," Paper Presented at the 79th Transportation Research Board Annual Meet, Washington, DC.

Livneh, M., Ishai, I., and Livneh, N.A. 1995 " Effect of Vertical Confinement on Dynamic Cone Penetrometer Strength Values in Pavement and Subgrade Evaluation," In Transportation Research Record 1473, TRB, National Research Council, Washington, D.C.

Norusis, M.J., 1993 "SPSS for Windows, Professional Statistics, release 6.0", Chicago, IL: SPSS Inc.

Rahim, A. and George, K.P. 2002 "Automated Dynamic Cone Penetrometer for Subgrade Resilient Modulus Characterization", Transportation Research Record No. 1806, 70-77, Washington, DC.

Thompson M.R., and Robnett, Q.L. 1976 "Resilient Properties of Subgrade Soils, Final Report—Data Summary", Transportation Engineering Series 14, Illinois Cooperative Highway Research and Transportation Program Series 160, University of Illinois at Urbaba-Champaign.

Soil profile interpretation based on similarity concept for CPTU data

F. Saboya Jr. & D.J.G. Balbi
Laboratory of Civil Engineering, State University of Norte Fluminense Darcy Ribeiro – UENF, Campos, RJ, Brazil

Keywords: piezocone test, clustering, similarity

ABSTRACT: Interpretation of soil profile from piezocone data is sometimes a difficult task specially, when the subsoil is highly stratified. For such a condition it seems interesting to use statistical treatment of the data in order to identify soil groups and the boundary amongst layers that could deserve special attention. Herein the cluster analysis is used as a classifying process in order to better define the description of the subsoil conditions at the geotechnical viewpoint. It is also shown the capability of this method which is of easy implementation in ordinary worksheets programs.

1 INTRODUCTION

Piezocone tests have become popular in past decade as a helpful and powerful tool in assessing the stratigraphic sequence of soil layers as well as in evaluating in-situ geotechnical parameters of sub-soil.

CPTU consists of static pushing of a probe into the soil where the point load, lateral friction and pore pressure are recorded in small intervals of depth.

The collected data are transformed into geotechnical parameters through charts and correlations proposed by several researches (Robertson, 1990; Robertson et. al., 1992; Senneset et. al., 1989)

In order to compare and classify the results, it is advisable to normalize the data. However the methodologies to do this are still open to debate. There are many proposals based on charts and empirical formulations.

Moreover, sometimes the main challenge resides in the precise assessment of layering of sedimentary deposits and the detection of lenses or mixed soil, which may play important role in the geotechnical behavior of the subsoil.

Therefore the main aim of this work is to use a statistical method based on similarity criteria of normalized values of point strength Q and Pore-Pressure B_q, as depicted and developed by Hegazy and Mayne (2002).

The CPTU tests were carried out in, Macae city, north of Rio de Janeiro State in a sedimentary basin of quaternary clayey soils.

2 SIMILARITY ANALYSYS

Similarity analysis is a statistical method to mining data that belong to a specific group that depicts similar mathematical characteristics (Everitt, 1974).

Several techniques have been proposed to analyze similarity. However, Hegazy and Mayne (2002) have suggested the single-link hierarchical method clustering for analyzing data from CPTU, due to its properties of being least affected by outliers and no prior information of data statistical distribution is necessary.

As reported by Hegazy and Mayne (2002) cluster analyses have as main target: (1) determination of groups similar of soils; (2) establishment of boundaries between layers and (3) location of lenses and mixed soils that cannot be classified in any pre-defined group of soil. Each group corresponds to a cluster number (N_c) and the number of clusters represents different types of soils or even the same soil with different properties.

Hegazy and Mayne (2002) point out that the development of clustering analysis requires some principal steps: (1) selection of variables to be analyzed; (2) standardization of the data; (3) determination of similarity matrix; (4) choice of clustering technique, (5) number of clusters and (6) interpretation.

The first step, variables selection, consists of the choice of most appropriated parameters obtained from CPTU data to represent the subsoil conditions. As explained by Hegazy and Mayne (2002) the normalized tip resistance and normalized parameters, given by Eq. 1 and Eq. 2 are most suitable for describing the soil profile and layering.

$$B_q = \frac{(u_2 - u_0)}{(q_t - \sigma_v')} \quad (1)$$

$$Q = \frac{(q_t - \sigma_v)}{\sigma_v'} \quad (2)$$

The next step is to normalize the data and identify those that are clearly outside of general trend in order to avoid influence on similarity matrix. This identification of abnormal data is done through the Z_{score}, which has zero mean and unitary standard deviation, defined by Eqs. 3 and 4.

$$\text{zscore} = z_{ij} = \frac{x_{ij} - E(X_j)}{\text{stddev}(X_j)} \quad (3)$$

$$\text{stddev}(X_j) = \sqrt{\frac{\sum_{i=1}^{n}[x_{ij} - E(X_{ij})]^2}{n-1}} \quad (4)$$

$E(X_j) = \text{mean}(X_j)$

x_{ij} = is the variable under consideration

n = number of data points

In order to assemble the similarity matrix, it is necessary to calculate the "distance" among a particular data (i) and the others subsequent (j).

This distance is given by, Eq. 5. For CPTU data the distance between B_q and Q for each depth considered is calculated.

$$d_{ij} = \frac{\sum_{i=1}^{n} x_{ik} x_{jk}}{\sqrt{\sum_{k=1}^{n}(x_{ik})^2 \sum_{k=1}^{n}(x_{jk})^2}} \quad (5)$$

The next step consists of choosing the number of cluster to be used in the analysis. This is the point where the reasoning is the key, because one can use as many cluster as data number. However, studies done by Hegazy (1998) have shown that, for CPTU tests, the ideal number of cluster should be between 2 to 100. But the author mentions that the maximum number of clusters is hardly higher than 15.

The last step is the interpretation of the results that are focused on the layering (soil profile), transition zones, lenses and anomalies.

Results obtained from similarity analysis shall be interpreted with caution, once a cluster with high similarity amongst data cannot be considered as a layer unless it shows thickness that characterizes it as a layer. The criteria adopted herein to consider an actual layer, is to show a set of data that corresponds to, at least, 50 cm in thickness. If the thickness is higher than 50 cm but less than 1.0m it can be considered as secondary layer or soil mix, according to established by Hegazy and Mayne (2002)

3 CPTU DATA

The data used herein were obtained from CPTU test carried out in Macae City, north of Rio de Janeiro State, Brazil.

No further considerations on the tests will be done because the main aim is to present the technique itself.

Initially the data were treated using the Z_{score} with depth in order to verify outliers and points with abnormal behavior, as shown in Fig. 1.

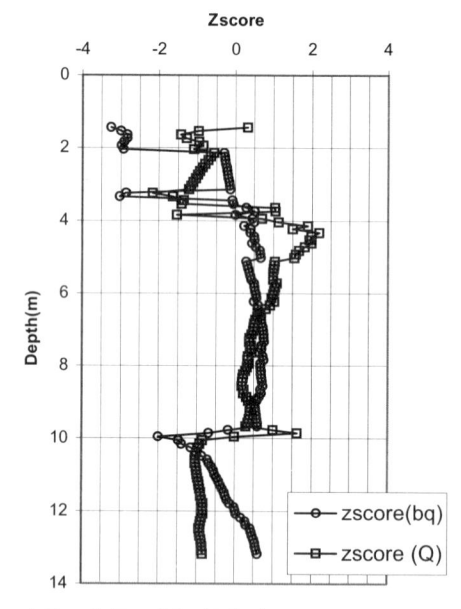

Figure 1. Z_{score} for B_q and Q with depth

It is very important to mention that, in this paper, the data interval used to classify the material was 0.10m in order to detect well-defined boundaries amongst layers as well as lenses and transitions. This has resulted in a similarity matrix of 118 rows and 118 columns.

To consider such a small interval it was necessary to adopt the following procedure to make the analysis feasible:

a) After assembling the similarity matrix, a summation of each column is done, i.e, each point has its own summation of similarity. The point that possesses higher value is taken as the reference for comparison to the others. Therefore a clear stratigraphy of the profile is obtained, once this point (reference) has the higher similarity compared to the others. However, some depth intervals are not fulfilled with this initial procedure and further steps are necessary until all intervals are fulfilled.

b) This next step consist of determining which point that belongs to not-fulfilled intervals has higher similarity summation, similar to item a.

This procedure is repeated until all columns of soil have been analyzed, as shown (a small interval due to the size of the data) in Table 1.

Table 1. Clustering Analysis for transition zones and mixed soils

Similarity(1)	Similarity(2)	Similarity (3)	Profile
1,000	0,985	0,980	Clay 1
1,000	0,988	0,976	
1,000	0,991	0,973	
1,000	0,993	0,970	
0,994	0,995	0,966	Mixed soil A1*
0,992	2,000	0,963	Clay 2
0,990	2,000	0,959	
0,988	2,000	0,956	
0,985	2,000	0,952	
0,983	2,000	0,948	
0,980	2,000	0,944	
1,000	0,983	3,000	Mixed soil A2*
0,975	0,919	0,993	
0,972	2,000	0,933	Clay 2
0,969	2,000	0,929	
0,993	0,949	3,000	Clay 3
1,000	0,956	3,000	
0,959	0,994	0,917	Transition S.1 - S.3
1,000	0,965	0,993	
0,994	0,953	3,000	Clay 3
0,976	0,919	3,000	
0,987	0,939	3,000	
0,975	0,919	3,000	

The number of clusters is determined by the necessity of the refinement. For a higher similarity degree, 0,99, sometimes it is necessary a higher number of cluster. However, there are not closed rules to establishing an ideal number of clusters, not an ideal similarity degree, either. Each case should be treated considering its own characteristics.

It is worthwhile to highlight that groups that have similarity with more that one cluster are considered transition layers. However some caution should be used when interpreting transition, taking into account the position and its relevancy for the interval under consideration.

Classification of layers with low or none similarity degree is considered to follow a special rule. If the layer cannot be classified in any cluster with more than 0.5m of relevancy, then this can be considered as transition, lenses or mixed soil according to the following:

a) Transition: the top of the non-similar layer shows high initial similarity with the immediately upper layer and the bottom of the layer shows similarity with the top of the immediately lower layer.

b) Mixed Soil: when the similar layer shows similarity with more than one cluster, and these do not show any boundary with this layer.

c) Lenses: classified as layers that do not possess any similarity with any other layer.

4 RESULTS

The results (Fig. 2) have shown that for a similarity of 0.99, the layers, transition zones and mixed soil are quite well defined. The resulting soil profile is shown in Figure 3 where it can be seen the great complexity of the soil profile with stratification, intercalation beds and transition zones.

This is typical of fluvial-lacustrine deposits where the inter-bedded layers play a crucial role in the mechanical behavior of subsoil.

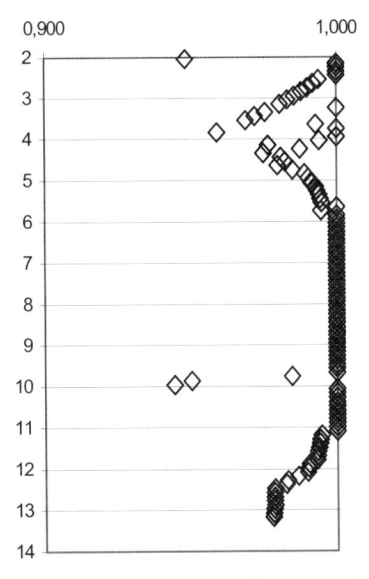

Figure 2 – Similarity for layers and transition zones with depth (similarity 0,99).

Soil Profile, Similarity 0.99

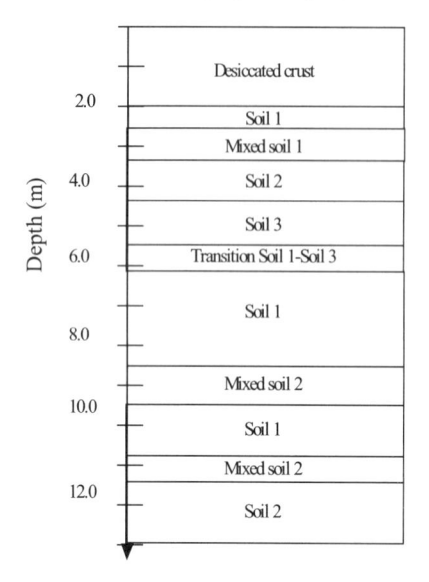

Figure 3 – Soil Profile obtained using similarity of 0,99.

5 CONCLUSIONS

Similarity analysis has become a helpful tool to assess soil profile due to its capacity in detecting changes that cannot be detected by sight or classification charts. This is particularly important when dealing with strongly layered soil profile where the presence of lenses is decisive for analysis involving consolidation.

It is worthwhile to detach that this method is capable to identify systematic errors as rod changes and casual pauses in the testing helping thus, a better understanding of mechanical properties of the subsoil. Another important feature of this method resides in the interaction with the interpreter by choosing the better number of cluster and/or the optimum similarity level to be used for a particular site or borehole.

REFERENCES

Everitt, B. (1974). *Cluster Analysis,* Halsted-Wiley, N.Y.
Hegazy, Y.A. ad Mayne, P.W. (2002) "Objective site characterization using clustering of piezocone data", *Journal of Geotechnical and Geoenvironmental Engineering,* 128 (12), 986-996
Hegazy, Y.A. (1998). "Delineating geostrratigraphy by cluster analysis of piezocone data." PhD Thesis, School of Civil and Environmental Engineering Dept. Georgia Inst. Of Technology, Atlanta.

Robertson, P.K. (1990). "Soil classification using the cone penetration test" *Canadian Geotechnical Journal.,* 27 (1), 151-158.
Robertson, P.K.; Sully, J.P.; Woelle, D.J.; Lunne, T.; Powell, J.J.M. & Gillespie, D.G. (1992). "Estimating coefficient of consolidation from piezocone tests" *Canadian Geotechnical Journal,* 29, 539-550.
Sennesset, K.; Sanven, R.; & Janbu, N. (1989) "Evaluation of soil parameters from piezocone testes", *Tranportation Research Record,* 1235, 24-37.

Centrifuge penetration tests in saturated layered sands

M.F. Silva & M.D. Bolton
Engineering Dept., University of Cambridge, Cambridge, UK

Keywords: centrifuge, piezocone, layered soil, sands, viscous fluid

ABSTRACT: In principle, the penetration resistance of saturated sands should be a function of relative density, relative crushability, and relative speed of penetration. Excess pore pressures due to high penetration velocities should be measurable, and related to soil permeability, and therefore grain size. Piezocone tests (PCPT) have been carried out in centrifuge models to investigate the sensitivity to layering and grain size. Models consisted of fine silica sand sandwiched between two layers of coarse silica sand. A 12-mm diameter model piezocone was pushed into the models at different penetration rates to observe the effect of layering under different loading conditions and permeabilities of the sand layers. Additionally, a more viscous pore fluid was used in a second series of penetration tests to correct the centrifuge modelling of penetration process and flow of fluid within soil. The data are used to discuss the methodology of characteristic grain size interpretation and the use of viscous fluid in centrifuge tests.

1 INTRODUCTION

Since its introduction as an *in situ* investigation device the cone penetration test (CPT) has been providing information which geotechnical practioners have used to identify soil stratigraphy. With the addition of pore pressure measurement the PCPT (or piezocone test) can lead, by careful interpretation, to good judgments about the characteristic grain size of soils.

However, the charts to identify the characteristic size of the soil grains are based on the uniformity of soil conditions. In reality, soils are the most heterogeneous of materials and the occurrence of layers of different grain size in a soil deposit can complicate the interpretation of PCPT data.

As the cone tip advances into different soil layers, the piezocone senses the effects of an approaching layer as well as being influenced by the soil properties behind its tip. This is due to the disturbance in both soil and pore fluid conditions surrounding a piezocone, which are associated with quantities q_c (cone resistance), f_s (sleeve friction) and u (pore pressure) measured on the probe. Figure 1 shows schematically the zones affected by the insertion of a cone. The disturbance zones interact with any inherent layering, changing the PCPT output.

Cone penetration data are used most directly in pile design. Following Kerisel's (1961) observations of embedment and size effects on full scale piles, De Beer (1963) used a classical bearing capacity solution to explain that the penetration resistance depends on the development of a static failure mechanism.

However, De Beer's (1963) theoretical explanation does not correctly reflect the steady penetration of a PCPT through different layer conditions. Investigations on the influence of layers on the penetration resistances were hence conducted mostly on model pile tests (e.g. Meyerhof and Valsangkar, 1977), calibration chamber tests (see Lunne *et al.*, 1997) and other special cases (e.g. Treadwell, 1976).

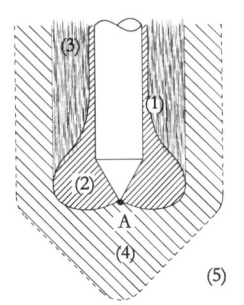

(1) Disturbed soil subject to shearing
(2) Direct disturbance zone when cone passes point A
(3) Previously disturbed soil
(4) Pre-compressed zone
(5) Undisturbed zone

Figure 1- Soil disturbance during cone penetration (adapted from Muromachi, 1981)

Table 1- Investigations on the effects of penetration rate in centrifuge models

References	Rate range (mm/s)	Probe diameter (mm)	g-level*	Remarks
Fine-grained soils				
Almeida and Parry (1985)	2-20	12.7	1g	Penetration rates between 2 and 20 mm/s have little effect on q_c in kaolin clay
Renzi et al. (1994)	2-60	11.3	100g	Increasing the penetration rate in silty clay, both cone resistance and pore pressure increased (q_c increased by 15%)
Coarse-grained soils				
Ferguson and Ko (1981)	1-19	8.0	4.5g	Moderate variations in the penetration rate would not influence q_c adversely. However, paper shows great scatter of data, q_c showing lower values at lower penetration speed
Corte et al. (1991)	0.55-2.2	20	6g	Saturated sand showed lower q_c than dry sand. Though neglected, the lowest speeds showed greater q_c values than the greater speeds
Van der Poel and Schenkeveld (1998)	1-250	11.2	35.6g	The cone resistance of the fastest CPT falls in between the slowest and medium CPT
Sharp et al. (1998)	1.0	4.0 8.0 12.0	9g 4.5g 3g	Excellent data agreement between the dry and saturated soil models

*g is the gravity acceleration

Recently, the layering effect has also been studied by numerical simulation (Ahmadi et al., 1999) and by centrifuge testing (Morley, 1997; Wright, 1998). The observed variability of soil behaviour type is inferred due to the occurrence of peaks and troughs on the cone penetration-graphs. The troughs are generally believed to be entirely due to more compressible soil layers, but the question arises whether they can alternatively be indicating layers of low permeability.

In this paper, centrifuge penetration tests into a three-layer silica sand system are described with particle size differences between successive layers as the prime variable. The intermediate layer consisted of fine sand to reduce the model permeability. To make the penetration resistance independent of changes on soil strength, the system consisted of sands of the same origin and the same relative density. To identify the fine sand layer the penetration tests were performed at different rates in the saturated soil models. A special 12-mm diameter model piezocone was designed to provide information as an *in situ* electric piezocone.

2 THE PENETRATION RATE EFFECT IN LAYERED SOILS

Soils create positive excess pore water pressures ahead of an advancing CPT probe. For the full range of possible penetration rates, undrained to partially or fully drained penetration can take place depending on the soil permeability and loading conditions. However, the excess pore water pressure in the vicinity of the piezocone filter element results from a combination of the physical displacement of soil and fluid during the penetration of piezocone, as well as from shear stresses generated in Zone 1, Fig. 1 (Burns and Mayne, 2002).

The common trend is that increasing the penetration rate, the dynamic pore pressure and cone resistance also increase (Danziger and Lunne, 1997, *apud* Lunne et al., 1997). On the other hand, measured negative excess pore water pressure in stiff clays, silts or dense fine sands indicates that soil relaxation or dilatancy is occurring, even though large positive pore pressures may exist just a few cone diameters beneath.

In the case of soil with layers of different permeabilities, the piezocone insertion modifies not only the properties of the surrounding soil from its initial condition but also the ambient total and effective stresses of the approaching layer due to pore pressure effects. The presence of a fine layer may hence restrict the water flow in the compressive zone ahead of the cone tip (Zone 4, Fig. 1). This could strongly influence the penetration resistance and it is more pronounced in fine-grained layers. Santoyo (1982), for example, observed consistent differences in *in-situ* penetration resistances at different speeds in soft clays interbedded with sand and hard silty clay lenses.

3 CENTRIFUGE MODELLING

CPTs in centrifuge models are useful to delineate layering, to demonstrate inhomogeneity within layers, and to validate comparisons with required prototype conditions (Bolton et al., 1999), and to investigate soil density changes provoked during testing, such as following earthquakes (Teymur, 2002).

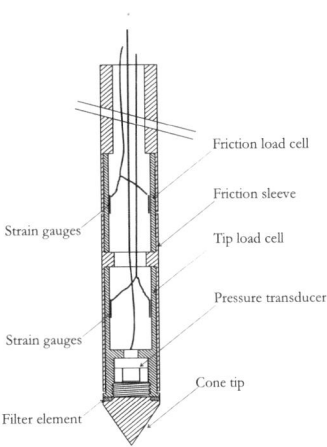

Figure 2- Model piezocone (not to scale)

The degree of drainage during a penetration process can best be inferred by changing the penetration rate, or by measuring ambient pore pressure within the soil, as well as on the piezocone. To date, most attempts to find the penetration rate effect were restricted to dynamic pore pressure measurements in fine-grained soils in the field. Few penetration testings to study the penetration rate effect using centrifuge facilities are found in the literature, and some are summarized in Table 1.

In *quasi*-static modelling at Ng, centrifuge tests are pleased to accept the N^2 factor on consolidation rate in accordance to Terzaghi's theory for a model with its drainage distances reduced by a factor N. In earthquake testing, centrifuge modellers must increase the frequency of cyclic loading by factor N to keep earthquake accelerations in proportion. They must therefore reduce the fluid viscosity by the model scale factor, so that the rate of creation of excess pore pressures is correctly balanced by the rate of dissipation to the model drainage boundaries.

This strategy of reducing soil permeability makes it more important to understand when the penetration speed of a penetrometer will itself induce excess pore pressures Δu during probing. Dimensional analysis of the penetration process reveals that

$$\Delta u \propto \eta D \bar{\varepsilon}_v v / K \qquad (1)$$

where η is the kinematic viscosity (e.g. 10^{-1} Nm^{-2}s for water at 20^0 C), D is the diameter of the penetrometer, $\bar{\varepsilon}_v$ is the mean volumetric compressive strain of the soil influenced by the passage of the penetrometer, v is the rate of penetration, and K is the intrinsic permeability of the soil.

It is clear from Eq. 1 that a model CPT probe of 12 mm diameter and 10 mm/s velocity, for example, should create smaller excess pore pressure by a factor 7.4 compared with a standard 35.7 mm diameter probe travelling at 25 mm/s in identical soil saturated with the same fluid and compressing similarly, i.e. under identical stresses. However, if the fluid viscosity is increased by a factor of 50, for example, then even the smaller, slower centrifuge CPT should create 7 times larger excess pore pressures than would a CPT in the same soil in the field.

In order to begin an assessment of this situation, two soil models were prepared for CPT sounding at different speeds with an identical fine sand layer, and saturated either with water or with a fluid whose viscosity was 50 times greater than water.

4 EQUIPMENTS AND INSTRUMENTATIONS

4.1 *Model piezocone*

A 60^0-conical tip piezocone with 12-mm outer diameter was designed as shown in Figure 2. The length dimensions were proportionally scaled down by a factor of three from specifications in the International Reference Procedure (IRPT, 2001). The conical tip and friction sleeve consisted of stainless steel so that consecutive penetration tests could be performed without replacement due to excessive abrasion.

Two load cells were designed to measure the cone resistance and sleeve friction. The adopted design also allowed the strain gauges to be glued to the load cell inner walls.

The dynamic pore pressure u_2 was measured by a commercial pressure transducer located inside the tip load cell. The pressure transducer was held in place using silicone rubber, which was also used as a sealing between the load cells and the piezocone shafts. The cone tip is removable so that the piezocone pressure system can be saturated. Sintered bronze was used as the filter element located at the cone shoulder.

The pressure transducer was calibrated inside a vessel filled with water. Compression was applied in the vessel, and both pressure and output readings were recorded. The process also allowed to verify the sealing against water leakage inside the piezocone, and to measure the cone area ratio a due to the effect of cone unequal area. A value of a equal to 0.75 was hence determined.

4.2 *Driving mechanism*

The mechanism for driving the model piezocone was designed to be used in soil models with high shearing resistance. The equipment design criterion consisted of a high load capacity up to 10 kN at any linear vertical speed (max. 10 mm/s). The actuator allows a maximum displacement of 300 mm, monitored continuously by a laser displacement transducer.

Figure 3 illustrates the driving mechanism. The model piezocone is attached to a carriage plate by a

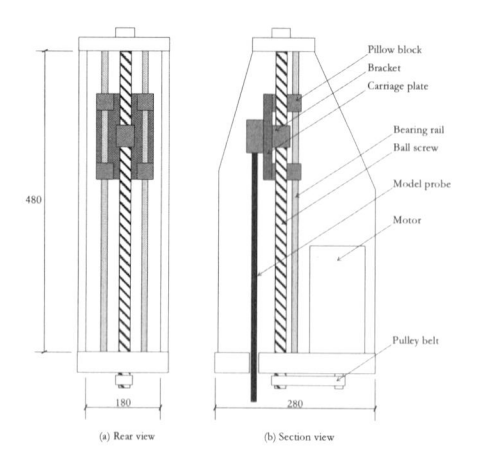

Figure 3- Schematic design of the driving mechanism (dimensions in mm, not to scale)

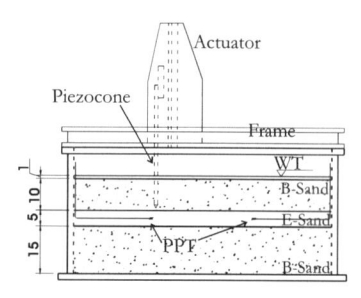

Figure 4- Section of the centrifuge model (units in cm, not to scale)

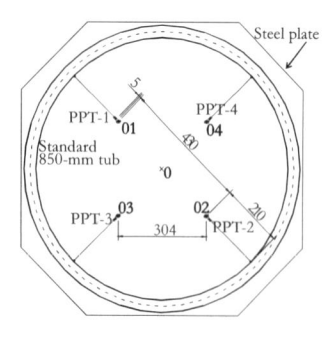

Figure 5- Location of the penetration tests and instrumentations in the 850 mm tub (units in mm, not to scale)

Table 2- Properties of the silica sands

Property	Fraction B	Fraction E
Grain size d_{10} (mm)	0.84	0.095
Grain size d_{50} (mm)	0.90	0.14
Grain size d_{90} (mm)	1.07	0.15
Specific gravity (G_s)	2.65	2.65
Maximum void ratio (e_{max})	0.820	1.014
Minimum void ratio (e_{min})	0.495	0.613
Friction angle at cons. volume (ϕ'_{cv})	36^0	32^0
Hydraulic conductivity (m/s)	0.7×10^{-2}	0.98×10^{-4}

series of aluminium connections. The carriage plate is driven by a ball screw and is guided by two slide rails. Four pillow blocks connect the guide rails to the carriage plate, providing linear vertical movement and thus not allowing rotation. The ball screw is connected to a 35 V DC servomotor by a timing belt on the bottom of the equipment. On the top of the screw system a spherical bearing supports the entire axial load which is generated during both upward and downward movement of the carriage plate. The load is then redistributed on the equipment frame. Also, the moments generated by the pillow blocks on the slide rails are supported by a vertical 20-mm thick aluminium plate.

4.3 Soil model

Commercially available silica sands of E and B British Standard fractions were used for the tests. The mechanical behaviour of the sands is well documented by high quality experimental data (e.g. Lee 1989, Tan 1990). Relevant geotechnical parameters are listed in Table 2.

The soil models consisted of three layers as shown in Figure 4. The sands were pluviated dry inside an 850-mm diameter steel tub to a given relative density (75%), instrumented with pore pressure transducers (PPTs), and saturated layer by layer. The first sand poured was the fraction B with 150 mm thickness. The 50-mm thick intermediate layer consisted of fraction E silica sand. The upper layer was again fraction B sand. The sand layers total is 300 mm in height.

4.4 Pore pressure transducers (PPTs)

The pore pressure transducers used for the centrifuge penetration tests were the Druck PDCR (see Konig et al., 1994). A sintered bronze metal with high air entry was used as a porous element to separate the soil from the PPT pressure diaphragm for measurement.

The pore pressure transducers were positioned in the fine layers during the sand pouring process. They were inserted halfway through the dry fine sand layers, at a horizontal distance of 11 mm from the location of the piezocone penetration (or 5 mm from the piezocone shaft, Fig. 5).

5 TEST PROCEDURES

5.1 Saturation techniques

The saturation processes took place before the models were transported to the centrifuge. The first soil model was saturated with de-ionised water. The water was inserted from four entrances located at the bottom of the tub at very low hydraulic gradient ($i \sim 0.1$). The process of saturation of the first model did not take more than 15 minutes. The model contained a total volume of 68 litres of water.

The second model was saturated with a 50-cP viscous fluid. The viscous fluid consisted of a solution of hydroxypropyl methylcellulose at a mass concentration of 2.2%. The concentration was selected according to the calibration curve shown in Figure 6. The curve was obtained following similar procedures adopted by Stewart et al. (1998). The equation is valid for the range of concentrations tested, between 0.8 and 3.6%. Additionally, a quantity of phenol equal to 0.5% of the mass of methylcellulose in the solution was added to minimize biological contamination.

In order to saturate the second model with the viscous fluid, a pressure lid was sealed over the top of tub and vacuum was then set up, creating a pressure gradient. The fluid feeding system was connected up to the four holes located at the bottom of the tub. Due to the high pressure gradient expected, the fluid rate of flow was then set at 0.7 l/h to avoid hydraulic failure by liquefaction. The saturation process took four full days to be completed. A total of 68 l of fluid was used to saturate the soil model.

5.2 Penetration tests

In each of the two soil models, four penetration tests were performed at different penetration rates. To avoid the boundary influence on the penetration resistances the piezocone was located according to the recommendation by Bolton et al. (1993): the penetration of a 12-mm diameter probe would not affect the properties of the soil outside a 480-mm diameter cylinder. The best suited locations and sequences of the penetration testings are shown in Figure 5, numbers 01 to 04. The cone tip was also stopped at 10 cm above the tub bottom to avoid the base boundary effect (Lee, 1989).

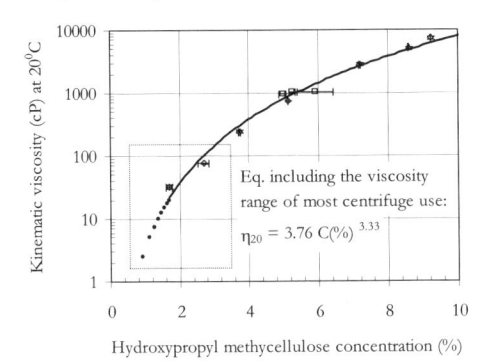

Figure 6- Kinematic viscosity at a temperature of 20^0 C (η_{20}) versus solution concentration of hydroxypropyl methycellulose

The centrifuge tests were performed at a centripetal acceleration of 50g. For each centrifuge penetration test, time was allowed after the centrifuge spinning-up for stabilization of the instrumentation readings. In each soil model, the first penetration test was performed at 0.4 mm/s. The second, third and fourth penetration test rates were respectively 0.8, 4.0 and 8.0 mm/s.

A fifth penetration test (control test) was also performed at the centre of each soil model. The objective of this test was to indicate any disturbance due to the four previous tests in the soil model as well as of the centrifuge stoppages to change the position of the driving mechanism. The penetration rate of the control test was the same as the first test: 0.4 mm/s. No significant changes of the cone resistance were observed when compared with the first penetration tests at 0.4 mm/s. Settlements were also measured during stoppage, but no changes were found.

6 RESULTS

The results of the first test series using the first soil model are shown in Figure 7. Only the four penetration tests at different rates on the locations specified in Fig. 5 are plotted. It can be observed that up to 60 mm model depth, the corrected cone resistances (q_t) at any rate increase almost linearly. The rate of increase then reduces over the next 40 mm (~ 3 cone diameters) in anticipation of the lower penetration resistance of the fine sand layer. The q_t value in the fine layer returns to the higher trend-line about 3 diameters above the lower layer of coarse sand. Ultimately, at a depth of 180 mm, i.e. at an effective overburden pressure corresponding to a depth of about 9 m in the field, the corrected penetration resistance is approximately 15 MPa.

Though noisier, the sleeve friction (f_s) curves of Fig. 7 show a similar trend with depth, including a dip as the cone tip approaches the intermediate layer. A value of about 300 kPa is measured at a depth of 180 mm. The friction ratio curves (R_f) show a steady average value of about 1.8% from 60 mm depth.

Figure 8 shows the penetration results in the second soil model. As in Fig. 7, the corrected cone resistances seem to increase linearly with depth at first. Differences are observed when the cone tip approaches the intermediate fine layer. As before, the corrected cone resistances approach the original trend curve when the cone has entered the lower coarse layer. The q_t values at 180 mm depth in the second model are seen to be about 20 MPa.

The sleeve friction curves in Fig. 8 also show increasing f_s with depth, but with much smaller frictional resistance compared to the first soil model results. The f_s value at 180 mm depth is only about 60 kPa. Consequently, the average friction ratio values are also reduced (R_f ~ 0.8%).

The dynamic pore pressures (u_2) measured behind the conical tip can be seen in Figs. 7 and 8, respectively for the first and second models. The piezocone recorded only the hydrostatic pressure u_0 at any penetration rate. Figure 7 only shows the dynamic pore pressure measurement at 8 mm/s penetration rate. The lower values of u_2 compared to u_0 above 30 mm are due to the process of re-saturation of the porous element during the penetration test. It is believed that the piezocone pressure system desaturated during the centrifuge spinning-up at position 4 (Fig. 5) while the cone tip was held above the model water table.

No perceptible changes of pore water pressure were read when the cone tip passed close to the pressure transducers located in the intermediate fine layer. Figure 9 shows the pore water pressure readings with cone tip displacement during penetration tests of the viscous fluid model. An average value of 64 kPa was read by all PPTs throughout the penetration processes.

Figure 10 shows the normalized cone resistance $Q_t = (q_t - \sigma_{v0})/\sigma'_{v0}$ with depth for the two soil models. In the first soil model an average Q_t-value of 110 was obtained. A greater scatter of Q_t at different penetration rates was however obtained in the second soil model, having an average value of 160.

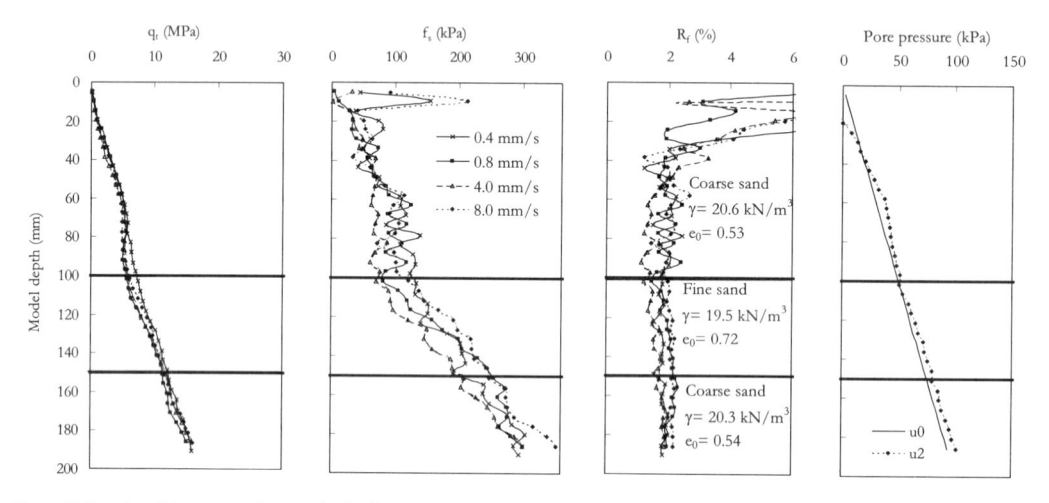

Figure 7- Results of the penetration tests in the first soil model saturated with de-ionized water

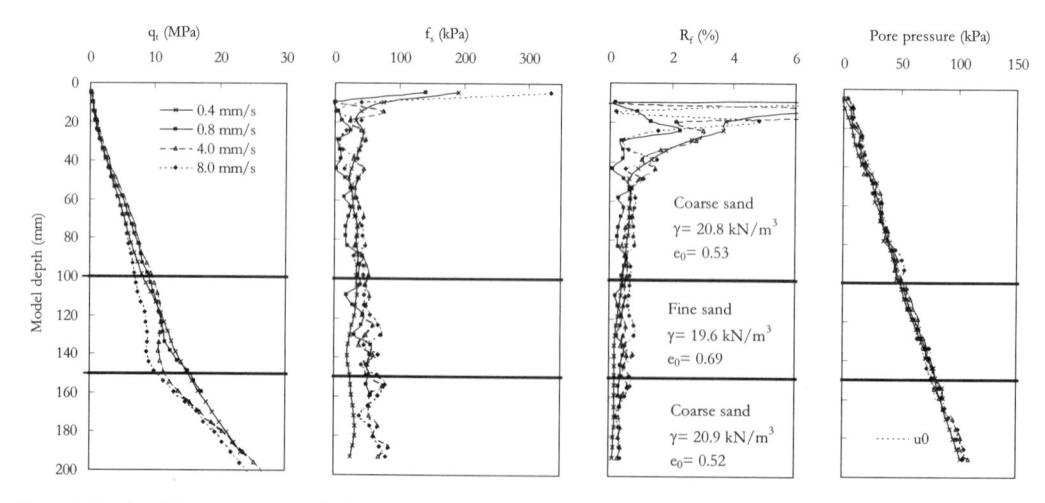

Figure 8- Results of the penetration tests in the second soil model saturated with 50-cP viscous fluid

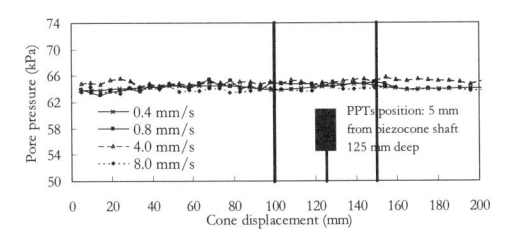

Figure 9- Pore pressure readings in the middle of the intermediate fine layer

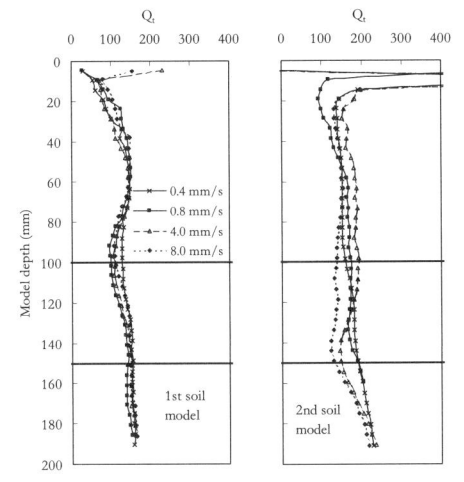

Figure 10- Normalized cone resistance with depth

The normalized pore pressure $B_q = (u_2 - u_0)/(q_t - \sigma_{v0})$ is zero with depth, because u_2 corresponds to the hydrostatic pressure u_0 in all tests. On the other hand, the normalized friction ratio $F_r = f_s/(q_t - \sigma_{v0})$ shows the same disparity of values as does the friction ratio R_f, because of the small values of total vertical stress σ_{v0} when compared to the corrected cone resistance q_t.

From the data shown above it is possible to use Robertson's (1990) soil behaviour type chart to verify its validity for centrifuge penetration tests. Analyzing both Q_t vs. F_r and Q_t vs. B_q charts, the soil models can be classified as silty to clean sands. The intermediate fine layer could not be identified by the interpretation of the piezocone parameters.

7 DISCUSSIONS

For the piezocone size and soil model geometry utilized, Figures 7 through 10 yield the following general discussions:
- The change in penetration resistance three cone diameters ahead of the layer boundaries can be due to the perceived difference in soil stiffness.
- After the cone tip crosses the interface of two layers the cone resistance is also dependent on the soil layer above the cone tip. However, the influence of the upper coarse layer is not so easily observed when penetrating the intermediate fine layer if the layer was of insufficient thickness, because the cone tip may be already "sensing" the lower coarse layer.
- Penetration rates up to 8 mm/s appear to have little influence on the cone resistance, even in fine sand with a pore fluid 50 times more viscous than water. The dynamic pore pressures (u_2) are also unaffected.
- The viscous fluid had the effect of reducing the sleeve friction by a factor of about 2.5 compared with the water-saturated model. Apparently, methylcellulose solution acts as a lubricant for sand sliding against stainless steel, though this needs further confirmation from direct shear tests.
- Although each of the soil layers in the second soil model is slightly denser than the corresponding layer in the first model, the 35% average increase in q_t seems surprisingly large. There is certainly no evidence that the methylcellulose solution reduced the angle of friction of sand on sand, therefore.
- In Fig. 8, the transitions between the intermediate fine and the coarser layers are more pronounced in the fast penetration tests than in the slow tests, though this could be coincidental. Further tests are required before cone resistance can properly be correlated with the relative density and crushability of various soil types.

8 CONCLUSIONS

Results of centrifuge penetration tests within saturated layered sands were presented. The penetration tests were performed at 50g with a 12-mm diameter piezocone and at different rates of penetration. It was expected that the identification of the intermediate fine layer would occur due to change of soil permeability. However, no sharp changes of cone resistance or generation of excess pore water pressure were measured by the piezocone. Only smooth changes of gradient on the penetration-graph curves were observed, which may be not easily associated with the penetration rate effect.

The effect of a different soil layer could be identified when the cone approached within a distance of three cone diameters. This distance of influence is observed only in soft soils in CPT tests (Treadwell, 1976; Lunne et al., 1997), but is within the range adopted by empirical methods for the prediction of pile end bearing capacity based on CPT results, which assume that the failure mechanism in sands extend to a depth around 0.7 to 4.0 pile diameters (e.g. Begemann, 1961; Te Kamp, 1977).

The difference in magnitude of the sleeve friction in the first and second soil models seems to be attributable to the lubricating properties of the hydroxypropyl methylcellulose. Further research on the subject is ongoing.

The differences in the penetration resistance seem not to be related to the change in pore fluid, but rather to variations in relative density. There is therefore no need to suppose that methylcellulose reduces the effective angle of friction of sand. Its use in dynamic centrifuge modeling to slow down pore fluid migration may therefore be acceptable if the apparent lubrication of metal interfaces can be overcome.

ACKNOWLEGDMENTS

The authors gratefully acknowledge the work of the staff members involved in the research, in particular Chris Collison, Steve Chandler, Jason Waters and Chris McGinnie. We would also like to thank Keith Wilkinson from Wilkinson Associates for the design of the centrifuge actuator. The first author is very grateful to the Brazilian agency CNPq for the financial support given during the research study at Cambridge University.

REFERENCES

Ahmadi, M.M., Byrne, P.M. and Campanella, R.G. (1999), "Numerical simulation of CPT tip resistance in layered soil." In: *Asian Institute of Technology, 40th Year Conference*, New Frontiers & Challenges, Nov.

Almeida, M.S.S. and Parry, R.H.G. (1985), "Small cone penetrometer tests in laboratory consolidated clays," *Geotechnical Testing Journal*, ASTM, 8(1): 14-24.

Begemann, H.K.S. (1961), "Previsions des fondations profondes a l'aide du penetrometre." In: *Proc. V ICSMFE*, Vol. 3, Paris.

Bolton, M.D., Gui, M.W., Garnier, J., Corte, J.F., Bagge, G., Laue, J. and Renzi, R. (1999), "Centrifuge cone penetration tests in sand," *Geotechnique*, 49(4): 543-552.

Bolton, M.D., Gui, M.W. and Phillips, R. (1993), "Review of miniature probes for model tests." In: *Proc. 11th Southeast Asian Geotechnical Conf.*, Singapore, May, 85-90.

Burns, S.E. and Mayne, P.W. (2002), "Analytical cavity expasion-critical state model for piezocone dissipation in fine-grained soils," *Soils and Foundations*, Japanese Geotechnical Society, 42(2): 131-7.

Corte, J.-F., Garnier, J., Cottineau, L.M. and Rault, G. (1991), "Determination of model soil properties in the centrifuge." In: *Centrifuge'91*, A.A. Balkema Ed., Boulder: 607-14.

De Beer, E. (1963), "The Scale Effect in Transportation of the results of deep sounding tests on the Ultimate Bearing Capacity of Piles and Caissons," *Geotechnique*, 13(1).

Fergunson, K.A. and Ko, H.Y. (1981), "Centrifuge model of the cone penetrometer." In: *Proc. of Cone penetration testing and experience*, ASCE, St. Louis: 108-27.

IRTP (2001), "International Reference Test Procedure (IRTP) for the Cone Penetration Test (CPT) and the Cone Penetration Test with pore pressure (CPTU)," Report of the International Society for Soil Mechanics and Geotechnical Engineering (ISSMGE).

Kerisel, J. (1961), "Fondations profondes en milieux sableux: Variation de la force portante limite en fonction de la densité, de la profondeur, du diamètre et de la vitesse d'enfoncement." In: *Proc. V ICSMFE*, Paris.

Konig, D., Jessberger, H.L., Bolton, M.D., Phillips, R., Bagge, G., Renzi, R., and Garnier, J. (1994), "Pore pressure measurement during centrifuge model tests: Experience of five laboratories." In: *Centrifuge'94*: 101-8.

Lee, S.Y. (1989), *Centrifuge modelling of cone penetration testing in cohesionless soils*, PhD Thesis, Cambridge University.

Lunne, T., Robertson, P.K. and Powell, J.J.M. (1997), *Cone Penetration Testing in Geotechnical Practice*, Blackie Academic and Professional.

Meyerhof, G.G. and Valsangkar, A.J. (1977), "Bearing capacity of piles in layered soils." In: *Proc. IX ICSMFE*, Tokyo, Vol. 1: 645-50.

Morley, P. (1997), *Centrifuge modelling of piles and anchors in calcareous sands*, Fourth Year Project, Cambridge University.

Muromachi, T. (1981), "Cone penetration testing in Japan." In: *Cone penetration testing and experience*, ASCE, St Louis, pp. 76-107.

Renzi, R., Corte, J.F., Rault, G., Bagge, G., Gui, M. and Laue, J. (1994), "Cone penetration tests in the centrifuge: experience of five laboratories." In: *Centrifuge'94*, A.A. Balkema Ed., Singapore: 77-82.

Robertson, P.K. (1990), "Soil classification using cone penetration test," *Canadian Geotechnical Journal*, Vol. 27, No. 1, pp. 151-158.

Santoyo, E. (1982), "Use of a static penetrometer in a softground tunnel." In: *ESOPT-II*, A.A. Balkema Ed., Amsterdam, 2: 835-9.

Sharp, M.K., Dobry, R. and Phillips, R. (1998), "Cone penetration modelling in sand for evalutation of earthquake-induced lateral spreading." In: *Centrifuge'98*, A.A. Balkema Ed., Tokyo: 161-6.

Stewart, D.P., Chen, Y.R. and Kutter, B.L. (1998), "Experience with the use of methylcellulose as a viscous pore fluid in centrifuge models," *ASTM Geotechnical Testing Journal*, Vol. 21, No. 4, December, pp. 365-9.

Tan, F.S.C. (1990), *Centrifuge and theoretical modelling of conical footings on sand*, PhD Thesis, Cambridge University.

Te Kamp, W.C. (1977), "Static cone penetration testing and foundations on piles in sand." In: *Fugro Sounding Symposium*, Utrecht, October.

Teymur, B. (2002), *The significance of boundary conditions in dynamic centrifuge modelling*, PhD Thesis, Cambridge University.

Treadwell, D.D. (1976), *The influence of gravity, prestress, compressibility, and layering on soil resistance to static penetration*, PhD Thesis, University of California, Berkeley.

Van der Poel, J.T. and Schenkeveld, F.M. (1998), "A preparation technique for very homogeneous sand models and CPT research." In: *Centrifuge'98*, A.A. Balkema Ed., Tokyo: 149-54.

Wright, L. (1998), *Centrifuge modelling of piles and anchors in calcareous sands*, Fourth Year Project, Cambridge University.

Proceedings ISC-2 on Geotechnical and Geophysical Site Characterization, Viana da Fonseca & Mayne (eds.)
© 2004 Millpress, Rotterdam, ISBN 90 5966 009 9

Sampling by drilling and excavations and groundwater measurements according to EN ISO 22475 for geotechnical investigations and site characterisation

F. Stölben
Stölben Bohrunternehmen GmbH, Zell/Mosel, Germany

V. Eitner
DIN Deutsches Institut für Normung e.V., Berlin, Germany

H. Hoffmann
Ingenieursozietät Prof. Dr.-Ing, Katzenbach GmbH, Darmstadt, Germany

Keywords: drilling, sampling, groundwater measurements, geohydraulic tests, qualification, conformity assessment

ABSTRACT: The Technical Committees of the European Committee for Standardisation (CEN) and the International Organisation for Standardisation (ISO) on geotechnical investigation and testing are preparing several common standards that deal with the direct investigation of soil, rock and groundwater. EN ISO 22475-1 "Geotechnical investigation and testing – Sampling by drilling and excavation and groundwater measurements – Part 1:Technical principles for execution" defines concepts and specifies requirements relating to exploration by excavation, drilling and sampling as well as groundwater measurements. It covers the general requirements for drilling rigs, ancillary equipment, methods for soil, rock and groundwater sampling, groundwater measuring stations (piezometers), groundwater measurements, handling, transport and storage of samples and reporting the results. EN ISO 22475-2 "Geotechnical investigation and testing – Sampling by drilling and excavation and groundwater measurements – Part 2: Technical qualification criteria for enterprises and personnel" specifies the technical qualification criteria for enterprises and personnel performing drilling and sampling services to ensure that both have the appropriate experience, knowledge and qualifications as well as the correct drilling and sampling equipment for the task to be carried out according to EN ISO 22475-1. EN ISO 22475-3 "Geotechnical investigation and testing – Sampling by drilling and excavation and groundwater measurements – Part 3: Conformity assessment of enterprises and personnel" deals with the conformity assessment of enterprises and personnel for ground investigation drilling and sampling as well as groundwater measurements according to EN ISO 22475-1 that comply with the technical qualification criteria according to EN ISO 22475-2. Every enterprise may individually choose a form of technical quality assurance according to these standards that is appropriate for its purpose (e.g. first, second or third party control). Moreover, there are several standards covering geohydraulic field tests such as permeability tests, water pressure tests and pumping tests.

1 INTRODUCTION

The Technical Committee of the European Committee for Standardization CEN/TC 341 "Geotechnical investigation and testing" prepares a set of standards in cooperation with the ISO/TC 182/SC 1 according to the Vienna Agreement (1991).

One of these standards – EN ISO 22475 "Geotechnical investigation and testing - Sampling by drilling and excavation methods and groundwater measurements" – deals with principles of the direct investigation of soil, rock and groundwater as subsoil and construction materials as part of the geotechnical investigation services. This standard consists of the following parts:
- Part 1: Technical principles for execution;
- Part 2: Technical qualification criteria for enterprises and personnel;
- Part 3: Conformity assessment of enterprise and personnel;

and defines concepts and specifies requirements relating to sampling by drilling and excavation and groundwater measurements.

The aims of such explorations are:
- to recover soil and rock samples of a quality sufficient to assess the general suitability of a site for geotechnical engineering purposes and to determine the required soil and rock characteristics in the laboratory
- to obtain information on the sequence, thickness and orientation of strata;
- to establish the type, composition and condition of strata and joint system and faults;

- to obtain information on groundwater conditions and recover water samples for assessment of the interaction of groundwater, soil, rock and construction material;
- to allow in-situ testing to be carried out.

EN ISO 22475 does not cover soil sampling for the purposes of agricultural and environmental soil investigation and water sampling for the purposes of quality control, quality characterization, and identification of sources of pollution of water, including bottom deposits and sludges.

Separate standards cover different geodyraulic test such as permeability tests, water pressure tests, pump tests etc.

2 TECHNICAL PRINCIPLES OF EXECUTION

2.1 Equipment

The drilling and sampling equipment selected shall of the appropriate size and type in order to produce the required quality.

If applicable, the drilling and sampling equipment shall be in accordance with ISO 3351-1, ISO 3352-1 and ISO 10097-1.

Drilling rigs with appropriate stability, power and equipment such as drill rods, casing, core barrels and bits shall be selected in order that the required sampling and borehole tests may be carried out to the required depth of the borehole and sampling categories. A selection of equipment which is currently used is given in an annex.

The drilling rig shall allow all drilling functions to be adjusted accurately. Where appropriate, the following drilling data should be measured and recorded against depth:
- Drill head torque;
- drill head rotational speed;
- feed thrust and pulling force;
- penetration rate;
- depth of hammering intervals;
- topographical depth;
- direction when inclined drilling;
- drilled length when inclined drilling;
- flushing medium pressure at the output of the pump;
- flushing medium circulation rate (input);
- flushing medium recovery rate.

2.2 Soil sampling

Techniques for obtaining soil samples can be generally divided in the following groups:
- sampling by drilling;
- sampling using samplers;
- block sampling;

Combinations of these sampling methods are possible and sometimes required due to the geological conditions and the purpose of the investigation.

There are three categories A, B and C of sampling methods. For given ground conditions, they are related to the best obtainable laboratory quality class of soil samples (defined in EN 1997-2) as shown in Tab. 1:
- category A sampling methods: samples of quality class 1 to 5 can be obtained;
- category B sampling methods: samples of quality class 3 to 5 can be obtained;
- category C sampling methods: only samples of quality class 5.

Table 1. Quality classes of soil samples for laboratory testing and sampling categories to be used (from EN 1997-2)

Soil properties / quality class	1	2	3	4	5
Unchanged soil properties					
particle size, particle size distribution	*	*	*	*	
water content	*	*	*		
density, density index, permeability	*	*			
compressibility, shear strength	*				
Properties that can be determined					
sequence of layers	*	*	*	*	*
boundaries of strata – broad	*	*	*	*	
boundaries of strata – fine	*	*			
Atterberg limits, particle density, organic content	*	*			
water content	*	*	*	*	
density, density index, porosity, permeability	*	*	*		
compressibility, shear strength	*				
Sampling category to be used	A				
		B			
					C

Samples of quality class 1 or 2 can only be obtained by using category A sampling methods. The intention is to obtain samples in which no or only slight disturbance of the soil structure has occurred during the sampling procedure or in handling of the samples. The water content and the void ratio of the soil correspond to that in-situ. No change in constituents or in chemical composition of the soil has occurred. Certain unforeseen circumstances such as varying of geological strata may lead to lower sample quality classes being obtained.

By using category B sampling methods, this will preclude achieving sampling quality class better than 3. The intention is to obtain samples containing all the constituents of the in-situ soil in their original proportions and the soil has retained its natural water content. The general arrangement of the different soil layers or components can be identified. The structure of the soil has been disturbed. Certain unforeseen circumstances such as varying of geological strata may lead to lower sample quality classes being obtained.

By using category C sampling methods, this will preclude achieving sampling quality class better than 5. The soil's structure in the sample has been totally changed. The general arrangement of the different soil layers or components has been changed so that the in-situ layers cannot be identified accurately. The water content of the sample may not represent the natural water content of the soil layer sampled.

2.3 *Rock sampling*

Techniques for obtaining rock samples can be divided in the following groups:
- sampling by drilling;
- sampling from trial pits, headings, shafts and from borehole bottom;
- integral sampling.

Combinations of these sampling methods are possible and sometimes required due to the geological conditions.
Rock samples are of the following types:
- cores (complete and incomplete);
- cuttings and retained returns;
- blocks.

The quality of the rock recovery is achieved by applying the following three parameters:
- total core recovery, *TCR*;
- rock quality designation, *RQD*;
- solid core recovery, *SCR*;

After recovery of the core barrels to the surface the rock recovery shall be assessed. In case the samples are extruded from the sampler and placed in a core box the sample shall be carefully logged. Core losses shall be filled with a dummy.

There are three categories of rock sampling methods depending on the best obtainable quality of rock samples under given ground conditions:
- category A sampling methods;
- category B sampling methods;
- category C sampling methods.

By using category A sampling methods it is intended to obtain samples in which no or only slight disturbance of the rock structure has occurred during the sampling procedure of the samples. The strength and deformation properties, water content, density, porosity and the permeability of the rock sample correspond to the in-situ values. No change in constituents or in chemical composition of the rock mass has occurred. Certain unforeseen circumstances such as varying of geological conditions may lead to lower sample quality being obtained.

By using category B sampling methods it is intended to obtain samples that contain all the constituents of the in-situ rock mass in their original proportions and the rock pieces have retained their strength and deformation properties, water content, density and porosity. By using category B sampling the general arrangement of discontinuities in the rock mass may be identified. The structure of the rock mass has been disturbed and thereby the strength and deformation properties, water content, density, porosity and permeability for the rock mass itself. Certain unforeseen circumstances such as varying of geological conditions may lead to lower sample quality being obtained.

By using category C sampling methods the structure of the rock mass and its discontinuities has been totally changed. The rock material may have been crushed. Some changes in constituents or in chemical composition of the rock material may occur. The rock type and its matrix, texture and fabric may be identified.

2.4 *Sampling of groundwater for construction purposes*

Groundwater sample methods shall be selected according to need. The quality of a groundwater sample is characterized by the extent to which it contains original constituents, such as suspended matter, dissolved gases and salts, or to which they have been contaminated during drilling. Groundwater can be sampled for the following purposes:
- to determine its aggressiveness to concrete;
- to determine its corrosive nature;
- to establish any risk to subsurface drainage systems and filters due to clogging and similar effects;
- to identify changes in groundwater quality or flow rate resulting from construction work;
- to determine the groundwater quality for recharging or infill in sewer lines;
- to determine its suitability as mixing water for construction material.

The number and location of collection points shall be specified in advance on the basis of the engineering problems involved and the local geological and hydrological conditions (see EN 1997-2). If a group of aquifers is encountered, it may be necessary to collect separate samples from each aquifer.

If it is intended to take water samples for chemical analysis, only air and clean water may be used as flushing medium. Flushing additives are not permitted.

2.5 *Groundwater measurements*

In order to obtain data on the magnitude, variation and distribution of the heads of groundwater or pore pressures in the ground, appropriate groundwater measuring stations shall be installed.

The type and arrangement of groundwater measurements shall be specified in accordance with EN 1997-2.

When drilling for piezometers, flushing media are allowed if:
- it is not clogging the filter pores; or

- an installation technique is used where the filter is protected until it reaches the borehole bottom and the piezometer then is driven into the borehole bottom.

Open or closed systems can be used to conduct groundwater measurements. The choice between open or closed systems should be made depending on the permeability of the ground, the rate of change in pore water pressure and the required precision and duration of the measurements.

In both systems a filter should be installed in the ground at the location at which the head of groundwater or the pore water pressure shall be measured. The filter shall prevent ingress of soil particles into the measuring system.

All components and equipment intended for installation in the ground shall be sufficiently resistant to mechanical loading and chemical attack by constituents in the groundwater. Any reactions between the materials used and the ground, in particular the formation of galvanic effects, shall be prevented.

Groundwater measuring stations shall be positioned and secured in such a way that third parties are not at risk. Appropriate measures shall be taken to avoid any risk to the groundwater measuring station due to contamination, flooding, traffic or frost.
Open systems can be divided in three groups as follows (see Fig. 1):
(a) observation borehole;
(b) open pipe;
(c) open pipe with inner hose.

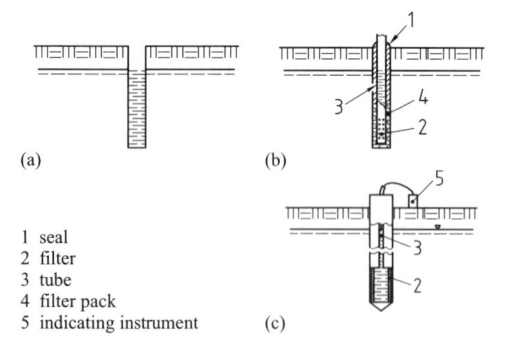

1 seal
2 filter
3 tube
4 filter pack
5 indicating instrument

Figure 1. Examples of open systems according to EN ISO 22475-1.

The piezometer in closed systems shall consist of a robust casing which is installed in the ground with a filter at the lower end (filter tip) and a water-filled chamber behind from where the water pressure is transmitted to the measuring device. Filters with sufficiently high air entry values shall be used. The pressure measurements can be performed as illustrated by Fig. 2:
(a) hydraulic measuring systems;
(b) pneumatic measuring systems;
(c) electrical measuring systems.

All measuring systems used shall be calibrated prior to commissioning the groundwater measuring station. This applies to both new and reused equipment. All parts of the measuring system that affect the accuracy of the measurements shall be calibrated.

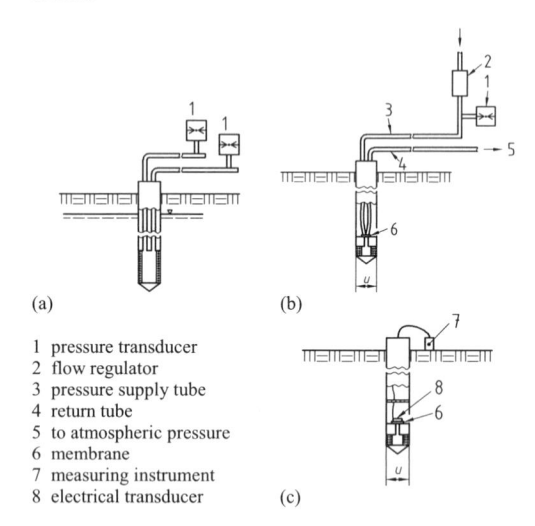

1 pressure transducer
2 flow regulator
3 pressure supply tube
4 return tube
5 to atmospheric pressure
6 membrane
7 measuring instrument
8 electrical transducer

Figure 2. Examples of closed systems according to EN ISO 22475-1.

The calibration results shall be documented in a report which, in addition to a description of the calibration procedure, shall include all information required to evaluate the measurements.

Measurements shall be checked if they represent effects of installation, time lag or groundwater fluctuations.

The results of the measurements shall be documented in a report which shall enable the values measured to be related to a particular stratum and interpreted unambiguously

2.6 *Geohydraulic testing*

EN ISO 22476-12 specifies requirements for the determination of the local permeability of soil below the groundwater level in a borehole by the Lefranc permeability test. The aim of the Lefranc permeability test is the quick determination of permeability parameters during drilling and sampling campaigns. This test is not comparable with pump tests. The determined permeabilities may considerably be different therefore.

EN ISO 22476-13 deals with water pressure tests in jointed rock mass as carried out during ground investigation drillings or to check whether injection work has been carried out successfully. The water pressure test gives a measure of the acceptance by in-situ rock of water under pressure. The test was originally introduced by Lugeon (1933) to provide a standard for measuring the impermeability of grouted ground; it is also widely used as a packer test to measure the permeability of dam foundations. In essence, it comprises the measurement of the volume of water that can escape from an uncased section of borehole in a given time under a given pressure. Flow is confined between known depths by means of packers. The flow is confined between two packers in the double packer test, or between one packer and the bottom of the borehole in the single packer test. The test is used to assess the amount of grout that rock accepts, to check the effectiveness of grouting, to obtain a measure of the amount of fracturing of rock, or to give an approximate value of the permeability of the rock mass local to the borehole.

EN ISO 22475-14 deals with pumping tests which extract water from an aquifer during a longer period (hours to weeks). Herewith, a measurable drawdown occurs at the extraction site and around it. The amount of drawdown depends on the extraction amount, the permeability of the tested ground and different further hydrogeological parameters.

2.7 *Handling, transport and storage of samples*

The conditions of soil and rock samples that were present after sampling according to sampling category A, B, or C, have to be preserved.

National laws or regulations have to be considered when transporting samples known or suspected to contain hazardous material.

A separate traceability record shall accompany each shipment so that the possession of the sample is traceable from collection to shipment to laboratory disposition.

When transferring the possession of samples the persons(s) relinquishing and receiving the samples shall sign, date, record the time and check completely the traceability record.

Every soil and rock sample has to be protected any time from direct sun light, heat, frost and rain.

2.8 *Reporting*

At the project site, a "Field Report" of sampling by drilling and excavation and groundwater measurements shall be completed which shall consists of the following, if applicable:
(a) summary log;
(b) drilling record;
(c) sampling record;
(d) back-filling record;
(e) record of identification and description of soil and rock;
(f) record of the installation of piezometers;
(g) record of groundwater measurements.

All field investigations shall be recorded and reported such that third persons are able to check and understand the results.

Additionally, a "Report of the Results" shall be completed which shall include the following essential information, if applicable:
(a) the field report (in original and/or computerised form);
(b) a graphical presentation of the record of the identification and description of soil and rock;
(c) a graphical presentation of the back-filling;
(d) a graphical presentation of the piezometer;
(e) a graphical presentation of the results of the groundwater measurements.

3 TECHNICAL QUALIFICATION CRITERIA

EN ISO 22475-2 specifies the technical qualification criteria for an enterprise and personnel performing sampling by drilling and excavations methods and groundwater measurement services so that both have the appropriate experience, knowledge and qualifications as well as the correct equipment for sampling by drilling and excavation methods and groundwater measurements for the task to be carried out according to EN ISO 22475-1.

4 CONFORMITY ASSESSMENT

EN ISO 22475-3 applies for the conformity assessment of enterprises and personnel performing sampling by drilling and excavation methods and groundwater measurements according to EN ISO 22475-1 that comply with the technical qualification criteria according to EN ISO 22475-2.

This part specifies the eligibility for conformity assessment, the application and assessment procedure as well as the re-assessment. The assessment report is given in an informative annex

5 CONCLUSIONS

ISO 22475 will harmonize drilling and sampling methods and groundwater measurements for geotechnical engineering purposes and will allow to compare these geotechnical services on common basis. This will lead to a fairer competition. They were also prepared to increase the quality world-wide on this field in geotechnical engineering.

Part 1 specifies the general principles of execution and Part 2 the technical qualification criteria for enterprises and personnel that perform sampling and groundwater measurements according to Part 1.

If an enterprise or personnel fulfils these technical qualification criteria of Part 2, they can prove their conformity by:

(1) a declaration of conformity by a contractor (first party control);
(2) a declaration of conformity by a client (second party control);
(3) a declaration of conformity by a conformity assessment body (third party control).

Every enterprise or personnel may decide individually, if they want to declare its conformity with Part 2 by first, second or third party control because none of these three documents requires such a declaration (see Fig. 3). A declaration of conformity may only be required by contract.

If an enterprise or personnel performing sampling and groundwater measurements according to Part 1 want to achieve a declaration of conformity by a conformity assessment body, the conformity assessment body is advised to use Part 3 which covers the conformity assessment of enterprises and personnel.

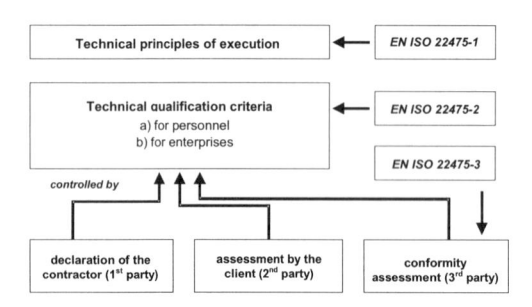

Fig. 3. Conformity assessment (after Stölben & Eitner, 2003)

REFERENCES

Lugeon, M. (1933): Barrages et Géologie. Dunod, Paris.
EN 791, Drilling rigs - Safety.
EN 1997-1, Eurocode 7: Geotechnical design - Part 1: General Rules.
EN 1997-2, Eurocode 7: Geotechnical design - Part 2: Ground investigation and testing.
ISO 3551-1, Rotary core diamond drilling equipment - System A - Part 1: metric units.
ISO 3552-1, Rotary core diamond drilling equipment - System B - Part 1: metric units.
ISO 5667 (all parts), Water quality - Sampling.
ISO 10097-1, Wireline diamond core drilling equipment - System A - Part 1: Metric units.
ISO 10381 (all parts), Soil quality - Sampling.
Stölben, F. & Eitner, V. (2003): Aufschluss- und Probenentnahmeverfahren und Grundwassermessungen im Rahmen von geotechnischen Erkundungen – Technische Grundlagen der Ausführungen, Qualifikationskriterien und Konformitätsbewertung. – Berichte von der 14. Tagung für Ingenieurgeologie, 26. bis 29. März 2003, Kiel; pp. 233 – 238.

Undrained shear strength and OCR of marine clays from piezocone test results

Fernando A. Trevor
Fugro Singapore Pte Ltd, Singapore

Paul W. Mayne
School of Civil & Environmental Engineering, Georgia Institute of Technology, Atlanta, GA, USA

Keywords: cavity expansion theory, cone penetrometer, critical state concept, piezocone, OCR, s_u

ABSTRACT: Significant improvements have been made in the interpretation of the piezocone penetration test (PCPT) results over the last decade. Herein, field PCPT data and laboratory test results from 12 deep boreholes advanced at 4 different sites in Southeast Asia have been investigated for the purpose of establishing an improved method to estimate overconsolidation ratio (OCR) and undrained shear strength (s_u) from PCPT results in these materials.
Cavity expansion theory and critical state concepts can be used in the interpretation of PCPT results. One of the reliable methods available with this theory and concept is presented along with recommendations to estimate OCR and s_u from PCPT results using a modified form with correction factors.
The case studies show piezocone predictions using the modified equations are in good agreement with the laboratory triaxial and consolidation test results. The authors have used this proposed method in marine clay soils at many other sites in Southeast Asia where PCPT and laboratory test results are available. The correction factors described in this paper have been determined on the basis of statistical evaluation of available laboratory test results. It is recommended to establish site-specific correction factors for other soils.

1 INTRODUCTION

In the ongoing and continuous search for a cost-effective method to derive soil parameters in the geotechnical industry piezocone penetration tests (PCPT) are increasingly carried out during soil investigations because at least three independent readings are acquired at one time. Documentation of estimated soil parameters from the PCPT data and the laboratory test results from reliable site investigation sites are thus of vital importance for further analysis.

One of the reasons for the rapid development and wide use of PCPT as a site investigation tool in clay soils is that, in theory, the boundary conditions are reasonably well defined in the test, allowing theoretical frame works to be used in the interpretation. However, in current practice, the evaluation of key soil parameters, such as the undrained shear strength and the overconsolidation ratio, are often based on empirical correlations. It will be of great value if the applicability of selected interpretation methods that are based on theoretical frameworks can be carefully evaluated with properly collected site investigation data.

This paper describes the PCPT interpretation method that was developed by Mayne (1991) based on cavity expansion theory and critical state concepts. This method was modified by Trevor F. (2001 MSc Thesis) incorporating correction factors to adjust soils of varying geologic origins and conditions.

2 PIEZOCONE EQUIPMENT AND TESTING

The basic components of piezocone testing equipment are the cone penetrometer, cone pushing equipment, and data acquisition system. The cone dimensions are described with the cone angle and the base area. The position of the cone filter can be at cone face (u_1), immediately behind the cone tip (u_2), or behind the friction sleeve (u_3). The International Reference Test Procedure (ISSMGE, 1999) suggests the filter to be located behind the cone tip (u_2) as the preferred cone, but it is not a requirement. The type of piezocone in this paper refers to u_2 type.

3 PIEZOCONE TEST RESULTS

The data acquisition system continuously records the test data acquired from the cone sensors while the piezocone is being pushed into the ground. The basic measurements are the cone resistance (q_c), sleeve friction (f_s) and the porewater pressure (u). These measurements are recorded with respect to time and corresponding depths. Corrected total cone resistance (q_t), friction ratio ($R_f = f_s/q_t$) and porewater pressure ratio [B_q = excess porewater pressure/(q_t-σ_{vo})] are usually computed from the measured piezocone data and are used in soil classification.

4 SOIL PARAMETERS IN CLAY

Shear strength parameters, overconsolidation ratio, deformation moduli, and permeability properties are the principal parameters of interest to a geotechnical design engineer. This paper presents the interpretation of undrained shear strength (s_u) and overconsolidation ratio (OCR), the two key engineering parameters from results of the piezocone penetration test in clay soils based on the theories of cavity expansion and critical state.

5 OVERCONSOLIDATION RATIO

Overconsolidation ratio (OCR) in one-dimensional compression is defined as the ratio between the maximum past vertical stress and in-situ effective vertical stress, or OCR = σ'_p/σ'_{vo}. The OCR is one of the important parameters in geotechnical engineering analysis. It provides the past stress history in a dimensionless and normalized form that is needed to evaluate the consolidation status of the soil. The measured magnitudes of cone resistance and porewater pressure from the piezocone test reflect the effect of soil stress history. Hence, OCR estimation from the piezocone tests may be feasible, if properly calibrated with oedometer data.

6 OCR ESTIMATION FROM PCPT RESULTS

A recently developed method for the interpretation of the overconsolidation ratio from piezocone data based on cavity expansion theory and critical state concept proposed by Mayne (1991) is reviewed below.

6.1 Mayne (1991) Method

It is generally accepted that the shearing mechanism around the cone tip may be approximately simulated by the expansion of a spherical cavity (Vesic, 1977; Teh and Houlsby, 1991), whereas that along the shaft of the piezocone is more consistent with the assumption of the expansion of a cylindrical cavity.

According to the spherical cavity expansion theory of Vesic (1977),

$$N_{kt} = 4/3 \, (\ln I_r + 1) + \pi/2 + 1 \quad (1)$$

where N_{kt} is the cone factor that is equal to q_{net}/s_u and I_r is the rigidity index that equals to G/s_u. Note that q_{net} is the net cone resistance that is equal to the difference between the corrected cone resistance and the total overburden pressure, or (q_t-σ_{vo}), and G is the shear modulus of the soil.

Rearranging the expression in Equation (1), the net cone resistance can be expressed as:

$$q_{net} = (q_t - \sigma_{vo}) = 1.33 s_u \ln I_r + 3.90 s_u \quad (2)$$

The excess porewater pressure ($\Delta u = u_2 - u_o$) generated during cone penetration may also be expressed in terms of relevant soil properties based on the cavity expansion and critical state concepts (Mayne and Bachus, 1988). The excess porewater pressure induced by an advancing probe is a result of a combination of changes in octahedral and shear stresses, and it can be expressed as:

$$\Delta u = \Delta u_{oct} + \Delta u_{shear} \quad (3)$$

Although the measured excess porewater pressure (Δu) is a combination of the octahedral and shear stresses, it is difficult to uncouple the two responses for direct field measurements. For spherical cavity expansion, Vesic (1972) indicated that the octahedral component of excess porewater pressure could be expressed simply as:

$$\Delta u_{oct} = (4/3) \, s_u \ln (I_r) \quad (4)$$

Using the strain path method, Baligh (1986) has shown that the porewater pressure measured just behind the cone tip (u_2) is dominated by octahedral stresses, with Δu_{shear} generally less than 20 percent of the total measured Δu. The magnitude of the induced porewater pressure depends on the relative distance between the total and effective stress paths, which are not truly known. If the mean effective stress path is assumed to be equal to the effective vertical stress for practical purpose, (Mayne and Bachus, 1988) the shear-induced component of the excess porewater pressure can be expressed as:

$$\Delta u_{shear} = \sigma'_{vo} \, [1-(OCR/2)^\Lambda] \quad (5)$$

where parameter Λ is the plastic volumetric ratio in critical state theory (Wroth, 1984).

For Type 2 piezocones, the measured excess pore pressures can be considered to be the sum combination of Equations (4) and (5), resulting in:

$$\Delta u = \Delta u_{oct} + \Delta u_{shear} \quad (6)$$

$$\Delta u = (4/3) s_u \ln I_r + \sigma'_{vo} [1-(OCR/2)^\Lambda] \quad (7)$$

Subtracting Equation (7) from (2) results in:

$$q_t - \sigma_{vo} - \Delta u = 3.9 s_u - \sigma'_{vo} [1- (OCR/2)^\Lambda] \quad (8)$$

or re-arranged to obtain:

$$(q_t - u_2) - \sigma'_{vo} = 3.9 s_u - \sigma'_{vo} [1-(OCR/2)^\Lambda] \quad (9)$$

Modified Cam Clay (MCC) is a simple critical state formulation that can be used to determine s_u from triaxial compression loading in terms of effective stresses and stress history effects (Wroth and Houlsby, 1985):

$$s_u = (M/2) (OCR/2)^\Lambda \sigma'_{vo} \quad (10)$$

where M is the slope of the critical state line in the mean effective stress-deviatoric stress (p':q) plane and it is equal to M = $6\sin\phi'_{cs}/(3-\sin\phi'_{cs})$. The effective friction angle at critical state ϕ'_{cs} is determined from conventional triaxial compression tests.

Combining Equation (9) and (10), OCR for Type 2 piezocones can be expressed as follows:

$$OCR = 2 \left[\frac{1}{1.95M+1} \left(\frac{q_t - u_2}{\sigma'_{vo}} \right) \right]^{1/\Lambda} \quad (11)$$

The parameter Λ is essentially constant for natural clays and averages about 0.75, 0.80 and 0.85 for compression, simple shear and extension modes, respectively (Mayne, 1988). Adopting a value $\Lambda=0.75$, corresponding to triaxial compression, the form of OCR for practical use becomes:

$$OCR = 2 \left[\frac{1}{1.95M+1} \left(\frac{q_t - u_2}{\sigma'_{vo}} \right) \right]^{1.33} \quad (12)$$

7 UNDRAINED SHEAR STRENGTH

The undrained shear strength is an important soil parameter that is needed for the analysis of geotechnical problems involving stability. Commonly, unconsolidated-undrained triaxial (UU) tests are performed in the laboratory to determine the undrained shear strength values. But in critical state theory, the undrained shear strength is referenced to that obtained from isotropically consolidated-undrained triaxial (CU) tests.

8 UNDRAINED SHEAR STRENGTH ESTIMATION FROM PCPT RESULTS

The undrained shear strength for triaxial compression mode is expressed in Equation (10) in terms of effective stresses and stress history effects. The OCR component in Equation (10) can be substituted by using Equation (11) in order to obtain the undrained shear strength directly from PCPT readings:

$$s_u = \frac{M}{2} \left(\frac{q_t - u_2}{1.95M + 1} \right) \quad (13)$$

9 CASE STUDIES

PCPT results and laboratory test results from 12 deep boreholes advanced at 4 different sites in Southeast Asia were used in the analysis. The OCR of the soils typically ranged between 1.0 and 3.5.

The available s_u values are from unconsolidated-undrained triaxial compression tests of high quality samples taken by pushing a thin-walled sampler with an internal diameter of 72 mm. The API (American Petroleum Institute) recommends UU tests in establishing shear strength profile for design of piles because of their consistency and repeatability. Few CU tests were carried out for the interest of this research study. Therefore s_u values from UU test results were used as the reference values. There could be considerable error in the UU tests because of sampling disturbance and omission of re-consolidation during testing. However, at offshore sites where stringent procedures are followed in soil sampling, it can be considered that the sampling disturbance is quite negligible.

The UU tests for case studies were performed on 72mm diameter undisturbed soil samples trimmed to a height of 144mm. Confining pressure equivalent to total overburden pressure was applied on the soil specimens. The specimens were then loaded to failure or 20% strain (whichever occurs first) at the rate of 1.5mm per minute in undrained condition.

For conversion among the tests, a correlation between UU and CU test results established by Chen and Kulhawy (1993) is presented below:

OCR	$s_u(UU) / s_u(CU)$
1.0~1.2	0.40~0.65
1.2~2.0	0.60~0.80
2.0~2.3	0.80~0.95

It should be noted that the ratio of $s_u(UU)$ over $s_u(CU)$ very much depends on the sample quality. The ratio of $s_u(UU)$ over $s_u(CU)$ indicates there will be a similar difference between the s_u values derived from PCPT results using Equation 13 and the reference s_u values.

Representative piezocone data from Site-1 are presented in Figure 1(a), showing cone resistance (q_c), sleeve friction (f_s), porewater pressure (u_2), and friction ratio ($R_f = f_s/q_c$) with depth. Figure 1(b) shows the net cone resistance and porewater pressure ratio with depth.

Soil properties and laboratory test results of the four sites are briefly described below.

9.1 Index Properties

Within the depth ranges cited, the average moisture contents, liquid and plastic limits of the four sites are summarized in Table 1.

Table 1. Index Properties

Location	Depth, m	Average Moisture Content, %	Average Liquid Limit, %	Average Plastic Limit, %
Site-1	0 ~ 3	80 ~ 90		
	3 ~ 110	25 ~ 50	>50	35 ± 9
Site-2	0 ~ 120	25 ~ 50	50 ~ 75	41 ± 9
Site-3	0 ~ 101	25 ~ 50	25 ~ 50	21 ± 7
Site-4	0 ~ 26	50 ~ 140		
	26 ~ 141	25 ~ 50	50 ~ 100	40 ± 12

The plasticity values generally plot close to the Casagrande A-line (Figure 2). The classification for clay soils at most sites is CH and for clay soils at Site-3 is CL based on the Unified Soil Classification System.

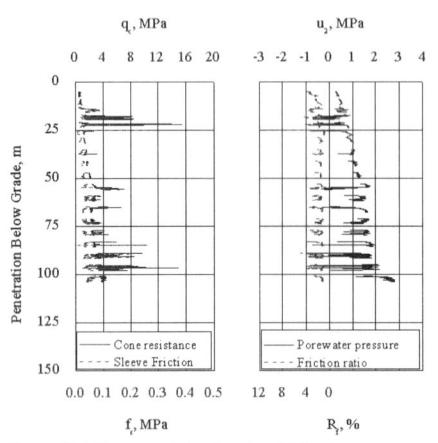

Figure 1(a) Piezocone Test Results of Site-1
(a) q_c, f_s, R_f & u_2 Profile (b) q_{net} & B_q Profile

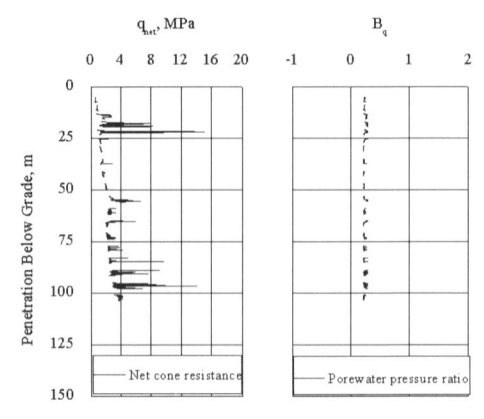

Figure 1(b) Piezocone Test Results of Site-1
(a) q_c, f_s, R_f & u_2 Profile (b) q_{net} & B_q Profile

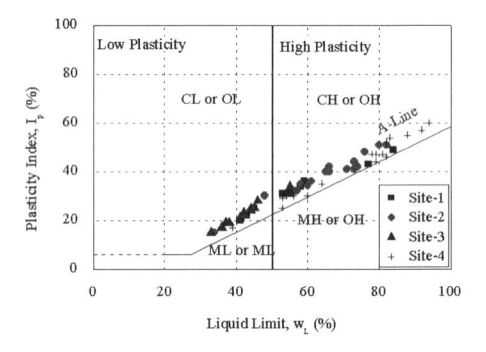

Figure 2. Plasticity Chart of Marine Clays

9.2 Consolidation Properties

The simplest laboratory test to evaluate consolidation properties of clay soil is the one-dimensional consolidation or oedometer test. The overconsolidation ratio (OCR) in one-dimensional compression test is defined as the ratio between the preconsolidation pressure and the existing effective vertical overburden pressure. The laboratory e/log p curve (void ratio vs Log-pressure) can be used to derive the preconsolidation pressure p_c and the overconsolidation ratio, using the procedure devised by Casagrande (1936). The OCR values also can be estimated based on index properties and undrained shear strength values. According to Ladd, et al. (1977), OCR can be estimated using the normalized undrained shear strength as follows:

$$OCR = \left[(s_u/\sigma'_v)/(s_u/\sigma'_v)_{nc}\right]^{1/\Lambda} \tag{14}$$

where $(s_u/\sigma'_v)_{nc}$ is equal to $(0.11+0.0037I_p)$ based on Skempton (1957) and Λ is assumed equal to 0.75 according to Mayne (1988).

9.3 Shear Strength Properties

Unconsolidated undrained and consolidated undrained triaxial compression tests are the most common test methods carried out in laboratories to determine the shear strength properties of clay soils.

The effective friction angle (ϕ') can be determined from Mohr circles corresponding to CU test results. The Mohr-coulomb failure envelope for normally consolidated clay passes essentially through the origin of the graph, and thus effective cohesion (c') can be considered equal to zero.

Both unconsolidated undrained and consolidated undrained triaxial compression tests were performed on undisturbed samples recovered from Site-1 and unconsolidated undrained tests were performed on undisturbed soil samples from the other 3 sites for the determination of the shear strength parameters. In addition, the effective friction angles were derived from the PCPT data using the Senneset et al. (1988) chart shown in Figure 3. The measured effective friction angles from the CU tests and the computed effective friction angles from the PCPT data are tabulated in Table 2.

The effective friction angles at critical state derived from the CU test results and the effective friction angles computed from the PCPT data are fairly comparable with a few noted exceptions. This shows that in the absence of effective friction angles at critical state, effective friction angles derived from other post-processing PCPT methods can be used as a substitute.

Table 2. Effective Friction Angles (Site-1)

Borehole	Depth m	Effective Friction Angle at Critical State (ϕ'_{cs}), deg	Effective Friction Angle from PCPT data (ϕ'), deg
BH-1	38.4	22	24
	62.5	30	25
	86.5	19	28
BH-2	25.3	26	20
	45.3	28	24
BH-3	29.4	30	22
	41.4	30	21
BH-4	42.3	29	25
	60.4	18	27
BH-5	69.4	26	27
	81.5	22	29

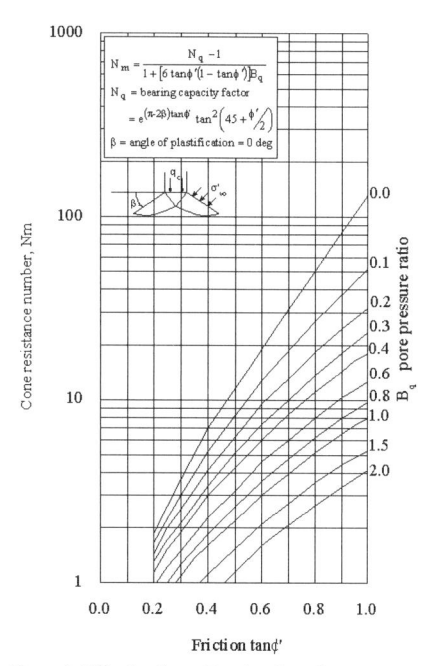

Figure 3. Effective Stress Bearing Capacity Factors (after Senneset et al., 1988)

Based on the CU tests results and the PCPT data, the average effective friction angle at critical state for Site-1 clay soils is 25 ± 5°. For Site-2, Site-3 and Site-4, effective friction angles were evaluated using the empirical correlation between effective friction angle (ϕ') and the plasticity index, Figure 4 (after U.S. Navy, 1971, and Ladd, et al., 1977).

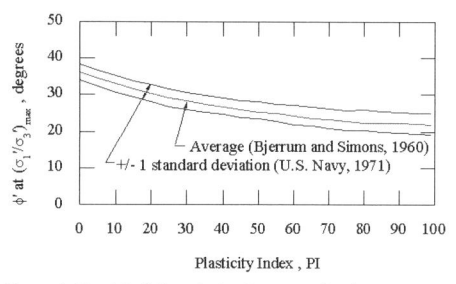

Figure 4. Empirical Correlation Between ϕ' and PI (after U.S.Navy, 1971 and Ladd, et al., 1977)

The estimated effective friction angles are presented in Table 3.

Table 3. Estimated Effective Friction Angles

Location	Average Plasticity Index	Average Effective Friction Angle at Critical State (ϕ'_{cs}) deg	Estimated Effective Friction Angle (ϕ') (from Figure 4) deg
Site-1	35 ± 9	25 ± 5	28
Site-2	41 ± 9	-	26
Site-3	21 ± 7	-	30
Site-4	40 ± 12	-	27

The slope of the critical state line (M) was determined using the laboratory test results that are summarized in Table 3 since M = 6sinϕ'/(3-sinϕ').

10 ESTIMATION OF s_u AND OCR

Based on Equations (12) and (13), values of s_u and OCR were interpreted from the piezocone test results. Figures 5 and 6 shows the results of borehole BH-1 at Site-1. Similar results were recorded for the rest of the boreholes as well.

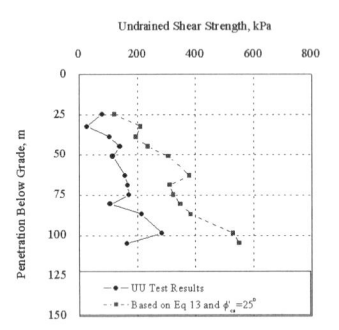

Figure 5. Undrained Shear Strength Profile (BH-1 at Site-1; using Equation 13)

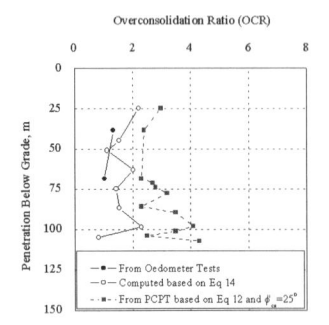

Figure 6. Overconsolidation Ratio Profile (BH-1 at Site-1; using Equation 12)

The comparison of predicted versus reference OCR and s_u values, show that the predicted values are higher than the reference values. However, the interpreted profiles and the laboratory test profiles have similar trends. Therefore, it is logical to calibrate the theoretical method discussed in this Paper to estimate overconsolidation ratio and the undrained shear strength parameters against the selected reference parameters in order to define correction factors. The correction factors are required in Equations (12) and (13) to improve the piezocone data interpretations.

11 CORRECTION FACTORS TO ESTIMATE s_u AND OCR FROM PCPT RESULTS

In the process of calibrating the cavity expansion-critical state method, interpreted profiles of OCR and s_u resemble measured profiles when the critical state parameter (M) was incorporated in the correction factor.

The correction factors derived by Trevor F. (2001 MSc Thesis) to predict OCR and s_u are designated as ξ_{OCR} and ξ_{su}, respectively. The OCR correction factor is ξ_{OCR}=(0.029+0.409M) and the s_u correction factor is ξ_{su}=(0.56-0.095M). Applying these derived correction factors Equations (12) and (13) can be rewritten as follows respectively:

$$\text{OCR} = 2\xi_{OCR}\left[\frac{1}{1.95M+1}\left(\frac{q_t - u_2}{\sigma'_{vo}}\right)\right]^{1.33} \quad (15)$$

$$s_u = \frac{M}{2}\xi_{s_u}\left(\frac{q_t - u_2}{1.95M+1}\right) \quad (16)$$

12 EVALUATION OF THE PROPOSED CORRECTION FACTORS

The proposed correction factors were evaluated using the piezocone data and the available laboratory test results from the four sites. The re-interpreted s_u and OCR profiles based on Equations (15) and (16) of BH-1 at Site-1 are presented in Figures 7 and 8, respectively. The re-interpreted s_u and OCR profiles of other boreholes are also similar to Figures 7 and 8 and the results agrees well with the laboratory test results.

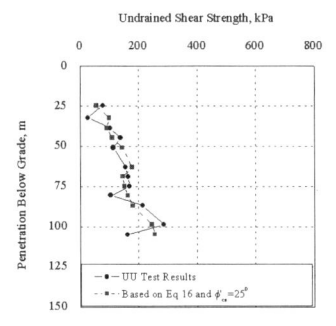

Figure 7. Undrained Shear Strength Profile
(BH-1 at Site-1; using Equation 16)

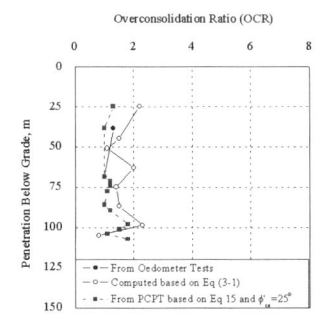

Figure 8. Overconsolidation Ratio Profile
(BH-1 at Site-1; using Equation 15)

To further evaluate the proposed correction factors, piezocone data from a new site designated as Site-5 where laboratory test results and field vane test (FVT) results are available were obtained. At Site-5, due to the lack of laboratory test results of plasticity indices and the consolidation test results, the evaluation was limited to interpretation of undrained shear strength values.

The undrained shear strength values from the FVT test must be multiplied by Bjerrum's correction factor to derive design shear strength values. The correlation between the Bjerrum's correction factor and the plasticity index established by Bjerrum (1972) is presented below:

Plasticity Index	Bjerrum's Correction Factor
20	0.99
30	0.91
40	0.84
50	0.78
60	0.73

Although plasticity index values for Site-5 is not available, Bjerrum's correction factor of 0.8 that corresponds to plasticity value of 45 was assumed to compute the undrained shear strength values.

At Site-5, a lower laboratory shear strength profile was recorded below 50m depth where expansive soils were encountered. The comparison of these three different test results shows that the proposed correction factors very well agrees with laboratory measured triaxial test results and with the in-situ vane test results.

The interpreted undrained shear strength values from the piezocone data (assuming $\phi'_{cs}=30°$), laboratory measured undrained shear strength values and the undrained shear strength derived from the field vane shear tests are shown in Figure 9.

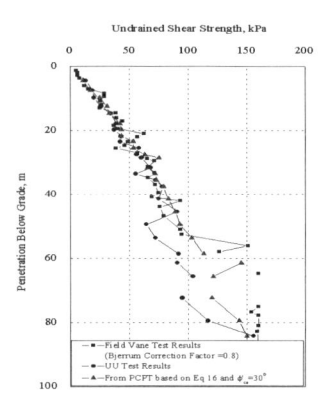

Figure 9. Undrained Shear Strength Profile
(BH-1 at Site-5; using Equation 16)

13 CONCLUSIONS

Relatively simple methods based on cavity expansion theory and critical state concepts are available to predict both the OCR and s_u parameters from piezocone penetration test (PCPT) results. However, these methods tend to over-predict OCRs of marine clay soils of Southeast Asia. Also, over predictions of s_u occurred when UU test results are used as the reference values. The UU tests for the case studies were carried out as described in Chapter 9.

Case studies demonstrate that, for expansive marine soils, the s_u values could be significantly underestimated by laboratory tests, even when high quality sampling and testing procedures are employed. The piezocone test is a valuable tool for recognising such discrepancies and for providing representative shear strength profiles with depth.

The correction factors derived to predict OCR and s_u are designated as ξ_{OCR} and ξ_{su} respectively. These correction factors are related to the critical state parameter M by $\xi_{OCR}=(0.029+0.409M)$ and $\xi_{su}=(0.56-0.095M)$. For site-specific calibration, modified equations that can be used to predict OCR and s_u from PCPT results are:

$$\text{OCR} = 2\xi_{\text{OCR}} \left[\frac{1}{1.95M+1} \left(\frac{q_t - u_2}{\sigma'_{vo}} \right) \right]^{1.33} \quad (15)$$

$$s_u = \frac{M}{2} \xi_{s_u} \left(\frac{q_t - u_2}{1.95M+1} \right) \quad (16)$$

ACKNOWLEDGEMENTS

The authors wish to thank ExxonMobil Exploration and Production Malaysia, JV Partners of the Highlands Gas Project PNG, TOTAL Indonesie, BP Vietnam and Premier Oil Natuna Sea BV for permission to publish laboratory and/or in-situ data in this paper. Authors also wish to express sincere thanks to the Managing Director of Fugro Singapore, Mr. Jeremy Paisley, for his generous support for the entire research study and to the academic staff of the Geotechnical Engineering department at Nanyang Technological University of Singapore for their assistance. Special thanks to Florence, Dion and Mondee for being a source of inspiration to make this paper a success.

REFERENCES

API, 2000. Recommended Practice for Planning, Designing and Constructing Fixed Offshore, Platforms -Working Stress Design, Edition 21, December 2000.
Baligh, M. 1986. Undrained Deep Penetrations and Pore Water Pressures, *Geotechnique 36 (4)*, 487-50.
Bjerrum, L. 1972. Embankments on soft ground, *Proceedings, Conference on Performance of Earth and Earth-Supported Structures,* ASCE, Vol. 2, 1-54.
Casagrande, A. 1936. The Determination of the Preconsolidation Load and its Practical Significance, *Proceedings of the First International Conference on Soil Mechanics*, Cambridge, Mass., Vol.3.
Chen, Y.J. and Kulhawy, F.H. 1993. Undrained Strength Interrelationships Among CIUC, UU, and UC Tests, *Journal of Geotechnical Engineering*, ASCE, Vol.119, No.11, 1732-1750.
ISSMGE, 1999. International Reference Test Procedure for the Cone Penetration Test with Pore Pressure, Report of the ISSMGE Technical Committee 16 on Ground Property Characterisation from In-situ Testing, Proceedings of the 12th European Conference on Soil Mechanics and Geotechnical Engineering, Amsterdam, Edited by Barends et al., Vol. 3, 2195-2222.
Ladd, C.C., Foott, R., Ishihara, K., Schlosser, F., and Poulos, H.G. 1977. Stress-Deformation and Strength Characteristics, State-of-the-Art Report, *Proceedings of the Ninth International Conference on Soil Mechanics and Foundation Engineering,* Tokyo, Vol. 2, 421-494.
Mayne, P.W. 1988. Determining OCR in clays from laboratory strength, *Journal of Geotechnical Engineering,* Vol.114, No.1, 76-92.
Mayne, P.W. and Bachus, R.C. 1988. Profiling OCR in clays by piezocone soundings, *Proc., First International Symposium on Penetration Testing*, ISOPT-1, Vol. 2, Orlando, 857-864.
Mayne, P.W. 1991. Determination of OCR in clays by piezocone tests using cavity expansion and critical state concepts, Japanese Society of Soil Mechanics and Foundation Engineering, *Soils and Foundations* – Vol.31, No.2, 65-76.
Senneset, K., Sandven, R., Lunne, T., By, T. and Amundsen, T. 1988. Piezocone Tests in Silty Soils, *Proceedings of the International Symposium on Penetration Testing*, ISOPT-1, Orlando, Vol. 2, 955-66, Balkema Pub., Rotterdam.
Skempton, A.W. 1957. Discussion on the Planning and Design of the New Hong Kong Airport, *Proceedings of the Institution of Civil Engineers,* Vol. 7, 305-325.
Teh C.I. and Houlsby, G.T. 1991. An Analytical Study of the Cone Penetration Test in Clay, *Geotechnique 41*, No.1, 17-34.
Trevor F. 2001. Evaluation of Soil Parameters from Results of Piezocone Tests in Clay, *MSc Thesis,* Nanyang Technological University, Singapore, 2001.
U.S.Navy. 1971, Soil Mechanics, Foundations, and Earth Structures, *NAVFAC Design Manual DM-7,* Washington, D.C.
Vesic, A.S. 1972. Expansion of Cavities in Infinite Soil Mass, *Journal of the Soil Mechanics and Foundations Division*, ASCE 98 (3), 265-290.
Vesic, A.S. 1977. Design of Pile Foundations, *Synthesis of Highway Practice 42*, Transportation Research Board, National Research Council, Washington, D.C., 68.
Wroth, C.P. 1984. The Interpretation of In Situ Soil Tests, *Geotechnique,* Vol. 34 (4), 449-489.
Wroth, C.P. and Houlsby, G.T. 1985. Property Characterization and Analysis Procedures, *Proceedings of the 11th International Conference of Soil Mechanics and Foundation Engineering,* San Francisco, Vol. 1, 1-55.

Results of a comparative study on cone resistance measurements

V. Whenham, N. Huybrechts, M. De Vos
Belgian Building Research Institute (BBRI), Belgium

J. Maertens
Jan Maertens bvba & Catholic University of Leuven (KUL), Belgium

G. Simon
Ministry of Equipment and Transport of the Walloon Region, D421 Geotechnical Direction

G. Van Alboom
Ministry of Flanders, AOSO, Geotechnics Division, Belgium

Keywords: mechanical, electrical, Cone Penetration Test, comparative study, standardization

ABSTRACT: Cone penetration tests (CPT) can be performed using a mechanical or an electrical cone. Because the electronic tip is recommended as a standard by the Eurocode 7, and as in Belgian very commonly mechanical CPTs are performed, a study is currently performed at the BBRI to quantify the ratios between the cone resistances as measured with four types of CPT tips - three mechanical tips (M1, M2 and M4) for discontinuous penetration, and the standard electric tip for continuous penetration. The data comes from 20 test sites where comparative CPTs recently have been performed in Belgium and covers various soil types, in particular stiff clays in which it has been experienced that the differences on cone resistance between electrical and mechanical measurements are significant.

1 INTRODUCTION

The three mechanical cone types used in Belgium are illustrated below in Figure 1: the simple cone with closing nut (CPT-M4), the mantle cone (CPT-M1) and the friction sleeve mantle cone (CPT-M2). The electrical cone (CPT-E1) is also illustrated in Figure 1. Electrical cones give more accurate readings than mechanical cones, but their use in Belgium is not standard due to the regular occurrence of hard inclusions in the soil, and the presence of very dense or cemented layers, which can cause damage to the rather expensive electrical cone.

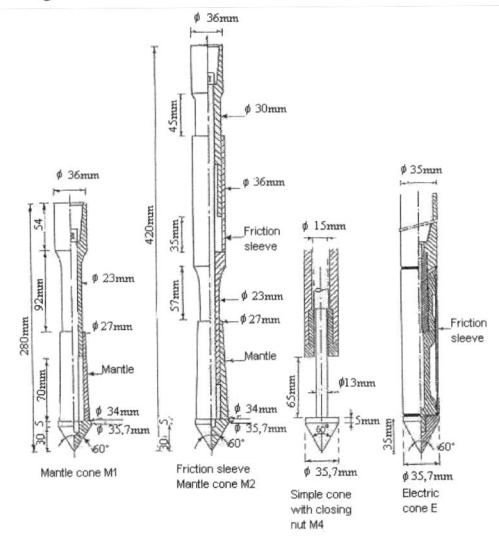

Figure 1. Cone penetrometer tips.

Systematic comparisons between tests with mechanical (CPT-M1, CPT-M2 & CPT-M4) and electrical (CPT-E1) cones show that the measured cone resistance may be affected by the type of cone (see Joustra K., 1974, Rol A.H., 1982).

Because the electrical tip is recommended as a standard by the Eurocode 7 (expected year of publication 2006) and also by the current draft of the National Annex (see De Vos, Bauduin, Maertens, 2003), and as in Belgium mostly mechanical CPTs are performed due to the soil conditions, a study is being undertaken at the BBRI in collaboration with the Geotechnical Division of the Flemish & Walloon Regional Ministries in order to investigate the difference between mechanical and electrical cone measurements and to summarise the conclusions into pragmatic conversion rules.

Figure 2. Location of the test sites in Belgium

2. THE COMPARATIVE STUDY

2.1. Collection of data

The cone penetration results from 20 test sites especially in the northern part of the country are gathered and analysed. The location of the test sites are represented in Figure 2.

2.2. Overview of the investigated sites and soil layer identification

For each investigated site, two approaches for soil classification are considered. The first one is based on the mechanical properties of the soil as deduced from the electrical CPTs (Robertson method, see Lunne & al., 1997), the second one is based on the nature of the soil as deduced from geological information and the results of borings performed on the test sites.

In Figures 3 to 8, typical electric and mechanical CPTs from some of the investigated sites together with the soil layer identifications based on the Robertson method (normal character) and the geologic informations (bold italic characters) are presented. Figures 3 and 4 correspond to sites where the subsoil consists of stiff tertiary clays (Rupelian clay or Boom clay and Ypresian clay). Because of the high cone resistance and the low friction ratio (which can be lower than 2% for the Ypresian clay) of the tertiary clays, they are often identified as silty or sandy soils when using soil classification methods based on CPT-E results. Based on the mechanical CPTs, the tertiary character of these soils is identifiable by an important increase in the total friction Q_{st}. The tertiary clays are also characterized by a regular increase in the cone resistance with depth.

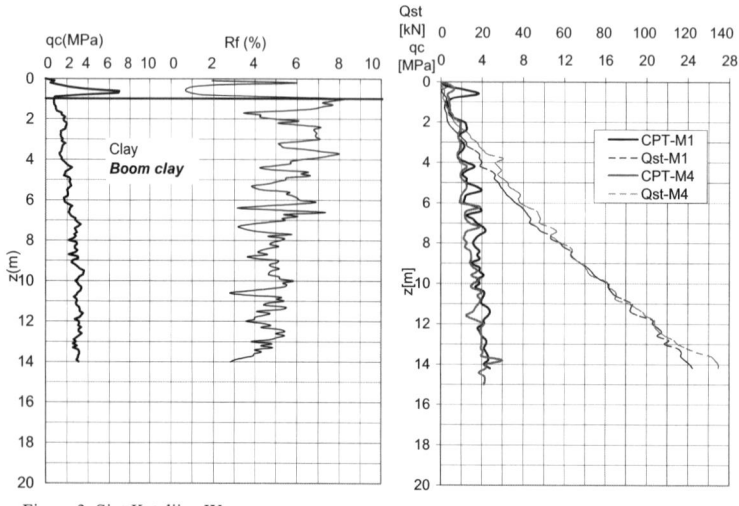

Figure 3. Sint-Katelijne Waver

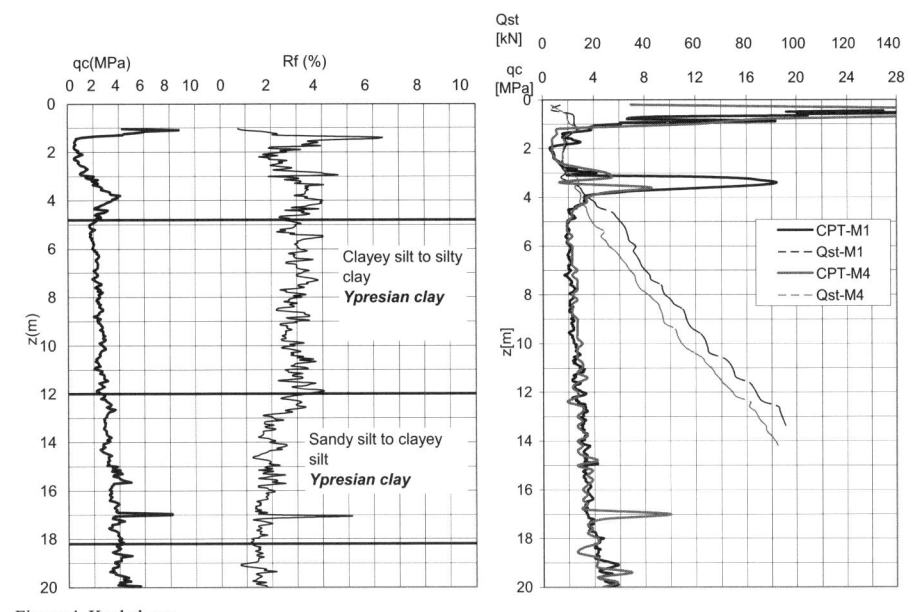

Figure 4. Koekelaere

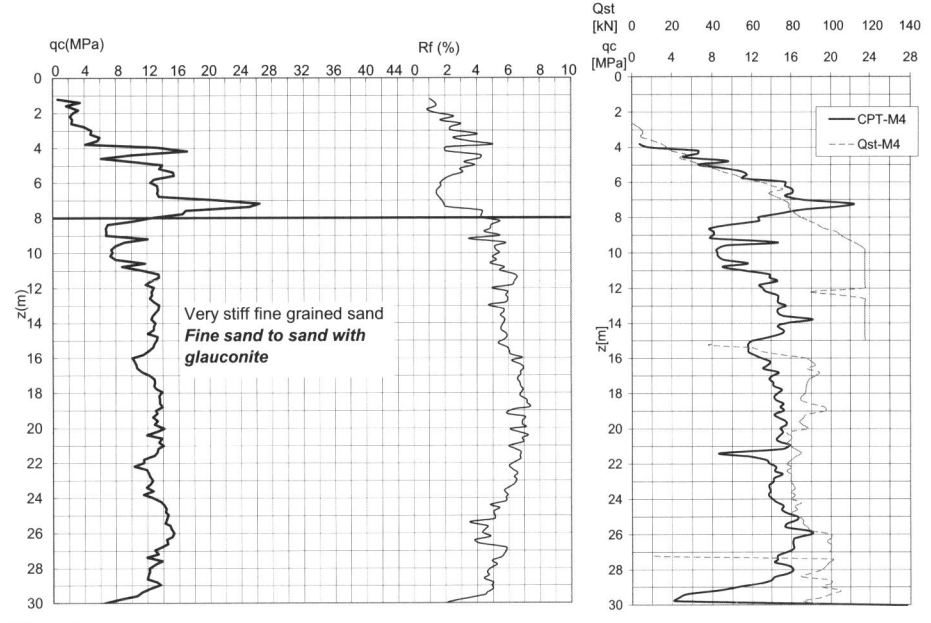

Figure 5. Antwerpen

In Figure 5, a site in Antwerpen where the subsoil consists of sand with a high content of glauconite is presented. As sands with a high content of glauconite have a high friction ratio, they are identified as clays or very fine grained soils following the CPT-E classification methods.

For the others types of soil (Fig.6 to 8), the classifications following both approaches are in good agreement. In Figure 6, the site of Loenhout is presented. The subsoil consists of tertiary sand covered by altering quaternary layers of silt, sand and clay.

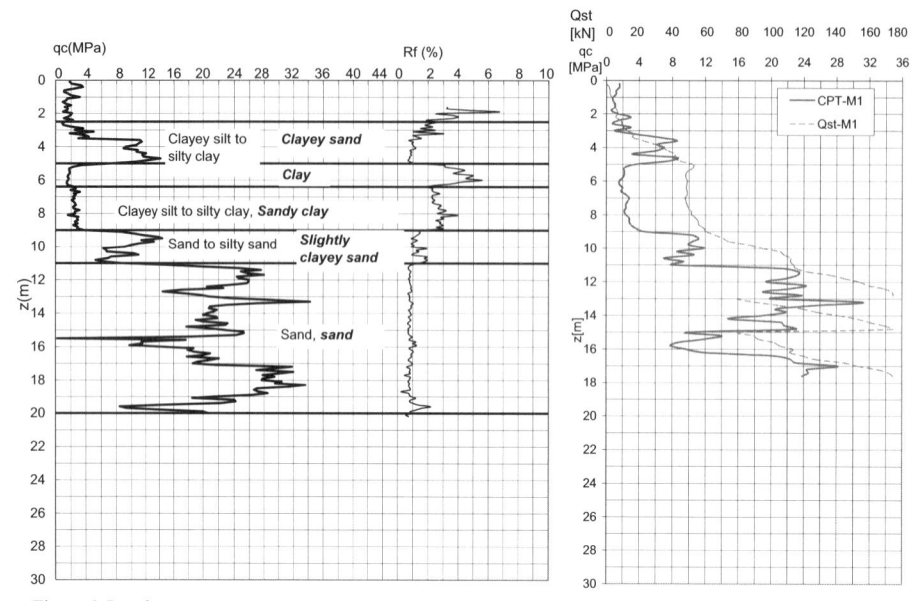

Figure 6. Loenhout

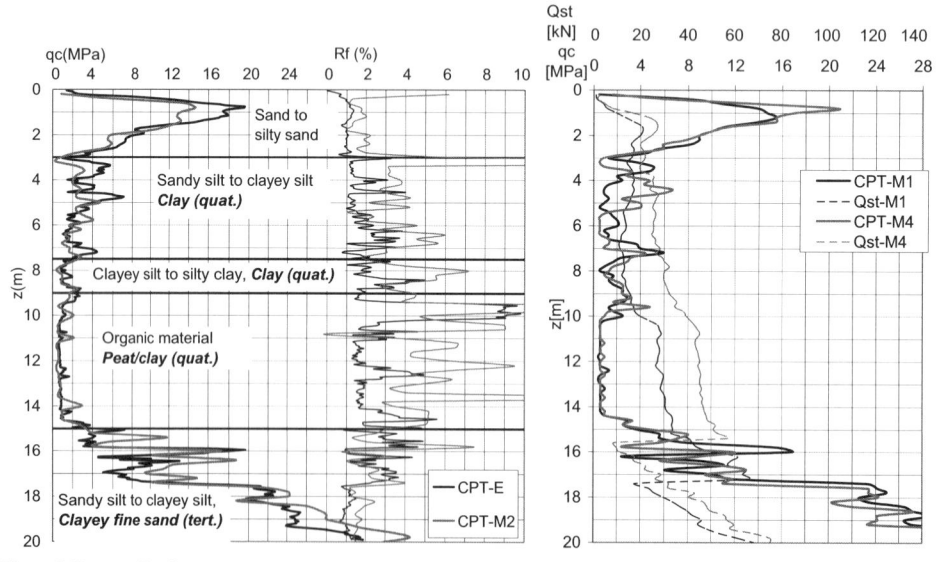

Figure 7. Beveren Doel

The subsoil in Beveren Doel (Figure 7) is characterized by weak soil conditions between 3 to 15m depth. For this site continuous CPT-M2 had been performed, from which it was possible to deduce a friction ratio as presented in Figure 7. It can be observed that the difference in friction ratio deduced from electric and mechanical CPTs is quite important. A general observation from the numerous CPTs performed in the framework of this study is that the repeatability and reliability of the mechanical friction measurements is quite not so good as for electrical CPTs. This is also the case for the discontinuous penetration measurements as can be seen by comparing the total side friction Q_{st} from the different types of cones.

The last example (Figure 8) is a loamy site in Ninove, which is representative for the quaternary subsurface of many sites in Belgium.

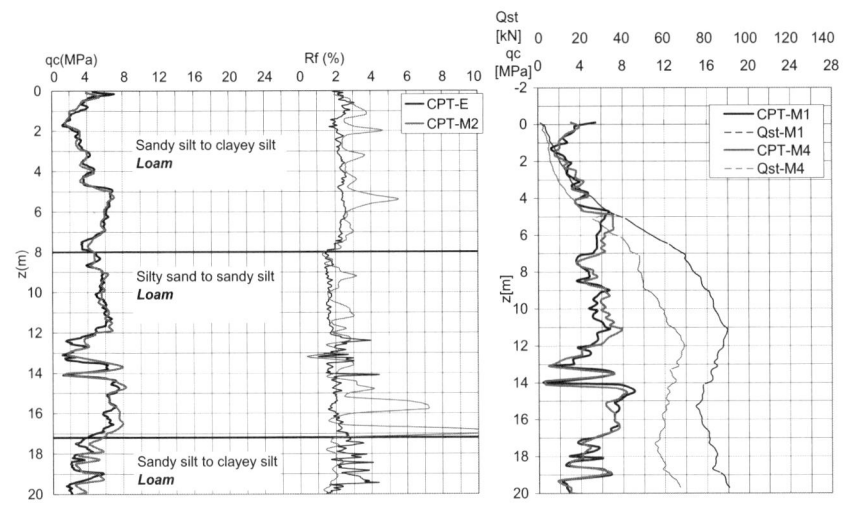

Figure 8. Ninove

2.3. Comparative analysis of q_c-values measured with different CPT tips

For comparing the influence of the different tips the mean values of cone resistance and friction ratio, calculated per meter of depth, are introduced. The cone resistances correspond to the 'q_c' value without any correction for the porewater pressure effects. Indeed the use of CPTU is not common in Belgium, and it is assumed that the influence of the porewater pressure on the ratio between electrical and mechanical cones is limited in most of the cases. Exception should be made for the soft clays, for which more investigation has to be performed. Based on these values, the ratios between the cone resistances obtained with the reference electrical cone and with the mechanical cones are compared. Examples are given in Figures 9 to 12 for different soil conditions.

2.4. Global statistical analysis of the ratios between electrical and mechanical cone resistance

For the global statistical analysis, four classes of soils are considered: "clay", "sand", "sand with glauconite" and "others". The "clay" soils cover the tertiary as well and the quaternary clays, as it has been observed by the analysis that for these soils the differences on cone resistance between electrical and mechanical measurements are quite similar. However the conversion factor for quaternary clays should be confirmed by more data for the whole range of very soft to stiff clays. Moreover the porewater pressure effects on the cone resistances and friction sleeve measurements have to be studied in more detail in the case of very soft clays. In the "sand" class the pure sand and the slightly clayey sands are gathered. A separated class is considered for the "sand with glauconite" because of the specific character of this sand (a.o. high Rf). Finally the "others" class covers all the types of soils between the clay and the slightly clayey sand.

Results are summarised in Table 1 and Figure 13. It can be observed that the average ratios between cone resistances as measured with various types of CPT tips are affected by the soil conditions. Clear trends are observable for clays; in particular the fact that measurements with the M1 and M2 mantle cones may give an increase of the cone resistance up to 30%, what can be explained by the friction of the soil on the mantle. For sands and intermediate soils, the results are more difficult to interpret. As a general rule, the ratios between the qc-values obtained with the mantle cones and the electrical cone are slightly below unity in the sand (0,97 and 0,90 respectively), and increase with the clay content. The ratio between the mechanical simple cone (M4) and the electrical cone is slightly above unity for all types of soils and is less influenced by the soil type. Statistically, for the other soils, the ratios are about one.

3 CONCLUSIONS

The difference between mechanical and electrical cone measurements is investigated in order to define conversion factors between the cone resistances as measured with various types of cones. Principal results are summarised in Table 1 and Figure 13. Because the ratios between cone resistances mainly depend on the soil types, the major difficulty is to define soil classification criteria for the choice of the conversion factors. In the "clay" class (Figure 13 & Table 1) the quaternary clay identified based on the CPT results (for example following the Robertson method) or based on the nature of the soil, and the tertiary clay identified based on the nature of the soil, are gathered. For these soils, the cone resistances measured with the mantle cones M1 and M2 are observed to be respectively 23 and 27% greater than the cone resistances measured with the electrical cone, especially for tertiary clays. For quaternary clays a same trend is observed, but should be confirmed by more data, covering the whole range of very soft to stiff clays. Also the porewater pressure effects have to be considered in the case of very soft clays. In the "sand" class, the sand, fine sand and slightly clayey sand as deduced from a CPT or boring identification are covered. In these soils, the mantle cones (M1 and M2) give cone resistances slightly below those obtained with the electrical cone, whereas the simple cone gives somewhat higher cone resistances, in particular in the sands with glauconite. The "others" class covers all the other types of soils. Statistically, the ratio between the cone resistances measured with mechanical and electrical cones are about the unity in these soils. This is the consequence of the averaging between soils with more or less sand or clay. As we can see in Figure 13 and Table 1, the uncertainty concerning the ratios for these soils is more important than for the pure clays or sands.

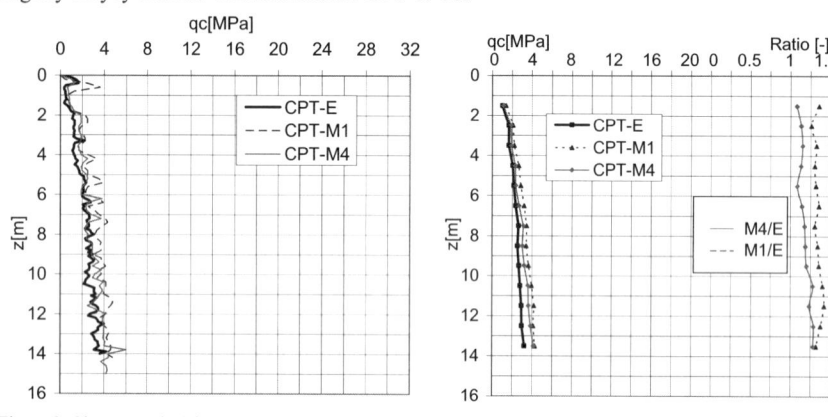

Figure 9. Clay example (Sint Katelijne Waver)

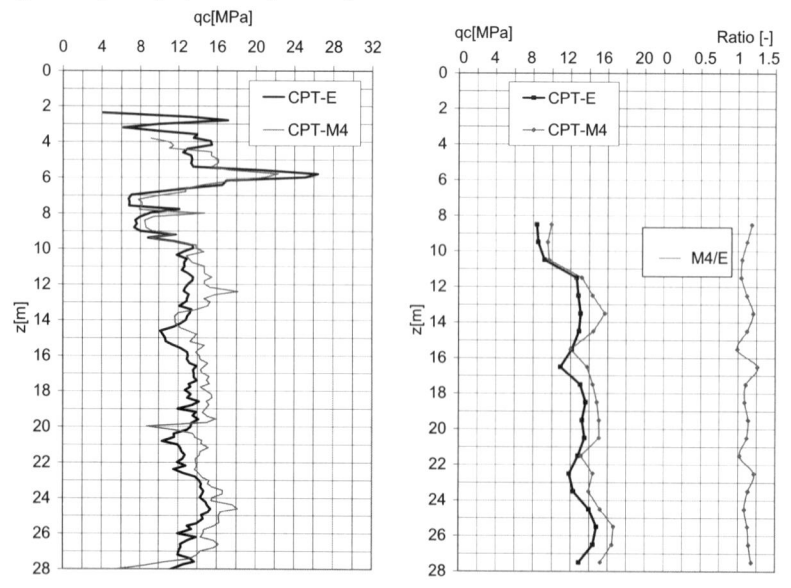

Figure 10. Sand with glauconite example (Antwerpen)

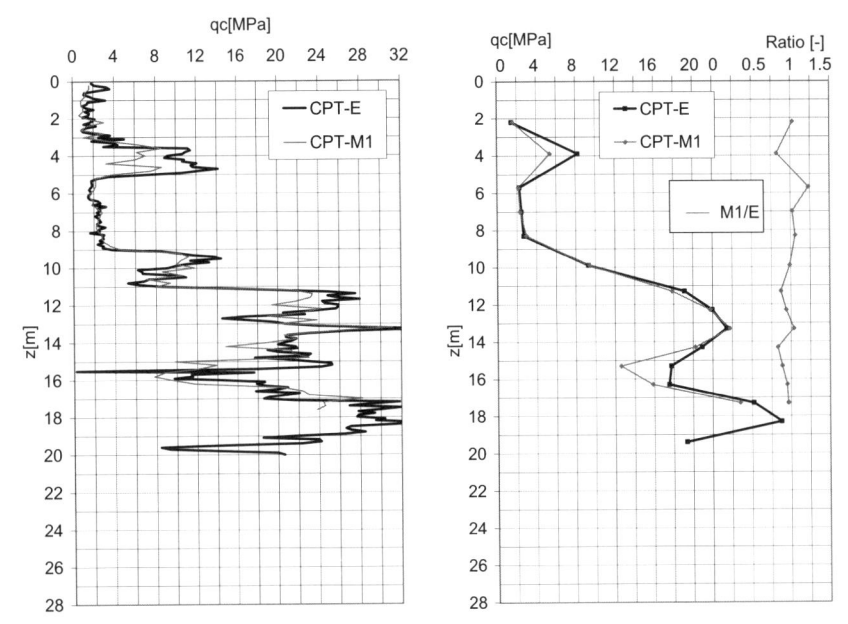

Figure 11. Sand example (Loenhout)

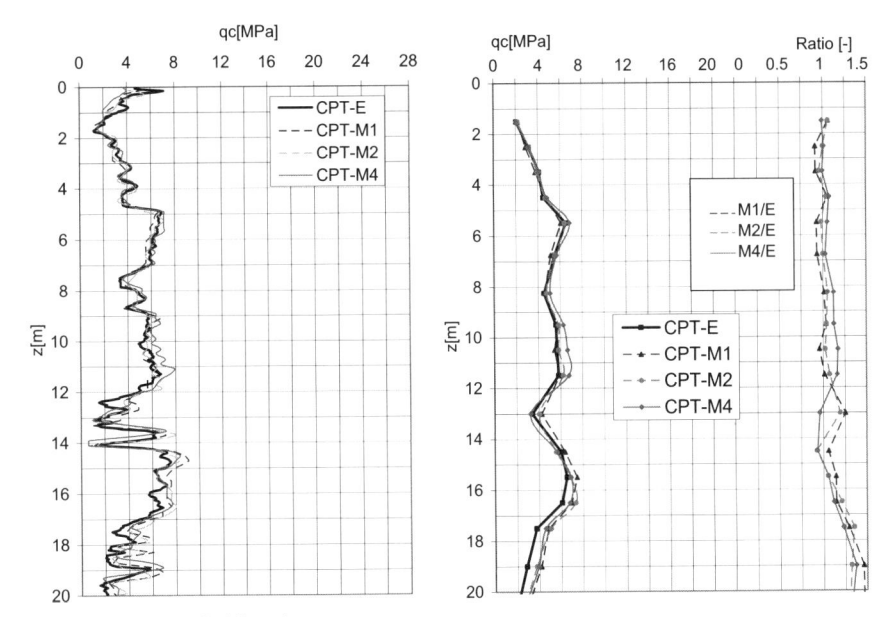

Figure 12. Loam example (Ninove)

Table 1. Ratios between cone resistances as measured with mechanical and electrical cones.

Clay CPT-M1/CPT-E	Ratios	1,23
	Standard deviation	0,09
	Coefficient of variation	8%
Clay CPT-M2/CPT-E	Ratios	1,27
	Standard deviation	0,25
	Coefficient of variation	20%
Clay CPT-M4/CPT-E	Ratios	1,08
	Standard deviation	0,14
	Coefficient of variation	13%

Others CPT-M1/CPT-E	Ratios	0,99
	Standard deviation	0,18
	Coefficient of variation	19%
Others CPT-M2/CPT-E	Ratios	1,01
	Standard deviation	0,28
	Coefficient of variation	18%
Others CPT-M4/CPT-E	Ratios	1,01
	Standard deviation	0,18
	Coefficient of variation	18%

Sand CPT-M1/CPT-E	Ratios	0,97
	Standard deviation	0,11
	Coefficient of variation	12%
Sand CPT-M2/CPT-E	Ratios	0,90
	Standard deviation	0,10
	Coefficient of variation	11%
Sand CPT-M4/CPT-E	Ratios	1,07
	Standard deviation	0,13
	Coefficient of variation	12%
Sand with glauconite CPT-M4/CPT-E	Ratios	1,07
	Standard deviation	0,13
	Coefficient of variation	12%

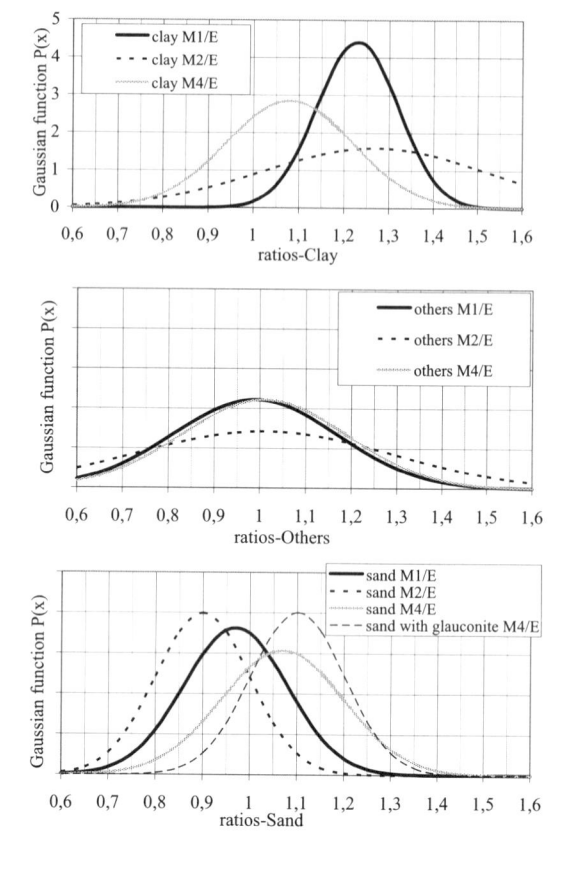

Fig 13. Ratios between cone resistances as measured with mechanical and electrical cones.

REFERENCES

De Vos M., Bauduin C., Maertens J. (2003) The current draft of the application rules of Eurocode 7 in Belgium for the design of pile foundations. Belgium Screw Pile Technology - design and recent developments; Proceedings, Brussels, May 7, pp303-322.

Joustra K. (1974) Comparative measurements on the influence of the cone shape on results of soundings. European Symposium on Penetration Testing ESOPT; Proceedings, Stockholm, June 5-7. Vol.2:2, pp:199-204.

Lunne T., Robertson P.K., Powell J.J.M.(1997) Cone penetration testing in geotechnical practice. Blackie Academic & Professional. London.

Rol A.H. (1982) Comparative study on cone resistances measured with three types of CPT tips. European Symp. On Penetration Testing; Proceedings, Amsterdam, May 24-27. Vol.2, pp 813-819.

2 Geophysical methods applied to geotechnical engineering

General report: Geophysical methods applied to geotechnical engineering

S. Foti
DISTR, Politecnico di Torino, Torino, Italy

A. (Tony) P. Butcher
BRE, Watford, UK

Keywords: geophysical tests, seismic tests, surface waves, site characterization, geophysics

ABSTRACT: The role of geophysical tests in geotechnical and geo-environmental engineering is constantly increasing. The related theme at this conference has attracted a large number of contributions mainly related to applications but also some containing methodological aspects and improvements in the interpretation processes. Many geophysical tests are based on the solution of an inverse problem, of which the one receiving most attention is probably surface wave tests. In this respect some aspects related to interpretation are explored in the present paper, together with a global overview of the papers presented for the theme. Also some key points for discussion are raised about the present and future of geophysics in geotechnical characterization.

1 INTRODUCTION

The potential of geophysical tests for engineering site characterization has been widely recognized, especially in recent decades. Their advantages are non invasiveness, the possibility of testing large volumes of soil, and their cost effectiveness. Conversely their interpretation is not always straightforward, requiring a high level of expertise and knowledge of the physical behaviour of soils and of the mathematical problems of the inversion of data in order to get meaningful results.

One of the main obstacles to wider use of geophysical techniques is often the unfamiliarity of geotechnical engineers regarding what they can expect from geophysical tests and the suitability of each technique in answering the question at hand on site (Wood, 1994). Interaction between engineers and geophysicists is essential in order to properly explore new applications of existing techniques and the development of new tools. A good contribution has been made recently in this direction from applied research from which results have been presented in a variety of conferences. In particular both the previous and the present conferences on site characterization included sessions dedicated to geophysical tests. Many contributions related to civil and environmental engineering applications of geophysical tests are reported at SAGEEP Meetings organized annually by the EEGS (Environmental and Engineering Geophysical Society) in the USA and at the annual conferences of the EAGE (European Association of Geoscientists and Engineers) Near Surface Geoscience Division. (formerly EEGS – European Section) in Europe.

The interest of the ISSMGE in geophysical testing is reflected in the activities of TC10, which in 1994 prepared a volume with the aim of introducing engineers to geophysical techniques and their capabilities and limitations (Wood, 1994). Their work is continuing with recommended test procedures for geophysical tests and their geotechnical applications.

The theme related to geophysical methods in the present conference has received a number of contributions related to both applications and new or improved interpretation techniques. It is important to emphasize that improvements in the interpretation are a key point in the success of application of existing technique to new problems, such as the ones posed by geotechnical and geo-environmental engineering. Often these improvements are directly linked to the application and to the purpose of the testing campaign; hence the interaction between the geotechnical engineer and the geophysicist is a critical aspect.

After a general overview of the papers of the present theme session and a short presentation of the most significant contributions, the paper will discuss some aspects related to the use and the interpretation of geophysical tests exemplified with surface wave applications. Finally some key points for discussion on this theme will be highlighted.

2 GENERAL OVERVIEW

27 papers on geophysical tests have been accepted for publication in the proceedings of the present conference, confirming the interest on the topic expressed in the previous venue of this specialty conference on site characterization (Atlanta, 1998), where 26 papers were included for the geophysical theme (Robertson & Mayne, 1998). However it must be noted that at the previous conference, a few of the papers classified as geophysical tests were related to special CPT devices developed for geophysical measurement at the tip, so that papers concerning geophysical testing have increased more significantly than these numbers show.

Papers presented in both editions make use of a wide variety of geophysical techniques, almost all classified as seismic and electrical measurements, with a few examples of electromagnetic or gravimetric methods. Several papers report results from different techniques applied on the same experimental site, but most of them just present the obtained results trying to assess the suitability of each technique at targeting a specific aspect. Little effort is spent in trying to use the different methods in a common interpretation in order to constrain their results or to obtain a single geophysical model of the subsurface. In the writers' opinion this is quite a pity since the possible synergies of different techniques are not fully explored. In general more detail on the physical ground model used to control or calibrate the geophysical model would enhance the case for application of the geophysical test. This aspect will be further discussed in section 3.3.

A wide variety of geophysical tests has been performed in the ISC'2 testing site. The results are reported in the papers by Lopes et al. (2004), Carvalho et al. (2004), Almeida et al. (2004).

Investigations of performance of different techniques for new field of applications are presented concerning the solution of different engineering problems: e.g. monitoring water saturation in sands (Kalantarian & Doser, 2004) and seepage risk (Comina et al., 2004)

The most popular techniques are by far the Surface Wave Methods (SWM) for seismic techniques and Electrical Resistivity Tomography (ERT) for electrical techniques.

Most applications are related to the determination of mechanical parameters of soil (basically from seismic techniques), the characterisation and monitoring of waste disposal, and the detection of voids and cavities. This latter application being the one for which the greater variety of techniques have been tried, as detailed in section 2.2.

2.1 *Surface Wave Methods*

Surface Wave Methods are present in 11 papers at the present conference, i.e. almost the 40% of the total number of papers for this theme, confirming the increasing interest both in the practicing and research communities. The growing interest in the topic and the spreading of such tests is also witnessed by the increase in the number of papers dealing with it in comparison to the previous ISC conference held in Atlanta, in which 8 papers contained surface wave topics (Robertson & Mayne, 1998).

Surface wave tests are essentially aimed at evaluating the small strain shear modulus profile at a site with an inversion process based on the geometrical dispersion of Rayleigh waves. The idea is borrowed from seismologists, who originally used dispersive properties of Rayleigh and Love waves for the characterization of the Earth's mantle (e.g. Dorman and Ewing, 1962). Early applications for engineering date back to the 1950s but the wide spread use of the technique in geotechnical engineering came after the introduction of the SASW (Spectral Analysis of Surface Wave) Method by the researchers of the University of Texas at Austin during the 1980s (Nazarian and Stokoe, 1984). The SASW method was originally developed using an impact source and a two-receiver scheme for the acquisition of experimental data and the name SASW since then has been associated to this two-station procedure, especially in geotechnical engineering literature. In parallel to the SASW method the Continuous Surface Wave (CSW) method has been developed (Abbiss, 1981 and Matthews et al, 1996) where the wave source is a variable frequency vibrator and acquisition using either 2 receivers or multi-receivers. Although the CSW technique requires a more sophisticated wave source it does allow specific frequency ranges to be targeted that is particularly useful for shallow exploration of complex layers particularly where stiffer layers overlie softer layers. The use of multiple receivers makes both acquisition and processing of surface wave data much faster and more accurate. Also in this case the analysis is performed in the frequency domain, so that the name SASW would be appropriate also for multistation tests, but often a different name is used in order to distinguish between different acquisition and processing schemes (e.g. MASW – Multistation Analysis of Surface Waves; Park et al. 1999). In general the basic principles are the same for any surface wave test independently from the acquisition and interpretation techniques, so that they can be generically addressed as surface wave methods (SWM).

Surface wave tests received little attention from geophysicists until very recently, surface waves being considered primarily as a source of unwanted noise (ground roll) in seismic records. Recently the interest in surface wave tests has greatly increased so that for example SAGEEP meetings now have a session entirely devoted to surface wave tests. Moreover both the EAGE-Near Surface Geophysics Journal and the EEGS Journal are preparing a special issue dedicated to SWM.

Back to the papers using SWM presented at this conference, it is noteworthy to mention that only 2 out of 11 papers use the two-station acquisition scheme, so that it is possible to state that multistation methods have become the most popular technique. Several papers present a comparison between the outcomes of different methods applied on the same site. Very often the results obtained from surface wave tests are compared to the results from borehole seismic tests (CHT; DHT), considered more accurate because they are based on a direct estimation of the travel time of body waves rather than on the solution of an inversion problem as surface wave tests. When looking at this comparison it is very important to bear in mind the different scale of application of the test, indeed surface wave test results are representative of the global behavior of a much larger volume of soil than that investigated with boreholes methods.

Here for example a comparison of the outcomes of borehole seismic measurements and SWM at the testing site specifically realized for this conference is reported (Figure 1). The tests have been performed independently (without the knowledge of each other results) and are reported in different papers (Carvalho et al., 2004; Viana da Fonseca et al., 2004; Lopes et al., 2004). The comparison appears very reasonable in the residual soils at the test site.

The contributions by O'Neill (2004a) and Ryden et al. (2004) are very significant because they propose very advanced techniques for the inversion of Rayleigh waves. The solution of the inversion problem is a central aspect of surface wave tests. Moreover it must be remarked that this aspect is still an open subject especially for situations in which the fundamental mode is not dominating the propagation phenomenon. This is not unusual since it is associated not only to inversely dispersive profiles (i.e. profiles with softer layers below stiffer ones) but also to marked jumps in impedance such those associated to the interface between soil and bedrock. This aspect has been explored in the 2-station SASW literature, where methods based on forward simulation of the complete wave field have been proposed (e.g. Ganji et al., 1998). For multistation tests this aspect still needs to be fully investigated. One of the main problems is related to the fact that the contribution of higher modes produces discontinuities in the apparent dispersion curve(s) so that the evaluation of partial derivatives (necessary for local search algorithms for the solution of the data inversion problem) faces some difficulties. For this reason global search methods, such as the Fast Simu-

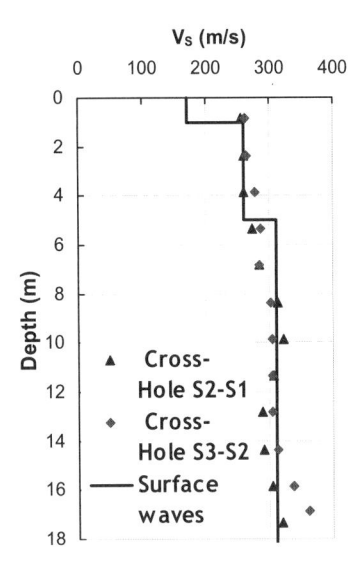

Figure 1 Comparison between Cross-Hole and Surface wave tests results at ISC'2 testing site

lated Annealing algorithm used by Ryden et al. (2004), look to be very promising.

Another contribution related to the solution of the data inversion problem is given by Comina et al. (2004). They propose a scheme of coupled inversion between Rayleigh waves and vertical electrical sounding data. The advantage is given by the strong analogy in the model used for the interpretation of these tests. The coupled inversion takes advantage from the use of common thickness unknowns in the two models so that the inversion process is globally better than separate inversion schemes. The final outcome is a better definition of the geometry. This is an example of possible synergies between different geophysical tests and a technique with wide potential application.

One appealing possibility related to surface wave testing is the analysis of Rayleigh wave attenuation aimed at estimating the damping of soils. In this respect it is very important to point out that reliable measurements can be obtained only using multiple receivers because of the very high sensitivity of such measurements to a variety of factors, ranging from calibration of the receivers to errors in their deployments (tilting, misalignments, etc.). The experimental data can be interpreted in the framework of viscoelastic layered media, accounting for geometrical attenuation and for the relationship between attenuation and dispersion (Aki and Richards, 1980). Applications of such approaches are reported in the literature (Lai et al., 2002; Xia et al., 2002; Foti, 2003).

2.2 Other seismic tests

The application of V_P seismic borehole tomography is reported in 3 papers, that show its effectiveness for the location of voids or highly weathered zones in rock mass (Oliveira & Coelho, 2004) and for the assessment of the performance of grouting techniques for soil improvement (Coelho et al., 2004; Yamaguchi & Satoh, 2004).

Classical geophysical tests such as seismic reflection or refraction tests from the ground surface and down-hole seismic tests are used in other papers. An interesting comparison between in situ results obtained with the seismic cone and laboratory measurements of shear wave velocity using bender elements is reported by Theron et al. (2004). As the role of geophysical measurements in the lab is steadily increasing, it is quite important to evaluate their relationship with in situ tests, as it will be discussed further in next sections.

An innovative method to locate voids in the ground by the analysis of seismic traces using wavelets is proposed by Gucunski et al. (2004). It is noteworthy to mention that the search for voids and cavities is one of the fields in which the potential of geophysical tests are most widely explored. This is due to the necessity of testing large volume of soils, which can be more easily and cost effectively pursued with non invasive methods from the ground surface. Another paper devoted to cavities location in karst terrane for bridge foundation projects makes use of Electrical Resistivity Tomography (Hiltunen & Roth, 2004). A fourth paper devoted to cavity location is based on the use of Ground Penetrating Radar and is aimed at assessing the risks associated to abandoned salt mines (Kulessa et al., 2004). The engineering interest in using geophysical tests for cavity detection is also witnessed by the fact that SAGEEP meetings have typically a session devoted to this topic.

2.3 Electrical Resistivity Tomography

As mentioned before Electrical Resistivity Tomography (ERT) is the non seismic geophysical technique receiving most attention in the papers presented at this conference. Indeed 8 papers report applications of ERT. This is most likely due to the improvements in multi-electrode equipment and to the availability of commercial software that makes this technique much easier to implement than in the past and obviously to the potential of these tests for mapping the underground layers quite accurately with fast non invasive measurements. Nevertheless it is very important that the test is interpreted by high level expert personnel to properly identify artifacts and unreliable data in order to have meaningful results.

The ERT is not able to give any information about the mechanical properties of the soil, but allows the mapping of layers with different composition and with different saturating fluid. The most significant applications are obviously related to environmental problems, indeed most of the papers presented at this conference use ERT for characterizing and monitoring waste disposal sites (Giacheti et al., 2004; Samsudin et al., 2004; Tibana et al., 2004).

Other interesting applications are related to the possibility of performing ERT in marine and fluvial sites to characterize submerged soil deposits (Brabers, 2004).

An interesting example of how additional a priori information can be profitably used to improve the reliability of resistivity imaging is reported by Wisen et al., 2004). They have used information from borehole logs in order to constrain the inversion of resistivity data. They have assembled a large number of surveys in order to develop a continuous stratigraphic model along the section of a proposed railway trench. The constraint given by borehole logs is very useful in determining sharp layer boundaries allowing the identification of geological structures, whereas traditional ERT inversion procedures, based on 2D smoothing algorithms, are not able to reconstruct sharp interfaces. As will be discussed in more detail in section 3.3, the use of constraints and a-priori data is essential for the calibration of geophysical test data, that is otherwise very often based only on the solution of a data inversion problem.

2.4 Other non-seismic tests

Some studies for the development of a new CPT probe to perform Time Domain Reflectometry measurements are reported by Lin et al. (2004). This cone would be able to measure dielectric properties as a function of frequency, which can be related to physical properties of soils. This technique can be especially interesting for hard-to-sample soils, for which the determination of natural porosity or water content is extremely difficult, yet it is of extreme importance for many engineering applications. The dielectric properties of the ground is important for the interpretation of GPR for example.

The applications related to other techniques are very limited in number, with a few applications of Vertical Electrical Soundings and one application concerning GPR for void detection (Kulessa et al., 2004). Probably alternative geophysical techniques would deserve more attention in relationship to their capabilities and potential for solving specific engineering problems.

3 SOME ADDITIONAL NOTES ON GEOPHYSICAL TESTS

In this section some general issues related to geophysical testing for engineering applications are discussed.

The interpretation of geophysical tests, especially those performed from the ground surface, is often based on the solution of an inversion problem. Indeed the identification of subsoil geometry and parameters is based on measurements of propagation and on the simulation of the forward problem of the propagation in an idealized medium. Two classes of problems arise: the appropriateness of the model considered in the associated forward problem and the intrinsic instability of inversion mathematical problems. A subsequent problem is given by the consequences of non-uniqueness of the solution in relation to the objective of the characterization campaign.

In the following some issues related to these topics will be discussed with reference to examples related to surface wave methods.

3.1 Model errors

The affinity between the model and the actual subsoil geometry and mechanical behaviour is of primary importance in order to get meaningful results from an inversion process.

For example a linear elastic layered medium is considered when interpreting surface wave data. The assumption of linear elastic behaviour is common to all geophysical seismic tests and it is very reasonable considering the very small strains induced in the ground by the propagation of seismic wave generated by the relatively light sources adopted for testing. On the other hand the hypothesis of horizontal layering is not always fulfilled by real deposits. The consequences of dipping layers or in general of lateral variations of properties along the profile can be very important. A systematic study of the effects of lateral variations using Finite Elements simulations has been reported by (Gucunski et al., 1996), showing that the dispersion curve is obviously strongly affected by the lateral variations, so that the inversion process can lead to misleading results.

Typically little attention is paid to this aspect in surface wave tests. In the 2-station SASW approach the strategy to cope with this problem was typically to repeat the test in each receiver setup with forward and backward shots following a common mid-point scheme for increasing receiver spacing. Afterwards the different branches of dispersion curve were averaged in order to get a single experimental dispersion curve to be inverted in order to obtain an average profile for the site. This strategy hides all the effects of lateral variations rather than giving a clear picture of the actual subsoil situation.

In multistation tests a simple strategy to investigate the existence of lateral variation can be again to repeat the test with forward and backward shots at the two of the receiver array. If the difference in the obtained dispersion curve is not very marked, it can be reasonable to invert the average dispersion curve.

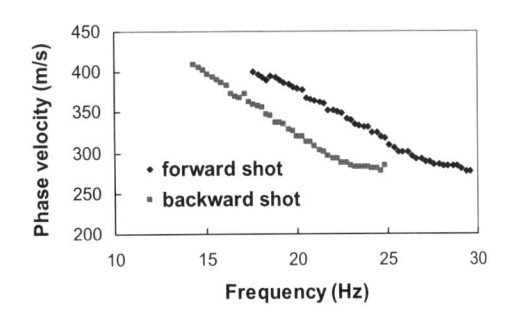

Figure 2 Experimental evidence of lateral variations in surface wave data (after Foti and Socco, 2000)

But in situation like the one reported in Figure 2, where the experimental dispersion curve is clearly different, an interpretation with a 1D model can lead to very misleading result, not reflecting the actual site condition. Some strategies for a 2D inversion of surface wave data have been proposed to map lateral velocity variations (Hayashi and Hikima, 2003).

A more rigorous approach to detect model errors in surface wave tests has been recently proposed (Strobbia & Foti, 2003). It is based on statistical analysis of phase versus offset data in frequency domain. Considering a single mode of propagation of a surface wave in a layered linear elastic medium, the theoretical phase versus offset relationship at a certain frequency is given by a straight line. Any deviation from this theoretical behaviour is associated to the violation of one of the fundamental hypotheses (Figure 3). Small deviations from the straight line are anyway to be expected due to experimental errors. Using a statistical criterion it is possible to effectively recognize model errors. This kind of approach has been proven to be able to recognize errors due to near field effects, lateral variations and higher mode contributions. Once these errors have been properly recognized, it is possible to adopt different strategies in order to obtain meaningful results (Strobbia & Foti, 2004).

Similarly, anisotropy of mechanical properties of the ground can also give misleading results. Work on stiff overconsolidated soils (anisotropic stress and mechanical structure) by Butcher and Powell, (1995) shows the shear wave velocity, derived from Rayleigh wave measurements, to be significantly lower than crosshole and downhole shear wave measurements as shown in Figure 4. This highlights the need for understanding and knowledge of the geotechnical properties of the ground when interpreting geophysical test data. Reliance on one testing or measurement technique will invariably lead to misleading outcomes.

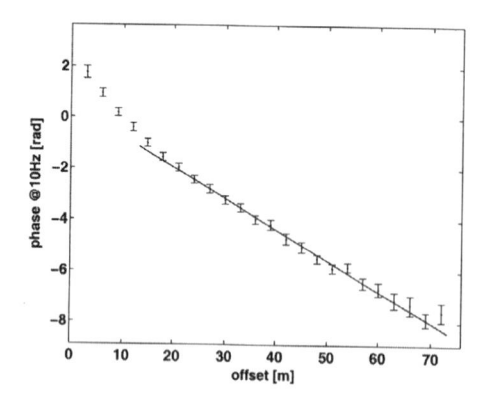

Figure 3 Phase-offset plot of surface wave data showing deviations from linearity (Strobbia and Foti, 2003)

3.2 Uncertainties quantification

Quantification of the uncertainties in the final results can give an important insight into the reliability of geophysical tests. Surface wave tests are affected by error in the measurements due to incoherent noise, bias introduced by the instrumentation, errors in the positioning of the receivers.

Uncertainties in surface wave tests have been extensively studied numerically by O'Neill (2004b). Following a Monte Carlo approach, the Author has explored the influence of a number of possible sources of error on synthetic dispersion curves.

Uncertainties introduced by incoherent noise at the site can be studied experimentally using several seismograms of the test at the same location and in the same testing configuration. Processing the different shots, mean and standard deviation of the phase velocity as a function of frequency can be obtained. Under the assumption of normal distribution of the experimental data, the uncertainty in the dispersion curve (Figure 5a) can then be mapped through the inversion process to the uncertainty in

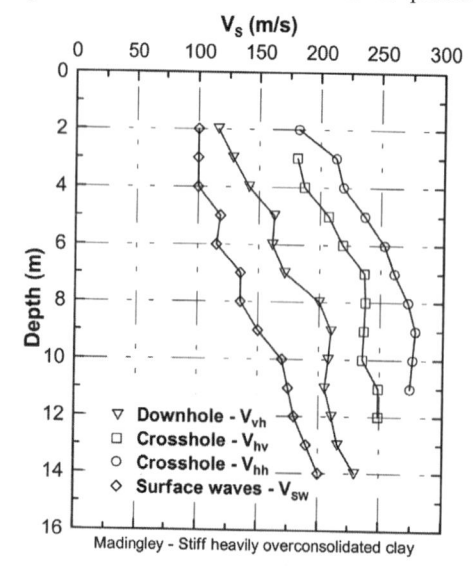

Figure 4 Comparison between Cross-Hole and Surface wave tests results for a stiff overconsolidated clay soil

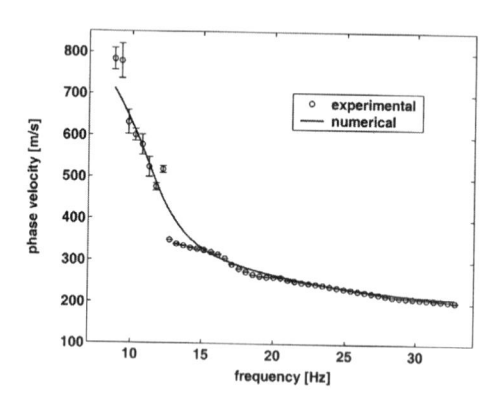

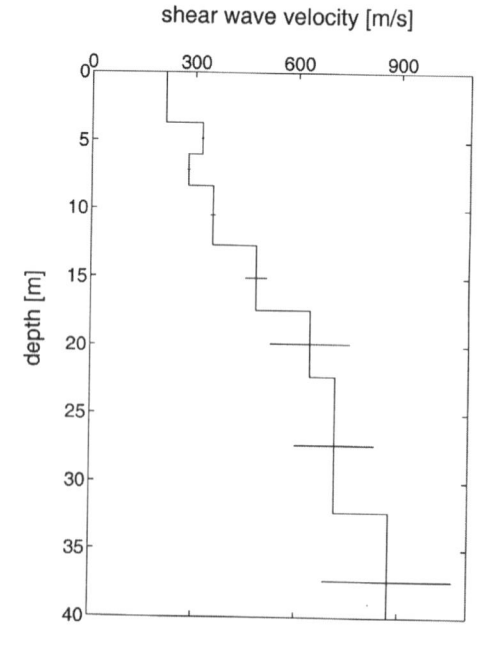

Figure 5 Uncertainties in the shear wave velocity profile obtained from surface wave tests

the shear wave velocity profile (Figure 5b). This information can give an important insight into the reliability of the final result. This aspect is particularly important considering the fact that the reliability of surface wave results are strongly dependent on depth. Shallower layers are typically well sampled because it is easier to get high frequency information, moreover the effect of ambient noise is greater in the low frequency range. However care should be exercised that this statistical type of data processing does not loose valuable information about layering and its variations.

3.3 Inversion process

Considering the instability of inversion problems, it is very important to use all the additional information available for the site. For example borehole logs, typically available in geotechnical projects, can give a very good constraint to the geometry during the inversion process, as shown in the paper at this conference by Wisen et al. (2004), regarding ERT. In surface wave tests, information from boreholes can be very useful in order to reduce the number of unknown in the inversion process, by assuming a-priori layer thicknesses.

Another practice very often adopted in the surface wave test to reduce the number of unknowns in the inversion process is to assume a priori values for the density and Poissons ratio of each layer. These values will change at the water table. Indeed the transition to full saturation is associated with a jump in Poissons ratio that must be carefully taken into account in order to get meaningful results (Foti and Strobbia, 2002). In this respect any information about water table position is of primary importance. When this information is not available from borehole logs, it can be convenient to execute at the same site some acquisitions for seismic refraction. Indeed P-wave refraction can be easily performed using the same equipment and testing configuration of surface waves, the only difference typically being the acquisition parameters in time. The presence of water table in soils is evidenced by the jump in P-wave velocity associated to the full saturation. Other similarities between P-wave refraction and surface wave tests can be shown (Foti et al., 2003) confirming the convenience of executing both tests whenever feasible. This shows that the combining of geophysical techniques can significantly improve the site characterization capabilities of geophysical techniques.

Another example of synergy between different geophysical techniques is represented by the coupled inversion of surface wave test and vertical electrical sounding presented in this conference by Comina et al. (2004).

3.4 Consequences of non-uniqueness of the solution

One of the critical aspects of all geophysical tests based on inversion processes is given by non-uniqueness of the solution. Indeed several combinations of model parameters can be associated to response of the forward problem very similar and practically equivalent with respect to experimental data. This is sometimes considered as the weak point of these techniques.

In reality the consequences of the non-uniqueness should be carefully evaluated with respect to the final object of the characterization campaign in order to explore their relevance.

The following example is related to surface wave tests for seismic local effects studies (Socco et al., 2003). In this case the geophysical tests are mainly devoted to building a geometrical model of the subsoil and the associated small strain shear modulus, while the non linear behaviour of soils is investigated with laboratory tests. In order to assess the influence of non uniqueness on the final objective of the study, shear wave velocity profiles that can be considered equivalent with respect to the experimental dispersion curve of surface waves have been obtained (Figure 6). The misfit between the corresponding numerical dispersion curve and the experimental dispersion curve for the site are substantially the same for the four profiles. These profiles have then been used for evaluating the site response using the code SHAKE (Schnabel et al., 1972). The response associated to three profiles is equivalent both in terms of frequency response (am

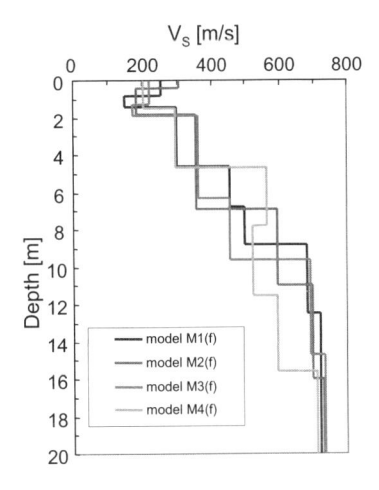

Figure 6 Shear wave velocity profiles obtained from the inversion of surface waves data (profiles giving equivalent dispersion curves with respect to experimental data)

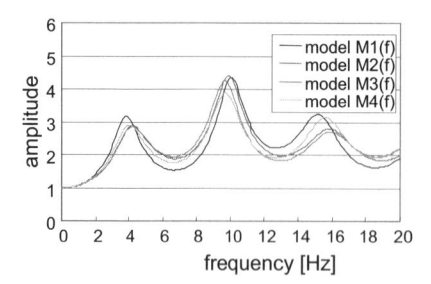

Figure 7 Amplification functions for the shear wave velocity profiles of Figure 6

plification factor) (Figure 7) and response spectrum for a sample recorded accelerogram corresponding to a real earthquake (Figure 8). Hence choosing any of the equivalent profiles in surface wave inversion would lead to essentially the same seismic response for the site. This final result is closely linked to the analogies in the two phenomena (surface wave propagation and seismic amplification in layered media).

4 KEY ASPECTS FOR DISCUSSION

Considering the ever increasing role of geophysical tests for geotechnical and geoenvironmental engineering and the contents of the papers presented at this conference, in the Authors' opinion the following aspects deserve a wide discussion in order to understand the actual perspective:

- Capabilities and limitations of geophysical tests are not always well understood, this is not helped by the lack of relationships, theoretical or empirical, between the geophysical measured properties and the geotechnical characteristics of the ground needed for design.
- The role of geophysical measurements in the laboratory. Is there a case for increasing the spectrum of techniques used in laboratory soil testing and then developing the relationship between laboratory and in situ geophysical tests?
- The role of geophysical testing in future developments. Ground investigation should be problem driven and therefore geophysical measurements should focus on problems that they can help to solve. How can the interaction of geotechnical and geophysical techniques be advanced, other than specialist CPT devices?
- The interaction between geotechnical engineers and geophysicists needs to improve with a better understanding of each others perspective. How can this be solved?
- The search for methods to solve the inversion of data from surface wave methods continues apace as we have seen from the papers at this conference and elsewhere. Are we in danger of generating techniques that are immensely powerful at inverting data but miss some details of the ground that are significant to the geotechnical characterisation?
- The growth in computational power has enabled the development of more complex data analysis and real time processing in the field. This has helped the geophysicist but has it helped the geotechnical characterisation of the ground?
- Topical applications. Void detection; geoenvironmental problems (especially the use of electrical techniques for plume detection); soil porosity. How successful are they and what are their limitations?
- Is the way ahead to encourage combined measurements of complementary techniques with coupled inversion for better definition of characteristics?

5 FINAL REMARKS

The increasing interest in geophysical tests for geotechnical applications is witnessed by the number of papers in this specific session. It would be worthy to expand this tendency because geophysical tests can give many physical insights relevant for geotechnical engineering and to solving geoenvironmental problems. In this respect a closer and closer collaboration between engineers and geophysicist has to be sought after in order to solve these problems. An interdisciplinary approach is required at an early stage of a project rather than later since it could lead also to significant contributions in the interpretation.

A significant contribution to help the wider use of geophysical tests and to make their results more useful could come from laboratory geophysical tests which constitute the natural counterpart of in situ testing. Gaining a better insight into what can be ob-

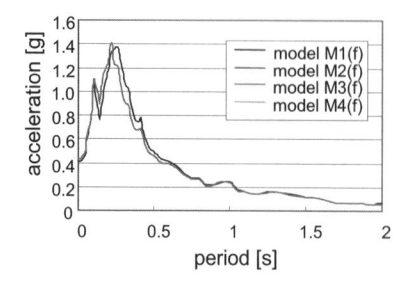

Figure 8 Response spectra for the shear wave velocity profiles of Figure 6

tained from these tests and the possible use of geophysical data in relationships to engineering applications.

Most of papers presented are related to just a couple of geophysical tests, showing the significance and potential of such tests but also highlighting the lack of attention to the full range of geophysical tests. Surface wave tests are present in a large number of contributions. One aspect to be noted is that almost all the papers reported work with the multistation acquisition scheme, with an impact source that has become standard. This may however not always be the best approach in complex stratigraphy. It must be remarked that surface wave tests are a very efficient tool to get the shear wave velocity profile in normally consolidated deposits, but special attention in processing and inversion is needed in order to have meaningful results.

Electrical Resistivity Tomography has also received great attention, especially due to its potential for use in geo-environmental characterization and monitoring, but significantly also combined with surface waves including coupled inversion, paving the way for future developments.

REFERENCES

Abbiss, C.P. (1981) "Shear wave measurements of the elasticity of the ground" Geotechnique, 31(1) 91-104.

Almeida, F., Hermosilha, H., Carvalho, J. M., Viana da Fonseca, A. & Moura, R. (2004) "ISC'2 Experimental Site Investigation and Characterization – part 2: from SH waves high resolution shallow reflection to shallower GPR tests", Proc. 2nd Int. Conf. on Site Characterization, Porto.

Butcher, A.P. and Powell, J.J.M. (1995) Practical considerations for field geophysical techniques used to assess ground stiffness. Proc. Int Conf. on Advances in Site Investigation Practice. ICE London, March. Thomas Telford.

Brabers, P. (2004) "Quantitative resistivity methods for marine and fluvial site investigations", Proc. 2nd Int. Conf. on Site Characterization, Porto

Carvalho, J.M., Viana da Fonseca, A., Almeida, F. & Hermosilha, H. (2004) "ISC'2 experimental site investigation and characterization - part 1: conventional and tomographic P and S waves refraction seismics vs electric resistivity", Proc. 2nd Int. Conf. on Site Characterization, Porto

Coelho, M.J., Salgado, F.M. & Fialho Rodrigues, L. (2004) "The role of crosshole seismic tomography for site characterization and grout injection evaluation on Carmo convent foundations", Proc. 2nd Int. Conf. on Site Characterization, Porto

Comina, C., Foti, S., Socco, L.V. & Strobbia, C. (2004) "Geophysical characterisation for seepage potential assessment along the embankments of the Po River", Proc. 2nd Int. Conf. on Site Characterization, Porto

Dorman J., Ewing M.. (1962) "Numerical inversion of seismic surface wave dispersion data and crust-mantle structure in the New York-Pennsylvania Area", J. Geophysical Research. 67 (13): 5227-5241

Foti S. (2003) "Small Strain Stiffness and Damping Ratio of Pisa Clay from Surface Wave Tests", Geotechnique, vol. 53 (5), 455-461

Foti S., Lai C.G., Lancellotta R. (2002) "Porosity of Fluid-Saturated Porous Media from Measured Seismic Wave Velocities", Geotechnique, vol. 52 (5), 359-373

Foti S., Sambuelli L., Socco L.V., Strobbia C. (2003) "Experiments of joint acquisition of seismic refraction and surface wave data", Near Surface Geophysics, EAGE, 119-129

Foti S., Socco V. (2000) "Analisi delle onde superficiali per la caratterizzazione dinamica dei terreni", La riduzione del rischio sismico nella pianificazione del territorio: le indagini geologico tecniche e geofisiche per la valutazione degli effetti locali, CISM, Lucca, 3-6 maggio, in italian

Foti S., Strobbia C. (2002) "Some notes on model parameters for surface wave data inversion", Proc. of SAGEEP 2002, Las Vegas, February 10-14, CD-Rom

Ganji V., Gucunski N., Nazarian S. (1998) "Automated inversion procedure for spectral analysis of surface waves", J. Geotech. and Geoenv. Eng. 124: 757-770

Giacheti, H.L., Elis, V.R. Peixoto, A.S.P., Mondeli, G. & Hamada, J. (2004) "The use of electrical resistivity to detection of leacheate plumes in waste disposal sites", Proc. 2nd Int. Conf. on Site Characterization, Porto

Gucunski N., Ganji V., Maher M.H.. (1996) "Effects of soil nonhomogeneity on SASW testing", Uncertainty in geologic environment: from theory to practice (Uncertainty 96). : 1083-1097

Gucunski, N., Shokouhi, P. & Maher, A. (2004) "Wavelets in surface wave cavity detection – theoretical background and implementation", Proc. 2nd Int. Conf. on Site Characterization, Porto

Hayashi K., Hikima K. (2003) "CMP analysis of multi-channel surface wave data and its application to near-surface S-wave velocity delineation", Proc. of SAGEEP2003. San Antonio. USA. April 6-10. CD-Rom

Hiltunen, D. R. & Roth, M. J. S. (2004) "Investigation of bridge foundation sites in karst terrane via multi-electrode electrical resistivity", Proc. 2nd Int. Conf. on Site Characterization, Porto

Jones R.B.. (1958) "In-situ measurement of the dynamic properties of soil by vibration methods" Geotechnique. 8 (1): 1-21

Kalantarian, E. & Doser, D. (2004) "Monitoring water saturation gradient in a silty sand site (West Texas) using geophysical techniques", Proc. 2nd Int. Conf. on Site Characterization, Porto

Kulessa, B., Ruffell, A. & Glynn, D. (2004) "Characterizing ground hazards induced by historic salt mining using ground-penetrating radar", Proc. 2nd Int. Conf. on Site Characterization, Porto

Lai C.G., Rix G.J., Foti S., Roma V. (2002) "Simultaneous Measurement and Inversion of Surface Wave Dispersion and Attenuation Curves", Soil Dynamics and Earthquake Engineering, Vol. 22, No. 9-12, pp. 923-930

Lin, C.P., Tang, S.H. & Chung, C.C. (2004) "Development of a TDR Dielectric Penetrometer", Proc. 2nd Int. Conf. on Site Characterization, Porto

Lopes, I., Strobbia, C., Almeida, I., Teves-Costa, P., Deidda, G.P., Mendes, M . & Santos, J.A. (2004) "Joint acquisition of SWM and other seismic techniques in the ISC'2 experimental site", Proc. 2nd Int. Conf. on Site Characterization, Porto

Matthews, M.C., Hope, V.S., Clayton, C.R.I. (1996). The use of surface waves in the determination of ground stiffness profiles. Proc. Instn Civ.Engrs Geotechnical Engineering. 119, Apr., 84-95.

Nazarian S., Stokoe II K.H. (1984) "In situ shear wave velocities from spectral analysis of surface waves", Proc. 8th Conf. on Earthquake Eng. - S.Francisco. 3: 31-38

Oliveira, M. & Coelho, M. J. (2004) "Decomposed rock mass characterization with crosshole seismic tomography at the

Heroísmo station site (Porto)", Proc. 2nd Int. Conf. on Site Characterization, Porto

O'Neill, A. (2004a) "Full waveform reflectivity for inversion of surface wave dispersion in shallow site investigations", Proc. 2nd Int. Conf. on Site Characterization, Porto

O'Neill, A. (2004b) "Shear velocity model appraisal in shallow surface wave inversion", Proc. 2nd Int. Conf. on Site Characterization, Porto

Park C.B., Miller R.D., Xia J. (1999) "Multichannel analysis of surface waves", Geophysics 64: 800-808

Robertson P.K., Mayne P.W. eds (1998) "Geotechnical site characterization", proc. of 1st Int. Conf. on Site Characterization ISC'98, Atlanta, Balkema, Rotterdam

Ryden, N, Park, C.B. & Ulriksen, P. (2004) "A framework for inversion of wavefield spectra in seismic non-destructive testing of pavements", Proc. 2nd Int. Conf. on Site Characterization, Porto

Samsudin, A. R., Hamzah, U. & Yaacob, W. Z. W. (2004) "Geoelectrical resistivity imaging of domestic waste disposal sites: Malaysian case study", Proc. 2nd Int. Conf. on Site Characterization, Porto

Schnabel P.B., Lysmer J., Seed H.B. (1972) "A computer program for earthquake response analysis of horizontally layered sites – User's manual", EERC, Berkeley, California, USA

Socco L.V., Foti S., Strobbia C., Giannone S. (2003) "Potenzialità e limiti di alcune tecniche di indagine sismica utilizzate per la valutazione della risposta sismica locale", Incontro Annuale GNGTS, in italian

Strobbia C., Foti S. (2003) "Statistical regression of phase difference in surface wave data", Proc. of SAGEEP 2003, San Antonio, USA, April 6-10, CD-Rom

Strobbia C., Foti S. (2004) "Multi-Offset Phase Analysis of Surface Wave Data (MOPA)", submitted for publication

Theron, M., Heymann G. & Clayton C. R. I. (2004) "The small strain stiffness of gold tailings", Proc. 2nd Int. Conf. on Site Characterization, Porto

Tibana S., Monteiro L.A.C., Santos Jr. E.L. & Almeida F.T. (2004) "An assessment of heavy metal contamination by an eletromagnetic survey in the solid waste disposal site of Campos dos Goytacazes (Brazil)", Proc. 2nd Int. Conf. on Site Characterization, Porto

Viana da Fonseca A., Santos J.A., Carvalho J., Ferreira C., Tuna C., Costa Santos, E. (2004) "Geotechnical characterization of a residual soil profile: the ISC'2 experimental site, FEUP", Proc. 2nd Int. Conf. on Site Characterization, Porto.

Wisén R., Dahlin T. & Auken E. (2004) "Resistivity imaging as a tool in shallow site investigation – a case study", Proc. 2nd Int. Conf. on Site Characterization, Porto

Wood R.D. ed. (1994) "Geophysical characterization of sites", volume prepared by ISSMFE TC10, Oxford & IBH Pub. Co., New Delhi

Xia J., Miller R.D., Park C.B., Tian G. (2002) "Determining Q of near-surface materials from Rayleigh waves", Journal of Applied Geophysics 51, 121-129

Yamaguchi, Y. & Satoh, H. (2004) "Evaluation of grouting effect with seismic tomography considering grouting mechanism", Proc. 2nd Int. Conf. on Site Characterization, Porto

ISC'2 experimental site investigation and characterization – Part 2: from SH waves high resolution shallow reflection to shallower GPR tests

F. Almeida & H. Hermosilha
Departamento de Geociências, Universidade de Aveiro, Portugal

J.M. Carvalho & A. Viana da Fonseca
Faculdade de Engenharia da Universidade do Porto, Portugal

R. Moura
Dep. Geologia, Faculdade de Ciências da Universidade do Porto, Portugal

Keywords: residual soils from granite, surface and borehole seismics, shallow SH Reflection, GPR

ABSTRACT: The in-situ investigation and characterization of ISC'2 experimental site comprised the application of several geophysical borehole, penetration and surface methods, namely S waves high resolution shallow reflection (SHRR), ground penetrating radar (GPR), P and S waves seismic cross-hole (CH), down-hole (DH), refraction, electrical resistivity, as well as geotechnical tests, namely SPT, CPT, DMT, among others. The site, located within the campus of Faculty of Engineering of the University of Porto (FEUP), is geologically formed by an upper layer of heterogeneous residual granite soil overlaying rather weathered granite contacting a gneissic migmatite. Direct and indirect results from some of the referred surveys are compared between them and with some of the available geological and geotechnical information, in particular hereby (Part II), those obtained from SHRR, GPR, and CH seismic tests. The integration of the available information was then used to build a tentative geological interpretation model. One of the objectives of this component of the global study, hereby presented, was to evaluate the possibility of obtaining valid information from a single technique, particularly, S waves shallow reflection.

1 INTRODUCTION

The ISC'2 experimental site is located in the surroundings of the Faculty of Engineering of the University of Porto (FEUP). In that place residual soils formed by the weathering of granite and metasediments on a complex environment. Figure 1 shows the local geology overlaying an infrared ortophoto and the test site location. FEUP north border was obtained from GPS measurements; the observation of the outcrop depicted on Figure 1 was of some help in figuring out the site geological complexity. The geological information obtained from the Porto geotechnical map (Carta Geotécnica do Porto, 1994), reveals contacts with N10°W bearing.

An intense planar anisotropy (foliation) can be observed in local metasediment borehole cores as well as at the referred outcrop (Figure 2). In this place the orientation of the local planar anisotropy is in accordance with the known 60° eastwards dipping of the local geological trends. This is interpreted as resulting from a metamorphic hercinian process leading to the preferential accumulation of Feldspar in bands (migmatite) and metasediments showing a gneissic texture.

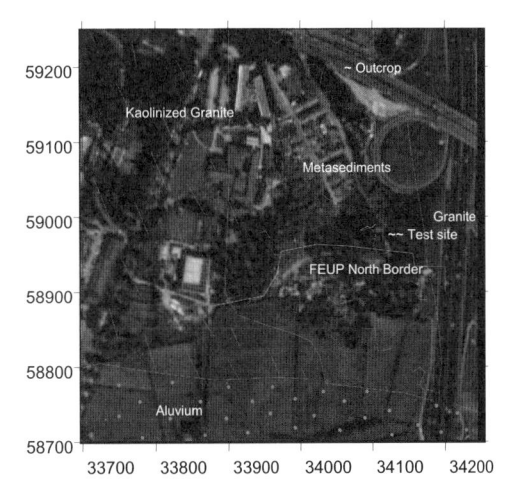

Figure 1 - Geology of ISC'2 experimental site area: Porto geological map overlaying an infrared ortophoto (UTM coordinates, datum EUROP50), from Carta Geotécnica do Porto, 1994.

Figure 2 - Geology of ISC'2 experimental site area: a) photo taken at the outcrop and b) gneiss-migmatite core from borehole S3.

Porto geological map show a clear contact line between the gneissic metasediment and the granite mass. In this complex structural and metamorphic zone it is admitted that the type of regional transition between the two bodies is not a single discontinuity plane but rather a gradual one consisting in an eastward "probabilistic" decreasing of feldspar bands maintaining the local trend of geological planar anisotropy. This interpretation assumes the same prevailing regional field of tectonic forces behind the resulting constant bearing and dipping of the geological anisotropy. Nevertheless, locally, there are zones of sudden lithologic changes.

The constitutive sub vertical structural/ metamorphic planar anisotropy of the rock masses can be seen as a determining factor to understand the geophysical response of the related residual soils resulting from the subsequent weathering process. The gradient orientation of this still active process differs greatly from the mentioned one related to the hercinian structural/metamorphic feldspar enrichment process that is believed to decrease towards east.

The weathering process will tend to transform the feldspar into kaolin (feldspar → kaolin) mainly in the geological contact zones between the two rock masses where the metamorphic/structural feldspar genesis was probably more active also where later fluid weathering action was more active.

In the referred geological model there are two factors (metamorphic planar anisotropy and weathering) which are believed to control the differential weathering processes between the diverse geological formations leading to a heterogeneous spatial distribution of the properties of the resulting soils.

The results from the geophysical surveys conducted in the experimental site depend, namely, on physical properties of the geological masses like the electrical conductivity related with the soil water content, and elastic properties which are theoretically related to the velocity of acoustic waves.

Considering that the Poisson ratio field values, derived from S and P seismic velocities (Part I), can be used as a lithologic or water content index (Almeida et al., 1999) and the S waves velocities can also be used as a low strain rigidity parameter, the interpretation of the geophysical data should be done in two layers separated by the water level.

Several geophysical, geological and geotechnical information was used to build the geological model shown in Figure 3.

The high resistivity anomaly above water level (Part I) is interpreted as being related to a lesser feldspar → kaolin band dipping 60° eastwards. The high S waves velocity from refraction tomography (Part I) is well correlated with that high resistivity anomaly. The lower S waves velocity zone in the middle of the profile is interpreted as a clayey band due to the fact that P waves velocity do not change significantly along the horizontal direction therefore, within this lower S waves velocity zones, Poisson ratio will increase to a clayey domain. The conventional refraction technique (Part I) show two interfaces (dotted lines in Figure 3) where velocity changes: the lower one is interpreted as being related to the seasonal water level and the upper one an interpretation (three layer model) very consistent namely with the tomographic and reflection velocity fields as well as with GPR results (Figures 16 and 17).

One of the objectives of the component of the global study hereby presented was to evaluate the possibility of obtaining valid information from a single technique particularly S waves shallow reflection. The rare opportunity of having data from other different methodologies (electrical imaging, tomographic and conventional refraction, cross-hole and borehole cores – Part I) was determinant for the present attempt.

It is important to note that seismic reflection is normally meant to be used in horizontally layered ground which is far from being the present case. One of the challenges was to figure out whether the

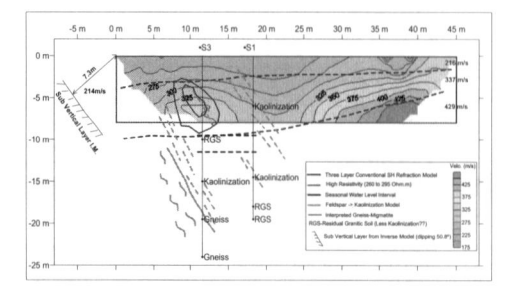

Figure 3 - 2D tomographic seismic refraction SH velocity model (m/s) overlaid by other geophysical results (Carvalho J. et al., 2004) and hypothetical geological model used for seismic reflection interpretation.

shallow reflection technique would detect sub vertical planar anisotropic structures.

2 SEISMIC WAVES - INITIAL CONSIDERATIONS

Seismic waves travelling through geological formations, namely residual soils, are influenced by changes in density and elastic properties, induced by variations in the water content; therefore, although in different ways, both compressional and shear seismic waves profiles are water content dependent.

A well known equation for S waves shows that their velocity equals the square root of the ratio between the shear strength and bulk density (Grant and West, 1965). As shear waves, contrary to compressional waves, are often though as being "blind" to air and water, their responses to soil fluid content vary accordingly. As an example, the increase in P waves velocity due to an increase in density caused by a raise in water content may produce either an increase in S waves velocity, if it occurs a predominant gain in cohesion due to moisture, or a decrease in velocity when the dominant factor is the loss of shear strength due namely to an increase of pore water pressure (West and Menke, 2000). Experiments with low and high tide shear velocity varying pattern with depth (West and Menke, 2000) in shallow unconsolidated sands concluded that: 1) shear wave velocity changes are controlled predominantly by variations in the shear strength of the matrix, not by small density changes that accompany fluid saturation; 2) shear velocities are sensitive to moisture content in the unsaturated zone, increasing by 10% due to cohesive attraction increment between grains and 3) in fully saturated sands pore fluid pressure lowers the matrix shear strength, decreasing velocity by at least 15%.

In a soil, the relation between media (solids plus water) shear modulus and matrix (solids) fluid bulk modulus can be described by Biot-Gassmann theory. Moreover, Myung W. Lee (Lee, 2003) formulation relates shear wave velocity and porosity based on the dependence among seismic waves velocity ratio (V_P/V_S), shear modulus and porosity/saturation. Biot's theory agrees with Gregory´s high porosity Class I grounds (Gregory, 1976). On the other hand, on Gregory's Class II ground, characterized by low porosity, the velocity increases with increasing water content (tested at low pressures with a frequency of 1MHz). The studied SH-Waves behaviour (Michaels and Barrash, 1996) across the water table in shallow environment (coarse grained highly permeable aquifer) indicates SH-wave velocities exhibiting predicted behaviour according to Gregory's low porosity Class II ground.

Velocity dependence of water content within the wave main travelling zone must be understood when interpreting shallow refraction data as much as effects of top soil moisture related with source and geophone-soil coupling. Shallow reflection experiments (Jefferson et al., 1998) conclude that water content of the upper 0.15m strongly affect the seismic records. These authors concluded that generally amplitude of the reflection signals increase with the increase in moisture showing a ringing character.

Shallow reflection SH wave techniques have been employed to geotechnical characterization of soils, namely recent experiments done in Portugal showed the successful application of this technique in sedimentary lagoon deposits (Ghose et al., 2002).

3 SITE INVESTIGATION

Shallow seismic methods/interpretations assuming sedimentary horizontal layered structures need to be adapted to this geological particular case where migmatite sub vertical residual structures are dominant.

A seismic line was planned (Figure 4) on a trench nearby three boreholes where respectively resistivity, refraction and cross-hole tests were previously carried out (J. Carvalho et al. 2004).

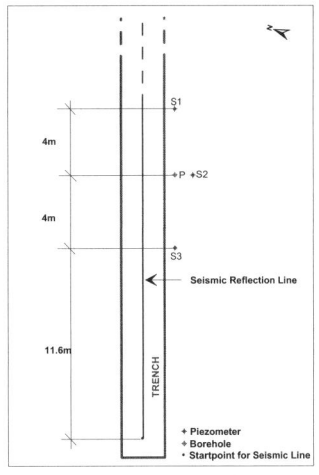

Figure 4 – Seismic line location on ISC'2 experimental site.

3.1 Shallow Seismic Reflection Test

In order to evaluate the response of the site to reflected shear waves a walk way noise test was done. The source shown on Figure 5 was designed specifically for the test site conditions. It is an impulse source with two wooden blocks coupled by steel peens, the upper block acting like a reaction mass; some stones were added to increase the reaction mass. The source was towed by a winch;

Figure 5 – Seismic shear wave impulsive source built for the specific site test location.

this process granted good coupling (increased surface contact) between source and top soil.

Acquisition took place in July, beginning of dry season, being the freatic layer around 10m deep.

3.2 Walk way noise test

The test was carried out during night time in order to avoid (as much as possible) the natural urban noise. For this test two Seistronix Ras24 seismographs were used, one with 24 channels and the other with 12 channels available. The 28Hz horizontal geophones spacing was 0.5m and minimum offset 1m with a sampling interval of 0.25ms. In order to obtaining a record with 72 channels the system was rolled over once, 5 stacks for each polarity were used and the shot was at the starting point of the seismic line (Figure 4).

Figure 6 shows the shots obtained for the two polarities, it is visible that some traces have a low frequency content (approximately 50HZ), this "noise" may be due to moisture condensed on the connections between geophones coupled to the wet soil on the bottom of the trench and the cables.

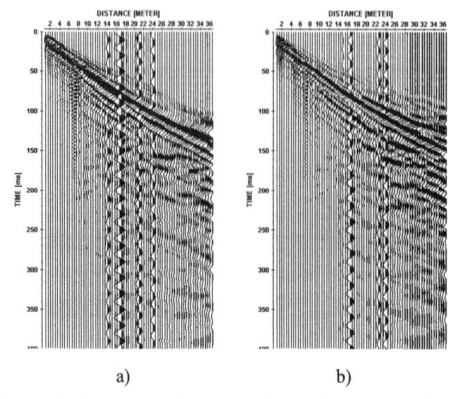

Figure 6 – Trace normalized shots obtained from the walk way noise test: a) Polarization A and b) polarization B.

On these two records was applied 45Hz to 55Hz Notch filter. Then, in order to minimize P wave events on the shots the two records on Figure 6 were subtracted without any other pre- processing. Figure 7 is the resultant record obtained for the subtraction; it shows that the signal has improved specially for larger offsets.

On the referred records, below Love wave cone zone, where SH reflection hyperbolas normally occur there are some nearly aligned crossed events. The adjusted velocity of these events around 210m/s to 250m/s are the velocities obtained from the refraction survey (Figure 3) for depths above 5m.

An interpretative inverse model of these events was considered based on the assumption that they were resultant from a dipping layer. The picked times for Event A on Figure 7-b) were plotted on Figure 8 overlaying the modeled reflection hyperbola dependent of the distance to reflecting layer, dipping angle and layer velocity (Dobrin, 1976); no constrains were imposed on this simple Solver inversion done with Excel©.

The event B on Figure 7-b) with negative apparent velocity (around 320m/s) are probably related to diffraction and back reflections events within the

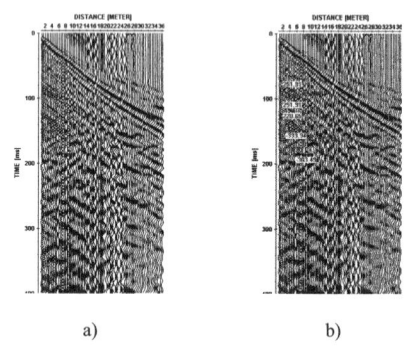

Figure 7 – Resulting shot record obtained after the subtraction of the two polarities. Records with trace normalized and visual AGC 50ms.

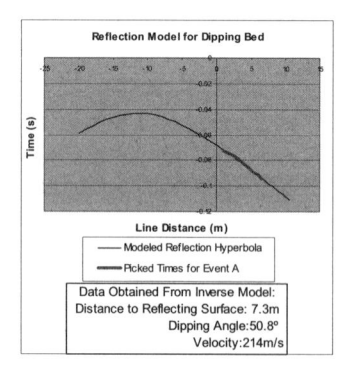

Figure 8 – Inverse model for event A on Figure 7.

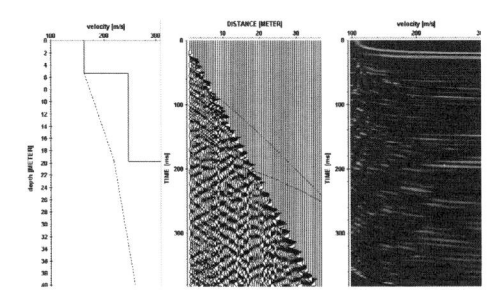

Figure 9 – Velocity analysis for the record obtained from the walk way noise test.

complex metamorphic sub vertical layering.

After the referred subtraction and previously to the velocity analysis shown on Figure 9, some pre-processing was carried out by applying a band pass Butterworth 55Hz/225Hz filter, a top mute and AGC 100ms gain.

Two reflection hyperbolas located below the described crossed events were then interpreted (Figure 9); the shallowest one with interval velocity of 162m/s and the deepest with 245m/s. Comparing the shallower interpreted reflector with cross-hole results (Figure 12-b) and Figure 9) it can be observed in both a velocity step increment around 4m depth; subsequently the lower velocity obtained from the walk way may result from the 4m depth stack velocity that has been admitted, since it was the first layer interval velocity.

From the walk way noise test it was possible to define a two way time limit of 200ms linked to a deep investigation around 20m, which is the approximate total depth of the available borehole cores. Geophone spacing used on this test was considered adjusted for the seismic reflection profile.

3.3 Seismic reflection profile

For the reflection profile we used the same parameters of the walk way noise test; an end-on-push scheme was used and the maximum number of channels was 24 to 32 per shot gather. The shot increment was 1m, with a total number of 12 shots and two polarizations. From this survey were obtained CMP's records spaced by 0.25m with variable fold coverage of 6 to 9 times along the distance comprised between 5m and 15m referred to the seismic line start point.

Figure 10 shows the shot gather at position 6m, after the subtraction of the two polarities, the same cross events occurs as well as a diffraction event not seen on the previous test. This supported the hypothesis that negative velocities can be due to diffraction events. These events have a low velocity (197m/s), meaning that the traveling waves are

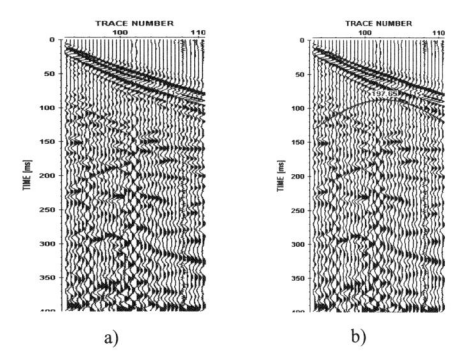

Figure 10 - Shot gather at 6m position, obtained from the reflection survey: a) the shot gather and b) same shot gather with interpreted diffraction. Records width trace normalized and AGC 100ms

associated with the propagation of seismic waves at shallow depths.

The next step was batch processing shot gathers:
- Band pass 24-200Hz;
- Notch filter 45/48/52/55 Hz;
- FK filtering to remove negative velocities;
- Top muting of first arrivals and Love waves;
- Spherical gain compensation;
- Composing CMP gathers.

After analyzing single CMP gather (example in Figure 11) a composed stack velocity section was obtained (Figure 13-a).

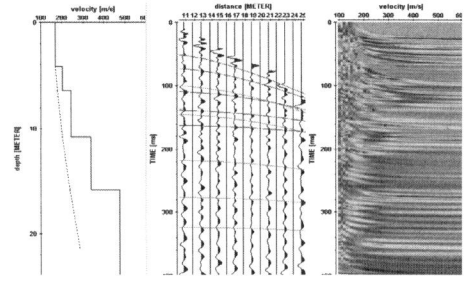

Figure 11 – Velocity analysis for CMP at 8m from starting point.

At this stage of processing two types of velocity fields were obtained: one from CMP velocity analysis and other from previous cross-hole measurements (J. Carvalho et al. 2004).

Figure 12 compares the vertical velocity distribution with depth obtained with cross-hole, CMP analysis, and shallow refraction tomography. It is possible to see consistent steeply changes with depth of the interval velocity models leading to the interpretation of three main velocity transition zones: the first one at 3m to 4m, the second at 6m to 7 m and the third at 10m to 12m.

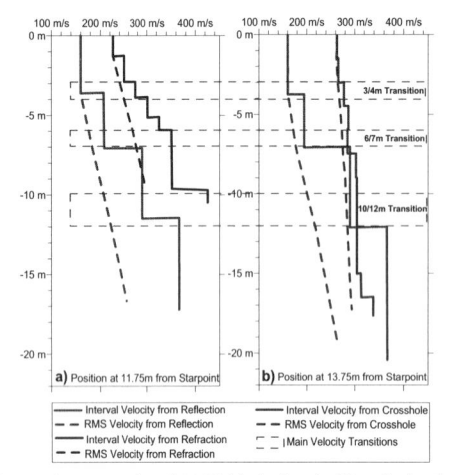

Figure 12 – Interval and RMS (dashed) velocities obtained: a) from tomographic refraction and reflection and on b) from cross-hole and reflection.

The RMS velocity gradient patterns with depth are similar except for cross-hole in the first 7m (Figure 12-b). This may be due to the effect of grout spreading around the boreholes causing the apparent raise in velocity nearby the surface where weathering is higher. Despite the very similar pattern of velocity change with depth between refraction and reflection, the refraction average velocity profile is higher; this difference may result from smoothing effects of the tomographic inversion process.

Two sections were then obtained for the seismic reflection profile, one using the adjusted velocities from the CMP´s (Figure 13), going from 4m to 18m on the transect line, and another with the stacking velocities obtained from the cross-hole survey (Figure 14), going from 13 to 17m meters on the transect line.

Despite the obtained cross-hole higher staking velocities the same behavior is observed along the sub vertical eastwards dipping direction between boreholes S3 and S1. On the reflection staking velocity model there is an high velocity zone (between 9m and 12m on the transect line); this maybe due to a geological sub vertical heterogeneity, common in Porto granite residual soils, that should be looked in terms of interval velocity.

Finally depth conversion was done using the 2D stacking velocity model from Figure 13-a). The 2D interval velocity model obtained is on Figure 15-a) and the depth section is on Figure 15-b).

To obtain the maximum shear modulus from the interval velocities ($G_0=\rho*V_s^2$) an average 18.1kN/m^3 density value was used (Figure 15-c).

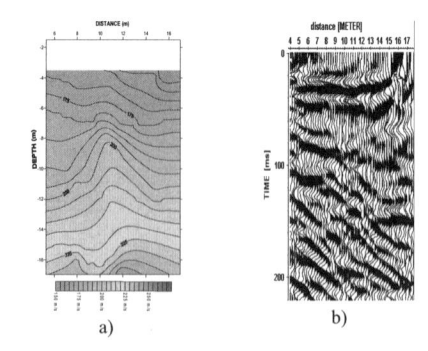

Figure 13 - a) 2D stacking velocity model from CMP reflection profile, b) the seismic NMO corrected and stacked.

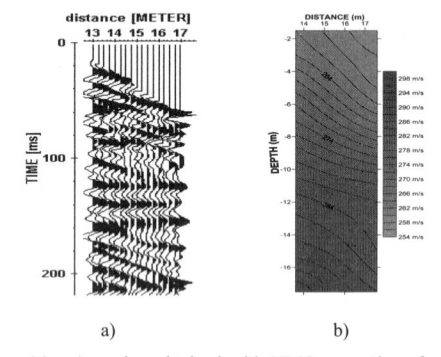

Figure 14 - a) section obtained with NMO corrections from cross-hole velocities and stacked b) 2D stacking velocity model from cross-hole.

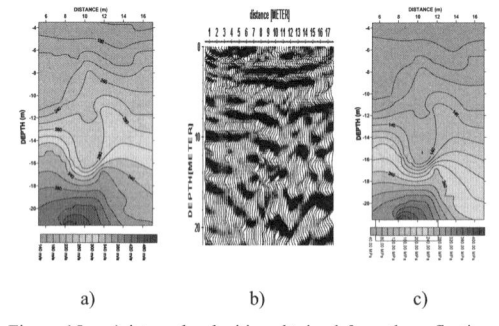

Figure 15 – a) interval velocities obtained from the reflection profile, b) seismic section converted to depth and c) maximum shear modulus (G_0) obtained from the interval velocities using an average density value of 18,1 kN/m^3.

From the maximum and minimum density values obtained on Part I the G_0 values can vary by $+/-$ 7%.

3.4 *GPR survey*

One GPR survey took place along the seismic line direction. For this profile was used a RAMAC GPR with a 100MHz shielded antenna. Despite the use of a shielded antenna there were some air diffractions possibly related with the trees above the trench. These inconvenient events were minimized by a migration process (0.3m/ns). The summarized following basic processing was done: Move Startime; DeWow 10ns; Average xy Filter; Band Pass Butterworth; Subtracting Average; Trace Amplitude Normalization.

Figure 16 shows the interpreted processed radargram. It is visible a reflector with very good agreement with the conventional refraction model first interface; the N-SPT values increase between 3m and 4m deep which supports the existence of a possible transition zone. There are some visible diffractions, and a possible less weathered zone between 0m and 5m along the transect line and from the surface to 3.5m depth.

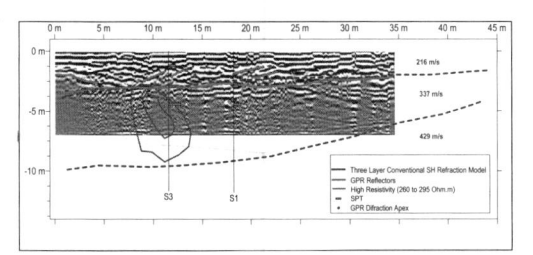

Figure 16 – Processed radargram with interpreted events overlaying the resistivity and conventional refraction model and N-SPT values in boreholes S3 and S1.

The next step was to overlay all the information interpreted (Figure 17) from seismic reflection, refraction, resistivity, GPR and from the geological information.

4 DATA INTEGRATION

Finally some of the more relevant geophysical interpreted data was overlaid, aiming at obtaining an overall conceptual model of the investigated geosystem. On Figure 17-a), the interval velocities obtained from CMP analysis are overlaid by the conventional refraction three layers model, the high resistivity contour lines and the hypothetical lateral dipping structure model (dip: 50.8°E) obtained from seismic inverted reflections. This model is believed to be related to the 60°E dipping gneissic-migmatite local structures.

On Figure 17-b) the seismic stacked section obtained is overlaid with additional information, namely the interpreted GPR reflectors and the interpreted discontinuities/transitions on the seismic section.

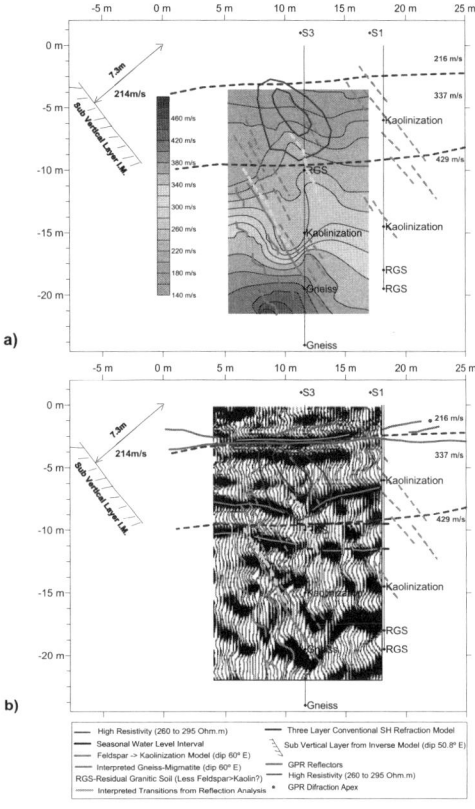

Figure 17 – Integration of the interpreted partial models: a) interval velocity model; b) stacked seismic reflection section.

5 CONCLUSIONS

The SH wave velocity fields (cross-hole, reflection and refraction) and resistivity model supports the local geological evidence.

Also, there is a very similar horizontal interface pattern common to interpretative models from seismic stacked section, conventional refraction and GPR radargram.

There are then three main interpreted considered transition zones: the first around 3/4m depth, giving a very good correlation with GPR and conventional seismic refraction; the second around 6/7m depth interpreted from CMP analysis, which is interrupted by a discontinuity and the third also obtained from CMP analysis from 10/12m with large discontinuity on the stacked section that can be correlated with the seasonal water level. The related interval velocities of these zones may be analyzed taking in account possible changes in the respective water content. The main behavior of these soils seems to fit Gregory Class II ground (low porosity) but westwards from S3, lowering of the velocity can be interpreted as saturates Gregory Class I ground (high porosity).

Between 3/4m and 6/7m there is a higher resistivity zone and a higher amplitude reflection zone from GPR that may be explained by lesser water content and porosity or less feldspar/kaolin on the matrix. Between 6/7m and 10/12m depth the increment of the velocity can be explained by the capillary moisture increase.

In a general way this combined challenging geophysical survey allowed several interpretative hypotheses concerning the characterization of the involved type of ground. At the same time brought several unsolved issues related with the behavior of shear waves on shallow weathered sub vertical heterogeneous structures with varying physical and mechanical properties. In particular, concerning seismic reflection, this type of ground should be tested on more controlled conditions, namely controlled source and more favorable locations. In addition some investigation should also be developed relating the Feldspar→Kaolin matrix index proprieties (porosity, permeability, water content, etc.) with seismic waves behavior in this geological environment.

ACKNOWLEDGMENTS

The authors are grateful to the Contractors that have sponsored this experimental site: MOTA-ENGIL; SOPECATE; TECNASOL-FGE and TEIXEIRA DUARTE. A special thank is due to Ing. Ricardo Andrade from Mota-Engil, and Ing. Nuno Cruz also from Mota-Engil, (formerly in CICCOPN, which has also supported this work)

Part of this work is financially supported by the research project: POCTI / ECM / 33796 / 1999: "*Management of sampling quality on residual soils and soft clayey soils. Comparative analysis of in situ and laboratory seismic waves velocities*".

We are also grateful to ICTE namely to Prof. Mendes Victor for the use of one Seistronix Ras24 seismograph.

REFERENCES

Almeida, F.E., Moura, R.M., Teves Costa, P., Oliveira, C.S., 1999. Caracterização das areias de Faro através de ensaios sísmicos: Sismo 99 2eme Rencontre en génie parasismique des pays méditerranés, Faro.
Carta Geotécnica do Porto, 1994. COBA/FCUP/CMP.
Carvalho, J.M., Viana da Fonseca, A., Almeida, F., Hermosilhe, H, 2004. ISC'2 experimental site investigation and characterization – Part I: Conventional and tomographic P and S waves refraction seismics vs. electrical resistivity – ISC'2 International Site Characterization – FEUP – Porto.
Dobrin, M.B., 1976, *Introduction to geophysical prospecting*: McGraw-Hill, Inc.
Ghose, R. and Goudswaard, J., 2002. Relating shallow S-wave seismic to cone penetration testing (CPT) in soft soil: a multi-angle, multi-scale approach: SEG conference, Calgary, Canada.
Ghose, R., Almeida, F., Hermosilha, H., Bonito, F. And Cardoso, C., 2002. Shallow S-wave Reflections over Lagoon deposits: 8th International Meting EEGS-ES – Aveiro, Portugal.
Grant, F.S., and West, G.F., 1965. Interpretation theory in applied geophysics: New York, McGraw-Hill Inc.
Gregory, A.R., 1976. Fluid saturation effects on dynamic elastic properties of sedimentary rocks: Geophysics, vol. 41 no. 5, p. 895-921.
Jefferson, R.D., Steeples, D.W., Black, R.A., 1998. Effects of soil-moisture content on shallow-seismic data: Geophysics Vol.63, Nº4, p.1357-1362.
Lee, M.W., 2003. Velocity ratio and its application to predicting velocities: U.S. Geological Survey Bulletin 2197.
Michaels, P. and Barrash, W., 1996. The anomalous behaviour of SH-waves across the water table: Proceedings of the Symposium on the Application of Geophysics to Engineering and Environmental Problems, EEGS:
West, M. and Menke, W., 2000. Fluid –Induced Changes in Shear Wave Velocity from Surface Waves: Proceedings, Symposium on the Application of geophysics to Engineering and Environmental Problems (SAGEEP), 21-28.
Viana da Fonseca, A., Carvalho, J., Ferreira, C., Tuna, C., Costa, E. & Santos, J., 2004. Geotechnical characterization of a residual soil profile: the ISC'2 experimental site, Porto. *Second Int. Conf. on Site Characterization – ISC'2*. Porto. Millpress, Rotterdam.
Viana da Fonseca, A., Matos Fernandes, M. & Cardoso, A.S. 1998. Characterization of a saprolitic soil from Porto granite by in situ testing, *First Int. Conf. on Site Characterization –ISC'98*. Atlanta: 2, 1381-1388. Balkema, Rotterdam.

Quantitative resistivity methods for marine and fluvial site investigations

Peteralv Brabers
Demco NV, Wintershoven, Belgium

Keywords: aquares, resistivity survey, geoelectrical, pipeline route, dredging, exploration

ABSTRACT: Geophysical methods are often being used in offshore site investigation programs as an exploration tool to define the general geological structure of the subsurface and it's vertical and horizontal variability. Geophysical results are used either to define in a more economically and geologically justified way geotechnical sampling locations or to get more continuous information between widely spaced borehole and CPT locations. Although earlier resistivity methods, as originally developed for land applications, have been used in offshore applications for more than 15 years now, they often tended to produce less quantitative information as compared to the classical seismic methods and as a consequence the offshore site investigation markets had lost their interest in them.
The Aquares resistivity or geoelectrical method was developed specifically for offshore applications. The last few years this technique has been applied successfully on various marine and fluvial site investigation projects and is more and more being used on port design & engineering projects, dredging reconnaissance projects, pipeline routes and gravel and sand exploration projects.
The resistivity data is acquired using a multichannel resistivity cable trailing behind the survey vessel on the seafloor. While the vessel is sailing electrical soundings are obtained every 3 seconds. During the fieldsurvey qualitative results are monitored on computer screen and allow already on site to define a number of subsequent CPT or borehole locations in an economically justified manner.
After processing the resistivity results are presented as colorcoded horizontal and vertical sections derived from a digital 3D model of the geological subsurface.
Quantitative marine resistivity methods can effectively be applied in conditions where "gas-masking", multiple reflections in shallow water, diffractions on coarse gravel, cap rock tend to reduce the effectiveness of classical seismic reflection methods.
A number of case studies are presented.

1 INTRODUCTION

As geotechnical investigations tend to be relatively expensive in offshore environments, offshore site investigators generally use geophysical methods to obtain a more detailed knowledge of the subsurface geology. Geophysical results were used either to define in a more economically and geologically justified way geotechnical sampling locations or to get more continuous information between the widely spaced borehole and CPT locations. For this purpose resistivity or geoelectrical methods, as originally developed for land applications, have been used in offshore applications for more than 15 years. As they had a reputation of producing less quantitative information compared to the more classical seismic methods offshore site investigation markets had lost their interest in them. At this moment these older methods are still doing useful work supplying formation factors and average resistivity values on marine cable routes with less importance attached to more quantitative information on sediment thicknesses and depths.

The Aquares resistivity method was recently developed specifically for quantitative offshore applications involving depths and thicknesses of geological structures. During the last several years this method has been successfully applied on various projects and is gradually again drawing the attention of port design engineers, dredgers, pipeline route designers and sand- and gravel markets.

2 PRINCIPLES

2.1 Land based applications

An electrical current is injected into the subsurface by means of two currrent electrodes. The voltage gradient associated with the electrical field of this current is measured between two voltage electrodes placed in between the current electrodes (see fig. 1a). Based on the measured values of current and voltage the average resistivity of the subsurface is calculated for a subsurface volume from the seafloor surface down to a certain penetration depth. The penetration depth depends on the distance between the current electrodes. Larger electrode distances are associated with increasing penetration depths.

If the measurements are repeated with progressively increasing current electrode distances information is obtained from progressively deeper geological structures (fig. 1a). As such, a fieldcurve is obtained showing the resistivity as a function of the horizontal distance between current electrodes. After computermodelling this fieldcurve is transformed into a geophysical subsurface section showing the resistivity as a function of depth. Various algorithms exist to carry out the resistivity curve inversion (M.H. Loke and R.D. Barker 1996, L.R. Lines and S. Treitel 1984, Zohdy A.A.R. 1989, Koefoed O. 1970).

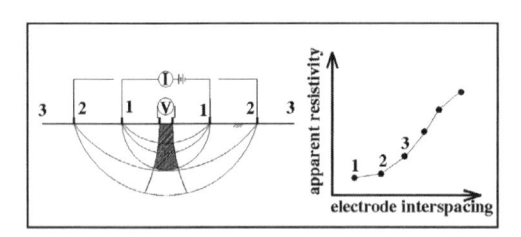

Figure 1a: Principles of Vertical Electrical Soundings
– On land

The resistivity of a geological structure depends on it's porosity, water saturation and the water resistivity. Gravel usually has a lower porosity than sand and it's resistivity thus is higher. Clay with generally very high porosities shows very low resistivities. Solid rock, on the other hand, has a low porosity and shows very high resistivities. Weathered rock tends to show relatively lower resistivities compared to solid rock. Every geological structure thus has it's own specific resistivity value.

2.2 Marine/fluvial applications

For water based applications the electrodes are placed on a multichannel cable trailing behind the survey vessel (fig. 1b).

During the fieldsurvey qualitative results are already shown on computer screen. This allows on site for a quick and well justified choice of subsequent CPT and borehole locations.

2.3 Data Processing

A complicated sequence of mathematical operations has to be followed before any interpretable results can be obtained.

First, the resistivity field data are edited and filtered to improve the signal/noise ratio. Geometrical corrections are applied to correct for the fact that the resistivity cable may show more or less significant curvature. Measurements made with a strongly curved cable are rejected. Other corrections are made to account for current losses into the watercolumn.

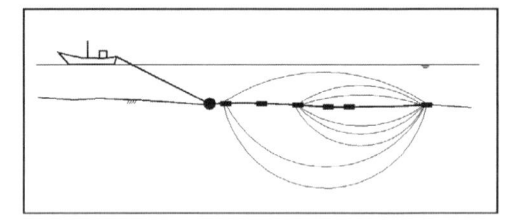

Figure 1b: Principles of Vertical Electrical Soundings
– On water

After interpolation of the resistivity information into a regular grid a digital 3D model of the subsurface is obtained. The results are visualised in color on vertical and horizontal cross sections showing the different geological structures in function of depth and geographical position.

The processing procedure described above is an interactive process. In order to extract a maximum of information out of the raw survey data the processing sequence has to be repeated several times to find the optimum processing parameters.

3 ADVANTAGES

At a speed of 1 sounding every 3 seconds the Aquares resistivity method is very fast compared to mechanical drilling or CPT's. It also has the advantage that a much larger volume is being sampled by a single sounding. While the subsurface volume sampled by a mechanically drilled borehole corresponds exactly with the borehole diameter (a few centimeters), the volume sampled by an electrical sounding may in some cases exceed 5 or 10 m in diameter. As a consequence, resistivity methods are quite suitable for determining various degrees of fracturing and weathering in rock.

Classical acoustical geophysical methods often have difficulties coping with geological situations limiting their effectiveness:

- The presence of gravel in the subsurface tends to obscure the information due to the appearance of diffractions.
- The application in shallow water causes multiple reflections to obscure the data.
- The application in areas with sediments rich in organic matter may cause "gas-masking" problems.
- The presence of "cap rock" on the seafloor tends to hide all geological structures underneath by reflecting most acoustical energy.

Thanks to the nature of geoelectrical methods, the abovementioned effects generally are not bothersome to resistivity methods. Furthermore, the results of a quantitative resistivity survey not only provide depths and thicknesses, it also results in a resistivity value for each geological structure which is an indication of it's porosity.

Thanks to the use of 3000 fold stacks, high electrical currents, signal enhancing electrode configurations, noise free electrode design, newly developed statistical techniques and appropriate processing algorithms signal/noise ratios have been improved significantly.

Rather than give an in-depth insight into the technical aspects of the method, this papers merely aims to present a number of case studies showing the possibilities of quantitative marine resistivity methods. For practical reasons the original colorcoded resistivity sections have been rendered in black-and-white by various ways of hatching.

4 CORRIB PIPELINE LANDFALL

In 2001 a resistivity survey was carried out at the landfall area of the Corrib gas pipeline along the West-Coast of Ireland. The proposed burial burial depth of the pipeline was 3m below seabed level. The geology is known to consist of sand, clay and gravel overlying a very hard metamorphic basement rock. About 10 survey lines were sailed parallel to the proposed piperoute, a 3D model was worked out including a number of horizontal and vertical sections to represent the geological subsurface. A horizontal section parallel to the seafloor at the proposed burial depth of 3 m is shown in figure 2 as well as a vertical resistivity section along the proposed pipe route.

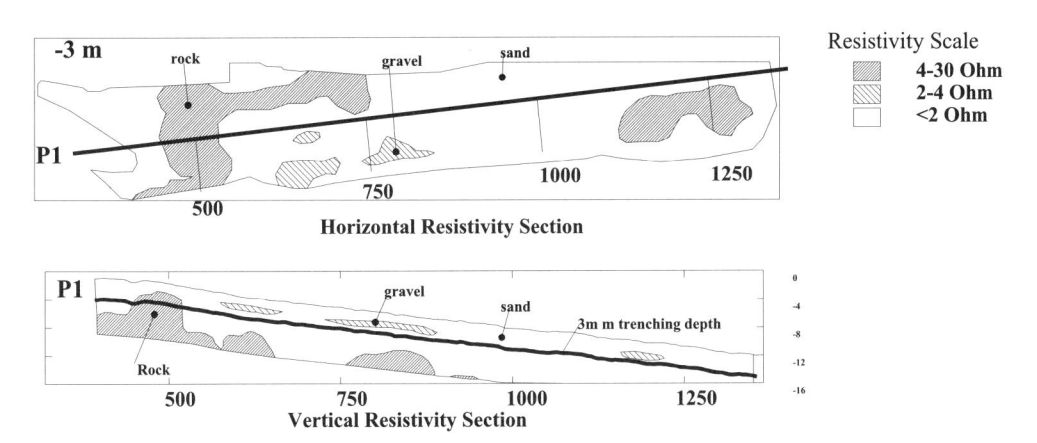

Figure 2: Corrib pipeline landfall resistivity survey

High resistivities (> 4 Ohmm) and intermediate resistivities (2-4 Ohmm) are hatched following the resistivity legend. High resistivity basement rocks reaching pipeline burial levels are visible at KP500. Other rockheads are located at deeper levels at KP850 and south of the route at KP1250. Intermediate resistivities correspond to gravel as confirmed by drilling. The lower resistivities (< 2 Ohmm) correspond to sand and clay.

Based on these results trenching costs were estimated and pipe burial operations were planned.

5 WEST-INDIA PORT DEVELOPMENT

In 2000 a resistivity survey was carried out along the West-Coast of India in view of the development of a new port. The exact location is not disclosed for commercial reasons. The geology along the West Coast of India is generally known to consist of mud and clay on basalt rock. Seismic exploration methods often fail because of multiple reflections in shallow water and effects of gas-masking in organic–rich muds.

A large number of resistivity lines were sailed in various directions following the accessibility of the survey area, a 3D model was worked out including horizontal and vertical resistivity sections as the one shown in figure 3.

The upper part of figure 3 shows a horizontal section at 11 m below Chart Datum while the lower part shows a vertical section along profile line P3 marked on the horizontal section. The vertical scale is exaggerated 100 times.

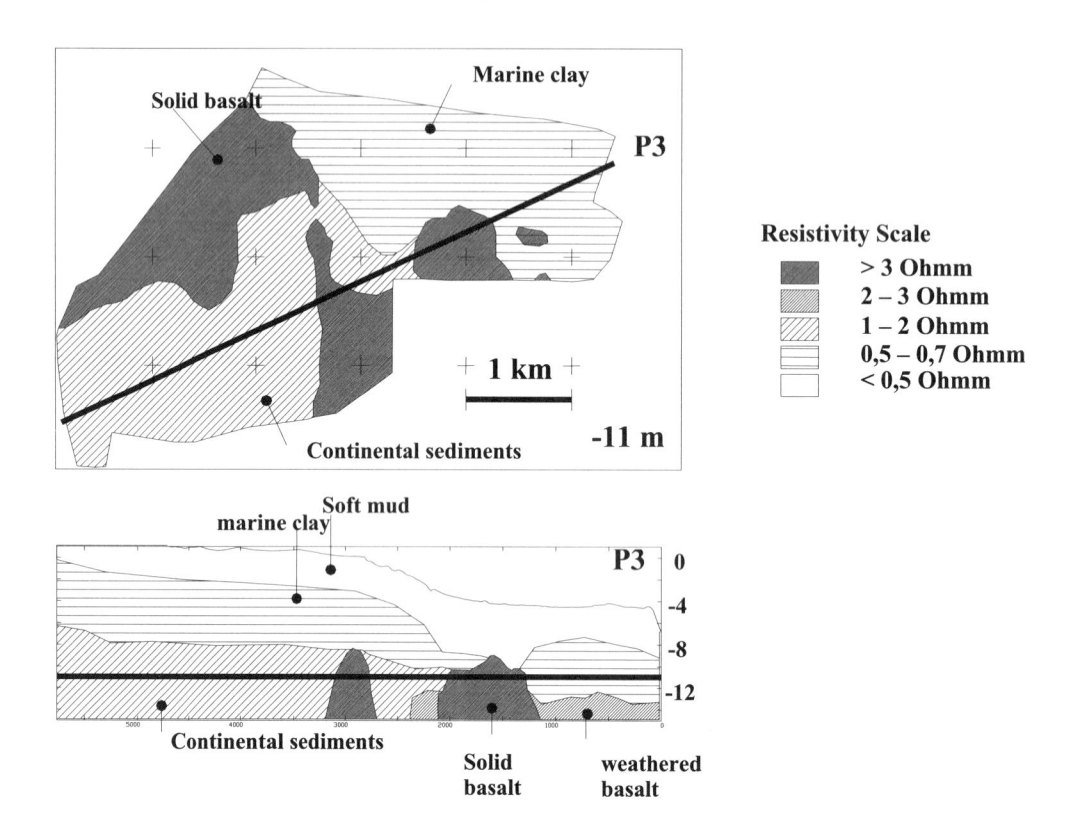

Figure 3: West-India port development resistivity survey

The vertical resistivity sections shows a low resistivity top structure of soft mud (< 0.5 Ohmm) and more consolidated marine clay (0.5-0.7 Ohmm) on top of a substratum consisting of high resistivity solid basalt (> 3 Ohmm) and weathered basalt with slightly lower resistivities (2-3 Ohmm) Intermediate resistivities (1-2 Ohmm) correspond to detritic sediments of continental origin deposited in valleys cut into the basaltic basement.

Based on above resistivity results, a drilling campaign was organised involving a limited number of (expensive) boreholes to confirm the geotechnical nature of each of above described structures. Based on the combined geophysical and geotechnical information turning basins, access channels and reclamation areas were designed in a most cost-effective way.

6 INN RIVER RESISTIVITY SURVEY

In 2003 a site investigation program was carried out on the river Inn in Austria and Germany along a 4 km long river section located downstream of a high dam. The geology in the survey area is known to consist of two gravel units separated by a clay intercalation. As the lower gravel unit is an important aquifer contributing to the region's drinking water supply local authorities were worried about man-induced effects of river erosion cutting through the impervious clay layer and exposing the lower aquifer to contaminated surface water.

In order to obtain more detailed information on the geological structure of the river bed a resistivity survey was carried out on the river before starting the drilling campaign. Seven lines were sailed at about 10 m interspacing, a 3D model was constructed and presented as various horizontal and vertical resistivity sections. Figure 4 shows some examples of such horizontal and vertical sections located between KP207.5 (downstream) to KP209.5 (upstream). The upper horizontal section is situated at NN458 (458 m above German Chart Datum) in the upper gravel unit while the lower horizontal section at NN454 is cutting through the lower gravel unit.

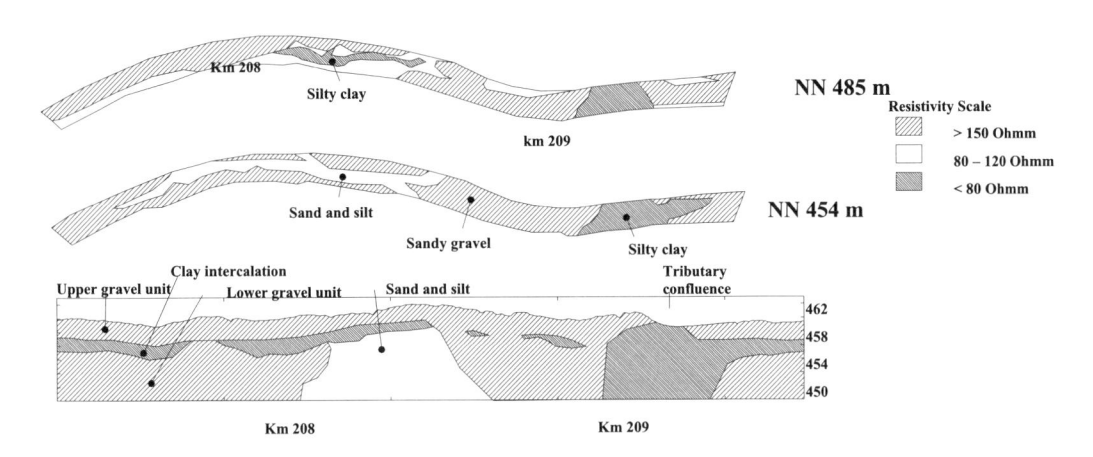

Figure 4: Inn river resistivity survey

Subsequent borehole information made clear that high resistivities (>150 Ohmm) shown in Figure 4 correspond to sandy gravel while lower resistivities (80-120 Ohmm) correlate with finer sediments such as sand and silt deposited as channel-fill sediments. Very low resistivities (< 80 Ohmm) are associated with the clay intercalation mentioned above. Very low resistivities are also associated with clay and silt deposits found at KP209.3 at the confluence of a small tributary on the right bank of the river. The vertical section below in Figure 4 shows this silt and clay deposit to be very thick. Apparently the vortex associated with the confluence of both the Inn and it's tributary must have been cutting through both gravel units as well as the clay intercalation. The idea that upper and lower gravel formations were two hydrologically separate units thus has to be abandoned. Even between KP208.5 and KP209.1 the low resistivity clay intercalation seems to be missing and mostly replaced by gravel.

7 MIDDLE EAST SAND SEARCH SURVEY

A resistivity survey was carried out on a location in the Persian Gulf with the aim of finding sand for reclamation purposes. The geology consisted of sand and silt sediments on top of a basement rock. A large surface area was overgrown with coral fields. Previous seismic and geotechnical surveys failed to find sand deposits in this area.

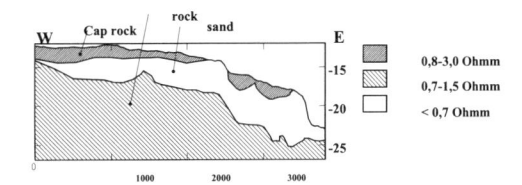

Figure 5: Sand search in caprock covered area

The western reaches of the resistivity section shown in Figure 5 show a high resistivity caprock layer (0.8-3.0 Ohmm) on top of a 4 to 5 m thick sand deposit (<0.7 Ohmm) resting on basement rock (0.7-1.5 Ohmm). Dredging operations in this area have indeed confirmed the existence of a 2 m thick caprock on top of the sand deposit.

8 CONCLUSIONS

The Aquares resistivity method is a quantitative geophysical tool that can be used in a large variety of site investigations involving dredging reconnaissance, port design, sand searches, pipe route surveys or sand and gravel exploration. It has been successfully applied in shallow water, gravel, in caprock covered areas and in organic rich sediments where seismic methods often tend to be less effective.

As each volume in the geological subsurface has a specific resistivity value defining the nature of the geological subsurface at each point in space, quantitative resistivity methods offer the possibility to develop digital 3D models of the geological subsurface which can be visualised as colored horizontal and vertical sections.

REFERENCES

Koefoed, O. 1970. A fast method for determining the Layer Distribution from the Raised Kernel Function in Geoelectrical Sounding. Geoph. Prosp. 18, 564-569.
Lines, L.R. and Treitel S. 1984. Tutorial. A review of least-squares inversion and its application to Geophysical Problems. Geoph. Prosp. 32, 159-186.
Loke, M.H. and Barker, R.D. 1996. Rapid Least-Squares Inversion of Apparent Resistivity Pseudosections by a quasi-Newtonian Method. Geoph. Prosp. 44, 131-152.
Zohdy, A.A.R. 1989. A new method for the automated interpretation of Schlumberger and Wenner sounding curves. Geoph. 54, 245-253.

# ISC'2 experimental site investigation and characterization – Part I: conventional and tomographic P and S waves refraction seismics vs. electrical resistivity

J.M.Carvalho & A. Viana da Fonseca
Faculdade de Engenharia da Universidade do Porto, Portugal

F. Almeida & H. Hermosilha
Departamento de Geociências, Universidade de Aveiro, Portugal

Keywords: residual soils from granite, surface and borehole seismics, electrical resistivity

ABSTRACT: The in-situ investigation and characterization of ISC'2 experimental site comprised the application of several geophysical borehole and surface methods, namely P and S waves seismic cross-hole (CH), down-hole (DH), refraction, electrical resistivity, S waves high resolution shallow reflection, ground penetrating radar (GPR), as well as geotechnical tests, namely SPT, CPT, DMT, among others. The site, located within the campus of Faculty of Engineering of the University of Porto (FEUP), is geologically formed by an upper layer of heterogeneous residual granite soil overlaying rather weathered granite contacting a gneissic migmatite.

Direct and indirect results from some of the referred surveys are compared between them and with some of the available geological and geotechnical information, in particular hereby (Part I), those obtained from P and S waves conventional (RC) and tomographic (RT) refraction, CH and DH seismic tests, as well as from two electrical resistivity traverses, conducted adjacent to three boreholes in which undisturbed soil samples were collected previously to seismic data acquisition. A regression exploratory analysis points out to the advantage of including both N-SPT values as well as depth values in modelling maximum shear modulus, G_0, behaviour.

1 INTRODUCTION

Residual soils from different types of granite are very common in the north-western part of Portugal where ISC'2 experimental site is located in the surroundings of the Faculty of Engineering of the University of Porto (FEUP). The site is geologically formed by an upper layer of heterogeneous residual granite soil of varying thickness, overlaying more or less weathered granite contacting an older gneissic migmatite with dominant sub-vertical foliation (Figures 1 and 3).

Upper crustal metamorphic processes involving the older gneissic migmatite formations seem to be the genesis, due to melting, of the existing embedded granitic bodies of various sizes and dimensions.

Differential weathering processes between the granitic and gneissic formations are significantly responsible for the very irregular weathering profile of the rock masses, which have themselves a quite irregular spatial distribution.

In general the thickness of these residual granitic soils (saprolitic formations) horizons vary a lot, between few meters to, sometimes, more than 20m.

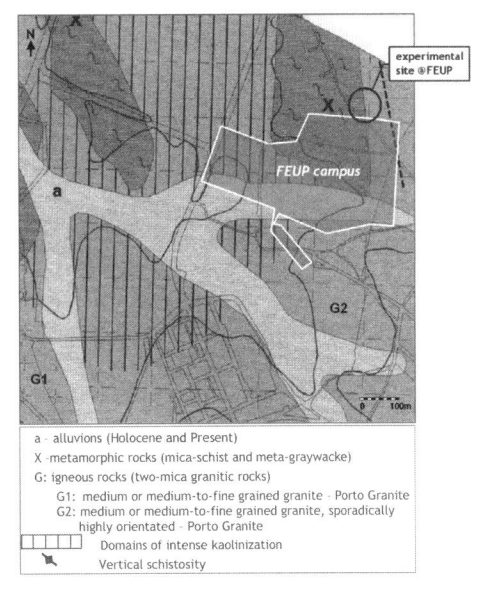

Figure 1. Geology of ISC'2 experimental site area: Porto geological map.

Although they often present strong heterogeneity, it is frequently observed an average gradual change of characteristics with depth, namely regarding their manifested mechanical properties.

The complexity of this kind of soils is often a hindrance for an accurate regional or local mapping of the spatial variability of the mechanical properties necessary for geotechnical design.

The data related to residual soils from granite compiled during the extensive in-situ and laboratorial investigation and characterization of ISC'2 experimental site, comprising the application of several geotechnical and geophysical surface and borehole techniques, namely seismic and electrical resistivity, is a valuable and rare opportunity to compare different methodologies and assess their relative advantages and limitations.

2 SITE INVESTIGATION

A seismic refraction survey, using P and S waves, was conducted adjacent to three boreholes (Figures 2 and 4), in which undisturbed soil samples were collected and some other geotechnical and geophysical tests were carried out, namely direct seismic transmission cross-hole and down-hole (CH and DH respectively). The treatment and interpretation of the acquired refraction data was done using conventional (RC) and tomographic (RT) methods (item 2.3).

Two pole-pole electrical resistivity sections separated by 1.5m were obtained paralelly to the sections resulting from the seismic survey (Figures 2 and 16).

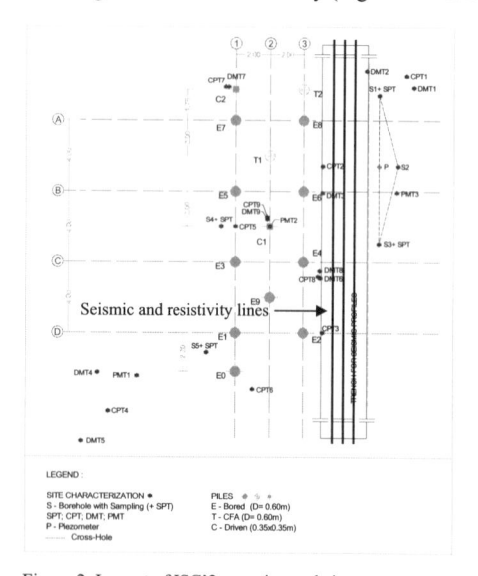

Figure 2. Layout of ISC'2 experimental site.

Direct and indirect results from the referred surveys are hereby compared between themselves and with some of the available geological and geotechnical information. In-situ and laboratory tests results as well as those resulting from using S waves high resolution shallow reflection and ground penetrating radar (GPR), are detailed in two other papers submitted to this Conference, (respectively, Viana da Fonseca et al., 2004 and F. Almeida et al., 2004).

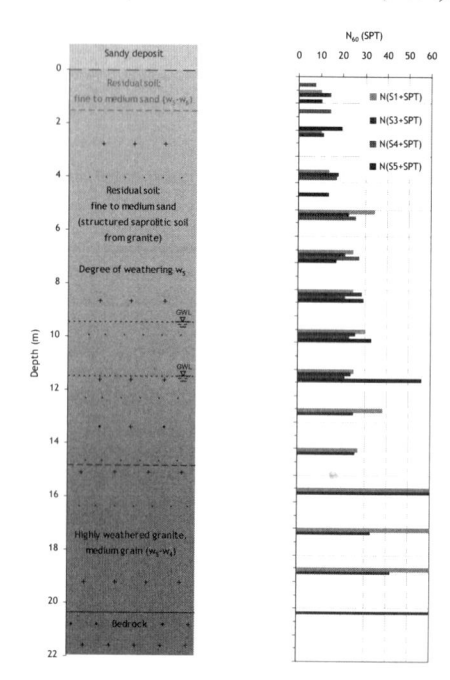

Figure 3. Schematic geological profile and NSPT values from boreholes S1, S2, S4 and S5.

2.1 Seismic cross-hole

Three boreholes, S1, S2 and S3 (Fig. 2), were used for an S and P waves seismic CH survey. The acquisition took place between boreholes S1-S2, S2-S3 and S1-S3, defining three sections, respectively, 4.25m, 4.35, 8.4m wide, going to a maximum depth of 18m.

Acquisition took place in July, in the middle of dry season, being the water level around 10m deep.

Data was collected using a Soil Dynamics bidirectional borehole vertical mechanical hammer and a Geo-Stuff BHG-3 triaxial geophone, coupled to a Geometrics Smartseis 12 channel seismograph.

Fig. 4 shows S and P waves velocities, V_S and V_P, obtained in the referred three sections.

Considering S waves results, it is evident the presence of a trend corresponding to an average gradual increase of velocity with depth, going from minimum values around 260 m/s to maximum values of around 350 m/s. There is also a very similar velocity pattern between all three sections, being each one very much representative of the others.

The resulting V_P and V_S values were used namely to estimate maximum shear modulus (G_0) and Poisson ratio (ν) values. These values are compared with those estimated by other mentioned seismic methods (Figures 21 and 22).

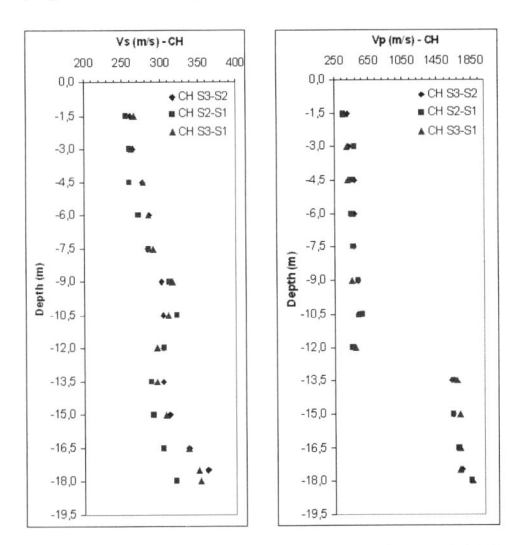

Figure 4. CH S and P seismic velocities, in sections S1-S2, S2-S3 and S1-S3.

The overall P waves variation in depth is quite distinctive, specifically for the clear "discontinuity" associated to the water level (w_L). Nevertheless, in each of the zones, above and below w_L, there is a trendily increase of V_P values similar to the one observed with S waves.

It is worth mentioning that apparently the presence of water was first clearly detected by P waves at 13.5m, being the measured water level at around 10m. With one of the shots located at 12m and the next at 13.5m deep, it could happen that an intermediary shot would previously detect the water. Anyway, the remaining apparent discrepancy may be explained by the seasonal change of the water level with a corresponding unsaturated soil layer of varying water content. On the other hand, the transitory decrease in both P and S waves velocity starting at 10.5m may be the water level presence distinctive sign.

2.2 *Seismic down-hole*

In borehole S3 was conducted a P and S waves DH seismic survey, with a 1.5m interval between shots, using the triaxial geophone referred in item 2.1. and a 10Kg hammer. Some of the preliminary results are presented, namely the obtained G_0 and ν values (Fig.21).

2.3 *Seismic refraction*

The P and S waves seismic refraction survey had three main objectives: to contribute for the site overall investigation/characterization; to compare the conventional (RC) and tomographic (RT) refraction interpretation methods; to study the correlations with other tests, in particular with data obtained from the electrical resistivity survey, described in item 2.4.

Seismic acquisition was performed along a 44.5m long traverse (Figures 2 and 5), having 1.5m meter spacing between geophones and nine shot points.

Figure 5. Seismic refraction P and S waves traverse.

Unfortunately, due to lack of free space, it was not possible to have a longer traverse, which has prevented attaining bedrock deeper formations.

The previously referred 12 channel seismograph was used with 14 Hz vertical and horizontal geophones.

Unfortunately, background noise from traffic and other sources was considerable during acquisition, namely the apparent refraction/reflection effect from the walls of the trench, in which the traverse was done. Figures 6 and 7 show the travel time curves, respectively, for S and P waves.

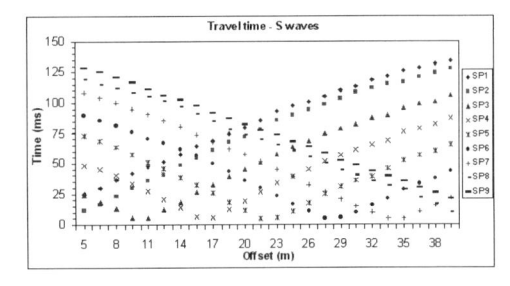

Figure 6. Refraction S waves travel time plots.

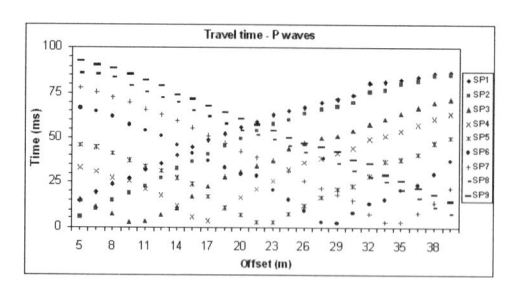

Figure 7. Refraction P waves travel time plots.

The generic pattern of the S and P waves travel time plots, points out to an average gradual increase of velocities with depth.

The interpretation resulting from the Time Delay and tomographic inversion methods are presented in the next figures. In Figures 8 and 9, two sections are presented for S waves and, respectively, for two and three layers models.

The conventional delayed time interpretation was done with software SIPT2 provided by Rimrock Geohysics® and the tomographic inversion was performed with the software SeisOpt22D provided by Optim software®.

The corresponding V_S values are consistent for both models and with those obtained from CH and RT. It can be noticed an apparent increase of velocity towards the right side of the graph (Figures 8 and 9) matching the observed variation in RT sections.

In Figures 10 and 11 the RT velocity sections respectively for S and P waves can be seen.

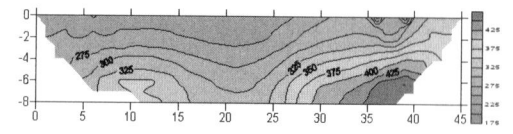

Figure 10. Refraction tomography S waves model.

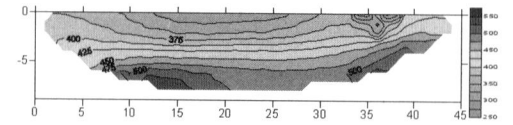

Figure 11. Refraction tomography P waves model.

The apparent increase of velocity referred in RC S waves sections is mapped in both RT sections. The overall pattern of velocity variability closely matches the one obtained with electric resistivity as it will be discussed later.

Two different types of graphs are presented in Figures 12 and 13, both showing the V_S and V_P variation along horizontal direction. This is done for different depths, according to the location of the centre of tomographic inversion cells. In Fig.13 it is obvious the very similar V_S and V_P pattern of variation along the horizontal direction, showing, namely, an increase of velocity in the section right side, more evident in case of V_S.

The scatter plot and regression line between V_S and V_P shown in Fig.14 confirm the good correlation between both velocities. This is also consequence of the fact this analysis is made above water level, where the soil matrix governs both velocities.

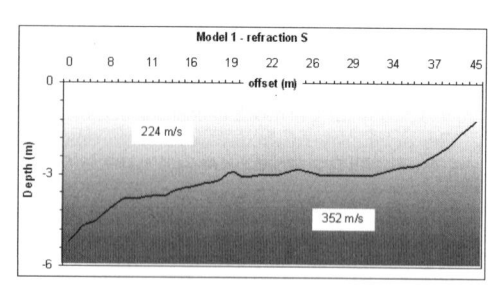

Figure 8. Refraction two layers model (RC interpretation).

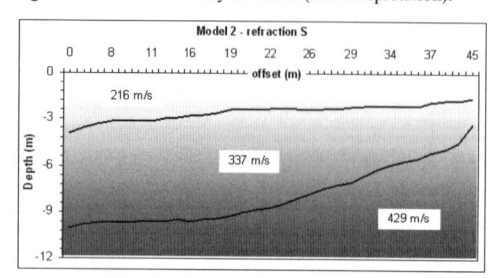

Figure 9. Refraction three layers model (RC interpretation).

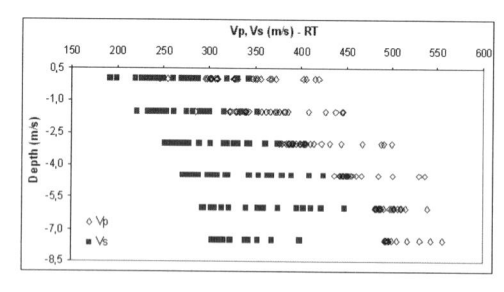

Figure 12. Seismic P and S waves from RT.

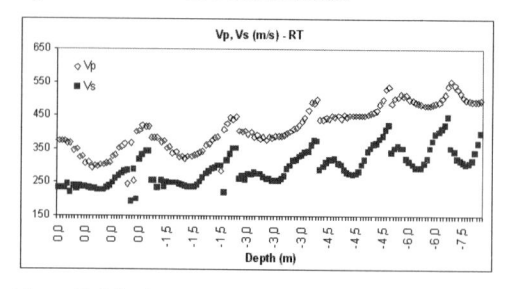

Figure 13. Seismic P and S waves from RT.

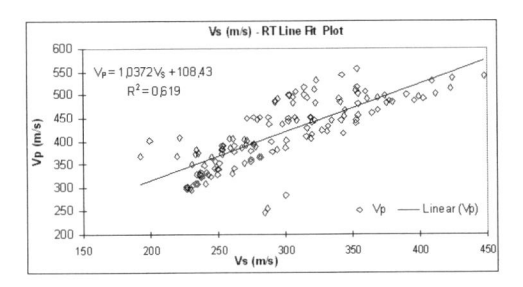

Figure 14. Seismic P and S waves regression from RT.

V_S and V_P variability with depth can be seen in Fig. 15 corresponding to CH sections S3-S2, S2-S1 and S3-S1 and to intersections of the RT sections along the vertical direction located approximately in S3, S2 and S1 horizontal coordinates.

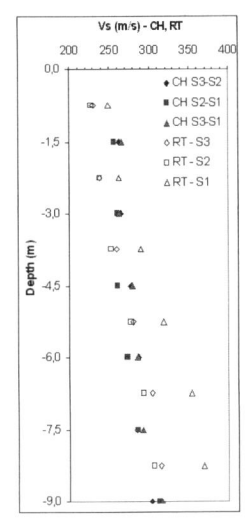

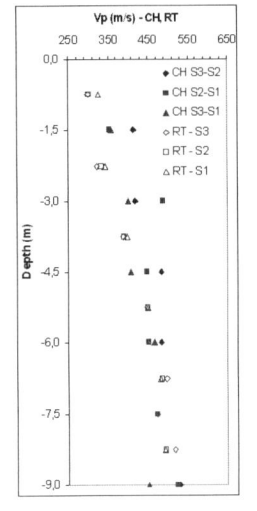

Figure 15. S and P waves seismic velocities taken directly from CH sections S1-S2, S2-S3 and S1-S3 and from refraction tomography "boreholes" located in the referred boreholes.

There is a very good generic agreement, except for the V_S wave RT "vertical seismic profile", VSP, centred in borehole S1. This fact is in accordance to the observed increase of velocity towards the right side of the RC and RT sections.

Fig. 16 presents the V_S and V_P variation with depth corresponding to CH sections S3-S2, S2-S1 and S3-S1 and to the corresponding RT "cross-hole" sections, CHRT. There is, also in this case, a good agreement between CH and CHRT sections.

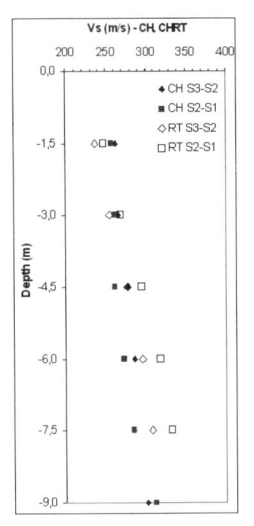

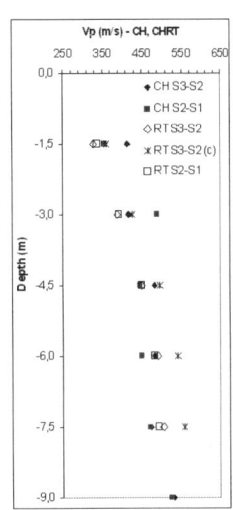

Figure 16. V_S and V_P from cross-hole (CH) and tomographic refraction CHRT sections.

2.4 *Electrical resistivity*

As previously mentioned, two electrical resistivity 33m traverses were developed using odd-even Pole-Pole spread electrode commutation for 3D surface imaging (Almeida F. *et al.*, 2001), parallel to the seismic refraction traverse but shorter than this one (Fig.17).

Figure 17. Electrical resistivity traverses.

The data was acquired with an automatic resistivimeter with 48 programmable channels. The electrodes where numbered from 1 to 48 (odd electrodes for current; $(2n-1) \times C$, and even electrodes, $2n \times C$, for potential), resulting on a $34.5 \times 1.5m^2$ rectangular grid. The potential distribution around each current electrode was analyzed in order to provide a quality control of data as shown for electrode 3C in Fig. 18.

The model sections resulting from the inversion of data using the software Res3Dinv© are shown in Figures 19 and 20.

Figure 18. Electric surface potential around electrode 3C (Fig. 17).

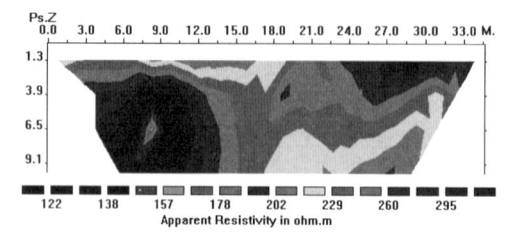

Figure 19. Electric resistivity section.

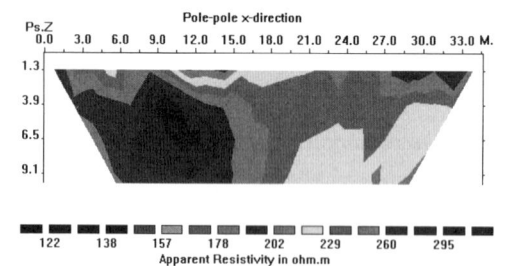

Figure 20. Electric resistivity sections.

Both models show zones of overall different apparent resistivities. The higher value zone, on the section left side, corresponds to a similar higher seismic velocity zone, observed in RT sections. On the other hand, the central/right lower value zone corresponds to a lower seismic velocity zone observed in RT sections. A tentative interpretation geological model is presented in Part II (F. Almeida et al., 2004), combining, namely, electrical and seismic information.

2.5 Derived fundamental results

Based in seismic velocities resulting from the various referred ways, maximum shear modulus, G_0, and Poisson ration, ν, values were calculated.

The density values used to calculate G_0 were taken from laboratory tests over undisturbed samples from borehole S2, ranging from $\gamma=16.8$ kN/m³ to $\gamma=19.4$ kN/m³.

Some of the obtained values are compared in Figures 21 and 22. In Fig. 21, G_0 and ν values were calculated with seismic V_S values obtained in CH sections S3-S2 and S2-S1 and DH in borehole S3.

As expected, G_0 and V_S variability with depth follows a similar pattern, smooth in general although more erratic in the case of DH based values.

The calculated G_0 values range from minimum values of around 90 MPa from DH tests and 120 MPa from CH tests, to a maximum value of 255 MPa.

The ν values show in general a bigger dispersion except in the saturated zone below 13.5m. While in the zone above 13.5m, the values vary around an average value of 0.25, below that level they are quite constant with values near 0.5 (around 0.48).This is an obvious sign of full saturation.

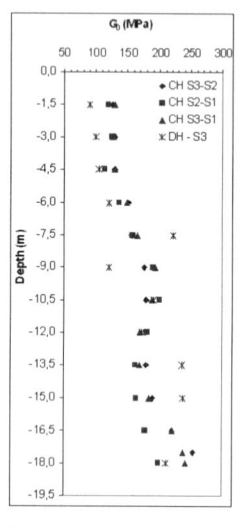

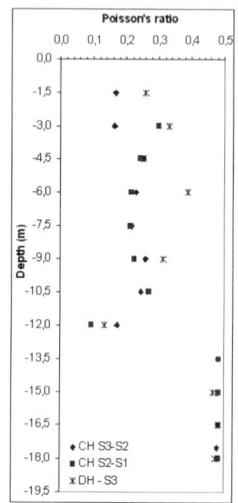

Figure 21. Shear modulus and Poisson ratio from CH sections and DH in borehole S3.

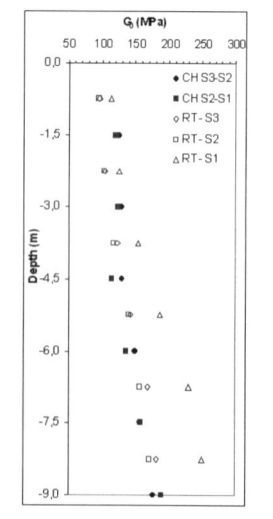

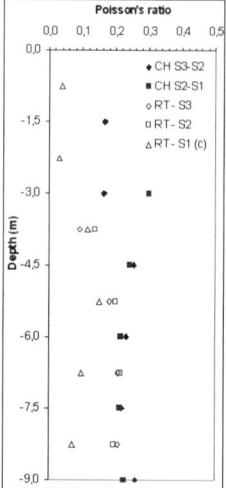

Figure 22 - Shear modulus and Poisson ratio from CH sections and "boreholes" RT.

G_0 and ν values plotted in Fig. 22, were calculated with seismic V_S values obtained in CH sections

S3-S2 and S2-S1 and RT "VSP" in "boreholes" S3, S2 and S1.

The V_S RT values used to calculate "VSP" RT-S1 (c) ν values had a previous 10% decrease correction in order to improve the results, otherwise too low.

Again and as expected according to seismic velocities results, G_0 values corresponding to borehole S1 location show higher values, with increasing divergence with depth.

The ν values corresponding to borehole S1 location show also a large dispersion with lower values. The remaining ν sets have a very similar trend mainly after 4.5m depth with values around 0.25.

3 EXPLORATORY REGRESSION ANALYSIS

Some of the calculated G_0 values were used in an exploratory uni- and bivariate regression analysis with N-SPT values and depth (associated with stress state level) as predictive variables. Both field N-SPT values, N_{SPT}, from boreholes S3 and S1, as well as the corresponding corrected ones, N_{60}, were used, being some of the results hereby discussed (Figures 23-28).

Figures 23, 24 and 25 show the projection of G_0, from CH section S3-S2, bivariate linear regresson models respectively on planes G_0-N_{SPT}, G_0-N_{60} and G_0-Depth, being the N-SPT values from borehole S3 and depths ranging from 1.5m to 15m.

Figures 26, 27 and 28 show similar results for G_0 values from CH section S2-S1 based on N-SPT values from borehole S1 and the same mentioned depths. Two of the N-SPT values were omitted in the graph due to the fact that they were rather out of the average overall trend. These values correspond to depths 6m and 13.5m.

In all the referred figures the trend lines based on calculated G_0 values are plotted.

The bivariate linear regression models coefficients used to calculate the predicted G_0 values are shown in Table 1, while in Table 2 the univariate power regression models coefficients used to calculate the G_0 trend lines are presented.

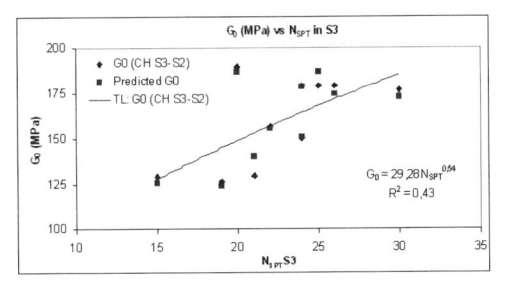

Figure 23. G_0 from CH S3-S2 vs. N_{SPT} in S3 and depth. Trend line for calculated G_0: power model $G_0=29.28N_{SPT}^{0.54}$.

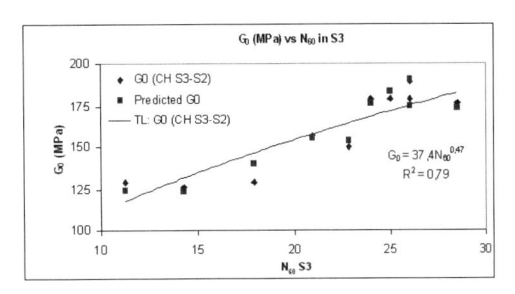

Figure 24. G_0 from CH S3-S2 vs. N_{60} in S3 and depth. Trend line for calculated G_0: power model $G_0=37.4N_{60}^{0.47}$.

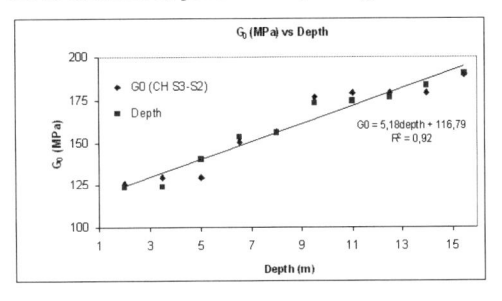

Figure 25. G_0 from CH S3-S2 vs. depth. Trend line for calculated G_0: linear model $G_0=5.18$depth$+116.79$.

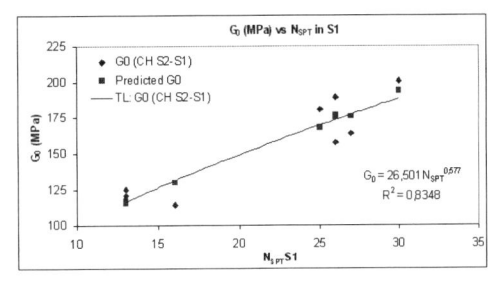

Figure 26. G_0 from CH S2-S1 vs. N_{SPT} in S1 and depth. Trend line for calculated G_0: power model $G_0=26.5N_{SPT}^{0.58}$.

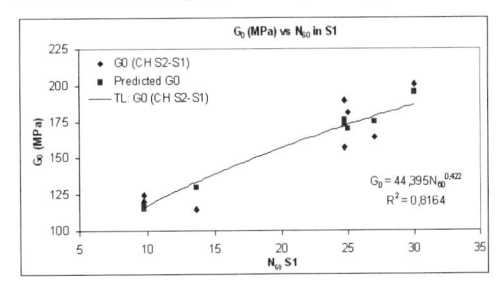

Figure 27. G_0 from CH S2-S1 vs. N_{60} in S1 and depth. Trend line for calculated G_0: power model $G_0=44.4N_{60}^{0.42}$.

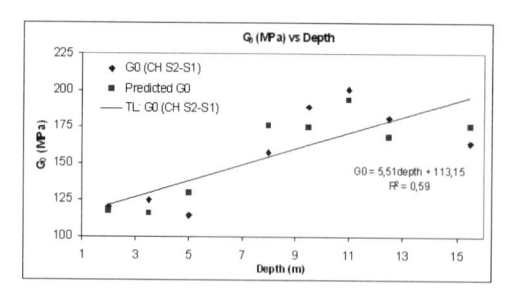

Figure 28. G_0 from CH S2-S1 vs. depth. Trend line for calculated G_0: linear model $G_0=5.51\text{depth}+113.79$.

Table 2 - G_0 power model trend lines.

G_0 univariate power models			
		N_{SPT}	N_{60}
CH S3-S2	Coef.	29.3	37.4
	power	0.54	0.47
	R^2	0.43	0.79
CH S2-S1	Coef.	26.5	44.4
	power	0.58	0.42
	R^2	0.83	0.82

The next equations are two bivariate linear regression models for G_0 estimation based both in N-SPT and depth values (Figures 24 and 26 and Table 1):

– For G_0 from CH S3-S2 and N_{60} in S3:

$$G_0 = 95.2 + 1.6 N_{60} + 3.5 depth \qquad (1)$$

– For G_0 from CH S2-S1 and N_{SPT} in S1:

$$G_0 = 54.3 + 4.9 N_{SPT} - 0.86 depth \qquad (2)$$

As referred in Table 1 the related multiple squared correlation coefficient values, M.R^2, are respectively 0.96 and 0.84 significantly higher than those obtained with univariate regression power models, specially for CH S3-S2 values.

The minus sign in models related to CH S2-S1 values (Table 1) may be explained by the non-linear overall correlation between G_0 and depth shown in Fig. 28.

Table 1 - G_0 bivariate linear regression models

G_0 bivariate linear regression models				
	Coefficients		Std. Error	M. R^2
CH S3-S2	Intercept	91.5	11.1	
	Depth	4.6	0.5	0.95
	N_{SPT}-S3	1.3	0.5	
	Intercept	95.2	8.7	
	Depth	3.5	0.7	0.96
	N_{60}-S3	1.6	0.6	
CH S2-S1	Intercept	54.3	23.6	
	Depth	-0.8	2.5	0.84
	N_{SPT}-S1	4.9	1.7	
	Intercept	75.6	17.2	
	Depth	-1.6	2.8	0.85
	N_{60}-S1	4.5	1.6	

4 CONCLUSIONS

The geophysical/geotechnical procedures described in this study, as well as the results obtained with the collected data, were part of the extensive ISC'2 experimental site investigation/characterization process.

This study concerns mainly the application of seismic borehole and surface methods as well as pole-pole electrical resistivity surface method.

Among the various aspects that have been previously summarized, one of the most relevant conclusions is the similarity in the spatial variability presented by seismic and electrical section models. In fact, the resulting models point out to the adequacy of both methods in mapping the local underground heterogeneities, both horizontally and vertically, inside the more or less weathered granitic mass as well as the boundary with the gneissic migmatite shown in Fig.1.

A tentative interpretation geological model, integrating also information from an S waves high resolution shallow reflection and GPR surveys, is presented in Part II, the referred paper submitted to this Conference (F. Almeida et al., 2004).

It was also interesting to observe the fact that each of the three obtained CH sections is very representative of the others, although one of them is around two times wider than the others.

The fact that the P waves CH detected clearly the presence of water around 3.5m under the piezometric measurement was associated to the high sensitivity of P waves to total saturation and in less degree to the presence of moist. Conversely, the water level presence distinctive sign may be the transitory decrease in both P and S waves velocity starting at 10.5m.

The exploratory regression analysis involving G_0, indicates the advantage of including depth (associated with stress state level) as predictive independent variable, besides N_{60} rather than N_{SPT}, in a linear bivariate model. The obtained multiple, M. R^2, and single, R^2, squared correlation coefficient values, are very indicative of the referred hypothesis.

ACKNOWLEDGMENTS

The authors are grateful to the Contractors that have sponsored this experimental site: MOTA-ENGIL; SOPECATE; TECNASOL-FGE and TEIXEIRA DUARTE. A special thank is due to Ing. Ricardo Andrade from Mota-Engil, and Ing. Nuno Cruz also from Mota-Engil, (formerly in CICCOPN, which has also supported this work)

Part of this work is financially supported by the research project: POCTI / ECM / 33796 / 1999: *"Management of sampling quality on residual soils and soft clayey soils. Comparative analysis of in situ and laboratory seismic waves velocities"*.

The authors would also like to acknowledge the valuable collaboration of Instituto Geológico e Mineiro (IGM) in particular to Prof. Mário Machado de Leite and Dr. Eurico Pereira.

REFERENCES

Almeida, F., Hermosilha, H., Carvalho, J.M., Viana da Fonseca, A. & Moura, R., 2004. *ISC'2 experimental site investigation and characterization – Part II: From SH waves high resolution shallow reflection to shallower GPR tests.* Second Int. Conf. on Site Characterization – ISC'2. Porto.

Almeida, F.; Carminé, P.; Gonçalves, L. and Daniel, L (2001) - *Odd-Even Pole-Pole Spread Electrode Commutation in Resistivity - 2D Cross Borehole and 3D Surface Imaging (Porto/Portugal Underground Tunnelling Case Studies).* Proceedings –Extended Abstracts - 7th Meeting of the Environmental and Engineering Geophysical Society (European Section), Birmingham, England, EGSI01 pp. 50-51.

ASTM (1993). *Standard test methods for cross-hole seismic testing. D4428/D4428M-91.* American Soc. Testing Mat., Standards Annual Book, Sec. 4. Vol. 04.08 pp.769-778. ASTM, Philadelphia.

Barros, J.M.C. (1997). *Dynamic shear modulus in tropical soils.* PhD. thesis, University of São Paulo, Brasil.

Dobrin, M.B., 1976, *Introduction to geophysical prospecting*: McGraw-Hill, Inc.

International Society for Rock Mechanics, 1988. *Suggested methods for seismic testing within and between boreholes.* Int. J. Rock Mech. Min. Sci. and Geomech. Abstr., 25 (6).

Pessoa, J.M.; Carvalho, J.M.C.; Carminé, P.;Viana da Fonseca, A. (2002). *Tomographic S and P waves velocities in a weathered granitic zone of Porto* (in Portuguese). 8º Congresso Nacional de Geotecnia, SPG, Lisboa.

Viana da Fonseca, A, Carvalho, J.M.C., Pessoa, J.M. 2002. *Cross-hole tests in weathering profiles of Porto granite. Elastic parameters and correlations* (in Portuguese). 8º Congresso Nacional de Geotecnia, SPG, Lisboa.

Viana da Fonseca, A., Carvalho, J.M., Ferreira, C., Costa, E., Tuna, C., Santos, J.A., 2004. *Geotechnical characterization of a residual soil profile: the ISC'2 experimental site, Porto.* Second Int. Conf. on Site Characterization – ISC'2. Porto.

Viana da Fonseca, A., Matos Fernandes, M. & Cardoso, A.S. 1998. *Characterization of a saprolitic soil from Porto granite by in situ testing,* First Int. Conf. on Site Characterization –ISC'98. Atlanta: 2, 1381-1388. Balkema, Rotterdam.

The role of crosshole seismic tomography for site characterization and grout injection evaluation on Carmo convent foundations

M.J. Coelho, F.M. Salgado & L. Fialho Rodrigues
LNEC - Laboratório Nacional de Engenharia Civil, Lisboa, Portugal

Keywords: seismic tomography, P-wave velocity, uncompressed soils, grout injection evaluation

ABSTRACT: The Carmo convent (XIV century) is founded on a silty-sand formation subjected to several negative impacts that have weakened the stiffness and strength characteristics of the local soils. A previous finite element analysis was carried out to assess the stability conditions for constructing a new underground railroad passing 20 meters below the convent foundations. The results indicated that upper soils might be, in some areas, highly uncompressed and very close to failure or in a plastic state. To confirm these and to try mapping the higher uncompressed zones, various instrumental measurements and in situ tests were carried out, including an extensive crosshole seismic tomography survey on 26 crosshole sections to obtain their P-wave velocity field. In general, the seismic tomographies obtained validated the finite element results.

A jet grout treatment test was carried out at a zone under the convent foundations where the seismic tomography had revealed very low P-wave velocity. Afterwards, the crosshole seismic test was repeated and the corresponding tomography showed that upper soil velocities had increased to values near the velocities for deeper (and less disturbed) soils. Additionally, three crosshole seismic tomographies were performed at a control site with a geological setting similar to the convent foundations, but not subjected to the same negative impacts. The P-wave velocities obtained for these undisturbed soils are about the same as the velocities for the convent foundations after the grout treatment, which confirms the test injection efficiency in improving soil characteristics below the Carmo convent foundations.

1 INTRODUCTION

The Carmo convent is a historical building from the XIV century located on the Carmo hill in Lisbon downtown. It is founded on miocenic terrains basically consisting of silty-sand soils, above the water table. These soils belong to "Areolas da Estefânia" formation (Almeida, 1986). In this formation, mainly at the surface and at deposit top, there are highly uncompressed soils (loose and very loose sands), which have already posed stability problems during the construction of several public works, like the Rossio railway tunnel (Almeida, 1991) and the Carmo convent itself (Fr. de Stª Anna, 1745) . Furthermore, at the convent site, this formation has been subjected to several negative impacts, like the 1755 Lisbon earthquake (M>8), which caused severe damages to the convent, due to the structure dynamic movements and due to the partial slope failure of Carmo hill. More recently, in the last decades, some shops on Carmo street, at the hill foot, have expanded their storage area by digging galleries inside this formation below the convent foundations. Since these galleries were excavated by artesian methods, the surrounding soils were significantly uncompressed. The impact from the earthquake and from diggings weakened considerably the stiffness and strength characteristics of local soils.

In 1996, due to the planned construction of a new underground railroad line, passing 20 meters below the convent foundations, it was necessary to study the site stability conditions. A finite element tension-deformational analysis for the convent foundation soils was carried out by Consórcio (1996a), based on available geological and geotechnical data (Consórcio, 1993). The results indicated that the soils between the galleries roof and the convent foundations might be in some areas highly uncompressed and very close to failure or in a plastic state (Consórcio, 1996a, Salgado, 1996 and Salgado & Coelho, 1996).

In view of these results, micro-piles were designed to reinforce the convent foundations. The micro-piles were to be founded on the "Areolas da Estefânia" underlying formation, named "Argilas dos Prazeres", in order to transfer the structure load to this deeper and stronger formation. However, this solution could pose other problems related with the uncompressed soils under the convent foundations. Micro-pile construction could dense the deeper sands and therefore cause additional movements on upper uncompressed soils. Another problem would be the possible negative friction on micro-piles caused by the creep or erosion of upper soils. To overcome these problems, the grout injection of these upper soils was proposed (Consórcio, 1996a, Salgado, 1996 and Salgado & Coelho, 1996).

To confirm the finite elements analysis and to map the higher uncompressed zones, an extensive crosshole seismic tomography testing was carried out by LNEC (*Laboratório Nacional de Engenharia Civil*) on 26 crosshole sections (see Fig. 1), using some of the micro-pile boreholes before their construction. The seismic tomographies obtained showed that soils above the galleries had much lower P-wave velocities than soils (of the same type) below the galleries, therefore validating the finite element results (Salgado & Coelho, 1996). Another objective of crosshole seismic tomography was the grout injection evaluation. The main-chapel zone, with an underground gallery below its foundations, in particular the F77-F92 crosshole section (gray zone in Fig. 1), was a critical zone from the stability point of view, where finite element analysis indicated a possible plastic state and the crosshole seismic tomography revealed very low velocities in the upper soils. Consequently, a jet grout treatment test was carried out in this zone to evaluate if injection could improve the soil characteristics (Consórcio, 1996b,c). After grout injection, the crosshole seismic test was repeated and the corresponding tomography showed that soil velocities had increased to values closer to the velocities for deeper and less disturbed soils.

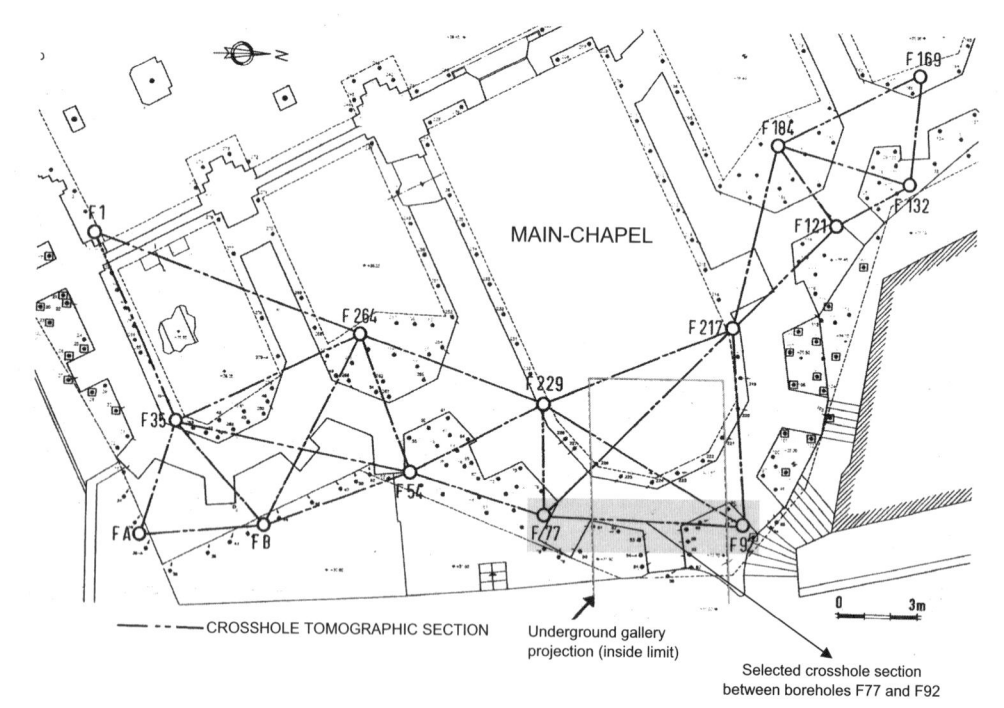

Figure 1. Carmo convent site: plane view of investigation area and tomographcic sections arrangement. The gray zone indicates the selected crosshole section (between boreholes F77 and F92), where grout injection was evaluated (adapted from Coelho, 2000).

To evaluate if this seismic velocity increment due the soils grout injection is acceptable, i.e., to evaluate if this grout injection type was sufficient to increase the soils strength and stiffness to adequate values, LNEC carried out a seismic tomographic test at a control site where the same geological formation ("Areolas da Estefânia") occurs, but where the upper soils are not or are less uncompressed. So, their P-wave velocities were expected to be representative values for soils with adequate stiffness and strength from a stability point of view.

This paper presents the seismic tomographies obtained at the crosshole section selected for the grout injection test, before and after injection, as well as the seismic tomographies at the control site, where the existence of no uncompressed soils was considered.

2 GEOLOGICAL SETTING

2.1 Carmo convent site

At the Carmo convent site, there is a sedimentary formation from miocenic age, named "Areolas da Estefânia". According to the previous geological-geotechnical study performed on this site (Consórcio, 1993), this formation consists mainly of fine silty-sand ("areolas"), medium sand, sandy-clay and interbedded calcarenite. Locally, this deposit is about 15 to 22 m thick with crossed stratification and frequent lateral *facies* variations. Under this formation, there is another miocenic sedimentary formation named "Argilas dos Prazeres" consisting basically of gray or green silty-clay. In general, there is a sandy-fill with variable thickness (0.5 to 10 m) above the convent foundations.

Fill soils are classified as loose, "Areolas da Estefânia" soils as loose to medium dense, and "Argilas dos Prazeres" soils as stiff to hard. The water table occurs near the interface between the two miocenic formations and therefore the upper formation ("Areolas da Estefânia") is unsaturated.

The sub-vertical boreholes F77 and F92 that were 32 and 34 m long define the crosshole section considered in this paper (see Figs. 1 and 3). They intercepted, from the top, 0.5 to 2 m of random sandy fill, 5.5 m of convent foundation material (probably consisting of a blend of rockfill and mortar), 15 to 15.5 m of silty-sand with interbedded calcarenite ("Areolas da Estefânia" formation) and under this, they intercepted silty-clay soils from "Argilas dos Prazeres" formation. This crosshole section also intercepts an underground gallery nearly transversal to it with about 11 m length and a vertical section of about 3.5 m height to 5 m width. This gallery was excavated from a shop on Carmo street, at Carmo hill foot, about 8.5 m below the foundations of the main-chapel of Carmo convent (Figs. 1 and 3). As Fig. 3 shows, this gallery was excavated inside "Areolas da Estefânia" formation and has a concrete lining of about 0.30 m.

2.2 Control site

A control site was chosen with a geological setting similar to the convent site but assumed to be an undisturbed site. This site is located on the crossroads of Anchieta and Capelo streets (in Lisbon) about 300m SW from Carmo hill crest (Carmo convent site), so the impacts from the 1755 earthquake are less severe (no slope failure occurred here). Additionally, no underground excavations or galleries exist at this site. Therefore, it was considered that, here, the original soils from "Areolas da Estefânia" formation would not be uncompressed and would have adequate stiffness and strength for site stability.

At this site, three sub-vertical boreholes F1 to F3 arranged in triangle were drilled to about 28 to 30 m depth (Fig. 4) and three crosshole seismic tomographies were performed at F1-F2, F2-F3 and F3-F1 sections. These boreholes intercepted from the surface, 1.5 to 2 m of sandy fill and 22.8 to 23.5 m of "Areolas da Estefânia" materials, where the upper 5 to 8 m was silty-sand with interbedded calcarenite or calacarenite with interbedded sand and the next 15.5 to 17.5 m was silty-sand. Under this formation they intercepted the "Argilas dos Prazeres" formation consisting here basically of silty-clay with interbedded marlstone.

3 SEISMIC TOMOGRAPHIES

3.1 Method guidelines

Over the last two decades, crosshole seismic tomography has been widely used to study the elastic properties of materials for a wide range of applications, from large engineering sites to ancient building surfaces (Pessoa, 1990 and Coelho, 2000).

For this particular crosshole seismc tomography, the objective is to reconstruct the P-wave velocity distribution in the crosshole section based on a multitude of seismic ray paths corresponding to crosshole seismic measures (Fig. 2). These measures can be the attenuation and velocity of both compressional and shear waves, but the most widely used is the P-wave velocity due to significant facility and quickness in data acquisition and unambiguity in time picking, in relation to S-wave velocity and attenuation measures.

For P-wave velocity tomography, the P-wave measured travel times are inverted into a velocity matrix, which comprises the area with seismic ray coverage. The most often used velocity inversion or reconstruction techniques are well described in the literature (Nolet, 1987, Pessoa, 1990 and Lo & Inderwiesen, 1994) and start with an initial velocity model for which the modeled travel times of seismic rays are calculated. The differences between modeled and measured travel times for seismic rays, called residuals, are then used for improving the initial model for the velocity distribution.

3.2 Selected section at Carmo convent site

At the main-chapel zone, below which there is an underground gallery about 8.5 m from its foundation (see Figs. 1 and 3), the seismic tomography of crosshole section F77-F92 (before the injection) showed very low P-wave velocities in the upper soils, between the foundation and the gallery, according to the finite elements analysis. This zone was selected to test a jet grout treatment to evaluate if injections would improve the elastic properties of these soils. After the grout injection, the crosshole seismic test was repeated and the corresponding

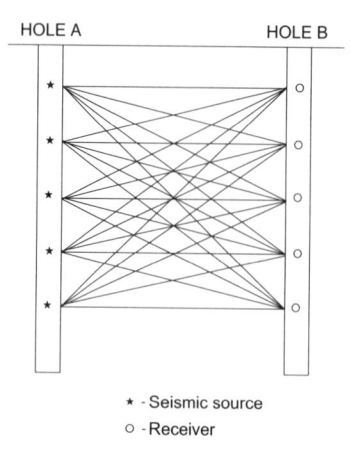

Figure 2. Field set-up for crosshole seismic tomography.

seismic tomography revealed velocity increments showing that upper soils velocity had increased to values closer to deeper soils velocity, namely to the ones below the underground galleries at the convent site.

At F77-F92 crosshole section, detonation caps were used as seismic source at borehole F92 and Oyo geophones were used as receivers at F77 borehole. The boreholes were steel cased. A twelve-channel digital data acquisition system (ABEM Terraloc MKIII) was used. Distance between boreholes is about 7.5 m. Both source and receiver intervals were, in general, 2 m, before and after injection. Seismic sources were located between 4 and 28 m along F92 borehole (additional shot at 29 m was performed after injection). Receivers were located along F77 borehole between 4 and 30 m before injection and between 2 and 28 m after injection (injection reduced the available length of F77 borehole). Therefore, the seismic ray coverage was not exactly equal before and after injection. Totals of 173 and 183 seismic rays (travel times) were used for tomographies, respectively before and after injection. Original seismic records were re-analysed and reprocessed for this paper but giving slightly differences in seismic tomographies comparatively with the ones presented in Salgado & Coelho (1996) and Coelho (2000).

The crosshole section was discretised by a grid of rectangular cells with 1m horizontal to 1.5m vertical and the same grid was used before and after injection tomographies. The center points of these cells (only those crossed by straight rays) are shown at seismic tomographies in Fig. 3. In the tomographic inversion processing the approximation of straight ray paths was used and a constant P-wave velocity for each cell was assumed. A SIRT (simultaneous iterative reconstruction technique) type algorithm implemented by Pessoa (1990) was used with minimum and maximum P-wave velocity constraints of 300 and 4000 m/s. The initial models were uniform velocity models equal to straight ray mean velocity (1024 m/s before injection and 1289 m/s after injection).

Fig. 3 shows the seismic tomographies corresponding, in both cases (after and before injection), to the fourth iteration. Even without numerical convergence, these models (with small number of iterations) were chosen to avoid the increment of velocity artefacts. The mean residual error (percentage value corresponding to the quotient between mean of absolute values of residuals and mean of measured times) is about 13% before injection and 15% after injection.

As Fig. 3 shows, the seismic tomography before injection revealed very low velocity in the upper soils, above the gallery, immediately below the convent foundations, where a zone with velocity less than 500 m/s occurs. Seismic tomography after injection shows a clear velocity increment in this zone evidencing a stiffness increment by means of soil grout injection. Apart from this and the two high velocity artefacts on bottom left corner and top right corner, seismic tomography after injection has a velocity distribution similar to the tomography before injection. Despite the complexity of the geological setting (with high velocity contrast structures) and the straight ray assumption, the seismic tomography revealed a reasonable concordance with it, except for the gallery location.

The relative maximum velocity centered about the gallery position can be explained by the existence of the gallery concrete lining, where certainly seismic rays suffer high bending because P-wave velocity is much higher in concrete than in sandy soils and in air surrounding the lining.

P-wave velocity difference obtained by subtracting cell velocity, before injection, from cell velocity, after injection, is mapped in Fig. 3(c). The most significant velocity increment (apart from the artefacts of bottom left corner and top right corner) occurs on upper soils between the convent foundations and the gallery evidencing good injection results.

3.3 Sections at control site

At this site, boreholes F1 to F3 were also steel cased and the distance between boreholes was about 8-9 m. Both source and receiver intervals were 2 m for all three sections. Seismic sources were located between 3 and 29 m along F1 and F2 boreholes. Receivers were located along F3 borehole between 2 and 26 m and along F2 borehole (for F1-F2 section) between 1 and 29 m. Totals of 205, 177 and 182 seismic rays (travel times) were used for F1-F2, F2-F3 and F3-F1 tomographies. The same acquisition equipment and tomographic processing as those

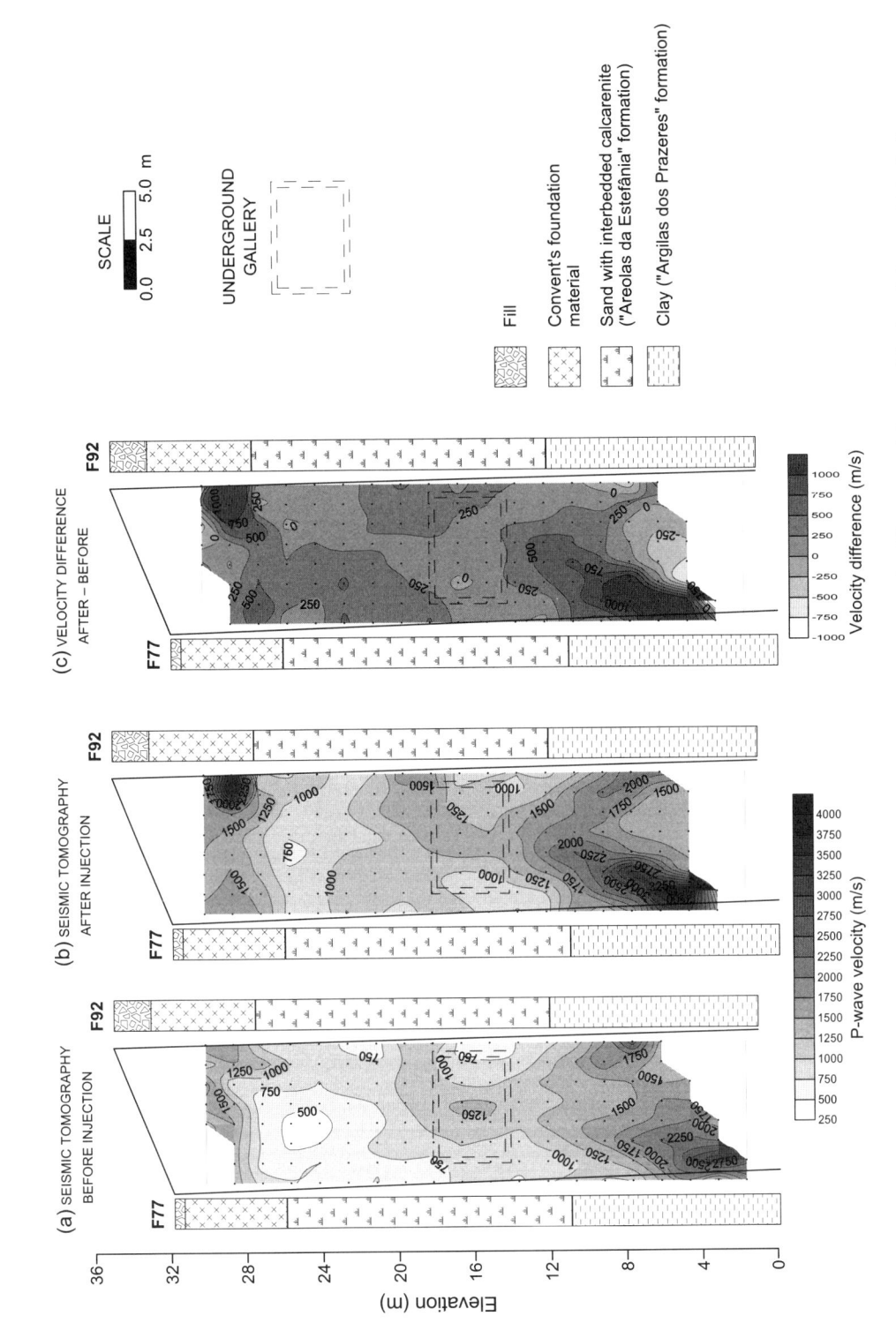

Figure 3. (a) and (b): Selected section seismic tomographies before (a) and after (b) grout injection. (c): Velocity difference map (after grout injection – before grout injection).

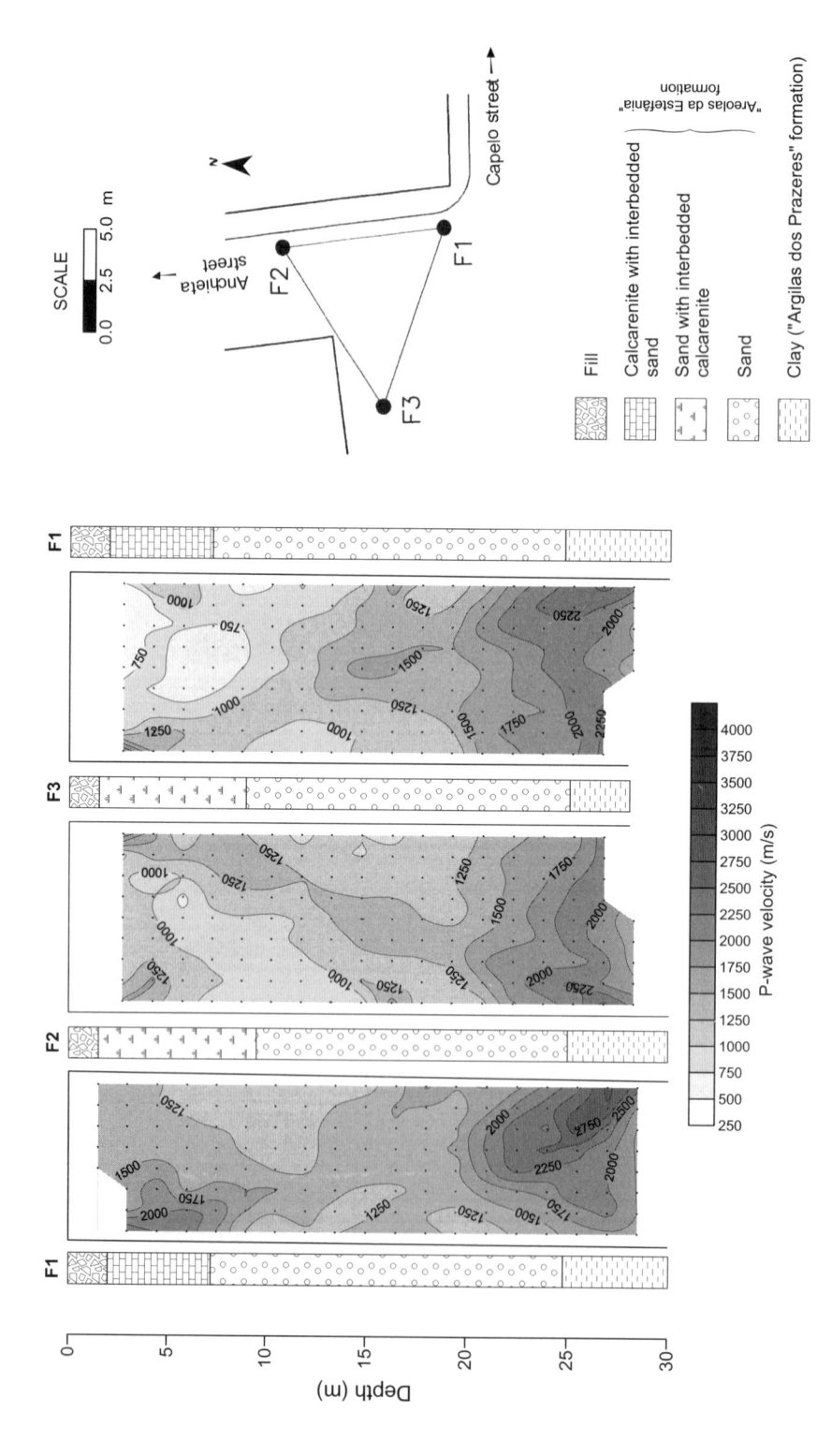

Figure 4. F1-F2, F2-F3 and F3-F1 seismic tomographies at control site.

used in convent site tomographies was also used here, including cell dimensions, velocity constraints and uniform velocity initial models.

The seismic tomographies in Fig. 4 were obtained at fourth iteration for all cases and have residual errors of about 6-10%. There can be observed that P-wave velocities at "Areolas da Estefânia" formation are higher than F77-F92 section velocities, before injection, but are about the same values as the ones attained at F77-F92 section after injection. This confirms that jet grout treatment performed at the selected zone of the convent foundations was sufficient to increment soils strength and stiffness to adequate values for site stability and for micro-pile construction with minimal adverse effects.

4 CONCLUSIONS

The Carmo convent is founded on a sedimentary formation where the soils are locally disturbed by past events that have weakened their stiffness and strength characteristics. This could endanger the stability of the convent during the construction of a planned underground railroad below the convent foundations.

Included in the studies for site stability conditions, a crosshole seismic tomography survey was carried out, which characterized the convent foundation soils, by means of P-wave velocity, evidencing some very low velocity zones.

After a jet grout injection test performed at a very low velocity zone, the corresponding crosshole seismic tomography was repeated revealing a P-wave velocity increment to values closer to the ones of deeper and less disturbed soils. To confirm if these higher velocities were characteristic of the same type of soils but with suitable strength and stiffness, an additional crosshole seismic tomography test was carried out at a control site where the same geological formation exists but where soils are undisturbed. The P-wave velocities obtained at the control site had about the same values as those obtained in the convent foundations after grout injection. Therefore, it was confirmed the grout injection test efficiency in improving the characteristics of soils under the Carmo convent foundations to acceptable values. Both set of results also showed that grout injection should be done, and how it should be done, in the remaining weak zones of the convent foundations.

In summary, crosshole seismic tomography significantly contributed to the characterization of the convent foundation soils and, in this case, it has also proved to be an indirect method to evaluate the grout injection efficiency.

ACKNOWLEDGEMENT

The authors thank METRO – Metropolitano de Lisboa, E.P. for their permission to publish this paper.

REFERENCES

Almeida, F.M. 1986. *Geologic map of the Lisbon region – scale 1:10000.* (in Portuguese). Serviços Geológicos de Portugal.

Almeida, I.M.B.M. 1991. *Geotechnical characteristics of the Lisbon soils.* (in Portuguese). Tese de Doutoramento (PhD Thesis), Universidade de Lisboa.

Coelho, M.J. 2000. *Crosshole seismic tomography for geotechnical investigation.* (in Portuguese). Trabalho de síntese (Monography), Provas de Assistente de Investigação do LNEC (Public examination for LNEC Research Assistant), LNEC, Lisboa.

Consórcio. 1993. *Restauradores - Baixa/Chiado and Rossio - Baixa/Chiado - Cais do Sodré: Geologic-geotechnical report.* (in Portuguese). Consórcio: BCP, CBPO, AGROMAN, SOMAGUE, PROFABRIL, KAISER, ACER e ACE, Report RT-00.0-104 for METRO – Metropolitano de Lisboa, E.P., Lisboa.

Consórcio. 1996a. *Carmo convent – Stress-strain model for micro-pile load evaluation.* (in Spanish). Consórcio: BCP, CBPO, AGROMAN, SOMAGUE, PROFABRIL, KAISER, ACER e ACE, Report RT-15.0-033 for METRO – Metropolitano de Lisboa, E.P. Lisboa.

Consórcio. 1996b. *Carmo convent – Grout injection test for "Areolas da Estefânia" – Results from "Ana Salazar" shop.* (in Portuguese). Consórcio: BCP, CBPO, AGROMAN, SOMAGUE, PROFABRIL, KAISER, ACER e ACE, Report RT-15.3-036 for METRO – Metropolitano de Lisboa, E.P., Lisboa.

Consórcio. 1996c. *Carmo convent – Grout injection on "Areolas da Estefânia" – "Ana Salazar" shop.* (in Portuguese). Consórcio: BCP, CBPO, AGROMAN, SOMAGUE, PROFABRIL, KAISER, ACER e ACE, Report RT-15.3-037 for METRO – Metropolitano de Lisboa, E.P. Lisboa.

Fr. J.P. de Sta. Anna. 1745. *Chronic of the Carmelitas.* (in Portuguese). Tomo Primeiro, Lisboa.

Lo, T. & Inderwiesen, P. 1994. *Fundamentals of seismic tomography.* SEG, Geophysical Monograph Series, N° 6, D.V. Fitterman Series Editor, USA.

Nolet, G. Ed. 1987. *Seismic tomography with applications in global seismology and exploration geophysics.* D. Reidel Publishing Company.

Pessoa, J.M.N.C. 1990. *Application of tomographic techniques to crosshole seismic exploration.* Trabalho de síntese (Monography), Provas de aptidão pedagógica e capacidade científica (Public examination for Assistant), Universidade de Aveiro, Portugal.

Salgado, F.M. 1996. *LNEC technical support for diagnosing the stability conditions and the reinforcement of the Carmo convent foundations and adjacent East hill.* (in Portuguese). Nota Técnica (Technical Report) 23/96 - NEGE, LNEC, Lisboa.

Salgado, F.M. & Coelho, M.J. 1996. *Geophysical survey with seismic methods on the Carmo convent site and on the crossroads of Capelo and Anchieta streets.* (in Portuguese). Nota Técnica (Technical Report) 33/96 - NEGE, LNEC, Lisboa.

Proceedings ISC-2 on Geotechnical and Geophysical Site Characterization, Viana da Fonseca & Mayne (eds.)
© 2004 Millpress, Rotterdam, ISBN 90 5966 009 9

Geophysical characterization for seepage potential assessment along the embankments of the Po River

C. Comina & S. Foti
DISTR, Politecnico di Torino, Torino, Italy

L.V. Socco & C. Strobbia
DIGET, Politecnico di Torino, Torino, Italy

Keywords: geophysical characterization, seepage, surface waves, inverse problems

ABSTRACT: The Po River is the longest Italian river. Seepage and flooding problems often occur along its embankments during periods of maximum flow and this is a major concern because the surrounding zones are highly populated and used for a variety of agricultural and production activities. Floods very often start with localized seepage that can degenerate causing inundations. The case of study presented is aimed at assessing the potential of different geophysical tests for the identification of geological sections that can be critical with respect to seepage. Geophysical tests are time and cost effective techniques and they can give good information in this respect. Electrical and seismic tests have been performed at a test site. In particular a novel procedure for the coupled interpretation of SWM and VES data has been used in order to obtain a more accurate reconstruction of the stratigraphies. This procedure takes advantage of the formal analogies of the two inversion processes in order to mitigate the non-uniqueness and ill-posedness inherent in all inverse problems.

1 SEEPAGE ALONG THE EMBANKMENTS OF THE PO RIVER

The Po river is the main and longest Italian river. The uncontrolled expansion of the urbanization during the last decades covered its alluvial planes. Local or extended defense systems (embankments, riverside protections) have been erected to contain the floods inside the main branch of the river at different stages.

In the present paper, the interest is focused on the system of external embankments, which protect the countryside during flooding. The embankments along the river, and in particularly at the site considered in the present study, have a trapezoidal shape. In order to contain the saturation line the face towards the river has a step slant while the face towards the countryside a lower one (in general embankments have a variable inclination between 1:4 and 1:7). The inclination of the embankment face towards the field, that is with a light slope, is also useful towards the risks of seepage, sliding and softening that mainly occurs within the foundation soil on the country side (Colombo, 1964).

Po embankments are generally 10 to 12 meters high and can be occasionally raised or even extended during periods of maximum flood of the river. The foundation soil can have generally very different characteristics depending on the location considering that the Po river flows for 650 Km in the "Padana" plain. Usually the soils are alluvial sediments transported over different ages by the river itself. A quite typical situation is depicted in Figure 1, in which a sequence of clayey and sandy layer is present.

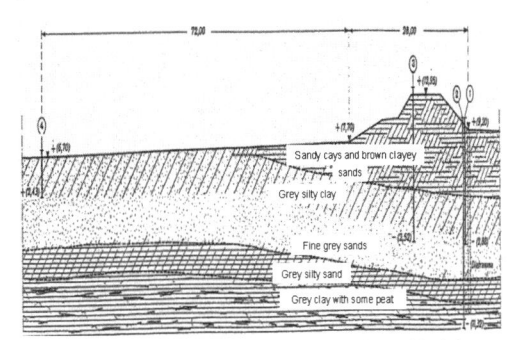

Figure 1. Example of section of the foundation soil of the Po river embankments; hydrographic left (Colombo, 1962).

One of the more frequent problems encountered in these river embankments is related to seepage phenomena. Indeed the presence of more permeable sandy layers below shallow clayey layers leads to excess of pore water pressures during flooding periods. Failures typically start with localized seepage events that can drive flows of water and large flooding in the surrounding countryside. The start and evolution of these phenomena are essentially linked to the geological features of the soil deposits along the embankments. Hence soil characterization plays a mayor role in the study of seepage.

Considering the large extension of the areas interested by this type of phenomena classical geotechnical techniques such as borehole logs and CPTs cannot be used for thorough investigations. Geophysical tests have the advantage that relatively large areas can be investigated with surveys carried out from the ground surface. In the present work, the potential use of geophysical tests has been assessed performing on the same site several seismic and electrical geophysical tests.

2 THE SITE

The site is located at Ponte Casati, in the municipality of Caselle Landi, district of Lodi, Lombardia (Northern Italy) (Figure 2). The site has undergone a general broken off in June 1917 and a large flood during one of the largest rain event in the 1950s. Moreover in the last years, especially during the autumn season, several localized seepage phenomena have developed along the embankments during the flooding of the Po River (Figures 3 and 4). In particular during the flood of October 2000 several such phenomena were detected along the section of embankment interested by this work. They were located as reported in the Figure 2. The site on which the tests reported in the present paper have been performed is located near to location E of Figure 2.

In this zone the Po river flows from north-west to south-est and just after the site there is a loop in the west-est direction; however in the location in which the tests were performed the river flows straight parallel to the embankment.

Figure 2. Position of seepage phenomena in October 2000 along the embankment.

The geology of the shallow formations in the zone is mainly constituted of recent alluvial deposits. The principal underground formations are made of more or less plain stratified stacks of homogeneous layers with lenses of different materials. The morphology of the zone is typical of the Po plain; embankments have the principal scope of defending the country and agricultural activities along them. Between the embankments and the river-bank there is the high-water bed zone where, when the river is in flood at its maximum level, the Po inundates. This zone is used for the cultivation of Poplar. The regular waterbed of the river is approximately 850m away from the embankment. The formation of seepage phenomena is due to excess water pressures during flood, when the river completely fills the high-water bed zone; the hydraulic gradient between the two sides of the embankment during flooding causes this localized failures.

Figure 3. Seepage phenomena and subsequent inundation of the countryside due to the huge rainfall of October 2000; this zone is indicated with the letter E in Figure 2.

Figure 4. Sand bags around a localized seepage phenomena in order to prevent flows that can start generalized failure.

The stratigraphy of the shallow subsurface at the site is representative of the geological conditions often associated to seepage events in the zone. Below a superficial sandy layer, a clayey layer underlain by sands is expected. The thickness of the clayey layer is the critical factor in this situation. Indeed when the river is in flood at its maximum level the water pressure in the underlying sand abruptly increases and the clayey layer prevents seepage in the short term. Thus, the objective of a characterization campaign for seepage potential is to locate, along the embankments, the zones where the thickness of the clayey layer is reduced. In particular the general scope of this campaign is to assess the potentiality of geophysical tests in locating the zones with maximum seepage risk. In this perspective the site was chosen because of his previous history in respect to this kind of phenomena.

3 GEOPHYSICAL TESTS

The objective of a soil characterization campaign for seepage potential detection is twofold. From one point of view areas in which the overall thickness of the clayey layer is reduced have to be detected. On the other hand, inside such areas, local variations of thickness have to be localized to identify the points in which the seepage can start.

In this perspective, a variety of seismic and electrical non-invasive tests have been carried out at the same site. Information about resistivity and seismic velocity variation with depth can be profitably used to identify the different materials. Five different profiles have been investigated in order to fully characterize the site (Figure 5). Three profiles are located in the area outside the embankment with respect to the river, where localized failures were observed in the past years (P1, P2 and P3); the other two profiles are located in between the embankments and the river banks (P4 e P5). The mid-point of each survey line has been kept fix when performing different tests in order to make reliable comparisons.

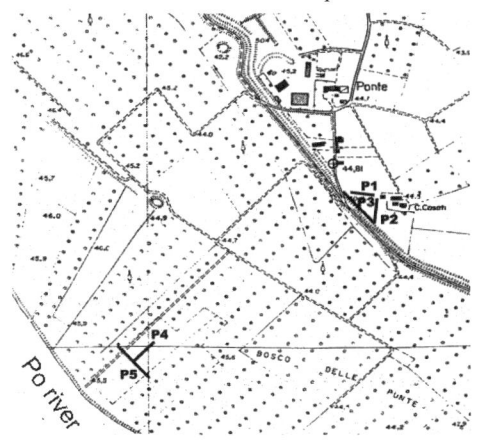

Figure 5. Localization of the profiles of the different geophysical techniques adopted.

The geophysical measurements were performed at two different stages. During a first campaign, which has been carried on in March 2001, the following tests were executed: P-wave refraction surveys, surface wave tests (SWM), vertical electrical sounding (VES) with Schlumberger configuration, electrical resistivity tomography (ERT) with Wenner configuration and horizontal electric profiling (HEP) with Ohm-mapper device. The second campaign, in September of the same year, was aimed at performing SH-waves seismic refraction tomographies. Table 1 reports a summary of the tests carried on in each survey profile.

Table 1. Surveys performed over each profile.

	S.E.V.	E.R.T.	H.E.P.	S.W.M.	P-wave Seismic Refr.	S-wave Seismic Refr.
P1	x	x		x	x	
P2	x	x	x	x	x	x
P3	x	x	x	x		x
P4		x				
P5	x	x		x	x	x

Important information with respect to the mechanical properties of the investigated soils can be obtained, in general, with seismic techniques. In particular SWM tests can give a good estimation of the small strain shear modulus of the soil, under the hypothesis of plain stratified formations. With the aim of mechanically characterize the investigated formations also P and S wave seismic profiles have been performed. In order to sample different depths two different impulsive sources have been used for SWM seismic tests: a sledge-hammer (6 Kg) and a heavy mass (130 Kg) mounted on a towing undercarriage, equipped with a winch of elevation that allows a 3 meters falling. The SH-wave source is instead a sledge-hammer impulsive source; the hammer rotates in the vertical plane perpendicular to the spreading and hits one side of the base of the chassis on which it is mounted. The base is filled up with heavy weights in order to improve the coupling with the ground.

On the other hand the information that can be obtained with geo-electrical techniques is the water table level and the characterization of the soils in terms of their electrical properties and stratigraphy. In particular with electrical techniques it is quite easy to localize the clayey formations (low values of the resistivity) with respect to the sandy ones. VES tests are interpreted under the hypothesis of horizontal stratification. Some ERT and HEP tests have been performed in order to investigate the lateral variations of the clayey formations and checking the above hypothesis. WIth the same perspective the SH-wave tomographies are used to assess lateral variations in terms of mechanical parameters. In most of the profiles investigated however the formations appear to be mainly horizontally stratified.

The VES technique has been performed also with the aim of developing a new joint inversion technique, with SWM data, which will be explained in details in the next paragraph.

A general description of the methods used in this campaign is outside the scope of the present paper, anyway some peculiar comments about HEP and S-wave refraction are reported in the following.

HEP have been performed with the Ohm-mapper Geonics® device that allows a very fast acquisition process. Indeed this device does not require conductive electrodes fixed in the ground but it entrusts the connection electrode-ground with capacitive high frequency electrodes that are dragged on the surface and that execute an high number of measures in fixed spacing dipole-dipole configuration. The energizing signal is managed from a transmitter connected to the current dipole, while the tension dipole is managed from a receiver synchronized on it. It is possible to investigate different portions of the subsoil executing several sets of measures along the same profile with increasing spacing between the centers of dipoles. In this way pseudosections of apparent resistivity are obtained. Such pseudosections are then inverted with algorithms analogous to those used for the resistivity tomographies. The advantage of this device is that HEP measurements are faster and can be performed also dragging the electrodes attached to a vehicle.

The SH-wave seismic refraction tomographies were performed with the use of new swyphones receivers (Sambuelli e Deidda, 1999). Every receiver is constituted of two conventional horizontal geophones mechanically and electrically coupled. The two geophones are mounted symmetrically on the same support and they are tilted in opposite direction with the same angle in respect to the vertical (this tilt allows to use them even if they are not perfectly horizontal). When the swyphone receiver is along the survey line the axes of both the geophones lays on the vertical plane perpendicular to the direction of the propagation. The two geophones are connected electrically in series, but with the polarity of one inverted with respect to the other one: the poles with positive sign are connected between them, while those with negative sign are both connected to the seismic cable. In such a way the signals generated from the two geophones, that have inverse polarity, are embezzled and they render a seismic trace in which the horizontal component (due to SH waves) is doubled, while that undesired P wave component is cancelled out. Using swyphone receivers it is therefore sufficient to hit the base of the SH-wave source from a single side, while using traditional geophones the energization must be repeated on the opposite side and stacked until the cancellation of P wave contribution. That remarkably increases the times and the costs of operation supplying a high quality "pure SH" signal.

4 JOINT INVERSION OF VES AND SURFACE WAVE DATA

A novel procedure for the coupled interpretation of SWM and VES data has been used in order to obtain a more accurate reconstruction of the stratigraphy. The results clearly show the potential of this technique in locating the clayey layer (due to the combination of shear wave velocity and resistivity information) and a good accuracy in determining its thickness. Both vertical electric sounding (VES) and seismic surface wave method (SWM) are time and cost effective surveying techniques widely used for engineering problems. Furthermore, the two techniques are both based on a monodimensional model and present many formal similarities in the interpretation of field data. A joint inversion could supply more reliable results, reducing interpretation ambiguities due to intrinsic ill-posedness of geophysical inverse problems. Previous studies developed the theoretical framework of the problem also proposing an inversion algorithm (Hering et al. 1995 and Misiek et al. 1997). A new, effective, joint inversion algorithm of Schlumberger VES data and fundamental Rayleigh mode dispersion curve has been implemented using a weighted damped least-square procedure. The layer thickness was chosen as the coupling parameter between the two sets of electrical and seismic parameters. A wide series of tests both on synthetic and field data were performed in order to evaluate the robustness of the proposed procedure even in situations in which some electric layer boundaries were not seismic boundaries and viceversa (Comina, 2001). All the results of the joint inversion showed a sensible improvement with respect to those of independent inversions. In some cases, the proposed algorithm was able to reconstruct stratigraphies not well resolved by the independent interpretation of the two techniques (Comina et al. 2002).

4.1 *Two Techniques: Analogies and Synergies*

VES technique is a well-established surveying technique, employed for decades thanks to the celerity of acquisition and to the relatively simple interpretation procedure. The Schlumberger array is one of the most commonly used testing configurations for engineering scale and the electrostratigraphy is inferred by the inversion of the apparent resistivity curve, assuming the soil as a stack of homogenous and isotropic layers. The vertical profile of the electric resistivity is usually interpreted in terms of lithology and water table.

On the other side the investigation based on Rayleigh wave propagation gained much popularity in recent years thanks to the possibility of estimating the stiffness profile of the investigated sites by a fast and effective test (Foti, 2000). The method is based on the dispersive characteristics of Rayleigh waves in a layered medium and assumes the soil as a stack of homogenous, isotropic and linear elastic layers. The final result is a vertical profile of the shear wave velocity by which the stiffness profile of the site can easily be determined.

The measuring technique is based on the acquisition of seismograms by using layouts of vertical low frequency geophones with regular spacing. Seismic sources could be either impulsive or controlled and should supply adequate energy in the frequency range of interest. The seismograms are analyzed in the fk domain, where the maxima of the spectra represent the apparent dispersion curve (Rayleigh phase velocity vs. frequency) which depends on the characteristics of the site and of the acquisition layout (Foti et al., 2000). The inversion of the dispersion curve supplies the vertical profile of SV-wave velocity.

The aforementioned techniques present many formal analogies both in the soil model and in the inversion algorithms (Comina, 2001):
- both techniques are based on layered models with a stack of homogenous and isotropic layers (monodimensional methods);
- the field curves have similar properties regarding resolution (both the techniques supply information about average properties of the investigated sites with resolution decreasing with depth) and problems concerning the non uniqueness of solution (the equivalence problem in VES);
- the forward problem solution can be obtained in both cases by assembling layer matrices.

These analogies make it possible to join the interpretation procedures in a coupled inversion algorithm by linking VES and SWM models by means of a coupling parameter.

Furthermore, the two techniques supply information which are partially complementary since they refer to different physical parameters: for instance, the evaluation of electric resistivity allows to describe the lithology providing indication about the presence of clayey soils and about the groundwater level, while the shear wave velocity furnishes information regarding the mechanical properties of the soil skeleton almost independently by the groundwater level.

4.2 The Joint Inversion Algorithm

The joint inversion algorithm, implemented in Matlab®, is based on the methods described by Hering et al. (1995) and Misiek et al. (1997), who proposed the joint inversion of electric and seismic data and the joint inversion of seismic data from different seismic techniques and of electric data from different electric techniques as well.

Both single and joint inversions involve the solution of the forward problems for seismic and electric. As the vertical electric soundings are concerned, the solution proposed by Koefoed (1979) has been adopted, while for the solution of the seismic problem the algorithm implemented by Rix and Lai (1998) was used, limiting the solution to the fundamental Rayleigh mode.

In order to link the two single inversion procedures in a joint inversion algorithm it is necessary to compare two different physical models and to select a coupling parameter. The first item is handled normalizing the data values with respect to measured data, while the layer thickness has been chosen as the coupling parameter. This choice implies that seismic interfaces and electric interfaces are assumed as coincident, but, since there are cases in which a variation of the electric properties does not necessarily corresponds to a variation of seismic properties (e.g. the water table sensibly modify the resistivity value while does not have a significant influence on shear wave velocity) the inversion algorithm should be able to manage also stratigraphies in which an interface exists just for one of the two models.

The solution of the joint inverse problem is structured similarly to each single inversion procedure: starting by an initial model which contains the parameters describing the electric and the seismic stratigraphies (with the possibility of introducing a variation of one of the physical characteristic without modifying the other) the forward problem is resolved and by means of an iterative process, the residuals between the experimental and the calculated data are minimized in order to obtain the final model which better approximates the field curves.

Analyzing the results of the joint inversion it should be taken into account that the two single inversions could supply a better approximation of the field curves because the joint one, operating on two different physical parameters, evaluates the solution which better satisfied contemporary the two models and that should be therefore much consistent with "reality".

After a series of test on synthetic data, a *weighted, damped least square* algorithm (Menke 1984, Tarantola 1987) has been chosen because of its stability and robustness. The algorithm solution is expressed as follows:

$$c^k = \left(G^T W_e G + \beta^2 W_m\right)^{-1} \cdot G^T W_e \cdot y^{k-1} \quad (1)$$

where: y is the residual vector normalized with respect to measured data; W_e is the weight matrix; W_m is the flatness matrix; β is a numerical damping factor (Menke 1984). In the joint inversion, the G matrix links the measured parameters ρ_{app} (apparent resistivity) and v_r (Rayleigh wave velocity) with ρ_i (true resistivity) and v_i (share wave velocity) of each layer and h_i (layer thicknesses). Hence the structure of the G matrix is the following:

$$G = \begin{bmatrix} \frac{\partial \rho_{app}^{(1)}}{\partial \rho_1} & \cdots & \frac{\partial \rho_{app}^{(1)}}{\partial \rho_n} & \frac{\partial \rho_{app}^{(1)}}{\partial h_1} & \cdots & \frac{\partial \rho_{app}^{(1)}}{\partial h_{n-1}} & 0 & \cdots & 0 \\ \vdots & & & & & & & & \\ \frac{\partial \rho_{app}^{(p)}}{\partial \rho_1} & \cdots & \frac{\partial \rho_{app}^{(p)}}{\partial \rho_n} & \frac{\partial \rho_{app}^{(p)}}{\partial h_1} & \cdots & \frac{\partial \rho_{app}^{(p)}}{\partial h_{n-1}} & 0 & \cdots & 0 \\ 0 & \cdots & 0 & \frac{\partial v_r^{(p+1)}}{\partial h_1} & \cdots & \frac{\partial v_r^{(p+1)}}{\partial h_{n-1}} & \frac{\partial v_r^{(p+1)}}{\partial v_1} & \cdots & \frac{\partial v_r^{(p+1)}}{\partial v_n} \\ \vdots & & & & & & & & \\ 0 & \cdots & 0 & \frac{\partial v_r^{(p+z)}}{\partial h_1} & \cdots & \frac{\partial v_r^{(p+z)}}{\partial h_{n-1}} & \frac{\partial v_r^{(p+z)}}{\partial v_1} & \cdots & \frac{\partial v_r^{(p+z)}}{\partial v_n} \end{bmatrix} \begin{vmatrix} x_j \\ Y_i \end{vmatrix}$$

(2)

where we find the partial derivatives of the measured quantities with respect of each layer characteristics; Y_i are the normalized differences between the values of the experimental and theoretical curves and x_i are the characteristic values calculated in the previous step of iteration. In the present study the calculation of partial derivatives has been performed numerically.

The structure of the G matrix point out that layer thickness is the only coupling parameter; hence the coupling is merely geometrical. A physical coupling would in fact involve relations between Rayleigh wave velocity and resistivity and between apparent resistivity and shear wave velocity, which would fill the zeros in the G matrix. The possibility of establishing this physical relation should be considered a future task of the research, investigating the relations between porosity and seismic velocity and between porosity and resistivity, as suggested by previous studies (Wempe 2000, Berryman 1995).

5 RESULTS

Results obtained along the various testing profiles analyzed in the present study lead to a stratigraphy composed, for the first 20 meters, of two different sandy layers, a superficial one till about 6-7 meters and a deeper one starting from 13-14 meters probably composed of coarser formations with some gravels. In between them there is a less permeable formation, probably of clayey nature, having thickness of approximately 6-8 meters. The water table is at very shallow depth, between 2 and 3 meters below the ground surface. From the quantitative point of view, the resulting differences in the values of resistivity can be interpreted with passages from saturated sands to saturated sands with percentages of clay around to 15%, as given by various semi-empirical relations. Such situation can lead to seepage problems during floods of the Po river and substantially it was the situation expected for this zone (in which one broken off generated in June 1917).

Very promising results have been obtained with the joint inversion of electrical resistivity soundings (VES) and the surface wave method (SWM). Results obtained from the joint inversion are: a profile of electrical resistivity and a profile of S-wave velocity. The combined analysis of the two profiles can give reliable indications about the type of material and moreover the joint inversion technique allows to obtain a more stable estimate of layer thicknesses.

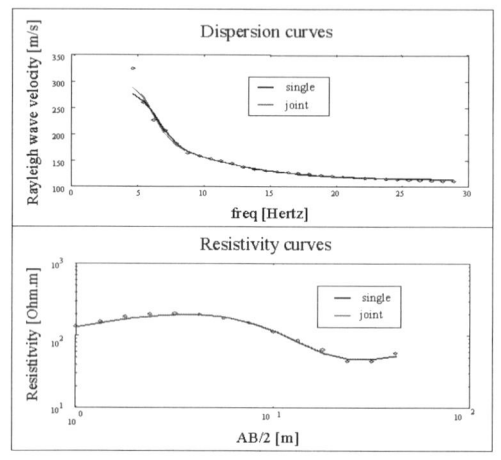

Figure 6. Comparison of single and joint inversion of dispersion and resistivity curves along P1 survey line (symbols = experimental data).

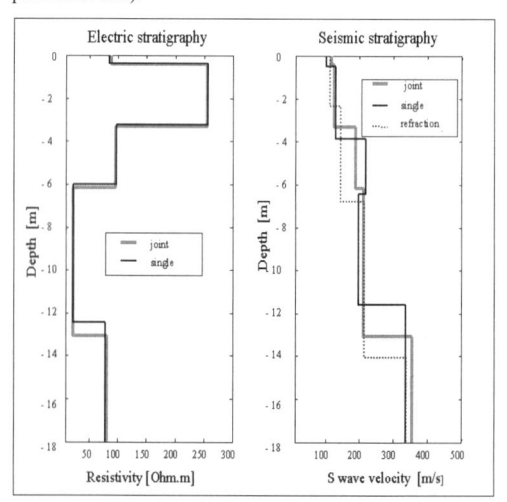

Figure 7. Comparison of single and joint inversion results along P1 survey line with a seismic refraction interpretation.

An interesting example of the improvement that a joint interpretation can lead with respect to the single inversions is reported in Figures 6 and 7. The final results are almost equivalent in terms of dispersion and resistivity curves (Figure 6) but the stratigraphies (Figure 7) show that the joint inversion supplies a seismic profile which is in better agreement with independent seismic refraction results, in particular concerning the depth of the last interface. The strong resistivity variation between 3 and 4 m depth is due to the water table. The information about water table position, inferred by electric stratigraphy, allows also a better estimate of the Poisson ratio values to be utilized for seismic inversion.

The limit of using SWM and VES is constituted by the restrictive hypothesis of horizontally stratified medium; hence their reliability is limited to relatively uniform formations with no marked lateral heterogeneities. This limitation can be overcome by electrical and seismic tomography, which allows locating variations of thickness in the layers. The tomographies have substantially confirmed the validity of the horizontal layering hypothesis for this site. The only remarkable difference to a horizontal stratified situation was noticed along profiles P1 and P2, which are more or less perpendicular to the embankment. In particular a rising of the clayey layer at a certain distance from the embankment has been detected.

A sample set of results for a single survey line from all the different techniques is reported in the following. In particular the line P3 runs parallel to the bottom line of the outer side of the embankment and it can be considered representative of the conditions leading to seepage failures.

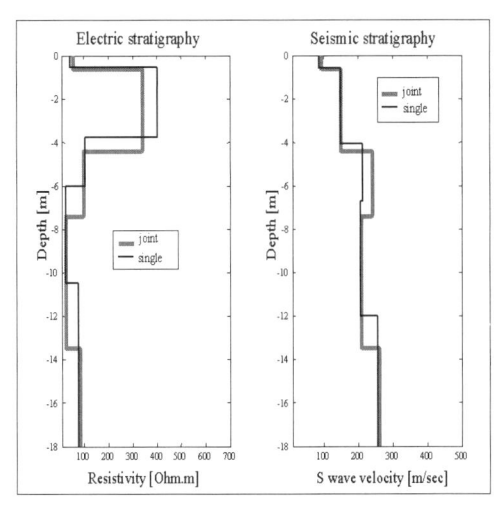

Figure 8. Comparison of single and joint inversion results along the P3 profile.

As it can be seen from the stratigraphies in Figure 8, the joint inversion supply a more defined interpretation than single inversion procedures, clearly locating the interfaces between layers. The joint inversion results are also well in agreement with the electric resistivity tomography both in terms of layer thickness and resistivity values (Figure 9).

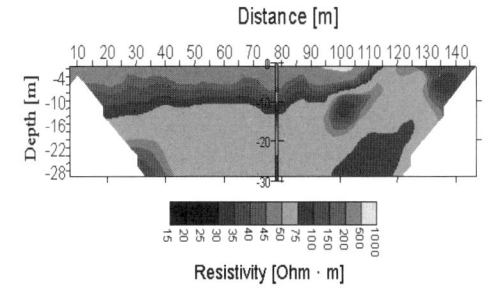

Figure 9. Electrical resistivity tomographies along the profile P3 with the VES executed in the center of the survey line (joint interpreted).

Even if with a bidimentional technique (such as the ERT) the lateral behavior of the formations can be distinguished (in particular the rising of the clayey layer) with respect to monodimensional methods the general trend of the section investigated is well confirmed with the VES (joint interpreted) reported in the center of Figure 9.

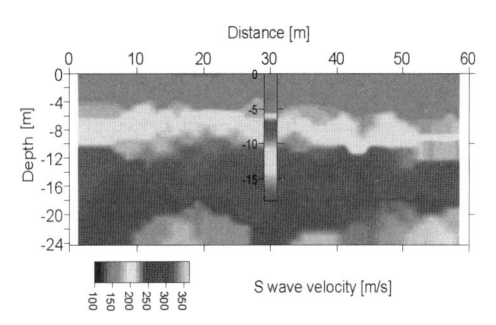

Figure 10. SH-waves seismic refraction tomography along the profile P3 together with the SWM invertion.

Figure 10 reports the comparison of SH-waves seismic refraction tomography with the SWM joint inversion (in the center of the figure). Also with respect to SH-waves seismic refraction tomographies the stratified nature of the site is confirmed and this can be considered another way in order to discriminate situations where the validity of a monodimensional approach is in doubt. It is noteworthy to mention that at depth greater than 10m the surface wave interpretation locates a slower (less stiff) layer. Such slower layer cannot be detected by the seismic refraction because of intrinsic limitations in the method itself, which is not able to resolve sequences characterised by velocity inversion.

With respect to the P-wave refraction surveys carried out at the site, because of the position of the water table (3 meters depth) and of the stiffness of the local formations, it is only possible to estimate the depth of the water level. For this reasons mainly S-wave measurements were performed. Anyway, the water table position is a useful element of validation the geo-electrical tests and constitutes an important element for the interpretation of SWM (Foti and Strobbia, 2003). In general since SWM and P-wave refraction utilize the same instrumentation (vertical geophones and impulsive sources) and testing configuration, they are easy to be performed together and it is convenient to get P-wave data to use the possible synergies (Foti et al., 2003).

Finally, preliminary results obtained with the Ohm-Mapper device on this site are not fully satisfactory if compared to ERT results. Still the potential of the technique in investigating large areas significantly reducing acquisition time is very high and further investigations in this direction are planned.

6 CONCLUSIONS

The aim of a characterization campaign for seepage potential is essentially the identification of layers with different permeability. Therefore the study has been finalized to check the ability of the various methods to discern the nature and the thickness of the various layers.

Geophysical techniques can be very useful in studying seepage potential. They have the advantage that relatively large areas can be investigated with surveys carried out from the ground surface. It is important to correlate electrical and seismic results in order to get more complete information. Geophysical techniques are based on the observation that various geotechnical materials are characterized by different characteristic values of electrical resistivity and stiffness (closely related to the velocity of propagation of the S-waves). In particular, the clayey materials have much lower values of electrical resistivity than those typical of sandy ones and analogously saturated materials introduce lower values of resistivity of the same dry materials. Moreover S-wave velocity constitutes a discriminating element that allows the differentiation of adjacent layers constituted from different materials.

The new joint inversion algorithm implemented can give more stable and accurate results, with the possibility of determining with greater accuracy the thickness of each layer, which is the coupling parameter of the joint inversion. The validity of such method is conditioned by the hypothesis of horizontally stratified medium, which is characteristic of both VES and SWM interpretation. However such hypothesis is fairly realistic for recent alluvial deposits along the embankments of the Po River. Anyway, it can be verified by electrical and seismic tomographies which supplies bidimensional information on the profiles, evidencing lateral heterogeneities if they are present.

ACKNOWLEDGEMENTS

This research has been sponsored by the Autorità di Bacino del Fiume Po (Po River Agency). Authors are grateful to the following individuals for help in performing and interpreting the tests: Prof. L. Sambuelli, Mr. R. Maniscalco, Mr. F. Bosco

REFERENCES

Berryman J.G. 1995 *"A Handbook of Physical Constants"* chapter 6, Mixture theories for rock properties, 205 - 228, American Geophysical Union, Washington D.C.

Comina C. 2001. *"Inversione congiunta di misure di resistività e di propagazione di onde superficiali"*. Master dissertation in Geotechincal Engineering, Politecnico di Torino, Torino.

Comina C., Foti S., Sambuelli L., Socco L.V. and Strobbia C. 2002. *"Joint inversion of VES and Surface Wave data"*. Proceedings of the SAGEEP, Las Vegas.

Colombo P. 1964. *"Studio delle caratteristiche dell'argilla limosa degli argini del Po"*. Istituto di Costruzioni Marittime, Università di Padova, Padova.

Foti S. 2000. *"Multistation methods for geotechnical characterization using surface waves"*. Phd dissertation in Geotechincal Engineering, Politecnico di Torino, Torino.

Foti S., Lancellotta R., Sambuelli L., Socco L.V. 2000. Notes on fk analysis of surface waves. Annali di Geofisica, vol. 43, n.6, 1199-1210

Foti S., Strobbia C. 2002. *Some notes on model parameters for surface wave data inversion*. Proc. of SAGEEP 2002, Las Vegas, USA, February 10-14, CD-Rom

Foti S., Sambuelli L., Socco L.V., Strobbia C. (2003) "Experiments of joint acquisition of seismic refraction and surface wave data", Near Surface Geophysics, EAGE, 119-129

Hering A., Misiek R., Gyulai A., Ormos T., Dobroka M. and Dresen L. 1995. *"A joint inversion algorithm to process geoelectric and surface wave seismic data. Part I: basic ideas"*. Geophysical Prospecting, 43, 135-156.

Koefoed O. 1979. *"Geosounding principles 1, resistivity sounding measurements"*. Elsevier Scientific Publishing.

Lai C.G. and Rix G.J. 1998. *"Simultaneous inversion of Rayleigh phase velocity and attenuation for near-surface site characterization"*. Georgia Institute of Technology, School of Civil and Environmental Engineering, report no. gitcee/geo-98-2, july.

Menke W. 1984. *"Geophisical data analysis: disctrete inverse theory"*. International Geophisics Series, Academic Press, San Diego.

Misiek R., Liebig A., Gyulai A., Ormos T., Dobroka M., Dresen L. 1997. *"A joint inversion algorithm to process geoelectric and surface wave sismic data. Part II: applications"*. Geophysical Prospecting, 45, 65-85.

Sambuelli L. and Deidda G. P. 1999. SWIPHONE^{TM}: a new seismic sensor with increased response to SH-waves. Proc. Of 5th Meet. Of EEGS, Budapest, SeP5.

Tarantola A. 1987. *"Inverse problem theory – methods for data fitting and model parameter estimation"*. Elsevier Scientific Publishing, New York.

Wempe W.L. 2000. *"Predicting flow properties using geophysical data: improving aquifer caracterization"*. Master dissertation, Stanford University, Stanford.

The use of multichannel analysis of surface waves in determining G_{max} for soft clay

S. Donohue, M. Long & K. Gavin
Dept. of Civil Engineering, University College Dublin, Ireland

P. O'Connor
Apex Geoservices, Killinierin, Gorey, Co. Wexford, Ireland

Keywords: multichannel, surface waves, soft clay, shear modulus, discrete particle scheme

ABSTRACT: The measurement of the small strain shear modulus, G_{max} is determined using the Multichannel Analysis of Surface Waves (MASW) method for two soft clay sites in Ireland. G_{max} profiles generated using the MASW method compare very well with values derived empirically from CPTU data and also with results of laboratory triaxial testing. A synthetic earth model generated using a Discrete Particle Scheme (DPS) was also used to evaluate the software, Surfseis. The MASW method compares well with both the conventional seismic methods and the synthetic model.

1 INTRODUCTION

The measurement of the small strain shear modulus, G_{max} of a soil is important for a range of geotechnical design applications. This usually involves strains of $10^{-3}\%$ and less. According to elastic theory G_{max} may be calculated from the shear wave velocity using the following equation:

$$G_{max} = \rho \cdot V_s^2 \qquad (1)$$

where G_{max} = shear modulus (Pa), V_s = shear wave velocity (m/s) and ρ = density (kg/m^3).

Several techniques are commonly used to measure V_s in both the field or in the laboratory. Intrusive field methods include cross-hole, down-hole and seismic cone methods. In these surveys seismic sources and receivers are located either between boreholes or between the surface and a point in a borehole or cone. Non intrusive field methods used to determine V_s include seismic reflection and refraction and surface wave surveys.

Laboratory methods used to compute V_s include the resonant column method and the bender element method where a shear wave is transmitted using a piezoceramic element from the top of the soil specimen and recorded with another piezoceramic element at the bottom. In this paper G_{max} is evaluated using the Multichannel Analysis of Surface Waves (MASW) method for two soft clay sites in the Irish midlands.

2 SURFACE WAVE ANALYSIS METHODS

The steady state Raleigh wave / Continuous Surface wave (CSW) technique was introduced by Jones (1958) into the field of geotechnical engineering. It has been developed further by others, such as Tokimatsu et al. (1991) and Mathews et al. (1996). The CSW method uses an energy source such as vibrator to produce surface waves.

In the early 1980's the widely used Spectral Analysis of Surface Waves (SASW) method was developed by Heisey et al (1982) and by Nazarian and Stokoe (1984). The SASW method uses a single pair of receivers that are placed collinear with an impulsive source (e.g. a sledgehammer). The test is repeated a number of times for different geometrical configurations.

The Multichannel Analysis of Surface Waves (MASW) technique was introduced in the late 1990's by the Kansas Geological Survey, (Park et al., 1999). The MASW method exploits proven multichannel recording and processing techniques that are similar to those used in conventional seismic reflection surveys. Advantages of this method include the need for only one shot gather and its capability of identifying and isolating noise. Donohue et. al. (2003) used the MASW method for determining G_{max} for very stiff glacial till. The MASW method was used for recording and processing of surface wave data for the two sites discussed in this paper.

3 SHEAR WAVE VELOCITIES FROM SURFACE WAVES

The type of surface wave that is used in geotechnical surface wave surveys is the vertically polarised Raleigh wave. In a non-uniform, heterogeneous medium, the propagation velocity of a Raleigh wave is dependent on the wavelength (or frequency) of that wave. The Raleigh waves with short wavelengths (or high frequencies) will be influenced by material closer to the surface than the Raleigh waves with longer wavelengths (or low frequencies), which reflect properties of deeper material. This is illustrated in Figure1.

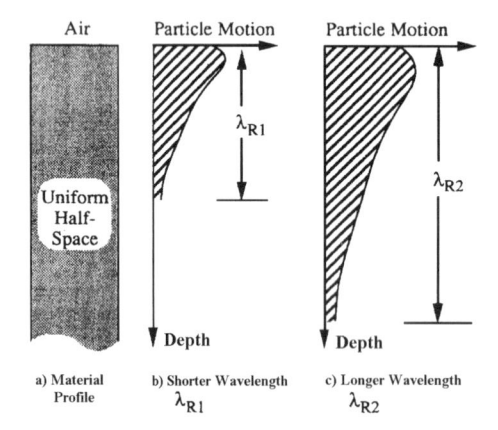

Figure 1. Approx. distribution of vertical particle motion with depth for two Raleigh Waves with different wavelengths (Stokoe et al., 1994)

This dependence of phase velocity on frequency is called dispersion. Therefore by generating a wide range of frequencies surface wave surveys use dispersion to produce velocity and frequency (or wavelength) correlations called dispersion curves.

After production of a dispersion curve the next step involves the inversion of the measured dispersion curve to produce a shear wave velocity – depth profile.

As an initial estimate, dispersion curves may be interpreted by assuming that the depth of penetration, z of a particular wave is a fraction of its wavelength, λ:

$$z = (\lambda/n) \qquad (2)$$

where n = a constant. The value of n is commonly chosen as either 2 or 3. Raleigh wave phase velocity, V_r, is then converted into shear wave velocity, V_s using equation (3).

$$V_s = (V_r/p) \qquad (3)$$

where p is a function of Poisson's ratio, ν. For $\nu = 0.2$, p =0.911 and for $\nu = 0.5$, p=0.955, therefore incorrectly approximating ν has minimal effect on V_s.

The software Surfseis performs the inversion procedure using a least-squares technique developed by Xia et al. (1999). Through analysis of the Jacobian matrix Xia et al. investigated the sensitivity of Raleigh wave dispersion data to various earth properties. S wave velocities are the dominant influence on a dispersion curve in a high frequency range (>5Hz). The inversion method produced by Xia et al. is an iterative method. An initial earth model (S wave velocity, P wave velocity, density and layer thickness) is specified at the start of the iterative inversion process. A synthetic dispersion curve is then generated. Due to its influence on the dispersion curve only the shear wave velocity is updated, after each iteration, until the synthetic dispersion curve closely matches the field curve. The Kansas Geological Survey produced the software Surfseis for use with the MASW method. Surfseis is evaluated in Section 6.

4 THE SITES

4.1 General

The two sites involved in this study have been used by researchers at UCD for several years and are generally well characterized. They are located at Athlone and Portumna towards the centre of Ireland. Full details of the Athlone site are given by Long and O'Riordan (2001). Conaty (2002) describes the sites at Portumna. The deeper soft soils at these sites are glacial lake deposits, which were laid down in a large pro-glacial lake, which was centred on the middle of Ireland, during the retreat of the glaciers at the end of the last ice age some 10,000 to 20,000 years B.P.

As the climate became warmer and vegetation growth was supported on the lake-bed, the depositional environment changed and the upper soils have increasing organic content. At the three sites the ground surface is underlain by two thin organic layers, calcareous marl and peat. The calcareous marl was formed when water super-saturated in calcium carbonate comes out of solution due to upward artesian flow.

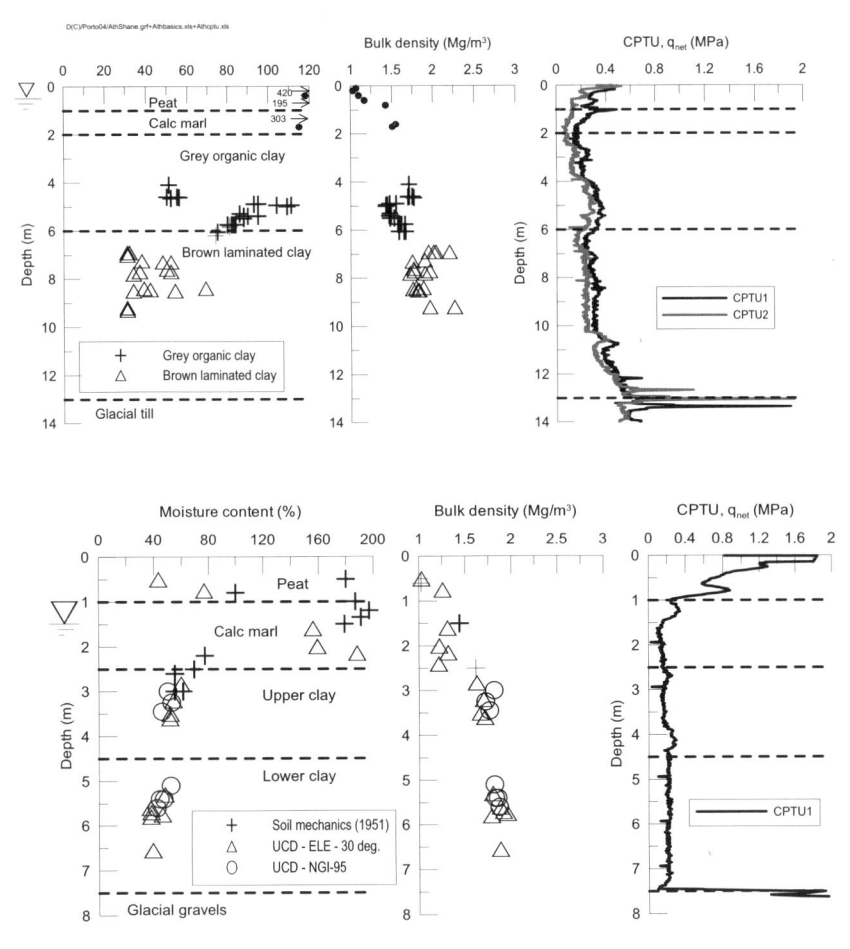

Figure 2a and b. Basic soil parameters Athlone and Portumna sites.

4.2 Athlone

At Athlone, two distinct strata were formed, as can be seen on Figure 2a. The lower soils are very soft brown horizontally laminated (varved) clays and silts with clearly visible partings typically 1 mm to 2 mm thick. These deposits are referred to as the brown laminated clay. They have average moisture content and bulk density of about 40% and 1.9 Mg/m³ respectively. Though there is some scatter in the data, there is no apparent trend in the parameters with depth. The brown laminated clay has a clay content of about 35% and has an average plasticity index (I_p) of about 18%.

For the brown laminated clay, laboratory strength data from high quality Sherbrooke block samples fall on or close to the $0.3\sigma'_{v0}$ line. However as detailed by Long and O'Riordan (2001) various vane tests and the cone pressuremeter tend to underestimate the strength, probably because of device insertion effects. Field vane sensitivity is about 2.5.

As the climate became warmer, the depositional environment changed and the upper soils show only some signs of varving and have an increasing organic content. The material deposited under these conditions is homogenous grey organic clay and silt.

As can be seen on Figure 2a, at about 4 m depth, its moisture content is about 55% and then increases to 110% before decreasing with depth to about 70%. Average bulk density is about 1.6 Mg/m³. In addition clay content and average I_p are of the order of and 25% and 40% respectively. These latter are higher than for the brown clay as a result of the increased organic content.

For the grey organic clay, strength data from both vane tests and laboratory testing shows considerable scatter. However values are generally greater than $0.3\sigma'_{v0}$ suggesting that some overconsolidation of the material has taken place. Average field vane sensitivity is about 8.5.

In the peat and calcareous marl material, the moisture content values are very high being consistently over 200%, with corresponding low bulk density values of the order of 1.2 Mg/m³ and 1.4 Mg/m³.

Piezocone (CPTU) q_{net} values are very low for all layers. They are slightly higher, however, in the peat and marl possibly due to the effects of fibrous inclusions. Values increase from about 0.15 MPa to 0.35 MPa in the grey organic clay, then fall back to about 0.2 MPa in the brown laminated clay followed by a gradual increase with depth, particularly below 10.5 m to about 0.6 MPa.

4.3 Portumna

As can be seen from Figure 2b, the Portumna clays are relatively uniform. In these deeper clay layers moisture content falls from about 50% in the upper clay layer to 40% in the lower layer. The corresponding bulk density values are 1.7 Mg/m³ and 1.85 Mg/m³. Additional data shows that the clay content is about 40% and I_p is 22%.

The peat layer has a very variable natural moisture content, which ranges between 45% and 180% and a corresponding bulk density of less than 1.2 Mg/m³. Equally the marl material has very high moisture content and a relatively low bulk density of about 1.25 Mg/m³.

Laboratory strength data from reasonable quality piston samples, suggests that the clay material is slightly overconsolidated with values falling in the range $0.4\sigma^1_{v0}$ to $0.5\sigma^1_{v0}$. Again vane strength values are lower, perhaps due to disturbance effects. Limited data on field vane sensitivity suggest that it is less than 5.

High CPTU q_{net} values were recorded in the peat, due probably to the effects of fibres. In the marl and in the layers below, values show almost no increase with depth, except perhaps in the upper clay layer and remain constant at about 0.25 MPa.

5 RESULTS OF MASW SURVEYS

5.1 General

The results of the MASW surveys for the two soft clay sites are discussed here. An impulsive source (sledgehammer) was used to generate the surface waves. The MASW results are compared to results from CPTU tests in terms of the cone tip resistance q_c. An empirical relation proposed by Mayne and Rix, (1993) was used to estimate G_{max} from the q_c data:

$$G_{max} = 99.5(p_a)^{0.305} * (q_c)^{0.695} / (e_0)^{1.13} \quad (4)$$

where q_c = the measured cone tip resistance (kPa) p_a = atmospheric pressure, e_0 = in situ void ratio.

In Section 6, the software, Surfseis, used to generate the shear wave velocity profiles before conversion to G_{max} profiles is evaluated using a discrete particle scheme.

5.2 Athlone

Three separate MASW survey lines were performed for the Athlone site to test the repeatability of the survey. The MASW lines were all parallel and located at two metre intervals. All tests were carried out at the same location (Profile D from Long and O'Riordan, 2001) as the CPTU and the Sherbrooke block sampling discussed above. The field set up for each of the Athlone survey lines consisted of 12 receivers (4.5 Hz geophones) at 1m intervals collinear with a chosen source location. A number of different source locations were chosen for each profile, at source receiver offsets of 1, 6, 13 and 25m. This was necessary to determine the optimum acquisition parameters thereby minimising the influence of near field and far offset effects (Park et al., 2002).

The depth of penetration of the MASW method for each of the profiles was 8.75 m, which was adequate for the site. This limitation resulted from the low Raleigh wave phase velocity (equation 5) of the very soft material under investigation, the limitation of 4.5 Hz geophones and the length of the receiver spread.

$$\lambda = (V_r/f) \quad (5)$$

where λ is the wavelength of a particular Raleigh wave with phase velocity V_r and frequency f. Therefore for a low velocity wave with a lowest recordable frequency (dependant on frequency on geophones) the wavelength and therefore the depth of penetration (eq. 2) of the survey will be smaller than for a higher velocity wave.

G_{max} values computed for the MASW survey at Athlone are presented in Fig. 3 along with the empirically derived profiles from the CPTU tests.

There is good agreement between the three MASW profiles for this site. There is a slight increase in the variation of the three profiles with depth. The difference in the top few metres is negligible and the maximum difference at 8.75 m depth is 2.4 MPa. There is also variation in G_{max} estimated from the two CPT cone tip resistance (q_c) profiles, indicative of ground variability.

The MASW results give similar and very low G_{max} values for the peat and marl, with the boundary between the marl and grey organic clay being clearly defined. It is also possible to define a boundary between the grey organic clay and the lower brown laminated clay.

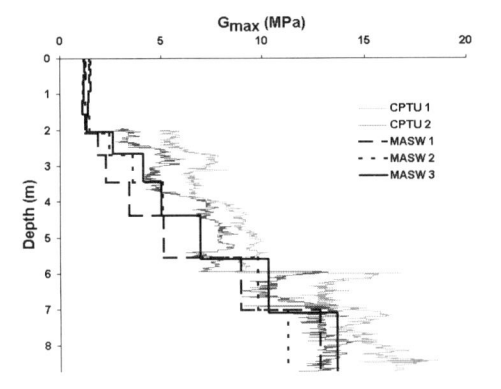

Figure 3. G_{max} from three MASW profiles compared with corresponding two CPT profiles for Athlone

CPTU data for the peat and calcareous marl has been ignored as the Mayne and Rix (1993) approach was never intended to be used for such materials. All the G_{max} values are very low for these very soft clays. While the first CPTU profile (CPTU1) gives a consistently higher result than the MASW profiles the second, CPTU2, gives a very similar result. In general the CPTU estimated profiles give higher G_{max} than the MASW results. Typically the CPTU approach gives values 30% higher than for the MASW survey for the grey organic clay and 20% higher for the brown laminated clay. However the values involved are so small that these differences are considered negligible.

5.3 *Portumna*

Two separate MASW survey lines were performed for the Portumna site to again test the repeatability of the survey. The MASW lines were parallel and located two metres apart and in the same location as the CPTU work and boreholes which yielded the piston samples described above. The field set up for the Portumna site also consisted of 12 receivers (4.5 Hz geophones) at 1 m intervals collinear with a chosen source location.

As with the Athlone site to determine the optimum acquisition parameters a number of different source locations were chosen for each profile, at source receiver offsets of 1, 6 and 13m. For the first profile, shot gathers were acquired at these locations on opposite sides of the receiver spread.

The same offsets were used for the second profile at only one side of the receiver spread. G_{max} values computed for the MASW surveys at Portumna are presented in Figure 4 along with a corresponding CPT profile. As explained above two profiles for the same set-up were acquired by switching the position of the source locations to the opposite side (labelled MASW 1a and MASW 1b (Opposite) in Fig. 4).

The depth of penetration of the MASW method for the survey lines, MASW 1a and MASW 1b (Opposite) was 8.75 m and for the MASW 2 profile the maximum depth was 10 m which were more than adequate for this site. Figure 4 only shows data to depth of 7.5 m because it is the range of interest for the soft clay material.

There is very good agreement between the three MASW profiles for this site. As in the Athlone site there is a slight increase in the variation of the three profiles with depth. Also as before the MASW survey clearly delineates the interface between the marl and the underlying clays. G_{max} calculated from the CPTU cone tip resistance (q_c) shows excellent agreement with the MASW profiles. A slight difference between the profiles occurs at the top of the clay layer at a depth of 2.5 m to 3.5 m where the CPTU derived G_{max} is slightly higher than any of the corresponding MASW profiles. As shown in Figure 4 the MASW method shows G_{max} increasing for the soft clay from 3 MPa at 2.5m depth to 16 - 19 MPa at a depth of 7.5m. The CPT estimated G_{max} for the clay increases from 7 MPa at the top of the stratum to 18.5 MPa at its base.

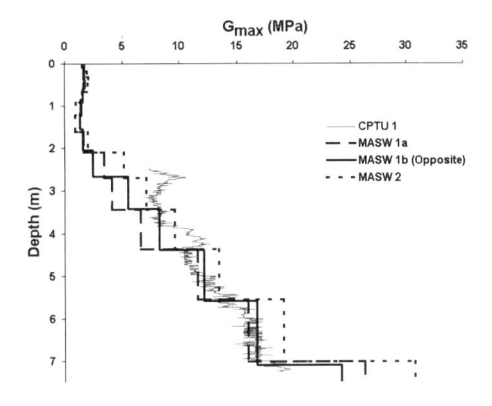

Figure 4. G_{max} from three MASW profiles compared with corresponding two CPTU profiles for Portumna

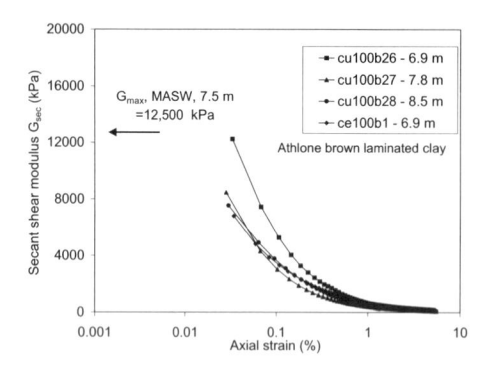

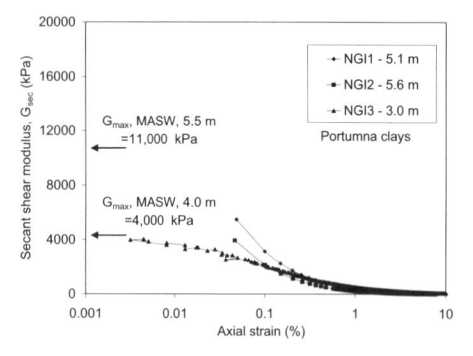

Figure 5. Comparison between laboratory test data and MASW output for (a) Athlone brown laminated clay and (b) Portumna clay.

5.4 Comparison with laboratory test data

A comparison between laboratory CAUC (anisotropically consolidated undrained compression) triaxial test data and MASW survey output is shown for Athlone brown laminated clay and Portumna clay on Figure 5a and 5b respectively. For Athlone the tests were carried out on high quality Sherbrooke block samples and for Portumna samples obtained using the NGI 95 mm diameter piston sampler was used. Strain resolution for Athlone is generally better as the axial displacement was measured using specimen mounted local gauges (Hall effect transducers).

It can be seen that the MASW G_{max} values relate well with the laboratory for these two lower plasticity clays. For the high plasticity Athlone grey organic clay (not presented here) the laboratory data yields higher stiffness values. The reason for this is not clear and warrants further study.

6 DISCRETE PARTICLE SCHEME

6.1 Overview

In order to evaluate the performance of the software, Surfseis, which is the main analysis tool in the MASW method, a Discrete Particle Scheme (DPS) was used. Developed in the Department of Geology, University College Dublin, Toomey and Bean (2000), the DPS allows the user to generate a synthetic earth model consisting of interacting particles. The particles are arranged in a closely packed isotropic hexagonal configuration (Figure 6) where each particle is assigned a density, diameter and P wave velocity.

V_s may be determined as the V_p to V_s ratio is fixed at 1.73. Also V_r is calculated, equation (3), using a value for Poisson's ratio of 0.25, which is fixed for the DPS. G_{max} was determined for the model using equation (1).

A geophysical experiment is set-up in the model with a source created and receivers (geophones) planted in the uppermost layer of particles. The output from this synthetic geophysical experiment is a seismogram. This synthetic seismogram is then converted to a format that is compatible with Surfseis and input into the software. As the input elastic moduli and wave velocities of the model are known the software was examined to see if it determines their correct values. A number of different models were tested, varying the number of layers, the layer thickness and stiffness. The results for two models are presented in Sections 6.2 and 6.3. The Shear wave velocity profiles for both models are very similar to the range of velocities that were observed in the field surveys of Athlone and Portumna. The reason for this was to test Surfseis for low velocity profiles. Donohue et al. (2003) tested the same software for higher velocities.

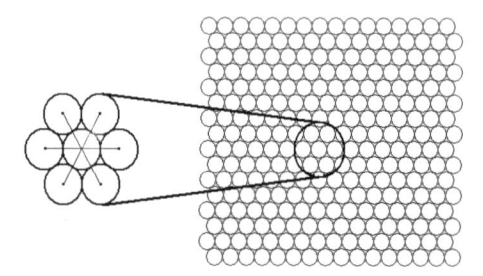

Figure. 6. The discrete particle scheme consists of particles arranged in a hexagonal geometry. Each particle is bonded to its six surrounding neighbours

Towards the end of the synthetic seismogram for the first DPS model a very small amount of noise was apparent. This was caused by reflections off the sides and bottom of the model. Although it is shown to have little or no impact on the MASW profile a second model was created (section 6.3) that was both wider and deeper so that any reflections would arrive later in the seismogram and so would not have an impact on the surface waves.

6.2 DPS Model 1

This Model is a 4-layer model where the first two layers are 1m thick, the third layer is four metres thick and the fourth extends to the base of the model. The particle diameter for this model is 0.1667m. The model is 510 particles wide (85m) and 501 particles deep (83.5m). There were 24 receivers selected at 1m intervals and the source to receiver offset was 1m. There is an increase in G_{max} with each deeper layer. The elastic properties and wave velocities of this model are listed in Table 1 below.

Table 1. Input parameters for DPS Model 1 where V_p, V_s and V_r are the P wave, S wave and Raleigh wave velocities, ρ = density and G_{max} = small strain shear modulus

	Depth (m)	V_s (m/s)	V_r (m/s)	V_p (m/s)	ρ (kg/m^3)	G_{max} (MPa)
Layer1	1	30	27.7	52	1200	1.08
Layer2	2	35.3	32.5	61	1300	1.59
Layer3	6	50.3	46.3	87	1600	4
Layer4	83.5	85	78.1	147	1800	13

The G_{max} profile for this DPS Model is shown in Fig. 7 along with the output MASW profile produced using the software, Surfseis. The depth of penetration for the MASW survey was 15m. This limitation was due to numerical constraints because the source to receiver offset and receiver spread could not be increased without the addition of noise to the end of the synthetic seismogram.

The Surfseis produced G_{max} profile compares well with the actual DPS G_{max} profile. The only difference of note between the two profiles is that the MASW method slightly underestimates G_{max} of the deepest layer by a maximum of 1.5 MPa. Also the MASW method shows the deepest layer beginning at 7m, an error of 1m. This was because the start of deepest layer in the DPS passes through the center of one of the inverted MASW layers. The MASW inversion then selected a G_{max} value in between the deepest layer and the layer directly above it.

The small amount of noise that was apparent on the synthetic seismic section appears to have no impact on the resultant G_{max} profile as shown below in Fig. 7.

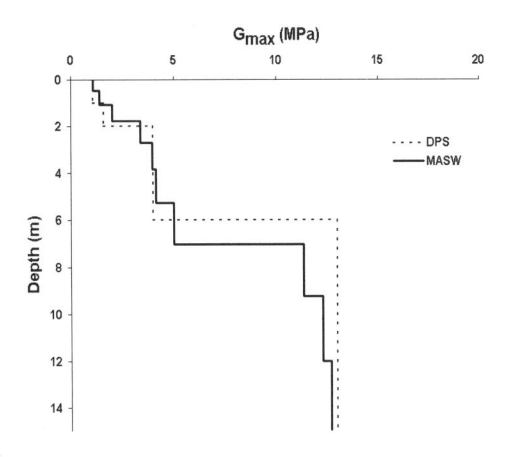

Figure 7. G_{max} profile for the first DPS model compared with the corresponding Surfseis (MASW) produced profile

6.3 DPS Model 2

A second larger DPS model was created to eliminate the small amount of noise that was observed in the previous model. An extra layer was also added to increase the complexity of the model. This Model is a 5-layer model where the first layer is 1m thick, the second 1.5m, the third and fourth are both 2m thick and the fifth extends to the base of the model. The particle diameter for this model is also 0.1667m. This model is 600 particles wide (100.02m) and 601 particles deep (100.19m). There were 24 receivers selected at 1m intervals and the source to receiver offset was 1m. There is an increase in G_{max} with each deeper layer. The elastic properties and wave velocities of this model are listed in Table 2 below.

Table 2. Input parameters for DPS Model 2 where V_p, V_s and V_r are the P wave, S wave and Raleigh wave velocities, ρ = density and G_{max} = small strain Shear Modulus

	Depth (m)	V_s (m/s)	V_r (m/s)	V_p (m/s)	ρ (kg/m^3)	G_{max} (MPa)
Layer1	1	34.7	31.9	60	1300	1.59
Layer2	2.5	39.8	36.7	69	1400	2.24
Layer3	4.5	50.3	46.3	87	1600	4
Layer4	6.5	60.1	55.3	104	1700	6.12
Layer5	100.19	75.1	69.1	130	1800	10.125

The G_{max} profile for this DPS Model is shown in Figure 8 along with two output MASW profiles produced using the software, Surfseis.

The only difference between the two MASW profiles was the center frequency of the source that was input into the DPS model. Center Frequencies of 7Hz and 12 Hz were selected for this.

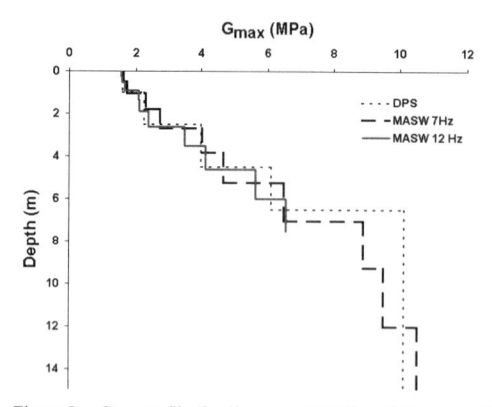

Figure 8. G_{max} profile for the second DPS model compared with the corresponding Surfseis (MASW) produced profile

Both Surfseis produced G_{max} profiles compare well with the actual DPS G_{max} profile. The 7Hz source detected all of the layers in the model and as with DPS model 1 the only difference of note between the DPS and the MASW profile is that the MASW method again slightly underestimates G_{max} of the deepest layer by a maximum of 1.2 MPa. The maximum depth using this frequency was 15m. This was again due to numerical constraints because the source to receiver offset could not be increased without the addition of noise to the end of the synthetic seismogram.

Due to its higher input center frequency (shorter wavelength) the12Hz source did not detect the deepest DPS layer (Section 3). The maximum depth using this frequency was 7.5m.

7 CONCLUSIONS

Shear wave velocity profiles were obtained in the field using the Multi Channel Analysis of Surface Waves (MASW) method at two soft clay sites in the Irish Midlands to determine the small strain shear modulus, G_{max} of this material and to compare the MASW derived stiffness profiles with corresponding CPT derived profiles.

At both sites the MASW produced profiles compared very well with values derived empirically from CPTU results.

The "depth of penetration" of the MASW profiles was limited to about 9 m, which was adequate for these sites. However if deeper profiles are required on soft material it is recommended that even lower frequency geophones are used along with a longer receiver spread (more geophones and/or wider geophone spacing).

A Discrete Particle Scheme (DPS) was then used to generate two separate layered earth models. A synthetic seismogram was produced from both models and was input in the software, Surfseis. As G_{max} for each of the models layers is known, Surfseis was examined to see if it determined their correct values. As shown the profiles compare very well.

ACKNOWLEDGEMENTS

The authors wish to thank Dr. C.J. Bean, Mr. M. Moellhoff and Apex geoservices for their assistance.

REFERENCES

Conaty, E. 2002. A study of the stability of the Shannon embankments. *M.Eng.Sc.* Thesis, University College Dublin.
Donohue, S., Gavin, K., O'Connor, P. 2003. G_{max} from Multichannel Analysis of Surface Waves for Dublin Boulder Clay. XIIIth *Euro. Conf. on Soil Mech. and Geotech. Eng.*, Prague.
Heisey, J.S., Stokoe, K.H and Meyer, A.H. 1982. Moduli of pavement systems from Spectral Analysis of Surface Waves. *Transportation Research Record, No. 852*, Washington D.C.: 22-31.
Jones, R. 1958. In-situ measurements of the dynamic properties of soil by vibration methods. *Geotechnique*, Vol. 8(1): 1-21.
Long, M.M. and O'Riordan, N.J. 2001. Field behaviour of very soft clays at the Athlone embankments. *Geotechnique*, 51 (4): 293-309.
Mathews, M.C., Hope, V.S., and Clayton, C.R.I. 1996. The use of surface waves in the determination of ground stiffness profiles. *Proc. Instn Civ. Engrs Geotech. Engng*, 119, Apr: 84-95.
Mayne, P.W., and Rix, G.J. 1993. G_{max}-q_c relationships for clays, *Geotechnical Testing Journal*, ASTM, 16 (1): 54-60.
Nazarian, S., and Stokoe, K.H. 1984. In situ shear wave velocities from spectral analysis of surface waves. *Proc. 8th world conf. On earthquake engineering*. Vol.. 3: 31-38.
Park, C.B., Miller, D.M., and Xia, J. 1999. Multichannel Analysis of surface waves. *Geophysics*, Vol. 64, No.3: 800-808.
Park, C.B., Miller, R.D., and Miura, H., 2002, Optimum field parameters of an MASW survey *[Exp. Abs.]: SEG-J*, Tokyo, May 22-23, 2002.
Stokoe, K.H., II, Wright, G.W., James, A.B., and Jose, M.R. 1994. Characterisation of geotechnical sites by SASW method. *Iin Woods, R.D., Ed., Geophysical characterization of sites*: Oxford Publ.
Tokimatsu, K., Kuwayama, S., Tamura, S., and Miyadera, Y. 1991. V_s determination from steady state Rayleigh wave method. Soils and Foundations, Vol. 31(2), 153-163.
Toomey, A., and Bean, C.J. 2000. Numerical simulation of seismic waves using a discrete particle scheme. *Geophys. J. Int.*, 141, 595-604.
Xia, J., Miller, R.D., and Park, C.B. 1999. Estimation of near surface shear wave velocity by inversion of Raleigh waves. *Geophysics*, Vol.64, No. 3, 691-700.

# The use of electrical resistivity for detection of leacheate plumes in waste disposal sites

V.R. Elis & G. Mondelli
University of São Paulo, USP – São Paulo, SP, Brazil

H.L. Giacheti, A.S.P. Peixoto & J. Hamada
São Paulo State University, Unesp – Bauru, SP, Brazil

ABSTRACT: Waste disposal in dumpsites or even in sanitary landfills can generate contaminants. The detection and delineation of the shape of the contamination plume can be assessed using monitoring systems or different site investigation techniques. This paper presents results of two case histories where vertical electrical and dipole-dipole electrical soundings were carried out in a deactivated dumpsite and in a sanitary landfill in Brazil. Test results were interpreted to assess physical properties of the ground and of the waste, as well as to detect and delineate the shape of leacheate plume. Test results show that the major advantage of using geophysical methods is to guide geotechnical *in situ* testing as well as to properly locate monitoring wells. Preliminary piezocone testing carried out at specific locations for both sites are presented and the advantages and limitations of this technology for site investigation of tropical soils is briefly discussed.

1 INTRODUCTION

In recent years there has been a steady increase in geoenvironmental engineering projects requiring enhancement in the site characterization techniques. Geoenvironmental characterization refers to the surficial and subsurficial representation that approximates the actual *in situ* conditions. This representation is typically developed by both surface and subsurface characterization (Campanella et al., 1998). A geoenvironmental site characterization requires stratigraphy, geotechnical, hydrogeological and some specific environmental parameters. Combining geophysical and geotechnical *in situ* techniques is the best way to achieve all the required information (Campanella et al., 1994).

Groundwater contamination is an excellent example of an area of interest within the scope of the geoenvironmental field. Geophysical methods, particularly electrical techniques, can be used to study different environmental characteristics that are important during the site characterization for waste disposal and monitoring migration of contamination plumes. The detection and delineation of the shape of the plume is an important component of geoenvironmental site characterization.

Two case histories are presented to show the use of conventional geoelectrical survey to detect leacheate plumes in waste disposal sites. The first one is a dumpsite over the watershed limit, on the recharge area of the Botucatu Aquifer. The other case history is a sanitary landfill located 1 km from a creek. At both sites the subsoil is mainly composed of sandstone. The dipole-dipole electrical profiling gave high quality information to assess groundwater contamination caused by a waste disposal for the two case histories. Physical properties of the ground (water table, groundwater flow and soil layer thickness) and the waste, as well as the interference with the surrounding media (detection and delimitation of plume and degree of contamination) were analyzed, using this technique.

The results showed that by selecting an appropriate geophysical technique and elaborating rational planning it is possible to assess the volume of waste, flow path and evaluate the impact caused by the waste disposal. The major advantage of using a geoelectrical method is to guide the drilling and sampling program as well as to define the location of the monitoring wells. This approach reduces the cost and produces much better results.

An appropriate geoenvironmental site characterization requires intrusive testing. Piezocone technology is not currently used for this purpose in tropical soils in Brazil. Preliminary results are presented and briefly discussed, indicating the advantages, limitations and the necessity of additional research to adjust this technology to the studied soils.

2 GEOENVIRONMENTAL SITE CHARACTERIZATION

2.1 Definition

Environmental, or geoenvironmental site characterization is a relatively new term for geotechnical engineers. So, there are several interpretations of what an environmental site characterization program is. Davies & Campanella (1995) define it as "*The field of study that links geological, geotechnical and environmental engineering, and the corresponding sciences, to form an area of interest that includes all environmental concerns within natural or processed geological media.*" According to the US EPA (1989), geoenvironmental site investigation includes stratigraphy, water level data, hydraulic conductivity and chemical distribution and sensors/receptors for potential and existing contaminants. The detection and assessment of groundwater contamination caused by waste disposal is an excellent example where geoenvironmental site characterization is necessary.

2.2 Geophysical Site Investigation

Geophysical methods, particularly the electrical techniques, can be used to study different environmental characteristics and they are very important during the site characterization for waste disposal and monitoring the migration of contamination plumes. The following geological, geotechnical, hydrogeological and environmental characteristics can be assessed using geophysical methods: (a) rock depth; (b) discontinuities; (c) changes on soil particle size; (d) groundwater level; (e) groundwater flow; (f) presence and three-dimensional distribution of the waste; (g) contaminated soil; (h) contaminated groundwater and plume shape. The first five characteristics are essential to assess any waste disposal sites. All characteristics presented are indispensable in the geoenvironmental monitoring of a waste disposal facility. All of the data obtained during geoelectrical profiling can be used to generate apparent resistivity maps at different depths and they can be used to assess the horizontal direction of contaminant flow.

2.3 Recent Technologies for Geoenvironmental Site Characterization

The modern site characterization approach is based on the cone penetration test (CPT). The standard piezocone (CPTU) measurements can be used as a tool for logging soils. Empirical and semi-theoretical correlations are available to estimate mechanical properties of the soils. The measurements of excess pore pressure, which are generated during penetration and their subsequent dissipation provided insight into the soil type and its hydraulics parameters (Campanella et al., 1994). As pointed out by the US EPA (1989) a geoenvironmental site characterization requires information of chemical distribution and source(s)/receptor(s) for potential or existing contaminants. The piezocone technology for environmental application included specific sensors for temperature, resistivity, pH, laser-induced fluorescence and ground penetration radar, among others. This technology also includes samplers to be used together with the piezocone for sampling soils, water and gas (Lunne et al., 1997).

3 RIBEIRÃO PRETO SITE

3.1 Site Description

The Ribeirão Preto site is a deactivated dumpsite located in Ribeirão Preto, São Paulo State, Brazil (Figure 1). Riberão Preto is a city in northeast São Paulo State, which has about 450,000 inhabitants. This dumpsite is over the surficial water divisor and at this site 600,000 m^3 of domestic waste was inappropriately disposed between 1975 and 1988. The landfill has two trenches, 300 meters long and 40 to 60 meters wide. The geology of the site is Sandstone from the Botucatu Formation covered with residual sandy soils. The topsoil is clayey material formed from Igneous Rock (Basalt) from the Serra Geral Formation.

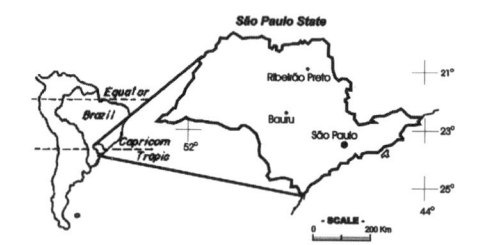

Figure 1: Cities where studied sites are located.

The traditional site characterization program carried out in this area included fieldwork to identify and assess the soil and rocks and to delimitate the zone affected by the domestic waste disposal. Laboratory tests to determine grain size distribution, cationic exchange capacity and specific surface of clay minerals were carried out on soil samples retrieved during Standard Penetration Tests (SPT).

There are six monitoring wells installed in this site. The water sampled in the saturated zone from these monitoring wells was used to assess contamination, comparing the values obtained with the background values. *In situ* hydraulic conductivity tests were carried out on the site using drilled boreholes, 100 mm in diameter. The coefficient of permeability for the topsoil varied between 10^{-7} to 10^{-8} m/s and it was equal to 10^{-6} m/s for the residual soil (Pejon & Zuquette, 1991).

3.2 Testing Results

3.2.1 Geoelectrical soundings

Several vertical electrical and dipole-dipole electrical profilings were carried out on the studied site. Dipole-dipole electrical profiling was interpreted with RES2Dinv (Loke, 1998) software for a better characterization of each cross section in terms of real resistivity, chargeability and investigation depth. Figure 2 shows a resistivity and chargeability cross-sections. The apparent resistivity map was elaborated based on all the dipole-dipole electrical profiling carried out on the studied site and the one for a 30 m theoretical depth is presented in Figure 3.

3.2.2 CPTU testing

After the interpretation of geoelectrical testing results, three piezocone tests were carried out to complement this investigation. As can be seen in Figure 3, CPT1 and CPT2 were carried out to cross the contamination plume generated from the waste deposit. CPT3 was pushed out of the plume. CPTU tests were not able to reach the groundwater table. Pore-pressure behind the tip was recorded using a slot-filter filled with automotive grease, as suggested by Larsson (1995) since groundwater level is deeper than the impenetrable layer for the cone. In this case, it was not possible to retrieve water to assess contamination. Soil samples were retrieved using direct-push technology at selected depths close to CPT1. Sampling depths were selected based on observed changes in q_c and R_f values to identify, classify and test soils. As piezocone testing does not provide soil samples, soil behavior type was identified using classification charts. CPTU testing results and soil classification based on the Robertson et al. (1986) approach are presented in Figure 4. This Figure also presents grain size distribution and electrical conductivity of the water leached through the sampled soils.

3.3 Discussion

3.3.1 Geoelectrical soundings

Based on the electrical resistivity cross sections it is possible to identify important characteristics, as soil layers, the trenches filled with waste, the saturated zone and the soil and water contaminated by the leaching from the waste. Figure 2 shows interpreted data for line C4, where the boundaries of two trenches can be seen, outlined by resistivity values lower than 25 ohm.m at the resistivity cross section. The chargeability cross section (Figure 2) also allows identification of trench location, since the waste shows higher chargeability values than the mass of natural soil (higher than 5.3 mV/V).

Figure 3 shows the map for a theoretical depth equal to 30 m, where the saturated zone is located. It permits visualization of the contamination plume (apparent resistivity values lower than 200 ohm.m) being generated from the waste deposit and moving from North (N) to Northeast (NE) and a less noticeable trend moving to the West (W). All geotechnical and geophysical testing data suggest that there is no burial waste at 30 m depth, however there is leaching from the waste reaching and spreading through the saturated zone. So, the map presented in Figure 3 shows the contamination plume moving and basically following the surficial water flow.

Another major aspect detected during the geoelectrical survey was to identify that the existing monitoring wells (P1 to P6 – Figure 3) are not properly located to detect the evolution of the contamination flow. Campanella et al. (1998) presented a mine tailings case history where the existing monitoring wells were not appropriately located. They used a portable surface geophysical tool called ground conductivity meter and resistivity piezocone tests to delineate an ionic rich plume and recommended new monitoring wells.

3.3.2 CPTU testing

The first step for piezocone interpretation is to determine if the penetration is drained or undrained. The partially saturated nature of tropical soils does not allow pore pressure measurements so it cannot be used to help identify the drainage conditions. For this reason, CPTU tests were carried out using a slot-filter filled with grease. Figure 4 shows that pore pressures are higher than zero above the groundwater level and these measurements helped to identify different soil behaviors.

For this site it is also observed that the Robertson et al. (1986) classification chart was not able to identify the soil based on grain size distribution, but the cone response was able to identify soils with distinct behaviors. As several researchers have stated, CPTU is an excellent tool for soil profiling. At the moment, for tropical soils, it cannot be used as an exclusive site investigation tool, since genetic characteristics affect soil behavior and soil sampling is required (Giacheti et al., 2003). The use of a soil sampler with direct-push technology is an interesting option to supplement site characterization. CPT test results can govern the depths from where to recover samples and the same truck or drill rig that pushes the cone will also push the soil sampler.

It is quite interesting to assess contamination by sampling groundwater. For this particular site, groundwater level is deeper than the layer impenetrable to the cone, so it cannot take advantage of using direct-push technology for water sampling.

For this reason water was leached through the sampled soils exactly where the contamination plume was identified. pH values were almost constant and electrical conductivity increased with depth, indicating that contamination also increases with depth (Figure 4), which is in good agreement with Elis (1998).

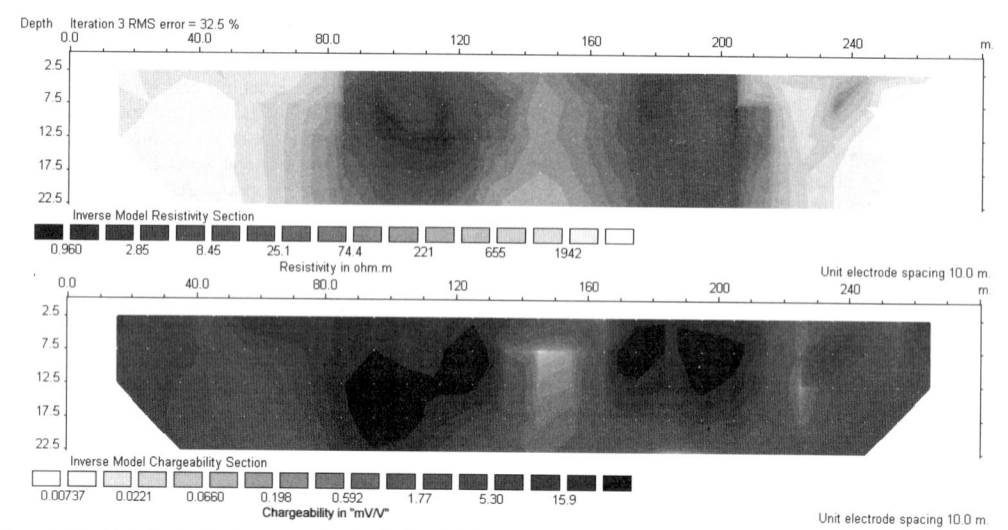

Figure 2: Electrical dipole-dipole cross section for the line C4, downstream from Ribeirão Preto Dumpsite.

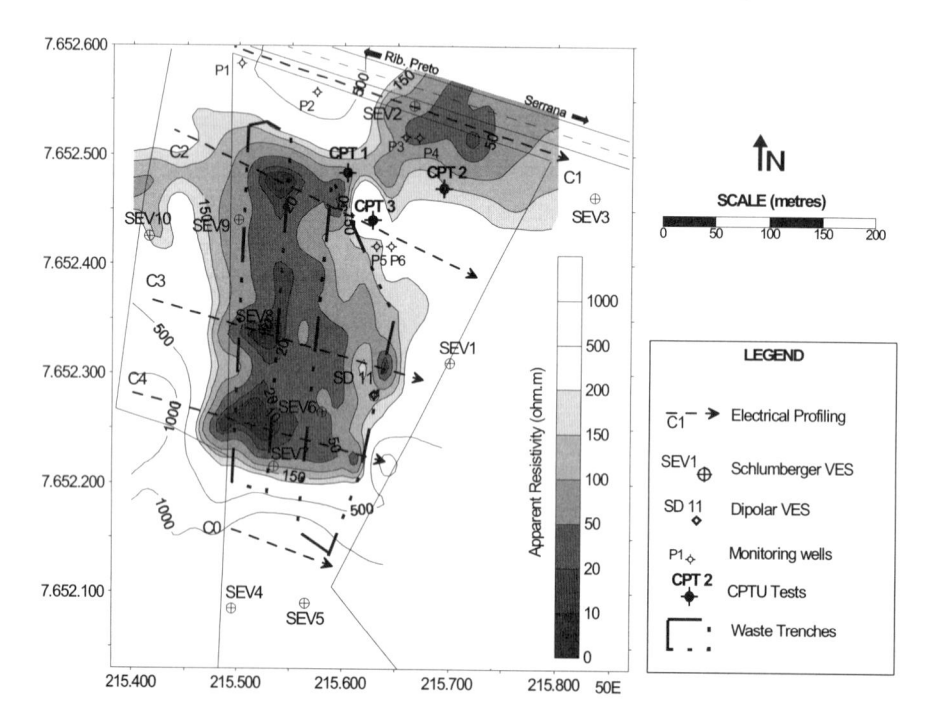

Figure 3: Apparent resistivity map at 30 m depth in the Ribeirão Preto Site. Low values (< 200 ohm.m) indicate the contaminated zone caused by leaching from the waste disposal is reaching the saturated zone.

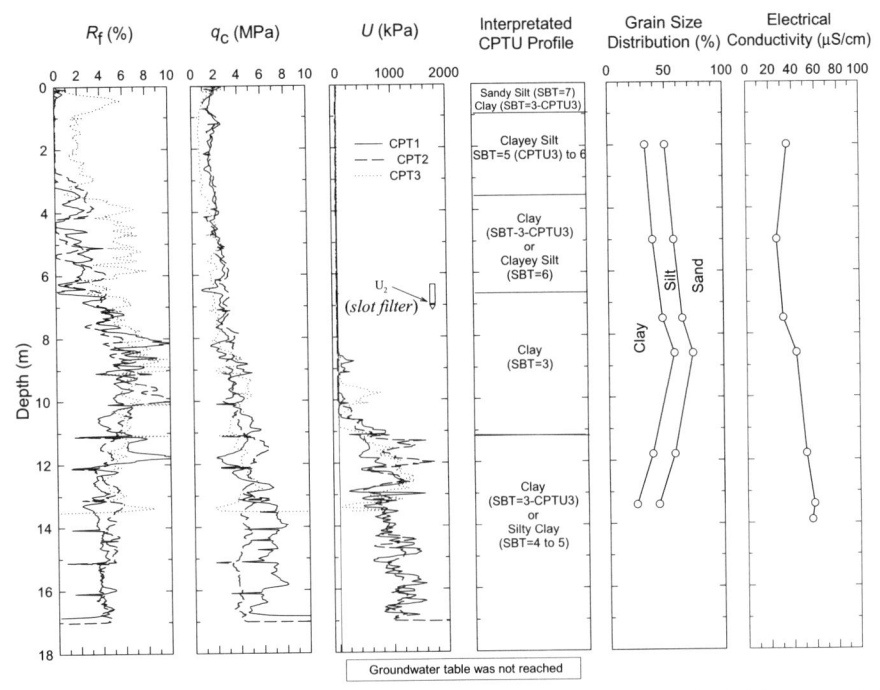

Figure 4: CPTU testing results, grain size distribution and electrical conductivity of the water leached through sampled soils from the Ribeirão Preto Site.

4 BAURU SITE

4.1 Site Description

This site is a sanitary landfill located in Bauru, São Paulo State, Brazil (Figure 1). Bauru is a city in central west São Paulo State and it has about 350,000 inhabitants. The sanitary landfill has an estimated volume equal to 958,000 m^3 and has been used for 10 years. The geology of the site is Sandstone from the Adamantina Formation, covered by alluvium sandy soils or colluvium clayey sands. Beneath these layers, there are residual soils from Sandstone. The depth of the groundwater level is always at least 5 m deeper than the landfill base.

The traditional site characterization program carried out at this site during the feasibility and design investigations included geological mapping, prepared based on aerial photos and fieldwork, to identify and assess the soil and rocks. Laboratory tests to determine grain size distribution and Atterberg limits were carried out on soil samples retrieved during Standard Penetration Tests (SPT). Laboratory hydraulic conductivity tests using the falling head procedure were carried out on undisturbed soil samples retrieved from this site. Hydraulic conductivity varied between 10^{-6} to 10^{-7} m/s.

4.2 Testing results

4.2.1 Geoelectrical soundings

Seven vertical electrical soundings (SEV1 to SEV7) with Schlumberger array at a maximum AB distance equal to 160 m were carried out close to the landfill area, but they did not cross it. Results can be used to identify groundwater level, layering of different materials, as well as to define a groundwater flow map. Two vertical electrical soundings (SEV8 and SEV9) were also executed over the dumpsite with the same array. Four dipole-dipole electrical profiles (C1, C2, C3 and C4) were carried out on the studied site. Two lines, C2 and C3, crossed the landfill: C1 is upstream and C4 is downstream from the sanitary landfill. Line C4 was carried out to identify anomalous zones, to characterize a probable contamination plume migration in the saturated zone. Dipole-dipole electrical profiling was also interpreted with the RES2Dinv (Loke, 1998) software. Figure 5 shows a resistivity and chargeability cross sections for line C4, located downstream from the sanitary landfill. The apparent resistivity map was elaborated based on all four dipole-dipole electrical profiles. The map for a 15 m theoretical depth is presented in Figure 6.

4.2.2 CPTU and RCPTU testing

Preliminary piezocone (CPTU) and resistivity piezocone (RCPTU) testing were carried out at selected locations (Figure 6), based on the interpretation of a conventional geoelectrical survey. CPT1 was carried out upstream from the landfill to be used as background testing but the groundwater level was deeper than the layer impenetrable to the cone, which is a similar problem to what occurred at the Ribeirão Preto site. RCPT14 and 15 were carried out to cross a probable contaminated plume. Two groups of piezocone tests were carried out close to temporary monitoring wells, located based on the low resistivity values of the saturated zone at 230 – 250 m (Figure 5). For RCPTU 5 pore-pressure was measured using conventional procedures to saturate the porous element with glycerin. The slot-filter filled with grease was used in the other tests. CPTU 4 and RCPTU 1, 2, 3 and 9 testing results are presented in Figure 7.

4.3 Discussion

4.3.1 Geoelectrical soundings

Vertical electrical profiling can be divided into two groups: soundings carried out over the dumpsite and soundings executed outside the dumpsite. The group of vertical electrical soundings carried out over the sanitary landfill permits identification of the dump thickness, where resistivity values are very low (2.4 to 11 ohm.m). The base of the trench can be identified by very high resistivity values. Resistivity values varied between 367 a 937 ohm.m for the soundings executed outside the dump. These results allowed characterization of important aspects of the soil mass as well as to estimate groundwater level. Resistivity values varied between 20 to 88 ohm.m below the groundwater level. Plotting the depth of groundwater for each vertical electrical sounding it is possible to generate a groundwater flow map, which shows that flow direction is Northwest (NW).

Figure 5 shows resistivity and chargeability cross sections for the line C4, downstream from the landfill. It can be used to identify saturated and unsaturated zones. Unsaturated soils have resistivity values higher than 150 ohm.m, associated to chargeability ranging from intermediate to low values. The saturated zone is characterized by the trend of lower resistivity with depth. It is located about 10 m deep in the borders and at about 5 m deep in the middle, because of local topography. The cross section shows low resistivity values on the saturated zone at the 120 – 145 m and 230 – 260 m locations, indicating contaminated regions. IP values show a vertical anomaly with high chargeability between 120 – 140 m, which suggests the presence of a discontinuity, conditioning a contamination flow (Figure 5).

The map for theoretical depth equal to 15 m presented in Figure 6 permits the characterization of soils below the water table. Based on this figure, it is also possible to identify the preferential direction of local flow, which is WNW (zones with resistivity values lower than 75 ohm.m).

4.3.2 CPTU and RCPTU testing

CPTU and RCPTU tests presented in Figure 7 were carried out to cross a probable contaminated zone, identified based on dipole-dipole soil profiling as well as to assess soil variability. As can be seen in this figure, point resistance (q_c), friction ratio (R_f), pore-pressure (U) and bulk resistivity (R) values are quite repeatable. Changes in q_c and R_f are related to different soil behaviors. The Robertson et al. (1986) classification chart was used to help identify the distinct soil behavior and to govern the depths from which to recover soil samples. It is interesting to emphasize that CPTU and RCPTU are different probes and the tests were carried out in different seasons. As several researchers have stated, CPTU is an excellent tool for soil profiling. RCPTU testing results presented in Figure 7 show that the soil profile is quite variable and SPT tests carried out at the site did not allow observing in detail the intercalated layers. Groundwater level was identified when bulk resistivity (R) suddenly dropped, at about 5 m depth. This information was confirmed with data from a temporary monitoring well installed close to the piezocone tests. Figure 7 shows that bulk resistivity (R) varies with soil texture and mineralogy. The interpretation of RCPTU tests to assess contamination in tropical soils is an ongoing research and it is not as straightforward as in sedimentary sands. More soil and water sampling is necessary to support interpretation. RCPT14 and 15 were carried out closer to the landfill, right over the contamination plume. The interpretation of these tests defined the depth to retrieve water with a sampler from direct push technology to assess contamination and it worked quite well for this purpose.

5 CONCLUSIONS

The case histories showed that geoelectrical survey gives high quality information to assess the groundwater contamination caused by a waste disposal. Both physical properties of the ground (water table, groundwater flow and soil layer thickness) and from the waste and the interference with the surrounding media (detection and delimitation of plume and the degree of contamination) can be assessed using this technique. It is also an important tool to determine the location of specific tests to complement a geoenvironmental site characterization as well as to define the best position for the monitoring system.

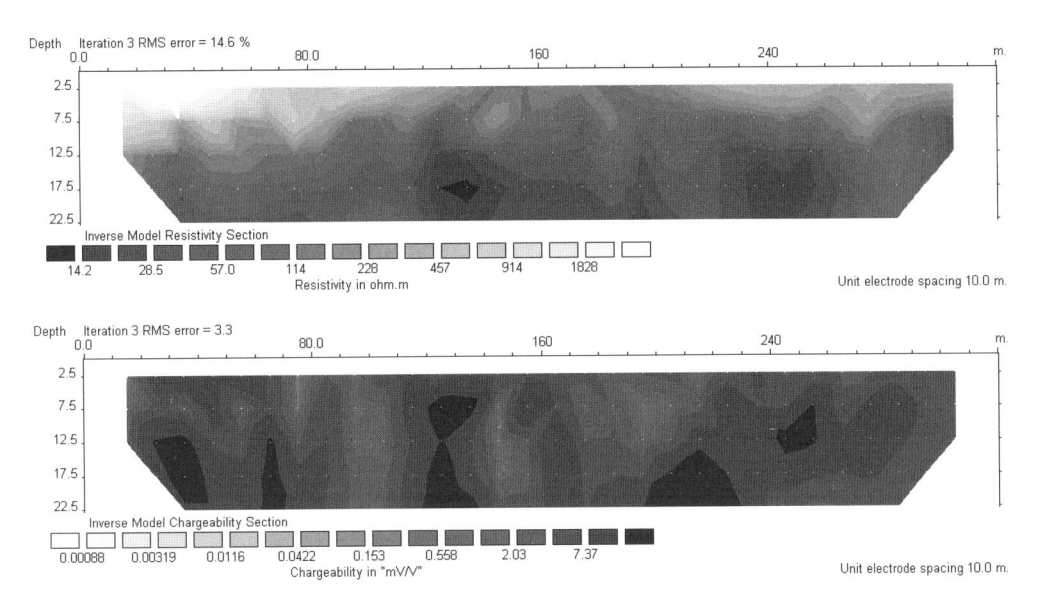

Figure 5: Dipole-dipole soil profiling for line C4, downstream from the Bauru Sanitary Landfill.

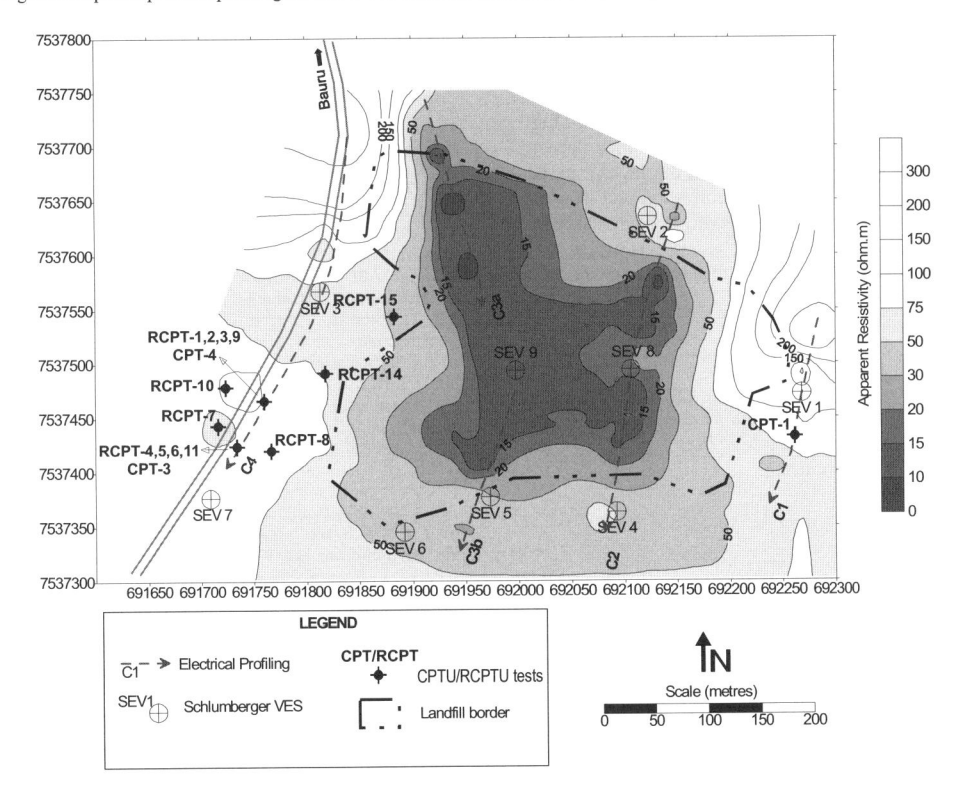

Figure 6: Apparent resistivity map at 15 m depth in the Bauru Site. Low values (< 75 ohm.m) indicate the contaminated zone caused by leaching from the waste disposal is reaching the saturated zone.

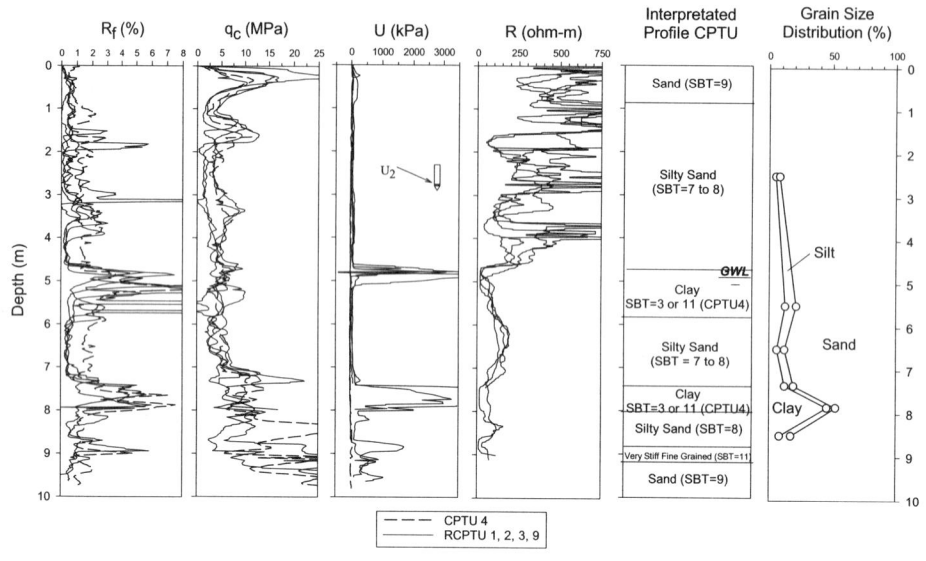

Figure 7: CPTU and RCPTU testing results and grain size distribution from the Bauru Site.

Piezocone technology has been increasingly used for a more efficient characterization of a geoenvironmental site. It reduces the contact between field personnel and contaminants and essentially produces no cuttings and little disturbance. This technology has some limitations for tropical soils, since the water table sometimes is deeper than the layer impenetrable to the cone. Another important aspect in tropical soils is that genesis affects soil behavior and soil sampling is required. The use of a soil sampler from the direct-push technology is an interesting option to support interpretation.

ACKNOWLEDGMENTS

The authors acknowledge the *Fundação de Amparo à Pesquisa do Estado de São Paulo* (FAPESP) for funding their researches.

REFERENCES

Campanella, R.G., Davies, M.P., Boyd, T.J. & Everard, J.L. (1994): "In-Situ Testing Methods for Groundwater Contamination Studies". *Symposium on Developments in Geotechnical Engineering, From Harvard to New Delhi, 1936-1994*, Balkema.

Campanella, R.G., Davies, M.P., Kristiansen, H. & Daniel, C. (1998): "Site characterization of soil deposits using recent advances in piezocone technology". *Proc. of the 1st Int. Conference on Site Characterization – ISC'98*, Atlanta, Georgia, Balkema, Roterdam, pp. 995-1000.

Davies, M.P. & Campanella, R.G. (1995): "Piezocone technology: Downhole geophysics for the geo-environmental characterization of soil". *SAGEEP 95*, Florida.

Elis, V.R. (1998): "Assessing the applicability of Geo-electrical methods for geophysical investigation of waste disposal sites". *Doctoral Thesis (in Portuguese)*, IGCE - Unesp, Rio Claro/SP/Brazil.

Giacheti, H.L.; Marques, M.E.M., & Peixoto, A.S.P. (2003): "Cone Penetration Testing on Brazilian Tropical Soils". *Soil Rock America 2003*, Cambridge, USA, V.1, pp. 397-402

Larsson, R. (1995): "The use of a thin slot as filter in piezocone tests". *International Symposium o Cone Penetration Tests*, CPT' 95, Sweden, V. 2, pp. 35-40.

Loke, M.H. (1998): "*RES2DINV ver. 3.3. for Windows 3.1 and 95– Rapid 2D resistivity and IP inversion using the least-squares method*". Penang: M. H. User's Manual, 35 p.

Lunne, T.; Robertson, P.K. & Powell, J.J.M. (1997): "*Cone penetration testing in geotechnical practice*". Blackie Academic & Professional, 311 p.

Pejon, O.S. & Zuquette, L.V. (1991): "Importance of geological-geotechnical studies for urban waste disposal". *II Symposium of Tailings Dams and Urban Waste Disposal*, (in portuguese), Rio de Janeiro/Brazil, pp. 367-377

Robertson, P.K.; Campanella, R.G; Gillespie, D. & Greig, J. (1986): "Use of piezometer cone data". *Proc. In-Situ-86*, ASCE Specialty Conference, pp. 1263-1280.

Shinn II, J.D. & Bratton, W.L. (1995): "Innovations with CPT for environmental site characterization". *Proceedings of CPT'95*, V. 2, pp. 93-98.

US EPA (1989): "Seminar on Site Characterization for Subsurface Remediation". *United States Environmental Protection Agency*, Technology Transfer, Report CERI-89-224, 350 p.

Wavelets in surface wave cavity detection – Theoretical background and implementation

N. Gucunski, P. Shokouhi & A. Maher
Dept. of Civil Engineering, Rutgers University, NJ, USA

Keywords: nondestructive testing, seismic techniques, SASW, wavelet transform, object detection, finite elements

ABSTRACT: A new approach based on representation of the surface response of a medium, in the time-frequency plane, is investigated in this study. Surface wave propagation in the soil medium is simulated by a transient response analysis on an axisymmetric finite element model. Cavities in a homogeneous half-space, as well as a pavement system with a variety of shapes and embedment depths, were considered. The continuous wavelet transform (CWT) has been introduced as a new analysis tool for extracting time and frequency information of nonstationary signals with good time and frequency resolutions. Different families of wavelets were examined for the analysis, and ultimately, Gaussian wavelets were chosen. It has been demonstrated that the surface response is of transient nonstationary nature with local time and frequency features. Using wavelet transform, the changing spectral composition of the surface transient response is presented in the time-frequency plane. Effects of different types of cavities on wavelet time-frequency maps (wavelet transform coefficients magnitude versus time and frequency for the response at different receiver locations) were analyzed. Time and frequency signatures of waves reflected from near and far faces of the cavity can be clearly observed in wavelet time-frequency maps. The pattern of changes of time-frequency maps at different receiver locations was reviewed. Based on these observations, a scheme for location of the cavities, and their size and depth estimation, using wavelet transform is suggested The proposed scheme has been used successfully to detect some known cavities in a pavement system. In addition, the applicability of the proposed method may be evaluated by a series of tests in a sand bin.

1 INTRODUCTION

Seismic techniques are typically used to measure and characterize a medium in terms of its elastic properties, i.e. elastic moduli, stiffness, and damping characteristics. The techniques may also be utilized in anomaly detection, and are advantageous where there is a significant rigidity contrast between the sought object and the surrounding medium. Presence of a cavity in a medium introduces a discontinuity in rigidity, which makes it an ideal target for seismic methods. The methods are based on either travel time or spectral analysis of elastic waves generated by different sources. Refraction and reflection methods are the most widely used travel time-based seismic techniques. The former usually fails to detect shallow cavities in a stratified medium. The latter is quite successful in detection of deep cavities, however, it is difficult to be utilized in near-surface cavity detection applications (Belesky and Hardy, 1986, Cooper and Ballard, 1988).

Application of surface waves in shallow cavity detection became especially plausible with the development of the Spectral Analysis of Surface Waves (SASW) technique. While the SASW method has been primarily used in evaluation of elastic moduli and layer thicknesses of layered systems, such as pavements (Nazarian and Stokoe, 1986), it has demonstrated to be reliable in detection of cavities in several numerical and experimental studies (Al-Shayea et al., 1994; Gucunski et al, 1996; Ganji et al., 1997). However, it has been discovered that the success in anomaly detection in this method depends on the receiver spacing and location, shear wave velocity, and damping of the surrounding soil medium (Ganji et al., 1997). When a shallow cavity is present due to damping, the reflections from cavity faces are of a limited duration. It has been also found that (Shokouhi et al., 2003) they have certain frequency bandwidths. In other words, presence of a cavity influences the frequency content of a limited portion of the response. This fact suggests that a time-frequency analysis of the response could be more efficient than analyzing it in merely one domain.

Wavelet transforms are fairly recent mathematical tools, which can be used to extract both local and global time and frequency information of a signal efficiently. Wavelet transforms were used herein to study simultaneous spatial and spectral changes in the near-surface wave pattern due to a presence of a cavity. Numerical simulations were utilized to simulate the seismic test performed on a soil medium and a pavement system. Using a wavelet transform, the theoretical response of a medium is transformed into a time-frequency presentation. Evaluating the features of this time-frequency presentation of the response is the basis for the proposed method for cavity detection and characterization.

The paper has two major parts. The first part focuses on the fundamentals of the theory of continuous wavelet transform (CWT) and the results of a wavelet analysis performed on the numerical results. The second part reflects the experimental implementation of the method in both laboratory and field conditions.

2 TIME-FREQUENCY ANALYSIS AND CONTINUOUS WAVELET TRANSFORMS (CWT)

The deficiency of the Fourier transform in analyzing nonstationary transient signals led to a definition of windowed Fourier transforms. Gabor (1946) introduced a short time window for extracting local information from the Fourier transform of a signal. In this case, spectral coefficients are computed just for this short length of the signal. The window is then moved to a new position and the calculation is repeated. This provides a local representation of the Fourier transform with a time resolution equal to the size of the window. One feature of the Gabor transform is that the width of the window function is constant. To have an adequate resolution, the windowed Fourier transform must be applied repeatedly with a varying window width each time. Wavelet transforms overcome this problem by relying on localized bases with varying bandwidths, in both time, and frequency domains.

Wavelet transforms are a new class of transformations, which can be used to present the changing spectrum of a nonstationary signal in a time-frequency plane. A 'mother wavelet' $\psi(\tau)$ is a well-localized function in both time and frequency domains. A wavelet $\psi_{a,t}(\tau)$, at scale a and time t, is obtained through scaling and translation of the mother wavelet function according to the following equation:

$$\psi_{a,t}(\tau) = \frac{1}{\sqrt{a}} \psi\left(\frac{t-\tau}{a}\right) \qquad (1)$$

At finer scales wavelets are shorter in the time domain, but have a broader frequency bandwidth, and vice versa. This fact implies a reciprocal relationship between the scale and the frequency. A wavelet transform is a convolution between a signal $x(\tau)$ and a set of wavelets over a finite range of scales. The continuous wavelet transform (CWT) is defined as,

$$W_\psi x_{a,t} = \int_{-\infty}^{\infty} x(\tau) \psi_{a,t}^*(\tau) d\tau \qquad (2)$$

where $W_\psi x_{a,t}$ is the wavelet coefficient at scale a and time t using wavelet function $\psi(\tau)$, and * indicates the complex conjugate. If a signal correlates well with the analyzing wavelet, the wavelet coefficient will be large; otherwise, it will be small. Plotting the wavelet coefficients for a range of scales (scale is a reciprocal function of frequency) and time produces a time-frequency representation of the signal.

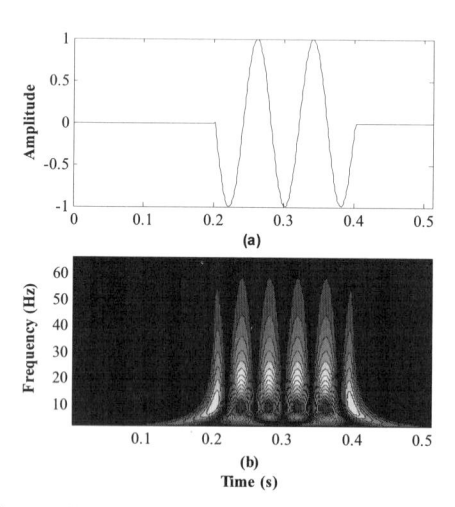

Figure 1. Illustration of continuous wavelet transform (a) a synthetic sinusoidal signal, (b) The corresponding Gaussian wavelet time-frequency map

The only constraint imposed on an integrable function wavelet is having a zero mean in the time domain, or a zero DC offset in the frequency domain. Therefore, many different wavelet functions have been developed for different analysis purposes. Among all types of wavelets examined, Gaussian real wavelets have proven to be the most appropriate wavelets for this study.

Wavelet transforms, by virtue of their varying bandwidth basis (wavelets), can capture local and global time and frequency characteristics of a signal very efficiently. It is useful to illustrate characteristics of the CWT through an analysis of a signal of known time and frequency properties. A synthetic signal is obtained from 512 evenly spaced samples of the function $\sin(25\pi t)$. The sine function is windowed using a unit square window of width 0.2s extended from 0.2s to 0.4s, and is shown in Fig. 1(a). The signal is analyzed using Gaussian wavelets, and the resulting time-frequency map is presented in Fig. 1(b). As can be clearly observed, the CWT map provides not only the fundamental frequency of 12.5 Hz, but also the time extent of the occurrence of this frequency component.

3 NUMERICAL SIMULATION OF WAVE FIELD BY FINITE ELEMENTS

3.1 Finite Element Model

A surface seismic test is simulated by a transient analysis of the response of a finite element model. Assuming that the wave propagation occurs in the vertical plane only, the soil medium is modeled as an axisymmetric model, with an impact source at its origin. Certain criteria are imposed to ensure that the finite element analysis accurately simulates the surface wave propagation (Ganji et al.,1997; Zerwer et al.,2002; and, Shokouhi et al.,2003).

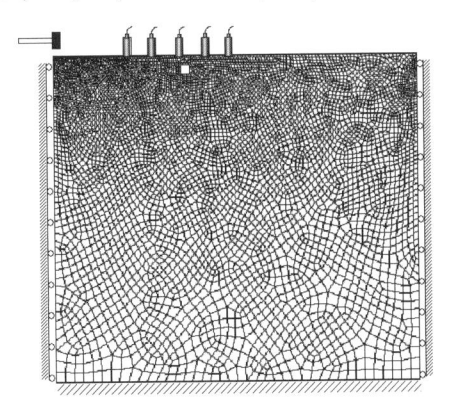

Figure 2. Finite element mesh.

The soil medium is described as a 40-m wide and 32-m deep model discretized by axisymmetric quadratic elements. The smallest element size is 0.25 m. The mesh is very fine between the source and the first receiver location. The element size gradually increases towards the boundaries. The soil is assumed to be linearly elastic with a shear wave velocity of 50 m/s. The impact is described as an impulse of a trapezoidal shape of 1-KN amplitude and 6-ms duration. A time step of 1 ms is chosen for the analysis. The finite element mesh is shown in Fig. 2.

3.2 Parametric study

To investigate changes in the time-frequency characteristics of the surface response caused by the presence of a cavity, the theoretical response of the medium at different surface points was transformed using Gaussian wavelets. Effects of the width, height, and embedment depth of the cavity were evaluated by considering cavities of different shapes and depth in a half-space and a pavement system (Fig. 3).

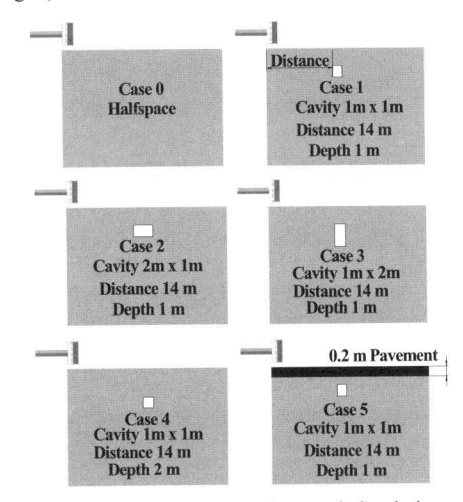

Figure 3. Cavity cases considered in the numerical analysis

The wavelet time-frequency maps of the response at a single receiver location (distance 11.7m from the source) for all of the cases are presented in Fig. 4. Peaks I, II and, III in Fig. 4(a) correspond to the arrivals of different components of incident waves, namely, compression, shear, and surface waves. Comparing Figs. 4(b)-4(f) to Fig.4(a) demonstrates the influence of the presence of a cavity on the response. Although the peaks of incident wave components (I, II, and III) are still present in these figures, a new sets of peaks (IV, and V) appear in Figs. 4(b) to 4(f). Peak IV in all these three figures marks the arrival of reflected waves from the boundaries of each of the cavities.

3.2.1 Effect of Cavity Width

Since the near face of the cavity in case 2 is at the same distance of 14m, peak IV starts at about the same time in this figure. However, the time bandwidth of the reflection peak is increased. The wider the cavity, the larger the time delay between the arrivals of reflected waves from near and far sides of the cavity. When the cavity is wide enough, the reflection peak will have two separate "hats"; each marking the arrival of the wave packet from one of the faces of the cavity. The two separate hats can be noticed for case 2 in Fig. 4(c). On the other hand, as the width of the cavity increases, the frequency of the reflection peaks decreases. Therefore, wider cavities cause two well-separated reflection hats of a lower frequency. Knowing the surface wave velocity of the medium, the time difference between these two hats (or equally the time bandwidth of the re-reflection peak) can be easily used to estimate the width of the cavity. Peak V signifies the end of the transient response, and does not have any physical significance.

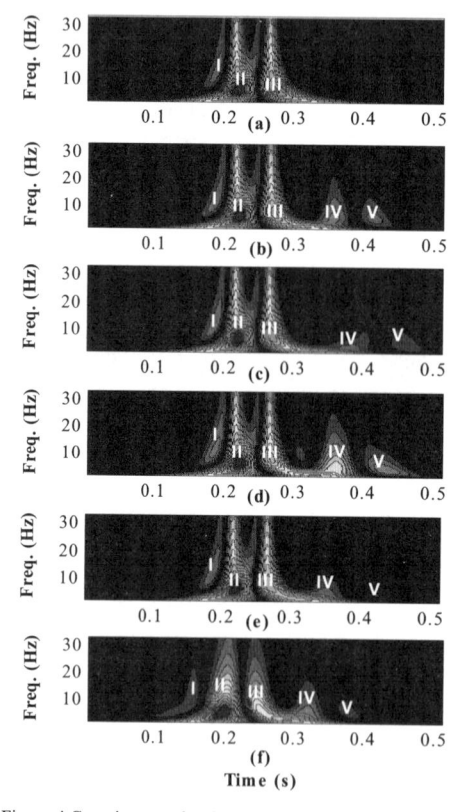

Figure 4. Gaussian wavelet time frequency map at 11.7 m for (a) case 0, (b) case 1, (c) case 2, (d) case 3, (e) case 4 and (f) case 5

3.2.2 *Effect of Cavity Height*

To study the effects of height of a cavity on wavelet time-frequency maps, the surface response at the receiver location 11.7 m for Case 3 is analyzed and presented in Fig. 4(d). Other than the three incident wave peaks (I, II and, III), two other peaks (IV, and V) are also illustrated in this figure. Peak IV corresponds to the arrival of reflected waves from the cavity faces. Given that the width of the cavity is the same as that of Case 1, the waves reflected from the near and far faces of the cavity are not well-separated, but rather form a single peak. However, the reflection peak in this figure is of a considerably higher energy. It can be concluded that increasing the height of the cavity results in reflection peaks of a broad frequency bandwidth and a higher energy level. However, the fundamental frequency of the reflections does not display a noticeable change.

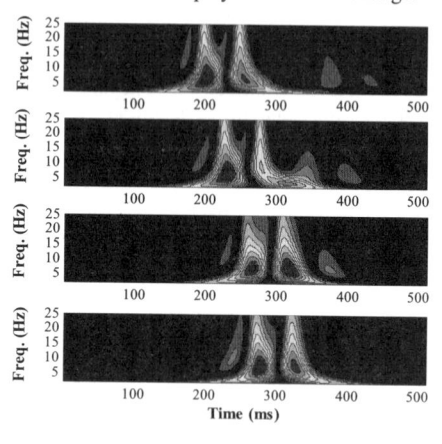

Figure 5. Gaussian wavelet transform for case 1 at receiver locations 10.6, 12.3, 13 and 14.8 m (top to bottom).

3.2.3 *Effect of Embedment Depth of the Cavity*

To study the effect of the embedment depth of the cavity on the wavelet transform maps, the response for Case 4, at distance of 11.7 m, is analyzed and presented in Fig. 4(e). As expected, the reflection peak for a deeper cavity is of a much lower energy level. Consequently, it was found that a cavity of an embedment depth larger than its height is hard to detect using this method.

3.2.4 *Effect of a Paving Layer*

In Case 5, the cavity is present under a 20 cm thick paving layer of a shear wave velocity of 250 m/s (five times the shear wave velocity of the soil). The response at a distance of 11.7 m is analyzed and presented in Fig. 4(f). Despite the high velocity layer on the top, the reflections from the cavity boundaries (peak IV) can still be clearly observed. The other characteristics of the reflection peak, such as the time width and peak frequency, are the same as those for Case 1. It can be concluded that shallow cavities in a pavement system (e.g. sinkholes and conduits) can be easily detected using this particular method.

3.2.5 *Effect of Receiver Location*

The results presented in Fig. 4 reflect the analysis of the surface response for different cases for a single receiver location (11.7m). Wavelet time-frequency maps for Case 1, and at different receiver locations in front of the cavity (10.6m, 12.3m, 13m and, 14.8m), are presented in Fig. 5. As can be observed, reflections from the cavity boundaries can be simply identified for the receiver locations in front of the cavity. Conversely, there is no noticeable signature of the cavity in the wavelet maps at the receiver location 14.8 m. Therefore, having the surface response at different points, the cavity can be located as being near receiver locations where the reflection peaks vanish. Afterwards, its size and shape can be approximated by studying the time and frequency characteristics of the reflection peak according to the discussion above.

4 CAVITY DETECTION SCHEME

Based on the wavelet analysis results, the following scheme for cavity detection is proposed:
1. Obtain and analyze the surface response for different radial directions, and a number of receiver positions along each direction, and present the results in the form of time-frequency maps.
2. Compare the pattern of changes in the time-frequency maps for each direction to those shown in Fig. 5. Choose the direction for which the changes in the maps resemble most in the presented pattern. If necessary, obtain the surface response in the selected direction at smaller receiver distances.
3. Approximate the location of the cavity based on the pattern of changes of time-frequency maps. The cavity is located near the receiver position where the reflection peaks vanish.
4. Calculate the distance of the near side of the cavity from the time position of the reflection peak in each time-frequency map (for each receiver position). The accuracy will increase if calculated for different locations, and then averaged.
5. Calculate the width of the cavity from the time bandwidth of the reflection peak. The accuracy will increase by averaging the time bandwidth for different receiver locations.
6. Determine the shape of the cavity from the time and frequency characteristics of the reflection peak. If the cavity is wide enough, the reflection peak will be recognized by two distinguished hats, a low peak frequency, and a narrow frequency bandwidth. The vertical elongation results in peaks of significantly higher energy levels and broader frequency ranges.

Cavities with an embedment depth of more than two times their size will be difficult to detect. In general, increasing the depth will result in reflection peaks of lower energy and narrower frequency bandwidths.

5 EXPERIMENTAL IMPLEMENTATION

5.1 *Detection of a pipe under a pavement*

A preliminary test setup was designed to evaluate the applicability of this analysis procedure in the field. An underground pipe with a 40 cm diameter, buried about 15 cm below the pavement surface perpendicular to the travel direction, was selected as the detection target. Data collection was conducted using a Seismic Pavement Analyzer (SPA) trailer (Nazarain *et al.* 1993). SPA is a device designed for evaluation of pavements and bridges using seismic techniques. This device is capable of performing an array of seismic tests, such as Impact Echo (IE), Impulse Response (IR), Ultrasonic Surface Waves (USW), and Spectral Analysis of Surface Waves (SASW). The SPA trailer collects time histories at different offsets from the source, which is either a low frequency or high frequency hammer. These time histories, along with the described analysis procedure using wavelet transform, were used in detection of the underground obstacle (i.e. pipe).

Figure 6. SPA trailer during testing (The exposed end of the pipe is marked)

The CWT analysis of the response at 0.3m, 0.6m and, 1.2m from the impact source is illustrated in Figure 7 (The SPA trailer during the testing of the selected pipe is shown in Figure 6). In this particular case, the buried pipe was located between 0.5 and 0.9m from the source. Since the receiver at 0.3m is relatively far from the pipe, and very close to the source, only the incident waves were observed at this location. However, the reflections from the pipe are clearly visible in the CWT of the receiver at 0.6m.
The position and characterization of the reflection peak is in conjunction with the results of the numerical analysis. The receiver at 1.2m is located after the pipe. It is believed that due to the shallow depth of the pipe, most of the energy of the incident wave is reflected. Therefore, this receiver has a much quieter CWT. This preliminary testing confirms the numeri

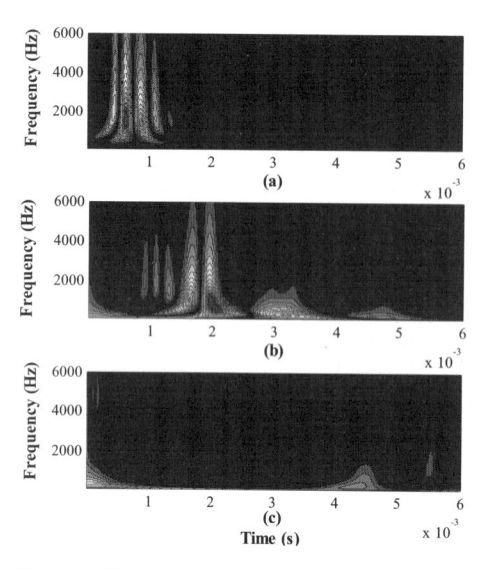

Figure 7. CWT maps for the response at (a) 0.3m, (b) 0.6m and, (c) 1.2m from the source.

cal simulation results. Coincidentally, a much more comprehensive laboratory testing is being carried out to quantify the advantages and limitations of this method in practice.

5.2 Laboratory detection of small cavities

For further evaluation of the potential of using the proposed method in practice, comprehensive laboratory testing is being conducted in a sand bin at Rutgers University. The bin, 4.2 m in diameter, and 1.8 m deep, is filled with fine to medium sand (Figure 8). There is a 30 cm thick, energy absorbing sawdust layer along the bottom and walls of the bin. The testing is being conducted using a Portable Seismic Soil Analyzer (PSSA), which is a portable seismic device. The device consists of an anvil type hammer and two accelerometers. The accelerometers are pneumatically coupled to the ground. The device has established that the source is near a receiver distance of 15 cm and a receiver spacing of 20 cm. PSSA is fully computer controlled. The device and the test setup are shown in Figures 9(a) and 9(b).

Detection of two obstacles is investigated in this set of experiments: a concrete obstacle and a cavity built out of Styrofoam panels. In both cases, the center of the buried object matches the center of the sand bin. The testing is conducted along six radial lines, 30 degrees apart (marked with yellow and pink lines in Figure 9). Testing along each line is conducted in 10 cm increments. While the laboratory testing is still in progress, preliminary results are promising with respect to verification of the results obtained from the numerical simulations.

Figure 8. Sand bin at Rutgers University

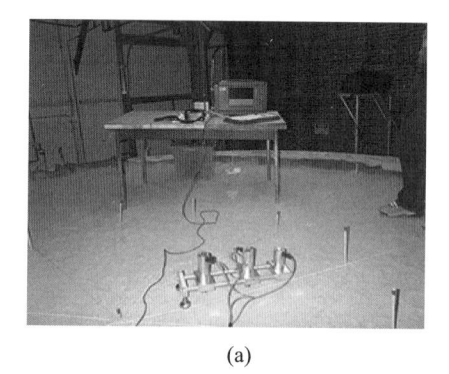

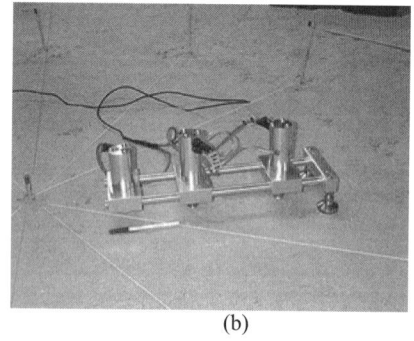

Figure 9. (a) Laboratory test setup in the sand bin, (b) PSSA.

6 CONCLUSIONS

Application of the continuous wavelet transform (CWT) for detection and characterization of underground shallows was investigated. Results of the numerical studies conducted were utilized to develop a proposed cavity detection scheme. This method was implemented in the field and demonstrated potential for detection of cavities and other anomalies in layered systems, such as soils and pavements.

REFERENCES

Al-Shayea, N.A., Woods R.D. & Gilmore, P. 1994. SASW and GPR to Detect Buried Objects. Proceedings Symposium on the Application of Geophysics to Engineering and Environmental Problems (SAGEEP). Wheat Ridge, CO, USA, 1994: 543-560.

Belesky, R.M. & Hardy H.R. 1986. Seismic and Microseismic Methods for Cavity Detection and Stability Monitoring of Near-Surface Voids. *Proceedings 27th US Symposium on Rock Mechanics, Rock Mechanics: Key to Energy Production, Univ. of Alabama, Tuscaloosa, AL, USA, 1986*: 248-258.

Cooper, S.S. & Ballard R.F. 1988. Geophysical Exploration for Cavity Detection in Karst Terrain. *Proceedings Symposium Geotechnical Aspects of Karst Terrains: Exploration, Foundation Des. And Pref., and Remedial Measures, Geotech. Spec., ASCE, New York, NY,USA*: Publication No. 14: 25-39.

Ganji, V., Gucunski, N. & Maher A. 1997. Detection of Underground Obstacles by SASW Method: Numerical Aspects. *Journal of Geotechnical and Geoenvironmental Engineering*, Vol. 123 (No. 3): 212-219.

Gucunski, N.,V. and M. Maher [1996]. "Effects of Obstacles on Rayleigh Wave Dispersion Obtained from the SASW Test", *Soil Dynamics and Earthquake Engrg.*, Vol. 15, No. 4, pp. 223-231.

Kaiser, G. 1994. *A Friendly Guide to Wavelets.* Birkhauser, Boston.

Nazarian, S. & Stoke II, K.H. 1986. Use of Surface Waves in Pavement Evaluation. *Transportation Research Record 1070*: 132-144.

Nazarian; S., Baker, M.R. & Crain, K. 1993. Development and Testing of a Seismic Pavement Analyzer, *Report SHRP-H-375, Strategic Highway Research Program, National research Council, Washington, D.C.,USA.*

Shokouhi, P. & Gucunski, N. 2003. Application of wavelet transform in detection of shallow cavities by surface waves. *Symposium on the Application of Geophysics to Engineering and Environmental Problems (SAGEEP). San Antonio, TX, USA.*

Zerwer, A., Cascante G. & Hutchinson, J. 2002. Parameter Estimation in Finite Element Simulation of Rayleigh Waves. *Journal of Geotechnical & Environmental Engineering* Vol. 128 (No. 3): 250-261.

Walker, J.S. 1999. *A primer on wavelets and their scientific applications*, CRC Press.

Investigation of bridge foundation sites in karst terrane via multi-electrode electrical resistivity

Dennis R. Hiltunen
The Pennsylvania State University, University Park, Pennsylvania, USA

Mary J.S. Roth
Lafayette College, Easton, Pennsylvania, USA

Keywords: bridge foundations, geophysics, karst terrane, electrical resistivity

ABSTRACT: Resistivity data was collected on several bridge foundation sites in Pennsylvania, in close proximity to geotechnical boring locations. Earth resistivity tomograms determined from multi-electrode measurements were plotted against drilling data and notable similarities were discovered. Resistivity profiles were found to match layering structure of drilling data in many cases, and resistivity values were indicative of materials observed in boring logs. In addition, resistivities could adequately predict transitions in material and top of rock profile. It appears that multi-electrode electrical resistivity is a useful tool for developing basic layer structure in karst terrane.

1 INTRODUCTION

Personnel of the Pennsylvania Department of Transportation (PennDOT) have described current site investigation practices for bridge foundations as inadequate, particularly in karst terrane where rock conditions are often highly variable. Bridges in Pennsylvania are largely founded on piles or other deep foundations. Highly variable subsurface conditions result in great uncertainty in deep foundation design and construction. PennDOT would greatly benefit from a better definition of subsurface conditions, including top of rock profile and quality of rock.

Typical PennDOT practice for bridges is to conduct two to three geotechnical borings per substructure unit in order to select and design a foundation. Both shallow and deep foundations are employed, and the decision is based primarily on the depth and quality of bedrock encountered in the borings. Shallow foundations are typically used where bedrock is shallow and of sufficient quality, while deep foundations are used for the remainder (and majority) of cases. For deep foundations, steel H-piles driven to refusal are typical. The design load capacity of these end-bearing piles is typically based upon an allowable stress in steel (6 to 9 ksi).

Due to uncertainty in subsurface conditions, experience with this process has led PennDOT to also conduct exploratory drilling at selected sites as part of the construction contract. Based upon judgments gathered from the original design borings, these exploratory borings are conducted on a small grid pattern over the area of a selected substructure unit, and they employ either traditional geotechnical boring equipment or air rotary drilling techniques. The intent is to reveal a more detailed subsurface model for the location prior to foundation construction, and the information is often able to provide this more detailed model. Because of significant lateral variability in karst terrane, this detailed model can be significantly different from that revealed by the two or three borings conducted for the original design. In this case, significant foundation design changes may be required after the construction contract has been awarded and begun, leading to significant overall cost increases for PennDOT. Also, because contractors are aware of the large uncertainty, contingencies can also be incorporated within the original agreement. It would appear that a more detailed subsurface model would be beneficial at the design stage. In addition, it would be desirable that these improved models not require the expense of exploratory geotechnical borings conducted on a small grid pattern.

2 VISION

Engineering geophysics is a promising solution to this problem. Engineering geophysics uses methods adapted from seismological and petroleum industries for characterization of shallow subsurface ground conditions, and enables "seeing" between boreholes or from instruments along the ground surface. Methods are analogous to medical techniques such as an

X-ray or MRI. The geophysical method explored in this study is multi-electrode earth resistivity, a non-destructive, in situ test procedure used to determine the subsurface's electrical resistivity profile. Electrical resistivity is an important physical parameter that can be used as a tool to characterize material changes with depth.

The vision of the research is to develop a characterization technique that will establish a credible subsurface model for a site for the design stage of a project. The model should be of sufficient detail such that significant and costly design changes during the construction contract are minimized, and contingencies for uncertain subsurface conditions are reduced. The methodology should be efficient with respect to both money and time, and capable of implementation in a wide variety of terrane and other surface and site conditions. The hypothesis is that images of electrical resistivity produced from geophysical measurements at a site can be used in conjunction with traditional geotechnical borings to establish a credible model.

The purpose of this paper is to document a study where multi-electrode earth resistivity was employed to characterize the subsurface in mantled karst terrane. Resistivity data was collected on several bridge foundation sites in Pennsylvania, in close proximity to geotechnical boring locations. Resistivity profiles determined from multi-electrode measurements were plotted against drilling data and notable similarities were discovered.

3 ELECTRICAL RESISTIVITY

The electrical resistivity method has been used in site characterization for about a century. Most earth materials are either good insulators or dielectrics, i.e., they do not conduct electricity very well. Rather, electricity is conducted through the subsurface via interstitial water. Rock typically has a significantly higher resistivity than soil because it has a smaller primary porosity and fewer interconnected pore spaces. It is thus drier. Earth materials such as clay tend to hold more moisture and generally conduct electricity much better; their resistivity values are typically much lower than that of rock. Thus, resistivity methods typically work well in characterizing karst terrane because of the high contrast in resistivity values between carbonate rock and moist, clayey residual soil overlying it.

3.1 Concept

Numerous articles and textbooks are available describing the resistivity method and its applications. For example, Dunscomb and Rehwoldt (1999) and Roth and Nyquist (2003) provide discussion of historical perspective, method summary, and current capabilities. Resistivity measurements (figure 1) are made by introducing current into the subsurface through two current electrodes (C1 and C2), and measuring the voltage difference with two potential electrodes (P1 and P2). From the magnitude of the introduced current, measured voltage, and a factor that accounts for the geometric arrangement of the electrodes, a resistivity value is calculated. The calculated resistivity value is not a true assessment of subsurface resistivity, but is instead an apparent value. Apparent resistivity is defined as the resistivity value that would be obtained if the subsurface were homogeneous. To obtain more accurate estimates for resistivities of inhomogeneous subsurface materials, measured values are compared with values calculated from an assumed model of the subsurface. This is typically an iterative procedure using inversion techniques in which the subsurface model is modified until a reasonable match is obtained.

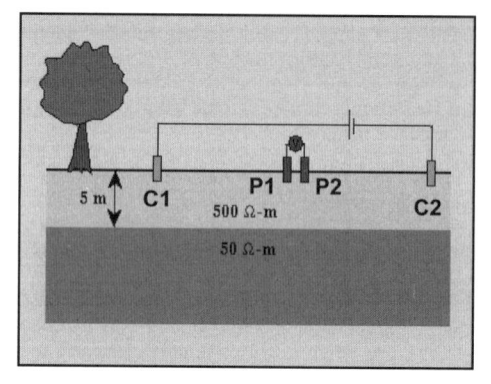

Figure 1. Four-Electrode Resistivity Test Schematic.

3.2 Traditional Four-Electrode Systems

A traditional four-electrode system consists of a power source, current meter, voltage meter, and four electrodes. Two survey methods are commonly used. First, in a sounding survey, the spacing between electrodes is increased between measurements, while the centerline of the electrode group remains fixed. As the electrodes are spread further apart, resistivites of deeper subsurface materials are obtained. The data from a sounding survey is typically interpreted by comparing the measured results to results calculated using a one-dimensional model of a layered subsurface system. The depth of investigation is governed by array type (the geometrical arrangement of the C1, C2, P1, and P2 electrodes), electrode spacing, and the specific subsurface materials present.

Second, in a profiling survey, the spacing between electrodes is fixed, and the electrode group is moved horizontally along a line between measurements. Resulting measurements can be used to locate variation in the subsurface along the measurement line. Interpretation of data obtained using this approach involves a simple plot of measurements as a function of distance along the line, followed by observation of variations of interest.

3.3 Multi-Electrode Systems

Recent development of multi-electrode earth resistivity testing has substantially improved investigation capabilities. Rather than moving equipment between data points, multi-electrode systems collect multiple data points with stationary equipment. These systems consist of multiple (usually 20 or more) electrodes (figure 2) connected to a switching device, a power source, a current meter, a voltage meter, and a data recorder (see figure 3 where these are combined within two units). Electrodes are spaced at equal distances along a survey line, and the switching device is used to automatically select combinations of four electrodes for each measurement. In the most recent devices now available, it is possible to apply current to two electrodes, and then simultaneously measure voltages across multiple pairs of electrodes, significantly reducing test time still further. The resulting data set consists of a combination of soundings and profiles, resulting in a two-dimensional survey of the subsurface materials. The depth of investigation using these methods is a function of line length, array type, and the subsurface materials present. Depths typically range from one-third to one-fifth of the length of the line.

A pseudo section is a simple plot of results from a multi-electrode test, where resistivity values are plotted at a horizontal location coinciding with the midpoint of the four electrodes responsible for the measurement, and at a depth proportional to the spread of the four electrodes. The process by which a geologic cross section of the subsurface is developed involves comparing pseudo sections of measured test data with pseudo sections of data calculated with an assumed model of the subsurface. This iterative "inversion" process continues until an acceptable match is found between the measured and theoretical pseudo sections. Once a match is found, the subsurface model producing the match is accepted as the best representation of actual subsurface conditions. This inversion process is automated via commercial computer programs, allowing for observation of approximate site conditions shortly following completion of test measurements.

4 PREVIOUS STUDIES IN KARST

Several studies reported in the literature have applied electrical resistivity techniques to site characterization of karst terrane. As shown in the following paragraphs, use of these techniques has proven quite satisfactory.

Figure 2. Resistivity Test Electrode

Figure 3. Resistivity Test System

Werner (1984) in a review of electrical resistivity methods for sinkhole prediction indicates that the problem of detection of sinkhole potential resolves to two aspects of local geology: the presence of a subterranean cavity and the presence of weakness, usually represented by fractures, in the overlying material. Neither condition is easily detected directly by the resistivity method unless the features are very large. However, increased weathering along zones of weakness and changes in ground-water drainage caused by cavities provide volumes of rock material with electrical resistivities different than that of the country rock, and these differences can be easily detected.

Lambert (1997) conducted a dipole-dipole survey in eastern Missouri to determine depth to top of bedrock and to locate voids within bedrock. He concludes that dipole-dipole surveying appears to be a viable tool for detecting and delineating karst features, particularly because of new equipment that allows for quick data collection and new modeling software that allows for more accurate interpretations. He cautions that the technique is a non-intrusive, indirect method of exploration, and should be used to supplement a subsurface exploration program that includes drilling or test pits.

Zhou, Beck, and Stephenson (1999) employed dipole-dipole electrical resistivity tomography during a site investigation on the Mitchell Plain of southern Indiana. They indicate that sinkholes are likely to form in depressions in the bedrock surface as a result of subterranean erosion of unconsolidated sediment by flowing groundwater. Electrical resistivity was used to identify such depressions, and they report that useful information was provided for identifying potential sinkhole collapse areas. They caution that their case study also demonstrates that data must be analyzed and interpreted cautiously, even with aid of boring data.

Dunscomb and Rehwoldt (1999) summarize the history and theory of electrical resistivity methods, including development of modern multi-electrode technology, and they present several case studies of two-dimensional resistivity profiling. They conclude that the method has very good resolving abilities in karst terrain to image geologic features such as pinnacled bedrock surfaces, overhanging rock ledges, fracture zones, and voids within the rock mass and within soil overburden. They caution that two-dimensional techniques only provide a slice of the subsurface, and karst features are generally highly variable in the third dimension. Thus, they recommend future experimentation with three-dimensional tomography. Also, although the technique has been shown to provide excellent information on subsurface features, there are resolution limits with respect to potentially significant but small karst features.

Roth, et al. (1999) present a case study of the reliability of multi-electrode resistivity testing for geotechnical investigations in karst terranes. In this case, comparisons between boring and resistivity results demonstrate that multi-electrode testing appears to be a valuable tool for geotechnical exploration in areas of karst overlain by clay soils. They suggest that by using intrusive methods to confirm interpretation of results, the test can predict depth to bedrock and determine trends in bedrock surface. However, they caution that reliability of the method is still in question with regard to locating and determining the size of possible voids. Three-dimensional variability and effects of line orientation both have significant influence on results and require further study. They suggest that use of three-dimensional resistivity tests may overcome some of these effects.

Kaufmann and Quinif (2001) used combined array two-dimensional electrical resistivity tomography to delineate cover-collapse sinkhole prone areas in southern Belgium. Simulations and field tests led to use of dipole-dipole and Wenner-Schlumberger arrays in a combined array inversion procedure to obtain a better image of the subsurface. A three-dimensional model of limestone bedrock was built using the two-dimensional results, and this model was compared to cone penetration (CPT) and boring results as well as locations and alignments of known cover-collapse sinkholes in and around the survey area. This comparison showed the validity of the proposed model and its usefulness to infer potential sinkhole collapse areas.

Roth and Nyquist (2003) evaluated multi-electrode resistivity testing in karst by comparing over 140 resistivity tests at two sites with results from 51 borings. The results demonstrated that multi-electrode resistivity can reliably map depth to bedrock with excellent repeatability, although there was some smoothing in areas where the true bedrock surface is highly irregular. Multi-electrode resistivity was not as effective at locating the size of voids, although resistivity results can provide useful information concerning void location. The likelihood of locating a void is increased if multiple orientations of lines are used and if lines are spaced at intervals of approximately 5 m or less.

5 KARST SITE RESULTS

Multi-electrode resistivity testing was conducted at five test sites in mantled karst terrane located along the new Interstate I-99 corridor between Bald Eagle and State College, Pennsylvania. Testing was conducted via a Sting R1/Swift system and manufactured by Advanced Geosciences, Inc. (AGI) of Austin, TX. The system employed 28 electrodes configured in a dipole-dipole array. By way of example, data from two structures are presented herein. Structures 318 and 319 will be two parallel two-lane bridges that carry I-99 over route US 322. Abutment 1 of Structure 318 will be supported on pile foundations, while Pier 2 of Structure 319 will be supported on shallow foundations bearing on rock.

Test results for Structure 318, Abutment 1 are shown in figures 4 and 5. Figure 4 is a resistivity tomogram for one line conducted over the planned location of the abutment. The line was 162 ft long and employed an electrode spacing of 6 ft. The resistivity contours outline a "pinnacle" of higher values between horizontal distances of 50 to 80 ft. Shown in figure 5 are material profiles determined from two geotechnical borings conducted along the same line as the resistivity test. Figure 5(a) logs the materials at a horizontal line distance of 67 ft, i.e., near the

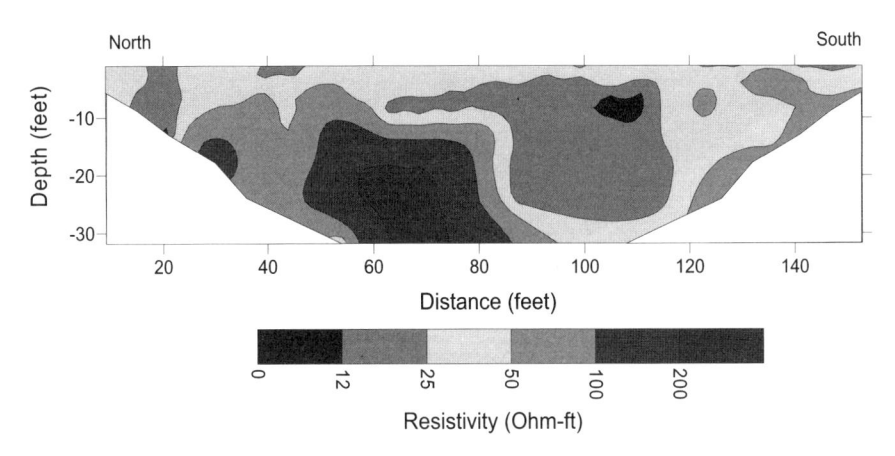

Figure 4. Resistivity Tomogram, Structure 318, Abutment 1

(a) Distance = 67 ft

(b) Distance = 85 ft

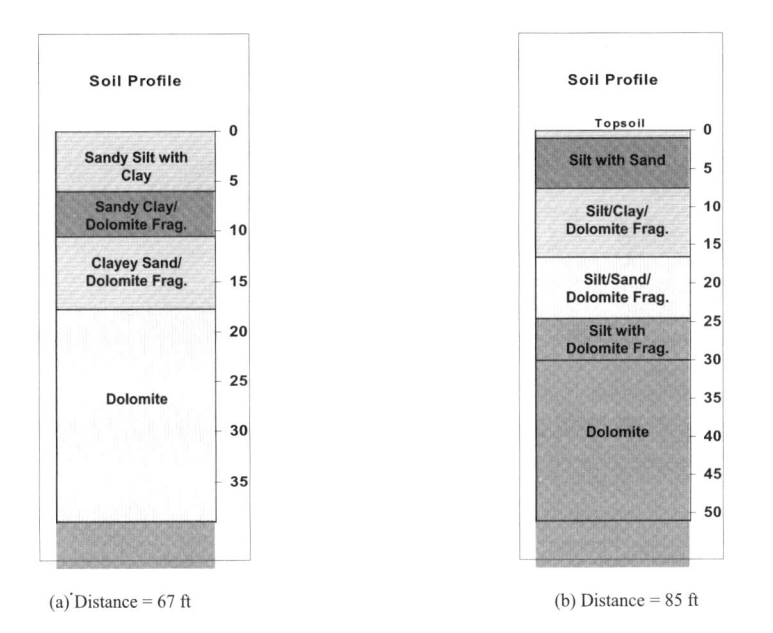

Figure 5. Boring Results, Structure 318, Abutment 1

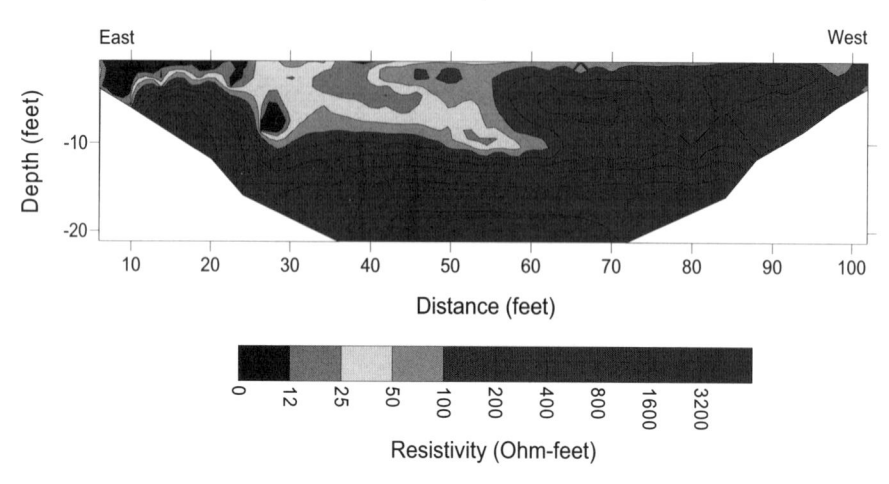

Resistivity Test #7

Figure 6. Resistivity Tomogram, Structure 319, Pier 2

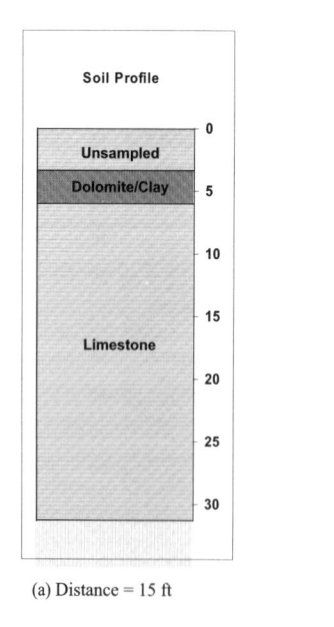

(a) Distance = 15 ft

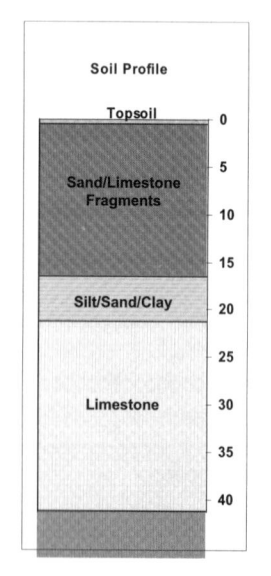

(b) Distance = 37 ft

Figure 7. Geotechnical Boring Results, Structure 319, Pier 2

"top" of the "pinnacle." Overburden soils are followed by a transition zone of soil and rock fragments, followed by dolomite at approximately a depth of 17 ft. Figure 5(b) logs the materials at a horizontal line distance of 85 ft, i.e., near the right "edge" of the "pinnacle." Here overburden soils are followed by a transition zone of soil and rock fragments, followed by dolomite at approximately a depth of 30 ft. The borings reflect similar materials at these two locations, but the depth to dolomite is considerably larger in the second boring. These data appear to verify at two points the pattern of behavior in resistivity values, i.e., there is a zone of shallower dolomite between line distances of 50 and 80 ft.

Test results for Structure 319, Pier 2 shown in figures 6 and 7 provide a scenario very similar to Structure 318. Figure 6 is a resistivity tomogram for one line conducted over the planned location of the pier. The line was 108 ft long and employed an electrode spacing of 4 ft. The resistivity contours outline a "pinnacle" of higher values between horizontal distances of 10 to 25 ft. Shown in figure 7 are material profiles determined from two geotechnical borings conducted along the same line as the resistivity test. Figure 7(a) logs the materials at a horizontal line distance of 15 ft, i.e., near the "top" of the "pinnacle." Overburden soils are followed by limestone at approximately a depth of 6 ft. Figure 7(b) logs the materials at a horizontal line distance of 37 ft, i.e., to the right "edge" of the "pinnacle." Here overburden soils and a mixture of soil and rock fragments are followed by limestone at approximately a depth of 21 ft. As with Structure 318, the borings reflect similar materials at these two locations, but the depth to limestone is considerably larger in the second boring. As before, these data appear to verify at two points the pattern of behavior in resistivity values, i.e., there is a zone of shallower limestone between line distances of 10 and 25 ft.

6 SUMMARY AND CONCLUSIONS

The purpose of this study was to employ multi-electrode resistivity testing to characterize the subsurface in mantled karst terrane. There is great need for a reliable testing method to map competent rock and estimate pile tip elevations during design of a structure. Data was presented for two bridge foundation sites. For each site, resistivity profiles were compared against material profiles available from geotechnical drilling. The resistivity profiles were found to match layering structure of the drilling data, particularly the top of rock profile. Based upon these findings, it can be concluded that multi-electrode resistivity is a useful tool for developing basic layer structure and top of rock profile in karst terrane.

ACKNOWLEDGEMENTS

The authors gratefully acknowledgement that the research described herein was funded by the Pennsylvania Department of Transportation via the University Based Research, Education, and Technology Transfer Program Work Order 82, Geotechnical Site Investigation for Bridge Foundations.

REFERENCES

Dunscomb, M.H. and Rehwoldt, E. (1999), "Two-Dimensional Profiling; Geophysical Weapon of Choice in Karst Terrain for Engineering Applications," *Hydrogeology and Engineering Geology of Sinkholes and Karst – 1999, Proceedings of the Seventh Multidisciplinary Conference on Sinkholes and the Engineering and Environmental Impacts of Karst,* Hershey, Pennsylvania, April 10-14, pp. 219- 224.

Kaufmann, O. and Quinif, Y. (2001), "An Application of Cone Penetration Tests and Combined Array 2D Electrical Resistivity Tomography to Delineate Cover-Collapse Sinkhole Prone Areas," *Geotechnical and Environmental Applications of Karst Geology and Hydrology, Proceedings of the Eigth Multidisciplinary Conference on Sinkholes and the Engineering and Environmental Impacts of Karst,* Louisville, Kentucky, April 1-4, pp. 359-364.

Lambert, D.W. (1997), "Dipole-Dipole D.C. Resistivity Surveying for Exploration of Karst Features," *The Engineering Geology and Hydrogeology of Karst Terranes, Proceedings of the Sixth Multidisciplinary Conference on Sinkholes and the Engineering and Environmental Impacts of Karst,* Springfield, Missouri, April 6-9, pp. 413-418.

Roth, M.J.S., Mackey, J.R., Mackey, C., and Nyquist, J.E. (1999), "A Case Study of the Reliability of Multi-Electrode Earth Resistivity Testing for Geotechnical Investigations in Karst Terrains," *Hydrogeology and Engineering Geology of Sinkholes and Karst – 1999, Proceedings of the Seventh Multidisciplinary Conference on Sinkholes and the Engineering and Environmental Impacts of Karst,* Hershey, Pennsylvania, April 10-14, pp. 247-252.

Roth, M.J.S. and Nyquist, J.E. (2003), "Evaluation of Multi-Electrode Earth Resistivity Testing in Karst." *Geotechnical Testing Journal*, ASTM, Vol. 26, pp. 167-178.

Werner, E. (1984), "Sinkhole Prediction - Review of Electrical Resistivity Methods," *Sinkholes: Their Geology, Engineering and Environmental Impact, Proceedings of the First Multidisciplinary Conference on Sinkholes,* Orlando, Florida, October 15-17, pp. 231-234.

Zhou, W., Beck, B.F., and Stephenson, J.B. (1999), "Application of Electrical Resistivity Tomography and Natural-Potential Technology to Delineate Potential Sinkhole Collapse Areas in a Covered Karst Terrane," *Hydrogeology and Engineering Geology of Sinkholes and Karst – 1999, Proceedings of the Seventh Multidisciplinary Conference on Sinkholes and the Engineering and Environmental Impacts of Karst,* Hershey, Pennsylvania, April 10-14, pp. 187-193.

Monitoring water saturation gradient in a silty sand site using geophysical techniques

Enayat Kalantarian
Environmental Science and Engineering program, University of Texas at El Paso, TX, USA

Diane Doser
Department of Geophysical Science, University of Texas at El Paso, TX, USA

Keywords: seismic techniques, experimental modeling, partial saturation, water table, instrumentation

ABSTRACT: We present the results of experimental investigations of the changes in water saturation of two silty sand sites in the vicinity of the Rio Grande near El Paso, Texas, using a combination of four geophysical techniques: seismic refraction, spectral analysis of surface waves (SASW), down-hole seismic and direct-current (DC) resistivity. These techniques were used to examine the periodic changes of the depth to the saturated zone (in Feb 2002, June 2002 & March 2003) in two study areas adjacent to two monitoring boreholes (~200 meters apart) within the Rio Bosque Wetland Park. Direct measurements of the water table level were taken during each field visit in two monitoring boreholes to verify the levels obtained by seismic surveys in the field. All the geophysical surveys used in this study independently indicated a three-layer model where layer thicknesses varied seasonally. Three techniques (SASW, DC resistivity and down-hole seismic) gave results for depth to saturation zones that were consistent with borehole observations. The refraction technique however, grossly overestimated (by a factor of 3 or more) the water table depth.

1 INTRODUCTION

The seasonal fluctuation of the water table depth in the vicinity of the Rio Grande Valley near El Paso is a function of the amount of rain, irrigation practices and river levels. The variation in soil moisture, salinity and grain size of the sediments within the Rio Grande valley near El Paso, Texas often makes it difficult to determine the saturation changes associated with the water table position using electrical resistivity or conductivity methods. Thus it is important to determine if other geophysical surveys could be used.

We conducted a series of geophysical surveys including seismic refraction, spectral analysis of surface waves (SASW), seismic down hole, and DC resistivity to determine how changes in the depth to the water table affected these surveys, which were conducted in February 2002, June 2002, and March 2003.

The surveys were conducted at two sites near boreholes located ~200 meters apart within the Bosque Wetland Park (Fig. 1). One borehole (RB-1A) is located about 20 meters from the Riverside Canal, which receives irrigation water from the Rio Grande from March to September. Another borehole (RB-3) is located near a diversion channel in the Bosque Wetland Park that receives peak discharge from November to February.

Our first set of seismic measurements was taken before the irrigation period in February 2002, when the average water table depth was about 2.8 meters in the RB-1A and 2.6 meters in the RB-3 borehole. A second set of seismic measurements was taken during the irrigation season in June 2002 when the average water table depth was about 2.35 meters in the RB-1A and 1.95 meters in the RB-3 borehole.

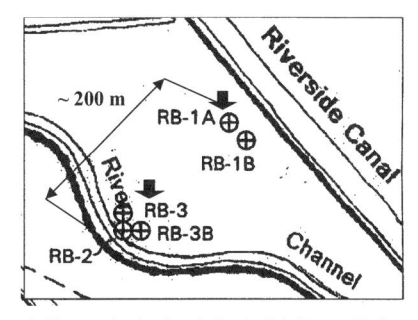

Figure 1. The monitoring boreholes the Rio Bosque Park

2 FIELD TECHNIQUES

In this section we describe our field measurements and discuss results and sensitivity of each geophysical technique.

2.1 Geophysical Measurements

The seismic refraction method utilizes P-waves, which travel at different velocities through different materials and are refracted at layer interfaces when the velocity of the lower layer is greater than that of the upper layer. The velocity of P-waves depends on the physical properties (i.e. rigidity, density, saturation) and degree of homogeneity of the medium. We used a 24-channel seismograph with 24 vertical component geophones and a sledge hammer/metal plate as an energy source. Two approximately 24 meter long (~ 79 feet) seismic lines with roughly 1-meter geophone spacing were recorded at the site. To verify this correlation between the geophone spacing and the corresponding wavelength, the SASW method was followed using the inequality of: $X/2 < L_{ph} < 3X$, (Nazarian 1984) where X and L_{ph} are geophone spacing and wavelength, respectively. Five shots (located at geophones 1, 7, 12, 18 and 24 m) were obtained for each line. Three to four hammer blows were performed at each source location to enhance signal quality by stacking the records to improve the signal-to noise ratio.

Since P-waves travel at the fastest speeds in a medium, the first seismic signal received by a geophone represents the P-wave arrival. Knowing the arrival times and distances from the source to each geophone, layer velocity can be obtained using a number of different analysis software. In this study, we used a computer program called "RefractSolve" by Burger (1992) to determine layer thickness and corresponding velocities. The resulting three-layer velocity models, with average P- velocities ranging from about 171 to 1330 m/s for both sites, are shown in figure 2. The 1330 m/s layer is considered to be saturated soil. However the depth to this layer (~ 7 m) is inconsistent with the depth to the saturated zone as measured by the other geophysical techniques and in the boreholes.

There are at least two reasons why the depth to the water table appears to be overestimated. First, there appears to be a low velocity zone (Nazarian and Desai, 1993) at 1-m depth that may contribute to an overestimation of the depth to the water table lying below it. (e.g. Dobrin, 1976). Second, if the upper portion of the saturated zone is a thin layer overlying a thicker, slightly higher velocity layer (such as a thick clay), the refracted arrivals from the thin layer will not be discernable because they always reach the surface layer than arrivals from the unsaturated zone or the deeper layer.

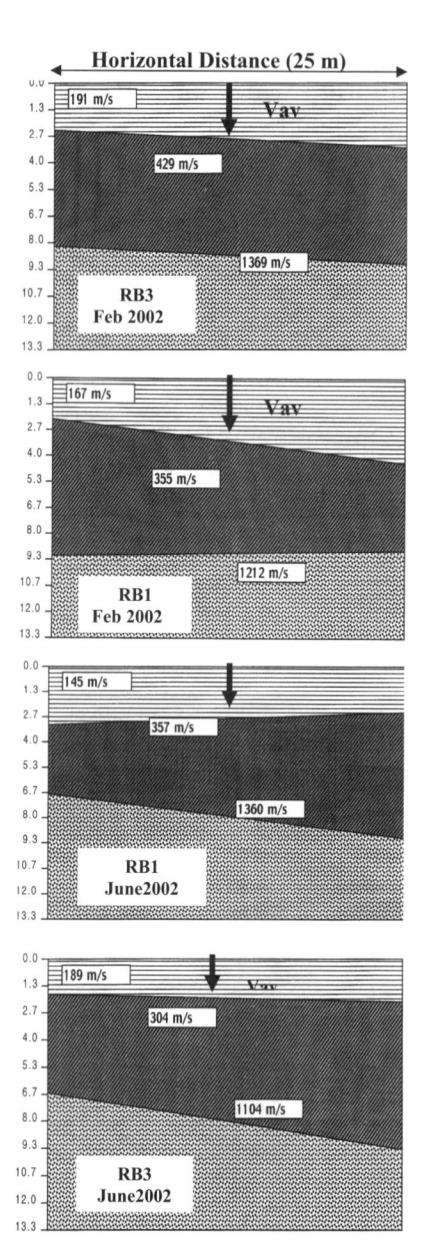

Figure 2. P- velocities models of the refraction data

This thin layer is termed "blind zone"., and it is difficult to recognize from seismic data alone (e.g. Soske, 1950).

We used the same 24-channel seismograph and geophones to record SASW data during the same field visits. A weight of about 50 lb (~ 24 kg) was dropped from a height of ~ 1.5 m to produce the seismic signal. We carried out three to five weight drops at each shot location along the line (at geophones 1, 7, 12, 18 and 24). During the recording process we selected those geophone channels that provided shot/detector distances, of 1, 2, 4, and 8 meters.

The wavelength/phase velocity relationships from all records were combined and used to generate dispersion curves. Nazarian and Stokoe (1984) have discussed this process in detail. The experimental dispersion curves were in our study constructed using a common receiver mid-point (CRMP) testing array (Nazarian and Stokoe 1986). In the CRMP array technique (Fig 3), an imaginary centerline for the receiver array was selected. Twenty-four receivers were then placed on the ground surface, and surface waves of various frequencies were generated in the medium. By selecting records from various geophones for various shot positions from the entire 24-channel data set, we were able to obtain the same mid-point detector/shot positions mentioned above without having to move our geophones during the data collection process. The records were monitored, captured, and saved for future reduction. After testing on one side of the array was complete, the receivers were kept in their original positions, but the source was moved to the opposite side of the imaginary centerline and testing was repeated.

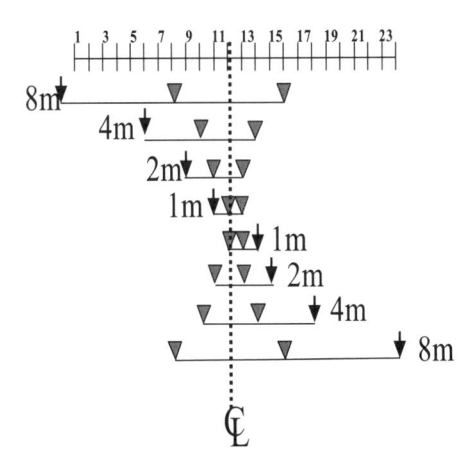

Figure 3. Common Receiver Mid Point (CRMP) array

The signals produced by the receivers were digitized and recorded in the time domain by the 24 channel seismograph. In order to gain more information from the extracted data, the recorded signals were transformed from the time domain o=into the frequency domain using Fast Fourier Transform algorithms.

The quality for signals for each test for all the frequency ranges (bandwidth) was verified using coherence function and the genuine signals were processed. In this selected range of coherence, phase information of the cross power spectrum was used to calculate phase velocities associated wavelengths for each frequency.

The iterative inversion software developed by Nazarian and Desai (1993) was used to determine the variation of the shear wave velocity profile at the sites. In this process the thickness of each layer was assigned by applying trial and error estimates to the program, and only the shear velocity of each layer was determined. The procedure was verified by comparison of the theoretical dispersion curve and the experimental dispersion curve obtained in the field. The experimental dispersion curve was constructed based on the data collected in the field in the form of phase information. The phase information was then analyzed using an unfolding process Nazarian (1984) to determine phase velocities and wavelengths for each frequency in order to build the theoretical dispersion curve. The RMS of experimental and theoretical dispersion curves was calculated to determine the best fit for the curves.

For layers with constant elastic properties, Raleigh wave (R-wave) velocity (V_R) and shear velocity (Vs) are related by Poisson's ratio (v).

$$V_R/V_S = [(1-v)/(0.5-v)]^{0.5} \quad (1)$$

Although the ratio of the R-wave to S-wave velocities increases as Poisson's ratio increases, the change in this ratio is not significant for shallow seismic studies. The effect of varying densities in the dispersion curve calculations is on the order of the effect of varying Poisson's ratios. Changing the values of the densities of the different layers does not significantly (< 1%) affect the shape of the dispersion curve Nazarian (1984). Thus, during the inversion process Poisson's ratios of 0.45, 0.40 and 0.35 (bottom to top layer of model) and a total unit density of 1800 kg/m^3 were assumed for our three-layer models.

The results obtained from the SASW studies also suggest a three- layer velocity model at the sites (Figure 4). Furthermore, the results indicate that the third layer has a lower shear velocity than the second layer for both sites. We interpret this low velocity layer as the fully saturated zone (water table). SASW analysis indicates that the average shear velocity of the saturated zones is about 140 m/s close to the borehole RB-1A, and 138 m/s near borehole RB-3.

Our third seismic technique used down-hole seismic methods to determine changes in compressional and shear wave velocity versus depth. In this technique, a seismic source is placed on the surface near a borehole (1 to 2 m away), and two geophones (14 Hz) are placed at selected depths (< 5 m) in the borehole. The source for P-waves is a hammer on a plate, for S-waves a railroad tie is hit by a hammer.

The railroad tie is held in place by the weight of a car. Multiple blows were made at both ends of the plank to generate SH-type waves with reversible first motions, as discussed by Crice (2001). Surface sources for P and S waves propagated elastic waves downward to an in-hole receiver, which was moved within the hole, thus providing a record section of waveforms for a series of depths. We used a Geometric R24 Seismograph for recording the data.

The raw data obtained from the down-hole surveys provided the travel times for compressional and shear waves from the source to the geophones. Note that data quality depends on borehole conditions. Brown et al., (2002) has indicated that the best results are obtained for an uncased borehole. However our down-hole logging method requires cased holes in order to clamp the receiver at various depths.

DC resistivity measurements were taken using the Wenner array (e.g. Sharma, 1997) with electrode spacings of 1, 2, 3, 4, 5, 7, 10, 15, and 20 m. An electrical current (I) was applied to the ground surface through two of the electrodes, while an additional electrodes measured variations in the potential of the electrical field V (voltage) that is set up within the earth by the current electrodes. The apparent resistivity is calculated from the current, voltage, and electrode geometry. The resistivity technique is sensitive to grain size and water salinity/water saturation.

In this work the apparent resistivity profiles show a drop in resistivity at a spacing of 4m (for RB1A February2002), 3m (for RB-3 June 2002) and 2 m (for RB1 July 2000) suggesting transition from unsaturated to saturated sediments. Observed apparent resistivity was modeled using the one-dimensional technique of Keller and Freschkencht (166). These results suggest water saturation at ~ 3 meter which is consistent with the SASW and down-hole survey results.

3 GEOPHYSICAL SURVEY RESULTS

We have summarized the measurements of all the geophysical surveys for both borehole sites in Figures 4a, 4b, 4c and 4d.

Our observations indicate that the average shear velocities measured by SASW (139 m/s) are about the same as the values determined from down-hole (133 m/s) measurements. The measured depths to the water table obtained from the two boreholes vary at both sites (Figures 4a through 4d). Part of this variation is due to the different layer interfaces obtained by the SASW and down-hole methods. Layer intervals for the down-hole profile were selected based on observed travel times and the down-hole lithology log. Lateral variability between surface array positions and down-hole measurements may also contribute to differences in shear wave velocities and the depth to interfaces. The down-hole method obtains velocities for a small volume close to the borehole, whereas the SASW method averages properties over horizontal distances related to the array length. Another possible explanation for the differences in shear velocities could be because of the Poisson's ratios (0.4, 0.35 and 0.30) assumed for initial analysis of the SASW technique.

Our results for the 4-meter SASW spacing suggest a sharp shear velocity decrease (spread out at almost a 2.4-meter depth), which is consistent with the direct measurements of water table depth in monitoring wells (2.5 m) and DC resistivity methods, while the layer velocities for the down-hole decrease at a 2 meter depth (Figures 4a, 4b, 4c and 4d).

The seismic refraction measurements give average P-wave velocities ranging from 171 m/s to 1330 m/s (a velocity consistent with saturated sediment). However layer thickness is not consistent with the resistivity, seismic down-hole and SASW model layer thicknesses or depths. This is similar to observation made along the banks of the Rio Grande northwest of El Paso (Hincapie et al. 2001). Note that depth to the 1330 m/s layer (saturated zone) is about 8.5 m deep in February and 6.5 m deep in June for both sites.

DC resistivity values obtained near the monitoring wells are consistent with our interpretation of partially and fully saturation layers obtained from the SASW and down-hole modeling (Figures 4a through 4d). The resistivity value of the second layer (~ 100 ohm-m) would correspond to a partially saturated layer, the second layer measured by the SASW and down-hole techniques. The third layer has resistivities that vary between 2-16 ohm-m, values consistent with saturated sediments. Depths to the third layer are consistent with depths to water table. The following table compares the seasonal results obtained by seismic techniques (refraction, SASW and down-hole) and DC resistivity.

Table 1. Seasonal water table fluctuation in the Park

	Borehole RB1A		Borehole RB-3	
Technique	Winter 2002	Summer 2002	Winter 2002	Summer 2002
	Depth (m)		Depth (m)	
Direct values	2.8	2.35	2.6	1.85
SASW	2.8	2.5	2.5	1.9
Down-hole	2	2	2	2
Refraction	9	8	8.7	8.2
Resistivity	3	3	None	3
Good results	SASW	SASW	SASW	SASW
	Resistivity	x	x	Down-hole

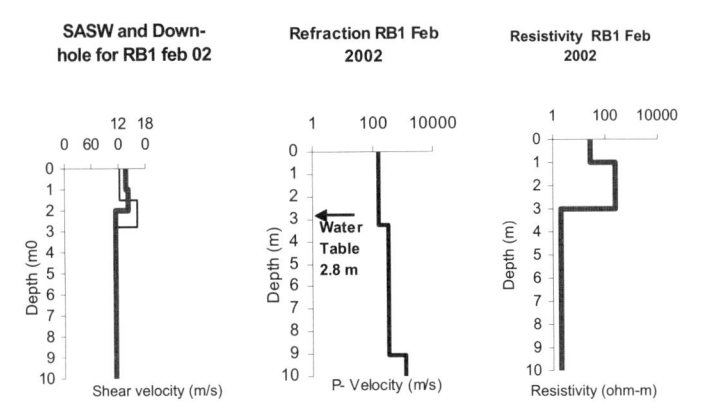

Figure 4a. Variations in geophysical parameters with depth at site in February 2002. SASW (thin line) and down-hole (thick line).

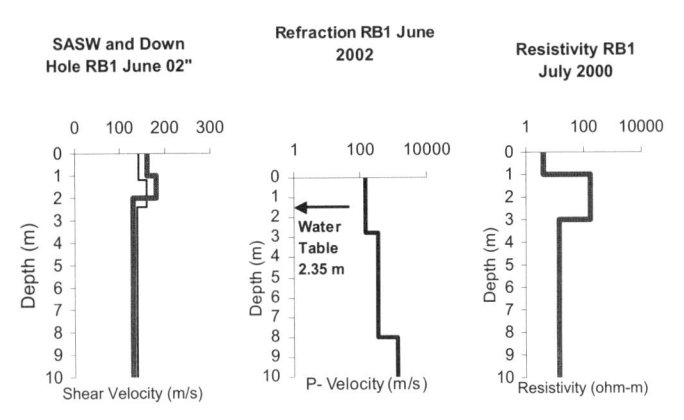

Figure 4b. Variations in geophysical parameters with depth in June 2002 and July 2000. SASW (thin line) and down-hole (thick line).

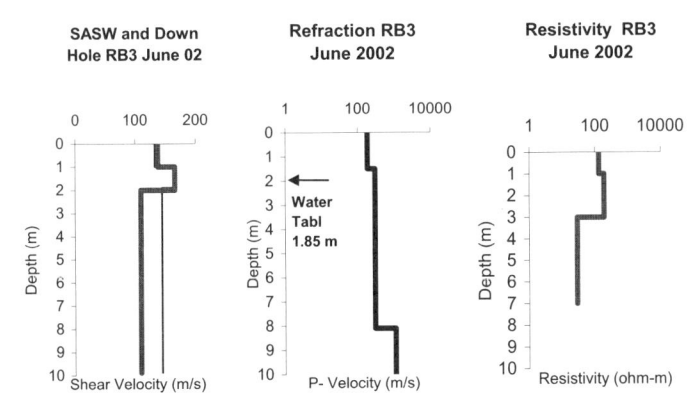

Figure 4c. Variations in geophysical parameters with depth at site in June 2002. SASW (thin line) and down-hole (thick line).

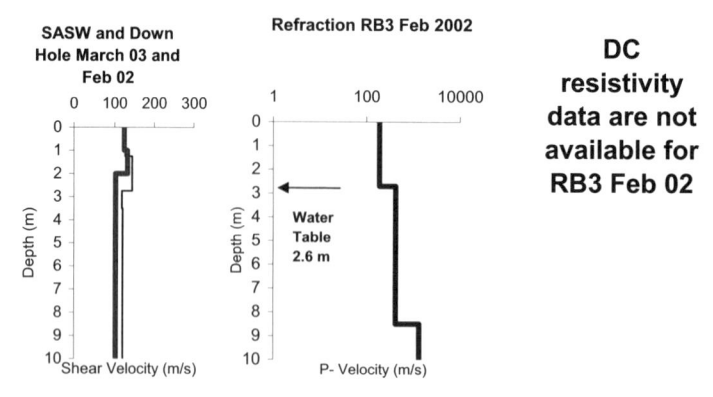

Figure 4d. Variations in geophysical parameters with depth at site in February 2002. SASW (thin line) and down-hole (thick line).

Our results suggest that the depth of the water table obtained by the down-hole method slightly underestimates the true depth, and that seismic refraction grossly overestimates the depth, but the refraction method show a decrease in the depth in the summer. DC resistivity overestimates the depth to the water table, mostly likely because we used water table beds in the modeling process. As a result, the SASW models both for the RB1-A and RB3 wells gave the best results in winter and summer 2002

4 CONCLUSIONS

SASW, seismic down-hole and resistivity studies at two sites in West Texas gave three-layer models consistent with known borehole and geologic information, including estimates of depth to the water table. SASW is the best at determining water table depth, while down-hole and resistivity gave depths within 1 m of the true water table. Although seismic refraction studies also gave three-layer models, depths to the water table (7-9 m) greatly overestimated the observed water table depth (2-3 m) similar to results from a previ-ous study by Hincapie et at. (2001). These results suggest that the seismic refraction technique may not be a suitable method for determining water table depth using at sites where the water table is expected to be < 5 m in depth. Notice that the verification technique of geophone spacing by the SASW method and the accuracy of the pervious investigators in the area suggested the scale of 1-m geophone spacing for the seismic refraction technique in this work.

ACKNOWLEDGMENTS

Special thanks to Soheil Nazarian and Deren Yuan for their modified inversion program to analyze the SASW tests, and Galen Kiap for his experimental support with down-hole data interpretation.

REFERENCES

Brown, L, T. D., M. Boore, and K, H. Stokoe II, 2002. Comparison of Shear-Wave Slowness Profiles at 10 Strong-Motion Sites from Noninvasive SASW Measurements and Measurements made in Boreholes, Bulletin of the Seismological Society of America. Vol. 92. No. 8. pp. 3116-3133.

Burger, H. R, 1992, "Exploration Geophysics of The Shallow Subsurface". Prentice Hall-Inc, pp. 20-145

Crice, D. 2001. Borehole Shear-Wave Surveys for Engineering Site Investigations, The first draft of Geostufuf pp 2-13

Doser, D.I, S Nazarian, D Yuan, and M.R.Baker 1998. Monitoring Water infiltration under an earth-fill levee with geophysical techniques, ISC ' 98 Proceed, First international Conf. on Site Characterization, PP 557 - 561

Hincapie, J, D. Doser, D. Yuan and M, Baker 2001. Detection of shallow water table fluctuation using the spectral analysis of surface waves (SASW) technique, Proceedings of the Symposium on the Applications of Geophysics to Environmental and Engineering Problems (SAGEEP 2001-Denver), on CD-ROM, paper ASP-4.

Keller, G,V, & FC, Freschknecht 1966. Electrical methods in geophysical prospecting. Pergamon Press.

Nazarian, S 1984. In Situ Determination of Elastic Moduli of Soil Deposits And Pavement Systems By Spectral-Analysis-Of-Surface-Waves Method, PhD Thesis, The University of Texas at Austin, Texas.

Nazarian, S., And M. Desai, 1993, Automated surface wave method: Field testing: Geotechnical Engineering Journal, Vol. 119. No. 7, pp. 1101-1111.

Stokoe, K. H, II, and S. Nazarian, (1985), "use of Rayleigh Waves in Liquefaction Studies", proceedings of a session sponsored by the Geotechnical Engineering Division of the American society of Civil engineering in conjunction with the ASCE convention in Denver, Colorado, pp 1-17.

Sharma, P. V., 1997, Environmental Engineering Geophysics: Cambridge Univ. Press, Cambridge, UK, 475 p.

Characterizing ground hazards induced by historic salt mining using ground-penetrating radar

Bernd Kulessa
School of Civil Engineering, Queen's University of Belfast, Belfast, BT9 5AG, UK

Alastair Ruffell
School of Geography, Queen's University of Belfast, Belfast, BT7 1NN, UK

Danny Glynn
Construction Service, New Works Division, Geotechnical Section, 4 Hospital Rd, Belfast, BT8 8JJ, UK

Keywords: ground hazards, historic mining, ground-penetrating radar

ABSTRACT: Abandoned and now buried mineshafts are a common hazard in many parts of the world. Their detection is a priority where they pose a threat e.g. to human life, infrastructure, or livestock. Often there are no surface expressions, and shaft detection with traditional geotechnical methods is expensive because often extensive, intrusive ground probing is required. Use of heavy probing equipment is unavoidable and increases the risk of collapse, therefore raising health and safety concerns. Significantly, geophysical methods are rapid, have the capacity to produce spatially complete ground images remotely from the surface, and require relatively lightweight equipment. They are therefore cost-effective, reduce the risk of failure in detecting shafts, and allow health and safety risks to be minimised. We report ground-penetrating radar (GPR) and electrical resistance (ER) surveys at two adjacent historic salt mines in Co. Antrim, Northern Ireland. At *French Park* pronounced brine seepage occurred in a field compromising sheep husbandry, and was suspected to originate in a buried mineshaft listed in historical records. The GPR surveys successfully detected the mineshaft and were used to define its location with metre-scale accuracy. These surveys further imaged several discrete features adjacent to the mineshaft including potential preferential drainage structures, and reflected collapsed ground beneath a near-by road which did not have a surface expression. The subsequent electrical resistance survey successfully delineated the exact contaminant-saturated area near the ground surface. Joint interpretation of the GPR and electrical resistance data suggests that the contaminants may spill out of the mineshaft and subsequently flow along the subsurface drainage structures until appearing at the surface. Remedial drainage structures have now been implemented at French Park. At *Maidenmount*, located only ~ 400 m uphill of the latter, GPR surveys were used to define the spatial extent of a large (likely >> 10^3 m^2) brine-filled underground cavity, the top of which has already collapsed over an area of ~ 50×30 m. The GPR data reflect the top of the cavity at ~ 30-35 m depth, suggesting an erosion rate of ~ 5 m year^{-1} since 1997. It therefore appears that collapse of the entire cavity is imminent, requiring urgent remedial action. We speculate that changes in the regional subsurface hydraulic regime caused by such collapse could further cause increased brine seepage out of the French Park mineshaft, potentially leading to further ground hazards at this site.

1 INTRODUCTION

The UK Department of Enterprise Trade and Industry (DETI) is responsible for the considerable number of abandoned mines and mineshafts in Northern Ireland (NI). DETI's remit is to ensure the safety and upkeep of these mines regardless of ownership or whether they are in public or privately owned ground. The present study focuses on investigating key characteristics of, and potential linkages between, two adjacent abandoned salt mines (*French Park* and *Maidenmount*, Figure 1) located north of Belfast Lough in Co. Antrim, NI. These two sites have been identified as being associated with particularly high health and safety risks. In the 1800's mining in the area focused on hand mining thick horizontal seams of salt. A combination of two vertical (for salt recovery and ventilation) and horizontal shafts were constructed, the latter being supported by regularly spaced salt pillars. Later commercial exploitation often involved pumping of water into one of the mineshafts, dissolving the underground salt deposits. The resulting solution was subsequently pumped to the surface for processing.

Dissolution of support salt pillars is suspected to cause continuous underground collapses in the survey area.

The present study aims to demonstrate that cost-effective, non-invasive geophysical methods, in particular ground-penetrating radar (GPR), have the capacity to provide spatially continuous information about mine-related hazards now buried in the subsurface. Such information is difficult to obtain using more conventional, invasive methods of geotechnical site investigation. The use particularly of microgravity methods for subsurface void detection is well documented, including mineshafts and salt dissolution structures (e.g. Sharma, 1997). Seismic reflection and electrical resistivity surveys have also been used for mineshaft and cavity detection (e.g. Miller and Steeples, 1994, 1995). GPR has been popularly used in the detection of sub-horizontal cavities such as caves, adits and passageways (e.g. Sharma, 1997), but surprisingly rarely for the location of vertical shafts (Leopold and Volkel, in press) or salt dissolution structures (Bornemann et al., 2001).

2 FIELD SITES AND METHODOLOGY

2.1 Field Sites

The abandoned mines at French Park and Maidenmount are located ~ 400 m apart (Figure 1). Since the-latter site is located uphill of French Park, it is likely that hydraulic processes at Maidenmount may influence those at French Park. In both areas, sheep and cattle grazing on poorly–drained soils is the main form of land use. A major satellite town of Belfast, Carrickfergus, is situated < 3 km southeast of the two sites. This town is undergoing expansion in housing and transport infrastructure. Thus the once isolated salt mining districts are now in places adjacent to housing and new roads. The area under study has a 5-15 m thick glacial till succession on top of Triassic silty mudstones and evaporites (Griffith et al., 1983).

2.1.1 French Park

At French Park continuous seepages of saline groundwaters have compromised sheep husbandry over the last five years. These seepages appear at the ground surface between a field access gate and a spoil heap, and subsequently flow downhill across the field (Figure 2). The brine spillage has been monitored by the Geological Survey (GSNI) for several years, and the adjacent roadway (Figure 2) was closed for health and safety reasons. A simple 'pipe and hardcore' remedial drainage design, supplemented by the inclusion of a herringbone drainage system to capture water seepages over a large area extending downhill from the projected mineshaft opening, was proposed by the consultants *W S Atkins*. It was speculated that trapped water within the abandoned, and now buried, *Marquis*

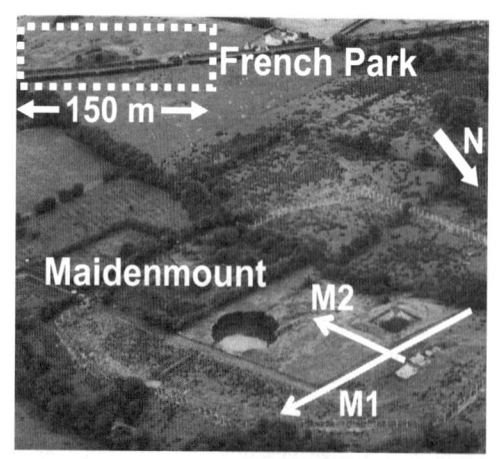

Figure 1. Aerial photograph showing the location of the two historic mining sites (French Park and Maidenmount), as well as the location of the two GPR lines (M1, M2) at Maidenmount. The crown hole is ~ 50 m in diameter (serving as scale).

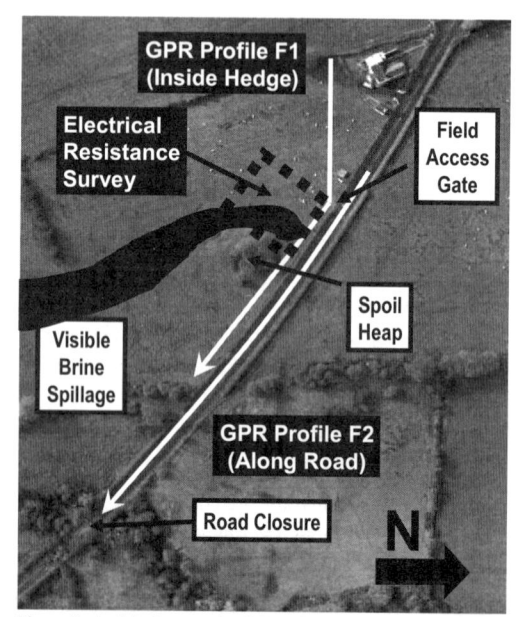

Figure 2. Aerial photograph of French Park, illustrating the two ground-penetrating radar profiles (*F1* and *F2*), the electrical resistance survey area (30×25 m, serving as scale), and site characteristics referred to in the text.

of *Downshire* mineshaft was hydraulically forced through the capped (conventionally by disused railway sleepers) top to the ground surface. Before instigating the work to provide the drainage system, NI *Construction Service* was asked to locate the actual mineshaft position. Initial thoughts of using intrusive investigation tools (such as drilling or exploratory excavation) to locate the mineshaft were abandoned amidst financial and health and safety fears that heavy mechanical equipment could induce collapse.

2.1.2 *Maidenmount*

400 to 500m north-west of the French Park site is a collapsed salt mine shaft at Maidenmount (Figure 1). The collapse comprises a ~ 30 m diameter, water-filled cylindrical chimney (or crown hole) of at least 20 m depth, above a ~ 100 m wide, ~ 30-40 m deep void, some 100m above the main salt-bearing horizon of Triassic mudstones. The void has been monitored until now by ultrasonic and borehole investigations, and has been migrating to the surface at several metres per year. The main void is now close to the base of its overlying chimney. Changes in the local drainage pattern occurred concurrently with the initiation of the seep downhill and French Park. The existence of a subsurface hydraulic link between the major collapse at Maidenmount and the initial study site must therefore be considered.

2.2 *Methods*

2.2.1 *Ground-Penetrating Radar: Concepts*

GPR is a popular geophysical survey technique that detects changes in dielectric permittivity in the ground. This is accomplished by sending a high-frequency (typically several MHz to > 1 GHz) electromagnetic wave into the ground using a transmitter antenna, and receiving the ground signal, reflected back to the surface after encountering a subsurface target, with a receiver antenna (Figure 3). The time between transmission and detection of the reflected signal may then be converted to depth if the velocity of GPR wave propagation is known. This time is commonly referred to as a two-way-travel time (TWT) because the signal travels down towards a subsurface target, and then back again to the ground surface. Radar data are commonly displayed using *radargrams*, which are plots of changes in received radar amplitude (either over time or with depth) against distance along the ground surface. Reflections caused at subsurface interfaces separating two regions of differing dielectric properties are then apparent as spatially distinct changes in radar amplitudes. Fresh water has a very high dielectric permittivity of ~ 80,

Table 1. Dielectric permittivities and GPR wave velocities of relevant geologic materials and fluids.

Material/Fluid	Permittivity	Velocity $m\,ns^{-1}$
Wet clay to gravel	7 – 40	0.05 – 0.11
Fresh/salt water	~ 80	~ 0.03
Hydrocarbon fluids	~ 2	~ 0.2
Air	1	~ 0.3

while air has a very low permittivity of 1. These and other dielectric permittivities of relevant subsurface fluids and materials are summarised in Table 1 (after Sharma, 1997), together with corresponding radar wave velocities. Wave velocity and signal attenuation respectively scale inversely and directly with dielectric permittivity. Based on expected results combined with independent information, radargrams may then be interpreted in terms of ground properties and buried materials. In the present case we expect the top of the mineshaft and other associated discrete features to show up as prominent reflections in the radargrams. In this context it is however important to recognize that different materials/fluids and any of their mixtures may have the same bulk dielectric properties, and thus equal velocities. It is therefore commonly not possible to determine the type of medium or pore fluid on the basis of GPR data alone.

2.2.2 *Ground-Penetrating Radar: Data Acquisition*

The GPR data at French Park were collected over two days in November 2002 using a MALÅ *RAMAC* GPR system with 100 MHz antennas (Figure 3). All data were collected on foot, with readings taken every 0.5 m along the profiles. Over 40 GPR lines, typically between several tens and > 100 m long, were collected on a variable grid broadly at and uphill of the location were the salt brine first appears on the ground surface (Figure 2). Two profiles (labeled *F1* and *F2*) are of particular importance to the present study (Figure 1).

The GPR data at Maidenmount were collected over seven days in July 2003 using the same MALÅ *RAMAC* system, however in this case 50 MHz antennas were used to achieve improved depth penetration. A total of 4725 m line length of GPR data were collected on foot around the crown hole using a variable grid, with readings taken every 1 m along the profiles. Two representative profiles are of particular importance to the present study, and are illustrated as lines *M1* and *M2* in Figure 1. These lines were collected perpendicular to each other on the northern side of the crown hole, where further collapse is expected. Both lines overlap (Figure 1), which may be used for verification purposes.

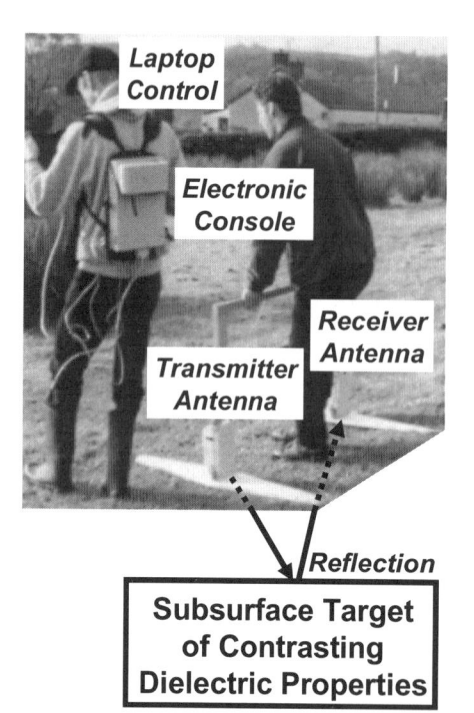

Figure 3. Instrumentation for ground-penetrating radar surveys, as explained in the text.

2.2.3 *Electrical Resistance Surveys*

Electrical resistance surveys involve injection of electrical current into the ground using two electrodes, and temporally coincident measurement of the resulting spatial voltage distribution in the subsurface using a second pair of electrodes (Figure 4). The current and voltage data may then be used to produce a 2-D map of the electrical resistance distribution across the survey area. The method works on the principle that the current will be concentrated where the ground is particularly conductive, and vice versa for relatively resistive areas. In these cases the voltages measured, and thus the electrical resistances, will respectively be comparatively low and high. Specifically, in the present case we expect the salt brine to show up as particularly conductive anomalies in the 2-D resistance map, while non-contaminated ground is expected to be characterized by relatively low conductivities.

The survey was conducted using an *RM-15* electrical resistance meter manufactured by GEOSCAN RESEARCH (Figure 4), which is more commonly used in archaeological applications. The depth penetration of the instrument is < 1 m. Measurements were made along 23 parallel lines spaced 1 m apart and 30 m long each, with readings taken every metre along the lines.

Figure 4. Photograph of the GEOSCAN RESEARCH *RM-15* electrical resistance meter. The reference electrodes at effective infinity are not shown for scaling purposes.

3 RESULTS

3.1 *French Park*

The raw radargrams along profiles F1 and F2 are illustrated in Figures 5a and 6a, plotted as TWT against distance along the profile. It is immediately obvious that there are strong reflections from the base of the overburden, which are disturbed or even absent in places (Figure 5a). There are also strong reflections from intact and disturbed internal layers in the road bed (Figure 6a). There are also numerous deeper, but less pronounced, reflections visible in these two figures. Raw voltage or absolute electrical resistance data are not shown here due to space considerations. The 2-D maps were characterized by strong conductive anomalies, which coincided well with the extent of the salt brine visible at the ground surface. The electrical resistance data are presented in terms of relative brine concentration in Figure 7, and interpreted in Section 4.2.

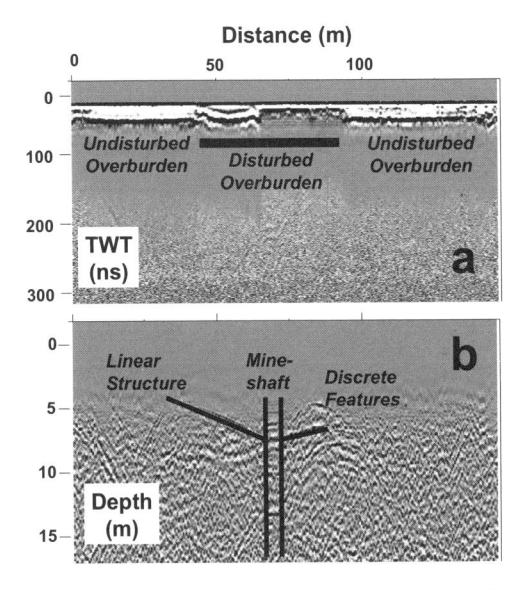

Figure 5. Raw (a) and processed (b) ground-penetrating radar data along profile F1 at French Park. The main features of interest are labeled.

3.2 Maidenmount

The raw radargrams from Maidenmount are also not shown here due to space considerations. We found that the noise level in the raw 50 MHz data was similar to that observed in the 100 MHz data from French Park. Notwithstanding, due to decreased signal attenuation subsurface reflectors were more readily apparent in the raw 50 MHz than in the raw 100 MHz data, as would be expected. The processed 50 MHz data are discussed in Section 4.3.

4 INTERPRETATION

4.1 Data processing and velocity estimation

All GPR data were processed with the INTERPEX software package *GRADIX*. Processing steps variably involved
- dewowing;
- drift removal;
- setting time zero;
- programmed gain control;
- band-pass frequency filtering;
- spectral trace balancing;
- velocity (f-k) filtering;
- conversion of TWT to depth.

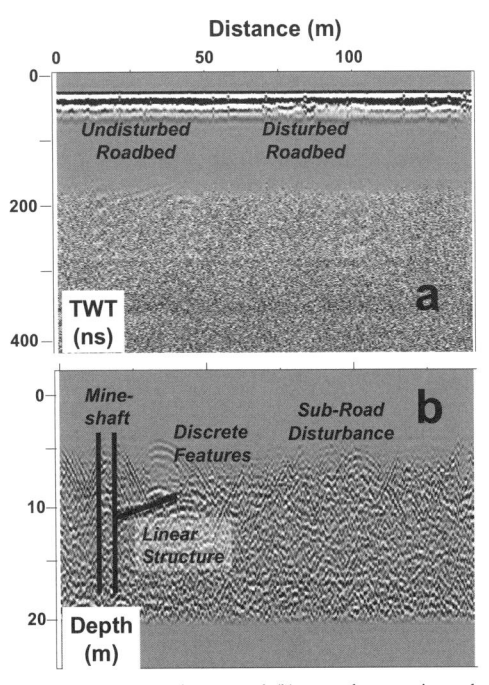

Figure 6. Raw (a) and processed (b) ground-penetrating radar data along profile F2 at French Park. The main features of interest are labeled.

A more detailed discussion of each processing step is beyond the scope of this paper. The effective till velocity necessary for depth conversion was obtained from move-out analysis of the diffraction hyperbolas in the French Park radargrams (Figures 5b and 6b). Such hyperbolas arise when a GPR wave strikes a discrete target, the target being located close to the apex of the hyperbola. This is because the TWT from the ground surface to the target is shortest when the GPR antennas are directly above the target, whereas the TWT increases when the antennas are located away from this closest location, thereby producing the two 'arms' of the hyperbola. Move-out analysis yields a till velocity of ~ 0.13 m ns^{-1}. Since both study areas were located close to each other, we considered it valid to assume that the same effective velocity applies to the till at Maidenmount, where diffraction hyperbolas were absent.

4.2 French Park

For convenience we have labelled all features in Figures 5 to 7, and individual analyses presented below are in specific reference to these figures without repetitive reference.

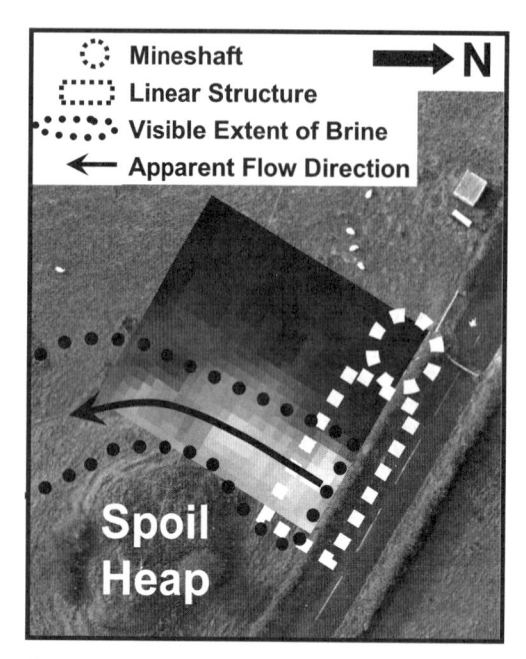

Figure 7. Summary of ground-penetrating radar results at French Park, together with electrical resistance data in relative units (lighter colours indicate higher brine concentration).

4.2.1 Mineshaft Location and Characteristics

We believe that the mineshaft shows up as a vertical series of reflections that all have approximately the same horizontal extent, and appear to be noticeably offset from surrounding reflections. Significantly in this context, the horizontal extents of these reflections agree well with the expected thickness of the mineshaft (~ 4 m). There are strong reflections near the top of the shaft, potentially indicating the presence of shaft capping. Multiple reflections originating inside the shaft could indicate the presence of some backfilled material.

The concave overburden disturbance corresponds to the access-way just behind the gate, and may therefore be interpreted as being closely related to either 'natural' (e.g. due to the regular heavy load) or 'artificial' (e.g. anthropogenic reinforcement of the access-way) tractor access to the field. To the right of this access-way disturbance is an area ~ 30 m in horizontal extent and clearly marked by missing reflections from the base of the overburden. This observation is consistent with disturbances due to mineshaft operation in this area, which is further confirmed by the apparent presence of associated buried objects (causing the diffraction hyperbolas). Unfortunately it is not possible to determine the exact nature of these objects from hyperbola analysis alone, although presence of abandoned work material or equipment is perhaps most likely. It could further be speculated that the access-way disturbance was once part of the excavation due to mineshaft installation, and was later intentionally backfilled to provide tractor access. If this is true then the mineshaft would once have been close to the centre of the excavation (and thus the overall overburden disturbance), which would appear plausible.

4.2.2 Salt Brine Seepage

The electrical resistance distribution across the survey area was transformed into equivalent relative brine concentration (Figure 7) by dividing all resistance data by the highest measured resistance value, yielding a concentration range from 0 to 1. Maximum concentrations occurred near the north-eastern corner of the survey area (Figure 2), giving the overall impression that brine concentration is low in the western part of the survey area, and high in the eastern part. The latter observation corroborates the clearly visible surface brine spillage, but also extends the area of brine contamination uphill. Brine concentrations decrease towards the south-west, and therefore directionally coincide with surface brine flow (Figure 8). We therefore conclude that the brine first appears at the surface in the north-eastern corner of the survey area (i.e. uphill from the spoil heap and towards the hedge; Figure 2), and subsequently appears to flow around the western end of the spoil heap, and then further downhill roughly parallel to the southern flank of the spoil heap (thereby causing the clearly visible surface spillage). We do however caution that the part of the plume could also flow downhill underneath the spoil heap (or indeed through it), and potentially also downhill along its eastern flank (i.e. along the corridor formed by the spoil heap and the hedge; Figure 2).

Figure 7 suggests that brine extrusion does, as such, not appear to be directly linked to the mineshaft (at least not at the < 1 m penetration depth of the resistivity meter deployed here). However, we observe that it appears to emanate near the eastern end of a strong linear GPR reflector inclined towards the mineshaft (Figures 5b and 6b respectively). Notably, there is also a corresponding inclined reflector on the western side of the mineshaft (Figure 5b). We speculate that these inclined reflectors may indicate the post-excavation bedrock surface, whose eastern (and thus downhill) expression could represent a preferential flow pathway routing salt brine flow from the mineshaft towards the point of surface extrusion. The associated buried objects also appear to be predominantly located in this eastern downhill area (Figures 5b and 6b) respectively), and could potentially provide further preferential pathways sustaining brine flow towards the ground surface.

4.2.3 Damage to Road Bed

Roadbed (Figure 5a) and sub-road (Figure 5b) disturbances are observed very close to an area fenced off beside the northern verge of the road where mine collapse had occurred previously. We therefore interpret these disturbances as being due to subsurface subsidence caused by mining-induced collapse. Since the sub-road subsidence has no obvious surface expression, it would have presented a hidden danger to traffic had the road not been wisely closed as a precautionary measure. Traffic loading would have increased the danger of further, and potentially sudden, subsidence or even collapse.

4.3 Maidenmount

A strong straight reflecting horizon is immediately apparent in the processed profile M1 at ~ 30 m depth (Figure 8a). This depth is consistent with what could reasonably be expected to be the top of the subsurface chamber. Closer inspection of the GPR reflections reveals that the first break has a positive phase, which commonly indicates an increase in dielectric constant. We therefore infer that the subsurface chamber is likely water or brine-filled rather than air-filled. If the top part of the chamber was air-filled we would instead expect to observe a negative first break because the dielectric constant of air is less than that of glacial till (Table 1). Profile M2 shows an inclined reflector dipping from ~ 24 m near the crown hole towards the east, reaching depths of > 30 m (Figure 8b). This observation is consistent with the top of the subsurface chamber increasing in depth away from the crown hole, which is expected. Encouragingly, the depth to the chamber top reflector is approximately the same in both profiles where they overlap (indicated by stars in Figures 8a and b).

Other profiles collected in the same area as lines M1 and M2 corroborate these findings, and previous evidence collected by GSNI suggests that the subsurface chamber is particularly extensive north of the crown hole. We are therefore confident that we may conclude that the subsurface chamber is water or brine-filled, and that the depth of the top of the chamber increases from > 20 m near the crown collapse to > 30 m several tens of metres away from it. Unfortunately, loss of signal strength due to attenuation with depth does not allow us to define the exact horizontal extent of the chamber. It could however be expected that it is horizontally more extensive than we may confidently infer here.

5 SYNTHESIS AND CONCLUSIONS

Our GPR surveys at French Park were successful in detecting and locating the mineshaft, and together with electrical resistance results these surveys are con-

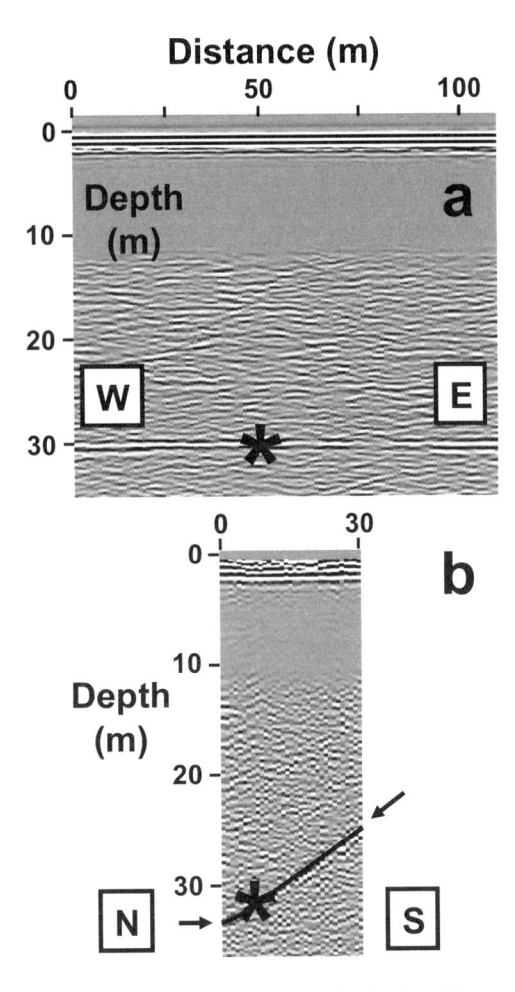

Figure 8. Processed ground-penetrating radar data from Maidenmount: (a) profile M1; (b) profile M2 (see Figure 1 for profile locations). The cross-over point between the two lines is marked with a star. Note that the deeper (northern) end of the reflector in (b) flattens out away from the crown hole.

sistent with salt brine seepage out of the mineshaft and subsequently upwards to the ground surface. We could further confirm that an extensive fluid-filled buried chamber extends northwards away from the crown hole at Maidenmount. Significantly, the depth to the top of the chamber is considerably less than it was six years ago (~ 60 m), which suggests an erosion rate of ~ 5 m per year, directed vertically upwards from the top of the chamber. Continuing collapse of the roof of the chamber is therefore evident.

Since the Maidenmount site is located only a few hundred metres away from and, importantly, uphill of the French Park mineshaft (Figure 1) it is important to elucidate whether potential hydraulic linkages exist between the two sites. It is particularly revealing to consider that only regional hydraulic potentials could force brine upwards out of the mineshaft. Such potentials are here expected to direct water flow from the upstream Maidenmount site to the downstream French park site. We hypothesize that continuing roof collapse at Maidenmount could have caused regional hydraulic potentials to increase in the direction of French Park, eventually forcing upward brine seepage out of the mineshaft to the ground surface. This would imply that further collapse at Maidenmount could enhance future seepage at French Park, which should be considered during design of appropriate remedial measures at French Park.

In conclusion, we have demonstrated that GPR surveys can yield spatially extensive information about subsurface hydraulic properties and processes at historic mining sites, which would be difficult to obtain using more conventional methods of geotechnical site investigation. Since GPR, or indeed many other geophysical methods, are cost-effective compared to drilling or invasive ground testing we recommend that application of geophysical methods should regularly be given serious consideration at the site investigation stage.

ACKNOWLEDGEMENTS

Our thanks to the landowners at French Park for allowing access. Help in the field from Paul McCarthy, Brendan McClean, Geoff Warke and Rachael McCallister is very gratefully acknowledged. Thanks also to Terry Johnson and Garth Earls (Geological Survey, N.Ireland) for advice, as well as John Meneely, John Davidson, and Barrie Hartwell (Queen's University Belfast) for kind equipment hire and support.

REFERENCES

Bornemann, O., G. Mingerzahn, and J. Behlau. 2001. Characterisation of Sites for Salt Caverns in the Middle European Zechstein Salt Basin Using Exploration Experiences of the Gorleben Salt Dome. In: *Technical Meeting Papers, SMRI, Fall 2001 Meeting, 7-10 September 2001, Albuquerque, New Mexico, USA*, 198-210.

Griffith, A. E., H. E. Wilson, and J. R. P. Bennett. 1983. Geology of the Country around Carrickfergus and Bangor: Sheet 29. *1983 Memoirs of the Geological Survey of Northern Ireland.*

Leopold, M. and J. Volkel. Detection of Neolithic flint mines in Bavaria, Germany. *Archaeological Prospection*, in press.

Miller, R. D. and D. W. Steeples. 1994. Applications of shallow high resolution seismic reflection experiments to various environmental problems. *Journal of Applied Geophysics*, **31**, 65-72.

Miller, R. D. and D. W. Steeples 1995. Applications of shallow, high-resolution seismic reflection. to various mining operations. *Mining Engineering*, **47**(4), 355-361.

Sharma, P. V. 1997. *Environmental and Engineering Geophysics.* Cambridge University Press, Cambridge.

Hydrogeological characterization by electrical resistivity surveys in granitic terrains

A.S. Lima & A.C.V. Oliveira
Departamento de Ciências da Terra, Universidade do Minho, 4710-057 Braga, Portugal

Keywords: hydrogeology, electrical resistivity, granite, mineral water

ABSTRACT: Dikes and veins are the most important hydrogeological structures found in granitic rocks which may act either as groundwater pathways or as impermeable barriers. In both cases there is a contrast between the resistivity of these structures and the resistivity of the host rock. In this paper we present the results of an electrical resistivity study carried out in Gerês area, Portugal. Geological setting of Gerês area is dominated by a biotite-rich coarse-grained porphyritic granite crossed by quartz veins and basic rock dikes. The Gerês – Lovios major fault crosses the area with a general NNE-SSW strike and it has a very important role in the deep flow of the Gerês hot springs. Thirty-three vertical electrical soundings were made using the Wenner alpha array. A 3D apparent resistivity model points out the main heterogeneities and the anisotropy of this hydrogeological framework. The lowest values of apparent resistivity belong to the southern sector of the study area, corresponding to the existing spring. In this place, there's another alignment of low resistivity values, probably related to a zone of quartz veins crossing the Gerês – Lovios major fault. The resulting node seems to have a decisive role in the ascending groundwater.

1 INTRODUCTION

In order to improve the mineral aquifer exploitation of the Gerês spa (Northern Portugal), a step-by-step hydrogeological exploration program began in 2000. General geology of this area has been already described by Medeiros et al. (1975). In our study, besides a detailed geological assessment, a geophysical survey was carried out, using electrical resistivity.

The studied area (Fig. 1) is largely occupied by biotite-rich coarse-grained porphyritic granite, crossed by quartz veins and basic rock dikes. The known Gerês-Lovios major fault, probably responsible for the groundwater deep flow, constitutes the most important tectonic structure of that region and is visible through an incised fracture valley along a NNE-SSW strike (Fig. 2). Thus, the geomorphology of the studied area corresponds to the Gerês river basin, a NNE-SSW elongated small catchment, tributary of Cávado river basin.

The Gerês thermal occurrence is a sodium bicarbonate water, with 330 µS/cm electric conductivity, corresponding to an electrical resistivity value of 20 Ωm at 47 °C (emergence temperature). It belongs to the sulphur group of Portuguese mineral waters, characterized by the presence of reduced sulphur species, alkaline pH and low values of redox potential, usually negatives.

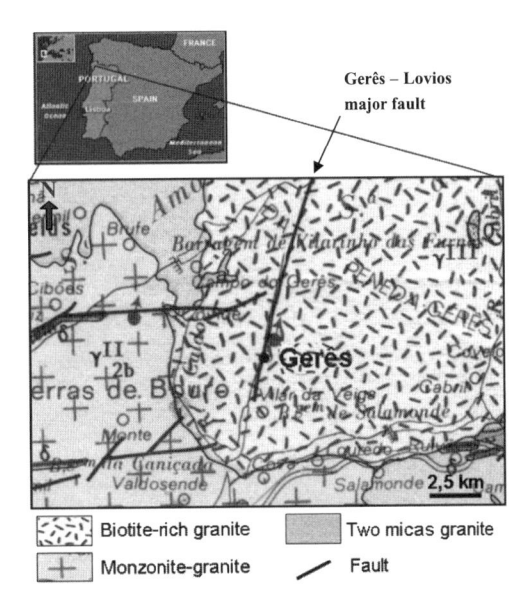

Figure 1. Location map and geology of the studied area. Geological map extracted from Carvalho (1992).

Earlier research developed by Lima et al. (1999), Lima and Silva (2000) and Lima (2001) established a conceptual model where the meteoric water infiltrates in the high relief East sector and moves Westward to the thermal springs of Gerês spa. Furthermore, according to those authors, water infiltrates through a set of fractures transverse to the Gerês-Lovios major fault and rises by free convection and also forced convection, helped by the higher permeability of the discharge zone.

Trying to verify that model, especially the groundwater pathways near the springs, an electrical resistivity survey was performed in the eastern sector of the already mentioned main fault (Fig. 2).

Evans (1999), is the most convenient arrangement for studying lateral resistivity variations.

The acquisition parameters of the resistivity survey are summarized in Table 1. All the VES starts with "a" = 2 m. Maximum values of "a" are very variable due to topographic restrictions. According to Loke (2000), the maximum depth of investigation in each VES is given by:

$$Z = 0,519 \, a \qquad (1)$$

where Z = depth of investigation, m; a = electrode spacing, m.

As shown in Table 1, the maximum depths of investigation, considering equation (1), range from 13 m, in VES3, to 70 m, in VES25. The mean and median values are, respectively, 44 m and 46.7 m. Modal depth is 52 m.

Table 1. Parameters of the thirty-three VES.

Sounding	Minimum "a" (m)	Maximum "a" (m)	Maximum depth (m)
VES1	2	110	57
VES2	2	28	15
VES3	2	26	13
VES4	2	54	28
VES5	2	54	28
VES6	2	88	46
VES7	2	90	47
VES8	2	66	34
VES9	2	130	67
VES10	2	42	22
VES11	2	34	18
VES12	2	42	22
VES13	2	70	36
VES14	2	90	47
VES15	2	64	33
VES16	2	50	26
VES17	2	78	40
VES18	2	84	44
VES19	2	90	47
VES20	2	88	46
VES21	2	102	53
VES22	2	110	57
VES23	2	114	59
VES24	2	126	65
VES25	2	134	70
VES26	2	100	52
VES27	2	110	57
VES28	2	100	52
VES29	2	100	52
VES30	2	100	52
VES31	2	100	52
VES32	2	100	52
VES33	2	100	52

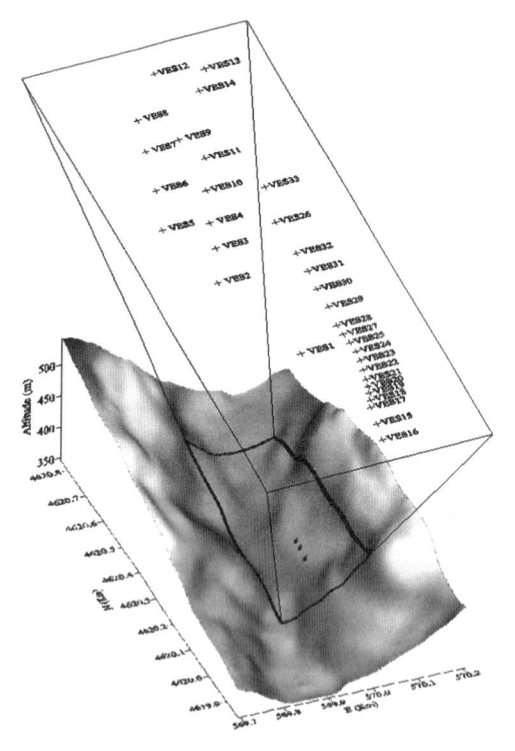

Figure 2. Morphology of the Gerês river valley and distribution of the thirty-three vertical electric soundings (VES#) in the studied area. VES1 is very close to the thermal springs (⑨).

2 METHODS

2.1 Resistivity survey

Resistivity survey was performed using an ABEM resistivemeter, model Terrameter SAS 300C. Thirty-three vertical electric soundings (VES) were made in the vicinity of spring's area (Fig. 2). Owing to geomorphologic features, there was no possibility to extend the studied area eastwards. The VES were made using the Wenner alpha array which, according to

2.2 Data processing and interpretation

To achieve the highest quality of field data, almost all measurements of resistance were taken after sixty-four readings. Resistance values acquired in the field were converted into apparent resistivity values according the equation:

$$\rho_a = 2\pi aR \qquad (2)$$

where ρ_a = apparent resistivity, Ωm; a = electrode spacing, m; R = resistance, Ω. As suggested by Loke (2000), unusual high or low apparent resistivity values were corrected manually, from graphical inspection, to obtain a smooth curve.

2D and 3D models were created with Rockworks software, an integrated geological data management, analysis, and visualization tool collection, by Rockware Inc. All built models were based in the inverse-distance/anisotropic algorithm.

3 RESULTS AND DISCUSSION

3.1 Vertical electrical soundings

As shown in Figure 3, the VES that is nearest to the thermal springs (VES1) shows high values of apparent resistivity near the topographic surface which, gradually, decreases as depth increases. Apparent resistivity reaches the lowest values (near 50 Ωm) between "a" = 20 m and "a" = 40 m and then goes back to higher values, but never reach the initial ones. Since the resistivity survey was made in the summer, the high values of apparent resistivity of the shallowest zone could be explained by the low degree of soil water saturation. Reaching the water table, apparent resistivity decreases to values little higher than the resistivity of thermal water. On the other hand, subsequent increase in apparent resistivity values is related to the influence of the bigger rock mass involved.

The VES9 is located around 250 m away from the thermal springs (Fig. 2) and shows a one-dimensional resistivity pattern similar to VES1 one. Nevertheless, there are three major differences between them: first, the mean value of apparent resistivity in the former is lower than in the latter; second, the lowest values of VES9 are reached deeper than VES1, which can be explained by the distance from the main anomaly area, near the VES1; third, in VES9, from "a"=100 m, values of apparent resistivity show a progressive decrease, reaching the lowest values at highest depths. Since the electrodes were spread out in the N-S direction, the decrease in apparent resistivity values with depth is probably related to the inclusion of the spring's area in the rock mass researched by electrical measurements. In fact, as one can see in Figure 3 (VES1), this area shows low apparent resistivity values, which induces the lowering of apparent resistivity values in VES9.

3.2 2D models

A 2D model based on apparent resistivity values for a = 10 m (Fig. 4) shows quite a subordination to terrain morphology. The lowest values are located near two incised streams, whose channels correspond to possible faults situated across the Gerês-Lovios major fault. The southern one is filled with a much crushed quartz vein. As showed by Pereira (1992), Pereira and Almeida (1994) and Lima (2001), this kind of veins is considered to be very important hydrogeological structures, due to their ability to conduct groundwater, which is supported by the electrical resistivity results.

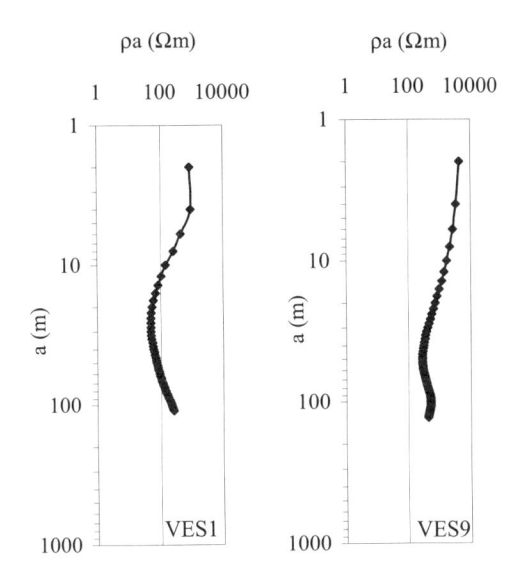

Figure 3. Variation of apparent resistivity with electrode spacing in VES1 and VES9.

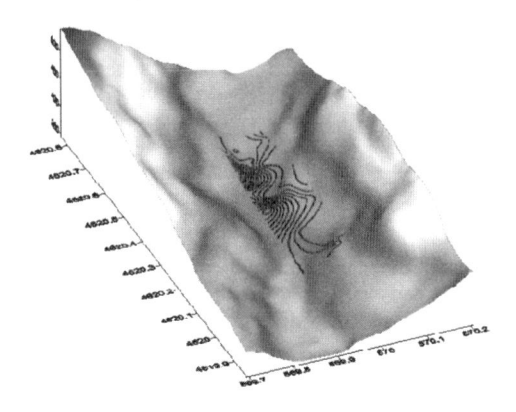

Figure 4. Overlapping of the two-dimensional distribution of apparent resistivity at a = 10 m and the morphology map.

The northern resistivity anomaly (north-eastern sector in Fig. 4) follows a possible NE-SW fracture zone. As shown in Figure 4, the apparent resistivity tends to increase westward, as a result of decreasing bedrock depth.

The general trend above mentioned is again present at "a" = 30 m and "a" = 50 m (Fig. 5). Two-dimensional model for "a" = 30 m suggests the presence of two fractures zones: one of them is located close to the mineral springs, being roughly W-E; the other, roughly NW-SE, is placed about 200 – 250 m apart from the former.

Comparing the models for "a" = 10 m and "a" = 30 m, there are evidences of a fracture zone in the NE sector but the results of these two models indicates different directions. This may be explained as follows: the apparent resistivity anomaly found in the "a" = 10 m model could be related to a shallow fracture, which supplies the tributary of Gerês river. The NW-SE fracture detected in the "a" = 30 m model, seems to be deeper and more persistent. This is confirmed by the "a" = 50 m two-dimensional model (Fig. 5 on the right).

The southern fracture (Fig. 5) is located in the same position both in the "a" = 30 m model and in the "a" = 50 m model, indicating a vertical fracture. The northern fracture is displaced northward from the "a" = 30 m model to the "a" = 50 m model, suggesting that the fracture dips towards north.

A two-dimensional model based in a north-south cross section, taken at the middle of the study area (AB profile, Fig. 5, left), shows the position of the water table (Fig. 6). The low degree of water soil saturation near the topographic surface is expressed by the highest values of apparent resistivity, reaching magnitudes of several thousands of Ωm. This model suggests that the south sector is less resistant than the northern one, being particularly promising for hydrogeological purposes. This is in agreement with the present knowledge of local hydrogeology, since the mineral aquifer discharge occurs in this site.

The CD cross-section (Fig. 7) shows the effect of the low weathered granite present in the NW sector of the studied area. In this sector there are several outcrops and some blocks of fresh granite, which account for the high values of apparent resistivity recorded. As in the AB cross-section, one can see in the south-eastern sector of the CD cross-section a low resistivity zone, probably related to the intersection between the major fault and the transverse one, where the mineral springs are located.

The EF and GH cross-sections are quite different, concerning both their absolute apparent resistivity values and spatial distribution of them. The EF model (Fig. 8), taken in the southern sector of the studied area across the Gerês River, shows a low resistivity anomaly near the mineral springs. This anomaly seems to spread towards the river (on the left). The GH model (Fig. 9), based in apparent resistivity values obtained in the northern sector (see Fig. 5) defines, as a whole, a zone of high resistivity values. The slight differences between these two cross-sections seem to correspond to different levels of soil water saturation. It should be noted that GH vertical profile doesn't cross the Gerês-Lovios fault zone.

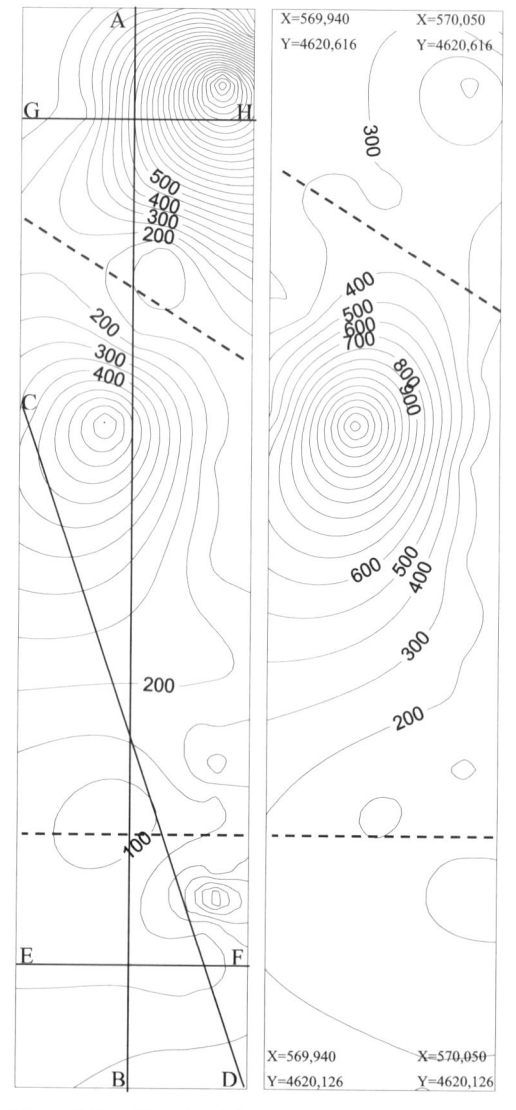

Figure 5. Two-dimensional (horizontal profiles) models of apparent resistivity for a = 30 m (left) and a = 50 m (right). Blue dashed lines show possible faults. AB, CD, EF and GH plain lines drawn in the left horizontal profile show the locations of vertical profiles plotted in Figs. 6, 7, 8 and 9, respectively.

3.3 3D model

Based in all apparent resistivity values acquired in the field surveys, and applying a upper filter obtained from altitude values of topography and an

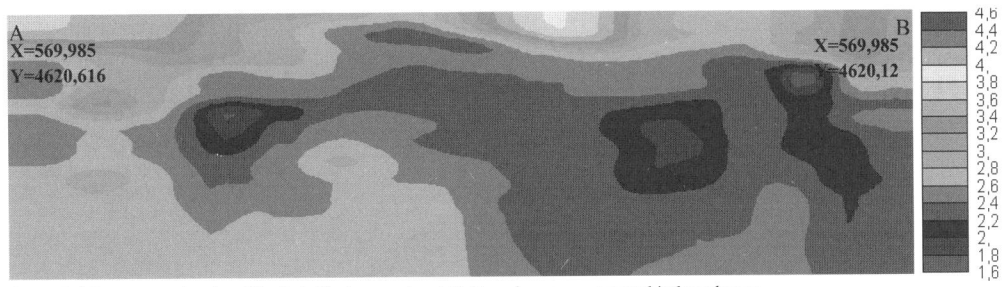

Figure 6. AB cross-section (see Fig. 5, left). Apparent resistivity values are expressed in log ohm.m.

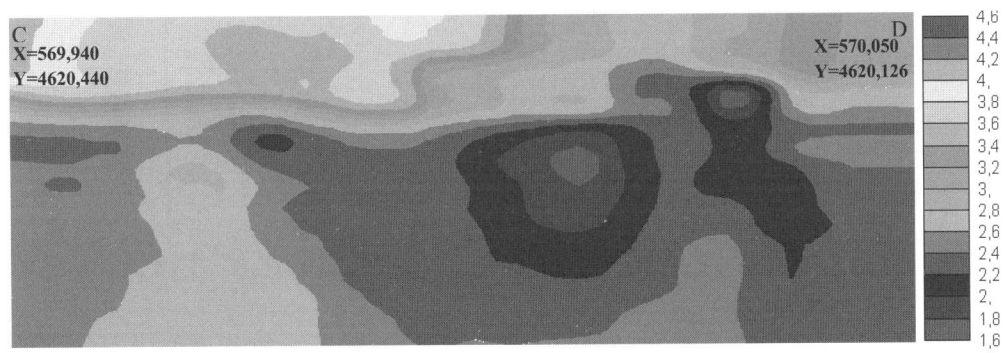

Figure 7. CD cross-section (see Fig. 5, left). Apparent resistivity values are expressed in log ohm.m.

Figure 8. EF cross-section (see Fig. 5, left). Apparent resistivity values are expressed in log ohm.m.

Figure 9. GH cross-section (see Fig. 5, left). Apparent resistivity values are expressed in log ohm.m.

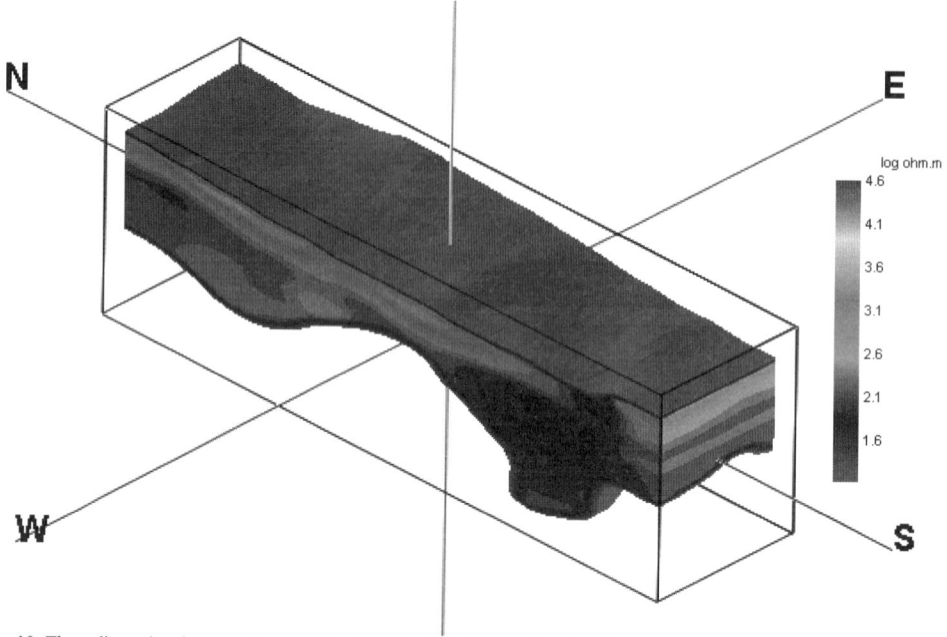

Figure 10. Three-dimensional model of apparent resistivity of Gerês area, showing the perspectives of southern and western plans. Color of upper and lower surfaces doesn't have any resistivity meaning.

lower filter based in the maximum "a" values reached in each vertical electric sounding, a 3D model was constructed to show a 3D perspective of the underground of the studied area (Figs. 10 and 11).

Figure 10 shows a perspective of southern and western plans. As a general trend, it could be noted the well defined dry upper soil zone, reaching very high values of apparent resistivity, usually in the magnitude of thousands of Ω.m.

As shown, the variation of maximum "a" values in the different soundings brings some difficulties to the interpretation of the deepest zone.

Considering that we are dealing with a heterogeneous medium it's not straightforward to extrapolate values from one known area to an unknown one. So, we don't know the behavior of the deepest areas of the middle and northern sectors western side of the studied area. Nevertheless, as shown in Fig. 10, the southern sector is characterized by low resistivity values from the surface to deeper zones. As already mentioned, this zone is coincident with the spring's discharge area. In addition, spring's discharge area shows low values of apparent resistivity, due to the influence of low resistivity of mineral water. Indeed, considering the mineralization of water (electrical conductivity = 330 µS/cm at 25 °C), as well as its emergence temperature (almost 47 °C), water resistivity is estimated in nearly 20 Ωm.

Figure 11 shows a perspective of eastern and northern plans of 3D apparent resistivity model. The low resistivity zones found in the previous perspective (Fig. 10) are also present here. The water table is close the topographic surface in the eastern sector, which is in accordance with field observations. Nevertheless, it's necessary to keep in mind that this water table corresponds to a non-mineral aquifer, that is, a phreatic aquifer, while the above mentioned water surface (seen in the western plan of Fig. 10) probably belongs to the mineral aquifer. The zone of high resistivity values found in the north-eastern sector of the studied area (Fig. 11) can be explained by two main reasons: the possible presence of fresh blocks of granites and the distance from the most important discharge area. It should be noted that the maximum deeps reached in the eastern sector is generally higher than those of western sector.

4 CONCLUSIONS

This study has enabled us to improve and confirm our understanding of the subsurface structures associated with the Gerês spa. The higher permeability of the present aquifer discharge area is confirmed by the low values of apparent electrical resistivity recorded there. As suggested by Lima and Silva (2000), the eastern sector of the Gerês mount seems

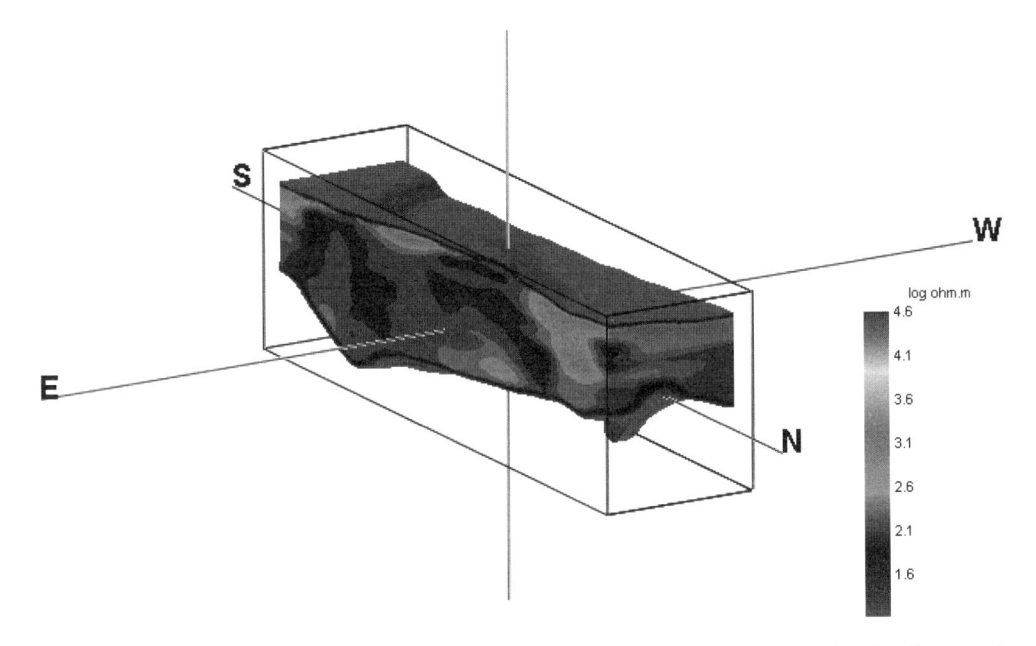

Figure 11. Three-dimensional model of apparent resistivity of Gerês area, showing the perspectives of north and eastern plans. Color of upper and lower surfaces doesn't have any resistivity meaning.

to play an important role in the groundwater flow, which is confirmed in our study. The western side of the Gerês river valley doesn't seem to contribute to the Gerês hydromineral occurrence, as was also proposed by Lima e Silva (2000). Thus, Gerês River is probably the mineral aquifer boundary. Limitations due to topographic impositions, didn't allow us to reach depths higher then 70 m (VES 25). Notwithstanding these restrictions related to topographic conditions, the low magnitude of apparent resistivity values suggests a promising hydrogeological environment, mainly in the southern sector of the studied area, close to the present springs. The usefulness of 3D resistivity imaging is proven despite the above mentioned unfavourable conditions.

ACKNOWLEDGMENTS

A. Lima acknowledges the support provided by the Empresa das Águas do Gerês, S.A. The authors are greatly indebted to Anabela Campos, Manuel Vieira, Fabíola Silva, Ana Paula Gonçalves and Rute Areal for their valuable help in the field surveys.

REFERENCES

Carvalho. D. 1992. *Carta Geológica de Portugal. Escala 1/500 000*. Serviços Geológicos de Portugal.
Evans, A. M. 1999. *Introduction to Mineral Exploration*. Oxford, U.K: Blackwell Science.
Lima, A.S. & Silva, M.O. 2000. Utilização de Isótopos Ambientais na Estimativa das Áreas de Recarga em Regiões Graníticas (Minho - NW de Portugal). *As Águas Subterrâneas no Noroeste da Península Ibérica. Textos das Conferências, Mesa Redonda e Comunicações, A Coruña, 3 a 6 de Julho de 2000*: 387-394.
Lima, A.S. 2001. *Hidrogeologia de Terrenos Graníticos. Minho - Portugal*. PhD thesis. Universidade do Minho, Braga, Portugal.
Lima, A.S., Silva, M.O., Carreira, P.M.M. & Nunes, D. 1999. An Isotopic Study of -Groundwater in Granitic Terrains (Northwest Portugal). *Ninth Annual V. M. Goldschmidt Conference, LPI Contribution n° 971, Lunar and Planetary Institute, Houston*: 171-172.
Loke, M.H. 2000. *Electrical Imaging Surveys for Environmental and Engineering Studies. A Parctical Guide to 2-D and 3-D Surveys*. Available at: http://www.georentals.co.uk
Medeiros, A.C., Teixeira, C. & Lopes, J.T. 1975. *Carta Geológica de Portugal na Escala 1/50 000. Notícia Explicativa da Folha 5-B (Ponte da Barca)*. Serv. Geol. Portugal, Lisboa.
Pereira, M.R.C. & Almeida, C. 1994. Captação de Águas Subterrâneas em Rochas Cristalinas: Factores que Influenciam a Produtividade. *Actas do 2° Congresso da Água: o Presente e o Futuro da Água em Portugal*, Lisboa, Vol II: 95-106.
Pereira, M.R.C. 1992. Importância dos Filonetes de Quartzo na Pesquisa de Água Subterrânea em Rochas Cristalinas. *Geolis* Vol VI, n° 1 e 2: 46-52.

Development of a TDR dielectric penetrometer

C-P. Lin, C-C. Chung & S-H. Tang
Dept. of Civil Engineering, National Chiao Tung University, Hsinchu, Taiwan

Keywords: dielectric permittivity, conductivity, Time Domain Reflectometry (TDR), CPT

ABSTRACT: Electrical properties of a soil include the electrical conductivity and dielectric permittivity. Conventional electrical probes measure only the resistivity (the reciprocal of the electric conductivity) of the soil. Interpretation of the resistivity alone for soil physical properties is difficult because it is sensitive to many factors, such as water content, soil types and ground water characteristics. The Time Domain Reflectometry (TDR) is a geophysical test method based on electromagnetic waves. It can be used to make simultaneous measurements of electrical conductivity, apparent dielectric constant, and dielectric constant at different frequencies of a soil. The dielectric properties provide extra information related to soil physical properties. Current TDR probes can only apply to soils at surface. The paper describes the development of a TDR cone penetrometer that is capable of providing continuous TDR measurements during the cone penetration. Various probe configurations were experimentally studied. The data reduction method for the determination of apparent dielectric constant and electrical conductivity has been formulated and calibrated. The region of influence around the probe and the effect of penetration on TDR measurements are also studied.

1 INTRODUCTION

Conventional in situ testing methods, such as SPT, CPT, and DMT, focus on mechanical response of the soil under test. The physical properties of a soil, such as water content, void ratio, and physical chemistry of pore water, are of greater concern in geo-environmental concerns. Soil-water interaction and soil microstructure are also important to the renewed focus on the fundamentals of soil behavior.

Physical properties of soils are estimated using laboratory techniques on samples retrieved from a borehole or test pit. While measurements can be made in the field, in general, these measurements are made after the field investigation and on a limited number of samples. Hence, questions arise as to how representative laboratory samples are of the actual field conditions. In particular, it is relatively difficult to obtain undisturbed sample in a sand deposit. It is also costly and time consuming to obtain physical properties using traditional drilling and laboratory techniques.

Field tests for characterizing soil physical properties are likely to be implemented using electrical methods. Electrical properties of a soil include the electrical conductivity and dielectric permittivity. Most electrical probes measure only the resistivity (the reciprocal of the electric conductivity) of the soil. Interpretation of the resistivity alone for soil physical properties is difficult because it is sensitive to many factors, such as water content, soil types and ground water characteristics. The Time Domain Reflectometry (TDR) is a geophysical test method based on electromagnetic waves. It can be used to make simultaneous measurement of electrical conductivity, apparent dielectric constant, and dielectric constant at different frequencies of a soil. The dielectric properties provide extra information related to soil physical properties.

Current TDR probes can only apply to soils near ground surface. A research to develop an efficient field method to estimate various physical properties of soils using time domain reflectometry was undertaken by the National Chiao Tung University in Taiwan. An initial objective of the research project was to better utilize the electrical properties for characterizing microscopic properties of soils. The main steps of the research are:
1. development of field probes suitable for TDR measurements of soils at various depths,
2. construction of homogenization models for soil electrical properties as a function of soil composition.

3. development of dielectric spectroscopy of soils using the field probe, and
4. development of theoretical or semi-empirical relations to extract soil physical properties from electrical properties.

As part of phase one, a TDR probe was developed to be used in conjunction with the Cone Penetration Test or deployed as a permanent sensor with the CPT. The probe was specifically designed to be directly inserted into soil without the need for digging, drilling, or other types of soil preparation. This paper describes details of the probe design and measurements that can be done.

2 BACKGROUND AND THEORY

2.1 Time Domain Reflectometry

The basic principle of time domain reflectometry (TDR) is the same as radar. But instead of transmitting a 3-D wave front, the electromagnetic wave in a TDR system is confined in a waveguide. Figure 1 shows a typical TDR measurement setup composed of a TDR device and a transmission line system. A TDR device generally consists of a pulse generator, a sampler, and an oscilloscope; the transmission line system consists of a leading coaxial cable and a measurement waveguide. The pulse generator sends an electromagnetic pulse along a transmission line and the oscilloscope is used to observe the returning reflections from the measurement waveguide due to impedance mismatches. Such instruments have been used since 1930's for cable testing prior to Fellner-Feldegg (1969) using them for measuring dielectric properties of liquids. The concept has been extended to measurements of electrical properties of soils in which TDR probes are embedded (Topp et al. 1980; Dalton et al. 1984; Heimovaara 1994; Lin 2003).

2.2 Measurements of electromagnetic properties

The electrical properties of a soil include dielectric permittivity (ε) and electrical conductivity (σ). The dielectric permittivity is in general a complex number and a function of frequency. The equivalent dielectric permittivity (ε^*), representing the total effect of the frequency-dependent complex dielectric permittivity (ε) and the conductivity (σ) of a soil, can be written as, (Ramo et al., 1994)

$$\varepsilon^*(f) = \varepsilon'(f) + j\varepsilon''(f) = \varepsilon'(f) - j\left(\varepsilon''(f) + \frac{\sigma}{2\pi f \varepsilon_0}\right) \quad (1)$$

where f is the frequency; j is $(-1)^{1/2}$; ε' and ε'' are the real and imaginary parts of dielectric permittivity, respectively; ε^{ii} is the imaginary part of the equivalent dielectric permittivity, and ε_0 is the dielectric permittivity of free space.

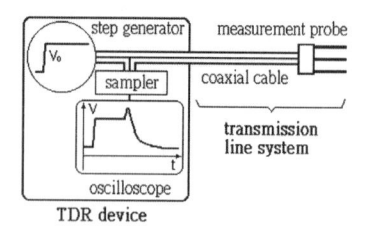

Figure 1. A typical configuration of a TDR measurement system.

The transmission line wave equation derived from Maxwell's equations governs the electromagnetic wave propagation in a transmission line. Propagation constant and characteristic impedance are two intrinsic parameters that can be defined in the general solution of the wave equation. The propagation constant, a function of the dielectric permittivity of the insulating material between conductors, determines the phase velocity and attenuation of the wave propagation. The characteristic impedance is a function of the cross-sectional geometry of the conductors as well as the dielectric permittivity of the insulating material between the conductors. Some electromagnetic wave is reflected and recorded by the TDR device if the impedance changes along the transmission line.

Since the dielectric permittivity of the insulating material depends on frequency, the propagation velocity is also a function of frequency. The TDR waveform recorded by the sampling oscilloscope is a result of multiple reflections and dispersion. A typical TDR output waveform is shown in Fig. 2. The experimental time-domain information may be treated in the frequency domain to obtain the dielectric permittivity as a function of frequency (Giese & Tiemann 1975; Heimovaara 1994; Lin 2003). This involves deriving the system function as a function of the impedance, propagation constant, and boundary conditions and it has different form depending on the configuration of the probe. The system function for the field probe to be developed involves much work in electromagnetics and is the main scope of phase three of the multiphase research project. However, simple methods are available for determining apparent dielectric constant and electrical conductivity.

The propagation velocity (v) of an electromagnetic wave that travels in a material with equivalent dielectric permittivity (ε^*) is a function of frequency since the dielectric permittivity depends on frequency. It can be written as, (Ramo et al., 1994)

$$v(f) = \frac{c}{\sqrt{\frac{\varepsilon'(f)}{2}\left(1 + \sqrt{1 + \left(\frac{\varepsilon''(f)}{\varepsilon'(f)}\right)^2}\right)}} \quad (2)$$

where c is the speed of light. The denominator in Eq 2 can be considered as the apparent dielectric permittivity of each frequency component. Topp et al. (1980) ignored the dielectric relaxation and loss and assumed the denominator to be a constant. Accordingly, the denominator in Eq 2 was replaced by the apparent dielectric constant (K_a) and the corresponding propagation velocity was called apparent velocity (v_a). K_a can be determined from the measured v_a to be,

$$\sqrt{K_a} = \frac{c}{v_a} = \frac{c\Delta t}{2L} \quad (3)$$

v_a is determined from the time difference between the arrivals of the two reflections (as shown in Fig. 2) and the round-trip length of the probe in the soil.

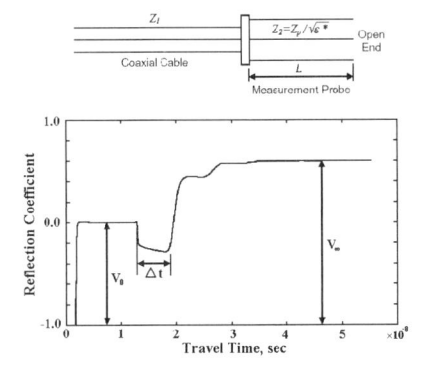

Figure 2. Interpretation of the TDR waveform to estimate apparent dielectric constant and electrical conductivity.

The electrical conductivity (σ) can be measured using the zero-frequency response, which is readily obtained from the reflected signal at long time, once all multiple reflections have taken place and equilibrium is reached (i.e. V_∞ in Fig. 2). According to the derivation of Giese & Tiemann (1975), the electrical conductivity can be written as

$$\sigma = \left(\frac{\varepsilon_0 c}{L}\right)\left(\frac{Z_p}{Z_s}\right)\left(\frac{2V_0}{V_\infty} - 1\right) = c_1 + \frac{c_2}{V_\infty} \quad (4)$$

where L is the length of the probe, Z_p is the impedance of the probe filled with air, Z_s is the output impedance of the TDR device (typically 50 ohm), V_0 is the amplitude of the signal coming from the TDR system, and V_∞ is the asymptotic value of the reflected signal. For probes of known characteristics, Z_p may be calculated from probe dimensions (Ramo et al., 1994). Alternatively, the lumped parameters (or called probe constants) c_1 and c_2 can be inferred from TDR measurements in media of known electrical conductivities.

3 PROBE DESIGN AND CALIBRATION

3.1 Probe design

Waveguides or probes for TDR measurements are primarily of two types: coaxial type and multiconductor type, as shown in Fig. 3(a) and 3(b). The coaxial type of probe is composed of a cylindrical cylinder (CC) acting as the outer conductor and a rod along the centerline of the cylinder acting as a central conductor. The multi-conductor type of probe is composed of one or more rods acting as the outer conductors and a center rod as the inner conductor. The coaxial type of probe is adopted for laboratory measurements such as in the compaction mold or in a Shelby tube, using the cylindrical cylinder as the outer conductor with the inner conductor being a rod inserted along the centerline of the soil in the mold. The multi-conductor probes can be used for in-place measurements. Conventional multi-conductor probes are 30 cm long and therefore difficult to insert at depths below a few feet. In order to adapt the TDR technique to a cone penetrometer application a new design is required for the probe. The multiple conductors are placed around a non-conducting shaft to form a TDR probe as shown in Fig 3(c).

Also shown in Fig. 3 are the electrical potential distributions corresponding to the cross-sections of different probe types. The electrical field is contained in the CC for a coaxial probe while it is open in multi-conductor probe. The material near the center conductor contributes more to the TDR response, and hence has higher spatial weighting to the dielectric properties measured from the TDR response. Baker & Lascano (1989) and Knight (1992) have studied the spatial sensitivity of the measured dielectric permittivity. It should be noted that the material inside the shaft of the TDR cone penetrometer is different from the surrounding material to be measured by design. Therefore, calibration procedures need to be developed for measurements of the apparent dielectric constant and electrical conductivity. In addition, probe should be designed to minimize the effect of the material inside the shaft and maximize the influence zone in the surrounding medium. A series of prototype probes were constructed in the lab to obtain the optimal configuration for the waveguide. The variables considered include the number of conductors and conductor width (or spacing). The PVC tubes were used as the shaft and copper strips as the waveguide conductors. The configurations of the prototypes were summarized in Table 1.

(a) (b) (c)

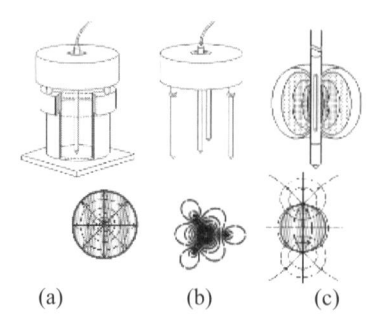

Figure 3. Configurations of types of transmission lines and illustrations of their associated electrical potential distribution.

Table 1. Probe types with different conductor configurations.

Type No.	Copper width (mm)	Copper length (mm)	No. of copper	Probe type
T1	20	200	4	
T2	30	146	3	
T3	20	200	2	
T4	10	200	2	
T5	3	200	2	

3.2 Calibration for dielectric constant and electrical conductivity

The dielectric constant measured by the penetrometer probe is a weighted average dielectric constant of the soil and the probe material between the conductors. A convenient homogenization model is based on Birchak's exponential model (Birchak et al. 1974), in which the effective (or measured) apparent dielectric constant ($K_{a,eff}$) is related to the soil dielectric constant ($K_{a,soil}$) and probe dielectric constant ($K_{a,probe}$) as

$$\left(K_{a,eff}\right)^n = a\left(K_{a,soil}\right)^n + (1-a)\left(K_{a,probe}\right)^n \quad (5)$$

in which n is an empirical constant that summarizes the geometry of the medium with respect to the applied electric field and a is a weighting factor of the surrounding soil. According to Birchak et al. (1974), the theoretical value for n is 1.0. The last term in Eq 5 can be lumped as an empirical parameter b, since the probe dielectric constant is a constant. The soil dielectric constant can be determined from the TDR penetrometer measurement as

$$\left(K_{a,soil}\right)^n = \frac{\left(K_{a,eff}\right)^n - b}{a} = \frac{\left(\frac{c\Delta t}{2L}\right)^{2n} - b}{a} \quad (6)$$

where n, a and b are calibration parameters for dielectric constant.

Similarly, the probe material between the conductors affects the effective electrical conductivity. Following the same reasoning for dielectric constant and assuming $n = 1$, the soil electrical conductivity can be determined from the TDR penetrometer measurement as

$$\sigma = \alpha + \frac{\beta}{V_\infty} \quad (7)$$

where α and β are calibration constants for electrical conductivity.

4 EVALUATION OF PROBE PEFORMANCE

The multi-conductor penetrometer waveguides may have different features in the TDR response depending on the number of conductors and conductor width. The optimum probe configuration should result in TDR waveforms in which travel time analysis can be easily performed. In addition, the effective dielectric constant should be as close to the soil dielectric constant as possible, and the probe have an influence zone around it as far as possible. These features associated with various probe types listed in Table 1 were evaluated.

4.1 TDR waveforms and effective dielectric constant

Time domain reflectometry measurements were made by attaching the TDR probe to a Tektronix 1502C (Tektronix, Beaverton, Or) via 2 m of 50-ohm coaxial cable fitted with 50-ohm BNC connectors at each end. The multi-conductor penetrometer waveguides were submerged in a big tank filled with tap water. Figure 4 shows the TDR waveforms in water for waveguides with different number of conductors. Similarly, the waveforms in water for 2-conductor waveguides with different conductor width are shown in Fig. 5. The waveform of a coaxial probe is also shown in Fig. 4 and Fig. 5 for comparison. The length of the coaxial probe is 116 mm. The length of the penetrometer waveguide is 200 mm except for probe T2. The TDR sends a step pulse down the cable and some of the wave energy is reflected from both the beginning and end of the probe as shown in Fig. 4 and Fig. 5. The first positive reflection is due to the connector between the cable and the probe. The sudden drop of the waveform resulting from the negative reflection occurs when the pulse enters the probe section. And the second positive reflection occurs at the end of the probe. As the number of conductors and conductor width increases, the impedance of the probe decreases and the negative reflection at the beginning of the probe increases, causing the waveform drops down to a lower level. In terms of waveform shape, the reflections in probe T1 is more apparent and can be easily identified.

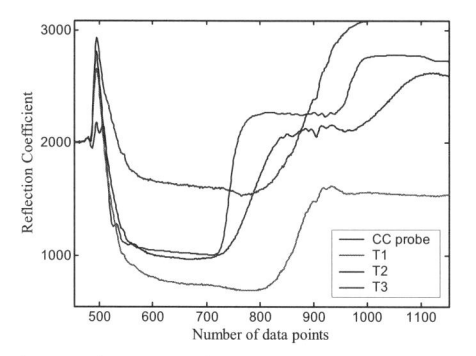

Figure 4. The TDR waveforms of probes with different number of conductors.

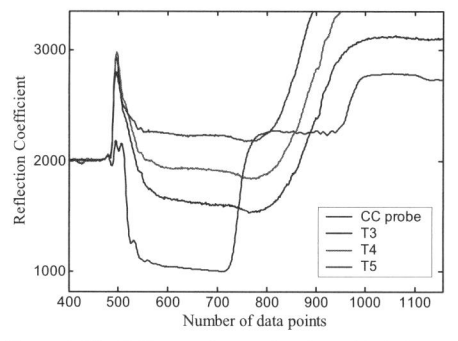

Figure 5. The TDR waveforms of probes with different conductor spacing (width).

Waveforms of the penetrometer probes are more dispersive (i.e. rise time of the step pulse is longer) than that of the coaxial probe. This is due to the connector between the BNC connector and the prototype probes. The traveltime Δt of the penetrometer probe is about 75% of that of the coaxial probe of the same length. All penetrometer probes perform similarly in this regard. The effective dielectric constants measured by the probes listed in Table 1 are all near 42, which is approximately equal to $(K_{a,water}+K_{a,probe})/2$, in which $K_{a,water} = 80$ and $K_{a,probe} \approx 4$. Considering the theoretical value $n=1.0$, Eq 5 can be simplified as

$$K_{a,eff} = \frac{K_{a,soil} + K_{a,probe}}{2} \qquad (8)$$

for TDR dielectric penetrometers shown in Table1.

4.2 *Radial sampling in TDR measurements*

The radial sampling in TDR measurements may be investigated using electromagnetic field theory. Alternatively, an experimental approach was taken since the theoretical derivation is too complicated and need to be experimentally verified. In order to investigate the radial sampling in TDR measurements using the dielectric penetrometers, the prototype probes were submerged in water-filled PVC tubes of different diameters. Since the dielectric constant of water and air are in two opposite extreme, $K_{a,wate} = 80$ and $K_{a,air} = 1.0$. The spatial weighting function may be defined experimentally as

$$F(r) = \frac{K_{a,r}}{K_{a,eff}} \times 100\% \qquad (9)$$

where $K_{a,r}$ is the effective dielectric constant measured in an water-filled PVC tube with inner diameter r and $K_{a,eff}$ is the effective dielectric constant measured in a big water-filled tank.

For different probe configurations, the spatial weighting function (F) can be plotted as shown in Fig. 6. The effective dielectric constant becomes asymptotic at a distance of 100 mm and greater. The majority of the electromagnetic response occurs within the first several centimeters in the radial direction. The spatial bias for the two-conductor probes (T3, T4, and T5) is slightly less than the three-conductor probe (T2) and four-conductor probe (T1); while Probe T1 and T2 have similar spatial weighting function. For the two-conductor configuration, the spatial bias is independent of the conductor width. Similarly, the weighting function for electrical conductivity is shown in Fig. 7. The material near the probe weights even more in conductivity than dielectric constant.

While probes T1 and T2 are more sensitive to near material in terms of dielectric constant; they are less sensitive in conductivity relative to 2-conductor probes (T3, T4, and T5). Observations from Fig. 6 and Fig. 7 raise the concern for the penetration (disturbance) effect on K_a and σ measurements in soils. The soil displaced by the penetrometer is likely to change the density of soil adjacent to the penetrometer. The nature of variation in density around the penetrometer due to cone penetration will influence the dielectric constant and electrical conductivity. This should be a common problem to all electrical probes that has been overlooked in the past.

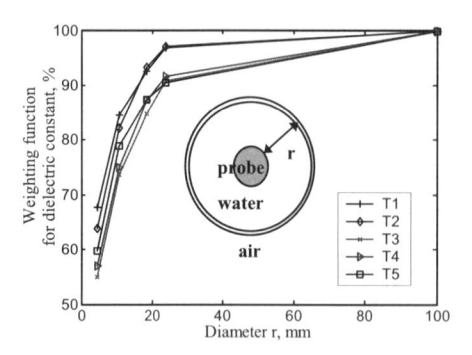

Figure 6. Spatial weighting function for dielectric constant.

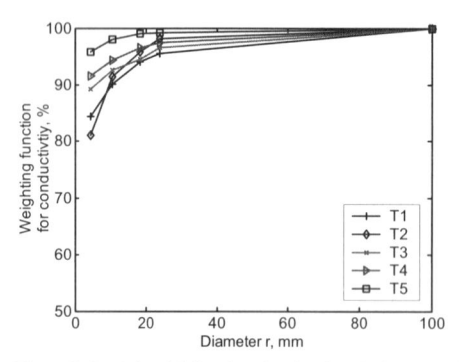

Figure 7. Spatial weighting function for electrical conductivity.

5 SIMULATED PENETRATION TEST

A TDR dielectric penetrometer was actually fabricated using the design similar to type T1. Type 1 may not be the optimum configuration as shown in Fig. 6. However it was selected at the time when the major concern was to have TDR reflection that can be identified most easily for all cases (i.e. from dry to wet soils). Figure 8 illustrates the design and picture of the probe. The probe consists of four arc-shape stainless steel plates and a delrin shaft. The thickness of the stainless steel was maximized to increase the axial strength of the probe. The stainless steel plates were fit into four grooves in the delrin shaft and fastened with screws. This probe was used to perform simulated penetration test in a calibration chamber.

5.1 Results of calibration

The TDR penetrometer shown in Fig. 8 differs from Type 1 probe in that thick conductors are embedded in a dielectric shaft instead of thin conductors bonded to the surface of the dielectric shaft. Calibration tests need to be carried out before it can be put into used for measurements of dielectric constant and electrical conductivity. Several liquids of known dielectric constants and electrical conductivities were used for calibrating the probe using Eqs 6 and 7. The materials used for calibrating dielectric constant were air, butanol, ethanol, and water; while water with different amount of added NaCl was used for calibrating electrical conductivity. Assuming theoretical value $n = 1.0$, the calibrated parameters $a = 0.34$ and $b = 1.91$, respectively. If n remained unknown during calibration, the calibrated parameters $a = 0.35$, $b = 1.78$, and $n = 0.96$. Note that the a value is smaller than 0.5, as suggested by Eq 8, because the thick conductor plates are embedded in the delrin grooves instead of stick to the surface. Using the calibrated parameters, the apparent dielectric constants of the calibrating liquids are plotted against their known values in Fig. 8. Both calibrated results provide fairly good fit. The theoretical value $n = 1.0$ is also verified in Fig. 9. For simplicity, $n = 1.0$, $a = 0.34$, and $b = 1.91$ are used. Similarly, the calibration constants for electrical conductivity were obtained as $\alpha = -0.04$ and $\beta = 145.71$. The estimated electrical conductivity using the calibrated parameters fits the known values extremely well, as shown in Fig. 10.

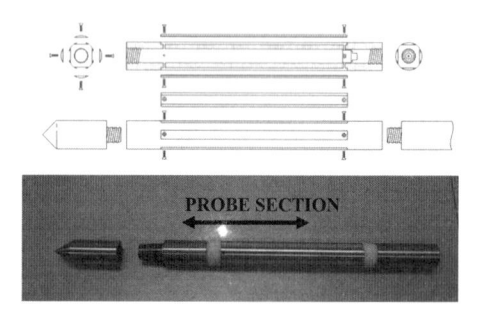

Figure 8. Prototype of the TDR penetrometer.

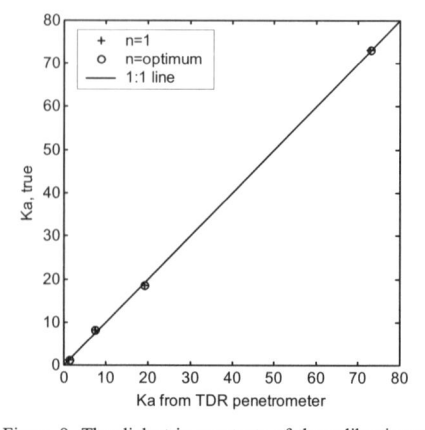

Figure 9. The dielectric constants of the calibrating materials vs. that estimated after calibration.

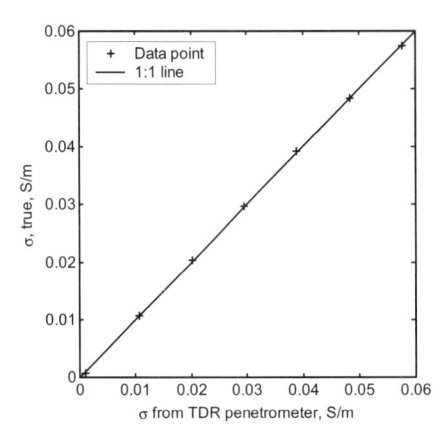

Figure 10. The electrical conductivities of the calibrating materials vs. that estimated after calibration.

5.2 Applications

A silty sand (SM) was used for the simulated penetration tests. Seven different gravimetric water contents were used to prepare samples in a calibration chamber. The soil and water were mixed thoroughly to obtain the desired water content. The mixed soil was sealed with plastic wrap and allowed to equilibrate for more than 24 h, to yield a uniform soil specimen. The soil was then compacted in the calibration chamber in layers and the total mass of the soil and chamber was measured. Two TDR measurements were taken, one with the TDR penetrometer and the other with a multi-rod probe (MRP) similar to Fig. 3(b). Then samples of the soil were oven-dried to determine the gravimetric water content. Tests were performed twice for each water contents to evaluate the repeatability.

The variation of TDR waveforms as the soil water content increases is shown in Fig. 11. The dielectric constant and electrical conductivity increases with water content, as can be inferred from Fig. 11. A good correlation between $\sqrt{K_a}$ and volumetric water content (θ) exists as shown in Fig. 12. The correlation between $\sqrt{\sigma}$ and θ shown in Fig. 13 also shows great linearity. The $\sqrt{K_a}$-θ relationship is relatively independent of soil type and electrical conductivity of pore water (reference). But the $\sqrt{\sigma}$-θ relationship greatly affected by pore water electrical conductivity. Therefore, apparent dielectric constant can be used for measuring volumetric water content (or void ratio when the soil is saturated). The volumetric water content and electrical conductivity can then provide extra information for determining the characteristic of the pore water. Further research involves the dielectric spectroscopy of soils using the TDR penetrometer. The dielectric spectrum may add another dimension of information to the apparent dielectric constant and electrical conductivity. De-

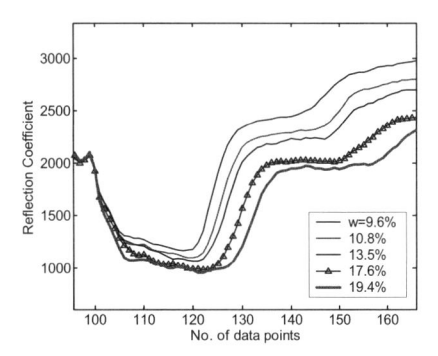

Figure 11. TDR waveforms for soils of different water contents.

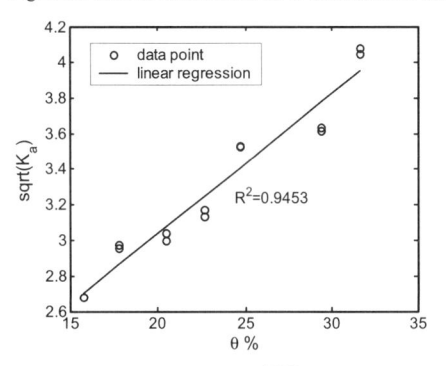

Figure 12. Correlation between $\sqrt{K_a}$ and θ.

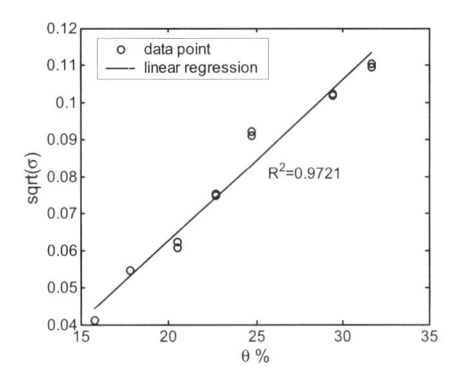

Figure 13. Correlation between $\sqrt{\sigma}$ and θ.

tailed discussion of the use of electrical properties is beyond the scope of this paper.

5.3 Effect of penetration

In addition to the measurements using the TDR penetrometers, TDR measurements were also performed using a MRP probe. The diameter of the multiple rods is 9.5 mm and the spacing between the center conductor and outer conductors is 65 mm. The effect of penetration on TDR measurements using the MRP is considered negligible (Siddiqui et al. 2000). Comparing the measurements of TDR penetrometer with that of MRP can reveal the effect of penetration. The comparison is shown in Fig. 14 and Fig. 15 for K_a and σ, respectively. No surcharge was added on top of the calibration chamber. Considering the low confining pressure, the soil in the simulated penetration test should be dilative. Hence, the void ratio increases due to probe insertion. The increase in void ratio results in a decrease in apparent dielectric constant and electrical conductivity, as verified in Fig. 14 and Fig. 15. The effect of penetration is much more pronounced for electrical conductivity than for dielectric constant. This can be explained by comparing the spatial weighting function for electrical conductivity to that for dielectric constant. The change in dielectric constant due to probe insertion is less than or comparable to the uncertainty associated with the $\sqrt{K_a}$-θ correlation in this case. This adds one more reason to why dielectric constant rather than electrical conductivity should be used for water content (or void ratio) measurements. The simulated penetration used a hammer to penetrate the TDR cone penetrometer. This may cause air gap betweenthe penetrometer and soil. More comprehensive study and refined penetration test may be necessary to quantify the effect of penetration in various cases.

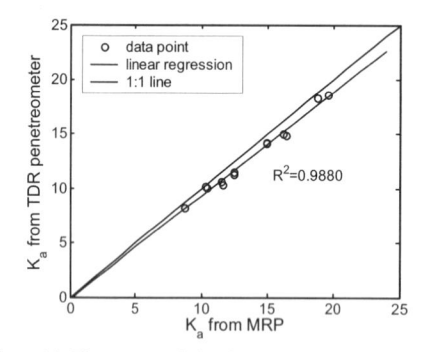

Figure 14. The apparent dielectric constant obtained from TDR penetrometer vs. that from MRP.

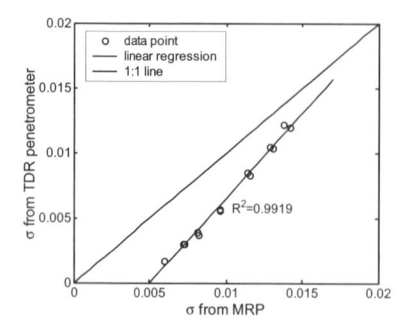

Figure 15. The electrical conductivity obtained from TDR penetrometer vs. that from MRP.

6 CONCLUSION

Time domain reflectometry is a promising technique for simultaneously measuring the dielectric constant and electrical conductivity of a soil in situ. Current TDR probes can only apply to soils near ground surface. The paper describes the development of a TDR cone penetrometer that is capable of providing continuous TDR measurements during the cone penetration. Various probe configurations were experimentally studied. The data reduction method for the determination of apparent dielectric constant and electrical conductivity has been formulated and calibrated. The region of influence around the probe and the effect of penetration on TDR measurements are also studied. Research is under way to develop dielectric spectroscopy using the TDR penetrometer and new applications of this new technique. More work must also be undertaken to quantify and minimize the effect of probe insertion.

ACKNOWLEDGEMENTS

The research was sponsored by the National Science Council of ROC under contract numbers 89-2218-009-100 and 90-2611-E-009-004.

REFERENCES

Birchak, J.R., Gardner, C.G., Hipp, J.E. & Victor, J.M., 1974. High dielectric constant microwave probes for sensing soil moisture. *Proceedings IEEE* 62:93-98.

Baker, J.M. & Lascano, R.J. 1989. The spatial sensitivity of time-domain reflectometry. *Soil Science* 147:378-384.

Dalton, F.N., Herkelrath, W.N., Rawlins, D.S. & Rhoades, J.D. 1984. Time-domain reflectometry: simultaneous measurement of soil water content and electrical conductivity with a single probe. *Science* 224:989-990.

Fellner-Felldegg, J. 1969. The measurement of dielectrics in the time domain, *Journal of Physical Chemistry* 73:616-623.

Giese, K. & Tiemann, R. 1975. Determination of the complex permittivity from thin-sample time domain reflectometry: improved analysis of the step response wave form. *Adv. Mol. Relax. Processes* 7:45-59.

Heimovaara, T.J. 1994. Frequency domain analysis of time domain reflectormetry waveforms: 1 measurement of the complex dielectric permittivity of soils. *Water Resources Research* 30:189-199.

Knight, J.H. 1992. Sensitivity of time domain reflectometry measurements to lateral variations in soil water content. *Water Resource Research* 28: 2345-2352.

Lin, C.-P. 2003. Analysis of a non-uniform and dispersive tdr measurement system with application to dielectric spectroscopy of soils. *Water Resources Research* 39(1): art. no. 1012.

Ramo, S., Whinnery, J.R. & Van Duzer, T. 1994. *Fields and Waves in Communication Electronics*. 3rd ed., John Wiley, New York.

Topp, G.C., Davis, J.L. & Annan, A.P. 1980. Electromagnetic determination of soil water content and electrical conductivity measurement using time domain reflectometry. *Water Resources Research* 16:574-582.

Joint acquisition of SWM and other seismic techniques in the ISC'2 experimental site

I. Lopes & I. Moitinho
Centro and Departamento de Geologia da Faculdade de Ciências da Universidade de Lisboa, Portugal

C. Strobbia
Politecnico di Torino, Dipartimento di Georisorse e Territorio, Italy

P. Teves-Costa
Centro de Geofísica, Departamento de Física da Faculdade de Ciências da Universidade de Lisboa, Portugal

G.P. Deidda
Università degli Studi di Cagliari, Dipartimento di Ingegneria del Territorio, Italy

M. Mendes
ICIST, Departamento de Física do Instituto Superior Técnico, Portugal

J.A. Santos
ICIST, Departamento de Engenharia Civil e Arquitectura do Instituto Superior Técnico, Portugal

Keywords: ISC'2 test site, surface wave method, SH-reflection, noise measurements

ABSTRACT: Different seismic techniques are jointly used for the characterization of the experimental test site prepared for the ICS'2 Conference: the Surface Wave Method, SH Reflection and Noise Analysis (H/V). Geologically the site is composed by a residual soil layer resultant of weathering of granites, about 15 meters thick, overlaying granite bedrock. This set of tests provide consistent data that allow comparison between the different types of non-intrusive seismic techniques, discussing the reliability and limitations for each particular technique applied for this case study.

1 INTRODUCTION

On the occasion of the International Site Characterization Conference ISC'2, the organizing committee set up a test site for geotechnical characterization and invited several groups of researchers to make a "scientific competition" carrying out experiments with different geophysical techniques, in order to find the most effective one. Each group, charged of testing a particular technique, received a little information about the site and so it experimented its technique without knowing if the technique itself would work or not.

The test site is located at the FEUP Campus (Figure 1) in Porto (Portugal). The ISC'2 test site was characterized by a pavemented area with trees and a slab, where it was opened a trench with 50×1.5×0.5m approximately (Figure 2). Our group were performed field tests in May 2003. The conditions were not the more appropriate for seismic testing existing: the presence of a building, a thick concrete plate (the slab), tree roots, a highway and a noisy machine working nearby make the site not suitable for seismic testing.

In the geological map (Figure 3) the site is located near a border zone between the Granites of Contumil and the Schist-Greywacke complex. The Contumil Granites are described as medium to coarse-grained granites with phenocrystals of feldspar, forming saprolitic soils, very weathered showing both arenization and kaolinitization. The cartography between these geological units in the contact zone is very difficult because here the materials are very weathered, the passage between materials is gradual and the deformation caused by the contact metamorphism resulting of ascent of the granitic magma, becoming sometimes very difficult to distinguish the rocks present (Costa & Teixeira, 1957).

Figure 1. Location of the ISC'2 Test site

Figure 2. ISC'2 test site

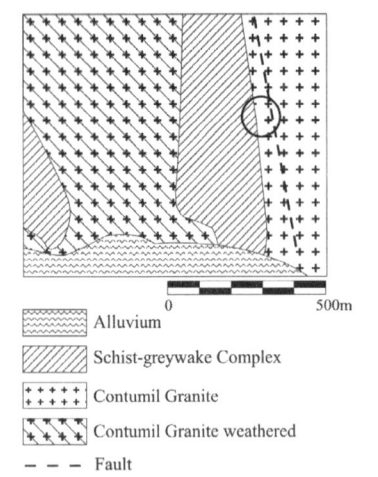

Figure 3. Geological map of the area of the ISC'2 test site (O).

The geological information that was given to our working group was only that the boreholes performed showed more or less 15 m of very weathered granites followed by granite bedrock.

Here, we refer to tests made using Surface-wave method (SWM), SH-wave seismic reflection method and Nakamura's noise spectral ratio method.

2 SEISMIC TECHNIQUES APPLIED

2.1 Surface Wave Method

The use of surface waves in geotechnical engineering applications was introduced with the SASW (spectral analysis of surface waves) method (Nazarian & Stokoe, 1984). In this method the Rayleigh waves are generated by an impulsive source and recorded in two receivers. More recently has been introduced an acquisition scheme with multiple receivers, usually 24 or more, which has a faster field procedure and more accurate results due to the higher spectral integrity of the acquired data (Gabriels et al., 1987; Park et al., 1999; Foti, 2000; Strobbia, 2003).

Generally the surface wave method can be divided in three different steps: acquisition; processing and inversion (Foti, 2000).

The acquisition (Figure 4) has the objective of recording the full waveform generated by an active or a passive source using a seismograph and low frequency transducers (usually below 10 Hz). To perform this step it is possible to use different kinds of equipment, for instance hammer or vibrating sources, geophones or accelerometers as receivers. The acquisition parameters, i.e. number of receivers, time sampling parameters, layout geometry are very important for the acquisition of good quality data.

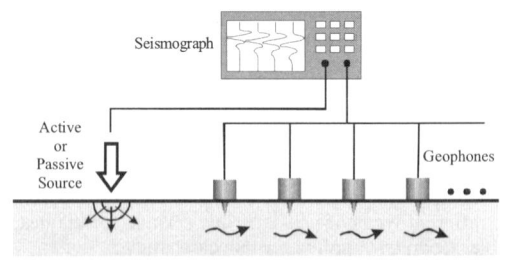

Figure 4. Multistation acquisition scheme

The processing consists of extracting from the raw data the dispersive characteristics of the Rayleigh waves, which are the more energetic events acquired (Richart et al., 1970), to obtain the dispersion curve, i.e. the phase velocity as a function of frequency (or wavelength). For instance the spectral analysis of data acquired with an impulsive source, with wide frequency content, can be performed using a Fourier Transform. The processing becomes difficult by the presence of other seismic events, lateral variations, attenuation of the signal with distance, and by the multi-modal nature of the Rayleigh wave propagation in vertically heterogeneous media.

The inversion is the last step of the surface wave method. The aim is to estimate the physical parameters of a soil model based on the dispersion curve obtained, i.e. transforming the properties of the wave propagation into the physical properties of a layered soil model. A set of mathematical relations is used to theoretically predict the physical behaviour of the system. When model parameters are known the output can be predicted in the forward problem. The opposite situation is when the output is known and the system unknown, being this the inverse problem, where the task is to estimate the model parameters.

2.1.1 Acquisition

The surface wave tests were performed using the multistation array configuration (Figure 4) with an impulsive source – 10 kg sledgehammer on metal plate. The data were acquired with a RAS 24 Seistronix station and 24 vertical low frequency geophones (Geospace, 4.5 Hz).

Several set-ups were used to acquire seismic data (Figure 5):
1. Line 1 has been acquired with three configurations within the trench, with at least two end-off shots on both sides of the line, one 48 m length with 2 m spacing between geophones, one with 24 m length with 1 m spacing and the last with 36 m length with 1 m spacing acquired in two different steps moving the geophones (Figure 6);
2. Line 2 was acquired above the slab, using a 12 m line with 0.5 m spacing between geophones;
3. Line 3 was acquired half in the slab half in the concrete pavement (N68°W), 12 m length with 0.5 m spacing between geophones.

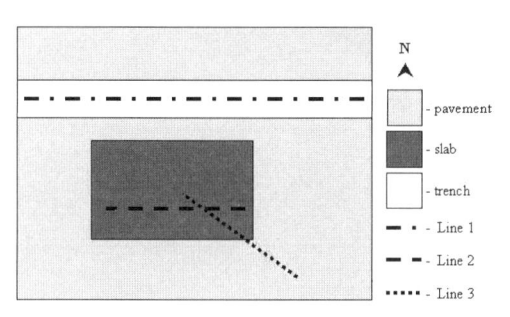

Figure 5. Scheme of the ISC'2 test site showing the locations of the acquisitions

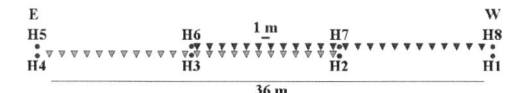

Figure 6. Scheme of the 36 m acquisition showing the several shot positions used (from H1 to H8).

Several shots were acquired and recorded separately for each line to allow a statistical evaluation of the data quality.

2.1.2 Processing and Inversion

The data were processed using the software package "PoliSurf" (Strobbia, 2003) created by the Politecnico di Torino surface wave workgroup. The raw data (distance-time) are transformed into the frequency-wavenumber (f-k) domain applying a high resolution f-k transform. The experimental dispersion curve is obtained in the f-k domain by searching the energy maximum for each frequency (Foti, 2000; Foti et al., 2001, Strobbia, 2003).

The set of data was processed to analyse which of the acquisitions gave better results, in terms of stability and uncertainty. Figures 7,8 and 9 present the processing of non filtered raw data for the best acquisition in the trench, the best acquisition in the slab and the acquisition with the direction N68°W (half in the slab and half in the low thickness pavemented soil). As it is possible to observe from the f-k spectrum, the main energetic event remains similar in all acquisitions even if the frequency content is different: the presence of the concrete on top allows the propagation of higher frequencies, and a wider frequency band is observed in spectra. The experimental dispersion curve are nevertheless similar, as it is possible to observe in Figure 10, and the observed differences can likely be due to heterogeneities of the site. In the following will be shown only the results of line 1 and line 2.

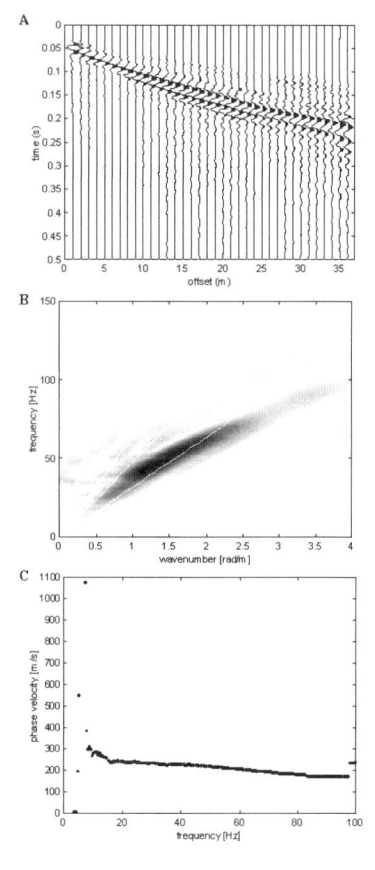

Figure 7. Processing of the best acquisition in the trench – line 1 (36 m): A – seismogram; B – f-k spectrum; C- experimental dispersion curve.

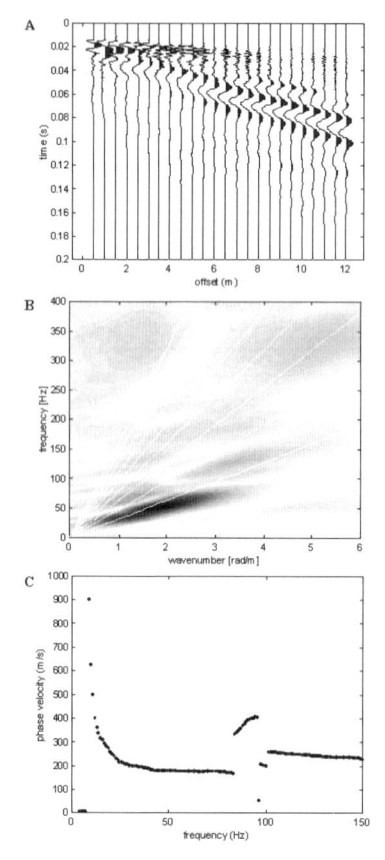

Figure 8. Processing of the best acquisition in the slab – line 2: A – seismogram; B – f-k spectrum; C - experimental dispersion curve.

The inversion was performed with a forward model, giving the model parameters and superimposing the experimental dispersion curve to the synthetic model curves by trial and error. The inversion has been performed using a low parameterisation, i.e. a small number of layers, for the parsimony principle (Tarantola, 1987).

The fitting for the best dispersion curve obtained in the trench (36 m), resulted in a low parameterised model with three layer plus bedrock with a gradual increment in velocity giving a 19 m thickness before bedrock (Figure 11A and 11C, Table 1).

With the same procedure the best acquisition in the slab resulted in a slightly different model also with four layers over bedrock but showing both slightly different velocities and thicknesses (Figure 11B and 11C, Table 1).

The velocities are influenced by the presence of the slab, whose effects decreases at low frequency: the first layer should be characterized or assumed a priori and used in the inversion. If the first stiff layer is simply neglected, a slight overestimation of the velocities is obtained. The depth of the model in the slab is limited by the short length of the array that increases the uncertainty at low frequencies.

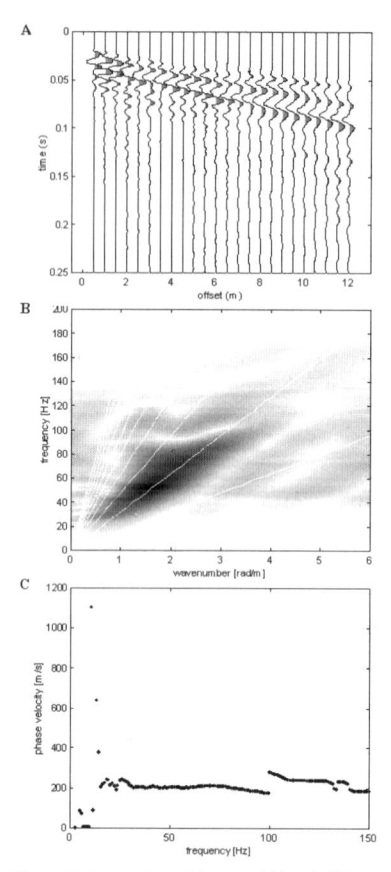

Figure 9. Processing of the acquisition half in and half out of the slab (N68°W) – line 3: A – seismogram; B – f-k spectrum; C- experimental dispersion curve.

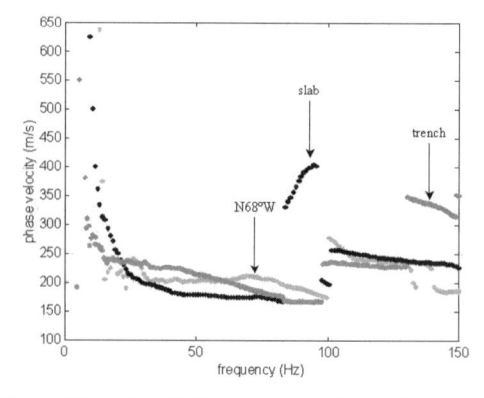

Figure 10. Experimental dispersion curves obtained for the three acquisitions presented.

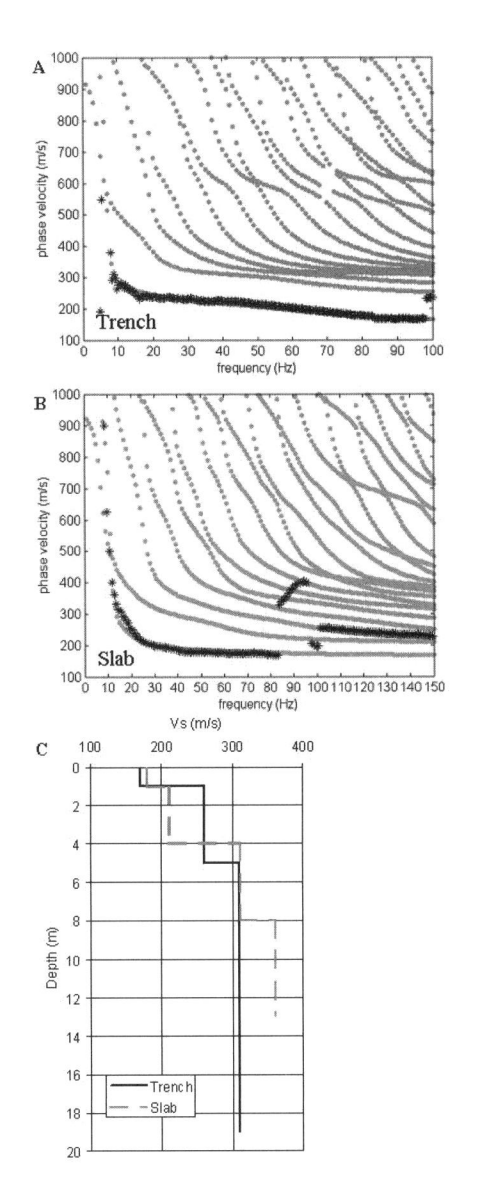

Figure 11. Fitting and low parameterised model for the acquisitions in the trench (A) and in the slab (B). In dots are the modal curves resulting from the forward algorithm for the model in C, and in asterisks are the experimental dispersion curves.

Table 1. Vs models obtained for the fitting in line 1 (36 m) and in line 2

	Line 1 (36 m)		Line 2 (slab)	
	V_S (m/s)	Thickness (m)	V_S (m/s)	Thickness (m)
Layer 1	170	1	180	1
Layer 2	260	4	210	3
Layer 3	310	14	310	4
Layer 4	-	-	360	5
Bedrock	1000	∞	1000	∞

One of the main limitations of the SWM is the simplicity of the model that is assumed (1D): it is hence important to check if a site fits this assumption.

The acquisition of multichannel data with different array configurations and repeated shots offers the possibility of processing redundant data, assessing the data quality, selecting the more reliable portions of the data set, checking the presence of lateral variations, and estimating the uncertainty.

The processing is based on the spectral analysis of data: in frequency domain the properties of the propagation of the different spectral components are estimated by means of two-dimensional transforms as the f-k transform, which give a robust estimate of the dispersive characteristics of the site. A pre-processing allows the identification of lateral variations and near field effects that have to be removed to avoid the data contamination by systematic errors.

Among the data that have been gathered in the trench, a simulated 36 channel acquisition has been selected as the best data set to be processed. The seismogram clearly shows the presence of a dominant dispersive event with a group velocity of more or less 200 m/s (Figure 7A).

The presence of lateral variations has to be assessed since it can strongly influence the inferred characteristics: besides the short length of the used array, the high variability of the geological environment produce lateral variations within the array. The lateral variations can be evidenced by means of the analysis of the phases as a function of the offset, with the MOPI procedure (Strobbia & Foti, 2003): the deviation from the linear behaviour of the phase can correspond to lateral variation and to near field effects. The experimental unwrapped phases at 45 Hz shows the change in the slope along the array (Figure 12): the same effect of velocity variation can be depicted splitting the data in 12 traces and analysing separately the 3 obtained data sets (Figure 13).

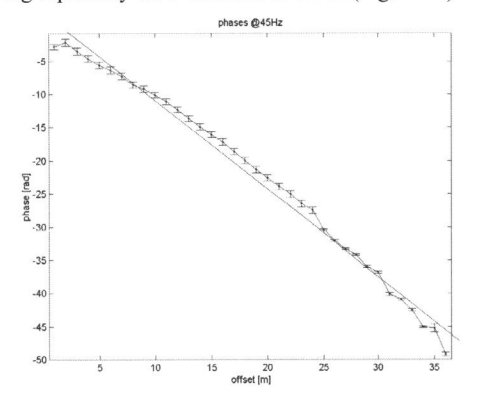

Figure 12. Phases at 45 Hz as a function of the offset, source in 37m: the slope increases with the offset, that is that the phase velocity decreases

The increase of the velocity in the last third of the array is confirmed by the dispersion curves (Figure 14) extracted from the three subset of traces.

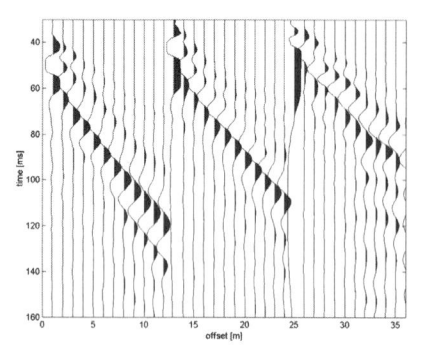

Figure 13. Seismograms of the three subsets of traces acquired in line 1 (36 m).

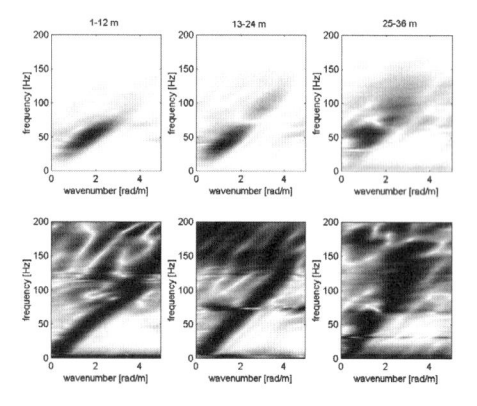

Figure 14. F-k spectrum and normalized spectrum resultant from the processing of the three subsets of traces in Figure 13.

The first part of the record is then analysed to estimate the dispersion curve: the multi offset phase inversion is applied to estimate the presence of near field effects. It can be shown that within a small distance from the source the propagation is influenced by near field effects, and the velocity is lower than the actual value. This could produce an underestimate of the phase velocity at low frequency, and this contamination can be stronger with a low number of receiver.

After having filtered this effects the dispersion curve is estimated by search of energy maxima: the curve that is obtained is however very similar to the one that has been inverted, and the differences are limited to the high frequency band. Lateral variations are hence not critical for the examined site.

2.2 Nakamura's method

Since the beginning of the XX century several scientists were investigating the usefulness of the seismic noise records. During the fifties the correlation between the seismic noise and atmospheric conditions, seismic sources, or other natural sources (as sea waves) has been widely studied and, up to the seventies, these correlations, in terms of noise nature, have been well established. Nowadays, individual seismic noise records are used to characterize the site conditions and array techniques are used to obtain the site vertical velocity profile. Between the different proposed techniques to analyse the seismic noise records, the methodology proposed by Nakamura, in 1989, has been widely used to characterize site conditions. It is well known that several earthquakes can produce different intensities, in a small area (for instance, within a small town). These differences in the observed intensities are due to different geological and geotechnical properties of the shallower soil formations. Nakamura (Nakamura 1989, 1996, 2000) proposed an easy and fast methodology to estimate a pseudo-transfer function for the soil layer and, in particular, its natural frequency.

Nakamura (1989) proposed the definition of a pseudo-transfer function obtained by the spectral analysis of the seismic noise: this function is given by the ratio between the spectra of the horizontal component and the spectra of the vertical component (noted as H/V, for simplification). To define this pseudo-transfer function for a surface soil layer, Nakamura considered the following assumptions: (i) the horizontal motion associated with the seismic noise is mainly composed by S waves whereas the vertical motion is composed by P waves; (ii) at the base of the surface layer the horizontal component is proportional to the vertical component; (iii) in the frequency range of interest the useful seismic noise is mainly the result of S wave reflections and refractions in the base of the surface layer. Taking into account these assumptions Nakamura approximated the true transfer function of the soil layer by the explained ratio H/V (detailed deduction can be found in Nakamura's papers).

If the noise records have the same amplitude than the earthquake records this pseudo-transfer function will be the same than the classical transfer function. However, due to the amplitude differences, this pseudo-transfer function estimates well the natural frequency of the soil deposit but gives a poor estimation of the correspondent amplification level.

2.2.1 Data acquisition and processing

In order to characterize the ISC'2 test site we performed 13 noise measurements in different points distributed along the entire site. The data were acquired with a CityShark seismic station equipped with a 3D 1 Hz Lennartz seismometer (LE-3D) with a sampling rate of 100 Hz. Each record has 10 minutes duration.

In the data processing, each record was divided in several windows of 20 seconds length, with 2 seconds of superposition, according to an algorithm performed to select the most "quiet" windows. The spectral H/V ratio was calculated for each window, and then the mean of all H/V ratios and the standard deviation was computed. Besides, all the records were processed together, in order to obtain a mean value for the whole site. This data processing was performed with the J-SESAME software, developed in the aim of the SESAME project (Sesame, WP03 Team, 2003).

2.2.2 Results

The analysis of the obtained H/V was performed from 0 to 20 Hz. Due to the similarities of the several H/V ratios obtained for each point, we present only the result of all records processed together - Figure 15. By the observation of this figure it is possible to suppose that this site is composed by a softer surface layer overlying a harder formation. This surface layer presents a mean natural frequency close to 4.5 Hz, in spite of some dispersion (values vary from 4.3 to 5.0 Hz, with one exception).

Taking into account that the natural frequency of a soil deposit is equal to Vs / 4 h, where Vs is the S wave velocity and h is the layer thickness, if we know one of these parameters, we can estimate the other.

2.3 SH-wave seismic reflection

Shallow seismic reflection has been widespread use in a variety of environmental, groundwater and engineering applications over the last 15 years. Abundant examples in the literature demonstrate that, in appropriate settings, seismic reflection can successfully contribute to the geologic, hydrogeologic and engineering characterization of the near surface.

Shallow seismic reflection, however, like all geophysical techniques, does not always work. In some cases the success or lack thereof is related to the unfavorable nature of the near surface. In other cases, survey design, equipment and processing techniques can cause failure. Effective use of the technique, even for applications that have classically met with success, strongly depends on the properties of the near surface and how they influence the seismic waves. The appropriateness of shallow seismic reflection must be evaluated for each distinct site and objective. For this concern the optimum test is a walkaway wavefield acquisition. Walkaway test results should guide decisions on how to adjust parameters and/or to modify equipment or, if no reflections are identified, even to terminate the survey.

A seismic walkaway noise test is a test to determine if seismic reflections can be recorded with various shot sources and densely spaced spread of geophones. Recording parameters (analogue filters, sample rate, recording time) are conservatively estimated to determine the optimum settings for a future seismic reflection profile. Once the raw data are collected, post-processing of the field records helps to identify reflections and define the optimum station spacing, source type, source size, source repetition, receiver type, and recording parameters.

In this work, since the available information was insufficient to appropriately design and execute a complete common-midpoint (CMP) seismic reflection profile, a SH-wave walkaway noise test was performed to get the necessary preliminary information about the site. The source position remained fixed as the geophone spread was moved away in order to span offsets from 1.2 m to 22.5 m with 0.3-m geophone interval. The source we used was a wooden plank that had been struck in a perpendicular direction to the geophone line with a 5-kg sledgehammer. The receivers were 20 Hz Swyphones^{TR} (Sambuelli et al., 2001), special detectors with increased sensitivity to horizontal motions. For each geophone spread position several shots were executed with different settings of the recording parameters. A complete set of seventy-two traces is shown in Figure 16.

The dominant energy on the seismogram is associated with direct and/or refracted SH-waves. No evidence for coherent hyperbolic events (reflections) is present through the section.

Since identification of reflection events on unstacked gathers is absolutely essential to the confident, appropriate, and ethical use of shallow seismic reflection profiling, we decided to stop the experiment because SH-wave reflection simply did not work on the ISC'2 test site.

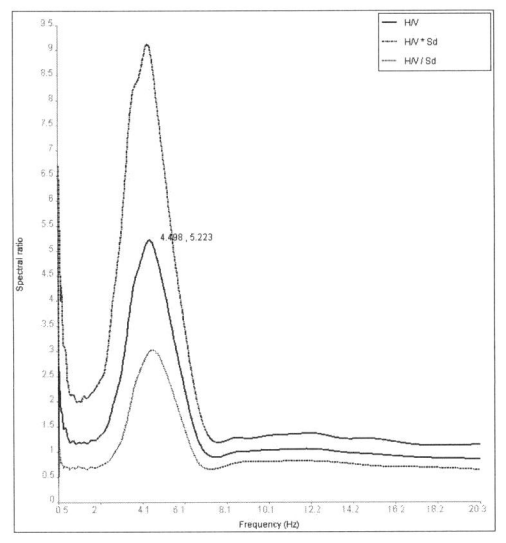

Figure 15. Mean H/V ratio obtained by the simultaneous processing of all the seismic noise records. Black thick curve represents the mean value, the upper dot-line curve represents the mean value multiplied by the standard deviation, and the dotted lower curve represents the mean value divided by the standard deviation.

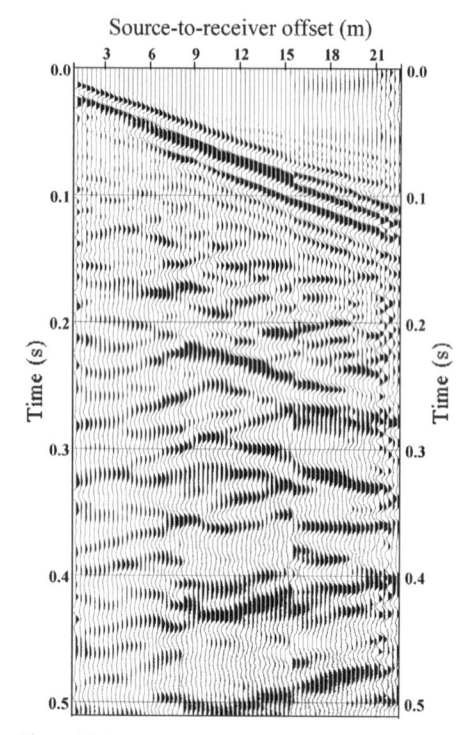

Figure 16. SH-wave walkaway noise test (100-ms AGC scaling). No evidence of reflections are present. The presence of linear low-frequency events below the first arrivals, that mimics the direct wave with the same apparent velocity and with the same or reversed moveout, suggests that vertical "walls" behind and in front of the line might be present.

In spite of the unsuccessful of SH-wave seismic reflection, we analysed the first arrivals of the acquired records to measure S-wave velocity in the near surface and to try to infer the attenuation properties of the near surface material by measurements of the quality factor Qs, related to damping ratio by

$$D\% = \frac{1}{2Q_S} \times 100 \quad (1)$$

2.3.1 Velocity estimation

To obtain values for velocities in the near surface, we picked two layers from the gather with velocities of 239 and 322 m/s (Figure 17). Intermediate refractors with intermediate velocities, however, were also possible to pick, suggesting a gradual increase in velocity with depth and/or a lateral variation.

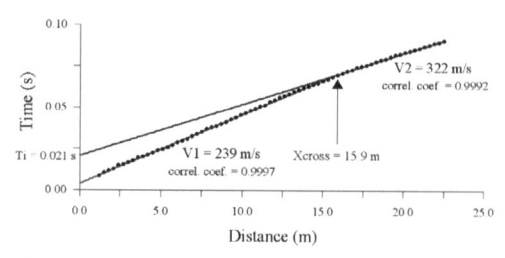

Figure 17. SH-wave travel-times for the first arrivals in the seismogram of Figure 16. In the case of horizontal layering, the values in the figure would give a model earth constituted by a surface layer, with S-wave velocity of about 240 m/s and thickness of about 3.5 m, over a half space characterized by a S-wave velocity of 320 m/s.

2.3.2 Qs estimation

For frequency-independent Qs in the bandwidth of interest, a spherical harmonic wave $A(R,\omega)$ propagating in an attenuating medium with angular frequency ω can be described by the expression

$$A(R,\omega) = \frac{1}{R} A_0(R_0,\omega) \cdot \exp(-\alpha R) \cdot \exp(-j\omega R/v) \quad (2)$$

where $\alpha(\omega) = \omega/2vQs$, $A(R,\omega)$ is the wave amplitude at distance R from the source, $A_0(R_0,\omega)$ is the amplitude at the source, and v is the phase velocity. The $1/R$ term accounts for geometrical spreading and is frequency independent. The exponential term $\exp(-\alpha R)$ accounts for the anelastic attenuation in the medium. The second exponential term $\exp(-j\omega R/v)$ is a time delay that does not enter into the amplitude, and therefore it can be omitted. Then, the spectral components for angular frequency ω at receivers 1 and 2, for example, are related by the expression

$$\frac{A_2(R_2,\omega)}{A_1(R_1,\omega)} = \frac{R_1}{R_2} \cdot \exp\left[-\frac{\omega}{2vQs} \cdot (R_2 - R_1)\right] \quad (3)$$

Converting to logarithms of the amplitude ratios, expression (3) can be written as

$$-\ln\left[\frac{A_2(R_2,\omega)}{A_1(R_1,\omega)}\right] = \ln\left(\frac{R_2}{R_1}\right) + \left[\frac{\omega}{2vQs} \cdot (R_2 - R_1)\right] \quad (4)$$

Hence, plotting the logarithm of the spectral ratio as a function of (angular) frequency should yield a linear trend whose intercept on the ordinate is a measure of elastic losses (geometrical spreading) and whose slope m is a function of Qs

$$Qs = \frac{R_2 - R_1}{2 \cdot m \cdot v} = \frac{\Delta t}{2 \cdot m} \quad (5)$$

where Δt is the travel-time difference between the signals at receivers 1 and 2.

For the present case, to estimate Qs from the first arrivals, these ones were windowed, by a Hamming window with length compatible with the duration of the event, before transforming to the frequency domain to avoid frequency distortions and to stabilize the Qs estimate itself. The dominant frequency of these data was below 170 Hz. Hence, the frequency bandwidth chosen for data analysis was between 20 Hz (the natural frequency of the receivers) and 170 Hz.

Spectral ratios were calculated for many configurations of two receiver locations. The slope is then determined by taking a weighted average of those spectral ratio data versus frequency as shown in Figure 18. Using the first fifty-two traces we obtain a mean value of $Qs = 19$ (D=2.6%), while using the other traces we obtain a mean value of $Qs = 55$ (D=0.9%). These values are appropriate for sands and silty sands.

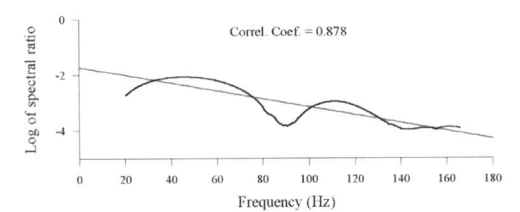

Figure 18. An example of spectral ratio (black curve) versus frequency graph obtained by spectral amplitude of signal from two traces, and used to estimate the quality factor Qs. The gray line is the line that best fits the spectral ratio. The slope of this line together with the travel-time difference of the used signals allow to estimate Qs.

3 CONCLUSIONS

It is well established in literature and also our experience that geophysical techniques allow accurate results in the characterization of a site, and are very helpful in geotechnical characterization helping to direct further work. However each technique has its limitations depending on the site conditions and the purpose of the work.

The test site has some peculiar characteristics that can be critical for some seismic techniques. Ambient noise can be a severe problem for some active seismic testing, as the presence of strong lateral variations of properties, artificial buried objects as for instance the presence of walls or slabs.

Moreover geophysical techniques are large-scale measurements and often the investigation depth depends on the length of the array, i.e. limitations in length are directly limitations in the achieved depth.

For the reasons above mentioned the ISC'2 test site was very challenging for the geophysical characterization. Using the several techniques in our means the workgroup tried to solve the challenge without having any previous idea about the site conditions or expected results. Several techniques were applied, obtaining different results as a function of the conditions and characteristics of the test site, and also as a function of the intrinsic limitations of each technique.

Surface waves are particularly robust because of the high energy of the wave analysed. The several sets of data were processed giving similar models. The results obtained are stable and the estimate obtained for the trench (line 1) seems reliable. The seismic soil profile obtained with the surface wave method is a three layers model with increasing velocity (Table 1) between 170 m/s and 310 m/s before a bedrock whose velocity is apparent and very difficult to determine with only this line length.

The Nakamura's methodology is mainly used to characterize different soil formations within a small area. It is devoted to identify relative soil properties and it did not allow a fine characterization of the soil structure. It is adequate to characterize the soil formations of a given site if it presents a simple structure, composed by a soil layer over a rock formation. In this case, H/V will give the natural frequency of the soil formation. If the soil structure is very complex, or composed by several layers, the interpretation of the H/V ratio is not simple, and needs more additional information.

We found for ISC'2 test site a natural frequency of 4.5 Hz. Taking into account the definition of the natural frequency of a soil deposit. If we considered the geological information given to our working group (15 m of very weathered granites overlaying the granite bedrock), it is possible to estimate a mean velocity of 270 m/s for the weathered layer. If we take the result obtained by the surface wave method, that identified a bedrock formation at 19 m depth, the mean velocity of the surface layer will be 342 m/s.

SH-wave walkaway noise test showed that ISC'2 test site is not appropriate for shallow seismic reflection. Unfavourable geological conditions together with a high noise level prevent to obtain useful SH-wave reflection data. However, first arrivals information allowed the determination of the S-wave velocities and quality factors for the near surface. The SH-wave walkaway noise test revealed very important noise sensitivity, probably associated to the trench geometry conditions, that make difficult to pick with confidence the arrival times of hyperbolic reflections. The undesired reflections from the lateral walls of the trench obscure the desired reflected arrivals from the depth interfaces. In addition, later processing procedures could not improve the signal-to-noise ratio in order to produce more useful seismograms. Therefore, turning our attention to the first arrival reaching the geophones, we could obtain information to estimate a simple structure with constant layer velocities. The thickness of first layer is about 3,5 m with Vs=239 m/s and D=2.6% overlying another layer with Vs= 322m/s and D=0.9%.

The SWM and SH-refraction results are in agreement for the depth and velocity model. The comparison with the Nakamura's method can not be done directly. In fact different combinations of velocities and layer thicknesses can give the same natural frequency.

The surface wave technique has a fast field procedure and is very robust, even if in general other seismic techniques, as the SH reflection, have a higher resolution. SWM proves to be very useful in urban conditions, giving a general idea about the subsoil properties and helping to redirect the characterization campaign.

ACKNOWLEDGMENTS

The authors acknowledge the ISC'2 organization committee for the invitation to do this work.

This work was developed under the research activities of the Centre of Civil Engineering (CEC) from the Faculty of Engineering of he University of Porto (FEUP) and supported by research project: POCTI / ECM / 33796 / 1999: *"Management of sampling quality on residual soils and soft clayey soils. Comparative analysis of in situ and laboratory seismic waves velocities"*, from FCT (Science and Technology Foundation).

This work was also developed under the research activities of CEGUL (Geology Centre of the University of Lisbon), CGUL (Geophysical Centre of the University of Lisbon) and ICIST (Institute for Structural Engineering, Territory and Construction of Instituto Superior Técnico) and was partially supported by pluriannual funding from FCT and by the scholarship SFRH/BD/2962/2000.

REFERENCES

Costa, J.C. & Teixeira, C. (1957) Geological Map of Portugal, scale 1:50.000. Descriptive report of the map 9-C Porto, Serviços Geológicos de Portugal, Lisboa. (in Portuguese).
Foti, S. (2000) Multistation Methods for Geotechnical Characterization using Surface Waves. Doutorato di Ricerca in Ingegneria Geotecnica. Politecnico di Torino, 229 p.
Foti S.; Lancellotta R.; Socco L.V. & Sambuelli L. (2001) Application of fk analysis of Surface Waves for Geotechnical Characterization. Proc. 4th Int. Conf. on Recent Advances in Geotechnical Earthquake Engineering and Soil Dynamics, CD-Rom.
Gabriels P.; Snieder R. & Nolet G. (1987) In situ measurement of shear wave velocity in sediments with higher-mode rayleigh waves, Geophysical prospecting, 35, 187-196
Nakamura, Y. (1989) A method for dynamic characteristics estimation of subsurface using microtremor on the ground surface, RTRI, 30-1, 25-33.
Nakamura, Y. (1996) Real-time information systems for seismic hazard mitigation UrEDAS, HERAS and PIC. QR of RTRI, 37 (3), 112-127.
Nakamura,Y. (2000) Clear identification of fundamental idea of Nakamura's technique and its applications, 12th World Conference on Earthquake Engineering.
Nazarian, S. & Stokoe, II K.H. (1984) In situ shear wave velocity from spectral analysis of surface waves. Proc. 8th Conf. on Earthquake Engineering, S. Francisco, vol. 3. Prentice Hall, pp. 31-38.
Park, C.B.; Miller, R.D. & Xia, J. (1999) Multichannel analysis of surface waves. Geophysics 64(3), pp. 800-808.
Richart, F.E.; Hall, J.R. & Woods, R.D. (1970) – Vibration of soils and foundations, Prentice-Hall.
Sambuelli, L.; Deidda, G.P.; Albis, G.; Giorcelli, E. & Tristano, G. (2001) "Comparison of standard horizontal geophones and newly designed horizontal detectors", Geophysics, vol. 66, n° 6, 1827-1837 pp.
Sesame - WP03 Team (2003) Multiplatform H/V processing software J-SESAME. Deliverable D09.03, June 2003.
Strobbia, C. (2003) Surface Wave Method. Acquisition, processing and inversion. PhD Thesis, Politecnico di Torino, 317p
Strobbia C. & Foti S. (2003) Statistical regression of the phase in surface wave, SAGEEP 2003, San Antonio.
Tarantola, A. (1987) Inverse Problem Theory: Methods for data fitting and model parameter estimation, Elsevier Science Publisher, Amsterdam, 614 p.

Decomposed rock mass characterization with crosshole seismic tomography at the Heroísmo station site (Porto)

M. Oliveira & M.J. Coelho
LNEC – Laboratório Nacional de Engenharia Civil, Lisboa, Portugal

Keywords: rock mass characterization, residual soil, seismic tomography, P-wave velocity

ABSTRACT: At the Heroísmo station site (of Porto underground) there is a granitic rock mass formation named "Granito do Porto", where the rock mass presents significant heterogeneity, with highly weathered and fractured rock zones and with frequent and large intercalations of residual (saprolitic) soil from granite. At the request of Transmetro-ACE, LNEC carried out crosshole seismic tomographies on several crosshole sections with about 30m depth in order to zone and to characterize these heterogeneities.
The tomographies obtained showed heterogeneous P-wave velocity distributions with frequent velocity inversions in depth according to highly weathered and fractured rock zones detected at the boreholes and even with the core recovery rates, especially for the case of residual soil. The lowest velocity zones occur typically on the surface region, above the water level, where the decomposition, weathering and fracture density are more accentuated and where the rock mass is more uncompressed.

1 INTRODUCTION

The Heroísmo station of the Light Rail Metro System, for the Metropolitan Area of Porto, is located at the intersection of Rua do Heroísmo and Rua António Carneiro, in the centre of Porto. This station belongs to Line C, which links the interface centres of Campanhã and Trindade. This line is a tunnel about 2.3km long excavated by a TBM-EPB tunnelling machine. The Heroísmo station was designed as an underground cavern structure with excavation from an access shaft.

Due to the heterogeneity of the granitic rock ("Granito do Porto"), in the area of the access shaft of the station, with various decomposed and fractured zones being found, seismic tests were carried out between boreholes in order to characterise the site in terms of P-wave velocity variations. These crosshole tests sought to obtain a broader coverage of the zone, as large volumes of rock were included in the seismic ray coverage between boreholes. The crosshole tests were based on a group of eleven boreholes (BH01 to BH11 on Fig. 1) until about 30m depth. For each of the tested crosshole sections (CRH01 to CRH10 on Fig. 1), the field set-up for P-wave crosshole seismic tomography was applied to obtain P-wave travel times along ray paths disposed in fans in the section. These multiple travel times measured along several directions allowed the bidimensional tomographic inversion of the times into the P-wave velocity field of the areas crossed by seismic ray paths (Coelho & Oliveira, 2001).

This paper summarises the local geological conditions, supported by the drilling investigation, as well as the method used for the seismic test and for the data processing, and the main results obtained.

2 GEOLOGICAL SETTING AND DRILLING INVESTIGATION

The rock mass found at the site, shown in Fig. 1, belongs to the formation known as "Granito do Porto", which is from the Hercynian age. In general, it is a very heterogeneous, medium-grained, two mica, granite. According to the International Society for Rock Mechanics classification (ISRM, 1981), the granite found ranges from moderately weathered (W3) to highly weathered (W4) and, sometimes, completely weathered, corresponding to a residual soil (W5, W6), even at considerable depths. Blocks of several dimensions of slightly weathered rock (corestones) can also exist, surrounded by granite that is almost decomposed and disintegrated into soil. The rock mass fracturing (discontinuity spacing) varies from moderately/lightly fractured (F3/2) to very fractured (F5).

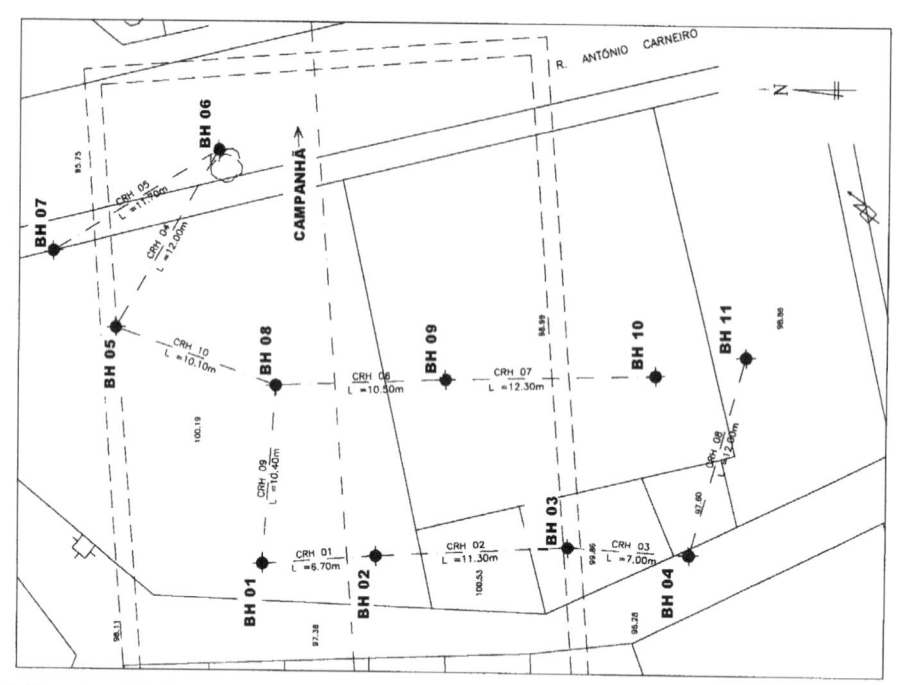

Figure 1. Plane view of the investigation site with the boreholes used (BH01 to BH11) and with the seismic crosshole sections carried out (CRH01 to CRH10).

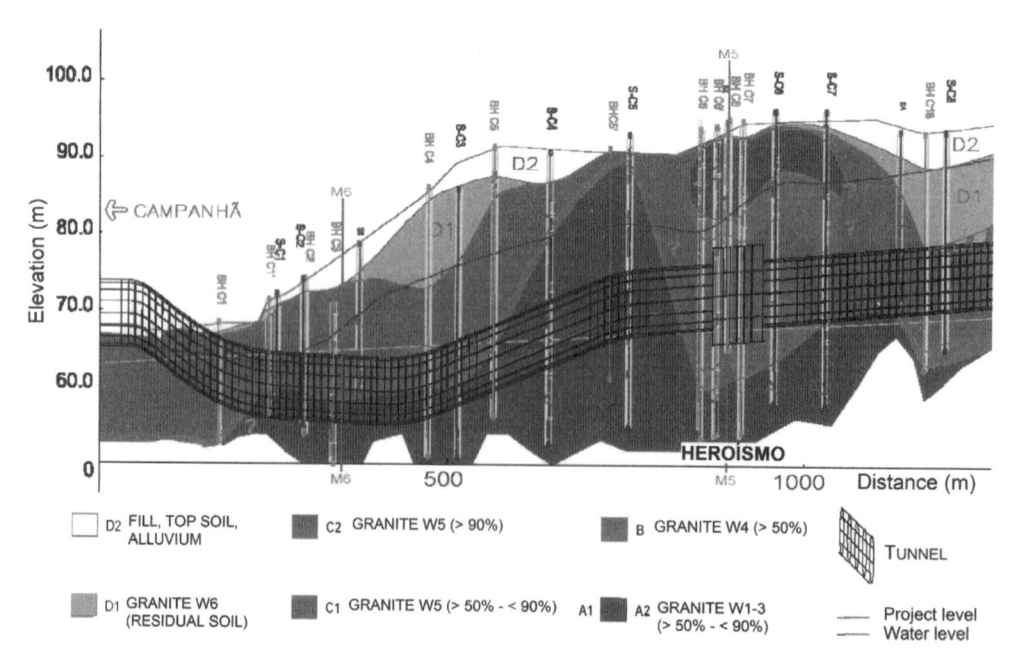

Figure 2. Geotechnical zoning for Line C profile, between Campanhã and Heroísmo station (adapted from Carminé, 2000 and Fruguglietti et al., 2000).

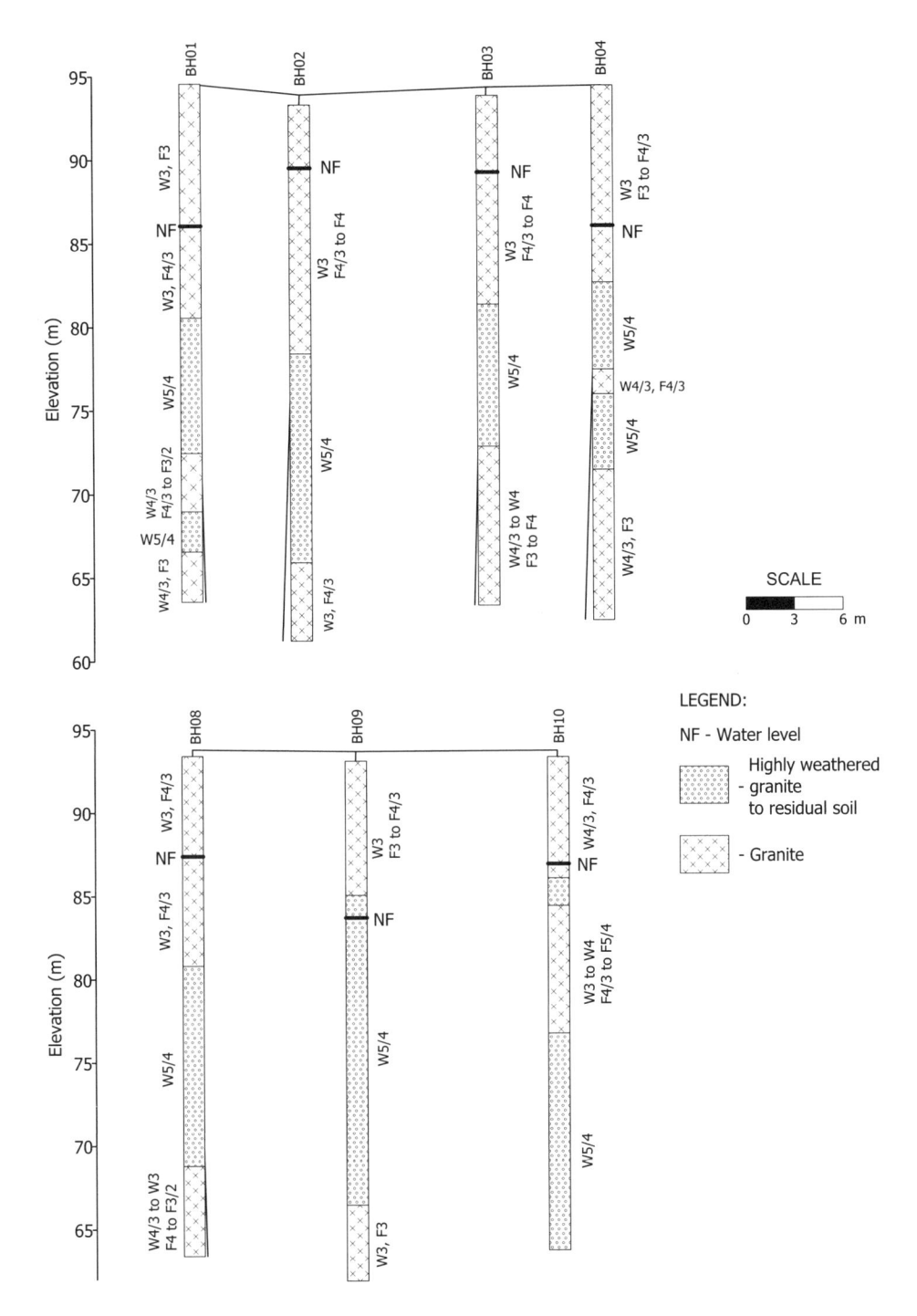

Figure 3. Borehole logs for BH01-BH02-BH03-BH04 and BH08-BH09-BH10 profiles (adapted from Transmetro-ACE).

The geological and geotechnical investigations carried out between 1995 and 1999, with a large number of mechanical tests and drilling being done, mainly along the underground sections, showed a very heterogeneous rock mass, both horizontally and in depth. Fig. 2 presents part of the geotechnical zoning along Line C (Carminé, 2000 and Frugugli-etti et al., 2000), between Campanhã and the Heroísmo zone.

It should be stressed that all the eleven boreholes (BH01 to BH11) drilled in the access shaft zone show high variation of granite weathering with depth. The boreholes show zones ranging from moderately to very weathered granite, with fracture density between F3/2 and F5, to a decomposed granite, and vice versa, as Fig. 3 shows for the two longest profiles (BH01-BH02-BH03-BH04 and BH08-BH09-BH10). At the zones described as completely to highly weathered granite (W5/4), the core recovery rates were, in general, very low. It should be pointed out that this "residual soil" exists at several depths and sometimes has significant extensions in depth (more than 15m height on BH09 borehole). In the area nearest to the surface, the granite is moderately to highly weathered (W3, W4/3) with fracture spacing varying from moderately spaced to closely spaced, with some intermediate sections (F3, F4/3, F4).

The water level at BH01 to BH11 boreholes was about 4.5 to 10m deep.

3 CROSSHOLE SEISMIC TOMOGRAPHIES

Crosshole seismic tomography tests were carried out by LNEC at CRH01 to CRH10 sections (Fig. 1) to obtain the P-wave velocity field (Coelho & Oliveira, 2001).

The seismic tomographies for the BH01–BH02–BH03–BH04 profile (formed by CRH01, CRH02 and CRH03 adjacent sections) and for the BH08–BH09–BH10 profile (formed by CRH06 and CRH07 adjacent sections) are here presented as the most significant examples of this site characterization.

3.1 Data acquisition

The field set-up for pure crosshole seismic tomography used at the Heroísmo station site consisted on placing the seismic source at successive depths in one borehole and recording the source-generated seismic waves at several receivers located along the adjacent borehole. This multiplicity of crosshole seismic measures corresponds to different ray paths along the crosshole section, such as those illustrated in Fig. 4 assuming straight ray paths. From these seismic records it is possible, at least theoretically, to pick the travel time (hence the velocity) or/and amplitude (hence the attenuation) for both P and S-waves. Nevertheless, in practice, due to the difficulties arising from data acquisition and from ambiguity in S-wave and amplitudes picking, the P-wave travel times are the most often used measurements.

For the crosshole tests at the Heroísmo station site, spacing along boreholes between successive receivers and seismic sources was about 1.5m. Borehole distances varied between 6.7m (CRH01 section) and 12.7m (CRH07 section). Electrical detonation caps were used for seismic sources and Oyo geophones were used for receivers. An ABEM Terraloc seismograph was used for data acquisition and recording. Seismic records had in general a high signal/noise ratio, making it possible to pick first break travel times without pre-processing. Higher noise records and records of which the first break travel times were beyond the expected velocity range were disregarded.

The steel cased boreholes used were sub-vertical and their (small) deviations to vertical were measured with an inclinometer. The mean values of these deviations were used for seismic data processing and for definition of the tomographic planes (Coelho & Oliveira, 2001 and Oliveira & Coelho, 2003). As BH01 to BH04 boreholes are almost coplanar, the CRH01, CRH02 and CRH03 sections were considered on a single tomographic plane (BH01-BH02-BH03-BH04 profile). A single tomographic plane (BH08-BH09-BH10 profile) was also considered for CRH06 and CRH07 adjacent sections.

Totals of 926 and 741 seismic rays (travel times) were used for BH01-BH02-BH03-BH04 and BH08-BH09-BH10 profiles. Fig. 4 shows these rays where straight ray paths were assumed.

3.2 Tomographic inversion

The crosshole seismic tomography here applied reconstructs the P-wave velocity field for a profile, from the P-wave travel times measured along a multitude of seismic ray paths at the crosshole sections considered for the profile. The measured travel times are inverted into a velocity matrix (grid of cells), which comprises the area with seismic ray coverage, by a tomographic inversion technique implemented and described by Pessoa (1990). This is a SIRT (Simultaneous Iterative Reconstruction Technique) type that uses the approximation of straight ray paths and assumes constant velocity for each cell of the grid. The process starts with an initial velocity model (matrix) for which the modeled travel times of (straight) seismic rays are calculated. Differences between modeled and measured travel times for seismic rays, called residuals, are then used for improving the initial model for the velocity distribution.

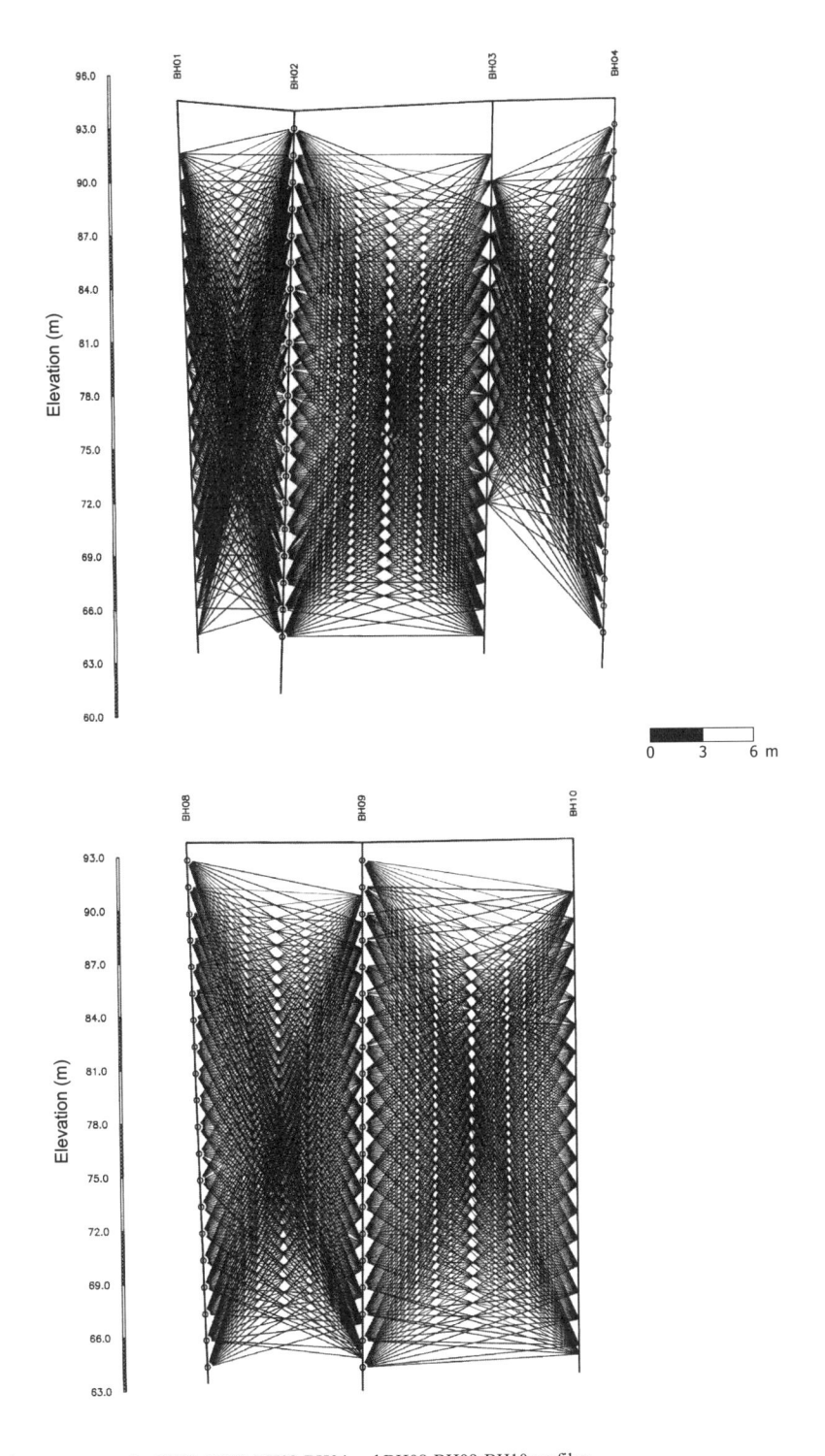

Figure 4. Seismic ray coverage for BH01-BH02-BH03-BH04 and BH08-BH09-BH10 profiles.

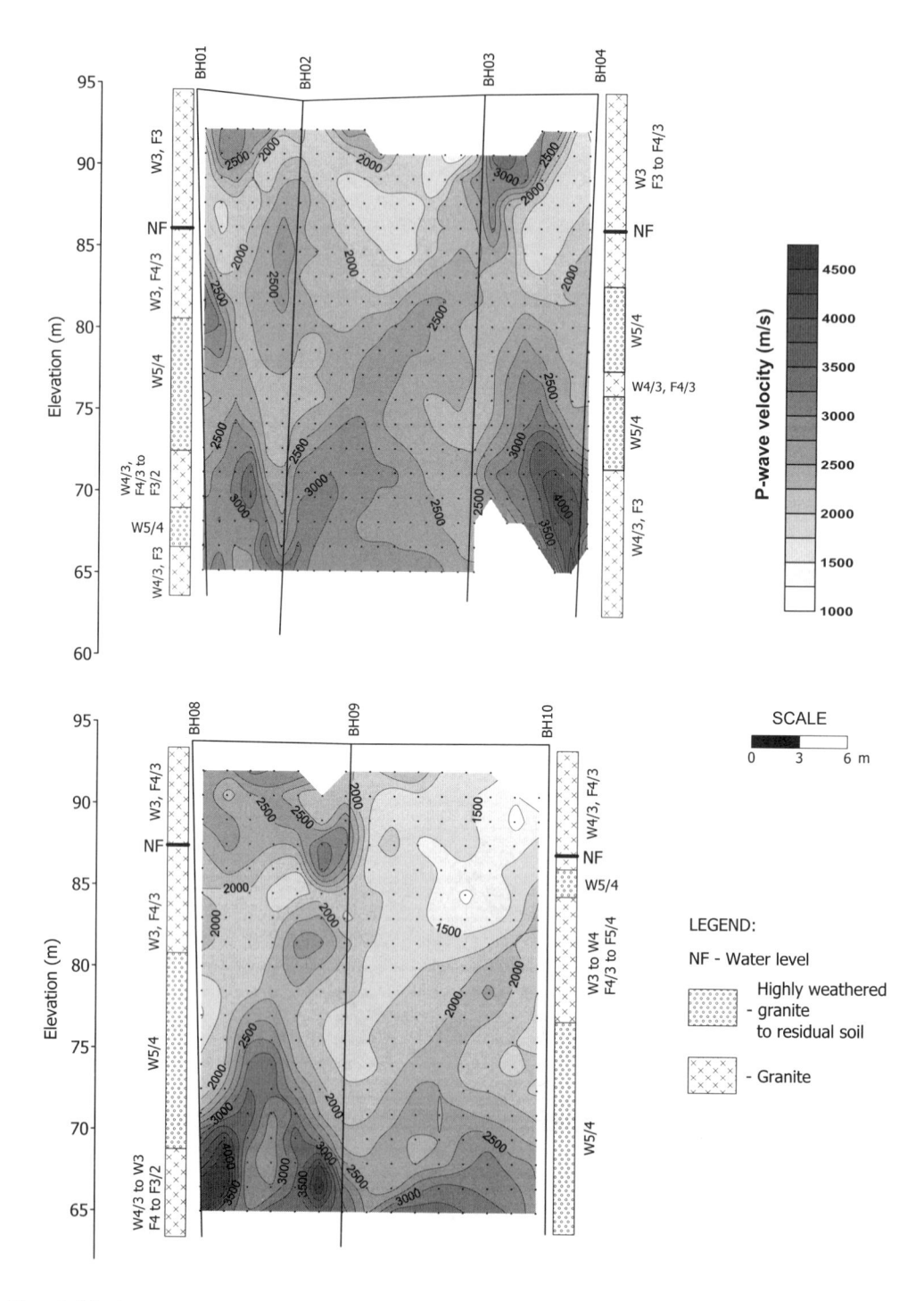

Figure 5. Seismic tomographies for BH01-BH02-BH03-BH04 and BH08-BH09-BH10 profiles.

The algorithm uses minimum and maximum velocity constraints for limiting the artefacts increment.

The BH01-BH02-BH03-BH04 profile was discretised by a grid of rectangular cells with 1m horizontal to 1.5m vertical. For the BH08-BH09-BH10 profile, 1.5m width square cells were used. The seismic tomographies of Fig. 5 show the center points of the cells crossed by straight rays. Initial models with uniform velocity (equal to mean of straight ray velocities) were used for both profiles, being 2406m/s for the BH01-BH02-BH03-BH04 profile and 2122m/s for the BH08-BH09-BH10 profile. Constraints of 340 and 4500m/s were used respectively for minimum and maximum P-wave velocity.

The seismic tomographies presented in Fig. 5, obtained by the tomographic inversion process described above, were chosen as the best estimations of the P-wave velocity field on the profiles. Even though without numerical convergence of the residuals, second iteration tomography, for the BH01-BH02-BH03-BH04 profile, and fourth iteration tomography, for the BH08-BH09-BH10 profile, were selected, to avoid the increment in velocity artefacts.

The mean residual error (percentage value corresponding to the quotient between the mean of the absolute values of the residuals and the mean of the measured times) for these tomographies is about 10% for the BH01-BH02-BH03-BH04 profile and 7% for the BH08-BH09-BH10 profile. These small values of residual error, together with the fact that the mean velocity of cells is very close to the mean velocity of straight rays for each profile, evidence high agreement (correlation) between the geophysical models and the measured travel times for seismic rays.

4 SEISMIC CHARACTERIZATION OF THE ROCK MASS

The seismic tomographies obtained at the site defined the main heterogeneities of the rock mass and characterized, by P-wave velocity, their different structures with high resolution. The tomographies showed significant variations and frequent velocity inversions, both laterally and in depth, evidencing a heterogeneous rock mass with irregular allocation of weathering and fracturing zones.

The highest velocity inversions and gradients are in general well correlated with the changes in the rock weathering and in the fracture density detected at the boreholes. This correlation makes it possible to deduce about the mechanical quality of the rock mass between boreholes.

In general, deeper zones have relatively higher velocities than shallower zones. This trend, despite being likely to reflect an increment in the mechanical quality of the rock mass, for some zones, is mainly related to the rock mass saturation in depth (in opposition to the unsaturated upper region), and can also be influenced by the rock mass compression increment with depth. The lowest velocity zones (less than 1500m/s) typically occur above the water level, on the uncompressed region, where the rock mass is certainly more weathered (or even decomposed) and more fractured. Simultaneously, some high velocity zones (above 2500m/s) occur on surface and unsaturated region, like the ones observed close to BH01, BH03 and BH09 boreholes in BH01-BH02-BH03-BH04 and in BH08-BH09-BH10 profiles. In these zones, the rock mass has locally a higher mechanical quality.

From the set of tomographies obtained at the site, it was possible to characterize quantitatively the rock mass, by means of the P-wave velocity range, as Table 1 shows. From the borehole logs, two kinds of materials were considered: granite and "residual soil" (highly weathered granite to residual soil, W5/4). This "residual soil" occurs below water level, for almost all of the boreholes used. Because saturation has a high influence on the P-wave velocity of rock mass, granite above and below the water level were considered separately.

Table 1. Characterization of rock mass at the Heroísmo station site, by means of P-wave velocity obtained from crosshole tomographies.

Material	Weathering grade	Fracture density	P-wave velocity m/s	P-wave velocity characteristic range m/s
Granite, above water level	W3	F3	2000 – 2750	1250 – 2750
		F4/3 to F4	1500 – 2500	
	W4/3 to W4	F4/3 a F5/4	1250 – 1750	
Granite, below water level	W3	F3	2000 – >3000	1750 – >3000
		F4/3 to F4	1750 – >3000	
	W4/3 to W4	F3/2 to F4	2000 – >3000	
"Residual soil" (highly weathered to decomposed granite), below water level	W5/4	—	1750 – 2500	1750 – 2500

5 CONCLUSIONS

The crosshole seismic tomographies carried out at the site of the access shaft to Heroísmo station provided, by means of P-wave velocities, a high resolution zoning of the rock mass. Furthermore, by correlation with borehole logs, the mechanical quality of the rock mass, namely between boreholes, can be inferred from those seismic tomographies.

The tomographies obtained are characteristic of a heterogeneous rock mass with highly variable mechanical quality, either in depth or laterally, and with irregular weathering (and fracture density) distribution.

This geophysical method produced useful additional information about the rock mass, in complement to the data from drilling and other mechanical investigations.

ACKNOWLEDGEMENTS

The authors wish to thank Metro do Porto, S.A., Normetro and Transmetro-ACE for their permission to publish this paper.

REFERENCES

Carminé, P. 2000. Geological-geotechnical conditions along underground sections of "Metro do Porto" project. (in Portuguese). *Proceedings 7° Congresso Nacional de Geotecnia. Porto, Portugal, April 10-13, 2000, Vol III*: 1377-1388.

Coelho, M.J. & Oliveira, M. 2001. *Crosshole seismic tomographies at "Heroísmo" station site.* (in Portuguese). Relatório (Report) 137/01-NP, LNEC, Lisboa.

Fruguglietti, A., Gasparini, M. & Centis, S. 2000. Influence of geotechnical conditions on the excavation method of "Metro do Porto" project. *Proceedings 7° Congresso Nacional de Geotecnia. Porto, Portugal, April 10-13, 2000, Vol III*: 1391-1400.

ISRM. 1981. *Rock characterization, testing and monitoring – ISRM Suggested Methods*. Editor: E. T. Brown. England: Pergamon Press Ltd.

Oliveira, M. & Coelho, M.J. 2003. Heroísmo station site. Crosshole seismic tomographies. (in Portuguese). *Proceedings Jornadas Hispano-Lusas sobre Obras Subterráneas. Madrid, Spain, September 15-16, 2003*: 477-486.

Pessoa, J.M.N.C. 1990. *Application of tomographic techniques to crosshole seismic exploration.* (in Portuguese). Trabalho de síntese (Monography), Provas de aptidão pedagógica e capacidade científica (Public examination for Assistant), Universidade de Aveiro, Portugal.

Shear velocity model appraisal in shallow surface wave inversion

Adam O'Neill
School of Earth and Geographical Sciences, The University of Western Australia, Perth, Australia

Keywords: Rayleigh wave, dispersion, error propagation, LVL/HVL resolution, Monte Carlo

ABSTRACT: Uncertainty of surface wave dispersion from numerical modeling and field repeatability tests shows a strong dependence on the spread length resolution limitation with frequency. High frequency uncertainty is about 1% and rises nonlinearly at low frequency to over 30%, where Gaussian distribution is invalid. The major source of dispersion error is additive noise to seismograms, static shifts and geophone placement, with geophone coupling and tilt effects being minimal. Monte Carlo ensemble analysis shows the half-space shear velocity to become unresolvable in the presence of realistic dispersion error. Shear velocity of a shallow LVL under a caprock is well resolved, but becomes poorly resolved when at depth underlying a buried HVL, as are the thicknesses of a LVL or HVL. Shear velocity 'smearing' between interfaces occurs at depth, where velocity gradients are more likely to be estimated instead of sharp contrasts, as well as an underestimate of half-space shear velocity. Gaussian error propagation is invalid, with shear velocity likelihood showing broad and/or bi-modal distribution. A 5 m thick, stiff layer at 20 m depth is found to be physically non-resolvable with dispersion measured from an array at the surface.

1 INTRODUCTION

The three-step process of acquisition-processing-inversion of surface wave dispersion provides a layered model *estimate* of shear wave velocity. This model must then be *appraised* for its accuracy and *interpreted* into a geotechnical framework. There are several factors which reduce the accuracy of an inverted model (Boschetti 1995):
1. Raw and processed data errors;
2. Inexact forward modelling;
3. Inappropriate parameterisation, and;
4. Detection of the global minimum.

Items 1 and 2 relate to the error propagation problem and 3 and 4 relate to the inherent nonuniqueness which is present even in noise-free, ideal cases and been intensively studied in linearised optimisation (Jackson 1973). While global optimisation methods more broadly sample model space and allow numerous layer parameterisations, surface wave dispersion calculation is still generally restricted to flat-layered models (Sambridge and Mosegaard 2002). However, even in a perfectly flat-layered site the assumption of fundamental mode dominance introduces a systematic error where dominant higher modes are known to be generated.

Model appraisal can be made by either deterministic (comparison to invasive tests results) or statistical (based on theoretical error propagation and non-uniqueness analysis) methods. Direct comparison of surface wave models to borehole results has been extensively reported (Brown et al. 2002; Lai et al. 2002). However, these do not quantify the source of error and cannot be taken as an absolute base, since invasive test results also suffer from systematic and random errors.

Statistical analysis of error propagation into the inverted model has been studied from limited repeated field testing in SASW (Tuomi and Hiltunen 1996) and MASW (Lai et al. 2002). The former showed dispersion repeatability of about 7% and normally distributed, but only over a high frequency range (60-100 Hz). The inverted shear velocity uncertainty was up to 8% and not normally distributed. The latter showed good dispersion repeatability (<1%) at high frequencies (>10 Hz), with poor repeatability at low frequency (<10 Hz) increasing nonlinearly to over 10%. Scatter in dispersion at all frequencies was shown to be normally distributed, leading to an average uncertainty in inverted shear wave velocity of 3%.

For model appraisal to be valid, data uncertainty must be quantitatively known. In shallow surface wave applications, measured dispersion dependency on field recording parameters has been intensively studied in both SASW (Hiltunen and Woods 1990)

and MASW (Park et al. 2002), usually for reducing 'near-field effects' and/or higher mode influence. In SASW, the scatter in observed dispersion occurs due to a number of various source-receiver layouts being employed for maximum bandwidth (Stokoe et al. 1994). Many studies on processing effects (Foti et al. 2002; Park et al. 2001), field repeatability (Beaty and Schmitt 2003), model parameterisation influences (Foti and Strobbia 2002; Rix and Leipski 1991) and dispersion sensitivity in forward (Socco and Strobbia 2003) and inverse (Luke et al. 2003) modelling have also been recently reported.

This paper is thus divided into two sections: (1) A thorough numerical error analysis of MASW dispersion, proven with repeated field measurements and (2) inverted model appraisal incorporating realistic data uncertainty. Unlike previous work, cases including dominant higher modes ('effective' phase velocity) are employed and individual error sources quantified. However, we will begin with the fundamental issue of how recording array aperture (spread length) affects resolution of multichannel dispersion measurement.

2 PHASE VELOCITY RESOLUTION

In group velocity analysis, poor resolution at low frequency is evident due to time-frequency tradeoff and in phase velocity measurement, similar smearing at low frequency occurs (Park et al. 1998). While dispersion error is usually estimated as a constant relative value, large uncertainty at low frequency has been reported in crustal (Pedersen et al. 2003), deep basin (Scherbaum et al. 2003) and shallow sediment (Louie 2001; Malagnini et al. 1997) investigations. In a MASW framework, Forbriger (2001) showed the phase slowness resolution dependency on spread length to be:

$$\Delta p(f) = \frac{1}{fL} \quad (1)$$

where $\Delta p(f)$ is the slowness resolution (s/m), f is the frequency (Hz) and L is the spread length (m). In other words, for best low frequency resolution (i.e. deep interpretation), longer spreads are required. Based on Equation (1), O'Neill (2003) showed that the maximum resolvable wavelength from a linear spread is $0.4L$. While, in general, irregular dispersion at low frequency is attributed to 'near-field effects', the dispersion curves of Lai et al. (2002) support the observation of larger low frequency uncertainty from shorter recording spreads.

3 SYNTHETIC DISPERSION UNCERTAINTY

The full-waveform PSV reflectivity method (O'Neill 2003) was employed with the shear velocity models (Cases 1 to 3) of Tokimatsu et al. (1992), shown in Figure 1.

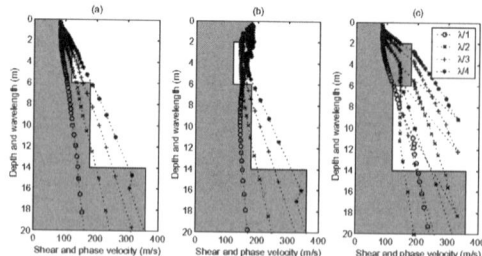

Figure 1. Shear velocity models of Tokimatsu et al. (1992): (a) Case 1, (b) Case 2 and (c) Case 3. Overlain are dispersion curves generated with the PSV method, scaled by 'approximate inversion' with four different wavelength reduction factors.

First, the effects of various acquisition parameters (geophone positioning / tilt / coupling, source type / frequency / depth, static errors and additive noise sources) on MASW dispersion were numerically tested using a 48-channel spread with 1 m geophone spacing and impact source at 5 m near offset. Raw seismograms were randomly perturbed and dispersion observed by the frequency-slowness (f-p) method for 30 trials. The standard deviations of phase velocity with frequency for Case 1 were summed and are shown ranked in Figure 2. It can be seen that geophone tilt and coupling have minimal influence, while positional error, static shifts and additive noise introduce larger error in observed dispersion. Explosive source depth errors occur due to higher modes being preferentially generated for gradually deeper sources in the upper layer, even though the model is normally dispersive.

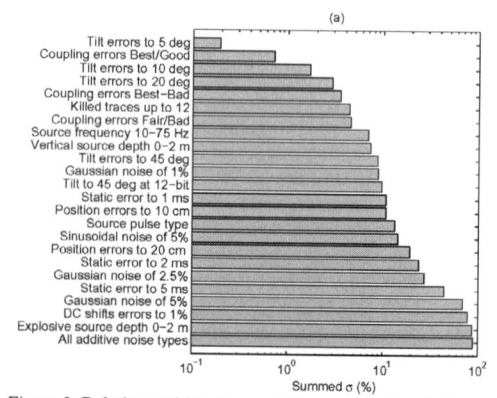

Figure 2. Relative ranking of error influences on Case 1 dispersion by summing relative standard deviations over 5-70 Hz.

The scatter in observed dispersion for all additive noise sources to the 48-channel gathers shows a Gaussian distribution at high frequency (Figure 3). However, at low frequency, larger outliers are better fitted with a Lorentzian distribution, caused by the poor phase velocity resolution.

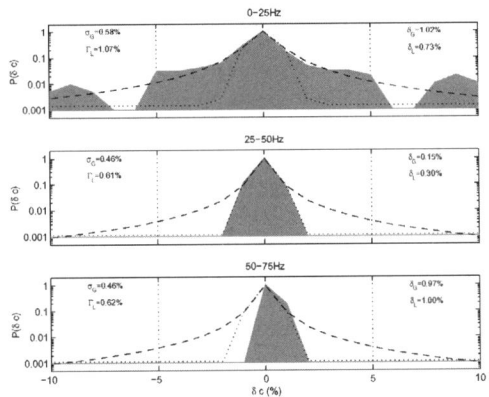

Figure 3. Statistical analysis of the dispersion error distribution caused by additive noise to synthetic seismograms averaged over 25 Hz bands, with Gaussian (dotted) and Lorentzian (dashed) distributions fitted.

A Monte Carlo repeatability analysis combining all the error sources of Figure 2 was run for 120 iterations. In addition, random spread layouts between 12-channels at 0.5 m spacing and 96-channels at 1 m spacing and near offsets of 0.5 m to 20 m were allowed. Analyses for Cases 2 and 3 are shown in Figures 4 and 5, where an f-p transform was used for the automatic dispersion picking, but the f-k transform gave similar results.

The nonlinear increase in error at low frequency exceeds 20% and mimics the theoretical trend of phase velocity resolution with frequency (Equation 1). The high frequency uncertainty is around 1-2%, but several times larger at, and around frequencies of transitions to, dominant higher modes. While the 'effective' dispersion segments correlate with the 'modal' curves, below about 5 Hz the observed dispersion does not show the expected high phase velocity. This will greatly limit the ability to interpret the half-space shear velocity, particularly in Case 3 (Figure 5). Essentially, the lower cutoff frequency of 5 Hz is the point where phase velocity is not resolvable by the f-p transform due to the spread length limitation.

The reduced χ^2 test was employed to test the assumption of Gaussian fit to the distribution of phase velocity scatter (Lai et al. 2002). shown in Figure 6 for the Monte Carlo dispersion repeatability tests of Cases 1, 2 and 3. It can be seen that below about 20 Hz in Cases 1 and 3 and for all frequencies in Case 2 that the χ^2 values exceed the acceptable level (approx. 6). At these frequencies, the scatter is better fitted with a Lorentzian distribution.

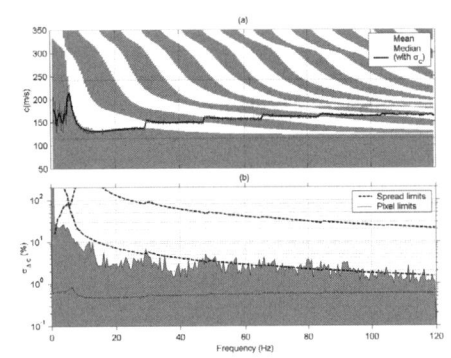

Figure 4. Monte Carlo repeatability of the Case 2 dispersion by f-p transform: (a) Median (black) and theoretical modal (grey) dispersion curves and (b) relative standard deviation (grey) and spread length resolution (dashed) limits with frequency.

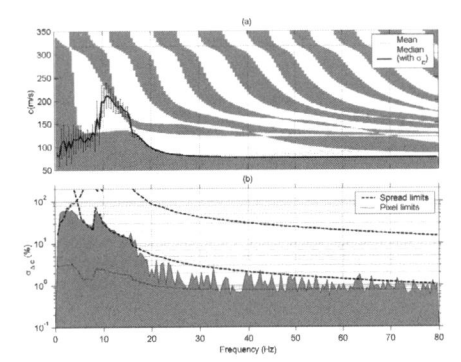

Figure 5. Monte Carlo repeatability of the Case 3 dispersion by f-p transform: (a) Median (black) and theoretical modal (grey) dispersion curves and (b) relative standard deviation (grey) and spread length resolution (dashed) limits with frequency.

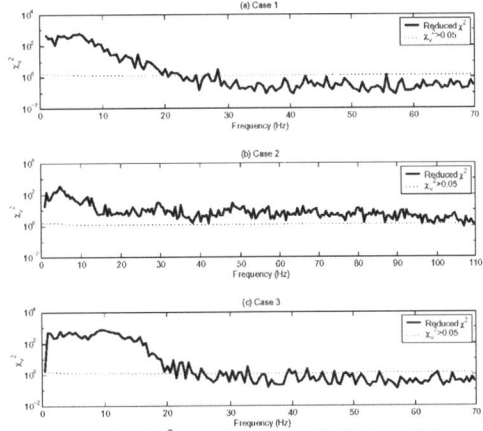

Figure 6. Reduced χ^2 test of Gaussian distribution of the synthetic Monte Carlo dispersion repeatability analysis: (a) Case 1, (b) Case 2 and (c) Case 3.

4 SYNTHETIC MODEL RESOLUTION AND APPRAISAL

An indication of how dispersion errors affect the model sensitivity and subsequent convergence in an inversion is shown in the RMS misfit planes in Figure 7. Considering the 1% misfit contour, layer thickness is poorly constrained in all cases, similar to the results of Mackenzie et al. (2001). The effect of realistic errors greatly lowers resolution of the half-space depth and shear velocity in Case 1, but has little effect on shear velocity resolution of the Case 2 LVL. Note the nonlinearity of the Case 3 HVL case by the undulating misfit at high shear velocity values, due to large variations in dominant higher modes.

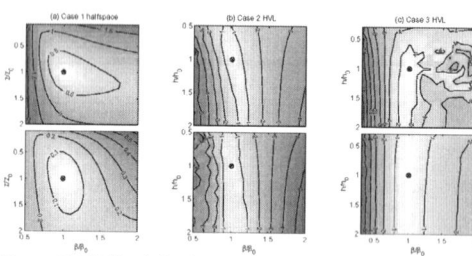

Figure 7. RMS misfit (%) subspaces for variations in layer thickness/depth and shear velocity: (a) Case 1 halfspace, (b) Case 2 low velocity layer and (c) Case 3 high velocity layer. For each, both constant 3% dispersion error (upper) realistic dispersion errors (lower) are used for the misfit weighting.

However, a better indication of model appraisal can only be made by a thorough, unconstrained Monte Carlo ensemble analysis. Considering the 4-layer synthetic models as a starting base, first a new layer stack is generated, comprising thicknesses either constant or geometrically increasing with depth. The shear velocities at each depth are perturbed by random amounts, proportional to the dispersion errors, scaled to approximate depths (see Figure 1) using a $\lambda/2.5$ reduction. The PSV method is used to forward model the dispersion curve, again with a 48-channel spread and 1 m geophone spacing and impact source at 5 m near offset. Unconstrained RMS misfit, weighted using the numerically modeled dispersion uncertainty, is measured at each iteration.

Figures 8 and 9 show the results of this analysis on the Case 2 and 3 models, where it can be seen that as the allowed data misfit increases, so too does the width of the acceptable shear velocity range. Of particular interest is the good constraint on the shallow layer shear velocities, in particular the LVL in Case 2 (Figure 8), similar to the results of Forbriger (2001), who also conversely showed the poor P-wave velocity resolution of a LVL. However, the LVL at depth in Case 3 (Figure 9) is very poorly constrained and both cases show 'smearing' between layers 3 and 4 (half-space), where the best-fitting models appear as a velocity gradient over 10-15 m depth, rather than a sharp interface. While this method is not as broad-sampling as genetic algorithm inversion (Mackenzie et al. 2001), the poor resolution at depth is similarly reproduced, in particular the higher likelihood of the inverted half-space shear velocity being an underestimate. Note there are no successful models in Case 3 under 0.5% misfit.

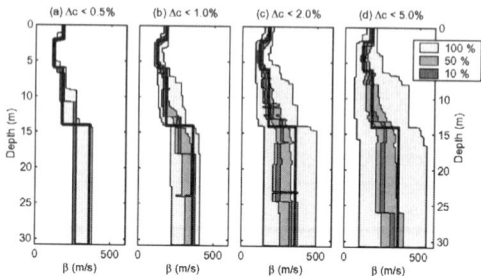

Figure 8. Monte Carlo ensemble analysis of Case 2, showing the top 10%, 50% and 100% of models from RMS data misfits within: (a) 0.5%, (b) 1%, (c) 2% and (d) 5%.

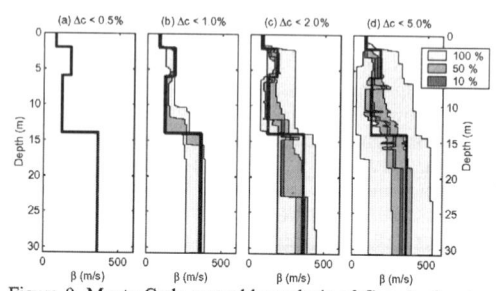

Figure 9. Monte Carlo ensemble analysis of Case 3, showing the top 10%, 50% and 100% of models from RMS data misfits within: (a) 0.5%, (b) 1%, (c) 2% and (d) 5%.

Figures 10 and 11 show the probability density functions of Figures 8 and 9 respectively, overlain with the shear velocity distributions averaged over the true layer depth ranges. Even though the Monte Carlo analysis employed normally distributed perturbations of shear velocity, Gaussian error propagation is invalid in these cases, where a skewed and/or bi-modal pattern emerges, especially for deeper layers. In a LVL below a buried shallow HVL (Figure 11), much larger uncertainty can be expected and is almost equally distributed, that is, near zero resolution. This type of Monte Carlo analysis is expected to show more realistic limitations than theoretical standard deviations of model parameters from the covariance matrix since there is no influence of the inversion regularisation.

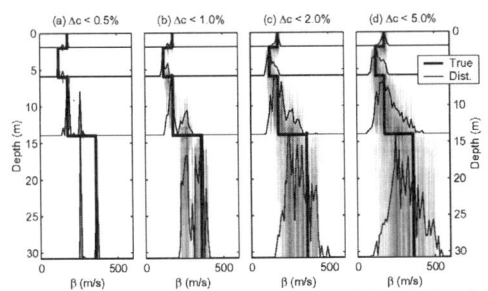

Figure 10. Monte Carlo ensemble analysis of Case 2, showing the probability density function image (grey) overlain with shear velocity distributions for each layer (black), from RMS data misfits within: (a) 0.5%, (b) 1%, (c) 2% and (d) 5%.

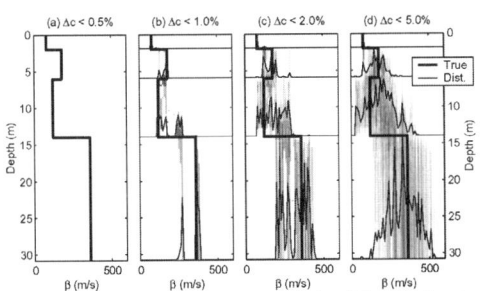

Figure 11. Monte Carlo ensemble analysis of Case 3, showing the probability density function image (grey) overlain with shear velocity distributions for each layer (black), from RMS data misfits within: (a) 0.5%, (b) 1%, (c) 2% and (d) 5%.

5 FIELD REPEATABILITY TESTING

The Perth Convention Centre site is on the Perth city foreshore in the Perth Basin, Western Australia, over land reclaimed from the Swan River in the 1950s. It was chosen as the location for the new Perth Convention Centre for its proximity to the CBD. However, special preparation was necessary to prevent subsidence due to a thick sequence of organic muds. The surface wave spread was about 100 m from the north shore of the Swan River and within several metres of the spread centre were several invasive tests, including a seismic cone penetrometer test (SC2), a logged geological borehole (CB5), and a cone penetrometer test (CC20). A summary of these is shown in Table 1. Below the artificial fill material, the soil is quite soft, and concrete piles driven hydraulically down to the siltstone 'basement' were to form the foundation for the construction.

A 24-bit, 8-channel National Instruments PC card was used to acquire the surface wave data, shot in May 2002, with both a 10 kg hammer on PVC plate and 50 kg weight drop as sources. To increase the number of channels, walkaway shooting was used. Here, the source was moved in multiples of a spread length while the spread remained planted, and this was repeated for both 1 m and 3 m geophone spacings. The hammer source was used for the close spacings and the heavier source for the wider geophone spacings. Typical composite shot gathers are shown in Figure 12.

Table 1. Summary of borehole tests near the Perth Convention Centre surface wave sounding point.

Test:	SC2	CB5	CC20
Depth (m)	Shear velocity (m/s)	Geological log	Tip resistance (MPa)
0–5	140	Sand (fill)	3–30
5–20	85–145	Silty-clay	<1
20–27	220–450	Silty-sand	2–12
27–36	135–200	Silty-clay	2–3
36	No data	Siltstone	No data

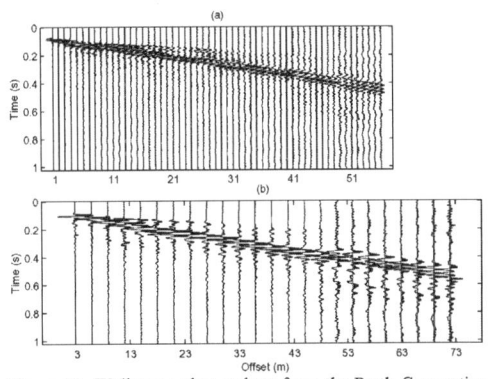

Figure 12. Walkaway shot gathers from the Perth Convention Centre site: (a) Hammer source with 56 channels at 1 m spacing and (b) weight drop source with 24 channels at 3 m spacing.

At least five records were saved from each shot position, thus a large number of possible combinations of 8-channel segments allowed a Monte Carlo dispersion error analysis, where the full trace window was employed in the f-p transform. The results of this from the 1 m and 3 m geophone spacing data are shown in Figures 13 and 14 respectively. Due to the longer spread length, the 3 m spacing data shows better repeatability at lower frequencies. However, at higher frequency, repeatability is poorer compared to the 1 m spacing data, possibly more affected by side-scattered wavefields.

The distribution of the 1 m geophone spacing dispersion is shown in Figure 15, averaged over 12.5 Hz bands. Since the near-offset was constant, the 'low-frequency effects' are confined to the 0-12.5 Hz range, where a Lorentzian distribution best fits the observed scatter. At higher frequency, the distribution is Gaussian, where the error influences were limited to geophone position, source variations and additive noise.

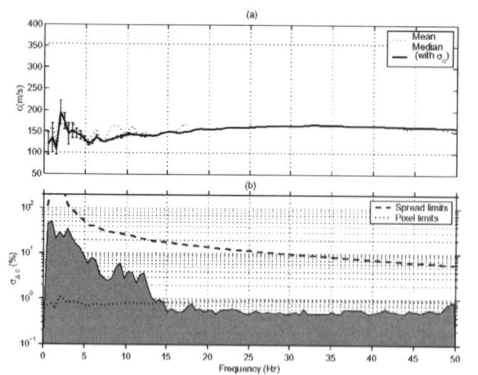

Figure 13. Monte Carlo repeatability of the 1 m geophone spacing Perth Convention Centre dispersion by f-p transform: (a) Median dispersion curve with error and (b) relative standard deviation (grey) and spread length resolution (dashed) limits with frequency.

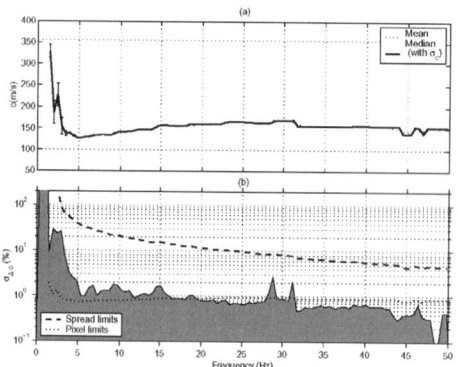

Figure 14. Monte Carlo repeatability of the 3 m geophone spacing Perth Convention Centre dispersion by f-p transform: (a) Median dispersion curve with error and (b) relative standard deviation (grey) and spread length resolution (dashed) limits with frequency.

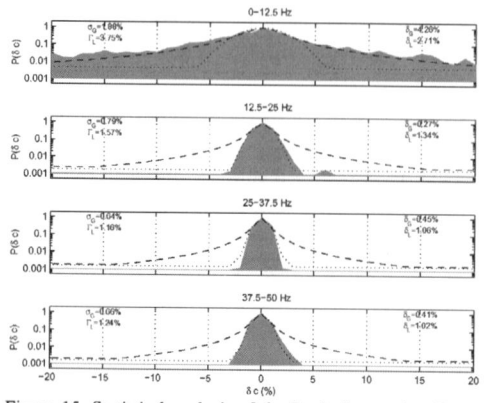

Figure 15. Statistical analysis of the Perth Convention Centre dispersion repeatability averaged over 12.5 Hz bands, with Gaussian (dotted) and Lorentzian (dashed) distributions fitted.

6 FIELD MODEL APPRAISAL

The median dispersion curve from the 3 m geophone spacing data (Figure 14) suggests that a high velocity basement has been detected, possibly up to 300 m/s, from the sharp rise in phase velocity at low frequency. For the inversion of this data, two different layer parameterisations were trialled: (1) 'Blind' inversion which employed 36 layers, each 1 m thick and (2) 'Borehole' inversion which used the layer interfaces identified in the SC2 log. Poisson's ratio was set at 0.4 for all layers and results of each inversion are shown in Figures 16 and 17 respectively, compared with 'true' shear velocities measured in the SC2 test.

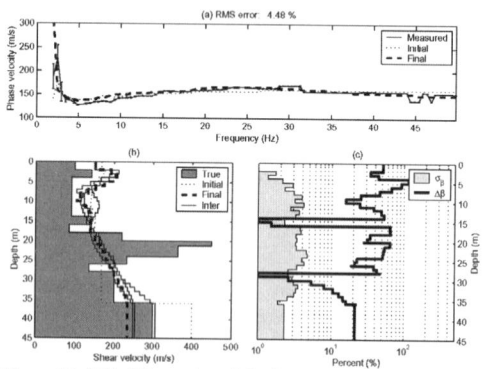

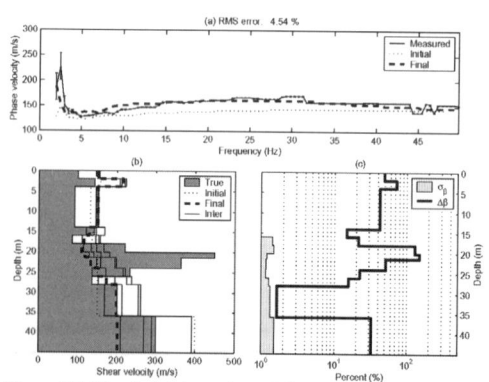

Figure 16. 'Blind' inversion of the 3 m geophone spacing dispersion at the Perth Convention Centre site, using a 36-layer model, each 1 m thick: (a) Dispersion curves with experimental error, (b) derived shear velocity models and (c) relative standard deviations (grey) and 'errors' (solid line) in inverted layer shear velocities.

Figure 17. 'Borehole' inversion of the 3 m geophone spacing dispersion at the Perth Convention Centre site, using the interfaces from the 'true' shear interval velocity from SC2: (a) Dispersion curves with experimental error, (b) derived shear velocity models and (c) relative standard deviations (grey) and 'errors' (solid line) in inverted layer shear velocities.

While the RMS error at the final iteration is quite large (approx. 4.5%), the three main points of note in the final model are: (1) Good correlation with the SC2 results in the trend of layer velocities between 1 and 15 m depth, except for a level shift, (2) good correlation at depths below 25 m (note that the 'true' siltstone velocity is an assumed value) and (3) no detection of the stiff HVL logged in the nearby invasive tests between about 20-27 m. The level shift in the recovered shear velocities in the upper 15 m supports the suspicion of error in the seismic cone penetrometer data. However, the inability to recover the silty-sand layer at 20 m depth was disappointing. Since the borehole was some 30 m from the spread centre it is possible that this layer has pinched-out, however, other boreholes all show this layer as pervasive over the entire site.

One quantitative test of the resolution in this case is to forward model the expected dispersion and invert the synthetic curve to check which layers can be detected under realistic field conditions. Figure 18 shows the results of this test using a 56-channel, 1 m geophone spacing synthetic shot gather generated from the SC2 shear velocity log. The models in Figure 18b show that the shallow layers are recovered well, but the existence of the hard layer at 20 m depth is not resolved. While the layers underlying this appear to be properly modelled, it is most likely coincidental due to the inherent 'smearing' and the method could only be guaranteed to provide reliable results to about 20 m depth.

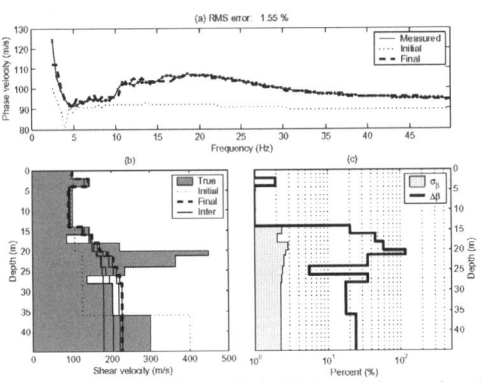

Figure 18. Inversion of a synthetic 56-channel data set based on the Perth Convention Centre seismic cone penetrometer log SC2, using the PSV method where the 'true' model is the shear interval velocity from this log (SC2), and these interfaces have been used for the inversion model: (a) Dispersion curves with experimental error, (b) derived shear velocity models and (c) relative standard deviations (grey) and errors (solid line) in inverted layer shear velocities.

Another resolution test is by Monte Carlo inference analysis about the 'true' model, to further see if the invisibility of the silty-sand layer is a physical or experimental limitation. Using the PSV method with a 24-channel spread at 3 m spacing, the starting model is randomly perturbed and dispersion misfit (relative to the starting model dispersion) measured, weighted by the uncertainties from the field repeatability test. The results for various allowed levels of RMS misfit are shown in Figure 19. While there are no successful models at the highest data likelihood, as the misfit is relaxed it is clear that the HVL at about 20 m depth is still not resolved in the most successful models. Only at larger allowed data misfit is the HVL evident, but only for the less successful models, where the nonuniqueness is much larger.

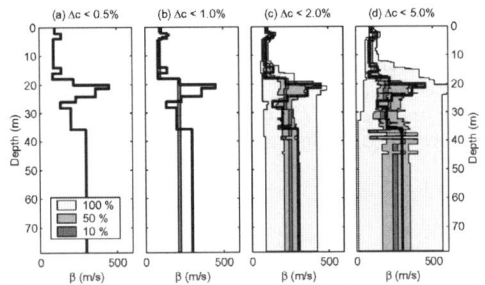

Figure 19. Monte Carlo ensemble analysis of the Perth Convention Centre shear velocity model from SC2, showing the top 10%, 50% and 100% of models from RMS data misfits within: (a) 0.5%, (b) 1%, (c) 2% and (d) 5%.

CONCLUSIONS

Dispersion uncertainty measured in shallow multichannel acquisition is important in the appraisal of models for engineering site investigation. From numerical modeling (incorporating dominant higher modes) a strong dependency on the spread length resolution limitation with frequency is exhibited. High frequency uncertainty is about 1% and rises nonlinearly at low frequency to over 30%, where Gaussian distribution is invalid. The major source of error is additive noise to seismograms, static shifts and geophone placement, with geophone coupling and tilt effects minimal. Allowing near offset and spread length as random variables, 'low frequency effects' can manifest over the entire recorded bandwidth.

Monte Carlo model ensemble analysis shows the half-space shear velocity to become unresolvable in the presence of realistic dispersion error, as is the thickness of a LVL and HVL. However, the shear velocity of a shallow LVL is well resolved, but poorly resolved when underlying a buried HVL. Shear velocity 'smearing' between interfaces occurs at depth, where velocity gradients are more likely to be estimated instead of sharp contrasts, as well as an underestimate of half-space shear velocity. Gaussian error propagation into the ensemble of shear velocity models is also invalid, where shear velocity likelihood shows a broad and/or bi-modal distribution.

Field repeatability tests verify the theoretical dispersion uncertainty analyses and error propagation. In particular, a 5 m thick, stiff layer at 20 m depth is found to be physically non-resolvable with dispersion measured from an array at the surface.

ACKNOWLEDGEMENTS

Klaus Holliger and Michael Roth are thanked for the FORTRAN reflectivity synthetic seismogram code. The Perth Convention Centre data was collected with Sebastiano Foti, using The University of Western Australia (UWA) Department of Civil Engineering seismograph. This work was conducted while the primary author (A. O'N) was in receipt of a University Postgraduate Award (UPA) from UWA.

REFERENCES

Beaty, K. S. and Schmitt, D. R., 2003, Repeatability of multi-mode Rayleigh-wave dispersion studies: Geophysics, 68, 782-790.

Boschetti, F., 1995, Application of genetic algorithms to the inversion of geophysical data: PhD thesis (unpublished), The University of Western Australia.

Brown, L. T., Boore, D. M. and Stokoe II, K. H., 2002, Comparison of shear-wave slowness profiles at 10 strong-motion sites from non-invasive SASW measurements and measurements made in boreholes: Bulletin of the Seismological Society of America, 92, 3116-3133.

Forbriger, T., 2001, Inversion flachseismischer Wellenfeldspektren: PhD thesis (unpublished), Universität Stuttgart [http://opus.uni-stuttgart.de/opus/volltexte/2001/861/]: Accessed 18 Sep, 2001.

Foti, S. and Strobbia, C., 2002, Some notes on model parameters for surface wave data inversion: Proceedings of SAGEEP.

Foti, S., Sabuelli, L., Socco, L. V. and Strobbia, C., 2002, Spatial sampling issues in fk analysis of surface waves: Proceedings of SAGEEP.

Jackson, D. D., 1973, Marginal solutions to quasi-linear inverse problems in geophysics: The edgehog method: Geophysical Journal of the Royal Astronomical Society, 35, 121-136.

Lai, C. G., Foti, S. and Rix, G. J., 2002, Analysis of uncertainty in surface wave testing: Journal of Geotechnical and Geoenvironmental Engineering (submitted).

Louie, J. N., 2001, Faster, better: Shear-wave velocity to 100 meters depth from refraction microtremor arrays: Bulletin of the Seismological Society of America, 91, 347–364.

Luke, B. A., Calderón-Macias, C., Stone, R. C. and Huynh, M., 2003, Non-uniqeness in inversion of seismic surface-wave data: Proceedings of SAGEEP.

Mackenzie, G. D., Maguire, P. K. H., Denton, P., Morgan, J. and Warner, M., 2001, Shallow seismic velocity structure of the Chicxulub impact crater from modeling of Rg dispersion using a genetic algorithm: Tectonophysics, 338, 97–112.

Malagnini, L., Herrmann, R. B., Mercuri, A., Opice, S., Biella, G. and de Franco, R., 1997, Shear-wave velocity structure of sediments from the inversion of explosion-induced Rayleigh waves: Comparison with cross-hole measurements: Bulletin of the Seismological Society of America, 87, 1413-1421.

O'Neill, A., 2003, Full-waveform reflectivity for modelling, inversion and appraisal of seismic surface wave dispersion in shallow site investigations: PhD thesis (unpublished), The University of Western Australia. [http://www.geol.uwa.edu.au/~aoneill/].

Pedersen, H. A., Coutant, O., Deschamps, A., Soulage, M. and Cotte, N., 2003, Measuring surface wave phase velocities beneath small broad-band arrays: Tests of an improved algorithm and application to the French Alps: Geophysical Journal International, 154, 903–912.

Park, C. B., Miller, R. D. and Xia, J., 1998, Imaging dispersion curves of surface waves on multi-channel record: Proceedings of SEG 68th Ann. Int. Mtg, 1377-1380.

Park, C. B., Miller, R. D. and Xia, J., 2001, Offset and resolution of dispersion curve in multichannel analysis of surface waves (MASW): Proceedings of SAGEEP.

Park, C. B., Miller, R. D. and Mura, H., 2002, Optimum field parameters of an MASW survey [http://www.terrajp.co.jp/OptimumFieldParametersMASWPark.pdf]: Accessed 18 Sep, 2002.

Rix, G. and Leipski, E. A., 1991, Accuracy and resolution of surface wave inversion: in Blaney, G. W. and Bhatia, S. K. (eds.), Recent advances in instrumentation, data acquisition and testing in soil dynamics: American Society of Civil Engineers, Geotechnical Special Publication No. 29, 17–32.

Sambridge, M. and Mosegaard, K., 2002, Monte Carlo methods in geophysical inverse problems: Reviews of Geophysics, 40, 3(1-29).

Scherbaum, F., Hinzen, K.-G. and Ohrnberger, M., 2003, Determination of shallow shear wave velocity profiles in the Cologne, Germany area using ambient vibrations: Geophysical Journal International, 152, 597–612.

Socco, L. V. and Strobbia, C. L., 2003, Extensive modeling to study surface wave resolution: Proceedings of SAGEEP.

Stokoe II, K. H., Wright, S. G., Bay, J. A. and Roesset, J. M., 1994, Characterization of geotechnical sites by SASW method: in Woods, R. D. (ed.), Geophysical characterization of sites: A. A. Balkema, 15-25.

Tokimatsu, K., Tamura, S., and Kojima, H., 1992, Effects of multiple modes on Rayleigh wave dispersion: Journal of Geotechnical Engineering, 118, 1529–1543.

Tuomi, K. E. and Hiltunen, D. R., 1996, Reliability of the SASW method for determination of the shear modulus of soils: in Uncertainty in geologic environments: from theory to practice (Uncertainty 96): American Society of Civil Engineers, 1225–1238.

Full waveform reflectivity for inversion of surface wave dispersion in shallow site investigations

Adam O'Neill
School of Earth and Geographical Sciences, The University of Western Australia, Perth, Australia

Keywords: synthetic seismograms, dominant higher modes, Occam's inversion, shear wave velocity

ABSTRACT: Dominant higher modes of surface waves are generated in sites with large stiffness contrasts and/or reversals between layers. A multichannel simulation employing full-waveform reflectivity properly models dispersion curve discontinuities and allows inversion for shear velocity structure. A fundamental-mode assumption causes low-velocity layers to be undetected and high-velocity layers to be interpreted both too deep and thick. Dominant higher modes are observed and inverted at a mine site over weathered sandstone and a compacted waste dump. Stacking dispersion curves provides a more unique dispersion for inversion.

1 INTRODUCTION

Surface waves have long been known to be affected by vertical variations in elastic parameters. Since the propagation of surface waves is dominated by the stiffness of the solid medium within a wavelength from the interface, in a layered system different frequency components travel at different velocity. The variation with frequency of phase or group velocity, the dispersion relationship, is the key property when modelling and interpreting surface wave data for shear velocity structure, which is an important parameter in geotechnical engineering (Whiteley 1990). In civil engineering applications (upper tens of metres), in situ imaging of shear velocity can be accomplished quickly through the inversion of surface wave dispersion (Cuéllar, 1997) and its use has become very common in recent years (Forbriger 2003a; Foti 2000).

Conventional surface wave surveying comprises three stages: acquisition, processing, and inversion (Gabriels et al. 1987). Seismograms acquired in the field are processed to extract the 'ground-roll' dispersion, the dominant component being surface waves. The inversion of the dispersion curve is conventionally based on a 1D plane-wave propagator matrix model (Buchen and Ben-Hador 1996). Differences between calculated and observed 'modal' dispersion curves are minimised for the mode of interest, based on an Earth model consisting of horizontal layers. Although all layer elastic parameters (compressional and shear velocities and density) and thicknesses may be unknown, dispersion is most sensitive to shear wave velocity, so it is this parameter which is sought (Beaty et al. 2002).

A major problem is that commonly used inversion methods may not provide reliable results in engineering cases where large stiffness contrasts and/or reversals occur, for example: caprock (laterite or calcarenite), soft layers (peats and clays) and shallow hard horizons (secondary cemented sediments or crystalline rock). Caprocks and buried hard horizons, in particular, can be problematic from both safety (onshore and offshore engineering foundation) and economic (mining overburden rippability) aspects. In these sites, an 'apparent' or 'effective' dispersion is observed, since a continuous, fundamental mode is not measured, but rather several modes combined into a discontinuous curve (Lai and Rix 1999; Tokimatsu et al. 1992).

While comparisons to invasive methods are generally good in 'normally dispersive' sites (Xia et al. 2002), discrepancies occur in more severely 'irregularly' and 'inversely dispersive' sites, due to the restriction to continuous, modal phase velocity dispersion observation and modelling (Brown et al. 2002). Mode mis-identification can also occur if the observed surface wavefield is assumed to be dominated by the fundamental mode. The following section introduces effective dispersion curves generated in synthetic cases, employing the new modelling procedure.

2 DISPERSION MODELLING METHODS

2.1 Plane-wave dispersion modelling

The conventional method for modelling dispersion curves is based on plane-wave propagator matrices (Buchen and Ben-Hador 1996). Since surface waves only propagate with specific wavenumbers, the roots of an implicit function in phase velocity–frequency space are the modal dispersion curves. The 'Fast Surface Wave' code (Schwab and Knopoff 1972) has been employed by other researchers in theoretical and field studies (Xia et al. 1999). However, it will be used in this paper (hereafter called the FSW method) to illustrate how plane-wave modal dispersion is unsuitable for inverting surface wavefields with dominant higher modes.

2.2 Full-waveform dispersion modelling

The new modelling method is a simulation of the multichannel analysis of surface waves (MASW) employed in modern field surveys. It comprises full-waveform shot gather generation followed by standard multichannel dispersion measurement, hereafter called the PSV method. It will be described starting with synthetic examples, the three shear velocity cases of Tokimatsu et al. (1992), shown in Figure 1. Case 1 is normally dispersive while Cases 2 and 3 are irregularly dispersive. In Case 2, the second layer is a low velocity layer (LVL) and in Case 3 it is a high velocity layer (HVL).

The kernel of the new method employs the reflectivity method (Müller, 1985), which belongs to the class of wavenumber or slowness integration methods for generating synthetic seismograms. Although the integrand is based on plane-wave matrices, and is still restricted to flat isotropic layering, summation of near- and far-field responses of all internal multiples over a well defined slowness range provides an accurate simulation of all P-SV wavefields. This allows modelling of Rayleigh waves, which are coupled P-SV interface waves, along with all other body wavefields (direct, refracted, reflected and guided). The synthetic shot gathers for a surface impact source for the models of Figure 1 are shown in Figure 2. These are trace-normalised, as is usual for displaying surface waves.

To extract dispersion curves from multichannel data, the frequency-slowness (f-p) method is used (McMechan and Yedlin 1981), with some practical aspects outlined in Louie (2001). Plane-wave transformation of the shot gather into the τ–p domain followed by FFT down the τ–axis, creates a dispersion image, shown in Figure 3. An equivalent procedure is to use a f-k transform (Gabriels et al. 1987).

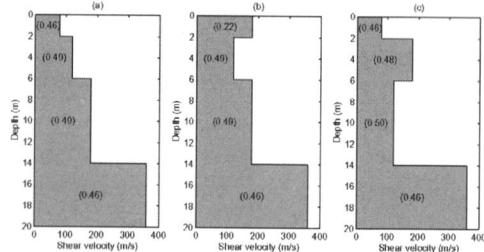

Figure 1. The three 1D shear velocity models of Tokimatsu et al. (1992): (a) Case 1, (b) Case 2, and (c) Case 3. The values in parentheses are the Poisson's ratios for each layer.

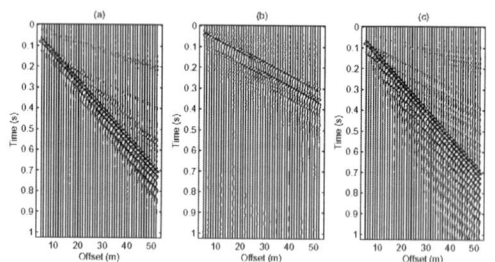

Figure 2. Three synthetic full-waveform P-SV common shot gathers corresponding to the models of Figure 2: (a) Case 1, (b) Case 2, and (c) Case 3.

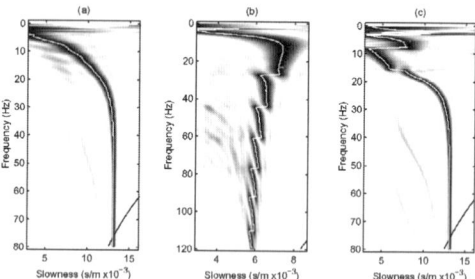

Figure 3. Frequency-slowness (f-p) transforms corresponding to the shot gathers of Figure 3 with the picked maxima: (a) Case 1, (b) Case 2, and (c) Case 3.

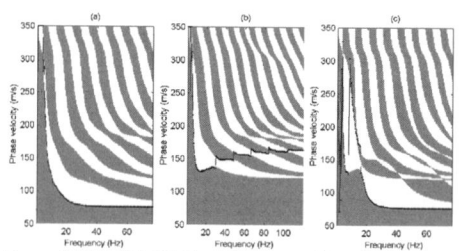

Figure 4. Picked P-SV dispersion curves from the f-p planes of Figure 4 (black lines) with modal FSW dispersion images (grey bands): (a) Case 1, (b) Case 2, and (c) Case 3.

The peak of the dominant lobe in f-p (or f-k) space is picked, and displayed as frequency-phase velocity (f-c), the black curves in Figure 4. The phase-velocity dispersion curve represents a path-average over the recording spread, the 'sounding point' to which the curve is assigned is conventionally assumed to be at the array centre. In engineering, a further transform is to present the dispersion as approximate depth-shear velocity, by calculating a scaled wavelength and phase velocity (usually 0.4λ and $1.1c$), which is known as 'approximate inversion' (Foti 2000).

In Figure 4, the grey bands show the theoretical FSW dispersion for each model, the edges of which are the modal dispersion curves. The PSV-generated dispersion for Case 1 matches the conventional FSW dispersion well, following the edge of the first grey band (fundamental mode). However, in Cases 2 and 3, the PSV dispersion 'jumps' up to higher phase velocities (higher modes), as is evident from the discontinuities in f-p space in Figure 3. While a propagating surface wave train is a superposition of 'normal modes', when higher modes are more energetic, they become known as 'dominant higher modes'. In Case 2 these occur at about 20 Hz intervals, starting at about 25 Hz, while in Case 3 only one dominant higher mode is evident, over 8-16 Hz.

Although in Figure 4 the FSW method correlates in segments, the onset frequencies of dominant higher modes cannot be predicted, so this method would be inadequate as an inversion kernel for the effective dispersion. Fundamental mode isolation filtering (Park et al. 2002) or multiple mode inversion (Xia et al. 2003) is not possible in these cases because of the limited bandwidths of each mode and illustrate why the FSW method is only suitable for inverting the continuous, well-separated dispersion curves (fundamental and/or higher modes) which occur in normally and weakly irregularly dispersive sites.

3 NEW INVERSION TECHNIQUE

3.1 Relation to previous work

The only reports to date which address the inversion of dominant higher modes by linearised optimisation are compared in Table 1. The forward calculation of Ganji et al. (1998) simulates an active source in a stiffness matrix formulation to reproduce the SASW test, while accounting for all modes and wavefields. A similar method was used by Hunaidi (1998) for modelling the 'mode jumping' associated with asphalt or road base cases but the inversion was based on a genetic algorithm. Lai and Rix (1998) showed that partial derivatives of phase velocity with respect to the layer velocities are valid for effective dispersion curves and provide code to compute these analytically. Forbriger (2003a,b) simulated a modified MASW test, where the reflectivity method was used to generate a dispersion (f-p) image, used in the inversion of the transformed data by numerical partial derivatives and was the only work where dominant higher modes were inverted in field data, even though they have been observed in other reports (Park et al. 1999; Rix et al. 2002) and inverted using 'modal' dispersion modelling.

In Table 1, the new scheme is described as a hybrid, in which the reflectivity seismograms are used to observe a dispersion curve, thus allowing rapid analytic partial derivatives calculation. Moreover, realistic data errors are incorporated on which the uncertainty propagation from field data to final model can be based.

Table 1. Comparisons between the methods of the new surface wave inversion procedure used in this paper and three recent methods which also account for dominant higher modes.

	Ganji et al (1998)	Lai and Rix (1998)	Forbriger (2003a,b)	Here
Forward engine	Active stiffness	Active propagator	Reflectivity	Reflectivity
Dataset	c-f curve	c-f curve	f-p plane	c-f curve
Dispersion method	SASW	f-k or SASW	Fourier-Bessel	f-k or f-p
Partial derivatives	Numerical	Analytic	Numerical	Analytic
Data errors	Neglected	Percentage	Statistical	True or modelled

3.2 Algorithm summary

The theoretical dispersion is modelled through an automation of the process illustrated in Figures 2 to 4. Firstly, the parameters as used in the acquisition of the field data are mimicked and a synthetic shot gather generated. Next, this is plane-wave transformed, again employing the same parameters as the field data. With carefully chosen frequency-slowness limits, the automatic ridge-picking (Figure 3) will not be affected by undesired wave fields, such as guided P-waves or aliased ground roll. The weighting for each frequency-phase velocity point is calculated as the inverse of the standard deviation measured from a Monte Carlo uncertainty analysis or simple noise model (O'Neill, 2003).

Using this forward kernel, the field dispersion is automatically inverted with a linearised optimisation based on Occam's inversion. This scheme and associated analytic partial derivatives are described in Lai and Rix (1998). It essentially provides the smoothest model which fits the data, the regularisation automatically rejecting unlikely structure, such as thin and highly-variable velocity layering. During inversion, only shear velocity is allowed to vary, while all other elastic parameters (Poisson's ratio, density and damping) remain fixed. In particular, the necessary a priori assumption of both the number and thicknesses of the layers is a major restriction to the linearised inversion method.

4 SYNTHETIC DATA INVERSION

4.1 Low velocity layer (LVL)

Results from inversions of the Case 2 effective dispersion (Figure 4b), using both the FSW and PSV methods, are shown in Figures 5 and 6. Here, the layer interfaces have been retained at their true depths with the starting shear velocity model a constant 150 m/s and Poisson's ratio maintained at its correct values for all layers. In Figure 5, the FSW method does not account for the dominant higher modes and the final dispersion curve is something of an average. Thus, the velocity reversal is not recovered and final shear velocity errors are of the order of 20% in the LVL and underlying layer. This result is very similar to the fundamental mode inversion of a LVL demonstrated in Lai and Rix (1999). The half-space velocity has not been plotted to preserve the LVL clarity, but it exceeds 700 m/s, or over 100% error. The final RMS error may appear acceptable and the theoretical uncertainties in final shear velocity (• :), calculated from the diagonal of the covariance matrix, are small. However, when compared to the actual error in inverted shear wave velocities (• :), it is apparent that the covariance matrix is strongly affected by the regularisation of the inversion at the last iteration and not representative of the actual final model accuracy. The range indicated by the intermediate iteration results ('Inter' in Figure 5b) are an indication of the degree of non-uniqueness of each inverted layer shear velocity, but still do not incorporate other parameterisation assumptions such as layer thickness.

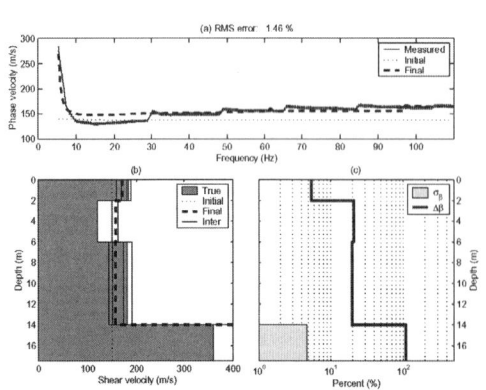

Figure 5. Inversion of the dispersion using the FSW method for Case 2, where the true depth interfaces have been used: (a) Dispersion curves with experimental error, (b) true and derived shear velocity models, and (c) relative standard deviations (grey) and errors (solid line) in inverted layer shear velocities.

When the PSV method is used (Figure 6), the dominant higher modes are properly modeled and after five iterations the RMS error is less than 1%. The shear velocity in the upper three layers is quite accurately obtained to within 1-2%. Poor recovery of basement parameters is normal because of the rapid loss of resolution at low frequency in the dispersion curve, which is where sensitivity is greatest to deeper layer shear velocity. Even though the range in intermediate LVL shear velocities is large, indicating a broad solution space and slower convergence for that parameter, is appears that the misfit minimum attained is near the global minimum.

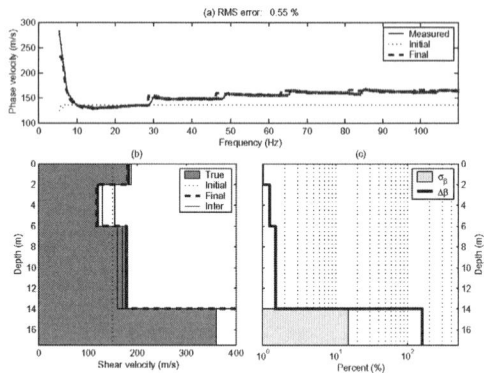

Figure 6. Inversion of the dispersion using the PSV method for Case 2, where the true depth interfaces have been used: (a) Dispersion curves with experimental error, (b) true and derived shear velocity models, and (c) relative standard deviations (grey) and errors (solid line) in inverted layer shear velocities.

In reality, the layer interfaces must be estimated from a priori data (e.g., nearby boreholes), or more commonly set as a stack of layers with defined thicknesses. Here 20 layers with thicknesses geometrically increasing from 0.5 m to 5 m to a half-space at 20 m depth are used and the starting shear velocity model automatically generated by the approximate inversion method, with Poisson's ratio set as 0.45 for all layers. The PSV inversion results are shown in Figure 7, with five iterations again reaching an RMS error less than 1%, and the shear velocity in the LVL is recovered to well within 10% on average. Although shear velocity standard deviations are well under 10% and range of intermediate shear velocities is low, the smoother model gives larger errors at depth and careful model appraisal must be made.

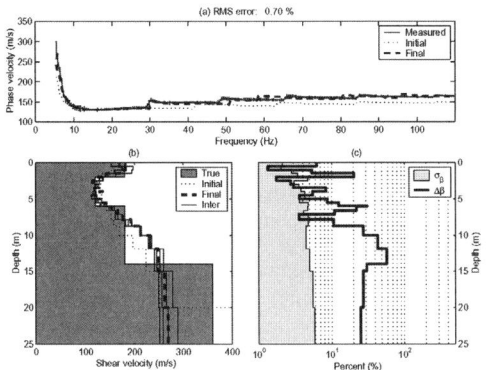

Figure 7. Inversion of the dispersion using the PSV method for Case 2, where a 20 layer model, with thicknesses geometrically increasing from 0.5 m to 5 to a half-space at 20 m depth, is used: (a) Dispersion curves with experimental error, (b) true and derived shear velocity models, and (c) relative standard deviations (grey) and errors (solid line) in inverted layer shear velocities.

4.2 High velocity layer (HVL)

In the literature to date, there are no reports of an inversion of a Case 3 style model (i.e., 'soft-hard-soft-hard'). Ganji et al. (1998) inverted a synthetic HVL dispersion, but dominant higher modes were not present. Rix et al. (2002) observed dominant higher modes at low frequencies in some field sites but inverted using a fundamental mode assumption. In Case 3, the dominant mode, which occurs at low frequency (8–16 Hz), causes a large discontinuity in the dispersion. Moreover, the signal to noise ratio at these low frequencies can be poor due to spread length and source spectrum limits.

Results employing the FSW method are shown in Figure 8. Here, 20 layers with thicknesses geometrically increasing from 0.25 m to 2.5 m to a half-space at 19 m depth are used, and the starting V_S model is a constant 150 m/s with Poisson's ratio set at 0.45 for all layers. The FSW method has converged to less than 1% and detected the general model trend, as shown in Figure 8b. However, by fitting the dominant higher mode as the fundamental in the 5–17 Hz band in Figure 8a, a large error is introduced into the HVL depth and thickness, as well as errors of up to 100% in the underlying LVL.

When the PSV method is used (Figure 9), the depth and thickness of the HVL and LVL are recovered well, to within 20% on average. Both the basement velocity and depth are better recovered than in Case 2, due to the slightly lower cutoff frequency. However, for the LVL underlying the HVL, there is somewhat more uncertainty than in Case 2.

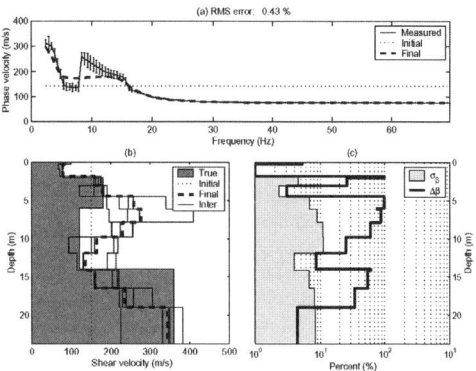

Figure 8. Inversion of the dispersion using the FSW method for Case 3, where a 20 layer model with thicknesses geometrically increasing from 0.25 m to 2.5 m to a half-space at 19 m depth is used: (a) Dispersion curves with experimental error, (b) true and derived shear velocity models, and (c) relative standard deviations (grey) and errors (solid line) in inverted layer shear velocities.

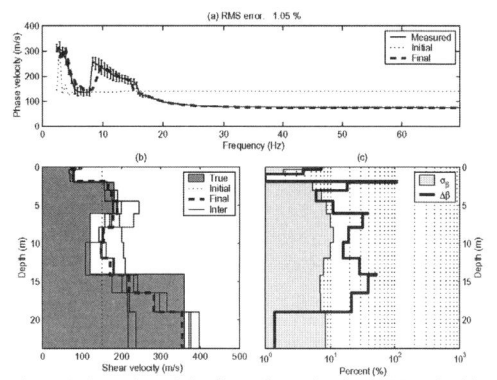

Figure 9. Inversion of the dispersion using the PSV method for Case 3, where a 20 layer model with thicknesses geometrically increasing from 0.25 m to 2.5 m to a half-space at 19 m depth is used: (a) Dispersion curves with experimental error, (b) true and derived shear velocity models, and (c) relative standard deviations (grey) and errors (solid line) in inverted layer shear velocities.

5 FIELD DATA INVERSION

5.1 Telfer mine site

The Telfer mine is a sediment-hosted gold deposit in the Great Sandy Desert in northwest Western Australia. Surface wave inversion was applied as an alternative to vertical seismic profiling (VSP) for engineering assessment of the in situ stiffness of both artificial fill material and the weathered and fresh sandstone basement. This is a vital parameter in choosing suitable sites for heavy vibrating machinery (such as milling equipment) and estimating any geotechnical ground improvement that might be needed. Even though the basement beds are known to be steeply dipping, it was anticipated that results over the sandstone would be more of an indication of the vertical weathering gradient. Sites over artificial fill material (such as the waste dump), depending on the degree of heterogeneity, would more likely resemble layered models.

A 24-channel Geometrics SmartSeis was used with 8 Hz geophones to acquire the data, in April 2003. As the minimum required investigation depth was 10 m, with 20 m preferred, spreads up to 50 m in length were necessary. To accomplish this, while maintaining 1 m geophone spacing to reduce spatial aliasing, walkaway shooting of two 24-channel spreads was performed. The source was a 10 kg sledge hammer on a thin steel plate, with the shot 5 m offset from the near geophone. It was discovered that strong lateral discontinuities prohibited the use of long spreads, so the near 24 channels were usually preferred. Figure 10 shows two shot gathers, where Line 9 was over weathered sandstone and Line 11 over a waste dump.

The f-p transforms of the shot gathers in Figure 10 are shown in Figure 11. Both lines show large discontinuities in the spectral ridge, suggestive of dominant higher modes. The character of the Line 9 dispersion (Figure 11a) is similar to Case 3 (Figure 3c), that is, a dominant higher mode, but over the 25–75 Hz band. Outside this range the dispersion is relatively smooth and appears to be of the fundamental mode. Line 11 (Figure 11b) shows the typical dominant higher modes which occur when a LVL is present, similar to Case 2 (Figure 3b).

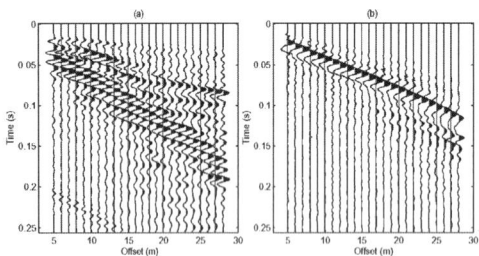

Figure 10. Raw 24-channel shot gathers from the Telfer mine site: (a) Line 9 and (b) Line 11.

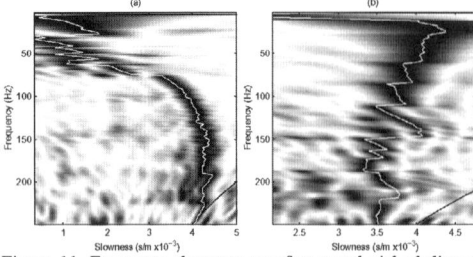

Figure 11. Frequency-slowness transforms and picked dispersion of the raw 24-channel shot gathers from the Telfer mine site: (a) Line 9 and (b) Line 11.

Inversion of the Line 9 dispersion is shown in Figure 12. Here 12 layers with thicknesses geometrically increasing from 0.25 m to 1 m to a half-space at 9 m depth are used. The starting shear velocity is model automatically generated by the approximate inversion method, and Poisson's ratio set at 0.4 for all layers, based on a V_P/V_S ratio around 2.5, estimated from first P-wave arrivals and high-frequency phase velocity. Although limited to the frequency range of 10–210 Hz, the PSV method accurately models the large discontinuity at about 30 Hz, which is a product of the large velocity contrast, modelled at a depth of about 2 m in Figure 12b. The final model suggests that a LVL may exist below this interface, between depths of about 4 and 8 m. While this LVL is geologically feasible, confident interpretation is questionable, considering the large range in the intermediate iterations and standard deviations in the final model, similar to the uncertainty in inversion of Case 3 (Figure 9). The interpreted weathering thickness is supported from nearby trenches where unrippable, fractured sandstone was revealed at shallow depth, about 1.5 m below the surface.

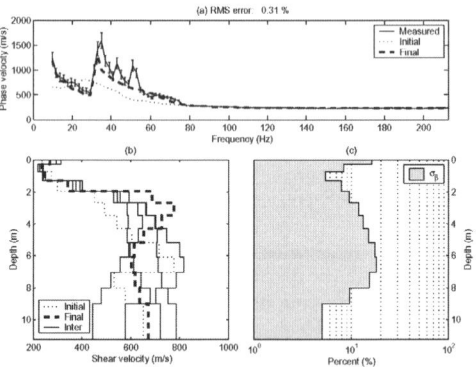

Figure 12. Inversion of the dispersion on Line 9 of the Telfer mine site, using the PSV method, where a model with 12 layers geometrically increasing in thickness from 0.25 m to 1 m to a half-space at 9 m depth is used: (a) Dispersion curves with experimental error, (b) derived shear velocity models and (c) relative standard deviations (grey) in inverted layer shear velocities.

Inversion of the Line 11 data is shown in Figure 13. Here 12 layers with thicknesses geometrically increasing from 0.25 m to 2 m to a half-space at 12 m depth are defined and the starting VS model automatically generated, with Poisson's ratio set at 0.4 for all layers, and the maximum usable frequency limited to about 120 Hz. Usually 100 Hz is more than sufficient to enable modelling of the shallowest layers. In Figure 13b, the PSV method has modelled the phase velocities of the dominant higher modes, but that their frequencies are somewhat displaced. This is similar to the Case 2 inversions of Figures 6 and 7, where even though the frequencies of transitions to dominant higher modes were not exactly modelled, the layer parameters are recovered well. The inversion has converged to a low RMS error, and standard deviations in the final model shear velocities are all less than 5%. This alone is not enough to quantify the accuracy of the inversion, but combined with the field observations of a hard, compacted, secondary cemented surface, and basement depth known to be at about 10 m, there is more confidence in the interpreted LVL shear velocity of about 250 m/s.

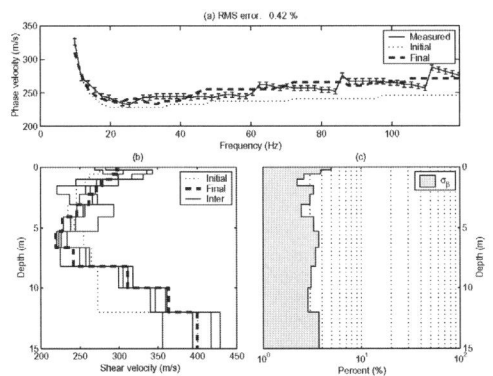

Figure 13. Inversion of the dispersion on Line 11 of the Telfer mine site, using the PSV method, where a model with 12 layers geometrically increasing in thickness from 0.25 m to 2 m to a half-space at 12 m depth is used: (a) Dispersion curves with experimental error, (b) derived shear velocity models and (c) relative standard deviations (grey) in inverted layer shear velocities.

5.2 Hyden fault scarp

The Hyden fault scarp, located about 340 km east of Perth, Western Australia, is the result of a prehistoric earthquake. A P-wave seismic reflection survey was shot in December 2001 using a 96-channel OYO DAS-1 receiver and 12-gauge shotgun blanks as a source, with the gun muzzle down a 0.3 m deep water-filled hole. The high-fold data proved useful for surface wave inversion, where better path-averaged dispersion could be achieved by stacking in the plane-wave transform domain. Trial and error revealed 50 channels to be the longest spread with minimal lateral discontinuity effects, where shot offsets offsets of 1 m to 10 m were used, the individual f-k transforms shown in Figure 14.

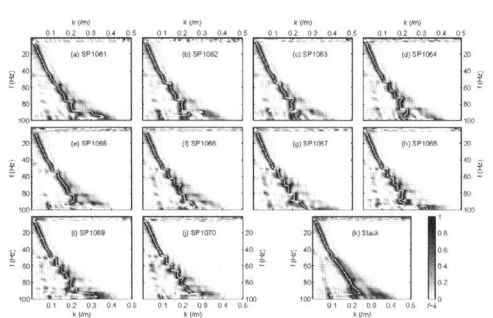

Figure 14. Method for dispersion stacking for a fixed, 50-channel geophone window from SP1011 to SP1060: (a) to (j) Individual f-k planes for shot offsets 1 m to 10 m and (k) the final summed image, from which dispersion can be picked.

The final stacked dispersion is shown in Figure 15a and was inverted by the PSV method. A model comprising layers starting from 1 m thick (at surface), below which they increase in thickness by the relation h=z/2, where h is the thickness and z is the depth to top, with a half-space below 20 m was used, the inversion results shown in Figure 15b. Although no borehole information is available to confirm these results, the final RMS error is low and the buried high velocity layering in the final model coincides with a GPR reflection at 4–5 m, which is anticipated to be a thick, laterite horizon. There is no indication of a hard basement and this overburden model was typical over the entire line, which helps to explain why reflections were absent in the P-wave shot gathers, due to the 'masking effect' of the shallow waveguide (Zahradnik and Bucha, 1998).

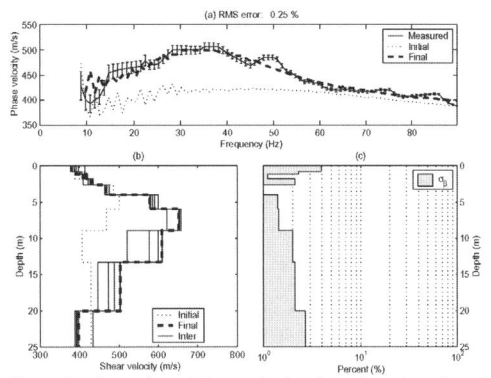

Figure 15. Inversion of the stacked-gather dispersion data at SP1035.5 at the Hyden fault scarp site, using the PSV method where a half-space at 20 m depth with layers linearly increasing in thickness starting with 1 m are used as the model: (a) Dispersion curves with experimental error, (b) derived shear velocity models and (c) relative standard deviations (grey) in inverted layer shear velocities.

6 CONCLUSIONS

Full-waveform P-SV reflectivity synthetic seismograms are the preferred basis for surface wave dispersion calculation, rather than plane-wave propagator matrix techniques. From synthetic examples it is clear that shallow layers, particularly low velocity ones, are recovered with better accuracy than deeper horizons. However, when 'masked' by a buried HVL, uncertainty in any deeper LVL is high. Use of the PSV method has been shown to be crucial when inverting structures that generate dominant higher modes. With plane-wave dispersion modelling, LVLs will not be detected and HVLs interpreted too stiff and deep. Poor recovery of deeper shear velocities and the need to assume a priori the depths to layer interfaces are major limitations to the accuracy of the linearised inversion method.

Dominant higher modes have been observed and inverted from field data with the new method, where conventional, fundamental mode methods would not be applicable. Although no direct ground-truthing was available, geological knowledge adds confidence to the accuracy of the interpreted shear wave velocity sections.

ACKNOWLEDGEMENTS

Klaus Holliger of the Swiss Federal Institute of Technology and Michael Roth of NORSAR provided the FORTRAN reflectivity synthetic seismogram code. The Telfer data was provided by Geoforce Pty Ltd. This work was conducted while the primary author (A. O'N) was in receipt of a University Postgraduate Award (UPA) from The University of Western Australia.

REFERENCES

Beaty, K.S., Schmitt, D.R., and Sacchi, M., 2002, Simulated annealing inversion of multimode Rayleigh wave dispersion curves for geological structure: Geophysical Journal International, 151, 622–631.

Brown, L.T., Boore, D.M., and Stokoe II, K.H., 2002, Comparison of shear-wave slowness profiles at 10 strong-motion sites from noninvasive SASW measurements and measurements made in boreholes: Bulletin of the Seismological Society of America, 92, 3116–3133.

Buchen, P.W. and Ben-Hador, R., 1996, Free-mode surface-wave computations: Geophysical Journal International, 124, 869–887.

Cuéllar, V., 1997, Geotechnical applications of the spectral analysis of surface waves: in McCann, D.M., Eddleston, M., Fenning, P.J., and Reeves, G.M. (eds.), Modern geophysics in engineering geology: Geological Society Engineering Geology Special Publication No. 12, 53–62.

Forbriger, T., 2003a, Inversion of shallow-seismic wavefields: I. Wavefield transformation: Geophysical Journal International, 153, 719–734.

Forbriger, T., 2003b, Inversion of shallow-seismic wavefields: II. Inferring subsurface properties from wavefield transforms: Geophysical Journal International, 153, 735–752.

Foti, S., 2000, Multistation methods for geotechnical characterization using surface waves: PhD thesis, Politecnico di Torino [http://www.polito.it/research/soilmech/foti/ index.htm]: Accessed 18 Apr, 2000.

Gabriels, P., Snieder, R., and Nolet, G., 1987, In situ measurements of shear-wave velocity in sediments with higher-mode Rayleigh waves: Geophysical Prospecting, 35, 187–196.

Ganji, V., Gucunski, N. and Nazarian, S., 1998, Automated inversion procedure for spectral analysis of surface waves: Journal of Geotechnical and Geoenvironmental Engineering, 124, 757–770.

Hunaidi, O., 1998, Evolution-based genetic algorithms for analysis of non-destructive surface wave tests on pavements: NDT&E International, 31, 273–280.

Lai, C.G., and Rix, G.J., 1998, Simultaneous inversion of Rayleigh phase velocity and attenuation for near-surface site characterization: Georgia Institute of Technology [http://www.ce.gatech.edu/~grix/surface_wave.html]: Accessed 18 Apr, 2000.

Lai, C. G. and Rix, G. J., 1999, Inversion of multi-mode effective dispersion curves: in Jamiolkowski, M., Lancellotta, R., and Lo Presti, D. (eds.), Pre-failure deformation characteristics of geomaterials: A. A. Balkema, 411–418.

Louie, J.N., 2001, Faster, better: Shear-wave velocity to 100 meters depth from refraction microtremor arrays: Bulletin of the Seismological Society of America, 91, 347–364.

McMechan, G.A., and Yedlin, M.Y., 1981, Analysis of dispersive waves by wave field transformation: Geophysics, 46, 869–874.

Müller, G., 1985, The reflectivity method: a tutorial: Journal of Geophysics, 58, 153–174.

O'Neill, A., 2003, Full-waveform reflectivity for modelling, inversion and appraisal of seismic surface wave dispersion in shallow site investigations: PhD thesis (unpublished), The University of Western Australia. [http://www.geol.uwa.edu.au/~aoneill/].

Park, C. B., Miller, R. D. and Xia, J., 1999, Multimodal analysis of high frequency surface waves: Proceedings of SAGEEP.

Park, C. B., Miller, R. D. and Ivanov, J., 2002, Filtering surface waves: Proceedings of SAGEEP.

Rix, G. J., Hebeler, G. L. and Catalina Orozco, M., 2002, Near-surface V_S profiling in the New Madrid seismic zone using surface-wave methods: Seismological Research Letters, 73, 380–392.

Schwab, F.A. and Knopoff, L., 1972, Fast surface wave and free mode computations: in Bolt, B.A. (ed.), Methods in computational physics: Academic Press, 87–180.

Tokimatsu, K., Tamura, S., and Kojima, H., 1992, Effects of multiple modes on Rayleigh wave dispersion: Journal of Geotechnical Engineering, 118, 1529–1543.

Whiteley, R.J., Fell, R., and MacGregor, J. P., 1990, Vertical seismic shear wave profiling (VSSP) for engineering assessment of soils: Exploration Geophysics, 21, 45–52.

Xia, J., Miller, R.D. and Park, C.B., 1999, Estimation of near-surface shear-wave velocity by inversion of Rayleigh waves: Geophysics, 64, 691–700.

Xia, J., Miller, R.D., Park, C.B., Hunter, J.A., Harris, J.B., and Ivanov, J., 2002, Comparing shear-wave velocity profiles inverted from multichannel surface wave with borehole measurements: Soil Dynamics and Earthquake Engineering, 22, 181–190.

Xia, J., Miller, R.D., Park, C.B. and Tian, G., 2003, Inversion of high frequency surface waves with fundamental and higher modes: Journal of Applied Geophysics, 52, 45–57.

Offshore geosciences for coastal infrastructure

J. Peuchen & W.M. NeSmith
Fugro, Netherlands

Keywords: coastal engineering, geophysics, in-situ testing, sampling, site investigation

ABSTRACT: Traditionally, land-type methods have been adapted to suit site characterisations in shallow water. In recent years, increasing water depths of coastal construction zones has meant that offshore technologies have become more appropriate. This paper describes recent experiences in the use of offshore geosciences for coastal projects. Particular characteristics for offshore investigation technologies are: (1) high efficiency and effectiveness of geophysics, (2) an excellent track record in Health Safety and Environment (HSE) management and (3) careful investigation planning. Continuous seismic refraction profiling and continuous electrical resistivity profiling have been found to be efficient for ground characterisation in shallow waters. For a coastal project, HSE considerations often include ship collision risk, adverse weather conditions and hazardous ground conditions. Any of these three conditions may preclude the use of fixed platforms such as jack-ups or anchored barges. Dynamically positioned vessels provide opportunities for reducing risks at their source. Only a limited number of dedicated vessels are globally available for geo-data acquisition. It can therefore be beneficial to participate in an offshore-vessel cruise by targeting a "pass-by" arrangement.

1 INTRODUCTION

The population of the earth is concentrated in coastal regions. Trade, scarcity of land, and the need for mineral resources and renewable energy are important drivers for expansion of facilities and infrastructure into deeper-water coastal zones. Traditionally, land-type methods have been adapted to suit site characterisations in shallow water. In water depths beyond about 15 m offshore geosciences can and do provide important contributions to coastal projects.

For the purposes of this paper "offshore geosciences" are summarised as:
- Geophysical survey providing an important contribution to site characterisation
- Ground investigation emphasis on in-situ testing
- Strong awareness and use of HSE systems

Regarding these subjects, it is the differences in the operational deployment from onshore to offshore sites that are primarily of interest. Measuring technologies typically originated onshore. Deployment techniques were further developed for offshore use. These have subsequently been expanded to the coastal zone.

This paper focuses on geotechnical site characterisation. It provides no coverage on geoscience topics such as met-ocean monitoring and environmental characterisation. Elliot and Stephens (2003) and Jean and Feld (2003) provide detailed information on these topics. Similarly, satellite- and airborne-based applications are excluded

2 ROLE OF GEOPHYSICS

2.1 *Comparison onshore – offshore – coastal*

The use of ground-based geophysical techniques for onshore site characterisation is not common. While there have been many technological advances in the area, a successful and efficient industry has yet to develop. Ground intrusion techniques dominate and are likely to do so for some time to come.

The role of geophysics in offshore scenarios is somewhat different. Marine geophysical surveying is a mature industry providing an important contribution to site characterisation, both technologically as well as in cost-efficiency. It is more than just a spin-off from oil and gas reservoir characterisation and ocean geology research. The principal systems, in order of business volume, are as follows:
1. Seismic reflection
2. Seismic refraction
3. Geo-electrical survey

The gradual increase in water depths of coastal construction areas has increased offshore vessel access to these sites and thus the ability to tow geophysical measurement systems through these areas has similarly increased. However, a coastal setting also introduces limitations including physical access constraints and, more particularly, uncertainties in geophysical data interpretation. The authors are aware of many successful and some unsuccessful coastal geophysical measurement projects. A majority of the unsuccessful ones are simply due to inadequate selection of tools and lack of expertise. Table 1 presents a summary of the role of marine geophysics in coastal infrastructure.

2.2 Seismic Reflection

Seismic reflection surveying dominates the geophysical testing market in volume and development efforts. Offshore technology in use ranges from "basic" tow systems to Autonomous Underwater Vehicles (AUVs) packed with advanced high-resolution equipment. The basic tow systems have largely remained unchanged for tens of years. However, the use of high-resolution technology is something of the past 10 years. The principal result of a seismic reflection survey is stratigraphic delineation. Recent developments in interpretation by inversion techniques show promise. However, they have yet to advance to a stage to provide adequate geotechnical correlations for extrapolation of ground properties. Figure 1 shows an example of fairly basic, low-cost results of a seismic reflection survey for a coastal site. The area is intended for land reclamation and subsequent high-capital development. The presence of the observed shallow gas is an important design consideration identified early on in project development.

Seismic reflection technology is available for coastal use within practically any global region.

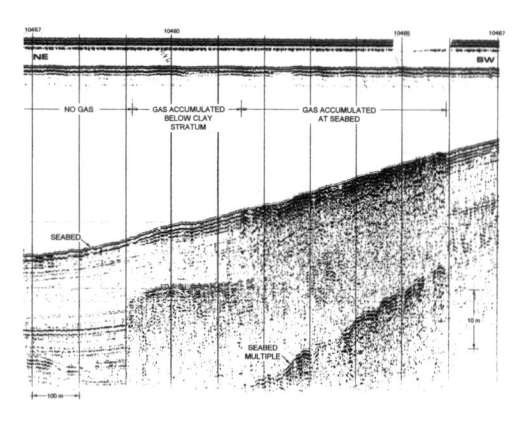

Figure 1. Example of Coastal Geohazard identified from Seismic Reflection Survey

2.3 Seismic Refraction

Seismic refraction is a well-known onshore investigation technique. Its traditional use is for characterisation of "overburden over rock" and for rock excavation potential. The use for estimation of excavation potential relates to seismic refraction providing a physical property of the ground, namely p-wave velocity. Sledge-towed coastal and offshore deployment systems have been successfully used since about 1997 (Puech and Tuenter, 2002). Systems without a sledge are also in use. Figure 2 presents the principle of the Gambas® seismic refraction system and an example of an integrated refraction profile.

The offshore/coastal track record stands at many thousands of kilometres of seismic refraction surveying. Yet it remains only a fraction of the volume of seismic reflection activities. The technology is available from a few global centres of expertise.

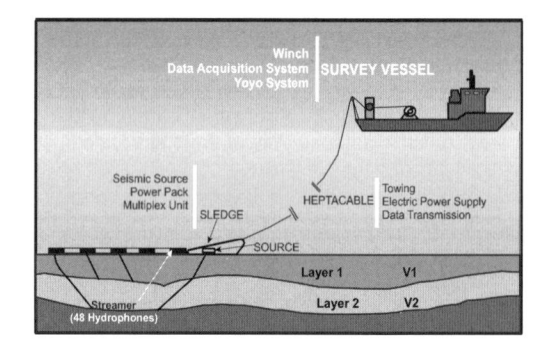

Figure 2a. Principle of Gambas® Seismic Refraction System
(after Puech and Tuenter, 2002)

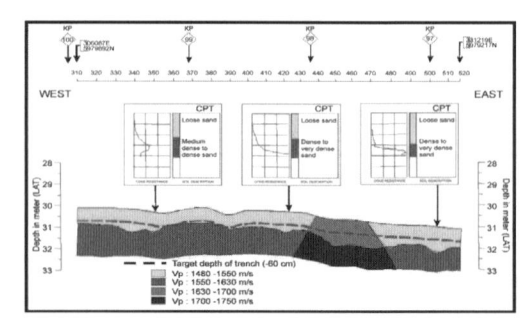

Figure 2b. Example of Seismic Refraction Value for Coastal Applications
(after Puech and Tuenter, 2002)

Table 1. Geophysics for Coastal Infrastructure

Feature	Seismic reflection	Seismic refraction	Electrical resistivity
Deployment			
Track record	From before 1970, mature industry	From 1997, limited global centres of expertise	From early 1990s, limited global centres of expertise
Mounting of measuring tool	Towed and hull-mounted systems from vessel suited to water depth; AUV may be economical in future	Seafloor-based sledge with trailing hydrophone cable; vessel tow is common but any towing device feasible	Seafloor-based cable; vessel tow is common but any towing device feasible; floating cable tow feasible for water depth of less than about 6 m
Interpretation			
Stratigraphy	Based on layering in acoustic impedance; possible masking of layering in shallow water because of seabed multiples, particularly in areas with coarse-grained sediments at seafloor; acoustic blanking in areas with shallow gas	Based on multi-layer stratigraphy with upper layers having lower seismic velocity than lower layer	Based on formation factor, limited strata delineation
Physical ground properties	None, possibly in future	p-wave velocity correlated to ground strength	Electrical resistivity correlated to ground unit weight

2.4 Electrical Resistivity

As with seismic refraction, electrical resistivity is a well-known onshore investigation technique. Its traditional use in geotechnics is for ground stratification and measurement of ground density. Electrical resistivity is a physical property of the ground/groundwater system and correlates mainly with ground porosity. Cable-towed coastal and offshore deployment systems have been in successful use for more than 10 years (e.g. Peuchen and Noort, 1996). However, as with seismic refraction, its use is a fraction of that of seismic reflection. The technology is available from a few global centres of expertise. An example of a bottom-dragged system for measuring electrical resistivity is presented in Figure 3.

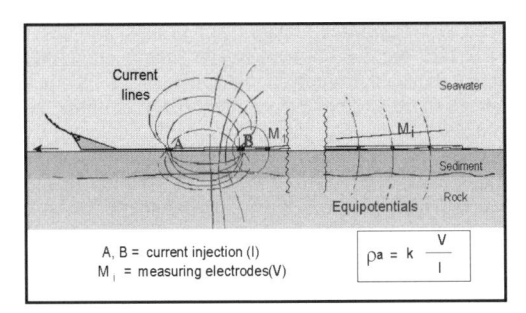

Figure 3. Principle of Rhobas Bottom-dragged Resistivity System
(after Puech and Tuenter, 2002)

3 GROUND INVESTIGATION

3.1 Comparison Onshore – Offshore - Coastal

Data collection technologies used in offshore intrusive ground investigations are as for onshore site investigations; i.e. a combination of (1) laboratory testing on recovered samples and (2) in-situ testing.

Interpretation of both laboratory and in-situ test data is tied in with databases shared with onshore applications. Of particular interest to offshore and coastal operations is any technique that is quick, robust and suited to remote operation.

There are no specific "offshore" laboratory test methods, other than, perhaps, multi-sensor core logging (MSCL). Figure 4 shows an example of a typical multi-sensor core logger. This technology is frequently used in ocean geology research and has now been exported to offshore geotechnical investigation. Available sensors include the following:
– Core diameter
– Gamma ray attenuation
– Magnetic susceptibility
– P-wave velocity
– Natural gamma
– Electrical resistivity
– Colour imaging and analysis.

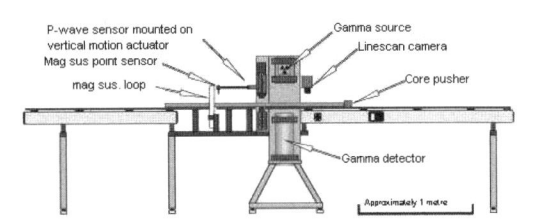

Figure 4. Typical Multi Sensor Core Logger

The general lack of specific offshore laboratory input is curious. One would expect a fast-track offshore industry to push for earlier delivery of test results and laboratory activities are notorious for extended time requirements. However, little research and development investment is currently targeted for novel laboratory technologies aimed at fast-track delivery. This is perhaps related to operators changing to a "commodity approach", giving focus to cost reduction rather than innovation.

Lunne (2001) presents an overview of common offshore in-situ test methods. The piezo-cone penetration test (CPTU or PCPT) dominates in-situ testing activities in offshore site characterisation. Three items are worthy of mention for the coastal zone, where onshore and offshore meet:
- Downhole geophysical logging
- Standard Penetration Test (SPT)
- Penetrometer devices.

Downhole geophysical logging has a long and successful track record in oil & gas reservoir characterisation. The mineral mining industry uses cheaper slimline versions of similar loggers. These are occasionally used in onshore geotechnical investigations. The slimline versions are also in use in offshore and coastal geotechnical investigations (Digby, 2002). Common downhole geophysical measurements are as follows:
- Natural gamma count
- Induced gamma count or gamma ray attenuation (using gamma source and receiver)
- Neutron density
- Lateral resistivity
- Magnetic susceptibility
- Sonic wave velocity.

These measurements all provide lithological indicators, as they react differently to variations in mineralogy. Cross plots of two or more measurements are used to assist with (1) stratigraphic determination, (2) detailed mineralogy, (3) in-situ porosity, density and moisture content determination and (4) estimation of seismic velocity parameters.

Recent developments in geophysical logging also include the addition of geophysical sensors to penetrometer-type tools to measure geophysical parameters whilst simultaneously performing an in-situ test such as a CPTU. In particular, the collection of natural gamma data in conjunction with CPTU performance allows correlating natural gamma measurements. A natural gamma sensor requires no radioactive source, thus easing HSE considerations. Natural gamma has shown some promise with regard to distinguishing between soil mineralogy (Digby, 2002).

The SPT is absent in the offshore investigation field, but is still the dominating in-situ test in some important coastal areas. One can perform SPTs in coastal waters but there are limitations, particularly: (1) the interpretation of results and (2) tool deployment safety. It is the authors' experience that SPT results can no longer be interpreted with confidence when rod length exceeds about 25 m. This value refers to rod length in borehole below seabed plus rod length covering water depth plus required rod stick-up above water level. This observation is based on confidential comparative studies between SPT, CPTU and Cone Pressuremeter Tests to beyond 100m depth (Fugro, 1994). Additionally, rod-based SPT activities generally fall outside acceptable personnel safety limits when carried out from a significantly heaving platform. HSE and efficiency considerations are likely to be the drivers for occasional use of a wireline SPT in coastal investigations. This technique is described with comments as follows:
- The sample tube, hammer and drop-height are as for the rod-type SPT. The driving mass is downhole and falls in an air-filled chamber.
- Energy delivery to the top of the sampler is affected by differential water pressure. No water pressure acts on the top of the downhole SPT tool where the rod enters the hammer chamber. This differential pressure increases with borehole fluid height, thus the mass of the housing must be high enough to stop the housing being forced up.
- The atmospheric chamber has a larger diameter than the sample tube in order to house the hammer. Penetration of the system requires the water below the atmospheric chamber to be displaced.
- Depth measurement is wireline-based rather than rod-based and thus requires a higher degree of observation to ensure accurate measurement.

The authors recommend use of CPTU technology and separate sampling as alternative to SPT. Equivalent SPT values can be obtained, if required, by application of correlations such as published by Robertson et al. (1983) and Kulhawy and Mayne (1990). This approach avoids time-consuming overwater SPT operations and suits practitioners familiar with SPT-based interpretation and design methods.

While the CPTU remains the dominant in-situ test, other penetrometer tools have been developed for offshore site characterisation. They are also applicable for coastal sites. These tools are typically for use in very soft normally consolidated marine clays. One common tool is the full rotation in-situ vane, which provides direct measurements of both peak and residual undrained shear strength (Figure 5). Output includes stress versus rotation information, as measured downhole.

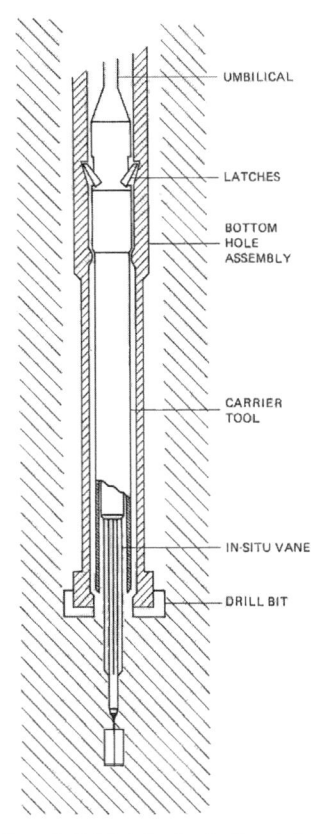

Figure 5. Full Rotation In-situ Vane (**after Geise et al., 1988**)

Two penetrometer-type tools include the T-bar and the spherical penetrometer (Lu et al., 2000; Figure 6). These tools provide direct measurement of in-situ undrained shear strength, according to simplified theoretical soil models. The authors note that data collection and research regarding the accuracy of T-bar and spherical penetrometers is still ongoing.

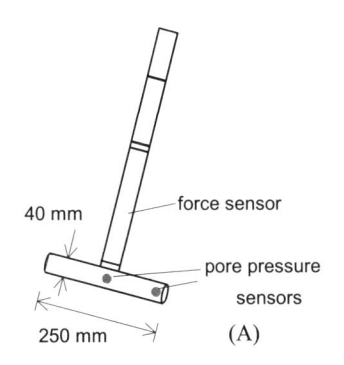

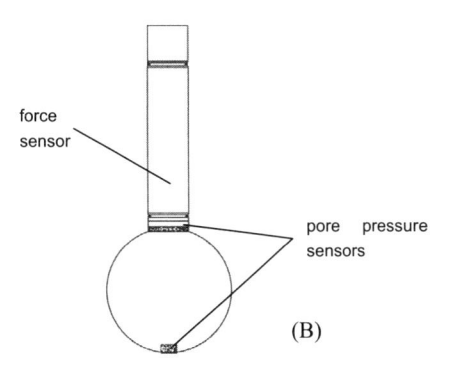

Figure 6. (A) T-bar Penetrometer (**after Randolph et al., 1998**) and (B) Ball Penetrometer

3.2 Deployment

Lunne (2001) describes offshore deployment systems, particularly so-called downhole systems and seabed-based systems. Figure 7 presents examples of these systems. Both systems find use in coastal areas with water depth exceeding 10 m to 15 m, depending on factors such vessel/barge draught, moonpool-based operations versus "over-the-side" operations and position stability (x, y, z).

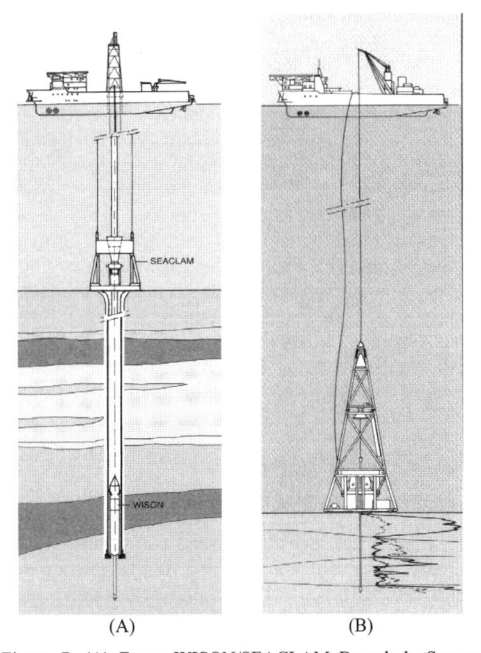

Figure 7. (A) Fugro WISON/SEACLAM Downhole System and (B) Fugro SEASPRITE Seabed System

A borehole is required for downhole sampling and in-situ test systems. Rotary drilling is the standard for borehole advancement. Outflow of drill cuttings is at seabed and not at drill platform level. Heave compensation systems allow controlled drilling from a heaving platform (vessel), typically up to a safe vessel heave of about 3 m at drill point. A stationary drill pipe for in-situ testing or sampling is obtained by a seafloor-based frame fitted with a pipe clamping system. Various rock core drilling systems are available, some of which are capable of recovering geotechnical drill core with "triple-tube" sample quality as obtained onshore (e.g. 102 mm diameter calcarenite core; Figure 8).

Figure 8. Example Recovery of Core with "Triple Tube" Sample Quality

Typically, seabed systems involve the deployment of a seafloor-based frame fitted with an in-situ testing or sampling tool advancement system. Tools are advanced in a continuous push mode similar to onshore in-situ testing techniques. Most systems include the ability to momentarily stop the push for performance of in-situ tests at discrete depths (e.g. the in-situ vane test). Depth limitations of seabed systems depend on the tool pushed and ground conditions. In soft soil, penetrometer-type devices can typically be pushed to a maximum of around 40 m below seabed.

Large diameter sampling devices that are deployed from a vessel and are self- or gravity-propelled into the seabed have also been developed for offshore site characterisations in the past decade (Borel et al., 2002). At present, longer length samplers (up to 20 m) such as the STACOR® typically require a greater water depth for use than is available for most coastal projects. Other similar but smaller length samplers (e.g. Gravity Corer or Vibro-Corer) with lengths of up to 6 m are routinely implemented in coastal site characterisations.

Seabed based drilling systems are remotely operated from the vessel from which they are deployed. These have not seen widespread use to date. Additionally, diver-operated seabed in-situ testing systems have seen some use in shallow coastal waters. These often fall far short of the HSE requirements of most operators.

4 HSE

Ground characterisation activities cannot and must not ignore HSE considerations. For a coastal project, such considerations often include ship collision risk, adverse weather conditions and hazardous ground conditions. Any of these three conditions may preclude the use of fixed platforms such as jack-ups and anchored barges taken into the coastal zone (Hodgson et al., 1995). Typical operations consist of equipment and personnel assembled to meet one-off project requirements. HSE statistics for these types of operations generally fall well short of society expectations.

Dynamically positioned vessels provide opportunities for reducing risks at their source. Figure 9 shows the dynamically positioned vessel, MV Bavenit during a recent coastal investigation. HSE for such vessels is based on tough oil and gas industry requirements. Only a limited number of dedicated vessels are globally available for geo-data acquisition. The vessels work on a 24-hour basis and numerous investigation systems are routinely available. Operational costs for sailing or stand-by are only marginally less than costs for actual data acquisition. It can therefore be beneficial to participate in an offshore-vessel cruise by targeting a "pass-by" arrangement.

Figure 9. Geotechnical vessel MV Bavenit at recent coastal project location

5 INVESTIGATION PLANNING

Site characterisation is about prediction of risk. Such prediction benefit must be balanced against price to be paid. Acceptability of risk is not only the domain of stakeholders but also that of society and other "third parties". Statutory HSE and other requirements are common.

Results of site characterisation activities must be available at appropriate key project dates. Obtaining a time slot in an offshore-vessel cruise often provides pass-by financial benefits. Planning should be well in advance to avoid disappointment in availability. In many situations, the notice period may well be less than the time required to obtain the necessary permits for coastal survey activities. A requirement for a three-month lead period is not uncommon.

Delays in investment planning decisions and long procurement times are known to upset time schedules (e.g. Peuchen and Noort, 1996). Recent software and communication developments offer some relief. It is now common to deliver a 90% complete report on field operations and data acquisition at the time of vessel demobilisation. Daily uploading of data in digital form from a vessel to a dedicated web site has been readily available. It has been in use for offshore and coastal projects since about 2000.

Dedicated vessels for ground characterisation generally include a basic testing laboratory for geotechnical classification. Results become available as data acquisition progresses. More advanced triaxial and oedometer testing can also be performed on board to cut report delivery times but this is generally not done in practice. Tight project deadlines occasionally require sharing of laboratory workload over multiple laboratories. This is a major logistical exercise and often requires consideration of across-border quarantine regulations designed to avoid spread of disease.

GIS (Geographical Information System) is a well-known tool for data presentation at various levels of sophistication (McNeilan et al., 1999). Unfortunately, project agreements frequently fail to set an appropriate GIS framework. Afterthoughts can lead to frustrations, if not failures, in data format exchange. Future high-level coastal projects are likely to attract use of expert-system visualisation and interpretation technology, such as those that are commonplace for oil and gas reservoir characterisation. Lattimer et al. (2000) present an example of the visualisation and analysis techniques.

6 WHAT WILL THE FUTURE HOLD?

Development of offshore site investigation technologies is typically driven by the fast-track oil and gas industry. Expectations of the industry with regard to efficiency will be particularly applicable in the areas of data analysis and presentation on site, at the completion of fieldwork. Alternatively, data presentation in real-time from remote sites to various offices allows project monitoring by various interested parties.

Similarly it is expected that there will be pressure from the oil and gas industry to develop novel laboratory technologies to address the issue of the extended time requirements currently required for laboratory test programs. Additionally, it is expected that efforts will continue to correlate in-situ test results with soil parameters typically obtained by laboratory testing.

The expected continuation of development in coastal areas with ever-increasing water depths can benefit from the opportunities offered by the off-shore geosciences. Particularly, the integration of geophysical and geotechnical investigation methods, stringent HSE practices and the specialist equipment and knowledge can provide important contributions. It is anticipated that these combinations of equipment and experience will be concentrated in a few global centres of expertise operating on a near continuous basis in coastal regions. Proper investigation planning to take advantage of availability will continue to increase in importance to ensure expediency.

REFERENCES

Borel, D., Puech, A., Dendani, H. and de Ruijter, M. (2002), "High Quality Sampling for Deep Water Geotechnical Engineering: the STACOR Experience", *Proceedings of the Ultra Deep Engineering and Technology Conference Brest, France.*

Digby, A. (2002), "Wireline Logging for Deepwater Geohazard Assessment", *Offshore Site Investigation and Geotechnics 'Diversity and Sustainability', Proceedings of the 5th International Conference of the Offshore Site Investigation Committee of the Society for Underwater Technology*, pp. 391-404.

Elliot, T. and Stephens, R. (2003), "Oceanographic Support for Deep Water Drilling and Field Development", *Proceedings of the PetroMin Deepwater Technology, Kuala Lumpur, Malaysia.*

Fugro (1994), Confidential Fugro Report No. K-2380/105, "Foundation Conditions", Bangladesh.

Geise, J.M., ten Hoope, J. and May, R.E. (1988), "Design and Offshore Experience with an In Situ Vane", *Proc. International Symposium on Laboratory and Field Vane Shear Strength Testing, ASTM STP 1014*, pp. 318-338.

Hodgson, A.J., Adam, C.H. and Sneddon, M. (1995), "Recent Developments in Planning and Execution of Nearshore Site Investigation", *Proceedings of the International Conference on Advances in Site Investigation Practice, Institution of Civil Engineers*, pp. 13-24.

Jean, G. and Feld, G. (2003), "The Measurement And Analysis Of Long-Period Waves To Support Coastal Engineering", *Long Period Waves Symposium, Proceedings of the XXXth IAHR Congress, Thessaloniki, Greece.*

Kulhawy, F.H. and Mayne, P.H. (1990), "Manual on Estimating Soil Properties for Foundation Design", *Electric Power Research Institute, EPRI.*

Lattimer, R.B., Davison, R. and van Riel, P. (2000), "An Interpreter's Guide to Understanding and Working with Seismic-Derived Acoustic Impedance Data", *The Leading Edge*, March, pp. 242-256.

Lunne, T. (2001), "In Situ Testing in Offshore Geotechnical Investigations", *In Situ 2001, Proceedings of the International Conference on In Situ Measurement of Soil Properties and Case Histories, Bali, Indonesia*, pp. 61-83.

Lu, Q., Hu, Y. and Randolph, M.F. (2000), "FE Analysis for T-Bar and Spherical Penetrometers in Cohesive Soil", *Proceedings of the Tenth International Offshore and Polar Engineering Conference, Seattle*, pp. 617-623.

McNeilan, T.W., Chacko, M.J., Dean, C.B., Reitman, J., Lam, I.P. and Buell, R. (1999), "An Integrated Approach to the Site Investigation and Earthquake Response Analysis for the San Francisco-Oakland Bay Bridge East Span Replacement", *Geotechnical Engineering for Transportation Infrastructure, Theory and Practice, Planning and Design,*

Construction and Maintenance, Proceedings of the 12th European Conference on Soil Mechanics and Geotechnical Engineering, Amsterdam, pp. 459-470.

Peuchen, J. and Van Noort, B.W. (1996), "Dredging Survey Strategy for an Intake Pipeline in Karstic Limestone", *HYDRO '96, 10th Biennial International Symposium of The Hydrographic Society, Port and Coastal Hydrography, Rotterdam, Netherlands, Proceedings, Special Publication No. 36*, pp. 141-151.

Puech, A. and Tuenter, H-J. (2002), "Continuous Burial Assessment of Pipelines and Cables: A State-of-Practice", *Offshore Site Investigation and Geotechnics 'Diversity and Sustainability', Proceedings of the 5th International Conference of the Offshore Site Investigation Committee of the Society for Underwater Technology*, pp. 391-404.

Randolph, M.F., Hefer, P.A., Geise, J.M. and Watson, P.G. (1998), "Improved Seabed Strength Profiling Using T-Bar Penetrometer", *New Frontiers, Proceedings of the International Conference on Offshore Site Investigation and Foundation Behaviour, Society for Underwater Technology, London*, pp. 221-235.

Robertson, P.K., Campanella, R.G. and Wightman, A. (1983), "SPT-CPT Correlations", *ASCE Journal of Geotechnical Engineering, Vol. 109, No. 11*, pp. 1449-1459

A framework for inversion of wavefield spectra in seismic non-destructive testing of pavements

N. Ryden & P. Ulriksen
Engineering Geology, Lund Institute of Technology, Lund University, Sweden

C.B. Park
Kansas Geological Survey, University of Kansas, KA, USA

Keywords: surface waves, guided waves, non-destructive testing, pavements, inversion, simulated annealing, MASW

ABSTRACT: A method for the inversion of wavefield spectra in non-destructive testing of pavements is outlined. Data acquisition and processing is based on the Multichannel Analysis of Surface Waves (MASW) method. The inversion of a vertical-shear wave velocity profile is conducted on the full wave field spectrum using the Fast Simulated Annealing (FSA) algorithm. This procedure avoids dealing with discrete dispersion curves and mode identification for the inversion of surface wave data. The viscoelastic properties of the asphalt layer are included in the forward model, resulting in an inverted mastercurve, showing the asphalt stiffness as a function of frequency. The outlined approach is demonstrated on both synthetic and field data. The main benefit is that the raw field data can be automatically processed and inverted without any subjective user input, while properly accounting for the interference of different modes and types of waves. The main disadvantage at this point of development is an extensive (several hours) computational time for the inversion process.

1 INTRODUCTION

Mechanistic and analytical models are the basis of modern pavement design. A prerequisite for using these models is that material properties, such as Young's modulus (E-modulus) and Poisson's ratio (v), can be measured and validated in the field. Stiffness properties obtained from seismic testing are low strain parameters that can be used, in conjunction with results from testing at higher strain levels, for calibration of non-linear material models (Tawfiq et al. 2000). Alternatively, seismic low strain stiffness properties can be adjusted with empirical relations and used directly in traditional pavement design criteria (Nazarian et al. 1999).

Since the early studies by Jones (1962) and Vidale (1964) surface wave methods have been recognized as a tempting approach for non-destructive testing (NDT) of pavements. The introduction of the Spectral Analysis of Surface Waves (SASW) method (Heisey et al. 1982; Nazarian 1984; Stokoe et al. 1994) has resulted in a widespread use of the method, and field protocols for pavement design and maintenance based on seismic methods are now being developed (Abdallah et al. 2003; Haegeman 2002). The evaluation of the compression (V_P) and shear (V_S) wave velocity of the top layer has proved to be most simple, accurate, and effective in seismic NDT of pavements (Roesset et al. 1990; Akhlaghi and Cogill 1994; Ryden 1999). Aouad (1993) and Wu et al. (2002) concluded that the SASW method is not very sensitive to the properties of the layer underlying the top asphalt layer (base layer). However, the potential of evaluating deeper embedded layers in a non-destructive manner has always been the main motivation for surface wave testing of pavements. For that reason a majority of recent publications have focused on the improvement of the evaluation (inversion) of the complete layered pavement system (Ganji et al. 1998; Gucunski et al. 2000; Wu et al. 2002).

In this paper we propose an alternative procedure, which differ in data acquisition, data processing and inversion. A detailed description of the procedure will be available in Ryden (2004). Data collection and processing is based on the Multichannel Analysis of Surface Waves (MASW) method (Park et al. 1999; Xia et al. 1999). The inversion of structural properties of the pavement system is conducted directly on the frequency phase velocity spectrum. A global optimization technique, Fast Simulated Annealing (FSA), is implemented for the inversion algorithm. With this procedure data reduction and the extraction of an experimental dispersion curve is avoided.

2 THEORETICAL CONSIDERATIONS

The proposed approach is based on a theoretical study of guided waves in pavement systems (Ryden 2004). A typical pavement structure consists of three different layers with different material and stiffness properties. Starting from the top, these layers are; asphalt layer, base layer (granular material), and subgrade (natural soil). The stiffness ratio between the layers is usually large and therefore the asphalt and base layer acts as effective wave guides. Wave propagation in the horizontal direction is similar to Lamb waves in free plates (Lamb 1917). The fundamental anti-symmetric mode (A0) is the dominating Lamb wave mode that can be excited with a vertical source at the surface of a horizontally aligned plate-like-structure (Wilcox et al. 2002). This is also the wave generated in the top asphalt layer in surface wave testing of pavements (Ryden and Park 2004). The lower velocity base layer is forced (tuned) to vibrate with the same frequency and horizontal wave number as the asphalt layer. Therefore higher modes of Lamb wave type are generated in the base layer. All these modes are leaky and energy is radiating down into the lowest velocity layer (subgrade). Rayleigh type of waves are only possible for wavelengths shorter than the thickness of the top asphalt layer, and also for very long wavelengths that are basically not affected by the pavement construction (Ryden 2004). It should also be noted that the plate like properties of the top layer dictates longitudinal waves to propagate as symmetric Lamb waves in this layer.

A theoretical layer model (Table 1) is used to exemplify the main features of guided waves in pavement systems. The first 120 modes (Fig. 1) of the theoretical layer model have been traced, in the complex wave number domain, using the dispersion curve software *Disperse* (Pavlakovic et al. 1997; *Disperse* 2001). For pavement systems where most modes are leaky, dispersion curves must be calculated in the complex wave number domain (Vidale 1964), where the imaginary part of the wave number represents the attenuation due to leakage (although all layer properties are elastic).

Table 1. Theoretical layer model.

Layer	Thickness (m)	V_s (m/s)	Density (kg/m^3)	Poisson's ratio
1	0.20	1400	2000	0.35
2	0.60	500	2000	0.35
3	∞	100	2000	0.35

In Fig. 1 all dispersion curves are gray-scale coded with respect to the attenuation due to leakage. Black parts represent low leakage (efficiently propagating guided waves) and gray parts represent higher leakage (strongly attenuating waves). The two lowest dark phase velocity asymptotes (a1 and a2) in Fig.1 correspond to interface waves with almost zero displacement at the surface. The overall trend of the black areas (a3) at higher phase velocities matches with free Lamb waves in the top layer (Ryden et al. 2004). This overall trend can be used for a simplified inversion of the stiffness and thickness of the top layer based on Lamb waves in a free plate (Jones 1962; Akhlaghi and Cogill 1994; Ryden et al. 2003).

Fig. 2 shows mode shapes at the corresponding frequencies and phase velocities marked as points (A) and (B) in Fig. 1. The mode shapes show the typical Lamb wave type of particle motion (Graff 1975) in both the asphalt and the base layer.

The nature of guided waves in pavement systems is more complicated than the traditional Rayleigh wave assumption (Ryden 2004). The theory briefly described above, affects all steps in the process of surface wave testing of pavements, and should be taken into account for the development of an improved seismic NDT technique for pavements.

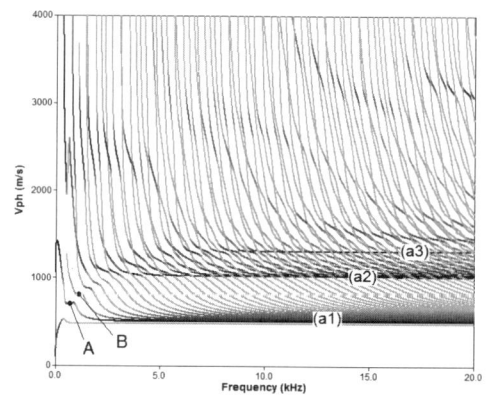

Figure 1. Dispersion curves from the theoretical layer model (Table 1).

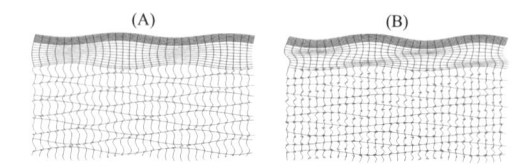

Figure 2. Mode shapes at (A) 500 Hz and (B) 1000 Hz as marked in Fig.1.

3 METHOD

The outlined method can be divided into three different parts, data acquisition, data processing, and inversion. The first two parts have been covered in Ryden et al. (2004).

3.1 Data acquisition and field set up

The Multichannel Simulation with One Receiver (MSOR) data acquisition technique is used to record data. In the MSOR data acquisition method a multichannel record is obtained with only one receiver. It is fixed at a surface point and receives signals $u(t)$ from several hammer impacts at incremental offsets (x) (Ryden et al., 2001). All recorded signals are then compiled to make an equivalent multichannel record $u(x,t)$ for dispersion analysis. A schematic description of the field set-up is shown in Fig. 3.

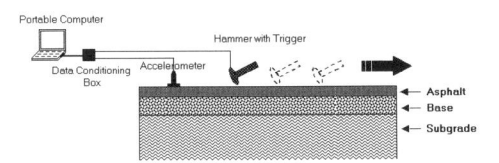

Figure 3. Schematics illustrating MSOR survey with fixed receiver and moving source.

The data acquisition system consists of a portable computer equipped with a A/D converter, source, receiver, and external signal conditioning. This configuration is called the Portable Seismic Acquisition System (PSAS) (Ryden et al. 2002).

Only one high frequency source, a 0.22 kg hammer, is used to cover all frequencies of interest. To improve source coupling and the precision of the source point, a steel spike is used as a source-coupling device. An accelerometer with a natural resonance frequency of 30 kHz is used as the receiver and is attached to the pavement with sticky grease.

3.2 Data processing

The multichannel equivalent record, $u(x,t)$, obtained from the MSOR method is automatically and objectively transformed to the frequency-phase velocity domain using the MASW processing scheme (Park et al. 1998; Park et al. 1999). The processing technique can be expressed with the integral:

$$S(\omega, c_T) = \int e^{-i(\omega/c_T)x} A(x, \omega) dx \quad (1)$$

where, $A(x,\omega)$ is the normalized complex spectrum obtained from the Fourier transformation of $u(x,t)$, ω is the angular frequency, and c_T is the testing phase velocity. $S(\omega,c_T)$ is the slant stack amplitude for each ω and c_T, which can be viewed as the coherency in linear arrival pattern along the offset range for that specific combination of ω and c_T. When c_T is equal to the true phase velocity of each frequency component there will be a maximum in $S(\omega,c_T)$. Calculating $S(\omega,c_T)$ over the frequency and phase velocity range of interest generates the wavefield spectrum where dispersion curves can be identified as high amplitude bands. In Park et al. (2001) a detailed parametric examination of Eq. 1 on its resolution in response to its parameters is presented. The MASW processing scheme is similar to the τ-p (time-slowness) method presented by McMechan and Yedlin (1981), but expressed in a simpler manner directly in the frequency-phase velocity domain, and by using the normalized amplitude spectrum $A(x,\omega)$ instead of the original true amplitude spectrum $U(x,\omega)$ (Forbriger 2003).

To illustrate the resolution that can be obtained with Eq. 1, over a realistic offset range (0.05 to 4.00 m), the theoretical layer model (Table 1) was analysed with the forward model (described in the following section). The resulting wavefield spectrum is shown in Fig. 4.

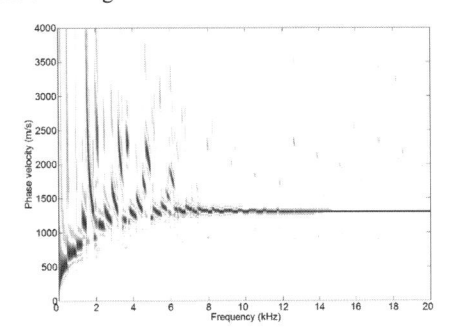

Figure 4. Frequency-phase velocity image obtained from the forward model used in the inversion algorithm, corresponding to the theoretical layer model (Table 1).

In this noise free case many modes can be identified (compare with Fig.1) but it should be noted that peak amplitudes in this image do not fully match with the normal modes in Fig. 1, especially not at lower frequencies (<1kHz). This is because the data processing scheme cannot fully resolve the exact phase velocities (normal modes) over a limited offset range due to mode interference (Ryden 2004). At this stage, one phase velocity at each frequency could be extracted from the peak amplitudes in Fig 4 and all points could be compiled and smoothed to an apparent dispersion curve. However, as evident from Fig 4 a lot of information should then be lost. Therefore we propose to preserve the data as a complete wavefield spectrum and conduct the inversion directly in this domain. By comparing Fig. 1 and Fig.4 it also becomes clear that the measured apparent dispersion curve is actually built up by many small portions of higher modes.

3.3 Inversion

Inversion of surface wave dispersion curves is usually defined as a nonunique and nonlinear inverse problem, where a direct solution is not possible. Nonunique means that different layer models may match the measured data equally well. Nonlinear means that small changes in the data can correspond to large changes in the layer model and vise versa.

The conventional inversion technique for surface wave methods is usually based on a linearized damped least squares solution, driven by partial derivatives of each inverted parameter (Jacobian matrix) (Menke 1989). The multimodal nature of surface wave data from pavements makes the partial derivatives approach difficult because the measurable apparent dispersion curve "jumps" between modes and the partial derivatives are not sufficiently stable (Foti et al. 2003; Ryden 2004). Another problem with linearized gradient based inversion techniques applied to nonlinear and nonunique problems, is that the solution is often dependent on the starting model and may not always converge to the global minima of the objective function (Luke et al. 2003). In the case of pavement systems the problem is further complicated by largely variable parameter sensitivities and correlated parameters (Ryden 2004). For example, the phase velocity of Lamb waves is strongly dependent on the bending stiffness of the plate. The bending stiffness is a function of both thickness and shear wave velocity, which makes these parameters correlated with respect to the phase velocity.

With the approach outlined here the inversion algorithm can be divided into two independent parts. The "forward model" which generates a frequency-phase velocity spectrum (S_{pred}) from a known layer model (**m**) with a corresponding mismatch (E) compared to the observed (measured) replica (S_{obs}), and the "global optimization" routine for minimizing E.

The forward model should ideally mimic the field set up and measured response as close as possible. The theoretically predicted response solution, $U_{pred}(\omega,k)$, from the stiffness matrix approach (Kausel and Roesset 1981) is used to find the dominating wave numbers (k_i) at each frequency for a given layer model (**m**) (Gucunski and Woods 1992; Ganji et al. 1999; Hunaidi 1998). Resonant wave numbers from the stiffness matrix method are superposed with their respective amplitude ratios over the measured offset range by using:

$$U_{pred}(x,\omega) = qR \int_0^\infty J_1(kR) J_0(kx) U_{pred}(\omega,k) dk \qquad (2)$$

where R represents the radius of the source point, q is the load intensity, and J_0 and J_1 are Bessel functions of the first kind of orders 0 and 1 respectively. In this case R and q can be omitted because the radius of the source point is much smaller than the wavelengths of interest and amplitudes are normalized in the subsequent processing of $U_{pred}(x,\omega)$. When $U_{pred}(x,\omega)$ has been calculated Eq. 1 is applied to generate the predicted wave field spectrum $S_{pred}(\omega,c_T,\mathbf{m}_j)$ (Ryden 2004). This theoretical wave-field spectrum is compared to the observed wave-field spectrum $S_{obs}(\omega,c_T)$ through the objective function (Ryden 2004):

$$E(m) = \left| 1 - \left(\frac{\sum_{i=1}^{M}\sum_{j=1}^{N} |S_{ij\,obs}| * |S_{ij\,pred}|}{\sum_{i=1}^{M}\sum_{j=1}^{N} |S_{ij\,obs}|^2} \right) \right| \qquad (3)$$

The mismatch (E) is then minimized with the Fast Simulated Annealing (FAS) algorithm (Szu and Hartley 1987). Simulated Annealing (SA) is a global direct Monte Carlo method (Laarhoven and Aarts 1987) that is especially developed for finding the global minima of challenging functions with many local minima. The method has been applied to surface wave inversion in previous studies, where discrete continuous dispersion curves have been successfully matched (Martínez et al. 2000; Beaty et al. 2002; Rambod and Gucunski 2003).

The basic concept of the SA method is to perform a random walk in the multidimensional parameter space intended to find the global minimum of the objective function (Eq. 3). The search strategy involves two control parameters, an initial temperature (T_0), and a cooling parameter (g). Each iteration involves a random perturbation of each parameter in **m** followed by a forward model and mismatch calculation. The initial model is randomly drawn from the predefined interval (Δm_i) of each parameter (i) to invert. If the calculated mismatch is smaller than that for the previously accepted model the new model is always accepted. However if the mismatch is larger than for the previous model the new model may also be accepted if:

$$r < e^{-\Delta E/T} \qquad (4)$$

where r is a random number from 0 to 1, ∇E is the mismatch difference between the current and the previously accepted model, T is the current temperature that is decreased with the number of accepted models "transitions" according to the employed cooling schedule. This second chance of acceptance (known as the Metropolis criteria) is the key concept of the SA algorithm, and can make the search escape from a local minima and move on in the search for the global minima.

The difference between standard SA and FSA lies in the way the new models are randomly perturbated for each new iteration. If the search range (∇m_i), assigned to each parameter to invert is too large, it will waste a lot of time at low temperatures where large perturbations are not likely. Therefore the FSA method involves a temperature scaled Cauchy distribution of the perturbation of each parameter (Sen and Stoffa 1995). The Cauchy distribution has a narrow central peak and heavy tails, providing concentrated local sampling in combination with occasional large perturbations. After each accepted transition each parameter (m_i) is pertubated to obtain a new testing value (m'_i) according to:

$$m'_i = m_i + \Delta m_i (T/T_0)(\eta_1 \tan(\eta_2 \pi/2)) \qquad (5)$$

where η_1 and η_2 are uniform random variables on [-1,1], and Δm_i are the predefined search interval for each parameter.

To test the outlined inversion scheme the synthetic data (Fig. 4) obtained from the forward model was inverted. A logarithmically sampled frequency range between 50 and 4000 Hz was used for the inversion to reduce the total number of data points presented in Fig. 4. The FSA schedule was set up with five transitions per temperature, and a linear cooling schedule ($T' = Tg$) with T_0=30, and g=0.98. The predefined interval (based on a reasonable range of material properties) for each parameter is given in Table 2. The inverted layer model is presented in Fig. 5 along with the target layer model.

Table 2. Parameter intervals used for the inversion of the theoretical layer model

Layer	Thickness (m)	V_S (m/s)	Density (kg/m³)	Poisson's ratio
1	0.10-0.5	1000-2000	2000	0.35
2	0.40-1.2	100-600	1500-2500	0.35
3	∞	50-300	2000	0.35

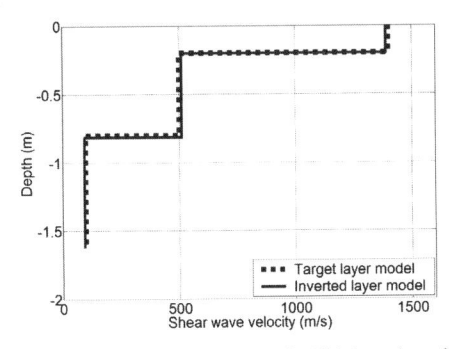

Figure 5. Inverted layer model from the FSA inversion scheme (solid line) applied to the theoretical layer model in Table 1 (dotted line).

Both the thickness and shear wave velocity of the first two layers are well resolved within 1% deviation from the target layer model (Table 1). The shear wave velocity of the halfspace shows the largest relative error of 5%.

Figure 6 shows the history of all iterations from the inversion. Each parameter value is plotted against the mismatch (E). This type of plot can be viewed as a multidimensional sensitivity analysis of each parameter. At large values of E parameter values are randomly distributed over the predefined intervals (Table 2). As the mismatch is decreased, the estimated parameter values approach the true values indicated with the dotted lines. The width of the peak at low E values indicates the resolution obtained from the inversion. The bulk density of the second layer is included in the inversion to exemplify a badly resolved parameter. Surface wave methods are not very sensitive to the bulk density of the materials and this parameter is usually not included in the inversion of surface wave dispersion curves (Xia et al. 1999).

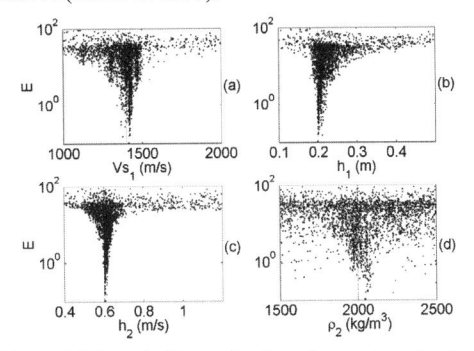

Figure 6. Mismatch (E) as a function of parameter value from all iterations used in the FSA inversion. The bulk density of the second layer (d) is not well resolved as indicated by the spread of the dots from each iteration.

4 FIELD DATA

The proposed method was applied to a recently constructed pavement, the Peab test site in Malmoe, Sweden. At this site, both thicknesses and materials for the different layers were well known (Table 3). This particular pavement construction should be a challenging target for seismic non-destructive testing because of the high velocity stabilized subgrade layer. The relatively thin base layer is embedded between two high velocity layers, the top asphalt layer and the stabilized subgrade.

Table 3. Pavement construction at the Peab test site.

Layer	Thickness (m)	Material	Density (kg/m³)	Poisson's ratio
1	0.12	Asphalt	2300	0.30
2	0.30	Base	2200	0.20
3	0.30	Stb. clay till	2100	0.20
4	1.00	Clay till	1900	0.35
5	∞	Clay till	1900	0.35

Following the MSOR method one accelerometer was located at zero offset (distance). While keeping the accelerometer at zero offset and by changing the impact points of the hammer from offset 0.50 m to 4.00 m with 0.05 m impact separation, data were collected with the PSAS system. The initial 0.50 m offset was used to reduce the near field effect at low frequencies. At each offset 4 impacts were stacked with the spike kept in a fixed position. The surface temperature was 23 degrees Celsius during data collection. The resulting observed wavefield spectrum is presented in Fig. 7a.

To account for the viscoelastic properties of the asphalt layer the shear wave velocity of the first layer is expressed as a power function of frequency:

$$V_S(f) = a_1 f^{a_2} \qquad (6)$$

where a_1 and a_2 are empirical coefficients defining the viscoelastic material properties. The dynamic E-modulus of the asphalt layer is usually defined by a "mastercurve" showing the E-modulus as a function of frequency at a reference temperature. Within the frequency range typically utilized in seismic non-destructive testing (~30 to ~20 000 Hz) the E-modulus frequency relation is known to be close to linear in a log-log scale (Aouad 1993), which motivates the chosen format of Eq. 6. The subsequent inversion can then work on the viscoelastic constants rather than on a constant shear wave velocity.

The matched theoretical wavefield spectrum from the inversion is plotted in Fig. 7b. First arrival longitudinal waves (at ~2800 m/s) are implicitly included for the inversion, since these longitudinal waves propagate as symmetric Lamb waves and are thus included in the stiffness matrix method. The corresponding inverted layer model is presented in Fig. 8 along with the frequency dependent shear wave velocity of the asphalt layer. It should be noted that the embedded low velocity layer (base) has been well resolved (see Fig 9b), as well as both the asphalt and the stabilized subgrade layer (Fig. 9a and 9c). The predefined search range of each parameter is equal to the x-axis range in Fig. 9.

In Fig. 7 both the observed and predicted wave field spectra are displayed with a linear frequency range. However, to reduce the computational time a logarithmic frequency scale from 50 to 5000 Hz was used in the inversion. The inversion scheme was run for 600 transitions using a total of 4000 trial layer models (see Fig 9). The total computational time was 4 hours on a 2 GHz PC.

5 CONCLUSIONS

The general framework for the inversion of wavefield spectra in non-destructive testing of pavements has been outlined. Data acquisition and processing is based on the MASW method. The inversion of a depth-shear wave velocity profile is conducted on the full wave field spectrum using the FSA algorithm. This procedure avoids dealing with discrete dispersion curves and mode identification for the inversion of surface wave data.

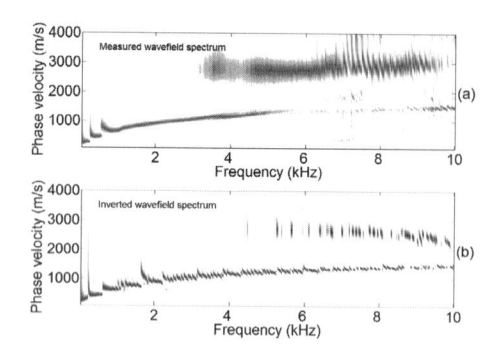

Figure 7. (a) Measured and (b) inverted wavefield spectrum from the Peab test site.

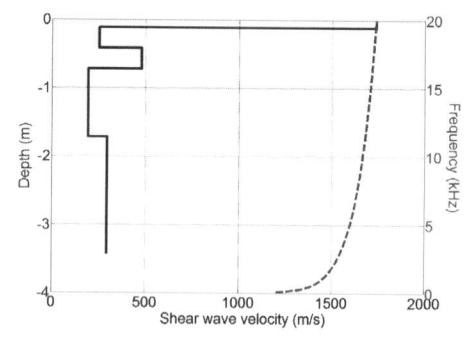

Figure 8. Final inverted layer model from the Peab test site. The dotted line shows the frequency dependent shear wave velocity of the asphalt layer.

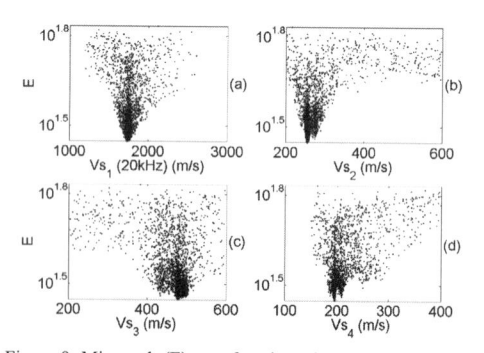

Figure 9. Mismatch (E) as a function of V_S from all iterations used in the FSA inversion of the data from the Peab test site. The spread of the dots indicates the multidimensional sensitivity for each inverted parameter.

The viscoelastic properties of the asphalt layer have been included in the forward model, resulting in an inverted "mastercurve", showing the asphalt stiffness as a function of frequency. With this procedure the asphalt design modulus at 30 Hz is obtained directly without out any empirical reduction factor.

The outlined approach is demonstrated on both synthetic and field data. The main benefit from this approach is that the raw field data can be automatically processed and inverted without any subjective user input, and by properly accounting for the interference of different modes and types of waves. The main disadvantage at this point is a long computational time for the inversion process.

It is believed that the resolution of the base layer (that has been recognized as difficult to resolve with traditional methods) has been improved with the outlined method. Higher resolution of the base layer properties is obtained because the full wavefield spectrum is utilized in the inversion method. Compared to the conventional approach where only one apparent dispersion curve is used, the amount of nonuniqueness may also be reduced by using the complete wave field spectrum in the inversion scheme.

ACKNOWLEDGEMENTS

We would like to give our sincere thanks to Peab Sverige AB, VINNOVA, and the Swedish road authority 'Vägverket', for financing this project. We also thank Professor Nenad Gucunski at Rutgers University for his assistance with the stiffness matrix method. The help from Mary Brohammer and Peter Jonsson with the preparation of this manuscript is also greatly appreciated.

REFERENCES

Abdallah, I., Yuan, D. & Nazarian, S. 2003. Validation of software developed for determining design modulus from seismic testing. Research Report 1780-5, Center for Highway Materials Research, the University of Texas at El Paso, TX.

Akhlaghi, B.T. & Cogill, W.H. 1994. Application of the free plate analogy to a single-layered pavement system, *INSIGHT* Vol. 36: 514-518.

Aouad, M.F. 1993. Evaluation of Flexible Pavements and Subgrades Using the Spectral-Analysis-of-Surface-Waves (SASW) Method. PhD thesis, Univ. of Texas at Austin, Texas.

Beaty, K.S., Schmitt, D.R. & Sacchi, M. 2002. Simulated annealing inversion of multimode Rayleigh wave dispersion curves for geological structure. *Geophys. J. Int.* Vol. 151: 622-631.

Disperse, 2001. A system for generating dispersion curves, User's manual version 2.0.11. Software version 2.0.15e. Guided Ultrasonics, London.

Forbriger, T. 2003. Inversion of shallow-seismic wavefields part 1: wavefield transformation. *Geophysical Journal International* Vol. 153: 719-734.

Foti, S., Sambuelli, L., Socco, V.L. & Strobbia, C. 2003. Experiments of joint acquisition of seismic refraction and surface wave data. *Near Surface Geophysics* Vol. 1: 119-129.

Ganji, V., Gucunski, N. & Nazarian, S. 1998. Automated inversion procedure for spectral analysis of surface waves. *Journal of Geotechnical and Geoenvironmental Engineering, ASCE* Vol. 124: 757-770.

Graff, K.E. 1975. *Wave motion in elastic solids*. Oxford University Press, London.

Gucunski, N., Abdallah, I. N., & Nazarian, S. 2000. ANN backcalculation of pavement profiles from the SASW test. In Geotechnical Special Publication (ASCE) No. 98, Pavement subgrade unbound materials, and nondestructive testing, 31-50.

Gucunski, N. & Woods, R.D. 1992, Numerical simulation of the SASW test, *Soil Dynamics and Earthquake Engineering* Vol. 11: 213-227.

Haegeman, W. 2002. In situ assessment of stiffness of a road sand embankment. Proceedings of the BCRA 2002, Lissabon, Portugal. 629-635.

Heisey, J.S., Stokoe, K.H. & Meyer, A.H. 1982. Moduli of pavement systems from spectral analysis of surface waves. *Transp. Res. Rec* 852, 22-31.

Hunaidi, O. 1998. Evolution-based genetic algorithms for analysis of non-destructive surface wave tests on pavements. *NDT&E International* Vol. 31: 273-280.

Jones, R., 1962. Surface wave technique for measuring the elastic properties and thickness of roads: Theoretical development, *British Journal of Applied Physics* Vol. 13: 21-29.

Kausel, E. & Roesset, J.M. 1981. Stiffness matrices for layered soils. *Bulletin of the Seismological Society of America* Vol. 71 (No. 6): 1743-1761.

Laarhoven, P.J.M. & Aarts, E.H.L. 1987. *Simulated Annealing: Theory and Applications*. Kluwer Academic Publishers. The Netherlands.

Lamb, H. 1917. On waves in an elastic plate. *Proc. Roy. Soc.* Vol. 93: 114-128.

Luke, B.A., Calderón-Macías, C., Stone R.C. & Huynh, M. 2003. Non-uniqueness in inversion of seismic surface-wave data. Proceedings of the Symposium on the Application of Geophysics to Engineering and Environmental Problems (SAGEEP 2003), San Antonio, Texas, SUR-05.

Martínez, M.D., Lana, X., Olarte, J., Badal, J. & Canas, J.A. 2000. Inversion of Rayleigh wave phase and group velocities by simulating annealing. *Physics of the Earth and Planetary Interiors* Vol. 122: 3-17.

McMechan, G. & Yedlin, M.J. 1981. Analysis of dispersive waves by wave field transformation. *Geophysics* Vol. 46 (No. 6): 869-874.

Menke, W. 1989. *Geophysical data analysis: discrete inverse theory*. Revised edition, Academic Press. New York.

Nazarian, S., Yuan, D. & Tandon V. 1999. Structural Field Testing of Flexible Pavement Layers with Seismic Methods for Quality Control. *Transp. Res. Rec.* 1654: 50-60.

Nazarian, S. 1984. In situ determination of soil deposits and pavement systems by spectral analysis of surface waves method. PhD thesis, Univ. of Texas at Austin, Texas.

Park, C.B., Miller, R.D. & Xia, J. 2001. Offset and resolution of dispersion curve in multichannel analysis of surface waves (MASW). Proceedings of the Symposium on the Application of Geophysics to Engineering and Environmental Problems (SAGEEP 2001), Denver, Colorado, March 4-7.

Park, C.B., Miller, R.D. & Xia, J. 1999. Multichannel analysis of surface waves. *Geophysics* Vol. 64: 800-808.

Park, C.B., Miller, R.D. & Xia, J. 1998. Imaging dispersion curves of surface waves on multi-channel records. Technical Program with biographies, SEG, 68th Annual Meeting, New Orleans, Louisiana, 1377-1380.

Pavlakovic, B. Lowe, M. Alleyne, D. & Cawley, P. 1997. Disperse a general purpose program for creating dispersion curves. *Review of Progress in Quantitative NDE* Vol. 16: 185-192.

Rambod, H. & Gucunski, N. 2003. Inversion of SASW dispersion curve using numerical simulation. Proceedings of the Symposium on the Application of Geophysics to Engineering and Environmental Problems (SAGEEP 2003), San Antonio, Texas, SUR-01.

Roesset, J.M., Chang, D.W., Stokoe II, K.H. & Auoad, M. 1990. Modulus and thickness of the pavement surface layer from SASW tests, *Transp. Res. Rec.* 1260: 53-63.

Ryden, N. & Park, C.B. 2004. Surface waves in inversely dispersive media. Submitted to: *Journal of Near Surface geophysics*.

Ryden, N. 2004. Ph.D. thesis in preparation.

Ryden, N., Park, C.B., Ulriksen, P. & Miller, R.D. 2004. A multimodal approach to seismic pavement testing. Accepted for publication in: *Journal of Geotechnical and Geoenvironmental Engineering, ASCE*.

Ryden, N. Park, C.B. Ulriksen, P. & Miller, R.D. 2003 Lamb wave analysis for non-destructive testing of concrete plate structures: Proceedings of the Symposium on the Application of Geophysics to Engineering and Environmental Problems (SAGEEP 2003), San Antonio, TX, April 6-10. INF03.

Ryden, N. Ulriksen, P. Park, C.B. & Miller, R.D. 2002. Portable seismic acquisition system (PSAS) for pavement MASW. Proceedings of the Symposium on the Application of Geophysics to Engineering and Environmental Problems (SAGEEP 2002), Las Vegas, NV, February 10-14, 13IDA7.

Ryden, N. Ulriksen, P. Park, C.B. Miller, R.D. Xia, J. & Ivanov, J. 2001. High frequency MASW for non-destructive testing of pavements-accelerometer approach. Proceedings of the Symposium on the Application of Geophysics to Engineering and Environmental Problems (SAGEEP 2001), Denver, Colorado, RBA-5.

Ryden, N. 1999. SASW as a tool for non destructive testing of pavements. MSc thesis, Univ. of Lund, Sweden.

Sen, M. & Stoffa, P.L. 1995. *Global Optimization Methods in Geophysical Inversion*. Elsevier, Amsterdam.

Stokoe, K.H., Wright, G.W., James, A.B. & Jose, M.R. 1994. Characterization of geotechnical sites by SASW method, in Geophysical characterization of sites. ISSMFE Technical Committee #10, edited by R.D. Woods, Oxford Publishers, New Delhi.

Szu, H. & Hartley, R. 1987. Fast simulated annealing. *Phys. Lett. A.* Vol. 122: 157-162.

Tawfiq, K., Sobanjo, J. & Armaghani, J. 2000. Curvilinear behavior of base layer moduli from deflection and seismic methods. . *Transp. Res. Rec.* 1716: 55-63.

Vidale, R.F. 1964. The dispersion of stress waves in layered media overlaying a half space of lesser acoustic rigidity. PhD thesis, Univ. of Wisconsin.

Wilcox, P., Evans, M., Diligent, O., Lowe, M. & Cawley, P. 2002. Dispersion and excitability of guided acoustic waves in isotropic beams with arbitrary cross section. *Review of progress in quantitative NDE* Vol. 21: 203-210.

Wu, H., Wang, S., Abdallah, I. & Nazarian, S. 2002. A rapid approach to interpretation of SASW results. Proceedings of the BCRA 2002, Lissabon, Portugal. 761-770.

Xia, J., Miller, R.D. & Park, C.B. 1999. Estimation of near-surface shear-wave velocity by inversion of Rayleigh wave: *Geophysics* Vol. 64: 691-700.

Geoelectrical resistivity imaging of domestic waste disposal sites: Malaysian case study

Abdul Rahim Samsudin, Umar Hamzah & Wan Zuhairi Wan Yaacob
School of Environmental and Natural Resource Sciences, Faculty of Science & Technology, Universiti Kebangsaan, Malaysia

Keywords: resistivity imaging, waste disposal site, contamination, leachate, groundwater

ABSTRACT: Geoelectrical imaging method is now frequently used for subsurface pollution studies. The method basically maps the distribution of the resistivity of subsurface materials. The resistivity image provides general information on subsurface stratification of the buried waste and soil and the depth to the bedrock below the lines of traverse. Underground soil and water that has been contaminated by leachate usually has a significantly lower resistivity value. The geoelectrical imaging method was used in this study to help delineating contaminated ground water and to assist in understanding the underground conditions at two domestic waste disposal sites in Malaysia. The quality and contaminated zone of the underground water was determined based on the measured geoelectrical resistivity value of subsurface materials. Two dimensional resistivity profiles and subsurface geological information from bore hole data were used to determine the extension and direction of the contaminant flow within the underground water system in the study areas.

1 INTRODUCTION

Solid waste in developing countries is generally disposed of in uncontrolled open dumps. The environmental consequences of such inadequate disposal sites are often quite evident, yet necessary improvements are seldom delt with. The most common practice for disposal of solid domestic wastes in Malaysia was depositing them in open dumping areas. These areas are vulnerable to ground and surface water pollution. Seriousness of the problem is still unknown and specific detailed study is generally needed.

Solid waste land disposal sites can be sources of groundwater contamination and the contamination problems are more likely to occur in humid areas, where the moisture available exceeds the ability of the waste pile to absorb water. In tropical countries like Malaysia which are characterized by high rainfall, contamination problems are expected to occur. The total pollutant load to the environment is dependent on the quantity and quality of the water that percolates through the disposal site and reaches the groundwater (Bengtsson et al., 1994). Waste disposal sites can seriously affect local wells and drilled holes used for public water supply and therefore, their locations must be planned and monitored carefully (Matias et al., 1994).

Assessing contaminated land poses many problems and is both difficult and hazardous. Major difficulties include locating the actual contamination and determining its spatial distribution. An accurate measurement of contaminant concentrations in highly heterogeneous soils is extremely difficult. Extensive drilling, probing and direct sampling of contaminated materials poses health and safety hazards and can be costly.

Non-invasive geophysical techniques can be used to delineate and identify pathways for contaminant migration. These techniques can be deployed rapidly and cost effectively, and have the ability to remotely locate significant contamination to determine its distribution and to monitor any change over time.

This paper describes results of geophysical surveys employing geoelectrical resistivity imaging to investigate the ground water contamination at two domestic waste dumping sites in Malaysia. The objectives of the surveys were to delineate and identify pathways for contaminant migration.

2 MATERIAL AND METHODS

The resistivity of the subsurface material can be measured by injecting a small current into the ground through two electrodes and the resulting voltage on the ground surface is measured at two potential electrodes (Reynolds, 1997). By varying the spacing between the electrodes, as well as the location of the electrodes, a 2-D electrical resistivity image of the subsurface can be obtained. The 2-D electrical imaging survey was carried out with a SAS1000 resistivity meter and Abem LUND ES464 electrode selector system. This system was connected to a total of 61 steel electrodes which were laid out on a straight line with a constant spacing via a multicore cable. The resistivity unit automatically selects the four active electrodes used for each measurement. The Wenner equal spacing electrode array was used for this survey. The measured resistivity data were interpreted using the RES2DINV inversion software (Loke and Barker,1995). Details about the survey and interpretation method can be found in the papers by Keller & Frischknecht (1966), Griffiths et al. (1990), Griffiths and Barker (1993), and Loke and Barker (1996).

The resistivity of the subsurface materials depend on several factors such as the nature of the solid matrix and its porosity, as well as the type of fluids (normally water or air) which fill the pores of the rock or soil. In general, rock and dry soil have high resistivities of several hundred or thousands ohm-meter. Unconsolidated water saturated soil and clay material have relatively low resistivity values of generally below 1000 ohm-m. The resistivity of sediments below water table is normally controlled by the resistivity of the groundwater which depend on the concentration of the electrolytes present. Fresh ground water generally have resistivity values of about 10 to 100 ohm-m.

The first dumping site is located about 10 km from Tampin Town in Negeri Sembilan. It occupies a small plot of generally flat land with a total area of about 15 square acres. The open dumping and burning operation startsed since 1981 with a total volume of 39,780 cubic metres of domestic waste being dumped at this site. Remains of partially buried and burnt old waste is found in the north east part of the dumping site whereas a relatively new waste is being dumped more towards south west side and adjacent to the existing pond (Figure 1).

The subsurface lithology of the area was determined from a total of twelve observation boreholes drilled to a maximum depth of 9 m to 15.0 m below ground level. The subsurface profile indicates that the aquifer layer which is made up of silty sand has variable thicknesses ranging from 5 to more than 10 metres. The water table along this line is relatively shallow with depth ranging from 0.66m to 2.00m below ground level.

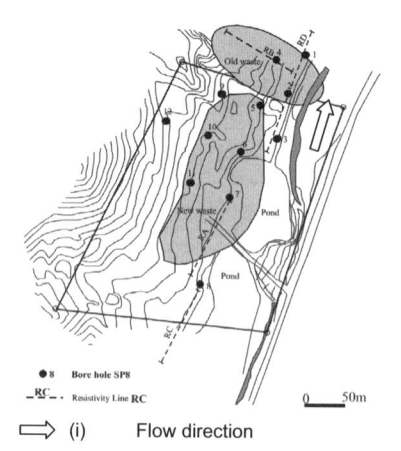

Figure 1. Topography map of first study site howing location of old and new wastes and resistivity lines

The local groundwater flow pattern within the vicinity of the dumping site has been based on the results of detailed ground and groundwater elevation measurement conducted by Malaysian Institute of Nuclear Technology (MINT). Observation of groundwater table at the boreholes indicated that the static groundwater level ranges from 1.45m to 2.44m. The groundwater appears to flow north east which follows the general trend of surface water flow in the area.

The second study site is located approximately 2km west of Dengkil Town in the state of Selangor. It is an open dumping site which caters for domestic waste from Sepang area and currently managed by Alam Flora Sdn. Bhd. It covers a flat lowland area close to a small river and underlain by an alluvial deposit which consists of mainly clayey soil.

3 RESULTS AND DISCUSSION

In the first study site, four resistivity imaging lines were established with three of them running parallel to the direction of local ground water flow of the study area and the fourth one (Line RB) is almost perpendicular to it (Figure 1). One of the line (Line RC) is located well outside the waste dumping area whereas 'Line RA' and 'Line RD' run across the new and old dumping sites respectively. The subsurface resistivity obtained from 'Line RC' was assumed as the standard resistivity value of the uncontaminated groundwater in the area. This resistivity value was used as a basis of interpretation in this study.

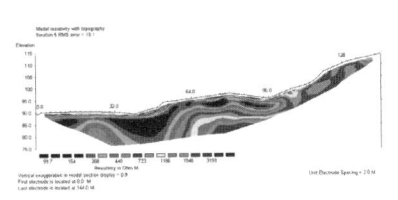

Figure 2. Resistivity depth section for line RC

Figure 2 shows resistivity depth section of Line RC. The low resistivity value of about 90 ohm.m was obtained for the aquifer layer which suggests that the groundwater underneath Line RC is relatively fresh and uncontaminated. Whereas the resistivity inverse sections for Line RA, Line RB and Line RD (Figure 3) have their minimum resistivity values ranging from 3.5 to 9.0 ohm.m.

These reasonably low resistivity values are interpreted to be related to the resistivity of the groundwater which has been contaminated by leachate produced by the disposed waste. The above interpretation is substantiated by the results of chemical analysis which indicate that the groundwater samples obtained from the corresponding boreholes have high chemical content (Tan, 1999; Bashilah Baharudin, 1999).

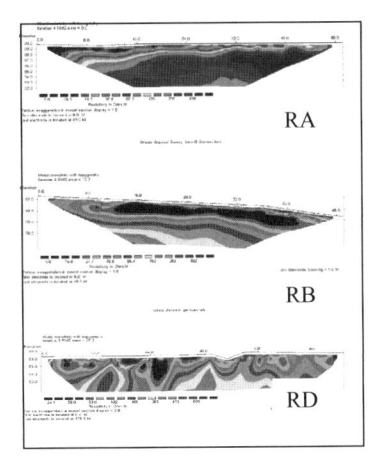

Figure 3: Resistivity depth section for resistivity lines RA, RB and RD

The resistivity and borehole results show that the contamination of the groundwater in the study area appears to be confined only within the vicinity of the dumping site. Water sample from borehole SP4 which is located in the old dumping site appears to have high concentration of chemical content and the mechanism by which the contamination arrived at that site was unclear. This will require further investigation and long term monitoring. The resistivity profiles also suggest a movement of contaminants towards northeast which follows a local trend of the groundwater flow in the area being studied.

For the second study site, a total of six lines of 2-D resistivity images were established with three of them located across the waste pile while the other three located outside the boundary of the waste disposal site (Figure 4).

Figure 5 shows the inversed resistivity models of the three resistivity lines measured on the top of the waste pile (Profiles 1, 5 & 6). The resistivity image of profile1 shows that the decomposed waste with highly conductive leachate has resistivity less than 2.5 ohm-m. The electrically conductive anomaly on the dumping site was interpreted as leachate which appears to have seeped through the underground soil at depth as far as 20m below surface (Profile 5). The 2D image of the resistivity also shows the thickness of the decomposing waste is slightly more than 10 meters (Profile 6) and this agrees well with the actual height of the waste pile observed in the field.

Three other resistivity sections (Figure 6) were measured around areas outside the boundary of the waste disposal site. The northern and western areas of the waste disposal sites are bounded by man-made drain. The 2-D resistivity sections show that, in general the soil surrounding the dumpsite has relatively high values (Profiles 2 & 3) which suggests that it is not much affected by the leachate runoff. However the leachate appears to have migrated outside into the eastern area of the dumpsite. This is illustrated by the near surface low resistivity layer observed on Profile 4.

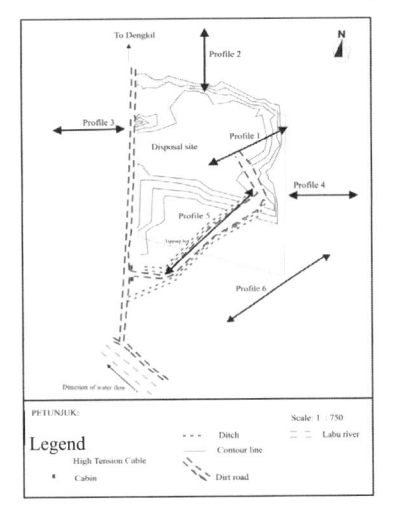

Figure 4. Location of waste disposal site and resistivity profiles

4 CONCLUSION

The results of the present study reveal that the two dimensional geoelectrical resistivity imaging can be used to investigate the subsurface migration of leachate at an open dumping site. The leachate migration was traced in form of low resistivity zones (with resistivity less than 2.0 ohm.m) of decomposing waste bodies saturated with highly conductive leachate. The resistivity imaging technique in conjunction with borehole information and hydrochemical analysis can be successfully used for pollution mapping of the ground water at a contaminated site.

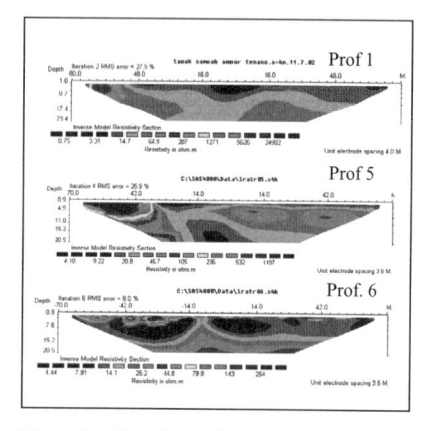

Figure 5. 2-D resistivity inversed models for profiles 1, 5 & 6 measured within the waste disposal site

The complexity of subsurface conditions beneath contaminated lands requires a multidiciplinary approach combining the systematic and careful application of hydrogeological, chemical and environmental geophysical techniques

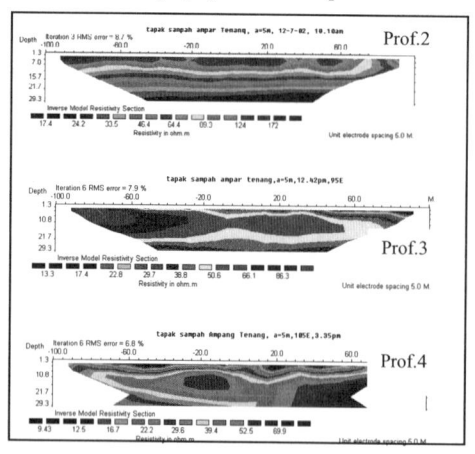

Figure 6: 2-D resistivity inversed models for profiles 2, 3 & 4 measured around the waste disposal site

ACKNOWLEDGEMENT

The authors wish to acknowledge the cooperation of Malaysian Institute of Nuclear technology (MINT) and Tampin District Council to enable geophysical field data to be collected as well as provide bore hole information for the first study area. The authors also thankful to Sepang City Council and Alam Flora Sdn. Bhd for permission to conduct resistivity survey in the second study area and the Malaysian Ministry of Sciences, Technology & Environment for financial assistance under IRPA(RM8) grant. 08-02-02-EA178 and IRPA(RM7) No.02-02-02-0010. Thank is also due to Universiti Kebangsaan Malaysia for partially supported the research through UKM short term Research Grant S/11/96.

REFERENCES

Bashillah Baharuddin, 1999. Penggunaan kaedah geofizik dalam pencemaran bawah tanah di Gemenceh, N.Sembilan, Tesis Sarjana Muda Sains & Kepujian, Jabatan Geologi, Universiti Kebangsaan Malaysia, Bangi, Selangor, sesi 1998/99 (Unpublished).

Bengtsson, L., Bendz, D., Hogland, W., Rosqvist, H. & Aksson, M., 1994. Water balance for landfills of different age, Journal of Hydrology, 158, 203 – 217.

Daniels,F. and Alberty,R.A., 1966. Physical chemistry, John Wiley & Sons,Inc.

Griffiths, D.H.,Turnbull, J. and Olayinka,A.I.,1990. Two dimensional resistivity mapping with a computer-controlled array, First Break 8, 121-129.

Griffiths, D.H. and Barker, R.D. 1993. Two-dimensional resistivity imaging and modeling in areas of complex geology. *Journal of Applied Geophysics*, 29 : 211 - 226.

Keller,G.V.and Frischknecht,F.C.,1966. Electrical methods in geophysical prospecting, Pergamon Press Inc., Oxford.

Loke, M.H. and Barker, R.D., 1995. Rapid least-squares inversion of apparent resistivity pseudosections using a quasi-Newton method. Geophysical prospecting.

Loke, M.H. and Barker, R.D. 1996. Rapid least-squares inversion of apparent resistivity pseudosections by a quasi-Newton method. *Geophysical Prospecting*, 44 : 131 – 152.

Matias, M.S., Marques da Silva, M., Ferreira, P. & Ramalho, E., 1994, A geophysical and hydrogeological study of aquifers contamination by landfill, Journal of Applied Geophysics, 32, 155 – 162.

Reynolds,J.M.,1997. An Introduction to Applied and environmental Geophysics, John Wiley &Sons, Inc.

Tan, C.A., 1999. Kajian bersepadu pencemaran air bawah tanah dengan Metoda geologi, geofizik dan hidrogeologi di tapak pembuangan sampah Gemenceh, N.Sembilan, Tesis Sarjana Muda Sains & Kepujian, Jabatan Geologi, Universiti Kebangsaan Malaysia, Bangi, Selangor, (Unpublished).

The small strain stiffness of gold tailings

M. Theron & C.R.I.Clayton
University of Southampton, Southampton, UK

G. Heymann
University of Pretoria, Pretoria, South Africa

Keywords: stiffness, shear wave velocity, gold tailings

ABSTRACT: Assessment of the stability of tailings dams, which are often located close to urban or mine developments, requires an estimate of the strength of the tailings material. The observed stability of such dams, even when built with relatively steep slopes, suggests that strength of tailings in situ may be higher than is typically estimated from laboratory tests on reconstituted specimens. Two schools of thought attempt to explain the stability of tailings dam slopes either on the basis of suction pressures, or of inter-particle bonding. This paper presents preliminary data from a research programme aimed at investigating the latter. Small strain stiffness measurements made in the field using a seismic cone are compared with laboratory results from bender element tests. The results suggest that the material found in gold tailings dams does not have high levels of bonding.

1 INTRODUCTION

The South African gold mining industry produces large quantities of waste material in the form of tailings slurry. These tailings consist of process water mixed with the rock flour generated by crushing, grinding and milling the gold bearing reef prior to extraction of the gold mineral. The tailings are disposed of in large tailings dam facilities. The ability of tailings to stand at relatively steep slope angles, relative to the angles of friction typically determined from triaxial compression tests on reconstituted specimens, suggests that there is a component of strength in the field that is not present in the laboratory specimens. In an attempt to better understand the in situ structure of the tailings an investigation was conducted to compare the small strain stiffness as measured by the seismic cone with the stiffness measured in the laboratory on remoulded material using bender elements.

Hardin and Drnevic (1972) investigated the effect of a number of parameters on the stiffness of young uncemented sands and clays using the resonant column tests. They found that the stiffness was strongly influenced by void ratio and stress condition and that other parameters such as degree of saturation, stress history, particle size, shape, gradation and mineralogy had a lesser effect. Subsequently, other workers have shown that soil stiffness is affected by bonding (Acar and El-Tahir 1986, Cuccovillo and Coop 1997) and ageing (Daramola 1980, Mitchell 1986 and Stokoe et al. 1995). These authors have demonstrated that stiffness increases with increased bonding as well as with ageing.

2 MATERIAL DESCRIPTION

The dam under investigation was 15 years old and situated near Welkom in the Freestate gold field of South Africa. The dam was built using the upstream semi-dry paddock method, whereby a perimeter wall known as a daywall is hydraulically constructed around the dam using the tailings material itself (eg. Wagener 1997). The daywall material is deposited in a series of paddocks constructed under controlled conditions during daytime. Sufficient time is allowed between depositions at each paddock to allow the material to dry. During the night the tailings material is discharged into the interior of the dam.

In situ seismic cone testing was conducted in the daywall of the dam. In addition a sample of material was taken from the daywall for laboratory bender element testing. Inter-bedded layers of coarse and fine material were visible in the undisturbed sample. The coarse and fine materials were separated and their particle size distributions determined. The average particle diameter (D_{50}) of the coarse material was 0.045mm and that of the fine material

0.015mm. This is within the range of average particle diameter between 0.008mm and 0.09mm as reported by Vermeulen (2001) based on work published by numerous authors. The specific gravity was 2.76 for both the coarse and fine material.

3 FIELD MEASUREMENT OF STIFFNESS

The in situ small strain stiffness of the tailings material was determined by measuring the elastic body wave velocities using a seismic cone.

The arrangement of the seismic cone test has been described by Heymann (2003) and is shown in Figure 1. The seismic energy source was at the surface and consisted of a wooden sleeper tied to the ground by an auger type anchor. The sleeper was hit with a heavy hammer to generate the seismic waves. Hitting the sleeper vertically downwards produced a seismic event rich in compression waves and hitting it horizontally from the side generated an event dominated by shear waves. In addition hitting the sleeper from opposite sides allowed shear waves of opposite polarity to be studied.

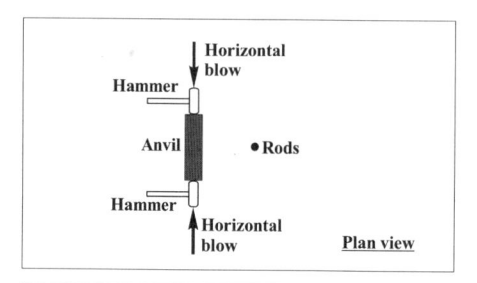

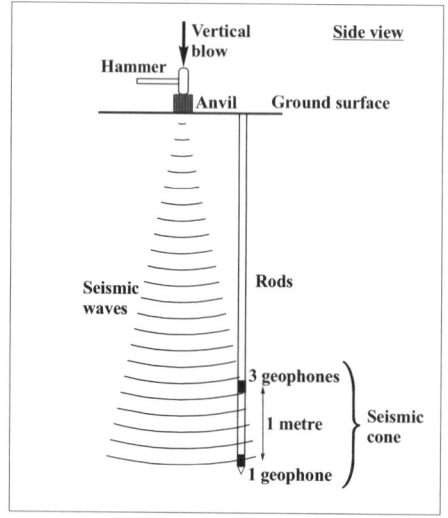

Figure 1: Arrangement of the seismic cone test.

The cone contained four geophones positioned at two levels. Three were positioned near the top of the cone in a triaxial configuration and one horizontal geophone was positioned near the bottom, at a distance of one metre below the top geophones. An external geophone was fixed to the sleeper at the ground surface to record the start time of the seismic event and initiate data logging. The system featured a pre-triggering facility making it possible to capture data from before the seismic event up to a specified time after the event. This enabled the full time history to be evaluated and avoided trigger timing errors that might occur if a switch mechanism were used to initiate data logging.

The cone was pushed into the tailings in increments of one metre. After penetrating a metre the probe rig motor was shut down and the seismic tests performed. Typically three measurements were recorded, two shear wave events obtained by hitting the sleeper horizontally from opposite ends, and one compression wave event by hitting the sleeper vertically. For each test the response of the soil at the geophones was digitally recorded.

Typical shear wave traces obtained when hitting the sleeper from opposite ends are shown in Figure 2. Clearly the time history recorded by the geophone was similar for both events but the polarity of the response was reversed. The traces in Figure 2 have a dominant frequency of about 50Hz, and coupled with the calculated shear-wave velocities (see later) suggest that except at shallow depth (less than about 4m) the nearest receiver will be more than 2 wavelengths from the source, thus putting both receivers in the far field. This provided confirmation that the observed reversing traces were from shear waves.

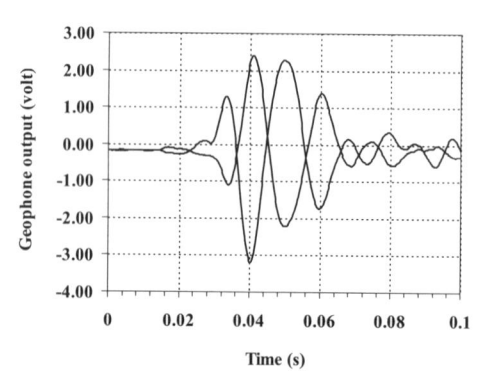

Figure 2: Shear wave reversal.

Figure 3 shows the response of the top and bottom horizontal geophones for the same seismic shear wave event. Provided (as was the case here) there is no pulse broadening as a result of high frequency attenuation, the travel time between the two geophones can be obtained either from the time offset between the first arrivals or the time offset between successive peaks. Since background noise and incoming compressional wave energy obscure the exact position of the first arrival, first peaks were used to obtain interval times between the top and bottom geophones. The wave velocity at each depth was determined by taking the travel time as the offset between the first peaks of the top and bottom geophone traces for the same seismic event and calculating the ray path length from the known positions of the geophones and the source.

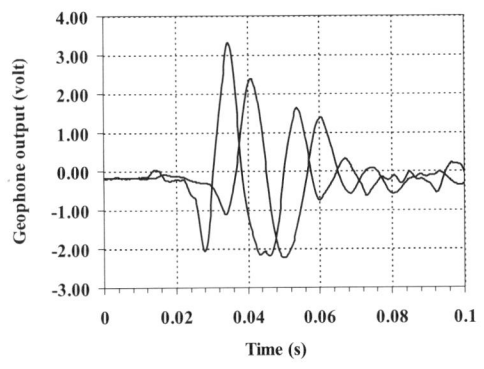

Figure 3: Shear wave arrival at the top and bottom geophones.

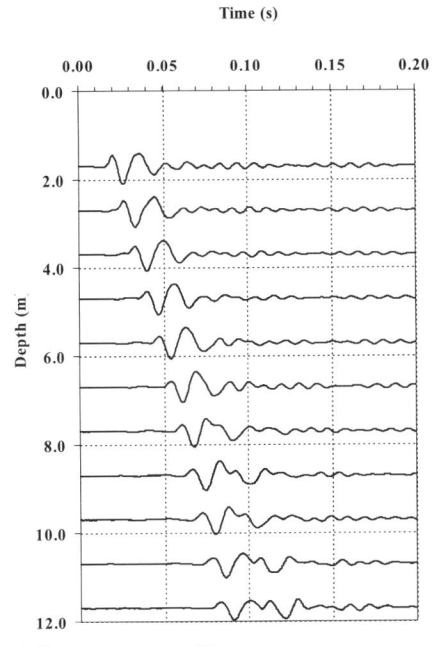

Figure 4: Shear wave trace profile.

Figure 4 shows the shear wave traces recorded by the lower of the horizontal geophones over the full depth of testing at the daywall of the tailings dam. The figure shows consistency in the time history of the wave events occurring at each depth. In addition it may be observed that the delay between the seismic event and the first arrival of the shear wave increased as the cone penetrated deeper, as a result of the increase in the distance between source and receiver. Further estimates of shear wave velocity were obtained from the shear wave traces in Figure 4, by calculating the slope of the first peak time vs. depth for successive groups of 3 traces.

The profiles of shear wave velocity with depth obtained as described above are shown in Figure 5. The shear wave velocity increased with depth from about 120m/s at shallow depths to over 200m/s at deeper levels.

The question arises as to whether the stiffness of the tailings material is due entirely to state effects (stress condition and void ratio) or whether bonding of the in situ material has occurred. Comparison of the undisturbed in situ shear wave velocity and stiffness with laboratory measurements on destructured material allows some judgement on this question.

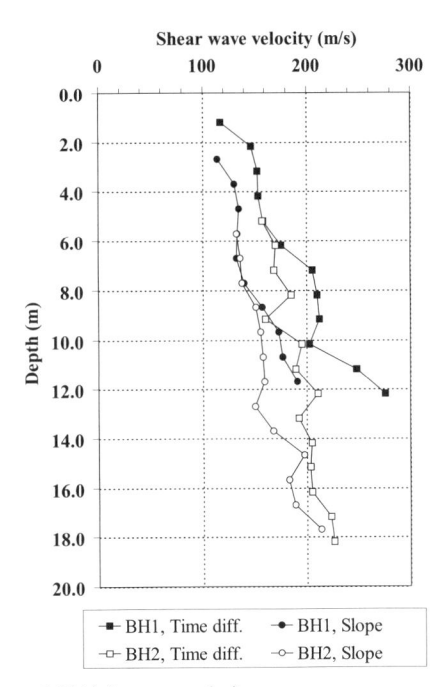

Figure 5: Field shear wave velocity.

4 LABORATORY MEASUREMENT OF STIFFNESS

The shear wave velocity of the tailings material was measured in the laboratory using bender elements. A bender element consists of two piezo-ceramic plates (bimorph elements), separated by a layer of high compliance material (Shirley and Hampton (1978)). Application of a voltage to a bimorph of this type causes it to bend and, when used in a suitable configuration, to generate shear waves in a soil specimen. The travel time of the shear wave between two points in a soil specimen can be used to calculate shear wave velocity as the ratio of the distance between the bender elements and the travel time. In the simplest methods the latter can be obtained as the time lag between the applied and received electrical signals, as displayed on an oscilloscope.

The basic laboratory configuration consisted of a conventional 100 mm diameter triaxial apparatus and 5-tonne loading frame. The cell and back pressure of the sample were applied by digital pressure controllers. Two bender element probes were inserted along the side of the sample at third height intervals (Figures 6 & 7). The bottom bender element probe acted as the transmitter whilst the top side bender element probe acted as the receiver. The signal to noise ratio was improved by the following actions:

- shielding around both bender element cables,
- power cables were separated from bender element cables,
- the triaxial apparatus and cable shielding was earthed to a common point provided on the oscilloscope.

Input signals were generated using a programmable function generator and the received signals were passed to a digital storage oscilloscope.

Tailing samples were prepared by reconstituting the material, and pouring it into a mould in such a way as to avoid segregation of the different particle sizes. A suction pressure of 20kPa was applied to the sample prior to removal of the mould and applying the cell pressure. Three samples with different void ratios were prepared. The initial void ratio of each specimen was determined from the measurement of the tailings mass and specimen dimensions. Each specimen was then isotropically consolidated under a series of effective confining pressures (25 to 600kPa) and swelled back in the same steps to the original effective confining pressure. Changes in the void ratio during the consolidation process were

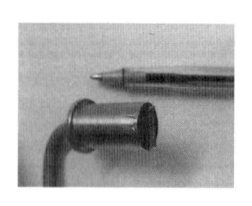

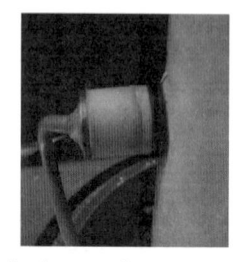

Figure 6: Photographs of side bender element probe

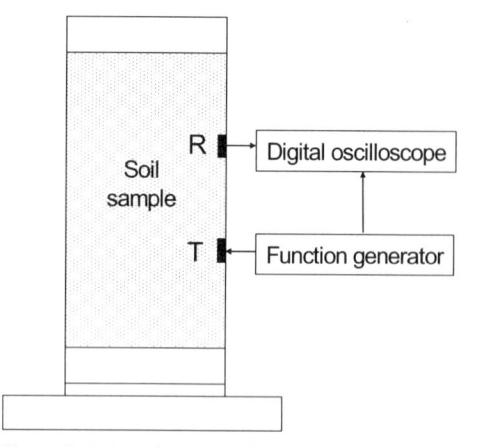

Figure 7: Schematic presentation of bender elements and instrumentation. T is the transmitter element and R the receiver element.

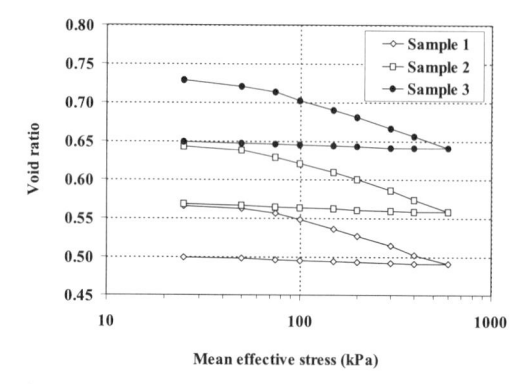

Figure 8: Void ratio changes caused by isotropic consolidation

monitored by the volumetric measurement of the pressure controllers. The void ratios of the three specimens under these consolidation loadings are shown in Figure 8.

The transmitter element (bottom side bender element) was excited with a 20V peak to peak single sine pulse at a frequency of 15kHz generated by the function generator. The internal trigger of the function generator was set to generate a single pulse every 4ms. The signals were recorded by a digital storage oscilloscope. At each effective pressure 128 traces were stacked to reduce noise. The arrival of the shear wave was taken as the first peak (as recommended by Viggiani and Atkinson (1995a)) and obtained from hand picking.

An input frequency of 15kHz was employed in order to reduce the influence of near field effects. Sanchez-Salinero et al. (1986), De Alba and Baldwin (1991) and Gohl and Finn (1991) suggest that a separation of 2 to 3 wavelengths (λ) between the transmitter and receiver should be sufficient for minimizing such effects. An input frequency of

15kHz resulted in a separation of at least 5λ. The shear stiffnesses of the samples were derived from the shear wave velocity as measured between the two side bender probes, and the bulk density of the specimen under its current effective confining pressure.

Figure 9 shows the small-strain shear stiffness (G_o) for the three samples at varying mean effective stress (p'). Neglecting the measurements at 25kPa effective confining pressure, which appear to have been influenced by stresses applied during specimen preparation, the relationship upon first loading is of the form:

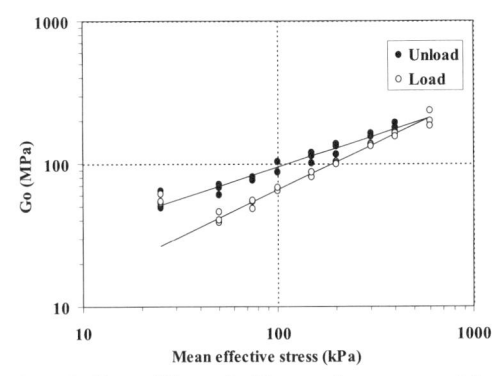

$$G_o = 3.3 \, p'^{\,0.65} \qquad (1)$$

Figure 9: Shear stiffness of tailing samples as measured by bender elements

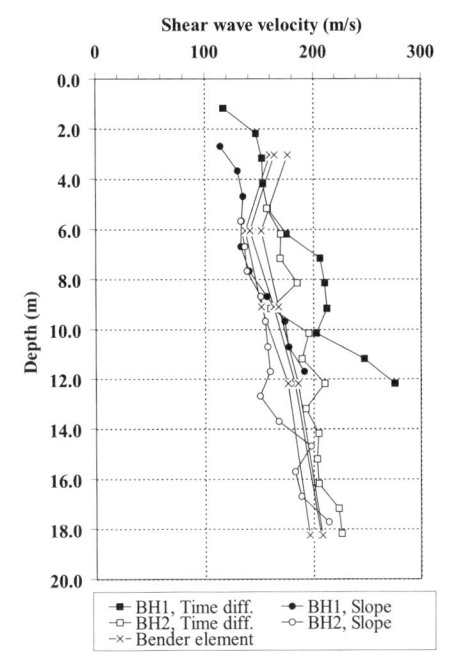

Figure 10: Comparison of laboratory measurements of shear wave velocity with values measured in the field.

5 DISCUSSION

It has been widely reported (e.g. Porovic, 1994, Lo Presti et al., 1993) that for rounded silica sands the exponent in Equation (1) is of the order of 0.5. Figure 9 shows that the tailings tested in this study the exponent was 0.65, which is close to the value of 0.653 reported by Viggiani and Atkinson (1995b) for reconstituted kaolinite. Clayton, Theron and Vermeulen (2004) show that the fine particles in gold tailings are platey, and that the introduction of even relatively small quantities of platy particles into rounded sand has the effect of suppressing dilation, and producing a clay-like behaviour.

Figure 10 compares the shear wave velocities determined from the bender element testing with values obtained from the two seismic cone tests in the daywall of the tailings dam. The depth / effective confining pressure relationship was calculated by assuming a K_o value of 0.4 (from an effective angle of friction of 37°, and Jaky's 1944 equation) with bulk densities obtained from undisturbed samples obtained from the daywall. The level of the water

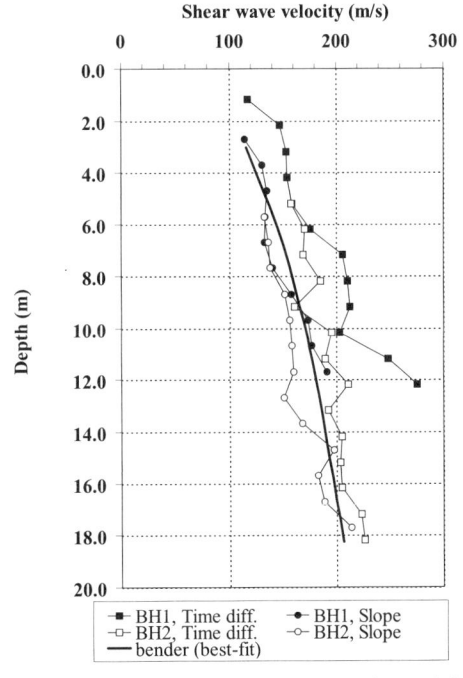

Figure 11: Comparison of best-fit laboratory estimate of shear wave velocity with values measured in the field.

table and the influences of seepage and anisotropy were obtained from piezocone data, following Rust, van der Berg and Jacobsz (1995).

It can be seen from Figure 10 that the shear wave velocities from BH 2 deviate very little from the profile predicted from bender element tests. Those in BH 1 show significant deviation, particularly at depths greater than 7m, perhaps suggesting some influence of bonding. Such an effect might also be produced at this location, however, by tailings containing a smaller proportion of platy fines, or by a locally lower groundwater regime.

Figure 11 superimposes a best-fit estimate of shear wave velocity obtained from Equation (1) and the depth / effective confining pressure relationship developed above on the field seismic velocity vs. depth profile. The agreement between these two estimates is encouraging.

6 CONCLUSIONS

Preliminary experiments comparing field and laboratory measurements of shear wave velocity suggest that these techniques may be of value in determining whether tailings are cemented *in situ*. For the gold tailings tested, the effects of bonding appear to have been limited.

REFERENCES

Acar, Y.B. and El-Tahir, A. 1986. Low strain dynamic properties of artificially cemented sand. *Journal of Geotechnical Engineering*, ASCE, Vol. 112, No.11, pp. 1001-1015.

Clayton, C.R.I., Theron, M. and Vermeulen, N.J. 2004. The effect of particle shape on the behaviour of gold tailings. *Proc. I.C.E. Skempton Memorial Conference*, London.

Cuccovillo, T. and Coop, M.R. 1997. Yielding and pre-failure deformation of structured sands. *Géotechnique*, Vol. 47, No. 3, pp. 491-508.

Daramola, O. 1980. Effect of consolidation age on stiffness of sand. *Géotechnique*, Vol. 30, No. 2, pp. 213-216.

De Alba P. and Baldwin K. 1991. Use of bender elements in soil dynamics experiments. Bhatia S.K. and Blaney G.W. (ed). *Recent Advances in Instrumentation, Data Acquisition and Testing in Soil Dynamics, Geotechnical Special Publication; Proc. of ASCE National Convention*, Florida, (12): 86-101.

Gohl W.B. and Finn W.D.L. 1991. Use of a piezocemraic bender element in soil dynamics testing. Bhatia S.K. and Blaney G.W. (ed). *Proc. of ASCE National Convention, Recent Advances in Instrumentation, Data Acquisition and Testing in Soil Dynamics, Geotechnical Special Publication*, Florida, (12):118-133.

Hardin, B.O. and Drnevich, V.P. 1972. Shear modulus and damping in soils: Design equations and curves. *Journal of the Soil Mechanics and Foundations Division*. ASCE. Vol 98, No. 7, p.667-692.

Heymann, G. 2003. The seismic cone tests. *Journal of the South African Institution of Civil Engineering*, Vol. 45, No. 2, pp.26-31.

Jaky, J. 1944. The coefficient of earth pressure at rest. *Journal of the Society of Hungarian Architects and Engineers*, Vol. 78, no. 22, pp.355-358.

Lo Presti, D.C.F., Pallara, O., Lancellotta, R., Armandi, M. and Maniscalco, R. 1993. Monotonic and Cyclic Loading Behaviour of Two Sands at Small Strains. Geotechnical Testing Journal, GTJODJ, Vol. 16, No. 4, pp. 409-424.

Mitchell, J.K. 1986. Practical problem from surprising soil behaviour. 20th Terzaghi lecture. *Journal of Geotechnical Engineering*, ASCE, Vol. 112, No.3, pp.259-389.

Porovic, E. 1994. Investigations of soil behaviour using resonant column torsional shear hollow cylinder apparatus. PhD Thesis, University of London.

Rust, E., van der Berg, J.P. and Jacobsz, S.W. 1995. Seepage analysis from piezocone dissipation results. *Proceedings of the International Symposium on Cone Penetration Testing*. Swedish Geotechnical Institute. Linköping, Sweden.

Sanchez-Salinero I., Roesset J.M. and Stokoe K.H. 1986. Analytic studies of body wave propagation and attenuation. *University of Texas*. GR86-15, Austin.

Shirley D.J. and Hampton L.D. 1978.Shear-wave measurements in laboratory sediments. *Journal of the Acoustic Society of America* 63 (2): 607-613.

Stokoe, K.H., Hwang, S.K., Lee, J.N.K and Andrus, R.D. 1995. Effects of various parameters on the stiffness and damping of soils at small to medium strains. *Proceedings of the first international conference on pre-failure deformation characteristics of geomaterials*, Vol. 2 pp.785-816. Balkema, Rotterdam.

Vermeulen, N.J. 2001. The composition and state of gold tailings. PhD thesis. *University of Pretoria*.

Viggiani G. and Atkinson J.H. 1995a. Interpretation of bender element tests. *Géotechnique* No. 45, Vol. 1, pp.149-154.

Viggiani, G. and Atkinson, J.H. 1995b. Stiffness of fine-grained soil at very small strains. *Géotechnique*, Vol. 45, No. 2, pp. 249-265.

Wagener, F. 1997. The Merriespruit slimes dam failure: Overview and lessons learnt. *Journal of the South African Institution of Civil Engineering*, Vol. 39, No. 3, pp.11-15.

An assessment of heavy metal contamination using electromagnetic survey in the solid waste disposal site of Campos dos Goytacazes, Brazil

S. Tibana, L.A.C. Monteiro, E.L. dos Santos Júnior & F.T. de Almeida
Laboratory of Civil Engineering of UENF Rio de Janeiro, Brazil

Keywords: dump site, heavy metal, geophysics, electromagnetic survey

ABSTRACT: This paper is the latest in a series published by the Civil Engineering Laboratory of the State University of North Fluminense (Rio de Janeiro / Brazil), reporting the contamination caused by the improper disposal of urban solid waste in the city of Campos dos Goytacazes (Brazil). A comprehensive research program has been carried out by this group in order to assess environmental impact of solid waste disposal in the industrial area of the city. The focused was heavy metals contamination. This work presents some results of a geophysics survey performed with the transient electromagnetic method. An electromagnetic device (SIROTEM MK-3) of the National Observatory of Rio de Janeiro was used. One hundred and eleven vertical profiles were performed. The results of this study conclude that there is a higher concentration of heavy metals in the areas of lower elevation of the landfill, what is reported in Santos et. al. (2003).

1 INTRODUCTION

The municipal landfill of Campos dos Goytacazes is located in its industrial area, about ten kilometers from downtown and spread over about 13 hectares (Araruna et. al., 2001). There have been some poor communities living close to the landfill, and, consequently, some people from these neighborhoods survive by recycling some solid waste. The topography of this area is level with some lagoons in the rainy season (Figure 1). The superficial water flows into the Vigário Canal and, then, to the Paraiba do Sul River (Santos Jr. et. al., 2002).

Due to the bad management of solid waste over the years, animals have free access to the landfill and become vectors of diseases.

All of these aspects create a public health and environmental problems. Since 2002, when the city council contracted a private company to manage the solid waste of the city, the landfill access has been controlled and the solid waste is being covered with layers of soil.

For decades, solid waste was disposed in the areas of lower elevation of the landfill, resulting in a higher concentration of heavy metals (Santos et. al., 2003). Nowadays, with a suitable management of solid waste the dumping site has different landscape but the residual pollution caused by years of unsuitable management cannot be eliminated easily.

Figure 1. Dumping site of Campos dos Goytacazes -RJ

Figure 2 presents a lay-out of the landfill. The school, the lagoon and the area where the solid waste is disposed are marked by black lines. The finer black lines are the level curves resulting from a topography survey before the site was used as a dump. The boundaries of the area considered in this work are shown with the thicker line that envelopes the school, the lagoon and the landfill area. It is easy to observe with old aerial photography that the lagoon area has diminished, in reality the solid waste was disposed into the lagoon diminishing the surface area of it. Nowadays, the area of the landfill extends into the lagoon more than that presented in Figure 2 and it is almost blocking the natural drainage canal.

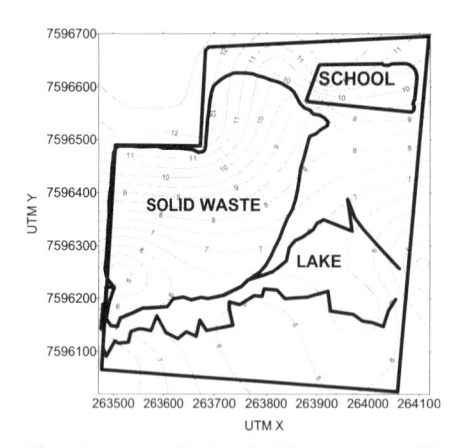

Figure 2. Lay out of the landfill (Monteiro et.al., 2002)

The flat relief showed by the topography survey and by Figure 1 before the start of landfill operations and the many lagoons are two good reasons not to dispose urban solid waste without protecting the sub-soil from contamination. This area needs a proper project to be used as a landfill.

Direct investigation shows that at about a depth of 4 meters in the lower area of the landfill there is a sand layer. It is hypothesized in this work that this layer is a result of a natural drainage for liquid from solid waste decomposition and from water infiltration, consequently, a way for heavy metals contamination to spread. Moreover, it is supposed in this study that the groundwater flows to the east (Santos et.al., 2002)

2 TRANSIENT ELETROMAGNETIC SURVEY

There are a wide variety of electromagnetic survey methods. Each method involves the measurement of one or more electric or magnetic fields induced in the subsurface by a primary field produced by a natural or an alternating current source (Sarma, 1997). In this research, the electromagnetic survey was performed by a transient electromagnetic method (TEM). Buselli et.al. (1990) show that the TEM can be applied effectively to the mapping of electrical conductivity changes associated with contaminated water near waste disposal sites.

A transmitting coil, a fine wire put down on the soil in a square of twenty-five meters side, induced an electromagnetic field. The electromagnetic field attenuations was monitored by a receiver coil that was installed the center of the squares (Monteiro, 2002). The result from each station makes a profile of soil resistivity. Examining all of the results, sections of resistivity can be defined by data interpolation. The geophysics survey was performed with the support of the National Observatory (BRAZIL).

The WinGLink 1.5 of Geosystem was used to interpret the profiles. A resistivity profile was obtained. The greater resistivity in depth, the lesser is the conductivity, consequently, there is less soil pollution.

When geophysics, specially the TEM method, is applied in dumping sites some problems with signals are expected because of the disturbance of solid waste composition. Usually, there is much noise that has to be eliminated before data interpretation.

3 EXPERIMENTAL PROGRAM

The geophysics survey consisted of one hundred and eleven stations carried out with an electromagnetic device. The positions of all stations and the boundaries of the landfill are presented in Figure 3 (Monteiro, 2002). Usually, the geophysics survey is performed in order to define sections, but in this study most of the station were done spread randomly inside of the landfill boundaries. The geophysics survey was programmed with disperse distribution of stations because inside the boundaries of the dump it was hypothesized to be more contaminated. Figure 3 shows the position of each station. The stations beginning with 1, right side of Figure 3, were positioned in the farm beside the dump considered as downstream from the landfill. The stations beginning with 2, left side of Figure 3, were located in another farm situated upstream of the landfill. And, finally, the stations beginning with 3 were placed inside the boundaries of the area above the solid waste disposal site. The area between stations beginning with 1 to the stations beginning with 3 does not have any station because of the energy transmission line.

One of the greatest difficulties was performing the investigation over the solid waste site. Figure 4 shows the group of technicians preparing one more station. The receiver coil and controller device are shown in Figure 5 and 6 respectively. The metal presence in the solid waste is a font of uncertainty.

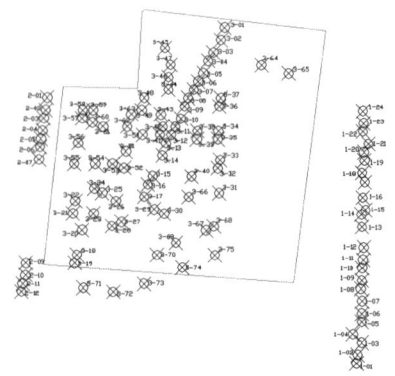

Figure 3. Position of each station during the geophysics survey

Figure 4. Geophysics survey in the landfill

Figure 5. The receiver coil in the landfill during an investigation

Figure 6. The controller device of SIROTEM MK-6

4 RESULTS AND ANALYSIS

In order to obtain better results from each station and, consequently, the conductivity and resistivity soils section, the results were screened eliminating the noise probably coming from the metal waste and from the transmission line of energy. Due to people recycling the solid waste much metal was pilled together resulting in high noise level.

The results of all stations were interpolated and maps of resistivity (ohm.m) were produced to investigate the spatial distribution of heavy metals.

In order to show the depth reached by heavy metals contamination, some maps were done with the geophysics survey at 5, 10 15, 20, 25, 30, 50 and 70 meters depth. The maps in Figure 7 to 10 show the spatial distribution of resistivity that vary from 25.0 ohm.m to 197.8 ohm.m. The blank space in the maps corresponds to the road with the electrical line as mentioned earlier.

The lower area of Figures 7 to 10 concern the lower area of the landfill where the solid waste is underwater, while the upper areas of the Figures concern the areas with higher elevations.

The Figure 7 shows the spatial resistivity at the depth of five. The darker tone of gray in the lower area of Figures indicates low resistiveity, about 25 to 30 ohm.m. Comparing Figure 2 with Figure 7, shows that the lower elevations area is more contaminated by heavy metals. In the middle of Figure 7 a medium level of contamination, about 90-100 ohm.m, was detected. In the area of higher elevation the tone of gray is very light and the resistivity was 120-140 ohm.m.

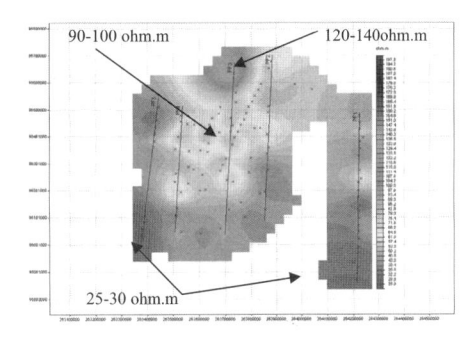

Figure 7. Map of five meters depth (Monteiro et.al., 2002)

Figures 8 to 10 show the spatial resistivity at the depth of ten, fifteen and thirty meters. The resistivity of higher elevation area increases from 120 to 200 ohm.m., suggesting that the heavy metal contamination, existing in this area is very low compared with the lower elevation area of the landfill.

It is easy to observe in figures 8 to 10 that as deeper as the cross section, higher is the resistivity in all areas, indicating that the concentration of heavy metals lessening. Moreover, another conclusion is that the area of high resistivity increases as the depth increases, suggesting that the heavy metal contamination decreases with depth.

Figure 11 shows a cross section at fifty meters depth. With this image it can be concluded that at this depth the pollutant is totally absent.

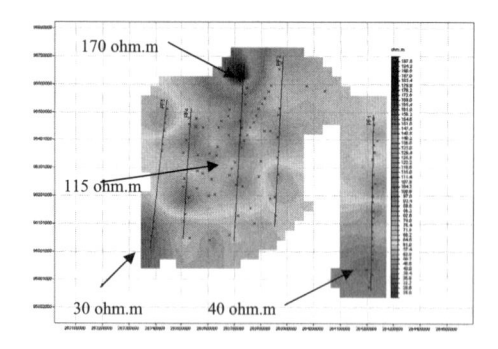

Figure 8. Map of ten meters depth (Monteiro et.al.2002)

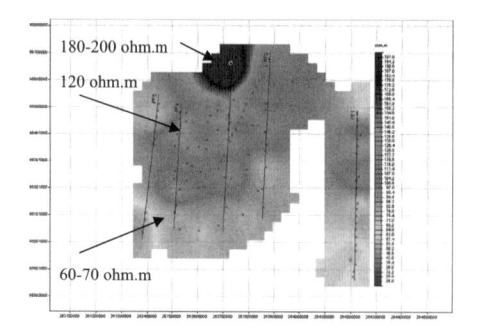

Figure 9. Map of fifteen meters depth (Monteiro et.al. 2002)

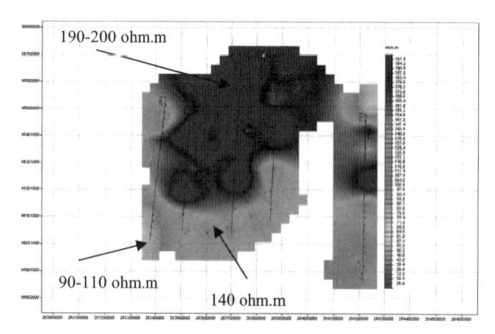

Figure 10. Map of thirty meters depth (Monteiro et. al. 2002)

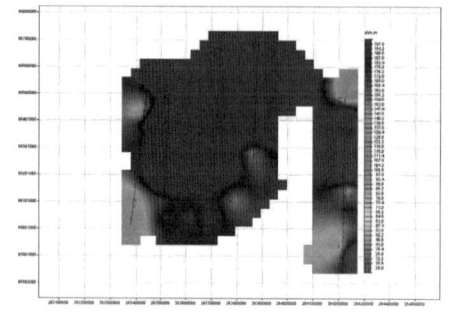

Figure 11. Map of fifty meters depth

Santos et al. (2002) found a concentration of pollution in the lower area also, even though it was a low concentration. The sand layer (Monteiro et.al., 2001) and the permeability (Monteiro et. al. 2002) of the cover layer are the reasons for lower values of concentration of heavy metals.

5 CONCLUSIONS

The results of the geophysics survey suggest that heavy metal contamination is concentrated in the lower area of the landfill, close to the lagoon, and that contamination reaches profound layers, about thirty meters deep

The results of this paper are in accordance with Santos et.al. (2003) and, considering the discussions presented in Santos et.al. (2002), the heavy metals have been transported to the east direction.

ACKNOWLEDGEMENTS

The authors of this paper are very gratefull to the geophysics technicians of National Observatory of Rio de Janeiro for their support during the geophysics survey and the interpretation of results

REFERENCES

Araruna Jr., J.T., Tibana, S., Monteiro, L.AC, Santos Jr, E.L.S. 2001. *Heavy metal contamination in a municipal dumping site in southeast Brazil*, Proc. XVII Int. Conf. on Solid Waste Technology Management 2001. V. 1, p. 886-901.

Buseli, G., Barber, C., Davis, G.B., & Salama, R.B. 1990. *Detection of groundwater contamination near waste disposal sites with transient electromagnetic and electrical methods*, Proc. 5° Investig. Geophysics - Geotechnical and Environmental Geophysics, ISBN 0-931830-99-0, v. II, p. 27-39.

Monteiro, L.AC, Santos Jr., E.L., Tibana, S., Araruna Jr., J.T. 2001. *Estudo da contaminação da área de disposição de resíduos sólidos urbanos (RSU) da cidade de Campos dos Goytacazes*. In: 21°. Congresso Brasileiro de Engenharia Sanitária e Ambiental, João Pessoa, Brazil. v. 1, p. 224-233.

Monteiro, L.AC. 2002. *Estudo da contaminação da área de disposição de resíduos sólidos da cidade de Campos dos Goytacazes / RJ pela técnica eletromagnética domínio do tempo*. MSc. Thesis presented in UENF – BRAZIL.

Monteiro, L.AC., Santos Jr., E.L., Tibana, S., Araruna Jr., J.T. .2003. *Physical dumping site characterization of Campos dos Goytacazes – RJ / Brazil* , Proc. of DMinUCE, London.

Santos Jr, E.L.; Tibana, S.; Souza, C. M. M.; Monteiro, L.A C.2003. *Heavy metal contamination in the dumping site of Campos dos Goytacazes - RJ / BRAZIL*. Proceedings of the 12th. Panamerican Conference on Soil Mechanics and Geotechnical Engineering, Boston. v. 1, p. 1-4.

Santos Jr, E.L., Monteiro, L.AC. & Tibana, S 2002. *Growdwater seepage in the dumping site of Campos dos Goytacazes / RJ – BRAZIL*. Proc. 8°. Cong. Nac. Geotecnia, Lisboa, Portugal v. 1, p. 1-6.

Sharma, P.V. 1997. *Environmental and engineering geophysics*, ISBN 0521 57632 6.

Geotechnical site characterisation using surface waves, case studies from Belgium and the Netherlands

V. van Hoegaerden, R.S. Westerhoff, J.H. Brouwer & M.C. van der Rijst
Netherlands Institute of Applied Geoscience TNO – National Geological Survey, Netherlands

Keywords: geotechnics, geophysics, surface wave, rayleigh wave, shear wave, cone penetration test (CPT), wave modelling

ABSTRACT: In the Netherlands, shallow (up to 30 meter below surface level) geotechnical site characterisation is normally carried out with cone penetrometers (CPT's). Although cost effective, a CPT has its disadvantages. First of all, the clients generally require 2D or 3D information, whereas the CPT delivers 1D information representative only for the direct vicinity of the penetrated hole. Secondly, CPT interpretation cannot be done without geological information, such as knowledge about the age of the sediments. Thirdly, since the instrument is pushed into the ground, it is not suited for sites where it is not allowed to use intrusive methods, e.g. at waste deposits. In 2001, a research project has been started to develop a geotechnical site characterisation method using seismic surface waves and multiple receivers. Having used the MASW-technique (Multi-channel Analysis of Surface Waves) on several projects, a number of (interpretation) improvements have been developed and implemented. In this article, case studies are presented from a site for possible nuclear storage (Dessel, Belgium) and a waste deposit (Zevenaar, The Netherlands). Both cases show that the proposed method (called ConsoliTest) is functionally operational and can play a significant role in geotechnical site characterisation.

1 INTRODUCTION

In 2001, a research project has started to improve geotechnical site characterisation using non-destructive seismic techniques. To focus on the depth range from surface till about 30 meter below.

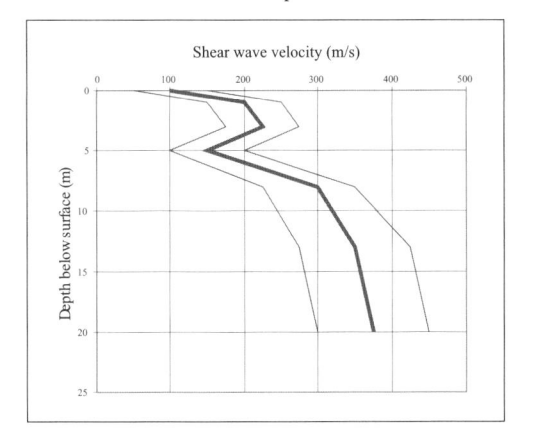

Figure 1. Ideal outcome (estimate curve plus standard deviation curves) of geotechnical site characterisation.

The data presentation in Fig. 1 is familiar to soil investigators and geotechnical engineers. It is the type of parameter, shear wave velocity, that needs further explanation. The shear wave velocity is the speed of propagation of a shear wave that travels through a medium (body) in the absence of reflections and refractions. Surface waves such as Rayleigh waves propagate at and below the boundary of the earth and air (a so-called free surface). Rayleigh waves travel with a velocity slightly less than the shear wave velocity at the free surface (see Russell (2001)). The value of information for shear wave velocity lies in the physical fact that the shear wave velocity is a measure of the ability of the sediment (or rock) to propagate shear stress. Generally said, low shear wave velocity hints at a low degree of consolidation and thus a high degree of fluidity. In fluids like water, the shear wave velocity is zero and therefore shear waves cannot propagate.

Overall, the shear wave velocity depends on the type of sediment, porosity, its pore content (water, brine or air), depth of burial and the geologic history (i.e. pre-loading etc.). Key to the perception of shear wave results is knowledge of the geology and the stratigraphy of the layers.

2 METHOD OF INVESTIGATION

In previous years, considerable progress has been made in the P-wave high-resolution reflection seismic technique, called HRS (Meekes (1992)). This technique works with multiple receivers and an energy source chosen especially for the area at hand. The area of investigation for HRS is however from about 25 till 500 meter below surface, i.e. below the geotechnical zone of interest.

The proposed technique for geotechnical site characterisation using surface waves builds on this expertise. It too uses multiple receivers with a constant receiver interval. Normally, only the vertical motion is recorded. The seismic energy source is a hammer or falling weight which impacts a metal plate or the ground directly (Fig. 2). The first step in the processing is similar to the MASW-method (Multi-receiver Analysis of Surface Waves, Gabriels et al. (1987), Park et al. (1998 and 1999) and Foti (2000)). The records are transformed from the (x,t)-domain into the (f,k)-domain and presented in a so-called dispersion plot (Fig. 3).

The dispersion plot shows the phase velocity as a function of frequency for the wave types recorded. Just as with acoustic waves, Rayleigh waves can exist in a normal (fundamental) mode as well as in higher modes. In a ground model with shear wave velocity increasing monotonically with depth, Rayleigh waves only propagate in the normal (or fundamental) mode. In that case, only one Rayleigh dispersion curve will be present in the dispersion plot with for each frequency the lowest phase velocity as possible. The occurrence of lower velocity layers in between higher velocity layers or the presence of a high velocity layer at the surface will lead to an attenuation of the normal mode on one hand and the presence of higher modes (and thus more dispersion curves) as well as other wave types such as Lamb waves (Park et al. (2002)) on the other.

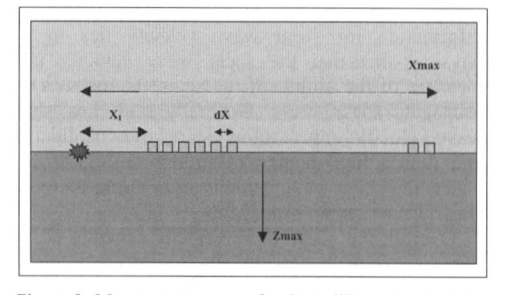

Figure 2. Measurement set-up for ConsoliTest. On the left is the energy source (red) and the squares represent the receivers.

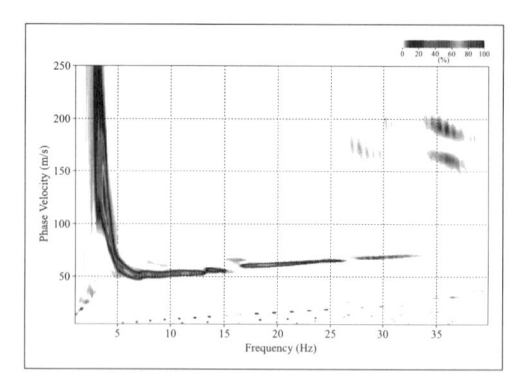

Figure 3. Example of a dispersion plot (phase velocity as a function of frequency).

The present implementation of the method only takes the dispersion curve of the fundamental Rayleigh wave mode into account for the inversion. This limitation causes the vertical resolution and accuracy to decrease with depth. Presently, the inversion of the dispersion curve is done with SurfSeis (Xia et al. (1999) and Park et al. (2001)). The 1D inversion algorithm uses the propagator matrix method to obtain theoretical dispersion curves for the fundamental Rayleigh mode (Schwab and Knopoff (1972), Haskell (1953) and Thomson (1950)). Important to keep in mind is the fact that the propagator matrix method can generate non-physical solutions for the fundamental Rayleigh wave in case of a ground model with low-velocity layer(s) in between higher velocity layers. One way to circumvent this, is to minimise the number of layers. The processing is concluded with verification and validation steps, as discussed below.

2.1 Verification and Validation Approach

Both for scientific as well as marketing reasons, verification and validation of the shear wave results is essential. The verification and validation has to honour the fact that the measurement method gives an averaged result over the line of measurement (receiver spread) instead of a 1D result in case of a CPT.

The most logical validation candidate is the so-called seismic cone penetration test (SCPT). This destructive measurement procedure uses a cone with three seismic receivers that is pushed stepwise into the (non-consolidated) sediments. Since the Consoli-Test measurement is a laterally averaging measurement of ground properties, at least three (S)CPT's should be placed equidistantly over the receiver spread length. Three or more (S)CPT's make it possible to rule out erroneous measurements and to signal lateral changes. For (qualitative) verification, a normal cone penetration test (CPT) can be used.

The second validation methodology is forward modelling of surface wave propagation and the creation of dispersion plots for a variety of (shallow) subsurface models, separate from the SurfSeis modelling and inversion process. In paragraph 3 and 4, examples will be shown of dispersion curves for any number of Rayleigh wave modes in one-dimensional (layer-cake) models. The 1D-implementation is based on Takeuchi and Saito (1972). Advantage of this approach over the propagator matrix method is its more realistic way of modelling the subsurface (with linear velocity trends in each depth interval, as in Fig. 1) as compared to the model of a stack of homogeneous layers in the Thomson-Haskell type of methods. In case the subsurface is not a layer-cake model or the limitations of the propagator matrix method are to be circumvented, wave modelling is done with finite-difference wave propagation methods. At present, a 2D implementation of P-SV wave propagation is used (Virieux (1986)). The theoretical seismic records are then analysed in SurfSeis to arrive at a dispersion plot. The 2D modelling can especially help the interpretation and inversion of complex subsurface models (Paragraph 4).

technology for geotechnical exploration of the 0-15 m depth range.

A geological description of the area can be found in Wouters and VandenBerghe (1994). The sediments till 20 m deep are pure quartz sands of Tertiary age (formation of Mol and possibly Kasterlee), locally with clay lenses. The sands vary in grain size. The line of interest is situated on a sand road, where the original relief (mild highs and lows) have been filled up with sand.

3.1 Results and Verification

Based on tests with various sledgehammers and varying source-to-first-receiver distances, the acquisition parameters as described have been chosen (Table 1). The ConsoliTest result for the line of interest is shown in Figure 4. The figure shows the colour contoured result from four 1D shear wave velocity profiles, situated at station 409, 427, 445, and 463. Each 1D shear wave velocity profile is representative of a section of 18 m length. Thus, the result for station 445 is the averaged result of ground properties from station 436 up to and including 454.

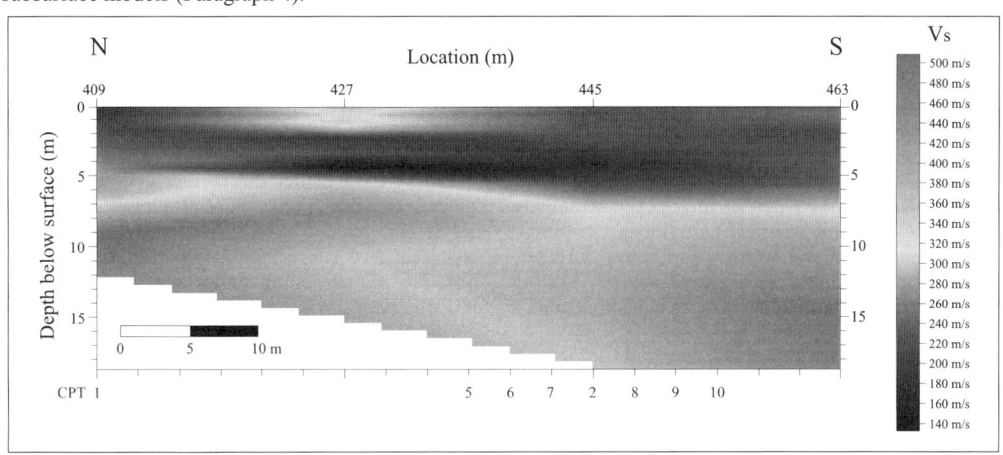

Figure 4. The 2D shear wave velocity profile for a line (Dessel, Belgium), constructed out of four 1D surface wave profiles at station 409, 427, 445, and 463.

3 CASE 1, A NUCLEAR STORAGE AREA IN DESSEL (BELGIUM)

In 2002, NIRAS/ONDRAF (the Belgian Agency for Radioactive Waste and Enriched Fissile materials) asked TNO to carry out a geotechnical exploration as part of a larger exploration project for a possible nuclear storage area in Dessel (Belgium). The objective was to test the feasibility of surface wave

Table 1. The acquisition parameters for the ConsoliTest measurements on the Dessel nuclear storage area.

Source	Sledgehammer of 10 kg
Receivers	4.5 Hz geophones
Number of geophones	36 (18 per analysis)
Geophone distance	1 m
Source – first receiver	5 m distance

The profile shows that the shear wave velocities are in general quite high, as can be expected from Tertiary sediments. Between 5 and 10 meter below surface, there is a contrast in shear wave velocity. From borings and laboratory soil tests, it is known

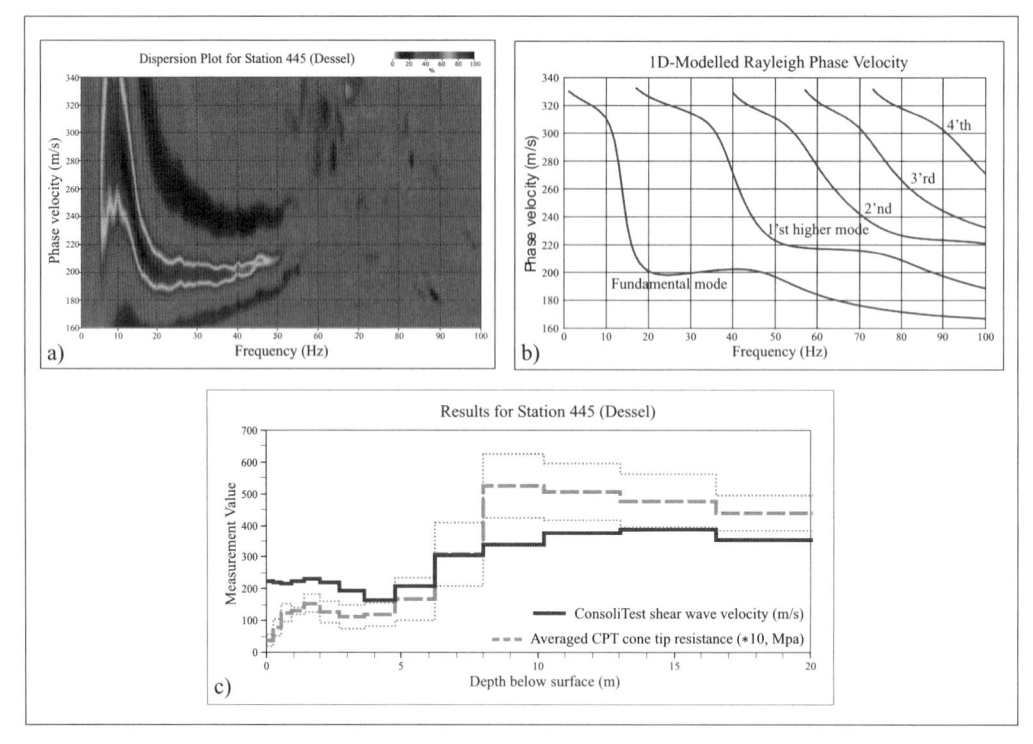

Figure 5. For station 445 (Dessel), 1D modelling shows that the picked dispersion curve is indeed a fundamental mode. The trends in the ConsoliTest result are verified by the CPT's.

that the shallow sediments (0 – about 8 m deep) are coarse quartz sands whereas the deeper sediments (about 8 - 20 m below surface) are fine quartz sands. In the zone of interest (0 till 10 m below surface), local variations in the shear wave velocity hint at grain size variations and/or changes in compaction.

3.2 Verification with 1D-modelling and CPT's

The verification of the Dessel results is done on the basis of 1D Rayleigh wave modelling and cone penetration tests (CPT's). In this paragraph, station 445 is taken as example. Figure 5.a shows the measured dispersion plot for station 445, with the coherent trend through $f = 20$ Hz and $v_f = 200$ m/s interpreted as fundamental Rayleigh mode.

Table 2. The subsurface model for the 1D Rayleigh wave modelling at station 445 (Dessel).

Layer	d (m)	v_s (m/s)	v_p (m/s)	ρ (g/cm^3)
1	2.6	240	700	1.6
2	3.4	160	1500	1.8
3	half-space	350	1600	2.0

To verify whether the correct coherent trend is picked for the fundamental Rayleigh mode, five Rayleigh modes have been modelled using a representative subsurface model (Table 2 and Fig. 5b). The 1D-modelling shows that the fundamental mode is clearly separated from the higher modes and confirms the choice made on the measured dispersion plot. Comparison with the measured dispersion plot also shows that there is no significant energy in the higher modes.

To check the inversion result in terms of trends and transitions (thick red line in Fig. 5c), seven CPT's have been placed on the 18 m spread of station 445 (annotated 5, 6, 7, 2, 8, 9, and 10 in Fig. 4). The comparison is done with the cone tip resistance of the CPT's, averaged over the same depth intervals as the inverted shear wave result. The thick dashed line (green) in Figure 5.c is the averaged cone tip resistance, with the thin dashed lines (green) indicating the spread (one standard deviation). The transition at about 6 m depth to higher stiffness is confirmed by the averaged cone tip resistance. Furthermore, the general trend in the shear wave result is supported by the trend in the cone tip resistance, as evidenced by a correlation coefficient (R^2) of 0.88 between the two curves.

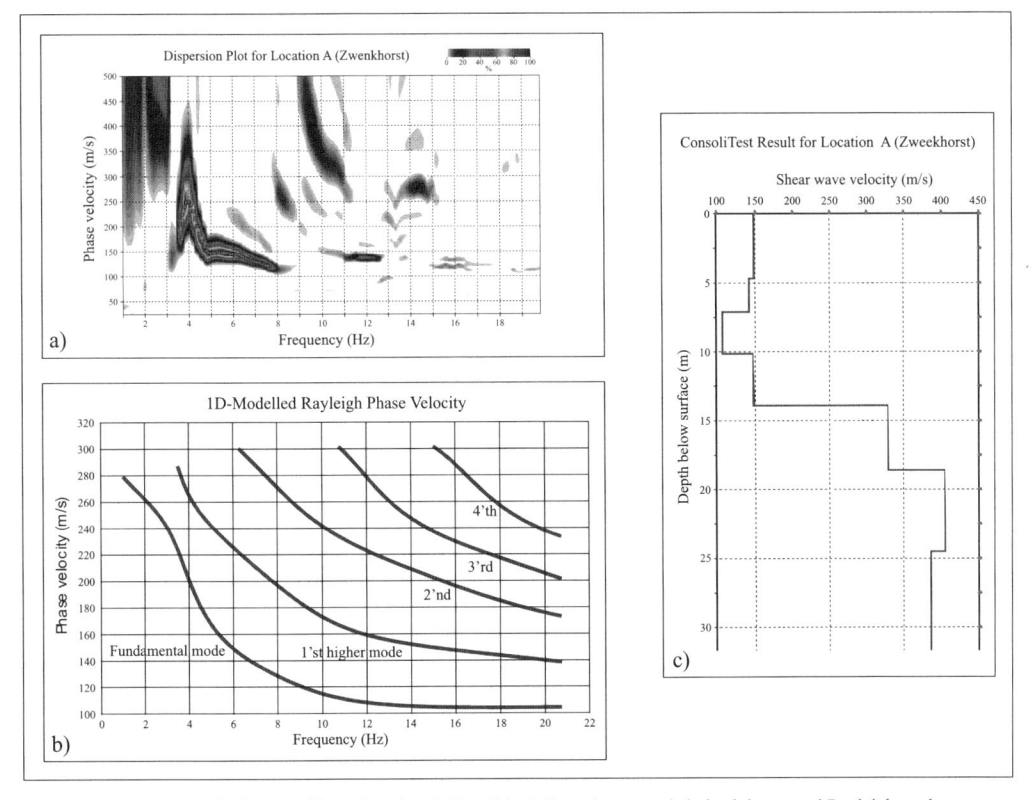

Figure 6. ConsoliTest results for Zweekhorst location A. The picked dispersion curve is indeed the normal Rayleigh mode.

4 CASE 2, A WASTE DEPOSIT IN ZEVENAAR (THE NETHERLANDS)

The second case is a waste deposit in Zevenaar (province of Gelderland) in the Netherlands. The waste deposit, called Zweekhorst, is owned by the Dutch environmental and waste company AVR. The goal is to determine the thickness of the waste deposit. Below the waste, there is a sand drainage layer of about 30 cm, a HDPE-foil and finally a sand-bentonite layer of about 20 cm. Below this, there is 1-2 meter sandy, silty clay (Formation of Echteld) and then a few tens of meters of coarse sand (Formation of Kreftenheye). Since it is not allowed or feasible to drill a hole or to penetrate the waste deposit with a CPT, this has to be solved with non-destructive methods. With the engineering company Grontmij as liaison, TNO has performed a feasibility study with the ConsoliTest technique at various locations on the waste deposit.

4.1 Zweekhorst Location A

Location A is situated on the highest point of the waste deposit, with an estimated waste thickness of about 12 meter. The waste consists of polluted soil, mixed with domestic and construction waste. To cater for the damping of energy in the waste, a

Table 3. The acquisition parameters for the ConsoliTest measurements on the Zweekhorst waste deposit.

Source	Falling weight (40 kg) on steal plate (maximum height 4 m)
Receivers	4.5 Hz geophones
Number of geophones	50
Geophone distance	1 m
Source – first receiver	1 m distance

falling weight has been used instead of a sledgehammer (Table 3). Both the dispersion plot, the inversion result as well as a verification using 1D Rayleigh wave modelling are shown in figure 6.

Based on the presence of a clear contrast in the shear wave velocity as well as a locally determined threshold value of 150 m/s for waste, the waste thickness is determined to be 12 ± 2 m.

4.2 Zweekhorst Location B

At location B, the thickness of the waste is known to be about 6 meter. The acquisition parameters are identical to those for location A (Table 3). Due to the presence of a hard, gravel rich top layer, the coupling of the geophones with the ground was not optimal. The measured dispersion plot is shown below (Fig. 7.a). Due to the lack of continuous features, a dispersion curve cannot be picked reliably for the fundamental Rayleigh mode.

To ease the interpretation, a 1D Rayleigh wave modelling is done. The model used for this is a hard top layer, a waste layer, the waste deposit bottom and then Kreftenheye sands (Table 4). The resulting theoretical Rayleigh dispersion plot shows a steep slope in phase velocity between 5 and 10 Hz and almost no dispersion above 15 Hz for the fundamental mode. The higher modes show a more gentle dispersion behaviour. The match with the measured dispersion plot remains difficult due to the absence of continuous features for frequencies higher than 20 Hz. Furthermore, the 1D-modelling does not give a measure of the relative energy of each Rayleigh mode.

Table 4. The subsurface model for the 1D and 2D wave modelling of Zweekhorst location B.

Layer	d (m)	v_s (m/s)	v_p (m/s)	ρ (g/cm^3)
1	0.2	250	1500	1.5
2	5.9	100	600	1.5
3	1.4	150	1500	1.7
4	half-space	250	1700	2.0

4.2.1 *Two-dimensional Wave Modelling*

To model as close as possible what has been measured in the field, 2D finite-difference wave modelling has been done. The program can use arbitrary 2D models or a SurfSeis generated 1D model and generates receiver records in SEG-2 format as well as a movie of the wave propagation (Fig. 8). The subsurface model used is identical to the one used for the 1D-modelling (Table 4).

The 2D wave modelling results are shown in Figure 9.b and d. The modelling is done for 24 receivers, with an offset from 5 till 28 meter with respect to the source position. The corresponding measurements in the field are shown next to it (Fig. 9.a and c). Both the receiver records as well as the dispersion plots show that the modelled data contains more high frequencies. This is due to the fact that damping is not (yet) incorporated in the wave modelling, other than attenuation due to geometric spreading. Especially the high frequencies are damped in the propagation from the hard top layer via the (least stiff) waste layer to the sediments and back again. In this case, the assumption of elastic wave propagation is thus questionable.

The phase velocities as modelled are similar to the measured phase velocities (Fig. 9.c and d). The measured dispersion plot however misses significant energy for frequencies higher than 20 Hz to make a full comparison. When the 2D-modelled dispersion plot (Fig. 9.d) is compared with the 1D-modelled dispersion plot (Fig. 7.b), it turns out that the fundamental and first higher mode are amalgamated in the 2D modelled dispersion plot. This raises the question whether this is really a fundamental Rayleigh wave, a Lamb (plate) wave or a mixture of Rayleigh and Lamb waves. The second and third higher Rayleigh wave mode however seem to match.

The constructive and destructive interference and damping of waves in the various layers can thus lead to complex wave behaviour that cannot be modelled in a physically correct way with 1D Rayleigh wave modelling. This example shows the advantages of 2D modelling in that it allows a realistic modelling of surface wave dispersion plots and a robust interpretation of wave modes. To improve the measurements, a vibrator should be used as energy source instead of a falling weight.

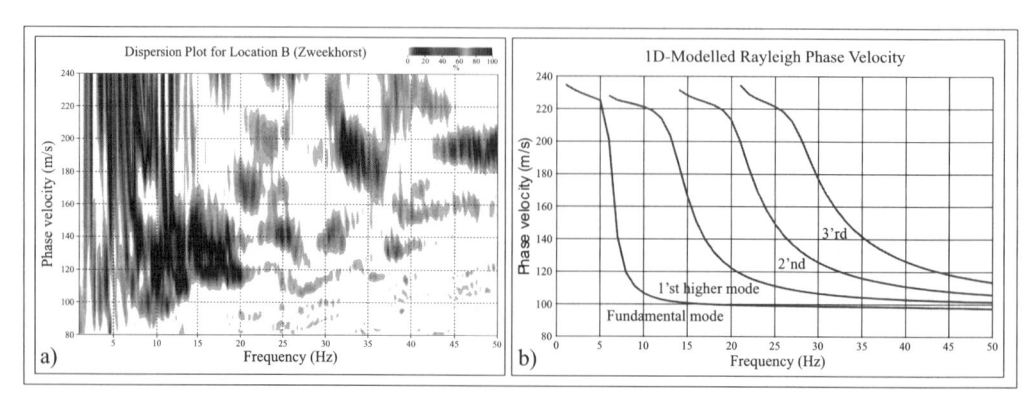

Figure 7. Measured and 1D-modelled dispersion plot for Zweekhorst location B.

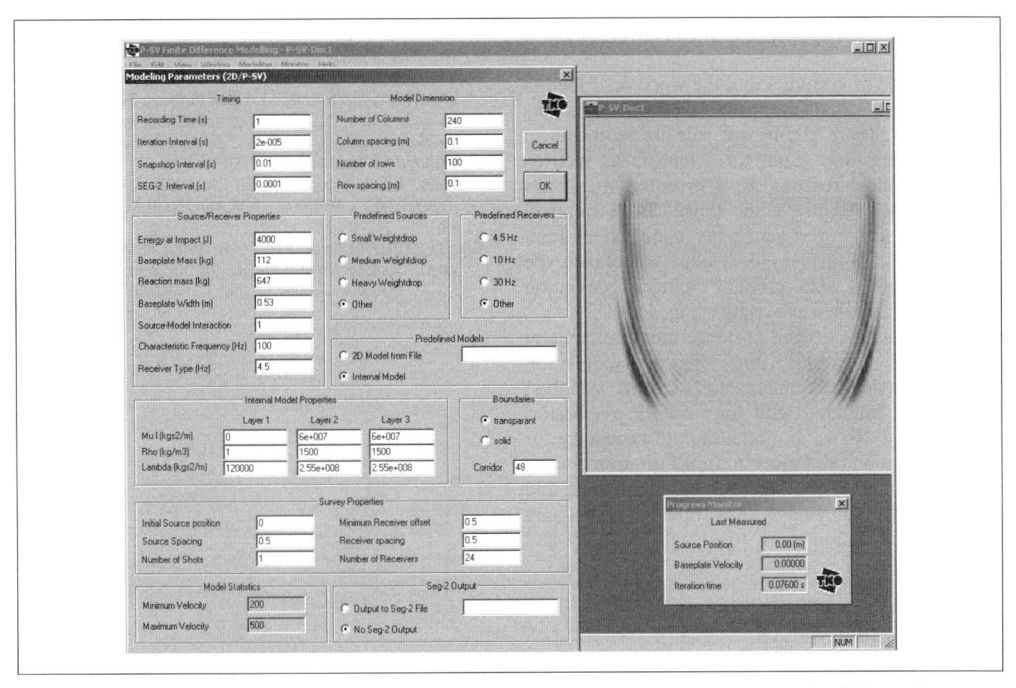

Figure 8. An impression of the 2D finite-difference wave modelling software (modelling parameters, movie window and progress monitor).

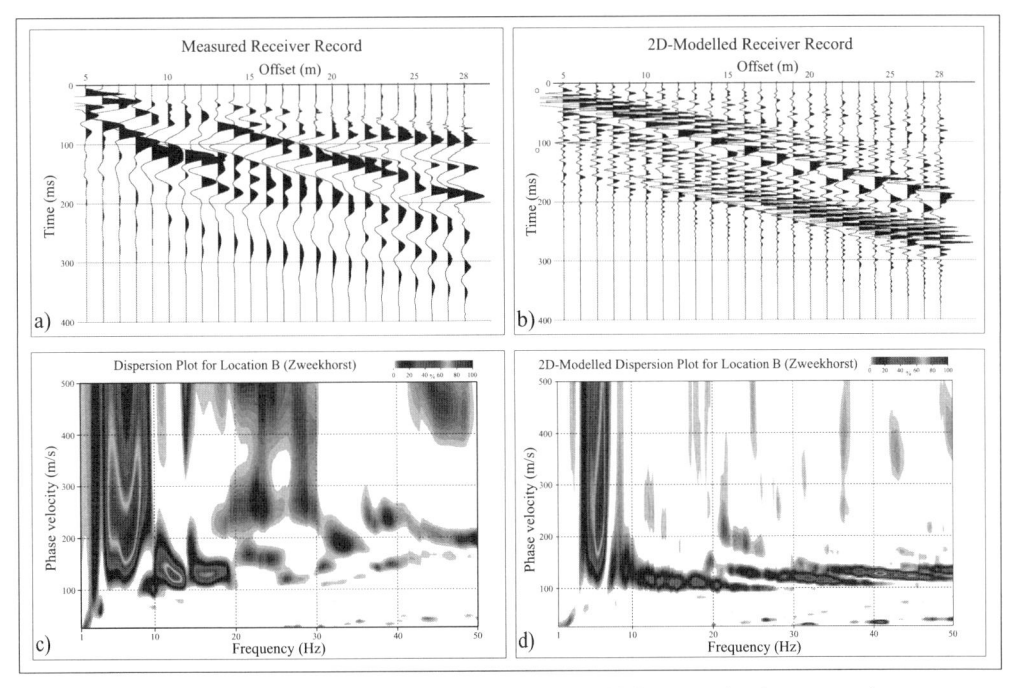

Figure 9. The 2D modelling results for Zweekhorst location B in comparison with the measured surface wave results.

5 CONCLUSIONS & RECOMMENDATIONS

The presented case studies and the work described in literature show that the proposed surface waves method can play a significant role in geotechnical site characterisation. Both for scientific as well as market technical reasons, it is essential to always combine the results with those from already established techniques, such as cone penetration tests (CPT's). To further increase the acceptance, the inversion should also produce standard deviation values for the shear wave results (Fig. 1).

To be able to perform accurate measurements at all onshore locations, the present acquisition equipment (sledgehammers and falling weights) needs to be extended with for instance a vibrator. Also, 1D and 2D wave modelling are necessary for the interpretation (and inversion) of the surface wave dispersion plots. In combination with geological and geomechanical knowledge, the proposed method can lift site investigation work to a higher quality and applicability level.

ACKNOWLEDGEMENTS

We would like to thank L. Wouters Bsc. (NIRAS/ONDRAF, Belgium) for permission to use the Dessel case study and J. van Erven (AVR, Rotterdam) for the opportunity to measure on the Zweekhorst waste deposit.

REFERENCES

Foti, S. 2000. *Multistation methods for geotechnical characterization using surface waves*, PhD thesis Politecnico di Torino.

Gabriels, P., Snieder, R. & Nolet, G. 1987. *In situ measurements of shear-wave velocity in sediments using higher mode Rayleigh waves*. Geophys. Prosp., 35, p. 187-196

Haskell, N.A. 1953, *The dispersion of surface waves in multilayered media*. Bull. Seism. Soc. Am., 43, p. 17-34.

Meekes, J.A.C. 1992. *Examples of the Application of Shallow Reflection Surveys in The Netherlands*. Quarterly Journal of Engineering Geology, 1992-IV.

Park, C.B., Ryden, N., Westerhoff, R.S. & Miller, R.D. 2002, *Lamb waves observed during MASW surveys* [Exp. Abs.], Soc. Expl. Geophysics, Salt Lake City, Utah

Park, C.B., Miller, R.D. & Brohammer, M. 2001. *SurfSeis MASW User's manual*. Kansas Geological Survey.

Park, C.B., Miller, R.D. & J. Xia 1999. *Multichannel analysis of surface waves*. Geophysics, v.64, n. 3, p. 800-808

Park, C.B., Miller, R.D. & Xia, J. 1998. *Imaging dispersion curves of surface waves on multi-channel record*. 1998 SEG Expanded Abstracts.

Russell, D.A. 2001. *Longitudinal and transverse wave motion*, http://www.gmi.edu/~drussell/Demos/waves/wavemotion.html

Schwab, F.A. & Knopoff, L. 1972. *Fast surface wave and free mode computations*: in Methods in Computation Physics, edited by B.A. Bolt, Academic Press, New York, p. 87-180

Takeuchi, H. & Saito, M. 1972. *Seismic surface waves*. In B.A. Bolt, Seismology: Surface Waves and Earth Oscillations (Methods in Computational Physics, Vol. 11). New York: Academic Press.

Thomson, W.T. 1950. *Transmission of elastic waves through a stratified solid*. Journal of Allied Physics, 21, p. 89-93

Virieux, J. 1986. *P-SV wave propagation in heterogeneous media: Velocity-stress finite difference method*. Geophysics, Volume 51, Number 4, p. 889-901

Wouters, L. & Vandenberghe, N. 1994. *Geologie van de Kempen (Een synthese)*. NIRAS/ONDRAF, NIROND-94-11

Xia, J., Miller, R.D. & Park, C.B. 1999. *Estimation of near-surface shear-wave velocity by inversion of Rayleigh waves*. Geophysics, vol. 64, no. 3, p 691-700.

Geophysical methods and identification of embankment dam parameters

O.K. Voronkov, A.A. Kagan, N.F. Krivonogova, V.B. Glagovsky & V.S. Prokopovich
B.E. Vedeneev Research Institute (VNIIG), St. Petersburg, Russia

Keywords: geophysical modeling, embankment dams, identification of parameters

ABSTRACT: The available experience of investigation of the state of embankment structures by geophysical methods shows the efficiency of those methods for detection and analysis of structural non-uniformity as well as the state and properties of soil masses. In cases of absence or failure of dam instrumentation, geophysical procedures prove useful for solving structural diagnosis problems. The Institute has developed a procedure permitting a full use of available information on the structure, particularly the data of geophysical survey, geological checking of site conditions and in-situ observations for construction of prediction models of the state of embankment dams. A system of methods recommended for geophysical diagnosis and control of the state of embankment dams and foundations includes various procedures - electrometric, seismoacoustic, thermometric, radio isotopic. The use of the recommendations developed for geophysical modeling, determination of physical-mechanical soil properties by geophysical investigations as well as identification method of model parameters of embankment structures and foundations serves as a basis for construction of efficient forecasting models.

1 INTRODUCTION

The evaluation of reliability and safety of hydro-technical constructions demands to fulfill computational investigations that are carried out on the base of modeling the system "foundation – structure". This modeling allows to find out peculiarities of texture, composition and properties of the ground.

The existing comparatively small experience of embankment structure state survey using geophysical methods indicates its efficiency at revealing and studying the inhomogeneity of structure, state and soil mass properties. When there is no control-and-measuring equipment (CME) or its failure on the dam the using of geophysical survey system is the method to solve the task of structure diagnosis.

At the Institute there has been developed a procedure that allows to apply quite fully all existing information about the structure including the data of geophysical investigations, geotechnical supervision and on-site observation to make up efficient predictive models of embankment dam state.

The procedure consists of the following:
– the choice of methods of geophysical diagnosis of embankment dams and foundations;
– the recommendations to determine physical-mechanical and seepage parameters of soils on the base of geophysical investigations;
– the procedure to create the design models of the dam - foundation system;
– the procedure to identify the parameters of mathematical models to evaluate the dam state and to predict its behavior.

2 GEOPHYSICAL METHODS FOR DIAGNOSIS OF EMBANKMENT DAMS STATE AND CHARACTERISTICS

Geophysical methods are used to solve the following tasks of the embankment dam diagnosis:
– detailed mapping of the depression surface in the downstream fill and detection of abnormal sites;
– determining the path of localised seepage;
– determining the sites of damages of the screen or defects of sheet piling;
– detecting, tracing, determining the depth and components of steeply-falling cracks and the areas of softening in the crest of the dam;
– determining the seasonal freezing (or thawing) depth of soils;

- estimation of soil layers inhomogeneity in the dam body according to the composition and state;
- estimation of mating quality of several soil and non-soil dam components;
- determining the soil density and porosity;
- determining the real seepage velocity and permeability coefficient;
- determining the static and dynamic characteristics of soils deformability;
- studying the tendencies of soil state and property changes under the long seasonal and extraordinary impacts.

The procedure of geophysical diagnosis includes the following methods:
- electrometrics (electrosounding, the method of natural electrical field, the method of charged body, resistance measurement, geological radiolocation; more seldom other methods of electrometrics can be used);
- seismoacoustical methods (seismic profiling and seismotranslucence using longitudinal and transversal waves, the more informative is the seismic tomography version of data observation and processing);
- thermometric methods (in wells and drills of any purpose, in springs, streams, in tail water; in a reservoir).

At express-diagnosis of embankment dams and their foundations we can confine ourselves to the methods of electrometry (3-4 methods) and thermometry.

The basic amount of geophysical investigations is carried out on the dam crest, downstream and upstream slopes (preferably on berms), from the ice or water surface of the reservoir, on the surface of sloping abutments. At making up the working program it should be necessary to specify a list, amount and procedure of investigations taking into account the peculiarities of the structure, the expected values of abnormalities, potential interference and also to formulate the demands for the results. When CME is not used data processing of geophysical investigations assumes to use as much as possible the information containing in design-construction and execution specification as well as in structure inspection statement.

To estimate the properties of dam and foundation soils for lack of CME the following geophysical methods are used:
- to study physical properties (density, porosity and et al.) – seismoacoustical, electrometric methods, and if there are wells – radioisotopic methods;
- to study mechanical properties (static and dynamic modules of elasticity, module of general deformation, Poisson coefficient and others) – seismoacoustical methods, more seldom – electrometric methods;
- to study seepage properties (permeability coefficient and seepage velocity, specific water absorbing) – electrometric, thermometric methods, more seldom – radioisotopic and seismoacoustical methods.

Depending on diagnosis tasks, constructive peculiarities of the structure and engineering-geological conditions the procedure recommends basic and subsidiary methods of the complex.

Among the possible types of geophysical investigations of the "dam - foundation" system (operating continuous, operating discrete and one-time observations) for the aim of diagnosis and monitoring it is recommended to use operating discrete and one-time observations that are much more cheaper than the operating continuous one and at the same time permit to find out tendencies and a character of state changing and properties of the system under study.

3 GEOPHYSYCAL MODELS OF EMBANKMENT DAMS

Geophysical modeling is a constituent of engineering-geological modeling. Under the geophysical model of the foundation or embankment structure it is understood the three-dimensional characteristic of distribution of physical properties (elastic, electrical, density properties and et al.) the values of which are determined by geophysical methods.

Engineering-geophysical modeling is necessary to solve diagnosis tasks of embankment dams and their foundations. According to the diagnosis stages it is useful to make up three types of models on every object:
- a priori model made up before on-site geophysical investigations of the object; it precedes the stage of direct monitoring and diagnosis of the dam state together with the foundation; making up a priori model allows to project and to prove the types, amount and the procedure of geophysical investigations;
- main model (main models) reflecting the state and properties of the "dam-foundation" system during all stages of geophysical investigations of the object on-site;
- a prediction model characterizing the state and physical properties of the structure and foundation in definite time intervals of maintenance period and based on the analysis of detected regularities of on-site geophysical study of the specific object, in particular the regularities of characteristic changing in time.

Besides the independent value connected with the necessity to develop rational complex of investigations on the corresponding stage the geophysical models can be transformed into the models according to the property types (deformative, strengthening, seepage, density and et al.) on the base of correlated or functional connections of geophysical characteristics with the corresponding property indices.

In practice the geophysical model of the embankment dam and the foundation represents a serious of geological sections, horizontal slices on different marks, specific maps, diagrams, tables and graphs. The most informative parts of the geophysical model of the embankment dam are the sections (lateral and longitudinal).

Inasmuch as the most informative methods at embankment dams diagnosis are the methods of electrometry, seismic prospecting and thermometry, there have to be developed geoelectrical (according to the values of specific electrical resistance ρ, polarizability η, dielectric constant ε), seismogeological (according to the values of the velocity of longitudinal V_p and lateral V_s waves) and geothermic (according to the temperature $t°$) models of the "embankment dam-foundation" system.

A priori models making up of which precedes the stage of on-site geophysical diagnosis are the base to develop the diagnosis project. The project includes estimation of possible differentiation of the section under study by geophysical characteristics, substantiation of particular investigation methods, a net of prospecting lines, observation systems and procedures, their periodicity, work amount and cost and et al.

The procedure of making up an a priori geophysical model of an embankment dam together with a foundation assumes the presence of the following information:
- about the type and project version of the dam, constructive elements, materials and soils of the dam body and antifiltration devices, about the state of elements (aerational, water-saturated, freezed);
- about geostructural diagram (model) of the foundation at the depth about 2H where H – the dam water height ;
- about values of geophysical characteristics (mean values and mean-square deviation or dispersion) of elements (soil and non-soil) of the dam and foundation obtained earlier according to the data of investigations of analogous embankment dams and also analogous rock and soil foundations.

On the base of generalization of on-site geophysical investigations data we made up a large table of mean values and mean-square deviations of characteristics of elastic (V_p, V_s) and electrical (ρ, η, ε) properties of soil and non-soil elements of dams and foundations being in different states: aerational, water-saturated, freezed ice-saturated. The table is to be published in the near future.

The procedure of models making up is illustrated by the example of the embankment dam of Belomor-Baltic Canal (Fig. 1 - 3).

4 ENGINEERING-GEOLOGICAL MODELS

The results of geophysical modeling along with the data of laboratory and field investigations of composition and properties of soils and underground water and along with the results of on-site observation and structure inspection are intended for making up engineering-geological and design models of the structure. The engineering-geological models are made up for different time periods and are revised according to the results of observation for structure maintenance. On such models it is obligatory to indicate the elements typical or abnormal by the composition, state, soil properties or by processes manifestation (seepage water outflows, suffusion carry-over of materials, landslides, rebounds, subsidence craters and so on). It is known that dangerous degree of such elements is different depending on their position in the body of the embankment dam or its foundation.

Using the models obtained in such a way there are carried out the calculated investigations that allows to find out the most dangerous sections (areas) in the body of the structure and foundation, i.e. the areas that are important from the point of view of the structure - foundation reliable operation. Then within the limit of such elements there are fulfilled engineering-geological investigations – drilling, boring pits, penetration tests , research-filtering work, laboratory soil tests, and sometimes when it is needed there are carried out research-on-site investigations of strength, deformability and water-permeability of these elements.

In the result of this work the abnormal areas found out by geophysical methods are provided with the detailed characteristics of soils composition, state, structure and properties. In some cases the abnormal area is not verified by direct methods of engineering-geological investigations or its boundaries are significantly being changed in comparison with the initial engineering-geological model.

After analysis of the obtained data the engineering-geological model is revised – the engineering-geological and design elements of the model are finally defined and there are specified soil property design characteristics within their limits. Let's note that according to Russian standards an engineering-geological element is some amount of soil of the same origin and type within which the values of soil properties are quasi-uniform and are changed irregularly, or the observing regularity can be practically

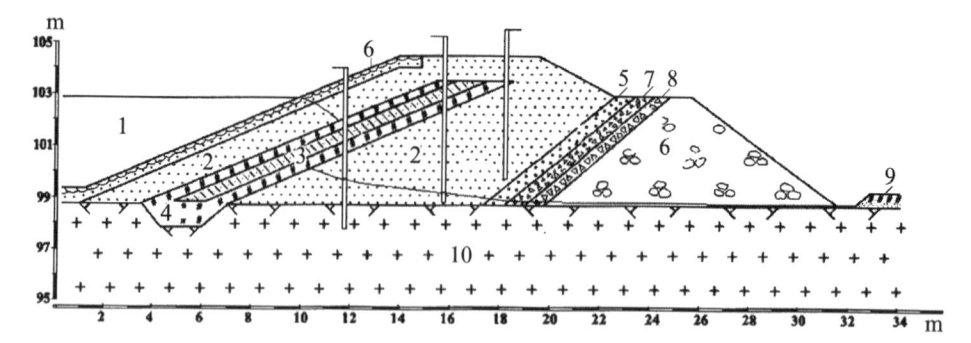

Figure 1. Geological scheme of cross-section of the embankment dam of Belomor-Baltic Canal and its foundation
1 – water, 2 – sand, 3 – clay sand, 4 – turf, 5 – coarse sand, 6 – stone, 7 – chipping, 8 – coarse crushed stone, 9 – turf, turfy sand, 10 – granite

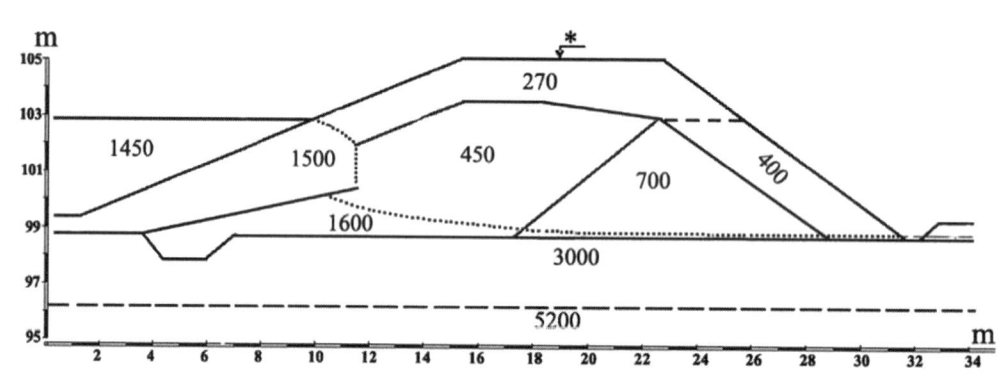

Figure 2. A priori seismogeological scheme of cross-section of the embankment dam of Belomor-Baltic Canal and its foundation; values of velocity V_p m/s are shown.

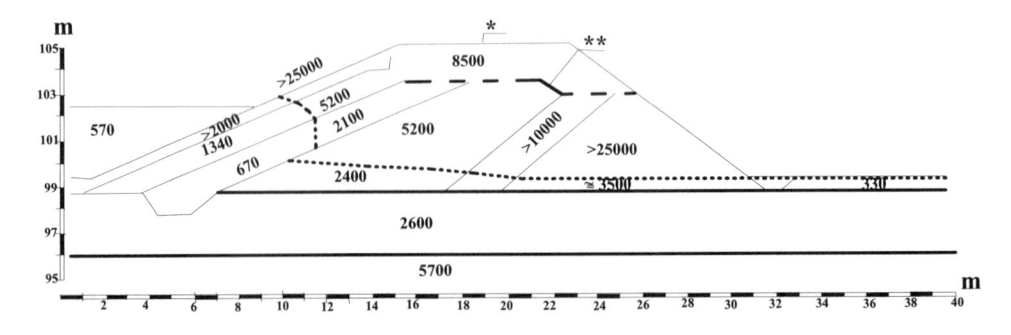

Figure 3. Geoelectrical scheme of cross-section of the embankment dam of Belomor-Baltic Canal and its foundation according to electrometry data; specific electrical impedance is shown (o/m), *-crest level after reconstruction of dam, ** - crest level before reconstruction of dam.

neglected. The design element is some amount of soil, not obligatory of the same origin and type, within which the standard and design values of characteristics, according to the peculiarities of using method for structure calculation, can be constant or changed by the definite law.

Engineering-geological models are subdivided into general (complex) and particular (specialized) ones.

General models characterize engineering-geological conditions of the object constructing and maintenance as a whole and are built using a complex of features. Particular models characterize specific performances of soil conditions and are built using one or limited set of features.

Specialized engineering-geological models are made up to calculate dam stability, settlements, seepage strength, water leakage and et al.

Accordingly to these tasks the complex engineering-geological model, soil property models – geomechanical, thermophysical, seepage ones that are built for the "foundation - structure" system or its elements (foundations, abutments and so on) are related to the main engineering-geological models.

Auxiliary models are used to make up main models. Geologo-structural, lithologo-genetic, geophysical models, soil properties models, models of geo-dynamic processes are related to the auxiliary models. In some cases the models of properties and processes are the main ones.

Geological structural and lithologo-genetic models are the base for development a complex engineering-geological model of foundation, but for an embankment structure – lithological models taking into account constructive peculiarities of a structure (upstream and downstream fills, cores and so on).

Geological-structural models are composed by the results of data analysis and generalization of engineering-geological investigations and study. At that there are indicated the elements differing by age, genesis, composition, structural-textural peculiarities, layering, erosion, and for rocks – by blocks, rock fracturing, natural stress state. Special consideration is given to the areas with increased rock fracturing, tectonic breaking, interlayers and insertions of dissoluble and karst soils.

At lithologo-genetic modeling of the foundation there are isolated the elements homogeneous by age, genesis and composition of soils. "Weak layers" are displayed in detail, i.e. such elements of the soil mass within which the composition, state and connected with them the soil properties are least favorable from the point of view of "dam-foundation" system work.

For embankment structures there are pointed out the elements, which are homogeneous or quasi-homogeneous by soils composition and properties depending on constructive peculiarities of the dam (screen, near-crest area, upstream and downstream fills and so on), and the technologies of their constructing (filling into water, dry, with soil compaction and so on).

The models of soils properties demonstrate three-dimensional distribution of foundation and structure elements according to their mechanical, seepage and thermo-physical properties. In this case the homogeneity is understood in statistical sense. According to Russian standards the element is considered to be homogeneous if the variation coefficient of the soil physical properties is not more than 15% and mechanical properties – 30%.

On preliminary stages of work the elements of the engineering-geological model of soils properties are pointed out by a composition and physical properties of soils. On later stages there are used the characteristics of mechanical, seepage and thermo-physical properties.

The models of geo-dynamic processes are also made up for the structure on the whole, its separate parts (near-crest area of the dam, abutments and so on) and shows types distribution, character and the degree of processes manifestation, intensity of their development, belonging to definite dam areas, to foundation, soils types according to composition, state and properties and so on. They can be composed also for several processes and phenomena.

Engineering-geological models can be of two types.

At first according to the results of engineering-geological investigations there is made up a diagnostic engineering-geological model illustrating engineering-geological conditions before dam constructing.

Taking into account changes supposing in the body and foundation of the structure at operation, there are made up prediction engineering-geological models showing distribution of elements within which the composition, state and soils properties change unfavorable for structure work. During structure operation there is carried out their monitoring the results of which are taken into account and for definite time moments there are made up engineering-geological models marking changes in the composition, state, in the soil structure and properties, in the boundary position of different elements, in the rate and character of geo-dynamic processes manifestation and so on.

Engineering-geological models are presented as cross-sections on typical and abnormal areas of the structure and foundation, maps-sections on different marks of the structure and foundation or three-dimensional models making up with the use of PC accompanied with tables, graphs and diagrams. These models can be disposable or constantly using for the investigated structure, can be periodically revised according to obtaining new data of engineering-geological monitoring.

5 MATHEMATICAL MODELS. PARAMETERS IDENTIFICATION

Mathematical models of the observed phenomena are used to predict the modes of embankment dams by data of on-site observations and geophysical investigations. The more comprehensive the selected model considers the processes occurring in the body of the dam and its foundation the more true and reliable results of predictions it will give and the deeper causes of the observed tendencies could be found out with the help of it.

Mathematical modeling of embankment dams and foundations includes solving the following problems:
- modeling of a seepage mode of embankment structures and foundations;
- modeling of temperature conditions of foundations and embankment structures, especially for ones situated in the north climate area;
- modeling of a stress-strain state of the "structure-foundation" system.

Quick development of computer facilities and numerical methods during last decades allows at present time to put and solve quite complicated tasks of statics and dynamics of embankment structures, hydromechanics, processes of heat and mass transferring et al. which are the base of mathematical modeling of the "embankment dam-foundation" system modes and states.

Validity of the receiving results is significantly determined by initial information of design model parameters. That is why parameters identification of the model is one of the important stages of mathematical modeling.

The aim of parameters identification of mathematical models describing states and modes of the "dam-foundation" system is such a kind of its parameters correction after which the modeling results correspond to the data of on-site observations.

For effective parameters identification it is needed to use data of on-site observation as much as possible. The validity of results of state estimation and prediction of embankment dam behavior directly depend on the amount and quality of information about the system.

To identify parameters of mathematical models there are chosen their initial values, then there is carried out systematic search of optimal values of model basic parameters. Optimization method includes choosing of special-purpose function, analysis of model sensitivity and search of optimal step in the direction of antigradient.

6 EXAMPLE OF PREDICTION MODEL CREATION

As an example of a prediction model creation it is considered the dam foundation model of complex of protection structures against flooding in Saint-Petersburg.

On the first stage to obtain more specific characteristics of soils there was modeled consolidation of the foundation on the dam site D-2 near the stake 22. Here in 1986 there was created a research field equipped with a complex of control-measuring equipment. On-site measurements on this site carried out by Saint-Petersburg State Architecture-Construction and Technical Universities, Bugrov et al. (1997), created a good base for development and identification of the model.

The design area for consolidation modeling of the foundation was the cross-section of the dam (Fig. 4a). The calculations were carried out using the method of finite elements with the help of software complex "DISK-Geomechanics", Bellendir et al. (1996), developed in VNIIG. The finite elements mesh consisted of 14985 triangular elements.

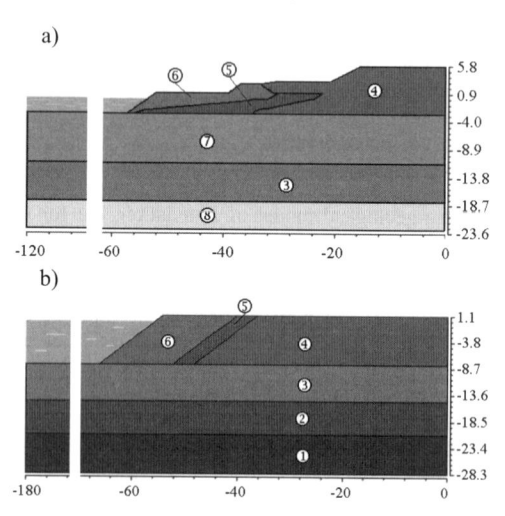

Figure 4. Design schemes: a) research ground, b) dangerous zone of dam;
1 – loam, 2 - plastic clay, 3 - flow loam, 4 – fine sand, 5 – sand-gravel soil, 6 - rockfill, 7 - flow clay, 8 – tough loam

Deformation modules and coefficients of horizontal seepage of the 3-d and 7-th layers were chosen as main parameters to identify the model. The values of the other parameters were taken according to the data of engineering-geological investigations. Identification was carried out in two stages for two groups of parameters determining stabilized and non-stabilized states that permits to divide the general extreme task into two tasks of smaller dimension.

The results of parameter identification are shown in Table 1.

Optimal values of varying parameters considerably differ from their initial values shown in brackets in Table 1 that is explained by considerable inhomogeneity of clayey soils of the dam foundation both by composition and by state (water content, consistency). Ground property parameters given in Table 1 were studied in laboratory conditions. It should be noted that during identification we could considerably improve the design model in the sense of corresponding the calculated results to the data of on-site investigations.

Table 1. Optimal values of design parameters

NN of layer	Deformation module (MPa)	Poisson's ratio	Permeability coefficient (m/day) horizontal	vertical
3	8.1 (2.7)	0.334	0.03 (0.00012)	0.00012
7	3.3 (2.0)	0.334	0.008 (0.000027)	0.000027

On the second stage of modeling the corrected parameters of the layer 7 were used to evaluate the state of the dam D-3 near the picket 62. Here in May, 1987 there had happened landslide of dam slopes in the direction of Neva bay.

The design area of the model is shown in Fig. 4b. At choosing values of the layer thickness there were taken into account the data of geophysical investigations fulfilled by the specialists of B.E.Vedeneev VNIIG.

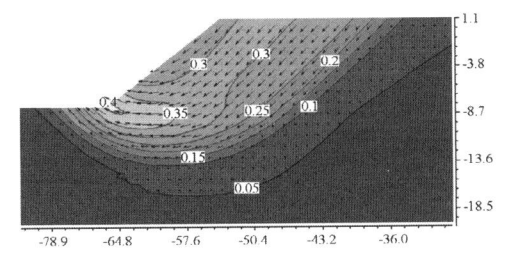

Figure 5. State of dam D-3 slope 10 days later after construction.

Finite elements mesh with thickening in landslide area included 12367 nodes and 24084 triangular elements. Consolidation processes and plastic deformation accumulation were considered in calculation. Constructing of primary embankment was suggested to be done promptly.

Having been obtained in the result of calculations, the areas of the limit state correspond to the most vulnerable area of the construction. In the design section such areas were situated near the slopes and had the width up to 15 meters by the dam crest. The view of the received fields of calculated displacements indicates the stability deficiency and demonstrates the position of the potential sliding surface (Fig. 5).

To estimate the dam state there were carried out calculations of slopes stability. Stability factors were determined using the computer program "STABILITY", Glagovsky et al. (1996), developed in B.E.Vedeneev VNIIG by searching the most dangerous round-cylindrical or broken-line surfaces of the sliding.

There was applied the method of stability evaluation according to the effective stresses in the body of the structure and foundation received at modeling the consolidation process for different moments of the constructing period.

The calculation results shown in Fig. 6 demonstrates that the slopes for some time after the dam had been constructed did not have sufficient reserve of stability. Structure stability factors begin to exceed their criterion values for the constructing period only after four months from the moment the dam had been constructed. Stability criteria for the period of operation start to be fulfilled only after seven months the dam had been constructed.

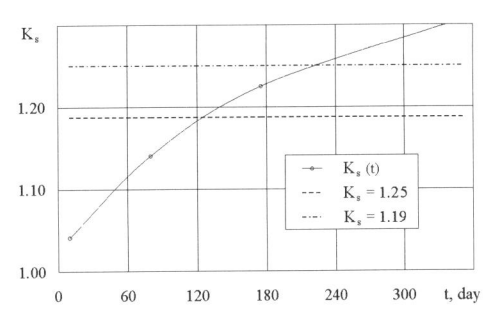

Figure 6. Stability factor depending on period after construction.

REFERENCES

Bugrov, A.K., Golly, A.V., Kagan, A.A., Kuraev, S.N., Pyrogov, I.A. & Shashkin, A.G. 1997. On-site investigations of stressed deformed state and consolidation of structure foundations of the flood protective complex in Saint Petersburg (in Russian). *Foundations and Soil Mechanics*, No 1: 14-20.

Bellendir, E.N., Glagovsky, V.B., Gotlif, A.A. & Prokopovich, V.S. 1996. Mathematical modeling of embankment structures and foundations (in Russian). *Proceedings of The B.E.Vedeneev VNIIG* Vol. 231: 272-286.

Glagovsky, V.B., Lipovetskaya, T.F. & Prokopovich, V.S 1996. Evolution of methods for estimating the "structure-foundation" system stability (in Russian). *Proceedings of The B.E.Vedeneev VNIIG* Vol. 231: 257-271.

Application of advanced geophysical technologies to landslides and unstable slopes

Robert J. Whiteley
Coffey Geosciences Pty. Ltd. Sydney, NSW, Australia

Keywords: geophysics, imaging, landfill, landslides, refraction, resistivity, seismic

ABSTRACT: Landslides and slope instability result from particular properties of soils, rocks and groundwater, their distribution and interaction. Quantification of these factors can be achieved by combining direct geotechnical testing and indirect geophysical investigation. Geophysical technologies assume increased importance where direct subsurface investigation by drilling is constrained by unfavourable and unsafe surface conditions. The important issues addressed are the lateral extent and depth of the affected area, the location of buried objects and density and moisture variations within and around the unstable mass. Consequently, the most successful geophysical technologies for detailed characterisation are seismic and electrical resistivity. Recently these methods have been advanced by improvements to digital acquisition equipment, tomographic imaging from boreholes and enhanced numerical analysis using personal computers. These advances allow for development of complex subsurface models that are required at landslide and unstable slope sites.
Case studies demonstrate the application of these advanced geophysical technologies to a variety of landslide and unstable slope problems. In central Thailand, seismic refraction was applied to delineate an unstable mass resulting from slump failures in siltstones at a proposed dam abutment. This method, enhanced by seismic ray tracing, was able to accurately locate the failed rock units. In Western Australia seismic refraction defined a major boundary fault in weathered granites that exerted controls over sliding failures of steep coal seams during open pit mining. In Malaysia, resistivity mapping located granite boulders and blocks in a failed slope that were posing hazards to remediation works. In Sydney, Australia seismic imaging from boreholes mapped the base of an unstable waste landfill.
Advanced geophysical technologies are powerful tools for the detailed characterisation of the complex subsurface conditions at landslides and unstable ground sites and are most effective when fully integrated with conventional site investigation methods.

1 INTRODUCTION

Once a landslide has developed or when unstable areas exhibit symptoms of past movement and incipient failure site investigations are normally undertaken to establish the factors affecting ground movement and to determine the appropriate remediation strategies for preventing or minimising future movement. It is widely recognised that landslides and slope instability result from particular properties of soils, rocks and groundwater, their distribution and interaction. Quantification of these factors can often be achieved by combined application of direct geotechnical testing and indirect geophysical investigation. Field work at landslides or on potentially unstable ground is difficult and risky. As a result geophysics is frequently considered to supplement drilling, however, there are widespread concerns that geophysical interpretations cannot cope with the complexity of subsurface conditions in these areas due to the apparent simplicity of many interpretative models.

Geophysically this complexity manifests itself as rapid variations in seismic velocity and electrical conductivity created by large changes in the elastic properties and groundwater conditions within displaced soil and rock masses that occur following substantial ground movements. These properties normally contrast strongly with the surrounding materials. As a consequence the geophysical technologies that have been most successful for detailed investigations at unstable sites are seismic and electrical resistivity (Bogoslavsky and Ogilvy, 1977). These methods have recently been advanced by improved digital acquisition equipment, the use of tomographic imaging from boreholes and enhanced numerical data processing algorithms using personal computers.

The major objective of this paper is to demonstrate, using a variety of case studies, the application of these advanced geophysical technologies to a variety of landslide and unstable sites where steep slopes are present.

2 CASE STUDIES

Table 1 lists the sites from which the case studies are taken, the nature of the instability, the major task for the geophysical work and the technologies that were used. These are discussed in the following sections.

Table 1. Geophysical case studies at unstable sites.

Site/ Location	Nature of Instability	Task	Geophysical Technology
Dam site, central Thailand	Slump failure of proposed abutment	Map unstable rock mass	Seismic refraction
Open pit coal mine, Western Australia	Sliding failures near unstable high wall	Locate major boundary fault	Seismic refraction
Road cut, Penang, Malaysia	Landslide in weathered granite	Locate buried granite boulders	Electrical resistivity
Waste landfill, Sydney, Australia	Soil movement on steep slope	Map base of fill	Borehole seismic imaging

2.1 Dam Site, central Thailand

During feasibility studies at a dam site on the Kwae Noi River in Central Thailand an extensive seismic refraction study was completed (Fell et al, 1992). This dam is currently under construction. Historic landslides had occurred on the proposed right abutment due to undercutting of the slope by the river. Landslide debris at this location produced a highly irregular ground surface strewn with large sandstone blocks. The seismic refraction study identified the shallow low velocity zone associated with this debris. This information and limited drilling led to inference of a rotational slide with a circular failure surface located close to the river level as shown in Fig. 1.
Re-interpretation of the seismic data in this region (Fig. 2) using seismic ray tracing (Whiteley,1994) improves definition of the slightly weathered rock surface and clearly showed the irregular surface of the displaced and slumped rock masses. The high Lugeon values obtained near the bottom of the inclined borehole DKN17 (Fig. 1) are located below the cusp points that represent the upper edges of slide surfaces between the displaced blocks of siltstone (Fig. 2).

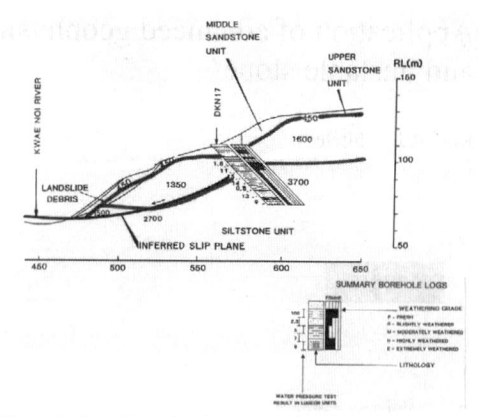

Figure 1. Seismic section from dam site (from Fell et al. ibid. Fig. 5.9).

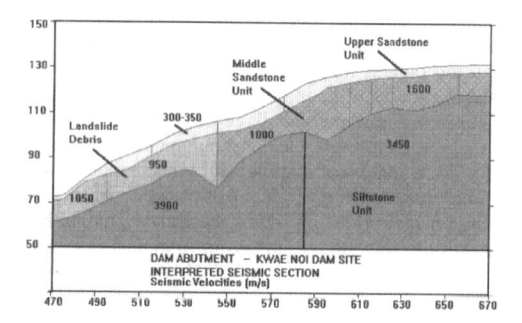

Figure 2. Re-interpreted seismic section from the Kwae Noi dam site.

2.2 Open Pit Coal Mine, Western Australia

The major coal resources of Western Australia occur in the intracratonic Collie Basin about 120 km south of Perth. A major north-south structural feature, the Muja Fault separates this basin from the Archaean basement of the Yilgarn Block, and forms the west wall of the Muja Open Cut coal mine. This area has long been recognised as being unstable (Joass, 1993) and various failures have occurred. At this location the coal sequence abuts the fault with seam dips a great as 60° due to fault drag. The coal is mined in a series of 120m wide strips advancing to the north that are backfilled with spoil as the mining proceeds.

The Fault, itself, is a normal fault dipping at an average angle of 80° into the basin. Adjacent to the fault plane is a zone of highly sheared basement rock consisting of chloritic schists and foliated gneiss. Beyond the fault zone moderately jointed gneiss is exposed. The shear zone is believed to represent an ancient fault zone along which the Muja Fault subsequently acted.

As the Muja Fault is concealed at many locations and has been found to wander, seismic refraction was chosen to map the fault zone as a lower cost alternative to detailed drilling. Early seismic refraction work over this fault (Peck and Yu, 1982) showed that substantial seismic velocity contrasts were present and indicated that the fault could be directly detected as a low velocity zone, however, later seismic work demonstrated that more reliable indicators of the fault were lateral structural and velocity variations at the freash and weathered bedrock levels.

Figure 3 shows an interpreted seismic section across the Muja Fault prior to mining with some of the seismic data. This section has been verified with computer ray-tracing and the geology from drilling.

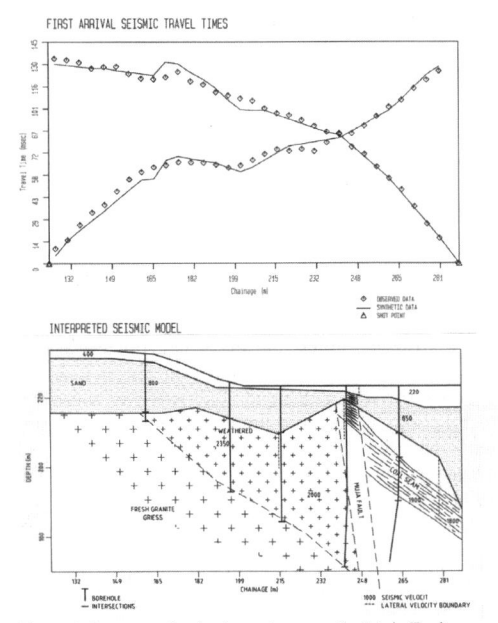

Figure 3. Interpreted seismic section over the Muja Fault

The seismic refraction work accurately mapped the weathered granitic and sedimentary bedrock across the fault. Substantial variations in bedrock levels occur across the fault with a displaced block of weathered granite correlating with moderately jointed gneiss extends from the fault zone for about 25 m (from about Ch. 215 to 240 m on Fig. 3). Movement of this block along the fresh bedrock surface is believed to represent the major source of the instability that manifests itself as sliding failures in the coal sequence followed by eventual failure of the high wall of the mine.

2.3 *Road Cut, Penang, Malaysia*

In late 1998 a steep slope of weathered granite beside a highway near Sun Moon City at Paya Terubong, Penang Malaysia failed. Large granite boulders were displaced and there were concerns that unstable buried boulders might remain in the area around the failed slope, posing a hazard during remediation works. The terrain in this region is highly irregular and natural slope of the hill prior to the landslide was between 45^0 and 65^0. As large electrical contrasts were expected between the completely weathered granite soils and the fresher less conductive granite boulders the electrical resistivity method was used to locate and determine approximate depth of any boulders.

The dipole-dipole sounding/profiling resistivity method was used along a number of profiles around the landslip with a dipole spacing of 5 m and n=1 to 5. The steep slopes made it necessary to compute measured earth resistances to apparent resistivities using actual electrode positions obtained from survey rather than the standard formula based on a flat earth model. These apparent resistivities were plotted as Edwards electrical pseudosections at the array midpoint and the "effective" or median depth (Edwards, 1977). The effective depth range using this method was between 2.1 and 7.4m. The apparent resistivity data were also contoured at each effective depth. This approach was preferred to 2D inversion (Loke and Barker, 1996) and insufficient data was obtained to permit accurate 3D inversion.

Figure 4 shows the resistivity profile locations (S1 to S9) and an apparent resistivity contour plan around the slide area at an effective depth of 2.1m. The areas of higher resistivity that have been marked are interpreted to represent the tops of shallow granite boulder concentrations.

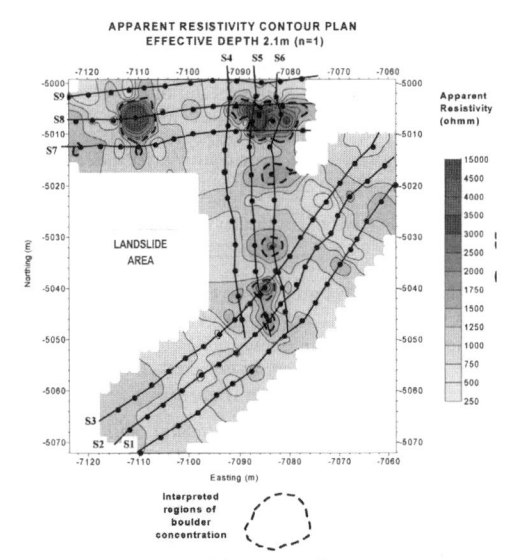

Figure 4. Apparent resistivity contour plan

An apparent resistivity pseudosection for Line S8 (Fig.4) is shown on Fig. 5 and Table 2 lists the interpretation of the anomalous resistivity highs on the pseudo-sections in terms of the presence of boulders from close to the surface to about 7m depth.

APPARENT RESISTIVITY PSEUDOSECTION LINE S8

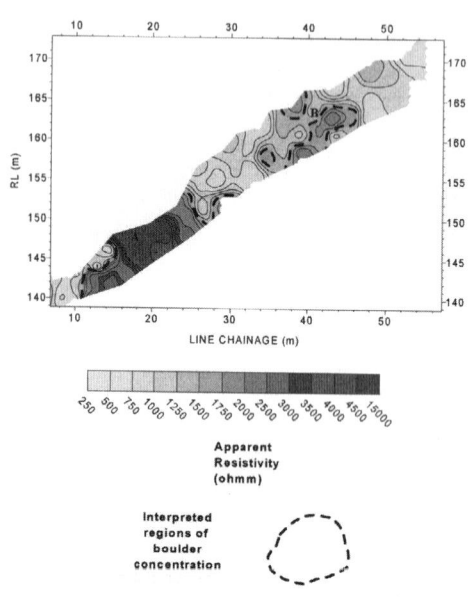

Figure 5. Apparent resistivity pseudosection with boulder areas marked.

Table 2. Interpretation of Anomalous Resistivity High Areas

	Description	Interpretation
A	Circular high enlarging with increasing depth	Large shallow boulder connected to or in close proximity with the rock mass or deeper boulders
B	Multiple complex highs decreasing with depth	Multiple boulders at shallow depth in weathered groundmass (grus)

The boulders in area B were considered the most hazardous. These were exposed and removed prior to the start of remediation works.

2.4 Waste Landfill, Sydney Australia

Uncontrolled dumping of domestic and building waste over a long period created an unstable fill site within an old sandstone quarry, adjacent to a river in south-western Sydney. The waste was believed to be at least 20 m thick. Although the site had been closed for many years and natural vegetation has returned, evidence of continuing instability was evident from visual inspection. There was concern about long stability of the waste and its potential to pollute the waterway should the slope fail. A key issue was the shape of the quarried rock surface beneath the potentially unstable fill.

Seismic imaging was completed between a pair of boreholes (BH101 and BH104) along a profile where surface subsidence had occurred near BH101. The crosshole imaging was supplemented by surface-to-borehole imaging either side of these holes with seismic refraction along the entire profile. Downhole and surface seismic sources were placed at 2 to 5 m intervals and the downhole hydrophone array had detectors at 2 m intervals. Fig. 6 shows the interpreted seismic incorporating all the seismic data and calibrated using the borehole logs.

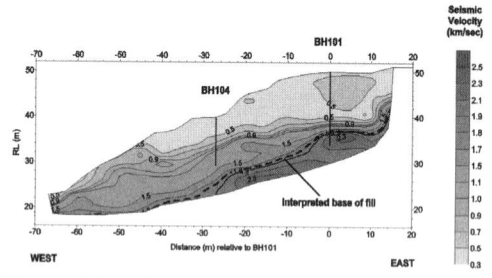

Figure 6. Seismic image of unstable waste fill site.

This image shows that the waste fill has a very low seismic velocity that rapidly increases near the bedrock surface. The rapid shallowing of these velocity contours on the right edge of the image represents the rise to the outcropping sandstone of the old quarry margin. The bedrock interface is close to the 1900 m/s (1.9 km/s) velocity contour. This contour indicates that the likely location of buried quarry benches occurs where the bedrock deepens abruptly or at distances of –5 m and –24 m from BH 101. The average bench height from the seismic image is about 4m.

The combination of limited drilling, installation of inclinometers and monitoring and this seismic imaging allowed an improved geotechnical model for the site to be derived that guided subsequent monitoring and remedial works.

3 CONCLUSIONS

The application of surface and borehole seismic and electrical technologies with the latest analysis and modelling methods provides powerful tools for the detailed characterisation of the complex subsurface conditions at landslides and unstable ground sites. These geophysical methods are able to deal with a wide variety of problems and are most effective when fully integrated with conventional methods for geotechnical site investigation.

REFERENCES

Bogoslovsky, V.A. and Ogilvy, A.A., 1977 Application of geophysical methods for the investigation of landslides. *Geophysics*, 42, 562-571.

Edwards, L.S. 1977 A modified pseudosection for resistivity and IP. Geophysics,42,1020-1036.

Fell, R, MacGregor, P and Stapledon, D. 1992 *Geotechnical Engineering of Embankment Dams*. Balkema, Rotterdam, 675p.

Loke, M.H. and Barker, R.D. 1996. Rapid least-squares inversion of apparent resistivity pseudosections by Quasi-Newton method. *Geophysical Prospecting*, 4,131-152.

Peck, W. and Yu, S.M. 1982 Seismic refraction studies for mine planning and design. *Coal Geology* 4,2,341-353.

Joass, G.G. 1993 Stability monitoring on the west wall of the Muja open cut. *Geotechnical Instrumentation and Monitoring in Open Pit and Underground Mining*, Szwedzicki (ed.), Balkema Rotterdam, 283-291.

Whiteley, R.J. 1994 Seismic refraction testing – a tutorial. in *Geophysical Characterization of Sites* ed. R.C. Woods, ISSMFE, New Delhi, 45-47

Resistivity imaging as a tool in shallow site investigation – a case study

R. Wisén & T. Dahlin
Department of Engineering Geology, Lund University, Sweden

E. Auken
HydroGeophysicsGroup, Department of Earth Sciences, University of Aarhus, Denmark

Keywords: applied, geophysics, resistivity, inversion, site investigation

ABSTRACT: Resistivity imaging in combination with drill hole information is a powerful tool in geotechnical site investigations. Traditionally the resistivity data is inverted with a 2D smooth inversion algorithm, resulting in a minimum structure model. In the geological interpretation it can be difficult to determine sharp layer boundaries in these models. An additional parameterized inversion of resistivity data, using few layer models can make the geological interpretation of the geophysical models easier and better.
The data presented was collected as part of a site investigation for a 2 km long and about 10 m deep railway trench. The geology in the area is sedimentary with Quaternary deposits overlaying limestone. The main issue at this stage of the project was to understand the hydrogeology in the area. The reason for this is to avoid inflow of water to the trench during and after construction. The main aim of this survey was mapping of the thickness and internal structure of the Quaternary deposits. An extensive dataset with geological and geotechnical data from auger- and core-drilling has been available for the geophysical processing and interpretation.
We recommend that the traditional 2D smooth inversion of resistivity data is combined with a parameterized layered inversion. In this case the Laterally Constrained Inversion from Aarhus University was used. The possibility to add discrete prior information in the LCI can limit the problem with ambiguity in geophysical inverse modeling. The final model based on both drilling and resistivity data is much enhanced compared to what is achieved using only one of the datasets alone.

1 INTRODUCTION

The adaptation of geophysical methods for engineering purposes has been one of the main contributions to the development of geotechnical site investigations in recent years (Morgenstern 2000). However, it is important to have a good conceptual geological model to function as a framework in which other types of data are placed (Sharp et al., 2000). Most investigations for infrastructural constructions require that the conceptual models are at least two-dimensional. Traditional geotechnical investigations often consist of sounding or sampling. In many cases it is difficult to create continuous models from this discrete data and therefore the combination of traditional geotechnical soundings and geophysical methods providing continuous models can be a powerful tool.

We will present a case where resistivity imaging has been used in combination with drilling information. Traditionally the resistivity data is inverted with a 2D smooth inversion algorithm, resulting in a minimum structure model. In the geological interpretation it can be difficult to determine sharp layer boundaries in these models. An additional parameterized inversion has therefore been utilized. The inversion is performed on few layer models. This automatically results in sharp layer boundaries. In the parameterized inversion it is also possible to use e.g. depth to layer boundaries from geotechnical investigations as prior information. This can limit the problem with ambiguity in geophysical inverse modeling.

A case from a site investigation for a railway trench in southern Sweden is presented. An extensive geotechnical dataset was available and used as prior information in the inversion of the resistivity data.

2 RESISTIVITY IMAGING

DC resistivity has developed tremendously during the last two decades and been used with good results.

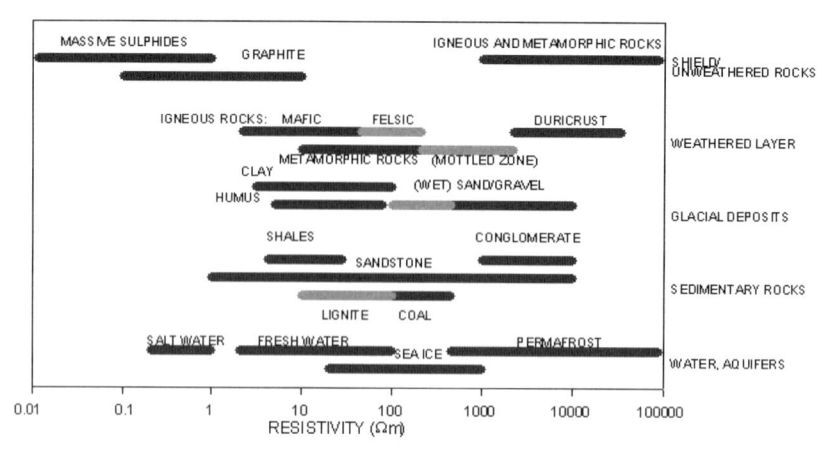

Figure 1 Typical resistivities in natural soil and rock materials (modified from Palacky 1987).

Continuous profiling to obtain a 2D image of the subsurface resistivity variation is a well-documented method (e.g. Griffiths and Turnbull, 1985; Overmeeren and Ritsema, 1988) based on the fact that different geological materials have different resistivities (Figure 1).

Resistivity measurements have many different applications in both environmental and engineering investigations (e.g. Dahlin 1996). Resistivity mapping is a common method for mapping of groundwater aquifers and their vulnerability (e.g. Sørensen et al., 2003; McGrath et al. 2002), delineation of landfill structures and leakage (e.g. Bernstone et al. 2000), site investigation for construction (e.g. Dahlin et al. 1999) and geological hazard assessment (e.g. Hack 2000; Suzuki et al. 2000).

2.1 Resistivity imaging systems

Resistivity data was collected as Continuous Vertical Electrical Soundings (CVES) data with a multi-electrode system that can perform roll-along measurements, i.e. continuous profiles with theoretically infinite length (Figure 2). The system used was a modified version of the ABEM Lund Imaging System (Dahlin et al. 2002).

Using this CVES system allows for a very flexible data collection. Based on the target and problem, any configuration and data density can be programmed and, if desirable, altered during measuring. The measurements in Sweden were performed with a combination of Wenner and Schlumberger configurations. In both cases the minimum spacing between current electrodes was 6 m and the maximum was 144 m. Today the system is further developed, now featuring more channels and IP measuring.

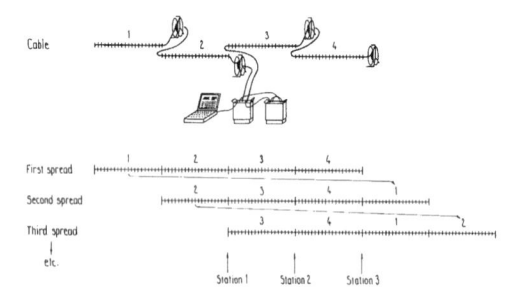

Figure 2 The principle of roll-along measuring with the system used for CVES measurements (modified from Overmeeren and Ritsema 1988).

2.2 Data processing – inversion algorithms

CVES data have generally been processed using 2D smooth inversion (e.g. Oldenburg and Li, 1994; Loke and Barker, 1996). A severe limitation of 2D smooth inversion is its inherited inability to determine sharp layer interfaces. This is to some extent improved by using the so-called robust inversion (Loke et al., 2003). For the 2D smooth inversion the commercially available software RES2DINV (Loke and Dahlin, 2002) was used in the robust inversion setting.

The LCI (Laterally Constrained Inversion) (Auken et al., 2003) was originally developed for inverting data from PACES (Pulled Array Continuous Electrical Sounding) system (Sørensen, 1996). The data quantities from the PACES system are extremely large, and 2D smooth inversion is therefore not practically possible on a routine basis. Because the PACES system is used in the Danish sedimentary geological environments with relatively smooth lateral resistivity variations, a layered inversion is desirable.

The LCI performs a parameterized inversion of many separate 1D models and data sets tying neighboring models together with lateral constraints on the model parameters. This is illustrated in Figure 3 b). The CVES data set in Figure 3 a) is divided into soundings and models. All models and corresponding data sets are inverted simultaneously, minimizing a common object function. The lateral constraints, data and constraints from prior information are all part of the inversion. Due to the lateral constraints, information from one model will spread to the neighbouring models. If the model parameters of a specific model are better resolved, due to e.g. prior information, this also spreads to neighbouring models. The result is a pseudo 2D lateral smooth model section with sharp layer interfaces. Furthermore, it is possible to add prior information by constraining model parameters, e.g. depth-to-layer interfaces based on lithology from drilling data.

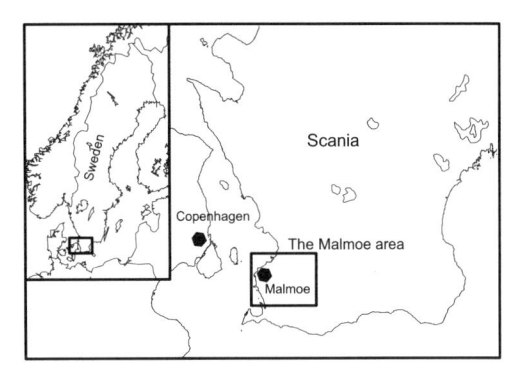

Figure 5 Map over Southern Scandinavia. The area in the box around Malmoe is enlarged in .

3 CASE STUDY

In year 1997 a parliamentary decision was made in Sweden to build the City Tunnel. With the increased person traffic from the Øresund fixed link[1] there began to be too much traffic for the existing railway system.

In the future all personal traffic going through Malmoe will use this connection, leaving the old tracks around Malmoe for freight trains.

The project is named The City Tunnel Project and construction is expected to begin during 2004 and to be completed in 2009. Totally, the project consists of 17 kilometers of railway, an expansion at Malmoe central railway station and the construction of two new stations. The connection between Malmoe central station and the Øresund fixed link is about 11 kilometers of double track railway, of which 6 kilometers will be built as double parallel tunnels under Malmoe city.

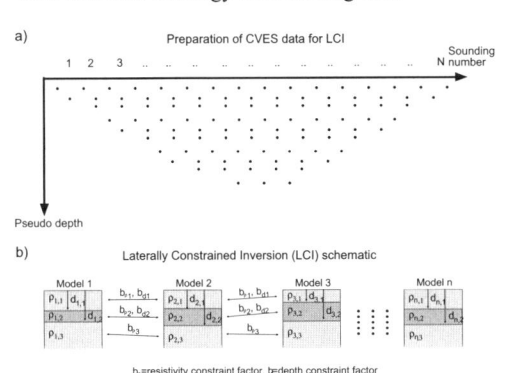

Figure 3 a) The CVES profile is divided into n data sets. b) The data sets are inverted simultaneously with a 1D model resulting in a pseudo-2D image. The models are constrained laterally, and each model allows prior constraints on the resistivities, thicknesses and depths.

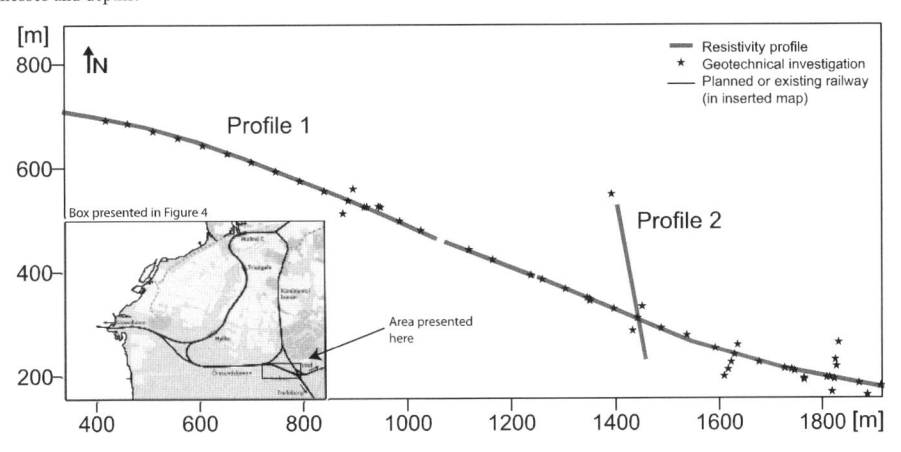

Figure 4 Map of Lockarp and the investigations.

The remaining 5 kilometers will be single track railway connecting the railway system around Malmoe with tracks heading east and southeast from Malmoe. The total expected cost of The City Tunnel Project is about 970 million €[2].

The case presented here describes part of the site investigation for a connection between the City tunnel and the main Swedish railway system. At this specific location (see Figure 5), a railway trench of about 2 km length and 10 m depth will be excavated.

3.1 Geotechnical investigations

The reference data in this area consists of almost 50 auger drillings and a few core and hammer drillings (see Figure 4 for positions). The results from the drilling consist of material classifications, levels for borders between different soil materials and level for the limestone. Most of the auger drilling and all core and hammer drilling were performed prior to the resistivity measurements. A handful of new auger drillings were made after the resistivity measurements as a direct consequence of these results.

3.2 Geology and hydrogeology in the area

The geological environment in the region is sedimentary and consists of Quaternary deposits underlain by Danian limestone; a generalization of the geology in Lockarp is shown in Figure 6. During different periods of the last glaciations, several ice fronts moving in different directions have influenced the geological environment in the area. This has resulted in a high geological variability.

Based on the geotechnical investigations a geological and hydrogeological model over the Lockarp area was created. Here the model has been extended with the expected physical property of resistivity. The model is divided into five units. From the surface and downwards they are:

Unit 1. Post- or late-glacial sediments, consisting of mainly sand and silt. Since this layer is situated above the primary groundwater surface, these sediments can be dry, but they can also be semi-dry, due to secondary aquifers, and/or mixed with the underlying clay till or organic material. The resistivity of this layer can vary significantly from about 100 Ωm to almost 1000 Ωm. The thickness is 0.5 – 2 m.

Unit 2. Clay till, alternating with thin sand and silt layers. The resistivity in the clay till is typically between 20 and 100 Ωm. The thickness is 2 - 5 m.

Unit 3. Sediments deposited between two clay tills, referred to as the inter-morainic sediments. This unit is found only in parts of the area. The inter-morainic sediments have been deposited on top of the lower clay till, Unit 4, and consist of mainly sand and silt layers. Sometimes the sediments can have layers of gravel or clay. The inter-morainic sediments are situated below the primary groundwater surface. The resistivity varies between 50 and 400 Ωm. The thickness can vary rapidly from 0 to 3 m.

Unit 4. Clay till containing silt and often sand. The resistivity of this clay till is slightly lower than the resistivity of the clay till in Unit 2, between 20 and 75 Ωm. The thickness ranges from 2 to 10 m.

Unit 5. Danian limestone. The limestone in the area undulates slightly and rises about 10 m from the western to the eastern part of the field area. The top of the limestone is often crushed and mixed with the lower clay till, unit 4. Sandy or very coarse local tills can be found directly on top of the limestone. The crushed and fissured upper part of the limestone is the primary aquifer in the area. It has a high hydraulic conductivity. The level of the groundwater pressure in the limestone can be found a few meters below the ground level. The resistivity of the upper part of the limestone and the coarse local tills vary between 100 Ωm – 600 Ωm.

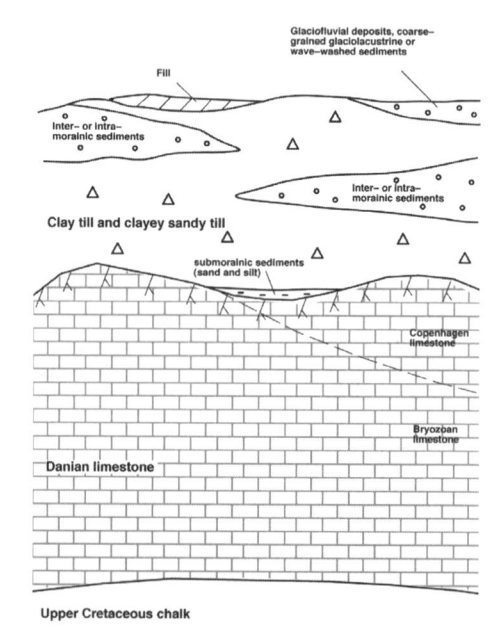

Figure 6 Generalized geological profile from the Lockarp area. Five units are used to describe the geology: Unit 1 – 4 consist of Quaternary deposits and unit 5 of Danian limestone.

---

[1] A bridge connecting Sweden with Denmark and the European continent. The bridge was opened in year 2000.
[2] At the time of writing.

It can be concluded that the limestone has high hydraulic permeability and makes up the main aquifer. The Quaternary deposits are generally thick enough to avoid contact between the trench and the main aquifer but it is not known if this is true for the entire distance. The Quaternary deposits are known to vary significantly in the area. The intermorainic sediments have high hydraulic permeability and makes up the secondary aquifer in the area.

3.3 Aim of investigation

From the geotechnical investigations performed prior to the geophysical investigations it is known that the planned bottom of the trench coincides with the level of intermorainic sediments along parts of the trench, unit 3, and in the eastern part of the area with the level of the limestone. The unit 3 sediments are found below the groundwater level, and possibly also below the groundwater pressure level in the limestone. The possibly large hydraulic conductivity in the limestone makes the groundwater situation, depth to limestone and soil composition and layering important questions for the environmental review, design and construction. Due to high variations in the layering and composition of the Quaternary deposits a continuous method of investigation was needed to identify areas with geological risks. Since the resistivity contrast between the different geological units is high, resistivity imaging was chosen.

3.4 Resistivity measurements

The resistivity measurements were performed in the summer of year 2000. Totally, about three km of CVES resistivity measurements were performed, see for positions. The resistivity data was originally inverted with 2D smooth inversion. It has now also been inverted with LCI. The main advantage lies in the co-interpretation of the two resistivity models. Five layers where used in the LCI and the layer boundaries are much easier to interpret in these models than in the minimum structure models from 2D smooth inversion. The lateral constraints in LCI force the model to become somewhat horizontally smeared and therefore the result from 2D smooth inversion is used for interpretation of vertical structures.

3.5 Results from resistivity imaging

In Figure 7 and in Figure 8 apparent resistivity data and inverted models from the investigated area are presented. A resistivity model with five layers that agree with the expected geological and geophysical model is identified. The high resistive bottom layer is interpreted as limestone. The thick low resistive layer on top is interpreted as the two clay-tills. The high resistive layer within this low resistive layer is interpreted as inter-morainic sediments that sometimes divide the two clay-tills. High resistive features in the top of the profile are interpreted as post- or late-glacial sediments.

The overall standard deviation of the residual error between measured data and model response is less than 1% for inversion results from Res2Dinv. For individual datasets it can be as low as 0.3%. This is very low and indicates good data. A big effort was made to have good electrode grounding and generally 100-200 mA was transmitted. The vicinity to an existing electrified railroad was not a problem due to the use of signal filters. These filters were set to minimize disturbances from 16.7 Hz that is the frequency of the Swedish railroad power lines.

The drilling data agree well with the resistivity data. Drilling data is divided into clay till, dark colour, and super- or inter-morainic sediments and limestone, light colour. All layer boundaries from the drilling data has been used as prior information in the LCI. This results in resistivity models that corresponds better to the drilling information. The models are not only altered in the positions where prior information has been added. The intermorainic sediments and depth to limestone in the resistivity models get better estimated and therefore the resistivities also get better estimated.

In Figure 8 d) it seems obvious that the resistivity data and the drilling data does not agree completely.

4 DISCUSSION

When combining the information on the lateral extension of the inter-morainic sediments from 2D smooth inversion and the layer interfaces from LCI inverted with prior information, it is clear that a better geological interpretation can be made than from any of the three data sets alone.

Due to high resistive equivalence in the inter-morainic sediments the properties of this layer can not be correctly determined. This results in that the thickness of the layer is exaggerated. The levels of all layers below are then suppressed and this gives rise to the impression of undulations in the limestone surface. When prior information on depth to layer boundaries are used in the LCI the thickness of the intermorainic sediments get better estimated and hence also the depth to lower layers.

If optimally positioned, a much smaller amount of prior information would be enough to achieve the result in Figure 7 d) (Wisén et al. 2002). The best way to do this is to perform the resistivity measurements early in the site investigation. The geotechnical investigations can then be placed based on the interpreted resistivity models. Finally the resistivity data can be re-inverted using the information from the drilling as prior information.

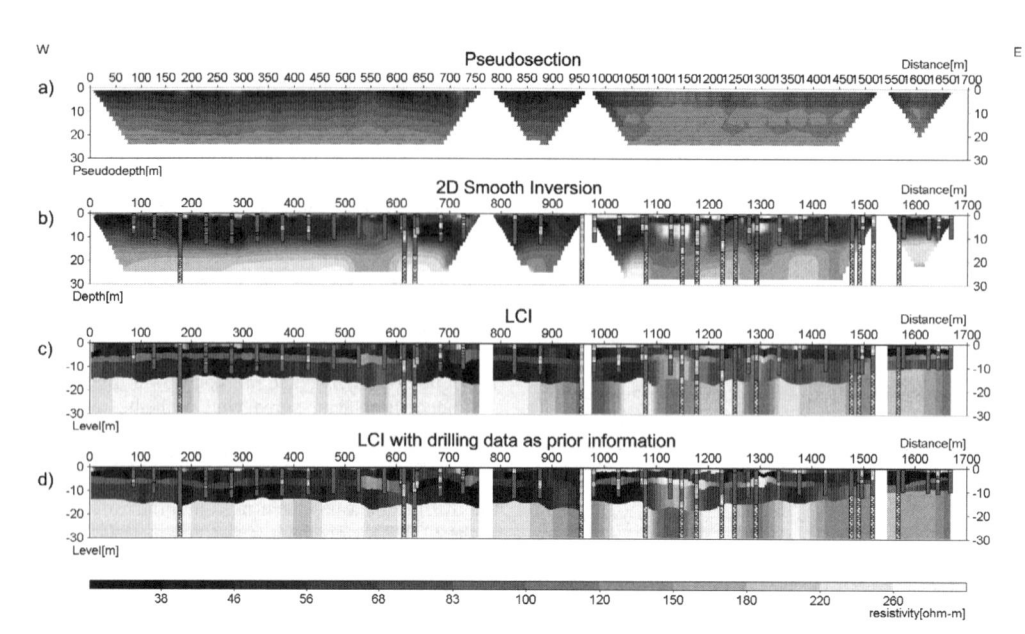

Figure 7 a) Pseudo section showing apparent resistivity for a measurement profile following the planned position of the railway trench in the eastern part of the investigated area. b) Resistivity model from inversion with Res2Dinv. c) Resistivity model from inversion with LCI. d) Resistivity model from inversion with LCI with lithological information from drillings as prior information in the inversion.

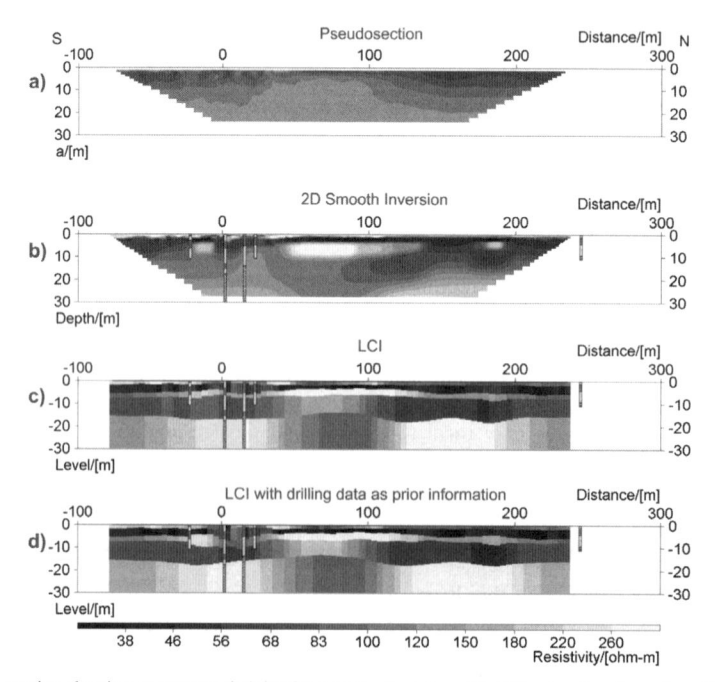

Figure 8 a) Pseudo section showing apparent resistivity for a measurement profile following the planned position of the railway trench in the eastern part of the investigated area. b) Resistivity model from inversion with Res2Dinv. c) Resistivity model from inversion with LCI. d) Resistivity model from inversion with LCI with lithological information from drillings as prior information in the inversion

The increased geological knowledge before positioning of expensive drillings will result in a more cost effective site investigation with a more reliable final geological model.

When prior information disagrees with the resistivity data, as in Figure 8 d), it is better to interpret the data sets separately, i.e. not using prior information in the LCI, but with the additional knowledge that inconsistencies exist between the drilling and resistivity data. The interpretations of resistivity and drilling data can be combined in the final geological/geotechnical model.

5 CONCLUSIONS

The results from 2D smooth inversion show a good horizontal resolution, but it does not describe the depth-to-layer interfaces well. The distinct layer interfaces from LCI make the resistivity model easier to interpret. Prior information improves the inversion result when used to solve ambiguity due to high-resistive equivalence. If the inversion where prior information is used show that resistivity data does not agree with the prior information the data sets should be interpreted separately. The information from inversion is then that there are inconsistencies between the two datasets.

It can be concluded that inversion results from the 2D smooth inversion and the LCI complement each other well. Together with geotechnical reference data, joint interpretation of resistivity models from these two methods increases the possibilities of precise geological interpretation.

ACKNOWLEDGEMENTS

We would like to thank the City Tunnel Project and Tyréns Infrakonsult AB for providing an interesting case study and high-quality geotechnical and geological data. We would also like to thank NORFA (The Nordic academy for Advanced Study) for a mobility scholarship that made the comparative study between the LCI and 2D smooth inversion possible.

REFERENCES

Auken, E., Christiansen, A.V., Jacobsen, B.H., Foged, N. and Sørensen, K.I., 2003, Piecewise 1D Laterally Constrained Inversion of resistivity data, submitted to *Geophysical Prospecting January 2003*.

Bernstone, C., Dahlin, T., Ohlsson, T. and Hogland, W. 2000, DC-resistivity mapping of internal landfill structures: two pre-excavation surveys, *Environmental Geology*, 39(3-4), p 360-371.

Dahlin, T., 1996, 2D resistivity surveying for environmental and engineering applications, *First Break*, 14(7), p 275-283.

Dahlin, T., Bjelm, L. and Svensson, C., 1999, Use of electrical imaging in site investigations for a railway tunnel through the Hallandsås Horst, Sweden, *Quarterly Journal of Engineering Geology*, 32(2), p 163-172.

Dahlin, T., Leroux, V. and Nissen, J., 2002, Measuring techniques in induced polarisation imaging, *Journal of Applied Geophysics*, 50(3), p 279-298.

Griffiths, D.H. and Turnbull, J., 1985, A multi-electrode array for resistivity surveying, *First Break*, 3(7), p 16-20.

Hack, R., 2000, Geophysics for slope stability, *Surveys in Geophysics*, 21, p 423-448.

Loke, M.H. and Barker, R.D., 1996, Rapid least-squares inversion of apparent resistivity pseudosections by quasi-Newton method, *Geophysical Prospecting*, 44, p 131-152.

Loke, M.H. and Dahlin, T., 2002, A comparison of the Gauss-Newton and quasi-Newton methods in resistivity imaging inversion, *Journal of Applied Geophysics*, 49(3), p 149-162.

Loke, M.H., Acworth, I. and Dahlin, T., 2003, A comparison of smooth and blocky inversion methods in 2D electrical imaging surveys, *Exploration Geophysics*, 34(3), p 182-187.

McGrath, R.J., Styles, P., Thomas, E. and Neale, S., 2002, Integrated high-resolution geophysical investigations as potential tools for water resource investigations in karst terrain, *Environmental Geology*, 42(5), p 552-557.

Morgenstern, N.R., 2000, Common Ground, *Proceedings of the GEOENG 2000 International Conference on Geotechnical and Geological Engineering, Melbourne, Australia, November 19-24, 2001*, ISBN: 1-58716-067-6, p 1-20.

Oldenburg, D.W. and Li, Y., 1994, Inversion of induced polarization data, *Geophysics*, 59, p 1327–1341.

Overmeeren, R.A. van and Ritsema, I.L., 1988, Continuous vertical electrical sounding, *First Break*, 6(10), p 313-324.

Palacky, G.J., 1987, Resistivity characteristics of geological targets, *Electromagnetic methods in applied geophysics*, ed. M.N. Nabighian, Society of Exploration Geophysics, Tulsa, p 53-130.

Sharp, J.C., Kaiser, P.K., Diedrichs, M.S., Martin, C.D. and Steiner, W. (2000) Underground works in hard rock tunnelling and mining, *Proceedings from GEOENG2000, Volume 1, Melbourne, Australia, November 19-24, 2001*, ISBN: 1-58716-067-6, p 841-926.

Sørensen, K.I., 1996, Pulled Array Continuous Electrical Profiling: *First break*, 14(3), 85-90.

Sørensen, K.I., Auken, E. and Jørgensen, F., 2003, Methodology for large-scale hydrogeophysical investigations, *submitted to Exploration Geophysics, March 2003*.

Suzuki, K., Toda, S., Kusunoki, K., Fujimitsu, Y., Mogi, T. and Jomori, A., 2000, Case studies of electrical and electromagnetic methods applied to mapping active faults beneath the thick quaternary, *Engineering Geology*, 56, p 29-45.

Wisén, R., Auken, E. and Dahlin, T., 2002, Comparison of 1D laterally constrained inversion and 2.5D inversion of CVES resistivity data with drilling data as apriori information, *Proceedings of 8th Meeting Environmental And Engineering Geophysics, Aveiro, Portugal, 8-12 September, 2002*, p 181-184.

Evaluation of grouting effect with seismic tomography considering grouting mechanism

Yoshikazu Yamaguchi & Hiroyuki Satoh
Public Works Research Institute, Japan

Keywords: dam, grouting, seismic tomography, permeability, cost-reduction

ABSTRACT: Geotomography is an exploration technique to image the detailed distribution of subsurface physical properties, and is positively applied to a geological investigation for dam foundations recently. If the applicability of this technique to the evaluation of grouting effect can be verified, it will become a relatively inexpensive way to evaluate the improved area by grouting remotely and on a large volume. Grouting test was conducted at the site composed of jointed quartz andesite. Seismic tomography was carried out to evaluate the grouting effect in addition to the conventional evaluation methods; e.g. permeability test and observation of drilled cores. Comprehensive analysis on test results was made, and the efficiency and applicability of seismic tomography as an evaluation method for grouting effect was also considered.

1 INTRODUCTION

Recently in Japan, development of rational grouting technology is one of the most important problems for the cost-reduction of dam construction. Geotomography is one of the newest subsurface exploration techniques, and it is frequently conducted to cost-efficiently prospect geological conditions over a wide area of ground in the civil engineering field. This technique allows one to reconstruct an image of the internal structure of the rock mass. This testing method is complex and expensive, but it is the only technique which can give a precise distribution of rock properties (Hudson, 1993). Under these circumstances, the rationalization of evaluation of grouting effects using geotomography instead of the conventional method has been studied (Hasui et al., 1992, Kawakami et al., 1994). This is not a direct evaluation based on the permeability, but an indirect evaluation based on other physical properties. Nevertheless, if the applicability of geotomography as a method of assessing grouting effects can be confirmed, it will be possible to improve and rationalize the assessment of grouting effects by appropriately combining this technology with conventional assessment methods. But previous researches (Hasui et al., 1992, Kawakami et al., 1994) suggests that the results of geotomography have been limited to considerations focused on qualitative changes in the distribution of physical properties before and after the grouting, and that quantitative analysis of geotomography results considering the grouting mechanism has not been made.

Considering the situation described above, seismic tomography was performed as a method of assessing the effectiveness of test grouting in jointed rock masses composed of dacite. Regarding these results, in addition to changes in the seismic wave velocity distribution caused by grouting as in the case of previous research, the quantitative relationship between these and changes in imperviousness was the object of an overall analysis accounting for the grouting mechanism. In addition, the applicability of seismic tomography as a grouting effects assessment method was discussed.

2 GEOLOGY OF TEST SITE

The test site consists of dacite in the Neogene period. The bedrock is overall hard and massive, and the observed results of core samples in the grouting holes at the test site described below have revealed that it is about 86% CM class and about 4% CH class. The remaining 10% is CL class or D class. The average physical properties of the rock in each rock mass class are summarized in Table 1.

Figure 1 shows the joint pattern of the bedrock in an equal-area map projected on a south hemisphere. This figure reveals that the joints in the bedrock are concentrated at N40° W75° NE and N70° E80° NS. Figure 2 schematically shows the relationship of the strike of these dominant joint groups with the plane location of the test site.

Table 1. Physical properties of rock in each rock mass class.

Rock mass class	Specific gravity in absolutely dry condition	Water absorption (%)	Effective porosity (%)	Unconfined compressive strength (N/mm^2)	Seismic wave velocity (km/s) Vp	Vs
CL	2.262	5.57	12.47	6.60	1.68	0.73
CM	2.318	4.63	10.57	15.17	2.22	0.99
CH	2.337	4.38	9.97	20.03	2.86	1.32

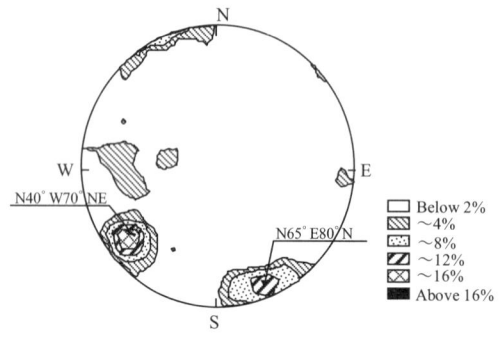

Figure 1. Joint pattern of bedrock.

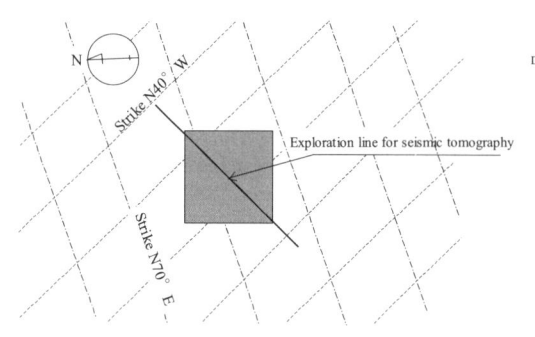

Figure 2. Schematic relationship of strike of dominant joint groups with plane location of test site.

(a) Plane arrangement

(b) Vertical arrangement in A-A' section

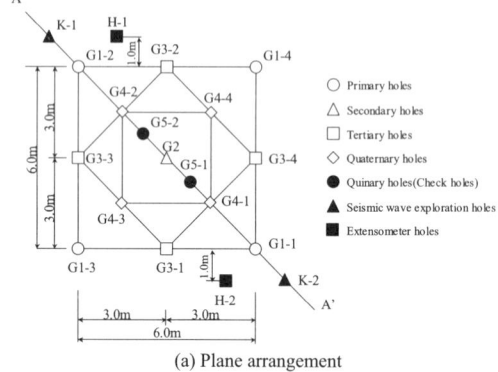

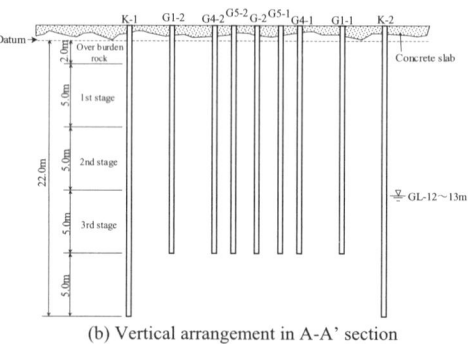

Figure 3. Grouting holes arrangement.

3 TEST GROUTING AND SEISMIC TOMOGRAPHY

3.1 Test Grouting

The test grouting was performed in accordance with the split spacing method and with the plane arrangement and the depths of the grouting holes shown in Figure 3.

The initial numerals in each hole number in the figure indicate the injection degree and the later numerals indicate the execution sequence of the holes in the same degree. On the excavated surface of the rock mass, a concrete slab at least 20 cm in thickness was placed, then slush grouting was performed on the 2 m thick overburden rock lying above the injection section. Later, the bottom surface of the concrete slab was assumed to be the datum plane GL - 0.0 m in the depth direction at the test site. The drilling was done using the rotary boring method with a diameter of 66 mm and cores were taken in order to observe the joints of the bedrock and the way that these joints were filled with cement grout. Details of the grouting specifications are shown in Table 2. To evaluate the permeability of the test site, water pressure tests were performed before each stage of the grout injection. In addition, to monitor bedrock displacement during the water pressure testing and the grouting, extensometers were installed at two locations as shown in Figure 3 (a). The depth of the anchor points of the extensometers was 5 m from the bottom edge of the injection section. The ground water level estimated from the water level inside the holes was located between GL - 12 to 13 m close to the boundary between stage 2 and stage 3.

Table 2. Test grouting specifications.

Grouting procedure	Down-stage grouting method (5m × 3stages)			
Grouting material	Normal portland cement(C) + Water(W)			
Water pressure test	stage	1st	2nd	3rd
	Maximum pressure (kPa)	294	490	980
Design injection pressure	stage	1st	2nd	3rd
	Design injection pressure (kPa)	294	490	980
Maximum injection discharge	20ℓ /min/stage (4 ℓ /min/m)			
Grout mix	Mixture (W/C)	8	6	4	2	1
	Quantity(ℓ)	400	400	400	400	1600
	Start mixture Lu < 20→ W/C=8					
	Lu≧ 20→ W/C=4 Lu : Lugeon value					
Closure criteria	When injection discharge is less than 1□/min/stage at design injection pressure, injection must be kept at the same condition in 30 minutes. If injection discharge doesn't increase, then injection will be stopped.					

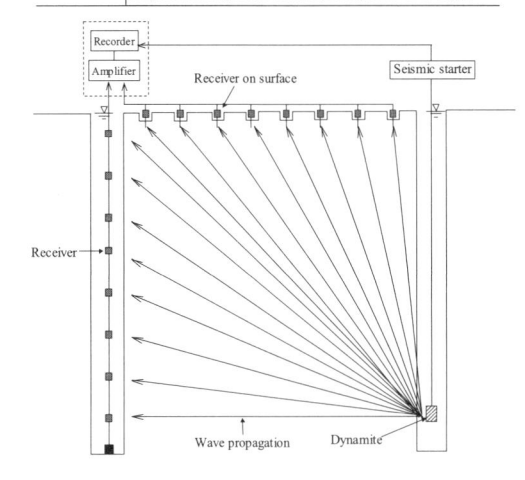

Figure 4. Schematic view of seismic tomography.

3.2 Seismic Tomography

Seismic prospecting was executed along the traverse line connecting hole K-1 and hole K-2 in Figure 3 three times: before grouting, three days after execution of the secondary holes, and after all the work was completed (ninth day after completion of the fourth degree holes). Specifically, seismic waves produced by hammers on the surface or dynamite detonations at shot points arranged inside hole K-2 at intervals of 1 m as shown in the outline in Figure 4, were received by geophones installed at intervals of 1 m on the ground surface and inside hole K-1. The length of the excavation of holes K-1 and K-2 is 22 m, and the holes were about 11 m apart for a total of 56 measurement points. The seismic wave velocity in this case is the P wave velocity.

The analysis was done by dividing the area enclosed by the measurement points into 242 unit blocks with sides of 1 m x 1 m, repeatedly performing calculations so that the theoretical travel-time of the seismic waves approached the observed travel time, and finding the final seismic wave velocity distribution model. The initial model of the velocity distribution was set using the BPT (Back Projection Technique) method and later analysis was based on the CG (Conjugate Gradient) method. The analysis did not use data if it was judged that the reading error of the arrival time of the initial motion in the data exceeded 0.5 ms. The number of data that were finally incorporated in the analysis were 686, 815, and 891 for the first, second, and third times respectively.

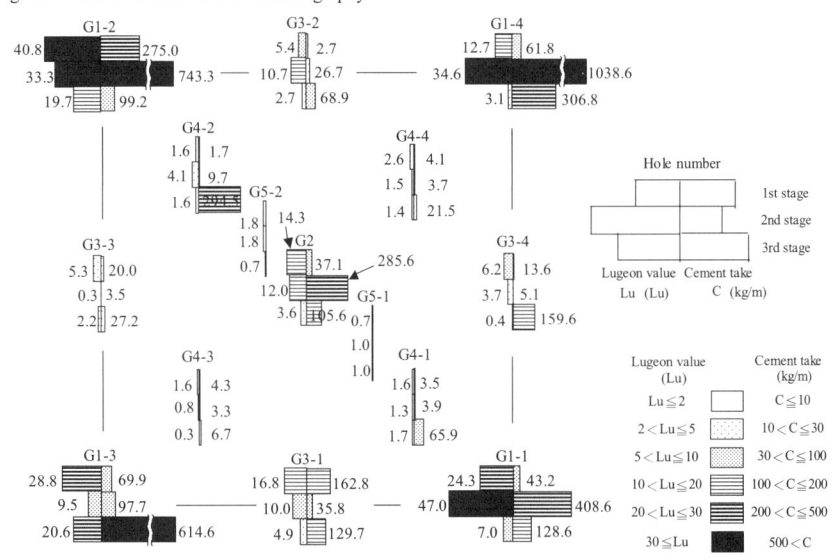

Figure 5. Summary of grouting test results.

(a) Injection without critical pressure

(b) Injection with critical pressure

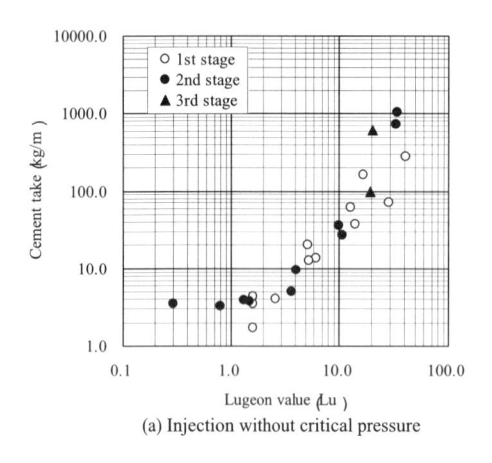

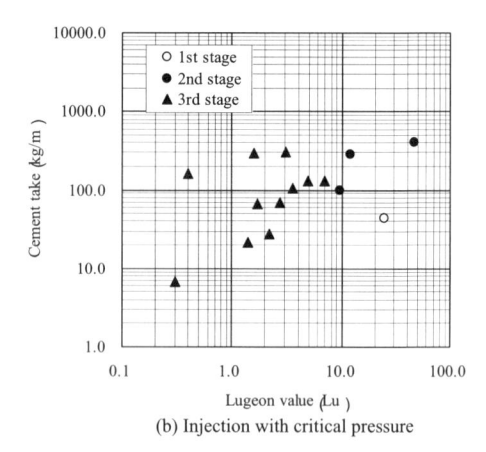

Figure 6. Relationship of Lugeon value with cement take.

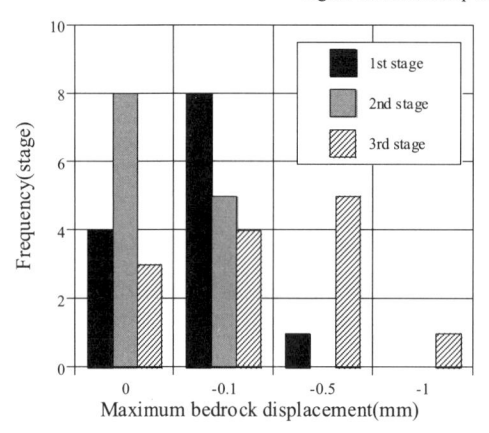

Figure 7. Frequency distribution of maximum bedrock displacement during grout injection.

4 TEST RESULTS

4.1 Test Grouting

Figure 5 shows the Lugeon values and the cement take during the test grouting. Judging from the figure, it can be concluded that the Lugeon value and the cement take decreased as the injection progressed and that the injection effectively improved the imperviousness of the bedrock. No grout leaks from the surface were observed during the injection at any stage.

Figure 6 presents an organization of the relationship of the Lugeon values with the cement take by dividing all grouting stages into those where the critical pressure appeared during water pressure testing and those where it did not. Figure 7 presents a frequency distribution found by organizing, for each stage, the larger maximum bedrock displacement of two values measured at two points during grout injection.

The critical pressure appeared during the water pressure testing in few cases at the first and second stages, but extremely often at stage 3. In cases where the critical pressure did not appear, the correlation of the Lugeon value with the cement take was high (coefficient of correlation = 0.918). The maximum bedrock displacement during injection at stage 1 and at stage 2 was smaller than 0.1 mm in almost all cases. Because the maximum injection pressure during the water pressure testing and the stipulated maximum pressure during grout injection were identical as shown in Table 2, if the critical pressure does not appear during water pressure testing, the critical pressure will also not appear during grouting. Consequently, it is assumed that during injection at stages 1 and 2, the cement grout filled the existing joints without increasing their apertures very much.

In cases where the critical pressure appeared during water pressure testing, the correlation of Lugeon value with the cement take is much lower than in cases where the critical pressure did not appear (coefficient of correlation = 0.484), and at the same time, the cement take for the same Lugeon value is greater than that in a case where the critical pressure did not appear. The maximum bedrock displacement during stage 3 injection is more than 0.1 mm in many stages. Consequently, it is assumed that during the third stage injection, as the cement grout filled the existing joints, it either expanded them slightly or formed new joints. Judging from the fact that at stages 1 and 2 adequate imperviousness was obtained by injection at an injection pressure lower than the critical pressure, the injection pressure at the third stage was slightly higher than the suitable injection pressure.

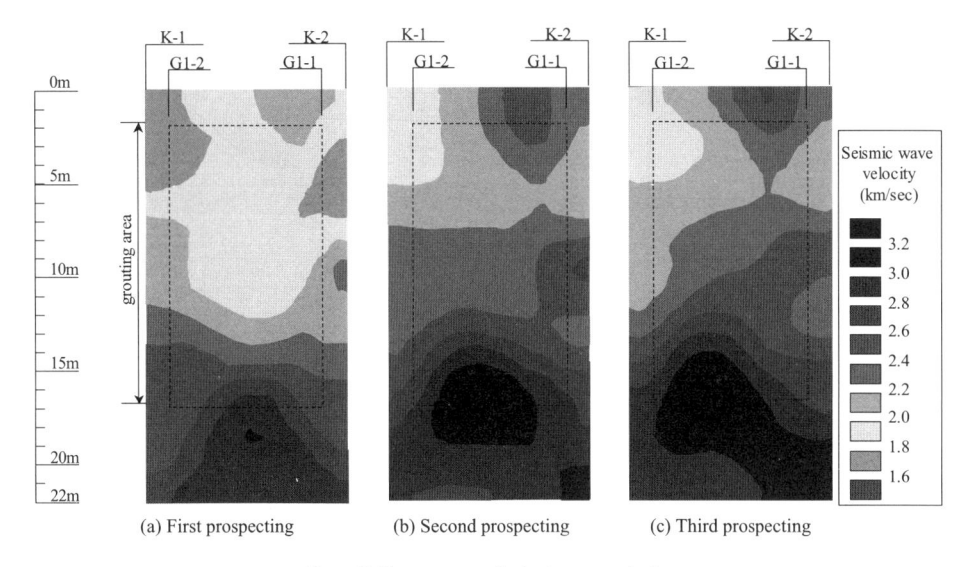

(a) First prospecting (b) Second prospecting (c) Third prospecting

Figure 8. Contour map of seismic wave velocity.

Table 3. Statistical quantities of seismic wave velocities for each grouting section.

Order of measurement	1st			2nd			3rd		
Stage	1st	2nd	3rd	1st	2nd	3rd	1st	2nd	3rd
Number of unit block	55	55	55	55	55	55	55	55	55
Mean value(km/s)	1.823	1.985	2.396	2.123	2.315	2.714	2.112	2.344	2.780
Standard deviation(km/s)	0.082	0.116	0.196	0.116	0.111	0.273	0.119	0.121	0.269
Coefficient of variation(%)	4.48	5.86	8.17	5.48	4.80	10.07	5.65	5.15	9.67
Maximum value(km)	2.03	2.26	2.87	2.45	2.64	3.29	2.39	2.55	3.31
Minimum value(km)	1.66	1.81	2.02	1.93	2.20	2.31	1.89	2.12	2.28

4.2 Seismic Tomography

Figure 8 shows the seismic wave velocity contour obtained from the analysis of the results of each seismic prospecting. Table 3 shows statistical quantities such as the average and standard differential of the seismic wave velocities for the unit blocks corresponding to the depth of each stage section for the grouting. These results reveal that the seismic wave velocity obtained from the second prospecting performed after the secondary hole execution is larger than that obtained from the first exploration performed before grouting throughout the entire exploration area. Although the seismic wave velocity detected by the third exploration performed after the grouting was completed was a little higher than that detected by the second exploration, there was little change in the overall distribution. It is believed that grouting reduces the scattering and contributes to the uniformity of the mechanical properties of bedrock (Kudo, 1963) But, according to Table 3, at this site test, the scattering of the seismic wave velocity represented by the coefficient of variation was smaller before grouting,

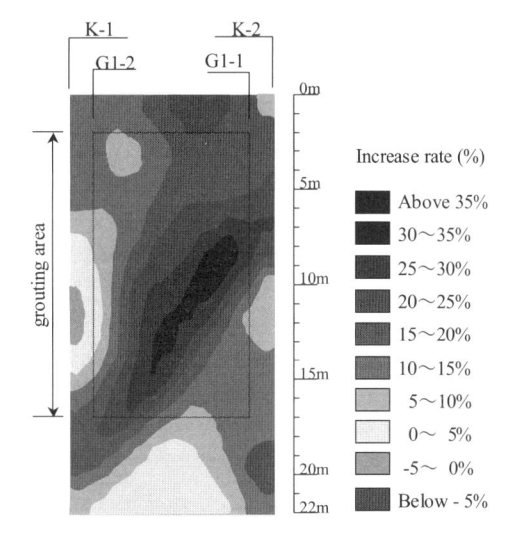

Figure 9. Contour map of increase rate in seismic wave velocity between first and third explorations.

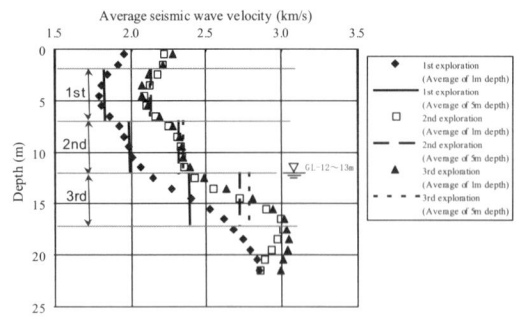

Figure 10. Distribution of seismic wave velocity in depth direction.

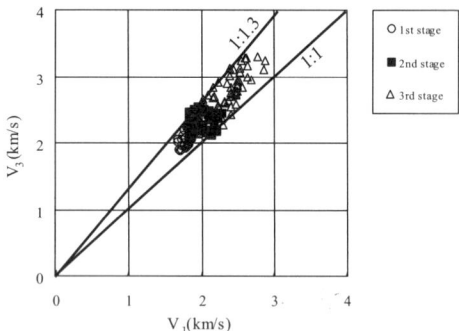

Figure 11. Relationship of seismic wave velocity from first exploration with that from third exploration.

indicating that the grouting did not tend to reduce scattering.

Figure 9 shows the contour of the rate of increase in the seismic wave velocity between the first exploration and the third exploration. This figure shows that the rate of increase tends to decline from the right to the left. Its incline generally conforms with the incline of the intersecting line of the N40° W75° NE joint group, one of the dominant joint groups formed at the site, and the seismic exploration section. Judging from this fact, it is presumed that this test grouting improved the imperviousness of the bedrock primarily by filling the existing joints with grout. The existence of one more dominant joint group, N70° E80° NS, did not influence the incline of the increase rate contour because in addition to the fact that the angle of intersection of this dominant joint group with the seismic wave exploration section is relatively small, the degree of concentration of the joints is lower below N40° W75° NE.

Figure 10 shows the depth direction distribution of the seismic wave velocity. The figure shows the average values for each 1 m that is the length of one side of the analysis unit blocks in the depth direction and the average value for each injection stage depth section with a length of 5 m. This figure reveals that regardless of the exploration period, the average seismic wave velocity increased in the depth direction. Because most of the bedrock in the grouting area is CM class regardless of its depth, as explained above, it is believed that the increase in the average seismic wave velocity in the depth direction might be a result of the effect of the confining pressure (Funato et. al 1987). But a close examination of the results of the first exploration reveals that the average seismic wave velocity of the third stage in the grouting injection section is far greater than that of the first and second stages. This is believed to be a result of the confining pressure dependency of the seismic wave velocity and the existence of the water surface near the depth boundary between stage 2 and stage 3. In addition, Figure 10 also reveals that although the seismic wave velocity increased sharply between first and second exploration, it did not change very much between the second and third exploration. But in parts deeper than the stage 3 depth, the third exploration revealed a rise in the seismic wave velocity. This is thought to be a result of a decline in the precision of the analysis resulting from the fact that near the bottom of the range of the seismic wave exploration, the velocity wave path is less dense than in other sections, but because, among the fourth degree holes, a relatively large quantity of cement was injected at stage 3 in holes G4-1 and G4-2 on the seismic wave exploration traverse line (see Figure 5), it may be a result of improvement in the part below stage 3.

Figure 11 shows the relationship of seismic wave velocity from the first exploration V_1 with seismic wave velocity from the third exploration V_3 in each unit block within the depth of the grouting section. This figure reveals that in almost all data, V_3 is greater than V_1 and that it is distributed within the range $V_3 = (1.0 \text{ to } 1.3) \times V_1$.

5 CONSIDERATIONS

The results of the test grouting have indicated that the first and second stage grouting filled the existing joints in the bedrock without expanding them and that the third stage grouting either expanded the existing joints or formed new joints as it filled the joints with grout. The results of the seismic tomography have demonstrated that the seismic velocity distribution is dependent on the confining pressure in the bedrock and that it is possible that the improvement effects by the grouting can be represented by an increase in the seismic wave velocity distribution. Based on these results, the possibility of performing quantitative evaluations of imperviousness

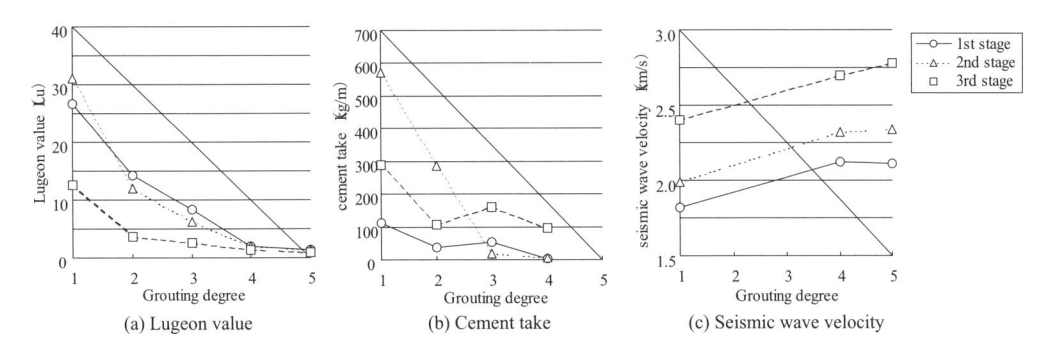

Figure 12. Changes with grouting degrees in Lugeon value, cement take and seismic wave velocity.

improvement achieved by grouting through the use of seismic tomography was discussed.

Figures 12 (a) and (b) show changes at each degree in the average Lugeon value and the average cement take at each stage, while Figure 12 (c) shows changes at each degree of the average value of the seismic wave velocity in the analysis unit blocks in the depth sections for each grouting stage shown in Table 3. In Figure 12 (c), the first, second, and third seismic wave exploration results correspond to the first, fourth, and fifth degree holes. Overall, the Lugeon value and the cement take declined at each degree, while the seismic wave velocity rose at each degree. These changes between degrees conform in that all reflect the results of grout gradually filling the joints in the bedrock. The tendency of the decline in permeability to converge after the fourth degree hole corresponds almost completely to the tendency of the increase in the seismic wave velocity to converge, suggesting a high degree of correlation between these factors.

Figure 13 shows the correlation of the seismic wave velocity with the Lugeon value for each stage. The seismic wave velocities corresponding to each Lugeon value were found as the average of a total of fifteen blocks, that is, five unit blocks (5 m) that correspond to the water pressure test stages in the vertical direction and 3 unit blocks (3 m) centered on the unit blocks including the hole axis in the horizontal direction, selected from among the seismic wave exploration results for the same degree. Because the results of this seismic wave tomography indicate the effects of the confining pressure dependency of the seismic wave velocity and the ground water conditions, the relationship between them has been organized for each stage. Judging from this figure, a relatively strong negative correlation was found between the seismic wave velocity and the Lugeon value at each stage. It also appears to be possible to classify the data from the first and second stages into one group and the data from the third stage where the groundwater conditions and injection mechanism

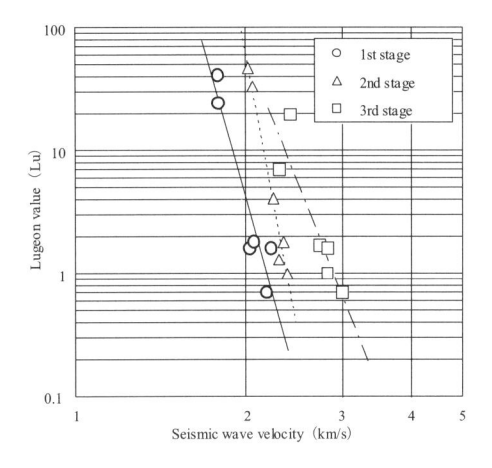

Figure 13. Correlation of seismic wave velocity with Lugeon value.

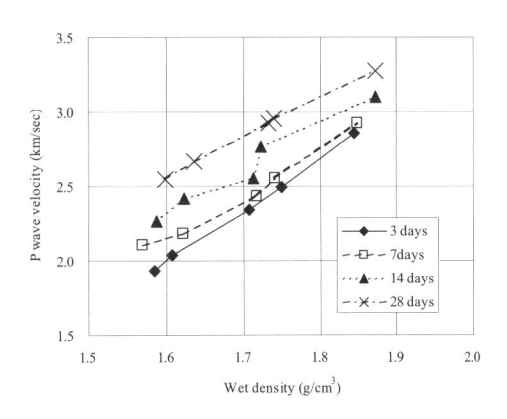

Figure 14. Seismic wave (P wave) velocity of hardened cement grouts (after Yamaguchi et al. 2000).

differed from those at the first two stages into another group.

Consequently, by considering the confining pressure of the seismic wave velocity, ground water conditions, and the injection mechanism, it is possible to quantitatively estimate to some extent the permeability of the bedrock based on the seismic wave velocity of the bedrock. But to perform a strict quantitative evaluation, it is necessary to study the effects on the seismic wave velocity of the grout that fills and hardens in the joints of a number of factors: the cement - water ratio of the injected grout is not constant, the density of the injected grout increases because of filtration caused by the action of the rock stress (Houlsby 1982, Ewert 1985, ISRM 1995), and during the seismic wave exploration, the age of the grout varied at each degree and stage. As one example, Figure 14 shows the wet density and the seismic wave velocity (P wave velocity) relationship under saturated surface-dry condition obtained as a result of ultrasonic wave-velocity measurements performed on 10 cm high cylindrical specimens with a diameter of 5 cm that were made by varying the water cement rate and the age among five and four values respectively (Yamaguchi et al. 2000). The cement used was Portland blast-furnace slag cement (Type B). This figure reveals that the seismic wave velocity varies considerably according to differences in density and age.

6 CONCLUSIONS

Test grouting and seismic wave tomography were executed in a jointed rock mass made of dacite to study the applicability of seismic wave tomography as a grouting effects evaluation method. The following conclusions were obtained.

(1) It was possible to gain a general understanding of the grout injection mechanism from the state of the occurrence of the critical pressure during the water pressure testing, the relationship between the Lugeon value and quantity of cement injected, and the state of the deformation of the bedrock during grouting.
(2) It was possible to judge the progress of the improvement achieved by the grouting as the rise in the seismic wave velocity obtained as a result of the seismic wave tomography.
(3) It has been concluded that by accounting for the injection mechanism of the grouting based on (1) and (2), seismic tomography is extremely applicable as a grouting effectiveness assessment method. But to do so, it is necessary to also study the confining pressure dependency of the seismic wave velocity and the ground water conditions.

REFERENCES

Ewert, F.K. 1985. *Rock grouting*, Springer-Verlag, pp.12-129.
Funato A. et. al. 1987. Measurements of ultra-sonic wave velocity of rock specimen under confining pressure, the 7th Japan Simposium on Rock Mechanics, pp.211-216.
Hasui, A. et al. 1992. On Evaluation of grouting effect for rock mass by crosshole seismic and borehole radar exploration, Journ. Japan Society of Dam Engineers, No.8, pp.35-44.
Houlsby, A.C. 1982. Optimum water : cement ratios for rock grouting, *Grouting in Geotechnical Engineering*, ASCE, pp.317-331.
Hudson, J.A. 1993. *Comprehensive rock engineering*, Vol.3, Pergamon Press, pp.635-650.
International Society of Rock Mechanics 1995. *Commission on Rock Grouting (Final Report)*.
Kawakami, T. et al. 1994. Evaluation of effects of rock grouting by Geo-tomography and in-situ rock deformation tests, the 9th Japan Simposium on Rock Mechanics, pp.337-342.
Kudo, S. 1963. On the investigation of grouting effetct by the seismic method, Report of the Public Works Research Institute, Ministry of Construction, Vol.114, pp.129-150.
Yamaguchi,Y., Yamamoto, S. and Abe, Y. 2000. Elastic wave velocity of hardened cement grouts, Journ. Japan Society of Dam Engineers, pp.56-63. (in Japanese with English summary)

3 Innovative technologies and equipment

General report: Innovative technologies and equipment

M. Long
Department of Civil Engineering, University College Dublin, Ireland

K.K. Phoon
Department of Civil Engineering, National University of Singapore, Singapore

Keywords: in situ testing, site investigation, soil, rock

ABSTRACT: A review and summary of all twenty five papers submitted to Theme 3 of ISC'2 on "Innovative technologies and equipment" is contained in this report. The papers confirm that significant and high quality research and development is taking place in this area. A range of materials from very soft sea bottom clays to hard rocks as well as contaminated material is dealt with. A particular concern is dealing with sites that have interbedded and random hard and soft zones. Four of the papers also deal with site characterization for seismic loading conditions.

1 INTRODUCTION

The objective of this report is to review all twenty five papers submitted to this theme and to attempt to draw general conclusions, trends and lessons learned on the overall subject of innovative technologies and equipment. As would have been expected, given the subject matter, this attracted interesting papers on a wide range of diverse topics. Therefore it is not easy to sub-divide or categorise the individual papers. However some attempts have been made to do this, as outlined on Table 1. Papers slotted under different topics could be cross-linked because the categories in Table 1 are non-unique. These papers are briefly discussed to highlight issues not covered under the broad categories in Table 1.

A feature of the papers presented is that four of them detail experience of full flow probes. Research in this area was re-commenced in Australia in the late 1990's having been started in the late 1930's at the Swedish Geotechnical Institute (SGI). Kallstenius (1961) describes the SGI Iskymeter, which is not unlike the T-bar penetrometer presented in the papers to this conference. Results from four different countries are given and therefore it is possible to undertake a more detailed review of the data given in these papers.

2 ADVANCED USE OF CPT

Four papers in this session (Harkes et al., Hebeler et al., Hryciw & Shin and Tan et al.) describe various physical augmentations of the standard CPT to widen its scope of applicability. The fifth paper (Areias et al.) describes an improvement in the procedure for conduct of SCPT. These papers are summarized briefly on Table 2.

3 INVESTIGATIONS OF HARD SOILS AND ROCKS

Möller et al. describe a method of using modern rock drilling machines provided with a number of transducers and a data acquisition system as a sounding tool in hard soils and rocks where normal penetrometers cannot be used. They illustrate that, by systematically measuring parameters such as rotational speed, thrust force, penetration rate, flush pressure and flush rate that very useful information can be obtained on the ground conditions. This information can then be used directly in design. The method is broadly known as Monitoring While Drilling (MWD).

The work by Möller et al. was undertaken in Sweden. Other similar methods are in use in other countries. For example Rygg and Andresen (1988) describe the rotary pressure sounding or "total sounding" approach used in Norway. It involves similar measurements, with particular emphasis being place on total thrust and penetration rate.

Table 1. Summary of papers – Theme 3 – Innovative technologies and equipment

Topic	No. of papers	Authors	Subject matter
Advanced use of CPT	5	Areias et al.	New shear wave source for SCPT
		Harkes et al.	Contaminated site assessment using electrical conductivity probe
		Hebeler et al.	Textured friction sleeves for CPT
		Hryciw and Shin	Thin layer characterization by VisCPT
		Tan et al.	Radioisotope cone to characterize lumpy fill
Investigation of hard soils and rocks	5	Baillot et al.	360° televising of drill holes
		Benoît et al.	Rock characterization using drilling parameters
		Möller et al.	Measuring while drilling in hard soil/rock
		Sacchetto et al.	CPT while drilling
		Shinn II and Haas III	Enhanced CPT penetration in highly contaminated conditions
Innovative in-situ testing devices	7	Bello et al.	Full displacement pressuremeter for municipal solid wastes
		Miller et al.	QA/QC of dynamic compaction using instrumented dilatometer
		Monnet and Senouci	In situ testing by the geomechameter
		Morii et al.	Permeability by pressure infiltrometer test
		Rahardjo et al.	Dual dilatometer for soft soils and peats
		Reiffsteck	In situ triaxial testing
		Wotschke and Friedel	Borehole tool for geophysical and geochemical testing
Full flow probes	4	Chung and Randolph	Different penetrometers in soft clay
		DeJong et al.	Full flow probes in soft layered clay
		Long and Gudjonsson	T-bar penetrometer in Irish soils
		Oung et al.	T-bar penetrometer in Dutch soils
Seismic site characterization	3	Apostolidis et al.	Use of microtremors
		Kim et al.	Shear wave velocity by SPT uphole test
		Park and Kim	Shear wave velocity by HWAW method
Liquefaction hazard assessment	1	Kuo et al.	Application of Iwasaki's P_L in Taiwan
Total number of papers	25		

Interpretation methods based on compound parameters that combine penetration rate, thrust, and rotation rate into a single parameter are also available. Benoît et al., Sacchetto et al., and Shin II & Haas III describe applications of similar systems in USA (New Hampshire), Italy (Parma, Po River, Venice) and USA (Washington State), respectively.

Benoît et al. describe the DPR (Drilling Parameter Recorder) system, which monitors a series of sensors to give data such as penetration rate, thrust, torque, rotation, water pressure and water inflow and outflow. It was designed with the objective of identifying fractures in contaminated rock. In addition they outline a number of useful compound parameters, which can be obtained from combinations of the measured ones. The "alteration index", which is indicative of relative hardness was useful in defining changes in lithology.

Table 2. Summary of reported work on advanced use of CPT

Paper	Motivation	Capabilities	Opportunities	Comments
Areias et al.	Increase the energy of shear wave in seismic cone penetration test to enhance signal-to-noise ratio and to extend testing depth.	Isolate hold-down loads (weight of CPT truck) on base plate in the vertical direction by a system of rollers so as to maximize transfer of energy from hammer to the ground.	Potential to generate shear wave signals 3 to 4 times stronger than those produced by conventional sources.	Method appears economical other than additional time needed to install rollers (?)
Harkes et al.	Augment CPT for detection of pollutants in contaminated sites.	Electrical conductivity probe.	Potential to map pollution distribution rapidly. Performance evaluated at a waste disposal in the Netherlands where distribution of chloride ions in the leachate has to be mapped.	Conversion from resistance to conductivity depends on temperature and conductivity depends on the number of cations and anions in solution. Thus, the electrical conductivity probe does not measure a specific pollutant directly and calibration with groundwater samples is necessary.
Hebeler et al.	CPT suffers from sensor overload in hard layers at cone tip. This limitation is less applicable to sleeve friction.	MSFA (Multi-sleeve friction attachment) consists of 4 friction mandrels with sleeves of varying roughness to determine interface shear vs. roughness relationship.	Potential to assess level of structure by comparing side friction and end bearing behaviour in structured and destructured soil. Performance evaluated at a uniform sand site in South Royalton, Vermont, USA.	Not fully tested in problematic soils such as conglomerates, gravels, cemented soils, stiff desiccated crusts and stiff layers overlying thick soft layers.
Hryciw & Shin	CPT resolution in stratigraphic classification is limited by tip resistance being influenced by soils with few diameters below tip and sleeve friction is only affected by soil adjacent to it.	VisCPT (vision cone penetrometer) provides a continuous vertical photo-log of soil. Image processing is used to extract a local homogeneity index that can be correlated to soil type.	Potential to resolve thin sand/silt partings (order of centimeters). Performance evaluated at a highly stratified hydraulic fill site on Treasury Island, San Francisco.	Local homogeneity index or other textural indices are sensitive to colour and illumination. It is difficult to establish general correlation between these indices and soil type. Baillot et al. used a video log of borehole wall for rock mass classification.
Tan et al.	Difficult to interpret CPT in random 3D matrix of stiff clay lumps and water/slurry filled voids (motivation is similar to Hryciw & Shin).	ND-CPT (nuclear density cone penetration test) measures in-situ wet density based on scattering of gamma rays.	Potential to detect water/slurry filled voids within a lumpy clay matrix (lump sizes order of cubic meters). Performance evaluated at a large land reclamation site off Pulau Tekong, Singapore.	Background count has to be determined to eliminate effect of natural radioactivity. This requires conducting ND-CPT twice in the same location with and without the radioactive source, which could be difficult in an offshore environment.

The equipment described Sacchetto et al. has the ability to record cone and drilling parameters at every 2 cm; thus making it quite ideal for sounding sites with alternating layers of hard and soft geomaterials. The equipment described by Shin II & Haas III has the ability to switch from a conventional CPT push system to a drill system to increase the depth of CPT sounding by overcoming gravel layers, cobbles or cemented and dense soils. In addition the EAPS (Enhanced Access Penetration) system supports an environmentally-safe drill spoils collection system for testing in highly contaminated sites.

Table 3. Summary of reported work on innovative in-situ testing devices

Paper	Device	Capabilities	Applications	Comments
Bello et al.	FDPM (full displacement pressuremeter)	Pressuremeter augmented to achieve larger radial expansion that is postulated to produce more realistic deformation parameters in municipal soil waste.	Field testing will commence in 2004 at the Muribeca landfill in the city of Recife, PE – Brazil.	It is possible to measure radial displacement up to 4 cm, but membrane can only be safely expanded to 2 cm at present (about 75% radial strain based on initial diameter of 5.4 cm.
Miller et al.	IDMT (instrumented dilatometer)	DMT augmented to record continuous pressure-membrane displacement curve; pore pressure & total pressure inside blade.	Unload-reload modulus appears to be a sensitive indicator of improvement produced by deep dynamic compaction in peat.	DMT correlations may not be applicable to IDMT indices
Monnet & Senouci	Geomechameter	Pressuremeter augmented with vertical hydraulic flow to vary vertical effective stress at the test level of the probe.	Measures friction angle, permeability, shear modulus & consolidation coefficient in reconstituted Hostun sand; limited field tests in Isère river sandy loam.	Applicable to soils with permeability higher than 10^{-8} m/s.
Morii et al.	GPI (Guelph pressure infiltrometer)	GPI augmented to measure both infiltration rate & volumetric moisture content.	Compute unsaturated hydraulic function of sand with gravel inclusions using genetic algorithm and FEM seepage analysis.	Initial matric suction in the soil mass likely to effect range over which the hydraulic function is valid.
Reiffsteck	LCPC in-situ triaxial test	Triaxial test at different stress paths and local small strain measurement using Hall effect transducers.	Tested in compacted sand and clayey silt	Useful to assess sample disturbance due to lack of internal clearance. Probably most useful in sand.
Rahardjo et al.	DDMT (dual dilatometer)	DMT with additional thicker blade attached at the top to produce larger strains that are postulated to improve sensitivity in soft soils.	Tested in soft clays & peats, Pelintung, Sumatera.	No clear improvement in interpretation is evident.
Wotschke & Friedel	Borehole tool for geophysical and geochemical testing.	Combines electrical resistivity tomography with in-situ soil gas analysis for investigating contaminated sites.	Tested in clayey lake sediments, overlain by a gravel aquifer, contaminated silt layer & gravel cover.	Able to detect two organic volatile compounds – trichloroethylene & tetrachloroethylene.

4 INNOVATIVE IN-SITU TESTING DEVICES

Seven papers in this session report on augmentation of existing devices to measure additional data (Miller et al., Morri et al.) or modification of existing devices: achieve larger radial expansion in pressuremeter (Bello et al.), induce vertical hydraulic flow around pressuremeter (Monnet & Senouci); attach additional blade above DMT (Rahardjo et al.); incorporate triaxial test setup with sampling tube (Reiffsteck) or combines electrical resistivity tomography with in-situ soil gas analysis (Wotschke & Friedel). A summary is given in Table 3 above.

5 FULL FLOW PROBES

5.1 Overall summary and general trends

Four papers in this session report on case histories of site characterization using full flow probes. These devices include T-bar, spherical ball and plate penetrometers. All of the authors state that the purpose of the use of the devices is to overcome the uncertainties in determining undrained shear strength (s_u) in soft clays, due to the usual corrections necessary with the CPT and the complicated failure mechanism around the cone. A summary of the reported work is given on Table 4.

Table 4. Summary of reported work on full flow penetrometers

Paper	Country	Devices used	Soil	General trends in results
DeJong et al.	US	T-bar, ball and plate	Connecticut valley varved clay (Amherst, Massachusetts)	In field CPTU q_t exceeds penetration resistance of T-bar, which in turn is about 38% greater than that of ball and plate.
Chung and Randolph	Australia	T-bar, ball and plate	Field – Burswood clay (Perth) Model – reconstituted Burswood clay	In field CPTU q_t exceeds that of all other devices with plate being marginally greater T-bar and ball. CPTU q_t increased more rapidly with depth. In contrast centrifuge model CPTU exhibits lowest q_t (probably reflecting higher rigidity index of natural clay). Model plate resistance lower than model T-bar or ball. Effect of surface roughness and aspect ratio marginal.
Oung et al.	Netherlands	T-bar only	Field – silty clay and peat in island of Marken, Ijsselmeer Model – Kaolinite	In 1g model tests CPTU q_t greater than T-bar resistance. In field CPTU and T-bar resistances similar. CPTU more "peaky". T-bar better at reflecting layer limits.
Long and Gudjonsson	Ireland	T-bar only	Soft glacial lake clays from Athlone, Portumna, Loughmore	For clay soils T-bar resistance only 50% of CPTU q_t. For organic and silty soils penetration resistance similar for both. CPTU q_t shows stronger increase with depth

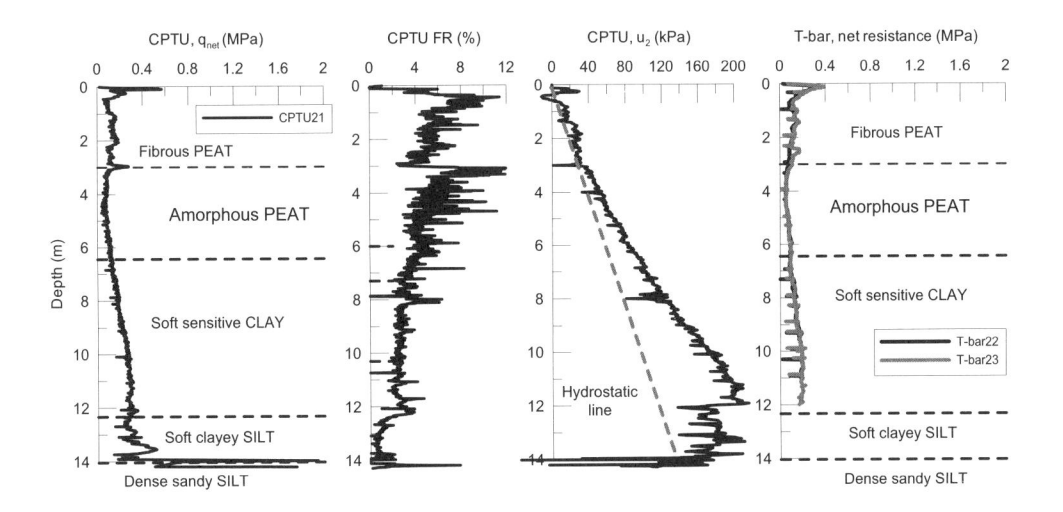

Figure 1a. Detailed T-bar and CPTU test results for Bundoran / Ballyshannon Bypass Project, Co. Donegal, Ireland

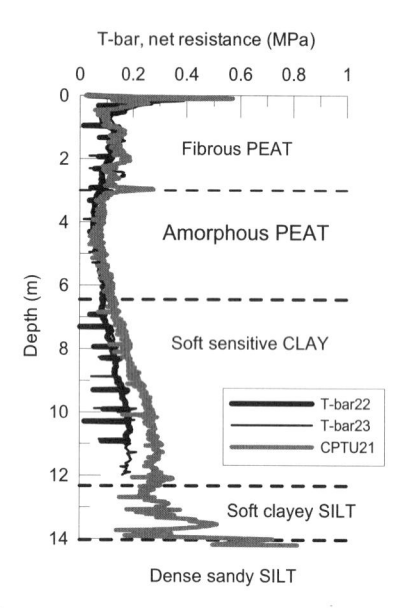

Figure 1b. Comparison of T-bar and CPTU test results for Bundoran / Ballyshannon Bypass Project.

5.2 *Differences in CPTU and T-bar penetration resistances*

From the table it can be seen that an important finding for all four studies is that, for clay soils, CPTU q_t generally exceeds the penetration resistance of the other devices. All of the authors have suggested this difference is due to the shape of the penetrometers, compared to that of the cone. These devices provide full flow (see Fig. 1 from DeJong et al.) in contrast to the CPTU which forces full and irrecoverable displacement of the soil, thus resulting in a greater penetration resistance. In contrast for the Dutch and Irish organic soils and the Irish silts the resistances measured by both devices are similar, possibly due to the "drained" nature of the flow mechanism of these material contrast to the "undrained" behaviour of the clay.

Another recent example of this can be seen in the data from the investigation of the Bundoran / Ballyshannon Bypass, Co. Donegal, Ireland. Full details of the tests are shown on Figure 1a and a direct comparison of T-bar and CPTU results is given on Figure 1b.

From the test results it can be seen that both CPTU net resistance and T-bar resistance show more or less identical and very low values in both the fibrous and amorphous peat but in the soft sensitive clay, the CPTU values exceed those of the T-bar.

Another important finding of the work reported above is that the difference in CPTU and T-bar resistance increases with depth. Randolph et al. (1998) also found similar differences between the CPTU and T-bar for a clay site, offshore Australia. They suggested that one possible explanation for the increasing divergence with depth is strength anisotropy. The "average" s_u deduced by the CPTU is most influenced by the soil response in triaxial compression (Baligh, 1996), where s_u is larger than in other shearing modes such as simple shear or triaxial extension. The symmetry of the T-bar deformation field, with full flow around the probe, is more likely to yield an average strength, which is generally close to that measured in direct simple shear (DSS). However, there are probably other contributing reasons because DeJong et al.'s data in highly anisotropic soil did not exhibit this effect clearly.

Lunne (2001) suggested that strain softening and rate effects may also be of importance in understanding the difference between the two devices.

5.3 *Undrained shear strength from full flow penetrometers*

The main objective of the use of full flow penetrometers is to derive undrained shear strength s_u. Also theoretically, one of the attractions of the full flow probes over the CPTU, is the existence of closely bracketed plasticity solutions for plane strain flow (T-bar) or axisymmetric flow (plate and ball) around the device. It was hoped that the range of resulting bearing capacity factors would be lower than the equivalent range for the CPTU. A summary of work reported here on the derivation of s_u and the related bearing capacity factors (N) is given on Table 5.

Unfortunately s_u varies with a large number of factors, most especially the soil type (e.g. plasticity), strain rate, anisotropy and the technique use to determine it and therefore it is not easy to make general conclusions. However some attempts have been made to relate T-bar penetration resistance to field vane s_u using the data from the four papers presented here in addition to some previously published data by Randolph et al. (1998) and Watson et al. (2000).

The resulting N values are presented on Figure 2 plotted against plasticity index (I_p) and the following could be concluded:

- On average the values are very close to the theoretical bearing capacity number of 10.5 from Randolph and Houlsby (1984).
- The overall scatter in the data is relatively small.
- In general values from the centrifuge tests on reconstituted clay are lower than for the intact clays.
- There does seem to be some indication of N decreasing with increasing I_p.

Table 5. Derivation of s_u and bearing capacity factors from full flow probes

Paper	Device	Technique used to determine s_u	Remarks
DeJong et al.	T-Bar Plate Ball	Field vane	• Large scatter in field vane data. • N lower for Ball and Plate than T-bar. • For weathered crust N average for T-bar, Ball and Plate = 9.7, 6.8 and 6.9. • For lightly overconsolidated zone, N average = 13.7, 10 and 10.
Chung and Randolph	T-Bar Plate Ball	Field vane	Average N values from full scale field tests • T-bar = 11.4 • Ball = 10.5 • Plate = 11.6
Chung and Randolph	T-bar	Lab vane	Average N values from lab. centrifuge tests lower than or field
Oung et al.	T-bar	Field vane, CIU	Full scale field tests N = 10.5 for CIU tests and in range 8 to 9.5 for vane tests.
Oung et al.	T-bar	UU, lab vane	1g model tests, N = 10.5 ± 1.4 for both
Long and Gudjonsson	T-Bar	CAUC	N average = 10.5. Range of N values lower than for CPTU in uniform soils but both are scattered for non-uniform organic soils.

Note: Some of the data was interpreted from the graphs

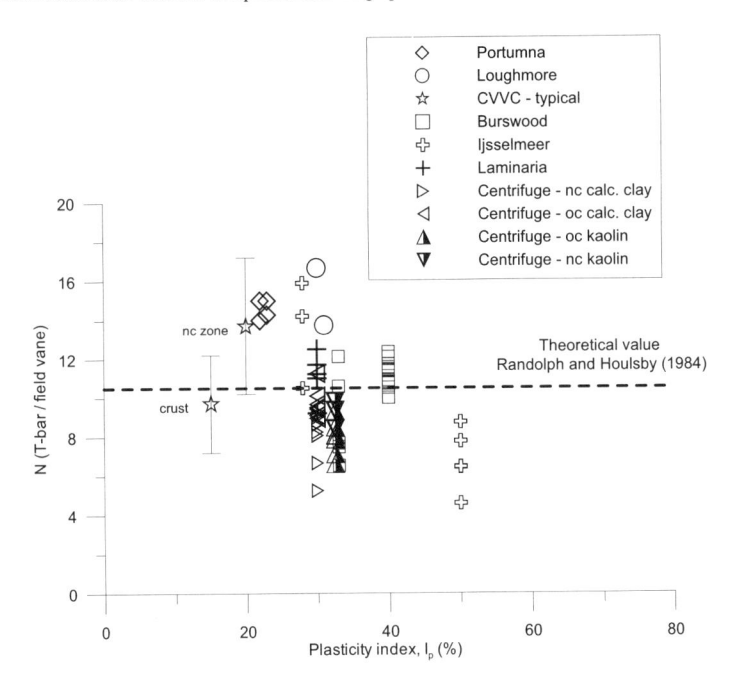

Figure 2. Bearing capacity factors from field vane and T-bar tests

Table 6. Summary of remoulded shear strength derivation from full flow probes

Paper	Device	No. of cycles to steady state	Remarks
DeJong et al.	T-Bar Plate Ball	Typically 13 Range 10 to 15	• T-Bar degrades fastest. • Rate of degradation in weathered crust higher than in lightly overconsolidated zone. • In lightly o-c zone, T-Bar $s_{ur} \approx s_{ur}$ field vane but others higher. • In crust all values similar. • S_t all devices lower than average S_t from field vane but within range of values.
Chung and Randolph	T-Bar Plate Ball	5 to 6	• Plate degrades fastest • After 5 to 6 cycles all devices gave degradation factor of 0.23 to 0.28 consistent with field vane $S_t = 3$ to 4.
Oung et al.			• No s_{ur} data reported.
Long and Gudjonsson	T-Bar only	5 to 6	• S_t similar to that from field vane

It is possible that effects of structure and ageing are contributory factors to the centrifuge test values being lower than those for the intact clays.

A trend with I_p relative to field vane s_u is quite possible due to combination of rate effects, vane insertion disturbance (followed by consolidation) and strain softening during T-bar penetration (Randolph, 2004). Normally it would be expected that strain rate effects would increase with increasing I_p therefore leading to higher N values, i.e. the opposite effect to that seen here.

Perhaps the higher I_p results in greater strain softening reducing the T-bar resistance (and therefore N) or alternatively the effects of insertion disturbance and partial reconsolidation decrease with increasing I_p resulting in higher s_u values (and therefore lower N). Clearly this is an area that warrants further work.

Nevertheless there is some promising evidence in the work presented here that the range of bearing capacity factors for the full flow probes may be relatively narrow and smaller than that of the CPTU. However there is an urgent need to verify this with further work by comparing results of full flow probe tests and s_u tests (using the same technique) for a range of different clays.

5.4 Remoulded shear strength from full flow penetrometers

Watson et al. (2000) proposed that remoulded shear strength (s_{ur}) and sensitivity (S_t) could be determined by these devices by applying several cycles of penetration and extraction over a range 20 to 40 times the diameter of the probe (approximately 1 m). The penetration resistance decreases with cycling until a steady state is achieved at s_{ur}. A summary of the reported data is given on Table 6.

From the table it can be seen that, despite some differences in rates of degradation an number of cycles required to reach steady state, the full flow devices can give useful estimates of s_{ur} and S_t.

5.5 Use of the T-bar in peat

The symmetry of the T-bar deformation field, with full flow around the probe, is more likely to yield an average strength, which is generally close to that measured in direct simple shear (DSS). The "average" s_u deduced by the CPTU is most influenced by the soil response in triaxial compression. Recently in Ireland there has been a number of catastrophic landslides in peat, where the mode of deformation was of simple shear. There is an urgent need to develop a technique for the determination of the shear strength of peat for the purposes of risk analyses of future slides.

It is possible that the T-bar may be useful for this purpose, as is illustrated on Figure 1a for the Bundoran - Ballyshannon Bypass, Co. Donegal and the following can be seen:

• Both CPTU and T-bar techniques characterize the peat as two separate layers: an upper fibrous zone and a lower amorphous mass.
• This lower layer is particularly weak and will have a strong influence on the stability of natural slopes, cuts or embankments constructed in this area. Therefore shear strength evaluation is critical. Due to the uncertainties involved in correcting the CPTU data, perhaps the T-bar will be the most useful tool to provide this data.
• However there is a need to correlate T-bar tests with results of high quality DSS lab tests in this difficult material.

Similar useful results have been found for two other Irish peat sites.

5.6 Other important findings

Some other important findings from the work, not dealt with in detail above, are as follows:

1. Corrections for CPTU much more significant than for other devices (Chung and Randolph, Long and Gudjonsson).
2. Only small difference between smooth and rough T-bar's. Difference consistent with theoretical formulation. (Chung and Randolph).
3. Little difference in results with various T-bar aspect ratios and any ratio in range 4 to 8 suitable (Chung and Randolph).

6 SEISMIC SITE CHARACTERIZATION

Three papers in this session report on geophysical prospecting using surface waves and shear waves. For seismic site characterization based on surface waves, it has been found that near field information is more suitable for exploring deep layer material properties and provides more spatially localized information (i.e., less affected by lateral heterogeneity between source and receivers). However, conventional Fourier transform techniques cannot exploit this advantage because the signal to noise ratio is too low in the near field.

The reason is that the signal to noise ratio at each frequency is computed as an average over the entire time domain, rather than over the duration where the signal is most significant. Park and Kim utilizes the harmonic wavelet transform to compute a local signal to noise ratio that is higher than the average one to improve the interpretation of near field information. Shear wave velocity profiles up to about 60 m deep were computed in field tests using short receiver spacing (about 2 m).

Apostolidis et al. were motivated by the same problem of exploring deep layers using surface waves. Their site is located in an urban area, which imposes further constraints on available space for deployment of receiver arrays and strength of energy sources. The authors demonstrated the use of microtremor readings to establish shear wave velocity profiles down to large depths between 100 and 300 m.

Kim et al. suggested another shear wave velocity interpretation technique based on an uphole test. The innovative aspect of their paper is to use the impact energy generated by SPT as a source, rather than borehole blasting. It is clearly economical to be able to obtain blow counts and shear wave velocity profiles using the same test. Shear wave velocity profiles up to depths between about 15 m and 30 m were computed in field tests.

7 OTHER TOPICS

Harkes et al., Shinn II and Haas III, and Wotschke & Friedel deal with contaminated sites. Harkes et al. discussed detection of chloride ions in solution using CPT that can measure electrical conductivity. Shinn II and Haas III described an environmentally-safe penetration system that can perform stratigraphic and contaminant profiling in a sealed casing using a CPT/Gas sampling probe. Detectors include a photo ionization detector (PID) for aromatics, flame ionization detector (FID) for hydrocardons, dry electrolytic conductivity detector (DELCD) for chlorinated and brominated solvents, and a photoacoustic IR analyzer for tetrachloride, chloroform, and water. Wotschke & Friedel uses a carrier tool that can conduct electrical resistivity tomography with in-situ soil gas analysis (trichloroethylene & tetrachloroethylene) in a plastic-cased borehole.

Baillot et al., Benoît et al., and Hryciw & Shin deal with borehole imaging systems. Baillot et al. used borehole video images to classify discontinuities of rock mass in their natural conditions. Benoît et al. briefly mentioned application of both optical and acoustic images to examine strike and dip of geologic structure. Interpretation appears to be manual in both papers. Hryciw & Shin combined imaging with the CPT for demarcation of thin sand/silt partings and analyzed the video logs automatically based on textural changes.

Kuo et al. assess the application of Iwasaki's liquefaction potential index (P_L) to Taiwan using data from the 1999 Chi-Chi earthquake. The results show that the P_L values as proposed by Iwasaki et al. appear to be too high for use in Taiwan and some revised values are suggested.

8 POINTS FOR DISCUSSION

8.1 Advanced use of CPT

Is there any possibility of applying VisCPT for detection of voids in a lumpy clay matrix? May not be possible to distinguish textural changes between clay lumps and slurry.

Is it possible to exploit the more localised influence zone of sleeve friction and making less variable f_s measurements further from the tip (> 0.35 m) to improve CPT interpretation in highly heterogeneous materials?

8.2 Investigation of hard soils and rocks

Is there merit in attempting to standardize MWD (measurement while drilling) techniques or are there too many variables involved, e.g. drilling equipment, parameters measured, ground conditions etc. to make the exercise worthwhile?

8.3 Innovative in-situ testing devices

Role of the dilatometer in ground investigation. Useful or not?

8.4 Full flow probes vs CPTU

Should full flow probes be considered for future or is CPTU adequate?

Standardisation of full flow probes (e.g. rate of penetration, smooth or rough, aspect ratio, number of cycles for remoulded strength tests etc.)?

Development of bearing capacity factors or full flow probes. Should in situ tests (e.g. vane) or lab tests be used? If lab tests, which? Triaxial maybe not same mode of shearing. Consider DSS?

8.5 Seismic site characterization

Effect of lateral heterogeneity in interpretation of shear wave velocity profile?

ACKNOWLEDGEMENTS

The authors are grateful to Lankelma Ltd., UK for assistance with the additional CPTU and T-bar testing reported here.

REFERENCES

Baligh, M.M. 1986. Undrained deep penetration, I: shear stresses; II: pore pressures. *Geotechnique*, 36, 4: 471 – 501.

Kallstenius, T. 1961. Development of two modern continuous sounding methods. Proc.5th ICSMFE, Paris, 1: 475 - 480

Lunne, T. 2001. In situ testing in offshore geo-investigations. *Proc. In situ 2001, Bali, Indonesia, International Conference on In Situ Measurement of Soil Properties and Case Histories*. Lunne and Rahardjo (eds.): 61-81.

Randolph, M.F., Hefer, P.A., Geise, J.M. and Watson, P.G. 1998. Improved seabed strength profiling using T-bar penetrometer. *Proc. Int. Conf. Offshore Site Investigation and Foundation Behaviour, SUT, London:* 221 – 236.

Randolph, M.F. and Houlsby, G. 1984. The limiting pressure on a circular pile loaded laterally in cohesive soil. *Geotechnique*, 34 (4): 613 – 623.

Randolph, M.F. 2004. Personal communication to authors, dated 17/3/04.

Rygg, N.O. and Andresen, A.A. 1988. Rotary-pressure sounding: 20 years experience. *Proc. 1st Int. Sym. on Pen. Testing, ISOPT-1, Orlando*, 1:453-457. *Also NGI Publication* No. 177.

Watson, P.G., Suemasa, N. and Randolph, M.F. 2000. Evaluating undrained shear strength using the vane shear apparatus. *Proc. 10th Int. Conf. on Offshore and Polar Eng., ISOPE 00, Seattle*, 2: 485 – 493.

The use of microtremors for soil and site categorization in Thessaloniki city, Greece

P. Apostolidis, D. Raptakis, M. Manakou & K. Pitilakis
Dept. of Civil Engineering, Aristotle University of Thessaloniki, Greece

Keywords: V_s velocity, Spatial Autocorrelation Coefficient, fundamental frequency, transfer function

ABSTRACT: Array measurements of microtremors at sixteen (16) sites in the city of Thessaloniki were performed to estimate the V_s velocity of soil formations for site effect analysis. The Spatial Autocorrelation Method (SPAC) was used to determine phase velocity dispersion curves. A Rayleigh wave inversion technique was subsequently applied to determine the V_s profiles at all the examined sites. The same measurements were used to determine the fundamental frequency of soil deposits at every site using the spectral ratio of the Horizontal to Vertical component of microtremor recordings (H/V technique). The total thickness of the soil deposits at all sites was estimated, using 1D theoretical transfer function of the V_s profiles, determined with SPAC method, and compared with H/V ratio. Finally, 2D soil cross sections were designed. The use of array measurements of microtremors is hence proved as a very efficient and cost –benefit tool for soil and site characterization for site effects and microzoning studies.

1 INTRODUCTION

Geometry, shear-wave velocity (V_s) of the alluvial deposits and depth of the bedrock are the key parameters, which control the site amplification.

The V_s velocity is usually determined in the field by using conventional seismic prospecting (e.g. reflection, refraction, borehole seismics) and in the laboratory through dynamic/cyclic tests (i.e. Resonant Column, Cyclic Triaxial), on intact and remolded soil samples. The use of conventional surface seismic methods presents practical difficulties in cases where deep sedimentary deposits are due to be investigated especially in urban areas. Moreover, the cost of large scale geophysical prospecting is very high. Additionally, the cost for implementing deep borehole seismics is also very high while the reliability of these seismic methods, such as crosshole and downhole, at large depths is often questionable, because of practical limitations (e.g. energetic sources, equipment management, etc). For these reasons in response numerical analysis engineers use as half-space a layer with V_s between 700m/sec and 1000m/sec.

In the present paper it is proposed to overcome in a reasonable degree the aforementioned problems by using array measurements of microtremors with both Spatial Autocorrelation Coefficient (SPAC) method and the spectral ratio of the Horizontal to Vertical component (H/V technique). The SPAC method is proved to allow the determination of V_s velocity at large depths (Apostolidis et al., 2004), while the H/V technique allows the estimation of the total depth of the rock basement with the combination of the fundamental resonant frequency with V_s velocity of the soil formations.

Microtremor measurements were performed at 16 representative sites from geological and geotechnical point of view, in the city of Thessaloniki (Fig.1). The city of Thessaloniki with a population of one million inhabitants has a long history of seismic catastrophes; it suffered severe damage during the destructive earthquake in 1978 (M_w=6.5). For the last two decades, a considerable research work has been conducted to investigate, together with the role of site effects by the damage distribution of the 1978 earthquake, the soil stratigraphy and the dynamic soil properties in order to assess future potential damage scenarios.

The basic target of microtremor measurements was the estimation of the V_s profiles for depths larger than 50 m, down to the top of the rock bedrock. These V_s profiles can be used not only for conventional 1-D ground response analysis but also for the design of 2-D cross-sections and 3-D mapping of the geological formations.

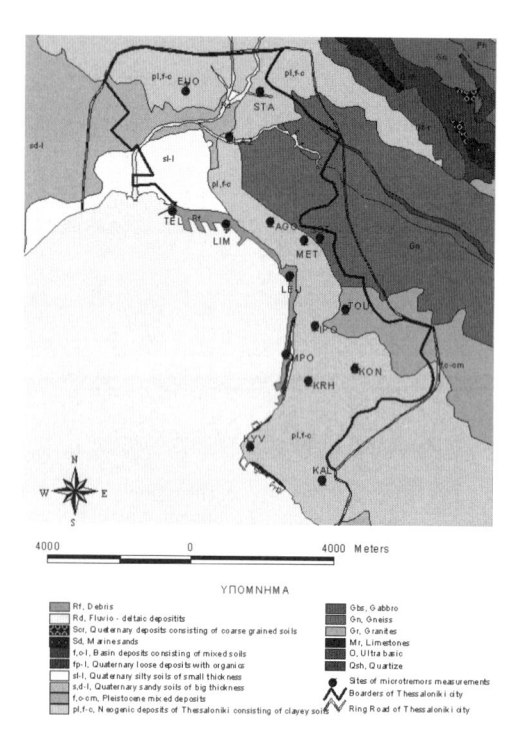

Figure 1. Geological map of Thessaloniki with the locations of the 16 sites of the microtremor measurements

2 V$_S$ PROILES WITH SPAC METHOD.

The fundamental assumption in SPAC method is that microtremor consists mainly of surface waves, from which it is possible to extract the dispersion curve of the phase velocity as a function of frequency in a next step. This dispersive curve is inverted to the shear-wave-velocity distribution with depths under the array which has been deployed. Aki (1957) gave a theoretical basis of the SPAC coefficient defined for ambient noise and developed a method to estimate the phase velocity dispersion of surface waves contained in microtremor using a specially designed circular array. Henstridge (1979) also introduced a "lucid" expression of the relationship between the spatial autocorrelation coefficient and the phase velocity of fundamental mode of Rayleigh waves. Okada (1997) extended it to an exploration method that is currently called the SPAC method. For a complete presentation of the method see also Okada (1999).

The measurements in Thessaloniki city were performed using circular arrays, which consists of three recording stations on the circumference envolving one common station. In this case different data sets were obtained at the same place, at the centre of the array. The recording instruments that were used consist of three-component broadband seismometers CMG-40T with standard frequency band from 0.033Hz to 50Hz and Reftek recorders of type 72A-07, in order to get period bandwidths adequate to explore V$_s$ velocity at layers as deep as possible. More detailed description of the recording status (day and time of measurements at every site, geometry of the arrays, etc) can be seen in a recent publication of Apostolidis et al (2004).

2.1 *Definition of the experimental dispersion curve*

The original signal trains were divided into multiple time windows, whose length varied depending on the wavelength associated with the penetration depth of interest (Apostolidis et al., 2004). The analysis procedure to estimate the SPAC coefficients and consequently the V$_s$ profiles is illustrated in fig. 2 for a single site, called KRH, as an example. Two circular arrays with radii of 10m and 35m were deployed, keeping the same central station for all measurements.

In the first stage of processing, we have examined the stationary nature of the microtremor recordings. If the microtremors could be considered as stationary, then stationary random function and space-correlation function between two stations can be computed as a function of the phase velocity of the surface waves. The frequency range in which microtremor can be considered stationary, was determined by the calculation of its power spectrum and the frequency coherence function (sometimes called the coherency squared function), estimated for different stations. At the specific site KRH and for the small array, the similar power spectra calculated for one time - window for all stations (Fig. 2a) and the coherency functions (Fig. 2b), imply that the recorded ambient noise is stable in the frequency range from 0.3 to 6Hz. In this frequency range, the SPAC method can give reliable results.

In the second stage, which is the main part of the analysis, the autocorrelation functions were calculated for each pair of stations of equal inter-stations distances. Figures 2c and 2d show the space correlation function calculated for each pair of equally distant stations and the spatial autocorrelation function (or the spatial autocorrelation coefficient). Figure 2e show the obtained SPAC coefficients at the finally selected frequencies between 2 and 6 Hz. The dispersion curve at lower frequencies was determined from the analyses of the larger array with the same procedure previously mentioned.

The experimental dispersion curve of phase velocity of Rayleigh waves obtained from the combination of the data of the two arrays is presented in Figure 3.

For the determination of the vertical V_s profiles an assumption of a complete initial soil model with thickness, V_p and V_s velocities, and density for each layer is needed. In case that theoretical dispersion curve (solid line in fig. 4) fits well the experimental one (circles in fig. 4), provided by SPAC method it is considered that the inverted soil model is accurate and finally acceptable (fig. 5).

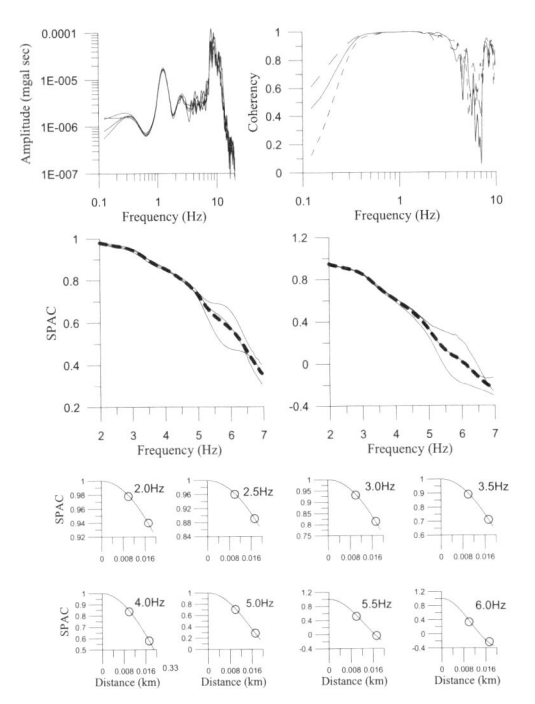

Figure 2. Analysis procedure for the estimation of the SPAC coefficient for the small array at site KRH (details in the text)

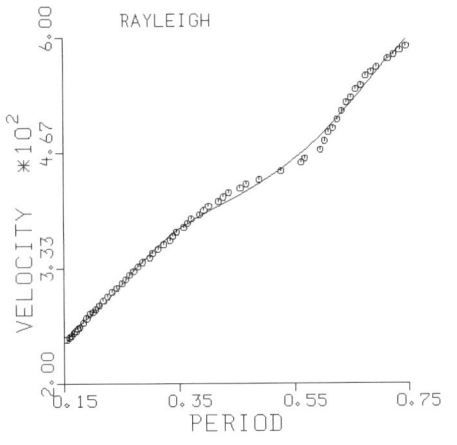

Figure 4. Fitting of the experimental and theoretical dispersive curve at site KRH

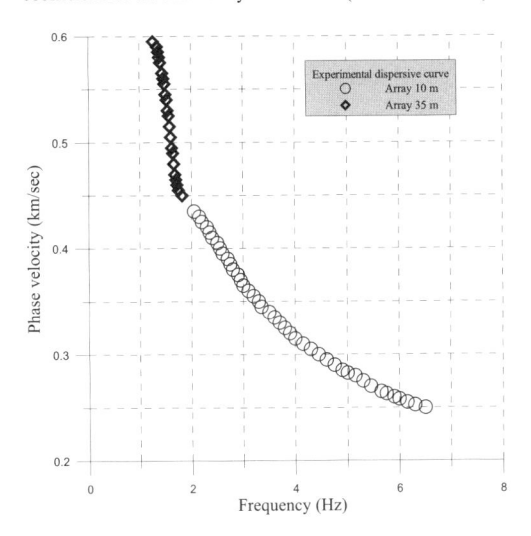

Figure 3. Dispersive curve of Rayleigh waves at site KRH

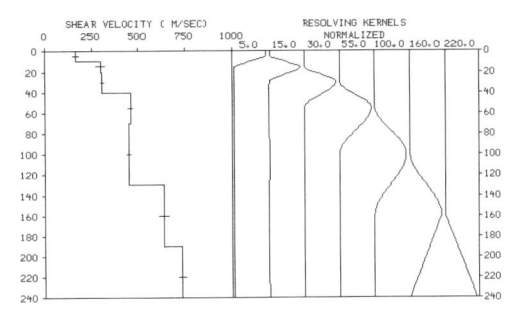

Figure 5. The final inverted Vs profile and the resolving kernels at site KRH

The reliability of the inverted V_s profile is confirmed by the shape of the resolving kernels, in form of a delta function that corresponds to the depth of the layers of the initial soil model (fig.5).

The same procedure used for site KRH was followed at all sites in Thessaloniki city and 16 V_s profiles were determined (Apostolidis et al., 2004).

2.2 Inversion procedure and inverted V_s profiles

The experimental dispersion curves defined with SPAC method at 16 sites were inverted using an iterative inversion algorithm introduced by Herrmann (1987), to reveal the V_s velocity structure below the circular arrays.

2.3 Comparison with Vs profiles from C-H tests

In order to validate the results of SPAC method, we compared the determined V_s profiles with the V_s profiles resulted from cross-hole measurements (Pitilakis et al., 1992; Raptakis et al., 1994) at many sites within the city. At site KRH, the V_s profiles are in good agreement, describing the stratigraphy and the V_s velocity of the main soil formations, down to a depth of 30 m, in which the borehole were terminated even if some local discrepancies due to the more detailed C-H measurements are observed.

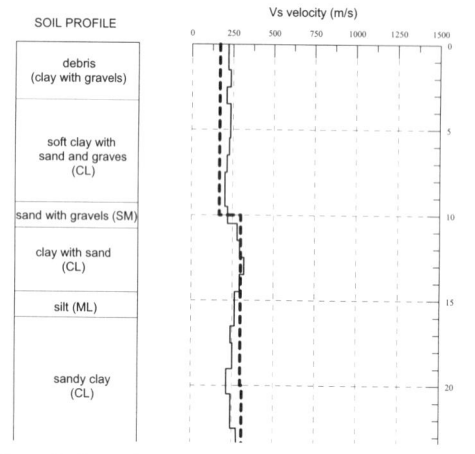

Figure 6. Comparison of the Vs model inferred from SPAC method (dashed line) and the model derived from C-H test (solid line) at site KRH.

The comparison at all available sites showed that SPAC method is reliable and fairly accurate for the estimation of both soil stratigraphy and V_s velocities in terms of site effect analyses.

3 FUNDAMENTAL FREQUENCY WITH H/V TECHNIQUE

Although the obtained penetration depths were sufficiently large, in some cases, the depth of the bedrock could not be estimated. For this, the H/V technique was applied on the same data sets in order to get the fundamental frequency of the sedimentary deposits. The fundamental frequency was used together with the inverted V_s profile to estimate the total thickness of the sediments down to the top surface of the bedrock, taking into account the theoretically defined site response based on the resolved SPAC V_s profile. For this reason and to compare 1-D theoretical transfer functions with empirical ones, H/V ratios were calculated at 16 sites within the city.

The Fourier spectra of each horizontal component and of the vertical component were calculated first. The sampling interval of the microtremor recordings was 0.008sec and the recording duration was 30min. The obtained waveforms were band-pass filtered from 0.1 to 25Hz and corrected for the baseline and the instrument response. The time window that was used was 20sec. The Fourier spectra were smoothed with the "weighted moving average method" (¼, +½, +¼) before calculating the H/V spectral ratio.

In figures 7 and 8 H/V spectral ratios of the microtremors components at site KRH are presented. The spectral ratios of both horizontal components remain constant at almost all the time windows. The resonant frequency is 0.5Hz.

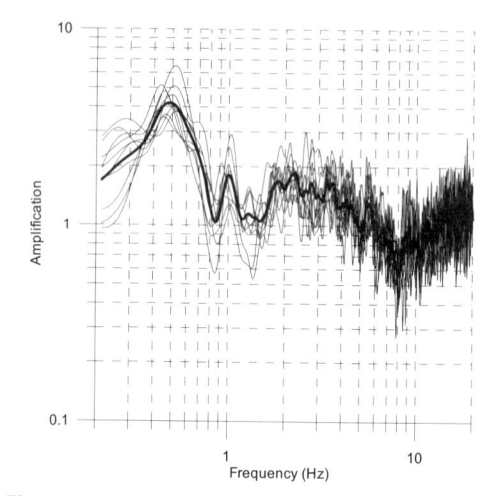

Figure 7. H/V spectral ratios and their average for N-S horizontal component at site KRH

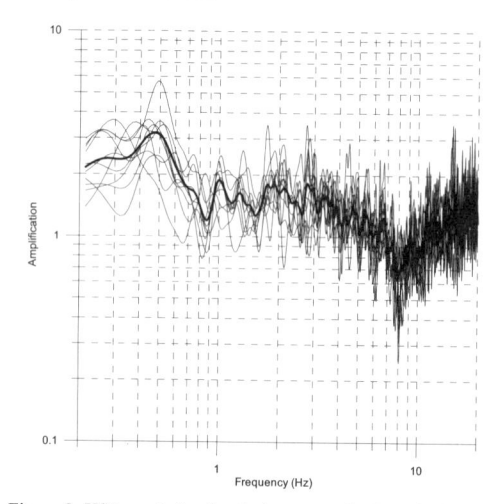

Figure 8. H/V spectral ratios their average for E-W horizontal component at site KRH

H/V spectral ratios at 16 sites are presented at fig. 9. The day and time of measurements, the final exploration depth determined with SPAC method and the fundamental resonant frequency are shown in Table 1.

Table 1. Day and time of measurements, exploration depths of Vs profiles determined with SPAC method and fundamental resonant frequency calculated with H/V technique

Site	Day	Time	Eploration. Depth (m)	Fund. F_o. (Hz)
KAL	Tuesday	10.0-16.0	280	0.85
KYV	Thursday	10.0-18.0	320	1.35
KRH	Sunday	4.30-11.0	240	0.5
MPO	Sunday	5.0-12.0	180	0.4
KON	Sunday	10-16.0	250	0.9
IPO	Sunday	5.0-9.0	110	0.97
TOU	Saturday	10.0-16.0	180	1.8
LEU	Saturday	4.30-11.0	130	0.87
MET	Wednes	10.0-16.0	35	3
AGO	Sunday	5.0-8.0	60	4
LIM	Saturday	7.0-13.0	180	0.9
TEL	Sunday	5.0-8.0	160	0.62
LAZ	Saturday	10.0-16.0	140	3
STA	Saturday	5.0-11.0	80	2.2
EUO	Wednes-	10.0-16.0	220	0.63
SST	Monday	12.0-15.0	80	-

The almost flat H/V spectral ratio at the rock site (Fig. 9) and the large variation of resonant frequencies at the other sites depict completely different soil conditions in the area of the city. All these results show that the geometry of the soil structure in which the city is located is rather complex suggesting that also complex site effects are expected (Raptakis et al., 2003 a and b).

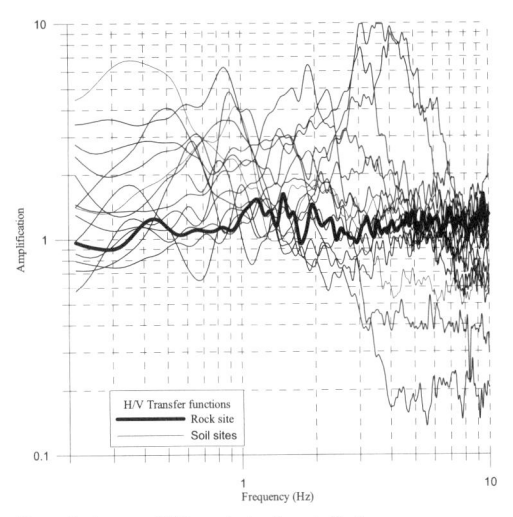

Figure 9. Average H/V spectral ratios at all sites

Moreover, to fully design 2-D cross-sections, the bedrock depth was estimated at all sites following a back analysis procedure presented in the next paragraph.

4 ESTIMATION OF THE BEDROCK DEPTH

The V_s profiles, determined with SPAC method as referred above, did not reach the bedrock at 9 sites (KYV, KRH, MPO, IPO, LEU, LIM, TEL, STA and EUO). At these sites, the total thickness of the sedimentary deposits was estimated using the V_s profile determined with SPAC method, the fundamental frequency calculated with H/V technique and 1-D theoretical transfer function computed at the same site where V_s profile is well known. Representative Q_s values for the soil formations used in these soil profiles were found by laboratory and geophysical surveys (Raptakis et al. 1994; Anastasiadis et al. 2001).

The basic idea is illustrated in figure 10 for the site KRH. The dashed lines show the H/V spectral ratios of E-W and N-S component.

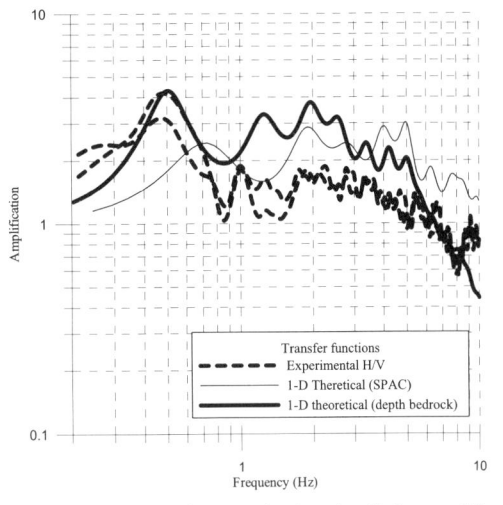

Figure 10. Empirical H/V spectral ratio and preliminary and final theoretical transfer function 1 at site KRH

The thin solid line is the theoretical transfer function based on V_s model determined with SPAC method. The difference of the fundamental resonant frequencies between the theoretical and the empirical transfer functions denotes that the sedimentary deposits at site KRH have larger thickness assuming that the V_s velocities are accurately defined.

In this sense, for the estimation of the thickness of the deepest soil formation two assumptions were made: a) the V_s velocity of this formation, remains stable down to the rock basement and b) the V_s value of the rock basement presents large contrast with the upper sediments. These two assumptions allowed estimating the thickness of this formation, which is 180m. Thus, the total thickness of the sedimentary deposits is 420m. To this end, the aforementioned defined and estimated final V_s model was used in order to compute the 1-D theoretical transfer function (bold solid line in Figure 10), which fits well with the empirical H/V spectral ratio (microtremors). We followed the same procedure for the other sites, in which the bedrock depth was not determined. All the final estimated depths of the bedrock in every site are presented in Table 2.

Table 2. Estimation of the bedrock depth with a back analysis procedure.

Site	Exploration. Depth (m) (SPAC)	Estimated. depth, (m) H/V-1D	Fund. Freq. (Hz)	V_s of the last formation SPAC, (m/sec)
KYV	320	790	0.35	1175
KRH	240	420	0.5	820
MPO	180	420	0.4	735
IPO	110	270	0.97	605
LEU	130	250	0.87	930
AGO	60	40	4	950
LIM	180	180	0.9	780
TEL	160	260	0.62	590
STA	80	90	2.2	765
EUO	220	350	0.63	940

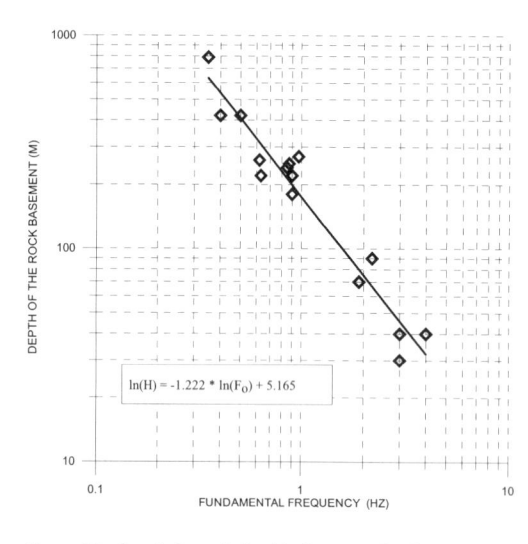

Figure 11. Correlation relationship between the fundamental frequency and the depth of the sedimentary deposits for the soil formations of Thessloniki.

These depths of the bedrock and the fundamental resonant frequency at the 16 sites were used to compute their correlation for the soils of Thessaloniki (Fig. 11). The empirical curve in fig. 11 could be successfully used for the estimation of the thickness of the sedimentary deposits at any site of interest in the frame of the microzonation study of Thessaloniki.

5 GEOTECHNICAL INTERPRETATION AND 2-D CROSS SECTIONS

Using all V_s profiles and the depth of the bedrock and the existing geotechnical data (Anastasiadis et al., 2001), the soils under the city of Thessaloniki were classified into five categories (A1, A, B, C and D).

Formation A1 is a sub-category of the basic formation A, composed of the softest soils, about 10m thick with V_S values of 130m/sec. Geotechnically it is classified as sandy silt (SM) and soft clay (CL). Formation A has V_s velocity values ranging from 150m/sec to 300m/sec with thickness from 5m to 35m. It is classified as clay with low to medium plasticity (CH-CL). In the eastern part of Thessaloniki, the thickness of formation A varies from 5m to 10m and V_s from 250m/sec to 300m/sec. It is composed of clay with gravels (CL-CG).

Formation B is underlying formation A; its V_s values are ranging from 300m/sec to 500m/sec and the thickness is varying from 20m to 130m. Formation C has V_s values from 500m/sec to 750m/sec and thickness ranging from 50m to 200m. Geotechnically, formations B and C are classified as stiff clays and stiff clays with sands (CL, CL-SC). Formation D is the deepest soil formation, overlying the real rock basement, having V_s values varying from 750m/sec to 1100m/sec. Based on all available data many reliable 2-D cross-sections of the subsoil stratigraphy in the city of Thessaloniki were proposed in terms of V_s velocities and proper geological – geotechnical judgment. Figure 12 presents a typical cross-section as an example.

6 CONCLUSIONS

Microtremor measurements within large densely urbanized cities, such as Thessaloniki, led to reliable V_s velocity profiles down to large depths (100m - 300m) using SPAC method. This is significant for accurate site response studies in big cities, where open free spaces suitable for the deployment of large arrays, are difficult to be found and high energy sources cannot be easily accepted. The same microtremor recordings have been used for the estimation of bedrock depths using both H/V technique and 1-D theoretical transfer function.

To this end, microtremors can be considered as a valuable complementary or alternative tool to determine V_s velocity and soil stratigraphy, which are the main requirements for soil and site characterization, site response analyses, microzonation studies, design of infrastructures and urban planning.

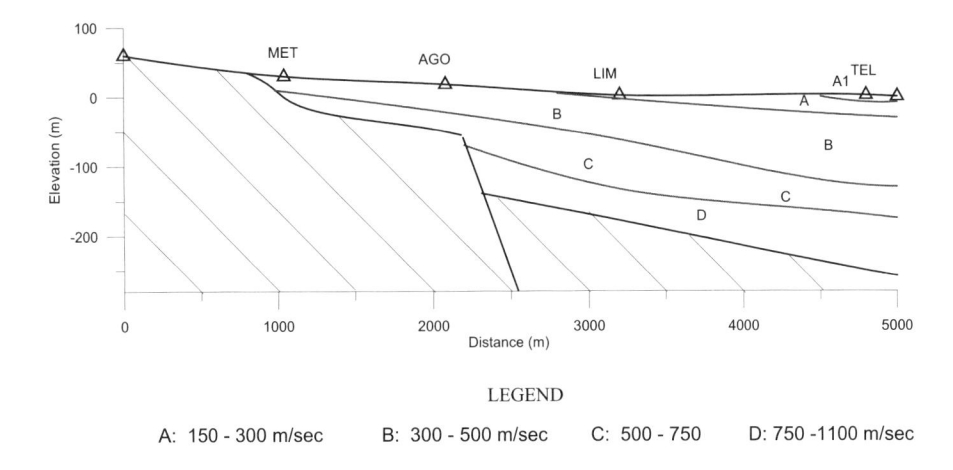

Figure 12. An example of a characteristic 2-D cross section that describes the basic soil formations of Thessaloniki city

REFERENCES

Aki K., 1957. *Space and Time Spectra of Stationary Stochastic Waves, with Special Reference to Microtremors*. Bull. Earthq. Res. Inst. Tokyo Univ. 25, pp. 415-457

Anastasiadis, A., D. Raptakis, and K. Pitilakis, 2001. *Thessaloniki's Detailed Microzoning: Subsurface Structure as Basis of Site Response Analysis*, PAGEOPH Special Issue on Microzoning, Vol. 158, N.11, pp. 2497-2533.

Apostolidis P., D. Raptkis, Z. Roumelioti & K. Pitilakis, 2004. *Determination of S-wave velocity structure using microtremors and SPAC method applied in Thessaloniki (Greece)*. Soil Dynamics and Earthquake Engineering 24, pp 49-67

Henstridge, J.D., 1979. *A signal processing method for circular arrays*, Geophysics, 44, pp. 179-184.

Herrmann, R., 1985. *Computer programs in seismology*, vol. III., Saint Louis University

Nakamura Y., 1989. *A method for dynamic characteristics estimation of subsurface using microtremors on the ground surface*.QR of RTR, 30-1, pp 25-33

Okada H., 1997. *A new method of underground structure estimation Using Microtremors*. Division of Earth Planetary Sciences, Graduate School of Science, Hokkaido University, Japan, Lecture notes.

Okada H., 1999. *A New Passive Geophysical Exploration Method Using Microtremors*. Division of Earth Planetary Sciences, Graduate School of Science, Hokkaido University, Japan, Lecture notes.

Pitilakis, K.D., Anastasiadis, A.I., Raptakis, D.G., 1992. *Field and Laboratory Determination of Dynamic Properties of Natural Soil Deposits*. Proc. 10th World Conference on Earthquake Engineering, Madrid, Vol. 3, pp. 1275-1280.

Raptakis, D.G., A.J. Anastasiadis, K.D. Pitilakis and K.S. Lontzetidis, 1994. *Shear wave velocities and damping of Greek natural soils*. Proc. 10th European Conf. on Earthq. Engng, Vienna, Austria, pp. 477-482.

Raptakis D., K. Makra, A. Anastasiadis & K. Pitilakis, 2003, *Complex site effect in Thessaloniki (Greece):I. Soil structure and confrontation of observations with 1D analysis*, Submitted in Bulletin of Earthquake Engineering

Raptakis D., K. Makra, A. Anastasiadis & K. Pitilakis, 2003, *Complex site effect in Thessaloniki (Greece): 2D SH Modeling and engineering insights*, Submitted in Bulletin of Earthquake Engineering

Raptakis, D.G., A.J. Anastasiadis, K.D. Pitilakis and K.S. Lontzetidis, 1994a. *Shear wave velocities and damping of Greek natural soils*. Proc. 10th European Conf. on Earthq. Engng, Vienna, Austria, 477-482.

New de-coupled shear wave source for the SCPT test

L. Areias, W. Haegeman & W.F. Van Impe
Department of Civil Engineering, University of Ghent, Belgium

Keywords: seismic source, shear waves, SCPT test, de-coupling

ABSTRACT: This paper describes a new seismic source to generate high-energy shear (S) wave signals for the seismic cone penetration (SCPT) test. The new source uses an innovative load de-coupling system that significantly reduces the effective overall mass of the source. Test results using the new source are presented and compared with results from a conventional source. The results show that S wave signals generated with the new system contain 3 to 4 times more energy than signals produced with existing conventional methods.

1 INTRODUCTION

The static cone penetration (CPT) test has been modified to measure seismic wave velocity in soils (Robertson et al, 1986). The SCPT test measures shear (V_s) and compression (V_p) wave velocities, from which the maximum shear (G_o) and elasticity (E_d) moduli are derived. G_o and E_d are related to the density of the medium (ρ) and Poisson's ratio (ν) by:

$$G_o = \rho V_s^2 \quad (1)$$

$$G_o = \frac{E_d}{2(1+\nu)} \quad (2)$$

Values of V_s and V_p are further related to ν by the equation:

$$\nu = \frac{\left(\frac{V_s}{V_p}\right)^2 - 0.5}{\left(\frac{V_s}{V_p}\right)^2 - 1.0} \quad (3)$$

SCPT tests are performed using a CPT truck. This allows both geotechnical and seismic data to be obtained simultaneously, in the same test. It also makes the SCPT test efficient and cost effective.

2 WAVE GENERATION

Shear waves in SCPT tests are generated at the surface using a beam-and-hammer type source. The beam or base plate is weighted down using the weight of the CPT truck to obtain firm and secure coupling with the ground. The weight of the CPT truck is placed directly on the beam. This method of weighting the beam is here called *coupled* loading.

The ends of the beam are struck horizontally to produce pairs of predominantly polarized S waves. Hitting the beam vertically generates predominantly compression (P) waves.

Generated seismic waves are detected by geophones (or accelerometers) installed in the cone. Typically, an array of 3 orthogonally-oriented geophones is located in the cone's housing, close to the cone's tip. Two of the geophones are installed horizontally to record the S waves. The third geophone is positioned vertically to capture the P signals.

3 OPTIMUM COUPLING FOR S WAVES

The amplitude of surface-generated S waves in the SCPT test partly depends on the level of coupling stress (Areias et al., 1999). Coupling stress is here defined as the normal static stress between the base plate and the ground.

Test results by the authors indicate that relatively low coupling stresses are required to generate maximum amplitude S wave signals, for given impact energies (Figure 1). These tests were performed using a specially designed 24 kg adjustable mechanical swing hammer to produce controllable and repeatable energy impacts. The results shown were obtained with 15° and 45° impact swing angles. The difference in impact energy between the two swing angles is approximately double.

It is common practice to weigh down the source using the weight of the CPT truck. Since a CPT truck can weigh up to 20 tons, a wide range of coupling stresses can be generated on the beam when it is loaded in this manner. As a result, SCPTs are being performed using a variety of coupling stresses. This can significantly reduce the energy and quality of the signals.

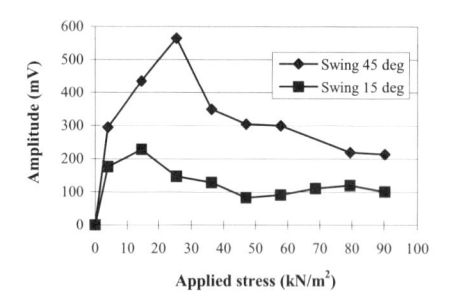

Figure 1. Variation of S wave amplitude with coupling stress in coupled-load tests (Areias et al., 1999)

4 MODELING OF S WAVE SOURCE

The base plate is modeled as a suspended simple beam to study its physical interaction with the ground. Assuming full elastic conditions during impact, and that the collision between the hammer and the beam is one dimensional, we can write the following expressions relating mass and velocity of the source system. From conservation of momentum we have:

$$m_h V_{hi} + m_b V_{bi} = m_h V_{hf} + m_b V_{bf} \quad (4)$$

Where:
m_h = mass of sledgehammer
V_{hi} = initial velocity of sledgehammer
m_b = mass of impact beam (base plate)
V_{bi} = initial velocity of impact beam
V_{hf} = final velocity of sledgehammer
V_{bf} = final velocity of impact beam

Since the collision is elastic, kinetic energy is conserved, by definition, and we obtain:

$$\frac{1}{2}m_h V_{hi}^2 + \frac{1}{2}m_b V_{bi}^2 = \frac{1}{2}m_h V_{hf}^2 + \frac{1}{2}m_b V_{bf}^2 \quad (5)$$

Rearranging (4) and (5), and noting that the base plate is initially at rest before collision, we can write the following expression for V_{bf}:

$$V_{bf} = \left(\frac{2m_h}{m_h + m_b}\right) V_{hi} \quad (6)$$

Equation (6) is only valid for a suspended beam, as noted. When the beam contacts the ground, a ground force (F_g) will act between the beam and the ground to resist movement of the beam. It is generally assumed that F_g is proportional to the particle velocity (V_x) of the ground just below the beam (Van der Veen et al., 1999; Miller and Pursey, 1954) and is expressed by:

$$F_g = V_x \, Z_{rad} \quad (7)$$

Where:
Z_{rad} = radiation impedance

Radiation impedance is a measure of the resistance of the ground to motion when a force is applied (Miller and Pursey, 1954). This represents the interaction of the source with the ground.

Assuming a purely physical interaction and perfect coupling between the beam and the ground, the particle velocity just below the beam will be equal to that of the beam itself, i.e. $V_x = V_{bf}$.

Noting that the mass of the base plate is much greater than that of the hammer ($m_b \gg m_h$), we can deduce from (6) and (7) that the ground force is inversely proportional to the mass of the plate, for given hammer impacts.

5 NEW S WAVE SEISMIC SOURCE

Conventional sources are weighted down by weights coupled directly to the impact beam, as shown in Figure 2. According to equations (6) and (7), this type of loading increases the overall mass of the beam-hammer system and should lead to a decrease in F_g and S wave energy.

A new seismic source system was designed to verify the results expressed by (6) and (7). Specifically, a new hold-down-load system was developed to reduce the overall mass of the source. This was done by isolating applied hold-down loads from the base plate.

The new system, herein referred to as *de-coupled* source system, uses a system of rollers (or other friction-reducing mechanism) to effectively decouple applied loads form the base plate along the horizontal direction. A simplified design of this system is shown in Figure 3.

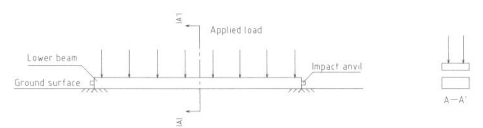

Figure 2. Typical conventional coupled source

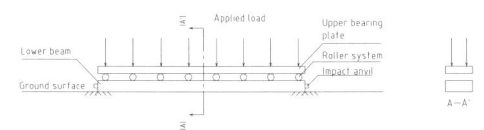

Figure 3. Simplified design of de-coupling mechanism for new seismic source

6 TEST RESULTS

Field tests were performed using both conventional and the new de-coupled seismic-source systems. These tests were carried out to evaluate the performance of the new de-coupled source system. The tests were performed using the adjustable mechanical swing hammer previously described to produce controllable and repeatable energy impacts. The tests were further performed using a solid steel beam as base plate.

Typical test results, showing amplitude as a function of applied coupling stress, are presented in Figures 4 and 5. Figure 4 shows results obtained using relatively low energy hammer impacts produced using 15° swings. Figure 5 presents results obtained using hammer swing angles of 45° for higher energy impact.

The results show that the amplitudes of S signals generated with the new de-coupled source are 3 to 4 times greater than those obtained using a conventional source. The tests also confirm the results of the physical modeling study, which suggested that F_g could be significantly increased by lowering the mass of the beam.

7 CONCLUSIONS

A newly designed de-coupled seismic source system for the SCPT test method has been briefly described. The new source uses an innovative load de-coupling system that effectively reduces the overall mass of the source to increase S wave energy.

Field tests show that amplitudes of S signals generated with the new system are 3 to 4 times greater than those obtained using conventional sources. The new system should improve SCPT testing in general by increasing S wave signal-to-noise ratios, extending test depth and enhancing accuracy of measured signals.

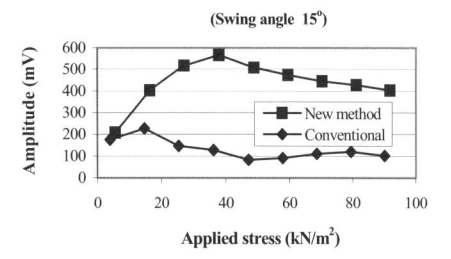

Figure 4. Comparison between new and conventional sources for hammer swings of 15°

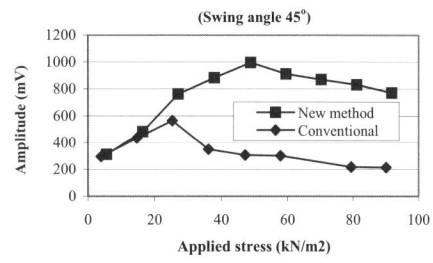

Figure 5. Comparison between new and conventional sources for hammer swings of 45°

ACKNOWLEDGEMENTS

The authors are grateful to Dr J. Brouwer for his helpful suggestions at the early stages of this research.

REFERENCES

Areias, Lou, W.F. Van Impe and W. Haegeman (1999), 'Variation of Shear Wave Energy with Coupling Stress in the SCPT method', *European Journal of Environmental and Engineering Geophysics*, Volume 4, Number 1, pp. 87-95.

Miller, D.F. and H. Pursey (1954), 'Field and Radiation Impedance of Mechanical Radiators on a Free Surface of a Semi-Infinite Isotropic Solid', *Proceedings of the Royal Society A 223*, pp.521-541.

Robertson, P.K., R.G. Campanella, D. Gillespie and A. Rice (1986). Seismic CPT to measure in-situ shear wave velocity, *Journal of Geotechnical Engineering*, Volume 112, nr. 8, pp. 791-803.

Van der Veen, Michiel, Jan Brouwer and Klaus Helbig (1999), 'Weighted Sum Method for Calculating Ground Force: an Evaluation by Using a Portable Vibrator System', *Geophysical Prospecting*, 47, pp. 251-267.

Comparing 360° televising of drill hole walls with core logging

R. T. Baillot & A. Ribeiro Jr.
Alphageos Tecnologia Aplicada S.A.

N. Barton
Nick Barton & Associates

R. Abrahão
FUNCATE - São Francisco Water Transfer System

Keywords: site investigation, core logging, rock mass characterization, televising boreholes

ABSTRACT: This paper presents a method of investigation of rock masses, used as a complementary or alternative process to diamond core drilling. A first version of the equipment for filming was developed in Japan and presented on the occasion of the ISC '98 (Atlanta, USA, 1998), but its production was not continued. The present version, treated in this paper, was developed in France by René Colas (1998). It was introduced into Brazil in November 2001, and into Germany in 2003. In the first part of this paper the fundamentals of the method are presented. Improvements made during its use are used to highlight the gains in productivity when investigating rock masses by these methods. There follows information about the equipment used (hardware and software), and the method of reporting results. Thereafter, data, results and possibilities of application of the method are summarized and discussed, especially concerning the São Paulo Metro – Line 4, and the San Francisco River Water Transfer project, maybe, one of the first times in which the televising of percussion boreholes has been used on a large scale in the world. As a basis for analysis of the images obtained in these site characterization campaigns, an application of the Q-system classification of rock masses was made, for comparing conventional core logging with the televising of percussion-drilled borehole walls. We were therefore comparing disturbed (core) samples with the relatively undisturbed borehole walls, making possible the identification of some factors that influence the disturbance of recovered core samples. The comparative qualities of the same rock masses are presented in the last part of the paper, using the six Q-parameters.

1 GENERAL CONCEPT

The method reported here is applied to the televising of the walls of bore-holes with diameters equal to or larger than 75 mm, and up to 250 mm, comprising two principal phases: image acquisition and image processing. For the first phase an image acquisition module is used, that is a segmented tube that houses a series of devices. In the lower segment, with transparent walls, lamps and a conical mirror are included. In the upper segment, with steel walls, are located a digital camera, a magnetic orientation system and the hardware necessary for the transmission of the camera images to the computer on the surface. When operating, the module is suspended on a cable and inserted into the borehole, moving through the target section. Thus the cylindrical wall of the hole is illuminated at the same time that its images, reflected by the conical mirror, are aimed at the digital camera. Due to the action of the conical mirror, the images obtained and sent by the camera are annular, from which the second phase is performed in the computer, using software especially developed for the restitution of the cylindrical form of the walls, pictured as a developed surface or as a virtual core sample. Details of the equipment are presented in Sec. 4.

In a typical form of application, the televising *begins* in diamond-drilled boreholes, permitting the technicians to familiarise themselves with the images of the various domains of the mass and the recovered core. From then on, the method can be used on larger numbers of cheaper and much faster percussion borings.

2 PRINCIPAL ADVANTAGES

The method is notable for improving productivity in the performance of the site investigation, leading to a reduction in time and costs, and gives operational simplifications, as follows:

Table 1. Relative cost between methods

CORE DRILLING

Description	Unit	Quant	USD/M	Total USD
Core drilling Ø 100mm	meter	100	150.00	15,000.00
			Total USD	15.000,00

TELEVISING, PERCUSSION AND CORE DRILLING

Activity option	Unit	Quant	USD/M	Total USD
Core drilling Ø 100mm	meter	40	150.00	6,000.00
Percussive drilling	meter	60	50.00	3,000.00
Televising services	meter	100	40.00	4,000.00
			Total USD	13.000,00

2.1 Productivity improvements

This is the result from the mobility of the equipment (mobile unit) and of the option to perform the televising in percussion holes, on a large scale. With the accesses prepared and depending on the distances between holes, the average daily production of a rotary-percussion borer, in the boring of holes 100 mm in diameter, is in the range of 60 to 100 meters. To the extent that they are readied, the holes can be televised at the rate of 60 meters per day per crew. The combination of percussive boring and televising covers 6 to 10 times more linear meters of hole per unit time, than diamond core drilling.

2.2 Reduction of time and cost

Greater productivity leads to reduction of the time schedule and the cost of the geotechnical investigation campaign. In the execution of 100 meters of rock boring with a diameter of 100 mm, the combination of percussive drilling and televising reduces the direct prices to about 86% of the price of diamond core drilling (Table 1). Bigger cost reduction is seen with more percussion holes.

2.3 Ease of visualisation and description of rock mass

Experience in Brazil demonstrates that televising reveals the rock mass close to its natural condition of occurrence, permitting the recognition of colour, granulation, texture, jointing and filled discontinuities. In general, the images indicate that the conditions of the mass are not as *unfavourable* as the core samples alone might suggest. This will be shown later on, in relation to the application of the Q-system for describing rock mass parameters. Among the improvements in visualisation and description of the mass are found the following:

2.3.1 Precise positioning of discontinuities in the mass

As the walls of the hole are pictured with a coverage of 100%, and thanks to the depth counter that registers the position of the sections pictured, the images locate the discontinuities in the mass exactly, while in core drilling the segments that compose the sample are not always completely recovered, thereby introducing inaccuracies in the evaluation (details below).

2.3.2 Automatic computing of discontinuities

In the images of the walls of the holes, formed as a developed cylindrical surface, the traces of dipping discontinuities appear as sinusoidal waves. The system software automatically computes the corresponding geological strike and dip angle, through the marking of three points in each trace, an operation performed by locating the cursor and clicking the mouse at each point. Linked to other computer applications, the system software provides a representation of the discontinuities in rose diagrams or lower hemisphere polar projections, or 'pap-logs' of dip and strike.

2.4 Ease of storage of bore profiles

With the elimination of most of the core samples of the rock mass, the televised profiles are filed electronically. A single CD, with 800 Mb memory, stores images equivalent to 500 meters of drilling. In core drilling, the boxes of samples must be preserved for a certain time, occupying physical space, and re-examination of samples implies visits to the storage location, manipulation of the boxes and correction of the profiles, requiring greater time in comparison with the time spent in re-examination of image profiles. If televising is used to complement drilling, photos of the real samples may be located with the virtual samples along the televised profile, permitting comparisons of the positioning and origin of discontinuities of the mass, either natural or induced, or those actually destroyed during drilling and core recovery.

2.5 Preparation of the report of results

In addition to the traditional printed format, the report of results of the televising of the bore-hole walls is furnished recorded on CD(s) that cannot be edited but permit viewing the entire profile of the hole (image of the hole wall surface and of the "core" sample), accompanied by the "pap-log" graph

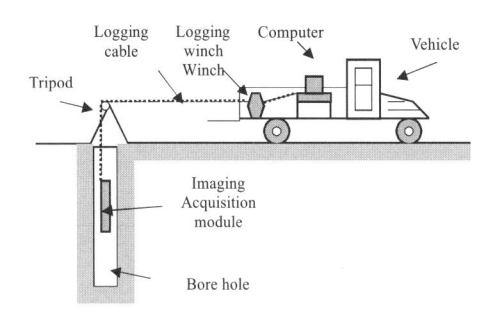

Figure 1. The basic equipment

indicating the angle of strike and dip of the discontinuities along the profile, with a legend of the respective classes (joints with and without fill, etc.). Polar projections and rose diagrams are options to be combined with the body of the report (see details of the software below in Item 4.2).

3 EQUIPMENT DETAILS

3.1 *Hardware*

The equipment used in the televising method is mounted and transported on a rubber-tyre vehicle, comprising a series of units (Figure 1).

3.1.1 *Image acquisition module*
This consists of a segmented tube (Ø=60mm) (see next, Figure 2); that is introduced into the borehole (minimum diameter 75mm) and connected to the computer installed on the surface near the hole. It houses several devices, particularly *lamps* to illuminate the wall of the bore-hole, a *conical mirror* to reflect and direct the wall images, a *digital video camera* to capture the images, and an *orientation system* so that the images will be referenced to the cardinal points. The focal distance of the camera may be adjusted. A conical device made of high-resistance plastic is connected to the lower end of the module to protect the apparatus from direct shocks in case of obstruction in the bore-hole. The upper cover of the module is provided with a central hole for passage of the support cable (see following item).

3.1.2 *Module suspension and displacement system*
This system consists of a *logging cable* Schlumberger 92.264 4.18P, linked to the interior of the *image acquisition module;* the cable, controlled

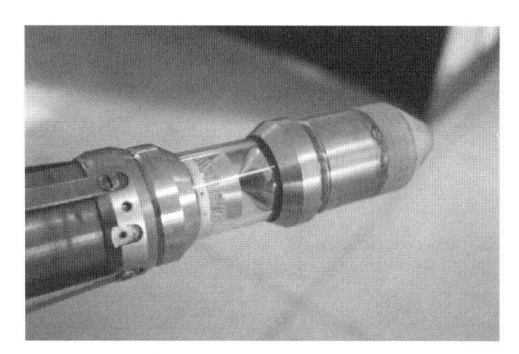

Figure 2. Imaging acquisition module

by a *logging winch* on the surface, also provides electric energy to the module and electronic transmission of the data collected. Provided with a *depth counter*, the system monitors the position of the module in the bore-hole in real time.

3.1.3 *Computers*
To acquire and process the images, the computer must have a RAM memory of 512 Mb, a HD of 40 Gb, an AGP video board with 64 Mb and a CD recorder. In addition, the image acquisition computer must be equipped with an additional mobile HD, CPI board specifically for the depth counter. For simple visualisation of the images, without editing, the computer must be provided with a Windows NT or XP (NTSF) system, a RAM memory of 128 Mb, and a drive for the CD-ROM.

3.2 *Software*

The annular and oriented images sent by the module are processed by the software, providing a quite ample view of the profile of the rock mass along the bore-hole, with control of the depth in each section. The final images can therefore be formatted as a developed cylindrical surface or as the virtual equivalent of a 'core sample' that is seen to rotate, or is held still, as the user requires. During the entire process, the computer works with three programmes, as follows

3.2.1 *Maki Cam*
The acquisition programme for the bore-hole wall image is actuated by a computer with characteristics described above. The image acquisition programme makes images of the wall about 30 cm long, through an optical arrangement consisting of a group of 12 lamps reflecting on the surface of a truncated cone of polished aluminium, which illuminates the wall, the image of which is reflected by a conical mirror and captured by a camera housed in the interior of the module. Each image of 30 cm or 624 pixels in

length, is composed of 156 slices of the image, taken at a speed of about 0.5 m/minute, with its depth and orientation registered.

3.2.2 *Maki Cad*

This is the image processing programme, which integrates the slices along the entire profile of the hole, composing a single image, including correction of possible distortions. The rose diagrams, stereograms and profiles of the 'pap-log' type, are produced automatically, at the termination of the plot of joints and filled fractures, being discrete according to the classification chosen for the discontinuities by the user.

3.2.3 *MakiVision*

A unique vision programme, especially created for the user to take advantage of all its resources with efficiency, without the necessity of a learning period, providing the opportunity to translate the images at various speeds and the reconstruction of "virtual samples" and their animation. The software is intended for the visualisation and analysis of the Image Profiles, Sinusoidal Profiles of the discontinuities and their Profiles, as well as the polar profile of the strike and dip of each of the discontinuities, and their Rose and Stereograph Projections, with an option for printing them. Each file sent to the client is accompanied by instructions on operation of the software, the simplicity of which assures its complete operation without any prior training, and in any type of computer, fixed or portable, that has a memory greater than 126 Mb.

4 BRIEF NOTES ON PROJECTS IN WHICH TELEVISING IS BEING USED MORE INTENSIVELY IN BRAZIL

The present work is based on televised data collected in the investigation of rock masses in projects of great importance, which will be constructed, now in development in Brazil. To these have been added others collected in experimental holes, as described below:

4.1 *São Paulo Metro – Line 4*

In the São Paulo Metro transport network, the pioneer 20,2 km North-South line; the 22,0 km East-West line, integrated with the commuter rail lines; and the 7,0 km Paulista Branch, linked to the North-South Line, have already been implemented. In sum, the existing system has capacity to transport 5,4 million passengers per day, but the population is 17 million. Now, work is beginning on the 6,7 km Line 4, linking the Centre to the Western Zone of the city.

Table 2. Site investigation résumé

São Francisco river water transfer system			
Site investigation	bore holes	length(m)	length %
Core drilling	17	447.66	31.6
Core drilling with televising	5	95.30	6.8
Percussive drilling and televising	48	874.00	61.6
Total	70	1,416.96	100.0

In general, Line 4 will be developed in tunnels, which may be excavated using either a Shield and TBM, or the NATM method, depending upon subsurface conditions.

4.2 *São Francisco River Water Transfer System*

By means of a partial diversion of the flow (average about 2.060 m^3/s) of the São Francisco River, the project proposes to irrigate arid areas in the Northeast of Brazil. To cross the mountain ranges between the contributory and beneficiary basins, as well as the use of natural stream beds, a series of channels, tunnels and pumping stations have been designed, involving excavation of rock and soil. The operational optimisation of the system will be achieved through implementation of reservoirs in the beneficiary basins and installation of hydroelectric power plants, with partial recuperation of the energy to be expended in pumping. The rock masses involved in the future construction of the project were investigated by means of televising core drilled holes and larger numbers of percussion drilled holes.

4.3 *Experimental bore-holes*

In addition to the cases in which televising was used in development of large projects, in others the method was employed in isolated holes, or in an experimental manner. In the initial situation, there was the necessity of complementing the data from core drilling, in gapped sedimentary rock, proposed for location of a dam on sandstone (the Furnas Formation). During the core drilling, to a depth of 15 meters, no recuperation of solid material was made, but its walls could be televised. In another case, a rotary boring was made in granite in the Barueri region, exclusively for tests of televising the walls in a section of rock at depths between 17 and 43 meters (or more precisely 17.46 and 42.60 m).

5 Q-SYSTEM FOR DESCRIPTION OF ROCK MASSES AND 360° TELEVISING OF HOLES

5.1 *Evaluation criteria for rock masses*

The proposed format for recording the statistical variation of the six Q-parameters has recently been described by Barton, 2002. A sheet contains six

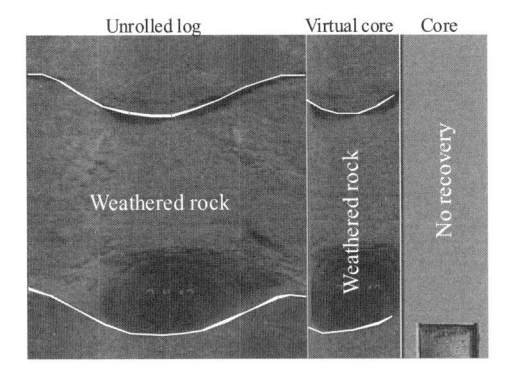

Figure 3. Weathered rock image

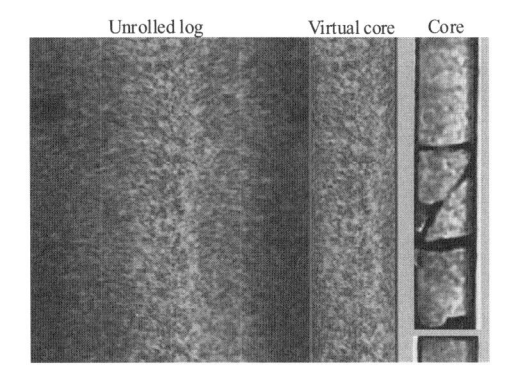

Figure 4. Incipient fractures image

main areas for recording the frequency of observations of the six Q-parameters, which are combined in the following way to calculate Q:

$$Q = \frac{RQD}{Jn} x \frac{Jr}{Ja} x \frac{Jw}{SRF}$$

(Note: Q=0.1-1= very poor, 1-4 poor, 4-10 fair, 10-40 good etc.)

The definitions of these six parameters is shown to the right-hand-side of each histogram, and the relevant descriptions and ratings are given above and below each histogram. Hand-written recording is followed by EXCEL plotting and calculation.

5.2 *Some typical conditions of masses observable by televising*

In recent rock mass investigative campaigns in Brazil, virtual and real samples were compared, indicating the utility of televised images in complementing data from rotary borings, revealing certain features of the mass, such as alteration and fracturing of the rock, as follows:

5.2.1 *Tension of weathered rock*
Along the televised profile, the extension of altered rock can be seen, generally as a function of the colours and reduced brightness that appear. Figure 3 shows images from the survey of the rock to be crossed by Line 4 of the São Paulo Metro.

5.2.2 *Incipient fractures*
In the case of core drilling in rock masses with incipient fractures, the forces and vibrations of the drilling induce an increased development and frequency of fractures in the samples, producing low RQD indices, possibly leading to underestimates of the quality or support capacity of the mass. The televised images presented in figure 4, show a granitic rock mass from the Barueri region, São Paulo State, with closed or poorly developed joints, the recognition of which required a careful and detailed examination of the developed cylindrical surface. The televising, proving the incipient character of the joints, and showing that their development is too small to cut through the virtual sample, suggests that some increase could be applied to the rock quality index, above that which examination of the real core sample would lead one to apply.

5.2.3 *Voids and erodible filling materials*
These occurrences produce low recovery from cored samples. As a consequence, the juxtaposition of segments that compose the sample does not help to determine the exact positioning of the discontinuities. Boring equipment with hydraulic jet does not correctly sample filled discontinuities of this type and, in consequence, in the core box, the filled discontinuities are not correctly represented. (With old borings, from mechanical drills operated manu--ally, a well-trained surveyor could detect such an occurrence, and place a wooden spacer.) In televised images, the discontinuities appear in their natural positions, making possible the determination of the individual thicknesses of the filled discontinuities, of the layers of materials, or the dimensions of the voids. Figure 5 refers to the detection of a horizontal fracture, with a 6 cm opening in the mass, partially filled with rock fragments. In the example, it can be verified that the juxtaposition of the segments of the real sample does not permit the definition of the position of the fracture with any assurance. RQD estimates are therefore changed.

5.2.4 *Vertical fractures flanked by incoherent materials*
This situation is pictured in Figure 6, in which the disturbance of the boring itself lead to washing of the incoherent materials (strongly weathered rock

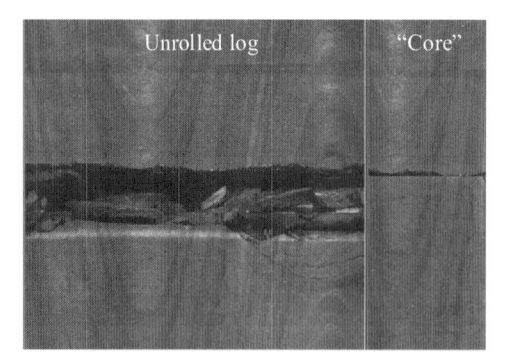

Figure 5. Voids and washable material image

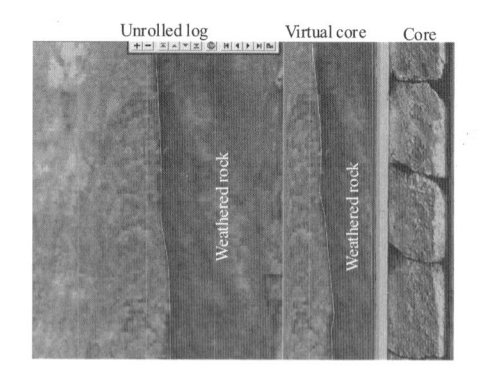

Figure 6 Voids and erodable material image

along one side of the vertical fracture) and intense fragmentation of the remaining rock, in spite of the greater coherence of this material. In this case the samples found in the core box are nothing more than an agglomeration of disconnected fragments, with a zero RQD index, it not even being feasible to conjecture the percentage recovery of the real sample. The televised data remain the only resource available for an evaluation of this particular rock mass location.

5.3 Q-index of some masses studied in the design of São Paulo Metro

For the study of the rock mass quality index, both by the examination of real samples and televised images, the profiles of rock mass borings were surveyed, in the course of the geotechnical investigation campaign for the final design of São Paulo Metro – Line 4. The criteria of selection were: 1) the study of diverse masses, not only in lithological terms but also as to the degree of alteration and fragmentation of the rock; 2) make possible the checking of the classification of the mass by different consultants who worked in the project development and/or the implementation of the works.

In accordance with these criteria, eight domains of rock were selected, crossed by borings SM-04 and SM-11 in the investigative campaign, of which the holes were also televised. At each domain possible, variations of basic parameters and respective quotients that define the rock mass quality index were evaluated, based on the examination of real samples and the analysis of images.

The variations having been identified, the results reflect the best and the worst values that can be indicated for the quality of the various domains of mass analysed. In the case of the televising, the lengths of the segments that compose the virtual samples, separated by visible fractures in the mass in its natural state, were verified, for assignment of equivalent RQD indices. The selected domains and the tabulation of results referring to the evaluation of the mass quality indices are presented in condensed form in Figure 7.

6 DISCUSSION AND CONCLUSIONS

Examination of results in the previous procedure confirms the expected tendency that, using televised images, higher quality classes may be assigned to rock masses, as compared with an evaluation based on diamond cored samples. To show better this result, the Q-index variation found for the eight domains of rock were located on Q-system classification scale (Barton, 2002) as indicated in Figure 7, in correspondence to the best and worst conditions of the mass in each domain studied. Figure 8 represents an exemple of the Q-parameter histogram method of logging (Barton,2002) with a conceptual image of likely bi-modal distribuition of RQD, J_n and J_a, that is expected to be obtained if television wall logging and core logging were combined on one sheet. In general, the results lead to the following discussion and conclusion:

6.1 Variation of the RQD/J_n Quotient

This quotient, where denominator represents the number of joint families, shows the tendency to assign to rock mass classifications relatively higher values by the televising method. If the best condition is considered, the quotient evaluated by televising is two or three time greater than that obtained by core logging. If, on the other hand, the worst condition is considered, the tendency is accentuated, with the quotient based on televising becoming three to six times greater than the other. This result is also

SÃO PAULO METRO: GEOMECHANICAL CLASSIFICATION OF ROCK MASSES

Bore-Hole n°	Depth m	Domain n°/ litology	Rock Mass Classes Related to Q Value								
			Ext.poor 0,01		Very poor 0,1	Poor 1	Fair 4	Good 10	Very good 40	Ext. good 100	Exc. good 400 1000
SM L4 04	35,90	1 milonite		0,17 - 0,3				11,0			
	37,30	2 q. vein					7,5 - 22,5	16,5			
	38,76	3		0,2			4,9				
	40,15	milonite						11 - 16,5			
SM L4 11	19,35	1 gnaiss					7,5 / 7,4	11,2 / 16,5			
	22,00	2 gnaiss		0,4			4,5 / 7,4	33,0			
	28,20	3 migmatite		0,4		3,8	7,4	33,0			
	33,82	4 schist	0,0				8,7	11,2	50,0		
	38,20	5 migmatite		0,4			4,9	16,5 / 33,0			
	53,24										

CORE / TELEVIEWER

Figure 7. Comparison of core drilling and televiewer logging

reflected in the relative positioning of the classifications indicated in Figure 7. The reason for this is evident in the conception and application of the televising method, in that the images obtained, with an *index of wall recovery* of 100%, picture the mass at the periphery of the boring, the zone least subject to damage induced by mechanical boring, and therefore in a condition much closer to natural. The segments of the video wall sample tend to be separated only by natural fractures, the incidence of segments with a length of less than 10 cm being quite rare, leading to RQD wall parameters greater than those produced by rotary borings.

Another reason for this result is the variation of the Jn parameter, the number of families of joints in the mass. At times, the evaluation of this parameter coincides in both the methods, while televising rarely indicates a lower value. However, the fragmentation of boring samples, greater than that of televised wall samples, results in worse values for this parameter, combining to reduce the quotient RQD/Jn. In this respect, it is important to also consider that in the televising method, the measured joint traces are easily transferred to a polar diagram in which the families are easily recognised.

6.2 J_r / J_a Quotient

In masses of sound rock, with high RQD indices, the examination of boring samples permit a very realistic evaluation of the Jr parameter, referring to the roughness of fractures. In the opposite case, with low RQD indices, the evaluation of this parameter is somewhat impeded due to the creation of fractures by the boring and to the reworking of the pieces themselves that make up the samples, disfiguring the natural characteristics of the surface of the joints. When the joints or discontinuities are open, not filled, a view of their walls can also be had by televising. On the other hand, so as not to harm the evaluation of the parameter Ja, referring to the alteration of the mass along the fractures, it is indispensable that careful attention be paid to the images of the first cored holes. A tendency to overevaluate this parameter may occur in televising, in rarer cases in which the images do not permit distinguishing altered or filled material from the rest of the mass. With all these possible tendencies, some of them prejudicial, the results of the quotient Jr / Ja may tend to one or the other side, and not indicate a consistent tendency for this or that method. The correct evaluation of the parameter Ja results from good technique in both methods being compared, with regard to recovery of core samples, and gauging borehole wall images in cored holes.

6.3 J_w / SRF Quotient

The parameters that define this quotient are difficult to evaluate, either by the conventional method of investigation of rock masses, with examination of boring samples, or by televising. The denominator of the quotient SRF, Stress Reduction Factor, that pictures the state of stress in the mass, is a parameter generally evaluated as a function of more ample knowledge of the mass, including considerations of prior experience in excavations, or results of tests in situ, in short, independently of isolated data obtained by core drilling or by televising. In the case in point, the rock mass in the São Paulo Metro – Line 4, the SRF parameter is evaluated as 1.0. The parameter Jw, in turn, refers to the value of water flows through the joints, which, in whatever hypothesis, always should be checked with especial attention during and after excavation. Divergences may arise in the evaluation of this parameter, to the extent that the televising reveals materials or fill in the fractures, reducing possible flow, at the same time that these materials are washed out in the process of core drilling, possibly increasing flow. Thus, in the specific case of the rock masses analysed in the present work, in the examination samples this parameter was assigned class C: high value flows or under high pressure along joints without fill, with a value equal to 0.5, while in televising the same parameter was assigned to class B: medium value flows or under medium pressure along the joints, with occasional out-washing of fill material. However, the experience acquired in excavations of the rock mass, may recommend an alteration of both evaluations.

6.4 Final Considerations

In the example shown in the spread-sheet (see Figure 8) we have imagined the combined result of conventional core logging using this system, together with the likely result of 360° borehole televiewer logging. For purposes of emphasis, we have shown how the recording of three of the parameters may be quite different, due to the better preservation of sensitive features in the borehole wall, as compared to the disturbed (or partly washed away) feature in the core. The bi-modal distributions of RQD, J_n and J_a show where the chief differences are likely to occur, due to reasons that will have become obvious, when studying the visual images of core-and-wall comparisons.

The advantage of recording observations in Q-system format is that many empirical links to rock mass parameters have been developed, for example deformation modulus and measures of rock mass strength, including the cohesive and frictional components of the rock mass – i.e. those requiring shotcrete and rock bolt support in the context of tunnelling. Q also correlates closely with the seismic velocity, making cross-checking (and extrapolation) between televiewer logging and remote sensing possible. Intuitively, the correlation between any televiewer logging of Q and velocity, should be more reliable than the correlation of any core logging of Q and velocity, due to the reduced disturbance. However the empirical correlation might change as a result.

The graph of Figure 7 shows, in general, that the most unfavourable classification of the mass, found by analysis of televised images, is situated in a category equal to, or immediately higher than the most favourable classification assigned to the same rock mass based on examination of core samples.

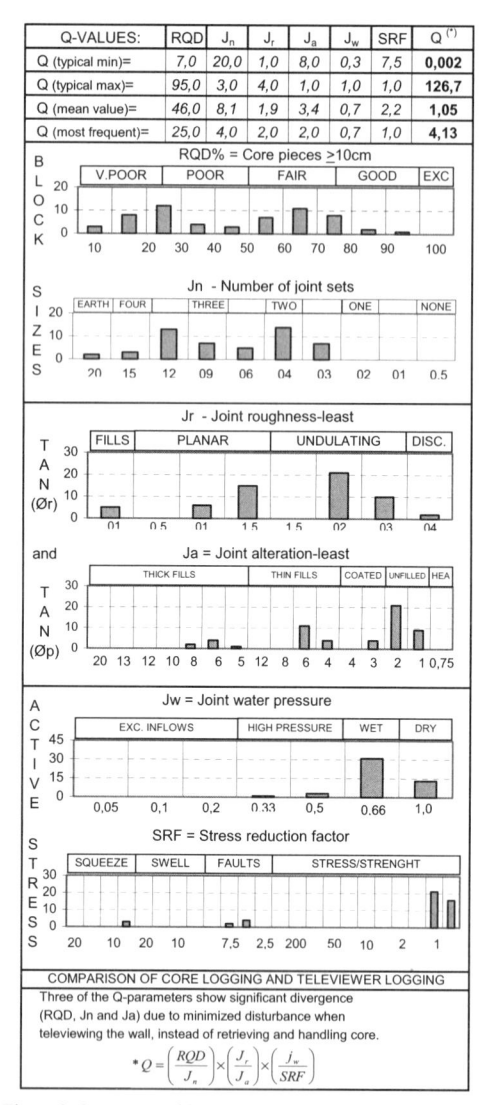

Figure 8. Q-parameter histogram

$$*Q = \left(\frac{RQD}{J_n}\right) \times \left(\frac{J_r}{J_a}\right) \times \left(\frac{J_w}{SRF}\right)$$

REFERENCES

Barton, N. 2002. *Some New Q-Value Correlations to Assist in Site Characterisation and Tunnel Design*, International Journal of Rock Mechanics & Mining Sciences 39, 2002,: 185-216.

Maki Vision version 3.13. 2001. *Software for visualization*; Colas Camera, May, 1998-April/2001.

User Guide MakiCad 5.0, 2001. *Software for analysis and visualization* Colas Camera.

User Guide MakiCam 5.0, 2001. *Acquisition software for MakiSystem*; Colas Camera.

ISC'98 1998. *First International Conference on Site Characterization*, April 1998, Atlanta, GA USA

Developments of a Full Displacement Pressuremeter for municipal solid waste site investigations in Brazil

Leonardo A.L. Bello
Pontifical University of Rio de Janeiro – PUC-Rio and Amazônia University – UNAMA, Brazil

Tácio M.P. De Campos & José T. Araruna Junior
Pontifical University of Rio de Janeiro – PUC-Rio, Brazil

Barry G. Clarke
University of Newcastle Upon Tyne, UK

Keywords: Full Displacement Pressuremeter, municipal solid waste, deformability, in situ techniques, field investigations

ABSTRACT: Settlements within municipal solid waste (MSW) landfills have normally been assessed by adapting the well known one dimension consolidation theory for soils to waste. By doing that, it is necessary to make use of waste primary and secondary compressibility indexes obtained by in and ex situ investigations, together with biological decay parameters. On the other hand, in situ determination of some MSW deformability parameters, such as shear and elasticity moduli and Poisson coefficient, is not yet a common practice nor a primary target of geotechnical engineering. More frequent is the assessment of such parameters originated from back analysis. This could be eventually comprehended due to waste highly heterogeneous constituents and some site operational difficulties, but never fully accepted. Only few direct investigation techniques are able to perform such task satisfactorily, and one of them is the pressuremeter. Therefore, this paper presents some steps towards the development of a full displacement pressuremeter, specially designed for MSW landfill investigation. The MSW-FDPM was conceived and designed at the University of Newcastle, UK, and assembled and calibrated at Pontifical University of Rio, PUC-Rio, Brazil. Special attention is given to its radial displacement system, which is made of three or four moving arms mounted along with Hall effect transducers (HET), located at different lengths on the probe, giving measures at tree to four different directions. The system allows measurements of up to 100% radial deformation, which means approximately 5cm radial, possibly reaching the undisturbed waste matrix. Physical positioning of the HET's and average radial deformation value, allow for anisotropy effect to be accounted. Probe's design characteristics are presented, as well as, some preliminary calibration results.

1 INTRODUCTION

Geotechnical problems associated with solid waste disposal have already been identified, although some mechanisms are yet in the process of full cognition. These are, for instance, the cases concerning MSW settlement analysis and deformation characteristics.

Generally, MSW settlement assessments have been carried out greatly by adaptations of the well known one dimension soil settlement theory, coupled with biological and creep parameter originated from decomposition and time processes, respectively. Developments in that field have produced numerous new mathematical approaches targeting the problem, which make use of fundamental parameters normally obtained from back analysis of long field monitoring campaigns and laboratory modified consolidation tests. On the other hand, MSW settlement assessments based on the peculiar waste's stress-strain behavior are not as usual as the classic approach. Perhaps this is due to the difficulties in determining in situ waste deformation parameters such as shear modulus, Young modulus and Poisson coefficient. Nevertheless, in situ soil investigation techniques have been employed within MSW cells in order to estimate such parameters. These include information based upon geophysics (e.g. Sharma et al., 1990; Singh & Murphy, 1990; Kavazanjian et al., 1994; Pereira et al., 2002), static penetration soundings (e.g. Sánchez-Alciturri et al., 1995; Jesberger & Kockel, 1995; Koda, 1997), and pressuremeter testing (e.g. Cartier & Baldit, 1983; Dixon & Jones, 1998; Wittle, 1998; Dixon et al, 2000)

Pressuremeter testing (PMT) is a well established in situ technique, and its advantages and applicability have been extensively discussed in literature reviews (e.g. Wroth, 1984; Robertson, 1986; Clarke, 1995), showing that self-boring pressuremeter (SBPM) possess the greatest potential to determine geotechnical parameters. Also, Clarke (1995) points out that PMT can be used in any type of subsoil and that its interpretation is based on a very simple theory.

PMT in MSW could be considered as a new geotechnical practice. In Brazil, although there are several MSW sites which have been intensively investigated, such adaptation hasn't been tested yet. This paper presents preliminary studies of a full displacement pressuremeter (FDPM) developed by the universities of PUC-Rio (Brazil) and Newcastle (United Kingdom) for this purpose.

2 THE MSW FULL DISPLACEMENT PRESSUREMETER

The use of SBPM in MSW has some drawbacks as it has been pointed out by Dixon & Jones (1998) and Dixon et al (2000) in terms of drilling velocity because of fiber-like characteristic of waste. They believe push-in pressuremeters (PIP) are more applicable, demanding low reaction forces (2 to 5ton). The FDPM presented here is a radial displacement monocell type of probe following the works of Withers et al. (1986) and Akbar (2001). It was conceived with the intention to have a low cost-benefit relation and to achieve grater cavity deformation than the usual commercial probes available, which, in turn, would provide basis for a more realistic evaluation of waste's deformation characteristics.

2.1 Main Probe

The main probe is made of high strength stainless steel. All other metallic parts which are in contact with the leachate (such as clamping and nut rings, conical tip, cable housing and rod adaptor) are made of steel coated chemically with a crust of nickel to protect against corrosion from the acid environment. The probe is 677mm long, with a testing length of approximately 500mm, and it has a 46mm diameter without the membrane on (Fig. 1).

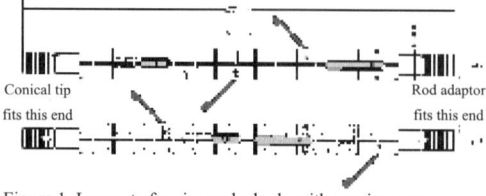

Figure 1. Lay-out of maim probe body with moving arms.

Four cavities were machined on its body to receive the systems of radial displacement measurements. Longitudinal and radial grooves were made on the outer diameter to allow instantaneous distribution of pressurized nitrogen between probe and membrane interface during expansion. An 8mm diameter longitudinal center hole was made to allow electric cable and gas passage. During assembly a conical tip is screwed on the left side of the probe (Figure 1), while the cable and rod adaptor are fitted on the right side of probe.

2.2 Radial Displacement Measurement System (RDMS)

During earlier phases of this research, great effort was devoted on the development of a system which would provide greater reading ranges than the usual ones available on other pressuremeter probes, at low costs. Such intention was justified considering the large deformation characteristics observed in MSW. Also, high heterogeneity and anisotropic behavior of waste could perhaps influence the cavity geometry formed during the test. As a result of that, expansion of a cylindrical cavity theory might not apply correctly. Therefore, in order to minimize this effect, one might get an average reading from different directions and positions on the probe and refer to it as the representative cylindrical cavity.

The sensor's choice was made on the basis of free space inside the probe and on cost-benefit relation of other types of transducers such as LVDT's. These aspects led to the selection of Hall Effect Transducers (HET) as displacement sensors. HET's are commonly employed in pressuremeter probes (eg. Allan, 1992; Akbar, 2001), and have the great advantage of being commercially offered at very low prices (£15) and small dimensions. They are fundamentally electronic devices that produce output voltage proportional to the intensity of the magnetic field to which they are exposed to, making it possible to one calibrate a system constructed with such sensors to whatever engineering units needed.

The system designed is very simple and consists of a movable L-shaped arm that rotates in relation to a pivoted pin, through the contraction and extension of a hair spring (Fig. 2). The arm's movement follows the membrane's expansion or contraction during the test. The L-shaped arm guards two button magnets (d=6mm, w=2mm) that are fitted and glued in two tight holes drilled at each side of the arm, near its bottom. Each magnet North Pole faces opposite directions, therefore, creating a specific magnetic field to produce the most linear output, as it was studied by Bello et al. (2002, 2003).

The HET employed is a 3 pin miniature (15.2mmx7.6mm) circuit chip SS94A1 LOHET II type, which has high temperature stability. This feature is important considering the high temperatures achieved inside MSW cells. The sensor is cemented on a thin acrylic plate, which, in turn, is screwed inside the probe's cavity, and remains somewhere in between the two magnets. It works with a regulated 9V supply and the output signal is amplified by a circuit from the electronic box before acquisition. As the arm moves, the relative position between HET and magnets changes and so does the magnetic field around the sensor, producing variation in output signal. The movement of the arm was limited to 40mm radial displacement, which for a 54mm initial cavity diameter (probe plus membrane) means approximately 148% cavity strain, defined as $\varepsilon_c = (r - r_i)/r_i \times 100\%$, where r represents cavity radius. Sensitivity, repeatability and hysteresis of each transducer were evaluated during calibration phases and a summary is presented on table 1.

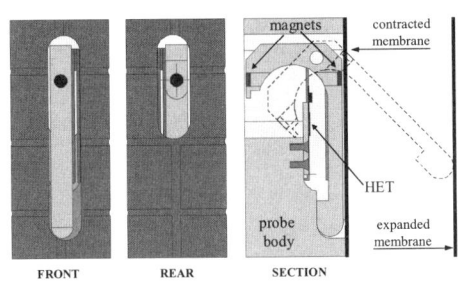

Figure 2. Details of probe's displacement measuring system.

Table 1. Characteristic of the RDMS built.

Sensor	Sensitivity (mV/mm)	Repeatability (%)	Hysteresis (% FS)
HET 1	102.3	99.47	0.37
HET 2	100.73	95.01	2.31
HET 3	112.20	97.14	1.99
HET 4	105.87	97.07	2.15

2.3 Temperature Sensor

A temperature sensor was adapted inside the probe to allow for corrections on the HET and membrane calibration curves due to temperature variation inside the MSW cells. Temperature could get as high as 60°C inside the cells originated by decomposition.

The sensor employed is a *National Semiconductor* LM35CH type. The LM35 series is a precision integrated-circuit temperature sensor, whose output voltage is linearly proportional to Celsius temperature. Its small size and location inside the probe can be notice on Figure 3 bellow.

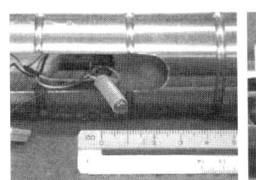

(a) sensor before assembly   (b) sensor mounted

Figure 3. Photos showing details of temperature LM35 sensor.

2.4 Rubber Membrane

The MSW-FDPM uses nitrile rubber tubes reinforced by one layer of nylon fibers as expansion membranes. Outer diameter is 50mm, whereas inner diameter is 42mm, providing a wall thickness of 4mm. They are normally manufactured in 10ft lengths and cut into pieces of approximately 600mm for testing purposes. With the membrane on, the diameter of the probe is approximately 53.5mm. The membranes being used by the time this paper was produced withstand radial displacement of only 25mm, which represents a 93% increase in respect to initial diameter. New types are being fabricated during the production of this paper that should go up to 40mm radial displacements, that is, should work to full range of the displacement measuring system described earlier.

2.5 Conical Tip

A 45° solid conical tip, with approximately 46.4cm², is attached to the probe's end in order to facilitate penetration. Its diameter is 5% grater than the main probe with the membrane on. According to Akbar (2001) this difference is enough to prevent damage to the membrane and to keep it from being dragged out of the clamping system during installation. By the time this paper was conceived preliminary studies were in progress to adapt a pressure cell and other sensor (e.g. ph meter and gas analyzer).

2.6 Pressure Control Unit (PCU)

The unit (Fig. 4) is built to control pressures of up to 58.6bar. It has two regulators, one giving fine control up to 10bar, and other allowing coarse control up to 58.6bar. Therefore, the unit is able to control tests in materials with a variety of stiffness. The box is equipped with quick release fittings, both for the inlet of gas pressure, and for the outlet where the gas-electric separator unit is connected.

Figure 4. Pressure control unit photo

2.7 Gas-Electric Separator Unit (GESU)

The GESU (Fig. 5) was built similar to the one developed by Akbar (2001) and it is made of two stainless steel tubes screwed together (parts 2 and 1). An 8 pin glass-metal seal (part 12) is located inside one end of the tube and is secured by screwing part 3, allowing continuity only of electronic signal. The unit holds the pressure transducer to measure the applied gas pressure. Metal cable glands (4 and 5) are used at both ends of part 1 to secure the cable and prevent damage to welded points. A loose nut adaptor (8) prevents any twisting of hose and cable while assembling. The electric cable inside the hose passes through the loose nut and to a 9 pole plug and socket (6 and 7). The GESU connects to PCU through a quick release fitting part.

On the surface the hose is attached to the GSU, which, in turn, goes in the pressure control unit.

A testing schematic drawing is presented on Figure 7.

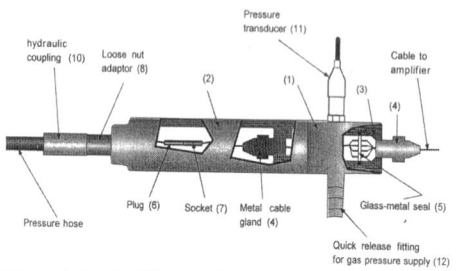

Figure 5. Detail of the gas-electric separator unit

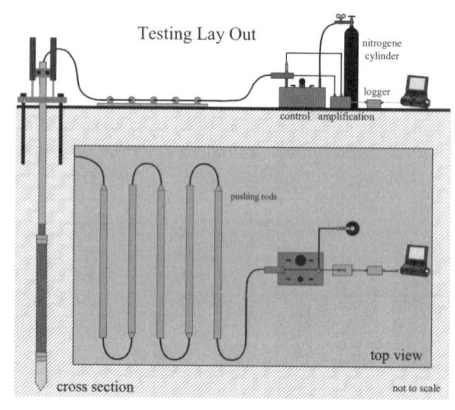

Figure 7. Schematic of the FDPM and its components.

3 MAIN INSTRUMENTS' CALIBRATION PROCEDURES AND RESULTS

2.8 *Electronic Control Unit (ECU)*

The ECU (Fig. 6) is responsible for supplying input voltage, for amplifying the output signals and to convert analog data to digital. The unit works with a 12V (2.3Ah) small rechargeable sealed battery. Inside the unit there is a printed electronic circuit that it was designed to supply regulated 9V to all the sensors and to amplify 5 output signals (4 HET and 1 pressure transducer). Individual signal offset and gain adjustments can be made prior to beginning of tests, by trimming potentiometer located on the board.

The ECU also holds inside the PICO ADC16 (*Pico Technology*), which is an 8 channels 16bits digital data logger. It converts 8 signals (4 HET, 1 pressure, 1 temperature and 1 reference) and sends them to a portable computer for acquisition.

3.1 *Pressure Transducer*

Pressure transducer calibration is a very common procedure in geotechnical laboratories (e.g. BS1377 Part 1:1990 Clause 4.4.4). Accurate dead weights resting on a precision made piston fitting into a matching cylinder generate a pressure in the oil of the system that can be used to calibrate the instrument. Here, a Bundenberg dead weight gauge was employed to calibrate a 15bar pressure transducer (part 11, Figure 5). For testing and calibration purposes the output signal is amplified by the electronic box, providing a sensitivity of 1.6mV/kPa. No hysteretic behavior was verified and the results and best fitting curve are illustrated on Figure 8 below.

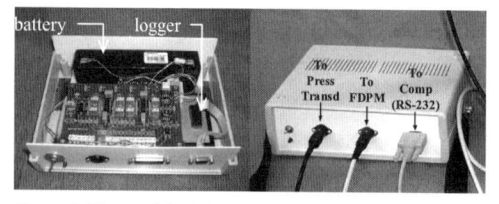

Figure 6. Photos of the ECU opened and closed.

2.9 *Other Parts*

Standard Cone pushing rods are used to push de pressuremeter into de MSW cells. A hydraulic hose with working pressure of 310bar carries pressurized nitrogen and electronic cabling down to de probe.

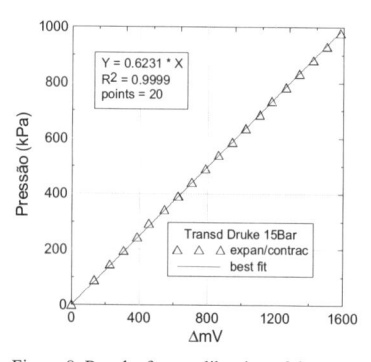

Figure 8. Results from calibration of the pressure transducer

3.2 Radial Displacement Measurement System (RDMS)

At room temperature (22°C) and having the whole pressuremeter assembled and on a horizontal position, calibration of the RDMS was made following the procedure suggested by Clarke (1995). The arm being calibrated was allowed to open or close in a controlled manner, while all the others were kept contract. A micrometer vertically mounted on a stand, adapted with a small acrylic plate at one end, was used to allow the arms to open and close, and, consequently, to measure radial displacements. Increments (expansion of the arm) of about 5% of the full range of each transducer were taken on the micrometer up to the maximum arm's aperture, following the same decrements steps (contraction of the arm). Corresponding readings were taken in the computer by the logger software, allowing 30seg for signal stabilization. This procedure was repeated several times for each arm transducer, and then continued to the next ones. The results found are presented through figures 9-12. There, one could identify 2nd degree calibration best fit curves similar to the ones observed by Akbar (2001). HET's are indexed in a crescent order from probe's top to bottom tip.

3.2.1 Temperature Effect

In order to evaluate temperature effect on the HET's the probe was assembled inside a stove adapted to run triaxial test with controlled temperature. A thermo gradient was applied to the probe with its arms all contracted and the readings were taken. Results illustrated on Figure 13 showed that there is small independent drift for each initial HET offset. Therefore, in real field testing situation, once testing depth is achieved, correction for zero reading output must be made according to temperature. This can be done either zeroing the output in the calibration equation on the software, or by trimming adjustments in the electronic box. The latter has shown better results, for temperature drift might force operational amplifiers to go beyond limits previously set, which, occasionally, are not possible to be undo with a merely equation changing.

3.3 Rubber Membrane Stiffness

Membrane stiffness is the pressure required to inflate the membrane and protective shaft (if applied), in air (Clarke, 1995). Calibrations affect interpretation of tests in soft soil and could be important in MSW. It should be performed at the same temperature as that in the ground. For MSW investigations, where temperatures can get as high as 60°C, it should be carried out inside a stove, for membrane behavior is temperature dependent.

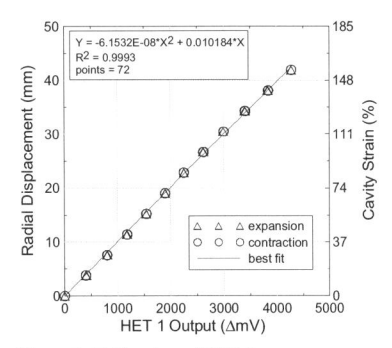

Figure 9. Calibration of HET 2

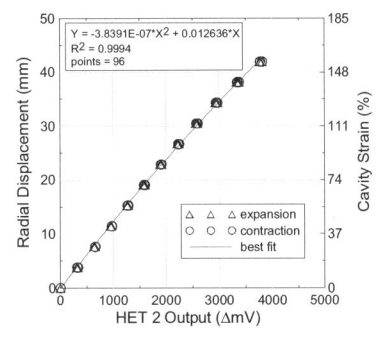

Figure 10. Result of calibration of HET 2.

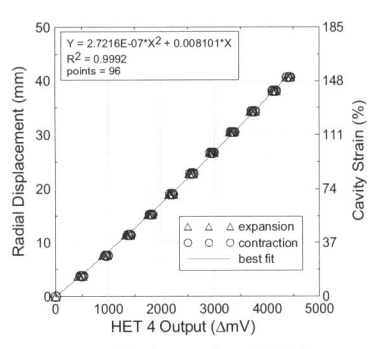

Figure 11. Calibration results of HET 4

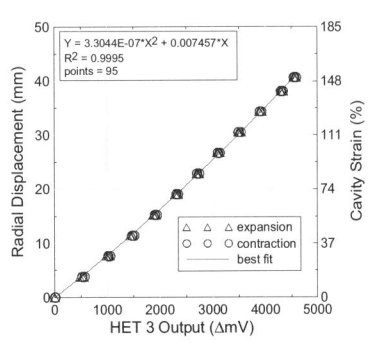

Figure 12. Calibration results of HET 3

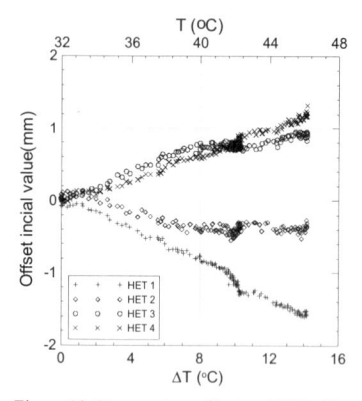

Figure 13. Temperature effect on HET with arms contracted

Calibration results presented here were taken at room temperature only, for adaptations to the stove were not finished in time. With the whole pressuremeter parts assembled and connected, the membrane was inflated and deflated following similar procedure used in a field test, that is, in 10kPa stress-controlled increments, held for 1min. Membrane burst was identified with radial displacements of about 25mm. So, to prevent membrane from being lost during calibration a 20mm limit was imposed when expansion in air was performed. At the moment, new types of membrane are being fabricated and tested to withstand greater displacements. Results found can be observed through Figure 14 to 17 for each transducer. According to Clarke (1995) the expansion curve is often non-linear and independent transducers show difference responses.

Third degree polynomial best fit curves are obtained from expansion and contraction phases separately. The lift-off pressure for such membranes is about 25kPa.

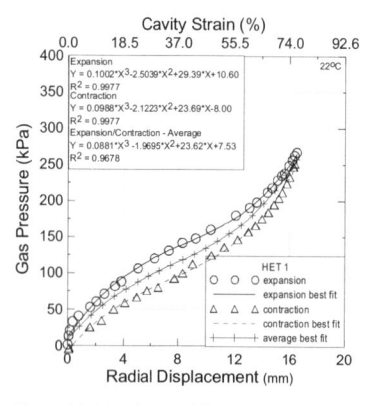

Figure 14. Membrane stiffness at HET1

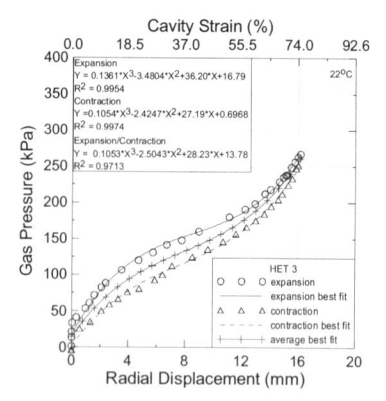

Figure 15. Membrane stiffness at HET3.

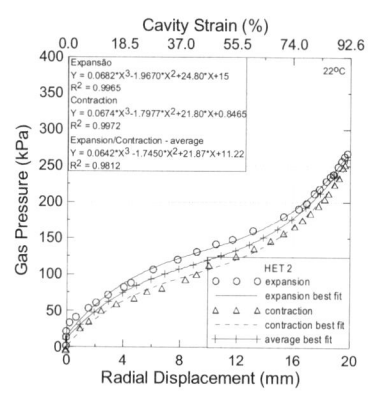

Figure 16. Membrane stiffness at HET2.

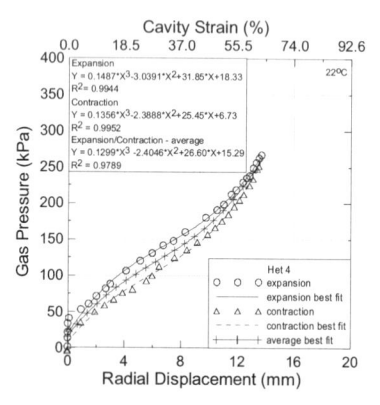

Figure 17. Membrane stiffness at HET4.

4 CONCLUSIONS

This paper presented preliminary studies towards the development of a low cost full pressuremeter to be primarily applied in MSW investigations. Main goal during earlier phases was to design a system to measure radial displacements with at higher values than the usual ones obtained in with PMT devices. The RDMS designed provided means to measure accurately up to 4cm (although it can effectively reach 5cm) radial expansions and it proved have a good cost-benefit relation. The use of Hall Effect Transducers proved to be efficient and cost-effective, fulfilling the main goal. Measurements at different directions and heights on probe could be a way to identify heterogeneity and anisotropy in MSW, but it needs to be better evaluated at field testing. Perhaps the membrane could be too stiff to identify such characteristics. Also, the rubber membrane applied in the studies was not satisfactory yet, for it cannot achieve compatible values with the RDMS. It bursts at lower values than expected (goes up to 25cm). New fabrications are being tested so far.

Field testing will be performed in the early months of 2004 at Muribeca MSW landfill in the city of Recife, PE – Brazil, where there has been extensively research going on. Campaign will involve SPT, CPT and PMT testing along with waste chemical and physical characterization to provide basis for a settlement analysis.

ACKNOWLEDGEMENTS

The authors would like to thank the financial support received from CAPES and CNPq/Pronex, and also to thank UNAMA, PUC-Rio and Newcastle University for their institutional support.

REFERENCES

Akbar, A. 2001. *Development of low cost in-situ testing devices*. PhD Thesis, Dept. Civil Eng. University of Newcastle Upon Tyne, UK, 369p.

Allan, P.G. 1992. Developments of a self-boring pressuremeter for the in-situ testing of weak rocks. PhD Thesis, Dept. Civil Engineering, University of Newcastle Upon Tyne, UK, 345.

Bello, L.A.L.. Clarke, B.G.; Araruna Jr., J.T. & De Campos, T.M.P. 2002. Developments of a cone pressuremeter to municipal solid waste deformation parameters assessments. In. *XII Brazilian Congress on Soil Mechanics and Geotecnical Engineering - XII COBRAMSEG*, São Paulo, Brazil, v. 2, pp. 835 – 845 (IN PORTUGUESE).

Bello, L.A.L.; De Campos, T.M.; Araruna Jr., J.T. & Clarke, B.G. 2003. Applicability of pressuremeter testing in MSW landfill investigations. In *V Brazilian Congress on Environmental Engineering – REGEO'2003*, Porto Alegre, RS, Brazil, May 20 – 23, Doc. DRF021, CD-ROM (IN PORTUGUESE).

Cartier, G. & Baldit, R. 1983. Comportment géotechnique des discharges de résidus urbains. *Bull. de Liason des Laboratoires des Ponts et Chaussées*, n.128, pp. 55 – 64.

Clarke, B.G. 1995. *Pressuremeter in geotechnical design*. Blackie Academic & Professional, Glasgow, 364 p.

Dixon, N. & Jones, D. R. V. 1998. Stress states in, and stiffness of, landfill waste. *Proc. Geotechnical Engineering of Landfills*. Eds. Dixon, N.; Murray, E. J. & Jones, D.R.V., Pubs. Thomas Telford, pp. 19 – 34.

Dixon, N.; Ng'Ambi, S.C; Jones, D.R.V. & Connel, A.K. 2000. The role of waste deformations on landfill steep side wall lining stability. *Proc. 38th Annual International Solid Waste Exposition – WASTECON2000*, The Institute of Waste Management: South Africa, v. 2, pp. 379 – 388.

Jesberger, H.L. & Kockel, R. 1995. Determination and assessment of the mechanical properties of waste material. In. *Waste Disposal by Landfill – GREEN'93, UK. Proc. Symp. On Geotechnics Related to the Environment – GREEN'93*, Balkema, Rotterdam, pp. 313 – 322.

Kavazanjian et al. 1995. Non-Intrusive Reyleigh wave investigations at solid waste landfills. *Proc. 1st Int. Conf. Environmental Geotechnics – 1st ICEG*, Edmonton, Alberta Bitech Publishers.

Koda, E. 1997. In-situ tests of MSW geotechnical properties. *Proc. 2nd Int. Symp. On Geotechnical Related to Environment – GREEN 2, Contaminated and derelict land*, R. W. Sarsby (ed), Krakov, Poland, Thomas Telford, London, pp. 247 – 254.

Pereira, A.G.H.; Sopeña, L. & Mateos, M.T. 2002. Compressibility of a municipal waste landfill. *Proc. 4th Int. Cong. On Environmental Geotechnics – 4th ICEG*. Rio de Janeiro, Brazil, pp. 201 – 206.

Robertson, P.K. 1986. In situ testing and its application to foundation engineering. *Can. Geotech. J.*, v. 23, pp. 573 – 594.

Sánches-Alciturri, J.M.; Palma, J.; Sagaseta, C. & Cañizal, J. 1995. Three years of deformation monitoring at Meruelo landfill. *Proc. Symp. On Geotechnics Related to the Environment, Waste Disposal by Landfill, GREEN'93*, Bolton, UK, Balkema, Rotterdam, pp. 365 – 371.

Sharma, H.D.; Dukes, M.T. & Olsen, D.M. 1990. Field measurements of dynamic moduli and Poisson's ratios of refuse and underlying soils at a landfill site. Geotechnics of Waste Fills – Theory and Practice, ASTM STP 1070, A. Landva e G.D. Knowles (ed), Philadelphia, pp. 57 – 70.

Singh, S. & Murphy, B. 1990. Evaluation of the stability of sanitary landfills. Geotechnics of Waste Fills – Theory and Practice, ASTM STP 1070, A. Landva e G.D. Knowles (ed), Philadelphia, pp. 240 – 258.

Withers, N.J.; Schaap, L.H. & Dalton, J.C.P. 1986, The development of a full displacement pressuremeter. *Proc. 2nd Int. :Symp. On Pressuremeter and its Marine Application*, Briaud e Audibert (eds), ASTM STP 950, Texan, USA, pp. 38 – 56.

Wroth, C.P. 1984. The interpretation of in situ soil tests. *Géotechnique*, v. 34, n. 4, pp. 449 – 489.

Whittle, R. 1998. Pressuremeter testing in waste. Cambridge Insitu Internal Report, site visited: http://www.cambridge-insitu.com/techpapers/Waste/Waste.htm (in march 2003).

Rock characterization using drilling parameters

Jean Benoît & Stanley S. Sadkowski
University of New Hampshire, Department of Civil Engineering, Durham, NH, USA

Wallace A. Bothner
University of New Hampshire, Department of Earth Sciences, Durham, NH, USA

Keywords: drilling parameters, fractured rock, bioremediation, geophysics

ABSTRACT: Identifying natural fractures in contaminated rock from conventional borehole investigations is an integral part of a bedrock bioremediation project in southeastern New Hampshire, USA. A series of boreholes have been used to develop protocols for bioremediation. Identification of connecting fractures between the boreholes allows monitoring of groundwater and evaluation of the effectiveness of the bioremediation techniques. This paper presents the results from a site characterization technique based on the recording of drilling parameters during the advance of each borehole. The drilling parameters are compared to downhole geophysical measurements such as videologging and acoustic televiewer monitoring. The use of the drilling parameters in this project helped improve the quality and efficiency of the drilling process and reduced the amount of fluid introduced and lost into the aquifer.

1 INTRODUCTION

Successful application of in situ bioremediation techniques to fractured rock aquifers is highly dependent on our ability to characterize geological rock units accurately and economically. An ongoing bedrock bioremediation project, funded by the USEPA, has allowed researchers from the University of New Hampshire Bedrock Bioremediation Center (BBC) to evaluate and develop innovative methods to detect fracture patterns and direction of groundwater flow in a metamorphic rock formation in southern New Hampshire, USA. A test site located at the Pease International Tradeport in Portsmouth, New Hampshire, has been selected for this study. The site, formerly part of the Pease Air Force Base, is contaminated with TCE (trichloroethylene) resulting from degreasing operations during equipment maintenance. Those contaminants migrated downward into the groundwater eventually finding their way to the weathered and competent bedrock. While TCE is the primary contaminant, other degradation products (dichloroethylene, DCE and vinyl chloride, VC) are present at different concentrations within the geologic profile. For this project, sets of paired boreholes approximately 10 m apart and 60 m deep were drilled to develop protocols for bioremediation. Identification of connecting fractures between the pairs of boreholes will allow monitoring of groundwater and evaluation of the effectiveness of the bioremediation techniques. This paper discusses the use of drilling parameters as a characterization technique for this bioremediation project.

2 SITE GEOLOGY

The subsurface conditions at the test site consist of approximately 10 to 20 m of Pleistocene glaciofluvial deposits overlying a relatively thin weathered bedrock zone followed by competent bedrock. Southeastern New Hampshire is underlain by Silurian and older metasedimentary rocks that are variably metamorphosed, tightly folded and faulted, and intruded by Paleozoic and Mesozoic igneous rocks. The bedrock in the area of Great Bay consists of biotite grade metasandstones and metashales of the Kittery and Eliot formations. The Kittery Formation is typically comprised of alternating beds of calcareous feldspathic metasandstone, some with well-preserved primary structures, and thin metashale intercalations. The Eliot Formation is a typically finer grained alternation of calcareous metasiltstone and metashale. Both units are tightly folded into asymmetric, shallow northeast plunging anticlines and synclines overturned to the southeast. Abundant Jurassic diabase dikes occur throughout southeastern New Hampshire and have been encountered in all of the BBC boreholes drilled thus far. They maintain a dominant northeast-southwest strike with steep dips to the northwest or southeast.

3 DRILLING PARAMETERS

Drilling parameter recorders (DPR) are computerized systems which monitor a series of sensors installed on standard drilling equipment. These sensors continuously and automatically collect data on all aspects of drilling, in real time, without interfering with the drilling progress. The DPR used for this project is a Jean Lutz CL88n, which has the capability of recording data on advance rate, downthrust and pull-up pressures, rod torque, rotation rate, mud/water pressure and flow, depth, and time. The data are displayed in real-time in digital form, as hard copy and are also stored on an electronic medium for further analysis.

Installation of the sensors for pressure measurements simply requires tee-connectors onto the appropriate hydraulic or water/mud lines. For the rotation speed, an electromagnetic sensor uses the bolt pattern on the rotating head to determine the revolutions per minute. The depth is measured using a pulley system mounted on the rig mast and attached to the drill head. For this project, water flow was measured using two flowmeters. One flowmeter measures the inflow while the second intercepts the return water, using a tee-connection, as it exits the borehole.

For this bedrock bioremediation project, a series of five cored boreholes, 150 mm in diameter, were drilled in bedrock using the DPR. A triple core barrel was used to advance the boreholes to depths of up to 60 m, resulting in continuous 100 mm rock cores. The inner barrel consisted of a lexan (plastic) liner designed to minimize microbial contamination and to help preserve the integrity of the rock cores. The core barrel had a length of approximately 1.5 m. The cores were retrieved by wireline. An additional borehole was advanced using a tricone roller bit.

A typical concatenated output from one of the cored borehole is shown in Figure 1 for depths of 29 to 35 m. The parameters are recorded as a function of depth rather than time. The recording depth interval is 5 mm. Each recorded interval is merged (concatenated) into a single profile. This operation requires a good understanding of drilling procedures and allows drilling artifacts and operator errors to be corrected as needed. Figure 1 shows the recorded drilling parameters as well as the lithology as logged by a geologist. Because thrust and torque are measured using pressure transducers, the values are shown in units of pressure as recorded by the DPR. The actual thrust and torque is easily calculated using the drill rig efficiency and piston configuration. For this project, drilling mud could not be used as it could interfere with the microbiological analyses. Drilling muds have been found to be highly contaminated with microorganisms. Consequently, the drilling fluid used was water pumped from an uncontaminated well. In addition, the return drilling fluid could not be re-circulated because of possible cross-contamination. Figure 1 shows the water pressure, the inflow of water as well as the return flow (outflow).

The use of a triple core barrel posed several challenges during the initial coring trials. Often the cores jammed into the inner barrel, essentially halting the drilling advance. Modifications to the inner diameters of the diamond impregnated coring bit and to the liner-retaining ring helped alleviate most of the core jamming. Several drilling parameter combinations were tried in an effort to maximize the core recovery and minimize blockage during drilling. Of major concern was also the quantity of drilling fluid used to advance the boreholes. Because of the TCE contamination, all return flow had to be contained and trucked to a water treatment facility.

Using individual DPR measurements, variations in the drilling parameters are interpreted to indicate the presence of fractures, changes in lithology, and competency of the bedrock. For example, under constant thrust and rotation rate, a variation in advance rate would suggest either a change in stratigraphy or the presence of an anomaly such as a cavity or a fracture. Using the torque measurements would further help determine the reason for the change in advance rate. At this site, a higher torque would be indicative of harder material or badly fractured rock while a lower torque would indicate the presence of a fracture. As shown in Figure 1, the lower torque at a depth of 32 m indicates the presence of a fracture while the higher torque at a depth of approximately 33. 8 m is as a result of a stratigraphic change. At the transition from the metasandstone to the stratified metashale, the driller increased the thrust and the rotation rate, which in turn increased the torque and advance rate. The penetration rate increased within the metashale. The water pressure and flow measurements can also be used to estimate the location of fractures. However, because of the core barrel configuration, fluctuations in water pressures cannot always be correlated to the presence of fractures. The fluid pressure and flow can be restricted due to jamming of the core in the barrel or, blockage of the drill bit and the return flow paths by cuttings. Nevertheless, the use of DPR while coring was immensely useful to the driller in delivering high quality cores with minimal water usage and, improving daily productivity.

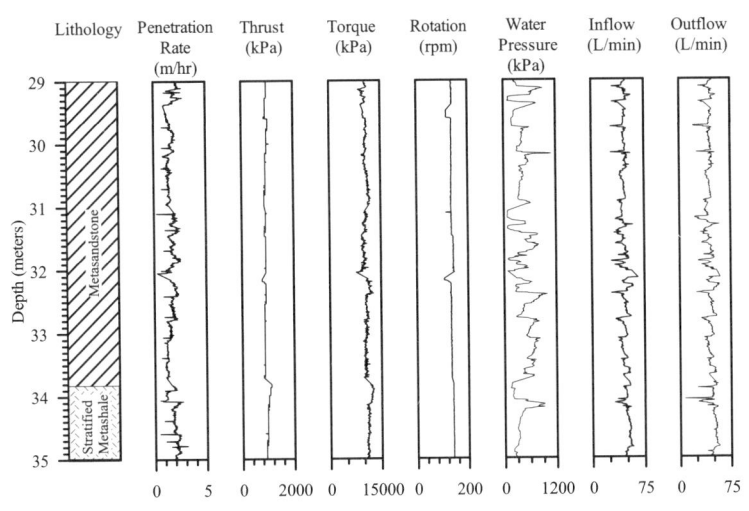

Figure 1: DPR Measured Parameters and Lithology for BBC-4 Using a Diamond Impregnated Coring Bit

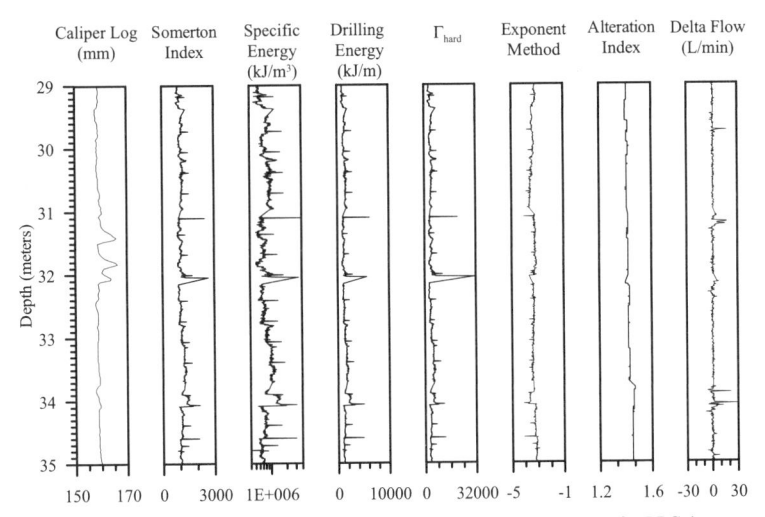

Figure 2: DPR Compound Parameters and Geophysical Caliper Log Data for BBC-4

Several methods of interpretation have also been developed using compound parameters. Compound parameters simply combine individual parameters into expressions of energy or empirical indices reflecting the resistance of the geological material to drilling. Table 1 presents the compound parameters currently used in this study. All methods use the penetration rate and the thrust or vertical force.

Except for the alteration index, all methods also use the rotation rate. Figure 2 shows the resulting compound parameters associated with the drilling parameters shown in Figure 1. Two additional parameters are shown. The caliper log was recorded after borehole completion and gives an indication of the condition of the borehole wall. The delta flow is the difference between the inflow and the outflow. A positive delta flow indicates a loss of fluid into the aquifer. For borehole BBC-4, more than 50,000 liters were used in the coring process from 23.4 to 55.2 m. Approximately 5.5% of that volume was lost into the aquifer through fractures in the bedrock formation. The example observations made at depths of 32 and 33.8 m using the standard drilling parameters are also reflected in the compound parameters. As the drilling advance slows, it requires more energy to destruct that particular layer.

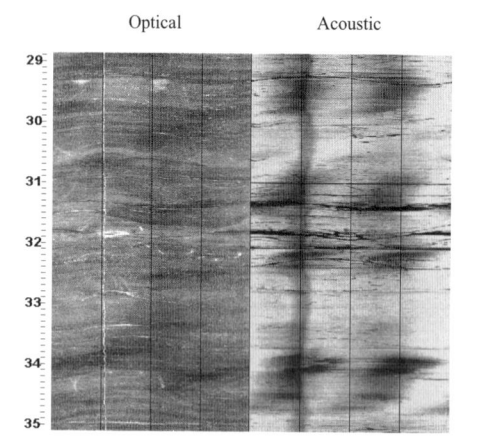

Figure 3: Geophysical Optical and Acoustic Televiewer Logs for BBC-4

Table 1: Summary of Compound Parameters used for DPR Data Analysis

Name	Equation	Units	Reference
Alteration index	$A_i = 1 + \dfrac{P}{P_0} - \dfrac{V}{V_0}$	---	Pfister, 1985
Drilling energy	$W = \dfrac{T\omega}{V}$	kJ/m	Pfister, 1985
Exponent method	$E = \dfrac{\log(\dfrac{V}{\omega D})}{\log(\dfrac{T_p D}{T})}$	---	Gui et al., 2002 (modification of Jordan and Shirley, 1978)
Somerton index	$S_d = P\sqrt{\dfrac{\omega}{V}}$	---	Somerton, 1959
Specific energy	$e = \dfrac{F}{A} + \dfrac{2\pi\omega T}{AV}$	kJ/m^3	Teale, 1964
Γ – hard parameter	$\Gamma_{hard} = \dfrac{\omega T_p D^2}{VT}$	---	Gui et al., 2002 (Girard, et al., 1986)

A= area removed by drill bit, D= bit diameter, F= vertical force, P= weight on bit, P$_0$= maximum theoretical weight on bit, ϖ= rotation speed, T$_p$= thrust, T= torque, V= penetration rate, V$_0$= maximum penetration rate.

The alteration index, indicative of relative hardness, also clearly shows a change in lithology at a depth of 33.8 m. It should be kept in mind that for non-horizontal fractures, only a portion of the fracture is intercepted as drilling occurs through each feature and thus changes in parameters are more progressive.

For the bedrock bioremediation project, a suite of downhole geophysical tests was carried out following borehole completion. Shown in Figure 3 are measurements using borehole optical televiewer and borehole acoustic televiewer. The optical televiewer produces a continuous 360-degree image of the borehole walls. The acoustic televiewer produces images of the borehole walls by sending an ultrasonic beam that is reflected back to the probe. The images are based on the amplitude and travel time of the reflected beam. Both televiewers provide accurate information on borehole deviation and orientation of the images. From the interpretation of these images it is possible to extract the strike and dip of each geologic structure. Using other geophysical tools such as natural gamma and caliper probes, the fractures were characterized as bedding planes, infilled fractures, open fractures or conductive fractures. Figure 3 clearly shows several interesting features at depths of 31 to 32 m. The horizontal feature at 31 m was initially interpreted as a horizontal fracture. However, the DPR measurements could not confirm this finding. After close examination of drilling procedures and the recovered cores, it was concluded that these types of features were the results of the drilling procedures at the beginning of each core run. Before starting a new core run, the driller would raise the core barrel approximately 0.2 m, slowly rotate the barrel with water flow until a good clean water return flow. During that process, a groove would be formed on the borehole wall and occasionally, depending on how the core broke at the end of the previous core run, the top of the core would also show a groove. Table 2 summarizes the geophysical interpretation for the log shown in Figure 3. Figure 4 is a photograph of the 100 mm diameter core illustrating the typical features shown in Figures 1 to 3. The photograph is annotated and shows an infilled fracture and two fractures. It should be noted that the geophysical measurements are made on the borehole walls and thus may not directly relate to the features observed on the actual core. For instance, drilling induced fractures will be logged on the core but will not be visible on the acoustic and optical televiewers. A drilling induced fracture is easily discerned in coarse-grained rocks but in the fine-grained rocks associated with this project, it is often impossible to differentiate them from a natural fracture or open bedding plane.

Fractures are observed on Figure 3 at depths of approximately 31.4, 31.8 and 32.1 m. As shown in Table 2, the dip of the fracture at about 32.1 m is nearly horizontal. For such an orientation, it is expected that a more significant response in drilling parameters would be visible. For that depth, the decrease in water pressure and the increase in water inflow were more apparent. For the other two depths (31.4 and 31.8m), the fractures are dipping more than 30 degrees and thus are intersected by the core bit more gradually. A slight increase in penetration rate is observed at those depths as well as a decrease in drilling fluid pressure. For all three of those depths the caliper log as seen in Figure 2 showed an increase in borehole diameter.

In an effort to understand better the DPR measurements during advance of the core barrel, an additional borehole was drilled using a tricone button bit. This non-coring borehole is a destructive method of advance and should provide a clearer picture of subsurface features unobstructed by the presence of the core barrel and core in the borehole. Figure 5 shows the drilling parameters for a section of BBC-7 borehole from 30 to 50 m. For this borehole it was possible to experiment with different drilling procedures. During this boring, the thrust and rotation rate were kept constant to observe resulting changes in torque and advance rate. Seen in Figure 5, from a depth of 30 to 38 m, the thrust was held constant at 3,500 kPa with a steady rotation of 120 rpm. From 38 to 42.9 m, using the same thrust, the rotation rate was increased to 150 rpm. Below 42.9 m, the rotation was returned to 120 rpm but the thrust was increased to 4,100 kPa. It can be observed that an increasing rotation coupled with a constant thrust directly translated into a doubling of the torque pressure without an apparent change in the penetration rate. Below 42.9 m, the increase in thrust also resulted in an increase in torque but not as significantly as the effect from increasing the rotation rate. The small increase in thrust did not result in a noticeable increase in penetration rate. However, since the thrust and rotation rate were kept constant, the variations in penetration rate and torque pressure provide a good indication of both fractures and lithologic changes. For example, at a depth of 32.2 m, the penetration rate increases by a factor of eight with an associated 35% increase in torque pressure. Those observations can be correlated to a fracture zone. During the borehole advance, the driller also maintained a constant inflow of water of 38 l/min. Variations in all drilling parameters will be further analyzed upon completion of the geophysical testing in BBC-7.

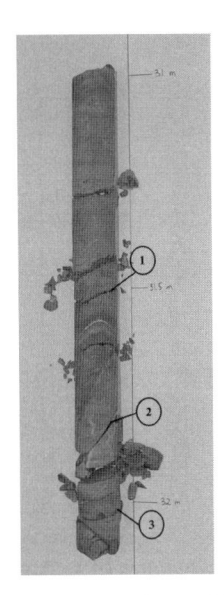

Figure 4: 100-mm Core Specimen from BBC-4

Table 2: Interpretation of Optical and Acoustic Logs (BBC-4)

Depth (meters)	Dip (degrees)	Dip Direction (degrees)	Comment
31.00	5.62	158.80	Bedding
31.19	19.78	195.19	Infilled fracture
31.29	19.68	188.60	Bedding
31.35	29.88	203.79	Fracture
31.38	39.96	227.03	Fracture
31.53	43.18	290.90	Infilled fracture (1 on Fig. 4)
31.65	62.04	237.64	Bedding
31.74	51.24	216.55	Infilled fracture
31.85	38.53	188.01	Fracture (2 on Fig. 4)
31.99	45.66	320.35	Bedding
32.05	9.23	70.97	Fracture (3 on Fig. 4)

Dip direction accounts for 16-degree declination of true north from magnetic north.

For both coring and non-coring holes, it also became apparent that bit wear significantly affected the drilling parameters. However, current interprettation methods do not properly take into account bit wear.

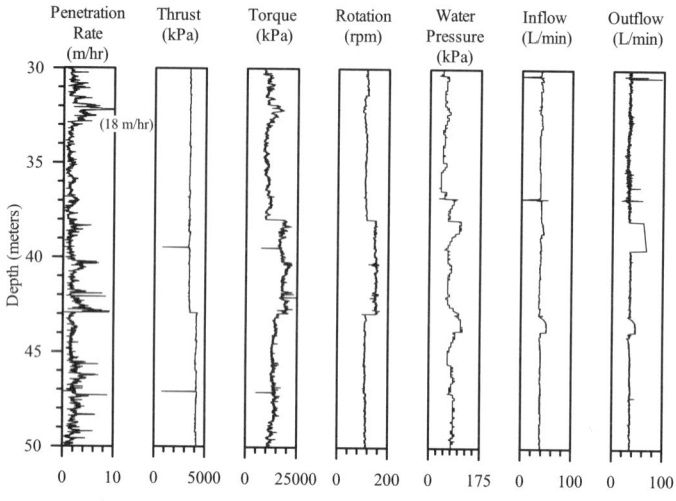

Figure 5: DPR Parameters for BBC-7 Using a Tricone Button Bit

4 CONCLUSIONS

For the bedrock bioremediation project in New Hampshire, the DPR has helped improve the quality and efficiency of the drilling progress and reduce the amount of fluid being introduced and lost into the aquifer. The use of both conventional and compound parameters accompanied by core logging and geophysical testing have improved our ability to locate fractures and changes in stratigraphy. Correlations between possible fracture zones and DPR data have been possible regardless of bit type and drilling procedure. However, bit wear should be properly integrated in the interpretation methods for effective analysis. In rock, bit wear can easily affect the measurements within a single borehole.

For destructive borehole advance, fracture zones are more readily identified using penetration rate and torque as indicators when thrust and rotation rate are held constant. Further comparisons of the two drilling methods to locate fractures will be made upon the completion of the geophysical testing.

The results of this investigation will be used to develop protocols for drilling and use of the DPR for future sites and other geological conditions. These protocols will provide in situ characterization rapidly and more cost effectively, thus limiting the need for expensive coring during rock investigations.

ACKNOWLEDGEMENTS

The Bedrock Bioremediation Center at the University of New Hampshire was funded by the U.S. Environmental Protection Agency, grant no. CR827878-01-0. This support is greatly appreciated.

REFERENCES

Girard, H., Morlier, P., Puvilland, O., Garzon, M. 1986. The Digital ENPASOL Method-Exploitation of Drilling Parameters in Civil Engineering. *Proceedings, 39th Canadian Geotechnical Conference. Ottawa, Ontario, August, 1986:* 59-68.

Gui, M.W., Soga, K., Bolton, M.D., Hamelin, J.P. 2002. Instrumented Borehole Drilling for Subsurface Investigation. *Journal of Geotechnical and Geoenvironmental Engineering.* Vol. 128 (No. 4): 283-291.

Jordan, J.R., Shirley, O.J. 1978. Application of Drilling Performance Data to Overpressure Detection. *Journal of Petroleum Technology.* Vol. 7: 987-991.

Pfister, P. 1985. Recording Drilling Parameters in Ground Engineering. *Journal of Ground Engineering.* Vol. 18 (No. 3): 16-21.

Somerton, W.H. 1959. A Laboratory Study of Rock Breakage by Rotary Drilling. *Journal of Petroleum Technology.* Vol. 216: 92-97.

Teale, R. 1964. The Concept of Specific Energy in Rock Drilling. *International Journal of Rock Mechanics, Mining Science.* Vol. 2: 57-73.

Proceedings ISC-2 on Geotechnical and Geophysical Site Characterization, Viana da Fonseca & Mayne (eds.)
© 2004 Millpress, Rotterdam, ISBN 90 5966 009 9

Penetration resistance in soft clay for different shaped penetrometers

S.F. Chung & M.F. Randolph
Centre for Offshore Foundation Systems, The University of Western Australia

Keywords: *In situ* penetration test, cyclic test, cone penetrometer, T-bar penetrometer, ball penetrometer, plate penetrometer, tip resistance, undrained shear strength, remoulded strength, sensitivity

ABSTRACT: This paper presents the results of *in situ* penetration tests and vane shear tests conducted in Western Australia, at the Burswood site near Perth. Penetrometers tested in the program included cone, T-bar, ball and plate. The paper will focus on the penetration and extraction resistance (q) profiles, and the derived undrained shear strength (s_u) profiles, obtained with the various penetrometers. The effects of surface roughness and aspect ratio of the T-bar on its resistance profile were also studied. Interestingly, the various penetrometers showed a narrow band of resistance profiles. Using a single value of bearing factor (N_k) for all penetrometers, the derived s_u profiles are comparable to the profiles measured directly from vane shear tests. Cyclic penetration and extraction tests were carried out at specific depths for each penetrometer (except for the cone). These tests comprised displacement cycles of ±0.5 m about the relevant depth, recording the penetration and extraction resistance over five full cycles. The results may be used to derive the remoulded strength and sensitivity of the soil. Also presented are the results of centrifuge tests carried out on reconstituted samples of clay material collected from the same site. Two boxes of soils were prepared and tested in-flight. In the first box, various types of penetrometers (cone, T-bar, ball and plate) were tested, while in the second box, four different lengths (hence aspect ratios) of T-bars were investigated. In addition, hand vane tests were conducted soon after completion of the penetration tests for each sample.

1 INTRODUCTION

The cone penetration test with pore pressure measurement (CPTU) is now used widely as the primary tool in a site investigation. The main advantages of the CPTU are that it provides reliable and continuous profiles of data of the soil with depth. Testing procedures and methods of interpreting the data of CPTU have been discussed extensively in the literature, and correlations developed from the three measurements (total cone tip resistance, q_t, excess pore pressure, Δu, and sleeve friction, f_s) to help identify the stratigraphic sequence (Robertson & Campanella, 1983; Lunne et al. 1997).

In clays, an estimate of shear strength is obtained from the net cone resistance, q_{cnet}, using a cone factor, N_{kt}:

$$s_u = \frac{q_{cnet}}{N_{kt}} = \frac{q_t - \sigma_{vo}}{N_{kt}} \quad (1)$$

where q_t is the total cone resistance and σ_{vo} the *in situ* vertical stress. In practice, errors in determining q_t from the measured resistance, q_c, and in estimating σ_{vo} and N_{kt} lead to quite large inaccuracies, particularly for soft sediments. As a result, in offshore practice it has become customary to use laboratory triaxial and simple shear test data to calibrate the value of N_{kt} for each new site.

The form of failure around the advancing cone entails that the value of N_{kt} will be affected by the soil stiffness (or rigidity index, I_r) and value of K_o (Teh & Houlsby, 1991). In an attempt to avoid the uncertainties involved with interpretation of the CPT in clays, while preserving the advantage of a continuous profile of resistance, the T-bar penetrometer was introduced (Stewart and Randolph, 1991). This was first tested in the centrifuge and later used in the field both onshore (Stewart and Randolph, 1994) and offshore (Randolph et al, 1998). The principle behind the T-bar, which consists of a cylindrical bar mounted at right angles to the push-rods, is that soil is able to flow around the cylinder from front to back, leading to a very localised plastic mechanism.

The relationship between net bearing resistance on the projected area of the T-bar and the shear strength of the soil is based on the plasticity solution of Randolph and Houlsby (1984). This gives a bearing factor that depends only on the surface roughness of the cylinder, varying between 9.1 (fully smooth) and 11.9 (fully rough). An average value of 10.5 was recommended for general use. Since the soil is able to flow around the T-bar, the overburden pressure is equilibrated above and below the T-bar, except at the shaft. However, since the projected area of the T-bar is considerably larger than the area of the shaft, the correction for overburden stress is relatively insignificant.

Alternative flow-round penetrometers consisting of either a spherical ball or circular plate have also been investigated recently, mainly through centrifuge tests (Watson et al. 1998) or numerical analysis (Lu et al. 2000).

This paper presents the results of in situ penetration tests for various penetrometers and vane shear tests performed in Western Australia, at the Burswood site near Perth. The results of centrifuge tests carried out on reconstituted samples of clay material collected from the same site are also presented.

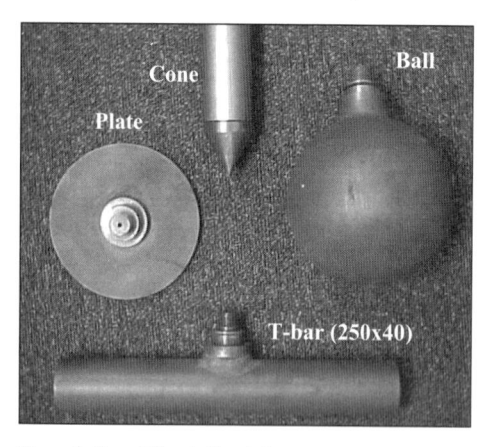

Figure 1. Cone, T-bar, ball and plate penetrometers.

2 FIELD TESTING

Penetrometers tested in the field included cone, T-bar, ball and plate. The cone penetrometer has an apex angle of 60° and a projected area of 10 cm². Two different sizes of T-bar penetrometers were tested, both of 40 mm diameter. The first T-bar was 250 mm long (aspect ratio, L/d = 6.25), giving a projected area of 100 cm², while the second T-bar was 160 mm long (L/d = 4), giving projected area of 64 cm². Both the ball and the plate penetrometers have a diameter of 113 mm, giving a projected area of about 100 cm². The longer T-bar was tested with both smooth and lightly sand-blasted conditions for the cylindrical surface, whereas the shorter T-bar,

the ball and the plate penetrometers were tested only with the sand-blasted surface condition.

All the penetrometers screw onto the standard cone rods, merely replacing the cone tip itself (see Figure 1). They all therefore measure the (u_2) pore pressure at the same position, immediately behind the penetrometer tip. The area ratio for the cone was measured by calibration as $\alpha = 0.70$.

The penetration tests started at ground level and extended to a depth of about 18 m. The clay layer of interest lies between depths of around 4 m and 17 m. The rate of penetration and extraction was 20 mm/s. Cyclic penetration and extraction tests were carried out at specific depths for each penetrometer, except the cone. These tests comprised displacement cycles of ±0.5 m about the mean depth, recording the penetration and extraction resistance over five full cycles.

3 CENTRIFUGE TESTING

Bulk samples of the Burswood clay were obtained from a depth of about 6 m. The clay was reconstituted at a water content of 125 % and used to prepare two boxes for the centrifuge. Both samples were consolidated in a press under a vertical stress of 35 kPa, and then transferred to the centrifuge for testing at 100g. A final sample thickness of between 200 to 230 mm was targeted, which would represent a prototype depth of 20 to 23 m.

Similar to field testing, in the first box, various penetrometers were tested, comprising the cone, ball, plate with diameters of 10 mm, 11.9 mm, 11.2 mm respectively, and the T-bar of 20 mm x 5 mm. In the second box, four T-bars of different lengths: 20 mm, 30 mm, 40 mm and 50 mm, all with the same diameter of 5 mm, were tested.

Unfortunately, it was not possible to measure pore pressures on the penetrometers during the centrifuge tests.

The rate of penetration and extraction was 1 mm/s. Hand vane tests were conducted soon after completion of the penetration tests for each sample.

4 PENETRATION AND EXTRACTION RESISTANCES

4.1 *Effect of corrections*

The effects of the corrections for unequal end area and overburden stress on the tip resistance are best illustrated in Figure 2, which shows the measured resistance and the corrected resistance profiles for field cone and (100 cm²) T-bar penetration tests. For the cone, the measured resistance (q_c) is first corrected for unequal end area, to total tip resistance (q_t) using the following relationship (Baligh et al, 1981; Campanella et al, 1982; Campanella, 1995):

$$q_t = q_c + u_2(1-\alpha) \quad (2)$$

where u_2 is the pore pressure measured immediately behind the cone tip and α is the unequal area ratio. Then the net tip resistance is calculated as below:

$$q_{cnet} = q_t - \sigma_{vo} \quad (3)$$

where σ_{vo} is the total overburden stress. Similarly, the measured tip resistances (q_m) of the T-bar and other penetrometers are corrected to net tip resistances (q_{net}) using the expression below:

$$q_{net} = q_m - [\sigma_{vo} - u_o(1-\alpha)]A_s/A_p \quad (4)$$

where u_o is the estimated hydrostatic water pressure, A_s is the cross-sectional area of the connection shaft and A_p is the projected area of the penetrometer.

For all these corrections, the *in situ* pore pressure has been taken as hydrostatic with a water table 1 m below the ground surface, and the average unit weight for the soil was 14 kN/m³ above the water table and 14.9 kN/m³ below the water table.

As may be seen in Figure 2, the correction for the T-bar is far less significant than that for the cone. This is mainly because the A_s/A_p term in Equation 4 is relatively small (generally around 0.1), and hence, reduces the weight of the bracketed term. Consequently, any uncertainty in estimating the unequal area ratio and overburden stress would have much less impact for the T-bar than for the cone.

Unless otherwise stated, all penetrometer resistance profiles presented hereafter are net values.

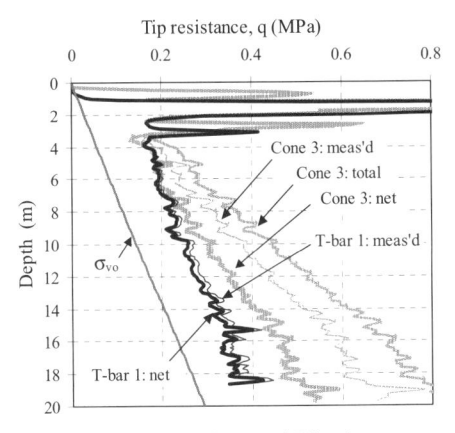

Figure 2. Correction of field cone and T-bar data.

4.2 *Effect of T-bar surface*

Figure 3 presents the penetration and extraction resistances of the smooth T-bar (Tests 1 and 2) and the lightly sand-blasted (rough) T-bar (Tests 3 and 4). All four tests show very similar resistance profiles, although the smooth T-bars tend to show penetration resistances slightly lower (~5 % on average) than the rough T-bars over the depth range 4 to 17 m. Such difference agrees with the theoretical T-bar resistance (Randolph and Houlsby, 1984), which varies by 5 to 7 % as the interface friction ratio increases from 0.2 to 0.4, or 0.3 to 0.5.

It should be noted that the sudden reductions in T-bar extraction resistances shown in the figure at depths of 9 m (Test 3) and 4 and 14 m (Test 4) are where cycles of penetration and extraction were applied during the initial penetration test (see later).

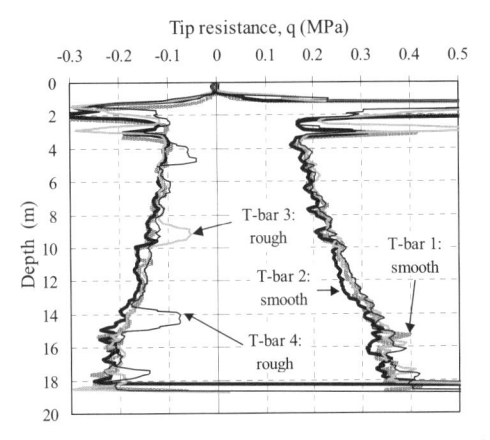

Figure 3. Penetration and extraction resistances for smooth and lightly sand-blasted T-bars (field results).

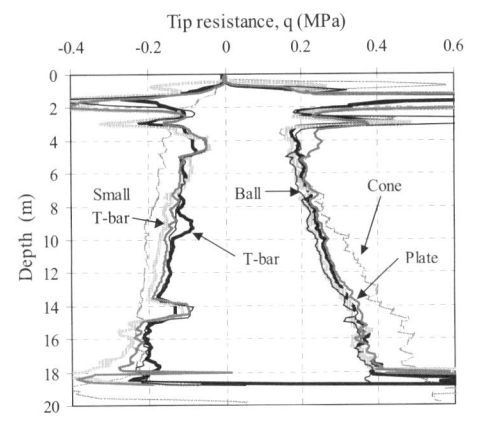

Figure 4. Summary of results from the different penetrometers tested in the field.

4.3 *Effect of various penetrometer shapes*

The penetration and extraction resistances (average values of two tests) for the various penetrometers tested in the field are plotted in Figure 4. During penetration, the cone resistance was found to increase more strongly with depth and exhibited the highest resistance below about 7 m. The other penetrometers showed a very tight band of resistances, with the plate penetrometer showing marginally higher resistance, particularly just below the surface crust at depths of 3 to 5 m.

During extraction, the cone penetrometer shows the highest resistance in the upper 15 m, followed by the smaller T-bar penetrometer over the same depth range. The latter penetrometer gave the highest extraction resistance below 15 m, possibly reflecting less disturbance, or remoulding of the clay, from the smaller penetrometer. A similar phenomenon was observed in the cyclic tests, as discussed later.

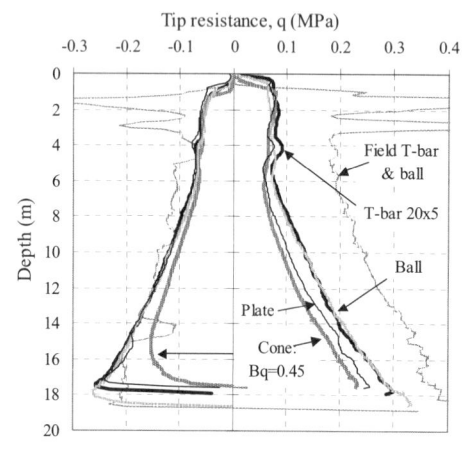

Figure 5. Summary of results from different model penetrometers in Box 1 of centrifuge tests.

Figure 5 presents the average resistance profiles for the different model penetrometers tested in Box 1 of the centrifuge tests. Average resistance profiles for the field T-bar and ball penetrometers are also included in the figure. Due to the inability to measure the pore pressure immediately behind the cone tip for the model cone penetrometer, the measured cone resistance was corrected to the net cone resistance using (Robertson and Campanella, 1983; Watson et al, 1998):

$$q_{cnet} = \frac{q_c - (\sigma'_v + u_o \alpha)}{1 - (1-\alpha) B_q} \quad (5)$$

where σ'_v is the estimated effective vertical stress and B_q is the ratio of the excess pore pressure to the net bearing pressure. The value of B_q was estimated from the field tests, where an average value of 0.45 was measured. The net cone resistance is relatively insensitive to B_q, since the area ratio for the model cone penetrometer is $\alpha = 0.86$.

The tip resistances for the model T-bar, ball and plate penetrometers were not corrected, since the areas of these penetrometers are much larger than the area of the penetrometer rod, so that the correction will be insignificant.

Note that the resistance profiles for all model penetrometers are plotted against their prototype depth, where 1 mm in the centrifuge would represent 0.1 m at prototype scale.

As may be seen in Figure 5, during penetration, the model cone penetrometer exhibits the lowest net resistance, in contrast to the field results. This finding may be affected by the uncertainty in correcting the cone resistance, but more probably reflects higher rigidity index for the natural clay, leading to higher cone resistance for a given shear strength (Teh & Houlsby, 1991). It was also interesting that the model plate penetrometer showed relatively lower resistance compared to the model T-bar and ball penetrometers.

During extraction, the model T-bar, ball and plate show very similar resistances, while the model cone exhibits lower resistance for most of the depth.

Of particular significance is the fact that the centrifuge results show broadly similar extraction resistances to the field values over the depth range 8 m to 16 m. This suggests that, for a given water content (these were similar in field and centrifuge test over that depth range), the remoulded strengths are similar, despite the peak strengths in the field being higher due to the innate structure of the undisturbed material.

4.4 Effect of aspect ratio of T-bar

Since the theoretical bearing factor for a T-bar is based on flow around an infinitely long cylinder (Randolph and Houlsby, 1984), it is important to investigate the effect, if any, of the finite length of the T-bar on the bearing factor. This was achieved by comparing the penetration resistances of T-bars with various aspect ratios, ranging from 4 to 10.

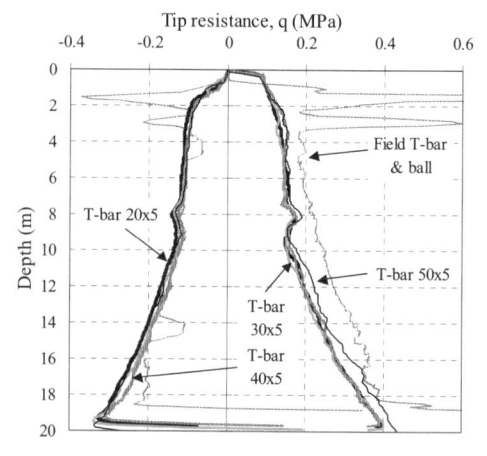

Figure 6. Summary of results from T-bars of different aspect ratio in Box 2 of centrifuge tests.

As was shown in Figure 4, the normal field T-bar with aspect ratio of 6.25 showed slightly higher penetration resistance than the smaller field T-bar (aspect ratio of 4), while the reverse was true during extraction. By contrast, the centrifuge test results, which are shown in Figure 6, indicate that the aspect ratio does not have any obvious effect on the resistance of the T-bar, at least for L/d of 4 to 10. Although the longest (50 mm x 5 mm) model T-bar does show slightly higher resistance than the other three T-bars over much of the depth, this is believed due to other effects, such as a locally stronger region of the soil sample or a slight bending moment on the load cell. During extraction, all four model T-bars showed very similar resistances.

The conclusion is that the small differences observed in the two field T-bars with different aspect ratios is not significant, and that any aspect ratio in the range 4 to 8 would be suitable.

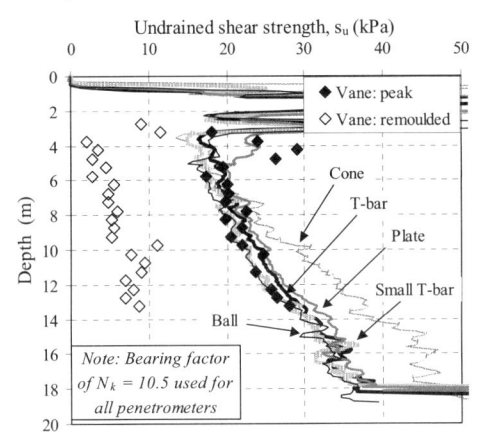

Figure 7. Undrained shear strength profiles derived from the field penetration tests.

5 UNDRAINED SHEAR STRENGTH

The undrained shear strength (s_u) may be estimated from the net penetration resistance through the following relationship (Campanella, 1995):

$$s_u = q_{net} / N_k \quad (6)$$

where N_k is the bearing factor for the penetrometer. A provisional N_k value of 10.5 has been used for all penetrometers to obtain profiles of undrained shear strength as plotted in Figure 7. Both peak and remoulded values of the shear strength measured from the *in situ* vane tests are also included for comparison in the figure. It may be seen that the peak s_u profile obtained from the field vane test is embraced within the band of s_u profiles derived from the field penetration tests, except for the cone penetrometer below 7 m, where it starts to exhibit higher shear strength. However, using an N_k value of 13 for the cone penetrometer for depth below 10 m, its s_u profile will be comparable to those of the other penetrometers. This may be an indication of variation of N_k value for the cone penetrometer with the soil properties.

The shear strength profiles derived from the standard (20 mm x 5 mm) model T-bar penetration tests and hand vane tests for both reconstituted samples are shown in Figure 8. It can be seen that the hand vane tests gave higher shear strength values for both boxes, especially in the region where the samples were normally consolidated. Nevertheless, the discrepancies in shear strengths measured by the hand vane and T-bar penetrometer are less than 20 %, which is consistent with data reported by Watson et al. (1998, 2000) for vane tests conducted in-flight in both normally consolidated calcareous clay and lightly overconsolidated kaolin clay. The reasons for these discrepancies may be attributed partly to strength anisotropy and partly to strain softening effects.

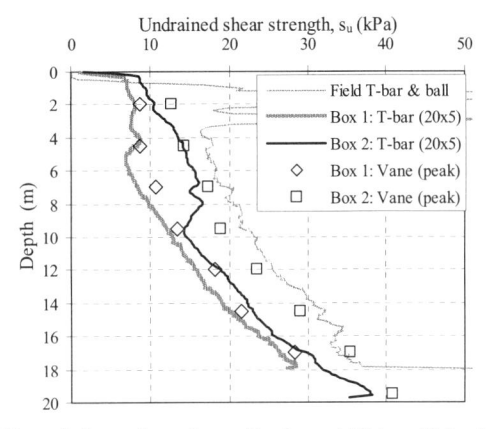

Figure 8. Comparison of s_u profiles for model T-bar with hand vane tests.

Figure 8 also shows clearly that Box 2 had a greater strength than Box 1, especially between depths 3 and 9 m, with a deeper transition depth from lightly overconsolidated to normally consolidated. These differences are due to the much longer laboratory-floor consolidation period for Box 2 (58 days compared with 7 days for Box 1), allowing more secondary consolidation to occur in addition to full primary consolidation. However, despite the longer consolidation period for Box 2, its strength profile is still lower than the field strength profile due to the innate structure of the undisturbed material in the field.

6 CYCLIC TESTS

Cyclic penetration and extraction tests were undertaken only in the field. Figure 9 presents result from a typical cyclic test performed during extraction at a depth of about 14.3 m for T-bar Test 3. It may be seen that both penetration and extraction resistances continue to degrade through the 5 cycles, but at a reducing rate, with the resistance stabilizing at a fully remoulded value.

The degree of degradation in resistance may be measured using a degradation factor, which is calculated by taking the mean (absolute) value of resistance during the half cycle of each 1 m stroke divided by the mean value for the initial penetration. A summary of the degradation factor against the number of cycles for the different penetrometers is shown in Figure 10.

Similar rates of degradation were found for all the penetrometers, although the plate penetrometer gave the most rapid degradation. The smaller field T-bar (ST-bar) showed the most gradual degradation, which is consistent with its higher extraction resistance compared to the other field penetrometers. All the penetrometers showed degradation factors converging to a final value of between 0.23 and 0.28 after 5 cycles. The average degradation factor of 0.25 is consistent with the sensitivity of 3 to 4 from the field vane tests. Hence, it appears that the cyclic test results may be used to measure the remoulded strength and sensitivity of the soil.

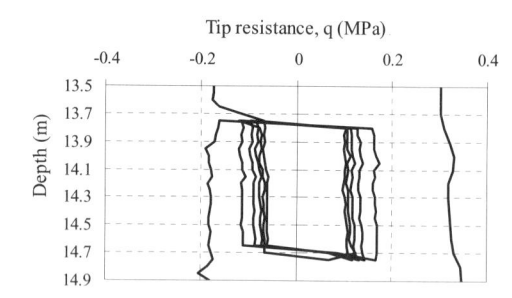

Figure 9. Cyclic test for field T-bar Test 3 at depth 14.3 m.

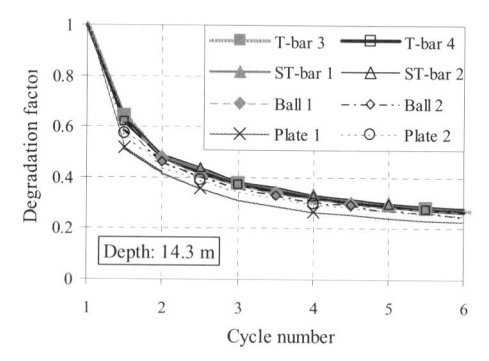

Figure 10. Degradation parameters for cyclic penetrometer tests at depth 14.3 m.

7 CONCLUSIONS

This paper has presented the results of in situ testing carried out at the Burswood site in Western Australia and centrifuge testing conducted on reconstituted samples of clay collected from the same site.

It has been shown that the corrections to derive the net penetration resistance from the measured resistance are significant for the cone penetrometer, but very small for flow-round penetrometers such as the T-bar, ball or plate. Hence, the net penetration resistance in soft clay may be estimated with greater confidence using a flow-round penetrometer.

The smooth field T-bar seems to show a slightly lower penetration resistance than the rough T-bar, in agreement with the predicted theoretical variation (Randolph and Houlsby, 1984).

The *in situ* tests showed a tight band of penetration resistance for the various shapes of flow-round penetrometers, implying similar values of bearing factor, although the net cone resistance increased more rapidly with depth. The model tests on reconstituted clay also gave similar penetration resistances, although this time the plate penetrometer unexpectedly gave lower penetration resistance than that for the T-bar and ball penetrometers, while the net cone resistance was the lowest.

The centrifuge tests on T-bars with aspect ratios between 4 and 10 did not seem to show any significant effect in the penetration resistance, although in field, the smaller T-bar (with L/d = 4) gave slightly lower resistance during penetration and higher resistance during extraction compared to the normal field T-bar (L/d = 6.25).

Using a provisional N_k value of 10.5 for all penetrometers, the shear strength profiles derived from the field penetration tests (except for the cone) are comparable to the results obtained from the field vane tests. However, the hand vane tests on the centrifuge samples gave higher s_u values compared to that derived from the model T-bar tests. The same phenomenon was observed by Watson et al (1998, 2000) for vane tests conducted in-flight in both normally consolidated calcareous clay and lightly overconsolidated kaolin clay.

In cyclic tests, all penetrometers showed degradation factors converging to an average value of about 0.25 after 5 cycles, which is consistent with the sensitivity of 3 to 4 obtained from field vane tests.

ACKNOWLEDGEMENTS

Special thanks are due to the sponsors of the joint industry project: Characterization of Soft Soils in Deep Water by In Situ Tests, undertaken jointly by NGI (project leader Mr Tom Lunne) and COFS, for permission to publish these data. Technical assistance from a number of people involved in the field and centrifuge tests is also greatly appreciated. This work forms part of the activities of the Centre for Offshore Foundation Systems (COFS), established and supported under the Australian Research Council's Research Centres Program.

REFERENCES

Baligh, M.M., Azzouz, A.S., Wissa, A.Z.E., Matyin, R.T. & Morrison, M.H. (1981). The piezocone penetrometer: Cone penetration testing and experience. *Proc. ASCE Conference on Cone Penetration Testing*, St Louis, pp. 247-263.

Campanella, R.G., Robertson, P.K. & Gillespie, D. (1982). Pore pressure during cone penetration. *Proc. 34th Canadian Geotechnical Conf.*, 5.2.1-14, Canadian Geotechnical Society.

Campanella, R.G. (1995). *Guidelines for Geotechnical Design Using the Cone Penetrometer Test and CPT with Pore Pressure Measurement, 5th Edition*, Civil Engineering Department, University of British Columbia, Vancouver, B.C., Canada V6T 1W5.

Lu, Q., Hu, Y. and Randolph, M.F. (2000). FE analysis for T-bar and ball penetration in cohesive soil. Proc. 10th Int. Offshore and Polar Engineering Conf. ISOPE 2000, Seattle, USA, 2, 617-623.

Lunne, T., Robertson, P.K. & Powell, J.J.M. (1997). *Cone Penetration Testing in Geotechnical Engineering*, Blackie Academic and Professional, London.

Randolph, M.F. & Houlsby, G.T. (1984). The limiting pressure on a circular pile loaded laterally in cohesive soil, *Geotechnique, Vol. 34*, No. 4, 613-623.

Randolph, M.F., Hefer, P.A., Geise, J.M. & Watson, P.G. (1998). Improved seabed strength profiling using T-bar penetrometer. *Offshore Site Investigation and Foundation Behaviour '98, Science and Underwater Technology*, London, 221-236.

Robertson, P.K. & Campanella, R.G. (1983). Interpretation of cone penetrometer tests. Part II: Clay, *Canadian Geotechnical Journal. Vol. 20*, 734-745.

Stewart, D.P. & Randolph, M.F. (1991). A new site investigation tool for the centrifuge, *Proc. Int. Conf. Centrifuge 1991*, Boulder, H.Y. Ko (ed.), Balkema, 531-538.

Stewart, D.P. & Randolph, M.F. (1994). T-bar penetration testing in soft clay, *J. of Geotech. Eng. Div., ASCE*, 120(12): 2230-2235.

Teh, C.I. & Houlsby, G.T. (1991). An analytical study of the cone penetration test in clay. *Geotechnique*, **41**(1), 17-34.

Watson, P.G., Newson, T.A. & Randolph, M.F. (1998). Strength profiling in soft offshore soils, *Geotechnical Site Characterization, Robertson & Mayne (eds), 1998*, Balkema, Rotterdam.

Watson, P.G., Suemasa, N. & Randolph, M.F. (2000). Evaluating undrained shear strength using the vane shear apparatus, *Proc. 10th Int. Conf. On Offshore and Polar Eng, ISOPE 00*, Seattle, **2**, 485-493.

Proceedings ISC-2 on Geotechnical and Geophysical Site Characterization, Viana da Fonseca & Mayne (eds.)
© 2004 Millpress, Rotterdam, ISBN 90 5966 009 9

Evaluation of the undrained shear strength profile in soft layered clay using full-flow probes

J.T. DeJong, N.J. Yafrate, D.J. DeGroot & J. Jakubowski
Department of Civil and Environmental Engineering, University of Massachusetts Amherst

Keywords: full-flow penetration probes, T-bar, Ball, Plate, CPTu, undrained shear strength, remolded undrained shear strength, in situ testing, soft clay, layered clay, varved clay

ABSTRACT: The undrained shear strength profile of soft clay deposits are routinely estimated with a piezocone (CPTu) and the field vane test (FVT). Though suitable for a majority of soil types, both devices require significant geometric and/or empirical corrections that are inherent to the device design and/or testing method. These corrections become problematic for soft soil and offshore deposits with extreme ambient pore pressures and/or overburden stress. Recently, "full-flow" probes, which provide a direct measure of the force required to induce soil flow, have been proposed to measure the undrained and remoulded undrained strengths of soft clays in situ. While analyses and tests have been performed with these probes, minimal full-scale field studies have been performed. Moreover, their suitability to anisotropic soft clays has not been investigated.

A series of full-flow soundings were performed at the University of Massachusetts Amherst test site, which consists of a continuous deposit of Connecticut Valley Varved Clay (CVVC) to a depth of 30 m. CVVC consists of alternating layers of sand-silt and clay, which together comprise a varve. Below the 6 m thick crust, the peak and remolded residual shear strength values determined with the FVT are uniform with depth and are approximately equal to 35 kPa and 4 kPa, respectively. The penetration resistance determined from the continuous penetration full-flow soundings are compared with that measured with the CPTu. Probe factors for both the overconsolidated crust and the lightly overconsolidated thick deposit are developed to estimate the undrained shear strength profile, using the FVT undrained shear strength as the reference. Finally, remolded undrained shear strength values determined from cycling the probes over selected intervals are compared with residual undrained shear strength values determined from the FVT.

1 INTRODUCTION

The state-of-practice for characterizing soft soil sediments is currently centered around the following three primary techniques: (1) high quality sampling and laboratory testing, (2) the field vane test (FVT), and (3) the piezocone penetration test (CPTu). While each has distinct advantages, they all pose challenges for characterization of soft clays, especially in deep water. Obtaining high quality undisturbed soil samples is difficult and expensive.

The field vane test (FVT) served as the primary in situ strength test to quantify the strength profile in situ prior to the regular implementation of the CPTu in the early 1990s. The field vane test has been modified minimally from its initial design (Figure 1) and provides estimates of the peak and residual undrained shear strength through the use of correction factors (Andresen and Bjerrum 1958). Though elegantly simple, a FVT test only provides strength data at one discrete position, requiring multiple tests to determine the strength profile with depth at one location. Furthermore, implementation of the FVT offshore at moderate to large water depths (>1000 m) can be problematic.

The piezocone penetrometer test (CPTu) (Lunne et al. 1997) is now generally preferred over the FVT due to the continuous measurement profile that is produced (Figure 1). However, the CPTu can have difficulty in the profiling of soft sediments, especially in deeper waters offshore, due to three primary factors. Firstly, due to the low strength of the soil the magnitude of the tip resistance (q_c) and sleeve friction (f_s) measurements are very small and often approach the resolution of the sensor load cells (Tanaka 1995, Lunne et al. 1997). Secondly, the CPT must be corrected for the differential pore pressure acting on the back of the cone tip, which may amount to several MPa, while simultaneously measuring soil strengths of 10 kPa or lower. Thirdly, due to the full displacement mode of the cone tip, the measured cone resistance must be corrected for the overburden pressure, which can be orders of magnitude larger than the net cone resistance value. Each of these factors introduces uncertainty, the combination of which limits accurate characterization of soft sediments with the CPTu.

2 BACKGROUND OF FULL-FLOW PROBES

Recognizing the limitations with the FVT and CPTu, full-flow probes were developed with the objective that a continuous *in situ* profile of strength and potentially other soil properties could be determined without the large uncertainties associated with the CPTu. In principle, the penetrometers are akin to a viscosity measurement as they induce plastic flow of the soil around a geometric probe during penetration. The three types of penetrometers that have been studied numerically and analytically and have been tested in the laboratory and centrifuge are the T-bar, Ball, and Plate penetrometers (Figure 1). For all three probes, the projected area is 100 cm^2, 10 times that of the standard 10 cm^2 CPTu. The flow mechanism around these probes is schematically shown in Figure 1. This full-flow mechanism provides advantages over the CPTu, which forces full and irrecoverable displacement of the soil. The load cell positioned directly behind the penetrometer measures the differential force acting on the geometric probe (e.g. T-bar, Ball, or Plate), thereby eliminating the need for adjustment of the measurement for the overburden stress and ambient pore pressure. This also enables improved measurement resolution through installation of a lower capacity load cell. Furthermore, the influence of incomplete saturation of the CPTu filter element is irrelevant to full-flow penetrometers.

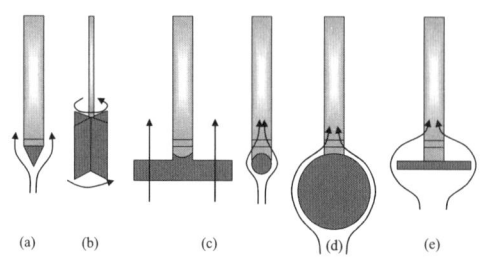

Figure 1. Schematic of in situ devices and soil displacement paths (a) CPTu, (b) FVT, (c) T-bar, (d) Ball, (e) Plate.

2.1 *Estimation of Undrained Shear Strength*

The measurements obtained from any *in situ* device must be adjusted to account for the respective geometry and deformation mechanism in order to estimate the undrained shear strength. The interpretation of the FVT differs from these devices and can be found elsewhere (Andresen and Bjerrum 1958, Terzaghi et al. 1996, ASTM D-2573 2003).

CPTu measurements must be corrected for various factors prior to using correlations to estimate the undrained shear strength of the soil. As detailed further elsewhere (e.g. Lunne et al. 1997), the undrained shear strength of clays can be estimated based on the following equation:

$$s_u = (q_t - \sigma_{vo}) / N_{kt} \quad (1)$$

where q_t is the corrected penetration resistance, $q_t = q_c + u_2(1-\alpha)$, where u_2 is the pore pressure measured on the shoulder of the CPTu and α is the cone ratio area, which typically varies from 0.55 to 0.95; σ_{vo} is the current total overburden stress; N_{kt} is the cone factor. Cone factors typically range from 7 to 15 (Robertson and Campanella 1983, Lunne et al. 1997), depending on the sensitivity and the degree of overconsolidation of the clay, although values as high as 30 have been reported for heavily overconsolidated fissured clays. The necessary correction for the pore pressure measurement, overburden stress, area ratio, and pore pressure contribute to uncertainty in estimating the undrained shear strength (Randolph et al. 1998). In combination, these factors can introduce a level of uncertainty that is unacceptable for stability and design calculations where the undrained shear strength is very low.

Due to the full-flow behavior of the soil around the T-bar, Ball and Plate probes during penetration, the pore pressure and overburden stress act on both the top and bottom surfaces of the probe and are thus assumed to be self-equilibrating. As a result, no corresponding correction factors are required for determining s_u for these probes. The undrained shear strength, s_u, is thus calculated using the following equation:

$$s_u = q_*/N_* \quad (2)$$

where q_* is the measured bearing resistance and N_* is the factor for a given full-flow probe. For simplicity, q_T and N_T will be used to denote the T-bar and similarly, q_B and N_B, and q_P and N_P, will be used to denote the Ball and Plate penetrometers, respectively.

The factors for full-flow penetrometers have been investigated with numerical and analytical analyses and through laboratory and centrifuge testing. Analytical and numerical analysis have provided upper and lower bound plasticity solutions for N_T for both a smooth and rough interface. For a smooth interface the lower and upper bound solutions are 9.14 and 9.21, respectively (Randolph and Houlsby 1984). The lower and upper bounds converge to a single value, 11.9, for a fully rough interface. Testing with scaled T-bar penetrometers in the centrifuge led to a recommendation of an N_T value of 10.5 (Stewart and Randolph 1991).

Factors for Ball and Plate penetrometers have subsequently been developed through similar studies. For the Ball penetrometer, the lower and upper bound solutions for a smooth and rough interface are 11.0 and 15.1, respectively (Randolph et al. 2001). The upper bound solution does not converge as the smooth and rough interface solutions are 11.8 and 15.5, respectively. While these analytical solutions are about 20% higher than for the T-bar, initial centrifuge scaled tests indicate that a Ball factor of similar magnitude to that recommended for the T-bar (10.5) may be appropriate. For the Plate probe, analytical plasticity lower and upper bound solutions for the smooth and rough interfaces are both coincident and are equal to 12.4 and 13.1, respectively (Martin and Randolph 2000).

2.2 Estimation of Remolded Undrained Shear Strength and Sensitivity

Beyond accurately predicting s_u, initial testing indicates that full-flow probes can provide a measure of the remolded undrained shear strength with a minor variation in the testing procedure (Watson et al. 2000). s_u is determined using the conventional approach during initial monotonic penetration. Following penetration to a specific depth, several cycles of penetration and extraction over a range 20 to 40 times the diameter of the probe (approximately 1 m) results in remolding of the soil within the flow zone. The penetration resistance decreases with cycling until a steady-state penetration resistance is achieved that reflects the remolded soil strength, s_{ur}. The sensitivity (S_t) of the soil may then be directly determined as follows:

$$S_t = s_u / s_{ur} \qquad (3)$$

where s_u is the undrained shear strength determined from the initial penetration and s_{ur} is the steady-state remolded undrained shear strength following repeating cycling.

The potential of full-flow penetrometers to assess the remolded undrained shear strength of the soil and subsequently the sensitivity has been focused on laboratory studies with the T-bar penetrometer (Watson et al. 2000). Preliminary tests show the remolded resistance to reduce rapidly within the first few cycles and then stabilize after 10 to 15 cycles at a resistance significantly less than the initial peak value.

3 LAYERED CLAY TEST SITE

Testing was performed at the US National Geotechnical Experimentation Site (NGES; Geocouncil 2001) located at the University of Massachusetts Amherst (UMass Amherst). An overview of the site conditions is presented here with further detail given in DeGroot and Lutenegger (2003).

3.1 Depositional History

The stratigraphy at the UMass Amherst NGES site is primarily composed of a 30 m thick deposit of Connecticut Valley Varved Clay (CVVC). This lacustrine soil was deposited approximately 15,000 years ago in glacial Lake Hitchcock during the retreat of the late Pleistocene ice sheet in New England, USA. The distinguishing feature of the deposit is alternating layers of clay and silt-fine sand. The resulting pair of layers, or varve, represents one year of deposition. The thickness of the varves varies considerably and ranges from a few millimeters to more than 1 m, with a general increase in varve thickness with depth. These variations depend on proximity of the glacier during deposition and also annual variations in sediment discharge and climate conditions. Typically, most of the variation in thickness of the varves is in the silt-sand layer whereas the clay layer changes minimally in thickness. There are about 1,389 varves at the UMass Amherst NGES site (Rittenour 1999).

During post glacial drainage of Lake Hitchcock, the Connecticut River formed and the surface sediments were exposed to an arid and cold climate. It is clear from visual inspection of samples and consolidation data that the upper few meters of the deposit has undergone significant changes as a result of desiccation, freeze/thaw cycles, possible permafrost conditions after drainage, and other weathering. This zone of the deposit, commonly known as a crust, extends to about 5 to 6 m below ground surface. The crust soil is typically brown and therefore oxidized as compared to the predominately reddish (local Triassic rock source) or gray colored (distant crystalline rock source), unoxidized lightly overconsolidated CVVC below the crust.

3.2 Classification and Stress History

State and index properties such as water content and Atterberg Limits for CVVC depend on the relative portions of the silt and clay layers in a test specimen. Figure 2 plots Atterberg Limits and water content versus depth based on bulk properties. The liquidity index varies from close to zero in the upper crust, increases significantly with depth in the crust, and averages 1.5 in the lightly overconsolidated zone.

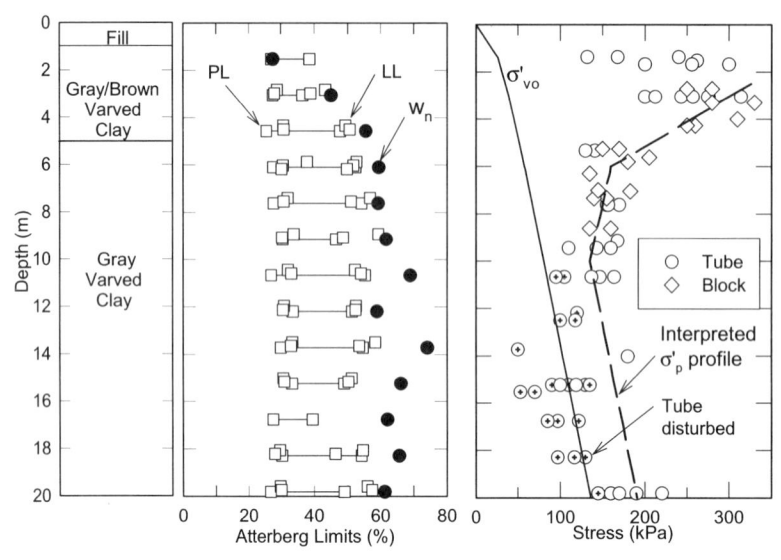

Figure 2. Summary of subsurface stratigraphy, Atterberg limits and natural water content, and stress history with depth at the UMass Amherst NGES test site.

The ground water table typically occurs in the upper 2 m below ground surface and varies by as much as ±2 m throughout the year, coinciding with changes in seasonal precipitation. Piezometer data show a slight artesian pressure that is consistent with local topography with a relatively uniform 0.87 kPa/m increase in pore water pressure with depth above hydrostatic conditions.

Figure 2 plots stress history data based on incremental load and constant rate of strain consolidation tests conducted on tube and block samples. The data clearly show evidence of the stiff crust in the upper five meters followed by light overconsolidation (low OCR) throughout the remainder of the deposit. Erosion is believed to have occurred at many locations in the Connecticut Valley during drainage of Glacial Lake Hitchcock, including at the NGES site. Aging and cementation are also possible contributing stress history mechanisms. For the "interpreted stress history" line fitted to the data in Figure 2, the OCR ranges from 9.3 at 2.5 m to 1.4 at a depth of 20 m.

The deposition mechanisms and subsequent stress history have resulted in a highly anisotropic deposit that varies substantially between the crust and lower deposit. For example, the strength anisotropy ratio $K_s = s_u(DSS)/s_u(CAUC)$ averages 0.63 and the hydraulic conductivity ratio, $r_k = k_h/k_v$, ranges between 7 and 80, depending on soil variability and laboratory versus field tests.

4 TESTING PROGRAM

The testing program consisted of soundings performed with a T-bar, Ball, and Plate probes with projected areas of 100 cm^2. The surfaces of all probes were roughened by sandblasting with the exception of the T-bar ends, which were machined smooth. Full-flow probe soundings were performed in predrilled holes to 2.5 m depth to avoid penetration through the very stiff upper crust. For comparison, four digital 10 cm^2 CPTu profiles were performed, with one representative profile presented herein. All penetrometer soundings were performed in accordance with ASTM 5778 (2003) during monotonic penetration. For the full-flow probes, repetitive cycling was performed during probe extraction at five depth intervals. It is noted that the time required for cycling at multiple depths during extraction significantly exceeded the time required for monotonic penetration.

Fourteen field vane test (FVT) profiles obtained previously were utilized as a benchmark in the analysis. FVTs were conducted using a Nilcon Vane Borer with a 130 mm x 65 mm vane with 1.9 mm thick rectangular blades and performed in accordance with ASTM 2573 (1994). The residual vane strengths were determined after 10 full revolutions of the vane were conducted. The s_u(FVT) values reflect high strength in the crust and rapidly decrease towards the bottom of the crust at approximately 6 m. Thereafter s_u(FVT) is approximately constant with depth with most values ranging between 30 and 40 kPa. Below a depth of 6 m, the data averages are s_u(FVT) = 35 kPa and s_{ur}(FVT) = 4 kPa, giving a sensitivity (S_t) based on these average values equal to 9 although individual test values vary between S_t = 5 and 25.

5 TEST RESULTS

Objective evaluation of full-flow probes to estimate the undrained shear strength (s_u) and the remolded undrained shear strength (s_{ur}) profile was referenced against FVT test s_u and s_{ur} (residual) data. Therefore, throughout the paper all N* factors (general term of N factors for all tests) presented are benchmarked to the s_u(FVT) profile. Explicit reference to the benchmark test procedure for determining N* values is critical and its importance cannot be overstated. For example, the value of N_{kt} for CVVC varies from 23, when referenced to DSS, to 13, when referenced to CAUC tests (DeGroot and Lutenegger 2002).

The penetration resistance pressure profile for the CPTu (q_t), T-bar, Ball and Plate probes is presented in Figure 3. All profiles show near identical trends with the CPTu pressure being the highest, which can be attributed to the full-displacement deformation of the penetrated soil. The T-bar mobilizes a penetration resistance that is about 38% larger than the Ball and Plate probe resistances, which are very similar. This implies that the plain strain deformation mode of the T-bar in the varved clay may provide greater resistance to mobilization than the axisymmetric deformation mode around the Ball and Plate probes. This is somewhat surprising considering that mobilization occurs over a larger number of varves for the Ball and Plate penetrometers, assuming that the approximate volume (depth) of soil engaged in the deformation envelope during steady state is a multiple of the probe cross-sectional diameter.

6 ANALYSIS OF UNDRAINED SHEAR STRENGTH

The undrained shear strength profile was analyzed by separating the profile into two distinct regions: a lower "lightly overconsolidated" zone that was considered to be between depths of 6-22 m and a "crust" zone between depths of 3-6 m. N* values for the CPTu (N_{kt}), the T-bar (N_T), the Ball (N_B), and the Plate (N_P) were determined using average values of the penetration resistance and the s_u(FVT) data within each depth increment. For the CPTu, N_{kt} was determined using $s_u = (q_t - \sigma_v)/N_{kt}$ while $s_u = q_*/N_*$ was used for the full-flow probes since correction for σ_v is not necessary due to the full-flow mechanism.

The N* values computed for the lightly overconslidated layer (6-22 m) are presented in Table 1. As evident and expected, the N* values for the axisymmetric probes (Ball, Plate) are lower than for the T-bar (plain strain). Implementation of N* and calculation of s_u with depth yields very similar profiles (Figure 4). All s_u(FVT) data between 6-22 m from 14 profiles are plotted for reference. The relatively large scatter in the FVT profiles, all of which were performed within 30 m of the full-flow soundings, reflect the natural variability of the deposit and implies a sensitivity of the FVT measurement to the relative percentage of silt versus clay engaged in shearing. Furthermore, differences in the shearing modes, with silt-silt and clay-clay shearing occurring with the FVT and mixed soil shearing occurring with the full-flow probes, may enable the probes to provide a more average response.

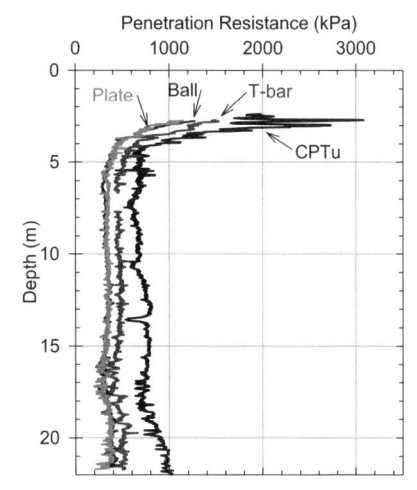

Figure 3. Penetration resistance profile for the CPTu (q_c), T-bar(q_T), Ball (q_b), and Plate (q_p).

The N* values for the lightly overconsolidated layer (OCR <2) (Table 1) are comparable to the analytically determined and recommended values. The N_T for this study exceeds both the analytical and

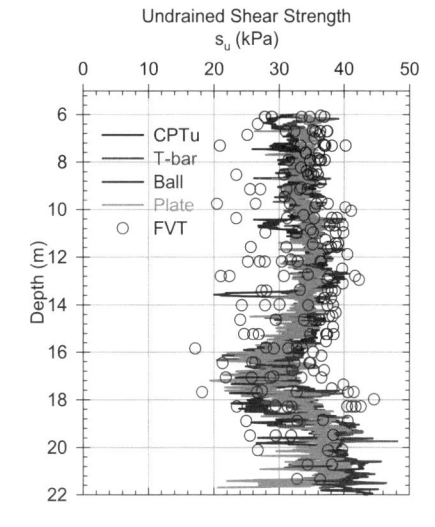

Figure 4. Undrained shear strength profile in the lightly overconsolidated zone between 3-6 m.

recommended values. In contrast, the N_B and N_P values agree with recommended values and are less than analytically determined values.

Table 1. Summary of N* values for probes.

Soil Layer (Depth)	CPTu $N_{kt}^†$	T-bar $N_T^†$	Ball $N_B^†$	Plate $N_P^†$
Lightly OC (OCR < 2) (6-22m)	15.0	13.7	10.0	10.0
Analytical Solutions	--	11.9[1]	15.1-15.5[2]	13.1[3]
Recommended Values	--	10.5[4,5]	10.5[5]	10.5[5]
Crust (OCR > 2) (3-6 m)	13.1	9.7	6.8	6.9

[†]Note: All values determined with reference to s_u values obtain from FVT data.
[1] Randolph and Houlsby, 1984; [2] Randolph et al, 2001; [3] Martin and Randolph, 2000; [4] Stewart and Randolph, 1991; [5] COFS, 2002.

A similar approach was implemented to obtain N* values in the overconsolidated crust zone between 3-6 m. The N* values are presented in Table 1 and the resulting s_u profiles are presented in Figure 5. The N* values in the crust for the full-flow probes are on average about 31% less than the corresponding value for the lightly overconsolidated layer (6-21m). In contrast, the N_{kt} value decreases by 12%. It is important to recognize that the values computed are averages and that the OCR in the crust layer decreases from about 8 to 2 between 3 and 6 m. As evident in Figure 5, the scatter in the FVT data is higher near 3 m and becomes more uniform with depth. Nonetheless, a predominant trend of lower N* values for the crust is evident.

Based on the above analysis, an accurate s_u profile with depth can be obtained with full-flow penetrometers. In addition, the N* values obtained for the lightly overconsolidated zone agree reasonably well with the previous analytical solutions and recommended values. Potential to determine s_u in the overconsolidated crust also shows promise.

7 ANALYSIS OF REMOULDED UNDRAINED SHEAR STRENGTH AND SENSITIVITY

Repeated cycling was performed over depth increments of 3.7-4.7 m, 6.7-7.7 m, 11.7-12.7 m, 15.7-16.7 m, and 20.7-21.7 m to evaluate the ability of full-flow probes to estimate the remolded undrained shear strength (s_{ur}). For efficiency cycling was performed during probe extraction following monotonic penetration. As a result, the first penetration segment of the cycling data was obtained prior to repeated cycling.

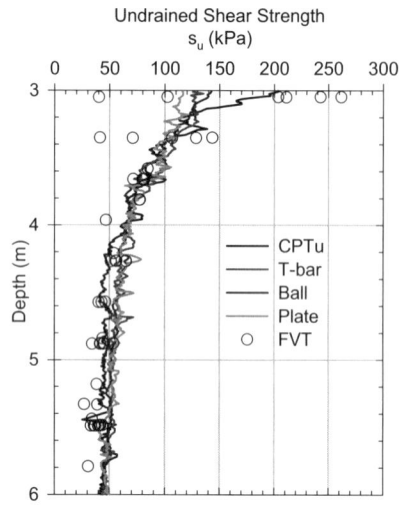

Figure 5. Undrained shear strength profile in the crust zone between 3-6 m using N* values from Table 1.

Two representative cycling tests are presented in Figure 6. T-bar cycling between 3.7-4.7 m (crust zone) and Ball cycling between 11.7-12.7 m (lightly overconsolidated zone) are presented in Figure 6 (a) and (b), respectively. For reference, the solid and dashed vertical lines are the average s_u(FVT) and s_{ur}(FVT) values for the specific depth increment, respectively. As evident, rapid decay in the shear strength occurs in the first few cycles, after which a consistent value of strength appears to be mobilized. A summary of the undrained shear strength mobilized during the penetration segment in each cycle at cycling intervals of 3.7-4.7 m (crust zone) and 11.7-12.7 m (lightly overconsolidated zone) for the three different full-flow probes are presented in Figure 7. In addition to the average s_u(FVT) and s_{ur}(FVT) reference lines, the maximum and minimum s_{ur}(FVT) values are referenced.

It is important to recognize that while the same symbolic notation, s_{ur}, is commonly used for the FVT and full-flow probes, the modes of shearing differ significantly. In the FVT test, s_{ur}(FVT) reflects a residual strength that exists on a distinct, defined shearing surface following very large strains. In contrast, the s_{ur} from full-flow probes reflect a remolded strength that exists after the soil has been tortuously mixed. Technically this latter strength measure may be more similar to remolded strengths measured in laboratory tests, such as the fall cone test. Due to these differences in the shearing mechanisms, it is possible that s_{ur}(FVT) values would differ from s_{ur} values obtained from T-bar, Ball, and Plate probe cycling. Nevertheless, short of laboratory test comparisons, s_{ur}(FVT) data provide an industry standard *in situ* estimate of high-strain strength.

The rate of degradation in the crust (3.7-4.7 m) is higher than in the lightly overconsolidated layer (11.7-12.7 m). In general, the T-bar degrades most rapidly. This may be due to the smaller cross-sectional area of soil that must be remolded. In the crust, all values converge following about 13 cycles, which agrees with previous observations that 10-15 cycles may be required to obtain a steady-state condition. In the lightly overconsolidated layer the shear strength mobilized by the Ball and Plate after 15+ cycles remains slightly higher than the corresponding T-bar values.

While the rates of degradation agree with general expectations, the undrained shear strength values to which the two tests converge differ. In the lightly overconsolidated layer (11.7-12.7 m), the T-bar approaches a value comparable to the s_{ur}(FVT). The Ball and Plate mobilized shear strength remains slightly higher with cycling. Further research is currently underway to understand whether the layered (varved) characteristics of the soil in combination with the larger cross-sectional area required for full remolding of the soil around the Ball and Plate can explain the observed differences in behavior. Nonetheless, the T-bar, Ball, and Plate probes estimate the sensitivity, S_t (= s_u/s_{ur}), to be about 7.0, 6.6, and 5.3, respectively. This slightly underestimates the average S_t(FVT) of 7.8, but lies within the range of values (5-25).

The undrained shear strength mobilized in the crust (3.7-4.7 m) approaches the minimum s_{ur}(FVT) obtained within the depth increment from among 14 FVT soundings. When the rods were released from the testing rig following cycling at this depth interval, the probe was observed to fall rapidly over a portion of the interval. With additional handling it was concluded that a cylindrical void space extending over much of the depth increment developed with cycling. This implies that the soil was gradually

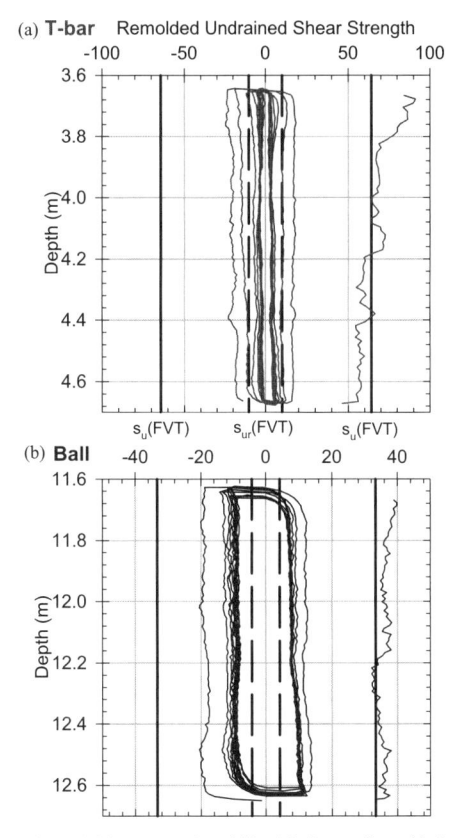

Figure 6. Shear strength mobilized during cycling with the (a) T-bar at 3.7-4.7 m and (b) Ball at 11.7-21.7 m. Determined using N∗ values from Table 1.

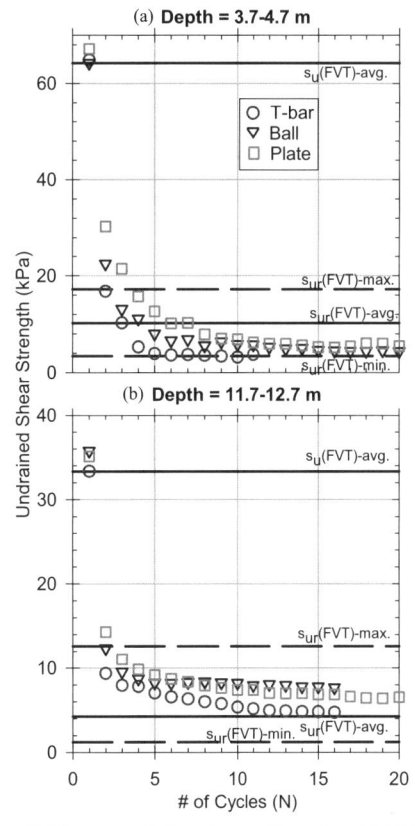

Figure 7. Shear strength degradation with cycling at (a) 3.7-4.7 m and (b) 11.7-12.7 m for full-flow probes. Determined using N∗ values from Table 1.

displaced rather than remolded during full-flow behavior. This deformation mechanism change during cycling may be due to a combination of minimal overburden stress, overconsolidation, and the layered soil characteristics.

With the behavior at all cycling intervals being similar to the trends presented, it is apparent that a measure of the remolded undrained shear strength can be obtained with the T-bar in lightly overconsolidated layered soils. With further investigation, it appears that the observed differences in Ball and Plate behavior compared to the T-bar and the different behavior within the crust zone can be explained in more detail.

8 CONCLUSIONS

A series of soundings performed with a CPTu and T-bar, Ball, and Plate probes and evaluation against field vane (FVT) data have enabled initial assessment of full-flow probes to characterize a highly layered (varved) anisotropic clay. The following observations have been made:
- The penetration resistance of the T-bar is higher than the Ball and Plate probes. The CPTu q_t resistance exceeds all probes. These differences are attributed to differences in displacement mechanisms (full vs. full-flow) and geometries of the full-flow probes.
- An accurate profile of the undrained shear strength as referenced to s_u(FVT) is attainable with all three probes in both the lightly overconsolidated zone (6-22 m) and in the overconsolidated crust zone (3-6 m) provided appropriate N_* values are obtained for each zone.
- The N_* values obtained for the T-bar, Ball, and Plate are generally similar to previous values obtained through analytical analyses and recommended values (Table 1). Specifically, the T-bar value is higher than previously reported values while the Ball and Plate values agree very well with the recommended value of 10.5. N_* values for the crust zone were on average 31% lower than for the corresponding value for the lightly overconsolidated layer.
- Cycling of full-flow probes shows potential to obtain the remolded undrained shear strength, especially in the lightly overconsolidated layer (6-22 m). In the crust, cycling may be problematic due to gradual soil displacement rather than remolding, which creates a cylindrical void.

Full-flow penetrometers provide a clear alternative to the CPTu for offshore applications where uncertainties in the correction for overburden stress can dominate determination of the undrained shear strength. Due to excessive costs associated with offshore testing, validation at land based sites that are highly characterized is preferable. Based on the first phase of testing presented herein, the applicability of full-flow probes may be extended to layered and anisotropic soils.

ACKNOWLEDGEMENTS

The work performed in this study has been supported by the National Science Foundation grant #CMS-0301448. This support is gratefully acknowledged.

REFERENCES

Andresen, A. and Bjerrum, L. (1958) "Vane testing in Norway", *Norwegian Geotechnical Institute*, Pub. No. 28.
ASTM (2003) *Annual Book of ASTM Standards*, Vol. IV, West Conshohocken, PA.
COFS (2002) *Australian Research Council Special Research Centre for Offshore Foundation Systems: Annual Report 2002*, University of Western Australia, Perth, Australia.
DeGroot, D.J. and Lutenegger, A.J. (2003) "Geology and Engineering Properties of Connecticut Valley Varved Clay", *Characterization and Engineering Properties of Natural Soils*, Tan et al. (eds.), Balkema, Vol. 1, pp. 695-724.
Geocouncil (2001) http://www.geocouncil.org/
Lunne, T. Robertson, P.K., and Powell, J.J.M. (1997) *Cone Penetration Testing in Geotechnical Practice*. Chapman & Hall, London, 312 p.
Martin, C.M. and Randolph, M.F. (2000) "Applications of the lower and upper bound theorems of plasticity to collapse of circular foundations", *Proc. Computer Methods and Advances in Geomechanics*, Vol. 2, pp. 1417-1428.
Randolph, M.F., and Houlsby, G.T. (1984) "The limiting pressure on a circular pile loaded laterally in cohesive soil", *Geotechnique*, Vol. 34, No. 4, pp. 613-623.
Randolph, M.F., Hefer, P.A., Geise, J.M., and Watson, P.G. (1998) "Improved seabed strength profiling using T-bar penetrometer", *Proc. Offshore Site Investigation and Foundation Behaviour 'New Frontiers'*, pp. 221-235.
Randolph, M.F., Martin, C.M., and Hu, Y. (2001) "Limiting resistance of a spherical penetrometer in cohesive material", *Geotechnique*, Vol. 50, No. 5, pp. 573-582.
Rittenour, T.M. (1999) "Drainage history of Glacial Lake Hitchcock, Northeastern USA", University of Massachusetts Amherst M.S. Thesis, 179 pp.
Robertson, P.K. and Campanella, R.G. (1983) "Interpretation of cone penetration tests, Part II: Clay", *Canadian Geotechnical Journal*, Vol. 20, pp. 734-745.
Stewart, D.P. and Randolph, M.F. (1991) "A new site investigation tool for the centrifuge", *Proc. Centrifuge 91*, pp. 531-538.
Tanaka, H. (1995) "National Report - The Current State of CPT in Japan", *Proc. CPT '95*, Vol. 1, pp. 115-124.
Terzaghi, K., Peck, R.B., and Mesri, G. (1996) *Soil Mechanics in Engineering Practice*, Wiley, New York, 549 pp.
Watson, P.G., Suemasa, N., and Randolph, M.F. (2000) "Evaluating undrained shear strength using the vane shear apparatus", *Proc. International Offshore and Polar Engineering Conference*, Vol. 2, pp. 485-493.

Rapid site assessment with the Electrical Conductivity Probe flexible approach for rapid and deep site investigation

M.P. Harkes, E.E. van der Hoek & J.J. van Meerten
GeoDelft, P.O. Box 69, 2600 AB Delft, Netherlands
m.p.harkes@geodelft.nl

Keywords: cone penetration, electrical conductivity, probe, rapid site assessment, pollutants, environment

ABSTRACT: The Electrical Conductivity Probe is a cone penetration technique for a rapid site assessment of the deeper underground (> 10 m - surface level). Measuring the electrical conductivity (EC) gives a good insight into the presence of specific pollutants.

1 INTRODUCTION

1.1 Project content

At a waste disposal site [area 83 ha] in the Netherlands waste has been dumped since 1963 [figure1]. The site was provided with an upper sealing layer during the mid-eighties, in accordance with government guidelines [Waste Disposal Regulations Soil Protection]. However, leachate had already filtered into the soil as a result of composting and from the site itself. The groundwater was therefore polluted. Salty debris from gas drilling holes in the deep subsoil had also been assimilated at several points in the main body of the site. This debris produces locally high concentrations of chloride in the leachate.

Figure 1: Activities at a waste disposal site

1.2 Innovation

In 1997, a groundwater extraction system was designed and installed that enables the protection of groundwater around the site against the run-off of pollutants from the site [Grondmechanica Delft, 1997]. The level of protection provided by the system is in accordance with the quality guidelines for ground- and surface water outside the present site, based on the current functions described in the Water Management Plan. This means that the concentration of chloride in the groundwater outside the site may not become higher than 300 mg/l.

1.3 Geohydrological management system

The Geohydrological Management System involves permanent groundwater extraction. The water is extracted on the north-west and south-west edge of the site, using 22 pump sumps and a drain measuring 375 m in length. Directional extraction takes place in the upper part of the aquifer. Between this aquifer and the waste disposal site a water retaining boulder clay layer is present. According to the design, the pollution already present in 1999 in the lower part of the water-transporting bed falls outside the scope of the management system.

1.4 Monitoring

Operation of the groundwater management system is checked regularly, by taking water samples from gauges positioned around the waste disposal site. In 2002, a significantly higher chloride concentration was measured in one of the gauges. This indicates that the situation deviates from the expected distribution [figure 2] due to the presence of density currents. As a possible hypothetical explanation it was stated that the pollutant might have sunk deeper due to local inhomogeneities in the boulder clay and due to the specific weight of salt water.

2 PROBLEM DEFINITION

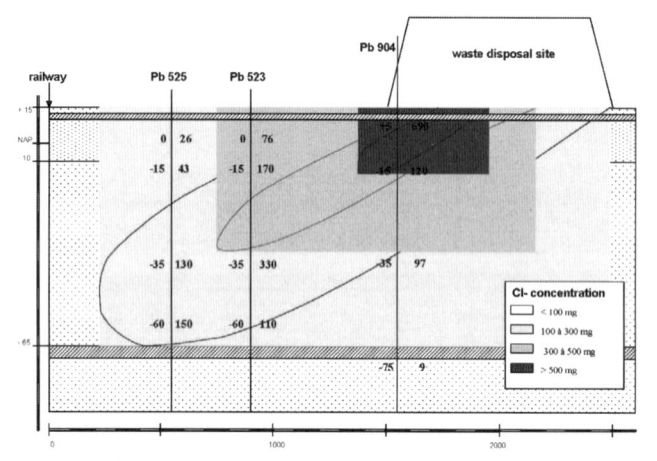

Figure 2: Assumed chloride distribution in the soil

2.1 Distribution mechanisms

Based on investigations performed in the past the assumed chloride distribution in the soil is as described in the above presented figure 2. At the site, [diluted] percolate [polluted mainly by chloride] flows through the boulder clay layer. High chloride concentrations have been measured at shallow depths close to the site. When the leachate infiltrates into the first aquifer, it was assumed that it mixes with fresh groundwater and flows in a south-west direction together with the groundwater. This leads to the assumed horizontal and vertical distribution of the chloride. The maximum concentrations are found at increasing depths as the distance from the site increases, however decreasing in absolute value. This distribution pattern is clearly seen in the above presented figure.

2.2 Urgency

Because the results from the monitoring showed concentrations conflicting with the assumed distribution the manager of the site requested GeoDelft to determine the extent of the chloride pollution, as well as the direction in which it had moved. This information was necessary to improve the groundwater management system. This management system is one of the elements incorporated in the environmental licence. As the municipal authority is able to restrict use of the waste disposal site, it was commercially important to ensure that the licence obligations were met as soon as possible.

3 SOLUTION

3.1 Flexible dynamic soil investigation

With GeoDelft's investigation concept known as flexible dynamic soil investigation [FDSI; Meurs, 2001], in-situ measurement methods are used and the strategy is adjusted throughout the investigation on the basis of the results obtained. The maximum number of analyses to be carried out is estimated in advance. The first measurement points are selected using hypotheses on the flow direction and the distribution of the pollution at depth.

In the field, analysis findings are used directly to pinpoint the following test location, so that the pollution contours for pre-defined threshold limits can be determined. In doing so for this case, chloride is the most important index parameter.

3.2 Supplementing FDSI

GeoDelft decided to use a probe to carry out conductivity measurements, and also to take several groundwater samples using the advanced multi-groundwater sampling probe. During the investigation, a depth of approximately 80 m beneath the surface level was reached in the aquifer consisting of coarse sand. This approach provided reliable results and was inexpensive. It also took considerably less time than the usual method for mapping pollution, which requires gauges to be positioned around the contaminated location during several series of measurement rounds.

Usually positioning of a deep gauge requires at least five working days. Once in position, a week must then pass before a test can be taken and analysed. This means that in the traditional approach a minimum of several weeks goes by before there is any insight into the distribution direction of the pollutants, and before the next series of gauges can be positioned for the following mapping round.

3.3 Converting resistance to conductivity

In the field, an extrudable probe that incorporates a cell for measuring electrical resistance is used to measure the resistance [Ohm] at a particular depth. Using the characteristic length of the cell [which varies per probe], the resistance is calculated per length-unit [Ohm.m].

The electrical resistance depends on the temperature. Once the extruded probe has reached the deeper underground, there is a pause in activity until the conductivity becomes constant. This shows that the effect of temperature increase caused by penetrating the soil has ceased. The natural temperature path may, however, vary per location. Only a few measurements of the groundwater temperature at this location are known. The only values found in the literature related to this location had been determined in gauges, and these give a groundwater temperature of 13°C.

Decomposition processes in the waste disposal site can result in increased temperatures. We can assume, however, that the influence of temperature increase in the soil beneath the site will be quickly reduced by the flow of groundwater, and that all measurements are carried out at 13°C.

In the laboratory, electrical conductivity measurements are usually carried out at 25°C, and all values were therefore converted to this temperature. To do so, the following formula [GeoDelft, 1983] was used:

$$\rho(t^{25°C}) = \rho(t^{13°C}) / (1 + 0{,}0025 * (13-25)) \qquad (1)$$

where:
$\rho(t)$ = electrical resistance (Ohm.m)

The electrical conductivity is the reciprocal of the resistance:

$$\rho(t) = 10000/\kappa(t) \qquad (2)$$

where: $\kappa(t)$ = the electrical conductivity (µS/cm)

3.4 From conductivity to chloride concentration

The electrical conductivity is determined by the number of different ions in solution. From the various measurement data available from around the waste disposal site since 1993, the analysed package of dissolved substances in the groundwater seemed to show as far as cations are concerned that there is a relatively large proportion of ammonium, sodium and potassium. As for anions, bicarbonate and sulphate occur alongside chloride. The contribution of other ions to the balance is negligible in comparison to these macro components.
It can now be stated that:

$$\Lambda(m) = \kappa/C \qquad (3)$$

where: $\Lambda(m)$ = molar conductivity
C = the concentration in mol/l

The contribution to conductivity differs per ion, and the total molar conductivity consists of the sum of cations and anions:

$$\Lambda(m) = \Sigma(V_+ . \lambda°) + \Sigma(V_- . \lambda°) \qquad (4)$$

where: $\Sigma(V_+)$ = proportion of cations [mol/l]
(ΣV_-) = proportion of anions [mol/l]
$\lambda°$ = specific conductivity

Using the known value for $\lambda°$, the electrical conductivity of a solution can therefore be calculated. The relationship is simplified somewhat because $\lambda°$ is dependent on the concentration.

The values for $\lambda°$ originate from Weingartner (1976):

$\lambda°$ (NH_4^+) = 73.4
$\lambda°$ (Na^+) = 50.11
$\lambda°$ (K^+) = 73.52
$\lambda°$ (Cl^-) = 76.34
$\lambda°$ (HCO_3^-) = 38.1
$\lambda°$ (SO_4^{2-}) = 160.0

For the chloride, bicarbonate and sulphate measurements carried out between 1995 and 2002, the electrical conductivity [EC] was calculated and plotted against the measured value.

3.5 Results

Figure 3 shows the indicative chloride concentrations measured using the electrical conductivity probe on the south side of the waste disposal site.

4 CONCLUSIONS

4.1 Conclusions about the distribution

After interpreting the measurements and comparing these with the intended situation that had been based on a previous forecast, the following conclusions can be drawn:
- The heaviest point of chloride pollution lies more to the south side of the site than was assumed in the past.
- In vertical direction, the distribution was over a larger area than previously predicted.

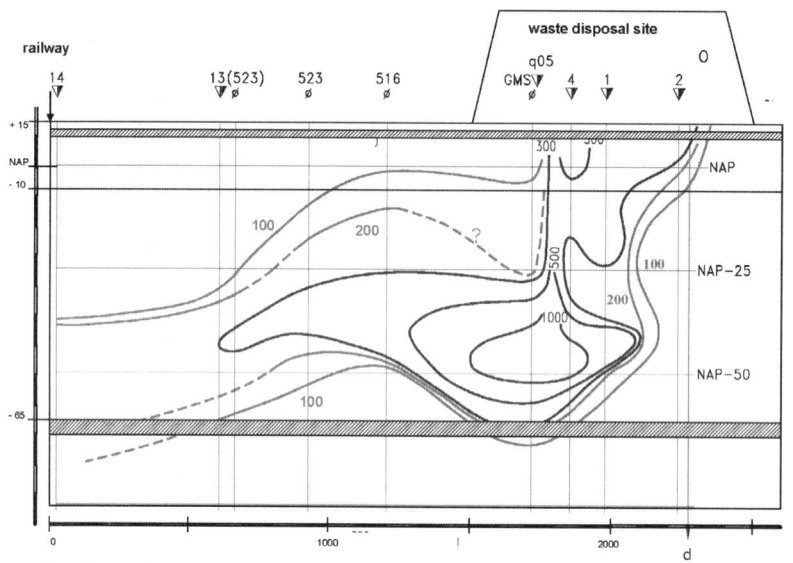

Figure 3: Actual chloride distribution
Contour line; 100, 200, 300, 500 en 1000 mg Cl⁻/l

- At a certain depth under the site (up to approximately NAP − 25 m) chloride concentrations are larger than predicted by a factor of 2.
- Infiltration of salt leachate from the site is probably greater than that used in the forecast calculations.
- On the south side of the waste disposal site, a local peak value of measured pollution is present deep in the subsoil. This indicates vertical transport due to a salt-water density current. This effect was not included in the forecast calculations.
- Extension of the groundwater management system with deep wells is necessary to prevent that leachate with a high chloride content passes the existing shallow extraction wells that were installed earlier based on the forecast of mixing.

4.2 Time savings

GeoDelft was able to carry out a 80 m deep cone penetration test measuring conductivity in less than a day, so providing immediate information about chloride levels. This made it possible to define the following measurement points for mapping the chloride pollution in an increasingly directed way. To guarantee that a measured conductivity agrees with an interpreted chloride level, GeoDelft collected calibration data by taking groundwater samples at various points and analysing them. The chosen working method ultimately led to time savings of approximately three months.

4.3 Innovative strategy

The approach used by GeoDelft for mapping the chloride pollution at this waste disposal site is an example of a new approach that can be described as a flexible and dynamic soil investigation [FDSI].

GeoDelft developed this innovative strategy in order to map groundwater pollution both quickly and reliably. Use of extrudable measuring- and sampling systems (such as the multi-groundwater sampling probe and the electrical conductivity probe) and 'on site' analyses are central to this approach. These techniques are extremely promising, and can be used in similar projects in Europe and throughout the world.

The investigation proved that storing highly concentrated brine in a waste dump can lead to density currents in groundwater. It was concluded that the following research on the spatial distribution should be supported with transport models based on the code HST3D rather than the more familiar MT3D.

ACKNOWLEDGEMENTS

Our special thanks go to Gerard van Meurs and Willem van der Zon of GeoDelft. We also thank the manager of this disposal site for his trust on the outcome of our innovative approach.

REFERENCES

Grondmechanica Delft (1997). Geohydrologisch beheerssysteem stortplaats VAM; Hoofdrapport: Basisontwerp CO-377270/16, Delft, augustus 1997.

Kirk Jr. K.L. (1987). HST3D: A computer code for simulation of heat and solute transport in three dimensional groundwater flowsystem. Water Resources Investigation, report 86-4095; Denver, Colorado, US.

Meurs, G.A.M. et al. (2001). Flexible and Dynamic Site Investigation. Second Int. Conf. and Ind. Exhibition, Field Screening Europe 2001. Strategies and Techniques for On-Site Investigation and Monitoring of Contaminated Soil, Water, Air and Waste, Karlsruhe 14 - 16 May 2001.

Weingartner, D. (1976). *Analytische Chemie; vraagstukken en uitwerking.* Amsterdam: Elsevier. ISBN: 901001584X

A framework for using textured friction sleeves at sites traditionally problematic for CPT

Gregory L. Hebeler & J.David Frost
Georgia Institute of Technology, USA

James D. Shinn II
Applied Research Associates, USA

Keywords: texture, friction sleeve, CPT, interface shear, surface roughness, multi friction sleeve attachment, problematic sites, penetrometer

ABSTRACT: The use of the cone penetration test (CPT), and other invasive in situ testing techniques, has not been feasible in certain geologic conditions due to difficulties, including: sensor overload, verticality, and large unsupported rod lengths. This manuscript presents a framework for using a multi-friction sleeve attachment (MFSA) to conduct investigations in stratigraphies typically problematic for conventional CPT testing. The recent development of a multi-friction sleeve attachment (MFSA) has shown that sleeve friction (f_s) measurements, obtained a larger distance behind the tip (> 0.35 m) allow for more accurate soil characterization and property determination. Additionally, testing with a series of variably textured friction sleeves allows for profiles of interface friction, soil friction, and end bearing (based on measurements derived from the use of textured sleeves) to be determined in a single sounding. Testing with the MFSA has been limited to predominantly granular soils at this stage in its development. However, its utility as a site characterization tool is very promising, and a number of previous hurdles to in-situ testing can now be removed or diminished. While the MFSA has not been extensively tested in problematic soils to date, preliminary analysis demonstrates its potential benefits in a variety of soil conditions, including the ability to determine levels of soil structure (e.g. cementation and diagenesis). Fundamental interface shearing concepts and example calculations are presented to relate MFSA results to conventional CPT results and to demonstrate how they can be used in place of conventional CPT techniques in problematic soil conditions.

1 INTRODUCTION

The cone penetration test (CPT) has become one of the most widely used and well-respected geotechnical testing tools. The use of the cone penetrometer, and other invasive in situ testing techniques, has not been feasible in certain geologic conditions due to difficulties with sensor overload, verticality, and large unsupported rod lengths. Geologic conditions that can lead to these penetration difficulties include conglomerates, gravels, highly cemented soils, stiff desiccated crusts, and stiff layers underlying thick soft layers typically found in coastal and/or alluvial regions (Lunne et al., 1997). The necessity to have cone tip (q_c) load cell resolution on the order of 1 N (~ 1 kPa for a 10 cm^2 CPT) limits the maximum tip load to 200 kN for most conventional CPT systems. As a result, sensor overloading at the tip can be common during testing in areas with natural gravels, desiccated crusts, and cemented layers. In many cases, these stiff layers are thin (on the order of 1-2 m) and occur near the surface, thereby obscuring the underlying soils from CPT classification. Consequently the use of CPT, CPTU and SCPT devices has been limited in these conditions, relegating engineers to the use of less robust in-situ techniques such as the standard penetration test (SPT) and visual classification. The sleeve friction (f_s) measurement is typically not subject to sensor overloading due to the much lower stresses associated with sleeve penetration and the associated shear loading mechanisms. However, the conventional CPT friction sleeve (f_s) measurement is considered highly variable and less reliable than the other CPT measures of tip stress (q_c) and pore pressure (u_2). The recent development of a multi-friction sleeve attachment (MFSA) (DeJong and Frost, 2002) has shown that f_s measurements made a larger distance behind the tip (> 0.35 m) using textured friction sleeves can allow for more accurate soil characterization and property determination.

In addition to geologic constraints, conventional CPT interpretation does not clearly differentiate between highly structured soils, such as cemented sands and quick clays, and other more stable unstructured materials showing similar CPT response. The current manuscript presents initial findings relating to the use of a multi-friction sleeve attachment for geologic conditions previously problematic for conventional CPT investigation, and to aid in the in situ quantification of soil structure. Fundamental concepts of interface shearing have been applied to early in situ results of MFSA soundings using textured friction sleeves, highlighting exciting new interpretation techniques to overcome the abovementioned limitations of conventional CPT testing.

2 FUNDAMENTAL INTERFACE SHEARING CONCEPTS

The geotechnical community has realized for many years the crucial role that interfaces play in the response of geotechnical systems. With interface shearing constituting a primary or secondary component in most geotechnical applications and testing methods, it is imperative to possess a strong fundamental understanding of interface shearing mechanisms. Along these lines, a number of researchers have made significant contributions to the understanding of soil – geomaterial interface behavior. The beginnings of modern interface research as it pertains to geotechnical engineering were initiated by Potyondy (1961) and Brummund and Leonards (1973). These early studies demonstrated the importance of the soil's moisture content, particle angularity, particle size, mineralogy, and normal load on interface strength. Brummund and Leonards (1973) further concluded that interface shear strength increased with counterface surface roughness reaching a limiting value at the internal shear strength of the adjacent soil, at which stage shearing was transferred away from the surface of the geomaterial into the soil mass. Modern compilations of factors affecting soil-geomaterial interfaces are shown in Tables 1 and 2, for coarse and fine-grained behaviors respectively.

Early conclusions regarding interface behavior were limited to qualitative assessments due to a lack of available methods to quantitatively characterize the surface roughness of the tested materials. Modern geotechnical research utilizes several methods to accurately characterize geomaterial surfaces including the popular surface profiling technique. The most commonly accepted parameters for characterizing surface roughness in the geotechnical community are R_{max} and R_a. R_{max} is the absolute vertical distance between the highest peak and lowest valley along the surface profile. Average roughness, R_a is defined as,

$$R_a = \frac{1}{L} \int_0^L |z| dx \quad (1)$$

where L is the sample length and z is the absolute height of the profile from the mean line. A thorough discussion of measurement techniques and international standards for surface roughness characterization can be found in Ward (1999), with a detailed discussion focused on geotechnical applications available in DeJong et al. (2002).

Table 1. Factors affecting soil-geomaterial interface behavior for soils exhibiting coarse-grained behavior (after Lee 1998)

Type	Factor	Significance
Soil	Angularity / Shape	High
	Cementation	Med
	Density	High
	Initial soil structure	Low
	Mean grain size (D_{50})	Medium
	Surface roughness	Low
	Uniformity coefficient (C_u)	Low
Geomaterial	Surface hardness	High
	Surface roughness	High
Testing	Normal stress	High
	Strain rate / Drainage	Low
	Test method	Low

Table 2. Factors affecting soil-geomaterial interface behavior for soils exhibiting fine-grained behavior

Type	Factor	Significance
Soil	Degree of saturation	High
	Diagenesis	Medium
	In situ water content	High
	Initial soil structure	Low
	Plasticity	High
	Specific surface	High
	Void ratio	Medium
Geomaterial	Surface hardness	Low
	Surface roughness	Medium
Testing	Normal stress	High
	Strain rate / Drainage	High
	Test method	Low

Quantification of the role of surface roughness on particulate-continuum interfaces was pioneered by Uesugi and Kishida (1986a, 1986b), in their laboratory work on sand-steel interfaces. The behavior qualitatively noted by Brummund and Leonards (1973) was quantified by a series of interface tests of sand sheared against steel plates of varying texture. The results were shown to be applicable over a range of particle sizes by the introduction of a normalized roughness parameter R_n,

$$R_n = \frac{R_{max}(L = D_{50})}{D_{50}} \quad (2)$$

Using R_n to characterize the surface roughness of the tested counterfaces, the surface roughness-interface shear relationship was found to be bilinear (Figure 1). The interface strength was shown to increase linearly, proportional to the increase in normalized roughness below a certain "critical" roughness value. Above the critical roughness, shearing was observed to transfer away from the interface into the adjacent soil mass with a measured interface strength equal to the internal shear resistance of the contacting soil.

The bilinear interface shear relationship observed by Uesugi and Kishida (1986b) has been shown to exist for a number of other geomaterial surfaces, including geomembranes, concrete, timber, and fiber reinforced polymers (Dove et al., 1997; Frost and Han, 1999; DeJong et al., 2000). Furthermore, the bilinear relationship has been shown to exist for a number of other surface roughness parameters (Lee, 1998). The existence of this fundamental particulate-continuum interface relationship, regardless of roughness parameter and geomaterial surface type, provides the opportunity to determine the full interface behavior of a soil deposit in situ without prior knowledge of the soil particle characteristics (De-Jong and Frost, 2002).

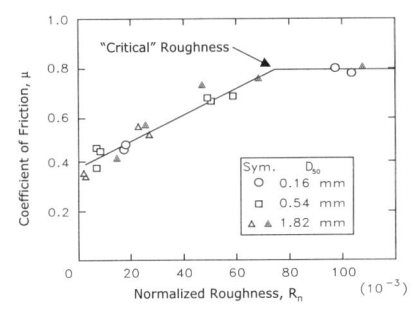

Figure 1. Surface roughness versus friction coefficient for sand-steel interfaces (after Uesugi and Kishida, 1986b).

3 THE MULTI-FRICTION SLEEVE PENETROMETER ATTACHMENT

A multi-friction sleeve penetrometer attachment has been developed by researchers at the Georgia Institute of Technology (DeJong and Frost, 2002) that allows for the direct in situ characterization of soil-interface strength during a conventional CPT sounding. The attachment utilizes four independent friction mandrels with replaceable sleeves of varying texture as shown in Figure 2. The MFSA was designed for use behind a conventional 15 cm² CPT device to allow for simultaneous determination of conventional CPT measurements (e.g., q_c, f_s, u_2) in conjunction with four individual interface shear measurements. The MFSA has an assembled length of 109 cm. When configured with a conventional digital CPT module (61 cm in length) the total instrument length becomes 170 cm.

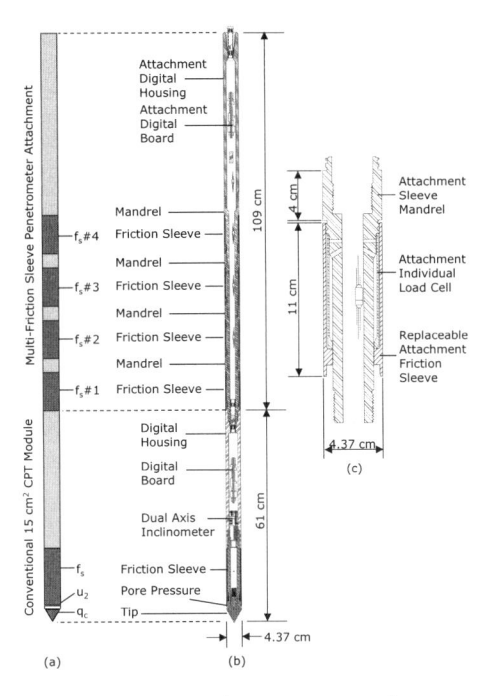

Figure 2. Multisleeve friction penetrometer attachment configured with conventional CPT Module. (a) schematic, (b) design detail, and (c) friction sleeve mandrel design detail (after DeJong, 2001).

The knowledge that both surface roughness and hardness are significant factors in the response of particulate–continuum systems guided the development of the MFSA and led to the use of replaceable textured friction sleeves of varying roughness. Wear of the testing materials is minimized by using hardened steel alloys throughout the device. The sleeve texture was designed to be "self-cleaning" to eliminate the possibility of soil particles clogging the texture, and changing the surface properties with depth during penetration. It was also found through extensive previous research and proof of concept testing that the texturing should consist of peak features extending beyond a base substrate to allow for better soil engagement across the range of particles sizes and shapes encountered in situ (DeJong, 2001). Additional considerations required that the sleeves induce internal shearing of the soil rather than only sliding along the surface at higher roughness values, and that the texturing pattern should be easily machinable. The resultant texturing pattern consists of an offset diamond texture (Figure 3) with variations in the height of the diamonds used to modify the magnitude of surface roughness.

4 PRELIMINARY MFSA RESULTS

A total of 94 soundings were conducted with the MFSA in the preliminary stages of device development at a relatively uniform sand site in South Royalton, Vermont, USA. The initial testing program was designed to fully evaluate the MFSA and the textured friction sleeves. The initial results have verified the proof of concept and established the reliability and robustness of the device. More details of the proof of concept testing and device development can be found in (DeJong, 2001; DeJong and Frost, 2002; and Frost and DeJong, 2004)

By configuring the MFSA with sleeves of increasing roughness in series, the entire interface shearing – surface roughness relationship of a site can be characterized in a single sounding. The results of such a sounding are shown in Figure 4, with $R_{max} = 0.25, 0.50, 1.00,$ and 2.00 mm for the four friction sleeves respectively. An overlay plot of the four friction sleeve profiles is shown in the rightmost plot of Figure 4 to highlight the effects of changing the sleeve surface texture.

5 MECHANISMS OF TEXTURED SLEEVE PENETRATION

As seen in Figure 3, the offset diamond texturing pattern used on the friction sleeves of the MFSA utilizes peak features above a constant substrate with a base diameter equal to that of a 15 cm² CPT. While the base pattern of the diamond texture is held constant, the height of the diamonds is varied to create patterns of varying surface roughness. The common diamond heights used in the preliminary testing

Figure 3. Schematic of MFSA textured sleeve pattern

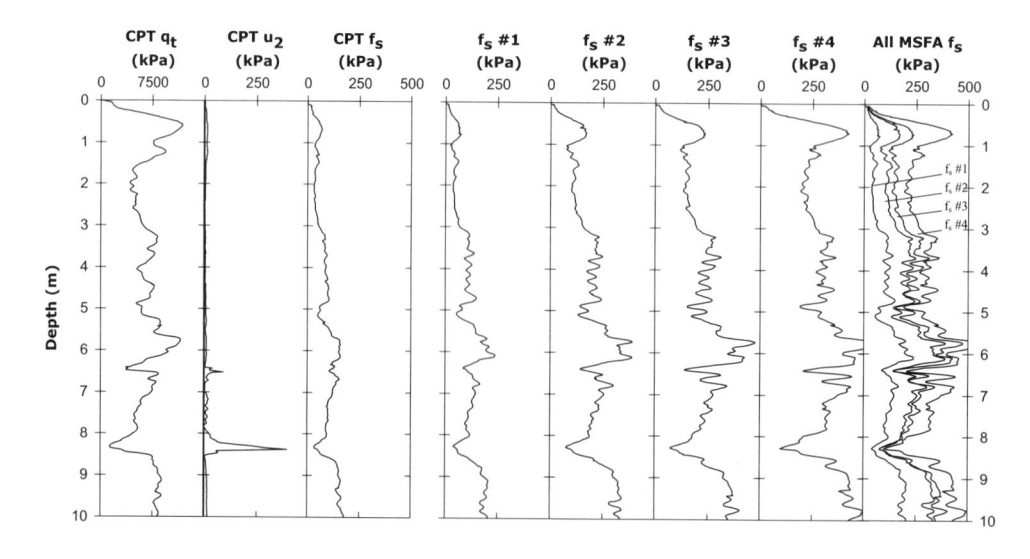

Figure 4. Results of a sounding with the multi-sleeve friction attachment configured with a conventional 15 cm² CPT and 4 sleeves of increasing roughness ($R_{max} = 0.25, 0.5, 1.0,$ and 2.0 mm).

sequence were R_{max} = 0.125, 0.25, 0.5, 1.0, and 2.0 mm in addition to "smooth" sleeves of roughness equal to conventional CPT friction sleeves, R_a = 0.50 µm.

While the offset diamond texture creates the desired combination of particle shearing and sliding to both prevent clogging and to engage the soil mass, a third mechanism exists during the penetration of textured sleeves. Due to the increased diameter of the diamond texture over the remainder of the CPT module, a punching shear or "bearing capacity type" failure zone exists along the leading edge of the diamond texture. Consequently, the measured sleeve stress consists of both interface shear resistance and an end bearing component herein termed the annular penetration force (APF). This mechanism was shown to exist by conducting soundings that varied the number of offset diamond rows placed in series. The results of each sounding were averaged over a common depth range of 1-10 m, resulting in a single representative value for each sounding. Figure 5 shows the averaged sleeve force plotted against the number of consecutive diamond rings, with the intercept at zero rings denoting the APF for a given diamond height. This relationship is represented by equation (3),

$$f_a = f_r * N_r + APF \qquad (3)$$

where f_a = average attachment sleeve force, f_r = average force per ring of texture (slope), N_r = # of consecutive rings of texture, and APF = annular penetration force (intercept). The APF mechanism is further supported by the trend of successive smooth sleeve forces resulting in an observed intercept, or APF, of zero.

While initially the existence of the APF force might seem to cause difficulty in interpretation of textured sleeve stresses, it in fact provides significant insight into the tested soil behavior when interpreted within the framework of known interface shearing behavior. In returning to the concept of a "critical roughness", above which the interface shear strength becomes constant and equal to the internal shear resistance of the soil, any divergence in measured sleeve stresses for textures above the critical roughness can be attributed to varying levels of APF. As such, by subtracting the measured stresses from sleeves with textures above the critical roughness value (e.g. R_{max} = 1.0 and 2.0 mm for the tested soil), the subtracted value becomes the difference in the annular penetration stress between the sleeves. Multiplying by the difference in circumferential area between the two textured sleeves gives the APF in the stable destructured zone behind the CPT module. This value can then be scaled by the differences in area between the textured annulus and the CPT tip to provide a measure of destructured "tip" stress. A similar calculation was previously conducted by Bloomquist et al. (1999) using an instrumented annular wedge of A_{proj} = 10 cm² placed approximately 1 m behind a conventional 10 cm² CPT in order to detect the presence of cementation in tested materials.

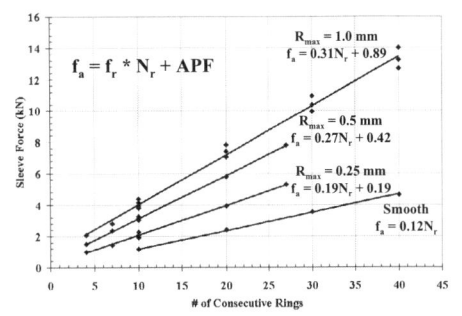

Figure 5. Trend in annular penetration force for a sampling of textured and smooth sleeves.

In a parallel manner, the measured values of two textured sleeves having roughness values below the critical roughness provide the ability to calculate a profile of calculated destructured smooth sleeve values. Because the trend below the critical roughness is known to be linear, the deviations in the measured sleeve stresses for roughnesses along this line can be used to estimate a destructured "smooth" sleeve response. With the MFSA configured with two sleeves above and two sleeves below the critical roughness value in order of increasing roughness, it is possible to calculate destructured end bearing ("tip") and side friction ("smooth sleeve") profiles from a single sounding. Figure 6 shows the destructured end bearing and side friction profiles calculated from the MFSA results, overlaid on the measured CPT q_c and f_s values for the same sounding. While a destructured smooth sleeve profile can be more easily obtained by simply using a smooth sleeve on the MFSA, the comparison shown in Figure 6 reinforces the validity of the presented concepts.

The calculated destructured end bearing and side friction values are not to be confused with attempted predictions of the CPT q_c and f_s measurements. The stress and soil conditions controlling the CPT response differ greatly from those affecting the MFSA. Additionally, the resolution of the textured friction sleeves is much finer than either the CPT tip or friction sleeve due to the vastly different stress conditions and geometric configurations controlling each sensor. The influence zone of the CPT tip has

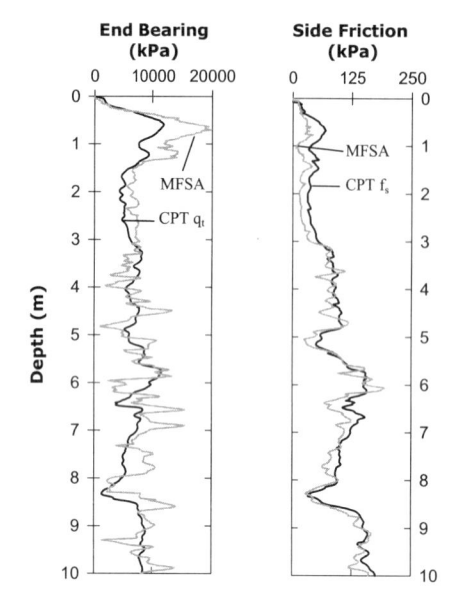

Figure 6. Comparison of CPT q_t and f_s measurements with the parallel destructured values estimated from the MFSA.

been extensively studied and while not fully understood has been found to be heavily dependent on the probe diameter, in addition to the in situ state of stress and soil conditions (e.g. Vesic, 1972 and Baligh, 1985). As such, the differences in the CPT tip area and the annular area of the textured friction sleeves can be associated with the size of the influence zone around each sensor. The CPT tip diameter is constant at $D_{tip} = 4.37\ cm$, with the maximum annular area of the textured friction sleeves resulting from the $R_{max} = 2\ mm$ sleeve, and equaling $D_{annular} = 0.4\ cm$. The MFSA is outfitted with shorter friction sleeves (11 cm) than the conventional CPT friction sleeve (16.4 cm), again resulting in a higher vertical resolution.

In addition to the differing levels of resolution between the various sensors, in situ stress and soil conditions are distinctly affected by the act of CPT penetration. The resultant soil conditions encountered by the MFSA are close to the residual state in a vastly different stress field to that experienced by the CPT tip and friction sleeve. Again looking at Figure 6, the noted differences in resolution and stress level become evident and comparing the two trends provides additional insight into the soil behavior. The results of Figure 6 correspond to a site of relatively uniform granular composition with minimal soil structure. Sites with significant levels of soil structure would show a large divergence in the trends calculated from the MFSA and measured by the conventional CPT sensors.

6 OPPORTUNITIES AT PROBLEMATIC SITES

A number of conventional penetration testing limitations are associated with sensor overload, verticality, and wear in abrasive soils. Some of these problems result from the placement of the CPT measurements at the front of the penetration zone, where abrasion and wear on devices is the highest, and the stress state is the most variable. The possibility of using a "dummy" tip with improved structural integrity ahead of a device like the MFSA provides a way to test in some intermediate conditions previously too harsh for conventional CPT use. This concept could be further extended for use in glacial tills and similar geologies, and for angled or horizontal pushes using modern drilling technologies and trailing the MFSA behind a drill head.

Additionally, as noted previously, having both a structured (CPT) and destructured (MFSA) measure of soil strength for multiple loading mechanisms: side friction (friction sleeves) and end bearing (tip, annular area of texture), can be effectively used to estimate the sensitivity, or level of soil structure, within a deposit. Soil encountered at a distance greater than 35 cm behind the CPT tip has been shown in previous research (Campanella and Robertson, 1981; DeJong and Frost, 2002; and Frost and DeJong, 2004) to be in a stable stress zone. Whereas the conditions encountered by the CPT tip and friction sleeve are known to be highly variable and more dependent on the intact strength and structure of the tested soils (Lunne et al., 1997).

7 CONCLUSIONS

This manuscript describes the use of a multi-friction sleeve penetrometer attachment (MFSA) to determine interface strength in situ over a full range of operational surface roughness values. Fundamental concepts of interface shearing of continuum and granular materials have been applied to the analysis of the MFSA, leading to a number of opportunities at sites traditionally problematic for conventional CPT devices. The salient conclusions are summarized below:

- Conventional "smooth" friction sleeves in contact with granular materials have been shown to only induce sliding failures, poorly modeling operational interface conditions in most geotechnical applications. The use of an offset diamond-texturing pattern has provided a means to effectively vary the surface roughness of friction sleeves, allowing for interface shearing mechanisms to be better controlled during in situ testing.
- The MFSA has been shown to be a viable in situ device, providing a means to determine the full

interface shear – surface roughness relationship in situ (Frost and DeJong, 2004)
- The penetration of textured friction sleeves consists of an annular penetration force (APF) in addition to the components of interface sliding and shearing.
- The APF, in combination with fundamental concepts of interface shearing determined previously in laboratory research, has allowed for destructured estimates of end bearing and side friction to be determined with the MFSA.
- Comparison of conventional CPT measurements with the destructured estimates calculated from the MFSA allow for improved characterization of soil structure and sensitivity.
- Use of the MFSA may be able to alleviate limitations at sites traditionally problematic to CPT use.

This paper represents an initial review of a number of concepts concerning the use of both textured friction sleeves and the MFSA at problematic sites. Research and testing with the MFSA is ongoing, and results from additional geologies will serve to provide further understanding of the presented concepts.

ACKNOWLEDGEMENTS

This research was partially supported by National Science Foundation grant #CMS-9978630, and the first author is supported by a National Defense Science and Engineering Graduate fellowship. Both sources of support are gratefully acknowledged. The assistance of Jason DeJong and Matt Evans during field testing, and the Vertek team at Applied Research Associates in South Royalton, Vermont for their collaboration in design and fabrication of the MFSA are also gratefully acknowledged. Finally, the tolerance of the Timian family in allowing almost 100 CPT sounding to be advanced in their front yard is most gratefully acknowledged.

REFERENCES

ASTM D3441-94 (1994) "Standard Test Method for Deep, Quasi-Static, Cone and Friction-Cone Penetration Tests of Soil," *ASTM*, West Conshohocken, PA.
ASTM D5578-95 (1995) "Standard Test Method for Performing Electronic Friction Cone and Piezocone Penetration Testing of Soils," *ASTM*, West Conshohocken, PA.
Baligh, M.M. (1985) "Strain Path Method," *ASCE Journal of Geotechnical Engineering*, Vol. 111, No. 9, pp. 1108-1136.
Bloomquist, D., Hand, R.S., and Anderson, J.B. (1999) "Development of a Cemented Sand Module for the Electronic Cone Penetrometer," *Transportation Research Record*, No. 1675, October, pp. 10-16.
Brummund, W.F. and Leonards, G.A. (1973) "Experimental Study of Static and Dynamic Friction Between Sand and Typical Construction Materials," *Journal of Testing and Evaluation*, Vol. 1, No. 2, pp. 162-165.
Campanella, R.G. and Robertson, P.K. (1981) "Applied Cone Research." *Proceedings, Symposium on Cone Penetration Testing and Experience: Geotechnical Engineering Division, ASCE*, October, pp. 343-362.
DeJong, J.T., Frost, J.D., Sacs, M. (2000) "Relating Quantitative Measures of Surface Roughness and Hardness to Geomaterial Interface Strength." *Proceedings of Geo-Eng 2000*, Sydney, AUS, CD-ROM.
DeJong, J.T. (2001) "*Investigation of Particulate-Continuum Interface Mechanisms and Their Assessment Through a Multi-Friction Sleeve Penetrometer Attachment,*" Ph.D. Thesis, School of Civil and Environmental Engineering, Georgia Institute of Technology, Atlanta, Georgia, USA, 360 pp.
DeJong, J.T and Frost, J.D., (2002) "A Multi-Friction Sleeve Attachment for the Cone Penetrometer," *ASTM Journal of Geotechnical Testing*, Vol. 25, No. 2, pp. 111-127.
DeJong, J.T., Frost, J.D., Saussus, D.R. (2002) "Relative Aspect of Surface Roughness at Particulate-Solid Interfaces," *ASTM Journal of Testing and Evaluation*, Vol. 30, No. 1, pp. 8-19.
Dove, J.E., Frost, J.D., Han J., and Bachus, R.C. (1997) "The Influence of Geomembrane Surface Roughness on Interface Strength," *Proceedings of Geosynthetics '97*, Vol. 2, pp. 863-876.
Frost, J.D., and DeJong, J.T. (2001) "A New Multi-Friction Sleeve Attachment," *Proceedings, 15th International Conference on Soil Mechanics and Geotechnical Engineering*, Istanbul, Vol. 1, pp. 91-94.
Frost, J.D., and DeJong, J.T. (2004) "In Situ Assessment of the Role of Surface Roughness on Interface Response," *ASCE Journal of Geotechnical and Geoenvironmental Engineering*, in press.
Frost, J.D. and Han, J. (1999) "Behavior at Interfaces between Fiber-Reinforced Polymers and Sands," *ASCE Journal of Geotechnical and Geoenvironmental Engineering*, Vol. 25, No. 8, pp. 633-640.
Jekel, J. W. A. (1988) "Wear of the Friction Sleeve and its Effect on the Measured Local Friction," *Proceedings, Penetration Testing 1988, ISOPT-1*, pp. 805–808.
Lee, S.W. (1998) "*Influence of Surface Topography on Interface Strength and Counterface Soil Structure*," Ph.D. Thesis, School of Civil and Environmental Engineering, Georgia Institute of Technology, Atlanta, GA, 336 pp.
Lunne, T., Robertson, P.K., and Powell, J.J.M. (1997) *Cone Penetration Testing in Geotechnical Practice*, Blackie Academic & Professional, New York, 312 pp.
Potyondy, J.G. (1961) "Skin Friction Between Various Soils and Construction Materials," *Geotechnique*, Vol. 11, pp. 339-355.
Robertson, P.K. (1990) "Soil Classification Using the Cone Penetration Test," *Canadian Geotechnical Journal*, Vol. 27, No. 1, pp. 151–158.
Uesugi, M. & Kishida, H. (1986a) "Influential Factors of Friction Between Steel and Dry Sands," *Soils and Foundations*, Vol. 26, No. 2, pp. 29-42.
Uesugi, M. & Kishida, H. (1986b) "Frictional Resistance at Yield Between Dry Sand and Mild Steel," *Soils and Foundations*, Vol. 26, No. 4, pp. 139-149.
Vesic, A.S. (1972). "Expansion of Cavities in Infinite Soil-Mass," *Journal of the Soil Mechanics and Foundations Division, ASCE*, Vol. 98, No. 3, pp. 265-289
Ward, H.C. (1999) Profile Characterization, *Rough Surfaces*, T.R. Thomas, Ed., Longman, London, 278 pp.

Proceedings ISC-2 on Geotechnical and Geophysical Site Characterization, Viana da Fonseca & Mayne (eds.)
© 2004 Millpress, Rotterdam, ISBN 90 5966 009 9

Thin layer and interface characterization by VisCPT

Roman D. Hryciw & Seungcheol Shin
Dept. of Civil and Environmental Engineering, University of Michigan, MI, USA

Keywords: cone penetration, vision cone penetrometer, soil stratigraphy, interfaces, thin layers

ABSTRACT: The vision cone penetrometer (VisCPT) was used to assess the reliability of conventional CPT-based site characterization in a stratified hydraulic fill. Thin layers and lenses which were often missed by the CPT were identified by the VisCPT. A model was also developed for CPT resistances at the interface of stronger, typically coarser soil overlying weaker material. The model predicted that the CPT tip resistance lags sleeve friction on average 10 cm at such interfaces. The result is an anomalous peak in friction ratio at, or just below the interface. A 10 cm downward shift in the tip resistances corrected the friction ratios and produced a CPT-based classification in excellent agreement with the VisCPT and the CPTU

1 INTRODUCTION

It is generally recognized that CPT tip resistance, q_c, is controlled by soils to several cone diameters below the tip. Meanwhile, CPT sleeve friction, f_s, is largely unaffected by soils beyond the contact area with the sleeve. The resulting effect on the computed CPT friction ratio ($FR = f_s/q_c \times 100$ %) and the implications to stratigraphic characterization can be significant yet unpredictable, particularly at the interfaces of soil strata. Development of the vision cone penetrometer (VisCPT) (Raschke and Hryciw, 1997) has made resolution of this problem attainable. Through digital processing of captured soil images, the VisCPT provides accurate information on soil types and the location of interfaces. Thus, previously unexplicable CPT trends can better be interpreted and geotechnical engineers can design based on less uncertainty regarding subsurface site conditions.

Presented in this paper is a model for CPT tip and sleeve resistance at the interface of a stronger soil overlying a finer-grained, weaker material. The model is used to correct and reinterpret a CPT test performed at the NSF National Geotechnical Experimentation Site (NGES) on Treasure Island in San Francisco, California. The VisCPT was also used to identify various thin soil layers that the CPT missed.

2 VISUAL DATA PROCESSING

The VisCPT records soil images at a field of view of approximately 1.5 cm (H) × 1.0 cm (V) at a pixel resolution of 720 × 480. The images are captured every 0.5 cm while the CPT advances at 2 cm/sec. The resulting overlap guarantees continuous coverage but requires digital trimming of images to produce a seamless vertical photo-log of the soil stratigraphy. The resulting visual data collected by the VisCPT is so large that manual interpretation of images is impractical, if not impossible. In this study, an image processing method based on 2nd order statistics of image pixel values is used to characterize the "texture" of soil images. The main goal of image processing is to extract textural indices from soil images which can be correlated to soil type. A study by Ghalib et al. (1998) showed that textural indices (Haralick et al., 1973) correlate well with soil particle size or soil type. Of the 16 textural indices defined by Haralick, *Local Homogeneity (LH)* which indicates the homogeneity of images, is adapted here. Generally, the textures of clay or silt images are more homogeneous than the textures of sand images so that the *LH* of fine materials will be larger than that of coarse materials.

3 DETECTION OF THIN LAYERS

The CPT lacks the ability to resolve thin layers. On occasion, a soil may be thick enough to be detected yet not thick enough to allow for full development of resistance at the tip or friction along the sleeve. In such cases, the CPT will incorrectly classify the soil. For instance, when a sand layer is present in clay but is not thick enough for the tip and sleeve resistances to fully develop the higher values that would be representative of the sand could not have enough chance to gain the full resistance and friction of the sand, the soil will be interpreted as finer and softer material than it actually is.

The VisCPT was utilized at the National Geotechnical Experiment site on Treasure Island in San Francisco. Since Treasure Island was built by hydraulic filling, the subsurface stratigraphy of the island is highly stratified with clay and silty clay seams interbedding silty sands or fine sands.

Figure 1 shows the CPT profile and the VisCPT-based *LH* of the soil column. The conventional CPT data was analyzed according to Robertson (1990). To classify the soil images, a certain range of *LH* was assigned to each type of soil as shown in Table 1. Since the magnification of the VisCPT camera was fixed at a 1.0 cm field of view, it was impossible for the VisCPT to classify soils that were finer than silt. Although, distinction between silty clay and clay was feasible.

Table 1. *Local Homogeneity* and Soil Description for VisCPT Classification

Soil Type	LH	Soil Description
2	< 0.28	Sand
3	0.28 – 0.32	Sand to silty sand
4	0.32 – 0.36	Silty sand to sandy silt
5	0.36 – 0.44	Sandy silt to clayey silt
6	0.44 – 0.52	Clayey silt to silty clay
7	0.52 – 0.60	Silty clay to clay
8	> 0.60	Clay
9		Organic material
10		Sensitive fine grained

Subfigures (c) and (d) in Figure 1 show the classification based on Robertson's chart and image processing, respectively. As observed in the figure, the CPT missed a 3 cm lens of sandy silt (soil type 5) at 3.1 m depth which was identified by the VisCPT and confirmed by high quality thin-walled tube sampling (Figure 1(f)). Also, silt seams (soil type 5 or 6) at 4.65, 4.75 and 5.15 m depth were not detected by the CPT. Apparently, the CPT is not able to detect soft thin layers interbedded in relatively thick harder layers.

Not only soft seams but also sand layers were not detected by the CPT when they were surrounded by silt or clay. A 10 cm silty sand layer at 5.85 m sandwiched between silt or clay material was missed. The soils between 5.7 m and 6.5 m were highly stratified with thin sands, silts and clays (soil types 4 to 8). However the soils were classified by the CPT as all silty or clayey materials (soil types 5 to 7).

In general, the CPT is unable to detect the presence of thin layers in fairly uniform soil profiles. In highly stratified deposits, the CPT inferred a somewhat averaged and more uniform stratigraphy than was actually present.

4 CPT MISINTERPRETATION AT INTERFACE

Another factor causing CPT misinterpretation of stratigraphy is that soil well ahead of the advancing cone affects, and possibly even controls the tip resistance. This is particularly true when the CPT advances through a hard soil towards a softer layer. The tip will begin to feel the presence of the soft material and the tip resistance will decrease before even contacting the softer soil. By contrast, when the CPT is advancing through a soft soil towards a harder layer, the tip must come very close to the interface of the two soils to sense the harder material. In the "soft to hard" transition, the CPT will require some penetration into the hard layer to develop the full resistance of the stronger soil. This is because the harder, less compressible soil is bulging upward into the softer overlying material.

Figure 2 illustrates a hypothetical CPT log of a stratified system where a soft sandy silt or clayey silt underlies a stronger clean sand. The q_c of the sand and the silt are assumed to have typical values of 20 MPa and 3 MPa respectively. The *FR* are 1 % for the sand and 2 % for the silt. As mentioned above, the q_c will begin to drop before reaching the silt. Virtually all of the tip transition will occur in the upper harder layer. Once in the weaker silt, the tip resistance rapidly stabilizes at a constant value representative of the silt. The length of the tip transition zone will depend on the relative soil strengths and stiffnesses of the two soils.

Meanwhile, the f_s will not begin to decrease until the bottom of the sleeve contacts the silt. It will then transition from 0.2 MPa (corresponding to the value in sand) to 0.06 MPa (the value observed in the silt) over a distance of 13 cm, that is, over the length of the sleeve. The center of the transition zone will coincide with the interface.

The differences in the two transition zones results in a characteristic *FR* profile with a peak coincident

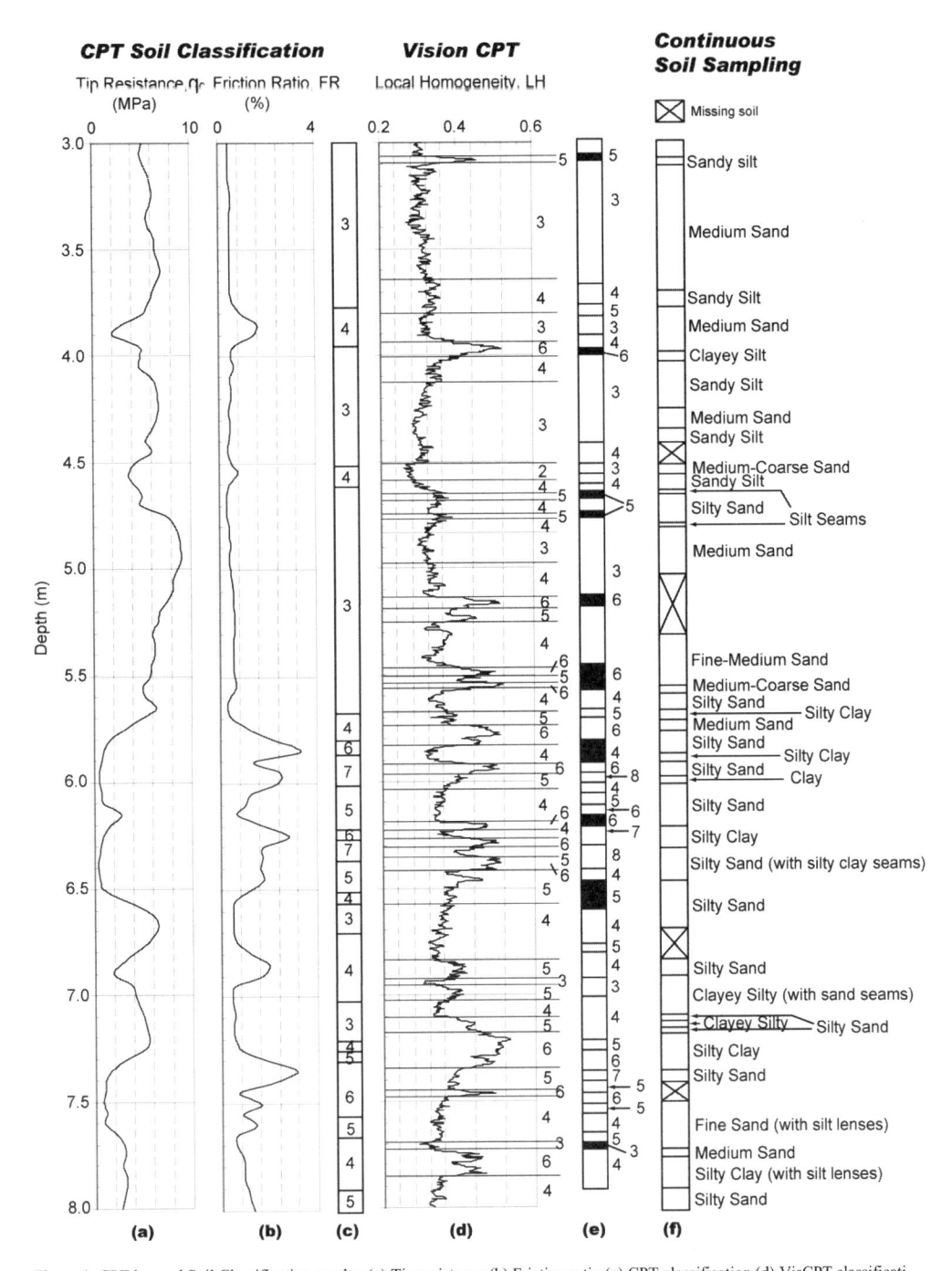

Figure 1. CPT log and Soil Classification results: (a) Tip resistance (b) Friction ratio (c) CPT classification (d) VisCPT classification (e) Corrected soil profile (f) Continuous soil sampling

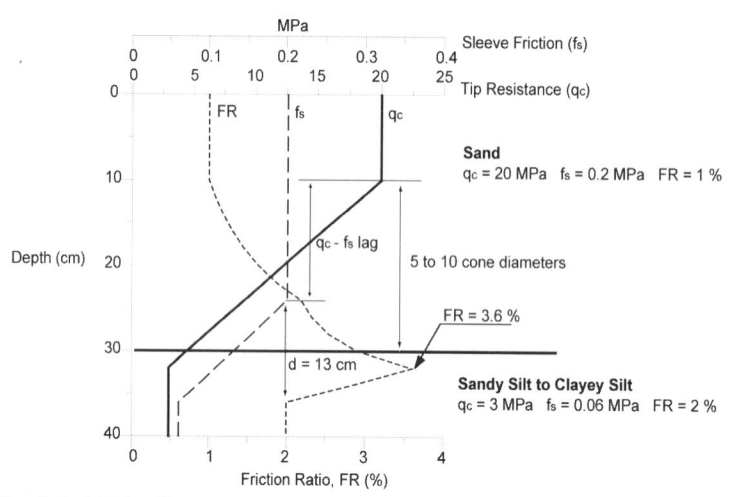

Figure 2. Hypothetical CPT readings at an interface

with the elevation at which the tip resistance reaches its lower limit. In the case of stronger soil over weaker or softer material, this peak occurs just below the interface. It should also be noted that the FR in the 13 cm surrounding the interface is larger than the FR in the softer soil itself.

The spike in the FR as shown in Figure 2 may imply that a thin (< 13 cm) clay seam exists whereas in reality, it is merely the CPT's response to an interface between a stronger soil above and a weaker one below.

The key observation is that while the CPT friction sleeve responds to the average frictional resistance over its 13 cm length, the tip resistance responds to soil further below the elevation of the cone. Thus, there is a "gap" or "lag" between responses. The q_c-f_s lag will be relatively small when the CPT advances from soft to hard soil but could be large when the CPT moves from hard to soft soil. As stated before, the actual q_c-f_s lag will be controlled by the relative stiffnesses and strengths of the two soils.

5 CPT CORRECTION

The above observations suggest that with the help of the VisCPT, the CPT readings may be corrected to account for the q_c-f_s lag. For the data shown in Figure 1, the VisCPT revealed that the CPT q_c was more representative of soil somewhat ahead of the tip than of the soil at the actual tip elevation.

Figure 3 adds two additional pieces of information to the CPT log shown in Figure 1, - the sleeve friction (f_s) and the pore water pressure (u) as measured by the CPTU. Peaks in q_c occurred at depths of 4.95, 5.65, 6.15, 6.65 and 7.25 m as summarized in Table 2. The corresponding peaks in f_s were at depths of 5.10, 5.75, 6.20, 6.80 and 7.30 m. The q_c-f_s lags were therefore 15, 10, 5, 15 and 5 cm. Or, on average, 10 cm. The FR corresponding to the first peak showed no distinct spike but on the remaining 4 occasions it spiked at 5.85, 6.25, 6.85 and 7.35 m depths. These spikes always occurred deeper than the elevations of the peaks of the associated f_s readings. Furthermore, in each case they corresponded very distinctly with the elevation at which q_c almost ceased to decrease. Finally, the five actual interface depths as observed by the VisCPT are listed in Table 2. They clearly show very good agreement with the observed spikes in the FR as had been predicted.

CPT-based classification of soils depends on coherent q_c and f_s records. In the vicinity of interfaces, the q_c-f_s lag destroys the coherence. The result is that spikes in FR such as shown by the solid line in the FR plot of Figure 3 do not correspond to finer-grained soils. These spikes also do not correspond to peaks of CPTU excess pore water pressure. Recognizing that the q_c is strongly controlled by an underlying softer soil, thereby creating the q_c-f_s lag, the authors shifted the q_c values in Figure 3 by the average observed q_c-f_s lag of 10 cm. The shifted q_c plot is shown by dashed lines superimposed over the original q_c data in Figure 3. The FR was then recalculated

Table 2. Depths of Peaks and Interfaces

Interface	Peak q_c Depth (m)	Peak f_s Depth (m)	Peak FR Depth (m)	q_c - f_s lag (cm)	Interface depth from VisCPT
1	4.95	5.10	None	15	5.12
2	5.65	5.75	5.85	10	5.90
3	6.15	6.20	6.25	5	6.20
4	6.65	6.80	6.85	15	6.85
5	7.25	7.30	7.35	5	7.20

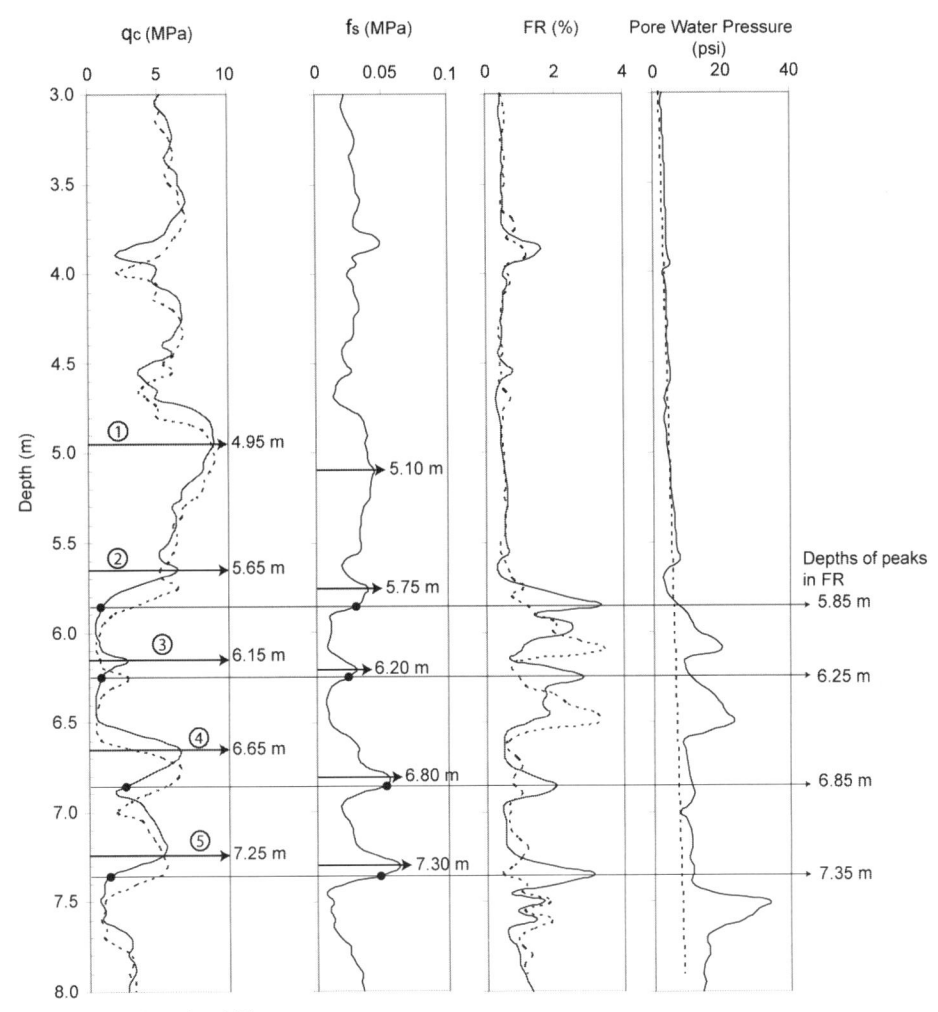

Figure 3. $q_c - f_s$ lag and q_c shifting

and the results are also shown on Figure 3 by dashed lines superimposed over the original FR plot. This corrected FR plot is clearly in better agreement with both the CPTU pore water pressures and the VisCPT logs.

A corrected stratigraphic soil column, following the 10 cm q_c shift is shown in Figure 1(e). The revised stratigraphic interpretation also includes thin soil layers that were missed by the CPT but revealed by the VisCPT as at elevations 3.1, 4.0, 4.65, 4.75, 5.2, 5.5 and 7.7 m. The corrected information and newly added layers are identified by the black areas in the Figure 1(e). Overall, the corrected soil profile shows more detail and agrees well with the profile based on the VisCPT and with the soil observed by sampling.

6 ADDITIONAL OBSERVATION AND DISSCUSSION

According to the CPT classification, two "silty sand to sandy silt" layers (soil type 4) were found at 3.8 m and 4.5 m. The thicknesses of the two layers were 15 cm and 10 cm respectively. The CPT suggests that the layers are sandwiched by "sand to silty sand" (soil type 3). However, based on both the VisCPT and collected soil samples, the soil at both of these depths is coarser than the soil above and below. The CPT classification of these two lenses as containing finer material was based on lower tip resistances and higher friction ratios. However, the f_s actually increased in these layers (see Figure 3) suggesting coarser material. The soil was actually coarse but very loose sand and was therefore mistakenly classified as being finer by the CPT.

In highly stratified regions such at depths 5.8, 6.2 and 6.5 m, a uniform shift in q_c to remove the q_c-f_s lag could be unproductive. In such regions, the VisCPT data was used to identify the strata elevations.

Textural indices are very sensitive to illumination and soil color. Thus, it is difficult to establish unique correlations between LH and soil types for classification that would be universally applicable. In this study, the range of LH for each soil type (Table 1) was determined by visually comparing soil images to computed LH values and CPT results.

It is interesting to note that f_s appears to be more reliable than tip resistance for stratigraphic delineation, particularly for locating interfaces, since f_s is only affected by soil adjacent to the sleeve and not by the soil beneath it. Shorter sleeve elements could conceivably result in shorter transition zones and therefore could better locate interfaces.

7 CONCLUSIONS

1. The vision cone penetrometer (VisCPT) provides higher resolution of thin soil layers than the CPT alone. The image textural soil index "*local homogeneity*" (*LH*) can be calibrated to provide soil classification with very good accuracy. Thin anomalous clay lenses and sand seams that may go undetected by the CPT are easily identified by the VisCPT.

2. The CPT tip resistance is controlled by soil up to several cone diameters below the tip while the CPT sleeve is affected only be soil adjacent to it. This leads to misclassification of soils in the vicinity of interfaces, particularly when the cone is pushed through stronger material into weaker soil, such as from sands into silts or clays.
3. Based on the results of a high quality CPT, VisCPT and CPTU tests performed on Treasure Island, it was observed that when the CPT passes from a coarser into a finer soil:
   a. A sharp increase in VisCPT *LH* reveals the elevation of the interface.
   b. The lag between peak q_c and peak f_s readings was observed to range from 5 to 15 cm (using a standard 10 cm^2 cone) with an average lag of 10 cm.
   c. The friction ratio displays a spike very close to the actual interface which also coincides with the elevation at which the tip resistance drops almost to the level representative of the lower soil.
   d. By shifting the tip resistance downward by 10 cm and recomputing the friction ratios, the soil classifications were in much better agreement with VisCPT and CPTU results.

REFERENCES

Ghalib, A., Hryciw, R. and Shin, S. (1998). "Image Textural Analysis and Neural Network for the Characterization of Uniform Soils". *In Proceedings of ASCE Congress on Computing in Civil Engineering*, pp 671 – 682. ASCE

Haralick, R., Shanmugam, K. and Dinstein, I. (1973). "Textural Features for Image Classification". *IEEE Transactions on System, Man and Cybernetics*. SMC-3(6): 610 – 621

Hryciw, R., Ghalib, A. and Raschke, S. (1998). "In-Situ Soil Characterization Using Vision Cone Penetrometer (VisCPT)". *In Proceedings of the 1st Int. Conf. on Site Characterization*. pp 1081 – 1086. ISC'98, A.A. BALKEMA

Raschke, S. and Hryciw, R. (1997). "Vision Cone Penetrometer (VisCPT) for Direct Subsurface Soil Observation". *ASCE Journal of Geotechnical and Geoenvironmental Engineering*. 23(11): 1074 – 1076

Roberson, P. (1990). "Soil Classification Using the Cone Penetration Test". *Canadian Geotechnical Journal*. 27(1): 151 - 158

Evaluation of shear wave velocity profile using SPT based uphole test

Dong-Soo Kim, Eun-Seok Bang & Won-Seok Seo
Department of Civil & Environmental Engineering, Korea Advanced Institute of Science and Technology, Daejeon, Korea

Keywords: SPT, uphole test, shear wave velocity, site characterization

ABSTRACT: A modified uphole test which is economical and reliable for obtaining shear wave velocity profile was introduced. Uphole test is a seismic field test using receivers on ground surface and a source in depth. In the proposed uphole test, SPT sampler which is common in site investigation, was used as a source and several surface geophones in line were used as receivers. The proposed uphole test can be performed to the deep testing depth without casing. Testing procedure and interpretation methods for obtaining interval times and shear wave velocity profile were introduced considering refracted ray path. Finally, uphole test were performed at several sites, and the applicability of the proposed uphole test was verified by comparing wave velocity profiles determined by the uphole test with profiles determined by downhole test, SASW test and SPT-N values.

1 INTRODUCTION

Determining reliable shear wave velocity (V_s) profile of a site has become important in the static deformation analysis as well as in the geotechnical earthquake engineering. Many seismic field tests such as crosshole, downhole, suspension PS logging, and SASW tests are now generally used for the evaluation of Vs profile. Each test has its own advantages and disadvantages and the results may not be coincident in many cases. Therefore, it is important to select adequate field testing technique considering site conditions and importance of projects to obtain the reliable V_s profile. The less have suppositions in testing and interpretation, the more have the trust of the result. But, it may not be efficient if it costs a lot in performing the test even if it can give the best result.

Bore-hole seismic methods such as downhole, in-hole and SPS logging tests are very attractive for several reasons. These methods require just one bore-hole to perform the tests and the interpretation procedure is very simple. However, in the layers of sedimentary gravels, weathered and fractured rocks which are frequently encountered in Korea, it is difficult to construct the bore-hole and to get a good coupling between surrounding soil and casing which is crucial in testing. In the downhole test, the wave propagation is hindered by these layers and the substantial amount of source energy is required to get a discernible signal (Bang 2001).

Uphole test is a seismic field test using receiver on surface and source in depth. It has been generally used for obtaining compression (P) and shear (S) wave velocity profiles using bore-hole blasting. Interpretation procedure is simple and it does not need a cased bore-hole. But, source generation is not cost-effective and can't be repeated at the same depth. And it is difficult to generate shear wave component using blasting and other in-hole sources.

Standard penetration test (SPT) is the most frequently used method in site investigation. The impact energy generated by SPT can be used as a source. The combination of split spoon sampler and hammer can be a good source which produces the significant amount of SV type shear waves inside the bore-hole. SPT is usually performed at every 1 or 1.5 meter intervals to the bottom of the bore-hole. If the receivers are placed on the ground surface, it would be feasible to perform the uphole test during SPT (Fig. 1). Because SPT is employed in almost every site investigations and the bore-hole casing is not required for the seismic test, the uphole test using SPT source would be an economical to obtain the additional shear wave velocity profile if the reliability of testing and data reduction method is secured. Because the distances between bore-hole and receivers are large, interpretation methods considering refracted ray path must be employed.

In this paper, the testing and data reduction procedures of uphole test using SPT source were investigated. Interpretation methods for obtaining interval times and for determining V_s profile were introduced considering refracted ray path. Finally, uphole test were performed at two sites, and the applicability of the proposed uphole test was verified by comparing wave velocity profiles determined by the uphole test with profiles determined by downhole test, SASW test and SPT-N values.

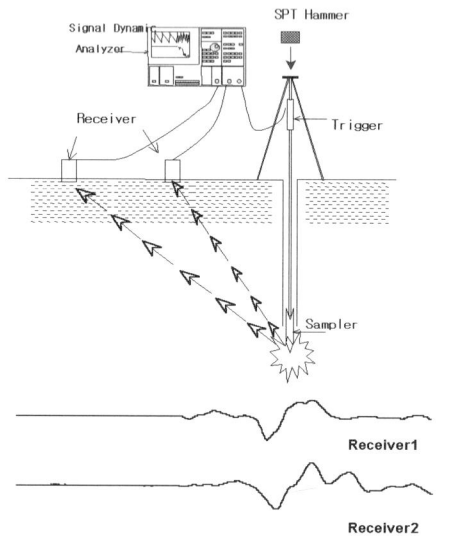

Figure 1. The schematic diagram of proposed uphole test based on SPT

2 PRELIMINARY FIELD TESTING

To understand the wave types generated by SPT sampler is crucial to the development of an SPT-based uphole test. Because the sampler moves downward during the impact, it can be postulated that the shear wave of the particle motion in vertical direction (SV wave) is generated and propagating horizontally and the compression wave is propagating in the vertical direction near the sampler as shown in Fig. 2. Therefore, the signals acquired by the receivers on the ground surface will be SV waves, except the receiver very close to the borehole. In general, the major direction of each wave motion will vary depending on the location of source and receiver. P-wave component will be mainly detected in the radial direction on the ground surface when SPT source is located at shallow depth but mainly detected in the vertical direction when located at deep depths. However, S-wave component will be mainly detected in both vertical and radial directions at any condition.

In order to investigate the types of waves generated by SPT sampler, the preliminary field testing was performed. Seven geophones were located at distances of 6m, 8m, 10m, 12m, 14m, 16m, 18m from the boring hole and the particle motions are measured during SPT. At a distance of 8m, 3-component (3D) geophones were installed to construct the particle orbits. Typical time domain signals in the radial, transverse, and vertical directions and the particle orbits obtained from 3D geophone at the SPT source depth of 9.0m are shown in Fig. 3 and Fig. 4, respectively. SV wave components generated by SPT sampler are dominant in the radial and vertical geophones as expected. Judging from the particle motions measured by 3D geophones, the arrival of SV wave is determined as indicated by arrow, where the substantial energy is arriving at radial and vertical receivers as shown in Fig. 3 and 4.

Exact determination of initial arrival of SV wave will be difficult because the polarity characteristics cannot be used in the measurements. However, for the interpretation of proposed uphole test, the only interval travel time information is required and it is not crucial to determine the exact arrival time at each receiver, as discussed later. In the proposed uphole test, only vertical geophones are used and the first peak (or trough) point in the vertical motion is decided as a SV wave arrival. An example of picking SV wave arrivals at each receiver is shown in Fig. 5. The arrivals of SV wave are clearly detected at each receivers of different distance from the borehole. It is also interesting to notice that P wave with higher frequency components is arrived earlier and separated by SV wave.

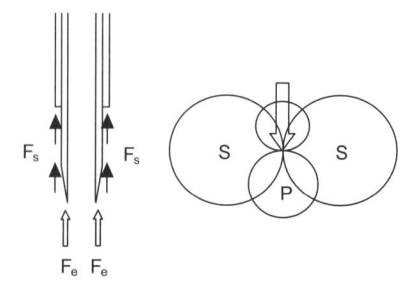

Figure 2. The assumed wave shape generated by SPT sampler source

3 TESTING PROCEDURE

Based on the preliminary field test results, the uphole testing procedure is proposed. To perform the seismic tests, the source, receivers and data acquisition system are usually required. The series of 1Hz vertical geophones (Mark Product L4-C) were used as receivers and the multi-channel signal analyzer (GRAPHTEC MA6000) was used to digitize and record the signals. As a source, the vibration generated at a depth by a split spoon sampler during the SPT hammer impact was used. The amount of energy generated is much larger than the surface plank wood source in the downhole test, but the reverse signal cannot be generated to capture the initial arrival time using the polarity characteristics. The schematic diagram of the proposed uphole test is shown in Fig. 1 and the testing procedures are as follows;

1) The receivers are placed on the ground at the selected intervals from the bore-hole. The spacing to the first receiver is usually selected away from the bore-hole to minimize the effect of compression wave.
2) After SPT test at a given depth, uphole test is performed. It would be better to drop the hammer manually after turning off the engine to reduce the noise. To trigger the source signal, the accelerometer or load cell is installed at the drilling rod near the anvil and the distance between trigger unit and bottom of the sampler is accurately measured to correct the trigger time.
3) The receiver signals are acquired and evaluated using the dynamic signal analyzer. Once the received signals are repeatedly similar as the previous triggered ones, the receiver signals are saved.

After drilling to the next depth and performing SPT test, the steps 2) and 3) are repeated to the final depth of the investigation.

4 UPHOLE DATA REDUCTION

4.1 *Determination of Arrival Time Difference*

In the analysis of uphole measurement, the determination of arrival time difference is crucial. Because it is difficult to use the polarity characteristics in the measurement of initial arrival time using SPT source, the arrival time difference between two signals is utilized in the analysis instead of measuring the direct arrival time from source to receiver.

Arrival time difference was determined either by peak-to-peak method or by cross-correlation method. In the peak-to-peak method, arrival time

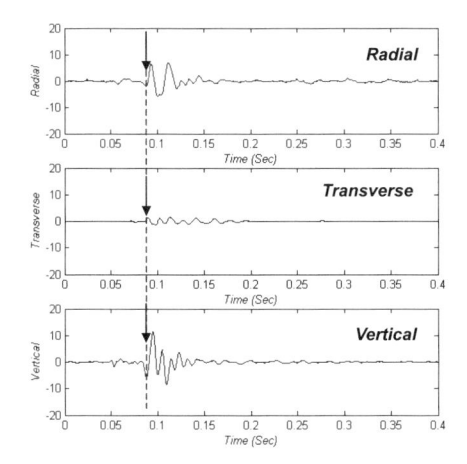

Figure 3. Typical time domain signals in three directions

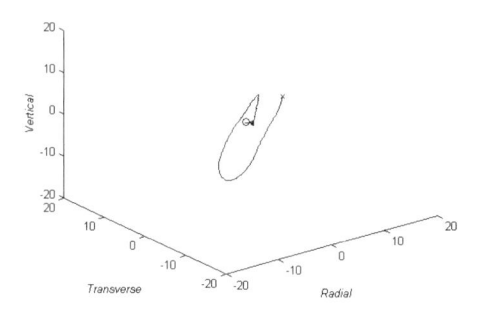

Figure 4. Typical particle motion orbits

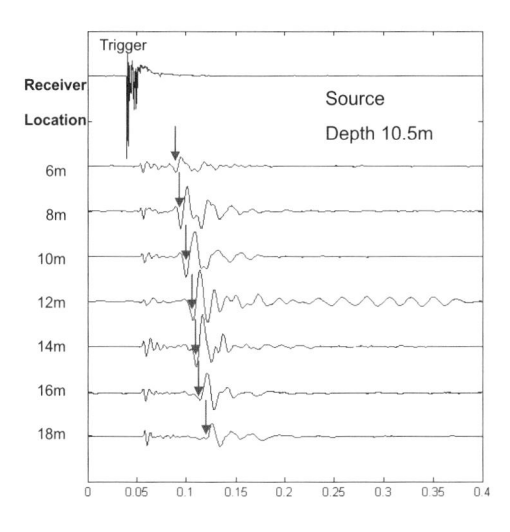

Figure 5. Determination of arrival of S wave

difference was evaluated by the time difference between first peaks (or troughs) of two receiver signals as shown in Fig. 6-a. In the cross-correlation method, the cross-correlation of two signals was determined by shifting one signal, relative to the other one, in steps equal to the digitized time interval. At each shift, the sum of the products of the signal amplitudes at each interval gives the correlation of that shift. After shifting through all of the time intervals, the cross-correlation can be plotted versus the time shift as shown in Fig. 6-b, and the time shift giving the greatest sum was taken as the arrival time difference (Campanella and Stewart 1992). Both methods can be used reliably when the shapes of two signals are not distorted, and the arrival time difference can be determined automatically without introducing the subjective opinion.

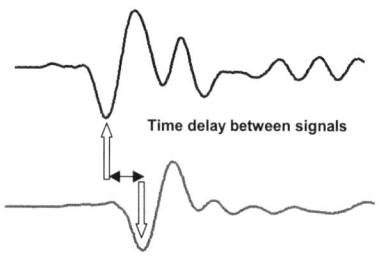

a) peak to peak method

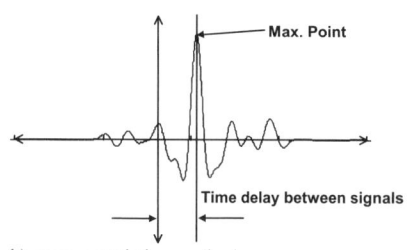

b) cross-correlation method

Figure 6. Determination of time delay between signals

4.2 Methods Obtaining V_s Profile

In order to determine the shear wave velocity profile using the proposed uphole test, interpretation methods that is similar to the downhole data reduction method (Kim et al. 2003 ; Joh and Mok 1998) was used in this study. The fundamental principle is that the velocity value of a layer is determined by the travel time delay between upper and lower point of a layer and the difference of each ray-path. So, testing depth, receiver distance and travel time information are required in the analysis. Because the distance from bore-hole to receivers are far, the refracted ray path based on Snell's law was used in the analysis.

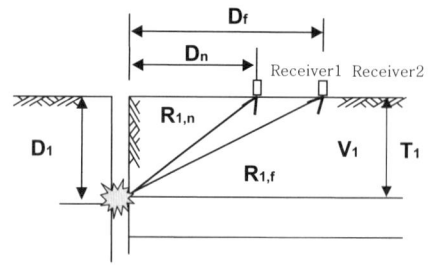

Figure 7. Interpretation method of uphole test at first layer

In this study, the shear wave velocity of first layer was obtained by using time delay between receivers as shown in Fig. 7 because the time delay at a given receiver between first and second testing depths is often negative. Therefore, Eq. 1 shows the time delay between receiver 1 (near) and receiver 2 (far). The wave velocity of 1st layer is obtained using Eq. 2.

$$DT_1 = T_{1,f} - T_{1,n} \tag{1}$$

$$V_1 = \frac{R_{1,f} - R_{1,n}}{DT_1} \tag{2}$$

In general case, at ith layer, travel times from source to each receiver in ith testing depth is given by Eq. 3 and the refracted ray path is determined considering two conditions, one is Snell's Law and the other is geometry criteria (Fig. 8).

$$T_i = \sum_{j=1}^{i} \frac{L_{ij}}{V_j} = \frac{L_{i1}}{V_1} + \frac{L_{i2}}{V_2} + ... + \frac{L_{ii}}{V_i} \tag{3}$$

$$\frac{\sin \alpha_{i1}}{V_1} = \frac{\sin \alpha_{i2}}{V_2} = ... = \frac{\sin \alpha_{ij}}{V_j} = ... = \frac{\sin \alpha_{ii}}{V_i} \tag{4}$$

$$T_1 \tan \alpha_{i1} + T_2 \tan \alpha_{i2} + ... + T_j \tan \alpha_{ij} + ... + T_i \tan \alpha_{ii} = D \tag{5}$$

From determined new refracted ray path(Eq. 6), the wave velocity of ith layer is determined using Eq. 7.

$$L_{ij} = T_j / \cos \alpha_{ij} \tag{6}$$

$$V_i = \frac{L_{ii}}{T_i - \sum_{j=1}^{i-1} \frac{L_{ij}}{V_j}} \tag{7}$$

where $T_{1,f}$: travel time of ith layer at receiver 2 (far); $T_{1,n}$: travel time of ith layer at receiver 1 (near); $R_{1,f}$:

the distance from source to receiver 2 (far) at 1st testing; $R_{1,n}$: the distance from source to receiver 1(near) at 1st testing; $α_{ij,n}$: incident angle from j^{th} layer to next layer of ith ray path; $L_{ij,n}$: the length of ray-path on jth layer of ith testing at receiver 1 (near); D_n : The distance between bore-hole and receiver 1; D_f : The distance between bore-hole and receiver 2; D_i : i^{th} testing depth

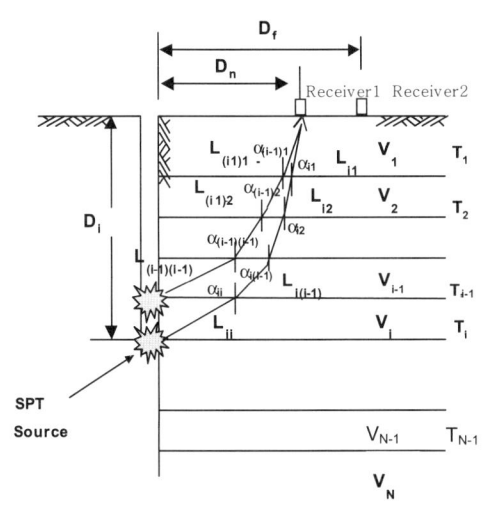

Figure 8. schematic diagram of interpretation uphole test

5 CASE STUDIES

Field tests were performed at three sites to verify the applicability of the proposed uphole test. Fig. 9 shows the typical signal traces with the source depth obtained from the receivers located at distance of 6m and 12m at Kim-je site. It is clear that the arrival of shear wave and time delay information can be obtained easily using peak to peak or cross-correlation method. From these time delay information, the shear wave velocity profiles at each receiver were determined by using the proposed interpretation method and compared with the results determined by downhole test and SPT-N value at the same location (Fig. 10). The shear wave velocity profiles determined by the signals of each receiver agreed well each other and the shape of profile is similar with the results obtained by downhole test and SPT-N value. The result obtained by the receiver 1 (6m) is not coincided with the other test results at deep testing depths. As previously mentioned, the signals obtained from near receiver can be easily interfered with some noise due to P wave component (Fig. 9a). Therefore, it is important to determine the adequate location of the near receivers considering the final testing depth because P wave component makes difficult to pick up the right S wave arrival.

Uphole tests were performed at two other sites; Gwangju and Yi-chon sites. Test results affected by noise signals due to P-wave component at near receivers were excluded in the analysis. Fig. 11 and 12 show the shear wave velocity profiles obtained by the proposed method and compared with SASW, downhole and SPT results. V_s profiles obtained by the proposed uphole tests matches reasonably with other test results, showing the potential of using in the site characterization.

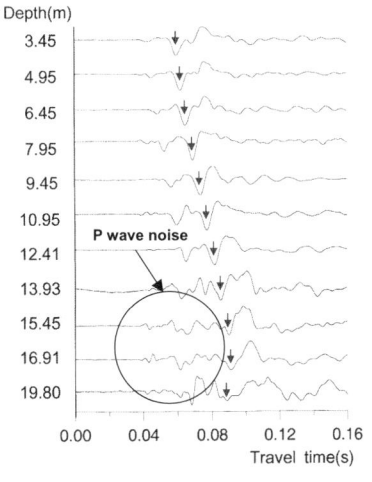

a) signal trace (Receiver 1 : 6m)

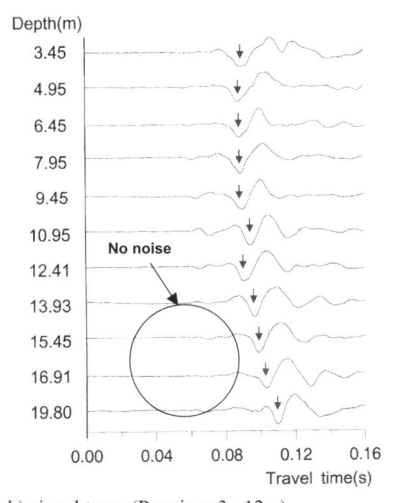

b) signal trace (Receiver 3 : 12m)

Figure 9. Signal traces at Kim-je Site

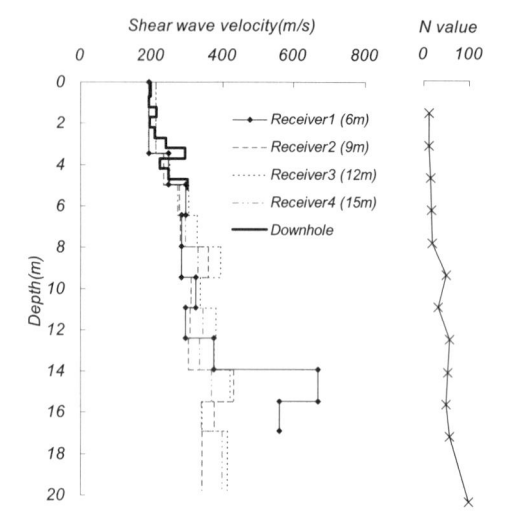

Figure 10. The result at Kim-je Site

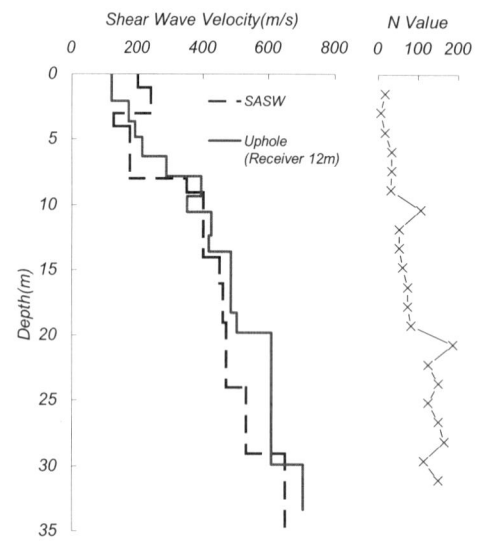

Figure 11. The result at Gwang-ju Site

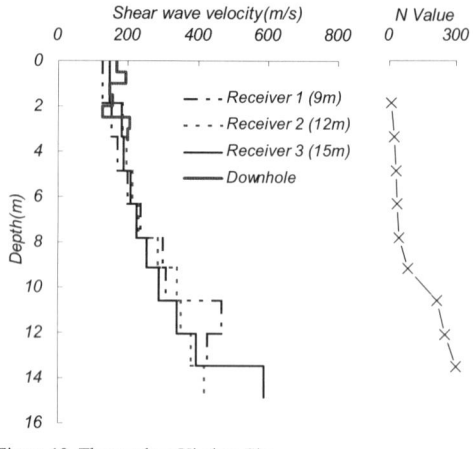

Figure 12. The result at Yi-chon Site

6 CONCLUSION

SPT based uphole test, which is simple and economical to determine shear wave velocity profile, was proposed. Testing procedure and interpretation method for obtaining shear wave velocity profile were introduced. The reliability and applicability of the proposed method was verified by performing field studies and comparing the results with those obtained by downhole test, SASW test and SPT-N values.

ACKNOWLEDGEMENTS

This study was sponsored by the SISTeC, which are gratefully acknowledged

REFERENCES

Bang, E.S. 2001. Evaluation of shear Wave Velocity Profiles using Dwonhole and uphole Tests. *Dissertation of Master of Science in Engineering*, KAIST, Daejon, Korea.

Campanella, R.G. and Stewart, W.P. 1992. Seismic cone analysis using digital signal processing for dynamic site characterization. *Can. Geotech. Journal. Vol 29*, pp.477-486.

Joh, S.H., Mok, Y.J. 1998. Development of an inversion analysis technique for downhole testing and continuous seismic CPT, *Jounal. of KGS, Vol. 14, No. 3, June*, pp. 95-108.

Kim, D.S., Bang, E.S., Kim, W.C. 2003. Evaluation of various downhole data reduction methods to obtain reliable V_s profile", *Geotechnical Testing Journal* (submitted)

Proceedings ISC-2 on Geotechnical and Geophysical Site Characterization, Viana da Fonseca & Mayne (eds.)
© 2004 Millpress, Rotterdam, ISBN 90 5966 009 9

Suitability of using Iwasaki's P_L in characterization of liquefaction damages during the 1999 Chi-Chi Earthquake, Taiwan

C.P. Kuo
Department of Construction Engineering, National Taiwan University of Science and Technology, Taipei, Taiwan

M. Chang, S.H. Shau & R.E. Hsu
Department of Construction Engineering, National Yunlin University of Science and Technology, Yunlin, Taiwan

Keywords: soil liquefaction, liquefaction damage assessment, liquefaction potential index, suitability

ABSTRACT: One of currently adopted methods in the Taiwanese Building Codes (ARBI 1998) for liquefaction damage assessment is based on Iwasaki's liquefaction potential index, P_L, which was also suggested by ISSMFE TC4 in 1993. The index was defined and calibrated by Iwasaki et al. (1982) using liquefaction and non-liquefaction incidents that had occurred in Japan. On September 21, 1999, a severe earthquake struck the center of Taiwan and resulted in significant liquefaction damages in sandy deposits of the Tzou-swei River, the largest alluvial fan to the mid-west of the island. An assessment of the liquefaction damages was conducted in this study based on the potential index approach and the results showed the computed damages appear to not completely agree with the observed situations. Original derivations of the liquefaction potential index by Iwasaki et al. were therefore reviewed. The results showed the threshold values of liquefaction potential index of 5 (insignificant potential) and 15 (significant potential), as proposed by Iwasaki et al. (1982), appeared to be sensitive to the database for analysis, and were too high (or unconservative) for use in Taiwan. Possible reasons might attribute to selections of the companion method for analysis of factor of safety against liquefaction, the depth-weighing function, as well as the sample space of liquefaction incidents. Based on liquefaction incidents of the 1999 earthquake, the aforementioned threshold values are suggested to be revised in order to be more consistent with the observations in Taiwan.

1 INTRODUCTION

Long before the 1999 Chi-chi earthquake, the works to establish a nationwide liquefaction potential map have been conducted by many universities and research institutes in Taiwan. The Liquefaction Potential Index, P_L or I_L, by Iwasaki et al. was adopted and became a normal method to estimate hazard by soil liquefaction in Taiwanese Building Code. The computing steps of this method are quite simple and clear for usage. However, this method would be influenced by completeness of boring holes logs and database. The phenomenon implicates subjectivities would be taken place while performing liquefaction potential analysis due to defects of databases. In fact, most liquefaction sites were successfully predicted prior to the earthquake in alluvial plain, river shoals, and reclaimed lands in mid-west Taiwan; even so, some liquefaction cases were predicted to be with low potentials, i.e. $P_L < 5$. In addition, some predictions were over-estimated, i.e. the $P_L > 15$ but no liquefactions observed.

Figure 1. Liquefaction sites in study area (modified from NCREE 2000)

Three problems need to be explained: (1) How well the estimation is suitable in Taiwan? (2) How to judge liquefactions were actually taken place? (3) How to get complete and correct database for analysis?

In fact, the method has to be characterized for some regions by its own database. An initial calibration by the earthquake was performed on Tzou-swei river alluvial plain, where most serious liquefaction cases were taken place during the quake. The liquefaction sites observed were shown in Figure 1. Some of them out of the region represent for ones in lands reclaimed from the sea. Around 20 liquefaction cases observed were used in this study.

2 DOUBTS ABOUT USAGE OF THE METHOD

Liquefaction hazard assessment using Iwasaki's P_L index is adopted into the Taiwanese Building Code as a standard to design or construct ground improvement works. However, this index was derived and defined based on a database with liquefaction and non-liquefaction cases in Japan; which might not suitable for sites in Taiwan. Some questions remains as to why compute a borehole data with only a depth shallower than 20m; how to integrate appropriate the P_L by a discontinuous SPT-N values? In addition, the threshold values of P_L of 5 (insignificant potential) and 15 (significant potential) are needed to be calibrated for use in areas other than in Japan.

3 SUMMARY FROM IWASAKI'S PREVIOUS WORKS

A brief review on Iwasaki's original research (1978, 1980, and 1982) was conducted and summarized as follows:

3.1 Review of Iwasaki et al. (1978)

- $P_L=0$ is considered to be the safest situation for site, i.e., no risk on liquefaction.

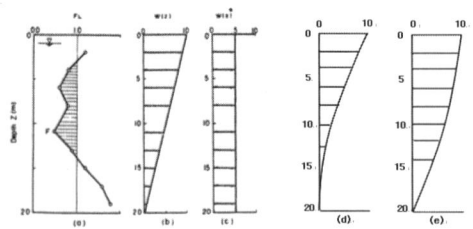

Figure 2. Different on weight functions for safety factors (modified from Iwasaki et al. 1978, Kuo et al. 2002)

- The weight function, $W(z)$, is used to calculate potentially different impacts of excess pore water pressures on soil layer at different depths. The function was set to be a reverse triangle distribution (Figure 2. (b)) from 0 to –20m in depth. It is interesting to know if other shapes of weight function (e.g. Figure 2. (c)-(e)) could provided more reasonable estimation?
- The computed P_L values for non-liquefaction cases were entirely less than 20; of which 70% cases were less than 5. Similarly, the P_L values for serious liquefaction cases were never less than 5; but more than 15 or even 20.

3.2 Review of Iwasaki et al. (1980)

- A comparison of safety factors (F_S or F_L) based on the estimation procedure by Seed and Idriss in 1971 and the procedure by Iwasaki indicated different P_L would come out when substituting F_S computed from Seed and Iriss's (1971) procedure with Iwasaki's. Accordingly the paper indicated that P_L should be computed using F_S from Iwasaki's procedure.
- Another concept was proposed that with more noticeable discrimination ability between liquefaction cases and non-liquefaction ones, the better method is.
- The computed P_L values for all non-liquefaction cases were less than 20; of which 65% of the cases were less than 5. Similarly, the P_L for 50% liquefaction cases were more than 15; only 10% of the cases with P_L less than 5.
- The method seemed to be more reliable due to completeness of database. However, the database was established by the cases in Japan.

3.3 Review of Iwasaki et al. (1982)

- The threshold values were defined herein:
  i. for non-liquefaction sites, most of the computed P_L were less than 15; and about 70% of them were less than 5.
  ii. for liquefaction sites, only 20% cases were computed with P_L less than 5; and 50% of them higher than 15 P_Ls.
- The research proposed the following:
  i. $P_L = 0$ stands for very low liquefaction risk;
  ii. $P_L \leq 5$ stands for low liquefaction risk;
  iii. $5 < P_L \leq 15$ stands for high liquefaction risk;
  iv. $P_L > 15$ stands for very high liquefaction risk.

4 CALIBRATION BY 1999 CHI-CHI EARTHQUAKE

In order to check the suitability of using P_L procedure in Taiwan, liquefaction incidents caused by the Chi-chi earthquake were to be used as a basis for comparison. The study herein collected a database with 28 liquefaction cases and 1101 Non-liquefaction cases.

Table 1. Quantities of Cases for Simulation

Round	Liquefactions	Non-liquefactions
In 2002	20	72
In 2003	28	1100

Table 2. Parameters for Simulation

Items	Values	Clarify
Earthquake Magnitude	7.3	In Richter Magnitude
SPT Energy Ratio(%)	73.5	Acceptable in Taiwan, suggested by NCREE
Peak Ground Accelerations (gal.)	See Figure 4.	Reasonable in Study Area with East West Component
Ground Water Level (m)	See Figure 5.	Contours Plus 3m; No More Than 1m Under Ground Surface

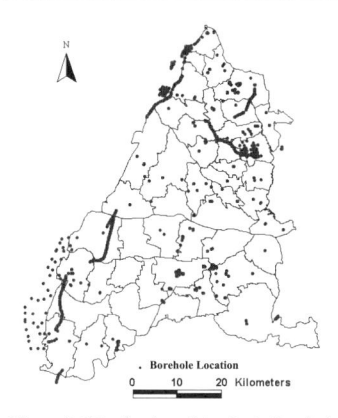

Figure 3. Distribution of Analysis Boreholes (Chang 2003)

The liquefaction analysis was based on simplified procedure by Seed and Iriss (Seed et al. 1971, Youd et al. 2001), Tokimatsu and Youshimi (1983), and New version JRA (JRA 1996). Three sets of analysis "Round 2002", "Round 2003#1", and "Round 2003#2" were performed in which "Round 2002" was computed precisely (Kuo et al. 2002, 2003), and the others were done in this study.

4.1 Conditions for Liquefaction Analysis

4.1.1 Liquefaction and Non-liquefaction Cases
As shown in Table 1, the case used in the 2002 and 2003 analyses were dramatically different; mainly because of the increase of non-liquefaction cases in 2003. Most of the non-liquefaction cases occurred at various places of the alluvial plain as shown in Figure 3.

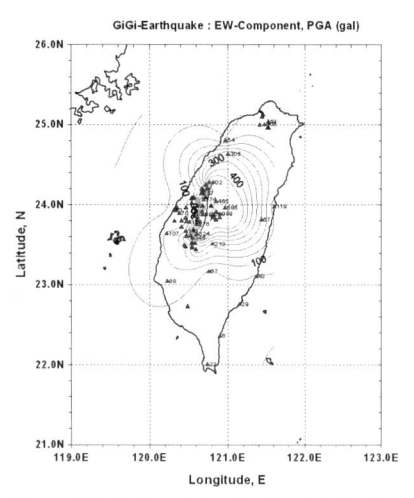

Figure 4. Peak Ground Accelerations Distribution During Chi-chi Earthquake, East West Component, in gal. (NCREE 2000)

4.1.2 Peak Ground Acceleration and Ground Water
Table 2. indicates values for the following analysis. The SPT energy ratio was decide to be 73.5%, different from the values (i.e. 60% and 72%) suggested by Seed et al. (1985) and used in Japan (Tokimatsu and Yoshimi 1983).

Peak ground acceleration distribution during the earthquake is shown in Figure 4. The east-west component was chosen because of the causative fault in the study area was in north-south direction.

Ground water table was one of major influence factors in the liquefaction analysis. Due to difficulty in acquiring exact data during the earthquake, a contour map interpolation based on borehole logs was used. In order to take into account the variation in ground levels among seasons, a long-term observation record of groundwater level in Tzou-swei River Alluvial Plain was used (Gea et al. 1996). The study indicated groundwater table would fluctuate in a range of ±3m. Accordingly, the groundwater level was adopted by an average contour as shown in Figure 5. and increased by 3m up, but not shallower than 1m below ground surface.

4.2 Depth-Weighted Integration for Computing P_L

Based on the original definition, function of P_L is in form of:

$$P_L = \int_0^{20} F(z) \cdot W(z) \cdot dz = \sum_0^{20m} P_{L(i)} \qquad (1)$$

where P_L = liquefaction potential index; $P_{L(i)}$ = liquefaction potential index increment corresponded to each SPT-N values in the range 0 to –20m; i depends on numbers of SPT performed in the borehole; $F(z)$

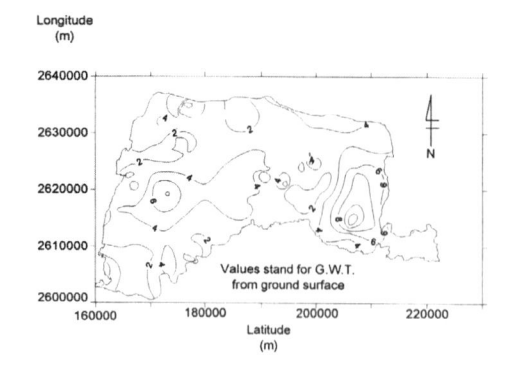

Figure 5. Average Ground Water Level based on borehole logs (Kuo 2001)

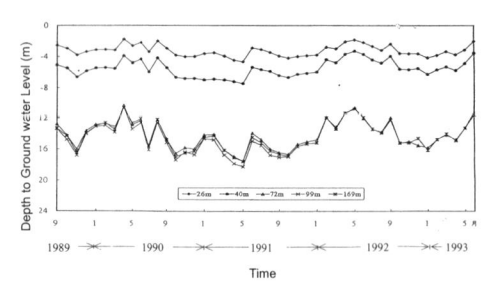

Figure 6. Variation in Groundwater Level (Gea et al. 1996)

f= $1 - F_{L(i)}$, and where $F_{L(i)}$ is the liquefaction safety factor at depth i; $W(z)$ = weight function of depth for safety function, which is in form of:

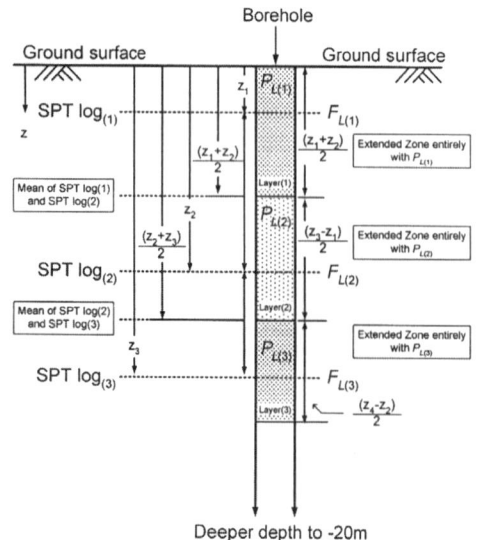

$$W(z) = 10 - 0.5z \qquad (2)$$

Figure 7. Definition of Symbols for Computing P_L

and $z = 0$ to 20(in meter, below ground surface). The following derivation is to clarify the computing of $P_{L(i)}$; definitions of the symbols are shown in Figure 7. For the first (computation) layer:

$$P_{L(1)} = \int_0^{\frac{(z_1+z_2)}{2}} F(z) \cdot W(z) \cdot dz$$

$$= \int_0^{\frac{(z_1+z_2)}{2}} \left(1 - F_{L(1)}\right) \cdot (10 - 0.5 \cdot z) \cdot dz$$

$$= \left(1 - F_{L(1)}\right) \cdot \left[10 \cdot \left(\frac{z_1 + z_2}{2}\right) - \frac{1}{4} \cdot \left(\frac{z_1 + z_2}{2}\right)^2\right] \qquad (3)$$

Same as the above, for the second (computation) layer:

$$P_{L(2)} = \int_{\frac{(z_1+z_2)}{2}}^{\frac{(z_2+z_3)}{2}} F(z) \cdot W(z) \cdot dz$$

$$= \left(1 - F_{L(2)}\right) \cdot \left[10 \cdot \left(\frac{z_2 + z_3}{2}\right) - \frac{1}{4} \cdot \left(\frac{z_2 + z_3}{2}\right)^2\right] -$$

$$\left(1 - F_{L(2)}\right) \cdot \left[10 \cdot \left(\frac{z_2 + z_1}{2}\right) - \frac{1}{4} \cdot \left(\frac{z_2 + z_1}{2}\right)^2\right] \qquad (4)$$

Hence, for Layer i, $P_{L(i)}$ becomes:

$$P_{L(i)} = \int_{\frac{(z_{i-1}+z_i)}{2}}^{\frac{(z_i+z_{i+1})}{2}} F(z) \cdot W(z) \cdot dz$$

$$= \left(1 - F_{L(i)}\right) \cdot \left[10 \cdot \left(\frac{z_i + z_{i+1}}{2}\right) - \frac{1}{4} \cdot \left(\frac{z_i + z_{i+1}}{2}\right)^2\right] -$$

$$\left(1 - F_{L(i)}\right) \cdot \left[10 \cdot \left(\frac{z_i + z_{i-1}}{2}\right) - \frac{1}{4} \cdot \left(\frac{z_i + z_{i-1}}{2}\right)^2\right] \qquad (5)$$

4.3 *Rules for the Threshold Value Determination*

To use a similar approach as Iwasaki (1982) for determination of threshold values, the study herein adopted the following procedures and the results are shown in Figure 8 to Figure 13.
- Rule1: Lower threshold value (i.e., 5 by Iwasaki), corresponding to accumulative percentage P_L = 70% from non-liquefaction cases
- Rule 2: Lower threshold value (i.e., 5 by Iwasaki), corresponding to accumulative percentage P_L = 20% from liquefaction cases
- Rule 3: Upper threshold value (i.e., 15 by Iwasaki), corresponding to accumulative percentage P_L = 90% from non-liquefaction cases

- Rule 4: Upper threshold value (i.e., 15 by Iwasaki), corresponding to accumulative percentage $P_L = 50\%$ from liquefaction cases

4.4 Results of Computation

The previous computation was made in 2002 (Kuo et al. 2002, 2003) and indicated in Table 3 and Figures 8 - 10. According to these data the simulation was to be thought to be in vain. Some of the lower threshold values could not be obtained. In order to improve the accuracy the sampling interval of P_L was further reduced as noted in Round 2003#2, where subdivision between $P_L = 0$ to 2.5 was made. The result was illustrated in Figures 11 to 13, and Table 3.

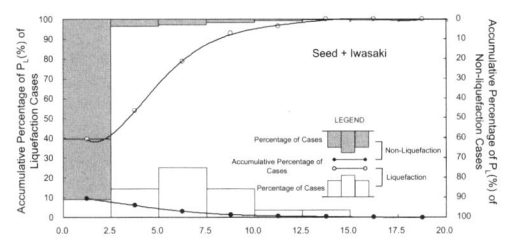

Figure 8. Result of Computation Round 2003#1 by Seed (1971, 2001)+ Iwasaki (1982) Method

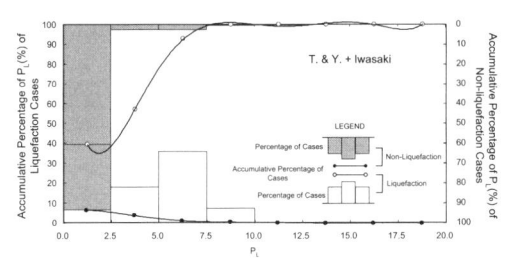

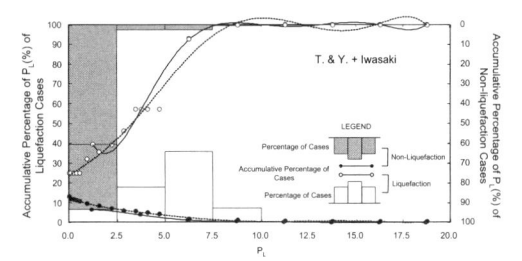

Figure 9. Result of Computation Round 2003#1 by Tokimatsu & Yoshimi (1983) + Iwasaki (1982) Method

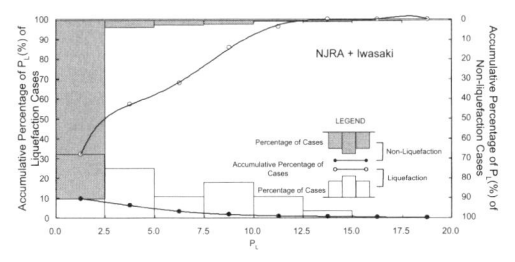

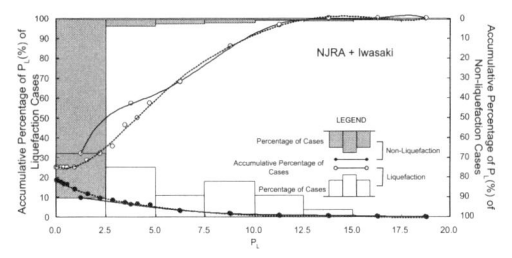

Figure 10. Result of Computation Round 2003#1 by NJRA (1996) + Iwasaki (1982) Method

Even through the threshold values could be found in the second round, however, all of them were zeros. An unreasonable situation showing the above rules may not be effective in defining the threshold values

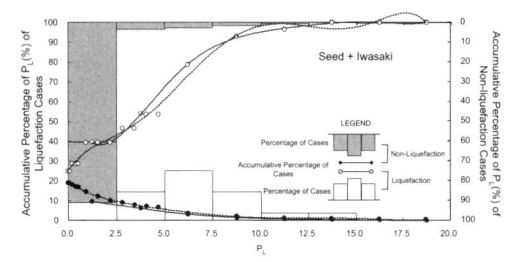

Figure 11. Result of Computation Round 2003#2 by Seed (1971, 2001) + Iwasaki (1982) Method

Figure 12. Result of Computation Round 2003#2 by Tokimatsu & Yoshimi (1983) + Iwasaki (1982) Method

Figure 13. Result of Computation Round 2003#2 by NJRA (1996) + Iwasaki (1982) Method

Table 3. Results of P_L Threshold Estimation

Round	Calibration Rules	Seed	T. & Y.	NJRA
In 2002	Rule 1	1.4	1.6	3.2
	Rule 2	0.5	0.4	0.1
	Average Value	0.95	1.0	1.7
	Rule 3	2.6	2.6	3.2
	Rule 4	6.6	4.6	7.8
	Average Value	4.6	3.6	5.5
In 2003#1	Rule 1	?	?	?
	Rule 2	?	?	?
	Average Value	?	?	?
	Rule 3	1.3	?	1.2
	Rule 4	2.9	3.3	2.5
	Average Value	2.1	?	1.9
In 2003#2	Rule 1	0	0	0
	Rule 2	0	0	0
	Average Value	0	0	0
	Rule 3	2.2	0.7	2.4
	Rule 4	2.8	3.2	4.1
	Average Value	2.5	2.0	3.3

for P_L. By the way, the fitting curves could be changed due to the subdivisions in scale of P_L.

4.5 Remarks

Previous study by the Authors (2002) mentioned the threshold values might be needed to reduce to 0.9-1.7 and 3.4-4.8, respectively, as compared to 5 and 15 suggested by Iwasaki et al. (1983). At that time the computation was based on relatively small database. However, the simulations herein indicated the increase in number of cases would not be helpful.

5 CONCLUSIONS

From the above discussions, some conclusions and suggestions can be drawn as follows. Iwasaki's Liquefaction Potential Index is commonly adopted by engineers, however, the method is based on database of Japan; the use of the method in other areas may require calibration. The calibration of the threshold values by 1999 Chi-chi earthquake seemed not to work well. Further study may be necessary. In comparison with the continuous information provided by CPT logs, the SPT data seems not to be suitably well for the integration for P_L. It would be suggested to calibrate the threshold values of P_L using CPT data.

ACKNOWLEDGEMENTS

Financial support from National Council for Research in Earthquake Engineering of Taiwan (NSC89-2711-3-319-200-28, NSC90-2711-3-319-200-12 and NSC91-2711-3-319-200-08) is very appreciated. Borehole information was provided by government departments, universities, schools, and private sectors, etc. Their kindly helps are very much appreciated. The Authors would like to thank T.M. Lin, Y.H. Chang, S.H. Chang, Y.Y. Chen, J.S. Hsu, G.D. Chu, and Y.A. Lin for assistances in setting up the database.

REFERENCES

Architecture and Building Research Institute, 1998, Building Codes for Fundamental Structures, *Newest Taiwanese Building Codes*, Taipei: 10-12 - 10-14

Chang, M., Kuo, C.P., Hsu, R.E., Shau, S.H. and Wu, T.F. 2003. Assessment on Liquefaction Potential of the Mid-West Alluvial Plain During the 1999 Chi-Chi Earthquake, Taiwan. *12th Panamerican Conference on Soil Mechanics and Geotechnical Engineering*. 237-244

Gea, Y.P., Lu, S.D., and Wang, Y.S. 1996. Hydrogeological Framework of the South Region of the Choshui River Fan. *Proc. Groundwater and Hydrogeology of Choshui River Fan*. 113-125. (in Chinese)

Hsu, R.E. 2002. *Evaluation on Liquefaction Potential of Alluvial Deposits of Chou-swei River Fan, Taiwan*. Master thesis. Department of Construction Engineering, National Yunlin University of Science and Technology. (in Chinese)

Iwasaki, T., Arakawa T., and Tokida, K. 1982. Simplified Procedures for Assessing Soil Liquefaction During Earthquakes. *Proceedings of Soil Dynamics & Earthquake Engineering Conference, Southampton*. 925-939.

Iwasaki, T., Tatsuoka, F., Tokida, K., and Yasuda, S.. 1978. A Practical Method for Assessing Soil Liquefaction Potential. *Proceedings of 2nd. International Conference on Microzonation for Safer Construction Research and Application*. 885-896.

Japanese Roads Association. 1996. 日本道路協會.道路橋示方書, 同解說, V 耐震設計編. (in Japanese characters)

Kuo, C.P. 2001. A Study on Liquefaction Potential of Alluvial Deposits at Yunlin County, Taiwan. Master thesis. Department of Construction Engineering, National Yunlin University of Science and Technology. (in Chinese)

Kuo, C.P., Chang, M., Hsu, R.E., Shau, S.H. and Lin T.M. 2002. A Study of Legitimacy of Using "Liquefaction Potential Index" in Taiwan. *Proceedings of Workshop on Evaluation Method for Liquefaction Potential and Map Illustration*. NCREE. (in Chinese)

Kuo, C.P., Chang, M., Hsu, R.E. and Shau, S.H. 2003. A Study on Localization of Iwasaki's Liquefaction Potential Index (P_L) in Taiwan. *Proceedings of the 10th Conference on Current Researches in Geotechnical Engineering in Taiwan*. 717-720. (in Chinese)

National Center for Research on Earthquake Engineering (NCREE), Taiwan. 2000. *Investigation Report on Geotechnical Damages during the 921 Chi-Chi Earthquake*. 111p. (in Chinese)

Seed, H.B. and Idriss, I.M. 1971. Simplified Procedure for Evaluating Soil Liquefaction Potential. *Journal of the Soil Mechanics and Foundations Division, ASCE* Vol. 97 (No. SM9): 1249-1273.

Seed, H.B., Tokimatsu, K., Harder, L.F. and Chung, R.M., 1985. Influence of SPT Procedures in Soil Liquefaction Resistance Evaluations. *Journal of Geotechnical Engineering, ASCE* Vol.111 (No.12):1425-1445.

Shau, S.H. 2003. *Preliminary Study on Liquefaction Resistance of Chou-swei River Sand by Low-Pressured Grouting*. Master thesis. Department of Construction Engineering, National Yunlin University of Science and Technology. (in Chinese)

Tatsuoka, F., Iwasaki, T., Tokida, K.I., Yasuda, S., Hirose, M., Imai, T. and Kon-no, M. 1980. Standard Penetration Tests and Soil Liquefaction Potential Evaluation. *Soils and Foundations, JSSMFE* Vol.20 (No.4): 95-111

Tokimatsu, K. and Yoshimi, Y. 1983. Empirical Correlation of Soil Liquefaction Based on SPT-N Value and Fines Content, *Soils and Foundations, JSSMFE* Vol.23 (No.4):56-74

Youd, T.L. and I.M. Idriss. 2001. Liquefaction Resistance of Soils: Summary Report from the 1996 NCEER and 1998 NCEER/NSF Workshops on Evaluation of Liquefaction Resistance of Soils. *Journal of Geotechnical and Geoenvironmental Engineering* Vol. 127(No.4): 297-313.

T-bar testing in Irish soils

Michael Long
Department of Civil Engineering, University College Dublin (UCD), Ireland

Gisli T. Gudjonsson
Statens Vegvesen, Bergen, Norway (Formerly research student UCD / NTNU Trondheim, Norway)

Keywords: in-situ testing, soft soils, CPTU, undrained strength, sensitivity

ABSTRACT: This paper investigates the application of the T-bar penetrometer for profiling the undrained shear strength and sensitivity of very soft soils at three sites in Ireland and compares the results to parallel CPTU tests. The work is based on comparisons with laboratory tests on high quality samples and field vane tests. It was found that, for the relatively uniform sites, the T-bar proved a very useful profiling tool and the range of calculated bearing capacity factors (N_{T-bar}) were less than for the CPTU (N_{kt}). The T-bar sensitivity values were also very close to those obtained from the field vane. With the possible exception of the CPTU bearing capacity factor $N_{\Delta u}$, both tools had difficulty in predicting the shear strength of the non-uniform organic soils.

1 INTRODUCTION

There are many engineering developments around the world that require accurate determination of the shear strength of soft sediments. Current state of practice, for in situ testing, is based primarily around piezocone testing (CPTU), to give a continuous strength profile, coupled with vane shear tests at discrete intervals. However, the corrections that need to be applied to the raw CPTU data and uncertainties over appropriate cone factors relating net cone resistance to shear strength have led to the pursuit of alternative shaped penetrometers.

Recently at the Centre for Offshore and Foundation Systems (COFS) in Australia (e.g. Stewart and Randolph, 1994 and Randolph et al., 1998, 2000) and at the Norwegian Geotechnical Institute (NGI) in Norway (Lunne, 2001) investigations have been performed using a cylindrical T-bar probe. This allows full flow around the probe (apart from a small region where the probe connects to the driving rods) reducing significantly the need to correct the measured bearing resistance for the ambient overburden stress. Also, theoretically, one of the attractions of the T-bar over the cone is the existence of closely bracketed plasticity solutions for plane strain flow around a cylinder.

Field experience in Australia and in Norway has confirmed the usefulness of the T-bar and has suggested that the range of bearing capacity factors determined (by correlation with good quality lab tests or in situ vane tests) is narrower than for the cone.

The objectives of this paper are to extend the Australian and Norwegian experience by examining the applicability of the T-bar to Irish soft soils. In particular, as the Norwegian and Australian work had been confined to clays with relatively low plasticity, it had been hoped to extend the database into high plasticity materials. Some work with the T-bar has been carried out on three Irish sites. All three sites are well characterised by a range of in situ test techniques together with high quality sampling (including block sampling).

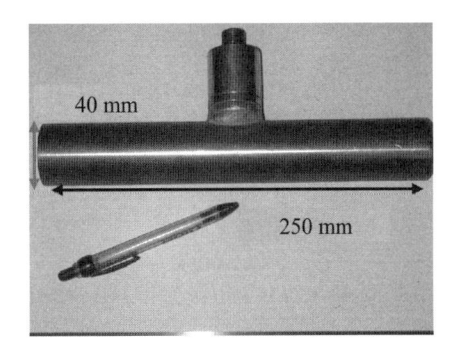

Figure 1. T-bar penetrometer

2 BACKGROUND TO T-BAR

Although it is generally accepted that the most useful tool for profiling shear strength in soft soils is the piezocone or CPTU, there are two sources of potential inaccuracy associated with the equipment itself. Firstly the measured cone resistance is very small and thus the accuracy of the readings is poor. In addition undrained shear strength (s_u) is normally determined via empirical relationships such as:

$$s_u = \frac{q_{net}}{N_{kt}} = \frac{(q_t - \sigma_{v0})}{N_{kt}} \quad (1)$$

$$q_t = q_c + (1-\alpha)u_2 \quad (2)$$

where q_t = corrected cone resistance; q_c = measured cone resistance; σ_{v0} = total vertical stress; α = net area ratio (0.8 in this case); u_2 = measured pore pressure behind the cone tip and N_{kt} = bearing capacity factor.

Thus, for the CPTU, the corrections to the measured resistance can be appreciable. Development of the T-bar and other similar devices has been driven by attempts to overcome these inaccuracies. Typically the bearing area of the T-bar is 10 times that of the cone, see Figure 1, thus improving the accuracy of the readings. In addition, the tip resistance of the T-bar is corrected for the unequal pore pressure effects using the expression:

$$q_{net} = q_c - [\sigma_{v0} - u_0(1-\alpha)]\frac{A_s}{A_p} \quad (3)$$

where u_0 is the ambient pore pressure and A_s/A_p is the ratio of the area of the connection of the shaft to the penetrometer and the area of the penetrometer in the plane normal to the shaft.

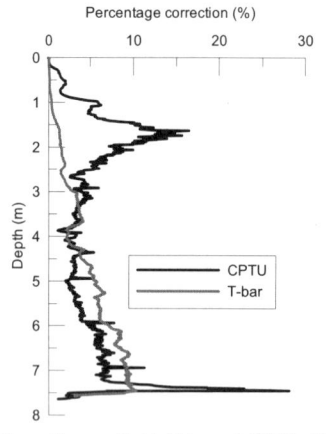

Figure 2. Corrections applied to T-bar and CPTU – Portumna

As this ratio is of the order of 0.1, the correction is far less significant than for the CPTU, where the ratio is unity. This is illustrated in Figure 2, which shows the percentage corrections between q_c and q_{net} for two tests at the Portumna site. As can be seen the adjustment to the cone data can be very significant, i.e. up to 17%, compared to an average correction for the T-bar of about 4%.

3 TEST PROCEDURE AND DATA ANALYSIS

3.1 CPTU

The piezocone CPTU used was of standard dimensions with a diameter of 35.7 mm and a projected area of 10 cm². For the cone penetration with pore pressure measurements the pore pressure transducer was located behind the cone (u_2 position). A standard penetration rate of 2 cm / sec. was used and readings were taken at 0.01 m. CPTU data was corrected and interpreted according to the recommendations of Lunne et al. (1997).

Undrained shear strength was determined from the net cone resistance (q_{net}) as described in Equation 1 and also from the effective cone resistance (q_e) and the excess pore pressure generated (Δu) as follows:

$$s_u = \frac{q_e}{N_{ke}} = \frac{q_t - u_2}{N_{kt}} \quad (4)$$

$$s_u = \frac{\Delta u}{N_{\Delta u}} = \frac{u_2 - u_0}{N_{\Delta u}} \quad (5)$$

3.2 T-bar – general and undrained shear strength

T-bar penetration tests were performed by removing the cone tip just below the load cell, as illustrated on Figure 1. The diameter of the T-bar was 40 mm and the length was 250 mm. It was roughened by light sandblasting. The actual T-bar used in these studies was on loan to UCD from NGI, who in turn had obtained it from COFS. Its projected area is 100 cm² (10 times that of the cone). Penetration rate was the same as for the piezocone at 2 cm/sec. and readings were taken with intervals of 0.01 m.

The following formulae were used to obtain s_u from the T-bar data:

$$s_u = \frac{q_{c(T-bar)}}{N_{T-bar}} \quad (6)$$

$$s_u = \frac{q_{net(T-bar)}}{N_{kt(T-bar)}} \quad (7)$$

where N_{T-bar} and $N_{kt(T-bar)}$ are bearing capacity factors.

3.3 Remoulded shear strength from T-bar

Remoulded shear strength (s_{ur}) can be interpreted from T-bar penetration when cyclic penetration and extraction is preformed. These tests comprised displacement cycles of ±0.5m about the relevant depth. Typically six cycles are performed The degradation factor for subsequent cycles (counting each reversal as a half cycle) is calculated by dividing the mean (absolute) resistance during the 1.0 m stroke by the initial penetration resistance.

4 SITES

4.1 General

The three sites involved in this study have been used by researchers at UCD for several years and are generally well characterized. They are located at Athlone and Portumna towards the centre of Ireland and at Loughmore, near Limerick in the Mid-West. Full details of the Athlone site are given by Long (2003) and by Long and O'Riordan (2001). Conaty (2002) and Gudjonsson (2003) described the sites at Portumna and Loughmore respectively. The soft soils at Athlone and Portumna are glacial lake deposits, which were laid down in a large pro-glacial lake, which was centred on the middle of Ireland, during the retreat of the glaciers at the end of the last ice age some 10,000 to 20,000 years B.P. At Loughmore there was a similar depositional environment, except that the lake involved was a local feature only and thus was much smaller. As the climate became warmer and vegetation growth was supported on the lakebed, the depositional environment changed and the upper soils have increasing organic content. At the three sites the ground surface is underlain by a layer of peat, typically 1 m thick.

For each site, basic parameters are given on Figures 3a to 3c and strength related parameters are given on Figures 4a to 4c.

4.2 Athlone

At Athlone, two distinct strata were formed, as can be seen on Figure 3a. The lower soils are very soft brown horizontally laminated (varved) clays and silts with clearly visible partings typically 1 mm to 2 mm thick. These deposits are referred to as the brown laminated clay. They have average moisture content, bulk density and clay content of about 40%, 1.9 Mg/m^3 and 35% respectively. The material has an average plasticity index (I_p) of about 18%. Though there is some scatter in the data, there is no apparent trend in the parameters with depth.

Laboratory strength data from high quality Sherbrooke block samples fall on or close to the 0.3σ'_{v0} line. However as detailed by Long and O'Riordan (2001) various vane tests and the cone pressuremeter tend to underestimate the strength, probably because of device insertion effects. Field vane sensitivity is about 2.5.

As the climate became warmer, the depositional environment changed and the upper soils show only some signs of varving and have an increasing organic content. The material deposited under these conditions is homogenous grey organic clay and silt. At about 4 m depth, its moisture content is about 55% and then increases to 110% before decreasing with depth to about 70%. Average bulk density and clay content is of the order of 1.6 Mg/m^3 and 25% respectively (Figure 3a). Average I_p is about 40%, being higher than for the brown clay as a result of the increased organic content.

Strength data from both vane tests and laboratory testing shows considerable scatter. However values are generally greater than 0.3σ'_{v0} suggesting that some overconsolidation of the material has taken place (Figure 4a). Average field vane sensitivity is about 8.5.

4.3 Portumna

As can be seen from Figure 3b, the Portumna clays are perhaps the most uniform of the materials under consideration. Moisture content falls from about 50% in the upper clay layer to 40% in the lower layer. The corresponding bulk density values are 1.7 Mg/m^3 and 1.85 Mg/m^3. Clay content is about 40% and I_p is 22%.

Laboratory strength data from reasonable quality piston samples, shown on Figure 4b, suggest the material is slightly overconsolidated with values falling in the range 0.4σ'_{v0} to 0.5σ'_{v0}. Again vane strength values are lower, perhaps due to disturbance effects. Limited data on field vane sensitivity suggest that it is less than 5.

4.4 Loughmore

The upper soils layers at Loughmore are highly organic with widely varying properties (Figure 3c). However the lower brown clay is relatively uniform with a moisture content of 40% to 50%, a bulk density of 1.7 Mg/m^3, a clay content of about 35% and an I_p of about 22% on average. Laboratory strength values, from samples taken with a conventional ELE piston sampler with a sharpened cutting edge angle are of the order of 0.3σ'_{v0}, with vane values being lower as before (Figure 4c). Field vane sensitivity is in the range 2 to 4.

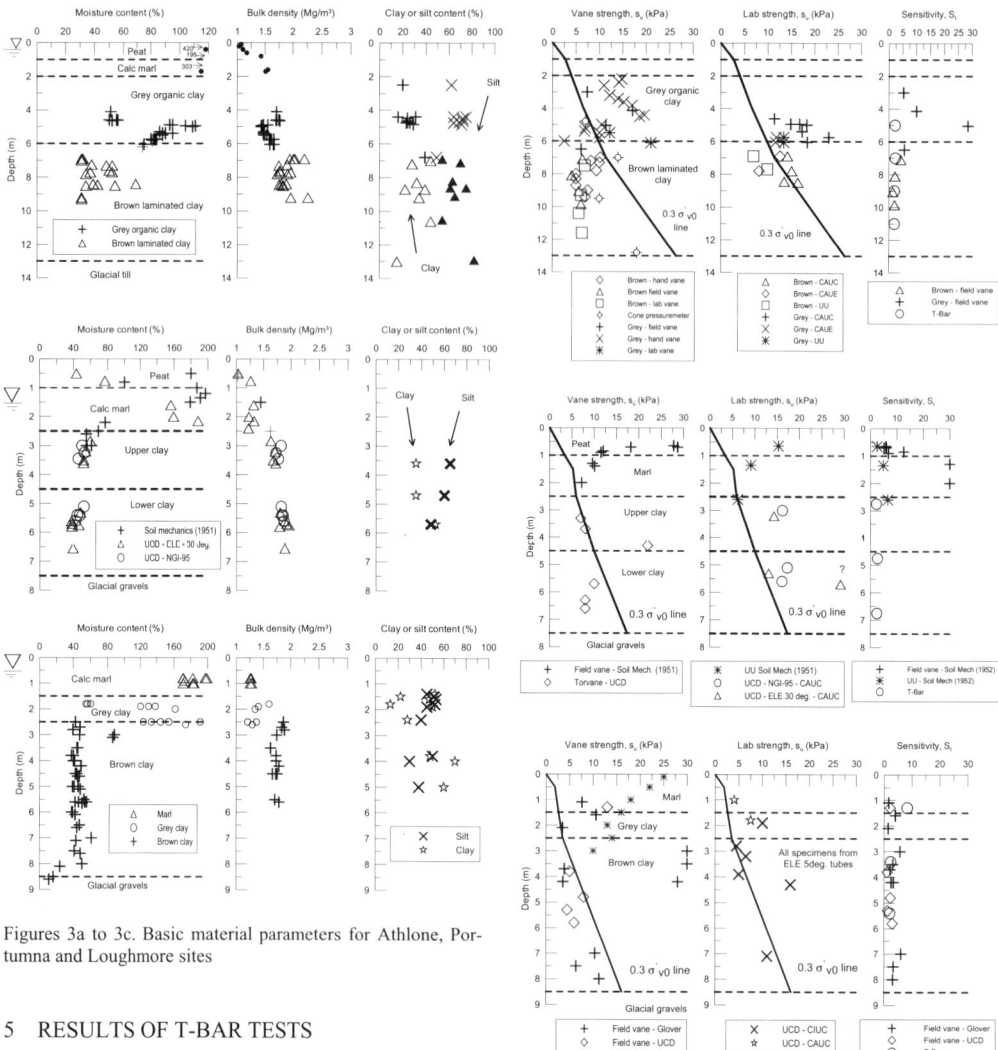

Figures 3a to 3c. Basic material parameters for Athlone, Portumna and Loughmore sites

5 RESULTS OF T-BAR TESTS

5.1 Athlone

The results of the T-bar tests are shown together with CPTU data for the three sites on Figures 5a to 5c respectively. All of the tests were located within about 0.5 m of one another. There is a reasonable agreement between the two CPTU tests and the two T-bar tests. All tests were carried out contemporaneously by Lankelma Ltd.

CPTU q_{net} values are very low and increase from about 0.15 MPa to 0.35 MPa in the grey organic clay, then fall back to about 0.2 MPa in the brown laminated clay followed by a gradual increase with depth, particularly below 10.5 m to about 0.6 MPa. Excess pore pressure (u_2) values for both tests are

Figures 4a to 4c. Strength data for Athlone, Portumna and Loughmore sites

practically identical and show a relatively uniform increase with depth, until about 10.5 m, below which there is a greater rate of increase.

T-bar net resistance values are also very low. In the grey organic clay they show a slight increase with depth from about 0.1 MPa to 0.25 MPa and these values are similar to those measured using the CPTU. However there appears to be practically no increase with depth in the brown laminated clay with values staying almost constant at about 0.2 MPa. These are approximately half the CPTU q_{net} values.

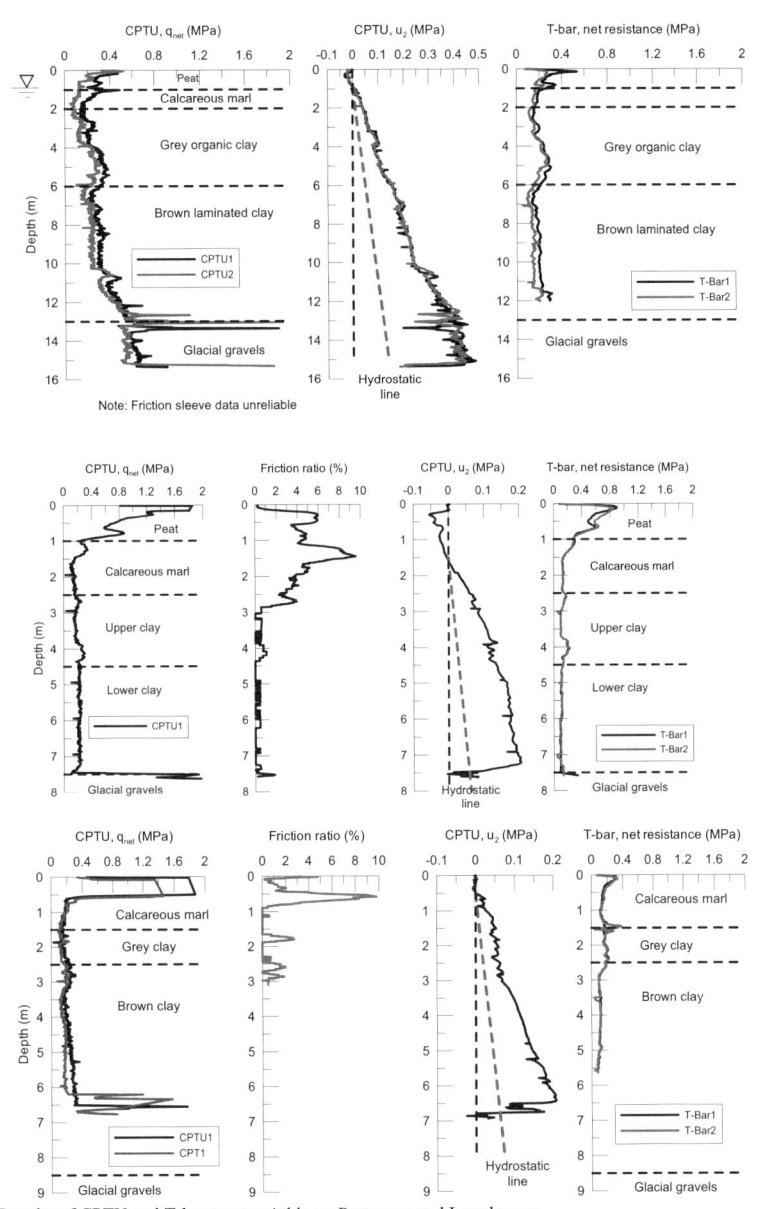

Figure 5a to c. Results of CPTU and T-bar tests at Athlone, Portumna and Loughmore

5.2 Portumna

At Portumna a single CPTU tests was carried out together with a pair of T-bar tests all of which were located very close to one another, see Figure 5b. The T-bar profiles are virtually identical. CPTU q_{net} values show almost no increase with depth, except perhaps in the upper clay layer and remain constant at about 0.25 MPa. Friction ratio falls close to zero confirming the materials are very soft. Excess pore pressure values show an increase with depth with the rate of increase falling off in the lower clay layer. Surprisingly the T-bar net resistance profiles show a slight decrease with depth from about 0.15 MPa in the upper clay to about 0.1 MPa in the lower layer. These values are again approximately 50% of those measured using the CPTU.

5.3 Loughmore

At Loughmore, a single CPTU test together with a CPT test was performed and they show reasonable agreement. Two T-bar tests were performed immediately adjacent to the cone tests. The pattern of behaviour is almost identical to that at Portumna with the T-bar net resistance values again showing a decrease with depth in this case to values less than 0.1 MPa. These values are again about 50% of those measured by the CPTU.

5.4 Upper organic layers – all sites

Similar to the finding made for the Athlone grey organic clay, it is interesting to notice that generally in the upper organic layers at all three sites, the net resistance from the cone penetration tests and the T-bar tests are in the same range.

6 UNDRAINED SHEAR STRENGTH

6.1 General

Values of N_{T-bar}, N_{kt}, N_{ke} and $N_{\Delta u}$ for all three sites are shown on Figures 6a to 6c respectively. The bearing capacity factors were obtained by correlation with results of CAUC triaxial tests on the best available samples as shown on the figures. These comprised Sherbrooke block samples at Athlone and either NGI 95 mm or ELE 100 mm fixed piston samples at Portumna and Loughmore. Therefore at the latter two sites sample disturbance effects would be expected to result in slightly higher values of the bearing capacity factors.

6.2 Undrained shear strength from T-bar results

Computed N_{T-bar} values have been compared with a theoretical value of 10.5 as recommended, for general use, by Randolph and Houlsby (1984). This analytical value was found to be dependant on the surface roughness of the cylinder described by its adhesion factor, α. The upper and lower bounds of the plasticity solution coincide at approximately 12 for a fully rough bar ($\alpha = 1$) and diverge slightly at lower values of α, with a minimum of about 9. It is unlikely that adhesion factors approaching 0 or 1 are achievable, despite the fact that the T-bar was sandblasted, therefore some intermediate bearing capacity factor must be chosen. This value of 10.5 was also adopted by Stewart and Randolph (1994). However it is likely that T-bar penetration resistance also depends on strength anisotropy, strain softening, clay remoulding and rate effects, all factors that are not normally included in the theoretical solutions. It is therefore necessary to develop empirical correlations for the T-bar, similar to the CPTU.

N_{T-bar} values for the Athlone brown laminated clay are very close, on average, to 10.5. Values for the Portumna clays and the Loughmore clay, below 4 m, are generally uniform but are lower than 10.5, being typically 7 to 8. N_{T-bar} values for the Athlone grey organic clay and the Loughmore materials above 4 m vary widely probably because of the varying organic content of these soils. In fact the Athlone grey organic clay N_{T-bar} values reduce uniformly with depth, reflecting the general reduction in organic content of the material with depth.

6.3 Undrained strength from CPTU results

In this case the bearing capacity factor values determined have been compared to the relatively well know limits given by Aas et al (1986), Senneset et al. (1982) and Lunne et al. (1997). For the Athlone grey organic clay N_{kt} and N_{ke} vary widely, again probably reflecting the variable organic content of the material. Only $N_{\Delta u}$ values, which are independent of cone resistance, seem to be reasonable. For the Athlone brown laminated clay all the values are relatively uniform with depth, albeit being on the high side of the published limits. For the Loughmore clays, the values are extremely variable, particularly above 4 m, the exception perhaps being $N_{\Delta u}$. In contrast the Portumna data are both uniform with depth and consistent with the published typical values.

6.4 Range of bearing capacity factors

One of the objectives of this work was to compare the range of bearing capacity factors obtained form the CPTU and T-bar to examine if, as found by the Norwegian and Australian research, that the range of N_{T-bar} was less than N_{kt}. For this exercise data for the deeper more uniform materials only was considered (i.e. Athlone brown laminated clay, Portumna clays and Loughmore clay below 4 m depth). On Figure 7, all N_{T-bar} and N_{kt} are plotted against I_p. It can be seen clearly that the range of N_{T-bar} is lower than that for N_{kt}. Even ignoring the high Loughmore N_{kt} value, the N_{T-bar} range is about 7 compared to 9 for the N_{kt} values.

It can also be seen form Figure 7 that there appears to be no clear relationship between N_{kt} and N_{T-bar} with I_p. A similar plot against pore pressure parameter, B_q, also shows no discernible trend. This finding is likely to be due to the low range of I_p values involved. Unfortunately the data for the higher plasticity (i.e. higher organic content) material had to be eliminated due to the high scatter.

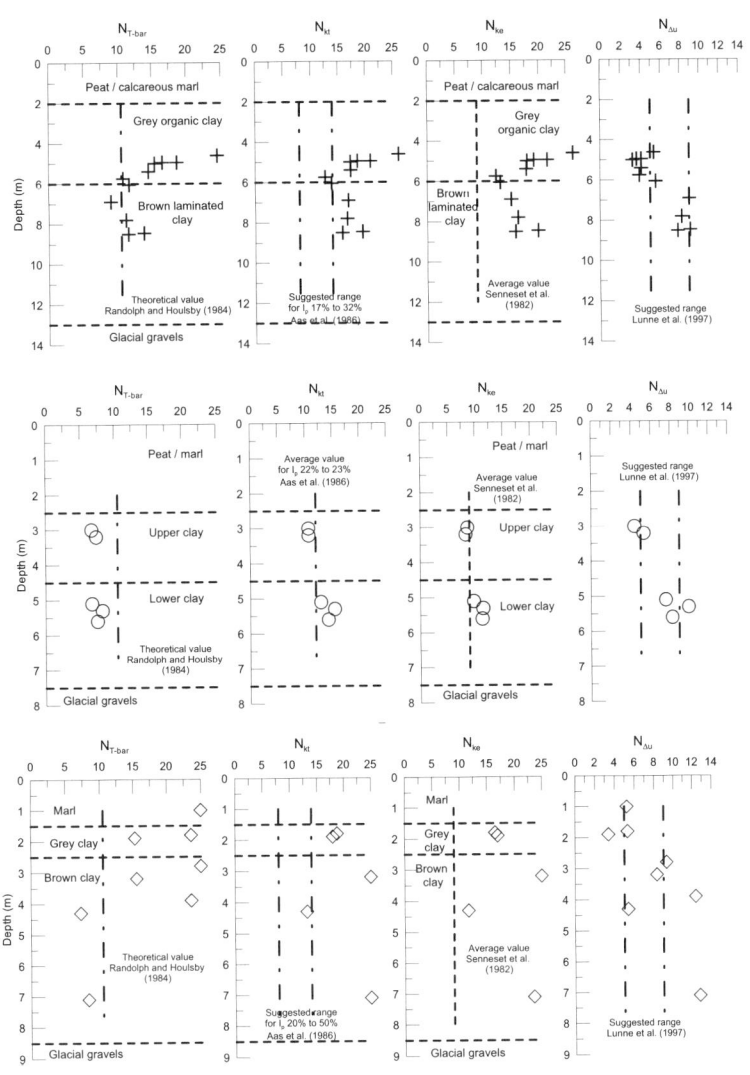

Figures 6a to 6c. Bearing capacity factors for Athlone, Portumna and Loughmore sites

7 REMOULDED SHEAR STRENGTH AND SENSITIVITY

Remoulded shear strength and sensitivity was determined from the T-bar data as described in Section 3.3. A typical cyclic test result for the Portumna site at 6.75 m is shown on Figure 8. The resistance during the initial penetration was about 0.1 MPa, while that during the first extraction cycle (following full penetration of the T-bar) was around –0.055 MPa, giving a degradation factor of 0.055/0.1 = 0.55.

The degradation factor for the subsequent cycles (counting each reversal as a half cycle) is calculated by dividing the mean (absolute) resistance during 1 m stroke by the initial penetration resistance of 0.1 MPa. Generally, each half cycle shows further degradation, although there is slight tendency in the test shown for the extraction resistance to be lower than the subsequent penetration resistance, giving a sawtooth effect in the degradation curve, illustrated in Figure 8a. The cyclic resistance appears to stabilise after a 5 to 6 cycles, to about 40% of the initial penetration resistance, see Figure 8b. Similar results were

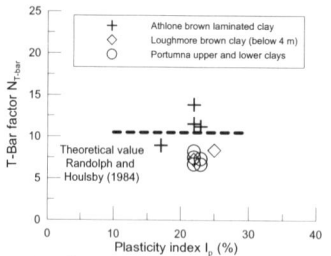

Figure 7. N_{kt} and $N_{T\text{-bar}}$ versus I_p

found for the other tests at Portumna and those at Athlone and Loughmore, suggesting that 6 cycles of penetration is adequate. Sensitivity values determined from the T-bar are compared to those from field vane on Figures 4a to 4c. It can be seen that the agreement is very good.

8 CONCLUSIONS

1. Results of pairs of T-bar tests carried out on the same site show a high degree of repeatability.
2. For the clay soils T-bar net resistance is very small and is approximately 50% that of the CPTU. For the organic and silty soils the net resistance from the two devices is of the same order of magnitude.
3. T-bar penetration resistance shows no increase with depth in contrast to the cone results.
4. For the uniform materials the bearing capacity factors are relatively uniform with depth and similar to the typical values suggested by others.
5. For the organic materials the parameters vary widely. Perhaps an exception to this is $N_{\Delta u}$, where the values are more reasonable, perhaps because it is independent of end resistance.
6. The range of $N_{T\text{-bar}}$ values are less than that for N_{kt}.
7. There is no discernible relationship between the bearing capacity values and I_p or B_q, possibly because the range of I_p values are too narrow.
8. Sensitivity determined from the T-bar results is very similar to that obtained from field vane tests.
9. The T-bar appears to be a very useful tool for profiling shear strength of very soft, relatively uniform soils. Further use of the device is encouraged.

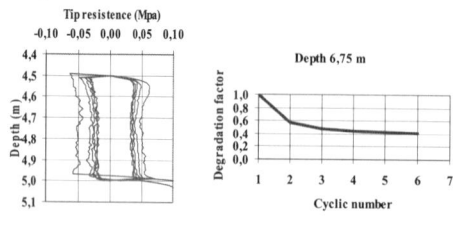

Figure 8. (a) Results of cyclic T-bar test, Portumna 6.75 m and (b) degradation number against number of cycles

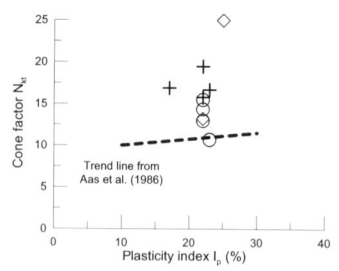

ACKNOWLEDGEMENTS

The authors are grateful for the support of NGI (Tom Lunne and Morten Sjursen) and Lankelma Ltd. (Brian Georgious and Dennis Hudson), who carried out the fieldwork

REFERENCES

Aas, G., Lacasse, S., Lunne, T. and Høeg, K. 1986. Use of in situ tests for foundation design on clay. *Proc. ASCE Conf. In situ '86*, Blacksburg:1 – 30.

Conaty, E. 2002. A study of the stability of the Shannon embankments. *M.Eng.Sc. Thesis*, University College Dublin.

Gudjonsson, G.T. 2003. Evaluation of the embankment failure on the Loughmore Link Road, Ireland. *Masters Thesis, Hovedoppgave 2003*, NTNU Trondheim, Norway.

Long, M. 2003. Characterisation and engineering properties of Athlone laminated clay. *Characterisation and Eng. Props. of Natural Soils*, Tan et al. (eds), Vol. 1: 757 - 790, Swets and Zeitlinger.

Long, M.M. and O'Riordan, N.J. 2001. Field behaviour of very soft clays at the Athlone embankments. *Geotechnique*, 51 (4): 293-309.

Lunne, T., Berre, T. and Strandvik, S. 1997. Sample disturbance effects in soft low plasticity Norwegian clay. *Proc. Sym. on Recent Developments in Soil and Pavement Mechanics, Rio de Janerio, June 1997:* 81 -92. Balkema.

Lunne,T., Robertson, P.K. and Powell, J.M. 1997. *Cone Penetration Testing in Geotechnical Practice*. Blackie Academic and Professional.

Lunne, T. 2001. In situ testing in offshore geo-investigations. *Proc. In situ 2001, Bali*, Lunne and Rahardjo (eds.): 61-81.

Randolph, M.F. and Houlsby, G. 1984. The limiting pressure on a circular pile loaded laterally in cohesive soil. *Geotechnique*, 34 (4): 613 – 623.

Randolph, M.F., Hefer, P.A., Geise, J.M. and Watson, P.G. 1998. Improved seabed strength profiling using T-bar penetrometer. *Proc. Int. Conf. Offshore Site Investigation and Foundation Behaviour, SUT, London:* 221 – 236.

Randolph, M.F., Martin, C.M. and Hu, Y. 2000. Limiting resistance of a spherical penetrometer in cohesive material. *Geotechnique*, 50, No. 5: 573 - 582.

Senneset, K., Janbu, N. and Svanø, G. 1982. Strength and deformation parameters from cone penetration tests. *Proc 2nd Eur. Sym. Pen Tes, ISOPT-1, Orlando,* 2: 995

Stewart, D. P. and Randolph, M. F. 1994. T-bar penetration testing in soft clay. *ASCE Journal of Geotechnical Engineering,* 120(12): 2230-2235.

Site characterization and QA/QC of deep dynamic compaction using an instrumented dilatometer

Heather J. Miller
Dept. of Civil Engineering, University of Massachusetts Dartmouth, North Dartmouth, MA, USA

Kevin P. Stetson
Sanborn Head & Associates, Westford, MA, USA

Jean Benoît
Dept. of Civil Engineering, University of New Hampshire, Durham NH, USA

Keywords: dilatometer, instrumented dilatometer, deep dynamic compaction, unload-reload modulus

ABSTRACT: An instrumented dilatometer (IDMT) was one of several in situ testing tools that were used on a major highway relocation project in Carver, Massachusetts (USA). Parts of the new highway span former cranberry bogs. Sheet piling was installed along both sides of the new highway alignment, and organic material was dredged from between the sheet pile walls. The area was then backfilled with sands. Since most of the sand was placed in a fairly loose state under water, liquefaction was a potential problem. Therefore, deep dynamic compaction (DDC) was used to densify the fill. An extensive in situ testing program was instituted to characterize the site conditions prior to densification, and to assess the sufficiency of the DDC after treatment. The results of this study suggest that the IDMT can be used to provide accurate and cost-effective stratigraphic profiles. The IDMT was particularly helpful in identifying pockets of organic soils (i.e., peat) that were not completely removed during the initial dredging operations. In terms of compaction assessment, the modulus values determined from the IDMT appear to be very sensitive indicators of densification effects.

1 INTRODUCTION

The state of Massachusetts Highway Department is in the process of relocating a section of US Route 44 from the existing Route 44 in Carver, MA to US Route 3 in Plymouth, MA. The study described herein was conducted at a section where mechanically stabilized earth (MSE) walls will eventually be constructed through former pond and cranberry bog areas. The native site stratigraphy consists of standing water and/or peat deposits of varying thickness that extend in depth up to a maximum of about 9.8 m. Glacial outwash deposits consisting of loose to dense, coarse to fine sands with lenses of silt, clay and gravel exist beneath the peat.

The construction project started with the installation of steel sheet piling through the pond/bog sections. The sheeting was located about 23.0 to 24.6 m off the proposed highway centerline. After removal of the peat deposits from within the sheet pile walls, granular fill was placed between the sheet piling by pushing the material forward (from the "land side") with a dozer. Fill was place from the dredged mudline (which varied widely in elevation) to approximately Elevation 34.5 m (roughly 1.6 m above the static groundwater table). A typical grain size distribution curve, as well as upper and lower limits of the range of grain size distribution of the fill material is provided in Figure 1. The fill is classified as poorly-graded sand according to the USCS classification system. The mean D_{50} is approximately 0.4 mm.

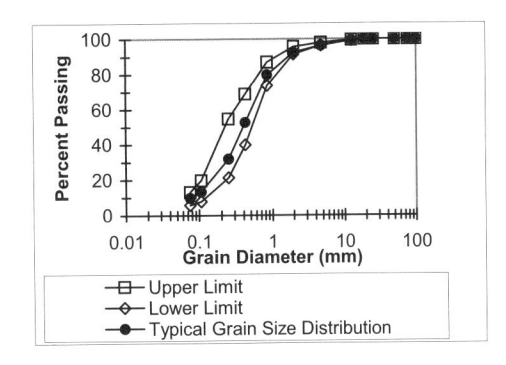

Figure 1. Grain Size Distribution of Hydraulic Fill Material

Since most of the sand was placed in a fairly loose state under water, the potential for liquefaction was a concern. Therefore, deep dynamic compaction (DDC) was used to densify the fill. In situ testing was conducted before and after compaction to obtain baseline soil parameters and to assess the sufficiency of the DDC treatment.

2 DEEP DYNAMIC COMPACTION PROGRAM

Deep Dynamic Compaction is a process whereby soil is densified by repeatedly dropping a massive weight from a crane to impact the ground. Dynamic energy is applied on a grid pattern over the site, typically using multiple passes with offset grid patterns. The DDC process, described in detail by Lukas (1995), is generally very effective in densifying loose granular deposits. The degree of improvement is a function of the applied energy per unit cross-sectional area, which is related to the tamper mass, the drop height, the number of drops and number of passes applied. The depth of improvement, which is a function of tamper mass and drop height, can be estimated using an empirical equation given by Lukas (1995). The maximum improvement resulting from DDC is predicted to occur within a zone from about 1/3 to 1/2 of the depth of improvement calculated using the equation proposed by Lucas (1995).

The DDC for this project was conducted using a tamper that weighed 15 Mg. The tamper was about 0.9 m high, with a hexagonal cross-sectional area of about 0.8 m on each side. Over the majority of the site, two passes of DDC were completed, each using a square grid pattern with a center-to-center spacing of 4.6 m (the grid pattern for the second pass was offset by about 2.3 m in each direction).

A drop height of 9.1 m was used for DDC within the vicinity of the in situ testing described in this paper. The number of drops applied at each drop location is shown in Figure 2. Based upon a 15 Mg tamper and a 9.1 m drop height, the depth of improvement computed using the empirical equation presented by Lucas is 5.9 m. The corresponding maximum improvement would then be predicted to occur within a zone between 1.9 m and 2.9 m below ground surface.

3 IN SITU TESTING PROGRAM

An extensive in situ testing program was carried out to provide baseline conditions of the hydraulic fill and to assess the degree of compaction resulting from the deep dynamic compaction. One of the field methods used in this testing program was a specially designed instrumented dilatometer (IDMT). A standard flat dilatometer was modified at the University of New Hampshire in an effort to better understand the mechanics and soil response during expansion of the dilatometer membrane (Stetson et al., 2003). This IDMT allows the continuous measurement of the complete membrane displacement range during the test, the pore pressure during insertion and testing and, the total pressure applied to the inside of the blade. These modifications were implemented without impacting the original blade design. Others similar probes have been previously designed and built for field testing and for use in calibration chambers (Motan and Gabr, 1985; Motan and Khan, 1988, Campanella and Robertson, 1991; Fretti et al., 1992, Kay and Chiu, 1993).

The testing procedure for the IDMT consists of hydraulically pushing the probe into the ground at a rate of 2 cm/s. Once the blade is at a testing depth, the downthrust is unloaded and the expansion of the membrane is initiated within the next 30 s. The rate of pressurization is designed to reach the A-reading within 30 to 60 s with the rate decreasing when approaching the A-reading to improve resolution at lift-off. For the remainder of the test, the pressure rate is kept nearly constant. To keep test times approximately constant, the average pressure rates during the pre-compaction and post-compaction profiles

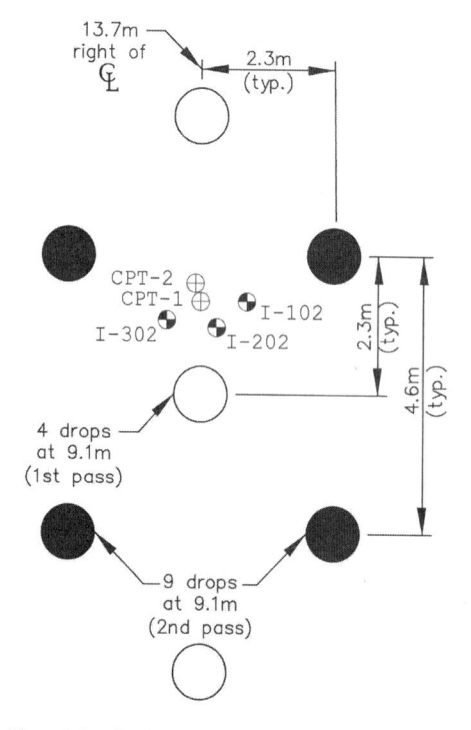

Figure 2. Details of Compaction Near IDMT Soundings

were 350 and 950 kPa/min, respectively. For each test, an unload-reload loop is conducted at a membrane displacement of approximately 0.6 mm. The final unloading rate is similar to the loading rate.

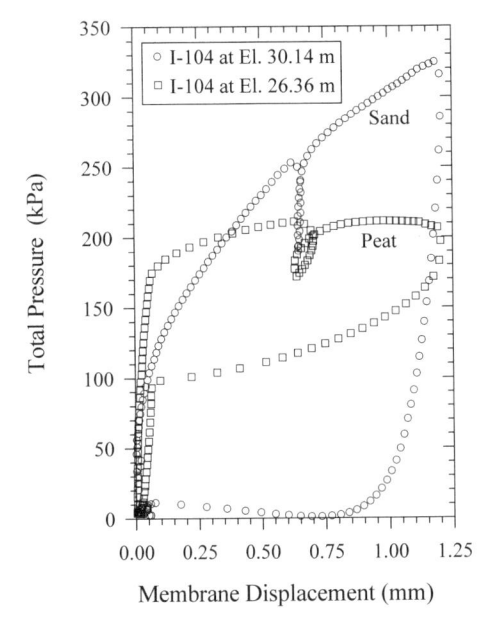

Figure 3. Typical pre-compaction IDMT tests in sand and in organic soil

Figure 3 shows two typical corrected pressure-displacement curves for that profile; test I-104 at El. 30.14 m was carried out within the hydraulic sands fill while test I-104 at El. 26.36 m was carried within a zone of soft organic material left in place prior to backfilling. Figure 4 shows the material index values, I_D, estimated from tests at I-104 and help confirm the type of soil in which the tests shown in Figure 3 were carried out. These IDMT test curves are corrected for membrane stiffness. The pressure-displacement curves are similar in appearance to self-boring pressuremeter curves. As the internal pressure approaches the lateral stress in the ground, the membrane starts lifting off. Because of soil disturbance due to blade penetration, excess pore water pressures are generated in the soft organic zone, leading to a substantial increase in lateral stress. That increase in lateral stress is reflected by the significantly higher lift-off pressure shown in Figure 3 for the test at 26.36 m. The response for the test in the soft zone is relatively flat following the unload-reload and actually shows a decrease in pressure with increasing displacement as the membrane stiffness becomes a significant component of the total pressure.

IDMT test profiles were carried out prior to and following deep dynamic compaction. Table 1 gives details relative to each sounding. Profiles I-102, I-202 and I-302 were carried out in the same vicinity, as shown in Figure 2, while profile I-104 was performed about 90 m away.

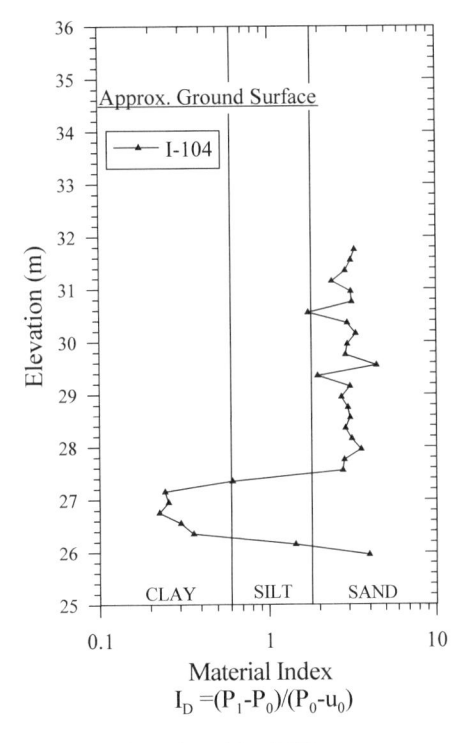

Figure 4. I_D values estimated from I-104

Table 1. Details of IDMT Soundings.

IDMT Profile	Surface Elevation (m)	Station I.D.[1] (ft)	Offset[2] (m)	Compaction Status	Date (M/D/Y)
I-102	34.48	156+00	14.5	Pre-DDC	12/10/02
I-202	34.58	155+98.6	14.0	Post-DDC	7/16/03
I-302	34.63	155+99	13.1	Post-DDC	8/15/03
I-104	34.56	159+04	1.1	Pre-DDC	12/11/02

[1] Station measurements in feet (1 foot ≈ 0.3 m)
[2] Distance to the right of centerline

Figure 5 shows two IDMT tests carried out at approximately the same depth, before and after deep dynamic compaction. The pressure-displacement test curves clearly depict the improvement from the DDC. The improvement is reflected in terms of higher lift-off and thus increased horizontal stress (or K) as well as increase in stiffness as indicated by the significantly larger pressure required to reach 1.1 mm expansion. Increases in lateral stress have also been reported by others using the DMT for QA/QC of deep dynamic compaction (Schmertmann et al., 1986, Marchetti et al., 2001). An enlarged view of the unload-reload loops for each of those two tests is shown in Figure 6. A straight line between the start of reloading and the loop closure is used to calculate the unload-reload modulus. It should be noted that the test curves in Figure 5 show every 5 data points recorded while the unload-reload loops in Figure 6 show every data point.

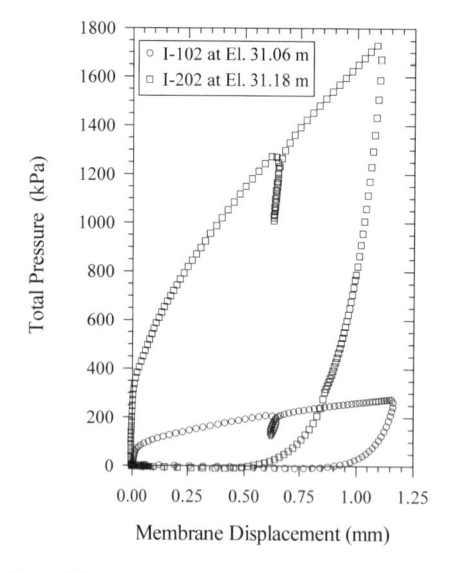

Figure 5. Pre- and post-DDC IDMT test curves

Figure 7 shows unload-reload modulus values for soundings I-102, I-202 and I-302. Those values were calculated according to Fretti et al. (1992). Average strain levels were calculated from the unload-reload loops for each of the four IDMT profiles. Prior to compaction, average strain levels for tests conducted in the hydraulic fill ranged from 3.2 to 3.6 x 10^{-4} mm/mm. After compaction, average strain values in that material ranged from 4.1 to 4.6 x 10^{-4} mm/mm.

Within the peat layer in I-104, the average strain level was 1.2 x 10^{-3} mm/mm.

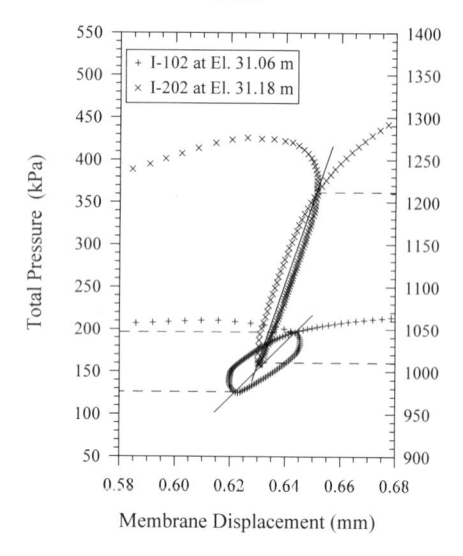

Figure 6. Pre- and post-DDC IDMT unload-reload loops

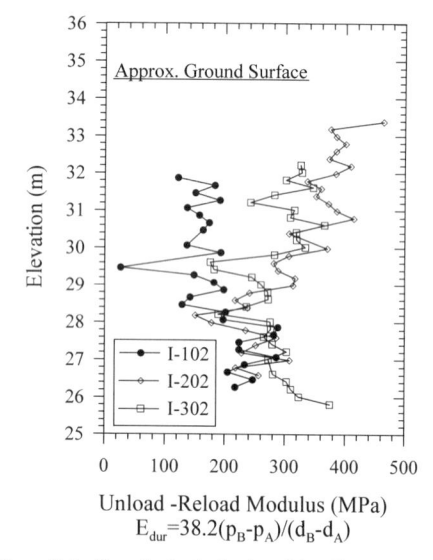

Figure 7. Profiles of unload-reload modulus, E_{dur}
Unload-Reload Modulus (MPa)
$E_{dur} = 38.2(p_B - p_A)/(d_B - d_A)$

As expected, the modulus values are greater following compaction with the most significant increases above Elevation 28. According to Lukas (1995), the maximum improvement should be approximately between Elevations 31.6 and 32.6. The results shown in Figure 7 indicate that the maximum improvement zone may extend somewhat deeper than those elevations. Post-compaction modulus values in the maximum improvement zone are about two times larger than the pre-compaction values. The native material is at an Elevation of about 28, and little improvement in modulus values has occurred below that elevation. The pre- and post-compaction modulus values seem to indicate the presence of a soft organic pocket at Elevation 29.5 m, and also a relatively soft zone at about Elevation 28.2 m.

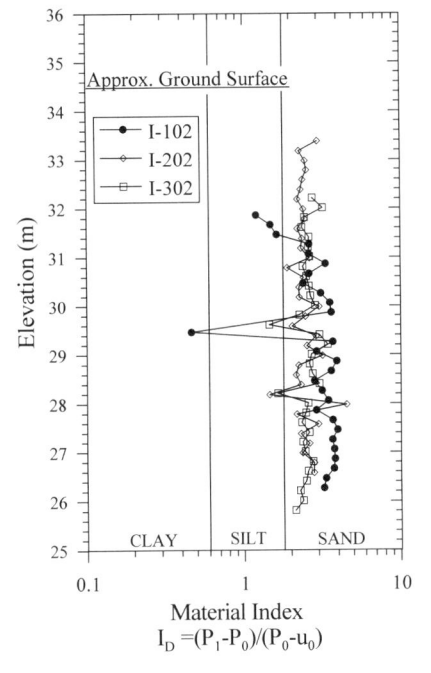

Figure 8. I_D values estimated from IDMT tests near Sta. 156

Figure 8 presents I_D values calculated from using the IDMT data and clearly shows that the soil at Elevation 29.5 m contains a clayey material. It should be noted that because of the impact of the unload-reload loop on the total time necessary to carry out an IDMT test, the IDMT indices might not be directly applicable in conventional Marchetti type correlations. On the average, each test took less than 5 minutes to carry out including one unload-reload loop and full unloading. With the compaction, the zone of organic soils was probably mixed with the surrounding sand and thus is identified as silt in profile I-302. The presence of the soft zone may also explain the lower degree of improvement to the native soil below that elevation. The organic material likely served as a damping layer, preventing full benefit of DDC below that zone.

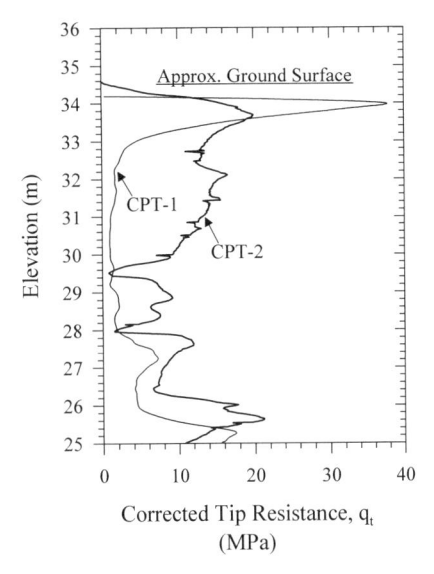

Figure 9. q_t values from CPT tests near Sta. 156

Table 2. Details of CPT Soundings.

CPT Profile	Surface Elevation (m)	Station I.D.[1] (ft)	Offset[2] (m)	Compaction Status	Date (M/D/Y)
1	34.48	156+00	13.7	Pre-DDC	1/15/03
2	34.58	155+01	13.6	Post-DDC	7/16/03

[1] Station measurements in feet (1 foot ≈ 0.3 m)
[2] Distance to the right of centerline

The results of the IDMT tests near Station 156 can also be compared with data from two cone penetrometer tests (CPT-1 and CPT-2) performed in the same general vicinity. Details relative to each sounding are given in Table 2. Plots of corrected tip resistance values for those CPT tests are shown in Figure 9. Interestingly, while the pre-compaction test (CPT-1) does not clearly identify the soft organic pocket at Elevation 29.5 m, the post-compaction CPT test does indicate significantly weaker zones at both Elevation 29.5 m and Elevation 28.2 m. It should be noted that the locations of these soft zones are erratic and of limited extent across the site. Similar to the trend with the IDMT unload-reload modulus values, the CPT tip resistance values following compaction show significant increases above Elevation 28, with only modest increases occurring in the native material below that elevation. The maximum improvement occurs approximately between Elevations 31 and 32.2, which is in close agreement with the maximum zone of improvement predicted by Lukas (1995).

4 CONCLUSIONS

The results of this study suggest that the IDMT is a very useful tool for providing stratigraphic profiles as well as parameters for QA/QC on in situ densification projects. The IDMT pressure-displacement curves are similar in appearance to self-boring pressuremeter curves, and enable a better understanding of the mechanics and soil response during expansion of the dilatometer membrane. During preliminary site investigations, the material index values estimated from the IDMT tests were particularly helpful in identifying pockets of soft organic soils (i.e., peat) that were not completely removed during the initial dredging operations. After compaction, the IDMT pressure-displacement curves and the unload-reload modulus values clearly depict the improvement that resulted from the DDC, with post-compaction modulus values in the maximum improvement zone of about two times larger than the pre-compaction values.

And finally, the IDMT proved to be helpful in understanding some of the factors that govern soil improvement resulting from DDC. IDMT unload-reload modulus values suggest that the maximum zone of improvement occurred approximately between Elevations 30 and 32, which is slightly deeper than the zone estimated using the equation proposed by Lucas (1995). In addition, the IDMT data indicate the presence of a very soft organic pocket at Elevation 29.5 m, and also a relatively soft zone at about Elevation 28.2 m. Those zones likely served as damping layers during the DDC, reducing the amount of energy transferred to the underlying material. The reduced effectiveness of the DDC in the material beneath those soft zones was confirmed by the minimal increases in IDMT unload-reload modulus values that resulted in that material.

ACKNOWLEDGEMENTS

The writers wish to acknowledge the Massachusetts Highway Department for their financial support for this research. Additionally, several MassHighway personnel provided much assistance in conducting the field testing for this project. Mr Peter Connors, Mr. Edward Mahoney, and the entire staff of the MassHighway Route 44 Field Office in Carver, MA are acknowledged in that regard. And finally, the writers appreciate help with field testing, data reduction, and preparation of figures that was provided by undergraduate research assistants Nicholas Yafrate and Tracy Willard.

REFERENCES

Campanella, R.G. and Robertson, P.K., 1991, "Use and Interpretation of a Research Dilatometer," *Canadian Geotechnical Journal*, Vol. 28, pp. 113-126.

Fretti, C., Lo Presti, D.C.F., and Salgado, R., 1992, "Development of the Research Dilatometer: In Situ and Calibration Test Results," *Rivista Italiana di Geotecnia*, Napoli, Italy, XXVI (4), pp. 237-243.

Kay, J.N. and Chiu, C.F., 1993, "A Modified Dilatometer for Small Strain Stiffness Characterization" *Proceedings of the Eleventh Southeast Asian Geotechnical Conference*, Singapore, pp. 125-128.

Lukas, R.G., 1995, *Geotechnical Engineering Circular No.1: Dynamic Compaction*, Report FHWA-SA-95-037, Federal Highway Administration, Washington, D.C.

Marchetti, S., Monaco, P., Totani, G. and Calabrese, M., 2001, "The Flat Dilatometer Test (DMT) in Soil Investigations", a report by the ISSMGE Committee TC16, *Proceedings of the International Conference on In Situ Measurements of Soil Properties*, Bali, Indonesia, May, 41 p.

Motan, S.E. and Gabr, M.A., 1985, "Flat Dilatometer and Lateral Soil Modulus," *Transportation Research Record 1022*, National Academy Press, Washington, DC, pp. 128-135.

Motan, S.E. and Khan, A.Q., 1988, "In-Situ Shear Modulus of Sands by a Flat-Plate Penetrometer: A Laboratory Study," *Geotechnical Testing Journal*, Vol. 11, No. 4, pp. 257-262.

Schmertmann, J., Baker, W., Gupta, R. and Kessler, K., 1986, "CPT/DMT QC of Ground Modification at a Power Plant", *Proceedings of In Situ '86*, Blacksburg, Virginia, June, ASCE Geotechnical Special Publication No. 6, pp. 985-1001.

Stetson, K.P., Benoît, J. and Carter, M.J., 2003, "Design of an Instrumented Flat Plate Dilatometer", *Geotechnical Testing Journal*, Vol. 26, No. 3, pp. 302-309.

Soil-Rock Sounding with MWD – a Modern Technique to Investigate hard Soils and Rocks

B. Möller
FmGeo AB, Malmö, Sweden

U. Bergdahl
Swedish Geotechnical Institute, Linköping, Sweden

K. Elmgren
Environmental Mechanics AB, Alingsås, Sweden

Keywords: soil-rock sounding, drilling equipment, MWD, sounding procedure, drilling parameters

ABSTRACT: This paper describes a method of using modern rock drilling machines provided with a number of transducers and a data aquisition system as a sounding tool in hard soils and rocks where normal penetrometers not can be used. The equipment as well as the sounding procedure are described. In addition a number of case records are presented to illustrate its use. The paper concludes that the improved drilling technique with registration of a number of drilling parameters can be used to find the transition between soil and rock, the content and size of boulders in the soils and to find different zones in rock with crushed material and various degree of induration.

1 INTRODUCTION

Modern structures, like tall buildings, long bridges and tunnels, often require also hard soils and rocks to be investigated with respect to their constitution, strength, deformation characteristics and water conductivity. Such ground in Sweden often contain moraine/till, a very dense soil deposited on the bedrock, coarse-grained sediments from the glacial rivers with sand, gravel, cobbles and boulders, and sedimentary and crystalline rocks with varying degrees of indurations and fissuring. For those soils and rock it is normally not possible to achieve borings to required depths using normal penetration testing methods like SPT-test or Dynamic probing. Therefore, the so called Soil-Rock Sounding (SRS) technique with MWD (Measuring While Drilling) was elaborated during the 1970's when the drilling rigs became motorized.

The principle of SRS is based on a hammer hitting a rod, producing a wave of energy conveyed to the bit, and thereby cuts the soil or rock material. The rod is thrusted at the top and rotated. The cavity at the drill-bit is cleaned by flushing through a channel in the rods and the drill-bit. This paper describes the SRS with respect to the equipment, reference test procedure and some case records where the method was used.

2 BACKGROUND

In order to investigate firm soil layers and rocks a special method of SRS was developed in Sweden and a first standard was proposed in 1974. This was an important step forward at that time as several types of equipment were used. The principle was that a rock drill bit was driven by a top hammer while the resistance of penetrated layers was measured manually in seconds for 0.2 m of penetration. In combination with the field engineer's observations of the behaviour of the drill rig and rod and the composition of the flushing water information about penetrated layers was obtained.

When equipment for automatic recording of different parameters became available in the 1980's, it became possible to record quick sequences and great amount of data. It became possible with different transducers to measure: rate of penetration or drilling resistance, the thrust force, torque, flush water volume, and flush water pressure. There was also a significant development of drill rigs and top hammers for different purposes which led the Swedish Geotechnical Society (SGF) to establish a new Reference Test Procedure for SRS in 1999. One aim was to make it possible to compare SRS carried out with different drill rigs.

3 DESCRIPTION OF EQUIPMENT

General: In order to obtain a standard procedure without specifying the drilling equipment in detail, the general outline is that different sizes of rigs and hammers should be acceptable, Figure 1a. Instead,

Figure 1 a) Typical Scandinavian drill rig used for soil- rock sounding.
b) Drillbits. Two common bits for Soil-Rock Drilling.

limitations in e.g. the penetration rate were stated to be between 0.2 and 0.6 meter per minute when drilling in hard rock (granite).

Drill bits: Both bits and button-bits can be used Figure 1 b. The diameter of the drill-bit will be related to the size of the hammer and the maximum rate of penetration in hard rock.

Rods: Normally rods with outside couplings (joints) size R32 or R38 are used. These rods have, however, a tendency to get stuck at redrawal in soils with boulders. Some operators therefore prefer a 44 mm diameter rod with inner joints.

Percussion hammer: A range of different percussion hammers can be used. The energy per blow can range from 200 to 900 Joule. If the calibration drilling of the rig indicates too high a penetration rate, the hydraulic pressure in the hammer can be reduced, or one can choose a larger drill bit for the job.

Flushing system: Water flushing is preferred but air flushing can be used, especially in winter conditions. When air flushing, no flow rate is measured. When water flushing, the flow rate and pressure are measured in a skid unit connected to the pump, but separated from the drill rig. The pump line diameter for water is normally 25 mm.

Table 1. The drilling parameters used for data collection in SRS.

Parameters	For short
Drilling head rotational speed (min^{-1})	Rotational Speed
Feed Thrust force load (kN)	Thrust Force
Hydraulic pressure in twisting machine (MPa) (function of Torque)	Pressure, Twisting Machine
Flush medium pressure at the pump (MPa)	Flush Pressure
Flushing medium circulation rate (l/min)	Flush Rate
Penetration rate (m/min)	Penetration Rate
Hammer pressure (On/Off parameter)	Pressure in Hammer

MWD: The MWD-technique was elaborated during 1980:s when the automatic data acquisition system on drill rigs became frequent and an essential part of the SRS. Figure 2 shows the principals of the system for automatic data aquisition used for SRS.

The sensors are mounted on the rig and connected with cables to a box for signal transformation. The box is connected to the logger with display, printing and storing facilities. The collected data can be stored on a diskette or directly sent to the office via modem and mobile phone.

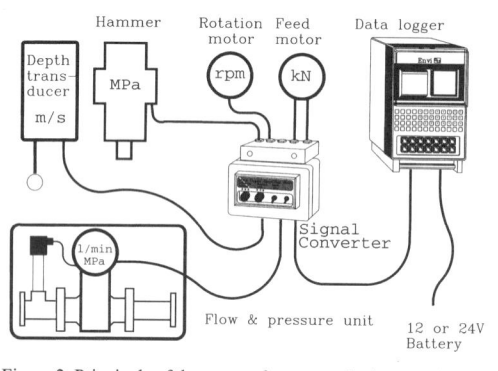

Figure 2. Principals of the system for automatic data acquisition in SRS.

4 APPLICATIONS

In soils the SRS is used for a relative indication of the stratigraphy of dense soils and the depth to bedrock. In addition, information on the thickness of e.g. old foundations in fills and size of boulders is obtained. To ensure that the bedrock is reached, the SRS is continued for another 3–5 m into the rock. The records are analyzed to find typical layers for sampling or in situ testing. Cutting samples indicate the type of soil penetrated. SRS is also widely used to penetrate dense soils overlying less dense e.g. in case of land reclamation, embankments and urban areas. In order to obtain a better resolution with respect to stratigraphy indications the size of the drill bit can be increased.

In rock, the SRS is also used to indicate different layers and especially fissured and crushed zones, where coring ought to be performed. In sedimentary rocks it is possible to indicate layers with different degrees of induration. The SRS can be combined with geophysical logging in order to increase the possibility of identifying different rock layers and pumping tests to estimate the conductivity of the fissured rocks.

The recording system could also be used together with core sampling giving an indication of the density variation in the rock as a complement to the core itself.

One SRS example is shown in Figure 3, where soundings were carried out to determine refusal levels for end-bearing piles driven for a library in Halmstad, on the Swedish west coast. Water was used for flushing and a button drill bit with 57 mm diameter was used. All parameters according to Table 1 were measured during drilling and are presented in Figure 3. The drilling resistance was calculated from the rate of penetration. Results indicate that the upper 31 m of the profile consists of soft soils and a layer of dense cohesionless soil from 31 to 34 m depth. That layer is underlain by bedrock.

Main parameters measured are the thrust force and the rate of penetration. For more detailed analysis the drilling resistance expressed in sec/0.2 m of penetration is used. Flush water volume and pressure are used as quality parameters e.g. to ensure that the drill bit is not clogged and that the flushing is sufficient to remove cuttings. In addition, these parameters can be used to detect more permeable layers primarily in rock mass. Torque, expressed as the hydraulic pressure in the twisting device, and the rate of rotation are kept constant at pre-set values throughout the entire sounding operation The hammer pressure is a so called on/off parameter and should be analysed together with e.g. the rate of penetration.

5 PROCEDURE FOR SOIL-ROCK SOUNDING WITH MWD

The idea behind SRS is, to base the standard on a "calibration drilling" where the rate of penetration must fall within certain limits. This is achieved by adjusting the drilling parameters and the size of the drill bit so

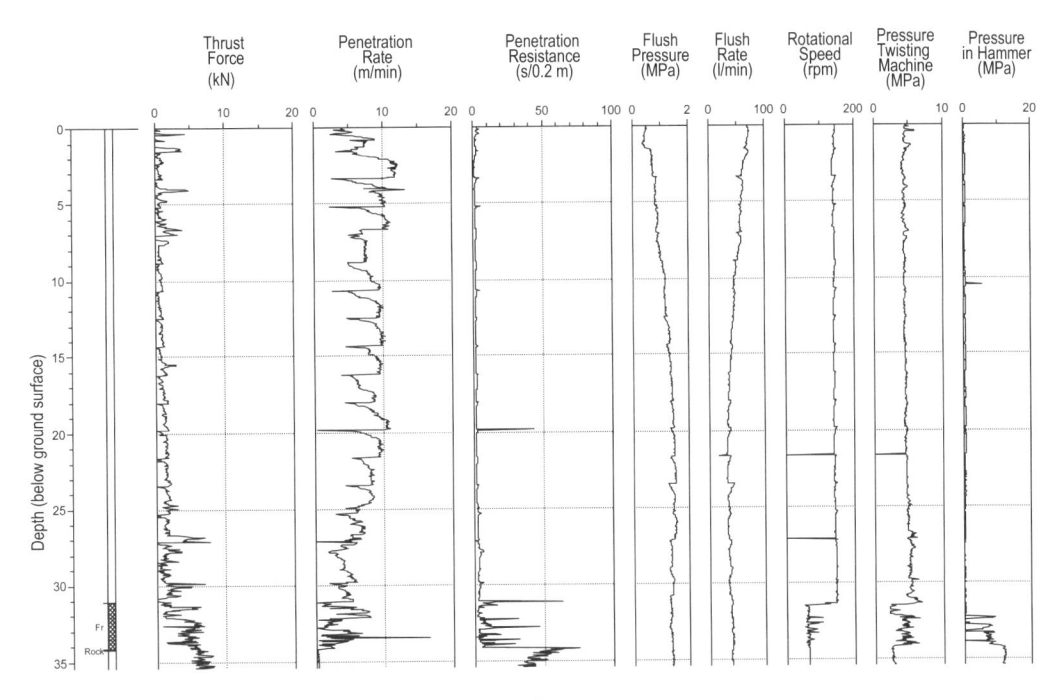

Figure 3. Exempele of results from soil-rock sounding in Halmstad.

that the rate of penetration falls between 0.2 and 0.6 meters per minute, corresponding to 60 to 20 sec/ 0.2 m of penetration.

Classification: SRS can be executed in three classes called SRS-1, SRS-2 and SRS-3, respectively. SRS-1 represents the old method used in Sweden before automatic readings were possible. In the Class SRS-2 and SRS-3 automatic registration are required. In Table 2 the requirements for registration of drilling parameters for the three classes are shown together with required accuracy.

Table 2. The requirements for registration in SRS type 1-3.

	SRS-1	SRS-2	SRS-3	Typical accurancy
Manual registration	yes	no	no	-
Automatic registration	Allowed	yes	yes	-
Depth	yes	yes	yes	0,1 m
Time, seconds/0,20 m	-	-		1 sec
Drilling resistance	yes	no[1]	no[1]	-
Rate of penetration	no	yes	yes	-
Bit load	no	yes	yes	10%
Hammer pressure, on/off	yes	yes	yes	5 %
Rotational pressure (on the hydraulic motor)	no	yes	yes	2 %
Revolutions per minute (RPM)	no	no	yes	2 %
Flushing pressure	no	no	yes	2 %
Flushing volume per minute	no	no	yes	5 %
Flushing media	Air/fluids	Air/fluids	Fluids	-
Sample cuttings	Optional	Optional	Optional	-
Notes by the operator during performance	yes	yes	yes	-

[1] The drilling resistance can afterwards be calculated from rate of penetration

Calibration: Transducers used for SRS shall be calibrated and approved at least every 6 months, prior to any major project, when the drilling rig is new or when there have been major changes in the drill rig which could affect the SRS results. Calibration procedures for all transducers are given in the Reference Test Procedure. Documentation from the calibration shall be available at the drill rig. In addition simple field checks of sensors on the drill rig should be performed frequently.

Preparations: Prior to each test, a number of items should be checked and recorded:
The drilling rig shall be steadily placed so it does not move during the test. Anchoring could be used The drilling are normally performed vertically. Maximum deflection from a given inclination is 2 degrees or 20mm/m. The drill-bit dimensions and edges should be within tolerances. The rods should be straight within tolerances. The drill rig should mechanically be in proper condition e.g. minimal friction. Drilling parameters to be used should be pre-set.

Execution of SRS: When sounding in rock, the pre-set parameters shall be the same as for the calibration drilling. In rock, the thrust force shall be kept as constant as possible. If, however, the rock quality is low, or the bit hits an inclined rock surface, the thrust force may be temporarely reduced.

In SRS-3, it is recommended to use casing through the soil layers. Data logging is used also during the casing installation. When SRS is used off-shore from a barge or a jack-up rig casing is recommended if the bending of the drill string exceeds tolerances. The straightness of the casing shall be checked before entering the drill string into the casing. In SRS-2 and −3 casing shall be used when the free length of the drill string is more than 3 meters. During drilling the thrust force, the rotational speed, the hammer pressure and the flush rate should be kept constant throughout the operation. Changes both within a borehole or between boreholes should be avoided.

Termination of SRS: Normally investigations of the rock surface for the design of normal foundations can be terminated after 3 meters drilling in solid rock. Investigations for heavy structures or underground structures the SRS could be terminated first after 5 meters drilling in solid rock or 5 metres below bottom of cavern or tunnel.

Data acquisition: Data should be collected at least every 50 mm for all drilling parameters but normally readings are taken every 20 mm.

In parallel to the automatic recording the driller shall record other observations during the SRS such as: Behaviour of the rig (shaking or stucking), type of cuttings, cutting samples, boulders penetrated or assumed rock surface.

6 CASE RECORDS

Investigation of limestone: For the construction of the Øresund Link between Sweden and Denmark where the ground mainly consist of limestone, indurated to various degrees and partly very fissured, efforts were made to find suitable investigation methods in order to:
- to find techniques for detecting the upper boundary of the Copenhagen limestone.
- to find the main features of the limestone layers and the occurrence of e.g. crushed zones or faults and unlithified layers.

A number of investigation methods were tested and the results were checked against a detailed mapping of the walls in a 13 m deep test pit in the limestone, Bergdahl et al (1995). Below some results from the SRS made are illustrated.

The SRS around the test pit were performed with hydraulic rigs provided with top hammers. Different kinds of drill bits were used. The reason for trying SRS was to find a penetration testing technique with high penetrability and capable of separating different limestone layers. Water as well as water with polymer additives were used as flushing medium. The soil-rock sounding parameters as mentioned above were automatically recorded by a Geoprinter. In addition the boreholes were used for borehole logging in order to improve the possibility of separating different layers in the limestone. The resistivity log was found to give the best resolution in separating layers with different water content down to 0.1 m in thickness.

The results of two such SRS expressed in penetration resistance sec/0.2 m are presented in Figure 4 together with the resistivity logging results in ohmm and the results of the mapping of the test pit walls close to the SRS. The mapping indicate the type of soil or limestone as well as the degree of induration** and fissuring* in the limestone.

The results from tests with different equipment indicated that the best resolution was achieved with an old type of X-bit with extra sharpened edges and with parameter readings taken every 10 mm. The penetration resistance curves clearly indicated where less indurated layers (H1-H3) and more indurated layers (H4-H5) were situated. When combining the penetration resistance with resistivity logging the water content (porosity) of the limestone layers could be clearly indicated. If the penetration resistance curve is combined with the bit force curve the evaluation of the SRS became more precise as the bit force normally drops in less indurated layers.

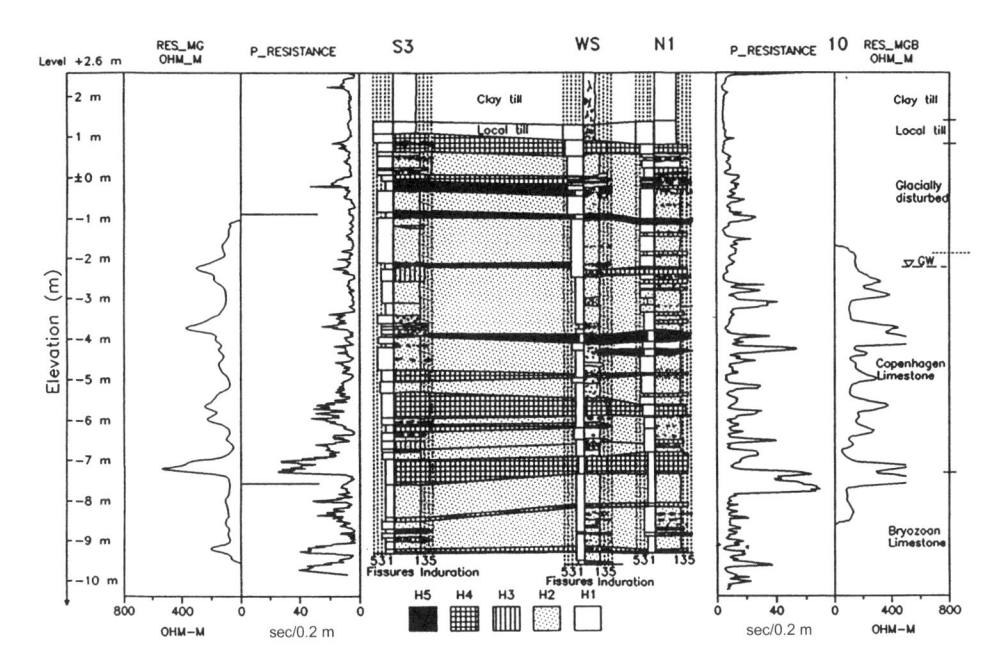

*) Degree of fissuring, according to Larsen (1988)
S1 Unfissured (no fissures)
S2 Slightly fissured (dist. between fissures >10 cm)
S3 Fissured (dist. between fissures 6-10 cm)
S4 Very fissured (dist. between fissures 2-6 cm)
S5 Crushed (dist. between fissures <2 cm)

**) Degree of induration, according to Larsen (1988)
H1 Unlithified (extremely weak R0)
H2 Slightly indurated (very weak R1)
H3 Indurated (weak R2)
H4 Strongly indurated (medium strong R3-strong R4)
H5 Very strongly indurated (very strong R5-extremely strong R6)

Figure 4. The stratigraphy of the test pit achieved from the mapping compared to the soil-rock soundings and the resistivity loggings in borehole Nos. S3 and 10 in Malmoe.

A correlation between results from density logging and the penetration resistance indicated that when the density was less than 2.0 g/cm^3 the penetration resistance was between 5 and 15 sec/0.2 m of penetration and from 2.0 to 2.7 g/cm^3 between 15 and 70 sec/0.2 m.

Investigations for driven end-bearing piles: When piles should be driven into dense layers of gravel and till with cobbles and boulders it is necessary to find the maximum depth (length) for the piles, normally the bedrock. Also the level from which an increased driving resistance for the piles could occur is important to find. Dynamic probing is normally used to find the length of end-bearing piles, however, it is sometimes too light to penetrate coarse and very dense layers. SRS is then often used to penetrate the soil to the bedrock.

For a new library in Halmstad a comprehensive investigation was carried out in order to determine pile length.

Figure 5 shows the results from investigations from the same site as the SRS presented in Figure 3. In addition Figure 5 contains results from a CPT test, a dynamic probing test and a simplified presentation of the SRS results as well as results from pile driving. The figure shows clearly the limitations in penetrability of the three methods.

The investigations show that the soil profile consists of a thin soil layer with organic content or road gravel. Beneath that a thin layer of sand underlain by clay and again a sand layer underlain by clay. The clay is underlain by cohesionless soil and till to the bedrock. The pile driving test indicated that all piles penetrated all the way to the bedrock or close to the bedrock.

Investigations of rock for foundation purposes: Investigations were carried out for a 620 m long bridge at Motala sound, founded on five piers and two abutments. Two of the piers were to be founded in water. SRS and core sampling were carried out for those piers.

The geology in that area is complex. The rock in the bridge alignment area consists of Ordovician limestone, shale and mudstone. The layers are dipping gently against north. The rock is covered by Quaternary soil to various depths.

Borehole 21, indicated that the rock consists of three layers of different rock types, namely reddish limestone overlaying (greenish) grey limestone, the latter containing a 2–3 m thick black shale layer. The grey limestone and the shale are often very fissured, with greenish or brownish clay infillings, while the reddish limestone generally appears more intact. Steeply inclined fissures filled with rock fragments and calcareous clay occured frequently. The limestone in Borehole 21 is more or less affected by glacial shearing and crushing.

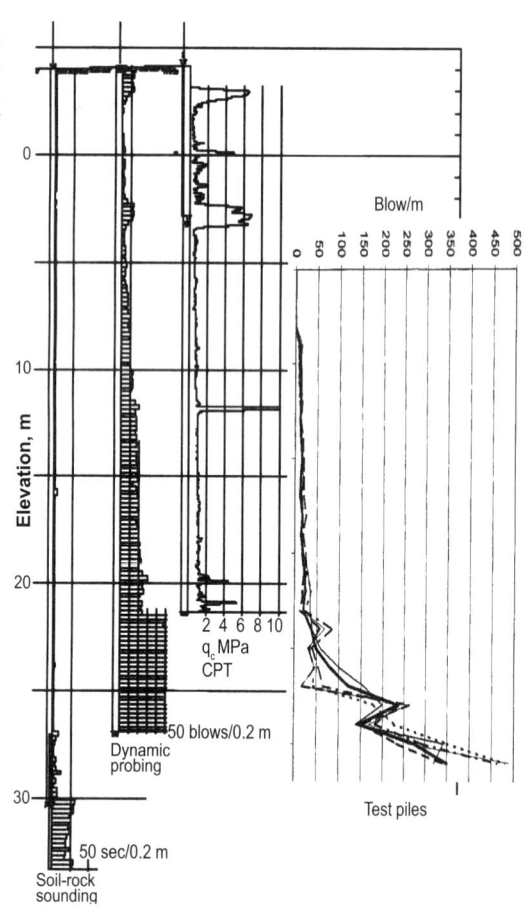

Figure 5. Results from site investigation and test pilings in Halmstad.

Figure 6 shows the results from SRS and core sampling in Borehole 21. The core sampling, see Figure 7 for example, was carried out with a wire-line tripple tubesampler (S-Geobor) with a core diameter of 102 mm.

On the basis of these results it was decided to found one of the piers on a shallow foundation and the other on a pile foundation. The foundation level for shallow foundation was recommended to be below 8 m depth (level –89).

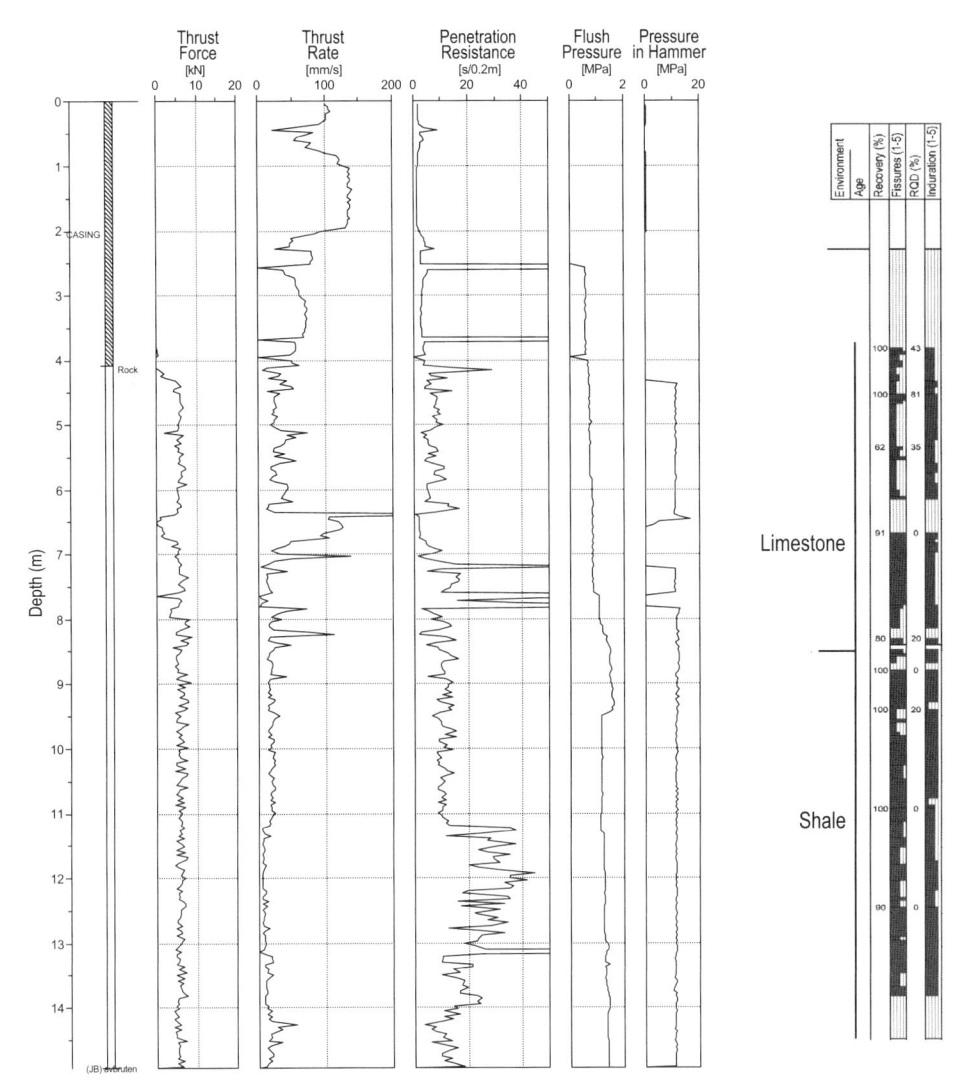

Figure 6. Results from soil-rock sounding and core drilling in Borehole 21.

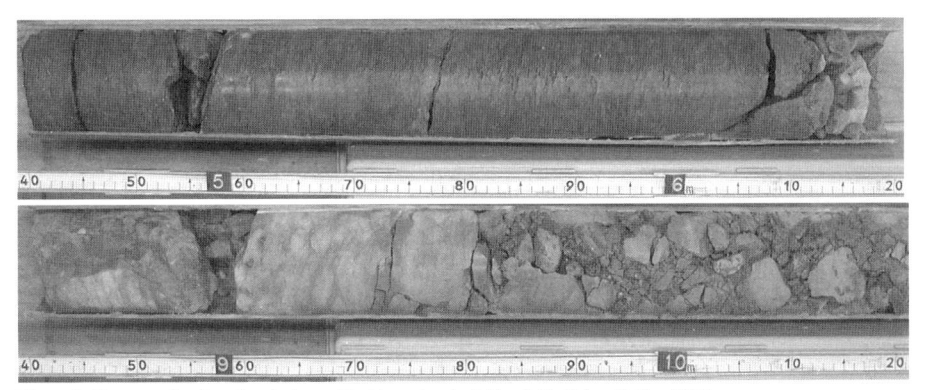

Figure 7. Core samples from borehole no. 21, Above levels 5.40 – 6.15. Below 9.40 – 10.20.

7. CONCLUSION

The equipment and technique of SRS as described above has proven to be a useful method for site investigations in dense soils and rocks. Many questions related to foundations or underground structures in such ground conditions can be solved by SRS. The SRS used in limestone had shown that it is a rapid technique with a high penetrability and good repeatability in indicating layers of high and low degrees of induration and fissured zones. By stabilizing the borehole with e.g. a polymer additive the borehole can be used also for logging of different kinds.

ACKNOWLEDGMENT

The authors would like to express their great and sincere appreciation to Håkan Garin, Scandiaconsult Sverige AB, for his kind assistance with providing us with documentation for both the library in Halmstad and the bridge across mainroad 50 in Motala.

REFERENCES

Bergdahl, U., Petersson, E. & Möller, B.,(1995). Site investigation in Limestone in Malmoe. European conference on soil mechanics and fondation engineering, 11, Copenhagen, May–June 1995. Proceedings, vol 5. The interplay between geotechnical and engineering geology. Danish Geotechnical Society. DGF Bulletin 11 p 5.149 – 5.158.

Colosimo, P.(1998).On the use of drilling parameters in rock foundations.International conference on site charaterization,ICS ′98,1,Atlanta,GA,19–22 April 1998.Proceedings,vol. 1 pp 347-352.

Nishi, K., Suzuki, Y. & Sasao, H. (1998).Estimation of soil resistance using rotary percussion drill.International conference on site characterization, ICS ′98,1, Atlanta,GA,19-2.2April1998. Proceedings, vol.1, pp393-398.

Larsen, G., Frederiksen, J., Villumsen, A., Fredericia, J., Gravesen, P., Foged, N., Knudsen, B. & Baumann, J. (1998). Vejledning i ingeniörgeologisk prövebeskrivelse. Danish Geotechnical Society. DGF Bulletin 1. Denmark (In Danish)

Reference Test Procedure for Soil-Rock Drilling (1999). Swedish Geotechnical Society. SGF Report 2:99. (In Swedish)

A new in situ apparatus: the geomechameter

J. Monnet
Lirigm, Université Joseph Fourier, BP53, 38041, Grenoble Cedex 9, France

S.M. Senouci
Laboratory SpA - Lab Serv. Res. Innovation, Via Vitorchiano, 165, 00189 Rome, Italy

Keywords: in-situ, apparatus, measurement

ABSTRACT: The difficulty or sometimes the impossibility to extract intact and representative samples to carry out laboratory tests, as well as the need to obtain fast results are important elements to choose in situ tests instead of laboratory tests. Consequently the in situ tests are more and more used in the prediction and the determination of soils and rocks properties. A new in situ testing apparatus has been thought up and developed. It is an evolution of the pressuremeter, using the forces generated by water flow around the measurement probe. The hydraulic flow allows the variation of the vertical stress, artificially, at the test level. The influence of this stress is taken into account in the interpretation of the test results. The apparatus allows its control and variation for a better evaluation of the soil mechanical characteristics as elastic shearing modulus, permeability and consolidation coefficients, cohesion and internal angle of friction.
 Hostun thin sand was chosen as a material to undergo the experimental study in laboratory. Isère loam was chosen to experiment the first model of the in situ apparatus. Results show the influence of the hydraulic stream flow on the expansion curves.

1 DESCRIPTION OF THE GEOMECHAMETER TEST

1.1 Introduction

The pressuremeter test can be considered as a shear test between the radial and circumferential stresses in plane strain condition, thus the vertical stress is the normal stress applied on the plane where shearing takes place (Baguelin et al. 1978). The theoretical limitation of the pressuremeter test is linked to the fact that the vertical stress is given by the unit weight of the soil. This test can be considered as a single shearing test, and it can only be used to determine either the internal angle of friction (Hughes et al. 1977) or the cohesion of soil (Baguelin et al. 1972). When the cohesion and the internal angle can both be determined, they are linked together so that the cohesion value depends on the internal angle of friction value (Baguelin et al. 1978). The principle of the geomechameter test is to create an hydraulic gradient to control the vertical stress, and to set its value to an appropriate range. This new apparatus allows the control of the three-dimensional state of stress around the probe, at the borehole wall. A series of three tests made with three different values of the vertical stress associated with the radial shearing stress, yields the cohesion and the internal angle of friction. This problem of local determination of the cohesion and the internal angle of friction is of great interest in Civil Engineering.

The pressuremeter can also measure the pore dissipation at the borehole when consolidation is reached, with a delay of one hour or more (Clarke et al. 1979). The geomechameter improves this delay by a simultaneous measurement of permeability and shear modulus and does not require waiting until final consolidation.

To validate the theoretical principles on which the geomechameter operates, it is necessary to use a standard soil which is well known and homogeneous, to control its relative density and its initial state of stress. All these parameters can be measured in a calibration chamber. On the other hand, during in situ experiments, the heterogeneity of the natural soil, its initial state of stress, and its density cannot be controlled. The testing of this new apparatus is realised in a quality procedure where each step is carefully controlled. The calibration chamber test is the first control operation that is needed for the validation of this apparatus. The last step will be the use of the Geomechameter in real conditions.

1.2 Geomechameter model probe

The geomechameter test (Fig. 1) consists of a monocell probe, which expands as a pressuremeter cell. In the soil around the probe, a water flow is applied. The experimental model probe is composed of an aluminium cylinder, 31 cm long and 6 cm in diameter. The measurement cell is 16 cm long and is placed between the injection cell upstream and the pumping cell downstream. The inside of the cylinder is massive and drilled by two boreholes to connect water used in the measurement cell and water, which comes from the pumping cell. The measurement cell is covered by a rubber membrane, which is inflated as soon as the geomechameter measurement begins. The pumping cell is covered by a 100 micrometer textile filter to prevent sand particles to enter the pump.

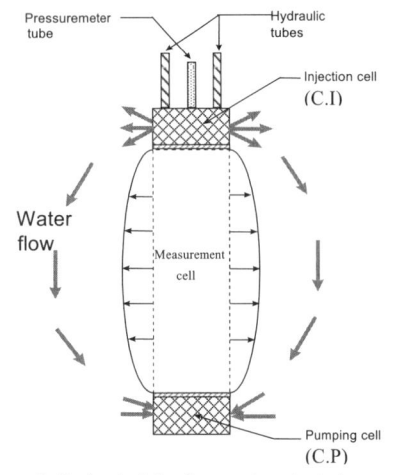

Figure 1: Basic principle of geomechameter test

1.3 Geomechameter in-situ probe

The prototype of the geomechameter probe (Fig. 2) is made of five different parts. The central part of the probe is identical to a pressuremeter probe with a measurement cell and two guard cells. Above the central cell, an injection cell is placed to inject water into the soil around the probe. Under the central cell, a pumping cell is placed to pump water from the soil around the probe. At each ends of the probe, there are two cells, which avoid the water flow to pour into the borehole. The in-situ probe is made so that:
- It is possible to disassemble the various parts for needs of maintenance
- The independence of the four circulation of fluid in the probe is insured (pumping, injection, air pressure, water pressure)

Figure 2: In-situ geomechameter probe

The preliminary calibration tests on the geomechameter probe was:
- The use of the apparatus with a free expansion of the probe (Fig. 4), to correct for pressure loss needs by the apparatus to inflate the probe
- The use of the apparatus with a probe placed in a thick metal tube (Fig. 5) until 1500kPa, to correct for volume loss produce by membrane compression and dilation of tube connexion
- The use of the apparatus with a probe placed in a water tank, to correct for the water head loss produced by the hydraulic circulation into the apparatus (Fig.6)

1.4 Control panel

The control panel (Fig. 3) is used for the measurements of the injection pressure and the pumping pressure are made by two glycerine-filled pressure gauges, graduated from - 0.1 to 0.15 MPa, and placed at the extremity of the pump. A volumetric gauge with a float is used for the measurement of the volumetric injected flow. It is controlled by a tap placed before and connected to the injection cell of the probe. Up stream from the pressure gauge, on the suction extremity of the pump, an electronic volumetric gauge gives the volumetric flow on suction. To prevent the failing of the pump, a valve-valve is disposed to deliver input water from a tank. If the pressure decreases too much, the valve-valve opens and the pump is fed by water from the tank, saturating the soil around the probe. As soon as the flow is established, the valve-valve closes and the water flow moves in a cyclic circulation.

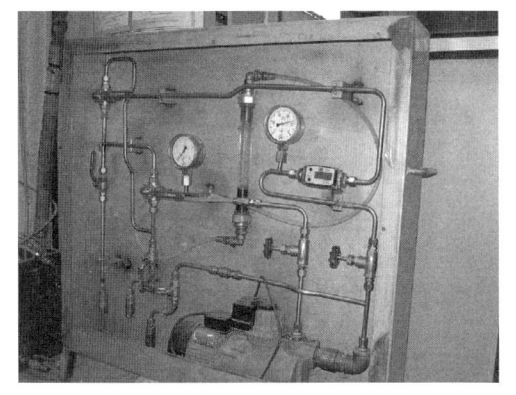

Figure 3: Control panel of the Geomechameter apparatus

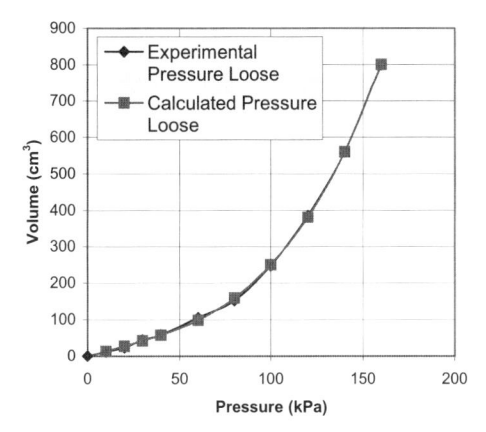

Figure 4: Free expansion test of the probe

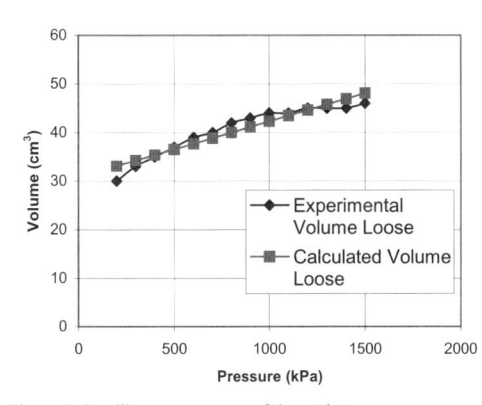

Figure 5: Swelling pressure test of the probe

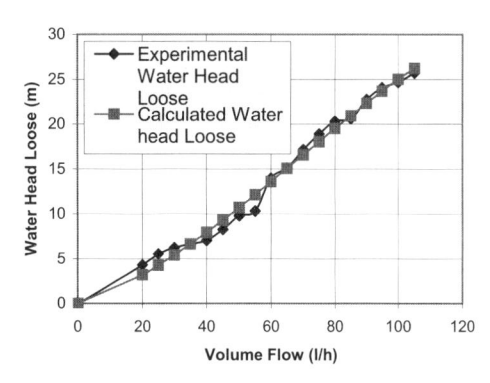

Figure 6: Water head loose with the probe put in a water tank

2 THEORETICAL STUDY

2.1 Notations

γ_w : Unit weight of water
γ' : Submerged unit weight
Φ' : Effective internal angle of friction
Φ_μ : Interparticle angle of friction
G : Shear modulus
H_i : Hydraulic head on injection
H_p : Hydraulic head on pumping
i : Hydraulic gradient
K_0 : Coefficient of earth pressure at rest
k : Permeability coefficient
l_e : Distance between injection - pumping cells
l_i : Length of the injection cell
N : Failure ratio equal to $(1-\sin\Phi')/(1+\sin\Phi')$
n : Dilatancy ratio equal to $(1-\sin\psi)/(1+\sin\psi)$
υ : Poisson's ratio
p : Pressure applied at the borehole wall
p_i : Pressure of the injected water
p_p : Pressure of the pumped water
p_0 : Initial hydraulic pressure for injection,
p_{lim} : Limit pressure of the pressuremeter test
ψ : Dilatancy angle
Q : Volumetric flow
r_0 : Radius of the borehole
σ_{vi}' : Effective vertical stress level injection cell
σ_{hi}' : Effective horizontal pressure at rest at the level of the injection cell
σ_v' : Effective vertical stress level middle probe
U_0 : Pore pressure at rest
z_{sim} : Depth simulated by the hydraulic gradient
z_i : Depth of the injection cell
z_p : Depth of the pumping cell

2.2 Equilibrium around the probe with the hydraulic flow

The geomechameter is an apparatus, which uses the hydraulic flow around an inflatable probe to increase the vertical effective stress at the level of the probe. When water is flowing into the soil, the power of the water decreases along trajectory. A force is applied on the soil particles in the direction of the stream by the water flow. In the geomechameter test, theses forces are similar to the gravity force in application and magnitude. When the permeability coefficient is higher than 10^{-8}m/s, the membrane expansion has no influence on pore pressure (Cambou and Bahar 1993, Frank and Nahra 1986) and the effective pressure can be used, but when permeability is lower than 10^{-10}m/s, the test cannot be performed because the soil behaves in undrained conditions. A numerical modelling of the test (Senouci and Monnet 1999) shows that for a unit volume element ΔV of the soil, the stream force is:

$$\vec{F} = \vec{i}.\gamma_w \Delta V \quad (1)$$

In the geomechameter test, water is injected into the soil by the injection cell and with a controlled injection pressure (p_i). After flowing into the soil, it is pumped through the pumping cell at a controlled pressure (p_p). The hydraulic gradient (2) is obtained by the difference between the injection and suction hydraulic head. The increase in effective stress (3) from the hydraulic water flow in the soil, at the level of the probe, is obtained by:

$$i = (H_i - H_p)/l_e \quad (2)$$

$$\delta\sigma'_v = (i\gamma_w)\frac{l_e}{2} \quad (3)$$

The increase in depth (5) is determined by:

$$\sigma'_v = \gamma.Z_{sim} = (\gamma + i.\gamma_w)\frac{l_e}{2} + \sigma'_{vi} \quad (4)$$

$$Z_{sim} = (1 + i\frac{\gamma_w}{\gamma})\frac{l_e}{2} + Z_i \quad (5)$$

A complete numerical modelling (Senouci and Monnet 1999) shows that the variation of the hydraulic gradient gives a variation of the vertical stress (3), which allows to determine an increase of the simulated depth of the probe. The differences between numerical finite element curves find by Plaxis program (Fig. 7) at the simulated depth (5) and the experimental measurements are very small.

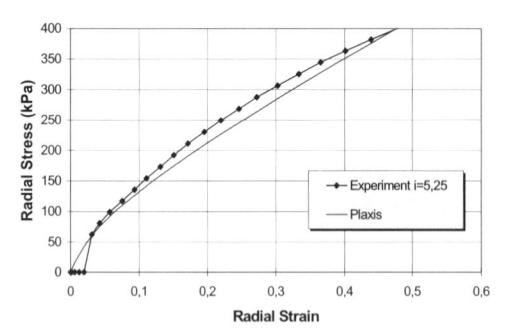

Figure 7: Comparison between the experiment and the finite element expansion of the probe for the simulated depth

2.3 Equilibrium around the probe with the expansion of the probe

We assume a non standard plasticity for a higher level of shear and a drained pressuremeter test (Monnet 1990, Yu and Houlsby 1991). This phenomenon is ruled by the angle of internal friction Φ' and the dilatancy angle Ψ. The dilatancy is a function of the interparticle angle of friction Φ_μ along the formula (Monnet and Gielly 1978):

$$\Psi = \Phi' - \Phi_\mu \quad (6)$$

The vertical stress is assumed to be equal to the modified value found by (4). As assumed before (Wood and Wroth 1977) plasticity may appear:
- into the horizontal plane between the radial stress σ_r and the circumferential stress σ_θ (first plastic area) in an area between the radii r_0 (borehole) and b (first plastic area).
- into the vertical plane between the vertical stress σ_z and the circumferential stress σ_θ in an area between the radii b and c (external radius of both plastic areas) in an area between the radii b (first plastic area) and c (second plastic area).

An elastic area extends beyond the radius c.

The global equilibrium of the soil around the probe takes into account the horizontal stresses:

$$\sigma'_\theta - \sigma'_r - r.d\sigma'_r/dr = 0 \quad (7)$$

and the vertical stresses:

$$d\sigma'_z/dz = \gamma \quad (8)$$

It was shown before (Monnet and Khlif 1994) that the theoretical relation between pressure applied to the borehole and radial strain is:

$$Ln\left[\frac{u_a}{a}(1+n) - C_1\right] = \delta.Ln(p) - \delta.Ln(\gamma.z) + Ln\left[(1-K_0)\gamma z.\frac{(1+n)}{2.\mu} - C_1\right] \quad (9)$$

with the small constant:

$$C_1 = \frac{n\left(\frac{u_a}{a}\right)(1+n)\left(\frac{\gamma z}{p}\right)^\delta + (1+n)(N-K_0)\frac{\gamma z}{2.\mu}}{1+n\left(\frac{\gamma z}{p}\right)^\delta} \quad (10)$$

and:

$$N = (1 - \sin \Phi') / (1 + \sin \Phi') \quad (11)$$
$$n = (1 - \sin \Psi) / (1 + \sin \Psi) \quad (12)$$
$$\delta = \frac{1+n}{1-N} \quad (13)$$

Formula (9) shows a linear relation between the logarithm of the pressure and the radial strain, which is a function of the internal angle of friction Φ' and the interparticle angle of friction Φ_μ. The measurement of the slope of this logarithmic relation and the knowledge of Φ_μ allows to find the friction angle with the pressuremeter test. The final value of Φ' is controlled by the superposition of the theoretical curve over the experimental curve (Monnet and Allagnat 2002).

3 EXPERIMENTAL STUDY

3.1 Experimental procedure with the model probe for the Hostun sand

Tests are carried out in a cylindrical tank, 60 cm in diameter and 100 cm in height. The probe is moulded in sand and is placed at the bottom of the tank under 0.8 m. of soil. Before filling the tank, water is circulated through the geomechameter to prevent any air lock, which might alter the measurement. The soil is placed in 10 cm thick layers and hand compacted with a 13 kg steel cylinder tamper of 15 cm diameter. The tamper has a falling course of 15 cm and 30 or 50 hammer blows are applied for each layer so that the design density is reached. Once the soil is compacted, it is saturated by the injection of water until the water table reaches the soil surface. Then the pump is powered on and the injection tap is open. The volumetric flow of water is controlled by the action on this tap. Water is injected into the soil through the injection cell and the pressure and the volumetric flow are measured. The hydraulic system has a cyclic flow and the injected water is pumped through the pumping cell. Water passes through pressure and volumetric gauges and comes back to the pump. A constant volumetric flow is reached after a few minutes. Then the probe is inflated with the normalised process (French Standard 1991). An unloading-reloading sequence is added before the swelling pressure is reached so that the reverse shear modulus can be measured (French Standard 1999). As soon as the test ends, the tank is dried by pumping, the sand is removed and the same experimental sequence is repeated.

Measurements are made with a precision of 1kPa on water pressure, 0.4l/mn on volumetric flow, 2cm³ on volume variation of the probe, and 2kPa on pressure applied inside the probe. As the vertical position of the probe is known within 1mm, precision on the increase of vertical stress (4) is 1kPa and simulated depth is known within 0.1m.

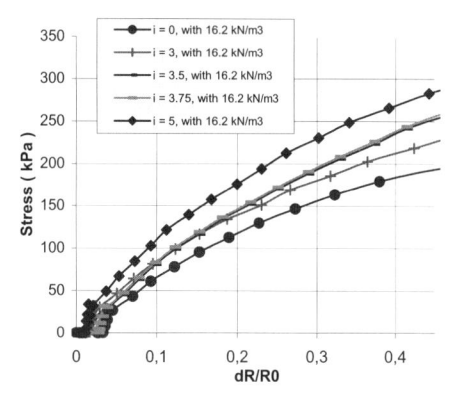

Figure 8: Influence of the hydraulic gradient on the Geomechameter experimental curves for the Hostun sand

3.2 Hydraulic gradient effect for the Hostun sand

The experimental curves, obtained when the hydraulic gradient varies, are shown in Fig. 8 at the depth of the probe (Z = 0.65 m). For all the curves, an unloading reloading sequence is performed at the beginning of the test so that the elastic modulus can be measured. The same density is used for each series of tests. The geomechameter curves show that the increase in hydraulic gradient leads to an increase of the inflate pressure of the cell for the same volume variation as predicted by the increase of the theoretical vertical stress in equation (4).

Table 1: Simulated depth of the Geomechameter test for the model probe

Test number	P_i (kPa)	P_p (kPa)	i Calculus	Z_{sim} (m)
16_1	1.5	-2.5	3	0.99
16_2	2	-3	3.5	1.05
16_3	2.5	-3	3.75	1.07
15_4	2	-4	4	1.10
16_4	3	-5	5	1.21
14_2	3	-5.5	5.25	1.24
14_3	5	-7.5	7.25	1.47

3.3 Simulated depth for the Hostun sand

The water flow around the probe creates artificial geostatic gravity. The hydraulic forces may be considered vertical and orthogonal to the horizontal plane where shearing occurs. These forces are linked to an increase in vertical stress. The influence of this artificial gravity is much more important around the probe over a distance equal to twice the radius measured from the edge of the borehole. In this zone, the theoretical calculations make with SEEP show that relation (5) can be applied. This equation is used to determine the controlled vertical stress and the simulated depth for each hydraulic gradient, with the at 0.65 m depth in the cylindrical tank. Table 1 gives the results of this study, and the correspondence between the hydraulic gradient calculated by (2) and the simulated depth obtained by (5). The water flow around the probe allows increasing the depth from 0.65 m, for no stream, to 1.47 m for an hydraulic drop of 7.25, which is 126% more than the initial depth of the probe.

3.4 Permeability and coefficient of consolidation for the Hostun sand

The determination of the permeability of the soil is made by formula (14) initially obtained for the pressuremeter-permeameter (Ménard 1957). This test is an injection of water in the soil from a borehole. Ménard assumes that near the injection cell, trajectories of water are normal to the edge of the borehole and the equipotential curves are cylinders coaxial with the borehole. This cylindrical behaviour expends along an area from the borehole where the trajectories are linear. Farther the flow seems to expend from a single point and the flow becomes spherical. Table 2 shows the results of this study.

$$k = \frac{\gamma_w Q}{2\pi l_i (p_i - p_0)} \left[\ln \frac{l_i}{r_0} + \frac{1}{2} \right] \quad (14)$$

Table 2 : Measured permeability of the sand

Q	P_i	k	γ
l/h	kPa	10^{-4} m/s	kN/m³
50	2	1.84	14.2
100	3	2.45	14.2
150	5	2.21	14.2
30	1.5	0.84	16.2
35	1.5	0.97	16.2
40	2	1.12	16.2
45	2.5	1.25	16.2
60	3	1.67	16.2

The measurements made on the constant head permeameter apparatus give 2.2 10^{-4} m/s for low density, and 1.2 10^{-4} m/s for high density. The values found with the geomechanical test are in the range of the laboratory measurements.

With this new apparatus we are able to infer the consolidation coefficient with the use of relation (15). The experimental consolidation of the sand cannot be observed, but this section is introduced only to show the ability of the apparatus to measure in situ such a coefficient by :

$$C_v = \frac{2.G.k}{\gamma_w} \frac{(1-\upsilon)}{(1-2\upsilon)} \quad (15)$$

Table 3 : Consolidation coefficient for the Hostun sand with the geomechameter

γ kN/m³	k_{moy} 10^{-4} m/s	G_{moy} kPa	Cv m²/s
14.2	1.5	4425	0.34
16.2	2.2	2020	0.10

This relation depends on the permeability coefficient, and on the Young modulus of the soil which is obtained from the unloading-reloading sequence at the beginning of the geomecanical curves. The simultaneous measurement of permeability and shear modulus of the soil on the unloading and reloading allows to find the consolidation coefficient without final consolidation of the soil. For soil with a consolidation coefficient of 14 m²/year, 90% of consolidation would be reached in 70 min. for strain holding test (Clarke et al. 1979), but other researchers find 55 hours (Baguelin et al. 1986).

We find the consolidation coefficient from the mean permeability given in Tables 2. The result of this study is shown on Table 3.

It can be seen that the consolidation coefficient for the Hostun sand is found between to 0.10m²/s to 0.34m²/s depending on the density level of the soil.

3.5 Measurement of the friction angle for the Hostun sand

The interpretation of the geomechametric curves is made with the program « GaiaPress » which uses an elasto-plastic theory in granular soil (Monnet and Khlif 1994). This theory is used with a simulated depth obtained by formula (4) which depends of the hydraulic gradient .

The determination of the elastic modulus of the soil is made on the unloading reloading cycle at the beginning of the test, or with the final unloading of the test.

The result of the interpretation of the 13 tests shows that the hydraulic gradient has no influence on the measurement of the friction angle. For the same density of the soil, the friction angle stay equal to the value found for the case without hydraulic gradient.

Fig. 9 shows that the friction angle depends on the density of the soil and varies between 31° for low density (14.2kN/m³) to 44° for high density (16.2kN/m³). These values are in the range of the expected values (Flavigny et al. 1990)

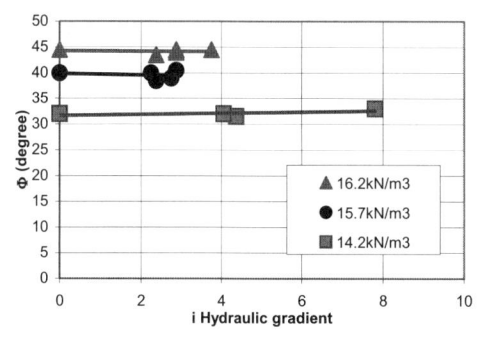

Figure 9: Non-influence of the hydraulic gradient on the friction angle measured with the Geomechameter

3.6 Measurement of the shearing characteristics of Isère river sandy loamy deposit with the in-situ probe

The geomechameter in-situ probe was tested in the Isère river sandy loamy deposit at a depth of 2.2m with a water table at 1.8m depth. Only a few tests were achieved but the preliminary results are in the range of the excepted ones. It can be seen (Fig. 10) that the theoretical results fit the experimental curve so that the control of mechanical parameters is valid with:
- Elastic modulus 5MPa,
- permeability coefficient $1.4 \cdot 10^{-6}$m/s,
- consolidation coefficient 9.10^{-4}m^2/s,
- cohesion 5kPa,
- friction angle 30.5°

The volumetric flow is about 40l/mn with 5h of test duration. This time is needed to reach a saturated stabilised flow, so that the vertical stress is modified by the geomechameter probe. The simulated vertical stress is 119kPa for an at rest vertical stress of 55kPa so that the increase of stress is 64kPa.

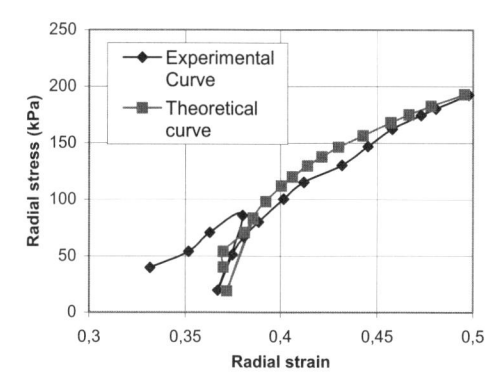

Figure 10: An example of in-situ Geomechameter test at 2,2m depth

Some triaxial tests were carried out on remoulded recompacted samples in drained condition of shearing. The results are shown on table 4. It can be seen:
- for the same level of stress (100kPa) than the geomechameter's ones, the Young modulus is closed to the in situ one with a difference of 10%,
- the cohesion 5kPa is not found in laboratory test
- the friction angle is 2° larger than the in situ one.

Table 4 : Results of triaxial tests on remoulded samples

σ'_3	E	ν	c'	ϕ
kPa	MPa		kPa	degree
60	2,7	0,47		
100	4,5	0,42		
200	10	0,3	0	33,5°
300	18,9	0,28		

These results can be explained by the way the sample is placed on the triaxial test, after remoulding and recompaction, where as the geomechameter test does not disturb the soil around the probe. Anyway the cohesion value is small and a definitive conclusion is not reached. Some tests in more cohesive soil are needed.

4 CONCLUSIONS

A new in-situ test was though out to test the soil into condition of three dimensional stress control. The geomechameter model probe shows its ability in the Hostun sand to increase the vertical stress around the probe and to measure the mechanical characteristics of the soil such as friction angle, permeability and consolidation coefficient with a good accuracy.

The in-situ geomechameter probe was used in the sandy loamy Isère river deposits at small depth and the results were compared with the triaxial test. It is found that the elastic and shearing characteristics are closed to the triaxial measured values so that the apparatus should be validated in this sandy loamy soil.

Further modifications are needed to improve the in-situ probe and some other soil should be tested to increase the use of this new apparatus.

REFERENCES

Baguelin F., Jézéquel J.F., Lemée E., Le Mehauté A., 1972, Expansion of cylindrical probes in cohesive soils, *ASCE, SM11, 1129-1142*.

Baguelin F, Jézéquel J.F., Shield D.H., 1978, The pressuremeter and foundation engineering, *Series on Rock and Soil Mechanics, Vol. 2, Trans. Tech. Publication*, 335-406.

Baguelin F., Frank R., Nahra R.., 1986, A theoretical study of pore pressure generation and dissipation around the pressuremeter, *Proc. 2nd Int. Symp. on Pressuremeter and its Marine Application, ASTM,* 169-186.

Cambou B., Bahar R., 1993, L'utilisation de l'essai pressiométrique pour l'identification de paramétres intrinsèques du comportement du sol, *Revue Française Géotechnique* (N°63): 39-53.

Clarke B.G., Carter J.P., Wroth C.P., 1979, In-situ determination of the consolidation characteristics of saturated clays, *Proc. 5th Europ. Conf. SMFE, Brighton,* Vol.2, 207-213.

Flavigny E., Desrues J., Palayer B., 1990, Note technique : Le sable d'Hostun RF, *Revue Française Géotechnique* (N°53): 67-70.

Frank R., Nahra R., 1986, Contribution numérique et analytique à l'étude de la consolidation autour du pressiomètre, *Rapport recherche LCPC* (N°137)

French Standard NF P 94-110, 1991,Essai pressiométrique Ménard, *AFNOR, Paris.*

French Standard NF XP 94-110, 1999,Essai pressiométrique Ménard, Partie 2 : Essai avec cycle , *AFNOR, Paris.*

Hughes J.M.O., Wroth C.P., Windles D., 1977, Pressuremeter tests in sand, *Geotechnique,* Jnl 27 (N° 4): 455-477.

Ménard L., 1957, Mesure in situ des propriétés physiques des sols, *Annales Ponts Chaussées* (N°14): 356-377.

Monnet J., Gielly J. 1978, Détermination d'une loi de comportement pour le cisaillement des sols pulvérulents, *Revue Française de Géotechnique* (N°7): 45-66.

Monnet J, 1990, Theoretical study of elasto-plastic equilibrium around pressuremeter in sands, *Proc. 3rd Int. Symp. Pressuremeter,* Oxford, 137-148.

Monnet J., Khlif J, 1994, Etude théorique de l'équilibre élastoplastique d'un sol pulvérulent autour du pressiomètre, *Revue Française Géotechnique* (N°67): 71-80

Monnet J., Allagnat D. 2002, Design of a large soil retaining structure with pressuremeter analysis, *Geotechnical Engineering 155,* Issue 1, 71-78.

Senouci S.M., Monnet J, 1999, Modélisation numérique du Géomécamètre, *Revue Française de Géotechnique* (N°88): 21-35.

Wood D.M., Wroth P.C. 1977, Some laboratory experiments related to the results of pressuremeter tests, *Geotechnique,* 27 (N°2): 181-201.

Yu H.S., Houlsby G.T. 1991, Finite cavity expansion in dilatant soils: loading analysis, *Geotechnique,* 41 (N°2): 173-183.

Permeability of sand sediment soil determined by the pressure infiltrometer test

T. Morii
Faculty of Agriculture, Niigata University, Niigata, Japan

Y. Takeshita
Faculty of Environmental Science and Technology, Okayama University, Okayama, Japan

M. Inoue
Arid Land Research Center, Tottori University, Tottori, Japan

Keywords: soil permeability, in-situ measurement, Guelph pressure infiltrometer, sand sediment soil, field-saturated hydraulic conductivity of soil, unsaturated moisture properties of soil

ABSTRACT: An integrated procedure to determine hydraulic conductivity of sand sediment soil that is characterized by an inclusion of gravel particles and cobbles in sand soil is proposed. Firstly the Guelph pressure infiltrometer (GPI) method to measure the soil permeability of the sand is introduced and extended so that it can estimate unsaturated moisture properties of the soil. Secondly a descriptive cylindrical soil model representing sand, gravel and cobbles, and voids within soil is assumed to determine the hydraulic conductivity of the sand sediment soil. A continuity law of flow discharge through the cylindrical soil model is introduced to derive theoretically a functional relationship of the hydraulic conductivity of the sand sediment soil with both the hydraulic conductivity of the sand measured by the GPI method and the gravel content of the soil. An accuracy of the functional relationship of the hydraulic conductivity is examined by laboratory permeability tests. Finally the GPI method and the functional relationship of the hydraulic conductivity are integrated to determine the soil permeability of the sand sediment soil. A numerical example is given to show an effect of the gravel content of soil on a prediction of storm runoff over the sand sediment soil.

1 INTRODUCTION

Permeability of sand sediment soil distributed in a valley watershed is a key parameter that may trigger a mountainous disaster such as a debris flow and a storm runoff over the soil. It is usually difficult to determine accurately and practically hydraulic conductivity of the sand sediment soil by using in-situ permeability test or laboratory permeability test of soil. This is because a value which can be measured by the in-situ or laboratory permeability test is only the hydraulic conductivity of sand which is merely a part of the sand sediment soil, and because the hydraulic conductivity of the sand sediment soil can be never characterized by this value of the sand permeability. This complicated and practically important problem restricts our accurate prediction of a water movement within the sand sediment soil or a flood runoff over the soil by using a numerical procedure based on a continuum theory.

In this study an integrated procedure to determine the hydraulic conductivity of the sand sediment soil that is characterized by an inclusion of large gravel particles and cobbles in the sand soil is proposed. In the integrated procedure, the hydraulic conductivity of the sand soil is measured by using an in-situ permeability test, and then an overall hydraulic conductivity of the sand sediment soil (in Section 3.1 this hydraulic conductivity is denoted by $K_{unified}$) is estimated by taking account of a continuity of water flow through soil and a gravel con-tent in soil. A Guelph pressure infiltrometer (GPI) method, which was developed by Reynolds and Elrick (1990) and Elrick and Reynolds (1992), is employed to measure the hydraulic conductivity of the sand soil. The GPI method is classified into a constant-head infiltration method and provides a simple in-situ permeability test. A field-saturated hydraulic conductivity of soil, K_{fs}, is determined by measuring a constant-head infiltration into soil from a single ring inserted into the soil surface in the GPI method.

In the following, firstly, a test procedure of the GPI method is outlined and extended so that the GPI method can determine unsaturated moisture properties of soil. Moisture content beneath the soil surface around the GPI ring is measured with time, and a genetic algorithm combined with a numerical saturated-unsaturated flow analysis is employed to estimate parameters describing the unsaturated moisture properties of soil from the in-filtration rate and the moisture content measured with time during the GPI test. Secondly a functional relationship is theoretically derived to evaluate the hydraulic conductivity of the sand sediment soil. A descriptive cylindrical soil model representing sand, gravel and cobbles, and voids within the soil is assumed and a continuity law of flow discharge through it is introduced to derive the functional relationship of the hydraulic conductivity of the sand sediment soil with the hydraulic conductivity of sand measured by the GPI method and the gravel content of the soil. The functional relation-ship which gives the hydraulic conductivity of the sand sediment soil is examined by a series of laboratory permeability test. Finally the integrated procedure that consists of the GPI method and the functional relationship of the hydraulic conductivity derived above is proposed to evaluate the soil permeability of the sand sediment. A numerical example is given to show a practical influence of the gravel content of soil on a storm runoff in the sand sediment soil. Some concluding remarks follow it.

2 IN-SITU PERMEABILITY TEST USING GPI

2.1 GPI method

The GPI consists of a single steel ring with a radius a inserted into the soil to depth d, a water supply tube and a water reservoir as shown in Figure 1 (Morii et al., 2000). The position of an air tube keeps the constant head of water H applied on the soil surface within the ring. Only the infiltration rate Q_s measured after the infiltration from the single ring into the soil reaches a quasi-steady state is required in the GPI method. Then the value of K_{fs} is calculated by the following equation (Reynolds and Elrick, 1990; Elrick and Reynolds, 1992):

$$K_{fs} = \frac{\alpha^* G Q_s}{\alpha^* a H + a + \alpha^* \pi a^2 G} \quad (1)$$

in which G is a dimensionless shape factor which takes account of the geometry of the infiltration surface within the ring. G is given by

$$G = 0.316 \frac{d}{a} + 0.184 \quad (2)$$

In Equation (1), α^* is a power describing an exponential relationship of unsaturated hydraulic conductivity with negative pressure head of soil, and is interpreted as an index of texture/structure component of soil capillarity. The GPI method requires that α^* be site-estimated by simple observation of soil. Values of α^* for various soil textures and structures are recommended by Elrick and Reynolds (1992).

2.2 In-situ permeability test

A practical effectiveness of the GPI method was examined by an in-situ permeability test conducted in the sand sediment soil about 50 m wide, 160 m long and 10 m thick. The sand sediment soil tested had been formed in the narrow valley of the mountainous watershed by the debris flow of soil. Five test points 10 to 20 m apart each were selected in the sand sediment soil. The soil contains a large number of gravel particles 30 to 70 cm in diameter and the sand to gravelly sand deposited among the gravel and cobbles as shown in Figure 2. The in-situ permeability tests using the GPI method were conducted on the surface soil at five test points. After completion of the tests, the soil was excavated about 1 m in depth by a back hoe and man-hands, and then

Figure 2. Sand sediment soil including gravel and cobbles.

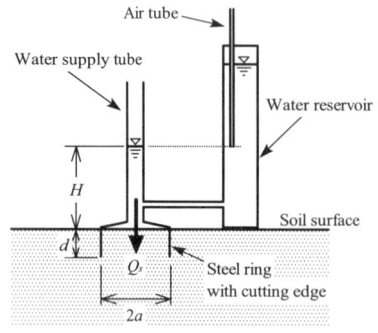

Figure 1. Schematic diagram of the GPI apparatus.

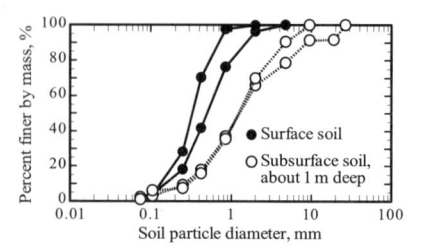

Figure 3. Grain size distribution of the sand tested.

the in-situ permeability tests using the GPI method were again carried out on the surface of the excavated soil. The steel ring with $a = 5.5$ cm was inserted about $d = 3.0$ cm into the soil, and 15 to 17 cm of H were applied on the soil surface in the tests. Figure 3 shows the results of grain size analysis of the sand. The soil near the sediment surface and the soil deposited in the deeper portion of the sediment are classified into sand and gravelly sand, respectively. In the GPI method, $a^* = 0.12$ cm^{-1} was adopted based on the site observation of the sand.

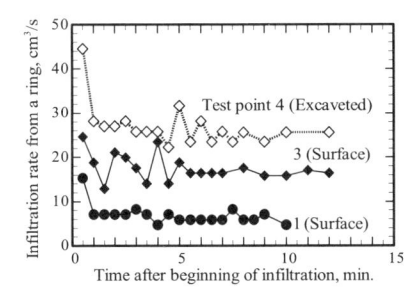

Typical examples of the infiltration rate measured during the in-situ permeability test using the GPI method are shown in Figure 4. As the soil tested is sand to gravelly sand, only 5 to 10 minutes were required to measure Qs irrespective of the preceding moisture condition of the soil. The total time required for setting the GPI apparatus, supplying water into the water reservoir and measuring the infiltration rate was a half to one hour.

K_{fs} of the soil determined by Equation (1) and corrected at 15 degree centigrade of water temperature are summarized in Table 1. Two measurements were repeated with slightly changing H at each test point. The in-situ tests on the excavated soil at Test point 2 were conducted in failure because of interference due to the small stone during the steel ring insertion into the soil. Sixteen values of K_{fs} given in Table 1 excluding those of Test point 2 are analyzed statistically to find out whether the values of K_{fs} are significantly different in the soil surface plane as well as along the soil depth. Table 2 shows the result of an analysis of variance based on a two-way layout method. Two factors describing the test point on the soil surface (that is Test point #1 to #5 on the soil surface and on the excavated soil surface) and the soil depth (that is the soil surface and the excavated soil surface) are selected, and both a variance ratio and a level of significance of these factors are calculated in the analysis of variance. It is found that the sand sediment soil is homogeneous in terms of per-

Table 1. Field-saturated hydraulic conductivity, K_{fs}[a], of the sand soil measured by the GPI method.

Soil depth	Surface		Excavated	
Repeat [b]	#1	#2	#1	#2
Test point 1	1.06×10^{-2}	1.48×10^{-2}	8.16×10^{-2}	8.32×10^{-2}
Test point 2	4.01×10^{-2}	5.96×10^{-2}	- [c]	-
Test point 3	4.05×10^{-2}	2.96×10^{-2}	1.86×10^{-1}	5.58×10^{-2}
Test point 4	1.65×10^{-2}	3.04×10^{-2}	4.85×10^{-2}	4.15×10^{-2}
Test point 5	2.66×10^{-2}	3.40×10^{-2}	4.13×10^{-2}	5.99×10^{-2}

a) K_{fs} in cm/s are corrected at 15 degrees in centigrade of water temperature.
b) Two measurements #1 and #2 were repeated with slightly changing H at each test point.
c) Tests on the excavated soil at Test point 2 were conducted in failure.

Table 2. Result of the analysis of covariance to examine statistical property of soil permeability.

Factors [a]	Degree of freedom	Variance ratio F_o	Level of Significance p [b]
Test point (A)	3	1.36	0.323
Test point (B)	1	8.76	0.018**
A×B	3	1.01	0.437
Error	8	-	-
Summation	15	-	-

a) A×B shows an interaction of the factors A and B on K_{fs}.
b) A set of two asterisks means that the factor has statistically a highly significant effect on K_{fs}.

meability in plane area as the probability of significance is not so small. But difference in K_{fs} measured at the surface soil and the excavated soil is statistically highly significant. From the result shown in Table 2, the sand sediment soil tested has non-uniform property of permeability along depth into the soil. This may be well explained by a process of sedimentation of sand and large gravel contained in the debris flow, where more massive gravel settles faster than smaller sand particles. This also can be explained by comparing the grading curves given in Figure 3. It may be right to conclude that a simplicity and rapidness of measurement provided by the GPI method realizes the statistical analysis of the soil permeability which has scarcely been tried in a soil investigation. This should be one of the special features of the GPI method.

2.3 Estimation of unsaturated moisture properties of soil

The GPI method was extended so that it could estimate the unsaturated moisture properties of soil. Figure 5 outlines the extension of the GPI method, where a volumetric moisture content beneath the soil surface near the GPI ring is measured by using a moisture sensor, Theta Probe type ML2x (Delta-Devices Ltd.), during the constant-head infiltration from the ring. Both the infiltration rate and the volumetric moisture content measured with time

Figure 4. Typical examples of the infiltration rate measured during the in-situ permeability test using the GPI.

during the GPI test were simulated iteratively by using the genetic algorithm (GA) combined with a FEM saturated-unsaturated flow analysis (GA+FEM) to estimate the unsaturated moisture properties of the soil (Takeshita and Morii, 2002). Functional relationships among volumetric moisture content, negative pressure head and unsaturated hydraulic conductivity of soil were assumed to be described by van Genuchten's equation (van Genuchten, 1980). The most optimal set of values of the soil parameters describing the functional relationships were determined by the GA so that it minimized a sum of squared deviations between the measurement and the FEM calculation of both the infiltration rate and the volumetric moisture content. An axisymmetric soil region, 50 cm in radius and 60 cm in depth, was selected for the FEM calculation as shown in Figure 5.

The measured data at Test point 3 shown in Table 1 were analyzed by the GA+FEM. Figure 6 shows comparison of the cumulative infiltration and the volumetric moisture content between the measurement and the FEM calculation which was obtained after the iterative estimations by the GA+FEM. Using the most optimal set of the soil parameters estimated in Figure 6, the van Genuchten's functional relationships of the unsaturated moisture properties of the sand are determined as shown in Figure 7. A relative hydraulic conductivity along the right y-axis of Figure 7 is defined as a ratio of the un-saturated hydraulic conductivity of soil to K_{fs}. A dotted line in Figure 7 is an exponential relationship of the unsaturated hydraulic conductivity of the sand that is drawn by using the recommended value of α^* mentioned in Section 2.1. A fairly good agreement of the unsaturated hydraulic conductivity between the van Genuchten's functional relationship and the dotted exponential line may mean that the introduction of the GA+FEM analysis into the in-situ GPI test provides practical estimation of the unsaturated moisture properties of soil.

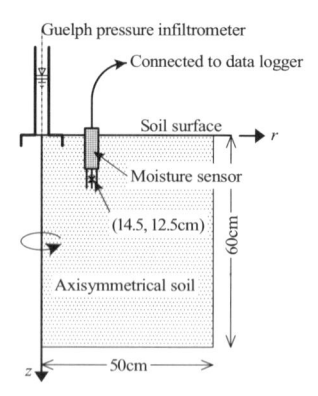

Figure 5. Moisture sensor, Theta ProbeML2x, inserted near the GPI ring to measure volumetric moisture content during infiltration.

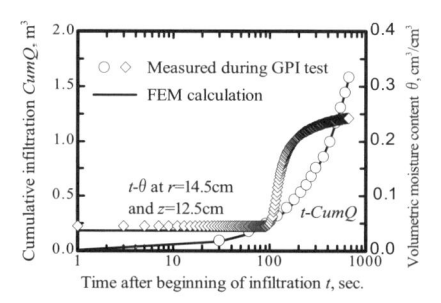

Figure 6. Comparison of cumulative infiltration and volumetric moisture content between the measurement and the FEM calculation after the GA+FEM iterations.

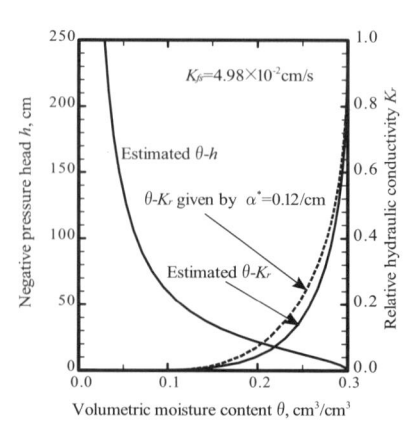

Figure 7. Unsaturated moisture properties of the sand estimated by the GA+FEM.

3 HYDRAULIC CONDUCTIVITY OF SAND SEDIMENT SOIL

3.1 *Hydraulic conductivity of sand sediment soil de-rived from cylindrical soil model*

In the numerical prediction of flow through the sand sediment soil as shown in Figure 2, both the sand and the gravel are totally unified into a porous continuum mass, and the value of the hydraulic conductivity of this unified porous mass is required. But the value that can be measured by the in-situ or laboratory permeability test is only the hydraulic conductivity of the sand which is merely a part of the sand sediment soil. Thus some procedure to evaluate the hydraulic conductivity of the unified mass of the sand and the gravel, $K_{unified}$, from the measurement of K_{fs} of the sand is needed.

To solve the problem mentioned above, a descriptive cylindrical soil model representing sand, gravel and voids is assumed as shown in Figure 8. A cross-sectional area A of the cylindrical soil model consists of the area of sand, void and gravel denoted by A_s, A_v and A_g, respectively. It is well understood that the value of the hydraulic conductivity which is measured by the in-situ permeability test represents the permeability of the region A_s+A_v. Denoting A_s+A_v by A_m and the hydraulic conductivity of A_m by K_m, then the flow discharge through A_m is given by $q_m = (K_m \cdot i) \cdot A_m$ in which i is a hydraulic gradient applied to the cylindrical soil model to move water along the axis as shown by a thick arrow in Figure 8. An amount of q_m should be equal to the flow discharge through A, that is $K_{unified} \cdot A \cdot i$, in the numerical calculation of flow because the gravel are completely impervious. Thus the hydraulic conductivity of the unified mass A is given by

$$K_{unified} = \frac{q_m}{A \cdot i} = K_m \left(1 - \frac{A_g}{A}\right) \qquad (3)$$

Assuming the same specific gravity G both for the sand and the gravel, and introducing a gravel content P that is defined as a ratio of mass of gravel to that of sand plus gravel, $A_g / (A_s + A_g)$, then Equation (3) can be rewritten as

$$\frac{K_{unified}}{K_m} = 1 - \frac{1}{1 + \left(\frac{1-P}{P}\right)\left(\frac{G \cdot \rho_w}{\rho_d}\right)} \qquad (4)$$

where ρ_d is a dry density of the sand and ρ_w is a water density. It should be noted that K_m in Equation (4) corresponds to K_{fs}.

To examine an accuracy of Equation (4), a series of laboratory one-dimensional permeability tests was conducted. Sand, 1 mm in maximum particle diameter and without fine particles, was mixed by P with river gravel sieved into 10 to 15 mm in diameter,

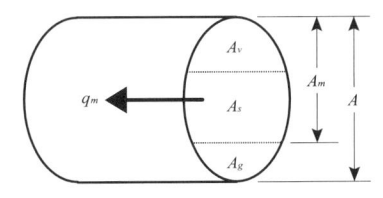

Figure 8. Descriptive cylindrical soil model assumed to derive the hydraulic conductivity of the sand sediment soil, $K_{unified}$.

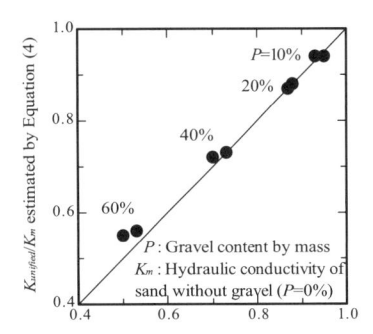

Figure 9. Comparison of the hydraulic conductivity of the sand-gravel mixture between the estimation by Equation (4) and the laboratory permeability test.

and compacted into an acrylic cylindrical column 10 cm in diameter and 100 cm long. P = 10, 20, 40 and 60 % were selected successively in the series of the laboratory one-dimensional permeability tests. A flow discharge through the soil specimen was measured at the top outlet of the column to determine the hydraulic conductivity of the sand mixed with the gravel, that is $K_{unified}$. The hydraulic gradient applied to the soil specimen was found from the measurement of total head along the soil column. The value of K_m was determined from the sand specimen without any gravel (P = 0 %).

Figure 9 shows a comparison of $K_{unified}/K_m$ measured in the laboratory one-dimensional test with the estimation by Equation (4) with a known value of G = 2.65. Two specimens were prepared in each test of P. A fairly good agreement of $K_{unified}/K_m$ between the measurement and the estimation shows a practical accuracy of Equation (4). $K_{unified}/K_m$ estimated by Equation (4) is slightly larger than the measurement at P = 60 %. This may be due to a non-uniform distribution of the gravel particles within the specimen, formed during pouring the sand-gravel mixture into the cylindrical column.

3.2 *Numerical example of storm runoff on sand sediment soil*

As shown in Equation (4), P is a key parameter that characterizes the permeability of the sand sediment soil. To show a practical influence of P on a prediction of storm runoff, a numerical sand sediment soil as shown in Figure 10 is selected and analyzed by the saturated-unsaturated flow FEM (Morii, 1999). The numerical sand sediment soil 100 m long and 10 m thick suffers from the storm 30 mm/h during 24 hours along its top surface and slope. K_{fs} = K_m = 3.0×10^{-2} cm/s, G = 2.65, and ρ_d = 1.35 g/cm³ were given according to some results of the in-situ test in Section 2.2 to calculate $K_{unified}$ by Equation (4). The unsaturated moisture properties of the soil were de-

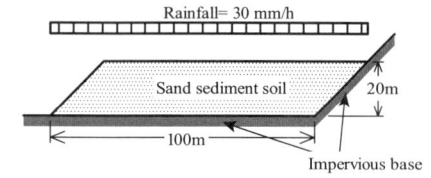

Figure 10. Numerical sand sediment soil to show practical influence of the gravel content on a prediction of storm runoff.

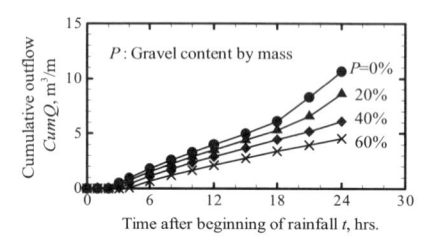

Figure 11. Cumulative outflow from the numerical sand sediment soil with time calculated by the FEM in which $K_{unified}$ estimated by Equation (4) is employed.

scribed by the van Genuchten's functional relationships estimated in Section 2.3 and given in Figure 7. An initial degree of saturation in the numerical sand sediment soil was assumed to be 60 %.

Figure 11 shows the numerical calculations of cumulative outflow through the soil slope from beginning to end of the storm. $P=$ 0, 20, 40 and 60 % were compared in the numerical calculations, in which $P=$ 0 % represents the sand soil without any gravel. It is found in Figure 11 that, if the value of the hydraulic conductivity of the sand measured by the in-situ permeability test is directly used in the numerical prediction of the flow in the sand sediment soil, the amount of the flow through the soil will be overestimated and, inversely, the flow over the soil surface is underestimated.

4 CONCLUSIONS

The integrated procedure to determine the hydraulic conductivity of the sand sediment soil that is characterized by an inclusion of the large gravel particles and cobbles in the sand was proposed. Firstly the Guelph pressure infiltrometer method was employed to measure the field-hydraulic conductivity of the sand, and extended so that it can determine the unsaturated moisture properties of the soil. Secondly the descriptive cylindrical soil model representing sand, gravel and voids was assumed to derive the functional relationship of the hydraulic conductivity of the sand sediment soil with the hydraulic conductivity of the sand measured by the GPI and the gravel content of the soil. The functional relationship derived was successfully examined by the laboratory permeability test. Finally a numerical example was given to show the practical influence of the gravel content on the prediction of storm runoff in the sand sediment soil.

The following is remarked:
a) The GPI method was effectively applied to the sand sediment soil to determine the field-saturated hydraulic conductivity of soil. It was found that the procedure to determine the field-saturated hydraulic conductivity of soil is consistently and is not time-consuming. It may be practically important to show that the permeability of soil could be statistically evaluated owing to simplicity and rapidness of measurement of the GPI method.
b) The GPI was extended to measure the volumetric moisture content near soil surface during the constant-head infiltration. Both the infiltration rate and the volumetric moisture content measured with time were successfully analyzed by the GA+FEM to estimate the soil parameters that describe the unsaturated moisture properties of soil.
c) The hydraulic conductivity of the unified soil mass, which represents the permeability of the sand sediment soil, was theoretically derived by applying the continuity law of flow to the descriptive cylindrical soil model, and effectively examined by the laboratory permeability test. The key parameter that characterizes the hydraulic conductivity of the sand sediment soil is the gravel content by mass. The numerical prediction of the storm runoff in the sand sediment soil showed the practical influence of the gravel content.

ACKNOWLEDGMENTS

The present study is supported by the Grant-in-Aid for Scientific Researches (B), No. 13556037 and No. 15360253, made by the Ministry of Education, Culture, Sports, Science and Technology, Japan. The study is also supported by the Joint Research Grant made by the Arid Land Research Center, Tottori University, Japan. The authors are grateful to Messrs. Satoshi Matsumoto and Takayuki Mori for their help to conduct the in-situ tests and to Mr. Tsuyoshi Yanagisawa for his help in the laboratory one-dimensional permeability tests.

REFERENCES

Elrick, D.E. & Reynolds, W.D. 1992. Infiltration from constant-head well permeameters and infiltrometers, in *Advances in Measurement of Soil Physical Properties: Bringing Theory into Practice* edited by G.C. Topp, W.D. Reynolds & R.E. Green, SSSA Special Publication 30, Madison, Wisconsin, Soil Science Society of America.

Morii, T. 1999. Prediction of water movement in soil by finite element method, *Bulletin of the Faculty of Agriculture, Nigata University, Japan*, 52(1): 41-54.

Morii, T., Inoue, M. & Takeshita, Y. 2000. Determination of field-saturated hydraulic conductivity in unsaturated soil by pressure infiltrometer method, in *Unsaturated soils for Asia* edited by H. Rahardjo, D.G. Toll & E.C. Leong, *Proceedings of the Asian Conference on Unsaturated Soils, Singapore, 18-19 May 2000*, Rotterdam, A. A. Balkema: 421-426.

Reynolds, W.D. & Elrick, D.E. 1990. Ponded infiltration from a single ring: I. Analysis of steady flow, *Soil Science Society of America Journal*, 54: 1233-1241.

Takeshita, Y. & Morii, T. 2002. Field measurement of unsaturated soil hydraulic properties using constant head permeability test, *Proceedings of the 3rd International Conference on Unsaturated Soils, 1, Recife, Brazil, 10-13 March 2002*: 417-423.

van Genuchten, M.Th. 1980. A closed-form equation for predicting the hydraulic conductivity of unsaturated soils, *Soil Science Society of America Journal*, 44: 892-898.

Proceedings ISC-2 on Geotechnical and Geophysical Site Characterization, Viana da Fonseca & Mayne (eds.)
© 2004 Millpress, Rotterdam, ISBN 90 5966 009 9

Adapted T-bar penetrometer versus CPT to determine undrained shear strengths of Dutch soft soils

O. Oung, J.W.G. Van der Vegt & L. Tiggelman
GeoDelft, Delft, Netherlands

H.E. Brassinga
Public Works Rotterdam, Rotterdam, Netherlands

Keywords: penetration testing, cone, CPT, T-bar, in-situ testing, soft soils, undrained shear strength

ABSTRACT: The objective of this work is to evaluate a new on-site investigation technique: the T-bar penetrometer to measure in-situ low undrained shear strengths. In laboratory conditions, model penetration tests in a prepared clay sample have been performed using a scaled cone penetrometer of the Dutch standard design and a jointless design where the sleeve is strain-gauged to measure the tip load using either a T-bar or a cone. The results obtained gave a better understanding of some effects affecting the tip load. In field conditions, penetration tests have also been performed using a T-bar and a cone. The undrained shear strengths evaluated from these penetration tests were compared mutually and to vane tests and triaxial tests. The capacity of the T-bar penetration tests to determine reliable values of undrained shear strength in-situ is discussed in comparison to the cone penetration tests.

1 INTRODUCTION

The soft peat and clayey soils in the Netherlands are often so soft that checks on the stability of a large number of constructions like dikes have to be carried out regularly and extensively. As in the case of dikes about 3 thousand kilometers are at stake and therefore the activities to evaluate the risks comprising soil investigation (typically soil sampling and triaxial tests) and calculations are very costly. In practice, because of limited budgets only limited amounts of tests can be carried out on sparse sites. To reduce the costs and to improve the quality of the predictions, alternative solutions need to be found.

The applications of in-situ and continuous techniques (like a Cone Penetration Test 'CPT') are generally of lower costs than multistage methods involving laboratory tests (like triaxial tests on bored samples). They also can provide more information than on-site discontinuous investigation methods like for instance vane shear tests, which are performed at chosen depths. Provided that the obtained parameters are reliable and accurate (limited corrections are needed) then it is worth promoting such techniques.

In this work, the parameter to be determined is the undrained shear strength c_u. In addition to the inherent advantages of using an in-situ and continuous method, the use of c_u in the predictions of the stability of dikes is new in the Netherlands. It is expected to produce less conservative predictions.

Moreover when the investigation technique to be investigated is approved then because of its low costs more sites can be investigated so that predictions can be more representative and of higher quality.

The objective of this paper is to evaluate the application of a T-bar penetrometer in comparison to the cone penetrometer to determine (low) undrained shear strengths in the typically Dutch soft soils. In the first part of this paper, tip resistances of a T-bar and a cone are investigated as well as some effects influencing the tip bearing in model tests. In the second part, penetration tests using T-bar and cone penetrometers on a dike in the island of Marken (the Netherlands) and complementary soil investigation methods are studied. The undrained shear strengths of each method are compared and the results are discussed.

2 MODEL TESTS

2.1 *Cones and T-bar for model tests*

The T-bar penetrometer was initially designed for centrifuge testing (Stewart & Randolph (1991)). The need for centrifuge testing to have a sensitive on-flight investigation technique was strong because clay consolidated in-flight changes drastic cally its strength when unloaded to 1-g conditions (Davies & Parry (1983)). In the centrifuge conditions, a sounding technique is required to be on one hand as reduced as possible in dimensions because of the limited size of the soil model and on the other hand as sensitive as possible. In this perspective, miniature cones have been built but their measurements suffer from the effect of water pressure occurring in soft cohesive soils affecting the global tip bearing (Tani & Craig (1995)). The lack of space in the shaft made it difficult, nearly impossible, to build in a (water) pressure transducer so that corrections were not possible. On the contrary, the T-bar penetrometer showed little effect of water pressure in its measurements. Later, it has been successfully tested in soft clay in field conditions (Stewart & Randolph (1994)) and offshore to investigate the seabed (Randolph & al. (1998)).

To determine the undrained shear strength c_u of soft clays sensitive penetrometers are required.

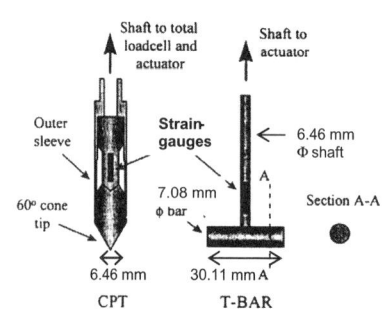

Figure 1. Miniature cone penetrometer and T-bar designed by the University of Western Australia for centrifuge testing.

Two types of sensitive penetrometers were used in the model tests:
- The T-bar and cone (6.47 mm in diameter) penetrometers of the University of Western Australia UWA. The design of these penetrometers differed from a standard CPT (Lunne & al. (1997)) by the measuring element. As shown in Fig. 1, strain gauges were glued along the inner shaft and the outer sleeve consists of a stiff plastic tube. The surface of the rod was smooth. The T-bar shown in Fig. 1 is not correctly scaled.
- The cone of 7-mm in diameter (without friction sleeve and pore pressure measurements) ordered by GeoDelft to AP van der Berg bv. (ApvdB). This cone was scaled down based on a standard cone.

The range of bearing capacity of the T-bar and cone of UWA was 200N. The accuracy on the full scale was 0.1%, which is 0.2N. For the cone_ApvdB, the range of bearing capacity was 250N and the accuracy full scale was 1N. Projected areas and range of bearing pressure are given in Table 1.

2.2 *Model tests*

Penetrations tests have been performed in a prepared clay cake. The clay was prepared by consolidating hydraulically kaolinite Speswhite clay slurry (100% water content) at a pressure of130kPa for ~30days in a cylinder of 900mm in diameter. This pressure was removed while performing the penetration tests. At the end of the consolidation, the clay was 585mm high. The clay was protected during the tests using a cling film. Free water left on top of the clay after consolidation was removed to prevent swelling and loss of strength in the upper part. The homogeneity of the clay cake was checked by determining water contents every 25mm in depth. The profile obtained is given in Fig. 2.

Table 1. Characteristics of the tips used in model tests

Type of head	T-bar UWA	Cone UWA	Cone ApvdB
Projected area [mm^2]	213.2	32.9	38.5
Range of bearing pressure [MPa]	0.938	6.09	18.2
N-factor	10.5 (+/-13%)	10-20 (15)	10-20 (15)

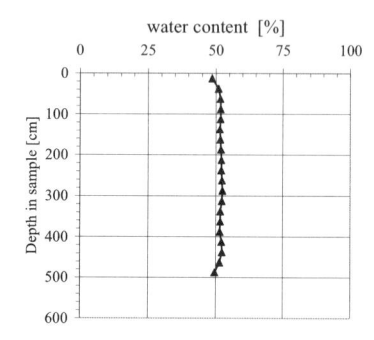

Figure 2. Profile of water contents 16 hours after penetration tests.

The penetration tests have been carried out at various penetration rates: 5, 0.5 and 0.05mm/s.

For comparison, each test was carried out by pushing away simultaneously 2 penetrometers into the clay sample. The 2 penetrometers consisted of on one side, the cone_ApvdB and on the other side either the cone_UWA or the T-bar_UWA. The minimum distance between two penetration sites was 100mm. Tests have been repeated to check their reproducibility.

The results of these tests are presented in Fig. 3. The following observations can be expressed:
1. The tip resistances of the T-bar and of the cone_ApvdB were reproducible (Fig. 3). Those of the cone_UWA were clearly not.

2. The cone_ApvdB presented rather stable measurement of the tip resistance in comparison to the cone_UWA. In the construction of the UWA penetrometer, the plastic skin was not directly mounted to the inner shaft at the shoulder of the tip but a distance of 2mm. This indicates that the UWA penetrometer could be sensitive to soil friction.
3. The cone_UWA and the cone_ApvdB penetrometers showed different tip resistances especially at the penetration rate of 5mm/s. The tip resistances profile of cone_UWA (1) varied very differently in comparison to the 3 other ones. Cone_UWA (1) is probably not representative (see Fig. 3). The tip resistance profile of cone_UWA (2) was larger than that of the cone_ApvdB, especially at 5mm/s. This difference can be explained by the effect of water pressure, which partly lowered the measured tip resistance of the cone_ApvdB in comparison to the cone_UWA. The UWA penetrometer was designed to be unsensitive to lateral water pressures. At the rate of 5mm/s, undrained conditions apparently ruled.
4. The T-bar_UWA showed clearly lower tip resistances than the two types of cones, independently from the penetration rates and types of constructions. The difference in the shape of the tip can be invoked here.

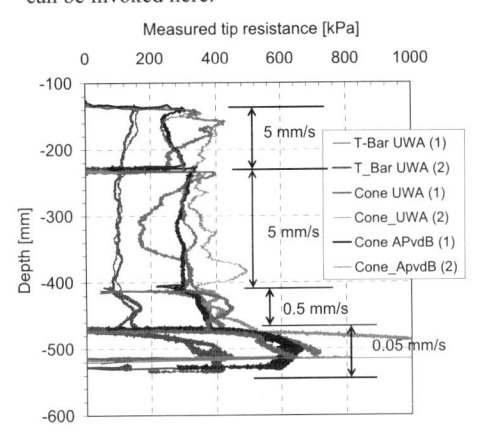

Figure 3. Profiles of measured tip resistances of penetration tests using 3 various tips, at various penetration rates

5. At a penetration rate of 0.05mm/s, a clear increase in the total tip resistance was observed in all the tests performed although passing from 5mm/s to 0.5 mm/s showed nearly no increase in the tip resistance. This point will be discussed in section 2.5.

2.3 Evaluation of total water pressure

Generally, at relatively large penetration rates, excess pore pressures are generated at the tip of penetrometer resulting in lower measured tip resistances. To correct from this effect, only water pressures at the tip of the penetrometer can be used. For the standard cone penetrometer, the measured bearing pressures should be corrected according to the following relation:

$$q_t = q_c + (1-a)u \qquad (1)$$

Where q_t is the corrected tip resistance, q_c is the measured tip resistance, a is net area ratio and u is the total water pressure acting behind the cone.

To estimate the total water pressure at 5mm/s, in the series of penetration tests performed, the penetrometer was held at a depth of ~-215mm. The tip resistances of the T-bar and the cone penetrometers varied accordingly (Fig. 4). Just after the standing still, the penetrometers were pushed away at a rate of 5mm/s. They then showed an immediate decrease followed by a steep increase, which may be related to relaxation of the soil around the penetrometer and dissipation of water pressure. The difference between these two measurements was relatively constant (~100kPa).

Moreover the bearing pressure of the T-bar was on average nearly zero and became even slightly negative at about the end of this dissipation/relaxation test. By holding the T-bar, the clay went to rest on the upper part of the bar. The soil stress pushing upwards the T-bar was compensated for only ~75%.

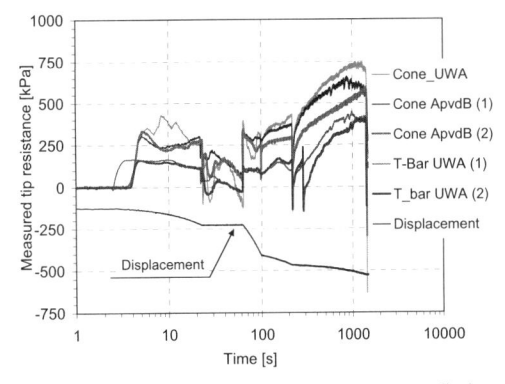

Figure 4. Measured tip resistances and penetrometer displacement versus time.

After 40s of water pressure dissipation, the penetrometers were pushed away at a rate of 5mm/s. This operation generated an immediate and sharp increase in bearing pressures (Fig. 4). Because of the undrained conditions, it was probable that the generated water pressures should have affected the total resistance of the cone penetrometers assuming that water pressure had nearly no effect on the T-bar measurement (Stewart & Randolph (1994)). A difference of ~100kPa between the two bearing pressures was reproducibly found. It is equivalent to a water pressure affecting the cone resistance with ~14kPa. The net area ratio of the cone_ApvdB was 0.86. Because of the small size and the homogeneity (Fig. 2) of the clay cake water pressures at the tip could be taken as nearly constant along the depth. For the evaluation of c_u, the value of 14kPa was used to correct the cone resistance from the effect of water pressure at the tip.

2.4 *Undrained shear strengths*

From the penetration tests, it is possible to evaluate the undrained shear strengths c_u of the soil. C_u is proportional to the total tip resistance:

For cone: $c_u = (q_t - \sigma_{vo})/N$ (2)

For T-bar: $c_u = q_t/N$ (3)

Where σ_{vo} is the total in-situ vertical stress and N a factor varying with the type of penetrometer. Eq.3 means that most probably soil cannot fill in immediately the room behind the T-bar while it is pushed away.

The N-factor of the T-bar penetrometers N_b derived from a plasticity solution, see Stewart & Randolph (1991). N_b has upper and lower limits corresponding to limits of roughness coefficients, which are in the range of 13% from the adopted value of 10.5. The cone-factor N_c varies between 10 and 20 and varies strongly with the type of soil encountered. Lunne & al.(1997) and Stewart & Randolph (1991) used a value of 15. In this work, the values of 15 and 20 were applied for N_c. The N-factor of the T-bar and the cone penetrometers are summarized in Table 1.

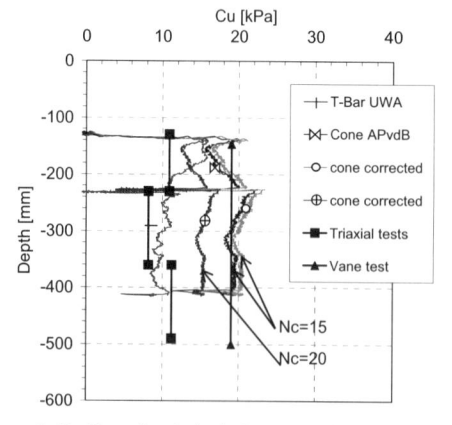

Figure 5. Profiles of undrained shear strengths (N-factor of cone: 15 and that of T-bar: 10.5). (for more details see text).

For comparison, triaxial tests (UU: unconsolidated undrained) have been performed on samples taken at three different depths. The tests were conform to the Dutch standards (NEN 5117) on a bored sample, which in turn was divided into 3 cylindrical samples of 67mm in diameter and 130 mm long. The results of these tests are presented in Fig. 5. In this figure, a line represents each triaxial test. Only one miniature vane test was performed at the a depth of -147mm. The measured s_u was multiplied by 0.8 according to Bjerrum's correction in order to obtain c_u (Bjerrum (1972)). This value was extrapolated to the rest of the clay assuming homogeneity (see line corresponding to the vane test in Fig. 5).

In Fig. 5, the c_u-values of the T-bar penetrometer was obtained from the measured tip resistances according to eq.2 whereas that of the cone penetrometer was corrected from the water pressure determined previously.

Correction of q_t should also take into account soil friction occurring behind the cone and the slit. Eq.1 would then be rewritten as follows:

$q_t = q_c + (1-a)u - bs_u$ (1bis)

b defines the length of the side of the cone tip (b = Lx4/D; D is shaft diameter). For the mini-cone_ApvdB, b = 1.1 and a = 0.86. As a consequence, the unwanted effect of soil friction (~15kPa) on q_t compensates that of the excess water pressure (~14kPa) in the tested soft clay as foreseen by Mesri (2001).

It can be concluded that the c_u-values of the T-bar penetration test agrees fairly with the UU triaxial tests. A good agreement is found between the vane test and the CPT when using not corrected cone resistances and an N-factor of 15 (Fig. 5). The c_u-values obtained from the CPTs were (~2 times) larger than those of the triaxial and of the T-bar penetration tests. This cannot be explained by the construction defaults but rather the interaction between soil and penetrometer for such soft clay and low penetration rates. More attention must be spent on this issue in the future.

2.5 *Effect of low penetration rates*

According to Campanella et al. (1983), at low penetration rates, the measured and the total (or effective) bearing pressures q_c and q_t increase as well as the measured sleeve friction. They found that this effect is mainly caused by the plasticity of the soil and is not related to total water pressures, which effect diminishes with the decrease of the penetration rates. In Fig. 3 however, the tip resistances at 0.05mm/s were substantially larger than those at 0.5 and 5mm/s independently of the types of penetrometer. Fig. 4 presents the measurements against the logarithm of time. The bearing pressures increased nearly linearly with log(time) and seemed to reach a maximum value, for both T-bar and cone_ApvdB. The reason for this effect is not clear. Possibly, the increase of the sleeve friction corresponded to an increase in a skin friction generated by the slow displacement of the shaft. This friction might be transferred by continuity to the tip of the penetrometer. Another possibility is the hardening of the soil created by the penetration in the initially soft clay.

Furthermore, it can be noticed that the bearing pressures of the cone_UWA at 0.05mm/s were much larger that those of the cone_ApvdB. This can be attributed to the construction of this cone, which makes it sensitive to soil friction as already mentioned.

2.6 Summary

Conclusions out of the model tests are:
- A T-bar penetrometer shows substantially lower tip resistance than a cone penetrometer.
- No correction on the T-bar resistance was needed to obtain c_u-values close to those of the triaxial tests.
- In the model tests performed, no correction on the cone resistance was necessary because the effect of soil friction acting on the cone substantially compensated that of water pressure acting behind the tip.
- The N-factor of 10.5 for the T-bar was acceptable. Acceptable values of the c_u coming from CPTs depend strongly on the choice of the cone N-factor. In the model tests, it should be large (15) to fit better the c_u-values of the vane test.

3 DETERMINATION OF c_U IN FIELD CONDITIONS

Undrained shear strengths in field conditions can be determined by applying the following characterization methods, see Lunne & al. (1997):
In-situ measurements:
- Field Vane Tests (FVT)
- Cone pressuremeter (CPM)
- Penetration tests (CPT)

Measurements upon undisturbed samples:
- Unconsolidated and Undrained UU tests at in-situ confining pressure
- Consolidated & Undrained CU tests at in-situ confining pressure

The field investigation has taken place in the island of Marken surrounded by the IJsselmeer (the Netherlands). This location presented soft soils like silty clay and Dutch peat. At two locations (HM 34.15K and HM 41.80) along the body and on the crown of the dike protecting the island from flooding, a series of field tests were performed with a small tractor. Testing at each location started with a CPT for soil stratification and classification. With the results of this CPT, the depths were chosen to perform the Field Vane Tests (FVT). At each of the 2 locations, a CPM test, a CPT (see section 3.1.2) and several samples were collected using the Begemann sampler. The triaxial tests were carried out on these samples. Each of these methods was operated at reasonable distance from each other as shown in Fig. 6. The results of the CPM test were not yet available at the time of writing.

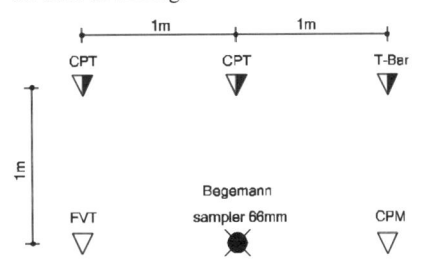

Figure 6. Relative distance of field tests.

3.1 Methods applied

In the following, a brief description of applied on-site investigation techniques is briefly given.

3.1.1 T-bar penetrometer

A T-bar penetrometer has been built according to the dimensions given in Stewart & Randolph (1994) for use in field conditions. This T-bar was constructed in such a way that it can be directly mounted on a 36mm shaft of the standard CPT equipment, at the location where a piezometer is usually mounted. A picture of the made T-bar and its dimensions are shown in Fig. 7. The penetration rate was 20mm/s.

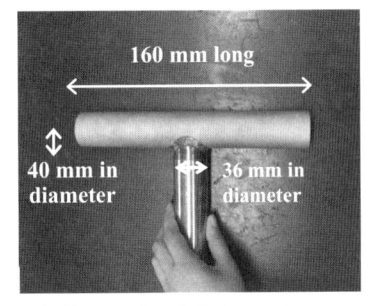

Figure 7. Photography of a T-Bar penetrometer

Figure 8. Mounting of the T-Bar penetrometer

The projected area of the T-bar was 6400mm². The accuracy was 0.01MPa. The procedure to perform the penetration tests is the same as that of a CPT. Fig. 8 shows how the adapted T-bar was mounted on a rig under a tractor.

3.1.2 Cone Penetration Test (CPT)

The standard and sensitive CPTs have been performed using a GeoDelft cone penetrometer conform the Dutch standard NEN 5140. The one used was a standard Dutch cone with a cone area of 1000mm², a friction sleeve of 150mm and a shaft of 36mm in diameter. The standard cone penetrometer was calibrated in pressures ranging from 0 to 50MPa. The sleeve was calibrated in the range of 0 to 0.7MPa. The sensitive cone penetrometer was treated in the same way as the standard one with the only difference that it was calibrated in a lower pressure range (0–14MPa) to obtain more accurate measurements (the accuracy was 0.02MPa). The sleeve friction was calibrated in the same range of pressures. The penetration rate was 20mm/s.

3.1.3 Field Vane Tests (FVT)

The Geonor field vane tests were performed conform the Dutch standard NEN 5106. The vane (65x 130mm²) was pushed (with a protection shield) into the soil and at a chosen depth the vane itself was pushed away 50cm further. The rotation rate of the vane was 0.1 degree per second. After measuring the maximum undrained shear strength, the vane was rapidly rotated at least 10 revolutions before the remoulded state was recorded.

3.1.4 The Delft continuous, or Begemann, sampler combined with triaxial tests

The Delft continuous or Begemann sampler produces a continuous profile of the soils encountered (Begemann (1966)). Both the sample diameter and its quality should be ideal for undisturbed laboratory testing. A cross-section through the sampler is shown in Fig. 9. The bottom 0.9m of the whole system consists of three tubes centered on one another as follows. The outside thick walled tube on which the compressive force is also exerted to push this part into the ground. Then a very thin walled tube of smaller diameter over which up to 23m of watertight pre-coated nylon stocking can be slid in folds. At the bottom end of the outside tube there is a cutting shoe so designed that a sample is obtained to the requisite diameter and the right length.

The triaxial tests were carried out in order to measure the stress-strain behavior and the shear strength parameters of a soil specimen under controlled stress conditions according to the Dutch NEN 5140 class 2. For this project, the samples were firstly anisotropically consolidated and all tests were performed in single stages in undrained conditions (CU).

3.2 Field penetration tests

Fig. 9 shows the results from the CPT and T-bar measurements on both locations.
The following observations can be made:

For both locations, the tip resistances measured were mainly comparable. The sleeve friction and consequently the friction ratio of the T-bar penetrometer were lower than that of the cone. The soil stratification interpreted from the q_t and R_f, of the CPTs agree with the Begemann samples. Although for the T-bar there are no empirical relations between q_t, f_s and R_f yet available, the measurements of Fig. 10 show that the same types of relations can be determined. It is not excluded that just a simple transposition of the empirical classification may be enough.

It can also be observed that the tip resistance of the cone shows a very peaked line in comparison to that of the T-bar. This can be attributed to the small projected area of the cone, which makes it sensitive to slight disturbance in the soil like fine heterogeneities. The T-bar however seems to present more average measurements of q_t as well as f_s.

Moreover at the site HM41.80, layer 5 consisting of peat ended at -5.82m for the T-bar whereas for the cone it was 0.38m deeper. This difference can be clearly observed on both the measurements of the tip resistance and the sleeve friction. The Begemann however shows the end of layer 5 at −5.47m. Assuming that the Begemann sampler provided with a nearly undisturbed sample and therefore the most re

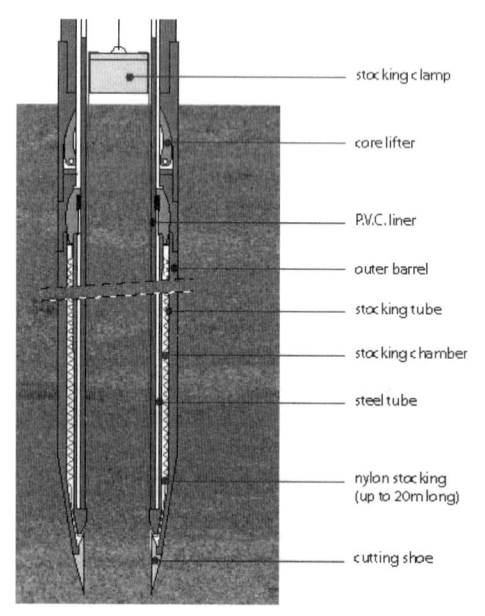

Figure 9. Begemann sampler of 66 mm in diameter

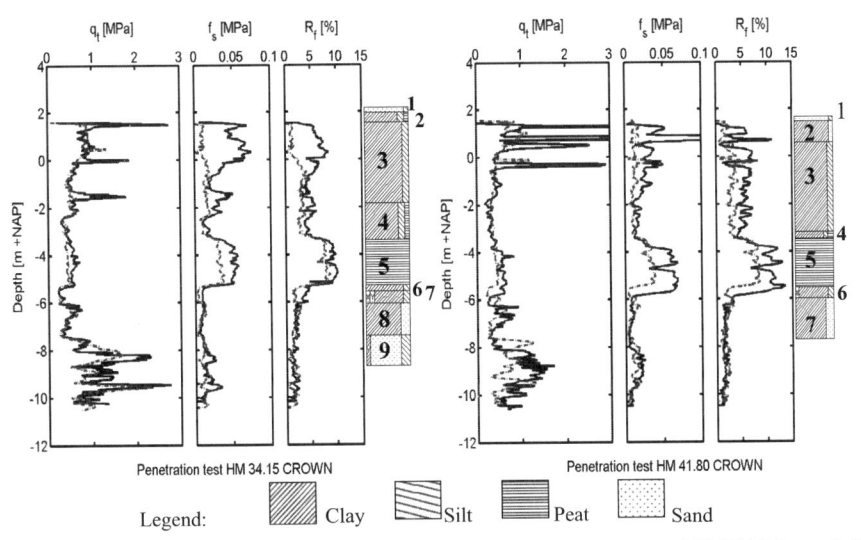

Figure 10. T-bar and Cone penetration tests and Begemann profiles: at site HM 40.80 Crown and HM34.15 Crown. Full line (blue): CPT. Dotline (red): T-bar penetrometer. Numbers in the profiles are related to soil layers.

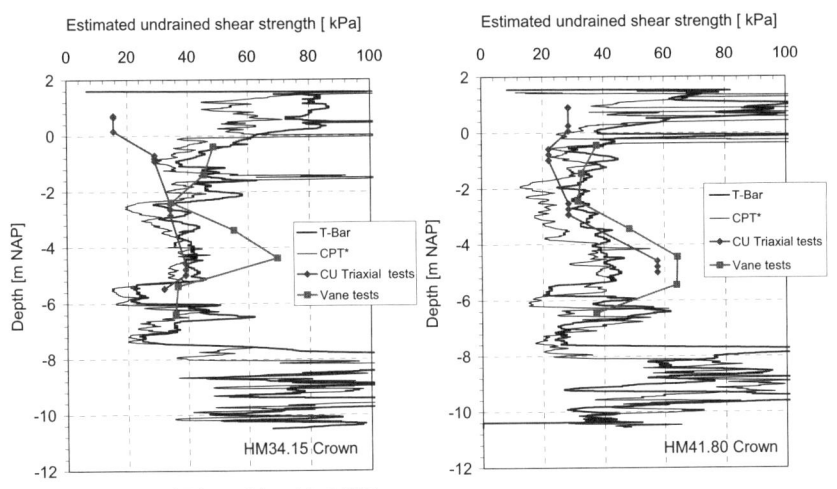

Figure 11. Profiles of Cu of T-bar and (sensitive) CPT.

liable method here, the difference in detecting the low limit of the peat layer between the cone and the Begemann was 0.73m. This is relatively large even if the Begemann sample was slightly disturbed and a drastic change in the soil layers can not be invoked because the sites of investigation were very close to each other. The reason for this large difference can be referred to the pattern of the soil flow around the penetrometer. The resolution to detect layers frontiers is nearly twice the diameter of the cone and it is about the double of that of the T-bar. In this case, the T-bar can reflect better the limits of the layers than the cone.

3.3 *Determination of c_u*

The undrained shear strengths of the penetration tests were estimated from the measured tip resistances. An N-factor of 10.5 was applied to the T-bar bearing pressures while 15 to the cone bearing pressures. In Fig. 11, the profiles of the undrained shear strengths of the penetration tests are plotted together with the results of the CU triaxial and field vane tests. The vane shear strength s_u has not been corrected. It is possible for clay to apply the correction on s_u using a factor depending on the plasticity index of the soil as suggested by Bjerrum (1972). This factor varies between 0.6 and 1.2. (Aas et al. (1986)). A good fit between vane and triaxial tests would require a Bjerrum's factor of 0.8.

In peat (layer 5 of Begemann samples), although the T-bar presented relatively lower tip resistances than the cone for both investigation sites, the c_u obtained were comparable. The shear strengths of the triaxial tests were close to those of the penetration tests only at the site HM34.15. The c_u of the vane tests (~60kPa) were very distinct from those of the T-bar and cone (~40kPa). These results showed that the vane shear strengths (both in clay and in peat) were larger than those of the triaxial and the penetration tests. The penetration tests either with the T-bar or the cone showed reproducible c_u. Their difference was about 5kPa. A good fit between the c_u-values would require applying an N-factor of 13 on the cone resistance.

In the silty clay layer, the tip resistances of the two penetrometers were rather similar, which means that either water pressure at the tip affected both measurements in the same way or this effect was negligible. The c_u values of the T-bar in the silty clay layers (layers 2 and 3) agreed with the vane shear strengths. The triaxial tests have provided with lower c_u- values than the other methods. However the undrained shear strengths calculated differed of ~15kPa on the site HM41.80 (Fig. 11). It seems that the effect of water pressures on the tip resistance was limited. This difference can be explained by the value of the N-factor chosen. An N-factor of 10 would have better fitted the c_u of the cone penetrometer than a value of 15. In this case, T-bar and cone had the same N-factor.

4 CONCLUSIONS

From the model and field tests performed, the following conclusions can be expressed:
- The model tests show that the measured tip resistance is rather dependent on the design and the construction of each individual model tip.
- The tip resistance measured in the model tests show large differences between the Australian T-bar and CPT, while in the field test there is a good agreement. Geometrical effects of the model penetrometers may cause this.
- The determination of the undrained shear strength from the T-bar in model tests is straightforward, while this determination from CPT tests depends on necessary correction due to pore pressures, soil friction and the N-factor.
- In the conditions of the field tests, the influence of local excess pore pressure on the measured cone resistance of a CPT to determine undrained shear strength could be neglected.
- The T-bar penetrometer as used in the field tests proved to be a practical and reliable tool for the continuous simultaneous determination of tip resistance, stratification and undrained shear strength. Whereas the combined effect of increased sensitivity of the measuring device and the small size of the cone penetrometer makes a sensitive CPT less reliable.
- In the Dutch engineering practice, the measured cone resistance is used as prescribed in NEN 6740. This prescription does not include correction for excess water pressure effect and the soil friction on the tip resistance although this might affect partly the measurements especially in cohesive soft soils. Nevertheless, it is of importance to investigate this issue to complete the evaluation of the T-bar penetrometer.
- Not experienced here but important to notify is the robustness of the T-bar penetrometer. Sensitive cone penetrometers can be definitively broken when it encounters hard things in the soil. This makes the T-bar penetrometer even more suitable for use in field conditions.

ACKNOWLEDGEMENTS

We would like to acknowledge Prof. Randolph and the University of Western Australia for providing us with their homemade penetrometers as well as AP vd Berg.bv for building a 7mm-cone.

REFERENCES

Aas, G; Lacasse, S., Tunne,T. and Hoeg, K; 1986. *Use of in situ tests for foundation design on clay*. Invited lecture ASCE Specialty Conf. "in-situ", Blacksburg.

Begemann, H.K.S.Ph. 1966. *The new apparatus for taking a continuous sample*. LGM-mededelingen. Vol. 10, N°4, Delft Geotechnics.

Bjerrum L., 1972. *Embankment in soft ground. Proc. Performance of earth and earth supported structures*. ASCE, Lafayette, Vol. 2, 1-54.

Campanella R.G., Robertson P.K. & Gillespie, D. 1983. *Cone penetration testing in deltaic soils*, Can. Geotech. J., vol. 20: 23-35.

Davies M.C.R. & Parry H.G. 1983. *Shear strength of clay in centrifuge models*. J. Geot. Eng. Vol. 109, N° 10, October: 1331-1337.

Lunne, T., Robertson, P.K. & Powel, J.M. 1997. *Cone penetration testing in geotechnical practice*. Ed. Blackie academic & professional, Chapman & Hall, London.

Mesri. 2001. *Undrained shear strength of soft clays from push cone penetration test*. Géotechnique. 2:167-168.

Randolph, M.F., Hefer, P.A. Geise, J. & Watson, P.G. 1998. *Improved seabed strength profiling using T-bar penetrometer*. In proceedings int. conf. offshore site investigation and foundation behaviour. 'New frontiers', Society for Underwater Technology. London: 221-235.

Randolph, M.F. & House, A.R. 2001. *The complementary roles of physical and computational modelling*. Int. J. Physical Modelling in Geotech. vol. 1: 01-08.

Randolph, M.F., Martin, C.M. & Hu, Y. 2000. *Limiting resistance of a spherical penetrometer in cohesive material*. Géotechnique, Vol. 50, N° 5, 573-582.

Stewart, D.A. & Randolph, M.F. 1991. *A new site investigation tool for the centrifuge.* Centrifuge '91. Ko (ed.), Balkema, Rotterdam. ISNB: 9061911931.

Stewart, D.A. & Randolph, M.F. 1994. *T-bar penetration testing in soft clay.* J. Geotech. Eng. Vol. 120, N° 12, December: 2230-2235.

Tani K. & Craig W.H. 1995. *Development of Centrifuge cone penetration test to evaluate the undrained shear strength profile of a model clay bed.* Soils and Foundations. Vol. 35, N° 2, June: 37-47.

Development of seismic site characterization method using harmonic wavelet analysis of wave (HWAW) method

Hyung-Choon Park
Department of Civil Engineering, University of Texas at Austin, Austin, Texas, USA

Dong-Soo Kim
Department of Civil and Environmental Engineering, KAIST, Daejeon, Korea

Keywords: HWAW, time-frequency analysis, dispersion curve, SASW, V_s profile

ABSTRACT: The new site characterization method using harmonic wavelet analysis of wave (HWAW) method was proposed. HWAW method based on time frequency analysis mainly uses the signal portion of the maximum local signal/noise ratio to evaluate the phase velocity and it can minimize the effects of noise. This method consists of three steps: field testing, evaluation of dispersion curve, and determination of V_s profile using single array inversion process. Field testing of this method is relatively simple and fast because one experimental setup which consists of one pair of receivers is needed to determine the dispersion curve of the whole depth. HWAW method use the near field information and can sample much deeper part of the site than the conventional phase unwrapping method. This method uses single array inversion which consider receiver location without increasing calculation time and complexities because the whole dispersion curve is determined from one experimental setup. To estimate the applicability of HWAW method, numerical simulations at layered soil and pavement profiles were performed. Field tests were also performed and shear wave velocity profiles obtained by HWAW method were compared with those by conventional SASW test and PS-suspension logging test. Through numerical simulations and field applications, the good potential of the proposed method is verified.

1 INTRODUCTION

The evaluation of shear modulus (or shear wave velocity) profile of the site is very important in the various fields of geotechnical engineering. To evaluate shear wave velocity profile, various in-situ seismic methods using surface waves have been developed (Nazarian and Stokoe 1984, Gabriels et. Al 1987, McMechan and Yedlin 1981). These surface wave based in-situ seismic methods have their own strength and weakness.

In this paper, new seismic site characterization method using the harmonic wavelet analysis of wave (HWAW) was proposed to overcome some of weaknesses in the existing surface wave based seismic site characterization methods. HWAW method is based on time-frequency analysis using harmonic wavelet transform. HWAW method mainly uses the signal portion of the maximum local signal/noise ratio to evaluate the phase velocity and it can minimize the effect of noise. The seismic site characterization method using HWAW method consists of three steps: field testing, evaluation of dispersion curve, and determination of V_s profile by single array inversion process. The field testing of this method is relatively simple and fast because one experimental setup which consists of one pair of receivers is needed to determine the dispersion curve of the whole depth of interest. The proposed method uses the near field information and can sample much deeper part of the site than the conventional phase unwrapping method. This method uses single array inversion which considers the variation of phase velocity with receiver location without increase of calculation time and complexities because the whole dispersion curve is determined from one experimental setup.

In this paper, the harmonic wavelet transform and the proposed method are briefly described. To estimate the applicability of the proposed method, numerical simulations at layered soil and pavement profiles were performed. Finally, field tests were performed, and the shear wave velocity profiles obtained by HWAW method were compared with those by conventional SASW test and PS-suspension logging test.

2 HWAW METHOD

2.1 Determination of Dispersion Curve

Wavelet analysis is fundamentally one of the correlation method (Newland 1999). Wavelet coefficient, a(t), provides information about the structure of input signal, s(t), and its relationship to the shape of the analyzing wavelet, w(t). The wavelet coefficient a(t) is defined by the correlation equation as follow;

$$a(t) = \int_{-\infty}^{\infty} s(t')w^*(t'-t)dt' \quad (1)$$

w*(t) is complex conjugate of w(t). When s(t′) correlates well with w*(t′- t), a(t) will be large, but when they do not correlate, a(t) will be small. Any wave shape can be used for wavelet if it is localized at a particular time. Harmonic wavelet is represented as follow;

In the frequency domain :

$$W_{m,n}(\omega) = \frac{1}{(n-m)2\pi} \quad \text{for } m2\pi \leq \omega < n2\pi$$
$$= 0 \quad \text{elsewhere} \quad (2)$$

In the time domain :

$$w_{m,n}(t) = \frac{e^{jn2\pi t} - e^{jm2\pi t}}{j(n-m)2\pi t} \quad (3)$$

where $j = \sqrt{-1}$. The harmonic wavelet is localized and has a harmonic characteristic in time domain. Harmonic wavelet coefficient, $a_{m,n}(t)$ which is defined by $W_{m,n}(\omega)$ can be represented as follow (Park and Kim 2001);

$$a_{m,n}(t) = s_f(t) + jH[s_f(t)] = x(t)e^{j\theta_{m,n}(t)} \quad (4)$$

where $s_f(t)$ is the output signal of ideal bandpass filtering operation where the magnitude of filter is $\frac{1}{2}|W_{m,n}(\omega)|$ and its bandpass is $m2\pi \leq \omega < n2\pi$; H represents Hilbert transform; x(t) is magnitude of $a_{m,n}(t)$; $\theta_{m,n}(t)$ is phase of $a_{m,n}(t)$. From Eq. (4), it can be noticed that the real part of $a_{m,n}(t)$ is output signal of bandpass filtering operation and the imaginary part of $a_{m,n}(t)$ is Hilbert transform of the real part of $a_{m,n}(t)$, namely, $a_{m,n}(t)$ is the analytic signal or Gabor's complex signal corresponding to $s_f(t)$. The harmonic wavelet transform functions as ideal band pass filter so that the harmonic wavelet coefficient, $a_{m,n}(t)$, for wavelet in the frequency band $m2\pi \leq \omega < n2\pi$ contains information only in the selective frequency band (m2π ,n2π).

When the signal passes through medium between receivers 1 and 2, the real part of $a_{m,n}^1(t)$ and $a_{m,n}^2(t)$ are $s_f^1(t) = y(t-t_g^1)\cos[(m+n)\pi(t-t_{ph}^1)]$ and $s_f^2(t) = y(t-t_g^2)\cos[(m+n)\pi(t-t_{ph}^2)]$, respectively, where the upper index indicates number of receiver. If the bandwidth of $W_{m,n}(\omega)$, (n-m)2π, is sufficiently narrow, then group delay t_g and phase delay t_{ph} have sensible meaning, and the envelop and phase of $s_f^1(t)$ and $s_f^2(t)$ are obtained from magnitude and phase of $a_{m,n}^1(t)$ and $a_{m,n}^2(t)$. The group and phase delays at receiver 1 and 2 are obtained from the magnitude and phase information of $a_{m,n}^1(t)$ and $a_{m,n}^2(t)$, and then the group and phase velocities are obtained from these delays. The procedure to determine the group and phase velocities at frequency (m+n)π, where (m+n)π is the center frequency of $W_{m,n}(\omega)$, is as follow (Park and Kim 2001);

1) Compute harmonic wavelet transform of signals obtained at receiver 1 and receiver 2.(Fig. 1)
2) Determine phase and group delays at frequency (m+n)π. which is the center frequency of arbitrary harmonic wavelet $W_{m,n}(\omega)$.
   a) the group delays at receiver 1 and 2 which are t_g^1 and t_g^2 are obtained. The group delay is a time corresponding to the maximum of magnitude of $a_{m,n}^1$ and $a_{m,n}^2$. (Fig. 2(a), (b))
   b) From phase information of $a_{m,n}^1$, θ_1 is taken as a phase corresponding to t_g^1 . (Fig. 2(c))
   c) The t_L and t_R are obtained from phase information of $a_{m,n}^2$. The t_L is the time corresponding to θ_1 which is the most close to t_g^2 on the left side of t_g^2 and t_R is the time corresponding to θ_1 which is the most close to t_g^2 on the right side of t_g^2 . (Fig. 2(d))
   d) t_{ph}^1 is defined as t_g^1 and t_{ph}^2 is either t_L or t_R depending upon which is closer to t_g^2. (Fig. 2(d))

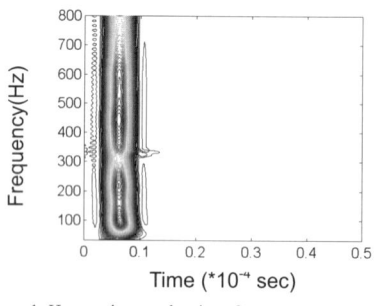

Figure 1. Harmonic wavelet time-frequency map

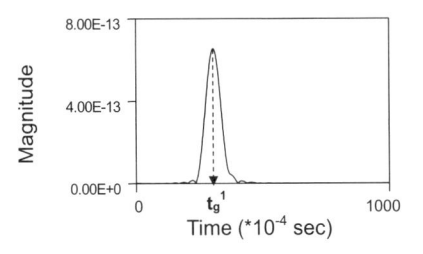

(a) Determination of group delay from magnitude of $a_{m,n}^1(t)$ at receiver 1

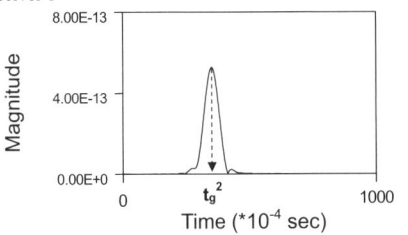

(b) Determination of group delay from magnitude of $a_{m,n}^2(t)$ at receiver 2

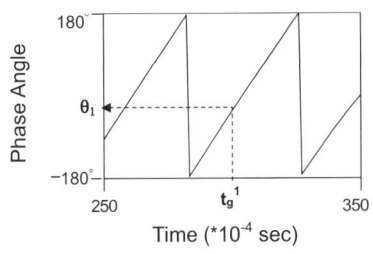

(c) Determination of θ_1 from phase of $a_{m,n}^1(t)$ at receiver 1

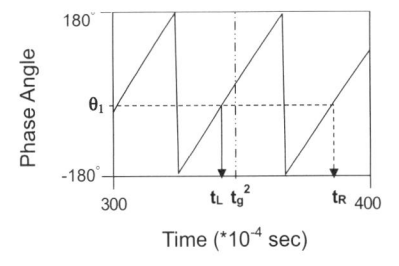

(d) Determination of phase delay from phase of $a_{m,n}^2(t)$ at receiver 2

Figure 2. Determination of the group and phase delays at frequency $(n+m)\pi$

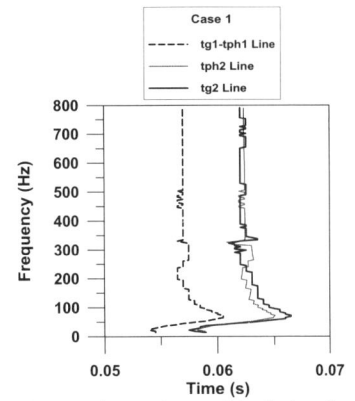

Figure 3. Phase and group delay in time-frequency domain

3) To determine the phase and group delays at whole frequencies, the procedure 2) has to be repeated for all harmonic wavelet coefficients. (Fig. 3)
4) If the distance between receiver 1 and receiver 2 is D, then the group velocity V_{gr} and the phase velocity V_{ph} at each frequency are obtained as follow;

$$V_{gr} = \frac{D}{t_g^2 - t_g^1} \qquad V_{ph} = \frac{D}{t_{ph}^2 - t_{ph}^1} \qquad (5)$$

In order that the above mentioned procedure is valid, the relative distortion of wave groups obtained by receivers 1 and 2 at each frequency component should be within some limit. The relative distortion can be determined by the period normalized time difference factor, Δt_T, as follow

$$\Delta t_T = -\frac{1}{V_{gr}} \cdot \frac{dV_{ph}}{d\lambda} \cdot D \qquad (6)$$

where λ is wave length. If Δt_T is N-0.5 < Δt_T < N+0.5 where N is integer, then the phase delay time at receiver 2 corresponding to real phase velocity is located (period)*N apart from t_{ph}^2 calculated by the procedure 2-d). Therefore, even if the absolute value of Δt_T is bigger than 0.5, t_{ph}^2 corresponding to the correct phase velocity can be determined from t_{ph}^2 of procedure 2-d). The procedure to determine the correct t_{ph}^2 from t_{ph}^2 of procedure 2-d) is called as data recovery process. HWAW method consists of preliminary calculation and data recovery process.

For non-stationary signals typically generated by impact source, the signal amplitude varies with time. In the conventional phase unwrapping method using Fourier transform, the signal to noise (S/N) ratio is measured at each frequency in the average sense. In the proposed method using time-frequency analysis, however, only information around t_g, where the signal energy is dominant, is utilized to evaluate the phase velocity at each frequency and the local S/N ratio around t_g is much greater than that in the average sense. This means that the proposed method is less affected by noise, and the phase velocity is obtained at each frequency independent of remaining frequency components so that the problem owing to spurious 360° cycle in the phase unwrapping method (Al-Hunaidi 1994) can be overcomed.

2.2 Test Setup and Inversion Process

For the site characterization, HWAW method can use two test setups; short receiver spacing setup and conventional test setup. In the short receiver spacing setup, source-receiver spacing is 6~12m and receiver spacing is 1~3m. In the conventional test setup, source-receiver spacing is over 10m and receiver spacing is same as source-receiver spacing. The proposed method also uses near field information to determine dispersion curve of whole depth from single test setup. It has been found that the near field dispersion curve includes long wave length component enough to explore deep layer and is more sensitive to deep layer material properties than far field dispersion curve. The dispersion curve determined by HWAW method and the theoretical dispersion curve using in the inversion process reflect the near field effect in the same way. Therefore, the near field dispersion curve can be used to evaluate deep soil profile. In general, for phase unwrapping method, average signal to noise (S/N) ratio is too low in the near field to determine the correct phase velocities. HWAW method can determine the correct phase velocities in the near field because the HWAW method uses local information where signal energy is dominant.

The phase velocities vary with receiver location, so test setup have to be considered for the inversion process. In the proposed method, the signal array inversion in which the theoretical dispersion curve is generated at receiver locations same as field test setup, is used determine the shear wave velocity profile. Because the proposed method uses just one test setup in the field, the single array inversion can be possible without increasing calculation time and complexity. If the proposed method uses short receiver spacing setup, the proposed method can minimize the possibilities of error due to lateral non-homogeneity and can determine detailed local soil profile along lateral direction by performing series of tests.

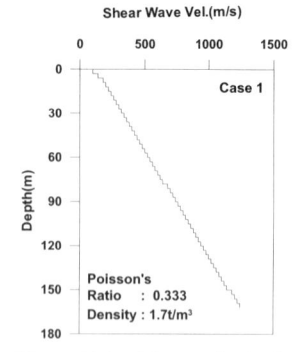

Figure. 4 Material properties for Case 1

Table 1. Material properties for Case 2

Case 2(Irregular Soil Profile)			
Thickness (m)	V_s (m/sec)	Density (t/m^3)	Poisson's Ratio
5	300	1.7	0.333
20	200	1.7	0.333
-	400	1.7	0.333

3 VERIFICATION

3.1 Numerical Simulation and Result

To verify the proposed method, numerical simulations were carried out using computer program FIT7 (Joh 1997) for soil profiles (Case 1, Case 2) and pavement profile (Case 3) as listed in Fig. 4, Table 1 and Table 2. Case 1 is regular soil profile in which stiffness increases with depth. Case 2 is irregular soil profile where stiffness varies irregularly with depth. Numerical test setups for soil profiles consist of two categories. The first category is short receiver spacing setup in which time signals generated at 6 and 7m from impact source and 12 and 13m from impact source were used as input signals. The second category is conventional test setup in which time signals are measured at 12 and 24m from impact source and used as input signals.

Fig. 5 shows the comparison of dispersion curve determined by the HWAW method and theoretical value for Case 1 and Fig. 6 shows the comparison for Case 2. It can be noticed that dispersion curve by proposed method coincide well with theoretical value in all test setup.

Table 2. Material properties for Case 3

Thickness (m)	Case 3(Pavement Profile)		
	V_s (m/sec)	Density (t/m^3)	Poisson's Ratio
0.1	1570	2.2	0.333
0.4	127	1.83	0.314
-	122	2.0	0.474

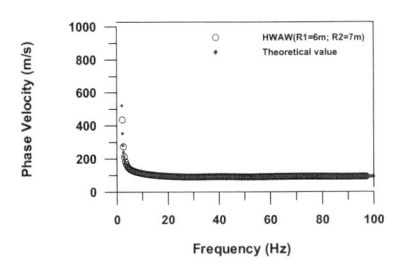

a) R1= 6m; R2=7m

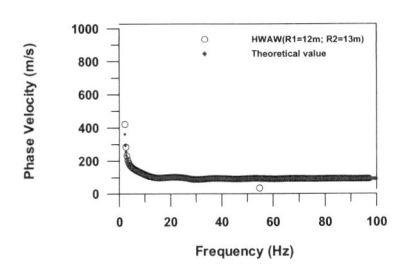

b) R1= 12m; R2=13m

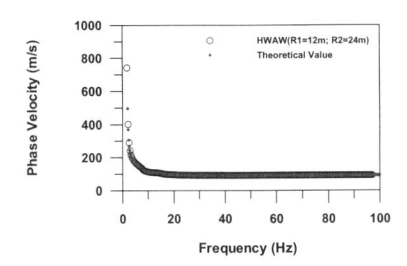

c) R1= 12m; R2=24m

Figure 5. Comparison of dispersion curves in Case1

Case 3 represents the pavement site in which stiffness of upper layer is much stiffer than those of lower layer. For this site, It is reported that conventional phase unwrapping method lead to incorrect

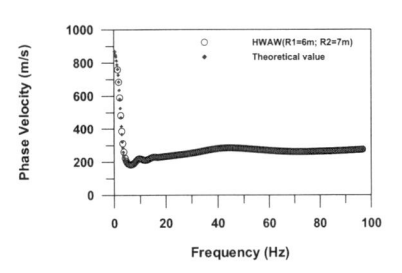

a) R1= 6m; R2=7m

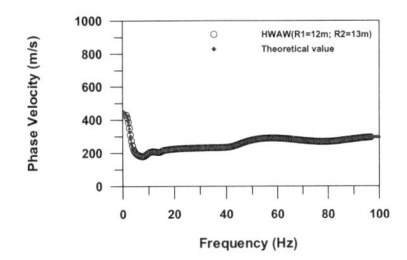

b) R1= 12m; R2=13m

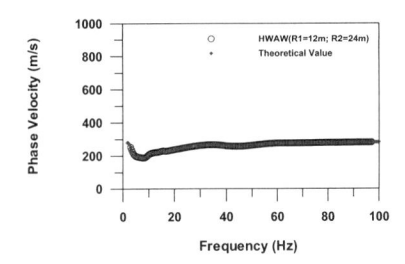

c) R1= 12m; R2=24m

Figure 6. Comparison of dispersion curves in Case2

dispersion curve (Al-Hunaidi 1994). Short receiver spacing setup in which time signals are measured at 6 and 6.1m from impact source were used for HWAW method. Fig. 7 shows the comparison of dispersion curve determined by HWAW method and theoretical value. It can be noticed that dispersion curve by proposed method coincide well with theoretical value over whole frequency range. The same pavement profile was studied by Al-Hunaidi (1994) who developed multiple filter/cross correlation method to overcome the problem of phase unwrapping method in pavement site and the dispersion curve obtaind by Al-Hunaidi was included in Fig. 7 for the comparison purpose. It can be noticed that the dispersion curve obtaind by multiple filter/cross correlation method was also agree well with those by HWAW method and theoretical value.

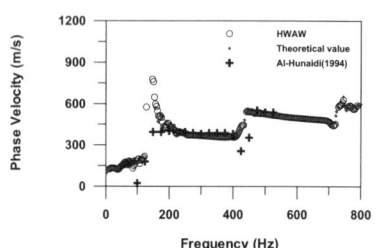

Figure 7. Comparison of dispersion curves in Case 3

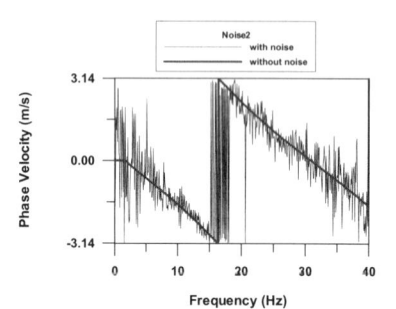

a) Phase spectrum of wave signal with and without noise

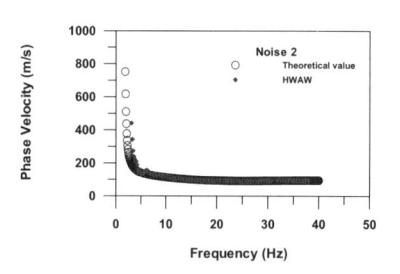

b) comparison of dispersion curve

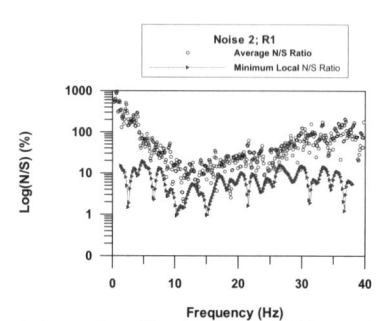

c) Comparison of local and average Noise to Signal ratio

Figure 8. Effect of noise in HWAW method

To evaluate the effect of noise in the proposed method, white random noise was added to time signals measured at 6m and 9m from the source for case 1. Fig. 8a) shows the phase spectrum of wave signals with and without noise and Fig. 8b) shows the comparison of dispersion curves by proposed method using signals with noise and theoretical value. In this case, phase unwrapping method may be difficult to determine dispersion curve from this phase spectrum, whereas the proposed method provide reasonable dispersion curve. Fig. 8c) shows the comparison of local and average noise to signal ratio of signal with noise at receiver 1. From the figure, it can be noticed that minimum local N/S ratio is generally much less than average N/S ratio. So, even though noise level which is defined by average noise to signal ratio is severe, HWAW method can determine correct dispersion curve.

3.2 *Field Application*

The proposed method was applied in the field. The short receiver spacing setup (R1=12.9m, R2=14.7m) and conventional test setup (R1=24m, R2=48m) were used to determine dispersion curves and the single array inversion was used to evaluate soil profile. The soil profiles by the proposed method were compared with those by SASW and PS-suspension logging tests.

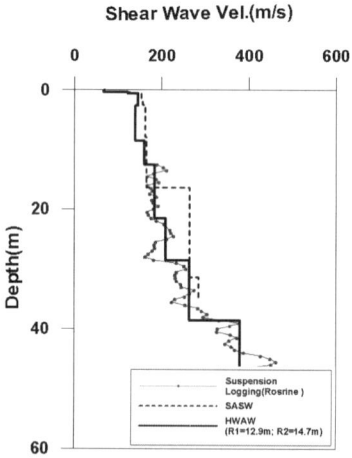

Figure 9. Comparisons of V_s profiles determined by short receiver spacing HWAW, SASW, and PS-suspension logging tests

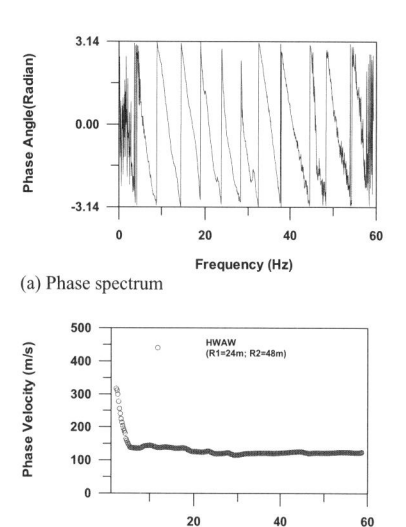

(a) Phase spectrum

(b) Dispersion curve determined by HWAW method

Figure 10. Phase spectrum and dispersion curve determined by HWAW method

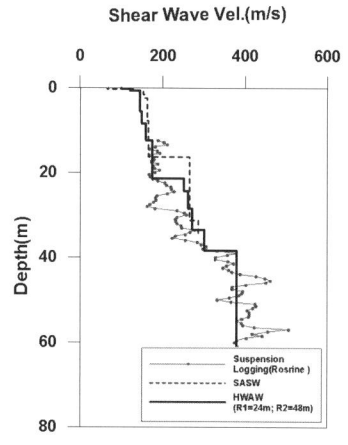

Figure 11. Comparisons of V_s profiles determined by long receiver spacing HWAW, SASW, and PS-suspension logging tests

Fig. 9 shows the comparison of shear wave velocity profiles determined by the proposed method using short receiver spacing setup, SASW test, and PS-suspension logging test. V_s profiles determined by three methods are very similar, but V_s profile by SASW test shows a little difference with those by the other methods. This difference can be explained by the lateral non-homogeneity of the site. In the SASW test, it is assumed that material between receivers is laterally homogeneous, and a long test line is needed to explore deep layer. Therefore, SASW test determine the soil profile of the site in the average sense when lateral non-homogeneity exists. But, HWAW method uses just short receiver spacing setup to determine dispersion curve of whole depth of interest. The short receiver spacing setup having receiver spacing of 1~3m can minimize the effect of lateral non-homogeneity and it has potentials to determine the detailed local V_s profile along the lateral direction such as two dimensional V_s map of the site.

Fig. 10a) shows the phase spectrum of wave signals recorded in the conventional test setup where receiver locations are 24m and 48m from impact source. Frequency range under 5Hz corresponding to wave length over 40m in this case is very important to explore the deep layer. However, as shown in the figure, signal quality is too low to determine phase velocities by phase unwrapping method because of the characteristic of impact source.

Fig. 10b) shows dispersion curve determind by HWAW method using conventional test setup. It can be noticed that HWAW method can determine reasonable dispersion curve over all frequency range including the range under 5Hz, even though signals have low average signal to noise ratio, because HWAW method use local information with maximum local signal to noise ratio. Fig. 11 shows comparison of V_s profiles determined by the proposed method using conviotional test setup, SASW test and PS-suspension logging test. V_s profiles determined by three methods are very similar, but at depth of about 20m, HWAW results provide the shear wave velocity values in the middle between SASW and PS loging results.

Fig. 12 shows comparison of V_s profiles determined by HWAW methods using short receiver spacing and long receiver spacing setup and SASW test. V_s profiles show a little difference at depths of about 20m. These differences between soil profiles may be due to two factors; The first factor is the variation of phase velocities with receiver locations and the second factor is lateral non-homogeneity of the site. Single array inversion was used for the proposed method and the theoretical dispersion curve was determined at the same receiver locations with field test setup. If site has no lateral non-homogeneity, V_s profiles determined by the proposed HWAW methods using short and long receiver spacing setups would be same. At this site, however, HWAW methods using different test setups are different each other, explaining that this site has some lateral non-homogeneity.It is also interesting to notice that V_s profile determined by HWAW method with long test setup is more similar to that by conventional SASW test and the longer the receiver spacing the more averaging the lateral non-homogeneity of the site.

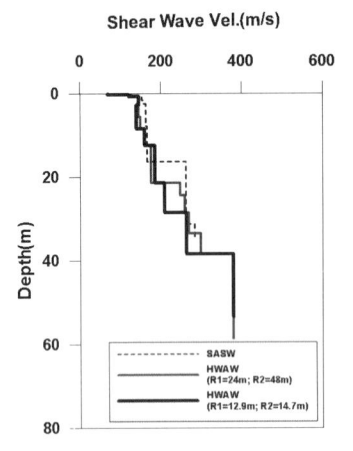

Figure 12. Comparisons of V_s profiles determined by HWAW methods using short and long spacing setup and conventional SASW test

4 CONCLUSIONS

In this paper, the new seismic site characterization method using the harmonic wavelet analysis of wave (HWAW) method was proposed. The HWAW method based on time-frequency analysis mainly uses the signal portion of the maximum local signal/noise ratio to evaluate the phase velocity and it can minimize the effects of noise. Field testing of this method is relatively simple and fast because one experimental setup which consists of one pair of receivers is needed to determine the dispersion curve of the whole depth. The proposed method uses the near field information and can sample much deeper part of the site than the conventional phase unwrapping method. The proposed method using short receiver spacing setup can determine the detailed local V_s profile along lateral direction. This method uses the single array inversion which considers the variation of phase velocity with receiver location without increasing calculation time and complexities. To estimate the applicability of the proposed method, numerical simulations at layered soil and pavement profiles were performed. Field tests were also performed and the shear wave velocity profiles obtained by the proposed method were compared with those by conventional SASW test and PS-suspension logging test. Through numerical simulations and field applications, the good potential of the proposed method is verified.

ACKNOWLEDGEMENTS

This study was sponsored by the SISTEC, which are gratefully acknowledged

REFERENCES

Al-Hunaidi, M.O. 1994. Analysis of disperded multi-mode signals of the SASW method using multiple filter/crosscorrelation technique. *Soil Dynamics and Earthquake Engineering* Vol 13: 13-24.
Nazarian S, & Stokoe KH. 1984. In situ shear wave velocities from spectral analysis of surface wave. *Proc. 8th Conf On Earthquake Eng. S.Francisco*: 31-38.
Newland DE. 1999. Ridge and phase identification in the frequency analysis of transient signal by harmonic wavelet. *J Vib And Acoustics* Vol 121: 149-155.
Gabriels P, Snider R, & Nolet G. 1987. In situ measurements of shear wave velocity in sediments with higher-mode Rayleigh waves. *Geophysical Prospecting* Vol 35: 187-196.
George A. McMechan, & Mathew J. Yedlin. 1981. Analysis of dispersive waves by wave field transformation. *Geophysics* Vol 46 (No. 6): 869-874.
Joh, S. H. 1997. *FIT&, software of Surface-Wave Forward Modeling, Inversion and Time-History Generation*, Chung-Ang University.
Park HC, & Kim DS. 2001. Evaluation of the dispersive phase and group velocities using harmonic wavelet transform. *NDT&E Int* Vol 34: 457-467.

# Use of dilatometer and dual dilatometer test for soft soils and peats

Paulus P. Rahardjo, Yunan Halim & Lento Sentosa
Parahyangan Catholic University, Bandung, Indonesia

Keywords: soft soil, peat, dilatometer, dual dilatometer

ABSTRACT: Measurement the properties of soft soils and peats in laboratory is very difficult task due to the difficulties in sampling and problems of disturbance. For these types of soils, in situ testing is preferable. This research uses the Marchetti dilatometer and dual dilatometer which was developed at UNPAR (Parahyangan Catholic University). Use of vane shear test is also common for soft soils, however not for peats, specially fibrous peats. Although the dilatometer is not a common tool in Indonesia, this device has been widely used in the world. The study area is located in Pelintung, Sumatera. In situ testing as well as laboratory tests were conducted and several correlations were established. The test result are promising especially new correlation are developed such as material index and the ratio of laboratory modulus and dilatometer modulus. The thrust of the blade during insertion is also measured and it is of interest that a correlation exists between the cone tip resistance and dilatometer thrust.

The Dual Dilatometer Test which was developed at UNPAR (UNPAR DDMT) has 2 membranes where the second one is mounted in the upper part of the blade with twice thickness than the standard one. The results of dilatometer index of both membranes were compared and it was concluded that for test on peats, a thicker blade gives a more reasonable result than the standard blade. This is probably due to the fact that the thicker blade displace the soil in larger strain and hence the measurement is more sensitive. In general, the use of UNPAR DDMT enhance the more suitable technology for peats.

1 INTRODUCTION

Flat Dilatometer Test (DMT) has been widely used and accepted as one of the in-situ testing equipment. The equipment measures soil responds against circular membrane inflated in lateral direction. To carry out the test, the blade is pushed to the depth of interest and the membrane is inflated outward by 1.1 mm. The pressures related to different position of the membrane is read from the manometer, i.e. the A, B, and C reading. By corrections, the readings are then used to calculate p_o, which is the lift off pressures closely related to "total horizontal pressures", p_1, the pressure of the soil when the membrane reach 1.1 mm and p_2 when the membrane is deflated. Several empirical corrections were developed for interpretation of soil shear strength, in-situ stress and soil moduli (Marchetti et al, 1980, Lutenegger, 1988).

The actual horizontal pressure is not p_o since the soil has been disturbed by insertion of the blade. Handy, et al. (1982) developed K_o Stepped Blade Test consisting four different thickness of blades to measure the lateral pressures of the soil for glacial till. By extrapolating the pressures based on blade thickness, the total horizontal pressures are interpreted at 'zero' thickness.

The idea of DMT and K_o-SBT lead the authors to develop Dual Dilatometer Test (DDMT) where an additional blade of double thickness to the standard blade is added. The measurement of standard DMT is repeated at the same depth by the second membrane. Prior to the development to UNPAR DDMT, Marchetti suggested that the use of dilatometer can be enhanced by various blade thickness or staging blade. By determining the soil pressure against several flat penetrometers of decreasing thickness one would obtain responses corresponding to a state of stress increasingly closer to the in situ state of stress. Marchetti state that the stress cannot be determined in correspondence of the intial state of stress, but one can reasonably expect to improve the accuracy of extrapolations towards the vertical axis (Marchetti, 1979).

2 DESCRIPTION OF UNPAR-DDMT

The width of Unpar DDMT is the same as Marchetti DMT (95 mm), however the blade is stepped with standard double thickness (15 and 30 mm) and two membranes of 60 mm diameter are mounted. Successive measurements are taken using each blade A Standard Gouda Penetrometer is used to push the DDMT into the ground and the thrust during insertion is read using the CPT manometer. The A, B, and C readings are taken for both membranes.

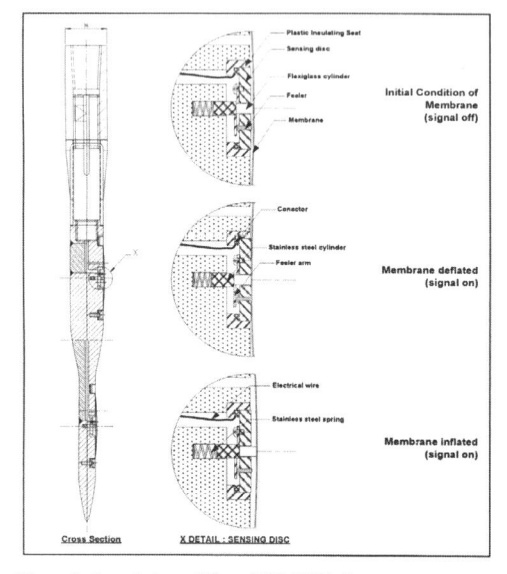

Figure 1. Description of Unpar DDMT Blade

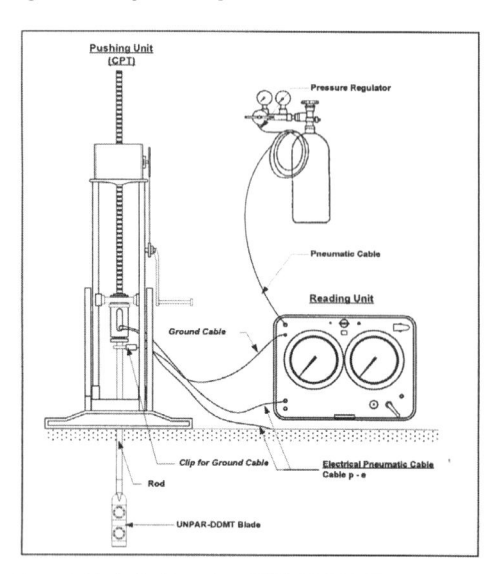

Figure 2. Method of Insertion of UNPAR DDMT

3 GEOLOGY OF THE SITE AND TESTING PROGRAMES

The study area is located in Pelintung, Sumatera, close to Bengkalis straits (between Malaka and Dumai) where an industrial site is being developed. Figure 3 shows the location and geology of the study area.

An access road about 4.0 km was constructed to connect the site to the existing Pakning – Dumai road. It is expected that the soil is under consolidated due to the fill placement which was placed about 1 years prior to testing. Based on the geological map, the area is described as coastal plain formed during Holocene period. The plain is dominated by soft clay sediment and vegetation forms the peats.

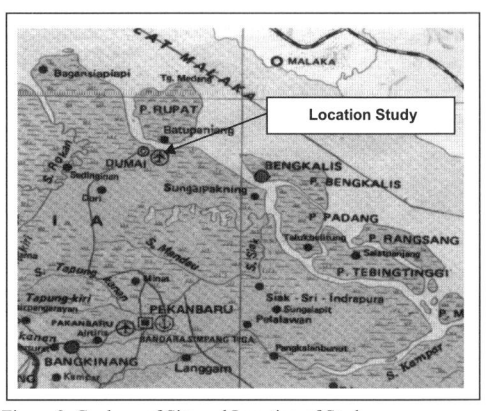

Figure 3. Geology of Site and Location of Study

The testing program consists of 4 drilling holes of 30 m depth with SPT and sampling, 3 DDMTs and 14 CPTs as shown in Figure 4.

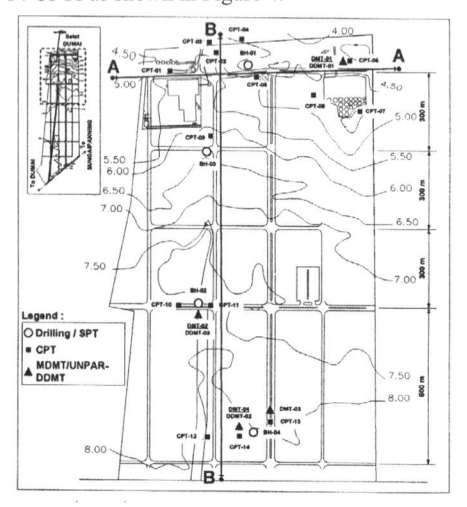

Figure 4. Testing Program and Locations

4 CHARACTERISTICS OF SOFT SOILS AND PEATS AT PELINTUNG

Based on data from drilling holes, the site may be characterized as soft recent deposits consisting of about 6 meters peats underlined by soft silty clays. Piston samplers were used to retrieve soil samples and characteristics of soft soils and peats were measured in laboratory and presented in the following.

The void ratio of peats are found as high as 3 – 16 and water content in the range of 250 – 900 %. The void ratio and water content of soft clays are subtantially lower than the peats as shown on Figure 5.

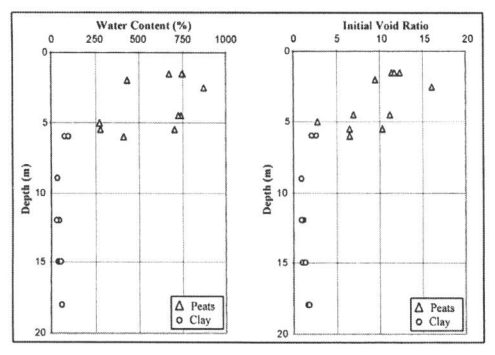

Figure 5. Water Content and Void Ratio vs Depth

As consequence of the high water content the peats is very compressible as depicted by compression index. The range of C_c is as high as 2-6 while the clay layer has much smaller range of 0.3-1.0 (Figure 6). The dependency of compression index to high water content is shown on Figure 7. Creep is also an issue to the deformation of the soils in long term period. And hence it is interesting to measure the creep parameter represented by C_α.

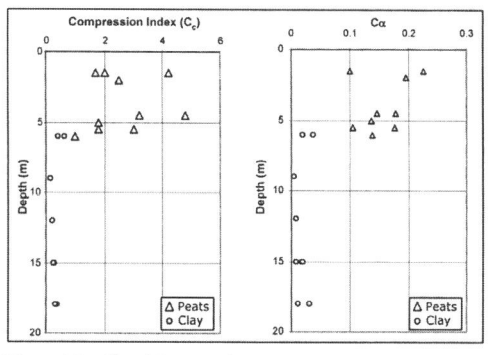

Figure 6. Profile of Compression Index and Values of C_α with Depth

Owing to the fact that the peat has much higher void than the clay, it is shown in Figure 8 that the coeficient of consolidation (and hence the permeability) could be 2-30 times higher. It is interesting that the value of C_v decrease rapidly by overburden pressure.

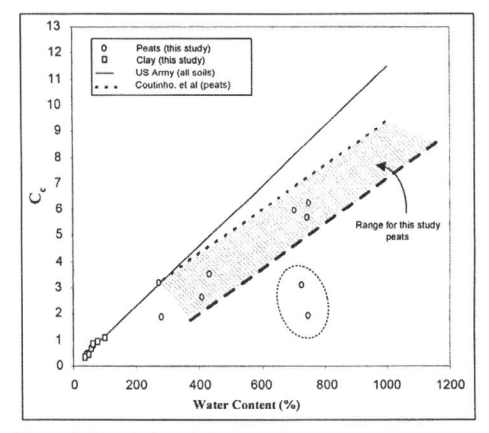

Figure 7. Correlation of Compression Index and Water Content

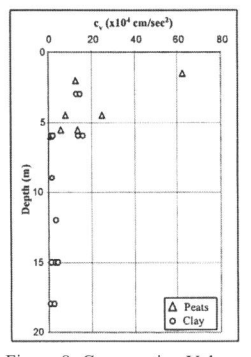

Figure 8. Comparative Values of Coefficient of Consolidation of Peats and Soft Clays

5 RESULTS OF DILATOMETER AND DUAL DILATOMETER TESTS

As defined by Marchetti (1980), the test results of DMT are presented in terms of material index, I_D, Horizontal Stress Index, K_D and Dilatometer Modulus, E_D, where

$$I_D = \frac{(p_1 - p_o)}{(p_o - u_o)}$$

$$K_D = \frac{p_o - u_o}{\sigma_v'}$$

$$E_D = 34.7(p_1 - p_o)$$

To distinguish the results of standard thickness blade (15 mm) that of the double thickness blade (30 mm), the index were written as $I_{D(30)}$, $I_{D(15)}$, $K_{D(30)}$, $K_{D(15)}$ and $E_{D(30)}$, $E_{D(15)}$. Three DDMT results are plotted in Figure 9, and it is shown that in general. Material Index measured by standard blade ($I_{D(15)}$) for peats is inconsistent and tend to give much higher values. This may be due to the existence of the fibers. The thicker blade however give more reasonable result. The measured modulus (E_D) is smaller for thicker blade.

It is shown that where compared to the results of drilling, the index material $I_{D(30)}$ is more consistent with the type of soil than $I_{D(15)}$. The dilatometer moduli of $E_{D(30)}$ are in the range of 30 % - 50 % of $E_{D(15)}$. But, this could be due to disturbance during insertion. The measurement of $E_{D(30)}$ is clearly conducted in a more disturbed condition, or it may be due to non linearity of the soil, where measurement of $E_{D(30)}$ is after much higher displacement.

6 INTERPRETATION OF TEST RESULTS AND DISCUSSION

A typical interpretation of soil parameter using DDMT is shown on Figure 9. A range of differences resulted in the interpreted parameter from the standard blade and the thicker blade. Of interest is the OCR profile where both show that the soil is under consolidated. In certain dense, this could be true since the test was conducted under fill placement, however the standard blade show that the clay at all depth are under consolidating. In general the pore pressure response as indicated by U_D shows that the peat layer is more permeable.

6.1 Interpretation of Material Type

The interpretation of material type is conducted based on Marchetti and Craps (1981). As shown on Figure 10, the peat is interpreted as sandy soil and the clay as silty soil if one use the standard DMT, while using the $I_{D(30)}$ shows better interpretation. Some data shows very low values of Dilatometer Moduli, specially for peats. The Marchetti and Craps chart may have to be extended to E_D value less than 0.5 MPA.

6.2 Interpretation of Undrained Shear Strength

The interpretation of undrained shear strength is based on empirical chart proposed by Marchetti (1980), Larsson & Eskillson (1989), Schmertmann (1991), and Kamei & Iwasaki (1995). Result of laboratory tests are also plotted as shown in Figure 11. Both the standard blade and the thicker blade are interpreted.

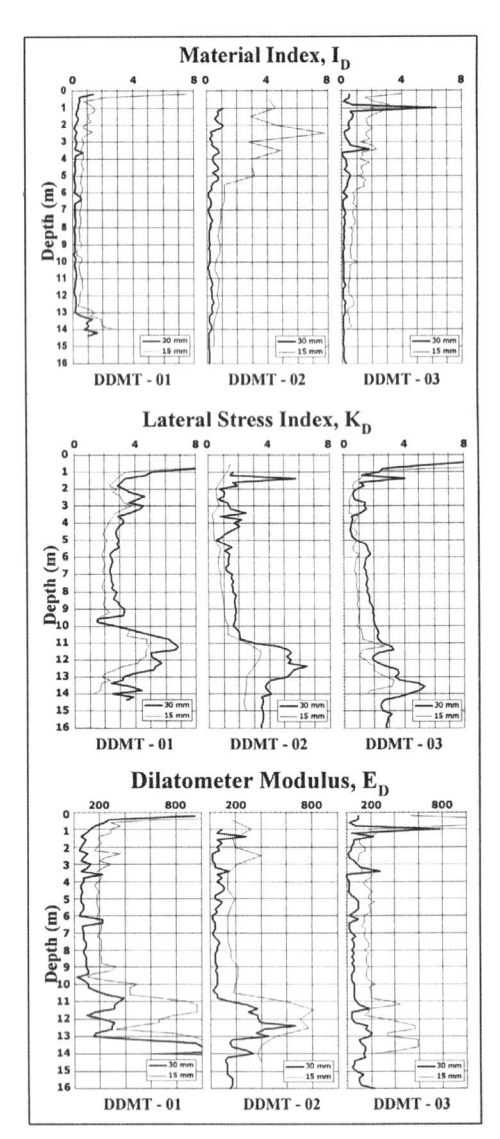

Figure 9. Dilatometer Index Profile

The Figure follow empirical equations as proposed in the following :
1. Marchetti,1980, : $s_u = f(K_D, \sigma_v')$
$$s_u = 0.22\, \sigma'_v\, (0.5\, K_D)^{1.25}$$

2. Larsson & Eskilson, 1989, : $s_u = f(p_1, u_o)$, based on Swedish organic clay.
$$s_u = \frac{(p_1 - u_o)}{9}$$

3. Schmertmann, 1991, : $s_u = f(p_o, u_o)$, based on data of soft clay

$$s_u = \frac{(p_0 - u_0)}{10}$$

4. Kamei & Iwasaki, 1995, : $s_u = f(E_D)$,
   $s_u = 0.018\ E_D$
5. Skempton (1957) proposed empirical correlation for normally consolidated clay where undrained shear strength ratio depend on plasticity index

$$\frac{s_u}{\sigma_v'} = 0.11 + 0.0037\ (PI)$$

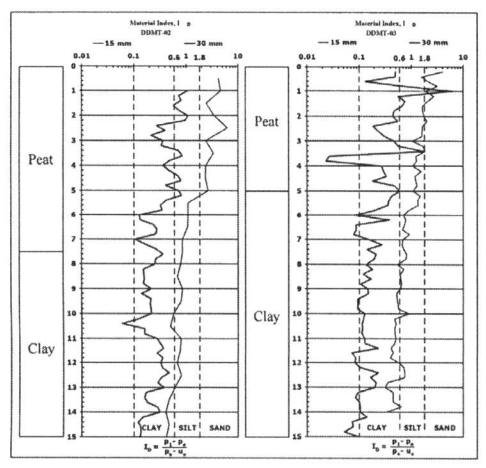

Figure 10. Comparative Interpretation of Soil Type

It is shown that every author uses different parameter for interpreting undrained shear strength. It is most reasonable the use of p_1 than other Dilatometer index. Although thicker blade give a more precise estimate, the method proposed by Larsson and Eskillson (1989) give the best result for the standard blade.

6.3 Interpretation of K_o

Figure 12 shown 3 method of empirical correlations:
1. Formula proposed by Marchetti, 1980 :

$$K_o = \left[\frac{K_D}{1.5}\right]^{0.47} - 0.6$$

2. Method by Larsson and Eskilson, 1989 :

$$K_o = 10^{(0.055(K_D - 3.5))} - 0.4$$

The above formula is valid for sensitive organic clay in Sweden.

3. Method proposed by Kulhawy and Mayne, 1990 :

$$K_o = \left[\frac{K_D}{2}\right]^{0.47} - 0.6$$

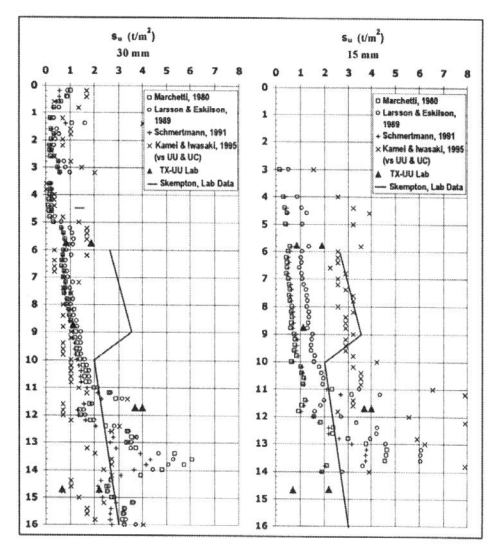

Figure 11. Interpretation of Undrained Shear Strength for DDMT-03

Kulhawy and Mayne (1990) basically identical to Marchetti recommendation except that the denominator is higher. This factor yield significant difference as shown in Figure 12. The interpretation of K_o give different values for standard blade and thicker blade. In general, the standard blade give much lower values in the order of 50 %. However method proposed by Larsson and Eskilson is less sensitive to K_D and both the standard blade and the thicker blade resulted in similar K_o values. This values is in the range of 0.3 – 0.5.

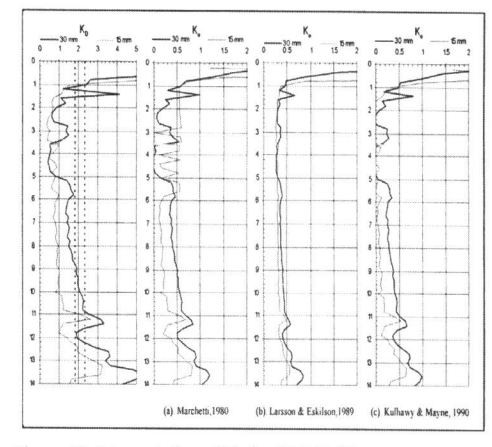

Figure 12. Interpretation of K_o for DDMT-03

6.4 Interpretation of Stress History and Preconsolidated Pressure

OCR is very important parameter for indication of stress history. Soft clay normally has OCR value close to 1.0 indicating normally consolidated clay. The upper clay layer may have OCR value higher than 1.0 due to fluctuation of water table or desiccation. Figure 13 is a plot of interpreted OCR based on different empirical formula:

1. Marchetti,1980
   $OCR = (0.5 K_D)^{1.56}$ for $I_D < 1.2$ (clay)

2. Larsson & Eskilson (1989),
   $OCR = 10^{0.16(K_D - 2.5)}$

3. Kamei & Iwasaki (1995),
   $OCR = 0.34 K_D^{1.43}$

4. Chang (1991a), for Singapore soft marine's clay
   $OCR = (0.5 K_D)^{0.84}$

It is interesting that the OCR interpreted by all of the above method yield OCR values less than 1.0. This could be indication of soil undergoing consolidation. Based on theoretical consolidation calculation, the soft soil is still consolidating due to fill placement of the road contruction. In such condition, the result of laboratory tests for OCR could be misleading. In general for K_D values less than 2.0, the OCR fall to values less than 1.0.

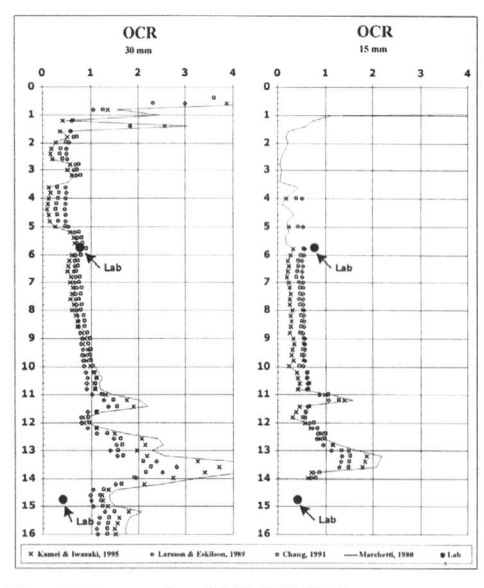

Figure 13. Interpretation of OCR (DDMT-03)

6.5 Interpretation of Moduli

The interpretation of soil moduli is based on method by Marchetti (1980) and Sue et al. (1993) :

1. Marchetti, 1980
2. Sue et al, 1993, proposed correlation based on soft marine clay in Taiwan for Material Index, $ID < 0.6$:
   $R_M = 0.5 + \log K_D$

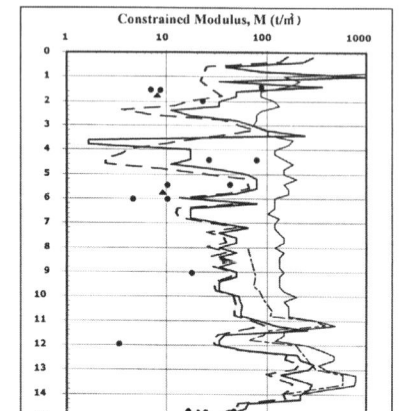

$M = R_M \cdot E_D$

Figure 14. Interpretation of Soil Moduli (DDMT-03)

The result of standard blade generally give higher value of constrained moduli, while the thicker blade yield closed values to laboratory test results. Empirical correlation of RM to Material Index can be established as shown on Figure 15. As the material is stiffer, RM is decreasing.

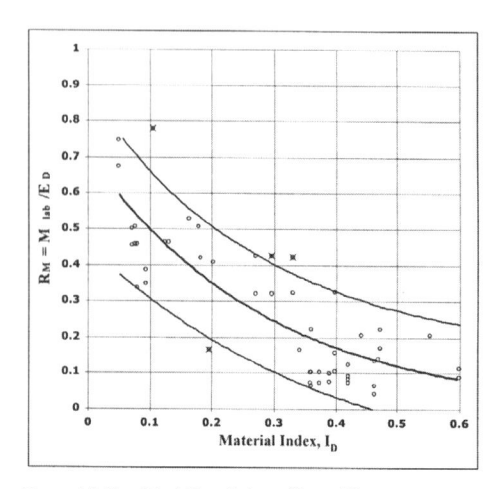

Figure 15. Empirical Correlation of I_D and R_M

6.6 Interpretation of Pore Pressure Response

Soil and pore pressure response is based on the value of p_0, p_1 and p_2. These values are normally higher than the hydrostatic pore water pressure. This means that additional pressure are derived from excess pore pressure and effective earth pressure. The value of p_2 is more representative of hydrostatic and excess pore pressure. As shown in the Figure, the peats do not yield significant excess pore pressure which means substantial excess pore pressure as been dissipated.

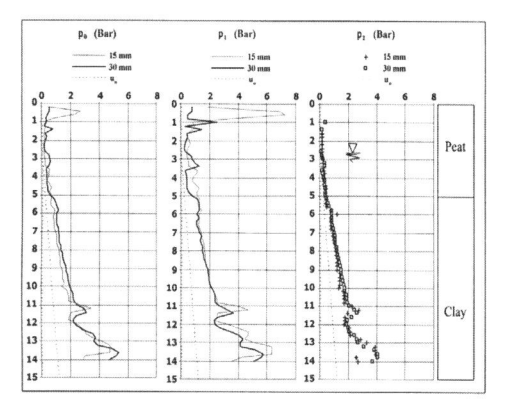

Figure 16. Measured Soil and Pore Pressure Response

6.7 Interpretation of DDMT Thrust

It is common to compare the DDMT thrust and the cone resistance of CPT. In this particular research, the results are plotted and an average of factor of 1.4 is obtained (Figure 17).

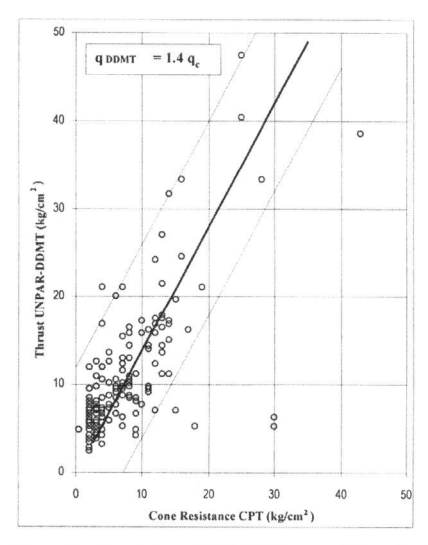

Figure 17. Correlation of DDMT Thrust and Cone Resistance

7 CONCLUSIONS

For test on peat and very soft soil, thicker blade give a more reasonable result than the standard blade. This is probably due to the fact that the thicker blade displace the soil in larger strain and hence the measurement is more sensitive.

Interpretation of K_o for standard blade and thicker blade give somewhat consistant values when using empirical formula by Larsson and Eskilson. Interpretation of OCR indicated that the subsoil is undergoing consolidation.

In general, the use of UNPAR DDMT enhance the more suitable technology for soft deposit and peats.

ACKNOWLEDGMENT

The author gratefully expressed appreciation to Prof. Marchetti of L'Quilla University, Italy for providing the dilatometer and to Kawasan Industri Dumai for permission to publish the data.

REFERENCES

ASTM Subcommittee D 18.02.10 - Schmertmann, J.H., Chairman (1986). *Suggested Method for Performing the Flat Dilatometer Test*. ASTM Geotechnical Testing Journal, Vol. 9, No. 2, June, 93-101.

Chang, M.F. 1991. *Interpretation of overconsolidation ratio from in situ tests in Recent clay deposits in Singapore and Malaysia*", Canad. Geot. Jnl.,Vol. 28.2 April : 210-225.

Coutinho, R.Q., Oliveira, J.T.R., Oliveira, A.T.J. 1998. *Geotechnical Parameters of Recife Organic Soft Soils-Peats*. Problematic Soils, Yanagisawa, Moroto & Mitachi (eds), Balkema, Rotterdam.

Handy, R.L., Remmes,B. Moldt,S., Lutenegger, A.J., Trott G. (1982). *In Situ Determination by Iowa Stepped Blade*.ASCE Jnl GED, GT 11, 1405-1422.

Kamey, T. & Iwasaki, K. (1995). *Evaluation of undrained shear strength of cohesive soils using a Flat Dilatometer*. Soils and Foundations, Vol. 35, No. 2, June, 111-116.

Kulhawy, F. & Mayne, P. (1990) "*Manual on Estimating Soil Properties for Foundation Design*" Report No. EL-6800 Electric Power Research Institute,. Cornell Univ. Ithaca, N.Y., 250 pp.

Larsson, R. & Eskilson, S. 1989. *DMT Investigations in Clay*. Swedish Geotechnical Institute, Publ. No. 243. Feb.

Larsson, R. & Eskilson, S. 1989. *DMT Investigations in Organic Soils*. Swedish Geotechnical Institute, Publ. No. 248. Aug.

Lutenegger, A.J. (1988). *Current status of the Marchetti dilatometer test*. Special Lecture, Proc. ISOPT-1, Orlando, FL, Vol. 1, 137-155.

Marchetti, S. (1979). *On The Determination of In Situ K_0 in Sand*. Contributo Panel Discussione Sessione N.7. Seventh Euruopean Conference on Soil Mechanics and Foundation Engineering. Brighton, England, Sept. 1979.

Marchetti, S. (1980). *In Situ Tests by Flat Dilatometer*. ASCE Jnl GED, Vol. 106, No. GT3, Mar., 299-321.

Marchetti, S. & Crapps, D. K. (1981). *Flat Dilatometer Manual*. GPE, Inc., Geotechnical.

Marchetti, S., Monaco, P., Totani, G. & Calabrese, M. (2001). *The flat dilatometer test (DMT) in soil Investigations. A report by the ISSMGE Committee TC 16*. Proceedings, International Conference on In-Situ Measurement of Soil Properties & Case Histories, Bali, Indonesia, 95-131.

Powell, J.J.M. & Uglow, I.M. (1988). *The Interpretation of the Marchetti Dilatometer Test in UK Clays*. ICE Proc. Conf. Penetration Testing in the UK, Univ. of Birmingham, July, Paper No. 34, 269-273.

Robertson, P.K., Campanella, R.G., Gillespie, D. & By, T. (1988). *Excess Pore Pressures and the Flat Dilatometer Test*. Proc. ISOPT-1, Orlando, FL, Vol. 1, 567-576.

Schmertmann, J.H. 1991. *The Mechanical Aging of Soils*. 25th Terzaghi Lecture ASCE Journal of Geotechn. Engineering, Vol. 117, No. 9: 1288-1330. Sept.

Schmertmann, J.H. 1991. *Discussions to Leonards and Frost* (1990). ASCE Journal of Geotechn. Engineering, Vol. 114, July 1988, in Vol. 117, No. 1: 172-188.

Skempton, A.W. and Bjerrum, L. (1957). *A Contribution to the Settlement Analysis of Foundations on Clay*. Geotechnique, vol. 7, p. 168.

US Army Corps of Engineers. 1990. Engineering and Design. *Settlement Analysis*. 30 September 1990: 3-34.

Field evaluation of the LCPC in situ triaxial test

Ph. Reiffsteck
Laboratoire Central des Ponts et Chaussées, 58 Bd Lefebvre 75732 Paris cedex 15 France

Keywords: in situ, apparatus, triaxial

ABSTRACT: A new in situ test method is proposed to investigate strength and deformation of soil masses. It is characterized as a field loading test realized in a tube placed by jacking. This communication will present the different stage of validation of this apparatus.

1 INTRODUCTION

Commonly in situ tests, like plate loading tests or borehole expansion tests, create an inhomogeneous stress field. Even if the relation between stress and deformations is well defined by theory of elasticity, the parameters needed by numerical models are obtained using corrections, correlations or empirical laws. So there is still a need for apparatus able to propose to numerical methods, rheological parameters in small deformation or plasticity directly measured in situ.

This communication will present in the principles of a specific in situ test called "triaxial in situ" and will more particularly illustrate the techniques used for the validation of the design of this testing device.

2 IN SITU TRIAXIAL TEST PRINCIPLE

The triaxial in situ testing device is an apparatus designed to reproduce the testing conditions of the laboratory triaxial test, in situ (AFNOR 1994a; Bishop and al.1975; LCPC, 1999). This type of apparatus has been already presented more extensively in Tani (1999) and Reiffsteck and Reverdy (2003). The test developed in LCPC consists of an open tube sampler where a membrane equipped with transducers is substituted to the liner and whose head comprises a piston activated by a jack (Figure 1). This device is forced into the soil by pushing. Once the apparatus reached the test depth, the testing program takes as a starting point the traditional triaxial test: application of horizontal pressure using a measuring cell then application of vertical stress using a piston, the hori-zontal stress being generally maintained constant. It creates a homogeneous stress field to the soil tested in the probe. Thanks to the control of the vertical piston and measuring cell, the apparatus makes it possible to impose various stress paths. Since special attention is paid to the characterization of deform-ability, and to study the behavior of soils at a small deformation level, the realization of local measure-ments are carried out by placing the transducers in contact with the soil: radial and axial displacement transducers and an pore pressure are fixed on the membrane.

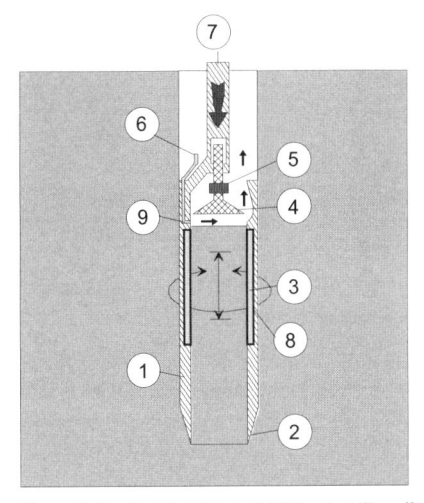

Figure 1. In situ triaxial test: LCPC patent (1: cell, 2: cutting shoe, 3: membrane, 4: cap, 5: force transducer, 6: connecting line, 7: axial load, 8: displacement transducer, 9: jetting tool)

Once the test is completed, the probe is retrieved from the borehole and the sample can be taken for further analysis and identification. Another possibility of this apparatus is to use it in a self-boring configuration, the device can be jacked steadily into the ground and there is no need to retrieve the probe between the tests. As in a self-boring pressuremeter a jetting nozzle placed in the driving module desegregates the soil that penetrates inside the device above the measuring cell (Baguelin and al., 1978; Benoît and al., 1995).

There is no inside clearance in the probe to avoid imposing deformations of extension and with its internal diameter of 100 mm and its 15 mm thickness, it presents an area ratio of 69% (AFNOR, 1995; Bat and al., 2000; Bigot and al., 1996; Hight, 2000; Hvorslev, 1949). This diameter and thickness are needed to limit the remoulding and the installation of the instrumentation but it also increases the force to insert the probe in the soil. For example, the pushing effort is estimated at 5 tons for a cohesion Cu>200kPa, case of Flandres clay (clay present in the North of France and similar to Boom clay).

Figure 2. In situ triaxial probe

A slenderness of two of the tested zone has been chosen to be close to the laboratory triaxial apparatus geometry; consequently the active membrane is 200 mm long.

2.1 Structure of equipment

The physical architecture of the machine is as follows: the actuators of the probe (membrane and jack) are connected to two pressure volume controllers being able to be controlled in pressure and in volume, the various transducers are connected to two data-acquisition card hosted in a notebook to collect measurements (see Figure 1 and 3).

The pressure controllers are connected to the computer by RS232 interface. A specific software analyzes the measurements, calculates the test parameters and sends the instructions to the controllers. The software is also programmed to carry out the various preliminary and posterior phases to the test.

Figure 3. Architecture of equipment ((1) probe, (2) Controller pressure volume, (3) notebook)

2.2 Measurement technique

Measurements of displacement use Hall effect semiconductors for vertical displacement on the membrane, and proximity transducers for radial displacement (Clayton et al., 1989). Measurements of pushing and vertical force and cell and pore pressure are taken in a traditional way. The other displacement transducers used for validation, are placed outside of the membrane to provide a reference.

3 FIRST TESTS

Before testing the apparatus on a real site, a period of tests, actually running, has to be made in well-controlled materials. It allows us to test the apparatus, check the transducers and validate the software.

Two sections of homogeneous soils have been built and tested by the Centre d'Expérimentation Routière of Rouen Laboratory in Normandy. The soils used were sand and clayey silt. In constructing the test section the soil was processed to uniform moisture, placed in 30 cm lifts, and compacted with a roller.

3.1 Preparation of the test

Before the test, the general set up and procedure are the preparation and saturation of the system: controllers, jack and measuring cell, gauging and calibration of the measuring cell and saturation of the pore water pressure measurement circuit. One proceeds to the installation of the porous stone, saturated and protected by a film of clay to prevent desaturation.

In the first phase of our study, we do not use the self boring module. Our first objective is to validate the concept of the probe. So our borehole is made in soils with a helical auger. On the level, which one wishes to test, a flat bottom auger is used to clean the borehole immediately before inserting the probe.

3.2 Procedure of test

Figure 4 shows a schematic layout of the testing set up also shown of Figure 5. The loads are applied by jacking against a steel frame loaded by dead weights.

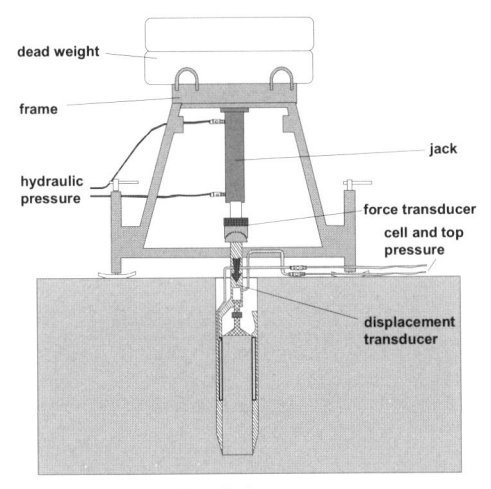

Figure 4. Principle of installation

The probe is inserted by pushing at the speed of a traditional sampler on a predetermined depth defines as the distance between the edge of the cutting shoe and the vents.

Figure 5. Picture of installation

During this phase the local transducers move upward until they are blocked (Figure 6). One can then put in contact the piston with the surface of the ground.

After that, it is necessary to proceed to a phase of relaxation: measure of pressures at constant volume.

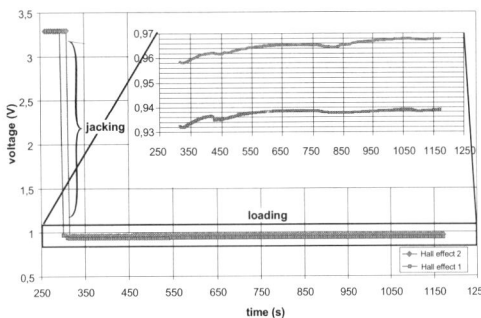

Figure 6. Evolution of vertical displacement during installation

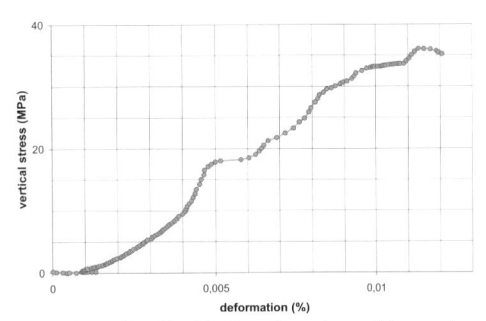

Figure 7. Results of loading in oedometric condition on clayey silt

It is then possible to carry out an isotropic or anisotropic consolidation by controlling the vertical force and the pressure in the measuring cell (Figure 7).

The phase of shearing can then be carried out according to the preset stress path (AFNOR, 1994b).

The tests actually running have shows the reliability of the apparatus and the pertinence of the design and particularly the type of transducers and the use of controllers. The first set of tests cannot be analyzed in soil mechanics framework. New tests investigating modulus and at rest earth pressure are planned for the beginning of 2004. As no ground water level exists in the fill, we were not able to test porewater pressures measurements.

Lastly, the procedure of end of test described below is applied.

3.3 Procedure of end of test

The end of test is done in a similar way than the one used for a sampler: the probe is retrieved from the pocket, the piston still in contact with the core, thus blocking the vents. Once retrieved, the probe lay out on a specific frame, the double effect action of the jack is used to retract the piston. After disconnection of tubing and cables of the transducers, and disassembling of the head containing the jack, one can eject the core by taking support with the thread of the cutting case.

4 FURTHER TESTS

Field evaluation will been performed in a over consolidated clay formation at a test site in Merville (Pas-de-Calais, France) and in sand at a second test site in Orléans (Loiret, France). Several in situ and laboratory tests have been made on this two sites. The results of pressio-penetrometers tests, self boring pressuremeter tests, cross-hole tests and triaxial tests with bender elements will be used to evaluate the potential of the testing device.

5 CONCLUSIONS

The design and laboratory evaluation of a in situ triaxial have been described. The following conclusions were reached.

The in situ triaxial test performed well during the first tests and gave measured lateral stresses in good agreement with those estimated for this type of fill material.

Further field trial tests are needed to fully investigate the performance of in situ triaxial test in well-documented ground.

REFERENCES

AFNOR 1994a Norme NF P 94-070 Essai à l'appareil triaxial de révolution - Généralités Définitions
AFNOR 1994b Norme NF P 94-074 Essai à l'appareil triaxial de révolution - Appareillage, Préparation des éprouvettes Essais UU, CU + u, CD
AFNOR 1995 Norme NF P 94-202 Sols: reconnaissance et essais - Prélèvement des sols et des roches – Méthodologie et procédures
Baguelin, F. Jézéquel, J.-F. Shield, D.H. 1978. The pressuremeter, Transtech publications
Bat A., Blivet J.-C. Levacher D. 2000. Incidence de la procédure de prélèvement des sols fins sur les caractéristiques géotechniques mesurées en laboratoire, RFG 91: 3-12
Benoît J. Atwood M.J. Findlay R.C. Hilliard B.D. 1995. Evaluation of jetting insertion for the self boring pressuremeter, Can. Geotech. J. 32 :22-39
Bigot, G. & Blivet, J.C. 1996. Prélèvement des sols et des roches, Bull. LPC 204: 113-117
Bishop, A.W. & Wesley, L.D. 1975. A hydraulic triaxial apparatus for controlled stress path testing, Géotechnique, 25(4): 657-670
Clayton, C.R.I. Khatrush, S.A. Bica, A.V.D. Siddique, A. 1989. The use of Hall effect semiconductors in geotechnical in-strumentation, GTJODJ ASTM, 12(1): 69-76
Hight, D.W., 2000. Sampling effects in soft clay : an update, The 4th Int. Geotec. Eng. Conf., Faculty of Engineering, Le Caire, Egypt
Hvorslev, J. 1949. Subsurface exploration and sampling of soils for civil engineering purposes, ASCE Report
LCPC, 1999. Procédé et dispositif d'essai triaxial in situ, Patent N°99.137.92, 9 pages
Reiffsteck Ph., Reverdy G., 2003. Presentation of a new soil investigation technique: the in situ triaxial test, In Natau, Fecker & Pimentel (eds) GeoTechnical Measurements and Modellling: 375-380. Rotterdam: Balkema

Tani, K. 1999. Proposal of new in-situ test methods to investigate strength and deformation characteristics of rock masses, In Jamiolkowski, Lancellotta & Lo Presti (eds), Prefailure deformation of geomaterials: 357-364. Rotterdam: Balkema

CPTWD (Cone Penetration Test While Drilling) a new method for deep geotechnical surveys

Massimo Sacchetto & Annalisa Trevisan
S P G drilling company, Adria (Ro) Italy

Kjell Elmgren & Kenth Melander
ENVI Environmental Mechanics, Alingsas Sweden

Keywords: cone penetration test, geotechnical survey, drilling, CPT

ABSTRACT: CPTWD represents integration of: wire-line drilling system, standard and modified piezocone, MDW-monitoring while drilling. During the CPTWD test the cone is protruding in front of the drill bit during drilling in the same way as a corer; CPTU data are stored in memory unit . At the same time as the CPTU data are logged, drilling parameters (MWD) are also recorded. The system allows for the change between CPT testing, continuous core drilling, down-hole testing and non coring drilling with MWD. The combination of CPT parameters and drilling parameters can be a very powerful basis for interpretation of the data. The advantages of this system compared to the other down hole type CPTU is that much longer strokes than the normal 3 m can be made. In addition, the information from the drilling parameters is very useful, especially in hard formations where CPTU cannot be performed. It is expected that future development work will improve the method in terms of the test procedures and in terms of capabilities of the system itself, by adding new tools (e.g. field vane test, Permeameter, fluid sampler, thin wall samplers, dilatometer DMT).

1 INTRODUCTION

The CPTWD is the acronymous of Cone Penetration Test While Drilling. Basically it can be considered as an integration between a wire-line system (core barrel, non coring device, cable recovery system, etc), piezocone and the MWD (Monitoring While Drilling) system.
Prior to the introduction of the CPTWD, it is necessary to briefly outline every part of the system: the wire-line system, the piezocone, the MWD.

2 DESCRIPTION OF THE SYSTEM

2.1 *Wire-line system*

The "recovery by cable or wire" system (hence the name "wire-line") is based on the use of a core barrel which, integrated with the casing, makes a unique body. Pushing and rotation are conveyed by the same casing.
The system is essentially composed of:
- Drilling rods (or "casing") inside which, at the bottom terminal part, there is an element (usually a slot) where the core barrel is lowered down by the wire and takes place inside, so allowing the drilling rods to rotate and push.
- Core barrel or drilling tool: this is placed inside (throughout the interior of) the drilling rods and has a hook up system.

In the following Picture 1, the particulars of a four hooks system are shown:

Picture 1

The drill mud will flow in the annular space between the external wall of the core barrel and the internal wall of the rods and "lubricates" the walls of the hole while removing the cuttings.
Most of the time the mud is recycled and stored on the surface before it is put back into circulation.
The following Picture 2 shows a typical installation for a wire-line c.c.drilling:

Picture 2

There are different types of core barrels as well as different types of tools which can be placed inside the drillstring (rods). The following Picture 3 shows the terminal part of a core barrel suitable for soft soils

Picture 3

The tool is lowered inside the drillstring (casing) and recovered by a cable or wire (therefore called "wire-line") moved by a special hoist. The wire has a "fishing-tool" (called "overshot") attached at its end. Such an overshot is designed in a way which allows the hooking/unhooking operation in an automatic way.

The wire line drilling operations are essentially executed as follows (referred to as c.c.drilling):
- The drill rig pushes and rotates the casing inside which the core barrel is placed (the shape and length of the core barrel depends on the type of soil). Normally, mud circulation inside the drillstring is used. The mud leaves the terminal part between the core barrel and the casing without interfering with the inside of the core barrel.
- When the drilling is completed, the rotary head of the drill rig is moved away laterally and the overshot is lowered down into the drillstring by the wire, driven by a "high-speed" hoist. The overshot hooks and the core barrel can be recovered.
- The core is taken out of the core barrel, which is cleaned and prepared for the next operation.
- The core barrel is lowered inside the drillstring again and one more rod is added at the top of the drillstring, the rotary head is connected to the rods, mud is injected and the drilling starts again.

The advantages of the wire-line system, compared to the traditional method are: better quality of the cores, higher percentage of recovery, faster execution, less power required from the drill rig and the possibility to easily change between different types of tools.

2.2 *Piezocone CPTWD*

For the experimental CPTWD system, a standard piezocone, having a memory data storage capability, has been used. (Ref. 1)

This type of cone, conforms with the following standard: "International Reference Test Procedure for CPT/CPTU from ISSMGE Rev.3". The accuracy class is class1.

In its' "standard" version it is a system which measures Q_c, F_s, U, (point resistance, local friction, pore pressure and inclination of the drillstring every 2 cm of penetration versus time in case of use in "memory mode", without cable)

During the sounding, two data files are produced: one in the internal microprocessor inside the CPTU cone, and the other in the data collector (microcomputer); Picture 4.

Picture 4

After recovering the cone, it is connected to the data collector and its CPTU data will be downloaded. Proper software in the data collector will allow for the synchronization between the memorized data as a function of time (inside the cone), and those as a function of depth (inside the data collector) . In this way, it is possible to obtain a normal data-file (Qc, Fs, U, incl.) versus depth.

In the special "CPTWD" version, Picture 5, an additional pressure sensor for the evaluation of the U in one more point of the piezocone (U_3 location according to the International Standards) and a rotation sensor have been added. The latter allowing one to determine whether the piezocone has been subject to rotation during the sounding, or not.

The reason to detect U_3 was to determine whether or not the U is affected by the overpressure caused by the injection of mud.

Besides this, the range of U has also been increased to 50 Bar in order to achieve test depths down to 300 m.

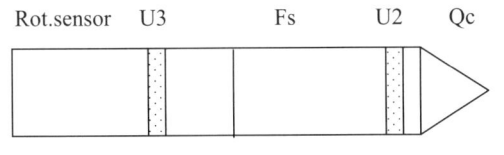

| Rot.sensor | U3 | Fs | U2 | Qc |

Picture 5

2.3 MWD (Monitor While Drilling)

It is a commonly and widely used method employed mainly in the deep oil drillings. It has been defined in several different ways, according to the company who built it, but is internationally known as MWD (Monitor While Drilling).

In order to monitor the reaction to the penetration of the soil in the drillrig, pressure sensors in the hydraulic oil circuitry, depth transducers, volume transducers, etc are applied. In the specific case of the CPTWD, the sensors signals are conditioned and amplified in the data collector, where they are stored and printed out in real time while drilling.

The measured parameters could be all the ones during the drilling. Still, they normally are:
- Bit-load (thrust pressure)
- Torque (rotation pressure)
- Mud pressure
- RPM (number of revolutions per minute)
- ROP (rate of penetration)
- Mud flow (volume of injected mud / min.)

2.4 The CPTWD Method

Basically, the CPTWD (CONE PENETRATION TEST WHILE DRILLING) system is an integration of the techniques described above. A wire-line core barrel has been modified, allowing it to keep a CPTU cone inside in the centre in such a way, that the cone is protruding from the bottom of the drill-bit by at least 35 cm.

A MWD data recording system is connected to the drill rig. In this case it is practical if the data collector is capable of handling both the drilling parameters and the CPTU data at the same time. Practically, the performance of such system are the following:
- Continuous core drilling, using the common core barrels, samplers, etc, but with the possibility of monitoring even the c.c.drilling with MWD; That is, to have a registration of all the implemented parameters even while carrying out a wire-line c.c.drilling.
- MWD with non coring drilling: A non coring tool, for example a tricone (always wire-line) is fixed inside the drillstring so that during the drilling the data collector records all the relevant drilling parameters every 2 cm. In this way, it is possible to (together with the monitoring of the cuttings) establish the stratigraphy.
- CPTU+MWD test at the same time: The CPTU cone-core-barrel is placed inside the drillstring. While the drillstring moves forward (at a speed as close as possible to the standard, 2 cm/s) the MWD system is always in function. After a substantial penetration, the CPTU cone is recovered, the data is downloaded to the data collector and subsequently synchronized with the corresponding MWD data. As a result, at the end of the drilling, it is possible to obtain a data matrix which includes (for every 2 cm of depth) for example: Qc, Fs, U_2, U_3, Bit load, Torque, Mud pressure, ROP, Mud volume and RPM. During an interruption of the sounding it is possible at any time in order to carry out dissipation tests. The evaluation of the drilling parameters in real time can be useful to the operator who eventually decides when to interrupt the penetration of the piezocone (i.e. because the soil is too dense). (Ref 3).
- It is useful to use the formula of the specific energy. This gives the Energy needed to penetrate one cubic meter of the soil material. (Ref. 2) For this reason (MWD), the CPTWD should not be considered a "blind" system. Several tests have been carried out and it has been noted that the operator can easily handle the change of stratigraphy (i.e. stop the drilling or decide when to change the tool) within a range of 10-20 cm.

In order to increase the safety of the cones, there are two alternative ways to go:
- alteration of the length of the piezocone in front of the drillstring system and
- a safety system which allows for the "return" or "recovery" of the piezocone back into the core barrel when the maximum value of resistance has been reached.

CPTU, as described above but without the detection of the MWD parameters, can also be used. In that case, the data collector connected to the drill rig only detects the depth versus time, in order to synchronise the CPTU data with the depth data, exactly like the "normal" CPTU tests in memory mode.
- DOWN-HOLE CPTU: the whole drillstring is lifted 1.5 – 3 m; a special core barrel having the memocone protruding for the same length (i.e. 1.5-3 m) is lowered and hooked inside the drillstring; then it is pushed down by the drill-rig (without rotation) for the same length. This allows to check and calibrate the CPTWD having

some "pure" (not affected by rotation and/or mud injection) CPTU data.
- DOWN HOLE tests of other sort: at any depth it is possible to lift the core-barrel and perform any type of test (SPT, Permeability test, Dilatometric, Vane test, etc).

The different options exposed above can be selected according to the type of soil as well as the scope of the bore-hole. The drillstring actually remains the same and the tools related to the three options are interchangeable any time the operator decides to do so.

2.5 Technical details of the CPTWD

The general scheme of the CPTWD prototype is shown below:

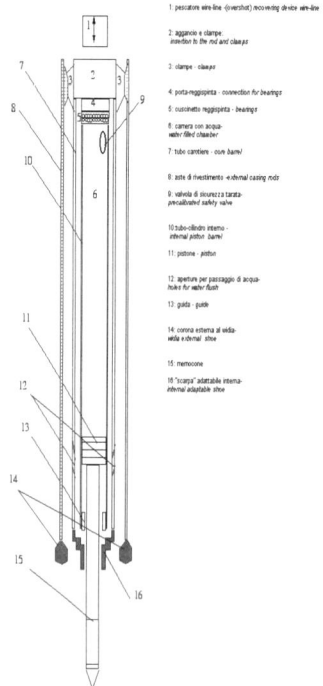

Picture 6

The system is basically made up of:
- Casing tubes. Those are of the same type used in c.c. drilling. For the prototype, the external diameter was 127 mm, which allows them to collect Ø 86 mm cores and Ø 88.9 mm (standard) undisturbed samples. The bottom part of the casing is equipped with a specially designed drill-bit.
- Hooking system. There are normally four hooks placed inside the core barrel, shaped in a way which allows them to get into (once they are unhooked from the overshot) the slot placed inside the casing, at a distance equal to the length of the core barrel.
- Core barrel. In this case it is not quite appropriate to use the definition core barrel because it is not used to carry out c.c.drilling operations, but it is basically a wireline barrel used to carry out drillings in gravel, which has been modified in a way to make it suitable to keep a CPTU cone, a security system, etc., inside itself. The so called core- barrel, is of the "fixed head" type. It rotates with the casing so that the bottom part breaks up the soil. It is similar to the type used in gravel. At the bottom, there is a drillbit of special design.

The following ,Picture 7, shows how the mud flows (at a maximum pressure) through the openings of the drill-bit and between the drillbit and the internal side of the casing:

Picture 7

The following scheme shows (left) the principle of the wire-line drilling system in general and of the CPTWD in particular:

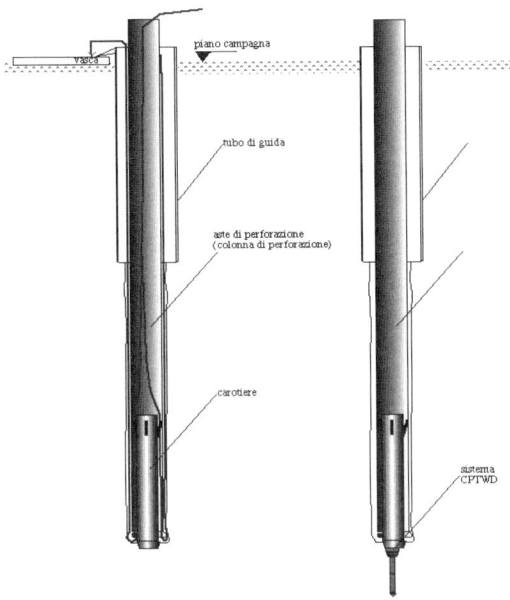

Picture 8

Safety device. Inside the "core barrel", the CPTU cone is assembled in a device which allows it to "return" into the core barrel when the resistance is greater than a pre-set maximum value.

Picture 9,.shows the CPTU cone after its retrieval at the end of the test.

Picture 9

3 RESULTS OBTAINED WITH THE CPTWD

3.1 Parma zone (July 2000)

The CPTWD system has been employed near Parma (Italy) in the geotechnical surveys for the project of some TAV (high speed train project) works on sites characterized by an alternance of fine soils (clay, silt, sand) and gravel layers often interlayed by clay intervals of little thickness (about a decimetre).

The execution of "traditional" static penetrometric tests would have been impossible due to the frequent presence of gravel intervals of different depth.

The geotechnical conditions in this site were well known due to the wide amount of tests which have been carried out; therefore it has been possible to compare the CPTWD results and assess its reliability.

The CPTWD test has been carried out as follows: The drilling starts with the use of the CPTU cone and MWD down to the top of the gravel; the core barrel with the CPTU cone was recovered at regular intervals (of about 1-2 m) in order to check the reliability of the data as well as the condition of the cone.

– As soon as the gravel was encountered, that is, as the Q_c was reaching 50 MPa, the "safety device" was trigged and the piezocone was retracted in to the core barrel. At the same time, the Operator had a way to realize not only the retraction of the cone, but also the variations of the drilling parameters, constantly monitored in real time, (every 2 cm), indicating a very dense layer.

– Once the cone was retracted, the core barrel with the piezocone was recovered and the data related to the last interval before getting to the gravel was down loaded and stored. After recovering the data, the CPTU cone was again installed in the core barrel and the cylinder of the "safety device" was again pressurized so that the CPTU cone could again be placed in position.

– After this, a non coring-tool (tricone) that made the non coring drilling possible, was installed and with simultaneous detection of the drilling parameters every 2 centimetres, drilling was carried out until another "penetrable" layer (clay, silt, sand) was found and then the CPTU+MWD started again.

Results are shown in the following graph:

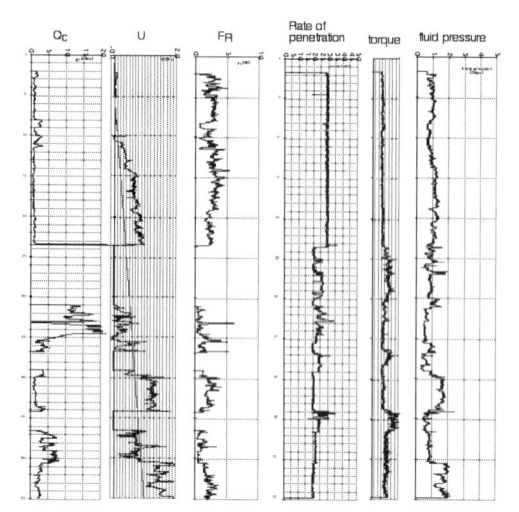

Picture 10

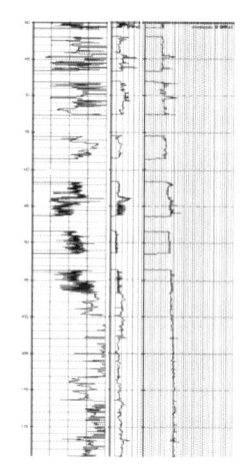

Picture 11

From left to right are Q_c, U, F_r, ROP (rate of penetration), Torque, Mud pressure.

The CPT graph is interrupted where there are layers of gravel. These are where the CPTU has been interrupted and MWD non coring drilling has been carried out.

3.2 *Piacenza zone (February-March 2001)*

The system has been used for the execution of two deep tests, carried out using a rig on a small jack-up barge, in order to investigate mainly sandy soils with gravel inter-layers. The drillings have been located on the two projected supports of the new tall bridge TAV (high speed train project) on the river Po.

Different from the work previously described, more advantage has been taken of the potentials offered by the wire-line recovery system.

In these soils, the CPTWD method has not been used continuously, but instead intermittent with continuous core. drilling, non coring drilling with MWD and the withdrawal of undisturbed samples, always using the same basic system and without changing the casing.

The withdrawal of the samples and the laboratory tests (along with the data coming from the other drillings made for the same project) allowed to assess the reliability of the CPTWD data, both in term of stratigraphy and in term of detecting the geotechnical properties (i.e. Ø in sandy soils).

Picture 11 is an example of a penetrometric graphic (Qc, Fs, U, Fr) from 60 to 120 m and one of the c.c. drilling operations in the same hole.

3.3 *Venice, treporti zone and mobile dams zone (July 2001-February 2004)*

Inside a project of monitoring the settlements of an experimental embankment, one vertical CPTWD, was carried out in June 2001 down to 110m, in mainly sandy soil, (very dense sand from -50 m).

After the CPTWD test, an experimental embankment (50 m diameter, 8 meters height) has been made, and some CPTU, DMT tests and bore-holes up to –60 m has been carried out in order to withdraw samples and to install the monitoring instrumentation (piezometers, assestimeters, inclinometers).

The monitoring job, or at least the interpretation of the data, is still in progress.

Due to very high resistance of the sand (sometimes Q_c greater than 50 Mpa), some intervals were carried out by using continuous core drilling (also in order to verify and calibrate the static test) and non coring drilling with MWD.

In order to keep a constant rate of penetration, having high density sandy intervals, the Operator sometimes increased the pressure of the mud injection at intervals in order to allow a 2 cm/s rate (Ref 5). Therefore, in order to evaluate the potential effects of such pressure in the penetrometric parameters, another pressure sensor has been added to the current model, just above the friction sleeve ("U3" according to the recent nomenclature).

The diagram below (Picture 12) shows the penetrometric graphic (Q_c, F_s, U, F_r) from 0 to 110 m.

The CPTWD data have been compared with the data of CPTU, DMT, bore-holes up to the depth of –60. The comparison shows that there is no relevant difference between CPTWD and CPTU data in the same range of depth (0 - -60). No comparison has been possible deeper than –60, due to high density of the sand not allowing deeper penetration to "normal" penetrometer.

At the present time, in the same area not far from the Jesolo site, in the same kind of soil, are in progress some comparative tests carried out with traditional method (use of the penetrometer inside pre-drilled holes) in order to compare CPTWD data with other kind of data, for the mobile dam project of Venice. For this project (mobile dams placed in the inlets of the Venice lagoon) have been carried out some CPTWD tests and dozens of deep bore-holes, CPTU and DMT tests, along with hundreds of undisturbed samples and laboratory tests.

All the tests have been carried out by operating two or three jack-up barges.

The job is still in progress, as well as the interpretation; the comparison of CPTWD data with other kind of data will be therefore very important for a validation of the results and as a matter of experimentation.

A bore hole up to –100 m by using CPTWD took around 4-5 working days; a bore hole carried out by using alternatively a drill rig (for cased pre-holes) and 300 kN static penetrometer took around 15 working days.

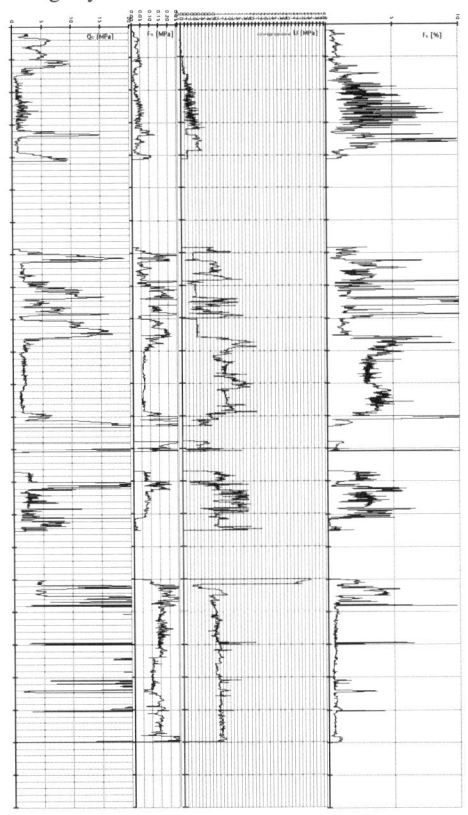

Picture 12

4 COMMENTS ON THE RESULTS

Actually, the total amount of current available CPTWD data is relatively small and at present, comparative tests with other testing methods have not been carried out in all kind of soils and for any range of depth.

Accordingly, it would therefore be necessary, to perform a more extensive validation of the obtained results with the following procedures:
- To execute comparative tests with CPTU and CPTWD at a depth which is achievable by the normal penetrometers. Such tests should be carried out in different types of soils in order to establish the correct testing procedures, in terms of the length of the cone, thrust calibration, pressure and volume of the injected fluid.
- To establish the cone's optimum length below the drill-bit by testing in different types of soil also at great depth, analyzing the variations of U_2 and U_3 as function of the pressure and the volume of the fluid injected at the bottom of the drillstring.
- To execute CPTWD tests at different penetration rate, within 1 and 2 cm/s, and compare the results
- Broad comparison between the penetrometric results and those obtained with other testing methodologies "in situ" (DMT, field vane test, SPT, etc) and in laboratory (where possible).
- To compare, when feasible, and economically possible, the penetrometric results obtained at great depth. For instance, comparing the CPTWD data with other penetrometric tests (wire line CPT down-hole used for offshore surveys) or penetrometric tests carried out with pre-drilled holes (as above mentioned).
- To evaluate not only with "in situ" tests, but also with laboratory tests and/or with the help of mathematic models, the influence of the rotation and fluid injection on the penetrometric data. In this way, it would be possible to calibrate, in a more scientific way, the MWD parameters.
- Since, for the first time, it is possible to compare results of MWD and CPTU from the same soil mass, there should be a possibility to carry out accurate calibrations of the drilling parameters and to get more objective and reliable data and possible correlations between the MWD and CPTU in soft soil.
- In case the CPTWD is evaluated in sufficiently documented and monitored cases, there is hope for the verification throughout "back-analysis", of the data obtained

5 POSSIBLE APPLICATIONS

The following are possible cases where the use of the CPTWD system could be competitive in relation to the traditional type of survey:
- Sites with alternating layers of non penetrable soil (i.e. gravel, cemented sands) and penetrable intervals of geotechnical interest (Ref. 4).
- Presence of compact overburden with subsequent penetrable layers.
- Geotechnical deep drilling (in particular the OFFSHORE drillings) where the cost/production ratio is important.
- Unavailability of the static penetrometer rig, providing the possibility to carry out static penetrometric tests with any kind of drillrig, at least at conventional depths.
- Geotechnical and environmental deep drillings: - at the end of the c.c.drilling/CPTWD, it is possible to install a piezometer and/or geotechnical instrumentation inside the bored hole.
- Deep surveys and main presence of sandy soils in which it is not possible to withdraw sufficiently undisturbed samples and the normal tests in situ (i.e. SPT) would not be economically efficient or reliable.
- The opportunity to have a more complete data matrix than with a normal penetrometer, which allows other kinds of interpretations. In particular, the fact of having in the same row (every 2 cm) the data of: Q_c, F_s, U_2, U_3, thrust, torque, rate of penetration, RPM, volume of injected fluid.

6 CONCLUSIONS

The CPTWD method lends itself to future improvements and developments, not only from a practical point of view, but also in a theoretical and interpretative area. The same general principle can be used for the implementation of other kind of tests with other kind of probes and sensors.

Still, with the idea of adapting other instruments to a core barrel wire-line, are in program some applications complementary to the CPTU test, for example:
- down hole wire-line field vane test.
- DMT (dilatometric) tests: the implementation of an application of a down-hole wire-line electronically operated Marchetti's dilatometer.
- execution of tests with Permeameter and sampling of fluids. Such applications has already been performed by NGI (Norwegian geotechnical institute) for the execution of offshore tests: a wire-line application of a Permeameter and gas sampler at great depth (D.G.S.: deep gas sampler) has been implemented by the NGI essentially for the petroliferous/oil drilling, but it is also applicable at shallow depth and is based on the same operating principle as the BAT/GEON.
- It is possible to adapt other types of sensors to the cone, i.e. sensors measuring chemical parameters.
- From a theoretical and interpretative point of view, the use of the CPTWD in soft soils, together with the MWD, will allow one to correlate, in a more efficient way, the mechanical characteristics of such soils with the data obtained by monitor while drilling tecnique.
- With the help of the right interpretative and mathematic models it might be possible to extend correlations which were found for soft soils also to coarse or rocky non penetrable soils.

The future improvements of the CPTWD system will include a combination between a drill rig and a pushing system of penetrometric type, possibly anchored in the ground (or to the floating craft in the case of offshore tests) for a higher capacity of pushing. This will allow a better control of the drilling parameters, (Ref. 5) leaving to the drill-rig the unique task of rotating the drillstring and injecting the mud.

REFERENCES

Elmgren, Kjell: *Slot type pore pressure CPTu filters. Behaviour of different filling media.* CPT 95, Linkoping 1995.

Fortunati, Fabio & Pellegrino, Guido: *The use of electronics in the management of site investigation and soil improvement works: Principles and applications.* ISC98, volume 1.

Larsson, Rolf: *Use of a thin slot as filter in piezocone tests.* CPT 95 Linkoping 1995.

Lunne et al. CPT: *Cone Penetration Testing in Geotechnical practice*

Peuchen, J.: *Commercial CPT profiling in soft rocks and hard soils.* ISC98, volume 2.

Tani, Kazuo: *Importance of instrumented drilling.* ISC98, volume 1.

te Kamp, W.G.B.: *The influence of the rate of penetration on the cone resistance qc in sand.* ESOPT II, Amsterdam 1982.

Proceedings ISC-2 on Geotechnical and Geophysical Site Characterization, Viana da Fonseca & Mayne (eds.)
© 2004 Millpress, Rotterdam, ISBN 90 5966 009 9

Enhanced access penetration system: a direct push system for difficult site conditions

J.D. Shinn II & J.W. Haas III
Applied Research Associates, Inc., VT, USA

Keywords: cone penetrometer, direct push, enhanced penetration, environmental site characterization

ABSTRACT: Access through gravel, cobble and cemented or dense soils is a limitation of the Cone Penetrometer Test (CPT) and other direct push techniques. Refusal in these layers has historically resulted in the need to mobilize a drill rig to penetrate the impervious layer, followed by continued investigation through a cased borehole. This approach increases cost and the detailed CPT profile of the site stratigraphy is lost. At highly contaminated sites, the cost of drilling and disposal of the drilling spoils can greatly increase the cost of characterization and extend the time required to conduct the investigation. This paper describes the Enhanced Access Penetration System (EAPS) which increases the depth to which CPT soundings can be conducted. EAPS consists of four major components: (1) a wireline CPT/Gas sampling probe and wireline soil and groundwater sampling system, (2) a small diameter air rotary drilling system, (3) environmental sensors that are used to detect and characterize contamination in both real and near-real time, and (4) an integral drill spoils collection and filtration system.

1 INTRODUCTION

The Cone Penetrometer Test (CPT) has proven to be a cost effective alternative to conventional drilling for environmental and geotechnical site investigation. Over the past decade, the U.S. Departments of Energy (DOE) and Defense (DOD) have developed CPT tools and sensors that enhance the investigative capabilities of the CPT.

While the CPT's investigative capabilities have continued to improve, the ability to achieve the depths of penetration required at many DOE sites such as Hanford, the Savannah River Site (SRS), and Paducah has remained a limitation.

The geology at these DOE sites can include very hard layers that will stop static penetration by conventional CPT and direct push sampling methods. These layers can be cemented soils, such as caliche, which is typically found in arid regions, coarse-grained formations (i.e., gravel and boulders), or volcanic flow fields. An outcrop of exposed caliche at the DOE Hanford site is depicted in Figure 1.

Under funding from the National Energy Technology Laboratory (NETL) and in cooperation with the DOE Hanford Site, Applied Research Associates, Inc. (ARA) has developed the Enhanced Access Penetration System (EAPS) that extends CPT penetration depth. EAPS consists of four major components: (1) a wireline CPT/Gas sampling probe and wireline soil and groundwater sampling system, (2) a small diameter air rotary drilling system, (3) environmental sensors that are used to detect and characterize contamination in both real and near-real time, and (4) an integral drill spoils collection and filtration system.

Figure 1. Hanford ringold formation showing cemented gravel and cobbles. Boulders up to 1 m in diameter can be encountered.

The wireline CPT/Gas sampling probe is used to determine soil stratigraphy and profile contaminants in real time. Once a refusal layer is encountered, the CPT/Gas sampling probe is withdrawn, leaving the push casing in place. A small diameter air rotary drill is then lowered through the casing and locked into the bottom of the casing. The drill is then used to penetrate the refusal layer. The return air and drill cuttings are routed through a series of filters to remove the drill cuttings. Volatile organic contaminants in the air stream are retained in a Granulated Carbon Trap ensuring that only clean air is emitted into the atmosphere. Once through the refusal layer, the wireline CPT/Gas sampling probe sounding is resumed. At any depth of special interest, the wireline CPT/Gas sampling probe can be removed and soil or water samples collected (again, without removing the casing) for either on-site or off-site testing. In highly impervious materials, the wireline-scale drill may prove ineffective at penetrating the refusal layer. For these situations, a larger diameter down-the-hole air hammer system was developed which is more energetic and can penetrate boulders and larger cobbles.

2 EAPS EQUIPMENT AND METHODS

The EAPS, photographed in Figure 2, consists of a CPT truck modified for both the small and large-diameter air rotary percussion and air rotary overburden drilling, and a support trailer housing an air filtration system. Additional equipment includes a 3-phase portable electric generator to power the vacuum system and a diesel engine driven 8.5-m^3/min air compressor with a capability of up to 1380 kPa air pressure. A Ford F-450 pickup is used to pull the support trailer.

Figure 2. EAPS at Umatilla chemical depot.

The major change to the CPT push system was integration of a combined push and drill system.
A photograph of the rotary drilling system integrated into the CPT push head is shown in Figure 3. The system supports both conventional CPT and drilling. During direct push operation, the drill head swings out of the way and allows the CPT rods to be added as with a conventional CPT push system. Once a refusal layer is encountered, the drill head swings into place and is locked to the push system. Drill pipes are lowered into the wireline casing rods and the drill bit locks into a taper lock built into a concentric bit attached to bottom of the wireline rods. Drilling can then commence. Electric solenoid switches are used to control the air flow to the drill and a manual lever is used to control the drill motor rotation. In practice, we normally apply only a moderate amount of down force while drilling. However, if required, the full push weight of the CPT truck can be applied to the drill and drill bit. On occasions this has proven useful to assist in moving cobbles and broken rock out of the way of the advancing drill.

Figure 3. Photograph of the EAPS air rotary drilling system.

2.1 Wireline CPT/Gas and water sampling systems

The EAPS CPT sampling tools are based on wireline soil and gas samplers that were developed by ARA for the DOE. Schematics of the wireline tools are shown in Figure 4.

The wireline system uses a 28.6-mm diameter piezocone. This size is not standard as per ASTM D3441, but a validation study reported by Farrington et al. (2000) examined the variability in differences between measurements taken with the same ASTM standard cone geometry as compared to the variability between measurements taken with a standard cone and the wireline cone. Statistical tests of hypothesis at 95% confidence confirmed that tip and sleeve stress measurements produced by the wireline cone do not differ from ASTM standard cone measurements.

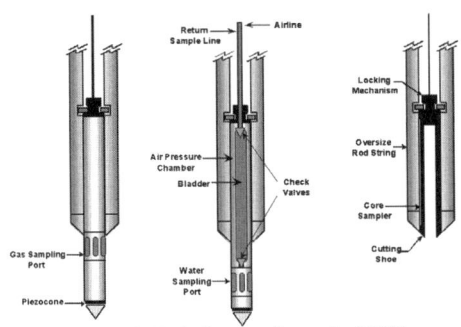

Figure 4. EAPS CPT wireline sampling tools. CPT/Gas, water, and soil samplers (left to right).

Table 1 lists the types of data that can be acquired with the EAPS data acquisition system.

Table 1. EAPS data acquisition capabilities.

CPT instrumentation	Drilling parameters	Other CPT probes available
Tip sleeve	Total push force	Total gamma
Pore pressure	Rotation speed	Spectral gamma
Total push force	Rotation torque	Video cone
Depth	Hammer blow rate	Soil moisture/resistivity
Soil gas flow rate	Supply air flow rate	Seismic
Temperature	Return air flow rate	*VOCs by GC detector or photo-acoustic IR

* Volatile Organic Compound (VOC)
  Gas Chromatograph (GC)
  Infrared (IR)

In operation, the wireline CPT/Gas sampling probe was used for *in situ*, direct push characterization of the site. Data from a typical sounding are shown in Figures 5 and 6. An air flow meter and two different analytical instruments were integrated into the EAPS data acquisition system. In practice, the airflow is monitored and recorded as the CPT push is being conducted. The analytical instruments included a gas detector manifold and a photo-acoustic spectroscopy infrared (PASIR) gas analyzer. Either of these instruments can be used, depending on the type and number of contaminants to be monitored. The manifold has three detectors: a photoionization detector (PID), a flame ionization detector (FID), and a discharge electron capture detector (DELCD). The PASIR analyzer was configured to monitor three gases: carbon tetrachloride, chloroform, and water. True continuous profiling is impractically slow, however, our practice is to monitor the gas readings semi-continuously at rod breaks (1-m intervals) or at locations likely to be contaminated (see Figure 5).

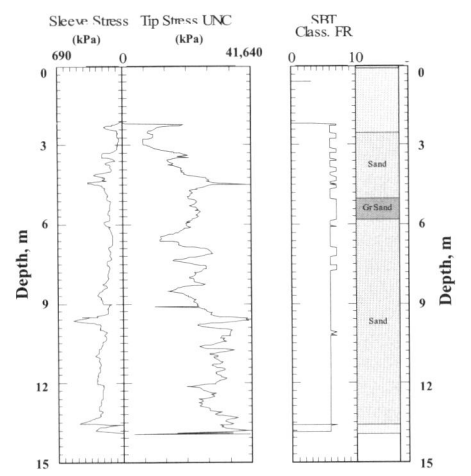

Figure 5. EAPS direct push data showing CPT characterization of soil behavior type (SBT) classified by friction ratio.

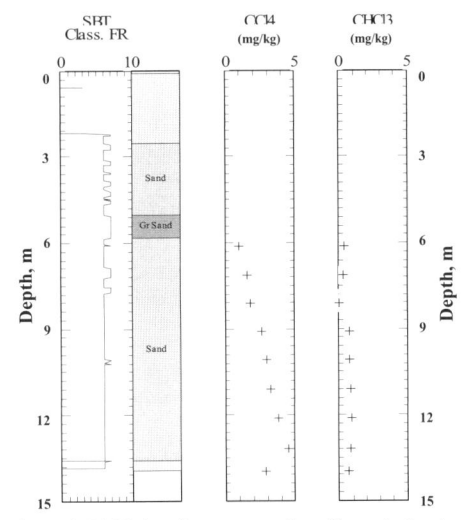

Figure 6. EAPS data from same push as Figure 5, showing gas analyzer results. Note contaminant concentrations increasing with depth below 6 m.

Detector output is displayed as a function of time and once the readings have stabilized, the CPT sounding can continue.

In some layers, fine-grained soils can clog the gas filter and reduce the flow rate. A system was developed for pressurizing the CPT probe to flush out fine-grained materials. The flushing system consists of a high-pressure breathable air tank with a regulator. Whenever the airflow from the CPT probe decreases to less than 0.5 L/m, the system is flushed with the breathable air. A test in the Hanford 200 West area confirmed that this approach returned the airflow to baseline values and could be used instead of pulling the wireline probe from the rod string for cleaning.

In regions of special interest, the wireline CPT/gas probe can be pulled from the rods and either a soil or water sampler inserted without pulling the rods. The soil sampler (see Figure 4) has proved to be highly effective for retrieving small volume samples for both on-site and off-site analysis. At many locations sample times as little as 2 to 3 minutes/sample have been routinely achieved.

Water samples can be obtained using the wireline bladder pump sampler (see Figure 4). As with the CPT probe and soil sampler, the bladder pump is locked into the push rods and can then be advanced into undisturbed soil. The bladder pump uses the breathable air tank on the push rig as a gas supply, eliminating the need for a separate clean gas source tank.

2.2 Particulate sampling

In addition to the *in situ* wireline samplers, an up hole system was developed to sample drilling spoils to either (1) conduct a cuttings analysis of the coarse fraction to classify the formation, or (2) collect a complete sample including coarse fraction down to particle sizes of 50 μm for laboratory analysis. The two particulate samplers are shown in Figure 7. A "Y" connector in the air return allows for a portion of the drilling spoils to be diverted from the main return line through the samplers. Once sufficient sample has been collected, two valves are closed to isolate the sample from the continuing main gas stream. Samples for geologic classification are safely collected in a glass jar that the geologist can log semi-continuously as material passes through it. The sample can also be analyzed to determine if the location is contaminated.

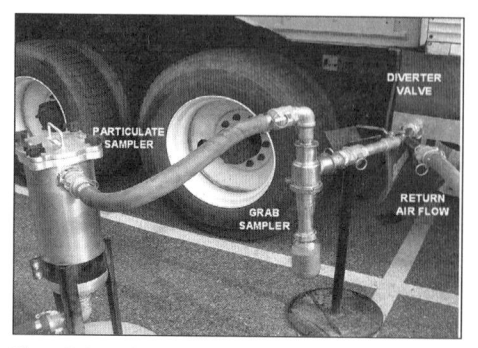

Figure 7. Jar and particulate drilling spoil samplers.

2.3 EAPS drilling options

In operation, wireline CPT probes are employed until refusal is encountered. The wireline probe is then withdrawn and a 25-mm diameter center bit is inserted into the ring bit attached to the bottom of the push rod string. A photograph of a selection of EAPS drill bits is shown in Figures 8 and 9. For penetration of boulders, a 73-mm OD casing with a 50-mm down-the-hole (DTH) hammer is used to advance the casing. The DTH hammer can be removed once the obstruction has been penetrated and the sounding continued with either the wireline CPT or a conventional CPT probe. This system can also be used when a larger diameter borehole is desired, (e.g., if high capacity 50-mm diameter pumps or other sampling systems which are too large for the wireline system are to be used).

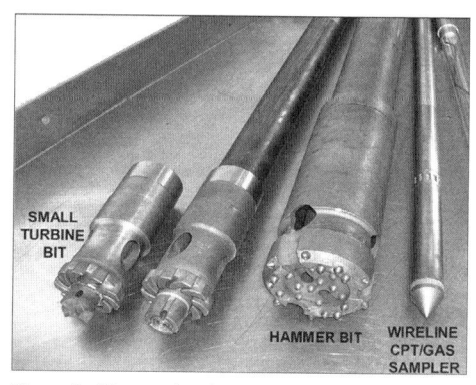

Figure 8. Photograph of EAPS drill bits and wireline CPT/Gas sampler.

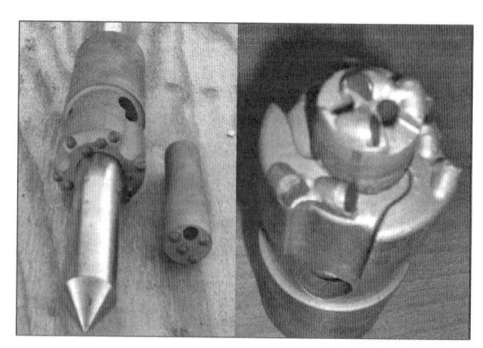

Figure 9. Photograph of EAPS CPT/hammer drills.

Testing of EAPS at the Hanford 200 W site has demonstrated that the system can routinely penetrate through cobbles and boulders and through the caliche layer located at a depth of 43 m. Without EAPS capability, CPT soundings were limited to a maximum depth of 18 to 21 m.

2.4 Drill spoils collection system

The EAPS drill spoils collection system is a unique feature that allows drilling operations to be performed at highly contaminated sites. The collection system consists of three major components. The first is the drill starter casing shown in Figure 10. The purpose of the starter casing is to seal the drill hole beneath the CPT truck and to prevent the escape of air from the borehole. The starter casing is pushed into the ground from inside the CPT truck to a depth of 3 m. A "dummy" point inside the starter casing is then withdrawn, leaving a clean, casing-lined hole. The CPT probe is lowered to the bottom of the starter casing and a centralizer is then lowered and installed to provide lateral support for the CPT rod. An airtight starter casing seal is then installed at the top of the casing. This seal consists of a square cross-section Teflon rope encircling the CPT rod and tightened against it using a compression fitting.

Figure 10. Drill starter casing and airtight seal.

Air and drilling spoils from the formation are returned up the annulus between the inner drill pipe and the outer CPT rod casing. The drill spoils are directed through an air swivel, visible in Figure 3, which is sealed to the top of the CPT push pipe. A packer similar to that used on the rod casing is tightened against the 25-mm OD drill pipe to again provide a seal. The drill spoils are directed through the 50-mm ID flexible hose to the air filtration system. Outside the CPT truck, a steel "Y" valve is used to either divert the drilling spoils directly to the filtration system or to a sampler, as previously discussed.

The air filtration system, shown in Figure 11, consists of four separate stages. The first is a cyclone separator, which separates all of the heavier material and much of the finer material from the air stream. The second stage consists of a bag filter system that separates all but the finest particles from the air stream. After the bag filter, the air stream is directed through a HEPA filter to remove all remaining particles and a carbon trap to remove any volatile compounds. A meter installed between the HEPA filter and carbon trap allows the operator to monitor airflow in the push rig with the EAPS data acquisition system. Continuous airflow monitoring has proven helpful in determining if the filter system is operating correctly and to indicate when the drill hole is becoming plugged.

Figure 11. Photograph of the EAPS air filtration system.

3 CONCLUSIONS

A system has been developed that integrates the ability of conventional CPT to provide a detailed stratigraphy and *in situ* characterization with the ability to penetrate through refusal layers such as cobbles, boulders and cemented or dense soil layers. The EAPS allows for rapid change from the CPT push to drilling configuration. At contaminated sites, the drill spoils are collected in a safe manner. In additional to the standard CPT data, EAPS provides the capability to acquire *in situ* information on the distribution of environmental contaminants and to collect physical samples for later testing in either an on-site or off-site laboratory.

ACKNOWLEDGEMENT

We gratefully acknowledge the Department of Energy and Hanford site technical team that has supported the development of the EAPS system. Dr. Karen Cohen facilitated the technical discussions with the Hanford site personnel. Dr. Scott Peterson facilitated the technical interchanges with Hanford personnel; and Grover "Skip" Chamberlain of DOE/EM50 also supported the project.

REFERENCES

American Society for Testing and Materials (ASTM). Standard Test Method for Deep, Quasi-Static, Cone and Friction-Cone Penetration Tests of Soil (D3441). November, 1994.

Farrington, S.P., Gildea, M.L., & Biachi, J.C. 2000. Development of a Wireline CPT System for Multiple Tool Usage. *Industry Partnerships for Environmental Science & Technology; Proceedings NETL Conference.* Morgantown, West Virginia, USA, October 17-19, 2000: 4.3.

Use of a radioisotope cone to characterize a lumpy fill

T.S. Tan, M. Karthikeyan, K.K. Phoon, & G.R. Dasari
Centre for Soft Ground Engineering, National University of Singapore, Singapore.

M. Mimura
Disaster Prevention Research Institute, Kyoto University, Japan.

Keywords: radioisotope cone, lumpy fill, site characterization, wet density

ABSTRACT: In Singapore, the uses of clays from marine dredging and land-based construction activities as fill for land reclamation provides a solution to the problem of disposal while turning such waste soils into material of economic value. A major problem with such fill is the very high degree of heterogeneity in the ground formed. The characterization of such ground at different stages of land reclamation works poses a major challenge as many traditional in-situ tests which provide point values are based on solutions essentially based on continuum mechanics. A major site characterization program was carried out as part of a research project. The Radioisotope cone penetration test is employed for site investigation to measure the continuous changes in density together with other usual cone parameters. The tests have been conducted on recently dumped lumpy fill prior to any soil treatment as well as on a newly formed lumpy fill, after a sand cap has been provided. Results from these tests shows that usual CPT parameters coupled with the additional density measurement will provide a much more complete picture of the fill formed. Problems faced in marine based investigations are also presented, and this has led to the development of a new cone.

1 INTRODUCTION

Dredging works in the coastal areas and excavations in urban areas produce large quantities of clay lumps. Use of these clay lumps for land reclamation is an environmental friendly solution to the disposal problem. When these big dredged clay lumps are used for reclamation, the land will have large initial inter-lump voids. If these large inter-lump voids do not close completely, even with surcharge, this will lead to excessive settlement when a structure is constructed on it. Characterization of the reclaimed fill must provide a clear idea of this. Characterization of a soil layer mainly relies on laboratory tests on undisturbed samples and / or in-situ tests. However for a lumpy fill, retrieval of undisturbed samples has been near impossible especially at the initial state when inter-lump voids are big. In-situ tests are generally based on the response of a continuum to a simple probe such as a cone penetrometer or a flat plate (dilatometer). The existence of large voids make meaningful interpretation of the obtained point values very difficult, if these are based simply on expected response of a continuum. Clearly, if the condition of the ground can be ascertained to some degree, this can greatly help in the interpretation.

One parameter that could provide some insight to the ground condition is the density. Therefore, it is necessary to have a reliable in-situ device to measure continuous subsurface profiles of water content or density of soil in addition to other point parameters such as cone resistance, pore pressure and sleeve friction.

In this paper, examples will be provided to demonstrate that for this particular type of ground, the addition of one additional parameter, the wet density, helps tremendously in the interpretation of results. This paper will describe the performance of a Gamma ray based radio-isotope (RI) cone penetrometer.

2 DESCRIPTION OF RADIOISOTOPE CONE

The in-situ wet density and water content of soil could be measured reliably using Nuclear-Density Cone Penetration Test (ND-CPT) and Neutron-Moisture Cone Penetration Test (NM-CPT) respectively (Shibata et al. 1993; Karthikeyan et al. 2001). These cone penetrometers will also measure cone resistance (q_c), sleeve friction (f_s) and pore pressure (u_2), providing more comprehensive subsurface information. If the soil is fully saturated, then only one test is needed, from which both the density and water content can be established. This was confirmed by Dasari et al. (2003) who had compared the water content measured directly by NM-CPT and the computed water content from the density measured by ND-CPT in the case of a saturated soil and found that both the data agreed with each other. The use of ND-CPT is recommended as this cone is influenced by a lesser number of environmental factors compared to NM-CPT.

dure for standard piezocone penetration tests specified by British Standard, BS 5930:1999. During testing, the ND-CPTs were pushed into the ground at a rate of approximately 2 cm/sec and cone resistance (q_c), sleeve friction (f_s), pore pressure (u_2) and ND count were recorded continuously. The detailed description and working procedure of ND-CPT was reported in Shibata et al. (1993) and calibration issues have been discussed by Shibata et al. (1993) & Dasari et al. (2003).

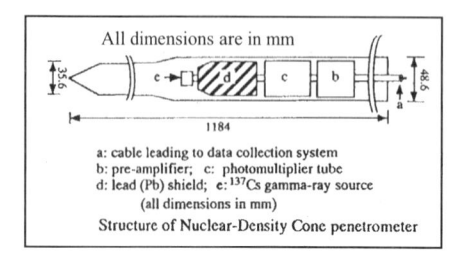

Figure 1 Diagram of Nuclear-Density Cone Penetrometer (ND-CPT) (after Shibata et al. 1993).

2.1 Nuclear-Density Cone Penetration Test (ND-CPT)

The Nuclear Density cone used in the present project is based on the design proposed by Shibata et al. (1993). Figure 1 shows major components of a ND-CPT. The lower part of the cone houses various sensors to measure the usual cone parameters, namely, cone resistance (q_c), sleeve friction (f_s) and pore pressure (u_2). The size of the lower part conforms to the standards recommended by the International Society for Soil Mechanics and Foundation Engineering (ISSMFE, 1989) for cone penetration testing. The diameter of the cone is 35.6mm, and the apex angle is 60°. The base area is 10 cm^2 and the area ratio, 'a' is equal to 0.75. A porous ceramic filter is located just behind the cone tip. The total length of the shaft housing the sensors is 258 mm. After this, the shaft tapers outwardly at an angle of 15°. The tapered portion of the shaft is 49 mm long and beyond this, the shaft has a constant diameter of 48.6 mm and extends for a total length of 896 mm. This upper part houses the radioisotope source, the detector, and a preamplifier.

The gamma ray source used in the ND-CPT is Cesium (Cs^{137}) with a half-life of 37.6 years, and the detector is sodium iodide activated with thallium (NaI (TI)) scintillator mounted on a photomultiplier tube. The length of the NaI scintillation detector is 10.2mm. The separation distance between the source and center of gamma detector is 255mm. The ND-CPTs were carried out according to the same proce-

Figure 2 Typical size of clay lump dredged from seabed.

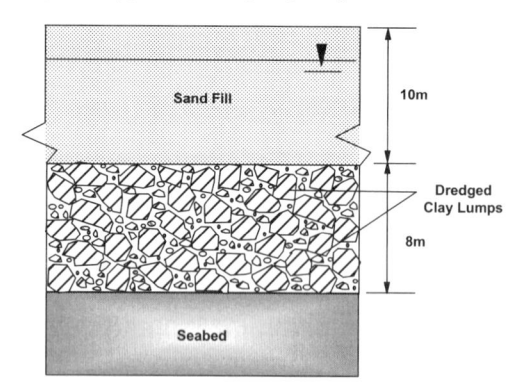

Figure 3 Schematic profile of reclamation with dredged clay lumps and sand surcharge (after Karthikeyan et al. 2004).

3 RECLAMATION USING LUMPY FILL

The use of dredged clay lumps for land reclamation is an attractive proposition because it provides an environmentally friendly solution to the disposal of these materials and recycles them as economically valuable fill. In coastal dredging, very large clamshells grab, with capacity ranging from 18m^3 to 25m^3 are used, and as a result, the clay lumps produced are usually quite big (1m^3-8m^3 in volume).

Figure 4 Photographic view of the Marine Cone Penetration System used in field investigation.

A typical clay lump dredged from the seabed is shown in Figure 2. These dredged clay lumps are placed in a barge and transported to a reclaimed site, where the lumps are discharged on to the seabed. Sand is usually used to cap the dredged clay lumpy fills. This is because the clay lumps can only be dumped up to a few meters below the sea level. Sand is then used to fill the reclaimed land above the sea level and to provide the surcharge to accelerate settlement.

Figure 3 shows a schematic section of the reclamation site with dredged clay lumps and sand surcharge. During initial stages of reclamation, there would be large voids between the big dredged clay lumps and these inter-lump voids (void between lumps) could be partly filled with small lumps and water. As a result of consolidation and compression, if the inter-lump voids (void between lumps) reduce to the size of intra-lump voids (void amongst particles within lump), then the lumpy fill is considered to have been "homogenized". From a construction perspective, the initial state of the lumpy fill needs to be established before contract can be called for the topping up. Subsequently, the change in the state in particular where the final state has reached a homogenized condition needs to be ascertained. This is clearly a major challenge and the results presented here is a first step towards achieving this understanding.

4 FIELD TESTS ON LUMPY FILL

The Pulau Tekong project was carried out by the Housing and Development Board to reclaim 1500 ha of land off the northeast coast of mainland Singapore. The project began in 1999 and 3 field pilot tests (TA 1, TA 2, & TA 3) are conducted in an on-going reclamation site. Clay lumps, excavated from the seabed using clam-shells to form sand keys for supporting the perimeter bunds, were then placed directly on the seabed from bottom-opening barges to form a lumpy fill layer, up to 3m below sea level (-3 mCD). Subsequently, capping sand was placed in several lifts up to +4 mCD. The site investigations were carried out using ND-CPTs to evaluate the state of the lumpy fill in two different stages. The first stage is immediately after placing the dredged materials in the reclamation area up to -3 mCD (Stage 1) and the second stage is immediately after dumping the capping sand, up to +4 mCD (Stage 2). A third stage is also planned and this will be after the end of consolidation.

In Stage 1, ND-CPTs were performed using a Marine Cone Penetration System, as shown in Figure 4. The control unit and the hydraulic power unit were placed on the barge. A hydraulic penetration system, which was mounted on steel frame or tower, was lowered on to the surface. The reaction force was provided by the dead weight of the steel tower. The barge is equipped with a four-point anchor mooring system to secure the barge in position and to maintain its stability during operations. In Stage 2, land is formed above the sea level and conventional CPT track was used to carry out the ND-CPTs.

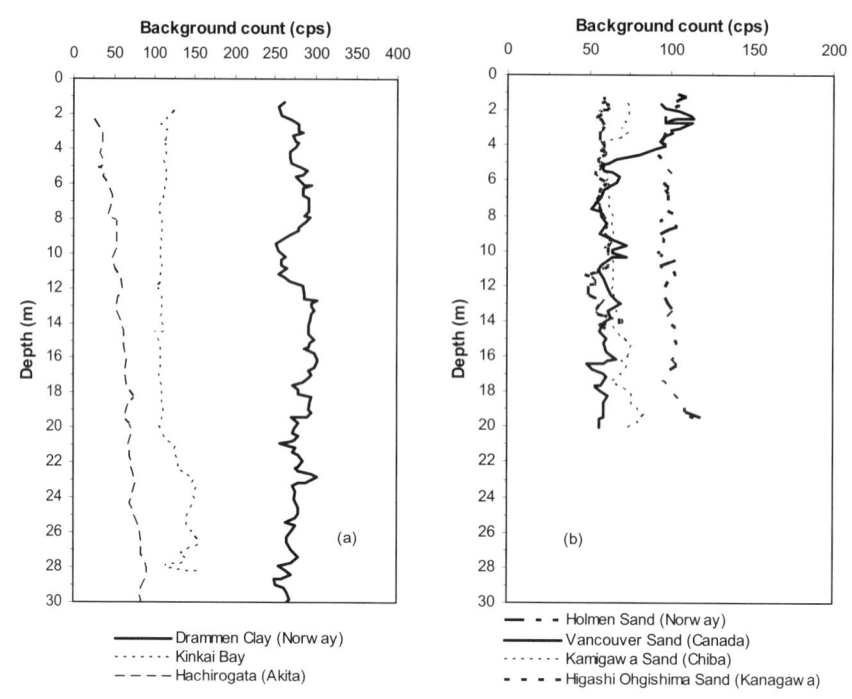

Figure 5 Typical background count profiles obtained for homogenous soils (a) Clay deposits (b) Sandy deposits (after Nobuyama, personal communication.).

4.1 *Problems faced in marine based investigation*

Problems in marine based investigation comes mainly from the fact that the current design of the RI cone requires two probing for each measuring point, one probing to obtain the background count of naturally occurring gamma photons and an another probing to measure the actual nuclear density (RI) count. For the measurement of density only gamma ray emitting from the gamma source and scattered in the soil medium should be used. Unfortunately, the RI count measurement from the gamma detector could not neglect the existence of the natural radioactivity in the environment from cosmic background. This natural radioactive (background) count is undesirable noise which must be subtracted from the count measured to give the actual nuclear density measurement. The background count is measured using a dummy cone, in which only the detector is placed to measure the naturally occurring gamma photons.

As the background gamma-ray comes from cosmic background, the intensity of background count measured at a point also varies greatly depending on soil type, mineral makeup, density of materials and its geological history. These profiles could possibly be used for a very preliminary classification of the soil type when other more decisive information is not available. Figure 5 shows the typical background count profiles obtained for homogeneous soils. As can be seen, the background count profile is weak in sand compared to that in clay. Thus, this can provide a rough guide to identify the boundary of clay and sand strata. It can be also seen in these background count profiles that there are little fluctuations with depth. Hence, if the soil is homogeneous and the variation in the background count profile is negligible within a test site, then double probing is not required for each measuring point. Only RI count needs to be measured at each point and an average background count within the test site is then used to determine the wet density of soil. This is especially important in marine measurement where each test is time consuming, and there will be great saving in cost if the number of probings can be reduced.

Figure 6 shows typical RI and background count profiles obtained for the lumpy fill site. Although all the six ND-CPTs are very closely spaced within a 5.0m x 5.5m square grid, significant variation in RI and background count profiles can still be observed, suggesting that the lumpy fill is highly heterogeneous. With this degree of variation in the background count, the use of an average background count profile within the test site needs to be examined.

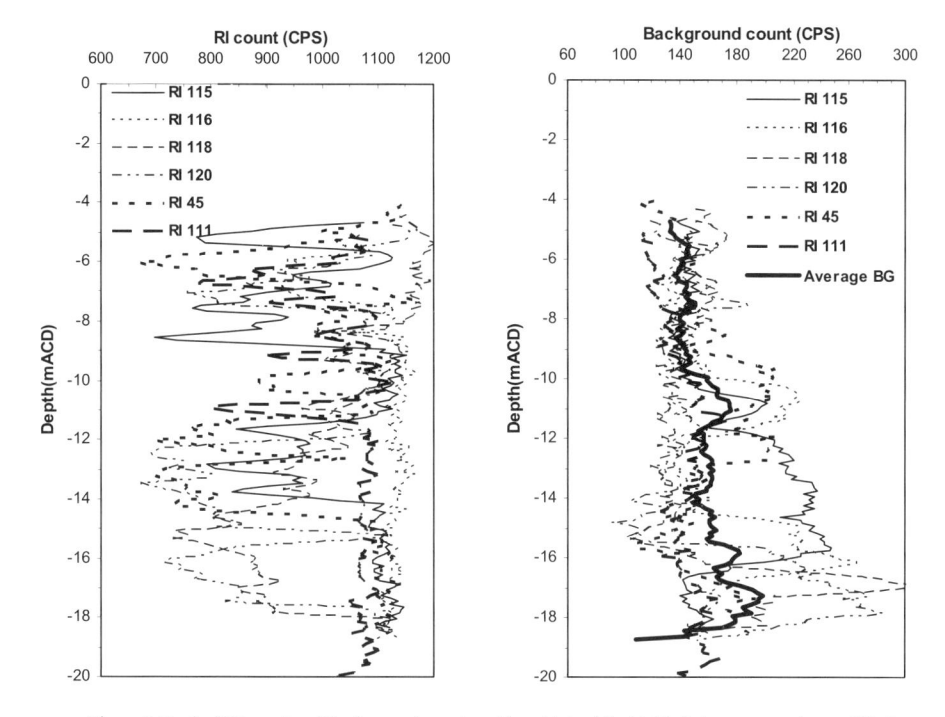

Figure 6 Typical RI count and Background count profiles obtained for highly heterogeneous lumpy fill site.

Figure 6 also shows the average background count profile obtained for the above six profiles. If the average background count profile is used for estimating the wet density, then there is an error of about 10% observed in the estimated wet density. Figure 7 shows the typical comparison of estimated wet density using actual measured background count and average background count. Therefore, for highly heterogeneous lumpy fill sites, two probings are required at each measuring point.

But in a marine based investigation (Stage 1), it is also near impossible to ensure that the two probes are done in the same hole due to practical difficulties. The variation in built-in coordinates for two different probing is about 0.15m to 1.5m. The non-correspondence of data obtained with two different holes will affect the accuracy of the measured density significantly for highly heterogeneous lumpy fill site. In order to avoid this discrepancy in the measured density profile, it is desirable to measure both the RI and background count with one probing. A new cone is currently being developed where the gamma ray section is extended so as to insert an additional detector that is outside of the gamma ray zone emitting from the source. With this, the cone is able to measure both the background and actual RI count during the same probing, thus removing the need to do two probings. Calibration works is in progress to establish the length needed to achieve this isolation.

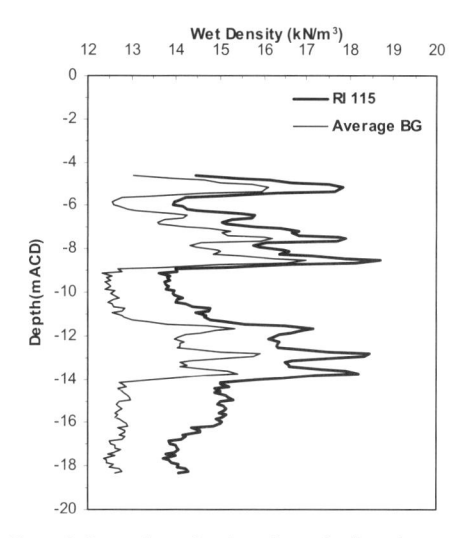

Figure 7 Comparison of estimated wet density using actual measured background count and average background count.

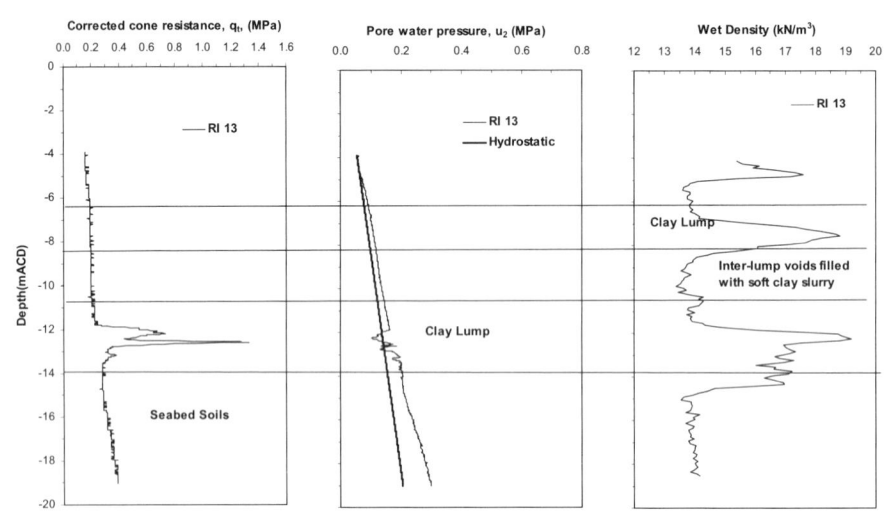

Figure 8 Typical cone resistance, pore pressure and wet density profiles at location RI 13 in TA 2.

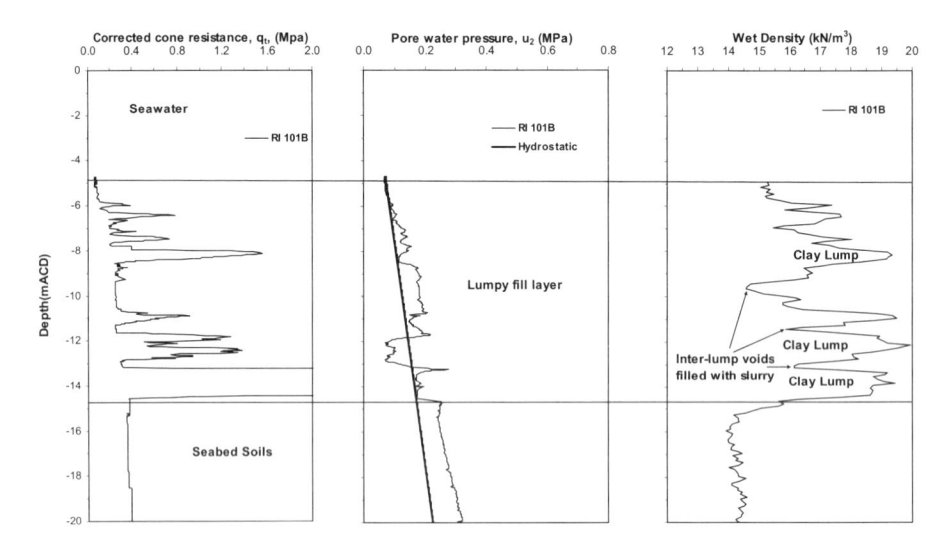

Figure 9 Typical cone resistance, pore pressure and wet density profiles at location RI 101B in TA 2.

4.2 Initial state of lumpy fill

Stage 1

Stage 1 is the investigation when the clay lumps have been placed but the sand surcharge has not been put on yet. For this, results from a pilot test area referred to as (TA 2) are discussed. This is one of 3 planned field pilot test areas involving the use of dredged clay lumps and sand surcharge as fill materials. The pilot test area is 100m long and 100m wide. The clay lumps were placed directly on the seabed which is about -13 to -14mCD from bottom-opening barges to form an 8m to 9m thick lumpy fill layer from June to July 2002. 120 numbers of marine ND-CPTs were conducted from August to November 2002. As discussed previously, when the dredged clay lumps were used for land reclamation, initially, there would be large voids between the big dredged clay lumps and these inter-lump voids could partly be filled with small lumps and water or clay slurry. Now, when the cone penetrates into this complicated soil, a number of probable soil profiles can result.

Figure 8 & 9 shows typical cone resistance, pore pressure and wet density profiles obtained using

ND-CPTs at location RI 13 & RI 101B. The changes in wet density, cone resistance and pore pressure together can be employed to identify the various sizes of clay lumps and inter-lump voids, as shown with horizontal lines in Figures 8 & 9. From the figures, it can be seen that there are appreciable changes in properties, confirming the heterogeneity of a lumpy fill.

From the density profile in Figure 8, two significant spikes could be observed at -7mCD and -12m CD. However, cone resistance and pore pressure profiles shows that there is only one spike at -12mCD. In a ND-CPT, the measuring sphere about the radioactive source is about 25 cm radius, and thus the density measured reflects the average around the central point of the radioactive source and detector configuration, in contrast with the essentially point-wise measurement in cone resistance and pore pressure. Thus Figure 8 suggests that the RI cone penetrates near to the edge of a clay lump from -7mCD to -9mCD, this is distinguished by the high average wet density and low cone resistance. However, from -12 to -14 mCD, it is penetrating into the interior of a clay lump, thereby measuring a high wet density and very high cone resistance at this depth. This vividly shows that using the cone resistance and pore pressure alone would have not picked up this detail.

At the same time, if there is no density measurement, then it is possible to infer from the usual cone parameter profiles that the cone at this location penetrates only one clay lump and this will mislead the interpretation of the actual ground. A similar observation could also be made in Figure 9. In this figure, it can be seen that there are significant spikes, and indicating the presence of lumps, as reflected by the combination of density and usual cone parameters.

From Figures 8 & 9, it can be deduced that the clay lumps are about 1-2m, consistent with that shown in Figure 2. The depth between these peaks is an indirect indicator of the size of the inter-lump voids. Studies are currently underway to interpret with greater details these "signatures". Also, in the inter-lump voids, the density is relatively high, generally above 14 kN/m³. This suggests that the voids have not been filled with sea-water, but slurry with significantly higher density than seawater. This is consistent with the observation that the seabed is covered with very soft clay, with density of about 14 kN/m³, or a water content of 110%, as can be observed. Therefore, for a highly heterogeneous lumpy fill site, the addition of one more parameter, the wet density, helps tremendously in the interpretation of results.

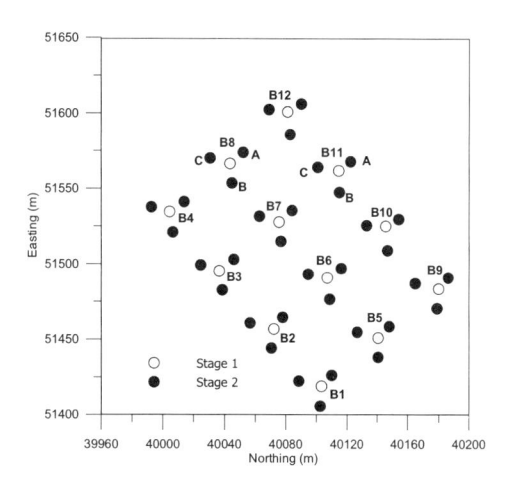

Figure 10 Location of ND-CPT after placing 3m to 4m dredged clays (Stage 1) and after placing 7m Hydraulic sand fill (Stage 2).

Stage 2
In Stage 2 investigation, the sand surcharge would have been placed on top of the lumpy clay fill. As the Stage 2 for TA2 has not been conducted yet, results from another pilot test area, (TA 3), are used. This pilot test area (TA 3) is 200m long and 150m wide and is subdivided into 12 grids, each of size 50m x 50m in plan. The clay lumps were placed directly on the seabed which is about -6 to -7mCD from bottom-opening barges to form a 3 m to 4 m thick layer from July to August 2001. Twelve marine ND-CPTs were conducted in the 1st week of September 2001 (denoted Stage 1 in Figure 10). Subsequently, 7 m of sand was placed in several lifts up to +4 mCD over a period of 6 months. Thirty-six ND-CPTs were conducted from 17th January to 6th March 2002 (denoted Stage 2 in Figure 10) and these tests were located in the triangular grid pattern, also shown in Figure 10.

Figure 11 shows typical cone resistance, pore pressure and density profiles obtained at pilot test area (TA 3) in Stage 1 and Stage 2. By comparing the four density profiles at each cluster, it is clear that the sand fill is above -3.5 mCD and the seabed clay below -7 mCD is homogeneous over this small lateral extent, but the dredged clay fill is heterogeneous. As can be seen in Figure 11, during Stage 1, two significant density spikes could be observed, and corresponding to this are two spikes in the pore pressure showing pore pressure significantly less than hydrostatic. This is the clearest evidence of the presence of lumps, the negative pore pressure is an indication that there have been tremendous load reliefs on these lumps excavated from some depth below seabed.

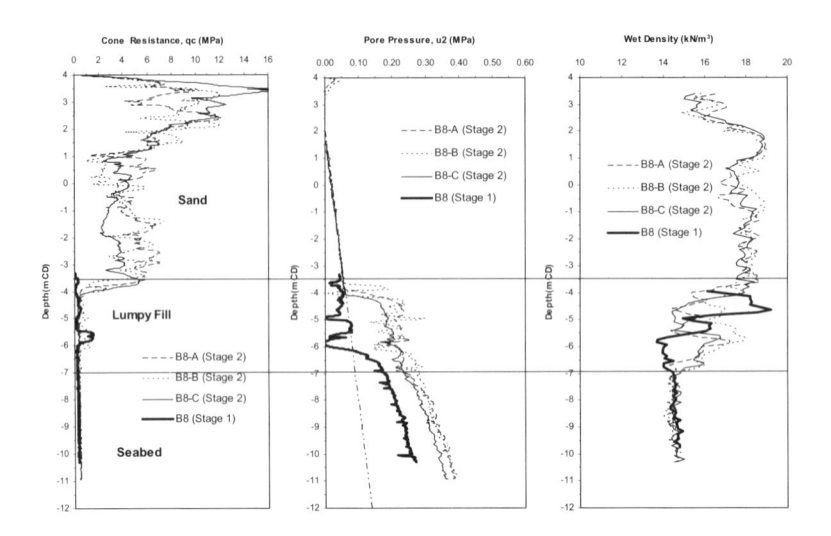

Figure 11. Typical con resistance, pore pressure and wet density at B8 (TA3) in Stage 1 & 2.

In Stage 2, with three density profiles measured at each cluster, the variation at each cluster reflects the heterogeneity of the lumpy clay fill formed. As can be seen in Figure 11, the wet density varying from 14 kN/m^3 to 18kN/m^3, indicating the presence of lumps, as reflected by the combination of density, pore pressure and sometimes cone resistance spikes. It is also observed that there is no clean sand pocket found in the lumpy fill, an indication that the sand used to cap the lumpy fill could not replace the slurry in the large inter-lump voids. The lumpy fill layer, after the placement of a sand surcharge, has zones of high density and low density, the denser zone is consistent with that of the original clay lump while the low density zone is due to the infilling of inter- lump voids with very soft clay slurry and also breakaway little lumps.

5 CONCLUSIONS

A comprehensive site investigation was carried out at a reclamation site where dredged clay lumps are used as fill materials. Radioisotope cone penetrometer was extensively used to evaluate the in-situ initial state of the fill formed by these dredged clay lumps. Results from these tests shows that usual CPT parameters coupled with the additional density measurement will provide a much more complete picture of the fill formed. However, the interpretation of the exact nature of the lumpy fill, which is a combination of lumps of soil formed with large inter-lump voids, is much more difficult. Problems faced while performing ND-CPTs in marine based investigation were also highlighted. Nevertheless, some aspects can still be discerned by examining more closely certain "signatures" present in the actual measured density profile.

ACKNOWLEDGEMENTS

The authors would like to thank the Housing and Development Board, Singapore, Toa–JDN (PUT) Joint Venture, Singapore and Kiso-Jiban Consultants Co., Ltd, Singapore for their support during site investigation and laboratory testing. The authors are also grateful to the National Science and Technology Board (NSTB) of Singapore for funding the present research work under grant NSTB/MCE/99/003.

REFERENCES

Dasari, G.R., Karthikeyan, M., Tan, T.S., Mimura, M., and Phoon, K.K. 2003. In-situ evaluation of radioisotope cone penetrometers in clays, *Geotechnical Testing Journal, ASTM*, Submitted for Publication.

ISSMFE, 1989. International reference test procedure for cone penetration test (CPT). *Report of the ISSMFE Technical Committee on Penetration testing of soils –TC 16, with references to test procedures*, Swedish Geotechnical Institute, Linkoping, Information, 7, 6 - 16.

Karthikeyan, M., Dasari, G.R., Tan, T.S., Lam, P.W., Loh, Y.H., Wei, J. and Mimura, M. 2001. Characterization of a reclaimed land site in Singapore, *Proceedings of the 3rd International Conference on Soft Soil Engineering*, 6-8 December 2001, Hong Kong, 587 –592.

Karthikeyan, M., Dasari, G.R., and Tan T.S. 2004. In-situ characterization of land reclaimed using big clay lumps, *Canadian Geotechnical Journal. (In print)*.

Shibata, T., M. Mimura and A.K Shrivastava. 1993. RI Cone Penetrometer experience in marine clays in Japan. *Proceedings of the 4th Canadian Conference on Marine Geotechnical Engineering*, Vol. 3: 1024 – 1033.

Development of a carrier tool for a multi-method investigation of brownfield sites

P. Wotschke & S. Friedel
Institute for Geotechnical Engineering, Swiss Federal Institute of Technology Zurich, Switzerland

Keywords: site investigation, multi-method approach, ERT, in-situ analysis, GPR

ABSTRACT: Efficient investigation of brownfield and contaminated sites requires economic acquisition of borehole information in the sense of data fusion. In this paper, we introduce a new technology in terms of a borehole tool capable of performing both geophysical and geochemical testing. In particular, the tool combines tomographical imaging techniques with in-situ analysis of soil gas. Essential parts of the new tool are 5 cm long metal spikes, pushed pneumatically through windows in conventional plastic filter casings. The spikes can serve as electrodes for in-hole, hole-to-surface und cross-hole electrical resistivity tomography, but also as hollow sampling noses for the extraction of pore air. Soil gas is analysed by a microchip sensor measuring the concentration of individual volatile organic compounds. The tool avoids any permanent installations of metal parts inside the borehole, allowing additional application of radar tomography. The modular design of the spike tool is open to be adapted to other in-situ analysis techniques, such as Laser Induced Fluorescence based on fibre optic transferred signals.

1 INTRODUCTION

Remediation and re-integration of potentially contaminated brownfield sites requires reliable and economic site investigation. In particular, existing boreholes should be used in the most economical way and the number of new ones should be minimised.

In this paper, we introduce a new borehole tool that can serve as a carrier for both geophysical and geochemical site technologies. The idea of this approach is to combine geophysical and pointwise sampled geochemical data to create a detailed image of the subsurface.

Tomographical imaging techniques such as electrical resistivity tomography (ERT) and Ground Penetration Radar (GPR) have now been used for several years to map and monitor contaminated sites and hydrogeological problems effectively (e.g. Daily & Owen 1991, Daily & Ramirez 1995, Binley et al. 2003a).

Especially the combination of both methods for cross-hole tomography greatly enhances information about the underground, because the two techniques deliver complementary information. GPR is more effective in high resistivity areas, and ERT proves to be more effective in low resistivity soils (Shinn 2000).

In practice, the combination is not as easy as just using them simultaneously. ERT requires a galvanic contact to inject currents and measure the resulting potential field. This is can be done in various ways: Slater et al. (2000), just as Binley et al. (2003b), constructed electrodes from stainless steel mesh and mounted it on insulating PVC pipes. As the permanent installation of cables would inhibit GPR due to unwanted reflections of the electromagnetic waves, additional holes for GPR are needed. An approach without permanent cables was suggested by Sauck (2000): a vertical array of electrodes mounted outside a PVC pipe is contacted by a set of sliders inside the pipe. However, only single-hole vertical resistivity profiles with this approach have been published yet. Vertical resistivity profiling, also possible with a set of electrodes mounted on a Cone Penetration Test (CPT) tool (Morey 1997), has a limited depth of investigation comparable to the separation of the electrodes involved.

Other approaches simply using cables lowered in uncased boreholes, work only if the borehole is a) stable (hard rock, otherwise one risks loss) and b) water saturated (to provide sufficient contact).

Our approach, in which pneumatic spikes are employed, is suitable for plastic cased boreholes in loose sediments and requires no permanent installations. Furthermore, a module of the tool is also capable of performing geochemical analysis of soil gas.

The most expensive and time consuming part of a site assessment is often the laboratory analysis of soil, water and gas samples. Effective field screening techniques can significantly reduce the quantity of samples going to the laboratory for analysis, thus controlling overall project costs (Hood et al. 2002). A screening sensor for a chemical in-situ analysis should comply with the following requirements:
Small size, and low cost,
- High sensitivity to the substances investigated well below legal environmental regulations,
- Repeatable and accurate results,
- Simple and robust operation in field,
- Safe handling to minimise an exposure to dangerous substances,
- Resource saving operation (e.g. water, electric power, equipment).

Within the frame of this study we restricted ourselves to a sensor system that is capable of detecting the most widespread and toxicologically relevant contaminants. The most abundant substances in the list of organic priority contaminants (Förstner et al. 1996) are Trichloroethylene (TRI) and Tetrachloroethylene (PER).

Both have a high vapour pressure and can be found in the liquid phase as Light Non Aqueous Phase Liquids LNAPL, and the gas phase as Volatile Organic Compounds (VOCs). In this study, we focused only on the gas phase.

Sampling systems that are capable of analysing VOCs quantitatively usually employ one of the following methods of sampling. In one widespread method, the sample is collected or accumulated in a storage medium e.g. in a glass ampul or on activated carbon (BUWAL 1998). A second approach transports the sample via a carrier gas to the surface (Hood et al. 2002). At the surface, the analysis is done with mobile laboratories, portable instruments or hand-held devices (Flachowsky 2003). Micro sensor systems for chemical analysis ("electronic noses") play a key role for the future of successful in-situ analysis and are currently still under going research (e.g. Barczewski et al. 2003).

In our approach we follow the technological perspective shown by Hagleitner et al. (2001), who proposes a highly sophisticated microchip sensor system and pattern recognition algorithm to determine the concentration of various organic compounds. The goal of our study was to assess the feasibility of such a system, to be integrated into a downhole measurement setup, to allow an in-situ real-time analysis of VOCs, at the depth of interest.

2 EQUIPMENT

2.1 Carrier Tool

The concept of the new downhole tool (patent pending, EPO Munich) that allows a combination of ERT and microchip in-situ soil gas analysis, while being open to other methods such as GPR, employs three functional elements as shown in Figure 1:
1) The stabilising and centralising element with the inflatable rubber hose,
2) The position detector,
3) The measuring sensor spike.

The modules of the probe have a diameter of 9 cm to operate in standardised 13.3 cm well pipes. Key parts are metal sensor spikes that can be pushed pneumatically through soft sections in the casing so that they are simultaneously in contact with the surrounding soil layers. The pressure used so far was 10 bar. The spikes can be used either as electrodes for ERT or as hollow sampling noses to extract pore air. The concept is open to be adapted to other sensoric principles.

Several spike modules can be combined to build a tool with between one to ten units. For this pilot study, the tool operated in a dedicated casing, a PVC pipe equipped every 0.25 m with weak ring areas of softer material (Styrodur®), which can easily be penetrated by the spikes. To detect the position of the rings reliably, we use a fine tuning winch and a magnetic position sensor at the top of the tool.

After completion of a measurement at a certain position, the spikes are pulled back into the probe. The tool can move to another position and start a new measurement. The operational sequence can be operated manually or, in a further project phase, automatically.

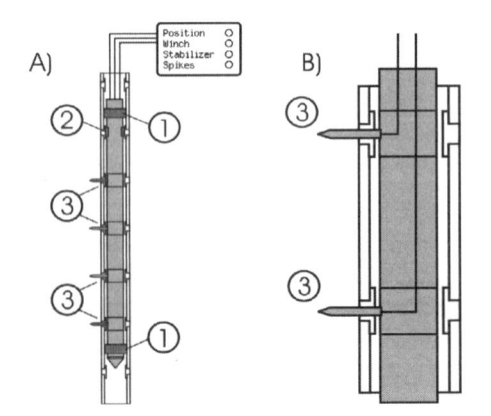

Figure 1. The multi-electrode probe design in the dedicated casing. A) Overview of the setup of the probe with its inflatable stabilisers (1), position detector (2) and sensor spikes (3). B) Detail of the spikes (3), pushed through the weak rings in the casing.

2.2 Multi-Electrode tool

One application for the massive stainless steel spikes in the new tool is using them as electrodes for electrical resistivity tomography (ERT). Subsurface electrodes can greatly enhance the resolving power of ERT with respect to surface only measurements. If only surface electrodes are used, reasonable resolution of structures can only be achieved at depths of about 25% of the largest electrode separation (e.g. Friedel 2003, Oldenburg & Li 1999).

For our field test, we assembled two probes of 1.60 m length and 16.5 kg weight each, comprising five spike units. The spikes ensure galvanic contact to the soil and allow currents to be injected and voltages to be measured. The electrodes are connected to an ERT device containing a power source, a selective high-resolution voltmeter and a multiplexer unit.

With the two ERT tools various measurements of single-hole (vertical profiling), hole-to surface and cross-hole were achieved during a field test on a brownfield test site in Zurich.

2.3 In-situ Analysis

The second sensor principle adapted to the spike tool was a chemical microchip sensor. Hagleitner et al. (2001) described a chemical sensor chip comprising three measuring principles: capacitive, mass-sensitive and calorimetric to measure quantitatively the concentration of various VOCs. The three micro system sensors are coated with selected polymers that allow preferential adsorption and show sensitivity to specific VOCs. Combining 3 sensors and 6 coatings and employing pattern recognition would allow a variety of VOCs to be distinguished and quantitatively analysed.

Since this sensor system is still in the laboratory stage and no field device is available, we restricted ourselves to an adaptation of the microchip unit to the spike probe, a feasibility study and laboratory calibration experiments for the capacitive sensor using several environmentally relevant compounds.

For laboratory testing, we used a chemical sensor setup, as indicated in Figure 2. The diameter of the sensor setup is 7 cm and fits smoothly into the probe. A micro pump (5) creates a constant gas stream of 120 ml per minute. It pumps the sample from outside the borehole, through the hollow spike into the probe. The gas is carried to the sensor (4) via chemically inert Teflon® pipes. The sensor sends its signals to the connected portable computer. Between measurements, a valve (3) leads the gas over a filter (2) to flush the sensor with clean gas. We used activated carbon as filter material.

The microchip unit presently contains the capacitive sensor coated by the polymer. As the gas sample flows over the sensor, diffusion of molecules into the polymer cause a change in dielectricity of the capacitor. This is measured as a frequency shift relative to a reference oscillator. In a single analyte system, the change of capacity is proportional to the concentration of a certain substance in the gas sample. A combination of various analytes and polymers was tested to calibrate the unit and assess the influence of environmental parameters such as humidity and temperature.

3 GEOPHYSICAL MEASUREMENTS

3.1 Results

The geoelectric multi-electrode tool was tested at a test-site in Zurich, where clayey lake sediments are overlain by a gravel aquifer, a potentially contaminated silty layer and a gravel cover. The ground water table was at 3.50 m below the surface. The principal layering was known from a previous investigation using 3-D surface ERT and georadar (Friedel et al. 2003). Five boreholes, at locations previously tested with a CPT, were also available to verify the results of the geophysical methods. In this paper, we restrict ourselves to a single 2-D ERT survey profile through a plane crossing two boreholes KB02-02 and KB02-03. The surface profile contained 25 electrodes at 1 m spacing. Similar conditions of limited space accessible for surface measurements (because of buildings, rubble, estate boundaries) are typical for many brownfield sites. To extend the depth of exploration of such a 25 electrode system beyond 5 m, the use of borehole electrodes is essential. Each of the five 10 m boreholes was equipped with the dedicated casing with 40 soft rings to be penetrated by the electrode spikes. For the sake of simplicity we show in this paper only the results of one ERT data set comprising 900 hole-to-surface pole-dipole measurements. A voltage dipole MN of 50 cm width was shifted down the borehole at 25 cm offsets (z-direction). For each dipole the current pole was shifted along the 25 surface elec-

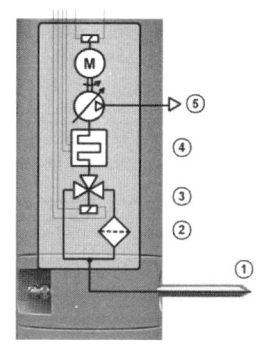

Figure 2. Scheme of the chemical sensor probe. A micro pump (5) takes the sample through the spike (1) over the microchip sensor (4). After the measurement, a valve (3) leads air to a filter (2) to flush the sensor. All parts are remote controlled by a portable computer.

trodes in the x-direction. As only surface electrodes were used for current injection, contact resistances of the spikes (up to 10 kOhm in the unsaturated gravel) did not cause problems. The whole set was measured twice; the deviation averaged around 1.5 percent and rarely exceeded 10 percent. The data were inverted with an iterative smoothness constrained Gauss-Newton and quasi-Newton inversion RES2DINV (Loke & Barker 1996, Loke & Dahlin 2002).

A peculiarity of the special data set appeared when the apparent resistivity ρ_a was plotted as a function of the positions of current and voltage electrodes $\rho_a(x_A, (z_M+z_N)/2)$. Where x_A is the position of the electrode A while z_M and z_N are the depths of the electrodes M and N in the borehole. This kind of data representation (pseudosection) already resembled, to a high degree, the principal layering known from the boreholes.

The pseudosection was therefore used as starting model for the inversion rather than a homogeneous halfspace. For both approaches, the data misfit was in the same order of magnitude. However, the final model obtained when starting from the pseudosection agreed much better with the borehole data.

The final resistivity image is shown in Figure 3 (bottom) and is compared with the borehole data and the results of cross-borehole radar tomography (centre) and vertical cross-hole radar profiling (top).

The ERT image is in very good agreement with the borehole logs. It shows resistivity of about 20 Ohmm for the bottom clay layer, around 100 for the water saturated aquifer, 200 to 500 Ohmm for the unsaturated aquifer, 50 Ohmm for the artificial filling, and again 500 Ohmm for the top gravel cover.

A particular feature appears inside the thick clay bed at the bottom where a narrow band of fine sand is embedded. This feature was previously unresolved by mere surface measurements. Using the borehole electrodes, this band could be resolved reliably as a layer along the whole profile.

Since no metal parts remained in the borehole after completion of the ERT measurements, a cross borehole radar tomography data set was measured using RAMAC 100 MHz antennae at 25 cm intervals. Although severe damping occurred in the clay beds, data processing allowed the detection of first breaks for nearly all 1500 tracks. The cross-borehole radar tomogram is very similar to the resistivity tomogram. However, the radar tomogram resolves the gravel aquifer only as one layer whereas ERT separates zones of different saturation within the gravel. Furthermore, ERT resolves the narrow sand band embedded in the clay between 7.20 m and 8.50 m better. Generally, the ERT resistivity image provided a higher resolution than the 100 MHz cross-hole GPR tomogram.

Besides the 100 MHz radar tomogram, we also tried to use RAMAC's 250 MHz borehole antennae. Due to the even higher attenuation of the shorter waves in the clay layers it was impossible to derive a complete cross-hole data set. Sufficient wave energy could only be transmitted between the boreholes in the sandy layers, as shown in Figure 3 (top). Here both antennae were moved parallel and simultaneously downward. The picture does not depict a special image but the arriving wave train for each antenna depth. It shows where transmission was possible and the time of the first arrival as a measure of wave velocity in the respective layer. Clearly the silty fill and the clay layer attenuate the signal completely. The gravel aquifer shows two distinct zones – unsaturated and water filled. And finally, the fine sand band seen in the ERT image and in the borehole logs is found also as a transmissive layer in 250 MHz parallel GPR profiling.

3.2 *Discussion*

The field test showed that the carrier tool employing pneumatically driven electrodes was stable mechanically and electrically under a variety of subsurface conditions from unsaturated gravel to clay, above and below the ground water table. Measurements were reliable and reproducible. The images delivered from the inversion of a specific pole-dipole data set showed very good agreement with borehole data and were cross-validated by subsequent cross-borehole GPR measurements within the same holes. Theoretically, the number of 5 electrodes at a distance of only one meter is the lowest limit suitable for tomographic measurements.

Ideally the complete borehole should be equipped with electrodes which would allow multiplexed switching of a much larger number of configurations. An extension of the probe to up to 10 spike electrodes seems possible, but the tool would increase in size and weight and become increasingly difficult to handle in the field.

The results suggest that such an extension might not be necessary. The images derived with our measurement setup show that even with a small number of electrodes, detailed images can be derived at low cost if appropriate configurations (e.g. pole-dipole, surface-to-hole) and a starting model (e.g. derived from a pseudosection) are used.

4 GEOCHEMICAL MEASUREMENTS

4.1 *Results*

During a laboratory study, four different polymers coatings were used for the capacitive microchip sensors: EC (Ethylcellulose), PDMS (Polydimethylsiloxan), PEUT (Polyetherurethan), and PCPMS (Polycyanopropylmethylsiloxan).

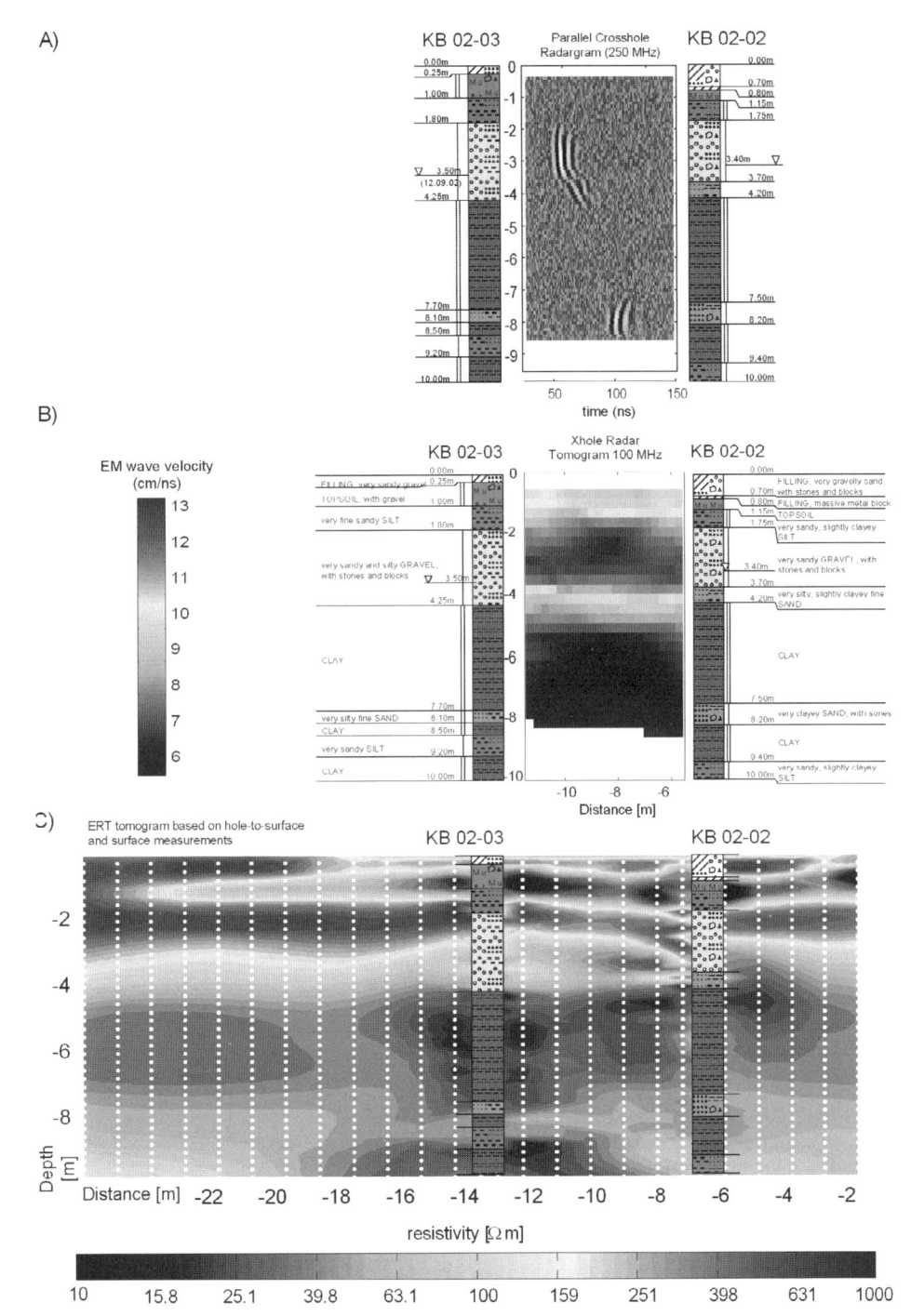

Figure 3. Comparison of different geophysical measurements. A) Radargram of a vertical cross-hole profiling, 250 MHz transmitter and receiver were lowered simultaneously. First arrivals on the time axis depict the travel time of the EM wave between the boreholes. Due to high attenuation no signal is transmitted in silty or clayey layers. B) 100 MHz cross-hole radar tomogram depicting average EM wave velocity, which is mainly a function of water content between the boreholes. C) ERT image derived from pole-dipole surface-to-hole measurements using a pseudosection starting model. Borehole data are included in all figures.

The sensors were exposed to small concentrations (compared to the vapour pressure) of TRI and PER. The measurement program flushes the sensors with clean synthetic air for about 600 s and exposes the sensor to 'contaminated' air for about 400 s. The concentration of the analyte were [20 35.6 63.2 112.4 200 200 112.4 63.2 35.6 20] Pa for TRI, and [16 28.5 50.6 90 160 160 90 50.6 28.5 16] Pa for PER. These concentrations were chosen to approach the order of magnitude of the Swiss legal limits (AltlV 1998). Figure 4 shows the result for TRI and PER on PDMS. The measurements for the two substances were taken one after the other, only displayed simultaneously in Figure 4.

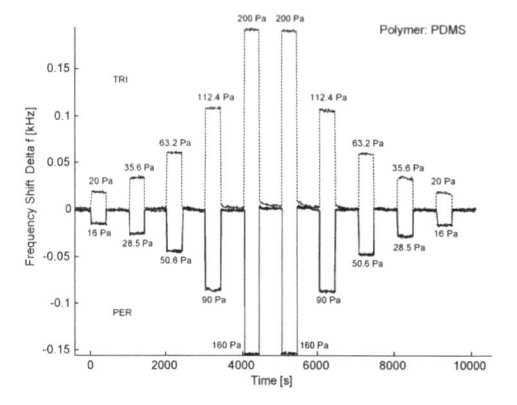

Figure 4. Reaction of a capacitive sensor coated with PDMS. TRI causes a positive frequency shift (dotted line), PER a negative shift (solid line). The shifts are spontaneous and stable.

The signals have the shape of stairs. Sharp signal flanks indicate that the diffusion of the analytes reach a steady stage quickly. Furthermore, this state is constant during the measurement period.

The dielectric constant of TRI (3.42), which is higher than the dielectricity of PDMS (2.6-2.8), causes a positive frequency shift.

In contrast to that, the dielectricity of PER (2.28) is lower than the dielectricity of PDMS. This causes a negative frequency shift.

The reliability of a chemical analysis can be expressed by three parameters: The sensitivity and the selectivity of the sensor to a certain analyte, and the reproducibility of the signal.

The sensor signal is measured for a variety of different analyte concentrations. We used a simple linear model to fit our data according to:

$$\Delta f^{cal}(c_i) = p_1 \cdot c_i + p_2 \qquad (1)$$

where c_i is the concentration of the analyte, Δf^{cal} is the calculated frequency shift of the capacitor, p_2 is an offset term and p_1 an approximation to the sensitivity as defined by the derivative of the signal with respect to the concentration:

$$p_1 = \frac{\partial(\Delta f)}{\partial c} \qquad (2)$$

A high p_1 value indicates a significant change of the sensor signal, caused by slightly different analyte concentrations. Optimally, the offset is zero, because the absence of the analyte should cause no signal.

Figure 5 shows the sensitivity regression curves for TRI and PER on PDMS for the data given in Figure 4. A mean value was calculated from each plateau of the observed signal Δf^{obs} and plotted

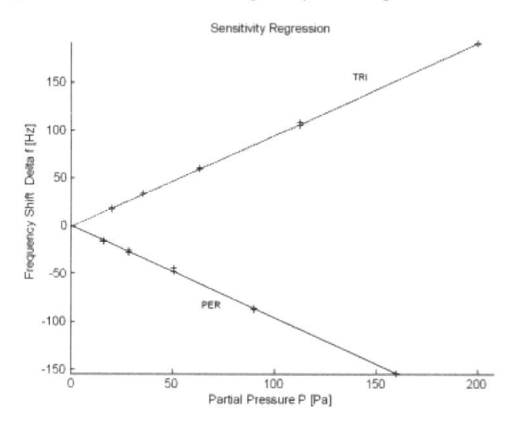

Figure 5. Sensitivity regression curves of TRI and PER on PDMS.

against the analyte concentration. The quality of the fit shows, that the linear calibration model (Eq. 1) is justified within the range of concentrations used.

To assess the quality of the calibration more quantitatively, the RMS error between measured and predicted data was calculated according to:

$$e_{rms} = \sqrt{\frac{1}{N} \cdot \sum_{i=0}^{N} \frac{\left(\Delta f^{cal}(c_i) - \Delta f^{obs}(c_i)\right)^2}{\left(\Delta f^{cal}(c_i)\right)^2}} \qquad (3)$$

Table 1. Summary of the sensitivity analysis for the analytes TRI and PER on the polymers EC, PCPMS, PDMS, and PEUT.

	TRI			PER		
	p_1 Hz/Pa	p_2 Hz	e_{rms} %	p_1 Hz/Pa	p_2 Hz	e_{rms} %
EC	4.47	68	12.8	-0.05	-18.7	41.0
PCPMS	0.74	11.2	16.7	-2.26	13.6	5.6
PDMS	0.96	-0.69	1.6	-0.96	0.74	4.6
PEUT	2.08	5.07	4.3	-4.45	14.1	6.5

In case of e_{rms} = 0, all measured data would perfectly fit the linear model. However due to measurement errors, such an ideal fit can hardly be achieved.

A summary of the results is given in Table 1. Sensitivity and accuracy of the sensors are very different for individual polymers and substances. To ensure the best possible performance on a TRI/PER analysis with a capacitor, a PDMS coating should be preferred to an EC coating. PDMS is more sensitive to TRI and PER, and the measurements are more accurate. Table 1 indicates that PCPMS and PEUT show very good results, too. However, their performance is less exact than the performance of PDMS.

With additional measurements, not included in this representation, we found the detection limits of the PDMS polymer sensor to be about 2.5 Pa for TRI and 2 Pa for PER. This below the Swiss legal limits of about 5 Pa each (AltlV 1998).

4.2 Discussion

The capacitive sensor analyses TRI and PER accurately and in a reproducible manner. The best performance we observed was with a PDMS polymer coating.

Up to now, we determined only singular analyte systems. Multi-component systems will provide information on polymer selectivity, cross-influences of other analytes and interactions between analytes. A single microchip sensor generates a single signal. In a singular system, the concentration of a known analyte can be determined. It is not possible to identify a contaminant with such a signal. Pattern recognition, as mentioned earlier in this article, is only possible with a variety of analyte specific signals. This requires a multi sensor chip. Such a system is still under research.

Flachowsky (2003) drew our attention to the high relative humidity that normally exists in soils. The strongly polar water molecule has a significant influence on the sensor capacity. The dielectric constant of water is very high (ca. 78). Hence, the presence of water covers all other analytes. With a $CaCl_2$ filter, it was possible to reduce selectively the water content to a negligible level. Such a filter has still to be included into the design of the pneumatic spike probe tool.

Other typical soil gas components, such as higher concentrations of CH_4 or CO_2 had no detectable influences on the sensor.

5 CONCLUSION

The aim of this work was to develop a tool that allows a combined geophysical and geochemical site investigation at low cost.

We developed a special design of a tool comprising as a key part stainless steel spikes that can be pushed pneumatically through weak zones in a plastic casing and can be used in various ways as sensors or sampling aids. So far we have adapted ERT and a specific chemical sensor to the design of the probe.

In a field test, we were able to prove that the subsurface resistivity models derived from measurements with the probe are of high resolution. Furthermore, it proved to be very consistent with borehole results and radar tomogram. The latter could only be performed in the same borehole because the probe design requires no metal parts to be permanently installed in the boreholes.

In a laboratory scale calibration experiment, we showed that the microchip sensor adapted to the probe is capable of detecting two of the most abundant VOCs in contaminated ground reliably and in the order of magnitude of Swiss environmental regulations. Future work will focus on widening the sensors spectrum of detectable contaminants by incorporating additional sensoric principles (mass sensitive and thermo sensitive) and pattern recognition software as suggested by Hagleitner et al. (2001).

Furthermore, the hollow spike of the carrier tool could also be adapted to various other chemical analysis sensors, e.g. optical geochemical analysis of soil gas or groundwater. For that purpose, the system could be equipped with a fibre optic cable, transmitting light of a laser source of a Laser Induced Florescence (LIF) device. Such an approach could help to identify hydrocarbons in-situ in the soil without taking a gas sample.

ACKNOWLEDGEMENTS

The authors would like to thank the reviewers for their valuable comments and suggestions on the manuscript. Thanks go to Rita Hermanns Stengele for her fund raising and her project support. Financial support for this research project is provided by a Swiss business company and the Swiss Commission for Technology and Innovation (6093.1 KTS). We would like to thank the colleagues from the ETH Physical Electronics Laboratory (PEL) for fostering our work, and making their sensors and measurement setup available to us. We appreciate the help of Hendrik Paasche and Holger Tetzlaff during the geophysical field work.

REFERENCES

AltlV 1998. Verordnung über die Sanierung von belasteten Standorten (Altlasten-Verordnung) vom 26. August 1998.
Barczewski, B., Baterau, K. & Müller, M. 2003. Physikalisch-chemische Sensoren. *Vor-Ort-Analytik für die Erkundung*

von kontaminierten Standorten. Erich Schmidt Verlag GmbH & Co., Berlin.

Binley, A., Cassiani, G., Middleton, R., & Winship, P. 2003a. Vadose zone flow model parameterisation using cross-borehole radar and resistivity imaging, in *Journal of Hydrology*, vol. 267, pp. 147-159.

Binley, A., Winship, P., West, J., Pokarb, M., & Middleton, R. 2003b. Seasonal variation of moisture content in unsaturated sandstone inferred from borehole radar and resistivity profiles, *Journal of Hydrology*, vol. 267, pp. 160-172.

BUWAL 1998. Arbeitshilfe Probenahme und Analyse von Porenluft. Bundesamt für Umwelt, Wald und Landschaft. Bern.

Daily, W. & Owen, E. 1991. Cross-borehole resistivity tomography, *Geophysics*, vol. 56, no. 08, pp. 1228-1235.

Daily, W. & Ramirez, A. 1995. Electrical resistance tomography during in-situ trichloroethylene remediation at the Savannah River Site, *Journal of Applied Geophysics*, vol. 33, pp. 239-249.

Flachowsky, J. 2003. Anwendung von Feldmesstechniken bei der Altstandorterkundung. *Vor-Ort-Analytik für die Erkundung von kontaminierten Standorten*. Erich Schmidt Verlag GmbH & Co., Berlin.

Förstner, U., Raudschus, M. & Reichert, J.K. 1996. *Chemie und Biologie der Altlasten*. Hrsg. Fachgruppe Wasserchemie der GDCh. VCH, Weinheim, New York, Basel, Cambridge, Tokyo.

Friedel, S. 2003. Resolution, stability and efficiency of resistivity tomography estimated from a generalized inverse approach. *Geophysics Journal International*, vol. 153, pp. 305-316.

Friedel, S., Wotschke, P. & Hermanns Stengele, R. 2003. 3-D Site Investigation Using GPR- and ER-Tomography with Pneumatic Borehole Electrodes; *EAGE 65th Conference & Exhibition*, Conference Proceedings, Stavanger, Norway.

Hagleitner, C., Hierlemann, A., Lange, D., Kummer, A., Kerness, N., Brand, O. & Baltes, H. 2001. Smart single-chip gas sensor microsystem. *NATURE*, Macmillan Magazines Ltd., vol. 414, pp. 293-296.

Hood, G.M., Allee, P.C. & Pucel, P.G. 2002. Evaluation of a cost effective SVOC field screening technique for use at petroleum contaminated sites. *Contaminated Soil Sediment & Water*, vol. 3/2002, pp. 8-21.

Loke, M.H. & Barker, R.D. 1996. Rapid least squares inversion of apparent resistivity pseudosections by a quasi-Newton Method. Geophysical Prospecting, vol. 44, 131-152.

Loke, M.H. & Dahlin, T. 2002. A comparison of the Gauss-Newton and quasi Newton methods in resistivity imaging inversion. *Journal of Applied Geophysics*. vol. 49, pp. 149-162.

Morey, R.M. 1997. Tomographic Site Characterization Using CPT, ERT and GPR. *Conference on Industry Partnerships to Deploy Environmental Technology*; Conference Proceedings. Morgantown, West Virginia, USA, October 21-23, 1997.

Oldenburg, D.W & Li, Y. 1999. Estimating depth of investigation in DC resistivity and IP surveys, *Geophysics*, vol. 64, pp. 403-416.

Sauck, W. A. 2000. A model for the resistivity structure of LNAPL plumes and their environs in sandy sediments, *Journal of Applied Geophysics*, vol. 44, pp. 151-165.

Shinn, J. 2000. Tomographic Site Characterization Using CPT, ERT and GPR. *Innovative Technology Summary Report DOE/EM-0517*, April 2000.

Slater, L., Binley, A., Versteeg, R., Cassiani, G., Birken, R., & Sandberg, S. 2002. A 3d ERT study of solute transport in a large experimental tank, *Journal of Applied Geophysics*, vol. 49, pp. 211-229.

4 New developments in interpretation of *in-situ* data

General report: New developments in the interpretation of *in situ* test data

Barry Lehane & Martin Fahey
School of Civil and Resource Engineering, The University of Western Australia

Keywords: *in situ* testing, parameter determinations

ABSTRACT: Most of the *in situ* test devices in popular use today have, with few exceptions, been developed more than 20 years ago. Although recent refinements have taken full advantage of on-going improvements in sensor design and data acquisition systems, the essential features of the tests have remained the same. Some modest changes in the way these *in situ* tests are interpreted have accompanied the newer technological developments and it is these changes that form the main focus of this discussion paper. The paper summarises particular features of papers submitted to this conference sub-theme and includes the authors' perspective on some recent general trends in this area.

1 BACKGROUND

Cone and standard penetration tests (CPTs and SPTs) are the most popular *in situ* tests employed for site characterisation and geotechnical engineering. In many countries, the CPT has long been the preferred test for most soils, primarily because of the severe difficulties associated with the interpretation of actual soil parameters from SPT blowcounts. However, despite ongoing theoretical developments in the modelling of the cone installation process, interpretative techniques that allow reliable assessment of soil strength parameters from the CPT end resistance (q_c) have remained elusive. It would appear that, although the boundary conditions associated with the CPT are simpler than those of the SPT, the complexities involved in modelling the installation process coupled with the need to employ sophisticated soil models in such analyses have made progress particularly slow. The analyses performed to date have shown that q_c depends on a wide range of soil parameters and that q_c cannot be generally related to any single soil parameter. For example, while it is common practice to assess a measure of undrained shear strength (s_u) from q_c using a cone factor N_{kt}, installation models such as the Strain Path Method, SPM (Baligh 1985) reveal (directly or indirectly) the importance of features such as clay structure and sensitivity, strain reversals, progressive failure and strain rate dependency on the mobilized cone end resistance. It is therefore not surprising that N_{kt} values correlated with a given measure of undrained strength (often a triaxial compression or vane strength) can vary by a factor of 2 between different clays.

Evidently, interpretation of soil parameters from our most popular *in situ* tests (i.e. CPT and SPT) is not straightforward and interpretative methods used in practice are empirical and therefore often of poor reliability. This is primarily because of the complex boundary conditions and the associated complex soil response that governs the test results. It should be noted, however, that the similarity of the CPT/SPT with the mode of installation of displacement piles has led to the development of relatively good direct correlations between q_c (and N) and displacement pile capacity. For the same reasons, the application of pressuremeter data to pile design in France has also met with reasonable success.

In situ tests that are designed to provide a direct measure of an *in situ* soil parameter are generally more desirable. Examples of popular *in situ* tests that fall into this category are (i) shear wave velocity measurement, and hence maximum shear stiffness (G_{max}) measurement, using a seismic CPT, (ii) vane shear tests to provide a measure of undrained strength and sensitivity corresponding to the vane mode of shearing, (iii) piezocone dissipation tests to allow inference of the (re-loading) lateral coefficient of consolidation and (iv) the self-boring pressuremeter test to measure the *in situ* lateral stress and shear stiffness as well as undrained and effective stress strength parameters. Interpretation using these *in situ* tests is relatively straightforward and has changed little in recent times. Unfortunately, however, *in situ* test devices that provide a direct measurement of a wide range of critical soil parameters do not exist and the practitioner needs to rely on empirical interpretative methods to deduce design values. Recent trends in the development of these methods are now discussed. Developments in the interpretation of *in situ* tests for environmental applications are not considered here.

2 GENERAL THEMES OF PAPERS ASSIGNED TO THIS SESSION[1]

The Authors were provided with copies of the 27 papers assigned to this session. These are listed at the end of this paper. Of these 27, some 16 deal with various aspects of the CPT, including interpretation or application of the seismic CPT (3 papers: *Baziw*; *Howie & Amini*; and *Schneider et al.*), and interpretation for shear strength (*Elmi & Favre*; *Liszkowski et al.*; *Tanaka & Tanaka*; and *Kim*) and other parameters (*Farouz et al.*; *Imre & Rózsa*; *Mlynarek & Tschuschke*; and *Roy*). The seismic test theme also includes one paper on interpretation of Love waves (*Guzina et al.*), and one on seismic cross-hole tomography (*Larsson & Mattson*). Another common theme was the use of various statistical interpolation or description methods to interpret or characterise CPT data – cluster and fuzzy methods, artificial neural networks, fractal analysis, and krieging-type methods (*Devincenzi et al.*; *Facciorusso & Uzielli*; *Failmezger et al*; *Kim*; and *Diaz-Rodriquez et al.*). In these, there is more of an emphasis on the CPT as a soil profiling tool, than as a method of determining soil parameters *per se*. There is also one paper (*Ahmadi & Robertson*) dealing with chamber size effects on measured q_c values in calibration chamber testing.

Other papers dealt with the vane shear test in clays (*Cerato & Lutenegger*), use of DMT stiffness in retaining wall design (*Monaco & Marchetti*). Interestingly, given that this conference series grew partly from the conferences on pressuremeter testing (at Oxford in 1990 and Sherbrooke in 1995), there are only two papers in this session on pressuremeter testing. One of these, by *Labrie et al.*, deals with the measurement of rock stiffness while, in the other, *Gomes-Correia et al.* illustrate the need to account for stiffness-strain non linearity when relating the Menard modulus with stiffness values inferred from Plate Load Tests and elemental values in triaxial tests. Other papers assigned to this session, but somewhat outside the theme of the session, include a paper by *Corrêa & de Souza* on laboratory testing of the contaminant sorption characteristics of residual soils, one by *Briaud* on foundation design methods, and one by *Vergauwen et al.* on a regional database for the soils of Flanders.

2.1 *Interpretation of seismic tests*

It is clear from papers presented to this conference that the measurement of shear wave velocities (usually in seismic cone penetration tests, SCPTs) and, to a lesser extent, compression wave velocities is gaining rapidly in popularity and already routine in many parts of the world. The anisotropic nature of *in situ* wave velocities (and hence very small strain stiffnesses) is now well established and recent efforts have focused on obtaining accurate and less subjective assessments of wave arrival times. The SCPT 'true interval' method of measurement of shear wave velocity, which involves two geophones 1m apart, is highly recommended by *Schneider et al.*, *Howie & Amini*, and others, and is likely to replace the 'pseudo-interval' technique as standard because of the relatively low cost of geophones. Signal processing techniques employed to analyse wave forms are complicated but greatly improve the reliability of the interpretation of arrival times. *Baziw*, for example, shows how deconvolution procedures can isolate the desired source wavelet and hence lead to an unambiguous assessment of arrival time.

Spectral analysis of surface (Rayleigh) waves is now a well established site characterisation technique, although it is known to run into difficulties when soil stiffness reduces with depth (Foti & Fahey 2003). At this conference, *Guzina et al.* present the computational framework for spectral analysis of surface Love wave and argue that the use of Love waves is preferred to Rayleigh waves because of their independence on the soil's Poisson ratio and the damping ratio for compressional waves. In addition to confirming the feasibility of this approach (as least from a theoretical standpoint), the same authors also show how application of the maximum-likelihood inverse theory (Tarantola, 1987) may be applied to treat noise polluted physical measurements.

Sand stiffness is understood to increase with the time after deposition or following significant disturbance ('remoulding'), although the mechanisms contributing to this increase are not well understood and are usually expected to take a considerable amount of the time to develop. However, *Howie & Amini* noticed that the shear wave velocity (V_s) measured using the pseudo interval technique increased by 8% when a wait time (between halting penetration and transmitting the shear wave) of 1 minute was used for the upper test in the interval, but wait times varying from 1 minute to 60 minutes were used for the

---

[1] Author names for papers to this Conference are in italics.

lower test(s) in the interval. Such increases were not obtained using the true interval SCPT, irrespective of the wait time. They attributed these results to the recovery of stiffness in a disturbed zone of sand around the penetrometer, as a function of wait time. The rapid increase in the sand stiffness of the disturbed zone has important implications for the shaft capacity of displacement piles in sand and suggests that the set-up of driven piles in sand observed by Chow et al. (1998), and others, may be related to the greater lateral stress increases during shear due to a higher level of constrained dilation.

Larsson and Mattsson also demonstrate how shear wave velocity measurements may be used effectively to infer soil stiffness variations beneath existing rail embankments (which are often inaccessible). The technique employed, which is referred to as seismic tomography (and is currently the subject of much research), involved locating a series of geophones in a borehole to one side of the embankment and a series of shear wave generators in a borehole on the other side of the embankment. Best-fit V_s values for a grid of discrete locations between the boreholes are then obtained to estimate the spatial V_s variation between the boreholes. As the solution to this inverse problem is not unique, the V_s profile that appears to be the most physically reasonable and compatible with the geological conditions is the one adopted. The authors show that the technique provides a credible estimate of *in situ* strengths beneath embankments that were broadly in line with vane strengths. It should be noted, however, that the analyses assumed incorrectly that the soil's very small strain stiffness (or velocity) was isotropic and that undrained strengths were proportional to this stiffness.

2.2 *Neural networks in soil parameter assessment*

Ding & Shang and *Kim* investigate the potential of artificial neural networks (ANNs) to deduce soil parameters from a series of independent measurements. *Ding & Shang* explore possible relationships between a soil's complex permittivity, ε_r^* (which is measured by transmission of electromagnetic waves through soil) and a soil sample's water content (w), density (ρ_b) and degree of saturation (S_r). The ANN approach does not presume any given relationships between input and output parameters and relies on a 'training procedure' for development of the (non-linear) relationships (networks) between these parameters. *Ding & Shang* show that, despite there being no well established relationship between ε_r^* and w, ρ_b and S_r, properly trained ANNs can predict these parameters with a typical R^2 correlation coefficient in excess of 0.8.

Kim also demonstrates the ability of ANNs to provide excellent predictions of triaxial compression undrained shear strengths (s_{utc}) using piezocone data. In addition to demonstrating the potential of ANNs for situations in which a reliable relationship between input and output parameters is not forthcoming, *Kim*'s work also shows that ANNs provide a more reliable prediction of s_{utc} using the corrected cone end resistance (q_t), the vertical total stress (σ_{v0}) and the pore pressure (u_2) than the following simple expressions:

$$s_{utc} = (q_t - \sigma_{v0})/N_{kT} \quad (1a)$$
$$s_{utc} = (u_2 - u_0)/N_{\Delta u} \quad (1b)$$

In other words, ANNs indicate that there remains scope for the development of an improved empirical relationship for s_{utc} from q_t, σ_{v0} and u_2. It follows that relationships such as those proposed by Lunne et al. (1985) between s_{utc}, ($q_t - u_2$) and the pore pressure ratio, B_q, should be pursued.

2.3 *Undrained strength/clay consistency*

While *Kim* demonstrated the potential of improved estimates of s_{utc} from piezocone data, a database from SE Asia compiled by *Tanaka & Tanaka* indicates that N_{kt} (see equation 1a), backfigured from s_u determined in unconfined compression, was independent of clay plasticity and varied from 4 to 23. The authors cite the influence of diatoms and geological history on the variability of measured s_u values, although it is clear that sampling disturbance and possible partial saturation of tested samples also had a major influence on the range of backfigured N_{kt} values. On the basis of *Kim*'s work, much less variability may be expected if both the q_t and u_2 values are included in a correlation with s_u.

Liszkowski et al. explore the relationship between q_t and a clay's liquidity index (I_L). The existence of such a relationship relies on an assumption that the CPT (undrained) end resistance is controlled by the clay's de-structured strength. These authors develop relationships of the following form:

$$(q_t - \sigma_{v0}) = \exp[(I_L - C_1)/C_2] \quad (2)$$

where C_1 and C_2 are constants that depend on the clay fraction. The expressions are seen to have much greater reliability than that of equation 1a, suggesting that q_t is controlled by operational strengths that are close to ultimate/critical-state strengths.

2.4 *Processing of in situ test data*

Our ability to record, store, process and analyse enormous quantities of data seems to be ever increasing. With such capabilities, it is not surprising to see the recent development of 'expert systems' to interpret field measurements. *Facciorusso & Uzielli* present one such example, which combined cluster analysis techniques with fuzzy classification to derive an objective means of delineating strata using mechanical cone penetration test data. The authors demonstrate the potential of such an approach, which may quickly find application in geophysical test interpretation. It should, however, be stated that an experienced geotechnicial engineer could use the same mechanical cone data to, very quickly, establish an arguably more precise delineation of the strata by visual inspection alone. 'Expert systems' of the future will need to incorporate such experience, but will never completely replace the need for physical examination and testing of the soil.

Our strong data processing capabilities are also in evidence in the paper presented by *Failmezger et al.* who show how these are now such that *in situ* test data may be analysed using a risk based approach with an assumed probability distribution of interpreted parameters.

Devincenzi et al. and *Vergauwen et al.* also show how data processing power can provide 3-D graphical representations of site stratigraphies, which may be included in databases for particular regions. An important feature of the former paper is that the interpretation of the data was carried out within the framework of a detailed knowledge of the geomorphology of the area of interest. This serves to emphasise the point that output from any automatic system used to construct stratigraphic sections from *in situ* test data must be checked for consistency with what is known about the geology or geomorphology of the site.

3 *IN SITU* TEST INTERPRETATION: FURTHER TRENDS

A number of further trends in *in situ* test interpretation that have not been highlighted in the foregoing are now summarised. These have become evident from other papers presented to this conference and following a review of recent publications in this area.

- The combination of the friction sleeve shear stress (f_s) with the CPT end resistance (q_c) into another parameter referred to as the friction ratio, R_f (= f_s/q_c) was a major step forward that enabled the soil type to be assessed from CPT data. As the number of *in situ* test parameters recorded increases, there has been a search for other such combined parameters to assist in soil characterisation and parameter determination. Examples include *(a)* the ratio of the very small strain stiffness (G_o) to either the CPT q_c or dilatometer E_d value to assess cementation and ageing effects for sand and *(b)* the ratio of the horizontal coefficient of consolidation (c_h) interpreted from piezocone dissipation tests to G_o to assess *in situ* permeability.

- The piezocone is acknowledged as being an excellent soil profiling tool. However, the complex mechanisms associated with cone penetrometer installation make the assessment of material parameters very difficult. For this reason, as described in detail by Randolph (2004), 'flow type' T-bar and ball penetrometers, which have better defined boundary conditions and are more amenable to analysis, are beginning to find wider use.

- The ability to easily vary the rate of penetration of various penetrometers is being explored as a means of assessing *in situ* consolidation characteristics (Randolph & Hope 2004).

- The centrifuge is beginning to be used as an alternative to calibration chamber experiments. For example, research at the University of Western Australia has employed the centrifuge and a model piezocone to examined effects of penetration rate in CPTUs in kaolin as well as partial drainage and rate effects in silts and clayey silts.

- Odebrecht et al. (2004), and others, have promoted idea of precise calibration of the energy input to SPT samplers and proposed that the applied dynamic energy can and should be quoted in place of the SPT N value.

- More precise numerical and analytical techniques for analysis of pressuremeter data, such as those proposed by Fahey & Carter (1993) and Mantaras & Schnaid (2002), are being employed. These may inspire practitioners to specify pressuremeter testing (particularly with the self-boring pressuremeter), which unfortunately appears to be in decline in some parts of the world, due to the perception that the test is difficult and expensive to perform, and to the scarcity of operators with the required skill in carrying out and interpreting the test.

- Renewed attempts to provide rational, less empirical means of interpreting vane and dilatometer test data are underway.

- Inversion techniques to interpret geophysical data are becoming more robust and reliable.

- The search for improved direct correlations between *in situ* test parameters and foundation performance continues. As the measurement of G_o and *in situ* stiffness anisotropy is becoming routine, the emphasis has been on combining such data with stiffness parameters measured at larger strain levels to deduce strain dependent moduli relevant to foundation design (Fahey & Lehane 2004). As postulated by Fahey (1998), combining the G_o value with the results of self-boring pressuremeter testing may be the ideal way of making progress in this area, but it is also likely that the stiffness measured by in the DMT, when combined with G_o, may provide the most economical means of achieving the same goal.

4 CONCLUDING REMARKS

There have, undoubtedly, been major developments in our ability to gather, analyse and process large quantities of *in situ* test data. *In situ* testing is now clearly a very powerful tool for site and soil characterisation. Techniques to assist interpretation of parameters from *in situ* tests for geotechnical designs have improved. There remains, however, a need to translate these developments to engineering practice and to convince the engineering profession of the merits of performing state-of-the-art *in situ* tests. For example, there is no point in telling a foundation designer of the need to measure G_o, if the designer has no simple means of performing a non-linear settlement analysis, or to measure undrained strength in a pressuremeter test in preference to a vane test if the effects of anisotropy and strain path dependence of undrained strength cannot be accommodated in a design procedure.

There is a now a common perception that methods of analysis in geotechnical engineering (for example, various commercially-available finite element packages that incorporate complex soil models) have outstripped our ability to measure the required input parameters by other than sophisticated laboratory tests on high-quality undisturbed soil samples. In soils that are difficult or impossible to sample, we are therefore left with nothing other than *in situ* test methods to determine the required parameters, and hence there is a need to continue to improve these methods.

5 PAPERS ASSIGNED TO THIS SESSION:

Ahmadi, M.M. & Robertson, P.K. Calibration chamber size and boundary effects for CPT qc measurements.
Baziw, E. State-space seismic cone minimum variance deconvolution.
Briaud, J.-L. Foundation engineering: recent contributions using in-situ testing.
Cerato, A.B. & Lutenegger, A.J. Disturbance effects of field vane tests in a varved clay.
Devincenzi, M.J., Colàs, S., Casamor, J.L., Canals, M., Falivene, O. & Busquets, P. High resolution stratigraphic and sedimentological analysis of Llobragat Delta nearby Barcelona from CPT and CPTU tests.
Diaz-Rodriquez, J.A., Moreno-Carrizales, P. & Springall, G. Fractal analysis of CPT data.
Ding, W. & Shang, J.Q. Neural networks in soil characterisation.
Elmi, F. & Favre, J.L. Cylindrical cavity expansion modelling for interpretation of cone penetration tests.
Facciorusso, J. & Uzielli, M. Stratigraphic profiling by cluster analysis and fuzzy soil classification from mechanical cone penetration tests in the habour area of Gioia Tauro, Italy.
Failmezger, R.A., Bullock, P.J. & Handy, R.L. Site variability, risk, and beta.
Farouz, E., Chen, J.-Y. & Failmezger, R.A. A case study using in-situ testing to develop soil parameters for finite element analyses.
Guzina, B.B., Madyarov, A.I., Osburn, R.H. Computational and physical basis for dynamic site characterization using Love waves.
Gomes Correia, A., Antao, A. and Gambin M. Using a non-linear constitutive law to compare Menard PMT and PLT E-moduli.
Howie, J.A. & Amini, A. Effect of ageing on shear wave velocity by seismic cone
Imre, E. & Rózsa, P. Dissipation test evaluation with a point-symmetrical consolidation model.
Kim, Y.-S. Feasibility of neural network application for determination of undrained shear strength of clay from piezocone measurements.
Labrie, D., Conlon, B. Anderson, T. & Boyle, R.F Measurement of in situ deformability in hard rock.
Larsson, R. & Mattsson, H. Estimation of shear strength increase beneath embankments by seismic cross-hole tomography.
Liszkowski, J., Tschuschke, M., Mlynarek, Z. & Tschuschke, W. Statistical evaluation of the dependence between the state parameters of choesive soils and CPTU data
Mlynarek, Z., Tschuschke, W. CPTU - precision and accuracy in post flotation sediments
Monaco, P. & Marchetti, S. Evaluation of the coefficient of subgrade reaction for design of multi-propped diaphragm walls from DMT moduli
Roy, D. Assessment of cyclic stability of cohesive deposits using cone penetration
Schneider, J.A., McGillivray, A.V & Mayne, P.M. Evaluation of SCPTU intra-correlations at sand sites in the Lower Mississippi River Valey, USA.
Strava Corrêa, A.C. & Moreira de Souza, N. Risk assessment for contaminated tropical soil.
Tanaka, M. & Tanaka. H. Examinations and considerations on the engineering properties and the cone factors of soils from the East Asia region.
Vergauwen, I., De Schrijver, P. & Van Alboom, G, The regional information system 'Databank Ondergrond Vlaanderen - DOV'. A powerful tool in geotechnical engineering.

REFERENCES

Baligh M.M.(1985). Strain Path Method. *Journal of the Geotechnical Engineering Division*, ASCE, 111, 1108-1136.
Chow F.C., Jardine R.J., Nauroy J.F. & Brucy F. (1998). Effects of time on capacity of pipe piles in dense marine sand. J. geotechnical & Geoenvironmental Engineering, ASCE, 124(3), 254-264.
Fahey M. & Carter J.P. (1993). A finite elemnt study of the pressuremeter test in sand using a non-linear elastic plastic model. *Canadian Geotechnical J.*, 30, 348-362.
Fahey M. & Lehane B.M. (2004) Stiffness parameters for settlement analysis of foundations from pressuremeter, dila-

tometer and seismic cone testing. *Proc 9th Australian- New Zealand Conf.*, Auckland, 2004, 2, 797-803.

Mantaras F.M. & Schnaid F. (2002). Cavity expansion in dilatant cohesive-frictional soils. *Geotechnique,* 52 (5), 337-348.

Lunne T. Christophersen H.P. & Tjelta T.I. (1985). Engineering use of piezocone data in North Sea clays. *Proc. 11th International Conf. Soil Mechanics and Foundation Engrg.*, 2, 907-912.

Odebrecht E., Schnaid, F., Rocha M.M. & Bernardes G.P. (2004). Energy efficiency for Standard Penetration Tests. (submitted for publication to *J. Geotech. and Geoenvironmental Engng*, ASCE)

Foti S. & Fahey M. (2003). Applications of multistation surface wave testing. *Proceedings of the 3rd International Symposium on Deformation Characteristics of Geomaterials (IS-Lyon)*, Lyon, France, 13–20, Balkema, Lisse.

Randolph M.F. (2004). Characterisation of soft sediments for offshore applications. Keynote lecture. *Proc. 2nd International Conference on Site Characterisation (ISC 04)*, Porto, Portugal, Sept. 2004, Balkema, Rotterdam.

Randolph M.F. & Hope S. (2004). Effect of cone velocity on cone resistance and excess pore pressures. *Proc. Int. Symp. On Engineering Practice and Performance of Soft Deposits*, Osaka (in press)

Tarantola A. (1987). Inverse problem theory. Elsevier: Amsterdam.

Discussion report: Benefit of new developments in interpretation of *in-situ* data

J. Peuchen
Fugro, Netherlands

Keywords: site investigation, *in-situ* testing

ABSTRACT: This note highlights points for conference theme discussion on New Developments in Interpretation of *In-Situ* Data.

1 INTRODUCTION

"New developments in interpretation of in-situ data" is the theme of this session. We share knowledge and technology, but what is important? What is to be highlighted? The following statements represent an attempt to set criteria to "what is important?".

As engineers and engineering geologists, we are "to create a sustainable natural and built environment for the benefit of future generations" (ICE, 2004).

This conference contributes to this mission. So we can rephrase the "what is important?" question to "what are the important contributions to the benefit of future generations?". Yet, opinions will differ. Some consensus can, perhaps, be derived from "management of risk", a buzzword in many societies and also in geotechnical practice. Eurocode EN1990:2002 - Basis of Structural Design, Clause 2.2 uses "reliability management". This term has a more positive connotation than risk. Particularly:

2.2 Reliability management
(1) P The reliability required for structures within the scope of EN 1990 shall be achieved:
   a) by design in accordance with EN 1990 to EN 1999 and
   b) by
   – appropriate execution and
   – quality management measures.

In further statements of Clause 2.2 on reliability management, we start to approach the theme of this conference and of this conference session:

e) *other measures relating to the following other design matters:*
   – *the extent and quality of preliminary investigations of soils and possible environmental influences;*
   – *the accuracy of the mechanical models used;*
   – *the detailing;*

It should be expected that most papers within the theme would be in the categories "the accuracy of the mechanical models used" and "the detailing". This is indeed the case. Many excellent papers provide a contribution "for the benefit of future generations".

However, papers within "the extent and quality of preliminary investigations of soils and possible environmental influences" are, perhaps, of greatest interest. The Clause 2.2 sequence appears to show a hierarchy in importance that is useful here. This hierarchy fits well with the recommendations of SISG (1993), as shown on Figure 1.

2 EXTENT AND QUALITY OF PRELIMINARY INVESTIGATIONS

Two papers within the theme show a good fit with "extent and quality of preliminary investigations":
- The regional information system "Databank Ondergrond Vlaanderen – DOV". A powerful tool in geotechnical engineering (Vergauwen et al., 2004)

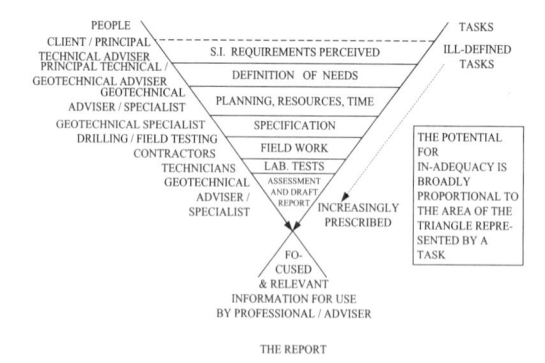

Figure 1: Geotechnical Quality for Site Investigation, adapted from SISG (1993).

- High resolution stratigraphic and sedimentological analysis of Llobregat delta nearby Barcelona from CPT and CPTU tests (Devincenzi et al., 2004).

Databank developments are not particularly new. Geological institutes have long been commissioned by governments to maintain geodatabanks. Also, any commercial geotechnical company would hold its own geodatabank, if not for maintaining a competitive edge. Its simplest form is probably a (city street) map with dots showing references to records. The advent of computers and internet changed the scene dramatically.

The MARIS GIS-based digital geo-databank (http://www.maris.nl) may well be the first ever with access by multiple users. MARIS started in 1985 as a Netherlands governmental initiative to improve overview and access to marine expertise, information and data related to the sea and its uses.

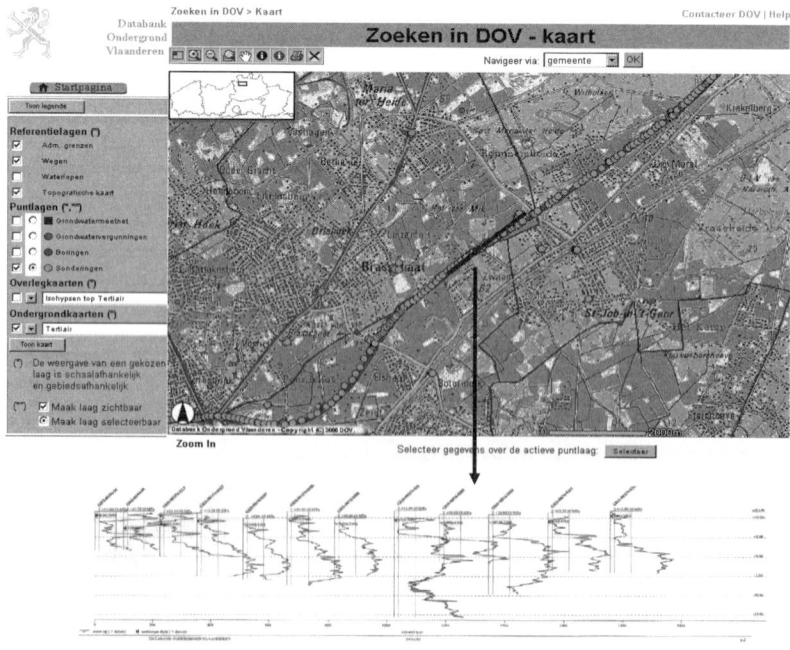

Figure 2. Databank Ondergrond Vlaanderen, geotechnical profile. After Vergauwen et al. (2004).

Since 1989, it operates as an independent foundation. Others followed suit, for example http://www.geonet.nl, http://eu-seased.net, http://dinoloket.nitg-tno.nl and, of course, http://dov.vlaanderen.be (Figure 2). Many of us have yet to benefit from this important development. Its importance is beyond technology. It may be regarded as, say, a "Factor-10" indicator for front-end decisions (see also Figure 1) that have most impact on reliability management. The value 10 is meant to indicate ten-fold potential benefit, for example in cost or in reliability (risk reduction). The assignment of a Factor-10 indicator to onshore geodatabank developments can be compared to site characterisation practice for front-end decisions for marine environment developments (Figure 3), as described by Campbell and Hooper (1993) and others. Here, the 3-D exploration seismic data provide much-wanted (Factor-10) information during early site appraisal evaluations. Regrettably, this conference theme has no such contributions from offshore practice.

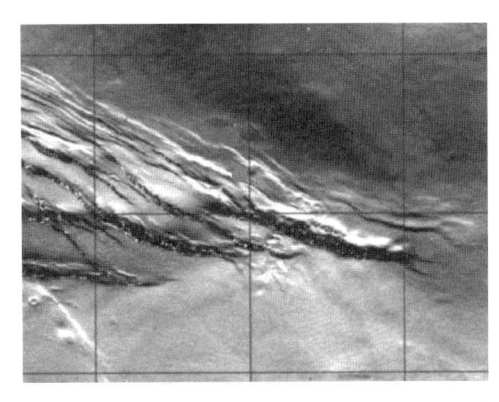

Figure 3: Seafloor image showing large fault scarps caused by active faulting. Seafloor relief is more than 80 m across some fault scarps. Water depth ranges from about 850 m to 1450 m, grid squares are about 5 km on a side. Image was generated from 3-D exploration seismic data (Campbell and Burrell, 2003).

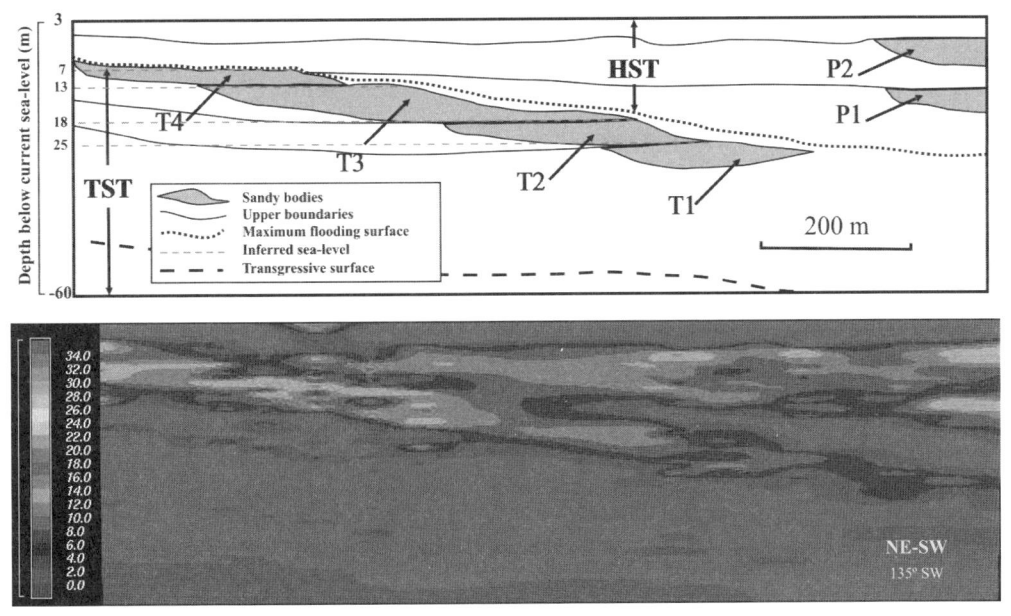

Figure 4. Stratigraphic cross-section Llobregat Delta, after Devincenzi et al. (2004).

The presented analysis of the Llobregat Delta may be regarded as a Factor-4 indicator. The analysis covers an area of 0.3 km² only, yet provides a good basis for interpretation of a 95 km² area of major economic value (Figures 4 and 5). A four-fold benefit is not unlikely, particularly as the improved geological and geotechnical understanding provides opportunities for enhanced interpretation of in-situ data.

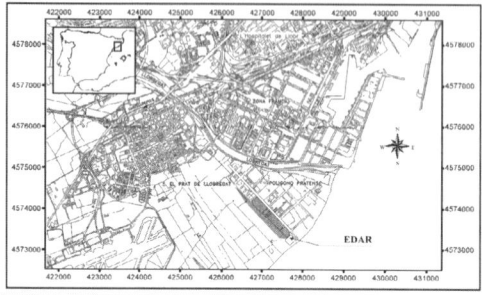

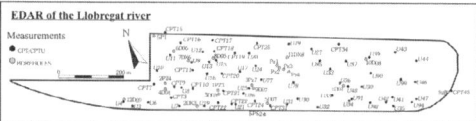

Figure 5. Study Llobregat Delta, after Devincenzi et al. (2004).

The remaining contributions to the theme can be assigned a Factor-2 indicator, albeit somewhat subjectively. The presented proposals and observations in this category are for more and better interpretation of specific results from available technologies for acquisition of in-situ test data. Improvement in interpretation by, say, a factor of two is a possible benefit.

3 CONCLUSION

The conference theme "new developments in interpretation of in-situ data" presents important contributions to the benefit of future generations. Much is yet to be learnt and made known to the wider profession. A challenge to us all.

REFERENCES

Campbell, K.J. and Burrell, R. (2003), "Deepwater Development Fast-tracking: The Critical Role AUV Surveys Play in Integrated Site Investigation and Geohazards Assessment", 7th Annual Offshore West Africa Conference and Exhibition 11-13 March, Windhoek, Namibia.

Campbell, K.J. and Hooper, J.R. (1993), "Deepwater Engineering Geology and Production Structure Siting, Northern Gulf of Mexico", in Advances in Underwater Technology, Ocean Science and Offshore Engineering, Offshore Site Investigation and Foundation Behaviour, Society for Underwater Technology, pp. 375-390.

CEN European Committee for Standardization (2002), "Eurocode - Basis of structural design", EN 1990:2002(E)

Devincenzi, M.J., Colas S., Casamor, J.L., Canals, M., Falivene, O. and Busquets, P. (2004), "High Resolution Stratigraphic and Sedimentological Analysis of Llobregat Delta nearby Barcelona from CPT and CPTU Tests", International Conference on Site Characterization ISC'2, Portugal.

Institution of Civil Engineers ICE (2004), "Our Vision", ICE Members' Handbook.

Site Investigation Steering Group SISG (1993), "Site Investigation in Construction 2: Planning, Procurement and Quality Management", Thomas Telford, London.

Vergauwen I., De Schrijver, P. and Van Alboom, G. (2004), "The Regional Information System "Databank Ondergrond Vlaanderen – DOV". A Powerful Tool in Geotechnical Engineering", International Conference on Site Characterization ISC'2, Portugal.

Calibration chamber size and boundary effects for CPT q_c measurements

M.M. Ahmadi & P.K. Robertson
Dept. of Civil and Environmental Eng., University of Alberta, Edmonton, Alberta, Canada

Keywords: cone tip resistance, calibration chamber, boundary condition, modeling, sand, size effect

ABSTRACT: A numerical modeling procedure for penetration mechanism is introduced to quantify chamber size effect and boundary conditions. Analyses are carried out for normally consolidated sand with K_0 equal to 0.5, and for loose, medium dense, and dense sand. In the numerical analysis, calibration chamber diameter and the boundary conditions are varied to investigate the effects of calibration chamber size and the boundary conditions on cone tip resistance. These analyses show that for loose sand a chamber diameter to cone diameter ratio of 33 or even smaller is sufficient for the boundaries not to influence the tip measurements. For dense sand however, numerical analysis shows that the chamber to cone diameter ratio should be more than 110 to ensure that boundaries are not affecting the tip measurements. The numerical analysis results obtained in this study are in agreement with the empirically based relations suggested by other researchers.

1 INTRODUCTION

The purpose of calibration chamber testing is to study the response of CPT in well defined and well controlled conditions of sand relative density, its stress state, OCR and boundary conditions. The results of these tests provide empirical relationships between the significant parameters affecting the cone penetration resistance namely sand relative density and stress state.

Calibration chambers have finite dimensions, and stress and strain conditions at the chamber boundaries need to be imposed. The boundaries imposed on a calibration chamber may not represent the real field situation. In order to simulate the semi-infinite soil mass in the field, the required boundary conditions are between two extreme limits of constant stress and zero displacement on both horizontal and vertical boundaries. This discrepancy may influence the cone tip resistance measured in calibration chambers.

Depending on whether stresses are kept constant or displacements are zero at the lateral and bottom sample boundaries, there are four different types of boundary conditions that can be applied in this type of calibration chamber (Lunne et al., 1997). These are listed in Table 1.

None of these four different boundary conditions simulate the field condition perfectly. The larger the chamber size, the less significant is the difference between results obtained in the chamber and the results obtained in the field. This means that the boundary conditions in the chamber can influence the results of penetration resistance if the chamber size is small.

Table 1: Boundary conditions available in Italian calibration chambers.

Type of boundary condition	Lateral and bottom boundary condition
BC1	Horizontal stress = Constant Vertical stress = Constant
BC2	Horizontal strain = 0 Vertical strain = 0
BC3	Horizontal strain = 0 Vertical stress = Constant
BC4	Horizontal stress = Constant Vertical strain = 0

The effect of chamber size and boundary condition on the tip resistance values measured in a calibration chamber has been recognized for many years. Parkin and Lunne (1982), based on penetration test results for two different chamber sizes and two different penetrometer sizes, have concluded that for loose sands, chamber size and boundary conditions do not have a significant effect on the cone tip resistance. For dense sands, on the other hand, the effects are considerable.

Lunne and Christophersen (1983), based on chamber test results on Hokksund sand, suggested that for a chamber to cone diameter ratio of 50, the difference in tip resistance obtained in the chamber and the field should be small.

Jamiolkowski et al. (1985) proposed a formula to relate the tip resistance obtained in the chamber to the tip resistance in the field.

$$q_{c,\text{field}} = q_{c,cc}(1 + \frac{0.2(DR\% - 30)}{60}) \quad (1)$$

In the above formula, $q_{c,cc}$ is the experimental value of tip resistance observed in the calibration chamber, DR is the relative density, and $q_{c,\text{field}}$ is the corrected tip resistance expected to be measured in the field for the same sand with the same relative density and the same in-situ stresses as in the chamber. The above formula is valid for a cone penetrometer with a projected cone area of 10 cm² in a 1.2 m diameter chamber; and it is based on the experimental results obtained on calibration chamber tests under BC1 type boundary condition.

Mayne and Kulhawy (1991), based upon examination of calibration chamber data, developed the following equation to relate the tip resistance measured in the chamber to the tip resistance in the field.

$$q_{c,\text{field}} = q_{c,cc}[\frac{(D_{cc}/d_{cone})-1}{70}]^{-DR(\%)/200} \quad (2)$$

In the above formula, D_{cc} is the diameter of the calibration chamber, and d_{cone} is the diameter of the cone. The parameters $q_{c,cc}$ and $q_{c,\text{field}}$ are the experimental value of tip resistance measured in the calibration chamber and the corrected tip resistance expected to be measured in the field respectively, and DR is the relative density as defined before.

As is noted in Eq. 2, Mayne and Kulhawy (1991) assumed that, regardless of the relative density and stress state, a chamber diameter to cone diameter ratio of 70 is sufficient to achieve the "free field" condition.

Salgado et al. (1998) proposed a penetration resistance theory to quantify chamber size effect, and to investigate the factors it depends on. They concluded that sand relative density, stress state, and intrinsic parameters are the most important factors to be considered in the magnitude of the chamber size effect.

In this paper the effect of chamber size and boundary will be presented based on numerical modeling procedure to analyze the penetration process suggested by Ahmadi (2000). A brief description of the constitutive modeling used in the analyses carried out in this paper is discussed first. The results of numerical analyses performed in this research program are presented afterwards. The analyses are carried out for dense, medium dense, and loose sand. Calibration chambers are modeled for all four different boundary conditions. All analyses were performed for a standard cone penetrometer having a conical tip with a base area of 10 cm² and 60 degree tip apex angle. The finite difference based computer program FLAC (2000) was used for all the analysis involved in this study.

2 CONSTITUTIVE LAW FOR SAND

The constitutive law used for the analysis of tip resistance in this paper was the model suggested by Ahmadi (2000). In the analysis it is assumed that the sand has the same mechanical properties as Ticino sand.

The Mohr-Coulomb elasto-plastic model was chosen for the analysis of cone tip resistance in sand. The values of stresses in close proximity to the cone tip are very much higher than those in the far field, and it is argued that the model parameters will therefore be different in the near and far field. Hence, the Mohr-Coulomb soil parameters are considered to be stress dependent. The stress dependent relations for shear and bulk modulus used in the Mohr-Coulomb soil model are:

$$G = K_G P_A (\frac{\sigma'_m}{P_A})^n \quad (3)$$

$$B = K_B P_A (\frac{\sigma'_m}{P_A})^m \quad (4)$$

In the above relations, σ'_m is the mean effective stress, P_A is the atmospheric pressure, or a reference stress, (equal to 1 kg/cm² =98.1 kPa), m and n are constants, and are both chosen to be 0.6, and K_G and K_B are constants that mainly depend on the relative density of the sand. The parameters used for K_G and K_B for Ticino sand are shown in Table 2. These values are in the range of values reported by Byrne et al. (1987).

Table 2. Parameters used for deformation and shear strength of Ticino sand.

DR%	K_G	K_B	ϕ'_0 (deg)	α (deg)
45	195	325	38.2	4.2
65	230	385	40.2	6.5
85	290	480	42.9	8.1

DR = average relative density of the tested specimens, at the end of consolidation.

The drained shear strength parameters of Ticino sand were obtained from triaxial tests carried out by ENEL/ISMES in Italy. Baldi et al. (1986) have summarized the results of these tests in terms of the curvilinear formula given by Baligh (1975):

$$\tau_{ff} = \sigma'_{ff}\left[\tan\phi'_0 + \tan\alpha\left(\frac{1}{2.3} - \log_{10}\frac{\sigma'_{ff}}{P_A}\right)\right] \quad (5)$$

Where τ_{ff} = shear stress on the failure surface at failure, σ'_{ff} = effective normal stress on the failure surface at failure, α = angle which describes the curvature of the failure envelope, and ϕ'_0 = secant angle of friction at $\sigma'_{ff} = 2.72\ P_A$.

Table 2 also shows the values of ϕ'_0 and α for Ticino sand for three different classes of relative density (Baldi et al., 1986).

The dilational characteristics of the sand are represented by the following relationship relating the dilation angle to the friction angle at failure and the constant volume friction angle:

$$\sin \psi = \sin \phi_f - \sin \phi_{cv} \qquad (6)$$

where: ψ = dilation angle, ϕ_f = friction angle at failure, ϕ_{cv} = constant volume friction angle. For Ticino sand, a value of 34.8 degress was assumed for the constant volume friction angle, as suggested by Salgado et al (1997).

The modeling procedure adopted in this study simulates the penetration process in a realistic way. The penetration modeling starts at the top of the grid, and progresses into the grid, and can end at any desired depth in the grid, meaning that the modeling process is simulating the cone moving downward in the ground. This is a moving boundary problem, and the large strain analysis was chosen to model the penetration mechanism. Because the problem has symmetry about the vertical axis, the axisymmetric option was used for this three dimensional problem in order to reduce the number of elements in the solution procedure. To study the boundary condition effects on cone tip resistance, the numerical grid is under different types of boundary conditions, similar to what is applied during calibration chamber testing, and described in Table 1. Details of the numerical approach to model the penetration mechanism and its verification are beyond the scope of this conference paper, and are fully presented in Ahmadi (2000).

3 ANALYSIS RESULTS

Figure 1 shows the result of analysis of penetration in dense sand for all four different boundary conditions. The sand relative density taken as input in this series of analysis is 90%, the effective vertical stress is 70 kPa, and K_0 is 0.5, giving a horizontal effective stress of 35 kPa. The high values of sand relative density together with low values of stress state selected as input for these analyses characterizes a very dense sand. Analyses are carried out for chambers of different diameters, namely 1.2 m, 1.8 m, 2.4 m, 3.0 m, and 4 m. The height of the numerical grid for all chambers is 1.5 m. The abscissa represents the ratio of calibration chamber diameter to cone diameter. The ratio of 33.6 corresponds to a chamber with a diameter of 1.2 meters and the ratio of 112.1 corresponds to a chamber with a diameter of 4 meters.

The ordinate represents the predicted tip resistance in MPa obtained for each analysis carried out under different boundary conditions and chamber diameters. The analysis predicts the tip resistance continuously for the whole height of the grid, and the tip resistance values taken in this paper is the predicted values of tip resistance at the middle height of the grid. This conforms to experimental procedure where the mid height tip resistance at the chamber is reported in the chamber data.

Figure 1 shows that the cone tip resistance is a function of boundary condition as well as the ratio of chamber diameter to cone diameter. For BC1 boundary condition where the chamber is under constant stresses in both horizontal and vertical directions, the tip resistance increases as the ratio of chamber to cone diameter increases from 33.6 to 112.1. This figure also shows the same trend for BC4 boundary condition. The numerical values of tip resistance obtained for both BC1 and BC4 boundary conditions for all values of chamber to cone diameter ratio are very close.

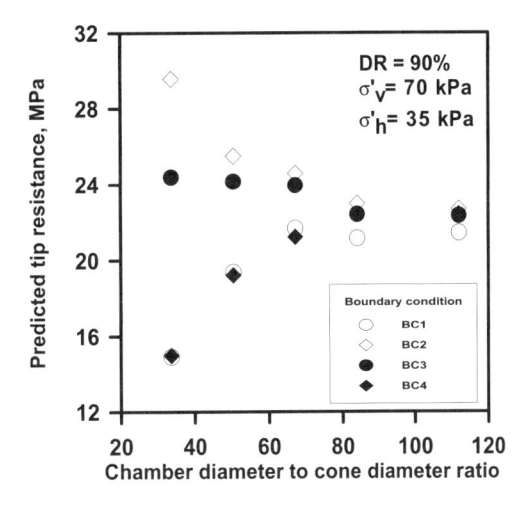

Fig. 1. Effect of chamber size and boundary condition on tip resistance (predictions for dense sand), [DR=90%, σ'_v = 70 kPa, K_0=0.5]

For BC2 boundary condition, the numerical analysis shows that the predicted tip resistance decreases as the chamber to cone diameter ratio increases. Figure 1 shows that as chamber diameter to cone diameter ratio increases from 33.6 to 112.1, the predicted tip resistance for BC2 boundary condition decreases from 29.6 MPa to 22.6 MPa. For BC3 type boundary condition, the value of tip resistance predicted for a 1.2 m radius chamber is 24.3 MPa. The figure shows that for chamber of larger diameter, the predicted value of tip resistance for BC3 boundary condition decreases slightly reaching to a value of 22.3 MPa for chamber diameter to cone diameter equal to 112.1. It can be seen that the BC3 results do change, but not significantly for the range of chamber diameter to cone diameter ratios studied in this paper.

Figure 1 shows that for values of chamber diameter to cone diameter ratio of 33.6, there is a large difference in cone tip resistance for different boundary conditions. This difference is especially noted between BC2 and BC1 (or BC4) boundary condition. This indicates that for dense sand, the results of cone tip resistance obtained in calibration chamber testing may be affected by the boundaries. In the case of dense sands, the numerical analysis indicates that a chamber to cone diameter ratio of 112 or more is needed to measure tip resistance during chamber testing which is not affected by the boundaries.

Figure 2 and Figure 3 show the results of a numerical analysis of penetration for all four boundary conditions and different ratios of chamber diameter to cone diameter for a medium dense sand and a loose sand respectively. The medium dense sand is characterized by a relative density of 70% and effective vertical stress of 300 kPa and effective horizontal stress of 150 kPa, resulting in the same value of K_0 as in Figure 1. The loose sand is characterized by a low value of relative density (DR=50%) and high values of stress state. In the analysis presented in Figure 3 for loose sand, the effective vertical stress is taken to be 700 kPa, and the effective horizontal stress to be 350 kPa. The K_0 value for these series of analysis is also 0.5. The high values of stress state together with a low value of sand relative density is chosen for loose sand to better show the contrast in response and analysis results with respect to dense sand.

Comparing Figure 1 and Figure 2, it is seen that that for a given chamber diameter to cone diameter ratio, the difference in predicted tip resistance for boundary conditions BC1 (or BC4) and BC2 obtained in the numerical analysis for medium dense sand is appreciably less than that obtained for dense sand. For medium dense sand and for a chamber to cone diameter ratio of 33.6, the predicted cone tip resistance for BC1 and BC2 boundary conditions is 18.9 and 21.8 MPa respectively. Figure 2 also shows that a chamber to cone diameter ratio of 67 is sufficient to produce penetration resistances that are not affected by the boundaries. This value of chamber diameter to cone diameter is also smaller than that shown in Figure 1.

For loose sand (DR=50%), the difference in predicted tip resistances for BC1 and BC2 boundary condition is even smaller. This trend is physically expected as Figure 3 shows that for any value of chamber to cone diameter ratio and any type of boundary condition, the predicted tip resistance is

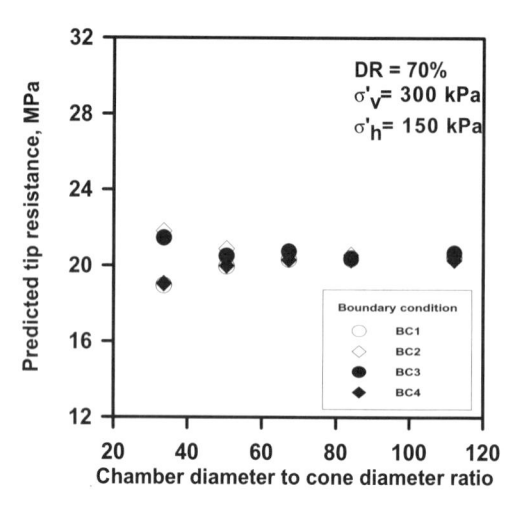

Fig. 2. Effect of chamber size and boundary condition on tip resistance (predictions for medium dense sand). [DR=70%, σ'_v = 300 kPa, K_0=0.5]

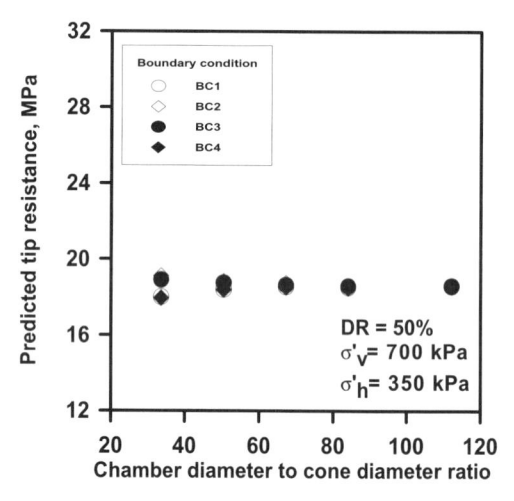

Fig. 3. Effect of chamber size and boundary condition on tip resistance (predictions for loose sand)., [DR=50%, σ'_v = 700 kPa, K_0=0.5]

about 18.6 MPa. This means that for loose sand, the size of the chamber or the type of the boundary condition imposed on the chamber walls during testing does not significantly affect the tip resistance values. For loose sand, a chamber to cone diameter ratio of 33.6 is sufficient to measure tip resistances that are not affected by the boundaries.

As discussed previously, calibration chamber measurements indicate that for loose sands under large confining stresses chamber size and boundary conditions do not influence the tip resistance measurements, whereas for dense sands under low stress states, chamber size and boundary condition may significantly affect the tip measurements. The numerical results obtained in this study are supported by experimental observation, and show good agreement with empirically based relationships suggested by other researchers.

4 CONCLUSION

The effect of chamber size and boundary condition on cone tip resistance is presented based on numerical modeling procedure to analyze the penetration mechanism. The analyses are carried out for dense, medium dense, and loose sand. Calibration chambers are modeled for all four different boundary conditions having different diameters. These analyses indicate that for dense sand, the results of cone tip resistance obtained in calibration chamber testing may be affected by the boundaries. In the case of dense sands, the numerical analysis indicates that a chamber to cone diameter ratio of 112 or more is needed to measure tip resistance during chamber testing which is not affected by the boundaries. The difference in predicted tip resistance for boundary conditions BC1 (or BC4) and BC2 obtained in the numerical analysis for medium dense sand is appreciably less than that obtained for dense sand, and for loose sand, the effect of calibration chamber size and its boundary on cone tip resistance is insignificant; and this difference is much smaller. The results of this study also indicate that boundary condition BC3 (i.e. zero horizontal strain plus constant vertical stress) appears to represent the best boundary condition with the least affect on the measured cone resistance.

Experimental observations indicate that for loose sands under large confining stresses chamber size and boundary conditions do not influence the tip resistance measurements, whereas for dense sands under low stress states, chamber size and boundary condition may significantly affect the tip measurements. The numerical results obtained in this study are supported by experimental observation, and show good agreement with empirically based relationships suggested by other researchers.

ACKNOWLEDGMENT

The authors are grateful to the Natural Sciences and Engineering Research Council of Canada (NSERC) for their financial support.

REFERENCES

Ahmadi, M.M. (2000). "Analysis of cone tip resistance in sand." PhD thesis, University of British Columbia, Vancouver, Canada.

Baldi, G., Bellotti, R., Ghionna, V., Jamiolkowski, M., and Pasqualini, E. (1986). "Interpretations of CPT's and CPTU's, 2nd Part: Drained penetration of sands." 4th Int. Conf. On Field Instrumentation and In-situ Measurements, Singapore, 143-156.

Baligh, M.M. (1975). Theory of deep site static cone penetration resistance. Publication No. R75-56, Department of Civil Engineering, Massachusetts Institute of Technology.

Byrne, P.M., Cheung, H., and Yan, L. (1987). "Soil parameters for deformation analysis of sand masses." Canadian Geotechnical Journal, 24(3), 366-376.

FLAC User Manual, version 4.0, (2000), Itasca Consulting Group Inc., USA.

Jamiolkowski, M., Ladd, C.C., Germaine, J.T., and Lancellotta, R. (1985). "New developments in field and laboratory testing of soils" Proc. 11th Int. Conf. on Soil Mechanics and Foundation Engineering, San Francisco.

Lunne, T., and Christophersen, H.P. (1983). "Interpretation of cone penetration data for offshore sands." Proc. 15th Annual Offshore Technology Conf., Houston, Texas, 181-192.

Lunne, T., Robertson, P.K., and Powell J.M. (1997). "Cone Penetration Testing in Geotechnical Practice", Blackie Academic & Professional.

Mayne, P.W., and Kulhawy F.H. (1991). "Calibration chamber data base and boundary effects correction for CPT data." Proc. 1st Int. Symposium on Calibration Chamber Testing, Potsdam, New York, 257-264.

Parkin, A.K. and Lunne, T. (1982). "Boundary effects in the laboratory calibration of a cone penetrometer in sand." Proc. 2nd European Symposium on Penetration Testing (ESOPT II), Amsterdam, Vol. 2, 761-768.

Salgado, R., Mitchell, J.K., and Jamiolkowski, M. (1997). "Cavity expansion and penetration resistance in sand." Journal of Geotechnical and Geoenvironmental Engineering, ASCE, 123(4), 344-354.

Salgado, R., Mitchell, J.K., and Jamiolkowski, M. (1998). "Calibration chamber size effects on penetration resistance in sand." Journal of Geotechnical and Geoenvironmental Engineering, ASCE, 124(9), 878-888.

# State-space seismic cone minimum variance deconvolution

Erick Baziw
Baziw Consulting Engineers Ltd., Vancouver, Canada

Keywords: Seismic cone penetration testing (SCPT), minimum variance, Kalman filter, smoothing, autoregressive moving average (ARMA)

ABSTRACT: Seismic Cone Penetration Testing (SCPT) is a geotechnical tool which facilitates the determination of low strain ($<10^{-4}$%) in-situ compression (P) and shear (S) wave velocities. The P-wave and S-wave velocities are directly related to the soil elastic constants of Poisson's ratio, shear modulus, bulk modulus, and Young's modulus. The seismic cone records the arrival of seismic waves generated at the surface using velocity or acceleration transducers installed in an electric piezocone. The in-situ P-wave and S-wave interval velocities are determined by firstly obtaining the corresponding time series arrival times or relative arrival times as the probe is advanced into the soil profile. Inversion analysis is then carried on the recorded P-wave and S-wave arrival times so that corresponding interval velocity profiles are obtained.

In SCPT there are site conditions which result in source wavelet multiples. These multiples complicate the recorded time series making the selection of interval arrival times a difficult task. This paper outlines a state-space smoothing Kalman Filter algorithm which deconvolves impedance structures from recorded source wavelet multiples. It is also demonstrated that the outlined algorithm is a highly beneficial tool in automating the determination of source wavelet arrival times when only a primary wavelet is recorded.

1 INTRODUCTION

There is considerable interest in methods of geotechnical *in-situ* engineering which enable shear (S) and compression (P) wave velocities in the ground to be accurately estimated. These measurements provide insight into the response of soil to imposed loads such as buildings, heavy equipment, earthquakes, and explosions. The S-wave and P-wave velocities are desired because they form the core of mathematical theorems which describe the elasticity/plasticity of soils and are used to predict settlement, liquefaction, and failure (Finn (1984); Andrus et al. (1999)). As such, accuracy in the estimation of shear and compression waves velocities is of paramount importance because these values are squared during the calculation of geotechnical parameters such as the Shear Modulus, Poisson's Ratio, and Young's Modulus, among others.

The Seismic Cone Penetration Test (SCPT) (an extension of the Cone Penetration Test (CPT)) was devised to measure seismic velocities directly through data obtained by installed seismic sensors in the cone penetrometer, in addition to the standard bearing pressure, sleeve friction, and pore pressure sensors (Campanella et al. 1986). As the cone penetrometer is advanced through the ground, using a pushing force, the advance is halted at one meter (or other such increment) intervals. When the cone is at rest, a seismic event is generated at the surface using a hammer blow or explosive charge, causing seismic waves to propagate from the surface through the soil to be detected by seismic sensors installed in the cone penetrometer. This event is recorded and the penetrometer is advanced another increment and the process is repeated. By determining the seismic arrival times interval seismic velocities are calculated over the depth increment under study (Baziw (1993 and 2002)).

In SCPT there are site conditions which may result in source wavelet multiples. These multiples complicate the recorded time series making the selection of interval arrival times a difficult task. Two examples of test environments which result in source wavelet multiples are remediation sites which contain concrete or stone columns and stratigraphic profiles which contain significant impedances between layering resulting in seismic source wavelet reflections.

As previously stated, the source wavelet multiples result in more complicated seismic time series. The recorded output is mathematically represented as the convolution of the source wavelet with the direct primary wavelet and the reflection coefficients between the different mediums present. The reflection coefficient for a normal incident wavelet is given as:

$$R = \frac{(\rho_2 V_2 - \rho_1 V_1)}{(\rho_2 V_2 + \rho_1 V_1)} \quad (1)$$

where R is defined as the reflection coefficient, ρ is the medium density and V is the medium velocity and it is assumed that the source wavelet travels from medium 1 and is reflected at medium 2. The mathematical representation of the convolution model is represented as follows:

$$z(k) = \sum_{i=1}^{k} \mu(i) S(k-i) + v(k), \quad k = 1, 2, \cdots, N \quad (2)$$

where $z(k)$ is the measurement, $\mu(i)$, $i = 1,2,...$, is the reflectivity sequence, $S(i)$, $i = 0, 1, ... I$ is a sequence associated with the seismic source wavelet, and $v(k)$ is the measurement noise. This paper outlines a seismic cone deconvolution algorithm where reflection coefficients are extracted from the raw seismic time series so that interval arrival times are more easily obtained for source wavelet multiples. In addition, when only a direct source wavelet is present in the recorded time series, seismic deconvolution can simplify and automate the determination of the source wavelet arrival times.

2 SEISMIC CONE DECONVOLUTION KALMAN FILTER

The Kalman Filter is an optimal unbiased minimum variance recursive filter which is based on state-space, time-domain formulations of physical problems. Application of this filter requires that the physical problem be modified by a set of first order differential equations which, with initial conditions, uniquely define the system behaviour. The filter utilizes knowledge of system and measurement dynamics, assumed statistics of system noises and measurement errors and statistical information about the initial conditions. Figure 1 illustrates the essential relation between the system, the measurements and the Kalman Filter.

Figure 1 indicates the scope of information the KF takes into account. As can be seen, the statistics of the measurement and state errors are essential components of the filter. The *a priori* information provides for optimal use of any number, combination and sequence of external measurements. The KF can be applied to problems with linear time-varying systems and with non-stationary system and measurement statistics. Problems with nonlinearities can be handled by linearizing the system and measurement equations or by implementing particle filtering. The Kalman Filter is readily applied to estimation, smoothing and prediction.

Mendel (1983) has carried out extensive work in fitting geophysical problems into state-space representations for the purpose of seismic deconvolution. By formulating the seismic cone deconvolution problem into a state-space representation allows for time variance of both the seismic source wavelet and ambient background noise, and modeling assumptions such as a minimum phase source wavelet are avoided. In addition, the KF has proven to be very robust in its ability to handle approximations to the source wavelet and perform well in high noise environments.

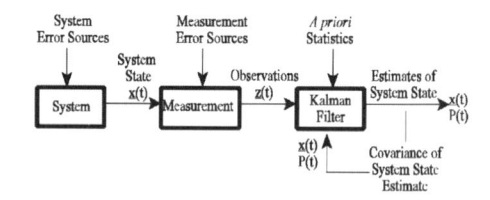

Figure 1. Block diagram of system, measurement, and Kalman Filter.

2.1 *Standard Kalman Filter Governing Equations*

In general terms, the Kalman Filter is a method for estimating a state vector $\underline{x}$ from measurement $\underline{z}$. The state vector may be corrupted by a noise vector $\underline{w}$ and the measurement vector is corrupted by a noise vector $\underline{v}$. The filter is applicable for systems that can be described by a first order differential equation in $\underline{x}$ and a linear (matrix) equation in $\underline{z}$. The filter can be described in both continuous and discrete form. The continuous state and measurement equations are given by eqs. (3) and (4) as follows:

$$\dot{\underline{x}}(t) = F(t)\underline{x}(t) + G(t)\underline{w}(t) \quad (3)$$
$$\underline{z}(t) = H(t)\underline{x}(t) + \underline{v}(t) \quad (4)$$

where $\underline{x}$ is an n-vector, $\underline{w}$ is a p-vector, and $\underline{z}$ and $\underline{v}$ are m-vectors. The random (vector) processes $\underline{w}$ and $\underline{v}$ are assumed to be zero mean, white noise processes. It is further assumed that $\underline{w}$ and $\underline{v}$ are statistically independent of each other. The corresponding discrete state and measurement equations are given by

$$\underline{x}_k = \Phi_{k-1}\underline{x}_{k-1} + \Gamma_{k-1}\underline{w}_{k-1}, \quad \underline{w}_k \approx N(\underline{0}, Q_k) \quad (5)$$
$$\underline{z}_k = H_k\underline{x}_k + \underline{v}_k, \quad \underline{v}_k \approx N(\underline{0}, R_k) \quad (6)$$

In eqs. (5) and (6), symbol N denotes a normal distribution with mean $\underline{0}$ and variance Q_k and R_k, respectively. In addition, Φ is defined as the *State Transition Matrix*, Γ is the *Input Transition Matrix*, and H is the *Measurement Matrix*. The discrete Kalman Filter estimation equations are outlined as follows:

State Estimate Extrapolation:
$$\hat{\underline{x}}_k(-) = \Phi_{k-1}\hat{\underline{x}}_{k-1}(+) \quad (7)$$

Error Covariance Extrapolation:
$$P_k(-) = \Phi_{k-1}P_{k-1}(+)\Phi_{k-1}^T + \Gamma_{k-1}Q_{k-1}\Gamma_{k-1} \quad (8)$$

The term $\Gamma_{k-1}Q_{k-1}\Gamma_{k-1}$ in eq. (8) is referred to as the *Noise Covariance Matrix*.

State Estimate Update:
$$\hat{\underline{x}}_k(+) = \hat{\underline{x}}_k(-) + K_k\left[\underline{z}_k - H_k\hat{\underline{x}}_k(-)\right] \quad (9)$$

Error Covariance Update:
$$P_k(+) = \left[I - K_kH_k\right]P_k(-) \quad (10)$$

I is the identity matrix in eq. (10).

Kalman Gain Matrix:
$$K_k = P_k(-)H_k^T\left[H_kP_k(-)H_k^T + R_k\right]^{-1} \quad (11)$$

Initial Conditions:
$$E[\underline{x}_0] = \hat{\underline{x}}_0, \quad E\left[(\underline{x}_0 - \hat{\underline{x}}_0)(\underline{x}_0 - \hat{\underline{x}}_0)^T\right] = P_0 \quad (12)$$

The computational sequence for the discrete Kalman Filter is outlined as follows:

A. At time index $k = 0$, specify initial conditions $\hat{\underline{x}}_0$, and P_0, and compute Φ_0 and Q_0.
B. At time index k=1, compute $\hat{\underline{x}}_1(-)$, $P_1(-)$, H_1, R_1, and the gain matrix K_1.
C. Using the measurement $\underline{z}_1$ at time index $k=1$, the best estimate of the state at $k=1$ is given by
$$\hat{x}_1(+) = \hat{x}_1(-) + K_1[z_1 - H_1\hat{x}_1(-)]$$
D. Update the error covariance matrix $P_1(+)$.
E. At time index $k=2$, a new measurement $\underline{z}_2$ is obtained and the computational cycle is repeated.

Smoothing is an off-line data processing procedure that uses all measurements between 0 and T to estimate a state at a time t where $0 \leq t \leq T$ (Gelb 1978). The smoothed estimate of $\underline{x}(t)$ based on all the measurements between 0 and T is identified as $\hat{\underline{x}}(t|T)$. There are three types of Kalman Filter smoothers which are identified as follows:

Fixed-interval smoothing: the time interval of measurements (i.e., the data span) is fixed, and we seek optimal estimates at some, or perhaps all, interior points.

Fixed-point smoothing: an estimate at a single fixed point in time is obtained, and the data span time T is assumed to increase indefinitely.

Fixed-lag smoothing: it is again assumed that T increases indefinitely, but in this case we are interested in an optimal estimate of the state at a fixed length of time in the past (i.e., $\hat{\underline{x}}(T-\delta|T)$ with δ held fixed).

In this paper we are interested in the implementation of a Fixed-Interval Smoother for seismic cone deconvolution. Mendel (1983) defines the discrete optimal fixed-interval smoothed estimate $\hat{\underline{x}}(k|N)$ (where $N = T/\Delta$ and Δ is the sampling rate) as follows:

$$\hat{\underline{x}}(k|N) = \hat{\underline{x}}_k(-) + P_k(-)\underline{r}(k|N) \quad (13)$$

where $k = N-1, N-2, \ldots, 1$, and $n \times 1$ vector $\underline{r}$ satisfies the backward-recursive equation

$$\underline{r}(j|N) = \Phi_{j+1}^{p'}\underline{r}(j+1|N) + H_j'\left[H_j P_j(-)H_j' + R_j\right]^{-1}\left[\underline{z}_j - H_j\hat{\underline{x}}_j(-)\right] \quad (14)$$

where $j = N, N-1, \ldots, 1$ and $\underline{r}(N+1|N) = 0$. In eq. (14), matrix Φ^p is defined as

$$\Phi_{k+1}^P = \Phi_{k+1}[I - K_k H_k] \quad (15)$$

The smoothing error covariance matrix $P(k|N)$ is defined as follows:

$$P(k|N) = P_k(-) - P_k(-)S(k|N)P_k(-) \quad (16)$$

where $k = N-1, N-2, \ldots, 1$, and $n \times n$ matrix $S(j|N)$, the covariance matrix of $\underline{r}(j|N)$, satisfies the following backward-recursive equation:

$$S(j|N) = \Phi_{j+1}^{p/} S(j+1|N) \Phi_p(j+1,j)$$
$$+ H'_j [H_j P_j(-) H'_j + R_j]^{-1} H_j \quad (17)$$

where $j = N, N-1, ..., 1$ and $S(N+1 | N) = 0$.

The computational sequence for the discrete fixed-interval Kalman Filter smoother can be thought of as a two pass process. In the first pass optimal real-time state estimates are obtained by implementation of previously outlines steps A to E. In the second pass results from the first-pass estimates (i.e., $\hat{x}$ and P) are reprocessed starting from time index N and utilizing eqs. (13) to (17).

2.2 Seismic Cone Deconvolution (SCD) Governing Equations

The seismic convolution model outlined in eq. (2) may be represented as an autoregressive moving average process (ARMA) (Mendel (1983)). The ARMA model is a combination of both an autoregressive (AR) process and a moving average (MA) process. An AR time series process is generated by a linear combination of past observations plus a Gaussian random variable. The MA process is generated by a finite linear combination of past and present inputs only. In the SCD algorithm the recorded source wavelet and any possible multiples are modeled as an ARMA process which is driven by a forcing function defined as the in-situ direct wavelet and reflection coefficients.

The first step in the SCD analysis is for the user to specify the order of the ARMA process. For example, the Z transform for a fourth order ARMA model is outlined as follows:

$$V(z) = \frac{b_1 z^4 + b_2 z^3 + b_3 z^2 + b_4 z^1}{z^4 + a_1 z^3 + a_2 z^2 + a_3 z^1 + a_4} = \frac{X(z)}{U(z)} \quad (18)$$

where the parameters $b_1, b_2, b_3,$ and b_4 define the MA process coefficients, while the parameters $a_1, a_2, a_3,$ and a_4 define the AR process coefficients. $X(z)$ is the Z transform of the seismic time series recorded and $U(z)$ is the Z transform of the direct wavelet and in-situ reflection coefficients.

In eq. (18), the output x_{k+4} is estimated on the basis of $x_{k+3}, x_{k+2}, x_{k+1}, x_k, \mu_{k+4}, \mu_{k+3}, \mu_{k+2},$ and μ_{k+1} according to the following equation:

$$x_{k+4} + a_1 x_{k+3} + a_2 x_{k+2} + a_3 x_{k+1} + a_4 x_k =$$
$$b_1 \mu_{k+4} + b_2 \mu_{k+3} + b_3 \mu_{k+2} + b_4 \mu_{k+1} \quad (19)$$

The SCD state-space formulation is based upon the technique utilized by Mendel (1983). Variable $y1_{k+1}$ whose Z transform is $Y1(z)$ is introduced into eq. (18) resulting in the following expression:

$$V(z) = \frac{b_1 z^4 + b_2 z^3 + b_3 z^2 + b_4 z^1}{z^4 + a_1 z^3 + a_2 z^2 + a_3 z^1 + a_4} \cdot \frac{Y1(z)}{Y1(z)} \quad (20)$$
$$= \frac{X(z)}{U(z)}$$

Equation numerator and denominator terms in eq. (20) gives the following two expressions:

$$x_k = b_1 y1_{k+4} + b_2 y1_{k+3} + b_3 y1_{k+2} \quad (21a)$$
$$+ b_4 y1_{k+1}$$
$$\mu_k = y1_{k+4} + a_1 y1_{k+3} + a_2 y1_{k+2} \quad (21b)$$
$$+ a_3 y1_{k+1} + a_4 y1_k$$

By choosing $x1_k = y1_k$, $x2_k = y1_{k+1}$, $x3_k = y1_{k+2}$, and $x4_k = y1_{k+3}$ we can fit eqs. (21a) and (21b) into a state-space formulation as follows:

$$\begin{bmatrix} x1_{k+1} \\ x2_{k+1} \\ x3_{k+1} \\ x4_{k+1} \end{bmatrix} = \begin{bmatrix} 0 & 1 & 0 & 0 \\ 0 & 0 & 1 & 0 \\ 0 & 0 & 0 & 1 \\ -a_4 & -a_3 & -a_2 & -a_1 \end{bmatrix} \begin{bmatrix} x1_k \\ x2_k \\ x3_k \\ x4_k \end{bmatrix} + \begin{bmatrix} 0 \\ 0 \\ 0 \\ 1 \end{bmatrix} \mu_k \quad (22)$$

where the direct source wavelet and reflection coefficients μ_k are defined as $E[\mu_k \mu_\tau] = Q_k \delta(k-\tau)$. μ_k is a Gaussian white noise processes with mean zero and time-variant variance of Q_k.

The discrete measurement equation is given as

$$z_k = \begin{bmatrix} b_4 & b_3 & b_2 & b_1 \end{bmatrix} \begin{bmatrix} x1_k \\ x2_k \\ x3_k \\ x4_k \end{bmatrix} \quad (23)$$

The computational sequence for the discrete SCD is outlined as follows:

F. Specify order of the convolution ARMA process and derive coefficients (e.g., $a_1, a_2, a_3, a_4, b_1, b_2, b_3,$ and b_4 - subsequently addressed in this paper).

G. Define state-space matrix equation (eq. (22)) and

measurement equation (eq. (23)).

H. Obtained estimates of the filtered states ($\hat{x}$) by implementing previously outlined Kalman Filter steps A to E.

I. Obtained smoothed estimates (eqs. (13) and (14)) by utilizing the values derived for $\hat{x}$ and P in step H and utilizing eqs. (13) to (17).

J. Derive reflection coefficients by implementing (Mendel (1983)):

$$\hat{\mu}(k|N) = Q(k)\Gamma' r(k+1|N) \quad (24)$$

The state-space formulation outlined in eq. (22) could also be modified to accommodate a more complicated ambient noise process as opposed to the assumed white measurement noise (Baziw and Weir-Jones (2002)). For example, a Gauss-Markov process can be used to describe many physical phenomena and is a good candidate to model possible seismic cone ambient measurement noise.

The Gauss-Markov process has a relatively simple mathematical description. As in the case of all stationary Gaussian processes, specification of the process autocorrelation completely defines the process. The variance, σ^2, and time constant, T_c (ie., $\beta = 1/T_c$), define the first-order Gauss-Markov process. The SCD state-space formulation is simply augmented with the discrete formulation of the Gauss-Markov process so that more structure is provided to the measurement noise.

2.3 Estimating the ARMA Parameters for a Seismic Cone Source Wavelet

As previously stated, the first step in the SCD algorithm is for the user to determine the order of the ARMA process and subsequently derive the necessary model parameters. This portion of the SCD analysis is referred to as *system identification*. The maximum-likelihood of approximating the true source wavelet with an ARMA model increases monotonically with increasing system order, while the computational cost of increasing the system order is proportional to n^3 where n is the ARMA model order (Mendel 1983). In general terms, the process of determining the ARMA model order is a trial and error approach. In this analysis, the investigator chooses a model which has the smallest number of parameters while meeting a performance index requirement that measures how well the ARMA model fits the actual in-situ model.

The technique utilized by the author in deriving the ARMA model parameters is based upon the work of Ogata (1987). In this approach to *system identification*, a least squares cost function which is defined as the difference between the ARMA model response and the corresponding experimental response is minimized.

2.3.1 ARMA Parameter Estimation by the Least Squares Method

The derivation of the ARMA model parameters by utilizing a least squares method is demonstrated by again considering the 4th order (i.e., $n = 4$) ARMA model given in eq. (18). If the numerator and denominator of eq. (18) is multiplied by z^{-4}, we obtain

$$\frac{X(z)}{U(z)} = \frac{b_1 + b_2 z^{-1} + b_3 z^{-2} + b_4 z^{-3}}{1 + a_1 z^{-1} + a_2 z^{-2} + a_3 z^{-3} + a_4 z^{-4}} \quad (25)$$

In eq. (25), the output x_k is estimated on the basis of x_{k-1}, x_{k-2}, x_{k-3}, x_{k-4}, μ_k, μ_{k-1}, μ_{k-2}, and μ_{k-3} according to the following equation:

$$\hat{x}_k = -a_1 x_{k-1} - a_2 x_{k-2} - a_3 x_{k-3} - a_4 x_{k-4} + b_1 \mu_k + b_2 \mu_{k-1} + b_3 \mu_{k-2} + b_4 \mu_{k-3} \quad (26)$$

where $\hat{x}_k$ is the estimated value of $\underline{x}_k$.

The error between the estimated output value and actual output value is defined as follows:

$$\varepsilon_k = x_k - \hat{x}_k \quad (27)$$

Since x_k depends on past data up to n sampling periods earlier, the error ε_k is defined only for $k \geq n$. By substituting eq. (26) and $k = n, n+1, ... N$ into eq. (27) and combining the resulting $N-n+1$ equations into vector-matrix equation, we obtain:

$$\underline{x}_N = C_N \underline{q}_N + \underline{\varepsilon}_N \quad (28)$$

where $\underline{x}_N = [x_4 \, x_5 \, ... \, x_N]$, $\underline{q}_N = [-a_1, -a_2, -a_3, -a_4, b_1, b_2, b_3, b_4]$, $\underline{\varepsilon}_N = [\varepsilon_4 \, \varepsilon_5 \, ... \, \varepsilon_N]$, and C_N is defined as:

$$\begin{bmatrix} x_3 & x_2 & x_1 & x_0 & \mu_4 & \mu_3 & \mu_2 & \mu_1 \\ x_4 & x_3 & x_2 & x_1 & \mu_5 & \mu_4 & \mu_3 & \mu_2 \\ . & . & . & . & . & . & . & . \\ . & . & . & . & . & . & . & . \\ . & . & . & . & . & . & . & . \\ x_{N-1} & x_{N-2} & x_{N-3} & x_{N-4} & \mu_N & \mu_{N-1} & \mu_{N-2} & \mu_{N-3} \end{bmatrix}$$

The least squares performance index is defined as:

$$J_N = \frac{1}{2}\sum_{k=4}^{N}\varepsilon_k^2 = \frac{1}{2}\varepsilon_N' \varepsilon_N \quad (29)$$

The least squares method involves minimizing eq. (29) such that the ARMA parameter values will best fit the observed data. In Ogata's formulation it is assumed that the input sequence $\{\mu_k\}$ is such that for $N>4$, $C_N'C_N$ is nonsingular. Ogata shows that the optimal estimate of $\hat{\underline{q}}_N$ is defined as:

$$\hat{\underline{q}}_N = [C_N'C_N]^{-1} C_N' \underline{y}_N \quad (30)$$

In eq. (30) it is required that $\{\mu_k\}$ is sufficiently time-varying so that $C_N'C_N$ is nonsingular.

Equation (30) is a first best estimate (in a least squares sense) of the ARMA parameters. Ogata presents a subsequent recursive formulation for the estimate of ARMA parameters utilizing eq. (30) as an initial estimate. The recursive least square estimation is defined as

$$\hat{\underline{q}}_{N+1} = \hat{\underline{q}}_N + K_{N+1}\left[y_{N+1} - \underline{c}_{N+1}\hat{\underline{x}}_N\right] \quad (31)$$

where

$$K_{N+1} = \frac{[C_N'C_N]^{-1}\underline{c}_{N+1}'}{1 + \underline{c}_{N+1}[C_N'C_N]^{-1}\underline{c}_{N+1}'}$$

and

$$\underline{c}_{N+1} = [y_N \mathbin{\vdots} y_{N-1} \mathbin{\vdots} y_{N-2} \mathbin{\vdots} y_{N-3} \mathbin{\vdots} \mu_{N+1} \mathbin{\vdots} \mu_N \mathbin{\vdots} \mu_{N-1} \mathbin{\vdots} \mu_{N-2}]$$

The method utilized by the author for determining the ARMA parameters for the source wavelet is to first convolve the isolated wavelet with a highly variable and known white noise process with mean zero and unity variance. This insures that $\{\mu_k\}$ is sufficiently time-varying so that $C_N'C_N$ is nonsingular. Initial estimates of the ARMA parameters are obtained by implementation of eq. (30), known $\{\mu_k\}$, and the convolved output sequence $\{y_k\}$. The initial estimates of $\hat{\underline{q}}_N$ are then feed into the recursive estimation equation defined by eq. (31) until the performance index (eq. (29)) reaches a predefined minimum.

3 EVALUATING THE SCD ALGORITHM WITH SIMULATED FINITE DIFFERENCE DATA

This section presents test bed simulation results when implementing the optimal estimation algorithm previously outlined. The first step in the simulation was to define a seismic source wavelet. Amini and Howie (2003) utilized a finite difference program (FLAC) to model the in-situ seismic cone wavelets. Figure 2 illustrates the simulated source wavelet generated by Amini and Howie (2003) obtained by personal communication. SCPT has the very beneficial feature in that the SH source wavelet is highly repeatable from site to site and its basic form is consistent through out a seismic profile except for the reduction in amplitude due to geometric spreading. The wavelet shown in Fig. 2 was generated by assuming a uniform halfspace with an in-situ shear wave velocity of 180 m/s and a sampling rate of 0.02 ms.

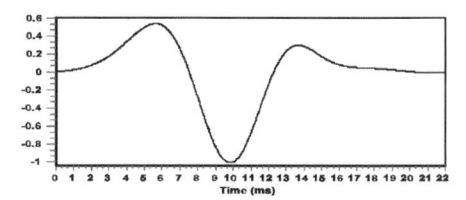

Figure 2. Finite difference simulate source wavelet.

The algorithm outlined in Section 2.3.1 was then implemented on the data illustrated in Fig. 2. In deriving the necessary ARMA parameters a 5th order AR and MA were utilized with a data compression ratio of 16 to 1 so that the sampling rate became 0.32 ms. In addition, it was found that ARMA model estimation algorithm worked best when the source wavelet was time reversed. The only impact that time reversing the source wavelet has on the SCD algorithm is that the recorded seismic time series must be time reversed when processing. Figure 3 illustrates the results of the ARMA estimation of the time reversed source wavelet shown in Fig. 2.

The estimated ARMA model is then convolved with the impedance structure illustrated in Fig. 4 to give the output shown in Fig. 5.

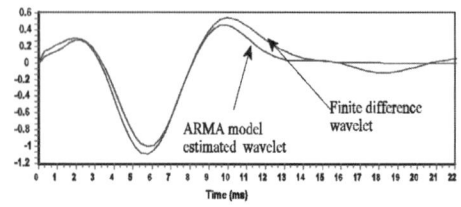

Figure 3. Estimating source wavelet in Fig. 2 with 5th order ARMA model

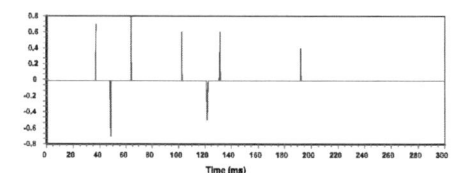

Figure 4. Reflection coefficients utilized to test the performance of the SCD algorithm.

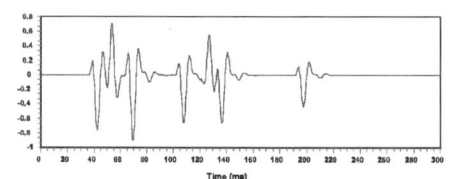

Figure 5. Output after convolving ARMA wavelet with reflection coefficients shown in Fig. 4.

If the SCD algorithm outlined in Section 2.2 is applied to the seismic data shown in Fig. 5, the reflection coefficients illustrated in Fig. 4 are recovered exactly.

Gauss-Markov noise is then added to the seismic data shown in Fig. 5 where a time constant of $T_c = 0.02$ ms and standard deviation of $\sigma = 0.002$ were specified. The SCD algorithm derived the output shown in Fig. 6. As is illustrated in Fig. 6 the arrival time location of the reflection coefficients is recovered exactly but there is some degradation in the estimation of the amplitudes. The SCD algorithm gives accurate estimates of the relative amplitudes of reflection coefficients.

The SCD algorithm was next tested for its ability to derive the reflection coefficients within a high noise environment. Figure 7 shows the seismic data of Fig. 5 with Gauss-Markov noise of $T_c = 0.02$ ms and $\sigma = 0.08$ added.

The SCD algorithm derived the estimates illustrated in Fig 8. As is shown in Fig. 8, the arrival time location of the reflection coefficients is recovered exactly but the algorithm fails to determine the amplitudes. The SCD algorithm gives accurate estimates of the relative amplitudes of the reflection coefficients which facilitates the investigator to recover the true amplitudes.

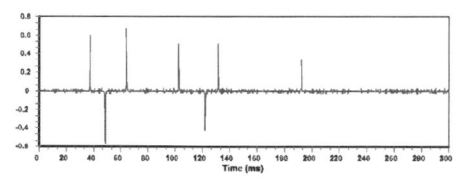

Figure 6. SCD estimated reflection coefficients.

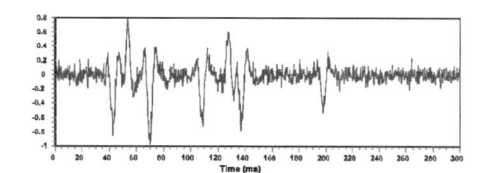

Figure 7. Seismic data of Fig. 5 with Gauss-Markov noise of $T_c = 0.02$ ms and $\sigma = 0.08$ added.

Figure 9 shows the SCD output for the data derived when the true time reversed wavelet of Fig. 2 is convolved with the reflection coefficients in Fig. 4 and Gauss-Markov noise of $T_c = 0.02$ ms and $\sigma = 0.08$ is added. The SCD algorithm provided very similar results to those illustrated in Fig. 8.

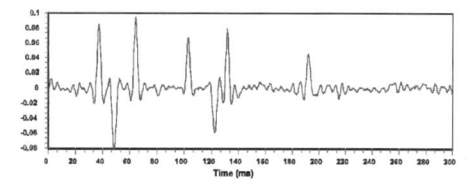

Figure 8. SCD estimated reflection coefficients for data shown in Fig. 7.

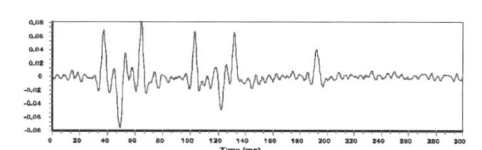

Figure 9. SCD estimated reflection coefficients when the source wavelet illustrated in Fig. 2 is utilized.

The SCD algorithm was next tested for its ability to extract the primary wavelet and the reflection coefficients when only a portion of the source wavelet is modeled. Figure 10 shows a 4th order AR and MA approximation to the portion of the time reversed source wavelet shown in Fig. 2. In this case the data was compressed at a 4 to 1 ratio resulting in a 0.08 ms sampling rate.

Figure 11 illustrates the SCD estimated reflection coefficients from the data shown in Fig. 7 (true time reversed wavelet utilized) and when only a portion of the time reversed source wavelet was modeled as illustrated in Fig. 10. As is shown in Fig. 11 the SCD algorithm recovered the exact time location (time shifted by constant amount due to truncated wavelet approximation) when only a portion of the source wavelet was modeled.

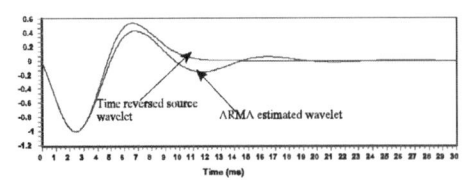

Figure 10. 4th order ARMA approximation to portion of source wavelet shown in Fig. 2.

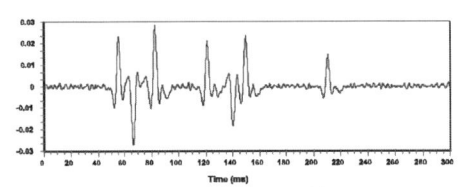

Figure 11. Estimated reflection coefficients when 4th order ARMA model utilized.

Figure 12 illustrates the ability of the SCD algorithm to simplify and automate the determination of the source wavelet arrival times when only a primary wavelet is present.

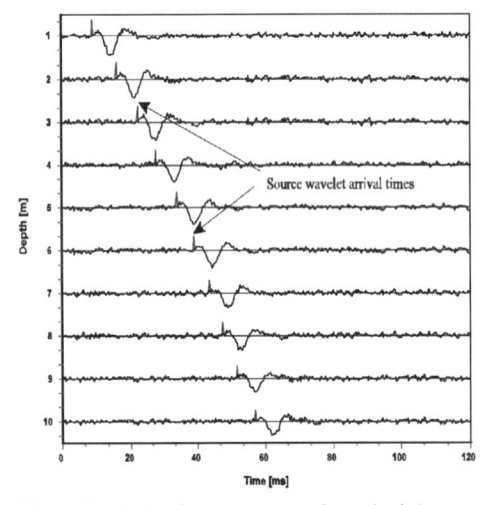

Figure 12. Estimating source wavelet arrival times from vertical seismic profile using SCD algorithm.

4 CONCLUSIONS

In SCPT there are site conditions which may result in source wavelet multiples. These multiples complicate the recorded time series making the selection of interval arrival times a difficult task. This paper outlined a state-space smoothing Kalman Filter algorithm which deconvolves impedance structures from source wavelets by modeling the convolution process as an ARMA model. The ability to obtain the in-situ reflection coefficients from multiples and/or reflections allows the investigator to map out under ground structures.

The SCD algorithm was demonstrated to be highly robust and accurate when utilizing approximations to the source wavelet and in high noise environments. In addition, when only a direct source wavelet is present in the recorded time series, the SCD algorithm significantly simplified and automated the determination of the source wavelet arrival times.

REFERENCES

Amini, A. and Howie, J.A. 2003. Numerical simulation of downhole seismic cone signals. 56th Canadian Geotechnical Conference, Winnipeg, Manitoba, Canada, September 28 to October 1, 2003.

Andrus, R.D., Stokoe, K.H. and Chung, R.M. 1999. Draft guidelines for evaluating liquefaction resistance using shear wave velocity measurements and simplified procedures, NISTIR 6277.National Institute of Standards and Technology, Gaithersburg, Md.

Baziw, E.J. 1993. Digital filtering techniques for interpreting seismic cone data. Journal of Geotechnical Engineering, ASCE, 119(6): 98-1018.

Baziw, E.J. 2002. Derivation of seismic cone interval velocities utilizing forward modeling and the downhill simplex method. Canadian Geotechnical Journal, 39: 1-12.

Baziw, E.J. and Weir-Jones, Iain 2002. Application of Kalman filtering techniques for microseismic event detection. Pure and Applied Geophysics, 159: 449-473.

Campanella, R.G., Robertson, FTC and Gillespie, D. 1986. Seismic cone penetration test. Proc. INSITU86. ASCE, Geot. Spec. Publ. No. 6, June: 116-130.

Finn, W.D. L. 1984. Dynamic response analysis of soils in engineering practice. In Mechanics of engineering materials. John Wiley & Sons Ltd., New York. Chapter 13.

Gelb, A. 1978. Applied optimal estimation. 4th Edition, MIT Press, Cambridge, Mass.

Mendel, J.M. 1983. Optimal seismic deconvolution an estimation-based approach. 1st ed., Academic Press.

Ogata, J. 1987. Discrete process control. 1st ed., Prentice- Hall, New Jersey. pp. 856-867.

The effects of vibration on the penetration resistance and pore water pressure in sands

J. Bonita
Weidlinger Associates, Washington D.C., USA

J.K. Mitchell & T.L. Brandon
Charles E. Via Dept. of Civil and Environmental Engineering, Virginia Tech, VA, USA

Keywords: cone penetrometer, calibration chamber, vibration, pore water pressure

ABSTRACT: Cone penetration tests performed in a calibration chamber reveal that both the penetration resistance and pore water pressure can be influenced by vibration, with the magnitude of influence depending on the confining pressure, density, and properties of the vibration. This influence of vibration is manifested in a reduction of penetration resistance and an elevation of pore water pressure within the zone of influence of the cone penetrometer, and is particularly evident when the applied frequency of vibration approaches the natural frequency of the soil mass. The influence of vibration on the penetration resistance and pore water pressure was quantified, resulting in estimations of the volume change potential of clean sands when exposed to vibration. A method that uses the tip resistance values from static and vibratory cone penetration tests to estimate the density of a clean sand soil is also introduced.

1 INTRODUCTION

The influence of vibration on the behavior of granular soils is often of interest to geotechnical engineers, particularly in potentially liquefiable areas, in areas sensitive to volume change during vibratory loading, or in areas where existing site conditions are deemed unsuitable to support shallow foundations. Pending the results of initial in-situ testing, a cost analysis is usually performed to compare the design implications, time, and financial incentives of ground improvement methods with the factors associated with construction of a deep foundation alternative. If deemed acceptable, the method used as a ground improvement technique is dependent on the type and extent of the soil deposit. Vibratory drum rollers are an inexpensive and efficient means of densifying loose granular soil deposits, but depths of penetration for conventional equipment is generally less than 2m (Sehn and Duncan 1990). Deep vibratory compaction methods are often used where thicker or deeper deposits of loose granular soil are present, but the true effect of the vibration is not recognized until the equipment is mobilized, ground improvement costs are incurred, and a post-compaction in-situ investigation is performed.

One alternative to this investigative approach is to perform a smaller scale vibratory based in-situ test that allows for estimations the volume change potential of the soil as it is exposed to dynamic loading. A vibratory cone penetrometer was developed for this purpose through joint research efforts at Virginia Tech and Georgia Tech. The results of calibration chamber tests using this vibratory cone penetrometer are presented, and a method for using the vibratory penetration test results to estimate the relative density (i.e. density index) and volume change potential of clean sand deposits is introduced. Pending field verification of the technique, it is estimated that it can be used to generate soil properties for static loading applications and also may provide useful information for ground improvement, dynamic loading, and liquefaction related evaluations.

2 TESTING SETUP AND EQUIPMENT

2.1 Calibration Chamber

The Virginia Tech calibration chamber is a flexible wall chamber with constant lateral and vertical stress boundary conditions. A schematic of the chamber system is presented in Figure 1. The chamber is 1.5m in diameter, 1.5m tall, and is cylindrical in shape. Sand was placed inside a 40-mil membrane liner inside the chamber by raining through an air pluviator. A perforated plate located at the bottom of the pluviator controls the placement density of the sand. The vertical stress was applied through air pressure bags acting on a free-floating base plate beneath the sample, and the confining pressure was applied to the sample independent of the vertical pressure using air in the annulus between the sample and the cylinder walls. The water pressure was applied through a third independent system. The saturation procedure involved an initial percolation of carbon dioxide through the soil followed by water inundation from the bottom to the top. Approximately four to six pore volumes of water were circulated through the sample to displace or dissolve any carbon dioxide entrapped in the void space. The sample was sealed and the pressures adjusted through the three independent systems to reach the desired vertical, horizontal, water pressure, and effective stress condition. B-values between 0.92 and 0.96 were measured using this procedure. These values were within the range of values reported by Bellotti et al. (1988) as typical for calibration chamber testing.

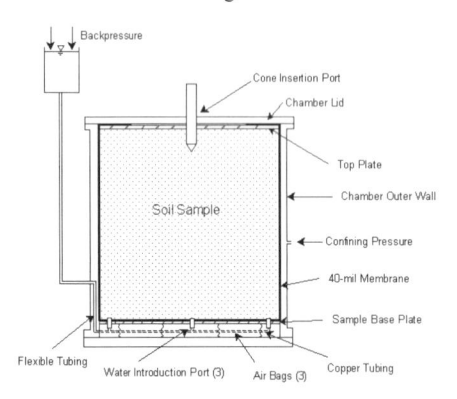

Figure 1. Virginia Tech Calibration Chamber (Not to Scale)

A hydraulic press was mounted onto the lid of the chamber to push the penetrometer at a constant rate of displacement (2 cm/sec) through sealed holes in the lid and top plate. Penetration tests were performed both statically and with a vibration applied at the top of the rod. During penetration, the tip resistance, sleeve friction, and pore water pressure were measured and recorded using an automated data acquisition system. Specific details of the Virginia Tech calibration chamber and testing methodology are given in Bonita (2000).

2.2 Testing Conditions

Over 115 cone penetration tests were performed in the calibration chamber in samples pluviated to relative densities (D_r) of 25% and 55% ($\pm 3\%$). Six penetration tests were also performed in miscellaneous samples ranging in density from very loose (D_r = 15%) to dense (D_r = 80%). Measurements of the tip resistance, sleeve friction, pore water pressure were recorded with depth during each penetration test. The force and frequency of vibration were also measured directly below the vibratory unit and at the penetrometer tip for all vibratory tests using a system of load cells and accelerometers.

The majority of the static (SCPT) and vibratory (VCPT) cone penetrometer tests were performed at vertical stress levels of 7.5 kPa, 55 kPa, and 110 kPa to simulate depths below the ground surface of about 0.8m, 6.4m, and 12.8m, respectively. Five penetration tests were performed at other stress levels. The confining and water pressures were adjusted on all tests to generate at rest effective stress horizontal earth pressure conditions (K_o=0.5).

2.3 Testing Materials and Equipment

Light Castle sand was used in all of the calibration chamber and laboratory testing. This sand has been used extensively by other researchers at Virginia Tech (e.g. Porter 1998), and is similar to Monterey 0/30 sand in physical properties and engineering behavior. A 15 cm^2 Fugro triple element piezocone and a 10 cm^2 single element piezocone were used in the testing. Pore water pressures were measured at both the midface (u_1) and shoulder (u_2) locations on the 15 cm^2 cone and at the shoulder (u_2) location on the 10 cm^2 cone. Both cones were robust and capable of withstanding the vibration induced through the vibratory loading portion of the testing.

Previous studies to link vibration with penetrometer testing were made in Japan (e.g. Sasaki and Koga 1982); however, the vibrators used in these investigations consisted of a horizontally rotating mass similar to that used in concrete vibrators. They found, however, that the horizontal vibration created gaps between the penetrometer and the soil, which obviously influenced the recorded penetration resistance values. More compete reviews of these investigations are given in Bonita (2000) and Bonita et al. (2000).

The vibratory unit used in our investigation consisted of a forced air system that rotated two eccentrically loaded masses opposite to one another (Figure 2). The vertical vibratory force generated by the unit was controlled by the placement of a an eccentric mass on each gear. A relatively constant maximum force of about 1.5 kN was measured at the load cells and computed at the tip of the cone penetrometer, indicating minimal energy loss through the cone rod system. The vibrator was designed so that the mass attached to the rotating gear system could be adjusted to maintain a constant peak dynamic force for different frequencies of rotation. The masses were positioned on the gears so that the horizontal component of motion was zero and the net vertical motion was twice that generated by a single rotating mass.

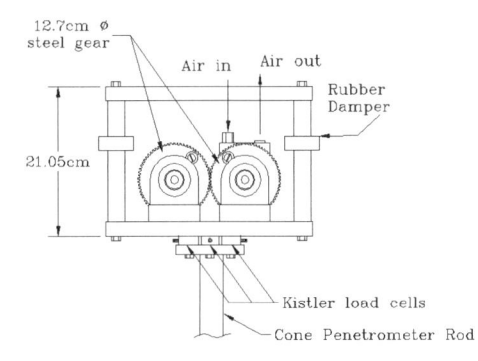

Figure 2. Counter Rotating Mass Vibrator

Vibration frequencies ranging from 5Hz to 135Hz were generated by the vibratory unit and measured at both the load cell and cone tip locations. Three secondary frequencies of vibration were consistently recorded at the load cells directly below the vibrator and at the cone tip. These frequencies matched well with the first three fundamental modes of vibration of the rod system estimated using an approach presented in Richart et al. (1970). These motions were significantly outside of the range of frequencies generated by the vibratory unit, and appeared to have little influence on the tip resistance or pore water pressure behavior during the penetration test.

3 TEST RESULTS

3.1 Penetration Resistance

The results of the calibration chamber tests indicated that it was possible to influence the measured cone penetration resistance, provided the magnitude of the force and frequency of vibration are large enough to induce a significant dynamic load into the soil. This influence of vibration is illustrated in Figure 3a, where the penetration resistance values measured in saturated loose samples at the low confining stress level are compared.

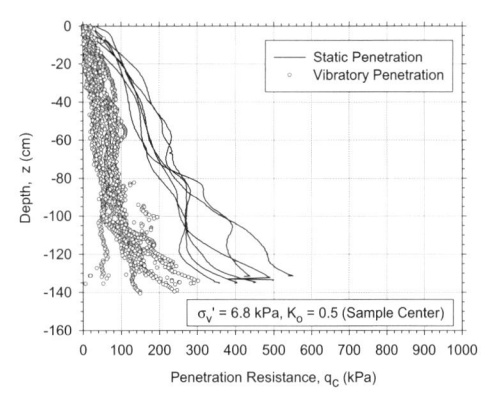

Figure 3a. Comparison of Static and Vibratory Penetration Resistance

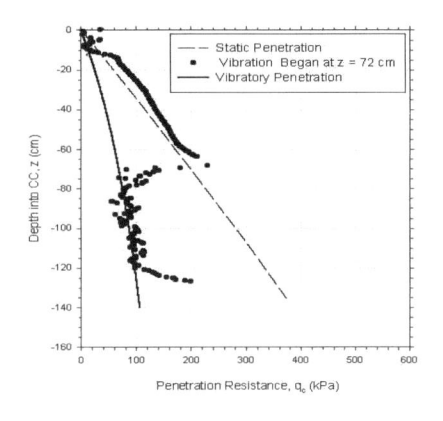

Figure 3b. Vibration Started Midway Through Penetration Test

A significant reduction in the penetration resistance at the center of the sample ($z = 75$ cm) due to the vibration (Figure 3a) was measured. This reduction is particularly evident in Figure 3b, where the penetration resistance is significantly reduced when the vibration was initiated when the penetrometer was at the midpoint of the sample.

The influence of vibration on the penetration resistance was quantified by generating a representative trend line for each testing condition and then comparing the trend lines for the static and vibratory tests. Examples of the regression curves generated for the static and vibratory penetration tests in the loose samples at the low stress conditions are presented in Figure 3b. This direct comparison is referred to as the reduction ratio (RR), as defined by:

$$RR = 1 - \frac{q_{cv}}{q_{cs}} \quad (1)$$

where RR is the reduction ratio, q_{cv} is the vibratory penetration resistance, and q_{cs} is the static penetration resistance. A version of this equation was proposed by Sasaki et al. (1986). A value of the reduction ratio near unity indicates that the vibratory penetration resistance was much less than that measured during static penetration, while a value near zero means that the static and vibratory penetration resistance values are approximately equal.

Plots showing representative trend lines for tests performed at each of the stress levels are presented in Figure 4 for both loose ($D_r = 25\%$) and medium dense ($D_r = 55\%$) samples tested using a vibration frequency of 135 Hz. A second abscissa was added to the plots to show the ratio of the vibratory to the static penetration resistance (q_{cv}/q_{cs}).

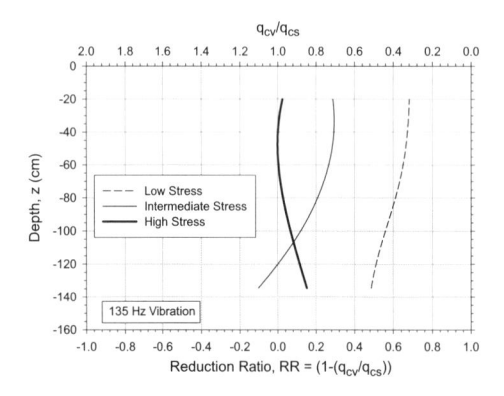

Figure 4a. Reduction Ratio at Different Density and Stress Levels – Loose Samples ($D_r = 25\%$)

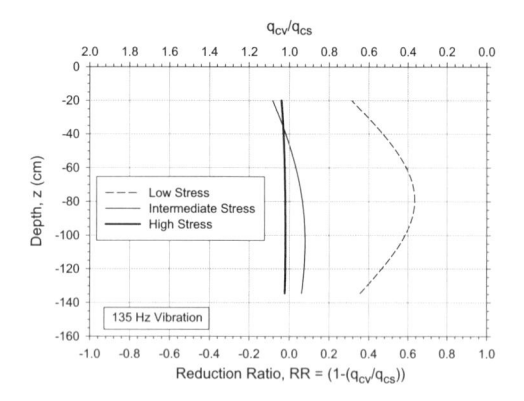

Figure 4b. Reduction Ratio at Different Density and Stress Levels – Medium Dense Samples ($D_r = 55\%$)

This comparison shows a RR of approximately 60% at the sample center in both the loose and medium dense soils at the low confining stress level. The effect of the vibration on the penetration resistance is reduced as the mean effective confining stress in the soil is increased, as indicated by RR values at the loose sample centers of approximately 25% and 7% at the intermediate and high stress levels, respectively. A negligible reduction in penetration resistance was observed at the sample center in the medium dense soils at the intermediate and high stress levels.

3.2 Pore Water Pressure

Statistical analyses of the data collected from the calibration chamber tests revealed a negligible difference in the magnitude of pore pressure increments (Δu) measured at the different transducer locations on the cone penetrometer (u_1 and u_2). This was noted for both static and vibratory penetration tests. The results of the testing also indicate that the induced pore water pressure ratio ($\Delta u/\sigma_{vo}$) measured at the cone reached a maximum value of only about 5% in all tests (Figure 5), which cannot by itself explain the considerable reduction in penetration resistance observed in the testing for certain conditions (Figure 4). However, pore pressure transducers placed in the soil deposit 0.35m away from a penetrating cone revealed a $\Delta u/\sigma_{vo}$ of about 44% at the low and intermediate stress states, which suggests that the pore pressure measured at the cone during vibratory penetration was not indicative of the actual pore water pressure at distances away from the penetrometer.

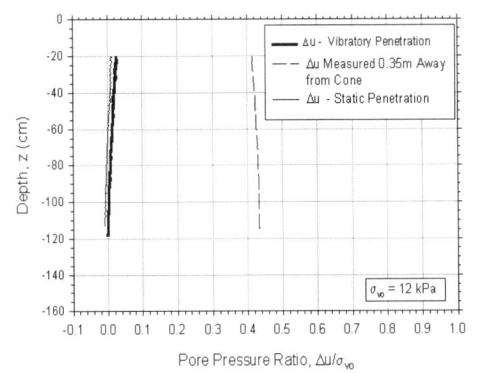

Figure 5. Pore Water Pressure At and Away from Cone Penetrometer

3.3 Influence of Frequency of Vibration

Figure 6 shows the tip resistance at the midpoint of the chamber in loose and medium dense samples at the low and intermediate stress level for tests at different vibration frequencies. As shown through the comparison, the vibratory penetration resistance is highly dependent on the frequency of vibration. Vibratory penetration tests at frequencies of 15 to 25 Hz did not give penetration resistance values that were significantly different from those obtained from static penetration (0 Hz). A maximum reduction in penetration resistance was measured during penetration with a frequency of vibration between 35 and 45 Hz, and a fairly constant reduction in penetration resistance was observed at frequencies above 55 Hz for all testing conditions.

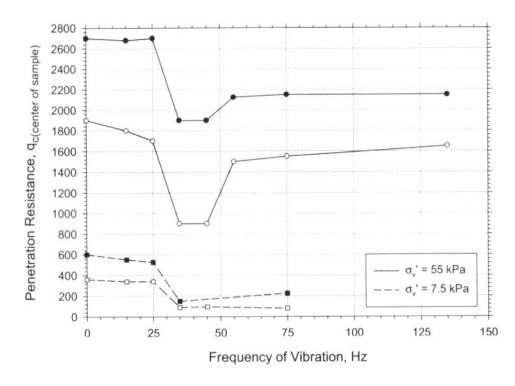

Figure 6. Influence of Frequency Vibration on Penetration Resistance Value

DISCUSSION OF RESULTS

3.4 Penetration Resistance and Pore Water Pressure

The intent of the vibrating penetrometer was to induce cyclic pore water pressure within the zone of influence of the cone while simultaneously measuring the penetration resistance. The influence of the vibration on the penetration resistance was identified by a reduction in the measured penetration resistance. Cavity expansion theory (e.g. Salgado et al. 1997) shows that the penetration resistance is controlled by plastic deformation of the soil adjacent to the penetrometer and by elastic deformation of the soil away from the penetrometer. A reduction of the penetration resistance through vibration requires that the effective stress in the zone of influence of the cone be reduced by an elevated pore water pressure under constant total confining stress conditions. This excess pore water pressure must occur adjacent to the cone penetrometer and at finite distances laterally and vertically away from the probe, and must be present during the shear-induced deformation associated with the penetration of the cone.

The application of cavity expansion theory also suggests that the zone of influence of the cone penetrometer increases as either the effective stress or density of the soil increases. In order to reduce the penetration resistance at the higher confining stresses or density, therefore, the total volume of soil influenced by the vibration must increase. Consequently, for soils at the higher stress levels or densities, the zone of influence of the vibration must increase in order to get a comparable decrease in penetration resistance.

The pore water pressure measured on the cone was significantly less than that measured 0.35m away from the penetrometer during vibratory penetration. The exact reason for this discrepancy is not clearly understood at this time. However, it does appear that elevated pore pressures were generated in the soil as a result of the vibration, whereas reduced pore pressures were present adjacent to the cone through the shear stresses generated by the penetration of the probe. Lunne et al. (1997) touched on this issue during discussions of the influence of shear stresses on pore pressure measurements for different transducer locations on the cone. These concepts may relate to the relationship of the in-situ void ratio and confining stress to the steady state line, and are most likely relate to the "B" and "$\bar{A}$" parameters defined by Skempton (1954). Further research in this area is currently in progress.

Pore pressure transducers placed in the loose and medium dense soil samples at low stress levels measured an induced pore pressure that was approximately equal to 88% of the total confining stress at a radial distance of 0.35m away from a vibrating penetrometer (Figure 5). It therefore appears that the zone of influence of the vibration has a minimum radius of at least 0.35m for the given densities and stress level. Estimations of the width of the elastic zone of influence of the cone penetrometer using relationships generated by Salgado et al. (1997) were less than or near 0.35m, suggesting the undrained conditions generated by the vibration encompassed the majority of the zone of influence of the cone penetrometer. Penetration tests performed under these conditions revealed significantly reduced tip resistance values, mainly due to the reduction in effective confining stress associated with the undrained conditions. However, the RR values noted for the loose samples at the intermediate and high stress level were only 25% and 5%, respectively, with a 135 Hz vibration, indicating that undrained condition induced by the vibration did not encompass the entire zone of influence of the penetrometer. Consequently, the effective stress in the soil was not dramatically reduced throughout the entire zone of influence of the cone, resulting in small to moderate reduction in the penetration resistance. Similar conclusions can be made for the tests performed in the medium dense samples tested at the intermediate and high stress levels.

3.5 Effects of Frequency of Vibration on the Penetration Resistance

The effect of the frequency of vibration on the measured penetration resistance and pore water pressure was evaluated as part of this investigation. Vibration frequencies of 15, 25, 35, 45, 55, 75, and 135 Hz were used in the testing. The peak dynamic force was adjusted for each of the tests to a fairly constant value of 1.5 kN.

An estimation of the natural frequency of the soil sample was generated using an expression developed from the wave equation for the motion of uniform soil on rock:

$$f = \frac{(2n-1)v_s}{4H} \quad n = 1, 2, ..., \infty \quad (2)$$

where f is the fundamental frequency, n is the mode of vibration, v_s is the shear wave velocity, and H is the thickness of the soil layer. The value of the shear wave velocity of the soil deposit used in this equation was estimated using a modification of an empirical expression developed by Hardin and Richart (1963), which takes into account the void ratio and effective confining stress in the soil:

$$V_s = [110.8 - (51)e] \cdot (\sigma_m')^{0.25} \quad (3)$$

where V_s is the shear wave velocity in m/sec, e is the void ratio, and σ_m' is the mean effective stress in the soil in kPa. The shear wave velocity using this expression was determined to be about 184m/sec for the loose (i.e. Dr = 25%) soil and 195m/s for the medium dense (i.e. Dr = 55%) soil. These values correspond to first fundamental frequencies of 31Hz and 33Hz, respectively, using Equation 2. Similar fundamental frequency values for soil were presented by Terzaghi et al. (1997). It should be noted that Equation 2 was developed based on a zero stress vertical boundary and a soil mass of infinite lateral extent. Since these boundary conditions are obviously different from those present in the calibration chamber, the estimation of the natural frequency of the calibration chamber sample using Equation 2 should be considered approximate.

As described earlier, the vibrations used in this investigation were generated from a pair of unbalanced rotating masses illustrated in Figure 2. The phase relationship associated with the rotation is such that the masses reach their top and bottom positions at the same time so that the addition of the vector components results in the cancellation of the horizontal component of motion. The dynamic analysis of the system shown in Figure 2 leads to a force of $Q = 2me\omega^2 sin(\omega t)$, which when introduced into Newton's Second Law of Motion yields the following equilibrium expression:

$$\sum F_y = m\ddot{x} = 2me\omega^2 \sin(\omega t) - kx - c\dot{x} \quad (4)$$

where m is the mass, ω is the circular frequency of motion, k is the stiffness, c is the damping coefficient, and x is the displacement. As indicated by Equation 4, the force generated by the rotating mass is proportional to the square of the frequency of oscillation. Using $\omega_n^2 = k/m$ as the natural frequency of the system and $D = c/c_c$, where c_c is the critical damping coefficient, Equation 4 is eventually reduced to:

$$M = \frac{A}{me} = \frac{(\omega/\omega_n)^2}{\left[1 - (\omega/\omega_n)^2 + (2D\omega/\omega_n)^2\right]^{\frac{1}{2}}} \quad (5)$$

The values of M in Equation 5 represent the dynamic magnification factor of the force during the vibration. The magnitude of the amplified excitation is the largest when the applied versus natural frequency ratio (ω/ω_n) approaches a value of 1.0, which indicates a resonance condition. The relationship also reveals that the magnification factors at ω/ω_n ratios above 1.0 that are greater than those below 1.0 for a defined difference in ω/ω_n away from $\omega/\omega_n = 1.0$.

The frequency of vibration associated with the largest reduction in penetration resistance was approximately equal to the natural frequency of the soil mass as determined using Equation 2. Penetration tests performed at frequencies greater than the natural frequency of the soil (i.e. 55 Hz) generated a larger reduction in the penetration resistance than those at the lower frequencies (i.e. 25 Hz). This behavior agrees with the amplification relationship in Equation 5 which suggests that the amplitude of the magnification factor is greater at the higher frequencies of motion for a defined difference in ω/ω_n away from $\omega/\omega_n = 1.0$. Similarly, the vibratory penetration resistance measured at the 75 Hz and 135 Hz frequencies were approximately equal, which agrees with the relationship that the ω/ω_n values are fairly constant above a value of about 2.0 (Figure 6).

Figure 7 gives a comparison of the normalized penetration resistance values (q_c/σ_v') for tests performed in loose and medium dense samples at the low and intermediate stress conditions with different frequencies of vibration. The normalized penetration resistance value reached a fairly constant value in the loose samples at both stress states and in the medium dense samples at the low stress state when the vibration frequency equaled or exceeded the natural frequency of the soil deposit. Although pore water pressures were not measured in the soil away from the cone in all of the tests, an elevated water pressure was recorded away from the cone penetrometer in those tests in the loose and medium dense samples at low stress states.

As noted in Figure 7, the q_c/σ_v' values generated at the intermediate stress level for the medium dense samples were significantly higher than those obtained for the loose samples at all frequencies of vibration. As noted above, cavity expansion theory shows that the zone of influence of the cone under these conditions is much larger than at the lower stress and density levels. Although not measured, it is estimated that the vibration induced pore water pressure was not elevated to a high enough level across the entire zone of influence of the penetrometer. Subsequently, the mean effective stress was not reduced and the corresponding q_c value was not influenced, as identified by the low RR and high q_c/σ_v' values. Thus, a measurement of the penetration resistance at or near a liquefied state

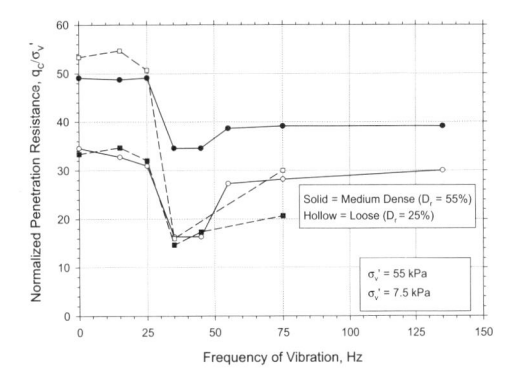

Figure 7. Influence of Frequency of Vibration on the Normalized Penetration Resistance Value

was not possible at these stress states using the given input vibration.

3.6 Estimation of the Relative Density Using Vibratory Penetration Resistance

The influence of vibration on the penetration resistance was quantified using the RR relationship in Equation 1. As noted in the previous section, the largest reduction in penetration resistance occurred when the frequency of vibration matched the natural frequency of the soil, which was about 35 Hz for the samples tested. The RR as a function of the effective stress is shown in Figure 8 for penetration tests performed in loose and medium dense samples under a 35 Hz vibration frequency. The values of the RR and effective stress at the center of the sample are used in the plot. It can be seen through this comparison that the highest RR value occurs at the lowest effective stress level. The RR decreases as the effective stress increases for a given relative density and decreases as the relative density increases for a given effective stress.

An expression was established for each of the density conditions that relates the value of RR to the vertical effective stress in the soil. The expression is based on a third order polynomial of the form:

$$RR = C_1 + C_2 \cdot \sigma_v' + C_3 \cdot (\sigma_v')^2 + C_4 \cdot (\sigma_v')^3 \quad (6)$$

where C_1, C_2, C_3, and C_4 are unknown constants and σ_v' is the vertical effective stress in the sample. The expression was written for each relative density using the established RR vs. σ_v' relationships, resulting in three equations and four unknowns. A fourth equation was generated by assuming the slope of the curve was zero at the high stress level, which implies that the derivative of Equation 6 is zero at this testing condition. Constants C_1 through C_4 could then be determined by simultaneous solution. The procedure was performed for both relative densities, which resulted in two separate best-fit curves that extended through the range of stress levels used in the testing.

The best-fit curves generated for both the loose and medium dense conditions are included in Figure 8. The RR data obtained from five sets of penetration tests outside the target stress level are also superimposed on this figure. As shown by the comparison, the penetration test data measured in tests outside of the target stress levels agree fairly well with the empirical approximation generated using Equation 6.

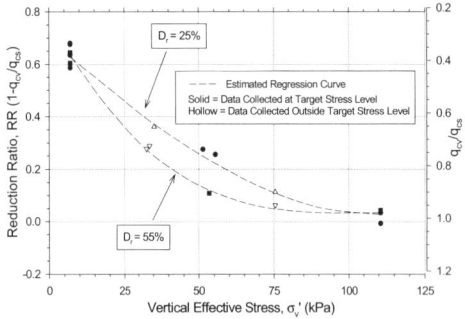

Figure 8. Measured vs. Estimated RR values

Equation 6 includes the independent variable of vertical effective stress and the constants C_1 through C_4. The RR value for a given effective stress varies for the two different densities because the values of the constants are different. An empirical expression was generated using the information from the curves for the loose and medium dense sand samples to generate an overall expression for RR that included both the independent variables of relative density and vertical effective stress. The procedure used to determine this expression involved plotting a given constant value determined from the curve fit versus the relative density (i.e. C_1 vs. D_r). A linear regression line was used to fit the data. The slope and ordinate of this line were then substituted back into the RR expression for the constant, resulting in an expression for RR that included new constants, relative density, and vertical effective stress. Using this procedure, a single empirical expression was generated that allowed for the computation of a family of curves for the different relative density and stress conditions. This empirical expression is noted in Equation 7 and the family of curves is illustrated in Figure 9:

$$RR = [C_1 \cdot D_r][C_2 + \sigma_v']^2[C_3 + \sigma_v'] + C_4[C_5 + \sigma_v'] \cdot [C_6 + (C_7 + \sigma_v') \cdot \sigma_v'] \quad (7)$$

where D_r is the relative density in units of percent, σ_v' is the vertical effective stress (in kPa), and C_1 through C_7 are empirical constants determined from the regression analysis.

The empirical family of curves generated using Equation 7 were compared to the RR values measured in tests performed in the six samples outside of the target stress range identified earlier (Figure 9). The few data points outside this stress range available from the testing seem to agree well with the values estimated from the curves, suggesting that it may be possible to use a direct comparison of the static and vibratory penetration resistance to estimate the relative density. However, only a few data points outside of the target relative densities have verified the technique, which implies that both additional calibration chamber testing and field verification is needed to quantify the validity of the proposed relationships.

4 CONCLUSIONS

The purposes of this investigation were to measure, evaluate, and quantify the influences of vibration on the penetration resistance and pore water pressure values measured during cone penetration tests. The results indicate that pore water pressure measurements on the cone penetrometer do not properly reflect the true magnitude of the pore water pressure generated in the surrounding soil. Under certain testing conditions, elevated pore water pressures were measured in the soil mass away from the cone penetrometer, suggesting the vibration was inducing structural collapse under an undrained condition. In tests performed in loose samples at all stress levels and medium dense samples at a low stress level, the magnitude of this induced pore water pressure approached the total confining stress in the soil, which in turn resulted in effective stress conditions within zone of influence of the cone that were at or near conditions of liquefaction.

It was also shown through this research that the penetration resistance values can be dramatically reduced if the volume of undrained soil encompasses the zone of influence of the cone penetrometer. Soil conditions at this state may be approaching flow liquefaction conditions, and the resulting reduced penetration resistance values may reflect the shear strength of the soil at or near a liquefied state. For a given testing condition, the magnitude of this reduction is largest when the frequency of vibration approximates the natural frequency of the soil deposit. Thus, the results of this investigation suggest possible approaches for estimation of both the shear strength of liquefied soil and the volume change of potentially liquefied soils in-situ; however, much additional testing and analysis is needed to properly test the concepts. Information related to the effect of vibration frequency on the soil behavior may also provide insight to vibration based soil improvement and pile driving applications.

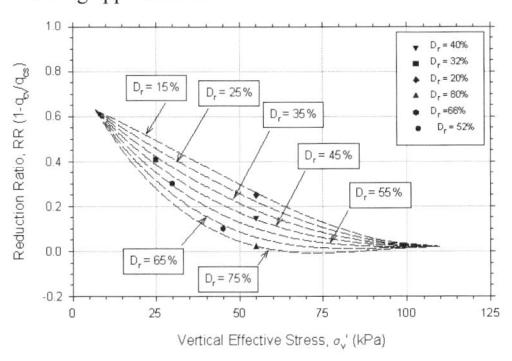

Figure 9. Estimation of Relative Density using RR values

ACKNOWLEDGEMENTS

The research described in this paper was supported by the National Science Foundation (Grant No. CMS-9810465) and the U.S. Geological Survey (Grant No. 1434-HQ-97-GR-03083). The authors also thank Paul Mayne and his graduate students at Georgia Tech and Jim Coffey, Jim Hawkins, J.T. McGinnis, Youngjin Park, David Plehn, and the other students and staff at Virginia Tech that participated in the project.

REFERENCES

Bellotti, R., Crippa, V., Pedroni, S., and Ghionna, V.N. (1988). "Saturation of Sand Specimen for Calibration Chamber Tests." *Proc. of the 1st Inter. Symp. on Penetration Testing (ISOPT-1)*, Orlando, FL, Balkema, p. 661-671.

Bonita, J.A. (2000). "The Effects of Vibration on the Penetration Resistance and Pore Water Pressure in Sands." Ph.D. Dissertation, Department of Civil Engineering, Virginia Tech, Blacksburg, VA.

Bonita, J.A., Mitchell, J.K. and Brandon, T.L. (2000). "In-Situ Liquefaction Evaluation Using a Vibrating Penetrometer," *Soil Dynamics and Liquefaction 2000, ASCE Geot. Specialty Publ.. No. 107*, p. 191-205.

Lunne, T., Robertson, P. K., and Powell, J. (1997). *Cone Penetration Testing in Geotechnical Practice*, Blackie Academic and Professional.

Porter, J.R. (1998). "An Examination of the Validity of Steady State Shear Strength Determination Using Isotropically Consolidated Undrained Triaxial Tests." Ph.D. Dissertation, Department of Civil Engineering, Virginia Tech, Blacksburg, VA.

Richart, F.E., Hall, J.R., and Woods, R.D. (1970). *Vibrations of Soils and Foundations*, Prentice Hall International Series, 414 p.

Robertson, P.K. (1994). "Suggested Terminology for Liquefaction: An Internal CANLEX Report." *University of Alberta*, Edmonton, Alberta.

Salgado, R., Mitchell, J.K., and Jamiolkowski, M. (1997). "Cavity Expansion and Penetration Resistance in Sand." *Journ. of Geot. and Geoenv. Eng.*, 123(4), p. 344-354.

Sasaki, Y. and Koga, Y. (1982). "Vibratory Cone Penetrometer to Assess the Liquefaction Potential of the Ground." 14th Joint Meeting of the U.S.-Japan Panel on Wind and Seismic Effects, Washington D.C., p. 541-555.

Sasaki, Y., Koga Y., Itoh, Y., Shimazu, T., and Kondo, M. (1986). "In Situ Test for Assessing Liquefaction Potential Using Vibratory Cone Penetrometer." 17th Joint Meeting of the U.S.-Japan Panel on Wind and Seis. Effects, Tsukuba, p. 396-409.

Sehn, A.L., and Duncan, M. (1990). "Instrumentation for Earth Pressures Due to Compaction." Trans. Res. Record 1277, p. 44-52.

Skempton, A.W. (1954). "The Pore Water Coefficients A and B." Geotechnique, Vol. 4, p. 143-147.

Terzaghi, K., Peck, R., and Mesri, G. (1997). Soil Mechanics in Engineering Practice. John Wiley and Sons, Inc. 549 p.

Proceedings ISC-2 on Geotechnical and Geophysical Site Characterization, Viana da Fonseca & Mayne (eds.)
© 2004 Millpress, Rotterdam, ISBN 90 5966 009 9

In situ testing and foundation engineering: recent contributions

Jean-Louis Briaud
Holder of the Buchanan Chair, Department of Civil Engineering, Texas A&M University, College Station, Texas 77843-3136, USA
(briaud@tamu.edu)

Keywords: shallow foundations, scale, embedment, load-settlement curve, horizontal load, piles, downdrag, bitumen coating

ABSTRACT: This article is an overview of contributions in foundation engineering and in situ testing by the author and his students at Texas A&M University. The topics include the scale and embedment effects for shallow foundations, the load-settlement curve method for shallow foundations, a simple method for horizontally loaded piles, and a method for calculating downdrag loads and for reducing them. Design methodologies are presented in each case. References are given for further details.

1 INTRODUCTION

Foundation engineering problems tend to be complicated geotechnical problems. Some 100 years ago the tendency for foundation engineering solutions was to resort to experimental correlations because theoretical soil mechanics was in its infancy. In the last 50 years, soil mechanics theory and the associated numerical simulations have made remarkable progress and have more than caught up with sampling and testing developments. There is a need to aim for solutions with a proper balance between theoretical and experimental considerations. In foundation engineering, this appears to give the best potential for reaching the threshold of optimum simplicity.

The methods presented here have been developed over the last 10 years by the author and his students. They include methods for shallow foundations, deep foundations. This is an excerpt of a more complete reference published previously (Briaud, 2002).

2 SHALLOW FOUNDATIONS: IS THERE SCALE AND EMBEDMENT EFFECT?

The students who worked on this project at various times are Philippe Jeanjean, Bob Gibbens, and Jayson Barfknecht. The sponsor was the Federal Highway Administration. The case considered is the one of a square footing in sand subjected to a vertical load applied at the center of the footing surrounded by a flat and horizontal ground surface. The average pressure under the footing is pf, the settlement is ρ, the footing width is B, the depth of embedment is D, and the average soil strength within the zone of influence of the footing is s_a. The question raised is: is there a scale effect and a depth of embedment effect on the load-settlement curve in this case?

The answer is based on theoretical considerations and on the results of experiments. The theoretical considerations included the bearing capacity equation and the theory of elasticity. The experiments performed for the study included five large footing load tests and over 30 plate load tests at the National Geotechnical Experimentation Site at Texas A&M University. The footings were 3x3m, 3x3m, 2.5x2.5m, 1.5x1.5m, and 1x1m; they were all embedded 0.75m into a medium dense silty sand. The results are presented in Briaud and Gibbens (1994, 1997, 1999). Figure 1 gives an example. The plates were square, the sizes were 0.1m, 0.2m, 0.3m, and 0.4m, and the embedment varied from 0m to 0.8m. The results are presented in Barfknecht and Briaud (1999). A review of the literature on this topic yielded four additional studies with footing load tests (Ismael, 1985, Pu and Ko, 1988, Khebib, Canepa, and Magnan, 1997, Lutenegger, 1995).

The answer to the question posed is limited by the evidence mentioned above and used to reach the following conclusions.
1. There is no scale effect and no embedment effect for the curve p_f/s_a vs. ρ/B regardless of the soil profile.
2. There is no scale effect and no embedment effect for the curve p_f vs. ρ/B when the soil strength profile is constant with depth.

853

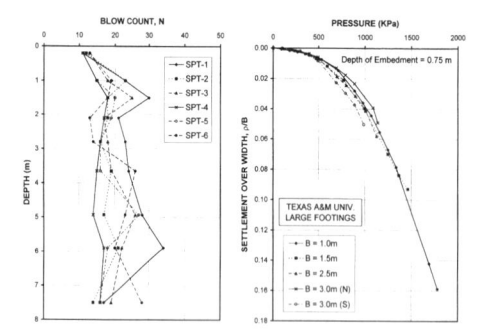

Figure 1: Results of the Large Footing Tets at Texas A&M University

3. There is a scale effect and an embedment effect for the curve p vs. ρ/B when the soil strength profile is not constant with depth. The effect is an increase or a decrease depending on whether the strength increases or decreases with depth. This effect disappears if the curve is normalized as p_f/s_a vs. ρ/B.

The general bearing capacity equation for sands assumes a soil strength profile which increases linearly with depth because ϕ and γ are constant. In this case, N_γ and N_q are constant and the equation gives the right influence of B and D. For any other strength profile, this equation does not represent the true variation of the bearing capacity because the assumptions no longer correspond to the strength profile.

3 SHALLOW FOUNDATIONS: THE LOAD SETTLEMENT CURVE METHOD

The students who worked on this project at various times are Philippe Jeanjean, Kabir Hossain, Bob Gibbens, Jayson Barfknecht, Jong Hyub Lee. The sponsor was the Federal Highway Administration. The load-settlement curve method is used to generate the complete load settlement curve for a footing. The Load Settlement Curve method (LSCM) replaces the calculations of bearing capacity and settlement which were done separately in the past. The LSCM was proposed by Briaud and Jeanjean (1994) for square footings in sand resting on a flat ground surface and subjected to a centered vertical load. This method is based on the point-by-point transformation of the pressuremeter curve (Briaud, 1992) into the load-settlement curve for the footing through the use of two equations.

$$\rho/B = 0.24 \, \Delta R/R_o \quad (1)$$
$$p_f = \Gamma \, p_p \quad (2)$$

where $\Delta R/R_o$ is the relative increase in pressuremeter radius, p_p is the pressure on the cavity wall applied by the pressuremeter probe, and Γ is a transformation function obtained experimentally and theoretically (Jeanjean, 1995) (Figure 2). As discussed in the previous section, the scale and embedment effect are directly tied to the soil strength profile within the zone of influence of the footing, and the p/s_a vs. ρ/B curve is independent of the foundation width B and the depth of embedment D. This indicates that the Γ function is independent of B and D.

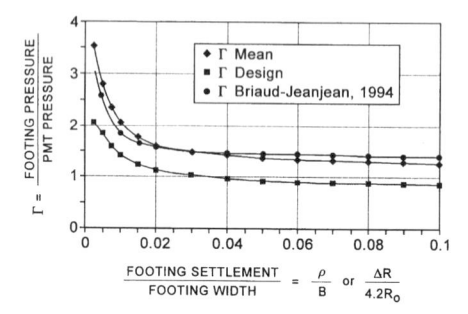

Figure 2: The Γ Function for the Load-Settlement Curve Method

This method was extended to the case of a rectangular footing (B wide, L long) subjected to an eccentric inclined load (eccentricity e and angle of inclination δ) at a distance d from the crest of a slope (Hossain, 1996). A number of correction factors are proposed based on numerical simulations calibrated against the large footing tests mentioned in the previous section. These factors are as follows:

Influence of the shape
$f_{L/B} = 0.8 + 0.2(B/L)$ (3)
Influence of eccentricity
$f_e = 1 - 0.33(e/B)$ center (4)
$f_e = 1 - (e/B)^{0.5}$ edge (5)
Influence of inclination
$f_\delta = 1 - (\delta/90)^2$ center (6)
$f_\delta = 1 - (\delta/360)^{0.5}$ edge (7)
Influence of a slope
$f_{\beta,d} = 0.8 \, (1 + d/B)^{0.}$ slope at 3 to 1 (8)
$f_{\beta,d} = 0.7 \, (1 + d/B)^{0.15}$ slope at 2 to 1 (9)

For the time being, the superposition of cases is taken into account by multiplying the influence factors as is common practice. There is some evidence that this approach is conservative (Hossain, 1996). More research is needed in this area.

The load-settlement curve method therefore consists of the following steps:
1. Perform preboring pressuremeter tests (PMT) within the depth of influence of the footing; usually at depths equal to 0.5B, 1B, and 2B.
2. Prepare the PMT curves and obtain the average curve (Briaud, Jeanjean, 1994).

3. Transform the average PMT curve point by point into the footing pressure vs. relative settlement curve:

$$\rho/B = 0.24 \, \Delta R/R_o \quad (10)$$
$$p_f = f_{L/B} \, f_e \, f_\delta \, f_{\beta,d} \, \Gamma \, p_p \quad (11)$$

where ρ is the settlement of the footing, B is the footing width, ΔR et R_o the increase in radius and the initial radius of the cavity in the PMT test respectively, p_f the footing pressure corresponding to ρ/B, $f_{L/B}$, f_e, f_δ, et $f_{\beta,d}$ the influence factors for shape, eccentricity, inclination, and proximity of a slope given by equations (3) to (9) above, Γ the function given in Figure 2 and which already includes the scale and depth of embedment effect, and p_p the pressure in the PMT test corresponding to $\Delta R/R_o$.

4. Prepare the load settlement curve for the footing once the p_f vs. ρ/B curve is known.

Figure 3 shows an example of the load settlement curve method.

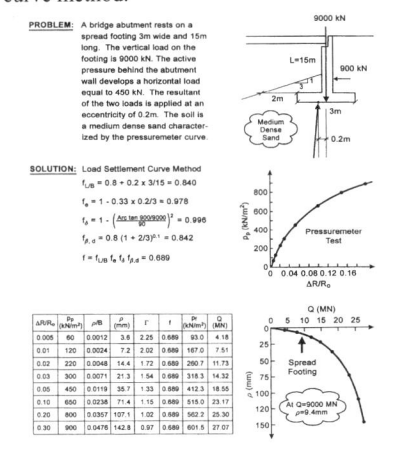

Figure 3: An Example of the Load-Settlement Curve Method

4 DEEP FOUNDATIONS UNDER HORIZONTAL LOADS: S.A.L.L.O.P.

The students and colleagues who worked on this project were Larry Tucker, Srini Donthireddy, and Marc Ballouz. The early part of the project was sponsored by the National Science Foundation. The problem is the one of a single pile subjected to a horizontal static load. This problem is often solved by assuming that the pile is an elastic member and that the soil can be represented by a series of non-linear horizontal springs called P-y curves. A method was developed at Texas A&M University to obtain the P-y curve directly from the pressuremeter curve. This method was simplified and lead to the Simple Analysis for Lateral Load On Piles (S.A.L.L.O.P.) (Briaud, 1997).

The following observation is the basis for the simplification. A conceptual plot of the soil resistance P per unit length of pile as a function of depth z is shown in Figure 4a. The sinusoidal nature of the P-z profile is such that the soil resistance alternates direction and essentially cancels itself out except for a shallow zone close to the ground surface which contributes most to the lateral resistance. More specifically there is a depth D_v where the shear force in the pile is zero (Figure 4b). The horizontal equilibrium of this shallow segment of pile is the basis of the SALLOP method.

The method consists of obtaining the lateral capacity H_{ou}, the horizontal movement at 1/3 of that load, and the maximum bending moment under $H_{ou}/3$. The lateral capacity H_{ou} is defined as the horizontal load corresponding to a horizontal movement at the pile head equal to B/10 where B is the pile diameter. The method was developed on a theoretical basis but was adjusted after studying a database of 20 full-scale pile load tests (Figure 5).

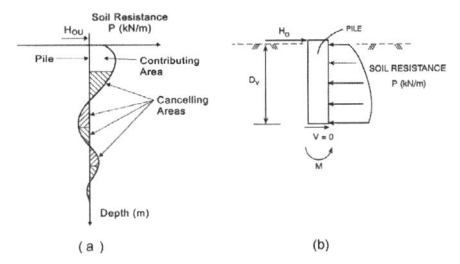

Figure 4: (a) Soil Resistance versus Depth, (b) Free Body Diagram of Shallow Segment

The steps of the method are as follows:
1. Perform preboring PMT tests within a depth corresponding to $2D_v$.
2. Reduce the data and obtain the profile of limit pressures p_l and the profile of first load modulus E_o. Select a design p_l value and a design E_o value from the profiles within the depth D_v. Use 1.5m if D_v is not known.
3. Calculate the zero shear depth D_v by using:

$$D_v = (\pi/4) \, l_o \quad \text{for } L > 3l_o \quad (12)$$
$$D_v = L/3 \quad \text{for } L < l_o \quad (13)$$
$$l_o = (4EI / K)^{1/4} \quad (14)$$

and L is the pile length, E the pile material modulus, I the moment of inertia of the pile, and K the horizontal soil modulus taken as $2.3E_o$ after studying the pile database. If the pile length is between l_o and $3l_o$ use linear interpolation.
4. Calculate the lateral capacity H_{ou} by using:

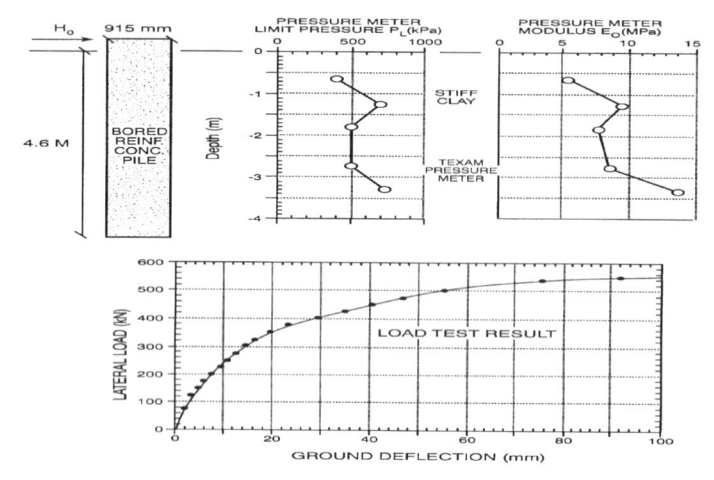

Figure 5: Example of Full Scale load Test in the Database

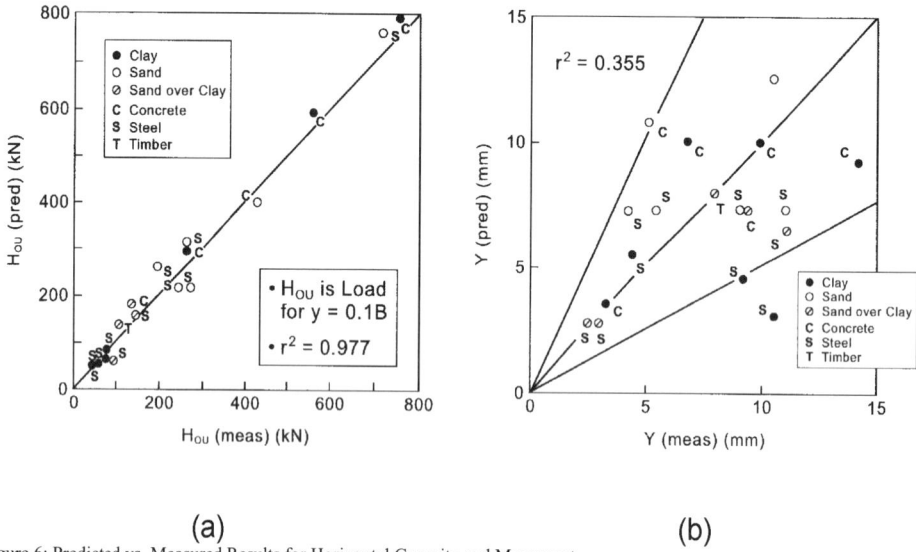

Figure 6: Predicted vs. Measured Results for Horizontal Capacity and Movement.

$$H_{ou} = \tfrac{3}{4}\, p_l\, B\, D_v \qquad (15)$$

5. Estimate the horizontal deflection yo under Ho (a safe fraction of H_{ou}) by

$$y_o = 2\, H_o / l_o\, K \quad \text{for } L > 3l_o \qquad (16)$$
$$y_o = 4\, H_o / L\, K \quad \text{for } L > 3l_o \qquad (17)$$

These equations are theoretically based and were used to find the best fitting value of K (K = 2.3 E_o).

The accuracy and precision of the SALLOP method can be evaluated on Figure 6. If a moment is also applied to the pile head, the reader is referred to Briaud (1997).

5 DEEP FOUNDATIONS: DOWNDRAG AND BITUMEN COATING

The students and colleagues who worked on this project were Larry Tucker, Randy Bush, Sangseom Jeong, Rajan Viswanathan, Mohamed Quraishi, and

Zaid Al Gurgia. The sponsor was the National Co-operative Highway Research Program. The project was aimed at developing a methodology to select bitumen coatings to reduce downdrag. It lead to publications outlining the procedure for uncoated piles and for bitumen-coated piles (Briaud, Tucker, 1997, Briaud, 1997), to a computer program called PILNEG (http://ceprofs.tamu.edu/briaud/pileneg.htm), and to a videotape on how best to coat the piles with bitumen.

The method to calculate the downdrag load and the allowable top load for an uncoated pile is outlined by working through an example by hand in this article. The computer program, the case of a pile group, and the bitumen selection process can be found in the publications mentioned. The example is the one of a single pile (Figure 7) driven in a soil deposit that will experience a settlement profile shown on the figure. The maximum shear stress that the soil can exert on the pile is taken as a constant equal to 25 kN/m^2 to simplify the calculations; it is assumed that the movement will be large enough to mobilize the full friction load in all cases. The point resistance is given by a load transfer curve as shown with a maximum point load of 1000 kN. The question is: how much load can be placed on top of the pile if the top settlement must be less than 14 mm; the problem is solved first for the uncoated pile and then for the coated pile.

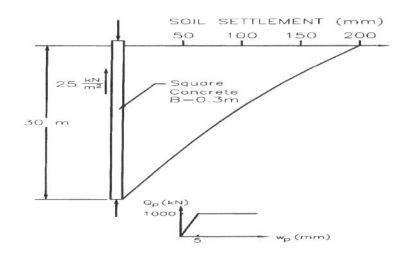

Figure 7: Example Problem for Downdrag Calculations

Uncoated Pile. The first step is to calculate the ultimate capacity of the pile in positive friction.
Q_u = (25 kN/m^2 x 1.2 m x 30 m) + 1000 kN = 1900 kN

Lets try a top load Q_t of 500 kN. The neutral point is found at a depth where the movement of the pile is equal to the movement of the soil. The calculations advance as a trial and error process.

If the neutral point is at a depth of 20 m, then according to the soil profile, the movement of the soil at that depth is w $_{NP(soil)}$ = 50 mm. The pile point carries a load Q_p of: Q_p = 500 kN + (20 m x 1.2 m x 25 kN/m^2) – (10 m x 1.2 m x 25 kN/m2) = 800 kN

For a point load of 800 kN, the point movement is given by the point load transfer curve as 4 mm.

Now it is possible to calculate the pile movement at the neutral point by adding the pile compression between a depth of 30m and a depth of 20m (depth of NP) to the 4 mm movement at the point.
w $_{NP(pile)}$ = 4 mm + (950 kN x 10^4 mm / 0.09 m^2 x 2x10^7 kN/m^2) = 9.3 mm

Since w $_{NP(soil)}$ ≠ w $_{NP(pile)}$ the initial guess of 20 m for the depth of the neutral point is incorrect and a new guess is required.

If the neutral point is at a depth of 29 m, then according to the soil profile, w $_{NP(soil)}$ = 5 mm and Q_p = 500 kN +870 kN – 30 kN = 1340 kN. This is not possible since the maximum point load is 1000 kN. The pile point will therefore reach the maximum load of 1000 kN and vertical equilibrium of the pile gives 500 kN + X = 1000 kN + (900 kN –X) or X= 700 kN

This downdrag value corresponds to 23.3 m of friction.

If the neutral point is at a depth of 23.3 m, then w $_{NP(soil)}$ = 35 mm and the movement at the top of the pile is: w$_{top}$ = 35 mm + (850 kN x 23300 mm / 0.09 m^2 x 2x10^7 kN/m^2) = 46 mm

This is more than the allowable movement of 14 mm. Therefore the top load must be reduced. Figure 8a gives the load distribution in the pile for a top load of 500 kN.

Lets try a top load Q_t of 100 kN, the same approach is taken. The final iteration is shown here.

If the neutral point is at a depth of 29 m, then according to the soil profile, w$_{NP(soil)}$ = 5 mm and Q_p = 100 kN +870 kN – 30 kN = 940 kN. The movement at the pile point is therefore w$_p$ = 4.7 mm

w $_{NP(pile)}$ = 4.7 mm + (955 kN x 103 mm / 0.09 m^2 x 2x107 kN/m^2) = 5.2 mm

In this case, w$_{NP(soil)}$ ≈ w$_{NP(pile)}$ and the neutral point position is indeed at a depth of 29 m. The top movement can now be calculated:
w$_{top}$ = 5 mm + (535 kN x 29000 mm / 0.09 m^2 x 2x107 kN/m^2) = 13.6 mm

This settlement is acceptable. The load distribution in the pile is shown on Figure 8(b). The distribution indicates that this pile which has a capacity of 1900 kN, can only be allowed to carry 100 kN (Figure 8b) because of the settlement criterion and of the downdrag. In addition this pile has a point resistance under working load which has a very low factor of safety against plunging. In this case, it becomes very advantageous to coat the pile with bitumen or other bond breakers as shown in the following.

Bitumen-Coated Pile. A bitumen coating which reduces the maximum shear stress that the soil can exert on the pile from 25 kN/m^2 to 2.5 kN/m^2 is selected. The coating however must only be applied to the part of the pile which will be subjected to downdrag. The neutral point is found in the same fashion as previously.

Lets try a top load Q_t of 500 kN.

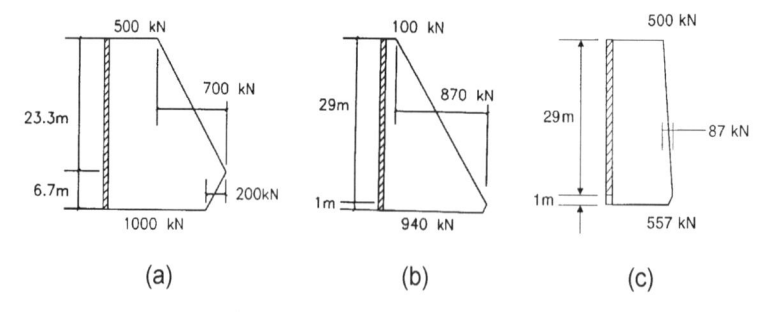

Figure 8: Load distribution in the pile for (a) the uncoated pile and a top load of 500 kN, (b) the uncoated pile and a top load of 100 kN, (c) the bitumen coated pile and a top load of 500 kN.

If the neutral point is at a depth of 29 m, then according to the soil profile, $w_{NP(soil)} = 5$ mm and $Q_p = 500$ kN $+ 87$ kN $- 30$ kN $= 557$ kN (Figure 8c). This corresponds to a point movement of 2.8 mm and a pile movement at the neutral point $w_{NP(pile)}$ close to $w_{NP(soil)}$

Therefore the neutral point is at a depth of 29 m. The settlement of the pile top is: $w_{top} = 5$ mm $+ (543.5$ kN $\times 29000$ mm $/ 0.09$ m$^2 \times 2 \times 10^7$ kN/m$^2) = 13.8$ mm

This is acceptable. The capacity of the pile is: $Q_u = (2.5$ kN/m$^2 \times 1.2$ m $\times 29$ m$) + (25$ kN/m$^2 \times 1.2$ m $\times 1$ m$) + 1000$ kN $= 1117$ kN

The factor of safety against plunging failure is therefore $1117 / 500 = 2.23$. Figure 8(c) shows the load distribution in the pile at working loads. By coating the pile with bitumen, the allowable load has been increased from 100 to 500 kN; coating the pile is estimated to increase the pile cost by 20 to 30 %.

6 NATIONAL GEOTECHNICAL EXPERIMENTATION SITES IN THE USA

There are 5 National Geotechnical Experimentation Sites (NGES) in the USA. They are located at Texas A&M University, on Treasure Island near the University of California at Berkeley, at the University of Massachussetts, at Northwestern University, and near Auburn University. The purpose of these sites is to investigate the behavior of geotechnical structures at full scale in a research environment; it was initiated in the early 1990s under the sponsorship of the Federal Highway Administration and the National Science Foundation. The soil at these sites is very well characterized and researchers throughout the USA and the world can come and conduct controlled research experiments. The results are organized in a database maintained by the Federal Highway Administration. The projects include shallow foundations, deep foundations, retaining walls, embankments, culverts as well as innovative in-situ testing and non-destructive detection methods. Information on these sites can be found at http://www.unh.edu/nges/

CONCLUSIONS

The summary of a number of projects has been presented in this article. For shallow foundations in sand, it is found that there is no effect of the depth of embedment and no effect of scale if the load settlement curve is plotted as a pressure over soil strength versus settlement over foundation width. For shallow foundations also, a method is proposed to obtain the complete load settlement curve for the footing using only the pressuremeter curve. For horizontally loaded piles, a hand calculation method using the pressuremeter limit pressure and modulus is proposed and evaluated. For downdrag on piles, a method is proposed to obtain the settlement of the pile and the downdrag load based on the PMT data collected near the pile tip. The National Geotechnical Experimentation Site at Texas A&M University is used for evaluation of in-situ testing in the USA.

REFERENCES:

Barfknecht, J., Briaud J-L., (1999), "Effect of Scale and Depth of Embedment for Footings in Sand," Research Report, Texas A&M University, Dpt. of Civil Engineering, College Station, Texas, USA.(can be obtained from: briaud@tamu.edu)

Briaud J-L., (1992), The Pressuremeter, A.A. Balkema, Rotterdam, The Netherlands. (http://balkema.ima.nl/Scripts/ cgi-Balkema.exe/author?AutNo=1367)

Briaud .-L., Gibbens R.M., (1999), "Behavior of Five Large Spread Footings in Sand," Journal of Geotechnical and Geoenvironmental Engineering, Sept. 1999, Vol. 125, No. 9, American Society of Civil Engineers, Reston, Virginia, USA.

Briaud, J-L., (1997) "Bitumen Selection for Downdrag on Piles," Journal of Geotechnical and Geoenvironmental Engineering, Vol. 123, No. 12, ASCE, New York,December 1997.

Briaud, J-L., (1997) "SALLOP: Simple Approach for Lateral Loads on Piles," Journal of Geotechnical and Geoenvironmental Engineering, Vol. 123, No. 10, pp. 958-964, ASCE, New York, October 1997.

Briaud, J-L., and Gibbens, R.M. (1997), "Large Scale Load Tests and Data Base of Spread Footings on Sand," Federal Highway Administration, Publication No. FHWA-RD-97-068, McLean, Virginia, USA, pp. 217.

Briaud, J-L., Gibbens, R.M., Editors (1994), "Predicted and Measured Behavior of Five Spread Footings on Sand," Geotechnical Special Publication No. 41, American Society of Civil Engineers, Reston, Virginia, USA, pp. 255.

Briaud, J-L., Jeanjean, Ph., (1994), "Load Settlement Curve Method for Spread Footings on Sand", Settlement '94 Specialty Conference, ASCE Specialty Publication No. 40, ASCE, Vol. 2, 1774-1804.

Briaud, J-L., Tucker, L.M., (1997) "Design and Construction Guidelines for Downdrag on Uncoated and Bitumen-Coated Piles", NCHRP Report 393, TransportationResearch Board, National Academy Press, Washigton DC, 1997.

Briaud, J-L., (2002) "Foundation Engineering: Some recent Contributions", Geotechnical Engineering, Vol. 18, No. 9, pp. 16-30, Korean Geotechnical Engineering Society, Seoul, Korea.

Hossain, K.M., (1996), "Load Settlement Curve Method for Footings in Sand at Various Depths, under Eccentric or Inclined Loads, and Near Slopes," Ph.D. Dissertation, Texas A&M University, Dpt. of Civil Engineering, College Station, Texas, USA.

Ismael, N.F. (1985), "Allowable Pressure from Loading Tests on Kuwaiti Soils," Canadian Geotechnical Journal, No. 22, Canadian Geotechnical Society, Ottawa, Canada, pp. 151-157.

Jeanjean, P., (1995), "Load Settlement Curve Method for Spread Footings on Sand From the Pressuremeter Test", Ph.D. Dissertation, Texas A&M University, Dpt. of Civil Engineering, College Station, Texas, USA.

Khebib, Y., Canepa, Y., and Magnan, M.-P. (1997), "Base de Donnees de Fondations Superficielles SHALDB: Essais des Laboratoires des Ponts et Chaussées," June 1997, Laboratoire Central des Ponts et Chaussées, Paris, France.

Lutenegger, A. (1995), personal communication by fax dated 9-27-95.

Pu, J-L., Ko, H-Y. (1988), "Experimental Determination of Bearing Capacity in Sand by Centrifuge Footing Tests," Proceedings of Centrifuge '88, J.-P. Corte Ed., A.A. Balkema Publishers, Rotterdam, Netherlands, p. 293-299.

Disturbance effects of field vane tests in a varved clay

A.B. Cerato & A.J. Lutenegger
University of Massachusetts, Amherst, MA, USA

Keywords: *in situ*, field vane, soil disturbance, varved clay, sensitivity

ABSTRACT: A field investigation was performed to evaluate the influence of disturbance on the measured undrained strength of a varved clay deposit using the Field Vane Test at the National Geotechnical Experimentation Site at the University of Massachusetts, Amherst. Four different field vanes of the same height and diameter but having different blade thickness were used to conduct FVT profiles over a depth of 3.6 m to 20.1 m. The geometry of the vanes used produced perimeter ratios ranging from 3.1% to 12.4%. The results show that the measured peak strengths are inversely proportional to the perimeter ratio of the vane. Extrapolation of the data to a vane of "zero" thickness may give a better indication of the in situ undisturbed strength for the use in design. In addition to giving lower peak strengths, the results indicate that thicker vanes do not give sufficient definition of the post-peak strength, a result of this behavior being "masked" by the soil disturbance. As the remolded strengths given by the vanes were essentially the same, the thicker vanes also give lower values of sensitivity. The results illustrate how field vane test results may easily be misinterpreted.

1 INTRODUCTION

The Field Vane Test (FVT) is still considered by many engineers as the preferred field method for providing reference values of undrained shear strength of soft and medium stiff clays. In addition to providing a measure of the peak undrained strength, the vane can also give an indication of the post-peak strength and the remolded (destructured) strength, and therefore soil sensitivity, s_t. It is the only in situ test with this capability at this time. Like nearly all in situ tests, however, the FVT is subject to disturbance effects resulting from the insertion of the vane into the ground. The study presented herein investigated the effects of vane thickness and resulting disturbance on undrained shear strength of Connecticut Valley Varved Clay (CVVC).

2 BACKGROUND

There are a number of factors that can affect the results obtained with the FVT and can therefore affect the interpretation of the undrained shear strength value obtained. These factors include both variations in the equipment used and variations in the test procedure. A number of the most important factors affecting the test results are given in Table 1.

Disturbance produced by inserting a vane should be related the geometry of the vane for the same soil. It is logical that the more soil that must be displaced to allow the vane to be inserted, the more the amount of disturbed soil. One might think that disturbance caused by inserting the vane might be related to the volume of the vane, however, Cadling and Odenstad (1950) suggested that the amount of disturbance could be described using the vane "perimeter ratio," defined as:

$$\alpha = \frac{4e}{\pi D} \quad (1)$$

where:

α = perimeter ratio (usually expressed as %)
e = thickness of vane blade (mm)
D = diameter of vane (mm)

Table 1 Factors that May Affect the Results of the FVT.

Factor	Reference
Vane Geometry	Cadling and Odenstad (1950)
	Osterberg (1956)
	Bazett et al. (1961)
Soil Fabric	Lo (1965)
	Lo and Milligan (1967)
	Loh and Holt (1974)
	Andrawes et al. (1975)
Rate of Shearing	Skempton (1948)
	Cadling and Odenstad (1950)
	Bazett et al. (1961)
	Aas (1965)
	Wiesel (1973)
	Perlow & Richards (1977)
Stress Distribution	Flaate (1966a)
	Donald et al. (1977)
	Menzies & Merrifield (1980)
	Wroth (1984)
Time Effects	Aas (1965)
	Torstensson (1977)

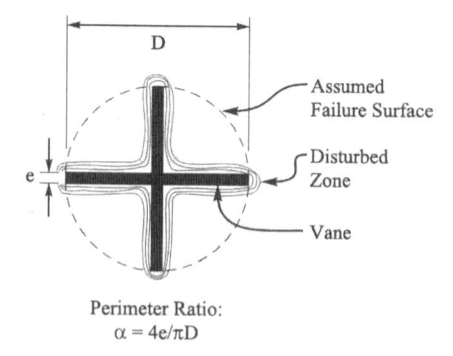

Figure 1 Disturbance caused by intrusion of the vane into clay.

Perimeter Ratio:
$\alpha = 4e/\pi D$

The parameter, α, is illustrated in Figure 1. If the zone of undisturbed soil adjacent to the vane blades is related to the blade thickness, then for the same diameter vane, a larger amount of "undisturbed" soil for testing will result from thinner blades. Thick blades on the vane will produce more disturbance for a vane with a constant diameter. Similarly, vanes with the same blade thickness, but different diameter should show less disturbance effects as the diameter increases. The perimeter ratio describes, in a rough sense, the amount of loss in strength that occurs along the cylindrical shearing surface produced by the vane rotation.

The undrained strength measurement should be more representative of "undisturbed" conditions. Equation 1 suggests that in order to reduce disturbance effects, either blade thickness must be reduced, or vane diameter must be increased, or both. There is a practical problem with machining vanes with blades that are too thin since they will have reduced strength and may bend or break during testing. Typical commercial vanes have perimeter ratios in the range of 4 to 8%.

LaRochelle et al. (1973) illustrated the importance of disturbance on the resulting undrained strength profiles obtained in sensitive Champlain Sea Clay ($s_t = 12$) by using vanes of the same diameter but different blade thickness. All other test procedures were held constant and the NGI vane device was used for all the tests. The resulting strength profiles are shown in Figure 2. These results show that for most of the tests in this soil, the measured undrained shear strength increases as the vane blade thickness decreases, i.e., as α decreases, which is expected.

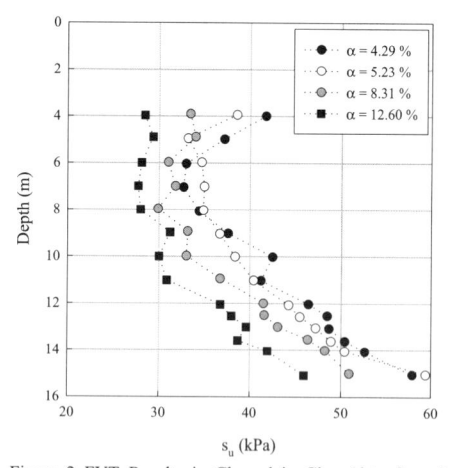

Figure 2 FVT Results in Champlain Clay (data from LaRochelle et al. 1973).

The results presented in Figure 2 were used by LaRochelle et al. (1973) to back extrapolate the measured undrained strength to a blade thickness of zero, providing a perimeter ratio of zero and therefore a condition of "zero disturbance." The results of this back extrapolation procedure for tests performed on Champlain Sea Clay indicated an approximate linear relationship between perimeter ratio and undrained shear strength for $\alpha = 4.3$ to 12.6 %. An increase in the extrapolated undisturbed strength of about 15% is suggested over the value obtained from the "standard" ($\alpha = 4.3\%$) vane. Similar results have been presented by Roy and Leblanc (1988) who used five different rectangular vanes that were produced with blades of different thicknesses and shaped in such a manner that their horizontal cross-sectional area, or volume, remained the same as the 130 mm by 65 mm standard vane manufactured by Nilcon. They showed that for sensitive marine clays at Saint-Alban and Saint-Louis in Canada, the linear extrapolation gave undrained strength values approximately 6.5 and 9 % higher than the standard vane ($\alpha = 4.3$ %) strength, which was less than the disturbance shown by LaRochelle (1973). These results show that disturbance may be reduced by using tapered blades.

Data from these previous studies, in which a perimeter ratio of about 5% was used as a reference, suggest that extrapolation to zero vane thickness can result in an increase in estimated undisturbed shear strength on the order of 6 to 15%; at least for sensitive clays. Some soils, such as clays with low sensitivity, may be relatively insensitive to blade thickness.

3 INVESTIGATION

In this study, the undisturbed strength was evaluated by conducting four parallel vane profiles with vanes of the same diameter but different blade thicknesses. This provided an estimate of the undisturbed strength of the Connecticut Valley Varved Clay (CVVC).

3.1 Site Geology

Tests were performed at the National Geotechnical Experimentation Site at the University of Massachusetts, Amherst. The site is situated within the Connecticut River Valley and within the extent of the former glacial Lake Hitchcock. The lacustrine sediment deposits were formed as a result of an ice-wall dam, which formed across the Valley in northern Connecticut creating seasonal deposition and settling of fine-grained particles over the coarser glacial till for a period of approximately 4000 years. This soil is locally known as Connecticut Valley Varved Clay (CVVC) and extends from Northern Vermont to Central Connecticut in the present Connecticut River Valley.

3.2 Field vane tests

The FVT measures the maximum torque applied to the vane through a rod system. This torque is then converted into the undrained shear strength assuming the distribution of the shear strength is uniform across the ends of the cylinder and around the perimeter (ASTM D2573). Four field vane profiles were conducted using vanes having different perimeter ratios ranging from 3.1 % to 12.5 %. The vanes were rectangular, four-bladed vanes with heights of 13.0 cm and diameters of 6.5 cm, providing height to diameter ratios of 2. The only difference then was the blade thickness, e, (1.58 mm, 3.18 mm, 4.76 mm and 6.35 mm). The blade thickness was varied in order to provide different values of α. The diameter of the vanes was held constant to give the same shear velocity. Table 2 gives the vane dimensions.

Table 2 Dimensions of Vanes.

Test #	Diameter (mm)	Height (mm)	Blade Thickness (mm)	Rod Diameter (mm)	Perimeter Ratio (%)
FVT-4	65	130	1.58	20	3.1
FVT-6	65	130	3.18	20	6.2
FVT-7	65	130	4.76	20	9.3
FVT-8	65	130	6.35	20	12.4

The Nilcon Vane Borer Model M-1000, which uses unprotected rods and an unprotected vane with a rod slip coupling, was used in the testing program (Figure 3). The rods had a diameter of 20 mm. The small diameter of the rods was to reduce the amount of rod friction, although the diameter of the rods had to be sufficient enough so that the elastic limit was not exceeded when the vane was stressed to its capacity.

Predrilled 76 mm diameter, 3.0 m deep holes were prepared prior to testing at each location to avoid damage to the vane during advancement through the stiff surficial crust and fill. The field vane, slip coupling and starting rod were placed through the center of the torque recording head. The slip coupling was located about 60 mm above the vane to permit 15° slip between the vane and the rods before the vane engaged. The torque head contained a self-recording device that utilized a steel pointer to scribe the corresponding torque on a waxed paper disc.

The vane was advanced in 0.6 m increments with testing starting at approximately 3.05 to 3.66 m below grade. Once the vane was in position, torque was applied to the vane within about 1 minute by a hand operated gear drive through 20 mm diameter torque rods extending to the surface. The applied rate of torque to the vane was approximately 0.1°/sec resulting in an average time to failure of 7 to 9 minutes.

Following failure of the soil at its peak strength, the test was continued at the same rate of torque to evaluate the post peak strength of the soil. Once this post peak strength was measured, the torque was removed and the rods manually rotated 10 complete revolutions to remold the soil in a cylindrical failure zone around the circumference of the vane. The remolded shear strength of the soil was

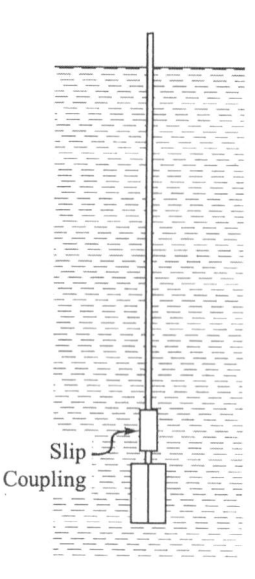

Figure 3 Unprotected Vane and Rods.

then determined. This measurement allowed the determination of the soil sensitivity, s_t. A sample recording of the measured torque showing the rod friction, peak strength, post peak strength and the remolded strength scribed onto the paper disc is shown in Figure 4.

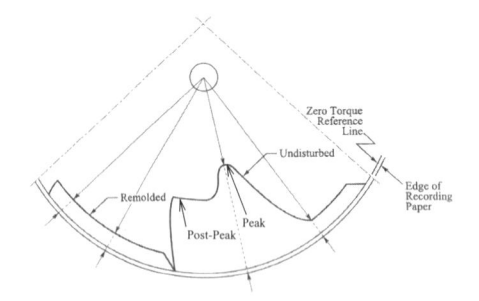

Figure 4 Sample Nilcon Field Vane Torque Trace.

4 RESULTS

The results from the FVT profiles are presented in Figure 5. These results show that undrained shear strength decreases as vane thickness increases. The extrapolated strength values shown in Figure 5 denote the "zero disturbance" shear strength values found by extrapolating the results of the vane tests back to a vane thickness of 0, or a perimeter ratio of 0. This extrapolation is illustrated in Figure 6, which shows results from three different depths in the profile. These results indicate an approximate linear relationship between perimeter ratio and undrained shear strength as has been previously noted. An increase in the estimated undisturbed strength of about 16 % is suggested over the value obtained from the "standard" vane, (α = 3.1%), which is slightly higher than other researchers have suggested, most likely because the reference vane used in this study had a slightly smaller perimeter ratio than previous studies. An additional effect of using a vane with thicker blades is that in a soft clay, higher pore water pressures will be generated during insertion. This leads to a reduction in measured strength. This effect can be quantified by waiting a sufficient time for pore water pressures to dissipate prior to shearing. Figure 7 shows expected response and extrapolated strength difference between immediate and delayed tests.

Post peak strength values are also lower with thicker vane blades and were difficult to interpret. The disturbance produced by the thicker blades tends to "mask" the post peak behavior. Since the remolded strength represents a completely "destructured" or fully disturbed condition, it should be expected that these values would be independent of vane blade thickness. That is, completely remolding the soil produces a minimum strength and therefore it should not make a difference on how the soil arrives at that condition on the resulting strength. The results shown in Figure 5 support this and do not show any consistent trend. The consequence of disturbance is really three-fold; 1.) peak values of undrained strength are lower as disturbance increases; 2.) post-peak strength is lower and poorly defined and 3.) soil sensitivity is lower.

The Undrained Brittleness Index, first suggested by Bishop (1971) may be used to define the reduction of undrained strength from peak to the residual (large strain) strength as:

$$BI = \frac{S_{up} - S_{upp}}{S_{up}} \qquad (4)$$

where:

s_{up} = Peak Undrained Shear Strength (kPa)
s_{upp} = Post Peak Undrained Shear Strength (kPa)

and may be related to the loss of soil structure. The sensitivity, s_t, of the soil was calculated using the peak strength and the remolded strength and is defined as:

$$s_t = s_{up}/s_{ur} \qquad (5)$$

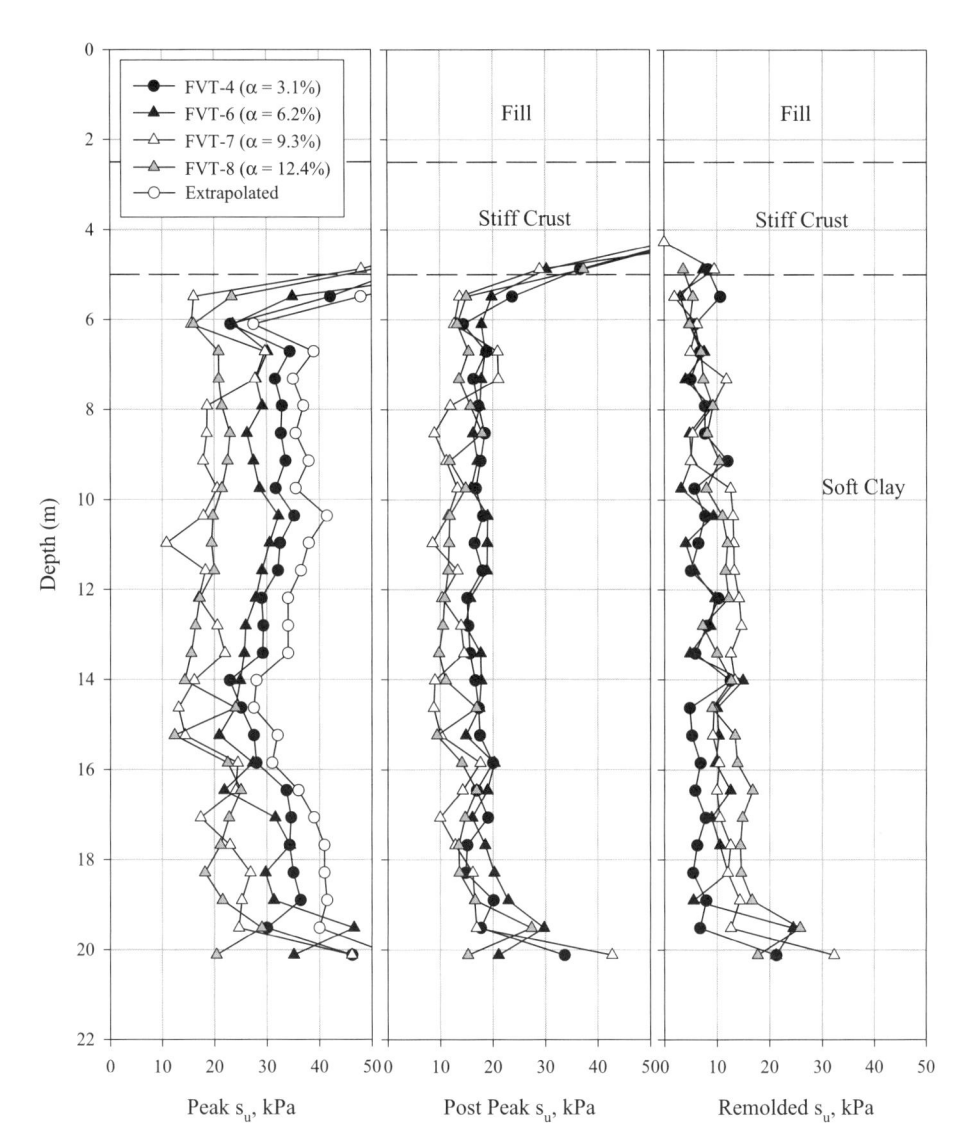

Figure 5 Results of FVT Tests in CVVC.

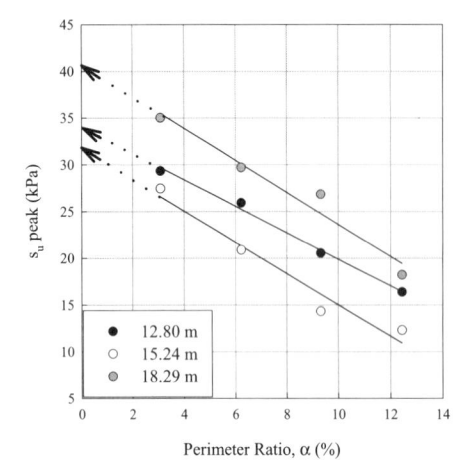

Figure 6 "Zero Disturbance" Extrapolation.

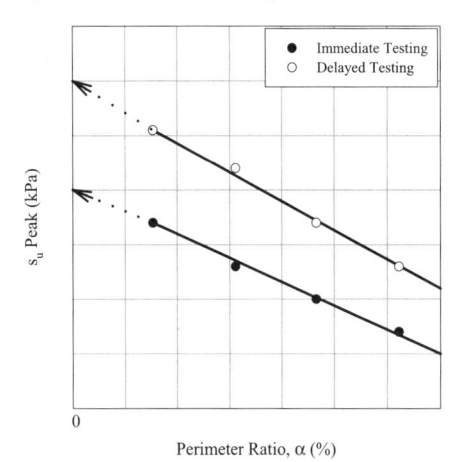

Figure 7 "Zero Disturbance" Extrapolation.

Figure 8 presents the results of the Brittleness Index (BI) and the sensitivity (s_t). The BI of this soil ranges from approximately 0.25 to 0.5 and the sensitivity ranges from approximately 1 to 10.

5 CONCLUSIONS

Measured Field Vane undrained shear strengths are directly related to the thickness of vane blades used within perimeter ratios ranging from 3 to 12 %. The results show that the measured peak strengths are inversely proportional to the perimeter ratio of the vane. In addition to giving lower peak strengths, the results indicate that thicker vanes do not give sufficient definition of the post-peak strength, a result of

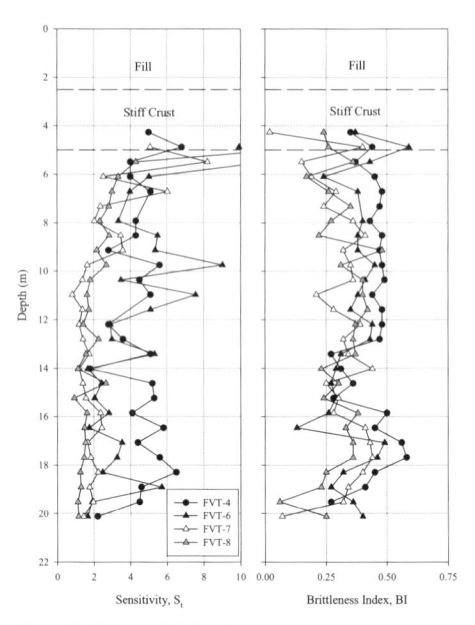

Figure 8 FVT s_t and BI Results.

this behavior being "masked" by the soil disturbance. As the remolded strengths given by the vanes were essentially the same, the thicker vanes also give lower values of sensitivity.

In a site investigation, it is important to know the exact type of equipment used in the testing program, since all vanes will not give the same strength values in a given soil. Therefore, the results of the testing must be reviewed with caution. For large projects, it would be helpful to perform two profiles with vanes of different geometries in order to understand disturbance effects in the particular soil and to allow an accurate prediction of shear strength for use in design.

Engineers may wish to consider the increase in undrained shear strength obtained by extrapolation to zero perimeter ratio over the value that is measured in the field. Extrapolation for a vane of "zero" thickness may give a better indication of the in situ undisturbed strength for the use in design. For some soils this should be accounted for in design as a real component of undrained shear strength, which is available but undetected in normal testing procedures. Ignoring this available strength is conservative in design.

REFERENCES

Aas, G. 1965. A Study of the Effect of Vane Shape and Rate of Strain in the Measured Values of In Situ Shear Strength of Clays. Proceedings of the 6th International Conference on Soil Mechanics and Foundation Engineering, Vol. 1, pp. 141-145.

Andrawes, K.Z. Krishnamurthy, D.N. and Barden, L. 1975. Anisotropy of Strength in Clays Containing Plates of Increasing Size. Proceedings of the 4th Southeast Asian Conference on Soil Engineering, Kuala Lumpur, Malaysia, Vol. 1, pp. 6-12.

Bazzet, D.J., Adams, J.I. and Matyas, E.L. 1961. An Investigation of Studies in a Test Trench Excavated in Fissured Sensitive Marine Clay. Proceedings of the 5th International Conference on Soil Mechanics and Foundation Engineering, Paris, Vol. 1, pp. 431-436.

Cadling, L. and Odenstad, S. 1950. The Vane Borer. Proceedings of the Royal Swedish Geotechnical Institute, No. 2, pp. 1-87.

Donald, I.B., Jordan, D.O., Parker, R.J. and Toh, C.T. 1977. The Vane Test – A Critical Appraisal. Proceedings of the 9th International Conference on Soil Mechanics and Foundation Engineering, Vol. 1, pp. 81-88.

Flaate, K. 1966. Factors Influencing the Results of Vane Tests. Canadian Geotechnical Journal, Vol. 3, No. 1, pp. 18-31.

LaRochelle, P., Roy, M. and Tavenas, F. 1973. Field Measurements of Cohesion in Champlain Clays. Proceedings of the 8th International Conference on Soil Mechanics and Foundation Engineering, Vol. 1.1, pp. 229-236.

Lo, K.Y. 1965. Stability of Slopes in Anisotropic Soils. Journal of the Soil Mechanics and Foundation Division, ASCE, Vol. 91, SM1, pp. 85-106.

Lo, K.Y. and Milligan, B. 1967. Shear Strength Properties of Two Stratified Clays. Journal of the Soil Mechanics and Foundation Division, ASCE, Vol. 93, SM1, pp. 1-15.

Loh, A.K. and Holt, R.T. 1974. Directional Variation and Fabric of Winnipeg Upper Brown Clay. Canadian Geotechnical Journal. Vol. 11, No. 3, pp. 430-437.

Menzies, B.K. and Merrifield, C.M. 1980. Measurements of Shear Stress Distribution on the Edges of a Shear Blade. Geotechnique, Vol. 22, No. 3, pp. 451-457.

Osterberg, J.O. 1956. Introduction to Vane Testing of Soil. ASTM Special Technical Publication, 193. pp. 1-7.

Perlow, M. and Richards, A.F. 1977. Influences of Shear Velocity on Vane Shear Strength. Journal of the Geotechnical Engineering Division, ASCE, Vol. 103, No. GT1, pp. 19-32.

Roy, M. and Leblanc, A. 1988. Factors Affecting the Measurement and Interpretation of the vane Strength in Soft Sensitive Clays. Vane Strength Testing in Soils: Field and Laboratory Studies, ASTM STP 1014, pp. 117-128.

Skempton, A.W. 1948. Vane Test in the Alluvial Plains of River Froth Near Grange Mouth. Geotechnique, Vol. 1, No. 2, pp. 111-124.

Torstensson, B.A. 1977. Time-Dependent Effects in the Field Vane Test. Proceedings of the International Symposium on Soft Clays, Bangkok, pp. 387-397.

Wiesel, C.E. 1973. Some Factors Influencing In Situ Vane Test Results. Proceedings of the 8th International Conference on Soil Mechanics and Foundation Engineering, Vol. 1.2, pp. 475-479.

Wroth, C.P. 1984. Interpretation of In Situ Tests. Geotechnique, Vol. 34, No. 4, pp. 449-489.

Assessment of cyclic stability of cohesive deposits using cone penetration

Roy Debasis
Indian Institute of Technology, Kharagpur 721302, WB, India

Keywords: cyclic stress ratio, cohesive soils, earthquake, cyclic stability

ABSTRACT: Unlike cohesionless soils, cohesive deposits are not susceptible to near-complete loss of shear strength under cyclic loading. Nevertheless, cohesive deposits are known to soften during earthquake-related cyclic loading that results in severe ground deformation and structural distress. Available frameworks for assessing cyclic stability of cohesive soils do not account for the intensity of cyclic shear stress. These procedures are based on very approximate indicators of cyclic soil strength. As a result, their performance is not always satisfactory. A framework has been proposed in this paper for assessing cyclic stability of cohesive soils utilizing a worldwide database of laboratory tests, earthquake-related ground failure case-histories and cone penetration tests in cohesive, fine-grained silts and clays. The proposed framework indicates that a Critical Stress Ratio (CSR) of less than 0.125 is unlikely to trigger large deformations in cohesive deposits even during great earthquakes. Nor are deposits characterized with a normalized total cone tip resistance of more than 70 susceptible to deform appreciably during such seismic events.

1 INTRODUCTION

Cohesive deposits are usually not susceptible to an almost complete loss of shear strength when subjected to cyclic simple shear simulating a strong earthquake. However, such deposits may soften during a cyclic undrained loading similar to an earthquake leading to the development of severe permanent ground deformation. Several case histories (Seed *et al.* 2003, Boulanger *et al.* 1999, Hyodo *et al.* 1994, Sasaki *et al.* 1980) can be found in the literature, which describe development of such distress in cohesive deposits during earthquakes.

Empirical guidelines are available for assessing cyclic stability of cohesive deposits. However, a number of published case histories (Seed *et al.* 2003, Boulanger *et al.* 1999, Hyodo *et al.* 1994, Sasaki *et al.* 1980) raise questions about the efficacy of these guidelines.

The problem basically arises because the indices upon which these guidelines are based are at best approximate measures of cyclic soil strength. Secondly, the assumption implicit in these guidelines that cyclic strength of cohesive soils can be assessed irrespective of the intensity and duration of the cyclic load is not in agreement with soil behavior observed in cyclic laboratory tests. Thus, it may be possible to assess the cyclic stability of cohesive soils using available guidelines if the Magnitude and ground acceleration level of the earthquake considered are similar to those based on which the guidelines were originally developed. However, use of these guidelines may not be appropriate for assessing the cyclic stability of the soils in an earthquake of different Magnitude and ground acceleration level.

Therefore there is a need for a framework for estimating cyclic strength of cohesive soils that reflect field performance and laboratory cyclic behavior. The framework should account for soil strength as well as the intensity and duration of cyclic load reasonably.

An attempt has been made in this study to develop such a framework utilizing a worldwide database of laboratory tests, earthquake-related ground failure case-histories and cone penetration tests in cohesive, normally- to lightly over-consolidated silts and clays obtained from the literature.

2 CYCLIC STABILITY OF COHESIVE SOILS

Saturated, dense, cohesionless soils and stiff, cohesive soils tend to dilate in drained simple shear. Loose or soft soils, on the other hand, tend to contract when sheared under drained conditions. During an earthquake a soil deposit is subjected to cycles of simple shear that are usually so rapid that an actual volume change cannot occur. Instead, the pore water pressure tends to decrease in dilative soils or increase in contractive soils. In loose cohesionless soils, the pore water pressure could increase to 100 % of the initial total stress causing a remarkable reduction in the shear strength. Such a reduction of shear strength could be so significant that the soil may start to behave like a viscous liquid. Typical stress-strain response of loose sand in a cyclic undrained direct simple shear test (DSS) with stress reversal is presented in Figure 1a and the corresponding pore water pressure development with the progress of cyclic simple shear is presented in Figure 1b.

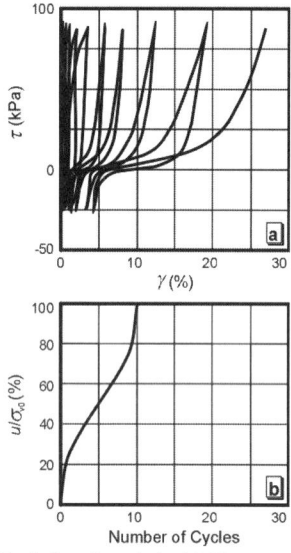

Figure 1. Typical cyclic undrained DSS response of loose sand (τ = shear stress, γ = shear strain, u = pore water pressure, and σ_{v0} = initial total vertical stress)

Like loose cohesionless soils, earthquake-related increase in pore water pressure in soft cohesive deposits causes a reduction in the frictional shear strength. However, the increase of pore water pressure within soft cohesive deposits during an earthquake is usually much smaller than that in a typical loose cohesionless deposit. Consequently, the shear strength reduction for soft cohesive deposits is usually less remarkable than loose cohesive soils. Nevertheless, soft cohesive soils do soften during earthquakes. As a result, severe deformation may develop within these deposits during an earthquake causing failure of structures and other facilities supported on them. Typical undrained DSS stress-strain response of soft cohesive soils in a cyclic test with stress reversal is presented in Figure 2a and the corresponding pore water pressure development is presented in Figure 2b.

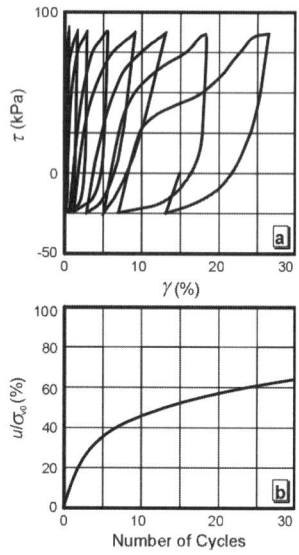

Figure 2. Typical cyclic undrained DSS response of soft clay

A state of zero effective stress is usually inferred when soil volcanoes are observed during or following an earthquake that eject liquefied soils. Development of a soil volcano requires a substantial increase of pore water pressure followed by its rapid dissipation. As discussed earlier, earthquake-related excess pore water pressure development in soft cohesive deposits is not as remarkable and its dissipation is not as rapid as those in loose cohesionless deposits. As a result, a soil volcano is rarely observed at sites underlain by soft cohesive deposits even when cyclic instability can be inferred from earthquake-related infrastructural damage. Therefore, it is difficult to assess whether cyclic instability has developed within a soft cohesive deposit based on post-earthquake field performance survey in undeveloped areas.

2.1 Factors Affecting Cyclic Stability of Soft Cohesive Soils

For a normally to lightly over consolidated, soft to firm, cohesive soil sample undergoing cyclic simple shear, deformation increases with the amplitude of cyclic shear stress and the number of shear stress cycles. Further, the amplitude of cyclic shear stress that results in an unacceptably large deformation within a certain number of cycles of shear stress application tends to increase with the static undrained strength of the soil and the effective geostatic stress. Thus Cyclic Resistance Ratio (CRR), defined as the ratio of single amplitude cyclic shear stress that results in a double amplitude shear strain of 5 % to the vertical effective stress at end of consolidation, can be used as an index of cyclic shear strength for a given amplitude and number of cycles of shear stress application.

Among other factors that affect cyclic strength of normally to lightly over consolidated, soft to firm, cohesive soils is cross anisotropy and soil fabric. Because of cross-anisotropy, soft clays usually exhibit about 40 % larger undrained cyclic strength in a triaxial (TX) test compared to a DSS test. Cyclic strength of soft cohesive soils also approximately depends on the liquidity and plasticity indices and sensitivity with soils of smaller liquidity index, greater plasticity index and less sensitivity tending to exhibit greater cyclic undrained strength. Further, it is a common practice to subject soil sampled to unidirectional cyclic simple shear in the laboratory. The cyclic strength of soils is approximately 10 % lower in multidirectional cyclic shear stress application typical of an earthquake ground motion in comparison with that in unidirectional cyclic simple shear.

3 DEFINITION

Cyclic instability is defined here as a phenomenon, in which a "large" deformation develops within a deposit of soil during a cyclic undrained shear loading. In this study, a double amplitude shear strain of 5 % is considered "large."

In contrast, the classical definition of liquefaction describes it as a state of zero effective stress that develops within soils during static or cyclic undrained shear loading due to pore water pressure rise.

4 REVIEW OF AVAILABLE GUIDELINES

Several guidelines are available for assessing susceptibility of cohesive deposits to liquefaction. These procedures and their field performance are briefly described in this section.

4.1 *The Chinese Criteria*

The most commonly used framework for assessing cyclic stability and liquefaction potential of cohesive soils, known as the Chinese Criteria (Seed and Idriss, 1982), originates from observed performance of clayey soils during earthquakes of moderate magnitude. These criteria use laboratory-based index measurements to infer whether or not a soil deposit is liquefiable. From a re-examination of the database upon which the Chinese Criteria are based, Andrews and Martin (2000) proposed a modified set of guidelines to assess liquefaction susceptibility of cohesive soils. Details of the Chinese Criteria and the modified guidelines proposed by Andrews and Martin are presented in Figure 3.

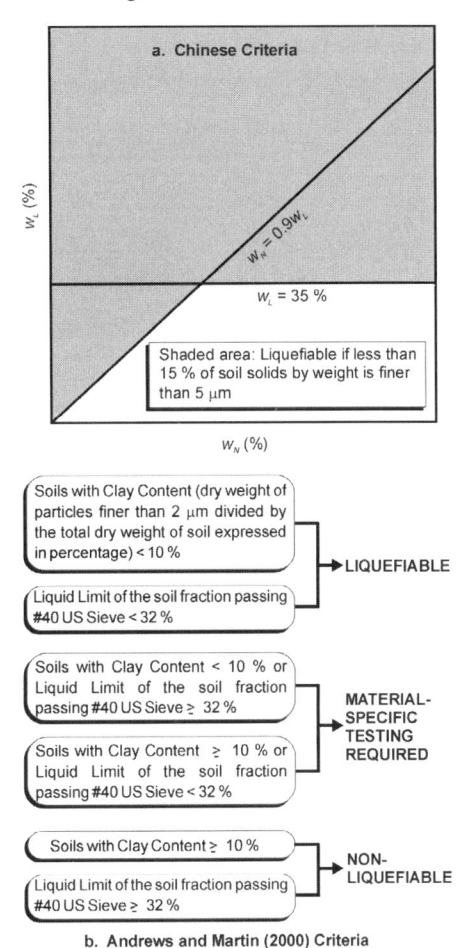

Figure 3. Chinese criteria and modified guidelines due to Andrews and Martin (2000)

4.2 *Other Guidelines*

Robertson and Wride (1998) consider soils with a Plasticity Index (I_P) of more than about 14 % to be too cohesive to liquefy. From laboratory testing of sands containing plastic fines, Polito (1999) concluded that clayey sands are likely to be susceptible to flow failure if I_P . 10 % and w_L . 25 %, where w_L is the Liquid Limit. To cover for the uncertainties in assessing the cyclic stability of clayey deposits, Robertson and Wride (1998) as well as Polito (1999) recommend that a laboratory testing program be undertaken for critical facilities located in areas underlain by extensive clayey deposits rather than relying on their screening procedures. This is an expensive proposition, especially in soft cohesive soil sites.

4.3 Field Performance of Available Guidelines

Laboratory-based research indicates that cohesive soils soften and deform under strong cyclic shaking regardless of their plasticity (Andersen, 1988, Boulanger et al. 1998, Dia Consultants 1998, Hyodo et al. 1994). These results, for instance, indicate that for six uniform sinusoidal cycles (considered equivalent to a Magnitude 6 earthquake) of reversible shear stress application, cohesive fine grained soils may sustain shear stress amplitude of about 0.20 to 0.25 times the initial vertical effective stress without large deformation. The corresponding cyclic shear stress amplitude for 26 cycles (considered equivalent to a Magnitude 6 earthquake) is 0.15 to 0.20 times the initial vertical effective stress. Since such ranges of amplitudes of cyclic shear stress is quite common especially for near-field earthquakes, large deformation is likely to develop at sites underlain by cohesive soils even in case of small earthquakes. The deformation is expected to be even more significant for great earthquakes of Magnitude 8 or greater.

Several field performance case histories (Seed et al. 2003, Boulanger et al. 1999, Hyodo et al. 1994, Sasaki et al. 1980) can be found in the literature, which describe development of liquefaction-like distress within cohesive deposits during earthquakes.

For instance, Boulanger et al. (1999) present a case history in which cohesive deposits at a site in California appeared to have liquefied during the 1989 Loma Prieta Earthquake. Boulanger et al. reached such a conclusion after finding ejecta of cohesive soils within soil volcanoes at several locations within this site. Also of interest is the fact that the ejecta are from cohesive layers deemed non-liquefiable according to the Chinese Criteria. Dia Consultants (1998) describe earthquake-related failure of embankments at clay sites in Japan. Some of these deposits are characterized with a Plasticity Index as large as 60 %.

More recently, liquefaction-like ground distress at several locations have been reported across the city of Adapazari in Turkey during the August 17, 1999 Kocaeli Earthquake (Mollamahmutoglu et al. 2002). Drill-hole soil samples from some of these sites indicate Fines Content of up to 92 %. Although the ejecta samples collected from the liquefied area had a Fines Content of up to 54 %, they were classified as non-cohesive. However, as indicated earlier, ejection of soil particles is unusual at sites underlain mainly by fine grained cohesive soils. Consequently, absence of cohesion in ejecta samples cannot confirm that cohesive soils at the sites where severe permanent ground deformation developed did not soften. A preliminary review of in-situ test data available at Pacific Earthquake Engineering Research Center of the University of California at Berkeley (http://peer.berkeley.edu/turkey/adapazari/) indicates that some of the building damage (e.g., at sites A and I in the vicinity of Cark Avenue) can probably be ascribed at least in part to softening of cohesive clayey soils.

Similar ground failure was also reported from several clay sites in Taiwan during the 1999 Chi-Chi Earthquake (Stewart et al. 2003). Cohesive deposits at these sites that liquefied in the classical sense were mainly of low plasticity and their performance were found to agree reasonably with Figure 1a except that in many occasions more than 15 % weight of these soils comprised of particles finer than 5 μm. The CPT based screening criterion did not perform satisfactorily at these sites.

Further, the guidelines discussed earlier essentially screen cohesive soil deposits for the likelihood of liquefaction using laboratory index measurements without considering the intensity and duration of cyclic loading. In contrast, review of cyclic undrained laboratory test data indicate that likelihood of development of large deformation within cohesive soils increases with amplitude of cyclic shear stress (i.e., intensity of shaking) and the number of cycles of stress application (i.e., duration of shaking).

In the following sections a framework has been proposed based on laboratory cyclic soil test, field performance case histories and cone penetration testing for assessing cyclic stability of cohesive deposits accounting for the intensity and duration of cyclic load application.

5 CYCLIC STABILITY FROM CONE TIP RESISTANCE

Cone tip resistance relates positively to the static undrained shear strength, effective geostatic stress (Robertson and Campanella, 1986) and the plasticity index (Mayne, 1998) for normally- or lightly overconsolidated cohesive soils. The influence of these factors on cone tip resistance and cyclic shear strength (discussed in Section 2.1) is thus comparable. It may therefore be possible to develop a relationship between cone tip resistance and cyclic shear strength of cohesive soils measured in laboratory DSS tests.

Similar to Robertson and Campanella (1986) procedure for assessing liquefaction potential of non-plastic granular deposits, attempt has been made in this study to identify a lower-bound relationship between the cyclic shear strength and normalized cone tip resistance. The Cyclic Resistance Ratio (CRR) determined from laboratory cyclic testing as well as Cyclic Stress Ratio (CSR) estimated from earthquake-related case histories in which severe ground deformation developed is used in this study as estimators of cyclic soil strength. The laboratory-based CRRs were corrected appropriately for multidirectionality of shear stress application and material anisotropy. The CRR is defined here as the ratio of cyclic shear stress to the vertical effective stress, at which a double amplitude shear strain of 5 % develops in 26 cycles in DSS or TX tests and the CSR is defined as:

$$\text{CSR} = 0.65 \times (a_{max}/g) \times (\sigma_v/\sigma'_v) \quad (1)$$

where a_{max} is the peak ground acceleration, g is the acceleration due to gravity, σ_v and σ'_v are the vertical total and effective stress prior to earthquake, respectively.

Normalized cone tip resistance q_{T1} given by

$$q_{T1} = q_T/\sigma'_v \quad (2)$$

is used in this study as the index of cyclic shear strength, where q_T is the total cone tip.

6 DATABASE

The composition of cyclic soil strength database used in this study is presented in Table 1. The corresponding cone tip resistances measured within these deposits were obtained from published records and other sources as listed in Table 2.

7 RESULTS

Laboratory CRRs and field CSRs for clayey soils listed in Table 1 are plotted against q_{T1} in Figure 4 for Magnitude 8.5 earthquakes. To convert the field performance data from smaller earthquakes, the Magnitude Scaling Factor (MSF) recommended by Youd et al. (2001) was used. A lower bound relationship between q_{T1} and CSR is also presented on Figure 4. Normally or lightly over consolidated deposits plotting above this line are likely to develop large shear deformations during earthquakes.

A review of Figure 4 indicates that a CSR of less than 0.125 is unlikely to trigger large deformations in cohesive deposits. Nor are deposits characterized with a $q_{T1} > 70$ susceptible to large earthquake-related deformations.

8 LIMITATIONS

The CSR - q_{T1} relationship proposed in this study is based on relatively few published laboratory test data and field performance records. Nor does the database considered in this study include cases representing non-occurrence of large earthquake-related deformations.

Secondly, this study uses a simple scheme for stress normalization of cone tip resistance described

Table 1. Data source and reference: cyclic laboratory testing and index properties

Deposit	Soil type, source for cyclic strength	OCR	I_P (%)	Reference
Adapazari, Turkey	Clay, Earthquake CSR	1.0	12 to 33	Mollamahmutoglu et al. 2002
Ariake, Japan	Clay, Laboratory CRR	1.0	40 to 60	Dia Consultants 1998
Cloverdale, BC, Canada	Sensitive clay, Laboratory CRR	1.0	27	Jitno 1990
Drammen, Norway	Clay, Laboratory CRR	1.0	27	Andersen et al. 1993
Gullfaks C, Norway	Clay, Laboratory CRR	1.3	23	Georgiannou et al. 1991
Moss Landing, California, USA	Clay, Earthquake CSR	1.0	15 to 25	Boulanger et al. 1998
St. Alban, Quebec, Canada	Sensitive clay, Earthquake CSR	2.0	20	Lefebvre et al., 1992
Troll A, Norway	Offshore, Clay, Laboratory CSR	1.45	37	By and Skomedal 1993
Troll B, Norway	Offshore, Sandy clay, Laboratory cyclic DSS test	1.45	20	By and Skomedal 1993

Table 2. Data source and reference: cone tip resistance

Site	Reference
Adapazari	Pacific Earthquake Engineering Research Center
Ariake	Dia Consultants 1998
Cloverdale	Sully and Campanella 1988
Drammen	Lunne, Personal Communication
Gullfaks C	Lunne et al. 1986, and Lunne anad Powell 1993
Moss Landing Marine Laboratories	Boulanger et al., 1998
St. Alban, Quebec	Demers, Personal Communication
Troll A and B	By and Skomedal 1993

by Equation 2 although some investigators (Wroth, 1988) prefer to use an alternative stress normali-

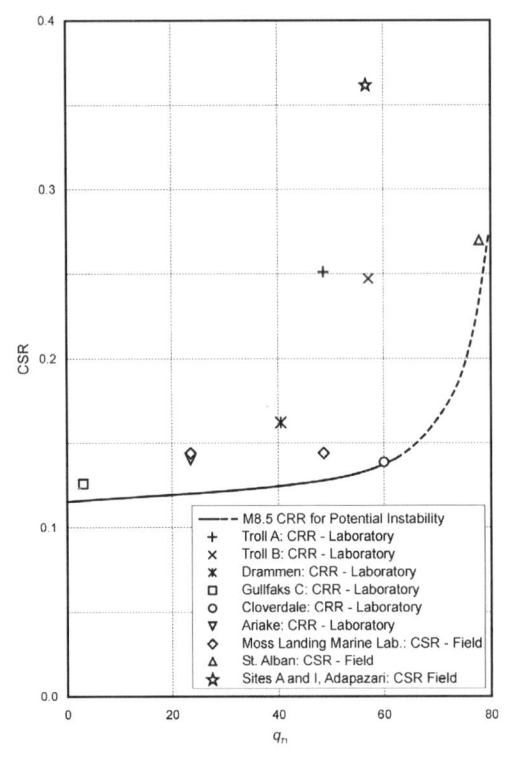

Figure 4. CSR for 5 % double amplitude shear strain for normally to lightly over consolidated cohesive soils for M 8.5 earthquakes

zation procedure, according to which the net total cone tip resistance is an appropriate index of shear strength of fine grained cohesive soils rather than q_{T1}.

Thirdly, the MSF used here is the same as those for cohesionless deposits.

9 CLOSURE

Cohesive fine grained soils soften and deform appreciably in laboratory cyclic simple shear tests that simulate moderate near field earthquakes. For laboratory undrained cyclic simple shear simulating larger earthquakes cohesive soil samples deform even more. Likelihood of cyclic softening and consequent permanent ground deformation at sites underlain by cohesive fine grained soils is often discounted because such soils rarely liquefy in the classical sense. Nevertheless, case histories documenting structural distress at sites underlain by clayey soils are not rare.

A tool has been proposed in this paper that uses normalized total cone tip resistance as an index of cyclic strength of cohesive soils for making a preliminary assessment of likelihood of development of large permanent ground deformation during earthquakes. Unlike similar tools currently available for this purpose, the procedure proposed in this study approximately accounts for the intensity and duration of ground motion.

ACKNOWLEDGEMENTS

The work presented in this paper would not have been possible without the technical help and encouragement of Mr. Keith Robinson and Mr. J. Martyn Bayne and the facilities extended to me by my former employer, Jacques Whitford and Associates Limited of Burnaby, British Columbia, Canada. Dr. Tom Lunne, Dr. Masayuki Hyodo, Dr. Richard Campanella and Mr. Denis Demers were kind enough to provide some of the piezocone penetration test and laboratory cyclic soil test data that form the basis of this study.

REFERENCES

Andersen, K. 1988. Properties of soft clay under static and cyclic loading. *Proceedings*, International Conference on Engineering Problems on Regional Soils. Elsevier. 7-26.
Andrews, D.C.A., and Martin, G.R. 2000. Criteria for liquefaction of silty soils. *Proceedings*, 12th World Conference on Earthquake Engineering, Auckland, New Zealand. Paper 0312.
Boulanger, R.W., Meyers, M.W., Mejia, L.H. and Idriss, I.M. 1998. Behavior of a fine-grained soil during Loma Prieta earthquake. *Canadian Geotechnical Journal* Vol. 35: 146-158.
By, T. and Skomedal, E. 1993. Soil parameters for foundation design, Troll Platform. *NGI Publications* Vol. 189. Norwegian Geotechnical Institute.
Dia Consultants. 1998. Investigations on the dynamic behavior of the Rokkaku River embankments. *Report* (in Japanese).
Georgiannou, V.V., Hight, H.W. and Burland, J.B. 1991. Behaviour of clayey sands under undrained cyclic triaxial loading. *Géotechnique* Vol. 41: 383-393.
Hyodo, M., Yamamoto, Y., and Sugiyama, M. 1994. Undrained cyclic shear behaviour of normally consolidated clay subjected to initial static shear stress. *Soils and Foundations* Vol. 34(4): 1-11.
Jitno, H. 1990. Stress-strain and strength characteristics of clay during post-cyclic monotonic loading. *M.A.Sc. Thesis*. University of British Columbia, Canada.
Lefebvre, G., Leboeuf, D., Hornych, P. and Tanguay, L. 1992. Slope failures associated with the 1988 Saguenay earthquake, Quebec, Canada, *Canadian Geotechnical Journal* Vol. 29: 117-130.
Lunne, T., Chistoffersen, H.P., and Tjetta, T.I. 1986. Engineering use of piezocone data in North Sea clays. *NGI Publications* Vol. 163. Norwegian Geotechnical Institute.
Lunne, T. and Powell, J.J.M. 1993. Recent developments in in-situ testing in offshore soil investigations. *NGI Publications* Vol. 189. Norwegian Geotechnical Institute.
Mayne, P.W., Robertson, P.K., and Lunne, T. 1998. Clay stress history evaluated from seismic piezocone tests. *Geotechnical Site Characterization* Vol. 2: 113-1118. Rotterdam: Balkema.
Mollamahmutoglu, M., Kayabali, K., Beyaz, T. and Kolay, E. 2002. Liquefaction-related building damage in Adapazari

during the Turkey earthquake of August 17, 1999. *Engineering Geology* Vol. 67: 297-307.

Polito, C.P. 1999. The effects of non-plastic and plastic fines on the liquefaction of sandy soils. *Ph.D.* Dissertation. Virginia Polytechnic Institute and State University, USA.

Robertson, P.K. and Wride, C.E. 1998. Evaluating cyclic liquefaction potential using the cone penetration test. *Canadian Geotechnical Journal* Vol. 35: 442-459.

Robertson, P.K. and Campanella, R.G. 1986. Guidelines for use, interpretation and application of the CPT and CPTU. *Soil Mechanics Series* No. 105, Department of Civil Engineering, University of British Columbia, Canada.

Sasaki, Y., Taniguchi, E., Matsuo, O. and Tateyama, S. 1980. Damage of soil structures by earthquakes. *Technical Note of PWRI* No. 1576. Public Works Research Institute, Japan (in Japanese).

Seed, H.B. and Idriss, I.M. 1982. Ground motion and soil liquefaction during earthquake. Monograph, Earthquake Engineering Research Institute, Oakland, California, USA.

Seed, R.B., Cetin, K.O., Moss, R.E.S., Kammerer, A.M., Wu, J., Pestana, J.M., Riemer, M.F., Sancio, R.B., Bray, J.D., Kayen, R.E. and Farris, A. 2003. Recent advances in soil liquefaction engineering. *Keynote Presentation*, 26th Annual ASCE Los Angeles Geotechnical Spring Seminar, Long Beach, California, USA.

Stewart, J.P., Chu, D.B., Lee, S., Tsai, J.S., Lin, P.S., Chu, B.L., Moss, R.E.S., Seed, R.B., Hsu, S.C., Yu, M.S. and Wang, M.C.H. 2003. Liquefaction and non-liquefaction from 1999 Chi Chi, Taiwan Earthquake. *Advancing Mitigation Technologies and Disaster Response*. TCLEE Monograph No. 25. ASCE, USA.

Sully, J.P. and Campanella, R.G. 1988. Interpretation of piezocone soundings in clay – a case history. *Penetration Testing in the UK*, 77-82. London: Thomas Telford.

Wroth C.P. 1988. Penetration testing – a more rigorous approach to interpretation. *Proceedings*, 1st International symposium on Penetration Testing, Florida, Vol. 1: 303-311.

Youd, T.L., Idriss, I. M., Andrus, R.D., Arango, I., Castro, G., Christian, J.T., Dobry, R., Finn, W.D.L., Harder, L.F., Jr., Hynes, M.E., Ishihara, K., Koester, J.P., Liao, S.S.C., Marcuson, W.F., III, Martin, G.R., Mitchell, J.K., Moriwaki, Y., Power, M.S., Robertson, P.K., Seed, R.B. and Stokoe, K.H., II. 2001. Liquefaction resistance of soils: summary report from the 1996 NCEER and 1998 NCEER/NSF workshops on evaluation of liquefaction resistance of soils. *Journal of Geotechnical and Geoenvironmantal Engineering* Vol. 127: 817-833.

Proceedings ISC-2 on Geotechnical and Geophysical Site Characterization, Viana da Fonseca & Mayne (eds.)
© 2004 Millpress, Rotterdam, ISBN 90 5966 009 9

High resolution stratigraphic and sedimentological analysis of Llobregat delta nearby Barcelona from CPT & CPTU tests

M.J. Devincenzi
IgeoTest, S.L., Figueres, Spain

S. Colàs, J.L. Casamor & M. Canals
GRC Geociències Marines, Dept. D'Estratigrafia, P. i Geociències Marines, Universitat de Barcelona, Spain

O. Falivene & P. Busquets
GRC Geodinàmica i Analisi de Conca, Dept. D'Estratigrafia, P. i Geociències Marines, Universitat de Barcelona, Spain

Keywords: CPT, CPTU, sequence stratigraphy, sediment facies, Llobregat delta

ABSTRACT: An interactive 3D model has been constructed for an area of 0.3 km^2 in the Llobregat delta based on soil parameters from CPTU. This 3D model shows, for the first time in the Llobregat delta, the high-resolution spatial distribution of sedimentary facies including beach ridge sands, floodplain silts and clays, delta front sands and silts and prodelta silts and clays. Each facies is described in terms of mechanical properties namely q$_c$ and FR% and is groundtruthed by means of continuous boreholes. Moreover, key surfaces for sequence stratigraphic interpretation, like maximun flooding surface and systems tract have been also identified. A geotechnical *in situ* test, CPT, proves therefore an excellent tool to investigate the geometry and evolution of sediment bodies from where Quaternary sea-level fluctuations can be inferred.

1 INTRODUCTION

This paper deals with the stratigraphic and sedimentological interpretation of a CPT/CPTU data set jointly with boreholes data from the deltaic plain of the Llobregat River within the municipality of El Prat de Llobregat, nearby Barcelona. Major infrastructures already existing in this area are being enlarged at present and new ones are under construction in the Llobregat delta-Barcelona area. These include doubling the operative surface of the Barcelona harbour, extending its airport, building a new sewage treatment plant (EDAR Riu Llobregat), deviating the Llobregat River lowermost course, setting up high speed railway connections, tunneling for a new subway line and restructuring one of the main industrial zones in Spain. Knowing the underground geological and geotechnical nature of this area is of prime importance to the development and economic growth of Catalonia and Spain.

Under the framework of the construction plan for the new sewage treatment plant on the delta plain of the Llobregat river, 70 CPT/CPTU tests and 20 coring boreholes were carried out within a 1.2 x 0.25 km^2 area on the right bank (Fig. 1). Despite of its small size, the study area is considered to be representative of the deltaic system because of its location near the current river mouth and orientation parallel to the river lowermost course.

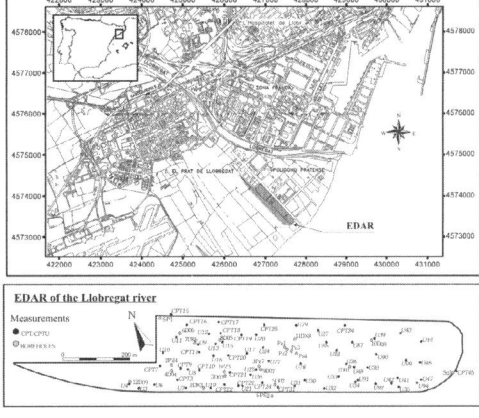

Figure 1. Study area and measurements location

The large amount of data available from the study area altogether with their high quality, facilitates the construction of robust 3D models which could be used to improve the interpretation of CPTU results for predictive purposes (c.f. section 5).

CPTU data for stratigraphy and sedimentary research in deltaic settings have been previously used by Amorosi & Marchi (1999) in the Po delta. Ventayol (2003) presented a general correlation amongst the previous mentioned main stratigraphic units in the Llobregat delta (cf. section 2.2) using CPT data

877

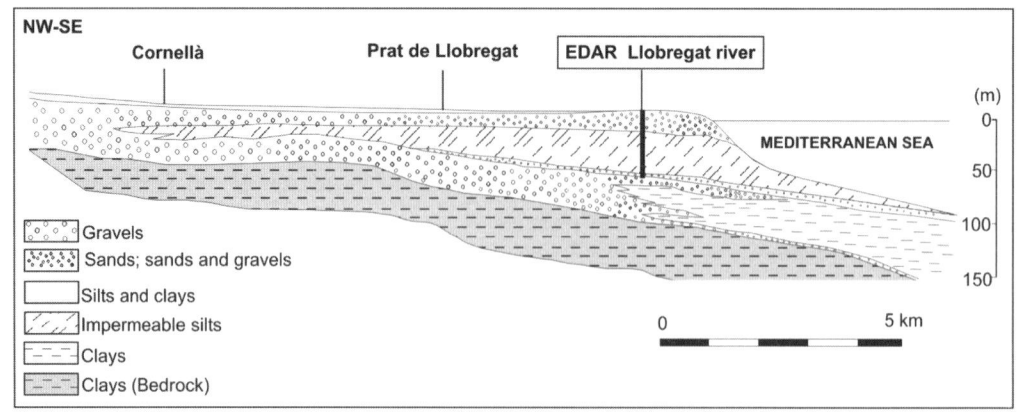

Figure 2. General architecture for the Llobregat Delta (modified from Marqués, 1974 and Bayó, 1985).

from the left river bank. Also, Gens & Lloret (2003) presented a stratigraphic cross-section made from CPTU and borehole data of the EDAR Riu Llobregat area.

2 GEOLOGICAL SETTING

2.1 *Deltaic sedimentary environments*

Deltas consist of three main sedimentary environments: 1) the delta plain, which is largely subaerial but contains subaqueous portions, 2) the delta front and 3) the prodelta. Each of these is characterized by its own assemblage of sediments and sedimentary structures.

1) The *delta plain* includes: i) distributary channels and their associated natural levees, point bars and crevasse splays (fluvial system) and ii) interdistributary deposits as bays and marshes. Distributary channels carry most of the coarse sediment that crosses the delta.
2) The *delta front* is the seaward part of the topset delta in the shallow subtidal zone. Typically this zone extends to depths of about 10 m or less, characterized by sand sediment with grain-size decreasing seaward, where fines become increasingly important.
3) The *prodelta* is the most widespread and homogeneous environment of the delta. It is dominated by silts and clays. Grain size fines seaward and grades into shelf muds.

2.2 *The Llobregat Delta*

The Llobregat delta is located south of Barcelona. The delta is formed on a 8 km wide continental shelf that displays inclinations of 0.3° to 0.7°. The delta plain spreads over 95 km^2 with 23 km of shoreline length. Towards the NW, the prodelta of the Llobregat interfingers with the prodelta of the Besós river.

The joint Llobregat-Besós prodelta system covers an area of 165 km^2 (Checa et al., 1988).

The general architecture of the Llobregat delta (Fig. 2) consists of five main units, which from base to top are: 1) a lower unit of Pliocene blue clays and shales with fragments of shells (bedrock), 2) fluvial gravels that form the lower aquifer probably younger than 18.000-15.000 y B.P., 3) prodelta deposits made of clayey silts, 4) the upper aquifer unit made of gravels, sands and some silts from delta front and delta plain environments and 5) the superficial level with flood plain fine sands, silts and clays and marsh clays (Marqués, 1974; Bayó, 1985).

The development of the modern Llobregat delta began with the Versilian transgression, about 10.000-11.000 y B.P. The sea level rose from around 65 m beneath the present sea level up to 3 m above it at 4500 yB.P., (Aloïsi & Monaco, 1978). According to several authors, the modern delta started growing at 6.000 y B.P. (Marqués, 1974; Serra & Verdaguer, 1983; Checa et al., 1988). Nevertheless, last century human activities such as river damming and water diversion greatly reduced sediment delivery to the delta, which has now segments being eroded along its shore.

3 FIELD WORK

The small size of the study area and the dense grid of geotechnical tests and boreholes resulted in a highly detailed view of the underground. Of the 70 total tests 60% were CPTU and 40% CPT. Depth investigation was in general 45-50 meters with some tests reaching 64 meters. Field work was performed by experienced technicians using a 20t truck. Equipment and procedures were in accordance with ASTM D 5778. Some DMT tests following ASTM D 6635 were also made with the same pushing rig and crew.

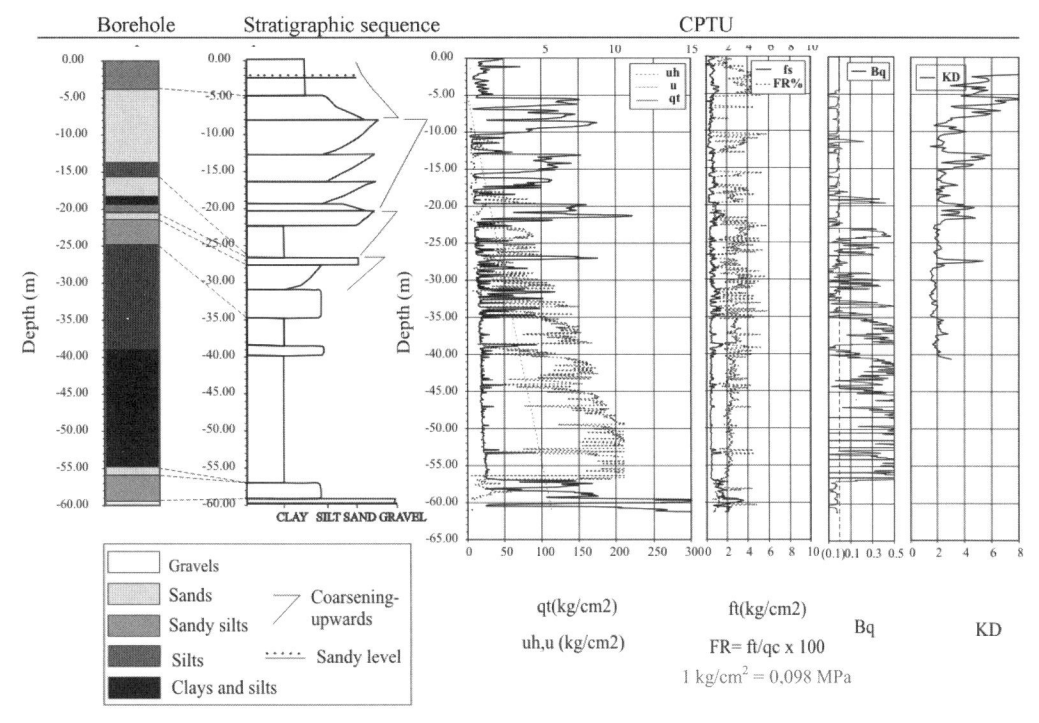

Figure 3. Example of CPTU 21 plot, borehole SPS2 and K_D from DMT 1.

The 20 continuously cored boreholes penetrated down to 60 meters. Undisturbed samples were taken and STPs were performed every 3 meters. Drills were also used to install extensometers for settlement monitoring.

Figure 3 shows a typical CPTU result in wich q_T, u_h, u, f_T, FR% and Bq are plotted in conjunction with a close borehole log and sedimentological interpretation. K_D from a nearby DMT test is also shown. Results clearly represent the generally assumed stratigraphic sequence of the delta edifice: upper sandy and lower gravel aquifers with an intermediate layer of prodelta clays and silts. The upper part of this intermediate layer having intercalations of silts or even very fine sands. This hole sequence is covered with floodplain and marsh sediments.

4 3D MODELLING

4.1 Input data

The variable used to construct the model was q_c. The main reasons for select q_c are 1) the possibility to include all CPT and CPTU tests in the same model, 2) its ability to distinguish between sedimentary facies with very contrasted mechanical response and, 3) the fact that the distinction between q_c and q_t does not bring more accuracy for soft facies distinction (e.g. flood plain silts and clays from marsh sediments). Other parameters could have been eventually used: FR, Bq, $K_{D\,DMT}$ or dissipation tests.

4.2 Interpolation technique

Previous work was developed in the same study area with the same dataset using kriging algorithms (Devincenzi et al., 2003). In this study, 3D interpolation has been developed using the minimum tension gridding algorithm implemented in the EarthVision® software of Dynamics Graphics, Inc.

The gridding technique represents the values of the input data as close as possible and calculates a plausible looking model grid with nodes that are or are not adjacent to input data points. This technique has two stages: 1) an initial estimate and, 2) biharmonic iterations with scattered data feedback.

The initial estimation calculates the q_c value for each grid node selecting the input data searched on the basis of octants. Once the estimates are complete, a number of iterations using a biharmonic cubic spline function begin to re-evaluate the grid nodes. The most useful characteristic of this function is the distribution of tension (second derivative or curvature) among the nodes so that the sum of the squares of the second derivatives is minimized. Simply stated, the curvature is distributed rather than

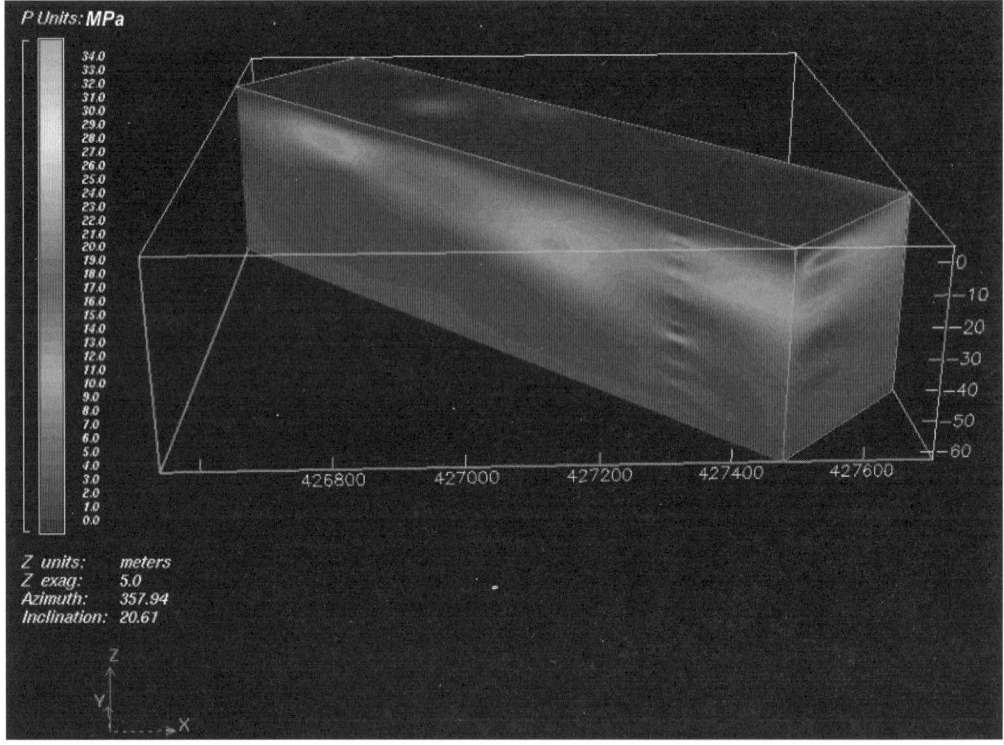

Figure 4. 3D grid for the study area.

concentrated at data points.

Grid spacing for the study area was 5x5x0.5 m with a grid rotation of 135° and a 0.01 vertical influence (Fig. 4). The grid rotation angle corresponds to the main directional trend of the sedimentation (approximately to the river course). Choosing the appropiate angle minimizes extrapolation where no data exist. Selecting a very small value for vertical influence implies that no weight is given to data above or below each interpolation node relative to data in the horizontal plane of that node (Flach et al., 2003).

4.3 Model Results

From the interpolated grid, an isosurface model has been constructed where q_c ranges of 1MPa are represented to improve identification of q_c changes.

This model differs from the grid since it has been cut under adjustments to the exact data location to avoid wrong interpolated values. Geometries of sediment bodies have been identified from the isosurfaces model.

Figure 5 displays two main levels: a coarse-grained level and a fine-grained level with high values and low values of q_c, respectively. These levels present an intercalation of low q_c values which thickness increases toward 135°.

5 RESULTS

5.1 Sedimentary facies characterization and SBT classification

Soil classification has been developed correlating tip resistance with sediment facies identified in boreholes. Plotting q_c versus friction ratio (FR) in the Robertson (1986) chart and adding sediment facies leads to q_c-sediment facies correlation. Particularly, four groups of facies have been identified in terms of grain size and spatial geometry that are clearly displayed by the 3D model: 1) flood plain facies, 2) beach facies, 3) delta front facies and 4) prodelta facies (Fig. 6).

Flood plain facies have a broad q_c range that corresponds to several sediment types inclusive of fluvial sands and gravels, leveé silts, crevasse sands silts and marsh clays. Crevasse sands and sandy silts and fluvial channel sands have been differentiated with q_c values of 6-10 MPa and 10-30 MPa respectively.

Beach ridge sands show values of 8-20 MPa, which are narrower than for fluvial channel sands since beach ridge sands are constituted by better sorted grain populations. These sediment bodies display coarsening-upwards trend.

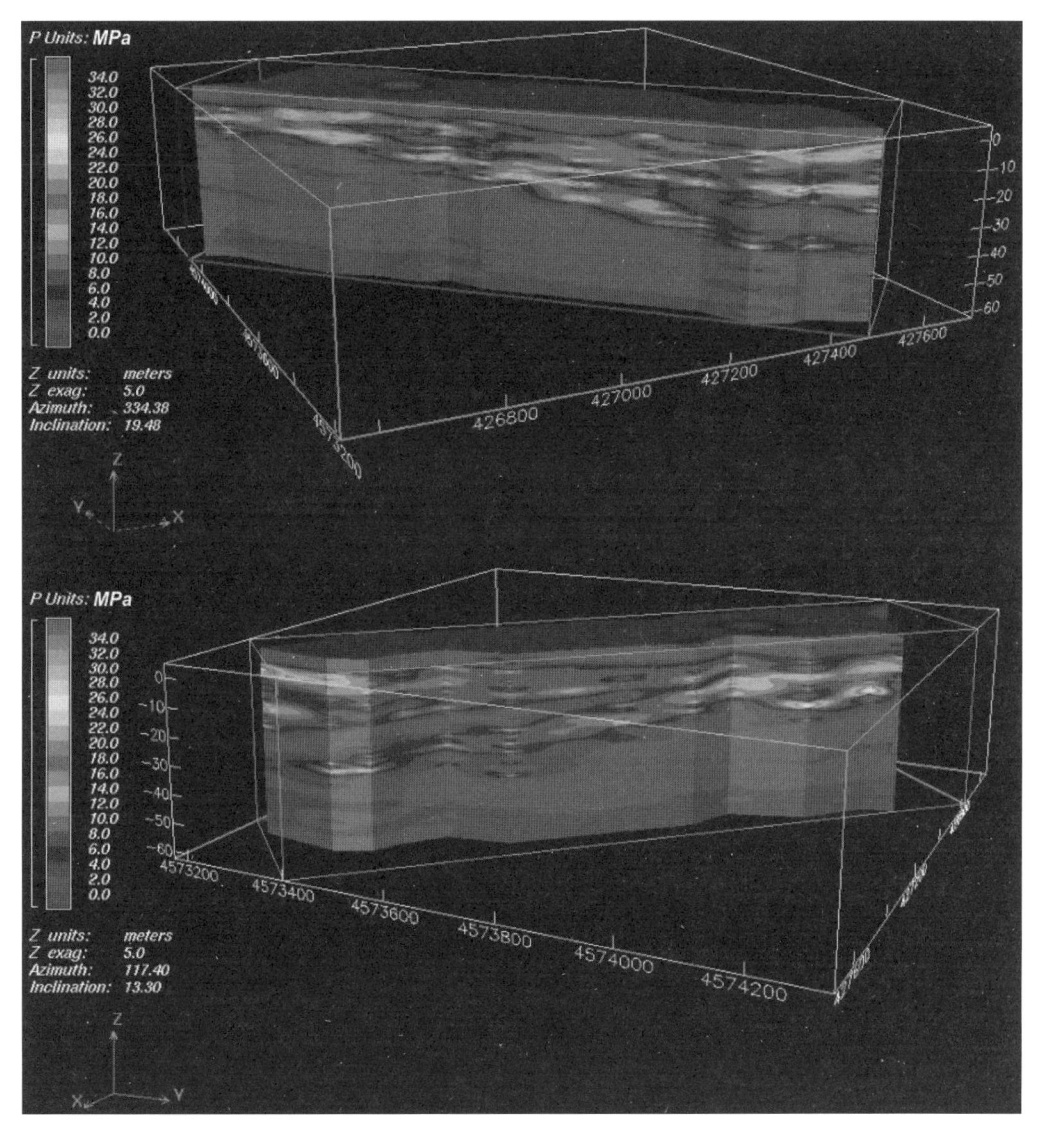

Figure 5. Isosurfaces model views. 3D grid model boundaries have been cut after adjustments based on data location.

Delta front deposits consists of silts, sandy silts and interbedded sands with q_c values around 1-2 MPa, wich tend to be slightly higher towards the top of the individual units.

Prodelta facies are most often made of silts but clayey silt and clay intervals could appear. Rare thin sandy intercalations have been occasionally observed near the top of these units close to the boundaries with the overlying delta front facies. Most of q_c values are, anyway, comprised between 1 and 2 MPa.

5.2 Evaluating the sedimentary evolution of the Llobregat delta

Relative sea-level positions have been identified from the location of landwards stepping beach ridge sands. Four beach ridge sandy units are observed in the 3D model (Fig. 7) at 7, 13, 18 and 25 m of depth below the current sea-level. Their distribution can be attributed to retrogradational deltaic sequences formed during periods of relative stillstand within the transgressive systems tract related to the last post-glacial sea-level rise as well as the subsequent shoreline migration (c.f. Glossary). The top of these

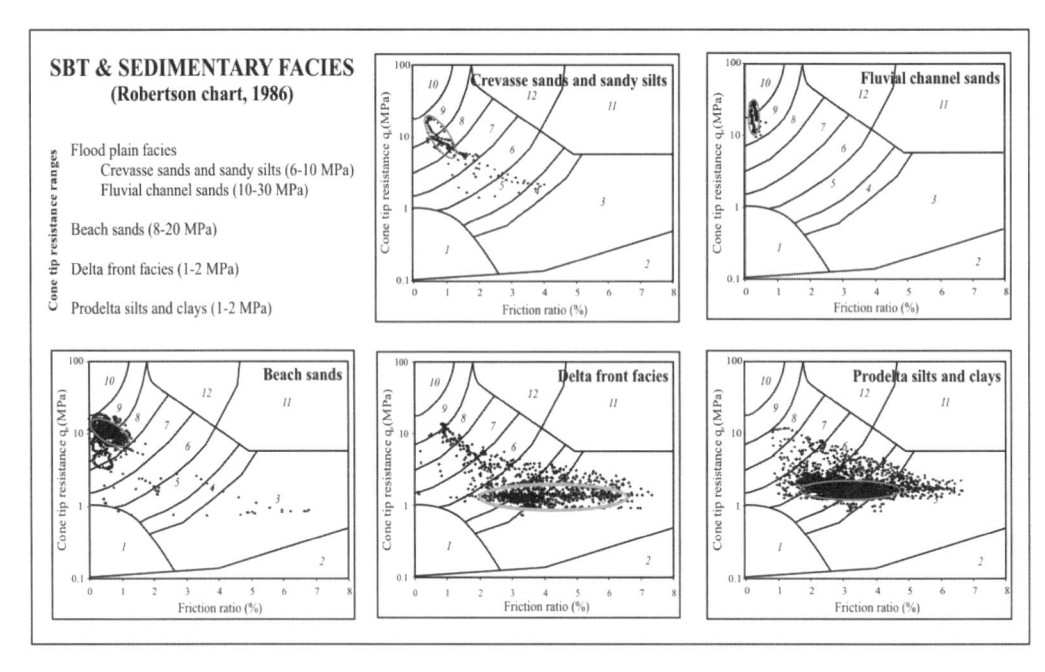

Figure 6. SBT and sedimentary facies on Robertson charts (1986).

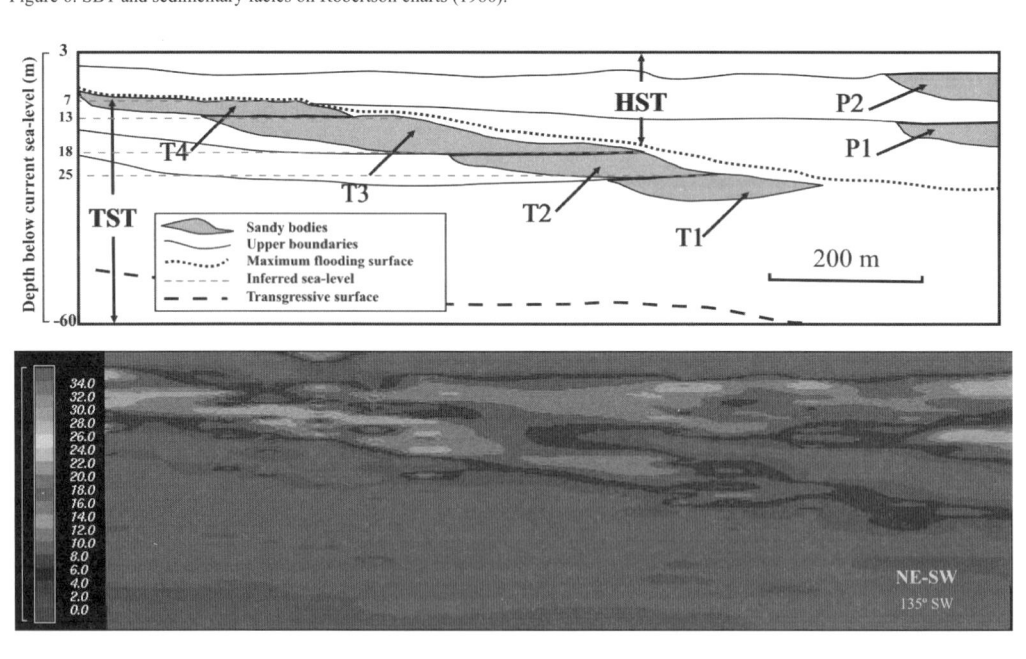

Figure 7. Stratigraphic cross-section showing the retrogradational sequences (T1, T2, T3, T4) as well as the progradational sequences (P1, P2) and the inferred sea-level stages at 7, 13, 18 and 25 m of depth. Maximum flooding surface corresponds to the top of the transgressive systems tract (TST) and the base of the high systems tract (HST). Transgressive surface corresponds the base of the transgressive systems tract. This interpreted cross-section is based on the lower cross-section from the 3D model (same vertical scale).

sandy bodies thus marks the boundary between the transgressive systems tract from the highstand systems tract (Fig. 7).This fact allows to infer four relative sea-level stages associated to the upper part of the Versilian transgression, likely from about 11000 to 6000 yBP.

Two additional, seaward shifted, uppermost beach ridge sands have been interpreted as representing two progradational phases associated to the highstand systems tract. This corresponds to a sustained delta growth and shoreline advance phase favored by the estabilization of the sea-level since 6000 y BP onwards.

Somoza et al. (1998) have identified similar retrogradational steps in the Ebro river. These authors associate this stacking pattern to decelerations in the rise of sea-level followed by stillstands or smooth falls in relative sea-level. Similar architectural configurations have been described along the Spanish margin by Díaz and Maldonado (1990) and Hernández-Molina et al. (1995, 2002) and the Adriatic margin by Amorosi and Marchi (1999), Amorosi and Milli (2001) and Ridente and Trincardi (2002).

6 CONCLUSIONS

CPT and CPTU measurements provide a high resolution dataset of the Llobregat delta, south of Barcelona, which is suitable for 3D modelling. Cone tip resistance constitutes a simple parameter that can be interpreted in terms of sediment facies if calibrations with boreholes are available.

3D modelling results unveil the internal structure of the delta by illustrating the spatial distribution of sedimentary units and q_c trends.

The minimum tension gridding technique constitutes an useful and practical tool to interpolate these data since its application is simple and no large calculation times are required. The modelling results obtained are supported by those from previous work in the same study area using kriging algorithms where the same trends were observed.

From the sedimentological viewpoint prodelta, delta front and delta plain sediments can be distinguished and characterized in terms of lithology, geometry and vertical grain size tendencies. Within the general architectural frame, our study allows differentiating fluvial channel sands, beach ridge sands, crevasse sands, delta front sands, silts and clays and prodelta silts an clays, jointly with key levels like the transgressive and flooding surfaces.

Four retrogradational sequences illustrate landward shoreline migration that likely took place during phases of relative sea-level fall or stillstands within the Versilian transgression.

The results achieved up to date are extremely promising in order to elucidate the internal structure and the construction of deltaic systems, as shown for a limited area in the Llobregat delta. Understanding how these systems formed and which are the resulting 3D geometries is also of prime importance for the geotechnical engineer since mechanical properties could be predicted. A joint research effort on deltaic systems involving academia and industry is particularly welcome because of the growing pressure these systems are supporting, as perfectly illustrated by the Llobregat Delta-Barcelona case study.

ACKNOWLEDGEMENTS

The authors gratefully acknowledge the permission of DEPURBAIX to use the data presented in this work.

GLOSSARY

A system tract is defined as a linkage of contemporaneous depositional systems that are defined by stratal geometries at bounding surfaces and internal parasequences stacking patterns. TRT (Transgressive Systems Tract) corresponds to retrogradational successions of shelf deposits. Its base is marked by the ravinement surface; its top is the Maximum Flooding Surface (MFS). The Highstand Systems Tract (HST), which consists of shelf to nonmarine deposits arranged in successive facies agradational and progradational, is eventually located over the MFS. (Miall, 1997).

REFERENCES

Aloïsi, J.C. & Monaco, A. 1976. *The Holocene transgression in the Golfe du Lion, southwestern France: Paleogeographic and Paleobotanical Evolution*. Gégr. phys. Quat., XXXII, N° 2: 145-161.

Amorosi, A. & Marchi, N. 1999. *High-resolution sequence stratigraphy from piezocone tests: an example from the Late Quaternay deposits of the southeastern Po Plain*. Sedimentary Geology, 128: 67-81.

Amorosi, A. & Milli, S. 2001. *Late quaternary depositional architecture of Po and Tevere river deltas (Italy) and worldwide comparison with coeval deltaic successions*. Sedimentary Geology, 144: 357-375.

ASTM D5778-95. 2000. *Standard Test Method for Performing Electronic Friction Cone and Piezocone Penetration Testing of Soils*. American Society for Testing & Materials, West Conshohocken, PA, USA.

ASTM D6635-01. 2002. *Standard Test Method for Performing the Flat Plate Dilatometer*. American Society for Testing & Materials, West Conshohocken, PA, USA.

Bayó, A. 1985. *Les aigües Recursos i riscos geològics. Història Natural dels Països Catalans, 3*. Enciclopèdia Catalana. Barcelona.

Checa, A., Diaz, J.I., Farran M. & Maldonado, A. 1988. *Sistemas deltaicos holocenos de los ríos Llobregat, Besós y Foix: modelos evolutivos transgresivos*. Acta Geológica Hispánica, 23: 241-225.

Devincenzi, M.J., Colàs, S., Falivene, O., Canals & M., Busquets, P. 2003. *Aplicación del piezocono para el estudio sedimentológico de detalle de los sedimentos cuaternarios del delta del Llobregat, Barcelona.* III Congreso Andaluz de Carreteras, Sevilla, España, Octubre 2003: 937-954.

Díaz, J.I. & Maldonado, A. 1990. *Transgressive sand bodies on the Maresme continental shelf, Western Mediterranean Sea.* Marine Geology, 91: 53-72.

Dynamic Graphics, Inc. 2002. *EarthVision® User's Guide.* Alameda, Canada.

Flach, P.G., Hamm, L.L., Harris, M.K., Thayer, P.A., Haselow, J.S. & Smits, A.D. 2003. *A maethod for characterizing hydrogeologic heterogeneity using lithologic data.* SEPM Concepts in hydrogeologic models of sedimentary aquifers. Pp. 119-137.

Gens, A. & Lloret, A. 2003. *Monitoring a preload test on soft ground.* Proc. Fiel Measurements in Geomechanics, Myrvoll (ed.): 53-59.

Hernández-Molina, F.J., Somoza, L., Rey, J. & Pomar, L. 1994. *Late Pleistocene-Holocene sediments on the Spanish continental shelves: Model for very high resolution sequences stratigraphy*, 120: 129-174.

Hernández-Molina, F.J., Somoza, L., Vázquez, J.T. & Rey, J. 1995. Estructuración de los prismas del Cabo de Gata: respuesta a los cambios climáticos-eustáticos holocenos. Geogaceta, 18: 79-82.

Hernández-Molina, F.J., Somoza, L., Vazquez, J.T., Lobo, F., Fernández-Puga, M.C., Llave, E. & Díaz-del-Río, V. 2002: 5-23. *Quaternary Stratigraphic stacking patterns on the continental shelves of the southern Iberian Peninsula: their relationship with global climate and palaeoceanographic changes.* Quaternary International, 92: 5-23.

Marquès, M.A. 1974. *Las formaciones cuaternarias del Delta del Llobregat.* Tesis doctoral. Universitat de Barcelona, Barcelona. Pp. 401.

Miall, A.D. 1997. *The geology of stratigraphic sequences.* Springler, Berlin: 433.

Ridente, D. & Trincardi, F. 2002. *Eustatic and tectonic control on deposition and lateral variability of quaternary regressive sequences in the Adriatic basin.* Marine Geology, 184: 273-293.

Robertson, P.K. 1986. *In situ testing and its application to foundation engineering.* Canadian Geotechnical Journal, 23: 573-594.

Serra, J. & Verdaguer, A. 1983. *La sedimentación holocena en el prodelta del Llobregat.* X Congreso Nacional de sedimentología, Mahón, September 1983.

Somoza, L., Barnolas, A., Arasa, A., Maestro, A., Rees, J.G. & Hernández-Molina, F.J. 1998. *Architectural stacking patterns of the Ebro delta controlled by Holocene high-frequency eustatic fluctuations, delta-lobe switching and subsidence processes.* Sedimentary Geology, 117: 11-32.

Ventayol, A. 2003. *Caracterización Geotécnica de sedimentos deltaicos mediante piezoconos. Aplicación al margen izquierdo del delta del Llobregat.* Ingeniería del Terreno. INGEOTER Vol. 2, C. López Jimeno (ed.): 413-433.

Fractal analysis of CPT data

J. A. Díaz-Rodríguez & P. Moreno-Carrizales
Department of Civil Engineering, National University of Mexico, Mexico

Keywords: CPT profiles, fractal analysis, fractal dimension, zoning Mexico City

ABSTRACT: This paper presents an application of the fractal analysis as a "descriptor" of the irregularity that present the soil profiles. The heterogeneity of a deposit is observed in the irregularity that present that graphs of some particular property versus the depth. Fractal principles are ideally suited to modeling a variety of problems in geomechanics. Soil profiles are generally described quantitatively, but the method describe here (the fractal approach) gives a quantitative description to characterize the CPT soil profiles.

1 INTRODUCTION

Site characterization is an important first step in the design a construction of many civil engineering projects. For sites with poor ground conditions and sensitive new facilities, foundation design and site development may be a significant factor in the overall project cost and viability.

The work required to properly characterize a project site may range from basic review of geological data to detailed site investigation.

The Cone Penetration Test (CPT) has become an important in-situ test for the characterization of soils where penetration is possible. The CPT offers a fast, economic and repeatable in-situ test especially in loose and soft sediments. As a logging tool CPT technique is unequalled with respect to delineation of stratigraphy and the continuous rapid measurement of soil parameters. Empirical and semi-theoretical correlations are available to estimate a full range of mechanical properties (Lunne et al. 1997).

In nature one finds many patterns that are irregular and fragmented that they cannot be described by classical Euclidean geometry. Scientists became more and more interested in trying to find order behind such complex, and seemingly chaotic, systems in nature. Natural objects, star constellations, coast lines, rivers, trees and soil profiles are composed of an apparent infinity of superimposed structures on all levels of detail.

Fractal geometry remains one of the most fascinating and useful areas of the physicals sciences at all levels. Fractals are used to model natural phenomena such as turbulent flow, Brownian motion, and porous surface.

Mandelbrot (1982) responded to the challenge to describe nature and so conceived and developed a new geometry. He identified the shapes of nature as fractals (from the Latin adjective *Fractus*, meaning fragmented)

Then, fractal geometry offers the opportunity of soil profiles quantification, and the possibility of relation soil profiles to specific soil properties and soil processes.

2 STRATIGRAPHICAL ZONING OF MEXICO CITY

The Basin of Mexico, where Mexico City is located, occupies a 9600 km^2 area. It is predominantly a flat lacustrine plain, placed at a 2250 m above sea level mean elevation, at the southern edge of the *Meseta Central*. During Pliocene-Holocene times, a segment of the Trans-Mexican Volcanic Belt was developed over this plateau. The basin is elongated in a NNE-SSW direction, over about 125 km length and 75 km width.

The basin remained open (exorrheic) until 700,000 years ago, when volcanic activity formed an enormous natural curtain: the Chichinautzin Range, that closed it and became an endorheic basin.

Based on boring performed at different times to depths of up to 100 m, Marsal & Mazari (1959) defined three zones in Mexico City: Lake zone, Transition zone and Hilly zone. Lake zone is a lacustrine zone, transition zone is predominantly sandy mate-

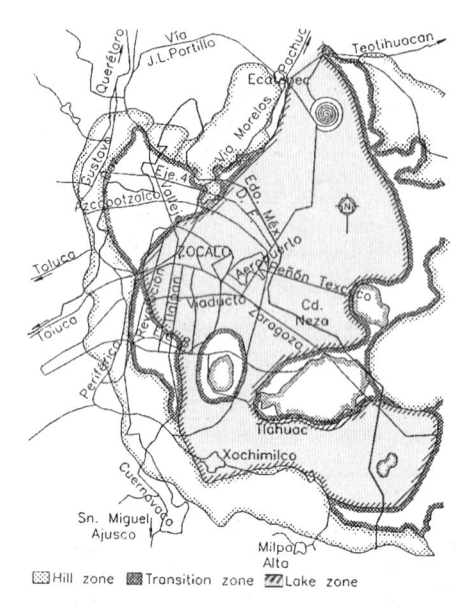

Figure 1. Subsoil zoning of Mexico City

rial, and the soils at Hill zone correspond to weathering of volcanic tuffs or eroding of such rocks. A current zoning of Mexico City is presented in Figure 1.

The sediments from Mexico City are complex mixture of crystalline minerals and amorphous material that challenge a simple nomenclature; they are heterogeneous volcanic and lacustrine sediments with microfossils (diatoms and ostracods) and organic matter. Physical, chemical, and mineralogical properties and also stratigraphy are described in Díaz-Rodríguez et al. 1998. Characterization and engineering properties of Mexico City are discussed in Díaz-Rodríguez 2003.

3 CONE PENETRATION TESTING

The characterization of soil materials through relatively inexpensive in situ testing is one of the geotechnical engineer's goals. The CPT profiles due to the continuous information provided and their easy execution are probably the most useful test for the definition of soil stratigraphy and following evaluation of mechanical properties.

The CPT has been used in Mexico City to define the stratigraphy of several places. Then, more than 104 soundings obtained by Geotec (2002) during the period 1985-2002 were analyzed. Figure 2 shows two typical records of CPT from Mexico City.

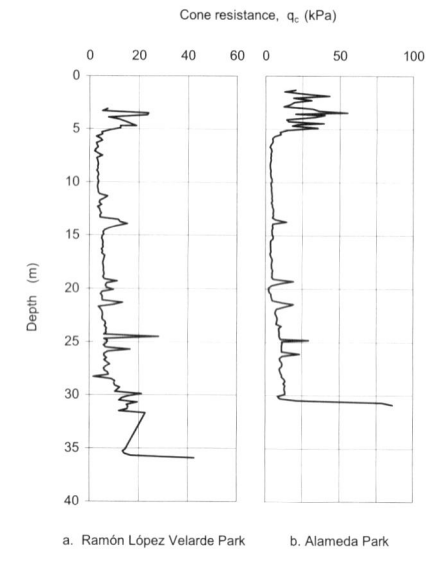

Fig. 2 Typical CPT profiles of Mexico City (Lake Zone)

The CPT equipment consisted of cone penetrometer, pushing equipment and data acquisition system. The equipment and penetration procedures fol-lowed ASTM D 3441-98. The cone has a 60° apex angle and a cross-sectional area of the tip of 10 cm^2. The friction sleeve, located above the conical tip, has a standard area of 150 cm^2 and the constant penetration rate was 2 cm/s.

4 FRACTAL MATHEMATHICS

Mandelbrot (1967) introduced the concept of fractals in a geological context. Noting that the length of a rocky coastline increased as the length of the measuring ruler size decreased according to a power law, he associated the power with a fractal (fractional) dimension. The fractal dimension is a measure of the roughness of the features. Mandelbrot (1982) has used fractal concepts to generate synthetic landscapes that look remarkably similar to actual landscapes. Mandelbrot has written a large number of scientific papers that deal with the geometry of the phenomena observed in many fields of science.

The central feature of a large class of fractal objects is self-similarity. Self-similarity implies that a structure is invariant to change of scale, i.e., it possesses dilatational symmetry and thus there exists no characteristic length associated with the structure.

Consider an irregular tortuous line. If it is fractal, then when a portion of this line is magnified it will look (statistically) similar to the whole line of which is a part. The rate of increase of measured length with ruler size is characterized by fractal dimension,

D. The implication is that D can be determined by measuring the length of the line over a range of measured scales. The most common method used for this proposes is known as the "box counting method" (and so the result is sometimes called the box-counting dimension).

In this method, a grid of squares is placed over the boundary or line profile and the number of squares through which any part of the line passes is counted (Fig. 3). This process is repeated with different grids having different size squares r. Using a range of values for r, yields the number of boxes, N, required to totally cover the line. N varies with r according to

$$N(r) = k / r^D = k (r)^{-D} \quad (1)$$

Where k is the constant of proportionality and D is the fractal dimension or box-counting dimension D_{BC}. A log-log plot of N versus r yields a line whose negative slope is the fractal dimension D (Fig. 4). Values of D close to the Euclidean value of 1 indicate that the line is relatively smooth and the estimate of the length is insensitive to ruler size. As the line becomes more rugged, D increases. Hence D is a quantitative measure of the roughness of a line or soil profile. The fractal dimension D of a geometric domain is a measure of the space-filling potential of the fundamental geometric shapes of which is composed. The value of D for a soil profile will be < 2, since 2 would be expected for a completely filled rectangle.

Interesting results have been presented for (Moore, Krepfl and Chuou 1991, Vallejo 1995, Díaz-Rodríguez & Moreno-Carrizales 1999).

Additionally, fractal geometry has the potential to be used in the modeling and scaling of soil physical processes and to provide insight into factors controlling soil profiling.

5 MATERIALS AND METHODS

As mentioned before, more than 104 CPT profiles (cone resistance q_c versus depth) were considered. All data are located in the Lake zone of Mexico City. The depths considered were from 7 to 30 m due to this is the most important stratum from the geotechnical point of view (Díaz-Rodríguez et al. 1998).

All profiles has 115 point registered on step of 20 cm of depth, however many of them showed big discontinuities in more than 25% of the selected range, for this reason those profiles were not considered.

The profiles selected were analyzed, and then a graphical representation was generated, all at same scale, depth (7-30 m), q_c (0.50 kg/cm²) and thickness line of 1 pixel.

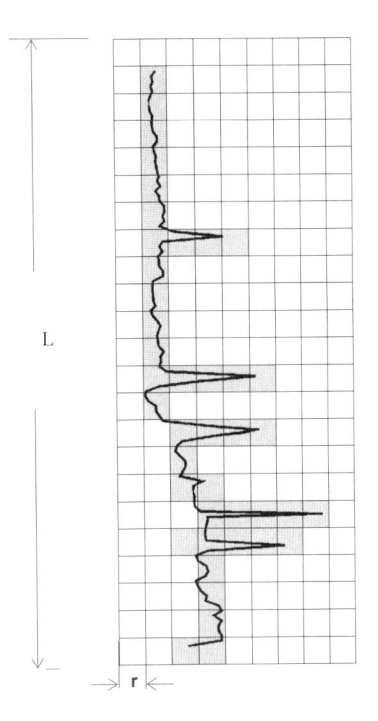

Figure 3. The box-counting method of determining fractal dimension.

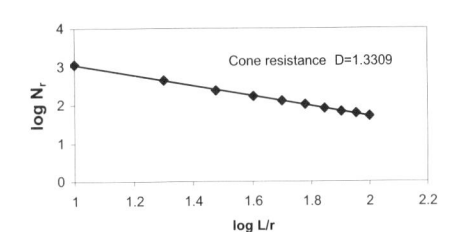

a. Ramón López Velarde Park

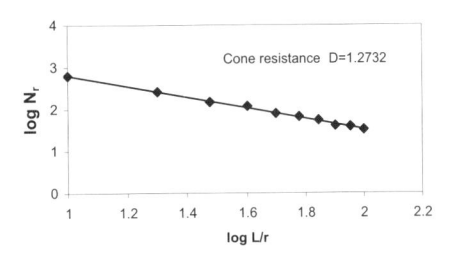

b. Alameda Park

Figure 4. A log-log plot showing the fractal dimensions.

Each graphical representation is an image file in format BMP (B/W) OF 2500X1900 PIXELS. To estimate the fractal dimension, D a computer program was developed in C++. The program "Fractal BC" implements the box-counting procedure analyzing each image file and calculate D according to equation (1). In total 38 profiles were analyzed.

This method is shown diagrammatically in Figures 3 and 4.

6 RESULTS

In all cases where the profiles were analyzed, it was concluded to be fractal. The fractal dimension for cone resistance, qc, for the Lake zone of Mexico City varied in value between 1.1090 and 1.3309 with and average of 1.2069 and a standard error of 0.0537.

The fractal dimension is a powerful and simple mathematical tool to measure the irregularity that present CPT profiles.

7 CONCLUSIONS

From the present investigation on typical Mexico City CPT profiling, the following main conclusions can be drawn.
1. This paper has shown the applicability of fractal theory to the analysis of CPT profiling.
2. The fractal geometry allows expressing the variability of CPT profiles numerically.
3. The fractal dimension for cone resistance was D = 1.2069 ± 0.0537.
4. The application of fractal geometry to soils is in its infancy.

REFERENCES

Díaz-Rodríguez, J.A. and Moreno-Carrizales, P. 2001. Estimating the soil fabric: a fractal approach, *XVth International Conference on Soil Mechanics and Geotechnical Engineering*, Istanbul, Turkey, Vol. 1: 27-31.

Díaz-Rodríguez, J.A., Lozano-Santa Cruz, R., Dávila-Alcocer, V.M., Vallejo, E. and Girón, P. 1998. Physical, chemical, and mineralogical properties of Mexico City: a geotechnical perspective. *Canadian Geotech. Journal* 35(4): 600-610.

Díaz-Rodríguez, J.A. 2003. Characterization and engineering properties of Mexico City lacustrine soils. in *Characterization and Engineering Properties of Natural Soils*. Tan et al. (eds). A.A. Balkema Publishers, The Netherlands.

Geotec. 2002. CPT files (1985-2002).

Mandelbrot, B.B. 1967. How long is the coast of Great Britain. Statistical self-similarity and the fractional dimension: *Science* 156: 636-638.

Mandelbrot, B.B. 1977. *Fractals: forms, chance and dimension*. San Francisco, Freeman.

Mandelbrot, B.B. 1982. *The fractal geometry of nature*. W.H. Freeman & Co., New York. 468 p.

Marsal, R.J. & Mazari, M. 1959. The Subsoil of Mexico City. Contribution to *First Panamerican Conf. on Soil Mechanics and Found. Eng.* Mexico City.

Moore, C.A., Krepft, M. and Chuou, H.A. 1991. Fractals in geomechanics. *Computer Methods and Advances in Geomechanics*, Beer, Booker & Carter (eds). A.A. Balkema: 371-376.

Lunne, T., Robertson, P.K., and Powell, J.J.M. 1997. *CPT in Geotechnical Practice*. Blackie Academic and Professional.

Vallejo, L.E. 1995. Fractal analysis of granular materials. *Géotechnique* 45 (1): 159-163.

Neural networks in soil characterization

W. Ding & J.Q. Shang
Dept. of Civil and Environmental Engineering, the University of Western Ontario, London, Canada

Keywords: Artificial Neural Network (ANN), complex permittivity, soil characterization

ABSTRACT: The complex permittivity of a soil recovered from a landfill site in Halton, Ontario, Canada, is measured using a custom developed apparatus in laboratory in the frequency range from 0.3 MHz to 1.3 GHz. The soil samples were compacted at various water contents, densities and degrees of saturation. A database consisting of the complex permittivity measurement of 122 soil specimens is established and three ANN models are trained, verified and tested to predict the soil water content, degree of saturation, and dry density. The results show that three ANN models perform well as evaluated in terms of RMS/Data Mean and correlation coefficients. The results of this study show promising potential for the further development of an in-situ soil property measurement system.

1 INTRODUCTION

The measurements of soil water content, density and degree of saturation are commonly practiced in geotechnical investigation and quality control of earthworks. The soil samples are collected from boreholes and their subsequent testing in the laboratory is often time consuming and expensive. The geophysics based techniques, such as electrical, electromagnetic, gravitational, magnetic, and radiometric methods, have shown promising potential in soil characterization. Among these techniques, the complex permittivity, which is measured through transmission and reflection of electromagnetic waves, is a relatively new concept in geotechnical engineering applications and has advantage over other methods such as the electrical conductivity (resistivity) and dielectric constant (e.g. resistivity survey, time-domain-reflectometry (TDR) etc.) mainly because the complex permittivity reflects the polarization and conduction behavior of soil in a broad frequency spectrum. A single measurement of the complex permittivity provides a set of data, typically in the order of several hundreds, in a specific measurement frequency range.

Artificial neural networks (ANNs) are used to capture the multi-variable relationship between the complex permittivity and soil properties. ANN is a massive distributed processor, which simulates the performance of biological neurons and the internal operation of human brains. ANNs have been applied in geotechnical and geoenvironmental engineering research for prediction, classification, approximation, and recognition.

In this study, ANNs are trained using a database, including complex permittivity, water content, density, degree of saturation, and pore fluid chemistry of 122 soil specimens from Halton Till, a site specific soil. Three ANN models are developed for soil property characterization based on the database, namely, the prediction of the soil water content, degree of saturation, and dry density. The performance of the models is evaluated using statistical tools.

2 BACKGROUND

The complex permittivity is an electrical property of materials. The complex permittivity ε^* (F/m) is expressed as:

$$\varepsilon^* = \varepsilon' - j\varepsilon'' \tag{1}$$

where ε' (F/m) is the real part of complex permittivity representing polarization of the material, or the ability to store electrical energy in an electric field, $j = \sqrt{-1}$, and ε'' (F/m) is the imaginary part of complex permittivity (the loss factor) describing the loss of the electrical energy in the material and is related to the electrical conductivity of the material (Kraszewski, 1997). The combination of these two parts reflects the response including polarization and conduction of constituents (atoms, molecules, etc) of the material in an applied electric field. The relative complex permittivity, which is the complex permittivity normalized by the permittivity of vacuum ε_0 (F/m), is commonly used in practice.

$$\varepsilon_r^* = \frac{\varepsilon^*}{\varepsilon_0} = \frac{\varepsilon'}{\varepsilon_0} - j\frac{\varepsilon''}{\varepsilon_0} = \varepsilon_r' - j\varepsilon_r'' \qquad (2)$$

where ε_r^* (dimensionless) is the relative complex permittivity, ε_r' (dimensionless) is the relative permittivity, ε_r'' (dimensionless) is the relative loss factor, and ε_0 is the permittivity of vacuum, where ε_0 = 8.854 ×10-12 F/m. In this study, the relative complex permittivity data are used throughout the ANN modeling process.

The research on the complex permittivity of soil has been conducted by a number of researchers. Santamarina and Fam (1997) measured the complex permittivity of bentonite and kaolinite in the frequency range from 0.2 GHz to 1.3 GHz at various soil moisture contents and solution concentrations. Klein and Santamarina (1997) discussed the method of complex permittivity measurement. Kaya and Fang (1997) reported the complex permittivity of bentonite and kaolinite mixed with methanol or aniline. Thevanayagam (1994) discussed the interaction effects of particle shape, orientation, porosity, and relative disparity between electrical parameters (conductivity and dielectric constant) of soil particles and pore fluid on electrical response of bulk soil. Chaudhari et al. (2001) measured dielectric relaxation on ethanol-nitrobenzene and ethanol-nitrotoluene binary mixtures at various concentrations and temperatures. Boyarskii et al. (2002) suggested a model of dielectric properties to describe dielectric and radiophysical properties of wet soils. Kaya (2002) used dielectric constant in a low frequency range to evaluate soil porosity.

The second author's research group has worked on the complex permittivity of soil. The work included the development of complex permittivity measurement systems for compacted soil (Shang et al. 1999; Scholte et al. 2002) and soil permeated with chemical solutions (Rowe et al. 2001); complex permittivity measured on soil samples with various water contents, densities, degrees of saturation, pore fluid compositions (including soil contaminated by heavy metals, organic compounds and landfill leachate), etc. (Shang and Rowe, 2003, Rowe et al. 2002, Shang et al. 2000, Scholte 1999, Xie 1999, Josic 2001). Shang et al. (2000) presented a work on using multiple linear regression analysis to link the complex permittivity and soil water content, density and salinity. However, the interactions between the complex permittivity and numerous influencing factors are non-linear and complicated, the quantitative characterization of soil is limited to a rather narrow range of soil properties. Therefore, a more advanced analysis tool - artificial neural network is introduced in this research.

Artificial Neural Networks (ANNs) are computer-based models that simulate neural structures of human brain neurons. ANNs learn from input information in the same way as human brains, where brain cells provide people with abilities to remember, think, analyze, and apply previous experiences to corresponding actions. The most distinguished property of ANNs is that they have abilities such as generalization, pattern recognition, image and speech recognition, classification, prediction, mathematical analysis, simulation of sophisticated physical processes, and quality control. A typical structure of an ANN consists of a number of processing units, or neurons, which are usually arranged in layers: an input layer, an output layer and hidden layers, as shown in Figure 1. The input layer is employed to introduce input variables. The hidden layers and output layer are fully or partially connected to the units in the input layer, which play internal roles and produce outputs in the execution of the network. For details on the mechanism of ANNs, readers are referred to Trajan (1999).

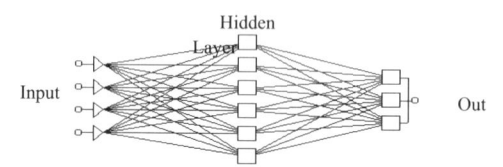

Figure 1. Typical Structure of an ANN

ANNs in soil characterization applications have been reported in literature. For example, Goh (1995) applied neural networks to analyze cone penetration test (CPT) data for foundation design parameters; Baldi et al. (1986) used the neural network to model the non-linear relationship between the cone tip resistance and the relative density of sand and the mean effective stress; Chang and Islam (2000) employed artificial neural networks to predict soil moisture measured by a remote sensing apparatus; Kurup and Dudani (2002) used artificial neural networks to profile the overconsolidation ratio (OCR) of clays using piezocone penetration test (PCPT) data; Penumadu and Zhao (1999) employed ANNs to model the stress-strain and volume change behavior of sand and gravel under drained triaxial compression test conditions; Najjar et al. (1996) employed artificial neural networks for predicting soil compaction parameters; Schaap and Leij (1998) applied artificial neural networks to predict soil hydraulic conductivity and soil water retention.

3 COMPLEX PERMITTIVITY OF HALTON TILL

In order to study the complex permittivity of a soil-water electrolyte system, it is essential to measure the complex permittivity in the frequency spectrum of interests, including the real and imaginary parts, on soil specimens with known properties, such as water content, degree of saturation, density and pore fluid chemistry. A laboratory measurement system was developed for measuring the complex permittivity of compacted or undisturbed soil specimens (Shang et al. 1999).

The Halton till, a soil recovered from Halton, Ontario, Canada, is used in this study. A total of 122 soil specimens were prepared in a rather broad range of physical and chemical properties. The water contents vary from 6.0% to 21.7%, with an average of 14.7 %; the dry densities vary from 1.60 Mg/m^3 to 2.08 Mg/m^3, with an average value of 1.88 Mg/m^3; the degrees of saturation vary from 36.7% to 100%, with an average value of 84.8%. The soil specimens were compacted under pre-determined water contents and compaction energy. The complex permittivity of soil specimens was measured in the frequency range of 0.3 – 1300 MHz. Details on the measurement procedure can be found in Shang et al. (1999) and Scholte et al. (2002). Figure 2 shows three typical traces of the real and imaginary parts of the complex permittivity versus frequency at various water contents, degrees of saturation, dry densities and pore fluid compositions. It may be seen from the figure that:
(1) Both real and imaginary parts of the complex permittivity are very high in the low frequency range (f < 100 MHz for the real parts and f < 200 MHz for the imaginary parts), which is mainly attributed to electrical double layer polarization, as discussed previously;
(2) (2) Dielectric dispersion begins from f ~ 400 MHz and approaches to ε_∞ ~ 2 before 800 MHz, after which all traces converge.

In addition, the static electrical conductivity and electrode polarization have considerable effects in the low frequency range. In the data analysis, therefore, the complex permittivity values in the frequency range of 200 MHz to 500 MHz are used based on the above considerations, including 47 pairs of real and imaginary parts at corresponding frequencies for each soil specimen. In the development of the ANN models, as will be discussed in the next section; the complex permittivities at certain specific frequencies are selected based on modeling needs.

The variations of the complex permittivity traces shown in Figure 2 as a function of soil properties illustrate the potential of using the complex permittivity for soil characterization. On the other hand, they also indicate the complexity of the problem as the traces are controlled by multiple factors simultaneously. Without a proper analytical tool, it is very difficult to distinguish a single soil property. It should be pointed out that in this study all measurements were made at room temperature with variations of less than 1 °C. Therefore the effect of temperature is not considered in the modelling process. With further enhancement of the database, he temperature effect can be incorporated into an ANN model as well.

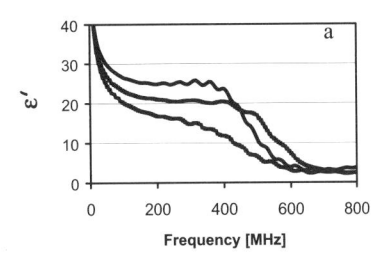

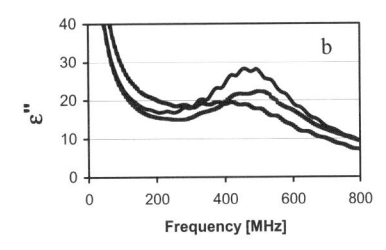

Figure 2. Typical complex permittivity traces. (a) Real part, (b) imaginary part

4 CHARACTERIZATION OF SOIL PROPERTIES USING ANN MODELS

The artificial neural network (ANN) is used as a modeling tool to relate the complex permittivity and soil properties by means of commercial software Trajan 4.0 (1999). The software allows the programmer to construct linear or non-linear models to solve multi-variable regression and classification problems. For characterization of soil properties of Halton Till, namely, the soil water content, degree of saturation, and dry density, a series of artificial neural networks with various architectures are designed and trained using experimental results on 122 soil samples, and the best 3 models are retained for the prediction of the three target properties. The complex permittivities of Halton Till measured from 200 MHz to 500 MHz and soil property parameters are used as inputs and outputs, respectively, to feed into the designed networks for supervised training. The trial and error method is used to design ANN architectures, select function parameters, and conduct input variable selection during modeling processes. The selections of network type (Multiplayer Perceptions (MLP), Radial Basis Function Networks (RBF) and Linear Networks (LN), Trajan, 1999), selection of function parameters (the learning rate, momentum, and number of epochs) and the architecture design are iteratively carried out and fine tuned during the modeling process.

The uniqueness in the ANNs modeling in this study is the input variable selection. This is because:
1. The large quantity of the input variables (94 complex permittivity (ε' and ε'') data for each soil specimen measured at 47 specific frequencies in the frequency range of 200 MHz to 500 MHz), corresponding to only 1 set of output variables, such as water content, dry density and degree of saturation for each measurement);
2. The limited training data (122 sets of complex permittivity measurement of soil specimens). Therefore, reducing the input variables and nework complexity becomes extremely important to improve the network performance. A case-specific modeling procedure is developed to select the input variables and network function parameters, design ANN architectures, and optimize the network performance. The input variable selection is performed following four steps to fit the designed networks (Ding 2002), i.e.,
   (1) Inputs (ε' and ε'') are selected in the frequency range from 200 MHz to 300 MHz, 300 MHz to 400 MHz, and 400 MHz to 500 MHz, respectively, to feed into the designed networks;
   (2) Based on the performance of networks, the important inputs as judged by the program are retained;
   (3) The retained inputs are divided into different combinations to fit the designed networks. Meanwhile, the network function parameters are fine adjusted to match the input combinations and network architectures, thus to optimize the network performance;
   (4) The trained networks are compared for their performance and architecture complexities, and the best networks are retained.

In the following sections, three ANN models for the prediction of soil water content, degree of saturation and dry density are presented and discussed.

4.1 ANN-M1- Prediction of Soil Water Content

The relationship between the real part of the complex permittivity (also known as the dielectric permttivity or dielectric constant) and soil water content has been long recognized and used in practice, e.g. Wang and Schmugge (1980) and Top et al. (1980). However, the studies are all based on a specific soil with property parameters such as density and pore fluid chemical composition kept relatively constant. To train the ANN model for prediction of soil water content of Halton Till with the density and pore fluid composition varying in a relatively broad range, 122 data sets are divided into 74 training, 24 verification and 24 testing data sets. The complex permittivity measured on the 74 soil specimens are used as inputs, and the soil water content data are used as target outputs to feed into designed networks for supervised training. Then 24 data sets are used to verify the network performance. Finally, the network predicts the soil water contents of 24 soil specimens in the testing phase. The predicted values by an ANN are compared with actual values from experiments, and the network is fine adjusted to reduce the error between the predicted value and actual value. The best network is an MLP network with nine neurons in the input layer, ten neurons in the hidden layer and one neuron in the output layer, as shown in Figure 3. The nine neurons correspond to nine inputs as the real parts of complex permittivities (ε') measured at frequencies 201 MHz, 472 MHz, 485 MHz, and 498 MHz, and the imaginary parts (ε'') measured at frequencies 201 MHz, 207 MHz, 311 MHz, 414 MHz and 427 MHz. The single output corresponds to the soil water content. Sensitivity analyses are performed to evaluate the importance of each input variable to the corresponding output. In Figure 3, the input neurons are arranged in the order of their relative importance. The imaginary part at frequency 311 MHz (ε''(311 MHz)) plays the most significant role among the nine inputs according to the sensitivity analysis. Details of the regression statistics are shown in Table 1. The soil water content (output) mean values for the training data sets, verification data sets and testing data sets are 14.8%, 14.6% and 14.7%, respectively. The Root Mean Squared (RMS) error function is defined as (Trajan, 1999):

$$RMS = \sqrt{\frac{\sum_{i=1}^{n}(X_{Pi} - X_{Mi})^2}{n}} \quad (4)$$

where X_{Pi} is the predicted output of the ith data set, X_{Mi} is the measured output of the ith data set, and n is the total number of all data sets. The ratio of RMS versus the data mean (RMS/Data-Mean) is used to evaluate the network performance. The RMS/Data-Mean values in the training, verification and testing data sets are 6.3%, 10.9% and 6.5% respectively. The correlation coefficients (R^2) for the training, verification and testing data sets are 0.94, 0.91 and 0.91, respectively. The results indicate that the model is able to capture the relationship between the inputs (complex permittivity) and outputs (soil water content) and can be used to predict the soil water content of Halton Till when other soil properties change in a relatively broad range.

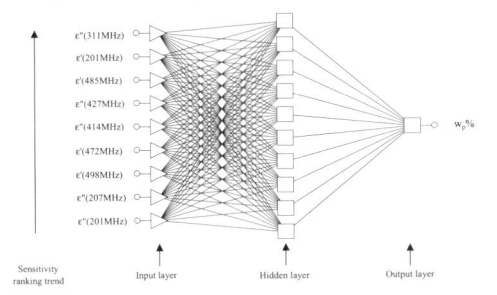

Figure 3. Architecture of ANN-M1

Table 1. Regression statistics of ANN-M1

	Training	Verification	Testing
Data Mean (%)	14.8	14.6	14.7
RMS Error (%)	0.93	1.60	0.95
RMS/Data Mean (%)	6.3	10.9	6.5
Correlation (R^2)	0.94	0.91	0.91

4.2 ANN-M2 - Prediction of Degree of Saturation

During the modeling of ANN-M2 for the prediction of the degree of saturation, the network architecture design and input variable selection are essentially the same as that in ANN-M1. The data sets are randomly divided to 67 training, 28 verification and 27 testing data sets. The best network has seven neurons in the input layer, six neurons in the hidden layer and one neuron in the output layer corresponding to the predicted degree of saturation. The architecture is shown in Figure 4. Among seven inputs, the real part at frequency 201 MHz (ε'(201MHz)) is the most distinguished input according to the sensitivity analysis.

The network regression statistics are summarized in Table 2. The RMS/Data-Mean values for the training, verification and testing data sets are 4.2%, 3.0% and 3.6% respectively, indicating the model performance is reliable and stable. The correlation coefficients for the training, verification and testing data sets are 0.97, 0.97 and 0.97, respectively. This further proves that the developed model is able to capture the essential association between the degree of saturation and complex permittivity with satisfactory performance.

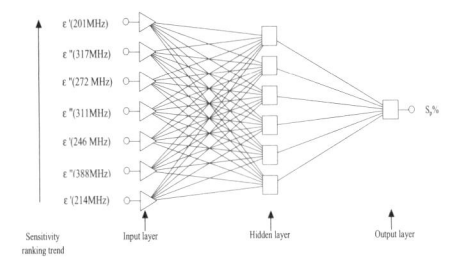

Figure 4. Architecture of ANN-M2

Table 2. Regression statistics of ANN-M2

	Training	Verification	Testing
Data Mean (%)	82.5	87.9	87.5
RMS Error (%)	3.46	2.65	3.13
RMS/Data Mean(%)	4.2	3.0	3.6
Correlation (R^2)	0.97	0.97	0.97

ANN-M2 has only 7 input variables and 6 hidden neurons. The simple architecture and less complexity result in fast execution and good network performance. Compared to ANN-M1 for the prediction of soil water content, ANN-M2 has higher correlation coefficients (R^2) in the training, verification and testing data sets and lower RMS/Data-Mean values. The soil water content represents the gravimetric ratio of water and dry solids, while the degree of saturation represents the volumetric ratio of water and the total voids. It is known that the complex permittivity is closely associated with the volumetric fractions of the water, air and solids in soil-water systems (Kraszewski 1997). Consequently, the complex permittivity is more sensitive to the changes in the degree of saturation than to the changes in the soil water content (the mass of water divided by the mass of dry solids based on soil mechanics definition), which results in higher correlation coefficients in ANN-M2 than those in ANN-M1. The independent development of ANN-M1 and ANN-M2 leads to the different ANN architectures, as shown in Figures 3 and 4. It is also shown that the most significant input and sensitivity rankings are quite different in ANN-M1 and ANN-M2.

4.3 ANN-M3 - Prediction of Soil Dry Density

Soil density is routinely measured in a geotechnical investigation. Scholte 1999, Xie 1999 and Josic 2001 prepared the soil specimens with relatively consistent dry densities. This was intended to observe the effects of other factors such as the soil water content and pore fluid composition on the complex permittivity. Shang et al. (2000) developed a number of linear regression models to relate three independent variables, namely, the water content, bulk density and salinity, to the soil complex permittivity at specific frequencies using data collected by Scholte (1999). It was found that the bulk density was significant only at a frequency of 200 MHz, and became insensitive at frequencies higher than 200 MHz; thereby it was dropped from the linear regression models. During the ANN modeling process, it is found that the complex permittivity data as the only inputs are not adequate to achieve reliable network performance, because the soil volumetric water content plays a predominant role in the soil complex permittivity, which overshadowed the effect of density (Ding, 2002). Therefore, a sub-model named as ANN-θ_v is developed to predict the volumetric water content as an aid in the prediction of dry density. Details of the modeling procedure and model performance, including input variable selection, function parameter selection, network architecture design, and statistics of network performance, can be found in Ding (2002). The output of the volumetric water content is used as one of the inputs in ANN-M3. It should be noted that the prediction capacity of ANN-M3 is not compromised by requiring the volumetric water content as an input. The volumetric water content values are generated as outputs from a sub-ANN model and are not directly measured from experiments.

The best network for the prediction of dry density is found to be an MLP network after a number of trials, as shown in Figure 5. The model consists of nine input neurons, five hidden neurons, and a single output corresponding to the dry density of Halton Till. According to the sensitivity analysis, the volumetric water content (θ_{vp}%) plays the key role in the prediction. Due to the lower sensitivity of the dry density to the complex permittivity and the effects caused by other factors such as the soil water content and pore fluid chemistry, a different approach is adopted in the training process of this model, including development of sub-model, the lower learning rate and larger number of epochs (Ding 2002). Following the similar procedure in ANN-M1 and ANN-M2, 122 data sets are randomly divided into 78 training, 24 verification and 20 testing data sets.

The regression statistics of this model are summarized in Table 3. The RMS/Data-Mean values of the training, verification and testing data sets are 1.6%, 1.6% and 2.5%, respectively, which indicates the relatively stable and satisfactory performance of the model. The correlation coefficients for the training data sets, verification data sets and testing data sets are 0.85, 0.92 and 0.75 respectively. The correlation coefficient of the testing data set is lower than those of the training and verification data set, which may be due to the rather narrow range of the dry density in the database, and the noise introduced by other influencing factors such as the soil water content and pore fluid composition. The performance of the network can be improved upon obtaining more representative data sets for training.

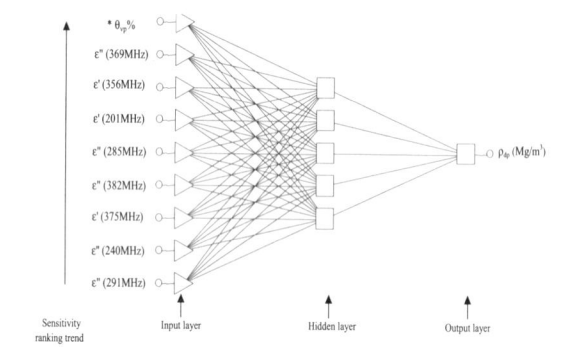

Figure 5. Architecture of ANN-M3

Table 3. Regression statistics of ANN-M3

	Training	Verification	Testing
Data Mean (Mg/m3)	1.88	1.87	1.91
RMS Error (%)	0.031	0.029	0.048
RMS/Data Mean (%)	1.6	1.6	2.5
Correlation (R^2)	0.85	0.92	0.75

5 CONCLUSIONS

The complex permittivity of Halton Till, a soil recovered from a landfill site in Halton, Ontario, Canada, was measured using a custom developed apparatus in laboratory in the frequency range from 0.3 MHz to 1.3 GHz. A database consisting of 122 soil specimen measurements is established for modeling purposes. Artificial Neural Networks (ANNs) are adopted and three MLP models named as ANN-M1 for predicting the soil water content, ANN-M2 for predicting the degree of saturation, and ANN-M3 for predicting the dry density are trained, verified, and tested. The models have appropriate architectures, reasonable complexities, and reliable performance. The models are able to predict soil properties with potential affecting factors changing in a relatively broad range. The experimental results used as the input and output data were measured by three researchers, which indicates that the random assignment of the data sets used for the training, verification and testing is independent.

For each soil specimen, there are 47 complex permittivity data (ε' and ε'') measured available as input variables, and there is only one set of output representing the soil water content, degree of saturation and dry density. Therefore, one of the keys to achieve good performance in the modeling process is to reduce the input variables (Ding 2002). The most representative and significant input variables are selected to relate the corresponding outputs, thereby to establish the most representative relationship between the inputs and outputs. The developed models perform well as judged in terms of the RMS/Data-Mean values and correlation coefficients. It is noted that the correlation coefficient in the testing data sets of ANN-M3 (R^2=0.75) is slightly lower than those of the training data sets (R^2=0.85) and verification data sets (R^2=0.92) due to the rather narrow range of the dry density of soil specimens. The performance of ANN-M3 model can be further improved when a more comprehensive database is available.

ACKNOWLEDGEMENT

The research is supported by the Natural Science and Engineering Research Council of Canada. Dr. R.K. Rowe was involved in the supervision of three graduate students involved in the research related to data collection.

REFERENCES

Baldi, G., Bellotti, R., Ghionna, V.N., Jamiolkowski, M., and Pasqualinin, E. (1986). Interpretation of CPTs and CPTUs- 2nd part: Drained Penetration of Sands. Proc. 4th International Geotechnical Seminar on Field Instrumentation and In Situ Measurements, School of Civil and Structure Engineering, Nanyang Technology Institute, Singapore, pp. 143-156.

Boyarshii, D.A., Tikhonov, V.V., and Komarova, N.Y. (2002). Model of Dielectric Constant of Bound Water in Soil for Application of Microwave Remote Sensing. Progress in Electromagnetic Research, PIER 35, pp 251-269.

Chang, D.H. and Islam, S. (2000). Estimation of Soil Physical Properties Using Remote Sensing and Artificial Neural Network. Journal of Remote Sensing of Environment, Vol. 74, No. 3, pp. 327-340.

Chaudhari, A., Das, A., Raju, G., Chaudhari, H., Khirade, P., Narain, N., and Mehrotra, S. (2001). Complex Permittivity Spectra of Binary Mixture of Ethanol with Nitrobenzene and Nitrotoluene Using the Time Domain Technique. Physical Science and Engineering. Vol. 25, No. 4, pp. 205-210.

Ding, W. (2002). Characterizing Halton Till Using Complex Permittivity and ANNs. M.E.Sc. Thesis, Civil and Environmental Engineering, the University of Western Ontario.

Goh, A.T.C. (1995). Modeling Soil Correlations Using Neural Networks. Journal of Computing in Civil Engineering, ASCE, Vol. 9, No. 4, pp. 275-278.

Josic, L. (2001). Complex Permittivity of Water and Soil Contaminated by Heavy Metals and Organic Compounds. M.E.Sc. Thesis, Civil and Environmental Engineering, the University of Western Ontario.

Kaya, A. (2002). Evaluation of Soil Porosity Using a Low MHz Range Dielectric Constant. Journal of Engineering and Environmental Science. Vol. 26, pp. 301-307.

Kaya, A., and Fang, H. (1997). Identification of Contaminated Soil by Dielectric Constant and Electrical Conductivity. Journal of Environmental Engineering, Vol. 123, No. 2, pp. 169-177.

Klein, K., and Santamarina, J.C., (1997). Methods for Broad-Band Dielectric Permittivity Measurements (Soil-Water Mixtures, 5 Hz to 1.3 GHz). ASTM Geotechnical Testing Journal, Vol. 20, No. 2, pp. 168-178.

Kraszewski, A. (1997). Microwave Aquametry. IEEE Press, New York.

Kurup, P.U., and Dudani, N.K. (2002). Neural Networks for Profiling Stress History of Clays from PCPT Data. Journal of Geotechnical and Geoenvironmental Engineering, Vol. 128, No. 7, pp. 569-579.

Mitchell, J.M. (1993). Fundamentals of Soil Behavior. 2nd Edition. John Wiley & Sons, Inc.

Najjar, Y.M., Basheer, I.A., and Naouss, W.A. (1996). On the Identification of Compaction Characteristics by Neuronets. Journal of Computers and Geotechnics, Vol. 18, No. 3, pp. 167-187.

Penumadu, D., and Zhao, R. (1999). Triaxial Compression Behavior of Sand and Gravel Using Artificial Neural Networks (ANN). Journal of Computers and Geotechnics, Vol. 24, pp. 207-230.

Rowe, R.K., Shang, J.Q., and Xie, Y. (2001). Complex Permittivity Measurement System for Detecting Soil Contamination. Canadian Geotechnical Journal, Vol. 38, No.3, pp. 498-506.

Rowe, R.K., Shang, J.Q. and Xie, Y. (2002) Effect of Permeating Solutions on the Complex Permittivity of Compacted Clay". Canadian Geotechnical Journal. Vol. 39, No. 5, pp. 1016-1025.

Santamarina, J.C., and Fam, M. (1997). Dielectric Permittivity of Soil Mixed with Organic and Inorganic Fluids. Journal of Environmental and Engineering Geophysics, Vol. 2, No. 1, pp. 37-51.

Schaap, M.G., and Leij, F.J. (1998). Using Neural Networks to Predict Soil Water Retention and Soil Hydraulic Conductivity. Soil and Tillage Research, Vol. 47, pp. 37-42.

Scholte, J.W. (1999). The Complex Permittivity of Compacted Halton Till. M.E.Sc. Thesis, Civil and Environmental Engineering, the University of Western Ontario.

Scholte, J.W., Shang, J.Q., and Rowe, R.K. (2002). Improved Complex Permittivity Measurement and Data Processing Technique for Soil-Water Systems. Geotechnical Testing Journal, Vol. 25, No. 2, pp. 187-198.

Shang, J.Q., Rowe, R.K., Umana, J.A., and Scholte, J.W. (1999). A Complex Permittivity Measurement System for Undisturbed/Compacted Soils, ASTM Geotechnical Testing Journal, Vol. 22, No. 2, pp.165-174.

Shang, J.Q. and Rowe, R.K. (2003) Detecting Landfill Leachate Contamination using Soil Electrical Properties. ASCE Practice Periodical of Hazardous, Toxic, and Radioactive Waste Management, Special Edition on Soil and Groundwater Remediation, Vol. 7, No.1, pp. 3-11.

Shang, J.Q., Scholte,J.W. and Rowe, R.K. (2000). Multiple Linear Regression of Complex Permittivity of a Till at Frequency Range from 200 MHz to 400 MHz. Subsurface Sensing Technologies and Applications. Vol.1, No,3, pp.337-356.

Thevanayagam S. (1994). Soil-Structure Characterization Using Electromagnetic-Waves. Particulate Science and Technology, 12 (3): 281-298.

Topp, G.C., Davis, J.L. and Annan, A.P. (1980). Electromagnetic determination of soil water content and electrical conductivity measurement using time domain reflectometry. Water Resources Research, Vol. 16, pp 574-582.

Trajan (1999). Manual for Trojan Neural Networks. Trojan Software Ltd, UK.

Wang, J.R., and Schmugge, T.J. (1980). An Empirical Model for the Complex Dielectric Permittivity of Soils as a Function of Water Content. Geoscience and Remote Sensing. Vol. 18, pp. 288-295.

Xie, Y. (1999). Complex Permittivity of a Clayey Till Permeated by Aqueous Ionic Solution. M.E.Sc. Thesis, Civil and Environmental Engineering, the University of Western Ontario.

Cylindrical cavity expansion modeling for interpretation of cone penetration tests

F. Elmi & J.L. Favre
Ecole Centrale Paris, Laboratory of Mechanics (MSSMat), France

Keywords: cylindrical cavity expansion, cone penetration test, pressuremeter test

ABSTRACT: The cavity expansion theory has been widely used in the analysis of geotechnical problems such as the bearing capacity of deep foundations, interpretation of pressuremeter tests and cone penetration test. In this paper, a method based on cavity expansion is used in order to determine the penetrometer tip resistance in soil. The cavity expansion method foresees two steps in order to predict the cone tip resistance (q_c). First, the limit pressure p_L is determined by numerical computation. During this step the cavity expansion is simulated by the finite element method with a non-linear elastoplastic constitutive model in order to find the limit pressure p_L.
The second step consists of relating the cavity expansion limit pressure to q_c. To this propose Salgado's slip modes have been used for the computation of q_c. This method differs from the prior cavity expansion method and extends it. It simulates the cavity expansion by the finite elements method and it uses a non-linear elastoplastic constitutive model.

1 INTRODUCTION

Cavity expansion analysis plays an important role in geotechnical engineering. It has been widely applied to geotechnical problems such as the bearing capacity of deep foundations, interpretations of pressuremeter tests, cone penetration tests, pile driving and detonation of explosive devices within soil deposits.

Both the spherical and cylindrical cavity expansions have been investigated by researchers. Published solutions vary because of differences in the constitutive models used to describe the stress-strain behaviour of the expanding body and the failure modes. Most analytical analysis have been based on elastic-perfectly plastic soil models (e.g. Hill, Chadwick, Salençon, Vesic & Yu).

2 DEEP PENETRATION MECHANISM

The analogy between cavity expansion and cone penetration was first pointed out by Bishop in connection with deep punching of metals, after observing that the pressure required to produce a deep hole in an elastic-plastic medium is proportional to that necessary to expand a cavity of the same conditions. It has been used successfully by different authors and for various types of materials, e.g. Gibson for clays, Skempton for sands and Ladanyi for sands, for sensitive clays and for rocks. (Ladanyi & Johnston, 1974)

2.1 Calculation of Penetration Resistance

When a penetrometer is pushed into the soil, it creates and expands a cylindrical cavity. Thus, there should be a relationship between penetration resistance and the pressure required to expand a cylindrical cavity in the soil from an initial radius. This initial radius can equal zero. Figure 1 shows the displacement field generated around the base of a pile when it is pushed down in to the soil. A rigid soil core with an approximately conical shape is formed at the base of the pile with theoretical apex angle equal ($\pi/4+\phi/2$). The displacement field immediately below the pile base, inside and along the surface of the conical soil core, is vertical. There is a rotation in the displacement field from vertical underneath the pile base to horizontal beyond the sloping line. The horizontal displacement field is compatible with the idea of an expanding cylindrical cavity.

Based on these observations, Salgado et al. (1997) have assumed the failure pattern of figure 2 for a penetrometer moving down through the soil mass. To the right of line BC, in zone P, the major principal stress and strain fields, and the displace

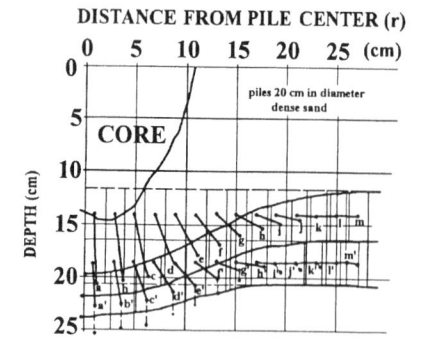

Figure 1. Displacements underneath tip of cylindrical penetrometer (after BCP 1971 in Salgado et al., 1997)

ment field are horizontal. The stress field in zone P is directly related to the pressure required to expand a cylindrical cavity from an initial radius equal to zero. In zone Q, and at its interface with zone T, the major principal stresses are vertical and directly related to penetration resistance. There is a 90° rotation from a vertical major principal stress in zone Q to a horizontal major principal stress in zone P. The angle ABC must accordingly be equal to 90° for the stress rotation to take place continuously in space. Penetration resistance is derived from the cylindrical cavity pressure using a stress rotation analysis.

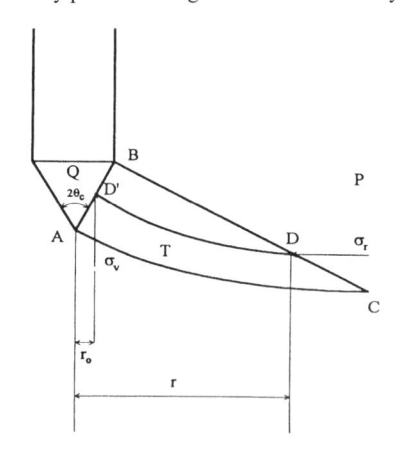

Figure 2. Slip pattern under cone penetrometer (Salgado et al., 1997)

2.2 Relation between p_L and q_c

Following Bolton, the major principal stresses in two different zones radiating from a same point, separated by an angle $\Delta\Psi$, for plane strain conditions, as shown in figure 3 are related through equation (1).

$$\sigma_1^A = \sigma_1^B e^{2\Delta\Psi \tan\phi} \tag{1}$$

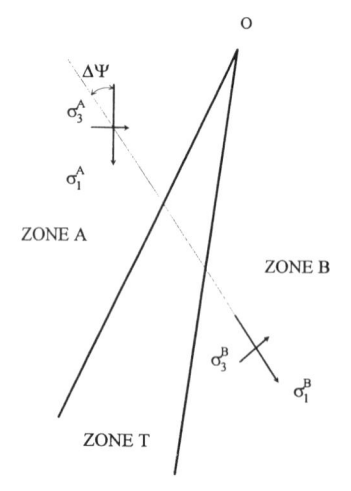

Figure 3. Stress rotation between zones A and B takes place in transition zone T

It can be applied to zones P and Q of figure 2, where $\Delta\Psi=\pi/2$, to give equation (2), or, equation (3).

$$\sigma_1^Q = \sigma_1^P e^{\pi \tan\phi} \tag{2}$$

$$\sigma_v = \sigma_r e^{\pi \tan\phi} \tag{3}$$

From this, Salgado et al. (1997) have shown the relation between σ_r and σ_v by equations (4), (5) and (6).

$$\sigma_v = p_L e^{\pi \tan\phi} \left(1 + C - 2C \frac{r_0}{d_c}\right)^{\beta-1} \tag{4}$$

$$C = e^{(\pi/2)\tan\Psi} \cot\theta_c \tag{5}$$

$$\beta = 1 - \frac{N-1}{N} \tag{6}$$

N is element's flow number; given by equation (7).

$$N = \tan^2\left(\frac{\pi}{4} + \frac{\phi}{2}\right) \tag{7}$$

The penetration resistance q_c resulting from integration on the cone surface is given by equation (8).

$$q_c = 2p_L e^{\pi \tan\phi} \frac{(1+C)^{1+\beta} - (1+\beta)C - 1}{C^2 \beta(1+\beta)} \tag{8}$$

3 PRACTICAL APPROACH TO DETERMINE THE PENETRATION RESISTANCE

A method based on cavity expansion is used in order to calculate the penetrometer tip resistance. Two steps are needed from the cavity expansion method in order to predict cone tip resistance (q_c):

First, one has to determine the limit pressure p_L by numerical computation. For this step the cavity expansion is simulated by finite element method with a non-linear elasto-plastic constitutive model in order to find the limit pressure p_L. The second step is to relate the cavity expansion limit pressure to q_c. In this paper we have used Salgado's slip mode presented above, for the computation of q_c.

3.1 Constitutive model

The soil behaviour is modelled by HUJEUX elasto-plastic model. This model is based on the theory of plasticity with hardening and the concept of critical state. It derives from the original Cam-Clay model. This model is a multimecanisms model, decomposed in three mechanisms of plane deformation in three orthogonal planes and one purely volumetric mechanism of consolidation.

The elastic part is supposed to obey a non-linear elasticity behaviour, where the bulk (K) and shear (G) modulus are functions of the mean effective stress (p') given by equations (9) and (10) where K_{ref} and G_{ref} are the bulk and shear modulus measured at the mean reference pressure (p_{ref}).

$$K = K_{ref} \left(\frac{p'}{p_{ref}} \right)^{n_e} \tag{9}$$

$$G = G_{ref} \left(\frac{p'}{p_{ref}} \right)^{n_e} \tag{10}$$

Adopting the soil mechanics sign convention (compression positive), the deviatoric primary yield surface of the k plane is given by equations (11), (12) and (13), where ϕ'_{pp} is the friction angle at the critical state.

$$f_k(\sigma, \varepsilon_v^p, r_k) = q_k + p'_k \cdot \sin \phi'_{pp} \cdot F_k \cdot r_k \tag{11}$$

$$F_k = 1 - b \ln(p'_k / p_c) \tag{12}$$

$$p_c = p_{co} \exp(-\beta \varepsilon_v^p) \tag{13}$$

The parameter b controls the form of the yield surface in the (p',q) plane and varies from $b = 0$ to 1 passing from a Coulomb type surface to a Cam-Clay type one. β is the plasticity compression modulus and p_{co} represents the critical state stress corresponding to the initial void ratio. The internal variable r_k, called degree of mobilized friction, is associated with the plastic deviatoric strain. This variable introduces the effect of shear hardening of the soil and permits the decomposition of the behaviour domain into pseudo-elastic, hysteretic and mobilized domains. It is given by equations (14), (15) and (16).

$$r_k = r_k^{el} + \frac{\int_0^t \bar{\varepsilon}^p dt}{a + \int_0^t \bar{\varepsilon}^p dt} \tag{14}$$

$$a = a_1 + (a_2 - a_1)\alpha_k(r_k) \tag{15}$$

$$\begin{cases} \alpha_k = 0 & \text{if } r^{ela} \langle r_k \langle r^{hys} \\ \alpha_k = \left(\frac{r_k - r^{hys}}{r^{mob} - r^{hys}} \right)^m & \text{if } r^{hys} \langle r_k \langle r^{mob} \\ \alpha_k = 1 & \text{if } r^{mob} \langle r_k \langle 1 \end{cases} \tag{16}$$

Where a_1, a_2 and m are model parameters and r_{hys} and r_{mob} designate the extend of the domain where hysteretic degradation occurs. The isotropic yield surface is given by equations (17) and (18).

$$f_{iso} = |p'| - d \cdot p_c \cdot r_{iso} \tag{17}$$

$$r_{iso} = r_{iso}^{elas} + \frac{\int_0^t |(\dot{\varepsilon}_v^p)_{iso}| dt}{c \cdot (p_c / p_{ref}) + \int_0^t |(\dot{\varepsilon}_v^p)_{iso}| dt} \tag{18}$$

Where d is a model parameter representing the distance between the isotropic consolidation line and the critical state line in the (e-$\ln p'$) plane and c controls the volumetric hardening.

3.2 Cylindrical cavity expansion modeling by FEM

The problem of cylindrical cavity expansion was investigated, through the HUJEUX model. In order to study the pressuremeter test and the effect of parameters on the limit pressure p_L and pressuremeter curves ($\Delta V/V_0$ against the pressure, p), we used the method of finite elements adopted in the GEFDYN code (Aubry et al. 1986).

The parameters of the model concern both the elastic and plastic behaviour of the soil. Table (1) represent the different parameters of HUJEUX model that we will use in this study. These parameters correspond to Flandre clay with $Ip = 58\%$, normally consolidated (OCR=1) and for the isotropic stress $p'=150$kPa.

All of our calculations in this study are made using an one-dimensional and axisymmetric problem in plane strain conditions (2D). The probe with radius r_0, applies the pressure on the cavity wall. A large variety of meshes have been examined and the mesh that is represented in figure 4.

The initial radius of cavity is equal $r_0 = 0.022$m. This corresponds to the CPMT test (Cone Pressuremeter Test). r is considered equal to $150r_0$ for satisfying the infinite conditions. The effect of mesh size is considerd in figure 5.

The results of FEM are shown in Figure 6. In this figure, we have drawn the curve of p against $\Delta V/V$ for different values of G_{max}. The effect of G_{max} on

Table 1 : Soil properties and constitutive model parameters

	Parameters	Values	Relations
Soil properties	w_L (%)	92	$Ip = 0.73 (w_L-13) = 58$
	C_c	0.36	Odometer test
	Cs	0.15	Odometer test
	e	0.95	Calculated
	$\rho(kg/m^3)$	1900	Measured
	σ'_0	150 kPa	$\rho.z$
	σ'_c	150 kPa	Odometer test
	n	0.487	$n = e/(1+e)$
	k_0	0.61	$k_0 = 1 - \sin\phi_{pp}$
Elastic	G_{ref}(MPa)	161	Measured
	K_{ref}(MPa)	350	$\nu = 0.33$ measured $K = 2G(1+\nu)/3(1-2\nu)$
	p_{ref}	1 MPa	
	n_e	0.8	Estimated [*]
Critical state and plasticity	ϕ'_{pp}	23	Estimated: $\phi'_{pp} = -8.38\text{Ln}(w_L) + 60.6$
	β	12	$\beta = 2.3(1+e)/C_c$
	p_{co} (MPa)	0.6	Graph of "relative arrangement" [*]
	d	2.0	$d = \exp(\Delta e/\lambda)$
	b	1	Estimated [**]
Flow rule and isotropic hardening	ψ_c	23	Estimated [**]
	α_ψ	1	Estimated [**]
	a_1	1e-4	Estimated [**]
	a_2	5e-3	Estimated [**]
	c_1	0.18	Estimated [**]
	c_2	0.09	$c_2 = 0.5c_1$
	m	2	Estimated [**]
Threshold domains	r^{ela}	0.025	Estimated [**]
	r^{hys}	0.1	Estimated [**]
	r^{mob}	0.5	Estimated [**]
	r^{ela}_{iso}	1e-4	Estimated [**]

[*] Favre et al. (2003)   [**] Modaressi et al. (2003)

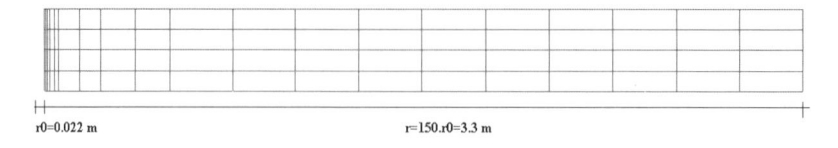

$r_0=0.022$ m $r=150.r_0=3.3$ m

Figure 4: Finite element mesh for modelling the cone pressuremeter test ($h=44$cm and $D=0.44$cm)

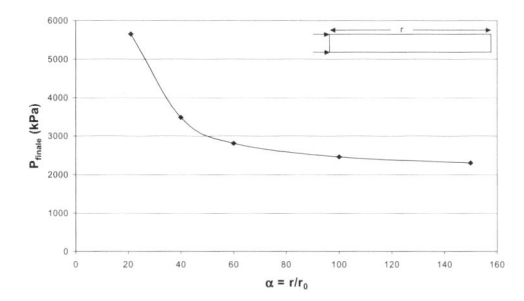

Figure 5 : Effect of mesh size on cavity expansion modelling

limit pressure is considerable. In this figure the limit pressure is very near one that has been measured *in-situ* but the form of the simulated curve in the beginning of the test is less steep than the reality. Biarez suggests (personal communication) that the fissuring and anisotropy in clays of Flandre can explain this phenomenon that one is not able to model by the numeric methods. It remains like a very strong hypothesis.

4 VALIDATION OF THE PROCEDURE

In order to validate this procedure, we used the resulats of a cone-pressuremeter carried out in experimental site of MERVILLE. The geological map of this site is presented in figure 7. The site is covered by Flandre clay. This clay is used for modeling of cylindrical cavity expansion in the previous step. The tip resistance is calculated by Salgado's method. Comparison between *in-situ* measurement of q_c and simulated ones is shown in figure 8. The very good agreement between two series of results is observed.

5 CONCLUSIONS

In this paper, a new procedure is used to find the bearing capacity of steady penetration of penetrometer in soils. This procedure couples the perfectly plastic solution of Salgado et al. (1997) and cylindrical cavity expansion modeling by finite elements method with the HUJEUX elasto-plastic multi-mechanism model with hardening. The advantage of this coupling procedure is to model the kinematics of penetration close to the tip, with perfect plasticity law and the kinematics far from the tip with an elasto-plastic law with hardening in conjunction with a finite elements method. This procedure reduces the calculation cost more than penetration methods with large deformation finite elements.

Once the procedure is validated, it will be used for modelling the CPT with the different characteristics of soil. For checking, the results are compared with literature correlations and a few results of in-situ tests. This will be a base for finding a robust correlation between q_c and the other parameters of soil.

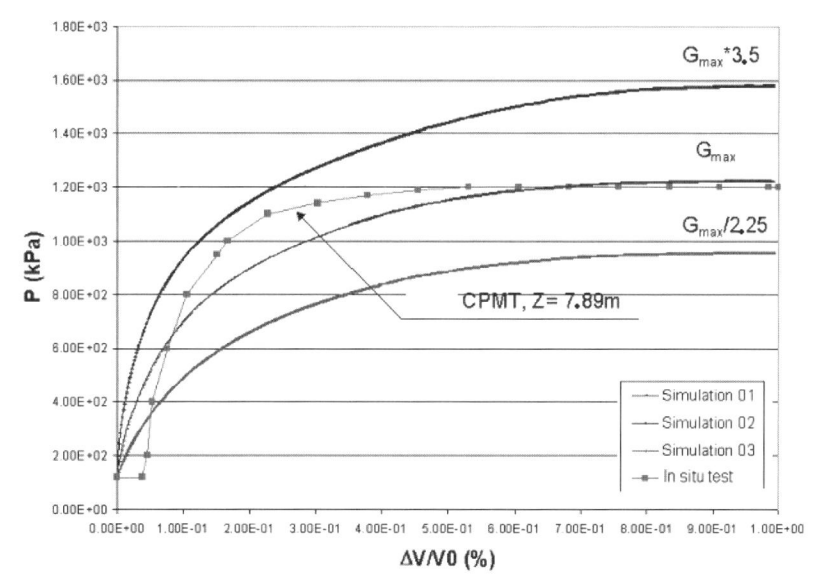

Figure 6: Cylindrical cavity expansion modelling in Flandre clay, z=7.89m

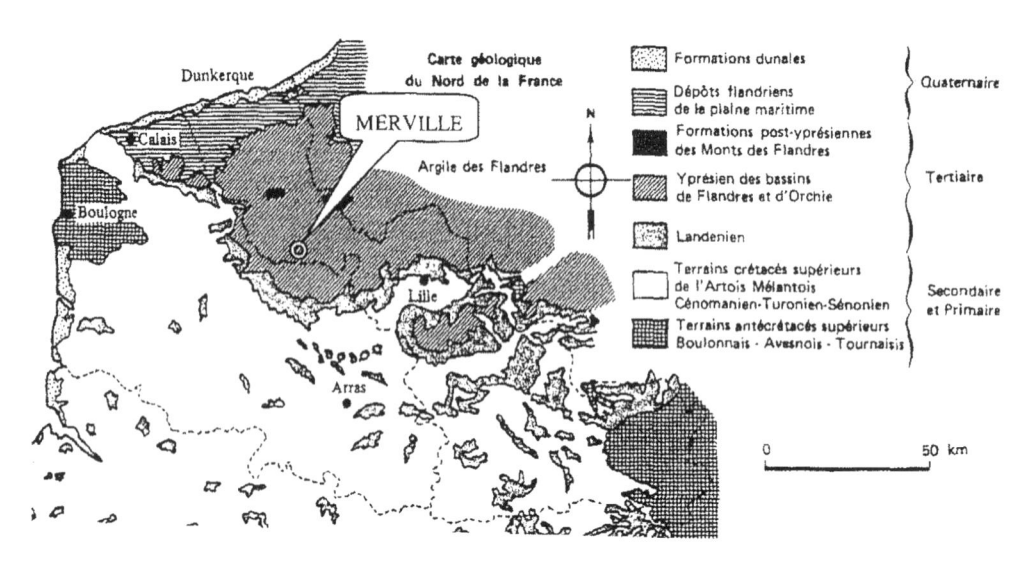

Figure 7: North of France geological map, experimental site of MERVILLE (after Ferber & Abraham, 2002)

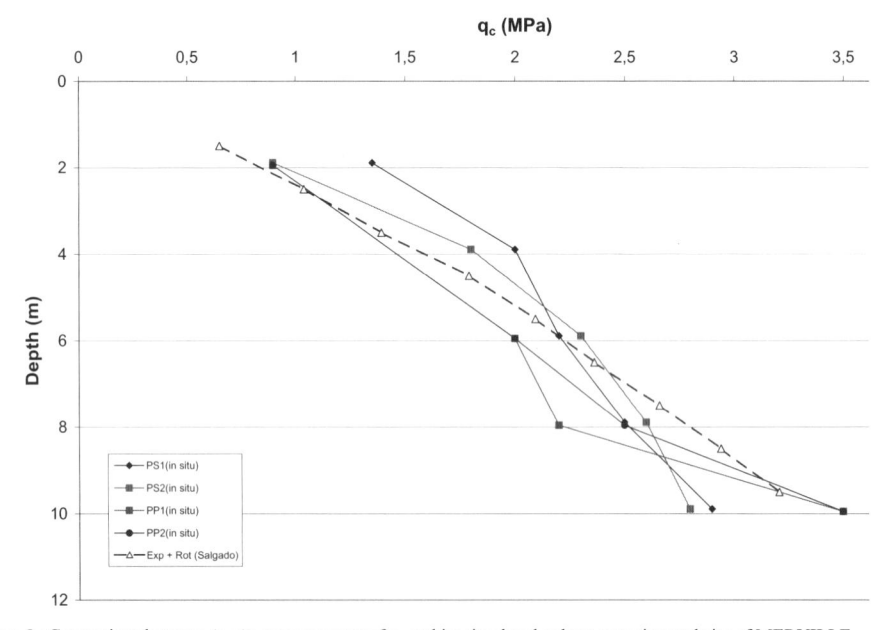

Figure 8 : Comparison between *in-situ* measurement of q_c and its simulated values, experimental site of MERVILLE

ACKNOWLEDGEMENTS

The Authors wishes to acknowledge the support provided by the FUGRO France Corporation. They would also like to thank Prof. Biarez J., Prof. Modaressi A. and Dr. Lopez-Caballero F. for stimulating discussions during this work.

REFERENCES

Aubry, D. Hujeux, J.C. Lassoudiere, F. Meimon, Y. 1982. *A double memory model with multiple mechanisms for cyclic soil behaviors*, International Symposium on Numerical Models in Geomechanics, Zurich,, vol. 1, 3-13.

AUBRY, D. CHOUVET, D. MODARESSI, A. & MODARESSI, H. 1986. *GEFDYN : Logiciel d'analyse de comportement mécanique des sols par éléménts finis avec prise en compte du couplage sol-eau-air,* Manuel scientifique, Ecole Centrale Paris, LMSS-Mat.

BCP Committee 1971. *Field tests on piles in sand,* Soil and Foundation, Tokyo, Japan, Vol. 11, No. 2 : 29-49.

Elmi, F. 2003. *Détermination des propriétés des sols marins par reconnaissance géophysique et géotechnique.* PhD thesis, Ecole Centrale Paris, France.

Elmi, F. & Favre, J.L. 2002. *Analogy between analyse of pile and penetrometer tip resistance by cavity expansion method,* Ninth International Conference on Piling and Deep Foundation, DFI2002, Nice (France).

Favre, J.-L., Biarez, J. & Mekkaoui, S. 2002. *Modèles de comportement en grandes déformations des sables et des argiles remaniées à l'œdomètre et triaxial,* Symposium international de l'identification et détermination des paramètres des sols et des roches pour les calculs géotechniques (PARAM02), Presses de l'ENPC, 369-384, Paris, 2-3 September.

Ferber, V. & Abraham, O. 2002. *Apport des méthodes sismiques pour la détermination des modules élastiques initiaux : Application au site expérimenta de Merville,* Paramètres de calcul géotechnique (PARAM 02), Magnan (ed.) : 41–48.

Hujeux, J.-C. 1985. *Une loi de comportement pour le chargement cyclique des sols, Génie Parasismique,* V. Davidovici, Presses ENPC, France : 278-302.

Ladanyi, B. & Johnston, G.H. 1974, *Behavior of circular footing and plate anchors embedded in permafrost,* Canadian Geotechnical Journal, Vol. 11 : 531-552.

Modaressi, A., Lopez-Caballero, F. & Elmi, F. 2003. *Identification of an elastoplastic model parameters using laboratory and in-situ tests.* Third International Syposium on Deformation Characteristics of Geomaterials (IS Lyon 03), Lyon (France), Septembre 22-24.

Salgado, R. Mitchell, J.K. & Jamiolkowski, M. 1997. *Cavity Expansion and Penetration Resistance in sand.* Journal of Geotechnical and Geoenviromental Engineering , Vol. 123, No. 4, ASCE.

Proceedings ISC-2 on Geotechnical and Geophysical Site Characterization, Viana da Fonseca & Mayne (eds.)
© 2004 Millpress, Rotterdam, ISBN 90 5966 009 9

Stratigraphic profiling by cluster analysis and fuzzy soil classification from mechanical cone penetration tests

J. Facciorusso & M. Uzielli
Department of Civil Engineering, University of Florence, Italy

Keywords: geologic uncertainty, stratigraphy, cone penetration testing, clustering, fuzzy

ABSTRACT: Cone penetration testing has gained increasing popularity in geotechnical site characterization due to its speed and economy, and to the good quality of its data in terms of precision, accuracy, repeatability and continuity of measurement when compared to other in-situ tests. In the present paper, cluster analysis and fuzzy soil classification from mechanical cone penetration tests are applied for stratigraphic delineation in the harbor area of the southern Italian town of Gioia Tauro. The main features of the clustering and fuzzy algorithms adopted are described. The results of stratigraphic profiling by cluster analysis and fuzzy classification for a number of soundings are shown and compared; the applicability of the adopted site characterization techniques is assessed through comparison with adjacent borehole logs and standard penetration tests.

1 INTRODUCTION

Cone penetration testing is increasingly employed in geotechnical site characterization due to the precision, accuracy, and repeatability of its output data. Data deriving from mechanical cone penetration testing, however, are affected by larger uncertainties than those from electrical cone penetration and piezocone testing; furthermore, the data sampling interval is greater and the distance between cone-tip resistance and sleeve friction measures is larger. Nonetheless, in Italy, at present, a considerable number of databases used for important geotechnical analyses - such as liquefaction susceptibility evaluation - consists of data from mechanical cone tests. Thus, it appears advisable to apply different techniques and compare their results to obtain reliable stratigraphic profiling from such data.

In the geotechnical literature, results of cone penetration tests have been interpreted to delineate stratigraphic profiles in one or more of the following ways: 1) visual examination of raw data; 2) empirical soil classification charts; 3) statistical methods; 4) fuzzy soil classification techniques; and 5) neural networks. Here, the applicability of cluster analysis and fuzzy soil classification to mechanical cone penetration test data for stratigraphic profiling is evaluated for the harbor area of the town of Gioia Tauro, in southern Italy; this study also assisted with liqufaction risk analysis in this area.. A description of the basic features of the algorithms employed is provided, as well as an overview and an assessment of the main results.

2 GEOLOGICAL FEATURES AND SOURCE DATA

The harbour area of Gioia Tauro, in the southern Italian region of Calabria, is located in a flat plain, originating from a depression, spreading along its length in a N-S direction and filled by continental sediments of Quaternary Age. The plain primariliy comprises granular saturated soils in its surficial layers (up to a depth ranging from 50 m to 70 m from ground level), overlying a layer of compacted clays and silty clays of considerable thickness (500m or more). The bedrock is located at variable depths of 500-600 m (Facciorusso and Vannucchi, 2003).

Figure 1 shows a representative cross-section from borehole log data and the locations of CPT and SPT tests. As may be observed, the first 20 meters (represented by a dashed area) of the cohesionless deposit - the maximum depth commonly investigated for liquefaction risk analyses - include a thick layer of made ground, overlying, with increasing depth:
- Soil A: coarse to medium loose aeolian sands, with a thickness varying between 3 and 5 m;

905

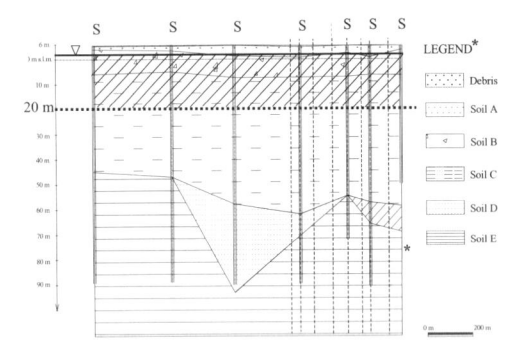

Figure 1 –Representative cross-section of soil stratigraphy in the Gioia Tauro harbor area with bore-hole (indicated as S) and CPT and SPT (represented by a dashed line) location.

- Soil B: coarse and medium to coarse sands with polygenic gravels or sandy polygenic alluvial gravels, with a thickness of about 10 m;
- Soil C: medium and fine to medium dense sands, having a thickness ranging from 30 m to 70 m, including a sequence of lenses and thin layers of sands, gravelly sands and fine silty sands; the top of layer C is found at depths ranging from 7 to 19 m from ground surface. While borehole logs frequently indicate the presence of heterogeneous layers in terms of composition, these are, nevertheless, indicated as single stratigraphic units.

The water table level is estimated at 2.3 m above sea level, corresponding to depths varying from 0.0 m to 3.2 m below the ground surface. The reference water table is the highest measured at the site (with daily fluctuations of 0.35-0.4 m).

Extensive geological and geotechnical surveys have been performed in the past (CPT and SPT tests, geotechnical boreholes, laboratory tests, etc.) to characterize the area, which hosts one of the most important trade port junction of southern Europe.

The results of 6 boreholes, of 25 profiles of mechanical CPT tests and 19 profiles of SPT tests were selected for the present study. The maximum investigated depth ranges between 40 and 91.3 m, for boreholes, between 20.5 m and 39.5 m, for SPT tests, from 20.5 m and 72 m, for CPT tests. The main characteristics of the CPT tests, such as investigated depth and spatial distance from other in situ investigations, are summarized in Table 1. CPT results, in terms of cone tip resistance, q_c, and friction ratio, R_f, in the upper 20m are plotted in Figure 2. The q_c values show a certain variability around an average value of 24 MPa, with a a maximum value of 44 MPa; the R_f values generally fall below 2%, though a peak of 7% is reached.

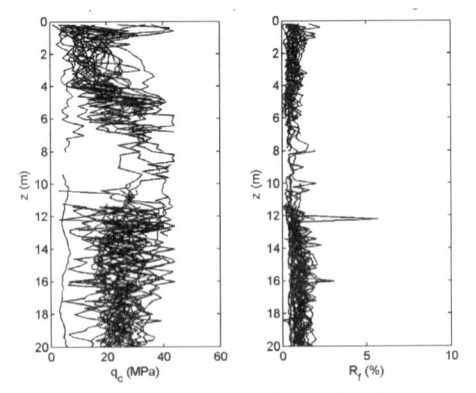

Figure 2 – Comprehensive plots of cone tip resistance, q_c, and friction ratio, R_f, for the 23 investigated soundings, limited to 20 m below ground level

3 CLUSTER ANALYSIS

In general terms, cluster analysis is the art of finding groups in data showing a certain degree of similarity in their mathematical description. The delineation of stratigraphies based on visual inspection of CPT profiles is a complex, subjective procedure, relying on expert judgement. Previous investigations (Hegazy 1998; Hegazy and Mayne 2002) have shown that the application of clustering to piezocone CPT results may allow for the objective detection of inherent correlations between data, and for the consequent assessment of the stratigraphic profile. Thus, in the context of the less accurate and reliable mechanical CPT testing, the attempt will be made herein to identify the presence and location of primary layers (arbitrarily defined as those having a thickness greater than 1 m), secondary layers (having a thickness between 0.5 m and 1 m), lenses, transitions layers, soil mixture, and other features, as defined by Hegazy and Mayne (2002), while excluding all data outliers deriving from accidental or systematic errors, as shown in Figure 3.

The clustering method adopted herein refers to the following succession of operations: 1) the data are arranged into n objects, (i.e. each set of measures performed at each depth investigated during CPT testing) and p attributes, (i.e. soil properties directly measured or derived during CPT testing) in a $n{\times}p$ matrix (in which each row represents all the considered properties at the corresponding depth); 2) two or more attributes are chosen to identify the structure present in the data and, if necessary, standardized to avoid their dependence on the choice of measurement units; 3) a set of proximity (distance) values between all possible pairs of objects is stored in a $n{\times}n$ matrix; 4) the objects are grouped in clusters on the basis of mutual distance.

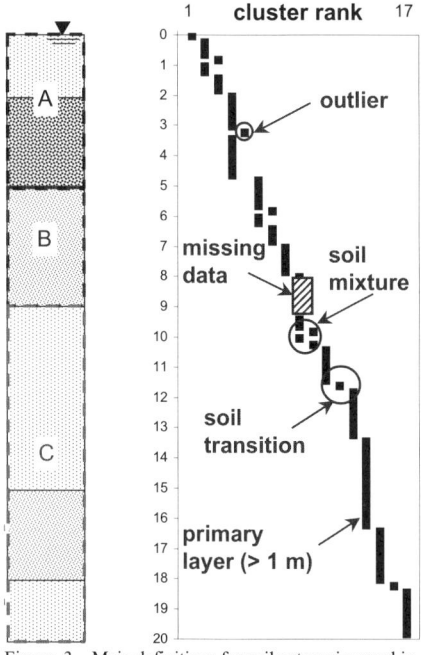

Figure. 3 – Main definitions for soil categories used in the clustering procedure of mechanical CPT data

Clustering can be performed by various algorithms which differ from each other in terms of: (a) the input data (e.g. type and number of variables, standardization method, etc.); (b) the mathematical function adopted to represent the distances between objects (e.g. euclidean, cosine, etc.); (c) the number of clusters (fixed in "partitioning methods", variable in "hierarchical methods"); (d) the procedure of grouping objects to generate clusters (merging objects by means of "agglomerative techniques" or splitting them by means of "divisive techniques"); and (e) the definition of the type of distance (e.g. minimum, maximum, average, etc.,)between clusters (e.g. Kaufman and Rousseeuw, 1990).

In the present study, the selection of the most suitable algorithm to perform clustering analysis was based on the nature of the data (continuous variables), their spatial distribution (generally irregular and widely spaced) and the past, still limited, experiences in the field of geotechnics (Młynarek and Lunne, 1987). A hierarchical agglomerative method was adopted, whereby clusters are merged using the average distance criterion. The algorithm, referred to by the acronym HAMAD, is shown schematically in Figure 4 and detailed hereinafter.

The Minkowski distance, selected as the proximity parameter, is given, between the i-th and the j-th object, by:

$$d(i,j) = \left(\sum_{k=1}^{p} (x_{ik} - x_{jk})^q \right)^{1/q} \quad (1)$$

where i and j vary between 1 and the number of objects, n; p is the number of variables considered and q is set equal to 2.

The HAMAD algorithm was applied to the nor-

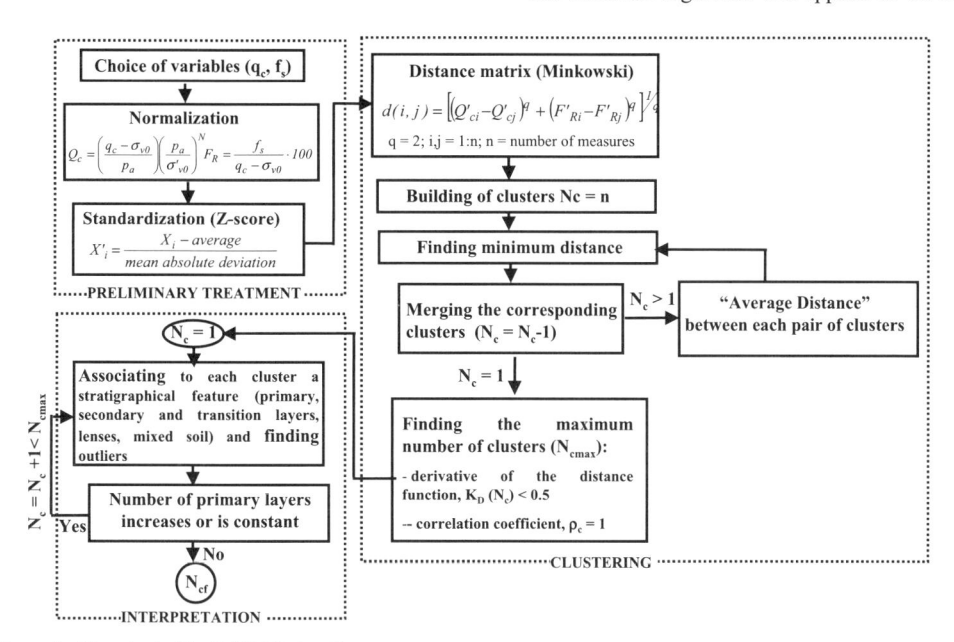

Figure 4 – Flow chart of the HAMAD algorithm

malized cone tip resistance, Q_c, and friction ratio, F_R:

$$Q_c = \left(\frac{q_c - \sigma_{v0}}{p_a}\right)\left(\frac{p_a}{\sigma'_{v0}}\right)^N \quad (2)$$

$$F_R = \frac{f_s}{q_c - \sigma_{v0}} \cdot 100 \quad (3)$$

where $p_a = 0.1$ MPa, σ_{v0} and σ'_{v0} are the vertical total and effective stresses, respectively, and N is an exponent depending on soil type (Robertson, 1990), ranging from 0.5 (for sands) to 1 (for clays). Q_c and F_R were selected in place of q_c and f_s as normalization allows to remove the influence of depth.

To avoid the dependence on the choice of measurements units, which may have a strong effect on the results of clustering, the two reference variables were subsequently standardized into the unitless input variables to the HAMAD algorithm, X_1' and X_2', using a modified Z-score method:

$$X'_i = \frac{X_i - m_X}{s_X} \quad (4)$$

where X_i is the variable to be standardized ($X_1=Q_c$, $X_2=F_R$) and

$$m_X = \left(\sum_{i=1}^n X_i\right)/n \quad (5)$$

$$s_X = \left(\sum_{i=1}^n |X_i - m_X|\right)/n \quad (6)$$

are the average and mean absolute deviation, respectively, of the sets of measurements to be standardized (i.e. X_1 and X_2). Kaufman and Rousseeuw (1990) suggested that using s_X instead of the standard deviation – the latter commonly used in the standard z-score method – allows for a more robust identification and treatment of outliers.

The Minkowski distance was calculated between pairs of standardized objects (X_{1i}, X_{2i}) and (X_{1j}, X_{2j}), measured at depths i and j, respectively, by means of Eq. 1.

Prior to implementing the HAMAD algorithm, each of the n objects initially constituted a single cluster (with the number of clusters, N_c, initially equal to number of objects, n). At the first step of the HAMAD algorithm, the two closest clusters were merged to form a new cluster. In each following step, the distance between all clusters was recalculated considering the average of the Minkowski distances between all pairs of objects in the two clusters (Figure 5); two clusters now closest were merged, and N_c decreased by 1.

While the procedure could be iteratively repeated until all data are merged into a single cluster ($N_c=1$), the algorithm was stopped for $N_c=N_{cmax}$. The thresh-

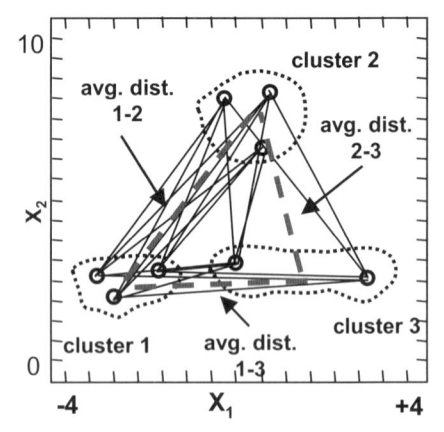

Figure 5 – Example of average distance clustering visual scheme

old value, N_{cmax}, was defined arbitrarily in terms of the following conditions: 1) the derivative, K_D, of the Minkowski distance function, $D(N_c)$, represented by the minimum distance between clusters at each step, was definitively less than 0.1 (Figure 6b); 2) the correlation coefficient, ρ_c, approached a value of 1 (Figure 6c) or was characterized, for N_{cmax}, by a significant variation (i.e. maximum relative peaks).

The correlation coefficient, ρ_c, is calculated between two adjacent cluster configurations (corresponding to two consecutive steps), as defined by Neter et al. (1990):

$$\rho_c = \frac{\sum_{i=1}^n (x_{i(j)} - m_{(j)})(x_{i(j+1)} - m_{(j+1)})}{\sqrt{\sum_{i=1}^n (x_{i(j)} - m_{(j)})^2}\sqrt{\sum_{i=1}^n (x_{i(j+1)} - m_{(j+1)})^2}} \quad (7)$$

where $x_{i(j)}$ is the cluster rank to which the i-th object

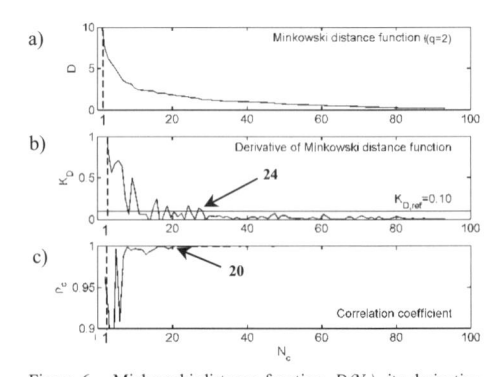

Figure 6 – Minkowski distance function, $D(N_c)$, its derivative, $K_D(N_c)$, and correlation coefficient, $\rho(N_c)$, for the selection of N_{cmax} ($N_{cmax} = 20$ in the example shown)

belongs at the j-th step and $m_{(j)}$ is a weigthed average of the number of objects included in each cluster at the j-th step.

All cluster configurations were then analyzed, and the resulting subdivision of data into clusters with depth (conceptually related to a possible soil stratigraphy) were interpreted visually on the basis of the criteria and definitions established by Hegazy (1998) (see, for example, Figure 3). The minimum number of clusters, N_{cf}, able to provide a reliable soil stratigraphy is defined as the value of N_c corresponding to the last step at which no primary layers (with thickness greater than 1.0 m) were formed as a result of HAMAD agglomeration.

4 FUZZY CLASSIFICATION

Zhang and Tumay (1999) proposed a possibilistic-fuzzy approach with the objective of addressing the observed uncertainty in the correlation between existing soil composition and the mechanical response to penetration in existing CPT-based classification charts, as investigated in previous works (Zhang and Tumay 1996a; Zhang and Tumay 1996b).

Zhang (1996a) observed two basic tendencies in existing soil behavior classification charts, as two almost orthogonal curve shapes: soil type changes in one direction, and in situ soil state (OCR, sensitivity, age, cementation, liquidity index, K_0, etc.) in the other.

The operation of soil chart simplification as proposed by Zhang and Tumay consists of the derivation of two independent indices (the soil classification index, U, and the in-situ state index, V) representing the two primary tendencies in the soil behavior classification chart by Douglas and Olsen (1981) shown in Figure 7, through the empirical superposition of a curvilineal orthogonal coordinate system along the principal tendencies in the original chart and, successively, the transformation of the curvilineal coordinate system into a cartesian coordinate system by conformal mapping (Zhang and Tumay 1996a), as shown in Figure 8.

Given q_c and R_f, from the Douglas and Olsen (1981) semi-logarithmic chart (R_f is in % and q_c is in tsf), the intermediate variables x, y, and $u(x,y)$ are calculated from the following relations:

$$x = 0.1539 R_f + 0.8870 \log(q_c) - 3.35 \quad (8)$$

$$y = -0.2957 R_f + 0.4617 \log(q_c) - 0.37 \quad (9)$$

$$u = -\frac{(a_1 x - a_2 y + b_1)(c_1 x - c_2 y + d_1)}{(c_1 x - c_2 y + d1)^2 + (c_2 x + c_1 y + d_2)^2} + \frac{(a_2 x + a_1 y + b_2)(c_2 x + c_1 y + d_2)}{(c_1 x - c_2 y + d1)^2 + (c_2 x + c_1 y + d_2)^2} \quad (10)$$

Finally the soil classification index $U=-u$ is defined, while the in-situ soil index, V, is not used in the fuzzy classification procedure.

In a subsequent paper, Zhang and Tumay (1999) introduced three fuzzy soil types: highly probable sandy soil (HPS), highly probable mixed soil (HPM) and highly probable clayey soil (HPC). The membership functions of the three fuzzy soil types, given in Eq. 11, Eq. 12 and Eq. 13, and shown in Figure 9, are functions of the soil classification index U:

$$\mu_s = \begin{cases} 1.0 & U > 2.6575 \\ \exp\left[-\frac{1}{2}\left(\frac{U - 2.6575}{0.834586}\right)^2\right] & U \le 2.6575 \end{cases} \quad (11)$$

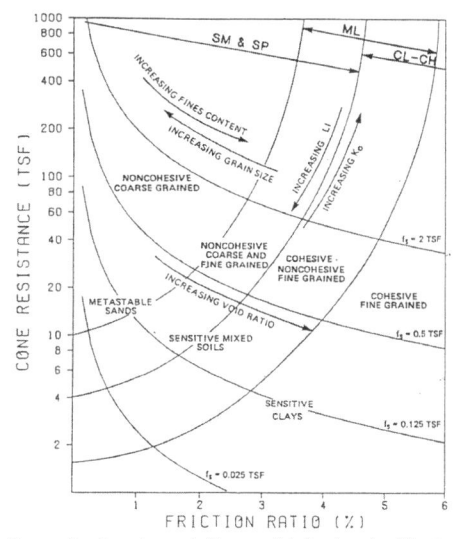

Figure. 7 - Douglas and Olsen soil behavior classification chart (1981)

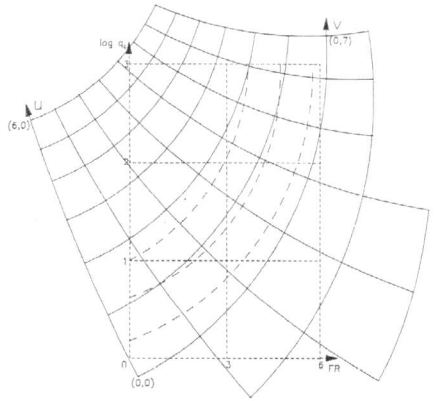

Figure 8 - Transformed Douglas and Olsen chart with superimposed U-V plane and boundary curves of soil classification criteria (Zhang 1994)

$$\mu_m = \exp\left[-\frac{1}{2}\left(\frac{U-1.35}{0.724307}\right)^2\right] \quad -\infty < U < \infty \quad (12)$$

$$\mu_c = \begin{cases} \exp\left[-\frac{1}{2}\left(\frac{U-0.1775}{0.86332}\right)^2\right] & U \geq 0.1775 \\ 1.0 & U < 0.1775 \end{cases} \quad (13)$$

They approximate the normal distributions for the three soil type groups as defined in a preceeding statistical analysis relating the average U value, representing soil behavior type, to the soil composition characteristics for eight sets of CPT sounding and boring data (Zhang and Tumay 1996b).

For any calculated value of U, by applying Equations 11, 12 and 13, it is thus possible to describe a continuous profile in terms of soil behavior, quantifying the degree of possibility that, at each depth, the penetrated soil behaves like a cohesionless, mixed or cohesive soil.

5 EVALUATION OF RESULTS

Clustering and fuzzy soil classification analyses were performed on the 25 mechanical CPT soundings selected for the study. For each sounding, the results were compared with the stratigraphic profiles of adjacent borehole logs and SPT profiles.

Poor to good agreement between the clustering outputs and the borehole logs was generally observed with a final number of clusters N_{cf} varying between 7 and 17. The agreement was assessed as being "good" if the main stratigraphic interfaces (dividing soils A, B, and C), and also the minor interfaces in the heterogeneous soil C were identified with sufficient accuracy, consistently with the limitations of the source CPT and SPT in-situ tests and with the subjectivity of borehole logs; "fair" if only the main units were identified; and "poor" if not even the main stratigraphic units were delineated accurately. In most cases, clustering also allowed to mark the presence lenses, soil transitions and heterogeneous layers, the latter being generally reported as single units in the borehole logs.

The results are summarized in Table 1, which reports, for each CPT considered: (a) CPT identification code from original data; (b) elevation of ground surface above sea level; (c) distance from the closest bore-hole or other in situ test; (d) final number of cluster adopted on the basis of the criteria treated previously; and (e) qualitative assessment of consistency with the stratigraphical interpretation from bore-hole and SPT results.

Figure 10d shows a typical example of a "good" final cluster configuration for one of the investigated CPT soundings, the results of which are reported in Figure 10a and 10b in terms of normalized cone tip resistance, Q_c, and friction ratio, F_R. In the example, N_{cf}=17. It may be seen that the main stratigraphic interfaces (continuous lines in Figure 10) as delineated by the borehole log (Figure 10 f) performed at a distance of 12.2 m (see Table 1) from the CPT vertical, are correctly detected by the cluster subdivision (at depth 4.95 m between the soils denoted as A and B, at depth 8.95 m, between soils B and C, respectively). Moreover, the soil previously indicated as C,

Table 1 - List of CPT considered in this study, with elevation of ground surface above sea level, Z, distance from the closest bore-hole or SPT test, d_B, final number of cluster adopted, N_{cf}, and qualitative assessment of agreement with bore-hole and SPT results

ID	Z (m)	d_B (m)	N_{cf}	Agreement
415	5.51	10.7	8	fair
436	5.56	8.1	12	good
444	5.50	8.3	10	poor
445	5.48	10.5	9	fair
447	5.34	8.4	8	fair
450	5.5	12.7	10	poor
452	4.11	3.5	10	poor
454	4.18	6.7	10	good
455	5.54	8.2	10	good
456	5.48	10.5	13	good
457	5.43	11.5	9	poor
458	5.49	6.0	10	fair
459	5.54	9.2	12	fair
460	5.55	10.4	11	fair
461	5.39	8.2	12	fair
462	5.47	9.5	7	good
465	5.55	8.8	17	good
468	5.46	10.5	8	fair
469	5.47	10.4	15	good
480	5.24	5.8	10	good
481	5.20	11.1	14	good
483	2.30	12.4	12	good
484	2.30	12.2	10	poor
488	2.89	8.8	9	fair
498	4.19	9.0	10	fair

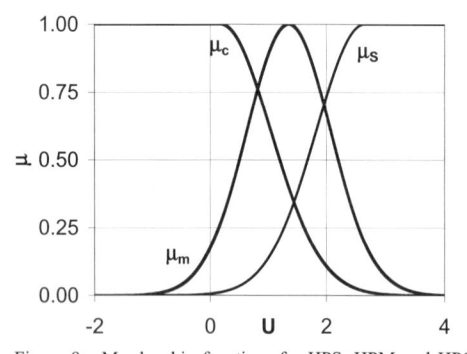

Figure 9 - Membership functions for HPS, HPM and HPC fuzzy soil types

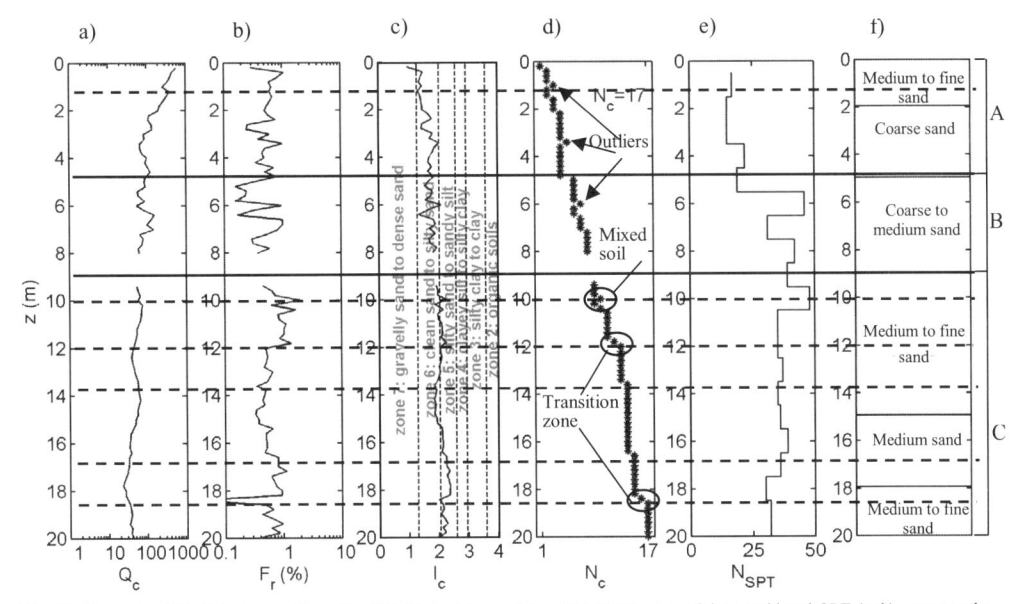

Fig. 10 - Example of final cluster configuration (d) obtained by applying HAMAD to one of the considered CPT (a, b), compared to Robertson (1990) soil classification and to SPT results (e) and bore-hole log interpretation (f) obtained at the nearest site

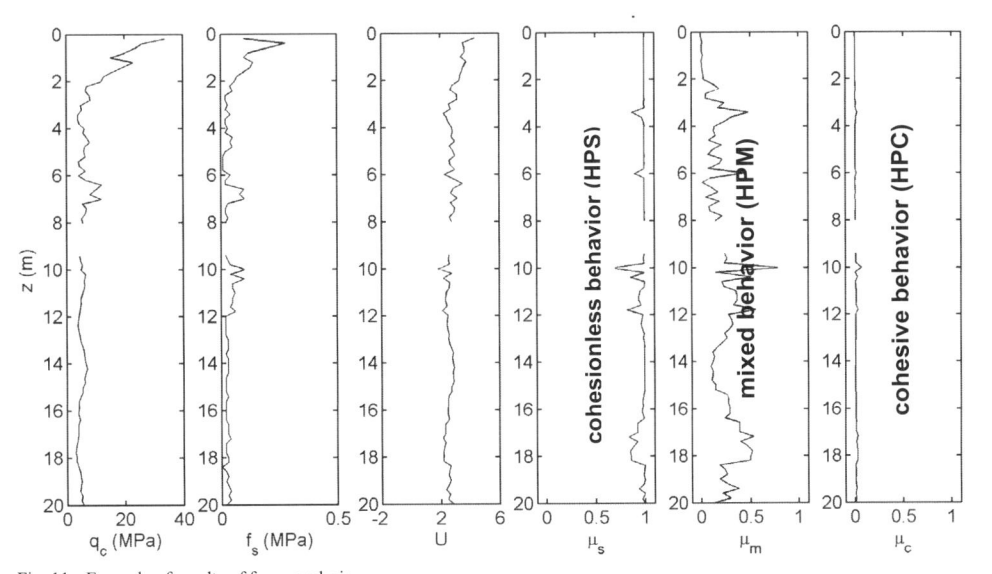

Fig. 11 - Example of results of fuzzy analysis

defined as being markedly heterogeneous by the geological description of the area and borehole logs (and whose heterogeneity may also be inferred by the SPT values reported in Figure 10e), is described by means of 6 distinct clusters identifying a set of secondary stratigraphic boundaries (dashed lines in Figure 10). It is also possible to recognize a transition zone (at depth of 11.9 m) and a mixed soil zone (at depth of 10.2 m). Finally, three outliers result from the cluster representation at various depths, as shown in Figure 10d. Shifts in the depths of the main stratigraphic interfaces between cluster profiles, borehole logs and SPT logs were acknowledged in some cases; these do not exceed 0.5 m in the soundings examined, and are most probably due to the well recognized uncertainty, existing in the true depth of CPT and SPT measurements.

Besides the expected, markedly cohesionless nature of the investigated soil layers, emphasized by the consistently high values of μ_s (seldom falling below 0.8) the fuzzy classification approach allowed to acknowledge a variable degree of "mixed" behavior with depth, with μ_m ranging between 0 and 0.6, with peaks around 0.8, and a very low degree of cohesive behavior, with μ_c rarely ranging beyond 0.1. The presence of intermediate-behavior soils in essentially cohesionless layers, which may be observed in Figure 11, in the example output of the same CPT previously considered, is consistent with the visual descriptions provided in the borehole logs, and with the variability of N_{SPT} values.

6 MAIN CONCLUSIONS

Clustering techniques and fuzzy soil classification were employed in an integrated approach for stratigraphic profiling. Even in a case of relatively homogeneous soil profiles in terms of soil compositions, and in presence of relatively less reliable and accurate input data - as are the results of mechanical CPT tests – it was possible to assess the capability of cluster analysis to identify the presence and position of stratigraphic discontinuities in the investigated soundings, and to gain a better comprehension of the variability of soil mechanical behavior with depth by the fuzzy membership functions.

The information gained from the successive application of the two techniques allows for the acknowledgement of the possible heterogeneity of layers, and for the direct quantification of the uncertainty in soil classification through the generally non-zero values of the fuzzy membership functions.

On the basis of the assessment of the agreement between the borehole logs, the results of adjacent SPT tests and the results obtained in the present study, it may be stated that the clustering-fuzzy integrated approach can provide an objective method for the subdivision of data on the basis of soil mechanical response to penetration, thus allowing to overcome the limitations and uncertainty, and lack of one-to-one correspondence between soil composition and mechanical behavior, inherent in a stratigraphic profiling based on composition criteria and visual inspection of CPT profiles.

Thus, it may assessed with sufficient confidence that clustering and fuzzy soil classification could be employed in the case examined, even in absence of numerous boreholes logs and other in-situ tests, for the delineation of soil stratigraphy from mechanical CPT data.

ACKNOWLEDGMENTS

This research has been carried out within the framework of a project financed by the MIUR (Ministry of Education University and Research) with the aim of performing geotechnical and seismological analyses in the Messina Strait area.

REFERENCES

Douglas, B.J., Olsen, R.S. (1981), "Soil Classification Using Electric Cone Penetrometer. Cone Penetration Testing and Experience." *Proceedings of the ASCE National Convention*, St. Louis, MI, U.S.A., 209-227.

Facciorusso, J., Vannucchi, G. (2003), "Liquefaction Hazard Maps of the Harbour Area of Gioia Tauro (Italy) by Geostatistical Methods", *Proc. SEE4*, Teheran (Iran)

Hegazy, Y.A. (1998), "Delineating Geostratigraphy by Cluster Analysis of Piezocone Data ", *Ph. D Thesis*, Georgia Institute of Technology, 464 pp.

Hegazy, Y.A., Mayne, P.W. (2002), "Objective Site Characterization Using Clustering of Piezocone Data", *Journal of Geotechnical and Geoenvironmental Engineering*, ASCE, 128(12): 986-996.

Kaufman, L., Rousseeuw, P.J. (1990), "Finding Group in Data: An Introduction to Cluster Analysis", John Wiley & Sons, Inc.

Młynarek, Z., Lunne, T. (1987), "Statistical estimation of homogeneity of a North Sea overconsolidated clay", Proc. Of 5th ICASP, Vancouver, Vol. 2, 961-968.

Neter, J., Wasserman, W., Kutner, M.H. (1990), "Applied Linear Statistical Models: Regression Analysis of Variance and Experimental Designs", Richard D. Irwin, Inc., Boston, Massachusetts, 1012 pp

Robertson, P.K. (1990), "Soil Classification Using the Cone Penetration Test", *Canadian Geotechnical Journal*, Vol. 27, 151-158.

Zhang & Tumay (1996a), "Simplification of Soil Classification Charts Derived from the Cone Penetration Test", *Geotechnical Testing Journal*, Vol. 19, No. 2, 203-216.

Zhang, Z., Tumay, M.T. (1996b), "The Reliability of Soil Classification Derived from Cone Penetration Test." *Uncertainty in the Geologic Environment: From Theory to Practice*, ed. C.D. Shackleford, P.P. Nelson and M.J.S. Roth, ASCE Geotechnical Special Publication No. 58, Madison, WI, U.S.A., 383-408.

Zhang, Z., Tumay, M.T. (1999), "Statistical to Fuzzy Approach Towards Soil Classification", *Journal of Geotechnical and Geoenvironmental Engineering*, ASCE, 125(3):179-186.

Site variability, risk, and beta

Roger A. Failmezger
In-Situ Soil Testing, L.C., 173 Dillin Drive, Lancaster, Virginia 22503,USA
insitusoil@prodigy.net

Paul J. Bullock
Department of Civil and Coastal Engineering, University of Florida, P.O. Box 116580, Gainesville, Florida 32611-6580, USA
pbull@ce.ufl.edu

Richard L. Handy
Handy Geotechnical Instruments, Inc., 1502 270th Street, Madrid, Iowa 50156, USA
rlhandy@iowatelecom.net

Keywords: beta probability, borehole shear, dilatometer, cone penetrometer, slope, settlement, pile

ABSTRACT: Geotechnical site investigations typically require a trade-off between the cost of the investigation and the designer's confidence in its results. Because cost restrictions usually limit the number of borings, soundings, and other tests, engineers tend to design conservatively to assure adequate performance. However, this conservatism increases the cost of the project without benefit to the owner. The engineer must discuss risk with the owner and design for the desired risk level chosen that the owner chooses. An accurate evaluation of the soil/rock properties at the site, combined with statistical risk analysis, can produce a more efficient and economical design with the desired confidence.
The Beta probability distribution provides a realistic and useful description of variability for geotechnical design problems. Site investigation methods that improve the accuracy of design parameters will reduce risk, and the design will then focus on the site's true soil variability without parasitic test variability. Three examples illustrate geotechnical risk analysis using the Beta distribution and emphasize the importance of minimizing testing variability.

1 INTRODUCTION

An economical site investigation leading to reasonable design safety usually requires two phases, the first using rapid, less expensive insitu tests to identify and map critical areas, and the second providing more detailed tests of selected soil strata and design analysis. The latter should replace any preliminary analyses and quantify risk. Site variability, soil test accuracy, and the accuracy of the design method all affect the reliability of the final design.

Probability distribution functions help define the relationship between variability and design performance risk. In the examples presented below, probability analyses using the Beta distribution indicate a nearly linear relationship between the nominal design factor of safety and the standard deviation (variability) for different probabilities of success. Note to improve client relations, we prefer the probability of "success" rather than its complement the probability of "failure".

After choosing a design factor of safety and computing its standard deviation, the engineer can use the figures below (or develop additional figures) to determine the probability of success. Adjustments to the design can then achieve the desired value. However, although greater safety minimizes the risk of sudden failure, excessive settlement, lateral movement, etc., it also results in higher construction costs.

2 OWNER INVOLVEMENT

The engineer must always design to avoid loss of life. However, all other design decisions are purely economical and the owner should decide the appropriate level of success. After all, it is the owner's money and soil. The owner is well versed with risk, for every financial decision he makes involves risk.

Because engineers typically do not discuss the possibility of failure with the owner, they assume liability that should remain with the owner. To cover the engineer's perceived liability, his designs are often overly conservative and costly and serve neither the owner nor the engineer. If the owner thinks the foundation costs will be too high, he will hire a second engineer and the first engineer may lose the project (perhaps, rightfully so). The owner and the engineer should mutually decide on the acceptable probability of success for the design. The owner's understanding and acceptance of the inherent risk help determine the feasibility of the project.

Our analyses below show results for success probabilities of 90%, 95%, 99%, and 99.9%. Choosing the most appropriate value depends on many factors such as the intended use and sensitivity of the facility, foundation redundancy, costs to repair, installation of performance monitoring instruments, and quality of the contractor. A structure with equipment sensitive to differential settlement should use a probability of success of 99 or 99.9%, whereas a warehouse that can tolerate more differential settlement and still function adequately can tolerate a lower 90% probability of success.

Pile supported structures that have some redundancy can also use lower probabilities of success, 90 or 95%. If one pile does not have its full desired capacity, a nearby pile may have additional capacity and provide the needed extra load capacity. Often pile groups need a whole number plus a fraction of a pile to carry the design load, but the additional pile is installed resulting in supplemental capacity (e.g. compute 8.2 piles, install 9 piles).

A slope's location may help decide its appropriate probability of success. Highway departments may construct slopes with lower level of success, choosing to save money by repairing the occasional failed slope rather than buying more right-of-way to build flatter slopes. However, on a heavily traveled road, a higher probability of success reduces the risk of a costly traffic delay.

Instruments can be installed to monitor the performance of construction. Unsatisfactory areas can be detected and stabilized. The owner can choose a lower probability of success (90% or perhaps lower) if he determines his savings from the less conservative approach will be greater than the remedial fixes that may occur in hopefully isolated areas.

The quality of the contractor and the quality of the engineering inspection may also influence the design probability of success. High quality contractors and inspectors will recognize and correct for unanticipated subsurface conditions, providing a better product less susceptible to damage. The engineer should work with the owner to initially pre-qualify contractors and later help the owner select a contractor that has submitted a responsive bid.

The engineer must educate the owner and explain why certain tests will be conducted and how that knowledge will be used for improved design. By being involved with the owner, the engineer will develop and improve their business relationship. The owner will not consider the engineer as a commodity service (hiring and selecting the engineer based on fee) but rather as a valuable contributor to his project. If the owner does not want to assume his risks and the engineer loses the project, the engineer has only lost a bad client.

3 EVALUATING STANDARD DEVIATION

The sources of standard deviation that affect risk assessment include the natural variability of the soil or rock, man-created variability added during the design process and intangible variability created by the owner. Man-created variability is how well the test/design predict what will occur. The designer should attempt to minimize the man-created variability by using a sufficient quantity of tests that accurately predict the design parameters and by choosing design methods that accurately predict performance. This value can be quantified from case study databases. The owner can create intangible variability by selecting a low bid contractor that is not pre-qualified or responsive, another firm to perform the inspection and not enough tests. If these sources of variability are considered to be independent of each other, then the overall standard deviation equals the square root of the sum of the individual standard deviations squared.

If the other sources of variability can be minimized, the engineer can focus on the site's geologic variability, defining areas of poor or favorable geologic conditions and possibly designing those areas of the site separately. A lower overall standard deviation results in a more efficient design with a lower factor of safety.

4 WHY USE THE BETA PROBABILITY DISTRIBUTION FUNCTION?

With sufficient data, several common population statistics can be calculated, such as the mean and standard deviation. Statistical analyses often use these values and assume normal or log-normal population distributions, which may not adequately characterize the soil test data. For a normal distribution the minimum and maximum limits are negative and positive infinity, respectively. The log-normal distribution uses limits of zero and positive infinity. In either case, these limits are unrealistic and often impractical. With the more versatile Beta distribution (of which the normal distribution is a specific subset) the engineer chooses the minimum and maximum limits. In our analyses we evaluated both 3 and 5 standard deviations from the mean as the minimum and maximum limits. Because there was little difference in the results, we concur with the recommendation by Harr (1977) to use minimum and maximum limits of 3 standard deviations away from the mean.

Steep and narrow Beta probability curves (with low standard deviations) describe homogeneous soil conditions, and flatter curves indicate imprecise or heterogeneous conditions. The following examples illustrate the evaluation of project risk for slope stability, vertical pile capacity, and settlement.

5 SLOPE STABILITY ANALYSIS

Many local and national codes specify a minimum factor of safety for earthen slopes. These specifications seldom consider the homogeneity of the subsurface conditions or the consequences of failure. The owner does not want to buy excess land to have overly conservative slopes nor does he want to have an expensive repair later. The Beta probability distribution allows engineers to consider the above conditions in their analyses. The area under the probability curve with a factor of safety less than 1.0 defines the probability of failure. Because the total area under the probability curve must equal 1.0, the probability of success equals 1.0 minus the probability of failure. Homogeneous subsurface conditions, with low uncertainty (standard deviation), will result in a sharply peaked and narrow Beta distribution curve at a given probability of success, with an average factor of safety slightly more than 1.0. Conversely, a heterogeneous subsurface, with high uncertainty, will result in a flat and wide Beta curve, with an average factor of safety much higher than 1.0 to achieve the same probability of success.

We performed parametric analyses with the Beta probability distribution for various factors of safety and standard deviations, choosing the minimum and maximum limits for the distribution as the average value ± 3 or ± 5 standard deviations. Figure 1 shows representative Beta distribution curves for the probability of success of 95% with limits equal to ± 3 standard deviations from the average value. Figure 2 shows the variation of the Beta value at the average safety factor for a range of safety factors and success probabilities. Figures 3 and 4, for Beta limits of ± 3 and ± 5 standard deviations respectively, show a nearly linear relationship between the average factor of safety and the standard deviation for a given success probability. The y-intercept was 1.00 and the coefficient of correlation was greater than 0.998. As the limit of standard deviation approaches zero, the beta curve becomes narrower and steeper, which results in the average factor of safety approaching 1.00.

The engineer may use design charts similar to Figures 3 and 4 (almost identical) to determine the probability of success for the average design safety factor required with a known (or assumed) standard deviation. The stability analysis methods presented by Christian (1997) or Duncan (2000) will help calculate the nominal (average) factor of safety and its tandard deviation for given slope conditions. To achieve a greater probability of success, the engineer should alter the design to increase the chosen safety factor, or decrease the variability (through better site characterization, more accurate analyses, ground modification, etc.).

5.1 *Slope Stability Example*

This example describes a hypothetical slope stability design using electric cone penetration tests, performed during a phase one subsurface investigation, that delineate three geologic strata at the site. The phase two investigation included five borehole shear tests performed in each stratum to estimate the average drained strength parameters and their standard deviations. The borehole shear test accurately measures the drained shear strength of the soil and compares well to laboratory strength tests (Handy, 1986).

We calculated the average and standard deviation of the slope's stability using the point estimate method (Christian, 1997), which performs multiple analyses using permutations of design variables by assigning a value of either the average plus one standard deviation or the average minus one standard deviation to each. Using the shear strength of each of the three strata and the groundwater level as the parametric variables, multiple runs with a Janbu stability analysis program provided a total of 16 permutations (2^n, where n = the number of variables = 4). For these permutations, the average factor of safety equaled 1.25 with a standard deviation of 0.15, and the design chart in Figure 3 indicates an acceptable 95% probability of success.

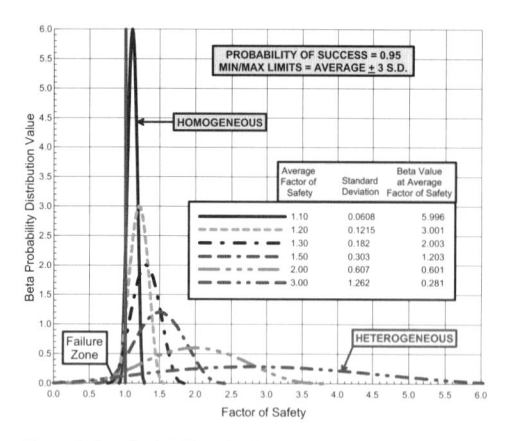

Figure 1: Beta Probability Distribution Curves for Slope Stability

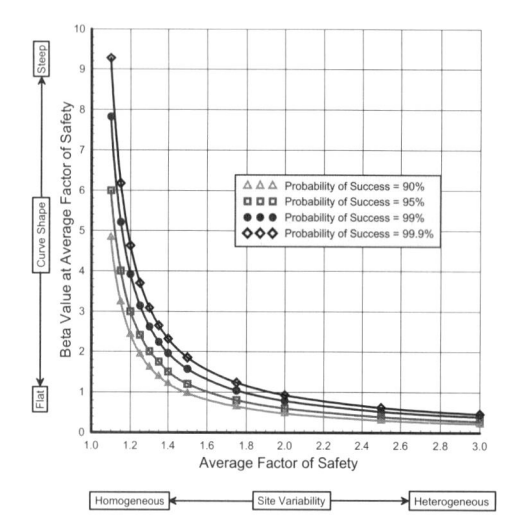

Figure 2: Beta Probability Analyses For Slope Stability with Min/Max Limits = Average $\pm$ 3 Standard Deviations

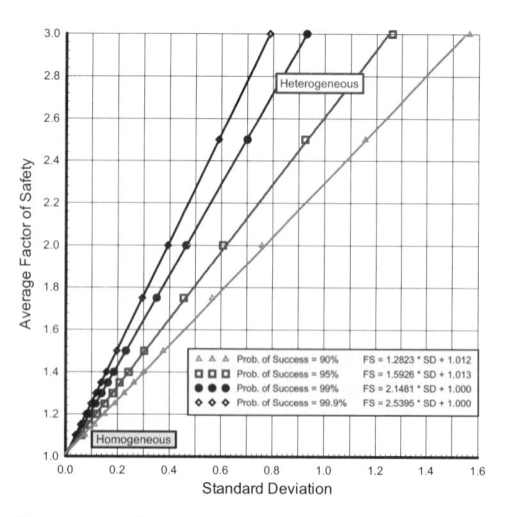

Figure 3: Design Chart of Beta Probability Distribution Analyses for Slope Stability with Min/Max Limits = Average $\pm$ 3 Standard Deviations

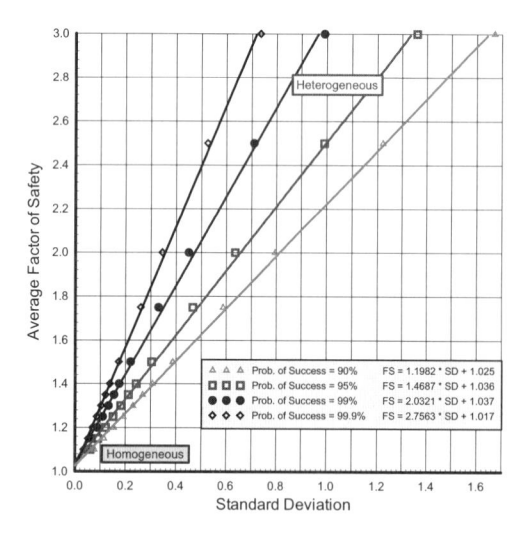

Figure 4: Design Chart of Beta Probability Distribution Analyses for Slope Stability with Min/Max Limits = Average $\pm$ 5 Standard Deviations

6 PILE CAPACITY

Beta probability distribution curves can be used to represent both the pile capacity and the load supported by the pile. Where these two curves intersect, the load exceeds the pile capacity and the intersecting area represents the probability of failure. Assigning the average applied load a value of 1.0, and calculating the pile capacity as a multiple of this applied load, leads to a unitless analysis convenient for design purposes. Standard deviations of 0.1 and 0.2 should adequately describe the normal variation of the actual load condition. As found above, minimum and maximum limits of the average value $\pm$ 3 standard deviations should adequately define the expected range of values, with a minimum value of at least zero. Figures 5 and 6 show representative Beta probability distribution curves for a probability of success equal to 95% with load standard deviations of 0.1 and 0.2 respectively.

For homogeneous sites the pile capacity has a low standard deviation, resulting in a narrow, peaked Beta curve with an average value close to the average load. For heterogeneous sites the predicted pile capacity is less accurate and its standard deviation is higher, resulting in a flat and wide Beta curve. For the same level of safety as the homogeneous case, the heterogeneous curve shifts farther to the right of the load curve and has a higher average value.

Figures 7 and 8 show the factor of safety, or pile capacity/load ratio, versus the standard deviation of the pile capacity for standard deviations of the load equal to 0.1 and 0.2, respectively. Again, we found a nearly linear relationship for a given probability of success! For the load standard deviation of 0.2, which contains more uncertainty, the probability of success lines shift up and to the left of the lines for a standard deviation of 0.1. Note that these two figures show a general relationship between the safety factor, the design variability, and the success probability that is valid for any pile design method. These design charts may also be used for tiebacks, soil nails, lateral capacity of piles and other applied load/soil capacity analyses.

6.1 Pile Capacity Example

This example considers twenty (20) cone penetrometer test soundings performed for the hypothetical design of a laboratory foundation supported by steel pipe piles. Column loads will require support of 200 kN per pile, with a standard deviation of 20 kN. For each sounding the LCPC pile capacity prediction method (see Campanella, et al., 1986) provided a pile designed to carry a load of 350 kN (nominal safety factor = 1.75). The design tip elevations for the different column loads in the foundation plan did not vary greatly, resulting in a standard deviation of only 35 kN due to the natural soil variability. Based on a database case study, Campanella, et al. (1986) indicate a coefficient of variation of 0.15 for the LCPC predicted capacity of driven steel pipe piles. Using this value, the standard deviation due to the LCPC method is 0.15 * 350 kN = 52.5 kN. The overall standard deviation equals the square root of the sum of the two individual standard deviations squared, or 63.1 kN. The columns were designed to exert a load of 200 kN per pile. Dividing by the 200 kN nominal applied load results in a unitless predicted pile capacity of 1.75, with a standard deviation of 0.32 and load standard deviation of 0.1. Because the building will contain sensitive laboratory equipment, the owner chose a 99% probability of success. However, Figure 7 indicates a probability of success of only 93% for the above parameters.

By increasing the pile diameter so that each pile will have a capacity of 400 kN, the natural standard deviation of 35 kN and the LCPC method standard deviation of 0.15 * 400 kN or 60 kN result in an overall standard deviation of 69.5 kN. Using the unitless values of 2.0 for the factor of safety and 0.35 for pile standard deviation, Figure 7 indicates an acceptable probability of success of 99%.

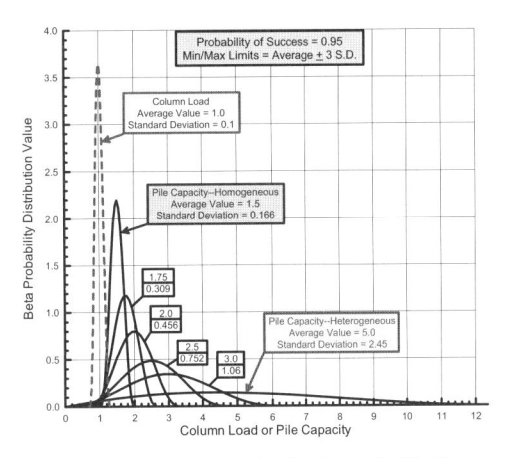

Figure 5: Beta Probability Distribution Curves for Pile Capacity Analyses, Column Load Standard Deviation = 0.1

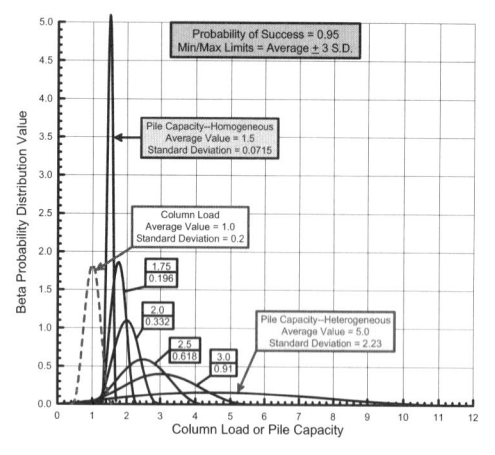

Figure 6: Beta Probability Distribution Curves for Pile Capacity Analyses, Column Load Standard Deviation = 0.2

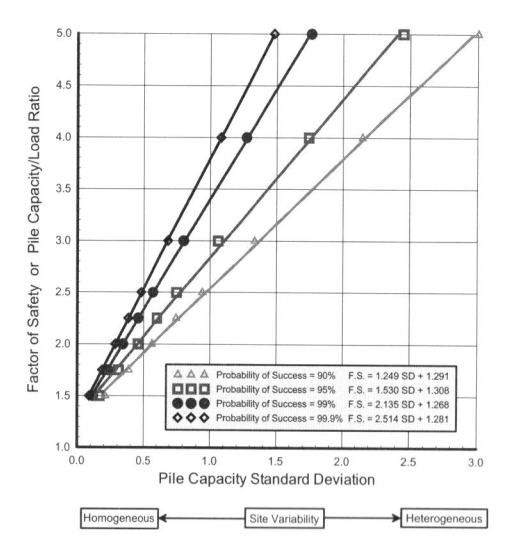

Figure 7: Design Chart for Beta Probability Distribution Analyses of Pile Design, Load Standard Deviation = 0.1

7 SETTLEMENT ANALYSIS

Engineers commonly consider total settlements exceeding 25 mm as unsatisfactory. Design approaches similar to that described below could use a different limit, or could alternatively seek to limit the differential settlement or angular distortion. Using

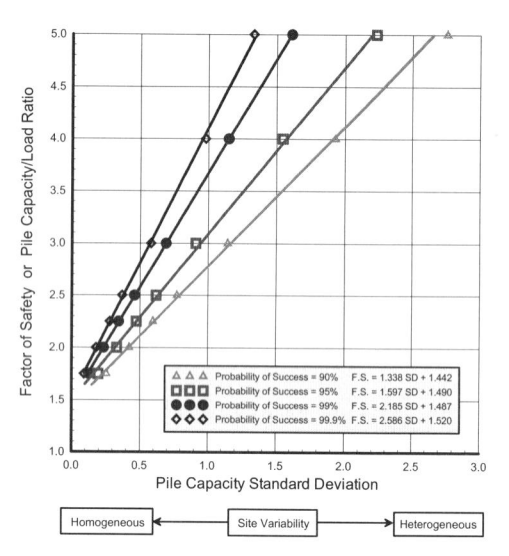

Figure 8: Design Chart for Beta Probability Distribution Analyses of Pile Design, Load Standard Deviation = 0.2

a Beta probability distribution, unacceptable settlement occurs in the zone where it exceeds 25 mm. Of course, the settlement distribution cannot start at less than zero, and again minimum and maximum limits of the average value $\pm$ 3 standard deviations provide reasonable bounds. Figure 9 shows representative Beta distribution curves for a probability of success of 95%, indicating that an increase in the standard deviation requires a decrease in the average settlement to obtain the same success. This requirement may result in a questionable reverse "J" or incorrect "U" shaped Beta distribution due to the high variability. If an unreasonable distribution of this type occurs, the engineer should reduce the variability through better quality testing or increase the allowable settlement threshold to more than 25 mm.

Figure 10 shows the relationship between the Beta distribution value and the average settlement for several success probabilities of 90, 95, 99 and 99.9%. Figure 11 provides a design chart for settlement analysis. When the beta distribution was "bell-shaped", Figure 11 shows there is a nearly linear relationship between average settlement and its standard deviation for a given probability of success. As the limit of standard deviation approaches zero, the beta curve becomes narrower and steeper, which results in the average settlement (y-intercept) approaching 25 mm.

To determine the standard deviation of settlement for a site, the engineer needs an accurate assessment of the soils' static deformation properties. For most projects requiring settlement assessments, Dilatometer tests (DMT) provide a satisfactory solution. Schmertmann (1986) presents a field-verified settlement calculation method for DMT data. Like traditional settlement predictions based on consolidation tests, this method divides the geologic sections into layers and computes the settlement of these layers. With DMT tests at 20 mm depth intervals, each layer may be as thin as 20-mm, and each DMT sounding provides a separate settlement estimate. By combining all of the settlement predictions, the engineer may compute an average and standard deviation, and then use Figure 11 to determine the probability of not exceeding a threshold limit of 25 mm. If the probability of success is too low, the engineer can perform additional DMT soundings, reduce the applied bearing pressure, or design footings individually.

The Standard Penetration Test (SPT) is also often used for the settlement design of spread footings in sands, particularly in the United States. From case study data, Burland and Burbridge (1985) show that settlement estimates based on the SPT N_{60} value have a coefficient of variation ≈ 0.67. This high value probably results from both the inherent variability of the SPT and the use of a dynamic penetration test to estimate static deformation properties (Failmezger, 2001). The Dilatometer, which is a calibrated static deformation test, more accurately predicts settlement and has a coefficient of variation of about 0.18 for all soils except quick silts (Failmezger, Bullock, 2004). In Table 1, the upper limits for average settlement are computed for DMT and SPT methods assuming that there is no site variability. At best, the SPT Beta probability distribution has a reverse "J" shape.

7.1 Settlement Example

Sixteen Dilatometer test soundings were performed for a hypothetical grocery store to depths of approximately 9 m. The soils below about 8 m were dense and no measurable settlement was expected below that depth. Schmertmann's method (1986) provided a settlement estimate for each sounding, with an average settlement of 18 mm and a standard deviation of 4.0 mm due to soil variability. Because the Dilatometer tests were pushed and no quick silts were present at the site, we assume a coefficient of variation for the test and prediction method of 0.18 , resulting in a standard deviation for the DMT prediction method of 0.18 * 18 mm = 3.2 mm. Using an overall standard deviation of 5.1 mm, from Figure 11 the probability of success is slightly more than 90% and is acceptable for the proposed building.

Table 1: Maximum Value of Average Settlement with Zero Site Variability for a Threshold Settlement of 25 mm

Probability of Success	Maximum Average Settlement (mm)	
	DMT Method in all soils with Coeff. of Variation = 0.18	SPT Method in only sands using N_{60} with Coeff. of Variation = 0.67
90%	20.17	12.5
95%	19.28	11.1
99%	18.03	9.5
99.9%	17.15	8.7

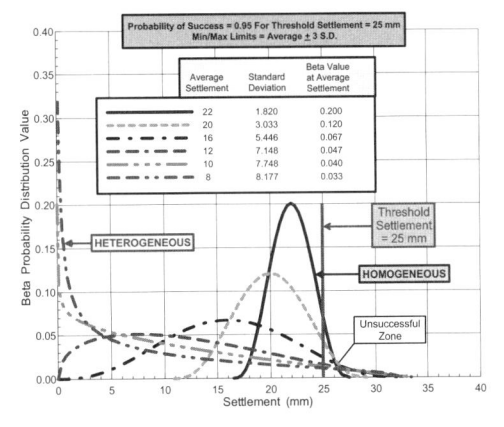

Figure 9: Beta Probability Distribution Curves for Settlement Analyses

CONCLUSIONS/RECOMMENDATIONS

Based on the above examples and our experience with the Beta probability distribution:
1. Good design requires the owner's acceptance and understanding of acceptable risk.
2. Risk analysis can provide more efficient and economical design.
3. Effective risk analysis requires the engineer to limit variability, as best possible, to that inherent in the geologic deposit.
4. A thorough and accurate site investigation helps to minimize design variability and improves design efficiency.
5. Soil tests that directly measure design parameters should reduce variability better than empirical correlations with indirect measurements.

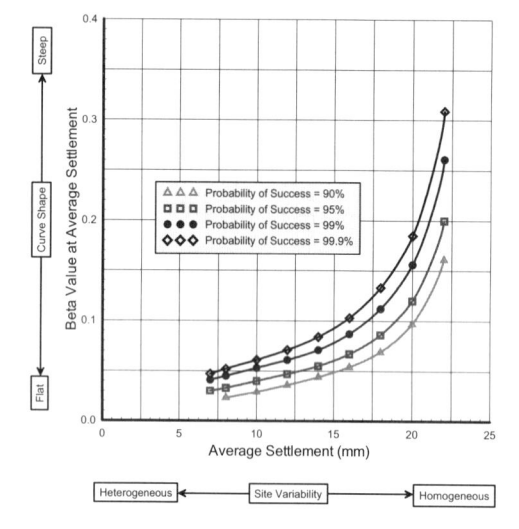

Figure 10: Beta Probability Distribution Analyses for Settlement with an Acceptable Settlement Threshold = 25 mm and Min/Max Limits = Average ± 3 Standard Deviations

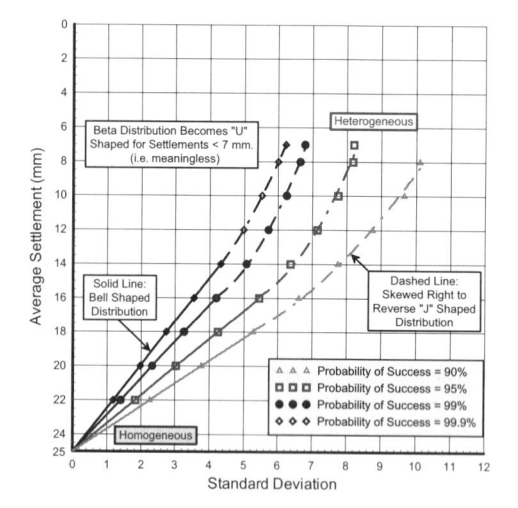

Figure 11: Design Chart for Beta Probability Distribution Analyses for Settlement with a Threshold Settlement = 25 mm and Min/Max Limits = Average ± 3 Standard Deviations

6. Design charts based on the Beta distribution provide a simple tool to choose safety factors appropriate to the desired success probability and to the combined level of variation inherent in the design method, site investigation method, and the site itself.
7. For a given probability of success, using the Beta probability distribution within common engineering limits provides a nearly linear relationship between the average value of the design parameter and its standard deviation.
8. The appropriate average safety factor of safety for slope stability should consider the site variability, consequences of failure and the necessary probability of success.
9. Risk analysis for pile capacity should consider the standard deviation of both the applied load and the soil capacity.
10. Previous database studies show that even in sands settlement analyses based on the Dilatometer test and design methods have much better accuracy than such analyses based on the Standard Penetration Test.

REFERENCES

Burland, J.B. and Burbridge, M.C., 1985, "Settlement of Foundations on Sand and Gravel", Proc., Inst. of Civ. Engrs, Part 1, 78, 1325-1381.

Christian, John T., Oct. 23, 24, 1997, "Reliability Approaches in the Design of Slopes", Excellence in Geotechnical Engineering, Central Pennsylvania Section, ASCE and Pennsylvania Department of Transportation.

Duncan, J. Michael, April 2000, "Factors of Safety and Reliability in Geotechnical Engineering", ASCE Journal of Geotechnical and Geoenvironmental Engineering, Vol 126, No. 4, pp. 307-316.

Failmezger, R.A., Rom, D., Ziegler, S.R., 1999, "SPT? A better approach of characterizing residual soils using other in-situ tests", Behavioral Characterics of Residual Soils, B. Edelen, Ed., ASCE, Reston, VA, pp. 158-175.

Failmezger, Roger A., 2001, Discussion of "Factors of Safety and Reliability in Geotechnical Engineering."

Failmezger, Roger A., Bullock, Paul J., 2004, "Individual foundation design for column loads", International Site Characterization '02, Porto, Portugal

Handy, Richard L., 1986, "Borehole shear test and slope stability" Proc., In-Situ '86: ASCE specialty conf. On use of in-situ tests and geotech engrg, ASCE, Reston, VA, pp. 161-175.

Harr, Milton E., 1977, Mechanics of Particulate Media: A Probabilistic Approach, McGraw-Hill, New York

Schmertmann, John H., 1986, "Dilatometer to compute foundation settlement", Proc., In-Situ '86: ASCE specialty conf. On use of in-situ tests and geotech engrg, ASCE, Reston, VA, pp. 303-321.

A case study using in-situ testing to develop soil parameters for finite-element analyses

E. Farouz & J.-Y. Chen
CH2M HILL, Inc., Herndon, Virginia, USA
efarouz@ch2m.com; jchen1@ch2m.com

R.A. Failmezger
In-Situ Soil Testing, L.C., Lancaster, Virginia, USA
insitusoil@prodigy.net

Keywords: dilatometer, cone penetrometer, finite-element analyses, slope stability

ABSTRACT: Finite-element analyses can accurately model soil's response to loading conditions. However, without realistic geotechnical parameters to model the stress-strain and strength characteristics of soils, its accuracy diminishes. This paper discusses the use of finite-element analyses with the computer program, PLAXIS, to evaluate the long-term performance of cut slopes at the Virginia Route 288 project, near Richmond, Virginia, USA. The 9-meter high cut slopes are located near an area with a history of slope failures. Limit-equilibrium slope stability analyses based on the conventional subsurface investigation approach using borings and overly conservative soil parameters derived from Standard Penetration Test results and back-analyses of historical slope failures near this area indicated that the cut slopes will be stable at a slope ratio of 5-horizontal to 1-vertical (5H:1V). Using the finite-element analyses with soil parameters developed based on the results of dilatometer tests (DMT) and piezo-cone penetrometer tests (CPTU), the cut slopes were found to be stable at a slope ratio of 3H:1V.

1 INTRODUCTION

The Virginia 288 PPTA (Public Private Transportation Act) project was approved for construction in December 2000 and construction started in April 2001. The project includes construction of approximately 17 miles of new highway with 23 bridges and overpasses. The design team on the project, led by CH2M HILL, was asked to reduce the cost of a cut slope within a segment of project designated as "Cut C." Cut C is located along the mainline of Virginia Route 288 immediately south of the James River. Documented historical slope failures near this area of the project led to conservative design of slopes in Cut C. The cut slopes were originally recommended to be at a slope ratio as flat as 5H:1V, including a drainage blanket. A proposal by the contractor initiated the study presented in this paper, to re-evaluate the stability of the cut slope. Results of this study led to a more reasonable and cost-saving design. The general location of this project is shown in Figure 1.

2 PROJECT GEOLOGY

The project is located in the Piedmont Physiographic Province of Central Virginia. The region is characterized by complexly folded and faulted igneous and

Figure 1. Site Location Map of the Virginia Route 288 Project

metamorphic rocks of Late Precambrian to Paleozoic age (Wilkes, 1988) below Triassic-aged coal measures, shales, and interbedded sandstones and shales. Geologic literature for the Midlothian Quadrangle of Virginia reports that a Tertiary-aged gravelly terrace deposit is present at the cut slope location, south of the James River flood plain and north of Bernard's Creek (Goodwin, 1970). This material is composed mostly of coarse gravel, with clayey sand beds inter-layered with the gravel. The matrix of the formation is predominantly sand with varying amount of clay.

3 PROJECT DESCRIPTION

The cut slope extends approximately between Virginia Route 288 mainline stations 158+20 and 161+00 and is entirely within the limits of Cut C, which extends from station 153+00 to station 163+00. The original designer of this roadway cut slope recommended a slope ratio as flat as 5H:1V at some cuts. The design included a drainage blanket. A schematic design cross-section is presented in Figure 2.

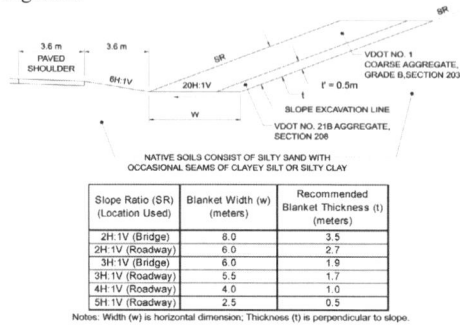

Figure 2. Original Schematic Design Cross-section of the Cut Slope (after HDR Engineering, Inc., 1999) [VDOT denotes Virginia Department of Transportation.]

Groundwater levels indicated by borings and monitoring wells in the Cut C area along Route 288 are summarized in Table 1. Generally, groundwater is observed to be near or above the finished grade between stations 154+00 and 163+00. At the maximum, groundwater is approximately 4 to 5 meters above the finished grade between station 155+00 and 160+00.

Table 1. Summary of Measured Groundwater Levels in Cut C Area (after HDR Engineering, Inc., 1999)

Station	Cut Depth (m)	Groundwater Elevation (m)	Groundwater Depth from Surface (m)	*Groundwater Height above Finished Cut (m)
153	2	Dry	3	-1
154	5	58	6	-1
155	8	61	3	5
156	10	62	5	5
157	8	60	4	4
158	9	60	5	4
159	8	60	3	5
160	6	59	2	4
161	4	56	1	3
162	5	54	3	2
163	2	52	3	-1

* Note that negative values indicate groundwater table below the finished cut.

Because geotechnical properties of soils are generally site-specific, even within the same geological formation, we performed in-situ testing and re-evaluated the slope stability, upon contractor's proposal to increase the slope ratio and avoid using drainage blanket to save valuable construction dollars. Based on our study presented hereafter, the cut slope is found to be stable at a slope ratio of 3H:1V.

4 IN-SITU TESTING

The in-situ testing program consisted of both piezo-cone penetrometer tests (CPTU) and dilatometer tests (DMT), which are continuous or at least near-continuous soil profiling techniques to delineate subsurface stratigraphy and soil properties. The CPTU data require a good estimate of correlation coefficients to determine strength and deformation parameters. These coefficients depend on the geologic formation and can be site-specific.

The Marchetti dilatometer test is a calibrated static deformation test. The lift-off pressure, p_0, and the pressure at full expansion, p_1, are measured. These two independent parameters are used to compute other soil parameters through triangulation (two variables to get a third variable). We used Marchetti's (1980) correlation to calculate the vertical constrained deformation modulus, M. This modulus is obtained after combining the dilatometer modulus, E_D, with the horizontal stress index, K_D, which is an indicator of stress history. We used Schmertmann's (1982) method for determining the drained friction angle in the cohesionless soils.

In our study, in-situ testing including three piezo-cone penetrometer tests, designated as PZ-1, PZ-2, and PZ-3, and four dilatometer tests, designated as DT-1, DT-2, DT-3, and DT-4, was performed at selected locations shown in Figure 3. DT-1, DT-2, and PZ-1 are located at the top of the cut slope on the south-bound-lane (SBL) side of the highway and DT-3, DT-4, and PZ-2 are located at the bottom of the cut slope on the SBL side. PZ-3 is an additional piezo-cone penetrometer test located at the top of the cut slope on the north-bound-lane (NBL) side of the highway. At the time of testing, the slope had already been cut close to the planned finished elevation, at a slope ratio of 3H:1V, without obvious distress.

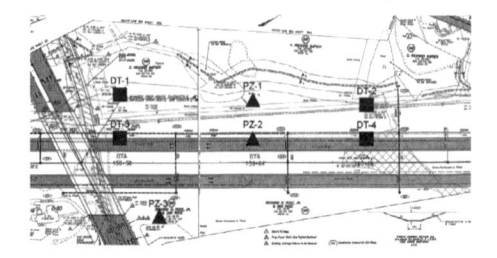

Figure 3. In-situ Testing Locations

Typical CPTU and DMT results from our study are presented in Figures 4 and 5, respectively. These results were obtained at testing locations PZ-1 and DT-1, shown in Figure 3. Interpreted DMT strength and deformation parameters from testing at DT-1 are presented in Figures 6 and 7, respectively. All the testing results consistently show that the soils within the cut slope are primarily sandy soils with occasional seams of clayey silt or silty clay, which correlates well with geological literatures (e.g. Goodwin, 1970).

From the DMT results obtained at DT-1, it is observed that a stiffer sandy soil layer exists at a depth between 0 and 2 meters below the top of slope, as indicated by the higher thrust required to push the dilatometer blade and the higher constrained modulus (M). Below a depth of 4 meters from the top of slope, the stiffness of sandy soils generally increases with increasing depth. For example, in DT-1, the constrained modulus (M) increases from 200 bars to 900 bars, between a depth of 4 m and 9 m. The drained friction angle (ϕ') of the sandy soils is generally greater than 37 degrees (ranging between 37 and 47 degrees) under the plane-strain condition. The drained friction angle under triaxial compression (ϕ'_{TC}) is averaging 38 degrees. Also, the sandy soil deposits within the slope is generally overconsolidated, with an overconsolidation ratio (OCR) decreasing with increasing depth.

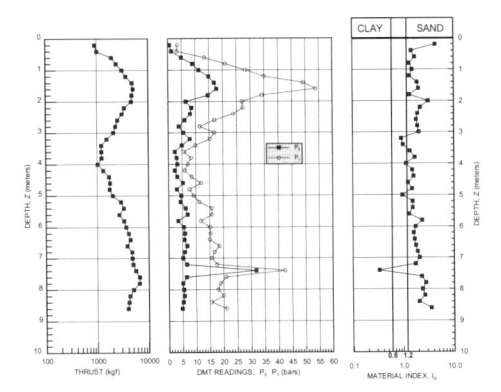

Figure 5. DMT Results Obtained at Test Location DT-1

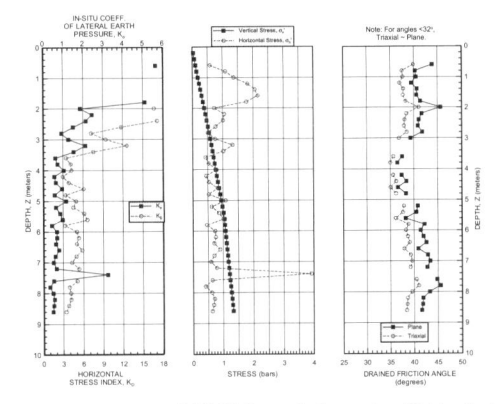

Figure 6. Interpreted DMT Strength Parameters Obtained at Test Location DT-1

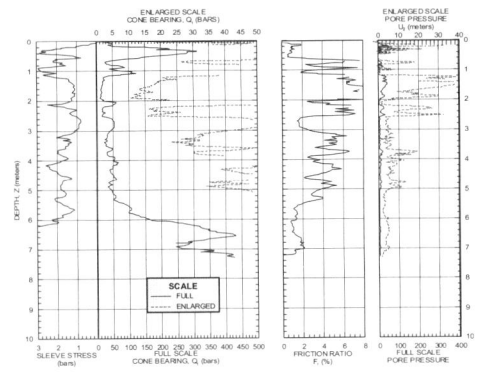

Figure 4. CPTU Results Obtained at Testing Location: PZ-1

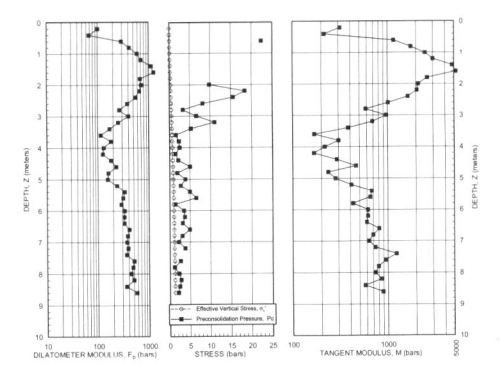

Figure 7. Interpreted DMT Deformation Parameters from Testing Results Obtained at DT-1

5 STABILITY ANALYSES

Slope stability analyses using a finite-element based computer program, PLAXIS (Brinkgreve and Vermeer, editors, 1998), were executed to evaluate the performance of the cut slope. A cross-section at the SBL side of Virginia Route 288 mainline station 158+20 was analyzed. This cross-section represents one of the deepest cut sections along this slope. The depth of this cut section is approximately 9 m with a revised slope ratio of 3H:1V. The top of the slope is at an elevation of 65 m above the mean sea level (MSL) and the bottom of the slope is at an elevation of 56 m above the MSL. The top of bedrock is at an

approximate elevation of 50 m above the MSL (6 m below the bottom of cut). A single soil type was used for soils above the rock, which is assumed as fixity in the model. This cut section was analyzed under the following two groundwater conditions:
1) The normal groundwater condition with the groundwater level at an elevation of 60 m above the MSL (4 m above the bottom of cut).
2) The worst-case groundwater condition with the groundwater level at an elevation of 65 m above the MSL (corresponding to a fully saturated cut slope).

In the model, the cut was excavated in three steps. Each cut step involved removal of soil of 3-m vertical thickness in accordance with the 3H:1V slope ratio, during a period of 2 months. Groundwater drawdown characteristics were able to be modeled during each cut step with the groundwater flow module in PLAXIS such that the effective stress within the cut slope can be estimated more accurately.

The hardening soil model with various strength, deformation, and groundwater flow parameters presented in Table 2 was used to model the soil behavior. Strength and deformation parameters were considered the most critical ones for this particular cut slope with regards to its stability and we relied on the DMT results to develop these parameters. CPTU results were used to confirm that variation of soil properties within the slope profile was small and a single soil type can reasonably represent the behavior of the slope. Sources or correlations where these parameters were developed are presented in Table 2 and discussed hereafter.
1) Moist and Saturated Unit Weights: The moist unit weight was estimated from the DMT results and it matched up well with the data in HDR Engineering, Inc. (1999). Therefore, both moist and saturated unit weights are the same as those in HDR Engineering, Inc. (1999).
2) Strength Parameters: The drained cohesion was assumed to be zero for a sandy soil. The drained friction angle was the minimum friction angle (37 degrees) under the plane-strain condition, indicated by DMT results. The correlation between friction angle and dilatancy angle was presented by Bolton (1986). As an order of magnitude estimate, the dilatancy angle was estimated to be: $\varphi = \phi' - 30$ degrees.
3) Deformation Parameters: The oedometer modulus was assumed to be the constrained modulus at a depth of 6 m. As a result, the reference pressure is the minor principal stress (effective horizontal stress) at a depth of 6 m, indicated by the DMT results. The Young's modulus (E) can be estimated from constrained modulus (M) and Poisson's ratio (υ) by: $E = M(1+\upsilon)(1-2\upsilon)/(1-\upsilon)$. The Poisson's ratio was determined to be 0.29 from the drained friction angle under triaxial compression (ϕ'_{TC}), using the relationship presented in Kulhawy and Mayne (1990): $\upsilon = 0.1 + 0.3$ ($\phi'_{TC} - 25$ degrees)/(20 degrees). The power (m) for stress-dependent stiffness was assumed to be 0.5 for a dense sand, according to Janbu (1963).
4) Hydraulic conductivity and void ratio: The hydraulic conductivity for dense sand with occasional seams of clayey silt or silty clay was interpreted from the guidelines in Terzaghi et al. (1996). Anisotropy was assumed in hydraulic conductivity such that the ratio between horizontal and vertical hydraulic conductivity is 1.5. The initial void ratio was assumed to be 0.5 for a typical dense sand matrix according to Terzaghi et al. (1996).

The ϕ-c reduction procedure in PLAXIS was performed to evaluate the stability of this cut slope. The factors of safety calculated from the ϕ-c reduction procedure under the normal and worst-case groundwater conditions are 2.2 and 1.2, respectively. Limit-equilibrium slope stability analyses were also performed to check the stability of the cut slope. The factors of safety calculated from limit-equilibrium analyses under the normal and worst-case groundwater conditions are 1.3 and 1.1, respectively. These factors of safety are lower than the ones obtained from finite-element analyses because a simple straight-line phreatic surface intercepted by the slope was assumed in the limit-equilibrium analyses while groundwater drawdown was modeled with assigned water heads (as the boundary conditions) and hydraulic conductivity of soils in the finite-element analyses. Groundwater drawdown in sandy soils increases the mean effective stress, as shown in Figure 8, and thus the shear strength of soils and factors of safety of the slope.

The incremental shear strain calculated from the ϕ-c reduction procedure is a good indication of the most critical failure surface of the slope. Under the normal groundwater condition, the incremental shear strain contours are presented in Figure 9. As shown in Figure 9, the most critical failure surface is influenced by groundwater drawdown and presence of the bedrock (assumed as fixity in the model). These two factors contribute to the overall stability of this cut slope.

As a result of the in-situ testing program and analyses using more realistic soil parameters from such testing, this cut slope was determined to be stable at a slope ratio of 3H:1V, without a drainage blanket. The saving of construction spending compared with an original 5H:1V slope with a drainage blanket, along both the NBL and SBL sides of the roadway is approximately half a million dollars, which was significantly more than the cost of the in-situ testing program and more refined analyses.

Table 2. Soil Parameters Developed from In-situ Testing and Used in the Finite-element Analyses

Soil Properties	Value	Unit	Source
Moist Unit Weight, γ	18.9	kN/m^3	Estimated from DMT results.
Saturated Unit Weight, γ_{sat}	20.2	kN/m^3	HDR Engineering, Inc. (1999).
Cohesion, c'	0	kPa	Assumed for the drained condition.
Drained Friction Angle, ϕ'	37	degrees	Estimated from DMT results.
Dilatancy Angle, φ	7	degrees	Bolton (1986).
Oedometer Modulus, E_{oed}	57000	kPa	Estimated from DMT results.
Secant Young's Modulus, E_{50}	45000	kPa	Estimated based on E_{oed} and Poisson's ratio.
Power, m	0.5	-	Janbu (1963).
Reference Pressure, p^{ref}	100	kPa	Estimated from DMT results.
Horizontal Permeability, k_x	1.5E-04	cm/sec	Terzaghi, Peck, and Mesri (1996).
Vertical Permeability, k_y	1.0E-04	cm/sec	Terzaghi, Peck, and Mesri (1996).
Initial Void Ratio, e_{init}	0.5	-	Terzaghi, Peck, and Mesri (1996).

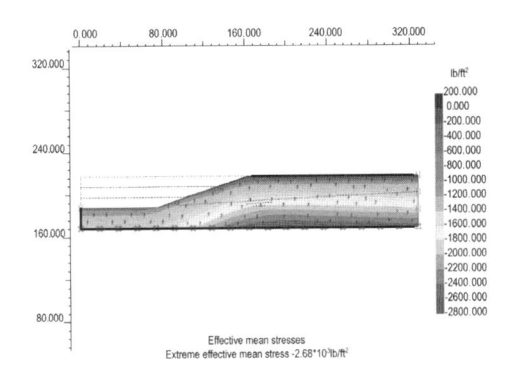

Figure 8. Influence of Groundwater Drawdown on the Mean Effective Stress Within the Slope [X-axis and y-axis show PLAXIS coordinates in feet.]

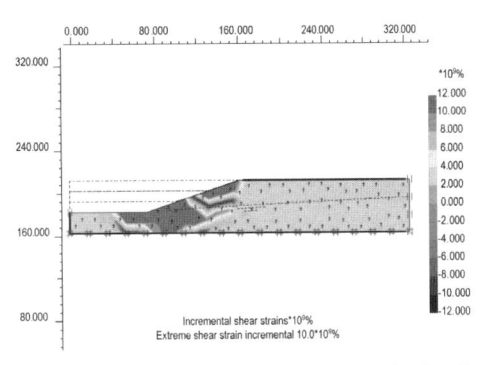

Figure 9. Incremental Shear Strain Contours Showing the Most Critical Failure Surface of the Slope [X-axis and y-axis show PLAXIS coordinates in feet.]

6 CONCLUSIONS

The following conclusions can be drawn from the project described herein.
1) Geotechnical properties of soils are site-specific and, under certain circumstances, in-situ testing offers the best measure to characterize various strength and deformation parameters of soils in place. The proper selection of geotechnical properties of soils can reduce the overall cost of the project.
2) In-situ testing is best performed by a specialist who has the knowledge of the geology and soil behavior of the site such that the soil parameters can be estimated more accurately.
3) The finite-element analysis can more accurately model the state of stress, stress-dependent deformability and strength, and groundwater characteristics within an earth structure. However, such an analysis requires more soil parameters than a conventional limit-equilibrium slope stability analysis. In-situ testing is considered the best way to obtain these soil parameters, especially within a sandy soil deposit where sampling and laboratory testing are more difficult and costly.
4) The slope modification presented herein has been implemented for a period of more than 2 years. Observations of the slope indicate that the performance has been satisfactory.

REFERENCES

Bolton, M.D. (1986). "The Strength and Dilatancy of Sands," Geotechnique, Vol. 36, No. 1, pp. 65-78.
Brinkgreve, R.B.J. and P.A. Vermeer, editors (1998). "PLAXIS Finite Element Code for Soil and Rock Analyses Version 7," Computer Program Manual, A.A. Balkema, Rotterdam, Netherlands.
Goodwin, B.K. (1970). "Report of Investigation 23 Geology of the Hylas and Midlothian Quadrangles, Virginia," Virginia Division of Mineral Resources, Charlottesville, VA.
HDR Engineering Inc. (1999). "Route 288 State Project 0288-072-104, PE101, Powhatan County, Virginia, Geotechnical Engineering Report for Roadway Design," Pittsburgh, PA.
Janbu, J. (1963). "Soil Compressibility as Determined by Oedometer and Triaxial Tests," Proc. ECSMFE Wiesbaden, Vol. 1, pp. 19-25.
Kulhawy, F.H. and P.W. Mayne (1990). "Manual on Estimating Soil Properties for Foundation Design," EL-6800 Research Project 1493-6, Final Report Prepared for Electric Power Research Institute, Palo Alto, CA.

Marchetti, S. (1980). "In Situ Tests by Flat Dilatometer," ASCE Journal of Geotechnical Engineering Division, March 1980, pp. 299-321.

Schmertmann, J.H. (1982). "A Method for Determining the Friction Angle in Sands from the Marchetti Dilatometer (DMT)," Proceeding of the Second European Symposium on Penetration Testing, Amsterdam, pp. 853-861.

Terzaghi, K.,R.B. Peck, and G. Mesri. (1996). "Soil Mechanics in Engineering Practice," John Wiley & Sons, Inc., New York, 549 pp.

VDOT (1997). "Metric Road and Bridge Specifications," Virginia Department of Transportation, Richmond, VA.

Wilkes, G.P. (1988). "Mining History of the Richmond Coalfield of Virginia," Virginia Department of Mines, Minerals & Energy, Charlottesville, VA.

Using a non linear constitutive law to compare Menard PMT and PLT E-moduli

António Gomes Correia
Universidade do Minho, Portugal

Armando Antão
Universidade Nova de Lisboa, Portugal

Michel Gambin
Apageo, France

Keywords: E-modulus, Menard PMT, plate load test, constitutive law

ABSTRACT: Engineers may now be bewildered by the various values of E-moduli they can get on the same soil if they are not aware of the variability of such a modulus with the strain level that the soil sustains. Today using simple softwares as Plaxis 7 may help a lot to understand the degradation of the E value when the strain level increase.
In this paper we compare the strain levels of the soil for the pressuremeter tests, the plate loading test and the tri-axial test. Using a non linear elastic plastic constitutive law (usually known as exhibiting isotropic strain hardening) we were able to find a good agreement between the various E-moduli obtained in these different tests.

1 INTRODUCTION.

At this time, there are more and more constitutive laws to model soil behaviour from micro-strains to failure. Research workers and engineers use numerical codes at various levels:
- Level 1: for routine estimation, assuming elastic parameters;
- Level 2: for advanced calculation, assuming non-linear soil stiffness and;
- Level 3: for research, using complex constitutive laws of material behaviour.

However, the use of these codes address a major practical difficulty which is related to the choice of the characteristic values to give to the relevant parameters of the constitutive laws.

These values are usually obtained from laboratory tests where distribution of stresses and strains are homogeneous and boundary conditions are well defined. However, it is always difficult to obtain undisturbed samples; moreover, the selection of samples and their size can lead to uncertainty regarding their representativity. Consequently, in-situ tests could then become an alternative means to obtain these values. In any case it is always interesting to be able to compare the field test results with laboratory tests, and, if possible, combine both results. Unfortunately it must be pointed out that the drawback of most routine in-situ tests lies in the fact that the stress and strain distribution necessary for the identification of constitutive laws is unknown.

In this respect, it is interesting to classify in-situ tests in three categories, regarding the measurement of soil stiffness:
- Category A: includes field measurements of shear waves velocity by geophysical tests;
- Category B: includes field tests such as pressuremeter tests and plate load tests which can yield E values using a simple theory;
- Category C: includes field tests for which the soil reaction cannot be modelled by a theory.

Tests of category A permit obtaining a value of the shear modulus representing the stiffness at very small strains (G_0), that is, in the true elastic domain of ground behaviour. G_0 is a function of the slope of the straight line tangent to the (q, dε/ε) curve at the origin (fig.1). It is a fundamental parameter of the ground, considered as a benchmark value, and often used to normalize the other values of G for different strains spans, under the ratio G/G_0.

The tests of category B are used to obtain in-situ stiffness characteristic values G for the stress and strain levels induced by the proposed structural loads on the tested ground. These test results are derived using the elastic theory in the following way: G is a function of the slope of either the tangent to point A or of the secant AB on the (q, dε/ε) curve. It can also be a function of the slope of OB. All these values can later be used taking into account the non-linear stiffness of the ground.

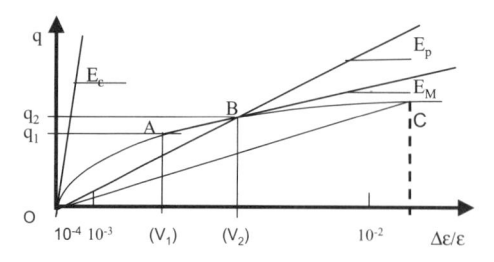

Figure 1. A typical stress-strain curve example showing tangent and secant moduli.

Tests of category C yield parameters closer to index tests; their results must be empirically correlated with the reference stiffness values obtained by the other in-situ test categories (A and B), or by laboratory tests.

This paper deals with the comparative analysis of category B field tests, here the Menard pressuremeter test (PMT) and plate load tests (PLT). Results are presented for two levels of calculations: level 1 and 2, which can be applied in engineering practice. Since it can be shown that level 1 approach leads to very different stiffness values for both types of tests, it can be of interest to move to level 2 to take into account the non-linear stiffness of soil. Further, the identification of parameters using level 2 permits to compare field results with tri-axial laboratory test results for the same level of stresses and strains.

2 PRESSUREMETER TESTS

2.1 Level 1 – Routine calculation and analysis

The pressuremeter originally developed by Louis Ménard in the late 50's (Ménard, 1955), is a very popular in-situ device used in various countries for most geotechnical projects, essentially for foundation design. In France, it has been standardized since the end of 60's (MEL, 1971), and in the USA since the mid 80's. The present standards are respectively AFNOR NF P 94-110-1 and ASTM D 4719-00. An ISO standard No. 22476-4 is under preparation on the same topic. The analysis of test results for routine calculations (level 1), leads to the Menard modulus E_M, among other stress-strain parameters. This modulus is determined in the pseudo-elastic zone of the curve relating the uniformly distributed pressure applied on the borehole wall to the volume change of the cavity. The theory of Lamé for thick cylinders is used assuming a linear-elastic, homogeneous and isotropic behaviour of the tested soil :

$$E_M = 2(1+\upsilon)V_0 \frac{dp}{dV} \tag{1}$$

where:
V_0 is volume of the probe measuring cell;
dp/dV is the slope of the secant AB on the pressuremeter curve in the pseudo-elastic zone;
υ is the Poisson's ratio.
The Menard modulus is a Young's modulus calculated for a Poisson's ratio of 0,33. Figure 2 presents a criterion to define the pseudo-elastic zone (Briaud, 1992). In Figure 3 another representation is proposed, which gives directly the tangent modulus G (Gambin and Jezequel, 1998).

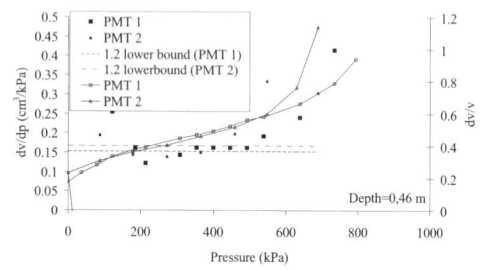

Figure 2. Definition of the pseudo-elastic zone for tests PMT1 and PMT2.

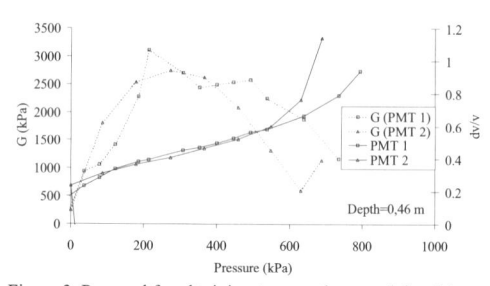

Figure 3. Proposal for obtaining tangent shear modulus G in the pseudo-elastic zone.

It must be pointed out that this modulus is related with the average stiffness exhibited by the ground associated with a determined strain level. Consequently the use of this value must be corrected by an empirical calibration factor to be applied in design, as for instance this is done in the French Code for foundation design (MEL, 1993, Clarke, 1995). Consequently Menard modulus must be considered as a test-specific design parameter.

For an unload-reload cycle a secant modulus can also be defined, which in our tests is around 2,2 times Ménard's modulus. If the cycle is selected in a cavity strain range of 0.1 – 0.2 %, the secant modulus can be used to predict settlements (Clarke, 1995).

2.2 Level 2 - Advanced Analysis

Since the pressuremeter test is an "in-situ" test where boundary conditions are known, the analysis of a test results has a sound theoretical basis. The procedure to identify parameters is essentially based on the nature of a "back analysis" requiring an appropriate modelling of the constitutive laws of the ground. Since the test analysis is an inverse problem, identification of parameters, analytically or numerically, depends entirely on the modelling technique (Gioda, 1985). Therefore, realistic model assumptions must be addressed, otherwise physical meaning of parameters can be loosed.

The advance analysis of test results is dealing with a non linear modelling analysis of soil behaviour. In this context three aspects must be considered:
- Elastic behaviour of ground for the domain of very small strains;
- Elasto-plastic behaviour for intermediate level of strains and;
- Non linear geometry for large strains.

Various constitutive laws were use in modelling the pressuremeter results: Cambou et Boubanga (1989), Shahrour et al. (1995), Biarez et al. (1998). In this paper we adopt a simple model developed by PLAXIS and called HSM (hardening soil model) which assumes a non linear elastic response of the soil during loading and a isotropic hardening during unloading. The geotechnical parameters which are necessary to carry out the study are well known by professionals. The main features of this model are:
- Stress dependent stiffness.
- Plastic straining due to both primary deviatoric and compression loading.
- Elastic unloading and reloading.
- Failure according to the Mohr Coulomb model.

The hyperbolic response due to tri-axial deviatoric loading, as well as the oedometer simulation, are shown in Figure 3, where the meaning of the stiffness parameters of the model is submitted.

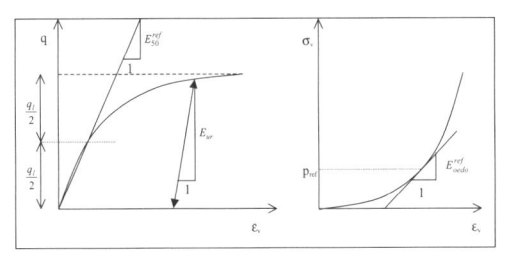

Figure 4. Schematic response for tri-axial and oedometer tests by Plaxis-HSM.

Furthermore, the model has parameters allowing changing strength and stiffness with depth, but this was not adopted in the calculations.

The identification of the parameters of this model has been done by fitting pressuremeter test data with the numerical results. However, a number of combinations of these parameters will produce the same sort of best fit, therefore it was necessary to exhibit some engineering judgement to ensure that parameters obtained are compatible with one another. Furthermore, laboratory results of the test material were also used, as the friction angle of tri-axial tests (Fleureau et al., 2002). This method of identification of parameters, known as computer-aided-modelling (CAM) is now seen as an alternative and a promising method in many geotechnical applications.

Figure 5 shows the best fit obtained for two pressuremeter test results (PMT1 and PMT2) obtained at the same depth in a embankment built with a silty sand; this soil is a residual soil from granite.

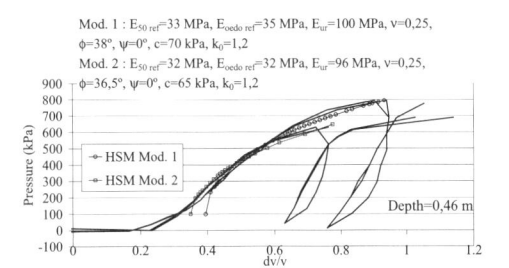

Mod. 1 : $E_{50\,ref}$=33 MPa, $E_{oedo\,ref}$=35 MPa, E_{ur}=100 MPa, ν=0,25, ϕ=38°, ψ=0°, c=70 kPa, k_0=1,2
Mod. 2 : $E_{50\,ref}$=32 MPa, $E_{oedo\,ref}$=32 MPa, E_{ur}=96 MPa, ν=0,25, ϕ=36,5°, ψ=0°, c=65 kPa, k_0=1,2

Figure 5. Best fitting curves of two pressuremeter tests using Plaxis-HSM.

3 THE PLATE LOAD TESTS

3.1 Level 1 – Routine calculation and analysis

The load plate test is also used in routine calculations of foundations (ASTM D1196), roads and airports (ASTM D1195, LCPC-CT2). Generally plates of large diameter are used with smaller plates on top to ensure a sufficient rigidity to obtain uniform settlements under the plate. As in the pressuremeter routine analysis, the modulus derived from the plate load test is valid only for a linear-elastic, homogeneous and isotropic ground. In this case the Boussinesq equation can be used to obtain the modulus for the first loading:

$$E = \frac{1,5 Q_{appl} D/2 (1-\nu^2)}{\delta} \qquad (2)$$

for the case of a rigid plate, as is the case of that used in our tests.

Figure 6 shows the results of plate load test carried out according to the LCPC procedure (LCPC-CT 2). When unload-reload cycles were realized, it is also possible, like in the pressuremeter, to define a secant modulus, which in our tests is around 1,8 times the modulus in the first loading. Figure 7 shows the increase of the secant modulus of reloading with the stress level. This is a typical behaviour of compacted unbound granular materials, where the strains under this level of stress are practically elastic (Gomes Correia and Biarez, 1999). In this domain, an increase of modulus with the increase of stress in the direction of the load is observed.

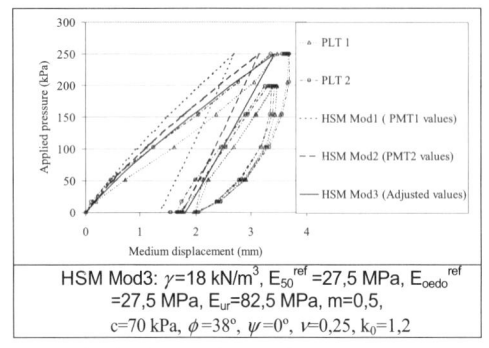

Figure 6. Plate load test results according to LCPC-CT2 procedure and best fitting curves.

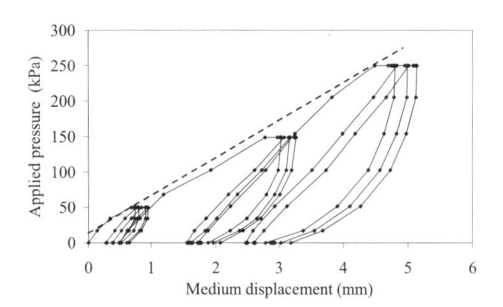

HSM Mod3: $\gamma=18$ kN/m^3, $E_{50}^{ref}=27,5$ MPa, $E_{oedo}^{ref}=27,5$ MPa, $E_{ur}=82,5$ MPa, m=0,5, c=70 kPa, $\phi=38°$, $\psi=0°$, $\nu=0,25$, $k_0=1,2$

Figure 7. Plate load test results according to ASTM-D1196 standard.

3.2 Level 2 - Advanced Analysis

Since the plate load test gives, like pressuremeter tests, a load-settlement curve, it might be expected that it would be possible to use an inverse technique to identify parameters of a non-linear model for the ground. So, the same numerical tools used for the pressuremeter analysis were applied: Plaxis with HSM constitutive law. In a first step, the same parameters obtained before were used (fig. 6), i.e. HSM Mod.1 and HSM Mod.2. One observes a ground response with a stiffer behaviour in loading as well in unloading. This can be a consequence of a less good adaptation of the model to simulate the field of stresses and strains of plate load test, which is different from that of the pressuremeter test. Indeed, plate load test procedure induces strain levels in the ground much lower than those used in the initial fitting of the pressuremeter curve. Another explanation can be the physical nature of the real behaviour of the material. In spite of the depth of 0,46 m, pressuremeter test results can be considered as representative of the plate load test with a 0,61 m plate diameter performed on surface. It is very likely in a granular material, compacted with a vibrating roller, that the state of density close to the surface is less dense, contributing to a global less stiff response of the ground under the plate load test. One better curve fitting to the plate load test (HSM Mod.3) is obtained reducing original modulus of pressuremeter simulations by about 15 % (fig. 6).

With these model parameters from the plate load test a numerical simulation was performed in order to observe the strain distribution in depth in relation to the total surface settlement for a given load. These results led to an almost unique relationship between the ratio of the settlement at a given depth and the total settlement, and the ratio between this depth and the diameter of the plate (fig. 8). This confirms that practically (about 90 %) settlement arises from strains up to a depth about twice the diameter of the loading surface.

It is also interesting to observe the results of numerical simulations of the variation of the secant modulus with depth on the axis of the load (fig. 9). These results can also be represented by replacing depth by the corresponding strain in the ground (fig. 10), which explains the increase of the modulus with depth, as a result of a decrease of the strain level. These results put in evidence the difficulty of identification of the parameters of a non linear law from a load-settlement curve of a plate load test without having other information. A possible solution to this is the measurement of deformations in depth under the plate with strain gauges, as it was proposed by Burland (1989).

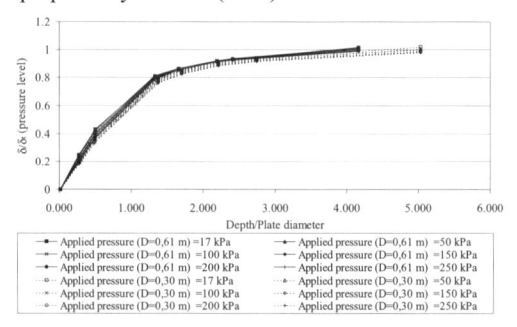

Figure 8: Ratio of the settlement at a given depth and the total settlement (δ/δ_t) as a function of the ratio between this depth and the diameter of the plate (z/D).

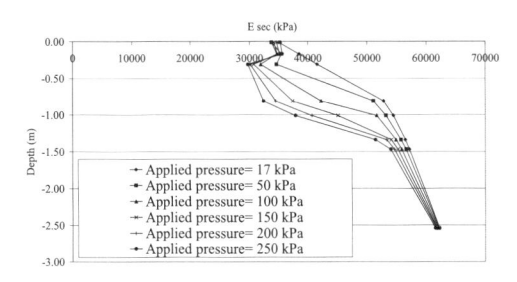

Figure 9. Variation of the secant modulus with the depth.

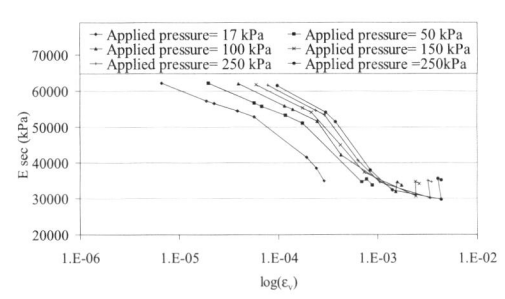

Figure 10. Variation of the secant modulus with strain.

4 COMPARISON BETWEEN MODULI FROM PRESSUREMETERS, PLATE LOAD TESTS AND TRI-AXIAL TESTS

The routine interpretations of PMT and PLT led to very different values of modulus. Besides, the Menard modulus is a tangent modulus, in the sense that it is obtained by the slope of the pseudo-elastic zone of the pressuremeter curve, while the modulus obtained by the plate load test is generally a secant modulus. In addition it is obvious that these modulus will be modified if test procedure or interpretation is modified. This is a consequence of the non-linear material behaviour, where the modulus depends on the level of stress and strain, among others. The main point is to know in practice how to use correctly these values. In fact, modelling geotechnical structures is being more and more popular, and consequently the results of category B tests must be more and more used for the identification of the model parameters. It is evident that the correctness of this identification is a function of the model adopted. So, the appropriateness of the model must be carefully analysed and confirmed.

The model chosen in this study is a compromise between a simple constitutive law of ground behaviour with the description of its more important aspects in terms of stress, strain and strength and the use of parameters having a physical meaning fully understood by practitioner engineers.

With this model, and by using the parameters deducted from pressuremeter and plate load tests, a numerical simulation of tri-axial tests is submitted on Figure 11. These simulations were done for different stress paths corresponding to those induced by the plate load test at various depths (0,1D; 0,5D; 1D and 2D). Figure 11 represents these results which express the secant and tangent modulus as a function of strain level. It can be observed that these results are very sensitive to the influence of stress paths and consequently to the depth.

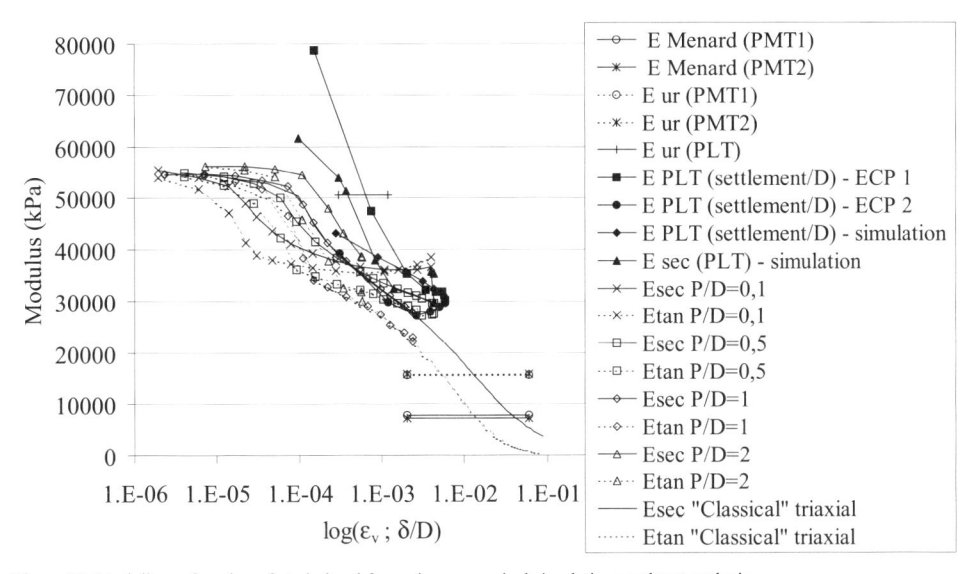

Figure 11. Moduli as a function of strain level for various numerical simulations and test analysis.

From Figure 11 it can also be noticed that the Menard modulus obtained in the routine analysis is associated to a strain level near 1 %, while the secant modulus of unload-reload of the plate load test is rather close to 0,1 %. This difference of strain levels leads to an unload-reload modulus of plate load test about three times the unload-reload modulus of pressuremeter, associated to a different strain level. On this Figure 11 the secant modulus of the plate load test (from the routine analysis) is submitted as a function of the ratio of settlement and plate diameter (tests PLT1 e PLT2). One of test results (PLT2) leads to a variation trend similar to the results of the simulation of the two tri-axial tests for the deeper stress paths. It is also interesting to observe that the ASTM D1195 procedure of plate load tests leads to levels of strains during loading variable between 0,01 % and 0,6 % (this last one for a stress of 250 kPa). Another interesting result comparable to the previous one is the variation of the secant modulus with the level of strain calculated in depth (result presented in Figure 10) for an average stress of 250 kPa applied by the plate.

5 COMPARISON BETWEEN RESULTS OF STRESS-STRAIN BEHAVIOUR FROM TRI-AXIAL AND PLATE LOAD TESTS

Hillier and Woods (2001) showed that for a non-linear elastic behaviour of the ground, the curve load-settlement of a plate load test:

$$Q = K\left(\frac{\delta}{D}\right)^n \quad (3)$$

follows the same power law as the constitutive model of the ground:

$$Q = K'\varepsilon^n \quad (4)$$

To validate this findings using our results, tri-axial test curves (vertical stress-vertical strain for different stress paths) were compared with similar plate load test results (applied vertical stress of PLT versus ratio of settlement to diameter δ/D of the plate) (fig. 12).

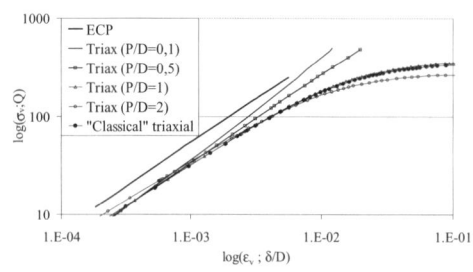

Figure 12. Comparison between tri-axial test strain and PLT strain

The analysis of these results shows a certain influence of the stress path. For the conventional stress path there is a domain of load where strains are 0,5 times the "relative strains" δ/D of the plate load test, whereas Hillier and Woods (2001) found, for a power of n=0,55, a value of 0,3. After a certain level of load it is evident that, due to the plasticity of the ground, this ratio is not valid any more.

6 CONCLUSION

Routine analyses of Menard PMT and PLT lead to E-moduli values which are quite different. Further these values would be more or less different if the way to carry out the tests is changed or the method of E values derivation is modified. In a compacted silty sand the first loading E-modulus for a PLT is 2.7 times that one of the Menard E modulus. Further the unload-reload modulus of the PLT is 3 times the unload-reloading E-modulus of PMT. This exemplifies the warning that these E-moduli values must only be used within specific rules of design.

It was shown that these apparent discrepancies between the E values stem from the fact that moduli are a function of the strain level. It is important to analyze these tests with relevant constitutive laws.

Using the previous modeling technique where the friction angle is derived from tri-axial tests, we can draw the following conclusions:
- the strain level associated with the Menard E modulus is of the order of 1 %, when the strain level associated with the PLT unload-reload modulus is about 0.1 %,
- the ASTM practice for the PLT leads to a strain level of 0.01 to 0.6 % during loading,
- these strain levels are in good agreement with the E-moduli values obtained for both types of tests,
- there is almost a single relationship between the ratio of the elementary settlement at a given depth over the total settlement and the ratio between this depth over the plate diameter. The full settlement arises within a depth equal to twice the plate diameter,
- in the non linear elastic behaviour domain of the soil, the curve which expresses the variation of the applied pressure during a PLT versus the ratio settlement over diameter is close to the curve which shows the function of the vertical stress versus the vertical strain in a tri-axial test. In the numerical modeling with a power law equal to 0.5, the ratio between the relative plate deformations or "relative strains" δ/D and the tri-axial strains was about 0.5.

REFERENCES

AFNOR, 2000, Essai pressiométrique Ménard. *Norme Française NF P94 – 110-1*

ASTM, 1993, Standard test method for repetitive static plate tests of soils and flexible pavements components, for use in evaluation and design of airport and highway pavements. *Annual Book of ASTM Standards, D 1195-93, American Society for Testing and Materials, West Conshohocken, Pensylvania*

ASTM, 1997, Non-repetitive static plate load test of soils and flexible pavement components for use in evaluation and design of airport and highway pavement, D1196-93, *American Society for Testing and Materials, West Conshohocken, Pensylvania*

ASTM, 2004, Standard test method for prebored pressuremeter testing in soils. *American Society for Testing and Materials, vol. 04.08, Soil and Rocks (I): D420-D5779, West Conshohocken, Pensylvania*

Biarez, J., Gambin, M., Gomes Correia, A., Flavigny, E.,Branque, D., 1998, Using pressuremeter to obtain parameters to elastic plastic models for sand, *Geotechnical Site Characterisation (ISC'98), Robertson and Mayne editors, Balkema, Rotterdam, Vol 2, pp.747-752*

Briaud, J-L., 1992, The pressuremeter, *Balkema, Rotterdam*

Burland, J., 1989, The 9th Bjerrum memorial lecture: Small is Beautiful, the Stifness of Soils at Small Strain, *Can. Geotech. Journal, Vol 26, pp.499-516*

Cambou, B., Boubanga, A., 1989, Press'ident, un logiciel d'aide à la caractérisation des sols,*Actes des Journées Universitaires de Génie Civil, Rennes*

Clarke, B., 1995, The pressuremeter in geotechnical Design,,*Blackie Academic and Professionnal, Glasgow.*

Fleureau, J-M.; Gomes Correia, A.; Hadiwardoyo, S.; Dufour-Laridan, E. & Langlois, V., 2002, Influence of suction on the dynamic properties of a silty sand. *Third International Conference on Unsaturated Soils. 10-13 de Março de 2002, Recife, Brasil*

Gambin, M., Jezequel, J., 1998, A new approach to the Menard PMT parameters, *Geotechnical Site Characterisation (ISC'98),Robertson and Mayne editors, Balkema, Rotterdam*

Gioda, G., 1985, Some remarks on back analysis and characterization problems. *Proc. 5th Int. Conf. on Num. Methods in Geomechanics, vol. 1, pp. 47-61*

Gomes Correia, A. and Biarez, J., 1999, Stiffness properties of materials to use in pavement and rail track design. *Proc. XIIth ECSMGE, vol.1, pp. 1245-1250*

Hillier, R.P. & Woods, R.I. (2001). Characterisation of non-linear elastic behaviour from field plate load tests. *Proc. XV th ICSMGE, Istanbul, Balkema, Rotterdam, vol. 1, pp. 421-424*

ISO/WD 22476-4, Geotechnical investigation and testing – field testing – part 4: Menard Pressuremeter test. *ISO Technical Programme: TC 182/SC1*

Ménard, L., 1955, Pressiomètre, brevet français d'invention, n° 1.117.983, 19.01.1955

LCPC, 1973, Essai à la plaque. *Mode operatoire CT-2, Laboratoire Central des Ponts et Chaussées, Paris*

MEL(Ministre de l'Equipement et du Logement), 1971, Essai Pressiométrique Normal, *Série des Modes Opératoires du LCPC, Dunod, Paris*

MELT, 1993, Fascicule 62, Titre V du CCTG, *Imprimerie des Journaux Officiels, Paris*

Shahrour, I., Kasdi, A., Abriak, N., 1995, Utilisation de l'essai pressiométrique pour la détermination des propriétés mécaniques des sables, *Revue Française de Géotechnique, N°73, Paris*

Computational and physical basis for dynamic site characterization using Love waves

Bojan B. Guzina, Andrew I. Madyarov and Robert H. Osburn
Dept. of Civil Engineering, University of Minnesota, MN, USA

Keywords: dynamic site characterization, layered half-space, Love waves, inverse problems

ABSTRACT: This paper deals with the computational and experimental framework for dynamic site characterization using the spectral analysis of Love waves. The methodology revolves around a model for the horizontally-polarized (SH) wave motion in a layered viscoelastic half-space due to action of a torsionally-vibrating disc. It is shown that the use of Love waves as a sounding tool reduces the number of elastic and damping parameters relevant to site characterization from five to three per soil layer. For practical purposes, solution to the inverse problem is reduced to the minimization of a functional quantifying the misfit between experimental observations and theoretical predictions of the surface ground motion. By virtue of a consistent viscoelastic analysis of the Love-wave dispersion and attenuation (as captured by the motion sensors), the proposed approach furnishes independent estimates of the shear modulus, S-wave damping, and mass density profiles that are not directly available from the Rayleigh-type spectral interpretations. To maintain physical relevance, the proposed computational developments are accompanied with the design of a compact, manually-operated device for generating axisymmetric Love-wave field.

1 INTRODUCTION

Over the past two decades, the Spectral Analysis of Surface Waves (SASW) method (Heisey *et al.*, 1982, Stokoe and Nazarian, 1983) has emerged as a powerful engineering tool for non-invasive site characterization. A general arrangement of the SASW test involves a vertically-oscillating source plate and a pair of remote sensors used to monitor the induced ground motion dominated by the surface, i.e. Rayleigh waves. Intrinsically, this technique makes use of the dispersion of (vertically-polarized) *Rayleigh waves* and their intimate dependence on site stratigraphy as a tool to resolve the subterranean variation of the elastic soil and rock properties, interpreted within the framework of lateral homogeneity (Ganji *et al.*, 1998). In a general isotropic setting where soil damping is also of interest (e.g. Rix and Lai, 1998), such material identification may involve as much as *six* parameters per soil layer, including the layer thickness, soil s (small-strain) shear modulus, Poisson s ratio, damping ratios for compressional and shear waves and mass density, which renders the data interpretation difficult.

To mitigate the latter problem, the focus of this investigation is the establishment of an alternative computational and physical framework for dynamic site characterization via the spectral analysis of (horizontally-polarized) *Love waves*, generated by a torsionally-oscillating source plate. Using a rigorous mathematical model for the horizontally-polarized (SH) wave propagation in a layered viscoelastic solid, it is shown that the use of Love waves as a probing tool rules out the influence of the soil s Poisson s ratio and damping ratio for compressional waves, thus reducing the maximum number of parameters relevant to seismic data interpretation from six to four per soil layer. To obtain the subsurface stiffness, damping and density profiles, the inverse problem is formulated in terms of a functional quantifying the misfit between experimental observations and theoretical predictions of the surface ground motion, synthesized in terms of the apparent surface wave velocity and amplitude decay of the ground motion. For an effective computational treatment of the gradient-based minimization employed to solve the problem, sensitivities of the visco-elastodynamic model with respect to layer thicknesses, shear moduli, damping ratios, and mass densities are derived analytically. For practical applications, the proposed back-analysis is complemented with a conceptual experimental setup aimed at generating the axially symmetric Love-wave field.

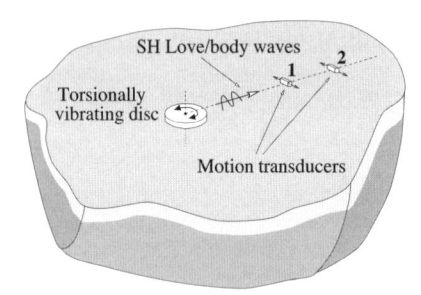

Figure 1. Conceptual setup for the spectral analysis of Love waves

2 DYNAMIC SITE CHARACTERIZATION

With reference to Fig. 1, the site characterization problem considered in this study assumes an experimental setup comprised of a torsionally-oscillating plate and a pair of tangentially-oriented geophones, used to monitor the surface vibration (see Guzina, 2000). To expedite both the measurements and spectral interpretation of the induced ground motion, torsional oscillations of the loading plate can be generated using an impact loading or an electromagnetic shaker excited by a white-noise random signal. The latter source has, for instance, been successfully used in resonant column (Al-Sanad et al., 1986) and small-scale centrifuge testing (Pak and Guzina, 1995).

On denoting the respective source-receiver distances corresponding to the nearer and farther geophone in Fig. 1 by r_1 and r_2, the *apparent phase velocity* of the horizontally-polarized (body and Love) waves propagating between the two transducers can be calculated as

$$c(\omega) = \frac{\omega(r_2 - r_1)}{\Delta\phi}, \tag{1}$$

where ω is the circular frequency of vibration, and $\Delta\phi$ is the *phase lag* between the response of the nearer transducer (denoted as "1") and the farther sensor (denoted as "2"), which can be expressed as

$$\Delta\phi = \arg\left[\frac{1}{w(r_1, r_2, \omega)}\right].$$

Here $w(r_1, r_2, \omega)$ is the *frequency response function*

$$w(r_1, r_2, \omega) = \frac{\hat{u}_\theta(r_2, \omega)}{\hat{u}_\theta(r_1, \omega)} \tag{2}$$

where $\hat{u}_\theta(r_k, \omega)$ is the (complex-valued) Fourier spectrum of the displacement record captured by the kth sensor $(k = 1, 2)$.

In what follows, a computational framework is developed that will allow estimation of the soil's subsurface profile in terms of layer thicknesses, shear moduli, S-wave damping ratios, and mass densities on the basis of observed i) dispersion relationship for the Love waves, $c(\omega)$ and ii) amplitude decay of the induced ground motion. As indicated by the preceding statement, term "Love waves" will hereafter be used in generalized sense as to denote both surface and (SH) body waves comprising the surface ground motion.

3 FORWARD MODEL

To establish a necessary computational framework for the spectral analysis of Love waves, it is useful to consider a viscoelastic ground model consisting of a uniform half-space overlaid by n parallel homogeneous layers. With the reference cylindrical coordinate system (r, θ, z) centered at the free surface $z = 0$, the kth layer (bounded by its upper and lower interfaces located respectively at depths z_{k-1} and z_k) is characterized by the thickness $h_k = z_k - z_{k-1}$, mass density ρ_k, Poisson's ratio ν_k, shear modulus G_k and damping ratios ζ_{pk} and ζ_{sk} for compressional and shear waves, respectively ($k = \overline{1, n}$). The bottom half-space is assumed to have density ρ_{n+1}, Poisson's ratio ν_{n+1}, shear modulus G_{n+1}, and damping ratios $\zeta_{p(n+1)}$ and $\zeta_{s(n+1)}$.

To simulate the testing configuration depicted in Fig. 1, the above multi-layered system is subjected to a time-harmonic torque of magnitude T generated by a torsionally-vibrating rigid circular plate with radius a, bonded to the top of the first layer. For practical purposes, action of the latter axisymmetric source can be approximated via linearly-varying tangential tractions

$$\tau_\theta(r, \theta, t) = T \frac{2rH(a-r)}{\pi a^4} e^{i\omega t}, \tag{3}$$

where $H(x)$ stands for the Heaviside step function. By virtue of the correspondence principle (Findley et al., 1989), the effects of material attenuation for this class of time-harmonic problems can be conveniently simulated by replacing G_k ($k = \overline{1, n+1}$) in the elastodynamic solution with the respective complex moduli

$$G_k^* = G_k \left(1 + i \frac{2\zeta_{sk}}{\sqrt{1 - 4\zeta_{sk}^2}}\right), \quad \zeta_{sk} < 0.5, \tag{4}$$

and effecting an analogous substitution for ν_k in terms of their complex counterparts ν_k^* ($k = \overline{1, n+1}$) (Guzina and Lu, 2002).

On employing the method of propagator matrices (Guzina and Pak, 2001) to derive the response of a layered viscoelastic solid due to time-harmonic torsional excitation, it can be shown that the *only* non-zero motion component at the surface of a layered half-space, induced by the torsional load (3), is the *axisymmetric* tangential displacement u_θ which permits the integral representation

$$u_\theta(r, z=0, t; \boldsymbol{p}) \equiv T\hat{u}_\theta(r, \omega; \boldsymbol{p})\, e^{i\omega t} =$$
$$= \frac{2Te^{i\omega t}}{\pi a^2 G_1^*} \int_0^\infty \mathcal{K}(\xi, \omega; \boldsymbol{p}) J_2(a\xi) J_1(r\xi)\, d\xi, \quad (5)$$

where J_m is the Bessel function of order $m = 1, 2$,

$$\boldsymbol{p} = (G_1^*, \rho_1, h_1, \ldots\ G_n^*, \rho_n, h_n, G_{n+1}^*, \rho_{n+1})^T$$

is the vector of material and geometric parameters of the half-space, and

$$\mathcal{K}(\xi, \omega; \boldsymbol{p}) = \frac{1}{\beta_1} \cdot \frac{\mathcal{B} + e_1 \mathcal{A}}{\mathcal{B} - e_1 \mathcal{A}}. \quad (6)$$

In (6), coefficients $\mathcal{A} \equiv \mathcal{A}(\xi, \omega; \boldsymbol{p})$ and $\mathcal{B} \equiv \mathcal{B}(\xi, \omega; \boldsymbol{p})$, which are functions of the transform parameter ξ, are calculated through a matrix product

$$\begin{pmatrix} \mathcal{A} \\ \mathcal{B} \end{pmatrix} = \boldsymbol{Q}_1 \boldsymbol{Q}_2 \ldots \boldsymbol{Q}_{n-1} \begin{pmatrix} \varkappa_n - \varkappa_{n+1} \\ \varkappa_n + \varkappa_{n+1} \end{pmatrix} \quad (7)$$

in terms of the propagator matrices ($k = \overline{1, n-1}$)

$$\boldsymbol{Q}_k = \begin{pmatrix} e_{k+1}(\varkappa_k + \varkappa_{k+1}) & (\varkappa_k - \varkappa_{k+1}) \\ e_{k+1}(\varkappa_k - \varkappa_{k+1}) & (\varkappa_k + \varkappa_{k+1}) \end{pmatrix}, \quad (8)$$

where

$$e_k = e^{-2\beta_k h_k}, \qquad k = \overline{1, n};$$
$$\varkappa_k = G_k^* \beta_k, \quad \beta_k = \sqrt{\xi^2 - \frac{\omega^2 \rho_k}{G_k^*}}, \quad k = \overline{1, n+1}. \quad (9)$$

Formula (7) is obtained by simplifying the recursive formulas for the *generalized* reflection coefficient, $\hat{R}_1^d = \mathcal{A}/\mathcal{B}$, for the downwardly propagating SH waves in the top layer impinging on the material interface at $z = h_1$. The latter coefficient, which includes multiple reflections and transmissions of horizontally polarized waves occurring at material discontinuities below depth $z = h_1$, can be derived by means of the propagator matrix approach (Guzina and Pak, 2001).

One may note that the horizontally-polarized displacement solution (5) does not depend on the Poisson s ratios ν_k nor the compressional-wave damping ratios ζ_{pk}, $k = \overline{1, n+1}$, of the layered half-space, thus reducing the total number of material parameters per soil layer relevant to dynamic site characterization from five to three, namely G_k, ζ_{sk} and ρ_k.

The semi-infinite integral in (5) can be effectively evaluated using the method of *asymptotic decomposition* wherein the leading asymptotic expansion of the kernel $\mathcal{K}$ is extracted and integrated in closed form so that the remaining part with strong decay can be evaluated numerically (see, e.g., (Guzina and Pak, 2001) in the context of visco-elastodynamic fundamental solutions).

4 INVERSE SOLUTION

To obtain the subsurface stiffness, damping and density profiles on the basis of (1), (2) and (5), solution to the inverse problem is sought by minimizing a functional quantifying the misfit between experimentally-observed and theoretically-predicted amplitude and phase information synthesizing the Love wave motion between receivers "1" and "2" (see Fig. 1).

4.1 Maximum-likelihood inverse theory

For an effective interpretation of noise-polluted field measurements, it is useful to employ the framework of the maximum-likelihood inverse theory (Tarantola, 1987) and postulate that all uncertainties present in the problem can be approximated as being Gaussian. To this end, an experiment resulting in the vector of measured data $\boldsymbol{d}^{obs}$ is assumed to provide an information about the *true* values $\boldsymbol{d}$ of observable parameters in the form of a conditional probability density

$$\chi(\boldsymbol{d}^{obs}|\boldsymbol{d}) = \alpha \cdot \exp\left[-\frac{1}{2}(\boldsymbol{d} - \boldsymbol{d}^{obs})^T \boldsymbol{C}_d^{-1}(\boldsymbol{d} - \boldsymbol{d}^{obs})\right],$$

where $\boldsymbol{C}_d$ is the data covariance operator and α is a normalizing constant. On neglecting the effect of modeling errors associated with the visco-elastodynamic prediction of ground motion measurements, $\boldsymbol{d} = \boldsymbol{g}(\boldsymbol{m})$, it can be shown (Tarantola, 1987) that the posterior information in the model space, which assumes no prior information on $\boldsymbol{m}$, is given by the Gaussian distribution $\sigma(\boldsymbol{m}) = const \times \exp[-\frac{1}{2}\mathcal{F}(\boldsymbol{m})]$ where

$$\mathcal{F}(\boldsymbol{m}) = [\boldsymbol{g}(\boldsymbol{m}) - \boldsymbol{d}^{obs}]^T \boldsymbol{C}_d^{-1}[\boldsymbol{g}(\boldsymbol{m}) - \boldsymbol{d}^{obs}]. \quad (10)$$

On the basis of (10), a maximum-likelihood solution to the featured inverse problem can be uniquely defined as a point $\boldsymbol{m} = \boldsymbol{m}_{ML}$ in the model space where $\sigma(\boldsymbol{m})$ is maximized, or equivalently where $\mathcal{F}(\boldsymbol{m})$ attains its minimum value.

4.2 Interpretation of the horizontally-polarized wave field

To apply the maximum-likelihood theory to inverse problems involving complex-valued vector $\boldsymbol{p} = (G_1^*, \rho_1, h_1, \ldots G_n^*, \rho_n, h_n, G_{n+1}^*, \rho_{n+1})^T$ of model parameters, it is convenient to introduce its real-valued, dimensionless counterpart

$$\boldsymbol{m} = \log[\mathrm{Re}(\bar{G}_1^*), \mathrm{Im}(\bar{G}_1^*), \bar{\rho}_1, \bar{h}_1, \ldots$$
$$\mathrm{Re}(\bar{G}_n^*), \mathrm{Im}(\bar{G}_n^*), \bar{\rho}_n, \bar{h}_n,$$
$$\mathrm{Re}(\bar{G}_{n+1}^*), \mathrm{Im}(\bar{G}_{n+1}^*), \bar{\rho}_{n+1}]^T \in \mathbb{R}^{4n+3},$$

where $\bar{G}_k^* = G_k^*/G_0$, $\bar{\rho}_k = \rho_k/\rho_0$, $(k = \overline{1, n+1})$ and $\bar{h}_k = h_k/a$, $(k = \overline{1, n})$, with G_0 and ρ_0 denoting the reference shear modulus and mass density, respectively.

Here one may note that the logarithmic representation of $\text{Re}(\bar{G}_k^*)$, $\text{Im}(\bar{G}_k^*)$, $\bar{\rho}_k$ and $\bar{h}_k$ is introduced to facilitate the imposition of the positivity requirement on G_k, ρ_k and h_k as well as the non-negativity restriction on ζ_{sk}, $k=\overline{1,n+1}$. In what follows, the components of $\boldsymbol{m}$ will be treated as *independent* material parameters instead of G_k, ζ_{sk}, ρ_k and h_k.

Assuming that the spectral measurements of the frequency response function (2) are performed at N discrete frequencies ω_j, the observed data and their predictions, synthesized in terms of the phase and amplitude information, can be suitably arranged into respective real-valued vectors as

$$\boldsymbol{d}^{obs} = (\bar{c}_1^{obs}, \ldots, \bar{c}_N^{obs}, |w_1^{obs}|^2, \ldots, |w_N^{obs}|^2)^T;$$
$$\boldsymbol{g}(\boldsymbol{m}) = (\bar{c}_1, \ldots, \bar{c}_N, |w_1|^2, \ldots, |w_N|^2)^T, \quad (11)$$

where $(\bar{c}_j \equiv c(\omega_j)/c_0, |w_j| \equiv |w(\omega_j)|)$ and $(\bar{c}_j^{obs} \equiv c^{obs}(\omega_j)/c_0, |w_j^{obs}| \equiv |w^{obs}(\omega_j)|)$ $(c_0 = \sqrt{G_0/\rho_0})$ are respectively the predicted (i.e. trial) and observed values of the (normalized) phase velocity and amplitude decay of Love waves between receivers "1" and "2", located at distances r_{1j} and r_{2j}. In (11), the phase-amplitude pairs $(\bar{c}_j, |w_j|)$ and $(\bar{c}_j^{obs}, |w_j^{obs}|)$ are computed using i) formulas (1), (2) and (5) (for a trial value of $\boldsymbol{m}$) and ii) experimental records, respectively. Owing to the fact that the ground motion is typically monitored using geophones, i.e. in terms of particle *velocity* (as opposed to particle displacement), it is important to note that formula (2) is equally applicable when $\hat{u}_\theta(r,\omega)$ is replaced by the corresponding Fourier spectrum of the ground velocity (i.e. geophone) record, $\hat{v}_\theta(r,\omega)$.

With the above definitions, the task of finding the maximum-likelihood solution to the inverse problem of dynamic site characterization associated with Fig. 1 can be reduced to the minimization of the cost functional $\mathcal{F}(\boldsymbol{m})$ given by (10) in the $(4n+3)$-dimensional parametric space.

4.3 Gradients of the objective functional

Owing to the appreciable computational cost of visco-elastodynamic solutions, functional $\mathcal{F}(\boldsymbol{m})$ given by (10) can be minimized most effectively by means of the gradient-based descent techniques such as the BFGS quasi-Newton method (Nocedal and Wright, 1999). For this class of solution algorithms, the central issue is an accurate and efficient evaluation of the gradient

$$\nabla_{\boldsymbol{m}}\mathcal{F}(\boldsymbol{m}) = 2[\boldsymbol{g}(\boldsymbol{m}) - \boldsymbol{d}^{obs}]^T \boldsymbol{C}_d^{-1} \frac{\partial \boldsymbol{g}(\boldsymbol{m})}{\partial \boldsymbol{m}}.$$

where $\partial \boldsymbol{g}(\boldsymbol{m})/\partial \boldsymbol{m}$ is the Jacobian operator of $\boldsymbol{g}(\boldsymbol{m})$. It can be shown that the complex-valued displacement field $\hat{u}_\theta(r,\omega;\boldsymbol{p})$ given by (5) is an *analytic function* of $\boldsymbol{p}$ within the parametric domain of interest so that the "parametric" differentiation of $\hat{u}_\theta$ associated with $\nabla_{\boldsymbol{m}}\mathcal{F}$ can be performed under the integral sign in (5). This fact, in turn, results in a semi-analytical expression for $\nabla_{\boldsymbol{m}}\mathcal{F}$ that can be computed with the same accuracy as $\mathcal{F}$ itself. For instance, a formula for the partial derivative of $\hat{u}_\theta$ with respect to $m_2 \equiv \log[\text{Im}(\bar{G}_1^*)]$, required (through (1)-(2) and (10)) for the calculation of $\nabla_{\boldsymbol{m}}\mathcal{F}$, can be derived as

$$\frac{\partial \hat{u}_\theta}{\partial \log[\text{Im}(\bar{G}_1^*)]} = i\,\text{Im}(G_1^*) \frac{\partial \hat{u}_\theta}{\partial G_1^*},$$

where

$$\frac{\partial \hat{u}_\theta}{\partial G_1^*} = -\frac{\hat{u}_\theta}{G_1^*} + \frac{2}{\pi a^2 G_1^*} \int_0^\infty \frac{\partial \mathcal{K}}{\partial G_1^*} J_2(a\xi) J_1(r\xi)\,d\xi, \quad (12)$$

$$\frac{\partial \mathcal{K}}{\partial G_1^*} = -\frac{1}{\beta_1}\left[\mathcal{K} + \frac{4h_1 e_1 \mathcal{AB}}{(\mathcal{B}-e_1\mathcal{A})^2}\right]\frac{\partial \beta_1}{\partial G_1^*} +$$
$$+ \frac{2e_1}{\beta_1(\mathcal{B}-e_1\mathcal{A})^2}\left(\mathcal{B}\frac{\partial \mathcal{A}}{\partial G_1^*} - \mathcal{A}\frac{\partial \mathcal{B}}{\partial G_1^*}\right),$$

with the partial derivatives of $\mathcal{A}(\xi,\omega;\boldsymbol{p})$ and $\mathcal{B}(\xi,\omega;\boldsymbol{p})$ given by

$$\frac{\partial}{\partial G_1^*}\binom{\mathcal{A}}{\mathcal{B}} = \frac{\partial \boldsymbol{Q}_1}{\partial G_1^*}\boldsymbol{Q}_2\ldots\boldsymbol{Q}_{n-1}\binom{\varkappa_n - \varkappa_{n+1}}{\varkappa_n + \varkappa_{n+1}}, \quad (13)$$

$$\frac{\partial \boldsymbol{Q}_1}{\partial G_1^*} = \begin{pmatrix} e_2 & 1 \\ e_2 & 1 \end{pmatrix}\left(\beta_1 + G_1^* \frac{\partial \beta_1}{\partial G_1^*}\right).$$

Consistent with the forward solution described earlier, quantities $\boldsymbol{Q}_k$, $\varkappa_k$, e_k and β_k in (13) are defined by (8) and (9). Expressions for their derivatives can be obtained in a straight-forward manner and are omitted herein for brevity. Sensitivities of the displacement field $\hat{u}_\theta$ with respect to other components of vector $\boldsymbol{m}$ can be obtained in a similar fashion. It should be noted, however, that each "propagator" matrix $\boldsymbol{Q}_k$ depends on the properties of layers k and $k+1$ *only*, thus significantly simplifying the parametric differentiation of matrix products in (7).

One may note that the kernel $\partial \mathcal{K}/\partial G_1^*$ in (12) has a faster rate of decay as $\xi \to \infty$ than its counterpart $\mathcal{K}$ featured in the forward solution (5). As a result, the improper integral in (12) can be computed without the aid of the asymptotic decomposition procedure that is used to evaluate (5). The same comment applies to other sensitivity functions as well.

5 PHYSICAL SETUP

To complement the preceding analytical and computational developments, a simple manual apparatus for generating axisymmetric Love waves is proposed in

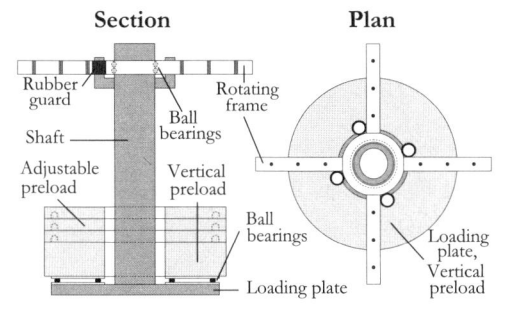

Figure 2. Manually-operated device for generating torsional impact

this study that consists of i) a hollow cylindrical shaft fixed to the bottom loading plate, ii) (primary and secondary) vertical preload, and iii) a rotational impact frame. To facilitate the transfer of angular momentum from the impact frame to the shaft and thus the loading disc, rubber guards are placed on small, yet stiff transfer "arms" as shown in Fig. 2. As indicated in the diagram, a slip between the loading plate and the soil is minimized using sufficient vertical preload. The primary preload rests on ball bearings which furnish a frictionless contact with the loading plate; the secondary preload, in the form of detachable solid annuli, is placed on top of the primary annulus. Similarly, the manually-operated impact frame slides on ball bearings that are supported by the shaft. An adjustable magnitude of the impact torque is provided by the set of mass attachments fixed to the rotating (impact) frame.

6 NUMERICAL RESULTS

In what follows, performance of the minimization algorithm is examined through numerical examples with synthetically-generated measurements. To this end, a two-layer-over-the-half-space viscoelastic system is taken as a test profile. In the configuration, each layer is 5 m thick, i.e. $h_1 = h_2 = 5$ m. Complex shear moduli of the layers and the half-space are taken as $G_1^* = (80 + 12.97i)$ MPa, $G_2^* = (180 + 14.44i)$ MPa and $G_3^* = (320 + 19.24i)$ MPa, respectively, which in accordance with (4) approximately correspond to the damping ratios $\zeta_{s1} = 8\%$, $\zeta_{s2} = 4\%$ and $\zeta_{s3} = 3\%$. For the minimization example, the mass density is assumed to be distributed uniformly throughout the half-space, i.e. $\rho_k = 2000 \, \text{kg/m}^3$ ($k = 1,2,3$), and to be known beforehand.

The synthetic dispersion curve, obtained from the forward solution for the test profile, is shown in Fig. 3a. To obtain the dispersion curve spanning a wide range of frequencies, three different receiver spacings (assuming $r_2 = 2r_1$, see Fig. 3) are used according to the following bounds, proposed by analogy to the Heisey s filtering criterion is SASW testing (e.g. Ganji et al., 1998):

$$\frac{2\pi c}{3} < \{\omega r_1 \equiv \omega(r_2 - r_1)\} < 2\pi c.$$

The above criterion can also be conveniently rewritten as a constraint on the phase lag $\Delta\phi$ between the records of two transducers, i.e.

$$\frac{2\pi}{3} < \Delta\phi < 2\pi.$$

Fig. 3b shows the corresponding phase lag dependencies on the excitation frequency for selected receiver spacings.

For the purpose of inversion, only a limited number of data points are needed to adequately define the dispersive properties (Fig. 3a) and attenuation characteristics (not shown) of Love waves. In this example, $N = 45$ frequency points were selected for the inversion as indicated in the Figure.

In the example described above, the minimization process involves 8 unknown parameters, namely the real and imaginary parts of the complex shear moduli G_1^*, G_2^* and G_3^*, and heights h_1 and h_2 of the two layers.

In the case of noise-free measurements, the data covariance matrix is $C_d = 0$, which also means that the components of d^{obs} are not correlated and should contribute to the cost functional $\mathcal{F}(m)$ with equal weights. Accordingly, one may take C_d^{-1} to be a unit matrix so that (10) simplifies to $\mathcal{F}(m) \equiv \mathcal{F}_1(m)$ where

$$\mathcal{F}_1(m) = \|g(m) - d^{obs}\|_2^2 \tag{14}$$

and $\|\cdot\|_2$ denotes the Euclidean norm in $\mathbb{R}^{2N}$.

Fig. 4 shows several steps of the iterative process minimizing $\mathcal{F}_1(m)$ in the featured (eight-dimensional) parametric space, starting from a *uniform* initial profile that assumes the correct number of layers (i.e. two layers over the half-space) with $G_k^{*initial} = (140 + 5i)$ MPa, $k = 1, 2, 3$, and $h_k^{initial} = 3$ m, $k = 1, 2$. As can be seen from the display, the final iteration resolves both geometric and material parameters of the test profile with reasonable accuracy. With regard to the shown distribution of Im(G^*), one should note that the damping ratio ζ_s is related to the imaginary part of the complex shear modulus through

$$\zeta_{sk} = \frac{\text{Im}(G_k^*)}{2|G_k^*|}, \quad k = \overline{1, n+1}. \tag{15}$$

On the basis of (15) and Fig. 4, the resolved damping ratios of the layers and the half-space can be calculated as $\zeta_{s1}^{est} = 8\%$, $\zeta_{s2}^{est} = 4.06\%$ and $\zeta_{s3}^{est} = 3.18\%$. These values should be compared to the "true" damping ratios $\zeta_{s1} = 8\%$, $\zeta_{s2} = 4\%$ and $\zeta_{s3} = 3\%$.

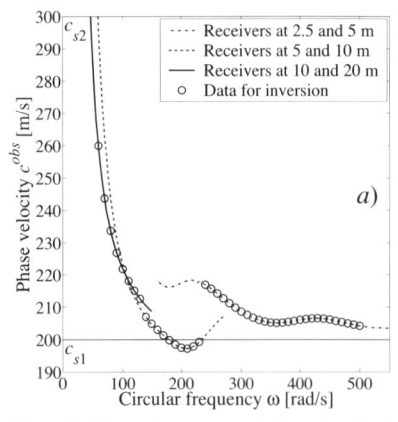

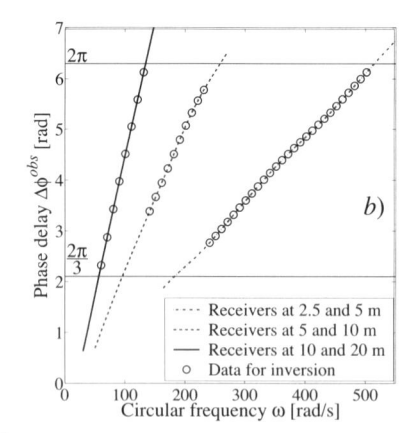

Figure 3. Dispersion curve and phase lag for the "true" profile

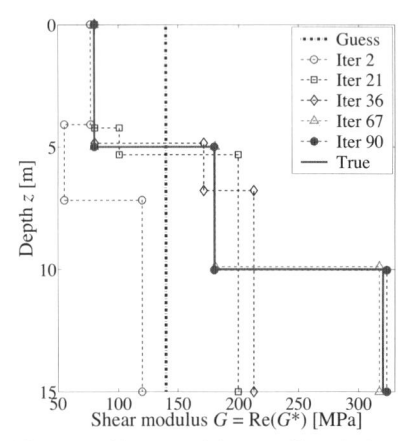

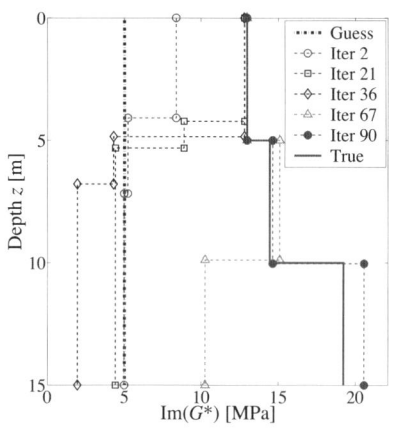

Figure 4. Shear modulus profiles during the iterative minimization process with noise-free measurements ($\rho_k = 2000\,\text{kg/m}^3$, $k = 1, 2, 3$)

To illustrate the utility of the maximum-likelihood theory, the same testing profile is considered as in the earlier example, but this time with noise-polluted displacement records. To this end, the noise level is set to 2.5% of the reference displacement (taken as the maximum displacement amplitude captured by the farthest receiver at $r = 20$ m) for all frequencies except those around 60 Hz, where it it increases in a bell-shaped manner up to 10% of the reference displacement. The value of $\boldsymbol{d}^{obs}$ used for the minimization was obtained by averaging through a number of realizations which helped reduce the effect of measurement noise. In terms of the data covariance matrix $\boldsymbol{C}_d$, the maximum-likelihood theory reduces the importance of data with poor signal-to-noise ratios, namely the data around 60 Hz and those captured by the farthest couple of receivers.

Several steps of the minimization procedure are shown in Fig. 5. The initial profile in this example is taken as $G_k^{*initial} = (140 + 5i)$ MPa, $k = 1, 2, 3$, $h_1 = 6$ m and $h_2 = 4$ m. On the basis of Fig. 5 and (15), the resolved damping ratios of the layers and the half-space can be calculated as $\zeta_{s1}^{est} = 8\%$, $\zeta_{s2}^{est} = 3.98\%$ and $\zeta_{s3}^{est} = 3.84\%$. These values should again be compared to the "true" damping ratios $\zeta_{s1} = 8\%$, $\zeta_{s2} = 4\%$ and $\zeta_{s3} = 3\%$. The comparison indicates that the properties of the bottom half-space and especially its damping ratio ζ_{s3} cannot be resolved very accurately since the low-frequency data (as captured by the farthest pair of receivers), that "illuminate" the properties of deep layers, are dominated by the noise and are therefore disregarded by the cost functional (10).

7 CONCLUSIONS

This paper reports a compact computational and physical framework for the in-situ evaluation of dynamic

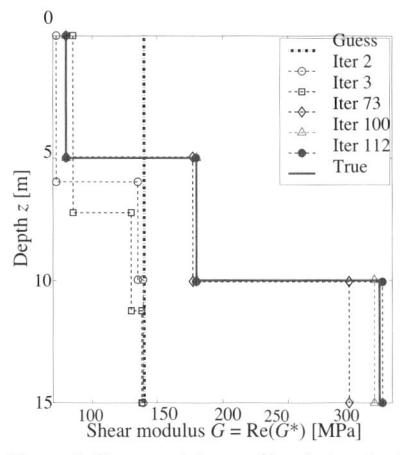

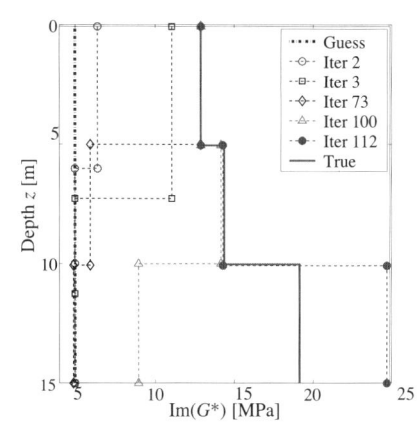

Figure 5. Shear modulus profiles during the iterative minimization process with noise-polluted measurements ($\rho_k = 2000\,\text{kg/m}^3$, $k=1,2,3$)

soil properties by means of the spectral analysis of (axisymmetric) Love waves. It is shown that the underlying forward solution rules out the effect of i) Poisson s ratios and ii) compressional-wave damping ratios of the multi-layered profile, thus rendering the special analysis of horizontally-polarized ground motion well suited to engineering applications where soil damping is important. Numerical examples with synthetically-generated measurements indicate that the dispersion and attenuation characteristics of Love waves, as captured by the discrete pairs of motion sensors, exhibit sufficient sensitivity on the stratigraphic arrangement to be used as a viable alternative (or complement) to their Rayleigh-wave counterpart in dynamic site characterization. The employment of the maximum-likelihood theory is shown to enhance the robustness of the minimization algorithm in dealing with noise-contaminated data. To provide an effective physical basis for its field applications, the proposed back-analysis is complemented with a conceptual experimental setup aimed at generating the axially symmetric Love-wave field.

ACKNOWLEDGMENTS

The support provided by the National Science Foundation through CAREER Award No. CMS-9875495 to B. Guzina and the University of Minnesota Supercomputing Institute during the course of this investigation is gratefully acknowledged.

REFERENCES

Al-Sanad, H.A., Aggour, M.S. and Amer, M.I. 1986. Use of random loading in soil testing, *Indian Geot. J.*, 16(2): 126-135.

Findley, W.N., Lai, J.S. and Onaran, K. 1989. *Creep and relaxation of nonlinear viscoelastic materials*, New York: Dover.

Ganji, V., Gucunski, N. and Nazarian, S. 1998. Automated inversion procedure for Spectral Analysis of Surface Waves, *J. Geotech. Geoenv. Eng.*, ASCE, 124(8): 757-770.

Guzina, B.B. 2000. Dynamic soil sensing via horizontally-polarized shear waves, *Use of Geophysical Methods in Construction*, Geotechnical Special Publication No. 108, ASCE: 95-110.

Guzina, B.B. and Lu, A. 2002. Coupled waveform analysis in dynamic characterization of lossy solids, *J. Eng. Mech.*, ASCE, 128(4): 392-402.

Guzina, B.B. and Pak, R.Y.S. 2001. On the analysis of wave motions in a multi-layered solid, *Q. Jl. Mech. Appl. Math.*, 54 (1): 13-37.

Heisey, J.C., Stokoe, K.H., II and Meyer, A.H. 1982. Moduli of pavement systems from spectral analysis of surface waves, *Transp. Res. Record*, 852: 22-31.

Nocedal, J. and Wright, S.J. 1999. *Numerical optimization*, New York: Springer.

Pak, R.Y.S. and Guzina, B.B. 1995. Dynamic characterization of vertically loaded foundations on granular soils, *J. Geot. Engrg.*, ASCE 121(3): 274-286.

Rix, G.J. and Lai, C.G. 1998. Simultaneous inversion of surface wave velocity and attenuation, *Geotechnical Site Characterization*, 2, A.A. Balkema, Rotterdam: 503-508.

Stokoe, K.H.I. and Nazarian, S. 1983. Effectiveness of ground improvement from spectral analysis of surface waves, *Proc. 8th Europ. Conf. on Soil Mech. and Found. Engrg.*, Helsinki: 31-38.

Tarantola, A. 1987. *Inverse problem theory*, Amsterdam: Elsevier.

Effect of ageing on shear wave velocity by seismic cone

John A. Howie & Ali Amini
Department of Civil Engineering, The University of British Columbia, Vancouver, BC, Canada

Keywords: seismic cone testing, sands, ageing, disturbed zone, shear wave velocity, small strain stiffness

ABSTRACT: During waiting periods of up to 60 minutes after a stop in penetration of a seismic cone in a sand deposit, shear wave travel times were observed to decrease. The paper discusses the causes of this effect and its practical implications. The effect can be explained by considering the effects of disturbance and the subsequent ageing of the zone of disturbance created by cone penetration. Laboratory and field evidence indicates that freshly deposited or recently disturbed sand experiences a time dependent gain in small strain stiffness. When cone penetration stops, the small strain stiffness of the disturbed zone around the cone will begin to increase. This results in an increase in V_s in the disturbed zone and in shorter arrival times when shear waves are generated at the ground surface. The possible errors in V_s determinations due to time effects can be avoided by observing consistent wait times after the stoppage of the penetration or by using a true interval seismic cone. The use of inconsistent wait times caused errors in V_s of up to 8%. It is also shown that the V_s interpreted from seismic cone data using interval techniques represents the V_s of the undisturbed soil.

1 INTRODUCTION

Time dependent changes in the properties of sand have been observed in field and laboratory testing. (Schmertmann 1991; Mitchell and Solymar 1984). Various mechanisms have been proposed for these ageing effects but the phenomenon is not well understood.

Studies on the long term set up of driven piles in sand have also shown an increase in pile capacity with time (e.g. Chow et al. 1998). This has been attributed to changes in stress conditions and ageing of the soil around the pile. Due to similarities between piling and cone penetration testing, the same phenomenon may be expected to occur around a cone used in site exploration. This paper presents the results of an investigation to determine whether time effects could be observed during seismic cone testing.

2 BACKGROUND

2.1 Ageing in sands

Much of the evidence for time effects on penetration resistance comes from in situ testing to assess the results of ground improvement. Both cone tip resistance, q_t, and shear wave velocity, V_s, have been observed to decrease in sands when measured soon after ground treatment and to then show an increase with time after treatment.

Penetration resistance has also been observed to change considerably with time after deposition of granular fill. Skempton (1986) documented increases in the Standard Penetration Test N-value of fills with time as discussed by Jamiolkowski et al. (1988). Robertson et al. (1995) found a linear trend between the change of normalized V_s and logarithm of age in sands.

Many laboratory studies have shown an increase in V_s with time after preparation of sand samples (Afifi and Woods 1971, Anderson and Stokoe 1978, Mesri et al. 1990 and Baxter 1999). Based on the theory of wave propagation in an isotropic, linear elastic medium, V_s can be used to estimate the small strain shear modulus of soil, G_o, using the expression:

$$G_o = \rho V_s^2 \qquad (1)$$

The effect of ageing on G_o obtained from resonant column tests or seismic wave velocities, may be approximated by the expression:

$$\frac{G_o(t)}{G_o(t_p)} = 1 + N_G \log\left(\frac{t}{t_p}\right) \qquad (2)$$

where t_p is the time to the end of primary compression (EOP), t is any time greater than t_p, and N_G is the slope of a plot of $G_o/G_o(t_p)$ versus the log of time. The reference time, t_p, is difficult to determine in sand but 1000 minutes is commonly used (Anderson and Stokoe 1978).

Daramola (1980) studied the effects of ageing on the stiffness of dense Ham River sand in conventional triaxial testing. The stiffness was observed to be a function of relative density, D_r, for fresh samples. However, after ageing, time was observed to have a great influence on stiffness and D_r was not the main factor controlling stress-strain response. Secant stiffnesses at strains less than 0.5% increased by 100% over three log cycles of time.

Shozen (2001) and Lam (2003) studied the effect of ageing periods of up to 10,000 minutes on the stress-strain response of very loose Fraser River sand. A period of ageing resulted in a much stiffer response during the initial portion of the stress-strain curve but the effect of ageing tended to disappear after increments of axial strain of about 0.05%. The curves coincided beyond the initial stages of loading. Fig. 1 shows data for ageing times of up to 1000 minutes after consolidation at a stress ratio of $\sigma'_1/\sigma'_3=2.0$ (Howie et al. 2001). The data suggest that ageing increases initial stiffness but has little effect on larger strain properties, including strength. Resolution of the strain measurement apparatus was not considered reliable for shear strains below about 0.02% and so the effect of ageing on G_o ($\gamma \sim 0.0001\%$) was not studied. While the above effects are only strictly applicable to samples aged for up to about 1 week, field evidence suggests that ageing continues over the long term.

The above observations suggest that tests responsive to the small strain stiffness of sand should be more sensitive to time effects than those tests that subject the soil to large strains. Thus, V_s should be very sensitive to time effects. This makes the seismic test ideal for studying the possible ageing effect. As the waves created in the seismic test result in strains in the range of only 10^{-5} to 10^{-6} (Stewart and Campanella, 1991), the testing is unlikely to alter the changes in soil stiffness induced by ageing. This is an advantage over periodic pile load testing to study time dependent changes in pile capacity, where the strains occurring during pile load testing will reduce or eliminate the effects of ageing.

2.2 *Seismic piezocone testing*

The standard piezocone, CPTU, has a conical tip with a 60° apex angle, is 10 cm² in cross-section, has a 150 cm² friction sleeve and pore pressure can be measured on or near the cone tip during penetration at the standard rate of 20 mm/sec. Tip resistance, q_t, sleeve friction, F_s, and pore pressure, u, are recorded at typical intervals of 2.5 or 5 cm.

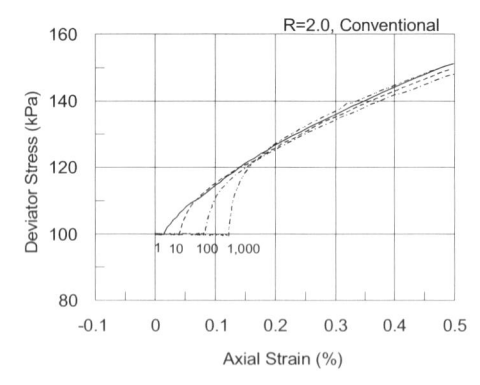

Figure 1. Effect of ageing on stress-strain behaviour of Fraser River Sand

The capability of the CPTU can be enhanced by the inclusion of accelerometers or geophones in a module mounted above the cone (Robertson et al. 1986). These can be used as receivers during downhole or crosshole seismic testing. Details of the seismic module in the cone at the University of British Columbia (UBC) are provided in Campanella and Stewart (1992). The combined tool is known as a seismic cone (SCPT) or seismic piezocone (SCPTU). SCPTU testing is commonly used to determine V_s.

At selected depth intervals (usually 1 m), the penetration is stopped and the rod string is unloaded. Seismic waves are generated at the surface and propagate to the cone as shown in Fig. 2. A pair of 12 kg swing hammers strikes either end of the front steel pad supporting the 13-tonne UBC cone truck. The pad is offset by 0.8 m from the vertical axis of the cone. The contact between steel hammer and steel pad triggers the data acquisition system, which records the horizontal particle motion that arrives at the accelerometer in the seismic module. The signal is displayed on the computer screen. The other end of the pad is struck in a similar manner to generate shear waves with inverse polarity and to obtain a mirror image response at the sensor as shown in Fig. 2. Two or more blows on each end of the pad are recorded to ensure repeatability of the signals. The cone is then advanced to the next depth interval and the procedure is repeated.

For a cone with seismic sensors at only one location, the above test procedure is called a pseudo interval technique. A true interval approach requires two sets of seismic sensors with a fixed distance between them. At each strike of the hammer, two signals are recorded at two depths from which the travel time interval can be obtained.

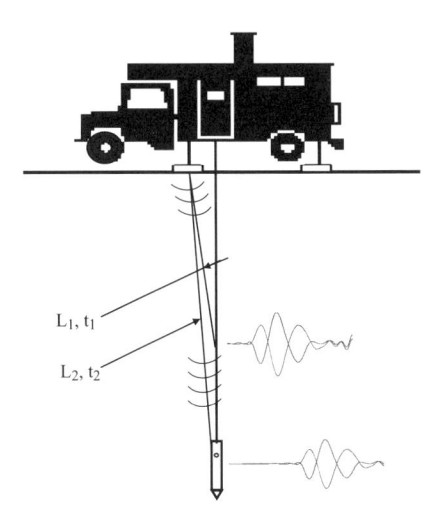

Figure 2. Seismic cone testing, UBC CPT truck

2.3 Interpretation of seismic signals from SCPT

The average V_s over a given depth interval can be calculated by an interval technique using Equation 3:

$$V_s = \frac{L_2 - L_1}{t_2 - t_1} = \frac{\Delta L}{\Delta t} \qquad (3)$$

where L_2 and L_1 are the slant distances between the sensor in the cone and the source beam taking account of the offset between the vertical axis of the cone string and the source beam (Fig. 2). The time interval, $\Delta t = (t_2-t_1)$, is the difference between the arrival times of shear waves at two successive depth intervals. The time interval, Δt, was determined by cross-correlation, a technique in which one signal is shifted by small time steps relative to another signal. At each shift, the sum of the product of the signal amplitudes is calculated and the time shift corresponding to the maximum cross-correlation is taken as the interval travel time between the two depths.

3 FIELD EVIDENCE OF CHANGES IN V_s DURING SCPT TESTING

Fig. 3 is a cone profile from the field test site. Two soundings were carried out. In the first, pseudo interval seismic tests were performed at 0.5m intervals in Zone 1, which extended from 10.45 m to 11.95 m. At each depth, the pad was struck at 1, 5, 10, 20, 40 and 60 minutes after cone penetration stopped. In the second sounding, conducted 3 m. from the first, a true interval cone was used to carry out testing in zone 2 (Fig. 3). Shear waves were generated every 5 minutes up to one hour after penetration stopped. Shear wave arrivals were recorded at upper and lower accelerometers for each hammer blow. For this research, the accelerometers were 0.975 m apart. All the seismic data were sampled at a frequency of 20 kHz (50 μsec per sample).

Fig. 4 shows the typical seismic signals obtained by the pseudo interval method at one depth at different wait times. These signals were windowed for the first complete cycle of the shear waves. As the wait time increased, the signals shifted to the left, indicating a faster travel time. This trend was observed at all test depths. The shift of the signals changed the interpreted V_s from 195 m/s to 210 m/s (8% increase) in 60 minutes as shown in Fig. 5. Fig. 6 illustrates how V_s was calculated. All the interval velocities for different wait times at 11.95 m were determined by cross-correlation with reference to a single record taken at the 1 minute wait time at the upper end of the interval, 10.95 m. Shear wave velocities were calculated for the left and right hits separately. Both confirmed a trend of increasing V_s values with longer wait times.

For signals recorded with the true interval cone, the same trend of faster arrival time with longer wait times was observed at each accelerometer location. However, Fig. 7 shows that the signal shift was almost identical at the upper and lower receivers. Consequently, the pairs of seismic signals at upper and lower interval depths with consistent wait times yielded a constant value of Δt. This results in a constant value of V_s, despite the faster arrival times at any one depth (Fig. 8). On the other hand, if the reference seismic signal at the upper level was fixed, the interpreted V_s would increase with time (Fig. 8). The results with the true interval cone suggest that if pseudo-interval data are gathered using a consistent wait time, a constant value of V_s should also be obtained.

In summary, the field data indicate the following apparent contradiction:
- A shift of about 500 μsec in seismic signals is observed over a wait time of 60 minutes, indicating that V_s increased with time;
- A constant Δt at any consistent wait time indicates that V_s did not change with time.

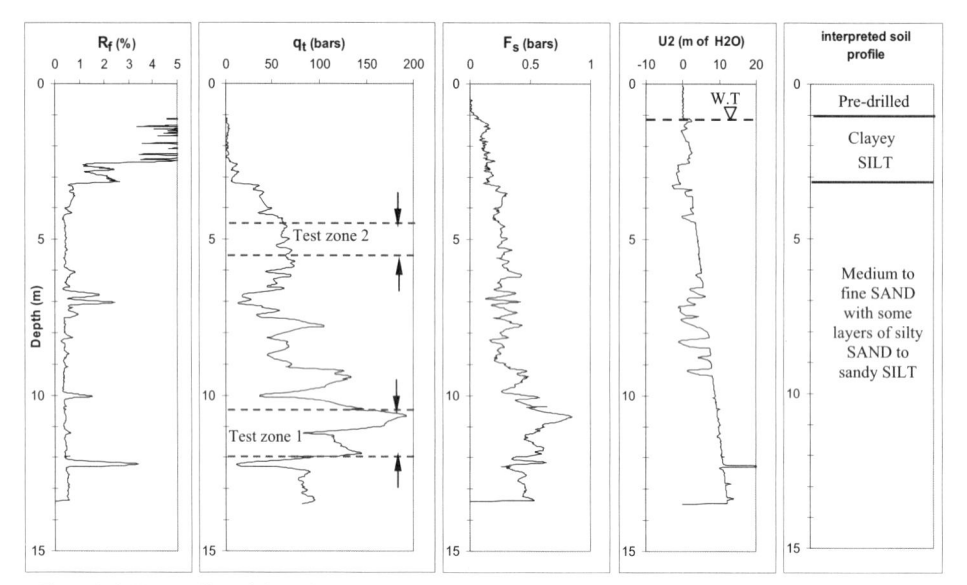

Figure 3. CPTU sounding, Richmond, B.C.

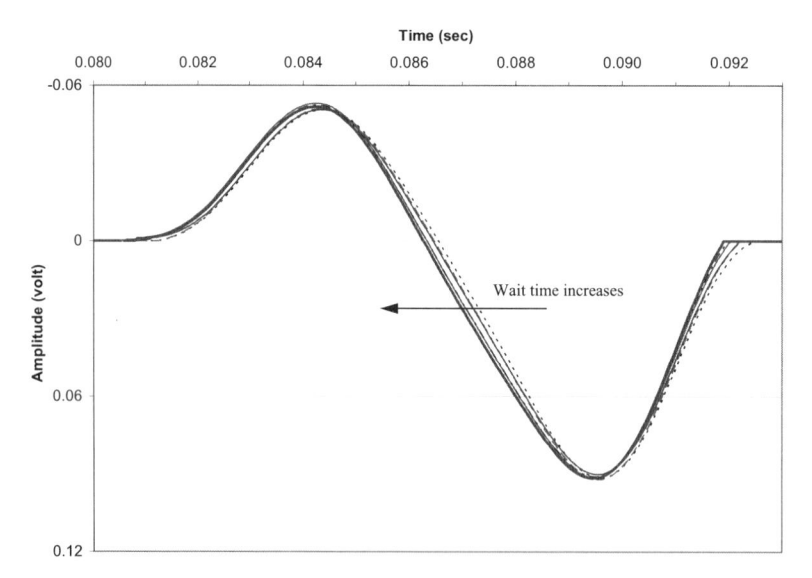

Figure 4. Windowed signals at 11.95m at wait times of 1, 5, 10, 20, 40 and 60 minutes after the stoppage of penetration-left hammer hit - low pass 250Hz- Richmond, B.C

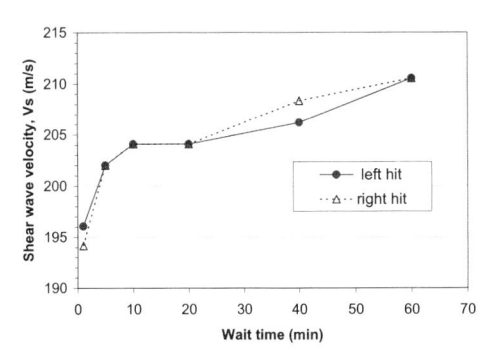

Figure 5. V_S at depth interval 10.95-11.95m, Richmond, BC.

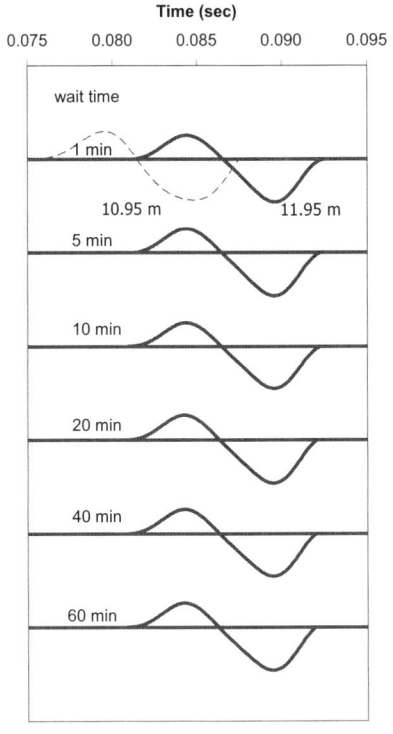

Figure 6. Signals used for cross-correlation at upper and lower interval depths (10.95 m and 11.95 m), Richmond, BC.

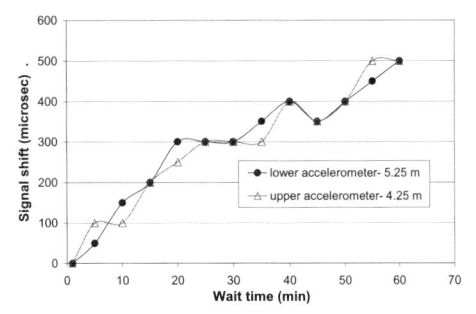

Figure 7. Shift of the seismic signals with wait time

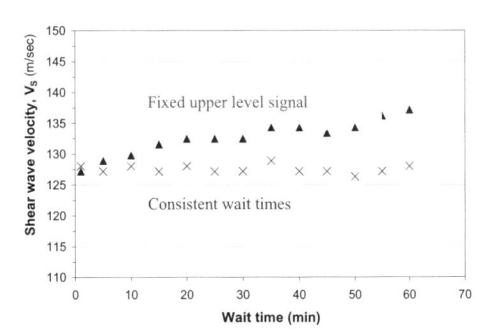

Figure 8. Effect of wait time and interpretation method on the calculated V_s

4 DISCUSSION

Attempts to resolve the observed contradiction focused on equipment performance, test and analysis procedures and on soil behaviour.

4.1 *Equipment performance*

Repeatability of the seismic testing was checked by repeatedly hitting the truck pad at various wait times while the seismic cone was held in a vice in the cone truck. Cross-correlation of the signals showed a maximum time shift of 50 µsec between signals. This corresponded to the data sampling rate or the resolution of the time records and indicates that the seismic testing system generates repeatable results. The observed seismic signal shift of 500 µsec in 60 minutes is much larger than the system resolution and is likely due to changes in soil conditions.

4.2 *Time dependency of shear wave arrival*

The change in shear wave travel time with wait time can be explained by consideration of disturbance and ageing effects in the soil affected by cone penetration. A zone close to the cone will experience large strains during cone penetration. The extent of the disturbed zone will depend on the strength and stiffness of the soil before penetration, and the magnitude of the strains will attenuate with distance from the cone.

Laboratory and field data suggest that the small strain stiffness of soil may be substantially reduced by disturbance and will increase logarithmically with time after disturbance. As a result, the small strain stiffness of the disturbed zone (Fig. 9) will begin to increase due to ageing immediately after cone penetration ceases. Increasing stiffness of the disturbed zone with time results in the observed faster arrival of the shear waves with wait time at any particular depth.

4.3 Calculation of V_s

If V_s is calculated relative to a fixed reference signal at the upper end of the depth interval, Δt decreases with wait time and V_s increases. To understand the finding of constant V_s with consistent wait time, it is necessary to consider the geometry of the problem. The conventional ray path model of seismic wave propagation is assumed as shown in Fig. 9.

When a ray encounters the disturbed zone, it will refract in accordance with Snell's Law. The angle of refraction depends on the stiffness ratio between undisturbed and disturbed ground. If both accelerometers at lower and upper interval depths are located in the same deposit, then the radius of the disturbed zone is approximately the same and the lengths of the travel paths in the disturbed zone are almost identical. This causes any change in V_s due to changes in the properties of the disturbed zone to be almost identical for both upper and lower accelerometers, as was observed in the field (Fig. 7). This results in the difference in arrival times at each of the accelerometers being constant and in a V_s that is independent of wait time.

A parametric study was conducted to investigate the implications of this model for the measurement of V_s by SCPTU. An idealized configuration is shown in Fig. 9. The variables in the analysis were the ratio of the stiffnesses in the disturbed and undisturbed zones, represented by the ratio V_{s2}/V_{s1}, and the radius of the disturbed zone, AB. Snell's law and the geometry of the model results in the following equations for a given travel path:

$$\frac{\sin(a_1)}{V_{s1}} = \frac{\sin(a_2)}{V_{s2}} \quad (3)$$

$$OA.\tan(a_1) + AB.\tan(a_2) = d \quad (4)$$

where d is the depth of the accelerometer, a_1 is the angle of incidence, a_2 the angle of refraction and OA is shown in Fig. 9. Theoretical arrival times are calculated by solving Equations (3) and (4) for the two unknown angles, a_1 and a_2, at each depth. ΔL and V_s are then calculated using the conventional interval method assuming straight travel paths as noted in section 2.3.

The results of the parametric study are shown in Table 1 and Fig. 10 for two depths: 4.25 and 5.25 m. For a constant radius of the disturbed zone, AB, an increased V_{s2}/V_{s1} ratio represents an increase in the stiffness of the disturbed zone with time. Therefore, an increased V_s ratio represents a longer wait time. As the stiffness ratio in the disturbed zone increases, the shear wave arrival time decreases but there is almost no change in the calculated V_s from consistent wait times (marked with ellipses in Table 1). For a consistent wait time, the measured V_s is very close to $V_{s1}=125$ m/s assumed for the undisturbed zone.

If the wait time is not consistent and a fixed reference signal is used (marked with rectangles in Table 1), then the interpreted V_s exceeds the undisturbed shear wave velocity. The error depends on the diameter of the disturbed zone and the magnitude of the increase in stiffness due to ageing.

These results are consistent with field observations and indicate that the V_s obtained from the seismic cone using interval methods of interpretation is a property of the undisturbed soil and is largely unaffected by disturbance, provided consistent field procedures are followed at each measurement depth.

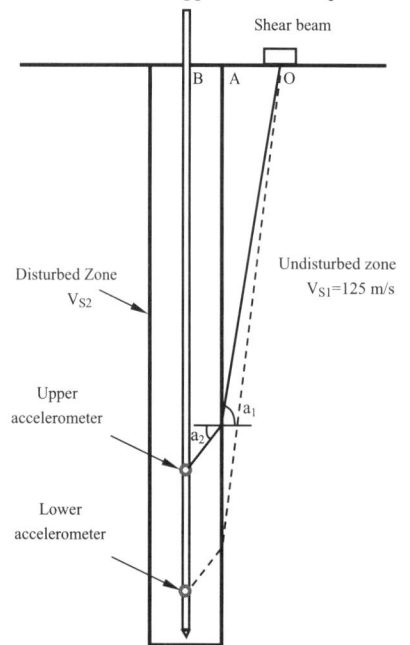

Figure 9. Schematic seismic wave ray path

In the true-interval measurements reported here, the cone was pushed directly to the test depth and measurements were taken at fixed times. Thus, the ageing effect was identical for top and bottom accelerometers and the above analysis is applicable. For a typical true interval probe, the upper receiver will be 1 metre behind the lower one. During a 1 metre push, the soil around the upper receiver will experience small additional disturbance compared to that which occurred during the initial penetration of the cone tip but the lower one will be in freshly disturbed soil. Hence, the lower receiver will likely experience more of an ageing effect than the upper one. This aspect of true-interval V_s measurements requires further study.

Table 1. Results of the parametric analysis

AB[I]	V_{S2}/V_{S1}	t[II]	Arrival time		Calculated V_s	
			4.25 m	5.25 m	Consistent[III]	Fixed[IV]
(m)	(-)		(sec)	(sec)	(m/s)	(m/s)
0.1	0.50	t_1	0.03585	0.04376	124.6	125
0.1	0.67	t_2	0.03536	0.04327	124.7	133.2
0.1	0.80	t_3	0.03507	0.04298	124.7	138.6
0.2	0.50	t_1	0.03712	0.04505	124.3	125
0.2	0.67	t_2	0.03614	0.04407	124.3	142.5
0.2	0.80	t_3	0.03556	0.04349	124.3	155.5
0.4	0.50	t_1	0.03970	0.04767	123.7	125
0.4	0.67	t_2	0.03774	0.04571	123.7	165.7
0.4	0.80	t_3	0.03657	0.04454	123.8	205.7

[I] Radius of the disturbed zone
[II] $t: t_3 > t_2 > t_1$ some arbitrary wait time associated with the V_S ratio
[III] Calculated V_S from signals with consistent wait times
[IV] Calculated V_S using a fixed upper interval depth signal

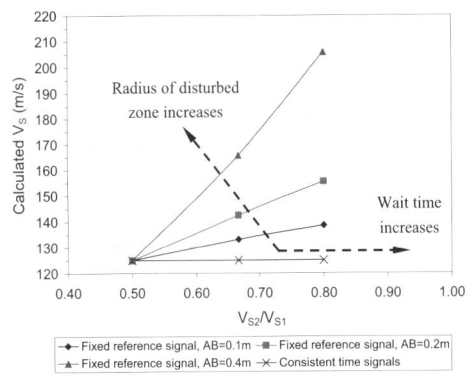

Figure 10. Effect of change of stiffness and size of disturbed zone on interpreted V_s- Undisturbed V_s =125m/s.

5 CONCLUSION

The results of this study show that the arrival time of shear waves during seismic cone testing in Fraser River sand is affected by the properties of the zone of disturbance created during cone penetration. The decrease in arrival time with wait time after penetration ceases is consistent with an increase in stiffness due to ageing. The ageing effect has been observed in laboratory and field tests on Fraser River and other sands. The results also indicate that the V_s values interpreted using interval techniques are very close to the shear wave velocities of the undisturbed soil, provided a consistent wait time is observed after penetration ceases.

The ageing effect could have practical implications for measurement of V_s by interval techniques if the time lag after the stoppage of cone penetration is not consistent or the ageing rate at each end of the depth interval is different due to a change in soil type. In the field testing reported here, a difference in wait time after penetration of 60 minutes between the two test depths resulted in an 8% error in V_s. Such an error translates to an error of 17% in G_o calculated by Equation (1). This effect can be largely eliminated by consistent field procedures. It should be noted that such a long wait time is not usual unless a technical problem occurs or a dissipation test is required.

ACKNOWLEDGEMENT

The authors acknowledge the financial support of the Natural Science and Engineering Research Council of Canada, the GREAT Award program of the Science Council of B.C. and the industrial collaborators: Klohn-Crippen Consultants Ltd., Conetec Investigations Ltd., Foundex Explorations Ltd. and Geopac West Ltd. The assistance of the technical support staff of the University of British Columbia, Department of Civil Engineering, especially Scott Jackson is greatly appreciated.

REFERENCES

Afifi, S.S., and Woods, R.D. 1971. Long-term pressure effects on shear modulus of soils. Journal of Soil Mechanics and Foundation Engineering, ASCE, 97(SM10), pp. 1445-1460.

Anderson, D.G. and Stokoe, K.H. 1978. Shear modulus, a time-dependent soil property. Dynamic Geotechnical Testing. ASTM Special Technical Publication 654, pp. 66-90.

Baxter, C.D.P. 1999. An experimental study on aging of sands. Ph.D. dissertation, Virginia Polytechnic Institute and State University, Blacksburg, Virginia.

Campanella, R.G., and Stewart, W.P. 1992. Seismic cone analysis using digital signal processing for dynamic site characterization. Canadian Geotechnical Journal, 29(3): pp. 477-486.

Chow F.C., Jardine R.J., Naroy J.F. and Brucy F. 1998. Effects of time on capacity of pipe piles in dense marine sand.

Journal of Geotechnical and Geoenvironmental Engineering, ASCE, 124(3), pp. 254-264.
Daramola, O. 1980. Effect of consolidation age on stiffness of sand. Geotechnique, 30: pp. 213-216.
Howie, J.A and Amini, A., Shozen, T. and Vaid, Y.P. 2001. Effect of time on in situ test results after ground improvement- new insights from laboratory and field data. In Proceeding of 54th Canadian Geotechnical Conference, pp. 467-474, Calgary, Alberta, Canada.
Ishihara, K. 1996. Soil behaviour in earthquake geotechnics. Oxford Science Publications.
Jamiolkowski M., Ghionna V.N., Lancellotta, R. and Pasqualini E. 1988. New correlations of penetration tests for design practice. Penetration Testing 1988. Edited by J. De Ruiter. In Proceedings of the 1st International Symposium on Penetration testing, Orlando, Florida, Balkema, Rotterdam.
Jamiolkowski, M. and Manassero, M. 1995. The role of in-situ testing in geotechnical engineering – thoughts about the future. In Proceedings of the International Conference on Advances in Site Investigation Practice. Thomas Telford Ltd., London, U.K., pp. 929-951.
Lam, C.K.K. 2003. Effects of Aging and Stress Path on Stress-Strain Behaviour of Loose Fraser River Sand. M.A.Sc. dissertation, University of British Columbia, Vancouver, BC.
Mesri, G., Feng, T.W. and Benak, J.M. 1990. Post-densification penetration resistance of clean sands. Journal of Geotechnical Engineering, ASCE, 116: pp. 1095-1115.
Mitchell, J.K. and Solymar, Z.V. 1984. Time-dependent strength gain in freshly deposited or densified sand. Journal of Geotechnical Engineering, ASCE, 110: pp. 1559-1576.
Robertson, P.K., Campanella, R.G., Gillespie, D., and Rice, A. 1986. Seismic CPT to measure in-situ shear wave velocity. Journal of Geotechnical Engineering Division, ASCE, 112(8): pp.791-803.
Robertson, P.K., Fear, C.E., Woeller, D.J., Weemees, I. 1995. Estimation of sand compressibility from seismic CPT. In proceeding s of the 48th Canadian Geotechnical Conference, Vancouver, Vol. 1, pp. 441-448.
Schmertmann, J.H. 1991. The mechanical aging of soils. 25th Karl Terzaghi Lecture. Journal of Geotechnical Engineering, ASCE, 117: pp. 1288-1330.
Shozen, T. 2001. Deformation under the constant stress state and its effect on stress-strain behaviour of Fraser River Sand. M.A.Sc. dissertation, University of British Columbia, Vancouver, BC.
Skempton, A.W. 1986. Standard penetration test procedures and the effects in sands of overburden pressure, relative density, particle size, ageing and overconsolidation. Geotechnique, 36: pp. 425-447.
Stewart, W.P., and Campanella, R.G. 1991. In situ measurement of damping of soil. In 2nd International Conference on Recent Advances in Geotechnical Earthquake Engineering and Soil Dynamics. St. Louis. University of Missouri-Rolla, Vol. 1, pp. 83-92.

Dissipation test evaluation with a point-symmetrical consolidation model

Emoke Imre
Geotechnical Research Group of the Hungarian Academy of Sciences at the Department of Geotechnics,
Budapest University of Technology and Economics, Hungary

Pál Rózsa
Department of Computer Science and Information Theory, Budapest University of Technology and Economics, Hungary

Keywords: dissipation test, inverse problem, analytical solution, coupled consolidation model

ABSTRACT: A point-symmetrical linear coupled consolidation model with constant displacement boundary condition is tested against measured dissipation test data. The initial condition and the size of the displacement domain (r_1) are identified. Results show that - similarly to the previously tested cylindrical model – very short data jets can successfully be evaluated if the sensor is well above the tip.

1 INTRODUCTION

The dissipation test can be used for the determination of the in situ coefficient of consolidation c on condition that proper model and non-linear inverse problem solver are available. Presently some numerical solutions of two dimensional models are used for the evaluation and, the non-linear inverse problem is solved approximately (e.g. Lunne et al, 1992). Neither the test nor the evaluation is fully standardized.

The test is performed in such a way that the steady penetration is stopped at any depth, the rod is kept in clamped position and, the pore water pressure is measured with time. The sensors are situated on the tip or along the shaft (Fig. 1). At the end of the steady penetration, the direction of the flow lines (being about perpendicular to the equi-pressure lines in Fig 2.) change from horizontal to vertical and the distance of the zero pressure line varies from about r_1 =37r_0 to about r_1 =23r_0.

Here a one dimensional point-symmetric coupled consolidation model with constant displacement boundary condition (Imre-Rózsa, 2002) is verified for five filter positions A, B, C, D, E (Fig. 1c). The non-linear inverse problem is solved with a precise, automatic (routinely applicable) method (Imre, 1996) entailing the reliability testing of the solution. The displacement domain r_1 is specified at the zero pressure line in the first step. Then r_1 is modified using a model law elaborated for this purpose.

Results show that the point-symmetrical model is slightly better than the previously tested cylindrical model (Imre-Rózsa, 1998; Imre, 2002) and both models are good if the sensor is well above the tip.

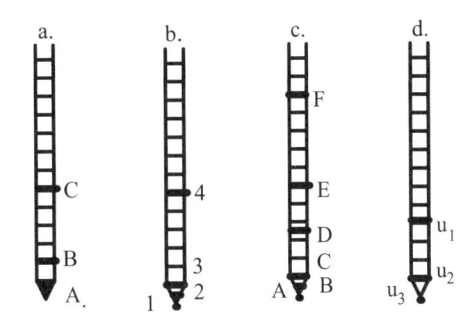

Figure 1. Some filter positions known from the literature. a. Torstensson (1977), b. Baligh and Levadoux (1986),c. Teh and Houlsby (1988), d. Kim, Lee and Kim (1997).

2 MODEL

The system of differential equations of the point-symmetrical coupled linear model:

$$E_{oed}\frac{\partial}{\partial r}\frac{1}{r}\frac{\partial}{\partial r}(rv)-\frac{\partial u}{\partial r}=0 \qquad (1)$$

$$-\frac{k}{\gamma_v}\frac{1}{r}\frac{\partial}{\partial r}(r\frac{\partial u}{\partial r})+\frac{\partial \varepsilon}{\partial t}=0 \qquad (2)$$

where u is pore water pressure [kPa], v is displacement [m], r is the space co-ordinate [m] and, t is time [s], k [m/s] is permeability, γ_v [kN/m³] is unit weight of water, E_{oed} [kPa] is oedometric modulus and, c [m²/s] is coefficient of consolidation:

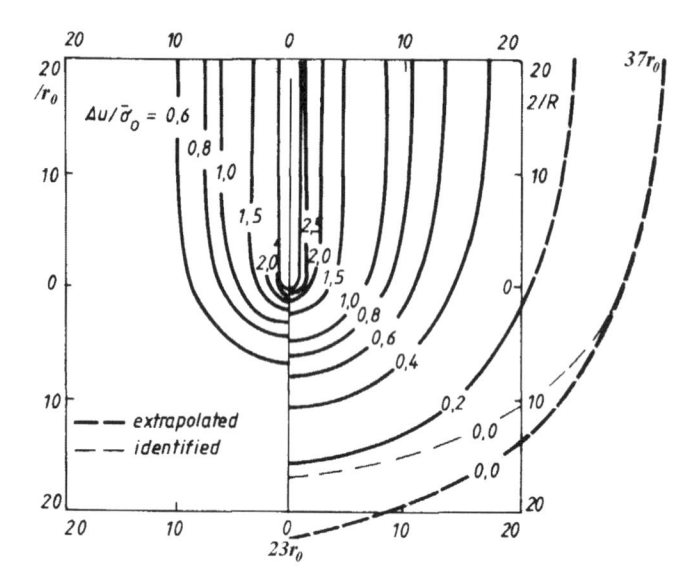

Figure 2. The pore water pressure results of the strain path method (after Baligh, 1986) indicating the approximated and identified zero pressure lines. Left and right: bilinear and hyperbolic modeling, thin dashed line is related to the point-symmetric model

$$c = \frac{k \cdot E_{oed}}{\gamma_v} \quad (3)$$

The space domain varies between the radius of the pile (r_0) and the distance of the zero pressure line (r_1). The homogeneous form of the boundary conditions are as follows:

$$u(t,r)|_{r=r_1} = 0 \quad (4)$$

$$\frac{\partial u(t,r)}{\partial y}\bigg|_{r=r_1} \equiv 0 \quad (5)$$

$$v(t,r)|_{r_0} \equiv v_0 > 0 \quad (6)$$

$$v(t,r)|_{r_1} \equiv 0 \quad (7)$$

The analytical solutions for u and for the transient part of v are as follows:

$$u(t,r) = \sum_{k=1}^{\infty} \frac{C_k E_{oed} e^{-\lambda_k^2 \cdot c \cdot t}}{\sqrt{r^3}} \left\{ \begin{array}{l} [3B_{1.5}(k,r) - (\lambda_k r) B_{2.5}(k,r)] - \\ [3B_{1.5}(k,r_1) - (\lambda_k r_1) B_{2.5}(k,r_1)] \end{array} \right\} \quad (8)$$

$$v^t(t,r) = \sum_{k=1}^{\infty} \frac{C_k e^{-\lambda_k^2 \cdot c \cdot t}}{\sqrt{r}} [B_{1.5}(k,r)] \quad (9)$$

$$B_{1.5}(k,r) = J_{1.5}(\lambda_k r) + \mu_k Y_{1.5}(\lambda_k r) \quad (10)$$

$$B_{2.5}(k,r) = J_{2.5}(\lambda_k r) + \mu_k Y_{2.5}(\lambda_k r) \quad (11)$$

where $J_{1.5}/J_{1.5}$ and $Y_{2.5}/Y_{2.5}$ are Bessel functions of the first and second kinds, with the order of 1.5 and 2.5, C_k ($k=1...\infty$) [kPa] are coefficients dependent on the initial transient displacement function $v'_0(r)$:

$$C_k = \frac{\int_{r_0}^{r_1} r \, v_0^t(r) [B_{1.5}(k,r)] dr}{\int_{r_0}^{r_1} r \, [B_{1.5}(k,r)]^2 dr} \quad (12)$$

The parameters λ_k [1/m] ($k=1...\infty$) are the roots of the following equation derived from (6) and (7):

$$J_{1.5}(\lambda_k r_1)/J_{1.5}(\lambda_k r_0) = Y_{1.5}(\lambda_k r_1)/Y_{1.5}(\lambda_k r_0) \quad (13)$$

Parameters μ_k [-] ($k=1...\infty$) can then be computed using equation (6) or (7).

3 MODEL LAW

At the end of the steady penetration, the direction of the flow lines (Fig. 2, being about perpendicular to the equi-pressure lines) change from horizontal to vertical for the filter positions indicated in Figure 1. The size of the displacement domain (r_1) – the zero pressure line - varies from about $r_1 = 37r_0$ to about $r_1 = 23r_0$. These facts indicate the problems of anisotropy and varying space domain.

No anisotropy effects were taken into account in this work. The difficulty related to the varying space domain was solved as follows.

No space normalization unit is included in the analytic solution of the model (equations (8) and (9)). Therefore, the analytical solutions related to different space domains cannot be transformed into each other.

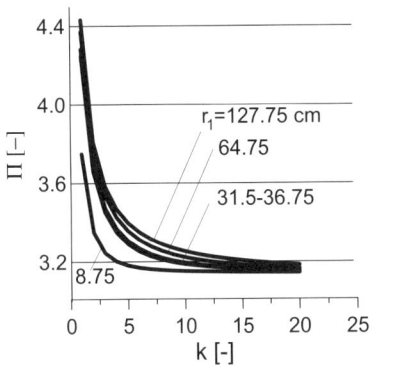

Figure 3 Results of numerical tests for $\Pi=(r_1-r_0)\,\lambda_k/k$ using various values for r_1 [cm] and $r_0=1.75$ cm

The roots λ_k ($k=1...\infty$) can be computed by solving the non-linear equation (13) for every value of r_1.

This can be avoided by using an approximate relationship for the roots λ_k ($k=1...\infty$) as follows:

$$\lambda_k \approx \frac{k\pi}{(r_1-r_0)} \qquad (14)$$

Inserting this into the analytic solution in the place of λ_k, the following new dimensionless variables appear in the arguments of the Bessel function and in the argument of the exponential function:

$$(r) = \frac{r}{r_1-r_0} \qquad (15)$$

$$T = \frac{ct}{(r_1-r_0)^2} \qquad (16)$$

Using the latter time factor, approximate model laws can be derived. Besides others, the following relation can be derived assuming that $T_1=T_2$ and $t_1=t_2$:

$$\frac{c_1}{c_2} = \frac{(r_{1,2}-r_{0,2})^2}{(r_{1,1}-r_{0,1})^2} \qquad (17)$$

The approximate relationship was numerically verified as follows. The roots λ_k ($k=1...\infty$) were determined for six values of r_1 (8.75 cm, 31.5 cm, 36.5 cm, 45.5 cm 64.75 cm, 127.75 cm) and one value of r_0 (1.75 cm). the quantity:

$$\Pi := \frac{\lambda_k(r_1-r_0)}{k} \qquad (18)$$

was plotted against k using the computed results. As it can be seen in Figure 4, value of Π is asymptotically equal to π.

4 INVERSE PROBLEM SOLUTION METHODS

The merit function was defined as follows:

$$F(p) = \frac{\sqrt{\sum_{i=1}^{N}u_m(t_i)-u(t_i,p)}}{N\,\underset{i}{max}(u_m(t_i))} \qquad (19)$$

where u is pore water pressure, N is number of data, m measured, p is parameter vector.

The merit function was geometrically explored in the space of the non-linearly dependent parameters by applying a sub-minimization for the linearly dependent parameters (Imre,1996).

The parameter vector consisted of the non-linearly dependent parameter coefficient of consolidation (c) and, the linearly dependent initial condition parameters (C_k, $k=1..n$).

The identified parameter p_i was considered reliable it was a unique solution of the inverse problem and, if its confidence interval was within its range. The uniqueness of the identified parameter was tested by the construction of the so called minimal sections of parameter p_i (Imre,1996) computable in parallel to the minimization.

The standard deviation σ_i was determined using the laniaries model (Press et al, 1985) The error of this estimation is twofold: (i) not only random but also deterministic (e.g. modeling, numerical) errors occur; (ii) the model is non-linear.

Therefore, the so computed σ_i were considered as error indicators. (It can be noted that σ_i can be determined from the minimal sections, too. In this work the latter method used as a qualitative check.)

The 88% confidence interval was computed from the Tschebiseff inequality:

$$P(|x_i-p_i| \geq a) \leq \frac{\sigma_i^2}{a^2}, a>0 \qquad (20)$$

where x_i and σ_i are expected (identified) value and standard deviation of the parameter p_i, respectively.

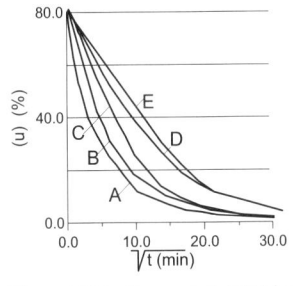

Figure 4. Data (Lunne et al. 1992) in u versus $t^{0.5}$ plot

5 RESULTS

The dissipation test data was measured in the Bothkennar soft clay, five filter positions E to A (Fig. 4).

Figure 2 shows that the extrapolated zero pressure line (r_1) varies from about $37r_0$ to $23r_0$ between filter positions E to A.

In the course of the model fitting, at first, value of r_1 was taken into account as $r_1 = 37r_0$ which is good for filter position E. From the so identified c the size of the displacement domain r_1 was back-computed for other filter positions using the approximate model law (Chapter 3). This result was verified again, the results of the latter verification are not presented here.

Although the direction of the flow lines changes from horizontal to vertical, no anisotropy effects were considered here. The number of the identified initial condition parameters was 20 as a maximum (C_k, $k=1..20$). Results only for $k=1$ are presented here since the identified functions were generally physically not admissible.

5.1 Long data set ($\kappa > 90\%$), $r1 = 37r0$

The coefficient of consolidation (c) and, its variance (σ_c/c) can be seen in Table 1. The fitted and measured data are shown in Figures 5 to 6. The minimal sections of the noise-free and the noise-polluted merit functions of c are shown in Figure 7.

According to the results, the solution of the inverse problem is unique. Both c and σ_c increases between filter positions E to A (except that the variance is greater in filter position D than in filter position C).

Accordingly, the noise-polluted merit function is steeper around the global minimum for filter position E than for filter position A. The fit is the best in filter position E and the worst in A. The lower boundary of the 88% confidence interval ($c-3\sigma_c$) is negative except for filter positions C and E.

5.2 Short data set ($\kappa < 90\%$), $r1 = 37r0$

In case of short data jets, the identified c can be seen in Table 2 and Figure 8, the minimal sections of the noise-free and the noise-polluted merit functions related to c are shown in Figure 12.

According to the results, in filter position E the identified c is independent of the length of the data jet.

Table 1. The fitting error, the identified coefficient of consolidation (c) and, its variance, point-symmetrical model

Filter position	Fitting error [%]	c [cm²/s]	σ_c/c [-]
A	9	0.09	0.57
B	8	0.06	0.47
C	5	0.04	0.28
D	6	0.02	0.35
E	4	0.02	0.21

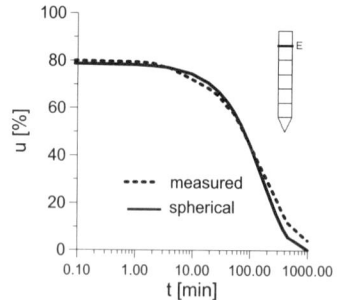

Figure 5. Measured versus fitted data, filter position E, point-symmetrical model.

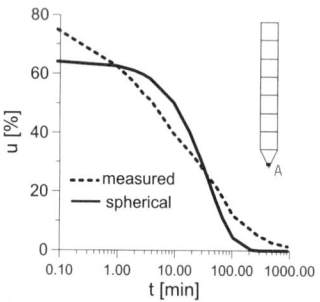

Figure 6. Measured versus fitted data, filter position A, point-symmetrical model.

The shape of the minimal section of the merit function is practically unchanged with decreasing t indicating that the solution is reliable.

However, in filter positions A to D the identified c is dependent on the length of data jet t (increases with decreasing t).

5.3 The case of varying r1

Based on the approximate model law (Equation (18)), r_1 was back-computed for filter positions A to D using the values of c shown in Table 1 being related to $r_1 = 37r_0$ assuming that the good value for c is equal to 0.02 cm²/s.

According to the results (Fig 2, Table 5), the back-computed r_1 varies from $37r_0$ to $17.44r_0$. The latter is somewhat under-predicted with respect to the results of the strain path method.

Table 2. The coefficient of consolidation c [cm²/s] from short data jets, point-symmetrical model

Duration [min]	Filter position				
	A	B	C	D	E
~2	0.6	0.4	0.14	0.12	0.02
~4	0.4	0.3	0.12	0.08	0.03
~18	0.2	0.13	0.07	0.04	0.03
~37	0.12	0.09	0.06	0.03	0.03
~68	0.1	0.07	0.05	0.03	0.02
~217	0.09	0.06	0.04	0.02	0.02

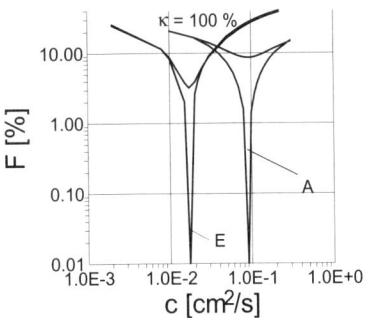

Figure 7. Minimal sections for the point-symmetrical model, long tests

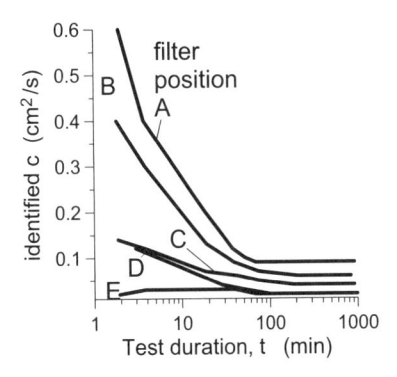

Figure 8. Variation of c with t, point-symmetrical model

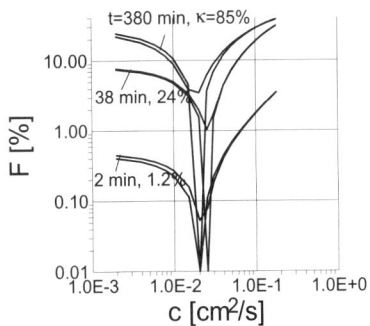

Figure 9. Minimal sections for the point-symmetrical model, short tests, filter position E.

Table 3. The relation between c and r_1 for the point-symmetrical model

Filter position	Identified c [cm2/s] assuming r_1=64.75cm c [cm²/s]	computed r_1 [cm] assuming c=0.02 cm²/s r_1 [cm]
A	0.09	30.52
B	0.06	37.38
C	0.04	45.78
D	0.02	64.75
E	0.02	64.75

Table 4. Approximate κ=50 % dissipation time (t_{50}) and approximate κ=90 % dissipation time (t_{90}) on the basis of the results shown in Figure 4

Filter position	t_{50} [min]	t_{90} [min]
A	8	200
B	22	250
C	60	300
D	90	>600
E	110	>600

6 DISCUSSION, CONCLUSION

6.1 Time factor concept

Although the dimensionless time factor:

$$T = \frac{ct}{r_0^2} \quad (21)$$

is not present explicitly in any solution or formulation, the analytical or numerical solutions of pile consolidation theories are generally presented in terms of it (e.g. Baligh and Levadoux, 1986; Teh and Houlsby, 1988; Lunne et al., 1992).

The applicability of the time factor concept was numerically tested for the cylindrical model in a previous paper (Imre, 2002b) and for the point-symmetrical model here. The following approximate relation for λ_k (k=1...∞):

$$\lambda_k \approx \frac{k\pi}{(r_1 - r_0)} \quad (22)$$

was numerically verified, then inserted into the analytical solution resulting in the following time factor:

$$T = \frac{ct}{(r_1 - r_0)^2} = \frac{ct}{r_0^2 \left(\frac{r_1}{r_0} - 1\right)^2} \quad (23)$$

in the case of both models. Being term $n=r_1/r_0$ constant in many cases (e.g. where the initial condition is given in terms of r_0, Fig. 2) the time factors (21) and (23) differs in a constant multipliers only.

Some further consequences are as follows. For a given normalized initial condition (i) the analytical solutions can approximately be transformed on the displacement domain, (ii) either c or r_1 can be identified not both and, the reliability of the identified c is not dependent on r_1.

6.2 Results of approximate identification methods

The approximate model fitting methods can be divided into two groups: (i) an estimation from a one point fit (e.g. the 50 % dissipation time t_{50}) or, (ii) an estimation from the slope of the first straight line

Proceedings ISC'2 on Geotechnical and Geophysical Site Characterization, Viana da Fonseca & Mayne (eds.)

portion of the u versus $t^{0.5}$ plot (e.g. Lunne et al, 1992).

The one point fitting at t_{50} provided a c of 0.15 cm^2/s for both filter positions B and C and, according to Table 4, t_{50} varied from 22 to 100 min. In this work 150 was used for the rigidity index.

The slope type approximate method resulted in values of c of 0.15 cm^2/s and, of 0.015 cm^2/s for filter positions B and C, respectively. The slope was determined from an at least 5 min long part of the test (Fig. 4).

The "true" value of c is 0.01 cm^2/s (Lehane and Jardine, 1994), or 0.0095 cm^2/s (Burns and Mayne, 1995).

6.3 Results of the cylindrical model

Some results related to the cylindrical model (Imre, 2002a) are summarized in Table 5-7 and compared with the results of the point-symmetrical model.

In filter position E both models gave good estimation for c (0.02 cm^2/s) even from a 2 min long test, the inverse problem solution did not depend on t (the true value of c is about equal to 0.01 cm^2/s).

In filter positions A to D both models gave worse estimation for c than in filter position E and, the estimations were dependent on the data jet length.

Table 5. The fitting error, the identified coefficient of consolidation (c) and, its variance, cylindrical model

Filter position	Fitting error [%]	c [cm^2/s]	σ_c/c [-]
A	9	0.13	0.56
B	8	0.09	0.45
C	5	0.06	0.28
D	6	0.03	0.37
E	4	0.02	0.23

Table 6. The coefficient of consolidation c [cm^2/s] from short data jets, point-symmetrical model

Duration [min]	Filter position				
	A	B	C	D	E
~2	0.8	0.6	0.2	0.2	0.03
~4	0.6	0.4	0.2	0.1	0.05
~18	0.2	0.2	0.1	0.06	0.04
~37	0.16	0.1	0.08	0.05	0.03
~68	0.12	0.1	0.06	0.04	0.03
~217	0.12	0.09	0.06	0.03	0.03
~380	0.11	0.09	0.06	0.03	0.02

Table 7. The relation between c and r_1 for the point-symmetrical model

Filter position	Identified c [cm^2/s] assuming r_1=64.75cm	computed r_1 [cm] assuming c=0.02 cm^2/s
	c [cm^2/s]	r_1 [cm]
A	0.13	25.38
B	0.09	30.52
C	0.06	37.38
D	0.03	52.86
E	0.02	64.75

7 CONCLUSION

The major part of the dissipation test evaluation procedures available over-predicts the value of the coefficient of consolidation c by about one order of magnitude. The necessary sampling time varies between about 5 min to 100 min.

The suggested evaluation procedures consist of two different one-dimensional analytical models and an automatic and "precise" inverse problem solution method. These procedures gave precise results for c on condition that the sensor was well above the tip. The last sampling time was equal to 2 min elapsed time. These methods gave results comparable with the existing methods for filter positions A to D.

It can be concluded that the suggested dissipation test evaluation procedures seem to be very economic for filter position E. In filter positions A to D the development of the two models is advisable.

REFERENCES

Baligh, M.M. (1986). Undrained deep penetration, II. pore pressures. *Geotechnique*, 36(4): 487-503.

Baligh, M.M.; Levadoux, J. N. (1986). Consolidation after undrained penetration. II. Interpretation. *Jl. of Geot. Eng. ASCE*, 112(7): 727-747.

Burns, S.E., Mayne, P.W (1995). Coefficient of consolidation (c_h) from type 2 piezocone dissipation test in overconsolidated clay. *Proc. of Int.Symp. on CPT*. 137-142.

Imre, E. and Rózsa, P. (1998). Consolidation around piles. *Proc. of 3rd Seminar on Deep Foundations on Bored and Auger Piles. Ghent* 385-391.

Imre, E. and Rózsa, P. (2002). Modelling for consolidation around the pile tip. *Proc. of the 9th Int. Conf. on Piling and Deep Foundations (DFI), Nizza*. 513-519

Imre, E. (2002a): Evaluation of "short" dissipation tests. *Proc. of the 12th Danube-European Conference*. 499-503.

Imre, E. (2002b). Pile consolidation models and scale effect. *Proc. of NUMGE 2002. Paris*.

Kim, Y.S., Lee, S.R., Kim, Y.T. (1997) . Application of an Optimum Design Technique for Determining the Coefficient of Consolidation by Using Piezocone Test Data. *Computers and Geotechnics*. 21:4:277-293.

Lehane, B. M.; Jardine, R. J. (1994) Displacement-pile behaviour in a soft marine clay. *Canadian Geotechnical Journal*. 31. 181-191.

Levadoux, J.; Baligh, M.M. (1986). Consolidation after undrained penetration. I. Prediction. *Jl. of Geot. Eng. ASCE* 112(7):707-727.

Lunne, T; Robertson, P.K.; Powell, J.J.M. (1992). *Cone Penetration testing*. Blackie Academic & Professional;

Teh, C.I. and Houlsby, G.T. (1988). Analysis of the cone penetration test by the strain path method. *Proc. 6th Int. Conf. on Num. Meth. in Geomechanics, Innsbruck*.

Torstensson, B. A (1977). *The pore pressure probe*. Paper No. 34. NGI

Feasibility of neural network application for determination of undrained shear strength of clay from piezocone measurements

Young-Sang Kim
Department of Ocean System Engineering, Yosu National University, Yeosu, Jeonnam 550-749, Korea

Keywords: undrained shear strength, piezocone test, neural network, tri-axial test

ABSTRACT: The feasibility of using neural networks to model the complex relationship between piezocone measurements and the undrained shear strength of clays has been investigated. A three layered back propagation neural network model was developed based on actual undrained shear strengths and piezocone measurements compiled from various locations around the world. Comparing model predictions of new piezocone data, not previously employed, with reference s_u values obtained from the CIUC and CAUC, validated the neural network model. It was also compared with conventional empirical and theoretical methods.
It was found that the neural network model is capable of inferring a relationship between piezocone measurements and the undrained shear strength of clays, which have various soil characteristics. It was also found that the developed neural network model could predict a more precise and reliable undrained shear strength than theoretical and empirical approaches.

1 INTRODUCTION

The piezocone penetrometer is widely used in onshore and offshore soil investigations at the present time. Existing devices provide information concerning the cone resistance (q_c), the sleeve friction (f_s), and penetration pore pressure (u). It is a common practice to correct the cone resistance and sleeve friction for pore pressure effects. The corrected values are designated q_T and f_T (Lunne et al., 1986). In parallel with equipment and hardware design and development, attempts have been made to theoretically and/or empirically correlate the test results to soil properties and especially to undrained shear strength (e.g., Lunne et al., 1985; Jamiolkowski et al., 1982; Keaveny and Mitchell 1986; Konrad and Law 1989; Chen and Mayne 1993). However, some of the procedures represent more or less local correlations and are not always applicable to different types of soil (Robertson et al., 1986). Rad and Lunne (1988) proposed a direct correlation between piezocone measurements and undrained shear strength s_u as an alternative to local correlations by subdividing 11 clayey soils with respect to the OCR value.
 The present paper examines the feasibility of neural networks to obtain a general relationship between the undrained shear strength of clay and piezocone measurements.

2 REVIEW OF THE THEORETICAL AND EMPIRICAL APPROACHES

2.1 *Theoretical Approaches*

The theoretical solutions available for the evaluation of undrained shear strength from piezocone measurements can be grouped into the following six classes - i.e., classical bearing capacity, cavity expansion theory, conservation of energy combined with cavity expansion theory, analytical and numerical approaches using linear and non-linear stress-strain relationships, strain path method, and hybrid method which combined a cavity expansion theory and Cam Clay Model. All the above said theories infer a relationship between cone resistance and undrained shear strength s_u as follows (Lunne et al., 1997):

$$q_c = N_c \cdot s_u + \sigma_i \quad (1)$$

where N_c= theoretical cone factor, σ_i =in situ total pressure depending on the theory used.

 Chen and Mayne (1993) proposed a simple piezocone model expressing the corrected cone tip resistance (q_T) and penetration pore pressure measured behind the tip (u_2) in formulations based on the cavity expansion and Modified Cam Clay model as follows.

$$(s_u)_{CIUC} = (q_T - u_2)/5.5 \quad (2.1)$$

$$(s_u)_{CAUC} = (q_T - u_2)/6.5 \quad (2.2)$$

where $(s_u)_{CIUC}$ = undrained shear strength obtinaed from isotropically consolidated triaxial compression test, $(s_u)_{CAUC}$ = undrained shear strength obtinaed from anisotropically consolidated triaxial compression test

Since theoretical solutions have limitations in modeling the real soil behavior under conditions of varying stress history, anisotropy, sensitivity, ageing and macro fabric, the empirical approach has been generally preferred although the theoretical model has provided a useful framework of understanding.

2.2 Empirical Approaches

The empirical interpretation of undrained shear strength s_u using piezocone measurements can be categorized into three groups as follows by the parameter used – i.e., net cone resistance ($q_T - \sigma_{vo}$), excess pore pressure ($u_2 - u_o$), and effective cone resistance ($q_T - u_2$):

$$s_u = \frac{q_T - \sigma_{vo}}{N_{kT}} \quad (3.1)$$

$$s_u = \frac{u_2 - u_o}{N_{\Delta u}} \quad (3.2)$$

$$s_u = \frac{q_T - u_2}{N_{ke}} \quad (3.3)$$

where q_T = corrected cone tip resistance = $q_c + (1-a)u_2$, q_c= measured cone tip resistance, a=net area ratio of cone geometry, u_2=pore pressure measured behind the cone tip, σ_{vo}=total overburden pressure, u_o=static pore pressure, N_{kT}, $N_{\Delta u}$, and N_{ke}=empirical cone factors.

Over the years, a large number of studies have been performed to evaluate undrained shear strength from site-specific empirical cone factors. However, it has been known that the range of the cone factors varies widely within the same cone factor – e.g., $N_{kT} = 5 \sim 30$, $N_{ke} = 1 \sim 13$, $N_{\Delta u} = 7 \sim 10$ (Lunne et al., 1997). It does not seem to be able to obtain the general cone factors, thus local correlation has always been preferred. However, places where no database is available, the set up of the local correlation takes too much time and extra efforts are required to obtain the reliable laboratory and field test data. Therefore, development of a model which can generalize the relationship between undrained shear strength and piezocone measurements is strongly required.

3 THE ARTIFICIAL NEURAL NETWORK MODEL

Neural networks have been found to be useful for analyzing complex relationships involving a multitude of variables in place of conventional mathematical models used in many geotechnical applications (Goh, 1994; Toll, 1996). Fig. 1 shows the architecture of a typical neural network consisting of three layers of interconnected nodes. Each node (neuron) is connected to all the nodes in the next layer. There is an input layer where data are presented to the neural network, and an output layer that holds the response of the network to the input. It is an intermediate layer, known as hidden layers that enable these networks to represent and compute complicated associations between patterns. Each hidden and output neuron processes its inputs by multiplying each input by its weight, summing the product, and then processing the sum using a nonlinear transfer function to produce a result. The S-shaped sigmoid curve is commonly used (Rumelhart et al., 1986).

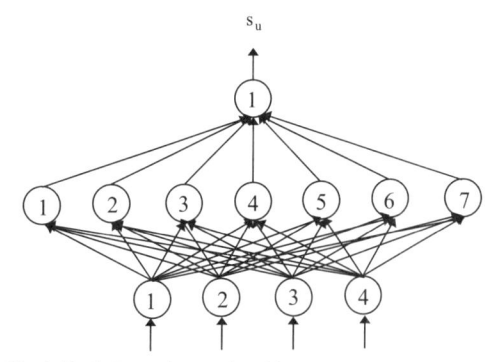

Fig. 1. Typical neural network architecture

In neural networks, the mathematical relationship between the variables does not have to be specified. Instead, the neural network 'learns' by adjusting the weights and biases between neurons in response to errors between the actual undrained shear strength and prediction values of the neural network model. At the end of this training stage, the neural network represents a model, which should be able to predict undrained shear strength for given new piezocone measurements.

3.1 Database and Training

The case records from Rad and Lunne (1988) were evaluated using the neural network. The data consisted of 11 clay sites (41 case records from Norway,

North Sea, UK, Brazil, and Canada). A total of 30 case records were used for the training phase and 11 for the testing phase that were randomly selected and had not been used for the training of neural network. Before the training and testing phase, all the input and target values were scaled to be fallen within the range [-1, 1] to make the training of the neural network more efficient.

In this study a three-layer back-propagation neural network, as shown in the Fig. 1, was used and programmed within MATLAB using a neural network toolbox, which provided an effective implementation of the neural networks. The back-propagation learning was done by continually changing the values of the network weights and biases in the direction of steepest descent with respect to error. Since, the simple back-propagation is normally very slow because it requires a small learning rate for stable learning, the the Levenberg-Marquardt (L-M) algorithm was used to improve speed and general training performance of the back-propagation. The mean squared error goal was 0.0005. The objective of training was obtained within a short time by use of the L-M algorithm.

Currently, there is no rule for determining the optimal model and the number of nodes in the hidden layer, except through experimentation. In this study, several neural network models were considered corresponding to input variables – i.e., overburden pressure σ_{vo}, corrected cone tip resistance q_T, pore pressure measured on cone tip u_1 and behind the cone tip u_2, and OCR. Model SuM3 was selected as a best model from the viewpoint of practicability and accuracy of prediction. Table 1 shows the details of the neural network model SuM3 and relative importance of each input neurons. The relative importance of each input neuron was obtained by partitioning the hidden-output connection weights into components connected with each input neuron (Garson, 1991; Goh, 1994).

4 RESULTS

4.1 Performance of ANN model for training and testing data

The prediction results of the neural network model SuM3 are compared with measured values in Fig. 2 and correlation coefficient R between prediction and measured s_u values are given in Table 1. In general, the neural network predictions from the testing phase as well as the training phase concur with measured triaxial compression undrained shear strengths, although the model was not trained explicitly for test data. This indicates that the neural network was successful not only in modeling the nonlinear relationship between the undrained shear strength and input variables but also in obtaining the generalized relationship among all data compiled from the various test sites.

Table 1. Details of neural network model SuM3 and relative importance of input neurons

Model	No. of hidden neuron	Input Variables and Relative Importance (%)			Correlation Coefficient, R (%)	
		σ_{vo}	q_T	u_2	Train	Test
SuM3	7	31.7	28.4	39.9	99.8	99.6

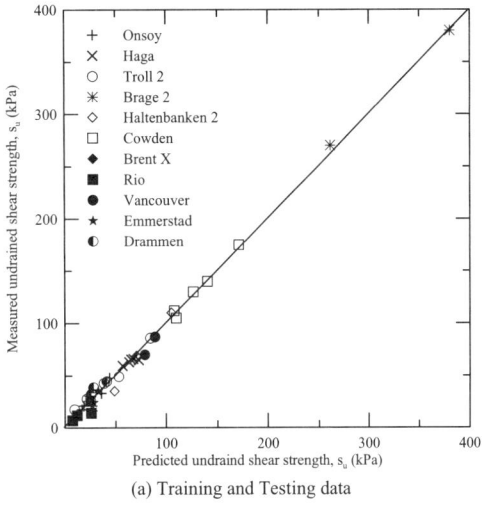

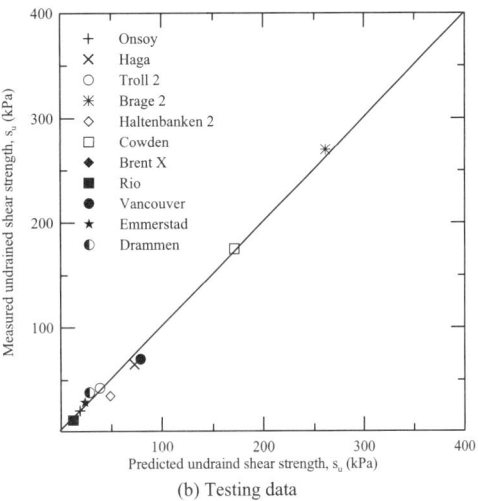

Fig. 2. Performance of the neural network model for training and testing data

4.2 Comparisons with conventional empirical methods

The prediction results for the test data using the conventional empirical methods were obtained from the

Eq. (3). Undrained shear strengths were normalized with those obtained from triaxal compression test and plotted with measured undrained shear strength in Fig. 3. Empirical cone factors N_{kT}, $N_{\Delta u}$, and N_{ke} were determined as 12.6, 4.78, and 9.42 from the training data.

As shown in Fig. 3, neural network model SuM3 gives less scatter results than the conventional empirical methods.

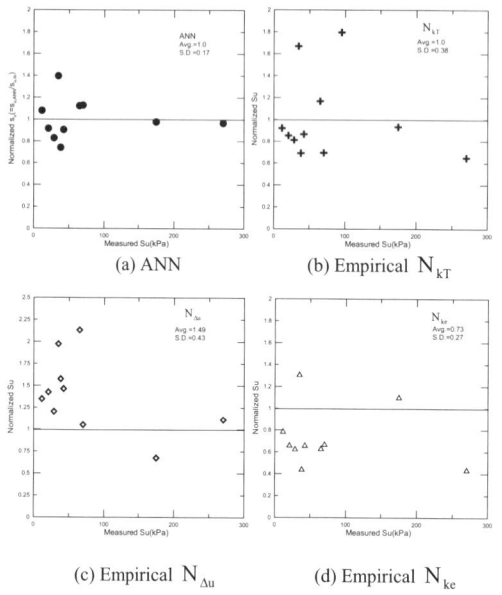

(a) ANN

(b) Empirical N_{kT}

(c) Empirical $N_{\Delta u}$

(d) Empirical N_{ke}

Fig. 3. Comparisons of prediction result with empirical models

4.3 Comparisons with theoretical method

Prediction results of hybrid theory (Chen and Mayne, 1993) were normalized with measured undrained shear strength and compared with those of neural network in Fig. 4. Since Eq. (2) were derived using the average soil parameters of $\phi' = 30$, $\Lambda = 0.75$, predicted undrained shear strength results show sometimes large deviations. While, neural network model gives more precise prediction results as shown in Fig. 3(a).

As Chen and Mayne (1993) pointed, Eq. (2.2) for undrained shear strength of ansiotropic case [Fig. 4(b)] shows slightly better results than those of isotropic case [Fig. 4(a)] because clays in nature are generally consolidated under anisotropic state of stress.

5 CONCLUSIONS

A simple back-propagation neural network model was used to estimate the undrained shear strength of clays from the piezocone measurements. After learn-

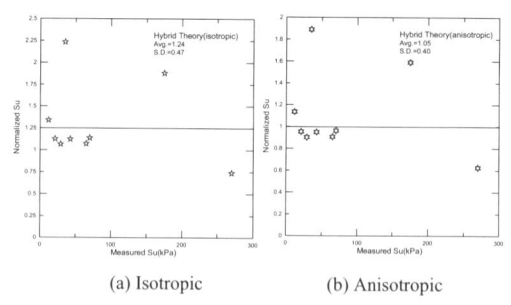

(a) Isotropic

(b) Anisotropic

Fig. 4. Normalized undrained shear strength of hybrid theory

ing from a set of actual undrained shear strengths obtained from the triaxial test (CIUC and CAUC), the neural network predictions of the undrained shear strength are found to concur with test data that were not previously known. Therefore, it can be concluded that neural network modeling is successful in generalizing the complex relationship between piezocone measurements and the undrained shear strength of clays. From comparisons with the conventional empirical methods and hybrid theory, it was found that the neural network predictions were more precise, consistent and reliable. Also, the neural network model is much simpler and practical. Another advantage of the neural network approach is the ease with which it can be retrained to improve its performance, as additional data are acquired.

ACKNOWLEDGEMENTS

The research was supported by the Basic Research Program (No. R05-2003-000-11073-0) of the Korea Science & Engineering Foundation.

REFERENCES

Chen, B.S.Y. and Mayne, P.W. (1993). Piezocone Evaluation of Undrained Shear Strength in Clays. *11th Southeast Asian Geotechnical Conference*, 4-8 May, Singapore: 91~98.

Garson, G.D. (1991). Interpreting neural-network connection weights. *AI expert*, 6(7), pp.47~51.

Goh, A.T.C. (1994). Seismic liquefaction potential assessed by neural-networks. *Journal of Geotechnical Engineering*, Vol. 120, No. 9, 1467~1480.

Jamiolkowski, M., Lancellotta, R., Tordella, L., and Battaglio, J.M. (1982). Undrained strength from CPT. *ESOPT*, Vol. 2, pp.599~606.

Keavney, M.J. and Mitchell, J.K. (1986). Strength of fine-grained soils using the piezocone. *Proceedings of Insitu '86, a specialty conference*, ASCE, Blacksburg, pp.668~685.

Konrad, J.-M. and Law, K.T. (1987). Undrained Strength from piezocone tests. *Canadian Geotechnical Journal*, Vol. 24 : 392~405.

Lunne, T., Eide, O., and de Ruiter, J. (1976). Correlation between cone resistance and vane shear strength in some

Scandinavian soft to medium stiff clays. *Canadian Geotechnical Journal*, Vol. 13, No. 4, pp430~441.

Lunne, T., Christophersen, H.P., and Tjelta, T.I. (1985). Engineering use of piezocone data in North sea clays. *Proceedings of 11th ICSMFE*, Vol. 2, San Francisco, pp.907~912.

Lunne, T., Eidsmoen, T., and Gillespie, D. (1986). Laboratory and field evaluation of cone penetrometers. *Proceedings of Insitu '86, a specialty conference, ASCE*, Blacksburg, pp.714~729.

Lunne, T., Robertson, P.K., and Powell, J.J.M. (1997). *Cone Penetration Testing*. Blackie Academic & Professional, pp57~62.

Rad, N.R. and Lunne, T. (1988). Direct Correlations between Piezocone Test Results and Undrained Shear Strength of Clay. *ISOPT-1*, Vol. 2:911~917.

Robertson, P.K., Campanella, R.G., Gillespie, D., and Greig, J. (1986). Use of piezometer cone data. *Proceedings of Insitu '86, a specialty conference, ASCE*, Blacksburg, pp.1263~1280.

Rumelhart, D.E., Hinton, G.E., and Williams, R.J. (1986). Learning Internal Representation by Error Propagation. *Parallel Distributed Processing: Foundations*, Vol. 1, MIT press, Cambridge, Mass.

Toll, D. (1996). Artificial Intelligence Applications in Geotechnical Engineering. *Electronic Journal of Geotechnical Engineering, Premiere Issue*.

Proceedings ISC-2 on Geotechnical and Geophysical Site Characterization, Viana da Fonseca & Mayne (eds.)
© 2004 Millpress, Rotterdam, ISBN 90 5966 009 9

Measurement of *in situ* deformability in hard rock

D. Labrie, B. Conlon, T. Anderson & R.F. Boyle
CANMET Mining and Mineral Sciences Laboratories, Natural Resources Canada, Ottawa, ON, Canada

© Minister of Natural Resources Canada, 2004 – Published with permission.

Keywords: *in situ* rock deformability, dilatometer testing, modulus of deformation, rock properties, scale effects, mine structures, hard rock

ABSTRACT: The analysis of a set of 84 dilatometer measurements made in three deep mines in northern Quebec revealed a correlation between the *in situ* modulus of deformation (E_d) and the *in situ* stress, suggesting that dilatometer measurements could be used as to assess stress levels and integrity of mine structures (Labrie et al., 1998). Since then more than a hundred measurements and three new sites have been added to the database, including a site less than 20 meters below surface. The database includes values for the intact rock modulus of deformation (E_l) determined in the laboratory using drill core recovered from the test sites, and estimates of the rock mass modulus of deformation (E_m) derived from E_l using the Rock Mass Rating. The enlarged database makes it possible to assess rigorously the correlation of E_d with *in situ* stress and to test the assumptions used to reduce the dilatometer measurements, in particular the use of the Lamé equation, which does not take *in situ* stress into account. Recent considerations to incorporate *in situ* stress in the computation are discussed.

1 INTRODUCTION

Strength and deformability are probably the two most important properties used to characterize the mechanical behavior of rock materials and to assess the stability of structures built in or founded on rock (Jaeger and Cook, 1969; Bieniawski, 1984; Brady and Brown, 1985; Yow, 1993). Goodman (1989) reviews many of the static and dynamic tests used to determine the deformability of rocks, describing the apparatus, field and laboratory techniques, and the interpretation of the results.

The dilatometer is one of the most versatile instruments used to determine the *in situ* modulus of deformation. The operating principle is quite simple: a known pressure is applied to the wall of a borehole and the displacement of the wall is measured. Depending on the particular instrument, dilatometers may be used in boreholes of various diameters with lengths of several hundred meters (Gill and Leite, 1995).

In this paper we describe a series of dilatometer tests that were carried out in six hard rock mines in northern Quebec, Canada, either to characterize rock structures (e.g., mine galleries, crown and sill pillars, stope abutments), estimate the level of damage occurring in rock masses in the vicinity of mine openings, or to provide data to support the interpretation of measurements made in boreholes. Field investigations were conducted to determine the geology, structure and stress regimes of the sites. Laboratory measurements of rock properties were made on core specimens. The significance of the dilatometer test results is assessed with respect to repeatability, correlation with deformability measurements of the rock mass and intact rock, and the influence of the *in situ* stress. Recently suggested methods to take stress effects into account are reviewed.

2 DILATOMETER TESTING

Dilatometers with a range of probe and membrane sizes and measuring devices are available (e.g., Cambridge Insitu, 2003; Interfels, 2003; OYO, 2000). The PROBEX-1 model (Roctest, 2002) for N-size boreholes (75.7 mm) was used for this series of tests. The probe consists of a high-pressure flexible membrane that can be inflated to 30 MPa, a hydraulic module comprising a dual piston and cylinder assembly to inflate and deflate the membrane, and a volume-change measuring device mounted inside the probe casing. A manual hydraulic pump is used to pressurize the system. The line pressure and the volume change are measured electronically with transducers mounted on the pump or built in the probe. A manually operated data acquisition module is mounted in the readout unit.

A standard testing procedure has been proposed for this type of dilatometer by the International Society of Rock Mechanics (IRSM, 1987), adapted by the manufacturer to respect the characteristics of its instrument (Roctest, 1992). A typical test is shown in Figure 1. All tests were conducted according to the procedure suggested by the manufacturer.

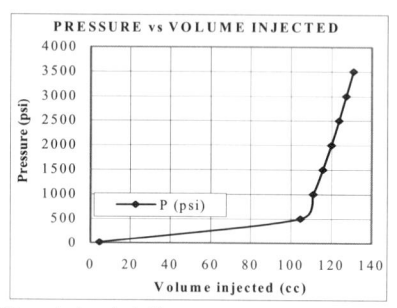

Figure 1. A typical dilatometer measurement.

The interpretation of the results is based on the well-known Lamé equation (Goodman, 1989):

$$E_d = r(1+v)(\Delta p / \Delta r), \quad (1a)$$

or in terms of volume, to respect the characteristics of the instrument used for the test program:

$$E_d = 2V(1+v)/((\Delta V / \Delta p)-c) \quad (1b)$$

where E_d represents the *in situ* modulus of deformation; v, the Poisson's ratio; r, the borehole radius; V, the borehole volume; p, the inflation pressure; and c, the constant of calibration of the probe.

3 MINE SITES INVESTIGATED

Six mines were involved in the testing program, with one to three sites of measurement investigated at each of these mines (Table 1). Tests were conducted in virgin ground, crown and sill pillars and stope abutments. Boreholes were drilled from the opening providing the best access to the structure to be investigated, usually haulage ways or mine galleries. The boreholes were usually drilled either vertically, or parallel to the main geological structures, or along the dip of ore bodies. If possible the boreholes were aligned parallel to a principal component of the general stress tensor. This is important when monitoring stresses in mine structures, as it limits the number of unknowns and facilitates the interpretation of results (Jaeger and Cook, 1969). Ground conditions and stress regimes determined or estimated for each test site are detailed in Table 1.

4 RESULTS

4.1 Induced stress levels

The magnitude and orientation of principal stresses at the test sites were determined for mine design purposes prior to the dilatometer test program (References are listed at the bottom of Table 1.) The orientation of the major principal stress component (σ_1) was usually perpendicular to the ore body or structure investigated, in accord with the trends for the Abitibi area (Corthésy et al., 1997).

4.2 Results of dilatometer tests

Dilatometer tests were conducted at twelve locations in six mines in the Abitibi and Saguenay areas of the Province of Quebec. The sites were located underground, except for the Pierre-Beauchemin mine, where boreholes were drilled from the surface to investigate the quality of the bedrock in the area of the crown pillars. These crown pillars are permanent structures usually left in place to protect the underground openings from intrusion of topsoil and are also a major element of ground support used to guarantee the general stability of a mine operation.

A total of 208 tests were done at these six mines during the last fifteen years: 128 were primary tests, being the first loading of the rock mass; and the remainder were secondary tests, i.e. reloading experiments carried out at the same position within the boreholes without moving the probe to verify the accuracy and the repeatability of measurements and identify the presence of strain-hardening, if any. Most (69%) of the reloading tests differed from the primary test by less than 20%. Reloading tests were done systematically at Joe Mann, Sigma, Francoeur and Niobec mines, the latter having the largest variability with seven of sixteen secondary test results showing a difference greater than 20%.

The distance between tests varied between 0.4 and 2 m, depending on the objectives of the various test programs, e.g., 0.4 m spacing was used at Mine Francoeur to map the depth of the damaged zone around a ventilation drift (Simon, 2002); 2 m spacing was used at Mine Pierre-Beauchemin to determine the conditions of crown pillars (Cockburn and Désormeaux, 1988). Elsewhere a 1 m interval was used, except where structural defects were detected in boreholes. The test point was moved to avoid perforating the membrane.

Table 1. List of mines and characteristics of dilatometer test sites.

MINE SITE	SITE (no.)	BOREHOLE	DEPTH (m)	ROCK-TYPE	$E_{LABORATORY}$ (GPa)	POISSON'S RATIO	RQD [1]	RMR [1]	σ_1 [2] (MPa)	σ_{mean} [2] (MPa)	REFERENCE
Mine Pierre-Beauchemin (former Eldrich) Evain (Abitibi, Qc)	1 1 2 3	87-E1 87-E1 87-E2 87-E3	7 – 20 7 – 20 10 – 22 8.5 – 21	Diorite (i) Diorite (a) Diorite (a) Tonalite	88.60 72.55 65.64 66.63	0.295 0.290 0.275 0.290	65 – 88 65 – 88 40 – 79 63 – 79	50 – 61 50 – 61 42 – 59 48 – 52	0.5 0.5 0.5 0.5	0.5 0.5 0.5 0.5	Cockburn & Désormeaux, 1988; Labrie, 1988
Mine Niobec Saint-Honoré (Saguenay, Qc)	1 2	600-F1 600-F3	190 – 200 185 – 200	Carbonatite Carbonatite	52.52 64.77	0.337 0.298	86 – 89 62 – 86	76 – 78 71 – 78	11.5 11.5	8.0 8.0	Labrie & Conlon, 1997a
Mine Joe Mann Chibougamau (Saguenay, Qc)	1 2	20-O-9 21-A-9	565 – 575 610 – 620	Gabbro Rhyolite	81.36 68.24	0.247 0.250	82 – 96 87 – 94	77 – 81 78 – 80	52.9 64.6	39.3 45.8	Labrie & Conlon, 1997b
Mine Sigma Val-d'Or (Abitibi, Qc)	1 1 2	17018 17018 17020	1460 – 1465 1465 – 1470 1573 – 1578	Andesite Diorite Quartz (+To)	62.05 65.69 64.13	0.300 0.309 0.320	74 – 97 74 – 97 100	75 – 81 75 – 81 94	66.4 66.4 70.8	43.8 43.8 46.8	Labrie & Conlon, 1997c
Mine Francoeur Arntfield (Abitibi, Qc)	1 1 1 1	99-2 99-2 V1 – V4 V5 – V8	135 – 137 135 – 137 135 – 137 135 – 137	Andesite Gabbro Andesite Gabbro	62.00 59.30 74.40 72.90	0.230 0.250 0.235 0.238	59 – 66 60 – 74 n/a n/a	53 – 54 58 – 61 n/a n/a	5.0 5.0 5.1 5.1	3.4 3.4 3.6 3.6	Simon, 2002
Mine Bell-Allard Matagami (Abitibi, Qc)	1 2 2	520-1 – 520-3 520-4 – 520-6 520-6	965 950 950	Rhyolite Rhyolite Rhyolite	64.52 63.86 63.86	0.240 0.237 0.237	93 – 99 85 85	78 – 84 79 79	62.2 34.8 34.8	39.0 23.0 23.0	Labrie et al., 2001

[1] References for description of rock masses and definition of quality indices: Rock Quality Designation (RQD) (Deere, 1964); Rock Mass Rating (RMR) (Bieniawski, 1984).
[2] References for calculation or estimate of stress regimes at sites of measurement: Major principal stress (σ_1); and Mean stress ($\sigma_{mean} = [\sigma_1 + \sigma_2 + \sigma_3]/3$). Mine Pierre-Beauchemin: Estimated by the authors; Mine Niobec: Yu et al., 1988; Mine Joe Mann: Yu et al., 1995; Mine Sigma: Aubertin et al., 1997; Mine Francoeur: Corthésy et al., 1997; and Mine Bell-Allard: Corthésy et al., 1999.

The probe was inflated to pressures up to 30 MPa at Mine Pierre-Beauchemin and 25 MPa at other sites, with increments of 3.45 MPa (500 psi). The pressure and volume of fluid injected to inflate the probe were recorded after each increment (Figure 1). Coefficients of linearity determined are mostly greater than 0.998, confirming the linear elastic behavior of rock types investigated here.

Only the primary test results are discussed here. Secondary test results are detailed by Labrie et al. (1998). Twenty of the 128 primary tests were discarded: four tests at Mine Pierre-Beauchemin were conducted in highly altered, soil-like material; four were single results within limited, specific rock units already covered by previous tests; eight were discarded because tests were done in unspecified ground conditions, close to the excavation and inducing a high stress gradient at the location of measurements (Aubertin et al., 2002); and four were rejected because their values exceeded the instrument limits (three of these were replaced by secondary results). All dilatometer tests carried out over the last fifteen years and considered in this article are summarized in Table 2.

4.3 Results of Laboratory Tests

An extensive laboratory test program was carried out on samples recovered from the boreholes used for dilatometer testing. Tests were carried out under uniaxial and triaxial compression modes according to ASTM (2002) standards. Specimens were loaded until failure while recording loads and confining pressures. Axial and circumferential deformations were measured with electrical strain gages and linear and circumferential transducers during uniaxial compression tests, but only with transducers during triaxial tests. Young's modulus and Poisson's ratio were determined (Table 1). Poisson's ratio was used for the reduction of the dilatometer measurements (Equation 1b).

Table 2. Results of dilatometer tests and comparison with results obtained in laboratory and field.

MINE SITE	BORE-HOLE	ROCK TYPE	RMR	$E_{DILATOMETER}$ Mean / N (GPa / n)	$E_{DILATOMETER}$ StDv / CV (GPa / %)	$E_{LABORATORY}$ (GPa)	E_{MASS} (GPa)	RATIO E_D / E_L	RATIO E_M / E_L	RATIO E_M / E_D
Mine Pierre-Beauchemin	87-E1	Diorite (i)	61	30.60 / 2	9.64 / 32	88.60	20.78	0.345	0.235	0.679
	87-E1	Diorite (a)	59	12.96 / 2	5.44 / 42	72.55	15.73	0.179	0.217	1.214
	87-E2	Diorite (a)	55	12.62 / 3	7.38 / 59	65.64	12.14	0.192	0.185	0.962
	87-E3	Tonalite (a)	50	18.47 / 4	3.76 / 20	66.63	10.03	0.277	0.151	0.543
Mine Niobec	600-F1	Carbonatite	77	26.72 / 8	4.57 / 17	52.52	22.52	0.509	0.429	0.843
	600-F3	Carbonatite	73	30.51 / 8	11.57 / 38	64.77	23.95	0.471	0.370	0.785
Mine Joe Mann	20-O-9	Gabbro	79	65.10 / 6	12.17 / 19	81.36	37.56	0.800	0.462	0.577
	21-A-9	Rhyolite	79	60.93 / 7	8.73 / 14	68.24	31.50	0.893	0.462	0.517
Mine Sigma	17018	Diorite	80	59.88 / 4	4.96 / 8	65.69	31.46	0.912	0.479	0.525
	17018	Andesite	81	71.65 / 5	9.68 / 14	62.05	30.83	1.155	0.497	0.430
	17020	Quartz+To	94	60.75 / 2	12.99 / 21	64.13	51.37	0.947	0.801	0.846
Mine Francoeur	99-2	Andesite	53	26.58 / 8	2.08 / 8	62.00	10.57	0.429	0.170	0.398
	99-2	Gabbro	59	32.77 / 2	0.50 / 2	59.30	12.86	0.553	0.217	0.392
	V1-V4	Andesite	53	35.31 / 10	7.53 / 21	74.40	12.94	0.475	0.174	0.367
	V5-V8	Gabbro	59	31.83 / 18	6.64 / 21	72.90	15.81	0.437	0.217	0.497
Mine Bell-Allard	520-1	Rhyolite	81	33.32 / 5	5.31 / 16	64.52	32.06	0.516	0.497	0.962
	520-5	Rhyolite	79	35.84 / 6	4.38 / 12	63.86	29.48	0.561	0.462	0.823
	520-6	Rhyolite	79	72.85 / 11	16.94 / 23	63.86	29.48	1.141	0.462	0.405

[1] Definition of abbreviations: Rock type: (i) intact rock; (a) altered rock. $E_{DILATOMETER}$ (E_D): Modulus of deformation determined with dilatometer in boreholes (Roctest, 1992); $E_{LABORATORY}$ (E_L): Modulus of deformation determined in the laboratory (ASTM, 2002); and E_{MASS} (E_M): Modulus of deformation determined for the rock mass (Nicholson and Bieniawski, 1990). StDv: Standard deviation; and CV: Coefficient of variation (StDv / Mean).

[2] References to original set of data and results: Mine Pierre-Beauchemin: Cockburn and Désormeaux, 1988; Mine Niobec: Labrie and Conlon, 1997a; Mine Joe Mann: Labrie and Conlon, 1997b; Mine Sigma: Labrie and Conlon, 1997c; Mine Francoeur: Labrie and Conlon, 2000; and Mine Bell-Allard: Labrie et al., 2002.

4.4 Deformability of rock masses

In order to analyze dilatometer test results objectively and compare them with results of laboratory tests carried out on rock specimens or the calculated deformability of rock masses, scale effects have to be considered (Bieniawski, 1984; Brady and Brown, 1985; Goodman, 1989; Jackson and Lau, 1990; Nicholson and Bieniawski, 1990; Pinto da Cunha, 1990; Pinto da Cunha and Muralha, 1990; Palmström and Singh, 2001; Asef and Reddish, 2002; Kayabasi et al., 2003).

Most practical formulations of scale effects are based on rock mass classifications. Rock mass indices are used in equations to reduce values determined in the laboratory on intact rock specimens and provide realistic values taking into account the inherent characteristics of rock masses (ISRM, 1978). We used the formulation proposed by Nicholson and Bieniawski (1990) to estimate the modulus of deformation for the rock mass:

$$E_m = (E_l/100)(0.0028 RMR^2 + 0.9 e^{(RMR/22.82)}) \quad (2)$$

where E_m and E_l represent the modulus of deformation of the rock mass and intact rock, respectively, and RMR is the Rock Mass Rating (Bieniawski, 1984). The derived values of E_m are listed in Table 2. The three moduli of deformation (E_d, E_m and E_l) are shown as a function of mean *in situ* stress in Figure 2.

5 ANALYSIS OF RESULTS

The series of field and laboratory measurements produced some interesting observations regarding the deformation moduli:

(i) The intact rock modulus (E_l or Young's modulus) was within quite a narrow range for all rock types tested: 60 to 75 GPa with an average of 67.4 GPa and a standard variation of 8.2 GPa;
(ii) The rock mass modulus (E_m) ranged from 10 to 40 MPa. It was lower near the surface where rock mass was more fractured and altered. The RMR was between 50 and 60 near the surface and about 80 at depth;
(iii) The dilatometer modulus (E_d) is strongly correlated with the observed *in situ* stress (Figure 3); and
(iv) The dilatometer modulus (E_d) is similar to the rock mass modulus (E_m) near the surface, at shallow depth and low stress; and close to the intact rock modulus (E_l) at depth and under high stress.

Regression lines fitted to the data (Figure 3) are described by the following equations:

$$E_m = 518.9 \sigma_{mean} + 14,244 \text{ (in } MPa\text{), and} \quad (3a)$$

$$R^2 = 0.7762 \quad (3b)$$

$$E_d = 857.1 \sigma_{mean} + 23,895 \text{ (in } MPa\text{), and} \quad (4a)$$

$$R^2 = 0.6767 \quad (4b)$$

where R^2 is the coefficient of correlation.

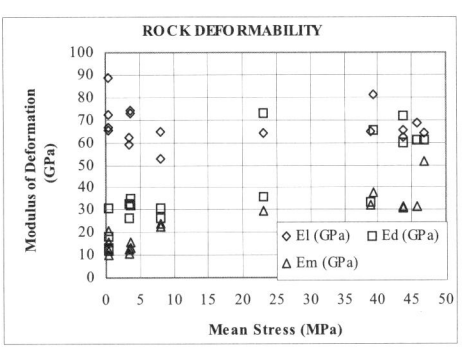

Figure 2. Deformation moduli as a function of mean *in situ* stress. The dilatometer modulus (E_d) is similar to the rock mass modulus (E_m) at low stress and similar to the intact rock modulus (E_l) at high stress.

In an attempt to improve the coefficients of correlation, the procedure was repeated with all results normalized by their Young's modulus (E_l), but no significant change was observed. In fact, the coefficients were slightly lower with values of 0.7424 and 0.6621 obtained for the rock mass (E_m) and the dilatometer (E_d) test results, respectively.

A last attempt was made to improve the quality of the correlation observed between dilatometer test results and the mean stress observed *in situ*, and therefore, propose an alternative interpretation of the data in Figure 3. An exponent-type regression was fitted to the data, similar to the approach proposed by Santarelli et al. (1986) and Santarelli and Brown (1987) to model the behavior of confined boreholes. The new relation is then given by the following expression:

$$E_d = 20,904 \sigma_{mean}^{0.2649} \text{ (in } MPa\text{)} \quad (5a)$$

and

$$R^2 = 0.7389 \quad (5b)$$

The coefficient of correlation of equation 5 shows a slight improvement compared to equation 4. More interesting however are the boundary values displayed by equation 5. At surface, unconfined, the modulus of deformation (E_d) is zero, and increases very rapidly to reach a value of 25 GPa under a stress level of 2 MPa, i.e. the average value of the modulus of the rock mass (E_m) under the same conditions. Above 10 MPa, the modulus of deformation increases slowly and reaches 58.9 GPa under an average stress level of 50 MPa. Within the range of stresses considered, both relations, 4a and 5a, are consistent with all results reported in the present paper.

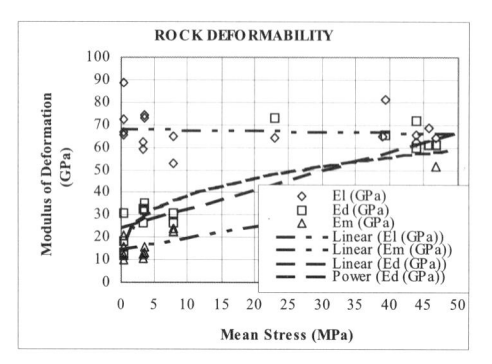

Figure 3. Correlation between deformation moduli and *in situ* stresses.

The modulus of deformation determined with the dilatometer (E_d) is very close of the modulus of the rock mass (E_m) at a low stress level, and similar to the modulus determined in the laboratory (E_l) on intact rock samples, at a high stress level. Intermediate stress levels would logically lead to intermediate values of deformability as well. Unfortunately, very few results are available for intermediate depths and stress levels. This is discussed below.

6 DISCUSSION

The discussion is limited to the issues of interpolation to intermediate stress levels and the effect of *in situ* stress on deformability measurements. Issues such as the comparison of results with similar results published in literature; the sensitivity of borehole dilatometers and limitations imposed by current instruments in practice; and the range and the meaning of rock properties determined with borehole dilatometers are addressed elsewhere (Labrie et al., 1998; Kanishiro et al., 1987; Goodman, 1989).

6.1 *Interpolation*

A fair estimate of the modulus of deformation at the scale of boreholes and for intermediate stress levels can be predicted by equations 4 or 5. However, the lack of experimental values at these stress levels remains a shortcoming. Additional measurements are needed to verify the proposed relationships. Nevertheless, our observations are consistent with those made by other researchers over the last twenty years (Koopmans and Hughes, 1986; Pinto da Cunha, 1990; Pinto da Cunha and Muralha, 1990).

6.2 *Influence of in situ stress*

Results shown in Table 2 and Figure 3 clearly indicate that the stress level has a direct influence on the modulus of deformation determined with dilatometer *in situ*, until its value becomes close to the one determined in the laboratory on the rock material.

The efforts of Santarelli et al. (1986) and Santarelli and Brown (1987) to model stresses and strains around boreholes and explain their behavior at failure has many similarities with our measurements at the rock mass scale and our attempts to characterize the effect of *in situ* stresses on the deformability of rock structures. Tests and observations by Santarelli et al. (1986) and Santarelli and Brown (1987) were conducted on non-linear porous materials that can only be modeled adequately through an intrinsic formulation that leads to a complete redefinition of the constitutive law of the material, e.g. the equivalent for porous non-linear materials of the well-known Hooke's law for elastic linear materials. Applied to the interpretation of dilatometer measurements, the implementation of a similar approach would mean a complete reformulation of the classic Lamé's solution, which was beyond the scope of the present article.

All rocks tested in the present program were hard rock materials, showing elastic linear behaviour, and therefore complying well with both Lamé's and Hooke's solutions (see Figure 1). Coefficients of linearity determined in the laboratory with compression tests and in the field with dilatometer tests are usually greater than 0.995 for stress levels discussed. Nevertheless, the results obtained show the importance of boundary conditions on the result of deformability measurements carried out with borehole dilatometers.

7 CONCLUSION

Over two hundred dilatometer tests were made in six mines of northern Quebec, Canada, over the last fifteen years. Mines were hard rock mines, operating at different depths and stress conditions, from surface to depths exceeding 1,800 meters. Tests were done in boreholes drilled from surface and underground openings, at sites providing the best access possible to structures to investigate. Results show a good correlation between stress levels observed at the location of measurements and the deformability of structures investigated, in accordance with results published by other searchers over the last twenty years (Koopmans and Hughes, 1986; Pinto da Cunha, 1990; Pinto Cunha and Muralha, 1990).

The main outcome of the test program is a data set that offers a wide range of deformability values that can be expected within rock structures under different stress conditions. Stress conditions are a critical factor to assess correctly the meaning of these values and decide on their application. Reasons for undertaking the test program and doing field measurements were detailed at the beginning of the article. These reasons cover most of the applications sought for dilatometer testing (Gill and Leite, 1995).

ACKNOWLEDGMENTS

The financial and logistical support provided by all mines participant in the test program is fully acknowledged. The work was funded in part through research contracts and grants assigned to *Department of Civil, Geological and Mining Engineering* of École Polytechnique de Montréal by the *National Sciences and Engineering Research Council of Canada (NSERC)* and the *Institut de recherche Robert-Sauvé en santé et en sécurité du travail (IRSST)*. The *Centre de recherches minérales (CRM)* of the Department of Natural Resources of Province of Quebec and CANMET Mining and Mineral Sciences Laboratories (CANMET-MMSL) of Natural Resources Canada also contributed to funding. Special thanks are offered to engineers, technicians and all personnel involved in the project and provided help and support, in particular: Jean Bastien at Mine Pierre-Beauchemin (formerly Mine Eldrich); Ghislain Pomerleau at Mine Niobec; Marc Huot and Alain Coulombe at Mine Joe Mann; Sylvie Poirier and Gilles Gagnon at Mine Sigma; François Girard at Mine Francoeur; and Denise Ouellet at Mine Bell-Allard. The help of former colleagues, consultants and suppliers is also acknowledged: Louis Marcoux and Richard Grenier at the *Centre de recherches minérales (CRM)*; Daniel Cockburn at Monterval; Michel Aubertin and Richard Simon, professors at École Polytechnique de Montréal; and Michel Blais and Jean-Pierre Perron at Roctest. The article was reviewed and edited by Ray Durrheim, Senior Scientist, CSIR-MiningTek (SA), and Michel Plouffe, manager, Ground Control Program, CANMET-MMSL. The attention and courtesy paid by the Organizers of the Conference to the authors are appreciated. The article is dedicated to the mentors of the first author: Denis E. Gill and Branko Ladanyi, Professors Emeritus at the *Department of Civil, Geological and Mining Engineering* of École Polytechnique de Montréal.

REFERENCES

Asef, M.R. & Reddish, D.J., 2002. The impact of confining stress on the rock mass deformation modulus. *Géotechnique*, Vol. 52, No. 4: 235-241.
ASTM, 2002. Standard Test Methods for Elastic moduli of intact rock core specimens in uniaxial compression (D 3148–96) and for Elastic moduli of undrained intact rock core specimens in triaxial compression without pore pressure measurement (D 5407–95). In *Annual Book of ASTM Standards 2002*, Section 4: Construction, Vol. 04.08: Soil and Rock (1): D 420 - D 5779: 345-349, 1211-1216. West Conshohocken (PA): ASTM International.
Aubertin, M., Li, L. & Simon, R., 2002. *Effet de l'endommagement sur la stabilité des excavations souterraines en roche dure*. Rapport R-312, Études et recherches, Institut de recherche Robert-Sauvé en santé et sécurité du travail (IRSST), Montréal (Québec).
Aubertin, M, Simon, R., Auer, L. & Gill, D.E., 1997. *Une étude sur le potentiel de coups de terrain à la mine Sigma en relation avec les effets du dynamitage de préfracturation*. Rapport CDT P1848, Centre de développement technologique, École Polytechnique de Montréal, Montréal (Québec).
Bieniawski, Z.T., 1984. Input parameters for design. In *Rock Mechanics Design in Mining and Tunnelling*: 55-96. Rotterdam: A.A. Balkema Publishers.
Brady, B.H.G. & Brown, E.T., 1985. Rock strength and deformability. In *Rock Mechanics for Underground Mining*: 86-134. London: George Allen & Unwin.
Cambridge Insitu, 2003. *Cambridge Insitu 73 & 95 mm Diameter High Pressure Dilatometer, Fact sheet*. Cambridge Insitu, Cambridge (England).
Cockburn, D. & Désormeaux, D., 1988. *Investigations géotechniques (pour la caractérisation des) piliers de surface à la mine Eldrich, Évain, Québec*. Rapport 1132-S, Monterval inc., Val-d'Or (Québec).
Corthésy, R., Leite, M.H. & Gill, D.E., 1997. *Élaboration d'un modèle de prédiction des contraintes in situ dans le Nord-Ouest québécois*. Rapport R-173, Études et recherches, Institut de recherche Robert-Sauvé en santé et sécurité du travail (IRSST), Montréal (Québec).
Corthésy, R. & Leite, M.H., 1999. *Mesures des contraintes aux Mines Matagami, Division Bell-Allard, Matagami*. Rapport final, Centre de développement technologique, École Polytechnique de Montréal, Montréal (Québec).
Deere, D.U., 1964. Technical description of rock cores for engineering purposes. *Rock Mechanics and Engineering Geology*, Vol. 1, No. 1: 17-22.
Gill, D.E. & Leite, M.H., 1995. Dilatometer testing of rock. In *The Pressuremeter and its New Avenues, Proc. 4th Int. Symp. On Pressuremeters*, Sherbrooke (Québec), G. Ballivy Ed.: 249-256. Rotterdam: A.A. Balkema Publishers.
Goodman, R.E., 1989. Deformability of rocks. In *Introduction to Rock Mechanics, 2nd Edition*: 179-220. New York: John Wiley & Sons.
Interfels, 2003. *Dilatometer, Fact sheet*. Boart Longyear Interfels GmbH, Bad Bentheim (Germany).
ISRM, 1978. Suggested methods for the quantitative description of discontinuities in rock masses. *Int. J. Rock Mech. Min. Sci.*, Vol. 15, No. 6: 319-68.
ISRM, 1987. Suggested methods for deformability determination using a flexible dilatometer. *Int. J. Rock Mech. Min. Sci.*, Vol. 24, No. 2: 123-134.
Jackson, R., Lau, J.S.O. (1990). The effect of specimen size on the laboratory mechanical properties of Lac du Bonnet grey granite. In *Proc. 1st International Workshop on Scale Effects in Rock Masses*, Loen (Norway), A.P. da Cunha Ed.: 165-174. Rotterdam: A.A. Balkema Publishers.
Jeager, J.C. & Cook, N.G.W., 1969. Underground measurements. In *Fundamentals of Rock Mechanics*: 361-383. London: Methuen & Co.
Kaneshiro, J.Y., Harding, R.C., Johannesson, P. & Korbin, G.E., 1987. Comparison of modulus values obtained from dilatometer testing with downhole seismic surveys and unconfined compressive tests at McKays Point Dam Site, California. In *Proc. 28th U.S. Symposium on Rock Mechanics*, Tucson (AR), Farmer, I.W. et al. Eds: 211-221. Rotterdam: A.A. Balkema Publishers.

Kayabasi, A., Gokceoglu, C. & Ercanoglu, M., 2003. Estimating the deformation modulus of rock masses: A comparative study. *Int. J. Rock Mech. Min. Sci.*, Vol. 40, No. 1: 55-63.

Koopmans, R. & Hughes, R.W., 1986. The effect of stress on the determination of deformation modulus. In *Proc. 27th U.S. Symposium on Rock Mechanics*, Tuscaloosa (AL), H.L. Hartman Ed.: 101-105. Littleton (CO): Society of Mining Engineers.

Labrie, D., 1988. *Résistance mécanique et modules de déformation élastique du matériau de la mine Eldrich, Évain, Québec*. Rapport de division MRL 88-48 (TR), Laboratoires de recherche minière, CANMET, Énergie, Mines et Ressources Canada, Ottawa.

Labrie, D. & Conlon, B., 1997a. *Caractérisation et instrumentation du pilier horizontal entre les niveaux 600 et 700 de la zone 102 à la mine Niobec, Saint-Honoré, Québec*. Rapport LMSM 96-083 (RC), Laboratoires des mines et des sciences minérales, CANMET, Ressources naturelles Canada, Ottawa.

Labrie, D. & Conlon, B., 1997b. *Caractérisation et instrumentation de deux piliers horizontaux à la mine Joe Mann, Chibougamau, Québec*. Rapport LMSM 96-084 (RC), Laboratoires des mines et des sciences minérales, CANMET, Ressources naturelles Canada, Ottawa.

Labrie, D. & Conlon, B., 1997c. *Caractérisation et instrumentation de trois structures minières à la mine Sigma, Val-d'Or, Québec*. Rapport LMSM 97-020 (RC), Laboratoires des mines et des sciences minérales, CANMET, Ressources naturelles Canada, Ottawa.

Labrie, D. & Conlon, B., 2000. *Vérification de l'endommagement dû au sautage et son effet sur la déformation du massif à la périphérie d'une galerie de ventilation à la mine Francoeur, Arntfield, Québec*. Rapport LMSM 99-069 (RC), Laboratoires des mines et des sciences minérales, CANMET, Ressources naturelles Canada, Ottawa.

Labrie, D., Conlon, B. & Anderson, T., 2001. *Essais de compression sur le matériau de la mine Bell-Allard, Matagami, Québec*. Rapport LMSM 01-018 (RC), Laboratoires des mines et des sciences minérales, CANMET, Ressources naturelles Canada, Ottawa.

Labrie, D., Conlon, B. & Boyle, R.F., 1998. In situ deformability of moderately to highly stressed mine structures in hard rock. *Int. J. Rock Mech. Min. Sci.*, Vol., 35, No. 4/5, Paper No. 123: 614-615.

Labrie, D., Conlon, B., Judge, K. & Anderson, T., 2002. *Observations et essais de déformabilité en forage à la mine Bell-Allard, Matagami, Québec*. Rapport LMSM 01-038 (RC), Laboratoires des mines et des sciences minérales, CANMET, Ressources naturelles Canada, Ottawa.

Nicholson, G.A., Bieniawski, Z.T. (1990). A non-linear deformation modulus based on rock mass classification. *Int. J. Min. Geol. Engng*, Vol. 8, No. 3: 181-202.

OYO Corporation, 2000. *(Dilatometer) Elastmeter-2, Fact sheet*. OYO Corporation, Ibaraki (Japan).

Palmström, A. & Singh, R., 2001. The deformation modulus of rock masses – Comparisons between in situ tests and indirect estimates, *Tunnelling and Underground Space Technology*, Vol. 16, No. 2: 115-131.

Pinto da Cunha, A., 1990. Scale effects in rock mechanics. In *Proc. 1st International Workshop on Scale Effects in Rock Masses*, Loen (Norway), A. Pinto da Cunha Ed.: 3-27. Rotterdam: A.A. Balkema Publishers.

Pinto da Cunha, A. & Muralha, J., 1990. About LNEC experience on scale effects in the deformability of rock masses. In *Proc. 1st International Workshop on Scale Effects in Rock Masses*, Loen (Norway), A. Pinto da Cunha Ed.: 219-230. Rotterdam: A.A. Balkema Publishers.

Roctest, 1992. *Borehole Dilatometer, Model Probex-1 Users' Manual*. Roctest Limited, Saint-Lambert (Longueuil), Québec.

Roctest, 2002. *Borehole Dilatometer, Model Probex-1, Fact sheet*. Roctest Limited, Saint-Lambert (Longueuil), Québec.

Santarelli, F.J. & Brown, E.T., 1987. Performance of deep wellbores in rock with a pressure-dependent elastic modulus. In *Proc. 6th ISRM Int. Congress on Rock Mechanics, Vol. 2*, Montréal (Québec), G. Herget & S. Vongpaisal Eds: 1217-1222. Rotterdam: A.A. Balkema Publishers.

Santarelli, F.J., Brown, E.T. & Maury, V., 1986. Analysis of borehole stresses using pressure dependent, linear elasticity, Technical Note. *Int. J. Rock Mech. Min. Sci.*, Vol. 23, No. 6: 445-449.

Simon, R., 2002. *Étude de l'effet du sautage adouci sur la fracturation des parois d'une excavation souterraine*. Rapport R-310, Études et recherches, Institut de recherche Robert-Sauvé en santé et sécurité du travail (IRSST), Montréal (Québec).

Yow, J.L., 1993. Borehole dilatometer testing for rock engineering. In *Comprehensive Rock Engineering, Volume 3*, J.A. Hudson Ed.: 671-682. London: Pergamon Press.

Yu, Y., Wong, A.S. & Toews, N.A., 1988. *Stability assessment of SL-102-19 and S-102-21 sill pillars and adjacent stopes at the Niobec Mine, Chicoutimi, Quebec – Part IV*. Division Report MRL 88-101 (TR), Mining Research Laboratories, CANMET, Energy, Mines and Resources Canada, Ottawa.

Yu, Y., Vongpaisal, S., Boyle, R.F. & Toews, N.A., 1995. *A strategic global mine stability assessment at Joe Mann mine, Chibougamau, Quebec*. Report MRL 95-015 (CL), Mining Research Laboratories, CANMET, Natural Resources Canada, Ottawa.

Estimation of shear strength increase beneath embankments by seismic cross-hole tomography

R. Larsson
Swedish Geotechnical Institute, Linköping, Sweden

H. Mattsson
GeoVista AB, Luleå, Sweden

Keywords: seismic test, cross-hole tomography, clay, undrained shear strength, strength increase, embankment

ABSTRACT: There is a great demand to utilise the shear strength increase resulting from consolidation beneath old embankments on soft soils when these structures are to be widened, raised or subjected to heavier or faster traffic loads. There are often no records of the loads and settlements, and access to the embankment for traditional geotechnical investigations is often restricted, particularly in the case of railway embankments. The method of seismic cross-hole tomography in order to estimate the increase in shear strength has therefore been tested beneath a number of well documented test embankments. The results have been shown to provide a good general picture of the shear strength beneath the embankments, in addition to fairly good estimates of the actual magnitudes of the shear strengths.

1 BACKGROUND

The Swedish railway system is currently being upgraded for faster and heavier trains. In this context, it will be of great benefit if the shear strength beneath old embankments on soft soil, many of which were constructed up to 100 years ago, can be utilised. However, this requires verification of the shear strength increase in relation to the soil outside the embankment, which involves certain problems. The size of the settlements that have occurred and the exact load that has been applied are rarely known. Traditional geotechnical investigations require access to the embankment and drilling through it. On the other hand, the railway authorities have set demands on uninterrupted traffic and prohibit equipment or personnel on the embankment for safety reasons. There is thus a considerable need for a method to estimate the shear strength increase that does not require access to the embankment itself.

2 HYPOTHESIS

Measurement of the shear wave velocity in the field is often used to estimate the in situ initial shear modulus G_0. It has been shown that the shear modulus in soft cohesive soils can be expressed as a function of the undrained shear strength (e.g. Larsson and Mulabdic, 1991). In principle, it should therefore be possible to estimate also the undrained shear strength from the measured shear wave velocity. However, both the shear modulus and the undrained shear strength are functions of the square of the shear wave velocity and very accurate measurements of the shear wave velocity are therefore required. Furthermore, the relation between the undrained shear strength and the shear wave velocity is also a function of the overconsolidation ratio (Andersen et al., 1988; Atkinson, 2000). However, a prerequisite for considerable consolidation settlements and shear strength increases beneath embankments is construction on normally consolidated or only slightly overconsolidated soft ground. The soil beneath the embankment will then remain in a normally consolidated or only slightly overconsolidated state also after the load application and throughout the consolidation process, unless a significant unloading is carried out. It should thus be possible to estimate the undrained shear strength beneath such embankments from the shear wave velocity.

3 SCOPE OF THE INVESTIGATION

The method of seismic cross-hole tomography has been used in an effort to estimate its usefulness to assess the increase in shear strength due to consolidation beneath three test embankments. These embankments are about 50 years old and loads, settlements, pore pressures and shear strength increases are well known. The study has been designed as a pilot project in order to determine whether the method can be used in practical applications and the accuracy with which the shear strength increase can be estimated.

4 THE TEST EMBANKMENTS

The measurements have been made in the test fields at Skå-Edeby and Lilla Mellösa, which are supervised by the Swedish Geotechnical Institute. A total of eight instrumented test embankments were constructed in these fields during the period 1945–1961 and have since been monitored continuously. Three of the embankments were built on natural ground and these have been used in this project.

The soft soils in the test fields consist of only slightly overconsolidated high-plastic clay on top of firm layers of sand or till. The average overconsolidation ratio beneath the thin crusts is about 1.15. The thickness of the clay layers vary between 12 and 15 metres and the top layers consist of organic clays. The liquid limits of the clays range from 130 to 55% and the plasticity indices from 85 to 30%.

The oldest embankment (test fill) was built in 1945 at Lilla Mellösa on top of 14 metres of clay. It had a height of 2.5 metres and a square outline with a base length of 30 metres. The settlements today amount to about 2.0 metres and are continuing at a current rate of approximately 10 mm/year. The compression of the soil layers is fairly evenly distributed over the depth of the soft soil profile. The undrained shear strength beneath the embankment has increased throughout the profile, but most significantly in the upper part, Fig. 1.

The test fill at Skå Edeby was constructed in 1957 on top of 12 metres of clay. It was 1.5 metres high and had a circular outline with a base diameter of 35 metres. In 2002, the total settlements amounted to 1.1 metres and the rate of ongoing settlements was 5–6 mm/year. Here, too, the settlements have with time become fairly evenly distributed with depth. The shear strength beneath the fill has been measured by field vane tests on two occasions after construction and the results in the latest investigation have also been checked by direct simple shear tests in the laboratory. Beneath the central parts of the fill, the shear strength has increased fairly evenly throughout the clay profile, Fig. 2.

A narrow embankment with a length of about 40 metres was also constructed at Skå-Edeby in 1961. It was similarly given a height of 1.5 metres but the

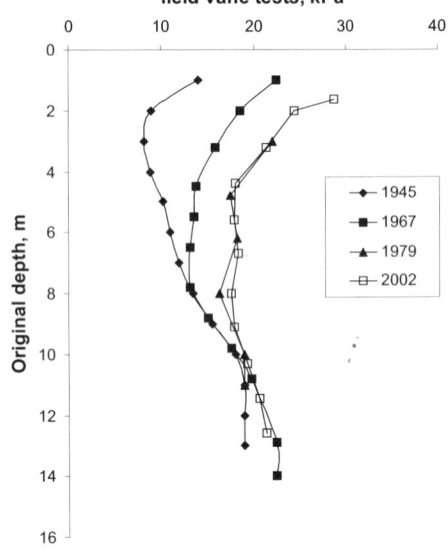

Fig. 1. Measured undrained shear strength beneath the test fill at Lilla Mellösa.

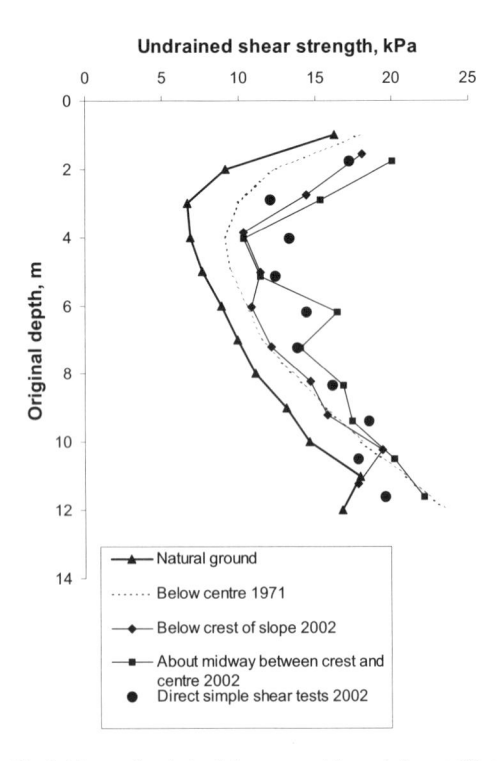

Fig. 2. Measured undrained shear strength beneath the test fill at Skå-Edeby.

crest was only 4 metres wide and the base 8.5 metres wide. The depth of the clay layers at this location was about 15 metres. The total settlements in 2002 amounted to about 1.1 metres, of which about 0.2 metre is related to horizontal movements in the subsoil. Due to the narrow embankment and the load distribution, the compression of the soil layers is mainly confined to the upper half of the soil profile. According to the estimated changes in water content, the compression of the soil is fairly evenly distributed down to an original depth of 8–9 metres, whereas results from oedometer tests and shear strength tests indicate that this limit is located at 7–8 metres depth, Fig. 3.

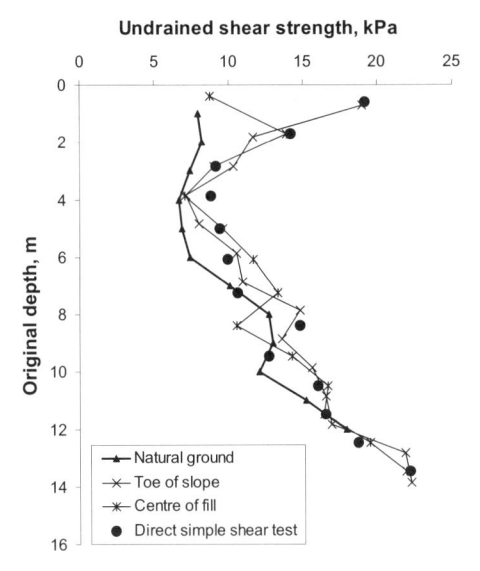

Fig. 3. Measured undrained shear strength beneath the test embankment at Skå-Edeby.

Further details of the soil conditions and the embankments at the test fields at Skå-Edeby and Lilla Mellösa can be found in Hansbo (1960), Chang (1981), Larsson (1986) and Larsson and Mattsson (2003), among other publications.

5 TEST METHOD

5.1 Seismic waves

Seismic investigations are based on the propagation of elastic waves in the ground. These waves are usually separated into compression waves, shear waves and surface waves. Shear waves have a particle motion that is perpendicular to the direction of propagation. The propagation velocities of shear waves are governed by the shear modulus and density of the ground. From the shear wave velocity, V_s, and the density, ρ, it is possible to calculate the shear modulus, G,

$$G = V_s^2 \rho$$

The shear modulus that is measured in most seismic tests is the initial shear modulus at very small strains, G_0. The relation between the initial shear modulus and the undrained shear strength, c_u, in Swedish normally consolidated or only slightly overconsolidated clays has been found to be (Larsson and Mulabdic, 1991)

$$G_0 \approx \frac{504 \, c_u}{w_L}$$

where w_L is the liquid limit.

The undrained shear strength in the correlation refers to values obtained in corrected field vane tests and direct simple shear tests.

5.2 Seismic tomography

Tomography is a well-known technique for creating images of projections (tomograms) of hidden objects by using X-rays, ultrasound or electromagnetic waves. The technique used in this project is termed *seismic crosswell direct wave traveltime tomography*, but is commonly called cross-hole tomography. Its basic principle is to estimate a velocity model of the ground by measuring the time it takes for elastic waves to propagate from a source to a receiver. To perform cross-hole tomography it is necessary to have at least two boreholes. An array of geophones is inserted in one hole and in the other an elastic wave is generated. A seismograph measures the time it takes for the wave to travel from the source point to the geophones. The source is then moved to another position in the hole and the procedure is repeated. The measurements will produce a number of arrival times of waves that have crossed the investigated area. The geophone distance and the wave frequency mainly govern the data resolution; the shorter the distance and the higher the frequency, the better the resolution. The spatial relation between the depth of, and distance to, the boreholes is also an important parameter since shallow boreholes and a large distance will lead to poor ray coverage, Fig. 4.

The measured first arrival times and the co-ordinates of the geophones and the source points are stored in a data file. The area between the boreholes is divided into a grid of velocity cells. Each cell is assigned an initial start value. A model program then calculates the time it takes for different rays to travel

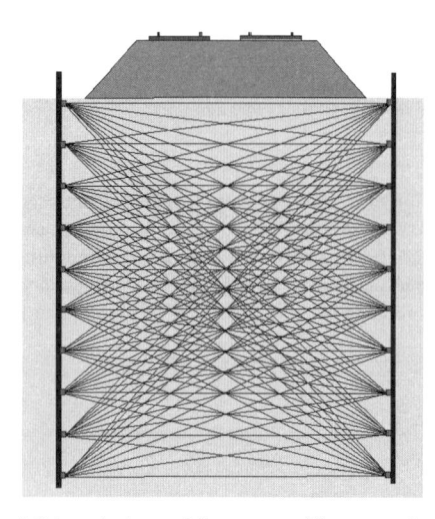

Fig. 4. Schematic picture of the coverage of the wave paths.

through the area between the boreholes. The calculated times are compared to the measured travel times, and the errors in the calculations are the differences between these two parameters. Different rays intersect each cell and the best-fit velocity is estimated by the least squares method. The procedure is repeated for a predetermined number of iterations or until a chosen acceptable difference between the measured travel time and the corresponding value calculated by the model is reached, the so-called RMS residual, which indicates the fit of the model. The size of an acceptable RMS value depends mainly on the measuring accuracy of the travel times for the shear waves. In this particular case, an RMS value of up to 25–30 ms is considered to indicate a good model fit. The velocity model does not provide a unique solution to the inversion problem, but with information about the geological conditions at the site it is possible to determine whether the established model is physically reasonable.

The software used in this project is called 3DTOM (Jackson and Tweeton, 1996). For inversion of travel time data, 3DTOM uses the SIRT method (Simultaneous Iterative Reconstruction Technique; Peterson et al., 1985). It is possible to model straight rays, crooked rays or combinations of these (hybrid modelling). The start model can be varied between a homogenous, a horizontally layered or a chequerboard model.

6   FIELD EQUIPMENT AND PROCEDURE

The equipment for collecting the data consisted of a TERRALOC MARK 3 (ABEM) seismograph and three 5-component 28 Hz sensor geophones (BG-K5) with pneumatic clamping devices. The vibration source was a screw plate attached to a hollow drilling pipe. A free-running inner rod system was inserted into the pipe. The rods were lifted and allowed to fall onto the top of the screw plate to generate vertically polarised shear waves. The triggering of the seismograph was carried out by a standard geophone (PE-3) attached to the drill pipe at its upper end. The geophones were set up to detect vertically polarised shear waves and were mounted at a distance of 1.0 metre from each other.

The geophones were lowered inside a vertical borehole with a casing and attached by inflating the pneumatic clamping devices. The casing consisted of a plastic bellows hose that is vertically elastic, which ensures good transmission of the signal from the clay to the geophones. In the first measurement position, the uppermost geophone was placed as close as possible to the ground surface, after which the second and third geophones were positioned at about 1 and 2 metres depth respectively. The screw plate was screwed 0.1–0.2 metre into the ground. The trigger geophone was attached to the pipe and a measurement was performed. The set of geophones was then lowered 3 metres and a new measurement performed. When the geophone array reached the bottom of the borehole, the screw plate was advanced downwards to 1 metre depth and the procedure was repeated with the geophones instead being lifted in 3-metre stages. This was repeated with the screw plate being advanced in 1-metre steps until it reached firm ground. Since the trigger geophone was mounted at the top of the drill pipe, a time delay was introduced as the wave had to travel upwards along the pipe before it reached the trigger geophone. Each data set was corrected for this delay.

The measurements beneath the square and circular fills were performed with one borehole positioned in the central part, but well outside the permanent instrumentation, and the other at the perimeter of the fill. The conditions in the other directions were assumed to be identical. The measurement beneath the embankment was preformed across it with one borehole at each side and the measurements in natural ground outside the fills were performed with geometries similar to those beneath the fills. The distance between the boreholes varied between 9 and 15 metres and the depth to solid ground varied in approximately the same way, which resulted in fairly square geometries and a ray coverage of about 0°–45°.

7 TEST RESULTS

Fig. 5 shows the tomograms from beneath the circular fill at Skå-Edeby and of the natural soil outside. There is a large velocity contrast between the gravel fill and the clay soil. Down to about 5 metres depth, there are still considerably higher velocities under the test fill than in the undisturbed soil. At 6–7 metres depth, there is a horizontal high-velocity sub-layer that cuts across the entire soil section. Beneath this layer, from 7 to 9 metres depth, the shear wave velocity is still higher under the test fill than in the natural soil outside. When the bottom of the clay layer is reached, the velocity rapidly increases, which indicates that a stiffer material underlies the clay. The dome shape of this layer is most probably an artefact created during the model inversion, which is caused by a combination of a fast velocity increase and a lack of data related to the bad ray coverage close to the boundary. A similar effect can be seen beneath the fill material and at the lower boundary of the section in natural soil. The circular anomalies appearing along a vertical line at the 13 m distance are also caused by the lack of coverage and true data due to the borehole being situated in this position. The RMS value of the test fill model is 21 ms and for the undisturbed soil 25 ms.

The tomograms from the road-like test embankment at Skå-Edeby are shown in Fig. 6. Here, the increase in shear wave velocity under the embankment is considerably less and can only be readily observed to a depth of about 5 metres. The RMS value of 39 ms is fairly high and indicates problems in fitting the model data to the measured travel times.

The results of the measurements at Lilla Mellösa greatly resemble those from the circular fill at Skå-Edeby, Fig. 7. The RMS values are low,: the level for the embankment model being 16 ms and for the natural soil 10 ms, which indicates that the models fit statistically well to the measured data. The boundary between the soft soil and the bottom is not as well-defined at the Lilla Mellösa site as at the Skå-Edeby site. At Lilla Mellösa, the clay layers are underlain by sand, whereas those at Skå Edeby lie on rock or till.

8 CORRELATION BETWEEN ESTIMATED AND MEASURED SHEAR STRENGTH

From a first glance at the tomograms in Figs 5–7, it is obvious that a considerable increase in shear wave velocity and undrained strength has occurred beneath the circular and square test fills in Skå-Edeby and Lilla Mellösa. It also indicates that any such increase beneath the road-like test embankment at Skå-Edeby is considerably smaller and is limited to the upper layers beneath the embankment. When making a more detailed evaluation, certain aspects have to be taken into account. These are:
- values close to the boreholes are more or less erroneous because of poor wave path coverage and are thereby misleading.
- values at the upper and lower boundaries of the section in a portion midway between the boreholes are more or less erroneous if the soft soil is overlain and underlain by considerably stiffer material. Even when this is not the case, the values are uncertain because of poor coverage in these parts.
- The relation between undrained shear strength and shear wave velocity is sensitive to the liquid limit (or plasticity index) of the soil.

Before evaluation, the data in the vertical strips with poor coverage close to the boreholes should be excluded. The width of these strips can be estimated from a sketch of the wave paths for the actual distances between the boreholes and depths between the measuring points, see Fig. 4. A similar estimate of uncertain zones at the upper and lower boundaries should also be made.

Since information on the distribution of the liquid limit beneath the loaded area under the present conditions is normally absent, this has to be estimated from the data in the natural soil outside or from investigations performed before the load application together with an estimate of the distribution of settlements with depth. The estimation of the total settlements beneath the fill material is fairly straightforward since the border between this material and the underlying clay is rather distinct. The estimation of the distribution with depth is more approximate and has to be made with consideration to the way in which the levels of different layers and stiffness borders beneath and outside the embankments are located in relation to each other. From the visual inspection of the tomograms beneath the large fills at Skå-Edeby and Lilla Mellösa, it is quite obvious that downward movements have occurred throughout the clay profiles beneath the fills. A more detailed distribution is difficult to interpret, but an assumption of an even distribution of compression with depth appears to be reasonable and will have to suffice. The estimation of the settlement distribution with depth beneath the road-like embankment is more difficult. However, the tomogram clearly indicates that the embankment has settled about 1 metre and that there are no significant settlements below 6–7 metres depth. A rough estimate is that the settlements are evenly distributed down to this depth.

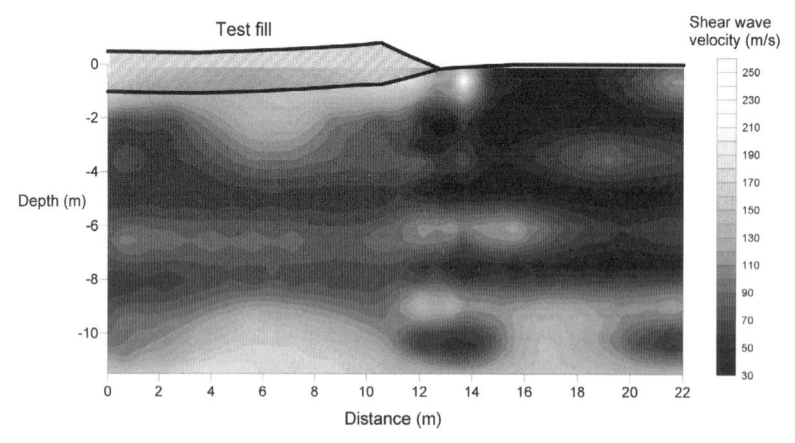

Fig. 5. Contour plot of the tomograms from the circular test fill and natural soil at the Skå-Edeby site.

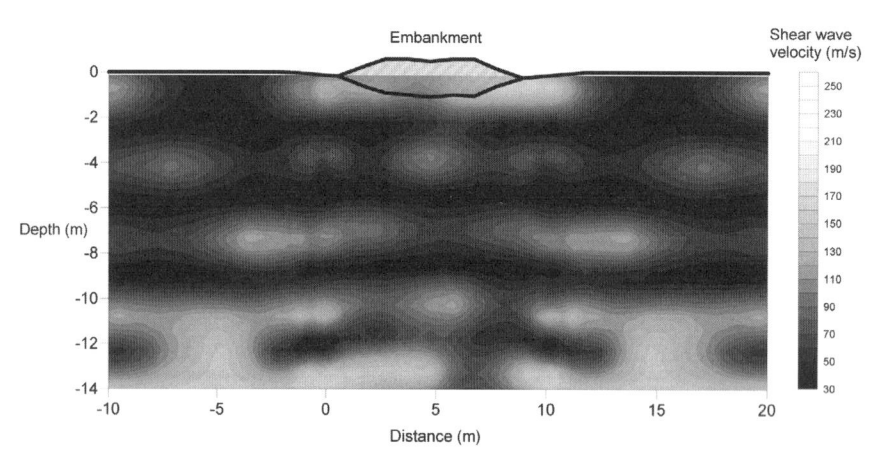

Fig. 6. Contour plot of the tomograms from the road-like test embankment and natural soil at the Skå-Edeby site.

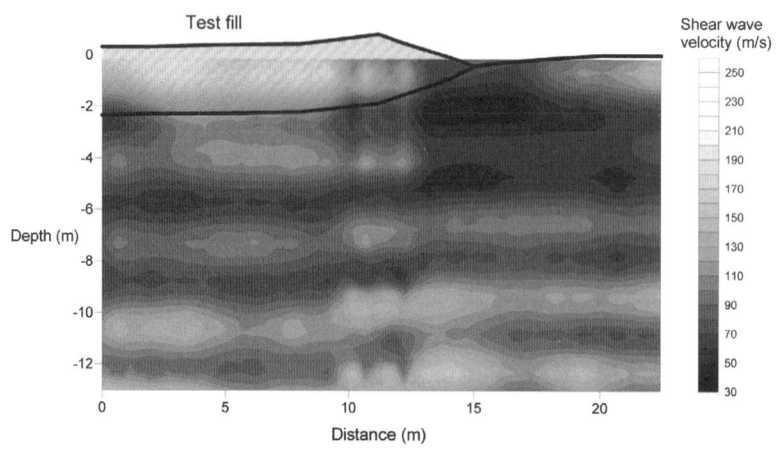

Fig. 7. Contour plot of the tomograms from the square test fill and natural soil at Lilla Mellösa.

The next step is to draw detailed tomograms with closely spaced contour lines for the shear wave velocities. For the tomograms of the natural soil, horizontal lines are drawn at selected evenly spaced depths. The zones with estimated erroneous data are excluded and the average velocity at each depth within the remaining zone is estimated. The same procedure is used for the tomograms beneath the loaded areas, but here the horizontal lines are adjusted while taking into consideration the estimated settlements at each level in such a way that lines corresponding to the original depths are produced. A relevant comparison can then be made between the estimated average velocities at the 'original' depths beneath the loaded areas and in natural ground.

There are then two ways of estimating the undrained shear strength beneath the loaded areas. The first is to use the empirical relation between shear wave velocity, density and liquid limit and the undrained shear strength. The estimated shear wave velocity beneath the fill is then used together with density and liquid limit at the original depth. In large settlements, there is also a certain increase in density, but this is of limited importance. However, there is always a certain spread in such relations. A more direct way is to use the undrained shear strength measured in the natural ground outside the loaded area. The shear wave velocities at each depth are then compared to velocities corresponding to the same original depths beneath the loaded area. The undrained shear strength beneath the loaded area is then calculated as the undrained shear strength in the natural soil multiplied by the square of the quotient between the shear wave velocities (see relations given in section 5.1)

$$c_{u2} = c_{u1} \left(\frac{V_{s2}}{V_{s1}} \right)^2$$

where
c_{u2} = undrained shear strength beneath loaded area
c_{u1} = undrained shear strength in natural ground
V_{s2} = shear wave velocity beneath loaded area
V_{s1} = shear wave velocity in natural ground

The evaluated shear wave velocities in natural ground and beneath the loaded areas suggest that an increase in shear wave velocity has occurred throughout the profiles beneath the large fills. This increase is large at the top but decreases with depth. Beneath the narrow embankment at Skå-Edeby, there is a significant increase in velocity down to 4 metres depth. The effect probably extends down to 6–7 metres depth, but an anomaly found in the shear strength determinations by the field vane tests at 4 metres depth also appears in the shear wave velocities. Below 6–7 metres depth, there is no indication of any increase in shear wave velocity.

The undrained shear strength calculated from the measured shear strengths in the field and the amplification $(V_{s2}/V_{s1})^2$ estimated from the measured shear wave velocities are shown in Fig. 8. This method provides the best estimate of the shear strength compared to using the empirical relation.

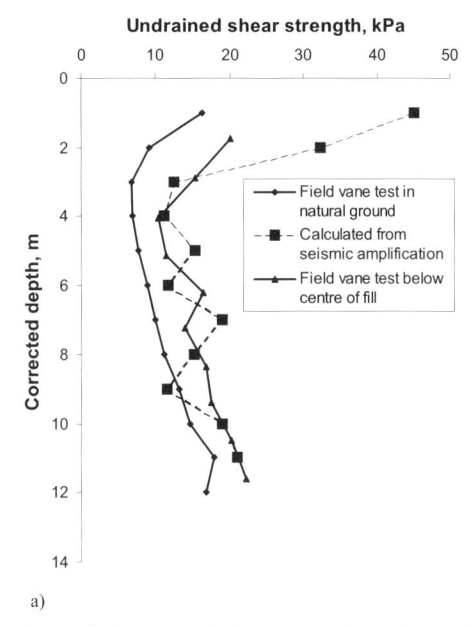

a)

Fig. 8. Evaluated undrained shear strength using the undrained shear strength in natural ground and the amplification estimated from the measured shear wave velocities.
a) At the circular fill at Skå-Edeby

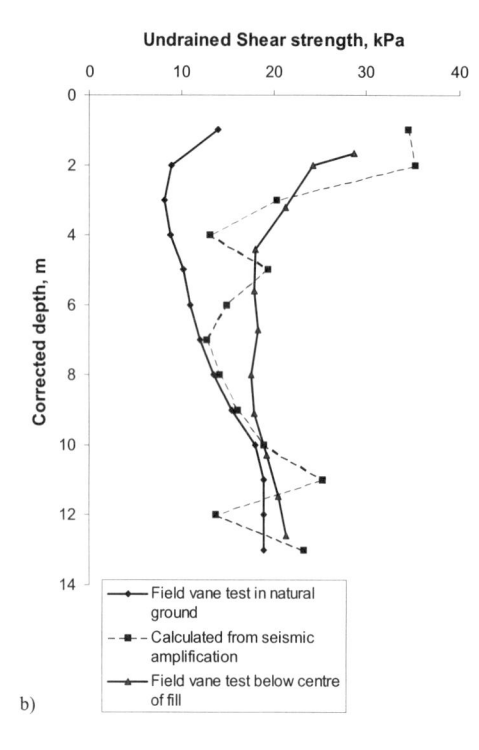

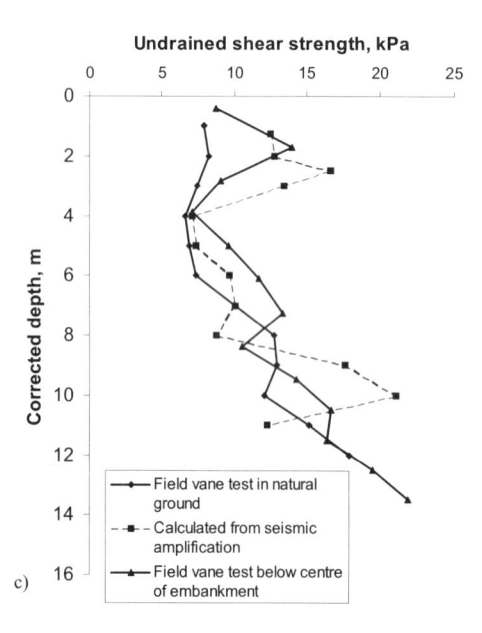

Fig. 8. Evaluated undrained shear strength using the undrained shear strength in natural ground and the amplification estimated from the measured shear wave velocities.
b) At the square fill at Lilla Mellösa
c) At the test embankment at Skå-Edeby

9 CONCLUSION

The possibility of estimating the increase in shear strength due to consolidation beneath embankments by seismic cross-hole tomography has been illustrated. The measurements were performed with readily available equipment and evaluation programs. The results have a certain degree of scatter, but the general pattern of the shear strength variation is obtained and an estimate of the operative strength on the basis of the tomograms would prove to be fairly close to the actual measured values.

ACKNOWLEDGEMENT

The project described in this paper was sponsored by the Swedish National Rail Administration, GeoVista AB and the Swedish Geotechnical Institute.

REFERENCES

Andersen, K.H., Kleven, A. and Heien, D. (1988). Cyclic Soil Data for Design of Gravity Structures. Norwegian Geotechnical Institute, Publication No. 175, Oslo.

Atkinson, J.H. (2000). Non-linear soil stiffness in routine design. Geotechnique, Vol. 50, No. 5, pp. 487-508.

Chang, Y.C.E. (1981). Long-term consolidation beneath the test fills at Väsby, Sweden. Swedish Geotechnical Institute, Report No. 13, Linköping.

Hansbo, S. (1960). Consolidation of Clay with Special Reference to Influence of Vertical Sand Drains. Swedish Geotechnical Institute, Proceedings No. 18, Stockholm.

Jackson, M.J. and Tweeton, D.R. (1996). 3DTOM: Three-dimensional geophysical tomography. Instruction manual. U.S. Geological Survey, Report of investigation 9617.

Larsson, R. (1986). Consolidation of soft soils. Swedish Geotechnical Institute, Report No. 29, Linköping.

Larsson, R. and Mattsson, H. (2003). Settlements and shear strength increase below embankments. Swedish Geotechnical Institute, Report No. 63, Linköping.

Larsson, R. and Mulabdic, M. (1991). Shear moduli in Scandinavian clays - Measurements of initial shear modulus with seismic cones - Empirical correlation for the initial shear modulus in clay. Swedish Geotechnical Institute, Report No. 40, Linköping.

Peterson, J.E., Paulson, B.N.P. and McEvilly, T.V. (1985). Applications of algebraic reconstruction techniques to cross-hole seismic data. Geophysics 50, pp. 1566-1580.

Statistical evaluation of the dependence of the liquidity index and undrained shear strength of CPTU parameters in cohesive soils

J. Liszkowski & M. Tschuschke
Institute of Geology, Adam Mickiewicz University, Poznań, Poland

Z. Młynarek & Tschuschke W.
Department of Geotechnics, August Cieszkowski Agricultural University Poznań, Poland

Keywords: cohesive soils, states of consistency, CPTU data

ABSTRACT: The paper presents simple empirical relationships enabling determination of the liquidity index from CPTU results. In construction of these dependencies the following factors were taken into account: grain size distribution of soils, their genesis, trend of penetration characteristics with depth, and the effect of stress history. The significance of the relationships was verified on basis of statistical criteria. The relationships between cone resistance and liquidity index and undrained shear strength are shown to be consistent.

1 INTRODUCTION

Research on cone penetration testing has allowed determination of a wide range of geotechnical parameters directly from analysis of penetration characteristics or, neglecting these parameters, designing shallow and deep foundations (Lunne et al. 1997). A specific exception from this rule is liquidity index. Poor literature on this issue is the result of limited application of the liquidity index in geotechnical subsoil description and replacing this parameter by undrained shear strenght. Recommendations of the Polish Standards (PN-B-02479, PN-81/B-03020) in case of uncomplicated constructions founded in simple geotechnical conditions, allow designing subsoil bearing capacity by strength and deformation parameters determined from correlation with the leading characteristics which, for cohesive soils, is the liquidity index. Hence, knowledge about the characteristics describing soil state has, in the above mentioned conditions, a significant practical importance, and the possibility of using state parameters determined directly from the results of cone penetration test would significantly facilitate geological and engineering recognition of subsoil and shorten design process. This idea is the main inspiration for an attempt at determination of empirical correlations between CPTU results and a characteristics describing consistency state of genetically differentiated cohesive soils.

2 IN-SITU INVESTIGATIONS

The area of the in-situ test investigations was limited to western Poland with subsoil consisting of soils of various geological formations. The field tests included CPTUs and taking soil samples from the layers selected on basis of the penetration characteristics. 35 test sites were selected for the study where depth of subsoil testing ranged from 18 m to 53 m. The cone penetration tests were carried out with a heavy 200 kN Track-Truck penetrometer by A.P. van den Berg, using a standard piezocone with base area 10 cm^3, and with a metal filter behind the cone. During deep penetrations a procedure protecting pushed in elements against buckling and breaking was applied (Tschuschke et al. 2001). According to The International Reference Test Procedure, changes of the following penetration parameters with depth were registered: cone resistance – q_c, sleeve friction – f_s and pore pressure measured behind the cone – u_2. On the basis of the registered values, corrected and normalized values of penetration parameters were determined:

- corrected cone resistance – q_t

$$q_t = q_c + (1-a) u_2 \quad (1)$$

- net cone resistance - q_n

$$q_n = q_t - \sigma_{VO} \quad (2)$$

- friction ratio - R_f

$$R_f = (f_s/q_t) \, 100\% \tag{3}$$

where: a – area ratio of the cone, σ_{vo} – total overburden stress. One metre long soil sample cores for laboratory studies were collected from the subsoil in the area of earlier penetrations with a Mostap sampler by A.P. van den Berg. From 35 soil profiles a total of over 200 soil cores were taken. Fig. 1 presents representative CPTU results and location of sampling places in soil profile.

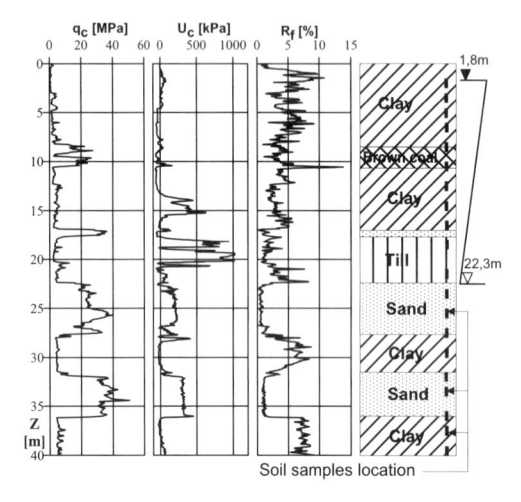

Fig.1 CPTU and soil profile with soil samples location.

3 LABORATORY TESTS

Laboratory studies were limited to those elements which facilitated identification of analyzed soils with respect to grain size distribution, consistency state and geological origin. Due to the fact that the studied issues concerned a given soil group, the laboratory tests were related exclusively to cohesive soils. Within the test scope the following determinations were carried out:
- grain size distribution by wet analysis
- water content - w_n
- plastic limit - w_p
- liquid limit - w_L
- calcium carbonate content - $CaCO_3$ and organic content - I_{om}

4 CLASSIFICATION OF THE ANALYZED SOILS

The main classification criterion for cohesive soils, according to the system used in Poland, is clay content – f_i and, additionally, silt and sand content (f_π, f_p). Due to clay content soils are grouped into classes:
- cohesionless ($f_i < 10\%$)
- mild cohesive (f_i 10-20%)
- cohesive (f_i 20-30%)
- very cohesive ($f_i > 30\%$)

Differentiation of grain size distribution is graphically presented by dispersion of the points in Feret's triangle (Fig. 2) which indicates that three groups of fine grained soils selected by the classification system with clay content over 10% were qualified for the problem analysis.

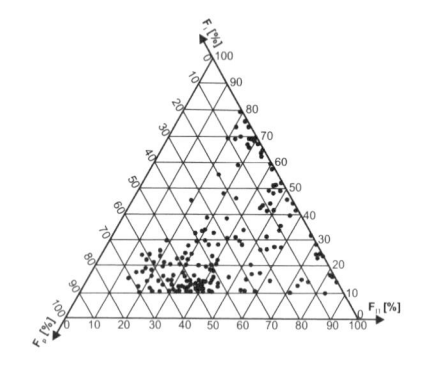

Fig.2 Feret triangle for three soil fraction contents determination.

The issue of adequacy of the penetration parameters can be best verified from evaluation of their degree of adjustment of a representative classification system worked out for the cone penetration method. Comparative analysis was carried out by means of placing points oriented by the penetration parameters on the CPTU classification chart (Fig.3). The points are at the same time graphically differentiated with respect to grain size distribution which was determined in laboratory for the soil samples collected from the subsoil. In the general evaluation of reliability of the results from the static penetration test it should be stated that location of the analyzed points on the CPTU classification chart is included within the areas selected in the system for the soils with the same grain size distribution. This fact allows assumption that also the consistency state of cohesive soils can be determined on basis of the penetration characteristics.

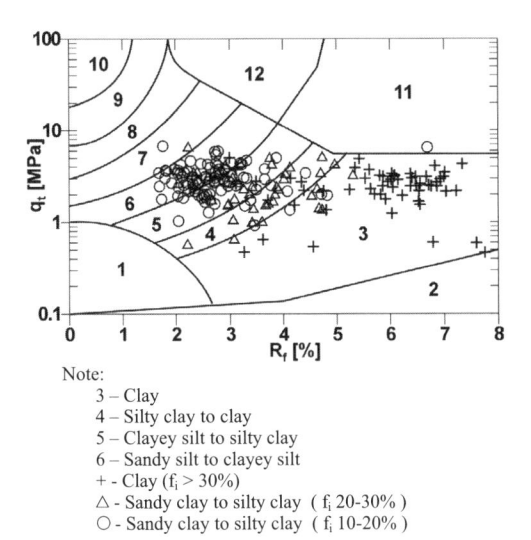

Note:
3 – Clay
4 – Silty clay to clay
5 – Clayey silt to silty clay
6 – Sandy silt to clayey silt
+ - Clay ($f_i > 30\%$)
△ - Sandy clay to silty clay (f_i 20-30%)
○ - Sandy clay to silty clay (f_i 10-20%)

Fig.3 CPTU chart for soil classification (Robertson et al. 1986).

5 STATISTICAL EVALUATION

For description of the variability of consistency state in cohesive soils a liquidity index - I_L is used which, according to the concept of this parameter, relates natural water content to Atterberg's limits (equation 4)

$$I_L = (w_n - w_p)/(w_L - w_p) \quad (4)$$

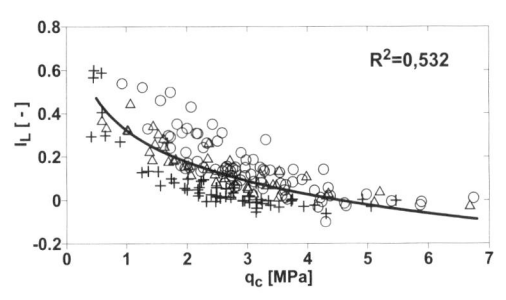

Fig.4 Relationship between liquidity index and measured cone resistance.

Statistical analysis of the empirical relationship between the state parameter – I_L and cone resistance indicated that the best estimation is obtained for a logarithmic type function (Fig. 5) described by the equation 5. Statistical significance of analyzed relationship (equation 5) is higher if the influence of pore pressure on cone resistance is included.

$$I_L = 0{,}323 - 0{,}214 \ln(q_t) \quad (5)$$

The correlation coefficient of the estimated function with the value $R^2 = 0{,}617$ indicates that even at the significance level $\alpha = 0{,}001$ zero hypothesis about lack of correlation between variables is rejected.

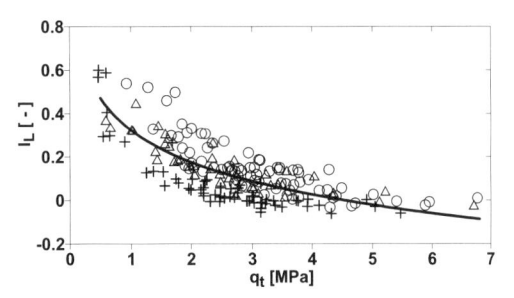

Fig.5 Relationship between liquidity index and corrected cone resistance.

Another stage in the analysis aiming at increasing significance of the correlation can be taking into account data on two important factors, namely: differentiation of soils with respect to grain size distribution, and considering, in the evaluation of the correlation relationship, the effect of stress state in the subsoil on the value of cone resistance. The first issue was verified in analysis of adequacy of a CPTU classification system and reliable assessment of soil kind carried out on its basis. From the analysis results that a good identifier of the soil kind is, in CPTU, friction ratio – R_f. Hence, indication of a statistically significant difference between mean values of the friction ratio in soil groups differentiated with respect to grain size distribution would indicate significance of this coefficient in constructing correlation dependence. The analysis proved the values of the statistics between the very cohesive ($f_i > 30\%$, $R_f = 5{,}83\%$) and cohesive ($f_i = 20-30\%$, $R_f = 3{,}53\%$) $t_{\alpha=0.05} = 9{,}95$, $t_{crit} = 1{,}99$, as well as between cohesive ($f_i = 20-30\%$, $R_f = 3{,}53\%$) and mild cohesive ($f_i = 10-20\%$, $R_f = 2{,}65$) soil groups $t_{\alpha=0.05} = 6{,}80$, $t_{crit} = 1{,}98$, lead to rejection of zero hypothesis justifying the thesis about statistically significant difference between the means ($t_\alpha > t_{crit}$). The second issue concerning taking into account stress state in subsoil in cone resistance evaluation, is related to observation of the increasing trend in the cone resistance distribution with the increase in the vertical component of geostatic stress. Releasing corrected value of cone resistance (equation 1) from the stress state in subsoil usually deals with replacing the former with, so called, net cone resistance (equation 2). Taking into account in evaluation of correlation relationship between cone resistance and liquidity index, of the factors presented above leads to the following shape of the dependence:

- for very cohesive soils – group 1 (Fig. 6)

$$I_L = 0,235 - 0,235 \ln(q_n) \quad ; \quad R^2 = 0,856 \qquad (6)$$

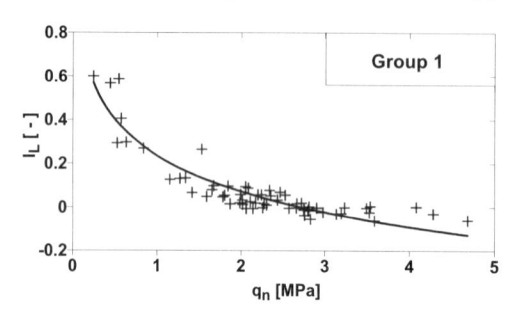

Fig.6 Liquidity index vs. net cone resistance for very cohesive soils.

- for cohesive soils – group 2 (Fig. 7)

$$I_L = 0,304 - 0,194 \ln(q_n) \quad ; \quad R^2 = 0,815 \qquad (7)$$

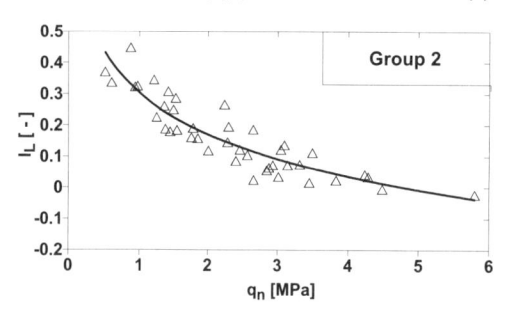

Fig.7 Liquidity index vs. net cone resistance for cohesive soils.

- for mild cohesive soils – group 3 (Fig. 8)

$$I_L = 0,416 - 0,284 \ln(q_n) \quad ; \quad R^2 = 0,779 \qquad (8)$$

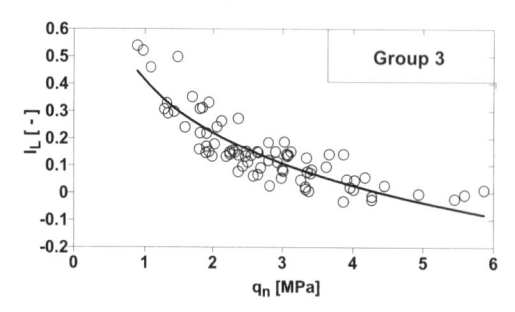

Fig.8 Liquidity index vs. net cone resistance for mild cohesive soils.

The correlation coefficients obtained for the relationships presented in equations 6-8 clearly increase significance of the correlation.

The final element of the analysis was taking into account genetical differentiation of the soils divided into groups with respect to grain size distribution. Considering the assumed division in subgroups the following estimation of the relationship and the correlation coefficient:

- for very cohesive soil ($f_i > 30\%$) (Fig.9)
  - Quaternary clays; $CaCO_3 > 0\%$ - subgroup 1.1

$$I_L = 0,244 - 0,240 \ln(q_n) \quad ; \quad R^2 = 0,806 \qquad (9)$$

- Tertiary clays $CaCO_3 = 0\%$, low plasticity $I_p < 50\%$ - subgroup 1.2

$$I_L = 0,265 - 0,213 \ln(q_n) \quad ; \quad R^2 = 0,969 \qquad (10)$$

- Tertiary clays $CaCO_3 = 0\%$ high plasticity $I_p > 50\%$

$$I_L = 0,230 - 0,247 \ln(q_n) \quad ; \quad R^2 = 0,825 \qquad (11)$$

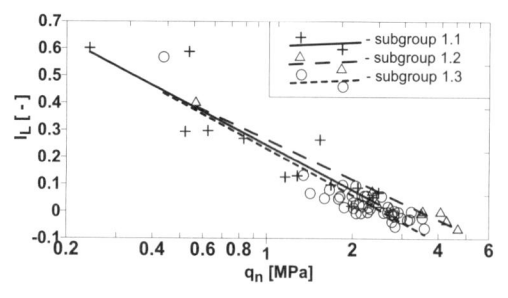

Fig.9 Liquidity index vs. net cone resistance for genetically differentiated very cohesive soils.

- for cohesive soils ($f_i = 20-30\%$) (Fig. 10)
  - solifluction tills, $CaCO_3 = 0\%$ - subgroup 2.1

$$I_L = 0,271 - 0,147 \ln(q_n) \quad ; \quad R^2 = 0,704 \qquad (12)$$

- the last glaciation tills; $CaCO_3 > 0\%$ - subgroup 2.2

$$I_L = 0,310 - 0,216 \ln(q_n) \quad ; \quad R^2 = 0,839 \qquad (13)$$

- earlier glaciations tills; $CaCO_3 > 0\%$ - subgroup 2.3

$$I_L = 0,375 - 0,254 \ln(q_n) \quad ; \quad R^2 = 0,796 \qquad (14)$$

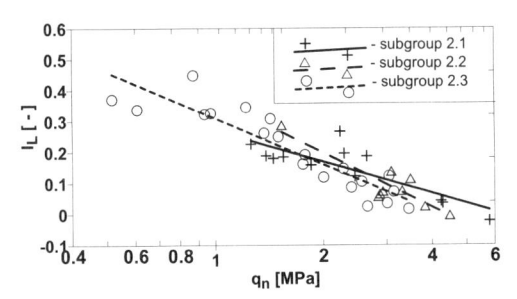

Fig.10 Liquidity index vs. net cone resistance for genetically differentiated cohesive soils.

- for mild cohesive soils (f_i=10-20%) (Fig. 11)
- solifluction tills, $CaCO_3$=0% - subgroup 3.1

$$I_L = 0{,}421 - 0{,}250 \ln(q_n) \; ; \; R^2 = 0{,}852 \quad (15)$$

- the last glaciation tills; $CaCO_3$>0% - subgroup 3.2

$$I_L = 0{,}500 - 0{,}333 \ln(q_n) \; ; \; R^2 = 0{,}837 \quad (16)$$

- earlier glaciations tills; $CaCO_3$>0% - subgroup 3.3

$$I_L = 0{,}344 - 0{,}238 \ln(q_n) \; ; \; R^2 = 0{,}791 \quad (17)$$

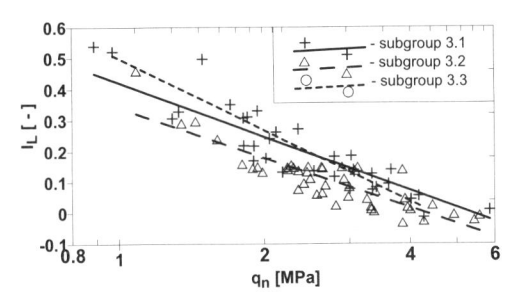

Fig.11 Liquidity index vs. net cone resistance for genetically differentiated mild cohesive soils.

The detailed equations for the genetically differentiated soils (equations 9-17) indicate great similarity of the function course, particularly within the soils of similar grain size distribution. Therefore, it would be justified to ask a question about statistical significance of additional divisions into subgroups. To answer this question a zero hypothesis was presented about lack of differences between the coefficients of regression. In order to verify this hypothesis the function should be transformed to linear shape by replacing the value of net cone resistance by a logarithmized value of this parameter, and then the regression coefficient of the estimation corresponds to the slope of a straight line. Statistical analysis indicated that in the case of very cohesive soils there is no basis for rejection of the zero hypothesis on the significance level α=0.05, which, in case of these soils leads to the conclusion about combining three analyzed subgroups into one common group with more favorable estimation of the regression coefficients. The most favorable estimation of the relationship is then given in the equation 6. In the group of cohesive soils the value of the statistics exceeded critical values of the test by comparing the first subgroup with the two remaining ones. Verification of the zero hypothesis indicated that for the first subgroup – solifuction tills, the best statistical solution is given in the equation 12, while the following equation for the remaining two subgroups is:

$$I_L = 0{,}312 - 0{,}210 \ln(q_n) \; ; \; R^2 = 0{,}853 \quad (18)$$

The idea of combining the second and the third subgroups is also supported by estimation of the correlation coefficient, more favorable than in the case when these groups were analyzed separately (equations 13 and 14).

In the group of mild cohesive soils the zero hypothesis was rejected at comparing regression coefficients between all three subgroups, proving statistical significance of discrimination from the group of: solifluction tills, the last glaciation tills and earlier glaciations tills. Therefore, a reliable solution for the group of mild cohesive soils was assumed to be division into subgroups (Fig. 11), for which the analyzed dependence is presented in the equations 15-17.

6 THE STATE OF CONSISTENCE OF THE SOILS WITH RESPECT TO UNDRAINED SHEAR STRENGTH

To determine undrained shear strength - S_u from CPTU results an empirical approach can be applied which is given in equation 19.

$$S_u = (q_t - \sigma_{V0}) / N_{kt} = q_n / N_{kt} \quad (19)$$

where: N_{kt} – empirical cone factor.

Comparison of functions which use cone resistance to determine liquidity index (equations 6-18) and undrained shear strength (equation 19) indicates that they are equal and in the case of a known value of cone factor - N_{kt}, it is possible to move from the state parameter to the strength one and the other way round. This idea seems justified on condition that the description of the consistency state of the soils includes, besides physical definition, criteria of strength division. Figure 12 presents such an idea, where to a traditional physical division into consistency states (Jumikis 1962) are attributed characteristic values of undrained shear strength (Anon 1995). The issue of mutual dependence between the discussed parameters and application to description of the consistency state of undrained shear strength is excellently presented by Figure 13. From the analysis of Fig. 13 results a very significant conclusion that cone factor is not a material constant and its value depends on many factors such as: stress history, plasticity index, grain size distribution, rigidity index, sensitivity. The determined dependencies describing liquidity index in the function of cone resistance are located within the variability of cone factor $N_{kt}=10 \div 25$, which corresponds to typical values of this parameter quoted in literature (Lunne et al. 1997).

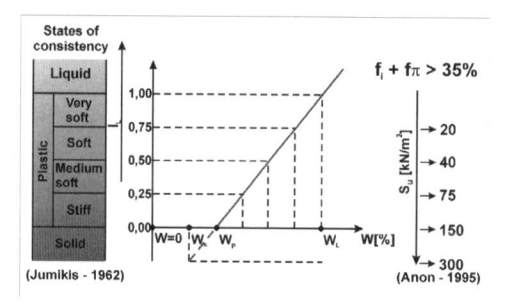

Fig.12 The idea of physical and strength description of consistency states of cohesive soils.

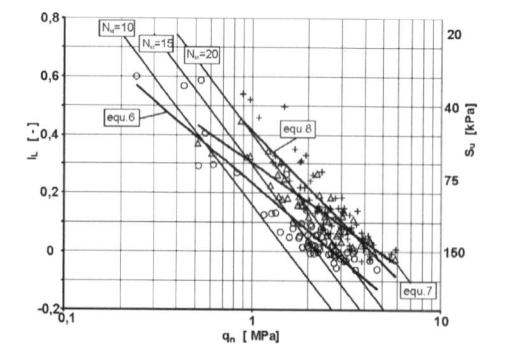

Fig.13 Relationship between parameters I_L and Su determined from cone resistance.

7 CONCLUSIONS

Detailed analysis of the study results enables giving a positive answer to the question about possibility of application of CPTU data to identify consistency state of cohesive soils. This statement implies a practical conclusion that on basis of CPTU results we obtain a favorable, continuous distribution of changes in the consistency state with depth determined at actual stress state in the subsoil, hence more realistic from evaluation of the consistency state from bore holes where this assessment has a local character, limited to the places of sampling. As a measure of the consistency state of soils can be assumed the liquidity index estimated from the value of the net cone resistance. The correlation dependence between these parameters is well described by a logarithmic function. The correlation significance is higher if in the evaluation of the cone resistance the effect of stress state in the subsoil and the effect of pore pressure on registered value of this parameter are taken into account.

The values of regression coefficients of the analyzed relationship also depend on grain size distribution and soil genesis. Statistical analysis indicated that the most favorable estimation of the dependence for very cohesive soils, irrespective of their origin, is given by equation 6, for cohesive soils, depending on their structure: equation 12 for solifluction tills, and equation 18 for the remaining tills. In case of mild cohesive soils, besides structural changes, a significant factor is also an overconsolidated ratio. In this group of soils a statistical significance of division into solifluction tills (equation 15), the last glaciation tills (equation 16) and earlier glaciations tills (equation 17) was found. Determined dependencies have high level of correlation whose measure is the correlation coefficient assuming the values from $R=0.84$ to $R=0.93$. In the case when the consistency state of the soils is defined by strength criteria, the dependencies describing the liquidity index and undrained shear strength are similar and at a given value of cone factor enable transfer from the state parameter to the strength one.

The results did not allow solution of all problems related to interpretation of penetration curves in cohesive soils, quite opposite, they exposed some issues which require further studies. Among them are:
- taking into account stress history of the subsoil in the evaluation by investigating the effect of OCR on statistical significance of the relationship
- working out an interpretation procedure of the penetration curves for anisotropic and non-homogeneous soils which would enable determi-

nation of adequate measures of mean penetration characteristics and which would constitute an unburdened, reliable parameter used in construction of correlation relationships
- supplementing determined correlation relationships with a group of cohesionless soils and organic ones.

REFERENCES

Anon, A. 1995. Identification and description of soils. *In situ Testing and Soil Properties Correlations; Proceedings International Conference on In Situ Measurement of Soil Properties and Case Histories. Bali, May 21-24, 2001*, Vol.2:2.1-5.

Jumikis, A.R. 1965. *Soil Mechanics*. New Yersey: D.Van Nostrand Company, Inc.

Lunne, T., Robertson P.K., Powell, J.J.M. 1997. *Cone Penetration Testing In Geotechnical Practice*. London: Blackie Academic and Professional.

PN-B-02479. Polish Standard. 1998. *Geotechnics – Ground investigation report – General rules*. Warsaw: PCN.

PN-81/B-03020. Polish Standard. 1981. *Building soils – Foundation bases – Static calculation and design*. Warsaw: PCN.

Robertson, P.K., Campanella, R.G., Gillespie, D., Greig, J. 1986. Use of piezometer cone data. *Use of In Situ Tests in Geotechnical Engineering; Proc. ASCE Speciality Conference In Situ'86. Blacksburg:* 1263-1280.

Tschuschke, W., Młynarek Z., Welling E. 2001. Interpretation of deep CPTU results. *Proc. International Conference on In Situ Measurement of Soil Properties and Case Histories. Bali, May 21-24, 2001*, Vol. 1: 655-659.

Volk, W. 1973. *Applied Statistics for Engineers*, 2nd Ed., New York: Mc Graw-Hill.

CPTU – replication test in post flotation sediments

Z. Młynarek & W. Tschuschke
Department of Geotechnics, August Cieszkowski Agricultural University Poznań, Poland

T. Lunne
Norwegian Geotechnical Institute, Oslo, Norway

Keywords: cone penetration test, statistical analysis, post flotation sediments

ABSTRACT: The paper presents the results of CPTU replication tests carried out on embankments of Żelazny Most dump. The precision of measured cone resistance, sleeve friction and pore pressure were evaluated. The analysis indicated that non homogeneity of post flotation sediments has greater effect on precision of friction sleeve and pore pressure than on cone resistance. The number of replications does not have a significant effect on precision of cone resistance or sleeve friction. In the replication test the change of precision of friction angle, evaluated from equations 4 and 5, is presented in Table 1.

1 INTRODUCTION

The static penetration method is at present used also in evaluation of strength and deformation parameters of soils which are known as " non textbook soils" (Lunne et al. 1997). A good example of this type of soils are mining wastes which are used in erecting earth constructions. Such is the Żelazny Most dump near Lubin, Poland, where post flotation tailings are transported to the dump crown and, after spreading on the beach, collected and built into embankments. High quality evaluation of geotechnical parameters of sediments constitutes a fundamental issue for calculations of embankment stability and making decisions about increasing its height. The problem is getting more important at present because the embankment height reaches locally 45.0 m and is planned to reach even 100 m.

The CPTU method has been used for prediction of strength and deformation parameters of post flotation sediments in the Żelazny Most dump since 1988 (Młynarek et al. 1995, 1997). By means of CPTU test, geotechnical parameters of the sediments are checked continually. A supplement of these tests are laboratory studies. Considerable height of the embankment, anisotrophy of the sediment macro-structure and great variability in grain size distribution (Pordzik et al. 2000) resulted recently in a question being raised about the precision of the parameters measured in CPTU test, which are later applied in determining shear strength and parameters of deformation characteristics (Młynarek, 2000).

The significance of this question is indicated by the fact that the Norwegian Geotechnical Institute and Department of Geotechnics, Agricultural University, Poznań, Poland, proved that precision of measuring CPTU parameters with various penetrometers is very differentiated (Lunne 2003). This paper presents an attempt at answering that question.

2 A PROCEDURE FOR DETERMINING PRECISION OF CPTU PARAMETERS

2.1 Study object

Replication penetration tests were carried out on a beach of the Żelazny Most dump. The tests were located in a square net at 1.5 m distance. Penetration was made to the depth of 25 m. Registration of the CPTU test parameter started from the depth of 8.0 m due to non homogenously compacted subsurface zone (Fig. 1) and according to the International Test Procedure for CPT and CPTU, Technical Committee 16 ISSMGE (2000).

2.2 Basis for theoretical analysis

Precision of the values measured in CPTU is decided by several factors, namely: the effect related to the quality of measuring equipment (cone, friction sleeve, system of registering pore pressure), the operator's effect, the effect related to interaction between the operator and measuring system of the

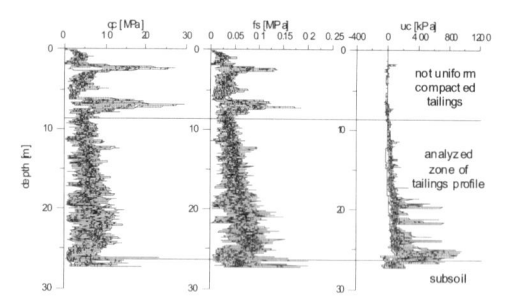

Figure 1. CPTU replication test profile.

penetrometer (Lee 1974, Młynarek 2003). Due to the fact that the tests were carried out with one type of penetrometer Hyson 20Tf by a.p. van den Berg (Holland) and one cone type, and that there was one operator, these factors were assumed to be constant. Variability of the parameters measured in the penetration process: cone resistance – q_c, friction sleeve – f_s, and pore pressure – u, in case of the post flotation sediments, is affected by the effect of trend related to change in stress state in the embankments with depth and variability of grain size distribution. Both factors appear together and can not be separated in the analysis.

To evaluate precision of the penetration parameters, a linear model was assumed:

$$Y_k(z) = T_k(z) + \varepsilon_k(z) \quad (1)$$

where:
$Y_k(z)$ – value of measured parameter with depth z,
$T_k(z)$ – value of trend at the depth z,
$E_k(z)$ – value of noise at the depth z,
while:
k – number of replication (k=1,...., 15).
With equation (1) it is possible to estimate trend of the parameters within the penetration depth range z=8, ..., 25m. To determine the trend measured parameters (q_c – every 2 cm) from all locations were used (Figs 2, 3, 4).

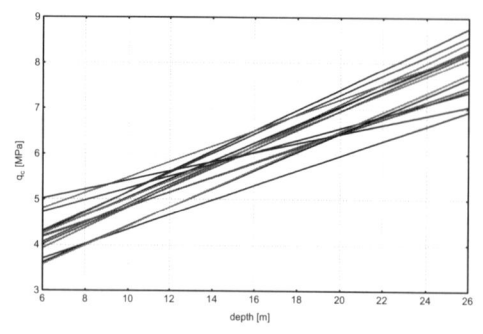

Figure 2. Regression functions for q_c at different locations.

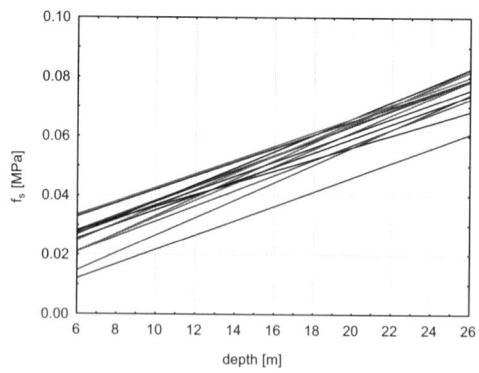

Figure 3. Regression functions for f_s at different locations.

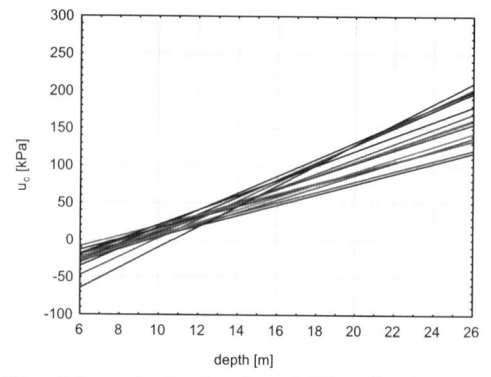

Figure 4. Regression functions for u_c at different locations.

As a measure of precision the following was assumed:

$$prec = mean\ (std(T_k(z))) \quad (2)$$

where:
$T_k(z)$ – value of trend at the depth z,
std ( ) – standard deviation,
mean ( ) – mean value.
The effect of variability of grain size distribution can be, in a sense, established by means of noise determination. Due to the fact that the sediments in the embankments can be assumed to be normally consolidated, the noise model was assumed to have identical form as the trend one.

The noise measure was determined from the dependence:

$$NOISE = \sqrt{1-p^2} \cdot \sigma \quad (3)$$

where:
ρ – autocorrelation coefficient,
δ – standard deviation from noise value.
The procedure of calculating the noise measure consists of several steps, namely: tapering of the

measured value of the parameter on every 2.0 m depth interval with a movable mean, noise determination, calculation of the standard deviation from the noise value, and determination of the autocorrelation function. An exponential function was assumed as a model of the autocorrelation function (Fig. 5).

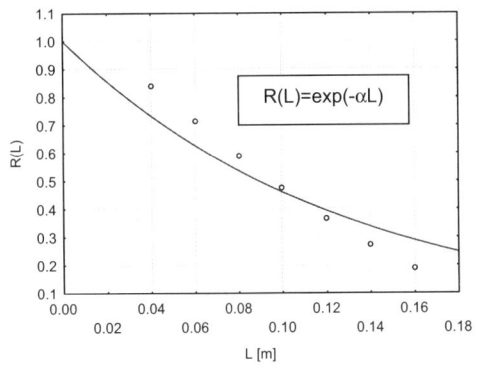

Figure 5. Autocorrelation function for q_c.

3 RESULT ANALYSIS

Fig. 5 presents adjustment of the autocorrelation function for cone resistance to assumed exponential model, while Figs 6, 7 and 8 give noise values for the CPTU test parameters. This figures show that the noise variability in each location is more non homogenous for f_s and u parameter than for q_c. This fact indicates that non homogeneity of grain size distribution in the sediments in each test point has greater effect on registering the friction sleeve and pore pressure than on registering cone resistance.

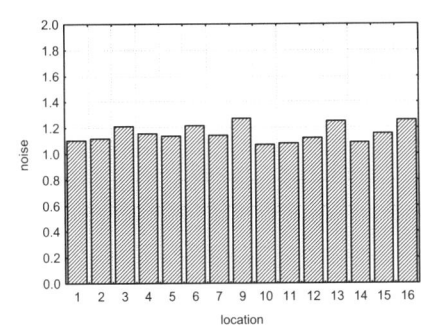

Figure 6. Noise for q_c at different locations.

Precision of the CPTU test parameters was determined as a global one for all replications of the CPTU. In order to determine the effect of stress state on the penetration parameters precision, the diagrams 9, 10 and 11 present a change in this precision with depth.

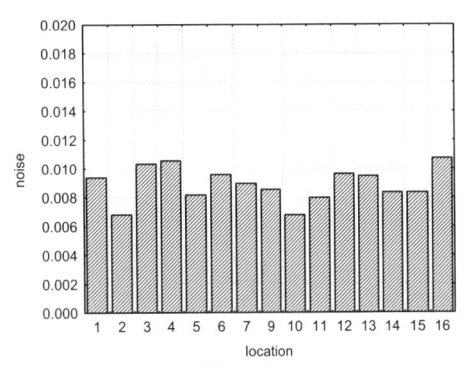

Figure 7. Noise for f_s at different locations.

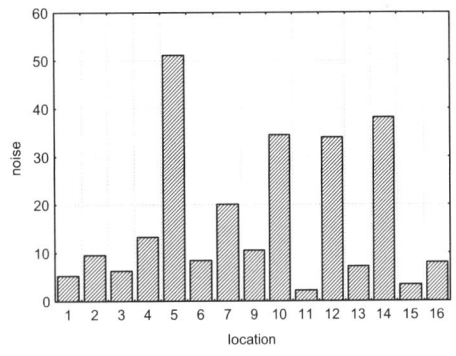

Figure 8. Noise for u_c at different locations.

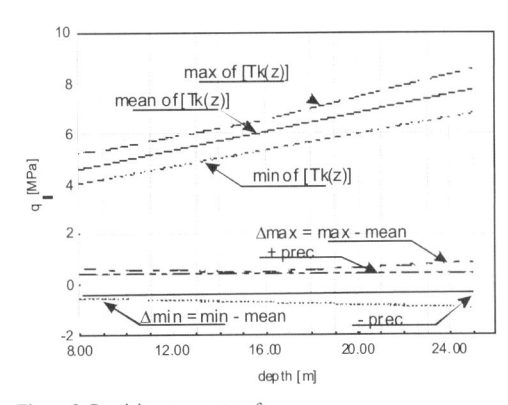

Figure 9. Precision parameters for q_c.

From economic point of view important is planning of the number of the CPTU tests to determine geotechnical parameters on a given or assumed precision level. This can be solved by determination of precision fluctuation depending on the number of

Table 1. The influence of number of replications on evaluation of precision of q_c, relative density and effective angle of friction of post flotation sediments.

Location	Mean (q_c) [MPa]	Precision of (q_c) [MPa]	Mean (I_D)	Precision of I_D	Mean (ϕ) [°]	Precision of ϕ [°]
1,4,13,16	6,40	0,362	0,631	0,011	34,57	0,108
6,7,10,11	5,92	0,397	0,616	0,013	34,42	0,132
13,14,15,16	6,05	0,458	0,620	0,015	34,46	0,151
1,3,5,7,10,12,13,15	6,12	0,447	0,622	0,014	34,48	0,143
1,2,3,4,5,6,7,9,10,11,12,13,14,15,16	6,14	0,387	0,622	0,012	34,49	0,123

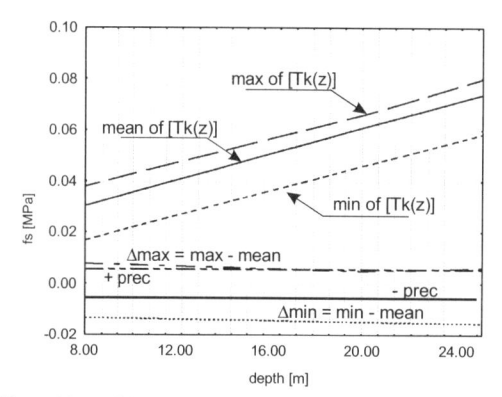

Figure 10. Precision parameters for f_s.

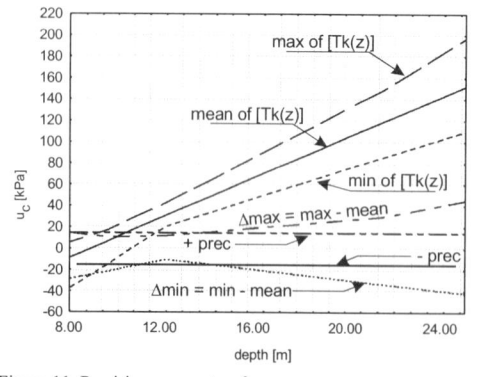

Figure 11. Precision parameters for u_c.

the CPTU tests. From Table 1 results that the number of replications does not significantly affect lowering or increasing precision of cone resistance. However, from this Table we can observe that location of replication tests can change the precision by about 20%.

4 EFFECT OF CONE RESISTANCE PRECISION LEVEL ON PREDICTION OF GEOTECHNICAL PARAMETERS OF THE SEDIMENTS

The strength and deformation parameters of post flotation sediments are obtained from the CPTU test parameters, as in structural soils, from empirical relationships. Therefore, as already it was pointed out, the quality of prediction of these parameters is affected by the precision of evaluation of the CPTU parameters..

In the zone of the CPTU replication tests the sediments with grain size distribution from fine sands to silt sands were built into the embankment. To design raising of the embankment the basic parameter is hence friction angle. Long term studies enabled construction of the empirical relationship which facilitates determination of friction angle in the sediments - ϕ' with the change in the relative density – I_D with depth (Młynarek, et al. 1998).

$$\phi' = 28{,}31 + 9{,}93 \cdot I_D \qquad (4)$$

$$I_D = -0{,}936 + 0{,}445 \log(q_c) - 0{,}10 \log(\sqrt{\sigma_{v0}}) \qquad (5)$$

where:
q_c – cone resistance (kPa),
σ_{v0} – overburden pressure (kPa).
This relationship is not given in dimensionless form, and therefore the units must be as shown. Table 1 gives the values and the change in the precision of the relative density (I_D) and friction angle (ϕ) determined with the equations 4 and 5. The effect of the number of replications and selection of the test point is similar as the changes in cone resistance precision. However, high precision of both relative density and friction angle from cone resistance and the dependencies 4 and 5, is noticeable. It is worthwhile noticing that the precision of the evaluation of the friction angle is very similar to that obtained for mineral sands (Jamiołkowski et al.1985).

5 CONCLUSIONS

The analysis indicated that the knowledge about precision of the measured CPTU test parameters during penetration of soils belonging to non text book group is of importance in predicting geotechnical parameters of the sediments. This importance should be considered on two planes, namely: assuming a safe level of the parameters for calculations of embankment stability, and reliability of the test itself as a method of studying so specific soil material. The precision level turned out to be very similar to that determined for this equipment at testing homogenous clay layer (Lunne et al. 2003), while the level of noise determined at each test point proved that non homogeneity and anisotrophy of the sediments are not as significant in evaluation of the geotechnical parameters as assumed before the analysis.

REFERENCES

Jamiolkowski M., Laad C.C., Germine J., Lancellotta R. 1985. *New developments in field and laboratory testing of soils.* State of Art Report. Proc. of 11th ICSMGE, San Francisco, Balkema, Rotterdam.

Lee J.K. 1974. *Soil Mechanics New Horizons*. Chap. Lumb P.: *Application of statistics in soil mechanics*. Butterworths.

Lunne T. Gauer P., Młynarek Zb., Wołyński W., Kroll M. (2003). *Quality of CPTU, Part II Statistical evaluation of differences between the CPTU parameters obtained from tests with various penetrometers*. Norwegian Geotechnical Institute Report, Oslo.

Lunne, T., Robertson P.K., Powell, J.J.M. 1997. *Cone Penetration Testing In Geotechnical Practice*. London: Blackie Academic and Professional.

Młynarek Zb. 2000. *Effectiveness of in situ tests in evaluation of strength parameters of postflotation sediments*. Geotechnics Conference 2000, Thailand.

Młynarek Zb. 2003. *Influence of quality of in-situ tests on evaluation of geotechnical parameters of subsoil*. Proc. 13th ECSMGE, Prague: 565-570.

Młynarek Zb., Tschuschke W., Lunne T. 1995. *Use of CPT in mine tailings*. Proc. of International Symposium On Cone Penetration Testing CPT'95. Linköping, Sweden.

Młynarek Zb., Tschuschke W., Welling E. 1998. *Control of strength parameters of tailings used for construction of reservoir dam*. Fifth International Conference on Tailings and Mine waste. Balkema, Colorado.

Młynarek Zb., Tschuschke W., Wierzbicki J. 1999. *Procedure for in situ evaluation of geotechnical parameters of high dams*. Fourth International Conference On Case Histories in Geotechnical Engineering. St. Louis, USA.

Pordzik P., Tschuschke W., Wierzbicki J. 2000. *Statystyczna ocena rozkładu uziarnienia osadów wbudowanych w zapory składowiska Żelazny Most*. Proc. of XII National Conference on Soil Mechanics and Foundation Engineering, Szczecin: 181-191.

Evaluation of the coefficient of subgrade reaction for design of multi-propped diaphragm walls from DMT moduli

P. Monaco & S. Marchetti
University of L'Aquila, Italy

Keywords: diaphragm walls, subgrade reaction method, flat dilatometer test

ABSTRACT: This paper presents the preliminary results of a numerical study on multi-propped walls retaining deep excavations. The study is based on the comparison between the results obtained by FEM (Plaxis) using the non linear Hardening Soil model and by the Subgrade Reaction Method, considering different wall / soil stiffness, excavation depth and prop spacing conditions. A tentative relation, "calibrated" vs FEM results, is proposed for deriving the coefficient of subgrade reaction Kh for design of multi-propped diaphragm walls from the constrained modulus M from DMT.

1 INTRODUCTION

Multi-propped or multi-anchored diaphragm walls are widely used today for retaining deep open pit excavations in urban areas (e.g. underground car parkings), often in combination with the "top-down" construction technique, in order to limit the deformations in the surrounding soil.

The design of multi-restrained retaining walls requires the use of methods of analysis which take into account the soil-structure interaction and permit to simulate the staged construction sequence. Though the Finite Element Method (FEM) approach is generally regarded today as the "way to the future", in common practice the simple and well-known Subgrade Reaction Method (SRM) or "spring method" is still widely used and often preferred to more sophisticated FEM analyses, particularly in the early stage of design. The SRM permits to model even relatively complex cases in a simple and quick way, providing in general sufficiently reliable values of stresses in the wall and supports. On the other hand the SRM has several drawbacks, deriving from the rough simplification assumed in simulating the response of the soil to wall movements. One critical shortcoming is the difficulty in evaluating the coefficient of subgrade reaction K_h on a rational base. K_h is by no means an intrinsic property of the soil. Its value depends not only on soil stiffness, but also on various "geometric-mechanical" factors (e.g. geometry and stiffness of wall/struts, excavation depth). Yet the influence of the above factors on K_h is not clearly understood. Hence indications for the selection of K_h values dependable for design may be helpful to many engineers who still rely on the "old" SRM for everyday practice.

This paper presents the preliminary results of a numerical study aimed at establishing tentative correlations for deriving K_h for design of multi-propped diaphragm walls from the constrained modulus M_{DMT} obtained from the flat dilatometer test (Marchetti 1980). Comparisons both in terms of M_{DMT} vs "reference" M and in terms of predicted vs measured settlements (see reference list in TC16 2001) have shown that, in general, M_{DMT} is a reasonably accurate "operative" modulus.

The study is based on the comparison between FEM and SRM results, according to the following procedure: (1) Analysis of the behavior of a multi-propped wall for various cases by FEM (Plaxis) using the non linear Hardening Soil model, with soil stiffness parameters estimated from M_{DMT}. (2) Analysis by SRM varying K_h until the results obtained by Plaxis for the same cases are appropriately reproduced. (3) Formulation of a tentative relation between backcalculated "best fit" K_h values (matching FEM results) and M_{DMT}.

The study is purely numerical. K_h values are "calibrated" based on FEM results, assumed herein to represent the "true" wall / soil behavior. No comparisons are shown between the behavior predicted by the models and observed in real cases.

2 SPRING MODEL VS CONTINUUM

In the spring model the soil is schematized by a set of independent horizontal springs, generally characterized by a bilinear elastic-plastic pressure-displacement relation (Fig. 1). The coefficient of subgrade reaction Kh (spring stiffness) is the initial slope of the curve until the limit pressure, active or passive, is reached. Hence a purely 2-D (plane strain) problem is converted into a 1-D problem. The soil behavior is "captured" by only one parameter, Kh. The simplest case of homogeneous isotropic linear elastic continuum requires a minimum of two parameters (E and ν, or G and K) to fully define it.

In general, it is not possible to establish a unique and straightforward correlation between K_h and soil stiffness E. Deriving K_h of the springs from E of the continuum involves trying to establish a link between parameters of different *models*. Yet engineers are familiar with moduli E, not with K_h, and many times even crude relations E to K_h may prove useful.

Typical ranges of K_h can be found in the literature, but great care is required owing to the problem-dependent nature of the parameter. For a given soil, values appropriate for strips, rafts, laterally loaded piles and flexible walls are all different (Clayton et al. 1993).

3 EXISTING K_h FORMULATIONS FOR WALLS

Various methods have been proposed for evaluating K_h for retaining walls (e.g. Terzaghi 1955, Ménard et al. 1964, Balay 1984, Becci & Nova 1987, Schmitt 1995, Simon 1995). Most formulations assume that K_h [F·L^{-3}] is directly proportional to the soil modulus E [F·L^{-2}]. In essence, almost all studies arrive to similar conversion

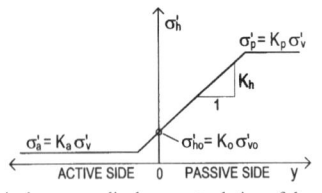

Fig. 1. Typical pressure-displacement relation of the springs

formulae $K_h \sim E/B$, where B is a dimension [L] which represents the width of the soil area "loaded" by the wall.

Most existing methods give indications for evaluating B for the simple mechanism of cantilever wall, generally suggesting to assume B proportional to the free cantilever height or embedded length. The behavior of multi-restrained walls is more complex and the estimate of B is uncertain, since the earth pressure distribution and the mode of deformation of the wall are not known *a priori*. E.g. multi-propped thick concrete diaphragm walls constructed by top-down technique, with basement floor slabs used as struts, generally exhibit a very stiff "box-type" behavior. The earth pressure may remain close to K_0 and multiple restraints permit only very small wall displacements. Hence K_h (ratio pressure/displacement) is expected to be higher, i.e. B lower, than for cantilever walls. FEM studies on propped and cantilever walls by Potts & Fourie (1984, 1986) and Fourie & Potts (1989) pointed out the enormous importance of the mode of deformation and displacement restraint on the earth pressure distribution and the resultant wall bending moments.

Various contributions on K_h given by French researchers are based on the original method by Ménard et al. (1964), which derives K_h over the embedded length of a cantilever wall from the pressuremeter modulus E_M:

$$K_h = E_M / [\alpha \cdot a/2 + 0.13 (9 a)^\alpha] \qquad (1)$$

This formula contains a dimensional parameter a (in m) related to wall geometry and a non dimensional factor α related to soil type. Ménard et al. (1964) assumed $a = 2/3$ of the embedded wall length (≈ distance between bottom of excavation and center of rotation at the toe of the wall). In practice a = height over which the soil is loaded by passive pressure. (Similar indications were given by Terzaghi 1955).

As reported by Amar et al. (1991), the pressuremeter modulus E_M is related to the oedometer modulus (in the same range of pressure) by the ratio $E_{oed} = E_M/\alpha$. For NC soils α varies between 1/3 in sands to 2/3 in clays (Ménard & Rousseau 1962). In principle, M_{DMT} (1-D modulus from DMT) can be used in Eq. 1 or derived formulae in place of E_M/α. (Various studies, e.g. Kalteziotis et al. 1991, Ortigao et al. 1996, Brown & Vinson 1998, have quoted similar ratios - generally ≈ 1/2 - between PMT and DMT moduli in different soils).

Balay (1984) adapted the Ménard formulation for evaluating K_h over the entire wall length, assuming $a = H$ (free cantilever height) above excavation level, while below the excavation a is related to the embedded length D and to the ratio D/H.

Schmitt (1995), based on the observation of different modes of deformation of stiff and flexible walls, adapted the Ménard formulation to take into account the flexural inertia of the wall EI, assuming $a \sim (EI/E_{oed})^{0.33}$ and $E_{oed} = E_M/\alpha$, thus obtaining:

$$K_h = 2.1 \, (E_{oed}^{4/3} / EI^{1/3}) \qquad (2)$$

According to the above relation, for a given soil modulus a stiff wall would have a lower K_h than a flexible wall. (Earlier studies on spread footings had indicated a lower influence of structure stiffness on K, e.g. Vesic 1961 found K inversely proportional to $EI^{1/12}$).

Simon (1995) extended the Ménard formulation adapted by Balay (1984) differentiating K_h for zones of "free" deformations (free height and embedded length of cantilever wall) and "restrained" deformations (zones between two props/anchors and behind a pretensioned anchor). For the zone between two props at distance L, assuming that a foundation of width B causes deformations over a length $L \approx 1.5\ B$, Simon (1995) proposed:

$$K_h = E_M / [0.13 (4.4 B)^\alpha + \alpha \cdot B/6] \quad (3)$$

The method by Becci & Nova (1987) differs from other spring methods for taking into account the non linear soil behavior, assuming the soil modulus E varying with stress level and loading path direction. K_h is evaluated as:

$$K_h = a \cdot E / L \quad (4)$$

E is assumed as the unloading-reloading modulus E_{ur} where the present stress level is lower than the maximum past stress level, as the virgin compression modulus when the maximum past stresses are exceeded. a is a non dimensional empirical factor (the Authors assume $a = 1$). As first approximation, L ($\approx$ width of soil zone involved by the wall movement) is assumed as an "average" width of the Rankine active and passive pressure wedges behind and in front of the wall (this leads to different K_h on the two sides).

4 FEM INPUT PARAMETERS

The geometry of the multi-propped wall selected for this study (Fig. 2) reproduces a typical configuration of common excavation works (e.g. car parking with 6 basement floors). The final depth of the excavation is 18 m. The total length of the wall is 24 m (embedded length 6 m). Six prop levels are equally spaced at 3 m intervals. The upper prop level is placed just at the top of the wall (ground surface). The half-width of the excavation is 20 m.

The finite element mesh used in Plaxis (plane strain analysis) is shown in Fig. 3. The wall was schematized as a "beam" element. The props were simulated by Plaxis "fixed-end anchor" elements (elastic springs of given axial stiffness, having one "movable" end connected to the wall and the other end "fixed" - zero displacement - at given longitudinal distance from the wall).

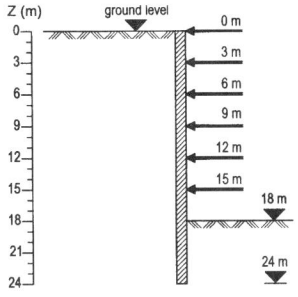

Fig. 2. Geometry of multi-propped wall

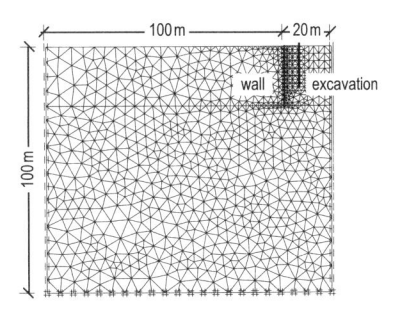

Fig. 3. Finite element mesh for multi-propped wall study

The soil behavior was simulated using the non linear Hardening Soil (HS) model of the Plaxis code. This model requires to input three stiffness moduli at the reference pressure $p^{ref} = 100$ kPa: the tangent oedometer modulus E_{oed}^{ref}, the triaxial modulus E_{50}^{ref}, the unloading- reloading modulus E_{ur}^{ref}. All moduli are stress-level dependent, according to the expressions $E_{oed} = E_{oed}^{ref} (\sigma'_1 / p^{ref})^m$, $E_{50} = E_{50}^{ref} (\sigma'_3 / p^{ref})^m$, $E_{ur} = E_{ur}^{ref} (\sigma'_3 / p^{ref})^m$, where σ'_1 and σ'_3 are the major and minor principal effective stresses. (Usually $m \approx 0.5$ to 1).

An intensive literature survey by Schanz & Vermeer (1997) indicated for remolded quartz sands (from very loose silty sands to very dense gravelly sands) $E_{50}^{ref} = 15$ to 75 MPa. The above range is remarkably similar to the range of M_{DMT} values found for sands of similar density. Schanz & Vermeer (1997) showed that E_{50}^{ref} is correlated to the 1-D modulus E_{oed}^{ref}. Hence, if one has data on the oedometer modulus, it can be used to estimate the triaxialmodulus. Based on the above considerations, in this study it was assumed $E_{oed} = M_{DMT}$, and all moduli required by the Plaxis HS model were estimated assuming M_{DMT} as the basic reference stiffness parameter. According to indications given by Plaxis researchers (Vermeer 2001), the following "typical" ratios between different moduli were adopted: $E_{50}^{ref} = E_{oed}^{ref}$, $E_{ur}^{ref} = 4 E_{oed}^{ref}$.

Two soil types were considered: a "soft soil" (e.g. a soft to medium clay) having $M_{DMT} = 4$ MPa and a "stiff soil" (e.g. a hard clay or a medium dense sand) having $M_{DMT} = 40$ MPa (at $p^{ref} = 100$ kPa). The above M_{DMT} values were selected to generally characterize a wide category of soft and stiff soils, and do not refer to any particular site. Having a real M_{DMT} profile from a specific site for a homogeneous soil, one could assume $E_{oed}^{ref} = M_{DMT}$ at $\sigma'_v = 100$ kPa. Also, the exponent m could be inferred from the rate of increase of M_{DMT} with depth.

The only difference between the "soft soil" and the "stiff soil" relates to stiffness. The "stiff soil" is simply a factor 10 stiffer than the "soft soil", but the relation $E_{50}/E_{oed}/E_{ur}$ is 1/1/4 for both soils. For both soils it was assumed $m = 0.5$ and $\nu_{ur} = 0.2$ (Poisson's ratio for unloading-reloading). All other parameters, in particular shear strength, were conveniently assumed equal for both soils: bulk unit weight γ = 19 kN/m³, friction angle Φ' = 30°, dilatancy angle ψ = 0, cohesion c' = 0.5 kPa, wall/soil friction angle δ = 20°. The wall was assumed to be installed without altering the in situ conditions ("wished in place"). The initial stresses were calculated assuming K_0 = 0.5.

The present study is restricted to fully drained conditions, with zero pore pressures everywhere. Seepage pressures have been ignored at this stage of the study, in an attempt to investigate the effects of earth pressures alone on the wall behavior.

Four different soil / wall stiffness combinations were considered (Table 1). The "rigid wall" is a 1 m thick reinforced concrete diaphragm wall, having Young's modulus E = 25000 MPa. The "flexible wall" is a steel sheetpile wall of similar structural capacity. Note that Case A and Case D have similar wall/soil stiffness ratios (EI/E_{oed}^{ref} = 58 and 52 respectively). For Case B EI/E_{oed}^{ref} = 521. For Case C EI/E_{oed}^{ref} = 5.8. Hence the wall/soil stiffness relation between the various cases is ≈ 1/10/100.

Very stiff props (30 cm thick concrete slabs, axial stiffness EA = 7500 MN/m) are associated to the "rigid wall" scheme. The "flexible wall" is supported by steel props of lower stiffness (EA = 657 MN/m). No prestress force was given to the props. The excavation sequence was simulated assuming that each prop level, starting from the uppermost one, is installed before excavating the 3 m soil layer immediately below, down to the next prop level (as in the top-down technique).

5 FEM RESULTS FOR MULTI-PROPPED WALL

The results of Plaxis analyses obtained for Cases A, B, C and D at the maximum excavation depth (18 m) are summarized in Fig. 4. Observations:

(a) The horizontal effective stresses developed on the back and on the front of the wall resulting from Plaxis calculation are shown in Fig. 4, compared to the distributions of at-rest, active and passive earth pressures (K_a and K_p calculated according to Caquot & Kerisel 1948). In general the full active condition behind the wall is not reached and σ'ₕ is intermediate between the initial K_0 and the K_a line. The passive pressure K_p is mobilized over ≈ 1/3 of the embedded wall length. Note that the earth pressure distribution is quite similar for Cases A and D, having similar wall/soil stiffness ratios.

(b) An "anomalous" earth pressure distribution results from the combination stiff soil/flexible wall (Case C). The progressive deflections of the wall towards the excavation promote the redistribution of σ'ₕ behind the wall in the embedded length and upper restrained zones (arching), while just above excavation level σ'ₕ decreases even below the "theoretical" minimum K_a. The maximum bending moment is much smaller than in all other cases, less than 50 % of the value calculated for the same soil in combination with the "rigid wall". A significant "fixity" moment at the toe of the wall is also observed in this case. (Similar results were obtained by Vermeer 2001 for a single-anchored wall of similar EI/E_{oed} ratio).

(c) The maximum horizontal displacements occur at ≈ 15-18 m depth, near the bottom of the excavation. The top of the wall does not move at all, due to the restraint provided by the upper props. This "deep" mode of deformation is opposite to the cantilever-type.

(d) The diagrams on the right in Fig. 4, inferred from Plaxis results, represent the variation of the earth pressure σ'ₕ on the back of the wall vs the horizontal wall displacement y at various depths (each curve refers to a given depth z, at 1.5 m intervals). To compare curves obtained at different depths, σ'ₕ is normalized to the vertical stress (earth pressure coefficient K = σ'ₕ / σ'ᵥ) and the horizontal displacement y is normalized to depth z (ratio y/z). The dashed horizontal lines represent the at-rest (initial) K_0 and the active K_a pressure coefficients. These curves correspond to the "active side" of the spring pressure-displacement curve in Fig. 1. (The attention is focused here on the "active side", more significant than the "passive side" for multi-propped walls, since their embedded length is generally small compared to the retained height). The figures show that, as the wall moves towards the excavation, σ'ₕ behind the wall varies differently from one case to another. In general K_a is not reached, except for Case C. In Cases A, C and D, after an initial decrease, σ'ₕ tends even to increase (well beyond K_0 in Case C) with wall displacement. (Again, the curves obtained for Cases A and D are quite similar). Only in Case B σ'ₕ decreases continuously. In general, however, the slope of the "normalized" curves obtained at various depths for each case is similar. Since the slope of these curves is equal to K_h of the springs (K_h = Δσ'ₕ /Δy) divided by the unit weight γ (assumed constant), this suggests that adopting K_h constant with depth in spring calculations should not involve a large error, at least as first approximation.

Table 1. Soil and wall stiffness parameters

	Soil stiffness E_{oed}^{ref} (MPa)	Wall stiffness EI (MNm²/m)
Case A soft soil/flexible wall	4	232
Case B soft soil/rigid wall	4	2083
Case C stiff soil/flexible wall	40	232
Case D stiff soil/rigid wall	40	2083

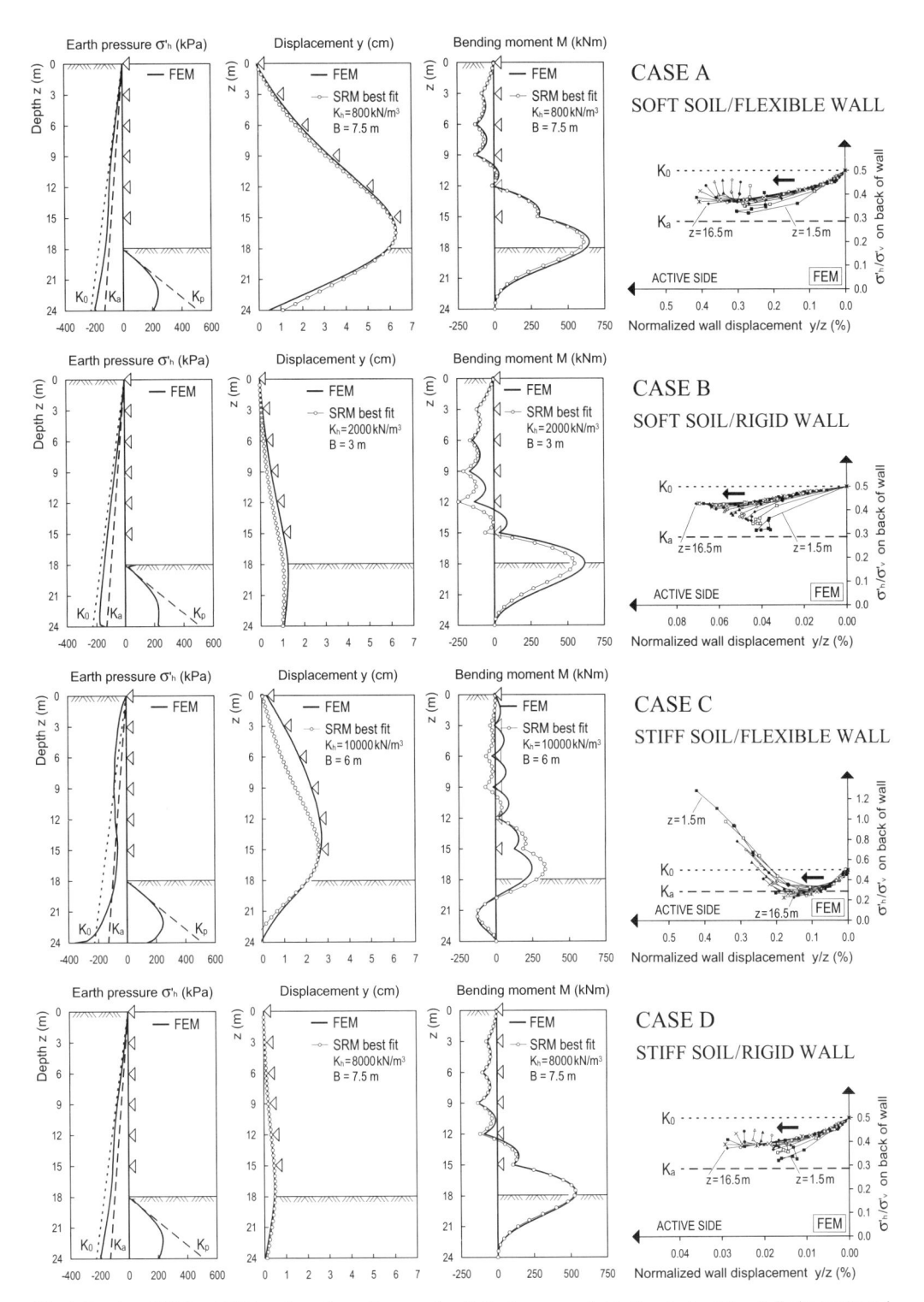

Fig. 4. Results of FEM and SRM analyses for multi-propped wall. Earth pressure distribution, horizontal wall displacements and bending moments at 18 m excavation depth. Normalized pressure–displacement curves on back of wall at 1.5 m depth intervals.

6 SPRING ANALYSIS RESULTS

The Subgrade Reaction Method was used to simulate the behavior of the multi-propped wall in Fig. 2, considering the same four cases analyzed by FEM. For each case SRM calculations were repeated varying Kh until the bending moments and the horizontal displacements of the wall were nearly equal to the values calculated by Plaxis.

As first approximation, K_h were assumed constant with depth and through all calculation steps, and equal on both sides of the wall. (A better estimate would involve K_h varying with depth and calculation sequence, e.g. K_h decreasing as excavation depth increases, and possibly different on the active and passive side).

The profiles of the horizontal displacements and bending moments calculated by SRM for the "best fit" K_h, i.e. the K_h values for which SRM results "match" FEM results (assuming the bending moment as "target" parameter), are shown in Fig. 4, superimposed to the Plaxis results. (The "best fit" K_h for matching horizontal displacements may differ slightly). Observations:

(a) The SRM reproduces well Plaxis results in Cases A and D, of "intermediate" wall/soil stiffness. The ratio between the "best fit" K_h calculated for the above two cases (K_h = 800 and 8000 kN/m³ respectively) is equal to the ratio between the corresponding soft/stiff soil moduli (1/10), confirming K_h proportional to soil stiffness.

(b) For Cases B and C, of "high" and "low" wall/soil stiffness respectively, the SRM is not able to reproduce Plaxis results with the same accuracy, even adopting different distributions of K_h with depth. This result is presumably a consequence of the limited ability of the spring model to simulate the soil behavior in particular conditions (e.g. in Case C arching cannot be simulated, since the spring would yield plastically as soon as K_a is reached and the earth pressure will never reach smaller values). The "best fit" backcalculated values are K_h = 2000 kN/m³ for Case B, K_h = 10000 kN/m³ for Case C.

The axial forces in the props calculated for the "best fit" K_h are generally in good agreement (± 10-20%) with the values calculated by Plaxis. Only in Case C the SRM largely underestimates (− 40-60 %) the forces in the upper prop levels.

The "best fit" K_h backcalculated from Plaxis results were compared to the values determined by various existing K_h formulations. The range of K_h obtained according to Eq. 4 by Becci & Nova (1987), assuming E = virgin compression modulus (not varying with stress, as first approximation), is similar to the "best fit" K_h. Also the K_h obtained according to Balay (1984) are not so far from the "best fit" K_h. The formulation by Schmitt (1995) overestimates K_h, giving K_h for the "rigid wall" less than 50 % of K_h for the "flexible wall", in contrast with the results of SRM-FEM comparison. The K_h obtained according to the formula by Simon (1995) for the inter-prop zone are also overestimated.

Hence, for the examined cases, existing K_h formulations taking into account wall/soil stiffness and restraint conditions do not reproduce the FEM-calculated behavior better than methods which do not take into account the above factors.

7 INFLUENCE OF EXCAVATION DEPTH AND PROP SPACING ON K_h

In order to investigate the influence of the excavation depth on K_h, FEM and SRM calculations for the multi-propped wall were repeated considering two different excavation depths, H = 12 m and H = 24 m. In both cases the embedded wall length was assumed equal to 6 m, as in the 18 m excavation previously considered.

For each excavation depth (H = 12, 18 and 24 m) the calculation was repeated assuming two different values of prop spacing, s = 3 m and s = 6 m.

This analysis was carried out only for the "rigid wall" (1 m thick concrete diaphragm wall).

Besides the "soft" and the "stiff" soil (E_{oed}^{ref} = 4 and 40 MPa respectively), a soil of "intermediate" stiffness (E_{oed}^{ref} = 16 MPa) was also considered. The above values of E_{oed}^{ref}, relative to a reference pressure p^{ref} = 100 kPa, correspond to average values of the constrained modulus M_{DMT} over the entire wall length equal to M_{DMT} ≈ 5-7 MPa for the "soft" soil, M_{DMT} ≈ 20-30 MPa for the "intermediate" soil and M_{DMT} ≈ 50-70 MPa for the "stiff" soil.

The "best fit" K_h resulting from SRM-FEM comparison for the examined cases are shown in Figs. 5 and 6. Fig. 5 shows that, for given soil stiffness M_{DMT} (average over total wall length) and prop spacing, K_h decreases significantly as the excavation depth increases from 12 to 18 m, but remains practically unchanged from 18 to 24 m. For given M_{DMT} and excavation depth, K_h decreases as prop spacing increases from 3 to 6 m, but such reduction is more pronounced for the lower depth (H = 12 m). In essence, K_h decreases as excavation depth and prop spacing increase, but the influence of both factors on K_h is smaller at higher excavation depths.

Fig. 6 shows the variation of K_h with M_{DMT} (average over total wall length) for different values of excavation depth and prop spacing. K_h increases with M_{DMT}, as expected. In most cases, for a given value of M_{DMT}, the values of K_h vary over a relative narrow range. Only in the case H = 12 m and s = 3 m (a nearly "zero displacement" restraint condition) K_h is significantly higher.

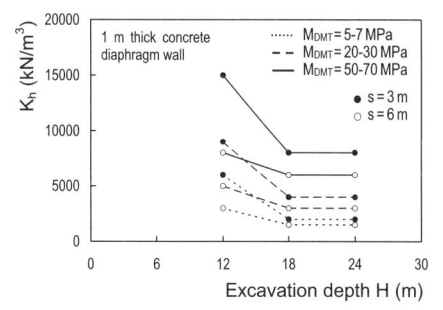

Fig. 5. "Best fit" K_h vs excavation depth H for different values of constrained soil modulus M_{DMT} and prop spacing s

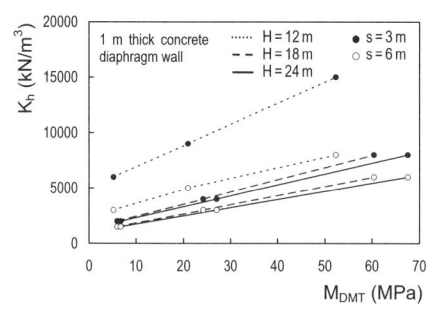

Fig. 6. "Best fit" K_h vs constrained soil modulus M_{DMT} for different values of excavation depth H and prop spacing s

8 TENTATIVE CORRELATION $K_h - M_{DMT}$

The results obtained for the examined cases indicate that K_h tends to decrease when the wall is subject to larger deformations as a consequence of deeper excavation or lower restraint degree, i.e. the width B of the soil zone involved by the wall movement increases. An attempt was made to interpret the above results based on a simple relation between K_h and the constrained soil modulus M_{DMT}, having the form (similar to existing K_h formulations):

$$K_h = M_{DMT}/B \quad (5)$$

The values of the "characteristic length" B obtained from Eq. 5 for the "best fit" K_h are plotted in Fig. 7 vs excavation depth, for three different ranges of M_{DMT} (average over total wall length) and two values of prop spacing ($s = 3$ m and $s = 6$ m). For a given M_{DMT}, B increases as excavation depth and prop spacing increase. For given excavation depth and prop spacing, B increases with M_{DMT}. For a "typical" wall configuration, say excavation depth $H = 18$ m and prop spacing $s = 3$ m, $B = 3$ m for $M_{DMT} = 5$-7 MPa, $B = 6$ m for $M_{DMT} = 20$-30 MPa, $B = 7.5$ m for $M_{DMT} = 50$-70 MPa. (The B values calculated for $H = 18$ m for Cases A, B, C and D are indicated in Fig. 4).

Fig. 7 may be used as a broad indication for selecting the values of B - hence K_h - for design of multi-propped diaphragm walls in cases of similar wall geometry / stiffness and prop spacing for different ranges of M_{DMT}.

For multi-propped walls it appears logical to link B to the retained excavation height H, not to the total or embedded length of the wall. In fact in this case, unlike for cantilever walls, the embedded length has a minor influence on the wall behavior and its value is not strictly dependent on wall equilibrium considerations. In the examined cases the ratio B / H was found ≈ 0.1-0.2 for $M_{DMT} = 5$-7 MPa, 0.2-0.4 for $M_{DMT} = 20$-30 MPa and 0.3-0.6 for $M_{DMT} = 50$-70 MPa.

9 COMPARISON WITH CANTILEVER WALL

To investigate the influence of restraints on K_h, the behavior of the "highly-restrained" multi-propped wall was compared to the behavior of a "fully-unrestrained" cantilever wall. FEM and SRM analyses were carried out for Cases A, B, C and D (Fig. 8), considering an excavation of 8 m depth. The embedded wall length (8 m) was established based on a conventional limit equilibrium analysis, assuming a factor of safety $Fs = 1.5$ on K_p. The excavation was simulated by 8 steps, each one corresponding to a 1 m excavation, to investigate in detail the initial part of the pressure-displacement curves. Observations:

(a) The full K_a condition is reached behind a large part of the wall in all cases, as expected, while σ'_h increases rapidly near the toe. K_p is mobilized over $\approx 1/4$ of the embedded length. This earth pressure distribution is consistent with the presence of a point of rotation at ≈ 6-6.5 m depth below excavation level.

(b) The "best fit" backcalculated values ($K_h = 800$-900 kN/m³ for the "soft soil", $K_h = 8000$-10000 kN/m³ for the "stiff soil") confirm the linear relation between K_h and soil modulus. For cantilever walls, the wall / soil stiffness has a minor influence on K_h. (In this case the "target" parameter of the SRM-FEM comparison was the wall displacement, since, as well known, for a cantilever wall the bending moments are practically not influenced by K_h).

(c) The K_h obtained for the cantilever wall for an excavation of 8 m depth are similar to the K_h found for the multi-propped wall for an excavation of 18-24 m depth.

(d) The "characteristic length" calculated based on Eq. 5 is $B \approx 5$-6 m, i.e. $\approx 2/3$-3/4 of the embedded length, very close to the values indicated by earlier K_h methods (e.g. Ménard et al. 1964). For cantilever walls, it appears appropriate to link B to the embedded wall length.

(e) The "normalized" pressure–displacement curves behind the wall at various depths (at 1 m intervals) calculated by Plaxis for the cantilever wall are significantly different from those obtained for the multi-propped wall (Fig. 4). σ'_h behind the wall decreases as wall displacement increases, and remains constant after K_a is reached. The initial part of the curves (shown in Fig. 8) is remarkably non linear. This behavior could be more closely simulated assuming for the springs a non linear pressure-displacement relation (e.g. similar to existing DMT-based formulations of $P–y$ curves for laterally loaded piles by Robertson et al. 1987, Marchetti et al. 1991), instead of the classical bilinear relation in Fig. 1. It is questionable, however, that this would lead to a substantially more accurate prediction of wall behavior, in view of the intrinsic approximation involved by the spring approach.

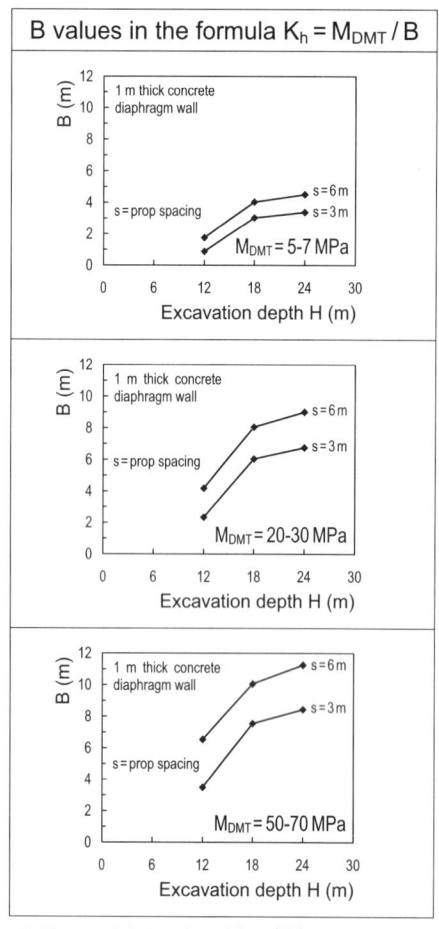

Fig. 7. Characteristic length B (M_{DMT}/K_h) vs excavation depth H for different values of M_{DMT} (average over total wall length) and prop spacing s

10 CONCLUSIONS

The Subgrade Reaction Method is widely used in current practice for design of multi-propped walls. Of course the crucial step is the selection of an appropriate value of the coefficient of subgrade reaction K_h. The numerical study illustrated in this paper investigates the influence of various factors on K_h (soil and wall stiffness, excavation depth, prop spacing). The possible K_h-DMT relationship is "calibrated" based on the results of FEM analyses carried out using the non linear Plaxis Hardening Soil model.

Several comparisons of DMT moduli vs "reference" moduli and DMT-predicted vs measured settlements (see e.g. TC16 2001) have shown that, in general, the constrained modulus M from DMT is a reasonably accurate "operative" modulus. Schanz & Vermeer (1997) indicated, for very loose to very dense sands, a range of values of the triaxial modulus E_{50}^{ref} (basic input parameter of Plaxis HS model, correlated to the 1-D modulus E_{oed}^{ref}) very similar to the range of M_{DMT} found for sands of similar density. In this study it was assumed $E_{oed} = M_{DMT}$, and all moduli required by the HS model were estimated assuming M_{DMT} as the basic reference stiffness parameter.

The results presented in this paper, though relative to a limited number of cases, indicate that K_h is proportional to soil stiffness and decreases as excavation depth and prop spacing increase, but its value tends to remain constant after a certain excavation depth is exceeded. The wall/soil stiffness largely influences the earth pressure distribution on the wall, but has a lower influence on K_h.

A tentative correlation between K_h and M_{DMT} is proposed, having the form $K_h = M_{DMT}/B$. Fig. 7 may be used as a broad indication for selecting the values of B - hence K_h - for design of multi-propped diaphragm walls in cases of similar wall geometry / stiffness and prop spacing for different ranges of M_{DMT}.

The B values ($B \approx 3$ to 8 m) obtained for "typical" multi-propped wall configurations (excavation depth 18-24 m, prop spacing 3 m) are similar to the values ($B \approx 5$-6 m) found for a cantilever wall for excavation depth 8 m.

Further investigations are needed to take into account different conditions and additional factors which may influence K_h (e.g. seepage pore pressures), in order to obtain indications for the selection of K_h in the general case.

Of course the results of this numerical study need to be validated based on field data from real cases.

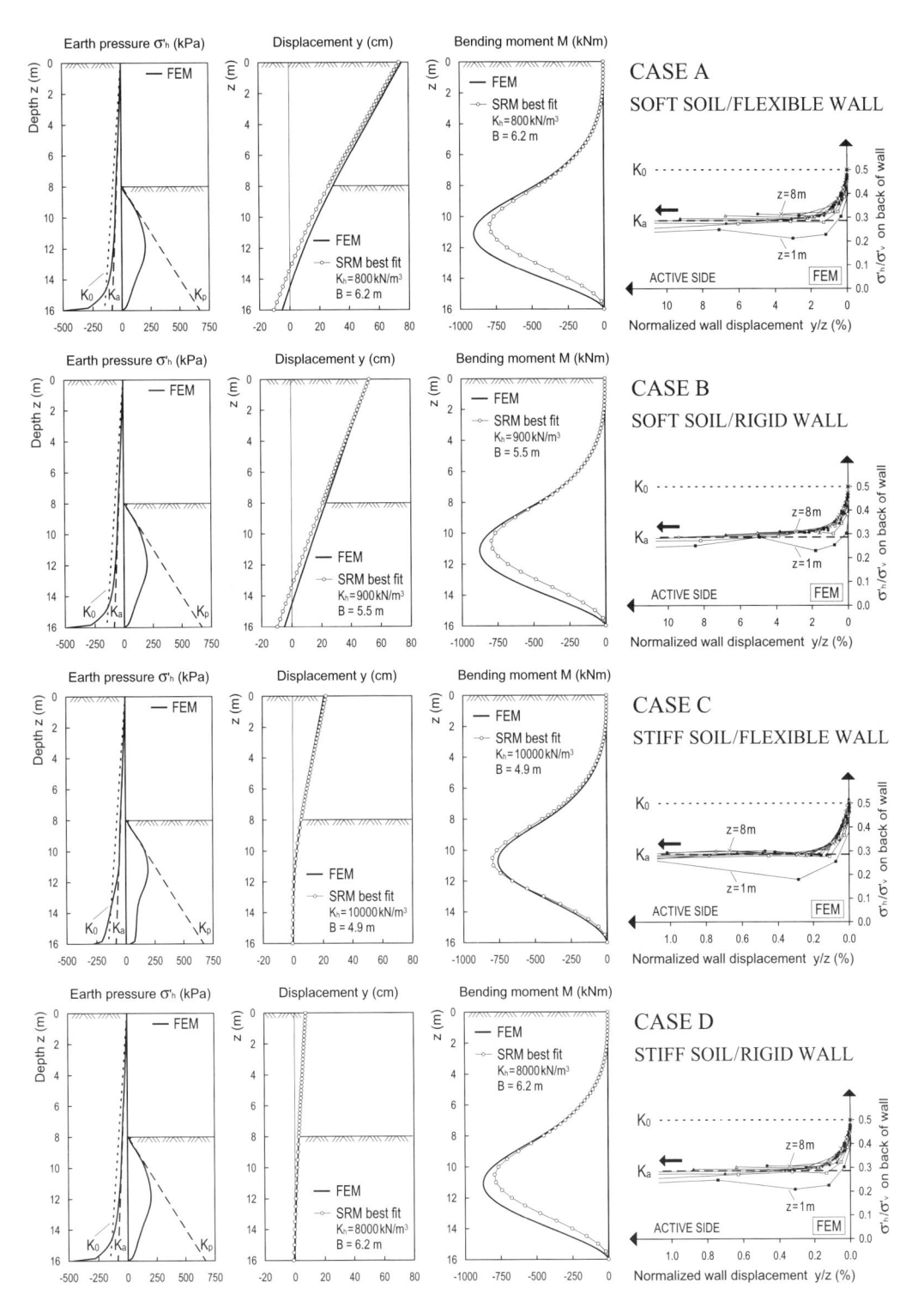

Fig. 8. Results of FEM and SRM analyses for cantilever wall. Earth pressure distribution, horizontal wall displacements and bending moments at 8 m excavation depth. Normalized pressure–displacement curves on back of wall at 1 m depth intervals.

REFERENCES

Amar, S., Clarke, B.G.F., Gambin, M.P. & Orr, T.L.L. 1991. The application of pressuremeter test results to foundation design in Europe. A state-of-the-art report by the ISSMFE European Technical Committee on Pressuremeters. Part 1: Predrilled pressuremeters and self-boring pressuremeters. A.A. Balkema.

Balay, J. 1984. Recommandations pour le choix des paramètres de calcul des écrans de soutènement par la méthode aux modules de réaction. *Note d'Information Technique, Laboratoire Central des Ponts et Chaussées.*

Becci, B. & Nova, R. 1987. Un metodo di calcolo automatico per il progetto di paratie. *Rivista Italiana di Geotecnica* 21, 1: 33-47.

Brown, D.A. & Vinson, J. 1998. Comparison of strength and stiffness parameters for a Piedmont residual soil. *Proc. 1st Int. Conf. on Site Characterization ISC'98, Atlanta,* 2: 1229-1234.

Caquot, A. & Kerisel, J. 1948. Tables for the Calculation of Passive Pressure, Active Pressure and Bearing Capacity of Foundations. Gautiers-Villars, Paris.

Clayton, C.R.I., Milititsky, J. & Woods, R.I. 1993. Earth pressure and Earth-retaining Structures. Chapman & Hall.

Fourie, A.B. & Potts, D.M. 1989. Comparison of finite element and limiting equilibrium analyses for an embedded cantilever retaining wall. *Géotechnique* 39, 2: 175-188.

Kalteziotis, N.A., Pachakis, M.D. & Zervogiannis, H.S. 1991. Applications of the Flat Dilatometer Test (DMT) in Cohesive Soils. *Proc. X ECSMFE, Florence,* 1: 125-128.

Marchetti, S. 1980. In Situ Tests by Flat Dilatometer. *ASCE Jnl GED,* 106, GT3: 299-321.

Marchetti, S., Totani, G., Calabrese, M. & Monaco, P. 1991. P-y curves from DMT data for piles driven in clay. *Proc. 4th DFI Int. Conf. Piling and Deep Foundations, Stresa,* 1: 263-272.

Ménard, L., Bourdon, G. & Houy, A. 1964. Étude expérimentale de l'encastrement d'un rideau en fonction des caractéristiques pressiométriques du sol de fondation. *Sols Soils* 9.

Ménard, L. & Rousseau, J. 1962. L'évaluation des tassements, tendances nouvelles. *Sols Soils* 1.

Ortigao, J.A.R., Cunha, R.P. & Alves, L.S. 1996. In situ tests in Brasilia porous clay. *Canad. Geot. Jnl,* 33, 1: 189-198.

Potts, D.M. & Fourie, A.B. 1984. The behaviour of a propped retaining wall: results of a numerical experiment. *Géotechnique* 34, 3: 383-404.

Potts, D.M. & Fourie, A.B. 1986. A numerical study of the effects of wall deformation on earth pressures. *Int. Jnl for Num. and Analytical Methods in Geomech.* 10: 383-405.

Robertson, P.K., Davies, M.P. & Campanella, R.G. 1987. Design of Laterally Loaded Driven Piles Using the Flat Dilatometer. *Geot. Testing Jnl,* 12, 1: 30-38.

Schanz, T. & Vermeer, P.A. (1997). On the Stiffness of Sands. *Proc. Symp. Pre-failure Deformation Behaviour of Geomaterials, ICE, London*: 383-387.

Schmitt, P. 1995. Méthode empirique d'évaluation du coefficient de réaction du sol vis-à-vis des ouvrages de soutènement souples. *Revue Française de Géotechnique* 71: 3-10.

Simon, B. 1995. Commentaires sur le choix des coefficients de réaction pour le calcul des écrans de soutènement. *Revue Française de Géotechnique* 71: 11-19.

TC16. 2001. The Flat Dilatometer Test (DMT) in Soil Investigations - A Report by the ISSMGE Committee TC16.

Terzaghi, K. 1955. Evaluation of coefficients of subgrade reaction. *Géotechnique* 4: 297-326.

Vermeer, P.A. 2001. On single anchored retaining walls. *Plaxis Bulletin* 10.

Vesic, A.B. 1961. Bending of beams resting on isotropic elastic solid. *ASCE Jnl Engineering Mech. Div.* 87, EM2: 35-53.

Evaluation of SCPTU intra-correlations at sand sites in the Lower Mississippi River Valley, USA

James A. Schneider
The University of Western Australia, Crawley, WA, Australia

Alec V. McGillivray and Paul W. Mayne
Georgia Institute of Technology, Atlanta, Georgia, USA

Keywords: sand, SCPTU, cone penetration test, shear wave velocity, maximum shear modulus, variability

ABSTRACT: A number of correlations exist between penetration resistance in sands and stiffness parameters, such as shear wave velocity, maximum shear modulus, and secant Young's Modulus. Those relationships between large- and small-strain properties have been shown to be influenced by overconsolidation ratio and ageing of a deposit, but also by grain characteristics. Since the seismic piezocone penetration test provides four independent measurements within a single sounding, including penetration resistance and downhole shear wave velocity, correlations between V_s or G_{max} and q_t can be evaluated from the same sounding. This study presents seismic piezocone penetration test data from relatively thick sand deposits of the Lower Mississippi River Valley in the United States. Those data are then compared to existing correlations between tip resistance and initial shear modulus to evaluate potential effects of density variations, particle characteristics, effective stress state, age of deposit, and parameter normalization.

1 INTRODUCTION

1.1 *Seismic piezocone testing*

The seismic piezocone test (SCPTU) measures the four parameters of tip resistance (q_t), sleeve friction (f_s), penetration pore pressures (in this study located immediately behind the tip, u_2), and a downhole shear wave arrival time for calculation of near-vertically propagating horizontally-polarized shear wave velocity ($V_{s,vh}$). The seismic cone has been in use for over two decades (i.e., Robertson, 1982), but is still typically not used in routine practice, even in areas of high earthquake activity where site-specific profiles of shear wave velocity are required for design, such as California, USA.

Comparison of SCPTU data from a single sounding eliminates additional variability in correlations induced by drilling, sampling, and disturbance prior to lab testing, as well as spatial variability of soils properties. In this study, the measurement of shear wave velocity will be compared to the parameter of cone tip resistance, but first the initial shear modulus (G_{max}) will be calculated as a function of shear wave velocity and total mass density (ρ_{tot}) as $G_{max} = \rho_{tot} \cdot V_s^2$. The parameters of G_{max} and q_t are both expressed in units of stress, so analyses can be normalized to eliminate the need for specific units tied to the equations.

1.2 *Controlling factors for G_{max} and q_t*

Many soil type-dependent correlations exist between G_{max} or V_s and cone tip stress q_t, but the strain incompatibility of measurements of the two parameters brings into question the reliability of those correlations.

Maximum shear modulus in sands is controlled by the number of contacts over a certain area, and the area of those contacts. As the void ratio decreases, the number of contacts within a certain volume increases, and thus the shear wave velocity and shear modulus should increase. As confining stress increases, the area of the particle contacts increase, and the shear modulus should increase as well (Santamarina & Aloufi, 1999). With ageing, the surface area of particle to particle contacts may creep and continue to increase in area, or possibly bond, thus increasing shear modulus. Alternatively, or in addition to the previously mentioned ageing effects, structural re-arrangement may occur, where weak force chains break down and are replaced by strong force chains (Bowman, 2002). Cementation effectively increases the area of the particle contacts, which should increase the shear modulus, but this is dependant on stress level, whether the soils is undergoing loading or unloading, and whether cementation occurred before or after loading and unloading periods (Rinaldi & Santamarina, 2003). Depending upon direction of wave propagation and particle

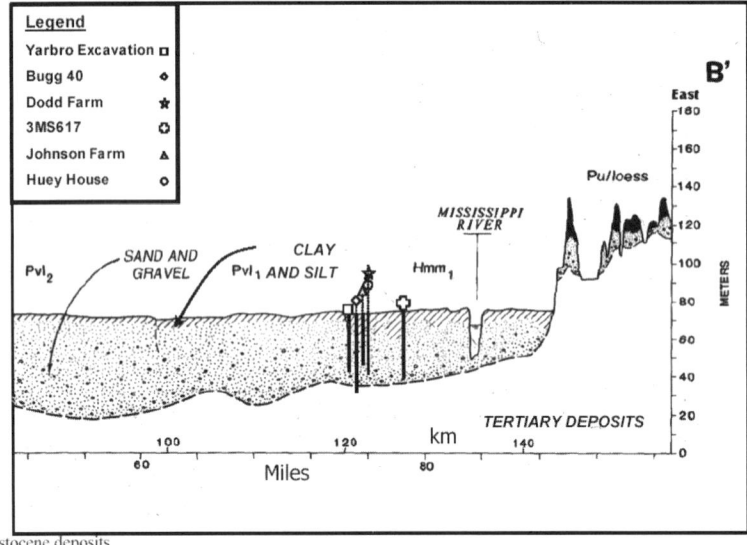

Figure 1. Generalized cross section through Mississippi River valley in NE Arkansas and SE Missouri with approximate locations and depths of seismic piezocone soundings [adapted from Saucier (1994)].

motion, anisotropic stress state and anisotropic soil fabric also influence measured shear wave velocities (Fioravante et al., 1998).

Cone tip resistance in sands is controlled by a complex interaction of phenomena as sand dilates, contracts, and rotates around a penetrating cone. These phenomena have been illustrated for pushed-in model piles, which are analogous to cone penetrometers, in siliceous and carbonate sands using image analysis techniques by White & Bolton (2004). Since cone penetration is a large strain process, particle behavior is no longer dominated by small strain behaviour at the contacts, and soil behavior is controlled by the lateral confinement and strength characteristics. Those characteristics are generally dependent upon relative density and horizontal stress state, which in turn influence shear modulus and a dilational component. That behavior has been assessed quantitatively using experimental data (e.g., Houlsby & Hitchman, 1988) and through analytical methods (e.g., Salgado et al., 1997).

Maximum shear modulus and cone tip resistance are controlled by very different phenomena. Since both parameters are influenced to some degree by the number of particles within a certain volume of soil and effective stress state, there could be some correlation between the two parameters, particularly in unaged normally consolidated deposits.

1.3 *Regional geologic overview*

The Mississippi River Valley has been formed by erosion of sediment by flow of glacial meltwater, and subsequent deposition of glacial outwash (Saucier, 1994). The upper 5 to 10 m of soil is predominantly recent deposits of clays and silts from the meandering path of the Mississippi River (Saucier, 1994). Consequently, formation of the alluvial valley has occurred in the Holocene.

Figure 1 shows a generalized cross section of the Mississippi River Valley, with the approximate location of six SCPTU soundings performed in Northeast Arkansas and Southeast Missouri. The cross section shows the upper 3 to 5 m filled in with clay and silt by recent meandering of the Mississippi River, and underlain by thick deposits of sand with some gravel to approximately 30 or 35 m depth. At 30 to 35 m depth, Tertiary deposits are encountered in the area where testing was performed. It is noted that a sounding at Bugg-40 penetrated into these Tertiary deposits, which were characterized as heavily overconsolidated clays for the final 5 m. Based on regional geology, it is inferred that the soil deposits tested are primarily Holocene age, but may be located near an interface between Holocene and Pleistocene deposits. Therefore, depending upon the thickness and lateral extent of scour during waning of the Wisconsin glaciation, some deposits tested may be of Pleistocene age. Since glaciation did not reach the lower Mississippi River valley, those sand deposits are considered to be predominantly normally consolidated.

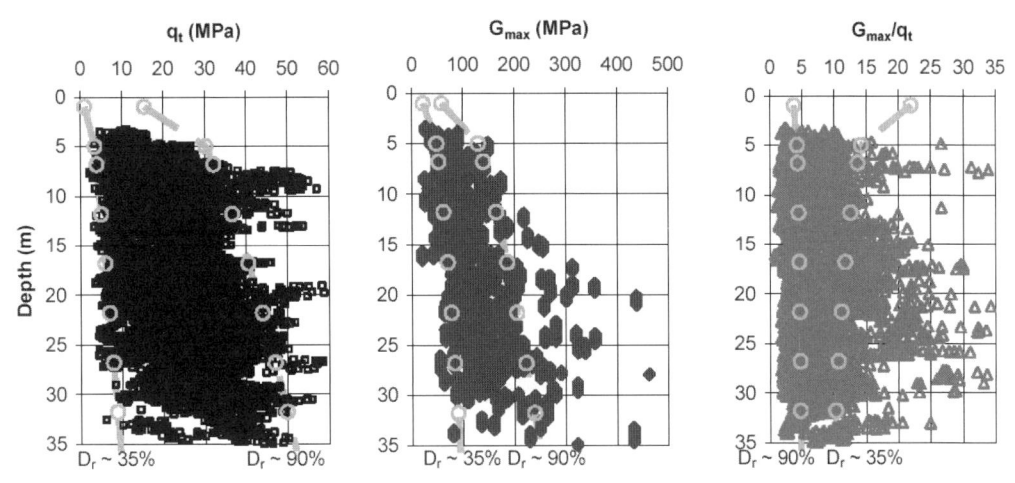

Figure 2. Profiles of all data with depth bounded by empirical relations to effective stress and relative density of 35 and 90 percent.

2 DATABASE DEVELOPMENT AND SOIL PROFILES

2.1 Overview of SCPTU data

SCPTU intra-correlations are reviewed relative to a database of layers selected from relatively thick (up to about 35 meter) Holocene sand deposits of the Lower Mississippi Valley, USA. The database consists of more than 90 layers from over 30 soundings at over 20 sites, within an area of approximately 5,000 km^2 along the Mississippi river. The layers are compiled from approximately 18,000 q_t measurements recorded at 0.05 or 0.025 m intervals, and approximately 500 corresponding shear wave velocity measurements recorded at 1 m intervals. Only saturated sands are included in the database, to eliminate the additional variability in shear wave velocity and cone tip resistance that may arise in partially saturated conditions.

All data points averaged into layer analyses of relationships between stiffness and cone tip resistance are shown on Figure 2, along with a profile of G_{max}/q_t with depth for each data point. The data cover a wide range of tip resistance values over all depths, indicating quite a heterogeneous depositional environment. Included in Figure 2 are the relative profiles from stress-level dependent empirical correlations corresponding to relative density values of approximately D_R = 35 percent and 90 percent, based on methods in Kulhawy & Mayne (1990) for q_t, and methods presented in Seed et al (1984), for G_{max}. The deposits were assumed to be normally consolidated (based on geologic history) for assessment of anisotropic stress state within the correlations.

2.2 Layer selection

When developing this database, consideration was given as to how to characterize each point. Including G_{max}/q_t in the database for every value of q_t is not appropriate, since that would associate one value of shear stiffness with approximately 40 values of q_t. This would bias the database towards the shape of the normalization function. It was also believed that averaging over the one-meter V_s intervals may become strongly influenced by layering of the deposit. Additionally, arrival time errors induced by analysis using the pseudo interval downhole method tend to average out over thicker layers. Therefore, representative layers, with a median thickness of approximately 4 m were used to develop the database.

Representative layer selection was based on tip resistance, friction ratio, excess pore pressure, and G_{max}/q_t ratio. A sharp discontinuity in any of those parameters, or noticeable change in slope of parameter versus depth, was considered as a new layer. An example profile separated into eight representative layers (dashed horizontal lines) is shown in Figure 3. Three saturated sand layers from that profile were added to the database; (i) 11.9 to 15.75 m; (ii) 15.75 to 23.4 m; and (iii) 23.4 to 32.4 m.

3 EVALUATION OF G_{max}-q_c CORRELATIONS

3.1 Previous studies

Correlations presented by Baldi et al. (1989), Rix & Stokoe (1991), Robertson et al. (1992), Fear & Robertson (1995), Hegazy & Mayne (1995), Andrus et al (2001), and Fahey et al. (2003) were reviewed under this study. Stress normalized behavior was

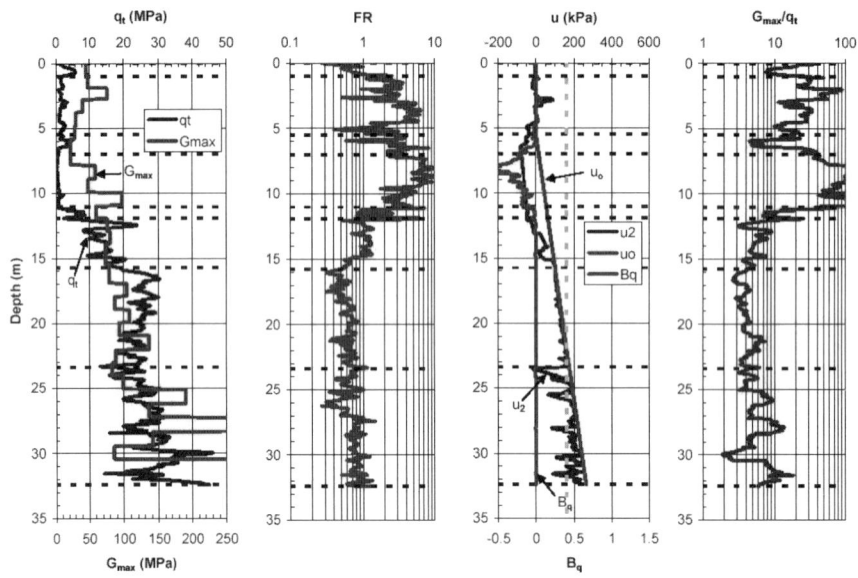

Figure 3. Profile of SCPTU parameters and layer selection for 3MS617 paleoliquefaction site

similar for each correlation, with slight changes in leading coefficients causing relatively linear offsets of the results. The work of Hegazy & Mayne (1995) also included CPT sleeve friction in the correlation to account for material differences.

Correlations in Rix & Stokoe (1991) and Fahey et al. (2003) present bounds of approximately ± 50 percent from the mean for G_{max}/q_t to q_t correlations, with a majority of field data in Holocene siliceous sands falling between the mean and mean minus 50 percent. Data lying between the mean and mean plus 50 percent trend typically includes aged and cemented sands, sands with high lateral stress, calcareous sands, and some siliceous sands. Andrus et al (2001) compare normalized parameters, but the selected normalization functions based on vertical effective stress do not account for the difference in how stress anisotropy influences q_t and $G_{max,vh}$. Since correlations discussed previously produced relatively similar values of variance and trends with increasing stress, field data from the Mississippi River Valley is presented relative to correlations of Rix & Stokoe (1991) and Andrus et al. (2001).

Rix & Stokoe (1991) compared G_{max}/q_t to normalized cone tip resistance. A leading coefficient, expressed here as K_G, was modified from the original form by a factor of 10 to incorporate q_{t1N}, $[q_{t1N}=(q_t/p_a)\cdot(p_a/\sigma'_{vo})^{0.5}]$ where p_a is atmospheric pressure in the same units as σ'_{vo} and q_t, rather than use the original units in kPa. The correlation as used is presented below:

$$G_{max}/q_t = K_G \cdot (q_{t1N})^{-0.75} \qquad (1)$$

where K_G is equal to 215, and was selected as the midpoint between the mean and mean minus 50 percent of the originally recommended trends, based on observations of larger field databases in Fahey et al. (2003).

The Andrus et al. (2001) correlation compares the stress-normalized shear wave velocity ($V_{s1} = V_s/(\sigma'_{vo}/p_a)^{0.25}$, with V_s in m/s)) to q_{t1N} as:

$$V_{s1} = K_V \cdot (q_{t1N})^{0.178} \cdot ASF \qquad (2)$$

where K_V is a leading coefficient equal to 77.4 and the ASF (age scaling factor) is equal to unity for Holocene soils and 1.41 for Pleistocene soils. An age scaling factor could be applied to the Rix & Stokoe (1991) correlation as well. Since G_{max} is a function of shear wave velocity squared, the ASF would also be squared, and equal to two.

3.2 G_{max} estimated from V_s

To calculate G_{max} from shear wave velocity, a correlation between V_s (in m/s), depth, z (in m), and total mass density, ρ_{tot} (in g/cc), was used (Mayne et al., 1999):

$$\rho_{tot} = 1+1/[0.614+58.7\cdot\{\log(z)+1.095\}/V_s] \qquad (3)$$

(n=727; $r^2 = 0.730$). This correlation performs quite well below the water table for deposits on land, but seems to be slightly shifted for use at offshore and nearshore sites, due to depth normalization rather than effective stress normalization.

3.3 Stress state and normalization

When evaluating the stress-normalized parameters of cone tip resistance and shear wave velocity in sands, often the only state parameter known with any degree of certainty is the effective vertical stress. So, it is convenient to normalize cone tip resistance in clean sands by effective vertical stress (divided by atmospheric pressure) raised to an exponent, n, that is often equal to 0.5. The resulting normalized quantity, q_{c1N} or G_{max1}, is primarily considered a function of relative density or void ratio, grain characteristics, and soil fabric, but what is often not discussed is the additional uncertainty induced by the normalization procedure itself. In natural soils, ageing, cementation, variation in horizontal stress state, stress history, and anisotropy, will all influence the actual normalized behavior.

The use of variable stress exponents has been proposed for normalization of CPTU data (Olsen & Mitchell, 1995). The results have been based on field data in sand and clay deposits that associate lower stress exponents (0.35 and 0.15) with overconsolidated and dense to very dense or stiff to hard deposits, and higher stress exponents (0.75 and 1.0) with very loose or soft deposits. In sands, the change in stress exponent is likely related to the decrease in dilation with increasing effective confining stress, as proposed by Houlsby & Hitchman (1988), but is also likely influenced by decreases in OCR and K_0 with depth, as well as cementation effects.

For normalization of small strain shear modulus, there is no influence of a dilation component, but cementation and stress history effects are likely to reduce the stress exponent to approximately 0.05 until the vertical stress exceeds the influence of the cementation (Rinaldi & Santamarina, 2003). Based on controlled laboratory experiments in sands, the stress exponent for shear modulus has been shown to be a function of the stress in the direction of wave propagation $(\sigma'_a)^{na}$ and the stress in the direction of particle motion $(\sigma'_b)^{nb}$, with the sum of na and nb typically equal to 0.5 and a ratio of na to nb of 1.0 to 1.5 (i.e., Fioravante et al., 1998). For the seismic cone test, σ'_a is equal to the vertical effective stress, and σ'_b is equal to the horizontal effective stress, so the normalization to $\sigma'_{vo}{}^{0.5}$ should be less influenced by a variation in K_0 than normalization for q_t, which may lead to a stress dependency in G_{max}/q_t ratios. That stress/depth dependency can be seen from the empirical trends in Figure 2, as G_{max}/q_t reduces from approximately 15 at 5 m to 10 at 30 m for the loose profile and increases from 4 at 5 m to 5 at 30 m for the dense profile.

Figure 4 presents variation in the q_t normalization function ($1/\sigma'_{vo}{}^n$) as a ratio to standard normalization based on $\sigma'_{vo}{}^{0.5}$. Potential errors between 5 and 20 m are generally up to about 20 percent, but may reach 50 percent or more at depth. If a constant K_0 is assumed, error is constant with depth, but soil deposits often do not have a constant K_0 with depth. With a constant error, no depth bias will be observed for individual measurements, and K_0 changes can be accounted for by modifying the leading coefficient, K, as shown for OCR from 1 to 10 by Baldi et al. (1989).

For an overconsolidated deposit modeled as a profile with a constant preconsolidation stress, p'_c, of 4 atmospheres, normalization error will change with depth. The influence of constant preconsolidation stress matches the trend of a stress exponent of 0.15 applied to vertical effective stress relatively well. Additionally, the use of the lower stress exponents approach is an appropriate normalization for lower values of K_0 at depth. Since denser soils are associated with lower values of stress exponents and lower values of K_0, this trend appears reasonable.

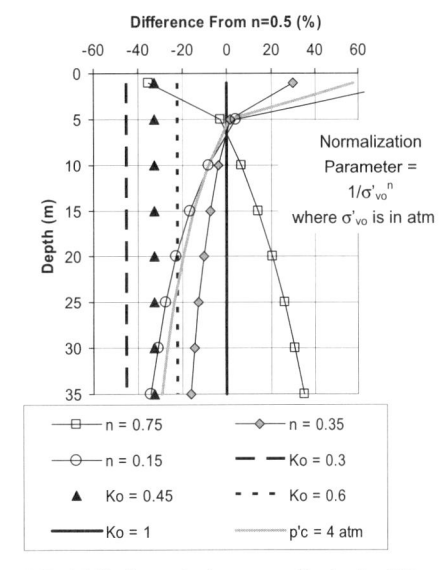

Figure 4. Variability in standard q_{t1N} normalization for different stress exponents and stress states

3.4 Data from the Mississippi River Valley

The Mississippi River Valley database has been separated into three groups by effective vertical stress at the midpoint of the layer, and plotted on Figures 5a and 5b in relation to the recommended trends of Equations 1 and 2, respectively. These correlations seem to generally fit the data within a variance of 25 percent, up to measured shear modulus values of 175 MPa. It was suggested in Section 1.3 that some tested soils in the Lower Mississippi River Valley may be of Pleistocene age, depending upon degree of scour induced by meltwater flow during waning glaciation. One characteristic of Pleistocene

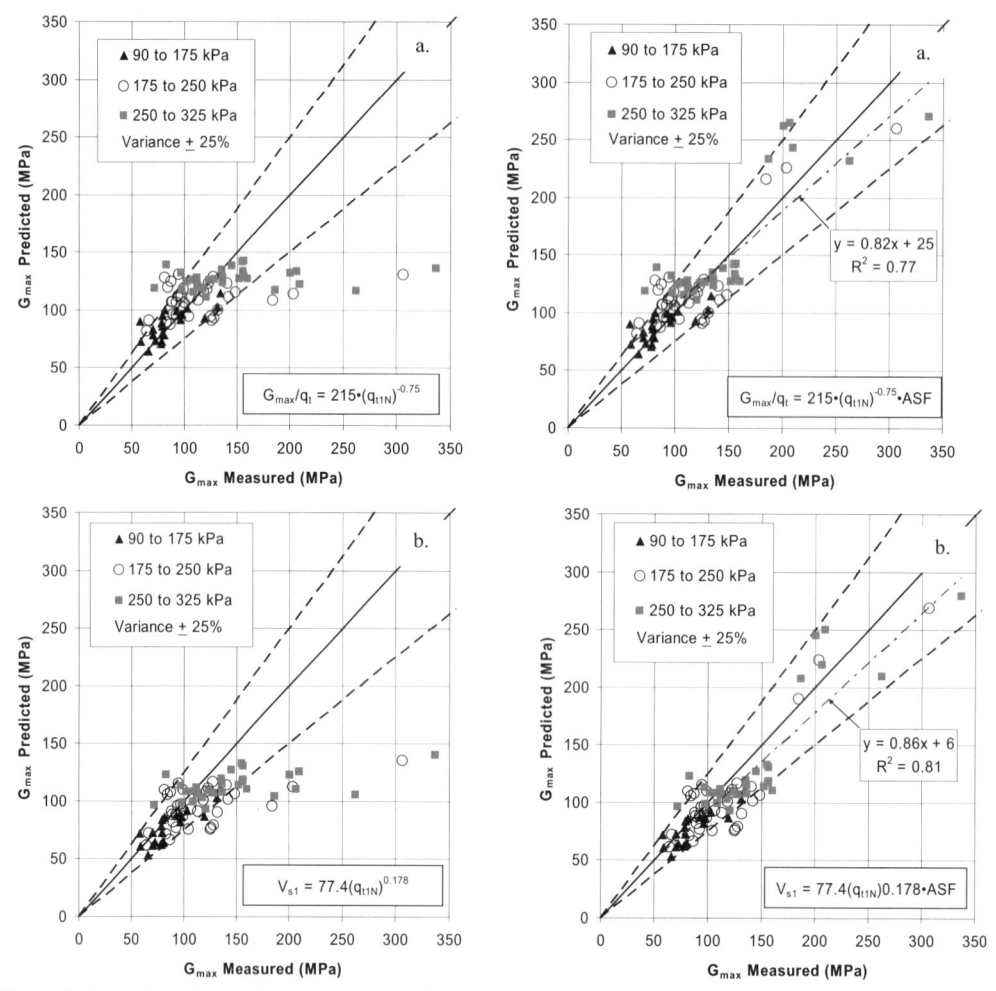

Figure 5. Comparison of cone tip resistance correlations to G_{max} calculated from shear wave velocity assuming Holocene deposits.

Figure 6. Comparison of cone tip resistance correlations to G_{max} calculated from shear wave velocity if layers with $G_{max,avg}$ greater than 175 MPa are considered Pleistocene.

deposits as compared to Holocene deposits would be a high shear modulus relative to cone tip resistance. Data from Figure 5 are re-plotted in Figure 6, assuming that soils with a shear modulus greater than 175 MPa are of Pleistocene age, and have an ASF relative to G_{max}/q_t of two.

When a Pleistocene ASF of two is applied to the layers with G_{max} greater than 175 MPa, the data matches trends within a variance of approximately 25 percent for G_{max} values between 50 MPa and 350 MPa. This is not meant to imply that all layers with a maximum shear modulus greater than 175 MPa are necessarily Pleistocene in age, but does imply measurement of V_s is necessary to minimize uncertainty in evaluation of shear modulus.

The coefficient of correlation, r^2, for Figure 6b is 0.81, as compared to the r^2 value for the original Andrus et al. (2001) correlation of 0.60. Since the Andrus et al. (2001) database consists of sites in the US, Canada, and Japan, while the Mississippi River Valley database in this paper has sites from one geologic origin, it is expected that a correlation coefficient should be higher in this case. It is also noted that the slope of the Andrus et al. (2001) relationship does not fall on the 1:1 trend line. If the leading coefficient is adjusted from 77.4 to 81.5 (or 11 percent change in G_{max}), the trend becomes closer to a 1:1 line, but the intercept moves further from the origin. Even though the data fit the trends relatively well in Figure 6, the increase in apparent correlation may be a function of compensating errors.

Figure 7 shows best fit linear trends between predicted and measured maximum shear modulus values for the Andrus et al. (2001) correlation. Similar trends were observed for all correlations evaluated. As stress level increases, these trends sharply deviate from a 1:1 relationship. Variance increase from approximately 15 percent for effective vertical stresses up to 175 kPa, to approximately 25 percent for effective vertical stress to 250 kPa, and to approximately 50 percent for effective vertical stress up to 330 kPa.

Since soil age generally increases with depth, there may also be a gradual increase in ASF with depth. Ageing effects are strongly controlled by stress state, density, and particle characteristics (Baxter, 1999), but it is very difficult to quantify the spatial variability of those parameters, as well as the absolute age of each layer, and the subsequent influence that the age has on G_{max} and q_t. Therefore, depth dependent ASF values will not likely increase correlation reliability.

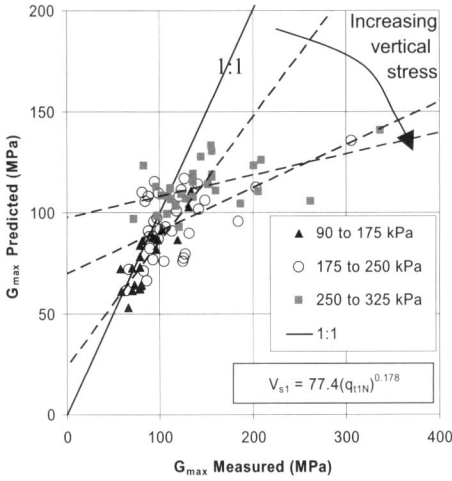

Figure 7. Assessment of stress influence on data trends with ASF assumed equal to one.

4 CONCLUSIONS & RECOMMENDATIONS

Cone tip resistance and maximum shear modulus are controlled by very different phenomena, but those parameters are also significantly influenced by stress state, void ratio, particle characteristics, and gradation. When maximum shear modulus is compared to cone tip resistance, it is inferred that additional insight may deduced about ageing, overconsolidation ratio, and/or horizontal stress state. Yet, since the number of controlling variables well exceeds the number of measured parameters, it is not possible to separate out the absolute role of each parameter, especially when dealing with averaged layers. Results of this study identified some potential Pleistocene layers that seem geologically feasible, but require additional study.

The variability of individual parameters that control G_{max}, q_t, and the G_{max}/q_t ratio are often on the order of 30 percent within an individual layer within the Mississippi River Valley. Variance in global G_{max}-q_t correlations is approximately 50 percent. A summary of this variability is provided in Table 1. Since these parameters may increase or decrease G_{max}/q_t relationships, and often produce somewhat compensating changes, use of G_{max}-q_t correlations without verification is prone to significant uncertainty. SCPTU provide a rapid and economical means for actual measurement of shear wave velocity profiles, and correlations should be used to infer other properties that significantly influence G_{max} relative to q_t rather than for evaluation of shear modulus. Ageing seems to influence G_{max}/q_t relationships in sands on the order of 100 percent when comparing Pleistocene to Holocene deposits. This is much larger than the 25 percent natural variability, and should be evident when comparing q_t to G_{max} for a specific deposit. On the other hand, since OCR has a minimal influence on G_{max}/q_t relationships in sands (10 percent) as compared to natural variability (typically 25 percent), reliable evaluation of OCR solely based on G_{max}/q_t relationships for sands is unlikely.

Table 1. Variability in Parameters that influence G_{max} and q_t in the Mississippi River Valley

Parameter	Variability (1 standard deviation)	Reference
q_t within layer	± 25% of mean	This Study
G_{max} / q_t within layer	± 30% of mean	This Study
Increase in G_{max} with increase in q_t	• Up to 40% over range of observed q_t values	This Study
G_{max} / q_t within local database for specific normalized q_t	• 50 % without deposit age consideration • Possibly 25 % with deposit age consideration	• Figure 5 • Figure 6
G_{max} / q_t within global database for specific normalized q_t	± 50 % in leading coefficient, K, for specific normalized q_t	Rix & Stokoe, 1991; Fahey et al., 2003
Age (Holocene to Pleistocene)	Possibly 100% increase in G_{max}	Andrus et al., 2001; Figures 5 & 6
OCR (1 to 10)	10% increase in leading coefficient, K, for specific normalized q_t	Baldi et al., 1989

While analyzing thicker (greater than 1 m) "layers" of soil are expected to minimize some error from thin layer corrections and pseudo interval processing techniques in the heterogeneous natural deposits of the Lower Mississippi River Valley, averaging the data is considered to be a major drawback in truly assessing the influence of maximum shear modulus on cone tip resistance in sands at a particular site. Use of true-interval seismic cone devices along with analysis of wave forms using digital signal processing methods shows great potential for generation of very detailed shear wave velocity profiles with less measurement induced uncertainty. To assess how variation in measured shear modulus with depth associates with variation in measured cone tip resistance with depth will be a major step forward in understanding cone tip resistance in layered natural deposits, since G_{max} is often an assumed input parameter for advance numerical models of penetration resistance in sands. Results of those studies would be beneficial for engineering analyses related to natural deposits and layered soils, such as liquefaction evaluation and load settlement behavior of deep and shallow foundations.

ACKNOWLEDGEMENTS

The authors would like to thank the staff members and graduate students and from Georgia Tech who helped collect data in the Mississippi River Valley, including: Ken Thomas, Tracy Hendren, Tom Casey, Billy Camp, Guillermo Zavala, and Tianfei Liao. Funding for aspects of this project was provided in part by the Mid-America Earthquake Center, U.S. Geological Survey, and the National Science Foundation. Conclusions, opinions, and recommendations herein are those of the authors and do not necessarily represent those of MAE, USGS, nor NSF. Appreciation is given to Dr. Roy Van Arsdale at the University of Memphis/CERI, Dr. Joan Gomberg and Dr. Buddy Schweig of the USGS/CERI, and Dr. Martitia Tuttle for their help in selecting and accessing these sites.

REFERENCES

Andrus, R.D., Piratheepan, P., Juang, C.H. 2001 Shear wave velocity – penetration resistance correlations for ground shaking and liquefaction hazards assessment, *Report*, Program Element 1, USGS Grant 01HQGR007, 6 pp.

Baldi, G., Bellotti, R., Ghionna, N., Jamiolkowski, M., and LoPresti, D.C.F. 1989 Modulus of sands from CPTs and DMTs, *Proc*, 12th Int. Conf. on Soil Mech. and Foundation Eng, Vol. 1, Balkema, Rotterdam, pp. 165–170.

Baxter, C.D.P. 1999 An experimental study on the aging of sands, *PhD Dissertation*, Virginia Tech, Blacksburg, Virginia, 303 pp.

Bowman, E.T. 2002 The ageing and creep of dense granular materials, *PhD Dissertation*, University of Cambridge, UK, 293 pp.

Fahey, M., Lehane, B., and Stewart, D. 2003 Soil stiffness and shallow foundation design in the Perth CBD, *Australian Geomechanics*, September, 30 pp.

Fear, C.E., and Robertson, P.K. 1995 Estimation of the undrained shear strength of sand: a theoretical framework, *Canadian Geotechnical Journal*, 32 (5), pp. 859-870.

Fioravante, V., Jamiolkowski, M., LoPresti, D.C.F., Manfredini, G., and Pedroni, S. 1998 Assessment of the coefficient of the earth pressure at rest from shear wave velocity measurements, *Géotechnique*, 48 (5), pp. 657-666.

Hegazy, Y. and Mayne, P.W. 1995 Statistical correlations between V_s and cone penetration test data for different soil types, *Proceedings*, CPT -95, Vol. 2, Linköping, pp. 173–178.

Houlsby, G.T., and Hitchman, R. 1988 Calibration chamber tests of a cone penetrometer in sand, *Géotechnique*, 38 (1), pp. 39–44.

Kulhawy, F.H., and Mayne, P.W. 1990 Manual on estimating soil properties for foundation design, *Report EL-6800*, Electric Power Research Institute, Palo Alta, CA, August, 306 pp.

Mayne, P.W., Schneider, J.A., and Martin, G.K. 1999 Small- and large-strain soil properties from seismic flat dilatometer tests, *Pre-Failure Deformation Characteristics of Geomaterials*, Vol. 1 (Torino), Balkema, Rotterdam, pp. 419–426.

Olsen, R.S., and Mitchell, J.K. 1995 CPT stress normalization and prediction of soil classification, *Proc., Int. Symp.on Cone Penetration Testing (CPT '95)*, Vol. 2, Swedish Geotechnical Society Report No. 3:95, Linköping, pp. 257-262.

Robertson, P.K. 1982 In-situ testing of soil with emphasis on its application to liquefaction assessment, *PhD Thesis*, Dept. Civil Engrg., University of British Columbia, Vancouver.

Robertson, P.K., Woeller, D.J., Kokan, M., Hunter, J., and Luternaur, J. 1992 Seismic techniques to evaluate liquefaction potential, *Proc. 45th Canadian Geotechnical Conference*, Toronto, Ontario, October 26-28, pp. 5:1-5:9.

Rinaldi, V.A., and Santamarina, J.C. 2003 Cemented soils: Behavior and conceptual framework, (in review), 39 pp.

Rix, G.J. and Stokoe, K.H. 1991 Correlation of initial tangent modulus and cone penetration resistance, *Calibration Chamber Testing*, Elsevier Science, New York, pp. 351–361.

Salgado, R., Mitchell, J.K., and Jamiolkowski, M. 1997 Cavity expansion and penetration resistance in sand, *J. of Geotechnical and Geoenvironmental Engineering*, 123 (4), pp. 344–354.

Santamarina, J.C. and Aloufi, M. 1999 Micro-scale interpretation of wave propagation in soils - fabric and fabric changes, *Pre-Failure Deformation Characteristics of Geomaterials*, Vol. 2, Balkema, Rotterdam, pp. 451–458.

Saucier, R.T. 1994 *Geomorphology and Quaternary Geologic History of the Lower Mississippi Valley*, U.S. Army Corps of Engineers Waterways Experiment Station, Vicksburg.

Seed, H.B., Wong, R.T., Idriss, I.M., and Tokimatsu, K. 1984 Moduli and damping factors for dynamics analyses of cohesionless soils, *Report No. UCB/EERC 84/14*, University of California, Berkeley, EERC, 37 pp.

White, D.J., and Bolton, M.J. 2004 Displacement and strain paths during pile installation in sand, *Géotechnique*, (accepted), 63 pp.

Risk assessment for contaminated tropical soil

Ana Cristina Strava Corrêa
Agência Nacional de Águas, Brasília, Brazil

Newton Moreira de Souza
Universidade de Brasília, Brasília, Brazil

Keywords: tropical soil, contamination attenuation, leachate, waste disposal, risk assessment

ABSTRACT: The results of our research on tropical soil attenuation capacity indicate that soil microstructure is modified due to leachate contact and biological attenuation plays an important role in the overall response of contaminant retention.
The study was based upon the observation, at laboratory scale, of three varieties of tropical residual soils. The soils studied came from potentials sites for solid waste disposal. Their natural bulk unit weights were found to vary from 16.3 to 17.0 kN/m^3 with a mean void ratio of 0,7. Mineralogical analysis indicated they were mainly formed by Kaolinite, Ilite and Quartz with traces of Goethite. The parameters found to enhance the soil attenuation capacity were porosity, structured clay and oxides contents.
These parameters were modeled for geo-environmental mapping of Brasilia. It resulted in a risk assessment tool for old waste disposal sites in this region. The final risk index is presented in a table obtained from GIS tools coupled with indexed layers.

1 INTRODUCTION

The attenuation capacity attributes considered in the literature, as CEC, low natural permeability, alkaline pH and clay content, do not match with tropical soil characteristics. Nevertheless, there are some reports about the good attenuation performance of tropical soils.

Santos (1996) inspected the unconsolidated 8 to 12 m deep, clayey highly permeable soil (10^{-3} cm/s) underneath Brasilia's landfill (without any kind of barrier) and found that heavy metal traces were concentrated just near the landfill base. After 25 years of continuous operation, receiving around 1200 ton/day of domestic waste dumped directly over the soil, the contamination traces were not found under 6 meters from the base.

Junqueira (2000) observed the same lateritic soil in a column leach test for leachate and found attenuation of ammonium, nitrates and COD after 5 pore volumes run. He concluded biological attenuation was the main process for attenuation. He also compared geotextile and sand drains and found that the sand effluent suffered attenuation.

Pohl (1996) studied tropical soils from two Brazilian regions. His column leach tests indicated attenuation of metals even under aggressive conditions (pH<5).

Boscov et al. (1999) concluded that lateritic clay may be used in sealing layers of waste sides after a series of diffusion and hydrodynamics dispersion and adsorption tests were performed.

Tropical residual soils have typical characteristics that made its interaction with contaminants very differentiated.

One of its properties is its usually high porosity, around 50%, but also, with high clay content. Those attributes together make what its known as structured soil. It also makes these soil excellent medium for bacterial fixation and development.

Others factors like high mean temperature and availability of electrons donors and acceptors makes biological process in tropical soil to be considered an important part of its containment capacity.

2 LITERATURE REVIEW

2.1 *Contaminant Transport and Fate Issues*

Contaminant flows through porous media are usually governed by advective-diffusive transport. Advection is the migration of contaminant due to bulk solution flow and diffusion is the migration due to a concentration difference in the absence of bulk solution flow. Mechanical dispersion is the spreading of contaminant during advection due to variations in pores and is considered negligible at low flow rates.

Shackelford & Rowe (1998) describe chemical and biological process affecting miscible contaminant transport as:
- Sorption (partitioning of contaminant between pore water and porous medium) - limiting case of ion exchange. Considering tropical soil, it might have low significance as CEC varies from 2 to5 meq/100g soil.
- Radioactive decay (irreversible decline in the activity of a radionuclide). Depends of type of residues disposed off. Its significance in attenuation is related to seepage velocity.
- Dissolution / precipitation (reactions resulting in release of contaminants from solids or removal of contaminants as solids) - dissolution is important at the migration front, while precipitation occurs particularly for pH above 7. Both processes are relevant for tropical residual soil typically enriched by Fe and Al minerals. Boscov et al. (1999) observed dissolution process of Fe and Al for pH range of 1 to 4 while testing tropical soils.
- Acid / base (reactions involving a transfer in protons - H^+) - important because controls other reactions like dissolution and precipitation. Hence, it also has significance for tropical soil contamination transport processes.
- Complexation (combination of anions and cations in a more complex form).
- Hydrolysis / substitution (reaction of a halogenated organic with water or a component ion of water - hydrolysis - or another anion - substitution). Boscov et al. (1999) registered a rapid decrease in Cr^{+3}, Fe^{+3} and As^{+3} in tropical soil during a diffusion test which was justified by hydrolysis process for pH under 4.
- Oxidation / reduction (reactions involving transfer of electrons) - important process controlling precipitation of metals. Oxidation also can occur though biological process.
- Biodegradation (reactions controlled by microorganisms) - important mechanism for organic compounds. Observed by Junqueira (2000) for tropical lateritic soil and in a sand drain for leachate collection.

2.2 Microbiology and Biodegradation

Tropical soils are predominantly well drained soils with high void ratio and the occurrence of soil aggregates results in the presence of numerous microenvironments within the soil (Baker & Herson, 1994).

Two of the known techniques for bioremediation in soil are biostimulation and bioaugmentation. The first is done by stimulating the indigenous microbial populations and the second is the introduction of specialized microbes for certain degradation job (Levin & Gealt, 1993). When leachate reaches the soil both processes may occur, since this sort of contaminant could be rich in nutrients and carries new species that may stay in soil aggregates. Baker & Herson (1994) propose that when evaluating the potential of using microbiological processes for remediation it is beneficial to check if some requirements for bacterial growth are present within the soil system. The following is a brief summary of each of these criteria:
- Soil humidity - Jaèckel et al. (2001) observed that one limiting aspect for microbial activity was soil humidity while monitoring methane production.
- Carbon and energy source - biodegradable organic compounds must be present to serve as carbon and energy source. Santos (1996) reported the ratio of biodegradable organic compound to COD varying from 80 to 50% for Brasilia landfill leachate.
- Electron acceptor - it is the substance that accepts electrons during an oxidation-reduction reaction. Oxygen is the electron acceptor in aerobic environments. Nitrate is the electron acceptor in anoxic environments. Sulfate, carbon dioxide, or reduced organics act as electron acceptors in anaerobic environments. In tropical structured soils there is possible the occurrence of both, aerobic and anaerobic environments. Moreover, leachate itself can provide some source of electron acceptor substances.
- Nutrients - substances that are taken in by the microorganisms and used in metabolic reactions. Baker & Herson (1994) point out that the preferable source of nitrogen is its reduced form of ammonium, which is abundantly found in leachate.
- pH - for most microbiological activities, the optimum pH range is 6.5 to 8.5.
- Temperature - perhaps the main responsible for the differentiated response of biodegradation within tropical soils. Metabolic reactions are more significant as temperature increases.
- Toxic materials - some microorganisms will respond to toxicity by limiting its growth, on the other hand, more resistant ones may develop and attenuate toxic materials.

Adequate contact - in order for the biodegradation of a contaminant occur, there must be sufficient interaction between the contaminate solute and the microorganisms. On the other hand, it is expected that the plume region would be richer in microorganisms because of the stimulation caused by leachate flow.

3 MATERIALS AND METHODS

3.1 Experimental materials

Samples of tropical lateritic and residual soil were collected from two distinct potential area for waste disposal. Those areas were selected by GIS tools as the ones that fulfill all requirements to be a new landfill site for Brasilia. The characteristics of those soils are presented in Table 1.

Table 1. Soil samples properties

Properties	SOB 2*	SAM 2	SAM3
Sand (%)	8	5	3
Silt (%)	50	69	77
Clay (%)	42	22	20
WL (%)	48	48	52
WP (%)	41	37	24
Bulk density (gf/cm^3)	1,02	1,40	1,71
Specific gravity (G_s)	2,73	2,84	2,88
Porosity (%)	64	50	40
Organic matter (%)	0,86	0,52	0,34
CEC (meq/100g)	3	4	5
Clay minerals	kaolinite	kaolinite	kaolinite
	gibbsite	illite	illite
(traces)	hematite	goethite	goethite

SOB 2 - Sobradinho - 2m deep;
SAM 2 - Samambaia - 2m deep;
SAM 3 - Samambaia - 3m deep.

The soils chosen were representative of three distinct horizons: "lateritic soil" (SOB 2), "transition zone" (SAM 2) and "saprolite horizon" (SAM 3).

Samples of methanogenic phase leachate were collected from Brasilia landfill experimental cells, filled with representative waste from the city. Table 2 presents the main characteristics of leachate collected.

Table 2. Characteristics of Collected Leachate

Parameter	Unit	Range Values
pH		8,3 - 8,7
Temperature	°C	22 - 25
COD	mg/L	536 - 629
Conductivity	mS	10 - 14
Salinity	g/L	6,5 - 8,5
Ammoniacal nitrogen	mg NH$_3$/L	616 - 945
Total Suspended Solids	mg/L	4500 - 8010
Total Dissolved Solids	mg/L	4400 - 7990
Total Volatile Solids	mg/L	3100 - 6580

Thornton et al. (2000) has shown there are systematic differences in the sorption and degradation of organic matter between acetogenic and methanogenic landfill leachate. On the other hand, Santos (1996) found that leachate reaching soil was at methanogenic phase even for recent deposits of Brasilia's landfill.

The air dried materials were packed at a moisture-density near the optimum obtained from Normal Proctor test. Columns had 4 to 6 centimeters of soil and were flushed with leachate at constant head. Table 3 summarizes the characteristics of modeled columns.

Table 3 - Characteristics of columns

Characteristic	SOB 2	SAM2	SAM 3
Optimum humidity (%)	31,5	26	23
Max. density (kN/m^3)	13,4	15,5	15,7
Moulding humidity (%)	34	27,5	25
Moulding density (kN/m^3)	13,2	15,3	15,5
Total volume (cm^3)	311	311	505
Pore volume (cm^3)	158	140	256

The columns were operated in upflow mode under a flow rate of 50 ml per day ($\approx$ 1/3 pore volume). Columns were saturated with the same fluid with which they were flushed. Columns flushed with saline solution of 1000ppm NaCl were used as blanks, and the soil packed for the same conditions of contaminated column.

3.2 Procedures for monitoring column leach test

The performance of soil was evaluated by monitoring indicators for physical retention, chemical adsorption and biological attenuation.

As physical indicators, suspended solids and turbidity were monitored.

Chemical reactions were observed by monitoring pH, total iron, electrical conductivity and salinity.

Biological processes were inferred by the observation of ammonium and COD breakthrough curves.

The hydraulic conductivity was also monitored for all columns.

3.3 Biomass determination

The determination of microbial activity was done following Jenkinson & Powson (1976) procedures. Soil was incubated, and the CO_2 produced is captured by KOH within the flasks with 20g of soil at field holding capacity humidity.

After 10 days, CO_2 produced is determined by titration with HCl. A second incubation period is followed for soil submitted to fumigation with chloroform. The biomass is then determined by:

$$BM = \frac{(X-Y)}{Kc} \quad (1)$$

where: X = C-CO_2 produced by fumigated soil (0-10 days); Y = C-CO_2 produced by non fumigated soil (0-10 days); and Kc = proportion of mineralised biomass in 10 days.

4 RESULTS

4.1 Solute breakthrough profiles

Soil physical fulfilment was monitored by solute breakthrough curves (BTC) of turbidity, that are shown in Figure 1 for soils SOB 2, SAM 2 e SAM 3.

Physical parameters BTC

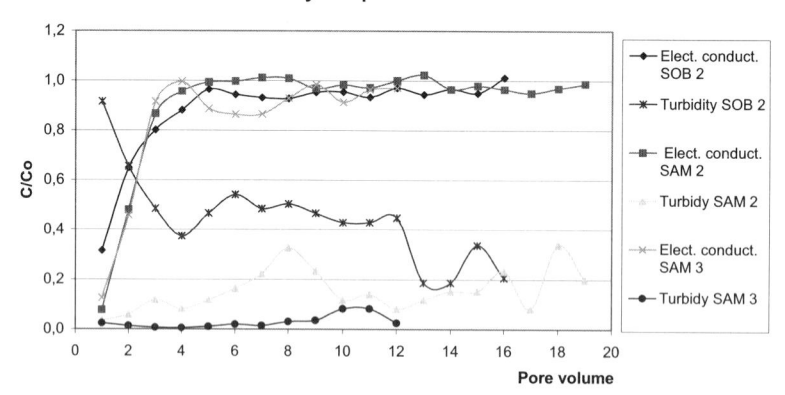

Figure 1 - Turbidity BTC for leachate flush experiment with soils SOB 2, SAM 2 and SAM 3.

Chemical parameters - SOB 2

Chemical parameters - SAM 2

Chemical parameters - SAM 3

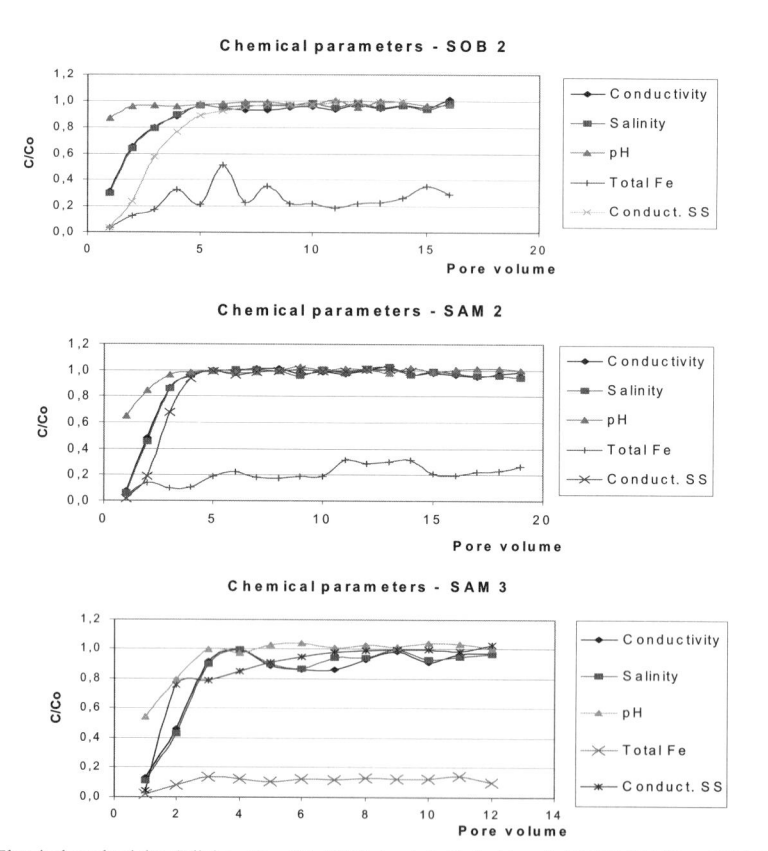

Figure 2 - Electrical conductivity, Salinity, pH and Fe BTC's leachate flushed in soils SOB 2, SAM 2 and SAM 3 columns. Electrical conductivity BTC for saline solution (SS) is also plotted as "blank" experiment.

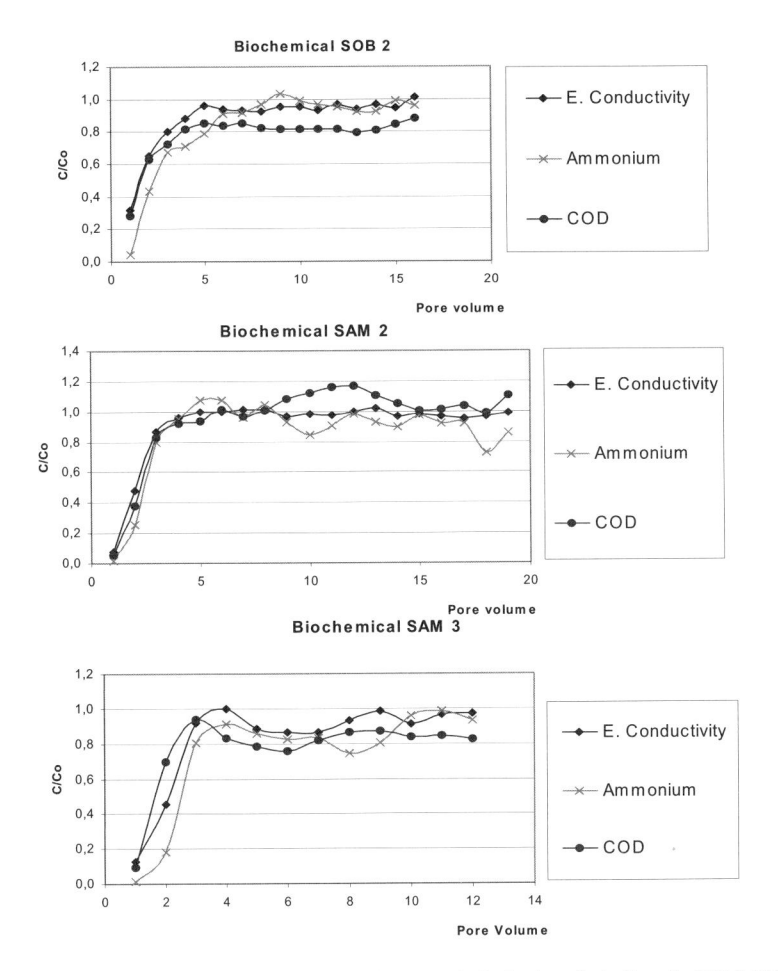

Figure 3 - Electrical conductivity, Ammonium and COD BTCs for leachate flushed in soils SOB 2, SAM 2 and SAM 3 columns.

The electrical conductivity BTC are plotted together in order to compare soil fulfilment after chemical equilibrium is reached.

Lateritic soil SOB 2 had an increasing removal performance for turbidity while saprolite SAM 2 and SAM 3 showed good turbidity removal, always over 60% of influent turbidity.

Results for solids showed that suspended solids were mainly dissolved salts. The mean result for effluent showed that best performance for solids retention was around 27% removal of TSS, obtained with saprolite soil SAM 3. On the other hand, volatile proportion of solids increased in soil SOB 2 effluent.

Chemical results are presented in Figure 2 for electrical conductivity, pH, salinity and Fe BTC's.

From electrical conductivity BTC's were plotted for saline solution and leachate. For all soils, electrical conductivity was slightly more retarded for saline solutions columns than the leachate ones.

Salinity and conductivity presented the same pattern indicating that either one could be used as an indicator for chemical behavior.

Buffering capacity of soils was observed through pH BTC's, to be coherent with natural soil pH of 5, 5.2 and 5.8 for SAM 3, SAM 2 and SOB 2, respectively. It reached the equilibrium for the same pore volume of chemical equilibrium for electrical conductivity.

All soil presented a good retention of total iron. Lateritic soil SOB 2 it was registered an increasing phase of effluent iron that reached maximum of 0,5 C/Co at the same pore volume chemical equilibrium was reached for electrical conductivity.

COD and NH_3 BTC's for leachate (Figure 3) were monitored for organics compounds retention behavior. COD BTC's had the same pattern for SAM 3 and SOB 2: (i) an initial phase of increasing values in terms of C/Co, (ii) a short second phase of maximum values of C/Co, and (iii) a third phase of equilibrium for 0,8 C/Co. The same pattern was also reported by Junqueira (2000) and Pohl (1996).

SAM 2 presented a differentiated behavior also observed by Junqueira (2000) while leaching a compacted lateritic clay. It seams that the higher density of this sample or smaller void spaces in soil pore reduced the biological process within the soil matrix.

4.2 Biomass determination

The microbial activity was determined for each sample of soils after column leach experiments were finished. Table 4 presents biomass expressed in terms of carbon.

Table 4 - Biomass (mg/kg) for different flushed fluid

Column leach	SAM 3	SAM 2	SOB 2
Saline solution (SS)	89,9	71,3	120,9
Leachate (LCH)	133,3	133,3	210,8
LCH/SS	1,5	1,8	1,7

Microbial activity increased in all three columns when flushed with leachate. On the other hand, the increment ratio, calculated as biomass within leachate over saline solution flushed columns, were roughly constant (1.5 to 1.8). It showed no relationship with the degree of soil laterization but could be related with leachate characteristics that enhance microbial activity at the same rate although for different media.

5 DISCUSSION

Considering that chemical equilibrium for BTC's were reached for all columns around the 4th volume of pores, it is reasonable to suppose that the attenuation observed for COD and NH$_3$ for soils SAM3 and SOB 2 were due to biochemical processes. Physical response could also be a consequence of microbial activity, as it could have happened for iron uptake and immobilization within soil columns.

Microbiological systems require certain conditions that tropical soils are able to provide.

This study has shown that for methanogenic phase leachate, there is potential degradation of organic compounds within tropical soils.

Leachate composition would affect the response of indigenous microorganisms from soil. The roughly constant proportion between biomass C before and after contamination showed that contamination process could be compared to a natural bioremediation technique of biostimulation, more than bioaugmentation.

Soil porosity, hence the degree of compaction, influence biological degradation probably because of the poor distribution of nutrients that are mainly physically retained at the first centimeters of columns. On the other hand, ions still migrate through diffusion, since it would be the predominant process as flow velocity decreases.

6 RISK ANALYSIS

The tested soil characteristics were located over the Brazil's Federal District map, as a layer of a Geographical Information System tool. Other environmental criteria were also considered for geology, hydrology, topography and vegetation.

For risk analysis it was considered the soil occupation as urban areas, roads distance and accessibility.

Risk assessment is proposed based upon the concept that it could be obtained from the product of environmental vulnerability (V) by the potential hazard of anthropogenic action (P), as shown in equation (2):

$$R = V * P \qquad (2)$$

In its turn, environmental vulnerability is the result of weighted mean values of selected layers. Equation (3) presents the general expression used for indexed overlay model supported by a GIS tool.

$$V_{f(x,y)} = \frac{\sum V_{i(x,y)} * P_i}{\sum P_i} \qquad (3)$$

where $V_{f(x,y)}$ is the final value of overlaid cell at (x,y) position, $V_{i(x,y)}$ is the environmental factor value considered in each layer at (x,y) position and P_i is the attribute weight.

Environmental vulnerability was then obtained mainly from three environmental factors: (i) geological / soil characteristics, (ii) hydro geological attributes and (iii) soil use / occupation. Tables 5 to 8 presents the correspondent values of attributes for each set of factors. All attributes were evaluated for contamination vulnerability within a 0 to 1 scale. More vulnerable attributes gets higher values, closer to 1.

Geological classes shown in Table 5 were valued considering the results obtained from the laboratory experiments described above. Typical tropical residual soils were evaluated as being less vulnerable to contamination.

Those tested values provide an approximation of possible results and can vary with the user priorities or will. The main objective of the proposed model is to provide a methodology for risk assessment considering important tropical residual soil characteristics.

Table 5 – Federal District geological classification

Litology	Texture	Value
Metargilito; Ardósia; Metamarga; Metassiltito	Clay / silt	0,2
	Concretionary-clay	0,4
	Gravel-clay	0,5
Calcário	Clay	1
Quartzito	Sandy	1
Dolomito	Clay	1
Metargilito; Metassiltito	Clay / silt	0,2
	concretionary-clay	0,4
	gravel-clay	0,5
Quartzito	Sandy	1
Quartzito	Sandy	1
Quartzito	Sandy	1
Metassiltito; Metargilito	Clay / silt	0,2
	concretionary-clay	0,4
	gravel-clay	0,5
Ardósia	Clay / silt	0,2
Metassiltito	concretionary-clay	0,3
	gravel-clay	0,4
Quartzito	Sandy	1
Metassiltito	silty	0,5
Quartzito	sandy	1
Filito Carbonoso	Clay / silt	0,2
	concretionary-clay	0,4
	gravel-clay	0,5
Filito Calcítico	gravel-clay	0,5
Calcixisto	Clay / silt	0,3
	Clay / silt	0,3
Micaxisto	concretionary-clay	0,5
	gravel-clay	0,6
Quartzo xisto	gravel-clay	0,7
	sandy	1
	clay	0,8
	gravel-clay	0,8
Others	gravel-sand	1
	intermediate	1
	gravel	1
	organic	1

Table 6 – Federal District soil depth

Unconsolidated material depth (m)	Value
0 - 2	1
2 - 5	0,8
5 - 10	0,6
10 - 20	0,4
> 20	0,2

Table 7 – Hydrogeology characteristics

Superficial net	Distance (m)	Value
	0 – 100	1
	100 – 250	0,8
	250 – 500	0,4
	> 500	0,1
Water table depth	(m)	
	0 – 3	0,8
	3 – 10	0,5
	> 10	0,1

On the other side, potential hazard was considered as the quantity of solid waste disposed over sites used at Federal District since the Brazilian capital creation, in 1960.

Table 9 summarizes the amount of solid waste disposed in each of the six sites used in Federal District since its creation, in 1960.

Relative potential hazard was calculated as a function of the rate of disposal over time and the total area used, for each site. Major hazard potentials get 1 value.

Table 10 shows the result of environmental risk index considering the location of each disposal area mentioned.

Table 8 – Soil occupation – distance from urban areas

Distance (m)	Value
0 – 200	1
200 – 500	0,8
500 – 1000	0,5
> 1000	0,1

Table 9 – Solid waste destination in Federal District

Nº	Site	Disposed waste (tons)	Relative potential hazard
1	Jockey Club landfill	1.504.718	0,57
2	Taguatinga/Ceilândia	674.763	0,73
3	Gama	28.845	1,00
4	Sobradinho	47.412	0,89
5	Planaltina	16.043	1,00
6	Brazlândia	6.782	1,00

Modified from Correa & Souza (2001)

Table 10 – Simulation of relative risk for disposal sites at Federal District

Nº	Site	Environmental vulnerability	Potential hazard	Risk level
1	Jockey Club landfill	0,51	0,57	0,29
2	Taguatinga / Ceilândia	0,79	0,73	0,57
3	Gama	0,46	1,00	0,46
4	Sobradinho	0,49	0,89	0,43
5	Planaltina	0,33	1,00	0,33
6	Brazlândia	0,36	1,00	0,36

7 CONCLUSION

Risk assessment is useful for decision making over priorities of remediation or soil recovery, considering sites vulnerability.

Table 10 presents the relative values of risks from the application of equation (2), for the same amount and toxicity of waste.

Higher risk obtained for Taguatinga / Ceilândia site was due to its location, over a porous quartz geological unit, very close to superficial water courses. The proximity of water courses was yet the main reason why Gama site got the second major relative risk. After 1990, it was indeed considered as a protected area for water supply use and requires intensive monitoring for water quality.

The risk offered by Jockey Club site was attenuated because of its tropical soil characteristics, in spite of the higher amount of waste received.

REFERENCES

Baker, K.H. & Herson, D.S. (1994). Introduction and Overview of Bioremediation. In Bioremediation, Baker & Herson (ed),. McGraw-Hill, Inc., New York, USA, pp 1-7.

Boscov, M.E.G.; Oliveira, E.; Ghilardi, M.P. and Silva, M.M. (1999). Difusão de metais através de uma argila laterítica compactada. *REGEO'99 - ABMS, Instituto Tecnológico de Aeronáutica*, São José dos Campos, S.P, pp 323 - 330.

Corrêa, A.C.S. & Souza, N.M. (2001). Áreas para disposição de resíduos sólidos urbanos no Distrito Federal: Teoria e prática. Revista Universa, v.9, n°2, Brasília, DF, pp 315-337.

Jaèckel, U.; Schnell, S. and Conrad, R. (2001). Effect of moisture, texture and aggregate size of paddy soil on production and consumption of CH4. Soil Biology & Biochemistry, Elsevier, 33, pp 965-971.

Jenkinson, D.S. and Powson, D.S. (1976). The effectsof biocidal treatments on metabolism in soil. V. A method for measuring soil biomass. Soil Biology biochemistry, 8: 209-213, 1976.

Junqueira, F.F. (2000). Análise do comportamento de resíduos sólidos urbanos e sistemas dreno-filtrantes em diferentes escalas, com referência ao Aterro do Jóquei Clube - DF. *Tese de Doutorado em Geotecnia, UnB*, Brasília, DF, publicação 06/2000, 261p.

Levin, M.A. and Gealt, M.A. (1993). Biotreatment of industrial and hazardous waste. Mc Graw-Hill,Inc.331p.

Pohl, D. H. (1996). Tropical Residual Soils: Applications in Solid Waste Containment. Tese submetida pela Universidade de Drexel, EUA, 325p.

Santos, P.C. (1996). Estudo da contaminação de águas subterrâneas de percolado de aterro de resíduos sólidos: caso Jóquei Clube de Brasília. Faculdade de Tecnologia, Departamento de Engenharia Civil, Pós-graduação em Geotecnia, Dissertação de Mestrado, Universidade de Brasília, DF, 145p.

Shackelford, C.D. and Rowe, R.K. (1998). Contaminant transport modeling. In Pinto, P.S.S.(ed) Environmental Geotechnics. III Int. Congress on Environmental Geothecnics, Lisboa, Portugal, pp. 939-956.

Thornton, S.F.; Bright, M.I.; Lerner, D.N. and Tellan, J.H. (2000). Attenuation of landfill leachate by UK Triassic sandstone aquifer materials: 2. Sorption and degradation of organic pollutants in laboratory columns. *J. of Contaminant Hydrology*, 43: 355-383.

An examination of the engineering properties and the cone factor of soils from East Asia

M. Tanaka & H. Tanaka
Port and Airport Research Institute, Yokosuka, Japan

Keywords: cone factor, engineering property, CPTU, unconfined compression test, diatom microfossils

ABSTRACT: General site investigations of soil sampling and cone penetration tests were carried out in Saga Ariake and Hachirogata(Japan), Pusan(Korea) and in some regions of Southeast Asian countries such as Bangkok(Thailand), Singapore and Hai phong(Vietnam). All the samplings were carried out using Japanese type samplers to remove the technical effect of the local sampling method. Using the clay samples obtained in the sampling, laboratory tests such as soil classification tests, unconfined compression tests and consolidation tests were subsequently performed. In this paper, the soil properties of these clays are compared and the cone penetration test results obtained from the six sites are also explained.

1 INTRODUCTION

Many Southeast Asian countries have been recently conducting civil engineering works for the development of infrastructure. As large portions of these works require a good knowledge of the soil characteristics, it is essential to possess simpler and faster methods of evaluating ground information in most geotechnical engineering fields.

One of the typical methods of evaluating the ground is to conduct the laboratory test on soil boring samples ; the other is to perform in-situ tests such as a cone penetration test (CPTU). It can easily been imagined, however, that ground-forming processes in this very wide region were largely different from area to area. The methods of investigating the ground vary significantly in each country depending on regional design standard and mechanical conditions. Therefore it would be necessary to use the same ground investigation method to compare the ground characteristics of various areas.

The authors conducted sampling and cone penetration tests on cohesive soils distributed in the following Asian regions: Saga Ariake and Hachirogata in Japan; Pusan in Korea; Bangkok in Thailand; Singapore; Hai Phong in Vietnam. Saga Ariake clay, Hachirogata clay, Pusan clay, and Bangkok clay belong to the Holocene cohesive soil ground; Singapore clay and Hai Phong clay belong to the Pleistocene ground. Engineering properties of these soils were evaluated with unconfined compression tests, and consolidation tests using the soil samples from above locations all taken by a Japanese-type sampling method. This paper describes the relationship between the engineering properties and the cone factors of the soil in each of the areas.

2 CONE FACTOR

The determination of the undrained shear strength of the ground from the results of a CPTU requires the cone factor (N_{kt}). The shear strength and N_{kt} of the soils in undrained condition are obtained from the following equation:

$$S_u = \frac{q_t - \sigma_v}{N_{kt}} \qquad (1)$$

where,
S_u: undrained shear strength from CPTU (kPa)
q_t: corrected cone resistance (kPa)
σ_v: overburden pressure, in total stress (kPa)
N_{kt}: cone factor

Lunne and Kleven (1981), Jamiolkowski et al. (1982), and Robertson et al. (1986) made detailed studies on N_{kt}. In Japan, Tanaka et al. (1992) performed CPTU on normally consolidated marine cohesive soils sampled from 7 locations, determining the ground characteristics and the values of N_{kt} from the results of the CPTU. In the determination of the N_{kt} values, they calculated undrained shear strengths from unconfined compression strengths. They compared the values of N_{kt} with those of I_p (plasticity index) to show that the values of N_{kt} of Japanese cohesive soil lie in the range of 8-16 regardless of the values of I_p.

Aas et al. (1986) calculated undrained shear strengths from triaxial compression strengths, simple shear strengths, and triaxial extension strengths to examine the relationship with I_p. The result showed the tendency for N_{kt} to increase with I_p. The experimental results of Tanaka et al. (1992) differed from those of Aas et al. (1986) in the interpretation of the N_{kt} versus I_p relationships. The difference was caused by the differences in the methods of obtaining undrained shear strengths and in the characteristics of the objective ground. The values of N_{kt} thus vary depending on ground-formation processes. The determination of undrained shear strengths from the results of CPTU is, however, significantly meaningful in practical applications.

This paper describes the relationships between the engineering properties and the cone factors of the ground in the East Asia. Figure 1 shows the locations in East Asia investigated by the authors. Sampling at each location was made in compliance with Japanese Geotechnical Society Standards "Method for obtaining Undistributed Soil Samples Using a Thin-walled Tube Sampler with a Fixed Piston" (JGS 1211). Bringing Japanese samplers to the test sites, the authors supervised local operators to take samples in accordance with Japanese standards.

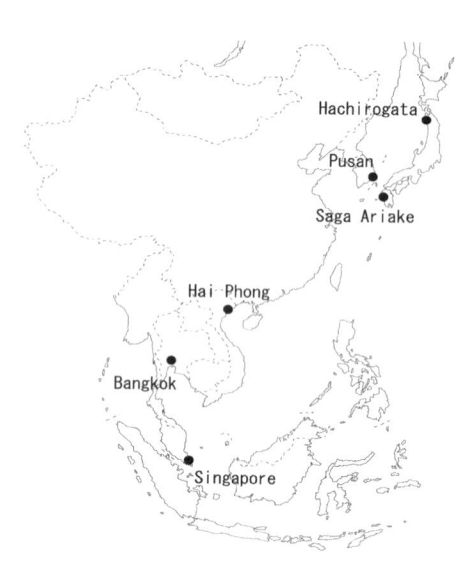

Figure 1. Investigation sites

3 RESULTS OF CASE STUDIES

3.1 *Japan*

3.1.1 *Saga Ariake*
Saga Ariake is located about 1,200km west of Tokyo. The site investigated was reclaimed some 300 years ago and is currently being used as a rice field. Figure 2 shows the results of soil property tests, unconfined compression tests, and CPTU on the site in Saga Ariake. The ground between the surface and a depth of about 18m consists of sensitive cohesive soil with high plasticity; the deeper ground is made of sandy soil. The values of soil grain density (G_s) ranged from 2.60 to 2.66g/cm^3, which were approximately the same as those of the G_s values of marine cohesive soils found in Japanese coastal areas. At depths greater than 13m, the values of natural water content (w_n) are lower due to the effect of the pumping of groundwater. The values of w_n were nearly the same as those of liquid limit (w_L) at all depths, indicating that the ground is susceptible to disturbance. At depths greater than 13m, the increase of the values of unconfined compression strength (q_u) tended to be larger than the increase of effective overburden pressures, as in the case of the distribution of water contents. The average value of N_{kt} was 10. It is assumed

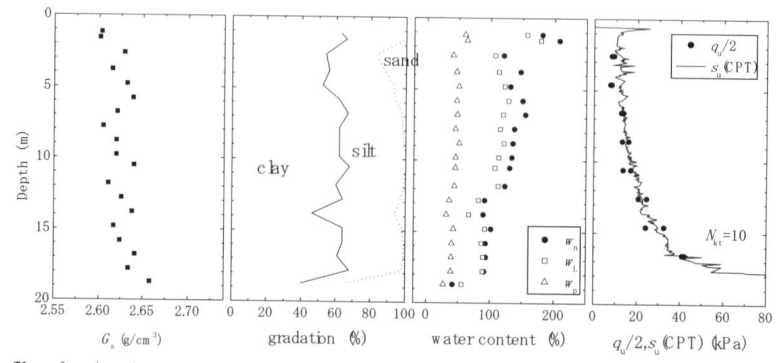

Figure 2. Profiles of engineering properties at Saga Ariake

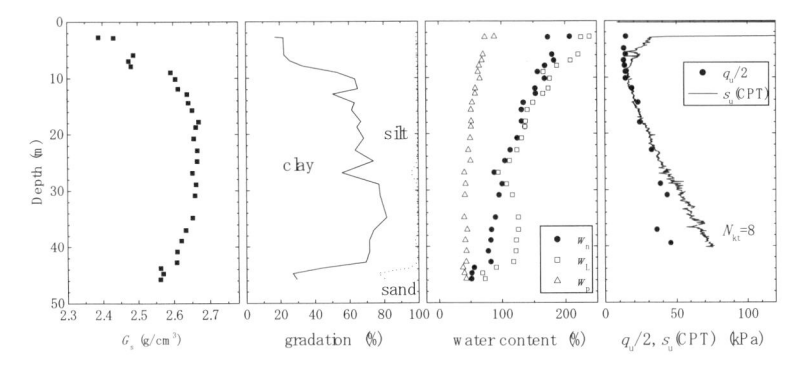

Figure 3. Profiles of engineering properties at Hachirogata

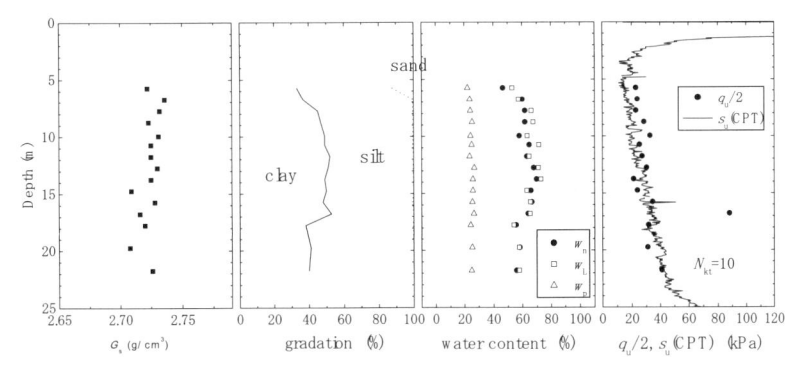

Figure 4. Profiles of engineering properties at Pusan

in this paper that clay fractions in grain size distributions have grains with diameters equal to or less than 5μm in accordance with the standard determined by JGS.

3.1.2 Hachirogata
Hachirogata is located in a northern part of Honshu. The target area is the land formed as a result of the reclamation of Lake Hachirogata implemented about 40 years ago. Figure 3 shows the results of physical tests, unconfined compression tests, and CPTU on the site in Hachirogata. The values of G_s showed extremely small values ranging from 2.40 to 2.62g/cm^3. The values of w_n in the ground from the surface to a depth of about -10m exceeded 150-200%. According to Tanaka and Locat (1999), this ground contains a large amount of diatoms microfossils entrapping a large portion of water inside the diatom shells and this will result in high water contents of the soil. Diatoms belong to the single-cell algae having porous shells mainly composed of silicic acids (SiO_2's). The fossilized remains of these silicic acids are known as diatomite. The mass of unit volume (ρ_t) of the ground rich in silicic acids is generally small. It ranged from 1.2 to 1.5g/cm^3 in this ground. The values of N_{kt} of this ground were about 8.

3.2 Pusan in Korea
Pusan is located 450km southeast of Seoul and near the mouth of the Nakdong River. The investigation was carried out in the site of Yangsan, a suburb of Pusan. The ground between depths of 6 and 22m consists of the deposit of marine cohesive soils; the deeper ground is made of stiff sandy soil. Figure 4 shows the results of physical tests, unconfined compression tests, and CPTU on the site in Yangsan. The values of G_s were about 2.72g/cm^3, which were smaller than those of G_s of Japanese cohesive soil. The values of ρ_t ranged from 1.62 to 1.74g/cm^3, which were larger than the average value of ρ_t of Japanese marine cohesive soil. In terms of clay contents, Pusan clay was relatively less plastic and less active. Contrary to smectite-dominant Japanese clay, the cohesive soil in this site contains illite, kaolinite, and quartz but no smectite. The values of N_{kt} of this ground were about 10.

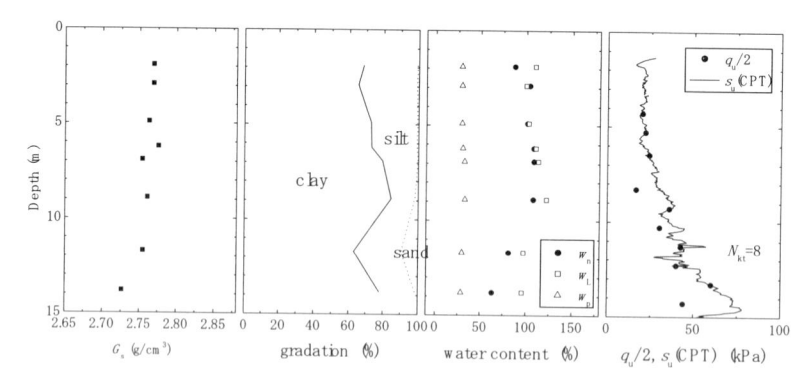

Figure 5. Profiles of engineering properties at Nong Ngoo Hao, in Bangkok

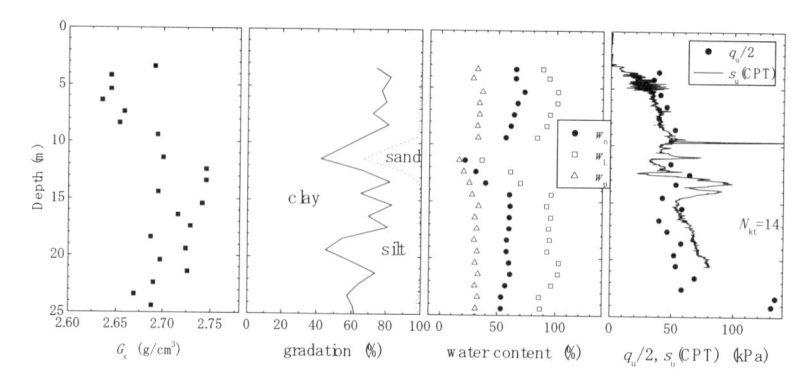

Figure 6. Profiles of engineering properties at Singapore

3.3 Nong Ngoo Hao in Thailand

The investigated site was Nong Ngoo Hao district located 10km east of downtown Bangkok. The ground of this district consists of the sedimentary deposit of soils carried by the Chao Phraya River. Figure 5 shows the results of physical tests, unconfined compression tests, and CPTU on the site in this district. The values of G_s were close to 2.76g/cm³. The clay contents of the soil were high. The values of w_n of the upper cohesive soil were almost the same as those of w_L; hence the soil was classified as susceptible to disturbance. The values of N_{kt} of this ground were about 8.

3.4 Singapore

The site is located about 5km from Changi International Airport. The tests were conducted at a point with a water depth of about 15m. Figure 6 shows the results of soil property tests, unconfined compression tests, and CPTU on the site. The ground consists of two marine clay deposit layers. In between the clay layers, there is a desiccated layer caused by a drop in the sea level during the ice age. The results of carbon isotope dating showed that the upper (ground level: 0.0-10.0m) and the lower (ground level: -15m or deeper) ones are sedimentary layers formed about 8,000 and about 25,000 years ago, respectively. The values of w_n were much lower than those of w_L; hence the soil was classified as stable. The values of N_{kt} were about 14 on average. σ_v in the calculation of $q_t - \sigma_v$ was calculated with the overburden pressure from the sea bottom surface.

3.5 Hai Phong in Vietnam

The investigation site is located in Hai Phong, the largest port city in Vietnam, about 100km away from Hanoi. The ground consists of the sediment of soil and sand carried by the Red River. The age determination using C14 (carbon dating method) indicated that it was the Pleistocene ground formed 20,000-40,000 years ago. It is currently covered by an about 7m-thick reclaimed layer. Figure 7 shows an overview of the soil. The values of w_L ranged from 45 to 67%, those of w_p (plastic limit) from 19 to 25%, and those of I_p from 24 to 42; hence the soil

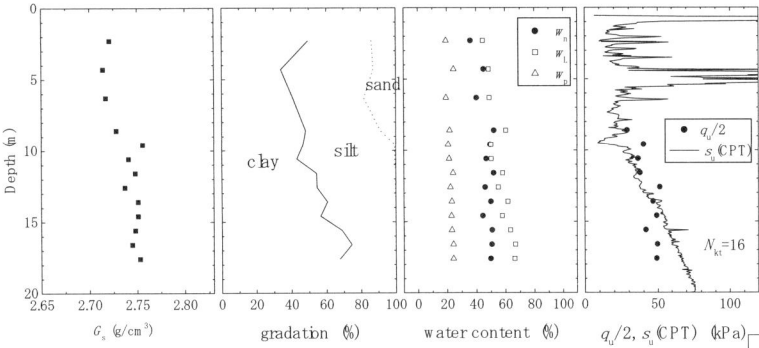

Figure 7. Profiles of engineering properties at Hai Phong

was mostly classified as a cohesive soil with low plasticity. The values of activity A_c of most samples were in the range of 0.6 to 0.9, which were smaller than those of the cohesive soil ground in Japan. The values of N_{kt} were around 16.

4 DISCUSSION

4.1 Engineering Properties

The characteristics of the ground measured at these 6 locations indicate that the G_s values of the Japanese ground are relatively smaller than those of the other East Asian areas by about 0.05–0.3g/cm^3, whereas the values of w_L and w_p are larger. Tanaka and Locat (1999) pointed out the existence of the diatoms microfossils (hereafter called diatoms) as one of the reasons for the findings above. Although many factors, such as sedimentary environments, clay minerals, and others, caused engineering characteristics to differ, this paper discusses the effects of diatom contents on engineering characteristics.

Figure 8(a) shows diatom content-G_s relationships, and Figure 8(b) the relationships between the diatom contents and the stress-strain curves obtained from the results of unconfined compression tests, where "B:D=75:25" means that the dry mass ratio of Bangkok clay is 75% and that of diatomite is 25% (Tanaka et al., 2003). It is evident from Fig. 8(a) that G_s decreases with increase in diatom content, reaching to the G_s values of diatoms; i.e., the increase of diatom contents cause G_s to decrease. The values of G_s for Japanese marine cohesive soil often show 2.65g/cm^3, suggesting that Japanese marine clay contains a dozen or so percent of diatoms. Shiwakoti et al. (2002) indicated that Japanese marine cohesive soil constituted several to nearly 50% of diatoms by mass. The results of our study support their findings.

The next discussion subject is the effect of diatom contents on the results of unconfined compression tests. As is obvious from Fig. 8(b), the unconfined compression strengths decreased with increase in diatom content. The stress-strain curves vary widely

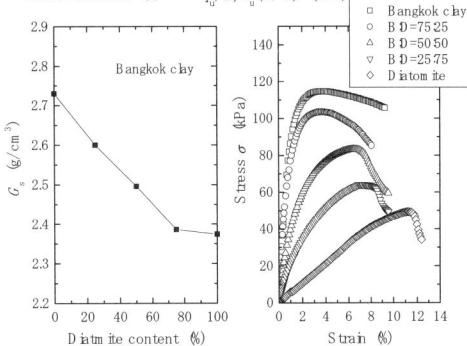

Figure 8. Effect of diatom content

with diatom content. To determine the undrained shear strengths of the cohesive soil ground such as the Japanese ground from unconfined compression strengths, it is required to take the effects of diatom contents into account.

4.2 Cone Factor

Table 1 shows the data of N_{kt} measured at the 6 survey sites classified with respect to principal clay minerals, geological ages, and overconsolidation ratios (OCRs). The Singapore clay deposit was divided into upper and lower layers because its sedimentary age was known. The values of N_{kt} of Holocene cohesive soil range between 8 and 12, which roughly agreed with the results shown by Tanaka et al. (1992). On the other hand, the values of N_{kt} of the Pleistocene ground were as large as 16 or 20. The values of N_{kt} of the Holocene cohesive soil were not influenced by principal clay minerals and OCRs.

Fig. 9 shows N_{kt}-I_p relationships. The values of N_{kt} of Saga Ariake clay, Hachirogata clay, Pusan clay, Bangkok clay, and upper Singapore clay, all of which belong to the Holocene cohesive soil, were in the range of 6-14, which is wider than the range shown by Tanaka et al. (1992). This is probably due to the fact that each clay deposit was formed in a different depositional environment, and that the undrained shear strengths of sand-containing samples were calculated based on the results of uncon-

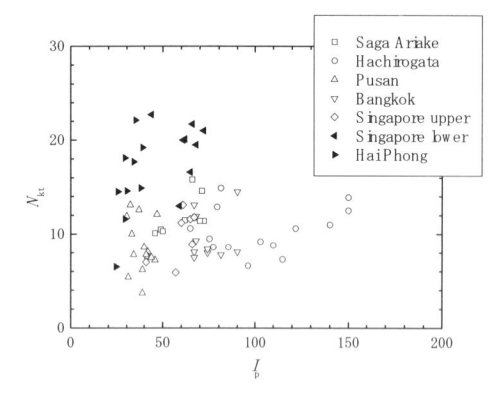

Figure 9. N_{kt}-I_p relationships

Table 1. Soil characteristics

site	clay mineral	geological epoch	OCR	N_{kt}
Saga Ariake	smectite	Holocene	1.0-1.3	10
Hachirogata	smectite	Holocene	1.0-1.1	8
Pusan	kaolinite	Holocene	1.0-1.3	10
Bangkok	smectite	Holocene	1.0-2.2	8
Singapore upper	kaolinite	Holocene	3.7-5.8	12
Singapore lower	kaolinite	Pleistocene	2.8-3.6	20
Hai Phong	kaolinite	Pleistocene	2.0-2.5	16

fined compression tests. The values of N_{kt} of the Pleistocene cohesive soil mostly range from 12 to 23, being large regardless of variations in I_p. This is because the undrained shear strengths of upper Singapore clay and Hai Phong clay were determined from the results of unconfined compression tests as it is usually done in the case of the Holocene cohesive soil. From the calculation of undrained shear strengths of such ground using simple shear strength, the values of N_{kt} are approximately 12. In the determination of N_{kt}, therefore, it is necessary to specify the method of obtaining undrained shear strength.

5 CONCLUSIONS

Boring and sampling in compliance with Japanese standards were conducted in the East Asia to investigate the engineering properties of the ground. Together with these, studies on the cone factors of the ground in the East Asia were also carried out from the viewpoints of geological ages and the method of determining undrained shear strength. The following are the conclusions of this investigation:
1. When determining the undrained shear strengths of the cohesive soil ground such as the Japanese ground from unconfined compression strengths, the effects of the diatom content need to be considered carefully.
2. From the fact that the unconfined shear strength for Pleistocene clays and clays with large OCRs is revealed in the range of relatively lower strengths than original values, it is recommended to set up a large value of cone factor for these clays.
3. When obtaining cone factors, it can be suggested to specify the method of evaluating undrained shear strength.

REFERENCES

Aas, G., Lacasse, S., Lunne, T. & Hoeg, K. 1986. Use of in situ tests for foundation design on clay", *Proceedings of ASCE Specialty Conference, In-situ '86, Use of in Situ Tests in Geotechnical Engineering*, 1-30.

Jamiolkowski, M., Lancellotta, R., Tordell, L. & Battaglio, M. 1982. Undrained Strength from CPT, *proceedings of the Second European Symposium on Penetration Testing*, 599-606.

Lunne, T. & Kleven, A. 1981. Role of CPT in North Sea Foundation Engineering, *Symposium on Cone Penetration testing and Experience, geotechnical Engineering Division, ASCE*, 49-75.

Tanaka, H., Sakakibara, M., Goto, K., Suzuki, K. & Fukasawa, T. 1992. Properties of Japanese Normally Consolidated Marine Clays Obtained from Static Piezocone Penetration Test, *Report of the Port and Harbour Research Institute*, Vol. 31, No. 4, 62-92. (in Japanese)

Tanaka, H. & Locat, J. 1999. A microstructural investigation of Osaka Bay clay, *Canadian Geotechnical Journal*, 36, 493-508.

Tanaka, M., Tanaka, H., Kamei, T. & Hayashi, S. 2003. Effects of Diatom Microfossil Contents on Engineering Properties of soils, *Proc. of The 13th Internationl Offshore and Polar Engineering Conference*, Vol. 2, 372-377.

Shiwakoti, D. R., Tanaka, H., Tanaka, M. & Locat, J. 2002. Influences of Diatom Microfossils on Engineering properties of soils, *Soils and Foundations, Japanese Geotechnical Society*, Vol. 42, No. 3, 1-17.

The regional information system 'Databank Ondergrond Vlaanderen - DOV'

I. Vergauwen, P. De Schrijver & G. Van Alboom
Ministry of Flemish Community, Department of Environment and Infrastructure, Geotechnics Division
Tramstraat 52, B-9052 Zwijnaarde, Belgium
ilse.vergauwen@lin.vlaanderen.be

Keywords: regional information system, subsoil, geotechnical database, Flanders

ABSTRACT: The Databank Ondergrond Vlaanderen - DOV (or database for the subsoil of Flanders) was created in 1996 as an initiative of three divisions within the Ministry of Flanders. DOV aims to collect, to interpret, to improve and to optimise access to geotechnical, geological and hydrogeological data of the Flemish subsoil. DOV has proved to be a valuable and powerful tool for the consultation of subsoil information. Engineers, architects, environmental specialists, and others, request data available in DOV on a daily basis to acquire reference material or as a basis to plan and work out more accurately further research in a specific area. Recently the DOV information system has been very valuable in several construction projects such as the extension of the ringway of Antwerp and the high-speed railway line in Flanders. DOV data could be used to calibrate, standardise and complete geotechnical data. Such was the case for the deepening of the canal Brussels-Charleroi at Halle where DOV data was used to trace the top of the bedrock and to determine the amount of rock to be dug out of the canal.

1 INTRODUCTION

In Flanders the demand for reliable and easy accessible data on the subsoil is constantly increasing. Although a lot of useful data are available within different institutes, organisations and companies, the consultation of these data is a difficult job as they are often too fragmented and non-digitised. The need for centralisation of the data was called for.

In 1996 a joint initiative of three departmental divisions within the Ministry of the Flemish Community resulted in the creation of the Database for the Subsoil of Flanders, commonly called 'Databank Ondergrond Vlaanderen' or 'DOV'. DOV intends to be an overall database compiling all geotechnical, geological and hydrogeological information on the subsoil of Flanders.

2 DOV INFORMATION SYSTEM

DOV is a co-operation within the Ministry of the Flemish Community between the Geotechnics Division and Water Division both in the Department of Environment and Infrastructure; and the Natural Resources and Energy Division in the Department of Economy, Labour, Internal Affairs and Agriculture (Boel et al, 2001, 2002, 2003).
The DOV information system is made up of 4 parts:

- DOV Input assures the correct intake of data into the database.
- DOV Output permits third parties to intensively explore and utilise the information.
- DOV Internet allows quick consultation of data. DOV Internet is freely accessible at http:\\dov.vlaanderen.be.
- DOV Report presents ready-made reviews with basic information of data requested in one of the previous modules.

The four DOV parts form a coherent entity. They are developed in a geographical orientated GIS environment so that any data is always linked to its geographic position within Flanders. Data consist of cartographic as well as alphanumeric information. The latter is depicted in DOV as a point at the plan co-ordinate of the testing point.

The amount of information is enormous. At present the processing of data available at the three partner divisions is in full process. At present DOV contains the alphanumeric point data of over 25,000 cone penetration tests; nearly 100,000 drilling points with bore hole descriptions and different types of geological interpretations; the administrative information of almost 25,000 groundwater licences and the piezometric groundwater level data and/or water quality analysis of 15,000 wells. Cartographic data integrated in DOV include the tertiary and quaternary geological maps, contour maps of several geological units and the groundwater vulnerability maps (De Ceukelaire et al, 2000; De Schrijver & Vergauwen, 2002).

The DOV information system is constantly improving. New and additional data are continuously added. The database structure and DOV parts are regularly revised, adapted and extended especially to the in- and output of new kind of data. The integration of data provided by other departments, public authorities and institutes is expected. First contacts with outside organisations have already been made in order to work out a methodology and framework for the intake of their additional information.

3 USING DOV

DOV has already proved to be a valuable and powerful tool for the consultation of subsoil information. Engineers, architects, environmental specialists and others request data available in DOV on a daily basis to acquire reference material or as a basis to plan and work out more accurately further research in a specific area.

3.1 *Extension of the ringway of Antwerp*

In order to relieve the busy traffic arteries of Antwerp, the project of extending the Antwerp ringway was put to study. This highway links the left and right bank of the river Scheldt and spans open sites as well as port areas and includes bridges, fly-overs and a tunnel construction.

The first object of the study was to collect all available geological and geotechnical information along the proposed highway extension. DOV revealed a density of information points in the area of interest (figure 1) and allowed an assessment of the overall soil layering. Gaps detected in the geotechnical data where additional testing was necessary could easily be pointed out.

In the investigated area soil conditions form a challenge as at the left bank the projected tunnel is crossing an area where the topsoil layers were eroded to great depth after an ancient dyke failure. The CPT-profile illustrates this (figure 2). The erosion reaches the Boom clay at its deepest point, resulting in a soft clay-silt layer of more than 20 m thickness.

From a hydrogeological point of view it was also necessary to discern the geological sequence over the site, up to the tertiary Boom clay. The principal geological units are:
- Sand fill layers
- Alluvial clay and peat
- Quaternary sands
- Tertiary sands (among others the glauconitic Antwerp sands)
- Tertiary Boom clay

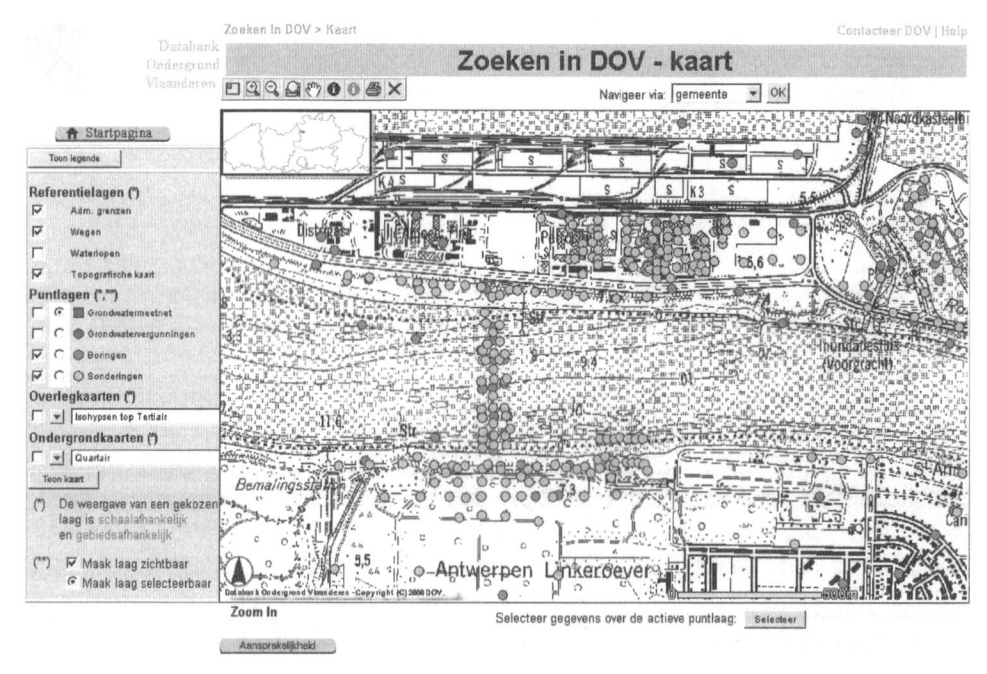

Figure 1. Example of DOV Internet showing a part of the area of interest where DOV data are indicated as light grey points for cone penetration test data and dark grey points for the drilling data. The topographic map of the area is displayed as a reference map.

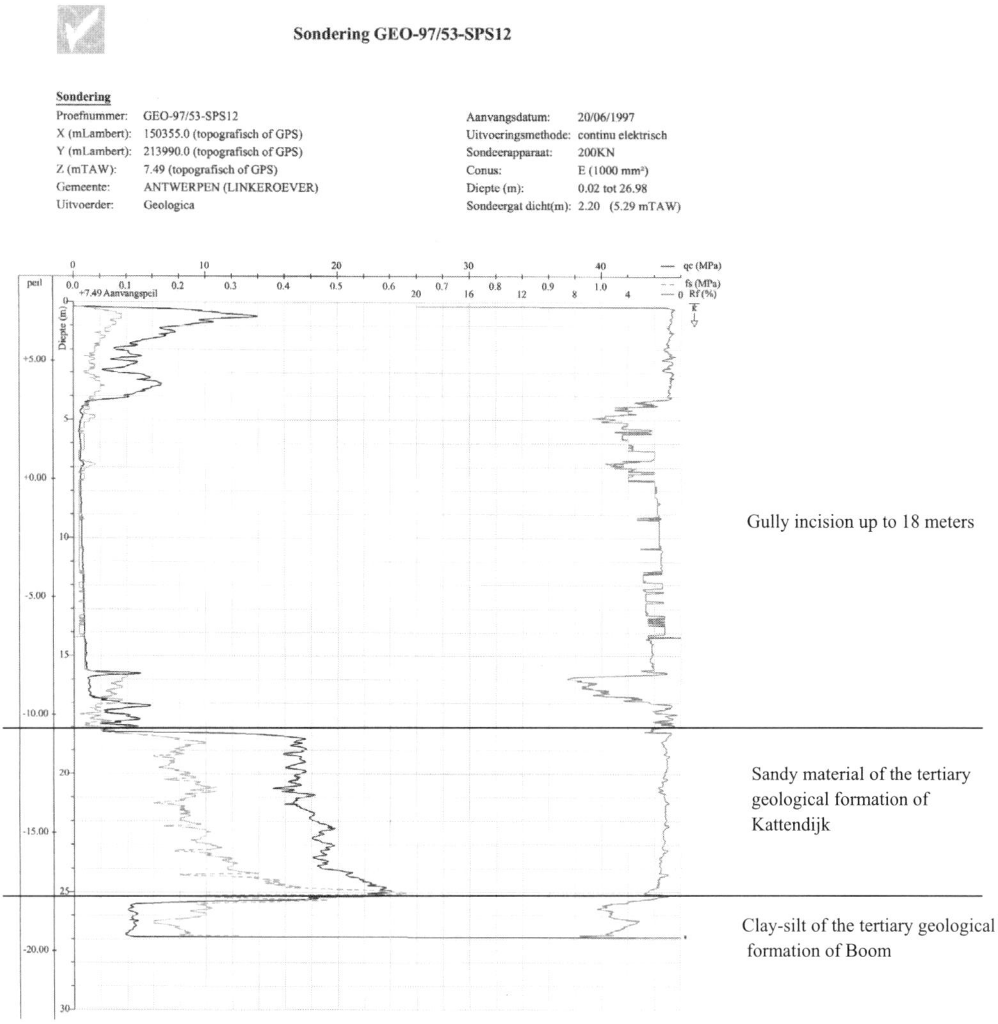

Figure 2. Typical CPT-profile where the topsoil was eroded by a deep gully after a major dyke failure.

From the information gained from of the DOV database a 3D-model of geotechnical soil units was created (figure 3). The model allowed assessment of the geotechnical impact of different planned routes for the ringway. Although traffic considerations did prevail on geotechnical grounds, a well-documented assessment could be made of pros and cons for each considered route.

Following the selection of the route of the ringway extension, a detailed soil investigation is being performed. These new data will form a new input in the DOV database and provide the project team with additional information.

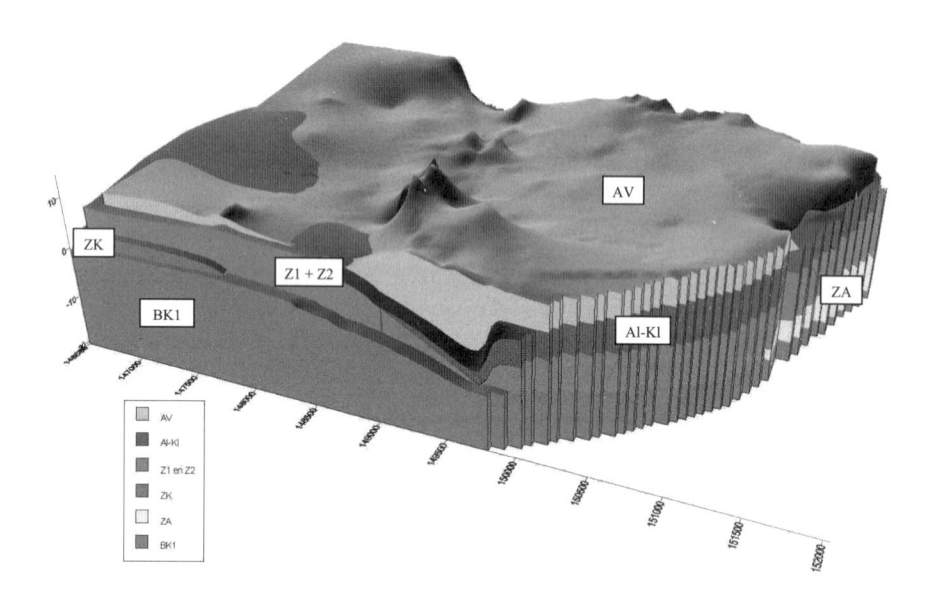

Figure 3. 3D-model of geotechnical units in the investigated area with sand fill layers (AV), alluvial clay and peat (Al-Kl), Quaternary sands (Z1 en Z2), Tertiary Kattendijk sands (ZK), Tertiary Antwerpiaan sands (ZA) and Tertiary Boom clay (BK), after Studiegroep Antwerpen Mobiel, 2003.

3.2 Deepening of the canal Brussels-Charleroi at Halle

In the study of deepening the canal Brussels-Charleroi at Halle DOV data are used to trace the top of the bedrock and to delimit the amount of rock to be dug out of the canal. In the investigated area the depth of the bedrock and the thickness of the upper mud layer vary a lot. At the start of the project plans were made to map out the bedrock and mud with the help of different geophysical methods. However, an initial study on the basis of the DOV drilling and cone penetration test data revealed extensive information on the exact depth of the bedrock throughout the area of interest. Three-dimensional interpolation of the data permitted a clear outlining of the bedrock. As a result of this initial investigation the engineers decided not to invest in a geophysical research. Areas with to few data could easily be filled in with additional drilling and cone penetration tests (figure 4).

3.3 The high-speed railway line in Flanders

The DOV information system has been very valuable in constructing the high-speed railway line in Flanders. DOV data could be used to visualise geological, geotechnical and hydrological data in cross sections. Data permitted the identification of the different geological layers, their composition, bearing capacity, stability and geophysical and hydrological parameters. On the basis of these data areas where additional research, such as additional drilling and cone penetration tests, was essential could easily be identified. Thus, the DOV data permitted an initial, fast and direct insight in the subsoil in the area of interest making a detailed investigation with a minimum of costs possible (figure 5).

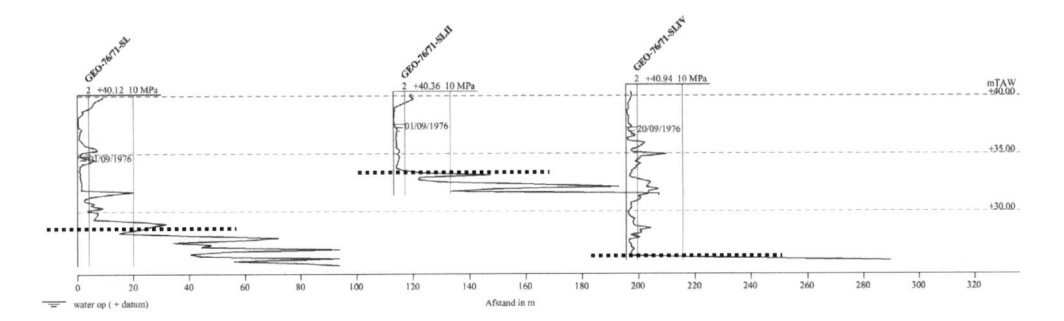

Figure 4. Extract of the geotechnical profile at Halle: the sudden increase in cone resistance indicates the transition zone between mud and bedrock. This transition occurs at different depths.

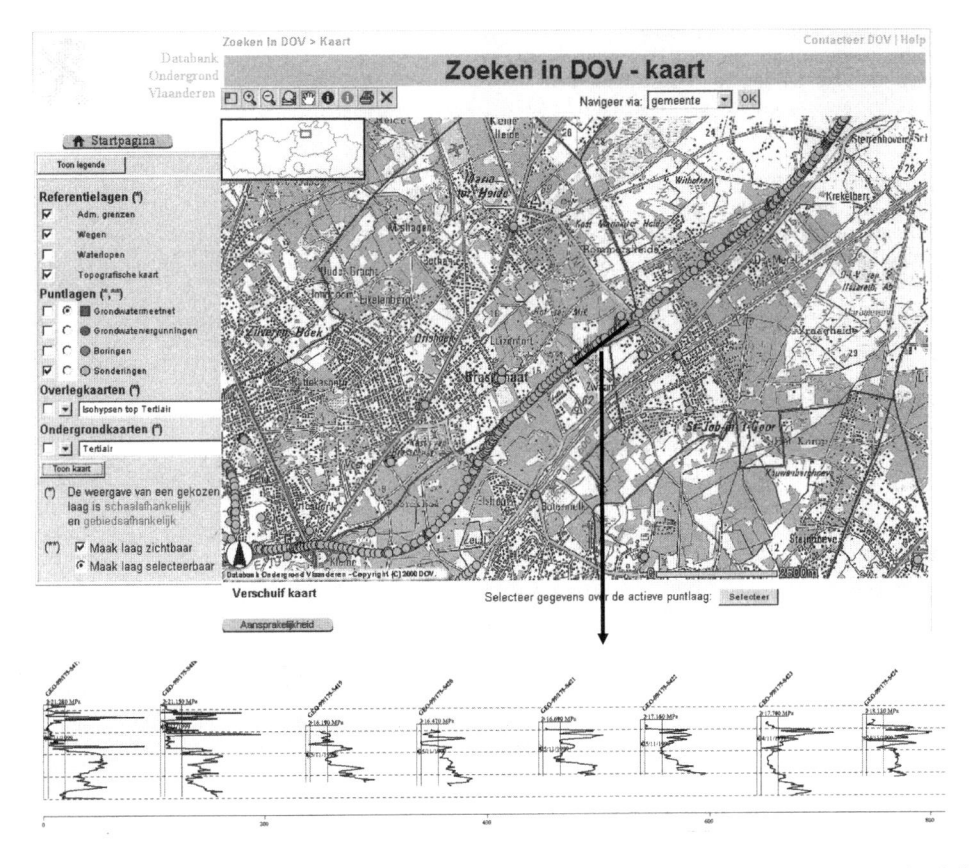

Figure 5. All along the high-speed railway line in Flanders huge amounts of geotechnical data are available in DOV (points indicate the presence of cone penetration tests). Data can be visualised in geotechnical profiles such as the illustrated profile along the indicated line.

4 CONCLUSIONS

DOV is a valuable and powerful database for the subsoil of Flanders. Since the activation of DOV Internet on May 22nd 2002 DOV is open to anyone. The DOV information system has been very valuable in several construction projects such as the extension of the ringway in Antwerp and the highspeed railway line in Flanders. DOV data could be used to calibrate, standardise and complete geotechnical data.

At the start-up of DOV quite some geotechnical companies and laboratories were rather sceptic. They feared that making data freely available to anyone would have a negative impact on their work orders. By now, the same companies are enthusiastic and intensive users of the DOV data. DOV is not interfering with their work; on the contrary, DOV forms an asset to their research and investigation. Almost any study undertaken by these companies is initiated by an exploration of the DOV data. Many companies are also willing to release their data to DOV in order to complete the database. At this moment the methodology and framework for the intake of this kind of additional data is worked out.

Through its central structure and direct accessibility of a large amount of geotechnical, geological and hydrogeological data, DOV is of scientific and economic importance for the use of soil and groundwater in Flanders.

REFERENCES

Boel, K., De Ceukelaire, M., De Schrijver, P. and Van Damme, M. (2001) Databank Ondergrond Vlaanderen -Jaarverslag 2000. MVG - LIN- AMINAL, Directoraat-generaal, Brussel, 24 pp.

Boel, K., De Ceukelaire, M., De Schrijver, P. and Van Damme, M. (2002) Databank Ondergrond Vlaanderen -Jaarverslag 2001. MVG - LIN- AMINAL, Directoraat-generaal, Brussel, 24 pp.

Boel, K., De Ceukelaire, M., De Schrijver, P. and Van Damme, M. (2003) Databank Ondergrond Vlaanderen -Jaarverslag 2001. MVG - LIN- AMINAL, Directoraat-generaal, Brussel, 24 pp.

De Ceukelaire, M., Jacobs, P., Sevens, E. and Mostaert, F. (2000) Recent initiatives in regional geological cartography in Flanders - Part 1: 'Databank Ondergrond Vlaanderen (DOV)': the geological information system for the subsoil of the Flemish Region. Proceedings of the Third Congress on Regional Cartography and Information Systems. Munich. 79-81.

De Schrijver, P. and Vergauwen, I. (2002) Databank Ondergrond Vlaanderen (DOV). Innovatieforum. Technologisch instituut - Genootschap Grondmechanica en Funderingstechniek. Antwerpen.

Studiegroep Antwerpen Mobiel (2003) Geotechniek. Infrastructuur en Kunstwerken L.O. Antwerpen.

Indices

Keyword index

accuracy 1035
active design 1415
ageing 943
airport 1125
analysis 3, 1543
analytical solution 951
angle of shearing resistance 75
anisotropy 209, 1717
ANSYS 1425
apparatus 741, 783
applied 607
aquares 427
Artificial Neural Network (ANN) 889
at rest stress state 1755
at-rest lateral earth pressure coefficient 1663
attenuation measurement 1883
autoregressive moving avarage (ARMA) 835

Ball 679
ball penetrometer 671
basaltic plateau 1177
bearing capacity assessment 1421
behaviour law 1425
beta probability 913
bioremediation 665
bitumen coating 853
block sample 1875
block sampling 1861
body and surface waves 1585, 1805
body waves 97
bonding 1869

bored piles 1207, 1431
borehole extensometers 1171
borehole shear 913
Borehole Shear Test 1663
boundary condition 829
bridge foundations 483
brittle flexes 1225
brown fields 1035
Bryozoan limestone 1813

Calabar 1125
calcareous 1191
calibration chamber 147, 829, 843
case history 97, 167, 1147
CASE method 299
case studies 1033
cavity expansion theory 391
cemented soil 1047
centrifuge 147, 377
CFA 1565
CH 1805
characterization 253, 315
clay 3, 133, 209, 271, 331, 971, 1191, 1319, 1775, 1827, 1869
clay seams 1543
clustering 373, 905
coal 1285
coarse sands 1843
coarse-grained alluvium of Tehran 1701
coastal engineering 555
coefficient of earth pressure at rest 1739
cohesion 1275

cohesive soils 869, 979
collapsible 1311
collapsible soils 1233, 1267
colluvium 1161
comparative study 399
complex permittivity 889
compressibility 1643
compression 1869
compression properties 1875
concrete 1481
conductivity 513
cone 345, 757
cone factor 1019
cone penetration 687, 701, 1597, 1695
cone penetration resistance 1687
cone penetration test 233, 399, 585, 787, 897, 987, 1003, 1035, 1185, 1507, 1663
cone penetration test/piezocone 1345
cone penetration testing 905
cone penetrometer 391, 671, 795, 843, 913, 921, 1439
cone resistance 315
cone tip resistance 829
conformity assessment 385
consolidation 287, 1739
consolidation coefficient 209
constitutive law 927
constrained modulus 345
contact 1425
contamination 571, 1339
contamination attenuation 1011
core logging 647
correlation 1075, 1089

cost estimation 1557
cost-reduction 615
coupled consolidation model 951
cover 1075
CPT 147, 315, 331, 513, 693, 757, 787, 877, 1075, 1119, 1275, 1421, 1499, 1627, 1731, 1799
CPT profiles 885
(CPT/CPTU) 1345
CPTU 315, 679, 719, 877, 1019, 1325, 1457, 1605, 1717, 1731, 1765, 1791, 1851
CPTU data 979
CPTU tests 1097
critical state concept 391
critical state soil parameters 1821
critical train speed 1651
cross-hole 1827
cross-hole test 1345, 1465, 1723
cross-hole tomography 971
cyclic stability 869
cyclic stress ratio 869
cyclic test 671
cylindrical cavity expansion 897, 1353

dam 615, 1285
damping ratio 1883
data interpretation 1061
DCP 367
de-coupling 643
deep dynamic compaction 727
deep mixing 1185
deformability 657, 1605
deformability moduli 1067
deformation moduli 279
deformation modulus of rock mass 265
deltaic 1605
depositional environment 1619
design optimisation 1089
diaphragm walls 993
diatom microfossils 1019
dielectric permittivity 513
dilatometer 345, 727, 775, 913, 921, 1439, 1597, 1673, 1695
dilatometer test 187, 265, 1663
dilatometer testing 963
direct push 795
discrete particle scheme 459

discussion guidelines 1391
dispersion 539, 1481
dispersion curve 767
dissipation test 951
dissipation testing 287
Distinct Element Method (DEM) 299
disturbance 271
disturbed zone 943
DMT 1275, 1605, 1717, 1731, 1765, 1799
dominant higher modes 547
downdrag 853
downhole tests 1695
DPL/M 1731
DPM 1219
DPSH 1731
dredging 427
drilled piles 1311
drilled shaft 1465
drilling 385, 787
drilling and *in-situ* testing 1533
drilling equipment 733
drilling parameters 665, 733
drilling process monitoring 1219
dual dilatometer 775
dump site 581
dynamic probe 1089
dynamic probing 325
dynamic site characterization 935
dynamic soil properties 1465
dynamic testing 133

E-modulus 927
earthquake 869, 1131
earthquake engineering 1199
edge failure 1081
effective stress 1041
elastic modulus 133, 345, 1681
elastic shear modulus 1869
elastic-plastic model 75
elasticity 167
electrical 399
electrical conductivity 687
electrical resistivity 433, 483, 505, 1147
electromagnetic survey 581
embankment 971
embankment dams 593, 1891
embankments 1605
embedment 853

empirical correlations 1843
energy 1565
energy efficiency 339
energy measurements 351
engineering geological investigation 1397
engineering geology 1813
engineering properties 1019, 1585
enhanced penetration 795
environment 687
environmental site characterization 795
Eocene 1081
equipment 271
error propagation 539
ERT 809
excavation 1415
expansive 331
experimental modeling 491
exploration 427

F^2 method 339
factor of safety 1575
failure costs 1747
fibre effect 271
field investigations 657
field testing 307
field vane 861
field vane test 271
field-saturated hydraulic conductivity of soil 749
fine-grained soils 133
Finite Element Method 1425, 1449
finite elements 475
finite-element analyses 921
fire 1481
Flanders 1025
flat dilatometer test 993, 1511
foundation design 1089
fractal analysis 885
fractal dimension 885
fractured rock 665
friction 1425
friction angle 1663
friction sleeve 693
Full Displacement Pressuremeter 657
full-flow penetration probes 679
fundamental frequency 635
fuzzy 905

FV method 339

gas pipeline 1161
general review 1391
geoelectrical 427
geohydraulic tests 385
geologic uncertainty 905
geological survey 1067
geological-geotechnical zoning 1067
geology 1105
geomembranes 1147
geophysical characterization 451
geophysical modeling 593
geophysical prospection 1443
geophysical surveying 1177
geophysical testing 1687
geophysical tests 409
geophysics 97, 409, 483, 555, 581, 585, 601, 607, 665, 1033, 1105, 1147, 1415, 1695
GeoQ 1747
geostatistics 1199, 1263
geotechnical 1105
geotechnical database 1025
geotechnical exploration 1397
geotechnical investigation 1111
geotechnical survey 787
geotechnics 97, 585, 1519
gold tailings 575
GPR 419, 809, 1075
GPS 1225
grain size effects 299
granite 39, 505
granitic residual soils 1279
granular soils 1139, 1191
gravel 299
grillages 1139
ground characterization 1219
ground deformations 233
ground hazards 497
ground improvement 1709
ground-penetrating radar 497, 1177
groundwater 571
groundwater measurements 307, 385
grout injection evaluation 443
grouting 615
Guelph pressure infiltrometer 749

guided waves 563

hard rock 963
hard soils-soft rocks 1293
hazardous waste 1075
heavy metal 581
heterogeneity 1035
heterogeneous soil 187
highways 1147
historic mining 497
hodograms 1611
horizontal load 853
HWAW 767
hydraulic conductivity 287
hydraulic fills 1111, 1331
hydrogeology 505

identification of parameters 593
IFR 1499
imaging 601
impedance 1465
impoundment 1285
inclinometer 1061, 1225
in-situ 741, 783, 861, 1213
in-situ analysis 809
in-situ and laboratory testing 1813
in-situ and laboratory tests 1361
in-situ cohesion 1047
in-situ ground monitoring 1061
in-situ measurement 749
in-situ penetration test 671
in-situ rock deformability 963
in-situ soil testing 3
in-situ techniques 657
in-situ test 167, 253, 265, 1421, 1619
in-situ testing 287, 555, 625, 679, 719, 757, 819, 825, 1041, 1233
in-situ testing, 1259
in-situ testing 1627, 1635, 1663, 1775
in-situ tests 97, 345, 351, 1033, 1053, 1139, 1301, 1331, 1709, 1843
in-situ void ratio 1835
in-situ tests 49
instrumentation 491
instrumented dilatometer 727
instrumented embankment 1511
instrumented load test 1551

interaction 1089
interface shear 693
interfaces 701
interpretation 3
inverse problems 451, 935, 951
inversion 563, 607, 1481
investigation 307
investigation programme 253
ISC'2 test site 521
iterative forward medeling (IFM) 1611

joint shear strength 1067

Kalman filter 835
karst terrane 483
K_0 1765, 1827

laboratory and field tests 1165
laboratory testing 307, 1161, 1775
laboratory tests 167, 1843, 1851, 1861
landfill 601, 1075
landslides 601, 1081
lateral skin friction of piles 359
layered clay 679
layered half-space 935
layered soil 377
leachate 571, 1011
leak detection 1147
lightweight truck penetrometer 1319
limestone 1207
limestone formation 1105
limestone rock mass classification 1397
limit pressure 1681
liquefaction 1131, 1199
liquefaction damage assessment 713
liquefaction potential index 713, 1199
Llobregat delta 877
load capacity 1499
load transfer 1431
load transfer curves 1525
load-settlement curve 853
Love waves 935
LPT 299
lumpy fill 801
LVL/HVL resolution 539

Mackintosh Probe 325
marine clay 1041, 1739
maritime work 1457
marl 1081
Maryland 331
MASW 563, 1111
material damping 1585, 1655, 1805
matric suction 1311, 1319
maximum shear modulus 1003
measurement 741
mechanical 399
mechanical behavior 187, 1325
mechanics 3
Menard PMT 927
metro 1119
mine structures 963
mineral water 505
minimum variance 835
modeling 829
moderator's questions 1391
modified DPL system 1519
moduli 75
modulus 1511
modulus of deformation 963
Mohr-Coulomb model 279
Mokattam city 1081
monitoring 1225, 1747
Monte Carlo 539
MSW landfill 1339
mudstone 331
multi friction sleeve attachment 693
multi-method approach 809
multichannel 459
municipal solid waste 657, 1233, 1259
municipal solid waste fills 1325
MWD 733

N-Value 339
neural network 957
Nigeria 1125
noise measurements 521
non-destructive testing 475, 563
non-engineered clayfill 1571
nonlinear deformability 1891
nonlinear response 1431
normalized cone tip resistance 1835
numerical analyses 1851

numerical modelling 1557
numerical simulation 147, 1891

object detection 475
Occam's inversion 547
OCR 391
oedometer tests 1791
Omega pile 1551
open pit 1415
open-pit mine sites 1035
opencast mines 1791
organic soil 1339
outcrop mapping 1533
overconsolidation ratio 1041, 1835

P-S logging 1473
P-wave velocity 443, 531
paleoliquefaction 1131
parameter determinations 819
partial drainage 49
partial saturation 491
pavement design 367
pavements 563
PCSV 1263
peat 775
penetration testing 209, 757
penetration tests 75
penetration tests: Standard Penetration Test (SPT) 1345
penetrometer 279, 693, 1519
penetrometer testing 1279
permeability 615, 1067, 1285
physical index 1619
physical modeling 147
piezocone 345, 377, 391, 1111, 1507, 1643
piezocone test 187, 373, 957
piezometers 1457
pile 913, 1425, 1439
pile capacity 1507
pile design 1519
pile group 1543
pile load test 1207
pile settlements 1525
piles 853, 1543
piling 1565
pipe piles 1489, 1499
pipeline route 427
piston sample 1875
piston tube sampling 1861

Plate 679
plate load test 75, 265, 927, 1047
plate loading 1701
plate penetrometer 671
PLT 1325
PMT 1473
polarization analysis 1611
pollutants 687
pore pressure 315, 1403
pore water pressure 843
porous 1311
porous clay 1301
portable dynamic cone penetration test 1155
Porto 1089
post flotation sediments 987
power 1565
prediction 1511
pressuremeter 1185, 1191, 1267, 1431, 1701, 1827
pressuremeter test 75, 897, 1681
pressuremeter tests interpretation 1353
probe 687
problematic sites 693
project-specific rock mass classification 1533
Pusan clays 1619

qualification 385
quality 1747

radioisotope cone 801
railroad embankment 1449
rain-induced slope failure 1155
rapid site assessment 687
rate of loading 133
Rayleigh wave 539, 585
RC tests 1805
refraction 601
refraction seismic 1533
refuse 1285
regional information system 1025
regression models 1687
relative compaction index 1371
relative density 1663
reliability 1575
remolded undrained shear strength 679

© 2004 Millpress, Rotterdam, ISBN 90 5966 009 9

remoulded strength 671
repeatability 325
residual 331
residual saprolitic soil 1755
residual saprolitic soils from granite 1361
residual soil profiles 1293
residual soils 345, 531, 1233, 1259, 1275, 1353, 1775, 1799
residual soils from granite 419, 433
Resilient Modulus 367
resistivity 601, 607, 1473
resistivity imaging 571
resistivity survey 427
resonant-column 1827
rheological test 1507
risk assessment 1011, 1155
risk management 1747
rock 625
rock mass characterization 531, 647
rock properties 963
rock socket 1207
rock support 1415
rockmass classification 39
rocks 1033

sample disturbance 1775, 1861, 1875
sampling 307, 385, 555, 1765
sand 3, 147, 829, 1003, 1489, 1663, 1775
sand & gravel piles 1709
sand sediment soil 749
sands 377, 943
saprolitic granite soils 75, 1345
SASW 475, 767, 1075, 1449, 1723, 1805
scale 853
scale effects 963
scattering 1891
SCPT 1655
SCPT test 643
SCPTU 1003
sediment facies 877
seepage 451
seismic 601
seismic cone 1185
seismic cone penetration testing (SCPT) 835, 1611

seismic cone penetrometer 1131
seismic cone testing 943
seismic CPT 1673, 1805
seismic profiling 1651
seismic refraction 1125, 1473
seismic source 643
seismic techniques 475, 491
seismic testing 97, 133
seismic tests 75, 409, 971
seismic tomography 443, 531, 615, 1067
sensitivity 209, 671, 719, 861
sequence stratigraphy 877
settlement 279, 913, 1165, 1191, 1439, 1511, 1543, 1571, 1605, 1673
settlement prediction. 1585
SH-reflection 521
shaft capacity 1565
shale 1081
shallow foundation design 1575
shallow foundations 853, 1165
shallow SH Reflection 419
shear modulus 133, 167, 459, 1687, 1701
shear strain 133
shear strength 49, 133, 209, 271, 287, 1041, 1047, 1875
shear wave 585, 1701
shear wave velocity 547, 575, 707, 943, 1003, 1263, 1655, 1687, 1755
shear waves 643, 1695
Shelby tube 367
silts 133, 187, 1643
similarity 373
simulated annealing 563
Singapore marine clay 1835
sink-hole 1403
site characterization 409, 707, 801, 1033, 1233, 1259, 1443, 1465, 1723, 1747, 1821
site investigation 233, 325, 555, 607, 625, 647, 809, 825, 1089, 1105, 1119, 1415, 1557, 1627
size effect 829
sleeve friction 315
sliding mechanism 1061
sliding micrometer 1225
slope 913
slope engineering 1219

slope failures 1177
slope stability 921, 1213, 1225, 1821
slurry 1285
small shear modulus 75, 1053
small strain 1717
small strain stiffness 49, 943
smoothing 835
soft clay 459, 679, 1165, 1457
soft rocks 1775, 1891
soft soils 719, 757, 775, 1651
soil 625, 1311
soil characterization 889
soil classification 359
soil disturbance 861
soil liquefaction 233, 713
soil nailing 1219
soil parameters 1791, 1851, 1861
soil permeability 749
soil profile 1585
soil stratigraphy 701
soil structure 49
soil-pile interaction 1431
soil-rock sounding 733
soil-structure 1089
soil-structure interaction 1635
soils 1033, 1655
solid waste landfill 1263
sounding procedure 733
Spatial Autocorrelation Coefficient 635
spectral ratio method 1883
spoil pile 1213
spread footing 1439
spring constant 133
SPT 299, 339, 351, 707, 1111, 1125, 1325, 1473
SPT instrumentation 293
SPT-T 1489
SPT-T tests 1311
stability 1791
stability assessment 1397
standard penetration 1597
Standard Penetration Test 293, 359, 1089, 1185, 1319, 1489, 1575, 1663, 1681
standardization 399
standards 253, 307
state 3
states of consistency 979
static loading 279

statistical analysis 987, 1371
statistics 1035
stiffness 3, 575, 1185, 1585, 1605, 1673, 1717, 1805
stiffness and strength 1361
stiffness non-linearity 49
strain rate 133, 209
stratigraphy 905
strenght increase 971
strength 3, 1185, 1285, 1869
stress-strain of soils 75
stress-strain relationship 1827
s_u 391
subgrade characterization 367
subgrade reaction 133
subgrade reaction method 993
subsidence 1171
subsoil 1025
suitability 713
surface and borehole seismics 419, 433
surface roughness 693
surface wave method 521
surface waves 97, 409, 451, 459, 563, 585, 1111, 1481
synthetic seismograms 547

T-bar 679, 757
T-bar penetrometer 671
T-Z curves 1525
tailings 1233, 1259, 1285, 1371
tailings densities 1331
tailings disposal systems 1097, 1331
tank 1191
technological characterization of tailings 1097

Tehran alluvium 1047
televising boreholes 647
test variability 1139
texture 693
thin layers 701
Time Domain Reflectometry (TDR) 513
time-frequency analysis 767
tip resistance 671, 1319
tomography 1611
topographic control 1171
topography 1105
torque 1489, 1519, 1565
torque measurements 359
transfer function 635
transported soils 75
transversely isotropic (TI) 1611
trial embankment 1571
triaxial 783
triaxial tests 957, 1791
tropical soils 1011, 1519, 1627
tunnel 1147, 1397, 1443, 1635
tunnel collapse 1473
tunnelling 39, 1557
types of hammer 339

ultimate capacity 1543
uncertainty 1575
uncompressed soils 443
unconfined compression test 1019
undisturbed samples 1843
undrained shear strength 671, 679, 757, 957, 971, 1739, 1851
undrained strength 719
unit skin friction 1489

unload-reload modulus 727
unsaturated flow 1403
unsaturated moisture properties of soil 749
unsaturated soil conditions 49
unsaturated soils 1233, 1267, 1301, 1551
uphole test 707

value engineering 1747
vane size 271
vane test 209, 1457
variability 1003, 1627
varved clay 679, 861
Venice 1511
Venice lagoon 187, 1643
vibration 843
viscous fluid 377
vision cone penetrometer 701
V_s profile 767
V_s velocity 635

waste disposal 1011
waste disposal site 571
water infiltration 1081
water level variations 1171
water table 491
wave attenuation 1655
wave modelling 585
wavelet transform 475
weathered granite 1089
weathering 1293
wet density 801

yield pressure 1681

zoning Mexico City 885

© 2004 Millpress, Rotterdam, ISBN 90 5966 009 9

Author index

Aboushook, M. 1081
Abrahão, R. 647
Agharazi, A. 265
Ahmadi, M.M. 829
Åhnberg, H. 271
Ajalloeian, R. 1397
Al Hamdan, W. 1035
Albuquerque Filho, L.H. 1097, 1331
Alicescu, V. 1403
Almeida e Sousa, J. 1301
Almeida, F. 419, 433
Amini, A. 943
Anderson, J.B. 1597
Anderson, N. 1723
Anderson, T. 963
Antão, A. 927
Apostolidis, P. 635
Araruna, J.T. 657
Arbanas, Ž. 1415
Arbaoui, H. 279
Areias, L. 643
Arroyo, M. 1457, 1605
Arulrajah, A. 287, 1041
Asghari, E. 1047
Assis, A.P. 1301, 1635
Auken, E. 607
Avsar, S. 1263
Azzam, R. 1035

Babendererde, S. 39
Baek, S.H. 1619
Baillot, R.T. 647
Balbi, D.J.G. 373
Balic, M. 1465

Bang, E-S. 339, 707
Barton, N. 647
Bates, L. 331
Baziw, E.J. 835, 1611
Beatrizotti, G. 1731
Bednarczyk, Z. 1791
Beira Fontaine, E. 1421
Bekkouche, A. 1425
Bello, L.A.L. 657
Benoît, J. 665, 727
Bergdahl, U. 733
Berglund, C. 271
Bernardes, G.P. 351
Bhadriraju, V. 1185
Bica, A.V.D. 1177
Black, J. 1851
Bo, M.W. 287, 1041
Bodé, L. 279
Boháč, J. 1571
Bolton, M.D. 377
Bonita, J. 843
Bothner, W.A. 665
Bouazza, A. 1263
Boyle, R.F. 963
Brabers, P. 427
Brandon, T.L. 843
Brassinga, H.E. 757
Bray, J.D. 1575
Bressani, L.A. 1177
Bressolette, Ph. 279
Briaud, J-L. 853
Brouwer, J.H. 585
Bullock, P.J. 913, 1439
Busquets, P. 877
Butcher, A.P. 409, 1717

Calabrese, M. 1511
Camapum de Carvalho, J. 1311
Campanella, R.G. 299
Campos, J.C. 1339
Canals, M. 877
Carvalho, J. 1361
Carvalho, J.M. 419, 433, 1755
Carvalho, P. 1293
Casamor, J.L. 877
Cavalcante, E.H. 293
Cavallaro, A. 1053
Cerato, A.B. 861
Chang, H-H. 1875
Chang, M. 713, 1061
Chang, M-F. 1431
Chen, J.W. 1061
Chen, J-Y. 921
Chien, S.C. 1061
Chiou, Y.F. 1061
Chung, C-C. 513
Chung, S.F. 671
Chung, S.G. 1619
Clarke, B.G. 657
Clayton, C.R.I. 575, 1565
Coelho, M.J. 443, 531
Colàs, S. 877
Comina, C. 451
Conlon, B. 963
Costa, E. 1361
Cotelo Neiva, J.M. 1067
Coutinho, R.Q. 1165, 1233, 1267
Cruz, N. 253, 1275, 1799
Cunha, R. 1519
Cunha, R.P. 1301, 1391, 1709

Cushing, A.G. 1285

Dahlin, T. 607
Daniel, C.R. 299
Danziger, B.R. 293
Danziger, F.A.B. 293
Dasari, G.R. 801
De, A. 1075
de Almeida, F.T. 581
de Arruda Dourado, K.C. 1233
de Campos, T.M.P. 657
de Carvalho, D. 359, 1421, 1551
De Mio, G. 1627
De Schrijver, P. 1025
de Souza Neto, J.B. 1233
De Vos, M. 399
Debasis, R. 869
Degrande, G. 1655
DeGroot, D.J. 679, 1775
Deidda, G.P. 521
DeJong, J.T. 679
Der Kiureghian, A. 1575
Devincenzi, M. 253, 877, 1457, 1605
Díaz-Rodríguez, J.A. 885
Ding, W. 889
Donohue, S. 459
Doo, G.C. 1061
dos Santos Jr., E.L. 581
Doser, D. 491
Dourado, K.C.A. 1267
Drijkoningen, G.G. 1449
Drossaert, F.H. 1449
Duarte, I.M.R. 1279
Dunn, R.J. 1075

Edet, A.E. 1125
Eitner, V. 307, 385
El-Sohby, M. 1081
Elis, V.R. 467
Elmgren, K. 733, 787
Elmi, F. 897
Endley, S.N. 1191

Facciorusso, J. 905
Fahey, M. 49, 819, 1673
Failmezger, R.A. 913, 921, 1439
Fairuz, S.M. 315
Fakher, A. 325, 1701
Falcão, P.R.F. 1311

Falivene, O. 877
Farias, M.M. 1635, 1709
Farouz, E. 921
Faure, R.M. 1481
Favre, J.L. 897
Fernández, A.L. 1525
Ferreira, C. 1361, 1755
Ferris, J.K. 1147
Fialho Rodrigues, L. 443
Figueiredo, S. 1799
Filz, G.M. 1525
Fioravante, V. 1585, 1805
Fityus, S.G. 331
Foá, S.B. 1635
Foged, N. 1813
Foti, S. 409, 451
Friedel, S. 809
Frost, J.D. 693
Fujii, T. 1155

Gaba, A.R. 1089
Galera, J.M. 1443
Gambin, M. 75, 927
Gardien, W. 1449
Gaspar, A. 1293
Gatto, R.L. 1339
Gavin, K. 459
George, K.P. 367
Ghionna, V.N. 1843
Giacheti, H.L. 359, 467, 1627
Glagovsky, V.B. 593
Glynn, D. 497
Gomes Correia, A. 75, 927
Gomes, R.C. 1097, 1331
Gómez-Escoubès, R. 1457, 1605
González Herrero, J.M. 1457
Gottardi, G. 1643
Gourvès, R. 279
Gucunski, N. 475, 1465
Gudjonsson, G.T. 719
Guimarães, R.C. 1311
Guirguis, N.Sh. 1105
Guo, J.Y. 1219
Gusmão, A.D. 1165
Guzina, B.B. 935

Haas III, J.W. 795
Haegeman, W. 643, 1655
Halim, Y. 775
Hamada, J. 467

Hammamji, Y. 1111
Hamzah, U. 571
Handy, R.L. 913
Harkes, M.P. 687
Hartlén, J. 1813
Hashemi, M. 1397
Hebeler, G.L. 693
Hegazy, Y.A. 1285
Herbschleb, J. 1119
Hermosilha, H. 419, 433
Heymann, G. 575
Hight, D. 1033
Hiltunen, D.R. 483
Hoek, E. 39
Hoffman, D. 1131
Hoffmann, H. 385
Hölscher, P. 1651
Hopman, V. 1651
Horta, E. 1597
Howie, J.A. 299, 943, 1259
Hryciw, R.D. 701
Hsu, H-H. 147
Hsu, R.E. 713
Huang, A-B. 147, 1473
Huybrechts, N. 399

Ilori, A.O. 1125
Imre, E. 951, 1507
Inoue, M. 749
Isetta, E. 1827

Jackson, P.G. 1813
Jadi, H. 1131
Jakubowski, J. 679
Jardas, B. 1415
Joh, S-H. 97

Kagan, A.A. 593
Kalantarian, E. 491
Karim, U.F.A. 1821
Karl, L. 1655
Karray, M. 1111, 1481
Karthikeyan, M. 801
Katzenbach, R. 307
Kavazanjian, E. 1263
Kelley, S.P. 1489, 1663
Khamehchian, M. 1701
Khodaparast, M. 325
Khoury, N. 1319
Kim, D-S. 339, 707, 767

Kim, S.W. 1619
Kim, Y-J. 1875
Kim, Y-S. 957
Kocsis, A. 1507
Kovačević, M.S. 1415
Krivonogova, N.F. 593
Kulessa, B. 497
Kulhawy, F.H. 1139
Kuo, C.P. 713

Labrie, D. 963
Ladeira, F.L. 1279
Larsson, R. 271, 971
LeBihan, J.-P. 1403
Lee, C.F. 1219
Lee, J-H. 1499
Lefebvre, G. 1111, 1481
Lehane, B. 49, 819, 1673
Leite, A.L. 1331
Lemos, L.J.L. 1345
Leroueil, S. 1403
Lewis, C.J. 1285
Li, D.J. 1869
Liao, J.J. 1473
Liao, T. 345
Lima, A.S. 505
Lima, C. 1067
Lin, C-P. 513, 1473
Lin, S.Y. 1061
Liszkowski, J. 979
Liu, W. 1723
Lo, K.F. 1883
Lo Presti, D.C.F. 167
Lombardi, G. 1827
Long, M. 459, 625, 719
Lopes, I. 521
López Carrasco, A. 1681
Low, H.E. 1835
Luna, R. 1131, 1723
Lunne, T. 315, 987, 1861
Lutenegger, A.J. 861, 1489, 1663

Madiai, C. 1687
Madyarov, A.I. 935
Maertens, J. 399
Maher, A. 475, 1465
Mahler, A. 1507
Manakou, M. 635
Mántaras, F.M. 1353

Marchetti, S. 993, 1511
Marinos, P. 39
Marques, E. 1293
Marques, F.E.R. 1301
Martínez, J.M. 1605
Mascarenhas, M.M.A. 1311
Massad, F. 1551
Massarsch, K.R. 133
Matasovic, N. 1075
Mateos, M.T. 1605
Mattson, H. 971
Maugeri, M. 1053
Mayne, P.W. 345, 391, 1003, 1033, 1131, 1695
Mazen, O. 1081
McGillivray, A. 1003, 1695
Melander, K. 787
Mendes, M. 521
Menkveld, A. 1821
Miller, H.J. 727
Mimura, M. 801
Minuto, D. 1827
Mitchell, J.K. 843
Młynarek, Z. 979, 987
Moghaddas, Sh. 1397
Moitinho, I. 521
Möller, B. 733
Monaco, P. 993, 1511
Mondelli, G. 467
Monnet, J. 741
Monteiro, L.A.C. 581
Moradi, M. 265
Moreira de Souza, N. 1011
Moreno-Carrizales, P. 885
Morii, T. 749
Moura, R. 419
Mulas, J. 1171
Mure, N. 1565
Murrieta, P. 1325

Nakamura, M. 1891
Nassif, H.H. 1465
NeSmith, W.M. 555
Nevels Jr., J.B. 1319
Neves, E. 1275
Neves, J. 1067
Ng, C.W.W. 1379
Nichol, D. 1147
Niedzielski, A. 1371
Nikraz, H. 287, 1041

Nilsson, T. 1519
Nunokawa, O. 1155

O'Connor, P. 459
O'Neill, A. 539, 547
Odebrecht, E. 351
Okada, K. 1155
Okereke, C.S. 1125
Oliveira, A.C.V. 505
Oliveira, D.A.F. 1325
Oliveira, H.R. 1161
Oliveira, J.T.R. 1165
Oliveira, M. 531
Oliveira, R. 1089
Ørbech, T. 1861
Osburn, R.H. 935
Oung, O. 757

Pahlavan, B. 1701
Paik, K.-H. 1499
Pais, R. 1533
Pan, Y.W. 1473
Pando, M.A. 1525
Park, C.B. 563
Park, H-C. 767
Passos, P.G.O. 1635, 1709
Peixoto, A.S.P. 359, 467
Pelli, F. 1827
Peral, F. 1171, 1443
Pereira, E.L. 1331
Pereira, F.M. 1097
Peuchen, J. 555, 825
Phoon, K.K. 625, 801, 1835
Piché, R. 1111
Pickles, A.C. 1089
Pinheiro, R.J.B. 1177
Pinho, A.B. 1279
Pitilakis, K. 635
Porbaha, A. 1185
Porcino, D. 1843
Pöschl, I. 1533
Poulos, H.G. 1543
Powell, J.J.M. 253, 1717
Prokopovich, V.S. 593
Pruška, J. 1225
Puppala, A.J. 1185

Quale, S. 1565

Ragusa, A. 1053

Rahardjo, P.P. 775
Rahim, A.M. 367
Randolph, M.F. 209, 671
Rao, P.M. 1191
Raptakis, D. 635
Ras, A. 1425
Reiffsteck, Ph. 783
Reynolds, J.M. 1147
Ribeiro Jr., A. 647
Ribeiro, L.F.M. 1097, 1331
Ritter, E. 1339
Rix, G.J. 1199
Robertson, P.K. 233, 829
Rocha de Albuquerque, P.J. 1551
Rocha, M.M. 351
Rodrigues, C.M.G. 1345
Rodríguez, A. 1171, 1443
Rodríguez Soto, A. 1681
Rohani, J. 315
Romero-Hudock, S. 1199
Roth, M.J.S. 483
Roxo, J.M.C. 1557
Rózsa, P. 951
Ruffell, A. 497
Rust, M. 1565
Ryden, N. 563
Ryu, C.K. 1619

Saboya Jr, F. 373
Sacchetto, M. 787
Sadkowski, S.S. 665
Sáenz de Navarrete, R. 1457
Salák, J. 1225
Salgado, F.M. 443
Salgado, R. 1499
Samsudin, A.R. 571
Sandven, R. 1775, 1791, 1851, 1861
Santos, C.B. 1301
Santos, J.A. 521, 1361
Satoh, H. 615
Schnaid, F. 49, 351, 1353
Schneider, J.A. 1003
Schröfel, J. 1225
Scott, J. 1565
Senouci, S.M. 741
Sentosa, L. 775
Seo, W-S. 339, 707
Seward, D. 1565

Shang, J.Q. 889
Shau, S.H. 713
Shibuya, S. 167, 1869
Shin, S. 701
Shin, Y-S. 1875
Shinn II, J.D. 693, 795
Shokouhi, P. 475
Silva Cardoso, A. 39
Silva, M.F. 377
Simon, G. 399
Simoni, G. 1687
Simonini, P. 187
Škopek, J. 1571
Socco, L.V. 451
Souza Neto, J.B. 1267
Sreerama, K. 1191
Steenfelt, J.S. 1207, 1813
Stephenson, R.W. 1723
Stetson, K.P. 727
Stewart, H.E. 1139
Stokoe II, K.H. 97
Stölben, F. 307, 385
Strava Corrêa, A.C. 1011
Strieder, A. 1177
Strobbia, C. 451, 521
Stuit, H.G. 1449
Sugawara, S. 1739
Sugiyama, T. 1155
Sy, A. 299

Takemura, J. 1765
Takeshita, Y. 749
Tan, T.S. 801, 1835
Tanaka, H. 1019, 1869
Tanaka, M. 1019, 1765
Tang, S.K. 1739
Tang, S-H. 513
Teves-Costa, P. 521
Tham, L.G. 1219
Theron, M. 575
Tibana, S. 581
Tiggelman, L. 757
Tinkler, J. 1473
Togliani, G. 1731
Tong, S.Y. 1739
Tonni, L. 1643
Torres, M. 1681
Totani, G. 1511
Tournier, J.P. 1111
Townsend, F.C. 1597

Trevisan, A. 787
Trevor, F.A. 391
Tschuschke, M. 979
Tschuschke, W. 979, 987
Tuna, C. 1361

Ulriksen, P. 563
Uzielli, M. 905

Vakili, J. 1213
Van Alboom, G. 399, 1025
van der Hoek, E.E. 687
van der Rijst, M.C. 585
Van der Vegt, J.W.G. 757
van Hoegaerden, V. 585
Van Impe, W.F. 643
van Meerten, J.J. 687
van Staveren, M.Th. 1747
Vannobel, P. 1111
Vennalaganti, K.M. 1191
Vergauwen, I. 1025
Viana da Fonseca, A. 75, 419, 433, 1275, 1293, 1361, 1755, 1799
Voronkov, O.K. 593

Waliński, M. 1371
Wang, Y.H. 1883
Watabe, Y. 167, 1765
Westerhoff, R.S. 585
Whenham, V. 399
Whiteley, R.J. 601
Wierzbicki, J. 1371
Wisén, R. 607
Wołyński, W. 1371
Woods, R.D. 97
Wotschke, P. 809

Yaacob, W.Z.W. 571
Yafrate, N.J. 679
Yamaguchi, Y. 615, 1891
Yamashita, S. 167
Yan, W.M. 1883
Yasrebi, S.S. 1047
Yu, H-S. 3
Yue, Z.Q. 1219

Záleský, J. 1225
Zekkos, D.P. 1575
Zhu, H. 1431